Radio Communication Handbook

SIXTH EDITION

Editor: Dick Biddulph, G8DPS

Radio Society of Great Britain

Published by the Radio Society of Great Britain, Cranborne Road, Potters Bar, Herts EN6 3JE.

First published 1938
Sixth edition 1994

ISBN 1 872309 24 0

Cover design: Geoff Korten Design.

Illustrations: Derek Cole and Bob Ryan (Radio Society of Great Britain), Ray Eckersley, and Jean Faithfull.

Production and typography: Ray Eckersley, Seven Stars Publishing.

Printed in Great Britain by The Bath Press Ltd, Lower Bristol Road, Bath BA2 3BL.

Acknowledgements

The principal contributors to this book were:

Chapter 1 – 'Principles'	Dick Biddulph, G8DPS
Chapter 2 – 'Passive components'	'Phosphor'
Chapter 3 – 'Semiconductors	Steve Price, G4BWE
Chapter 4 – 'Electronic tubes and valves'	George Jessop, G6JP
Chapter 5 – 'Building blocks'	Erwin David, G4LQI Peter Saul, G8EUX
Chapter 6 – 'HF receivers'	Pat Hawker, G3VA
Chapter 7 – 'HF transmitters and transceivers'	Mike Grierson, G3TSO
Chapter 8 – 'VHF/UHF receivers, transmitters and transceivers	Robin Hewes, G3TDR Steve Thompson, G8GSQ
Chapter 9 – 'Microwaves'	Mike Dixon, G3PFR
Chapter 10 – 'Telegraphy and keying'	George Benbow, G3HB
Chapter 11 – 'Propagation'	Ray Flavell, G3LTP
Chapter 12 – 'HF antennas'	Les Moxon, G6XN
Chapter 13 – 'VHF/UHF antennas'	Roy Powers, G8CKN
Chapter 14 – 'Power supplies'	Dick Biddulph, G8DPS
Chapter 15 – 'Measurements and test gear'	Clive Smith, G4FZH
Chapter 16 – 'Construction and workshop practice'	Tom Kirk, G3OMK
Chapter 17 – 'Electromagnetic compatibility'	Robin Page-Jones, G3JWI
Chapter 18 – 'Amateur satellites and space communications'	John Branegan, GM4IHJ
Chapter 19 – 'Image techniques'	Mike Wooding, G6IQM
Chapter 20 – 'Data communications'	Ian Suart, GM4AUP
Chapter 21 – 'Operating technique and station layout'	John Allaway, G3FKM Dick Biddulph, G8DPS Harold Fenton, G8GG
Chapter 22 – 'General data'	Dick Biddulph, G8DPS Erwin David, G4LQI

Acknowledgement is also made to the authors of articles published in *Radio Communication* from which extracts have been made. Other acknowledgements are made in the text where appropriate.

Contents

Foreword

by HRH Prince Philip, Duke of Edinburgh, KG, Patron of the Radio Society of Great Britain

WINDSOR CASTLE

The Radio Communication Handbook has become a standard reference work for amateurs and professionals across the world and this, the 6th edition, brings the book fully up to date.

Radio has come a very long way since Marconi first demonstrated its potential nearly a hundred years ago. Professional research and development have naturally been responsible for most of the advances, but amateurs have played a significant part in pioneering new techniques and in solving some of the mysteries and anomalies in the propagation of radio waves.

Amateur radio has long been a popular hobby for people of all ages and in all parts of the world. I know that many young people take it up under the Skill Section of the Duke of Edinburgh's Award and similar Schemes in other countries. In 1988, I launched the Radio Society of Great Britain's Project YEAR (Youth into Electronics via Amateur Radio) and I am glad to know that it has also persuaded the Radio Communications Agency to introduce a Novice Licence, with no age limit.

I hope that all radio enthusiasts will find this new edition of the Handbook useful, informative and an encouragement to experiment.

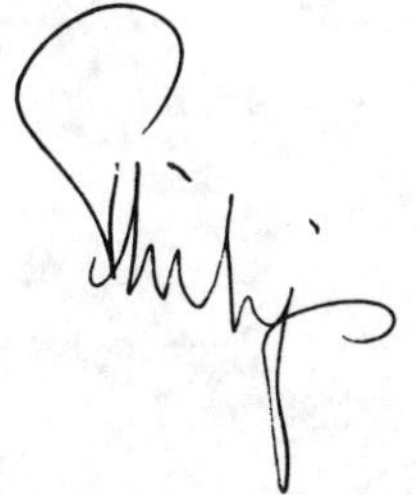

Preface

SINCE the fifth edition of this book, amateur radio has moved on apace, especially in the hardware which is now available to us. Integrated circuits of a complexity not dreamed of then are not only available but reasonably cheap. This means that all chapters have been revised and those dealing with such hardware have been totally rewritten.

Some new features have been introduced. Notably, a chapter on 'Building blocks' has been designed to encourage you to get out your pencil and paper and design something. The new chapter on 'Construction and workshop practice' is there to help you realise your project! The separate chapters in the previous edition on 'VHF/UHF receivers' and 'VHF/UHF transmitters' have been combined to reflect the modern tendency to use transceivers.

The chapters on hardware ('HF receivers', HF transmitters and transceivers', 'VHF/UHF transceivers' etc) all contain practical designs which are up-to-date at the time of writing.

Another innovation is the inclusion of full-sized PCB layouts at the back of the book. These may be copied for your own use but not for commercial applications without the written permission of the RSGB.

Some chapters have been omitted altogether or combined with others. Mobile and portable equipment is now virtually all commercial and enough details in the 'hardware' chapters are there for the home constructor. 'Noise' is included in the chapters on 'HF receivers' and 'VHF/UHF receivers, transmitters and transceivers'.

With a multitude of authors, it is inevitable that there is some duplication and there may also be some errors. If you notice any of the latter, please let me know.

Lastly, as Editor, I gratefully acknowledge the kindness of our Patron, HRH Prince Philip, Duke of Edinburgh, KG, in writing the Foreword, the work of all the contributors, Ray Eckersley, G4FTJ, who was Production Editor, and Derek Cole, Bob Ryan and freelance illustrators who did all the drawings. Thanks also to many amateurs who were consulted by letter and telephone in order to clear up errors and ambiguities.

Dick Biddulph, G8DPS

1 Principles

A KNOWLEDGE of the fundamental physical principles underlying radio communication is just as important to the amateur as to the professional radio engineer. It forms the starting point of all design work and is essential for proper maintenance and successful operating.

In the limited space available here it is not possible to discuss these basic principles in great detail, but a number of references are included in the bibliography at the end of the chapter as a guide to those readers who would like to make a further study of any particular part of the subject. Most public libraries will be only too glad to obtain the books listed if they are not already on their shelves.

ATOMS AND ELECTRONS

All matter is composed of *molecules*, which is the name given to the smallest quantity of a substance which can exist and still display the physical and chemical properties of that substance. There is a very great number of different sorts of molecules. Further study of a molecule discloses that it is made up of smaller particles called *atoms* and it has been found that there are about 102 different types of atoms. All molecules are made up of combinations of atoms selected from this range. Examples of different atoms are atoms of hydrogen, oxygen, iron, copper and sulphur, and examples of how atoms are combined to form molecules are (a) two atoms of hydrogen and one of oxygen to form one molecule of water, and (b) two atoms of hydrogen, one of sulphur and four of oxygen to form one molecule of sulphuric acid.

Atoms are so small that they cannot be seen even under the most powerful optical microscopes. Their behaviour, however, can be studied and from this it has been discovered that atoms are made up of a positively charged relatively heavy core or *nucleus* around which are moving a number of much lighter particles each negatively charged, called *electrons.* Atoms are normally electrically neutral; that is to say, the amount of positive electricity associated with the nucleus is exactly balanced by the total amount of negative electricity associated with the electrons. One type of atom differs from another in the number of positive and neutral particles (called *protons* and *neutrons)* which make up the nucleus and the number and arrangement of the orbital electrons which are continually moving around the nucleus.

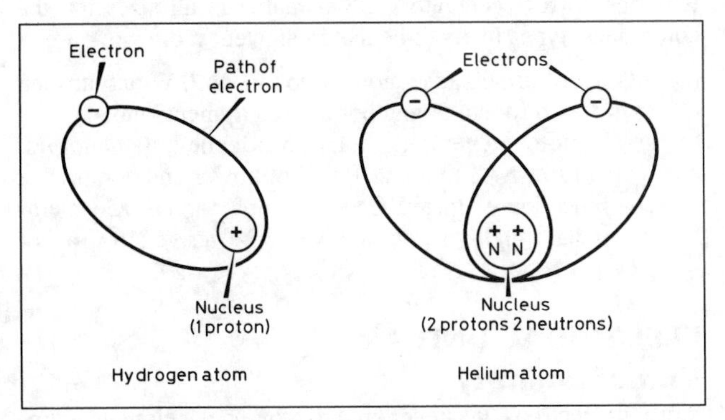

Fig 1.1. Structure of hydrogen and helium atoms

Large atoms, such as those of uranium, are complex, but small atoms, such as those of hydrogen and helium, are relatively simple, as can be seen in Fig 1.1.

The important fact which emerges from the preceding paragraphs is that all matter is made up of positively and negatively charged particles, which means that electricity is latent in everything around us.

CONDUCTORS AND INSULATORS

The ease with which the electrons in a substance can be detached from their parent atoms varies from substance to substance. In some substances there is a continual movement of electrons in a random manner from one atom to another and the application of an electrical pressure or voltage (for example from a battery) to the two ends of a piece of wire made of such a substance will cause a drift of electrons along the wire called an *electric current;* electrical *conduction* is then said to take place. It should be noted that if an electron enters the wire from the battery at one end it will be a different electron which immediately leaves the other end of the wire.

By arbitrary convention, the direction of current now is said to be *from positive to negative*.

Materials which exhibit this property of electrical conduction are called *conductors.* All metals belong to this class. Materials which do not conduct electricity are called *insulators,* and the following list gives a few examples of commonly used conductors and insulators.

Conductors	Insulators
Silver	Mica
Copper	Quartz
Aluminium	Glass
Brass	Ceramics
Steel	Ebonite
Mercury	Plastics
Carbon	Air and other gasses
Solutions of salts	Oil
or acids in water	Pure water

SOURCES OF ELECTRICITY

When two dissimilar metals are immersed in certain chemical solutions, or *electrolytes,* an *electromotive force* (EMF or *voltage*) is created by chemical action within the cell so that if the pieces of metal are joined externally to the cell there will be a continuous flow of electric current. This device is called a *simple cell* and such a cell, comprising copper and zinc rods immersed in diluted sulphuric acid, is shown in Fig 1.2(a). The flow of current is from the copper to the zinc plate in the external circuit; ie the copper forms the positive (+) terminal of the cell and the zinc forms the negative (–) terminal.

In a simple cell of this type hydrogen forms on the copper electrode, and this gas film has the effect of increasing the internal resistance of the cell and also setting up within the cell a *counter* or *polarising* EMF which rapidly reduces the effective EMF of the cell as a whole. This polarisation effect is overcome in practical cells by the introduction of chemical agents surrounding the anode for the purpose of removing the hydrogen by oxidation as soon as it is formed. Such agents are called *depolarisers*.

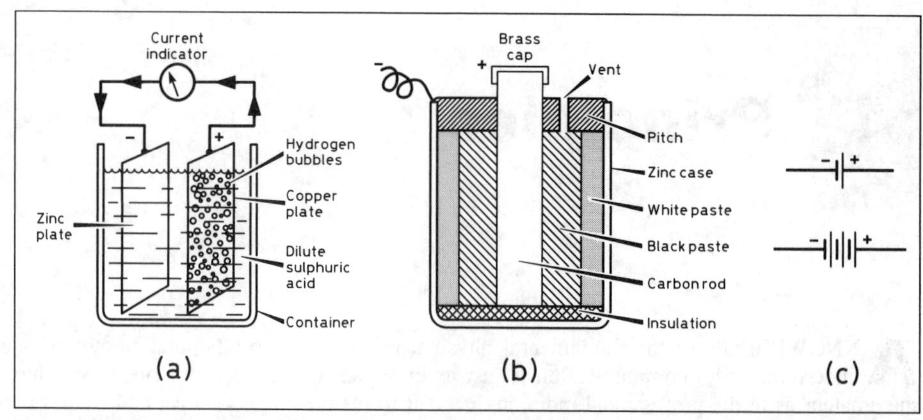

Fig 1.2. The electric cell. A number of cells connected in series is called a *battery*. (a) A simple electric cell consisting of copper and zinc electrodes immersed in dilute sulphuric acid. (b) Sectional drawing showing construction of a dry cell. (c) Symbol used to represent single cells and batteries in circuit diagram

Primary cells

Practical cells in which electricity is produced in this way by direct chemical action are called *primary cells;* a common example is the Leclanché cell, the construction of which is shown diagrammatically in Fig 1.2(b). The zinc case is the negative electrode and a carbon rod is the positive electrode. The black paste surrounding the carbon rod may contain powdered carbon, manganese dioxide, zinc chloride, ammonium chloride and water, the manganese dioxide acting as depolariser by combining with hydrogen formed at the anode to produce another form of manganese oxide and water. The remainder of the cell is filled with a white paste which may contain plaster of Paris, flour, zinc chloride, ammonium chloride and water. The cell is sealed with pitch except for a small vent which allows accumulated gas to escape. The EMF developed by a single dry cell is about 1.5V and cells may be connected in series (positive terminal to negative terminal and so on) until a battery of cells, usually referred to simply as a *battery,* of the desired voltage is obtained. The symbol used to denote a cell in a circuit diagram is shown in Fig 1.2(c). The long thin stroke represents the positive terminal and the short thick stroke the negative terminal. Several cells joined in series to form a battery are shown; for higher voltages it becomes impracticable to draw all the individual cells involved and it is sufficient to indicate merely the first and last cells with a dotted line between them with perhaps a note added to state the actual voltage. The amount of current which can be derived from a dry cell depends on its size and the life required and may range from a few milliamperes to an ampere or two.

Secondary cells

In primary cells some of the various chemicals are used up in producing the electrical energy – a relatively expensive and wasteful process. The maximum current available also is limited. Another type of cell called a *secondary cell, storage cell* or *accumulator* (different names for the same thing) offers the advantage of being able to provide a higher current and is capable of being charged by feeding electrical energy into the cell to be stored chemically, and be drawn out or discharged later as electrical energy again. This process of charging and discharging the cell is capable of repetition almost indefinitely.

The most common type of secondary cell is the *lead-acid cell* such as that used in motor-car batteries. It consists of two sets of specially prepared lead plates immersed in a mixture of sulphuric acid and water. Briefly, the action of the cell is as follows: in the discharged state the active material on each plate is lead sulphate. During the charging process the lead sulphate in the positive plate is changed to lead dioxide and sulphuric acid, and the lead sulphate in the negative plate to a form of lead called *spongy lead* and sulphuric acid. During discharge the reverse action takes place and lead sulphate forms again on both plates. The state of the charge of the cell may be checked by measuring the specific gravity of the electrolyte with a hydrometer since the concentration of sulphuric acid increases during charging and decreases again as the cell is discharged. Typical values of specific gravity are 1.250 for a fully charged cell and 1.175 for a discharged cell. The terminal voltage of a single lead-acid cell when fully charged and left standing is about 2.05V. In the discharged condition the voltage falls to about 1.85V.

To obtain a long life cells should not be overcharged since this causes flaking and buckling of the plates. Cells should not be left in a discharged state because the lead sulphate may undergo a physical change which it is difficult to break down in the recharging process and the capacity of the cell will be impaired; cells in this condition are said to be *sulphated*, a condition which may sometimes be partially eradicated by prolonged charging at a low current.

Mechanical generators

Mechanical energy may be converted into electrical energy by moving a coil of wire in a magnetic field. Direct-current or alternating-current generators are available in all sizes but the commonest types likely to be met in amateur radio work are:

(a) AC petrol-driven generators of up to 1 or 2kW output such as are used for supplying portable equipment; and
(b) small motor generators, sometimes called *dynamotors,* which furnish up to about 100W of power and comprise a combined low-voltage DC electric motor and a DC generator so that a high-voltage supply may be derived from a 6 or 12V car battery.

ELECTRICAL UNITS

Unit of quantity

Since all electrons, whatever kind of atom they belong to, carry the same charge, the amount of electricity associated with the

POWERS, EXPONENTIALS AND LOGARITHMS

It is tedious to put in all the noughts in very large or very small numbers so they are often expressed as powers of 10 or exponentials.

100 = 10^2 (say "10 squared") or 10 exp 2

The latter is sometimes shortened further to 10e2.

0.001 = 10^{-3} ("ten to the minus three") or 10 exp–3 or 10 e–3

where 'exp' or 'e' is shorthand for 'exponential'. When the exponential is based on 10 as above, then the exp part is the *logarithm* of the number (properly it should be called the *common logarithm* since there is another kind; see below.) So we have:

log 100 = 2 and log 0.001 = –3.

The main properties of logarithms ('logs') is that to *multiply* two numbers, add their logs and to *divide* two numbers, subtract their logs. So:

$1000 \times 10000 = 10^3 \times 10^4$

The logs are 3 and 4 so the product is 10^7 which is 10,000,000. This is a very simple example to illustrate the point.

There is a different kind of log based on the 'irrational' (not completely known) number 'e' (not to be confused with exponential, above) which has a value of 2.718282 approximately. It is used in certain calculations such as those on time constants (see later).

electron could be used as the unit of quantity of electricity. It is, however, too small for use as a practical unit, and a more convenient unit is called the *coulomb;* this is equivalent to the charge on approximately 6×10^{18} (six million million million) electrons. The analogy here between the molecule of water as a unit of quantity of water and the practical unit, the litre, is obvious.

A quantity of electricity or number of coulombs is usually denoted by the symbol *q*.

Unit of current flow

Continuing with the water analogy, whereas a flow of water is spoken of as *x* litres per second, the flow of electricity can be expressed as *x* coulombs per second. A current of one coulomb per second is called an *ampere* (abbreviated to *amp)* and the strength of a current is said to be *x* amperes. A current flow is usually denoted by the symbol *I*. The currents used in radio are often very small fractions of an ampere and for convenience the two smaller units *milliampere* (meaning a thousandth of an ampere and abbreviated to 'mA') and *microampere* (meaning a millionth of an ampere and abbreviated to 'µA') are used. Thus a current of 0.003A is written as 3mA. See Table 1.1 for abbreviations.

The relation between the total quantity of electricity (*q*) which has passed a point in a wire, the time of flow (*t*) and the rate of flow (*I*) is therefore:

Quantity (coulombs) = Current flow (amperes) × Time (seconds)

or in symbols:

$$q = I \times t$$

Unit of electric pressure

In order to make electricity flow continuously through a circuit it is necessary to have some device which can produce a continuous supply of electrons. This may be a battery in which the supply of electrons is produced by chemical action or a dynamo or generator in which mechanical energy is turned into

Table 1.1. Units and symbols

Quantity	Symbol	Unit	Abbreviation
Charge	*q*	coulomb	C
Current	*I*	ampere (amp)	A
Voltage*	*E* or *V*	volt	V
Time	*t*	second	s or sec
Resistance	*R*	ohm	Ω
Capacitance	*C*	farad	F
Inductance	*L*	henry	H
Mutual inductance	*M*	henry	H
Power	*P*	watt	W
Frequency	*f*	hertz†	Hz
Wavelength	l	metre	m

* 'Voltage' includes 'electromotive force' and 'potential difference'.
† The unit 'hertz' was formerly called 'cycle per second'.

Since the above units are sometimes much too large (eg the farad) and sometimes too small, a series of multiples and sub-multiples are used:

Unit	Symbol	Multiple
Microamp	µA	1 millionth (10^{-6}) amp
Milliamp	mA	1 thousandth (10^{-3}) amp
Microvolt	µV	10^{-6}V
Millivolt	mV	10^{-3}V
Kilovolt	kV	10^{3}V
Picofarad	pF	10^{-12}F
Nanofarad	nF	10^{-9}F
Microfarad	µF	10^{-6}F
Femtosecond	fs	10^{-15}s
Picosecond	ps	10^{-12}s
Microsecond	µs	10^{-6}s
Millisecond	ms	10^{-3}s
Microwatt	µW	10^{-6}W
Milliwatt	mW	10^{-3}W
Kilowatt	kW	10^{3}W
Gigahertz	GHz	10^{9}Hz
Megahertz	MHz	10^{6}Hz
Kilohertz	kHz	10^{3}Hz
Centimetre	cm	10^{-2}m
Kilometre	km	10^{3}m

Note: The sub-multiples abbreviate to lower case letters. All multiples or sub-multiples are in factors of a thousand except for the centimetre.

electrical energy. As we have seen, the battery or generator produces an electrical pressure, sometimes called an *electromotive force* or EMF, which may be used to force a current through a circuit. The unit of electrical pressure is the *volt*, and voltages are usually denoted in formulae by the symbol *E* or *V*.

Unit of electrical resistance

The ease with which an electric current will flow, or can be conducted, through a wire will depend on the dimensions of the wire and the material from which it is made. The opposition of a circuit to the flow of current is called the *resistance* of the circuit and is usually denoted by the symbol *R* in formulae. The resistance of a circuit is measured in *ohms* (abbreviated to the Greek letter Ω), and a circuit is said to have a resistance of one ohm if the voltage between its ends is one volt when the current flowing is one amp. For convenience, because the resistances used in radio equipment may be up to 10 million ohms, two larger units called the *kilohm* (meaning 1000 ohms) and the *megohm* (meaning 1 million ohms) are used. Thus 47,000 ohms (shortened to 47kΩ) may be referred to as '47 kilohms'. See Table 1.1 for abbreviations.

RESISTANCE AND CONDUCTANCE

Different materials may be compared as conductors by measuring the resistance of samples of the materials of the same size and shape. The resistance (in ohms) of conducting material of

Table 1.2. Specific resistance of commonly used conductors [1]

Conductor	Specific resistance (Ω-cm × 10^6)
Silver	1.6
Copper	1.7
Aluminium	2.7
Brass	7
Iron	10.1
Special high-resistivity alloys:	
Manganin	48
Eureka	42
Nichrome	108

one centimetre in length and one square centimetre in cross-sectional area is called the *specific resistance* or *resistivity* of the material. Specific resistance is quoted as *x* ohms per centimetre cube or more simply as *x* ohm-cm. If the specific resistance of a material is known, the actual resistance of, say, a piece of wire made from that material can be calculated since the resistance of a wire increases proportionally with its length and inversely with its cross-sectional area. If the length of a wire is *l* cm, its cross-sectional area *a* sq cm, and its specific resistance *S* ohm-cm, its actual resistance will be given by the formula:

$$R = \frac{S \times l}{a} \text{ ohms}$$

Approximate specific resistances of typical materials used in radio equipment are given in Table 1.2. At radio frequencies the resistance of a wire may be very much greater than its direct-current value because radio-frequency currents only travel along the surface shell of the wire and not through the whole body as in the case of direct current. This is called the *skin effect* and is important in making high-quality radio coils.

Sometimes, instead of speaking of the resistance of a circuit the *conductance of* the circuit is referred to: this is simply the reciprocal of the resistance. The unit of conductance is the *siemens (*sometimes called the *mho),* and conductance is usually denoted in formulae by the symbol *G*.

The relation between resistance and conductance is therefore:

$$G = \frac{1}{R} \quad \text{or conversely } R = \frac{1}{G}$$

Thus a resistance of 10Ω has a conductance of 0.1 siemens.

Ohm's Law

Ohm's Law states that the ratio of the voltage applied across a resistance to the current flowing through that resistance is constant. This ratio is equal to the value of the resistance. Writing this as a formula:

$$\frac{\text{Voltage}}{\text{Current}} = \text{A constant} = \text{Resistance}$$

or in symbols:

$$\frac{E}{I} = R$$

Note that if *E is* in volts and *I* in amperes, then *R* is in ohms. It will be seen that this formula relates current, voltage and resistance, so that if two of these quantities are known the third may be calculated by suitably rearranging the formula in the following ways:

$$E = I \times R \quad \text{(giving } E\text{, knowing } I \text{ and } R\text{)}$$

$$I = \frac{E}{R} \quad \text{(giving } I\text{, knowing } E \text{ and } R\text{)}$$

$$R = \frac{E}{I} \quad \text{(giving } R\text{, knowing } E \text{ and } I\text{)}$$

The application of Ohm's Law is illustrated by the following example.

Example. Consider the circuit shown in Fig 1.3 which consists of a battery E of voltage 4V and a resistance R of 8Ω. What is the magnitude of the current in the circuit?

Here $E = 4$V and $R = 8\Omega$. Let *I* be the current flowing in amperes. Then from Ohm's Law:

$$I = \frac{E}{R} \text{ amperes}$$

$$= \frac{4}{8} = \frac{1}{2} = 0.5\text{A}$$

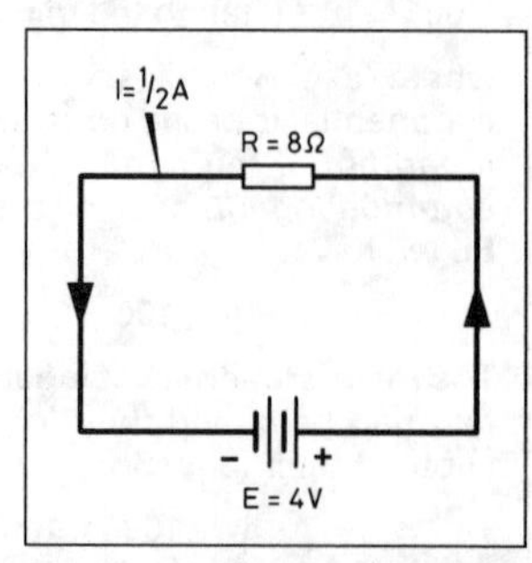

Fig 1.3. Application of Ohm's Law

It should be noted that in all calculations based on Ohm's Law care must be taken to ensure that *I, E* and *R* are in consistent units, ie in amperes, volts and ohms respectively, if errors in the result are to be avoided.

Electrical power

When a current of electricity flows through a resistor, eg in an electric fire, the resistor gets hot and electrical energy is turned into heat. The actual rise in temperature depends on the amount of power dissipated in the resistor and the shape and size of it. Sometimes the power dissipated is so small that the temperature rise is not very noticeable, but nevertheless whenever an electric current flows through a resistor power is dissipated therein. The unit of electrical power is the *watt,* usually denoted in formulae by the symbol *W*. The amount of electrical power dissipated in a resistor is equal to the product of the voltage across the resistor and the current flowing through the resistor. Thus:

$$\text{Power (watts)} = \text{Voltage (volts)} \times \text{Current (amperes)}$$

or in symbols:

$$W = E \times I$$

Since from Ohm's Law:

$$E = I \times R \quad \text{and} \quad I = E/R$$

the formula for the power dissipated in a resistor may also be written in two further forms:

$$W = E^2/R$$

$$W = I^2 \times R$$

These formulae are useful for finding, for example, the power input to a transmitter or the power dissipated in various resistors in an amplifier so that suitably rated resistors can be selected. To take a practical case, consider again the circuit of Fig 1.3. The power dissipated in the resistor may be calculated as follows:

Here $E = 4$V and $R = 8\Omega$. Let *W* be the power in watts. Then:

$$W = \frac{E^2}{R} = \frac{4 \times 4}{8} = 2\text{W}$$

It must be stressed again that the beginner should always see that all values are expressed in terms of volts, amperes and ohms in this type of calculation. The careless use of megohms or milliamperes, for example, may lead to an answer several orders too large or too small.

Decibels

In radio and line transmission engineering power ratios are often expressed in terms of units proportional to the logarithm of their ratio. In this way gains and losses in amplifiers, networks and transmission paths can be added and subtracted instead of having to be multiplied together. The *bel* is the logarithm to the base 10 (common logarithm) of the ratio of two powers. It is too large for normal use so the unit in common use is the *decibel* (one tenth of a bel) abbreviated 'dB', and widely used. The *neper*, which is the logarithm to the base e (2.718282) of the same ratio is occasionally used.

If there are two power levels, P_1 and P_2, for example at the input and output of an amplifier, the power ratio in decibels, N, is given by the expression

$$N = 10 \log_{10} \frac{P_1}{P_2} \text{ dB}$$

Expressed in nepers:

$$N = 0.5 \log_e \frac{P_1}{P_2} \text{ nepers}$$

Nepers can be converted easily into decibels, and vice versa, since from the two equations above it follows that 1 neper = 8.69dB.

Curve A in Fig 1.4 can be used for determining the number of decibels corresponding to a given power ratio and vice versa.

Since power can be expressed in terms of current in (or voltage across) a resistance, it follows that, for a given value of resistor, voltage and current ratio can also be expressed in decibels provided they are associated with the *same* resistance.

If $$P_1 = I_1^2 R = \frac{E_1^2}{R} \quad \text{and} \quad P_2 = I_2^2 R = \frac{E_2^2}{R}$$

then for a current ratio of I_1/I_2:

$$N = 20 \log_{10} \frac{I_1}{I_2} \text{ dB}$$

and for a voltage ratio of E_1/E_2:

$$N = 20 \log_{10} \frac{E_1}{E_2} \text{ dB}$$

Fig 1.4, curve B, shows the relationship between decibels and voltage and current ratios, and Table 1.3 gives some useful approximate power, voltage and current ratios in terms of decibels.

It should be noted that *strictly speaking* current and voltage ratios should only be expressed in decibels when the resistance at which they are measured is the same. In practice, however, the voltage gains of amplifiers having very different input and output resistances are often quoted in this way; it is an arbitrary use of the decibel which, although sometimes convenient in practice, often leads to confusion.

Resistors used in radio equipment

As already mentioned, a resistor through which a current is flowing may get hot. It follows therefore that in a piece of radio equipment the resistors of various types and sizes that are needed must be capable of dissipating the power as required without overheating.

Generally speaking, radio resistors can be divided roughly into two classes, (a) low power up to 3W and (b) above 3W.

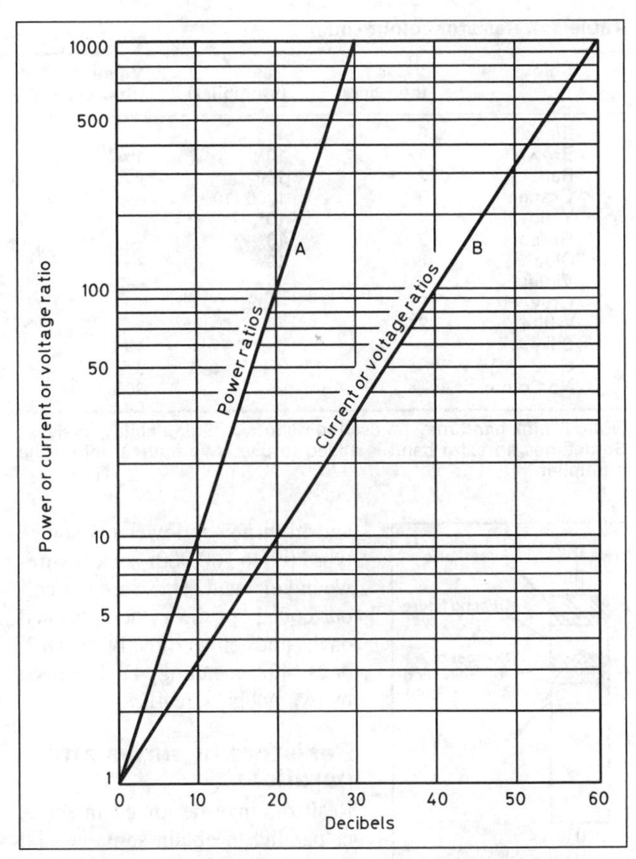

Fig 1.4. Graph relating decibels to power or current or voltage ratio

The low-power resistors are usually made of carbon or metal film and may be obtained in a wide range of resistance values from about 10Ω to 10MΩ and in power ratings of 0.1W to 3W. Typical carbon resistors are shown in Chapter 2 – 'Passive components'. For higher powers, resistors are usually wire-wound on ceramic formers and the very fine wire is protected by a vitreous enamel coating. They are also shown in Chapter 2.

Resistors, particularly the small carbon types, are usually colour-coded to indicate the value of the resistance in ohms and sometimes also the tolerance or accuracy of the resistance. The standard colour code is shown in Table 1.4.

The colours are applied as *bands* at one end of the resistor as shown in Fig 1.5. As an example, what would be the value of a resistor with the following colour bands: yellow, violet, orange, silver?

The yellow first band signifies that the first figure is 4, the violet second band signifies that the second figure is 7, while the orange third band signifies that there are three zeros to follow; the silver fourth band indicates a tolerance of ±10%. The value of the resistor is therefore 47,000Ω ±10% (47kΩ ±10%).

So far only fixed resistors have been mentioned. Variable resistors, sometimes called *potentiometers* or *volume controls*, are also used. These are usually panel-mounted by means of a threaded bush through which a ¼in or 6mm diameter spindle protrudes and to which the control knob is fitted. Low-power high-value variable resistances use a carbon resistance element

Table 1.3. Decibel conversion

Decibels	1	2	3	6	10	20	40
Power ratio	1.25	1.5	2	4	10	100	10,000
Voltage or current ratio	1.12	1.26	1.41	2	3.16	10	100

Table 1.4. Resistor colour code

Colour	Value (numbers)	Value (multiplier)	Value (tolerance)
Black	0	1	—
Brown	1	10	1%
Red	2	100	2%
Orange	3	1000 (10^3)	—
Yellow	4	10^4	—
Green	5	10^5	—
Blue	6	10^6	—
Violet	7	—	—
Grey	8	—	—
White	9	—	—
Silver	—	0.01	10%
Gold	—	0.1	5%
No Colour	—	—	20%

Note: A pink band may be used to denote a 'high-stability' resistor. Sometimes an extra band is added to give *two* figures before the multiplier.

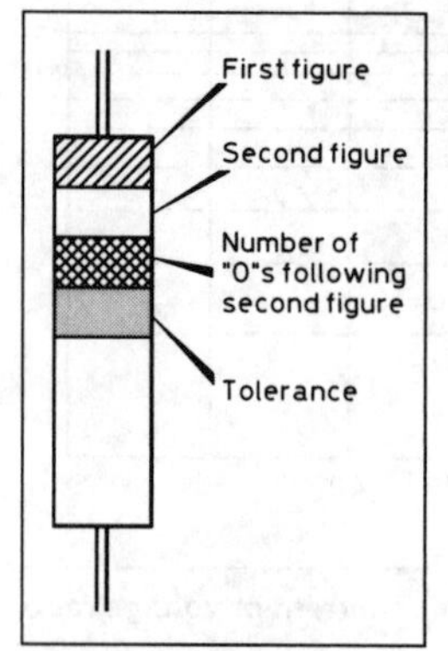

Fig 1.5. Standard resistance value markings

and high-power lower-resistance types (up to 100,000Ω) use a wire-wound element. These are not colour coded but the value is printed on the body either directly eg '4K7' or as '472' meaning '47' followed by two noughts, ie 4K7.

Resistors in series and parallel

Resistors may be joined in series or parallel to obtain some special desired value of resistance. When in series, resistors are connected as shown in Fig 1.6(a) and the total resistance of the resistors is equal to the sum of the separate resistances. The parallel connection is shown in Fig 1.6(b), and with this arrangement the reciprocal of the total resistance is equal to the sum of the reciprocals of the separate resistances.

If R is the total resistance, these formulae can be written as follows:

Series connection

$$R = R_1 + R_2 + R_3 + \text{etc}$$

Parallel connection

$$\frac{1}{R} = \frac{1}{R_1} + \frac{1}{R_2} + \frac{1}{R_3} + \text{etc}$$

Considering the case of only two resistors in parallel:

$$\frac{1}{R} = \frac{1}{R_1} + \frac{1}{R_2} \quad \text{or} \quad \frac{R_1 + R_2}{R_1 \times R_2}$$

which by inversion gives:

$$R = \frac{R_1 R_2}{R_1 + R_2}$$

This is a useful formula since the value of two resistors in parallel can be calculated easily.

To illustrate these rules two examples are given below.

Example 1. Calculate the resistance of a 30Ω and a 70Ω resistor connected first in series and then in parallel.

In series connection:

$$R = 30 + 70 = 100\Omega$$

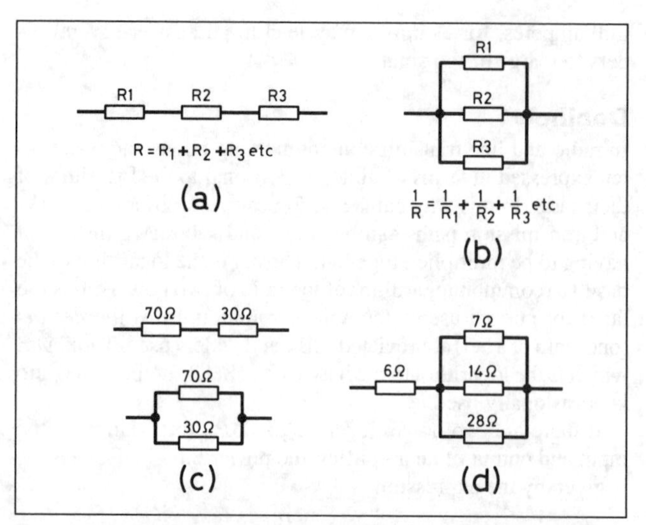

Fig 1.6. Resistors in various combinations: (a) series, (b) parallel, (c) series and parallel, (d) series-parallel. The calculation of the resultant resistances in (c) and (d) is explained in the text

In parallel connection, using the special formula for two resistors in parallel:

$$R = \frac{R_1 R_2}{R_1 + R_2}$$

$$= \frac{30 \times 70}{30 + 70} = 21\Omega$$

These two calculations are illustrated in Fig 1.6(c).

Example 2. Three resistors of 7, 14 and 28Ω are connected in parallel. If another resistor of 6Ω is connected in series with this combination, what is the total resistance of the circuit?

The circuit is shown in Fig 1.6(d). Taking the three resistors in parallel first, these are equivalent to a single resistance of R ohms given by:

$$\frac{1}{R} = \frac{1}{7} + \frac{1}{14} + \frac{1}{28} = \frac{4 + 2 + 1}{28} = \frac{1}{4}$$

Therefore $R = 4\Omega$

This in series with the 6Ω resistor gives a total resistance of 10Ω for the whole circuit.

CAPACITORS AND CAPACITANCE

Capacitors have the property of being able to store a charge of electricity. They consist essentially of two conducting plates or strips separated by an insulating medium called a *dielectric.* When a capacitor is charged there is a voltage difference between its plates, and the larger the plates and/or the smaller their separation, the greater will be the charge that the capacitor holds for any given voltage across its plates.

The unit of *capacitance* is called the *farad* and is the capacitance of a capacitor which holds a charge of one coulomb when the voltage across its plates is one volt. A farad is far too large a unit for practical purposes and capacitors are usually measured in *microfarads* (μF), one millionth of a farad, and in *picofarads* (pF), one million-millionth of a farad.

The capacitance of a capacitor depends on the area of its plates, on the distance by which they are separated and the material between them. When the space between the plates of a capacitor is occupied by some other insulating medium than air the

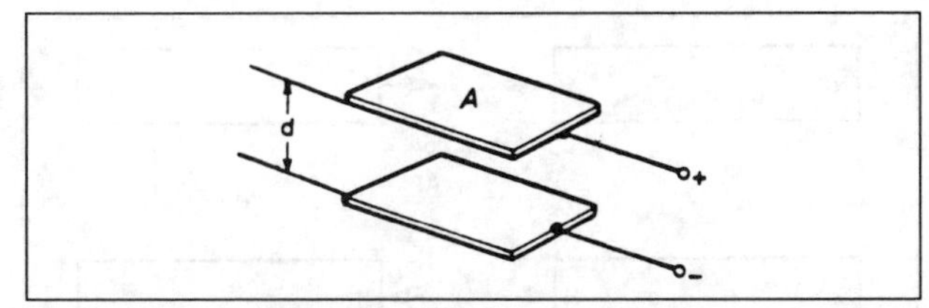

Fig 1.7. Parallel-plate capacitor. The capacitance is proportional to the area *A* and inversely proportional to the spacing *d*

capacitance of the capacitor is increased, the area and spacing being assumed the same in both cases. The factor by which the dielectric increases the capacitance compared with air is called the *dielectric constant* of the material. This factor is sometimes called the *permittivity* of the material and is denoted by the symbol *K*. Typical values of *K* are: air 1, paper 2.5, glass 5, mica 7. Certain ceramics have much higher values of *K* of 10,000 or more. If the dielectric is a vacuum, as in the case of the inter-electrode capacitance of a valve, the same value of *K* as for air may be assumed. (Strictly, $K = 1$ for a vacuum and is very slightly higher for air.) The voltage at which a capacitor breaks down depends on the spacing between the plates and the type of dielectric used. Capacitors are often labelled with the maximum working voltage which they are designed to withstand and this figure should not be exceeded.

Fig 1.7 shows an elementary capacitor made of two parallel metal plates of area *A* sq cm separated by a distance *d* cm. If the plates are of unequal area, only the smaller one should be taken into account since it is actually the cross-sectional area of the active dielectric which determines the capacitance. If the dielectric constant of the material between the plates is *K*, the capacitance of the capacitor will be:

$$C = \frac{0.0885\,KA}{d}\ \text{pF}$$

where *A* is in square centimetres and *d is* in centimetres.

As an example, suppose two metal plates each 1 sq cm in area are spaced 0.004cm apart in air. The capacitance would be:

$$C = \frac{0.0855 \times 1 \times 1}{0.004}\ \text{pF} = 22\text{pF}$$

If mica (for which the value of *K* is 7) had been used as a dielectric the capacitance would be:

$$C = \frac{0.0855 \times 1 \times 7}{0.004}\ \text{pF} = 155\text{pF}$$

Capacitors used in radio equipment

The values of capacitors used in radio equipment extend from below 1pF to 100,000µF. They are described in detail in Chapter 2.

Capacitors in series and parallel

Capacitors can be connected in series or parallel, as shown in Fig 1.8, either to obtain some special capacitance value using a standard range of capacitors, or perhaps in the case of series connection to obtain a capacitor capable of withstanding a greater voltage without breakdown than is provided by a single capacitor. When capacitors are connected in parallel, as in Fig 1.8(a), the total capacitance of the combination is equal to the sum of the separate capacitances. When capacitors are connected in series, as in Fig 1.8(b), the reciprocal of the equivalent capacitance is equal to the sum of the reciprocals of the separate capacitances.

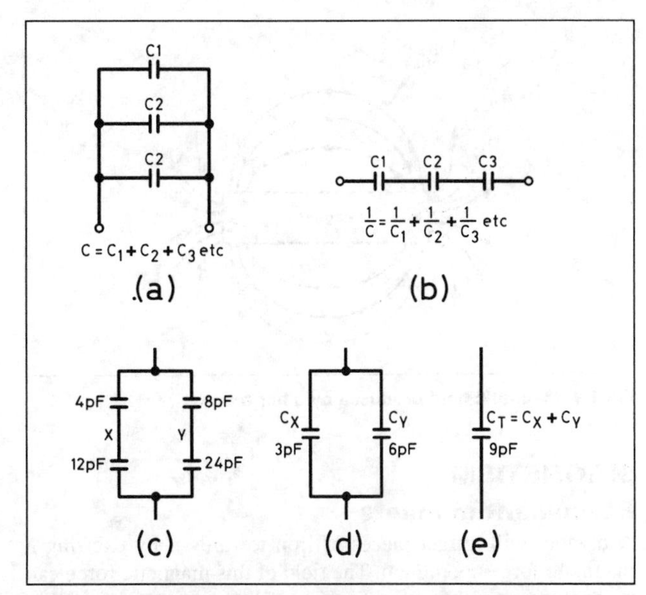

Fig 1.8. Capacitors in various combinations: (a) parallel, (b) series, (c) series-parallel. The calculation or the resultant capacitance of the combination shown in (c) required first the evaluation of each series arm X and Y as shown in (d). The single equivalent capacitance of the combination is shown in (e)

If *C* is the total capacitance these formulae can be written as follows:

Parallel connection

$$C = C_1 + C_2 + C_3 \text{ etc}$$

Series connection

$$\frac{1}{C} = \frac{1}{C_1} + \frac{1}{C_2} + \frac{1}{C_3} + \frac{1}{C_4} \text{ etc}$$

Similar to the formula for resistors in parallel, a useful equivalent formula for two capacitors in series is

$$C = \frac{C_1 C_2}{C_1 + C_2}$$

The use of these formulae is illustrated by the following example:

Example. Two capacitors of 4 and 12pF are connected in series; two others of 8 and 24pF are also connected in series. What is the equivalent capacitance if these series combinations are joined in parallel?

The circuit is shown in Fig 1.8(c). Using the formula for two capacitors in series, the two series arms X and Y can be reduced to single equivalent capacitances C_X and C_Y as shown in Fig 1.8(d). Thus:

$$C_X = \frac{4 \times 12}{4 + 12} = \frac{4 \times 1}{16} = 3\text{pF}$$

$$C_Y = \frac{8 \times 24}{8 + 24} = \frac{8 \times 24}{32} = 6\text{pF}$$

These two capacitances may now be compounded in parallel to give the total effective capacitance represented by the single capacitor C1 in Fig 1.8(e).

$$C = C_X + C_Y = 3 + 6 = 9\text{pF}$$

The total equivalent capacitance of the four capacitors connected as described is therefore 9pF.

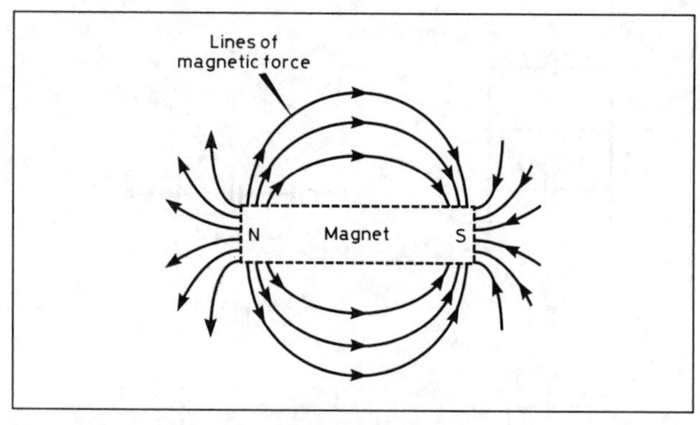

Fig 1.9. Magnetic field produced by a bar magnet

MAGNETISM

Permanent magnets

A magnet will attract pieces of iron towards it by exerting a magnetic force upon them. The field of this magnetic force can be demonstrated by sprinkling iron filings on a piece of thin cardboard under which is placed a bar magnet. The iron filings will map out the magnetic field as sketched in Fig 1.9 and the photograph Fig 1.10. It will be seen that the field is most intense near the ends of the magnet, the centres of intensity being called the *poles,* and *lines of force* spread out on either side and continue through the material of the magnet from one end to the other.

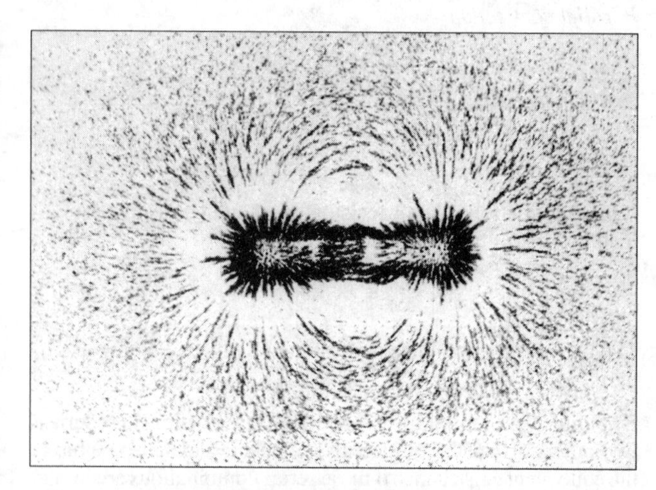

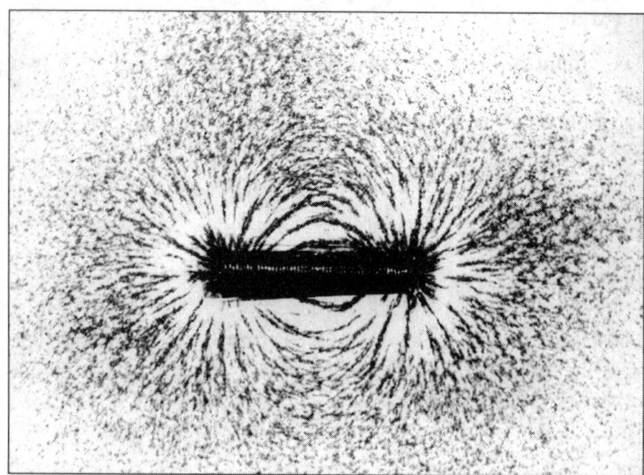

Fig 1.10. Iron filings mapping out the magnetic field of (top) a bar magnet, (bottom) a solenoid carrying current

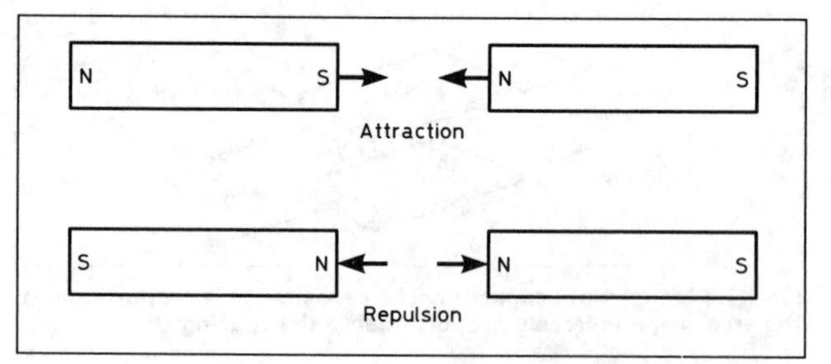

Fig 1.11. Attraction and repulsion between bar magnets

If such a magnet is suspended so that it can swing freely in a horizontal plane it will always come to rest pointing in one particular direction, namely towards the Earth's magnetic poles, the Earth itself acting as a magnet. A compass needle is simply a bar of magnetised steel. One end of the magnet (N) is called the *north pole,* which is an abbreviation of 'north-seeking pole' and the other end (S) a *south pole* or south-seeking pole. It is an accepted convention that magnetic force acts in the direction from N to S as indicated by the arrows on the lines of force in Fig 1.9.

If two magnets are arranged so that the north pole of one is near the south pole of another, there will be a force of attraction between them, whereas if similar poles are opposite one another, the magnets will repel one another: see Fig 1.11.

Magnets made from certain kinds of iron, nickel and cobalt alloys and certain ceramics, the hard ferrites (see Chapter 2) retain their magnetism more or less permanently, and find many uses in radio equipment, such as loudspeakers, polarised relays, headphones, cathode-ray tube focusing arrangements and magnetron oscillators.

Other types of iron and nickel alloys and some ceramics (the soft ferrites), eg soft iron, are not capable of retaining magnetism, and therefore cannot be used for making permanent magnets. They are effective in transmitting magnetic force, however, and are used as cores in electromagnets and transformers. These materials concentrate the magnetic field by means of a property called *permeability.* To a first approximation, the permeability is the ratio of the magnetic field with a core of, for example, soft iron, to that without it.

Electromagnets

A current of electricity flowing through a straight wire exhibits a magnetic field, the lines of force of which are in a plane perpendicular to the wire and concentric with the wire. If a piece of cardboard is sprinkled with iron filings, as shown in Fig 1.12, they will arrange themselves in rings round the wire, thus illustrating the magnetic field associated with the flow of current in the wire. Observation of a small compass needle placed near the wire would indicate that for a current flow in the direction illustrated the magnetic force acts clockwise

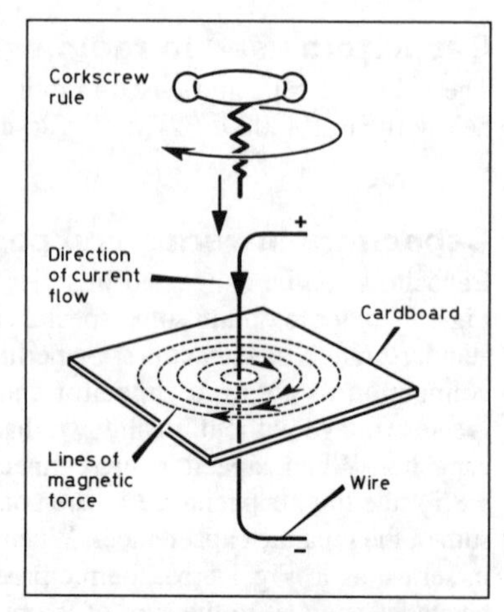

Fig 1.12. Magnetic field produced by current flowing in straight wire

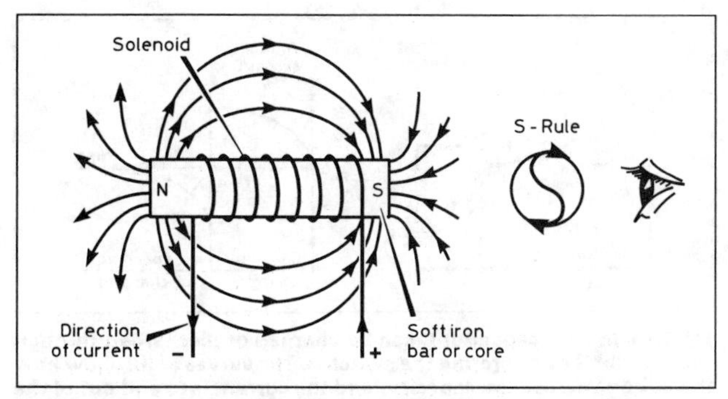

Fig 1.13. The S rule for determining the polarity of an electromagnet

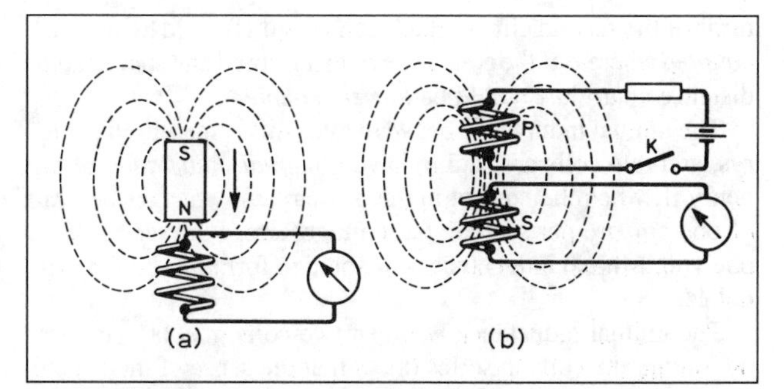

Fig 1.14. Electromagnetic induction. (a) Relative movement of a magnet and a coil causes a voltage to be induced in the coil. (b) When the current in one of a pair of coupled coils changes in value, current is induced in the second coil

round the wire. A reversal of current would reverse the direction of the magnetic field.

The *corkscrew rule* enables the direction of the magnetic field round a wire to be found. Imagine a right-handed corkscrew being driven into the wire so that it progresses in the direction of current flow; the direction of the magnetic field around the wire will then be in the direction of rotation of the corkscrew.

The magnetic field surrounding such a straight wire is relatively weak, but a strong magnetic field can be produced by a current if instead of a straight wire a coil of wire or *solenoid is* used: moreover, the field can be greatly strengthened if a piece of soft iron, called a *core*, is placed inside the coil. The strength of the solenoid magnet is increased by the introduction of the core (the *permeability effect*, cf dielectric constant). Fig 1.13 and the photograph Fig 1.10(b) shows the magnetic field produced by a solenoid, which it will be seen is very similar to that of a bar magnet as shown in Fig 1.9. A north pole is produced at one end of the coil and a south pole at the other. Reversal of the current will reverse the polarity of the electromagnet. The polarity of a solenoid can be deduced from the *S rule*, which states that the pole which faces an observer looking at the end of a solenoid is a south pole if to him the conventional current is flowing (ie from positive to negative) in a clockwise direction; see Fig 1.13.

The strength of a magnetic field produced by a current is directly proportional to the current, a fact made use of in moving coil meters (see Chapter 15 – 'Test equipment'). It also depends on the number of turns of wire, the area of the coil, and the permeability of the core.

Interaction of magnetic fields

In a similar way as the attraction and repulsion of permanent magnets, there can be interaction between the fields of electromagnets or between a permanent magnet and an electromagnet, and in just the same way the interaction is manifest as a force causing relative motion between the two. This interaction forms the basis of several types of electromechanical devices which are used in radio such as loudspeakers, earphones, and moving-coil meters. These will be described in detail in Chapter 2.

ELECTROMAGNETIC INDUCTION

If a bar magnet is plunged into a solenoid coil as indicated in Fig 1.14(a), the moving-coil microammeter connected across the coil will show a deflection. The explanation of this phenomenon, known as *electromagnetic induction*, is that the movement of the magnet's lines of force past the turns of the coil causes a voltage to be induced in the coil which in turn causes a current to flow through the meter. The magnitude of the effect depends on the strength and rate of movement of the magnet and the size of the coil. Withdrawal of the magnet causes a reversal of the current. No current flows unless the lines of force are moving relatively to the coil. The same effect is obtained if a coil of wire is arranged to move relatively to a fixed magnetic field. Dynamos and generators depend for their operation on the principle of electromagnetic induction.

Consider a pair of coils of wire are arranged as shown in Fig 1.14(b). When the switch K is open there is no magnetic field from the coil P linking the turns of the coil S, and the current through S is zero. Closing K will cause a magnetic field to appear due to the current in the coil P. This field, as it builds up from zero, will induce a voltage in S and cause a current to flow through the meter for a short time until the field due to P has reached a steady value, when the current through S falls to zero again. The effect is only momentary and is completed in a small fraction of a second. The change in current in the circuit P is said to have *induced* a voltage in the circuit S. The fact that a changing current in one circuit can induce a voltage in another circuit is the principle underlying the operation of transformers.

Self-inductance

If a steady current is flowing through a coil there will be a steady magnetic field due to that current. A current change will tend to alter the strength of the field which in turn will induce in the coil a voltage tending to oppose the change being made. This process is called *self-induction.* A coil is said to have *self-inductance*, usually abbreviated to *inductance.* It will have a value of one *henry* (H) if, when the current through the coil changes at a rate of one ampere per second, the voltage appearing across its terminals is one volt. Inductance is usually denoted by the symbol L in formulae. As the inductance values used in radio equipment may be only a very small fraction of a henry, the units *millihenry* (mH) and *microhenry* (μH) meaning one thousandth and one millionth of a henry respectively are commonly used.

The inductance of a coil depends on the number of turns, on the area of the coil and the permeability of the core material on which the coil is wound. The inductance of a coil of a certain physical size and number of turns can be calculated to a fair degree of accuracy from formulae or they can be derived from coil charts.

Mutual inductance

A changing current in one circuit can induce a voltage in a second circuit: see Fig 1.14(b). The strength of the voltage induced in the second circuit depends on the closeness or tightness of the magnetic coupling between the circuits; for example, if both coils are wound together on an iron core practically all the lines of force or magnetic flux from the first circuit will link with the turns of the second circuit. Such coils would be said to be *tightly coupled* whereas if the coils were both air-cored and spaced some distance apart they would be *loosely coupled.*

The mutual inductance between two coils is measured in henrys, and two coils are said to have a *mutual inductance* of one henry if, when the current in the primary coil changes at a rate of one ampere per second, the voltage across the secondary is one volt. Mutual inductance is denoted in formulae by the symbol M.

The mutual inductance between two coils may be measured by joining the coils in series (a) so that the sense of their windings is the same and (b) so that they are reversed. The total inductance is then measured in each case.

If L_a and L_b are the total measured inductances, L_1 and L_2 are the separate inductances of the two coils and M is the mutual inductance:

$$L_a = L_1 + L_2 + 2M$$

$$L_b = L_1 + L_2 - 2M$$

therefore: $$L_a - L_b = 4M$$

ie $$M = \frac{L_a - L_b}{4}$$

The mutual inductance is therefore equal to one-quarter of the difference between the series-aiding and series-opposing readings. For a more detailed description, see Chapter 2.

Inductors in series and parallel

Provided that there is no mutual coupling between inductors when they are connected in series, the total inductance obtained is equal to the sum of the separate inductances. When they are in parallel the reciprocal of the total inductance is equal to the sum of the reciprocals of the separate inductances.

If L is the total inductance (no mutual coupling) the relationships are as follows:

Series connection

$$L = L_1 + L_2 + L_3 \text{ etc}$$

Parallel connection

$$\frac{1}{L} = \frac{1}{L_1} + \frac{1}{L_2} + \frac{1}{L_3} \text{ etc}$$

For the special case of two inductors in parallel:

$$L = \frac{L_1 \times L_2}{L_1 + L_2}$$

It may be noted that this formula is of the same type as that relating to two resistors in parallel or capacitors in series. The use of these series and parallel inductance formulae is illustrated by the following example:

Example. Two inductors of 10 and 20μH are connected in series, and two others of 30 and 40μH are also connected in series. What is the equivalent inductance if these series combinations are connected in parallel? Assume that there is no mutual induction.

The 10 and 20μH coils in series are equivalent to (10 + 20) = 30μH. The 30 and 40μH coils in series are equivalent to (30 + 40) = 70μH.

These two equivalent inductances of 30μH and 70μH respectively are in parallel and will therefore be equivalent to one single inductance of:

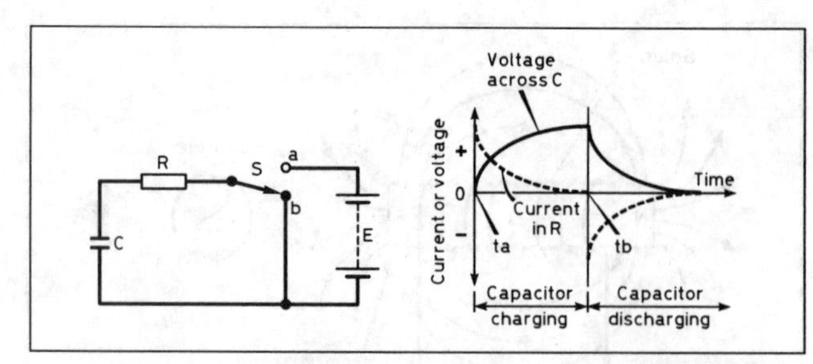

Fig 1.15. In (a) a capacitor C can be charged or discharged through the resistor R by operating the switch S. The curves of (b) show how the voltage across the capacitor and the current into and out of the capacitor vary with time as the capacitor is charged and discharged. The curve for the rise and fall of current in an LR circuit is similar to the voltage curve for the CR circuit

$$\frac{30 \times 70}{30 + 70} = \frac{30 \times 70}{100} = 21\mu\text{H}$$

This is the value of the equivalent inductance of the four coils in this series-parallel arrangement.

TIME CONSTANT

Fig 1.15(a) shows a circuit in which a capacitor C can either be charged from a battery of voltage E, or discharged through a resistor R, according to whether the switch S is in position a or b.

If at some instant, t_a, the switch is thrown from b to a, current will start to flow into the capacitor with an initial value E/R. As the capacitor charges the voltage across the plates increases and the current through the circuit therefore falls away, as shown in the charging portion of Fig 1.15(b); when fully charged to the voltage E the current will have dropped to zero.

When, at some time t_b, the switch is thrown back to b, the capacitor will discharge through the resistor R, the current being in the opposite direction to the charging current, starting at a value $-E/R$ and dying away to zero. As the capacitor discharges the voltage across its plates falls to zero as shown in the discharge portion of Fig 1.15(b). Graphs of this shape are called *exponential* curves.

It is evident that an infinite time is theoretically required to charge and discharge a capacitor through a resistor; the larger R or C, the longer the time taken. This time is specified by the *time-constant* of the CR circuit, being the time taken for a capacitor to discharge to 37%, ie (1/e)th (where e is the base of natural logarithms), of its initial voltage, or to charge to 63%, ie (1 – 1/e)th, of its final voltage. The time constant, in seconds, is given by the product of the capacitance in farads and the resistance in ohms. Thus:

$$t = CR \text{ seconds}$$

As an example, the time constant of a capacitance of 0.01μF (10^{-8}F) and a resistance of 47kΩ (4.7×10^4Ω) is:

$$T = 10^{-8} \times 4.7 \times 10^4$$

$$= 4.7 \times 10^{-4}\text{s}$$

$$= 0.47\text{ms}$$

Examples in which the time constants of CR circuits are important are to be found in radio receiver circuits associated with the second detector and AGC rectifier. The time constant of the signal detector diode load and reservoir capacitor is chosen to be long compared with the period of the IF signal to give good smoothing or filtering, but not so long that the voltage across the load resistor cannot follow the highest-speed audio-amplitude variations likely to be encountered. On the other hand, the time constant of the AGC circuit is chosen to be long compared with the period of the lowest audio-frequency amplitude variations to avoid suppression, due to negative feedback, of the modulation in the AGC-controlled stages, but at the same time short enough to follow the fastest fading likely to be encountered.

Time constants are very important in pulse circuits which are expected to handle fast transient waveforms.

The rise and fall of current in an inductance in circuit with a resistance also takes a definite time depending on the values of L and R in question. The larger the value of L and the smaller the value of R, the longer it takes for the current to rise or decay. The shape of the current curves for an LR circuit are similar in shape to the voltage curves for the CR circuit shown in Fig 1.15(b).

The time constant of an LR circuit is defined as the time taken for the current to reach 63%, $(1 - 1/e)$th, of its final value, or to decay to 37% of its initial value. The time constant in seconds is given by the inductance in henrys divided by the resistance in ohms. Thus:

$$t = \frac{L}{R} \text{ seconds}$$

ALTERNATING CURRENT

Previous sections have been concerned mainly with current flowing in one direction through a circuit, such a current being produced, for example, by a battery, ie *direct current* (usually abbreviated to *DC*).

There is another important type of current flow in which current flows backwards and forwards alternately through a circuit. Such currents are called *alternating currents (AC)*.

An alternating current could be obtained by connecting a resistor to a battery through a reversing switch as shown in Fig 1.16(a). On operating the switch S backwards and forwards from a to b, the current from the battery would flow in alternate directions through the resistor R; the graph in Fig 1.16(b) shows how the current would vary with the passage of time and the reversals of the switch. Such a current waveform, in which the current flows at a steady value for equal times in both directions, is called a *square wave.*

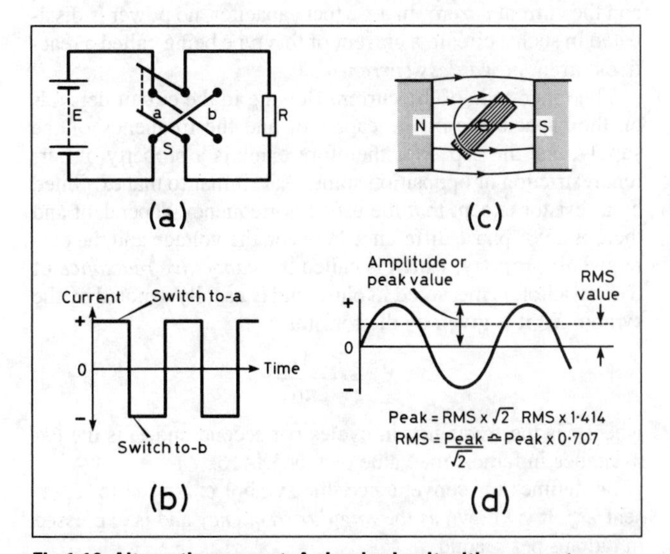

Fig 1.16. Alternating current. A simple circuit with a current-reversing switch shown at (a) produces a square-wave current through the resistor R as shown in (b). When a coil is rotated in a magnetic field as in (c) the voltage induced in the coil has a sinusoidal waveform (d)

The most common alternating current is that used for the electricity supply mains, in which in the UK and most of Europe the direction of current flow is continuously and smoothly changing from positive to negative and back again at a rate of 50Hz. In the USA and Canada the rate, or *frequency*, is 60Hz. In each cycle the direction changes twice, but for the purpose of most calculations it is the complete cycle which is more significant than the two half-cycles in opposite directions.

The alternating voltage of the electricity supply does not have a square waveform but fluctuates with a gradual and smooth variation like the swinging of a pendulum. Such a current would be produced by the continuous and steady rotation of a coil in a magnetic field as indicated in Fig 1.16(c), when the circuit is completed by some device such as a resistor. This is the basic principle used in the *alternator,* a machine for generating alternating current. The waveform of this current has a shape which is defined mathematically as a *sine wave,* shown in Fig 1.16(d), and the current is said to vary *sinusoidally*, ie it follows the same pattern as the trigonometrical sine of an angle plotted against the angle.

Alternating currents of much higher frequency are used in radio communication, the frequencies commonly used in shortwave amateur work ranging between 1,800,000 and 30,000,000Hz (18–30MHz). In the VHF/UHF bands frequencies used are up to 432,000,000Hz (432MHz) and much higher for microwaves.

The sine wave is fundamental because all other regular waveforms can be generated from a series of sine waves added together. For example, a square wave, in which the voltage changes abruptly, is the sum of the fundamental frequency as a sine wave plus one-third of the amplitude of the third harmonic (three times the fundamental) plus one-fifth of the fifth harmonic plus one-seventh of the seventh harmonic etc. The more terms used, the nearer it is to a true square wave. Going up to the ninth harmonic gives a square wave which is good enough for most purposes [2]. Note that in this case only odd harmonics are used.

Characteristics of alternating currents

From the preceding paragraph it can be seen that one feature which distinguishes one alternating current from another is the rate at which the complete cycles of current-reversal take place; this is called the *frequency* of the alternating current and is measured in *hertz* (cycles per second).

A second distinguishing feature of an alternating current is its magnitude or *amplitude,* by which is meant the maximum value reached during one cycle or alternation: see Fig 1.16(d).

It is possible to have two alternating currents whose frequency and amplitude are exactly equal but with a time lag or phase difference between the two so that they are not performing the same part of their cycles at the same instant. This is illustrated in Fig 1.17 where the solid and dotted lines are of the same amplitude and frequency but differ in phase by one quarter of the time taken for one cycle as indicated by the time x in the diagram [3].

Thus an alternating current or any other alternating quantity has *three* properties: frequency, amplitude and phase.

Another term used with reference to alternating current is the *period* of a wave, which is the time taken to perform one cycle. The period of a wave is the reciprocal of the frequency, thus:

$$\text{Frequency} = \frac{1}{\text{Period}}$$

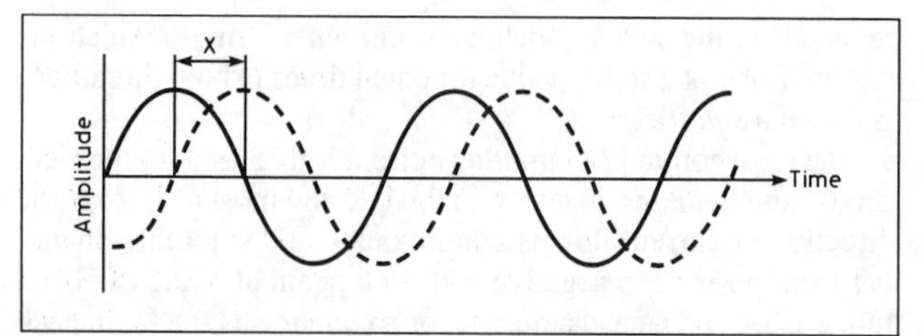

Fig 1.17. Phase difference between two alternating currents. Provided that they both have the same frequency the phase difference will remain constant

$$\text{Period} = \frac{1}{\text{Frequency}}$$

Frequency is usually denoted by the symbol *f* and period by the symbol *t*. As an example, the period of the 50Hz mains is given by:

$$t = \frac{1}{f} = \frac{1}{50} \text{ seconds}$$

The value of an alternating voltage or current can be specified by its amplitude. This is often called the *peak* voltage or *peak* current, being the highest value reached during a cycle. Thus the peak value of any alternating voltage or current is equal to its amplitude (in volts or amperes) as shown in Fig 1.16(d).

The peak value, however, is not the most common way of specifying an alternating voltage or current. The usual value adopted is the *root-mean-square* or *RMS* value which is equal to the direct voltage or current which would produce the same amount of power in a resistive load as the alternating voltage or current would produce. The RMS value is less than the peak value, the two being related if the waveform is sinusoidal as follows:

$$\text{RMS value} = \frac{\text{Peak value}}{\sqrt{2}} \quad 0.707 \times \text{Peak value}$$

or conversely:

$$\text{Peak value} = \text{RMS value} \times \sqrt{2} = 1.414 \times \text{RMS value}$$

An AC mains supply of 240V RMS thus has a peak voltage of 240 × 1.414 or approximately 340V.

AC circuit containing resistance only

Ohm's Law can be used to find the current flowing through a resistor when a certain alternating voltage is applied across its terminals. The current will flow backwards and forwards through the resistor under the influence of the applied voltage and will be in phase with it. The voltage and current waveforms for the resistive circuit of Fig 1.18(a) are shown in Fig 1.18(b).

The power dissipated in the resistor can be calculated direct from the small power formulae provided that RMS values for voltage and current are used, viz:

$$W = EI \qquad W = I^2 R \qquad W = \frac{E^2}{R}$$

If peak values *E* for voltage and *I* for current are used these formulae become:

$$W = \frac{EI}{2} \qquad W = \frac{I^2 R}{2} \qquad W = \frac{E^2}{2R}$$

AC circuit containing capacitance only: reactance of capacitor

When an alternating voltage is applied to a capacitor as shown in Fig 1.18(c) the capacitor will be alternately charged in one

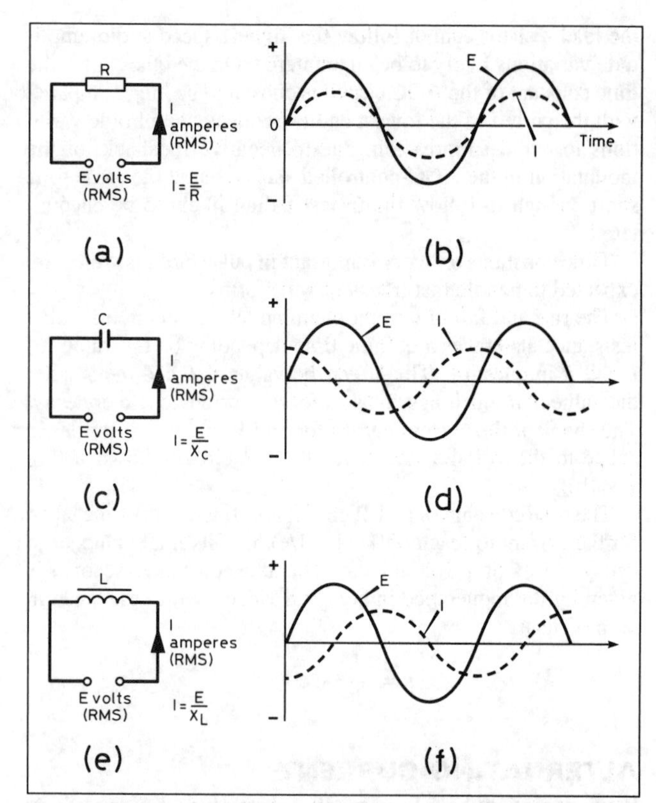

Fig 1.18. Voltage and current relationships in AC circuits comprising (a) resistance only, (c) capacitance only, (e) inductance only

direction and then charged in the opposite direction as the supply alternates. At each reversal it becomes momentarily discharged. An observer watching the flow of current in the wires connecting the capacitor to the AC supply would see an alternating current flowing backwards and forwards through the wires. This current flows into and out of the capacitor and not actually through it. A careful study would show that the current is one-quarter period (or 90°) out-of-phase with the voltage, and in advance of it, as shown in Fig 1.18(d)). This must be so, for when the capacitor is fully charged the voltage is a maximum and the current is zero. In a perfect capacitor, no power is dissipated in such a circuit, a current of this type being called a reactive current or *wattless current.*

The magnitude of the current flowing in the circuit depends on the capacitance of the capacitor and the frequency of the supply, and the capacitor therefore exhibits a property of current restriction or opposition somewhat similar to that exhibited by a resistor except that the effect is frequency-dependent and there is a 90° phase difference between the voltage and the current. This property, which is called the *capacitive reactance* of the capacitor, is measured in ohms and is usually denoted by the symbol X_c; it is given by the formula

$$X_c = \frac{1}{2\pi fC}$$

where *f* is the frequency in cycles per second and *C* is the capacitance in farads: the value of π is 3.1416.

Sometimes for convenience, the symbol ω is used to represent $2\pi f$. It is known as the *angular frequency* and is expressed in radians per second.

The following example shows how this formula is applied.

Example. Calculate the reactance offered, at frequencies of 50Hz and 50kHz respectively, of a capacitance of 2μF.

In the formula for capacitive reactance:

$$X_c = \frac{1}{2\pi fC}$$

Before any calculation is made it must be remembered to express f and C in the correct units.

Considering first the frequency of 50Hz:

$$X_c = \frac{1}{2\pi \times 50 \times 2 \times 10^{-6}}$$

$$= 10{,}000/2\pi = 1592\Omega$$

Considering next the frequency of 50kHz (ie 50,000Hz):

$$X_c = \frac{1}{2\pi \times 5 \times 10^4 \times 2 \times 10^{-6}}$$

$$= 10/2\pi = 1.592\Omega$$

It is worth noting that the reactance is inversely proportional to the frequency, so that it would be possible to derive the reactance at 50kHz directly from the reactance at 50Hz simply by dividing by 1000.

The current flowing through a capacitor can be calculated by using Ohm's Law, regarding the reactance of the capacitor X_c as replacing the more usual resistance R, viz:

$$I = E/X_c \quad (\text{cf } I = E/R)$$

AC circuit containing inductance only: reactance of coil

The opposition of an inductance to alternating current flow is called the *inductive reactance* of the coil: see Fig 1.18(e). It is proportional to the inductance of the coil and also to the frequency and is denoted by the symbol X_L. It may be calculated from the formula:

$$X_L = 2\pi fL$$

As in the case of a capacitor, the current in an inductance and the voltage across it are one-quarter period (or 90°) out-of-phase and no power is dissipated in a purely inductive circuit. In an inductive circuit, however, the phase of the current is 90° behind the voltage, ie exactly opposite to the case of the capacitive circuit: see Fig 1.18(f).

Example. Calculate the inductive reactance of a coil of 2H inductance at frequencies of 50Hz and 50kHz respectively.

The formula for inductive reactance is:

$$X_L = 2\pi fL$$

Since f is 50Hz and L is 2H:

$$X_L = 2\pi \times 50 \times 2\Omega$$

$$= 200 \times 3.14\Omega$$

$$= 628\Omega$$

Because inductive reactance is directly proportional to frequency, the reactance of the 2H coil at 50kHz will be one thousand times its reactance at 50Hz, namely 628,000Ω.

The current through an inductance and the voltage across it are connected by the modified Ohm's Law formula:

$$I = E/X_L \quad (\text{cf } I = E/R)$$

Note that pure capacitance and pure inductance do not exist. In practice they are both 'contaminated' by resistance which is that of the leads in the case of a capacitor or of the winding in the case of an inductor.

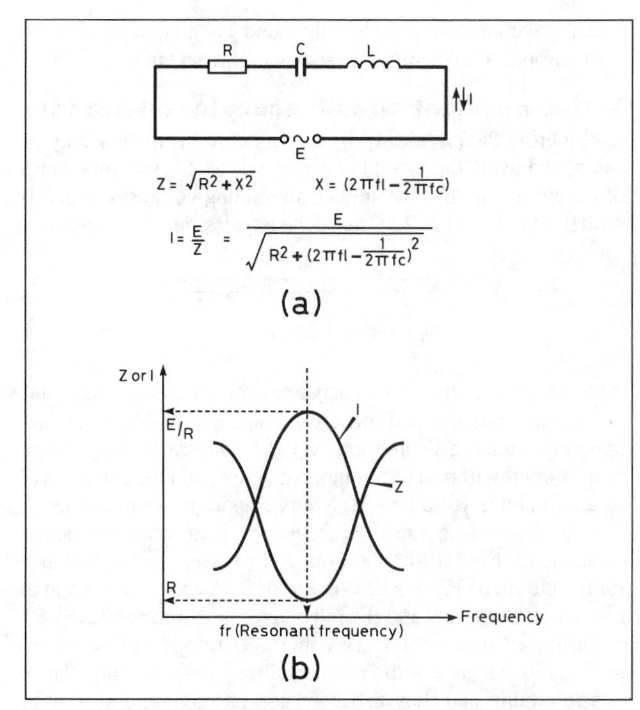

Fig 1.19. The series-resonant circuit. The curves shown at (b) indicate how the impedance and the current vary with frequency in the type of circuit shown at (a)

Susceptance

A term sometimes used in calculations is the *susceptance* of a coil or a capacitor. Susceptance is simply the reciprocal of reactance in the same way that conductance is the reciprocal of resistance. Susceptance is in general denoted by the symbol B and is measured in *siemens* (abbreviated to 'S'). Thus

$$B_L = 1/X_L \qquad B_C = 1/X_C$$

AC circuit containing L, C and R: impedance

An AC circuit may contain resistance, inductance and capacitance in series, as shown in Fig 1.19(a), each of which opposes the flow of current in a different way as mentioned in the previous sections. The opposition of the composite circuit to alternating current flow is called the *impedance* of the circuit and is denoted by the symbol Z.

The impedance of the whole circuit cannot be found by simply adding the reactances and resistances together, since the relative current and voltage phases associated with each component are not the same. The inductive and capacitive reactance have exactly opposite phase effects and consequently they must first be subtracted from one another to find the total reactance in the circuit, thus:

$$X = X_L - X_C$$

The impedance is then found by compounding the resistance with the total reactance of the circuit. This has to be done by taking the square root of the sum of the squares of these two quantities; thus:

$$Z = \sqrt{(R^2 + X^2)}$$

The current flowing through the circuit may then be found using Ohm's Law but with Z replacing R as follows:

$$I = E/Z$$

If the current in the circuit is known, the voltage across each individual component can be calculated using Ohm's Law with the appropriate value of the resistance or reactance.

Series-resonant circuit: acceptor circuit [4]

Considering the circuit of Fig 1.19(a), current flowing may be calculated using the formula $I = E/Z$, where Z is the impedance of the circuit and E *is* the applied alternating voltage. When Z is written in terms of R, L, C and f the expression of I becomes:

$$I = \frac{E}{\sqrt{R^2 + \left(2\pi fL - \frac{1}{2\pi fC}\right)^2}}$$

At one particular frequency, depending on the exact values of C and L, the capacitive and inductive reactances will be equal and opposite. Under this condition, called *resonance*, the impedance of the circuit will be a minimum, equal to R, and the current will be a maximum. Resonance is a very important phenomenon as it is used to select a desired frequency from other unwanted frequencies. Fig 1.19(b) shows how the current and impedance for the circuit of Fig 1.19(a) vary with frequency, the current at resonance being E/R and the impedance at resonance being R.

Series-resonant circuits are sometimes referred to as *acceptor circuits* because at the resonance frequency the impedance is a minimum and they therefore accept maximum current at this frequency.

The frequency at which a certain coil and capacitor resonate when connected together can be found by equating the inductive and capacitive reactances, thus:

$$2\pi fL = \frac{1}{2\pi fC}$$

Solving for f:

$$f = \frac{1}{2\pi\sqrt{LC}}$$

If L is in henrys and C is in farads, f will be in hertz.

This formula, next to Ohm's Law, is probably the most used of all in radio work as it permits the calculation of the inductance of a coil which will tune to a desired frequency with a given capacitance. Its use, together with other formulae relating to the resonant circuit, is illustrated by the following worked examples:

Example 1. What value of inductance is required in series with a capacitor of 500pF for the circuit to resonate at a frequency of 400kHz? (Assume no resistance.)

From the resonance formula:

$$f = \frac{1}{2\pi\sqrt{LC}}$$

the inductance is:

$$L = \frac{1}{4\pi^2 f^2 C}$$

Expressing the frequency and the capacitance in the proper units ($f = 4.0 \times 10^5$Hz and $C = 5.0 \times 10^{-10}$pF):

$$L = \frac{1}{4\pi^2 \times (4 \times 10^5)^2 \times (5 \times 10^{-10})} \text{ henrys}$$

Taking $\pi^2 = 10$, this becomes:

$$L = \frac{1}{3200} \text{ H}$$

$$= 312.5\mu\text{H}$$

Example 2. If an inductance of 50μH is in series with a capacitance of 500pF, what is the resonant frequency? (π^2 may be taken as 10.)

At resonance:

$$f = \frac{1}{2\pi\sqrt{LC}}$$

or

$$f^2 = \frac{1}{4\pi^2 LC}$$

Expressing the inductance and the capacitance in the proper units ($L = 5.0 \times 10^{-5}$H and $C = 5.0 \times 10^{-10}$F):

$$f^2 = \frac{1}{4 \times 10 \times (5.0 \times 10^{-5}) \times (5.0 \times 10^{-10})}$$

$$= 10^{12}$$

Therefore $f = 10^6\text{Hz} = 1\text{MHz}$.

Example 3. If the effective series inductance and capacitance of a vertical antenna are 20μH and 100pF respectively and the antenna is connected to a coil of 80μH inductance, what is the approximate resonant frequency?

In this example the antenna and coil together will resonate at a frequency determined by the capacitance and the sum of the antenna effective inductance and the loading coil inductance. At resonance:

$$f = \frac{1}{2\pi\sqrt{LC}}$$

Here the relevant values of inductance and capacitance expressed in the proper units are:

$$L = (2.0 + 8.0) \times 10^{-5}\text{H}$$

$$C = 1.0 \times 10^{-10}\text{F}$$

Therefore

$$f = \frac{1}{2\pi \times \sqrt{(1.0 \times 10^{-4}) \times (1.0 \times 10^{-10})}}$$

$$= \frac{1}{2\pi \times 10^{-7}} \text{ Hz}$$

$$= 1.6\text{MHz approximately.}$$

Magnification factor *Q* [5]

At resonance the voltage across the coil (or the capacitor) in the circuit of Fig 1.19(a) can be considerably greater than that supplied to the circuit by the generator. The current at resonance is determined by the value of the resistor R whereas the voltage across the coil (or the capacitor) is given by the product of the current and the appropriate reactance which may be many times the value of R. The ratio of the voltage across the coil (or the capacitor) to that across the resistor is called the *magnification factor* or Q of the circuit. If I is the current at resonance:

$$Q = \frac{IX_L}{IR} = \frac{2\pi fL}{R} = \frac{\omega L}{R}$$

or

$$Q = \frac{IX_C}{IR} = \frac{1}{2\pi fCR} = \frac{1}{\omega CR}$$

where $\omega = 2\pi f$.

The Q of a tuned circuit is determined mainly by the coil since good-quality capacitors have negligible losses. The Q of a

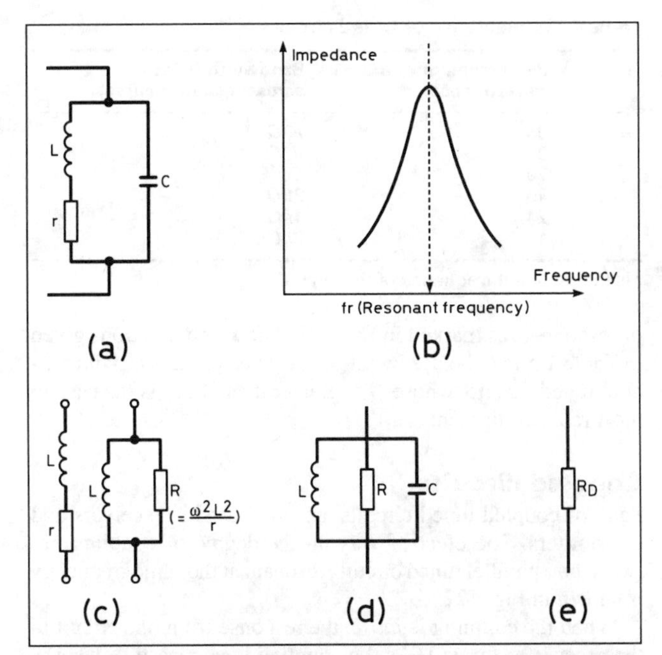

Fig 1.20. The parallel-resonant circuit. The curve shown at (b) indicates how the impedance varies with frequency in the type of circuit shown at (a). A series-to-parallel transformation to replace the series resistance by an equivalent parallel resistance can be made as shown at (c). The circuit shown at (a) can therefore be replaced by that shown at (d) in which a perfect coil and capacitor are shunted by a parallel resistance. At resonance the impedance of this circuit reduces to a fictitious resistance as shown at (e)

tuned circuit determines its *selectivity,* ie its ability to pick out a wanted signal from a number of unwanted signals on adjacent frequencies. A high Q-value corresponds to good selectivity.

Parallel-resonant circuit: rejector circuit

If a coil and capacitor are connected in parallel, as in Fig 1.20(a), resonance effects appear when the current is of such a frequency that the reactance of the coil and capacitor are equal in magnitude and opposite in sign. The resonant frequency of a parallel-tuned circuit is, for practical purposes, given by the same formula as for the series circuit, viz:

$$f = \frac{1}{2\pi\sqrt{LC}}$$

At resonance the impedance of a parallel-tuned circuit is resistive and is a maximum as shown in Fig 1.20(b). If the coil and capacitor were perfect reactive components, the impedance at resonance would be infinite. In practice the capacitor can be looked upon as a nearly perfect reactance, but there is always appreciable resistance associated with the coil which limits the resonant impedance of the circuit. The value of the impedance of a parallel-tuned circuit at resonance is called the *dynamic resistance* of the circuit. This is a fictitious resistance and appears to exist only for alternating currents of the resonant frequency; the DC resistance of the circuit is, of course, relatively very low.

Parallel-tuned circuits are sometimes called *rejector circuits* because at resonance they have a high impedance and therefore reject current at their resonant frequency.

Dynamic resistance

There is a series-to-parallel transformation which can be made between the two circuits of Fig 1.20(c) in which the series resistance r can be replaced by an equivalent shunt resistance R. Providing the value of r is small compared with the reactance of the coil, as is usually the case in radio circuits, the value of R is given by the formula:

$$R = \frac{X_L^2}{r} = \frac{\omega^2 L^2}{r}$$

where $\omega = 2\pi f$.

It follows that the parallel-tuned circuit of Fig 1.20(a) can be replaced by that shown in Fig 1.20(d) in which a perfect coil and capacitor are shunted by a resistor R. At resonance the capacitive and inductive reactances cancel out and the impedance of the circuit reduces to the fictitious resistance R_D shown in Fig 1.20(e): this is the *dynamic resistance* of the circuit.

Since at resonance $\omega L = 1/\omega C$, the value of the dynamic resistance of a parallel-tuned circuit is given by:

$$R_D = \frac{L}{Cr}$$

Further, since $Q = \omega L/r$, the dynamic resistance can also be expressed as:

$$R_D = \frac{Q}{\omega C}$$

From this it can be seen that for a high dynamic resistance (which is necessary for high gain in an RF amplifier when a parallel-resonant circuit is used as an anode load), the ratio of L to C should be high and r should be small (ie the Q should be high). The gain of an FET radio-frequency or intermediate-frequency amplifier is roughly equal to the dynamic resistance of the tuned-circuit load multiplied by the mutual conductance of the FET: see Chapter 3 – 'Semiconductors'.

L/C ratio

Examination of the formula for the resonant frequency of a tuned circuit:

$$f = \frac{1}{2\pi\sqrt{LC}}$$

shows that the resonant frequency is determined by the product of the inductance and capacitance, LC. It follows that there is a wide range of L and C values which can be chosen to resonate at a particular frequency. Put another way, for a particular resonant frequency there is flexibility in choice of the L/C ratio, ie a large L and small C (or alternatively a small L and a large C) may be used, provided that the LC product gives the required resonant frequency.

The choice of L/C ratio is determined by practical considerations connected with the particular application of the tuned circuit.

In HF receivers it is usual to employ circuits with a high L/C ratio as this leads to circuits with a high dynamic resistance and therefore high stage gain, the minimum possible value of C being determined by the stray circuit capacitance. On the other hand, a low L/C ratio, ie a high C value, is usually employed in variable frequency oscillator circuits associated with transmitters in order to swamp out the effect of the active device and other stray circuit capacitances which tend to make the frequency of the oscillator unstable.

In the case of HF transmitter output tank circuits the L/C ratio is specially chosen so that when damped by the load (or antenna) the effective Q of the tank circuit has a value which gives a good compromise between efficiency and harmonic suppression. This matter is dealt with in Chapter 7.

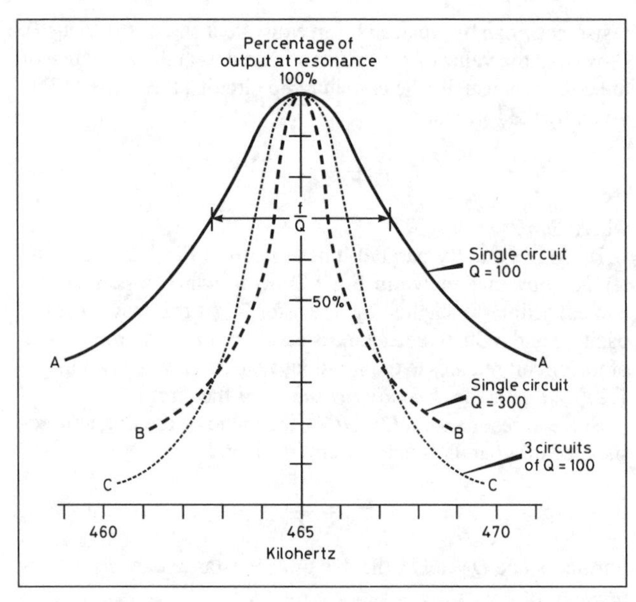

Fig 1.21. Selectivity curves of single and cascaded tuned circuits

Resonance curves and selectivity [4]

The curves of Figs 1.19(b) and 1.21(b) show how tuned circuits are more responsive at their resonant frequency than at neighbouring frequencies. Curves of this nature are called *resonance curves* and show how a tuned circuit can be used to pick out, or select, a wanted signal from a number of unwanted signals on other frequencies. Fig 1.21 shows a resonance curve A for a single parallel-tuned circuit having a resonant frequency of 465kHz and a Q of 100. When such a circuit is connected as the anode load in a simple single-stage intermediate-frequency amplifier circuit, the curve shows how the gain varies with frequency. For instance, it will be seen that an unwanted signal at 461kHz (4kHz off tune) would give only about 50% of the output of the wanted signal (of equal input amplitude) at 465kHz. The dashed curve B is for a similar circuit but with a Q of 300 and shows clearly how a higher Q gives greater selectivity.

There is a practical upper limit to the value of Q which can be obtained with a coil of reasonable size and, since much greater selectivity is required in a receiver than can be obtained with a single-tuned circuit, several amplifying stages are commonly used in cascade, each with a resonant circuit as its anode load. The dotted curve C in Fig 1.21 shows the response of three circuits in cascade, each with a Q of 100. The response at resonance has been rated at 100% in each case, regardless of the additional gain due to the extra stages. It will be seen that the curve is wider at the peak or nose than that of a single circuit with a Q of 300 (a desirable feature for telephony reception) and gives more selectivity down the skirts than the single circuit.

Another method of obtaining better selectivity than can be provided by a single tuned circuit is to apply *reaction,* or *positive feedback,* a method commonly used in straight receivers. The application of reaction (controlled positive feedback, see below) artificially increases the Q of the circuit, a value of several thousand being readily obtainable, with a consequent increase in selectivity and sensitivity.

Table 1.5 is useful for deriving approximate selectivity curves from the Q of the circuit and the operating frequency. Thus, as an example, a circuit having a Q of 100 and a resonant frequency of 465kHz would have a *bandwidth* (ie width across the resonance curve) of $f/Q = 465/100 = 4.65$kHz between points where the output had fallen to 70% of the resonant value on either side

Table 1.5. Selectivity of tuned circuits

Percentage of output at resonance	Bandwidth (width across response curve)
95	$f/3Q$
90	$f/2Q$
70	f/Q
45	$2f/Q$
24	$4f/Q$
12	$8f/Q$

f is the resonant frequency of the circuit.

of resonance, as marked in Fig 1.21. The approximations given in Table 1.5 may be used with series-tuned circuits or with parallel-tuned circuits whose Q is greater than 10, as is the case in most radio equipment.

Coupled circuits

Pairs of coupled tuned circuits are often used in receivers and transmitters. The efect of varying the degree of coupling between two parallel tuned circuits resonant at the same frequency is shown in Fig 1.22.

When the coupling is *loose*, the response from one circuit to the other is as curve I. As the coupling is increased to what is known as *critical coupling*, the output at resonance increases to curve II; here the mutual coupling between the coils is $1/Q$ of the inductance of either coil. Further increase (*tight coupling*) results in the formation of the double-humped characteristic shown in curve III, where the output at resonance is decreased.

Two tuned circuits are often mounted in a screening can, the coils generally being wound the necessary distance apart on the same former to give the required coupling. The coupling is then said to be *fixed.*

Some various arrangements for coupling tuned circuits are shown in Fig 1.23. The practical design of coupled tuned circuits is dealt with in Chapter 5 – 'Building blocks'.

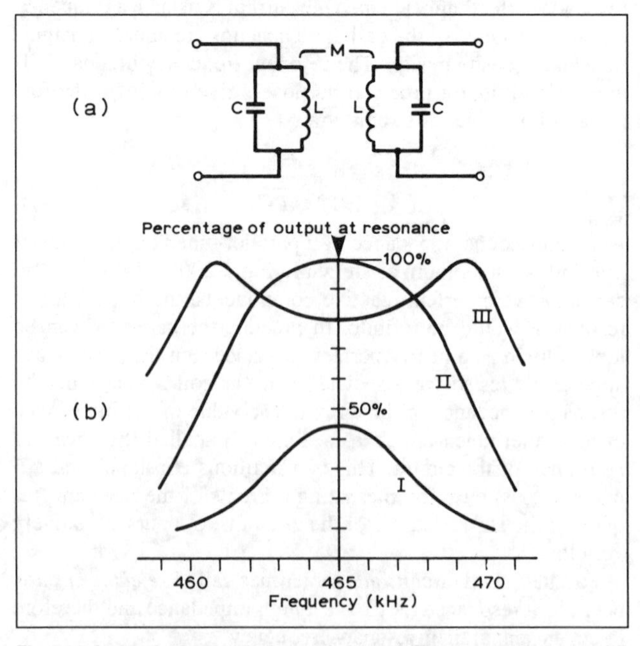

Fig 1.22. Inductively coupled tuned circuits. The curves shown at (b) represent the various frequency response characteristics of the coupled circuit shown at (a) for different degrees of coupling. The degree of coupling between two tuned circuits determines the overall selectivity. Close coupling gives a broad frequency response while loose coupling gives a narrow one

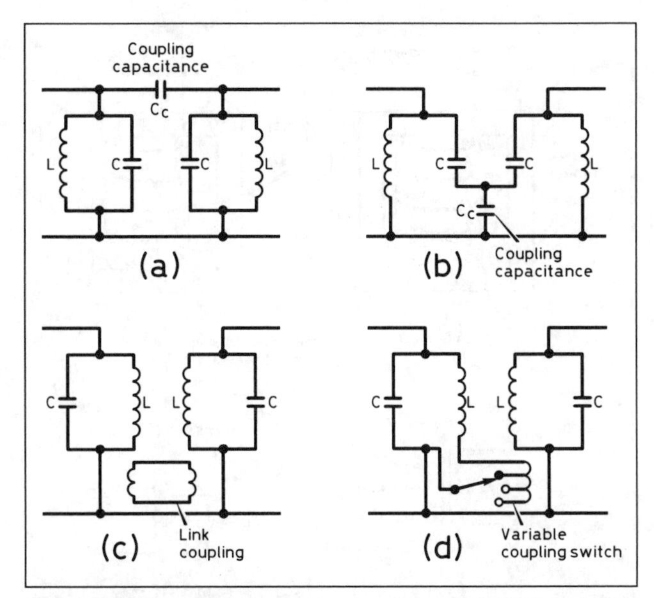

Fig 1.23. Various arrangements for coupling tuned circuits. (a) Top-capacitance coupling. (b) Bottom-capacitance coupling. (c) Link coupling. (d) Variable inductive coupling for variable-bandwidth IF amplifier transformers

FILTERS

Wave filters, usually referred to simply as *filters*, are of two types: (a) *passive* which are networks of reactive components, ie capacitors and inductors and (b) *active* which contain a gain-providing (amplifying) stage and usually just resistors and capacitors. Both exhibit certain characteristics as the frequency of a signal applied to them is varied. They are classed according to their frequency response and there are three types which are commonly used in amateur radio work, as follows:

1. *Low-pass filters* which have the characteristic of passing all frequencies below a specified frequency, called the *cut-off frequency*, f_c, and attenuating all frequencies above f_c.
2. *High-pass filters* which have the characteristic of passing all frequencies above a specified cut-off frequency, f_c, and attenuating all frequencies below f_c.
3. *Band-pass filters* which pass all frequencies between two specified cut-off frequencies f_1 and f_2, and attenuate all frequencies outside these two limits.

Filters are described in terms of their cut-off frequency(s), their pass-band ripple (ie how nearly the pass-band response is to being flat) and their stop-band attenuation. A further useful number is the steepness of the transition between pass and stop-bands.

The various types of filters are described in Chapter 5 – 'Building blocks'.

TRANSFORMERS

The fact that a changing current in one circuit can induce a current in a second circuit, as in Fig 1.14(b), is the basis of transformer action.

Transformers are useful for transferring electrical energy from one circuit to another without direct connection, for example from the collector circuit of one transistor to the base circuit of another transistor. In the transfer process it is possible to change the relative voltages and impedances of the primary and secondary circuits, examples being the supply of a low voltage to operate semiconductor devices from the high-voltage supply

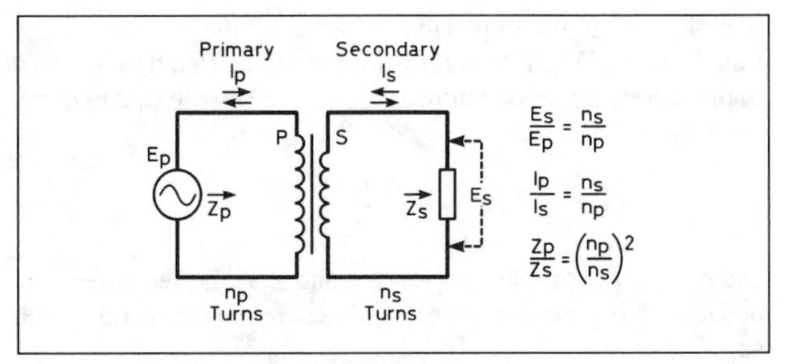

Fig 1.24. The low-frequency transformer

mains and the impedance matching of a low-resistance loudspeaker to a transistor collector circuit.

The coupled circuits referred to in the discussion on Fig 1.22 are examples of transformers with very loose coupling between the primary and secondary windings. In audio-frequency and power supply transformers, tight coupling is required, and the primary and secondary windings are therefore wound on a core of iron laminations with a construction similar to that of the low-frequency choke. Ferrite cores are also used for some applications such as for higher frequencies (10–100kHz)

The size of core used in the transformer depends on the amount of power to be handled.

Fig 1.24 shows a simple transformer with a primary winding P and a secondary winding S. Since both windings are in the same alternating magnetic field, the induced voltages will be in proportion to the number of turns on each coil. Thus:

$$E_s = \frac{E_p}{n_s/n_p}$$

where n_p is the number of turns on the primary; n_s is the number of turns on the secondary; E_p is the voltage across the primary; and E_s is the voltage across the secondary.

The ratio n_s/n_p is called the *turns ratio* and a transformer may step a voltage up or down according to whether n_s/n_p is greater or less than unity.

As an example, consider a transformer which supplies valve heaters having 1200 turns in its primary and 32 turns in its secondary. When connected to a 240V mains supply the secondary voltage will be:

$$E_s = 240 \times \frac{32}{1200} = 6.4\text{V}$$

As long as there is no load connected to a transformer the primary current should be very small. This current is called the *magnetising current* and can be neglected in most calculations when compared with the primary current due to the secondary load. Since the intensity of the magnetic field set up by the primary and secondary current is the same, and since the field intensity is proportional to the number of ampere-turns of each winding, it follows that the primary current will be equal to the secondary current multiplied by the turns ratio. Thus:

$$I_p = I_s \frac{n_s}{n_p}$$

where I_p is the current in the primary and I_s is the current in the secondary.

Taking the previous example again, if the heater current is 6A the primary current would be:

$$I_p = \frac{6 \times 32}{1200} \text{ amps} = 160\text{mA}$$

A transformer has the property of being able to transform

impedances. If the impedance offered by the primary of the transformer of Fig 1.24 is measured it will be found to be equal to the secondary or load impedance divided by the square of the turns ratio. Thus:

$$Z_p = Z_s \left(\frac{n_p}{n_s}\right)^2$$

where Z_p is the primary impedance and Z_s is the secondary impedance. It is probably more convenient to remember this result as:

$$\frac{Z_p}{Z_s} = \left(\frac{n_p}{n_s}\right)^2$$

which means that *the impedance ratio is equal to the square of the turns ratio.*

The transformation of impedance is a valuable property and transformers are widely used for matching a load to a source of power, the turns ratio being calculated to give maximum transfer of power to the load.

For example, the load into which an audio frequency amplifier transistor delivers its maximum undistorted output may be 64Ω. However, the loudspeaker which it is desired to use with the transistor may have an average impedance of only 4Ω and a matching transformer will therefore be required. The transformer turns ratio can be calculated from the formula just mentioned by taking the square root of the impedance ratio, thus:

$$n_p/n_s = \sqrt{(64/4)} = 4/1 = 4$$

This means that a transformer with four times as many turns on the primary as on the secondary will give the required impedance match.

Transformers are often referred to by their primary-to-secondary winding turns ratio. Thus a 4-to-1 (or 4:1) transformer would have four turns on the primary for each turn on the secondary, and would be a step-down transformer suitable for matching a primary-to-secondary impedance ratio of 4^2:1 or 16:1. A transformer with a 1:2 ratio would have two turns on the secondary for each turn on the primary, and would be a step-up transformer. It would match a primary-to-secondary impedance ratio of 1:4.

Auto-transformers

It is not always necessary to have isolation between the primary and secondary circuits of a transformer and in this case an *auto-transformer* can be used.

An auto-transformer can be looked upon as an ordinary transformer in which part of the primary or secondary winding is common to the other. Fig 1.25(a) shows the magnetic core of an ordinary step-up transformer with a two-turn primary, shown dotted, and a four-turn secondary. The primary is shown lying alongside the corresponding turns of the secondary; such a transformer would have a voltage step-up of 2:1 and its circuit symbol is as shown in Fig 1.25(b).

In Fig 1.25(c) the two-turn primary has been merged (or made common) with the secondary to form an auto-transformer. The two turns between the common point A and the tap B form the primary and the whole four turns from A to C form the secondary. The circuit symbol for such an auto-transformer is shown in Fig 1.25(d).

The voltage ratio of an auto-transformer (Fig 1.25(d)) is calculated from the ratio of the number of turns n_p between the common point A and the tap B to the total number of turns between A and C, thus:

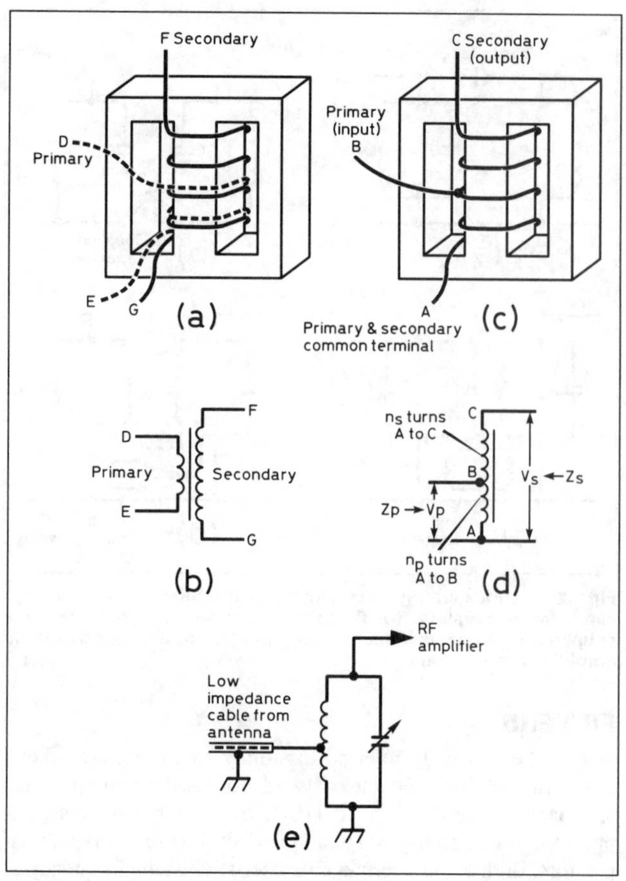

Fig 1.25. Derivation of the auto-transformer is shown at (a), (b) and (c). The circuit symbols is shown at (d) and a typical use at radio frequency is shown at (e)

$$\frac{V_p}{V_s} = \frac{n_p}{n_s}$$

As with ordinary transformers, auto-transformers can be used to provide impedance transformation or matching, the impedance ratio being equal to the square of the turns ratio. Thus in Fig 1.25(d):

$$\frac{Z_p}{Z_s} = \left(\frac{n_p}{n_s}\right)^2$$

Auto-transformers can be used to step up or down depending on whether the tap is the primary or secondary connection.

For AC power work, auto-transformers have the advantage that the windings and their insulation take up less space than in an ordinary transformer. They are commonly used to step-down or boost an AC supply voltage to some other value. For example a step-down auto-transformer, with a suitable tap, could be used to run 110V AC equipment from 240V AC mains.

The tapped tuned circuit is an example of a tuned auto-transformer which is often used at radio frequency. A typical application is shown in Fig 1.25(e) in which a tapped tuned circuit is used to match and provide a voltage step-up between a low-impedance antenna feeder and the input of a RF amplifier stage.

SCREENING

When two circuits are near one another, unwanted coupling may exist between them due to stray capacitance between them or due to stray magnetic coupling.

Stray capacitance coupling can be eliminated by placing an

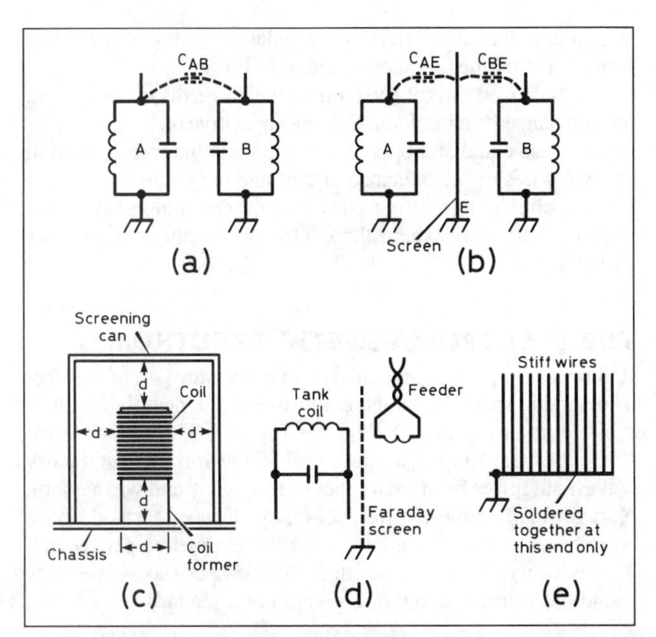

Fig 1.26. (a) Stray capacitance coupling C_{AB} between two circuits A and B. The introduction of an earthed screen E in (b) eliminates direct capacitance coupling, there being now only stray capacitance to earth from each circuit C_{AE} and C_{BE}. A screening can (c) should be of such dimensions that it is nowhere nearer to the coil it contains than a distance equal to the diameter of the coil *d*. A Faraday screen between two circuits (d) allows magnetic coupling between them but eliminates stray capacitance coupling. The Faraday screen is made of wires as shown at (e)

earthed screen of good conductivity between the two circuits in question as shown in Fig 1.26(b). There is then only stray capacitance from each circuit to earth and no direct capacitance between them. A useful practical rule is to position screens so that the two circuits are not visible from one another.

Stray magnetic coupling can occur between coils and wires due to the magnetic field of one coil or wire embracing the other. At radio frequency, coils can be screened by placing them in closed boxes or cans made from material of high conductivity such as copper, brass or aluminium. In practice, eddy currents are induced in the screening can which in turn set up a field which opposes and practically cancels the field due to the coil beyond the confines of the can.

If a screening can is too close to a coil the performance of the coil, ie its Q and also its inductance, will be considerably reduced. A useful working rule is to use a screening can of such a size that nowhere is it nearer to the coil than a distance equal to the diameter of the coil; see Fig 1.26(c).

At low frequencies where screening due to eddy currents is not so effective it may be necessary to enclose a low-frequency choke or transformer in a box of high-permeability magnetic material such as Mumetal in order to obtain satisfactory magnetic screening. Such measures are not often required but a sensitive component such as a microphone transformer may be enclosed in such a screen in order to make it immune from hum pick-up.

It is sometimes desirable to have pure inductive coupling between two circuits with no stray capacitance coupling. In this case a *Faraday screen* can be employed between the two coils in question as shown in Fig 1.26(d). This arrangement is sometimes used between an antenna and a receiver input circuit or between a transmitter tank circuit and an antenna. The Faraday screen is made of stiff wires (Fig 1.26(e)) connected together at one end only, rather like a comb. The screen is transparent to magnetic fields because by its very construction there is no continuous conducting surface in which eddy currents can flow and thereby give magnetic screening. However, because the screen is connected to earth it acts very effectively as an electrostatic screen, eliminating stray capacitance coupling between the circuits.

UNBALANCED AND BALANCED CIRCUITS

Unbalanced circuits

An unbalanced circuit is shown in Fig 1.27(a). A generator or signal source is shown connected to a load resistor through a single conductor; the return path for current is through what is termed an *earth return*. In practice, although the term 'an earth return' is used, the return path is usually through the chassis of the equipment or the screening of a coaxial cable.

Unbalanced circuits of this type are very commonly used in radio equipment and are perfectly satisfactory provided leads

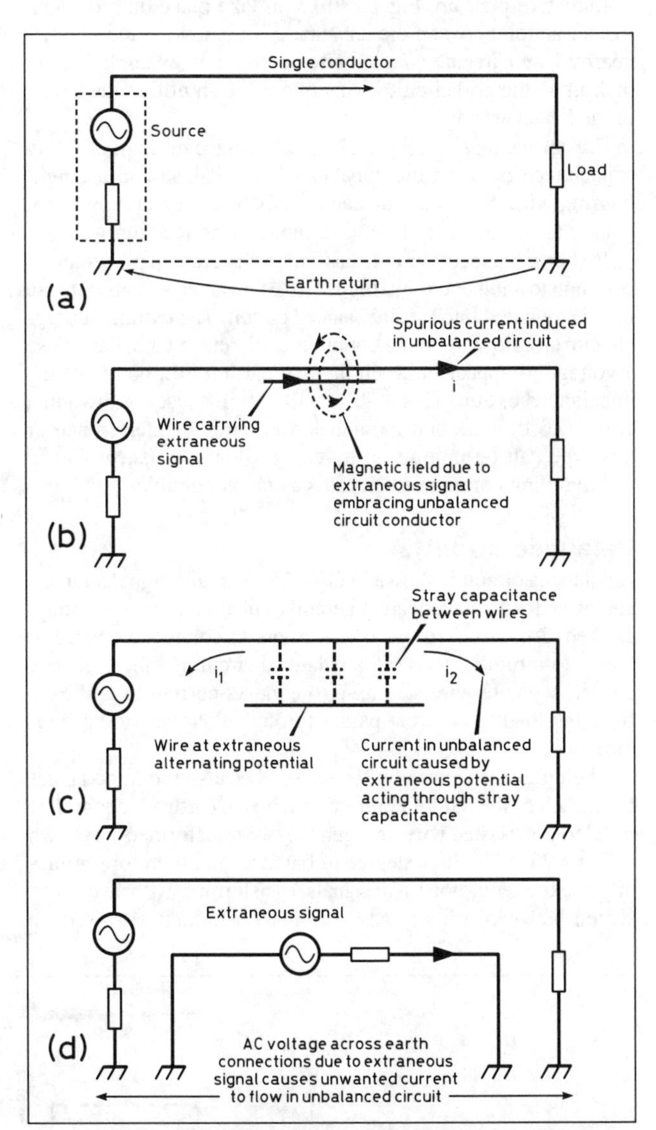

Fig 1.27. The unbalanced circuit. (a) The basic unbalanced circuit showing earth return path. (b) How extraneous signals and noise can be induced in an unbalanced circuit by magnetic induction. (c) Showing how extraneous signals can be induced by stray capacitance coupling. (d) Showing how extraneous signals can be induced due to a common earth return path

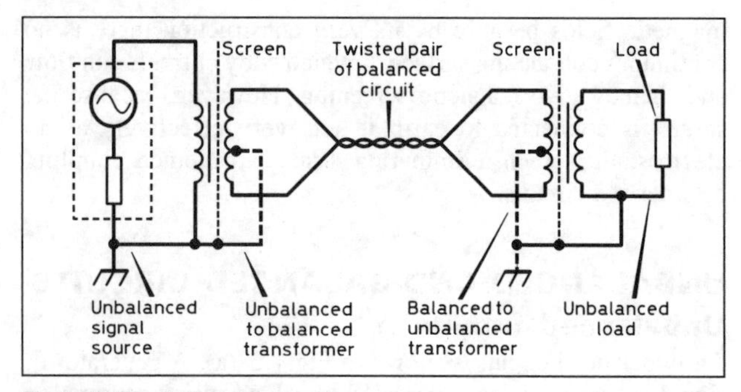

Fig 1.28. The balanced circuit

are kept short and are spaced well away from other leads. The circuit is, however, prone to the pick-up of extraneous noise and signals from neighbouring circuits by three means: inductive pick-up, capacitive pick-up and through a common earth return path.

Inductive pick-up, Fig 1.27(b), can take place due to transformer action between the unbalanced circuit wire and another nearby wire carrying an alternating current; an example is hum pick-up in the grid circuit wiring of a valve amplifier due to AC in the heater wiring.

Capacitive pick-up, Fig 1.27(c), takes place through the stray capacitance between the unbalanced circuit lead and a neighbouring wire. Such pick-up can usually be eliminated by introducing an earthed metal screen around the connecting wire.

If the unbalanced circuit has an earth return path which is common to another circuit, Fig 1.27(d), unwanted signals or noise may be injected into the unbalanced circuit. The extraneous signal current, flowing in the common earth return path, can cause a voltage to appear between the two earth return points of the unbalanced circuit. This voltage will in turn cause an unwanted current to flow around the unbalanced circuit. Interference of this type can be minimised by using a low-resistance chassis and avoiding common earth paths as far as possible.

Balanced circuits

A balanced circuit is shown in Fig 1.28. As many signal sources, and often loads as well, are inherently unbalanced (ie one side is earthed) it is usual to use transformers to connect a source of signal to a remote load via a balanced circuit. In the balanced circuit, separate wires are used to conduct current to and back from the load; no current passes through a chassis or earth return path.

The circuit is said to be 'balanced' because the impedances from each of the pair of connecting wires to earth are equal. It is usual to use twisted wire between the two transformers as shown in Fig 1.28. For a high degree of balance, and therefore immunity to extraneous noise and signals, transformers with an earthed screen between primary and secondary windings are used. In some cases the centre taps of the balanced sides of the transformers are earthed as shown dotted in Fig 1.28.

The balanced circuit overcomes the three disadvantages of the unbalanced circuit. Inductive and capacitive pick-up are eliminated since equal and opposite currents are induced in each of the two wires of the balanced circuit and these cancel out. The same applies to interfering currents in the common earth connection in the case where the centre taps of the windings are earthed.

THE ELECTROMAGNETIC SPECTRUM

Radio waves are electromagnetic waves forming part of the electromagnetic spectrum which comprises radio, heat, light, ultraviolet, gamma rays, and X-rays: see Fig 1.29. The various forms of electromagnetic radiation are all in the form of oscillatory waves, but differ from each other in frequency and wavelength. Notwithstanding these differences they all travel through space with the same speed, namely, 3×10^{10} centimetres per second. This is equivalent to about 186,000 miles per second (ie once round the world in about one-seventh of a second).

Frequency and wavelength

The distance travelled by a wave in the time taken to complete one cycle of oscillation is called the *wavelength.* It follows that wavelength, frequency and velocity of propagation are related by the formula

$$\text{Velocity} = \text{Frequency} \times \text{Wavelength}$$

or

$$c = f \times \lambda$$

where c is the velocity of propagation (3×10^{10}cm/s), f is the frequency of oscillation (Hz), and λ is the wavelength (cm).

This formula enables the wavelength to be calculated if its frequency is known, and vice versa. The following example shows what is meant:

Example. What are the frequencies corresponding to wavelengths of (i) 150m, (ii) 2m and (iii) 75cm?

From the formula $c = f\lambda$, the frequency is given by:

(i) $\lambda = 150\text{m} = 1.5 \times 10^4\text{cm},\ c = 3 \times 10^{10}\text{cm/s}$

Therefore $$f = \frac{3 \times 10^{10}}{1.5 \times 10^4} = 2 \times 10^6\text{Hz}$$

$$= 2.0\text{MHz}$$

(ii) $\lambda = 2\text{m} = 2 \times 10^2\text{cm}$

Therefore $$f = \frac{3 \times 10^{10}}{2 \times 10^2} = 1.5 \times 10^6$$

$$= 150\text{MHz}$$

(iii) $\lambda = 75\text{cm}$

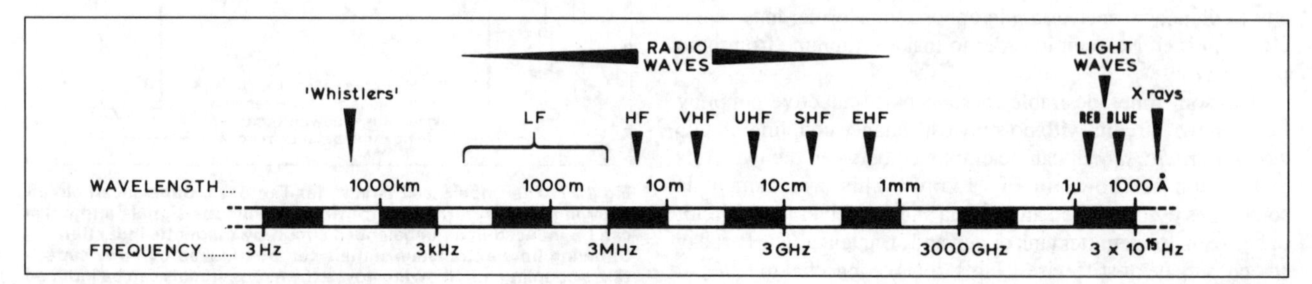

Fig 1.29. The electromagnetic spectrum

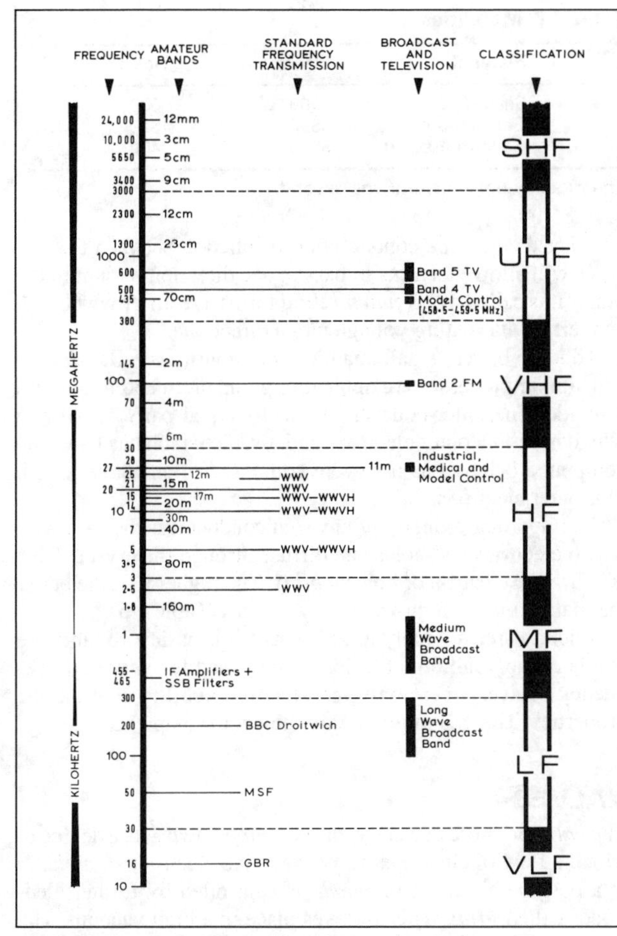

Fig 1.30. Amateur bands in relation to other services

Therefore
$$f = \frac{3 \times 10^{10}}{75} = 4 \times 10^{8}$$
$$= 400\text{MHz}$$

Fig 1.30 shows how the radio spectrum may be divided up into various bands of frequencies, the properties of each making them suitable for specific purposes. Amateur transmission is permitted on certain frequency bands in the HF, VHF and UHF ranges.

ANTENNAS

An *antenna* (or *aerial*) is used to launch electromagnetic waves into space or conversely to pick up energy from such a wave travelling through space. Any wire carrying a high-frequency alternating current will radiate electromagnetic waves and conversely an electromagnetic wave will induce a voltage in a length of wire. The problem in antenna design is to radiate as much transmitter power as possible in the required direction or, in the case of a receiver, to pick up as strong a signal as possible, very often in the presence of local interference.

An antenna may be considered as a tuned circuit consisting of inductance, capacitance and resistance, and for maximum radiation an antenna is usually operated so that it is naturally resonant at the operating frequency, maximum radiation corresponding to maximum high-frequency current flowing in the antenna.

There are two basic types of antenna used by amateurs: (a) the Hertzian antenna, known as a *dipole* or *doublet,* which is self-resonant by virtue of its length in relation to the wavelength of the signal, and (b) the Marconi antenna which may be of any

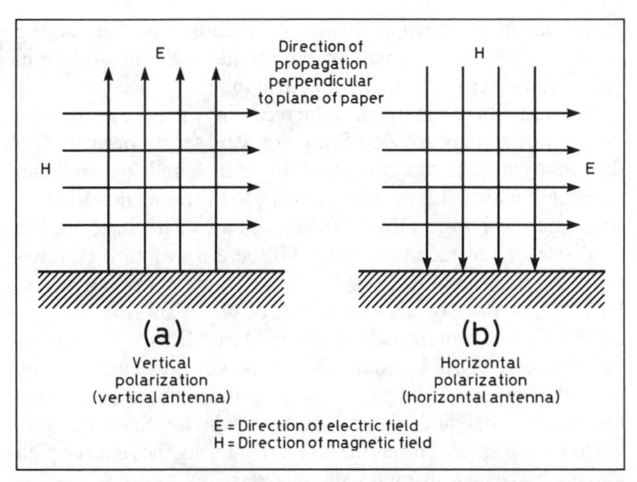

Fig 1.31. Diagram illustrating the polarisation of electromagnetic waves: (a) represents vertical polarisation and (b) represents horizontal polarisation, the distinction being in terms of the electric vector rather than the magnetic vector.

convenient length and relies on an earth connection; it is tuned to resonance by means of a coil or capacitor. The simple Hertzian antenna is one half-wavelength long and is therefore well suited to the shorter waves (higher frequencies) whereas the Marconi antenna is mainly used at longer wavelengths, eg 160m and 80m, where it is often impracticable to erect a Hertzian dipole. More detailed explanations of antenna theory and practical design will be found in Chapters 12 and 13.

PROPAGATION OF RADIO WAVES

Radio waves, on leaving an antenna situated in a uniform dielectric such as air at ground level, travel through space in straight lines with the velocity of light. The waves consist of interdependent electric and magnetic fields which act in directions mutually at right-angles and also at right-angles to the direction of propagation of the wave.

If the electric field acts in a vertical direction the wave is said to be *vertically polarised* and such waves are launched and best picked up by vertical antennas: see Fig 1.31(a). If the electric field is horizontal the waves are said to be *horizontally polarised* and horizontal antennas are then used: see Fig 1.31(b).

Horizontal polarisation is most commonly used by amateurs on the HF bands for long-distance communication. Vertical polarisation is frequently used for mobile work and local work on VHF and UHF bands where the all-round or *omnidirectional* characteristics of simple vertical antennas are an advantage.

The propagation of radio waves to remote parts of the Earth depends on reflection of the waves from layers of ionised gas situated between 70 and 200 miles or so above the Earth in a region called the *ionosphere.* As the reflecting properties of this region depend on the frequency used and are continually changing with night and day, the time of year, and sunspot activity, it can be imagined that the mechanism of long-distance short-wave propagation is complex and very variable. However, transmission conditions to distant points can be forecast with some accuracy.

The propagation of VHF and UHF signals rarely depends on the conditions in the ionosphere but on meteorological conditions in the lower levels of the atmosphere called the *troposphere.*

Fading

It is common for HF signals which have arrived at a receiver from a distant transmitter to vary continually in strength. This phenomenon is known as *fading* and is caused by the varying relationship between signals arriving at the receiver by different paths whose relative lengths are changing.

The strength of the signal at the receiving antenna terminal is the sum of signals arriving from a transmitter by perhaps two different paths. If the two paths differ in length by an exact number of wavelengths, corresponding to the frequencies in use, the signals arriving by the two paths will arrive in phase and the total strength of received signal will be a maximum. If, however, the path lengths differ by an odd number of half-wavelengths, the two signals will be out of phase and the total received signal strength will be a minimum. Thus if a signal is being received both by ground-wave and sky-wave paths, or by two different sky-wave paths, fading will be experienced when the effective height of the ionised layer which reflects the sky-wave varies, since this has the effect of altering the relative path lengths between transmitter and receiver.

Fading may be slow or very rapid: if it is rapid, the fading appears as an audible low-frequency flutter. During fading, speech transmitted by AM sometimes becomes badly distorted because the carrier and various sideband frequencies (see below) do not all fade equally at the same time.

The effect of fading is combated by applying automatic volume control (better referred to as *automatic gain control)* to receivers so that the overall gain of the set varies inversely with the strength of the received signal. Other methods used are:

(a) highly directive antenna systems, particularly in the vertical plane to restrict the possibility of multi-path transmission; and
(b) *diversity reception* in which two or more spaced antennas or two or more different frequencies are used for a transmission. Diversity reception depends for its operation on the fact that, statistically, fading is unlikely to occur at two different points or on two different frequencies at the same time. Two receivers are used with an automatic change-over arrangement for selecting the stronger of the two signals, a considerable reduction in fading being achieved.

Propagation is dealt with in detail in Chapter 11.

SEMICONDUCTORS

Certain elements and compounds are intermediate in electrical conductivity between metals and insulators. They are called *semiconductors*. The best known examples are silicon, germanium and the compound gallium arsenide. In all cases, the pure material is practically an insulator. This is an *intrinsic* semiconductor. Nearly all semiconductors are used in the form of near-perfect crystals which have been grown from a molten pool of pure material.

Silicon and germanium have an atomic structure in which there are four electrons in the outer shell (the electrons are arranged in so-called *shells*). Silicon has 14 electrons surrounding its nucleus divided up into shells of 2, 8 and 4. Phosphorus has 2, 8 and 5, while boron has 2 and 3. Only the outermost shells are involved in the conduction process.

Very small amounts of, say, phosphorus or boron in silicon increase the conductivity by several orders of magnitude (an order of magnitude is by a factor of 10, ie 10 times). Since phosphorus has one extra electron in its outer shell, this can diffuse through the crystal of silicon. This causes the conductivity which in this case is called *N-type*. On the other hand, boron has one fewer electron than silicon so it creates a space or *hole* where an electron should be which can also diffuse through the crystal.

Table 1.6. Mobilities

Material	Electrons	Holes
Silicon	1900	500
Germanium	3800	1820
Gallium arsenide	8800	400

The mobility units are cm²/volt second.

This is called *P-type* conductivity. A junction between the two types will allow electrons to pass in one direction but not in the other. It is called a *P-N junction* and forms a rectifier which will convert an alternating voltage into a direct one.

Addition of very small quantities of impurity is called *doping* and similar processes are applied to germanium and to gallium arsenide. The latter consists of exactly equal parts of gallium which has three outer electrons and arsenic which has five. The compound behaves as if it were composed of an element with four outer electrons.

An important property of any semiconductor is the ease with which electrons and holes can diffuse through the crystal. This is called the electron or hole *mobility*. The higher it is, the better the material is for high-frequency use (see Table 1.6).

Silicon is most widely used because it is easiest to fabricate and is cheap. Gallium arsenide is only used for very high frequency devices where only the *electron* mobility controls the properties. There is more on this subject in Chapter 3.

VALVES

The *valve* or, more correctly, the *thermionic valve*, is a device in which a flow of electrons from a negative electrode or *cathode* to a positive electrode or *anode* is controlled by further electrodes called *grids*. This all takes place in a high vacuum. The electrons are generated by a hot cathode. The ease with which the electrons are liberated from the cathode depends on the *work function* of the material. As far as amateurs are concerned only two materials are used: oxide coatings (a mixture of barium and strontium oxides) and thoriated tungsten (tungsten wire with thorium oxide mixed with it). See Table 1.7.

Valves have almost become obsolete for low-power purposes such as in receivers. They are still used for high-power stages in transmitters especially at VHF, UHF and in the microwave region. There is more specific information in Chapter 4.

QUARTZ CRYSTALS

Quartz is one of a group of crystals which are *piezoelectric* – when they are subject to strain, a charge develops across them. This only applies to certain faces or cuts at certain angles to the faces. The corollary is that, if a voltage is applied, a strain (movement) takes place. If a plate of quartz is cut in the right manner, it can ring like a bell, but at a much higher frequency, if struck

Table 1.7. Work functions

Material	Work function
Barium/strontium	2.35eV
Thorium/tungsten	3.71eV

eV = electron volts. The smaller the work function, the lower the temperature the cathode will emit electrons. This emission is governed by the Richardson equation:

$$I = AT^2 \exp(-p/kT)$$

where A and k are constants, p is the work function and T is the absolute temperature (degrees C + 273).

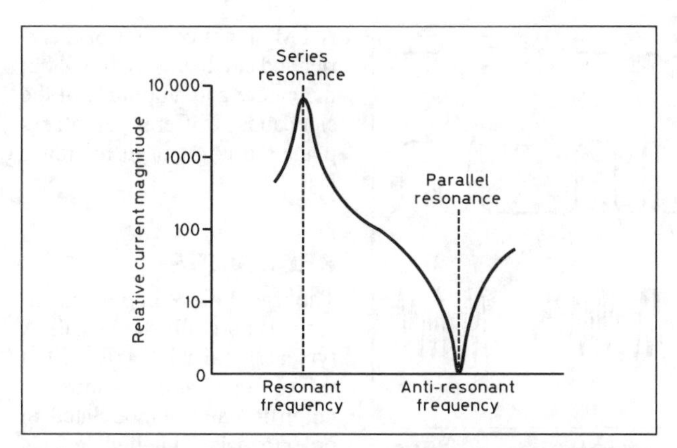

Fig 1.32. Typical variation in current through a quartz crystal with frequency. The resonant and anti-resonant frequencies are normally separated by about 0.01% of the frequencies themselves (eg a few hundred hertz for a 7MHz crystal)

or otherwise stimulated. This property is very similar to the resonance of a tuned circuit and can be used to control an oscillator at a fixed frequency. Compared with a conventional tuned circuit, a quartz crystal has a very high Q (values of up to a million are possible) and its frequency is fixed. It has *two* resonant points, a series-resonant point where the effective resistance is very low, and a parallel-resonant point where the effective resistance is very high. See Fig 1.32.

There are more details in Chapter 5 – 'Building blocks'.

AMPLIFIERS

Any circuit in which a signal is increased in magnitude is called an *amplifier*. The essential part is an active device such as a transistor or valve. They are classified by their frequency range, selectivity and size (power) of their output.

At one end of the scale there are the audio amplifiers used to bring the output of a microphone (a few millivolts) up to that which is needed for modulating a transmitter (see below). At the other end, there is the power amplifier of a transmitter which provides all the power needed to communicate with the world.

Details of the various sorts are given in Chapter 5.

Feedback

If some of the signal from the output of an amplifier is fed back to the input, there are two possibilities: first, the overall gain is reduced because the feedback signal is out of phase with the input; this is called *negative feedback*. Second, the signal is increased because the feedback signal is in phase with the input; this is *positive feedback*.

Negative feedback is used to improve the fidelity of an amplifier since any distortion generated in the amplifier will be reduced by more than the reduction of the signal [6].

Positive feedback is used in certain types of receiver to increase the signal without having to add further amplifying stages. At its extreme, when the feedback signal is greater than the input signal, the circuit oscillates. This can happen even if there is no apparent input signal because slight noise builds up. When a circuit oscillates, the frequency depends on the circuit elements. A tuned circuit will allow oscillation on its resonant frequency. The amplitude of oscillations increases until either the gain is reduced by some factor or one of the circuit elements saturates, ie it cannot accept any higher signal.

TRANSMITTERS

A transmitter consists of an oscillator and an amplifier with some means of imposing intelligence on the signal. The last function is called *modulation*. Fig 1.33 shows block diagrams of transmitters and much more detail is given in Chapters 7 and 8. The oscillator may be fixed in frequency, eg by a crystal, or variable (VFO) using a conventional tuned circuit, or it may be a combination of both, a VFO mixing with several crystal oscillators to produce the required final frequencies.

Modulation

A simple sine wave signal conveys no information except that it is there and perhaps the direction of its origin.

The simplest form of modulation is on-off (see Fig 1.34) which is used for morse code, radio teleprinter (also called 'RTTY') and, more exotically, for pulse modulation at microwave frequencies. This is done by switching one of the stages of the transmitter on and off as necessary. The switching must be done slowly, ie the square waveform must be rounded off. This is because a true square wave has odd harmonics up to a very large number (see above – they are theoretically present to infinity!) and these would cause sidebands (see below) to spread over a

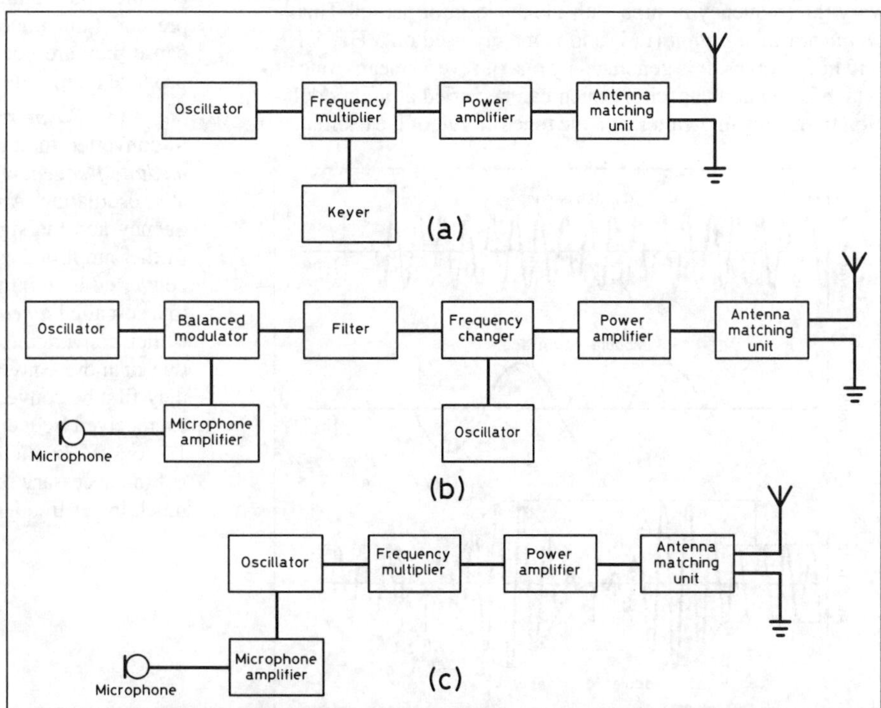

Fig 1.33. Block diagrams of transmitters: (a) for CW morse telegraphy; (b) for single sideband telephony; (c) for frequency modulated telephony (here the oscillator contains a variable reactance device for modulation). Note that in all cases the antenna matching unit must contain a low-pass filter with a cut-off frequency just above the operating frequency

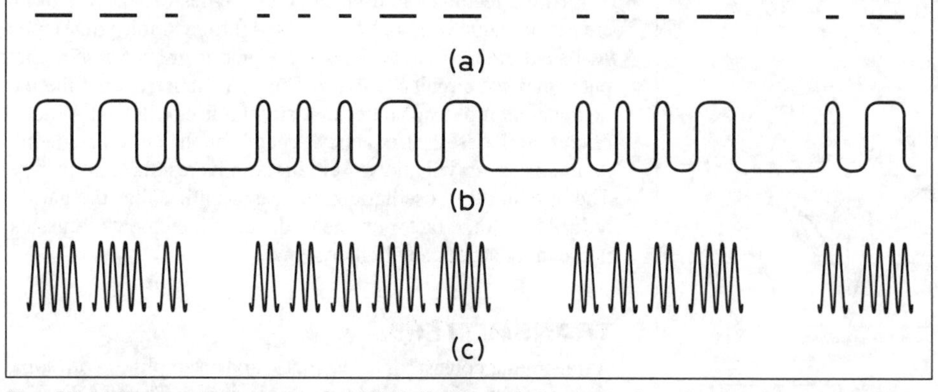

Fig 1.34. (a) The morse characters for G3VA. (b) The switching waveform for the transmitter. (c) The output of the transmitter

wide frequency range. This wastes spectrum space and causes key clicks which interfere with other band users.

With amplitude modulation, the amplitude of the high frequency carrier is varied as the audio signal varies (see Fig 1.35). This can be analysed into three parts, a carrier wave, the original radio frequency generated in the transmitter, and two *sidebands* which are higher and lower than the carrier by the audio modulating frequency. A simplified picture of this process is shown in Fig 1.36 where the modulation is a single audio frequency. The audio information is carried in both sidebands so one is redundant and could be removed without reducing the information content. This gives single sideband full carrier transmission. For speech, the carrier is also unnecessary and can be left out, giving single sideband suppressed carrier transmission, commonly called *single sideband* (SSB). The techniques for generating this mode are found in Chapter 7 – 'HF transmitters and transceivers'.

Another method of putting information on the carrier is to vary the frequency in time with the audio information. This is *frequency modulation* (FM) and is mainly used on VHF, UHF and microwaves. It is generated by one of several means, one of which is to put a capacitor which can be varied at the modulation frequency in parallel with the tuned circuit of the oscillator.

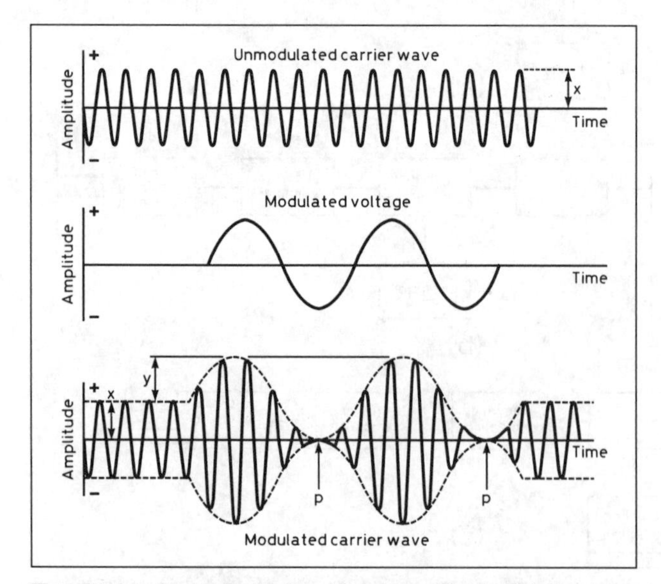

Fig 1.35. Amplitude modulation of a carrier. The modulation envelope may be displayed on an oscilloscope and represents 100% modulation by a sinusoidal signal. It shows graphically how the sum of the carrier and sideband voltages varies with time

A FM signal consists of a carrier and sidebands as in AM but the nature and amplitude of the sidebands is different. A full explanation is given in reference [7].

RECEIVERS

The signal picked up by an antenna is normally very small. A typical signal at HF will be from 0.5 to 10μV and this must be amplified and demodulated in order to make it intelligible. This is done in a *receiver*. The receiver needs not only to provide the gain but also to provide selectivity so that only the signal required is amplified and demodulated. Also, it should not respond to very large signals close to the required one.

There are several kinds of receiver but they are of two basic types:

(a) *Straight* receivers in which the received signal is amplified at signal frequency, demodulated and the resulting audio signal further amplified before being applied to a loudspeaker or headphones. A variant of this is the *direct-conversion* receiver where the signal is demodulated by mixing with a locally generated carrier of the same frequency as the signal. This is mainly used for SSB when the carrier has the frequency of the suppressed carrier of the transmitter. A second variant is the *super-regenerative* receiver. In this, the demodulator is an oscillating mixer in which the oscillations are interrupted at a moderate frequency (this is called *quenching* and is generally done at 30 to 500kHz). The rate at which they build up depends on the size of any signal present at the input. These are not often used nowadays because they are not well understood and they are liable to produce interference with other stations.

(b) *Superheterodyne* receivers (*superhets*) in which the signal in converted to another, often fixed, frequency (the *intermediate frequency* or IF) by mixing with a locally generated oscillation. Amplification takes place at the fixed frequency and the signal is then demodulated and the audio further amplified as above. A variant is where the signal is converted by mixing with one of several fixed frequencies to a so-called *tuneable IF* which is common to all bands. It is then converted to a fixed IF as before. There may, too, be two or more conversions of frequency. As an example, it may first be converted to a frequency well above the highest received frequency, eg for an HF receiver covering from 1.7 to 30MHz, the first IF may be as high as 45MHz but to get the necessary selectivity it has to be changed again to a much lower frequency such as 450kHz.

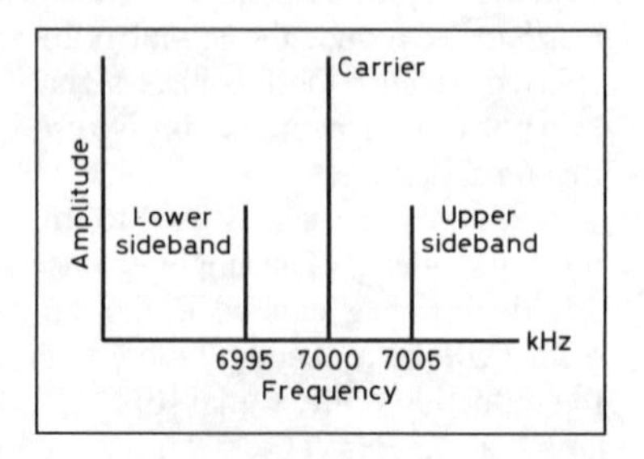

Fig 1.36. Sideband spectrum of a 7MHz carrier which is amplitude modulated by a 5kHz audio tone

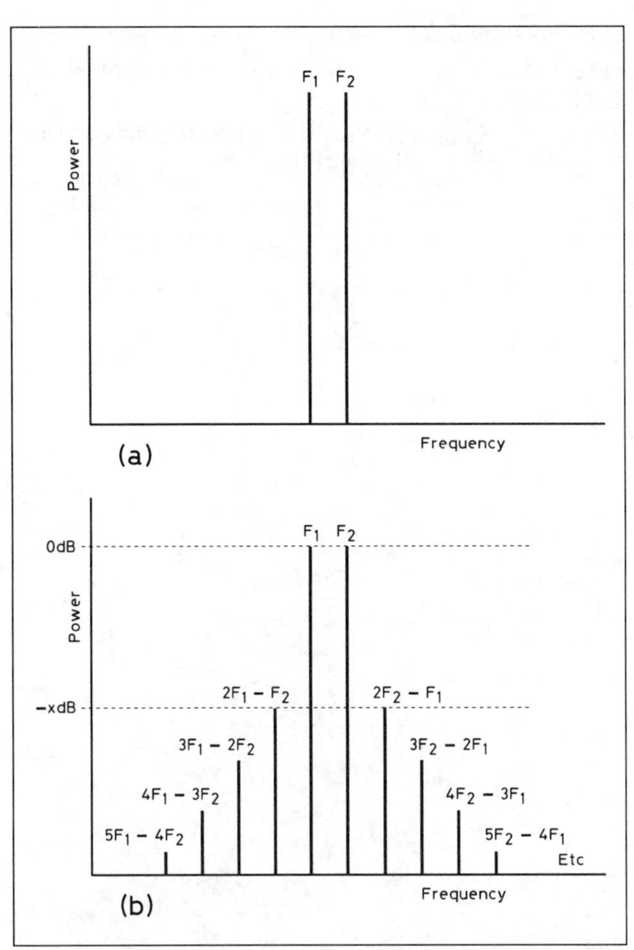

Fig 1.37. Spectrum of two frequencies (a) before and (b) after passing through a non-linear device

TRANSCEIVERS

These are simply receivers and transmitters in the same box. They have the same tuning control for both and so are normally employed to work *co-channel*, ie to transmit and receive on the same frequency. To do this, they usually share the same local oscillator and, in the case of SSB transceivers, will often share other components between the transmitter and the receiver.

There is much more detailed information on these topics in Chapters 6, 7 and 8.

INTERMODULATION

Whenever two radio frequencies are applied together to a non-linear device, the output from that device will contain the original frequencies and several others which are combinations of the originals and their harmonics. This is called *intermodulation* and the new frequencies are called *intermodulation products*.

The most important case is when the device is a power amplifier in a single sideband (SSB) transmitter. If this is not truly linear (and it never is!), intermodulation products are formed and distortion results.

Consider an SSB transmitter modulated by two audio tones, f_1 and f_2. These should produce only two radio frequencies, $f_0 + f_1$ and $f_0 + f_2$ (assuming we are generating the upper sideband and f_0 is the suppressed carrier frequency). Let us call these F_1 and F_2 respectively and plot a spectrum of them, ie a graph of power against frequency. It will be as shown in Fig 1.37(a).

Note that, since f_1 and f_2 are audio frequencies and f_0 is a radio frequency, then F_1 and F_2 will be very close together due to the relationship:

$$F_1 - F_2 = f_1 - f_2$$

If this signal is passed through a non-linear amplifier, the spectrum changes as intermodulation products appear: see Fig 1.37(b). The nearest products, $2F_1 - F_2$ and $2F_2 - F_1$, are called the *third-order products* and are always larger than the fifth ($3F_1 - 2F_2$ etc) and higher-order products. The power in these relative to the fundamentals, xdB in Fig 1.37(b), should be at least 30dB down and preferably more.

Intermodulation can also occur in other systems. In receivers it can occur in the 'front-end' if this is non-linear and, in a crowded band such as 7MHz with high-powered broadcast signals close to (and even in!) the amateur band, intermodulation between these can occur. This degrades the receiver performance if the products are within the pass-band of the receiver. The use of high-gain RF amplifiers in receivers can be a major source of this problem and may be countered by reducing the gain by an input attenuator. Alternatively, only just enough gain should be used to overcome the noise introduced by the mixer. For frequencies up to 14MHz, it is not necessary to have any pre-mixer gain provided the mixer is a good low-noise design (see Chapter 6 – 'HF receivers').

Intermodulation can also occur in unlikely places such as badly constructed or corroded joints in metalwork. This is the so-called *rusty-bolt effect* and can result in intermodulation products which are widely separated from the original transmitter frequency and which cause interference with neighbours' equipment (see Chapter 17 – 'Electromagnetic compatibility').

REFERENCES

[1] *Tables of Physical and Chemical Constants*, Kaye and Laby, 14th edn, Longman, London, 1972, pp102–105.
[2] M G Scroggie, *Wireless World* November 1977, pp79–82.
[3] *Second Thoughts on Radio Theory*, 'Cathode Ray', Iliffe, London, 1955, pp71–77.
[4] *Second Thoughts on Radio Theory*, 'Cathode Ray', Iliffe, London, 1955, pp109–113.
[5] *Second Thoughts on Radio Theory*, 'Cathode Ray', Iliffe, London, 1955, pp125.
[6] *Second Thoughts on Radio Theory*, 'Cathode Ray', Iliffe, London, 1955, pp161–173.
[7] *VHF/UHF Manual*, G R Jessop, G6JP, 4th edn, RSGB, pp5.24–5.29.

BIBLIOGRAPHY

Foundations of Wireless, M G Scroggie, Iliffe, London.

Radio Engineers' Handbook, F E Terman, McGraw-Hill, New York.

Radio Data Charts, R T Beatty (revised by A L M Sowerby), Iliffe, London.

Radio Engineering, F E Terman, McGraw-Hill, New York.

The Radio Amateur's Handbook, ARRL.

Shortwave Wireless Communication, Ladner and Stoner, Chapman and Hall, London.

Radio Receiver Design, K G Sturley, Chapman and Hall, London.

Radio Engineering, E K Sandeman, Chapman and Hall, London.

The Radio Amateurs' Examination Manual, George Benbow, G3HB, RSGB.

A Guide to Amateur Radio, Pat Hawker, G3VA, RSGB.

Outline of Radio and Television, J Pat Hawker, Newnes, London.

Solid State Design for the Radio Amateur, W Hayward, W7ZOI, and D DeMaw, W1FB, ARRL.

The Art of Electronics, P Horowitz and W Hill, 2nd edn, Cambridge University Press, Cambridge, 1989.

2 Passive components

RELIABILITY is just as essential for home constructors as it is for the professional, perhaps even more so owing to the frustration caused by trying to locate a breakdown with inadequate instrumentation. High-quality components tend to be expensive, but with care can be bought at rallies or reclaimed from surplus, providing one knows what to choose. This chapter gives some guidance.

To avoid the requirement for fault-finding, it is always advisable to test such components before use, and to derate where possible. In 1889 a Swedish chemist, Arrhenius, established an exponential temperature law for the rate of chemical reactions, and as a consequence of this it can be predicted that lowering the temperature of components greatly reduces the failure rate due to destructive reactions, providing that freezing of liquid contents does not take place. Similarly, operating below the maker's maximum voltage or current rating favours long life.

Many components are now made for surface mounting.

WIRE

Copper wire is the most often used conductor, being of low resistivity and easily worked. PVC covering is usually used for equipment wiring, and Kynar for wire-wrap, but there are higher-temperature coverings and sleeving available. Table 2.1 gives the characteristics.

For inductors and transformers the oleo enamel, cotton, regenerated cellulose and silk of yesteryear have given way to polyesterimide or similar plastic. These have better durability and come with a variety of trade names. They are high-temperature coatings, being generally designated by the maximum operating temperature or trade name. They also have very good resistance to abrasion and chemical attack. Such coatings are difficult to strip mechanically, especially on fine wires.

There are also polyurethane coatings, usually pink or, for identification, green, purple or yellow in colour. These are suitable up to 130°C. Polyurethane-covered wires are self-fluxing if a sufficiently hot iron is used, but remember that poisonous fumes are created.

Dual coatings of polyesterimide with a lower-melting-point plastic layer outside are made. When a coil has been wound with these, it is heated by passage of current, and the outer layers fuse, forming a solid bond. Dual coatings are also found where the outer coating is made of a more expensive material than the inner.

Table 2.1. Equipment wire and sleeving insulation

Material	Temperature guide (maximum °C)	Remarks
PVC	70–80	Toxic fume hazard
Cross-linked PVC	105	Toxic fume hazard
Cross-linked polyolefine	150	
PTFE (Tefzel)	200	Difficult to strip
Kynar	130	Toxic fume hazard. $\varepsilon_r \approx 7$
Silicone rubber	150	Mechanically weak
Glassfibre composite	180–250	Replaces asbestos

If temperature range is critical, check with manufacturer's data.

Table 2.2. Resistivity and temperature coefficient for commonly used metals

Metal	Resistivity at 20°C (μΩ-cm)	Temperature coefficient of resistivity per °C from 20°C (ppm/°C)	Notes
Annealed silver	1.58	4000	
Annealed pure copper	1.72	3930	
Aluminium	2.8	4000	(1)
Brass	9	1600	(1)
Soft iron	11	5500	
Cast iron	70	2000	
Stainless steel	70	1200	(1)
Tungsten	6.2	5000	
Eureka	50	40	(1, 2)
Manganin	43	±10	(3)
Nichrome	110	100	(1)

Notes:
(1) Depends on composition and purity. Values quoted are only approximate.
(2) Copper 60%, nickel 40% approx. Also known as 'Constantan', 'Ferry' and 'Advance'. High thermal EMF against copper.
(3) Copper 84%, manganese 12%, nickel 4%. Low-temperature coefficient only attained after annealing at 135°C in an inert atmosphere. Cannot be soft-soldered.

Oleo enamel and early polyesterimide crack with age and kinking, so old wire should be treated with suspicion. The newer synthetic plastic coatings can resist abrasion to a remarkable degree, but kinked wire of even this type should not be used for inductors.

Wire size can be expressed variously in millimetres diameter of conductor, Standard Wire Gauge (swg) or American Wire Gauge (awg). The sizes and recommended current-carrying capacity are given in Chapter 22. Precise current-carrying capacity is difficult to quote as it depends on the allowable temperature rise and the nature of the cooling.

Since RF currents only penetrate to a limited depth ($6.62/(F_{Hz})^{0.5}$ cm for copper), coils for higher frequencies are sometimes silver plated; silver has a marginally better conductivity than copper (see Table 2.2) and is less liable to corrosion, though it can tarnish if not protected. Generally the cost is not justified. Where high currents are involved, tubes may be used to reduce cost.

For some purposes low resistivity is not required, and a low temperature coefficient of resistivity is preferred. Table 2.2 gives the values of resistivity and temperature coefficient of commonly used materials.

FIXED RESISTORS

These are probably the most widely used components and are now among the most reliable if correctly selected for the purpose.

As the name implies, they resist the passage of current, and in fact the unit 'ohm' is really the 'voltage per amp'.

For many years resistors were made of a moulded mixture of carbon and binder, either painted or enclosed in an insulating tube. They are cheap and are still found, but suffer from a large negative temperature coefficient of −150 to −1500 parts per million per degree Centigrade (ppm/°C), and an irreversible change with age or heat due to soldering. Despite these defects, moulded carbon resistors are still used as they are less inductive than other common types. 20% is a common tolerance – closer tolerance cannot be maintained in use.

Due to the granular nature of the carbon and binder, more than thermal noise is generated when current flows, so moulded carbon composition resistors should not be used where low-level signals are present with appreciable current flow.

Tubes of carbon composition in various diameters with metallised ends are made for high-wattage dissipation. These may bear the trade name Morganite, and are suitable for RF loads.

The Second World War produced a need for closer tolerances, and carbon films were deposited on ceramic substrates by 'cracking' a hydrocarbon. To adjust the value, a helix was cut into the film, making the component more inductive than the moulded one. Protection is either by varnish, conformal epoxy or ceramic tube. Tolerance down to 5% is advisable; the large negative temperature coefficient makes the value of closer tolerance doubtful. The liability to total failure is greater than with moulded carbon composition.

In the 'thirties, Dubilier made a metal-film resistor of great bulk and stability, but it never became popular. In the 'fifties metal-oxide film resistors were introduced, soon to be replaced by metal-film devices. These were more stable, less noisy and more reliable than carbon film, and are now only very slightly more expensive. They use the same helical groove technique as carbon film and so are slightly inductive. Little difficulty is to be expected on this account up to, say, 50MHz. Protection is either by conformal or moulded epoxy coating.

Tolerances down to 0.1% and temperature coefficients of ±15ppm/°C are commonly available, and values range from 0.1Ω to 1MΩ at power ratings from 1/8W to 6W. More precise tolerances can be made to order. These small components are limited to around 200V, but higher-voltage versions are made.

These resistor values are quoted from a series which allows for the tolerance. The series is named according to the number of members in a decade. For example, E12 (10%) has 12 values between 1 and 10 (including 1 but not 10). In this case each will be the 12th root of 10 times the one below, rounded off to two significant figures (Table 2.3). E192 allows for 1%, the values being rounded off to three significant figures in this case. Unfortunately E12 is not a subset of E192 and some unfamiliar numbers will be met if E12 resistors are required from 1% stock. Values are usually marked by the colour code shown in Chapter 1 but some manufacturers print the value on their products.

Metal film resistors can also be made into integrated circuit packages, both single and dual-in-line being available in several different configurations. Using these can save space and tracking on printed circuit boards.

For higher powers, wire-wound resistors are used. The cheaper variety are wound on a fibre substrate protected by a rectangular ceramic tube, or rarely, moulded epoxy. More reliable ones are wound on a ceramic substrate protected by vitreous or silicone enamel, and for greater heat dissipation then encased in an aluminium body which can be bolted onto the chassis or heatsink. Vitreous enamelled wire-wound resistors can be made with a portion of the winding left free from enamel, so that one or more taps can be fitted. Wire-wound resistors are of no use at RF owing to their inherent reactance.

Table 2.3. 'E' range of preferred values, with approximate tolerance values from next number in range

E3 ±40%	E6 ±20%	E12 ±10%	E24 ±5%
1.0	1.0	1.0	1.0
—	—	—	1.1
—	—	1.2	1.2
—	—	—	1.3
—	1.5	1.5	1.5
—	—	—	1.6
—	—	1.8	1.8
—	—	—	2.0
2.2	2.2	2.2	2.2
—	—	—	2.4
—	—	2.7	2.7
—	—	—	3.0
—	3.3	3.3	3.3
—	—	—	3.6
—	—	3.9	3.9
—	—	—	4.3
4.7	4.7	4.7	4.7
—	—	—	5.1
—	—	5.6	5.6
—	—	—	6.2
—	6.8	6.8	6.8
—	—	—	7.5
—	—	8.2	8.2
—	—	—	9.1

The values shown above are multiplied by the appropriate power of 10 to cover the range.

Unless an aluminium-clad resistor is used, care should be taken to ensure that the heat generated does not damage the surroundings, particularly if the resistor is mounted on a printed circuit board. Marking is by printed value which often includes the wattage.

Fig 2.1 includes photographs of many of the fixed resistors that have just been described.

VARIABLE RESISTORS

Often a resistor has to be made variable, either for control or preset adjustment purposes, for which latter it is usually called a *trimmer*. For versatility, variable resistors are made with a moveable tapping point, and the arrangement called a *potentiometer*, or *pot* for short, the name being derived from a laboratory instrument for measuring voltage which used the same configuration. If only a variable resistor or trimmer is required, it is advisable to connect one end of the track to the slider.

Carbon composition, cermet, conductive plastic and wire-wound construction are all used. Rotary tracks with both multi-turn and single turn are all available, both for pots and trimmers, in varying degrees of accuracy. The hi-fi world uses linear motion versions for controls which are seldom suitable for the radio amateur. Wire wound has the disadvantages of limited resolution and higher price, but the advantages of higher wattage and good stability. It is not always possible to tell the type from the external appearance, as Fig 2.1 shows.

Single-turn rotary pots for volume control and many other uses usually turn over 250°, with log and anti-log law as well a linear law being available. Precision components are linear and turn over some 300°. For greater setting accuracy, multiturn pots are used, with turn-counting dials made to fit if required.

Values are frequently only in the E3 range to 20% and are printed on the component. Some makers use an exponent system of marking.

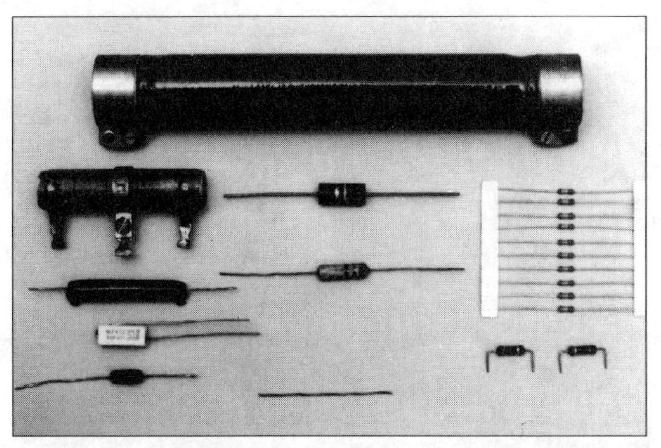

Fig 2.1(a). Fixed resistors. Top: 50W wire wound, vitreous-enamel covered. Left, top to bottom: 10W wire wound with adjustable tapping, 12W vitreous enamelled, 4W cement coated and 2.5W vitreous enamelled. Centre: 2W carbon composition, 1W carbon film. Right, top: ten ¼W metal film resistors in a 'bandolier' of adhesive tape, two carbon film resistors with preformed leads. Bottom: striped marker is 50mm long

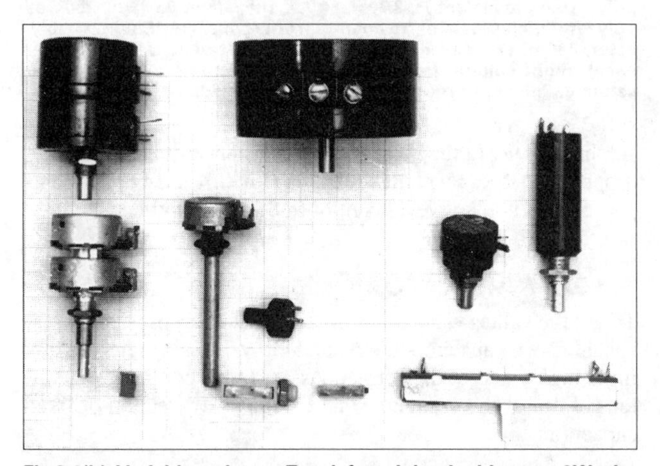

Fig 2.1(b). Variable resistors. Top, left to right: double-gang 2W wire wound, 5W wire wound. Middle, left to right: two independent variable resistors with coaxial shafts, standard carbon track, single-turn preset, standard wire wound and 10-turn variable resistor. Bottom, left to right: 10-turn PCB vertical mounting preset, 10-turn panel and PCB horizontal mounting preset. Right: a slider variable resistor

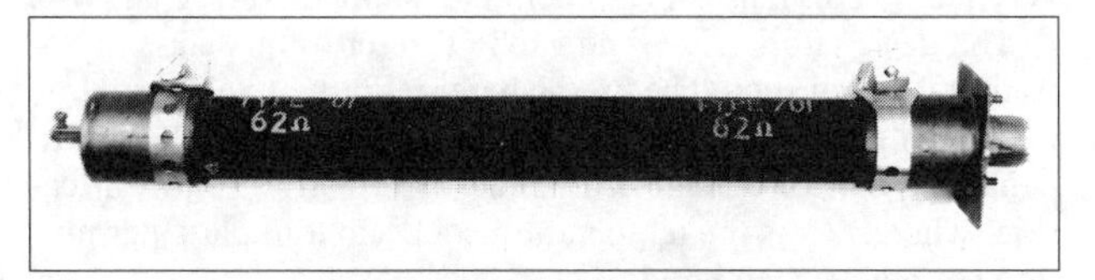

Fig 2.1(c). A 62Ω, 50W dummy load with UHF coaxial connector. Amateur made. Suitable for 200W for short periods

NON-LINEAR RESISTORS

All the above resistors obey Ohm's Law closely, that is to say that the current through them is proportional to the applied voltage (provided that the physical conditions remain constant and no work is done). Resistors except metal film and wire wound do have a slight voltage coefficient of resistance, which is so low that it does not affect amateur radio usage. For some purposes constancy of resistance is not wanted, and there are devices available which have this property.

Thermistors are made with sintered oxides or sulphides of various metals which have a large temperature coefficient of resistance, and are deliberately allowed to get hot. Both positive (PTC) and negative (NTC) temperature coefficient thermistors are made. Of all the forms of construction, uninsulated rods or discs with metallised ends and lead wires are the most interesting to the amateur. Bead thermistors are occasionally met.

Disc PTC thermistors are used for over-current protection, either in air or enclosed in the device to be protected. Self-generated heat due to excess current or rise of ambient temperature will raise the resistance and so prevent further rise of current. PTC thermistors should not be used in constant-current circuits because that can result in ever-increasing power dissipation and therefore temperature (thermal runaway), so destroying the device.

PTC thermistors, unlike metals, do not exhibit a positive temperature coefficient over the whole range of temperatures likely to be encountered. Fig 2.2(a) shows this behaviour with temperature for a particular device.

One type of 'thermistor', not usually classed as such, is the tungsten filament lamp. Tungsten has a temperature coefficient of about 5000ppm/°C (see Table 2.2) and lamps operate with a temperature rise of some 2500°C. So an increase to about 12 times the cold resistance can be expected if the lamp is lit to full brilliance. These lamps are relatively non-inductive, and make good RF loads; however, the ohmic value may be uncertain.

NTC thermistors find a use for limiting in-rush current with capacitor-input rectifier circuits, as they start with a high resistance which decreases as they heat up: Fig 2.2(b). (Remember that while still hot after a short break, the in-rush current will not be limited on restart.)

Use is made of the self-heating property in bead thermistors for the stabilisation of RC oscillators. Also they find a use in temperature compensation of transistor bias circuits and temperature-compensated crystal oscillators (TCXOs), where self-heating is arranged to be negligible and the response to ambient temperature wanted.

The ohmic value of thermistors is specified at a low temperature, usually 20 or 25°C, and again at some higher temperature. Because of the dependence of resistance on temperature, thermistors can be used as temperature sensors, providing that there is a closely defined relation between temperature and resistance and self-heating negligible.

When self-heated, the thermistor's temperature may rise considerably, and allowance made for this in the mounting arrangement.

Non-ohmic resistors exist in which the slope resistance (change of voltage divided by change of current) is not constant, these being called *voltage-dependent resistors* (VDRs). Symmetrical VDRs are known by the trade name (Metal Oxide) Varistor, and exhibit no rectifying properties. The metal oxide now used is zinc oxide, replacing silicon carbide (Atmite, Metrosil or Thyrite), as its change of resistance with voltage is greater: Fig 2.2(c). There is a certain voltage for a particular type below which the current is negligible, and above this voltage the current through the device increases rapidly.

Varistors are much used for protection of susceptible circuits from over-voltage but, owing to their capacitance being of the order of nanofarads, they cannot be used for RF circuits. Their suitability for any particular purpose must be found from the catalogues. (Special zener diodes are also made for protection purposes.)

While the subject of light is still fresh, it is time to mention the *light-dependent resistor* (LDR). The resistance of cadmium sulphide and selenium shows great dependence on light. Perhaps the original Mullard ORP12 is the best-known cadmium

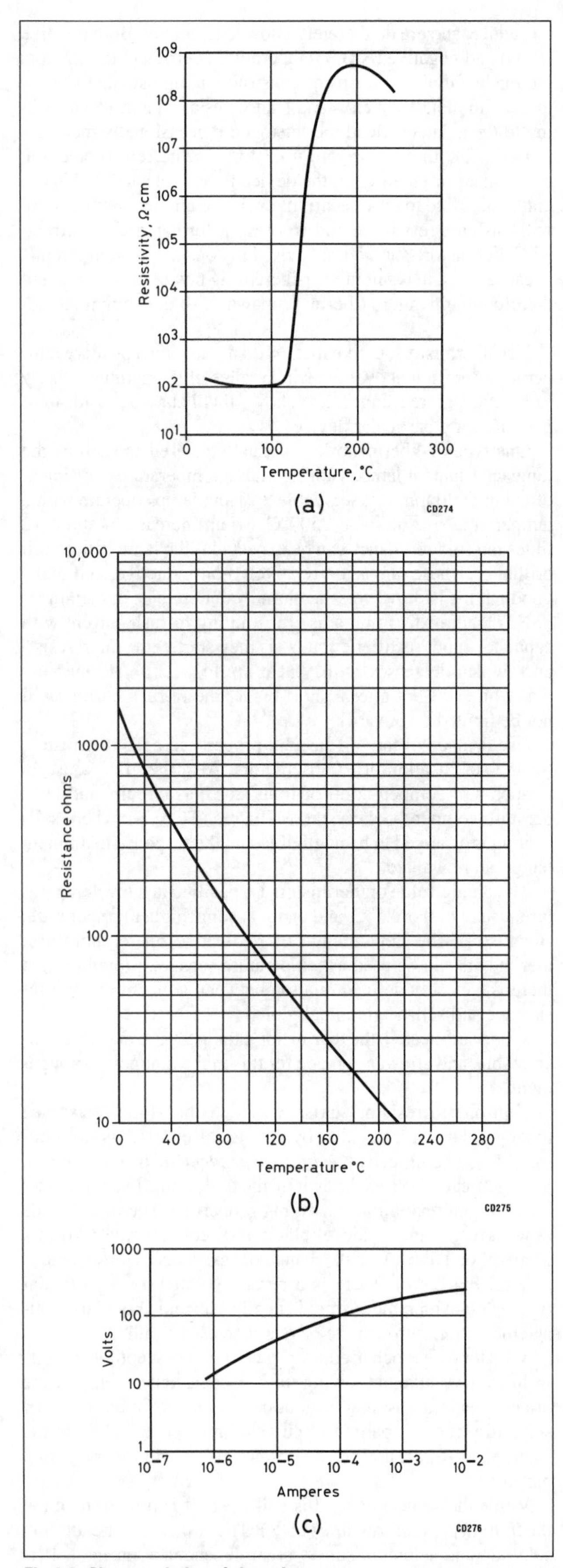

Fig 2.2. Characteristics of thermistors

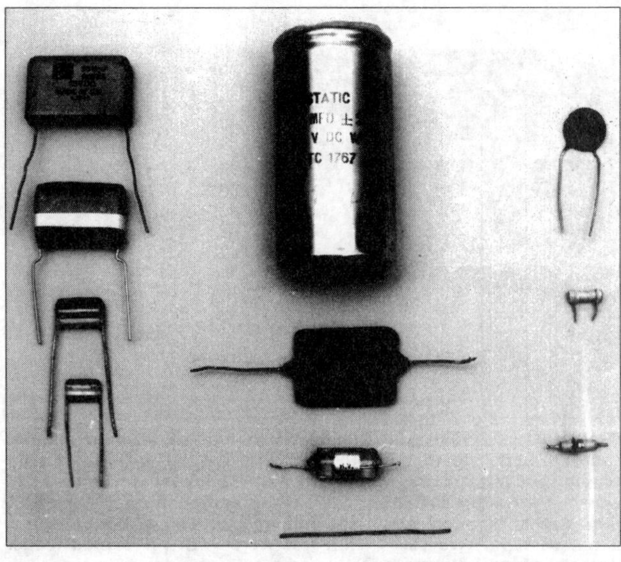

Fig 2.3. Fixed capacitors. Left column (top to bottom): 10µF, 63V polycarbonate dielectric, 2.7µF 250V, 0.1µF, 400V, 0.033µF, 250V all polyester dielectric. Centre column from top: 2.5µF 150V ±2% polyester, 240pF ±1% silver mica, 1nF ±2% polystyrene (all close tolerance). Right column from top: 4.7nF disc ceramic, 220pF tubular ceramic, 3nF feedthrough. Striped marker is 50mm long

sulphide LDR. In the dark, it has a resistance of at least 1MΩ, dropping to 400Ω when lit with 1000 lux. It has been used as part of an automatic level control in SSB transmitters.

FIXED CAPACITORS

To get the values wanted for amateur radio purposes into a reasonable size, capacitors are made with various dielectrics having a high value of dielectric constant (ε_r) according to the intended use of the capacitor. Fig 2.3 shows a selection of fixed capacitors.

Ceramic capacitors

These are the smallest and have a dielectric in one of three classes. Class 1 has a low ε_r and therefore a larger size than the other two classes for a given capacitance value. The loss factor is very low, comparable with silvered mica, and the value stays very nearly constant with temperature, applied voltage and life.

The designators 'COG' and 'NPO' refer to this class of capacitors, which should be used where stability of value and low loss are of greater importance than small size. Temperature compensating units are made with either N*** or P*** identification, where *** is the temperature coefficient in parts per million per degree Centigrade.

Class 2 comprises medium- and high-ε_r capacitors. They both have a capacitance value dependent on temperature, applied voltage and age, the effects being much greater for the high-ε_r material. (The capacitance lost by ageing can be restored by heating above the curie temperature, the value of which must be obtained from the manufacturer.)

'X7R' refers to a medium-ε_r material, 'Z5U' and 'Y5U' to the high-ε_r ceramic. Fig 2.4 shows the performance of these types with respect to age, temperature and voltage. Note the effect of voltage particularly. Leadthrough types are made in these materials for effective RF bypassing.

Class 3 have barrier-layer dielectric, giving very small size, poor loss factor and wide tolerance. Z5U dielectric has made these obsolete, but they may be found in surplus equipment.

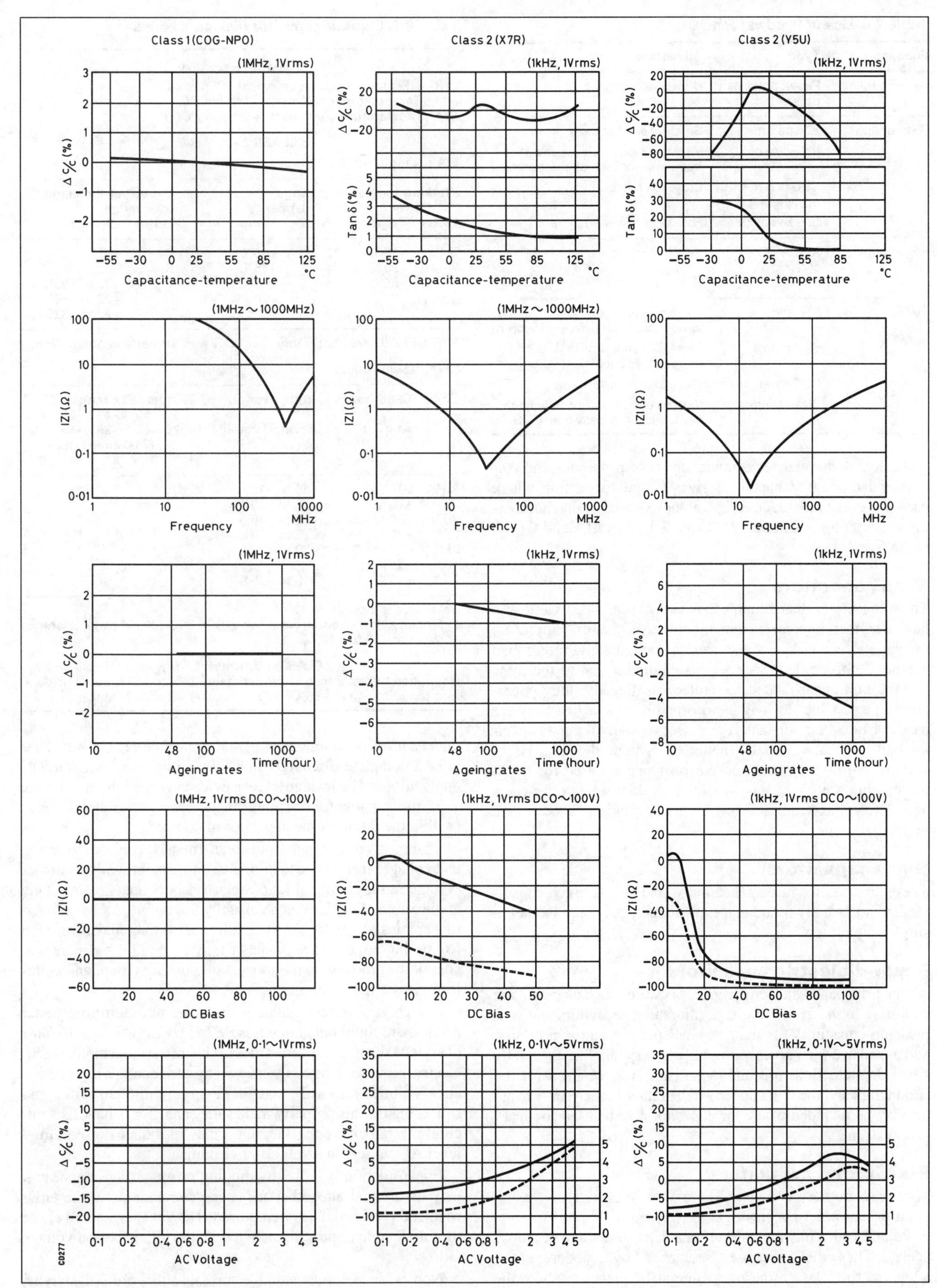

Fig 2.4. Ceramic capacitor characteristics

Table 2.4. Use of fixed capacitors

Function	Type	Advantages
AF/IF coupling	Paper, polyester, polycarbonate	High voltage, cheap
RF coupling	X7R ceramic	Small, cheap but lossy
	Polystyrene	Very low loss, low leakage but bulky and not for high temperature
	COG ceramic or silver mica	Close tolerance
	Stacked mica	For use in power amplifiers
RF decoupling	X7R or Z5U ceramic disc or feedthrough	Very low inductance
Tuned circuits	Polystyrene	Close tolerance, low loss, negative temperature coefficient (–150ppm)
	Silver mica	Close tolerance, low loss, positive temperature coefficient (+50ppm)
	COG ceramic	Close tolerance, low loss
	Class 1 ceramic	Various temperature coefficients available, more lossy than COG

Table 2.4 summarises ceramic capacitor properties and recommended usage. Chapter 22 gives the colour coding where used, but some manufacturers print the exponent value numbering system on the component. Table 2.5 gives EIA and CECC coding.

Table 2.5. Ceramic capacitor dielectric codes

CLASS 1 – ε_R < 500

COG, NPO temperature coefficient ±30ppm/°C
N*** Temperature coefficient: –***ppm/°C
P*** Temperature coefficient: +***ppm/°C

CLASS 2 – ε_r > 500

EIA coding

Working temperature range Lower Letter	Temp	Upper Figure	Temp	Capacitance change over range Letter	Change
Z	+10°C	2	+45°C	R	±15%
Y	–30°C	4	+65°C	S	±22%
X	–55°C	5	+85°C	T	+22 to –33%
		6	+105°C	U	+22 to –56%
		7	+125°C	V	+22 to –82%

X7R and Z5U are commonly met. X7R was formerly denoted W5R.

CECC 32100 coding

Code	Capacitance change over range At 0V DC	At rated voltage	Temperature range (°C) –55 +125 (Final code figure 1)	–55 +85 (2)	–40 +85 (3)	–25 +85 (4)	+10 +85 (6)
2B*	±10%	+10 to –15%		*	*	*	
2C*	±20%	+20 to –30%	*	*	*		
2D*	+20 to –30%	+20 to –40%				*	
2E*	+22 to –56%	+22 to –70%		*	*	*	*
2F*	+30 to –80%	+30 to –90%		*	*	*	*
2R*	±15%		*				
2X*	±15%	+15 to –25%	*				

Reference temperature +20°C.
Example: X7R (EIA code) would be 2R1 in CECC 32100 code, tolerance ±15% over –55 to +125°C.

CLASS 3 BARRIER LAYER

Not coded but tolerance approximately +50% to –25% over temperature range of –40 to +85°C. Refer to maker for exact details.

Mica capacitors

These are larger than ceramic capacitors for a given value and have a temperature coefficient of between +35 and +75ppm/°C, the variation being due to the material being natural rather than synthetic. Silver electrodes are plated directly on to the mica and the unit encapsulated for protection. Because they generally remain stable for long periods and have low loss they are used for tuned circuits and filters. Occasionally a mica capacitor will jump in value and this effect is unpredictable. High-voltage and high-current stacked-foil types are used for transmitters, but are not so easy to obtain as they were some years ago. The voltage rating and value are normally printed on the body.

Glass capacitors

These are used for rare applications up to 200°C. Such capacitors are not usually found in the catalogues, but could be met in surplus equipment.

Paper-dielectric capacitors

At one time paper-dielectric capacitors were much used, but are now only found as high-voltage smoothing capacitors and suppression capacitors for mains use. Paper capacitors are large and expensive for a given value, but are very reliable, especially when derated. Both foil and metallising may be used for the conductors and the units do tend to be rather inductive. A resistor may be included where the capacitor is to be used for spark suppression.

Plastic film capacitors

These are are very common. There are two basic types, using polar or non-polar plastics.

Polar plastics include polycarbonate, polyester and cellulose acetate. These dielectrics are characterised by a moderate loss, which can increase to a large value at frequencies where the plastic's molecule exhibits resonance. They also suffer from *dielectric absorption*, meaning that some charge reappears later after a complete discharge, and this is important in some DC applications. The insulation resistance is very high, making the capacitor suitable for coupling AC across a potential difference (within the rating of the capacitor of course).

Common values range from about 1nF to several microfarads, and voltages from 63 up to kilovolts, components with the higher voltage ratings being able to replace paper capacitors in many applications. Tolerance is not usually important as these capacitors are unsuitable for tuned circuits but, as they may be used in RC timing circuits, 5% or better may be bought for increased cost. Polycarbonate is the most stable of this group, and cellulose acetate the cheapest.

Polyethylene terephthalate is the material used for polyester and is more commonly known as *PETP*, *Mylar* (USA) or *Melinex* (Terylene when used for textiles). This material exhibits piezoelectric properties, as can be shown by connecting a high-resistance voltmeter to a dipped PETP high-value capacitor and squeezing it. This property makes the capacitor behave as a microphone, albeit a poor one, but it may introduce noise in low-level AF circuits in a vehicle environment.

Encapsulation is either by dipping or moulding in epoxy; it may be omitted altogether for cheap components where environmental protection is not required. The value is marked either by colour code or printing, along with the tolerance and voltage rating.

Non-polar materials suitable for capacitors are polystyrene, polypropylene and polytetrafluorethylene (PTFE). Capacitors

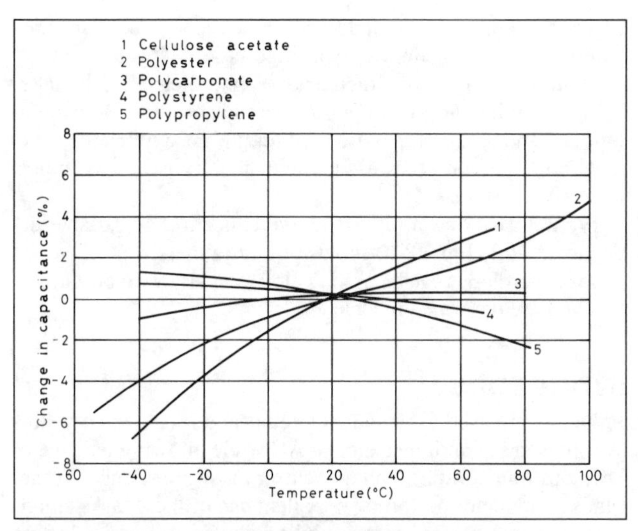

Fig 2.5. Temperature coefficients of plastic capacitors

made of these all have very low loss independent of frequency, but do tend to be bulky. Polystyrene capacitors in particular have a temperature coefficient around –150 ppm/°C, which makes them able to compensate for the similar positive temperature coefficient of ferrite-cored coils used in tuned circuits. Unfortunately polystyrene capacitors are limited to +70°C, and care must be taken when soldering them lest melting of the dielectric takes place. The outer foil of these may be marked with colour, and should be used for the more nearly grounded electrode.

Polypropylene is satisfactory up to +85°C and is recommended for pulse applications. Again, take care with the soldering.

Values, tolerance and voltage rating are usually written on the body. Small values are mainly of tubular axial construction, but radial and single-ended versions may be met with, of either tubular or boxed construction.

Fig 2.5 shows the effect of temperature on five types of plastic capacitor.

PTFE capacitors are not generally available for fixed values, but the material is used for trimmers, to be described later. Small lengths of PTFE (or if unobtainable, polystyrene) insulated coaxial cable make very good low-loss capacitors for use as tuning elements in multiband HF antenna systems. Cutting to an exact value is easy, as the capacitance per metre is quoted for types in common use.

Electrolytic capacitors

These are much used for values larger than about 1μF. The dielectric is a thin film of oxide on either aluminium or tantalum, and an electrolyte is used to contact the other plate. The electrolyte may be either solid or liquid (Fig 2.6).

These capacitors are usually made *polarised*, that is to say they have a positive and negative terminal. For some purposes non-polarised versions are made with an oxide film on both plates. Alternatively back-to-back polarised types can be connected as shown in Fig 2.7, which also gives some other hints on using electrolytics. However, tantalum capacitors can only withstand approximately 0.3V reverse bias, and it is not good practice to allow such reverse bias. Aluminium capacitors should never be reverse biased.

Electrolytics are rated for a voltage at a specified maximum temperature, and it is here that derating is most advisable. At full rating, some electrolytics have only 1000h life. There is always an appreciable leakage current, worst with liquid-electrolyte aluminium types. With non-use, gradual deformation of

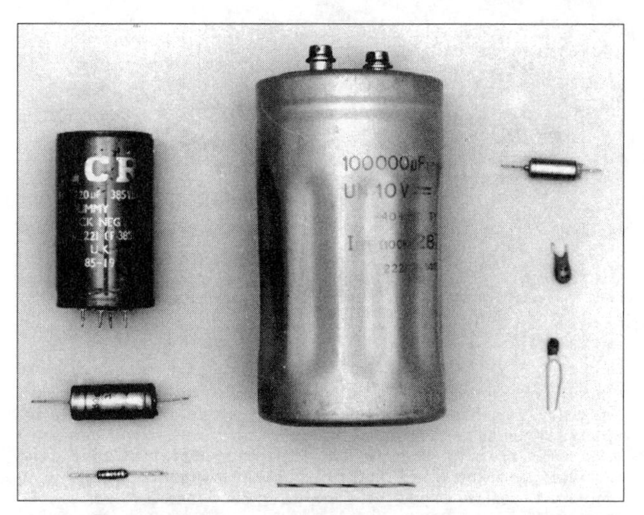

Fig 2.6. Electrolytic capacitors. Left, from top: 220μF 385V, 150μF 63V, 2.5μF 15V, electrodes. Centre: 100,000μF, 10V, all with aluminium. Right, from top: 68μF 15V, 22μF 25V and 22μF 6.3V, all tantalum types. Striped marker is 50mm long

the oxide layer takes place and to reform it a low voltage should be applied, being slowly raised to the rated voltage as the leakage current decreases.

Tantalum electrolytics should be protected from in-rush current, as this can lead to quick failure to a short-circuit condition. For all types, a ripple current is quoted and should not be exceeded to prevent abnormal rise of temperature.

Although capacitive reactance decreases with frequency, electrolytics are unsuitable for RF use. The innards may have considerable inductance and the equivalent series resistance (ESR) increases with frequency. So a small ceramic capacitor should always be parallelled with an electrolytic for RF use; the value depends on the frequency involved, 0.1μF to 1nF being common values. Special considerations apply for switch-mode converters, details of which are given in Chapter 14.

VARIABLE CAPACITORS

These are most often used in conjunction with inductors to form tuned circuits, but there is another use in RC oscillators (see Chapter 5). Small values (say less than 100pF) with a small capacitance swing are used as trimmers to adjust circuit capacitance to the required value.

Capacitors with vacuum as a dielectric are used for withstanding large RF voltages and currents. The capacitance is adjusted through bellows and the motion is linear. The range of adjustment generally only allows one amateur band to be tuned at a

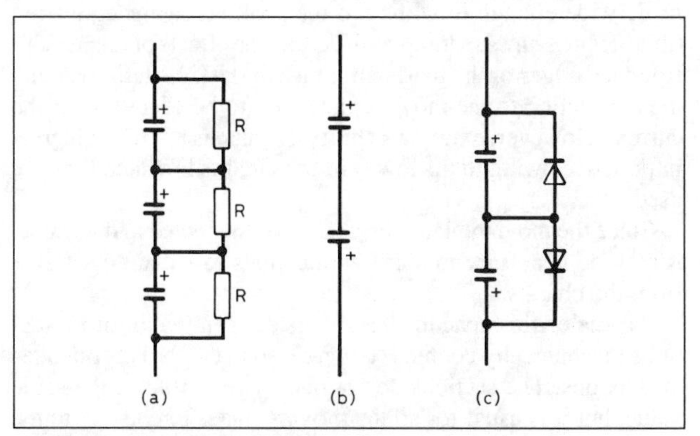

Fig 2.7. Circuits for electrolytic capacitors

Fig 2.8(a). Variable capacitors, Left: 500pF maximum, four-gang air-spaced receiver capacitor. Right: 125pF max transmitter capacitor with wide spacing. Bottom, left to right: 500pF max twin-gang receiver capacitor. Solid dielectric receiver capacitor. Single-gang 75pF max receiver capacitor. Twin-gang 75pF max receiver capacitor

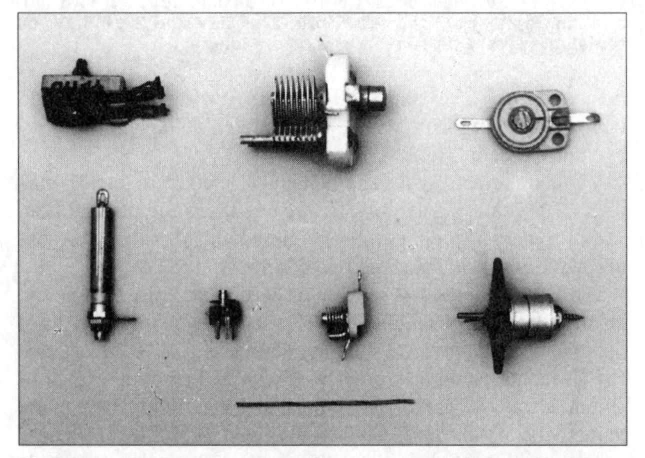

Fig 2.8(b). Trimmer capacitors. Top, from left: 50pF max 'postage stamp', 70pF max air dielectric, 50pF max flat ceramic. Bottom, from left: 27pF max tubular ceramic, 30pF max plastic film, 12pF max miniature air dielectric, 30pF max 'beehive' air dielectric

time. For the power levels allowed to amateurs, the expense of new vacuum capacitors is not justified.

Air-spaced parallel-plate variable capacitors are widely used, both for tuning and trimming. For precise tuning, for example, in a variable frequency oscillator (VFO), the construction must be such that the capacitance does not vary appreciably with temperature or vibration. This implies high-grade dielectric supports and robust, backlash-free mechanical design, not factors which make for cheapness.

For valve power amplifier use, the plate spacing has to be made wide enough to withstand the peak voltage encountered (this includes any superimposed DC and the effects of mismatch). Rounded edges of the plates minimise the risk of flash over and the connection to the moving plates has to be adequate for the current. However, extreme stability of value is not of such great importance owing to the low Q of the circuits in which they are used.

Since the moving plates are generally connected to the frame, it may be necessary in some applications to isolate the frame from the chassis.

Air-dielectric capacitors can be made with two or more sections mechanically coupled (*ganged*), so that one shaft operates all sections. The sections do not necessarily all have the same value, but it is usual for all the moving plates to connect to the frame. Some very small capacitors made for tuning transistor radios have plastic-film dielectric; these are not recommended where precise capacitors are specified.

Trimmers may have either air or a solid dielectric, the latter being a low-loss non-polar plastic, ceramic or mica. Both parallel-plate and tubular types are used for precise applications, and mica compression types used where stability is of less importance than cost or size.

Fig 2.8(a) shows a representative collection of variable capacitors, while Fig 2.8(b) shows some trimmers.

Junction diodes can be used as electronically variable capacitors and are described in Chapter 3.

INDUCTORS

Inductors are used for resonant circuits and filters, energy storage, presenting an impedance to AC while allowing passage of DC, and transforming voltage, current and impedance. Sometimes an inductor performs more than one of these tasks simultaneously, but it is easier to describe the wide range of inductors by these categories.

Resonant circuits

Either air- or magnetic-cored coils are used for tuned circuits. In the latter case provision may be made for some variation by altering the position of the core. The criterion of goodness for a tuning coil is Q, the ratio of reactance to resistance. With air-cored coils, the ratio of diameter to length should be greater than 1 for maximum Q, but Q only falls off slowly as the length increases. It is generally more convenient to use a long thin coil rather than a short fat one, accepting the slight loss of Q.

Calculation of the inductance of air-cored coils is difficult, and usually done by the use of charts. Fig 2.9 shows some different air-cored coils that may be used, including a variable type much sought after for linear amplifiers and antenna tuning units.

Magnetic cores may be either ferrite for the lower radio frequencies and iron powder for VHF. Screw cores can be used for adjustment in nominally air-cored coils or in closed magnetic circuits. Copper or brass slugs are used to reduce the inductance for tuning purposes. For closed magnetic circuits, a factor called A_L is quoted for the extreme positions of the core. A_L is the number of nanohenrys that one turn would have. As inductance is proportional to the square of the number of turns, the inductance of a wound coil of n turns would be n^2A_L. A number of different magnetic core assemblies are shown in Fig 2.10.

There are two common types of ferrite material: manganese zinc (Mn/Zn) and nickel/zinc (Ni/Zn). Mn/Zn has lower resistivity than Ni/Zn and lower losses due to *hysteresis*. This latter effect is caused by the magnetic flux lagging behind the magnetising field, so making a resistive component to the alternating current in any coil wound round the material. As the applied frequency is increased, eddy currents and possibly dimensional resonance play a greater part, and the higher resistivity Ni/Zn ferrite has to be used. Unfortunately Ni/Zn has greater hysteresis loss. Both types of ferrite saturate in the region of 400mT, this being a disadvantage in high-energy uses (*saturation* is when increasing the magnetic field ceases to produce the same increase of flux as it did at lower values).

There are many varieties of ferrite cores on the market, for whose properties it is best to consult the catalogues. There are also microwave ferrites such as yttrium iron garnet.

Inevitably coils will possess not only inductance but undesired resistance and distributed capacitance. The latter confuses the measurement of inductance, and limits the tuning range with a variable capacitor. However, when winding a tuning coil it

Fig 2.9(a). Air-cored inductors. Top: high-power transmitting on ceramic former. Left: VHF inductors moulded in polythene, the bottom one in a shielding can. Centre: small air-cored HF inductor (could have an iron dust core). Right: HF inductor wound on polystyrene rod. The striped marker is 50mm long

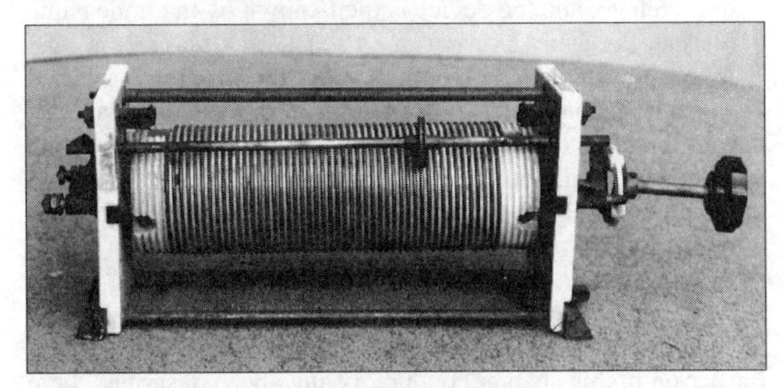

Fig 2.9(b). 'Roller-coaster' variable inductor

may be necessary to use a coating for mechanical reasons. All 'dopes' have a dielectric constant (ε_r) greater than one and so will increase the self-capacitance. Polystyrene, with an ε_r of only 2.5, dissolved in toluene (hazardous vapour) is recommended.

Energy-storage inductors

The type of core used will depend on the frequency. At mains supply frequency, silicon iron is the only choice. To minimise eddy current loss caused by the low resistivity of iron, thin laminations are used. A tape of grain-orientated silicon steel (GOSS), whose grains lie along the major direction in which the magnetising field also lies, can also be used. To prevent core saturation, a gap is left in the magnetic circuit. The calculation of core size, gap and turns is outside the scope of this book.

At the higher frequencies used in switch-mode converters, ferrites are widely used in the form of E-shaped half-cores used in pairs. As with silicon iron, a gap is required to prevent saturation, and because of the low saturation flux, most of the energy is actually stored in this gap. The ferrite manufacturers' data books give most of the information required for design.

Chokes

While energy-storage inductors are often called 'chokes', the term is used for inductors which pass DC but impede AC. The materials used are as for energy-storage inductors at the lower frequencies, but at higher frequencies where ferrites are too lossy,

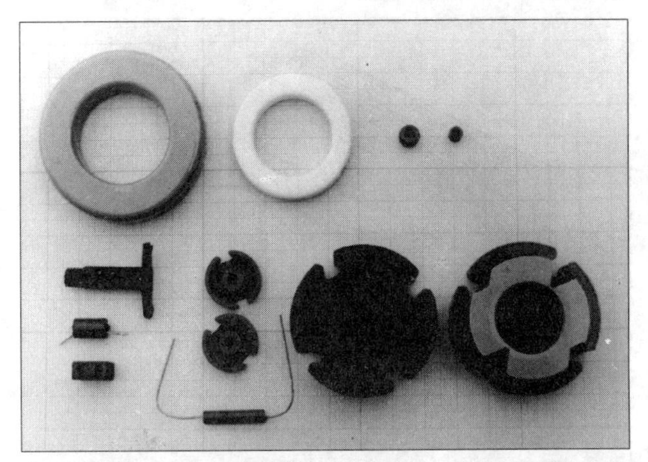

Fig 2.10. RF magnetic materials. Top, left to right: large, medium and small ferrite rings, and a ferrite bead with a single hole. Below, left top: tuneable RF coil former; middle: six-hole ferrite bead with winding as an RF choke; bottom: ferrrite RF transformer or balun former. Centre: small pot cores and an RF choke former with leads moulded into the ends. Right: large pot core with former

powdered-iron alloys are used. At even higher frequencies (in the VHF region), air-cored chokes must be used.

There is a problem with self-resonance in RF chokes. It is impossible to avoid self-capacitance and at some frequency this will resonate with the inductance to give a very high resistance. This is ideal but at frequencies higher than this the so-called inductor will behave as a capacitor whose impedance decreases with frequency. The situation is complicated by the fact that self-capacitance is distributed, and there may be multiple resonances. For feeding HT to valve anodes in linear amplifiers, a choke is required which must not have resonances in any band to be used. Where available, a wave-winder is used to wind 'pies' of differing sizes which are put in series. A less-effective, but simpler alternative is to sectionalise a single-layer solenoidal winding. Fig 2.11 shows a variety of chokes.

Transformers

Chapter 1 described *mutual inductance*, and this property enables transformers to be made. The magnetic flux from one coil is allowed to link one or more other coils where it induces an EMF. Two types of transformer are met, *tightly coupled* and *loosely coupled*. With tight coupling, the object is to make all the flux from the first (primary) coil link the other (secondary)

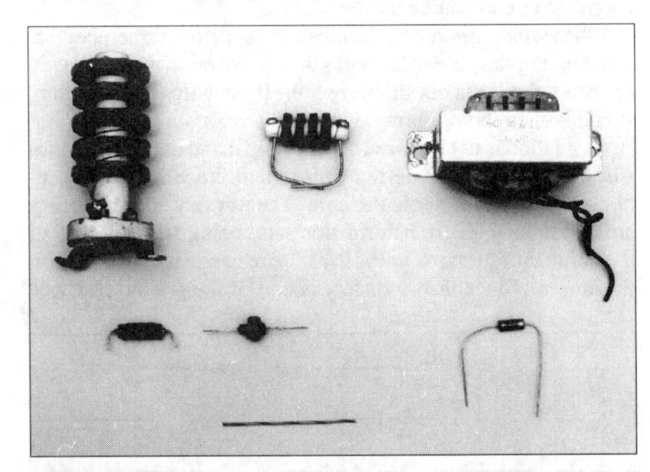

Fig 2.11. RF and AF chokes. Top, left to right: 1mH large, 1mH small RF chokes and 15H AF type. Bottom, left to right: 10µH single layer, 300µH 'pile' wound, 22µH encapsulated. Striped marker is 50mm long

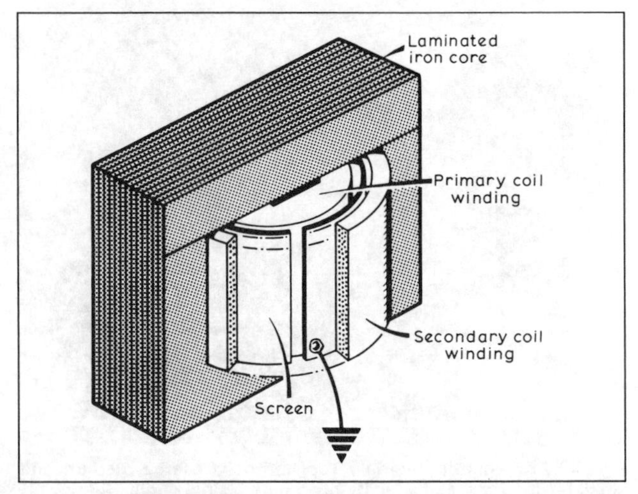

Fig 2.12. A Faraday screen is an earthed sheet of copper foil between the primary and secondary windings. For clarity, a gap is shown here between the ends, but in reality they must overlap while being insulated from one another to avoid creating a shorted turn. As copper is non-magnetic, mutual induction is not affected. The screen prevents mains-borne interference from reaching the secondary winding by capacitive coupling. Some transformers may have another screen (shorted this time) outside the magnetic circuit to prevent stray fields

coils. In this case, the ratio of the induced EMF to the applied EMF is that of the turns ratio of secondary to primary. Magnetic cores can be used, the type being chosen in the same way as for the inductors previously described. Formulae for calculation of the required turns are given in Chapter 22 with the wire data.

Leakage inductance, caused by all the flux from the primary not linking the secondary, is minimised by interleaving the primary and secondary windings; however, this does increase the inter-winding capacitance. Where insulation between primary and secondary is important, a divider is built into the bobbin, the primary being on one side and the secondary on the other or, alternatively, two concentric bobbins may be used. These methods of construction increase the leakage inductance, and are responsible for the poor regulation of many commercial 'safe' transformers.

If capacitive coupling between primary and secondary must be minimised, a Faraday screen is inserted: see Fig 2.12 for the way this is done. It is essential to avoid creating a shorted turn. Where stray field is to be reduced, a shorted turn outside the magnetic circuit can be used.

When winding cored transformers or inductors, remember that some part of the bobbin's cheeks will be covered by the core: do not bring the leads out in this region! If the wires are very thin, skeining them before bringing them out makes a sounder job.

If isolation is not required between input and output, the *auto-transformer* will be preferred. This will increase the power handling capacity of a particular core. The primary and secondary are continuous, the transformation ratio being still the ratio of the number of turns, as in Fig 2.13. Wire gauge is chosen to suit the current. Since the secondary is part of the primary (or vice

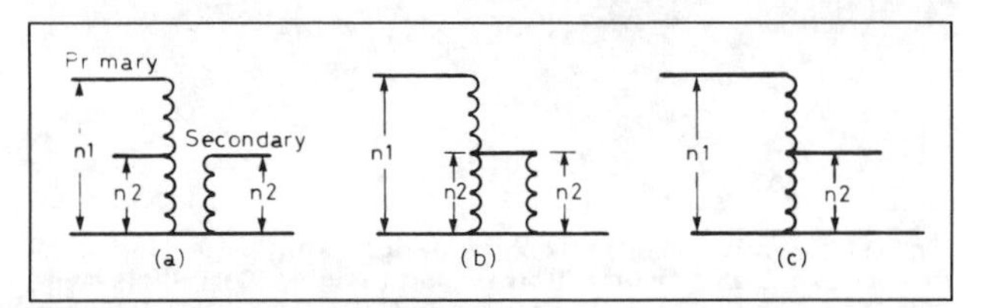

Fig 2.13. The auto-transformer

Fig 2.14. A Variac

versa), more effective use is made of the winding window. The tapping point can be made variable to allow adjustment of the output voltage, and the device is then known by the trade name of Variac (Fig 2.14).

In loosely coupled transformers where the windings are tuned by capacitors, the degree of coupling controls the bandwidth of the combination (Fig 2.15). Such transformers are widely used in IF stages. Again, ferrite, iron-dust or brass cores may be used in the appropriate frequency range. Loose coupling is not used in mains transformers.

MATERIALS

Earlier on in this chapter conductors have been described; here properties of some insulators used in construction will be considered. These will be grouped into three categories: inorganic, polar and non-polar plastics. Permanent magnets will also be mentioned.

Mica (mainly impure aluminium silicate), glass and ceramics are the most likely inorganics to be used (asbestos is out of favour at the moment!). Vitreous quartz is one of the best insulators but it is not generally available to the amateur. Glass and ceramics are used for antenna insulators. Glass has a tendency to collect a film of moisture on its surface, but the leakage will not be serious for general use; if it is then a coating of silicone varnish will stop moisture formation.

Ceramic coil formers leave little to be desired, but are becoming increasingly difficult to obtain.

Non-polar plastics such as polystyrene, polypropylene and PTFE are very good insulators at all amateur frequencies. Polypropylene rope, because of its electrical properties, avoids the use of insulators at the end of an antenna which is convenient for portable operation.

Polar plastics such as bakelite (Tufnol), nylon, Perspex and PVC have higher losses which are frequency dependent. These should not be used to withstand high RF stress. Some protective varnishes come into this category, being made for mains frequency use, and only the polystyrene dope previously described should be used on RF coils.

Permanent magnets may either be a special steel, ceramic or rare-earth alloy, being discovered in that order. Steels such as

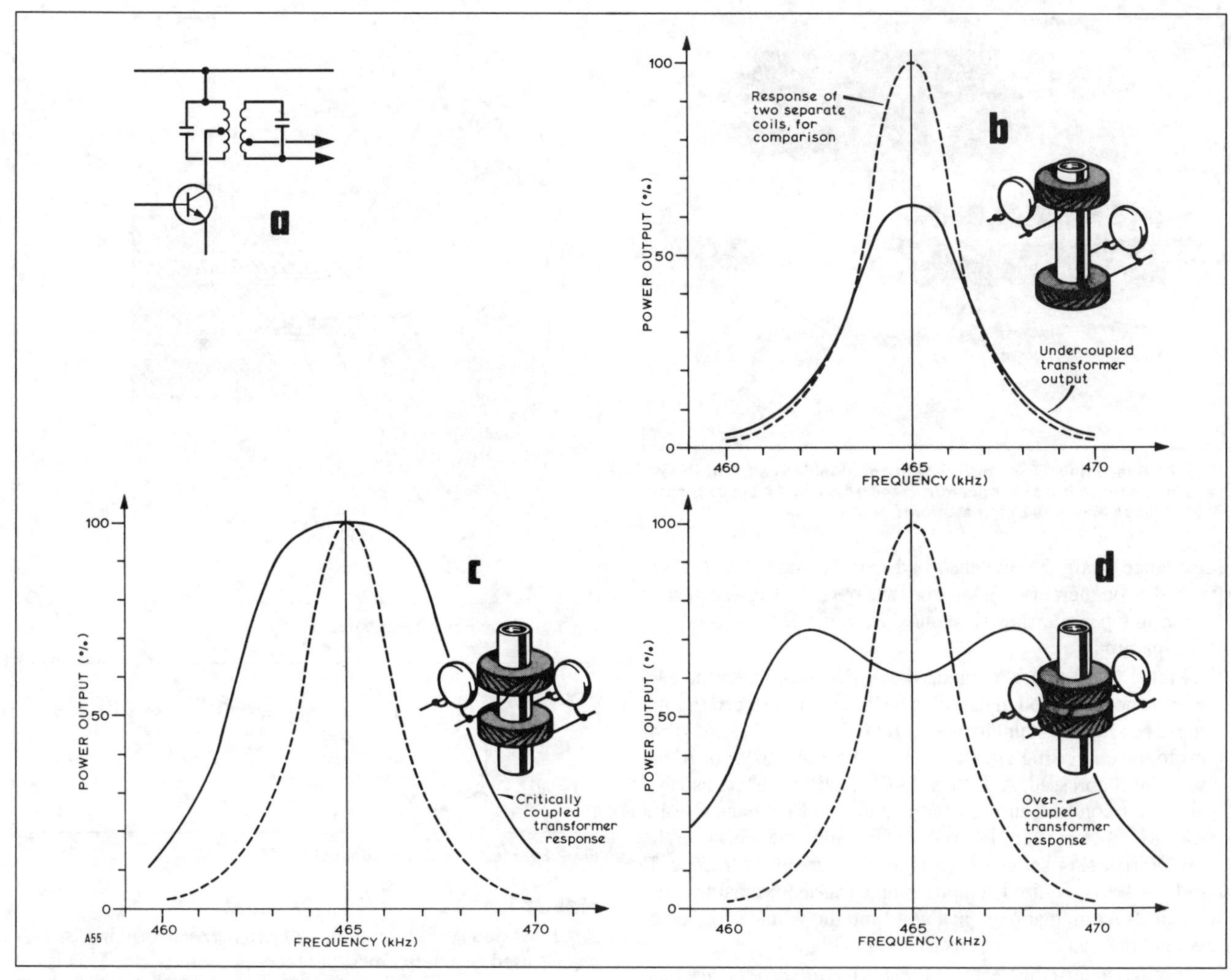

Fig 2.15. Bandwidth of coupled circuits

Alnico and Ticonal were much used in loudspeakers and DC machines, but now have been replaced by the other two materials. Ceramic magnets are cheap and withstand being left without a keeper better than steel. Rare-earth magnets (samarium cobalt or neodymium-boron-iron) store more energy than the other two, are very good on magnetic open-circuit and are expensive.

RELAYS

The Post Office 3000 and 600 type relays are still widely available in amateur circles (Fig 2.16). However, their openness is a disadvantage and many different sealed types are now available, most with lower operating power requirements. Different contact materials suit differing needs, such as very high or very low currents. The abbreviations 'NO', 'NC', and 'CO' stand for 'normally open', 'normally closed' and 'change-over', the first two referring to the state when not energised. Some manufacturers use 'Form A' for 'NO', 'Form B' for 'NC' and 'Form C' for 'CO'. Multiple contacts are indicated by a number in front of the abbreviation.

Latching relays are made which are stable in either position and only require set and reset momentary energisation through separate coils or a pulse of appropriate polarity on a single coil. These relays use permanent magnet assistance, meaning that correct polarity must be observed.

It is possible to get relays for operation on AC without the need for rectifiers, the range of operating voltages including 240V, 50Hz.

If RF currents are to be switched, reed relays (Fig 2.17) are suitable, or alternatively coaxial relays which maintain a stated

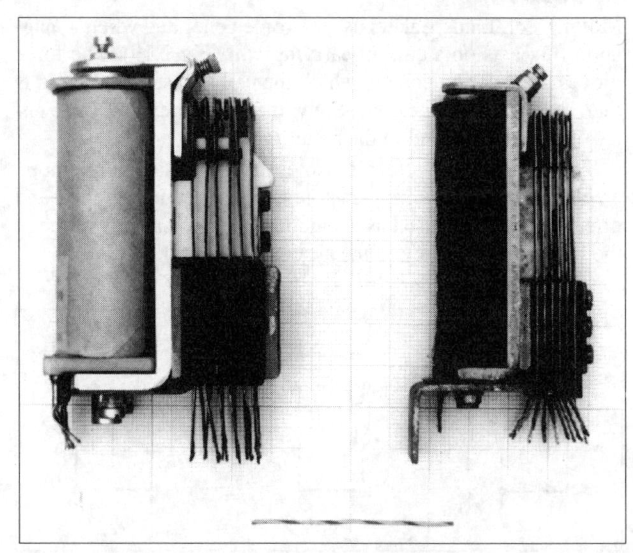

Fig 2.16. Post Office relays. Left: type 3000; right: type 600

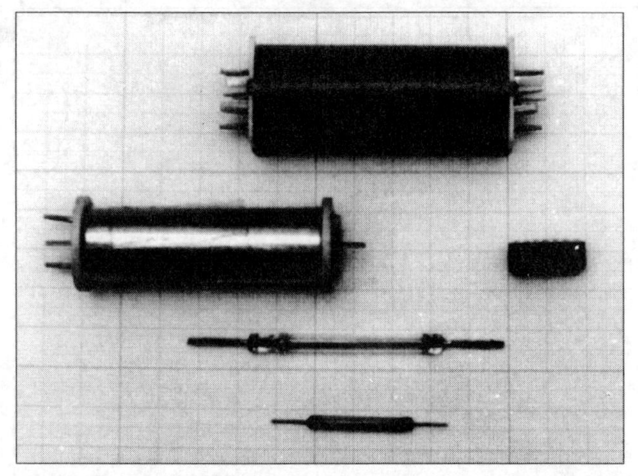

Fig 2.17. Reed relays. Top: four reeds in one shielded coil; middle left: reed in open coil; right: miniature reed relay in 14-pin DIL case. Below: large and small reed elements

impedance along the switched path can be used. Reed relays (with dry or mercury-wetted contacts) are among the fastest operating types and should be considered if a relay is used in a break-in circuit.

A relay is by nature an inductive device and, when the current is turned off through its coil, a back EMF is induced in such a sense as to try to maintain the current. This could cause problems to the energising circuit, in particular a transistor or IC, so it must be suppressed. A diode across the coil, connected so that it does not conduct during energisation, will conduct when a back EMF is created but in so doing will delay the release of the relay. Varistors or series RC combinations are also effective. If rapid release is required, a high-voltage transistor must be used with an RC combination which will limit the voltage applied to less than the maximum of this transistor (Fig 2.18).

Another undesirable feature of relays is *contact bounce* which may introduce false signals in a digital system. Mercury-wetted reed relays solve this problem, but some will only work in a particular orientation.

ELECTRO-ACOUSTIC DEVICES

Loudspeakers

Moving-coil loudspeakers use the force generated when a magnetic flux acts on a current-carrying coil (Fig 2.19). The force moves the coil and cone to which the coil is fastened. Sound is then radiated from the cone. Some means of preventing the entry of dust is provided in the better types.

The piezoelectric effect is also used to move a diaphragm by an electric field. The frequency response may not be good and often the device is used as a sounder to radiate a single tone at the resonant frequency of the system (Fig 2.20).

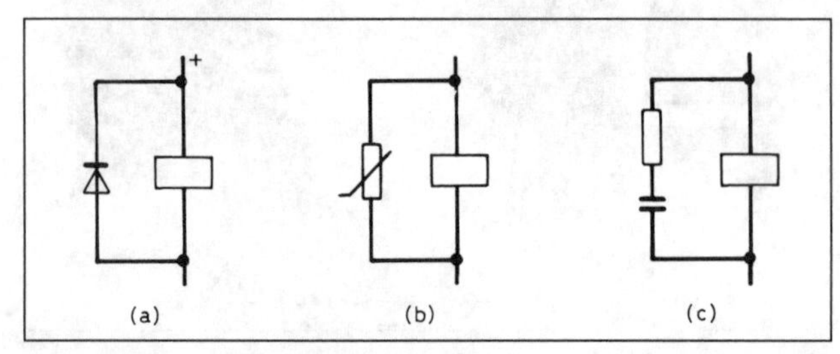

Fig 2.18. Relay coil suppression circuits

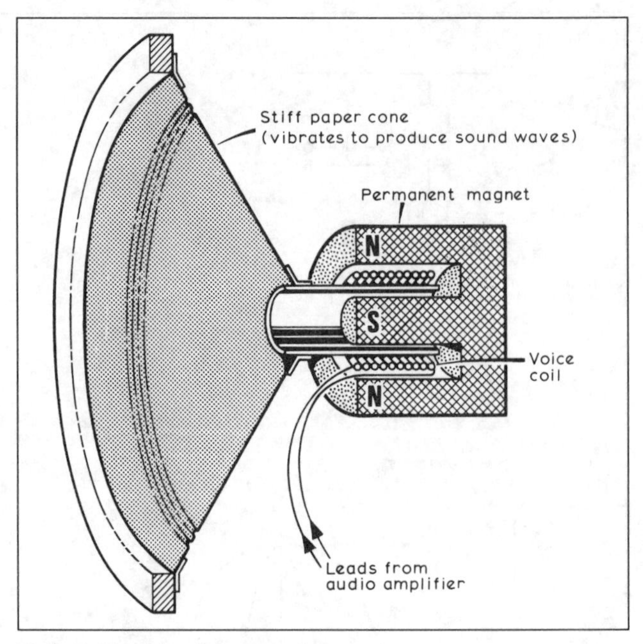

Fig 2.19. Moving-coil loudspeaker

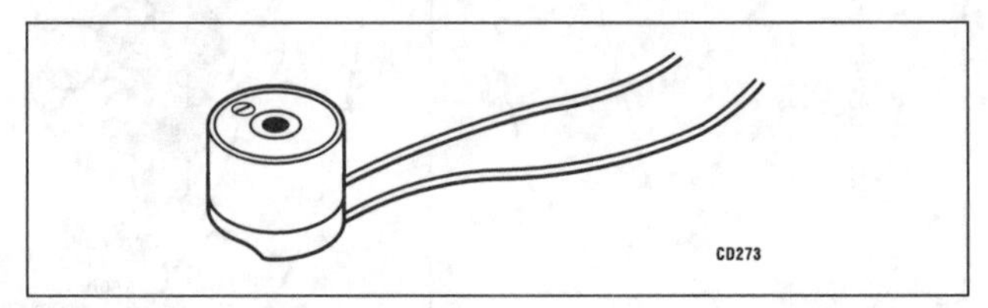

Fig 2.20. Piezoelectric loudspeaker

Headphones

Small versions of the moving-coil and piezoelectric loudspeakers are used as headphones, but there is another sort. This is the *diaphragm* type, where a magnetic diaphragm is attracted by a permanent magnet with an audio field superimposed on it (Fig 2.21). The permanent field is necessary because magnetic attraction is proportional to the square of the current, and would give rise to second-harmonic production if an over-riding steady field were not present.

Microphones

The moving-coil loudspeaker in miniature forms the *dynamic microphone*. It is possible to use the same unit both as a loudspeaker and microphone, often in hand-held equipment. The impedance is generally below 100Ω, and suitable matching must be performed.

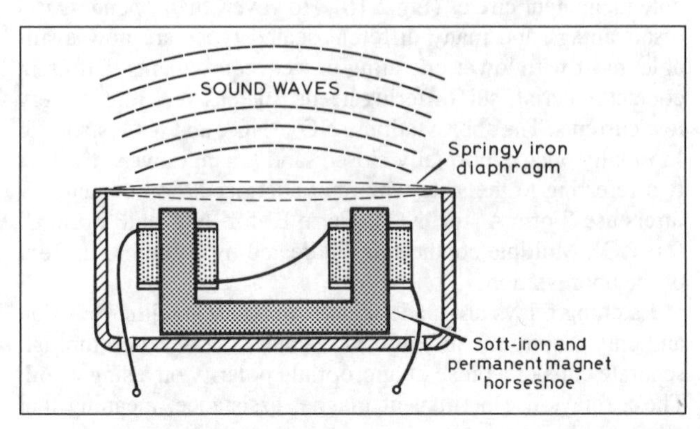

Fig 2.21. Moving-iron headphone

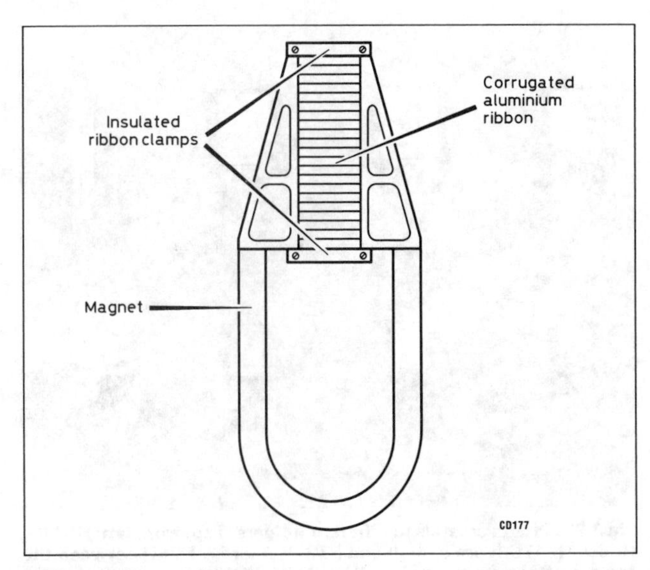

Fig 2.22. Ribbon microphone

A variant, the *velocity* or *ribbon microphone*, uses a thin aluminium ribbon suspended between the poles of a permanent magnet. The ribbon moves with the velocity of incident sound waves and so induces a small EMF. Like a dipole antenna it has a figure-of-eight polar diagram. If spoken to at close range, low frequencies are emphasised. As the ribbon impedance is only a fraction of an ohm, a matching arrangement must be used. Due to its bulk it is not much in vogue today, but there are some small American velocity microphones on the surplus market (Fig 2.22).

The piezoelectric loudspeaker of Fig 2.20 is reversible, and a smaller version forms the basis of the crystal microphone. Its equivalent circuit is a high capacitive reactance in series with a small resistance. The effect of this is to cause a loss of low frequencies if the load has too low a resistance; FETs make a very suitable load. Moisture should be avoided, as some crystals are very prone to damage.

It is possible to use the diaphragm headphone as a microphone, since the effect on which it is based is reversible, and some ex-service microphone capsules are made this way.

Probably the carbon microphone is the oldest still in common use, one form of construction being shown in Fig 2.23. Sound waves alternately compress and relieve the carbon granules, altering the current when energised by DC. It amplifies, but is basically noisy and subject to making even more unpleasant blasting noises if overloaded acoustically. The impedance is in the hundreds of ohms region.

The above types can be made 'noise cancelling' by allowing noise to be incident on both sides of the diaphragm with little effect. Close speaking will influence the nearer side of the diaphragm more, improving the voice-to-ambient-noise ratio.

The so-called *condenser microphone* has received a new lease of life under the name of *electret microphone*. The condenser microphone had a conductive diaphragm stretched in front of another electrode and a large DC potential difference was maintained across the gap. Variations in position of the diaphragm by sound waves cause a change of capacitance and hence an AF current flows. A large series resistance was included in the circuit, and this AF current created a voltage. Normally a valve amplifier was included in the assembly to avoid long high-impedance leads.

In the electret microphone, the DC bias is replaced by the field from an *electret*, which is the electrostatic equivalent of a

Fig 2.23. Microphone inserts. Left: carbon mic. Centre: dynamic mic. Right: electret mic

permanent magnet. The valve amplifier is replaced by an FET, making a very small unit possible.

QUARTZ CRYSTALS

Quartz, a mineral composed of silica (the main constituent of sand) is a *piezoelectric* material, ie it generates an electric charge between two faces when it is strained, eg when it is compressed, bent, sheared or twisted. Also, if subjected to an electric charge, deformation takes place.

The basic quartz crystal is hexagonal in section and has hexagonal points. The piezoelectric effect only takes place at certain angles to the main crystal axis. If a slice is cut at the correct angle to the axis, it can act like a bell and ring if struck. Ringing is of course oscillation and, with small crystal slices, will take place in the megahertz region. By adding an amplifier circuit with positive feedback in the correct phase, these oscillations can be maintained (see Chapter 5 – 'Building blocks') and a stable oscillator results where the frequency of oscillation is determined almost entirely by the size of the crystal, ie by the mechanical characteristics of the quartz and the type of strain involved; very small changes can be made by *pulling* by means of a capacitor or capacitor/inductor combination. The charge is communicated to the crystal by means of electrodes plated on to the surface.

The type of deformation, stretching, bending, shearing or twisting, depends on the angle of cut and detailed discussion of this is out of place here. Details can be found in reference [1].

The quartz crystal behaves as a tuned circuit with a very high Q (see Chapter 1 – 'Principles'). As an example, a 7MHz crystal in a modern holder may have a Q of 100,000 and have the equivalent circuit shown in Fig 2.24(a) where C_s, L_s and R_s represent the properties of the crystal and C_p the shunt capacitance of the electrodes and the stray capacitance of the holder. The measured values of these parameters are strange when compared with a normal LC circuit for the same frequency as Table 2.6 shows.

Note the very large equivalent inductance of the crystal. Also a Q of 300 for a coil/capacitor tuned circuit is very good.

If the reactance of a crystal is plotted against frequency (Fig 2.24(b)), the resulting curve shows two resonances, a low-impedance or series resonance and a high-impedance or parallel resonance. The frequencies are very close together and which is excited will depend on the oscillator circuit. In a crystal of, say, 7MHz, the difference between these two frequencies is of the order of a kilohertz. It depends on the exact size and shape of the electrodes.

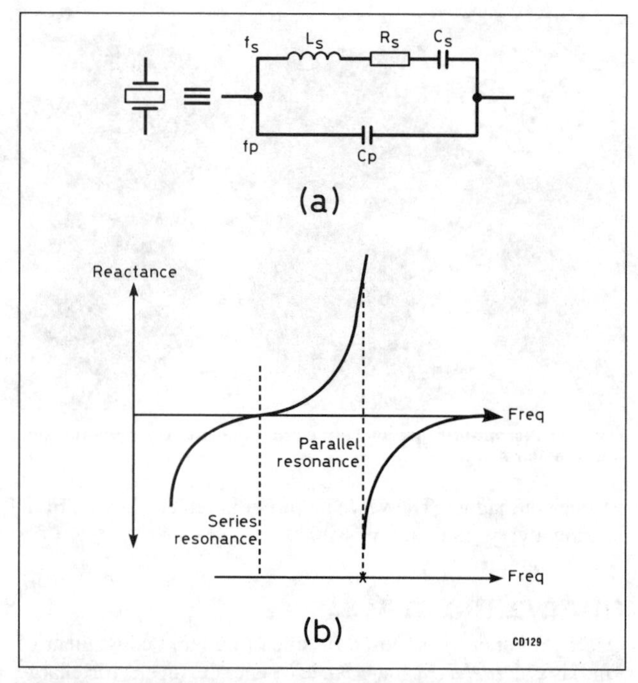

Fig 2.24. (a) Equivalent circuit of a piezoelectric crystal as used in RF oscillators or in filter circuits. Typical values for a 7MHz crytal are given in Table 2.6. C_p represents the capacitance of the electrodes and the holder. (b) Typical variation of reactance with frequency. The series and parallel reactances are normally separated by about 0.01% of the frequencies themselves (eg about a kilohertz for a 7MHz crystal)

Table 2.6. Parameters of a crystal compared with an LC circuit

Parameter	7MHz crystal	7MHz LC
L_s	42.5mH	12.9µH
C_s	0.0122pF	40pF
R_s	19Ω	0.19Ω
Q	98,000	300

The stability of the frequency of a crystal oscillator depends on the temperature of the crystal and, to some extent, on its age. The frequency-temperature relationship depends greatly on the cut of the crystal and manufacturers' data should be consulted for details. Probably the most useful for amateurs is the type which has a flat region (ie no change of frequency) over the range 20 to 30°C. All crystals change frequency as they age. The amount is small and depends on the quality of the crystal and on whether or not it has been subject to abuse such as dissipating too much power. In an oscillator, there can be power dissipated in the crystal by having too large a feedback. In general, crystals behave best if loaded as lightly as possible commensurate with them providing the output required.

Quartz crystals can normally be cut to resonate at frequencies up to about 30MHz; above this, they are too fragile to be of practical value. An overtone crystal is therefore used which oscillates at a frequency which is nearly equal to an odd multiple of the fundamental frequency. It will still try to oscillate at its fundamental but can be forced into its overtone by the external circuit. Most crystals can be forced into overtone mode but those specially designed for it are best.

Since quartz crystals act as tuned circuits, they can also be used as filters. The use of cheap, readily available computer or colour TV crystals in filters has been well documented [2, 3].

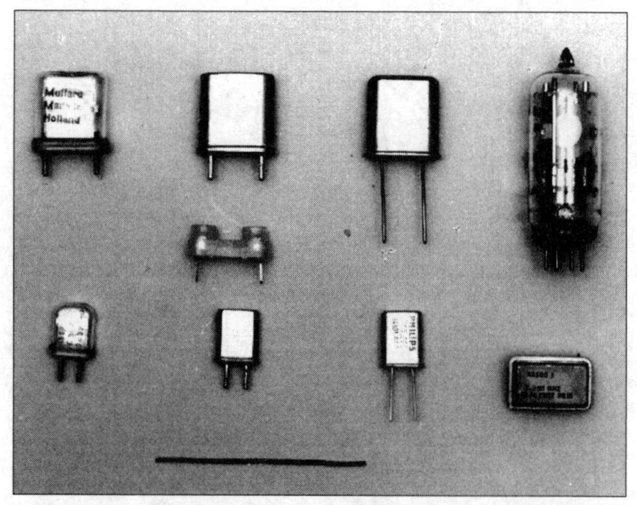

Fig 2.25. Quartz crystals in different holders. Top, from left: HC27U, HC6U, HC47U (wire ended) and B7G. Holder for HC6U between the rows. Bottom, from left: HC29U, HC25U, HC18U (wire ended) and a packaged crystal oscillator. The striped bar is 50mm long

Ceramic resonators

Certain ceramics, notably those containing titanium and/or zirconium oxides, have similar properties to quartz and are used as resonators for oscillators and filters. They have a lower *Q*, a lower maximum frequency and a higher temperature coefficient but these are offset by a lower price. They are also only available in a limited range of frequencies.

There is another type of ceramic resonator which works at microwave frequencies and is widely used in satellite TV receivers. These may become available in the amateur market eventually.

A further type of resonator is the *surface acoustic wave* (SAW) filter. In this the ceramic (usually a single crystal of lithium niobate) has printed on its surface a pattern of electrodes with which acoustic energy (sound but very much ultrasonic) can be launched and received. The precise pattern determines the frequency response. These are widely used in TV receivers at IF, say around 35–45MHz. They are produced for much higher frequencies.

Mechanical design

Quartz crystals must be protected from the environment and are mounted in holders; these are usually evacuated. There are several types and sizes (Fig 2.25) but basically they are either in metal or glass. If in metal, the cheapest holder is solder sealed, next is resistance welded and the most expensive is cold welded. Fig 2.25 shows the types of case usually met in amateur equipment, together with a crystal in a B7G valve enclosure.

REFERENCES

[1] *Shortwave Wireless Communication,* Ladner and Stoner, 4th edn, Chapman and Hall, London, 1946, pp308–338. Note that this is an old reference and it does not show modern crystal designs but is very good in showing the 'cuts' of crystals.

[2] G3JIR, *Radio Commmunication* 1976, p896; 1977, pp28, 122 and 687.

[3] G3OUR, *Radio Communication* 1980, p1294; 1982, p863.

3 Semiconductors

THE development of semiconductor technology has had a profound effect on all areas of electronic engineering and telecommunications. Since the first transistors were made in the late 'forties, an amazingly diverse range of semiconductor devices has appeared, ranging from the humble silicon rectifier diode used in power supplies to specialised GaAsFETs (gallium arsenide field effect transistors) for low-noise microwave amplification, and also digital LSICs (large-scale integrated circuits), which have made powerful desktop computers a reality, and involve the fabrication of many thousands of transistors on a single chip of silicon.

Within amateur radio, there are now virtually no items of electronic equipment that cannot be based entirely on semiconductor, or 'solid-state', engineering. However, thermionic valves may still be found in some high-power linear amplifiers used for transmission (see Chapter 5).

SILICON

Although the first transistors were actually made using the semiconductor germanium (atomic symbol Ge), this has now been replaced almost entirely by silicon (Si). The silicon atom, shown pictorially in Fig 3.1, has a central nucleus consisting of 14 protons, which carry a positive electrical charge, and also 14 neutrons. The neutrons have no electrical charge, but they do possess the same mass (atomic weight) as the protons. Arranged around the nucleus are 14 electrons. The electrons, which are much lighter particles, carry a negative charge. This balances the positive charge of the protons, making the atom electrically neutral and therefore stable.

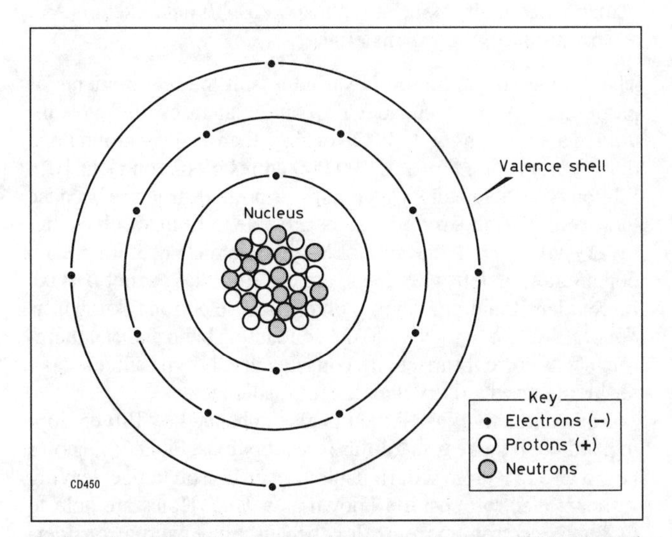

Fig 3.1. The silicon atom, which has an atomic mass of 28.086. 26% of the Earth's crust is composed of silicon, which occurs naturally in the form of silicates (oxides of silicon), eg quartz. Bulk silicon is steel grey in colour, opaque and has a shiny surface

The 14 electrons are arranged within three groups, or *shells*. The innermost shell contains two electrons, the middle shell eight, and the outer shell four. The shells are separated by forbidden regions, known as *energy band gaps*, into which individual electrons cannot normally travel. However, at temperatures above absolute zero (0 degrees Kelvin or minus 273 degrees Celsius) the electrons will move around within their respective shells due to thermal excitation. The speed, and therefore the energy, of the excited electrons increases as temperature rises.

It is the outermost shell, or *valence band*, which is of importance when considering whether silicon should be classified as an electrical insulator or a conductor. In conductors (see Fig 3.2(a)), such as the metals copper, silver and aluminium, the outermost (valence) electrons are free to move from one atom to another, thus making it possible for the material to sustain electron (current) flow. The region within which this exchange of electrons can occur is known, not surprisingly, as the *conduction band*, and in a conductor it effectively overlaps the valence band. In insulating materials (eg most plastics, glass and ceramics), however, there is a *band gap* (forbidden region) which separates the valence band from the conduction band (Fig 3.2(b)). The width of the band gap is measured in electron volts (eV), 1eV being the energy imparted to an electron as it passes through a potential of 1V. The magnitude of the band gap serves as an indication of how good the insulator is and a typical insulator will have a band gap of around 6eV. The only way to force an insulator to conduct an electrical current is to subject it to a very high potential difference, ie many thousands of volts. Under such extreme conditions, the potential may succeed in imparting sufficient energy to the valence electrons to cause them to jump into the conduction band. When this happens, and current flow is instigated, the insulator is said to have *broken down*. In practice, the breakdown of an insulator normally results in its destruction. It may seem odd that such a high potential is necessary to force the insulator into conduction when the band gap amounts to only a few volts. However, even a very thin slice of insulating material will have a width of many thousand atoms,

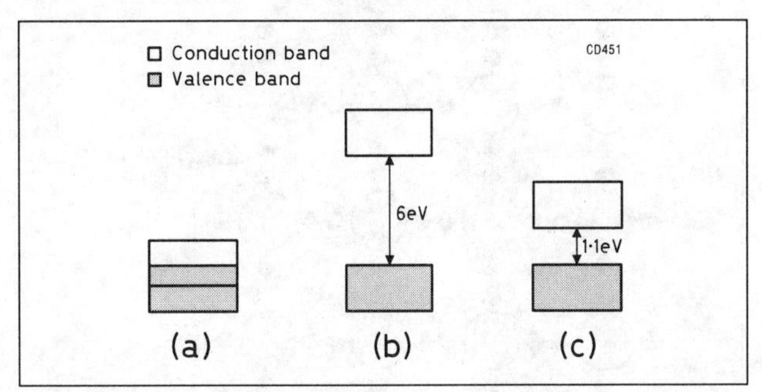

Fig 3.2. Band gaps of conductors (a), a typical insulator (b) and silicon (c)

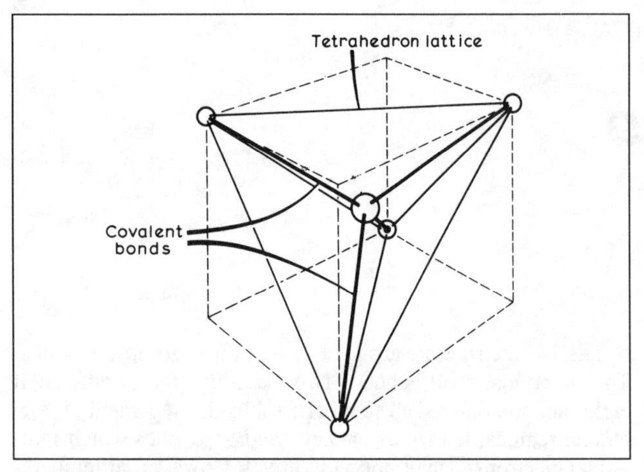

Fig 3.3. A three-dimensional view of the crystal lattice

and a potential equal to the band gap must exist across each of these atoms before current flow can take place.

Fig 3.2(c) shows the relationship between the valence and conduction bands for intrinsic (very pure) silicon. As can be seen, there is a band gap of around 1.1eV at room temperature. This means that intrinsic silicon will not under normal circumstances serve as an electrical conductor, and it is best described as a *narrow band-gap insulator*. The relevance of the term 'semiconductor', which might imply a state somewhere between conduction and insulation, or alternatively the intriguing possibility of being able to move from one to the other, is partly explained by the methods that have been developed to modify the electrical behaviour of materials like silicon.

Doping

The atoms within a piece of silicon form themselves into a criss-cross structure known as a *tetrahedral crystal lattice* – this is illustrated three-dimensionally by Fig 3.3. The lattice is held together by a phenomenon called *covalent bonding*, where each atom shares its valence electrons with those of its four nearest neighbours by establishing electron pairs. For greater clarity, Fig 3.4 provides a simplified, two-dimensional, representation of the crystal lattice.

It is possible to slightly alter the crystal lattice structure of silicon, and through doing so modify its conductivity, by adding small numbers of atoms of other substances. These substances, termed *dopants* (but also, although rather erroneously, referred to as 'impurities'), fall into two distinct categories:

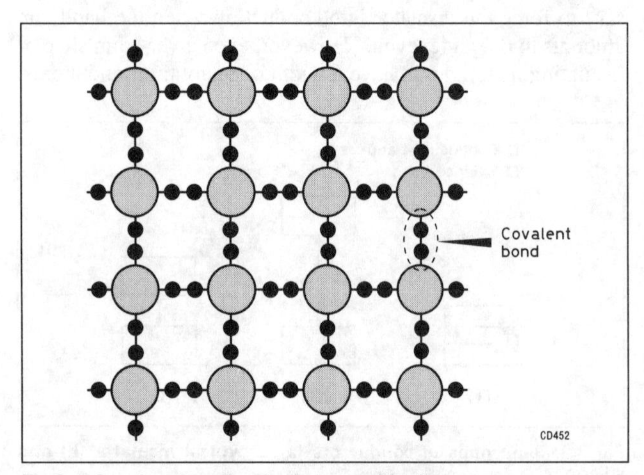

Fig 3.4. Covalent bonding in the crystal lattice

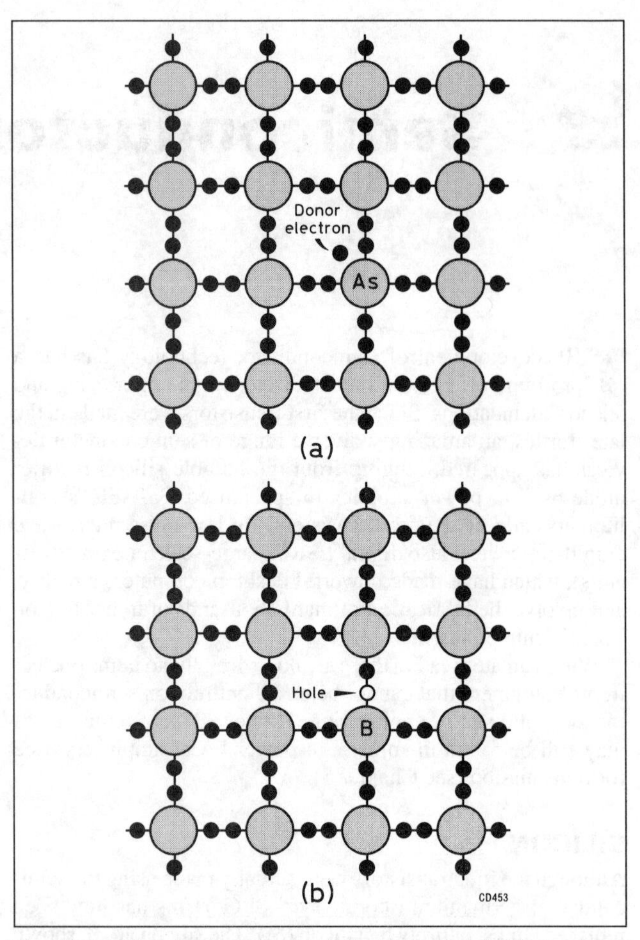

Fig 3.5. (a) The introduction of a dopant atom having five valence electrons (arsenic) into the crystal lattice. (b) Doping with an atom having three valence electrons (boron) creates a hole

- Group III – these are substances that have just three valence electrons: one less than silicon. Boron (atomic symbol B) is the most commonly used. Material doped in this way is called *P-type*.
- Group V – substances with five valence electrons: one more than silicon. Examples are phosphorus (P) and arsenic (As). This produces *N-type* material.

The doping concentration is varied to suit the requirements of particular semiconductor devices, but in all cases the levels involved are amazingly small. Typically, there will be around one dopant atom for every 100,000,000 atoms of silicon (1 in 10^8). Silicon modified in this way retains its normal structure because the Group III and Group V atoms are able to fit themselves into the crystal lattice. In the case of N-type silicon (Fig 3.5(a)), each dopant atom is left with one spare electron that cannot partake in covalent bonding. These 'untethered' electrons, known as *donors*, are free to move into the conduction band and can therefore act as current carriers. In consequence, N-type silicon has a far higher conductivity than intrinsic silicon.

The conductivity of silicon is also enhanced by P-type doping, although the reason for this is less obvious. Fig 3.5(b) shows how a Group 3 atom will fit itself into the crystal lattice, leaving a vacant electron position known as a *hole*. Holes are able to facilitate electron (current) flow because they exert a considerable force of attraction for any free electrons. Semiconductors which utilise both free electrons and holes for current flow are known as *bipolar* devices. Electrons, however, are more mobile than holes and so semiconductor devices designed to operate at the highest frequencies will usually rely on electrons as their main current carriers.

THE PN JUNCTION DIODE

The PN junction diode, or rectifier, is the simplest bipolar device. Fig 3.6 shows that the junction consists of a sandwich of P- and N-type silicon. This representation suggests that a diode can be made by simply sticking together small blocks of P- and N-type material. In practice, however, such a diode would hardly work at all. The surfaces to be mated will not be smooth enough to facilitate good ohmic contact, and it is difficult to prevent the formation of a thin layer of silicon dioxide (SiO_2) caused by interaction with the air. This is bound to cause problems because silicon dioxide is a very good insulator. Furthermore, the lattice structure at the surface is bound to be imperfect, with many partial ('dangling') atomic bonds.

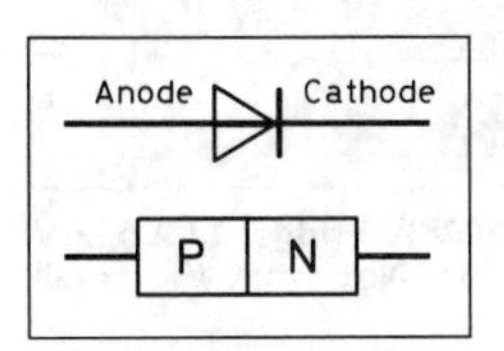

Fig 3.6. The PN junction and diode symbol

The process used to fabricate working junction diodes is shown in Fig 3.7, and variations of this technique are employed in the manufacture of nearly all other semiconductor devices. The starting point is a piece of N-type silicon which is called the *substrate* (see Fig 3.7(a)). An insulating layer of silicon dioxide is intentionally grown onto the top surface of the substrate in order to protect the silicon underneath, and the next step (Fig 3.7(b)) is to etch a small hole, referred to as a *window*, into the oxide layer – possibly using hydrofluoric acid. A P-type region is now formed by injecting the dopant boron through the window using a process known as *diffusion*, or alternatively *ion implantation* (Fig 3.7(c)). This means, of course, that a region of the N-type substrate has been converted to P-type material. However, the process does not involve removing any of the N-type dopant, as it is simply necessary to inject a higher concentration of P-type atoms than there are N-type already present. *Ohmic* contacts (ie contacts in which the current flows equally in either direction) are now bonded to the substrate and window (Fig 3.7(d)). In practice, large numbers of identical diodes will be fabricated on a single slice, or *wafer*, of silicon which is then cut into *chips* containing a single diode. Finally, metal leads are bonded onto the ohmic contacts before the diode is encapsulated in either epoxy resin, plastic or glass.

The diode's operation is conditioned by what happens at the junction between the P-type and N-type silicon. As Fig 3.8(a) shows, in the absence of any external electric field or potential difference, holes (represented by '+' symbols) will migrate by attraction from the P-type material into the N-type. Conversely, free electrons (represented by '–' symbols) will migrate from the N-type material into the P-type. The migration of carriers in this way gives rise to a potential difference across the junction

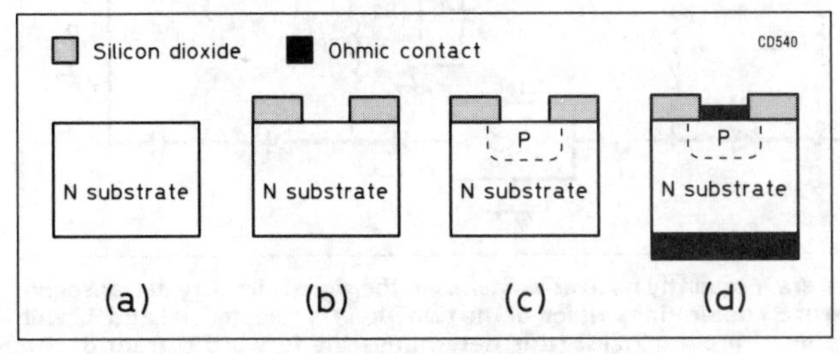

Fig 3.7. Fabrication of a PN junction diode

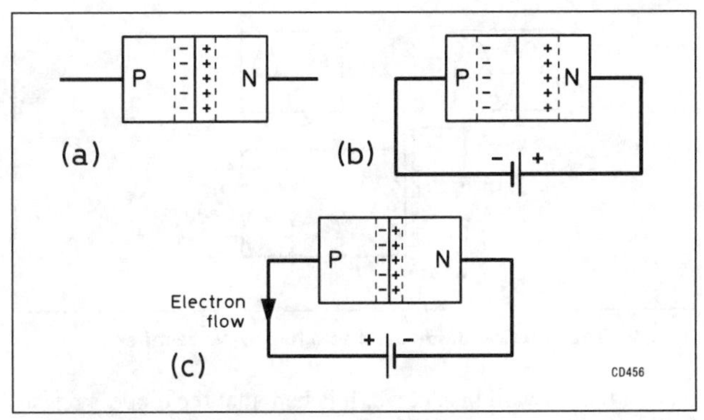

Fig 3.8. Behaviour of the junction diode under zero bias (a), reverse bias (b) and forward bias (c)

within an area known as the *depletion region*. The magnitude of this potential (called the *barrier height*) depends on the doping concentration, but for a typical silicon diode it will be around 0.7V. It is important to realise, however, that the potential cannot be measured by connecting a voltmeter externally across the diode because there is no net flow of current through the device.

If a battery is connected to the diode as in Fig 3.8(b), the external potential will simply reinforce the charge across the depletion region, causing it to widen (because the polarity of the external potential is the same as that of the barrier). Under these conditions practically no current will flow through the diode, which is said to be *reverse biased*. However, if the external potential is raised above a certain voltage the barrier may break down, resulting in reverse current flow.

Reversing the polarity of the battery (Fig 3.8(c)) makes the external voltage oppose the barrier potential and 'squashes' the depletion region, which therefore becomes narrower. The barrier height has therefore been reduced and under this condition, known as *forward bias*, current is able to flow through the device by diffusion of current carriers (holes and electrons) across the depletion region.

The diode's electrical characteristics are summarised graphically in Fig 3.9. Region A shows that where only a small forward bias voltage is applied, the depletion region's barrier height causes significant resistance to current flow. Nevertheless, as the forward bias is voltage is increased, the current rises exponentially (ie for each doubling of voltage the current rises by a factor of four). Region B commences at a point often termed the

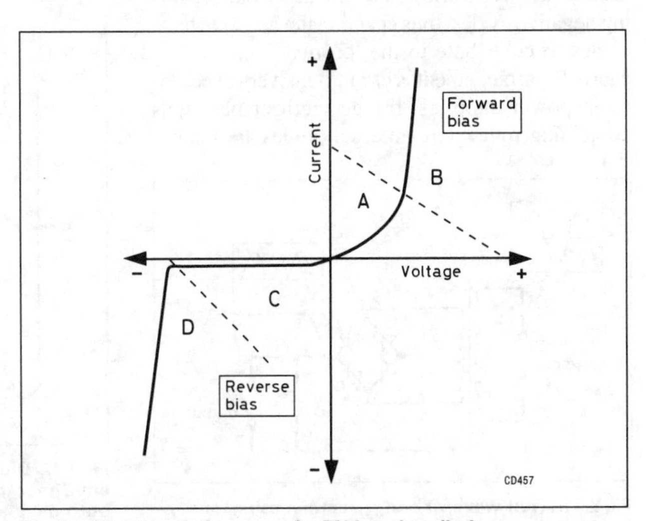

Fig 3.9. Characteristic curve of a PN junction diode

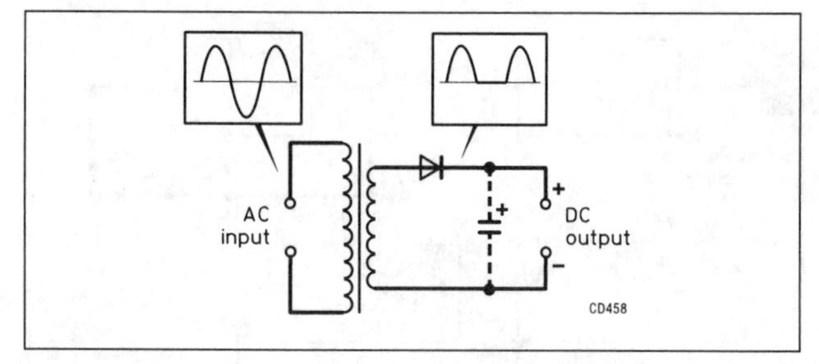

Fig 3.10. The junction diode used as a half-wave rectifier

knee of the forward bias curve. It is here that the diode's resistance to further current flow has been reduced to insignificance, and external resistance becomes the major factor in determining current. There remains, however, a small voltage drop which, in the case of an ordinary silicon diode, is around 0.7V .

Under reverse bias (region C) only a very small current flows, far less than 1μA for a silicon diode at room temperature. This current is due to *minority carriers* – particles that find a way of moving in the opposite direction to that determined by the potentials established within and outside the device, eg because of thermal excitation – and it does not vary with reverse voltage. Region D shows that if the reverse bias is increased above the diode's breakdown voltage, resistance to current flow dramatically falls.

Diodes have a very wide range of applications in radio equipment, and there are many types available with characteristics tailored to suit their intended use. Fig 3.10 shows a diode used as a rectifier of alternating current in a simple power supply. The triangular part of the diode symbol represents the connection to the P-type material and is referred to as the *anode*. The single line is the N-type connection, or *cathode* (Fig 3.6). Small diodes will have a ring painted at one end of their body to indicate the cathode connection. The rectifying action allows current to flow during positive half-cycles of the input waveform, while preventing flow during negative half-cycles. This is known as *half-wave rectification* because only 50% of the input cycle contributes to the DC output. Fig 3.11 shows how a more efficient power supply may be produced using a bridge rectifier, which utilises four diodes. Diodes A and B conduct during positive half-cycles, while diodes C and D conduct during negative cycles. Notice that C and D are arranged so as to effectively reverse the connections to the transformer's secondary winding during negative cycles, thus enabling the negative half-cycles to contribute to the 'positive' output. Because this arrangement is used extensively in equipment power supplies, bridge rectifier packages containing four interconnected diodes are readily

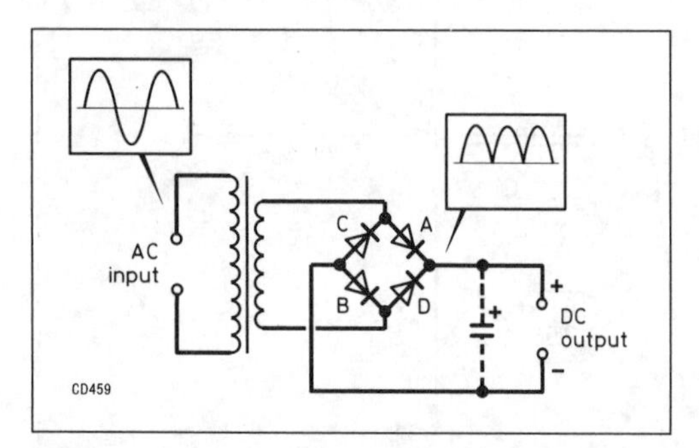

Fig 3.11. A full-wave, or bridge, rectifier using four junction diodes

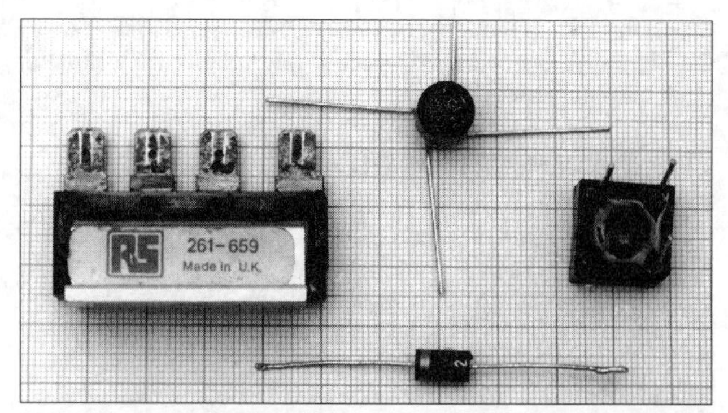

Rectifiers. From the left, clockwise: 35A, 100PIV bridge, 1.5A 50PIV bridge, 6A 100PIV bridge, 3A 100PIV diode. Note that the small squares are 1mm across

available. When designing power supplies it is necessary to take account of the voltages and currents that the diodes will be subjected to. All rectifiers have a specified maximum forward current at a given case temperature. This is because the diode's forward voltage drop gives rise to power dissipation within the device – for instance, if the voltage across a rectifier diode is 0.8V at a current of 5A, the diode will dissipate 4W ($0.8 \times 5 = 4$). This power will be converted to heat, thus raising the diode's temperature. Consequently, the rectifier diodes of high-current power supplies may need to be mounted on heatsinks. In power supplies especially, diodes must not be allowed to conduct in the reverse direction. For this reason, the maximum reverse voltage that can be tolerated before breakdown is likely to occur must be known. The term *peak inverse voltage* (PIV) is used to describe this characteristic (see Chapter 14).

Miniature low-current diodes, often referred to as *small-signal* types, are very useful as switching elements in transceiver circuitry. Fig 3.12 shows an arrangement of four diodes used to select one of two band-pass filters, depending on the setting of switch S1. When S1 is set to position 1, diodes A and B are forward biased, thus bringing filter 1 into circuit. Diodes C and D, however, are reverse biased, which takes filter 2 out of circuit. Setting S1 at position 2 reverses the situation. Notice how the potential divider R1/R2 is used to develop a voltage equal to half that of the supply rail. This makes it easy to arrange for a

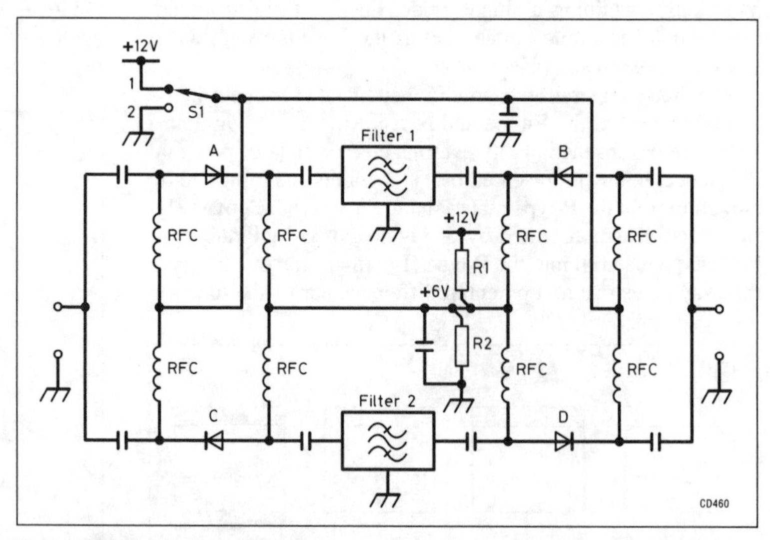

Fig 3.12. Diodes are frequently used as switches in the signal circuitry of transceivers. The setting of S1 determines which of the two filters is selected. R1 and R2 will both have a value of around 2.2kΩ (this determines the forward current of the switching diodes). For HF applications the RFCs are 100μH

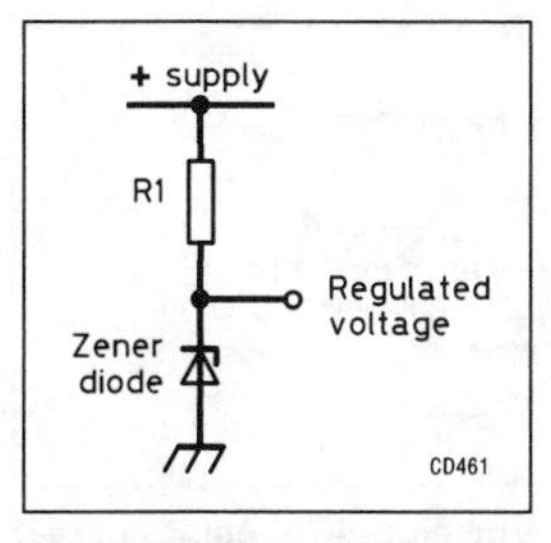

Fig 3.13. A zener diode used as a simple voltage regulator

forward bias of 6V, or a reverse bias of the same magnitude, to appear across the diodes. Providing that the peak level of signals at the filter terminations does not reach 6V under any circumstances, the diodes will behave as almost perfect switches. One of the main advantages of diode switching is that signal paths can be kept short, as the leads (or PCB tracks) forming connections to the front-panel switches carry only a DC potential which is isolated from the signal circuitry using the RF chokes and decoupling capacitors shown.

Zener diodes

These diodes make use of the reverse breakdown characteristic discussed previously. The voltage at which a diode begins to conduct when reverse biased depends on the doping concentration. As the doping level is increased, the breakdown voltage drops. This fact can be exploited during the manufacture of the diodes, enabling the manufacturer to specify the breakdown voltage for a given component. Zener diodes with breakdown voltages in the range 2.7V to over 150V are available and can be used to provide reference voltages for power supplies and bias generators.

Fig 3.13 shows how a zener diode can be used in conjunction with a resistor to provide voltage regulation (note the use of a slightly different circuit symbol for the zener diode). When power is initially applied the zener diode will start to conduct as the input voltage is higher than the diode's reverse breakdown value. However, as the diode begins to pass current, an increasingly high potential difference will appear across resistor R1. This potential will tend to rise until the voltage across R1 becomes equal to the difference between the input voltage and the zener's breakdown voltage. The net result is that the output voltage will be forced to settle at a level close to the diode's reverse breakdown potential. The value of R1 is chosen so as to limit the zener current to a safe value (the maximum allowable power dissipation for small zener diodes is around 400mW), while ensuring that the maximum current to be drawn from the regulated supply will not increase the voltage across R1 to a level greater than the difference between the zener voltage and the minimum expected input voltage (see Chapter 14).

Varactor diodes

The varactor, or variable capacitance, diode makes use of the fact that a reverse-biased PN junction behaves just like a capacitor with two plates. The diode capacitance is produced by the charge across the depletion region. As the reverse bias is increased the depletion region becomes wider. This produces the same effect as moving the plates of a capacitor further apart – the capacitance is reduced (see Chapter 1). The varactor can therefore be used as a voltage-controlled variable capacitor, as demonstrated by the graph at Fig 3.14. The capacitance is governed by the diode's junction area (ie the size of the plates) and also the width of the depletion region for a given value of reverse bias, which is a product of the doping concentration. Varactors are available covering a wide range of capacitance spreads, from around 0.5–10pF up to 20–400pF. The voltages at which the stated maximum and minimum capacitances are obtained will be quoted in the manufacturer's literature, but they normally fall in the range 2–20V. The maximum reverse bias voltage should not be exceeded as this could result in breakdown.

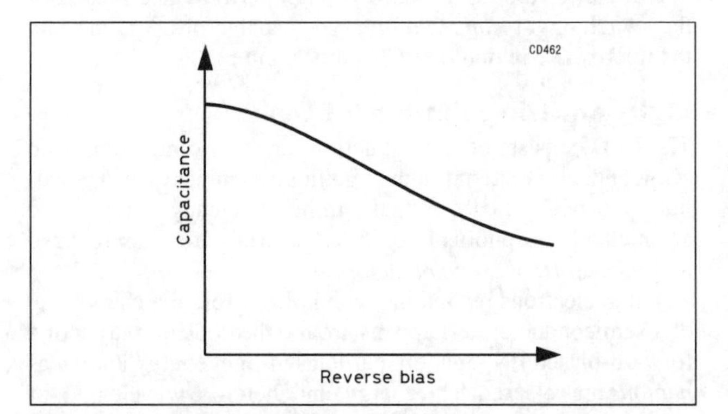

Fig 3.14. Relationship between capacitance and reverse bias of a varactor (variable capacitance diode)

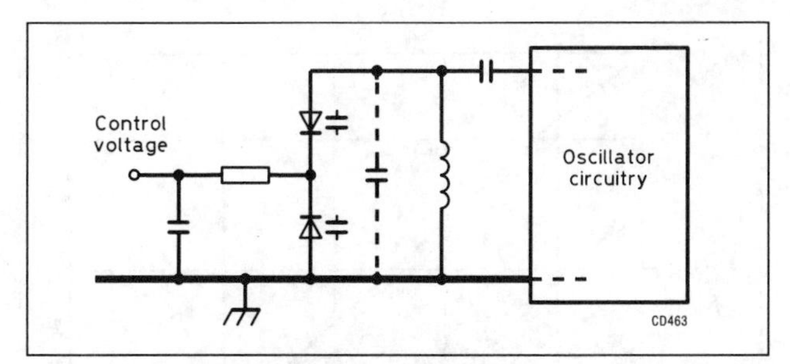

Fig 3.15. Two varactor diodes used to tune a voltage-controlled oscillator. The capacitor drawn with dotted lines represents the additional component which may be added to the tank circuit in order to modify the LC ratio and tuning range

Varactors are commonly used to achieve voltage control of oscillator frequency in frequency synthesisers. Fig 3.15 shows a typical arrangement where two varactors are connected 'back to back' and form part of the tank circuit of a voltage-controlled oscillator (VCO). The use of two diodes prevents the alternating RF voltage appearing across the tuned circuit from driving the varactors into forward conduction, which is most likely to happen when the control voltage is low. Because the varactor capacitances appear in series, the maximum capacitance swing is half that obtainable when using a single diode. Three-lead packages containing dual diodes internally connected in this way are readily available. It is also possible to obtain multiple-diode packages containing two or three matched diodes but with separate connections. These are used to produce voltage-controlled versions of two-gang or three-gang variable capacitors.

When used as a capacitive circuit element, the varactor's Q may be significantly lower than that of a conventional capacitor. This factor must be taken into account when designing high-performance frequency synthesisers (see Chapter 5).

PIN diodes

The PIN diode differs from a conventional PN junction diode in that it is fabricated with a region of almost pure (*intrinsic*) silicon sandwiched between the normal P- and N-layers (Fig 3.16). When forward bias is applied, current carriers will diffuse into the I-region, thus lowering the diode's resistivity and permitting current flow. As the graph at Fig 3.17 shows, the PIN diode's resistance is inversely proportionate to forward bias current, and this relationship is maintained over a surprisingly wide range. This characteristic is exploited to good effect in voltage-controlled RF attenuators.

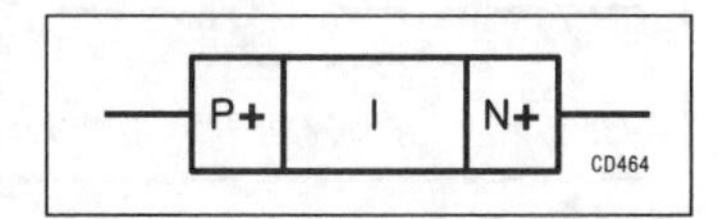

Fig 3.16. The PIN diode

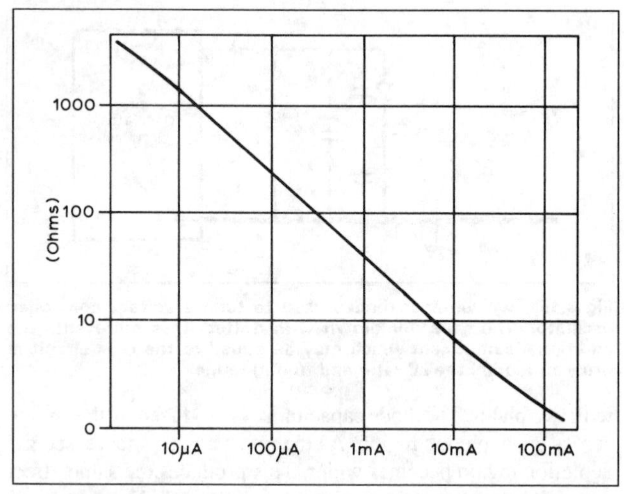

Fig 3.17. Relationship between forward current and resistance for a PIN diode

For instance, a PIN diode may be placed between the output of the first band-pass filter and the input of the RF amplifier in a receiver. A potentiometer is then employed to adjust the diode's forward current, thus providing a very effective RF gain control.

Under zero or reverse bias, the I-layer reverts to its high-resistance state and the diode functions as a low-value capacitor. PIN diodes can therefore be used as RF switches and large types, which have high reverse breakdown voltages and are capable of carrying forward currents of a few amps, are frequently employed as solid-state substitutes for transmit/receive changeover relays in transceivers. PIN diodes used in this way provide faster, more reliable switching than that obtainable with electromechanical relays, and are of course absolutely silent in operation.

Germanium point-contact diodes

Ironically, one of the earliest semiconductor devices to find widespread use in telecommunications actually pre-dates the thermionic valve. The first broadcast receivers employed a form of envelope detector (RF rectifier) known colloquially as a *cat's whisker*. This consisted of a spring made from a metal such as bronze or brass (the 'whisker'), the pointed end of which was delicately bought into contact with the surface of a crystal having semiconducting properties, such as galena, zincite or carborundum.

The germanium point-contact diode is a modern equivalent of the cat's whisker and consists of a fine tungsten spring which is held in contact with the surface of an N-type germanium crystal (Fig 3.18). During manufacture, a minute region of P-type material is formed at the point where the spring touches the crystal. The point contact therefore functions as a PN diode. In most respects the performance of this device is markedly inferior to that of the silicon PN junction diode. The current flow under reverse bias is much higher – typically 5μA, the highest obtainable PIV is only about 70V and the maximum forward current is limited by the delicate nature of the point contact.

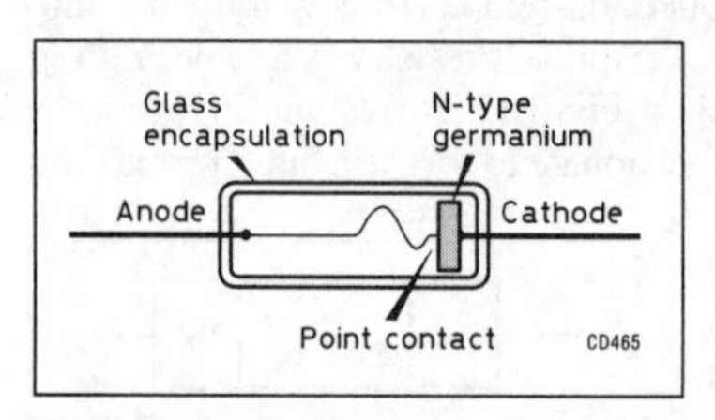

Fig 3.18. The germanium point-contact diode

Nevertheless, this device has a number of saving graces. The forward voltage drop is considerably lower than that of a silicon junction diode – typically 0.2V – and the reverse capacitance is also very small. There is also an improved version known as the *gold-bonded* diode, where the tungsten spring is replaced by one made of gold. Fig 3.19 shows a simple multimeter probe which is used to rectify low RF voltages. The peak value can then be read with the meter switched to a normal DC range. The low forward voltage drop of the point-contact diode leads to more accurate readings.

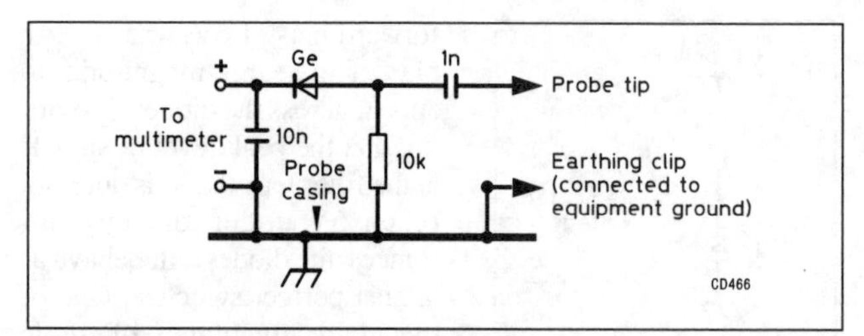

Fig 3.19. A simple RF probe using a germanium (Ge) point-contact diode

The hot-carrier diode

Ordinary PN junction diodes suffer from a deficiency known as *charge storage*, which has the effect of increasing the time taken for a diode to switch from forward conduction to reverse cut-off when the polarity of the applied voltage is reversed. This reduces the efficiency of the diode at high frequencies. Charge storage occurs because holes, which are less mobile than electrons, require a finite time to migrate back from the N-doped cathode material as the depletion region widens under the influence of reverse bias. The fact that in most diodes the P-type anode is more heavily doped than the N-type cathode tends to exacerbate matters.

The hot-carrier, or *Schottky barrier*, diode overcomes the problem of charge storage by utilising electrons as its main current carriers. It is constructed (Fig 3.20) in a similar fashion to the germanium point-contact type, but there are a few important differences. The semiconductor used is N-type silicon which is modified by growing a layer (the *epitaxial region*) of more lightly doped material onto the substrate during manufacture. The device is characterised by its high switching speed and low capacitance. It is also considerably more rugged than the germanium point-contact diode and generates less noise.

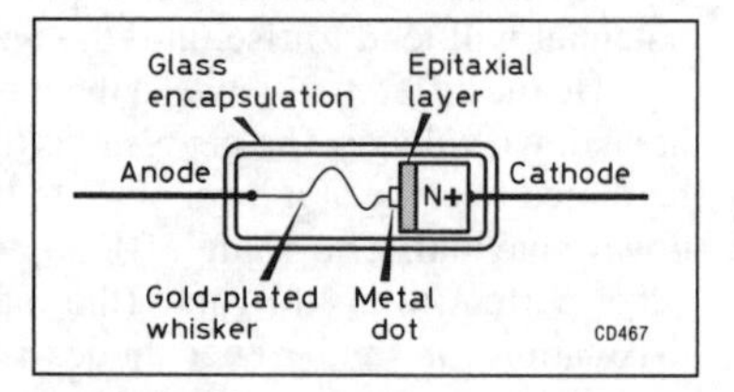

Fig 3.20. The hot-carrier (Schottky) diode

Hot-carrier diodes are used in high-performance mixers of the switching, or *commutating*, type capable of operating into the microwave region (see Chapters 5 and 6).

Light-emitting diodes (LEDs)

The LED consists of a PN junction formed from a compound semiconductor material such as gallium arsenide (GaAs) or gallium phosphide (GaP). As gallium has a valency of three, and arsenic and phosphorus five, these materials are often referred to as *Group III-V semiconductors*.

When electrons recombine with holes across the energy gap of a semiconductor, as happens around the depletion layer of a forward-biased PN junction, particles of light energy known as *photons* are released. The energy, and therefore wavelength, of the photons is determined by the semiconductor band gap. Pure gallium arsenide has a band gap of about 1.43eV, which

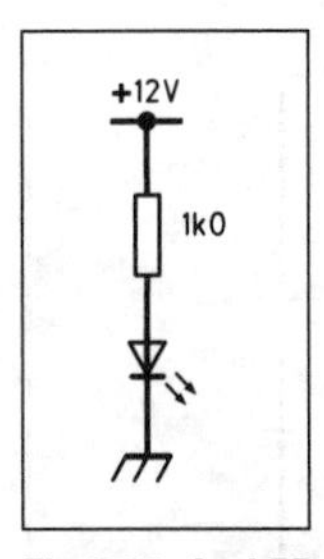

Fig 3.21. An LED emits light when forward biased. The series resistor limits the forward current to a safe value

produces photons with a wavelength of 880 nanometres (nm). This lies at the infrared end of the spectrum. Adding aluminium to the gallium arsenide has the effect of increasing the band gap to 1.96eV which shortens the light wavelength to 633nm. This lies in the red part of the visible spectrum. Red LEDs can also be made by adding phosphorus to gallium arsenide. Green LEDs (wavelength 560nm) are normally made from gallium phosphide.

The LEDs used as front-panel indicators consist of a PN junction encapsulated within translucent plastic. At a current of 10mA they will generate a useful amount of light without overheating. The forward voltage drop at this current is about 1.8V. Fig 3.21 shows a LED operating from a 12V power rail (note the two arrows representing rays of light which differentiates the LED circuit symbol from that of a normal diode). The series resistor determines the forward current and so its inclusion is mandatory. LEDs are also used in more complex indicators, such the numeric (seven-segment) displays employed in frequency counters (see Chapter 15).

The Gunn diode

The Gunn diode, named after its inventor, B J Gunn, comprises little more than a block of N-type gallium arsenide. It is not, in fact, a diode in the normally accepted sense because there is no P-N junction and consequently no rectifying action. It is properly called the *Gunn effect device*. However, 'diode' has become accepted by common usage and merely explains that it has two connections.

It consists of a slice of low-resistivity N-type gallium arsenide on which is grown a thin epitaxial layer, the active part, of high-resistivity gallium arsenide with a further thicker layer of low-resistivity gallium arsenide on top of that. Since the active layer is very thin, a low voltage across it will produce a high electric field strength. The electrons in gallium arsenide can be in one of two conduction bands. In one they have a much higher mobility than in the other and they are initially in this band. As the electric field increases, more and more are scattered into the lower mobility band and the average velocity decreases. The field at which this happens is called the *threshold field* and is 320V/mm. Since current is proportional to electron velocity and voltage is proportional to electric field, the device has a region of *negative* resistance. This odd concept is explained by the definition of resistance as the slope of the voltage-current graph and a pure resistor has a linear relationship (Fig 3.22). On the other hand, the Gunn diode has a roughly reversed 'S'-shaped curve (Fig 3.23) and the negative resistance region is shown by

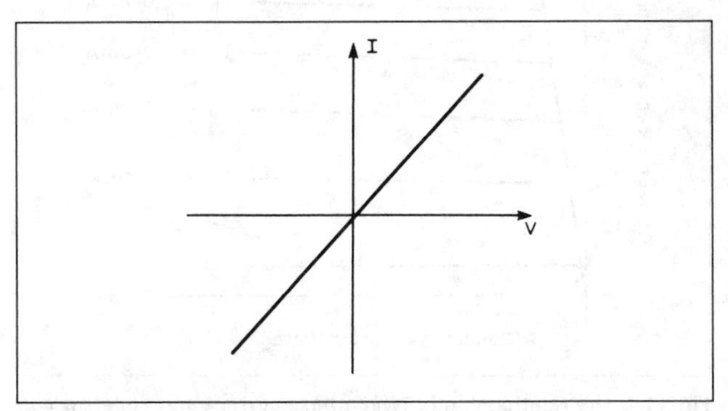

Fig 3.22. Voltage-current through a resistor

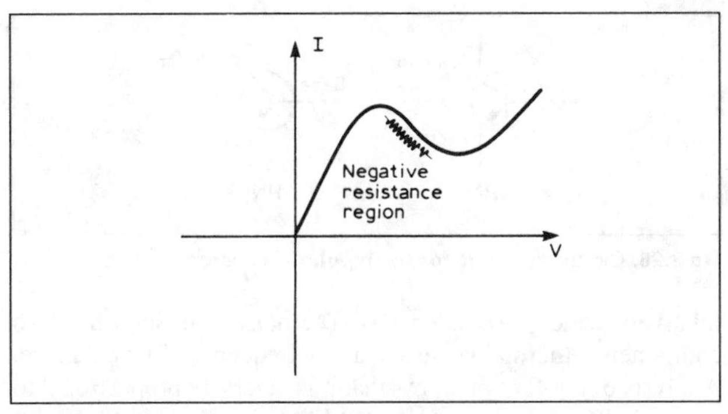

Fig 3.23. Voltage-current through a Gunn diode

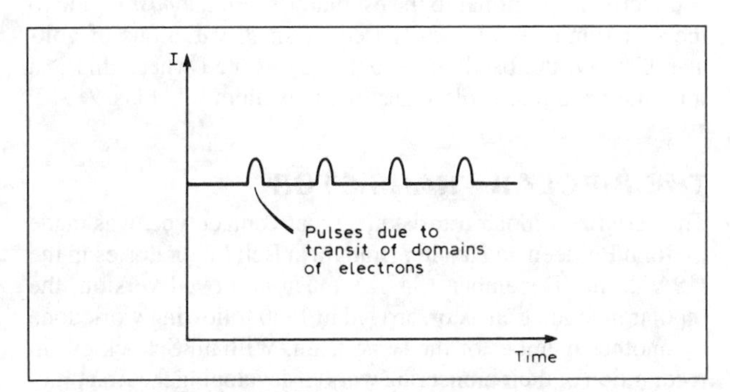

Fig 3.24. Current in Gunn diode

the hatching. The current through the device takes the form of a steady DC with superimposed pulses (Fig 3.24) and their frequency is determined by the thickness of the active epitaxial layer. As each pulse reaches the anode, a further pulse is generated and a new domain starts from the cathode. Thus the rate of pulse formation depends on the transit time of these domains through the epitaxial layer. In a 10GHz Gunn diode, the layer is about 10µm thick, and a voltage of somewhat greater than 3.5V gives the high field state and microwave pulses are generated. The current through the device shows a peak before the threshold voltage followed by a plateau. See Fig 3.25. The power output reaches a peak at between 7.0 and 9.0V for 10GHz devices.

The Gunn diode is inherently a wide-band device so it is operated in a high-*Q* cavity (tuned circuit) and this determines the exact frequency. It may be tuned over a narrow range by altering the cavity with a metallic screw, a dielectric (PTFE or Nylon) screw or by loading the cavity with a varactor. With a 10GHz device, the whole of the 10GHz amateur band can be covered with reasonable efficiency. Gunn diodes are not suitable for narrow-band operation since they are of low stability and have

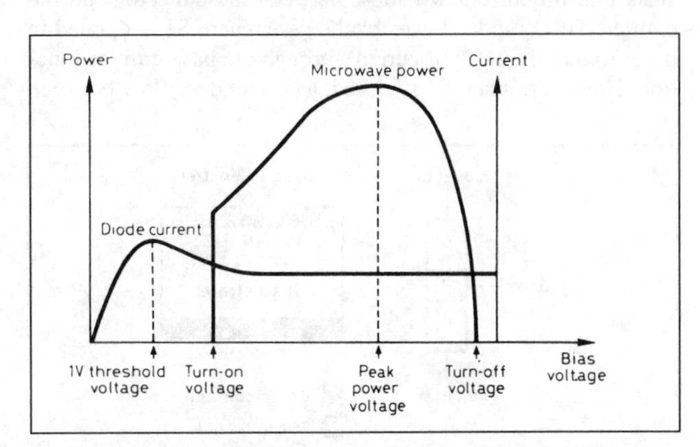

Fig 3.25. Characteristic shape of a Gunn oscillator's bias power curve

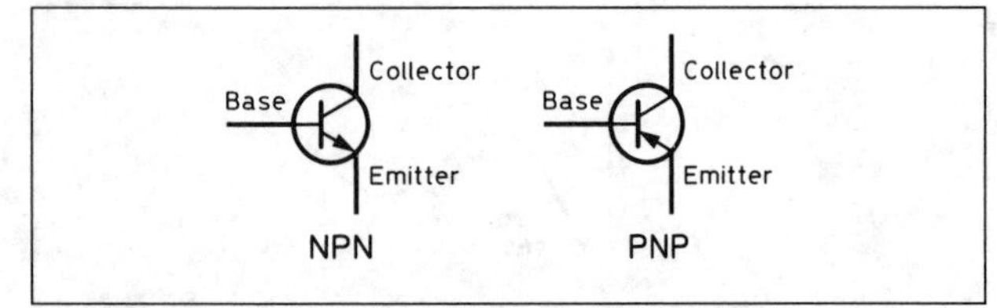

Fig 3.26. Circuit symbols for the bipolar transistor

relatively wide noise sidebands. The noise generated has two components, thermal noise and a low frequency 'flicker' noise. Analysis of the former shows that it is inversely proportional to the loaded Q of the cavity and that FM noise close to the carrier is directly proportional to the oscillator's voltage *pushing*, ie to the variation of frequency caused by small variations of voltage. Clearly, the oscillator should be operated where this is a minimum and that is often near the maximum safe bias.

THE BIPOLAR TRANSISTOR

The very first bipolar transistor, a point-contact type, was made by John Bardeen and Walter Brattain at Bell Laboratories in the USA during December 1947. A much-improved version, the bipolar junction transistor, arrived in 1950 following work done by another member of the same team, William Schockley. In recognition of their pioneering work in developing the first practical transistor, the three were awarded the Nobel Prize for physics in 1956.

The bipolar transistor is a three-layer device which exists in two forms, NPN and PNP. The circuit symbols for both types are shown in Fig 3.26. Note that the only difference between the symbols is the direction of the arrow drawn at the emitter connection.

Fig 3.27 shows the three-layer sandwich of an NPN transistor and also how this structure may be realised in a practical device. The emitter region is the most heavily doped.

In Fig 3.28 the NPN transistor has been connected into a simple circuit which allows its behaviour to be analysed. The PN junction between the base and emitter forms a diode which is forward biased by battery B1 when S1 is closed. Resistor R1 has been included to control the level of current that will inevitably flow, and R2 provides a collector load. A voltmeter connected between the base and emitter will indicate the normal forward voltage drop of approximately 0.6V typical for a silicon PN junction.

The collector-base junction also forms the equivalent of a PN diode, but one that is reverse biased by battery B2. This suggests that no current will flow between the collector and the emitter. This is indeed true for the case where S1 is opened in order to halt the flow of current through the base-emitter junction. However, when S1 is closed, a current does flow between

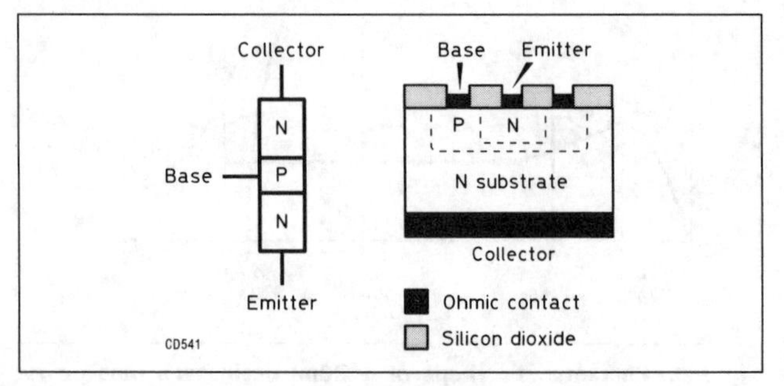

Fig 3.27. Construction of an NPN transistor

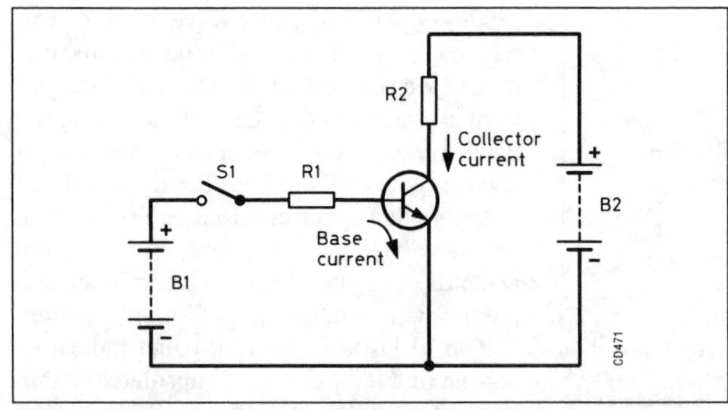

Fig 3.28. Applying forward bias to the base-emitter junction of a bipolar transistor causes a much larger current to flow between the collector and emitter. The arrows indicate conventional current flow, which is opposite in direction to electron flow

the collector and emitter. The reason for this is that the base region is made sufficiently thin to allow some of the electrons injected into the base from battery B1 to diffuse across the base and into the collector base junction. Electrons that find their way across the base will then be attracted to the positive collector. The resultant collector current is therefore flowing through a reverse-biased junction, but this has been made possible by the availability of electrons able to migrate across the base region. As the emitter region is heavily doped, there are far more electrons available than are required to maintain the base-emitter current dictated by the value of R1. These 'surplus' electrons are available to increase the current flow into the collector. Indeed, the collector current will be significantly larger than the base current, and the ratio between the two (known as the transistor's β) is related to the doping concentration of the emitter divided by the doping concentration of the base. The PNP transistor operates in a similar fashion except that the polarities of the applied voltages, and also the roles of electrons and holes, are reversed. The graph at Fig 3.29 shows the relationship between base current and collector current for a typical bipolar transistor.

The circuit shown in Fig 3.28 therefore provides the basis for an amplifier. Small variations in base current will result in much larger variations in collector current. The DC current gain (β), or h_{FE}, of a typical bipolar transistor at a collector current of 1mA will be between 50 and 500, ie the change in base current required to cause a 1mA change in collector current lies in the range 2 to 20μA. The value of β is temperature dependent, and so is the base-emitter voltage drop (V_{BE}), which will fall by

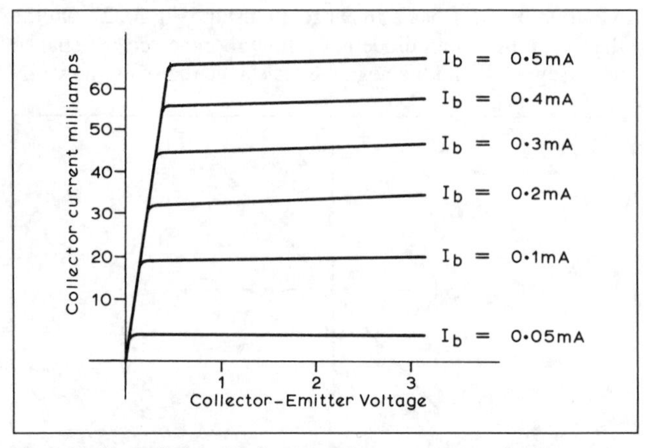

Fig 3.29. The relationship between base current and collector current for a bipolar transistor

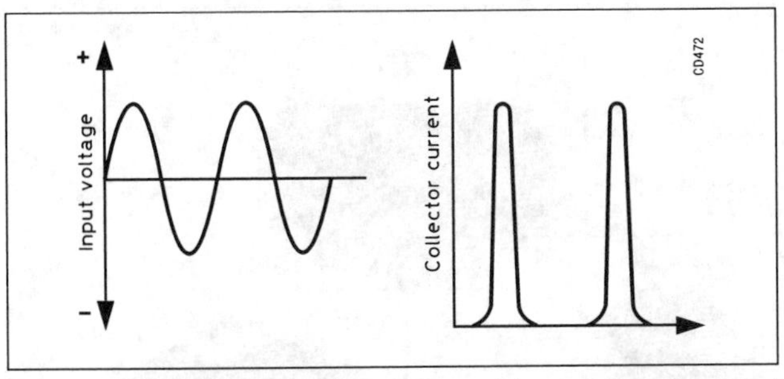

Fig 3.30. The circuit shown in Fig 3.28 would not make a very good amplifier

approximately 2mV for every 1°C increase in ambient temperature.

A transistor amplifier

If the battery (B1) used to supply base current to the transistor in Fig 3.28 is replaced with a signal generator set to give a sine-wave output, the collector current will vary in sympathy with the input waveform as shown in Fig 3.30. A similar, although inverted, curve could be obtained by plotting the transistor's collector voltage. Two problems are immediately apparent. First, because the base-emitter junction only conducts during positive half-cycles of the input waveform, the negative half-cycles do not appear at the at the output. Second, the barrier height of the base-emitter junction results in the base current falling exponentially as the amplitude of the input waveform drops below +0.6V. This causes significant distortion of the positive half-cycles, and it is clear that if the amplitude of the input waveform were to be significantly reduced then the collector current would hardly rise at all.

The circuit can be made far more useful by adding bias. Fig 3.31 shows the circuit of a practical *common-emitter* amplifier operating in Class A. (The term 'common emitter' indicates that the emitter connection is common to both the input and output circuits.) Resistors R1 and R2 form a potential divider which establishes a positive bias voltage at the base of the transistor. The values of R1 and R2 are chosen so that they will pass a current at least 10 times greater than that flowing into the base. This ensures that the bias voltage will not alter as a result of variations in the base current. R3 is added in order to stabilise the bias point, and its value has been calculated so that a potential of 2V will appear across it when, in the absence of an input signal, the desired standing collector current of 1mA (0.001A) flows (0.001A × 2000Ω = 2V). For this reason, the ratio between

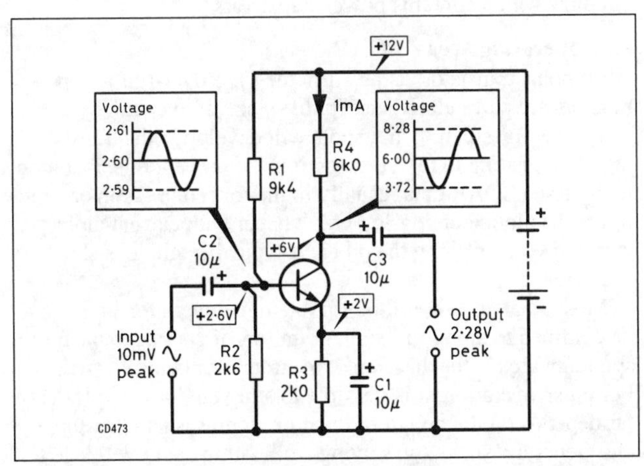

Fig 3.31. A practical common-emitter amplifier

the values of R1 and R2 has been chosen to establish a voltage of 2.6V at the base of the transistor. This allows for the expected forward voltage drop of around 0.6V due to the barrier height of the base-emitter junction. Should the collector current attempt to rise, for instance because of an increase in ambient temperature causing V_{BE} to fall, the voltage drop across R3 will increase, thus reducing the base-emitter voltage and preventing the collector current rising. Capacitor C1 provides a low-impedance path for alternating currents so that the signal is unaffected by R3, and coupling capacitors C2 and C3 prevent DC potentials appearing at the input and output. The value of the collector load resistor (R4) is chosen so that in the absence of a signal the collector voltage will be roughly 6V – ie half that of the supply rail. Where a transistor is used as an RF or IF amplifier, the collector load resistor may be replaced by a parallel tuned circuit (see Chapter 5).

An alternating input signal will now modulate the base voltage, causing it to rise slightly during the positive half-cycles, and fall during negative half-cycles. The base current will be similarly modulated and this, in turn, will cause far larger variations in the collector current. Assuming the transistor has a β of 100, the voltage amplification obtained can gauged as follows:

In order to calculate the effect of a given input voltage, it is necessary to develop a value for the base resistance. This can be approximated for a low-frequency amplifier using the formula:

$$\frac{26 \times \beta}{\text{Emitter current in milliamperes}}$$

The emitter current is the sum of the base and collector currents but, as the base current is so much smaller than the collector current, it is acceptable to use just the collector current in rough calculations:

$$\frac{26 \times 100}{1} = \frac{2600}{1} = 2600\Omega$$

Using Ohm's Law, and assuming that the peak amplitude of the input signal is 10mV (0.01V), the change in base current will be:

$$\frac{0.01\ (V)}{2600\ (R)} = 0.0000038\text{A}\ (3.8\mu\text{A})$$

This will produce a change in collector current 100 times larger:

$$\beta \times 3.8\mu\text{A} = 100 \times 3.8\mu\text{A} = 380\mu\text{A}$$

The change in output voltage (again using Ohm's Law) is:

$$380\mu\text{A} \times 6\text{k}\Omega\ (\text{R4}) = 2.28\text{V}$$
$$(38 \times 10^{-5} \times 6 \times 10^{3} = 2.28)$$

The voltage gain obtained is therefore:

$$\frac{2.28\ (\text{output voltage})}{0.01\ (\text{input voltage})} = 228\ (47\text{dB})$$

Note that the output voltage developed at the junction between R4 and the collector is phase reversed with respect to the input. As operating frequency is increased the amplifier's gain will start to fall. There are a number of factors which cause this, but one of the most significant is charge storage in the base region. For this reason, junction transistors designed to operate at high frequencies will be fabricated with the narrowest possible base width. A guide to the maximum frequency at which a transistor can be operated is given by the parameter f_T. This is the frequency at which the β falls to unity (1). Most general-purpose

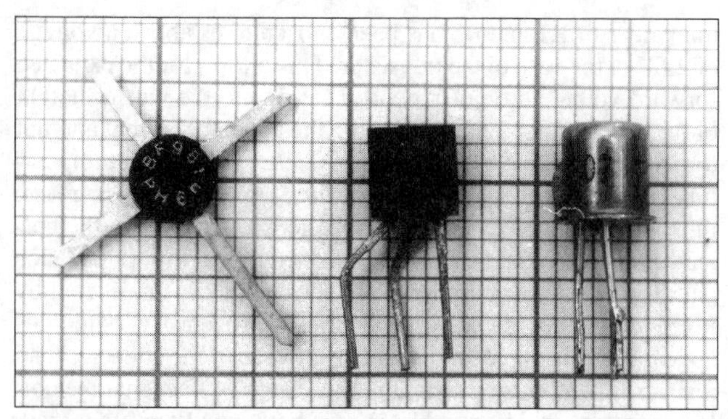

Low-power devices. From the left: UHF dual-gate MOSFET, HF bipolar transistor (plastic cased), AF bipolar transistor (metal cased)

junction transistors will have an f_T of around 150MHz, but specialised devices intended for use at UHF and microwave frequencies will have an f_T of 5GHz or even higher. An approximation of a transistor's current gain at frequencies below f_T can be obtained by:

$$\beta = f_T \div \text{operating frequency}$$

For example, a device with an f_T of 250MHz will probably exhibit a current gain of around 10 at a frequency of 25MHz.

Maximum ratings

Transistors may be damaged by the application of excessive voltages, or if made to pass currents that exceed the maximum values recommended in the manufacturers' data sheets. Some, or all, of the following parameters may need to be considered when selecting a transistor for a particular application:

V_{CEO} – The maximum voltage that can be applied between the collector and emitter with the base open-circuit (hence the 'O'). In practice the maximum value is dictated by the reverse breakdown voltage of the collector-base junction. In the case of some transistors, for instance many of those intended for use as RF power amplifiers, this rating may seem impracticably low, being little or no higher than the intended supply voltage. However, in practical amplifier circuits, the base will be connected to the emitter via a low-value resistance or coupling coil winding. Under these conditions the collector-base breakdown voltage will be raised considerably (see below).

V_{CBO} – The maximum voltage that can be applied between the collector and base with the emitter open-circuit. This provides a better indication of the collector-base reverse breakdown voltage. An RF power transistor with a V_{CEO} of 18V may well have a V_{CBO} rating of 48V. Special high-voltage transistors are manufactured for use in the EHT (extra high tension) generators of television and computer displays which can operate at collector voltages in excess of 1kV.

V_{EBO} – The maximum voltage that can be applied between the emitter and base with the collector open-circuit. In the case of an NPN transistor, the emitter will be held at a positive potential with respect to the base. Therefore, it is the reverse breakdown voltage of the emitter-base junction that is being measured. A rating of around 5V can be expected.

I_C – The maximum continuous collector current. For a small-signal transistor this is usually limited to around 150mA, but a rugged power transistor may have a rating as high as 30A.

P_D – The maximum total power dissipation for the device. This figure is largely meaningless unless stated for a particular case temperature. The more power a transistor dissipates, the hotter it gets. Excessive heating will eventually lead to destruction,

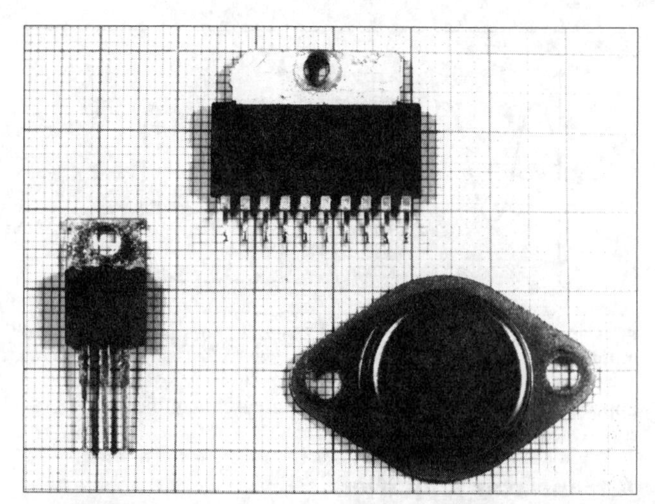

Power devices. From the left, clockwise: NPN bipolar power Darlington transistor, power audio amplifier IC, NPN bipolar power transistor

and so the power rating is only valid within the safe temperature limits quoted as part of the P_D rating. A reasonable case temperature for manufacturers to use in specifying the power rating is 50°C. It is unfortunate that as a bipolar transistor gets hotter, its V_{BE} drops and its β increases. Unless the bias voltage is controlled to compensate for this, the collector current may start to rise, which in turn leads to further heating and eventual destruction of the device. This phenomenon is known as *thermal runaway*.

The possibility of failure due to the destruction of a transistor junction by excessive voltage, current or heating is best avoided by operating the device well within its safe limits at all times (see below). In the case of a small transistor, junction failure, should it occur, is normally absolute, and therefore renders the device useless. Power transistors, however, have a more complex construction. Rather than attempting to increase the junction area of an individual transistor in order to make it more rugged, power transistors normally consist of large numbers of smaller transistors fabricated on a single chip of silicon. These are arranged to operate in parallel, with low-value resistances introduced in series with the emitters to ensure that current is shared equally between the individual transistors. A large RF power transistor may contain as many as 1000 separate transistors. In such a device, the failure of a small number of the individual transistors may not unduly affect its performance. This possibility must sometimes be taken into account when testing circuitry which contains power transistors.

Safe operating area – SOAR

All bipolar transistors can fail if they are over-run and power bipolars are particularly susceptible since, if over-run by excessive power dissipation, hot spots will develop in the transistor's junction, leading to total destruction. To prevent this, manufacturers issue SOAR data, usually in the form of a graph or series of graphs plotted on log-log graph paper with current along one axis and voltage along the other. A typical example is shown in Fig 3.32.

Most amateur use will be for analogue operation and should be confined to area I in the diagram (or, of course, to the corresponding area in the diagram of the transistor being considered). For pulse operation, it is possible to stray out into area II. How far depends on the height of the current pulses and the duty cycle. Generally speaking, staying well within area I will lead to a long life.

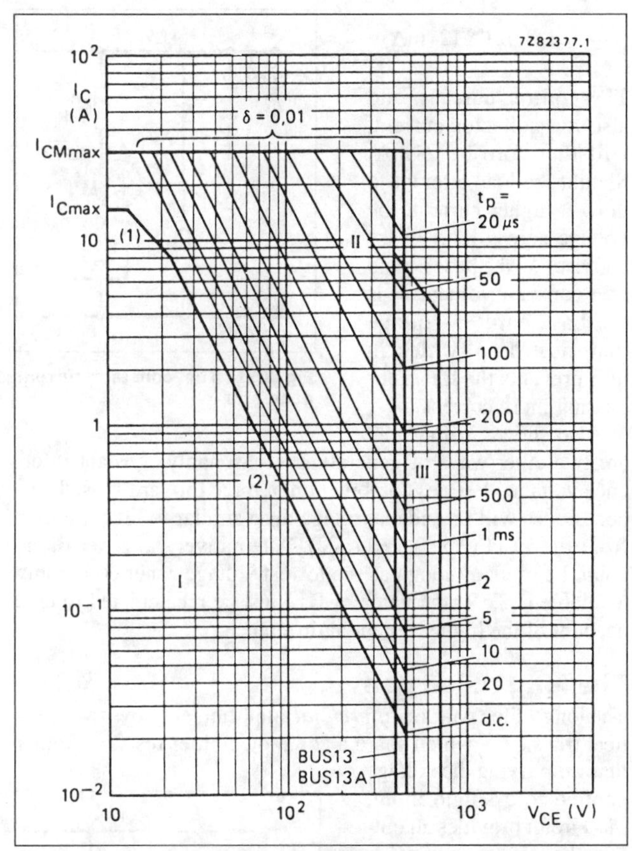

Fig 3.32. The safe operating (SOAR) curves for a BUS13A power bipolar transistor. When any power transistor is used as a pass device in a regulated PSU, it is important to ensure that the applied voltage and current are inside the area on the graph marked 'I – region of permissible DC operation'. Failure to do so may result in the device having a very short life because of *secondary breakdown*. This is due to the formation of *hot spots* in the transistor's junction. Note that power FETs do not suffer from this limitation. *Key to regions*: I – Region of permissible DC operation. II – Permissible extension for repetitive pulse operation. III – Area of permissible operation during turn-on in single-transistor converters, provided $R_{BE} \le 100\Omega$ and $t_p \le 0.6\mu s$. (Reproduction courtesy Philips Semiconductors)

All bipolar transistors have this SOAR but, in the case of small devices, it is not often quoted by the manufacturers and, in any case, is most unlikely to be exceeded.

Common-base and common-collector configurations

The common-emitter configuration is usually preferred because it provides both current and voltage gain. The input impedance of a common-emitter amplifier using a small, general-purpose transistor will be around 2kΩ at audio frequencies but, assuming that the emitter resistor is bypassed with a capacitor, this will drop to approximately 100Ω at a frequency of a few megahertz. The output impedance will be in the region of 10kΩ. Chapter 5 provides more information on the design of common-emitter amplifiers.

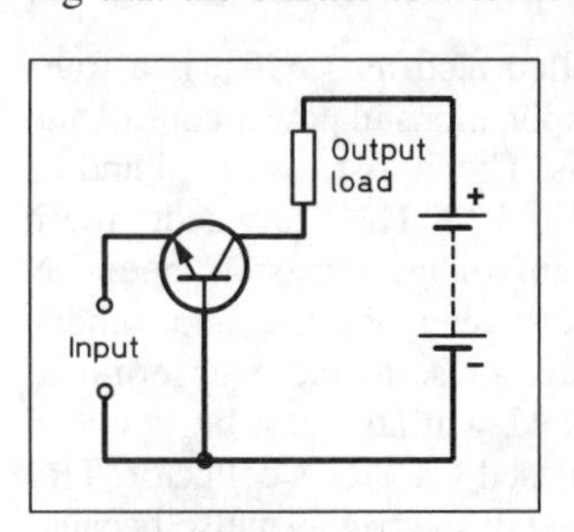

Fig 3.33. The common-base configuration

The common-base configuration is shown in Fig 3.33. The input and output coupling and bias circuitry has been omitted for the sake of clarity. As the emitter current is the sum of the collector and base currents, the current gain will be very slightly less than unity. There is, however, considerable voltage gain. The input impedance is very low, typically between 10 and 20Ω, assuming a collector current of 2mA. The output impedance is much higher, perhaps 1MΩ at audio frequencies. The output voltage will be in phase with the input. The common-base configuration will sometimes be used to provide amplification where the transistor must operate at a frequency close to its f_T.

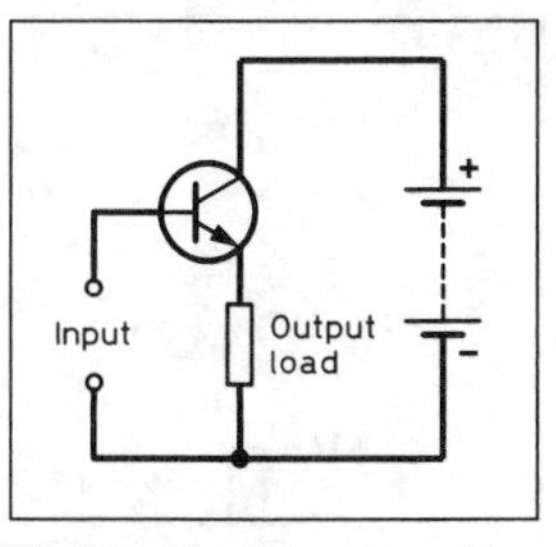

Fig 3.34. The common-collector, or emitter-follower, configuration

The common-collector configuration is normally referred to as the *emitter follower* (see Fig 3.34). This circuit provides a current gain equal to the transistor's β, but the voltage gain is very slightly less than unity. The input impedance is much higher than for the common-emitter configuration and may be approximated by multiplying the transistor's β by the value of the emitter load impedance. The output impedance, which is much lower, is normally calculated with reference to the impedance of the circuitry which drives the follower, and is approximated by:

$$\frac{Z_0}{\beta}$$

For example, if the transistor used has a β of 100 at the frequency of operation and the emitter follower is driven by circuitry with an output impedance (Z_0) of 2kΩ, the output impedance of the follower is roughly 20Ω. As with the common-base circuit, the output voltage is in phase with the input. Emitter followers are used extensively as 'buffers' in order to obtain impedance transformation, and isolation, between stages (see Chapter 5).

The Darlington pair

Fig 3.35 shows how two transistors may be connected to produce the equivalent of a single transistor with extremely high current gain (β). A current flowing into the base of TR1 will cause a much larger current to flow into the base of TR2. TR2 then provides further current gain. Not surprisingly, the overall current gain of the Darlington pair is calculated by multiplying the β of TR1 by the β of TR2. Therefore, if each transistor has a β of 100, the resultant current gain is 10,000.

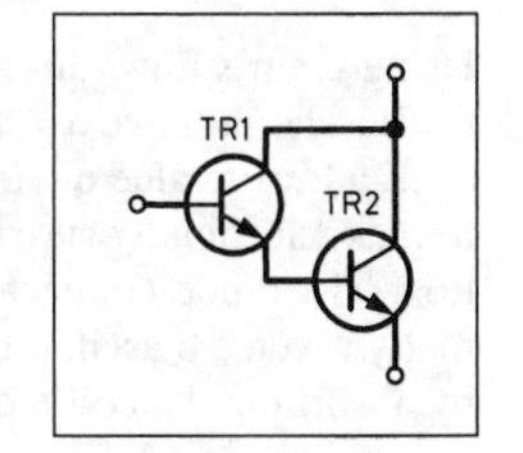

Fig 3.35. A Darlington pair is the equivalent of a single transistor with extremely high current gain

As an alternative to physically connecting two separate devices, it is possible to obtain 'Darlington transistors'. These contain a pair of transistors, plus the appropriate interconnections, fabricated onto a single chip.

The transistor as a switch

There is often a requirement in electronic equipment to activate relays, solenoids, electric motors and indicator devices etc using control signals generated by circuitry that cannot directly power the device which must be turned on or off. The transistor in Fig 3.36 solves such a problem by providing an interface between the source of a control voltage (+5V in this example) and

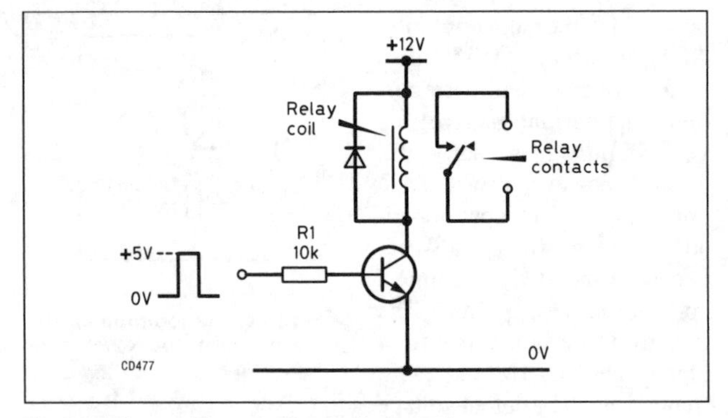

Fig 3.36. A transistor used to switch a relay

a 12V relay coil. If the control voltage is absent, only a minute leakage current flows between the collector and emitter of the switching transistor and so the relay is not energised. When the control voltage is applied, the transistor draws base current through R1 and this results in a larger current flow between its collector and emitter. The relay coil is designed to be connected directly across a potential of 12V and so the resistance between the collector and emitter of the transistor must be reduced to the lowest possible value. The desired effect is therefore the same as that which might otherwise be obtained by closing a pair of switch contacts wired in series with the relay coil.

Assuming that R1 allows sufficient base current to flow, the transistor will be switched on to the fullest extent (a state referred to as *saturation*). Under these conditions, the voltage at the collector of the transistor will drop to only a few tenths of a volt, thus allowing a potential almost equal to that of the supply rail (12V) to appear across the relay coil, which is therefore properly energised. Under these conditions the transistor will not dissipate much power because, although it is passing considerable current, there is hardly any resistance between the emitter and collector. Assuming that the relay coil draws 30mA when energised from a 12V supply, and that the β of the transistor is 150 at a collector current of this value, the base current will be:

$$\frac{30 \times 10^{-3}}{150} = 200\mu A$$

Using Ohm's Law, this suggests that the current limiting resistor R1 should have a value of around 25kΩ. In practice, however, a lower value of 10kΩ would probably be chosen in order to make absolutely sure that the transistor is driven into saturation. The diode connected across the relay coil, which is normally reverse biased, protects the transistor from high voltages by absorbing the coil's back EMF on switching off.

Constant-current generator

Fig 3.37 shows a circuit that will sink a fixed, predetermined current into a load of varying resistance. A constant-current battery charger is an example of a practical application which might use such a circuit. Also, certain low-distortion amplifiers will employ constant-current generators, rather than resistors, to act as collector loads. The base of the PNP transistor is held at a potential of 10.2V by the forward voltage drop of the LED (12 – 1.8 = 10.2V). Note that although LEDs are normally employed as indicators, they are also sometimes used as reference voltage generators. Allowing for the base-emitter voltage drop of the transistor (approximately 0.6V), the emitter voltage is 10.8V. This means that the potential difference across R1 will be held at 1.2V. Using Ohm's Law, the current flowing through R1 is:

$$\frac{1.2}{56} = 0.021A\ (21mA)$$

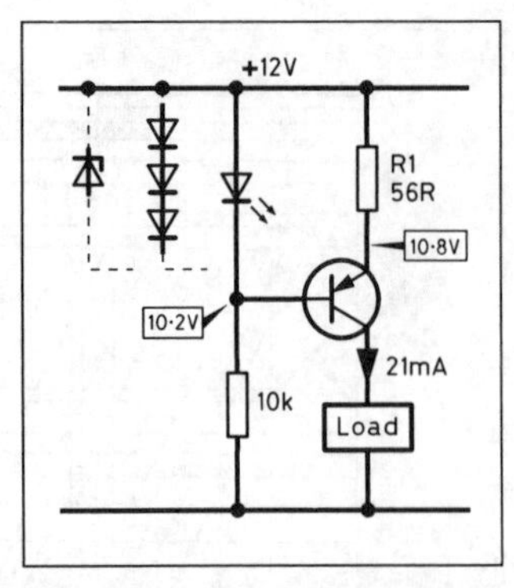

Fig 3.37. The constant-current generator

The emitter current, and also the collector current, will therefore be 21mA. Should the load attempt to draw a higher current, the voltage across R1 will try and rise. As the base is held at a constant voltage, it is the base-emitter voltage that must drop, which in turn prevents the transistor passing more current.

Also shown in Fig 3.37 are two other ways of generating a reasonably constant reference voltage. A series combination of three forward-biased silicon diodes will provide a voltage drop similar to that obtained from the LED (3 × 0.6 = 1.8V). Alternatively, a zener diode could be used, although as the lowest voltage zener commonly available is 2.7V, the value of R1 must be recalculated to take account of the higher potential difference.

The long-tailed pair

The long-tailed pair, or *differential* amplifier, employs two identical transistors which share a common emitter resistor. Rather than amplifying the voltage applied to a single input, this circuit provides an output which is proportional to the difference between the voltages presented to its two inputs, labelled 'a' and 'b' in Fig 3.38.

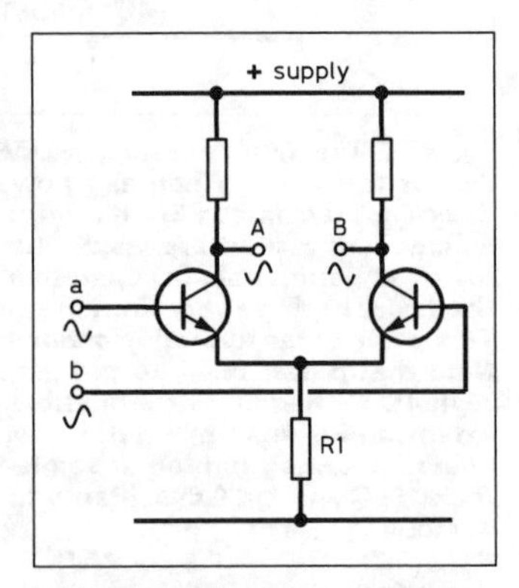

Fig 3.38. A differential amplifier or long-tailed pair

Providing that the transistors are well matched, variations in V_{BE} and β will cause identical changes in the potential at the two outputs, A and B. Therefore, if both A and B are used, it is the voltage difference between them that constitutes the wanted output. The long-tailed pair is very useful as an amplifier of DC potentials, an application where input and output coupling capacitors cannot be used. The circuit can be improved by replacing the emitter resistor (R1) with a constant-current generator, often referred to in this context as a *current source*. The differential amplifier is used extensively as an input stage in operational amplifiers (see p3.17).

Thyristors

The thyristor, or silicon controlled rectifier (SCR), is a four-layer PNPN device which has applications in power control and power supply protection systems. The thyristor symbol and its equivalent circuit is shown in Fig 3.39. The equivalent circuit consists of two interconnected high-voltage transistors, one NPN and the other PNP. Current flow between the anode and cathode is initiated by applying a positive pulse to the gate terminal, which causes TR1 to conduct. TR2 will now also be switched on because its base is forward biased via TR1's collector. TR1 continues to conduct after the end of the trigger pulse because collector current from TR2 is available to keep its base-emitter

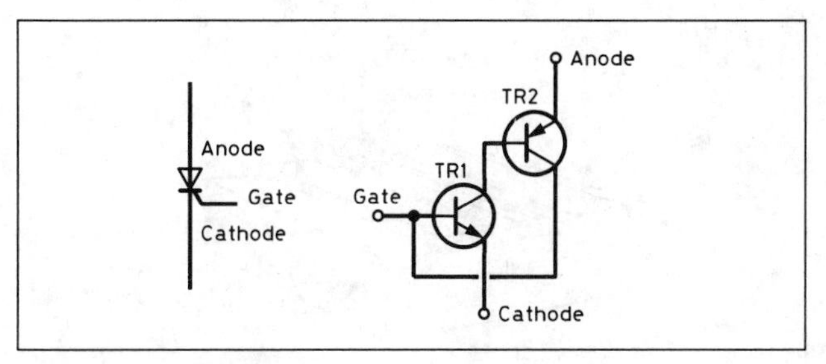

Fig 3.39. The symbol for a thyristor (silicon controlled rectifier) and its equivalent circuit

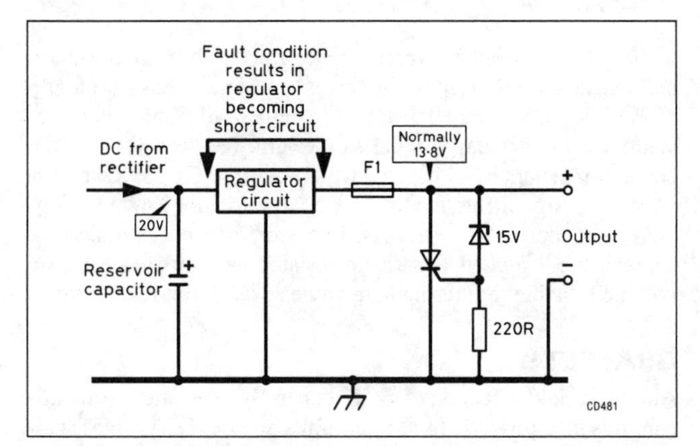

Fig 3.40. Crowbar over-voltage protection implemented with a thyristor

junction forward biased. Both transistors have therefore been latched into saturation and will remain so until the voltage between the anode and cathode terminals is reduced to a low value.

Fig 3.40 shows an over-voltage protection circuit for a power supply unit (PSU) based on a thyristor. In the event of regulator failure, the PSU output voltage rises above the nominal 13.8V, a situation that could result in considerable damage to any equipment that is connected to the PSU. As this happens, the 15V zener diode starts to conduct, and in doing so applies a positive potential to the thyristor gate. Within a few microseconds the thyristor is latched on, and the PSU output is effectively short-circuited. This shorting, or *crowbar* action, will blow fuse F1 and, hopefully, prevent any further harm.

The thyristor will only conduct in one direction, but there is a related device, called the *triac* (see Fig 3.41 for symbol), which effectively consists of two parallel thyristors connected anode to cathode. The triac will therefore switch currents in either direction and is used extensively in AC power control circuits, such as the ubiquitous lamp dimmer. In these applications a trigger circuit varies the proportion of each mains cycle for which the triac conducts, thus controlling the average power supplied to a load. Having been latched on at a predetermined point during the AC cycle, the triac will switch off at the next zero crossing point of the waveform, the process being repeated for each following half cycle.

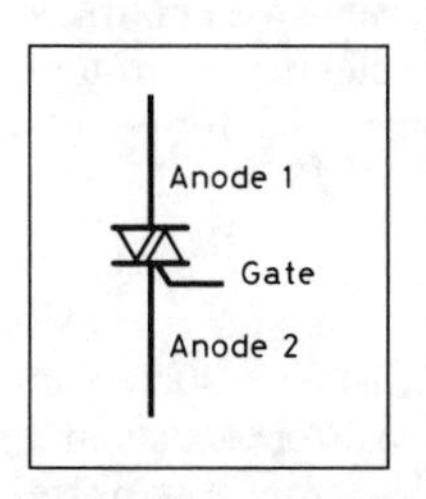

Fig 3.41. The triac or AC thyristor

THE JUNCTION FIELD EFFECT TRANSISTOR

Like its bipolar counterpart, the junction field effect transistor (JFET) is a three-terminal device which can be used to provide amplification over a wide range of frequencies. Fig 3.42 shows the circuit symbol and method of construction of an N-channel JFET. This device differs from the bipolar transistor in that current flow between its drain and source connections is controlled by a voltage applied to the gate. This means that in normal operation current does not flow into the gate at all, and the concept of current gain, as applied to bipolar devices, is meaningless. The current flow is by majority carriers, ie electrons in N-type devices. Therefore, holes have no part in the process and this allows very good high-frequency performance since hole diffusion is the slower conduction mode in bipolar devices.

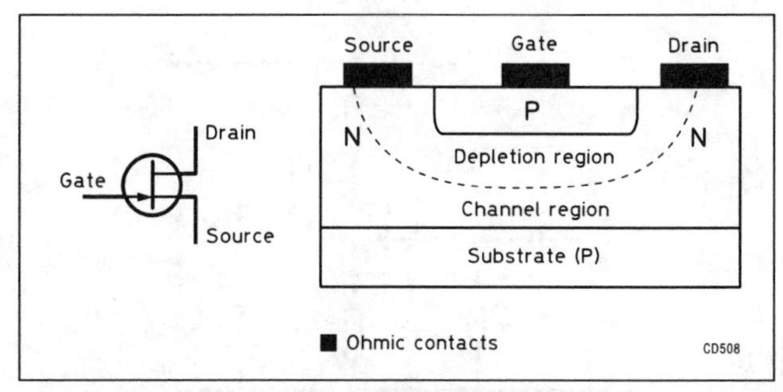

Fig 3.42. The circuit symbol and construction of the JFET (junction field effect transistor)

A negative voltage applied to the P-type gate sets up an electric field which intrudes into the N-type channel and establishes a depletion region (see Fig 3.42). This reduces the electron flow (current) between source and drain by effectively constricting the channel. If the negative potential applied to the gate is made large enough, the depletion region extends right across the channel, thus cutting off the drain current (I_D) completely. This condition is known as *pinch-off*. Therefore, providing that the gate is negatively biased at a point below pinch-off, a varying signal voltage applied to the gate will cause proportional variations in drain current. The gain, or *transconductance* (G_m or Y_{fs}), of a field effect device is expressed in siemens (see Chapter 1). A small general-purpose JFET will have a G_m of around 5 millisiemens. Fig 3.43 shows the characteristic curves for such a device.

The circuit of a small-signal amplifier using a JFET is shown in Fig 3.44. The potential difference across R2 provides bias by establishing a positive voltage at the source, which has the same effect as making the gate negative with respect to the source. R1 serves to tie the gate at ground potential (0V), and in practice its value will determine the amplifier's input impedance at audio

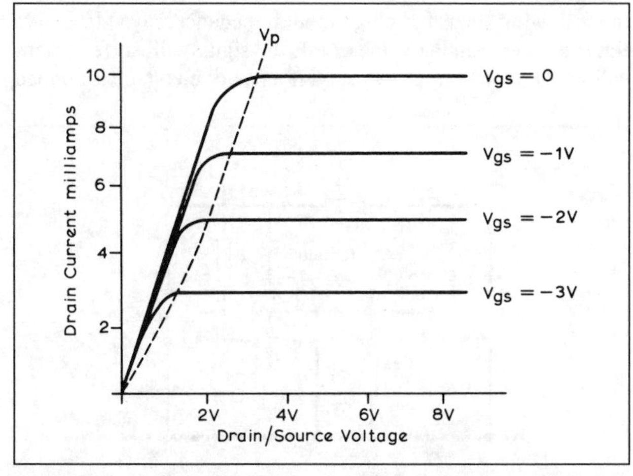

Fig 3.43. The relationship between gate voltage and drain current for a JFET

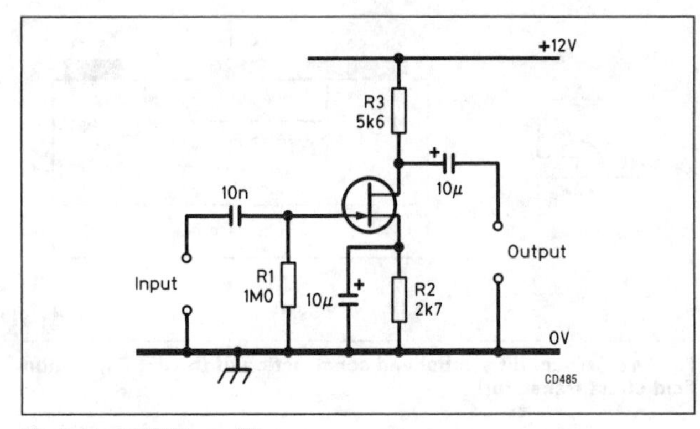

Fig 3.44. A JFET amplifier

frequencies. The inherently high input impedance of the JFET amplifier is essentially a result of the source-gate junction being reverse biased. If the gate were to be made positive with respect to the source (a condition normally to be avoided), gate current would indeed flow, thus destroying the field effect. The value of R3, the drain load resistor (R_L), dictates the voltage gain obtained for a particular device transconductance (assumed to be 4 millisiemens in this case) as follows:

$$\begin{aligned}\text{Voltage gain} &= G_m \times R_L \\ &= 4 \times 10^{-3} \times 5.6 \times 10^3 \\ &= 4 \times 5.6 \\ &= 22.4 \text{ (27dB)}\end{aligned}$$

The voltage gain obtainable from a common-source JFET amplifier is therefore around 20dB, or a factor of 10, lower than that provided by the equivalent common-emitter bipolar amplifier (see p3.9). Also, the characteristics of general-purpose JFETs are subject to considerable variation, or 'spread', a fact that may cause problems in selecting the correct value of bias resistor for a particular device. However, the JFET does offer the advantage of high input impedance, and this is exploited in the design of stable, variable frequency oscillators (see Chapter 5). JFETs are also employed in certain types of RF amplifier and switching mixer (see Chapters 5 and 6).

The JFET is often used as a voltage-controlled variable resistor in signal gates and attenuators. The channel of the JFET in Fig 3.45 forms a potential divider working in conjunction with R1. Here R2 and R3 develop a bias voltage which is sufficient to ensure pinch-off. This means that the resistance between the source and drain will be around 10MΩ and so, providing that the following stage has a high input impedance, say at least five times greater than the value of R1, the signal will suffer practically no attenuation. Conversely, if a positive voltage is applied

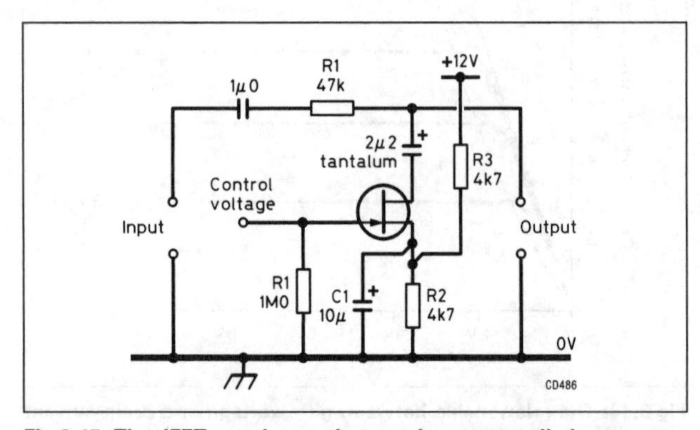

Fig 3.45. The JFET may be used as a voltage-controlled attenuator

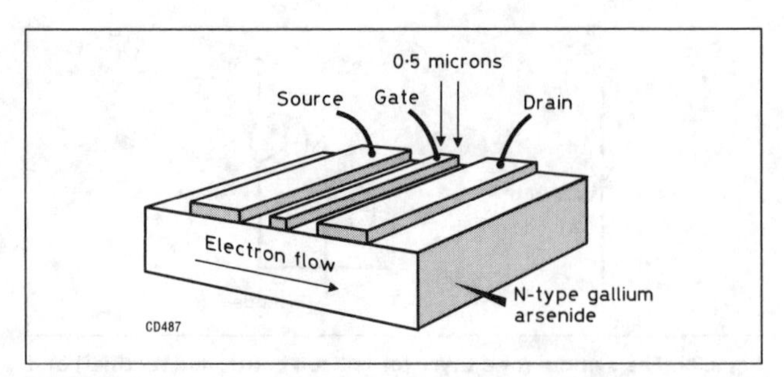

Fig 3.46. Construction of a GaAsFET (gallium arsenide field effect transistor)

to the gate sufficient to overcome the effect of the bias, the channel resistance will drop to the lowest possible value – typically 400Ω for a small-signal JFET. The signal will now be attenuated by a factor nearly equal to the ratio between R1 and the channel resistance – 118, or 41dB (note that C1 serves to bypass R2 at signal frequencies). The circuit is not limited to operation at these two extremes, however, and it is possible to achieve the effect of a variable resistor by adjusting the gate voltage to achieve intermediate values of channel resistance.

GaAsFETS

Although field effect devices are generally fabricated from silicon, it is also possible to use gallium arsenide. GaAsFETs (gallium arsenide field effect transistors) are N-channel field effect transistors designed to exploit the higher electron mobility provided by gallium arsenide (GaAs). The gate terminal differs from that of the standard silicon JFET in that it is made from gold, which is bonded to the top surface of the GaAs channel region (see Fig 3.46). The gate is therefore a Schottky barrier junction, as used in the hot-carrier diode (see p3.6). Good high-frequency performance is achieved by minimising the electron transit time between the source and drain. This is achieved by reducing the source drain spacing to around 5 microns and making the gate from a strip of gold only 0.5 microns wide (note that this critical measurement is normally referred to as the *gate length* because it is the dimension running parallel to the electron flow).

The very small gate is particularly delicate, and it is therefore essential to operate GaAsFETs with sufficient negative bias to ensure that the gate source junction never becomes forward biased. Protection against static discharge and supply line transients is also important.

GaAsFETs are found in very-low-noise receive preamplifiers operating at UHF and microwave frequencies up to around 20GHz. They can also be used in power amplifiers for microwave transmitters (see Chapters 8 and 9).

MOSFETS

The MOSFET (metal oxide field effect transistor), also known as the *IGFET* (insulated gate field effect transistor), exists in a number of forms, with applications ranging from low-noise preamplification at microwave frequencies to high-power amplifiers in HF and VHF transmitters. Large scale integrated circuits (LSICs), including the latest microprocessors, also make extensive use of MOS transistors.

The MOSFET differs from the JFET in having an insulating layer, normally composed of silicon dioxide (SiO_2), interposed between the gate and channel. The insulation prevents current flowing into, or out of, the gate under all conditions. This makes the MOSFET easier to bias and guarantees an extremely high input resistance under all conditions. The insulating layer

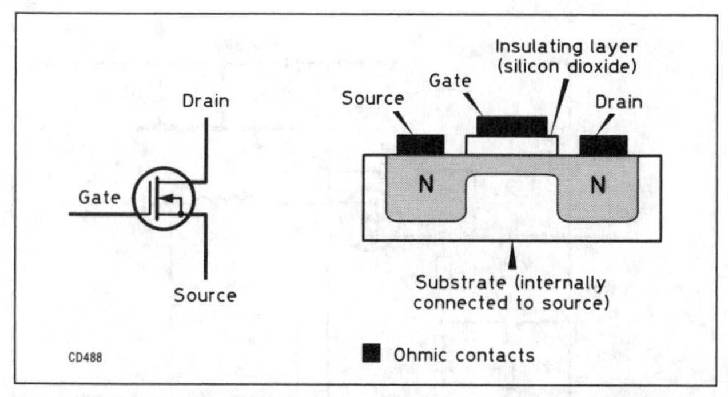

Fig 3.47. Circuit symbol and method of construction of an N-channel MOSFET (metal oxide semiconductor field effect transistor). The insulating layer prevents the flow of gate current but does not effect the creation of an electric field, and therefore a depletion layer, within the channel region. The circuit symbol for the P-channel type is the same, except that the direction of the arrow is reversed

behaves as a dielectric, with the gate forming one plate of a capacitor. Gate capacitance varies depending on the area of the gate itself and its general effect is to lower the impedance seen at the gate as frequency rises. The main disadvantage of this form of construction is that the very thin insulating layer can be punctured by high voltages appearing on the gate. Therefore, in order to protect these devices against destruction by static discharges, internal zener diodes are normally incorporated. Unfortunately, the protection provided by the zener diodes is not absolute, and all MOS devices should therefore be handled with care.

MOSFETs are available with channel regions fabricated from either N-type or P-type material, conduction being provided by electrons in N-type, and holes in the P-type. However, as electrons have greater mobility than holes, the N-channel device is normally favoured because it promises better high-frequency performance. There is a further subdivision into *depletion-mode* and *enhancement-mode* types.

Depletion-mode devices have a channel that is normally conducting, and gate bias must be applied in order to vary, or stop the flow. In enhancement-mode devices the opposite applies. With zero gate bias no drain current will flow, but when bias is applied the channel resistance will drop.

Adding further to the variety of MOSFETs available, there are also dual-gate types. To summarise by example, an N-channel enhancement-mode MOSFET will not pass drain current in the absence of gate voltage (or, to put it more correctly, if the gate is at the same potential as the source). To initiate drain current, it is necessary to apply positive bias to the gate, thereby making the gate positive with respect to the source.

Fig 3.47 shows the circuit symbol and construction method for a small single-gate MOSFET, and Fig 3.48 features the dual-gate equivalent. Dual-gate MOSFETs perform well as RF and

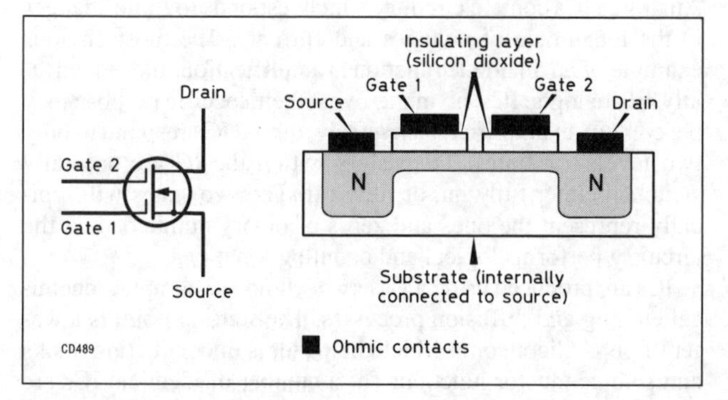

Fig 3.48. A dual-gate MOSFET

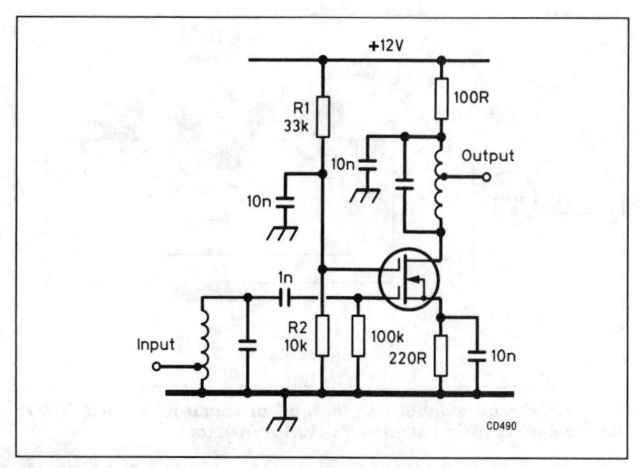

Fig 3.49. An RF amplifier using a dual-gate MOSFET

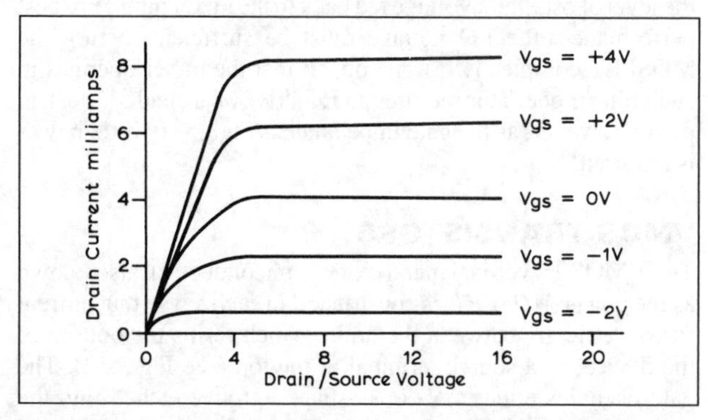

Fig 3.50. The relationship between gate voltage and drain current for a MOSFET

IF amplifiers in receivers. They contribute little noise and provide good dynamic range. Transconductance is also higher than that offered by the JFET, typically between 7 and 15 millisiemens for a general-purpose device.

Fig 3.49 shows the circuit of an RF amplifier using a dual-gate MOSFET. The signal is presented to gate 1, and bias is applied separately to gate 2 by the potential divider comprising R1 and R2. Selectivity is provided by the tuned circuits at the input and output. Care must be taken in the layout of such circuits to prevent instability and oscillation (see Chapter 5). A useful feature of the dual-gate amplifier is the ability to control its gain by varying the level of the gate 2 bias voltage. This is particularly useful in IF amplifiers, where the AGC voltage is often applied to gate 2 (see Chapters 5 and 6). The characteristic curve at Fig 3.50 shows the effect on drain current of making the bias voltage either negative or positive with respect to the source.

Dual-gate MOSFETs may also be used as mixers. In Fig 3.51, the signal is applied to gate 1 and the local oscillator (LO) drive to gate 2. There is a useful

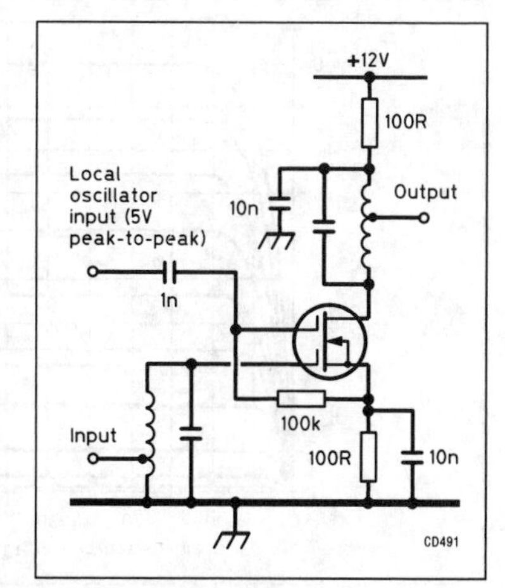

Fig 3.51. A mixer circuit based on a dual-gate MOSFET

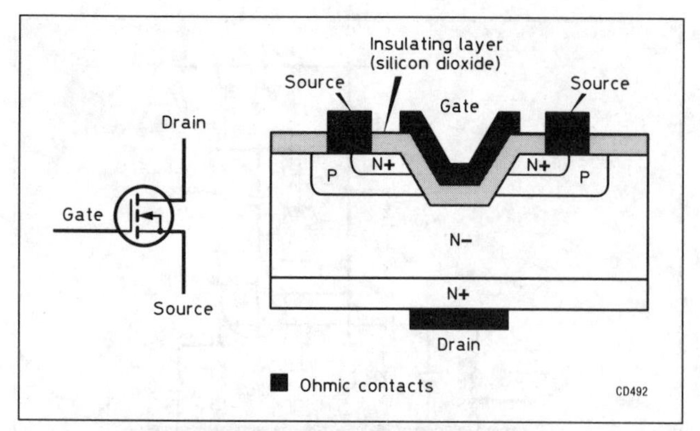

Fig 3.52. Circuit symbol and method of construction of a VMOS (vertical metal oxide semiconductor) transistor

degree of isolation between the two gates, and this helps reduce the level of oscillator voltage fed back to the mixer input. For best performance, the LO voltage must be sufficient to turn the MOSFET completely off and on, so that the mixer operates in switching mode. This requires an LO drive of around 5V peak to peak. However, as the gate impedance is high, very little power is required.

VMOS TRANSISTORS

The VMOS™ (vertical metal oxide semiconductor), also known as the *power MOSFET*, is constructed in such a way that current flows vertically between the drain, which forms the bottom of the device, to a source terminal at the top (see Fig 3.52). The gate occupies either a V- or U-shaped groove etched into the upper surface. VMOS devices feature a four-layer sandwich comprising N+, P, N– and N+ material and operate in the enhancement mode. The vertical construction produces a rugged device capable of passing considerable drain current and offering a very high switching speed. These qualities are exploited in power control circuits and transmitter output stages. Fig 3.53 shows the characteristic curves of a typical VMOS transistor. Note that the drain current is controlled almost entirely by the gate voltage, irrespective of drain voltage. Also, above a certain value of gate voltage, the relationship between gate voltage and drain current is highly linear. Power MOSFETs fabricated in the form of large numbers of parallel-connected VMOS transistors are termed HEXFETs™.

Although the resistance of the insulated gate is for all intents

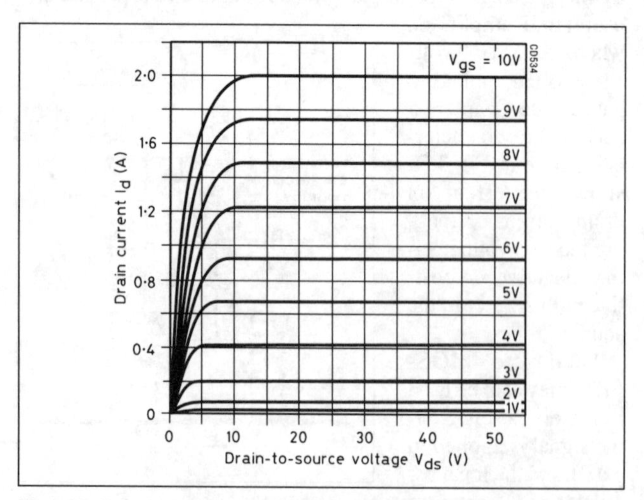

Fig 3.53. Relationship between gate voltage and drain current for a VMOS power transistor

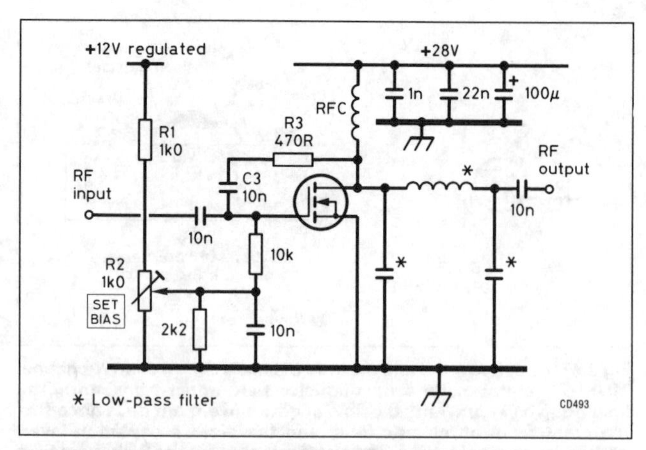

Fig 3.54. A linear amplifier utilising a VMOS power transistor. Assuming a 50Ω output load, the RFC value is chosen so that it has an inductive reactance (X_L) of approximately 400Ω at the operating frequency

and purposes infinite, the large gate area leads to high capacitance. A VMOS transistor intended for RF and high-speed switching use will have a gate capacitance of around 50pF, whereas devices made primarily for audio applications have gate capacitances as high as 1nF. A useful feature of these devices is that the relationship between gate voltage and drain current has a negative temperature coefficient of approximately 0.7% per degree Celsius. This means that as the transistor gets hotter, its drain current will tend to fall, thus preventing the thermal runaway which can destroy bipolar power transistors.

Fig 3.54 shows the circuit of a simple HF linear amplifier using a single VMOS transistor. Forward bias is provided by the potential divider R1, R2 so that the amplifier operates in Class AB. In this circuit R3 and C3 provide a small amount of negative feedback to help prevent instability. More complex push-pull amplifiers operating from supplies of around 50V can provide RF outputs in excess of 100W (see Chapters 5 and 7).

INTEGRATED CIRCUITS

Having developed the techniques used in the fabrication of individual semiconductor devices, the next obvious step for the electronics industry was to work towards the manufacture of complete circuits on single chips of silicon. The first integrated circuit (IC) was made in 1958 by Jack Kilby of Texas Instruments, just eight years after the birth of the bipolar junction transistor. The earliest ICs contained less than 50 components, but the technology has now advanced to a point where it is possible to mass-produce microprocessor chips containing over three million individual transistors.

ICs fall into two broad categories – *analogue* and *digital*. Analogue ICs contain circuitry which responds to finite changes in the magnitude of voltages and currents. The most obvious example of an analogue function is amplification. Indeed, virtually all analogue ICs, no matter what their specific purpose may be, contain an amplifier. Conversely, digital ICs respond to only two levels, or states. Transistors within the IC are normally switched either fully on, or fully off. The two states will typically represent the ones and zeros of binary numbers, and the circuitry performs logical and counting functions.

ICs are produced using a variety of photo-lithographic, chemical etching and diffusion processes. The starting point is a wafer of pure silicon, onto which the patterns of conducting tracks and semiconductor junctions for a number of identical ICs are photo-etched. Doping is normally carried out by diffusion at

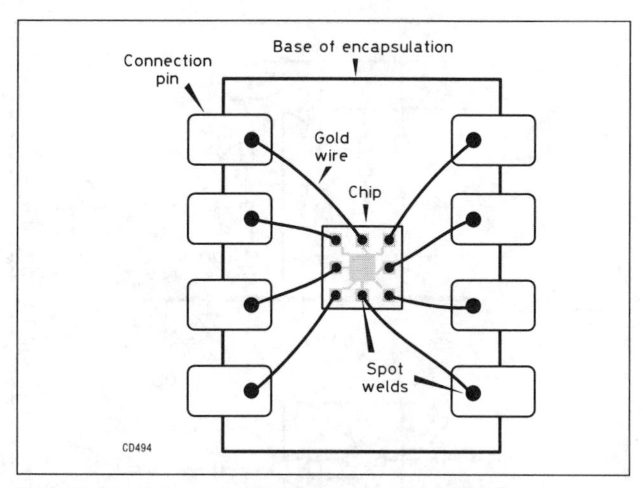

Fig 3.55. Internal construction of an integrated circuit

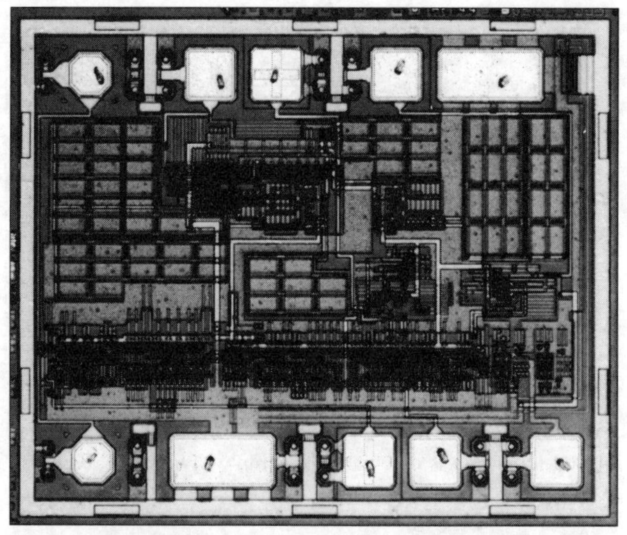

Photograph of the SP8714 integrated circuit chip showing pads for bonding wires (Plessey Semiconductors)

high temperatures. The processes are often repeated a number of times in order to fabricate different layers. The result, hopefully, is a wafer containing a large number of correctly functioning circuits. The wafer is now cut into individual chips, and any faulty ones discovered during an automated testing procedure are discarded. Each remaining chip is now affixed to the base of its encapsulation, and very fine gold wires are spot welded between pads located around the chip's periphery and the metal pins which serve as external connections – see Fig 3.55. Finally, the top of the encapsulation is bonded to the base, forming a protective seal. Most general-purpose ICs are encapsulated in plastic, but some expensive devices that must operate reliably at high temperatures are housed in ceramic packages.

An important advantage of integrated, or 'monolithic', construction is that because all the components are fabricated under exactly the same conditions, the operational characteristics of the transistors and diodes, and also the values of resistors, are inherently well matched.

LINEAR INTEGRATED CIRCUITS

Operational amplifiers

The operational amplifier (op-amp) is a basic building-block IC that can be used in a very wide range of applications requiring low-distortion amplification and buffering. Modern op-amps feature high input impedance, an open-loop gain of at least 90dB (which means that in the absence of gain-reducing negative feedback, the change in output voltage will be at least 30,000 times greater than the change in input voltage), extremely low distortion, and low output impedance. Often two, or even four, separate op-amps will be provided in a single encapsulation. The first operational amplifiers were developed for use in analogue computers and were so named because, with suitable feedback, they can perform mathematical 'operations' such as adding, subtracting, logging, antilogging, differentiating and integrating voltages.

A typical op-amp contains around 20 transistors, a few diodes and perhaps a dozen resistors. The first stage is normally a long-tailed pair and provides two input connections, designated *inverting* and *non-inverting* (see the circuit symbol at Fig 3.56). The input transistors may be bipolar types, but JFETs or even MOSFETs are also used in some designs in order to obtain very high input impedance. Most op-amps feature a push-pull output stage operating in Class AB which is invariably provided with protection circuitry to guard against short-circuits. The minimum value of output load when operating at maximum supply voltage is normally around 2kΩ, but op-amps capable of driving 500Ω loads are available. Between the input and output circuits there will be one or two stages of voltage amplification. Constant-current generators are used extensively in place of collector load resistors, and also to stabilise the emitter (or source) current of the input long-tailed pair.

In order to obtain an output voltage which is in phase with the input, the non-inverting amplifier circuit shown at Fig 3.57 is used. Resistors R1 and R2 form a potential divider which feeds a proportion of the output voltage back to the inverting input. In most cases the open-loop gain of an op-amp can be considered infinite. Making this assumption simplifies the calculation of the closed-loop gain obtained in the presence of the negative feedback provided by R1 and R2. For example, if R1 has a value of 9kΩ and R2 is 1kΩ, the voltage gain will be:

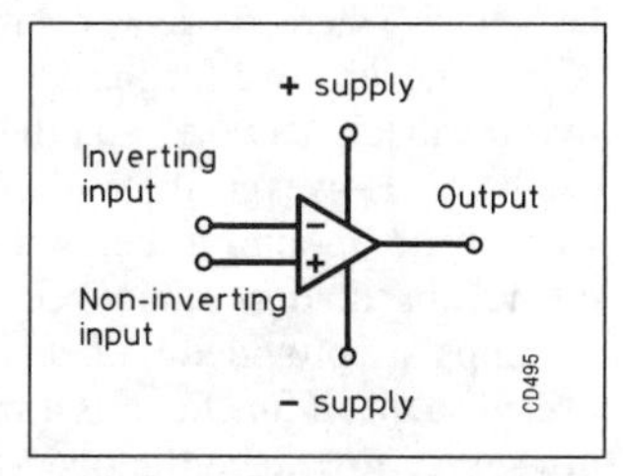

Fig 3.56. The circuit symbol for an operational amplifier (op-amp)

$$\frac{R_1 + R_2}{R_2} = \frac{9000 + 1000}{1000} = 10 \text{ (20dB)}$$

If R2 is omitted, and R1 replaced by a direct connection between the output and the inverting input, the op-amp will function as a unity gain buffer.

Fig 3.58 shows an inverting amplifier. In this case the phase of the output voltage will be opposite to that of the input. The

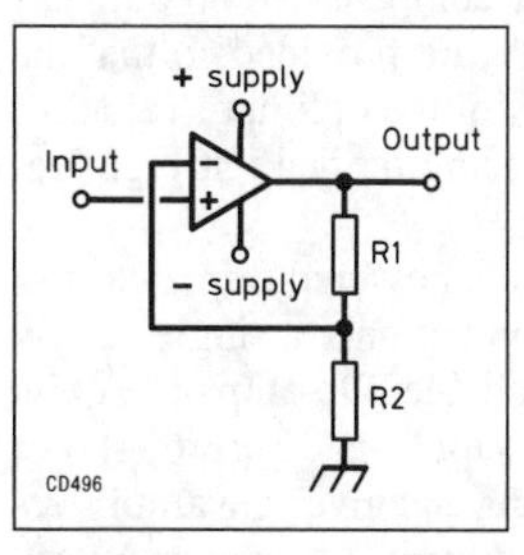

Fig 3.57. A non-inverting amplifier

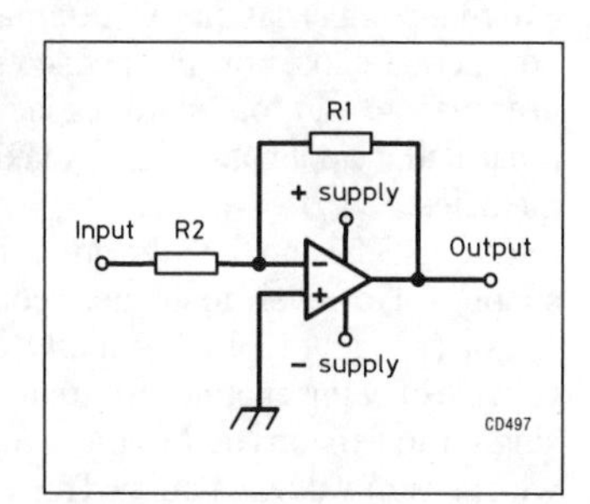

Fig 3.58. An inverting amplifier

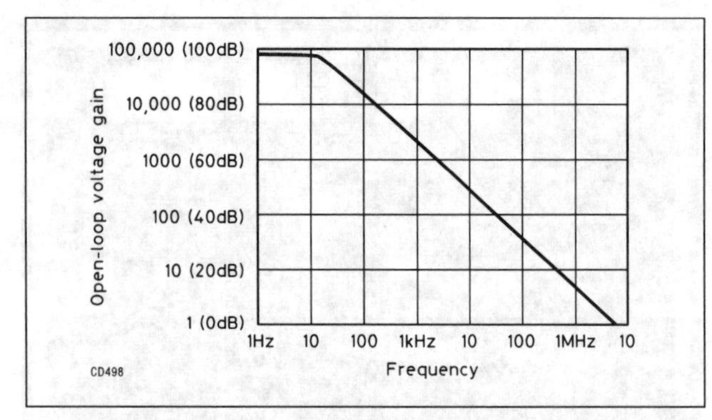

Fig 3.59. The relationship between open loop gain and frequency for a typical internally compensated op-amp

closed-loop gain is calculated by simply dividing the value of R1 by R2. Therefore, assuming the values are the same as used for the non-inverting amplifier:

$$\frac{R_1}{R_2} = \frac{9000}{1000} = 9 \text{ (19dB)}$$

At low frequencies, the operation of the negative feedback networks shown in Figs 3.57 and 3.58 is predictable in that the proportion of the output voltage they feed back to the input will be out of phase with the input voltage at the point where the two voltages are combined. However, as the frequency rises, the time taken for signals to travel through the op-amp becomes a significant factor. This delay will introduce a phase shift that increases with rising frequency. Therefore, above a certain frequency – to be precise, where the delay contributes a phase shift of more than 135° – the negative feedback network actually becomes a positive feedback network, and the op-amp will be turned into an oscillator. However, if steps are taken to reduce the open-loop gain of the op-amp to below unity (ie less than 1) at this frequency, oscillation cannot occur. For this reason, most modern op-amps are provided with an internal capacitor which is connected so that it functions as a simple low-pass filter. This measure serves to reduce the open-loop gain of the amplifier by a factor of 6dB per octave (ie for each doubling of frequency the voltage gain drops by a factor of two), and ensures that it falls to unity at a frequency below that at which oscillation might otherwise occur. Op-amps containing such a capacitor are designated as being 'internally compensated'.

Compensated op-amps have the advantage of being absolutely stable in normal use. The disadvantage is that the open-loop gain is considerably reduced at high frequencies, as shown by the graph at Fig 3.59. This means that general-purpose, internally compensated op-amps are limited to use at frequencies below about 1MHz and they will typically be employed as audio frequency preamplifiers, and in audio filters. There are, however, special high-frequency types available, usually featuring external compensation. An externally compensated op-amp has no internal capacitor, but connections are provided so that the user may add an 'outboard' capacitor of the optimum value for a particular application, thus maximising the gain at high frequencies.

In Figs 3.57 and 3.58 the op-amps are powered from dual-rail supplies. However, in amateur equipment only a single supply rail of around +12V is normally available. Op-amps are quite capable of being operated from such a supply, and Fig 3.60 shows single-rail versions of the non-inverting and inverting amplifiers with resistor values calculated for slightly different gains. A mid-rail bias supply is generated using a potential divider (R3 and R4 in each case). The decoupling capacitor C1 enables the bias supply to be used for a number of op-amps (in simpler circuits it is often acceptable to bias the op-amp using a potential divider connected directly to the non-inverting input of the op-amp, in which case C1 is omitted). The value of R5 is normally made the same as R1 in order to minimise the op-amp input current, although this is less important in the case of JFET op-amps due to their exceptionally high input resistance. C2 is incorporated to reduce the gain to unity at DC so that the output voltage will settle to the half-rail potential provided by the bias generator in the absence of signals.

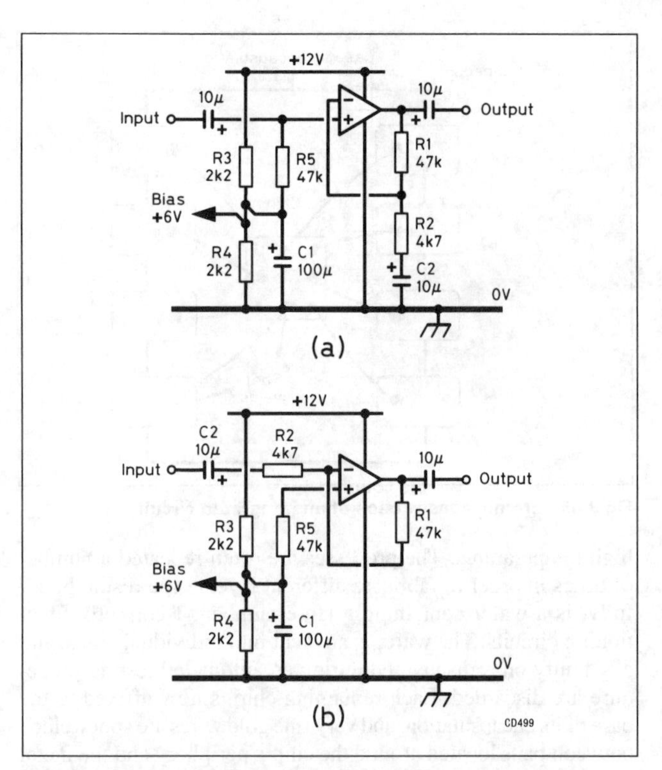

Fig 3.60. Single supply rail versions of the non-inverting (a) and inverting (b) amplifier circuits

Fig 3.61 shows a differential amplifier where the signal drives both inputs in antiphase. An advantage of this balanced arrangement is that interference, including mains hum or ripple, tends to impress voltages of the same phase at each input and so will be eliminated by cancellation. This ability to reject in-phase, or *common-mode*, signals when operating differentially is known as the *common-mode rejection ratio* (CMRR). Even an inexpensive op-amp will provide a CMRR of around 90dB, which means that an in-phase signal would have to be 30,000 times greater in magnitude than a differential signal in order to generate the same output voltage.

Further information on op-amps and their circuits is given in Chapters 5 and 22.

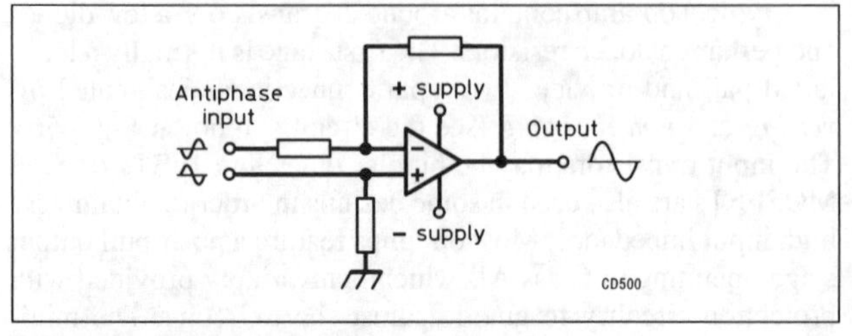

Fig 3.61. A differential amplifier

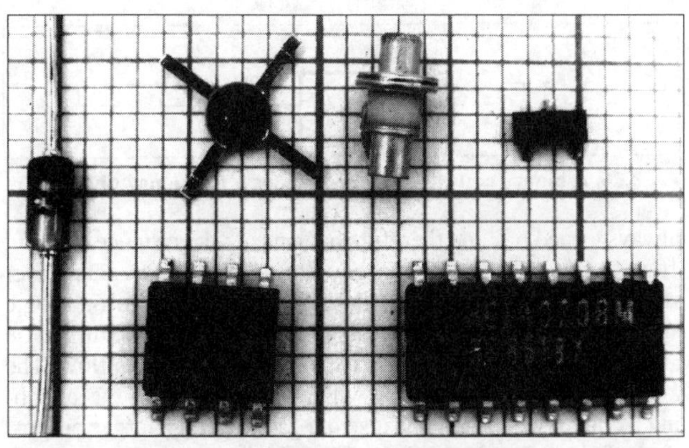

Small semiconductor devices. From the left, clockwise: signal diode, hot-carrier diode 'quad' for a double-balanced mixer, 10GHz Gunn effect device, surface-mounting bipolar transistor, surface-mounting 17-stage ripple counter, surface-mounting dual op-amp. The small squares in the background are 1mm

Audio power amplifiers

The audio power IC is basically just an op-amp with larger output transistors. Devices giving power outputs in the range 250mW to 40W are available, the bigger types being housed in encapsulations featuring metal mounting tabs (TO220 for example) that enable the IC to be bolted directly to a heatsink. Fig 3.62 shows a 1W audio output stage based on an LM380N device.

The LM380N has internal negative feedback resistors which provide a fixed closed loop voltage gain of approximately 30dB. A bias network for single-rail operation is also provided, and so very few external components are required. Not all audio power ICs incorporate the negative feedback and single-rail bias networks 'on-chip', however, and so some, or all, of these components may have to be added externally.

Voltage regulators

These devices incorporate a voltage reference generator, error amplifier and series pass transistor. Output short-circuit protection and thermal shut-down circuitry are also normally provided. There are two main types of regulator IC – those that generate a fixed output voltage, and also variable types which enable a potentiometer, or a combination of fixed resistors, to be connected externally in order to set the output voltage as required. Devices capable of delivering maximum currents of between 100mA and 5A are readily available, and fixed types offering a wide variety of both negative and positive output voltages may be obtained. Fig 3.63 shows a simple mains power supply unit

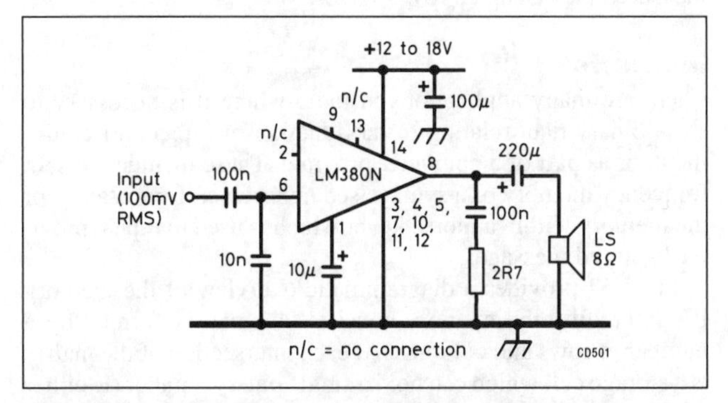

Fig 3.62. An audio output stage using the LM380N IC. The series combination of the 2.7Ω resistor and the 100nF capacitor at the output form a Zobel network. This improves stability by compensating for the inductive reactance of the loudspeaker voice coil at high frequencies

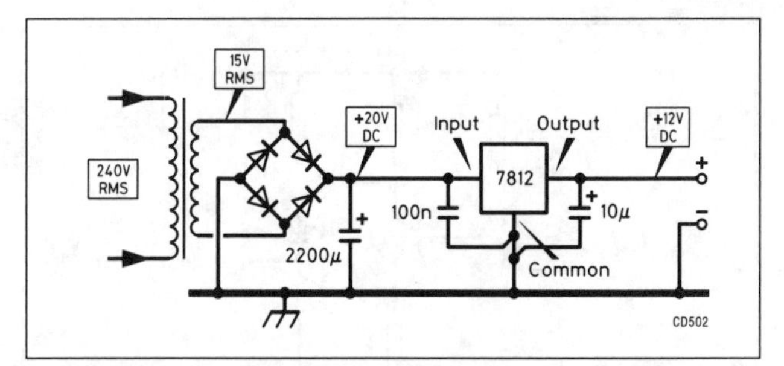

Fig 3.63. A simple 12V PSU (power supply unit) utilising a fixed voltage IC regulator type 7812

(PSU), based on a type 7812 regulator, providing +12V at 1A maximum.

Switched-mode power regulator ICs, which dissipate far less power, are also available.

Further information on regulator ICs is given in Chapter 14.

RF building blocks

Although it is now possible to fabricate an entire broadcast receiver on a single chip, this level of integration is rarely possible in amateur and professional communications equipment. To achieve the level of performance demanded, and also provide a high degree of operational flexibility, it is invariably necessary to consider each section of a receiver or transceiver separately, and then apply the appropriate technology to achieve the design goals. In order to facilitate this approach, ICs have been developed which perform specific circuit functions such as mixing, RF amplification and IF amplification.

Examples of IC mixers based on bipolar transistors are the Philips/Signetics NE602 (featuring very low current consumption as demanded in battery operated equipment) and the Plessey SL6440, a high-level device with programmable intercept point, which may be employed in the front-end of high-performance HF receivers. A popular IF amplifier is the Plessey SL1612. There are also devices offering a higher level of integration – often termed *sub-system ICs* – which provide more than one block function. For instance, the Plessey SL6700 contains two IF amplifiers, an automatic gain control (AGC) generator and a balanced mixer. For more information on how to use these ICs, see Chapters 6 and 7.

Amplification at UHF and microwave frequencies calls for special techniques and components. A wide range of devices, known generically as *MMICs* (microwave monolithic integrated circuits) are available.

DIGITAL INTEGRATED CIRCUITS

Logic families

In digital engineering, the term 'logic' generally refers to a class of circuits that perform relatively straightforward gating, latching and counting functions. Historically, logic ICs were developed as a replacement for computer circuitry based on large numbers of individual transistors, and, before the development of the bipolar transistor, valves were employed. Today, very complex ICs are available, such as the microprocessor, which contain most of the circuitry required for a complete computer fabricated on to a single chip. It would be wrong to assume, however, that logic ICs are now obsolescent because there are still a great many low-level functions, many of them associated with microprocessor systems, where they are useful. In amateur radio, there are also applications which do not require the processing capability of a digital computer, but nevertheless depend

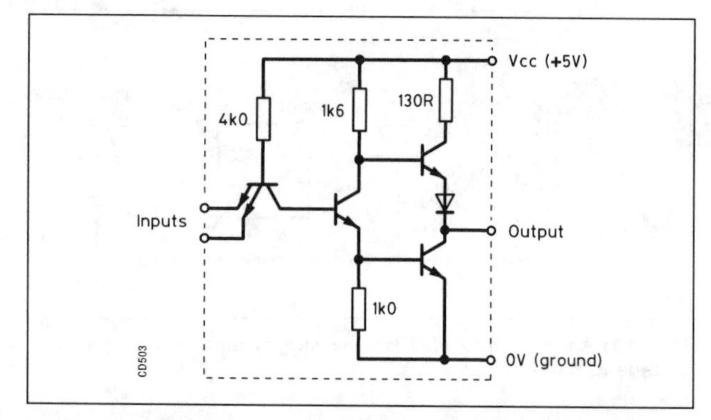

Fig 3.64. The internal circuit of a standard TTL two-input NAND gate. 74LS (low-power Schottky) TTL logic uses higher-value resistors in order to reduce power consumption, and the circuitry is augmented with clamping diodes to increase the switching speed of the transistors

upon logic – the electronic morse keyer (see Chapter 10) and certain transmit/receive changeover arrangements (see Chapter 7), for example.

By far the most successful logic family is *TTL* (transistor, transistor logic). Originally developed in the 'sixties, these circuits have been continuously developed in order to provide more complex functions, increase speed of operation and reduce power consumption. Standard (type 7400) TTL requires a 5V power rail stabilised to within ±250mV. Logic level 0 is defined as a voltage between zero and 800mV, whereas logic 1 is defined as 2.4V or higher. Fig 3.64 shows the circuit of a TTL NAND gate ('NAND' is an abbreviation for 'negative AND' and refers to the gate's function, which is to produce an output of logic 0 when both inputs are at logic 1). Note the use of a dual-emitter transistor at the input which functions in the same way as two separate transistors connected in parallel. One of the first improvements made to TTL was the incorporation of Schottky clamping diodes in order to reduce the turn-off time of the transistors – this enhancement produced the 74S series. Latterly, a low-power consumption version of Schottky TTL, known as *74LS*, has become very popular. The latest versions of TTL are designed around *CMOS* (complementary metal oxide silicon) transistors. The 74HC (high-speed CMOS) series is preferred for general use but the 74HCT type must be employed where it is necessary to use a mixture of CMOS and 74LS circuits to implement a design. A 74HC/HCT counter will operate at frequencies up to 25MHz.

The 'complementary' in 'CMOS' refers to the use of gates employing a mixture of N-channel and P-channel MOS transistors. Fig 3.65 shows the simplified circuit of a single CMOS inverter. If the input of the gate is held at a potential close to zero volts (logic 0), the P-channel MOSFET will be turned on, reducing its channel resistance to approximately 400Ω, and the N-channel MOSFET is turned off. This establishes a potential very close to the positive supply rail (V_{dd} – logic 1) at the gate's output. An input of logic 1 will have the opposite effect, the N-channel MOSFET being turned on and the P-channel MOSFET turned off. As one of the transistors is always turned off, and therefore has a channel resistance of about 10,000MΩ, virtually no current flows through the gate under static conditions. However, during transitions from one logic state to another, both transistors will momentarily be turned on at the same time, thus causing a measurable current to flow. The average current consumption will tend to increase as the switching frequency is raised because the gates spend a larger proportion of time in transition between logic levels. The popular CD4000 logic family is based entirely on CMOS technology. These devices can operate with supply voltages from 5 to 15V, and at maximum switching speeds of between 3 and 10MHz.

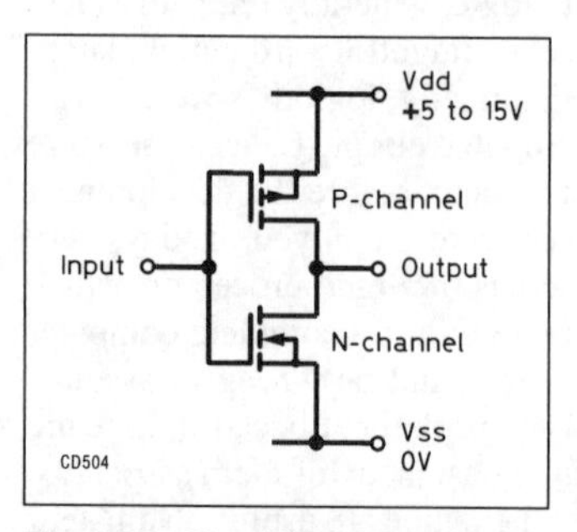

Fig 3.65. Simplified circuit of a CMOS inverter as used in the CD4000 logic family

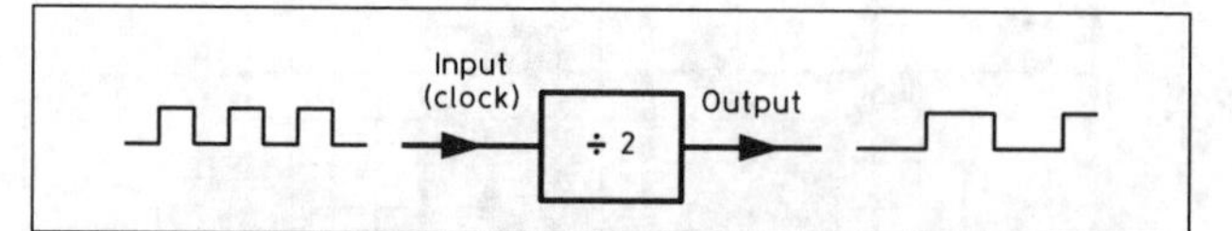

Fig 3.66. A binary counter may be used as a frequency divider

One of the most useful logic devices is the *counter*. The simplest form is the binary, or 'divide-by-two' stage shown at Fig 3.66. For every two input transitions the counter produces one output transition. Binary counters can be chained together (*cascaded*) with the output of the first counter connected to the input of the second, and so on. If four such counters are cascaded, there will be one output pulse, or count, for every 16 input pulses. It is also possible to obtain logic circuits which divide by 10 – these are sometimes referred to as *BCD* (binary coded decimal) counters. Although originally intended to perform the arithmetic function of binary division in computers, the counter can also be used as a frequency divider. For instance, if a single binary counter is driven by a series of pulses which repeat at a frequency of 100kHz, the output frequency will be 50kHz. Counters can therefore be used to generate a range of frequencies that are sub-multiples of a reference input. The crystal calibrator (see Chapter 15) uses this technique, and frequency synthesisers employ a more complex form of counter known as the programmable divider (see Chapter 5).

Counters that are required to operate at frequencies above 100MHz use a special type of logic known as *ECL* (emitter coupled logic). ECL counters achieve their speed by restricting the voltage swing between the levels defined as logic 0 and logic 1 to around 1V. The bipolar transistors used in ECL are therefore never driven into saturation (turned fully on), as this would reduce their switching speed due to charge storage effects (see p3.9). ECL logic is used in frequency synthesisers operating in the VHF to microwave range, and also in frequency counters to perform initial division, or *prescaling*, of the frequency to be measured (see Chapter 15).

Memories

There are many applications in radio where it is necessary to store binary data relating to the function of a piece of equipment, or as part of a computer program. These include the spot frequency memory of a synthesised transceiver, for instance, or the memory within a morse keyer which is used to repeat previously stored messages.

Fig 3.67 provides a diagrammatic overview of the memory IC. Internally, the memory consists of a matrix formed by a number of rows and columns. At each intersection of the matrix is a storage cell which can hold a single binary number (ie either a zero or a one). Access to a particular cell is provided by the memory's address pins. A suitable combination of logic levels, constituting a binary number, presented to the address pins will instruct the IC to connect the addressed cell to the data pin. If

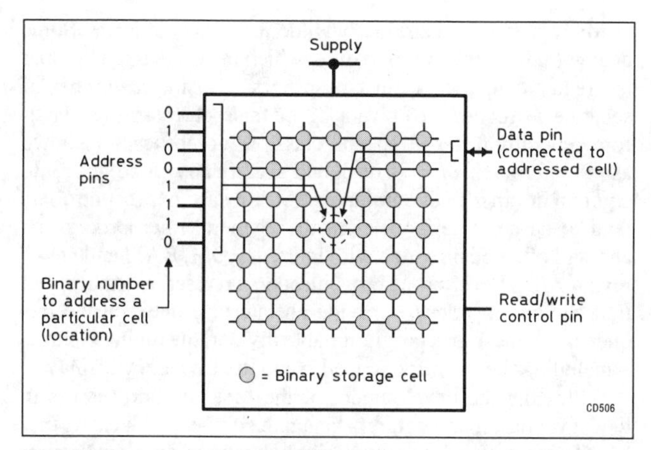

Fig 3.67. Representation of an IC memory

the read/write control is set to read, the logic level stored in the cell will appear at the data pin. Conversely, if the memory is set to write, whatever logic level exists on the data pin will be stored in the cell, thus overwriting the previous value. Some memory ICs contain eight separate matrixes, each having their own data pin. This enables a complete binary word (*byte*) to be stored at each address.

The memory described above is known as a *RAM* (random access memory) and it is characterised by the fact that data can be retrieved (read) and also stored (written) to individual locations. There are two main types of RAM – *SRAM* (static RAM) in which data is latched within each memory cell for as long as the power supply remains connected, and *DRAM* (dynamic RAM) which uses the charge, or the absence of a charge, on a capacitor to store logic levels. The capacitors within a DRAM cannot hold their charge for more than a few milliseconds, and so a process known as *refreshing* must be carried out by controlling circuitry in order to maintain the stored levels. The method of creating addresses for the memory locations within a DRAM is somewhat complex, in that the rows and columns of the matrix are dealt with separately. DRAM memory chips are used extensively in desktop computers and related equipment because the simple nature of the capacitor memory cell means that a large number of cells can easily be fabricated on a single chip (there are now DRAMs capable of storing 16 million binary digits, or *bits*). SRAMs fabricated with CMOS transistors are useful for storing data that must be retained while equipment is turned off. The low quiescent current consumption of these devices makes it practicable to power the memory from a small battery located within the equipment, thus providing an uninterrupted source of power.

The *ROM* (read-only memory) has data permanently written into it, and so there is no read/write pin. ROMs are used to store computer programs and other data that does not need to be changed. The ROM will retain its data indefinitely, irrespective of whether it is connected to a power supply. A special form of ROM known as the *EPROM* (eraseable programmable ROM) may be written to using a special programmer. Data may later be erased by exposing the chip to ultra-violet light for a prescribed length of time. For this reason, EPROMs have a small quartz window located above the chip which is normally concealed beneath a UV-opaque protective sticker. The *EEPROM* (electrically eraseable programmable ROM) is similar to the EPROM, but may be erased without using UV light. The *PROM* (programmable ROM) has memory cells consisting of fusible links. Assuming that logic 0 is represented by the presence of a link, logic 1 may be programmed into a location by feeding a current into a special programming pin which is sufficient to fuse the link at the addressed location. However, once a PROM has been written to in this way, the cells programmed to logic 1 can never be altered.

Analogue-to-digital converters

Analogue-to-digital conversion involves measuring the magnitude of a voltage or current and then generating a numeric value to represent the result. The digital multimeter works in this way, providing an output in decimal format which is presented directly to the human operator via an optical display. The analogue to digital (A/D) converters used in signal processing differ in two important respects. Firstly, the numeric value is generated in binary form so that the result may be manipulated, or 'processed' using digital circuitry. Secondly, in order to 'measure' a signal, as opposed to, say, the voltage of a battery, it is necessary to make many successive conversions so that amplitude changes occurring over time may be captured. In order to digitise speech, for instance, the instantaneous amplitude of the waveform must be ascertained at least 6000 times per second. Each measurement, known as a *sample*, must then be converted into a separate binary number. The accuracy of the digital representation depends on the number of bits (binary digits) in the numbers – eight digits will give 255 discrete values, whereas 16 bits provides 65,535.

Maximum *sampling frequency* (ie speed of conversion) and the number of bits used to represent the output are therefore the major parameters to consider when choosing an A/D converter IC. The fastest 8-bit converters available, known as *flash* types, can operate at sampling frequencies of up to 20MHz, and are used to digitise television signals. 16-bit converters are unfortunately much slower, with maximum sampling rates of around 100kHz. It is also possible to obtain converters offering intermediate levels of precision, such as 10 and 12 bits.

Having processed a signal digitally, it is often desirable to convert it back into an analogue form – speech and morse are obvious examples. There are a variety of techniques which can be used to perform digital-to-analogue conversion, and ICs are available which implement these.

Microprocessors

The microprocessor is different from other digital ICs in that it has no preordained global function. It is, however, capable of performing a variety of relatively straightforward tasks, such as adding two binary numbers together. These tasks are known as *instructions*, and collectively they constitute the microprocessor's *instruction set*. In order to make the microprocessor do something useful, it is necessary to list a series of instructions (write a program) and store these as binary codes in a memory IC connected directly to the microprocessor. The microprocessor has both data and address pins (see Fig 3.68), and when power is first applied it will generate a pre-determined start address and look for an instruction in the memory location accessed by this. The first instruction in the program will be located at the starting address. Having completed this initial instruction, the microprocessor will fetch the next one, and so on. In order to keep track of the program sequence, and also provide temporary storage for intermediate results of calculations, the microprocessor has a number of internal counters and *registers* (a register is simply a small amount of memory). The manipulation of binary numbers in order to perform arithmetic calculations is carried out in the *arithmetic logic unit* (ALU). A *clock oscillator*, normally crystal controlled, controls the timing of the program-driven events. The microprocessor has a number

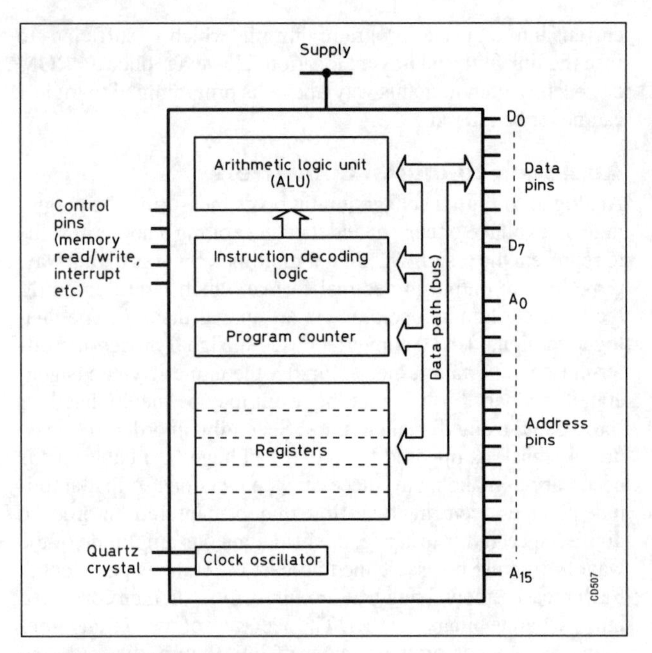

Fig 3.68. An 8-bit microprocessor

of control pins, including an interrupt input. This allows normal program execution to be suspended while the microprocessor responds to an external event, such as a keyboard entry.

Microprocessors exist in a bewildering variety of forms. Some deal with data eight bits at a time, which means that if a number is greater than 255 it must be processed using a number of separate instructions. 16-bit and 32-bit microprocessors are therefore generally faster. A special class of microprocessor known as the *microcontroller* is designed specifically to be built into equipment other than computers. As a result, microcontrollers tend to be more self-sufficient than general microprocessors, and will often be provided with internal ROM, RAM and possibly an A/D converter. Microcontrollers are used extensively in transceivers in order to provide an interface between the frequency synthesiser's programmable divider, the tuning controls – including the memory keypad – and the frequency display.

Following the development of the first microprocessors in the 'seventies, manufacturers began to compete with each other by offering devices with increasingly large and ever-more-complex instruction sets. There has been something of a backlash against this trend in recent years, with the emergence of the *RISC* (reduced instruction set microprocessor). The rationale behind the RISC architecture is that simpler instructions can be carried out more quickly, and by a processor using a smaller number of transistors.

DSP (digital signal processing) ICs are special microprocessors which have their instruction sets and internal circuitry optimised for fast execution of the mathematical functions associated with signal processing – in particular the implementation of digital filtering (see Chapter 5).

4 Electronic tubes and valves

MODERN electronic tubes and valves have attained a high degree of reliability and are still available in many forms for a wide variety of common and specialist applications. They have, however, been superseded for virtually all low-power purposes by semiconductors and thus there will be no mention of the use of valves in receivers in this chapter. Readers who need information on this topic are referred to earlier editions of this book.

FUNDAMENTALS

Emission

In most types of evacuated electronic tube the emission of electrons is produced by heating the cathode, either directly by passing a current through it, or indirectly by using an insulated heater in close proximity. The quantity of electrons emitted is governed by the construction and surface coating of the cathode and the temperature to which it is heated. This is known as *thermionic emission*.

Emission may also be produced when electrons impinge on to a surface at a sufficient velocity. For example, electrons emitted from a hot cathode may be accelerated to an anode by the latter's positive potential. If the velocity is high enough, electrons will be released from the anode. This is known as *secondary emission*.

The emission of electrons from metals or coated surfaces heated to a certain temperature is a characteristic property of that metal or coated surface. The value of the thermionic emission may be calculated from Richardson's formula:

$$I_s = A_1 T^2 e^{-b_1/T}$$

where I_s is the emission current in amperes per square centimetre; A_1 is a constant of the emitting substance; T is absolute temperature in kelvin (K); and b_1 is a constant depending on the material of the emitting surface.

Alternatively it may be calculated from a similar formula developed by Dushman.

Electron flow

Electrons are negatively charged. When in an evacuated tube an electron leaves a parent molecule, as for example during emission from a cathode, the molecule becomes more positively charged. If an electrode such as an anode is placed near to the cathode and is charged positively with respect to it, the electrons released by emission from the cathode will be attracted to the anode. As the electrons traverse the space between one electrode and another they may collide with gas molecules (because no vacuum can be perfect) and such collisions will impede their transit. For this reason the residual gas left inside the evacuated envelope must be minimal. An electronic tube which has been adequately evacuated is termed *hard*.

However, if a significant amount of gas is present, the collisions between electrons and gas molecules will cause it to ionise. The resultant blue glow between the electrodes indicates that the tube is *soft*. This blue glow should not be confused with a blue haze which may occur on the inside of the envelope external to the electrode structure: this is caused by bombardment of the glass, and in fact indicates that the tube is very hard.

Space charge

When electrons travel from cathode to anode in useful quantities they form a cloud in the space between the electrodes. The electric charge associated with this cloud is known as the *space charge*. It tends to repel the electrons leaving the cathode because it carries the same polarity. However, if the anode potential is sufficiently high, the effect of the space charge will be overcome and electrons will flow from the cathode, the flow being completed by an external circuit back to the cathode. As the anode potential is raised, the electron flow or current will increase to a point where the space charge is completely neutralised and the total emission from the cathode reaches the anode. The flow can be further increased only by raising the cathode temperature.

Cathodes

Although several types of cathode are used in modern valves, the differences are only in the method of producing thermionic emission. The earliest type is the *bright emitter* in which a pure tungsten wire is heated to a temperature in the region of 2500–2600K. At such a temperature emission of 4–40mA per watt of heating power may be obtained. Bright emitters are still employed in high-power transmitting valves for broadcasting but the only common amateur use is in diodes for applications such as noise generators. The life of a pure tungsten filament at full operating temperature is limited by evaporation of the tungsten, failure occurring when about 10% has been evaporated.

Dull emitters are directly heated thoriated tungsten cathodes which produce greater emission than bright emitters and require less heating power. In a dull emitter, a small quantity of thorium oxide is introduced into the pure tungsten wire. A process known as *carburisation* is used to create an outer skin of tungsten carbide on the wire which facilitates the reduction of the thorium oxide to metallic thorium, stabilises the emission and increases the surface resistance of the cathode to gas poisoning. Typical emission efficiency is in the region of 30–100mA per watt of heating power at an operating temperature of 1900–2100K. This type of cathode is relatively fragile and valves should not be subjected to shocks or sharp blows. Long life may be expected, provided the operating temperature is correctly maintained. In particular, the rated voltage or current should be closely controlled. Operation at constant filament power will give the longest life.

Oxide-coated cathodes are the most common type of thermionic emitter found in both directly and indirectly heated valves. In this type, the emissive material is usually some form of nickel ribbon, tube or thimble coated with a mixture of barium and strontium carbonate, often with a small percentage of calcium. During manufacture, the coating is reduced to its metallic form and the products of decomposition removed during the exhaustion process. The active ingredient is the barium which provides much greater emission than thoriated tungsten at lower heating powers. Typically, 50–150mA per watt is obtained at temperatures of 950–1050K.

Although the emission efficiency of oxide-coated cathodes is high and large currents may be drawn, they are less able to resist the poisoning effects of gas or ion bombardment. This type must not be operated under temperature-limited conditions.

In certain valves that are subject to back-bombardment of the cathode, such as magnetrons, some form of protected cathode coating/material is necessary. Such cathodes are known as *impregnated*, the active coating material being mixed with nickel or tungsten powder; this mixture is then coated on the cathode surface.

An *indirectly heated cathode* is a metal tube, sleeve or thimble shape, having a coating of emissive material on the outer surface. The cathode is heated by radiation from a metal filament, called the *heater*, which is mounted inside the cathode. The heater is electrically insulated from the cathode. The emissive material is generally the same as that employed for filamentary oxide-coated cathodes and operates at about the same temperature. The cathode may be made of pure nickel or of special alloys, depending upon the purpose of the valve. The heater is normally made of tungsten or molybdenum-tungsten alloy.

The life of valves with oxide-coated cathodes is generally good provided the ratings are not exceeded. Occasionally there is some apparent reduction in anode current due to the formation of a resistive layer between the oxide coating and the base metal, which operates as a bias resistor.

In *cold-cathode* valves, such as gas stabilisers, the cathode is an activated metal or coated surface.

The heater or filament voltage should be accurately measured *at the valve base* and adjusted to the correct value as specified by the makers. This needs great care as it must be done with the stage operating at its rated power.

Anodes

In most electronic tubes the anode takes the form of an open-ended cylinder or box surrounding the other electrodes, and is intended to collect as many as possible of the electrons emitted from the cathode; some electrons will of course be intercepted by the grids interposed between the cathode and the anode.

The material used for the anode of the small general-purpose type of valve is normally bright nickel or some form of metal coated black to increase its thermal capacity. Power dissipated in the anode is radiated through the glass envelope, a process which is assisted when adequate circulation of air is provided around the glass surface. In some cases a significant improvement in heat radiation is obtained by attaching to the valve envelope a close-fitting finned metal radiator which is bolted to the equipment chassis so that this functions as a worthwhile heatsink.

Higher-power valves with external anodes are cooled directly by forced air, by liquid, or by conduction to a heatsink, as follows:

Forced-air cooling requires a blower, preferably of the turbine type, capable of providing a substantial quantity of air at a pressure high enough to force it through the cooler attached to the anode.

Liquid cooling calls for a suitable cooler jacket to be fitted to the anode; this method is generally confined to large power valves. If water is used as the coolant, care must be taken to ensure that no significant leakage occurs through the water by reason of the high voltage used on the anode.

4CX250B showing its air-cooled vanes

Conduction cooling carries the heat from the anode to a suitable heatsink via a heat-conducting insulator forming part of the tube envelope or by attachment of a heat-conducting block connecting the anode to the heatsink. The latter is preferable, for the heat conductor block forms part of the external equipment, but a heat conductor that is part of the tube envelope is lost when failure occurs and the tube is discarded.

In certain UHF disc seal valves a different form of conduction cooling is used, the anode seal being directly attached to an external tuned-line circuit that functions as the heatsink radiator. Needless to say, it must be suitably isolated electrically from the chassis.

Whatever the type of electron tube and whatever method is used to cool it, the limiting temperatures quoted by the makers, such as bulb or seal temperatures, should never be exceeded; under-running the device in terms of its dissipation will generally greatly extend its life.

Grids

The electron flow from cathode to anode may be controlled by the introduction of one or more electrodes known as *grids*, the

CCS1 conduction-cooled tetrode (equivalent electrically to 4CX250B) mounted in a HC1 heat-conduction block

Table 4.1. Classification of electronic tubes and valves

Number of electrodes	Generic title	Number of grids
2	Diode	None
3	Triode	1
4	Tetrode	2
5	Pentode	3
6	Hexode	4
7	Heptode	5
8	Octode	6

number of such electrodes depending on the purpose for which the tube or valve is required. Electronic tubes are classified by a generic title based on the number of active electrodes they contain, as shown in Table 4.1.

Mechanically, the grid electrode takes many forms, dictated largely by power and the frequency of operation. In small general-purpose valves the grids are usually in the form of a helix (molybdenum or other suitable alloy wire) with two side support rods (copper or nickel) and a cross-section varying from circular to flat rectangular, dependent on cathode shape.

In some UHF valves the grid consists of a single winding of wire or mesh attached to a flat frame fixed directly to a disc seal.

In high-performance tubes where close clearance between cathode and grid is required for high mutual conductance, the grid winding is made on to a frame consisting of support rods and metal straps, the helix being wound under considerable tension (about half its breaking strain). This method of construction is generally confined to Grid No 1 (the control grid), where the minor axis of cross-section is decided by the support rod diameter.

In beam tubes such as klystrons, travelling-wave tubes and cathode-ray tubes, the electron flow is concentrated through a single hole in the electrode plates, the potential applied to a plate having the effect of varying the tube characteristics. For higher power tubes the grid may be in the form of a 'squirrel cage'.

Primary emission from a grid due to heating must be minimised to avoid affecting the tube's operation. To inhibit emission at the normal operating temperature various types of coating or plated surface perform both this function and at the same time increase the working surface of the grid. It is particularly important to minimise grid emission in transmitting valves, where the grid may be operating in a positive mode. Similar operating conditions apply also to Grid No 2 (the screen grid), more especially in audio output and transmitting types.

Both No 1 and No 2 grids are sometimes fitted with radiation fins to cool them in the interests of holding primary grid emission at a satisfactory level.

Where a variable gain facility is required, the No 1 (control) grid is given a variable helix pitch to enable it to handle changes from high gain to relatively low gain. Pitch variation is conveniently introduced by providing suitable gaps in the winding around the centre of the cathode system.

TUBE TYPES

Diodes

The simplest form of electronic tube is the two-electrode diode, consisting simply of an emitting surface (heater-cathode or filament) and an anode. The current which can be drawn from the anode is governed by the type of emitting cathode employed, its temperature and its spacing from the anode.

For a given operating cathode temperature a point of saturation is reached beyond which no further increase in current can be obtained unless the cathode temperature is increased.

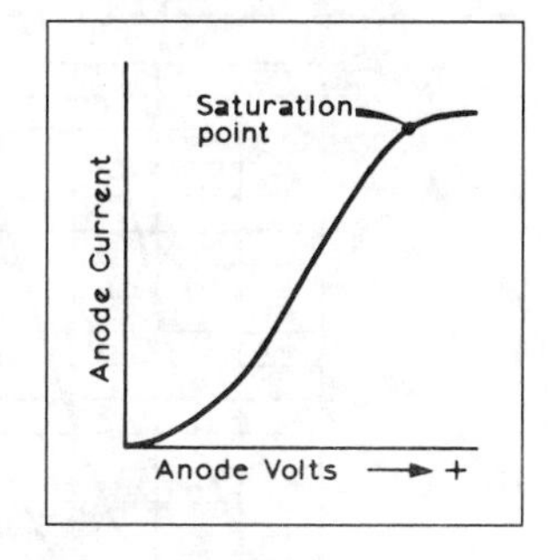

Fig 4.1. Typical diode characteristic, showing emission limitation for a given filament temperature

Diodes have a wide variety of applications, from low-level RF signal detection to power rectification up to very high voltages. To provide a low-impedance characteristic, where the voltage drop across the diode is virtually independent of the current drawn, gas filling is used, either with mercury vapour or an inert gas such as xenon. The resultant low voltage drop helps to reduce the anode dissipation.

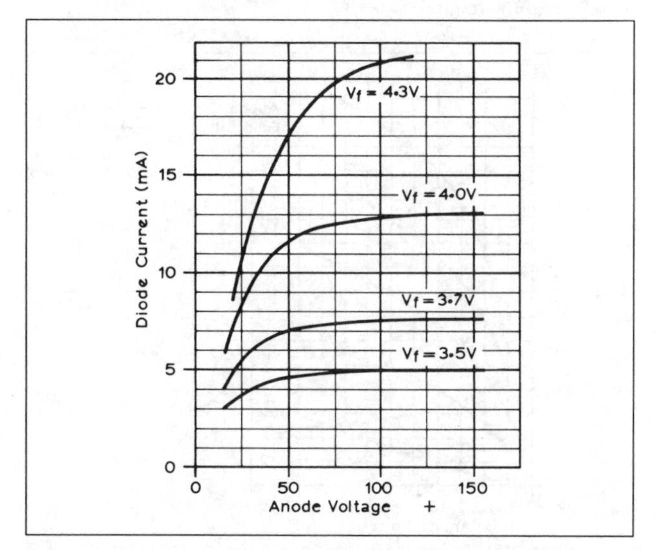

Fig 4.2. Diode saturation curve showing the effect of different filament voltages and hence temperatures

Triodes

By introducing a grid between the cathode and anode of an electronic tube the electron flow may be controlled. This flow may be varied in accordance with the voltage applied to the grid, its value being decided by the geometry of the grid and in particular the amplification factor and mutual conductance required from the valve.

Varying the potential applied to the grid modifies the space charge, but because the grid has an open mesh only partial interception of the electrons occurs, the majority of them remaining available for acceleration to the high-potential anode. Electron collection by the grid is low if its potential is low, or negative, but increases significantly as the voltage is made more positive.

It is important to recognise that a small general-purpose triode or pentode used as an audio amplifier with a negative potential on its grid may, if operated as a positively driven amplifier, be called upon to withstand more grid dissipation than the designer intended, and could be less reliable than a valve designed specifically for the purpose.

The grid voltage (grid bias) for a small general-purpose valve may be obtained by one of several methods:

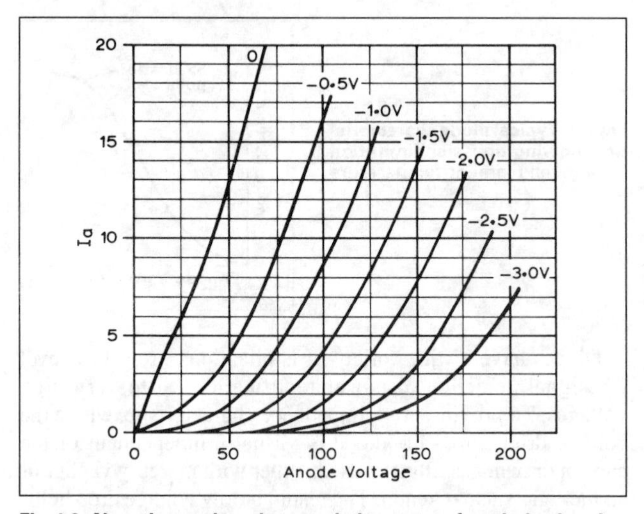

Fig 4.3. Negative-region characteristic curves of a triode showing the reduction in anode current which occurs with increase of negative grid voltage

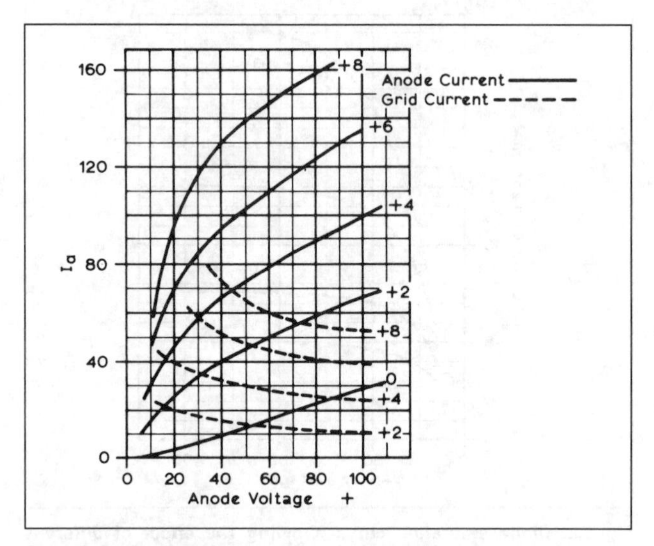

Fig 4.4. Positive-region triode characteristic curves showing how enhanced values of positive grid voltage increase anode current flow

1. *A separate battery.*
2. *A resistor connected between the cathode and the chassis (earth)* so that when current flows the voltage drop across it renders the cathode more positive with respect to the chassis (earth), and the grid circuit return becomes negative with respect to cathode.
3. *A resistor connected between the grid and the chassis (earth).* When the grid is so driven that appreciable current flows (as in an RF driver, amplifier or multiplier), the grid resistor furnishes a potential difference between grid and chassis (earth), and with cathode connected to chassis a corresponding negative voltage occurs at the grid. A combination of grid resistor and cathode resistor is good practice and provides protection against failure of drive, which a grid resistor alone would not give.
4. *Contact potential.* A high value of resistance (1–10MΩ) is used as the grid resistor. A small current flow through it will provide sufficient negative voltage for small-signal applications (eg an input audio amplifier). Characteristic curves which show graphically the relationship between anode current and anode voltage for various values of grid voltage are given, one illustrating the curve produced when negative grid voltages are applied, the other those for positive grid voltages, together with the corresponding grid currents.

As power amplifiers, triodes have the virtue of simplicity, especially when used in the earthed-grid (*grounded-grid*, see later) mode. At VHF and UHF the electrons can have a very short transit time if the valve is built on the planar principle. In this, the cathode, grid and anode are all flat and it is possible to have only a very short distance between them. Also the connection to each electrode has a very low impedance. Both factors promise good operation at VHF/UHF. From this design, it is clear that earthed-grid operation is best when the grid with its disc-shaped connector acts as a screen between the input (cathode) and the output (anode) circuits. (See the section on 'Disc seal valves' later in this chapter. The 2C39A illustrated there is a good amplifier up to at least 2.3GHz.)

Tetrodes

A tetrode ('four-electrode') valve is basically a triode with an additional grid mounted outside the control grid. When this additional grid is maintained at a steady positive potential a considerable increase in amplification factor occurs compared with the triode state; at the same time the valve impedance is greatly increased.

The reason for this increased amplification lies in the fact that the anode current in the tetrode valve is far less dependent on the anode voltage than it is in the triode. In any amplifier circuit, of course, the voltage on the anode must be expected to vary since the varying anode current produces a varying voltage-drop across the load in the anode circuit. A triode amplifier suffers from the disadvantage that when, for instance, the anode current begins to rise due to a positive half-cycle of grid voltage swing, the anode voltage falls (by an amount equal to the voltage developed across the load) and the effect of the reduction in anode voltage is to diminish the amount by which the anode current would otherwise increase. Conversely, when the grid voltage swings negatively, the anode current falls and the anode voltage rises. Because of this increased anode voltage the anode current is not so low as it would have been if it were independent of anode voltage. This means that the full amplification of the triode cannot be achieved. The introduction of the screen grid, however, almost entirely eliminates the effect of the anode voltage on the anode current, and the amplification obtainable is thus much greater.

A screen functions best when its voltage is below the mean value of the anode voltage. Most of the electrons from the cathode are thereby accelerated towards the anode, but some of them are unavoidably caught by the screen. The resulting screen current serves no useful purpose, and if it becomes excessive it may cause overheating of the screen.

If in low-voltage applications the anode voltage swings down to the screen voltage or lower, the anode current falls rapidly

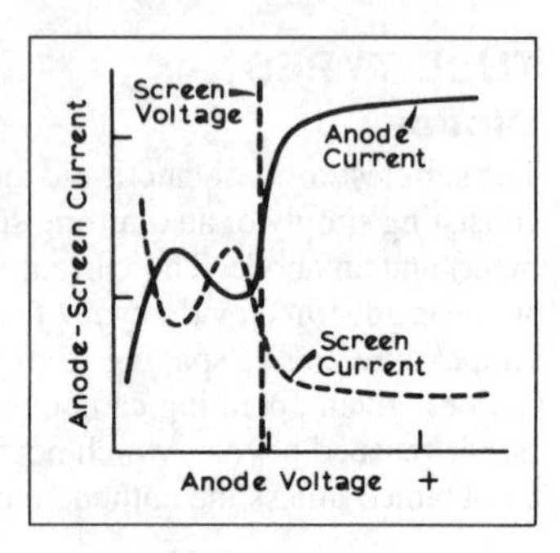

Fig 4.5. Characteristic curves of pure tetrode (often termed *screened grid*) showing the considerable secondary emission occurring when no suppression is used

while that of the screen rises due to secondary emission from the anode to the screen. It should be noted that the total cathode current is equal to the sum of the screen and anode currents.

Another important effect of introducing the screen grid (No 2) is that it considerably reduces the capacitive coupling between the input (control) grid and the anode, making possible the use of stable, high-gain RF amplification. To utilise this facility additional shields are added to the grid (electrically connected) so that the input connection cannot 'see' the anode or its supports. With such a structure it is possible to reduce the unit's capacitance by a factor of almost 1000 compared with the triode. Adequate decoupling of the screen at the operating frequency by the use of a suitable external bypass capacitor is essential.

In another type of tetrode, known as the *space-charge grid tetrode*, the second grid is positioned between the usual control grid and the cathode. When a positive potential is applied to this space-charge grid, it overcomes the limiting effect of the negative space charge, allowing satisfactory operation to be achieved at very low anode potentials, typically 12–24V.

Pentodes

To overcome the problem presented by secondary emission in the pure tetrode, a third grid may be introduced between the screen and the anode, and maintained at a low potential or connected to the cathode. Anode secondary emission is overcome and much larger swings of the anode voltage may be realised. This third grid is known as the *suppressor grid* (G3). Other methods which achieve the same effect are:

(a) increasing the space between screen grid and anode, as in the Harries critical anode-distance valve;
(b) fitting small fins to the inside surface of the anode; or
(c) fitting suppressor plates to the cathode to produce what is known as the *kinkless tetrode*, which is the basis of the beam tetrode suppression system.

In some special types of pentode where it is necessary for application reasons to provide two control grids, the No 3 (suppressor) grid is used as the second and lower-sensitivity control for gating, modulation or mixing purposes. Units of this type need to have a relatively high screen grid (No 2) rating to allow for the condition when the anode current is cut off by the suppressor grid (No 3).

Beam tetrodes

A beam tetrode employs principles not found in other types of valve: the electron stream from the cathode is focused (*beamed*) towards the anode. The control grid and the screen grid are made with the same winding pitch and they are assembled so that the turns in each grid are in optical alignment: see Fig 4.6. The effect of the grid and screen turns being in line is to reduce the screen current compared with a non-beam construction. For example, in a pentode of ordinary construction the screen current is about 20% of the anode current, whereas in a beam valve the figure is 5–10%.

The pair of plates for suppressing secondary emission referred to above is bent round so as to shield the anode from any electrons coming from the regions exposed to the influence of the grid support wires at points where the focusing of the electrons is imperfect. These plates are known as *beam-confining* or *beam-forming* plates.

Beam valves were originally developed for use as audio-frequency output valves, but the principle has been applied to many types of RF tetrodes, both for receiving and transmitting. Their superiority over pentodes for AF output is due to the fact that

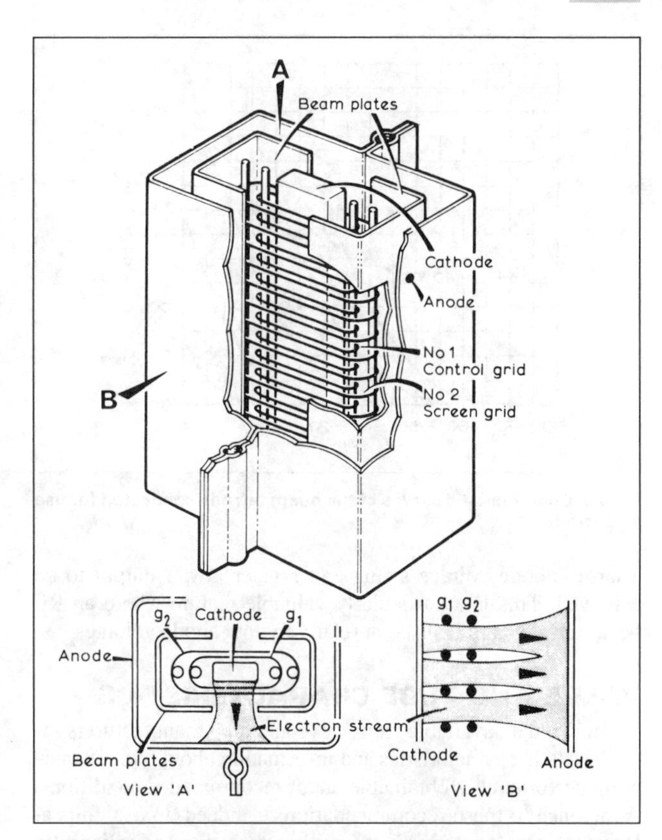

Fig 4.6. The general arrangement of a modern beam tetrode showing the aligned grid winding and the position of the beam forming plates. View 'A': looking vertically into a beam tetrode. View 'B': showing how the aligned electrode structure focuses electrons from cathode to anode

the distortion is caused mainly by the second harmonic and only very slightly by the third harmonic, which is the converse of the result obtained with a pentode. Two such valves used in push-pull give a relatively large output with small harmonic distortion because the second harmonic tends to cancel out with push-pull connection.

Fig 4.7 shows the characteristic curves of a beam tetrode. By careful positioning of the beam plates a relatively sharp 'knee' can be produced in the anode current/anode voltage characteristic, at a lower voltage than in the case of a pentode, thus allowing

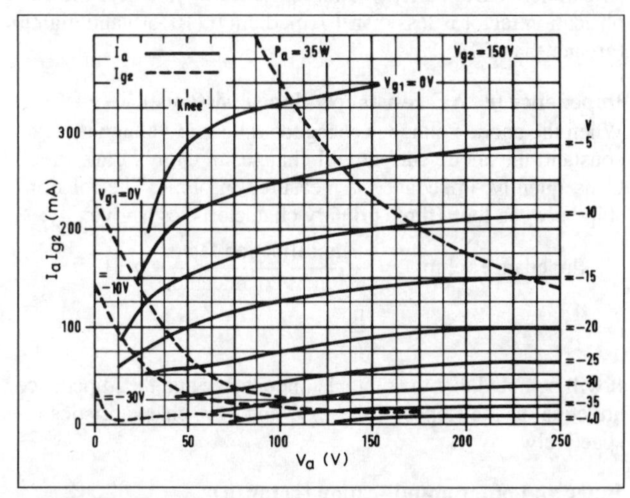

Fig 4.7. Characteristic curves of a beam tetrode. Anode secondary emission is practically eliminated by the shape and position of the suppressor plates

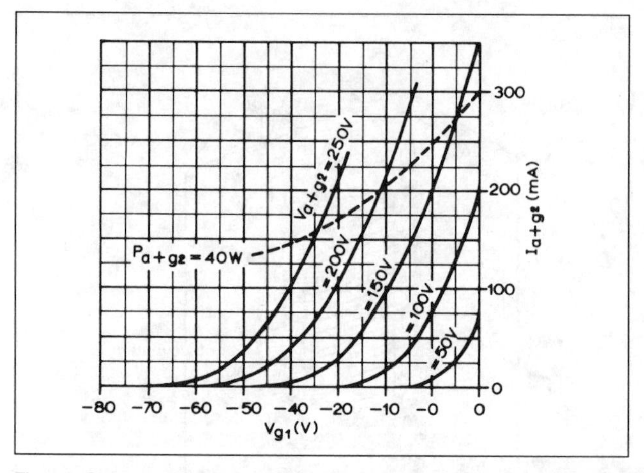

Fig 4.8. Characteristic curves of the beam tetrode connected for use as a triode

a larger anode voltage swing and greater power output to be achieved. This is a particularly valuable feature where an RF beam tetrode is to be used at relatively low anode voltages.

VALVE AND TUBE CHARACTERISTICS

Technical data available from valve and tube manufacturers includes static characteristics and information about typical operating performances obtainable under recommended conditions. Adherence to these recommendations – indeed, to use units at lower than the quoted values – will increase life and reliability, which at higher than quoted values can easily be jeopardised. In particular, cathodes should always be operated within their rated power recommendations. The following terms customarily occur in manufacturers' data:

Mutual conductance (slope, g_m, transconductance)

This is the ratio of change of anode current to the change of grid voltage at a constant anode voltage. This factor is usually expressed in milliamperes per volt or micromhos (1mA/V = 1000 micromhos).

Amplification factor (μ)

This is the ratio of change of anode voltage to change of grid voltage for a constant anode current. In the case of triodes classification is customarily in three groups, low μ, where the amplification factor is less than 10, medium μ (10–50) and high μ (greater than 50).

Impedance (r_a, AC resistance, slope resistance)

When the anode voltage is changed while grid voltage remains constant, the anode current will change, an Ohm's Law effect. Consequently, impedance is measured in ohms. The relationship between these three primary characteristics is given by:

$$\text{Impedance (ohms)} = \frac{\text{Amplification factor}}{\text{Mutual conductance}} \times 1000$$

or

$$r_a = \frac{\mu}{g_m}$$

It will be noted that the mutual conductance and the impedance are equal to the slopes of the I_a/V_g and I_a/V_a characteristics respectively.

Inner and outer amplification factor (μ)

In tetrodes and pentodes it is customary to quote the amplification factor of the device as if it were a triode, using the screen (G2) as the anode, or screen and anode connected together. This is known as *inner* μ or *inner amplification factor*. Similarly, where the screen (G2) is considered as a control grid in conjunction with the anode, the ratio of change of anode volts to the change of screen voltage for a constant anode current is known as the *outer* μ or *outer amplification factor*.

Electrode dissipation

The conversion from anode input power to useful output power will depend upon the tube type and the operating conditions. The difference between these two values, known as the *anode dissipation*, is radiated as heat. If maximum dissipation is exceeded overheating will cause the release of occluded gas, which will poison the cathode and seriously reduce cathode emission. The input power to be handled by any valve or tube will, in the limiting case, depend on the class of operation. Typical output efficiencies expressed as percentages of the input power are:

Class A	33%
Class AB1	60–65%
Class AB2	60–65%
Class B	65%
Class C	75–80%

Considering a valve with a 10W anode dissipation, the above efficiencies would give outputs as follows (assuming there are no other limiting factors such as peak cathode current):

Class A	5W
Class AB1 and AB2	15–18.5W
Class B	18.5W
Class C	30–40W

Control and screen grid dissipations

The control grid (No 1) and screen grid (No 2), due to their more fragile construction, have a lower thermal capacity than the anode and their maximum ratings must therefore not be exceeded. The control grid is heated considerably by radiation from the cathode and therefore its dissipation rating is quite small; any significant overheating will cause grid primary emission (*grid emission*). The screen grid, being shielded to a large extent from the cathode by the control grid, can accept a higher temperature rise before primary emission ensues. In an aligned-grid tetrode any grid wire which becomes distorted by overheating may become red hot by drawing excessive current.

Hum

When a cathode is heated by AC the current generates a magnetic field which can modulate the electron stream, and a modulating voltage is injected into the control grid through the interelectrode capacitance and leakages: additionally there can be emission from the heater in an indirectly heated valve (see below).

When operating directly heated valves such as transmitting valves with thoriated tungsten filaments, the filament supply should be connected to earth by a centre tap or a centre-tapped resistor connected across the filament supply (a *hum-bucking* resistor). The hum is usually expressed as an equivalent voltage (in microvolts) applied to the control grid. Valve hum should not be confused with hum generated in other circuit components.

Electrode primary emission

In an indirectly heated valve it is possible for the heater to become contaminated with emissive material and so emit electrons. This is known as *heater emission*. The emitted electrons will be attracted to any electrode which is positive with respect to the heater, and such an electrode may be the cathode or the screen grid or anode; it can even be the control grid which, although generally negative to the cathode, can become sufficiently positive to one end of the heater during at least part of the AC cycle of the heater supply. This explains why it is sometimes recommended that the centre-tap of the heater supply should be connected to a positive point in the circuit to reduce hum. The grid can thereby be maintained at a negative potential with respect to the heater.

Grid primary emission is a condition in which a grid commences to emit electrons itself and 'competes' with the cathode. The effect is produced by the heating of the grid which may be caused by an excessive flow of grid current, by the close proximity of the hot cathode or by radiated heat from the anode. The effects are accentuated if the grid becomes contaminated with active cathode material, which can happen if the valve is appreciably over-run even for a relatively short time.

The control grid, the screen and the suppressor grid are all subject to these effects. They are avoided by keeping the grid-cathode resistance low and by avoiding excessive heater, anode or bulb temperature (ie by not over-running the valve).

Two examples of the effects of control-grid primary emission can be given:

1. In a small output valve the anode current rises steadily, accompanied by distortion due to grid current flowing in the high-resistance grid leak in such a direction as to oppose the grid bias.
2. In a power amplifier or frequency multiplier in a telegraphy transmitter the drive diminishes when the key is held down, accompanied by rising anode current.

Primary screen emission: in an oscillator or amplifier when the screen voltage is removed, for example by keying, the output is maintained often for quite long periods if the valve is already hot.

Anode primary emission is similar to grid emission but occurs when the anode attains a sufficiently high temperature to emit electrons. This effect occurs mostly in rectifiers and causes breakdown between anode and cathode.

Secondary emission

When an electron which has been accelerated to a high velocity hits an electrode such as a grid or anode, electrons are dislodged and these electrons can be attracted to any other electrode having a higher potential. This effect is termed *secondary emission*. Under controlled conditions one electron can dislodge several secondary electrons, and a series of secondary-emitting 'cathodes' will give a considerable gain in electrons. This principle is used in the electron-multiplier type of valve.

Cathode-interface impedance

When a valve is operated for long periods, particularly with low cathode current or at complete cut-off, the mutual conductance steadily falls and so also does the available peak emission. This effect is due to the growth of a film between the metallic cathode and its emissive coating. This film possesses an impedance (the *cathode interface impedance*) which may be represented by a resistance with capacitive shunt connected in series with the cathode and acting as an automatic bias resistor. The rate of growth of interface resistance is considerably affected by the material of the cathode and is accelerated by high temperatures resulting from excessive heater voltage. Since the cathode-interface resistance is normally of the order of a few hundred ohms, it has a most serious effect on valves having a high slope and a short grid base because the normal cathode resistor is likely to

Typical valves used in many linear amplifiers. Left: 6146B transmitting type. Right: 6JS6 'sweep tube' originally designed for TV line output circuits

be comparable with this value. The effect of the parallel capacitance is to make the drop in performance less noticeable as the frequency is increased.

VALVE APPLICATIONS

Amplifiers

When an impedance is connected in series with the anode of a valve and the voltage on the grid is varied, the resulting change of anode current will cause a voltage change across the impedance. The curves in Fig 4.9 illustrate the classifications of valve amplifier operating conditions, showing anode current/grid voltage characteristics and the anode current variations caused by varying the grid voltage.

Class A

The mean anode current is set to the middle of the straight portion of the characteristic curve. If the input signal is allowed either to extend into the curved lower region or to approach zero grid voltage, distortion will occur because grid current is caused to flow by the grid contact potential (usually 0.7–1.0V). Under Class A conditions anode current should show no movement with respect to the signal impressed on the grid. The amplifier is said to be *linear*.

Class AB1

The amount of distortion produced by a non-linear amplifier may be expressed in terms of the harmonics generated by it. When a sine wave is applied to an amplifier the output will contain the fundamental component but, if the valve is allowed to operate on the curved lower portion of its characteristic, ie running into grid current, harmonics will be produced as well. Harmonic components are expressed as a percentage of the fundamental. Cancellation of even harmonics may be secured by connecting valves in push-pull, a method which has the further virtue of providing more power than a single valve can give, and was once widely found in audio amplifiers and modulators.

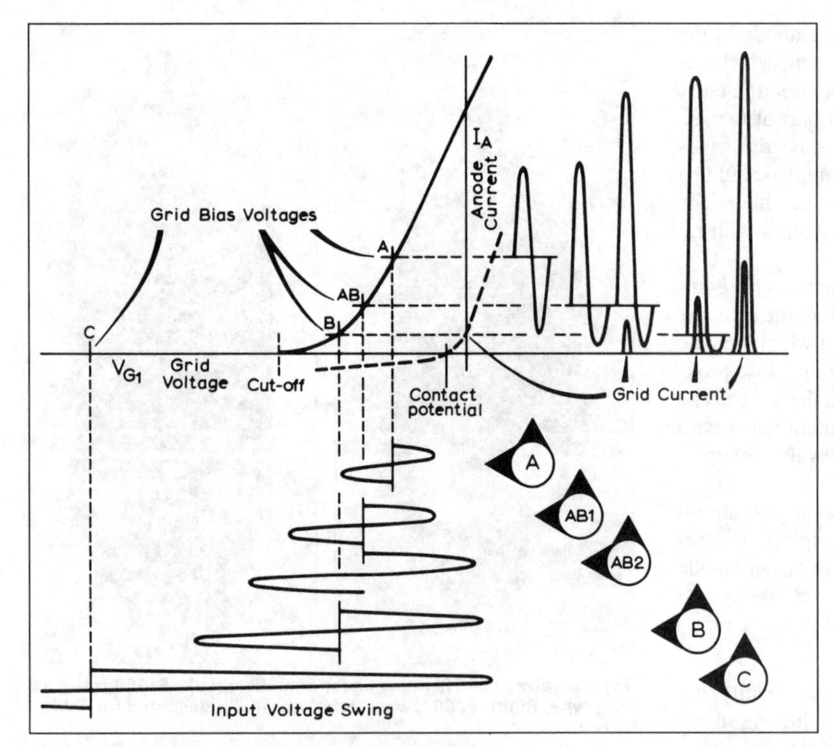

Fig 4.9. The valve as an amplifier: the five classes of operation

Class AB2

If the signal input is increased beyond that used in the Class AB1 condition peaks reaching into the positive region will cause appreciable grid current to be drawn and the power output to be further increased. In both the Class AB1 and AB2 conditions the anode current will vary from the zero signal mean level to a higher value determined by the peak input signal.

Class B

This mode, an extension of Class AB2, uses a push-pull pair of valves with bias set near to the cut-off voltage. For zero signal input the anode current of the push-pull pair is low, but rises to high values when the signal is applied. Because grid current is considerable an appreciable input power from the drive source is required. Moreover, the large variations of anode current necessitate the use of a well-regulated power (HT) supply.

Class C

This condition includes RF power amplifiers and frequency multipliers where high efficiency is required without linearity, as in CW, AM and NBFM transmitters. Bias voltage applied to the grid is at least twice, sometimes three times, the cut-off voltage, and is further increased for pulse operation. The input signal must be large compared with the other classes of operation outlined above, and no anode current flows until the drive exceeds the cut-off voltage. This could be for as little as 120° in the full 360° cycle, and is known as the *conduction angle*. Still smaller conduction angles increase the efficiency further, but more drive power is then needed. Pulse operation is simply 'super Class C'; very high bias is applied to the grid and a very small angle of conduction used.

Grid driving power

An important consideration in the design of Class B or Class C RF power amplifiers is the provision of adequate driving power. The driving power dissipated in the grid-cathode circuit and in the resistance of the bias circuit is normally quoted in valve manufacturers' data. These figures frequently do not include the power lost in the valveholder and in components and wiring, or the valve losses due to electron transit-time phenomena, internal lead impedances and other factors. Where an overall figure is quoted, it is given as *driver power output*. If this overall figure is not quoted, it can be taken that at frequencies up to about 30MHz the figure given should be multiplied by two, but at higher frequencies electron transit-time losses increase so rapidly that it is often necessary to use a driver stage capable of supplying 3–10 times the driving power shown in the published data.

The driving power available for a Class C amplifier or frequency multiplier should be sufficient to permit saturation of the driven valve, ie a substantial increase or decrease in driving power should produce no appreciable change in the output of the driven stage. This is particularly important when the driven stage is anode-modulated.

Passive grid

In linear amplifiers the driver stage must work into an adequate load, and the use of the passive grid arrangement is to be recommended. A relatively low resistance (typically 1kΩ) is applied between grid and cathode with a resonant grid circuit where appropriate. This arrangement helps to secure stable operation but should not be used as a cure for amplifier instability.

Grounded cathode

Most valves are used with the cathode connected to chassis or earth, or where a cathode-bias resistor is employed it is shunted with a capacitor of low reactance at the lowest signal frequency used so that the cathode is effectively earthed. In modulated amplifiers two capacitors, one for RF and the other for AF, must be used.

Grounded grid

Although a triode must be neutralised to avoid instability when it is used as an RF amplifier, this is not always essential if an RF type of tetrode or pentode is employed. However, above about 100MHz a triode gives better performance than a tetrode or pentode, providing that the inherent instability can be overcome. One way of achieving this is to earth the grid instead of the cathode so that the grid acts as an RF screen between cathode and anode, the input being applied to the cathode. The capacitance tending to make the circuit unstable is then that between cathode and anode, which is much smaller than the grid-to-anode capacitance.

The input impedance of a grounded-grid stage is normally low, of the order of 100Ω, and therefore appreciable grid input power is required. Since the input circuit is common to the anode-cathode circuit, much of this power is, however, transferred directly to the output circuit, ie not all of the driving power is lost.

Grounded anode

For some purposes it is desirable to apply the input to the grid and to connect the load in the cathode circuit, the anode being decoupled to chassis or earth through a low-reactance capacitor.

Such circuits were employed in cathode followers and infinite-impedance detectors.

Neutralising amplifiers

Instability in RF amplifiers results from feedback from the anode to the grid through the grid-to-anode capacitance and is minimised by using a tetrode or pentode. At high frequencies, particularly if the grid and/or anode circuit has high dynamic impedance, this capacitance may still be too large for complete stability. A solution is to employ a circuit in which there is feedback in opposite phase from the anode circuit to the grid so that the effect of this capacitance is balanced out. The circuit is then said to be *neutralised.*

A typical arrangement is shown in Fig 4.10. Here the anode coil is centre-tapped in order to produce a voltage at the 'free' end which is equal and opposite in phase to that at the anode end. If the free end is connected to the grid by a capacitor (C_n) having a value equal to that of the valve grid-to-anode capacitance (C_{g-a}) shown dotted, any current flowing through C_{g-a} will be exactly balanced by that through C_n. This is an idealised case because the anode tuned circuit is loaded with the valve anode impedance at one end but not at the other; also the power factor of C_n will not necessarily be equal to that of C_{g-a}.

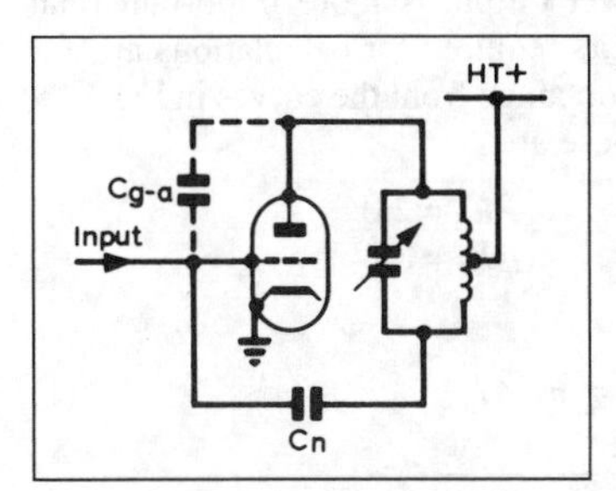

Fig 4.10. Neutralising a grounded-cathode triode amplifier. The circuit is equally suitable for a tetrode or a pentode

The importance of accurate neutralisation in transmitter power amplifier circuits cannot be overstressed, and will be achieved if the layout avoids multiple earth connections and inductive leads; copper strip is generally preferable to wire for valve socket cathode connections.

Double tetrodes for VHF/UHF service

Early designs of double tetrode contained two complete electrode systems in one envelope. In current designs there is a common cathode and a common screen grid, with separate control grids and anodes positioned on each side of the cathode. In the design illustrated at Fig 4.11 crossover neutralising capacitors are built in. The form of construction shown helps to reduce the effects of circulating currents and lead inductance, especially in cathode and screen.

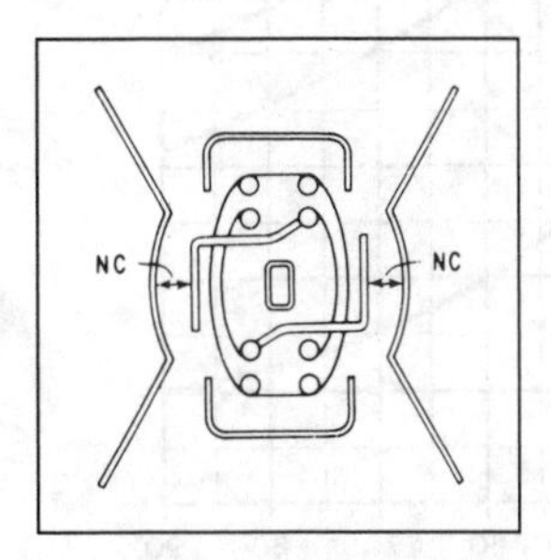

Fig 4.11. A cross-section of a VHF double tetrode such as the QQV03-20A

Suitably designed Class C amplifier and frequency multiplier valves will give satisfactory operation up to about 600MHz.

The QQV03-20A and QQV06-40A double tetrodes

CALCULATION OF OPERATING CONDITIONS FOR RF AMPLIFIERS

In a tuned amplifier the anode and grid voltages are of sine-wave form and in phase opposition. The anode current does not flow continuously, but in a series of pulses whose duration varies from 40° to more than 180° of each complete cycle of 360°. The grid current flows for a shorter duration, since this only occurs when the grid is positive relative to the cathode.

Fig 4.12. Basic circuit of a tuned amplifier

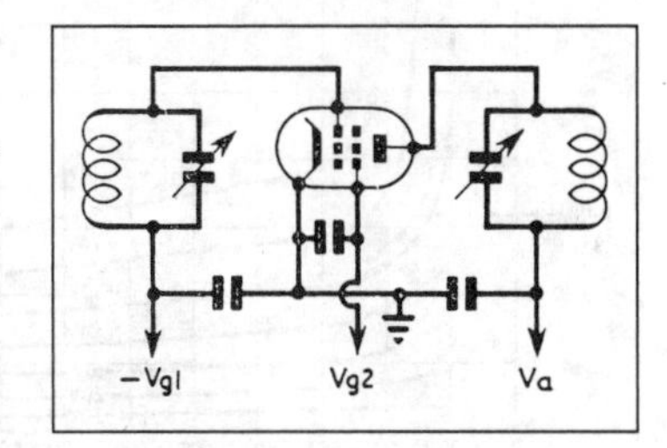

Figs 4.12 and 4.13 show the basic circuit and phase relationships respectively. It will be seen that the peak values of anode and grid currents occur when the anode voltage is low and the

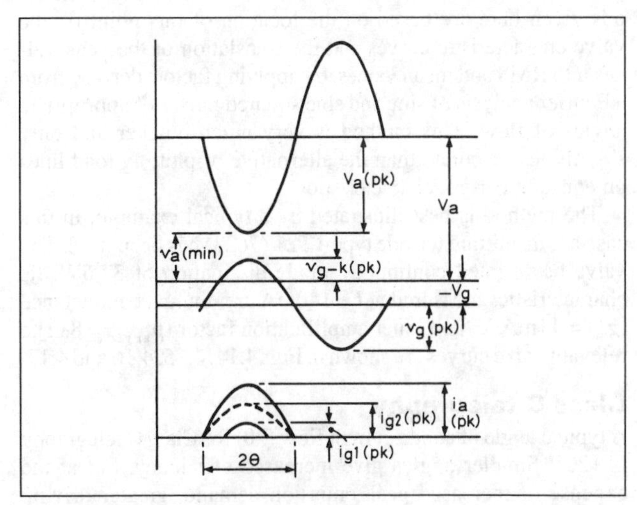

Fig 4.13. Anode/grid phase relationship in the tuned amplifier

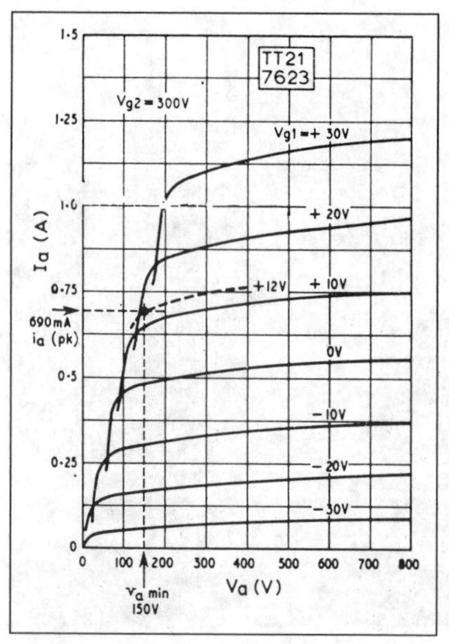

Fig 4.14. I_a at various values of V_{g1} for a TT21 (7623) transmitting valve

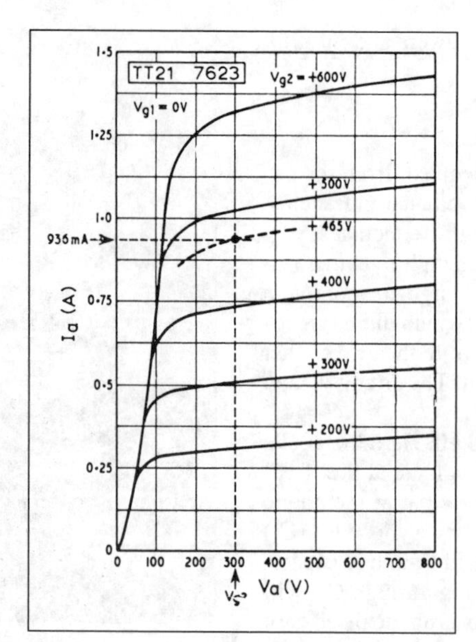

Fig 4.15. I_a at various values of V_{g2} for a TT21

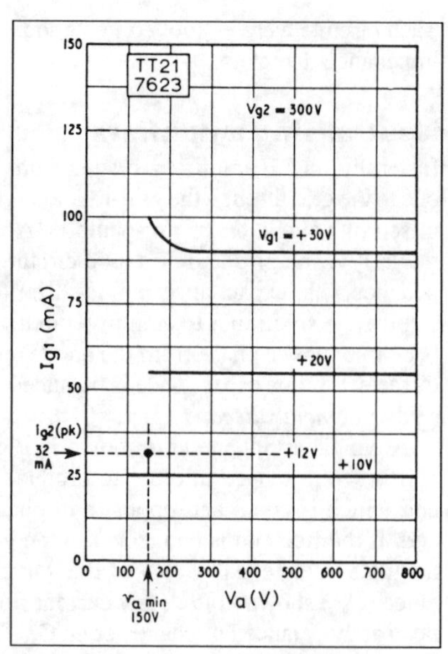

Fig 4.16. I_{g1} at various values of V_{g1} for a TT21

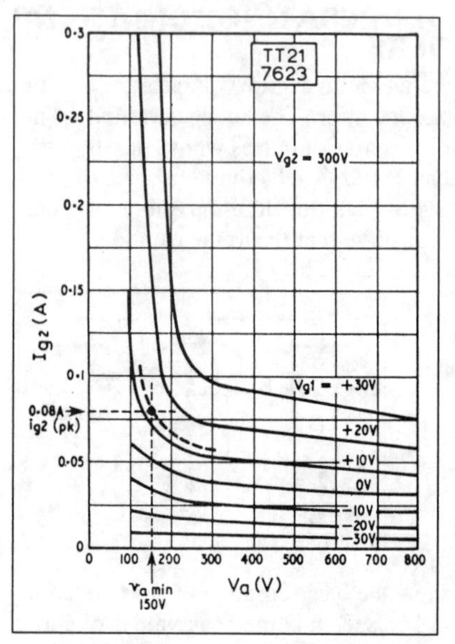

Fig 4.17. I_{g2} at various values of V_{g2} for a TT21

grid voltage is at its maximum positive value. The design methods given here are based on the location of this point on the valve characteristic curves and the translation of the peak values into RMS and mean values, by applying factors derived from a Fourier analysis of sine and sine squared pulses of appropriate angles of flow. This method is very much quicker and only slightly less accurate than the alternative of plotting load lines on constant current characteristics.

The method is best illustrated by a typical example; in this case a transmitting tetrode type TT21 (7623) has been used. The valve has a rated continuous anode dissipation of 37.5W. Its characteristics measured at I_a= 140mA are: mutual conductance (g_m) = 11mA/V, and inner amplification factor ($\mu_{g1\text{-}g2}$) = 8. The relevant valve curves are shown in Figs 4.14, 4.15, 4.16 and 4.17.

Class C telegraphy

A typical angle of anode current flow (2θ) for Class C telegraphy is 120°. Smaller angles give increased efficiency, but at the expense of increased peak emission demand, greater driving power and possibly shorter valve life. Larger angles are sometimes used when power output is more important than efficiency. The design factors required for calculations are F_1, F_2, F_3 and F_4. These can be obtained from the curves in Fig 4.18 for an angle of θ of 60°. These are:

$$F_1 = 4.6 \qquad F_3 = 2.0$$
$$F_2 = 1.8 \qquad F_4 = 5.8$$

The design formulae are:

Peak anode current $\quad i_{a(pk)} = F_1 \times I_a \qquad (1)$

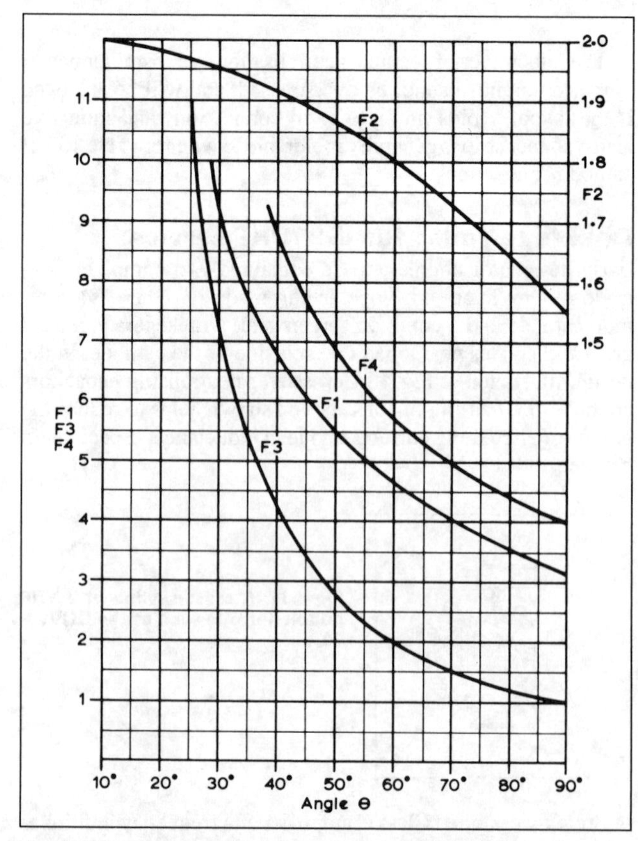

Fig 4.18. Design factors for Class C telegraphy (see text)

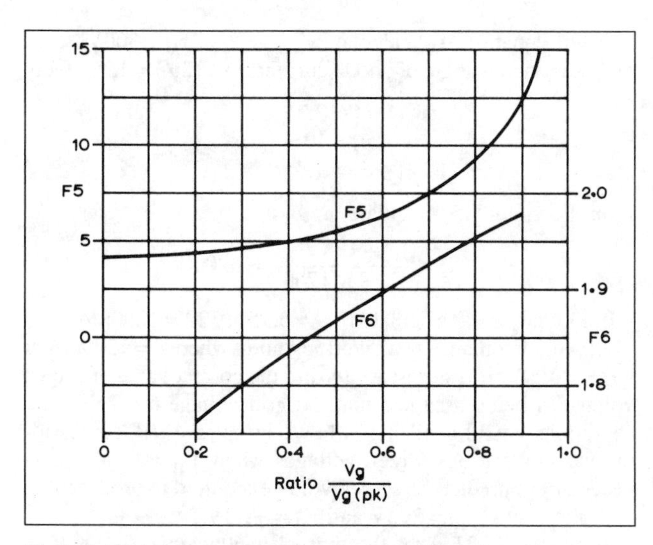

Fig 4.19. Design factors for Class C telegraphy (see text)

Peak anode voltage $v_{a(pk)} = V_a - v_{a(min)}$ (2)

Power output $P_{out} = \frac{F_2}{2} \times I_a \times v_{a(pk)}$ (3)

Grid voltage (triodes)

$$-V_g = \frac{V_a \times F_3}{\mu} + (v_{g\text{-}k(pk)}) \times (F_3 - 1) \quad (4a)$$

Grid voltage (tetrodes)

$$-V_g = \frac{V_{g2} \times F_3}{\mu_{(g1\text{-}g2)}} + (v_{g1\text{-}k(pk)}) \times (F_3 - 1) \quad (4b)$$

Peak grid voltage $v_{g1(pk)} = V_{g1} + (v_{g1\text{-}k(pk)})$ (5)

Calculate ratio $V_g/v_{g(pk)}$ and from curve in Fig 4.19 read F_5 and F_6.

Grid current $I_g = \frac{i_{g(pk)}}{F_5}$ (6)

Grid dissipation $p_{g1} = \frac{I_g \times F_6 \times (V_{g1\text{-}k(pk)})}{2}$ (7)

Driving power $P_{dr} = p_{g1} + (V_g \times I_g)$ (8)

Screen current $i_{g2} = \frac{i_{g2(pk)}}{F_4}$ (9)

Screen dissipation $p_{g2} = V_{g2} \times I_{g2}$ (10)

Output impedance $Z_a = \frac{v_{a(pk)}}{F_2 I_a}$ (11)

In order to choose a value for anode input which will exploit the ratings of a chosen valve, an estimated efficiency may be assumed. Alternatively, the input may be fixed by other considerations such as available power supplies or licence regulations.

A reasonable efficiency for a Class C amplifier at frequencies up to 30MHz is 75%. Hence, for the valve chosen, which has an anode dissipation rating of 37.5W:

$$\text{Anode input} = \frac{37.5}{1 - 0.75} = 150\text{W}$$

At an anode voltage of 1000 this corresponds to a DC anode current of 150mA.

From Equation (1) calculate $I_{a(pk)} = 4.6 \times 150 = 690$mA. Next locate the current on the valve's anode current (I_a), anode voltage (V_a) characteristic (Fig 4.14) at a low value of anode voltage, just inside the knee of the curve; this corresponds to an anode voltage of 150V and a grid voltage of +12V.

From Equation (2), calculate $v_{a(pk)} = 1000 - 150 = 850$V.

From Equation (3) calculate:

$$P_{out} = \frac{1.8}{2} \times 0.15 \times 850 = 115\text{W}$$

The anode dissipation is the difference between anode input and power output.

$$p_a(\text{dissipation}) = 150 - 115 = 35\text{W}$$

This dissipation is sufficiently close to the maximum rating and can be accepted for the rest of the calculation. If the figure had been greater or considerably lower than the rated maximum, a new design should be made using a different power input, angle of flow or minimum anode voltage $V_{a(min)}$.

The chosen valve is a tetrode and from Equation 4(b) calculate grid voltage:

$$-V_g = \frac{300 \times 2}{8} + 12 \times 1 = -87\text{V}$$

From Equation (5) calculate $v_{g(pk)} = 87 + 12 = 99$V.

Calculate:

$$\frac{V_g}{v_{g(pk)}} = \frac{87}{99} = 0.88$$

and from Fig 4.19 read values of F_5 and F_6. These are 11.7 and 1.975 respectively.

From the grid current (I_g), anode voltage (V_a) curves of the TT21 (7623) a peak grid current of 32mA occurs at $V_a = 150$V and $V_{g1} = +12$V.

From Equation (6) calculate $I_g = 32/11.7 = 2.75$mA.

From Equation (7) calculate:

$$p_{g1} = \frac{2.75 \times 1.975 \times 12}{2} = 32.5\text{mW}$$

From Equation (8) calculate:

$$P_{dr} = 32.5 + (2.75 \times 87) = 273\text{mW}$$

The driver stage should produce considerably more than this minimum power in order to allow for losses in the coupling system.

From the screen grid current (I_{g2}), anode voltage (V_a) curves of the TT21 (7623), a peak screen current of 80mA occurs at $V_a = 150$V and $V_{g1} = +12$V.

From Equation (9) calculate $I_{g2} = 80/5.8 = 13.8$mA.

From Equation (10) calculate $p_{g2} = 300 \times 13.8 = 4.15$W. This dissipation is within the maximum rating of 6W and is acceptable.

From Equation (11) calculate $Z_a = 850/(150 \times 1.8) = 3.16\text{k}\Omega$.

It is now possible to design a pi-coupler to match 3.16kΩ to the impedance of the load.

Anode-modulated amplifiers

Anode-modulated amplifiers are designed in a similar manner to that given for Class C telegraphy, but checks must be made to ensure that the required conditions at the modulation crest are met.

At the modulation crest, the anode and screen voltages will be increased but the bias will be unchanged; hence the angle of anode current flow will increase. Typical values are between 150° and 180°. In making a design, it is necessary to assume an angle and later check the accuracy of the assumption. In the following equations, values at the crest of modulation are indicated by ($'$), thus θ' may be between 75° and 90°. Since the amplifier is assumed to be linear, then:

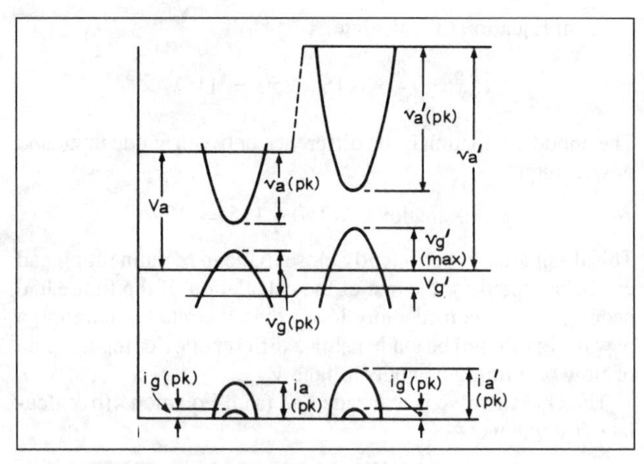

Fig 4.20. Phase relationship at the carrier and modulation crest for an anode-modulated Class C amplifier

$$P'_{out} = 4P_{out} \quad (12)$$

$$v'_{a(pk)} = 2v_{a(pk)} \quad (13)$$

Hence

$$v'_{a(min)} = 2v_{a(min)} \quad (14)$$

By using Equation (3) rearranged, the anode current at modulation crest can be calculated from:

$$I'_a = \frac{P'_{out} \times 2}{F'_2 \times v_{a(pk)}} \quad (15)$$

and from Equation (1):

$$i'_{a(pk)} = F'_1 \times I'_a$$

Normally, the positive grid voltage may be assumed to have the same value as calculated at the carrier. The peak working point corresponding to $i'_{a(pk)}$, $v'_{a(min)}$ and $v_{g1\text{-}k(pk)}$ must be located on the anode current (I_a), anode voltage (V_a) curves. In the case of a tetrode, a value of the screen voltage must be found which satisfies these conditions. In triodes, it may be found that a different (usually greater) value of $v_{g\text{-}k(pk)}$ is required to satisfy $i'_{a(pk)}$ and $v'_{a(min)}$.

The grid current at the modulation crest is usually significantly less than at the carrier. By using some grid leak bias, the angle of flow can be increased to 180°, requiring less bias, and hence making available an increased positive grid excursion. An alternative is to supply sufficient modulation to the driver stage to provide the required positive excursion.

For convenience of illustration, it will be assumed that the foregoing Class C telegraphy design is now to be modulated, but it should be noted that this will not necessarily give a practical result, since the anode dissipation rating may be exceeded during modulation.

It is usual practice to quote anode dissipation ratings at carrier (unmodulated) conditions of two-thirds of the maximum valve rating. This is based on the assumption that the average power dissipation will be increased by 1.5 times when modulation is applied. In the valve used for the example, the anode dissipation under modulation must be reduced to 37.5/1.5 = 25W.

In practice, however, with speech waveforms of relatively high peak-to-mean ratio, it is satisfactory to use a rather higher dissipation rating. When speech compression is used, or continuous 100% tone modulation is applied, it is important to ensure that the actual anode dissipation under modulation conditions is within the maximum rating. Returning to the previous design:

From Equation (12) calculate $P'_{out} = 4 \times 115 = 460$W.

From Equation (13) calculate $v'_{a(pk)} = 2 \times 850 = 1700$V.

From Equation (14) calculate $v'_{a(min)} = 2 \times 150 = 300$V.

Assuming an angle of anode current flow (2θ′) = 150°, then:

$$F'_1 = 3.75$$
$$F'_2 = 1.69$$
$$F'_3 = 1.35$$

From Equation (15) calculate

$$I'_a = \frac{460 \times 2}{1.69 \times 1700} = 320\text{mA}$$

From Equation (1) calculate $i'_{a(pk)} = 3.75 \times 320 = 1200$mA.

In order to obtain a peak working point where $i'_a = 1200$mA at $v'_{a(min)} = 300$V, it is necessary to find the correct value of screen voltage, it being assumed that the grid voltage for the carrier conditions is still available (+12V). From the I_a/V_a curves for the valve at various screen voltages when $V_{g1} = 0$, it is now necessary to predict the screen voltage required to produce $i'_{a(pk)} = 1200$mA at $v'_{a(min)} = 300$V and $v'_{g1} = +1.2$V.

From the TT21 data, the mutual conductance (g_m) at I_a = 140mA is 11mA/V, therefore, at $i'_{a(pk)} = 1200$mA, the mutual conductance will increase by:

$$\left(\frac{1200}{140}\right)^{1/3} = 2.046 \text{ which gives 22mA/V approx}$$

From this it follows that the anode current at $V_{g1} = +12$V will be $12 \times 22 = 264$mA greater than the value at $V_{g1} = 0$V.

The point on the curve that now has to be found is for V'_a = 300V, $I_a = 1200 - 264 = 936$mA. This corresponds to a screen voltage of 465V.

The screen voltage should therefore be increased by slightly more than 1.5 times when the anode voltage is doubled by modulation. The modulation transformer should be designed to provide this screen modulation point either by a tap on the main winding or by additional winding.

The assumed angle of flow can be checked to see if it is realistic, by calculation of the bias from Equation (4b).

$$-V'_{g1} = \frac{465 \times 1.35}{8} + 12 \times 0.35 = -82.5\text{V}$$

This is close enough to the original value of –87V for a practical design.

In practice, the regulation of the driver source, the change of grid current when the screen voltage is raised, and the method of obtaining the bias, will modify the available positive grid voltage at the crest, but the calculation gives sufficient guide as a practical starting point.

Class AB and Class B linear amplifiers

In Class AB and Class B linear service, the amplifier is required to handle modulated waveforms without distortion. The amplification of single sideband suppressed carrier signals is the most usual example.

In a Class B amplifier, the angle of flow of anode current is close to 180°. An acceptable design can be made using the procedure given for Class C telegraphy but with θ = 90°. In practice, however, such amplifiers are operated with some standing anode current ($I_{a(0)}$) in the absence of a signal, as a means of improving the linearity.

Class AB amplifiers invariably operate at significant standing anode current. Design curves based on angle of flow are therefore inconvenient, and curves based on the ratio of mean anode current under driven conditions to standing anode current are more useful.

The curves given in Fig 4.21 are suitable. In these, F_7

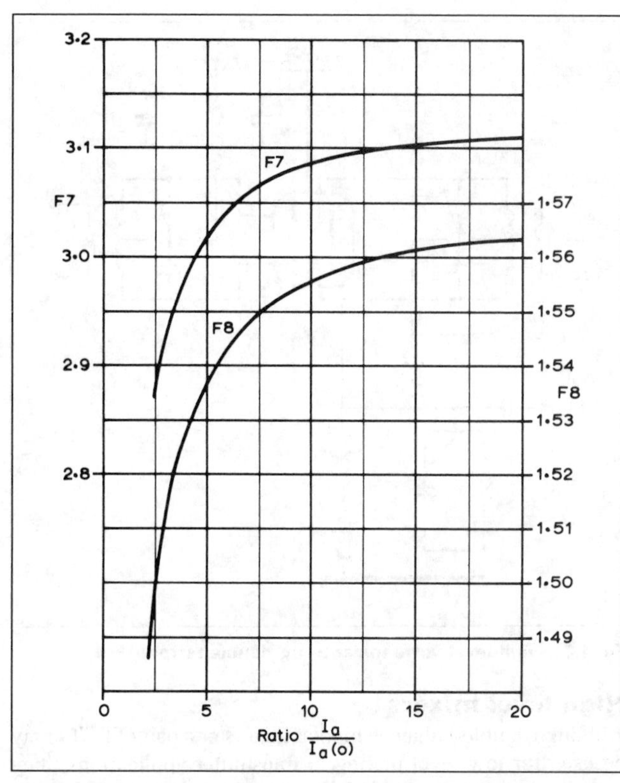

Fig 4.21. Design factors for Class AB and Class B linear amplifiers (see text)

corresponds to F_1 and F_8 to F_2; from which, under these new conditions:

Peak anode current $\quad i_{a(pk)} = F_7 \times I_a \quad (16)$

Power output $\quad P_{out} = \frac{F_8}{2} \times I_a \times v_{a(pk)} \quad (17)$

In a typical Class AB amplifier driven to maximum peak envelope power the valve will have an anode efficiency of about 70%. The anode dissipation is a maximum at some value of drive less than the maximum. The anode dissipation at maximum drive must therefore be less than the maximum rating, say, 80%.

Taking the same example as used for the Class C calculations, the TT21 (7623), an anode dissipation of 30W is a suitable starting point. In a final design, the values must be chosen so that, taking into account the peak-to-mean ratio of the modulation waveform, excessive anode dissipation does not occur. Taking anode dissipation as 30W and anode efficiency of 70%, then:

Anode input $\quad P_{in} = \frac{30}{1-0.7} = 100\text{W}$

Decide on the anode voltage; in this case, taking $V_a = 1000\text{V}$, the anode current $I_a = 100\text{mA}$.

Next, it is necessary to decide the zero signal (standing) anode current $I_{a(0)}$; this depends on a compromise between efficiency and intermodulation distortion. Generally a current corresponding to about 66% of the rated anode dissipation is typical, from which:

$$I_{a(0)} = (2/3) \times 37.5 = 25\text{mA}$$

Then:

$$I_a/I_{a(0)} = 4$$

From Fig 4.21:

$$F_7 = 2.99 \quad \text{and} \quad F_8 = 1.53$$

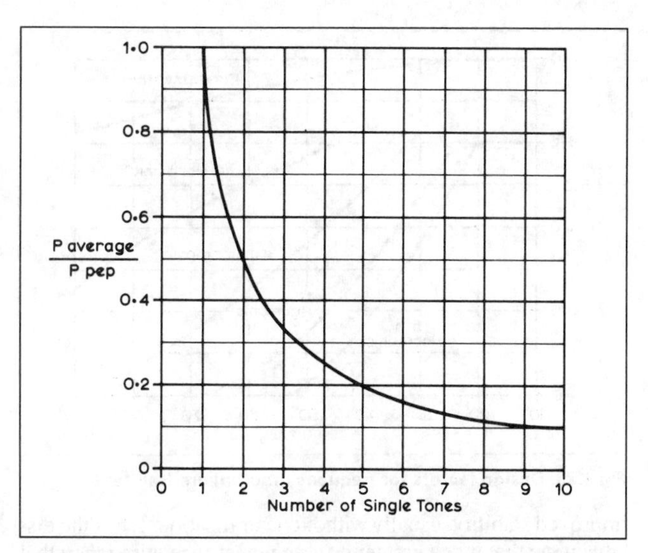

Fig 4.22. The assessment of intermodulation in linear amplifiers

and from Equation (16):

$$i_{a(pk)} = 2.99 \times 100 = 299\text{mA}$$

Locate this current on the I_a/V_a characteristic curve to find the value of $v_{a(min)}$. To preserve linearity it is important that this point shall not be in the curved part of the knee characteristic. From the curve a value of 100V is suitable.

Hence:

$$v_{a(pk)} = 1000 - 100 = 900\text{V}$$

and from Equation (17)

$$P_{out} = (1.53/2) \times 0.10 \times 900 = 69\text{W}$$

Anode dissipation $\quad p_a = 100 - 69 = 31\text{W}$

The calculation of driving power (if any) and anode load impedance follows the same procedure as for Class C telegraphy. The bias will, however, be decided by the chosen value of $I_{a(0)}$. The approximate value can be taken from the characteristic curve, but in practice should be set to give the required value of $I_{a(0)}$. The intermodulation of linear amplifiers is frequently assessed by using a test signal consisting of two or more signals (tones) of equal amplitude. The average power output will decrease as the number of tones is increased in the test signal as shown in Fig 4.22.

In the usual case of a two-tone test signal, and assuming ideal linear characteristics, the relation between single and two-tone conditions is:

$$I_{a\,(two\text{-}tone)} = \frac{2}{\pi} I_{a\,(single\text{-}tone)}$$

Average input power:

$$P_{in\,(two\text{-}tone)} = V_a \times I_{a\,(two\text{-}tone)}$$

Average output power:

$$P_{out\,(two\text{-}tone)} = {}^1/_2\, P_{out\,(single\text{-}tone)}$$

Grounded-grid operation

All the preceding designs are based on the assumption that the signal is applied to the grid and the cathode earthed (grid drive or common cathode connection). Sometimes the signal is applied to the cathode and the grid earthed (grounded grid or cathode drive connection). This arrangement has the advantage of

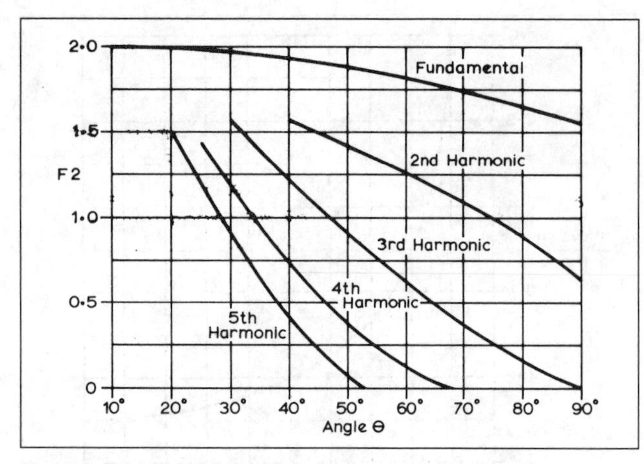

Fig 4.23. Design factors for frequency multipliers (see text)

improved stability, usually without neutralisation. It has the disadvantage that much greater driving power is required than that needed for grid drive connection, although some of the driving power is recovered in the output circuit.

The driving power:

$$P_{dr} = (V_g \times I_g) + p_{g1} + \left(\frac{v_{g1(pk)} \times F_2 \times I_a}{2}\right)$$

The drive power which appears in the output is:

$$\frac{v_{g1(pk)} \times F_2 \times I_a}{2}$$

In the case of a tetrode, there is a small additional driving power which is not recovered in the output; this occurs due to the product of peak drive voltage and the fundamental component of the screen current. It is usually sufficiently small to be ignored.

Frequency multipliers

Frequency multipliers are Class C amplifiers in which the anode circuit is tuned to a harmonic of the drive frequency, and may be designed in the same way as a Class C amplifier. In general, smaller angles of flow are used, as this tends to increase the harmonic output. The factor F_2, which in the amplifier design gives the ratio of peak fundamental to DC anode current, is replaced by a factor giving the ratio for peak harmonic to DC anode current. These factors for harmonics up to the fifth are shown in Fig 4.23.

Factors

F_1 and F_7 $\quad \dfrac{\text{Peak anode current}}{\text{DC anode current}}$ (assuming sine waveform)

F_2 and F_8 $\quad \dfrac{\text{Peak fundamental component of anode current}}{\text{DC anode current}}$ (assuming sine waveform)

F_3 $\quad \dfrac{1}{1 - \cos\theta}$

F_4 $\quad \dfrac{\text{Peak screen current}}{\text{DC screen current}}$ (assuming squared sine waveform)

F_5 $\quad \dfrac{\text{Peak grid current}}{\text{DC grid current}}$ (assuming squared sine waveform)

F_6 $\quad \dfrac{\text{Peak fundamental component of grid current}}{\text{DC grid current}}$ (assuming squared sine waveform)

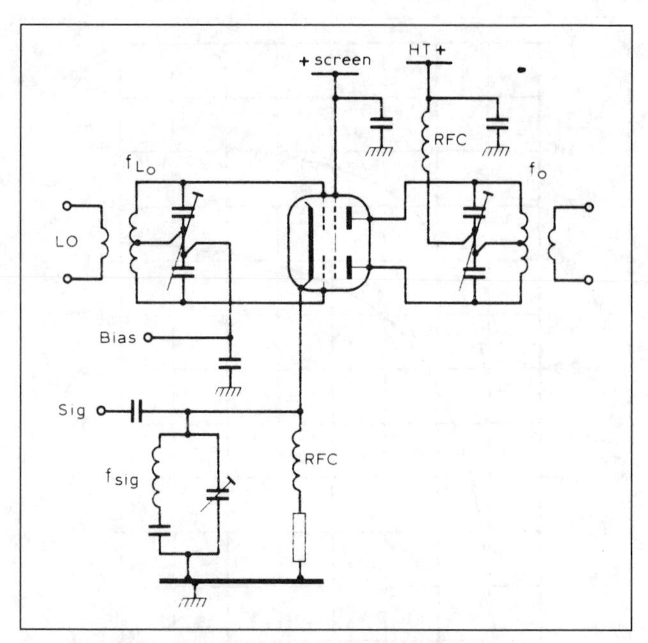

Fig 4.24. High-level valve mixer using double tetrode

High-level mixers

Although double-balanced bipolar mixers or a pair of FETs may be used for low-level mixing in transmitter applications, they really do need a considerable increase in the number of linear amplifier stages to raise the output to an appropriate level, and this may be inconvenient at UHF.

Another approach to the problem is to use some form of high-level mixer. For VHF and UHF up to the 432MHz band the double tetrode offers a convenient circuit, using such valve types as the QQV02-6 or 3-20. In these the local oscillator input is normally injected into the two control grids (in push-push) and the other signal input into the cathode circuit (in parallel). Care is necessary of course to preserve as good a balance as possible in the push-push circuits to ensure maximum attenuation of the unwanted signals (see Fig 4.24).

For UHF (particularly for the 1.3GHz band) one or other of the grounded-grid triodes such as the 8255 or EC88 for low power, or 2C39A for high power, can be used. In this case the two input signals to be mixed are either injected into the grid and cathode or, more commonly, the grid is grounded and both are injected into the cathode (see Fig 4.25). In the latter case some care is needed to isolate the two signals at their input and at the same time to adequately ground the local oscillator input tuned circuit. The value of C in Fig 4.25(b) must be small to avoid excessive absorption of lower-frequency signal and yet be large enough to effectively ground the local oscillator tuned input circuit.

With all types of mixer in which the cathode has an RF input, the heater must be isolated by RF chokes in each lead.

VALVES AND TUBES FOR SPECIAL PURPOSES

Disc seal valves

In the disc seal triode (Fig 4.26), characteristically of high mutual conductance, the electrode spacing is minimal. The 'top hat' cathode contains the insulated heater, one side of which is connected to the cathode and the other brought out coaxially through the cathode sleeve connection. The fine-wire grid stretched across a frame emerges through the envelope by an annular connection. Because the clearance between grid and

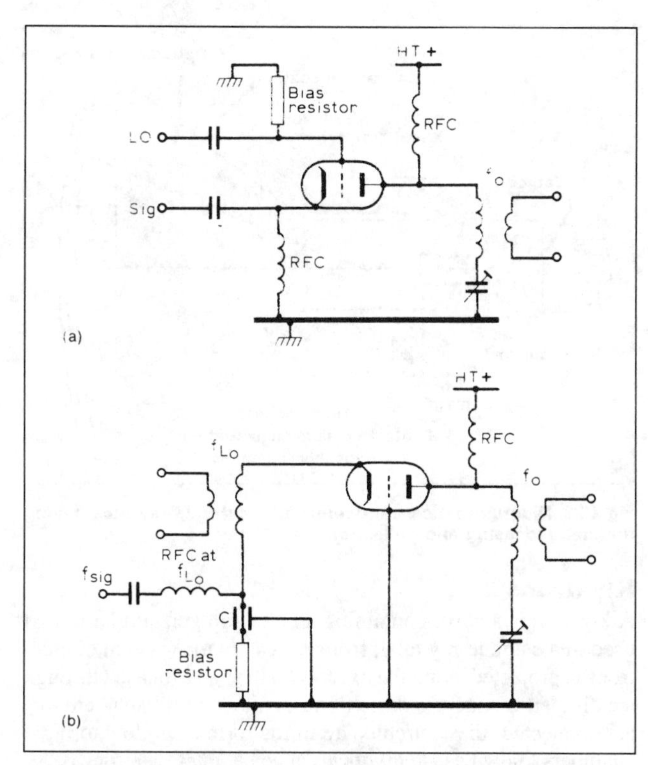

Fig 4.25. Grounded-grid valve mixers. (a) Grid and cathode injection. (b) Cathode injection

cathode is very small, the cathode surface is shaved during construction to provide a plane surface. The anode also emerges via an external disc for coaxial connection. On larger disc seal valves the anode may form part of the valve envelope.

Disc seal valves are available for power dissipations of a few watts to 100W with forced air cooling and outputs in the frequency range 500–6000MHz. It should be noted that maximum power and frequency are not available simultaneously.

Disc seal valves, although intended for coaxial circuits, may be effectively employed with slab-type circuits. Important points to be observed are (a) only one electrode may be rigidly fixed, in order to avoid fracture of the seals (which is more likely to occur in the glass envelope types); and (b) except in the case of forced-air anode cooling the anode is cooled by conduction into its

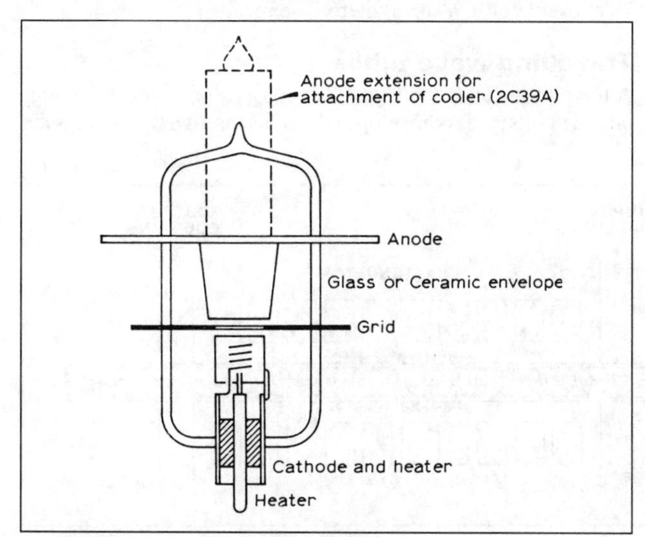

Fig 4.26. General form of a disc seal valve

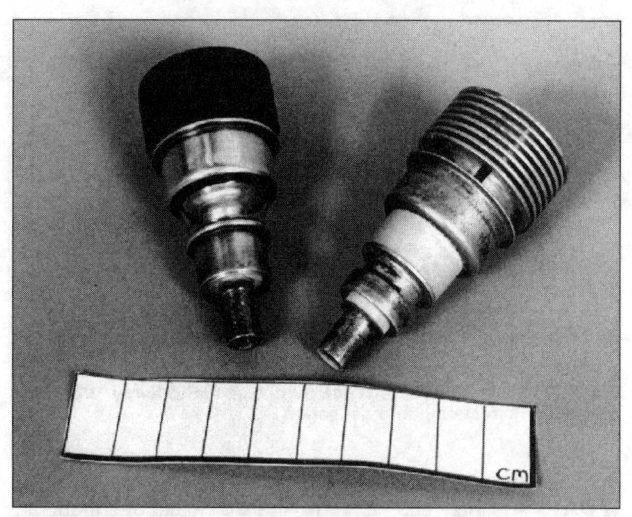

The 2C39 disc-seal triode, showing glass version (left) and ceramic version (right). The anode cooling fins on the ceramic version are removable, clamped in placed with Allen bolts

associated circuit. With a shunt-fed circuit thin mica insulation will function both as a capacitance and a good transmitter of heat.

Certain sub-miniature metal ceramic envelope types such as the 7077, although of the generic disc seal form, require special sockets if they are used in conventional circuits. Many of them give significant output at the lower SHF bands.

Cathode-ray tubes

A cathode-ray tube contains an electron gun, a deflection system and a phosphor-coated screen for the display. The electron gun, which is a heated cathode, is followed by a grid consisting of a hole in a plate exerting control on the electron flow according to the potential applied to it, followed in turn by an accelerating anode or anodes.

The simplest form of triode gun is shown at Fig 4.27. The beam is focused by the field between the grid and the first anode. In tubes where fine line spots and good linearity are essential (eg in measurement oscilloscopes), the gun is often extended by the addition of a number of anodes to form a lens system. Beam focusing may be by either electrostatic or electromagnetic means.

Oscilloscope tubes

Electrostatic focusing is used in oscilloscope tubes, the deflection system consisting of pairs of plates to deflect the beam from its natural centre position, depending on the relative potentials applied. Interaction between the two pairs of deflection plates is prevented by placing an isolation plate between them (Fig 4.28).

After its deflection the beam is influenced by a further

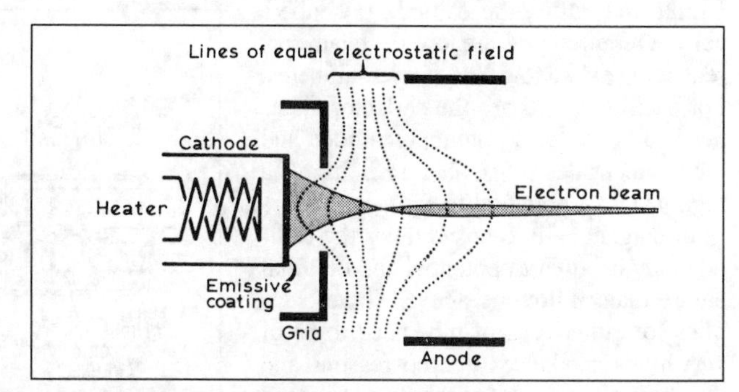

Fig 4.27. Diagram of the electron gun used in cathode-ray tubes, travelling-wave tubes and klystrons

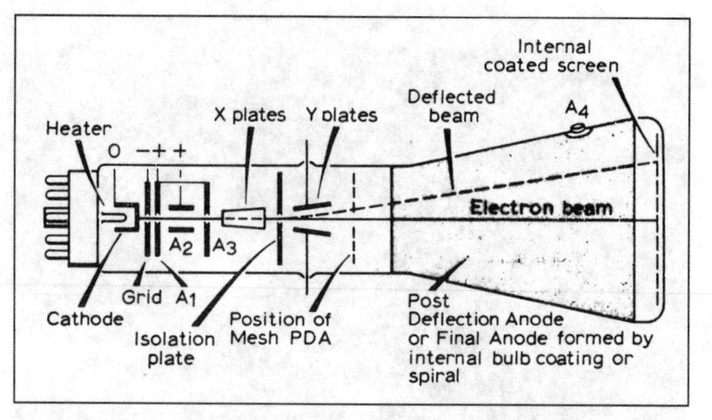

Fig 4.28. Diagrammatic arrangement of a cathode-ray tube with electrostatic focusing and deflection

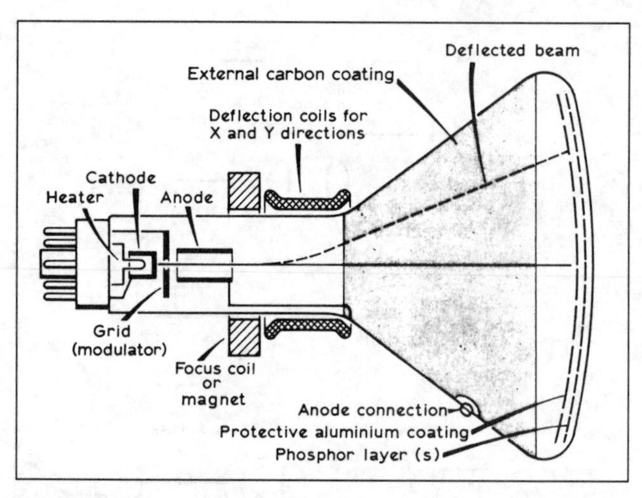

Fig 4.29. Diagrammatic arrangement of a cathode-ray tube having magnetic focusing and deflection

accelerating electrode known as the *post-deflection accelerator* (PDA) which may take the form of a wide band of conducting material on the inside of the cone-shaped part of the bulb, or of a close-pitch spiral of conducting material connected to the final anode. For some purposes when it is important to maintain display size constant irrespective of the final anode voltage, a mesh post-deflection accelerator is fitted close to the deflecting plates.

If a double beam is required the electron flow is split into two and there are two sets of deflection plates. Alternatively, two complete systems are enclosed in one tube. Although a common X-deflection plate may be fitted, the advantage of two complete systems lies in providing complete alignment of the timing (horizontal) deflection. By setting two systems at an angle to one another adequate overlap of each of the displays is provided.

Radar and picture tubes

In radar and television tubes magnetic focusing and deflection are common (Fig 4.29), the deflection angle being vastly greater than with an oscilloscope tube. Such tubes employ a simple triode or tetrode electron gun with an anode potential of 20kV or more.

Screen phosphors

Many types of phosphor are used for coating the screens of cathode-ray tubes, their characteristics varying according to the application. In all of them the light output is determined by the final anode voltage used, but where this exceeds about 4kV the phosphor is protected against screen burn by a thin backing layer of evaporated aluminium. Oscilloscope tubes require a phosphor with a wide optical band to give a bright display for direct viewing. This phosphor, yellow-green in colour, extends into the blue region to enable direct photographs to be taken from the display.

CR tube power supplies

Unlike the valve, the cathode ray tube's current requirements are low, the beam current being only a few tens of microamperes. For oscilloscope work the deflector plates need to be at earth potential, cathode and other electrodes consequently being at a high negative potential to earth.

In magnetically focused tubes the cathode may be at earth potential and the final anode many kilovolts above. Power supplies for either type of tube need to be of very high impedance for safety reasons, and the short-circuit current should not exceed about 0.5mA.

Klystrons

A klystron is a valve containing an electron gun similar to that used in a cathode-ray tube, from which a narrow beam of electrons is projected along the axis of the tube and focused through small apertures across which one or more oscillatory circuits are connected; these circuits are in the form of hollow toroidal chambers known as *rhumbatrons*. The beam of electrons is velocity-modulated by the application of an RF field, for example by passing the beam through small apertures having an RF voltage between them. If this velocity-modulated beam is now passed through a field-free space, known as a *drift space*, the faster-moving electrons will begin to overtake the slower ones so that alternate regions of high and low electron density will exist at some point along the beam.

The rhumbatrons are located at points of maximum density. If two rhumbatrons are coupled together by means of a feedback loop, or if a reflector electrode maintained at a slightly negative potential with respect to the cathode is mounted at the end of the tube, the energy will be reflected back into a single rhumbatron. The electrons become bunched due to their velocity modulation and, since the energy is in phase by reason of the distance of travel to the reflector and back again, sustained oscillation will take place. Klystrons are employed as oscillators, the reflector type being preferred for low-power work, eg as a local oscillator in a superheterodyne receiver.

Travelling-wave tubes

A travelling-wave tube (Fig 4.30) consists of a wire helix supported in a long glass envelope which fits through two waveguide

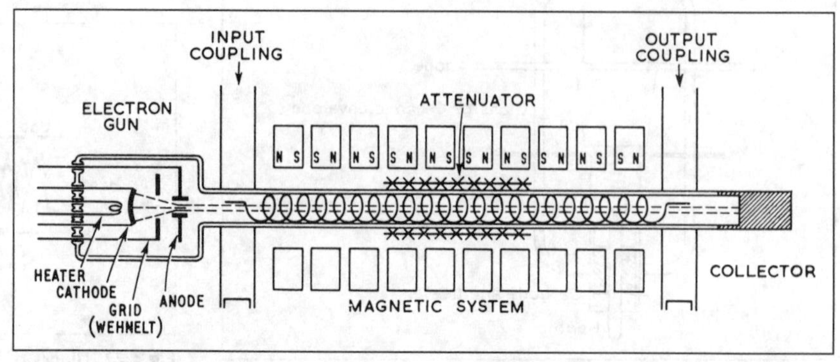

Fig 4.30. Travelling-wave tube

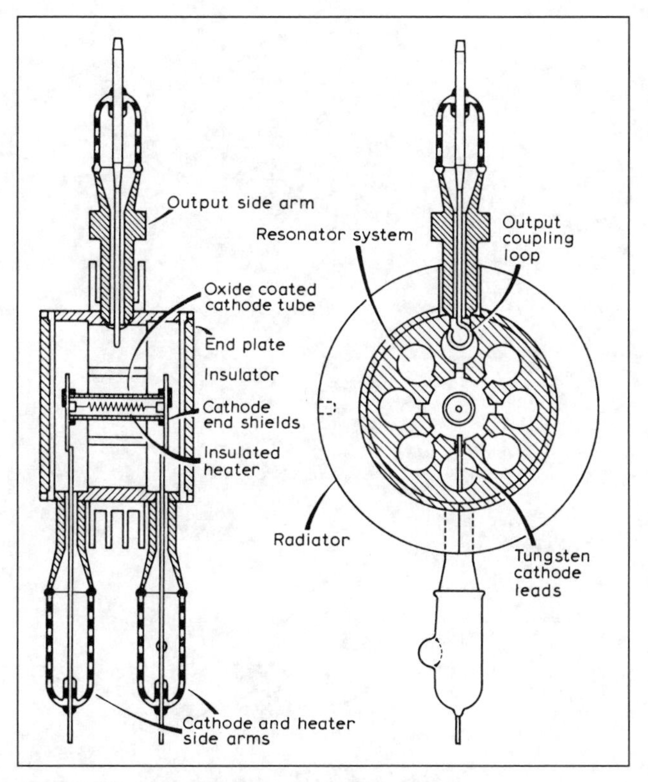

Fig 4.31. General arrangement of a magnetron

stubs or coaxial couplers used as input and output elements. At one end of the tube is a conventional electron-gun assembly which directs an electron beam to the collector at the other end. The RF signal input is coupled into the helix and will run round the turns of the helix at roughly its normal velocity, but its axial velocity may be only about a tenth of that value, depending on the pitch of the helix. If the electron beam is sent along the centre of the helix and focused by a magnet it will be charge-density modulated by the voltage pulse (ie amplitude-modulated) owing to the interaction between the magnetic field of the current in the helix and the electrons. This modulation grows in amplitude along the length of the helix roughly according to the square of the axial distance from the beginning of the helix. By suitable design the output waveguide coupling will extract more energy than was put in by the input coupling and a considerable power gain can be achieved. For example, travelling-wave tube amplifiers with a bandwidth of 500MHz centred on frequencies in the range 1700–8500MHz are available. Outputs of 5–10W and power gains of 30–40dB are possible.

To prevent self-oscillation, it is necessary to introduce some attenuation between the input and output, and to match the input and output impedances to the tube. Serious mismatch at the output is likely to damage the tube.

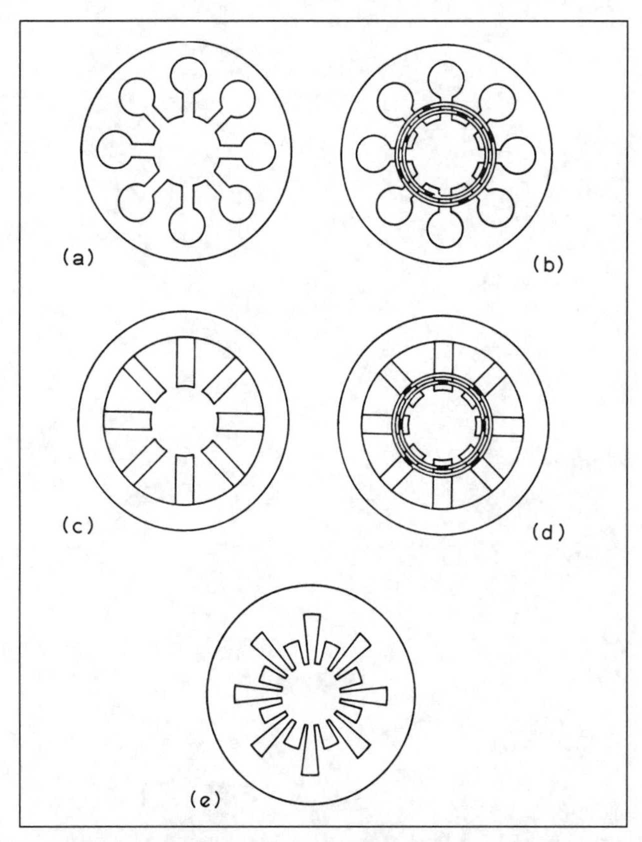

Fig 4.32. Various forms of anode cavity in the magnetron: (a) unstrapped cavity, (b) strapped cavity, (c) unstrapped vane-block, (d) strapped vane-block, and (e) rising-sun block

The magnetron

A magnetron is a diode with a cathode and anode in a cylindrical assembly. In a simple magnetron the anode is split into two parts; more commonly it comprises multiple cavities which resonate at the operating frequency. The whole assembly is mounted in a magnetic field parallel to the axis of the electrodes. The magnet may be an integral part of the valve (the *package magnetron*) or a close-fitting separate assembly.

The axial magnetic field causes the path of the electrons leaving the cathode to be curved. At a certain field intensity the electrons fail to reach the anode and return to the cathode in circular orbits. The time taken for an electron to complete its orbital journey decides the frequency of oscillation. The energy associated with the moving electrons is given up in the space around the cathode, transferred to the resonant cavities and coupled out into an associated waveguide by loop and probe or by a windowed slot cut in the back of one of the cavities. The anode cavity can take a number of different forms and these are illustrated in Fig 4.32.

5 Building blocks

OSCILLATORS

This section is concerned with oscillator design. Oscillators are a fundamental topic in radio and are used in all types of equipment. The principles of design can be easily determined and, as in all circuit design, it is important to determine the objectives before starting the task.

An oscillator would ideally generate a pure sine wave. If a spectrum of voltage against frequency were to be plotted, it would consist of a single line at the required frequency: Fig 5.1(a). This figure shows the spectrum of the 'ideal' oscillator. Spectral plots like this from a spectrum analyser are much used in professional electronics, but are rarely available to the amateur. A more practical approach is to tune across the frequency band of interest with a reasonably good receiver, preferably on single sideband (SSB). The ideal oscillator would then produce a single clean beat note, of amplitude related to the amplitude of the oscillations, and of course the coupling into the receiver.

No oscillator produces this ideal response, and there are always harmonics, noise and often sidebands in the response. A more practical plot is shown in Fig 5.1(b). Tuning a receiver across this would show where the problems are! The *noise floor* (the background level of noise which is equal at all frequencies) is apparent, as is the *'1/f' noise*, which is an increase in noise close to the centre (wanted, or *carrier*) frequency. The harmonics also show these effects, as do the sidebands, spurious oscillations caused usually in synthesisers by practical limitations in loop design.

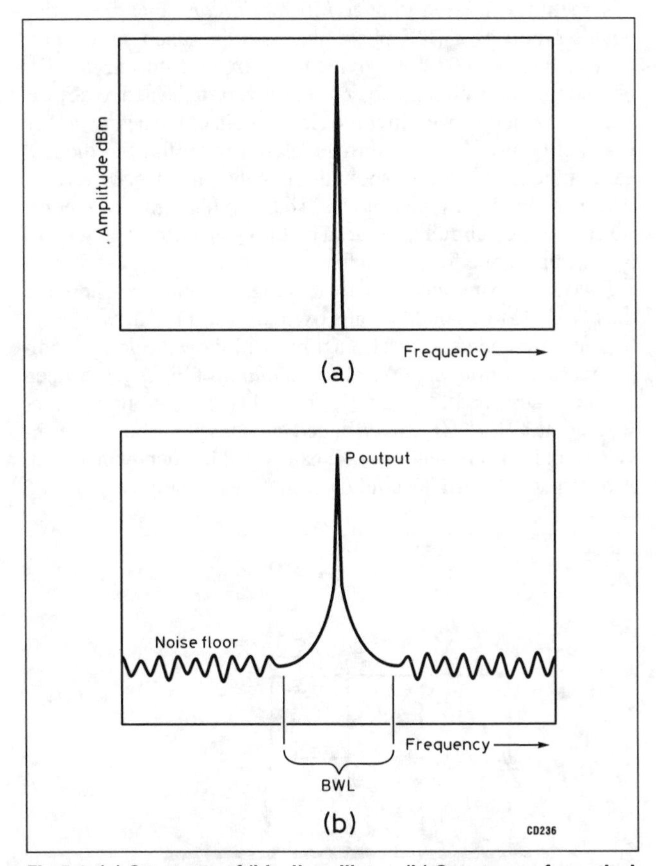

Fig 5.1. (a) Spectrum of 'ideal' oscillator. (b) Spectrum of practical oscillator showing noise

Noise and sidebands are almost always undesirable, and can be very difficult to remove once generated, so one purpose of this chapter is to indicate ways of minimising them. Harmonics are less of a problem. They are a long way in frequency terms from the wanted signal, and can often be simply filtered out. In some cases, they may even be wanted, since it is often convenient to take an output frequency from an oscillator at a harmonic instead of the fundamental. Examples of this are shown below for crystal oscillators.

Oscillators may be classed as *variable frequency oscillators* (VFOs), *crystal oscillators* (COs), including the class of *variable crystal oscillators*, (VXOs), *phase-locked loop synthesisers* (PLLs), which include VFOs as part of their system, and more recently *direct digital synthesisers* (DDSs). Any or all of these may be used in a particular piece of equipment, although recent trends commercially are to omit any form of 'free-running' (ie not synthesised) VFO. The name is retained, however, and applied to the sum of the oscillators in the equipment. Since these are digitally controlled, it is possible to have two (or more) 'VFOs' in a synthesised rig, where probably only a single variable oscillator exists physically. It can be retuned in milliseconds to any alternative frequency by the PLL control circuit.

All oscillators must obey certain basic design rules, and examination of preferred designs will show that these have been obeyed by the designer, whether deliberately or by empirical methods.

The first essential for an oscillator is a gain element. This will be an active device such as a bipolar transistor (sometimes called a 'BJT' for *bipolar junction transistor*), a junction field-effect transistor (JFET), a metal-oxide semiconductor FET (MOSFET) or a gallium arsenide FET (GaAsFET). There are lesser-used devices such as gallium arsenide bipolars, which at the time of going to press are just coming on to the market, and gallium arsenide HEMTs (high electron mobility transistors), a variant of the GaAsFET with lower noise and usually biased exactly as the latter. Operational amplifiers are rarely used at RF, since very few are capable of high frequency operation. Some integrated circuits do contain oscillator-maintaining circuits, but they are very often variants of standard discrete circuits. Valves, now long obsolete in the context of oscillator design, will be neglected in this text, but it is worth observing that most of the fundamental circuit configurations owe their origins to valve designs from the early part of the 20th century. For those interested in valve design, reference should be made to earlier editions of this handbook, and to references [1] and [2]. A more recent and readily available source is Section 4 of reference [3].

The basic requirements for an oscillator are:

1. There must be gain over the whole frequency range required of the oscillator.
2. There must be a feedback path such that the product of the forward gain and the feedback attenuation still leaves a net *loop gain* greater than 1.
3. The feedback must occur in such a way that it is in phase with the input to the gain stage.
4. There must be some form of resonant circuit. This is most often an inductor and capacitor (L-C). More precise frequency control can be achieved using electromechanical structures such as *crystals*, *ceramic resonators* or *surface acoustic wave* (SAW) *resonators*. This latter category is only really feasible in specialist applications where the high cost can be justified.

   The resonant circuit may not be the one apparent because, as in many non-intended oscillations, a feedback around many paths including the parasitic components in the circuit can control the oscillations, which can of course be at more than one frequency. At low frequencies, a resistor-capacitor (RC) or inductor-resistor (LR) network may be used for frequency control. These cannot be said to be truly resonant in themselves but, with active gain stages, resonant peaks which are sometimes of very high Q can be achieved.

   A potential successor to the crystal and ceramic resonator is the electrochemically etched mechanical resonator, usually in the same silicon chip as the oscillator circuit. These devices are still in the research stage, but may well come into common use soon.

PLL synthesisers all contain oscillators, usually at least two, one being a VFO and the other a crystal reference frequency. In the special case of the DDS, described below, there is no oscillator at the output frequency since this is a truly synthesised output.

An oscillator therefore needs gain, feedback and a resonant circuit. However, there are more criteria which can be applied:

5. For stable oscillations, the resonant circuit should have high Q. This tends to rule out RC and LR oscillators for high frequencies, although not completely since they are often used for audio applications. More information on low-frequency oscillators can be found in audio texts, especially reference [4].
6. For stable oscillations, the active components should also show a minimum of loading on the resonant circuit, ie the loaded Q should be as high as possible. Although an oscillator could be said to be the ultimate in Q-multipliers, high inherent Q gives better stability to the final product.
7. The choice of active device is critical in achieving low-noise operation. This will be covered extensively below, since in almost all cases low-noise oscillations are required, and the difference between 'noisy' and 'quiet' oscillators may be only in the choice of a transistor and its matching circuit.
8. The output stage of an oscillator can be very important. It must drive a following stage adequately, without *load pulling*, ie variations in the load impedance should not change the oscillator frequency.
9. The power supply to an oscillator should be very carefully considered. It can affect stability, noise performance and output amplitude. Separate supply decoupling and regulation arrangements are usually essential in good oscillator design.

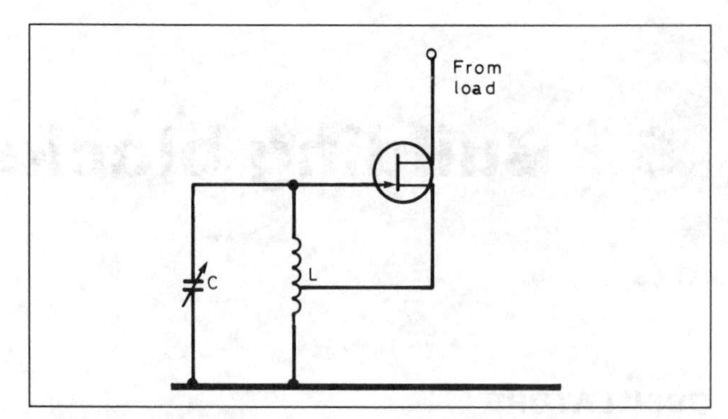

Fig 5.2. Hartley oscillator

To many people, the above comments may be acceptable for VFOs, but may seem excessive in the design of crystal oscillators or synthesisers. While it is certainly true that all but the poorest crystal oscillators will give acceptable results, good circuit design can make the results even better, and for synthesisers, good, stable, voltage-controlled oscillator (VCO) design is essential for a well-behaved loop with low phase noise. A VCO is the special case of a VFO where the tuning capacitance is replaced by a voltage-variable element, for example a *varactor* or *tuning diode*. This is a class of diode optimised for wide capacitance variation with applied voltage. Other means of electronic tuning are possible in some circuits, but varactor tuning is most common; it is not without problems, as described below.

VFOs

Most, but not all, VFOs consist of a single active device with a tuned circuit and feedback network to cause the oscillation. This is sometimes referred to as a *180° oscillator*, since the active device produces a 180° phase shift and the tuned circuit produces a further 180° shift, giving the 360° total shift needed for operation. Some circuits do not even invert in the active device, but use it as a follower. In this case, the gain of the active device may not be used, or may only be used as a buffer, but the real gain of the circuit comes about through the impedance transformation occurring in the follower, so that a high transformer ratio at the tuned circuit, produced by the Q, is buffered down to a lower impedance for the feedback network.

Two such examples are shown in Figs 5.2 and 5.3. These are the classic Hartley and Colpitts oscillators in FET-driven form.

Some comments on these diagrams will show the design compromises in action. In the *Hartley oscillator* of Fig 5.2, the tuned circuit is designed for high Q, preferably using tinned copper wire or, at VHF/UHF, silver-plated copper wire. The active device is an FET, chosen for good gain at the frequency involved. A dual-gate MOSFET could be used or, especially at UHF and

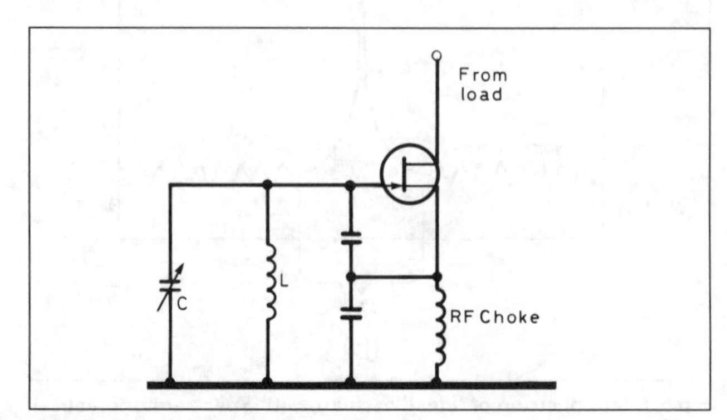

Fig 5.3. Colpitts oscillator

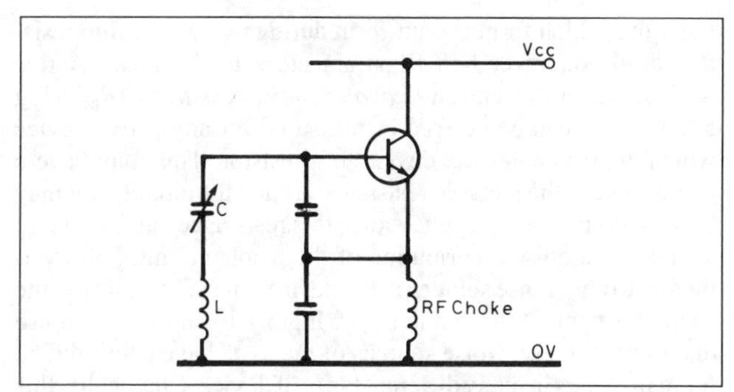

Fig 5.4. Clapp oscillator

microwave frequencies, a GaAsFET. The high input impedance of the FET puts little loading onto the tuned circuit, while the FET source provides feedback at a low impedance. The feedback voltage is transformed up to a higher impedance and appropriate phase, so the oscillation loop is complete. This is one of the examples where the active device provides only impedance matching to the oscillator loop; no phase shift in the FET is involved.

Fig 5.3 shows the capacitively tapped version of basically the same circuit. Capacitive transformers are less easy to understand intuitively, but if one thinks of the whole as a high-impedance resonant circuit, then a capacitive tap is a reasonable alternative. This circuit is known as the *Colpitts oscillator*. While in principle the two circuits are very similar, in practice the additional capacitors of the Colpitts make the circuit less easy to tune over a wide range. However, in the Hartley, with careful design and a low-input-capacitance FET, a very wide range of total circuit capacitance (and hence frequency range) can be achieved. Of course, in a particular application there may be no need for a wide range of frequencies, and in this case it is usual to add capacitance to the input to reduce the overall swing. One aspect of most active devices is that the input capacitance is slightly voltage variable, so that, if the FET has a small variable input capacitance and the circuit has a much larger fixed capacitance, then the total variation is reduced. The Colpitts circuit is also preferred at higher frequencies, since the transformer action of the inductor is much reduced where the 'coil' is in the form of a straight wire, say at 500MHz or above.

Sometimes described as a variant of the Colpitts circuit is the *Clapp oscillator*: Fig 5.4. This shows a series-resonant circuit which is more easily matched into the bipolar transistor shown. This circuit is especially suitable to UHF and low microwave work, where it is capable of operating very close to the cut-off frequency (F_t) of the transistor.

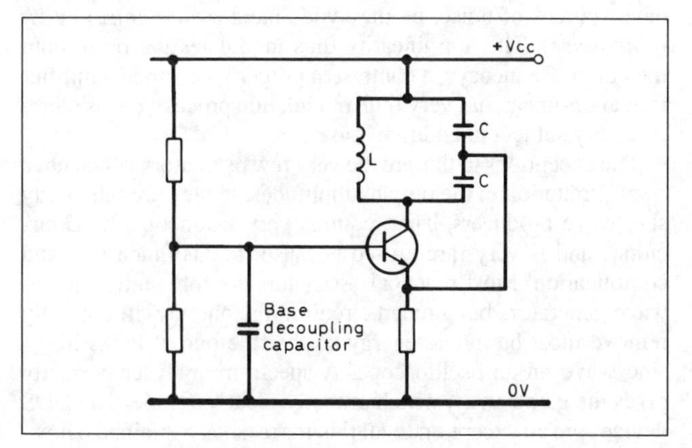

Fig 5.5. 180° phase-shift oscillator

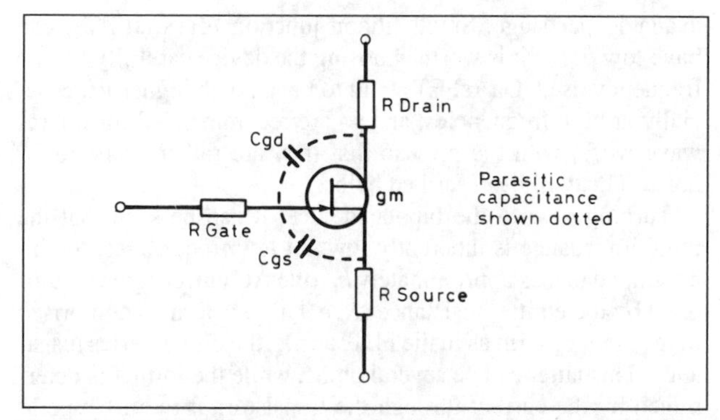

Fig 5.6. FET input characteristics. Input resistance is very high. Input capacitance is approximately equal to C_{gs} + (voltage gain) × C_{gd}. Input equivalent noise resistance is approximately equal to R_{GATE} + R_{SOURCE} + $1/g_m$. Typically R_{GATE} is less than 10Ω and R_{SOURCE} is approximately equal to R_{DRAIN} and less than 100Ω. Power devices may be less than 1Ω

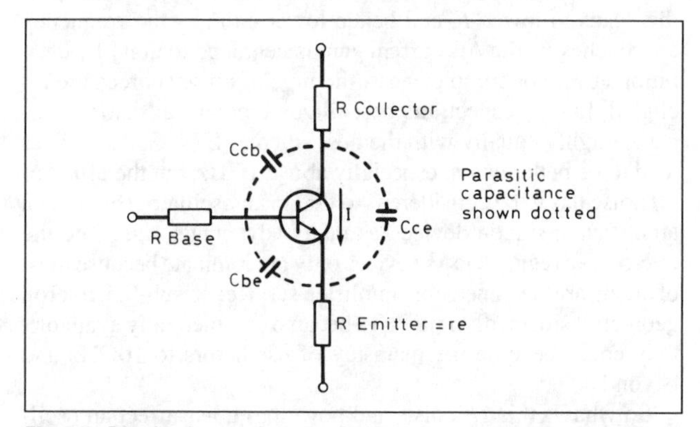

Fig 5.7. Bipolar input characteristics. Input resistance is approximately $R_{BASE} + h_{fe} \times (r_e + R_e)$ where $R_e = 1/g_m$ and $g_m = (q/kT) \times I$. Input capacitance is very approximately equal to $C_{be} + V_{gain} \times C_{cb}$ but transistor effects will usually increase this dynamically. Input equivalent noise resistance (R_{IN}) is equal to $R_{BASE} + (R_e + r_e)/2$. Typically, for a small device, R_{BASE} is about 100Ω, R_e is 26Ω at 1mA, and r_e is 3Ω. Therefore R_{IN} is approximately 129Ω

Variants of both the Hartley and Colpitts circuits have been derived for feedback around the active device including a 180° phase shift. An example is shown in Fig 5.5 and this is again more suited to a bipolar transistor.

In the past, there was much discussion on whether the best choice for the active device in an oscillator was a bipolar transistor or an FET. There are many factors involved in this choice, but some simple rules can be derived. First, since the loaded Q must be maintained as high as possible, an FET is attractive for its high input impedance. A bipolar transistor could be used as an emitter follower with an undecoupled emitter resistor, but a noise analysis of the circuit would put all of that resistance in the noise path. As an example, Figs 5.6 and 5.7 compare noise sources in oscillator (or amplifier) input circuits. The FET, provided it is operated well within its frequency range, has a high input impedance. Any input capacitance is absorbed into the tuned circuit fairly directly, since it looks just like a capacitor with at most an ohm or two of input series resistance and perhaps 1 or 2nH of series inductance from the bondwire on the chip. The output impedance at the source, ie in a follower, is $1/(g_m)$, (plus a small ohmic resistance term) where g_m is the mutual conductance at the operating frequency. As a gain stage, the output impedance of most FETs is very high, so the gain is determined by g_m and the load impedance. An exception to this is the GaAs MESFET, where the output impedance is typically a few hundred ohms, so there is a serious limit to the gain available per stage. Small silicon junction FETs (JFETs) can have low g_m, so it is worth choosing the device carefully for the frequency used. GaAsFETs tend to have much higher g_m, especially at high frequencies, and so are recommended for microwave work, with the proviso that they are rather prone to $1/f$ noise. This will be described below.

Turning now to the bipolar device, it can be seen that the input impedance is inherently low. At low frequencies, the input impedance is approximately h_{fe} (the AC current gain) multiplied by the emitter resistance ($R_e + r_e$). This term is comprised of R_e, the $1/g_m$ term as in the FET, and r_e, the ohmic series resistance. This latter can be several ohms, while the former is determined by the current through the transistor; at room temperature, $R_e = 26/I_e$, where I_e is the emitter current in milliamps. Thus at, say, 10mA, the total emitter resistance of a transistor may be an ohm or two, and the input impedance say 50 to 100Ω. This will severely degrade the Q of most parallel-tuned circuits. There is no advantage in using a smaller emitter current, since this leads to lower F_t and hence lower gain; as the frequency approaches F_t, the AC current gain is degraded to unity by definition at F_t. For completeness, the bipolar noise sources are included. In this respect, the best silicon bipolar transistors compare roughly equally with the best silicon FETs. GaAs devices tend to be better again, especially above 1GHz, but the effect of $1/f$ noise has to be considered, so that in an oscillator (but not an amplifier) a silicon device has many advantages right into the microwave region. GaAs devices only predominate because most of them are designed for amplifier service; if sub-0.5 micron geometry, silicon discrete FETs became commercially available, they could become the mainstay of oscillators to 10GHz and beyond.

So what is this $1/f$ noise, and how does noise affect an oscillator? Noise is more familiarly the province of the low-noise amplifier builders, but the principles are the same for oscillators. Where noise comes into the receiver context is that if the signal to be generated is the local oscillator (LO) in a receiver, then most or all of the local oscillator noise is modulated onto the wanted signal at the intermediate frequency (IF). There are several mechanisms for this, including the straightforward modulation of the wanted signal, and the intrusion in the IF of *reciprocal mixing* (see Chapter 6), where a strong but unwanted signal, which is close to the wanted one within the front-end passband, mixes with LO noise. A low-noise oscillator is therefore a major contributor to a low-noise receiver. On transmit, the effects may not be so obvious to the operator, but the transmission of noisy sidebands at potentially high powers is inconvenient to other band users, and may in extreme cases cause transmissions outside the band. In practice, an oscillator with low enough noise for reception is unlikely to be a problem on transmit.

All devices, active and passive, generate noise when not at absolute zero temperature. Inductors and capacitors only do so through their non-ideal resistive terms, so can be neglected in all practical cases. Resistors and transistors (FET and bipolar) are the real sources of noise in the circuit. The noise power generated by a resistor is:

$$\text{Noise power} = 4kTBR$$

where k is Boltzmann's constant, a fundamental constant in the laws of physics; T is the absolute temperature (room temperature is usually approximated to 300K, ie 27°C); B is the measurement bandwidth and R is the resistance value.

This equation is accurate for most resistors, although some older types, notably carbon-composition types, do have additional noise due to the many miniature semiconducting junctions which form and un-form during use. Metal film resistors and modern carbon films are close to the ideal. Active devices are less so, and an *equivalent noise resistance* (R_{IN} in Fig 5.6 and 5.7) can be derived or measured for any active device which approximates the device to a resistor. This may have a value close to the metallic resistance around the circuit, or it may be greater. For many applications, this resistance can be used to estimate the noise contribution of the whole circuit. However, there are other noise sources in the device, most noticeably in the oscillator context those due to $1/f$. Fig 5.1(b) showed $1/f$ noise diagrammatically. Noise sources of this type have been studied for many years in many different types of device. Classically, this noise source increases as frequency is reduced. It was thought that there had to be a turnover somewhere in frequency, but this need not be the case. If we take a DC power supply, then it has a voltage at the output, but it has not always been so. There was a time before the supply was switched on, so the 'noise' is infinite, ie $1/f$ holds good at least in qualitative terms. This gives rise to the concept of noise within a finite bandwidth, usually 1Hz for specification purposes. Practical measurements are made in a sensible bandwidth, say a few kilohertz, and then scaled appropriately to 1Hz. The goodness of an oscillator is therefore measured in noise power per hertz of bandwidth at a specified offset frequency from the carrier. A good crystal oscillator at 10MHz would show a noise level of −130dBc (decibels below carrier) at 10kHz offset. A top-class professional source might be −150dBc. There are several possible causes of $1/f$ noise and, when examining the output of an oscillator closely, there is invariably a region close to the carrier where the noise increases as $1/f$. This is clearly a modulation effect, but where does it come from?

The answer is that almost all (but not all) oscillators are very non-linear circuits. This is inherent in the design. Since it is necessary to have a net gain around the oscillation loop (which is the gain of gain stage(s) divided by the loss of feedback path), then the signal in the oscillator must grow. Suppose the net gain is 2. Then the signal after one pass round the loop is twice what it was, after a further pass four times and so on. After many passes, the amplitude should be 'infinite'. There clearly must be a practical limit. This is usually provided by voltage limitations due to power supply rails, or gain reduction due to overdrive and saturation of the gain stage(s). Occasionally, in a badly designed circuit, component breakdown can occur where breakdown would not be present under DC conditions. This can be disastrous as in a burn-out or, more sinister, as in base-emitter breakdown of bipolar transistors, which will lead to permanent reduction of transistor gain and eventual device failure in service.

So, almost all practical oscillators are highly non-linear and have sources of noise, be they wide-band (*white noise*) or $1/f$ (*pink noise*). The non-linearity thus modulates the noise onto the output frequency. In contrast, a properly designed amplifier is not non-linear, has very little modulation process, and is therefore only subject to additive noise.

The exceptions to this are the very few oscillators which have a soft limitation in the output amplitudes, ie they are inherently sine wave producers. This requires very fast-acting AGC circuitry, and is very rarely used because it adds much cost and complication. Most practical oscillators are inherently square-wave generators but a tuned circuit in the output will normally remove most harmonic energy so that the output looks like a sine wave on an oscilloscope. A spectrum analyser normally gives the game away, with harmonics clearly visible. The DDS devices are an exception to all this for reasons explained below.

Generally, a silicon device will have very much lower levels

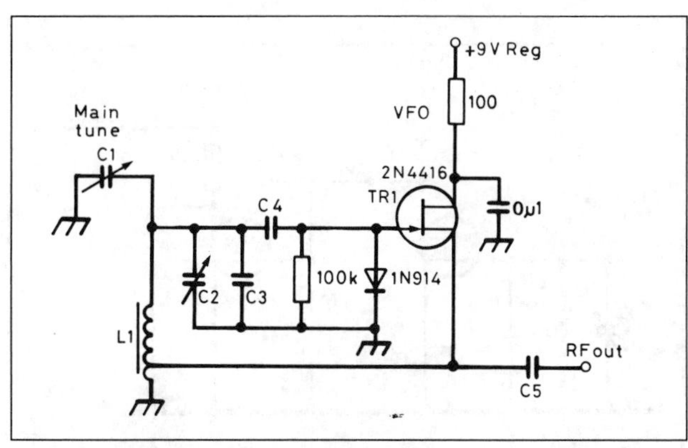

Fig 5.8. Hartley VFO *(W1FB's QRP Notebook)*

of 1/*f* noise than a gallium arsenide device. This is because silicon is a very homogeneous material, of very high purity, and in modern processes with very little surface contamination or *surface states*. Gallium arsenide is a heterogeneous material which is subject to surface states. The *1/f knee* is the frequency below which the noise levels associated with the surface states start to increase, typically at 6dB/octave of frequency. In, say, 1980 this frequency would have been 100MHz or more for most gallium arsenide processes. In 1990 the figure was as low as 5MHz in some processes. Silicon typically has a 1/*f* knee of less than 100Hz, sometimes less than 10Hz, so 1/*f* noise is unimportant for most practical cases. For gallium arsenide, however, that does mean that there is a possibility of increased close-to-carrier noise in GaAs-based oscillators. Whether this makes the oscillator better or worse than a silicon design depends on the exact choice of devices and circuits. More information on these topics can be found in reference [5].

VARIABLE FREQUENCY OSCILLATORS (VFOs)

The science of VFO design has been described many times in the amateur literature. The design should start with a set of clear objectives, such as the required frequency range, the exact means of tuning, and the requirement for stability. In VFO design, as distinct from voltage-controlled oscillator (VCO) design which will be described below, the tuning is carried out by a passive component such as a mechanically variable capacitor, or occasionally by a variable inductor. The most usual requirement is for an oscillator having SSB quality stability over a reasonable period of time, ie the frequency should not shift by more than 100Hz in an hour, and should always be repeatable to within 300Hz or so. The usual frequency is 5MHz or so, with typically 500kHz frequency shift available. The major application is in tuning of SSB receivers or transceivers where a synthesiser is not used. The VFO is used either in a second IF conversion or in an analogue phase-locked loop. Alternatively, VFOs are used in single-conversion receivers with free-running oscillators. This puts a great deal of emphasis on VFO stability, and a really stable VFO for 30MHz SSB operation is very difficult to achieve. VFOs for operation at, or multiplication to, VHF frequencies are not practical for transmission, although they may be adequate for FM reception if carefully designed.

Figs 5.8 to 5.10 show a range of practical VFO circuits. The circuit in Fig 5.8 is a conventional Hartley oscillator as outlined above. The feedback tap should be arranged to be about 0.25 of the coil from the earthy end. The main tuning capacitor should be mechanically very robust and mounted in such a way as to

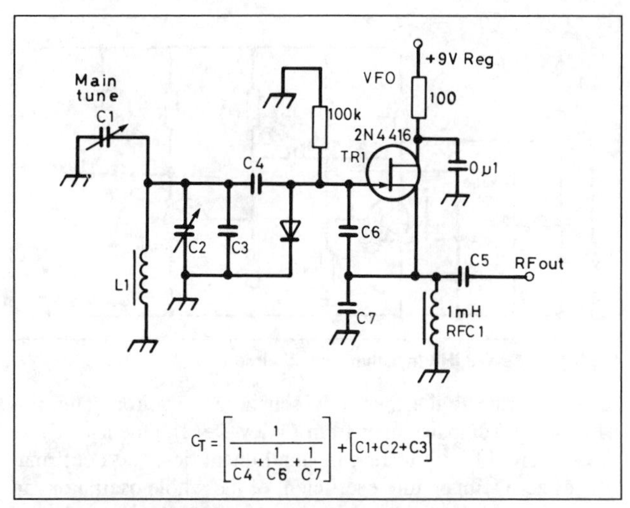

$$C_T = \left[\frac{1}{\frac{1}{C4}+\frac{1}{C6}+\frac{1}{C7}}\right] + \left[C1+C2+C3\right]$$

Fig 5.9. Colpitts VFO *(W1FB's QRP Notebook)*

avoid the possibility of vibrations. All fixed capacitors should be silver mica types; ceramics are notorious for changing their values arbitrarily. Very-small-value ceramics are better in this respect, but should be avoided where possible. Although a bipolar transistor could be used in this circuit with bias modification, it will tend to reduce circuit *Q* and hence be less stable. The FET should be chosen for low noise and, although JFETs are usually preferred, this circuit will work well with modern dual-gate MOSFETs such as the BF981. The second gate should be connected to a bias point at about 4V.

Fig 5.9 shows the Colpitts version of the circuit. The additional input capacitance restricts the available tuning range but this is rarely important; it is more relevant to avoid voltage variable capacitance in the device. Again, a dual-gate MOSFET version is possible. A varactor-tuned version of the Hartley is shown in Fig 5.10. Many other variants on these basic circuits have been published over the years; some examples are shown in Figs 5.11 and 5.12. Other classes of oscillator have been suggested, most notably the design due to P G Martin, G3PDM (Fig 5.13) and described more fully in reference [6]. In this design, which includes a buffer stage applicable to most of the oscillators described in this section, extreme precautions were taken to reduce drift. Many of these are simply applied to any oscillator, such as building the oscillator in a separate enclosure within the main unit, careful decoupling and stabilisation of power supplies, ensuring cleanliness of all components and the use of silver-plated copper wire for the tank coil. More difficult

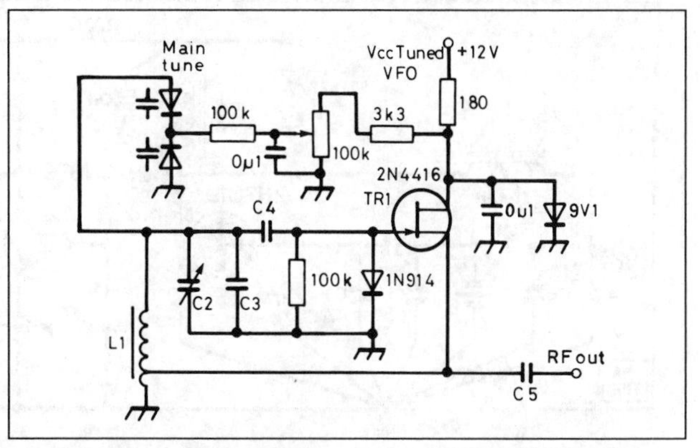

Fig 5.10. Varactor-tuned Hartley VFO. Suggested values for reactances are: X_{C1} 1200Ω; X_{C2} 3500Ω; X_{C3}, X_{C4}, X_{C5} 880Ω; X_{C6}, X_{C7} 90Ω; X_{L1} 50Ω *(W1FB's QRP Notebook)*

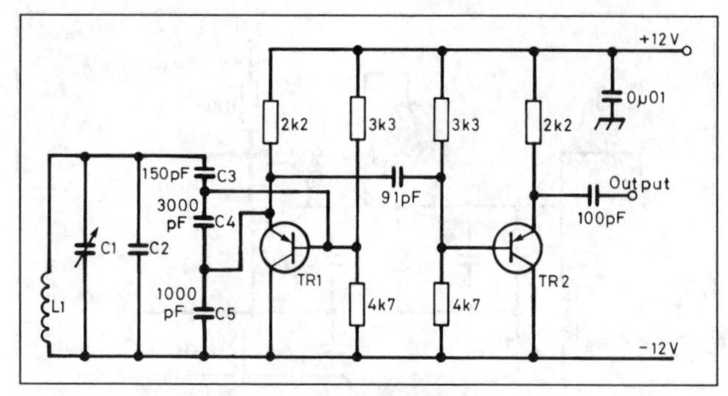

Fig 5.11. The W3JHR 'synthetic rock' circuit

to incorporate is the thermally sensitive capacitor (Thermotrimmer or Tempatrimmer from Oxley Developments). These devices are difficult to find in small quantities, and confirmation of zero temperature coefficient of the whole oscillator can be a time-consuming business. Again, an updated version of this circuit would probably use a dual-gate MOSFET, which should offer marginally better performance. The buffer stage could also use more easily obtainable devices such as the 2N3904 or the BF199.

One further class of oscillator, relatively rarely used, but having some advantages in special cases, is the *Franklin*. This is

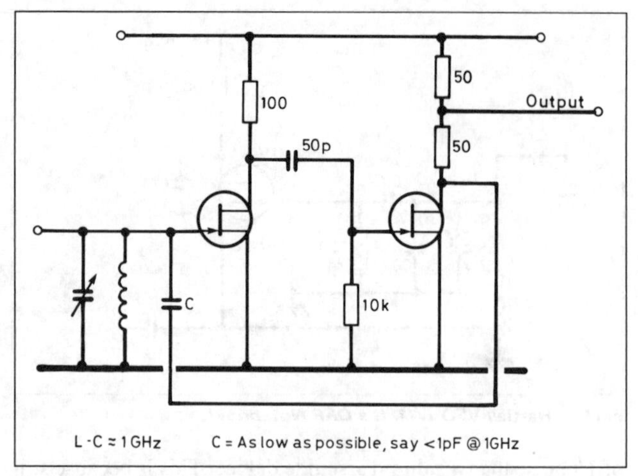

Fig 5.14. The Franklin oscillator

shown in Fig 5.14. In this example, the active devices are GaAs MESFETs. As mentioned above, the MESFET tends to have relatively low gain per stage, so there are some advantages in using two gain stages in series to provide positive gain over a very wide frequency range. The devices are shown running with self-bias on the gates. Ideally, they should be run close to I_{dss} where the gain is usually greatest. The availability of GaAsFETs is still rather variable, so no specific types are suggested. Surplus 'red spot' and 'black spot' devices have been successfully tried in this circuit. The feedback path is arranged to give 360° phase shift, ie two device inversions. The feedback is then directed into the tuned circuit. Although the output impedance is relatively low, it is prevented from damping the resonant circuit by very light coupling. The high gain ensures oscillation even in this lightly coupled state, so that very high circuit *Q* is maintained, and the oscillation is stable and relatively noise-free. A particular feature of this configuration is the insensitivity of the circuit to supply variation; but the supply should still be well decoupled at the operating frequency. A disadvantage of the circuit is the total delay through two stages, which limits the upper frequency, but with gallium arsenide devices this can still yield 5GHz oscillators.

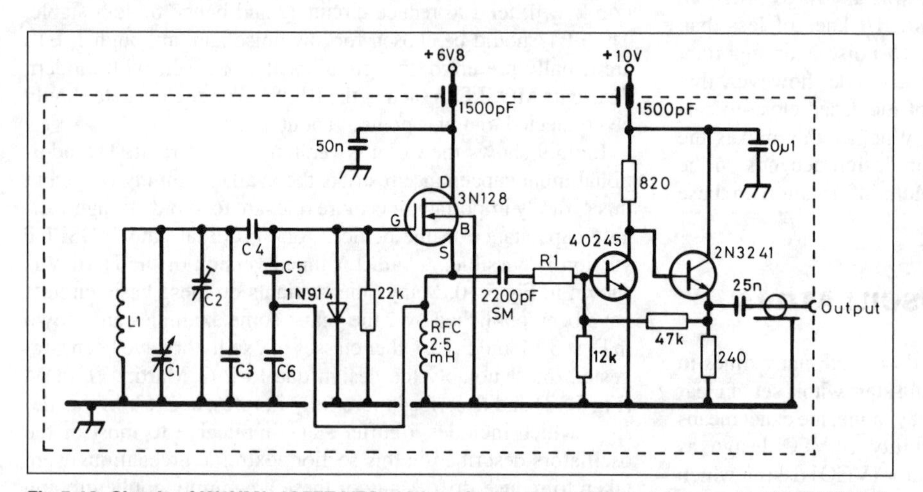

Fig 5.12. Circuit of W2YM's IGFET VFO. Values for the 3.5–4MHz version are L1, 17t, 20 B & S, 16tpi, 1in diam; C1, 100pF; C2, 25pF, C3, 100pF silver mica; C4, 390pF sm; C5, C6, 680pF sm; R1, between 12kΩ and 47kΩ selected for 2V peak output level at input to transmitter. More readily available transistors are BF981 (both gates strapped together) and 2N3904

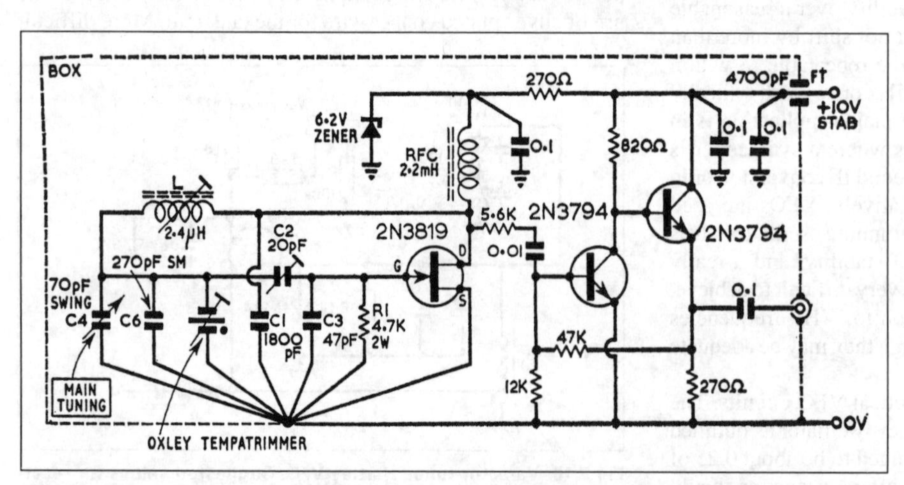

Fig 5.13. The high-stability FET Vackar oscillator developed by G3PDM to cover 5.88 to 6.93MHz for the Mark 2 version of his *Radio Communication Handbook* (5th edn) receiver

CRYSTAL OSCILLATOR CIRCUITS

This section describes a range of fixed-frequency, as distinct from VXO, crystal oscillators with practical circuit details. Most circuits shown will benefit from the addition of a buffer amplifier to avoid loading of the circuit, and some circuits for buffers are shown in this and the section on VFOs.

Many crystal oscillator circuits have been described in the past, each with some claimed advantage over all the others. Actually, crystal oscillator circuits are not hard to build, and most reasonable designs work fairly well since the crystal itself provides extremely high Q, and in such a way that it is difficult to degrade the Q without deliberately setting out to do so.

Fundamentally, as has been shown above, an oscillator requires positive feedback to oscillate. A signal must pass from the output of the amplifier stage to its input, with sufficient amplitude and suitable phase shift that oscillations are maintained. Most commonly, a single active device is used with a 180° phase shift, so that a further 180° is needed in the tuned circuit. This is usually easily achieved. It is also usually obvious which sense is needed in the feedback circuit, since in COs this is usually capacitive. If an inductive feedback arrangement is used, then it is possible to get the phase wrong; in this case, reverse the sense of the feedback winding.

The amplifier circuit needs to have sufficient gain at the operating frequency to overcome losses in the feedback path. For fundamental-mode crystals, any small HF transistor is adequate, even a BC109 type, but for overtones, more care in choice is needed, so a BF series is recommended, eg BF199. Alternatives in Jedec 2N types are 2N3904 or 2N918.

Excessive feedback should be avoided, primarily because it tends to reduce stability. Excessive oscillation levels at the crystal can lead to damage, but this is very unlikely in 12V operated circuits. Sufficient feedback is needed to guarantee starting of the oscillator; some experimentation may be needed, especially if the crystal is an old type. The old series of surplus FT243 types cannot really be recommended now, since it is so many years since their manufacture. The best results will be obtained with modern crystal types in HC6/U, HC18/U, HC25/U or similar holders. All of these should give good results.

Most crystals come in sealed metal cans. These really are 'no user serviceable parts inside', but it is interesting to open up an old one carefully to examine the construction. The crystal plate is held in fairly delicate metal fingers, so shocks to the crystal should be avoided. The higher-frequency crystals are very delicate indeed; until recently, fundamental modes were limited to 20MHz. Very recently, fundamental mode crystals have been offered on the professional market to 250MHz. These use expensive ion beam milling techniques to produce local areas of thinning in otherwise relatively robust crystal blanks, so maintaining some strength while achieving previously unattainable frequencies.

Harmonic modes of crystal operation are available from 20MHz to about 70MHz on third overtones and to 130MHz on fifth overtones. Crystals have been made for higher overtones, but circuit design and tolerancing is impractical; even fifth overtone at 116MHz can be difficult. A better approach for high harmonics is a separate harmonic generator.

Crystals made for commercial use are similar in specification to amateur devices; they may be additionally aged or more commonly ground slightly more precisely to the centre frequency. For most purposes this is not significant unless multiplication into the microwave region is required. Similarly, crystals are sometimes seen in glass packaging. This can mean a higher degree of hermeticity, but again for amateur purposes this is not significant. Occasionally, crystals are specified at a temperature other than room temperature (see Fig 5.15). Ideally, the crystal should have a zero temperature coefficient around the operating temperature, so that temperature variations in the equipment or surroundings have little effect. In some cases, the crystal is placed in a crystal oven at 70°C. It is important to note, however, that

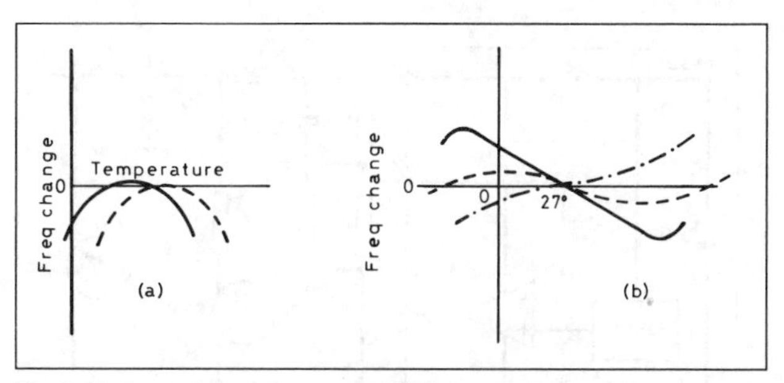

Fig 5.15. Crystal temperature coefficients. Frequency/temperature curves of 'zero-temperature coefficient' crystals. (a) Typical BT-cut crystals. (b) Typical AT-cut crystals

the best result will only be obtained if the crystal is cut for the oven temperature. Although any temperature stabilisation is useful, a room-temperature-cut crystal may have little advantage in an oven. Circuits for thermal stabilisation of crystals are shown in Figs 5.16 and 5.17.

Figs 5.18 to 5.21 show four basic crystal oscillator circuits. The first (Fig 5.18) is a FET-based *Pierce oscillator*, with the crystal in the feedback path from drain to gate. The only frequency setting element is the crystal. For this to be the case, the RFC must tune with the circuit parasitic capacitances to a frequency lower than the crystal. Most small JFETs will work satisfactorily in this fundamental-mode circuit. Note that the FET source is grounded for maximum gain. This also implies that a depletion-mode FET is essential, ie a type that passes current at zero gate-source voltage. This is true of most JFETs. MOS devices could be used with a forward bias on the gate, or GaAsFETs with a small reverse bias. However, in most cases the JFET will show the best noise performance and for probably the lowest cost.

Fig 5.19 shows an overtone version of basically the same

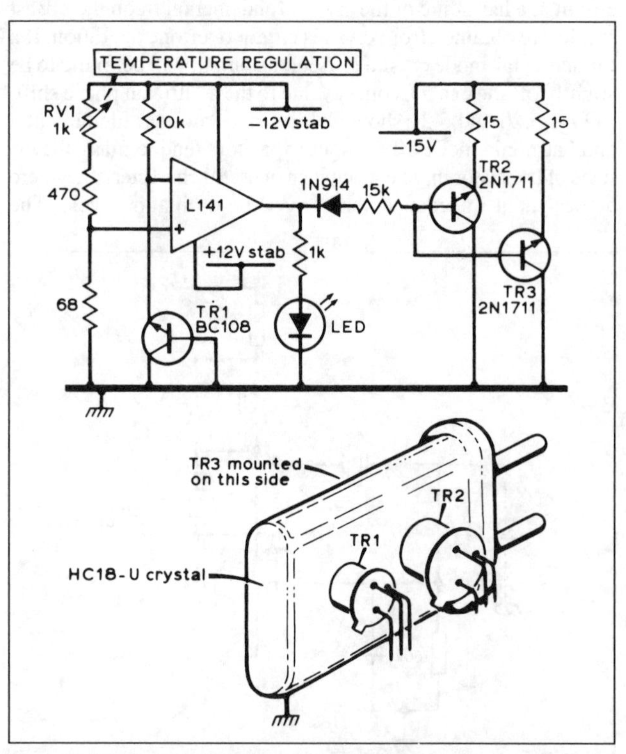

Fig 5.16. Proportional temperature control system for HC/16U crystals as proposed by I6MCF and using a pair of transistors as the heating elements and a BC108 transistor as the temperature sensor

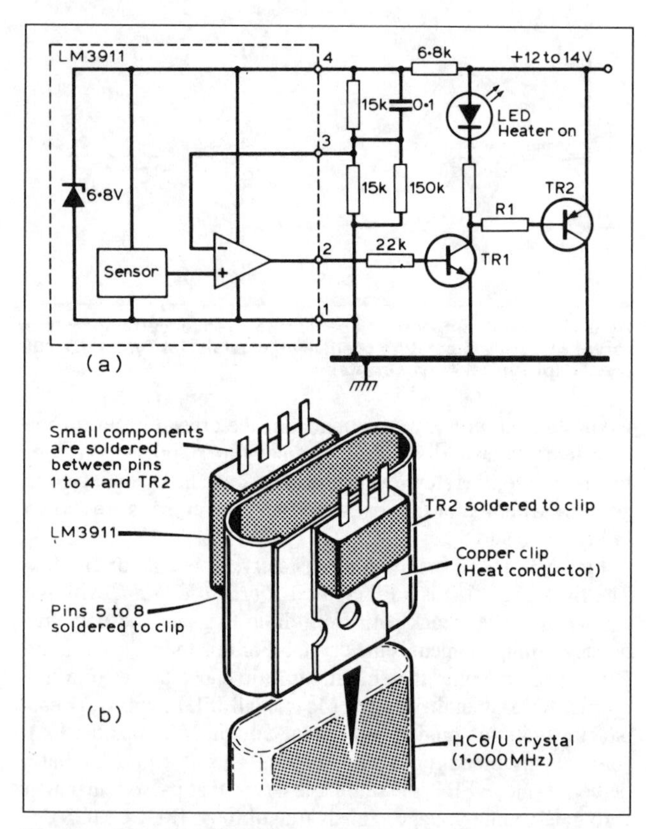

Fig 5.17. Circuit diagram (a) and mechanical details (b) of the temperature control system used by G3SEK, based on the LM3911 IC. Collector of TR2 and pins 5 to 8 of LM3911 device are soldered to the copper mounting clip. TR1 can be almost any NPN silicon transistor; TR2 is a tab-mount PNP silicon power transistor

circuit. In this case, feedback is only permitted by the tuned circuit at a harmonic of the crystal fundamental frequency. Best results are obtained from crystals cut for overtone operation. If a fundamental-mode crystal is used, the output will be found to be slightly off the exact frequency due to the additional phase shift.

Figs 5.20 and 5.21 show Colpitts oscillators in fundamental and harmonic modes. The input capacitors tend to mask the effects of transistor input capacitance; a useful characteristic where device input impedance varies over the oscillation cycle. The

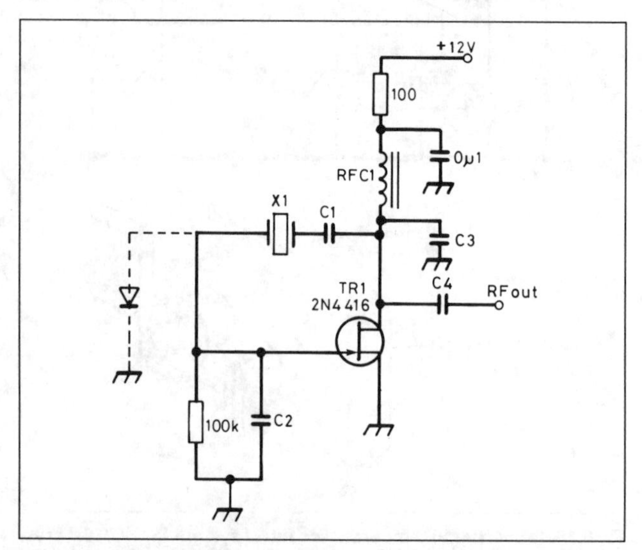

Fig 5.18. Pierce oscillator. Reactance values are: X_{C1}, X_{C3} 230Ω; X_{C2}, X_{C4} 450Ω. *(W1FB's QRP Notebook)*

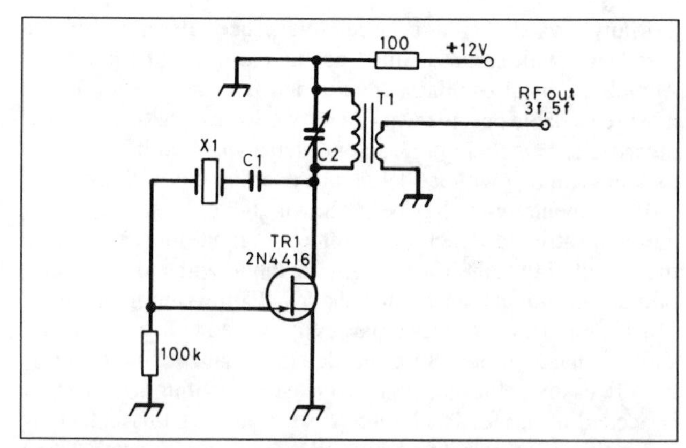

Fig 5.19. Overtone oscillator. Reactance values are: X_{C1} 22Ω, X_{C2} 150Ω at resonance. *(W1FB's QRP Notebook)*

chief differences in these circuits are the use of bipolar transistors with the different biasing arrangements shown, and the capacitive transformer feedback at the crystal. While the feedback voltage may appear to be slightly less than unity, this is multiplied by the Q of the tuned circuit, ensuring oscillation. An advantage of this configuration is that the crystal is grounded at one end, so facilitating switching between crystals, which is much less easily achieved in the Pierce circuits. Crystal switching of overtone oscillators can be achieved, but is not recommended, since it can lead to loss of stability or even *moding* of the crystal, ie some crystals can oscillate at an unintended harmonic, or even another frequency altogether.

The buffer amplifiers of Figs 5.22 and 5.23 are also shown to indicate that buffering is usually needed. The output stages of Fig 5.13 could also be used.

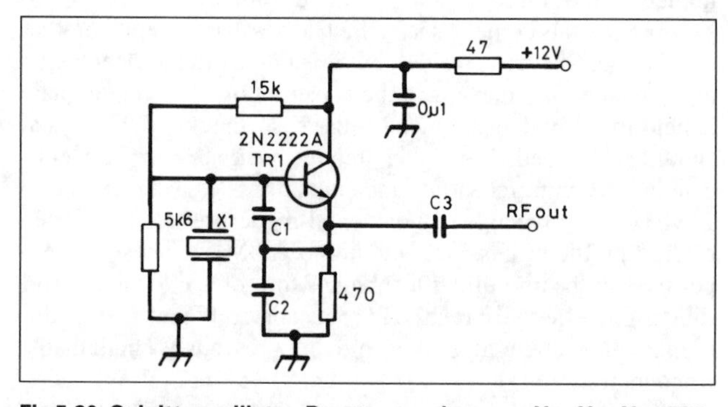

Fig 5.20. Colpitts oscillator. Reactance values are: X_{C1}, X_{C2}, X_{C3} 450Ω. *(W1FB's QRP Notebook)*

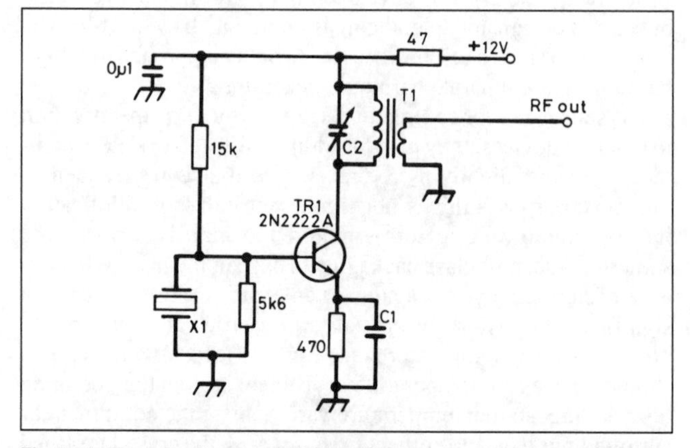

Fig 5.21. Tuned collector oscillator. Reactance values are: X_{C1} 225Ω; X_{C2} 150Ω. *(W1FB's QRP Notebook)*

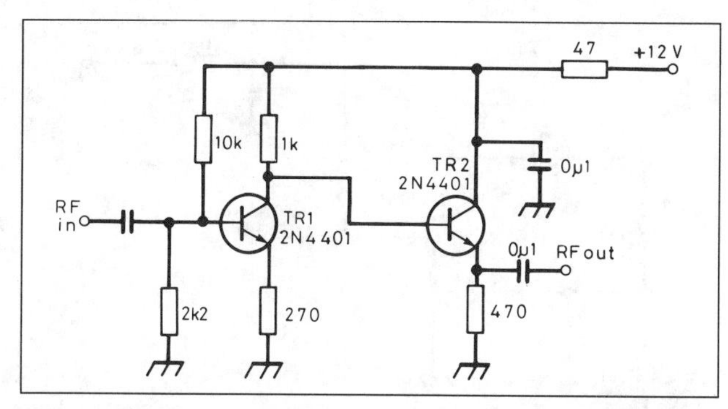

Fig 5.22. Direct-coupled buffer circuit. *(W1FB's QRP Notebook)*

Relatively few circuits for integrated circuit oscillators have been described. One such is shown in Fig 5.24. This uses a standard CMOS gate. Almost any of the CMOS logic families will work in this circuit. The primary frequency limitation is the crystal. This oscillator will generate a very square output waveform, and with the faster CMOS devices can be a rich source of harmonics to be used, for example, as marker frequencies. Of course, a square wave contains only odd harmonics.

Variable crystal oscillators (VXOs)

The VXO has been called 'the poor man's synthesiser', with some justification. Where there is a need for a very stable oscillator with only a small tuning range, the VXO is a candidate. The principle can be seen in Fig 5.25, where the equivalent circuit of a crystal is shown with a series inductance and capacitance to allow tuning. The whole circuit has a resonant frequency which can be varied by change of the inductive reactance in series with the crystal; for convenience, tuning is usually carried out with a variable capacitor. All the usual precautions in VFO design should be used, but VXOs are less sensitive since they contain one very high *Q* element, the crystal. The two problems of VXOs are the very limited tuning range, typically of 0.1–0.2%, and the tendency to instability if this range is exceeded. A good VXO should be very close to crystal stability, with the added ability to change frequency just a little. Of course, multiplication to higher frequencies is possible, and several of the older types of commercial equipment (ICOM IC-202, Mizuho) successfully used VXOs on the 2m band. Several VXO circuits are shown in Figs 5.26 to 5.30. As before, FETs are often the best choice of active device: either the JFETs shown, or MOSFETs such as the BF981, BF982 or BF960. Similar Japanese types are the 3SK88; US types such as the 40673 may also be used.

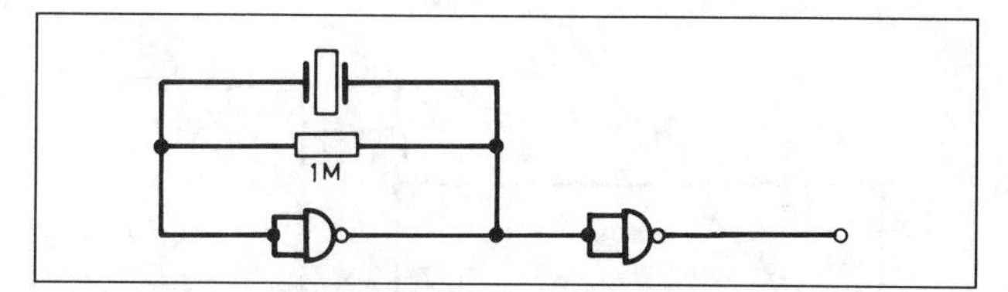

Fig 5.24. CMOS oscillator circuit. With modern 'HC' CMOS, this will oscillate to over 20MHz with a good square-wave output

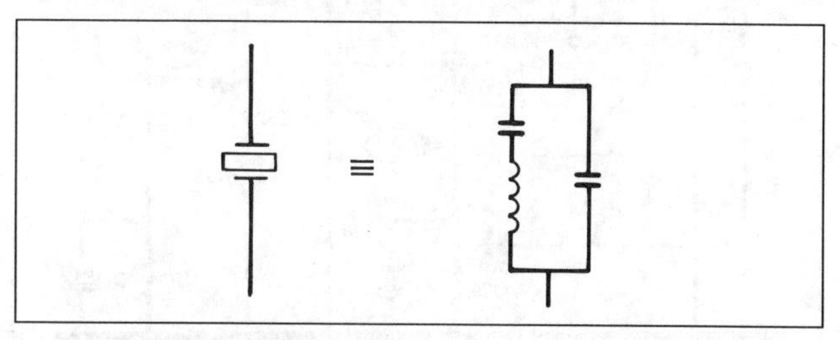

Fig 5.25. Crystal equivalent circuit

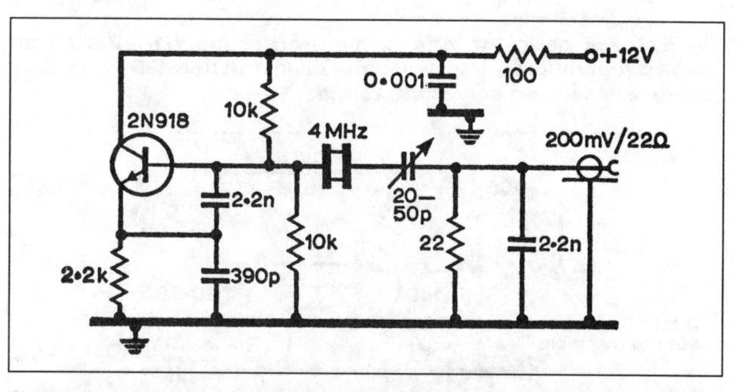

Fig 5.26. Low noise and low harmonic output is claimed by DJ2LR for this crystal oscillator in which the crystal not only acts as the frequency-controlling element but also forms a low-pass filter. High spectral purity is particularly impertant for such applications as the master oscillator of a frequency synthesiser or for reducing reciprocal mixing in high-performance receivers

Excessive stray capacitance should be avoided in the circuit. Ideally, the VXO should be able to shift from below to just slightly above the nominal crystal frequency. This will be easier with the circuit of Fig 5.30 than Fig 5.29, since the feedback capacitors limit the minimum capacitance. Choice of crystal is also important. Fundamental-mode crystals can be used, but overtone types will usually be better in pulling range since they have smaller plated areas on the crystal and therefore less capacitance. Some crystals are made specifically for VXO work, although these are rarely available to the amateur. If a crystal is to be ground for a specific frequency, it is safest to specify the operating frequency as a fundamental, although the requirement for VXO operation could also be specified. All VXOs should operate from a well-regulated and decoupled supply, and a separate box is very desirable. Buffering should also be provided. Crystal switching in VXOs should be avoided, although the commercial rigs referred to above did employ switching with 14MHz crystals. Non-linear operation of capacitance against frequency should also be expected, although some linearisation is possible using combinations of series and parallel trimmer capacitors with the crystal. A more modern approach is to fit a low-cost LCD frequency readout, thus avoiding the need for a linear scale.

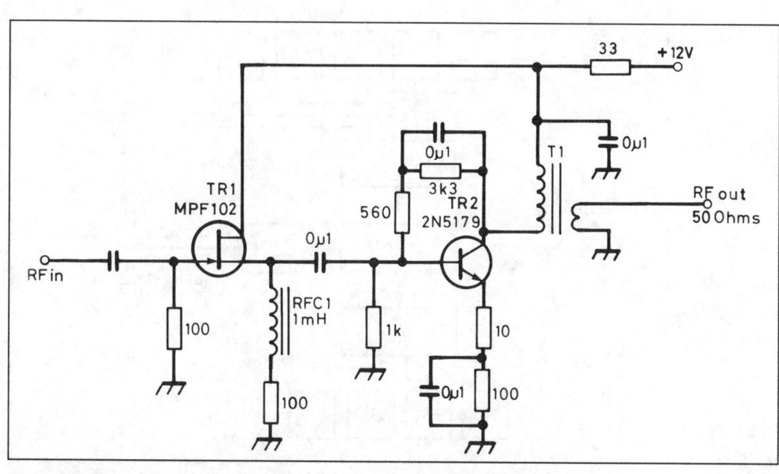

Fig 5.23. Transformer output buffer circuit. *(W1FB's QRP Notebook)*

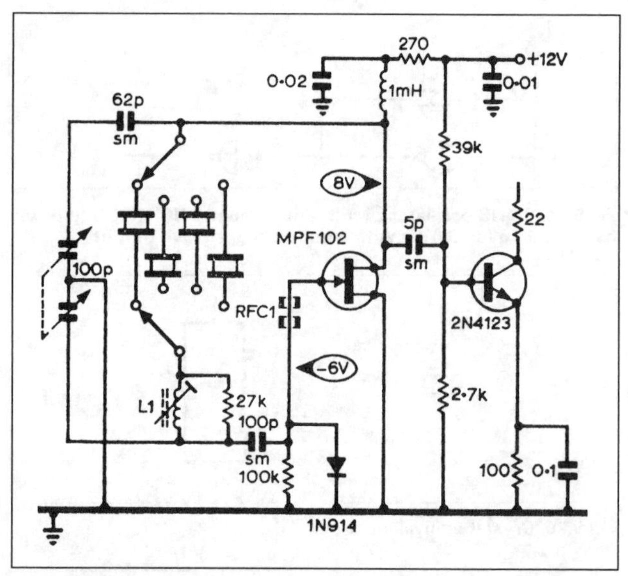

Fig 5.27. FET oscillator forming the heart of the W1CER VXO for 144MHz operation. L1 of medium-Q construction (38–85μH slug tuned). Subsequent stages use toroids

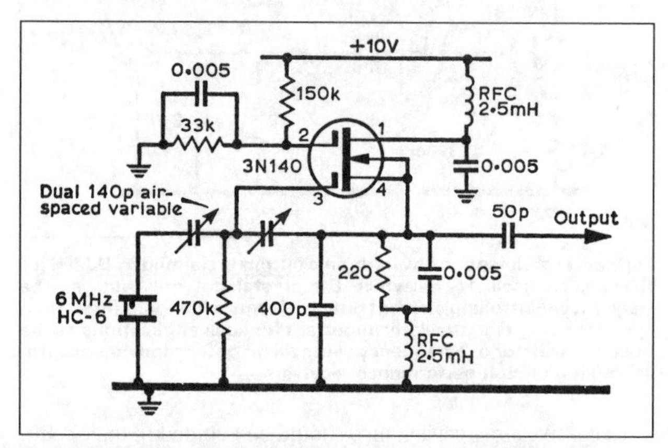

Fig 5.28. A capacitive VXO that is claimed to achieve up to 5kHz deviation with a standard 6MHz AT-cut HC6 crystal using a dual 140pF capacitor (eg Hammarlund MCD-140-D). Layout should keep stray capacitances affecting crystal as low as possible. Other types of dual-gate MOSFET could be used

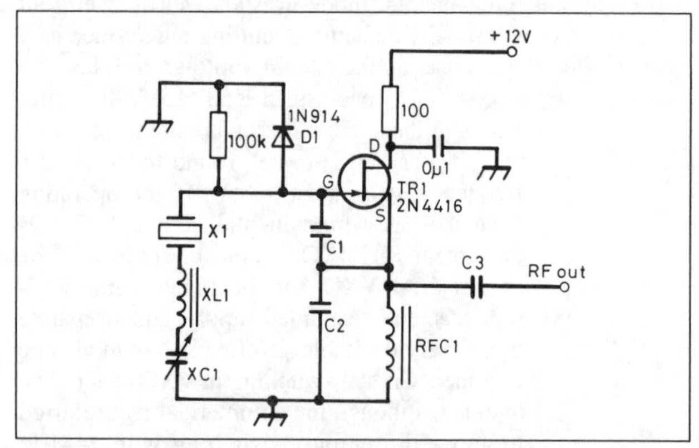

Fig 5.29. Colpitts VXO. $L_{\mu H} = X_L/(2\pi f)$. Reactance values are: X_{C1} 110Ω; X_{L1} 1.3Ω; C1, C2, C3 same as Fig 5.9; RFC1 (X_L) = 12.5kΩ

FREQUENCY SYNTHESIS

Phase-locked loop frequency synthesisers

This section and the subsequent one will address the basics of frequency synthesiser design. The split into two sections

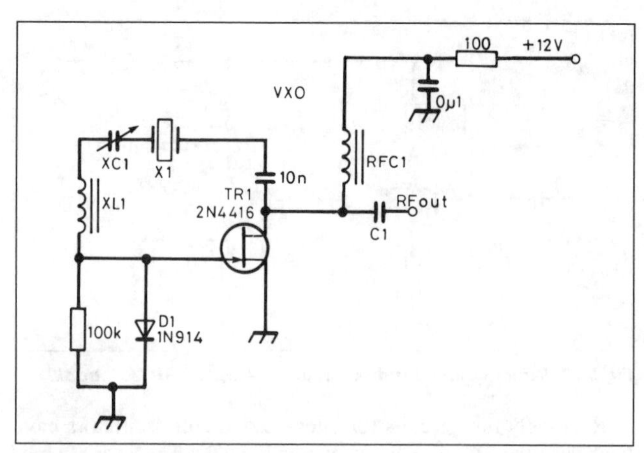

Fig 5.30. Pierce VXO. Reactance values are: X_{C1} 110Ω; X_{L1} 1.3Ω; C1 (X_C) 450Ω; RFC1 (X_L) 12.5kΩ

distinguishes classical phase-locked loop (PLL) synthesisers from the newer direct digital synthesisers (DDS). However, since the history of synthesis has common aspects from both areas, it is discussed first in order to give a perspective to the design of frequency synthesisers in general.

Direct synthesis (but not direct digital synthesis) has been known for many years, and simply means the production of an output frequency, either a transmitter output or a local oscillator for a receiver, by the addition of several (sometimes many) oscillators, usually by a combination of mixing, multiplying, re-mixing etc with many stages of filtering. This was a truly analogue technique and expensive to carry out really well. Examples were primarily limited to specialist, often military, synthesisers in the 'fifties. The output spectrum could be made very clean, especially in respect of close-to-carrier phase noise, but only by significant amounts of filtering. Direct synthesisers are still being designed in specialist areas, because they can give very fast frequency hopping and low spurious levels well into the microwave region.

A simple version did go into quantity production, in the early US-market CB radios. This was the *crystal bank synthesiser* shown in Fig 5.31. This arrangement used two arrays of crystals, selected in appropriate combinations to give the required coverage and channel spacing. It worked very well over a restricted frequency range, could offer very low phase noise since crystal oscillators are inherently 'quiet', and could be made acceptably (for the time) compact and low power. A familiar

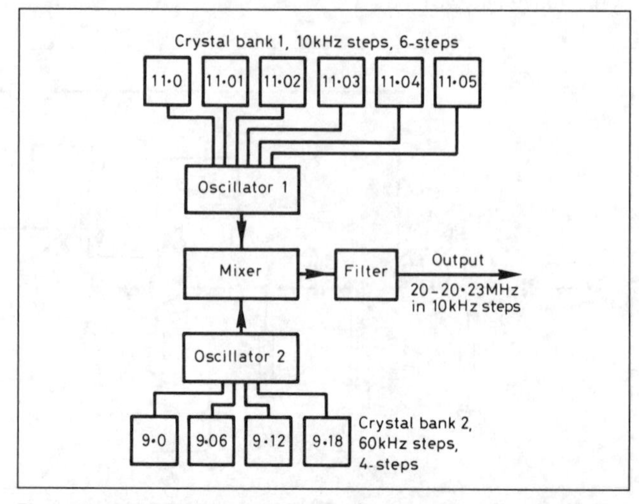

Fig 5.31. Simplified crystal bank synthesiser

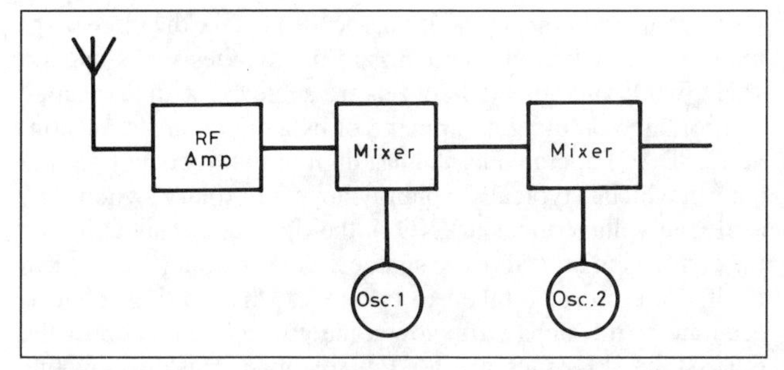

Fig 5.32. Multiple oscillators/mixers, an alternative to frequency synthesis

example on the UK market some 20 years ago was the very well known Liner 2. This used crystal bank synthesis [7] with a front-panel-tuned VXO on a further crystal oscillator for the conversion up and down from the effective 'tuneable IF' at 30MHz to the working frequency of 145MHz. The need for the VXO illustrates the weakness of this type of synthesiser; close channel spacing is essentially prohibited by the very large number of crystals which would be needed. Triple bank versions were described in the professional literature, but were not produced for amateur purposes. The other disadvantage of this class of synthesiser is the high cost of the crystals, but it remains an interesting technique capable of very high performance with reasonable design. The well-documented problems of the Liner Two were not connected with the synthesiser but were in a potentially overdriven mixer.

A popular alternative to synthesis for many years used a carefully engineered VFO and multiple mixer format. Examples are in the KW range, the Yaesu/Sommerkamp FT-DX series [8] and, in homebrew form, the G2DAF (see earlier editions of this handbook). This technique survived in solid-state form, with among others, the FT101 [9]. The technique is illustrated in Fig 5.32. However, in parallel with these came the early attempts at combining the best VFO characteristics with synthesis in order to build a single-conversion receiver, with the objective of achieving exceptionally wide dynamic range. An example of this was G3PDM's classic phase-locked oscillator, as used in his receiver and described in earlier editions of this handbook [6]. This was a true synthesiser, albeit analogue, based primarily on valves, with a VFO using an FET and two bipolar transistors which is still regarded as the standard to beat. This circuit is shown in Fig 5.13. The author was present at a demonstration of this receiver where a signal below 1μV suffered no apparent degradation from a 10V signal 50kHz away. This amazing >140dB dynamic range was made possible by the extremely low phase noise of this synthesiser, and of course the superb linearity of a beam-deflection mixer run with 100V push-pull local oscillator drive! About this time (1969), the first truly digital (but not direct) synthesisers were beginning to appear, chiefly in military equipment. The outstanding example is the UK-sourced Clansman series of radios, still in service with UK forces at the time of going to press.

A basic PLL digital synthesiser is shown in Fig 5.33. Actually, this is very much simplified. The VCO output is divided in a programmable divider down to a frequency equal to that of the crystal-controlled reference source; this may be at the crystal frequency or more usually at some fraction of it. The phase comparator compares the two frequencies, and then produces a 'DC' voltage which is proportional to the difference in phase. This is applied to the VCO in such a way as to drive it towards the wanted frequency. When the two frequencies at the phase comparator input are identical, the synthesiser is said to be *in lock*.

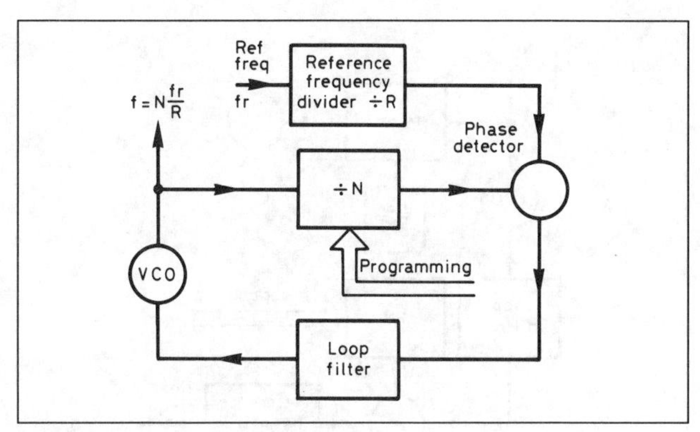

Fig 5.33. Basic PLL synthesiser

Ideally, the phase comparator is actually a phase and frequency comparator, so that it can bring the signals into lock from well away. The 'DC' term referred to is a varying voltage determined by the frequency and phase errors. The bandwidth available to this voltage is said to be the *loop bandwidth* and it determines the speed at which lock can be achieved. There is a (usually) wider bandwidth called the *lock-in* bandwidth, which is the range of starting frequencies from which lock can be achieved.

Fully programmable dividers of appropriate frequency range were not immediately available to the early PLL designers. Two solutions were proposed for this, the *dual modulus prescalers* and a variant of the crystal mix scheme. The dual modulus counter is a particularly clever use of circuit tricks to achieve the objective of high-speed variable ratio division. It was found that dividers could be built which could be switched very rapidly between two different division ratios, for example 10 and 11. If the divider is allowed to divide by 11 until the first programmable counter (the A-counter in Fig 5.34) reaches a preset count (say, A) and then divides by 10 until the second counter reaches M, the division ratio is:

$$(11 \times A) + 10 \times (M - A) \text{ or } (10M + A)$$

The advantage in the system is that the fully programmable counters M and A need only respond, in this example, to one-tenth of the input frequency, and are therefore easier to manufacture. The disadvantage is that there is a minimum available count of $A \times (11)$, where A is the largest count possible in the A-counter. This is rarely a problem in practice.

The scheme of a dual modulus based PLL is shown in Fig 5.34. It consists of a VCO, usually in discrete component form,

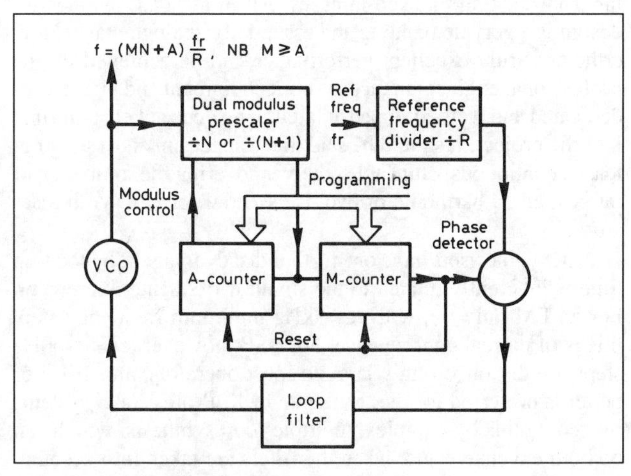

Fig 5.34. Dual modulus prescaler

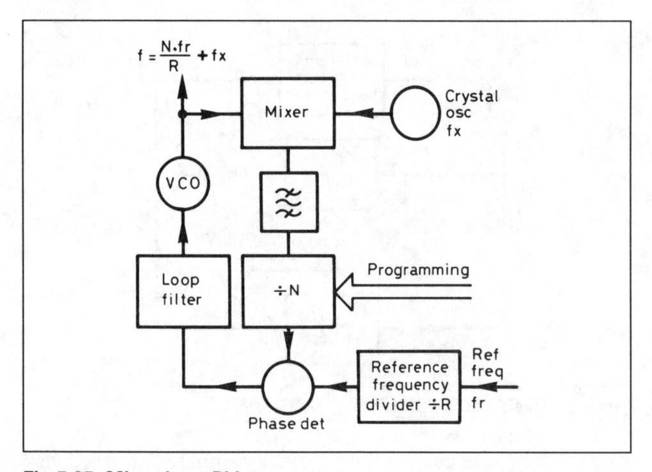

Fig 5.35. Mixer-loop PLL

the dual modulus counter, a main counter, a reference oscillator and divider chain, a phase detector and a loop amplifier/ filter. This type of synthesiser has been described many times, both in the professional and amateur fields [10]. In some variants, it probably represents the current status of PLL frequency synthesis, especially for very-wide frequency range applications such as TV receivers and satellite tuners. The second method, that of crystal mixing, is probably the most widely used technique in amateur and PMR equipment, where wide operating frequency range is not needed. It offers advantages of simplicity and low component count, especially when the digital functions of the system can be contained in a single chip. The scheme is shown in Fig 5.35. Unlike the dual modulus scheme, no prescaler is needed, but a high-frequency mixer is necessary, usually with two crystals, one for mixing down and one for the reference.

Almost all the commercially available PLL synthesiser chips use an external VCO. Despite the stabilisation offered by the PLL, it is essential to build a very-good-quality VCO, since VCO phase noise and jitter are potential sources of reciprocal mixing in a receiver. While the lock loop can take care of phase noise close to the carrier, a badly designed circuit or a badly laid out PCB can cause a synthesiser to 'feature' sidebands on the output spectrum at the comparison frequency and sometimes harmonics of it. The VCO dominates the noise away from the nominal frequency, ie outside the PLL bandwidth, which may be only a kilohertz or so, less in the case of very narrow frequency increments. Phase detector design is another critical area; the most recent PLL chips offer two or more phase detectors to cover the far-from-lock and locked-in cases. All in all, PLL synthesiser design is a very demanding and essentially analogue task. Nevertheless, truly excellent performance can be achieved by the professional engineer in factory-built equipment and by the very dedicated and well-equipped amateur who does not cost his time into the project. It should be said that a spectrum analyser, or at least a continuously tuneable receiver covering the frequency of interest and a harmonic or two, is essential for any synthesiser work.

PLL synthesisers have one further disadvantage. The lock-up time is inherently related to the smallest frequency increment. For an FM-only rig, with say 5kHz minimum frequency step, this is of no real consequence. For SSB, however, even 100Hz steps are disconcertingly large to most operators, and 10Hz or better is preferred for the 'analogue feel'. Professional systems overcome this by complex, multiple-loop synthesis, which can be both expensive and, when the filters are taken into account, bulky.

Amateur equipment practice has tended to take the more pragmatic route. Minimum step size in the synthesiser is usually 1kHz, while increments between are achieved with a separate control knob on older equipment, or by a digital analogue converter (DAC) operated by the last digit of the frequency setting control which is typically a photo-chopper or rotary switch. The analogue voltage tunes a VXO in the rig, which may either be the conversion crystal in the synthesiser or the reference crystal itself. Care must be taken to ensure that the pulling range is accurate or the steps will show a jump in one direction or the other at the 1kHz increments. Actually, many rigs do show this if observed carefully; the reason is that until recently the DAC was a fairly simple affair consisting of a resistor array and switching transistors. More recently, commercial DACs have been used, but this is not a complete solution, since the VXO is unlikely to be linear to the required degree, so some step non-linearity is inevitable. At least if the end points are not seriously wrong, this should not be a problem. Some rigs do tune in 10Hz steps, especially on HF; this is an extension of the technique to 100 steps instead of just 10. Again, there is the issue of the 1kHz crossover points; but with 10Hz steps this can be made less noticeable on a well-designed and adjusted rig.

Properly designed, PLL synthesis works very well indeed, and most commercial amateur equipment uses some form of PLL synthesiser. However, the problems of providing narrow frequency increments, with the demand for complete HF band coverage at least on receive have lead to rather compromised synthesiser designs. This has been seen in particular in reviews of otherwise excellent radios limited in performance by synthesiser noise and spurious outputs causing reciprocal mixing. One solution being used by commercial companies, and now falling in price to 'amateur' levels, is direct digital synthesis, and this is the subject of the next section.

One variant on PLL synthesis which appeared some years ago, and which is particularly suitable to amateur construction, is the so-called *huff-and-puff VFO* [11]. This consists (Fig 5.36) of a VFO, which must be built to the very highest standards, and a digital locking circuit. This behaves like a PLL, having the same frequency reference chain and phase-sensitive detector, but the divider chain is largely omitted. The VFO is tuned in the normal way, and the loop, such as it is, locks the VFO frequency to the nearest multiple of the reference frequency possible. The reference is typically below 10Hz, so that analogue feel is retained in the tuning. The loop of course can take a significant time to lock up, so various provisions to speed this can be employed. The loop must not jitter between different multiples of the reference, hence the need for very high initial stability in the loop. As a commercial proposition, this scheme is unattractive compared to a true PLL, but as an amateur approach it can overcome the VFO tuning problem in an elegant way.

Direct digital synthesis (DDS)

Direct digital synthesis has been frequently mentioned in *Radio Commmunication*, and is increasingly a feature of advanced amateur equipment [12–14]. The basic direct digital synthesiser consists of an arrangement to generate the output frequency directly from the clock and the input data. The simplest conception is shown in Fig 5.37. This consists of a digital accumulator, a ROM containing the pattern in digital form of a sine wave, and a digital-to-analogue converter. Dealing with the accumulator first, this is simply an adder with a store at each bit. It adds the input data word to that in the store. The input data word only changes when the required frequency is to be changed. In the simplest case the length of accumulator is the clock frequency

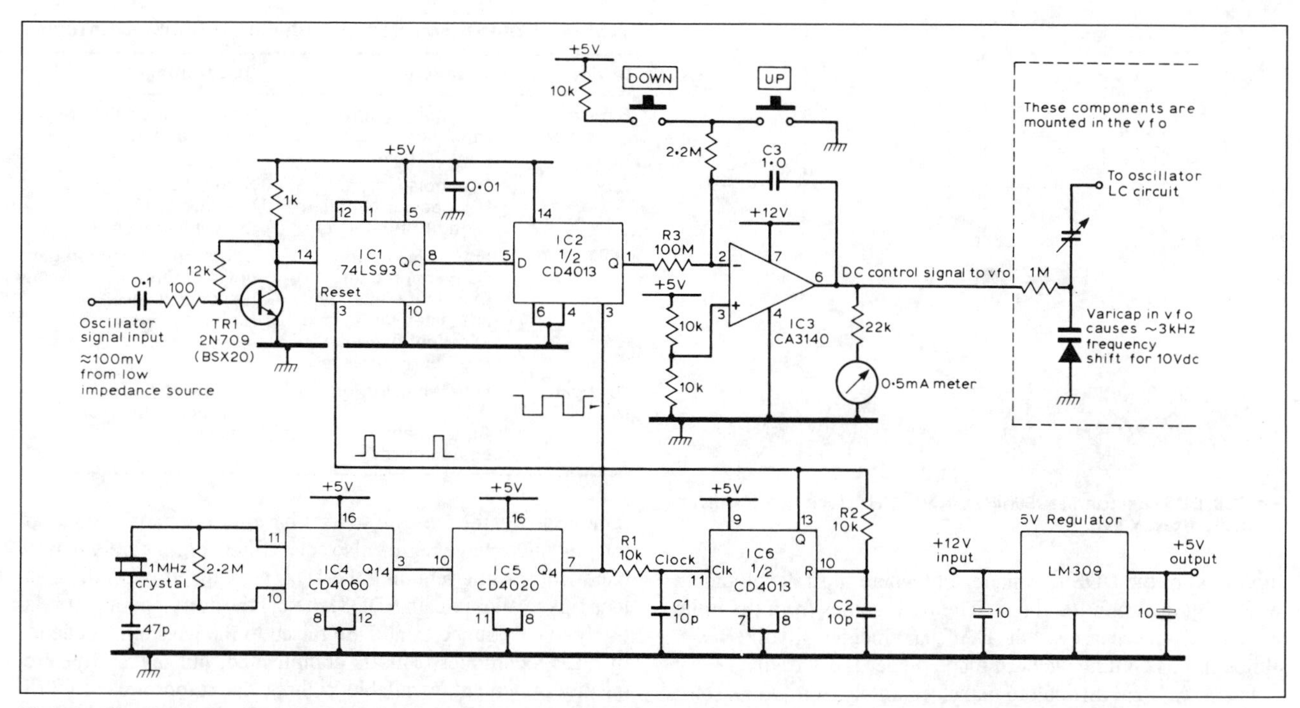

Fig 5.36. Circuit details of the CMOS form of the 'huff-and-puff' VFO stabilising system – a version of PA0KSB's system as described in *Ham Radio*

divided by the channel spacing, although it is usually calculated the other way round, ie if a 5kHz channel spacing up to 150MHz is needed, and since at least two clock pulses are required per output cycle, a clock frequency of at least 300MHz will be required. More conveniently, a 16-bit accumulator gives 65,536 steps. If the step size is to be 5kHz, then a clock frequency to the accumulator of 65,536 × 5kHz is needed, ie 327.68MHz. This makes programming particularly easy, especially as most DDS devices have a parallel-input data format. Such a DDS would produce any frequency in the range covered, ie 5kHz to over 100MHz in a single range without any tuned circuits, although of course it does depend for stability on the clock source, which would normally be crystal controlled. Other frequency increments are available; any multiple of 5kHz by selection of input data, and others by choice of clock frequency, eg 6.25kHz requires a clock at 204.8MHz with a two-channel (2 × 3.125kHz) program word. Of course, frequency multiplication or mixing can be used to take the output to higher frequencies.

A more advanced device is shown in Fig 5.38 [15]. This has on-chip DACs and facilities for square, triangle and sine outputs. Two DACs are needed because both phase and quadrature signals are available in true and complement form. The most significant bit (MSB) from the accumulator feeds the square-wave output buffer direct. In parallel, the next seven bits from the accumulator, which digitally represent a sawtooth waveform, feed a set of XOR gates, under control of the MSB, so that a triangle output is generated, in digital form, at this point in the circuit. Actually, two triangles in quadrature are generated. These can be digitally steered to the output DACs or can be used to address a ROM containing data for sine and cosine waves; only 90° is needed, since all four quadrants can be generated from one. Finally, if selected, the digital sine/cosine is fed to the DACs for conversion to analogue form.

This device was designed to operate at up to 500MHz output frequency with 1Hz steps, so the full range is 1Hz to 500MHz, again in a single 'range' with no means of or requirement for tuning. The clock frequency is necessarily very high; the quadrature requirement adds a further factor of two, so nominal clock frequency is 2^{31}Hz, ie 2.147483648GHz. This must be supplied with crystal-controlled stability, although since phase noise is effectively divided down in the synthesiser and the frequency is fixed, it can be generated by a simple multiplier from a crystal source.

Output spectra are shown in Figs 5.39 and 5.40. Fig 5.39 shows a clean output at exactly one quarter of the clock frequency. Fig 5.40 shows a frequency not integrally related to the clock frequency. Here the spurious sidebands have come up to a level about 48dB below the carrier. This is the fundamental

Fig 5.37. DDS concept

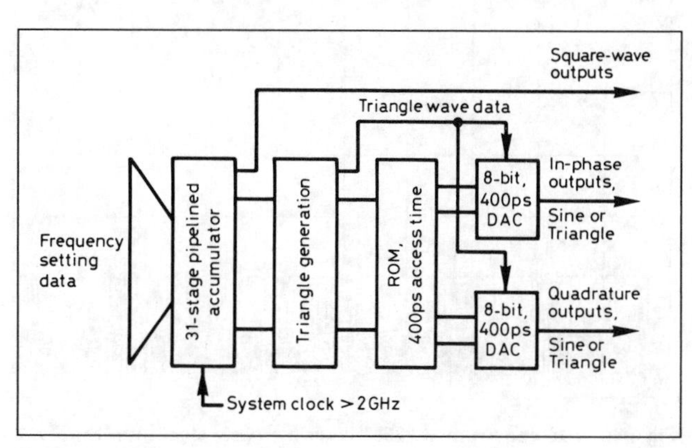

Fig 5.38. More complex DDS

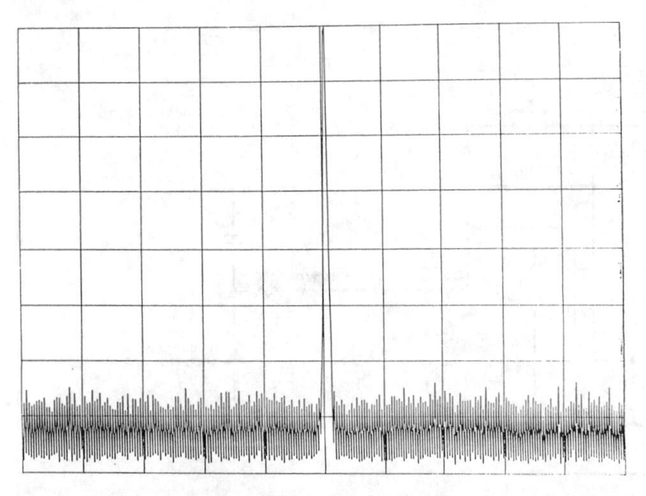

Fig 5.39. DDS spectrum at 250MHz with a 1GHz clock. 10MHz/div, X axis; 10db/div, Y axis

limitation of the DDS technique, and where most development work is going on in the future. The limit comes from the finite word size and accuracy of the DAC, and incidentally the ROM, although this could have been made bigger fairly easily.

Fast DACs are difficult to make accurately, for two reasons. The first is technological. IC processes in general have a limit to a component matching accuracy of about 0.1%, ie 9 or 10 bits accuracy in a careful design. Other techniques, such as laser trimming of the resistors, can be used on some processes. However, this tends to use older, slower processes, and especially requires large resistors for trimming which slow the DAC settling time. The second problem is simply the requirements on fast settling: the DAC is required to get to the final value quickly and this will in general happen with the smallest number of bits. An 8-bit system as in this example was chosen to fit the process capabilities and the device requirements, but this leads to a limitation in the high level of spurious signals present. Basically, although some frequencies are very clean, in the worst case the spurious level is $6N$ dB below the carrier, where N is the number of effective DAC bits. For an 8-bit system, this gives –48dB. Oversampling, ie running the clock at more than twice the output frequency, gives some improvement (at 6dB per octave) by improvement of the DAC resolution, but only up to a limit of the DAC's accuracy. Typically this is about 9 bits or –54dB. In some applications, this may not be serious, since filtering can

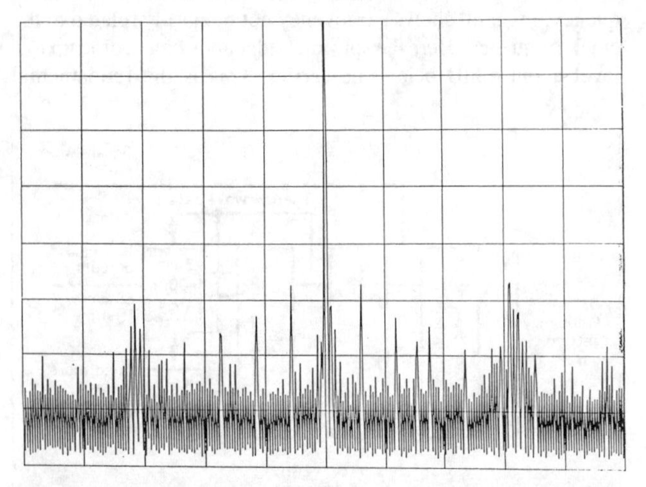

Fig 5.40. DDS spectrum at 225MHz with a 1GHz clock. 10MHz/div, X axis; 10dB/div, Y axis

Table 5.1. Comparison of semiconductor performance in mixers

Device	Advantages	Disadvantages
Bipolar transistor	Low noise figure High gain Low DC power	High intermodulation Easy overload Subject to burn-out
Diode	Low noise figure High power handling High burn-out level	High LO drive Interface to IF Conversion loss
JFET	Low noise figure Conversion gain Excellent square-law characteristic Excellent overload High burn-out level	Optimum conversion gain not at optimum square-law response level High LO power
Dual-gate MOSFET	Low IM distortion AGC Square-law characteristic	High noise figure Poor burn-out level Unstable

remove all but the close-in spurs. Phase-locked translation loops into the microwave region also act as filters of relatively narrow bandwidth, while retaining the 1Hz step capability. In the amateur rigs available using DDS, the synthesisers operate at relatively low frequencies and are raised to the working frequency by PLL techniques. This is complicated, but makes fine frequency increments available without the compromise of PLL design. The devices used are CMOS types, with DAC accuracies of 8 and 10 bits. This gives spurious signals of theoretically –60dB referred to the carrier, which is adequate with filtering from the PLL.

Although 'discrete' DDS designs using TTL and CMOS have been published, they have been limited to very low (usually audio) frequencies. Newer IC processes, and especially the reduced cost of DACs, will soon lead to real use by amateurs of the DDS technique. Dedicated DDS chips are becoming available at low cost, and it is likely that this trend will continue. Eventually, we may see faster, more accurate DACs and new filtering techniques combine to produce a 100dB clean synthesised oscillator, with no tuning or set-up procedures at all and with extremely narrow channel spacing at low cost.

MIXERS

Terms and specifications

A mixer is a device of which the desired output frequency is the sum or difference of two input frequencies. Mixers have many functions, sometimes going by another name.

The *front-end* of a superheterodyne receiver contains a mixer or *first detector*, in which the incoming signal (of varying frequency) is mixed with a *local*, ie in-receiver, oscillator or frequency synthesiser, to yield a fixed *intermediate frequency* (IF). After IF filtering and amplification, this IF is mixed with a *beat frequency* in the *demodulator, product detector* or *synchronous detector,* to obtain as the difference frequency an audio output. In a single-sideband transmitter, the amplified audio from the microphone is mixed, in a *modulator*, with a *carrier*.

Any device with a non-linear voltage/current characteristic can serve as a mixer. However, the output amplitude of an ideal mixer bears a linear (proportional) relationship to the amplitude of one input, the *signal*, if the amplitude on the other input, eg from the *local oscillator*, is held constant. Diodes, bipolar transistors, junction, single and dual-gate MOSFETs, as well as their valve equivalents, are used as mixers (see Table 5.1).

Of the two inputs to a mixer, one, f_s, contains the intelligence, the second, f_o, is specially generated to shift that intelligence to (any positive value of) $\pm f_s \pm f_o$, of which only one is the desired

Table 5.2. Mixing products in single, balanced and double-balanced mixers

Unbalanced mixer					
	f_o	$2f_o$	$3f_o$	$4f_o$	$5f_o$
f_s	$f_o \pm f_s$	$2f_o \pm f_s$	$3f_o \pm f_s$	$4f_o \pm f_s$	$5f_o \pm f_s$
$2f_s$	$2f_s \pm f_o$	$2f_o \pm 2f_s$	$3f_o \pm 2f_s$	$4f_o \pm 2f_s$	$5f_o \pm 2f_s$
$3f_s$	$3f_s \pm f_o$	$3f_s \pm 2f_o$	$3f_o \pm 3f_s$	$4f_o \pm 3f_s$	$5f_o \pm 3f_s$
$4f_s$	$4f_s \pm f_o$	$4f_s \pm 2f_o$	$4f_s \pm 3f_o$	$4f_o \pm 4f_s$	$5f_o \pm 4f_s$
$5f_s$	$5f_s \pm f_o$	$5f_s \pm 2f_o$	$5f_s \pm 3f_o$	$5f_s \pm 4f_o$	$5f_o \pm 5f_s$
Balanced mixer – half the number of mixer products					
	f_o	$2f_o$	$3f_o$	$4f_o$	$5f_o$
f_s	$f_o \pm f_s$	$2f_o \pm f_s$	$3f_o \pm f_s$	$4f_o \pm f_s$	$5f_o \pm f_s$
$3f_s$	$3f_s \pm f_o$	$3f_s \pm 2f_o$	$3f_o \pm 3f_s$	$4f_o \pm 3f_s$	$5f_o \pm 3f_s$
$5f_s$	$5f_s \pm f_o$	$5f_s \pm 2f_o$	$5f_s \pm 3f_o$	$5f_s \pm 4f_o$	$5f_o \pm 5f_s$
Double-balanced mixer – one quarter the number of mixer products					
	f_o	$2f_o$	$3f_o$	$4f_o$	$5f_o$
f_s	$f_o \pm f_s$	—	$3f_o \pm f_s$	—	$5f_o \pm f_s$
$3f_s$	$3f_s \pm f_o$	—	$3f_o \pm 3f_s$	—	$5f_o \pm 3f_s$
$5f_s$	$5f_s \pm f_o$	—	$5f_s \pm 3f_o$	—	$5f_o \pm 5f_s$

f_o is the local oscillator. Note that a product such as $2f_o \pm f_s$ is known as a *third-order product*, $3f_s \pm 3f_o$ as a *sixth-order product* and so on.

output. In addition, the mixer output contains both input frequencies, their harmonics, and the sum and difference frequencies of any two of all those. If any one of these many unwanted mixer products almost coincides with the wanted signal at some spot on a receiver dial, there will be an audible beat note or *birdie* at that frequency. Similarly, in a transmitter, a *spurious* output may result.

In general-coverage receiver and multiband transmitter design this problem can be made more manageable by the use of *balanced* mixers. By applying a signal to a balanced input port of a mixer, the signal, its even harmonics and their mixing products will not appear at the output. If both input ports are balanced, this applies to both f_o and f_s, and the device is called a *double-balanced mixer*.

Table 5.2 shows how balancing reduces the number of mixer products. Note that products such as $2f_o \pm f_s$ are known as *third-order* products, $3f_o \pm 3f_s$ as *sixth-order* and so on. Lower-order products are generally stronger and therefore more bothersome than higher-order products.

A mixer can have *conversion gain* or *conversion loss*. It is expressed, in decibels, as the output level over the signal input level. Positive decibel figures mean gain, negative numbers loss.

All mixers have limitations.

Isolation between the input ports of a mixer refers to the input applied to one port affecting whatever is connected to the other input port. In the simplest of receivers using a mixer without much isolation, peaking the antenna tuning circuit connected to the signal input may 'pull' the frequency of an unbuffered free-running oscillator connected to the LO port. Conversely, the LO signal may be transmitted from the antenna!

Noise is generated in all mixers. It is quantified as a *noise figure*, expressed in decibels over the noise generated by a resistor of the same value as the impedance of the mixer port at the prevailing temperature, eg 50Ω at 17°C. If mixer noise is significant as compared to the smallest signal to be processed, the signal-to-noise ratio will suffer. The same noise level or *noise floor*, in terms of dBm, is shown in Fig 5.41 [16]. Below 10MHz, however, atmospheric noise will override receiver noise and most mixers will be adequate in this respect.

Overload occurs if an input signal exceeds the level at which the output is proportional to it, always assuming that the local oscillator level is adjusted for best performance. The overloading input may be at a frequency other than the desired one, as happens in the mixer preceding the crystal filter in a receiver.

Compression is the conversion gain reduction which occurs

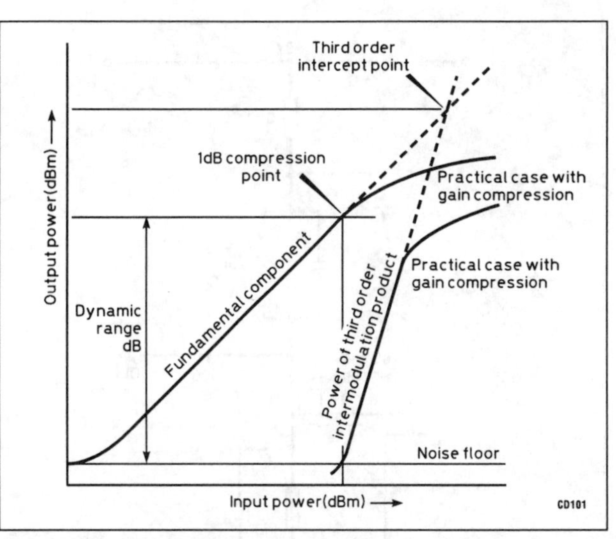

Fig 5.41. Noise floor, 1dB compression point, dynamic range and 3rd-order intercept indicated on a mixer output vs input plot *(GEC-Plessey Professional Products IC Handbook)*

when the signal input exceeds the maximum the mixer can handle in a linear manner. The input level for 1dB of conversion gain reduction is often specified. Note the bending of the 'fundamental component' line in Fig 5.41. The signal input level should be kept below the one causing this bending.

Intermodulation products are generated when the input contains two or more strong unwanted signals, from which a mixing product falls within the wanted-output bandwidth. Intermodulation performance is often specified as the power level in dBm of the *third-order intercept point*. This is the fictitious intersection on a signal input power vs output power plot, Fig 5.41, of the extended (dashed) fundamental (wanted) line and the third-order intermodulation line. Note that the third-order line rises three times steeper than the fundamental line.

Practical examples of mixers

A single-diode mixer is frequently used in microwave equipment. As a diode has no separate input ports, it provides no isolation between inputs. The mixer diode D1 in Fig 5.42,

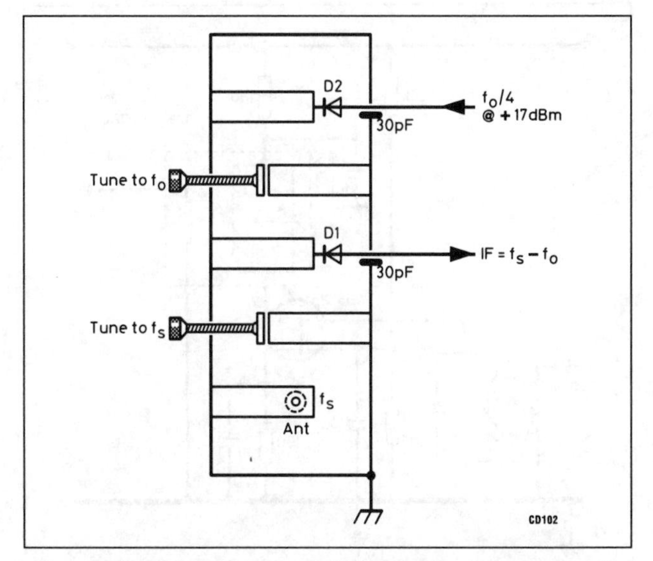

Fig 5.42. Single-diode mixer for 1296MHz. An interdigital filter provides isolation between ports. D1 (HP5082-2817) is the mixer and D2 (HP5082-2853) is the last multiplier in the LO chain *(QST)*

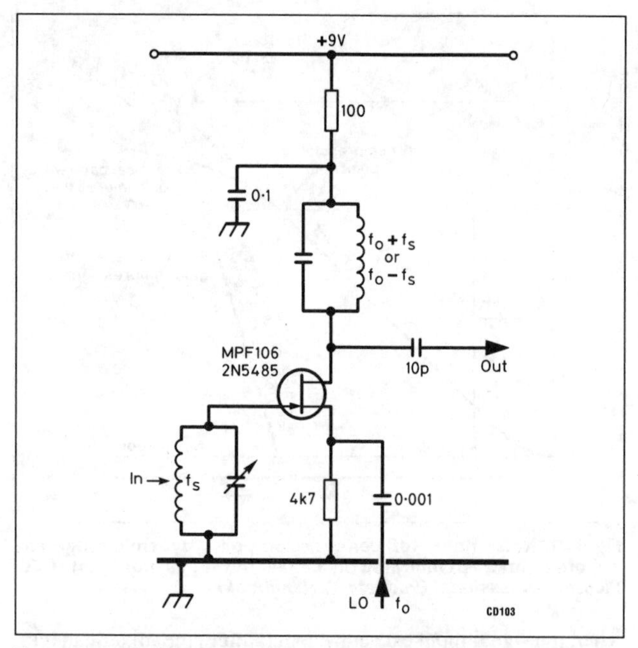

Fig 5.43. A single JFET makes a simple, inexpensive mixer. The LO drives a low-impedance port, requiring it to supply some power

however, is used in an interdigital filter [17]. Each input frequency readily passes by its high-Q 'finger' to the lower-Q (diode-loaded) finger to which D1 is connected but cannot get beyond the other high-Q finger which is tuned to the other input frequency. In this design f_s = 1296MHz and f_o = 1268MHz. The 30pF feedthrough capacitor at the mixer output terminal acts as a short to earth for both input frequencies, but is part of a tuned circuit (not shown) at the output frequency of 1296 − 1268 = 28MHz. Diode D2 is not part of the mixer, but is the last link in a local oscillator chain. D2 operates as a quadrupler from 317 to 1268MHz. To get a good noise figure, the mixer D1 is a relatively expensive microwave diode, HP5082-2817. In the multiplier spot, the cheaper HP5082-2853 suffices.

Fig 5.43 shows a junction FET in what is probably the simplest general-purpose mixer circuit. It provides some isolation

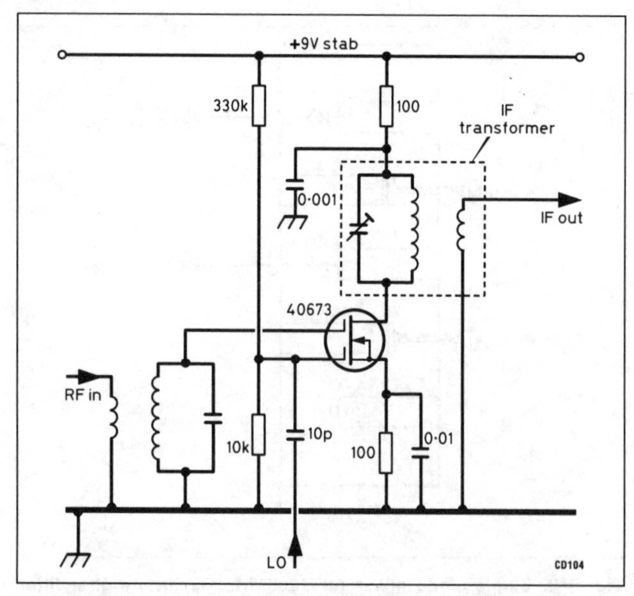

Fig 5.44. The gates of a dual-gate MOSFET provide two high-impedance input ports but the noise figure is high

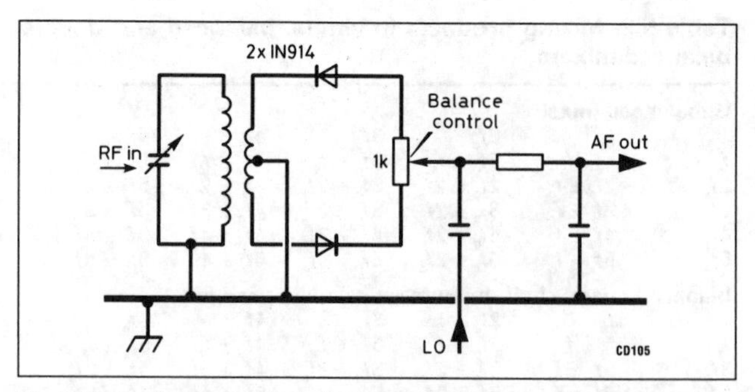

Fig 5.45. In this two-diode balanced mixer for direct-conversion receivers, balancing helps to keep LO drive from reaching the antenna

between ports, low noise, high conversion gain and a reasonable dynamic range. The latter can be improved by the use of a power FET such as the U310, E310 or CP640 series biased for high source current, say 40mA, depending on the device. JFETs require careful adjustment of bias (source resistor) and local oscillator input for best performance.

The signal is applied to the high-impedance gate; hence, damping on the resonant input circuit is low and gain is high. The local oscillator, however, feeds into the low-impedance source; this implies that *power* is required from the local oscillator, which may require buffering. One could reverse the inputs, thereby reducing the local oscillator power requirements, but that would reduce the gain. The output is taken from the drain with a tuned circuit selecting the sum or difference frequency.

Fig 5.44 shows a dual-gate MOSFET in a mixer circuit in which both inputs feed into high-impedance ports. Though the dynamic range of these devices is somewhat limited, they are used because of their simplicity and low cost.

Fig 5.45 is a balanced version of the diode mixer. It is popular for direct conversion receivers where the balance reduces the amount of local oscillator radiation from the receiving antenna. The RC output filter reduces local oscillator RF reaching the following AF amplifier.

Figs 5.46 and 5.47 are balanced versions of JFET and dual-gate MOSFET mixers. The notes on these devices in unbalanced mixers apply equally to the balanced versions.

Fig 5.48 represents a high-level balanced valve mixer as used in early VHF transverters. It is still useful for converting an obsolete VHF transmitter into a transverter.

As with all high-level mixers, there is a danger of the local oscillator frequency being radiated. To prevent this, at least four but preferably six tuned circuits should be used between the mixer and the antenna.

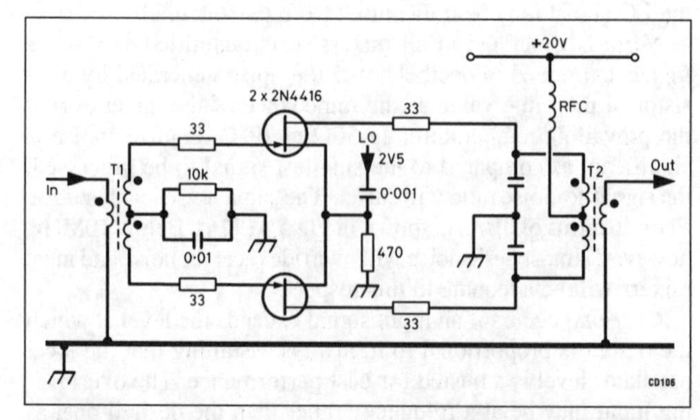

Fig 5.46. Two JFETs in a balanced mixer are capable of the performance required from all but the most expensive communications receivers

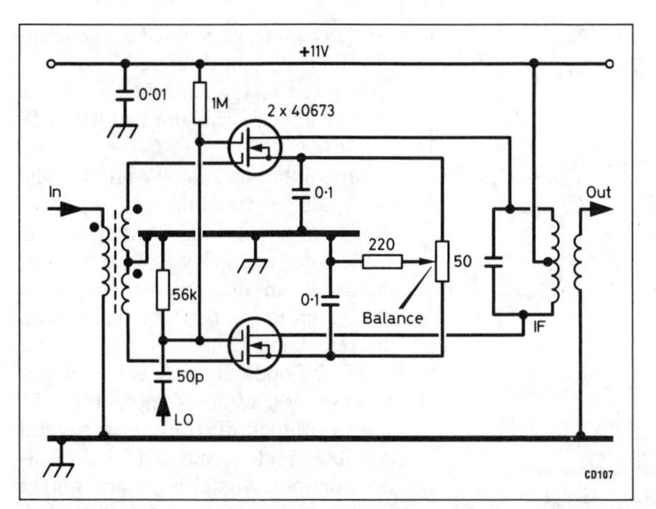

Fig 5.47. Two dual-gate MOSFETs in this balanced mixer require low LO power. At the lower frequencies, where noise is not a major consideration, they perform as well as JFETs

The QQV03-10 mixer anode current with no local oscillator or signal input should be about 15mA, and with only local oscillator drive about 30mA. Applying signal input to the cathode should drive this up by another one to two milliamps.

The following two double-balanced mixers can be used where low noise and the greatest possible dynamic range must be combined with maximum suppression of unwanted outputs, eg in continuous-coverage receivers and multiband transmitters. Three different approaches are presented.

Fig 5.49 shows a diode-ring mixer. It has been used for many decades and its performance can be second to none. Going around the diode ring, all diodes point in the same direction; this differs from a rectifier bridge. The diodes should be matched for both forward and back resistance. The inexpensive 1N914 is a favourite, but Schottky barrier types such as the BA481 (for UHF) and BAT85 (for lower frequencies and large-signal applications) are capable of higher performance. The wide-band transformers are wound on small ferrite toroids. For HF and below the Amidon FT50-43 is suitable, and 15 trifilar turns of 0.2mm diameter enamelled copper wire are typical. The dots indicate the same end of each wire. Complete ring mixers are available, eg the popular model SBL-1. They contain four matched diodes and the two transformers but are more expensive than the sum of their components.

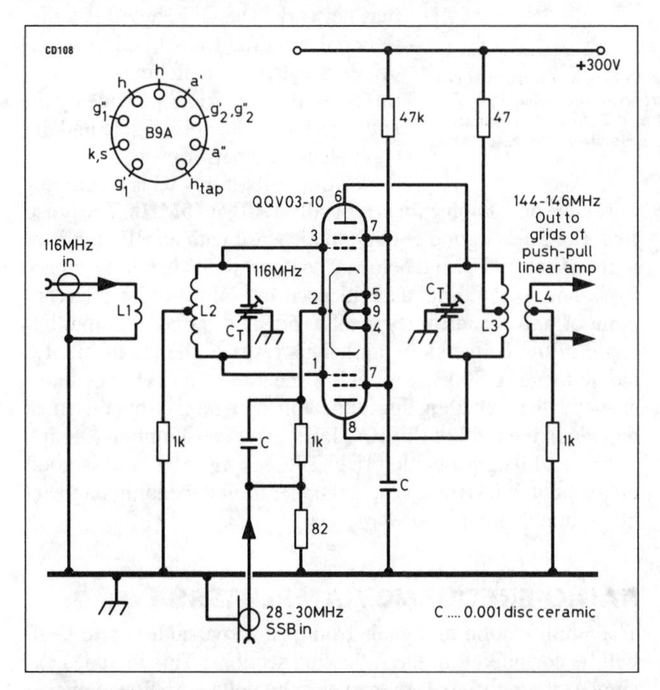

Fig 5.48. Though high-level mixing in transmitters is generally discouraged because of the danger of spurious frequencies being transmitted, this balanced valve mixer is useful in the conversion of valved PMR equipment into VHF transverters. Good shielding and four to six high-*Q* tuned circuits at the output frequency can keep spurious radiation under control

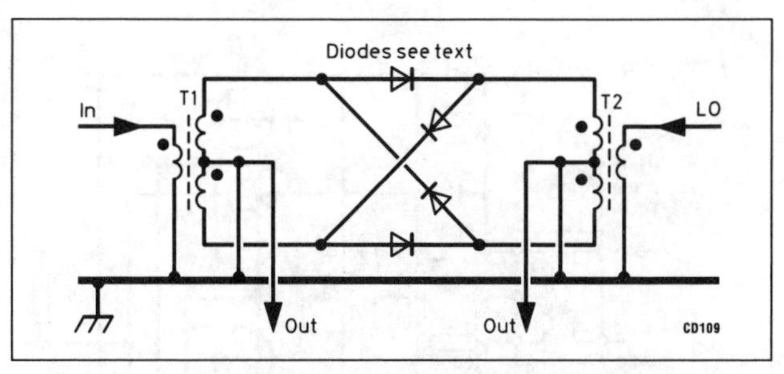

Fig 5.49. Diode ring mixers are capable of very high performance at all the usual signal levels but several sometimes costly precautions have to be taken to realise that potential; see text. Further improvement of the balance can be obtained by the use of push-pull feed to the primaries of T1 and T2 and feeding the output from between the T1 and T2 centre-taps into a balanced load

Diode mixers are low-impedance devices, typically 50Ω. This means that the local oscillator must deliver *power*, and 5mW (7dBm) or more, depending on components, is required for best dynamic range.

Passive mixers, ie those not containing amplifying devices such as transistors, have a conversion loss, ie the wanted-frequency output is always less than the signal input, –7dB being typical for diode ring mixers. In receiver first mixers, this loss must be made up in the following IF amplifier. Where noise figure is important, say above 25MHz, this IF amplifier must be a low-noise type.

The LO and output ports of a diode ring mixer must be resistively terminated. Failing to do so at the LO port may increase third-order intermodulation. As it is difficult to predict the output impedance of an oscillator circuit over a wide frequency range, it is best to generate more LO power than that required by the mixer and insert a resistive attenuator, 3–6dB being common. Reactive termination of the output port can increase conversion loss, spurious responses and third-order IMD. Most mixers work into a filter to select the desired frequency. At that frequency, the filter may have a purely resistive impedance of the proper value but at the frequencies it is designed to reject it certainly does not. If there is more desired output than necessary, an attenuator is indicated; if not, the filtering may have to be done after the following amplifier, which may be the one making up for the conversion loss. A cascode amplifier is sometimes used as it can have a good noise figure and, over a wide frequency range, a predictable resistive input impedance.

Another double-balanced mixer capable of high performance uses four JFETs. The one shown in Fig 5.50 is used in the Yaesu FT-1000 HF transceiver. The circuit offers some conversion gain and the local oscillator needs to supply little power as it feeds into the high-impedance gates. Loading the output with a crystal filter apparently does no harm.

Bipolar ICs can also be configured as single-balanced or double-balanced mixers for use in the VHF range and below. ICs developed for battery-powered instruments such as portable and cellular telephones have very low power consumption and are

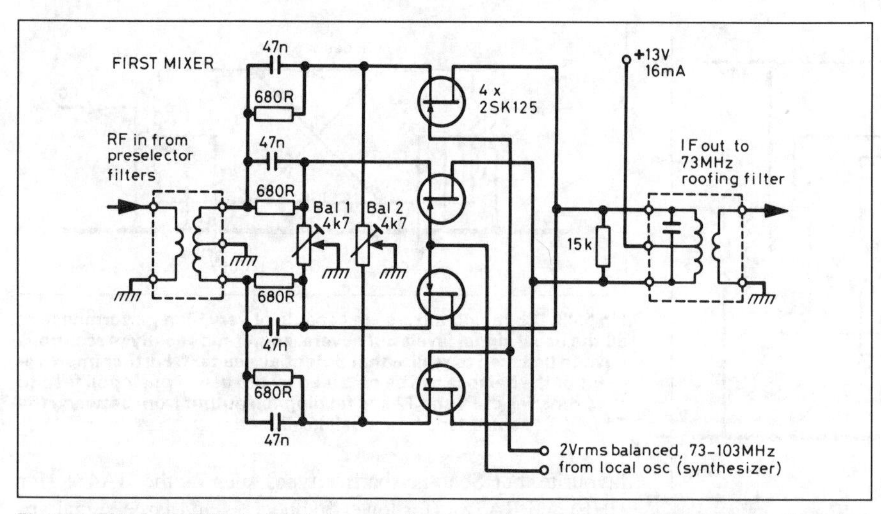

Fig 5.50. This four-JFET double-balanced mixer is as good as any seen in amateur receivers but does not have the problems of conversion loss, termination and high LO drive associated with diode ring mixers

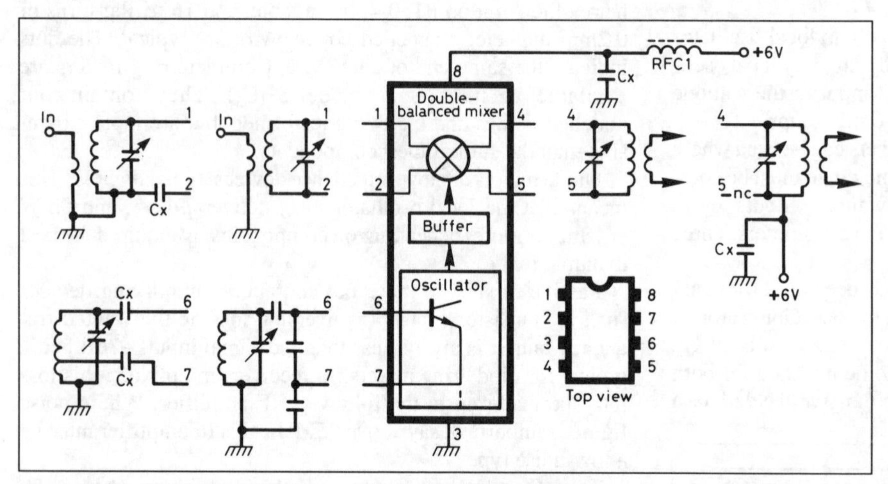

Fig 5.51. The NE602N double-balanced-mixer/oscillator is shown here in block diagram form with some options for external circuitry. This IC combines good performance with low DC requirements and a reasonable price. The option on the right for pins 1 and 2 should preferably use a split-stator capacitor. The option on the left for pins 6 and 7 is for a Hartley oscillator, the one on the right for a Colpitts

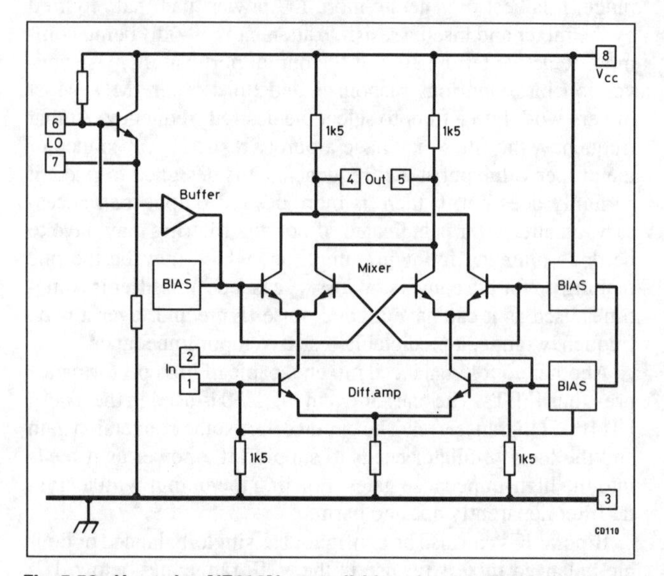

Fig 5.52. How the NE602N monolithic mixer/oscillator works is described in the text with reference to this equivalent circuit (*Signetics RF Communications Handbook*)

used to advantage in QRP amateur applications. Often, a single IC contains not only a mixer but other functions such as a local oscillator and RF or IF amplifiers.

Monolithic technology relieves the home constructor of the task of matching components and adjusting bias while greatly simplifying layout. Also, the results are more predictable, if not always up to the best obtainable with discrete components.

The NE602N mixer/local oscillator IC may serve as an example. Fig 5.51 shows a block diagram with several possible input, output and local oscillator options. Signal input and output can be either balanced or single-ended. Balanced inputs enhance the suppression of unwanted mixing product and reduce LO leakage into the signal input. Hartley and Colpitts connections are shown for the local oscillator. In the latter, a crystal can be substituted for the parallel-tuned LC circuit, or an external LO signal can be fed into pin 6. All external connections should be blocked for DC by capacitors as all biasing is done within the IC.

Fig 5.52, the equivalent circuit of the IC and the following description of how it works were taken from reference [18]. The IC contains a Gilbert cell (also called a *transistor tree*), an oscillator/buffer and a temperature-compensated bias network. The Gilbert cell is a differential amplifier (pins 1 and 2) which drives a balanced switching cell. The differential input stage provides gain and determines the noise figure and the signal handling performance.

Performance of this IC is a compromise. Its conversion gain is typically 18dB at 45MHz. The noise figure of 5dB is good enough to dispense with an RF amplifier in receivers for HF and below. The large-signal handling capacity is not outstanding, as evidenced by a third-order intercept point of only −15dBm (typically +5dBm referred to output because of the conversion gain). This restricts the attainable dynamic range to 80dB, well below the 100dB and above attainable with the preceding discrete-component mixers, but this must be seen in the light of this IC's low power consumption (2.4mA at 6V) and its reasonable price. The on-chip oscillator is good up to about 200MHz, the actual upper limit depending on the *Q* of the tuned circuit.

RADIO-FREQUENCY AMPLIFIERS

The amplification of signals from just above audio up to UHF will be considered in the following sections. This includes examples of amplifiers for receiver input voltages below a microvolt in 50Ω, ie about a femtowatt (10^{-15}W!) up to the legal power limit for UK amateurs (400W PEP), especially those which might be considered for home construction projects. Included are small-signal amplifiers with discrete semiconductors and ICs and

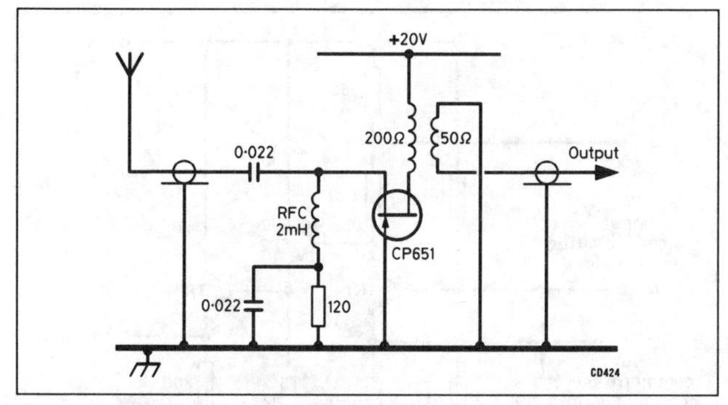

Fig 5.53. 10dB broad-band RF amplifier using a power FET in the common-gate mode. It will handle 0.5 to 40MHz signals from 0.3µV to almost 3V p-p with a noise figure of 2.5dB. The drain current is 40mA

power amplifiers with discrete semiconductors, hybrids (semi-integrated modules) and valves.

Low-level discrete-semiconductor RF amplifiers

Discrete semiconductors are indicated when the function requires no more gain than that which can be realised in one stage, generally between 6 and 20dB, or when the required noise level or dynamic range cannot be met by an IC. In receiver input stages, ie ahead of the first mixer, both conditions may apply.

Modern design of the first stage in a receiver demands filters rejecting strong out-of-band signals, a wide dynamic range, low noise IF at VHF and above, and only enough gain to overcome the greater noise of an active first mixer or the conversion loss of a passive mixer, typically a gain of 10dB. Bipolar, field-effect and dual-gate FET transistors can provide that. AGC can easily be applied to FET amplifiers, but the trend is to switch the RF amplifier off when not required for weak-signal reception.

In most of the following low-level RF circuits inexpensive general-purpose transistors such as the bipolar 2N2222A, FETs BF245A, MPF102 and 2N3819, and dual-gate MOSFETs 40673 and 3N201 can be used. However, substitution of one device by another or even by its equivalent from another manufacturer frequently requires a bias adjustment. In each of the following examples, the device which was originally designed in is shown, regardless of its current availability or price.

For exceptional dynamic range, power FETs are sometimes used at currents up to 100mA. Similarly, for the best noise figure above 100MHz, gallium-arsenide FETs (GaAsFETs) are used instead of silicon types.

A simple untuned amplifier with a power MOSFET in a grounded-gate circuit is shown in Fig 5.53. The Teledyne-Crystalonics CP651 was specially made for this purpose but the Philips BLF221 may be a suitable substitute. This circuit is suitable for use in an HF antenna distribution amplifier, ie where several receivers tuned to different frequencies must work off one antenna. Beware, however, of operating any amplifier that does not have the dynamic range of a power FET without preselector filtering; a strong out-of-band signal could overload it and block your receiver.

Grounded-gate FETs are also used as buffers between diode mixers and crystal filters. The mixer then works into the non-reactive load it requires and the crystal filter input is properly terminated as well.

The amplifier in Fig 5.54 uses a dual-gate enhancement-mode MOSFET, Signetics type SD301. On 144MHz, the advantage of this device is that input circuit tuning for maximum gain and for best noise figure coincide. If a noise generator is available

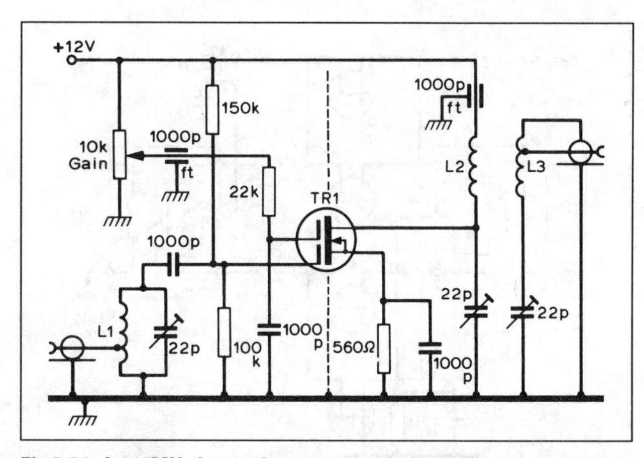

Fig 5.54. A 144MHz low-noise preamp using an enhancement-mode dual-gate MOSFET SD301. If a depletion-mode FET such as the 40673 is used, the 150kΩ resistor can be omitted. L1: 5t 0.3mm tinned copper tapped 1t from earthy end, 10mm long, 6mm dia. L2, L3: 6t 1mm tinned copper, 18mm long, 8mm dia. L3 tapped 1t from earthy end. L2, L3 mounted parallel, 18mm between centres

for tune-up, or on a lower-frequency band, a less-expensive, general-purpose, depletion-mode, dual-gate MOSFET can be used. The 150kΩ bias resistor can then be left out.

The input tank circuit and the output band filter represent good practice for single-band operation and should be tuned for flat response over the desired band and rapid roll-off outside it. If no sweep generator and 'scope are available, all three trimmers should be peaked on a weak centre-band signal with the coupling of the output filter loosened (greater spacing between L2 and L3); then, without touching the trimmers, L2 and L3 should be brought together just enough to flatten the response over the band. On a narrow band such as 70MHz this is easy but on a wide band such as 28MHz this may take some effort.

Fig 5.55 represents a *cascode* amplifier. The tuned input circuit is only lightly loaded. The amplifier is very stable, and smoothly changes from gain to attenuation as the AGC voltage causes the bipolar transistor to reduce the gain of the input FET.

Four FETs in a push-pull parallel grounded-gate circuit, Fig 5.56, are used in the RF amplifiers of some top-grade HF receivers. Push-pull reduces second-order intermodulation, several smaller FETs in parallel approach the wide dynamic range

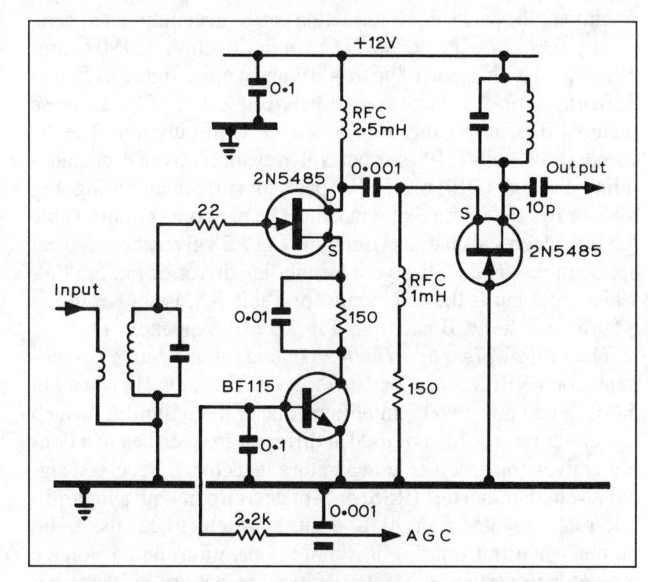

Fig 5.55. A cascode RF amplifier using two JFETs. The bipolar transistor controls the gain in response to the applied AGC voltage

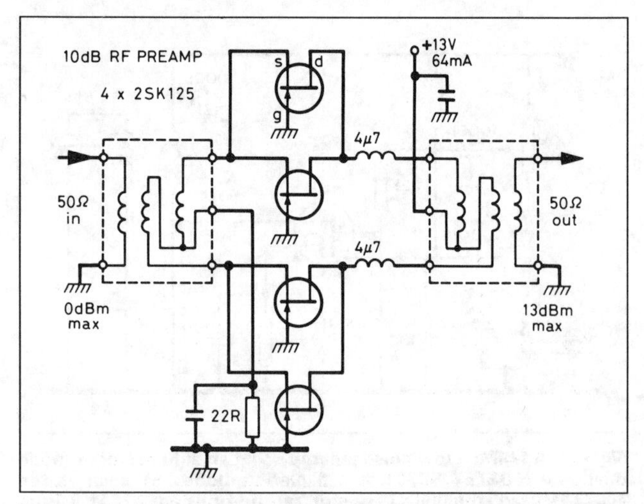

Fig 5.56. Four grounded-gate FETs in push-pull parallel running high drain currents give this RF stage of the Yaesu FT-1000 HF transceiver its excellent dynamic range, suppression of second-order intermodulation, and proper termination of preceding and following bandpass filters

of a power FET, and the source input provides, through a simple transformer, proper termination for the preselector filters.

Low-level IC amplifiers

Monolithic integrated-circuit RF amplifiers come in a great variety, sometimes combined on one chip with other functions such as a mixer, detector, oscillator or AGC amplifier. A linear IC, though more expensive than its components in discrete form, has the advantage of well-specified performance in the proven PCB layouts given in the manufacturer's data sheet.

Some ICs are labelled 'general purpose', while others are optimised for a specific application. If intended for hand-portable telephones, low power consumption might be a key feature, but linearity might be unimportant in such an NBFM system. In an IC for fixed-station SSB, the reverse might apply. Any potential user would do well to consult the data sheets or books on several alternative ICs before settling on any one. New devices are being introduced frequently and the price of older ones, often perfectly adequate for the intended function, is then reduced. The following elaborates on a few popular devices, but only a fraction of the data sheet information can be accommodated here.

The *GEC-Plessey SL560C* is a single-function 300MHz amplifier [16] with a gain of up to 40dB and a noise figure as low as 2dB; the supply voltage can be between 2 and 15V; all these features depend on the user-selectable configuration. The IC comes in an 8-pin TO-5 case but is also available as an 8-pin dual-inline package (DIP) or as a flat pack for surface mounting. Fig 5.57 shows the internal diagram and the pin connections, Table 5.3 lists electrical characteristics and Fig 5.58 gives three typical applications. The PCB layout for the line driver of Fig 5.58(a), which doubles as the test circuit for Table 5.3, is shown in Fig 5.59(a) and the PCB pattern in Fig 5.59(b) (Appendix 1).

The *Philips-Signetics NE604A* contains all the active components for a NBFM voice or data receiver IF system [18] (see Fig 5.60). It contains two IF amplifiers, one of them limiting, which can either be cascaded or used at different frequencies in a double-conversion receiver, a quadrature detector, a 'received signal-strength indicator' (RSSI or S-meter) circuit with a logarithmic range greater than 90dB, a mute switch (to cut the audio output when transmitting in simplex operation) and a voltage regulator to assure constant operation on battery voltages between 4.5V (at 2.5mA) and 8V (at 4mA). The IC is packaged in

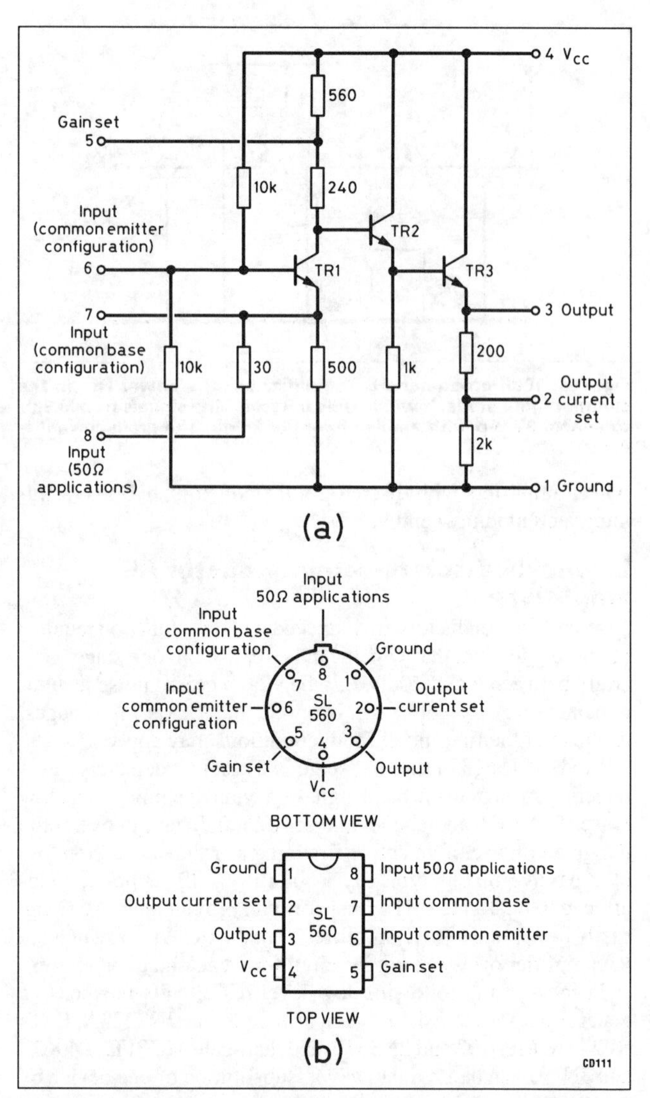

Fig 5.57. The SL560 low-noise VHF IC amplifier. (a) Internal diagram. (b) Pin connections *(GEC-Plessey Professional Products Handbook 1991)*

a 16-pin DIP (604AN) but is also available in a flat-pack for surface mounting (604AD). The operating temperature range of the NE604A is 0 to 70°C. The –40 to 85°C version is called SA604A; skiers please note! Table 5.4 gives operating characteristics and Fig 5.61 shows the test circuit, a typical application at 455kHz; Table 5.5 is the components list [18].

Table 5.3. The SL560C electrical characteristics

Characteristic	Value Min	Typ	Max	Unit	Conditions
Small-signal voltage gain	11	14	17	dB	
Gain flatness	—	±1.5	—	dB	10–220MHz
Upper cut-off frequency	—	250	—	MHz	
Output swing	+5	+7	—	dBm	V_{cc} = 6V
	—	+11	—	dBm	V_{cc} = 9V
Noise figure, common emitter	—	1.8	—	dB	R_s = 200Ω
	—	3.5	—	dB	R_s = 50Ω
Supply current	—	20	30	mA	

Test conditions (unless otherwise stated): Frequency = 30MHz; V_{cc} = 6V; R_S = R_L = 50Ω; T_{amb} = 25°C; test circuit: Fig 5.58(a). (Taken from *Plessey Professional Products Handbook 1991*).

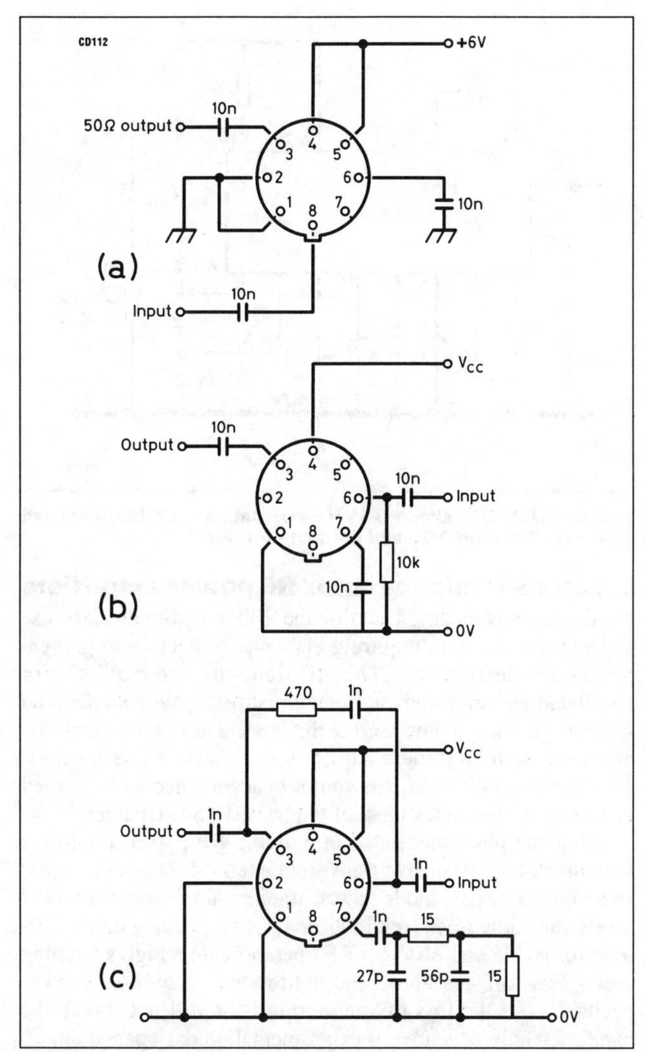

Fig 5.58. The SL560 IC in three different configurations. (a) 50Ω line driver and test circuit. Gain is 14dB, bandwidth is 220MHz at P_{out} = 1mW, 50Ω or 200MHz at P_{out} = 5mW, 50Ω. (b) Low-noise preamplifier. Voltage gain is 32dB at 6V or 35dB at 10V. Noise figure is 1.8dB (R_s = 200Ω). Supply current 6mA at 6V or 12mA at 10V. Bandwidth is 75MHz. (c) Wide-band amplifier. Gain is 13dB at V_{cc} = 9V, –1dB at 6MHz and 300MHz. ***(GEC-Plessey Professional Products Handbook 1991)***

The *GEC-Plessey SL6700A* is also an IF system for battery operation, but aimed at AM, CW or SSB reception. The block diagram (Fig 5.62) shows two IF amplifiers, an AM detector

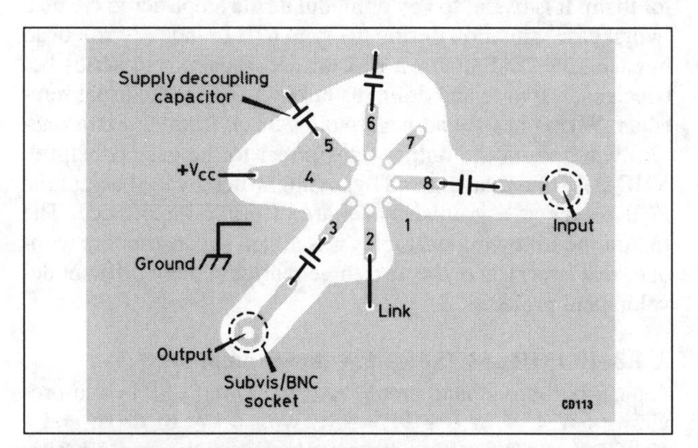

Fig 5.59(a). The SL560 50Ω line driver and test circuit layout ***(GEC-Plessey Professional Products Handbook 1991)***

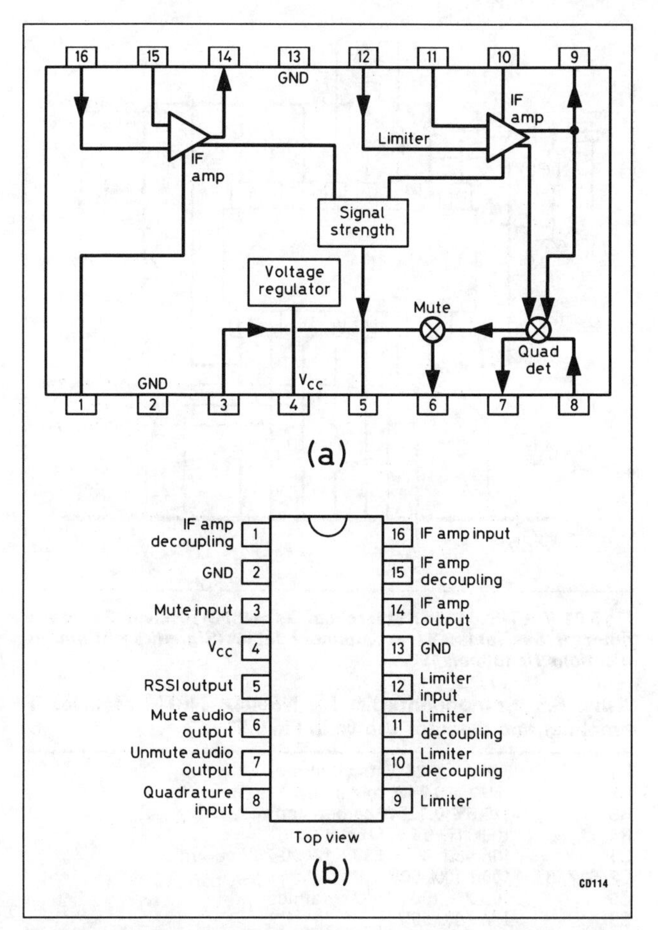

Fig 5.60. The NE604A FM receiver IF IC. (a) Block diagram. (b) Pin connections of NE604AN ***(Signetics RF Communications Handbook 1989)***

with delayed AGC which is applied to the internal IF amplifiers but is also available for an external RF amplifier, a noise blanker, and a double-balanced modulator which can be used as the mixer in a single-conversion AM receiver, as a first-to-second IF converter or, as is done in Fig 5.63, as a product detector for CW or SSB. This IC comes in a 18-pin DIP and operates over the whole avionics temperature range of –55 to 125°C.

Mini-Circuits and Avantek make several types of *MMIC* (microwave monolithic integrated circuit) which provide wide-band gain up to the 2.3GHz band with very few external components. They have a 50Ω input, can be cascaded without interstage tuning, have noise figures as low as 3.2dB at 1GHz and are capable of 10–20mW of output into 50Ω. Packaging is for

Table 5.4. The most important NE604A electrical characteristics

Parameter	Test conditions	NE604A Min	Typ	Max	Unit
Input limiting – 3dB	Test at pin 16	—	–92	—	dBm/50Ω
AM rejection	80% AM 1kHz	30	34	—	dB
Recovered audio level	15nF de-emphasis	110	175	250	mV_{RMS}
Recovered audio level	150pF de-emphasis	—	530	—	mV_{RMS}
SINAD sensitivity	RF level –97dBm	—	16	—	dB
Total harmonic distortion		–35	–42	—	dB
Signal-to-noise ratio	No modulation for noise	—	73	—	dB
RSSI output	RF level = –118dBm	0	160	550	mV
	RF level = –68dBm	2.0	2.65	3.0	V
	RF level = –18dBm	4.1	4.85	5.5	V
RSSI range	R4 = 100kΩ pin 5	—	90	—	dB

Taken from *Signetics RF Communication Handbook 1989.*

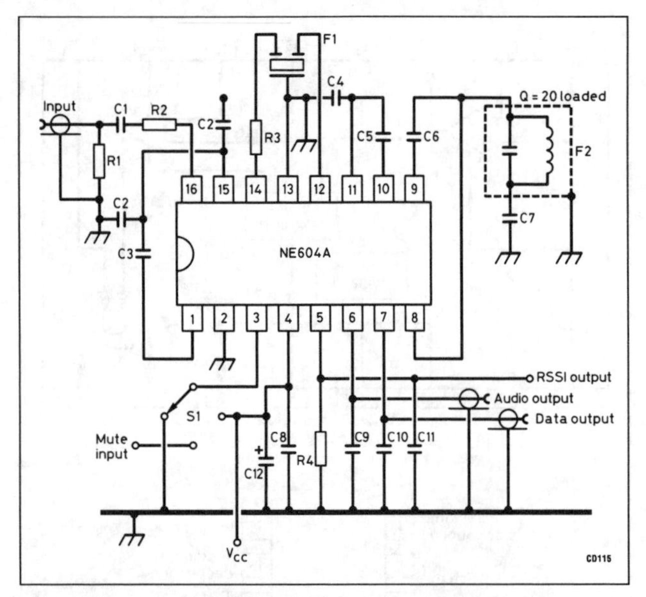

Fig 5.61. The NE604A in its test circuit as a NBFM receiver IF amp and detector. See Table 5.5 for component details *(Signetics RF Communications Handbook 1989)*

Table 5.5. Components list for NE604A NBFM receiver IF amplifier and detector shown in Fig 5.61

R1	51R 1% 0.25W metal film
R2	1k5 1% 0.25W metal film
R3	1k5 5% 0.125W carbon comp
R4	100k 1% 0.25W metal film
C1	10n +80 –20% 63V K10000-Z5V ceramic
C2–5, 7, 8	100n 10% 50V
C6	10p 2% 100V NPO ceramic
C9	15n 10% 50V
C10	150p 2% 100V N1500 ceramic
C11	1n 10% 100V K2000-Y5P ceramic
C12	6µ8 20% 25V tant
F1	455kHz ceramic filter, Murata SFG455A3

direct soldering to PCB tracks. The only external components required are a choke and series resistor in the 12VDC supply lead and DC blocking capacitors at their input and output. As new models appear frequently, no listing is given here. Data sheets are available in the UK from Mainline Electronics, Leicester, who stock the devices.

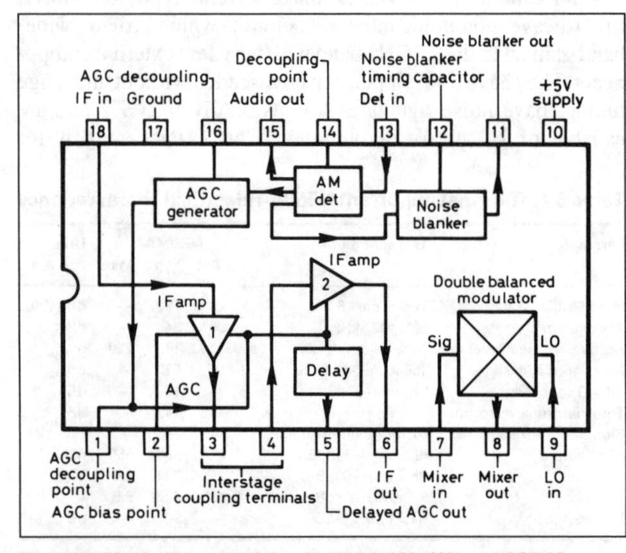

Fig 5.62. Block diagram of the SL6700A AM, CW and SSB IC

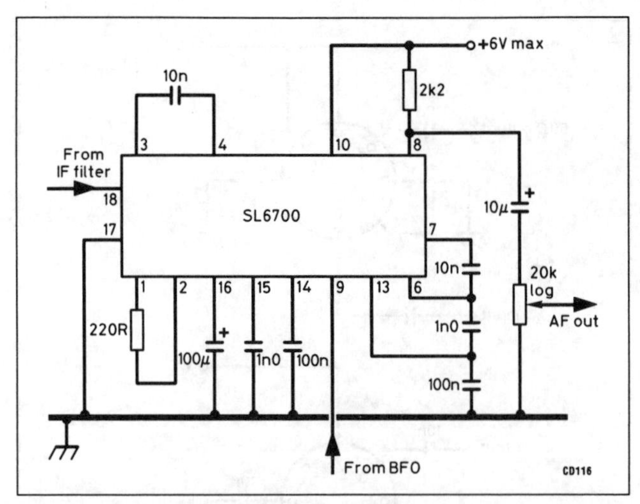

Fig 5.63. In G3TSO's simple 3.5MHz superhet, the SL6700A was used as IF amplifier with AGC and product detector

Discrete-semiconductor RF power amplifiers

Solid-state power amplifiers of the 100W-and-over class use expensive transistors requiring elaborate protection to prevent premature destruction. The RF transformers required for multiband HF amplifiers at the high currents involved demand specialised components and techniques. These constraints do not make such amplifiers attractive to amateur experimenters, but their assembly from kits and/or in accordance with detailed instructions eliminates most of the hazards. See Chapter 7.

Amateurs have succeeded in building RF power amplifiers with much less expensive transistors intended for audio, digital switching or switch-mode power supplies, albeit at lower power levels and sometimes with reduced upper frequency limits. Criteria for stable and efficient RF operation are a high switching speed, low 'on' resistance and not-too-high gate-to-source capacitance. As the loss of such a transistor will not 'break the bank', a whole new area of experimentation has opened up.

Here follow some other results published in recent years. Note that most use push-pull circuits. This suppresses even harmonics and thereby reduces the amount of output filtering required. With any wide-band amplifier a proper harmonic filter for the band concerned *must* be used; without it even the third or fifth harmonic is capable of harmful interference.

Parasitic oscillations have also dogged experimenters. They can destroy transistors before they are noticed. A spectrum analyser is the professional way to find them but with patience a continuous-coverage HF/VHF receiver can also be used to search for them. It is useful to key or modulate the amplifier to create a 'worst case' situation during the search. If found, a ferrite bead in each gate lead and/or a resistor and capacitor in series between each source and drain (as in Fig 5.66) are the usual remedies. W1FB has found negative feedback from an extra one-turn winding on the output transformer to the gate(s) helpful. VHF-style board layout and bypassing, whereby the back of the PCB serves as a ground plane, are of prime importance. The first of the following examples was aimed at constructors without great experience; the next three represent more difficult development projects.

A 1.8–10.1MHz MOSFET 5W power amplifier

A push-pull broadband amplifier using DMOS FETs and providing 5W CW or 6W PEP SSB with 0.1W of drive and a 13.8VDC supply has been described by Drew Diamond, VK3XU [20]. He used two inexpensive N-channel DMOS switching FETs

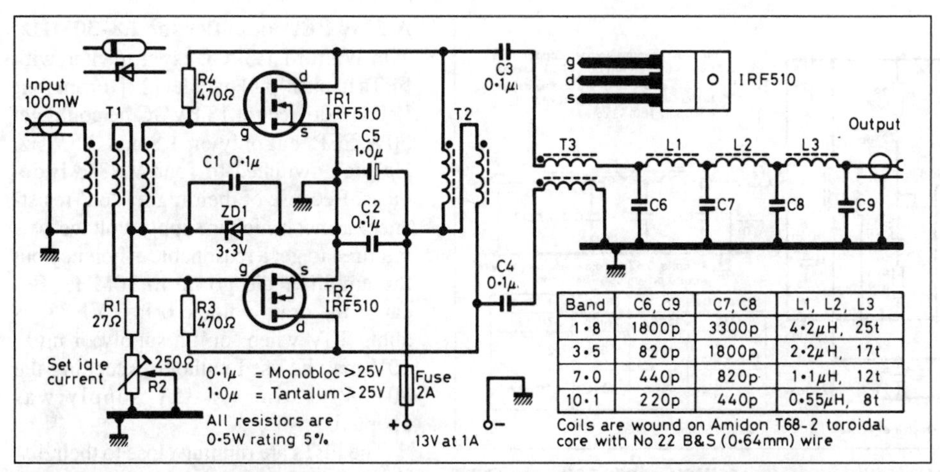

Band	C6, C9	C7, C8	L1, L2, L3
1·8	1800p	3300p	4·2µH, 25t
3·5	820p	1800p	2·2µH, 17t
7·0	440p	820p	1·1µH, 12t
10·1	220p	440p	0·55µH, 8t

Fig 5.64. Circuit diagram of the 1.8–10.1MHz amplifier providing 5W CW or 6W PEP SSB using switching FETs. T1 comprises three 11t loops of 0.5mm enam wire on Amidon FT50-43 core. T2, T3 are three 11t loops of 0.64mm enam wire on Amidon FT50-43. * Indicates start of winding

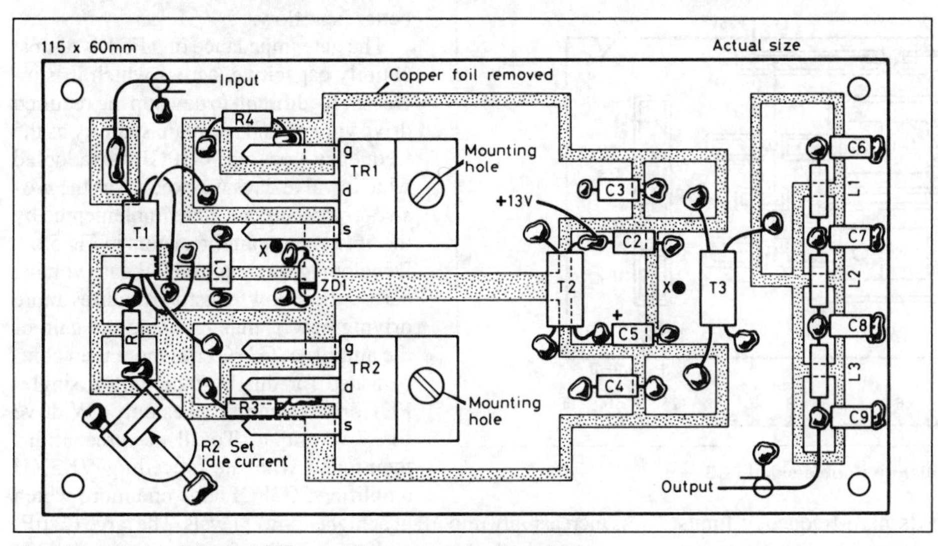

Fig 5.65. Layout of the 5W FET amplifier on double-sided PCB

which make useful RF amplifiers up to about 10MHz. VK3XU's amplifier had a two-tone IMD of better than –30dBc (–35dBc typical) and, with the output filter shown in Fig 5.64, all harmonics were better than –50dBc. The amplifier should survive open or shorted output with full drive and remain stable at any load SWR.

The drain-to-drain impedance of the push-pull FETs is 2 × 24 = 48Ω so that no elaborate impedance transformation is needed to match into 50Ω. T3 serves as a balun transformer. T2 is a balanced choke to supply DC to the FETs. Negative RF feedback is provided by R3 and R4, stabilising the amplifier and helping to keep the frequency response constant throughout the range. The heatsink of the bias zener ZD1 is positioned against the heatsinks of TR1 and TR2 with a small blob of petroleum jelly so that it tracks the temperature of the FETs, causing the bias voltage to go down when the temperature goes up. The amplifier enclosure must have adequate ventilation.

The complete amplifier, with an output filter for one band, can be built on a double-sided 115 × 60mm PCB. For stability, the unetched 'ground plane' should be connected to the etched-side common/earth in at least two places marked 'X' in Fig 5.65. Drill 1mm holes, push wires through and solder top and bottom.

If multiband operation is required, the filter for the highest band should be accommodated on the amplifier board and kept in the circuit on all bands. Lower-frequency filters can then be mounted on an additional board.

Polystyrene or silvered-mica capacitors should be used in the filters. Hard-to-get values can be made up of several smaller ones in parallel.

When setting up, with R2 at minimum resistance the desired no-signal current is 200–300mA. With 100mW drive and a 50Ω (dummy) load, the supply current should be about 1A. After several minutes of operation at this level and with suitable heatsinks, the latter should not be uncomfortably hot when touched lightly. While 100mW drive should suffice on the lower bands, up to 300mW may be needed at 10.1-MHz, which is about the limit for these FETs. Overdriving will cause flat-topping. With larger heatsinks and higher supply voltage, more output would be possible, as is shown below.

A MOSFET 1.8–7MHz 25W power amplifier

VK3XU later described a more powerful version of his 5W amplifier using Motorola MTP4N08 MOSFETs (80V at 4A) made for switch-mode power supplies [21]. This is shown in Fig 5.66. VK3XU lists performance with MTP4N08 devices as: frequency range 1.8–7MHz (with reduced output up to 14MHz); output power nominally 25W, typically 30W PEP or CW with 1W drive; input SWR less than 1.2:1; two-tone IMD about –35dBc; harmonic output, depending on low-pass output filter, –50dBc; No output protection is required; the amp will withstand any output SWR, including short- and open-circuit at full drive; DC supply is 25V at 2A (no regulation required) or 13.8V at reduced output.

The IRF510, having the same pin-out and voltage/current ratings and lower input and output capacitances, should give better performance above 7MHz; proper output filters must be used for each band, eg those shown in Table 5.6, which was taken from reference [19].

One might be tempted to use, in HF amplifiers, VHF transistors salvaged from retired AM PMR transmitters. Because of the very high power gain of VHF transistors used way below their design frequency, stability problems must be expected.

The construction, Fig 5.67, is very much like that of the 5W version. The MOSFETs must each be fitted with an adequate heatsink. If this amplifier were to be operated at its rated 50W input and 25W output, each transistor has to dissipate 12.5W; To hold the FET tab temperature to 125°C at an in-cabinet ambient of 50°, ie a rise of 75°, heatsinks rated less than 75/12.5 = 6°/W would be required; this would be correct in duplex FM or RTTY service but when using morse, AMTOR or SSB, the average dissipation would be roughly half that and 10°/W heatsinks

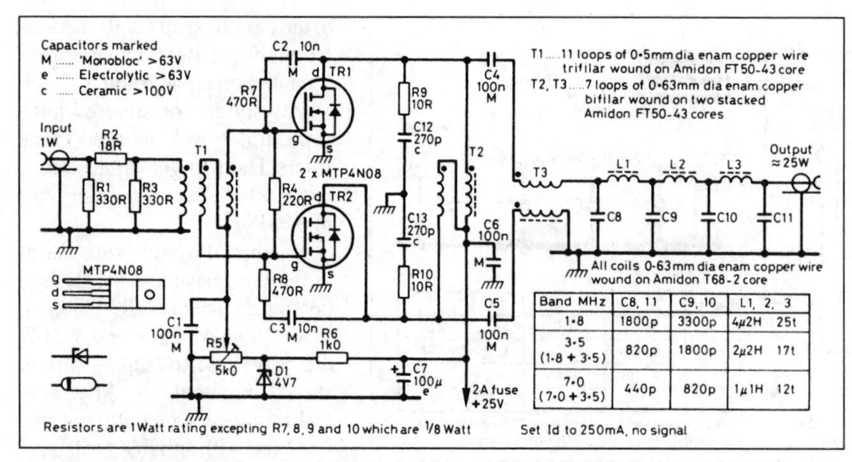

Fig 5.66. Circuit diagram of the 1.8–7MHz amplifier providing 25W CW or PEP SSB using FETs intended for switching power supplies

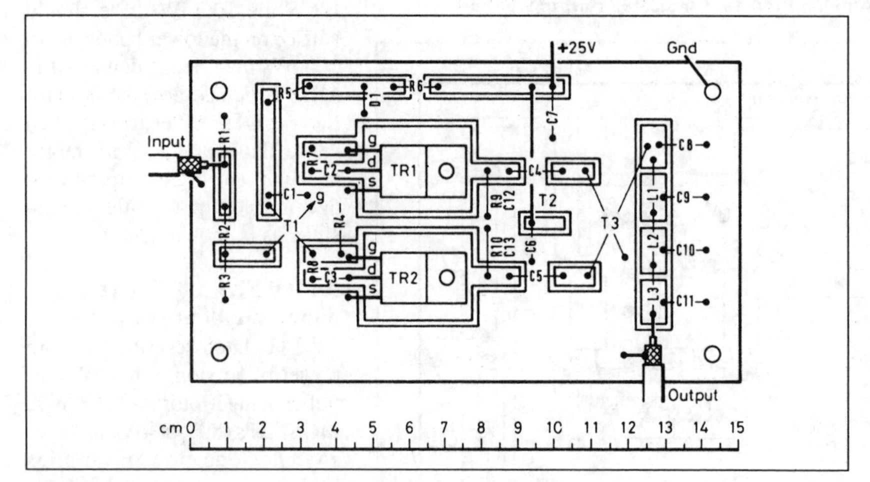

Fig 5.67. Layout of the 25W FET amplifier on double-sided PCB

would do. Pushing the amplifier to its high-frequency limits, however, would reduce efficiency and require larger heatsinks. If in doubt, use a small blower. Do remember that the tabs of these FETs are connected to the drains, so they are 'hot' both for DC and RF.

Table 5.6. 50Ω low-pass output filters for HF amplifiers up to 25W

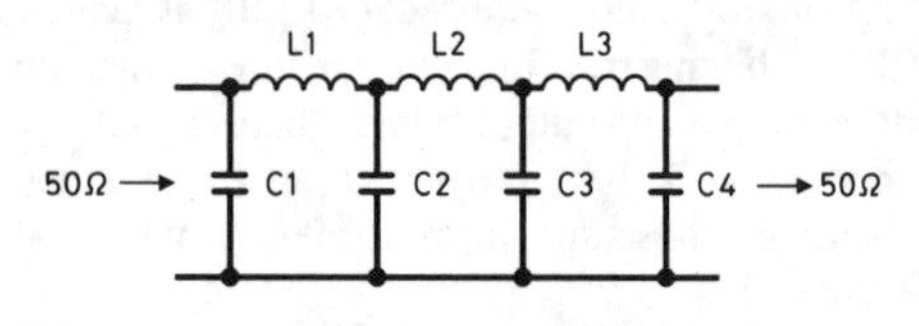

Band (MHz)	C1, C4 (pF)	C2, C3 (pF)	L1, L3 (µH)	(turns)	L2 (µH)	(turns)	Wire (Cu enam)	Cores
3.5	560	1200	2.46	21	2.89	23	0.4mm, 26awg	T-50-2
7	470	820	1.4	17	1.56	18	0.5mm, 24awg	T-50-2
10.1	220	470	0.96	15	1.13	17	0.5mm, 24awg	T-50-6
14	110	300	0.6	11	0.65	14	0.5mm, 24awg	T-50-6
18	100	250	0.52	11	0.65	13	0.6mm, 22awg	T-50-6
21	110	240	0.48	11	0.56	12	0.6mm, 22awg	T-50-6
24	120	270	0.54	12	0.63	13	0.6mm, 22awg	T-50-6
28	56	150	0.3	8	0.38	10	0.8mm, 20awg	T-50-6

Notes. Various cut-off frequencies and ripple factors used to achieve preferred-value capacitors. Coil turns may be spread or compressed with an insulated tool to peak output. Cores are Amidon toroids. Capacitors are silver mica or polystyrene, 100V or more.

A 50W PEP amplifier for 1.8–30MHz

Tim Walford, G3PCJ, experimented with FETs in push-pull parallel [22]. From four IRF510 devices at 13.8VDC he could get 50W PEP, but only on 1.8 and 3.5MHz. He then switched to four VN88AF devices. Because of their higher 'on' resistance, a much higher supply voltage was required to get a reasonable efficiency but the amplifier did go up to 30MHz. Because the drain voltage of the VN88s is about 10V when 'on', a supply of up to 40V may be used without exceeding the 80V V_{ds} rating. A 35V supply was adopted.

The FETs are running close to their dissipation limits, so a large heatsink is required. The VM88AFs were eventually replaced by VM88AFDs because of their better heat flow.

The gate impedance of a FET is almost entirely capacitive, across which it is increasingly difficult to develop the required drive voltage and maintain stability as the frequency goes up; G3PCJ has adopted what in valve days was known as the *passive-grid technique*, here implemented by the 56Ω 'swamping' resistors in Fig 5.68. Developing the required RF gate voltage across these low resistors requires more driving power, thus reducing the gain of the amplifier. G3PCJ has more than compensated for this by including a single-FET driver stage; it takes only 1W drive for 50W output. To all the precautions mentioned with the preceding VK3XU amplifiers, G3PCJ adds one more which is increasingly important at higher power levels. The large (22µF) capacitor in the bias supply allows time for the antenna change-over relay to close before forward bias reaches the FET gates; the diode in the PTT line 'kills' that bias as soon as the PTT switch is released, ie before the antenna relay opens (the PTT switch is connected between +12V and the relay coil to earth). These measures assure that there is no output from the amplifier unless the antenna relay is connecting the load, thus eliminating a source of damage to the relay or the FETs.

50W of RF from one low-cost FET

David Bowman, G8DPW, investigated inexpensive audio transistors at RF [23]. He ended up with the Hitachi 2SK413, an N-channel MOSFET made for ultrasonic power applications. Its specifications are: V_{ds} 140V maximum, V_{gs} ±20V maximum, channel dissipation 100W maximum, rise time 15ns, fall time 50ns, case TO-3P.

For the 7MHz amplifier using this device, Fig 5.69, G8DPW chose an LCL T-output network with a Q of 12; this, in contrast to a wide-band transformer, suppresses harmonics, including the second which could become bothersome in this single-ended design. When it was found on test that for stability a capacitance of 200pF (X_C about $9R_L$) was required between drain and source, the T-network became a pi-L network, but that does not affect the principle. The variable capacitor peaks the output at the operating frequency. G8DPW believes his amplifier would work at much higher frequencies given the proper output tuning/

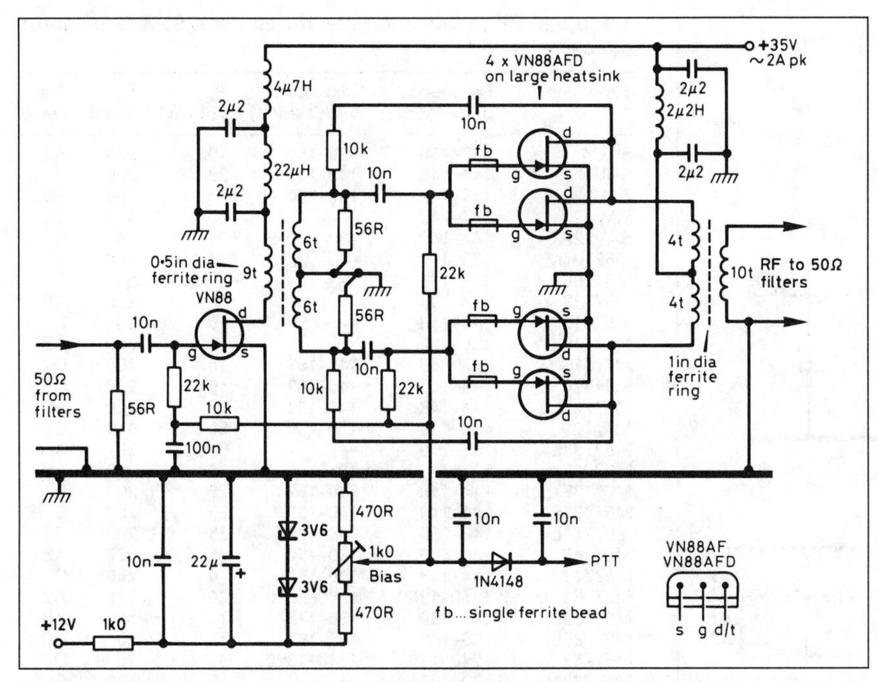

Fig 5.68. 50W PEP from push-pull parallel VN88AFDs. G3PCJ's amplifier works on all bands, 1.8–28MHz

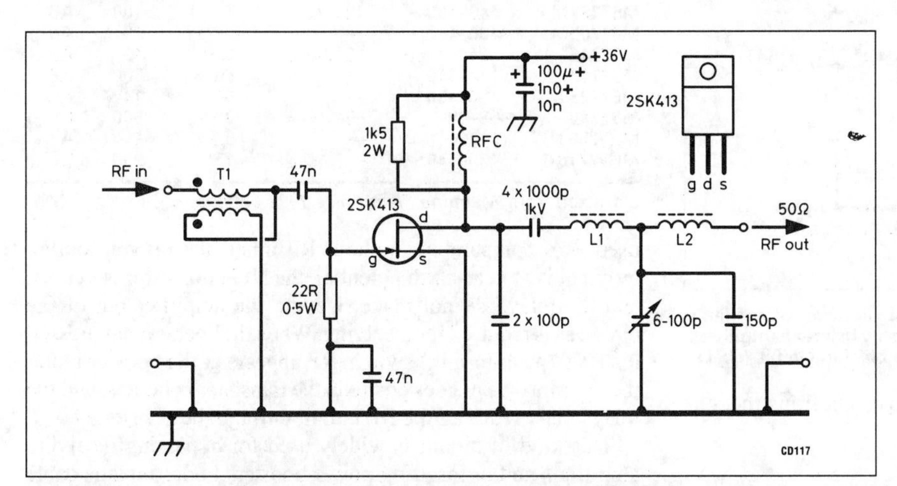

Fig 5.69. A single-ended 50W 7MHz amplifier with an ultrasound power FET. The coil details for 7MHz are as follows. T1, 4:1 broad-band transformer, 8 bifilar turns of 0.5mm enam copper twisted 5t per inch on ferrite core OD 12.4mm, μ = 125 (Bonex No 530090). L1, 18t 1.2mm on T-130-6 toroid. L2, 23t 1.2mm on T-130-6 toroid. Capacitors rated at 350V. RFC, 14t 1mm enam copper on 1in OD ferrite toroid. Performance details are given in Table 5.7

matching network. He tried 14MHz; it worked, but not enough drive was available to establish the efficiency. The passive gate, biasing and heatsinking techniques are as for the G3PCJ amplifier described above.

Considering the higher power level, the external low-pass filters require more substantial inductors that those shown in Table 5.6.

Table 5.7. Performance of 50W single-FET amplifier

Frequency (kHz)	Supply voltage (V DC)	CW output power (W)	Efficiency (%)
7100	36	52	73–75
7100	43	73	73–75
14,330	43	38	—

Drive power at 7 and 14MHz was approximately 1.5 and 1W respectively.

VHF linear amplifiers

For medium-power VHF linear amplifiers there is another cost-saving option. Because transistors intended for Class C (AM, FM or data) cost a fraction of their linear counterparts, it is tempting (and feasible) to use them even for linear applications. If simple diode biasing alone is used, however, the transistors, such as the MRF227, go into thermal runaway. Two solutions to this problem were suggested in the ARRL lab for further experimentation [24]. Both methods preserve the advantages of a case earthed for DC.

The current-limiting technique of Fig 5.70(a) should work with any device, but is not recommended where current drain or power dissipation are important considerations. The approach is simply to use a power supply which is current limited, eg by a LM317 regulator, and to forward bias the transistor to operate in Class B. Forward bias is chosen by the values of R_{B1} and R_{B2}.

The active biasing circuit of Fig 5.70(b) is not new but has not been much used by amateurs. The collector current of the transistor is sampled as it flows through a small-value sensing resistor R_S; the voltage drop across it is amplified by a factor R_F/R_I, the gain of the op-amp differential amplifier circuit. If the collector current goes up, the base voltage is driven down to restore the balance. The gain required from the op-amp circuit must be determined empirically. While experimenting, the use of a current-limiting power supply is recommended to avoid destroying transistors.

50–1296MHz power amplifier modules

Designing a multi-stage UHF power amplifier with discrete components is no trivial task. More often than not, individually-tuned interstage matching networks are required. Duplicating even a proven design in an amateur workshop or on the factory floor has its pitfalls. One solution, though not the least expensive, is the use of a sealed modular sub-assembly. These are offered by several semiconductor manufacturers, including Mitsubishi, Motorola and Toshiba. For each band, there is a choice of output power levels, frequency ranges, power gain and class of operation (linear in Class AB for all modes of transmission or non-linear in Class C for FM or data modes only). Table 5.8 is a partial list from the 1992 Mainline Electronics (Leicester) catalogue. All require a 13.8V DC supply, sometimes with separate external decoupling for each built-in stage, and external heatsinking. In general, their 50Ω input provides a satisfactory match for the preceding circuitry, and a pi-tank (with linear inductor on the higher bands) is used for antenna matching and harmonic suppression. They are designed to survive and be stable under any load, including open- and short-circuit. As each model has

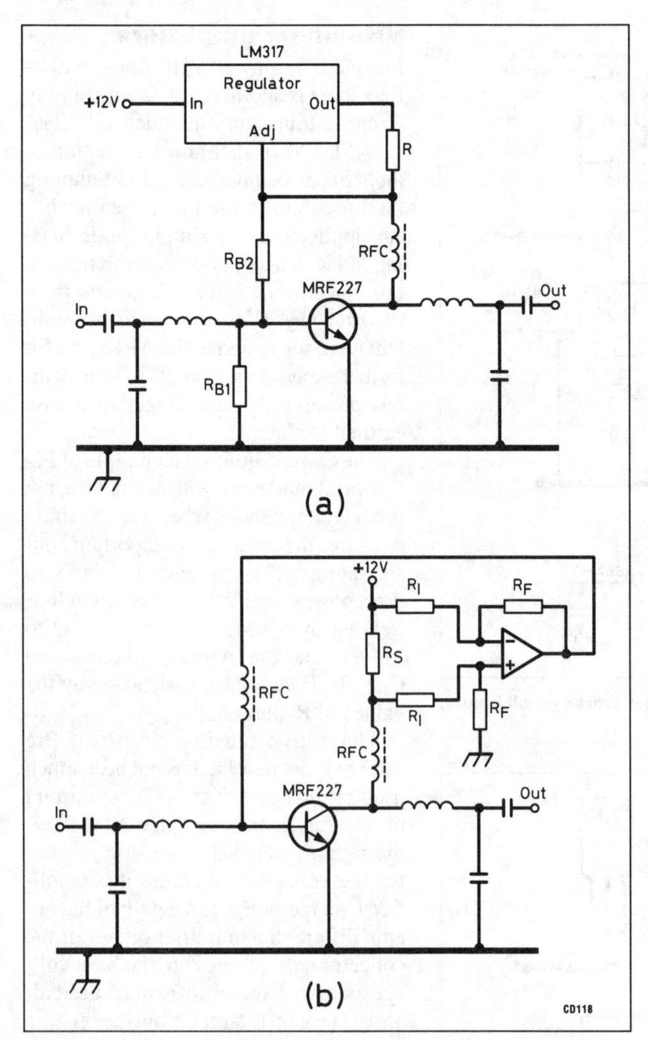

Fig 5.70. Two methods to prevent thermal runaway of medium-power VHF transistors designed for Class C only in linear amplifiers. (a) A supply current regulator. (b) An active biasing circuit *(QST)*

its peculiarities, it is essential to follow data sheet instructions to the letter.

A 435MHz 15W mobile booster

PA0GMS built this 70cm power amplifier with RF VOX to boost the FM output from his hand-held transceiver to a respectable 15W for mobile use [25]. The circuit is shown in Fig 5.71. Its top half represents the amplifier proper, consisting of (right to left) the input T/R relay, input attenuator to limit input power to what the module requires for rated output, the power module with its power lead decoupling chokes (ferrite 6 × 3mm beads with three turns of 0.5mm diameter enamelled copper wire) and bypass capacitors (those 0.1µF and larger are tantalum beads with their negative lead earthed; lower values are disc ceramics, except for the 200pF which is a feedthrough type). The output tank (two air trimmers and a 25mm long straight piece of 2mm silvered wire) feeds into the output T/R relay.

Both relays are National model RH12 shielded miniature relays not designed for RF switching but adequate for UHF at this power level. The choke between them is to stop RF feedback through the relays. The relays are activated by what is sometimes misnamed a *VOX* (voice-operated control) but is in fact an RF-operated control. Refer to the diagram at the bottom of Fig 5.71. A small fraction of any RF drive applied to the amplifier input is rectified and applied to the high-gain op-amp circuit here

Table 5.8. RF power modules for all 50–1296MHz amateur bands

Type	Frequency band (MHz)	Max band-width (MHz)	P_{out} (min W)	P_{in} (mW)	Bias class
SAU4	430–450	430–450	10	200	AB
SAV7	144–148	143–149	28	200	C
SAV12	144–148	130–160	5	150	AB
SAV17	144–148	—	50	400	C
SAV22A	144–148	130–160	7	15	C
M57704M	430–450	—	13	200	C
M57706	145–175	—	8	200	C
M57713	144–148	—	10	200	AB
M57715	144–148	—	10	200	C
M57716	430–450	—	10	200	AB
M57716N	142–163	140–165	10	200	C
M57726	144–146	140–152	40	200	C
M57727	144–148	140–150	25	200	AB
M57729	430–450	430–450	30	600	C
M57732L	135–160	—	5	20	AB
M57735	50–54	48–56	10	200	AB
M57737	144–148	140–152	25	200	C
M57737R	144–148	140–152	25	200	C
M57741L	148–160	140–170	25	200	C
M57745	430–450	430–450	25	300	AB
M57747	144–148	140–150	10	200	C
M57762	1240–1300	1240–1300	20	1000	AB
M57783L	135–160	125–175	7	50	AB
M57785L	135–150	127–163	7	50	AB
M57787	1240–1300	1240–1300	3	10	C
M57788M	430–450	—	45	400	C
M57796H	150–175	140–190	7	200	AB
M57796MA	144–148	135–160	7	200	AB
M57797MA	440–450	420–460	7	100	AB
M67705M	430–470	—	7	20	AB
M67715	1240–1300	1220–1320	1.5	10	AB
M67727	144–148	—	65	500	AB
M67728	430–450	—	65	14W	AB
M67742	68–88	60–100	30	500	C
MHW591	1–250	—	0.7	1	A
MHW710-1	400–440	—	15	250	C

Extracted from *Mainline Electronics 1992 Catalogue.*

used as a comparator. Without RF input, the op-amp output terminal is at near-earth potential, the NPN transistor is cut off, and the relays do not make, leaving the amplifier out of the circuit as is required for receiving. When the operator has pressed his PTT (push-to-talk) switch, RF appears at the booster input, the op-amp output goes positive, the transistor conducts, and the relays make, routing the RF circuit through the booster.

This kind of circuit is widely used for bypassing receiving amplifiers and/or inserting power boosters in the antenna cable when transmitting. At VHF or UHF, this is often done to make up for losses in a long cable by placing these amplifiers at masthead or in the loft just below. Though not critical, the 1pF capacitor may have to be increased for lower frequencies and lower power levels, and reduced for higher frequencies and higher input power. The NPN transistor must be capable of passing the coil currents of the relays. While not an issue with 0.5W in and 15W out, at much higher power levels proper relay sequencing will be required. See G3PCJ's four-FET 50W HF amplifier and Fig 5.68 above.

No PCB is used for the amplifier proper. It is built into a 110 × 70 × 30mm tin-plate box, Fig 5.72. The module is bolted through the bottom of the box to a 120 × 70mm heatsink which must dissipate up to 30W. Heat transfer compound is used between the module and the box and also between the box and the heatsink; both must be as flat as possible. Care is required when tightening the bolts as the ceramic substrate of the module is brittle. The other components are soldered in directly. The RF-operated relay control is a sub-assembly mounted on a 40 × 30mm PCB.

Adjustments are simple. Set the input attenuator to maximum

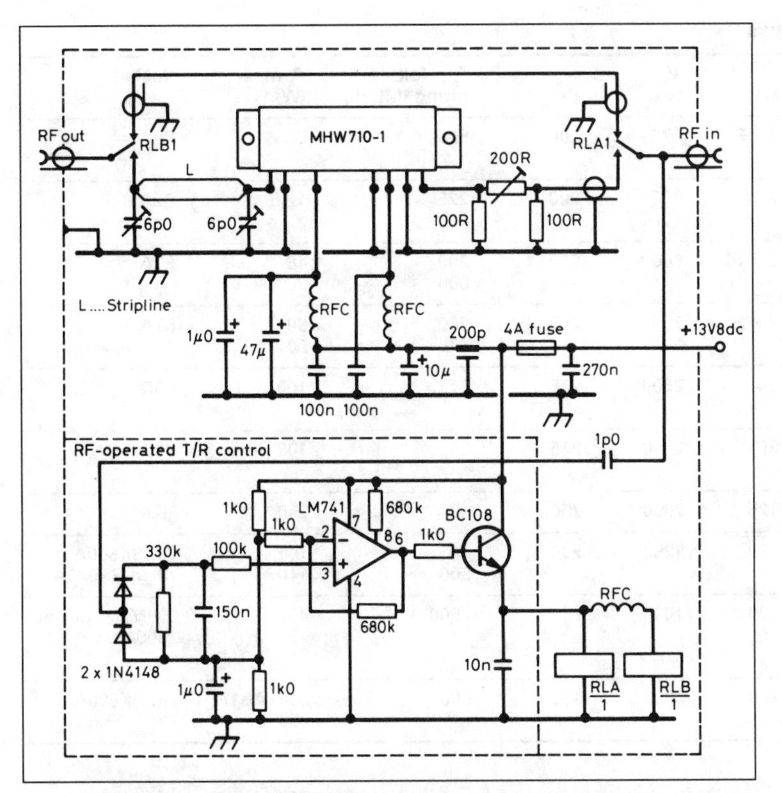

Fig 5.71. Circuit of the PA0GMS 15W 435MHz mobile booster using a power module and a universal RF-operated T/R control

resistance and the two air trimmers to half-mesh. Apply 0.6 or 0.7W input from the hand-held and verify that both relays make. A few watts of output should be generated at this stage. Adjust the two air trimmers to peak the output, then reduce the input attenuation until the output reaches 15W with a supply current of about 3A. This completes the adjustment. Harmonics were found to be –60dBc and intermodulation –35dBc.

Valve power amplifiers

Warning – all valve power amplifiers use lethal voltages. Never reach into one before making *sure* that the HT is off and HT filter capacitors discharged; do not rely on bleeder resistors – they can fail open-circuit without you knowing it. If there is

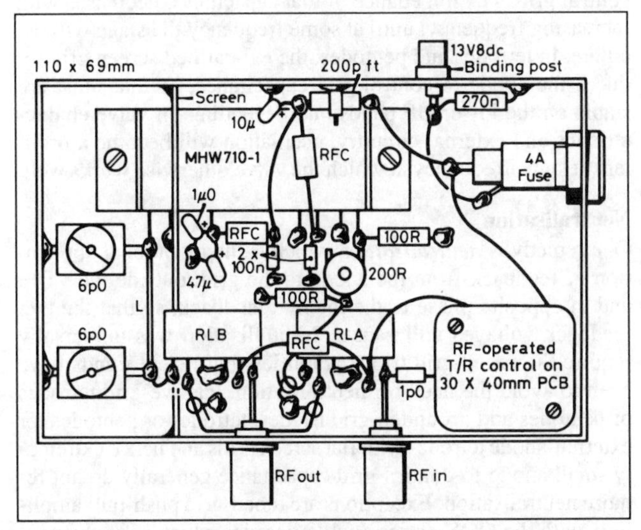

Fig 5.72. Construction of the PA0GMS 15W 435MHz mobile booster. The RF-operated T/R control is a sub-assembly on a PCB

mains voltage in the amplifier enclosure, be sure that the plug is pulled before working on it.

Virtually all current factory-made HF amplifiers up to the legal power limit (400W = 26dBW PEP output in the UK) are solid state. The same is true for VHF and UHF equipment up to 50W. For home construction, however, the criteria are different. Power transistors are expensive and can be instantly destroyed, even by minor abuse; some components, such as a transformer for a 50V/20A power supply, are not easy to find at a reasonable price. By contrast, valves do tolerate some abuse and those who consider the search for inexpensive components an integral part of their hobby find rich pickings at rallies and surplus sales. Table 5.9 gives rough operating conditions (PEP or CW) for some valves still popular with amateur constructors.

All high-power amplifiers require an RF-tight enclosure and RF filtering of supply and control leads coming out of them. Doing otherwise invites unnecessary interference with any of the electronic devices in your and your neighbours' homes.

Technical alternatives will now be considered and several economical ways of implementing them are given. All are single-stage linear amplifiers intended to boost the output power of a transceiver.

Choosing valves

The first choice concerns the valves themselves; one to four of these must be capable of delivering the desired power gain and output power at the highest intended operating frequency at reasonable filament and anode voltage and, where required, control-grid bias and screen-grid voltage. The availability and cost of the valve(s), valve socket(s), power supply transformer(s) and anode tank capacitor will frequently restrict the choice.

The only valves of recent West-European production which are of interest to the economy-minded constructor of HF amplifiers were not designed for radio transmitters but to drive deflection yokes and high-voltage transformers at the horizontal sweep frequency in CTV receivers; accordingly, the manufacturers' data sheets are not much help to the designer of RF circuits. However, amateurs have published the results of their experiments over the years. For the popular (originally Philips) pentode PL519, which can produce 100W PEP at HF and reduced output up to 50MHz, some information is given in Tables 5.9 and 5.10. With fan cooling, long valve life can be expected.

Another HF favourite is the (originally GE) 'beam power' tetrode type 813. This valve, first made more than 50 years ago, is now available new from Eastern European and Chinese manufacturers at a reasonable price but there are good buys in used American JAN-813s available; these can be tested by comparison with an 813 which is known to be good in the type of amplifier and at the approximate frequency and ratings of intended usage. This valve is rated at 100W anode dissipation; with 2500V on the anode and forced-air cooling a single valve can provide an output of 400W PEP, but it is better practice to use two valves in push-pull or parallel to produce this output without such cooling. One disadvantage is the requirement for a hefty filament transformer; each 813 requires 10V at 5A, AC or DC. A detailed description by G3ISD of a 400W HF amplifier using two 813s appeared in reference [26].

For VHF and UHF two series of valves stand out. One comprises the double-tetrodes (originally Philips) QQV02/6,

Table 5.9. Ratings of some commonly used PA valves

Type	Base	Heater (V)	Heater (A)	P_a (W)	V_a (V)	V_{g2} (V)	F_{max} full rating (MHz)	P_o max (W)	Socket
QQV02-6*	B9A (Fig 5.48)	6.3 12.6	0.8 0.4	2 × 3	275	200	500	5	B9A
QQV03-10* 6360	B9A (Fig 5.48)	6.3 12.6	0.8 0.4	2 × 5	300	200	225	12.5	B9A
QQV03-20A* 6252	B7A (Fig 5.73)	6.3 12.6	1.3 0.65	2 × 10	600	250	200 600	48 20	B7A
QQV06-40A* 5894	B7A (Fig 5.73)	6.3 12.6	1.8 0.9	2 × 20	750	250	200 475	90 60	B7A
PL519/40KG6A 40KD6A	B9D (Fig 5.74) 9RJ (Fig 5.74)	42	0.3	35	2500	275	21	100	B9D
EL519/6KG6A 6KD6A	B9D (Fig 5.74) 9RJ (Fig 5.74)	6.3	2.0	35	2500	275	21	100	B9D
813	5BA	10	5.0	125	2500	750	30	250	5BA
4X150A QV1-150A	B8F Special	6.0	2.6	150	1250	200	165 500	195 140	2m SK600A 70cm SK620A
2C39A 7289 3CX100A5	Disc seal	6.3	1.05	100	1000	—	2500	40 17	500MHz special 2500MHz
4CX250B QE61-250	B8F Special	6.0	2.6	250	2000	400	500	300 (AB1) 390 (C)	2m SK600A 70cm SK620A

* Double tetrode with built-in neutralisation.

QQV03/10, QQV03/20A and QQV06/40A; the other is a family of (originally Eimac) 'external-anode' valves including the beam-power tetrodes 4X150A and 4CX250B. Though expensive when new, these valves can be found 'good used' as a result of the military and avionics practice of replacing valves on a 'time expired' rather than 'when worn out' basis. Test them as recommended for the 813 above.

On 144MHz and below, a QQV06/40A in Class AB1 can produce up to 100W PEP on a plate voltage of 1kV. In 432MHz operation, the efficiency will be lower, but 60W can be obtained with reduced input; fan cooling will increase the life expectancy of valves run near their dissipation limits.

Obsolete mobile and base station transmitters using these double-tetrodes are sold at very low prices. Their power amplifier sub-assemblies often can be converted into effective amateur amplifiers with little or no mechanical work.

External-anode valves can produce 'maximum legal' power on VHF and UHF but they always require forced-air cooling and the construction of the amplifiers is mechanically demanding. Convertible surplus equipment with them is not often found but several proven designs are detailed in Chapter 8 and the RSGB *VHF/UHF Manual*.

For high power in the 23 and 13cm bands, the disc-seal triode 2C39A is universally used. Amplifier designs appear in Chapter 9 and the RSGB *Microwave Handbook*.

Table 5.10. Two-tone measurements by G4DTC on his PL519 grounded-grid amplifier with fan cooling

Test frequency	3.7MHz
Anode voltage	710V
Anode current (no signal)	20mA
Anode current (max signal)	162mA*
Control-grid bias	–5V set for minimum cross-over distortion on 'scope
Average RF current into 70Ω	1.15A
PEP RF output	185W*
PEP RF input	20W* (estimated)
Anode load resistance	2kΩ (estimated)
Valve inter-electrode capacitances	
Anode to all other electrodes	22pF measured cold
Anode to control grid	2.5pF from data sheet
Control grid to cathode	20pF measured cold

* At this level, the solder of the anode caps would melt even with fan cooling! It is reassuring to know that overdriving the amplifier will not distort the output, ie cause splatter, or instantly destroy the valve but it obviously is not for routine operation.

Valve configuration

There are three different ways to drive an amplifier valve: tuned grid, passive grid and grounded grid; the advantages and disadvantages of each are shown in Table 5.11. One disadvantage listed for tuned-grid amplifiers is the requirement for neutralisation; this may need explanation. The anode-to-control-grid capacitance within a valve feeds part of the RF output back to the control grid. The impedance of that capacitance decreases with increasing frequency, until at some frequency the stage will oscillate. In tetrodes and pentodes, the RF-earthed screen grid reduces the anode-to-control-grid capacitance, so that most are stable on the lower HF bands but, depending on valve characteristics and external circuitry, oscillation will become a problem at some frequency at which the valve otherwise works well.

Neutralisation

One remedy is *neutralisation*, which is the intentional application of feedback from the anode to the grid, outside the valve and in opposite phase to the internal feedback, so that the two feedback voltages will cancel. Neutralisation was universally required in tuned-grid triode amplifiers but recent trends have been to avoid the need for neutralisation. Passive-grid tetrodes or pentodes and grounded-grid triodes, tetrodes or pentodes, or external-anode tetrodes with flat screen grids and hence extremely small anode-to-control-grid capacitance generally do not require neutralisation. Exceptions are tuned-grid push-pull amplifiers with the QQV-series double-tetrodes which have the neutralisation capacitors built-in.

Table 5.11. Advantages and disadvantages of tuned-grid, passive-grid and grounded-grid PA valves

TUNED GRID, CLASS AB1

Advantages
(a) Low driving power.
(b) As there is no grid current, the load on the driver stage is constant.
(c) There is no problem of grid bias supply regulation.
(d) Good linearity and low distortion.

Disadvantages
(a) Requires tuned grid input circuit and associated switching or plug-in coils for multiband operation.
(b) Amplifier must be neutralised.*
(c) Lower efficiency than Class AB2 operation.

TUNED GRID, CLASS AB2

Advantages
(a) Less driving power than passive grid or cathode-driven operation.
(b) Higher efficiency than Class AB1.
(c) Greater power output.

Disadvantages
(a) Requires tuned grid input circuit.
(b) Amplifier must be neutralised.
(c) Because of wide changes in input impedance due to grid current flow, there is a varying load on the driver stage.
(d) Bias supply must be very 'stiff' (have good regulation).
(e) Varying load on driver stage may cause envelope distortion with possibility of increased harmonic output and difficulty with TVI.

PASSIVE GRID

Advantages
(a) No tuned grid circuit.
(b) Due to relatively low value of passive grid resistor, high level of grid damping makes neutralising unnecessary.
(c) Constant load on driver stage.
(d) Compact layout and simplicity of tuning.
(e) Clean signal with low distortion level.
(f) Simple circuitry and construction lending itself readily to compact layout without feedback troubles.

Disadvantages
(a) Requires higher driving power than tuned grid operation.

CATHODE DRIVEN

Advantages
(a) No tuned grid circuit.
(b) No neutralising (except possibly on 28MHz).
(c) Good linearity due to inherent negative feedback.
(d) A small proportion of the driving power appears in the anode circuit as 'feedthrough' power.

Disadvantages
(a) High driving power – greater than the other methods.
(b) Isolation of the heater circuit with ferrite chokes or special low-capacitance-wound heater transformer.
(c) Wide variation in input impedance throughout the driving cycle causing peak limiting and distortion of the envelope *at the driver.*
(d) The necessity for a high-*C* tuned cathode circuit to stabilise the load impedance as seen by the driver stage and overcome the disadvantage of (c).

* Neutralisation is built into the QQV-series of double tetrodes.

A double-tetrode linear VHF amplifier

The amplifier shown in Fig 5.73 [27] is popular for 50, 70 and 144MHz; differences are in tuned circuit values only. With 1.5–3W PEP drive, 70–90W PEP output can be expected, depending on frequency. For CW and SSB, convection cooling is sufficient. For FM or data, the input power should be held to 100W or a little more with fan cooling. The anodes should not be allowed to glow red. The circuit exemplifies several principles with wider applicability. Typical values for this particular amplifier are given below in *(italics)*.

- It is a push-pull amplifier, which cancels out even harmonics; this helps against RFI, eg from a 50MHz transmitter into the FM broadcast band.
- It also makes neutralisation easier; the neutralisation achieved by capacitors from each control grid to the anode of the other tetrode (here within the valve envelope) is frequency independent as long as both grid and anode tuned circuits are of balanced construction and well shielded from each other.
- Control-grid bias *(30V)* is provided from a fairly low-impedance source set for a small *(35mA)* 'standing' (ie zero signal) current between 10 and 20% of the peak signal current *(250mA)*; the latter must be established by careful adjustment of the RF input. This is different from Class C amplifiers (in CW, AM and FM transmitters) with large (tens of kilohms) grid resistors which make the stage tolerant of a wide range of drive levels.
- The screen-grid voltage is stabilised; this also differs from the practice in Class C amplifiers to supply the screen through a dropping resistor from the anode voltage; that could cause non-linear amplification and excessive screen-grid dissipation in a Class AB1 linear amplifier.

  Where very expensive valves are used, it is wise to arrange automatic removal of the screen-grid voltage in case of failure of the anode voltage; a valve with screen-grid but no anode voltage will not survive long. This can be combined with the function of relay RLC, which removes screen-grid voltage during receive periods, thereby eliminating not only unnecessary heat generation and valve wear but also the noise which an 'idling' PA sometimes generates in the receiver.
- Tuning and coupling of both the grid and plate tanks of a linear amplifier should be for maximum RF output from a low but constant drive signal. This again differs from Class C amplifiers, which are tuned for maximum 'dip' of the DC anode current.
- This amplifier circuit has been used on 50, 70 and 144MHz. Table 5.12 gives tuned-circuit data; there is no logical progression to the coils from band to band because the data were taken from the projects of three different amateurs. Using a 'dip' oscillator, ie before applying power to the amplifier, all tuned circuits should, by stretching or squeezing the air-wound coils, be pre-adjusted for resonance in the target band, so that there is plenty of capacitor travel either side of resonance.
- Note the cable from the power supply to the amplifier: the power supply has an eight-pin chassis socket and the amplifier an eight-pin chassis plug; the cable has a plug on the power supply end and a socket on its amplifier end. Disconnecting this cable at either end never exposes dangerous voltages.

A grounded-grid 200W linear HF amplifier

The amplifier shown in Fig 5.74 was designed by G3TSO to boost the output of HF transceivers in the 5 to 25W class by about 10dB [28]. It uses a pair of CTV sweep valves in parallel; originally equipped with the American 6KD6, the European EL519 or PL519 are equivalent and the pin numbers in the diagram are for them. The choice depends only on the availability and price of the valves, sockets and filament supply; the 6KD6 and EL519 have a 6.3V, 2A filament, while the PL519 requires 42V, 0.3A.

The two filaments can be connected in series, as shown in the diagram, or in parallel. Table 5.13 provides details of some other components.

A few of the design features are:

- It may seem that the pi-input filter, L1 and C1–C11, is superfluous; it is if the driving stage is a valve with a resonant anode circuit. The cathodes do not, however, present to a wide-band semiconductor driver a load which is constant over the RF cycle, causing distortion of the signal *in the driver stage.* This is avoided by inclusion of a tuner which does double duty as as an impedance transformer from 50Ω. The pi-filter

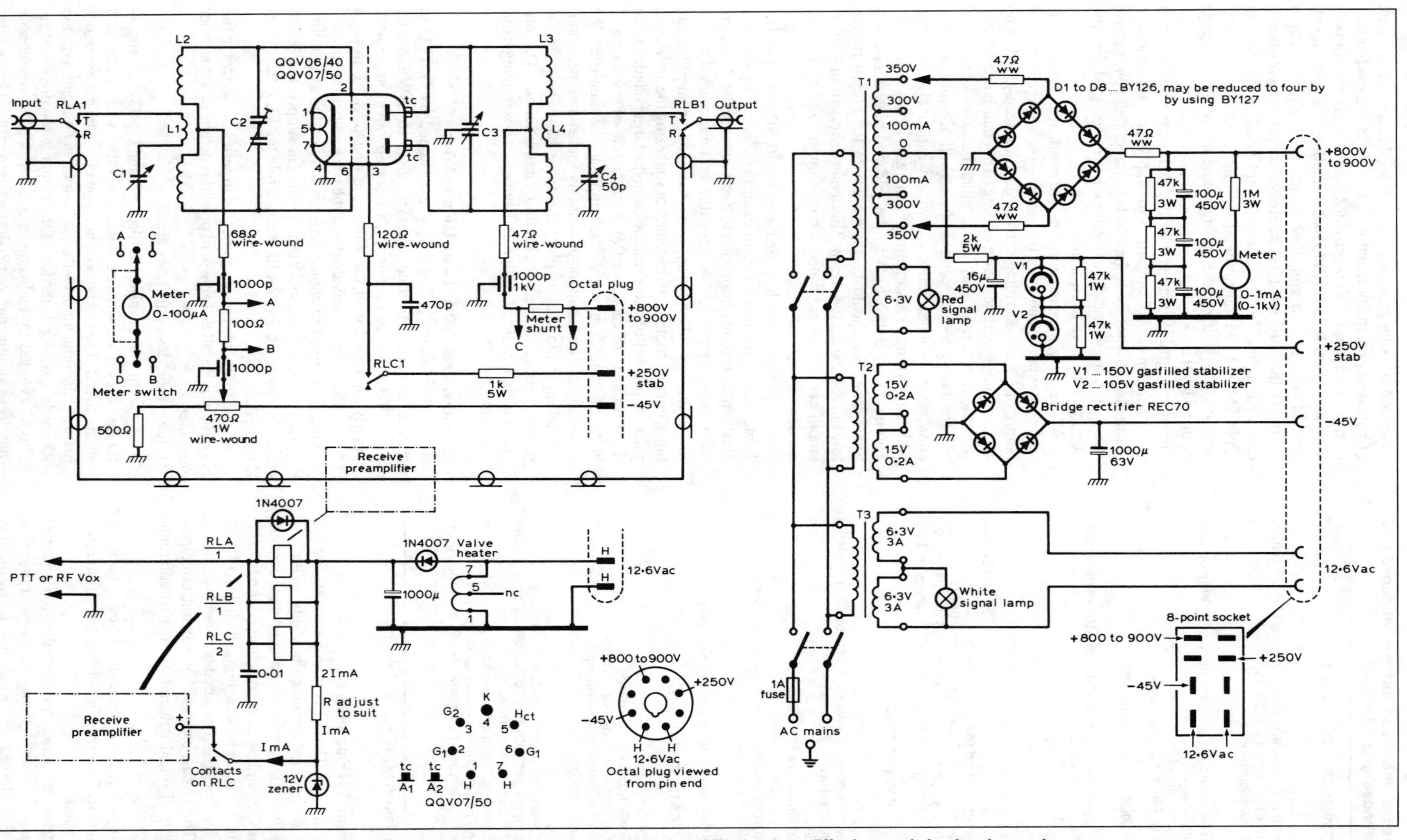

Fig 5.73. A tuned-grid double-tetrode linear VHF amplifier and its power supply. 50, 70 and 144MHz versions differ in tuned-circuit values only

Table 5.12. Tuned circuits for the double-tetrode linear VHF amplifier

Band	C1	L1	L2	C2	C3	L3	L4
144MHz	30pF	1½t, 8mm ID, 2mm dia insulated	2 + 2t, 8mm ID, 2mm dia, length 28mm, gap 8mm	20 + 20pF	12 + 12pF	2 + 2t, 16mm ID, 3mm dia Cu tubing, length 25mm, gap 8mm	1t 16mm ID, 2mm dia well-insulated
70MHz	50pF	2t 9.5mm ID, 0.9mm dia insulated	8 + 8t, 9.5mm ID, 1.2mm dia	20 + 20pF	20 + 20pF	8 + 8t 25mm ID, 1.2mm dia	1t 25mm ID, 1.2mm dia well-insulated
50MHz	50pF	2t 15mm ID, 0.8mm dia insulated 7cm to C1 twisted	3 + 3t 15mm ID, 0.8mm dia	50 + 50pF	30 + 30pF	6 + 6t 15mm ID, 1mm dia	2t 15mm ID, 0.8mm dia well-insulated, 10cm to C4 twisted

C2 and C3 are split stator. C4 is 50pF.

is sufficiently flat to cover each band without retuning, but not to include any of the WARC bands.

- The valves, with all three grids earthed, ie connected as *zero-bias triodes* and without RF drive, would remain within their dissipation rating; with separately adjustable bias to each control grid, however, it is possible not only to reduce the no-signal anode current to the minimum necessary for linearity *(20mA each valve)* but also, by equalising the two idling currents, assure that the two valves will share the load when driven. This is particularly important when valves with unequal wear or of different pedigree are used together. It may make the purchase of 'matched pairs' unnecessary.
- When valves are operated in parallel, the wires strapping like electrodes together, along with valve and socket capacitances, will resonate at some VHF or UHF frequency. If the valves have sufficient gain at that frequency, the stage may burst into parasitic oscillation, especially at the modulation peaks when power gain is greatest. Even in a single-valve or push-pull stage this can happen as HF tuning capacitors are short-circuits at VHF or UHF. In this amplifier, several precautions are taken. Both pins of each screen and suppressor grid are earthed to chassis by the shortest possible straps; the control grids are by-passed to earth through C13 and C14 in a

Table 5.13. Components for the G3TSO grounded-grid 200W linear HF amplifier (Fig 5.74)

L1	16t, 1.6mm Cu-tinned, 25mm dia, 51mm long, taps at 3, 4, 6, 10t
L2	18t, 1.2mm Cu-tinned, 38mm dia, first 4t treble spaced, remainder double spaced. Taps at 3, 4, 6, 9t
L3, 4	5t 1.2mm Cu-enam wound on 47Ω 1W carbon resistor
RFC1	40t 0.56mm Cu-enam on ferrite rod or toroid (μ = 800)
RFC2	40t 0.56mm Cu-enam on 12mm dia ceramic or PTFE former
RFC3	2.5mH rated 500mA
RFC4	15 bifilar turns, 1.2mm Cu-enam on ferrite rod 76mm × 9.5mm dia
C1–4, 9, 10, 12–14	Silvered mica
C5–8	Compression trimmers
C15	Two 500p ceramic TV EHT capacitors in parallel
C16	4.7n, 3kV or more ceramic, transmitting type or several smaller discs in parallel
C18	3 × 500p ganged air-spaced variable (ex valved AM receiver)
C19	1n 750V silver mica, transmitting type or several smaller units in parallel
R1, 2	10R 3W carbon (beware of carbon film resistors, which may not be non-inductive)
RV1, 2	10k wirewound or linear-taper cermet potentiometer
Coaxial cable	RG58

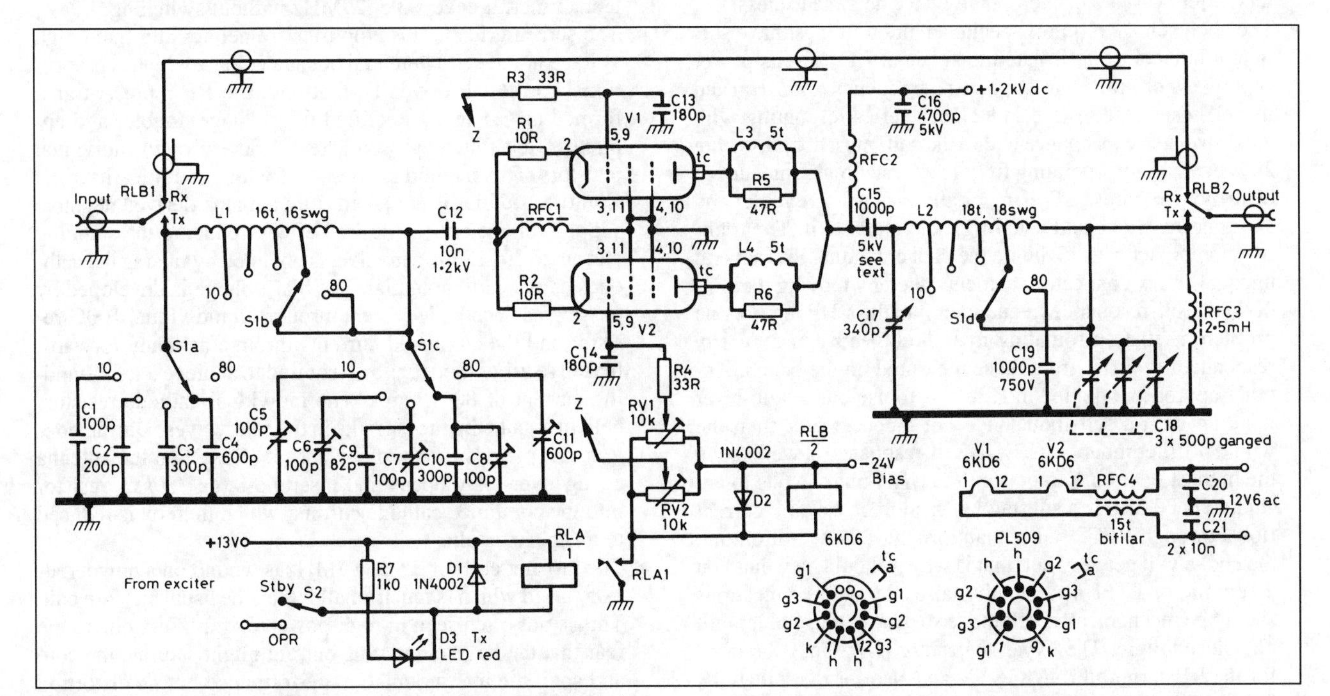

Fig 5.74. A grounded-grid 200W linear HF amplifier built by G3TSO, mostly from junk-box parts. No screen-grid supply is required

like manner; some valve sockets have a metal rim with four or more solder lugs for that purpose. Carbon stopper resistors R5 and R6 are used to lower the Q of parasitic resonances. Mere resistors would dissipate an intolerable fraction of the output, especially on the highest operating frequency where the valve's anode to earth capacitances are a large part of the total tank capacitance and carry high circulating currents; small coils L3 and L4 are therefore wound on the stoppers. These coils are dimensioned to have negligible impedance at all operating frequencies but not to short the stopper resistor at a potential parasitic frequency. The resistor with coil is called a *parasitic suppressor* and considerable experimentation is often required; sometimes one even can get away without them.

- The insulation between valves' cathodes and filaments was never made to support RF voltages. The cathodes of grounded-grid amplifiers, however, must be 'hot' for RF; the filaments, therefore, should be RF-insulated from earth; this is the purpose of RFC4. The bypass capacitors C20 and C21 prevent RF getting into the power supply. The filament wiring, including that on RFC4, must carry the filament current (4A for two parallel EL519 filaments) without undue voltage drop. Figure-8 hi-fi speaker cable is suitable.
- The anode tuning capacitor C17 is a 'difficult' component. Not only must it be rated for more than twice the off-load DC plate voltage *(3.5mm spacing)*, but its maximum capacitance should be sufficient for a loaded Q of 12 at 3.5MHz, while its minimum capacitance should be low lest the Q gets too high on 29MHz. If an extra switch section S1e is available, C17 can be 100pF maximum with fixed capacitors of 100pF and 250pF respectively being switched across it on 7 and 3.5MHz. These fixed capacitors must be rated not only for high RF voltages *(3.5kV)* but also for the high circulating currents in a high-Q tuned circuit. Several low-capacitance silvered-mica capacitors in parallel will carry more current than one high-capacitance unit of the same model. High-voltage, high-current 'transmitting type' ceramic capacitors are available. Beware of Second World War surplus moulded mica capacitors; after 50 years or more, many have become useless.
- The anode choke requires special attention. It must have sufficient inductance to isolate the anode at RF from its power supply, even at the lowest operating frequency *(3.5MHz)* and its self-resonance(s) (the inductance series-resonating with the distributed capacitance within the coil) must not fall within 20% or so of any operating frequency lest circulating current destroys the choke. Self-resonances can be revealed by shorting the choke and coupling a dip meter to it. The traditional approach was to divide the choke winding into several unequal series-connected segments, thereby moving their individual self-resonance frequencies into the VHF region; one problem is that surrounding metallic objects cause *in situ* resonances to differ from those measured on the bench. Another approach, adopted in RFC2, is to use one single-layer winding with a self-inductance not much greater than the whole pi-filter inductor *(L2)*; its self-resonance is well above the highest operating frequency *(29MHz)* but one has to accept that at 3.5MHz a substantial fraction of the tank current flows through RFC2 rather than through L2. At full output, the choke will get very hot; that is why it should be wound on a ceramic or PTFE rod or tube; also, the wire connections should be mechanically well made so that melting solder will not cause failure. The bypass capacitor *(C16)* must be rated for the RF current through RFC2 and also for more than the DC supply voltage.
- The bandswitch in the pi-output filter *(S1d)* must withstand high RF voltages across open contacts and high circulating currents through closed contacts; above the 100W level, ceramic wafers of more-than-receiver size are required.
- The 'safety choke' *(RFC3)* is there to 'kill' the HT supply (and keep lethal voltages off the antenna!) in case the capacitor *(C15)* should fail short-circuit. RFC3 must be wound of sufficiently thick wire to carry the HT supply's short-circuit current long enough to blow its fuse.
- At HF, coaxial relays are unnecessary luxuries; power switching relays with 250VAC/5A contacts are adequate. In some relays with removable plastic covers all connections are brought out on one side with fairly long wires running from the moving contacts to the terminals; if so, remove the cover and those long wires and solder the (flexible) centre conductors of input and output coaxial cables directly to the fixed ends of the moving contact blades. In this amplifier, one relay, RLB, switches both input and output. This requires care in dressing RF wiring to avoid RF feedback from output to input. A relay with three changeover contacts side by side is helpful; the outer contact sets are used for switching, the spare centre set's three contacts are earthed as a shield. The foolproof way is to use two relays.

A passive-grid 400W linear HF amplifier

PA0FRI's Frinear-400 is shown in Fig 5.75 [29]. It has several interesting features.

- Being a passive-grid amplifier, most of the input power is dissipated in a hefty carbon resistor. The voltage across it is applied to the control-grids of the valves and, considering the low value of the resistor (50 or 68Ω), one might expect this arrangement to be frequency-independent; however, the capacitances of the four grids, sockets and associated wiring add up to about 100pF which is only 55Ω at 29MHz! This capacitance must be tuned out if what is adequate drive on 3.5MHz is to produce full output on the higher-frequency bands. PA0FRI does this with a dual-resonant circuit *(L3 and ganged tuning capacitors)* similar to the well-known E-Z-Match antenna tuner; it covers 3.5–29MHz without switching.
- The screen grids in this amplifier are neither at a fixed high voltage nor at earth potential but at a voltage which is proportional to the RF drive. To that end, the RF input is transformed up 3:1 in *T1*, rectified in a voltage doubler and applied to the four bypassed screen grids through individual resistors. This method is consistent with good linearity.
- Control-grid bias is not taken from a mains-derived negative supply voltage but the desired effect, reducing the standing current to 20–25mA per valve, is obtained by raising the cathodes above earth potential. The bias voltage is developed by passing each cathode current through an individual 100Ω resistor and the combined currents through as many forward-biased rectifier diodes as are required to achieve a total standing current of 80–100mA. The individual cathode resistors help in equalising the currents in the four valves. During non-transmit periods the third contact set *(RLA3)* on the antenna changeover relay opens and inserts a large *(10kΩ)* resistor into the combined cathode current, which thereby is reduced to a very low value.
- The pi-filter coil for 3.5 and 7MHz is wound on a powdered-iron toroid which is much smaller than the usual air-core coil. This is not often seen in high-powered amplifiers due to the fear that the large circulating current might saturate the core and spoil the intermodulation performance but no distortion was discernible in a two-tone test.

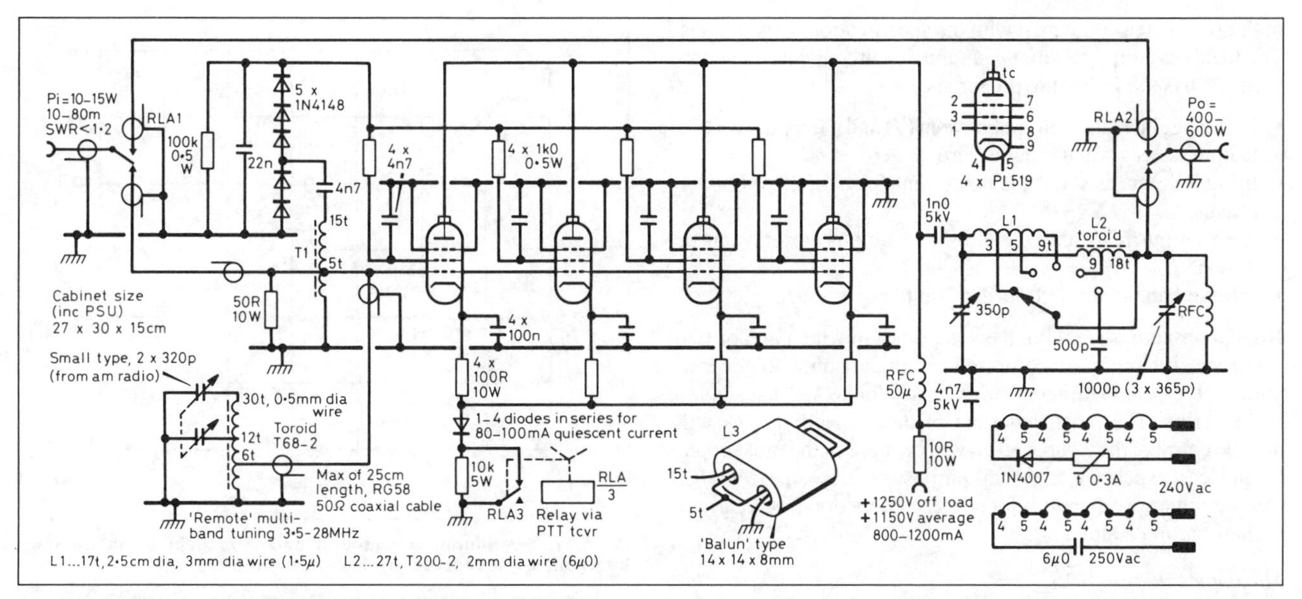

Fig 5.75. A passive-grid 400W linear HF amplifier, the PA0FRI Frinear. Screen-grid voltage is derived from the RF input and no grid bias supply is required

- In Fig 5.75, the 42V filaments of the four valves and a capacitor are shown series connected to the 240V mains. This 0.3A chain is the way these valves were intended to be used in CTV sets and it does save a filament transformer, but this method is not recommended for experimental apparatus such as a home construction project. Besides, a 6μF 250VAC capacitor is neither small nor inexpensive, and generally not available from component suppliers. Also, with lethal mains voltage in the amplifier chassis, the mains plug must be pulled every time access to the chassis is required and after the change or adjustment is made there is the waiting for filaments to heat up before applying HT again. It is much safer and more convenient to operate the filaments in parallel on a 42V transformer (3 × 12.6 + 5V will do), or to use EL519 valves in parallel, series-parallel or series on 6.3, 12.6 or 25.2V respectively.

DC AND AF AMPLIFIERS

This section contains information on the analogue processing of signals from DC (ie zero frequency) up to 5kHz. This includes audio amplifiers for receivers and transmitters for frequencies generally between 300 and 3000Hz as well as auxiliary circuitry, which may go down to zero frequency.

Operational amplifiers (op-amps)

An op-amp is an IC which serves to drive an external network of passive components, some of which function as a feedback loop. The name *operational amplifier* comes from their original use in analogue computers where they were used to perform such mathematical operations as adding, subtracting, differentiating and integrating. In analogue computing, as in the many different current applications, the *transfer function* of the complete circuit (the output versus input relationship) is exclusively described in terms of the external components; if the properties of the IC itself appear in the transfer function, the op-amp is inadequate for the job.

Why use IC op-amps rather than discrete components? Op-amps greatly facilitate circuit design. Having established that a certain op-amp is adequate for the intended function, the designer can be confident that the circuit will reproducibly respond as calculated without having to worry about differences between individual transistors or changes of load impedance or supply voltages; furthermore, most op-amps will survive accidental short-circuits of output or either input to earth or supply voltage(s). General-purpose op-amps are often cheaper than the discrete components they replace and most are available from more than one maker. Manufacturers have done a commendable job of standardising the pin-outs of various IC packages. Many devices are available as one, two or four in one DIL package; this is useful to reduce PCB size and cost but does not facilitate experimenting.

Large books have been written about the use of op-amps; some of these can be recommended to amateur circuit designers as most applications require only basic mathematics; several titles are found in the Cirkit and Maplin catalogues. In the following, no attempt is made to even summarise these books, but some basics are explained and a few typical circuits are included for familiarisation.

The symbol of an op-amp is shown in Fig 5.76. There are two input terminals, an inverting or (–) input and a non-inverting or (+) input, both with respect to the single output, on which the voltage is measured against earth or 'common'. The response of an op-amp goes down to DC and an output swing both positive and negative with respect to earth may be required; therefore a dual power supply is used, which for most op-amps is nominally ±15V. Where only AC (including audio) signals are being processed, a single supply suffices and blocking capacitors are used to permit the meaningless DC output to be referred to a potential halfway between the single supply and earth established by a resistive voltage divider. In application diagrams, the power supply connections are often omitted as they are taken for granted.

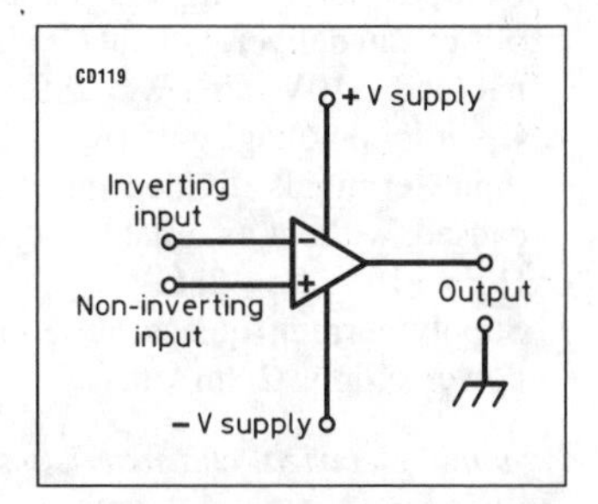

Fig 5.76. Symbol of an operational amplifier (op-amp)

To understand the use of an op-amp and to judge the adequacy of a given type for a specific application, it is useful to define an *ideal* op-amp and then compare it with the specifications of real ones. The ideal op-amp by itself, *open loop*, ie without external components, has the following properties:

- Infinite gain, ie the voltage between (+) and (–) inputs is zero.
- Output is zero when input is zero, ie zero *offset*.
- Infinite input resistance, ie no current flows in the input terminals.
- Zero output resistance, ie unlimited output current can be drawn.
- Infinite bandwidth, ie from zero up to any frequency.

No real op-amp is ideal but there are types in which one or two of the specifications are optimised in comparison with general-purpose types, sometimes at the expense of others and at a much higher price. The most important of these specifications will now be defined; the figures given will be those of the most popular and least expensive of general-purpose op-amps, the µA741C. Comparative specifications for a large variety of types are given in the Maplin catalogue.

Maximum ratings

Values which the IC is guaranteed to withstand without failure include:

- Supply voltage: ±18V.
- Internal power dissipation: 0.5W.
- Voltage on either input: not exceeding applied positive and negative supply voltages
- Output short-circuit: indefinite.

Static electrical characteristics

These are measured at DC (V_s = ±15V, T = 25°C).

- Input offset voltage (V_{oi}): the DC voltage which must be applied to one input terminal to give a zero output. Ideally zero. 6mV max.
- Input bias current (I_b): the average of the bias currents flowing into the two input terminals. Ideally zero. 500nA max.
- Input offset current (I_{os}): the difference between the two input currents when the output is zero. Ideally zero. 200nA max.
- Input voltage range (V_{cm}): the *common-mode* input, ie the voltage of both input terminals to power supply common. Ideally unlimited. ±12V min.
- Common-mode rejection ratio (CMRR): the ratio of common-mode voltage to differential voltage to have the same effect on output. Ideally infinite. 70dB min.
- Input resistance (Z_i): the resistance 'looking into' either input while the other input is connected to power supply common. Ideally infinite. 300kΩ min.
- Output resistance (Z_o): the resistance looking into the output terminal. Ideally zero. 75Ω typ.
- Short-circuit current (I_{sc}): the maximum output current the amplifier can deliver. 25mA typ.
- Output voltage swing (±V_o): the peak output voltage the amplifier can deliver without clipping or saturation into a nominal load. ±10V min (R_L = 2kΩ).
- Open-loop voltage gain (A_{OL}): the change in voltage between input terminals divided into the change of output voltage caused, without external feedback. 200,000 typ, 25,000 min. (V_o = ±10V, R_L = 2kΩ).
- Supply current (quiescent, ie excluding I_o) drawn from the power supply: 2.8mA max.

Dynamic electrical characteristics

- Slew rate is the fastest voltage change of which the output is capable. Ideally infinite. 0.5V/µs typ. (R_L ≥ 2kΩ).

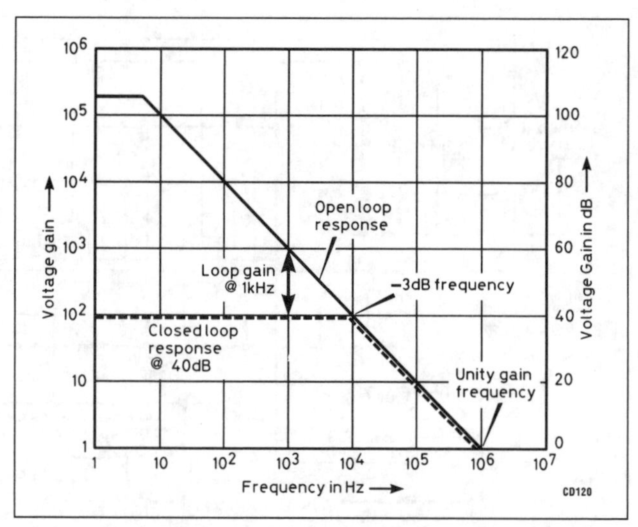

Fig 5.77. Open-loop, closed-loop and loop gain of an op-amp, internally compensated for 6dB/octave roll-off

- Gain-bandwidth product: the product of small-signal open-loop gain and the frequency (in hertz) at which that gain is measured. Ideally infinite. 1MHz typ.

Understanding the gain-bandwidth product concept is basic to the use of op-amps at other than zero frequency. If an amplifier circuit is to be unconditionally stable, ie not given to self-oscillation, the phase shift between inverting input and output must be kept below 180° at all frequencies where the amplifier has gain. As a capacitive load can add up to 90°, the phase shift within most op-amps is kept, by *internal frequency compensation*, to 90°, which coincides with a gain roll-off of 6dB/octave or 20dB/decade. In Fig 5.77, note that the open-loop voltage gain from zero up to 6Hz is 200,000 or 106dB. From 6Hz, the long slope is at –6dB/octave until it crosses the unity gain (0dB) line at 1MHz. At any point along that line the product of gain and frequency is the same: 10^6. If one now applies external feedback to achieve a signal voltage or *closed-loop* gain of, say, 100 times, the –3dB point of the resulting audio amplifier would be at $10^6/100$ = 10kHz and from one decade further down, ie 1kHz to DC, the closed-loop gain would be flat within 1%. At 3kHz, ie the top of the communication-quality audio range, this amplifier would be only 1dB down and this would just make a satisfactory 40dB voice amplifier. If one wished to make a 60dB amplifier for the same frequency range, an op-amp with a gain-bandwidth product of at least 10^7 = 10MHz should be used or one without internal frequency compensation.

Op-amp types

The specifications given above apply to the general-purpose bipolar op-amp µA741C. There are special-purpose op-amps for a variety of applications; only a few are mentioned here, each with a reasonably priced example and the relevant specification.

Fast op-amps

The OP-37GP has high unity gain bandwidth (63MHz), high slew rate (13.5V/µs) and low noise, (3nV/√Hz) which is important for low-level audio.

FET-input op-amps

The TL071-C has the very low input currents characteristic of FET gates (I_{os} ≤ 5pA) necessary if very-high feedback resistances (up to 10MΩ) must be used, but it will work on a single battery of 4.5V, eg in a low-level microphone amplifier, for which it

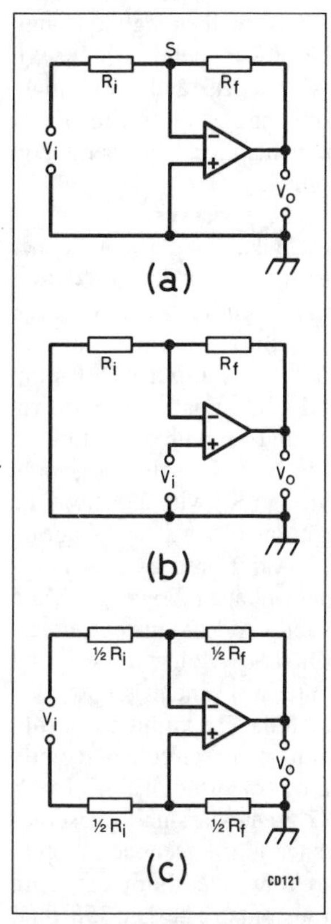

Fig 5.78. Three basic op-amp circuits: (a) inverting; (b) non-inverting; and (c) differential amplifiers

has the necessary low noise. The input impedance is in the teraohm ($10^{12}\Omega$) range.

Power op-amps

The LM383 is intended for audio operation with a closed-loop gain of 40dB and can deliver up to 7W to a 4Ω speaker on a single 20V supply, 4W on a car battery, and can dissipate up to 15W if mounted on an adequate heatsink.

Basic op-amp circuits

Fig 5.78 shows three basic configurations: *inverting* amplifier (a), *non-inverting* amplifier (b) and *differential* amplifier (c).

The operation of these amplifiers is best explained in terms of the ideal op-amp defined above. In (a), if V_o is finite and the op-amp gain is infinite, the voltage between the (–) and (+) input terminals must approach zero, regardless of what V_i is; with the (+) input earthed, the (–) input is also at earth potential, which is called *virtual earth*. The signal current I_s driven by the input voltage V_i through the input resistor R_i will, according to Ohm's Law, be V_i/R_i. The current into the ideal op-amp's input being zero, the signal current has nowhere to go but through the feedback resistor R_f to the output, where the output voltage V_o must be $-I_sR_f$; hence:

$$V_o = -V_i\frac{R_f}{R_i} \quad \text{in which} \quad -\frac{R_f}{R_i} = A_{cl}, \text{ the closed-loop gain}$$

The signal source 'sees' a load of R_i; it should be chosen to suitably terminate that signal source, consistent with a feedback resistor which should, for general purpose bipolar op-amps, be between 10kΩ and 100kΩ; with FET-input op-amps, feedback resistors up to several megohms can be used.

The junction marked 'S' is called the *summing point;* several input resistors from different signal sources can be connected to the summing point and each would independently send its signal current into the feedback resistor, across which the algebraic sum of all signal currents would produce an output voltage. A summing amplifier is known in the hi-fi world as a *mixing amplifier*; in amateur radio, it is used to add, not mix, the output of two audio oscillators together for two-tone testing of SSB transmitters.

For the non-inverting amplifiers, Fig 5.78(b), the output voltage and closed-loop gain derivations are similar, with the (–) input assuming V_i:

$$V_o = V_i\frac{R_i + R_f}{R_i} \quad \text{in which} \quad \frac{R_i + R_f}{R_i} = A_{cl}, \text{ the closed loop gain}$$

The signal source 'sees' the very high input impedance of the bare op-amp input. In the extreme case where $R_i = \infty$, ie left out, and $R_f = 0$, $V_o = V_i$ and the circuit is a *unity-gain voltage follower*, which is frequently used as an impedance transformer having an extremely high input impedance and a near-zero output impedance. Non-inverting amplifiers cannot be used as summing amplifiers.

A *differential amplifier* is shown in Fig 5.78(c). The differential input, ie V_i, is amplified according to the formulae given above for the inverting amplifier; the common-mode input, ie the average of the input voltages measured against earth, is rejected to the extent that the ratio of the resistors connected to the (+) input equals the ratio of the resistors connected to the (–) input. A differential amplifier is useful in 'bringing to earth' and amplifying the voltage across a current sampling resistor of which neither end is at earth potential; this is a common requirement in current-regulated power supplies; see also Fig 5.70.

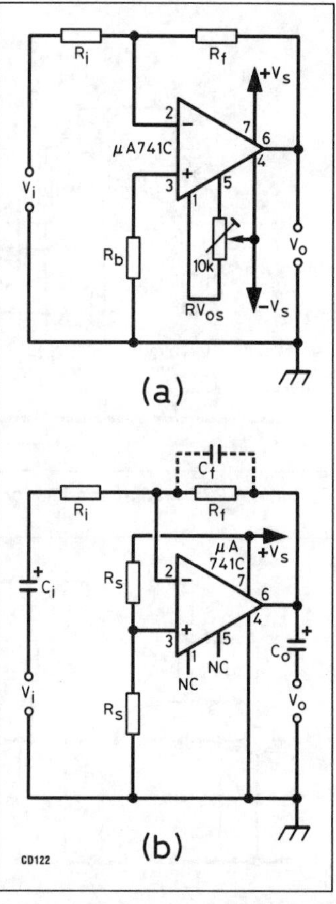

Fig 5.79. DC to audio amplifier (a) requires dual power supplies; it has input bias current and voltage offset compensation. The audio-only amplifier (b) needs only a single power supply

Power supplies for op-amps

Where the DC output voltage of an op-amp is significant and required to assume earth potential for some inputs, dual power supplies are required. Where a negative voltage cannot be easily obtained from an existing mains supply, eg in mobile equipment, a *voltage mirror* can be used. It is an IC which converts a positive supply voltage into an almost equal negative voltage, eg type Si7661CJ. Though characterised when supplied with ±15V, most op-amps will work off a wide range of supply voltages, typically ±5V to ±18V dual supplies or a single supply of 10V to 36V, including 13.8V. One must be aware, however, that the output of most types cannot swing closer to either supply bus than 1–3V.

In audio applications, circuits can be adapted to work off a single supply bus. In Fig 5.79, the AC-and-DC form of a typical inverting amplifier with dual power supplies (a) is compared with the AC-only form (b) shown operating off a single power supply.

The DC input voltage and current errors are only of the order of millivolts but if an amplifier is programmed for high gain, these errors are amplified as much as the signal and spoil the accuracy of the output.

In Fig 5.79(a), two measures reduce DC errors. The input error created by the bias current into the (–) input is compensated for by the bias current into the (+) input flowing through R_b, which is made equal to the parallel combination of R_i and R_f. The remaining input error, $I_{os} \times R_b + V_{os}$, is trimmed to zero with RV_{os}, which is connected to a pair of amplifier pins intended for that purpose. Adjustment of RV_{os} for $V_o = 0$ must be done with

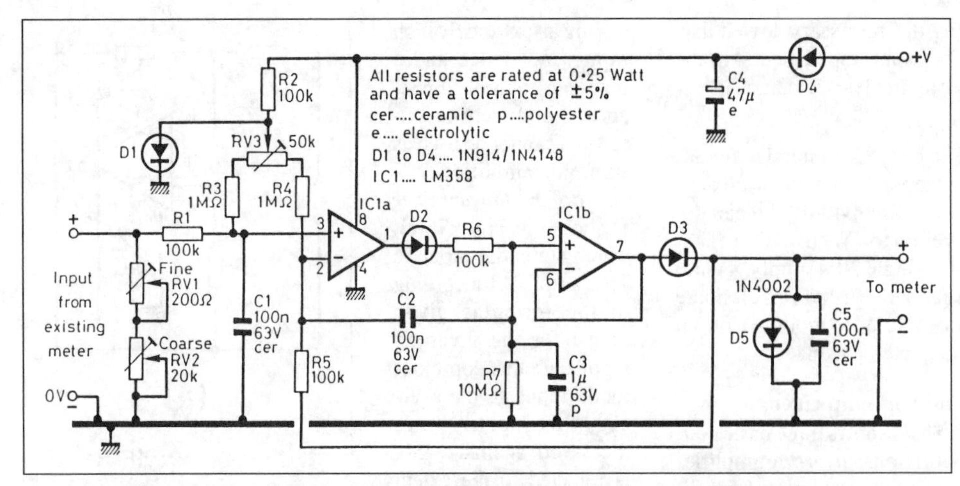

Fig 5.80. A PEP-reading module for RF power meters using a dual op-amp (GW4NAH)

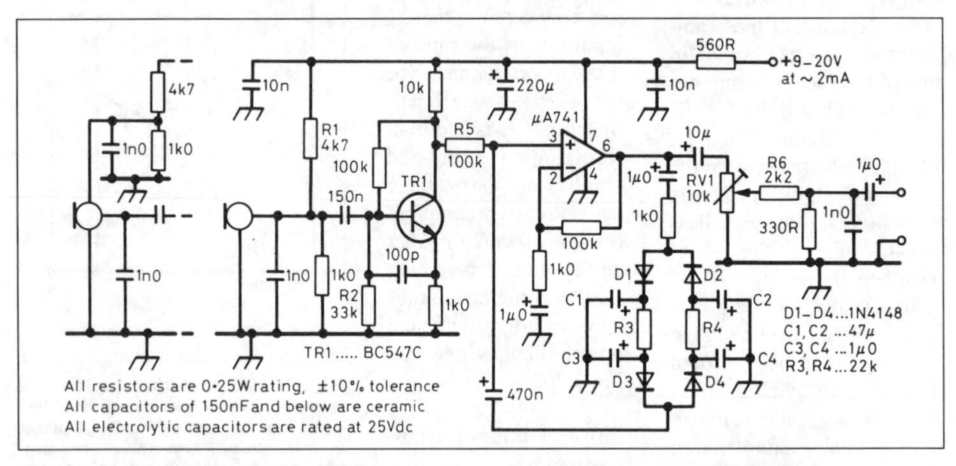

Fig 5.81. A microphone amplifier with op-amp dynamic range compressor (PE1CXO)

the V_i terminals shorted. Note that other amplifier designs use different offset arrangements such as connecting the slider of the offset potentiometer to $+V_s$. Consult the data sheet or catalogue.

In audio-only applications, DC op-amp offsets are meaningless as input and output are blocked by capacitors C_i and C_o in Fig 5.79(b). Here, however, provisions must be made to keep the DC level of the inputs and output about half-way between the single supply voltage and earth; this is accomplished by connecting the (+) input to a voltage divider consisting of the two resistors R_s. Note that the DC closed loop gain is unity as C_i blocks R_i.

In the audio amplifier of Fig 5.79(b), the high and low frequency responses can be rolled off very easily. To cut low-frequency response below, say 300Hz, C_i is dimensioned so that at 300Hz, $X_{Ci} = R_i$. To cut high-frequency response above, say 3000Hz, a capacitor C_f is placed across R_f; its size is such that at 3000Hz, $X_{Cf} = R_f$.

For more sophisticated frequency shaping, see the section on active filters.

A PEP-reading module for RF power meters

The inertia of moving-coil meters is such that they cannot follow speech at a syllabic rate and even if they could, the human eye would be too slow to follow. The usual SWR/power meter found in most amateur stations, calibrated on CW, is a poor PEP indicator.

GW4NAH designed an inexpensive circuit [30], Fig 5.80, on a PCB small enough to fit into the power meter, which will rise to a peak and hold it there long enough for the meter movement, and the operator, to follow.

The resistance of RV1 + RV2 takes the place of the meter movement in an existing power meter; the voltage across it is fed via R1 and C1 to the (+) input of op-amp IC1a. Its output charges C3 via D2 and R6 with a rise time of 0.1s, but C3 can discharge only through R7 with a decay time constant of 10s. The voltage on C3 is buffered by the unity-gain voltage follower IC1b and is fed via D3 to the output terminals to which the original meter movement is now connected. The input-to-output gain of the circuit is exactly unity by virtue of R5/R1 = 1. C2 creates a small phase advance in the feedback loop to prevent overshoot on rapid transients. The LM358 dual op-amp was chosen because, unlike most, it will work down to zero DC output on a single supply. The small voltage across D1 is used to balance out voltage and current offsets in the op-amps, for which this IC has no built-in provisions, via R3, R4 and RV3. D4 protects against supply reversals and C4 is the power supply bypass capacitor. D5 and C5 protect the meter movement from overload and RF respectively.

A microphone amplifier with dynamic-range compressor

In this design by PE1CXO [31], Fig 5.81, the BC547C transistor serves as a 20dB preamplifier for the signal from the microphone; its collector voltage also sets the DC level for the op-amp inputs at roughly half the supply voltage.

Disregarding for now the diode circuit at its output, the op-amp is connected as a non-inverting ×101 (40dB) audio voltage amplifier, for a total gain of 60dB. Audio output from the op-amp is rectified in diodes D1 and D2, which charge capacitors C1 and C2 positive and negative respectively. The voltage difference between C1 and C2 discharges via R3, D3, D4 and R4. The capacitors C3 and C4 serve to make the top of D3 and D4 'cold' for AC, which has a dual purpose. It removes AC ripple from the current through D3 and D4 and it provides an earth reference for a voltage divider consisting of R5 and the impedance of the diodes (D3 and D4, back-to-back, in parallel). This latter impedance depends on the C1 and C2 DC discharge through D3 and D4. The larger this current, the lower the impedance of the diodes and the lower the input voltage to the op-amp's non-inverting (+) input.

As the signal voltage at the op-amp input is small, the non-linearity of the diodes creates but little distortion. 2.5V p-p is available at the op-amp output.

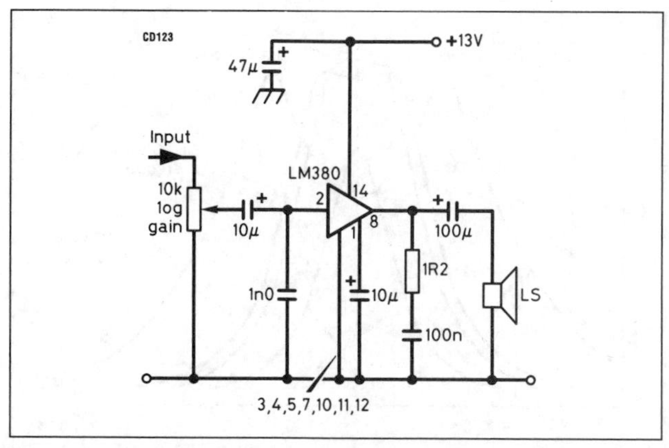

Fig 5.82. This simple receiver audio IC provides up to 40dB voltage gain and over a watt of low-distortion output power to drive an 8Ω loudspeaker

Receiver audio

The audio signal obtained from the demodulator of a radio receiver generally requires filtering and voltage amplification; if a loudspeaker is to be the 'output transducer' (as distinct from headphones or an analogue-to-digital converter for computer processing) some power amplification is also required. For audio filtering, see the section on filters. For voltage amplification, the op-amp is the active component of choice for reasons explained in the section on them; there is also a great variety of ICs which contain not only the op-amp but also its gain-setting resistors and other receiver functions such as demodulator, AGC generator and a power stage dimensioned to drive a loudspeaker. A fraction of a watt is sufficient for a speaker in a quiet shack but for mobile operation in a noisy vehicle several watts are useful. Design, then, comes down to the selection from a catalogue or the junk box of the right IC, ie one that offers the desired output from the available input signal at an affordable price and on available power supply voltages. The data sheet of the IC selected will provide the necessary details of external components and layout. Fig 5.82 is an example of voltage and power amplification in an inexpensive 14-pin DIL IC; the popular LM380 provides more than 1W into an 8Ω speaker on a single 13.8V supply. Even the gain-setting resistors are built-in [32].

Transmitter audio

With the virtual disappearance from the amateur scene of high-level amplitude modulation, the audio processing in amateur as in commercial and military transmitters consists of low-level voltage amplification, compression, limiting and filtering, all tasks for op-amp circuitry as described in the section on them. The increase in the consumer usage of radio transmitters in cellular telephones and private mobile radios has given incentive to IC manufacturers to integrate ever more of the required circuitry on to one chip. As an example, the SL6270C gain-controlled microphone preamplifier/VOGAD [16], Fig 5.83, requires only a few external resistors and capacitors to do all that the circuit of Fig 5.81 does and that in an 8-pin DIL package! For those amateurs who wish to construct and use their equipment, rather than experiment with it, these ICs are the obvious choice.

FILTERS AND LC COUPLERS

Filters are circuits designed to *pass* signals of some frequencies and to *reject* or *stop* signals of others. Amateurs use filters ranging in operation from audio to microwaves. Applications include *preselector* filters which keep strong out-of-band signals from overloading a receiver, *IF* (intermediate frequency) filters which provide adjacent-channel selectivity in superheterodyne receivers, and *audio* filters which remove bass and treble not essential for speech communication from a microphone's output to restrict the bandwidth taken up by a transmitted signal. The transmitter output is filtered to keep harmonics from being radiated.

Filters are classified by their main frequency characteristics. *High-pass* filters pass frequencies above their *cut-off* frequency and stop signals below that frequency. In *low-pass* filters the reverse happens. *Band-pass* filters pass the frequencies between two cut-off frequencies and stop those below the lower and above the upper cut-off frequency. *Band-reject* (or *band-stop*) filters stop between two cut-off frequencies and pass all others. *Peak filters* and *notch* filters are extremely sharp band-pass and band-reject filters respectively which provide the frequency characteristics their names imply.

A *coupler* is a unit that matches a signal source to a load having an impedance which is not optimum for that source. An example is the matching of a transmitter's transistor power amplifier requiring a 2Ω load to a 50Ω antenna. Frequently, impedance matching and filtering is required at the same spot, as it is in this example, where harmonics must be removed from the output before they reach the antenna. There is a choice then, either to do the matching in one unit, eg a wide-band transformer with a $1:\sqrt{(50/2)} = 1:5$ turns ratio and the filtering in another, ie a 'standard' filter with 50Ω input and output, or to design a special filter-type LC circuit with a 2Ω input and 50Ω output impedance. In multiband HF transceivers, transformers good for all bands and separate 50Ω/50Ω filters for each band would be most practical. For UHF, however, and for the high impedances in RF valve anode circuitry, there are no satisfactory wide-band transformers; the use of LC circuits is required.

Ideal filters and the properties of real ones

Ideal filters would let all signals in their intended pass-band through unimpeded, ie have zero *insertion loss,* suppress completely all frequencies in their stop-band, ie provide infinite *attenuation,* and have sharp *transitions* from one to the other at their *cut-off* frequencies (Fig 5.84). Unfortunately such filters

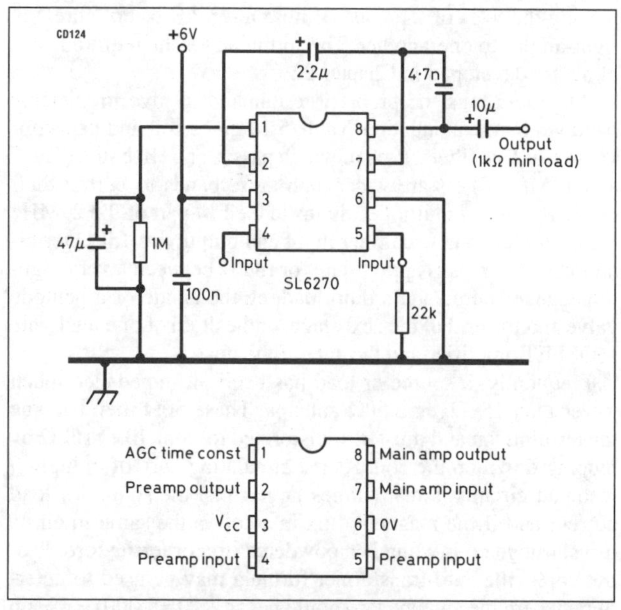

Fig 5.83. A VOGAD IC provides microphone gain and dynamic range compression with only a few external passive components *(GEC-Plessey Professional Products Handbook)*

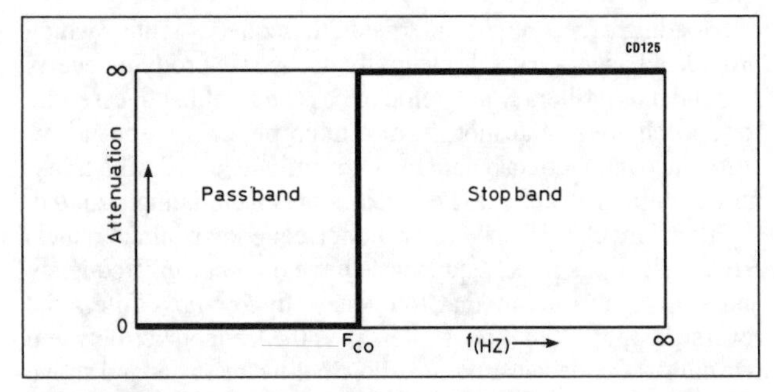

Fig 5.84. Attenuation vs frequency plot of an ideal low-pass filter. No attenuation (insertion loss) in the pass-band, infinite attenuation in the stop-band, and a sharp transition at the cut-off frequency

do not exist. In practice, the cut-off frequency, ie the transition point between pass-band and stop-band, is generally defined as the frequency where the response is –3dB (ie down to 70.7% in voltage) with respect to the response in the pass-band; in very sharp filters, such as crystal filters, the –6dB (half-voltage) points are frequently considered the cut-off frequencies.

There are several practical approximations of ideal filters but each of these optimises one characteristic at the expense of others.

LC filters

If two resonant circuits are coupled together, a band-pass filter can be made. The degree of coupling between the two resonant circuits, both of which are tuned to the centre frequency, determines the shape of the filter curve. See Fig 5.85. Undercoupled, critically coupled and overcoupled two-resonator filters all have their applications.

Four methods of achieving the coupling are shown in Fig 5.86. The result is always the same and the choice is mainly one of convenience. If the signal source and load are not close together, eg on different PCBs, placing one resonant circuit with each and using link coupling is recommended to avoid earth loops. If the resonant circuits are close together, capacitive coupling between the 'hot ends' of the coils is easy to use and the coupling can be adjusted with a trimmer capacitor. Stray inductive coupling between adjacent capacitively coupled resonant circuits is avoided by placing them on opposite sides of a shield and/or placing the axes of the coils (if not on toroids or pot cores) at right-angles to one-another. The formulae for the required coupling are developed in Chapter 22.

All filters must be properly terminated to give predictable bandwidth and attenuation. Table 5.14 gives coil and capacitor values for five filters, each of which passes one HF band. (The 7 and 14MHz filters are wider than those bands to permit their use in frequency multiplier chains to the FM part of the 29MHz band.) Included in each filter input and output is a 15kΩ termination resistor, as is appropriate for filters between a very high-impedance source and a ditto load, eg the anode of a pentode valve and the grid of the next stage, or the drain of one dual-gate MOSFET amplifier and the gate of the next.

Frequently, a source or load has itself an impedance much lower than 15kΩ, eg a 50Ω antenna. These 50Ω then become the termination and must be transformed to 'look like' 15kΩ by tapping down on the coil. As the circulating current in high-Q resonant circuits is many times larger than the source or load current and if the magnetic flux in a coil is the same in all its turns, true in coils wound on powdered iron or ferrite toroids or pot cores, the auto-transformer formula may be used to determine where the antenna tap should be: at $\sqrt{(50/15,000)} \approx 6\%$ up from the earthy end of the coil. In coils without such cores, the flux in the end turns is less than in the centre ones, so the tap

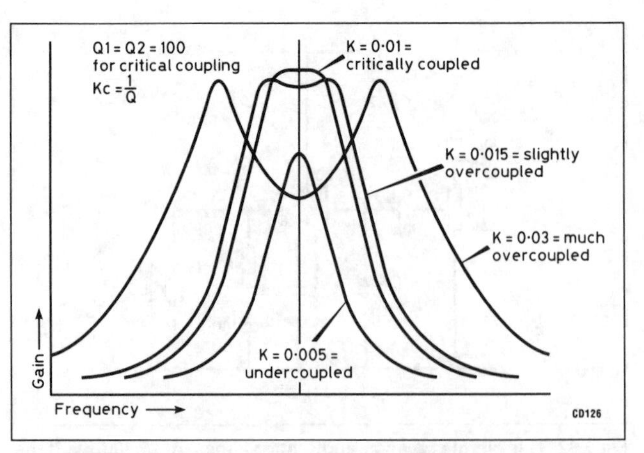

Fig 5.85. Frequency response of a filter consisting of two resonant circuits tuned to the same frequency as a function of the coupling between them (after Terman). Under-coupling provides one sharp peak, critical coupling gives a flat top, mild overcoupling widens the top with an acceptably small dip and gross overcoupling results in peaks on two widely spaced frequencies. All these degrees of coupling have their applications

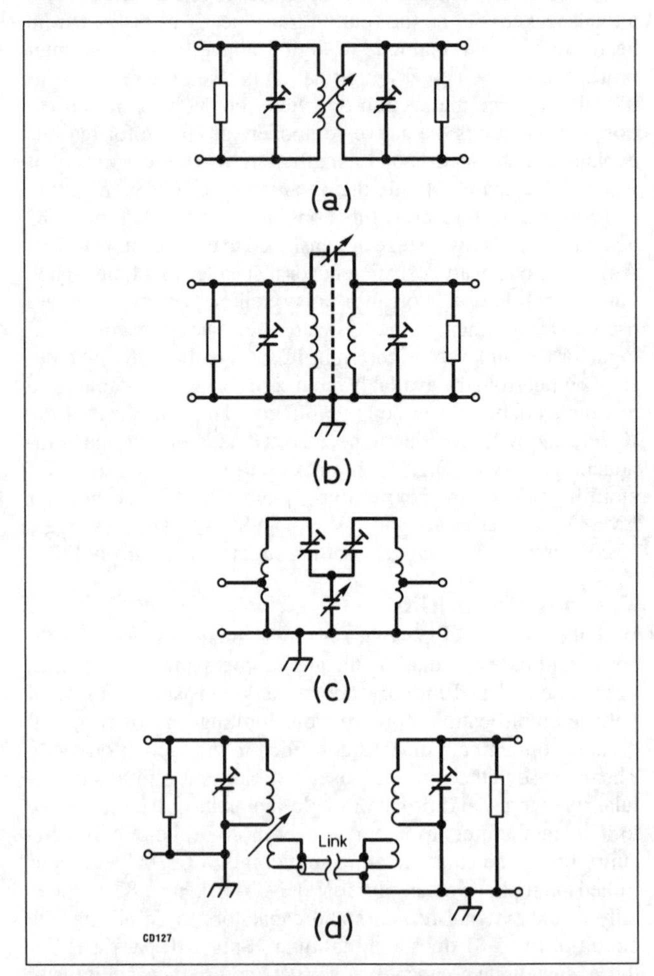

Fig 5.86. Four ways of arranging the coupling between two resonant circuits. Arrows mark the coupling adjustment. (a) Direct inductive coupling; the coils are side-by-side or end-to-end and coupling is adjusted by varying the distance between them. (b) 'Top' capacitor coupling; if the coils are not wound on toroids or pot cores they should be shielded from each other or installed at right-angles to avoid uncontrolled inductive coupling. (c) Common capacitor coupling; if the source and/or load have an impedance lower than the proper termination, they can be 'tapped down' on the coil (regardless of coupling method). (d) Link coupling is employed when the two resonant circuits are physically separated

Table 5.14. Band-pass filters for five HF bands and 15kΩ input and output terminations as shown in Fig 5.86(b)

Lowest frequency (MHz)	Centre frequency (MHz)	Highest frequency (MHz)	Coupling (pF)	Parallel capacitance (pF)	L (μH)	Winding details (formers ¾in long, 3/8in dia)
3.5	3.65	3.8	6	78	24	60t 32swg close-wound
7	7.25	7.5	3	47	10	40t 28swg close-wound
14	14.5	15	1.5	24	5	27t 24swg close-wound
21	21.225	21.45	1	52	1	12t 20swg spaced to ¾in
28	29	30	0.6	10 primary	3	21t 24swg spaced to ¾in
				30 secondary	1	12t 20swg spaced to ¾in

The use of tuning slugs in the coils is not recommended. Capacitors can be air, ceramic or mica compression trimmers. Adjust each coupler to cover frequency range shown.

must be experimentally located higher up the coil. Impedances lower than 50Ω can be accomodated by placing that source or load in series with the resonant circuit rather than across all or part of it.

DC operating voltages to source and load devices are often fed though the filter coils. If properly bypassed, this does not affect filter operation. To avoid confusion, DC connections and bypass capacitors are not shown in the filter circuitry in this chapter.

Several more sophisticated LC filter designs are frequently used. All can be configured in high-pass, low-pass, band-pass and band-reject form. *Butterworth* filters have the flattest response in the pass-band. *Chebyshev* filters have a steeper roll-off to the stop-band but exhibit ripples in the pass-band, their number depending on the number of filter sections. *Elliptic* filters have an even steeper roll-off, but have ripples in the stop-band *(zeroes)* as well as in the pass-band *(poles)*: see Fig 5.87. Chebyshev and elliptical filters have too much *overshoot*, particular near their cut-off frequencies, for use where pulse distortion must be kept down, eg in RTTY filters.

The calculation of component values for these three types would be a tedious task but for filter tables normalised for a cut-off frequency of 1Hz and termination resistance of 1Ω (or 1MHz and 50Ω where indicated). These can be easily scaled to the desired frequency and termination resistance. See Chapter 22.

M-derived and *constant-k* filters are older designs with less-well-defined characteristics but amateurs use them because component values are more easily calculated 'long-hand'. The diagrams and formulae to calculate component values are contained in Chapter 22.

From audio frequency up to, say, 100MHz, filter inductors are mostly wound on powdered-iron pot cores or toroids of a material and size suitable for the frequency and power, and capacitors ranging from polystyrene types at audio, to mica and ceramic types at RF, with voltage and current ratings commensurate with the highest to be expected, even under fault conditions.

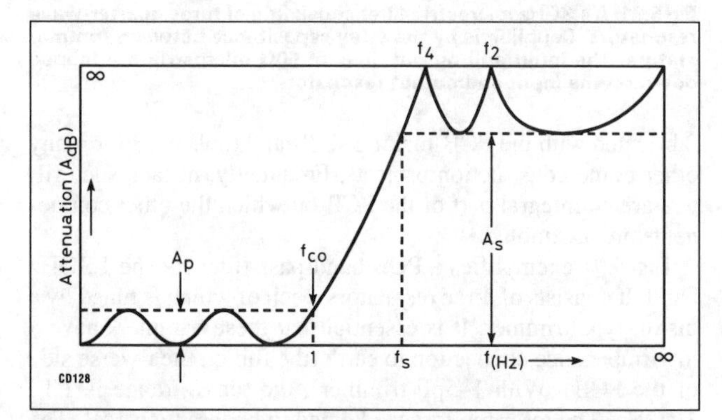

Fig 5.87. Attenuation vs frequency plot of a two-section elliptic low-pass filter. *A* is the attenuation (dB); A_p is the maximum attenuation in the pass-band or ripple; f_4 is the first attenuation peak; f_2 is the second attenuation peak with two-section filter; f_{co} is the frequency where the attenuation first exceeds that in the pass-band; A_s is the minimum attenuation in the stop-band; and f_s is the frequency where minimum stop-band attenuation is first reached

Fig 5.88 shows an audio filter to provide selectivity in a direct-conversion receiver. It is a three-section elliptic low-pass filter with a cut-off frequency of 3kHz, suitable for voice reception.

To make any odd capacitance values of, say, 1% accuracy, one starts with the next lower standard value (no great accuracy required), measures it precisely (ie to better than 1%), and adds what is missing from the desired value in the form of one or more smaller capacitors which are then connected in parallel with the first one. The smaller capacitors, having only a small fraction of the total value, need not be more accurate than, say, ±5%.

Filtering at VHF and above

At HF and below, it may be assumed that filters will perform as designed if assembled from components which are known to have the required accuracy, either because they were bought to tight specifications or were selected or adjusted with precise test equipment. It was further assumed that capacitor leads have negligible self-inductance and that coils have negligible capacitance.

At VHF and above these assumptions do not hold true. Though filter theory remains the same, the mechanics are quite different. Even then, the results are less predictable and adjustment will be required after assembly to tune out the stray capacitances and inductances. A sweep-generator and oscilloscope provide the most practical adjustment method. A variable oscillator with frequency counter and a voltmeter with RF probe, plus a good deal of patience, can also do the job.

At VHF, self-supporting coils and mica, ceramic or air-dielectric trimmer capacitors give adequate results for most applications. For in-band duplex operation on one antenna, as is common in repeaters, bulky and expensive very-high-*Q* cavity resonators are required, however.

The band-pass filter of Fig 5.89 includes four parallel resonant circuits [27]. Direct inductive coupling is used between the first two and the last two; capacitive coupling is used between

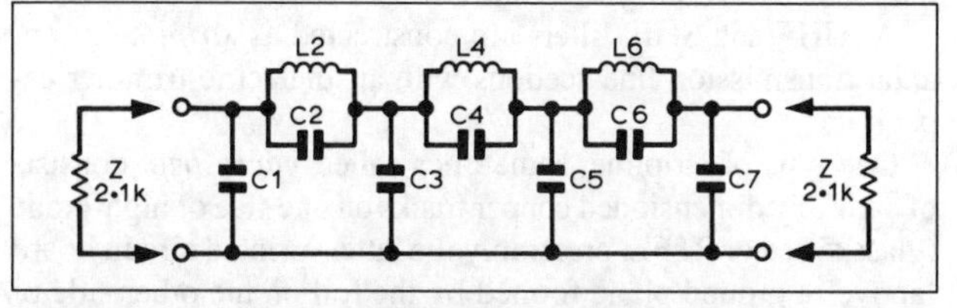

Fig 5.88. A three-section elliptic low-pass filter with 3kHz cut-off. C1 = 37.3nF, C2 = 3.87nF, C3 = 51.9nF, C4 = 19.1nF, C5 = 46.4nF, C6 = 13.5nF, C7 = 29.9nF, mica, polyester or polystyrene. L2 = 168mH, L4 = 125mH, L6 = 130mH are wound on ferrite pot cores

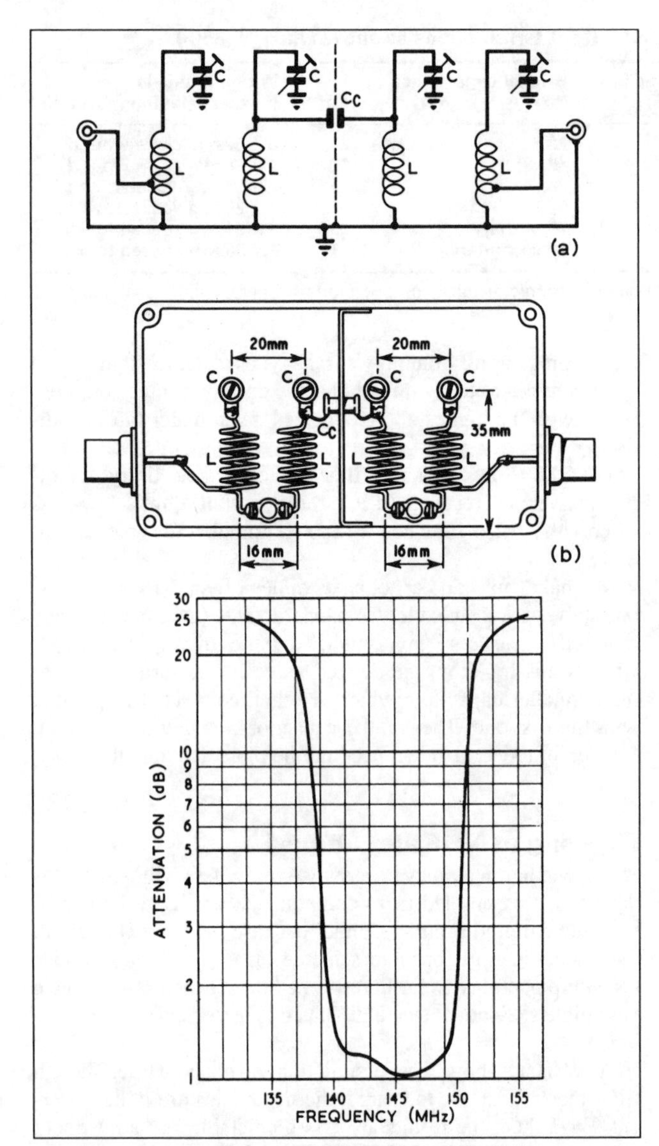

Fig 5.89. A four-section band-pass filter for 145MHz. (a) Direct inductive coupling between sections 1–2 and 3–4, top capacitor (0.5pF) coupling between sections 2–3; both input and output are tapped down on the coils for 50Ω terminations. (b) The filter fits into an 11 x 6 x 3cm die-cast box; coils are 3/8in (9.5mm) inside dia, 6½ turns bare 18swg (1.22mm dia) spaced 1 wire dia; taps 1t from earthy end; C are 1–6pF piston ceramic trimmers for receiving and QRP transmitting; for higher power there is room in the box for air trimmers like those in Fig 5.92. (c) Performance curve

the centre two, where a shield prevents stray coupling. The input and output connections are tapped down on their respective coils to transform the 50Ω source and load into the proper terminations. This filter can reduce harmonics and other out-of-band spurious emissions when transmitting and suppress strong out-of-band incoming signals which could overload the receiver.

At UHF and SHF, filters are constructed as *stripline* or co-axial transmission line sections with air-dielectric trimmer capacitors.

One type of stripline, sometimes called *microstrip,* consists of carefully dimensioned copper tracks on one side of high-grade (glass-filled or PTFE, preferably the latter) printed circuit board 'above' a ground plane formed by the foil on the other side of that PCB. The principle is illustrated in Fig 5.90 and calculations are given in Chapter 22. Many amateurs use PCB strip lines wherever very high Q is not mandatory, as they can be

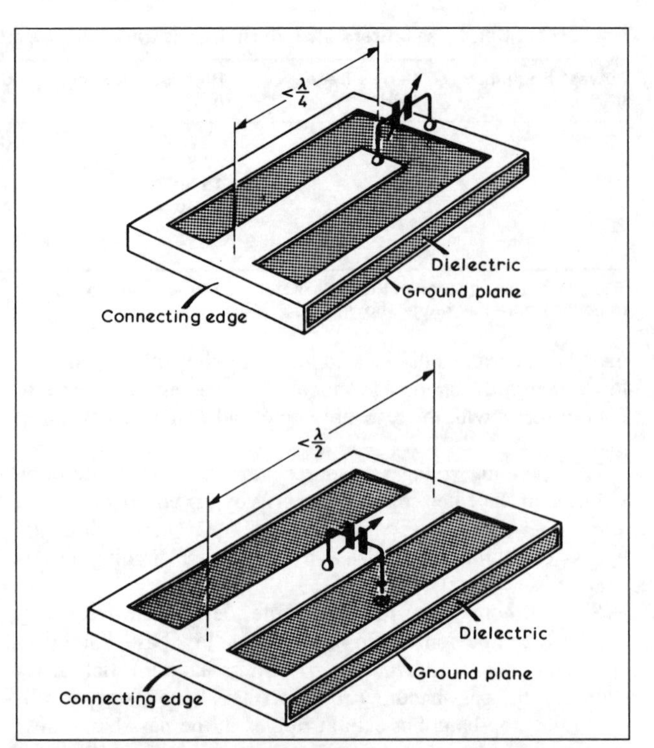

Fig 5.90. Basic microstrip quarter-wave (top) and half-wave resonators. The electrical length of the strips is made shorter than their nominal length so that trimmers at the voltage maxima can be used to tune to resonance; the mechanical length of the strips is shorter than the electrical length by the velocity factor arising from the dielectric constant of the PCB material

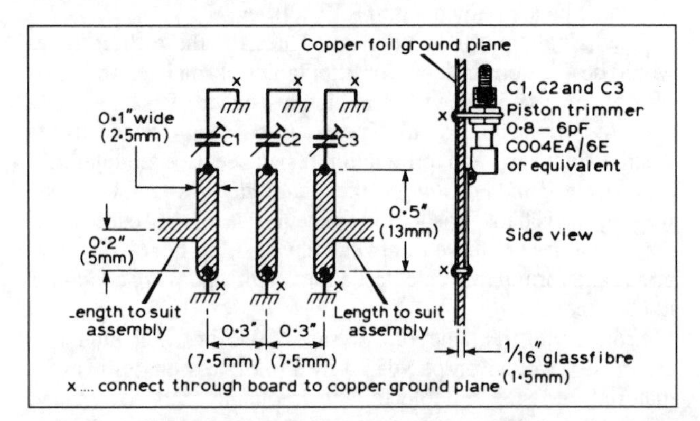

Fig 5.91. A 1.3GHz microstrip filter consisting of three quarter-wave resonators. Coupling is by the stray capacitance between trimmer stators. The input and output lines of 50Ω microstrip are tapped down on the input and output resonators

fabricated with the PCB-making skills and tools used for many other home-construction projects. Frequently, in fact, such filters are an integral part of the PCB on which the other components are assembled.

Fig 5.91 exemplifies a PCB band-pass filter for the 1.3GHz band. It consists of three resonators, each of which is tuned by a piston-type trimmer. It is essential that these trimmers have a low-impedance connection to earth (the foil on the reverse side of the PCB). With 1–5pF trimmers, the tuning range is 1.1–1.5GHz. The insertion loss is claimed to be less than 1dB. The input and output lines, having a characteristic impedance of 50Ω, may be of any convenient length. This filter is not intended for high-power transmitters.

For higher power, the resonators can be sheet copper striplines, sometimes called *slab lines*, with air as the dielectric. Fig 5.92

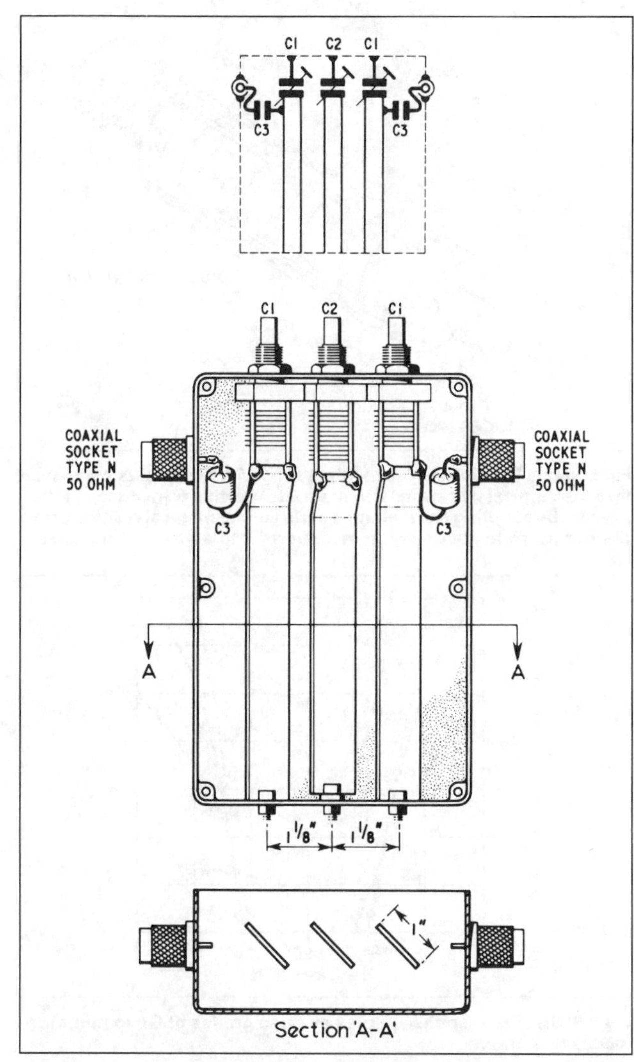

Fig 5.92. The circuit and layout for a 100W 145MHz slab-line filter. The strips are 1 × 1/16in (25 × 1.5mm) sheet copper, offset at 45° to avoid overcoupling. Input and output lines are 6½in (165mm) long; the centre resonator is slightly shorter to allow for the greater length of C2 and a rib in the cast box. C1 = 50pF, C2 = 60pF, C3 = 4.4pF

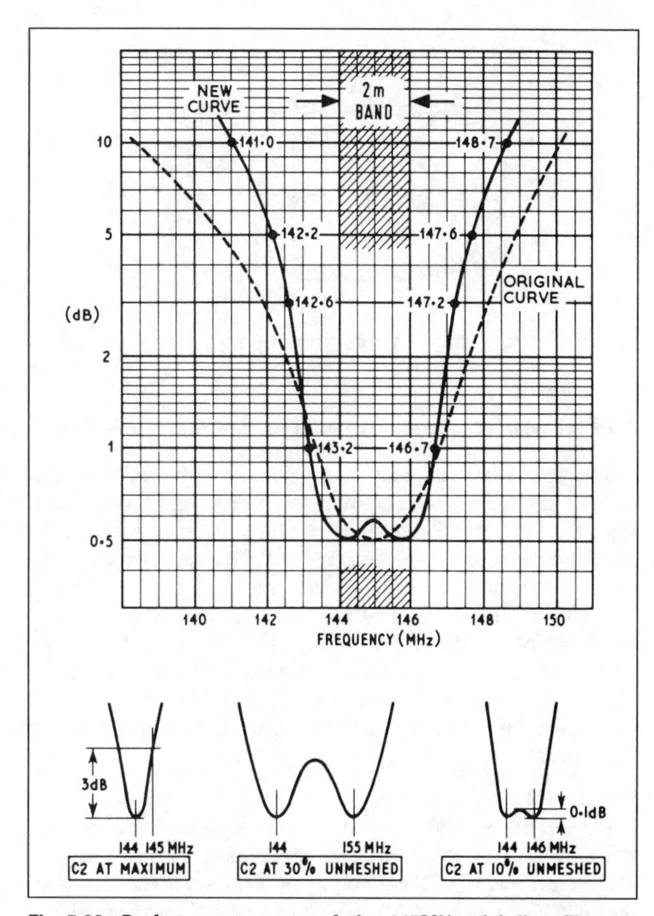

Fig 5.93. Performance curves of the 145MHz slab-line filter as affected by the setting of C2. Note that the insertion loss is only 0.6dB. Also, compare the −10dB bandwidth of this filter with that of the filter of Fig 5.89: 7.7MHz for this filter when tuned to 'new curve', 13MHz for the four-section filter with small, lower-*Q* coils

gives dimensions for band-pass filters for the 144MHz band [27]. The connections between the copper fingers and the die-cast box are at current maxima and must have the lowest possible RF resistance. In the prototype, the ends of the strips were brazed into the widened screwdriver slots in cheesehead brass bolts. The 'hot' ends of the strips are soldered directly to the stator posts of the tuning capacitors. After the input and output resonators are tuned for maximum power throughput at the desired frequency with the centre capacitor fully meshed, the latter is then adjusted to get the desired coupling and thereby pass-band shape (Fig 5.93).

For the highest Q at UHF and SHF, and with it the minimum insertion loss and greatest out-of-band attenuation, coaxial cavity resonators are used. Their construction requires specialised equipment and skills, as brass parts must be machined, brazed together and silver plated. The *Microwave Handbook* (RSGB) elaborates on these techniques.

LC matching circuits

LC circuits are used to match very low impedances, such as VHF transistor collectors, or very high impedances, such as valve anodes, to the 50 or 75Ω coaxial cables which have become the standard for transporting RF energy between 'black boxes' and antennas. The desired match is valid only at or near the design frequency. The calculations for L, pi and L-pi circuits are given in Chapter 7.

High-*Q* filter types

The *shape factor* of a band-pass filter is often defined as the ratio between its –60dB and –6dB bandwidth. In a professional receiver, a single-sideband filter with a shape factor of 1.8 could be expected while 2.0 might be more typical in good amateur equipment. It is possible to make LC filters with such performance, but it would have to be at a very low intermediate frequency (10–20kHz), have many sections, and be prohibitively bulky, costly and complicated. The limited Q of practically realisable inductors, say 300 for the best, is the main reason. Hence the search for resonators of higher Q.

Several types are used in amateur equipment, including mechanical, crystal, ceramic, and surface acoustic wave (SAW) filters. Each is effective in a limited frequency range and *fractional bandwidth* (the ratio of bandwidth to centre frequency in percent). See Fig 5.94.

Mechanical filters

There are very effective SSB, CW and RTTY filters for intermediate frequencies between 60 and 600kHz based on the mechanical resonances of small metal discs (Fig 5.95). The filter comprises three types of component: two magnetostrictive or piezo-electric transducers which convert the IF signals into

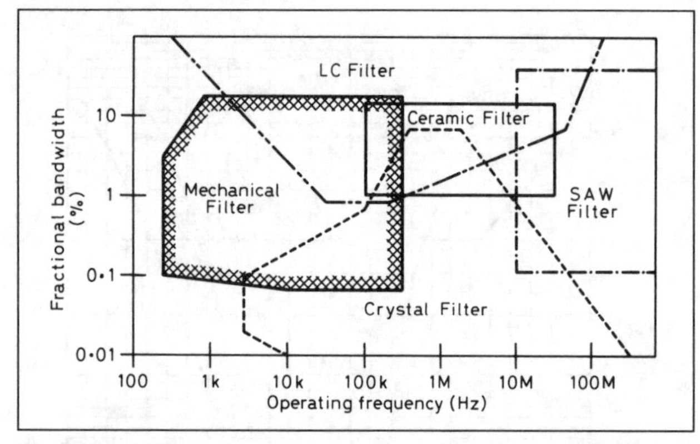

Fig 5.94. Typical frequency ranges for various filter techniques

mechanical vibrations and vice versa, a number of resonator discs, and coupling rods between those disks. Each disc represents the mechanical equivalent of a high-Q series resonant circuit, and the rods set the coupling between the resonators and thereby the bandwidth. Shape factors as low as 2 can be achieved. Mechanical filters were first made and used in amateur equipment by Collins Radio (USA). Goyo (Japan) still make the mechanical filters formerly made by Kokusai. The latter range includes a plug-in unit intended to replace a 455kHz IF transformer in a valve receiver, thereby achieving selectivity compatible with SSB reception.

Crystal filters

Piezo-electric crystals, cut from man-made quartz bars, are resonators with extremely high Q, tens of thousands being common. Their electrical equivalent, Fig 5.96(a), shows a very large inductance L_s, an extremely small capacitance C_s and a small loss resistance R_s, in a series-resonant circuit. It is shunted by the parallel capacitance C_p, which is the real capacitance of the crystal electrodes, holder, socket, wiring and any external *load capacitor* one may wish to connect across the crystal. At frequencies above the resonance of the series branch, the net impedance of that branch is inductive. This inductance is in parallel resonance with C_p at a frequency slightly above the series resonance: Fig 5.96(b). The essentially mechanical series resonance, the *zero*, may be considered user-immovable but the parallel resonance or *pole* can be pulled down closer to the series resonant frequency by increasing the load capacitance, eg with a trimmer.

Though their equivalent circuit would suggest that crystals are linear devices, this is not strictly true. Crystal filters can therefore cause intermodulation, especially if driven hard, eg by a strong signal just outside the filter's pass-band. Sometimes, interchanging input and output solves the problem. Very little is known about the causes of crystal non-linearity and the subject is never mentioned in published specifications.

The two most common configurations are the *half-lattice* filter, Fig 5.97, and the *ladder* filter, Fig 5.98.

Half-lattice filter curves are symmetrical about the centre frequency, an advantage in receivers, but they require crystals differing in series resonant frequency by somewhat more than half the desired pass-band width and an RF transformer for each two filter sections. Two to four sections, four to eight crystals, are required in a good HF SSB receiver IF filter. Model XF-9B in Table 5.15 is a high-performance filter which seems to be of this type. Half-lattice filters make good home construction projects only for the well-equipped and experienced. Fig 5.99 shows a filter by G2DAF which is about as simple as a half-lattice filter can be.

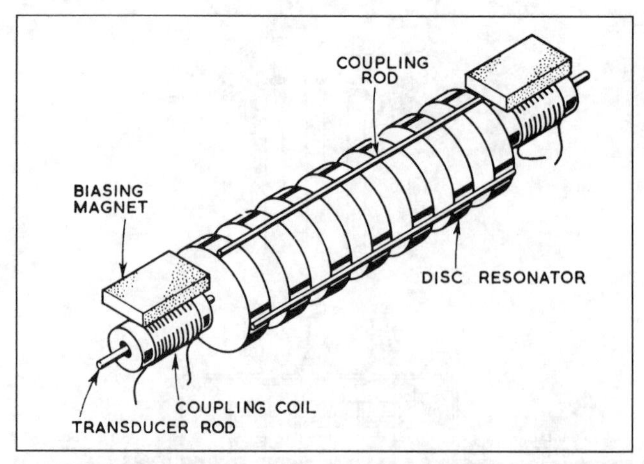

Fig 5.95(a). The Collins mechanical filter. An IF signal is converted into mechanical vibrations in a magnetostrictive transducer, is then passed, by coupling rods, along a series of mechanical resonators to the output transducer which reconverts into an electrical signal

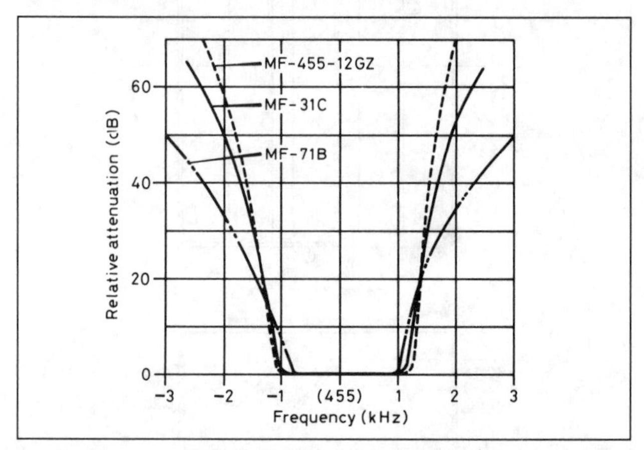

Fig 5.95(b). The response curves of three grades of Goyo miniature mechanical filters

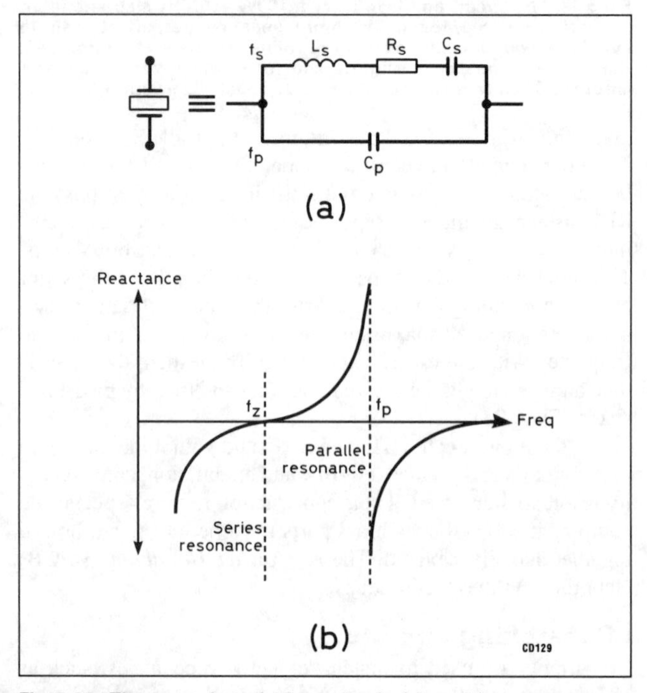

Fig 5.96. The crystal equivalent circuit (a) and its reactance vs frequency plot (b)

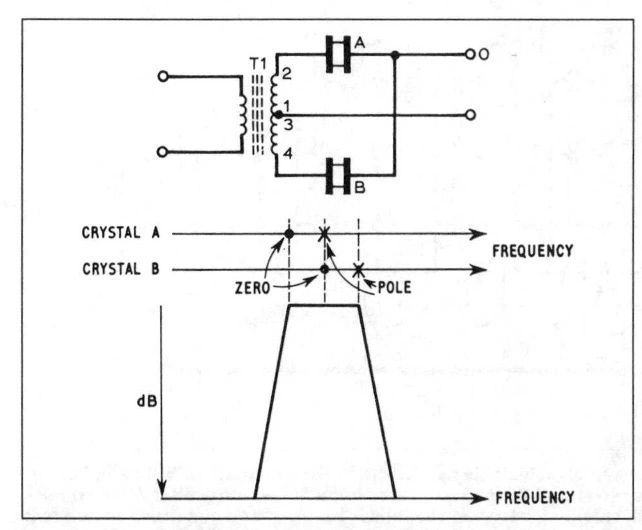

Fig 5.97. The basic single-section half-lattice filter diagram (top) and its idealised frequency plot (bottom). Note the poles and zeros of the two crystals in relation to the cut-off frequencies; if placed correctly, the frequency response is symmetrical. The bifilar transformer is required to provide balanced inputs

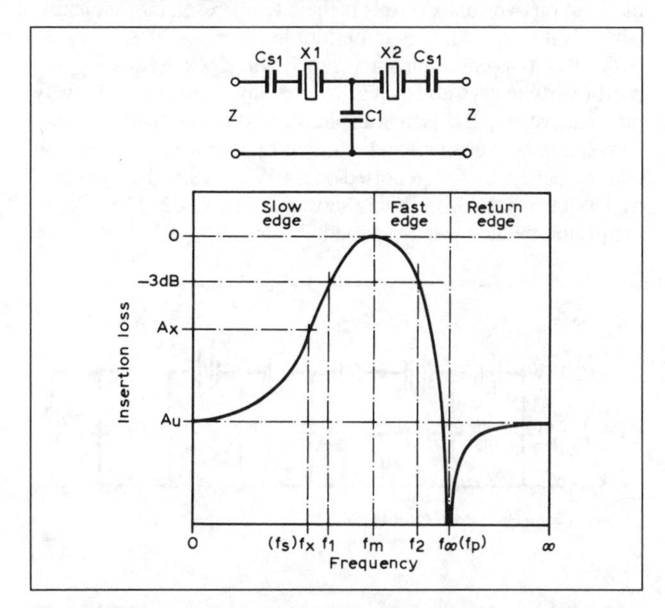

Fig 5.98. The basic two-pole ladder filter diagram (top) and its generalised frequency plot (bottom). Note that the crystals are identical but that the resulting frequency response is asymmetric

Ladder filters require no transformers and use crystals of only one frequency but they have asymmetrical pass-band curves; this creates no problem in SSB generators but is less desirable in a receiver with upper and lower sideband selection. With only five crystals a good HF SSB generator filter can be made. Model XF-9A in Table 5.15 seems to be of this type.

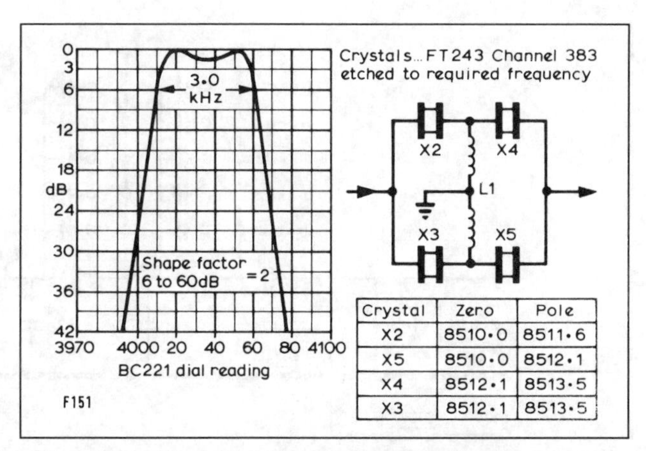

Crystal	Zero	Pole
X2	8510·0	8511·6
X5	8510·0	8512·1
X4	8512·1	8513·5
X3	8512·1	8513·5

Fig 5.99. G2DAF made this seemingly simple two-section half-lattice SSB filter by etching up to the desired frequencies originally identical quartz plates from Second World War surplus pressure-mounted crystal units; modern crystals have plated-on electrodes and are hermetically sealed, defying DIY frequency changes. Note that X5 has much greater pole-zero spacing than the other crystals, resulting in a slight asymmetry of the frequency curve. L1 consists of nine bifilar turns on a small ferrite toroid. The nominal termination impedance is 2kΩ

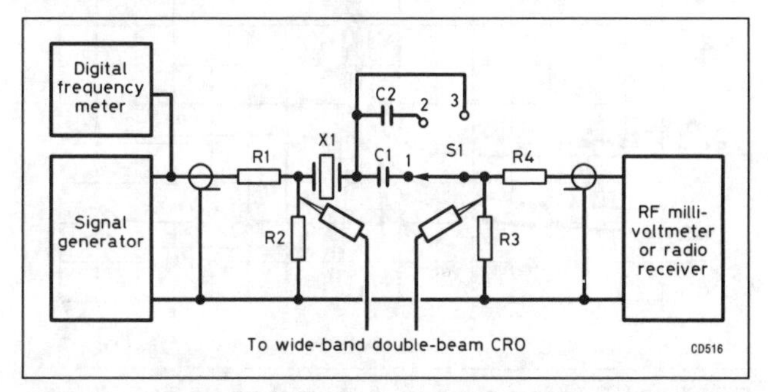

Fig 5.100. G3JIR's crystal test circuit. At resonance, both series (switch position 3) and parallel (switch positions 1 and 2), the two 'scope traces are in-phase. For 9MHz filter crystals, R1 = R4 = 1kΩ, R2 = R3 = 220Ω. The signal generator and frequency meter should have a frequency resolution of 10Hz

Ladder crystal filters using inexpensive consumer (3.579MHz telephone and 4.433 or 8.867MHz TV colour-burst) crystals have been successfully made by many amateurs. A word of warning: these crystals were made for parallel-resonant oscillator service; their parallel resonant frequencies with a given load capacitance, typically 20pF, were within 50ppm or so when new. Their series-resonant frequencies, unimportant for their original purpose but paramount for filter application, can differ by much more. It therefore is very useful to have many more crystals than one needs so that a matched set can be selected.

To design a filter properly around available crystals one must know their characteristics. They can be measured with the test circuit of Fig 5.100, where the three positions of S1 yield three equations for the three unknowns L_s, C_s and C_p. R_s can be

Table 5.15. KVG 9MHz crystal filters for SSB, AM, FM and CW bandwidths

Filter type	XF-9A	XF-9B	XF-9C	XF-9D	XF-9E	XF-9M
Application	SSB TX	SSB TX/RX	AM	AM	NBFM	CW
Number of poles	5	8	8	8	8	4
6dB bandwidth (kHz)	2.5	2.4	3.75	5.0	12.0	0.5
Passband ripple (dB)	<1	<2	<2	<2	<2	<1
Insertion loss (dB)	<3	<3.5	<3.5	<3.5	<3	<5
Termination	500Ω/30pF	500Ω/30pF	500Ω/30pF	500Ω/30pF	1200Ω/30pF	500Ω/30pF
Shape factor	1.7 (6–50dB)	1.8 (6–60dB) 2.2 (6–80dB)	1.8 (6–60dB) 2.2 (6–80dB)	1.8 (6–60dB) 2.2 (6–80dB)	1.8 (6–60dB) 2.2 (6–80dB)	2.5 (6–40dB) 4.4 (6–60dB)
Ultimate attenuation (dB)	>45	>100	>100	>100	>90	>90

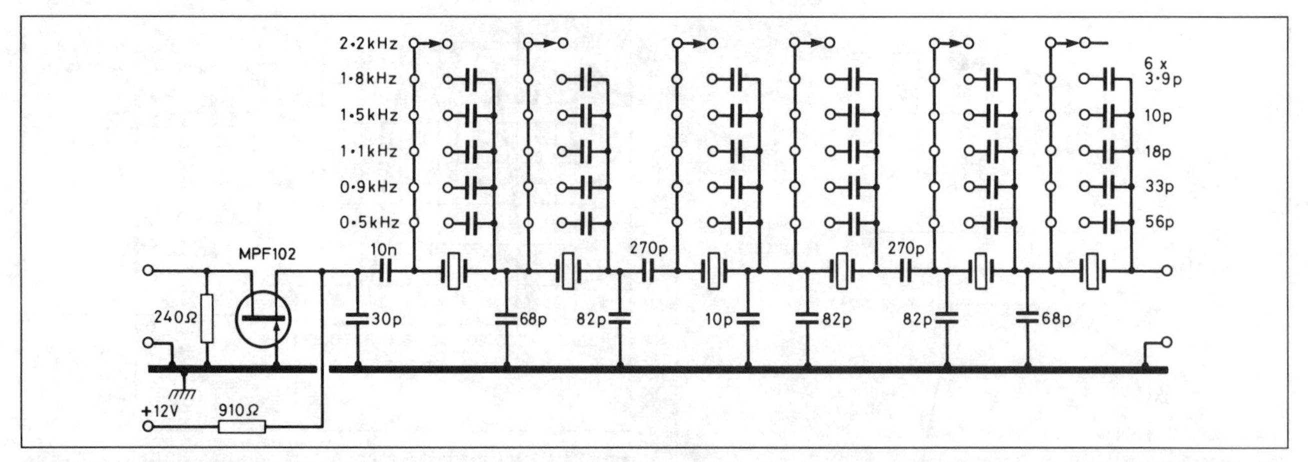

Fig 5.101(a). Y27YO described this six-section 4.433MHz ladder filter with switch-selectable bandwidth in *Funkamateur* 1/85. The FET buffer is included to present a stable, non-reactive load to the preceding diode mixer and the proper source impedance to the filter. While it uses inexpensive components, including PAL TV colour-burst crystals, careful shielding between sections is required and construction and test demand skill and proper instrumentation

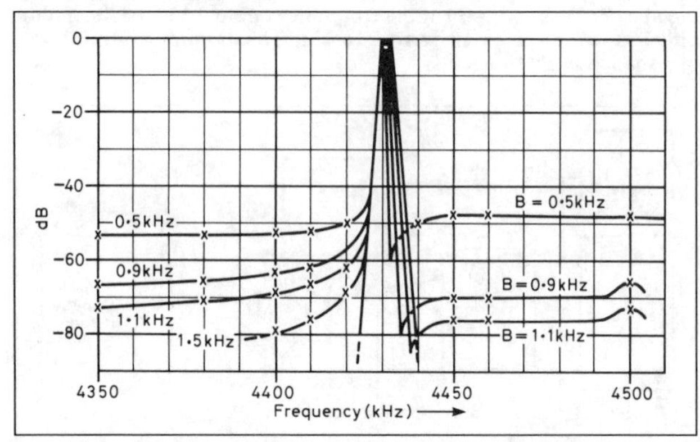

Fig 5.101(b). Response curves of the filter shown in Fig 5.101(a)

measured after establishing series-resonance with S1 in position 3 by substituting non-inductive resistors for the crystal without touching the frequency or output level of the signal generator. A resistor which gives the same reading on the output meter as the crystal is equal to R_s. Note that the signal generator must be capable of setting and holding a frequency within 10Hz or so. All relevant information and PC programs to simplify the calculations are given in reference [33].

If a sweep generator and 'scope are used to adjust or verify a completed filter, the sweep speed must be kept very low: several seconds per sweep. Traditionally this has been done by using 'scopes with long-persistence CRTs. Sampling 'scopes, of course, can take the place of long-persistence screens.

The home-made filter of Fig 5.101, described by Y27YO, uses six 4.433MHz PAL colour TV crystals [34]. The filter bandwidth can be changed by switching different load capacitors across each crystal. Note that at the narrower bandwidths the ultimate attenuation is reduced because of the greater load capacitors shunting each crystal. Each crystal with its load capacitors and switch wafer should be in a separate shielding compartment.

The input impedance of any crystal filter just above and below its pass-band is far from constant or non-reactive; therefore it is unsuitable as a termination for a preceding diode-ring mixer, which must have a fixed purely resistive load to achieve the desired rejection of unwanted mixing products. The inclusion in this filter of a common-gate FET buffer solves this problem.

Virtually all available crystals above about 24MHz are *overtone* crystals, which means that they have been processed to have high Q at the third or fifth mechanical harmonic of their fundamental frequency. Such crystals can be used in filters. The marking on overtone crystals is their series resonant frequency, which is the one which is important in filters.

48MHz is a common microprocessor clock frequency and third-overtone crystals for it are widely available and relatively inexpensive. 48MHz also is a suitable first IF for dual- or triple-conversion HF receivers, which then require a *roofing filter* at that frequency. PA0SE reported on a 48MHz SSB filter designed by J Wieberdink [35]. This is shown in Fig 5.102. The following prototype data were recorded:

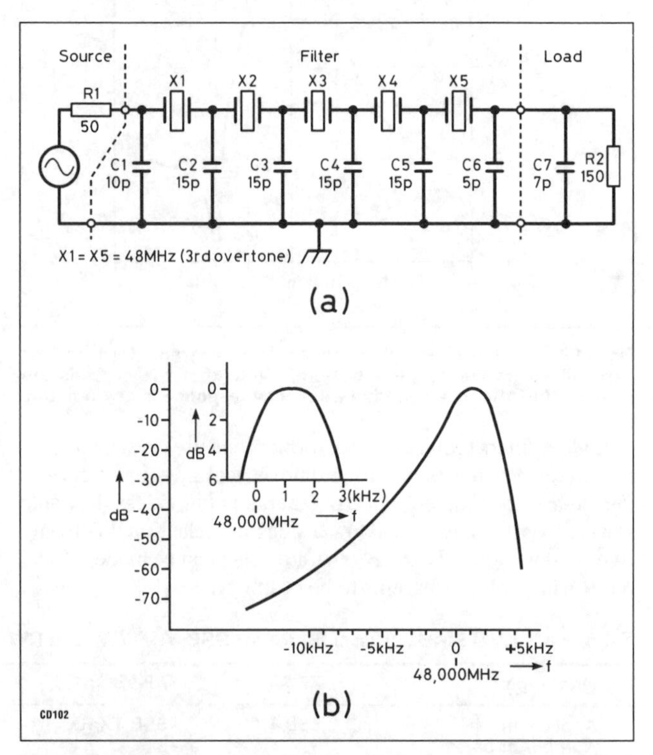

Fig 5.102. Overtone crystals, always marked with their series-resonant frequency, are used in this 48MHz ladder filter. This design by J Wieberdink from the Dutch magazine *Radio Bulletin* 10/83 makes a good roofing filter for an HF SSB/CW receiver, but its top is too narrow to pass AM or NBFM and its shallow skirt slope on the low side requires that it be backed up by another filter at a second or third IF

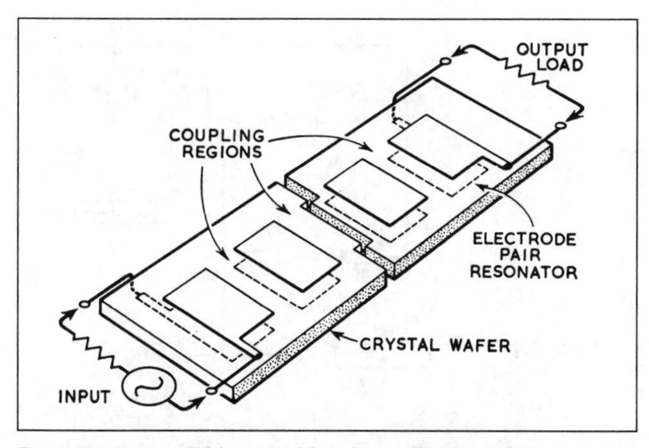

Fig 5.103. A monolithic crystal band-pass filter consists of several resonators, ie pairs of electrodes, on a single quartz plate. The coupling between the resonators is essentially mechanical through the quartz. Monolithic filters are mounted in three or four-pin hermetically sealed crystal holders. At HF, they are less expensive though less effective than the best discrete-component crystal filters, but they do make excellent VHF roofing filters

f_0	48.0012MHz
–3dB bandwidth	2.6kHz
–6dB bandwidth	3kHz
–40dB bandwidth	9kHz
–60dB bandwidth	15.1kHz
Spurious responses	Less than –70dB
Pass-band ripple	Less than 0.2dB
Insertion loss	2.1dB

These data were measured with the 50Ω source and 150Ω//7pF load shown in the diagram, these giving the best results. Adequate shielding of the whole filter and between its sections is essential at this high frequency.

Monolithic crystal filters

Several pairs of electrodes can be plated onto a single quartz blank as shown in Fig 5.103. This results in a multi-section filter with the coupling between elements being mechanical through the quartz. Monolithic crystal filters in the 10–100MHz range are used as IF filters where the pass-band must be relatively wide, ie in AM and FM receivers and as roofing filters in multimode receivers.

Ceramic filters

Synthetic (ceramic) piezo-electric resonators are being made into band-pass filters in the range of 400kHz to 10.7MHz. Monolithic, ladder and half-lattice crystal filters all have their ceramic equivalents. While cheaper, ceramic resonators have lower Q than quartz crystals and their resonant frequency has wider tolerance and is more temperature dependent. While ceramic band-pass filters with good shape factors are made for bandwidths commensurate with AM, FM and SSB, they require more sections to achieve them, hence their insertion loss is greater. Care must be taken that BFOs used with them have sufficient frequency adjustment range to accommodate the centre frequency tolerance of a ceramic filter, eg 455±2kHz at 25°C. Input and output matching transformers are included in some ceramic filter modules. Fig 5.104 shows a muRata ladder filter and a Toko monolithic filter.

Active filters

When designing LC audio filters, one soon discovers that the inductors of values one would wish to use are bulkier, more

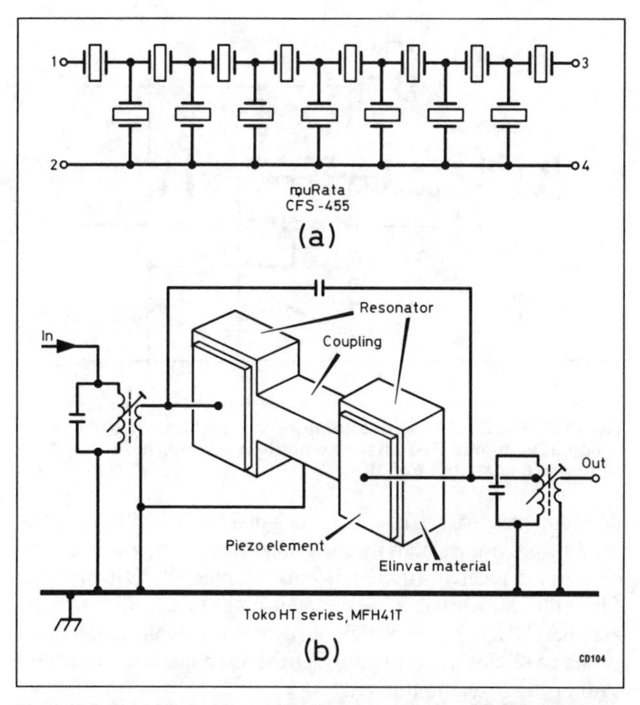

Fig 5.104. Ceramic band-pass filters are made in several configurations resembling those of crystal filters. While cheaper than the latter, they require more sections for a given filter performance, hence have greater insertion loss. Centre-frequency tolerances are 0.5% typical. Shown here are a ladder filter (a) and a monolithic filter (b)

expensive and of lower Q than the capacitors. Moreover, this lower Q requires the use of more sections, hence more insertion loss and even greater bulk and cost. One way out is the *active filter*, a technique using an amplifier to activate resistors and capacitors in a circuit which emulates an LC filter. Such amplifiers can be either single transistors or IC operational amplifiers, both being inexpensive, small, miserly with their DC supply, and capable of turning insertion loss into gain. Most active audio filters in amateur applications use two, three or four two-pole sections, each section having an insertion voltage gain between one and two. Filter component (R and C) accuracies of better than 5% are generally adequate, polystyrene capacitors being preferred.

The advantage of op-amps over single transistors is that the parameters of the former do not appear in the transfer (ie output vs input) function of the filter, thereby simplifying the calculations. Note that most IC op-amps are designed to have a frequency response down to DC. To allow both positive and negative outputs, they require both positive and negative supply voltages. In AC-only applications, however, this can be circumvented. In a single 13.8V DC supply situation, one way is to bridge two series-connected 6.8kΩ 1W resistors across that supply, each bypassed with a 100μF/16V electrolytic capacitor; this will create a three-rail supply for the op-amps with 'common' at +6.9V, permitting an output swing up to about 8V p-p. The input and output of the filter must be blocked for DC by capacitors.

A complete active filter calculation guide is beyond the scope of this book but some common techniques are presented in Chapter 22. Here, however, follow several applications, one with single transistors and others using op-amps.

A discrete-component active filter, which passes speech but rejects the bass and treble frequencies which do not contribute to intelligibility, is shown in Fig 5.105, together with its

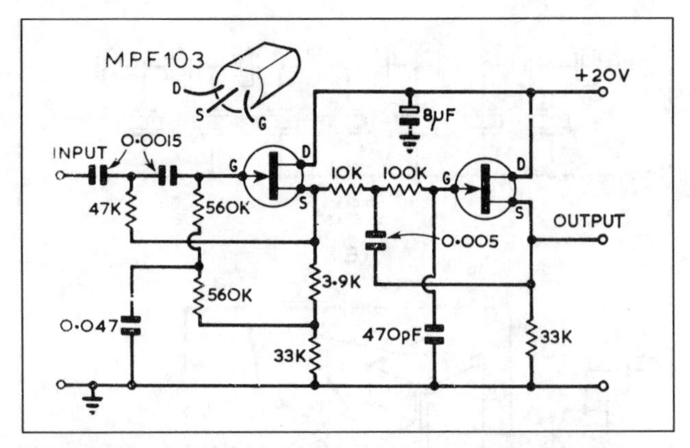

Fig 5.105. The ZL2APC active band-pass audio filter using FETs and a single DC supply. The filter attenuations are referred to the 1kHz response and an input of 1V RMS

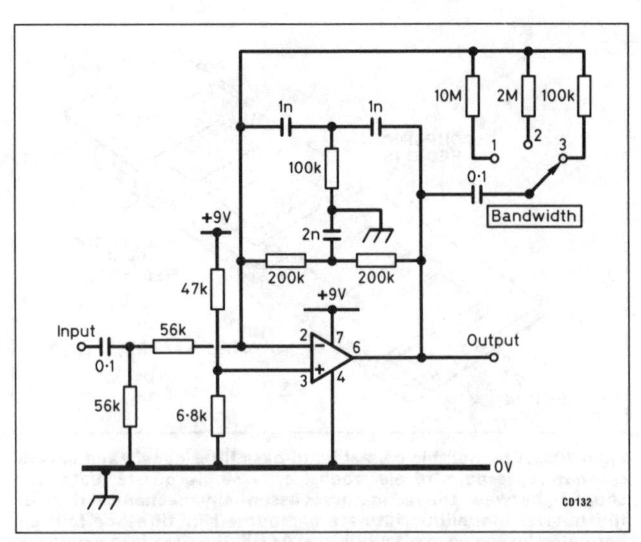

Fig 5.106. G3SZW put an 800Hz twin-T filter in the feedback loop of an op-amp to obtain a peak filter, then widened the response to usable proportions with switch-selectable resistors: to 60 and 180Hz (positions 1 and 2) for CW or 300–3500Hz (position 3) for voice. This filter does not require dual DC supplies

frequency response. This filter, designed by ZL2APC, might be used to provide selectivity for phone reception with a direct-conversion receiver (though it must be pointed out that active filters do not have sufficient dynamic range to do justice to those very best DC receivers, which can detect microvolt signals in the presence of tens of millivolts of QRM on frequencies which the audio filter is required to reject).

A *twin-T* filter used in the feedback loop of an op-amp is shown in Fig 5.106. A twin-T filter basically is a notch filter which rejects one single frequency and passes all others. With the R and C values shown, that frequency is about 800Hz. Used in the feedback loop of an op-amp, as in this design by G3SZW, the assembly becomes a peak filter which passes only that one frequency, too sharp even for CW. G3SZW broadened the response by shunting switch-selectable resistors across the twin-T; to 60Hz with 10MΩ, 180Hz with 2MΩ, and from 300Hz to 3500Hz, for phone, with 100kΩ.

Twin-T filters require close matching of resistors and capacitors. That would, in this example, be best accomplished by using four identical 1nF capacitors and four identical 200kΩ resistors, using two of each in parallel for the earthed legs.

This circuit also demonstrates another technique for the use of op-amps on a single supply, here 9V. The DC level of both inputs is set by the voltage divider to which the (+) input is connected:

$$(6.8/(6.8 + 47)) \times 9 = 1.14\text{V}$$

The op-amp's DC output level is set by its DC input voltage and inverting gain:

$$1.14 + ((200 + 200)/(56 + 56)) \times 1.14 = 5.2\text{V}$$

roughly half-way between +9V and earth. The capacitor in series with the bandwidth switch is to prevent the lowest bandwidth resistor from upsetting the DC levels. The input and output blocking capacitors have the same purpose with respect to any DC paths through the signal source or load.

A CW filter with two 741 or 301A-type op-amps in a multiple-feedback, band-pass configuration was described by LA2IJ and LA4HK [36], and is shown in Fig 5.107. It would be a worthwhile accessory with a modern transceiver lacking a narrow crystal filter.

The first stage is fixed-tuned to about 880Hz, depending partly on the value of R2. The corresponding resistor in the second stage is variable and with it the resonant frequency can be adjusted to match that of the first stage, or to a slight offset for a double-humped band-pass characteristic. In the first state the filter has a pass-band width of only 50Hz at –6dB (about 640Hz at –50dB). When off-tuned the effective pass-band can be widened to about 200Hz (1550Hz). Note that a bandwidth as narrow as 50Hz is of value only if the frequency stability of both the transmitter and receiver is such that the beat note does not drift out of the pass-band during a transmission, a stability seldom achieved with free-running home-built VFOs. Also, no filter can compete with the ears of a skilled operator when it comes to digging out a weak wanted signal from among much stronger QRM.

A scheme to provide second-order CW band-pass filtering or a tunable notch for voice reception was described by DJ6HP [37]. It is shown in Fig 5.108, and is based on the three-op-amp so-called *state variable* or *universal* active filter. The addition of a fourth op-amp, connected as a summing amplifier, provides the notch facility.

The resonant Q can be set between 1 and 5 with a single variable resistor and the centre frequency can be tuned between 450 and 2700Hz using two ganged variable resistors.

Switched-capacitor filters

The switched-capacitor filter is based on the digital processing of analogue signals, ie a hybrid between analogue and digital signal processing. It depends heavily on integrated circuits for its implementation. It pays to study the manufacturers' data sheets of devices under consideration before making a choice.

While switched-capacitor filters had been introduced to amateurs before, eg the Motorola MC14413/4 (superseded by MC145414) by W1JF [38] and the National Semiconductor MF10 by AI2T [39], it was an article by WB4TLM/KB4KVE [40] featuring the AMI S3528/9 ICs in the AFtronics SuperSCAF that led to the BARTG R5 design by G3ISD [41] shown in Fig 5.109 and Tables 5.16 and 5.17. A kit for this instrument is currently available from BARTG.

WB4TLM/KB4KVE describe the operation of switched-capacitor filters as follows: "The SCF works by storing discrete samples of an analogue signal as a charge on a capacitor. This charge is transferred from one capacitor to another down a chain of capacitors forming the filter. The sampling and transfer operations take place at regular intervals under control of a precise frequency source or clock. Filtering is achieved by combining the charges on the different capacitors in specific ratios and by

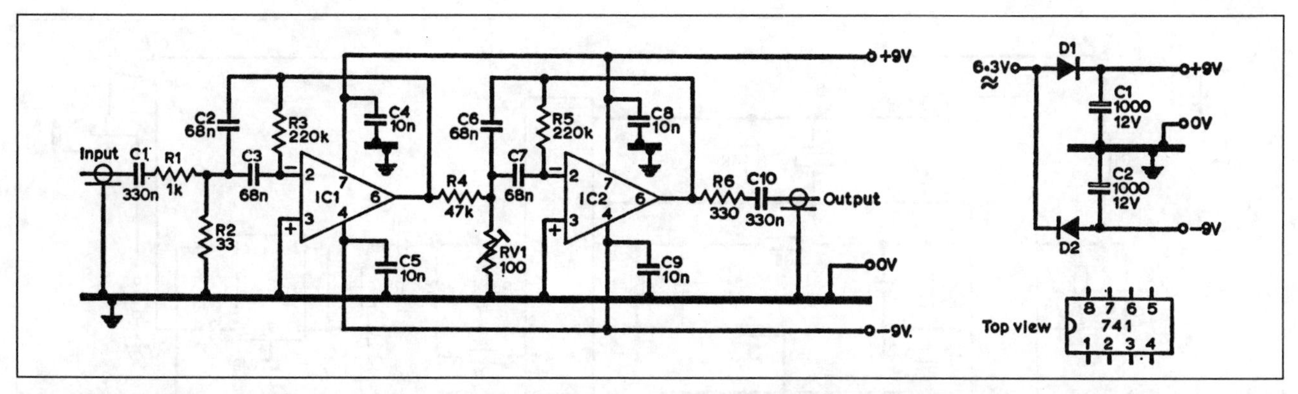

Fig 5.107(a). This CW band-pass filter using two multiple-feedback stages comes from LA2IJ and LA4HK. The centre frequency of the second stage can be equal to or offset from the first. If the two frequencies coincide, the overall bandwidth is 50Hz at –6dB (640Hz at –50dB), if staggered 200Hz (1550Hz). The dual power supply is derived from 6.3VAC, available in most valved receivers or transmitters

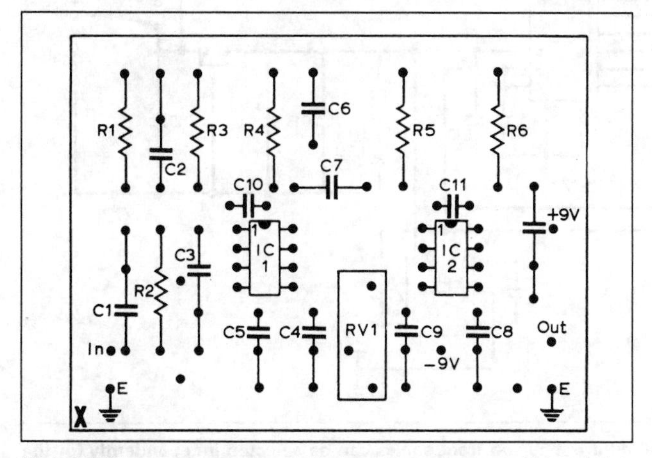

Fig 5.107(b). Component layout for the filter shown in Fig 5.107(a). The PCB pattern is shown in Fig 5.107(c) in Appendix 1

feeding charges back to the prior stages in the capacitor chain. In this way, filters of much higher performance and complexity may be synthesised than is practical with analogue filters".

The BARTG R5 filter includes a seven-pole high-pass and a seven-pole low-pass SCAF of which the cut-off frequencies can be chosen by selecting one of 40 positions of two switches according to Table 5.17. If used as an audio filter behind a receiver, one might listen to CW with the filter switches set to H7 and L9 for a pass-band of 635–904Hz, while for voice reception H3 and L27, 273–2711Hz would be appropriate. Other selections would be useful for RTTY tones. In addition to the filters, the kit contains ICs for audio amplification to speaker level and ±5VDC from the 12VDC power input, as well as the necessary passive components.

ATTENUATORS

Attenuators are resistor networks which reduce the signal level in a line while maintaining its characteristic impedance. Table 5.18 gives the resistance values to make up 75Ω and 50Ω unbalanced RF T- and pi-attenuators; the choice between the T- and pi- configurations comes down to the availability of resistors close to the intended values, which can also be made up of two or more higher values in parallel; the end result is the same. Here are some of the applications.

Receiver overload

Sensitive receivers often suffer overload (blocking, cross-modulation) from strong out-of-band unwanted signals which are too close to the wanted signal to be adequately rejected by preselector filters. This condition can often be relieved by an attenuator in the antenna input. In early solid-state receivers, which were very overload-prone, there sometimes was a choice of two levels of attenuation, eg 20 or 40dB. In more recent receivers, one reduction of sensitivity is usually provided for by a switch which removes the RF amplifier stage from the circuit; only one attenuator is then provided, usually 20dB.

There is an additional advantage of an attenuator. However carefully an antenna may be matched to a receiver at the wanted

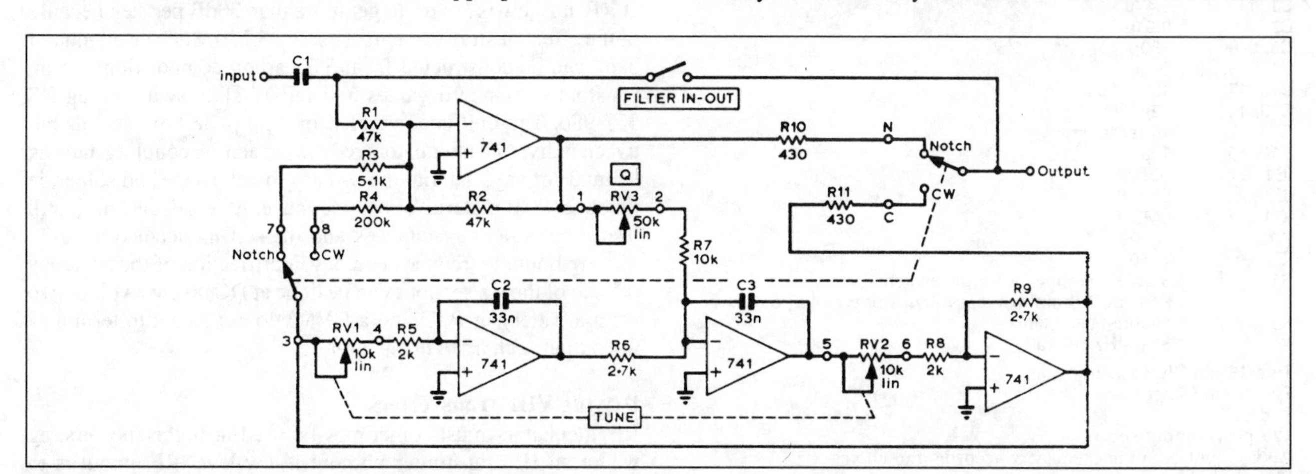

Fig 5.108. A universal filter scheme described by DJ6HP provides variable-*Q* band-pass filtering for CW or a tunable notch for 'phone. Dual DC supplies between ±9V and ±15V are required

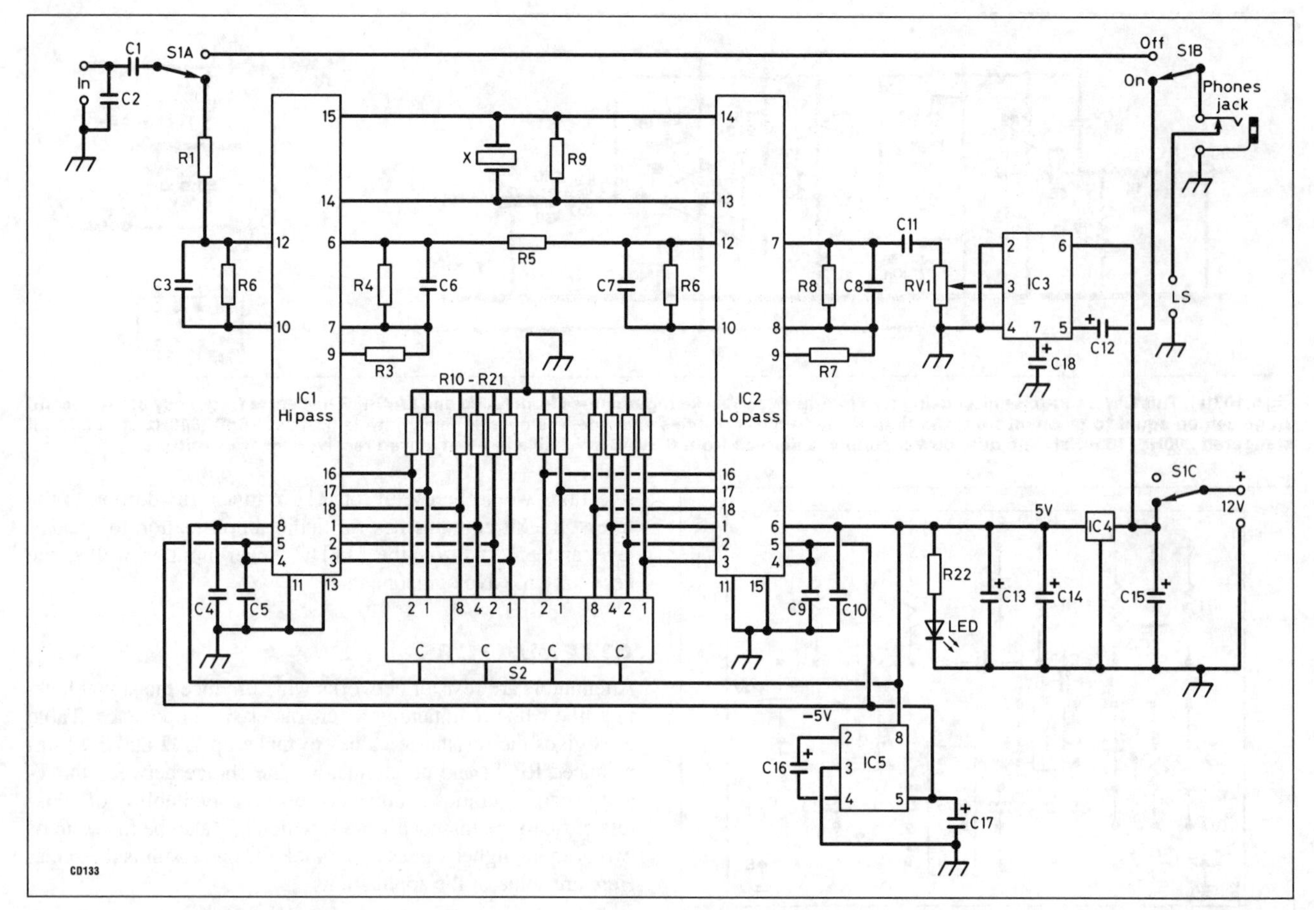

Fig 5.109. The schematic of the BARTG R5 switched-capacitor filter. 40 different cut-off frequencies can be selected independently for the high-pass and the low-pass filter to configure the optimum pass-band for CW, speech or data *(Datacom)*

frequency, it is likely to be grossly mismatched at the interfering frequency, leaving the receiver's preselector filter poorly terminated and less able to do its job. A 10dB attenuator keeps the nominally 50Ω termination of the preselector filter between 41 and 61Ω under all conditions of antenna mismatch.

Table 5.16. R5 filter components list

R9	10M, 0.25W
R22	330R, 0.25W
RV1	4k7 preset
C1, 11	0.47μ
C2	0.01μ
C3, 6–8	680p
C4, 5, 9, 10	0.1μ
C12, 17	100μ, 16V
C13, 15	22μ, 25V
C14	1μ tantalum
C16, 18	10μ
IC1	S3529
IC2	S3528
IC3	LM386
IC4	78L05
IC5	7660
S1	3-pole changeover toggle switch
S2	Four BCD thumbwheel switches or two 40-pos CB channel switches
X	3.58MHz crystal

Two 18-pin DIL sockets
Two 8-pin DIL sockets
Red LED
12V power input socket
Audio input/ouput sockets/jacks according to choice
Phones jack

All resistors except R9 and R22 are 47k, 0.25W.

An S-meter as a field strength meter

When plotting antenna patterns, the station receiver is often used as a field strength indicator. As the calibration of S-meters is notoriously inaccurate, it is better to always use the same S-meter reading, say S9, and to adjust the signal source or the receiver sensitivity to get that reading at a pattern null. All higher field strengths are then reduced to S9 by means of a calibrated switchable attenuator.

If six attenuators of 1, 2, 3, 5, 10 and 20dB can be individually switched in and out, a range of 0–41dB in 1dB steps results; at RF, it is unwise to try to get more than 20dB per step because of the effect of stray capacitances: Fig 5.110. Such an attenuator bank can be constructed from 5% carbon composition or film resistors of standard values and DPDT slide switches, eg RS 337-986. The unit is assembled in a tin-plate box; shields between individual attenuators reduce capacitive coupling and can be made of any material that is easy to cut, shape and solder, ie tin plate, PCB material or copper gauze. At VHF, the unit is still useful but stray capacitances and the self-inductance of resistors are bound to reduce accuracy. Verification of the accuracy of each of the six sections can be done at DC, using a volt or two from a battery or PSU and a DVM; do not forget to terminate the output with a 50Ω resistor.

Driving VHF transverters

RF attenuators must sometimes be used at higher powers, eg where an HF transmitter without a low-level RF output is to drive, but not to overdrive, a VHF transverter. Here is a case history.

Table 5.17. BARTG R5 filter. Cut-off frequencies selected by each of the 40 high-pass and low-pass switch positions

Switch position	High-pass	Low-pass
00	40	44
01	91	100
02	182	200
03	273	300
04	363	399
05	455	500
06	546	601
07	635	699
08	726	799
09	822	904
10	914	1005
11	1005	1105
12	1099	1209
13	1179	1297
14	1271	1398
15	1355	1491
16	1453	1598
17	1535	1688
18	1627	1790
19	1731	1904
20	1808	1989
21	1892	2081
22	1985	2183
23	2086	2295
24	2198	2418
25	2260	2486
26	2392	2632
27	2465	2711
28	2543	2797
29	2625	2887
30	2712	2983
31	2805	3086
32	2905	3196
33	3013	3314
34	3129	3442
35	5423*	5965*
36	3254	3579
37	3389	3728
38	5811*	6392*
39	3537	3891

* Note these frequencies. Table taken from *Datacom*.

A Ten-Tec Corsair II HF transceiver was to drive a Microwave Modules 28/144 VHF transverter. The output from the HF rig could be limited, by means of its ALC control, to between 100 and 25W PEP. The transverter could accommodate inputs between 0.005 and 0.5W PEP. The proper solution was to bring out the low-level drive to the HF rig's PA and feed it into the transverter; this has been successfully implemented, but not until after the warranty on the HF rig had expired. Meanwhile, an attenuator had been used, calculated and implemented asfollows.

The power incompatibility was at least 10 log (25/0.5) = 17dB; a 20dB attenuator would allow for calibration and component errors and had to be dimensioned to withstand an input of 25W, PEP on SSB and for short periods of CW testing. Table 5.18 showed pi-attenuator values of 61Ω shunt and 248Ω series resistors but no great precision was required. A 50Ω 100W Cantenna dummy load was available for the input resistor; it obviously could handle the power, and the load of 43Ω, ie less than the nominal 50Ω seen by the HF rig, would not matter as the SWR would be well below 1.5:1. For the other resistors, 2W carbon film types were the biggest non-inductive types found in component distributors' catalogues. The voltage across the Cantenna would be $\sqrt{(25W \times 43\Omega)}$ = 33V of which about 30V would be across the series resistor. 30V across 248Ω = 3.6W. The series resistor was made up of two 470Ω 2W resistors in parallel. For the output shunt resistor, a 56Ω 1W unit was handy. These three resistors were shielded in a small tin box to avoid 28MHz radiation. The assembly is sketched in Fig 5.111.

Table 5.18. Resistor values for 50Ω and 75Ω T- and pi-attenuators

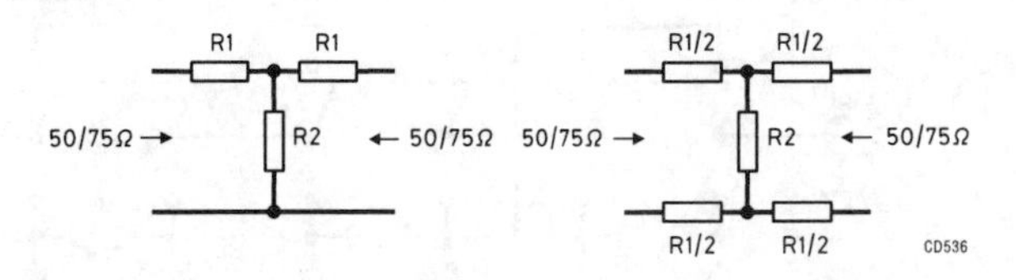

T-pad

Attenuation (dB)	50Ω R1	50Ω R2	75Ω R1	75Ω R2
1	2.9	433	4.3	647
2	5.7	215	8.6	323
3	8.5	142	12.8	213
4	11.3	105	17.0	157
5	14.0	82	21.0	123.4
6	16.6	67	25.0	100
7	19.0	56	28.7	83.8
8	21.5	47	32.3	71
9	23.8	41	35.7	61
10	26.0	35	39.0	52.7
11	28.0	30.6	42.0	45.9
12	30.0	26.8	45.0	40.2
13	31.7	23.5	47.6	35.3
14	33.3	20.8	50.0	31.2
15	35.0	18.4	52.4	25.0
20	41.0	10.0	61.4	15.2
25	44.7	5.6	67.0	8.5
30	47.0	3.2	70.4	4.8
35	48.2	1.8	72.4	2.7
40	49.0	1.0	73.6	1.5

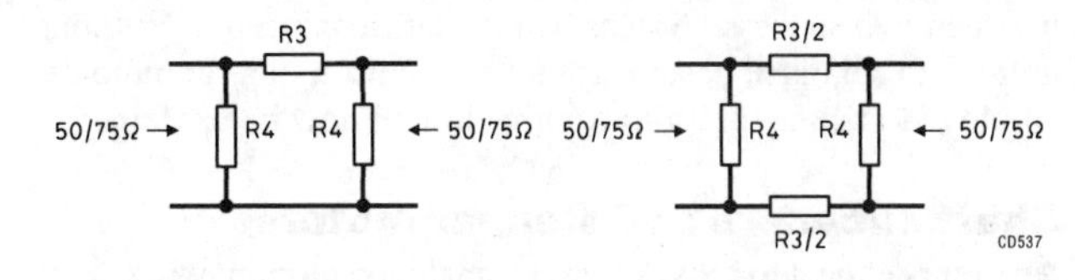

π-pad

Attenuation (dB)	50Ω R3	50Ω R4	75Ω R3	75Ω R4
1	5.8	870	8.6	1305
2	11.6	436	17.4	654
3	17.6	292	26.4	439
4	23.8	221	35.8	331
5	30.4	179	45.6	268
6	37.3	151	56.0	226
7	44.8	131	67.2	196
8	52.3	116	79.3	174
9	61.6	105	92.4	158
10	70.7	96	107	144
11	81.6	89	123	134
12	93.2	84	140	125
13	106	78.3	159	118
14	120	74.9	181	112
15	136	71.6	204	107
20	248	61	371	91.5
25	443	56	666	83.9
30	790	53.2	1186	79.7
35	1406	51.8	2108	77.7
40	2500	51	3750	76.5

As the transverter has a separate 28MHz receive output and the HF rig a separate receive antenna input, no relay was required to bypass the attenuator on receive.

ANALOGUE-DIGITAL INTERFACES

Most real-world phenomena are *analogue* in nature, meaning they are continuously variable rather than only in discrete steps; examples are one's height above ground going up a smooth

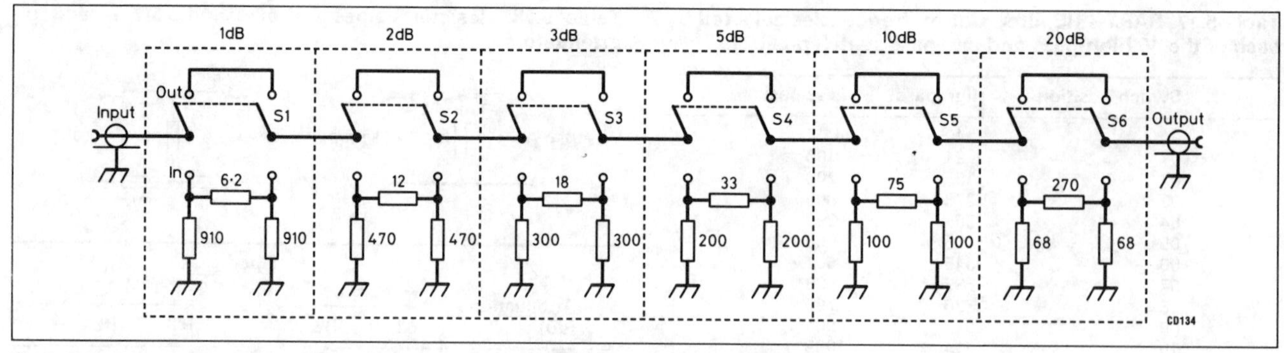

Fig 5.110. 50Ω attenuator bank with a range of 0–41dB in 1dB steps *(ARRL Handbook)*

wheel-chair ramp, the shaft angle of a tuning capacitor, the temperature of a heatsink and the wave shape of one's voice. However, some are *digital* in nature, meaning that they come in whole multiples of a smallest quantity called *least significant bit* (LSB); examples are one's height going up a staircase (LSB is one step), telephone numbers (one cannot dial in between two numbers) and money (even when the LSB is the farthing of blessed memory, any sum is a whole multiple thereof).

Frequently, it is useful to convert from analogue to digital and vice versa in an *analogue-to-digital converter* (ADC) or a *digital-to-analogue converter* (DAC). If anti-slip grooves with a pitch of, say, 1cm are cut across the wheelchair ramp, it has, in fact, become a staircase with minuscule steps (LSBs) which one can count to determine how far up one is, with any position between two successive steps being considered trivial. Without going into details of design, a number of conversions commonly employed in amateur radio equipment will now be explained.

Shaft encoders and stepper motors

Amateurs expect to twist knobs, an analogue motion, when they want to change frequency. Traditionally, that motion turned the shaft of a variable capacitor or screwed a core into or out of a coil. Now that frequencies in many radios are generated by direct digital synthesis, a process more compatible with keyboard entry and up/down switches, amateurs not only still like to twist knobs, they also like them to feel like the variable capacitors of yesteryear; equipment manufacturers comply but what the knob actually does drive is an optical device called an *incremental shaft encoder*: Fig 5.112 [42].

The encoder is a disc divided into sectors which are alternately transparent and opaque. A light source is positioned at one side of the disc and a light detector at the other. As the tuning knob is spun and the disc rotates, the output from the detector goes on or off when a transparent or an opaque disc sector is in the light path. Thus, the spinning encoder produces a stream of pulses which, when counted, indicate the change of angular position of the shaft. A second light source and detector pair, at an angle to the main pair, indicates the direction of rotation. A third pair sometimes is used to sense the one-per-revolution marker seen at the right on the disc shown.

Available encoder resolutions (the number of opaque and transparent sectors per disc) range from 100 to 65,000. The SSB/CW tuning rate of one typical radio was found to be 2kHz per knob rotation in 10Hz steps; this means an encoder resolution of 200. The ear cannot detect a 10Hz change of pitch at 300Hz and above, so the tuning feels completely smooth, though its

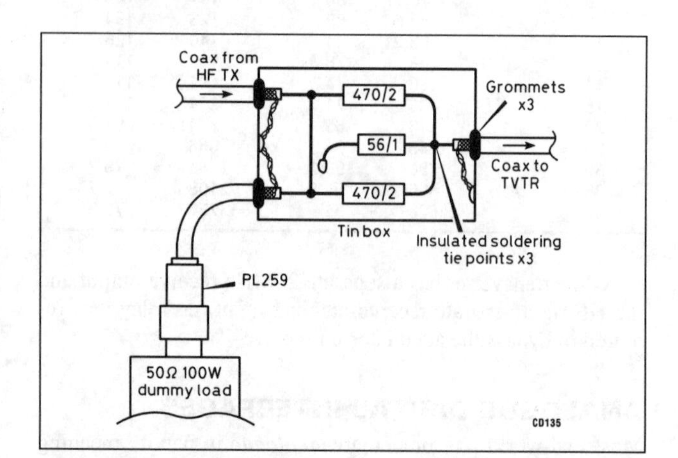

Fig 5.111. Attenuator to reduce 25W HF transceiver output to less than 0.5W for a VHF transverter

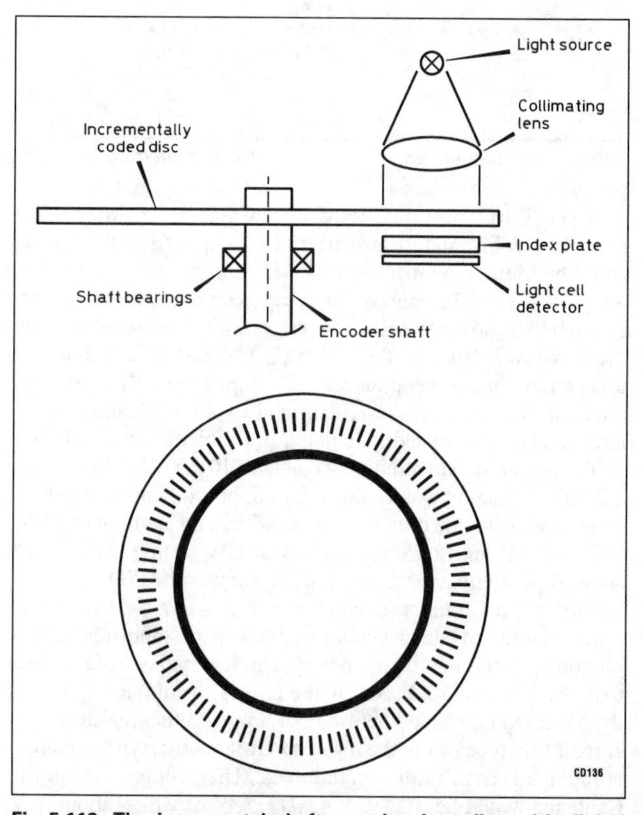

Fig 5.112. The incremental shaft encoder. In radios with digital frequency synthesis, the tuning knob turns the shaft which turns the disc which alternatingly places transparent and opaque sectors in the light path; each pulse from the light detector increases or decreases the frequency by one step. (Taken from *Analog-Digital Conversion Handbook*, 3rd edn, Prentice-Hall, 1986, p444, by permission. Copyright Analog Devices Inc, Norwood, MA, USA)

output is a digital signal in which each pulse is translated into a 10Hz frequency increment or decrement.

Stepper motors do the opposite of shaft encoders; they turn a shaft in response to digital pulses, a step at a time. Typical motors have steps of 7.5 or 1.8°; these values can be halved by modifying the pulse sequence. Pulses vary from 12V at 0.1A to 36V at 3.5A per phase (most motors have four phases) and are applied through driver ICs for small motors or driver boards with ICs plus power stages for big ones. The drivers are connected to a control board which in turn may be software-programmed by a computer, eg via an RS232 link (a standard serial data link). In amateur radio they are used on variable capacitors and inductors in microprocessor-controlled automatic antenna tuning units and on satellite-tracking antenna azimuth and elevation rotators.

Digital frequency meters

To pretune a receiver to an as-yet-unheard SSB transmitter of known frequency, one has to be spot-on, ie within 100Hz. In a receiver 'band' 500kHz wide, and allowing half of the total, ie 50Hz for the dial reading error alone, a dial resolution of one part in 10,000 would be required; with analogue dials this is impractical, if not impossible. The solution is a digital frequency read-out. On AM or NBFM, the required accuracy is an order of magnitude less but even there a digital read-out is an operating convenience.

In recent factory-made receivers or transceivers, all frequencies as well as the read-out are derived by digital synthesis from a single crystal-controlled master clock oscillator; no conversion is required.

In older factory-made equipment and in most home-built radios the frequency generating system (for the non-channelised bands) includes a free-running VFO and often one or more crystal oscillators; their combined drift with time and temperature is unpredictable. The solution is a digital frequency counter.

A radio may have a digital frequency meter built in, one may be retrofitted or a separate instrument may be used; they all work on the same principles: a crystal-controlled timebase oscillator, through a frequency divider, alternately opens and closes a gate at precise intervals; the signal of unknown frequency passes through that gate to a digital counter for as long as the gate is open. When the gate closes, the number of cycles counted is held in a latch and applied to a decoder which converts the binary count into decimal digits and each digit into the appropriate segments of the familiar seven-segment numerals of the LED or LCD display. During the gate-closed period, the counter is reset to zero and the display holds until refreshed after the next count.

Professional-grade frequency counters have a choice of gate times, eg 0.1, 1.0 and 10s and 8- or 9-digit resolution. Their absolute accuracy is the accuracy of the timebase plus or minus one count. The timebase accuracy depends on how little the crystal oscillator drifts with time and temperature from the frequency to which it was last set during calibration against an atomic (caesium) clock or a standard radio station such as MSF at Rugby. Assuming three-monthly calibrations, ambient temperatures between 15 and 35°C and pre-aged crystals in glass (the best but needlessly expensive) or in resistance or cold-welded metal cases, one can expect 1ppm for a plain oscillator (XO), 0.1ppm for a temperature-compensated oscillator (TCXO) and 0.01ppm for an oven-controlled oscillator (OCXO). TCXOs or even OCXOs sometimes are options for expensive amateur transceivers. Oscillators aimed at the amateur market may be an order of magnitude less stable when new, but they usually settle down as

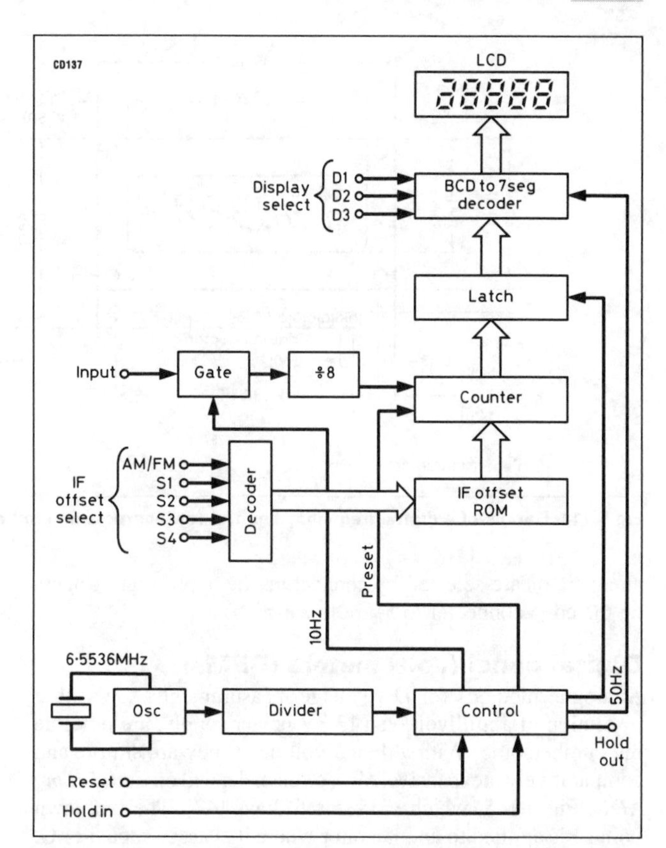

Fig 5.113. Block diagram of a digital frequency counter, Cirkit model FC177 *(Cirkit)*

their crystals age. Solder-sealed crystals, however, are distinctly inferior in this respect and epoxy-sealed crystals should not be used in frequency counters at all.

Amateurs can find cheap crystals of almost-frequency-meter grade in quartz clocks from wrecked cars. These usually are 4.194,304MHz (= 2^{22}Hz), easy to divide down to a one-second gate; the divider IC is in the clock. Because they must keep time in spite of wide temperature variations, car clock crystals usually are of much better quality than those sold for computer clock oscillators. Car clocks with a second hand can be checked with just patience instead of test equipment: adjust the clock for best timekeeping and then monitor it for a while to see how much that adjustment changes with time. A change of 1s/day/day = 12ppm; 1s/day/week = 1.65ppm; 1s/day/month = 0.38ppm.

There is a reasonably priced frequency counter module with five LCD digits for incorporation in amateur construction projects; it is sold by Cirkit, their model FC177. It will count frequencies up to 3999.9kHz, ie including the 1.8 and 3.5MHz amateur bands, with a resolution of 100Hz. It will read transmitter frequencies and also receiver frequencies; to be able to do that, the module counts the local oscillator frequency, starting not from zero but from the IF which can be one of several common IFs pre-programmed in the unit. Designed for broadcast receivers, the unit can read VHFs which have been processed by a ÷100 prescaler with a choice of IF offsets near 10.7MHz. The resolution then is 10kHz, suitable only for the FM part of the 50MHz band (20kHz channel spacing) and the air band but not for the 145MHz band which has 25 or 12.5kHz channel spacing. The block diagram is shown in Fig 5.113 and an outline in Fig 5.114.

The module requires a stabilised supply voltage of 5V at 4mA. The IF offset, place for the decimal point and MHz/kHz

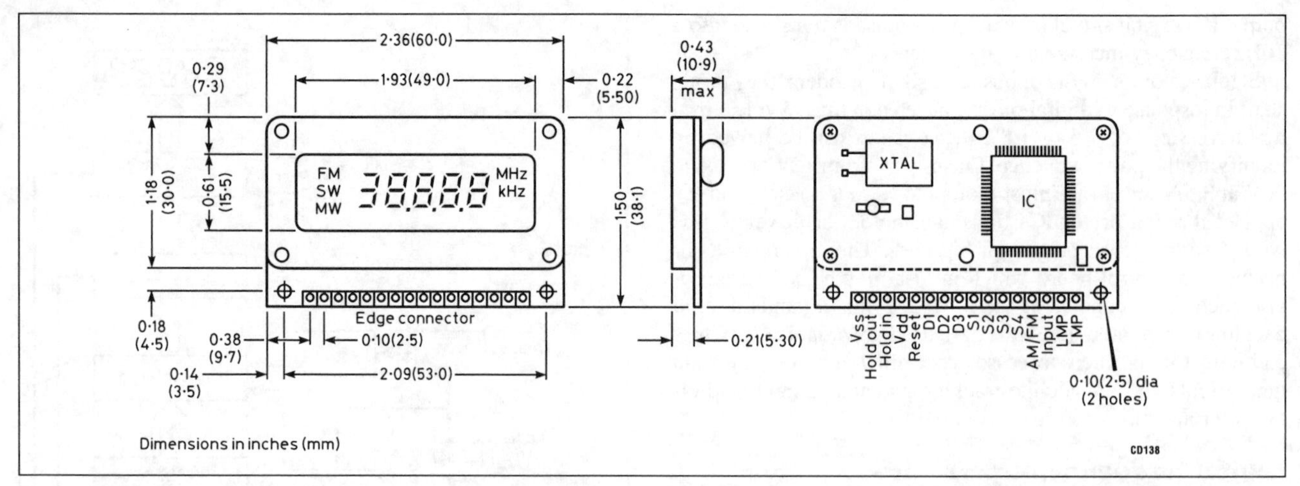

Fig 5.114. Details of a digital frequency counter for incorporation in home-built radios or for retrofitting. Cirkit model FC177 *(Cirkit)*

annunciator are selected by connecting the appropriate fingers on the edge connector to 5V or to earth.

Digital panel (volt) meters (DPMs)

Analogue methods for DC voltage measurements, say with a resolution of a millivolt on a 13.8V power supply, are possible but cumbersome. With a digital voltmeter they are simple and comparatively inexpensive. Most of them depend on a *dual-ramp ADC*, Fig 5.115, which works as follows [42]. The unknown voltage is applied to an integrator where it charges capacitor C; at the same time a counter is started, counting clock pulses. After a predetermined number of counts (a fixed interval of time, T), the signal input is switched off and a reference voltage of opposite polarity is applied to the integrator. At that instant, the accumulated charge on C is proportional to the average input over the time interval T. The integral of the reference is an opposite-going ramp with a slope V_{REF}/RC. During that time, the counter is again counting from zero. When the integrator output reaches zero, the count is stopped and the analogue circuitry is reset. Since the charge gained is proportional to $V_{IN}T$, and the equal amount of charge lost is proportional to $V_{REF}\delta t$, then the number of counts relative to the full count is proportional to $\delta t/T$, or V_{IN}/V_{REF}. If the output is a binary number, it will be a binary (digital) representation of the input voltage. However, if the input is attenuated and offset by $V_{REF}/2$, the output will be in *offset binary*, suitable for input to a computer. Simultaneously, the output is converted into BCD (binary coded decimal) form and applied to an LED or LCD digital display.

Compared with other ADCs, dual-slope integration has great advantages if its slow speed does not matter. In addition to excellent linearity and, by virtue of the averaging of the signal over its input interval, good immunity against HF noise, the slow speed can be turned into an advantage: if the conversion time is made equal to a mains period, ie 20ms, common-mode hum is cancelled out.

The whole dual-slope ADC, except for the integrating capacitor C, can be bought in a single DIP IC which operates on a single supply voltage, usually 5V.

Most DPMs have 3½ digits (counting up to ±1999) and are accurate to ±0.1% ±1 count ±0.01%/°C. Their basic voltage range usually is ±199.9mV, though range resistors for ±1.999V, ±19.99V and ±199.9V may be built in. The much-more-expensive 4½-digit models would be specified ±0.01% ±1 count ±0.003%/°C.

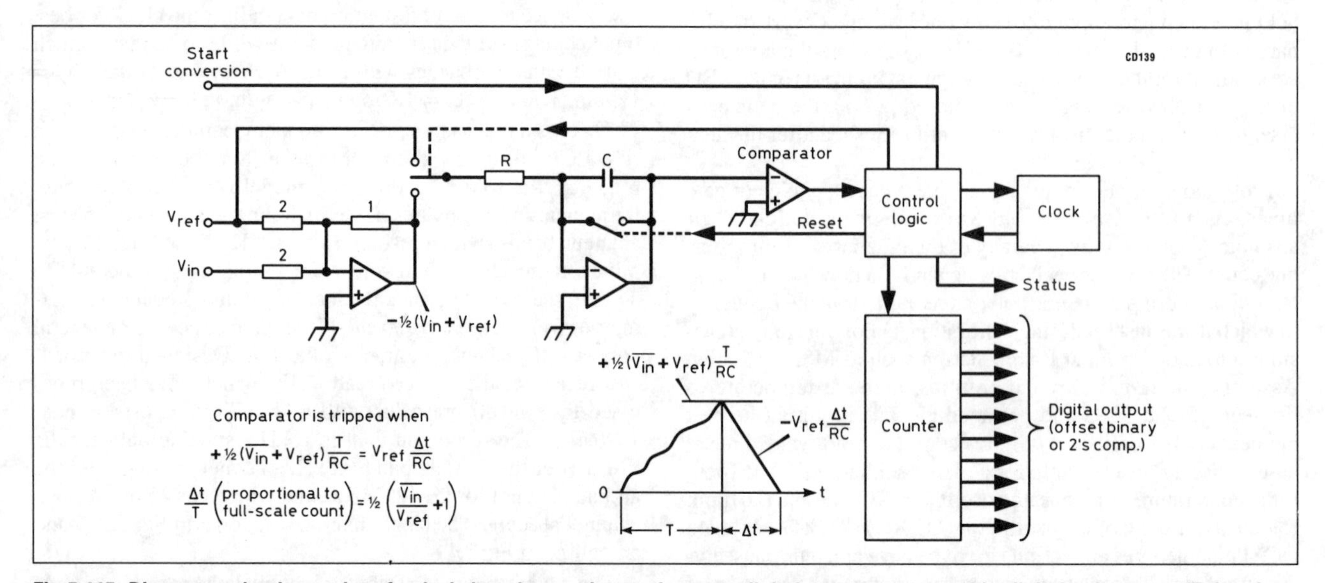

Fig 5.115. Diagram and voltage plot of a dual-slope integrating analogue-to-digital converter as used in digital voltmeters. (Taken from *Analog-Digital Conversion Handbook*, 3rd edn, Prentice-Hall, 1986, p215, by permission. Copyright Analog Devices Inc, Norwood, MA, USA)

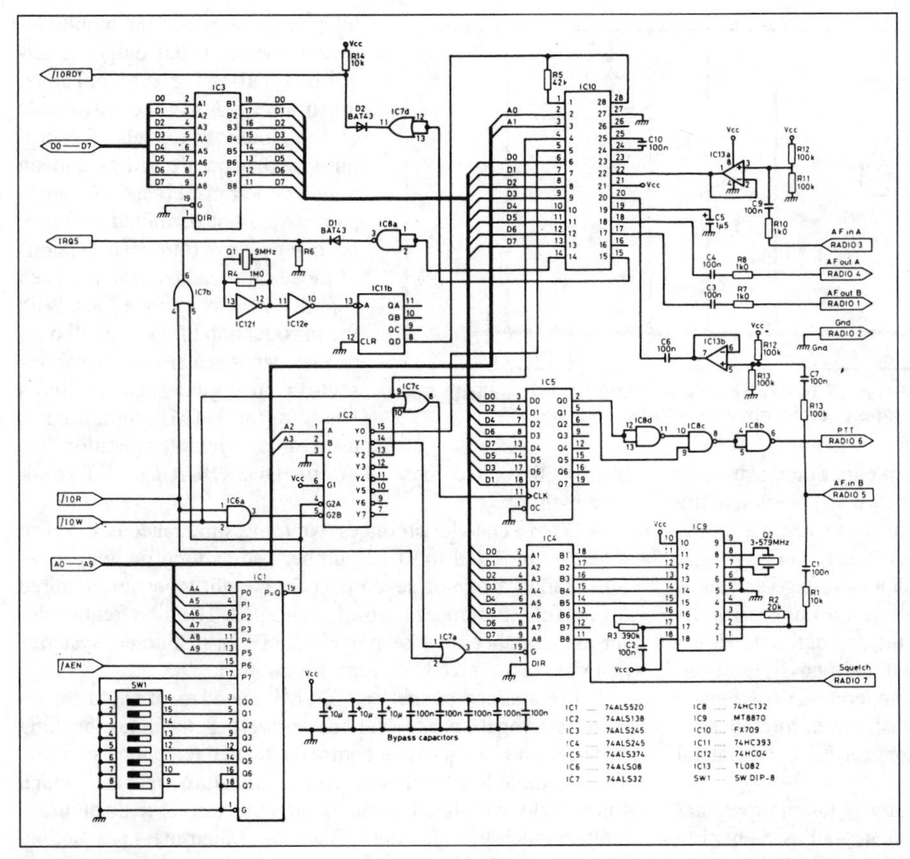

Fig 5.116. Speech encoder/decoder designed by DG3RBU and DL8MBT for voice mailboxes on UHF FM repeaters

Digitising speech

While speech can be stored on magnetic tape, and forwarded at speeds a few times faster than natural, each operation would add a bit of noise and distortion and devices with moving parts, ie tape handling mechanisms, are just not suitable for many environments. Digital speech has changed all that. Once digitised, speech can be stored, filtered, compressed and expanded by computer, and forwarded at the maximum speed of which the transmission medium is capable, all without distortion; it can then be returned to the analogue world where and when it is to be listened to.

If speech were to be digitised in the traditional way, ie sampled more than twice each cycle on the highest speech frequency, eg at 7kHz for speech up to 3kHz, and 10-bit resolution of the amplitude of each sample were required for reasonable fidelity, a bit rate of 70kbit/s would result, about 20 times the 3.5kHz bandwidth considered necessary for the SSB transmission of analogue speech. Also, almost a megabyte of memory would be required to store each minute of digitised speech. Differential and adaptive techniques are used to reduce these requirements.

DG3RBU and DL8MBT developed hardware and software for the digital storage and analogue retransmission of spoken messages on normal UHF FM voice repeaters and the digital forwarding of such messages between repeaters via the packet network. They described their differential analogue-digital conversion as follows.

"The conversion is by *continuously variable slope decoding* (CVSD), a form of delta modulation; in this process it is not the instantaneous value of an analogue voltage that is being sampled and digitised but its instantaneous slope at the moment of sampling. Binary '1' represents an increasing voltage, '0' a decreasing voltage. Encoding the slope of the increase or decrease depends on the prior sample. If both the prior and present samples are '1', a steeper slope is assumed than when a '1' follows a '0'. Decoding does the same in reverse. This system of conversion is particularly suitable for speech; even at a relatively modest data rate of 16kbit/s it yields good voice quality. An FX709 chip is used.

"The analogue-to-digital converter (ADC) for speech input and digital-to-analogue converter (DAC) for speech output are assembled on a specially designed plug-in board for an IBM PC computer: Fig 5.116 [43]. It provides all the required functions, starting with the address coding for the PC (IC1 and 2). In the FX709 (IC10), the signal passes through a bandpass filter to the one-bit serial encoder; after conversion to an eight-bit parallel format the data pass to the PC bus via IC3; in the other direction, voice signals pass through a software-programmable audio filter; registers for pause and level recognition complete the module. The FX709 has a loopback mode which permits a received and encoded speech signal to be decoded and retransmitted, an easy way to check the fidelity of the loop consisting of the radio receiver-encoder-decoder-radio transmitter.

"The built-in quartz clock oscillator (IC12f) and divider (IC11) permit experiments with different externally programmable clock rates. The maximum length of a text depends on the data rate. We used 32kbit/s, at which speed it is hard to tell the difference between the sound on the input and an output which has gone through digitising, storage, and reconversion to analogue. The maximum file length is 150s. The reason for this time limit is that the FX709 has no internal buffer; this requires that the whole file must be read from RAM in real time, ie without interrupts for access to the hard disk."

TONE SIGNALLING DEVICES

Tone signalling is extensively used for the control of remote FM stations and repeaters. Three different applications are of interest.

Repeater access by 1750Hz toneburst

To activate the transmitter in an IARU Region 1 (Europe and Africa) VHF or UHF repeater, its receiver must not only receive a carrier (and some of them speech) but also a 1750Hz ±5Hz toneburst with a deviation of 2.5kHz. The duration of the toneburst required differs from repeater to repeater but is generally between 200 and 500ms, sometimes more. Factory-made VHF and UHF FM transceivers for the European amateur market include a tone generator but home constructors and users of ex-PMR transceivers can build one.

G3VEH's Super-Bleep is inexpensive to build; it includes a twin-T op-amp oscillator and another op-amp as a timer. Fig

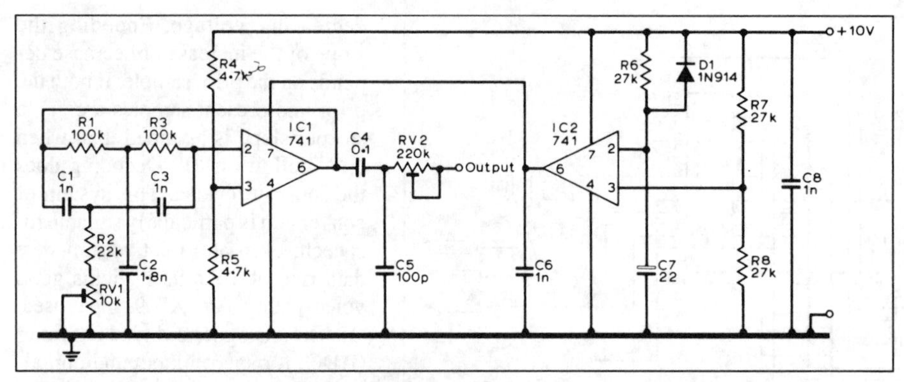

Fig 5.117. The Super-Bleep 1750Hz toneburst generator and timer (G3VEH). RV1 adjusts the frequency and RV2 adjusts the level. The supply can be anywhere between 5 and 15V

5.117. It is keyed on by PTT switching in either the positive or the negative supply lead. The oscillator then will start sending the tone and the voltage at IC2 pin 2 will rise; when it exceeds the voltage at pin 3 (one-half the supply voltage), pin 6 will flip from high to low, depriving IC1 of its supply voltage and stopping oscillation. The values of R6 and C7 set the duration of the toneburst. When the PTT is released at the end of the transmission, C7 discharges through D1. This circuit's possible disadvantage may be that in the extremes of temperature experienced in mobile operation its frequency may drift out of tolerance, in spite of the use of polystyrene capacitors for C1–3 and metal film resistors for R1–3.

A crystal-controlled toneburst generator is less temperature dependent and does not require a frequency adjustment: Fig 5.118. The 7168kHz oscillator output is divided by 4096 (2^{12}) in a CMOS divider IC to provide 1750Hz. The timing circuit uses the reset facility of the IC, pin 11. When C5 has charged to above half the supply voltage (this takes about 400ms) the IC resets and mutes the tone though the oscillator continues to function. D1 provides a fast discharge path for C6. The output from the divider is a square wave which is changed to a triangular wave by R4-C7. The output is set for a deviation of 2.5kHz by RV1. Power for the unit may be taken from any +10V regulated source in the transmitter; power consumption is minimal. If no such regulated source is available, a simple zener diode with series resistor can be used; good regulation is not required but protection against voltage spikes is.

Dual-tone multi-frequency (DTMF) signalling

Since its US introduction in the 1950s as a replacement for the rotary telephone dial, the *touch-tone pad* has found world-wide acceptance, not only for keying telephone numbers but for many remote control applications as well.

Many American repeaters have *auto-patches*, ie landline telephones on which the mobile repeater user can dial outgoing telephone calls; for that purpose, touch-tone pads have been available on the American models of microphones for mobile radios and on hand-held transceivers. As auto-patches are not permitted on European repeaters, European versions of those same radios have not been supplied with touch-tone pads. With the introduction of voice mailboxes on German repeaters (see the above section on digital speech) this is changing as DTMF signalling is used for their remote operation. Table 5.19 lists the commands for the DG3RBU/DL8MBT prototype mailbox.

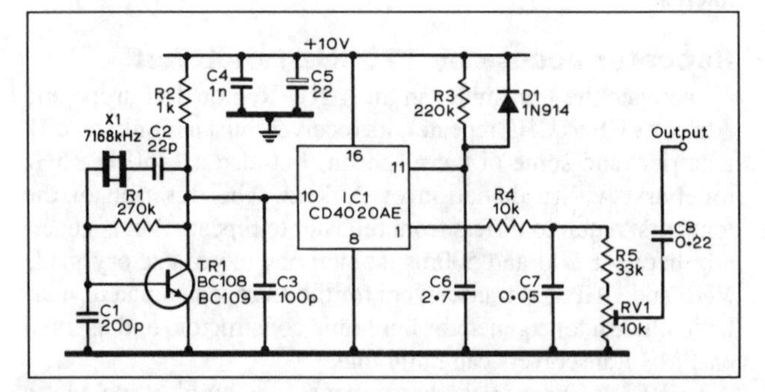

Fig 5.118. A 1750Hz toneburst is generated by dividing a 7168kHz crystal frequency. The IC also works as a timer

A DTMF encoder produces two tones simultaneously. Seven tones are required for a 12-number pad as used on telephones (also called a two-of-seven encoder); eight tones are required for the full 16-number (two-of-eight) pad. The tone frequencies for each number are shown in Fig 5.119. In 12-number systems, column 4 and its 1633Hz tone are omitted.

Self-contained pocket-size DTMF encoders are sold for remote interrogation of telephone answering machines, banking by phone, and a variety of computer-related telephone services. Their audible tone output is from a miniature 'speaker' which can be held over the microphone of a transceiver to do all that a built-in touch-tone pad does. Because of their mass production, they are inexpensive.

DTMF decoders can be assembled from one IC, eg ITT model 3201, which produces a 'data valid' signal and a hexadecimal character corresponding to any valid received pair of DTMF tones. The hex output can then be connected to a logic array which can be programmed for the intended task. In the UK, DTMF decoders are used in repeaters for remote control by their managers but for few other amateur radio projects.

Table 5.19. DTMF commands for the DG3RBU/DL8MBT prototype voice mailbox for UHF repeaters

Command	Function
999	Log-on for a new user and, in reply, issue of a user number
1xx	Log-on with previously assigned user number
01	List of all available commands
02	Reads off a group of 10 user numbers with associated calls. Repeating 02 identifies the next 10 user numbers etc
021	Renewed entering of own callsign (if not secured by prior 023)
022	Listen to one's own callsign
023	Secures own callsign against unauthorised tampering. Can be overridden only by sysop (system operator)
03	Speaking clock
04	List of own messages
041–049	Read own messages (041 first message, 042 second etc). After each message the user number of its sender is given
051–059	Erases own messages (051 first message etc)
06	News/info bulletin
07xxx	Sending a message; xxx is the addressee's user number. Key the five digits without pause, then release PTT and wait for prompt before entering a message; do not release PTT while sending a message
09	Exit from mail box
*	Abort partly keyed DTMF sequence and words spoken immediately thereafter; those words will be read back upon release of PTT

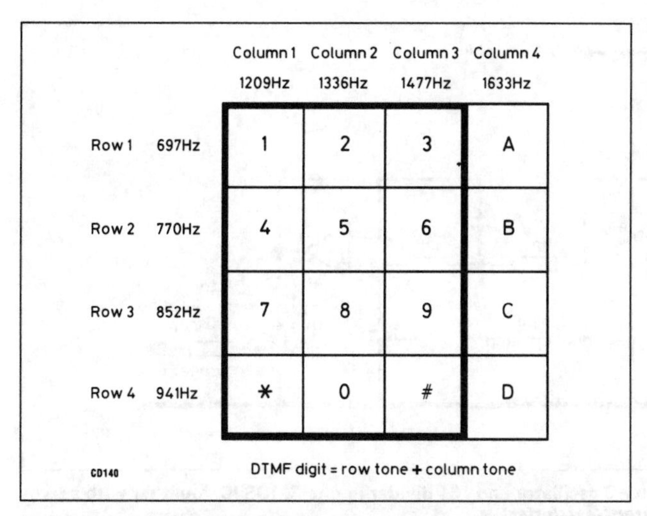

Fig 5.119. The DTMF (dual-tone multifrequency) key frequencies. 12-number (two-of-seven) pads only have three columns; 16-number (two-of-eight) pads have all four columns

Continuous-tone coded squelch systems (CTCSS)

If an FM receiver requires, for its squelch to open and remain open, not only a carrier of sufficient strength but also a tone of a specific frequency, much co-channel interference can be avoided.

If a repeater transmits such a tone with its voice transmissions but not with its identification, a CTCSS-equipped receiver set for the same tone would hear all that repeater's voice transmissions but not its idents. Its squelch would not be opened either, eg during lift conditions, by a repeater on the same channel but located in another service area and sending another CTCSS tone. Conversely, a mobile, positioned in an overlap area between two repeaters on the same channel but which have their receivers set for different CTCSS tones, could use one repeater without opening the other by sending the appropriate tone. Similar advantages can be had where several groups of stations, each with a different CTCSS tone, share a common frequency. With all transceivers within one group set for that group's CTCSS tone, conversations within the group would not open the squelch of stations of other groups monitoring the same frequency.

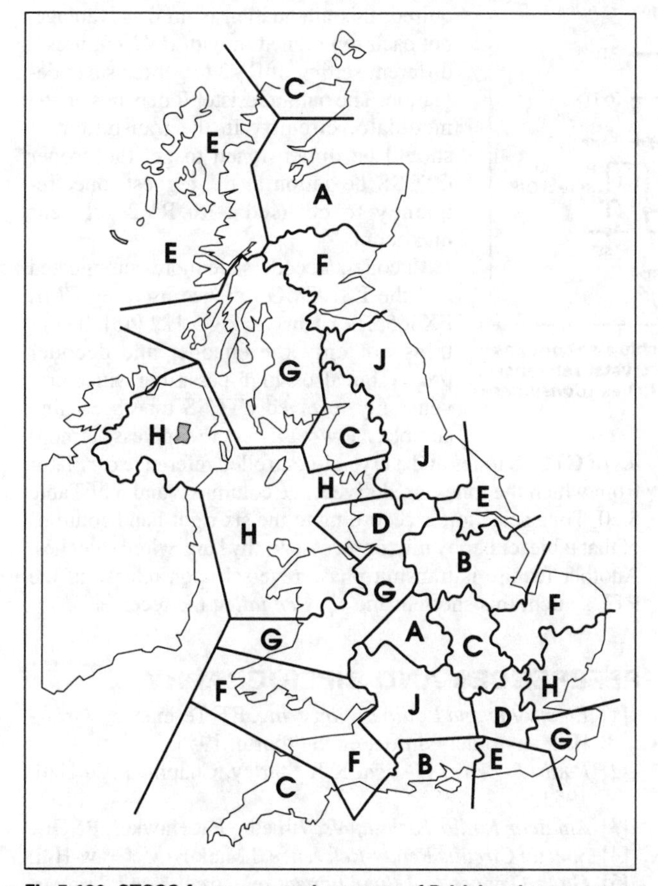

Fig 5.120. CTCSS frequency assignments of British voice repeaters by area

Table 5.20. CTCSS (continuous tone coded squelch system) EIA-standard frequencies. The letters behind some frequencies refer to British areas where that frequency has been assigned to voice repeaters as per Fig 5.120. The eight right-hand columns refer to Fig 5.122

Nominal frequency (Hz)		FX365 frequency (Hz)	Δf_o (%)	D_0	D_1	D_2	D_3	D_4	D_5
67.0	A	67.05	+0.07	1	1	1	1	1	1
71.9	B	71.90	0.0	1	1	1	1	1	0
74.4		74.35	–0.07	0	1	1	1	1	1
77.0	C	76.96	–0.05	1	1	1	1	0	0
79.7		79.77	+0.09	1	0	1	1	1	1
82.5	D	82.59	+0.10	0	1	1	1	1	0
85.4		85.38	–0.02	0	0	1	1	1	1
88.5	E	88.61	+0.13	0	1	1	1	0	0
91.5		91.58	+0.09	1	1	0	1	1	1
94.8	F	94.76	–0.04	1	0	1	1	1	0
97.4		97.29	–0.11	0	1	0	1	1	1
100.0		99.96	–0.04	1	0	1	1	0	0
103.5	G	103.43	–0.07	0	0	1	1	1	0
107.2		107.15	–0.05	0	0	1	1	0	0
110.9	H	110.77	–0.12	1	1	0	1	1	0
114.8		114.64	–0.14	1	1	0	1	0	0
118.8	J	118.80	0.0	0	1	0	1	1	0
123.0		122.80	–0.17	0	1	0	1	0	0
127.3		127.08	–0.17	1	0	0	1	1	0
131.8		131.67	–0.10	1	0	0	1	0	0
136.5		136.61	+0.08	0	0	0	1	1	0
141.3		141.32	+0.02	0	0	0	1	0	0
146.2		146.37	+0.12	1	1	1	0	1	0
151.4		151.09	–0.20	1	1	1	0	0	0
156.7		156.88	+0.11	0	1	1	0	1	0
162.2		162.31	+0.07	0	1	1	0	0	0
167.9		168.14	+0.14	1	0	1	0	1	0
173.9		173.48	–0.19	1	0	1	0	0	0
179.9		180.15	+0.14	0	0	1	0	1	0
186.2		186.29	+0.05	0	0	1	0	0	0
192.8		192.86	+0.03	1	1	0	0	1	0
203.5		203.65	+0.07	1	1	0	0	0	0
210.7		210.17	–0.25	0	1	0	0	1	0
218.1		218.58	+0.22	0	1	0	0	0	0
225.7		226.12	+0.18	1	0	0	0	1	0
233.6		234.19	+0.25	1	0	0	0	0	0
241.8		241.08	–0.30	0	0	0	0	1	0
250.3		250.28	–0.01	0	0	0	0	0	0
No tone		No tone	—	0	0	0	0	1	1

Taken from Consumer Microelectronics Ltd *IC Data Book*, 1st edn.

The Electronic Industries Association (EIA) has defined 38 CTCSS standard *sub-audible* tone frequencies. They are all between 67 and 250Hz, ie within the range of human hearing but outside the 300–3000Hz audio pass-band of most communications equipment and their level is set at only 10% of maximum deviation for that channel; hence sub-audible.

The left column of Table 5.20 gives the frequency list. Frequencies followed by a letter have been assigned to British repeaters; the same letters appear on the map of Fig 5.120.

CTCSS encoders (tone generators) and also decoders (tone detectors) are offered as options for many current mobile and hand-held VHF and UHF FM transceivers. They are very small but expensive.

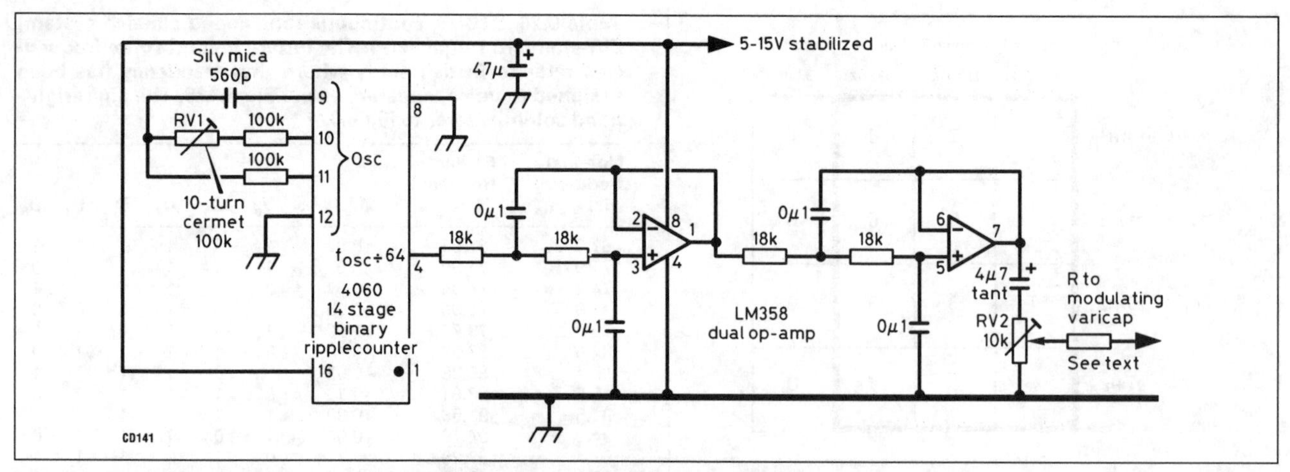

Fig 5.121. The Tuppenny simple CTCSS encoder (G0CBM) consists of an RC oscillator and ÷64 divider in one CMOS IC, together with a two-stage active LP filter built around a dual op-amp (G8HLE) *(Kent Repeater Newsletter)*

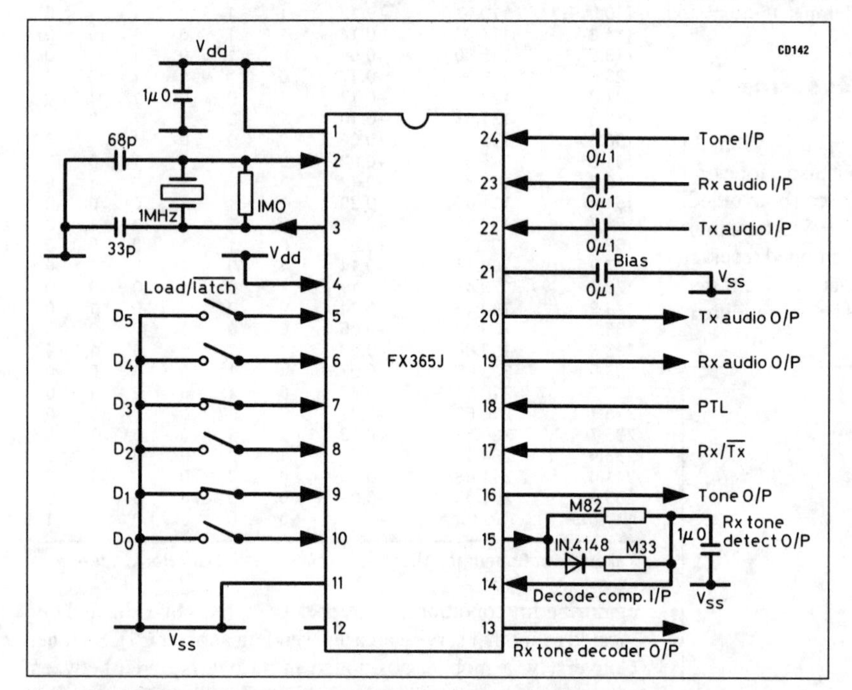

Fig 5.122. CTCSS encoder/decoder using a CML CMOS LSI device. The chip also contains 300Hz-cut-off HP and LP filters to separate tones and speech and a crystal reference oscillator. Tone selection can be by microprocessor or hard-wired switches *(Consumer Microcircuits IC Data Book)*

Commercial standards require encoder frequencies to be within 0.1% of the nominal tone frequency under all operating conditions, attainable only with crystal control; most amateur repeaters are more tolerant, however, and RC-oscillators have been used successfully. The encoder tone output must be a clean sine wave lest its harmonic content above 300Hz becomes audible; this requires good filtering.

G0CBM's very simple Tuppenny CTCSS tone generator was designed for retrofitting in a surplus PMR transmitter to access local repeaters: Fig 5.121 [44]. It is built around the inexpensive CMOS oscillator-frequency divider IC 4060. The parts connected to pins 9, 10 and 11 are the frequency-determining components of the RC oscillator; they have a tolerance of 5% but parts with the lowest possible temperature coefficient should be chosen to obtain adequate frequency stability, especially under mobile operation. RV1 allows frequency adjustment over the range 4288–7603Hz, which, after dividing by 64, yields at pin 4 any CTCSS frequency between 67 and 119Hz; this includes tones A–J assigned to UK repeaters.

A two-stage active low-pass filter was designed by G8HLE to get sufficient suppression of any harmonics above 300Hz of all tones. The LM358 dual op-amp was chosen because it is small, cheap and works on a low, single supply voltage. The –6dB frequency was chosen at 88Hz, ie the higher tones to be passed fall outside the pass-band. As the tone amplitude is far greater than required, this attenuation is no disadvantage, but each tone tuned in with RV1 requires a different setting of RV2 to get the same deviation. The output resistor R depends on the modulator circuitry in the transmitter. It should be dimensioned to get the proper CTCSS deviation at the highest tone frequency to be used with RV2 set near maximum.

Encoders/decoders are more complicated and the LSI CMOS device used, eg CML FX365, is expensive: Fig 5.122 [45]. It contains not only the encoder and decoder proper, but also a high-pass filter which prevents any received CTCSS tone becoming audible, a low-pass filter to suppress harmonics of CTCSS tones and a crystal-controlled reference oscillator from which the tones are derived; see columns 2 and 3 of Table 5.20. Tone selection is according to the six right-hand columns of that table, either by microprocessor or by hard-wired switches. Another feature is transmit phase reversal upon release of the PTT switch; this shortens the *squelch tail* at the receiver.

REFERENCES AND BIBLIOGRAPHY

[1] *Electronic and Radio Engineering*, F E Terman, McGraw-Hill, any older edition, eg the fourth, 1955.

[2] *Radio Receiver Design*, K R Sturley, Chapman and Hall, 1953.

[3] *Amateur Radio Techniques*, 7th edn, Pat Hawker, RSGB.

[4] *Special Circuits Ready-Reference*, J Markus, McGraw-Hill.

[5] *GaAs Devices and their Impact on Circuits and Systems*, ed Jeremy Everard, Peter Peregrinus.

[6] 'A receiver with noise immunity and frequency synthesis', Peter Martin, G3PDM, *Radio Communication Handbook*, 5th edn, RSGB, 1976, pp10.104–10.108.
[7] *Owners Manual, Belcom Liner 2.*
[8] See for example, the instruction manuals for the FT-400 series, especially the FT-DX401, FR-DX400.
[9] *Maintenance Service Manual, FT-101 series*, Yaesu Musen Co Ltd.
[10] The Plessey Company, now part of GEC, have published several applications books describing PLL synthesis and any of these are well worth acquiring, although the older ones are only obtainable second-hand. Examples are *Radio Communications Handbook*, 1977; *Professional Radio Communications*, 1979; *Radio Telecoms IC Handbook*, 1987; *Professional Data Book*, 1991; and *Frequency Dividers And Synthesisers IC Handbook.*
[11] Klaas Spaargaren, PA0KSB, in *Ham Radio* December 1977, but see also reference [3] above, Chapter 4.
[12] 'Technical Topics', Pat Hawker, G3VA, *Radio Communication* December 1988, pp957–958.
[13] 'ICOM IC-725 HF Transceiver', P J Hart, G3SJX, *Radio Communication* September 1989, pp56–58.
[14] 'Direct digital synthesis, what is it and how can I use it?', P H Saul, *Radio Communication* December 1990, pp44–46.
[15] *Plessey Semiconductors Data Sheet SP2002.*
[16] *GEC-Plessey Professional Products IC Handbook.*
[17] W2CQH, *QST* January 1974.
[18] *Signetics RF Communications Handbook.*
[19] Doug DeMaw, W1FB, *QST* April 1989, pp30–33.
[20] Drew Diamond, VK3XU, *Amateur Radio* October 1988.
[21] Drew Diamond, VK3XU, *Amateur Radio* January 1991.
[22] Tim Walford, G3PCJ, *Radio Communication* May 1990, p30.
[23] David Bowman, *Radio Communication* December 1989, pp36–37.
[24] Zack Lau, KH6CP, *QST* October 1987.
[25] *Radio Communication* December 1992, p49.
[26] E J Hatch, G3ISD, *Radio Communication* May 1982 and March 1984.
[27] *VHF/UHF Manual*, ed G R Jessop, G6JP, 4th edn, RSGB, 1983.
[28] Mike Grierson, G3TSO, *Radio Communication* March 1990, p35.
[29] *Radio Communication* August 1992, p39.
[30] GW4NAH, *Radio Communication* January 1989, p48.
[31] *Radio Communication* November 1992, p28.
[32] *Radio Communication* June 1991, p45.
[33] 'Computer-aided ladder crystal filter design', J A Hardcastle, G3JIR, *Radio Communication* May 1983.
[34] H R Langer, Y27YO, *Funkamateur* January 1985; *Radio Communication* June 1985, p452.
[35] J Wieberdink, *Radio Bulletin* (NL), October 1983.
[36] LA2IJ and LA4HK, *Amator Radio* November 1974.
[37] DJ6HP, *cq-DL* February 1974.
[38] W1JF, *QST* November 1982, July 1984 and January 1985.
[39] AI2T, *CQ* January 1986.
[40] WB4TLM and KB4KVE, *QST* April 1986.
[41] E J Hatch, G3ISD, *Datacom* (British Amateur Radio Teledata Group) Autumn 1989 and January 1990.
[42] *Analog-Digital Conversion Handbook*, 3rd edn, Prentice-Hall, 1986, p215.
[43] *Radio Communication* May 1992, p62.
[44] *Kent Repeater Newsletter* January 1993.
[45] *IC Data Book*, 1st edn, Consumer Microcircuits Ltd, p2.40.

Bibliographic notes

Reference [11] above lists several of the titles of applications books which have been produced over recent years by Plessey Semiconductors and by GEC-Plessey Semiconductors. These are normally available free to electronics professionals, but may be more difficult to obtain by the amateur. They do, however, appear at rallies and radio car boot sales, and are well worth a small sum in purchase. Other very useful sources are reference [3], from which many circuits have been abstracted, and Doug DeMaw's *W1FB's QRP Notebook* (ARRL) which was the source of Figs 5.8–5.9, 5.18–5.22, 5.28 and 5.29.

6 HF receivers

AMATEUR HF operation, whether for two-way contacts or for listening to amateur transmissions, imposes stringent requirements on the receiver. The need is for a receiver that enables an experienced operator to find and hold extremely weak signals on frequency bands often crowded with much stronger signals from local stations or from the high-power broadcast stations using adjacent bands. The wanted signals may be fading repeatedly to below the external noise level, which limits the maximum usable sensitivity of HF receivers, and which will be much higher than in the VHF and UHF spectrum.

Although the receivers now used by most amateurs form part of complex, factory-built HF transceivers, the operator should understand the design parameters that determine how well or how badly they will perform in practice, and appreciate which design features contribute to basic performance as HF communications receivers, as opposed to those which may make them more user-friendly but which do not directly affect the reception of weak signals.

Ideally, an HF receiver should be able to provide good intelligibility from signals which may easily differ in voltage delivered from the antenna by up to 10,000 times and occasionally by up to one million times (120dB) – from less than 1μV from a weak signal to nearly 1V from a near-neighbour. To tune and listen to SSB or to a stable CW transmission while using a narrow-band filter, the receiver needs to have a frequency stability of within a few hertz over periods of 15min or so, representing a stability of better than one part in a million. It should be capable of being tuned with great precision, either continuously or in increments of at most a few hertz.

A top-quality receiver may (or may not) be required to receive transmissions on all frequencies from 1.8MHz to 30MHz (or even 50MHz) to provide 'general coverage' or only on the bands allotted to amateurs. Such a receiver may be suitable for a number of different modes of transmission – SSB, CW, AM, NBFM, data (RTTY/packet) etc – with each mode imposing different requirements in selectivity, stability and demodulation (decoding). Such a receiver would inevitably be complex and costly to buy or build. On the other hand, a more specialised receiver covering only a limited number of bands and modes such as CW-only or CW/SSB-only, and depending for performance rather more on the skill of the operator, can be relatively simple to build at low cost.

As with other branches of electronics, the practical implementation of high-performance communications receivers has undergone a number of radical changes since their initial development in the mid-'thirties, some resulting from the improved stability needed for SSB reception and others aimed at reducing costs by substituting electronic techniques in place of mechanical precision.

However, it needs to be emphasised that, in most cases, progress in one direction has tended to result in the introduction of new problems or the enhancement of others: "What we call progress is the exchange of one nuisance for another nuisance" (Havelock Ellis) or "Change is certain; progress is not" (A J P Taylor). As late as 1981, an Australian amateur was moved to write: "Solid-state technology affords commercial manufacturers cheap, large-scale production but for amateur radio receivers and transceivers of practical simplicity, valves remain incomparably superior for one-off, home-built projects." The availability of linear integrated circuits capable of forming the heart of communications receivers combined with the increasing scarcity and hence cost of special valve types has tended to reverse this statement. It is still possible to build reasonably effective HF receivers, particularly those for limited frequency coverage, on the kitchen table with the minimum of test equipment.

Furthermore, since many newcomers will eventually acquire a factory-built transceiver but require a low-cost, stand-alone HF receiver in the interim period, the need can be met either by building a relatively simple receiver, or by acquiring, and if necessary modifying, one of the older valve-type receivers that were built in very large numbers for military communications during the second world war, or those marketed for amateur operation in the years before the virtually universal adoption of the transceiver.

Even where an amateur has no intention of building or servicing his or her own receiver, it is important that he or she should have a good understanding of the basic principles and limitations that govern the performance of all HF communications receivers.

BASIC REQUIREMENTS

The main requirements for a good HF receiver are:

1. Sufficiently high *sensitivity,* coupled with a wide *dynamic range* and good *linearity* to allow it to cope with both the very weak and very strong signals that will appear together at the input; it should be able to do this with the minimum impairment of the *signal-to-noise ratio* by receiver *noise, cross-modulation, blocking, intermodulation, reciprocal mixing, hum* etc.
2. Good *selectivity* to allow the selection of the required signal from among other (possibly much stronger) signals on adjacent or near-adjacent frequencies. The selectivity characteristics should 'match' the mode of transmission, so that interference susceptibility and noise bandwidth should be as close as possible to the intelligence bandwidth of the signal.
3. Maximum freedom from *spurious responses* – that is to say signals which appear to the user to be transmitting on specific frequencies when in fact this is not the case. Such spurious responses include those arising from image responses, breakthrough of signals and harmonics of the receiver's internal oscillators.
4. A high order of *stability,* in particular the absence of short-term frequency drift or jumping.

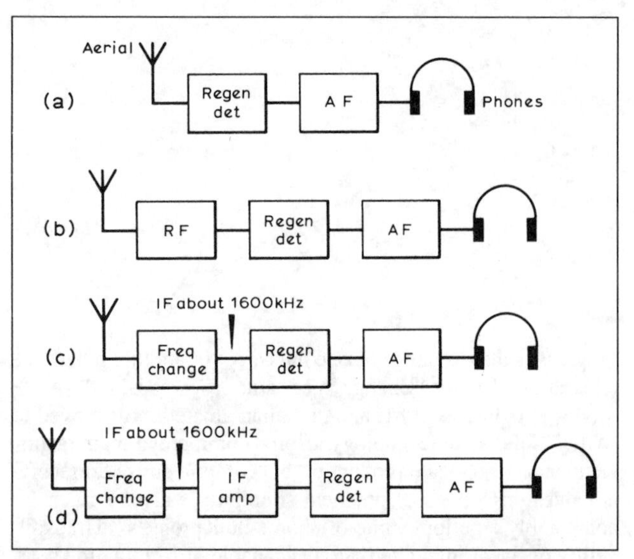

Fig 6.1. Simple receivers intended primarily for morse reception. (a) Two-stage 'straight' receiver with high-gain regenerative detector. (b) The addition of a tuned RF stage improves sensitivity and selectivity (this arrangement is often known as a *TRF receiver*). (c) The simplest form of superhet. (d) The addition of an IF stage will greatly increase the gain and selectivity

5. Good read-out and calibration of the frequency to which the set is tuned, coupled with the ability to reset the receiver accurately and quickly to a given frequency or station.
6. Means of receiving SSB and CW, normally requiring a stable beat frequency oscillator preferably in conjunction with product detection.
7. Sufficient amplification to allow the reception of signals of under 1μV input; this implies a minimum voltage gain of about one million times (120dB), preferably with effective automatic gain control (AGC) to hold the audio output steady over a very wide range of input signals.
8. Sturdy construction with good-quality components and with consideration given to problems of access for servicing when the inevitable occasional fault occurs.

A number of other refinements are also desirable: for example it is normal practice to provide a headphone socket on all communications receivers; it is useful to have ready provision for receiver 'muting' by an externally applied voltage to allow voice-operated, push-to-talk or CW break-in operation; an S-meter to provide immediate indication of relative signal strengths; a power take-off socket to facilitate the use of accessories; an IF signal take-off socket to allow use of external special demodulators for NBFM, FSK, DSBSC, data etc.

In recent years, significant progress has continued to be made in meeting these requirements – although we are still some way short of being able to provide them over the entire signal range of 120dB at the ideal few hertz stability. The introduction of more and more semiconductor devices into receivers has brought a number of very useful advantages, but has also paradoxically made it more difficult to achieve the highly desirable wide dynamic range. Professional users now require frequency read-out and long-term stability of an extremely high order (1Hz stability is needed for some applications) and this has led to the use of frequency synthesised local oscillators and digital read-out systems; although these are effective for the purposes which led to their adoption, they are not necessarily the correct approach for amateur receivers since, unless very great care is taken, a complex frequency synthesiser not only adds significantly to the cost but may actually result in a degradation of other even more desirable characteristics.

So long as continuous tuning systems with calibrated dials were used, the mechanical aspects of a receiver remained very important; it is perhaps no accident that one of the outstanding early receivers (HRO) was largely designed by someone whose early training was that of a mechanical engineer.

It should be recognised that receivers which fall far short of ideal performance by modern standards may nevertheless still provide entirely usable results, and can often be modified to take advantage of recent techniques. Despite all the progress made in recent decades, receiver designs dating from the 'thirties and early 'forties are still capable of being put to good use, provided that the original electrical and mechanical design was sound. Similarly, the constructor may find that a simple, straightforward and low-cost receiver can give good results even when its specification is well below that now possible. It is ironical that almost all the design trends of the past 30 years have, until quite recently, impaired rather than improved the performance of receivers in the presence of strong signals!

BASIC TYPES OF RECEIVERS

Amateur HF receivers fall into one of two main categories:

(a) *'straight'* regenerative and *direct-conversion* receivers in which the incoming signal is converted directly into audio by means of a demodulator working at the signal frequency;
(b) single- and multiple-conversion *superhet* receivers in which the incoming signal is first converted to one or more intermediate frequencies before being demodulated. Each type of receiver has basic advantages and disadvantages.

Regenerative detector ('straight' or TRF) receivers

At one time valve receivers based on a *regenerative* (*reaction*) detector, plus one or more stages of AF amplification (ie 0-V-1, 0-V-2 etc), and sometimes one or more stages of RF amplification at signal frequency (1-V-1 etc) were widely used by amateurs. High gain can be achieved in a correctly adjusted regenerative detector when set to a degree of positive feedback just beyond that at which oscillation begins; this makes a regenerative receiver capable of receiving weak CW and SSB signals. However, this form of detector is non-linear and cannot cope well in situations where the weak signal is at all close to a strong signal; it is also inefficient as an AM detector since the gain is much reduced when the positive feedback (*regeneration*) is reduced below the oscillation threshold. Since the detector is non-linear, it is usually impossible to provide adequate selectivity by means of audio filters. Only by careful design is it possible to design a solid-state receiver of this type which is able to cope with modern band conditions, although for CW operation it can still provide a useful low-cost receiver for portable operation.

Better performance can be achieved by the application of (regenerative) Q-multiplication to the RF tuned circuit followed by an infinite impedance detector.

Simple direct-conversion receivers

A modified form of 'straight' receiver which can provide good results, even under modern conditions, becomes possible by using a linear detector which is in effect simply a frequency converter, in conjunction with a stable local oscillator set to the signal frequency (or spaced only the audio beat away from it). Provided that this stage has good linearity in respect of the signal

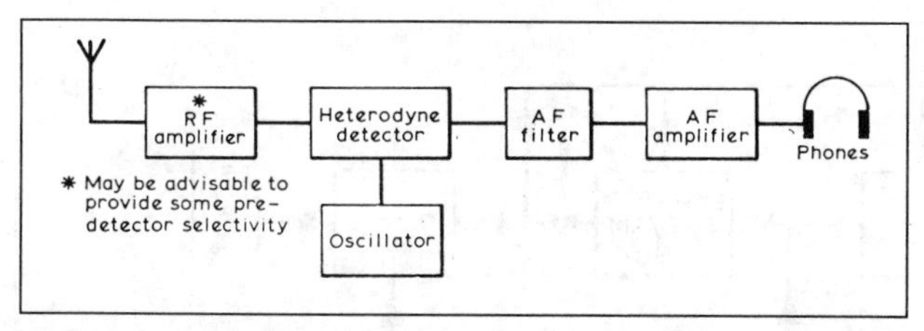

Fig 6.2. Outline of simple direct-conversion receiver in which high selectivity can be achieved by means of audio filters

path, it becomes possible to provide almost any desired degree of selectivity by means of audio filters. This form of receiver (sometimes termed a *homodyne*) has a long history but only in the past few decades has it been widely used for amateur operation since it is more suited (in its simplest form) to CW and SSB reception than AM. The *direct-conversion* receiver may be likened to a superhet with an IF of 0kHz or alternatively to a straight receiver with a linear rather than a regenerative detector. In a superhet receiver the incoming signal is mixed with a local oscillator signal and the intermediate frequency represents the difference between the two frequencies; thus as the two signals approach one another the IF becomes lower and lower. If this process is continued until the oscillator is at the same frequency as the incoming signal, then the output will be at audio (baseband) frequency; in effect one is using a frequency changer or translator to demodulate the signal. Because high gain cannot be achieved in a linear detector, it is necessary to provide very high AF amplification. Direct-conversion receivers can be designed to receive weak signals with good selectivity, but in this form do not provide true single-sideband reception (see later); another problem often found in practice is that very strong broadcast signals (eg on 7MHz) drive the detector into non-linearity and are then demodulated directly and not affected by any setting of the local oscillator.

A crystal-controlled converter can be used in front of a direct-conversion receiver, so forming a superhet with variable IF only. Alternatively a frequency converter with a variably tuned local oscillator providing output at a fixed IF may be used in front of a direct-conversion receiver (regenerative or linear demodulator) fixed tuned to the IF output. Such a receiver is sometimes referred to as a *supergainer* receiver.

Two-phase and 'third-method' direct-conversion receivers

An inherent disadvantage of the simple direct-conversion receiver is that it responds equally to signals on both sides of its local oscillator frequency, and cannot reject what is termed the *audio image* no matter how good the audio filter characteristics; this is a serious disadvantage since it means that the selectivity of the receiver can only be made half as good as the theoretically ideal bandwidth. This problem can be overcome, though at the cost of additional complexity, by phasing techniques similar to those associated with SSB generation. Two main approaches are possible: see Fig 6.3.

Fig 6.3(a) shows the use of broad-band AF 90° phase-shift networks in an 'outphasing' system, and with care can result in the reduction of one sideband to the extent of 30–40dB. Another possibility is the polyphase SSB demodulator which does not require such critical component values as conventional SSB phase-shift networks.

Fig 6.3(b) shows the *'third method'* (sometimes called the *Weaver* or *Barber* system) which requires the use of additional balanced mixers working at AF but eliminates the need for accurate 90° AF networks. The 'third method' system, particularly in its AC-coupled form *(The Radio & Electronic Engineer* Vol 43, No 3, March 1973, pp209–215) provides the basis for high-performance receivers at relatively low cost, although suitable designs for amateur operation are rare. Two-phase direct-conversion receivers based on two diode-ring mixers in quadrature (90° phase difference) are capable of the high performance of a good superhet.

Super-regenerative detector receivers

The high gain of a regenerative detector can be further increased by introducing a voltage at a supersonic frequency in such a way that RF oscillation ceases every half cycle of this second frequency, termed the *quench frequency*. This quenching voltage can be generated in a separate stage or, more commonly, within the regenerative detector stage. Although extremely high gain in a single stage is possible, this form of reception introduces high interstation noise and has poor selectivity, while additional complications are needed for CW reception and to avoid the emission of a rough signal capable of causing interference over a wide area. Super-regenerative receivers are suitable for both AM and FM reception but are today seldom used on HF; however, the principle remains of interest.

Superhet receivers

The vast majority of amateur receivers are based on the *superhet* principle. By changing the incoming signals to a fixed frequency (which may be lower or higher than the incoming signals) it becomes possible to build a high-gain amplifier of controlled selectivity to a degree which would not be possible over a wide spread of signal frequencies. The main practical disadvantage

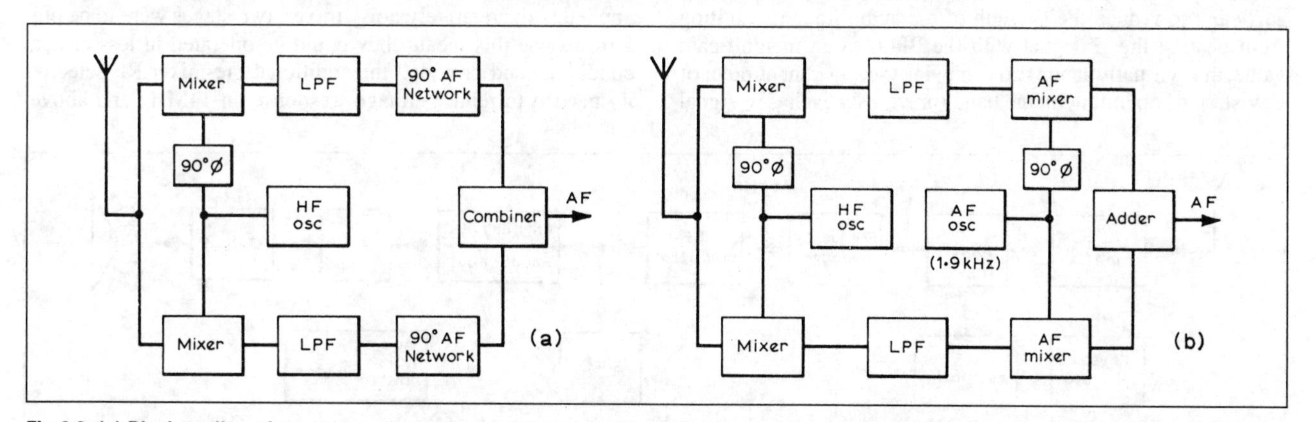

Fig 6.3. (a) Block outline of two-phase ('autophasing') form of SSB direct-conversion receiver. (b) Block outline of 'third method' (Weaver or Barber) SSB direct-conversion receiver

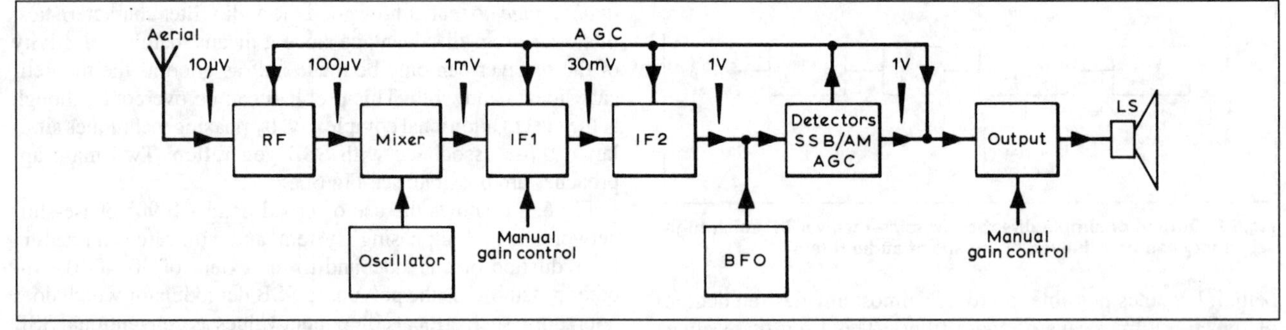

Fig 6.4. Block outline of representative single-conversion superhet receiver, typical of many used for amateur operation and showing typical levels of signal at various stages

with this system is that the frequency conversion process involves unwanted products which give rise to spurious responses, and much of the design process has to be concentrated on minimising the extent of these spurious responses in practical situations.

A *single-conversion* superhet is a receiver in which the incoming signal is converted to its intermediate frequency, amplified and then demodulated at this second frequency. Virtually all domestic AM broadcast receivers use this principle, with an IF of about 455–470kHz, and a similar arrangement but with refinements was found in many communications receivers. However, for reasons that will be made clear later in this chapter, some receivers convert the incoming signal successively to several different frequencies; these may all be fixed IFs: for example the first IF might be 9MHz and the second 455kHz and possibly a third at 35kHz. Or the first IF may consist of a whole spectrum of frequencies so that the first IF is variable when tuning a given band, with a subsequent second conversion to a fixed IF. There are in fact many receivers using double or even triple conversion, and a few with even more conversions, though unless care is taken each conversion makes the receiver susceptible to more spurious responses. The block diagram of a typical single-conversion receiver is shown in Fig 6.4. Fig 6.5 illustrates a double-conversion receiver with fixed IFs, while Fig 6.6 is representative of a receiver using a variable first IF in conjunction with a crystal-controlled first local oscillator (HFO).

Many modern factory-built receivers up-convert the signal frequency to a first IF at VHF as this makes it more convenient to use a frequency synthesiser as the first HF oscillator: Fig 6.7(a).

As the degree of selectivity provided in a receiver increases, it reaches the stage where the receiver becomes a single-sideband receiver, although this does not mean that only SSB signals can be received. In fact the first application of this principle was the single-signal receiver for CW reception where the selectivity is sufficient to reduce the strength of the audio image (resulting from beating the IF signal with the BFO) to an insignificant value, thus virtually at one stroke halving the apparent number of CW stations operating on the band (previously each CW signal was heard on each side of the zero beat). Similarly double-sideband AM signals can be received on a set having a carefully controlled pass-band as though they were SSB, with the possibility of receiving either sideband should there be interference on the other. This degree of selectivity can be achieved with good IF filters or alternatively the demodulator can itself be designed to reject one or other of the sidebands, by using phasing techniques similar to those sometimes used to generate SSB signals and for two-phase direct-conversion receivers. But most receivers rely on the use of crystal or mechanical filters to provide the necessary degree of sideband selectivity, and then use heterodyne oscillators placed either side of the nominal IF to select upper or lower sidebands.

It is important to note that whenever frequency conversion is accomplished by beating with the incoming signal an oscillator lower in frequency than the signal frequency, the sidebands retain their original position relative to the carrier frequency, but when conversion is by means of an oscillator placed higher in frequency than the carrier, the sidebands are inverted. That is to say, an upper sideband becomes a lower sideband and vice versa (see Fig 6.8).

DESIGN TRENDS

After the 'straight' receiver, because of its relatively poor performance and lack of selectivity on AM phone signals, had fallen into disfavour in the mid-'thirties, came the era of the superhet communications receiver. Most early models were single-conversion designs based on an IF of 455–470kHz, with two or three IF stages, a multi-electrode triode-hexode or pentagrid mixer, sometimes but not always with a separate oscillator valve. This approach made at least one RF amplifying stage essential in order to raise the level of the incoming signal before it was applied to the relatively noisy mixer; two stages were to be preferred since this meant they could be operated in less critical conditions and provided the additional pre-mixer RF selectivity needed to reduce 'image' response on 14MHz and above.

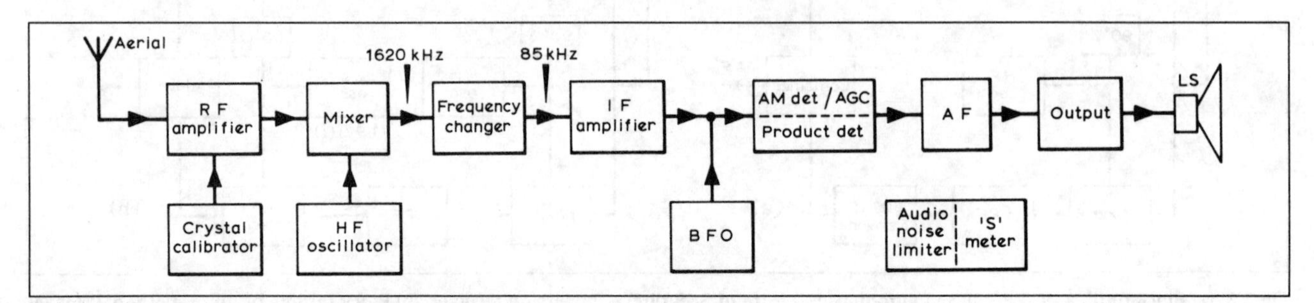

Fig 6.5. Block outline of double-conversion communications receiver with both IFs fixed

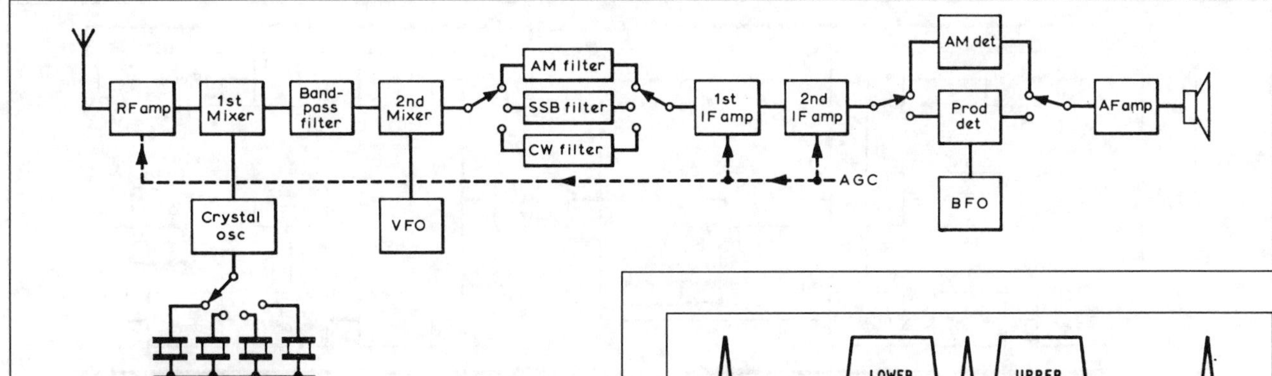

Fig 6.6. Block diagram of a double-conversion receiver with crystal-controlled first oscillator – typical of many current designs

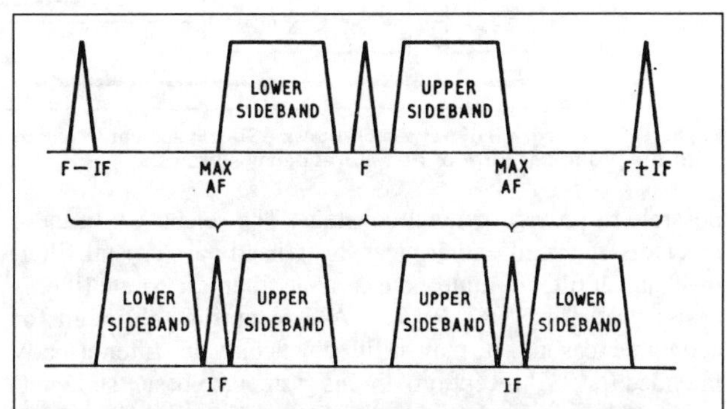

Fig 6.8. A local oscillator frequency lower than the signal frequency (ie *f* − IF) keeps the upper and lower sidebands of the intermediate frequency signals in their original positions. However, when the local oscillator is placed higher in frequency than the signal frequency (*f* + IF), the positions of the sidebands are transposed. By incorporating two oscillators, one above and the other below the input signal, sideband selection is facilitated (this is generally carried out at the final IF by switching the BFO or carrier insertion oscillator)

Usually a band-switched LC HF oscillator was gang-tuned so as to track with two or three signal-frequency tuned circuits, calling for fairly critical and expensive tuning and alignment systems. These receivers were often designed basically to provide full coverage on the HF band (and often also the MF band), sometimes with a second tuning control to provide electrical band-spread on amateur bands, or with provision (as on the HRO) optionally to limit coverage to amateur bands only. Selectivity depended on the use of good-quality IF transformers (sometimes with a tertiary tuned circuit) in conjunction with a single-crystal IF filter which could easily be adjusted for varying degrees of selectivity and which included a phasing control for nulling out interfering carriers.

Later, to overcome the problem of image response with only one RF stage, there was a trend towards double- or triple-conversion receivers with a first IF of 1.6MHz or above, a second IF about 470kHz and (sometimes) a third IF about 50kHz.

With a final IF of 50kHz it was possible to provide good single-signal selectivity without the use of a crystal filter.

The need for higher stability than is usually possible with a band-switched HF oscillator and the attraction of a similar degree of band-spreading on all bands has led to the widespread adoption of an alternative form of multi-conversion superhet; in effect this provides a series of integral crystal-controlled *converters* in front of a superhet receiver (single or double conversion) covering only a single frequency range (for example 5000–5500kHz) This arrangement provides a fixed tuning span (in this example 500kHz) for each crystal in the HF oscillator. Since a separate crystal is needed for each band segment, most receivers of this type are designed for amateur bands only (though often with provision for the reception of a standard frequency transmission, for example on 10MHz); more recently some designs have eliminated the need for separate crystals by means of frequency synthesis, and in such cases it is economically

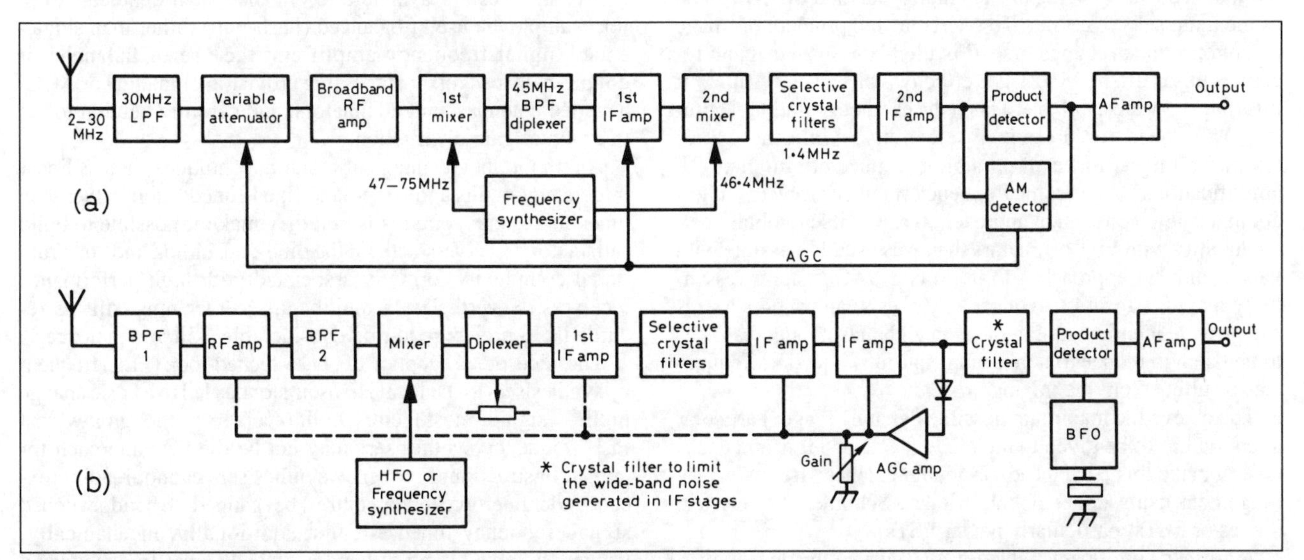

Fig 6.7. Representative architectures of modern communications receiver designs. (a) General-coverage double-conversion superhet with up-conversion to 45MHz first IF and 1.4MHz second IF. (b) Single-conversion superhet, typically for amateur bands only, with an IF in the region of 9 or 10MHz

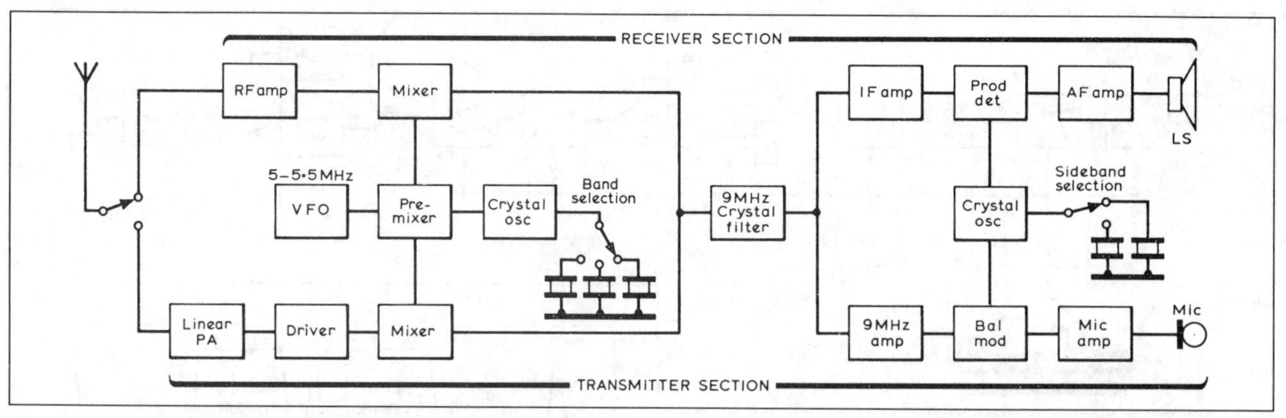

Fig 6.9. Block diagram of a typical modern SSB transceiver in which the receiver is a single-conversion superhet with 9MHz IF in conjunction with the pre-mixer form of partial frequency synthesis

possible to provide general coverage. The selectivity in these receivers is usually determined by a band-pass crystal filter, mechanical filter or multi-pole ceramic filter, a separate filter is being used for SSB, CW and AM reception (although for economic reasons sets may be fitted with only one filter, usually intended for SSB reception). In this system the basic 'superhet' section forms in effect a variable IF amplifier.

In practice the variable IF type of receiver provides significantly enhanced stability and lower tuning rates on the higher frequency bands, compared with receivers using fixed IF, though it is considerably more difficult to prevent breakthrough of strong signals within the variable IF range, and to avoid altogether the appearance of 'birdies' from internal oscillators. With careful design a high standard of performance can be achieved; the use of multiple conversion (with the selective filter further from the antenna input stage) makes the system less suitable for semiconductors than for valves, particularly where broad-band circuits are employed in the front-end and in the variable IF stage.

There is now a trend back to the use of fixed IF receivers, either with single conversion or occasionally with double conversion (provided that in this case an effective *roofing filter* is used at the first IF). A roofing filter is a selective filter intended to reduce the number of strong signals passing down an IF chain without necessarily being of such high grade or as narrow-band as the main selective filter. To overcome the problem of image reception a much higher first IF is used; for amateur band receivers this is often 9MHz since effective SSB and CW filters at this frequency are available. This reduces (though does not eliminate) the need for pre-mixer selectivity; while the use of low-noise mixers makes it possible to reduce or eliminate RF amplification. To overcome frequency stability problems inherent in a single-conversion approach, it is possible to obtain better stability with FET oscillators than was usually possible with valves; another approach is to use mixer-VFO systems (essentially a simple form of frequency synthesis) and such systems can provide identical tuning rates on all bands, though care has to be taken to reduce to a minimum spurious injection frequencies resulting from the mixing process.

To achieve the maximum possible dynamic range, particular attention has to be given to the mixer stage, and it is an advantage to make this a balanced, or double-balanced (see later) arrangement using either double-triodes, Schottky (hot-carrier) diodes or FETs (particularly power FETs).

A further significant reduction of spurious responses may prove possible by abandoning the superhet in favour of high-performance direct-conversion receivers (such as the Weaver or 'third-method' SSB direct-conversion arrangement); however, such designs are still only at an early stage of development.

Most modern receivers are built in the form of compact transceivers functioning both as receiver and transmitter, and with some stages common to both functions (Fig 6.9). Modern transceivers use semiconductor devices throughout the receiver stages, although valves may continue to be used in the power stages of the transmitter. Dual-gate FET devices are generally found in the signal path of the receiver. Most transceivers have a common SSB filter for receive and transmit; this may be a mechanical or crystal filter at about 455kHz but current models more often use crystal filters at about 3180, 5200, 9000kHz or 10.7MHz, since the use of a higher frequency reduces the total number of frequency conversions necessary.

One of the fundamental benefits of a transceiver is that it provides common tuning of the receiver and transmitter so that both are always 'netted' to the same frequency. It remains, however, an operational advantage to be able to tune the receiver a few kilohertz around the transmit frequency and vice versa, and provision for this *incremental tuning* is often incorporated; alternatively some amateurs prefer to use a separate external VFO so that the two frequencies may be separated when required.

The most critical aspect of modern receivers is the signal-handling capabilities of the early (*front-end*) stages. Various circuit techniques are available to enhance such characteristics: for example the use of balanced (push-pull) rather than single-ended signal frequency amplifiers; the use of balanced or double-balanced mixer stages; the provision of manual or AGC-actuated antenna-input attenuators; and careful attention to the question of gain distribution.

An important advantage of modern techniques such as linear integrated circuits and wide-band fixed-tuned filters rather than tuneable resonant circuits is that they make it possible to build satisfactory receivers without the time-consuming and constructional complexity formerly associated with high-performance receivers. Nevertheless a multiband receiver must still be regarded as a project requiring considerable skill and patience.

The widespread adoption of phase-locked-loop (PLL) frequency synthesisers as the local HF oscillator has led to a basic change in the design of most factory-built receivers, although low-cost PLL frequency synthesisers may not be the best approach for home-construction. Frequency synthesisers cannot readily (except under microprocessor control) be 'ganged' to band-switched signal-frequency tuned circuits; additionally, mechanically-ganged variable tuning of band-switched signal-frequency and local oscillator tuned circuits as found in older communications receivers would today be a relatively high-cost technique.

These considerations have led to widespread adoption in factory-built receivers and transceivers of 'single-span' up-conversion multiple-conversion superhets with a first IF in the VHF range, up to about 90MHz, followed by further conversions to lower IFs at which the main selectivity filter(s) are located.

In such designs, pre-selection before the first mixer or pre-amplifier (often arranged to be optionally switched out of circuit) may simply take the form of a low-pass filter (cut off at 30MHz) or a single wide-band filter covering the entire HF band. Higher-performance receivers usually fit a series of sub-octave band-pass filters, with electronic switching (preferably with PIN diodes).

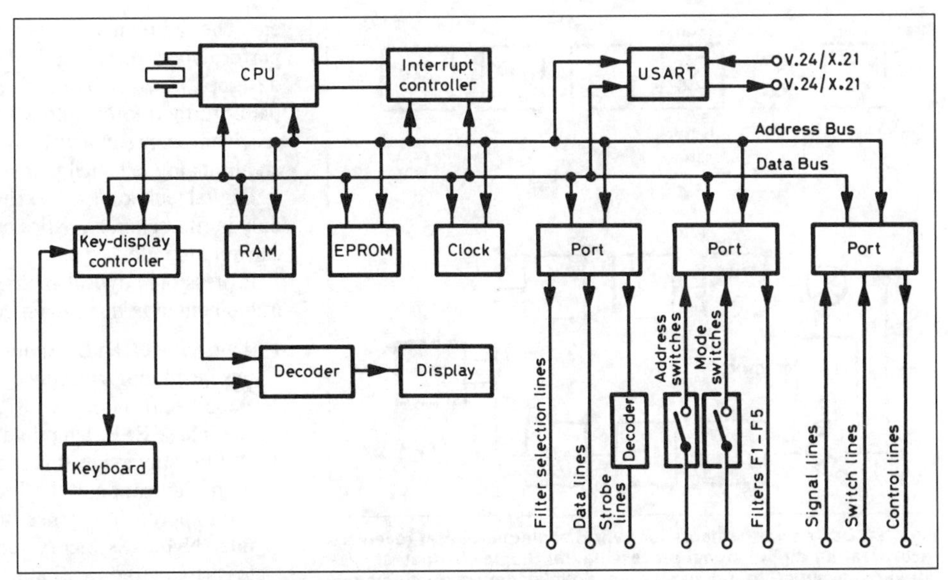

Fig 6.10. Architecture of the elaborate internal computer system found in a modern professional fully synthesised receiver (DJ2LR)

With fixed filtering, even of the sub-octave type, very strong HF broadcast transmissions will be present at the mixer(s) and throughout the 'front-end' up to the main selectivity filter(s). To enable weak signals to be received free of intermodulation products, this places stringent requirements on the linearity of the front-end. The use of relatively noisy low-cost PLL frequency synthesisers also raises the problem of 'reciprocal mixing' (see later). For home-construction of high-performance receivers, the earlier design approaches are still attractive, including the 'old-fashioned' concept of achieving good pre-mixer selectivity with high-Q tuned circuits using variable capacitors rather than electronic tuning diodes. Diode switching rather than mechanical switching can also significantly degrade the intermodulation performance of receivers.

DIGITAL TECHNIQUES

The availability of general-purpose, low-cost digital integrated circuit devices made a significant impact on the design of communication receivers although, until the later introduction of digital signal processing, their application has been primarily for operator convenience and their use for stable, low-cost frequency synthesisers rather than their use in the signal path.

By incorporating a digital frequency counter or by operation directly from a frequency synthesiser, it is now normal practice to display the frequency to which the receiver (or transceiver) is tuned directly on matrices of light-emitting diodes or liquid crystal displays. This requires that the display is offset by the IF from the actual output of the frequency synthesiser or free-running local oscillator. Such displays have virtually replaced the use of calibrated tuning dials.

PLL synthesisers are commonly 'tuned' by a rotary shaft-encoded switch which can have the 'feel' of mechanical capacitor tuning of a VFO, but this may be supplemented by push-buttons which enable the wanted frequency to be punched in. Common practice is for the frequency change per knob revolution to be governed by the rate at which the knob is spun, to speed up large changes of frequency. With a synthesiser the frequency may change in steps of 100Hz, 10Hz or even 1Hz, and it is desirable that this should be free of clicks and should appear to be almost instantaneous. Many of the factory-built equipments incorporate digital memory chips which can be programmed with frequencies to which it is desired to return to frequently. It should be noted that a phase-locked oscillator has an inherent jitter that appears as phase and amplitude noise that can give rise to reciprocal mixing and an apparent raising of the noise floor of the receiver. Digital direct frequency synthesis (DDS) can reduce phase noise and is being increasingly used.

Digital techniques may be used to stabilise an existing free-running VFO by continuously 'sampling' the frequency over pre-determined timing periods and then applying a DC correction to a varactor forming part of the VFO tuned circuit. The timing periods can be derived from a stable crystal oscillator and the technique is capable of holding a reasonably good VFO to within a few hertz. Here again some care is needed to prevent the digital pulses, with harmonics extending into the VHF range, from affecting reception.

Microprocessor control of user interfaces can include driving the tuning display, memory management (including BFO and pass-band tuning), frequency and channel scanning; data bus to RS-232 conversion. As stressed by Dr Ulrich Rohde, DJ2LR, several key points need to be observed:

1. Keypad and tuning-knob scanning must not generate any switching noises that can reach the signal path.
2. All possible combination of functions such as frequency steps, operating modes, BFO frequency offset and pass-band tuning should be freely and independently programmable and storable as one data string in memory.
3. It is desirable that multilevel menus should be provided for easy use and display of all functions. This includes not only modes (USB, LSB, CW etc) but also AGC attack and decay times. Such parameters should be freely accessible and independently selectable.

It should be understood that most of the above microprocessor functions are for user convenience and do not add to the basic performance (apart from frequency stability) of a receiver; and unless care is exercised may in practice degrade performance.

Digital signal processing

Digital techniques have been used in HF receivers for several decades, notably in the form of PLL-frequency synthesisers,

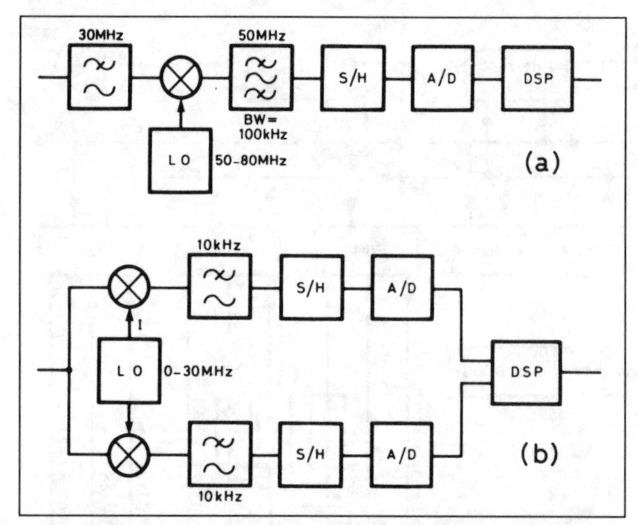

Fig 6.11. Basic arrangements for hybrid analogue/digital receivers incorporating digital signal processing. (a) Single-channel, single-conversion superhet (in practice a double-conversion analogue front-end is more likely to be employed to bring the IF signal down to 1.6 or 3MHz). (b) Dual-channel homodyne approach: S/H, sample and hold; A/D, analogue/digital converter; DSP, digital signal processing

digital frequency displays, frequency memories, microprocessor controlled mode-switching, variable tuning rates etc. By the mid-'eighties, professional designers were engaged in further extensions in the use of digital rather than analogue approaches in the main signal path of HF receivers.

Digital signal processing (DSP) is seen as opening the way to the 'all-digital receiver' in which the incoming signals would be converted directly into digital form and then processed throughout the receiver as a digital bitstream. In time, this approach is expected to become practical and provide high performance at lower than analogue costs, but this remains a long-term aim due to the limited resolution and sampling frequencies of existing VLSI analogue-to-digital converters (ADC).

However, hybrid analogue/digital receivers incorporating DSP, with digitisation of incoming signals at audio frequency (baseband frequency) or at a relatively low IF have been developed for professional, military and maritime applications, providing performance characteristics comparable to good analogue receivers. An advantage of DSP is that such processing can provide accurately shaped selectivity characteristics at a lower cost than a full complement of analogue (crystal) band-pass filters as fitted in high-grade, multimode professional receivers.

Potentially, digital technology offers a lower component count, easier factory assembly, higher reliability and the lower costs that can ensue from using the standard digital devices developed for use in computers etc. The radio amateur, however, should not expect an overall performance significantly better than is possible in a well-designed receiver based on analogue technology in the signal path; furthermore, unless the designer exercises great care, receivers using sophisticated DSP may suffer in comparison with an all-analogue signal path.

Digital technology does, however, open new possibilities not only in filter characteristics but also in such areas as AGC, noise cancellation etc.

Professional hybrid analogue-digital HF receivers have been implemented in three main configurations:

1. Two-phase (SSB demodulation) direct-conversion with baseband DSP filtering.
2. Superhet front-end with 'zero-IF' (super-gainer type) with two-phase SSB demodulation and baseband DSP.
3. Multi-conversion superhet with a low final IF (LF or MF), with digitisation at the low IF of up to about 500kHz. This third approach may use 'undersampling' below the Nyquist rate (Nyquist's theory requires that sampling should be at least twice the maximum frequency of the sampled waveform) with the aliasing products eliminated by means of relatively low-grade analogue filtering (eg ceramic filter). For example, a sampling frequency of 96kHz may be used to digitise signals at an IF of 456kHz.

Amateur-radio transceivers have been marketed with DSP filtering as an optional extra, in this case at baseband frequencies behind a conventional SSB front-end with a product detector after analogue SSB filters.

DSP at other than audio frequency has the problem that the dynamic range is adversely affected by the need that much of the receiver gain must be ahead of the ADC, and therefore must be under automatic gain control. With effective AGC the dynamic

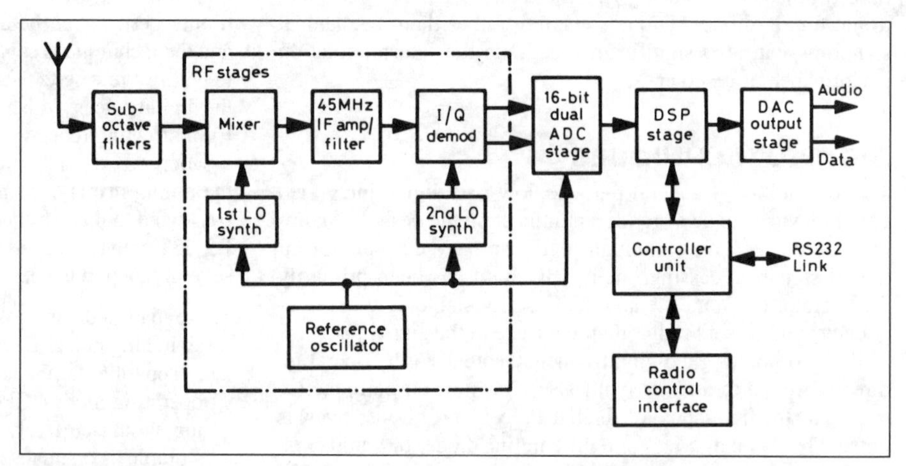

Fig 6.12. Outline of prototype high-performance analogue/digital professional communications receiver with baseband digitisation following two-phase (I/Q) demodulation (Roke Manor Research Ltd)

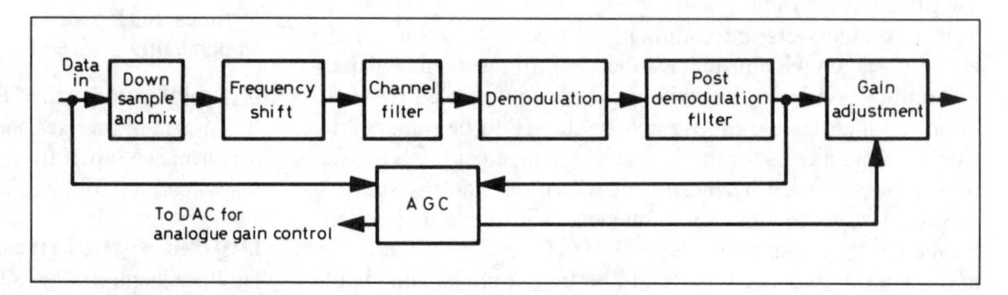

Fig 6.13. Software structure of the STC receiver

range may be specified as better than 120dB but it should be appreciated that the instantaneous dynamic range will be much lower than this, since current ADCs cannot cope with large dynamic ranges. It should be appreciated that a large instantaneous dynamic range is highly desirable when trying to copy a weak narrow-band signal immediately alongside a very strong signal. The linearity and resolution of the ADC is thus a key requirement for a high-performance receiver. An ADC with a resolution of 12 or more bits per sample is required.

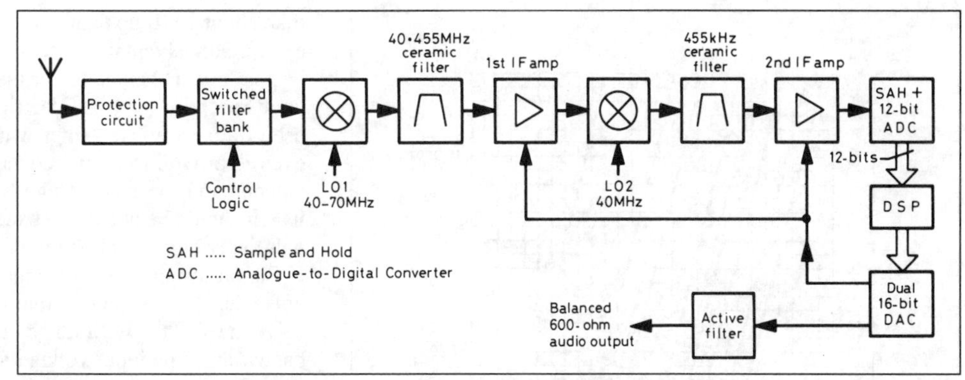

Fig 6.14. Block diagram of the STC marine HF band analogue/digital receiver with sub-Nyquist sampling of 455kHz IF

There is no doubt, however, that DSP can provide excellent filter characteristics programmed to match the required bandwidth of different modes. For example an SSB filter's attenuation at a 500Hz offset from the upper band edge can be over 60dB without any adjustments required and is insensitive to external component and environmental changes, and without the phase delays that can affect performance of data transmissions. The same DSP filter can be programmed to provide a series of band-pass filters (eg 500Hz, 1kHz, 2.7kHz, 6kHz etc), each having shape factors superior to a complete set of high-cost crystal filters. In practice, many of the advantages of DSP filtering may be lost due to limitations in the analogue front-end and in the ADC as noted above. For home-construction, the difficulties that still exist in programming a DSP filter may greatly outweigh the advantages; nevertheless, as current difficulties are overcome, it seems certain that DSP will come to play an increasingly important role in HF receivers and many other aspects of amateur radio.

RECEIVER SPECIFICATION

The performance of a communications receiver is normally specified by manufacturers or given in equipment reviews or stated in the various constructional articles. It is important to understand what these specifications mean and how they relate to practical requirements in order to know what to look for in a good receiver. It will be necessary to study any specifications with some caution, since a manufacturer or designer will usually wish to present his receiver in the most favourable light, and either omit unfavourable characteristics or specify them in obscure terms. The specifications do not tell the whole story: the operational 'feel' may be as important as the electrical performance; the 'touch' of the tuning control, the absence of mechanical backlash or other irregularities, the convenient placing of controls, the positive or uncertain action of the band-change switch and so on will all be vitally important. Furthermore, there is a big difference between receiver measurements made under laboratory conditions, with only locally generated signals applied to the input, and the actual conditions under which it will be used, with literally hundreds of amateur, commercial and broadcasting

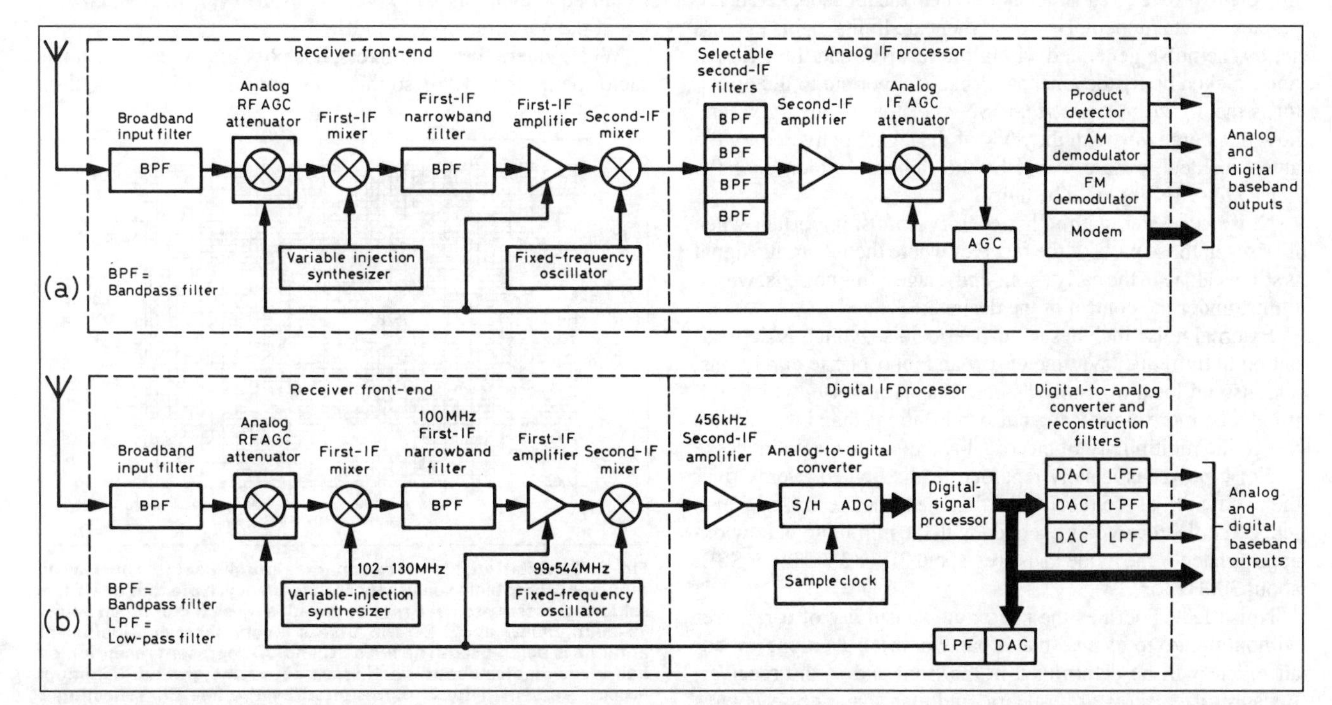

Fig 6.15. (a) Outline of typical professional analogue communications receiver with several band-pass filters and demodulation circuits for the different modes. (b) Use of a software-reconfigurable digital signal processor at the 455kHz second IF potentially reduces cost by eliminating the multiple band-pass filters etc

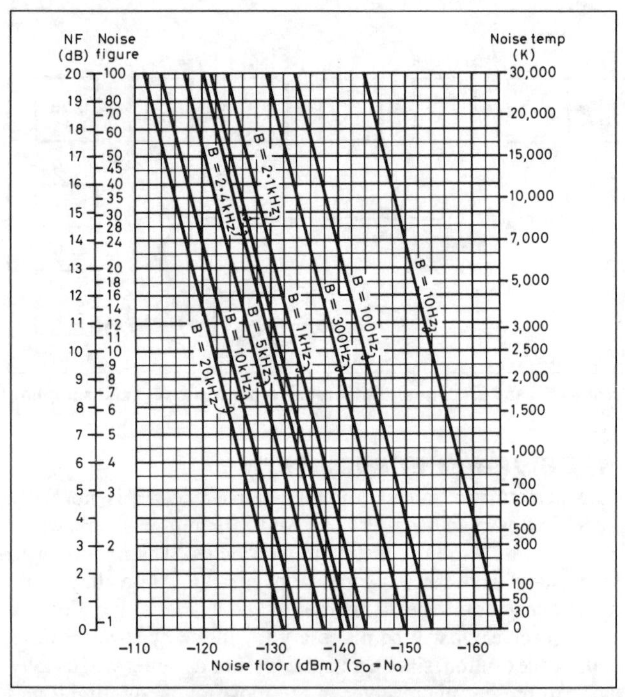

Fig 6.16. Relationship between the sensitivity of a receiver as defined in terms of noise factor (dB), noise figure or noise temperature and the noise floor (noise = signal) in dBm for various receiver bandwidths. Note how the minimum detectable signal reduces with narrower bandwidths for the same noise factor

signals being delivered by the antenna in the presence of electrical interference and possibly including one or more 'blockbusting' signals from a nearby transmitter.

Sensitivity

Weak signals clearly need to be amplified more than strong ones in order to provide a satisfactory output to the loudspeaker, headphones or data modem. However, there are limits to this process set by the noise generated within the receiver and the external noise picked up by the antenna. What is important to the operator is the *signal-to-noise ratio* (SNR) of the output signal and how this compares with the SNR of the signal delivered by the antenna. Ideally these would be the same, in which case the *noise factor* (NF) would be unity.

Noise generated within the receiver is most important when it arises in those parts of the receiver where the incoming signal is still weak, ie in the early (front-end) stages; this noise is, within limits, under the control of the designer.

External noise includes atmospheric noise, which is dependent on both frequency, time of day and ionospheric conditions, and also on local man-made electromagnetic noise, which will usually be more significant in urban locations than in rural sites due to the multiplicity of electrical and electronic appliances.

Noise power is usually regarded as distributed evenly over the bandwidth of the receiver so that the effective noise level is reduced if the receiver is operated with the minimum bandwidth appropriate for the mode in use (eg about 2.1 or 2.7kHz for SSB, about 300Hz for CW)

Noise factor defines the maximum sensitivity of a receiver without regard to its pass-band (bandwidth) or its input impedance, and will be determined in the front-end of the receiver. Because of galactic, atmospheric and man-made noise always present on HF there is little need for a receiver noise factor of less than 15–17dB on bands up to about 18–20MHz, or less than about 10dB up to 30MHz, even in quiet sites. It may, however, be an advantage if the first stages (preamplifier or mixer or post-mixer) have a lower noise figure since this will permit good reception with an electrically short antenna or allow the use of a narrow-band filter, which attenuates the signal power even if providing an impedance step-up, to be used between antenna and the mixer stage (ie improved pre-mixer selectivity). It should be noted that excessive sensitivity is likely to impair the strong-signal handling capabilities of the receiver.

Signal-to-noise ratio (or more accurately signal-plus-noise to noise ratio) gives the minimum antenna input voltage to the receiver needed to give a stated output SNR with a specified noise bandwidth. The input voltage for a given input power depends on the input impedance of the receiver; for modern sets this is invariably 50 or 75Ω but for older (though still useful) receivers this may be 400Ω and such sets may thus appear wrongly to be less sensitive unless the impedance is taken into account. The signal delivered by the antenna from a weak incoming signal may well be of the order of −130dBm, representing 0.14μV across a matched 50Ω line. It should be noted that where sensitivity is defined in terms of SNR, this depends not only on input impedance but also on the output SNR, which while usually 10dB may occasionally be 6dB; it will also depend as noted above on the noise bandwidth of the receiver.

It is useful to note also the *minimum discernible signal* (MDS) which is defined as where the signal is equal to the noise voltage, and thus 10dB less than the minimum signal for a SNR output of 10dB, although normally given in dBm. The MDS is thus the *noise floor* of the receiver and represents the power required from a signal generator to produce a 3dB (S+N)/N output, ie where the input signal equals the background noise. MDS (dBm) = −174dBm + 10 log BW + NF where NF is in decibels.

The noise floor of HF receivers is conventionally given in −dBm in a 3kHz noise bandwidth and in practice may range from about −120dBm to about −145dBm. Alternatively, receiver noise in 3kHz bandwidth at 50Ω impedance may be represented by an equivalent signal expressed in dBμV EMF, which will equal the noise factor less 26dB.

With modern forms of mixers, it is possible to achieve a noise factor better than 10dB so that no preamplification is required

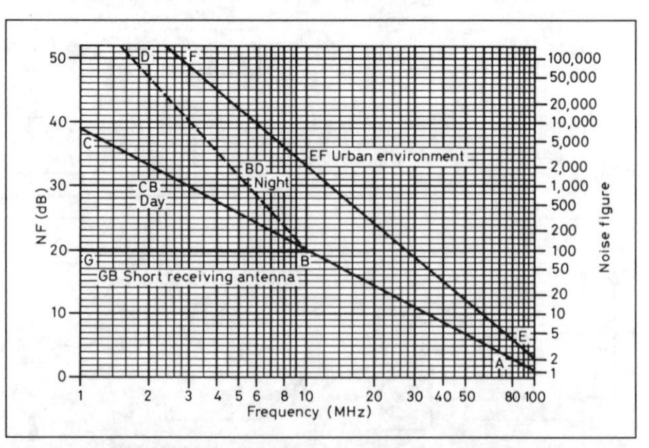

Fig 6.17. Relationship between maximum usable sensitivity, minimum acceptable sensitivity and frequency. Note that on HF it is seldom necessary to have receivers with a noise factor much below 15–20dB, below about 21MHz, unless a very short and inefficient antenna is being used. The lines BC and BD represent reception on half-wave dipoles about a half-wave above the ground. Minimum acceptable sensitivity is represented by the line EF, reflecting a noisy urban environment. An HF receiver with a noise factor of about 10–15dB can usually be designed to have better strong-signal performance than the typical figure of 6dB of many modern receivers

to achieve maximum usable sensitivity on HF up to about 21MHz; however, low-gain amplification may still be advisable in order to minimise radiation of the local oscillator output from the antenna and to permit the use of narrow-band or resonant input filtering.

Strong-signal performance

The ability of a receiver to receive effectively weak signals in the presence of much stronger signals at frequencies not far removed from the desired signal is today recognised as more important than extreme sensitivity. The strong signals may come from local or medium-distance amateur stations (particularly during contest operation) or at any time from high-power (typically 500kW) broadcast stations in bands adjacent to the 7, 14 and 21MHz amateur bands. Strong signals can result in desensitising (*blocking*) the receiver, cross-modulation and/or the generation of spurious intermodulation products. The effects arise from non-linearity in the receiver and can be specified in terms of the receiver's *dynamic range*, although it is important to note that the dynamic range needs to be specified separately for blocking, cross-modulation and intermodulation. The most useful and most critical specification is given by the *third-order intercept point* or as the *spurious-free dynamic range*.

Strong-signal performance and a wide dynamic range have become widely recognised as important and highly desirable characteristics of receiver performance, although it can be argued that, at least for the amateur service (particularly CW), accepted methods of laboratory measurement may not provide a realistic specification. It is conventional practice to make dynamic range measurements using two signal generators with frequencies spaced 20kHz apart; however, where receivers incorporate relatively low-cost PLL synthesisers, the spacing between the two signals may have to be increased to 50 or even 100kHz in order to overcome the effects of synthesiser phase noise (made even more significant by the trend to up-conversion to a VHF intermediate frequency because VHF oscillators are generally worse than HF ones in this respect). It would be more useful if instantaneous dynamic range measurements were made with frequencies spaced 2 or 5kHz apart. It should also be remembered that in practice, the receiver will be required to cope not with just two locally generated signals but with dozens of strong broadcast carriers which will reach the vulnerable mixer stage unless filtered out (or at least reduced) by pre-mixer selectivity. Multiple carriers produce a multitude of IMPs which may resemble noise, thus raising the noise floor and decreasing the sensitivity of the receiver.

In specifying receiver dynamic range use is made of the concept of *intercept points*, although it is important to realise that these are purely graphic presentations based on two-signal laboratory measurements and do not represent directly the signal-handling properties of a receiver. Third-order intermodulation products are measured in the laboratory with the aid of a spectrum analyser and two high-performance signal generators. The intercept point is found with the aid of a graph with axes representing output power in dBm plotted against input power in dBm, on which the wanted signal and the *N*-order intermodulation product outputs are extended to the points where the IMPs intercept the wanted signal plot. This will occur at signal inputs well above those that can in practice be handled by the receiver.

With both axes logarithmic, the second-order IMPs will have twice, and the third-order IMPs three times, the slope of a wanted signal. For third-order IMPs, $N = 3$ and the IMP power is proportional to the cube of the input powers. In other words, for every 1dB increase in the input powers, the IP3 output power increases by 3dB. When plotted graphically there will thus be a point at which the IP3 output crosses and overtakes the plot of a wanted signal.

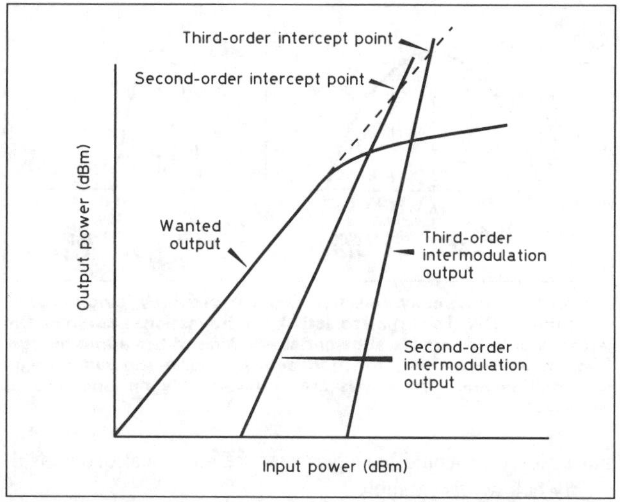

Fig 6.18. Graphical representation of receiver dynamic range performance

This shows that with very strong off-frequency signals, IMPs would completely block out the wanted signal. But for specification purposes we are more concerned to know the point at which the IMPs rise above the noise floor and can just be heard as unwanted interference.

A useful guide to the signal-handling performance of a receiver is given by the *spurious-free dynamic range* (SFDR). The upper end is defined by the signal level applied to the receiver input which produces third-order IMD products equal to the noise floor of the receiver under test. The lower end is defined as the minimum input signal at the noise floor of the receiver, ie the *minimum discernible signal* (MDS).

The SFDR is given by 2/3 × (IP3 – No) where SFDR is in decibels, IP3 is the third-order intercept point in dBm and No is the receiver noise floor in dBm. In the laboratory the receiver noise floor will be that of the internally generated noise and this will usually be significantly below the noise floor with the receiver connected to an antenna. The full SFDR may thus not be usable in practice, and it may be more useful to know the maximum input signal level, Pi(max) that produces third-order IMD products equal to the receiver noise floor:

$$\text{Pi(max)} = 1/3\,(2\text{IP} + \text{No})$$

where Pi(max) is the maximum input signal in dBm, IP is the third-order intercept point in dBm and No the receiver noise floor in dBm.

SFDR is sometimes termed the *two-tone dynamic range*, but note that receiver dynamic range may be defined differently, for example as the *blocking dynamic range* (BDR) representing the difference between the 1dB compression point and the noise floor. With double-balanced mixers, a rule of thumb puts the third-order intercept point some 10 to 15dB above the 1dB compression point but, since the spurious-free IMD range is restricted to where the IMPs rise above the noise floor, the blocking dynamic range will be significantly greater than the SFDR. For example, an extremely-high-performance front-end with a third-order input intercept point of +33dBm, a 1dB input compression point of +14.3dBm and a minimum discernible signal of –133.4dBm would have a spurious-free dynamic range of 111dB but a blocking dynamic range of 147.7dB. In practice

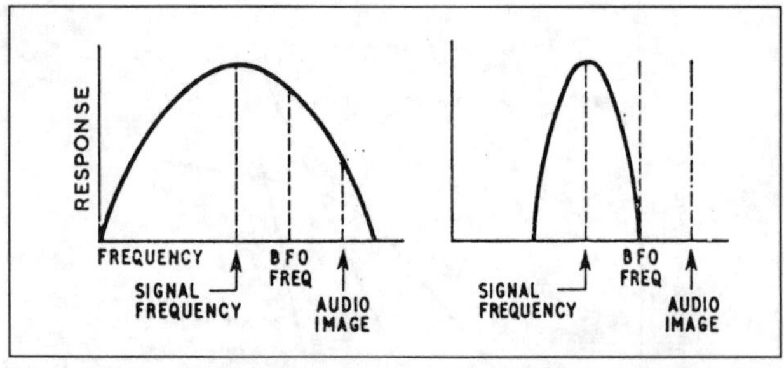

Fig 6.19. How a really selective receiver provides 'single-signal' reception of CW. The broad selectivity of the response curve on the left is unable to provide substantial rejection of the audio 'image' frequency whereas with the more selective curve the audio image is inaudible and CW signals are received only on one side of zero-beat

most receivers would have performance specifications considerably below this example.

It needs to be stressed that in order to achieve state-of-the-art meaningful dynamic performance of a receiver, extreme care must be taken not only with the vulnerable mixer stages but throughout the receiver, up to and including the final selectivity filters. On the other hand, budget-conscious constructors/purchasers can take heart from the fact that receivers with far less stringent specifications may still prove entirely adequate for many forms of amateur activity not involving the reception of very weak signals under the most hostile conditions such as international HF contests, or, for example, 7MHz night-time operation in the presence of strong broadcast carriers.

Selectivity

The ability of a receiver to separate stations on closely adjacent frequencies is determined by its *selectivity.* The limit to usable selectivity is governed by the *bandwidth* of the type of signal which is being received.

For high-fidelity reception of a double-sideband AM signal the response of a receiver would need ideally to extend some 15kHz either side of the carrier frequency, equivalent to 30kHz bandwidth; any reduction of bandwidth would cause some loss of the information being transmitted. In practice, for average MF broadcast reception the figure is reduced to about 9kHz or even less; for communications-quality speech in a double-sideband system we require a bandwidth of about 6kHz; for single-sideband speech about half this figure or 3kHz is adequate, and filters with a nose bandwidth of 2.7 or 2.1kHz are used, providing, in the case of a 2.7kHz filter, audio frequencies from

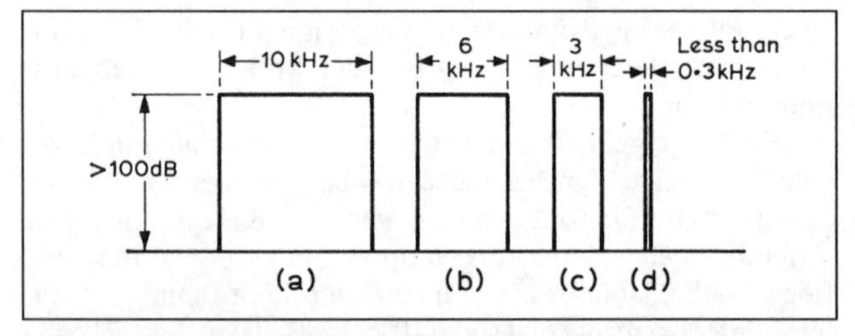

Fig 6.20. The ideal characteristics of the overall band-pass of a receiver are affected by the type of signals to be received. (a) This would be suitable for normal broadcast reception (DSB signals) permitting AF response to 5kHz. (b) Suitable for AM phone (AF to 3kHz). (c) For SSB the band-pass can be halved without affecting the AF response (in the example shown this would be about 300 to 3300Hz). (d) Extremely narrow channels (under 100Hz) are occupied by manually keyed CW signals but some allowance must usually be made for receiver or transmitter drift and a 300Hz bandwidth is typical – by selection of the BFO frequency any desired AF beat note can be produced

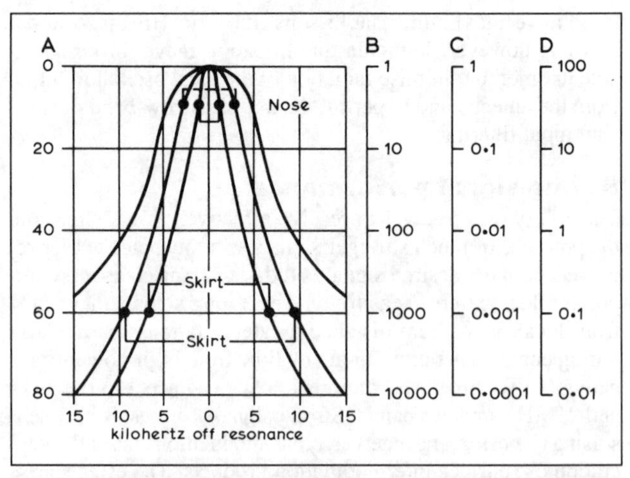

Fig 6.21. The ideal vertical sides of Fig 6.20 cannot be achieved in practice. The curves shown here are typical. These three curves represent the overall selectivity of receivers varying from the 'just adequate' broadcast curve of a superhet receiver having about four tuned IF circuits on 470kHz to those of a moderately good communications receiver. A, B, C and D indicate four different scales often used to indicate similar results. A is a scale based on the attenuation in decibels from maximum response; B represents the relative signal outputs for a constant output; C is the output voltage compared with that at maximum response; D is the response expressed as a percentage

300 to 3000Hz with little loss of intelligibility. For CW, at manual keying speeds, the minimum theoretically possible bandwidth will reduce with speed from about 100Hz to about 10Hz for very slow morse; in practice, however, the stability of the receiver or transmitter will seldom allow a bandwidth of much less than 100–300Hz to be used. Excessively narrow filters with good shape factors make searching difficult – and many operators like to have some idea of signals within a few hundred hertz of the wanted signal. Ideally, again, we would like to receive just the right bandwidth, with the response of the receiver then dropping right off as shown in Fig 6.20, to keep the *noise bandwidth* to a minimum. Although modern filters can approach this response quite closely, in practice the response will not drop away as sharply over as many decibels as the ideal.

To compare the selectivity of different receivers, or the same receiver for different modes, a series of curves of the type shown in Fig 6.21 may be used. There are two ways in which these curves should be considered: first the bandwidth at the *nose,* representing the bandwidth over which a signal will be received with little loss of strength; the other figure – in practice every bit as important – is the bandwidth over which a powerful signal is still audible, termed the *skirt* bandwidth.

The nose bandwidth is usually measured for a reduction of not more than 6dB, the skirt bandwidth for a reduction of one thousand times on its strength when correctly tuned in, that is 60dB down. These two figures can then be related by what is termed the *shape factor,* representing the bandwidth at the skirt divided by the bandwidth at the nose. The idealised curves of Fig 6.20, which have the same bandwidth regardless of signal strength, would represent a shape factor of 1; such a receiver cannot be designed at the present state of the art, although it can be approached by digital filters and some SSB filters; the narrower CW filters, although much sharper at the nose, tend to broaden out to about the same bandwidth as the SSB filter and thus have a rather worse shape factor, although this may not be a handicap. Typically a high-grade modern receiver might have an SSB shape factor of 1.2 to 2 with a skirt bandwidth of less than 5kHz.

It should be noted that such specifications are determined when applying only *one* signal to the input of the receiver; unfortunately in practice this does not mean that a receiver will be unaffected by very strong signals operating many kilohertz away from the required signal and outside the IF pass-band; this important point will be considered later in this chapter.

It therefore needs to be stressed that the effective selectivity cannot be considered solely in terms of static characteristics determined when just one test signal is applied to the input, but rather in the real-life situation of hundreds of signals present at the input: in other words it is the *dynamic selectivity* which largely determines the operational value of an amateur HF receiver.

Automatic gain control (AGC)

The fading characteristics of HF signals and the absence of a carrier wave with SSB make the provision of effective AGC an important characteristic of a communications receiver, although it should be stressed that no AGC may be preferable to a poor AGC system that can degrade overall performance of the receiver. Unfortunately, unwanted dynamic effects of an AGC system cannot be deduced from the usual form of receiver specification which usually provides only limited information on the operation of the AGC circuits, indicating only the change of audio output for a specified change of RF input. For example, the specification may state that there will be a 3dB rise in audio output for an RF signal input change from 1μV to 50mV. This represents a high standard of control, provided that the sensitivity of the receiver is not similarly reduced by strong off-tune signals, and that the control acts smoothly throughout its range without introducing intermodulation distortion etc.

Basically AGC is applied to a receiver to maintain the level of the wanted signal output at a more or less constant value, while ensuring that none of the stages are overloaded with consequent production of IMPs etc. The control voltages, derived usually at the end of the IF amplifying stages, are applied to a number of stages in the signal path while usually ensuring that the IF gain is reduced first, and the RF/first mixer later, in order to preserve the SNR.

When AGC is applied to an amplifier this shifts the operating point and may affect both its dynamic range and the production of intermodulation distortion. One partial solution is the use of an AGC-controlled RF attenuator(s) ahead of the first stage(s).

All AGC systems are designed with an inherent delay based on resistance-capacitance time-constants in their response to changes in the incoming signal; too-rapid response would result in the receiver following the audio envelope or impulsive noise peaks. The delays are specified as the *attack time*, ie the time taken for the AGC to act, and the *decay time* for which it continues to act in the absence or fade of the wanted signal.

For AM signals, now only rarely encountered in amateur radio, the envelope detector may be used directly to generate the AGC voltage. In this case the attack and delay times need to be fast enough to allow the AGC system to respond to fading, but slow enough not to respond to noise pulses or the modulation of the carrier. Typically time-constants are about 0.1 to 0.2s.

For amateur SSB signals there is no carrier level and the AGC voltage must be derived from the peak signal level. This is sometimes done by using the AF signal from a product detector but it is more satisfactory to incorporate a dedicated envelope detector to which a portion of the final IF signal is fed. For SSB, a fast attack time is needed but the release (decay) time needs to be slow in order not to respond to the brief pauses of a speech transmission. Such a system is termed *hang AGC* and can be used also for CW reception. Typically the attack time needs to

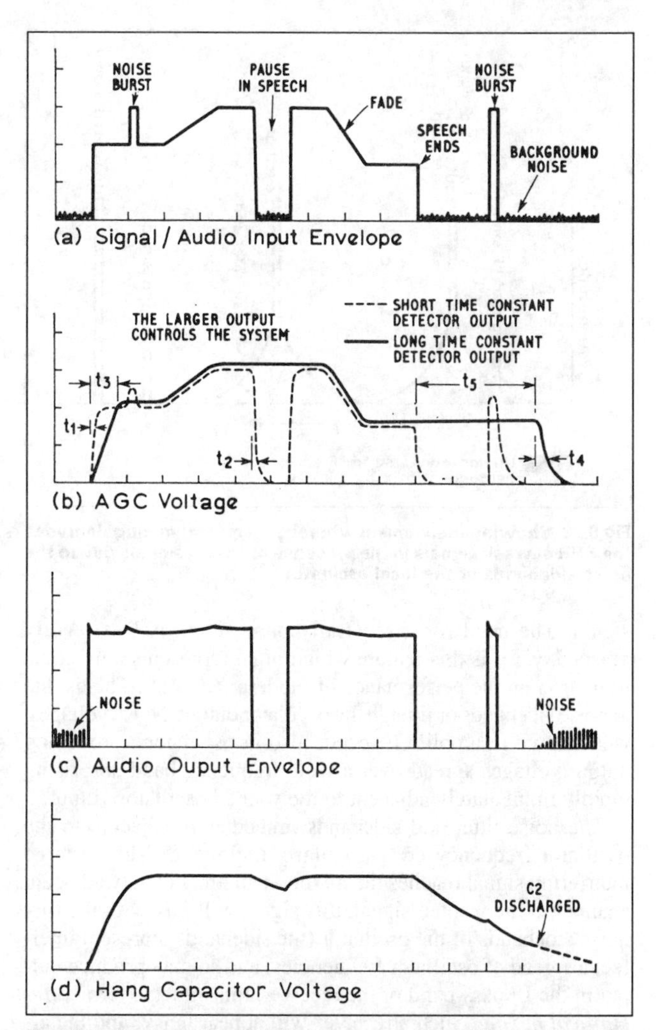

Fig 6.22. The behaviour of dual time-constant AGC circuits under various operating conditions, where t_1 is the fast detector rise time, t_2 the fast detector decay time, t_3 the slow detector rise time, t_4 the slow detector fall time, and t_5 the hang time. This approach is used in the SL621 IC or can also be implemented using discrete devices

be less than 20ms and the release time some 200 to 1000ms; with a receiver intended also for AM reception it is useful to have a shorter release time (say 25ms) available for *fast AGC*. Hang AGC systems are often based on two time-constants to allow the system to be relatively unaffected by noise pulses while retaining fast attack and hang characteristics.

For both SSB and CW reception, it is important that the BFO or carrier insertion oscillator should not affect the AGC system and for many years this problem was sidestepped by turning the AGC system off during CW reception. Even with modern AGC systems, it may be useful to be able to turn the system off, particularly where final narrow-band CW selectivity depends on post-demodulation filtering, with the result that the AGC system will react to signals within the bandwidth of the final IF system which may be about 3kHz unless narrow-band IF crystal filters are fitted.

Oscillator noise and reciprocal mixing

A single-conversion superhet requires a variable HF oscillator which should not only be stable but should provide an extremely 'pure' signal with the minimum of noise sidebands. Any variation of the oscillator output in terms of frequency drift or sudden 'jumps' will cause the receiver to detune from the incoming

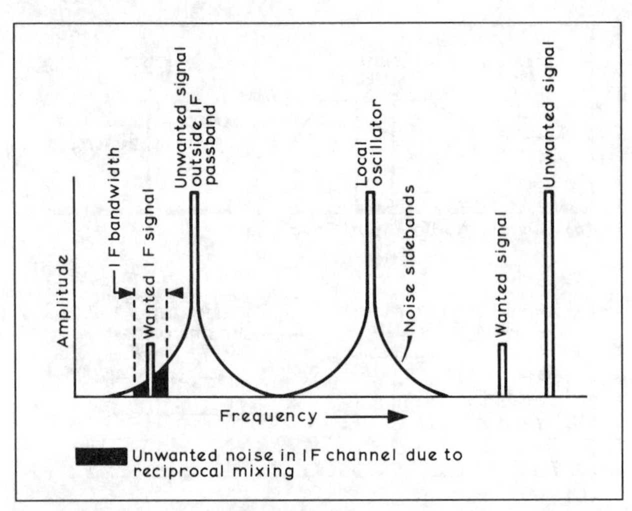

Fig 6.23. Showing mechanism whereby reciprocal mixing degrades the SNR of weak signals in the presence of strong signals due to the noise sidebands of the local oscillator

signal. The need for a spectrally pure output is less readily grasped, yet it is this feature which often represents a practical limitation on the performance of modern receivers. This is due to noise sidebands or jitter in the oscillator output. Noise voltages, well known in amplifiers, occur also in oscillators, producing output voltages spread over a wide frequency band and rising rapidly immediately adjacent to the wanted oscillator output.

The noise jitter and sidebands immediately adjacent to the oscillator frequency are particularly important. When a large interfering signal reaches the mixer on an immediately adjacent channel to the wanted signal, this signal will mix with the tiny noise sidebands of the oscillator (the sidebands represent in effect a spread of oscillator frequencies) and so may produce output in the IF pass-band of the receiver: this effect is termed *reciprocal mixing*. Such a receiver will appear noisy, and the effect is usually confused with a high noise factor.

At HF the 'noise' output of an oscillator falls away very sharply either side of the oscillator frequency, yet it is now recognised that this noise may be sufficient to limit the performance of a receiver. Particular care is necessary with some forms of frequency synthesisers, including those based on the phase-locking of a free-running oscillator, since these can often produce significantly more noise sidebands and jitter than that from a free-running oscillator alone.

Synthesisers involving a number of mixing processes may easily have a noise spectrum 40 to 50dB higher than a basic LC oscillator. A tightly controlled phase-locked oscillator with variable divider might be some 20 to 30dB higher than an LC oscillator, but possibly less than this where the VCO is inherently very stable and needs only infrequent 'correction'.

For LC and crystal oscillators, it appears that field-effect transistors provide minimum noise sidebands; valves next; with bipolar transistors third. This is another reason why the FET is a good choice for an oscillator.

Fig 6.23 shows in simplified form the basic mechanism of reciprocal mixing. Because of the noise sidebands (or 'skirts') of the local oscillator, some part of the strong unwanted signal is translated into the receiver's IF pass-band and thus reduces the SNR of a weak wanted signal; further (not shown) a small fragment of the oscillator noise also spreads out to the IF, enters the IF channel and reduces the SNR. In practice the situation is even more complex since the very strong unwanted carrier will itself have noise sidebands which spread across the frequency of the wanted signal and degrade SNR no matter how pure the output of the local oscillator.

Noise generated by an oscillator is specified in terms of dBm in a 1Hz bandwidth at a specified frequency from its carrier frequency. At some frequency offset the noise spectrum begins to rise out of the noise plateau (−174dBm/Hz) and increases by four times each time the offset from the carrier is halved. A very-high-performance receiver with a 2kHz noise bandwidth requires that the oscillator phase noise needs to be of the order of −150dBc/Hz at, say, 20kHz off-tune, a very stringent requirement.

Fig 6.24 indicates the practical effect of reciprocal mixing on high-performance receivers; in the case of receiver 'A' (typical of many high-performance receivers) it is seen that the dynamic selectivity is degraded to the extent that the SNR of a very weak signal will be reduced by strong unwanted signals (1 to 10mV or more) up to 20kHz or more off-tune, even assuming that the front-end linearity is such that there is no cross-modulation, blocking or intermodulation. Reciprocal mixing thus tends to be the limiting factor affecting very weak station performance in real situations, although intermodulation or cross-modulation characteristics become the dominant factors with stronger signals.

It is thus important in the highest-performance receivers to pay attention to achieving low noise-sidebands in oscillators, and this is one reason why the simpler forms of frequency synthesisers, which often have appreciable jitter and noise in the output, must be viewed with caution despite the high stability they achieve. The three major forms of basic oscillator noise are:

(a) low frequency (LF) noise which predominates very close to carrier but is insignificant beyond about 250Hz;
(b) thermal noise which predominates between about 250Hz to about 20kHz from carrier; and
(c) *shot noise* attributable to noise current and more or less evenly spread at all frequencies.

Since the shot noise of an oscillator spreads across the IF channel, care is needed with low-noise mixers to limit the amount of this *injection-source noise* that enters the IF amplifier. Balanced and double-balanced mixers provide up to about 30dB rejection of oscillator noise. Another technique is to use a rejector *trap* (tuned circuit) resonant at the IF between oscillator and mixer.

Optimum oscillator performance calls for the use of a FET with a high forward transconductance and for a high unloaded tank-circuit *Q*.

For switching-mode mixers and product detectors the optimum

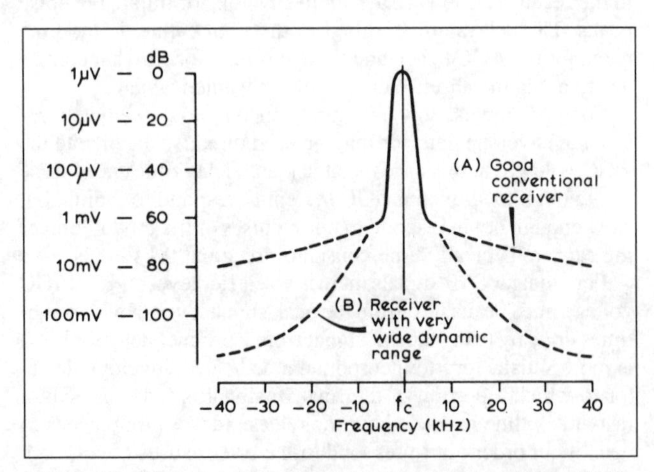

Fig 6.24. Reciprocal mixing due to oscillator noise can modify the overall selectivity curve of an otherwise very good receiver

oscillator output waveform would be a square wave but this refinement is comparatively rare in practice.

Stability

The ability of a receiver to remain tuned to a wanted frequency without *drift* depends upon the electrical and mechanical stability of the internal oscillators. The primary cause of instability in an oscillator is a change of temperature, usually as the result of internally generated heat. With valves, even in the best designs, there will usually be steady frequency variation of any oscillator using inductors and capacitors during a period of perhaps 15min or more after first switching on; transistor and FET oscillators reach thermal stability in a few seconds, provided that they are not then affected by other local sources of heat. However, the stability of a solid-state oscillator may be more affected than a valve oscillator by changes in the ambient temperature.

After undergoing a number of heat cycles, some components do not return precisely to their original values, making it difficult to maintain accurate calibration over a long period. Many receivers include a crystal-controlled oscillator of high stability providing marker signals (for example every 100kHz) so that receiver calibration can be checked and brought into adjustment.

Receiver drift can be specified in terms of maximum drift in hertz over a specified time, usually quoting a separate figure to cover the warming-up period. For example a high-stability receiver might specify drift as "not worse than 200Hz in any five hour period at constant ambient temperature and constant mains voltage after one hour warm-up".

Mechanical instability, which may appear as a shift of frequency when the receiver is subjected to mechanical shock or vibration, cannot easily be defined in the form of a performance specification. Sturdy construction on a mechanically rigid chassis can help. The need for high stability for SSB reception has led to much greater use of crystal-controlled oscillators and various forms of frequency synthesis.

Spurious responses

A most important test of any receiver is the extent to which it receives signals when it is tuned to frequencies on which they are not really present, so adding to interference problems and misleading the operator. Every known type of receiver suffers from various forms of 'phantom' signals, but some are much worse than others. Unfortunately, the mixing process is inherently prone to the generation of unwanted 'products' and receiver design is concerned with minimising their effect rather than their complete elimination.

Spuriae may take the form of:

(a) external signals heard on frequencies other than their true frequency;
(b) carriers heard within the tuning range of the receiver but stemming not from external signals but from the receiver's own oscillators (*birdies*);
(c) external signals which cannot be tuned out but are heard regardless of the setting of the tuning knob.

In any superhet receiver tuneable signals may be created in the set whenever the interfering station or one of its harmonics (often produced within the receiver) differs from the intermediate frequency by a frequency equal to the local oscillator or one of its harmonics. This is reflected in the general expression:

$$mf_u \pm nf_0 = f_i$$

where m, n are any integers, including 0, f_u is the frequency of the unwanted signal, f_0 is the frequency of the local oscillator, and f_i is the intermediate frequency.

An important case occurs when m and n are 1, giving $f_u \pm f_0 = f_i$. This implies that f_u will either be on the frequency to which the set is correctly tuned (f_s) or differs from it by twice the IF ($2 \times f_i$), producing the so-called 'image' frequency. This is either $f_u = f_s + 2f_i$ (for cases where the local oscillator is higher in frequency than the wanted signal), or $f_u = f_s - 2f_i$ (where the local oscillator is lower in frequency than the wanted signal).

For an example, take a receiver tuned to about 14,200kHz with an IF of 470kHz and the oscillator high (ie about 14,670kHz). Such a receiver may, because of 'image', receive a station operating on 14,200 + (2 × 470) = 15,140kHz. Since 15,140kHz is within the 19m broadcast band, there is thus every likelihood that as the set is tuned around 14,200kHz strong broadcast signals will be received.

To reduce such undesirable effects, pre-mixer selectivity must be provided in the form of more RF tuned circuits or RF bandpass filters, or by increasing the Q of such circuits, or alternatively by increasing the frequency difference between the wanted and unwanted image signals. This frequency separation can be increased by increasing the factor $2f_i$, in other words by raising the intermediate frequency, and so allowing the broadly tuned circuits at signal frequency to have more effect in reducing signals on the unwanted image frequency before they reach the mixer.

It should also be noted from the general formula $mf_u \pm nf_0$ that 'image' is only one (though usually the most important) of *many* possible frequency combinations that can cause unwanted signals to appear at the intermediate frequency, even on a single-conversion receiver. The problem is greatly increased when more than one frequency conversion is employed.

Even with good pre-mixer selectivity it is still possible for a number of strong signals to reach the mixer, drive this or an RF stage into non-linearity and then produce a series of intermodulation products as spurious signals within an amateur band (see Fig 6.25).

The harmonics of the HF oscillator(s) may beat against incoming signals and produce output in the IF pass-band; strong signals may generate harmonics in the receiver stages and these can be received as spuriae. Most such forms of spuriae can be reduced by increasing pre-mixer selectivity to decrease the number of strong signals reaching the mixer, by increasing the linearity of the early stages of the receiver, or by reducing the amplitude of signals within these stages by the use of an antenna attenuator.

Very strong signals on or near the IF may break directly into the IF amplifier and then appear as untuneable interference. This form of interference (although it then becomes tuneable) is particularly serious with the variable IF type of multiple-conversion receiver since there are almost certain to be a number of very strong signals operating over the segment of the HF spectrum chosen to provide the variable IF. Direct breakthrough may occur if the screening within the receiver is insufficient or if signals can leak in through the early stages due to lack of pre-mixer selectivity. For single-conversion and double-conversion with fixed first IF, it is common practice to include a resonant 'trap' (tuned to the IF) to reject incoming signals on this frequency. The multiple-conversion superhet contains more internal oscillators, and harmonics of the second and third (and occasionally the BFO) can be troublesome. For amateur-bands-only receivers every effort should be made to choose intermediate and oscillator frequencies that avoid as far as possible the effects of oscillator harmonics (*birdies*).

Because of the great difficulty in eliminating spurious responses in double- and triple-conversion receivers, the modern designer tends to think more in terms of single conversion with high IF (eg 9MHz). Potentially the direct-conversion receiver is even more attractive, though it needs to be fairly complex to eliminate the audio-image response. It must also have sufficient linearity or pre-detector selectivity to reduce any envelope detection of very strong signals which may otherwise break through into the audio channel, regardless of the setting of the heterodyne oscillator. The direct-conversion receiver can also suffer from spuriae resulting from harmonics of the signal or oscillator and this needs to be reduced by RF selectivity.

Cross-modulation, blocking and intermodulation

Even with a receiver that is highly selective to one signal down to the –60dB level, there remains the problem of coping with numbers of extremely strong signals. When an unwanted signal is transmitting on a frequency that is well outside the IF passband of the receiver, it may unfortunately still affect reception as a result of cross-modulation, blocking or intermodulation.

When any active device such as a transistor or valve is operated with an input signal that is large enough to drive the device into a non-linear part of its transfer characteristic (ie so that some parts of the input waveform are distorted and amplified to different degrees) the device acts as a 'modulator', impressing on the wanted signal the modulation of the strong signal by the normal process of mixing. When a very strong signal reaches a receiver, the broad selectivity of the signal frequency tuned circuits (and often any tuned circuits, including IF circuits, prior to the main selective filter) means it will be amplified, along with the wanted signal, until one or more stages is likely to be driven into a non-linear condition. It should be noted that the strong signal may be many kilohertz away from the wanted signal, but once this *cross-modulation* has occurred there is no means of separating the wanted and unwanted modulation. A strong CW carrier can reduce the amplification of the wanted signal by a similar process which is in this case called *desensitisation*, or in extreme cases *blocking*.

In these processes there need be no special frequency relationship between wanted and unwanted signals. However, a further condition arises when there are specific frequency relationships between wanted and unwanted signals, or between two strong unwanted signals; a process called *intermodulation*: see Fig 6.25.

Intermodulation is closely allied to the normal mixing process but with strong signals providing the equivalent of local oscillators, unwanted intermodulation products (IPs) can result from many different combinations of input signal.

It should be appreciated that cross-modulation, blocking and intermodulation products can all result from the presence of extremely strong unwanted signals applied to any stage having insufficient linearity over the full required dynamic range. The solution to this problem is either to reduce the strength of unwanted signals applied to the stage, or alternatively to improve the dynamic range of the stage.

Clearly the more amplifying stages there are in a receiver before the circuits or filters which determine its final selectivity, the greater are the chances that one or more may be overloaded unless particular care is paid to the *gain distribution* in the receiver (see later). From this it follows that multiple-conversion receivers are more prone to these problems than single-conversion or direct-conversion receivers.

It is often difficult for an amateur to assess accurately the performance of a receiver in this respect; undoubtedly many multi-conversion superhets and many receivers based on bipolar transistors fall far below the desired performance. Even high-grade valve receivers are likely to be affected by S9 + 60dB signals 50kHz or more away from the wanted signal.

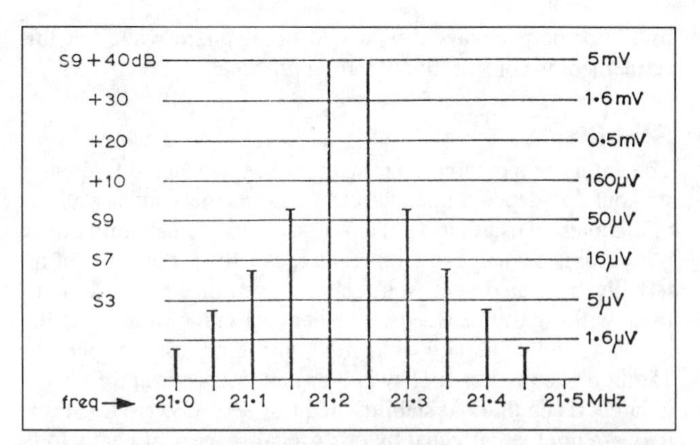

Fig 6.25. Intermodulation products. This diagram shows the effect of two very strong signals on 21,200kHz and 21,250kHz reaching the front-end of a typical modern transceiver and producing spurious signals at 50kHz intervals. Note the S9 signals produced as the third-order products $f_1 + (f_1 - f_2)$ and $f_2 - (f_1 - f_2)$. Three very strong signals will produce far more spurious signals, and so on

The susceptibility of receivers to these forms of interference depends on a number of factors: notably the type of active devices used in the front-end of the receiver; the pattern of gain distribution through the receiver and how this is modified by the action of AGC. On receivers suffering from this problem (often most apparent on 7MHz where weak amateur stations may be sought alongside extremely strong broadcast stations) considerable improvement often results from the inclusion of an attenuator in the antenna feeder, since this will reduce the unwanted signals to an extent where they may not cause spuriae before reducing the wanted signal to the level where it cannot be copied.

If sufficient dynamic range could be achieved in *all* stages before the final selectivity filter, then selectivity could be determined at any point in the receiver (for example at first, second or third IF or even at AF).

In practice some semiconductor receivers are unable to cope with undesired signals more than about 20–30dB stronger than a required signal of moderate strength, though their 'static' selectivity may be extremely good. The overall effect of limited dynamic range depends upon the actual situation and band on which the receiver is used; it is of most importance on 7MHz or where there is another amateur within a few hundred yards. Even where there are no local amateurs, an analysis of commercial HF signals has shown that typically, out of a total of some 3800 signals logged between 3 and 29MHz at strengths more than 10dB above atmospheric noise, 154 were between 60–70dBµV, 72 between 70–80dBµV, 36 between 80–90dBµV and 34 between 90–100dBµV. If weak signals are to be received satisfactorily the receiver needs to be able to cope with signals some 100dB stronger (ie up to about S9 + 50dB). This underlines the importance of achieving front-end stages of extremely wide dynamic range or alternatively providing sufficient pre-mixer selectivity to cut down the strength of all unwanted signals reaching the mixer (preferably to well under 100mV).

Another form of spurious response is where a strong signal breaks through into the IF stages of the receiver. For example, with a simple single-conversion receiver having an IF of 470kHz, strong coastal stations and ships operating around 500kHz may be heard; these signals can usually be eliminated by providing additional pre-mixer tuned circuits. With receivers having a high (first) IF or with the first IF tuneable over a range of frequencies, it becomes increasingly difficult to keep strong HF signals from breaking through into the IF stages. Whereas ideally one would like an IF rejection (compared with the wanted signal) of almost 120dB, most amateurs would be well satisfied with 80–100dB of protection; in practice many receivers with variable first IF do not provide more than about 40–60dB protection.

Choice of IF

Choice of the intermediate frequency or frequencies is a most important consideration in the design of any superhet receiver. The lower the frequency, the easier it is to obtain high gain and good selectivity and also to avoid unwanted leakage of signals round the selective filter. On the other hand, the higher the IF, the greater will be the frequency difference between the wanted signal and the 'image' response, so making it simpler to obtain good protection against image reception of unwanted signals and also reducing the 'pulling' of the local oscillator frequency. These considerations are basically opposed, and the IF of a single-conversion receiver is thus a matter of compromise; however, in recent years it has become easier to obtain good selectivity with higher-frequency band-pass crystal filters and it is no longer any problem to obtain high gain at high frequencies. The very early superhet receivers used an IF of about 100kHz; then for many years 455–470kHz was the usual choice – many modern designs use between about 3 and 9MHz, and SSB IF filters are now available to 40MHz. Where the IF is *higher* than the signal frequency the action of the mixer is to raise the frequency of the incoming signal, and this process is now often termed *up-conversion* (a term formerly reserved for a special form of parametric mixer). Up-conversion, in conjunction with a low-pass filter at the input, is an effective means of reducing IF breakthrough as well as image response.

A superhet receiver, whether single- or multi-conversion, must have its first IF outside its tuning range. For general-coverage HF receivers tuning between, say, 1.5 and 30MHz, this limits the choice to below 1.5MHz or above 30MHz. To reduce image response without having to increase pre-mixer RF selectivity (which can involve costly gang-tuned circuits) professional designers are increasingly using a first IF well above 30MHz. This trend is being encouraged by the availability of VHF crystal filters suitable for use either directly as an SSB filter or more often with relaxed specification as a *roofing filter.*

The use of a very high first IF, however, tends to make the design of the local oscillator more critical (unless the Wadley triple-conversion drift-free technique is used). For amateur-bands-only receivers the range of choice for the first IF is much wider, and 3.395MHz and 9MHz are typical.

A number of receivers have adopted 9MHz IF with 5.0–5.5MHz local oscillator: this enables 4MHz–3.5MHz and 14.0–14.5MHz to be received without any band switching in the VFO; other bands are received using crystal-controlled converters with outputs at 3.5 or 14MHz. With frequency synthesisers, up-conversion is commonly found, with the first IF between 40–70MHz.

Gain distribution

In many receivers of conventional design, it has been the practice to distribute the gain throughout the receiver in such a manner as to optimise signal-to-noise ratio and to minimise spurious responses. So long as relatively noisy mixer stages were used it was essential to amplify the signal considerably before it reached the mixer. This means that any strong unwanted signals, even

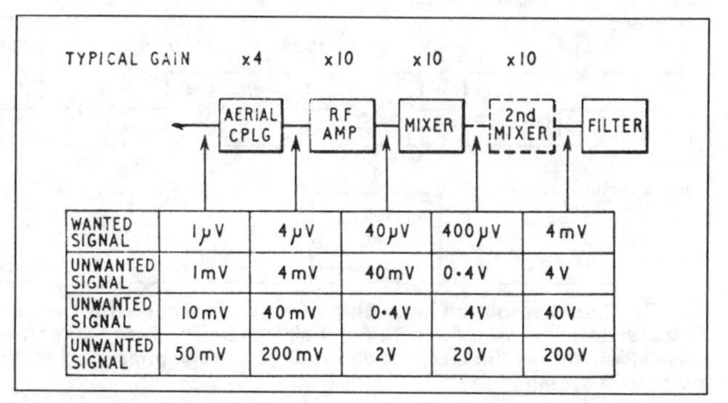

WANTED SIGNAL	1µV	4µV	40µV	400µV	4mV
UNWANTED SIGNAL	1mV	4mV	40mV	0·4V	4V
UNWANTED SIGNAL	10mV	40mV	0·4V	4V	40V
UNWANTED SIGNAL	50mV	200mV	2V	20V	200V

Fig 6.26. This diagram shows how unwanted signals are built up in high-gain front-ends to levels at which cross-modulation, blocking and intermodulation are virtually bound to occur

when many kilohertz from the wanted signal, pass through the early unselective amplifiers and are built up to levels where they cause cross-modulation within the mixer: see Fig 6.26. Today it is recognised that it is more satisfactory if pre-mixer gain can be kept low to prevent this happening. Older multi-grid valve mixers had an equivalent noise resistance as high as 200,000Ω (representing some 4–5µV of noise referred to the grid). Later types such as the ECH81 and 6BA7 reduced this to an ENR of about 60,000Ω (about 2.25µV of noise) while the ENR of pentode and triode mixers was lower still (although these may not be as satisfactory for mixers in other respects).

For example, the 6J6 had a mixer ENR of under 2000Ω, while the 7360 beam-deflection mixer had extremely wide dynamic range and an ENR of only 1500Ω. When the ENR is less than about 3000Ω no pre-mixer amplification is needed in an HF receiver.

The noise contribution of semiconductor mixers is also low – for example an FET mixer may have a noise factor as low as 3dB – so that generally the designer need no longer worry unduly about the requirement for pre-mixer amplification to overcome noise problems. Nevertheless a signal frequency stage may still be useful in helping to overcome image reception by providing a convenient and efficient method of coupling together signal-frequency tuned circuits, and when correctly controlled by AGC it becomes an automatic large-signal attenuator. Pre-mixer selectivity limits the number of strong signals reaching the mixer.

Fig 6.27 shows a typical gain distribution as found in a modern design in which the signal applied to the first mixer is much lower than was the case with older (valved) receivers.

The significant conversion losses of modern diode and FET ring mixers (6–10dB), the losses of input band-pass filtering and the need for correct impedance termination means that in the highest performance receivers it is desirable to include low-gain, low-noise, high dynamic range pre- and post-mixer amplifiers and a diplexer ahead of the (main or roofing) crystal filter to achieve constant input impedance over a broad band of frequencies.

If the diplexer is a simple resistive network this will introduce a further loss of some 6dB. Stage-by-stage gain, noise figure, third-order intercept, 1dB compression point and 1dB desensitisation point performance of the N6NWP high-dynamic-range MF/HF front-end of a single-conversion (9MHz IF) receiver (*QST* February 1993, and see later) is shown in Fig 6.28. More complex diplexers may be used to divert the local oscillator feedthrough signal from entering the IF strip.

For double- and multiple-conversion receivers, the gain-distribution and performance characteristics of all the stages

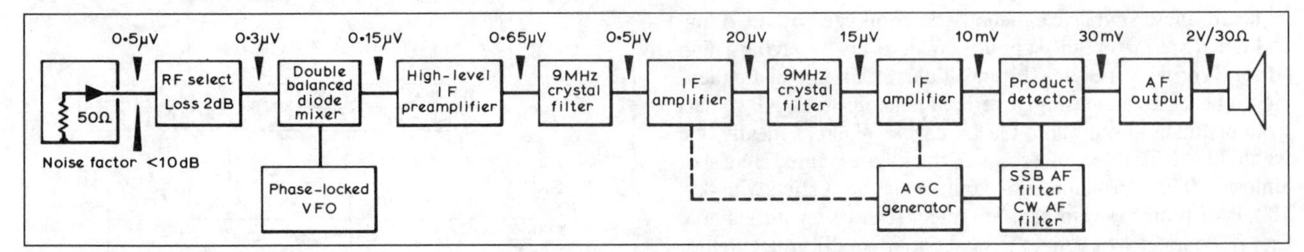

Fig 6.27. Gain distribution in a high-performance semiconductor single-conversion receiver built by G3URX using seven integrated circuits, 27 transistors and 14 diodes. 1μV signals can be received 5kHz off-tune from a 60mV signal and the limiting factor for weak signals is the noise sidebands of the local oscillator, although the phase-locked VFO gives lower noise and spurious responses than the more usual pre-mixer VFO system

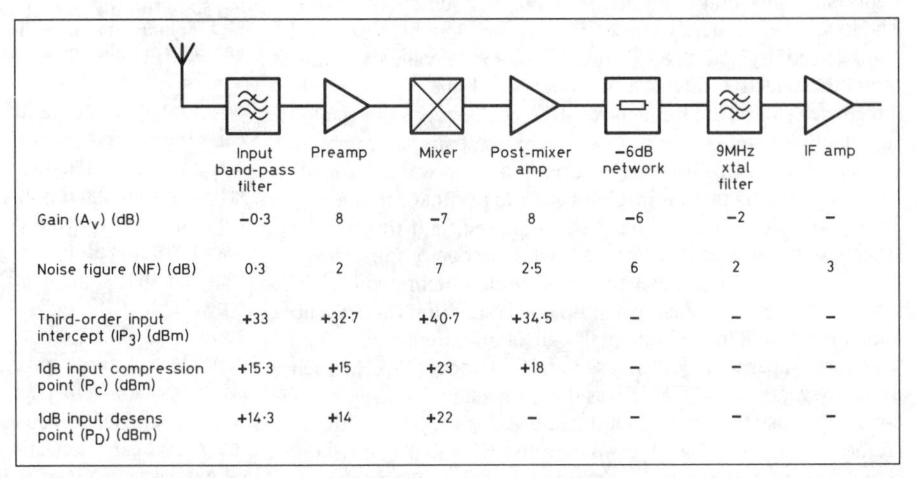

	Input band-pass filter	Preamp	Mixer	Post-mixer amp	−6dB network	9MHz xtal filter	IF amp
Gain (A_V) (dB)	−0·3	8	−7	8	−6	−2	–
Noise figure (NF) (dB)	0·3	2	7	2·5	6	2	3
Third-order input intercept (IP_3) (dBm)	+33	+32·7	+40·7	+34·5	–	–	–
1dB input compression point (P_C) (dBm)	+15·3	+15	+23	+18	–	–	–
1dB input desens point (P_D) (dBm)	+14·3	+14	+22	–	–	–	–

Fig 6.28. Stage-by-stage gain and characteristics of the N6NWP front-end of single-conversion 14MHz HF receiver with 9MHz IF

preceding the selective filter need to be considered. In general the power loss in any band-pass filter will decrease as the bandwidth increases: for example a 75MHz, 25kHz-bandwidth filter might have a power loss of 1dB whereas a 75MHz, 7kHz-bandwidth filter might have a power loss of 3.5dB.

STABILITY OF RECEIVERS

The resolution of SSB speech and the reception of a CW signal requires that a receiver can be tuned to, and remain within, about 25–30Hz of the frequency of the incoming signal. At 29MHz this represents a tolerance of only about one part in a million. For amateur operation the main requirement is that this degree of stability should be maintained over periods of up to about 30min. Long-term stability is less important for amateurs than short-term stability; it will also be most convenient if a receiver reaches this degree of stability within a fairly short time of switching on.

It is extremely difficult to achieve or even approach this order of stability with a free-running, band-switched variably tuned oscillator working on the fundamental injection frequency, although with care a well-designed FET oscillator can come fairly close. This has led (in the same way as for transmitting VFO units) to various frequency-synthesis techniques in which the stability of a free-running oscillator is enhanced by the use of crystals. The following are among the techniques used:

1. *Multi-conversion receiver with crystal-controlled first oscillator and variable first IF.* This very popular technique has a tuneable receiver section covering only one fixed frequency band; for example 5.0–5.5MHz. The oscillator can be carefully designed and temperature compensated over one band without the problems arising from the uncertain action of wavechange switches, and can be separated from the mixer stage by means of an isolating or buffer stage. For the front-end section a separate crystal is needed for each tuning range (the 28MHz band may require four or more crystals to provide full coverage of 28.0–29.7MHz).
2. *Partial synthesis.* The arrangement of (1) becomes increasingly costly to implement as the frequency coverage of the variable IF section is reduced below about 500kHz, or is required to provide general coverage throughout the HF band. Beyond a certain number of crystals it becomes more economical (and offers potentially higher stability) if the separate crystals are replaced by a single high-stability crystal (eg 1MHz) from which the various band-setting frequencies are derived (Fig 6.29). This may be done, for example, by digital techniques or by providing a spectrum of harmonics to one of which a free-running oscillator is phase-locked. It will be noted that with this system the tuning within any band still depends on the VFO and for this reason is termed *partial* synthesis.
3. *Single-conversion receiver with heterodyne (pre-mixer) VFO.* In this system of partial synthesis, the receiver may be a single-conversion superhet (or dual-conversion with fixed IFs) (Fig 6.30). The variable HF injection frequency is obtained using a heterodyne-type VFO, in which the output of a crystal-controlled oscillator is mixed with that of a single-range VFO, and the output is then filtered and used as the injection frequency. The overall stability will be much the same as for (1) but the system allows the selective filter to be placed immediately after the first mixer stage. However, to reduce spurious responses the unwanted mixer products of the heterodyne-VFO must be reduced to a very low level and not reach the mixer. As with the tuneable IF system, this arrangement results in equal tuning rates on all bands. The system can be extended by replacing the series of separate crystals with a single crystal plus phase-locking arrangement as in (2).
4. *Fixed IF receiver with partial frequency synthesis.* An ingenious frequency synthesiser (due to Plessey) incorporating an interpolating LC oscillator and suitable for use with single- or multiple-conversion receivers having fixed intermediate frequencies is outlined in Fig 6.31. The output of the VFO is passed through a variable-ratio divider and then added

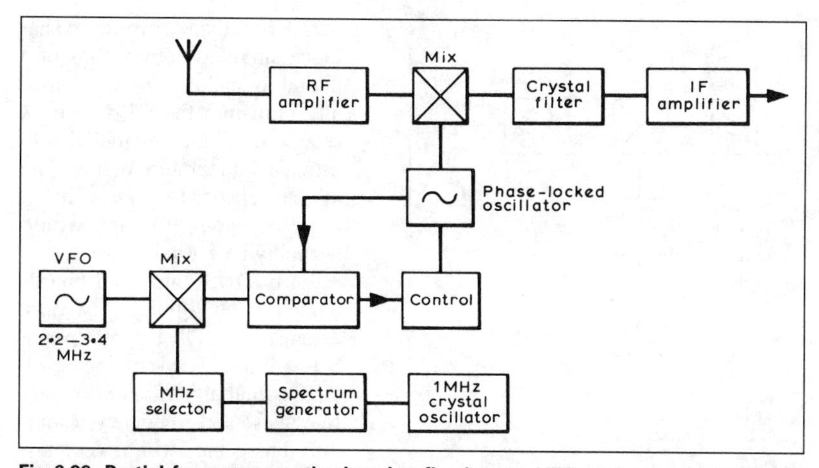

Fig 6.29. Partial frequency synthesis using fixed-range VFO with megahertz signals derived from a single 1MHz crystal

to the reference frequency by means of a mixer. The sum of the two frequencies applied to the mixer is selected by means of a band-pass filter and provides one input to a phase comparator; the other input to this phase comparator is obtained from the output of the voltage-controlled oscillator after it has also been divided in the same ratio as the interpolating frequency. The phase comparator can then be used to phase lock the VCO to the frequency $mF_{ref} + f_{VFO}$ where m represents the variable-ratio division. If for example f_{ref} is 1MHz, f_{VFO} covers a tuning range of 1–2MHz and the variable ratio dividers are set to 16, then the VCO output can be controlled over 17,000–18,000kHz; if the ratio divider is changed to 20 then the tuning range becomes 21,000–22,000kHz and so on. The use of two relatively simple variable ratio divider chains thus makes it possible to provide output over the full HF range, with the VFO at a low frequency (eg 1–2MHz).

5. *VXO local oscillator.* For reception over only small segments of a band or bands, a variable-crystal oscillator (VXO) can be used to provide high stability for mobile or portable receivers. As explained in Chapter 5, the frequency of a crystal can be 'pulled' over a small percentage of its nominal frequency without significant loss of stability. The system is attractive for small transceivers. Oscillators based on ceramic resonators can be 'pulled' over significantly greater frequency ranges.

6. *Drift-cancelling Wadley loop.* A stable form of front-end for use with variable IF-type receivers is the multiple-conversion Wadley loop which was pioneered in the Racal RA17 receiver (Fig 6.32). By means of an ingenious triple-mixing arrangement a variable oscillator tuning 40.5 to 69.5MHz and a 1MHz crystal oscillator provides continuous tuning over the range 0.5 to 30MHz as a series of 1MHz segments. Any drift of the variable VHF oscillator is automatically corrected. Although the system has been used successfully in home-constructed receivers, it is essential to use a good VHF band-pass filter (eg 40MHz ± 0.65MHz) if spurious responses are to be minimised.

7. *PLL frequency synthesisers.* In this system there is only one crystal reference oscillator which determines the stability of the system by 'locking' to it the output of a voltage-controlled oscillator (VCO) by means of a phase-locked-loop. The widespread use of relatively inexpensive PLL synthesisers as the MF/HF local oscillator of amateur-radio receivers and transceivers became possible when the basic PLL was integrated into a single IC device used with an external VCO. The term 'PLL' refers to a feedback loop which tracks small differences in phase between the input and feedback signals. A phase detector (comparator) measures the phase difference with its output filtered by a low-pass filter and applied to a VCO so as to change its frequency in the the direction that reduces the phase difference between the input signal and the local oscillator. The loop is in phase lock when the phase difference is zero. Fig 6.33 shows a PLL and Fig 6.34 shows the usual way of implementing a simple digital PLL frequency synthesiser by means of a programmable variable divider.

For the home-constructor, the building of a good PLL synthesiser is not so easy as might be suggested by the simplified description of the basic arrangement. It has been underlined that the VCO is critical to achieving good performance, especially in respect of phase noise, and hence reciprocal mixing in a receiver. Phase detector design is also critical and some PLL chips now offer two or more phase detectors to reduce noise both close in and away from the locked situation. Minimum step size often found in amateur equipment

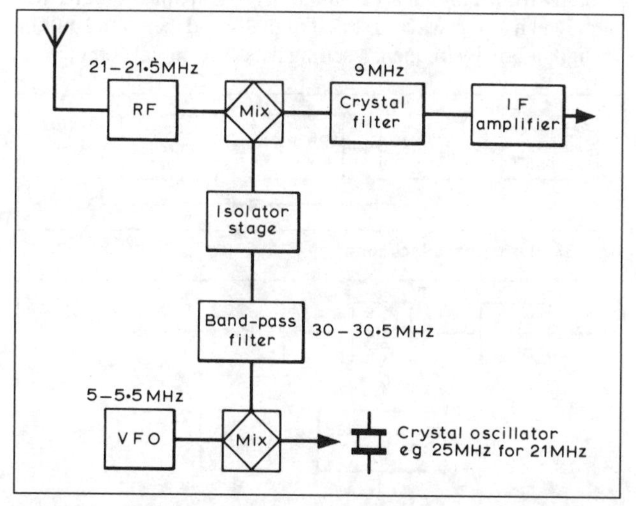

Fig 6.30. Pre-mixer heterodyne VFO system provides constant tuning rate with single-conversion receiver. Requires a number of crystals and care must be taken to reduce spurious oscillator products reaching the main mixer

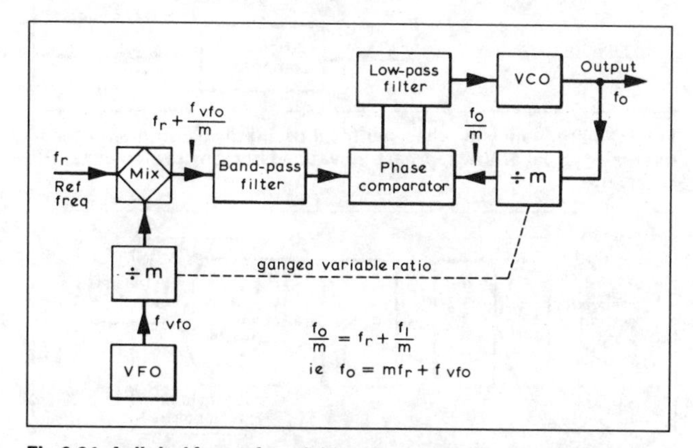

Fig 6.31. A digital form of partial synthesis developed by Plessey and suitable for use in single-conversion receivers

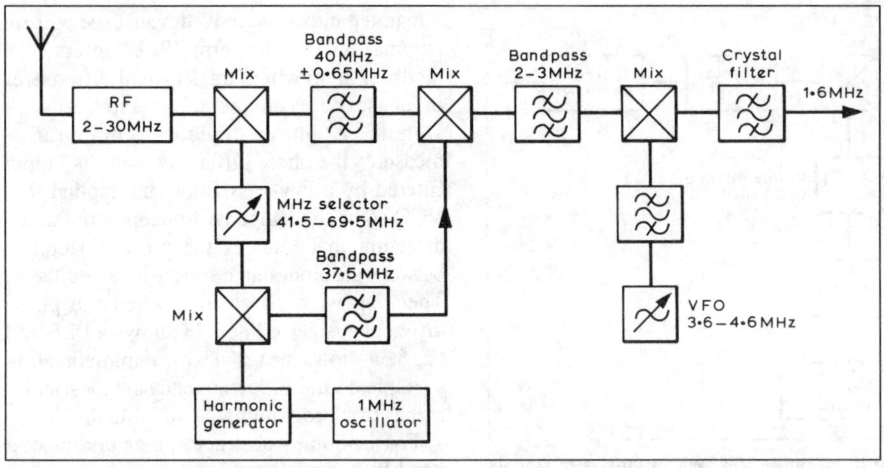

Fig 6.32. The Wadley drift-cancelling loop system as used on some early Racal HF receivers but requiring a considerable number of mixing processes and effective VHF band-pass filters

is 1kHz, often with a second tuning knob (eg 'clarifier') to provide fine tuning between the 1kHz steps. Alternatively by a digital analogue converter (DAC) operated by the last digit of the frequency set.

8. *Direct digital synthesis (DDS).* A later development has been the DDS in which the required frequency is generated directly from a digital accumulator (read-only memory, ROM) in which a sine-wave is stored digitally and then 'read out' in digital form by frequency-setting data. The digital output from the ROM is then converted to analogue sine waves by means of a digital-to-analogue converter usually built into the DDS chip. A second DAC may be included to provide a quadrature output. The system depends for its stability on the clock source which would normally be crystal controlled.

Early DDS chips and boards tended to be significantly more expensive than PLL synthesisers, but costs have been coming down. DDS potentially offers extremely fine phase and frequency resolution, broad bandwidth, very fast switching speed, and excellent phase-noise performance.

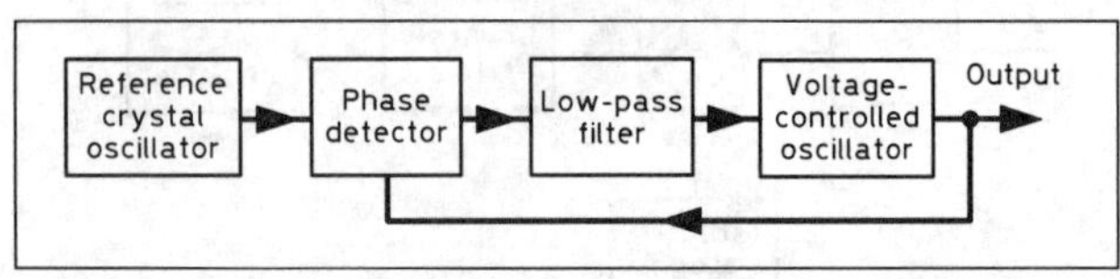

Fig 6.33. Basic phase-lock loop

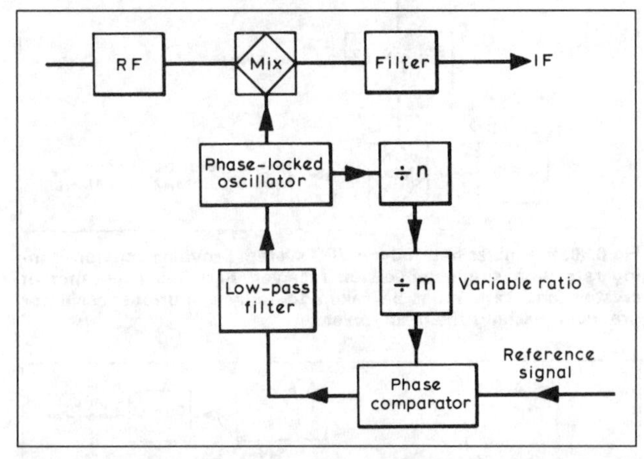

Fig 6.34. Frequency synthesised local oscillator using digital techniques – typical of the approach now used in many professional HF receivers

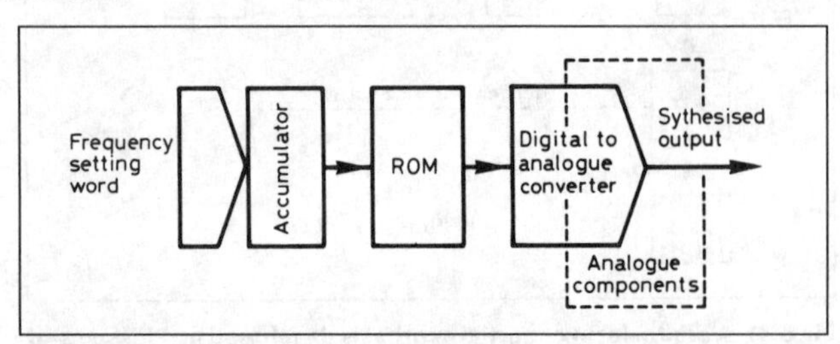

Fig 6.35. Direct digital frequency synthesiser

ACTIVE DEVICES FOR RECEIVERS

The amateur is today faced with a wide and sometimes puzzling choice of *active devices* around which to design a receiver: valves, bipolar transistors, field-effect transistors including single- and dual-gate MOSFETs and junction FETs; special diodes such as Schottky (hot-carrier) diodes, and an increasing number of *integrated circuits,* many designed specifically for receiver applications.

Each possesses advantages and disadvantages when applied to high-performance receivers, and most recent designs tend to draw freely from among these different devices.

Valves

In general the valve is bulky, requires additional wiring and power supplies for heaters, generates heat, and is subject to ageing in the form of a gradual change of characteristics throughout its useful life. On the other hand it is not easily damaged by high-voltage transients; was manufactured in a wide variety of types for specific purposes to fairly close tolerances; and is

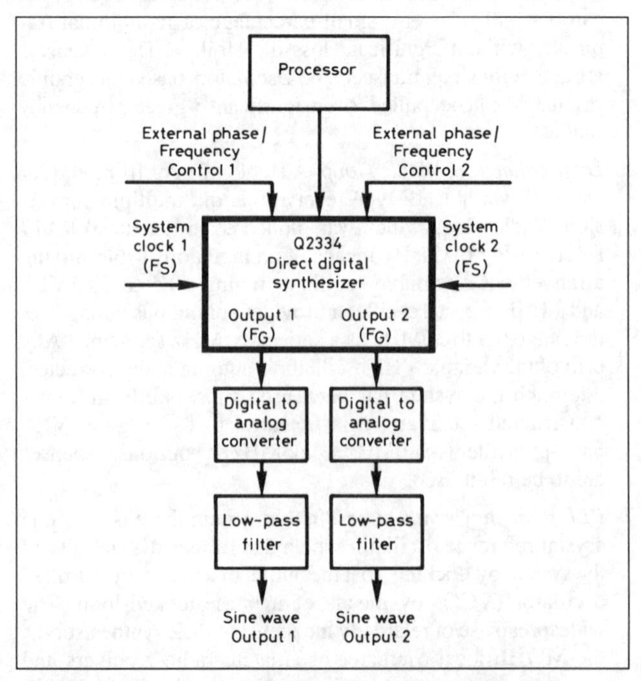

Fig 6.36. Block diagram of the Qualcomm Q2334 DDS system

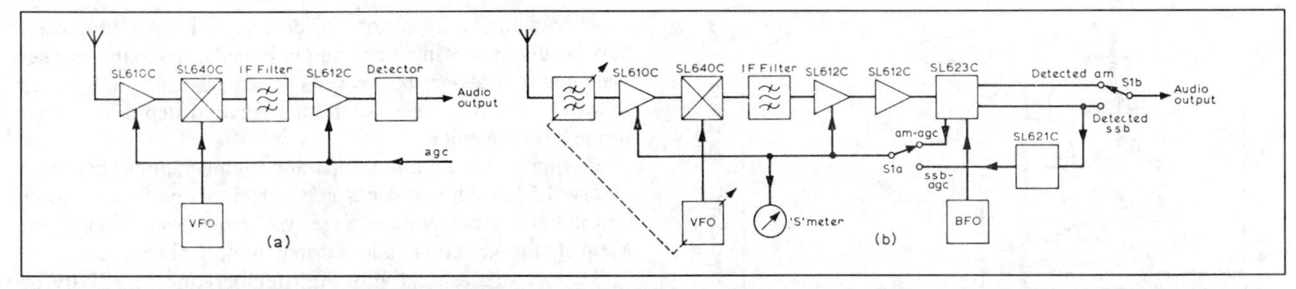

Fig 6.37. (a) Basic superhet receiver based on integrated circuits. (b) A more typical communications receiver for AM and SSB reception using the SL600 range of integrated circuits manufactured by Plessey Ltd

capable of handling small signals with good linearity (and special types can cope well with large signals). It is only fairly recently that semiconductor receivers have approached or even surpassed the dynamic range of a good valve receiver; nevertheless the future now lies with the semiconductor.

Bipolar transistors

These devices can provide very good noise performance with high gain, are simple to wire and need only low-voltage supplies, consuming very little power and so generating very little heat (except power types needed to form the audio output stage). On the other hand, they are low-impedance, current-operated devices, making the interstage matching more critical and tending to impose increased loading on the tuned circuits; they have feedback capacitances that may require neutralising; they are sensitive to heat, changing characteristics with changing temperature; they can be damaged by large input voltages or transients; and provide significant noise sidebands when used as an oscillator. Their main drawback in the signal path of a receiver is the difficulty of achieving wide dynamic range (except multi-emitter RF power types) and satisfactory AGC characteristics. On the other hand, bipolar transistors are suitable for most AF applications. They are not recommended for RF/mixer/oscillator/IF applications, unless care is taken to overcome their limitations, for example by using relatively high-power types. The bipolars developed for CATV (wire distribution of TV) such as the BFW17, BFW17A, 2N5109 etc, used with heatsink, can form excellent RF stages or mixers.

Field-effect transistors

These devices offer significant advantages over bipolar devices for the low-level signal path. Their high input impedance makes accurate matching less important; their near square-law characteristics make them comparable with variable-mu valves in reducing susceptibility to cross-modulation; and they can readily be controlled by AGC systems. The dual-gate form of device is particularly useful for small-signal applications, and forms an important device for modern receivers. They tend, however, to be limited in signal-handling capabilities. Special types of high-current field-effect transistors have been developed capable of providing extremely wide dynamic range (up to 140dB) in the front-ends of receivers. A problem with FET devices is the wide spread of characteristics between different devices bearing the same type number, and this may make individual adjustment of the bias levels of FET stages desirable. The good signal handling capabilities of MOSFET mixers can be lost by incorrect signal or local oscillator levels.

Integrated circuits

Special-purpose integrated circuits use large numbers of bipolar transistors in configurations designed to overcome many of the problems of circuits based on discrete devices. Because of the extremely high gain that can be achieved within a single IC they also offer the home-constructor simplification of design and construction. High-performance receivers can be designed around a few special-purpose linear ICs, or one or two consumer-type ICs may alternatively form the 'heart' of a useful communications receiver. It should be recognised, however, that their RF signal-handling capabilities are less than can be achieved with special-purpose discrete devices and their temperature sensitivity and heat generation (due to the large number of active devices in close proximity) usually make them unsuitable for oscillator applications. They also have a spread of characteristics which make it desirable to select devices from a batch for critical applications.

Integrated-circuit precautions

As with all semiconductor devices it is necessary to take precautions with integrated circuits, although if handled correctly high reliability may be expected.

Recommended precautions include:

1. Do not use excessive soldering heat and ensure that the tip is not at significant potential to earth (due to mains leakage).
2. Check and recheck all connections several times before applying any voltages.
3. Keep integrated circuits away from strong RF fields.
4. Keep supply voltages within ±10% of those specified for the device from well-smoothed supplies.

Integrated circuit amplifiers can provide very high gains (eg up to 80dB or so) within a single device having input and output leads separated by only a small distance: this means that careful layout is needed to avoid instability, and some devices may require the use of a shield between input and output circuits. Some devices have earth leads arranged so that a shield can be connected across the underside of the device.

Earth returns are important in high-gain devices: some (eg SL610) have input and output earth returns brought out separately in order to minimise unwanted coupling due to common earth return impedances, but this is not true of all devices.

Normally IC amplifiers are not intended to require neutralisation to achieve stability; unwanted oscillation can usually be traced to unsatisfactory layout or circuit arrangements. VHF parasitics may generally be eliminated by fitting a 10Ω resistor or ferrite bead in series with either the input or output lead, close to the IC.

With high-gain amplifiers, particular importance attaches to the decoupling of the voltage feeds. At the low voltages involved, values of series decoupling resistors must generally be kept low so that the inclusion of low-impedance bypass capacitors is usually essential. Since high-Q RF chokes may be a cause of RF oscillation it may be advisable to thread ferrite beads over one lead of any RF choke to reduce the Q.

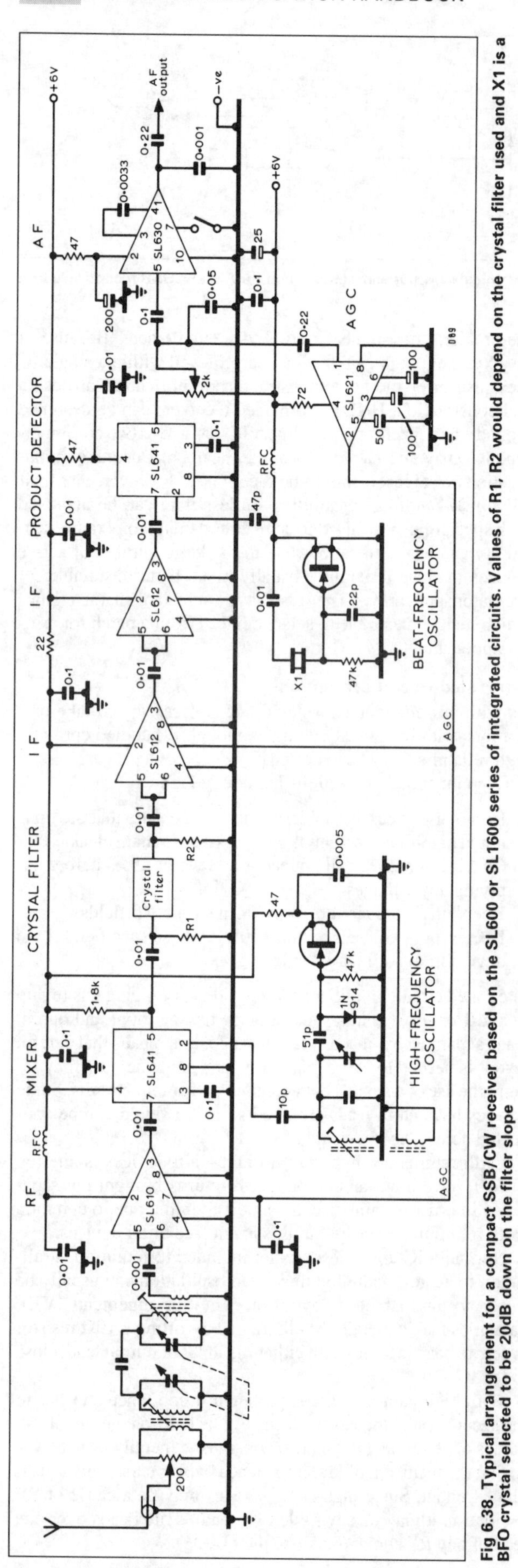

Fig 6.38. Typical arrangement for a compact SSB/CW receiver based on the SL600 or SL1600 series of integrated circuits. Values of R1, R2 would depend on the crystal filter used and X1 is a BFO crystal selected to be 20dB down on the filter slope

As with bipolar transistors, IC devices (if based on bipolars) have relatively low input and output impedances so that correct matching is necessary between stages. The use of a FET source follower stage may be a useful alternative to step-down transformers for matching.

Maximum and minimum operating temperatures should be observed. Many linear devices are available at significantly lower cost in limited temperature ranges which are usually more than adequate for operation under normal domestic conditions.

Because of the relatively high temperature sensitivity of bipolar-type integrated circuits, they are not suitable as free-running oscillators in high-performance receivers.

For the very highest grade receivers, discrete components and devices are still required in the front-end since currently available integrated circuits do not have comparable dynamic range. The IC makes possible extremely compact receivers; in practice miniaturisation is now limited – at least for general-purpose receivers – by the need to provide easy-to-use controls for the non-miniaturised operator.

SELECTIVE FILTERS

The selectivity characteristics of any receiver are determined by filters: these filters may be at signal frequency (as in a straight receiver); intermediate frequency; or audio frequency (as in a direct-conversion receiver). Filters at signal frequency or IF are usually of band-pass characteristics; those at AF may be either band-pass or low-pass. With a very high first IF (112–150MHz) low-pass filters may be used at RF.

A number of different types of filters are in common use: LC (inductor-capacitor) filters as in a conventional IF transformer or tuned circuit; crystal filters; mechanical filters; ceramic filters; RC (resistor-capacitor) active filters (usually only at AF but feasible also at IF).

The significance of 'nose' and 'skirt' bandwidths, and 'shape factor', was noted earlier. In many receivers, the overall shape factor is about 3 to 10. Very high-grade filters can have shape factors better than 1.1. A filter for SSB reception should have a nose bandwidth of about 2.2–2.5kHz, for CW about 300–500Hz.

LC filters

By using high-Q inductors it is possible to construct effective SSB filters only up to about 150kHz, but LC filters (IF transformers) are generally used only in support of other forms of filter. Fig 6.39 shows the conventional use of LC filters to form inter-stage coupling in valve receivers. The slope and shape of the characteristics are affected by the degree of coupling between the resonant circuits. The effective Q of filters of this type may be increased considerably by the technique of *Q-multiplication:* this depends on the high Q obtainable from a tuned circuit when it is operated near the point of oscillation. Some modern amateur receivers use Q-multiplication to sharpen up the selectivity for CW while using an SSB filter; this is a more economical approach than using separate SSB and CW crystal/mechanical filters.

Crystal filters

A quartz crystal is the equivalent of a series/parallel tuned circuit of extremely high Q. During the period 1935 to about 1960 single-crystal filters were commonly fitted in communications receivers, with a sharply peaked resonance usually between 455 and 470kHz, but occasionally about 1600kHz. Such filters can have a nose bandwidth of only a few hundred hertz, though the shape factor is not very good; variable selectivity can be achieved

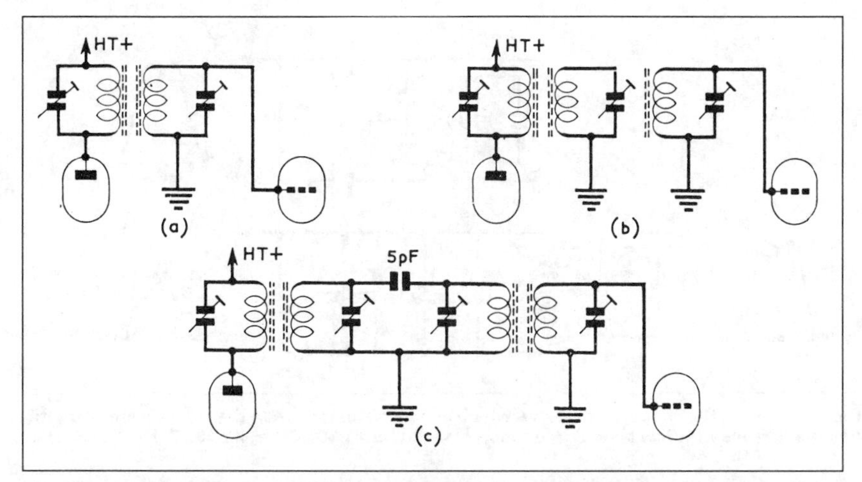

Fig 6.39. Typical inter-valve IF transformer couplings. (a) The conventional double-circuit IF transformer. (b) A triple-tuned IF transformer provides a useful improvement in skirt selectivity. (c) If a triple-tuned IF transformer is not available, two conventional IF transformers can be used to provide four tuned circuits per stage

by varying the impedance into which the filter operates, a low impedance providing extremely narrow bandwidths. Although this form of filter is still very well suited to CW reception, it is not ideal for SSB, although satisfactory results can be achieved by the use of a treble-rise AF network (*'stenode'* reception). Largely as a result of the increasing use of SSB, it is now more common to use multiple-crystal filters providing band-pass characteristics with centre frequencies up to 9MHz or 10.7MHz. Crystals may be grouped to form one filter or distributed over a number of stages.

Roofing filters

If the main selective filter is placed early in the signal path (eg immediately following a diode mixer or low-gain, low-noise, post-mixer amplifier) where the signal voltage is very low, subsequent amplification (of the order of 100dB or more) will introduce considerable broad-band noise, unless further narrow-band filtering is provided (which may be AF filtering). Another answer is to use an initial roofing filter with the final, more-selective filter(s) further down the signal path.

Crystal filters

The selectivity of a tuned circuit is governed by its frequency and by its Q (ratio of reactance to resistance). There are practical limits to the Q obtainable in coils and IF transformers. In 1929, Dr J Robinson, a British scientist, introduced the quartz crystal resonator into radio receivers. The advantages of such a device for communications receivers were appreciated by James Lamb of the American Radio Relay League and he made popular the IF *crystal filter* for amateur operators.

For this application a quartz crystal may be considered as a resonant circuit with a Q of from 10,000 to 100,000 compared with about 300 for a very-high-grade coil and capacitor tuned circuit. From Chapter 1, it will be noted that the electrical equivalent of a crystal is not a simple series- or parallel-tuned circuit, but a combination of the two: it has (a) a fixed series resonant frequency (f_s) and (b) a parallel resonant frequency (f_p). The frequency f_p is determined partly by the capacitance of the crystal holder and by any added parallel capacitance and can be varied over a small range.

The crystal offers low impedance to signals at its series resonant frequency; a very high impedance to signals at its parallel resonant frequency, and a moderately high impedance to signals on other frequencies, tending to decrease as the frequency increases due to the parallel capacitance.

While there are a number of ways in which this high-Q circuit can be incorporated into an IF stage, a common method providing a variable degree of selectivity is shown in Fig 6.40. When the series resonant frequency of the crystal coincides with the incoming IF signals, it forms a sharply tuned 'acceptor' circuit, passing the signals with only slight insertion loss (loss of strength) to the grid of the succeeding stage. The exact setting of the associated parallel resonant circuit, at which the crystal will offer an extremely high resistance, is governed by the setting of the *phasing control* which balances out the effect of the holder capacitance. Such a filter can provide a nose selectivity of the order of 1kHz bandwidth or less, while the sharp rejection notch which can be shifted by the phasing control through the pass-band can be of the order of 45dB. Fig 6.41 shows the improvement which can be obtained by switching in a filter of this type in a good communications receiver (a simple method of switching the filter is to arrange the phasing trimmer to short-circuit at

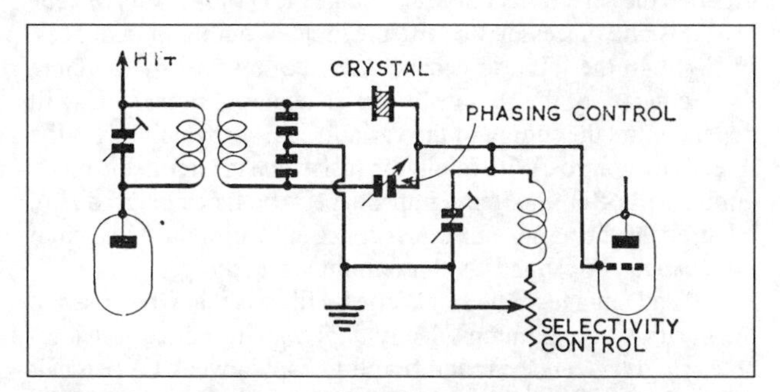

Fig 6.40. Variable-selectivity crystal filter. Selectivity is greatest when the impedance of the tuned circuit is reduced by bringing the variable resistor fully into circuit. For optimum results there must be adequate screening to prevent stray coupling between the input and output circuits which would permit strong off-resonance signals to leak round the filter. This type of filter in found mainly in the older valve receivers

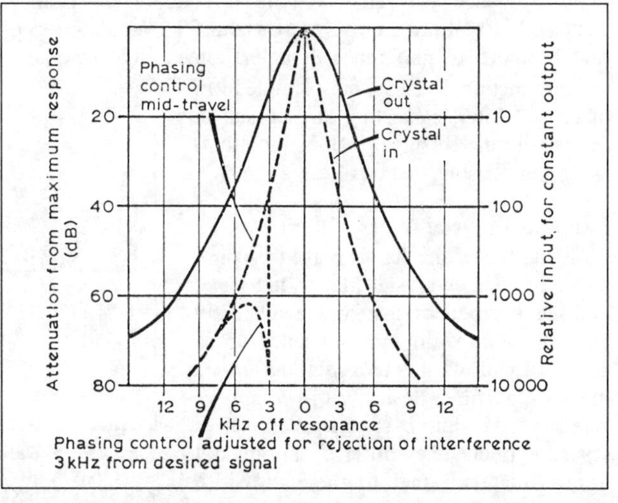

Fig 6.41. A graph showing the improvement in selectivity which can be obtained by the use of a crystal filter of the type shown in Fig 6.40

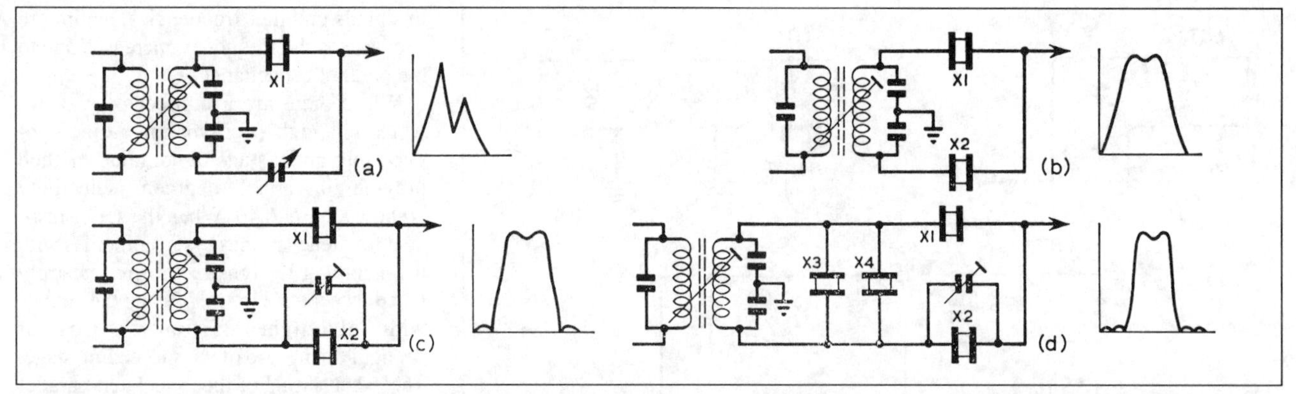

Fig 6.42. Half-lattice crystal filters showing the improvement in the shape of the curve which can be obtained when the crystals are correctly balanced or when extra crystals are used to reduce the 'humps'. Typical crystal frequencies would be X1 464.8kHz, X2 466.7kHz, X3 463kHz, X4 468.5kHz

one end of its travel). Inspection of the response curve will show that the improvement in the skirt selectivity is not so spectacular as at the nose, and even with a well-designed filter may leave something to be desired in the presence of strong signals.

The degree of selectivity provided by a single crystal depends not only upon the Q of the crystal and the IF but also upon the impedance of the input and output circuits. The lower these impedances are, the greater will be the effect of the filter, though this will usually be accompanied by a rise in insertion loss. To broaden the selectivity curve and make the Q of the filter *appear* less, it is only necessary to raise the input or output impedances. In Fig 6.40 the input impedance may be lowered by detuning the secondary of the IF transformer. The output impedance will depend upon the setting of the variable resistor which forms the selectivity control. With minimum resistance in circuit, the tuned circuit will offer maximum impedance, which can be gradually lowered by bringing more resistance into circuit. Maximum impedance corresponds with minimum selectivity.

A disadvantage of the single crystal filter is that when used in the position of maximum selectivity, it may introduce considerable 'ringing', rendering it difficult to copy a weak CW signal: this is due to the tendency of a high-Q circuit to oscillate for a short period after being stimulated by a signal, producing a bell-like echo on the signal.

Since the minimum nose bandwidth of a single crystal filter may be as low as 100–200Hz it is not possible to receive AM signals through the filter unless its efficiency is degraded by a high-impedance or by 'stenode' tone correction. If this is done, such a filter will often prove most useful for telephony reception through bad interference, though at some cost to the quality of reproduction. Even the most simple form of crystal filter, consisting of a crystal in series with the IF signal path, without balancing or phasing, can be of use.

Band-pass crystal filters

As already noted, the sharply peaked response curve of a single-crystal filter is not ideal, and has by modern standards a relatively poor shape factor: improved results can be achieved with what is termed a *half-lattice* or band-pass filter. Basically, this comprises two crystals chosen so that their series resonant frequencies differ by an amount approximately equal to the bandwidth required; for example about 300Hz apart for CW, 3–4kHz apart for AM telephony or 2kHz apart for SSB, see Fig 6.42(b). This form of filter, developed in the 'thirties, has in recent years come into widespread amateur use. Although it has a much improved slope over the single crystal filter, in its simplest form there will still be certain frequencies, just outside the main pass-band, at which the attenuation is reduced. Unless a balancing trimmer is connected across the higher-frequency crystal, the sides of the response curve tend to broaden out towards the bottom. As the capacitance across the crystal is increased, the sides of the curve steepen but the side lobes tend to become more pronounced. Capacitance across the lower frequency crystal broadens the response and deepens the trough in the centre of the pass-band.

Examples of band-pass filter response characteristics are shown in Fig 6.42.

To eliminate the 'humps' in the response curve, additional crystals may be included in the filter; these may be up to about six in number, Fig 6.43.

Alternatively, additional filter sections may be incorporated in the IF section in cascade. An advantage of using several cascaded filters is that less critical balancing and adjustment are needed. Provided that there is no leakage of IF signals around the filter due to stray capacitances or other forms of unwanted coupling (an important consideration with all selective filters) extremely good shape factors of the order of 1.5 can be achieved with about three cascaded filters on about 460kHz. There will be an insertion loss of the order of 6dB per section.

To provide a sharp variable rejection notch, it is advantageous to incorporate some form of bridged-T filter or Q-multiplier in order to be able to eliminate steady heterodyne interference.

To reduce susceptibility to cross-modulation or blocking,

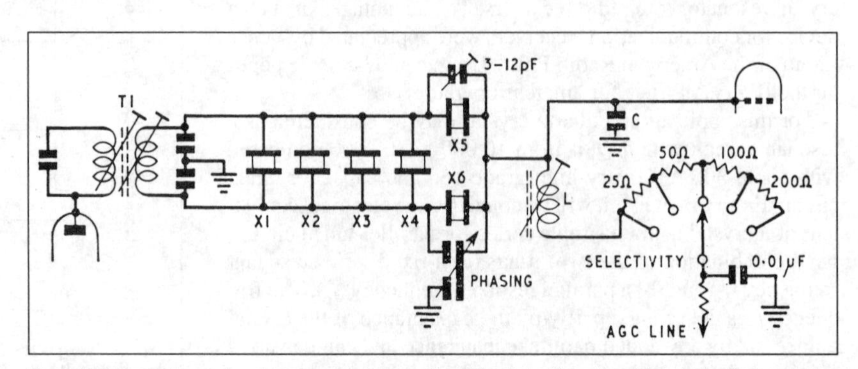

Fig 6.43. Up to about six crystals may be used in a single half-lattice filter. Here is a variable selectivity unit for 465kHz using FT241 crystals. X1 461.1kHz (49); X2 462.9kHz (50), X3 468.5kHz (53); X4 470.4kHz (54); X5 464.5kHz (51); X6 466.7kHz (52). Numbers in brackets refer to channel numbers for FT241 crystals. T1 and L/C should be tuned to mid-filter frequency

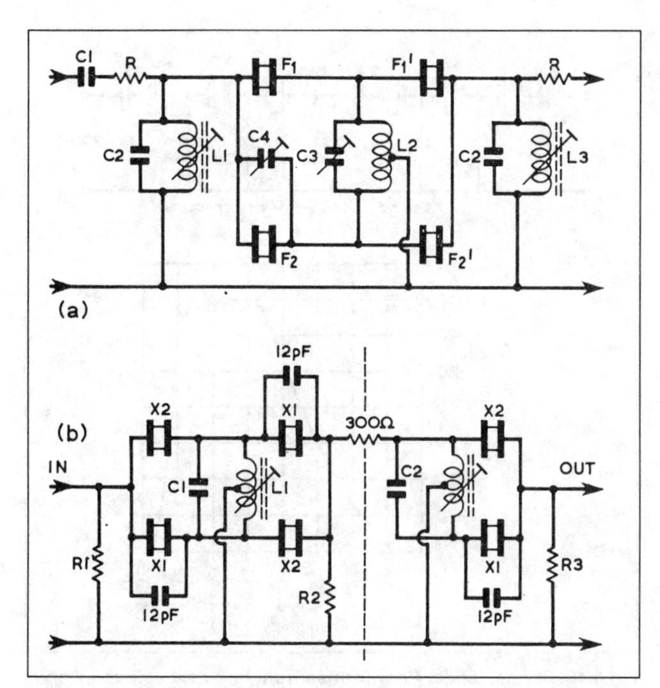

Fig 6.44. Typical HF crystal filters using FT243 crystals. (a) Filter using four crystals (F1 = F1'; F2 = F2' = F1 + 1500 to 2500Hz); C1, 2000pF; C2, 47pF; C3, 3–30pF; C4 3–10pF. L1 and L3 to resonate at filter frequency with C2. L2 to resonate with about 15pF setting of C3. R should be 2000Ω. (b) Filter using six crystals and capable of nose pass-bands of 2.4–3kHz and –60dB skirt bandwidth of 6–7kHz. The sets of X1 and X2 crystals should be separated by 1.7kHz in series-resonant mode. R1, R2 and R3 are 560 to 820Ω

crystal and other selective filters should be placed as near as possible to the front end of the receiver.

Where the first fixed IF is too high for a crystal filter to be completely effective, it is nevertheless advantageous to include a roofing filter at this point, with the more selective filter later in the receiver (but still at the earliest practicable stage).

High-frequency crystal filters

At high frequencies it becomes increasingly more difficult to obtain entirely satisfactory results with home-construction, although excellent factory-built SSB band-pass filters are available (at appreciable cost) at 9 and 10.7MHz. Filters for use as roofing filters are similarly available at VHF.

Effective HF crystal filters using FT243 crystals between 5.5 and 6.5MHz have been built by a number of amateurs; typical designs are shown in Fig 6.44.

It should be noted that crystal filters are not always linear and passive and can give rise to intermodulation products when subject to high-level signals; IMD performance can sometimes be improved by interchanging the input and output connections. The IMD performance of crystal filters may impose a limit to the spurious-free dynamic range of a high-performance receiver.

Typically good SSB band-pass crystal filters use four to six crystals. A more recent form of HF and VHF filter is the compact *monolithic crystal filter* (MXF). Such filters consist of a quartz wafer on which pairs of metal electrodes are deposited on opposite sides of the plate (Fig 6.45). Complete filters may occupy only a TO-5 transistor capsule, and can be designed for resonant frequencies up to the UHF region.

Ladder crystal filters

For home-construction a particularly useful alternative to the half-lattice arrangement is the *ladder filter,* which can provide excellent SSB filters at frequencies between about 4–11MHz.

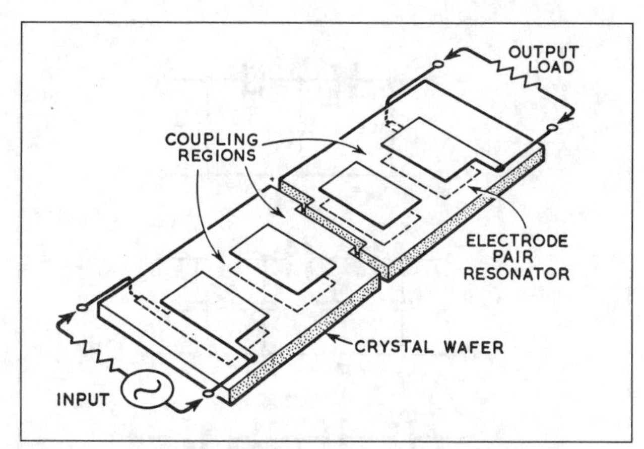

Fig 6.45. Monolithic form of HF crystal band-pass filter (MXF)

This form of crystal filter uses a number of crystals of the same (or nearly the same) frequency and so avoids the need for accurate crystal etching or selection. Further, provided it is correctly terminated, it does not require the use of transformers or inductors. Plated crystals such as the HC6U or 10XJ types are more likely to form good SSB filters, although virtually any type of crystal may be used for CW filters.

A number of practical design approaches have been described by J Pochet, F6BQP, ('Technical Topics', *Radio Communication* September 1976 and *Wireless World* July 1977) and in a series of articles 'Some experiments with HF ladder crystal filters' by J A Hardcastle, G3JIR *(Radio Communication* December 1976, January, February and September 1977).

Fig 6.46 outlines the F6BQP approach. By designing for lower termination impedances and/or lower frequency crystals excellent CW filters can be formed. A feature of the ladder design is the very high ultimate out-of-band rejection that can be achieved (75–95dB) in three- or four-section filters. For SSB filters at about 8MHz a suitable design impedance would be about 800Ω with a typical 'nose' band-pass of 2.0–2.1kHz.

Intercept points of SSB and roofing filters can be over +50dBm with low insertion losses, and over +45dBm for narrow-band CW filters.

The ladder configuration is particularly attractive for home-built receivers since they can be based on readily available, low-cost 4.43MHz PAL colour-subcarrier crystals produced for use in domestic colour-TV receivers to P129 or P128 specifications. In NTSC countries, including North America and Japan, 3.58MHz TV crystals can be used, although this places the IF within an amateur band and would be unsuitable for single-conversion superhet receivers. Low-cost crystals at twice the PAL sub-carrier frequency (ie 8.86MHz) are also suitable.

Virtually any combination of crystals and capacitors produces a filter of some sort, but published equations or guidance should be followed to achieve optimum filter shapes and desired bandwidth. If this is done, ladder filters can readily be assembled from a handful of nominally identical crystals (ideally selected with some small offsets of up to about 50 or 100Hz) plus a few capacitors, yet providing SSB or CW filters with good ultimate rejection, plus reasonably low insertion loss and pass-band ripple. Ladder filters have intercept points significantly above those of most economy-grade, lattice-type filters.

Figs 6.48 and 6.49 show typical ladder filter designs based on 4.43MHz crystals.

A valuable feature of the ladder configuration is that it lends itself to variable selectivity by changing the value of some, or preferably all, of the capacitors. The bandwidth can be varied

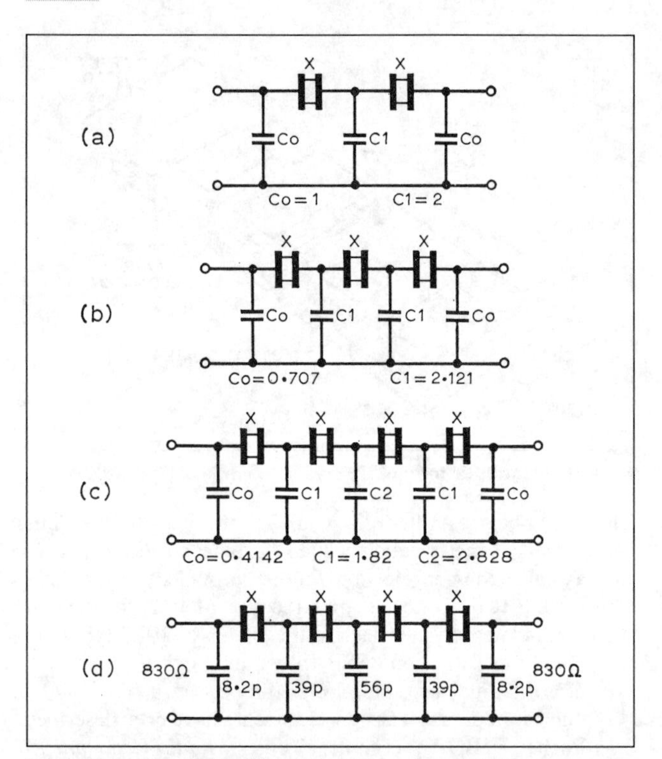

Fig 6.46. Crystal ladder filters, investigated by F6BQP, can provide effective SSB and CW band-pass filters. All crystals (X) are of the same resonant frequency and preferably between 8 and 10MHz for SSB units. To calculate values for the capacitors multiply the coefficients given above by $1/(2\pi fR)$ where f is frequency of crystal in hertz (MHz by 10^6), R is input and output termination impedance and 2π is roughly 6.8. (a) Two-crystal unit with relatively poor shape factor. (b) Three-crystal filter can give good results. (c) Four-crystal unit capable of excellent results. (d) Practical realisation of four-crystal unit using 8314kHz crystals, 10% preferred-value capacitors and termination impedance of 820Ω. Note that for crystals between 8 and 10MHz the termination impedance should be between about 800 and 1000Ω for SSB. At lower crystal frequencies use higher design impedances to obtain sufficient bandwidth. For CW filters use lower impedance and/or lower frequency crystals

over a restricted but useful range simply by making the middle capacitor variable, using a mechanically variable capacitor or electronic tuning diode (which may take the form of a 1W zener diode). Fig 6.50 shows an 8MHz filter which has a 'nose' bandwidth that can be varied from about 2.8kHz down to 1.1kHz.

Fig 6.51 illustrates a 4.43MHz nine-crystal filter built by R Howgego, G4DTC, for AM/SSB/CW/RTTY reception; the 3dB points can be varied from 4.35kHz down to 600Hz. In development, he noted that the bandwidth is determined entirely by the 'vertical' capacitors. If, however, these are reduced below about 10pF, the bandwidth begins to narrow rather than widen. The maximum bandwidth that could be achieved was about 4.5kHz. This could be widened by placing resistors (1kΩ to 10kΩ) across the capacitors but this increases insertion loss. Terminating impedances affect the pass-band ripple, not the bandwidth.

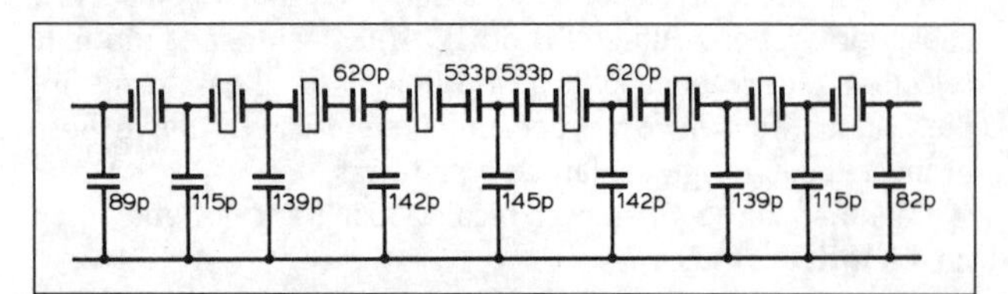

Fig 6.47. Circuit of eight-pole SSB filter. Nominal frequency 9511kHz. Input and output impedance 240Ω. –6dB bandwidth 2470Hz, –60dB bandwidth 4440Hz

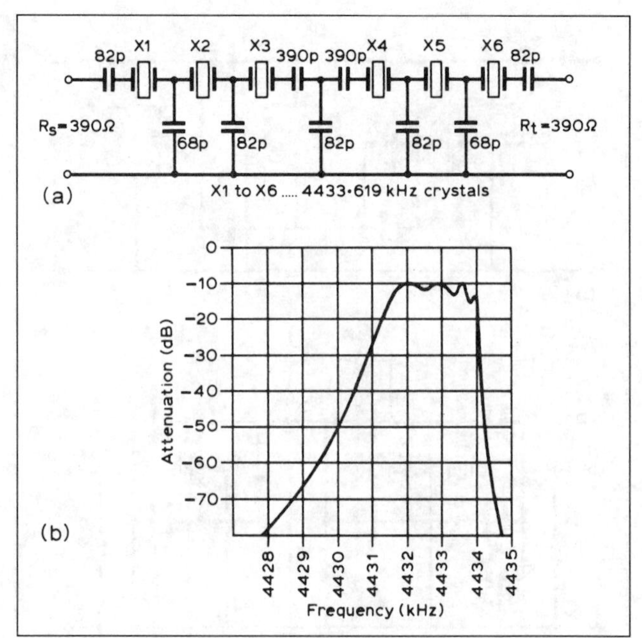

Fig 6.48. Crystal ladder filter formed from low-cost colour-TV crystals, providing effective filter for a lower-sideband generator (crystals for NTSC 525-line TV are 3579.545MHz.)

The filter of Fig 6.51 gives continuously variable selectivity yet is relatively easy to construct. It is basically a six-pole roofing filter followed by a variable three-pole filter. It was based on low-cost Philips HC18-U type crystals and these were found to be all within a range of 80Hz. Crystals in the large case style (eg HC6-U) tend to be about 200Hz lower. C1, C2 is a 60 + 142pF miniature tuning capacitor as found in many portable broadcast receivers. The integral trimmers are set for maximum bandwidth when capacitor plates are fully unmeshed. Set R1 for best compromise between minimum bandwidth and insertion loss, 1.2kΩ nominal.

The following specification should be achievable: C1, C2 plates unmeshed, 3dB points at 4437.25kHz and 4432.90kHz, bandwidth 4.35kHz. C1, C2 plates half-meshed, 3dB points at 4434.0kHz and 4432.90kHz, bandwidth 1.10kHz. C1, C2 plates meshed 3dB points at 4433.5kHz and 4432.90kHz, bandwidth 600Hz. Insertion loss in pass-band: maximum (R1 2.5kΩ, R2 1.2kΩ) 10dB, minimum (R1 0kΩ, R2 1.2kΩ) 6dB. Pass-band ripple 1–3dB (dependent on R1). Ripple reduces with bandwidth. Stop-band attenuation better than 60dB. –20dB bandwidth typically 1kHz wider than the –3dB bandwidth.

The filter shown in Fig 6.52, designed by D Gordon-Smith, G3UUR, provides six different bandwidths suitable for both SSB and CW operation, switching the value of all capacitors, and ideally preceded by a roofing filter. In order to reduce the number of switched components, the terminating resistors remain constant and the ripple merely decreases with bandwidth. This also reduces the variation in insertion loss. The 2.4kHz position has a 1dB ripple Chebyshev response, and the 500Hz position represents a Butterworth response. A 5:1 bandwidth change is possible if the maximum tolerable ripple is 1dB. This range is fixed by design constraints: the ratio of 1dB to 0dB ripple response terminating resistance is approximately 5:1 for the same bandwidth, and therefore the same terminating resistance satisfies bandwidths that have a ratio of about 5:1. The main disadvantage is that the pass-band moves low in frequency as it is narrowed. This could be compensated for by moving the carrier crystals down in sympathy with the filter centre frequency; a

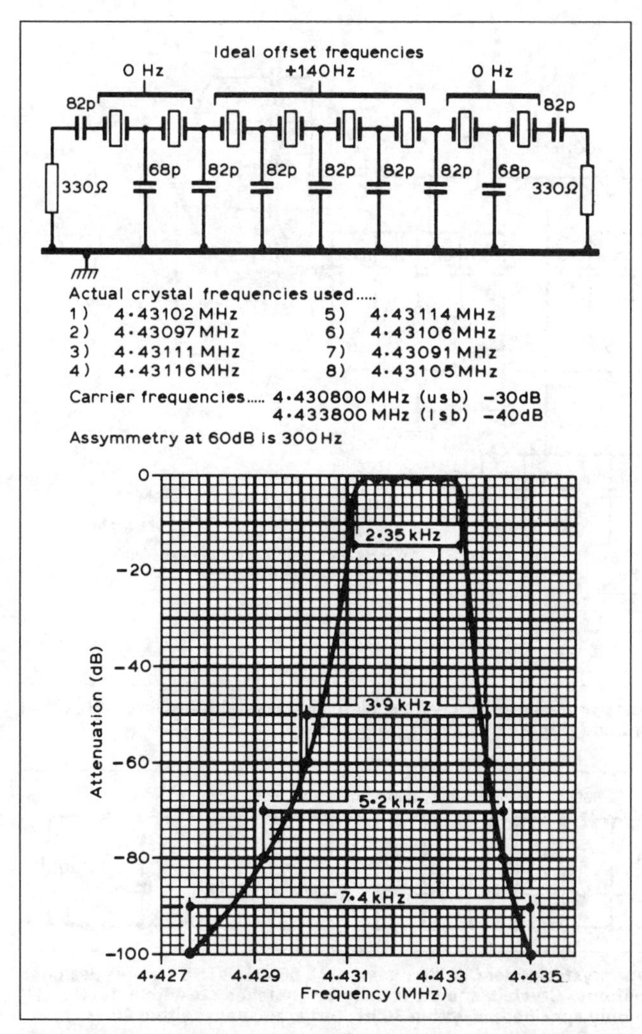

Fig 6.49. Ladder crystal filter using P-129 specification colour-TV 4.43MHz crystals. It provides a performance comparable with the ladder filter in the Atlas 180 and 215 transceivers. Insertion loss 4–5dB, shape factor (6/60dB) 1.66. Note that the rate of attenuation on the low-frequency side of the response is as good as an eight-crystal lattice design. On the HF side it is better

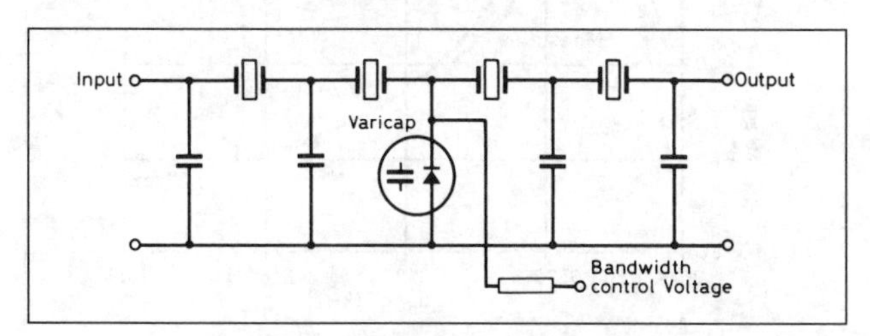

Fig 6.50. Simple technique for varying the bandwidth of an 8MHz crystal ladder filter from about 2.8kHz down to 1.1kHz

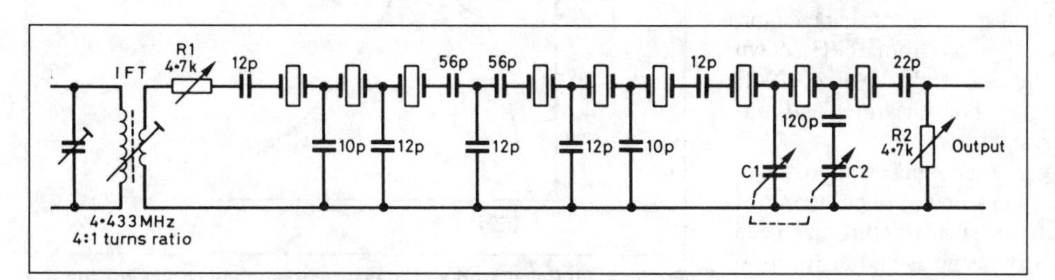

Fig 6.51. Variable-selectivity ladder filter using low-cost PAL colour-TV crystals for AM/SSB/CW/RTTY reception

total shift of 1kHz or less would probably be adequate, and could be corrected at the VFO by the RIT shift control.

A different approach to ladder filters is to put the crystals in shunt with the signal rather than in series. The filter then has its steeper slope on the low-frequency side instead of the high-frequency side, and this may be preferable for narrow-bandwidth filters (eg CW filters), particularly when using relatively low-Q plated crystals in HC-18 holders. Fig 6.53 shows a shunt-type crystal filter designed by John Pivnichy, N2DCH.

Ceramic filters

Piezoelectric effects are not confined to quartz crystals; in recent years increasing use has been made of certain ceramics, such as lead zirconate titanate (PZT). Small discs of PZT, which resonate in the radial dimension, can form economical selective filters in much the same way as quartz, though with considerably lower Q. Ceramic IF *transfilters* are a convenient means of providing the low impedances needed for bipolar transistor circuits. The simplest ceramic filters use just one resonator, but numbers of resonators can be coupled together to form filters of required bandwidth and shape factor. While quite good nose selectivity is achieved with simple ceramic filters, multiple resonators are required if good shape factors are to be achieved. Some filters are of 'hybrid' form using combinations of inductors and ceramic resonators.

Examples of ceramic filters include the Philips LP1175 in which a hybrid unit provides the degree of selectivity associated with much larger conventional IF transformers; a somewhat similar arrangement is used in the smaller Toko filters such as the CFT455C which has a bandwidth (to –6dB) of 6kHz. A more complex 15-element filter is the Murata CFS-455A with a bandwidth of 3kHz at –6dB, 7.5kHz at –70dB and insertion loss 9dB, with input and output impedances of 2kΩ and centre frequency of 455kHz. In general ceramic filters are available from 50kHz to about 10.7MHz centre frequencies.

Ceramic filters tend to be more economical than crystal or mechanical filters but have lower temperature stability and may have greater pass-band attenuation.

Mechanical and miscellaneous filters

Very effective SSB and CW filters at intermediate frequencies from about 60 to 600kHz depended on the mechanical resonances of a series of small elements usually in the form of discs: Fig 6.56. The mechanical filter consisted of three basic elements: two magneto-striction transducers which convert the IF signals into mechanical vibrations and vice versa; a series of metal discs mechanically resonated to the required frequency; disc coupling rods. Each disc represents a high-Q series resonant circuit and the bandwidth of the filter is determined by the coupling rods. 6–60dB shape factors can be as low as about 1.2, with low pass-band attenuation. The limitation of mechanical filters to frequencies of about 500kHz or below has led to their virtual disappearance from the amateur markets and they are now found only in older models despite their excellent performance below 500kHz.

Other forms of mechanical filters have been developed which include ceramic piezoelectric transducers with mechanical coupling: they thus represent a combination of ceramic and mechanical techniques.

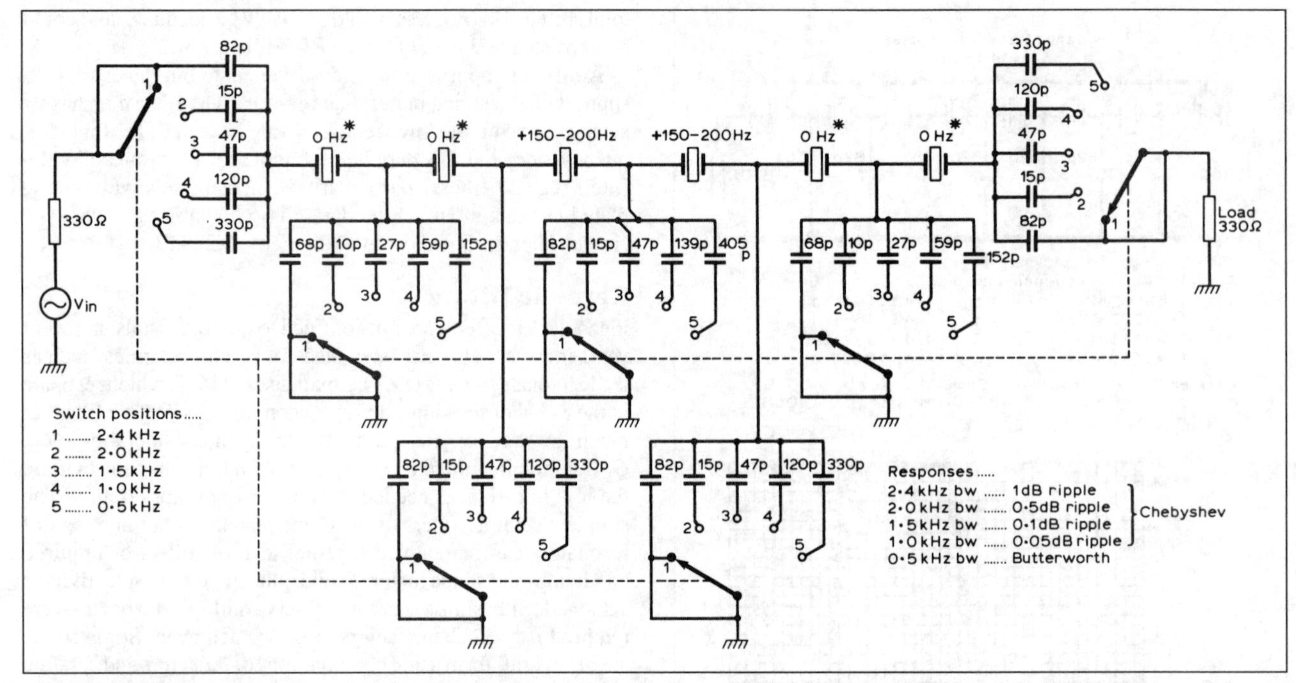

Fig 6.52. G3UUR's design for a switched variable-bandwidth ladder filter using colour-TV crystals. Note that crystals shown as 0Hz offset can be in practice ±50Hz without too detrimental an effect on the pass-band ripple

These filters may consist of an H-shaped form of construction; such filters include a range manufactured by the Toko company of Japan. Generally the performance of such filters is below that of the disc resonator type, but can still be useful.

Surface acoustic wave filters are available for possible band-pass filter applications where discrete-element filters have previously been used, including IF filters.

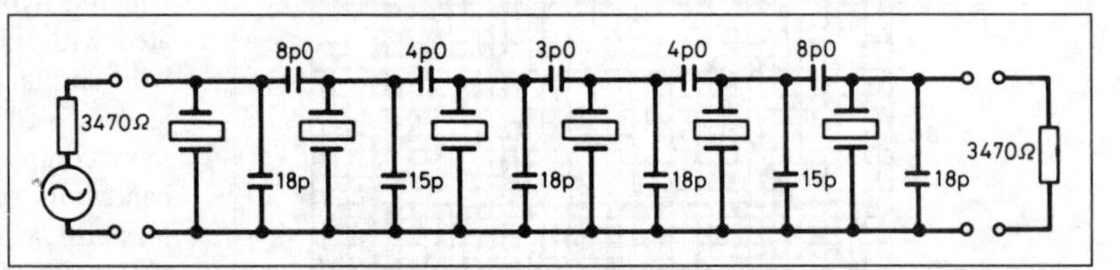

Fig 6.53. Shunt-type crystal ladder CW filter using six 3.58MHz NTSC crystals designed for 3470Ω terminations. Crystals should usually be matched to within 100Hz. (TV crystals are often only specified as within 300Hz but are usually within 200Hz.)

Variable IF pass-bands

The use of two mixers both controlled from the same variable oscillator can provide *pass-band tuning* (*IF shift*) and *variable bandwidth tuning* (VBT), techniques outlined in Fig 6.57. IF shift positions the IF pass-band of the signal passing through the SSB filter but does not change the overall selectivity. With VBT two filters are used, with the second filter variably aligned with the first filter and thus providing variable bandwidth tuning. A more ambitious version of VBT requiring four mixers and two ganged oscillators tuning in opposite directions has been used by Rohde & Schwarz in a professional receiver developed in the late 'sixties. In this form, two high-grade low-pass filters at 30kHz, using inductors rather than crystals, were arranged so that they provided a band-pass filter acting on both upper and lower sidebands: Fig 6.58.

As shown in Fig 6.59 this filter had an excellent shape factor, giving a bandwidth continuously adjustable from ±6kHz down to ±150Hz with substantially similar slope right down to −70dB at all settings, without introducing the non-linearities and limited dynamic range inherent in crystal filters.

A more recent, and potentially more economical, approach to variable selectivity is now possible with a single pre-programmed digital signal processing (DSP) filter; such filtering has been implemented at audio (baseband) frequencies and at IF up to about 455kHz. The limiting factor tends to be the dynamic range and resolution of the analogue-to-digital converter. Such filters

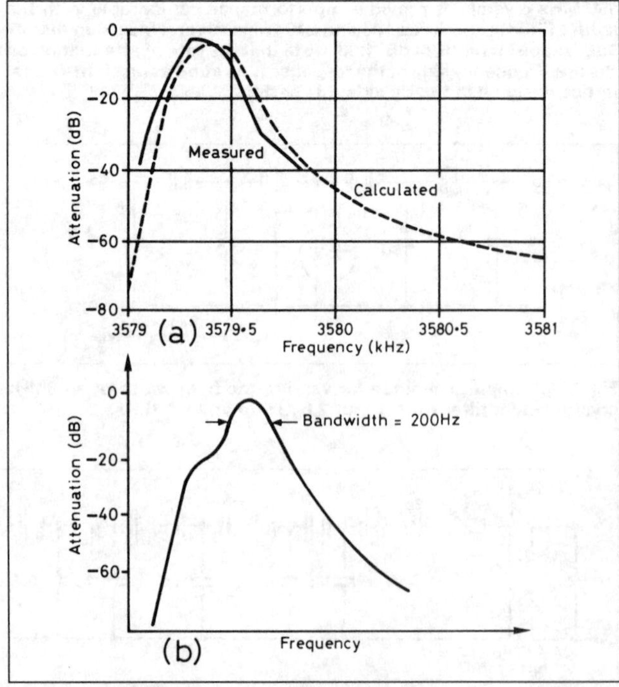

Fig 6.54. (a) Calculated and measured response curves for the CW filter. (b) Filter response resulting from use of poorly matched crystals

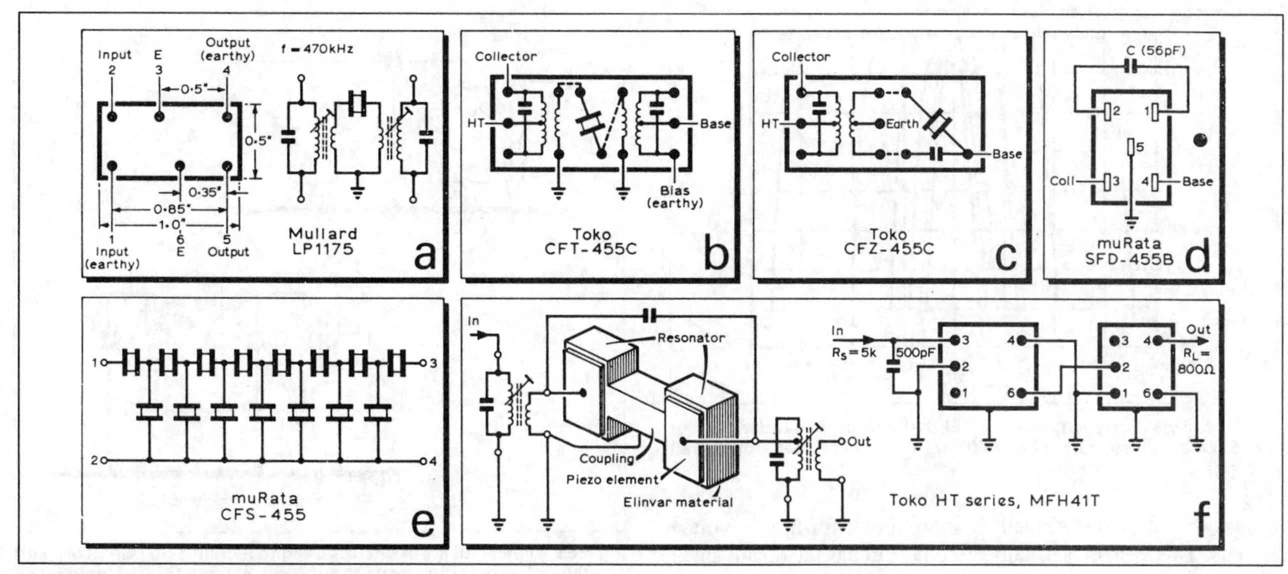

Fig 6.55. Representative types of ceramic filters

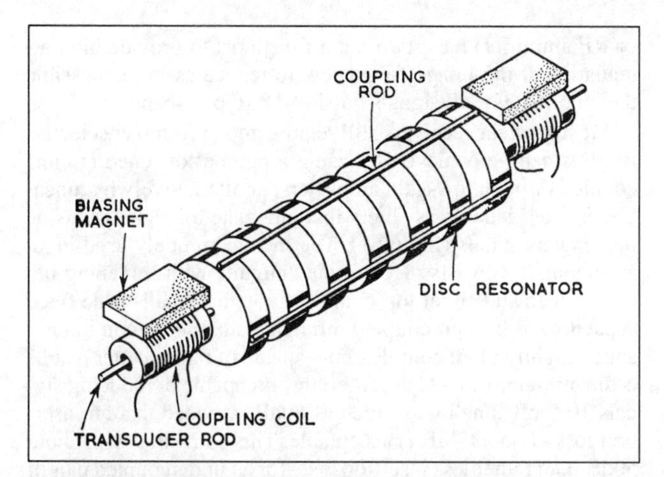

Fig 6.56. The Collins mechanical filter. IF signals are converted into mechanical vibrations by a magneto-strictive transducer and passed along a series of resonant discs, then finally reconverted into electrical (IF) signals by a second magneto-strictive transducer. The bandwidth of the filter is governed by the number of resonant discs and the design of the coupling rods. Very good shape factors can be achieved but the maximum frequency of such filters is usually about 500kHz

can also provide multiple notches in the pass-band that automatically suppress unwanted carriers.

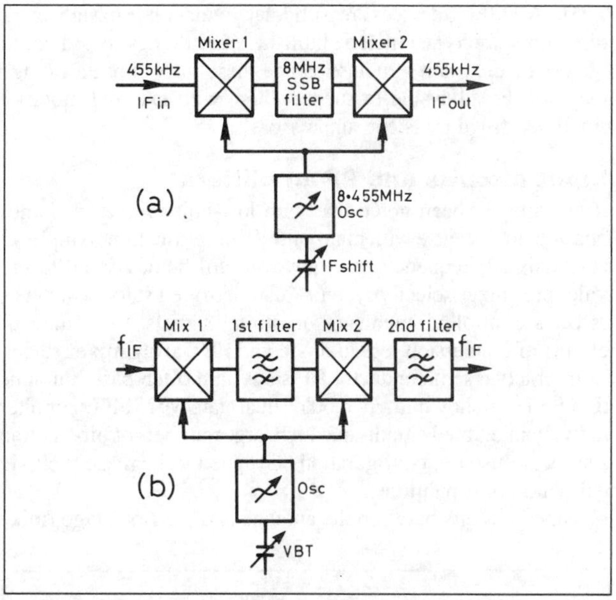

Fig 6.57. (a) IF shift positions the IF pass-band but does not change the overall selectivity. (b) The use of a second filter variably aligned with the first filter provides variable-bandwidth tuning

CIRCUITRY

Receiver protection

Receivers, particularly where they are to be used alongside a medium- or high-powered transmitter, need to be protected from high transient or other voltages induced by the local transmitter or by build-up of static voltages on the antenna. Valve receivers may suffer burn-out of antenna input coils; semiconductors used in the first stage of a receiver are particularly vulnerable and invariably require protection. The simplest form of protection is the use of two diodes in back-to back configuration. Such a combination passes signals less than the potential hill of the diodes (about 0.3V for germanium diodes, about 0.6V for silicon

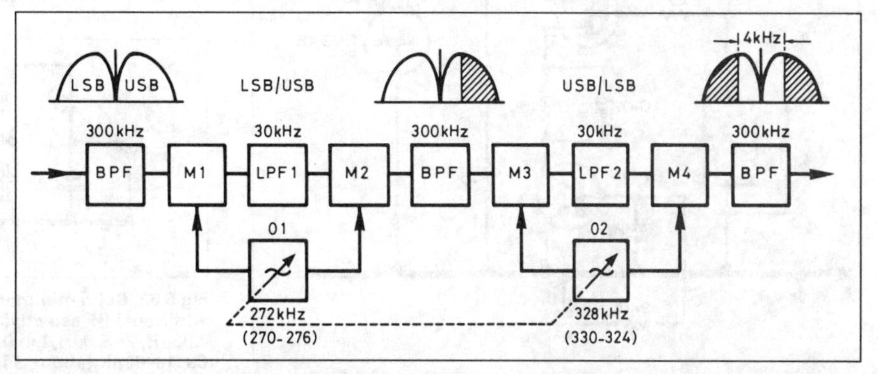

Fig 6.58. The basic principles of the 1969 Rohde and Schwarz EKO7-80 filter, based on two low-pass filters using inductors and not crystals to provide continuously variable bandwidth

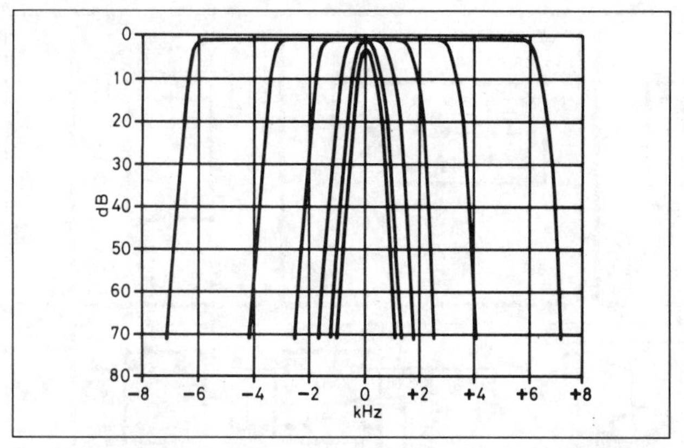

Fig 6.59. Selectivity curves of the EK07-80 filter at bandwidths of ±0.15, ±0.30, ±0.75, ±1.5, ±3.0, ±6.0kHz. Note the similar slope at all settings down to –70dB

diodes) but provides virtually a short-circuit for higher-voltage signals. This system is usually effective but has the disadvantage that it introduces non-linear devices into the signal path and may occasionally be the cause of cross- and inter-modulation.

The MOSFET devices are particularly vulnerable to static puncture and some types include built-in zener diodes to protect the 'gates' of the main structure. Since these have limited rating it may still be advisable to support them with external diodes or small gas-filled transient suppressors.

Input circuits and RF amplifiers

It has already been noted that with low-noise mixers it is now possible to dispense with high-gain RF amplification. Amplifiers at the signal frequency may however still be advisable to provide: pre-mixer selectivity; an AGC-controlled stage which is in effect a controlled attenuator on strong signals; to counter the effects of conversion loss in diode and FET-array mixer stages.

In practice semiconductor RF stages are often based on junction FETs as shown in Fig 6.60 or dual-gate MOSFETs, or alternatively integrated circuits in which large numbers of bipolar transistors are used in configurations designed to increase their signal-handling capabilities.

Tuned circuits between the antenna and the first stage (mixer

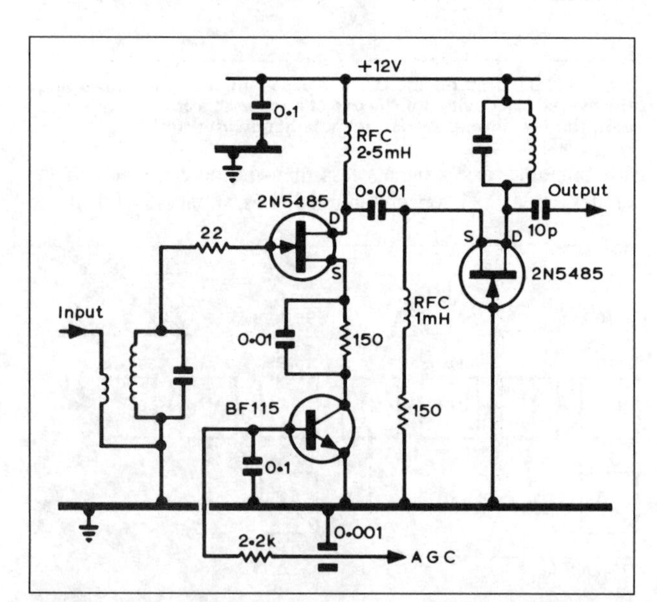

Fig 6.60. Cascode RF amplifier using two JFETs with transistor as AGC

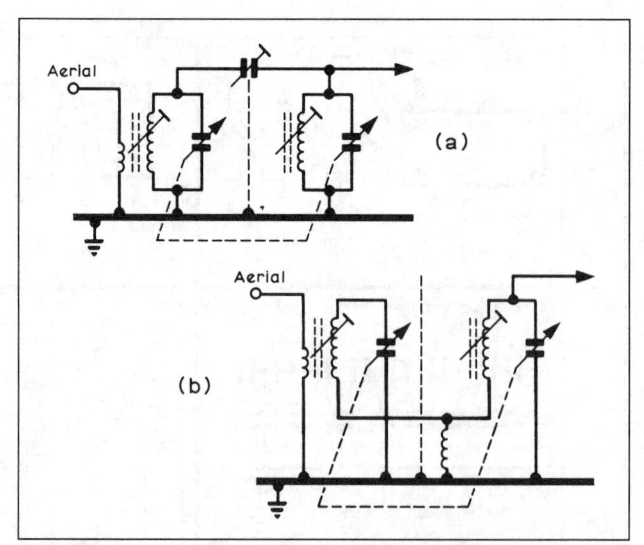

Fig 6.61. Typical RF input circuits used to enhance RF selectivity and capable of providing more than 40dB attenuation of unwanted signals 10% off-tune

or RF amplifier) have two main functions: to provide high attenuation at the image frequency; to reduce as far as possible the amplitude of all signals outside the IF pass-band.

Most amateur receivers still require good pre-mixer selectivity; this can be achieved by using a number of tuned circuits coupled through low-gain amplifiers, or alternatively by tuneable or fixed band-pass filters that attenuate all signals outside the amateur bands (Fig 6.61). The most commonly used input arrangement consists of two tuned circuits with screening between them and either top-coupled through a small-value fixed capacitor, or bottom-coupled through a small common inductance. Slightly more complex but capable of rather better results is the minimum-loss Cohn filter; this is capable of reducing signals 10% off-tune by as much as 60dB provided that an insertion loss of about 4dB is acceptable. This compares with about 50dB (and rather less insertion loss) for an undercoupled pair of tuned circuits. The Cohn filter is perhaps more suited for use as a fixed band-pass filter which can be used in front of receivers having inadequate RF selectivity. Fig 6.62 shows suitable values for 3.5 and 14MHz filters.

Broad-band and untuned RF stages are convenient in construction but can be recommended only when the devices used in the front-end of the receiver have wide dynamic range. An example shown in Fig 6.63 is a power FET designed specifically for this application and operated in the earthed-gate mode suitable for use on incoming low-impedance coaxial feeders.

Unless the front-end of the receiver is capable of coping with

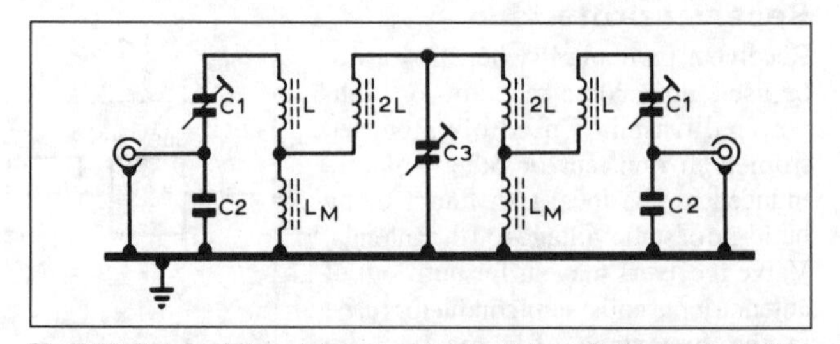

Fig 6.62. Cohn minimum-loss band-pass filter suitable for providing additional RF selectivity to existing receivers. Values for 14MHz: L 2.95µH, 2L 5.9µH, Lm 0.27µH, C1 22pF with 25pF trimmer, C2 340pF, C3 10–60pF (about 34pF nominal). Values for 3.5MHz: L 8µH, 2L 16µH, Lm 2.4µH, C1 150pF + 33pF + 5–25pF trimmer, C2 1nF, C3 150pF + 10–60pF trimmer

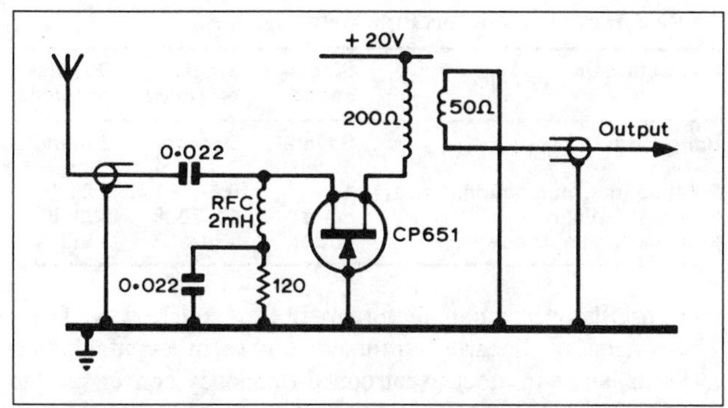

Fig 6.63. Broad-band RF amplifier using power FET and capable of handling signals to almost 3V p-p, 0.5 to 40MHz with 2.5dB noise figure and 140dB dynamic range. Drain current 40mA. Voltage gain 10dB. An alternative device would be a 2N5435 FET

the full range of signals likely to be received, it may be useful to fit an attenuator working directly on the input signal. Such attenuators are particularly useful in front of integrated-circuit and MOSFET amplifiers. Fig 6.64 shows simple techniques for providing manual attenuation control; Fig 6.65 is a switched attenuator providing constant impedance characteristics.

A wide-band amplifier placed in front of a mixer of wide dynamic range must itself have good dynamic range. Dynamic range can be defined as the ratio of the minimum detectable signal (10dB above noise) to that signal which gives a barely noticeable departure from linearity (eg 1dB gain compression).

Amplifiers based on power FETs can approach 140dB dynamic range when operated at low gain (about 10dB) and with about 40mA drain current. This compares with about 90 to 100dB for good valves (eg E810F), 80 to 85dB for small-signal FET devices; 70 to 90dB for small-signal bipolars. The dynamic range of an amplifier can be increased by the operation of two devices in a balanced (push-pull) mode.

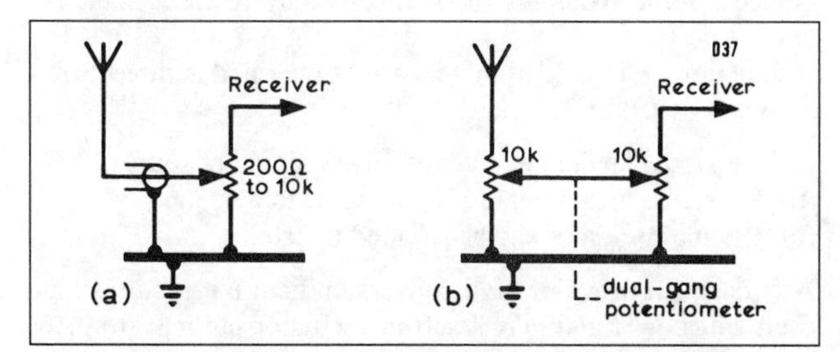

Fig 6.64. Simple attenuators for use in front of a receiver of restricted dynamic range. (a) No attempt is made to maintain constant impedance. (b) Represents less change in impedance

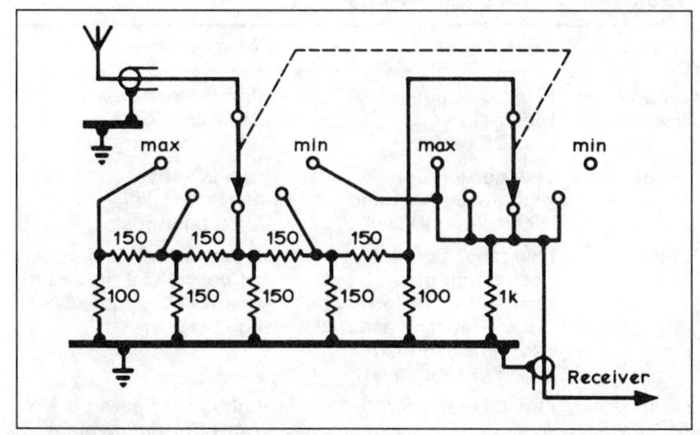

Fig 6.65. Switched antenna attenuator for incorporation in semiconductor receivers

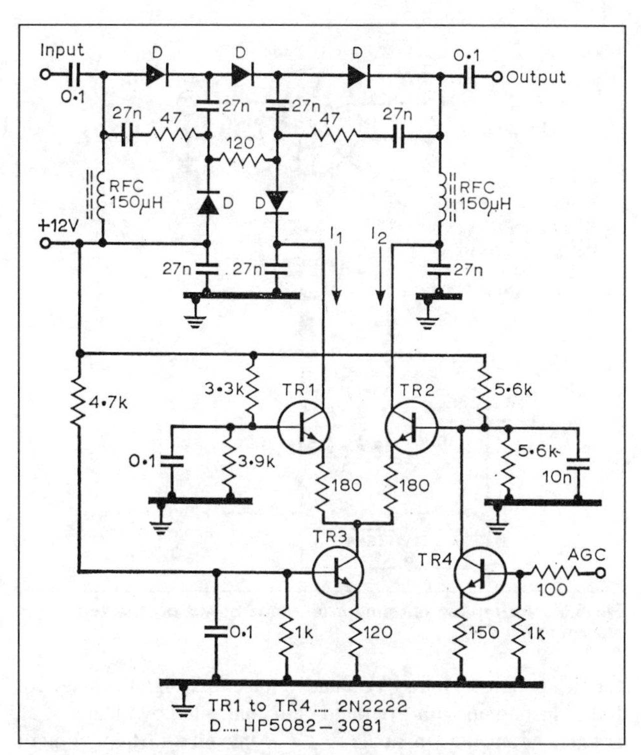

Fig 6.66. Five P-I-N diodes in a double-T arrangement form an AGC-controlled attenuator. The sum of the transistor collector currents is maintained constant to keep input and output impedances constant

Various forms of attenuators controlled from the AGC line are possible. Fig 6.66 shows a system based on P-I-N diodes; Fig 6.67 is based on toroid ferrite cores and can provide up to about 45dB attenuation when controlled by a potentiometer. Fig 6.68 shows an adaptation with MOSFET control element for use on AGC lines although the range is limited to about 20dB.

The tuned circuits used in front-ends may be based on toroid cores since these can be used without screening with little risk of oscillation due to mutual coupling.

It is important to check filters and tuned circuits for non-linearity in iron or ferrite materials. Intermodulation and cross-modulation can be caused by the cores where the flux level rises above the point at which saturation effects begin to occur.

Fig 6.63 shows an amplifier with a dynamic range approaching 140dB and suitable for use in front of, or immediately behind, a double-balanced Schottky diode mixer. It should be appreciated, however, that power FETs are relatively expensive devices, although these are some lower-cost devices such as the Siliconix E310.

Since the optimum dynamic range of an amplifier is usually achieved when the device is operated at a specific working point

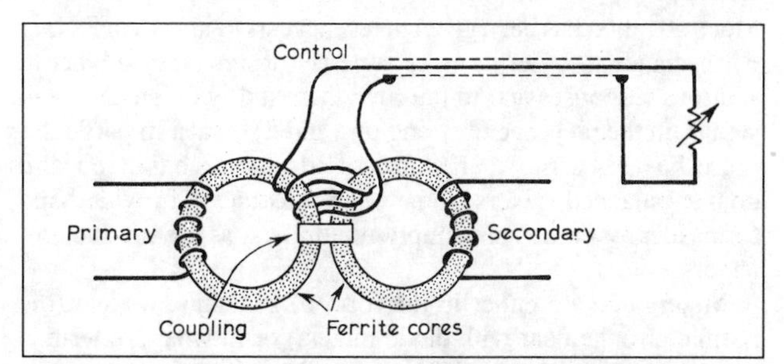

Fig 6.67. Basic form of RF level control using two toroidal ferrite cores

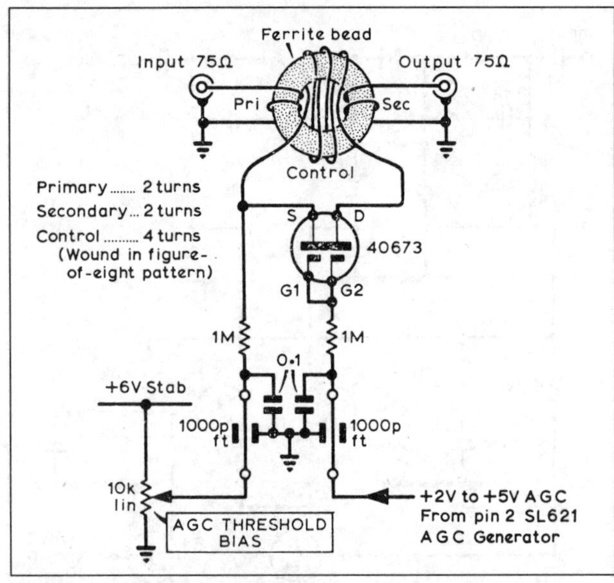

Fig 6.68. Automatic antenna attenuator based on the technique shown in Fig 6.67

(ie bias potential) it may be an advantage to design the stage for fixed (low) gain with front-end gain controlled by means of an antenna attenuator (manual or AGC-controlled). Attenuation of signals ahead of a stage subject to cross-modulation is often beneficial since 1dB of attenuation reduces cross-modulation by 2dB.

For semiconductor stages using FETs and bipolar transistors the grounded-gate and grounded-base configuration is to be preferred. Special types of bipolar transistors (such as the BF314, BF324 developed for FM radio tuners) provide a dynamic range comparable with many junction FETs, a noise figure of about 4dB and a gain of about 15dB with a collector current of about 5mA. Generally, the higher the input power of the device, the greater is likely to be its signal-handling capabilities: overlay and multi-emitter RF power transistors or those developed for CATV applications can have very good intermodulation characteristics.

For general-purpose, small-signal, unneutralised amplifiers the zener-protected dual-gate MOSFET is probably the best and most versatile of the low-cost discrete semiconductor devices, with its inherent cascode configuration. If gate 2 is based initially at about 30 to 40% of the drain voltage, gain can be reduced (manually or by AGC action) by lowering this gate 2 voltage, with the advantage that this then *increases* the signal-handling capacity at this stage. This type of device can be used effectively for RF, mixer, IF, product detector, AF and oscillator applications in HF receivers.

Mixers

Much attention has been given in recent years to improving mixer performance in order to make superhet designs less subject to spurious responses and to improve their ability to handle weak signals in the presence of strong unwanted signals. In particular there has been increasing use of low-noise balanced and double-balanced mixers, sometimes constructed in wide-band form. Ideally mixers are simply multipliers, as are product detectors.

Mixers operate either in the form of switching mixers (the normal arrangement with diode mixers) or in what are termed *continuous non-linear* (CNL) modes. Generally switching mixers can provide better performance than CNL modes but require more oscillator injection, preferably in near-square-wave form. The concept of 'linearity' in mixers may seem a contradiction in terms, since in order to introduce frequency conversion the device must behave in a highly non-linear fashion in so far as the oscillator/signal mixing process is concerned, and the term 'linearity' refers only to the signal path.

Table 6.1. Basic mixer arrangements

Characteristic	Single-ended	Single-balanced	Double-balanced
Bandwidth	Several decades	Decade	Decade
Relative intermodulation density	1	0.5	0.25
Interport isolation	Little	10–20dB	>30dB
Relative oscillator power	0dB	+3dB	+6dB

For valve receivers a breakthrough was the appearance of the *beam-deflection mixer* (7360, 6JH8) which could be used in unbalanced or balanced arrangements as a switching-type mixer, with a noise performance that makes it suitable for use as the first stage of an HF receiver. As for all switching mixers there is a requirement for appreciable oscillator power. The 7360 can provide exceptional performance in its ability to handle input signals of the order of 2.5V, with conversion gain of 20dB and noise figure of only 5dB. Beam-deflection mixer valves are no longer manufactured, with the result that they have become rare and expensive, and, although capable of handling very strong signals, their IMD performance is less striking.

Because of their near square-law characteristics field effect devices make successful mixers provided that care is taken on the oscillator drive level and the operating point (ie bias resistor); preferably both these should be adjusted to suit the individual device used.

Optimum performance of a mixer requires correct levels of the injected local oscillator signal and operation of the device at the correct working point. This is particularly important for FET devices. Some switching-mode mixers require appreciable oscillator power.

Junction FETs used as mixers can be operated in three different ways:

(a) RF signal applied to gate, oscillator signal to source;
(b) RF signal to source, oscillator signal to gate; and
(c) RF and oscillator signals applied to gate.

Approach (a) provides high conversion gain but requires high oscillator power and may result in oscillator pulling; (b) gives good freedom from oscillator pulling and requires low oscillator power, but provides significantly lower gain; (c) gives fairly

Table 6.2. Device comparisons

Device	Advantages	Disadvantages
Bipolar transistor	Low noise figure High gain Low DC power	High intermodulation Easily overloaded Subject to burnout
Diode	Low noise figure High power handling High burn-out level	High LO drive Interface to IF Conversion loss
JFET	Low noise figure Conversion gain Excellent in performance Square-law characteristic Excellent overload High burn-out level	Optimum conversion gain not possible at optimum square law response High LO power
Dual-gate MOSFET	Low IM distortion AGC Square-law characteristic	High noise figure Poor burn-out level Unstable

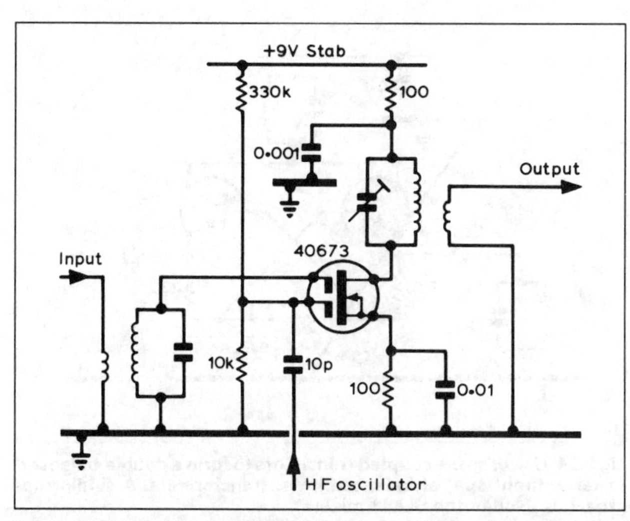

Fig 6.69. Typical dual-gate MOSFET mixer – one of the best 'simple' semiconductor mixers providing gain and requiring only low oscillator injection, but of rather limited dynamic range

high gain with low oscillator power and may often be the optimum choice. For all FET mixers careful attention must be paid to operating point and local oscillator drive level. For most applications the dual-gate MOSFET mixer (Fig 6.69) probably represents the best of the 'simple' single-device arrangements.

During the 'eighties and 'nineties, much attention has been given to improving the input intercept point of receivers in order to cope better with the range of strong signals that reach low-noise broad-band mixers with a minimum of pre-mixer selectivity. A number of high-performance, double-balanced mixers have been developed and are available in package form for home-construction. It has been noted that a double-balanced mixer offers advantages over single-device and single-balanced mixers in reducing the number of IMPs, and also the oscillator radiation from the antenna and the oscillator noise entering the IF channel.

Double-balanced mixers using four hot-carrier diodes when driven at a suitable level and with correct output impedance can provide third-order intercept points (IP3) of up to about +40dBm, representing a useful margin over that required for high-performance amateur operation (about +20–25dBm), a figure that is more likely to be achieved with packaged DBMs at reasonable cost. Conversion loss is likely to be at least 6dB, and optimum performance requires a high-level of oscillator injection (up to about +20–30dBm) with a near-square waveform.

Table 6.3. Mixing products in single, balanced and double-balanced mixers

	f_o	$2f_o$	$3f_o$	$4f_o$	$5f_o$
Unbalanced mixer					
f_s	$f_o \pm f_s$	$2f_o \pm f_s$	$3f_o \pm f_s$	$4f_o \pm f_s$	$5f_o \pm f_s$
$2f_s$	$2f_s \pm f_o$	$2f_o \pm 2f_s$	$3f_o \pm 2f_s$	$4f_o \pm 2f_s$	$5f_o \pm 2f_s$
$3f_s$	$3f_s \pm f_o$	$3f_s \pm 2f_o$	$3f_o \pm 3f_s$	$4f_o \pm 3f_s$	$5f_o \pm 3f_s$
$4f_s$	$4f_s \pm f_o$	$4f_s \pm 2f_o$	$4f_s \pm 3f_o$	$4f_o \pm 4f_s$	$5f_o \pm 4f_s$
$5f_s$	$5f_s \pm f_o$	$5f_s \pm 2f_o$	$5f_s \pm 3f_o$	$5f_s \pm 4f_o$	$5f_o \pm 5f_s$
Balanced mixer – half the number of mixer products					
f_s	$f_o \pm f_s$	$2f_o \pm f_s$	$3f_o \pm f_s$	$4f_o \pm f_s$	$5f_o \pm f_s$
$3f_s$	$3f_s \pm f_o$	$3f_s \pm 2f_o$	$3f_o \pm 3f_s$	$4f_o \pm 3f_s$	$5f_o \pm 3f_s$
$5f_s$	$5f_s \pm f_o$	$5f_s \pm 2f_o$	$5f_s \pm 3f_o$	$5f_s \pm 4f_o$	$5f_o \pm 5f_s$
Double-balanced mixer – one quarter the number of mixer products					
f_s	$f_o \pm f_s$	—	$3f_o \pm f_s$	—	$5f_o \pm f_s$
$3f_s$	$3f_s \pm f_o$	—	$3f_o \pm 3f_s$	—	$5f_o \pm 3f_s$
$5f_s$	$5f_s \pm f_o$	—	$5f_s \pm 3f_o$	—	$5f_o \pm 5f_s$

f_o is the local oscillator. Note that a product such as $2f_o \pm f_s$ is known as a *third-order product*, $3f_s \pm 3f_o$ as a *sixth-order product* and so on.

An active form of double-balanced ring modulator can be based on four symmetrical medium-power FETs, such as the Siliconix U350 devices, or the SD5000 quad-FET package. Such mixers are more sensitive than diodes to changes in termination, particularly reactive components. A less costly approach is the use of a push-pull arrangement with N-junction FETs or dual-gate MOSFETs (which provide slightly higher conversion gain).

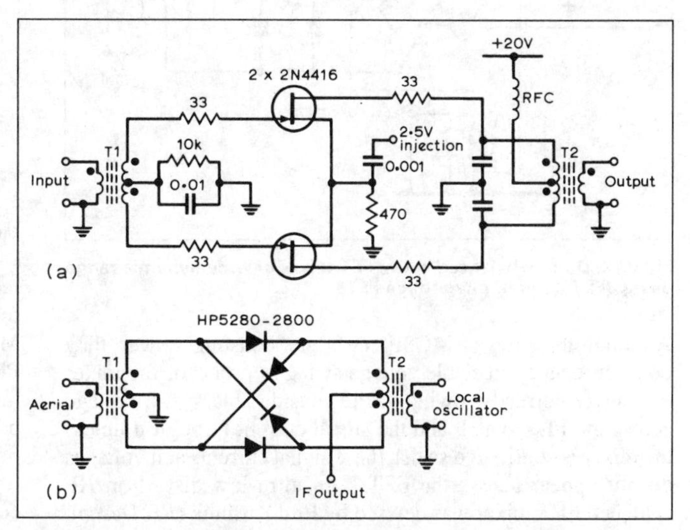

Fig 6.70. Low-noise mixers of wide dynamic range. (a) Balanced FET mixer (preferably used with devices taking fairly high current). (b) Diode ring mixer using Schottky (hot-carrier) diodes

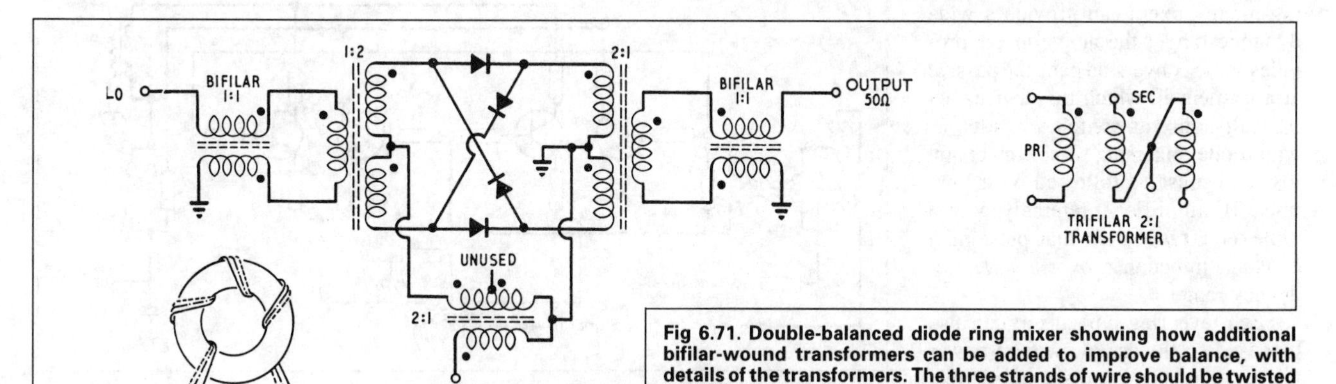

Fig 6.71. Double-balanced diode ring mixer showing how additional bifilar-wound transformers can be added to improve balance, with details of the transformers. The three strands of wire should be twisted together before winding; each winding consists of 12 to 20 turns (depending on frequency range) of No 32 enamelled wire. Injection signal should be 0.8 to 3V across 50Ω (4–12mW)

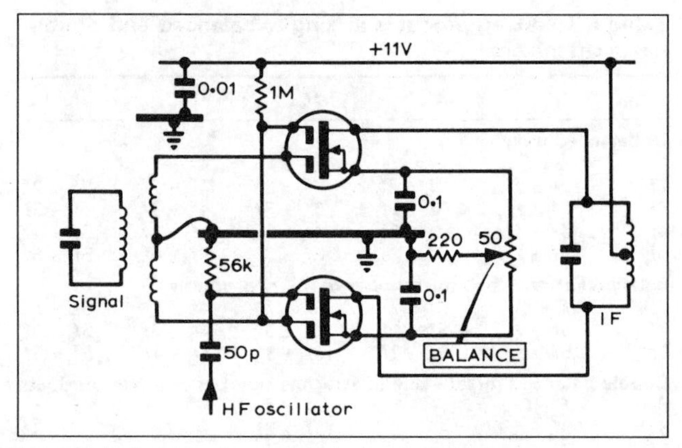

Fig 6.72. Balanced mixer using dual-gate MOSFETs

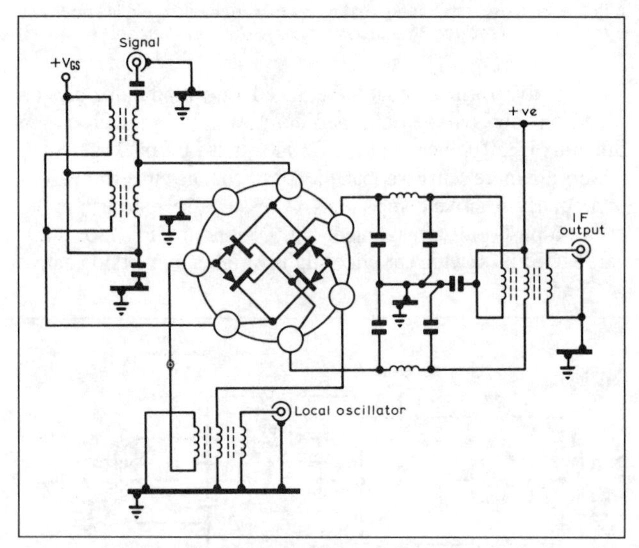

Fig 6.73. Double-balanced active FET mixer of wide dynamic range using JFET quad of power-type FETs

Although the gates of MOSFETs do not consume power, they do require an appreciable voltage swing in order for the mixer to operate correctly, owing to the considerable signal voltage across the FET switch and the signal current through it unless the mixer is configured so that these signal currents and voltages do not appear across the FETs. An ultra-low-distortion HF switched FET mixer was devised by Eric Kushnik (*RF Design*, September 1992) in which an IP3 of +25dBm has been measured with –3dBm local oscillator power.

Both active and passive FET switching-mode mixers can provide a wide dynamic range; the active mixer provides some conversion gain, the passive arrangement in which the devices act basically only as switches results, as with diode mixers, in some conversion loss and must be followed by a low-noise IF amplifier, preferably with a diplexer arrangement that presents a constant impedance over a wide frequency range.

High-level ring-type mixers can also be based on the use of medium-power bipolar transistors in cross-coupled 'tree' arrangements. With such an arrangement there need be no fundamental

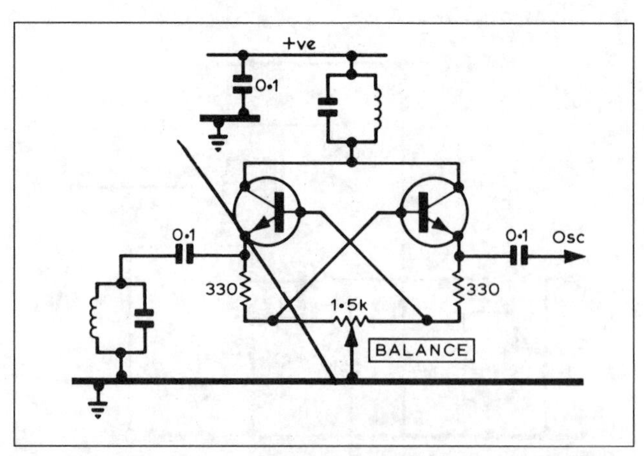

Fig 6.74. Use of cross-coupled transistors to form a double-balanced mixer without special balanced input transformers. A similar approach is used in the SL641 mixers

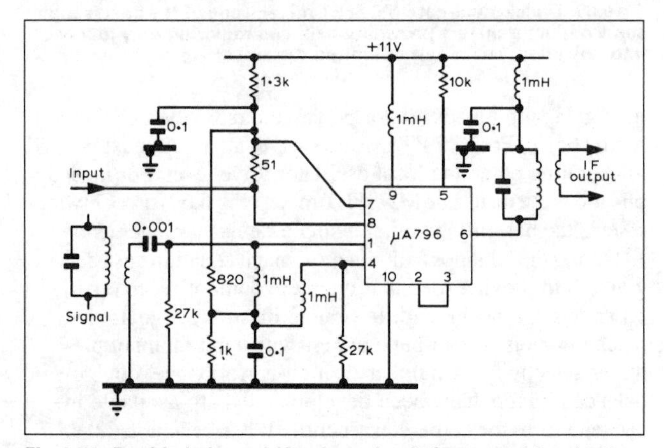

Fig 6.75. Double-balanced IC mixer circuit for use with µA796 or MC1596G etc devices

requirement for push-pull drive or balanced input/output transformers. This approach is used at low-level in the Motorola MC1596 and Plessey SL640 IC packages, and at high-level in the Plessey SL6440. The SL6440 was specifically developed as a high-level, low-noise mixer and is capable of +30dBm intercept point, +15dBm 1dB compression point with a conversion 'gain' of –1dB.

Fig 6.76 shows the basic arrangement of this class of mixer, comprising a pair of transconductance mixers with emitter resistors added for IM improvement. Resistors in the base and

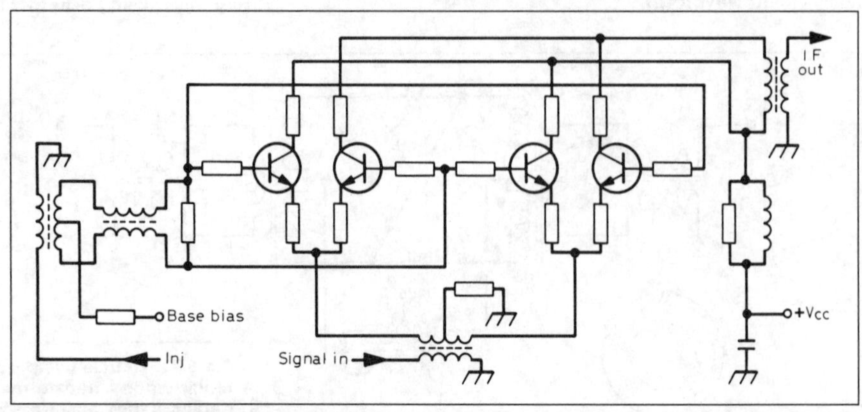

Fig 6.76. A double-balanced active mixer using bipolar transistors in a degenerated version of the balanced transconductance mixer. This approach is used in the Plessey SL6440 high-level IC and in the Motorola low-level MC1596 IC (*Radio Receivers* – Gosling (ed))

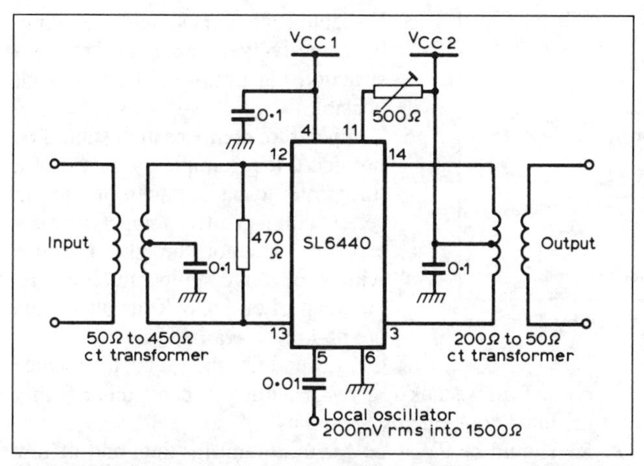

Fig 6.77. Typical application of the SL6440 in HF communications receiver as mixer with good dynamic range

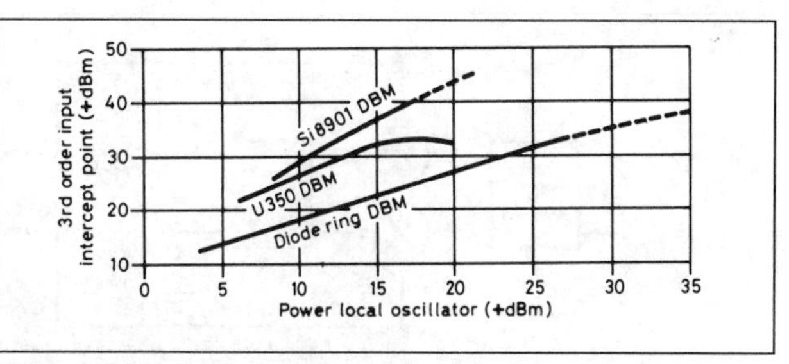

Fig 6.79. Performance comparison between Si8901 DBM, U350 active FET DBM and diode ring DBM

collector leads add loss of ultra-high frequencies to suppress parasitic oscillations caused by resonances formed by circuit and transistor capacitances together with the leakage reactance of the associated transformers. Injection is via a balancing transformer to the bases of the bipolar transistors which are overdriven, resulting in signal switching action. Collector supply voltage is applied to the output transformer centre-tap through a parallel resistor-inductor to further suppress oscillation. With medium-power transistors, this active mixer can give a 3dB gain, 9dB noise figure and +25dBm input intercept over the 2 to 30MHz band as an up-converter to the 100MHz range. This device can be used with or without preamplification and with little pre-mixer selectivity in high-performance home-built receivers.

Common to both diode and switching-mode FET mixers is their square-law characteristics, an important factor in maintaining low distortion during mixing. Equally important for high dynamic range is the ability to withstand overload that can be a major cause of distortion. Some designs with passive ring mixers use paralleled diodes to provide greater current handling; the penalty attached to this approach is the need for a very large increase in local-oscillator power. A form of high-performance mixer using monolithic quad-ring double-diffused MOSFETs developed by Ed Oxner, KB6QJ (*Siliconix Application Note AN85-2*) used a resonant-gate drive transformer to provide sufficient switching voltage without a corresponding increase in switching power, an effective technique but not easily implemented with broad-band mixers.

Switching mixers, if they are to achieve a high third-order intercept point, require a drive that must:

(a) approach ideal square wave;
(b) ensure a 50% duty cycle; and
(c) have sufficient amplitude to switch the devices fully 'on' and 'off', and, in the case of FET devices, offer minimum reduced resistance when 'on'.

Further, to maintain good overall performance in terms of minimum conversion loss, maximum dynamic range (ie taking into account the noise figure) and maximum strong signal performance, it is desirable to incorporate image-frequency termination.

With any mixer operating directly on the incoming RF signals without pre-mixer amplification, the IF amplifier that follows the mixer, either directly or after a roofing filter, must have a low noise figure and a high intercept figure, preferably with a diplexer arrangement to achieve constant input impedance over a broad band of frequencies.

Image-rejection mixer

In recent years, there has been a marked trend towards multi-conversion general-coverage receivers with the first IF in the VHF range since this facilitates the use of frequency synthesisers covering a single span of frequencies with a broad-band up-conversion mixer. It has long been considered that subsequent down-conversion mixer stages should not change the frequency by a factor of more than about 10:1 in order to minimise the 'image' response. Thus a receiver with a first IF of the order of 70MHz or 45MHz and a final IF of, say, 455kHz or 50kHz (to take advantage of digital signal processing) normally requires an intermediate IF of, say, 9 or 10.7MHz and possibly a further IF of about 1MHz. With triple or even quadruple conversions, it becomes increasingly difficult to achieve a design free of spurious responses and of wide dynamic range.

It is possible to eliminate mid-IF stages by the use of an image-rejecting two-phase mixer akin to the form of demodulator used in two-phase direct conversion receivers; this is best implemented using two double-balanced diode-ring mixers or the equivalent. Fig 6.81 outlines the basic arrangement of an image-rejection

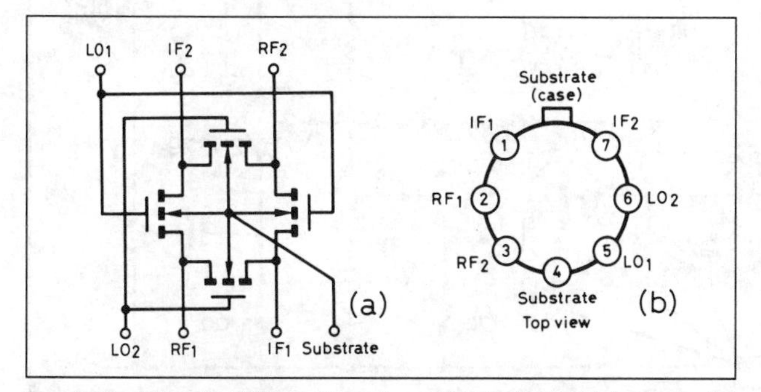

Fig 6.78. Siliconix Si8901 ring demodulator/balanced mixer. (a) Functional block diagram. (b) Pin configuration with Si8901A in TO-78 and Si8901Y as surface-mounted So14 configuration

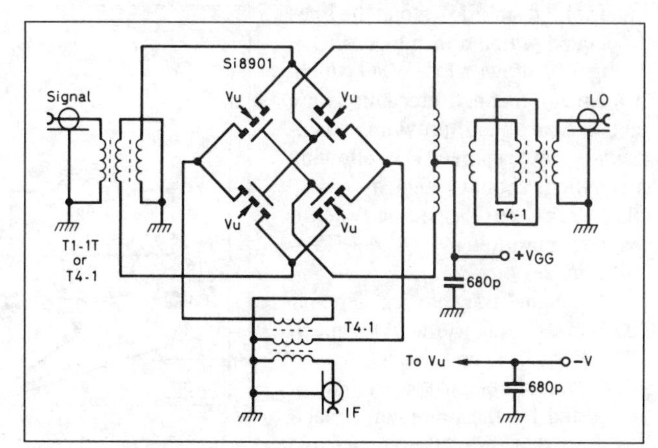

Fig 6.80. Prototype commutation double-balanced mixer as described in the Siliconix application notes

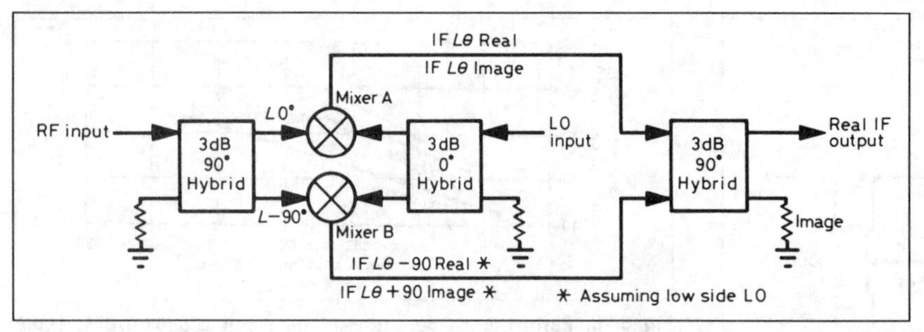

Fig 6.81. Image-rejecting mixer from *Microwave Solid-state Circuit Design* by Bahl and Bhartia (1988)

mixer. Such a mixer can convert a VHF signal directly to, say, 50kHz while maintaining image response at a low level.

The HF oscillator(s)

The frequency to which a superhet or direct-conversion receiver responds is governed not by the input signal frequency circuits but by the output of the local oscillator. Any frequency variations or drift of the oscillator are reflected in variation of the received signal; for SSB reception variations of more than about 50Hz will render the signal unintelligible unless the set is retuned. Some of the design techniques by which stability can be improved have already been outlined, but for many amateur receivers stability is still largely determined by one or more free-running oscillators.

The overall stability of a single-conversion model thus depends on the high frequency oscillator (HFO) and to a lesser extent (since it is usually at a lower frequency) on the BFO; that of a double-conversion model on the stability of all three oscillators, and so on. Generally it is more difficult to achieve good stability as the frequency increases, which is one reason why many receivers use a crystal-controlled HFO, while designing the variable oscillator at lower frequency.

The prime requirements of an oscillator used for heterodyne conversion are: freedom from frequency changes resulting from mechanical vibration or temperature changes; sufficient output for maximum conversion efficiency; low harmonic output (particularly important in second and third oscillators); no undue variation of output throughout the tuning range. The methods and circuits used to achieve these results are substantially the same as for transmitter VFOs.

Where a VFO is to be used in conjunction with the variable IF or pre-mixer technique it need cover only a single range; for example 1MHz or commonly 500kHz. The absence of range switching makes it possible to design this for high stability, along the lines associated with a transmitter VFO.

Fig 6.82 shows a FET VFO which, if all recommended precautions are taken, is capable of providing an extremely stable source. The following precautions should be noted:

1. Design with genuine Vackar configuration, ie C1/(C4 + C6) ≈ C3/C2 ≈ 6.
2. Mount in strong box (eg die-cast).
3. Use high-quality variable capacitor (eg Jackson U101)
4. C2 should be air-spaced and adjusted for the minimum capacitance that allows the circuit to oscillate freely.
5. Variable capacitors should preferably be effectively cleaned before construction (eg ultrasonic bath if at all possible).
6. Temperature compensation should be provided, for example by means of a differential capacitor with one leg in series with a positive temperature coefficient capacitor, the other in series with a negative temperature coefficient capacitor (Oxley Tempatrimmers are no longer available).
7. C1, C3 and C6 should be silver-mica types firmly stuck (eg by Araldite) to convenient firmly mounted components or chassis.
8. R1 should be 2W type for minimum heating, and of low inductance.
9. The use of good buffer/isolating amplifier is essential; the two-stage arrangement shown should prove satisfactory.
10. Use in conjunction with a well-stabilised power supply (for example using a zener diode with constant-current FET). Disc ceramic bypass capacitors should be used liberally along the supply rail to prevent feedback.
11. L, C1, C2, C3, C4, C6, R1 and FET source connection should have a common earthing point (for example one of the fixing screws of C4).
12. Ceramic coil formers preferred but iron dust-cores can be used to facilitate calibration adjustment; ferrite cores should be avoided for this application.
13. Keep wiring leads short and use stiff wire (16 or 18swg) for all interconnections affecting the oscillator tank coil.

IF amplifiers

The IF remains the heart of a superhet receiver, for it is in this section that virtually all of the voltage gain of the signal and the selectivity response are achieved. Whereas with older superhets having significant front-end gain, the IF gain was of the order of 70–80dB, today it is often over 100dB.

Where the output from the mixer is low (possibly less than 1μV) it is essential that the first stage of the IF section should have low-noise characteristics and yet not be easily overloaded. Although it is desirable that the crystal filter (or roofing filter) should be placed immediately after the mixer, the very low output of diode and passive FET mixers may require that a stage of amplification takes place before the signal suffers the insertion loss of the filter.

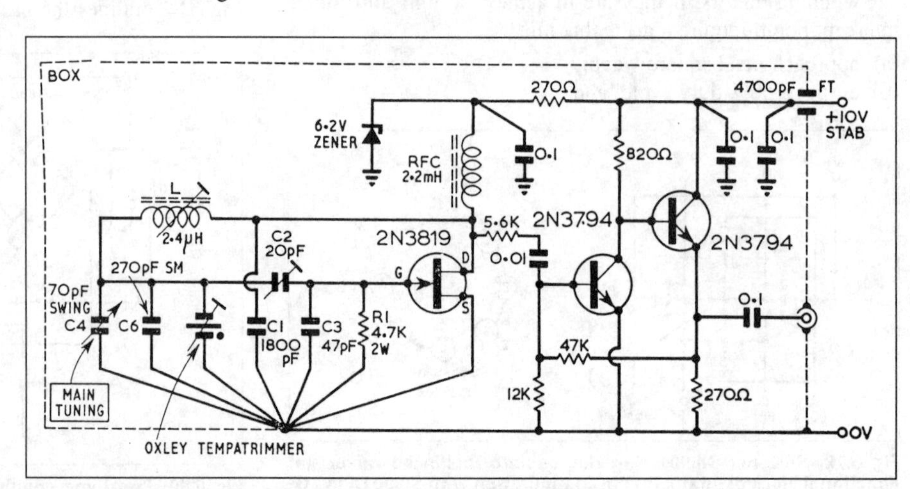

Fig 6.82. High-stability FET Vackar VFO covering 5.88 to 6.93MHz

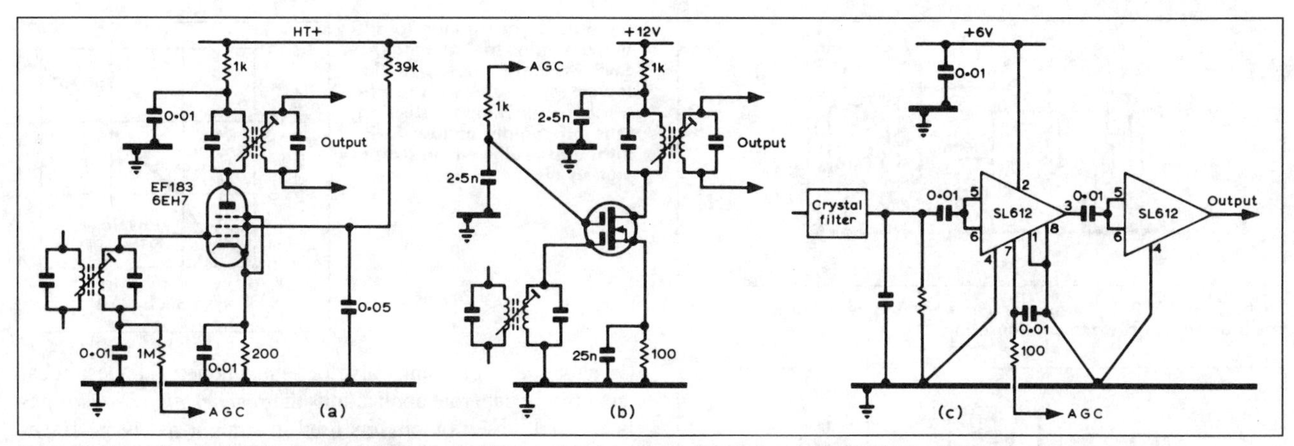

Fig 6.83. Typical automatic gain-controlled IF amplifiers using (a) valves, (b) dual-gate MOSFETs and (c) integrated circuits

Similarly it is important that where the signal passes through the sideband filter at very low levels the subsequent IF amplifier must have good noise characteristics. Further, for optimum CW reception, it will often be necessary to ensure that the *noise bandwidth* of the IF amplifier *after* the filter is kept narrow. The noise bandwidth of the entire amplifier should be little more than that of the filter. This can be achieved by including a further narrow-band filter (for example a single-crystal filter with phasing control) later in the receiver, or alternatively by further frequency conversion to a low IF.

To achieve a flat AGC characteristic it may be desirable for all IF amplifiers to be controlled by the AGC loop, and it is important that amplifier distortion should be low throughout the dynamic range of the control loop.

The dual-gate MOSFET (Fig 6.83(b)) with reverse AGC on gate 1 and partial forward AGC on gate 2 has excellent cross-modulation properties but the control range is limited to about 35dB per stage. Integrated circuits with high-performance gain-controlled stages are available (Fig 6.83(c)).

For valve amplifiers frame-grid valves such as the EF183 provide a control range of over 50dB and cope well with large signals (Fig 6.83(a)).

Where a high-grade SSB or CW filter is incorporated it is vital to ensure that signals cannot 'leak' around the filter due to stray coupling; good screening and careful layout are needed.

In multiple-conversion receivers, it is possible to provide continuously variable selectivity by arranging to vary slightly the frequency of a later conversion oscillator so that the band-pass of the two IF channels overlap to differing degrees. For optimum results this requires that the shape factor of both sections of the IF channel should be good, so that the edges are sharply defined.

With double-tuned IF transformers, gain will be maximum when the product kQ is unity (where k is the coupling between the windings). IF transformers designed for this condition are said to be *critically coupled*; when the coupling is increased beyond this point (*over-coupled*) maximum gain occurs at two points equally spaced about the resonant frequency with a slight reduction of gain at exact resonance: this condition may be used in broadcast receivers to increase bandwidth for good-quality reception. If the coupling factor is lowered (*under-coupled*) the stage gain falls but the response curve is sharpened, and this may be useful in communications receivers.

FETs as small-signal amplifiers

Field effect transistors make good small-signal RF or IF amplifiers and offer advantages when used properly. Although vulnerable to electrostatic discharge when out of circuit, unless protected by an internal diode(s), once wired in circuits having an easy, low-resistance path between their gate(s) and earth they can provide low-noise, rugged amplifiers. However, despite being, like thermionic valves, voltage-controlled devices, they should not be used as replacements to valves in similar circuits.

FETs have very high slope of up to 30mA/V (much higher than most valves) but also much greater drain-to-gate capacitance than the inter-electrode capacitances of triode valves. This is a sure prescription for self-oscillation if connected in a typical pentode-type amplifier circuit.

However, the FET has very low output impedance and can be used as a stable low-gain device, yet providing excellent stage gain from the voltage step-up that can be readily achieved with a resonant input transformer. For example, the 21MHz preamplifier shown in Fig 6.84 using the 2N3819 FET (10mA/V) with a 330Ω resistor as load has a device gain of only three, but the input tuned circuit can provide a voltage gain of about seven, resulting in a stable voltage gain of about 21, with the FET's very high input impedance presenting only light loading of the input transformer, providing an unconditionally stable stage gain of over 20dB. Figs 6.85, 6.86 and 6.88 show typical FET and dual-gate FET small-signal amplifiers.

The most critical amplifier in a modern, high-performance receiver is usually the *post-mixer amplifier*, ie the IF amplifier that follows the mixer, either directly or after a crystal filter. It needs to be of low noise in order to cope with the conversion loss of a ring mixer, and if preceding the crystal filter will have to cope with large off-frequency signals, requiring a high intercept characteristic. For the highest-performance receivers, push-pull bipolar transistors with a noise figure of about 2dB and a third-order intercept point equal or better than that of the mixer are required.

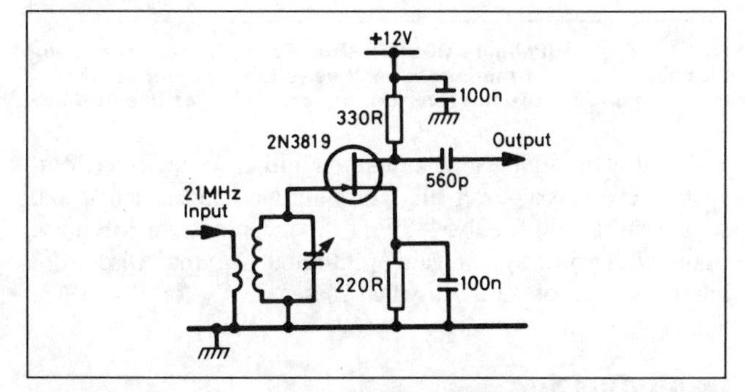

Fig 6.84. Stable FET preamp with low-impedance output load and with the main part of the gain coming from the step-up input transformer

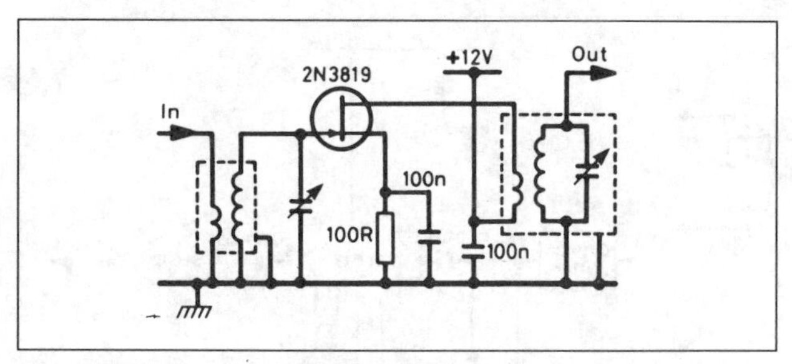

Fig 6.85. Stable FET IF amplifier using two bipolar transistor-type IF transformers in reverse configuration

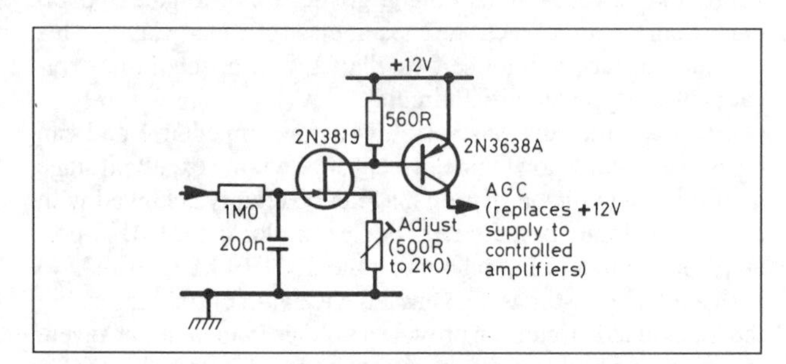

Fig 6.86. Dual-gate FET IF amplifier

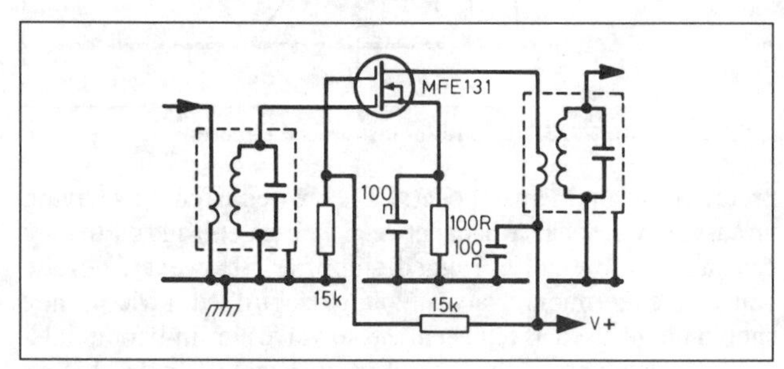

Fig 6.87. AGC amplifier to provide varying supply voltage to FET RF/IF stages

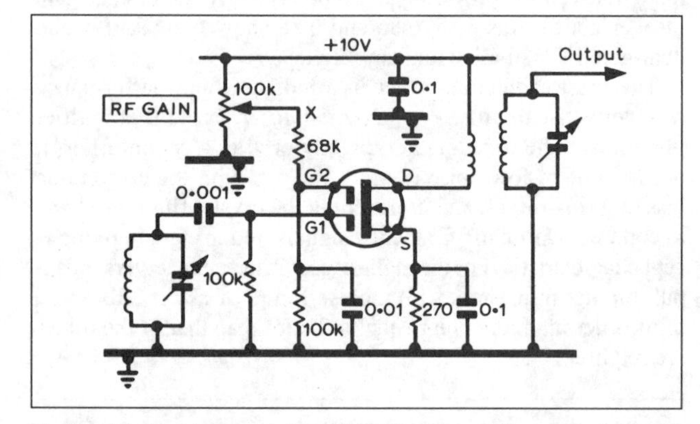

Fig 6.88. Typical dual-gate MOSFET RF or IF amplifier. G2 is normally biased to about one-third of the positive voltage of drain. In place of manual gain control point X can be connected to a positive AGC line

Possibly the simplest post-mixer amplifier for high-performance receivers is a power-FET common-gate stage, such as that shown earlier. With a 2N5435 FET this can provide a 2dB noise figure with a 50Ω system gain of 9dB and an output third-order intercept point of +30dBm when biased at V_{DD} of 12–15V at 50mA.

Demodulation

For many years, the standard form of demodulation for communications receivers, as for broadcast receivers, was the

Fig 6.89. Synchronous (product) detection maintaining the SNR of signals down to the lowest levels whereas the efficiency of envelope detection falls off rapidly at low SNR, although as efficient on strong signals

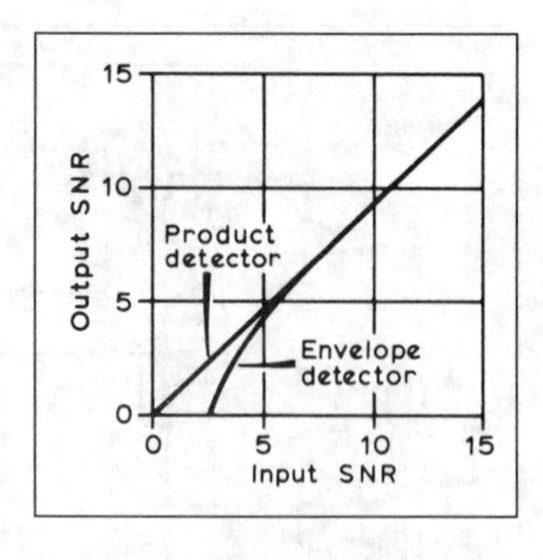

envelope detector using valve or semiconductor diodes; occasionally for superhet applications the *regenerative detector* has been used, based on circuits used in straight receivers. Envelope detection is a non-linear process (part mixing, part rectification) and is inefficient at very low signal levels. On weak signals this form of detector distorts or may even lose the intelligence signals altogether. On the other hand, *synchronous* or *product detection* preserves the signal-to-noise ratio, enabling post-detector signal processing and audio-filters to be used effectively (Fig 6.89). Synchronous detection is essentially a frequency conversion process and the circuits used are similar to those used in mixer stages. The IF or RF signal is heterodyned by a carrier at the same frequency as the original carrier frequency and so reverts back to the original audio modulation frequencies (or is shifted from these frequencies by any difference between the inserted carrier and the original carrier as in CW where such a shift is used to provide an audio output between about 500 and 1000Hz).

It should be noted that a carrier is needed for both envelope and product detection: the carrier may be radiated along with the sidebands, as in AM, or locally generated and inserted in the receiver (either at RF or IF – usually at IF in superhets, at RF in direct-conversion receivers).

Synchronous or product detection has been widely adopted for SSB and CW reception in amateur receivers; the injected carrier frequency is derived from the beat frequency oscillator, which is either LC or crystal controlled. By using two crystals it is possible to provide selectable upper or lower sideband reception.

The use of synchronous detection can be extended further to cover AM, DSBSC, NBFM and RTTY but for these modes the injected carrier really needs to be identical to the original carrier, not only in frequency but also in phase: that is to say the local oscillator needs to be in *phase coherence* with the original carrier (an alternative technique is to provide a strong local carrier that virtually eliminates the original AM carrier – this is termed *exalted carrier detection).*

Phase coherence cannot be achieved between two oscillators unless some effective form of synchronisation is used. The simplest form of synchronisation is to feed a little of the original carrier into a local oscillator, so forcing a phase lock on a free-running oscillator; such a technique was used in the *synchrodyne* receiver. The more usual technique is to have a *phase-lock loop.* At one time such a system involved a large number of components and would have been regarded as too complex for most purposes; today, however, complete phase-lock loop detectors are available in the form of a single integrated circuit, both for AM and NBFM applications.

Apart from the phase-lock loop approach a number of alternative forms of synchronous multi-mode detectors have been developed. One interesting technique which synthesises a local phase coherent carrier from the incoming signal is the *reciprocating detector.*

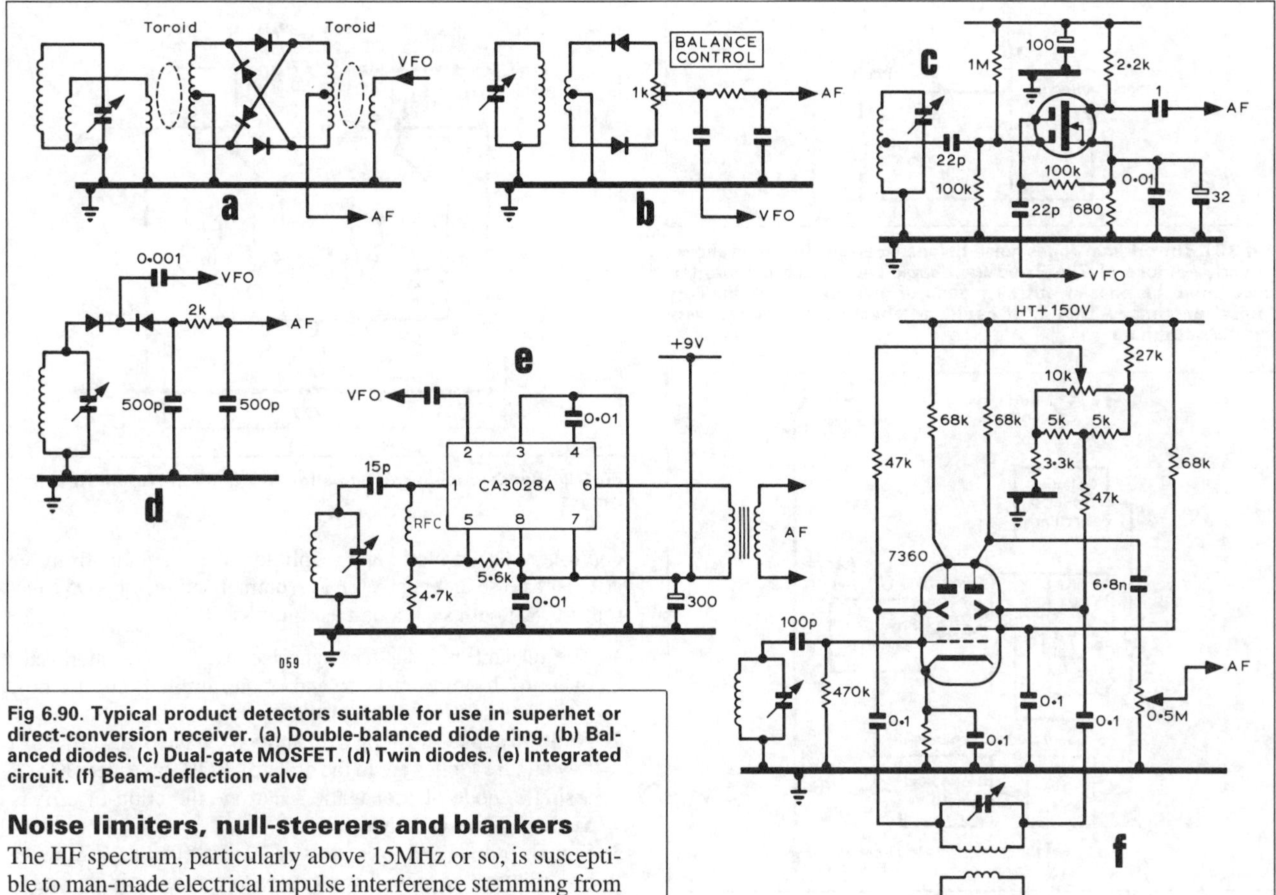

Fig 6.90. Typical product detectors suitable for use in superhet or direct-conversion receiver. (a) Double-balanced diode ring. (b) Balanced diodes. (c) Dual-gate MOSFET. (d) Twin diodes. (e) Integrated circuit. (f) Beam-deflection valve

Noise limiters, null-steerers and blankers

The HF spectrum, particularly above 15MHz or so, is susceptible to man-made electrical impulse interference stemming from electric motors and appliances, car ignition systems, thyristor light controls, high-voltage power lines and many other causes. Static and locally generated interference from appliances, TV receivers etc can be a serious problem below about 5MHz, and may be maximum at LF/MF.

These interference signals are usually in the form of high-amplitude, short-duration pulses covering a wide spectrum of frequencies. In many urban and residential areas this man-made interference sets a limit to the usable sensitivity of receivers and may spoil the reception of even strong amateur signals.

Because the interference pulses, though of high amplitude, are often of extremely short duration, a considerable improvement can be obtained by 'slicing' off all parts of the audio signal which are significantly greater than the desired signal. This can be done by simple AF limiters such as back-to-back diodes. For AM reception more elegant noise limiters develop fast-acting biasing pulses to reduce momentarily the receiver gain during noise peaks. The ear is much less disturbed by 'holes of silence' than by peaks of noise. Many limiters of this type have been fitted in the past to AM-type receivers.

Unfortunately, since the noise pulses contain high-frequency transients, highly selective IF filters will distort and broaden out the pulses. To overcome this problem, *noise blankers* have been developed which derive the blanking bias potentials from noise pulses which have not passed through the receiver's selective filters. In some cases a parallel broadly tuned receiver is used, but more often the noise signals are taken from a point early in the receiver. For example, the output from the mixer goes to two channels: the signal channel which includes a blanking control element which can rapidly reduce gain when activated; and a wide-band noise channel to detect the noise pulse and initiate the gain reduction of the signal channel. To be most effective it is necessary for the gain reduction to take place virtually at the instant that the interference pulse begins. In practice, because of the time constants involved, it is difficult to do this unless the signal channel incorporates a time delay to ensure that the gain reduction can take place simultaneously with or even just before the noise pulse. One form of time delay which has been described in the literature utilises a PAL-type glass ultrasonic television delay line to delay signals by 64μs. It is, however, difficult to eliminate completely transients imposed on the incoming signal.

One possible approach, which has been investigated at the University College, Swansea, is to think in terms of receivers using synchronous demodulation at low level so that a substantial part of the selectivity, but not all of it, is obtained after demodulation. This allows noise blankers to operate at a fairly low level on AF signals.

A control element which has been used successfully consists of a FET gate pulsed by signals derived from a wide-band noise amplifier. The noise gate is interposed between the mixer and the first crystal filter, with the input signal to the noise amplifier taken off directly from the mixer.

Noise limiters and noise blankers are suitable for use only on pulse-type interference in which the duty cycle of the pulse is relatively low. An alternative technique, suitable for both continuous signals and noise pulses, is to null out the unwanted signal by balancing it with anti-phase signals picked up on a short 'noise antenna'. This has led to the revival of the 'thirties Jones noise-balancing technique in which local interference could be phased out by means of pick-up on an auxiliary noise antenna. This

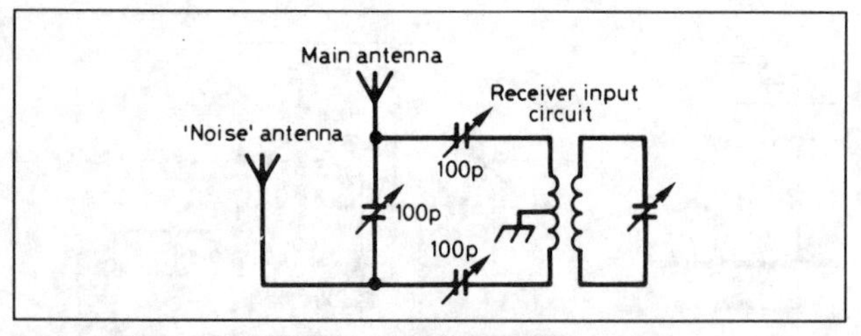

Fig 6.91. The original Jones noise-balancing arrangement as shown in early editions of *The Radio Handbook*. Local electrical interference could be phased out by means of pick-up on the auxiliary 'noise' antenna. Although it could be effective it required very careful setting up

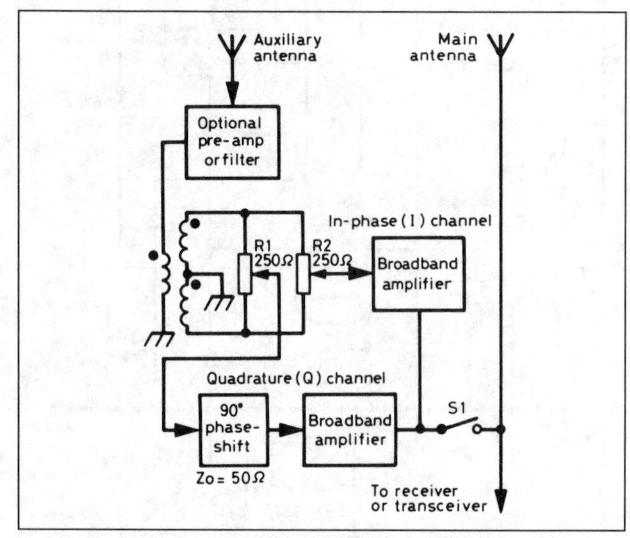

Fig 6.92. Functional diagram of the electronic null steering unit for use in conjunction with an HF receiver or transceiver (S1 is a relay contact to disconnect the system during transmission) as described by John Webb, W1ETC, of the Mitre Corporation in 1982

system (Fig 6.91) was capable of reducing a specific unwanted source of interference by several S-points while reducing the wanted signal by only about one S-point, but required critical adjustment of the controls.

In the early 'eighties John K Webb, W1ETC, developed a more sophisticated method of phasing out interference using coiled lengths of coaxial cable as delay lines to provide the necessary phase shifts: Fig 6.92. This included a compact null-steerer located alongside the receiver/transceiver, capable of generating deep nulls against a single source of interference, resembling the nulls of an efficient MW ferrite-rod antenna. The two

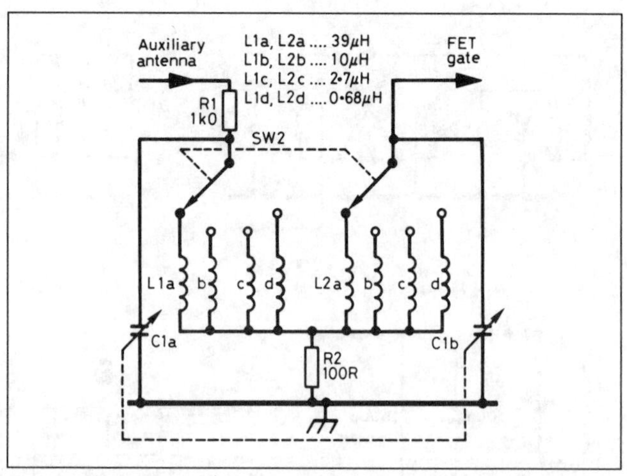

Fig 6.94. Bandswitching modification to provide 1.8 to 30MHz in four ranges

controls adjusted phase and amplitude of the signals from the auxiliary noise antenna. W1ETC summarised results (*QST* October 1982) with such a unit as follows:

1. The available null depth in signals propagated over short paths of up to 20 miles is large and stable, limited only by how finely the controls are adjusted.
2. Nulls on signals arriving over short skywave paths of up to a few hundred miles are in the order of 30dB, provided there is a single mode of propagation and one direction of arrival. Such nulls are usually stable.
3. Signals propagated over paths of 10 to 100 miles may arrive as a mixture of ground-wave and skywave. A single null is thus ineffective.
4. Signals propagated by skywave over long distances frequently involve several paths, each having a different path length so that a single null has little effect on what is usually the 'wanted' signal.
5. Broad-band radiated noise can be nulled as deeply as any radio signal. This seems to be a more effective counter to noise than blanking or limiting techniques with local electrical noise deeply nulled, and with little effect on wanted long-distance signals.

To meet result (5) the interference has to be directly radiated to the receiver antennas and not enter the receiver in a less-directional manner (for example re-radiation from mains cabling etc). Various methods of implementing null-steering have been described: Figs 6.93–6.95 show a design by Lloyd Butler, VK5BR, utilising the phase shifts of off-tune resonant circuits.

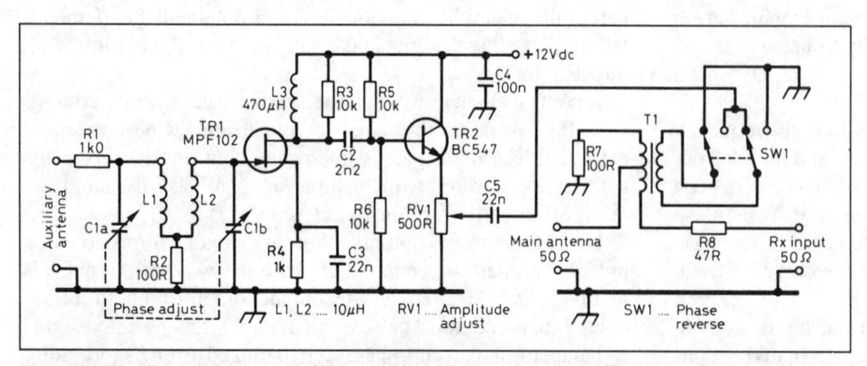

Fig 6.93. VK5BR's Mk 2 interference-cancelling circuit as described in the January 1993 issue of *Amateur Radio*. As shown this covers roughly 3.5 to 7MHz. C1 ganged 15–250pF variable capacitor or similar. L1, L2 miniature 10μH RF chokes. T1 11 turns quadfilar wound on Amidon FT-50-75m toroidal core

AF stages

The AF output from an envelope or product detector of a superhet receiver is usually of the order of 0.5 to 1V, and many receivers incorporate relatively simple one- or two-stage audio amplifiers, typically using an IC device and providing about 2W output. On the other hand the direct-conversion receiver may require a high-gain audio section capable of dealing with signals of less than 1μV.

Provided that all stages of the receiver up to and including the product detector

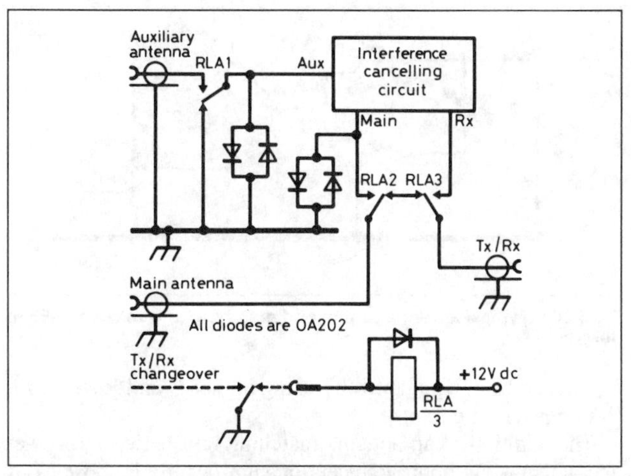

Fig 6.95. Transmit-receive switching with protection diodes for use of the VK5BR interference-cancelling circuit with a transceiver

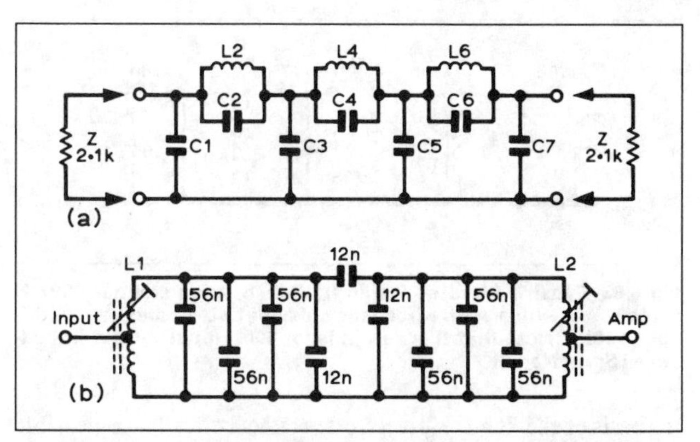

Fig 6.96. (a) Phone and (b) CW AF filters suitable for use in direct-conversion or other receivers requiring very sharply defined AF responses. The CW filter is tuned to about 875Hz. Values for (a) can be made from preferred values as follows: C1 37.26nF (33,000 + 2200 + 1800 + 220pF); C2 3.871nF (3300 + 560pF); C3 51.87nF (47000 + 4700 + 150pF); C4 19.06nF (18,000 + 1000pF); C5 46.41nF (39,000 + 6800 + 560pF); C6 13.53nF (12,000 + 1500pF); C7 29.85nF (27,000 + 2700 + 150pF). All capacitors mica or polyester or styroflex types. L2 168.2mH (540 turns), L4 124.5mH (460 turns); L6 129.5mH (470 turns) using P30/19 3H1 pot cores and 0.25mm enam wire. Design values based on 2000Ω impedance

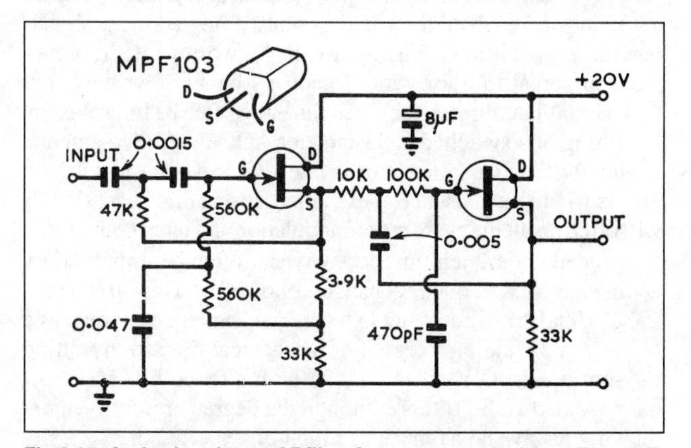

Fig 6.97. Active band-pass AF filter for amateur telephony. The –6dB points are about 380 and 3200Hz, –18dB about 160 and 6000Hz

are substantially linear, many forms of post-demodulation signal processing are possible: for example band-pass or narrow-band filtering to optimise signal-to-noise ratio of the desired signal, audio compression or expansion; the removal of audio peaks, AF noise blanking, or (for CW) the removal by gating of background noise. Audio phasing techniques may be used to convert a DSB receiver into an SSB receiver (as in two-phase or third-method SSB demodulation) or to insert nulls into the audio pass-band for the removal of heterodynes. Then again, in modern designs the AGC and S-meter circuits are usually operated from a low-level AF stage rather than the IF-derived techniques used in AM type receivers.

It should be appreciated that linear low-distortion demodulation and AF stages are necessary if full advantage is to be taken of such signal processing, since strong intermodulation products can easily be produced in these stages. Thus, despite the restricted AF bandwidth of speech and CW communications, the intermodulation distortion characteristics of the entire audio section should preferably be designed to high-fidelity audio standards. Very sophisticated forms of audio filtering, notching and noise reduction are now marketed in the form of add-on digital-signal-processing (DSP) units.

Audio filters may be passive using inductors and capacitors, or active, usually with resistors and capacitors in conjunction with op-amps or FETs (Figs 6.96–6.99). Many different circuits have been published covering AF filters of variable bandwidth, tuneable centre frequencies and for the insertion of notches. The full theoretical advantage of a narrow-band AF filter for CW reception may not always be achieved in operational use: this is because the human ear can itself provide a 'filter' bandwidth of about 50Hz with a remarkably large dynamic range and the ability to tune from 200 to 1000Hz without introducing 'ringing'.

MODIFICATIONS TO RECEIVERS

While the number of amateurs who build their own receivers from scratch is today in a minority, some newcomers or those with limited budgets buy relatively low-cost models or older second-hand receivers and then set about improving the performance. Old, but basically well-designed and mechanically satisfactory, valved receivers can form

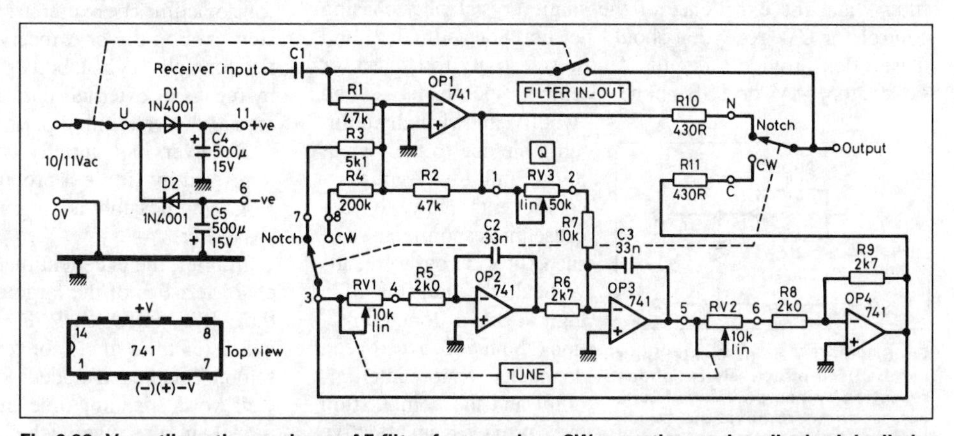

Fig 6.98. Versatile active analogue AF filter for speech or CW reception as described originally by DJ6HP in 1974 and which continues to represent an effective design. It provides a CW filter tuneable over about 450 to 2700Hz with the *Q* (bandwidth) variable over a range of about 5:1. For speech the filter can be switched to a notch mode. Although modern digital audio filters could provide more precisely shaped tuneable filtering, this analogue filter has received many endorsements over the years

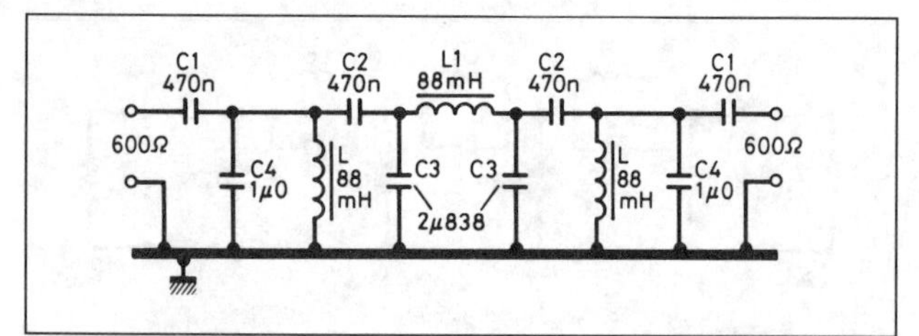

Fig 6.99. Passive AF filter design by DJ1ZB using standard 88mH toroids and with a centre frequency of about 420Hz and bandwidth about 80Hz. Note that this design is for 600Ω input/output impedance (*Sprat* No 58)

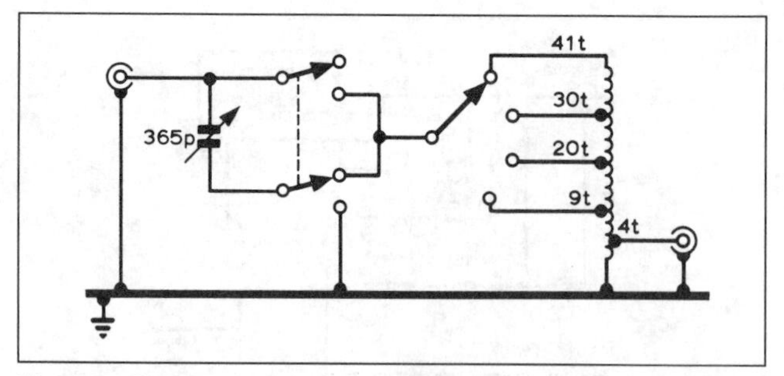

Fig 6.101. Typical antenna tuning and matching unit to cover 0.55 to 30MHz

the basis of excellent receivers, often rather better than is possible by modifying some more recent low-cost receivers. The main drawback of the older receivers is their long warm-up period, making it difficult to receive SSB signals satisfactorily until the receiver has been switched on for perhaps 15 or 20min.

Some of the older models using relatively noisy mixer stages may be improved on 14, 21 and 28MHz by the addition of an external *preamplifier* and such a unit may also be useful in reducing image and other spurious responses. However, high-gain preamplifiers should not be used indiscriminately since on a low-noise receiver they will seriously degrade the signal-handling capabilities without providing a worthwhile improvement of signal-to-noise ratio. Receivers having low noise but poor signal-handling capabilities can more often be improved by the fitting of a switched, adjustable or AGC-controlled antenna attenuator. Such an attenuator is likely to prove of most use on 7MHz where the presence of extremely strong broadcast signals will often result in severe cross-modulation and intermodulation.

A receiver deficient in selectivity can often be improved by adding a second frequency changer followed by a low-frequency (50 to 100kHz) IF amplifier (a technique sometimes known as a *Q5-er*); or by adding a crystal or mechanical filter, or by fitting a *Q*-multiplier. CW reception can be improved by the use of narrow-band audio filters, although the degree of improvement may not always be as much as might be expected theoretically because of the ability of an experienced operator to provide a high degree of discrimination.

Older receivers having only envelope detection may be improved for SSB and CW operation by the fitting of a product detector; or for NBFM reception by adding an FM discriminator.

Many older receivers use single rather than band-pass crystal filters (and the excellence of the single crystal plus phasing control for CW reception should not be underestimated) and these often provide a degree of nose selectivity too sharp for satisfactory AM or SSB phone reception: speech may sound 'woolly' and virtually unintelligible due to the loss of high- and low-frequency components. However, because the response curve of such filters is by no means vertical, the addition of a high degree of tone correction (about 6dB/octave) can do much to restore intelligibility and the combination then provides an effective selectivity filter for SSB reception. The tone correction circuit shown in Fig 6.100 is suitable for high-impedance circuits and can be adapted by using higher C and lower R for low-impedance circuits.

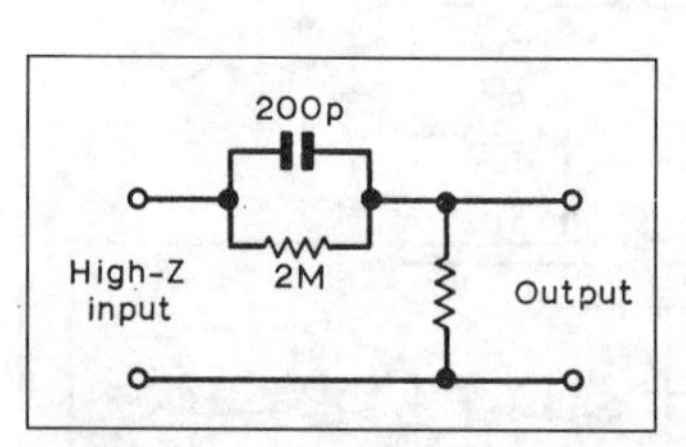

Fig 6.100. The single-crystal filter can be used effectively for phone reception by incorporating AF tone correction to remove the 'woolliness' of the heavily top-cut speech. A simple network such as the above provides top lift that restores intelligibility when used with the response curve of a typical single-crystal filter

The addition of an antenna matching unit between receiver and antenna can improve reception significantly in those cases where appreciable mismatch may exist (for example when using long-wire antennas with receivers intended for use with a 50 or 70Ω dipole feeder) (Fig 6.101).

A common fault with older receivers is deterioration of the Yaxley-type wave-change switch and/or the connection to the rotor spindle of the variable tuning capacitors; such faults may often cause bad frequency instability and poor reset performance. Improvement is often possible by the careful use of modern switch-cleaning lubricants and aerosols.

A simple accessory for older receivers (or those modern receivers not already incorporating one) is a crystal calibrator providing 'marker' signals derived from a 100kHz or 1MHz crystal. While a simple 100kHz oscillator will usually provide harmonics throughout the HF range, the availability of integrated circuit dividers makes it practicable to provide markers which are not direct harmonics of the crystal. For example 10kHz or 25kHz or even 1kHz markers can be provided using TTL decade divider logic or divide-by-two devices.

A receiver deficient in HF oscillator stability on the higher frequency bands may still form the basis of a good tuneable IF strip when used on a low frequency band in conjunction with a crystal-controlled converter. Again, when the basic problem is oscillator drift due to heat, this can sometimes be reduced by fitting silicon power diodes in place of a hot-running rectifier valve or by adding temperature compensation to the HFO. A more drastic modification is to replace an existing valve HFO with an internal or external FET VFO. Excessive tuning rate can sometimes be overcome by fitting an additional or improved slow-motion drive. A receiver with a good VFO can be modified for really high-stability performance (better than about 20Hz) by means of external 'huff and puff' digital stabilisation using crystal-derived timing periods.

Receivers not initially designed for SSB operation can be improved by fitting a product detector with crystal-controlled BFO, and possibly adding a good mechanical or crystal band-pass filter.

In brief, the excellent mechanical and some of the electrical characteristics of the large and solidly built receivers, such as the AR88, HRO, SUPER-PRO and some Eddystone models which featured single conversion with two tuned RF stages, are seldom equalled in modern 'cost-effective' designs. It may prove well worth spending time and trouble to up-grade these vintage models into receivers which can be excellent even by modern standards. Post-war Collins and Racal RA17L valved receivers remain highly regarded HF receivers.

The following summary indicates some common faults with older models and ways in which these can be overcome.

1. *Poor sensitivity.* Due to atmospheric noise this usually only degrades performance on 21 and 28MHz and then only on older valve models. Sensitivity can be improved by the addition of a preamplifier, but gain should not be more than is necessary to overcome receiver noise. Note that the sensitivity of a receiver may have been impaired by poor alignment, or by mismatched antennas, or due to the ageing of valves.
2. *Image response.* This can be reduced by additional pre-mixer selectivity, often most conveniently by means of a low-gain preamplifier with two or more tuned circuits. It is also possible to use a pre-tuned filter such as the Cohn minimum-loss filter for particular bands.
3. *Stability.* This is a direct function of the oscillators within the receiver. Excessive drift and frequency 'jumping' may be due to a faulty valve or band-change switch, or to incorrect adjustment of any temperature-compensation adjustments. Drift can sometimes be reduced by reducing the amount of heating of the oscillator coil by fitting heat screens, or by the addition of temperature compensation. But often with older receivers it will be found difficult to achieve sufficient stability on the higher frequency bands. In such cases considerably greater stability may be achieved by using the receiver as a variable IF system on one of the lower frequency bands, with the addition of one or more crystal-controlled converters for the higher frequency bands. It is worth noting that all oscillators (not only the first 'HFO') may be the cause of instability (eg second or third frequency conversion oscillator or even the beat frequency oscillator).
4. *Tuning rate.* The tuning rate of some older but still good receivers tends to be too fast for easy tuning of SSB and CW signals. Often this problem can be overcome by the fitting of an additional slow-motion drive on the main tuning control. Alternatively the receiver may be used, as mentioned above, as the variable IF section with a converter since the tuning rate on lower frequency bands may be satisfactory. Performance on SSB may be improved also by fitting a product detector where only envelope detection is built-in.
5. *Selectivity.* It is possible to improve the selectivity of a receiver by fitting an external low IF section, or by fitting a (better) crystal filter, or a *Q*-multiplier. Many SSB receivers make little provision for narrow-band CW reception and in such cases it may be possible to include a single-crystal filter with phasing control in one of the later IF stages, or to add a *Q*-multiplier or audio filter.
6. *Blocking and intermodulation.* Performance of many semiconductor (and some valve) receivers can be improved by the addition of even a simple antenna attenuator for use on 7MHz in the presence of extremely strong broadcast signals.

BUILDING RECEIVERS

For many years the percentage of home-built HF receivers in use on the amateur bands has been very low and increasingly has been confined to specialised sectors of the hobby such as compact, low-power (QRP) portable operation, often on one specific band and for a single mode of operation (often CW only) or as introductory receivers for those to whom the rather daunting cost of factory-built receivers or transceivers can represent a deterrence to HF operation.

There remain, however, valid reasons to encourage home-construction, not only for those with limited budgets but also as an ideal form of 'hands-on' learning process. A major advantage of building or modifying older receivers is that the constructor can then be confident of maintaining it in good trim. It is a substantial advantage for the amateur, particularly if located a long way away from the suppliers, to use equipment that he or she feels capable of keeping in good condition, and carrying out his or her own repairs when necessary. Increasingly factory-built equipment for the amateur does not lend itself to home-servicing.

Nor should it be forgotten that for those with the necessary practical experience, the availability of complex integrated circuits developed for consumer electronics and ceramic resonators that do not require skilled alignment has eased the construction of both simple and high-performance receivers. The factory designer must usually cater for all possible modes and bands, whereas the constructor can build a no-compromise receiver to suit his or her own particular interests. The amateur can still provide himself with a station receiver or transceiver, or a portable receiver or transceiver that can bear comparison with the best available factory models, and in doing so prove that the communications receiver is not a 'black box' or 'consumer appliance' of which the technology remains a largely unknown quantity.

It has traditionally been a feature of the hobby of amateur radio that the enthusiast strives to understand the technology; the home-construction and home-maintenance of equipment are not assets that should be surrendered lightly despite the undoubted attractions of factory-designed equipment.

Simple receivers, converters and single-band QRP transceivers can be built in an evening or two, but an advanced receiver – or receiver section of a multiband transceiver – may take several months of work and adjustment.

This section provides circuit diagrams of a number of relatively simple receivers, including direct-conversion (both heterodyne and regenerative models) and superhets that do not depend on initial up-conversion to VHF or frequency-synthesisers.

Although the homodyne-type direct conversion receiver and the 'zero-IF' form in which the output from a first mixer is fed directly into a product detector make possible low-cost HF receivers which are simple to build, care must be taken if optimum performance is to be achieved.

Because the necessary high overall gain (about 100dB) is achieved virtually entirely in the AF amplifier, these stages are very sensitive to hum pick-up at the 50Hz mains supply frequency and its harmonics. The AF stages should be well decoupled with the mixer and first AF stage in close proximity to enable the connections between them to be as short as possible; power supplies need to be well smoothed with care taken not to introduce 'earth loops' and/or direct pick-up from the magnetic field surrounding mains transformers and inductors. AF hum can usually be avoided altogether with battery-operated receivers.

The high AF gain may also introduce 'microphony' with components acting as microphones. A loudspeaker tends to make components vibrate and headphones are preferable, with the added advantage of requiring less AF output and overall gain. Ceramic capacitors, which exhibit piezo-electric characteristics, tend to introduce microphony; moulded polycarbonate capacitors are much to be preferred. Similarly, ferrite toroid cores used for AF filters can also introduce microphony. For inductors with values greater than 0.1mH, screened air-cored inductors are preferable.

Direct-conversion receivers of all types, including those with regenerative detectors, tend to suffer from local oscillator radiation from the receiving antenna and can cause interference in the locality, unless RF leakage through even a double-balanced mixer/demodulator is reduced, eg by the use of a broad-band

Table 6.4. Guide to coil windings

Diameter of former	Number of turns: Main tuned winding	Low-impedance antenna	Medium-impedance antenna	Reaction	SWG (main winding)	Approximate frequency range (MHz): Minimum	Maximum	Remarks
1½in ribbed air-core	3	1	2	2	20	13.5	31	Turns spaced two wire diam
	5	1	2 or 3	2 or 3	20	11.5	23	Turns spaced one wire diam
	9	2	4	3	22	6.5	14	Slight spacing
	17	3	5	4	24	3.4	6.8	Close wound
	42	6	10	10	30	1.6	3.3	Close wound
7/8in ribbed air-core	8	2	3	3	24	1.6	30	Close wound
	18	3	5	5	24/26	7	16	Close wound
	40	6	10	10	30	3.5	8	Close wound
3/8in dust-iron core	8	2	3	3	26	13.5	31	Closewound
	14	3	5	4	28	7.0	15	Close wound
	26	5	8	6	30	3.2	7	Close wound
	40	6	10	9	32	1.6	3.6	Slightly pilewound

Figures are given as a guide only and are based on a tuning capacitor with a maximum capacitance of 160pF. The maximum frequency limit will depend largely on the value of stray capacitances, while such factors as closeness of turns, lengths of lead, position of dust core (where used) will materially affect the frequency coverage. The reaction winding for TRF receivers should be close to the lower end of the main winding. The antenna coupling coil, if of low impedance, should be wound over the earthy end of the main winding. If of medium impedance or for inter-valve use, the coupling coil should be spaced a little way from the lower end of the main winding. Where an HT potential exists between windings, care should be taken to see that insulation is adequate. Small departures from the quoted wire gauge will not make any substantial difference. Generally, reaction and coupling windings can be of moderately fine wire. The number of turns for intermediate ranges can be judged from the figures given

isolator. A low-gain resonant or broad-band RF stage reduces oscillator radiation.

Microphony may also arise in the signal-frequency components, with the mixer acting as a phase detector, reflecting the interaction between the LO radiation or leakage and the incoming signal in a high-Q tuned circuit. Remember that the LO voltage is very much stronger than the incoming signal and it may be difficult to eliminate RF microphony altogether without effective screening and isolation.

Oscillator radiation can also result in RF hum that appears as 50Hz hum in the headphones. This arises from the oscillator signal becoming hum modulated in the mains wiring/rectifiers and then being re-radiated back to the receiver antenna. The cure is to stop the LO signal from radiating; all connections to the receiver, such as power supply and headphones, should be RF grounded to the receiver case using decoupling capacitors.

Toroid cores

Increased use is being made of small toroid cores; since the magnetic field is virtually closed, inductors made in this way are largely self-screening and can be used in close proximity to the chassis or other components. To preserve the closed magnetic field the inductors should be wound symmetrically and it is not good practice to trim the coils by spreading or compressing the turns.

Approximate windings can be determined from: turns = $1000\sqrt{(L/A_l)}$ where L is the wanted inductance in millihenrys and A_l is the factor 'millihenrys per thousand turns' which has to be obtained from manufacturer's data (occasionally stamped on the core) or from trial windings. Conventional GDO checking is not usually possible due to the closed magnetic loop. Fig 6.102 shows a technique providing adjustment where it is not important to preserve a completely closed loop.

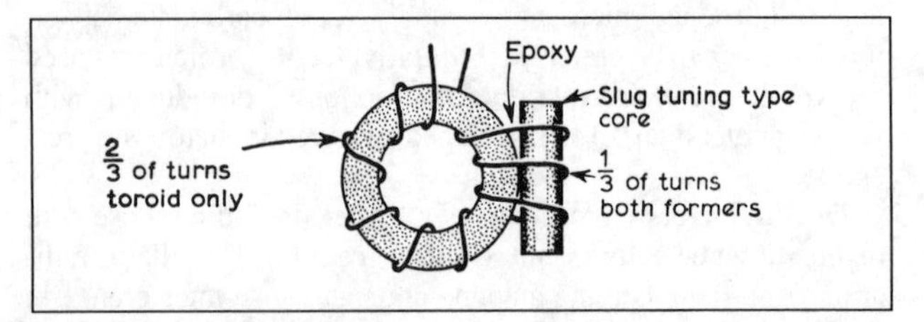

Fig 6.102. Tuneable toroid technique. About 10% variation in inductance can be achieved

When trimming by subtracting turns a useful accessory is a small crochet hook. Resonant circuits using toroids plus disc-ceramic capacitors may be trimmed by reducing the value of the capacitor by grinding the capacitor on a sanding disc or grinding wheel (up to 50% reduction in capacitance is usually possible provided that the grinding is longitudinal and does not smear metal dust between the plates). Exposed plate edges should be sealed by wax or polystyrene dope.

One cannot tap part of a turn on a toroid form; any time the wire passes through the central hole, it is effectively *one* turn. Try to wind turns just tight enough to keep them sliding around the core, but not too tight. Coils have a higher figure of merit if wound loosely but crossovers should be avoided. Avoid the use of the epoxy-type adhesive to hold windings in position on small coils since this increases distributed capacitance and reduces the number of turns, lowering the figure of merit. Nylon screws are the best mounting medium and will hold turns in place. Low-impedance link turns can be wound over other windings but space the link winding evenly around the core for optimum results.

Building stable HF oscillators

Most home-built receivers still tend to rely for stability on a tuneable HF oscillator, since the alternative in the form of a digital frequency synthesiser can pose problems in the amount of noise produced by the simpler synthesisers, and the use of direct digital synthesis (DDS) adds significantly to the cost of a receiver, although may well come to be the standard form of synthesiser for future high-performance receivers.

Construction of a stable free-running oscillator demands a similar approach to that used for transmitter VFOs. A band-switched oscillator that is adequately stable for SSB and data transmission is not an easy task, particularly where the fundamental frequency is above about 10MHz. Single-range oscillators can be used in double-conversion superhets where the first HFO is crystal controlled. Alternatively an oscillator covering

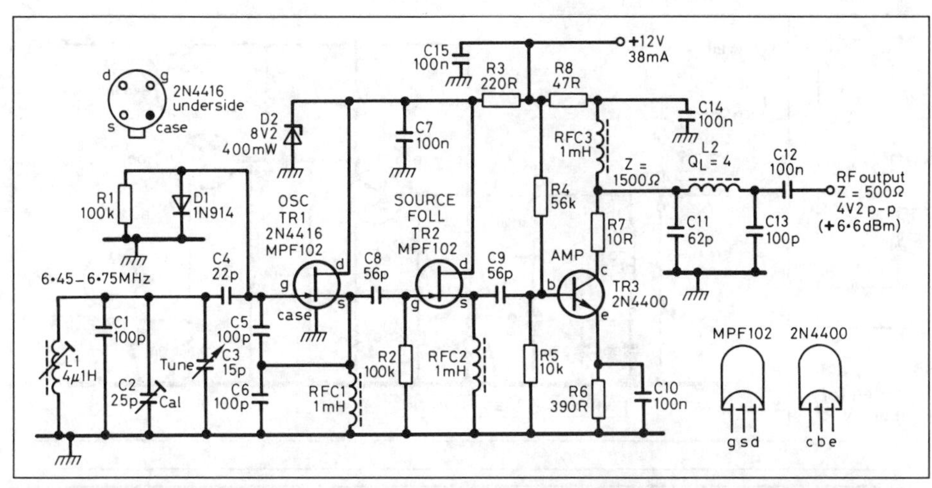

Fig 6.103. The VFO described by W1FB in *CQ* covering 6.45 to 6.75MHz but suitable for other frequency ranges by appropriate changes to L1/C1. C1, C4, C5 and C8 are NP0 ceramic or polystyrene types. C2 is a ceramic trimmer and C3 a miniature 15pF air variable. L1 is 32 turns of No 28 enamel wire on Amidon T-50-6 (yellow) toroid. L2 has 25 turns (No 28 enam) on Amidon FT-37-61 ferrite toroid

5.0 to 5.5MHz can be used for a two-band, 3.5 and 14MHz receiver with a (first) IF of 9MHz.

Some construction hints, based on advice given by Doug DeMaw, W1FB, include: select a suitable oscillator device, preferably a JFET or MOSFET since bipolar transistors tend to show greater changes of internal resistance and capacitance with temperature, and are relatively noisy. It is preferable to use a good mechanical tuning capacitor rather than an electronic tuning diode (which changes capacitance with temperature); use double-bearing capacitors that turn freely, avoid capacitors with aluminium vanes – plated brass or iron vanes are better. Use 0.5W carbon-film or carbon-composition resistors since they tend to be more stable than the small 0.25 or 0.125W types. Two or more fixed capacitors in parallel (eg two 50pF NP0 ceramic capacitors to provide 100pF) provide more surface area than a single capacitor, minimising internal heating from RF current. Use single-sided G-10 glass epoxy board material for VFO printed circuit boards, rather than double-sided PCBs or phenolic-base board. Double-sided boards may form unstable low-*Q* capacitors. Remember that PCBs are not the only form of construction. Physically separate and shield the oscillator from other heat-producing components. Use the lowest practical oscillator voltage (eg 6 to 8V) to reduce internal heating, regulated by a zener diode or, better still, its own small three-terminal IC regulator. Filter all DC leads entering the VFO box with an RF choke and 1000pF feedthrough capacitor to keep RF energy entering the VFO circuit. Do not use low-cost plastic trimmer capacitors in a VFO; use miniature air trimmers or NP0 ceramic units.

Recognise that a high-dynamic-range mixer requires substantial injection power, often more than 100mW, so that isolator/amplifier stages are usually required following the oscillator device. Push-pull (kallitron) oscillators can be an advantage: Fig 6.104.

Designs for home construction

The simplest types of HF receiver suitable for home construction 'on the kitchen table' without the need for high-grade measuring equipment etc are undoubtedly the various forms of 'straight' (direct-conversion) receivers using either a regenerative detector, or *Q*-multiplier plus source-follower detector, or the now more popular homodyne form of direct conversion, preferably using a balanced or double-balanced product detector. In practice, to avoid the complications of band-switching, most of the simple models tend to be either single-band receivers or may still use the once-popular 'plug-in' coils.

Figs 6.105-6.110 show a representative selection of circuit diagrams for simple receivers, suitable for use directly with high-impedance headphones, or with output transformers permitting the use of modern low-impedance headphones.

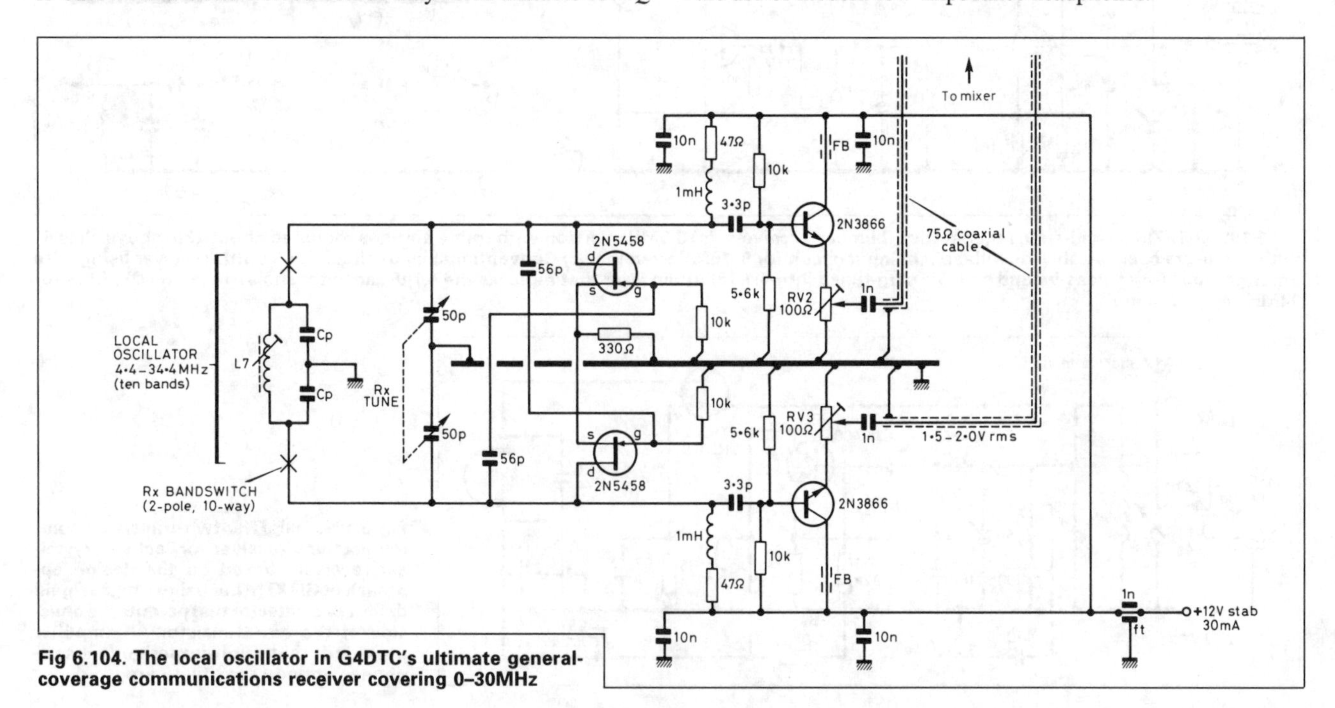

Fig 6.104. The local oscillator in G4DTC's ultimate general-coverage communications receiver covering 0–30MHz

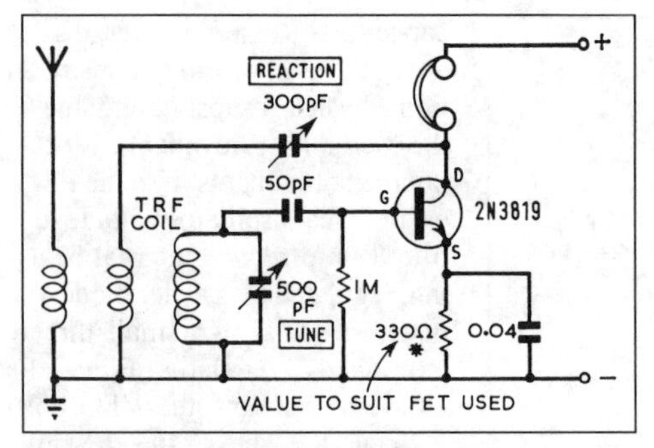

Fig 6.105. A single-FET receiver

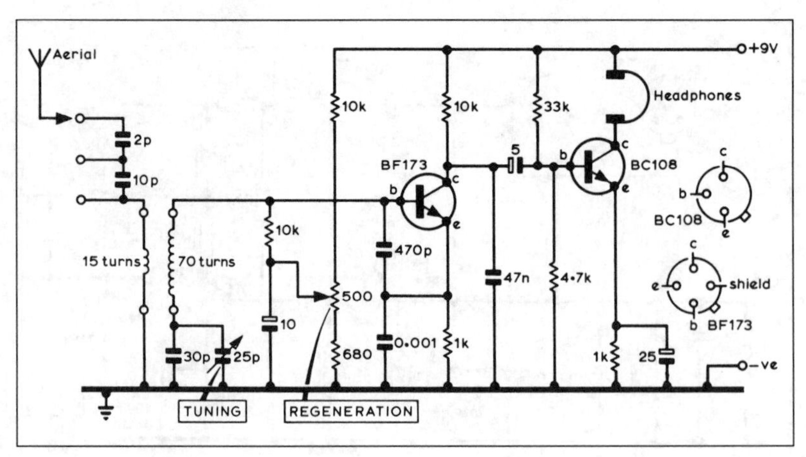

Fig 6.106. A simple 'straight' receiver intended for 3–5MHz SSB/CW reception and using Clapp-type oscillator to improve stability

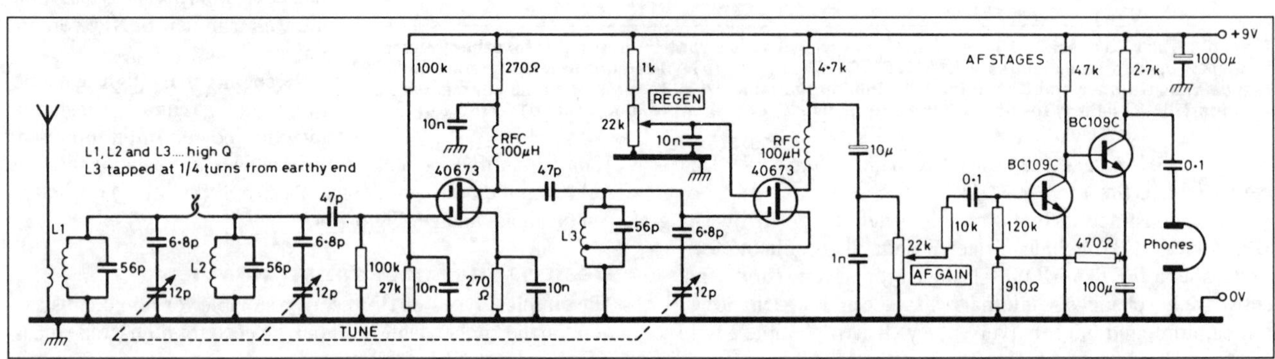

Fig 6.107. Solid-state 14MHz TRF 'straight' receiver originally described by F9GY in *Radio-REF* in the 'seventies and intended primarily as a monoband CW receiver

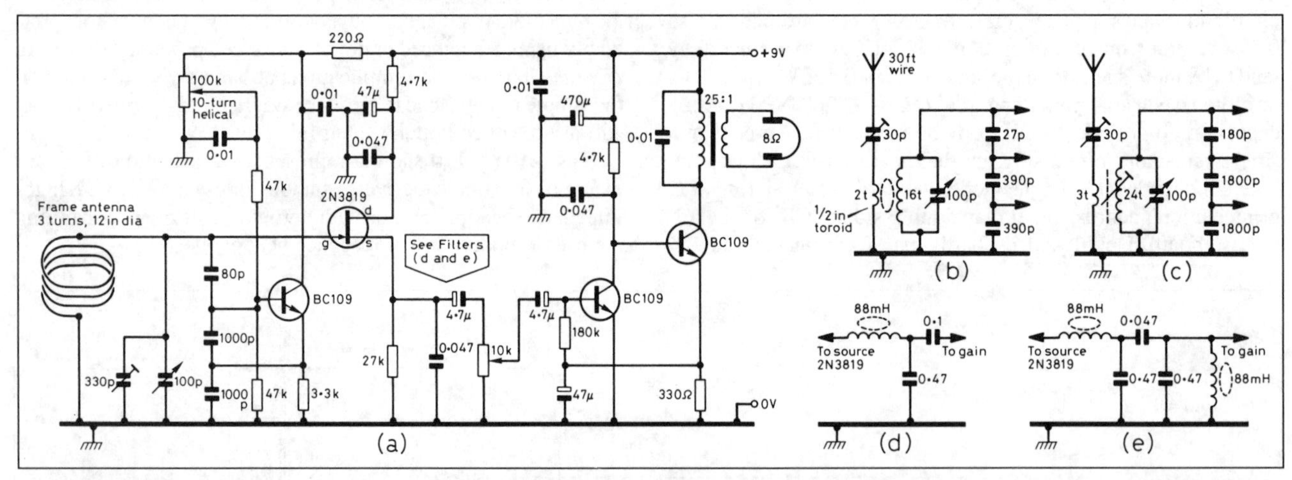

Fig 6.108. GI3XZM's solid-state regenerative 'blooper' receivers. (a) 3.5MHz version with frame antenna mounted about 12in above chassis with miniature coaxial-cable 'downlead'. (b) Input circuit for 9–16MHz version. (c) Conventional input circuit for 3.5MHz receiver using wire antenna (coil 26swg close-wound on 0.5in slug-tuned former). (d) Audio filter that replaces the 4.7μF capacitor shown in (a). (e) CW filter for 14MHz Mk 2 version

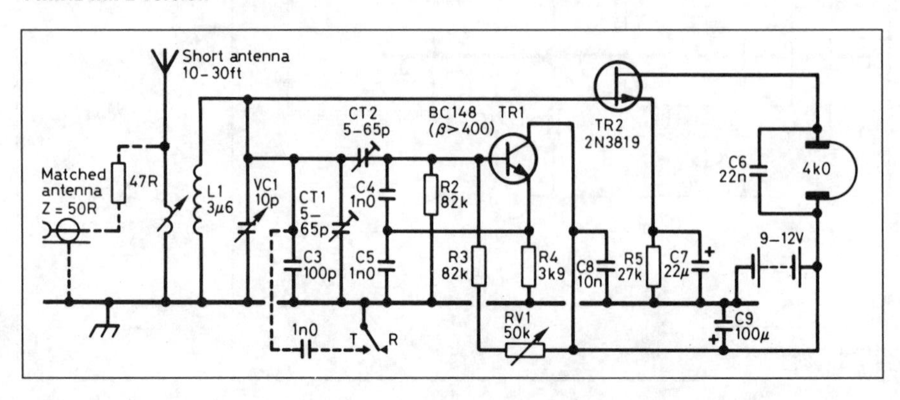

Fig 6.109. G3RJT's 'two-transistor communications receiver' or 'active crystal-set receiver', based on the design approach of GI3XZM but using a higher-gain drain-bend detector that permits the omission of the two-transistor AF amplifier provided that high-impedance headphones of good sensitivity are used

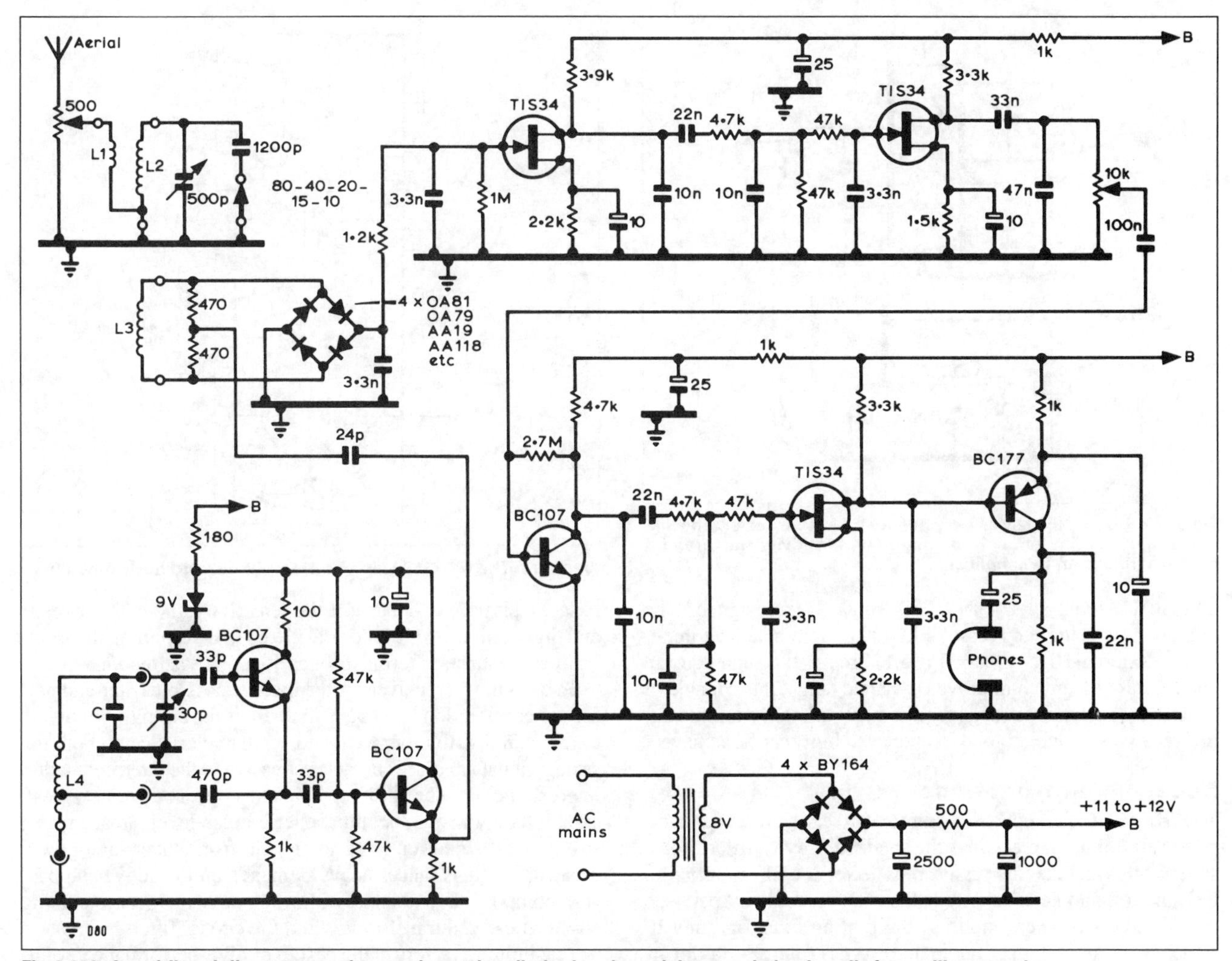

Fig 6.110. A multiband direct-conversion receiver using diode ring demodulator and plug-in coils for oscillator section

For simple receivers where a very high dynamic range is not sought, construction can be simplified by the use of NE602 or NE602A-type IC devices which contain a double-balanced mixer, oscillator and isolator stages. This device, originally developed for VHF portable radiophones, has been widely adopted by amateurs for use as frequency converters, complete front-ends for direct-conversion receivers, and as product detectors etc. Fig 6.111 outlines the NE602 and some ways it can be used, while Fig 6.112 shows its use as the front-end of a 28MHz direct-conversion receiver which could be adapted for lower HF bands.

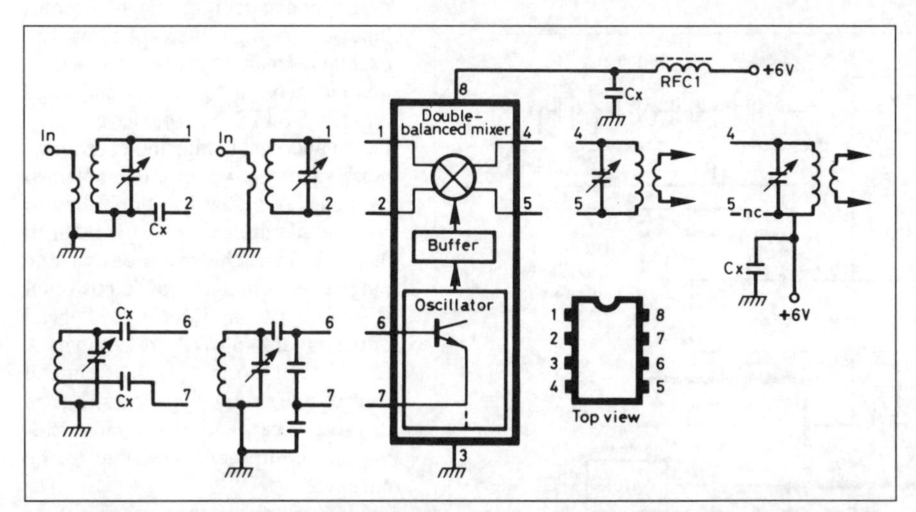

Fig 6.111. Typical configurations of the NE602N. Balanced circuits are to be preferred but may be more difficult to implement. Cx blocking capacitor 0.001 to 0.1μF depending on frequency. RFC1 (ferrite beads or RF choke) recommended at higher frequencies. Supply voltage should not exceed 6V (2.5mA). Noise figure about 5dB. Mixer gain 20dB. Third-order intercept 15dBm (do not use an RF preamplifier stage). Input and output impedances are both 2 × 1.5kΩ. Drop-in replacement type NE602A has an extra 5dB or so of dynamic range

The NE602 is equally suitable for use as a crystal-controlled frequency converter to provide extra bands in front of an existing receiver or for the 'super-gainer' form of simple superhet or as the mixer/oscillator stages of a conventional superhet, possibly using a second NE602 as a product-detector.

Receivers based on standard IC devices have particular application for 'listening' or as the receiver section of compact low-power transceivers.

Fig 6.113 shows how a Motorola MC3362 IC can form the complete front-end of a single-band superhet including IF and detector stages as part of a compact transceiver.

Fig 6.115 is a 3.5MHz superhet design based on NE602/SL6700/

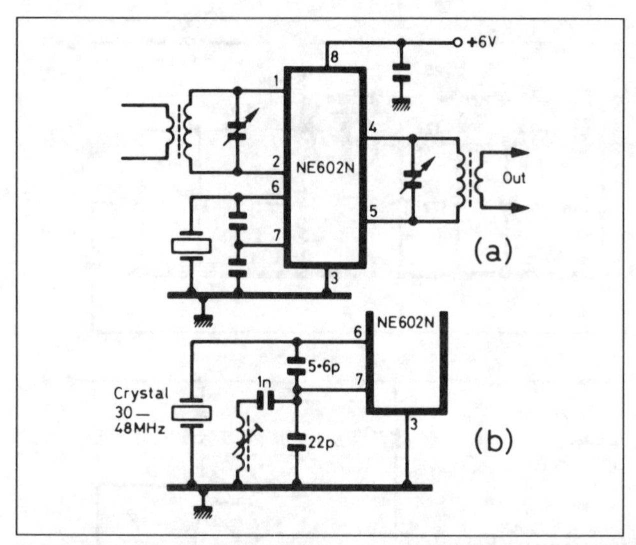

Fig 6.112. Use of NE602N as a crystal-controlled converter. The 5dB noise figure is low enough for optimum sensitivity up to about 50MHz without an RF amplifier

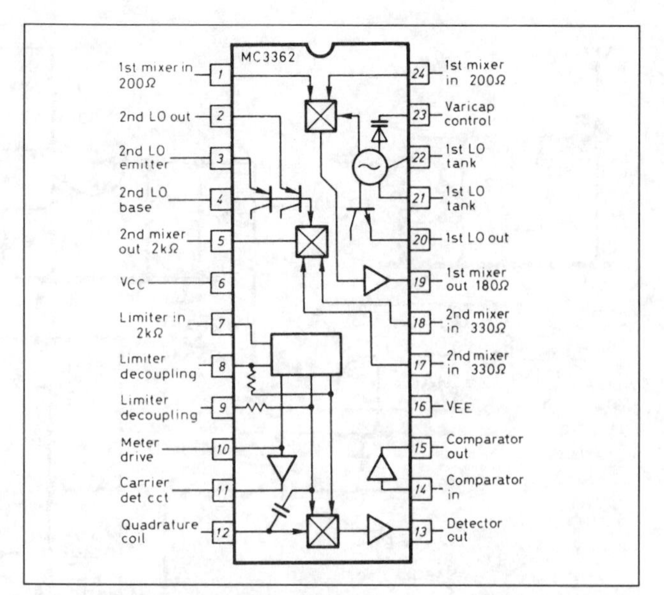

Fig 6.114. The MC3362 chip showing pin-out and basic functions

LM380 ICs plus an 2N3819 FET BFO. This design can be upgraded by using the Plessey SL6440 double-balanced mixer with an external HF oscillator. Fig 6.117 represents such a design and can provide much improved dynamic range; it is capable of satisfying the dynamic range requirements of main communication receivers, provided good frequency stability can be achieved.

Super-linear front-ends

The *front-end* of a superhet or direct-conversion receiver comprises all circuitry preceding the main selectivity filter. For a superhet this includes the passive preselector, the RF amplifier(s), the mixer(s) and heterodyne oscillator(s), the diplexer between mixer and post-mixer amplifier, the roofing filter, and any IF stages up to an including the main (crystal) filter. For a direct-conversion receiver, the front-end comprises all stages up to the selective audio-filter(s), including the product detector (which for a high-performance receiver may be of the two-phase, audio-image-rejecting type). For any receiver, superhet or direct-conversion, in which digital signal processing at IF or audio baseband is used to determine the selectivity, the A/D converter and digital filtering must be considered in determining the front-end performance in terms of linearity and dynamic range.

In designing a receiver for the highest possible front-end performance, attention must be paid to all of the circuitry involved, to the gain distribution, to the noise characteristics, to both the strong-signal handling characteristics and to the intermodulation intercept points. The ability to hold and copy an extremely weak signal, barely above the atmospheric noise level, adjacent to a strong local signal or close to signals from super-high-power broadcast signals places heavy demands on the active and passive components available within amateur budgets, including any ferrite-cored transformers and the crystal filters.

The limiting factors in the design of high-performance, solid-state HF receivers remain the spurious-free dynamic range (SFDR) of the mixer-stage, the noise and stability of the associated oscillator and strong-signal performance of the filters. Jacob Makhinson, N6NWP (*QST* February 1993) believes, by applying known design principles, radio amateurs can construct a high-performance front-end which combines a very high intercept point with excellent sensitivity. Used with a low-noise local oscillator, a front-end based on a DMOS FET quad device as a double-balanced switching mixer and low-noise square-wave injection, obtained by means of a dual flip-flop followed by a simple diplexer network at an IF of 9MHz, can achieve a wide dynamic range even when a suitable push-pull RF low-noise amplifier is used ahead of the mixer. N6NWP stresses that "A receiver incorporating such a front-end can provide strong-signal performance that rivals or exceeds that of most commercial equipment available to the amateur."

Fig 6.122 shows the essentials of the N6NWP 14MHz mixer based on a Siliconix/Calogic Si8901/SD8901 DMOS FET quad device together with a 74HC74 dual flip-flop to provide

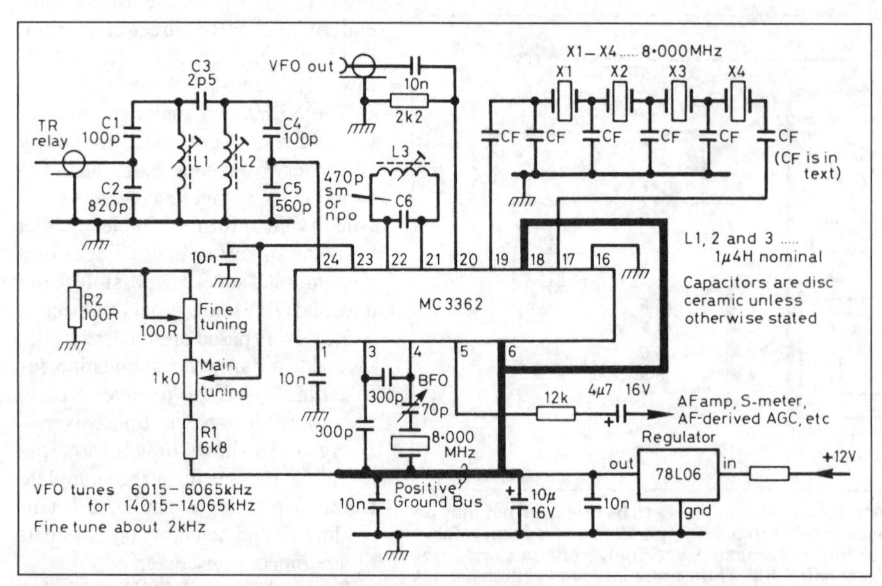

Fig 6.113. How K9AY uses the MC3362 as the complete front-end of the 14MHz superhet receiver section of his 5W QRP transceiver

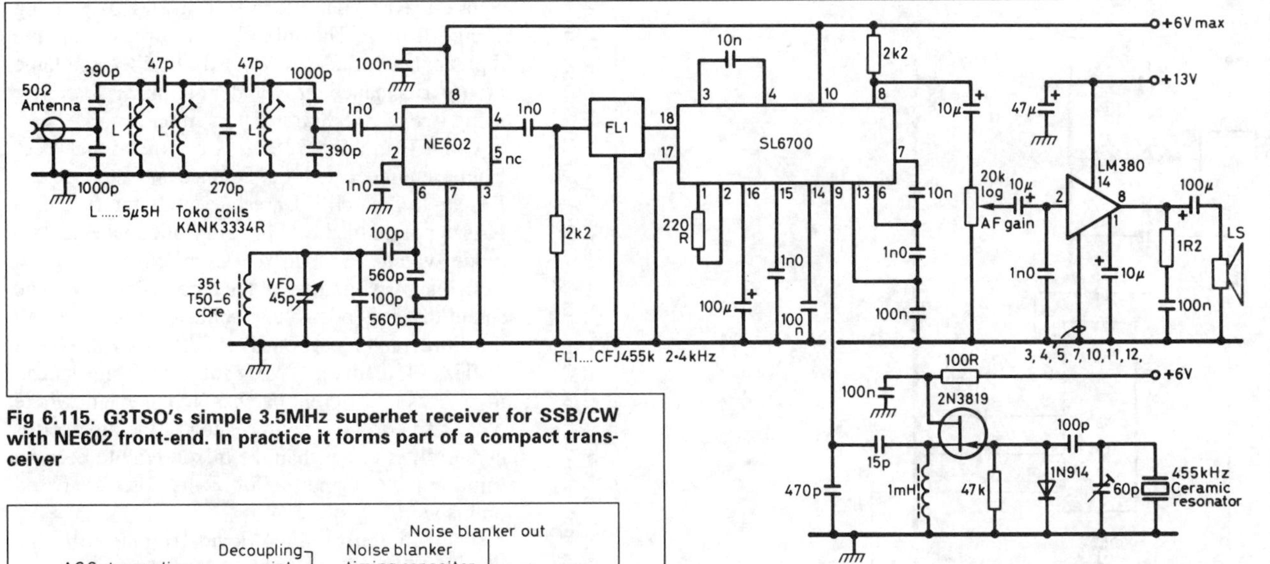

Fig 6.115. G3TSO's simple 3.5MHz superhet receiver for SSB/CW with NE602 front-end. In practice it forms part of a compact transceiver

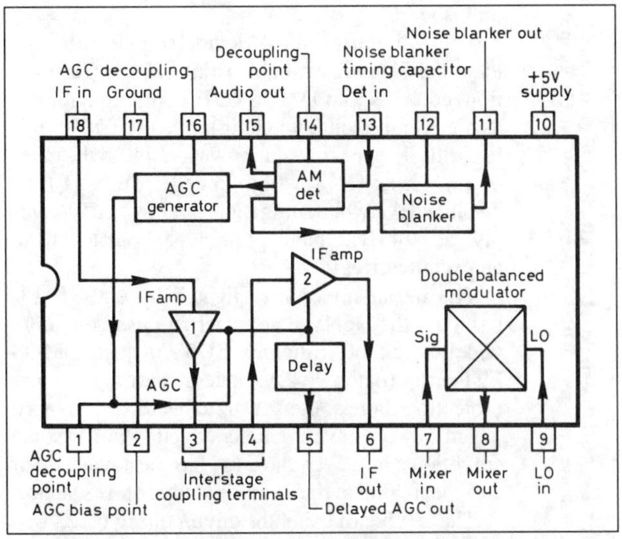

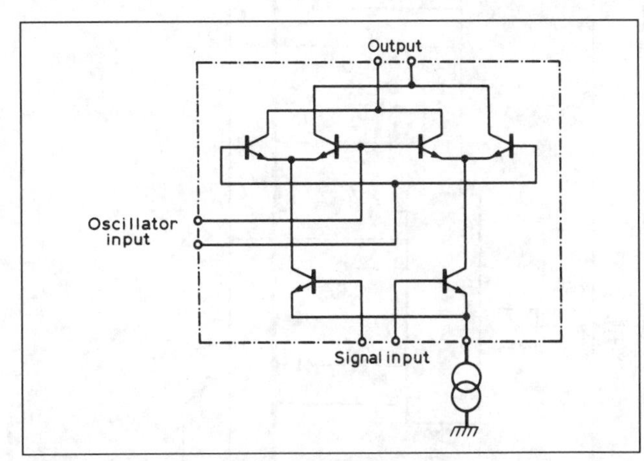

Fig 6.118. The basic transistor 'tree' mixer/modulator arrangement which forms the basis of the SL6440 mixer IC

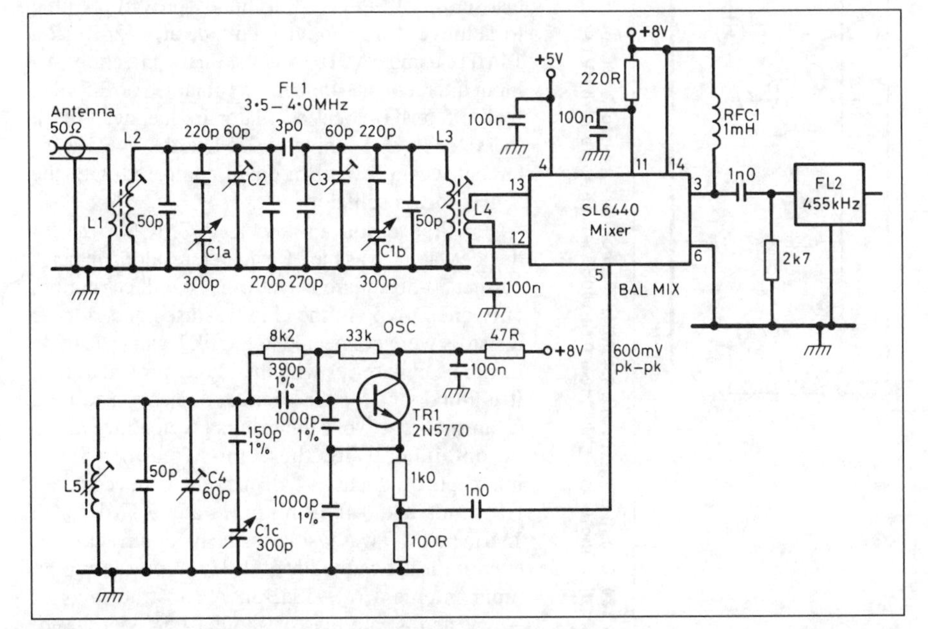

Fig 6.117. Circuit diagram of the front-end of the high-dynamic-range receiver based on the Plessey SL6440 double-balanced mixer and SL6700 subsystem IC

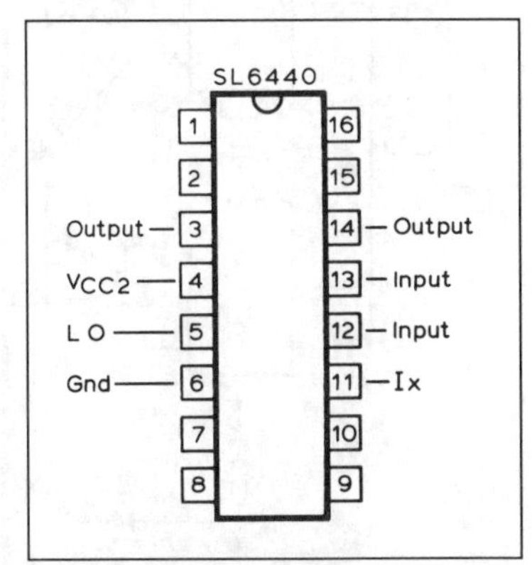

Fig 6.119. Pin connections for the plastic or ceramic SL6440A/C device. Pin connections viewed from above

Fig 6.116. Block diagram of the Plessey SL6700 subsystem IC used as the heart of the simple superhet and the high-dynamic-range 3.5MHz receiver

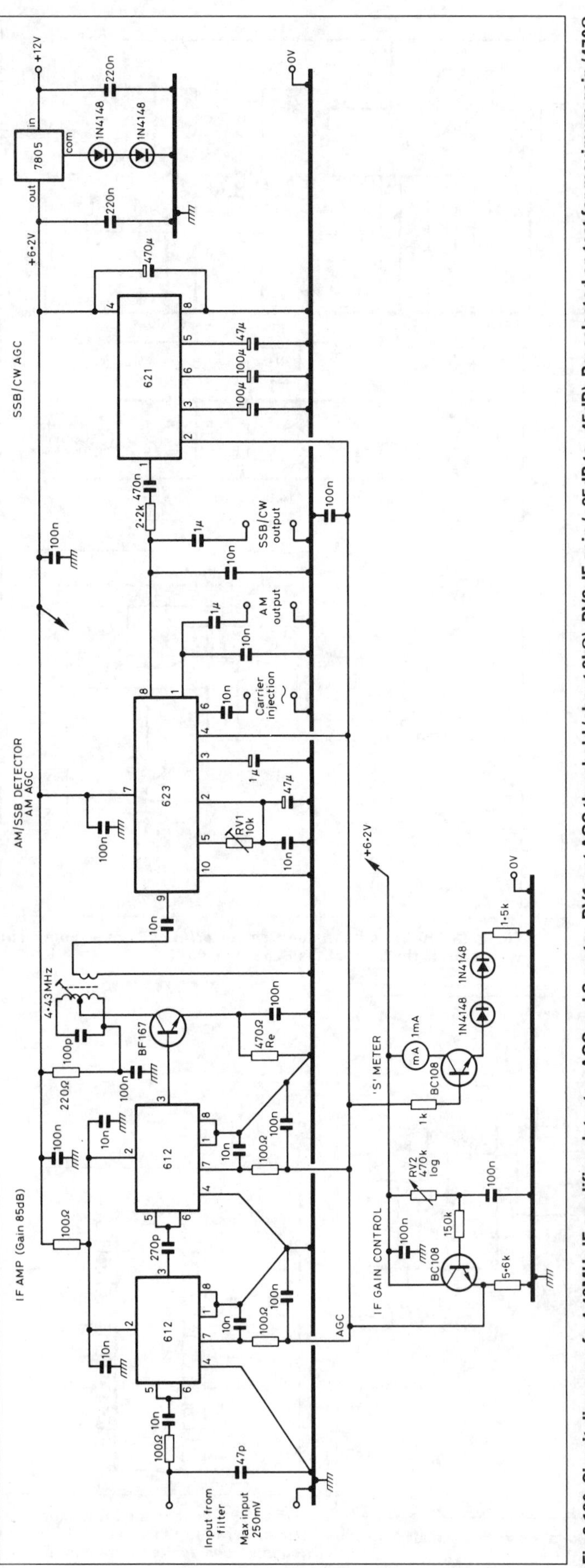

Fig 6.120. Circuit diagram of a 4.43MHz IF amplifier, detector, AGC and S-meter. RV1: set AGC threshold (about 3kΩ). RV2: IF gain (+85dB to –45dB). Rc: selected on test for maximum gain (470Ω nominal). S-meter is highly linear 0 to 140dB above noise with RV1 set so that noise almost triggers AGC. AM output 55mV for 10µV input. SSB output 30mV for 2µV input. Carrier about 200mV RMS

square-wave 5MHz injection from a VFO operating about 10MHz. The mixer is followed by simple diplexer network. N6NWP also described a suitable high-performance pre- and post-mixer amplifier using two MRF586 transistors in push-pull.

Colin Horrabin, G3SBI, is a professional electronics engineer with the Science and Engineering Research Council's Daresbury Laboratory which has supported his investigative work on the new H-mode switched FET mixer described later.

In his investigations of the N6NWP front-end he found that it is possible to achieve extremely high third-order intercept points (+50dBm on 1.8, 3.5 and 7MHz, +45dBm on 14MHz subject to some spread in devices in different batches) to the extent where other parts of the circuitry, such as the diplexer or crystal filter rather than the mixer, tend to become limiting factors, and the following notes are based on his work.

The Siliconix Si8901 device (replaced by the identical SD8901 available from Calogic Corporation) contains four DMOS FETs configured for use as a commutation (switching) mixer. N6NWP utilises this device with square-wave drive to the gates of the FETs from a high-speed CMOS D-type bistable device operating unusually from a 9V supply. The 14MHz intercept point is about +39dBm, an excellent figure.

The initial mixer investigated by G3SBI was based on the N6NWP approach but used the more widely available Siliconix SD5000 quad DMOS FET array (batch 9042). Since this array has gate-protection diodes, the substrate needs to be biased negatively to prevent gate conduction under some conditions; however, the array has the advantage of close matching of the drain-to-source on-resistance.

The test board using the circuit of Fig 6.123 was made using earth-plane construction, with all transformers and ICs fitted into turned-pin DIL construction, so that they could be changed easily. With the test set-up of Fig 6.124, it initally proved possible to achieve a true input intercept of +42dBm on 14MHz using 5MHz local oscillator injection. An input intercept of +45dBm was obtained on 3.5MHz with a 5.5MHz local oscillator frequency. With a local oscillator running at 23MHz the 14MHz intercept was a few dBm down compared with the 5MHz local oscillator.

For this reason an advanced CMOS 74AC74 device was used as the LO squarer, resulting in near-perfect 50-50 square waveform. To reduce ringing, only one D-type in the chip was used and stopper resistors were connected to the FET gates; a single ferrite bead in series with the V_{cc} pin proved useful. It is important for the oscillator injection to be a clean square wave if the results given above are to be obtained. With these modifications input intercepts of at least +42dBm were achieved on all HF bands and +46dBm on 1.8 and 3.5MHz. On 7MHz no IMD was visible on the spectrum analyser, even with a bandwidth of 10Hz, representing an input intercept of +50dBm. A substrate bias of –7.5V and a gate bias of about +4.5V were used; conversion loss was 7dB.

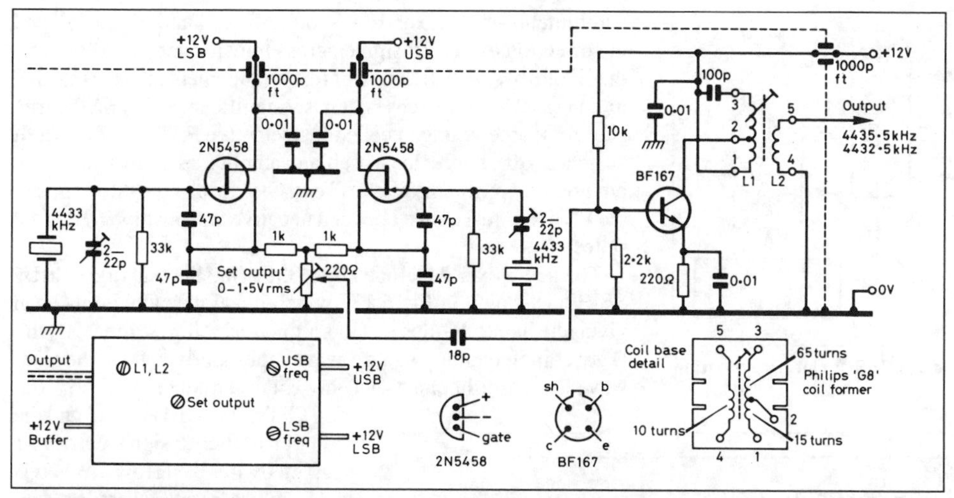

Fig 6.121. Circuit diagram of the carrier injection/BFO. L1 80ft 36swg on 6mm former tapped at 15t. L2 10ft 36swg over 15t of L1. Set output to 200mV (150mV) peak into 1kΩ. Crystal frequencies pulled: USB 4435.5kHz, LSB 4432.5kHz

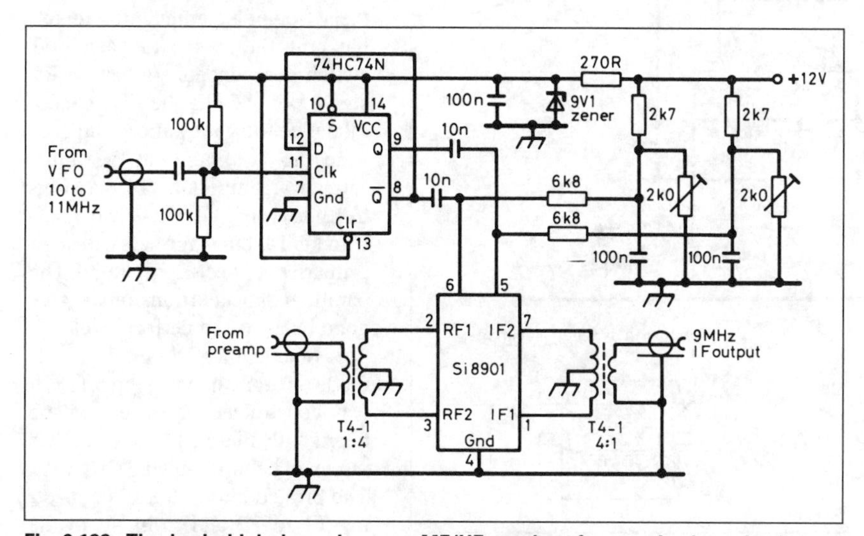

Fig 6.122. The basic high-dynamic-range MF/HF receiver front-end mixer circuitry as developed by N6NWP

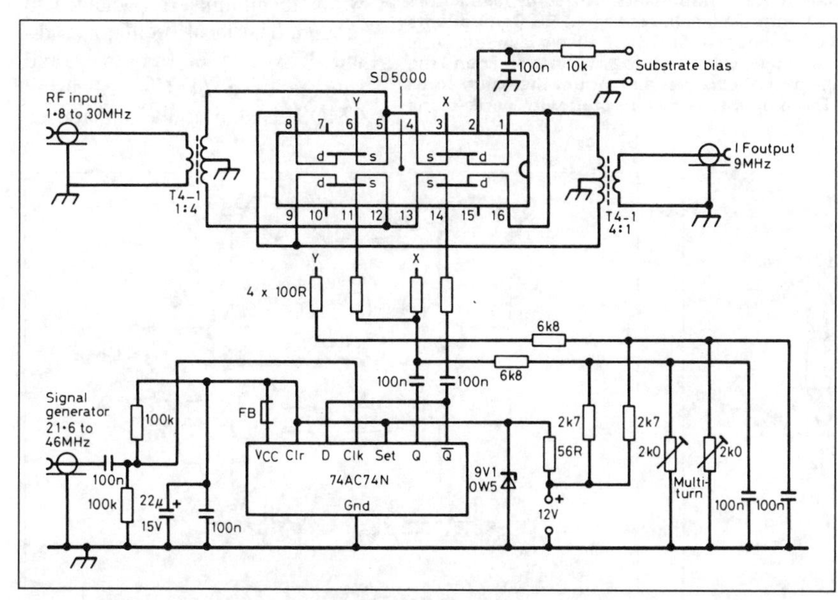

Fig 6.123. G3SBI's modified N6NWP-type mixer test assembly using the SD5000 FET array and 74AC74N to provide square-wave injection from a high-quality signal generator source at twice the required frequency

However the intercept point was found to degrade sharply as soon as the input signal exceeds +7dBm (0.5V); the situation can be recovered by dropping the gate bias voltage, but it is then no longer possible to achieve intercept points above +45dBm.

Fig 6.125 shows a 9MHz post-mixer amplifier developed by G3SBI, again based on the N6NWP approach but with changes that provide improved performance in terms of gain and output intercept point, with a noise figure of 0.5dB. Whereas N6NWP used the MRF586 device, G3SBI used the MRF580A device, giving a lower noise figure at a collector current of 60mA. Measured performance showed a gain of 8.8dB, output intercept +56dBm, noise figure 0.5dB. However, a crystal filter driven by the amplifier would present a complex impedance, particularly on the slope and near the stop-band, and would seriously degrade performance of the amplifier.

G3SBI also investigated the performance of quadrature hybrid 9MHz crystal-filter combinations. He found that the performance of budget-priced crystal lattice-type filters is a serious limitation (home-made ladder filters appear to have higher intercept points and lower insertion loss although the shape factor may not be quite so good). The problem with budget-priced lattice filters can be reduced by eliminating the post-mixer amplifier with the mixer going immediately to a quadrature hybrid network 2.4kHz-bandwidth filter, followed by a low-noise amplifier. The 2.4kHz filter is then used as a roofing filter. Although this is not an ideal arrangement, it can result in an overall noise figure of about 13.5dB (5dB noise figure due to the filter and amplifier, another 7dB from mixer loss, and a further 2.5dB loss due to the antenna input band-pass filter). This would be adequate sensitivity on 7MHz without a pre-mixer amplifier.

The N6NWP-type mixer followed basically accepted practice in commutation (switching mixers) achieving a +50dBm input intercept point on 7MHz when used with a precise square-wave drive, but the performance fell off on bands lower and higher in frequency than 7MHz. Results could be improved on the lower frequency bands by altering the capacitive balance of the RF input and above, but this had no significant effect on 14MHz and above.

Fig 6.126 illustrates a conventional

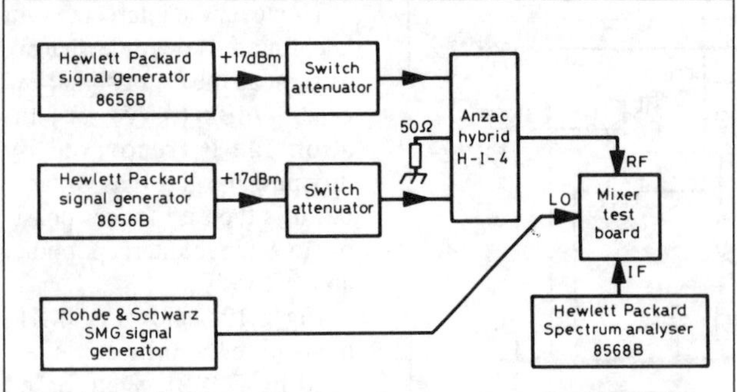

Fig 6.124. The test instrumentation used by G3SBI for intermodulation tests on the mixer

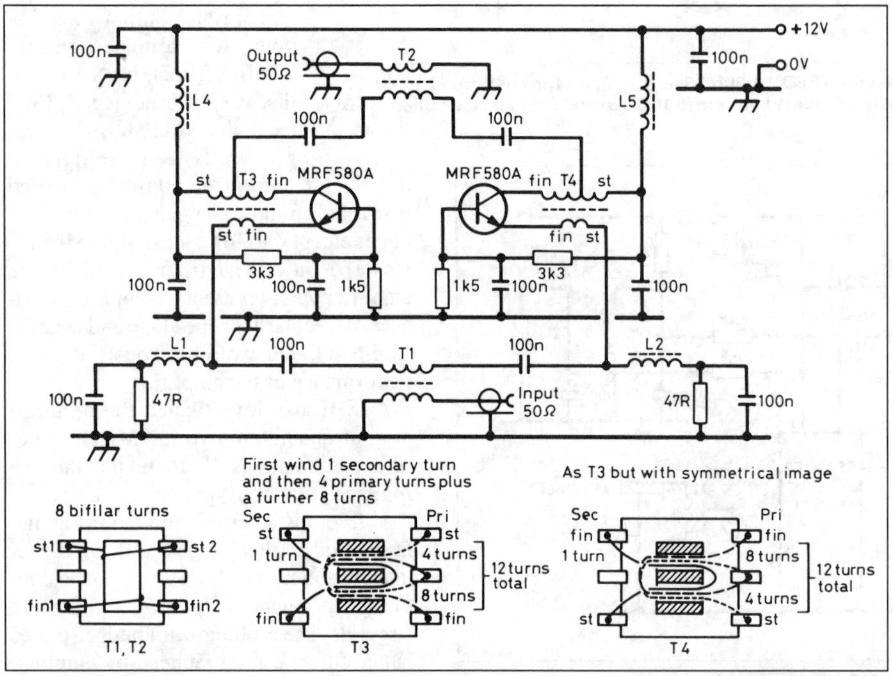

Fig 6.125. G3SBI's modified post-mixer test amplifier adapted from the N6NWP design but using the MRF580A devices. All resistors 0.25W metal-film RS Components. All 0.1μF capacitors monolithic ceramic RS Components. L4, L5 4t of 0.315mm dia bicelflux wire on RS Components ferrite bead. L1, L2 5t 0.315mm dia bicelflux wire (RS Components). T1–T4 use 40swg bicelflux wire. Take two glassfibre Cambion 14-pin DIP component headers, cut each into two parts and bend the tags 90° outwards. Stick a piece of double-sided tape onto the header and mount the balun cores on this. Wind the transformers as shown above. The amplifier is constructed with earth-plane layout

commutation ring mixer. If A is 'on', FETs F1 and F3 are 'on' and the direction of the RF signal across transformer T2 is given by the 'F' arrows. A deficiency of this arrangement is that as the RF input signal level increases, it has a significant effect on the true gate-to-source voltage needed to switch the FET 'on' or keep it switched 'off'. Larger local oscillator amplitudes are then required, but linearity problems may still exist because of the difference in the FET 'on' resistance between negative and positive RF signal states.

The new mixer developed by G3SBI (intellectual title held by SERC) is shown in Fig 6.127 which illustrates why it has been given the name 'H-mode'. Operation is as follows: Inputs A and B are complementary square-wave inputs derived from the sine-wave local oscillator at twice the required frequency. If A is 'on' then FETs F1 and F3 are 'on' and the direction of the RF signal across T1 is given by the 'E' arrows. When B is 'on', FETs F2 and F4 are 'on' and the direction of the RF signal across T1 reverses (arrows 'F'). This is still the action of a commutation mixer, but now the source of each FET switch is grounded, so that the RF signal switched by the FET cannot modulate the gate source voltage.

In this configuration the transformers are important: T1 is a Mini-Circuits type T4-1; T2 is two Mini-Circuits T4-1 transformers with their primaries connected in parallel. The parallel-connected transfomers give good balance and perform well.

A practical test circuit of the H-mode mixer is shown in Fig 6.128. It was constructed on an earthplane board with all transformers and ICs mounted in turned-pin DIL sockets. The printed circuit tracks connecting T1 to T2 and from T2 to the SD5000 are kept short and of 0.015in width to minimise capacitance to ground. The local oscillator is divided by two in frequency and squared by a 74AC74 advanced CMOS bistable similar to that used

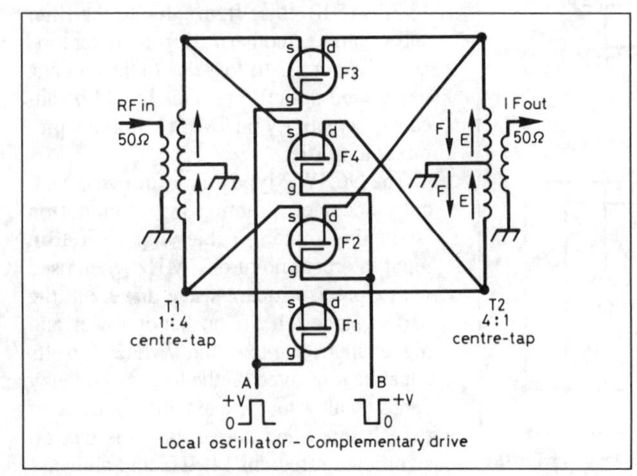

Fig 6.126. Conventional commutation ring mixer

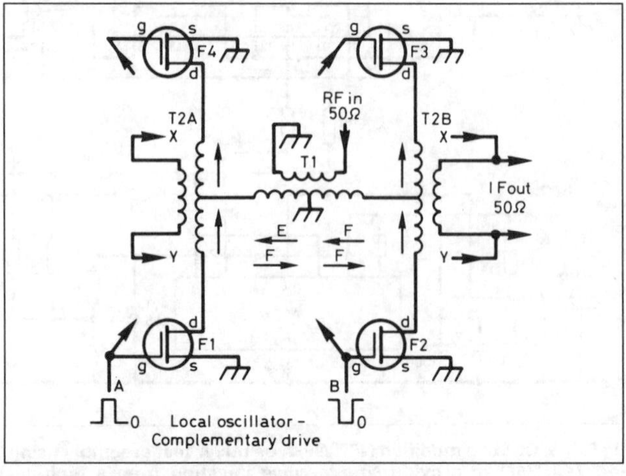

Fig 6.127. G3SBI 'H-mode' mixer

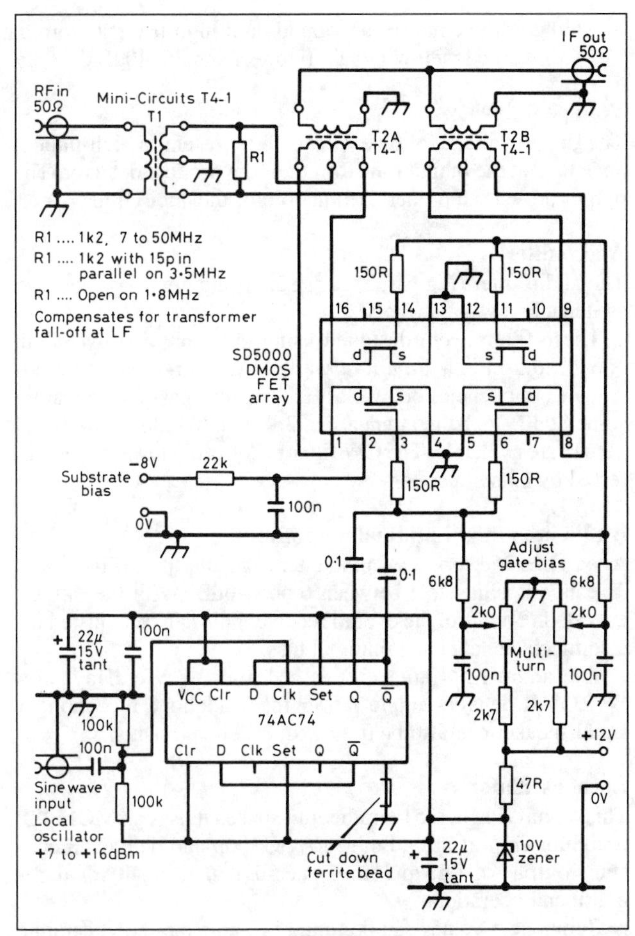

Fig 6.128. Test assembly for 'H-mode' mixer

in the N6NWP-type mixer. However, the bistable is run from +10V instead of +9V and a cut-down RS Components ferrite bead is inserted over the ground pin of the 74AC74 to clean up the square wave.

The preferred method of setting the gate-bias potentiometers with the aid of professional-standard test equipment is as follows. One potentiometer is set to the desired bias voltage for a specific test run, the other is then set by looking at the RF-to-IF path feedthrough on the spectrum analyser at 14MHz, and

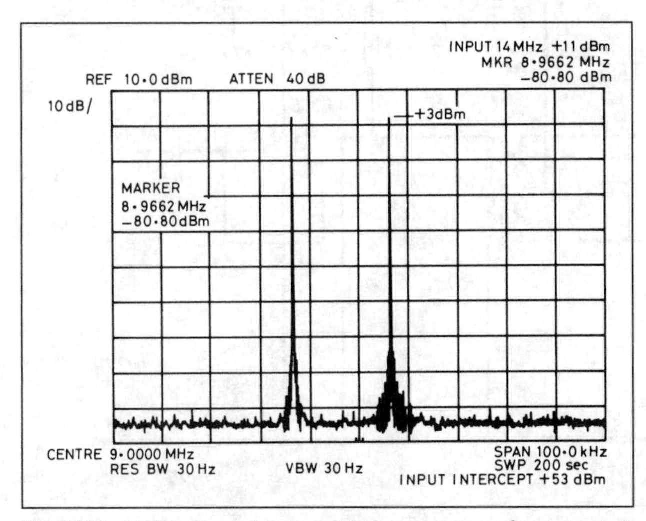

Fig 6.129. 14MHz input intermodulation spectrum for output at 9MHz

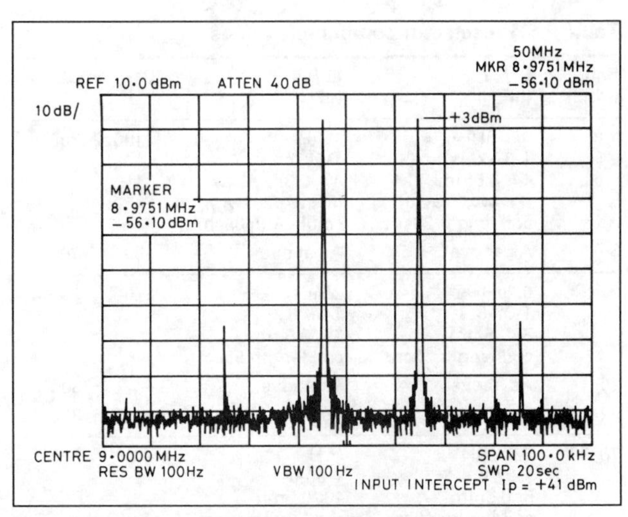

Fig 6.130. 50MHz input, output spectrum at 9MHz

adjusting the potentiometer for minimum IF feedthrough. The setting is quite sharp and ensures good mixer balance. An RF test signal of 11dBm (0.8V RMS) was used for each test signal for the two-tone IMD tests. The gate bias level chosen enabled an input level of +12dBm to be reached before the IMD increased sharply.

The performance of the H-mode test mixer was as follows. With an input RF test level of +11dBm (spaced at 2kHz or 20kHz) the conversion loss was 8dB; RF to IF isolation −68dB; LO to IF isolation −66dB. Input intercept points: 1.8 to 18MHz +53dBm; 21 to 28MHz +47dB, or better; 50MHz +41dB. These results were achieved with a gate-to-source DC bias of +1.95V and −8V substrate bias, a square-wave local oscillator amplitude of 9V and IF at 9MHz. It seems likely that a good performance could be achieved with an H-mode mixer transformer-driven from a sine-wave source provided the injection is via capacitors so that bias pots could still be used. There seems no reason why an H-mode mixer should not be used in an up-conversion arrangement rather than for a 9MHz IF.

PRACTICAL DESIGNS

A high-performance direct-conversion receiver

This design by Roelof Bakker, PA0RDT, first appeared in the Summer 1993 issue of *Sprat*, and is reprinted by kind permission of the G-QRP Club. It follows the normal DC-receiver format: preselector, RF amplifier, product detector, audio filter, audio interstage and final amplifier, and of course a local oscillator.

Preselector

The preselector (Fig 6.131) is patterned after an idea offered by Wes Hayward, W7ZOI, in a *Ham Radio* article. The insertion loss is smaller than 3dB and varies very little over a 12-frequency range. Component values are given in Table 6.5.

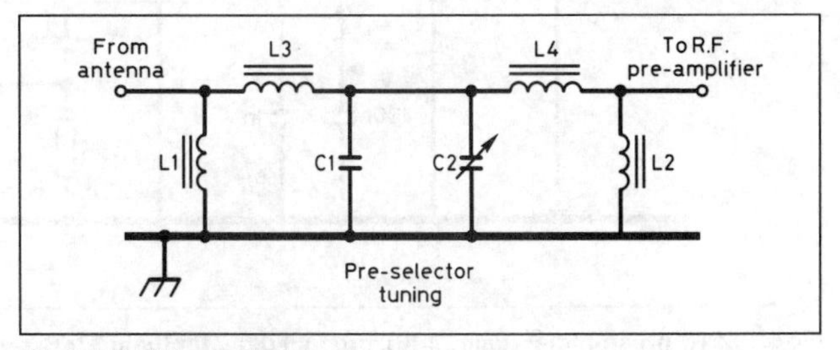

Fig 6.131. RF preselector

Table 6.5. Preselector component values

Band (MHz)	L1/L2 (pF)	L3/L4 (pF)	C1	C2 var
1.8	0.82μH T-37-2 14t 0.5mm (26swg/25awg) coil length 20cm	10μH T-50-2 45t 0.3mm (30swg/28awg) coil length 85cm	1100	300
3.5	As above	As above	180	300
7	0.2μH T-37-6 8t 0.5mm coil length 12cm	2.5μH T-50-6 25t 0.5mm coil length 55cm	270	300
10	As above	As above	—	300
14	As above	As above	—	300
18–30	0.082μH T-37-6 5t 0.5mm coil length 10cm	1μH T-50-6 16t 0.5mm coil length 40cm	—	300

RF amplifier

A monolithic microwave integrated circuit type MAR6 is used (Fig 6.132). It provides 20dB gain. These devices are unconditionally stable and easy to apply. If better IMD performance is needed a MAV11 can be used at the cost of a substantially higher current drain. In that case the 560Ω resistor should be changed to 100Ω.

Product detector

This is built around a Plessey SL6440 (Fig 6.132). High gain is achieved by intentionally not terminating this mixer. This has little effect on the IMD performance. T2 is an audio interstage transformer from an old AM radio. Various types were tried, including a home-made one wound on a high-μ ferrite toroid. They all worked well with gain figures from 35–40dB.

Voltage follower

The LF356 voltage follower (Fig 6.132) presents a high-impedance load at the output transformer of the product detector. The output provides a proper termination for the audio filter.

Audio filter

The audio filter (Fig 6.133) consists of one high-pass and two switchable low-pass filters.

These filters were designed using data supplied by Stefan Niewiadomski in a *Ham Radio* article. The high-pass filter attenuates frequencies below 350Hz. The low-pass filters feature some 80dB stopband attenuation. 3dB bandwidths of 1600 and 300Hz are available. The CW filter benefits the lower pitch preferred by many operators.

Audio interstage and final amplifier

Another LF356 serves as audio interstage amplifier (Fig 6.134). The gain is adjustable between 6 and 40dB. With the SET RECEIVER GAIN control the overall receiver gain can be adjusted for a comfortable level and minimal hiss.

The audio final amplifier is a design by Wes Hayward, W7ZOI. It delivers ample power for headphone reception. A small speaker can also be used with excellent results.

Local oscillator

This circuit (Fig 6.135) is rather unusual as it uses an MC1648P oscillator chip. It has a built-in AGC loop and delivers a constant output of 0dBm (1mW) at 50Ω, making it ideal for multiband operation.

Component values can be found in Table 6.6. NPO ceramic

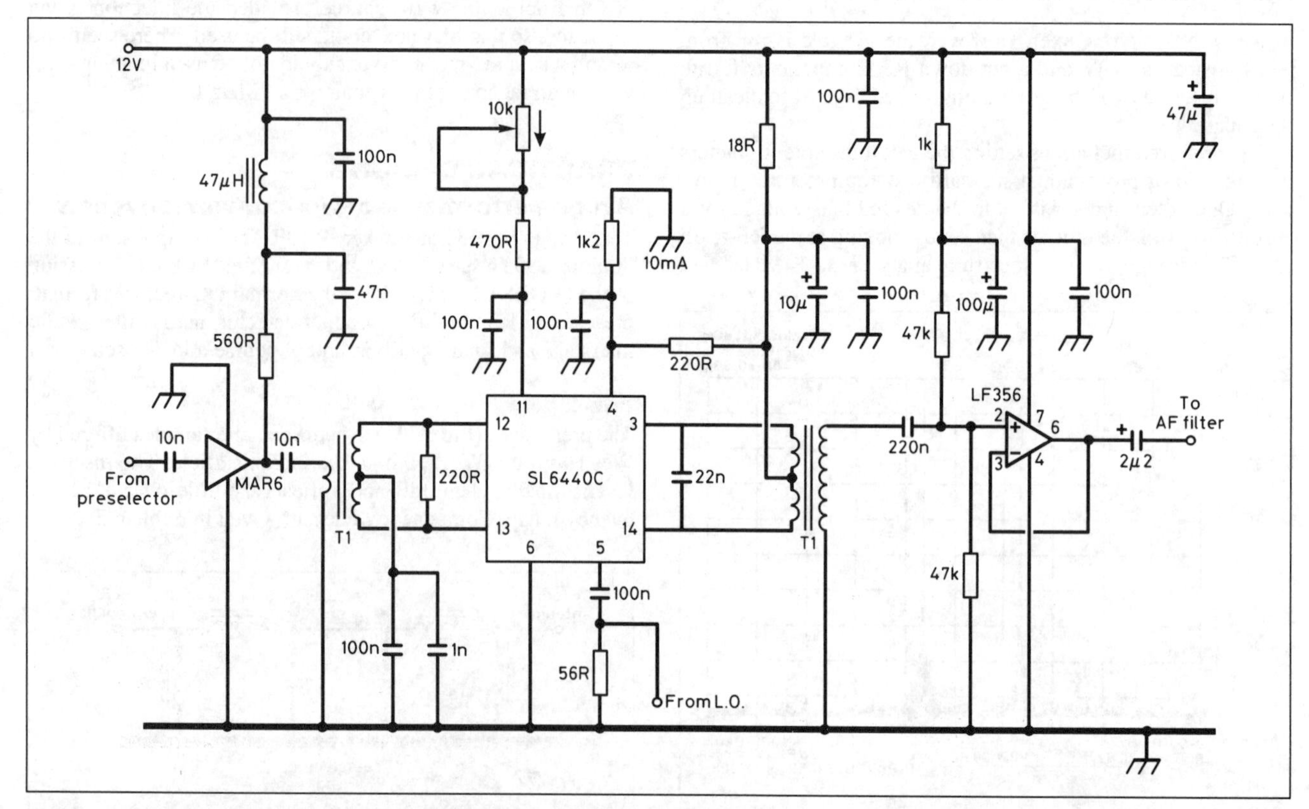

Fig 6.132. RF preamplifier (gain 20dB), product detector (gain 37dB) and voltage follower (gain 0dB). T1 is 10t trifilar wound on FT-37-43 ferrite core. T2 is an audio interstage transformer (see text)

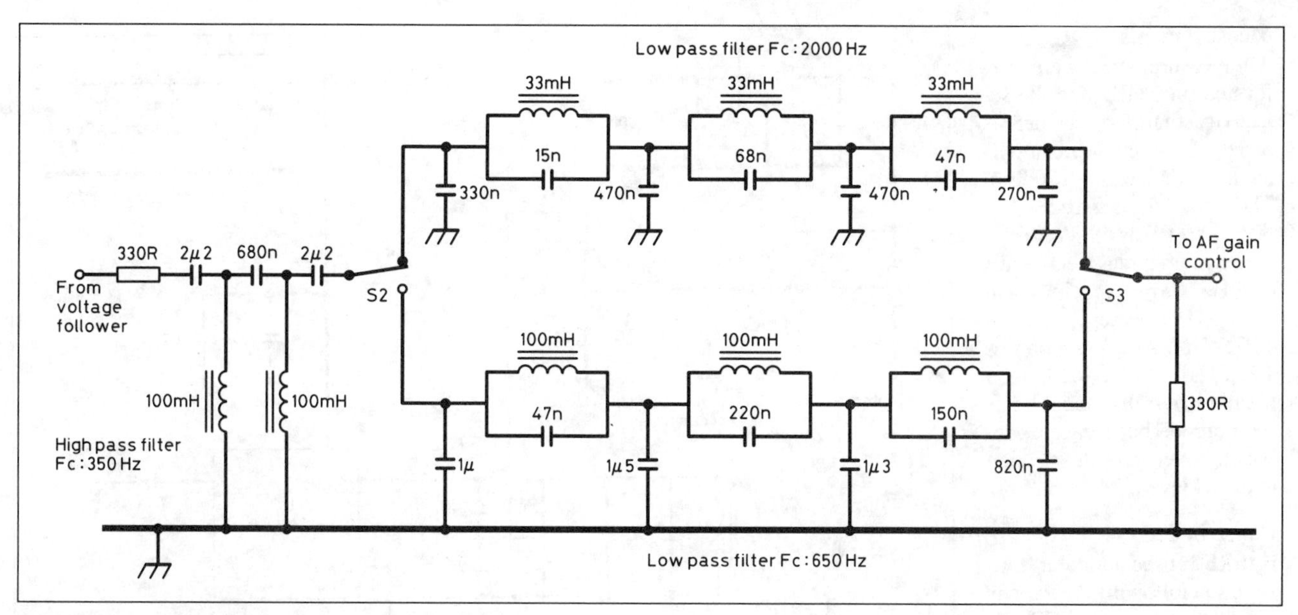

Fig 6.133. Audio filters. All coils Toko 10RB series

and polystyrene capacitors were used. After warm-up, frequency stability was no major problem on any band 160–10m. C5/6 set the frequency. C9/10 and C7 control the bandspread. T-50-6 toroids were used for L5 but suitable Toko or Neosid pre-wound coils can be used as well. Trimmer capacitor C5 can then be omitted and the frequency set by the adjustable core of L5.

The FREQUENCY SHIFT circuit is a great help in dodging QRM. If a CW signal is tuned from the high side and an interfering signal appears on the channel, switch S1 to the DOWN position. For a beat note of 500Hz the VFO shifts 1000Hz down so the audio image can be received. More often than not this channel is not disturbed by QRM. The 20kΩ and 50kΩ trimmer potentiometers should be adjusted for a frequency shift of twice the frequency of the preferred beat note.

Any other VFO can be used as long as it delivers 0dBm or 0.225V RMS into a 50Ω load.

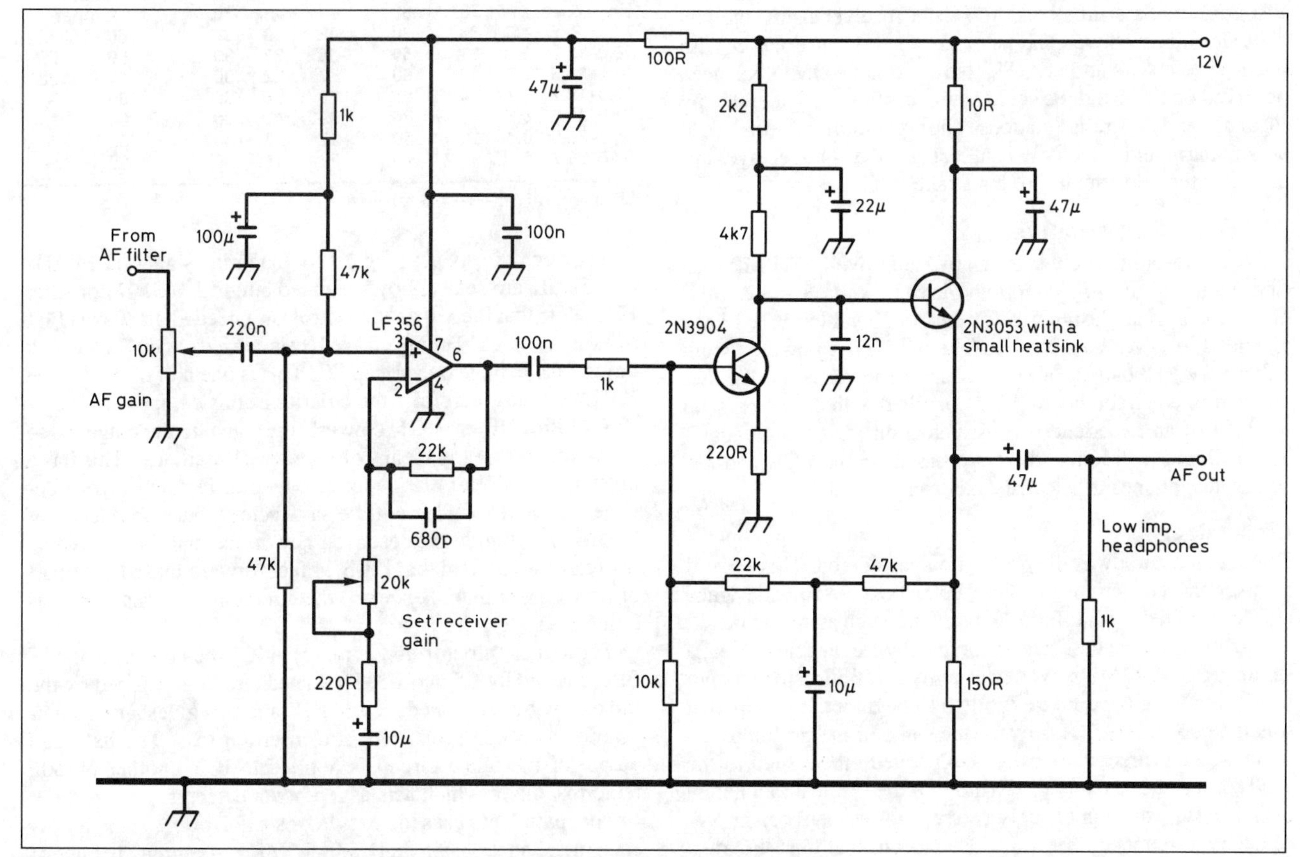

Fig 6.134. Audio interstage amplifier (gain 6–40dB) and audio final amplifier (gain 20dB)

Measurements

All measurements were performed on 7MHz and the selectivity control was set for SSB reception. Without preamplifier the MDS was −126dBm. The sensitivity for 10dB (S + N)/N ratio was 25µV. Two-tone dynamic range was 95dB (20kHz signal spacing) and third-order intercept point was +16.5dBm. AM detection was −35dB or 4mV. Without a preamplifier this receiver is 'bombproof' but using one enables a better gain distribution to be achieved.

Construction

PA0RDT used modular construction following the various circuit diagrams. No PCB patterns are available because none were used. Instead, neat 'ugly' construction was used with home-made tin-plate boxes with glass feedthroughs and RG174 coaxial cable for interconnections. VFO construction should be as sturdy as possible. An Eddystone 898 dial was used, which provides the luxury of a tuning rate of 4kHz per revolution.

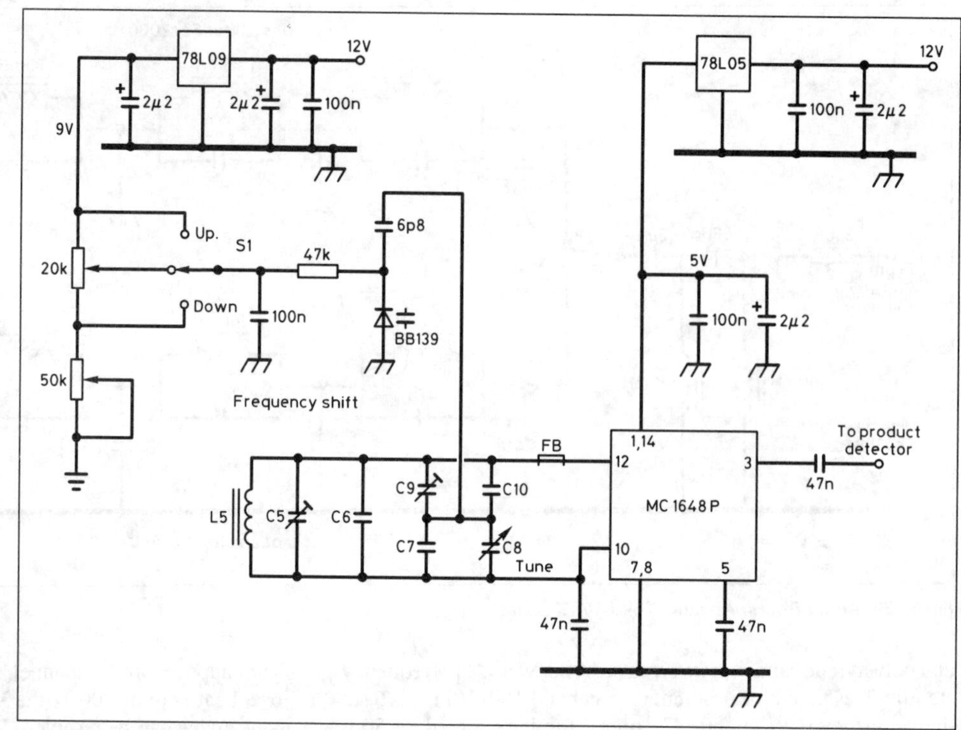

Fig 6.135. Local oscillator

Performance

PA0DRT made a nine-band 160–10m receiver using the circuits described above. The antenna was a 150ft doublet with open-wire feeders and an ATU. No overload or IMD has been observed on any band: he could listen on 40m at night and copy Japanese and Australian stations. The preamplifier seems not to do any harm and selectivity is amazing good. The receiver approaches the performance of his Drake R4C!

Table 6.6. Local oscillator component values

Band (MHz)	L5 (µH)	Turns	C5 var (pF)	C6 (pF)	C7 (pF)	C8 tuning (pF)	C9 var (pF)	C10 (pF)
1.8–2.0	66	128	60	—	—	30	-jumper-	
3.5–4.0	21.5	73	60	—	—	30	-jumper-	
7.0–7.2	4.3	33	60	22	33	30	60	82
10.0–10.2	1.8	21	60	68	33	30	60	56
14.0–14.5	0.79	14	60	82	22	30	60	68
18.0–18.2	0.77	14	60	22	56	30	60	—
21.0–21.5	0.51	11	60	39	47	30	60	33
24.8–25.0	0.46	10	60	22	56	30	60	—
28.0–28.7	0.35	9	60	22	47	30	60	—

L5 is wound on a T-50-6 toroid.

The Yearling receiver

This easy-to-build receiver design by Paul Lovell, G3YMP, was first published in *D-i-Y Radio* and later in *Radio Communication*. The original design was for the 20m band but it has been extended to cover 80m as well. The receiver is powered from either a 9V PP3 battery or mains adaptor and can be built either on the printed circuit board (PCB) supplied with the kit or (with the help of an experienced constructor) on a prototype board. The building instructions here are based on the PCB and kit. Either headphones or a loudspeaker can be used.

Circuit design

Direct conversion was the first option considered. This type of receiver has the merit of simplicity but, unless carefully designed and constructed, can suffer from problems such as strong breakthrough from broadcast stations on nearby frequencies. Also, a stable VFO at 14MHz, while certainly not impossible, is not something to be undertaken lightly by a beginner. The superhet was then considered. This overcame some of the problems associated with direct conversion, but created others such as the need for a rather expensive IF filter. So the result was a happy compromise, which in effect is a direct conversion receiver preceded by a frequency converter. This means that the VFO runs at about 5MHz instead of 14MHz, so stability is much better.

The circuit is given in Fig 6.136. Incoming signals at 14MHz or 3.5MHz are selected by the tuned circuit L1/C1/D1 or filter FL1. Note that the ANT TUNE control isn't needed on 80m as FL1 does all the work! Tuning on 14MHz is carried out by RV2 which adjusts the voltage on varicap D1. This is one half of diode type KV1236 – note carefully the polarity of this component.

A Philips NE602 (IC1) converts the signal to the range 5.0 to 5.5MHz (approx) by means of its internal oscillator. This has a crystal (X1) working at about 8.9MHz. In fact any crystal between 8.8 and 9.0MHz will be satisfactory, but a frequency of 8.95MHz will give greatest accuracy on the dial. More observant readers will note that D1 is in fact forward biased over part of its voltage range. However, the circuit as it stands performs quite adequately.

The signal output from the mixer in IC1 then passes to the IF filter formed by C5 and L2. The tuned circuit is damped by the rather low output impedance of IC1, and this gives a nice compromise between selectivity and insertion loss. The balanced output of the tuned circuit is applied to IC2, another NE602 mixer/oscillator which acts as a product detector.

The main VFO uses the oscillator section of IC2, which covers a range of approximately 5.0–5.5MHz. Assuming the use of a 9MHz crystal for X1, the 20m band will track within the range

5.00–5.35MHz and the 80m band from 5.2–5.5MHz. Note that the LF ends of the respective bands will be at opposite ends of the dial, since 20m makes of the sum of the receiver's two oscillator frequencies, and 80m uses the difference between them.

Main tuning is carried out by RV4, and RV5 provides the bandspread control. Tuning is carried out by means of the voltage on varicap D2 which, in association with C9, C10 and L3, determines the frequency of oscillation. The varicap is a dual type – cut in two with a sharp knife. Voltage regulator IC3 in the supply lines to the early stages makes stability surprising good.

The audio output from IC2 is amplified by IC4a and filtered by low-pass filter IC4b, before being further amplified to speaker level by IC5.

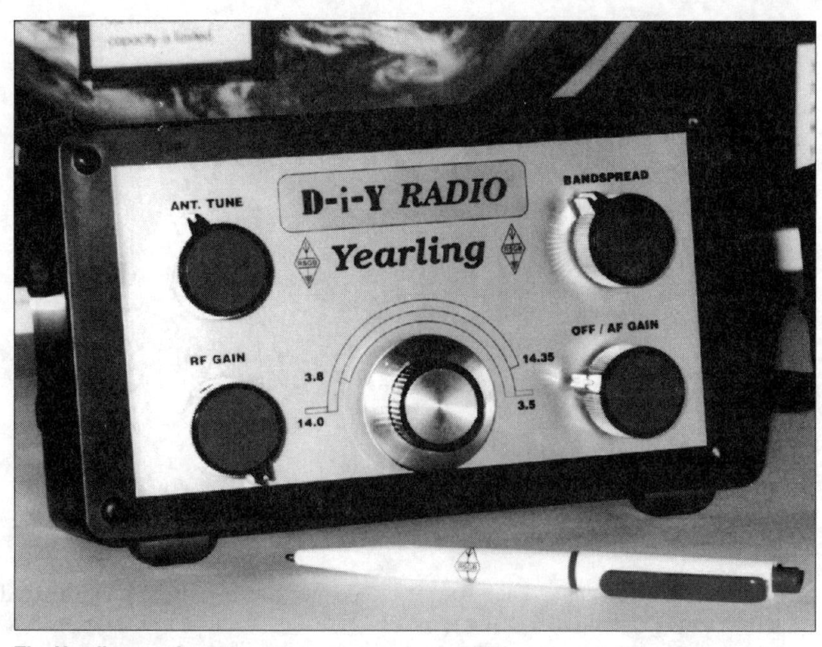

The Yearling receiver

Construction

Many Yearlings have been built from kits without problems, and the designer knows of a number of constructors who have successfully used a prototype board instead of the PCB. The prototype was built using just such a method and is illustrated in the photograph. No special precautions are needed but, as with any radio, neat wiring makes the tracing of faults a much easier process.

Fig 6.137 shows the connections to the gain and tuning controls. Screened cable should be used for the leads to the volume control, but stranded bell wire should be satisfactory elsewhere. Incidentally, there was no problem with using IC sockets for all the 8-pin devices. The coils are colour coded, with L1 having a pink core and L2 a yellow one.

It is rather easy to wire the varicaps incorrectly but, if you are using the PCB, the lettering on D1 should be next to coil L1 and the lettering on D2 should be facing resistor R7. Fig 6.138 shows the bandchange switch and the 80m filter which is glued to the side of the case. Holes for the five controls are 10.5mm diameter, and the speaker and power connectors have 6.3mm and 11mm holes respectively. The antenna and earth sockets need 8mm holes.

Setting it up

Connect a 9V battery and a reasonable antenna, and, when you switch on, you should hear some stations – or at least some whistles! It is suggested that you start with the 20m band, and make the adjustments before fitting the controls and sockets to the case.

A signal generator is useful, of course, but not essential to get the receiver working. Carry out the following steps and your Yearling should burst into life.

1. Set the core of L2 to mid-position.
2. Set RV1, RV2 and RV4 to mid-position and rotate the core of L1 until you hear a peak of noise. Now adjust L2 for maximum noise.
3. Tune carefully with the main tuning control RV4 until you hear amateur signals. Adjustment of the bandspread may be needed to clarify the speech.
4. Switch off the receiver and fit the controls and sockets to the case.

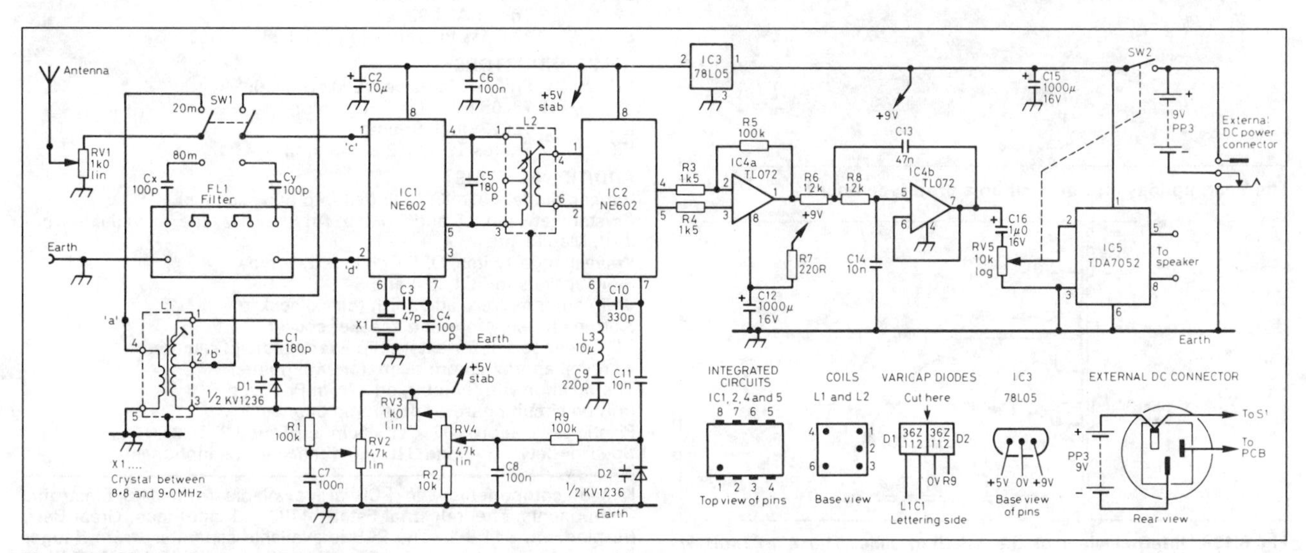

Fig 6.136. Circuit diagram of the Yearling

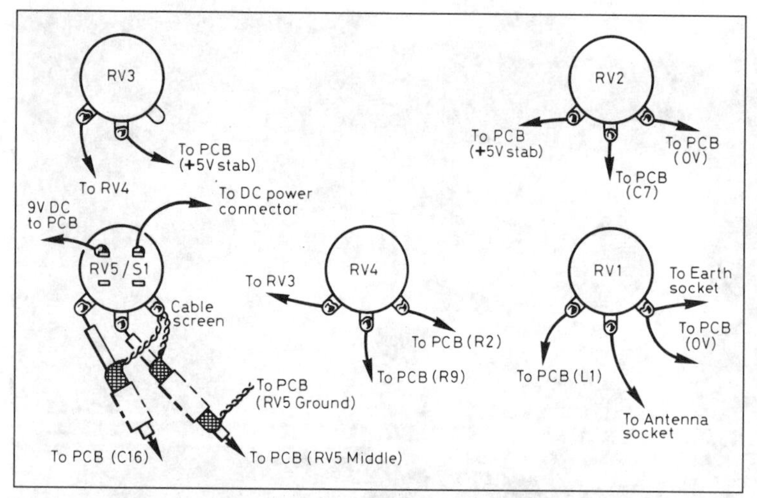

Fig 6.137. Rear view of the variable resistors. Check the connections carefully to make sure that the wires fit the correct holes on the board

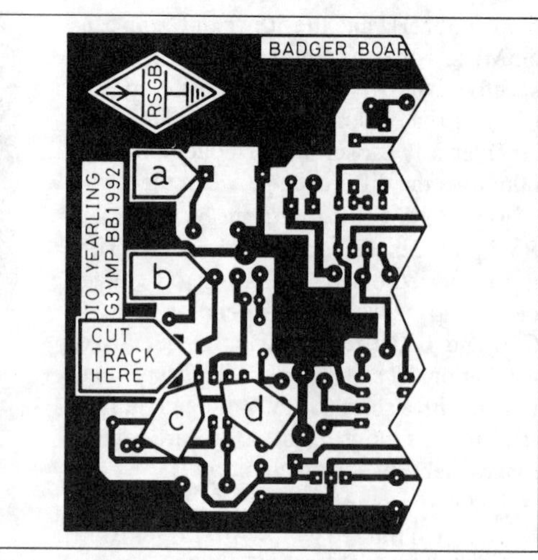

Fig 6.139. The underside of the PCB. Wires are connected from the switch and filter as shown

5. Finally, adjust the tuning knob so that the pointer roughly agrees with the dial. Due to the spread of varicap capacitance values, you may find the tuning a little cramped. This is easily fixed by adding a resistor (try 22kΩ to start with) in series with RV3 or adjusting the value of R2.

The Yearling may also be built on a prototype board

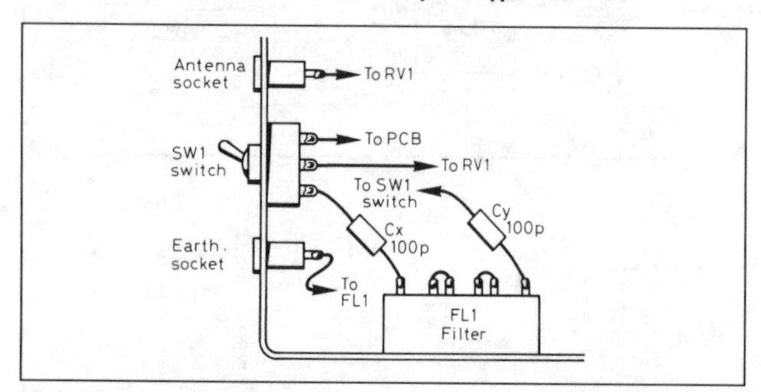

Fig 6.138. Internal view of the Yearling case. The 80m filter is attached to the base with glue

Table 6.7. Components list for the Yearling receiver

CAPACITORS	
All rated at 16V or more	
C1, 5	180p polystyrene, 5% or better tolerance
C2	10μ electrolytic
C3	47p polystyrene, 5% or better tolerance
C4, Cx, Cy	100p polystyrene, 5% or better tolerance
C6–8	100n ceramic
C9	220p polypropylene, 2% or better tolerance
C10	330p polypropylene, 2% or better tolerance
C11, 14	10n ceramic
C12, 15	1000μ electrolytic
C13	47n, 5% polyester
C16	1μ electrolytic
RESISTORS	
All 0.25W 5%	
R1, 5, 9	100k
R2	10k
R3, 4	1k5
R6, 8	12k
R7	220R
RV1, 3	1k linear
RV2, 4	47k linear
RV5	10k log with switch (SW2)
INDUCTORS	
L1	Toko KANK3335R
L2	Toko KANK3334R
L3	10μ, 5% tolerance (eg Toko 283AS-100)
SEMICONDUCTORS	
IC1, 2	Philips/Signetics NE602 or NE602A
IC3	78L05 5V 100mA regulator
IC4	TL072 dual op-amp
IC5	Philips TDA7052 audio amp
ADDITIONAL ITEMS	
Varicap diode, Toko KV1236 (cut into two sections)	
Crystal, between 8.8 and 9.0MHz. An 8.86MHz type is available from JAB, Maplin, etc	
Wavechange switch, DPDT changeover type	
8-pin sockets for IC1, 2, 4 and 5	
4mm antenna (red) and earth (black) sockets	
3.5mm chassis-mounting speaker socket	
DC power socket for external power supply (if required)	
4 knobs, approx 25mm diameter with pointer	
Tuning knob with pointer, eg 37mm PK3 type	
Printed circuit board or prototype board	
Plastic case, approx 17 × 11 × 6cm, eg Tandy 270-224.	
Speaker between 8 and 32Ω impedance (or headphones)	

Kits of components and PCB are available from JAB Electronic Components, The Industrial Estate, 1180 Aldridge Road, Great Barr, Birmingham B44 8PE. The PCB is available separately from Badger Boards, 87 Blackberry Lane, Four Oaks, Sutton Coldfield, B74 4JF.

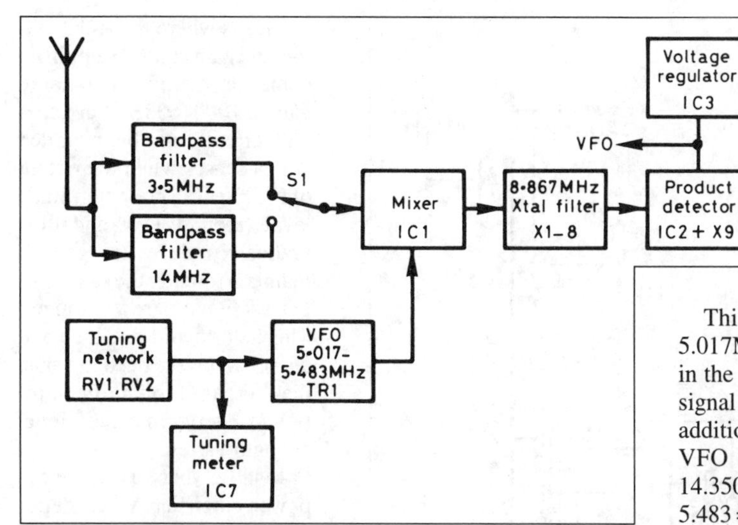

Fig 6.140. Block diagram of the two-band superhet. Coverage of either 20m or 80m is obtained simply by selecting the appropriate input band-pass filter with S1

6. Check the 80m band – this should work without further adjustments to the coils. Fig 6.139 shows the additional connections for 80m as the Yearling was originally designed for 20m only.
7. Finally, fix the PCB inside the case (double-sided sticky tape works well).

Conclusion

It is hoped that you get as much enjoyment from building the Yearling as the designer did. It certainly goes to show that amateur construction doesn't have to be complicated to be effective.

The *D-i-Y Radio* team designed a self-adhesive front panel for the Yearling which gives it a very smart appearance. This is included with the kit, or is available free from the *Radio Communication* office at RSGB HQ on receipt of an A5-size, stamped, self-addressed envelope.

A two-band superhet

This design by Steve Price, G4BWE, first published in the September and October 1993 issues of *Radio Communication*, covers two of the most popular HF bands – 20 and 80m. It boasts a sensitivity of around 0.5µV and features an eight-pole crystal filter which gives excellent SSB selectivity of approx 2.5kHz at –6dB points. Despite its high performance, component costs have been kept to a minimum, and all items should be readily available. Also, in order to make construction as painless as possible, printed circuit boards plus detailed assembly instructions are available from G4BWE.

Fig 6.140 shows a block diagram of the receiver. Signals picked up by the antenna are routed through two band-pass filters – one covering 3.5 to 3.85MHz (80m), and the other 14 to 14.35MHz (20m). Because these filters have a bandwidth of at least 350kHz, there is no need to provide tuning controls for them. S1 selects the desired band by connecting the output of the appropriate filter to the input of the first mixer, IC1. Amazingly, this is all that is required to switch between 80 and 20m.

The receiver is a single-conversion type with an intermediate frequency of 8.867MHz. IC1, working in conjunction with the VFO, has the job of converting signals to this frequency so that they may pass through the eight-pole crystal filter. The VFO is tuned between 5.017 and 5.483MHz by RV1 and RV2 to provide coverage of 3.384 to 3.85MHz and 13.884 to 14.35MHz.

This works as follows. Imagine that the VFO is tuned to 5.017MHz. The mixer will now produce an output of 8.867MHz in the case of two signal frequencies: first by subtraction for a signal at 13.884MHz (13.884 – 5.017 = 8.867) and second by addition for a signal at 3.85MHz (3.85 + 5.017 = 8.867). If the VFO is set at its HF limit the signal frequencies become 14.350MHz (14.350 – 5.483 = 8.867) and 3.384MHz (3.384 + 5.483 = 8.867) respectively. These calculations reveal that choosing an IF of 8.867MHz results in unnecessary coverage, extending 116kHz below the lower edge of both 20m and 80m (13.884 to 14MHz and 3.384 to 3.5MHz), but in practice this matters little. Of course, satisfactory performance hinges on the ability of the band-pass filters to provide sufficient attenuation of signals on the unselected band. If this condition is not met, S1 will be only partially effective as a band change switch.

Filter crystals

The reason for choosing an IF of 8.867MHz, rather than the nearest round figure (9MHz – which would work just as well), is that 8.867MHz quartz crystals are manufactured in large quantities for use in domestic TV receivers (some TVs use a 4.433MHz crystal, but that simply means that both types are readily available). In consequence 8.876MHz crystals are normally a lot cheaper than 9MHz types and so it makes sense to construct the IF filter using these.

After passing through the IF filter, the selected signal is demodulated by a product detector. This stage is simply another mixer, and has the job of down-converting the 8.867MHz IF to audio. It does this by mixing the IF signal with the output of a crystal oscillator operating close to the IF. The next stage is a low-noise audio preamplifier. This produces an output that varies from approximately 1mV, for a signal of strength S1 (ie 1µV at the antenna), to around 250mV if the signal is S9. There clearly needs to be further voltage amplification in order to make S1 signals loud enough to drive an 8Ω loudspeaker – 1mV would produce only 0.125µW! There is a problem, however, in that by providing additional amplification for weak signals, we run the risk of the much stronger S9 signals grossly overloading the receiver and causing distortion.

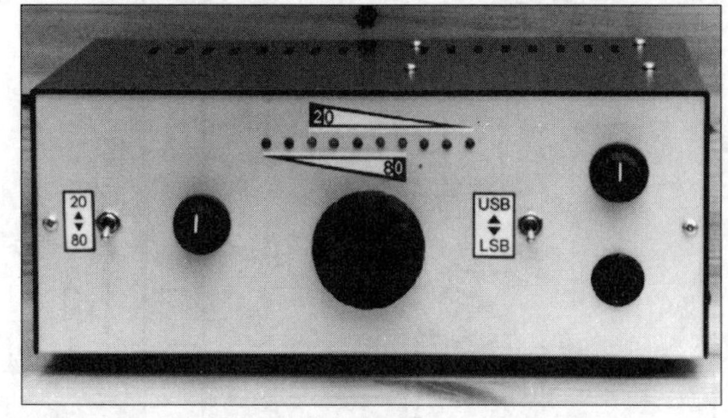

Front view of the two-band superhet

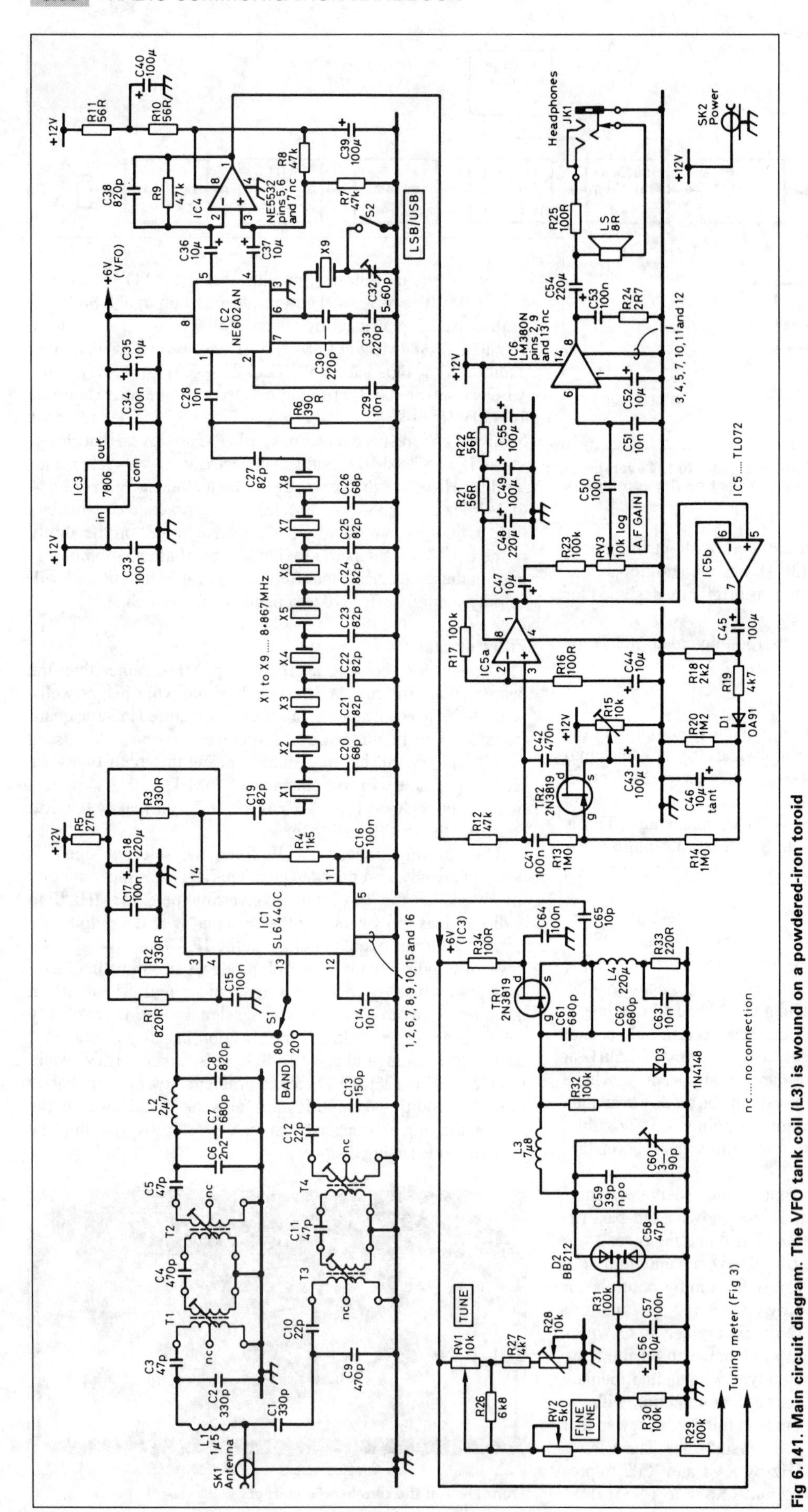

Fig 6.141. Main circuit diagram. The VFO tank coil (L3) is wound on a powdered-iron toroid

This is where the AGC (automatic gain control) amplifier comes in. IC5 has a maximum gain of 1000 (60dB), but a special circuit has been added which senses when the output of IC5 rises to a pre-determined level (around 1V), and then gradually reduces the gain to maintain roughly the same output as signals grow stronger. This helps avoid overload and also obviates the need to adjust the AF gain (volume) control (RV3) every time the signal strength varies.

Finally, there is an audio power amplifier which delivers a maximum output of around 500mW.

A closer look

Fig 6.141 shows the main circuit diagram, excluding the tuning meter, which will be discussed later. Signals from the antenna are routed to the band-pass filters via L1 and C1. These components form a simple diplexer which works rather like the crossover network in a hi-fi loudspeaker. They direct the lower frequencies to the 3.5MHz filter (C2–8, T1 and 2 plus L2) and higher frequencies to the 14MHz filter (C9–13 plus T3 and 4). RF transformers (T1–4) have adjustable ferrite tuning slugs which enable the filters to be aligned following construction.

The 3.5MHz filter has an additional low-pass section at its output, known as a *pi-network*, consisting of C6, 7, 8 and L2. This helps to block very strong broadcast signals at around 7MHz (eg from stations on the 41m broadcast band). It is necessary to attenuate these because the mixer (IC1) partially functions on the third harmonic of the VFO. But how can this cause interference? Well, imagine that the VFO is set at 5.362MHz in order to listen on 3.505MHz (5.362 + 3.505 = 8.867MHz). The third harmonic of the VFO will be 5.362 × 3 = 16.086MHz.

Now, if the 3.5MHz band-pass filter has insufficient attenuation at 7.219MHz, a strong signal on this frequency may be audible (16.086 – 7.219 = 8.867MHz). It would be wrong to pretend that the additional attenuation above 3.85MHz provided by the pi-network is always sufficient to completely eliminate this form of interference, or 'spurious response', but in practice it is reduced to a very low level. Luckily, the simpler 14MHz band-pass filter has proved just as effective at rejecting 7MHz signals.

The output from the appropriate band-pass filter is selected by the band change switch, S1, a single-pole double-throw (SPDT) miniature toggle type mounted on the front panel. IC1 is a high-performance bipolar mixer with an on-chip VFO buffer amplifier. The VFO utilises an FET (TR1), and is tuned with a varicap D2. This component functions as a voltage-controlled variable capacitor, the control voltage being determined primarily by RV1, a 10-turn wirewound potentiometer. RV2 is the FINE TUNE control and enables the frequency set by RV1 to be altered by plus or minus 2kHz. C60 determines the VFO upper frequency limit (5.483MHz), and R28 fixes the lower limit (5.017MHz).

Inside view of the two-band superhet

The mixer output (IC1, pin 14) feeds an eight-pole crystal ladder filter formed by a network of 8.867MHz quartz crystals (X1–8), which work in conjunction with C19–27. In essence, the capacitors spread out the resonant frequencies of X1 to X8, and give a band-pass response over approximately 2.5kHz. This bandwidth is fine for SSB, and also allows reception of CW and data modes. The product detector, IC2, is a mixer complete with on-chip oscillator transistor.

Sideband selection

Crystal X9 determines the oscillator frequency, which – when S2 is open – is increased slightly by C32. S2 selects the appropriate carrier frequency for either LSB (normally employed by amateur stations on 80m) or USB for 20m SSB. IC2 requires a 6V supply. This is generated by the voltage regulator IC3, which is also used to power the VFO and tuning network.

IC4, a low-noise op-amp, has the job of raising the level of the demodulated signals appearing at the outputs of IC2 (pins 4 and 5) by 36dB (ie the voltage gain is approximately 65). Using both the product detector outputs (which are antiphase) helps promote stability and increases noise rejection. The AGC amplifier, IC5, generates a gain control voltage as follows. After passing through IC5a, signals are routed via a simple buffer stage comprising IC5b. IC5b has no gain and is simply used to isolate the output of IC5a. D1 rectifies the signal by allowing only the positive half-cycles of the audio waveform to reach C46. The voltage across C46 is therefore proportionate to the average signal level and can be used to control the receiver's gain.

TR2 acts as a voltage-controlled variable resistor and works in conjunction with R12 to form a potential divider at the input of IC5a. As the gate voltage of TR2 (derived from C46 via R14) rises, the resistance between its source and drain falls. This will attenuate the signal by a factor dependent on the ratio between the value of R12 and the source/drain resistance. For instance, if the source/drain resistance is the same as R12 (47kΩ), the signal level will be halved (ie attenuated by 6dB). This action reduces the input to IC5 and so prevents overload. A very strong signal will cause the source/drain resistance of TR2 to drop to only a few hundred ohms, thus attenuating the signal by over 40dB (a factor of 100). R15 determines the gate voltage at which TR2 begins to operate. It is set to prevent the AGC affecting signals that are weaker than about S5. R20 provides a discharge path for C46, thereby allowing the receiver to gradually recover gain when signal levels drop.

IC6 is the audio power amplifier. This drives an internal 8Ω loudspeaker or alternatively headphones via JK1. A stereo jack socket is preferred because it enables ordinary hi-fi headphones to be used without having to change their plug. Readers who are more familiar with the design of superheterodyne receivers may be surprised that there is no IF amplifier (although it should be noted that both IC1 and IC2 contribute significant gain). To make up for this, the AF gain is much higher than in most other superhets. Steps have been taken, however, to maintain audio stability and, for instance, both IC4 and IC5 are 'double-decoupled' by C39, C40 plus R10, R11 and C48, C49 plus R21, R22 respectively.

Tuning meter

An analogue tuning scale could be made by arranging for a pointer to move across a strip of card, suitably marked. As the shaft of RV1 rotates 10 times, it is not possible to fix the pointer directly to this and so a reduction drive or a system of pulleys must be considered. The alternative is an electronic indicator which works by measuring the tuning voltage presented to the varicap (D2).

Fig 6.142 shows the circuit of such a meter designed for the two-band superhet. It uses an LM3914 bargraph driver (IC7) and, as can be seen, requires only a few additional components. The LM3914 measures the tuning voltage fed via R29 (Fig 6.141) to pin 5 and illuminates one of 10 light-emitting diodes (LED1–10) depending on its magnitude. The upper scale limit (LED10)

Close-up view of the circuit board

is determined by the voltage on pin 6, which is equal to the maximum tuning voltage (6V), minus a small drop introduced by R38. The lower limit (LED1) is set by adjustment of R36. LED brightness is controlled by R35.

The tuning meter, including the 10 LEDs, is built onto a separate PCB. This gives greater freedom in positioning the display on the front panel. Small round LEDs (3mm diameter) are most convenient as these can simply be pushed into a row of equally spaced holes drilled through the front panel. Fig 6.143 and the photographs show how the display looks. Different coloured LEDs were used to highlight the bottom 100kHz of each band where CW and data transmissions are normally found.

Because the relationship between VFO frequency and display indication is slightly non-linear, the LEDs highlighting the CW/data segment of 80m are one position further in from the scale end than those for 20m. The arrow markings above and below the display act as a reminder that the receiver tunes 'backwards' on 80m.

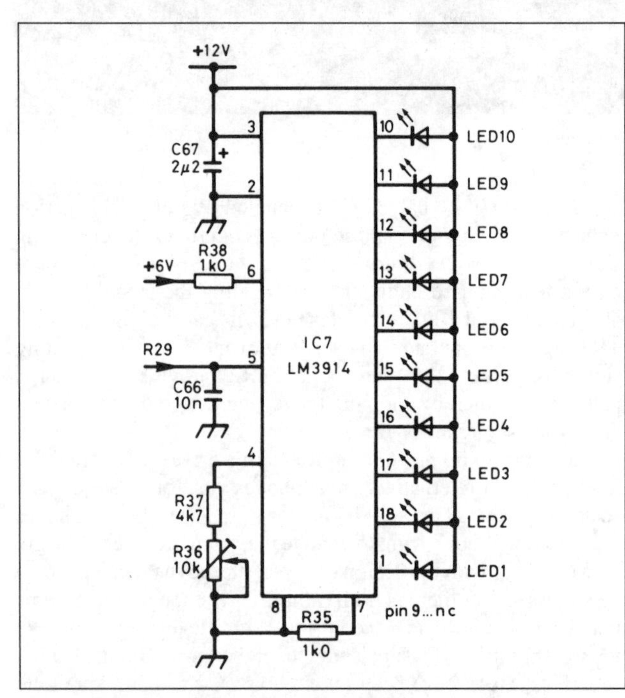

Fig 6.142. The LED bargraph tuning meter

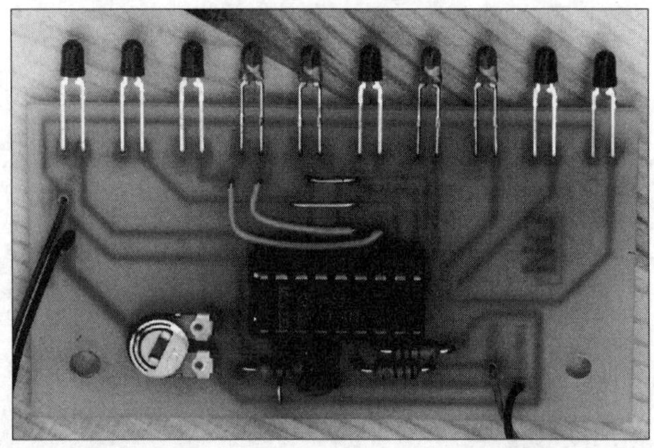

The LED bargraph circuitry

Construction hints

The first prototype of the two-band superhet was built on Veroboard (stripboard) and worked perfectly well. Nevertheless, most constructors will no doubt wish to make use of the two printed circuit boards (PCBs) now available.

The PCBs are supplied with layout drawings, which show the positions of all components; a parts identification and mounting guide, plus notes on the finer points of assembly and testing. It is recommended that all the ICs except IC3 and IC6 are mounted in sockets.

Components may be obtained from a number of well-known suppliers by mail order (see the end of the components list for ordering details of kits and PCBs), or you could go bargain hunting at the next mobile rally! Clubs eager to promote home construction might consider ordering some, or all, of the parts in bulk as a service to their members.

Once all the components and flying leads have been soldered onto the PCBs, the next task is to drill the selected case (this must be metal for screening reasons – see components list for a recommendation). The exact positions of the mounting holes for the PCBs, controls, LEDs, sockets and loudspeaker are partly a matter of choice. The dimensions of these components and the size of the chosen case must also be taken into account.

The photographs give a good idea of the layout which G4BWE

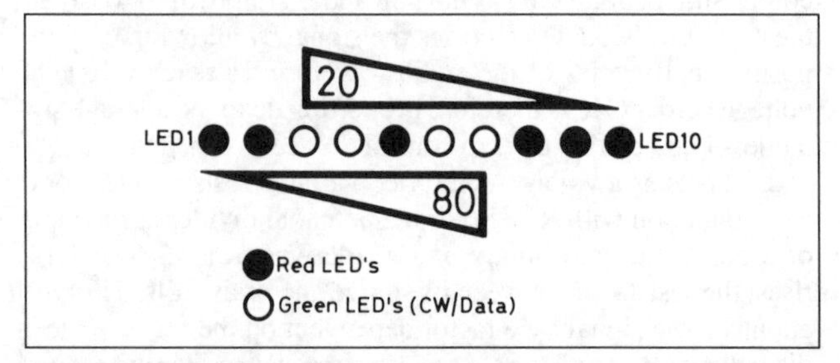

Fig 6.143. Suggested scale markings to enhance the LED tuning display

Table 6.8. Components list for the two-band superhet

RESISTORS

All fixed resistors are 0.25W, 5% carbon film.

R1	820R
R2, 3	330R
R4	1k5
R5	27R
R6	390R
R7, 8, 9, 12	47k
R10, 11, 21, 22	56R
R13, 14	1M
R15, 28, 36	10k preset (horizontal)
R16, 25, 34	100R
R17, 23, 29, 31, 32	100k
R18	2k2
R19, 27, 37	4k7
R20	1M2
R24	2R7
R26	6k8
R30	680k
R33	220R
R35, 38	1k0
RV1	10k lin, 10 turn (wirewound)
RV2	5k lin
RV3	10k log

CAPACITORS

C1, 2	330p ceramic plate or polystyrene
C3, 5, 11	47p ceramic plate or polystyrene
C4, 9	470p ceramic plate or polystyrene
C6	2n2 polystyrene
C7	680p ceramic plate or polystyrene
C8, 38	820p ceramic plate or polystyrene
C10, 12	22p ceramic plate or polystyrene
C13	150p ceramic plate or polystyrene
C14, 28, 29, 51, 63, 66	10n disc ceramic
C15, 16, 17, 33, 34, 41, 50, 57, 64	100n disc ceramic
C18, 48, 54	220µ 16 or 25V radial electrolytic
C19, 21, 22, 23, 24, 25, 27	82p ceramic plate or polystyrene
C20, 26	68p ceramic plate or polystyrene
C30, 31	220p ceramic plate
C32	5-60p foil trimmer
C35, 36, 37, 44, 47, 52, 56	10µ 25V radial electrolytic
C39, 40, 43, 45, 49, 55	100µ 16 or 25V radial electrolytic
C42	470n Siemens style polyester (10mm lead spacing)
C46	10µ 16 or 25V tantalum bead electrolytic
C53	100n sub-miniature polyester (5mm lead spacing)
C58	47p polystyrene
C59	39p NPO plate ceramic
C60	3-90p foil trimmer
C61, 62	680p polystyrene
C65	10p plate ceramic
C67	2µ2 35V tantalum bead electrolytic

INDUCTORS

L1	1µ5 Toko 7BS (part no 283AS-1R5)
L2	2µ7 Toko 7BS (part no 283AS-2R7)
L3	7µ8 41 turns of 26SWG enamel on T68-6 powdered iron toroid
L4	220µ Toko 7BS (part no 283AS - 221)
T1, 2	Toko KANK3333R
T3, 4	Toko KANK3334R

SEMICONDUCTORS

D1	OA91
D2	BB212 varicap-
D3	1N4148
LED1–10	3mm round (colour(s) of choice)
TR1, 2	2N3819
IC1	SL6440C
IC2	NE602AN
IC3	7806
IC4	NE5532
IC5	TL072
IC6	LM380N
IC7	LM3914

ADDITIONAL ITEMS

JK1	Headphone jack socket (see text)
LS1	8R loudspeaker (eg Maplin YT25C)
SK1	SO239 socket
SK2	2.5mm DC power socket
S1	SPDT miniature toggle
S2	SP or SPDT miniature toggle

CRYSTALS

X1-9	8.867MHz

MISCELLANEOUS

3 × 8-pin DIL IC sockets
16-pin DIL IC socket
18-pin DIL IC socket
Metal case, eg Maplin 2108, size: 225mm(w), 175mm(d), 89mm(h), order code XJ30H is suitable)
Knobs for RV1–3, insulated cable for flying leads, short length of RG174A/U miniature coax for connection between SK1 and PCB, 6BA nuts and bolts and 6mm spacers for PCB mounting.

The two printed circuit boards and further constructional details for this project may be ordered from the designer: Steve Price, 9 Spurcroft Road, Thatcham, Berks RG13 4XX. Contact him for further details.

A full kit of parts is available from JAB Electronic Components. This includes the case and all components. JAB Electronic Components, The Industrial Estate, 1180 Aldridge Road, Great Barr, Birmingham B44 8PE. Tel: 021-366 6928

Most of the components, including Toko inductors and the foil trimmers, are also available from Cirkit Distribution Ltd, Park Lane, Broxbourne, Herts EN10 7NQ.

The NE602AN (IC2) may be obtained from MACRO Ltd, Burnham Lane, Slough SL1 6LN. Alternatively, it should be possible to substitute the older NE602N in this design.

8.867MHz quartz crystals are available from Maplin Electronics, PO Box 3, Rayleigh, Essex SS6 2BR.

adopted – note the central position of the main tuning control (RV1) and the row of LEDs directly above this. The BAND-CHANGE switch (S1) and FINE TUNE control (RV2) are mounted to the left of the front panel, and the LSB/USB switch (S2) plus the AF GAIN control (RV3) and headphone socket (JK1) to the right. The main PCB is mounted on the base using 6BA nuts and bolts with 6mm spacing pillars – when deciding on the position of this PCB, remember to check that there is sufficient clearance for the panel-mounted components.

The antenna and power sockets (SK1 and SK2) are fixed to the rear panel. Finally, the loudspeaker (LS1) is bolted to the case lid underneath some pre-drilled ventilation slats (the slats are not really large enough to act as an efficient vent for LS1 and so a pattern of round holes has been drilled in the rear panel to allow some sound to escape from the rear of the speaker).

Testing and calibration

The receiver requires a regulated power supply of 12V DC. Quiescent current consumption is around 75mA, rising to approximately 300mA in the presence of a strong signal with the AF gain fully advanced. Constructors who do not already possess a suitable mains power supply unit may wish to consider building one. A simple fixed-voltage design using a type 7812 12V 1A regulator IC is quite adequate.

First set RV2, RV3, R15, R28, C32 and C60 at mid-travel. After connecting the power supply, check with a multimeter (negative test probe to ground) that there is around 12V on pins 8 of both IC4 and IC5 and also on pin 14 of IC6. There should be a slightly lower voltage on pins 3 and 14 of IC1 and 6V on pin 8 of IC2. Initially there will be very little sound from the loudspeaker, but after a few seconds the background noise will gradually fade up as the AGC voltage settles.

After allowing 5–10min for the VFO to warm up, it may be calibrated using a digital frequency meter (DFM). Connect the DFM probe to the junction of C65 and IC1, pin 5. Constructors who do not have access to a DFM could use a general-coverage receiver in SSB/CW mode. A short length of cable should be connected to the test receiver's antenna socket and the other end allowed to dangle inside the superhet's case.

Proceed as follows:

1. Set RV1 fully clockwise and check that there is 6V at the junction of RV2, R29 and R30 – this confirms that the track of RV1 has been connected the right way round.
2. Adjust C60 using a brass or plastic trimming tool to obtain a VFO frequency as close as possible to 5.483MHz (if you are using a general-coverage receiver instead of a DFM, simply tune the receiver to 5.483MHz and listen for a heterodyne as the VFO is tuned onto frequency).

3. Set RV1 fully anti-clockwise and adjust R28 to obtain a VFO frequency of 5.017MHz.

The VFO is now adjusted for coverage of 3.384 to 3.85MHz and 13.884 to 14.350MHz plus or minus the 2kHz latitude provided by the FINE TUNE control (RV2). This explains why RV2 must be set at mid-travel during calibration. The next task is to align the input band-pass filters:

4. Connect the antenna which will normally be used with the receiver. A single wire of between 25 and 40ft, reasonably elevated, is fine. An earth connection is beneficial, but if this is already provided via the mains power supply unit then there is no need to bother further.
5. Set S1 to 20m and S2 to USB (ie the contacts of S2 should be closed). With RV1 near mid-travel, tune around to locate an amateur SSB signal. Having found one, use the trimming tool to adjust the core of T3 for maximum level. Now adjust T4 in the same way (T4 may not 'peak' quite as sharply as T3; this is OK).
6. Tune away from any signals so as to obtain a steady background hiss. Change to LSB by opening S2 and adjust C32 so that the tonality, or 'pitch', of the background noise is the same as when S2 is set to USB.
7. Set S1 to 80m and find an amateur signal near the centre of this band. Now adjust the cores of T1 and T2 in the same way as you did for 20m.

If you have difficulty in finding a signal on either band, remember that this may be due to poor propagation conditions. Be prepared to wait a few hours and try again. If possible, use a second receiver to check. Alternatively, constructors who have access to an RF signal generator may use this instead.

Preset R15 is now set so that the AGC starts to operate when the signal level reaches approximately S5 to S7 (signals of this level will sound about five times louder than the normal background noise on 20m). There will be a degree of trial and error involved in finding the best setting for R15, but it should be possible to arrive at a point where the AGC is effective at preventing overload for signal strengths up to around S9 + 30dB. Signals weaker than S5 to 7 will not, of course, sound as loud as those which are stronger, but this should not prove a problem – simply increase the AF gain using RV3!

Finally, adjust R36 on the frequency meter board so that LED1 is illuminated when RV1 is fully anti-clockwise.

Finishing touches

The two-band superhet works particularly well within the limitations of such a simple design and should provide many hours of listening pleasure.

Possible additions include an ATU – this will improve antenna matching and also provide extra rejection of out-of-band signals – and a crystal calibrator to give a more accurate indication of frequency.

7 HF transmitters and transceivers

THE PURPOSE of a transmitter is to generate RF energy which may be keyed or modulated and thus employed to convey intelligence to one or more receiving stations. This chapter deals with the design of that part of the transmitter which produces the RF signal, while the methods by which this signal may be keyed or modulated are described separately in other chapters. Transmitters operating on frequencies between 2 and 30MHz only are discussed here; methods of generating frequencies higher than 30MHz are contained in other chapters. Where the frequency-determining oscillators of a combined transmitter and receiver are common to both functions, the equipment is referred to as a *transceiver*; the design of such equipment operating in the HF spectrum is also included in this chapter.

One of the most important requirements of any transmitter is that the desired frequency of transmission shall be maintained within fine limits to prevent interference with other stations and to ensure that the operator remains within the allowed frequency allocation. Spurious frequency radiations capable of causing interference with other services, including television and radio broadcasting, must also be avoided. These problems are considered in Chapter 17 – 'Electromagnetic compatibility'.

The simplest form of transmitter is a single-stage, self-excited oscillator coupled directly to an antenna system: Fig 7.1(a). Such an elementary arrangement has, however, three serious limitations:

(a) the limited power which is available with adequate frequency stability;
(b) the possibility of spurious (unwanted) radiation;
(c) the difficulty of securing satisfactory modulation or keying characteristics.

In order to overcome these difficulties, the oscillator must be called upon to supply only a minimum of power to the following stages. Normally an amplifier is used to provide a constant load on the oscillator: Fig 7.1(b). This will prevent phase shifts caused by variations in the load from adversely affecting the oscillator frequency.

A FET source follower with its characteristic high input impedance makes an ideal buffer after a VFO. A two-stage buffer amplifier is often used, Fig 7.1(c), incorporating a source follower followed by a Class A amplifier.

Transmitters for use on more than one frequency band often employ two or more oscillators; these signals are mixed together to produce a frequency equal to the sum or difference of the original two frequencies, the unwanted frequencies being removed using a suitable filter. This *heterodyne* transmitter, Fig 7.1(d), is very similar in operation to the superhet receiver. More recently, frequency synthesisers have become popular for the generation of RF energy, Fig 7.1(e), and have almost entirely replaced conventional oscillators in commercial amateur radio equipment. Synthesisers are ideally suited for microprocessor control applications and multi-frequency coverage.

Oscillators are covered in detail in Chapter 5, but as they form the basis of any transmitter they cannot go unmentioned in the description of transmitters.

THE CRYSTAL OSCILLATOR

The simplest method of achieving a high degree of frequency stability is by using a quartz crystal to control the frequency of the oscillator. Such an oscillator, when correctly designed and adjusted, remains the most frequency-stable device available to the amateur.

The quartz crystal behaves like a tuned circuit of exceptionally high Q-value and may therefore be connected in the oscillator circuit as would a normal frequency-determining tuned circuit. Recently ceramic resonators have become available for a limited selection of frequencies; they offer frequency stability approaching that of a quartz crystal with the advantage of considerably lower cost.

Two oscillator circuits commonly used are the Colpitts, Fig 7.2(a), and the Pierce, Fig 7.2(b). Both are suitable for *fundamental-frequency operation*, that is to say when the oscillator

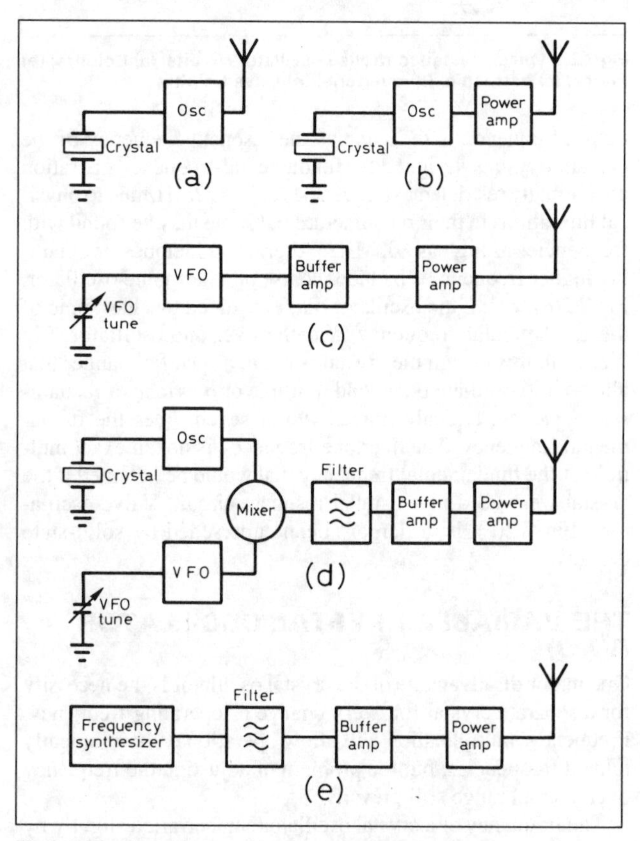

Fig 7.1. Block diagrams of basic transmitter types

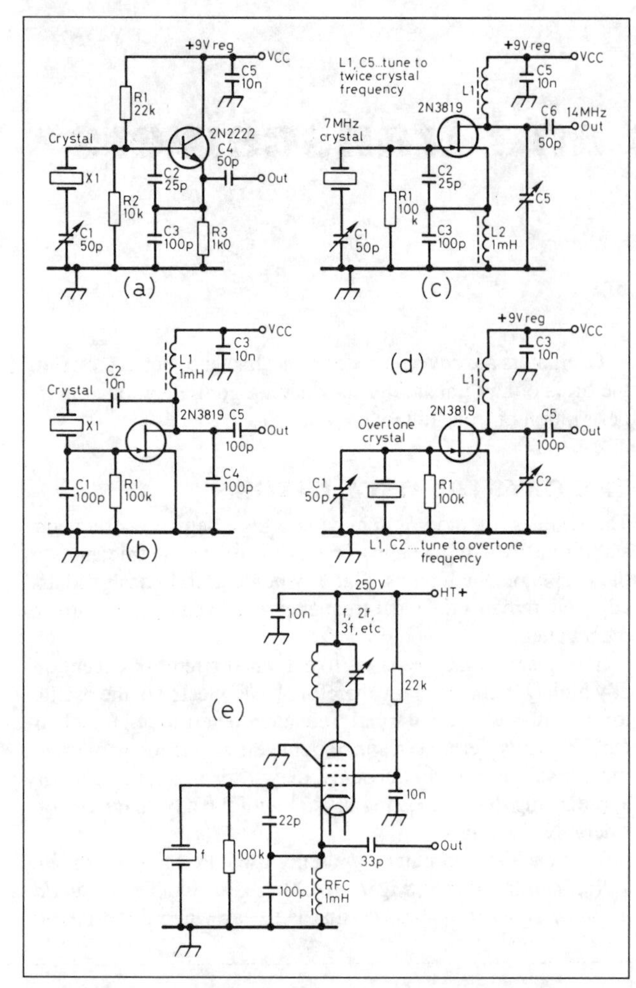

Fig 7.2. Typical crystal-controlled oscillator circuits. (a) Colpitts; (b) Pierce; (c) harmonic; (d) overtone; (e) valve Colpitts

output frequency is the same as the resonant frequency of the crystal. Crystals designed for fundamental-frequency operation are normally restricted to frequencies below 22MHz due to physical limitations in their manufacture but some may be found with frequencies as high as 100MHz. Alternative methods of obtaining higher frequencies include the use of a harmonic oscillator, Fig 7.2(c), where the oscillator output is tuned to a harmonic of the fundamental frequency, and the overtone oscillator, Fig 7.2(d). In this design the crystal is cut in a specific manner that allows it to oscillate on an odd multiple or *overtone* of its natural frequency, typically three, five or seven times the fundamental frequency. The overtone frequency is not an exact multiple of the fundamental frequency that would be obtained if the crystal were used in a parallel-resonant circuit. Valve oscillators, Fig 7.2(e), have largely been superseded by solid-state designs.

THE VARIABLE CRYSTAL OSCILLATOR (VXO)

One major disadvantage of the crystal oscillator is the necessity for a separate crystal for every change of operating frequency. Frequency multiplication will allow operation on harmonically related frequencies, but the problem of adjusting the frequency over a small range still prevails.

The frequency of a crystal oscillator can be varied slightly by changing the circuit in which it operates; this process may be

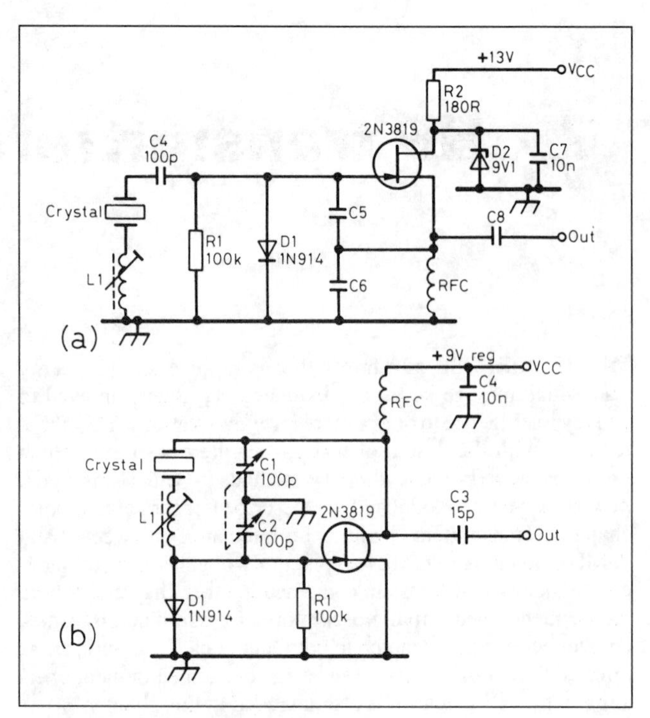

Fig 7.3. VXO circuits. (a) Colpitts; (b) Pierce

referred to as *pulling* or *rubbering* the crystal. AT-cut crystals in HC-6/U holders seem to give the greatest range of frequency change when used in a variable crystal oscillator (VXO) circuit. To obtain the maximum frequency shift it is important to reduce stray circuit capacitances to a minimum by using low-loss components and preferably avoiding the use of crystal holders. The higher the fundamental frequency, the greater will be the range of frequency movement. Typically a 3.5MHz crystal may only move 2 to 3kHz, whereas a 7MHz crystal may move as much as 10kHz. Ceramic resonators can usually be pulled more readily than quartz crystals.

Fundamental-frequency crystals below 22MHz are suitable for VXO operation. The Colpitts oscillator, Fig 7.3(a), is tuned by adjusting the value of the series inductor L1 but only limited frequency movement is possible. In the Pierce oscillator, Fig 7.3(b), a greater range of adjustment is possible, and the value of XL is adjusted to give the maximum frequency coverage throughout the operating range of C1. XL is not adjusted again during normal operation. The VXO is particularly suited to portable transmitting equipment where full band coverage is sacrificed in order to achieve stability and simplicity.

VARIABLE FREQUENCY OSCILLATOR (VFO)

Variable frequency oscillators represent the simplest practical way of providing frequency control of amateur radio transmitters when continuous coverage of a band of frequencies is required. An acceptable degree of stability can be achieved using VFOs at frequencies up to 10MHz, but lower-frequency oscillators tend to be more stable. Traditionally, higher frequencies were achieved by frequency multiplication of a highly stable low-frequency oscillator. This technique proved advantageous because of the harmonic relationship between amateur bands. With the decline in use of stand-alone AM and CW transmitters this technique has been superseded by the heterodyne process where the VFO output is mixed with a crystal oscillator to generate a higher-frequency signal: Fig 7.1(d).

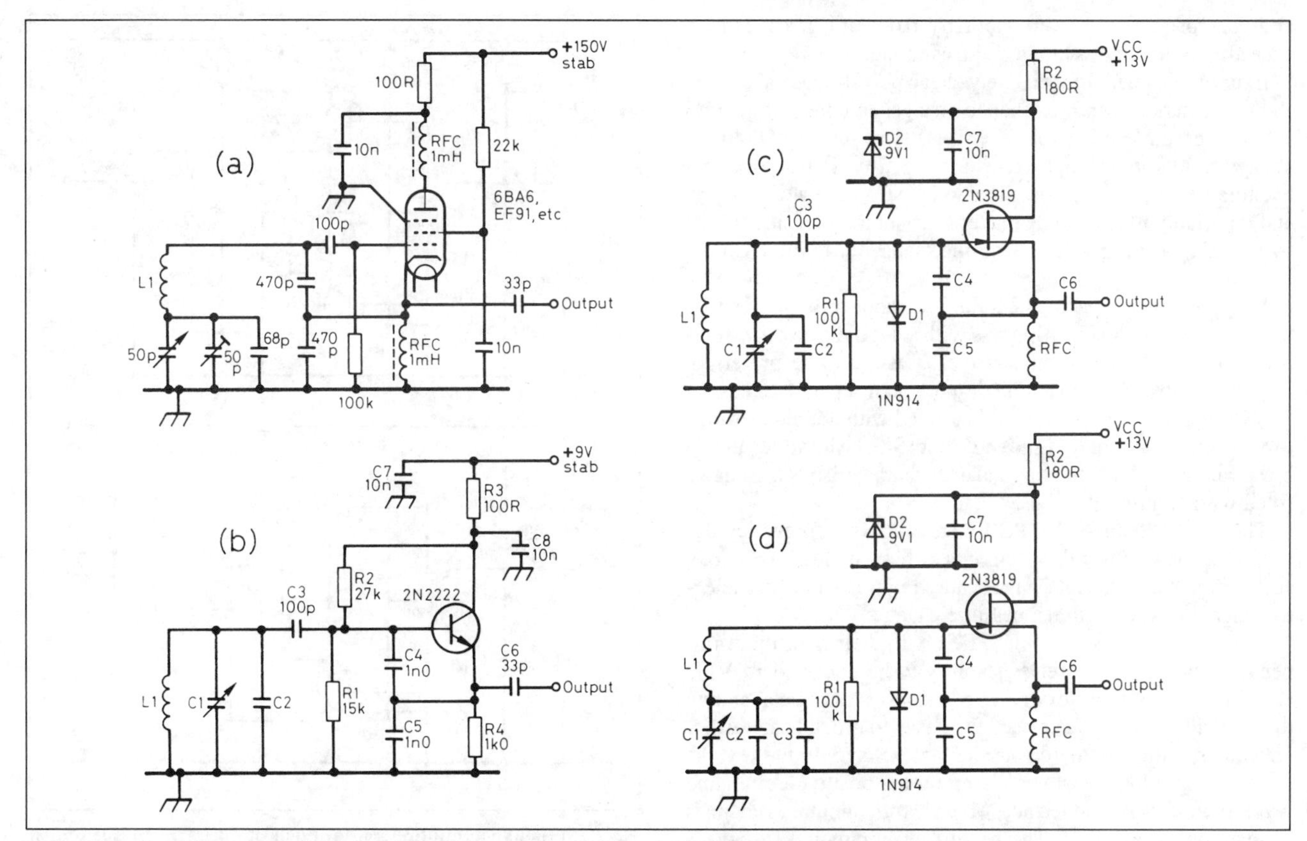

Fig 7.4. Popular VFO circuits. (a) Clapp; (b) Colpitts; (c) parallel-tuned Colpitts; (d) series-tuned Colpitts

VFO design recommendations

The rules for the avoidance of frequency drift in a VFO have remained largely the same, even though valves have given way to bipolar transistors and more recently FETs. They may be summarised as follows.

1. Ensure complete rigidity and strength of the mechanical structure and wiring.
2. Protect the tuned circuit from the effect of heat sources.
3. Construct the tuned circuit and associated oscillator components of materials having the lowest RF losses.
4. Avoid draughts across the tuned circuits.
5. Operate the oscillator at the lowest possible power level.
6. Provide voltage-stabilised power supplies.
7. Use an emitter or source follower to buffer the oscillator to minimise the loading effects of subsequent stages.
8. Screen circuits to prevent transfer of RF energy to and from other stages.

The decline in use of the thermionic valve as a VFO, Fig 7.4(a), provided solutions to such problems as hum modulation, associated with an AC heater supply, and the long-term drift caused by a continuous heat source.

The bipolar transistor, Fig 7.4(b), brought with it its own problems. The major source of drift associated with such devices is caused by inductance change in the parallel-tuned circuit, and the low input impedance of bipolar devices necessitates a high C to L ratio, with the result that a relatively small change in a low value of inductance can give rise to a major shift in frequency. Another disadvantage of a low-L circuit is that the connecting leads related to the coil degrade the tuned circuit Q by virtue of being parasitic inductances in series with the desired inductance.

The introduction of FET devices, Figs 7.4(c) and 7.4(d), with their high input impedance, has solved some of the problems associated with bipolar devices. Both types are still commonly found in oscillator circuits.

Often the desire to miniaturise equipment results in the use of less-than-ideal coils in oscillator circuits. The large-diameter, air-cored coil constructed from thick silver-plated wire wound on a ceramic former is often impractical and replaced by the smaller slug-tuned oscillator coil. If there is no alternative, it is good practice to use a slug-tuned coil which has a low-permeability core, and this should only just enter the coil at resonance. This will minimise thermal drift caused by changes in the properties of the core. Powdered iron cores are more stable than ferrite and are recommended if a core must be used. The slug mechanism must be mechanically stable in order to prevent frequency changes caused by vibration.

Considerable care must be taken if toroidal cores are used in VFO circuits as changes in core characteristics may occur with changes in ambient temperature. In general, toroids should only be used where the ambient temperature is likely to remain constant. Amidon (dash 7) mixture (colour-coded white) powdered-iron toroids offer improved thermal characteristics which are more suitable for oscillator applications.

Ceramic or fused silica formers are preferred for VFO tuned circuits, while plastic materials should only be used when nothing else is available.

Traditionally, silver or dipped-mica capacitors in the frequency-determining part of VFO circuits have always been regarded as essential to aid stability. They are good but seldom good enough on their own. The drift characteristics of mica capacitors differ greatly from one example to another. Much of the heating is due to relatively high RF currents flowing through the capacitors in an oscillator circuit. Ceramic compensating capacitors often have to be used to achieve ultimate stability, and this process of adjustment by trial and error can be exceedingly time consuming and frustrating.

Experiments with the relatively cheap polystyrene capacitors show a remarkable indifference to changes in temperature, and substituting these types for silver mica types in a VFO almost always results in an improvement in stability. Polystyrene capacitors drift in the opposite direction to silver mica capacitors, and invariably it is a simple matter to counter any residual drift by selecting a small-value silver mica capacitor to finalise drift cancellation.

Some trimmer capacitors exhibit very poor drift characteristics: ceramic and plastic trimmers with solid dielectric are notably poor. Air dielectric on ceramic bases are to be preferred, especially those that have milled rather than separate plates.

VFO tuning capacitors should be fitted with double bearings to aid mechanical rigidity, and capacitors fitted with silver-plated brass vanes exhibit better thermal drift characteristics than those fitted with aluminium vanes.

The use of double-sided PCB materials in the frequency-determining part of oscillator circuits is best avoided as it may suffer from capacitance changes caused by variations in dielectric characteristics with temperature change.

The use of JFETs and MOSFETs with their high input impedance has allowed greater stability to be achieved in VFO circuits. In the search for improved stability it has been shown that a small silicon diode placed from gate to ground can further enhance stability by regulating the bias voltage. In the reverse direction the diode acts as a clamp on the positive-going sine wave to limit transconductance, which in turn minimises changes in junction capacitance. The benefit of employing a diode is most noticeable when a source bias resistance is employed. Some designers claim that the use of dual-gate MOSFETs with both gates strapped together in oscillator circuits affords improved stability over JFET devices.

Load isolation

An important consideration in VFO design is how to extract the RF signal from the oscillator without imposing an excessive load upon it, and it is also necessary to protect it against load variations. The most common method employed is to loosely couple a FET source follower to the source of the oscillator. The coupling capacitor should be a thermally stable type such as polystyrene, and should be as low a value as possible. The use of a series resistance should be avoided as it will degrade the noise performance. Both oscillator and source follower can be conveniently operated from a common, stabilised supply voltage.

It has been long established that pulling caused by load changes can be minimised by operating the VFO at half the output frequency. This necessitates a frequency-multiplying stage, once common practice in valve transmitters. The use of frequency multiplication is still recommended for use in simple 'straight through' CW transmitters where the VFO may be prone to frequency changes caused by stray coupling with higher-powered output stages.

A frequency multiplier is a non-linear amplifier which produces an output having a high harmonic content. The output tuned circuit is selected to tune to the desired harmonic frequency and serves to attenuate the fundamental signal as well as the unwanted harmonics. Frequency multipliers operate in Class C. Single-ended multipliers can multiply by odd or even numbers but the signal level diminishes with the value of multiplication.

The use of a twin device doubler, Fig 7.5, offers a considerable improvement in efficiency over a single-ended type but a true push-pull circuit, Fig 7.5(a), is only capable of producing

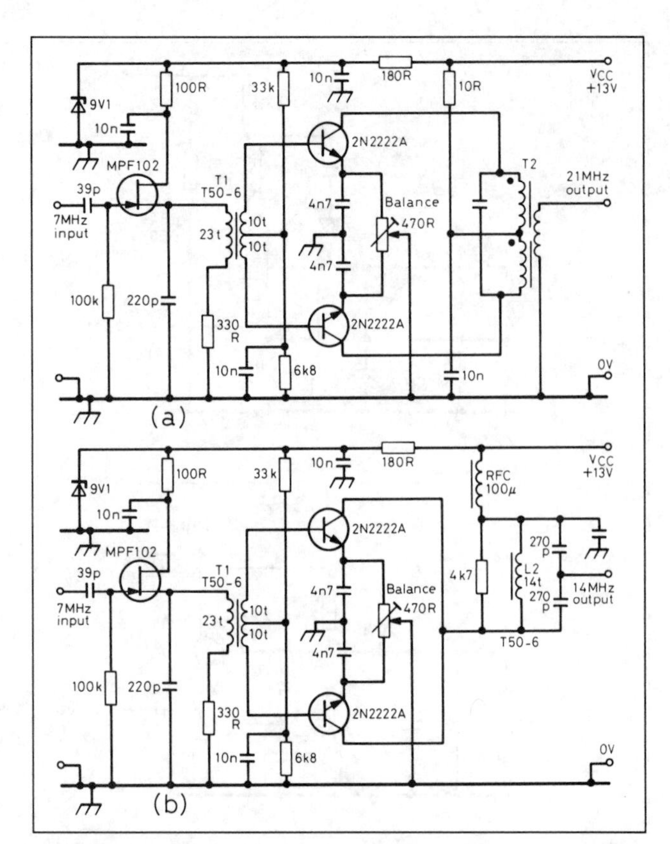

Fig 7.5. Frequency multipliers. (a) Push-pull tripler; (b) push-push doubler

odd-order harmonics. The circuit can be modified by connecting the collectors in parallel while the bases are connected in push-pull, Fig 7.5(b), resulting in a *push-push* doubler. This eliminates odd-order harmonics and passes only even-order ones. Frequency multiplication techniques, once common in most amateur band transmitters, are now seldom used.

Low-level RF amplifiers

Source followers

The source follower, Fig 7.6(a), is the modern-day equivalent to the cathode follower, Fig 7.6(b), of the valve era. It is usually a JFET, though MOSFETs are occasionally used. Operating in common-drain mode, it has a very high input impedance combined with a low output impedance. The source follower can be used to isolate loads as well as provide impedance transformation with unity gain. Emitter followers, Fig 7.6(c), have the same function but use bipolar devices, these having a much lower input impedance with consequent loading of the preceding circuit. It is always good practice to isolate an oscillator from its load by using either a source or emitter follower.

Voltage amplifiers

In order to preserve the highest possible levels of stability, oscillators are specifically designed to produce low RF output levels. Often this is apparent when a newly constructed oscillator fails to oscillate when initially switched on. Unless an oscillator is to be connected directly to a mixer requiring a low input level, such as a modern IC mixer, the RF voltage produced by the oscillator will have to be amplified to a more suitable level. In the case of a CW transmitter, several amplifier stages will have to be cascaded before the final power level is achieved. A diode ring mixer circuit, found in many HF transceivers, will typically

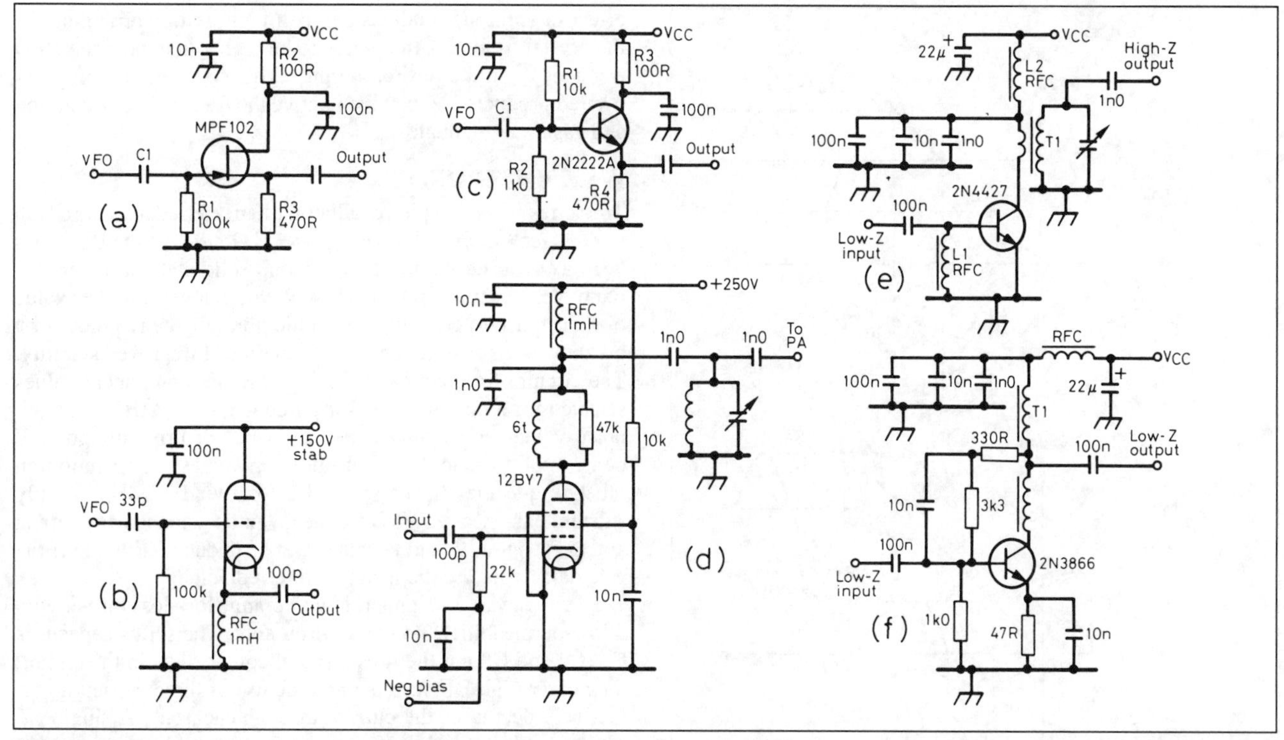

Fig 7.6. Low-level amplifiers. (a) Source follower; (b) cathode follower; (c) emitter follower; (d) and (e) Class C amplifiers; (f) Class A amplifier

require an oscillator injection in the order of +7dBm, which corresponds to 500mV RMS into 50Ω. This can normally be achieved with a single amplifier stage. It is important not to introduce harmonics or spurious signals during the amplification process, and for this reason it is normal to operate low-level amplifier stages in a linear manner, ie in Class A.

Fig 7.6(d) illustrates a typical valve amplifier stage used to drive the power amplifier in a CW transmitter; it is operated in Class C for maximum efficiency. Fig 7.6(e) shows a comparable solid-state bipolar amplifier for use in a similar transmitter driving a valve PA. Fig 7.6(f) illustrates a Class A (linear) amplifier suitable for increasing the power level in an oscillator chain or for amplification in a SSB transmitter.

The various amplifier operating classifications are as follows:

Class A: RF drive and DC bias are set so that the device is conducting continuously. The conduction angle is 360°: Fig 7.7(a). The output current swing is a direct replica of the input current swing, resulting in linear amplification. Efficiency is low, theoretically 50% but typically 25–30%.

Class AB: The device is biased so that output current flows for less than 360°, but more than 180°. Efficiency is greatly improved, typically 50%, but at the expense of linearity. However, it is still very acceptable for the most rigorous SSB applications.

Class B: Bias and RF drive are set to a level where the device is just cut off and output current flows for exactly one half of the cycle (180°). Efficiency is improved further (65%) while linearity is maintained at an acceptable level: Fig 7.7(b).

Class C: The device is biased off such that only peaks in the drive current cause output current pulses. Efficiency rises to around 80%, but the linearity is now extremely poor. The output waveform must be restored in a tank circuit: Fig 7.7(c).

Choice of VFO frequency

Traditionally a VFO would be operated on either the desired transmitting frequency or a direct sub-multiple of it. A VFO operating on the 160m band (1.8MHz) can be frequency doubled to operate on 80m (3.6MHz) by selecting its second harmonic; this can be further doubled again for operation on 40m (7.2MHz). The VFO frequency will of course have to be lowered to 1.75MHz to place the signal in the 40m band. Similarly, a 7MHz VFO can be doubled to 14MHz, tripled to 21MHz, or quadrupled to 28MHz. Thus, using just two VFOs, coverage of all the traditional HF amateur bands is possible. This technique has the advantage of allowing a relatively stable low frequency oscillator to be used on higher frequencies where a comparable level of stability would be difficult to achieve. Each time the frequency is multiplied, the tuning range and the drift also increase by the same multiplication factor.

If the output from a VFO is mixed with the output from a crystal oscillator, then it is possible to achieve a signal on either the sum or difference of the two oscillator frequencies. The tuning bandwidth is controlled by the frequency range of the VFO, while the drift will be the sum of the drift of the two oscillators, which can either add or subtract from one another. This heterodyne or mixing frequency generator exhibits better frequency stability than the multiplying type, but it has a number of disadvantages. The unwanted frequency from the mixing process, together with the two fundamental frequencies, has to be removed using suitable filters. Additional frequencies are generated by the mixing of multiples of the original oscillator harmonics, producing spurious products. By careful selection of the oscillator frequencies it is possible to arrange for the spurii to fall on frequencies where they are least likely to cause any adverse affects. A popular frequency of operation for VFOs for amateur equipment is in the region of 5.0MHz with a tuning range of approximately 500kHz. By very slight adjustment of component values, the same VFO circuit can be adapted for use directly on 1.8, 3.5 or 7MHz, and with care on 10MHz. The

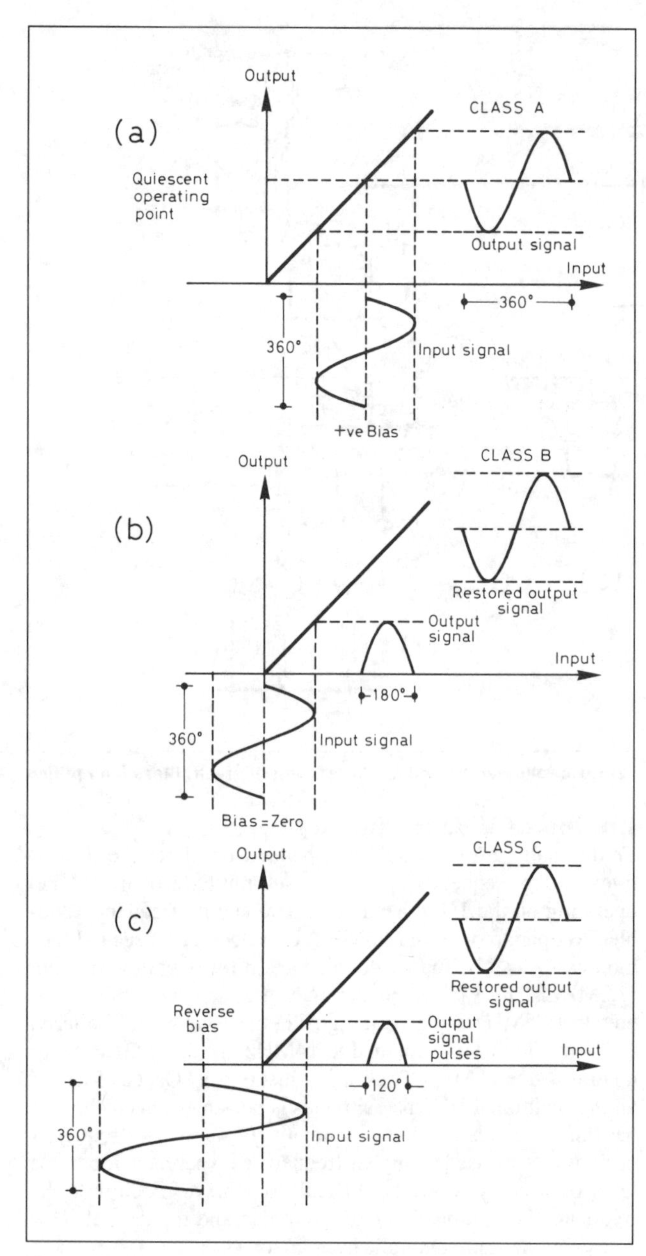

Fig 7.7. Bias arrangements for amplifiers

effect of spurious products can be minimised by arranging for the crystal oscillator frequency to be higher than both the VFO frequency and the desired output frequency. The resulting unwanted products are then well above the frequency of operation and easier to attenuate.

A practical VFO

The series-tuned Colpitts oscillator, often referred to as the *Clapp* or *Gouriet-Clapp* and devised by G C Gouriet of the BBC, has been a favourite with both valve and solid-state designers for many years. The series-tuned oscillator enables a higher value of inductance to be used than would normally be required for a parallel-tuned design, resulting in claims of improved stability. The circuit of a practical VFO, Fig 7.8, has component values selected for a nominal operating frequency of 5MHz, but reactance values are tabulated for frequency-determining components to allow calculation of optimum values for operation on other frequencies in the range 1.8 to around 10MHz. Simply substitute the desired VFO frequency into the reactance formula and the required *L* and *C* values can be deduced. Round off the calculated value to the nearest preferred value.

L1 uses a variable inductor on a ceramic low-loss former with a low-permeability, powdered-iron core. The series capacitors C1, C2 and C3 are the most critical components in the circuit; they carry high RF currents and the use of three capacitors effectively decreases the current through each one. A single capacitor in this location may cause frequency-jumping brought about by dielectric stress as a result of heat generated by the RF current. C3 is selected to counteract the drift of the circuit. In most circuits the tuning capacitor C6 will be located in series with the inductor and in parallel with C1–C3; a trimmer capacitor may also be placed in parallel with the main tuning capacitor. By placing the tuning capacitor in parallel with C5, a smaller tuning range is possible. This may be desirable for a 40m VFO where the frequency range required is only 100kHz.

The use of a JFET such as a 2N3819 or MPF102 ensures minimum loading of the tuned circuit. TR2, another JFET, acts as a buffer amplifier lightly coupled to the source of TR1. Both transistors operate from a zener-stabilised 9V power supply. TR3 is a voltage amplifier operating in Class A and uses a bipolar device such as the 2N2222A. Bias for Class A operation is established by the combination of the emitter resistor and the two base bias resistors which are also connected to the stabilised 9V supply. The collector impedance of TR3 is transformed to 50Ω using a pi-network which also acts as a low-pass filter to

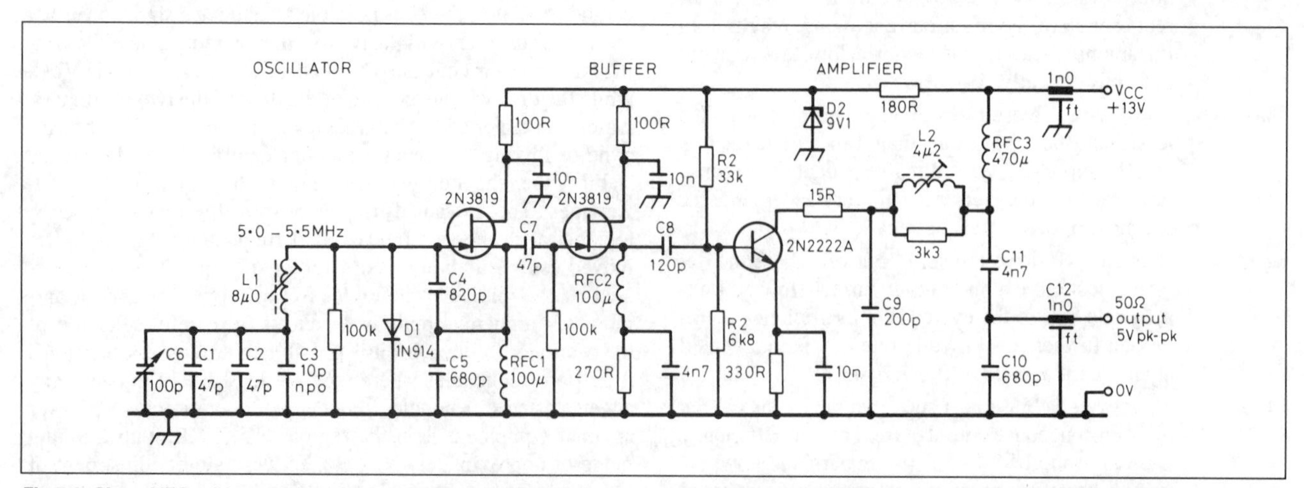

Fig 7.8. Clapp VFO with amplifiers for 5.0–5.5MHz. Reactance values are L1 = 265Ω, L2 = 140Ω, C1 = 690Ω, C2 = 690Ω, C3 = 2275Ω, C4 = 33Ω, C5 = 48Ω, C6 = 303Ω min, C7 = 690Ω, C8 = 227Ω, C9 = 152Ω, C10 = 48Ω, C11 = 4.5Ω, C12 = 23Ω, RFC1 = 4400Ω, RFC2 = 4400Ω

attenuate harmonics. The loaded Q of the output network is reduced to approximately 4 by adding a parallel resistor across the inductor, and this broadens the bandwidth of the amplifier. For optimum output the network must be tuned either by adjusting L2 or by altering the value of C9. The 15Ω series resistor is included to aid stability.

The VFO described in Fig 7.8 is capable of driving a solid-state CW transmitter, and it is ideally suited to provide the local oscillator drive for a diode ring mixer. It also has sufficient output to drive a valve amplifier chain in a hybrid design. A suitable push-pull frequency-doubler circuit is shown in Fig 7.5 for use in such applications as straight through, TRF (tuned radio frequency) transmitters for CW operation.

Evaluation of a newly constructed VFO

The VFO is the heart of the transmitter and will determine its ultimate stability. It is therefore essential that it is properly tested before being used to transmit on the air. The following points may be used as a guide to testing VFO circuits.

1. Locate the oscillator inside the chassis or casing in which it will ultimately be operated.
2. Operate the oscillator, together with its associated buffer amplifiers, into a suitable dummy load.
3. Monitor the current drawn by the VFO – it should not exceed 10–20mA. A higher current will be drawn by any additional Class A amplifiers.
4. Check that the unit is oscillating by using an oscilloscope or frequency meter. The use of a receiver is possible, provided that the frequency of oscillation can be located. Ensure that the waveform is sinusoidal; if it is not it will contain unwanted products.
5. Adjust the oscillator to cover the desired frequency range. This may require adjustment of the relative L/C ratio as well as the initial tuning of the oscillator to the desired frequency. If the tuning range is excessive, reduce the capacitance swing or increase the C to L ratio. If the bandspread is insufficient it will be necessary to either increase the size of the tuning capacitor, or alternatively the C to L ratio must be reduced to make the tuning range greater. Changing the L/C ratio over too greater range may have an adverse affect on stability.
6. Stability measurements take a great deal of time and patience. The VFO should be allowed to operate in its intended location for a period of time, during which frequency measurements are made at regular intervals. A graph can be plotted to determine the drift characteristics. Temperature effects can be simulated by the application of a hair drier to the outside of the unit.
7. Substitution of one or more capacitors in a VFO circuit, followed by a further drift checks, will allow the selection of the most suitable component to neutralise any drift experienced. In the case of a series-tuned Colpitts oscillator, drift cancellation is best performed by selection of the fixed-value series capacitors. Always allow at least 30min for any heat generated by soldering to dissipate after the addition or removal of a component.
8. Particular attention should be paid to the mechanics associated with the slow-motion drive attached to a VFO. The drive should have a suitable reduction ratio between 6:1 and 100:1, with no backlash or stiff spots. The VFO enclosure and drive mechanism should share a common mounting, in order to minimise frequency movement caused by expansion or flexing of the unit.

INTERSTAGE COUPLING

Correct impedance matching between a stage and its load provides maximum transfer of power. The load may be an antenna or a succeeding stage in a transmitter, and correct matching between either is therefore essential for efficient operation. A variety of matching networks are available for both valve and solid-state designs, and the choice will depend on a number of conditions such as: drive power versus tolerable mismatch, selectivity and impedance matching ratio. Output impedance can be calculated approximately by the formula:

$$Z = \frac{V_{cc}^2}{2P_{out}}$$

where Z is in ohms and P_{out} is the power output from the stage.

Determination of the base input impedance of a stage is difficult without sophisticated test equipment, but if the output of a stage is greater than 2W its input impedance is usually less than 10Ω and may be as low as 1 to 2Ω. For this reason some kinds of LC matching do not lend themselves to this application. With the precise input impedance of a stage unknown, adjustable LC networks often lend themselves best to matching a wide range of impedances. A deliberate mismatch may be introduced by the designer to control power distribution and aid stability, in which case excess power must be available compared with that required for the matched condition.

In the interest of stability it is common practice to use low-Q networks between stages in a solid-state transmitter, but the penalty is poor selectivity and little attenuation of harmonic or spurious energy. Most solid-state amplifiers use loaded Qs of 5 or less compared to Qs of 10 to 15 found in valve circuitry. All calculations must take into account the input and output capacitances of the solid-state devices, which must be included in the network calculations. The best source of information on input and output capacitance of power transistors is the manufacturers' data sheets. Impedances vary considerably with frequency and power level, producing a complex set of curves, but input and output capacitance values do not vary with frequency or power level.

Transformer coupling has always been popular with solid-state devices: Fig 7.9(a). Toroidal transformers are the most efficient and satisfactory up to 30MHz, the tapping and turns ratio of T1 being arranged to match the collector impedance of TR1 to the base impedance of TR2. In some circumstances only one turn may be required on the secondary winding. A low value of resistor is used to slug this secondary winding to aid stability.

An alternative to inductive coupling is to use a capacitive divider network which resonates L1 as well as providing a suitable impedance tap: Fig 7.9(b). Suitable RF chokes wound on high-permeability ($\mu = 800$) ferrite beads, adequately decoupled, are added to aid stability. This circuit is less prone to VHF parasitics than the one in Fig 7.9(a), especially if C2 has a relatively high value of capacitance.

When the impedance values to be matched are such that specific-ratio broad-band transformers can be used, typically 4:1 and 16:1, one or more fixed-ratio transformers may be cascaded as shown in Fig 7.9(c). The system exhibits a lack of selectivity, but has the advantage of offering broad-band characteristics. Dots are normally drawn on to transformers to indicate the phasing direction of the windings. (All dots start at the same end.)

Simple LC networks provide practical solutions to matching impedances between stages in transmitters. It is assumed that normally high output impedances will be matched to lower-value input impedances. However, if the reverse is required, networks can be simply used in reverse to effect the required transformation.

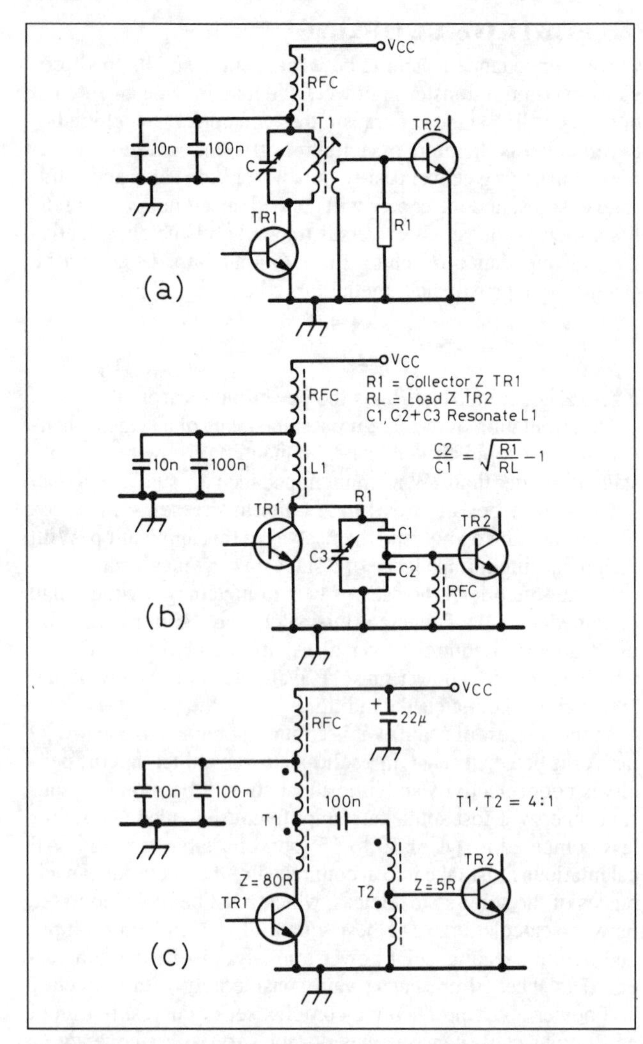

Fig 7.9. Interstage coupling. (a) Transformer coupling; (b) capacitive divider coupling; (c) 16:1 broad-band transformer matching

Networks 1 and 2, Figs 7.10(a) and 7.10(b), are variations on the L-match and may used fixed or with adjustable inductors and capacitors. Network 3, Fig 7.10(c), is used by many designers because it is capable of matching a wide range of impedances. It is a low-pass T-network and offers harmonic attenuation to a degree determined by the transformation ratio and the total network Q. For stages feeding an antenna, additional harmonic suppression will normally be required. The value of Q may vary from 4 to 20 and represents a compromise between bandwidth and attenuation.

Network calculations are as follows:

Network 1:

1. Select Q_1
2. $X_{L1} = Q_1 R_S + X_{Cs}$
3. $X_{C2} = Q_L R_L$
4. $X_{C1} = \dfrac{R_V}{Q_1 - Q_L}$

where:

$$\text{Load } Q = Q_L = \sqrt{\left(\frac{R_s(1+Q_1^2)}{R_L}\right)} - 1 = \sqrt{\frac{R_V}{R_L}} - 1$$

$$R_V = R_S(1 + Q_1^2) = \text{Virtual resistance of network}$$

Network 2:

When $R_S < R_L$

1. Select Q_1
2. $X_{C1} = Q_1 R_S$
3. $X_{C2} = R_L \sqrt{\dfrac{R_S}{R_L - R_S}}$
4. $X_{L1} = X_{C1} + \left(\dfrac{R_S R_L}{X_{C2}}\right) + X_{Cs}$

Network 3:

1. Select Q_1
2. $X_{L1} = (R_S Q_1) + X_{Cs}$
3. $X_{L2} = R_L Q_L$
4. $X_{C1} = \dfrac{R_V}{Q_1 + Q_L}$

where:

$$R_V = R_S(1 + Q_1^2)$$

$$Q_L = \sqrt{\left(\frac{R_V}{R_L}\right)} - 1 = \text{Load } Q$$

Conventional broad-band transformers are very useful in the construction of solid-state transmitters. They are essentially devices for the transformation of impedances relative to the ratio of the transformer and are not specific impedance devices, eg a solid-state PA with a nominal 50Ω output impedance may employ a 3:1 ratio broad-band output transformer. The impedance ratio is given by the square of the turns ratio and will be 9:1. The output impedance will only be 50Ω when the collector output impedance is:

$$\frac{50}{9} = 5.55\Omega$$

This will only occur at one specific power output determined using the formula:

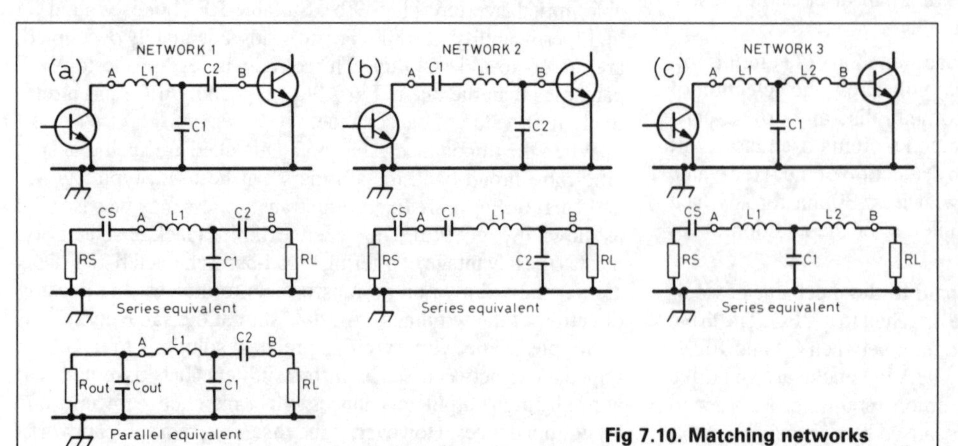

Fig 7.10. Matching networks

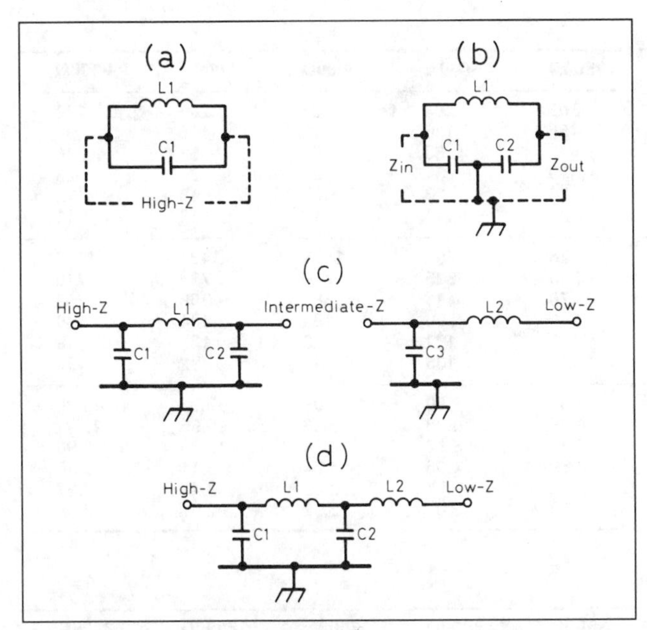

Fig 7.11. Pi-network and derivation. (a) Parallel-tuned tank circuit; (b) parallel-tuned tank with capacitive tap; (c) pi-network and L-network; (d) pi-L network

$$Z = \frac{V_{cc}^2}{2P_{out}}$$

Another type of broad-band transformer found in transmitting equipment is the *transmission-line transformer*. This acts as a conventional transformer at the lower frequencies but, as the frequency increases, the core becomes less 'visible' and the transmission-line properties take over. The calculations are complex, but it is well known that a quarter-wavelength of line can exhibit impedance-transformation properties. When a quarter-wave line of impedance Z_0 is terminated with a resistance R_1, a resistance R_2 is seen at the other end of the line.

$$Z_0^2 = R_1 R_2$$

Transmission-line transformers are often constructed from twisted pairs of wires wound onto a ferrite toroidal core having a initial permeability (μ) of at least 800. This ensures relatively high values of inductance with a small number of turns. It should be borne in mind that with high-power solid-state transmitters some of the impedances to be transformed are very low, amounting to only a few ohms.

ANODE TANK CIRCUITS

Pi-network

While the use of valves in amateur equipment has declined rapidly over the last decade, there are still numerous hybrid designs using valves in the final amplifier. Additionally, high-power valve linear amplifiers are likely to be in use for some considerable time. The *pi-tank* (Fig 7.11) has been the most popular matching network since the introduction of multiband transmitters in the 'fifties. It is easily bandswitched and owes its name to its resemblance to the Greek letter π.

The anode tank circuit is required to meet the following conditions.

1. The anode circuit of the valve must be presented with the proper resistance in relation to its operating conditions to ensure efficient generation of power.
2. This power must be transferred to the output without appreciable loss.
3. The circuit Q must be sufficient to ensure good flywheel action in order to achieve a close approximation of a sinusoidal RF output voltage. This is especially important in Class C amplifiers where the output is in the form of a series of pulses of RF energy which must be restored to the original shape.

Tank circuit Q

In order to quantify the ability of a tank circuit to store RF energy (essential for flywheel action), a quality factor Q is defined. Q is the ratio of energy stored to energy lost in the circuit.

$$Q = 2\pi \frac{W_S}{W_L} = \frac{X}{R}$$

where W_S is the energy stored in the tank circuit; W_L is the energy lost to heat and the load; X is the reactance of either the inductor or capacitor in the tank circuit; R is the series resistance.

Since both circulating current and Q are inversely proportional to R, then the circulating current is proportional to Q. By Ohm's law, the voltage across the tank circuit components must also be proportional to Q.

When the circuit has no load the only resistance contributing to R are the losses in the tank circuit. The unloaded Q_U is given by:

$$Q_U = \frac{X}{R_{loss}}$$

where X is the reactance in circuit and R_{loss} is the resistance losses in circuit.

To the tank circuit, a load acts in the same way as circuit losses. Both consume energy but only the circuit losses produce heat energy. When energy is coupled from the tank circuit to the load, the loaded Q (Q_L) is given by:

$$Q_L = \frac{X}{R_{load} + R_{loss}}$$

It follows that if the circuit losses are kept to a minimum the loaded Q value will rise.

Tank efficiency can be calculated from:

$$\text{Tank efficiency} = 1 - \left(\frac{Q_L}{Q_U}\right) \times 100$$

where Q_U is the unloaded Q and Q_L is the loaded Q.

Typically the unloaded Q for a pi-tank circuit will be between 100 and 300 while a value of 12 is accepted as a good compromise for the loaded Q. In order to assist in the design of anode tank circuits for different frequencies, inductance and capacitance values for a pi-network with a loaded Q of 12 are provided in Table 7.1 for different values of anode load impedance.

Pi-L network

The *pi-L network*, Fig 7.11(d), is a combination of the pi-network and the L-network. The pi-network transforms the load resistance to an intermediate impedance, typically several hundred ohms, and the L-network then transforms this intermediate impedance to the output impedance of 50Ω. The output capacitor of the pi-network is in parallel with the input capacitor of the L-network and is combined into one capacitor equal to the sum of the two individual values.

The major advantage of the pi-L network over a pi-network is considerably greater harmonic suppression, making it particularly suitable for high-power linear amplifier applications. A table of values for a pi-L network having a loaded Q of 12 for different values of anode load impedance is given in Table 7.2. Both

Table 7.1. Pi-network values for selected anode loads (Q_L = 12)

	MHz	1500Ω	2000Ω	2500Ω	3000Ω	3500Ω	4000Ω	5000Ω	6000Ω	8000Ω
	1.8	708	531	424	354	303	264	229	206	177
	3.5	364	273	218	182	156	136	118	106	91
C1	7.0	182	136	109	91	78	68	59	53	46
(pF)	14.0	91	68	55	46	39	34	30	27	23
	21.0	61	46	36	30	26	23	20	18	15
	28.0	46	34	27	23	20	17	15	13	11
	1.8	3413	2829	2415	2092	1828	1600	1489	1431	1392
	3.5	1755	1455	1242	1076	940	823	766	736	716
C2	7.0	877	728	621	538	470	411	383	368	358
(pF)	14.0	439	364	310	269	235	206	192	184	179
	21.0	293	243	207	179	157	137	128	123	119
	28.0	279	182	155	135	117	103	96	92	90
	1.8	12.81	16.60	20.46	24.21	27.90	31.50	36.09	39.96	46.30
	3.5	6.59	8.57	10.52	12.45	14.35	16.23	18.56	20.55	23.81
L1	7.0	3.29	4.29	5.26	6.22	7.18	8.12	9.28	10.26	11.90
(μH)	14.0	1.64	2.14	2.63	3.11	3.59	4.06	4.64	5.14	5.95
	21.0	1.10	1.43	1.75	2.07	2.39	2.71	3.09	3.43	3.97
	28.0	0.82	1.07	1.32	1.56	1.79	2.03	2.32	2.57	2.98

Table 7.2. Pi-L network values for selected anode loads (Q_L = 12)

	MHz	1500Ω	2000Ω	2500Ω	3000Ω	3500Ω	4000Ω	5000Ω	6000Ω	8000Ω
	1.8	784	591	474	397	338	297	238	200	152
	3.5	403	304	244	204	174	153	123	103	78
C1	7.0	188	142	114	94	81	71	57	48	36
(pF)	14.0	93	70	56	47	40	35	29	24	18
	21.0	62	47	38	32	27	23	19	16	12
	28.0	48	36	29	24	21	18	15	13	9
	1.8	2621	2355	2168	2026	1939	1841	1696	1612	1453
	3.5	1348	1211	1115	1042	997	947	872	829	747
C2	7.0	596	534	493	468	444	418	387	368	337
(pF)	14.0	292	264	240	222	215	204	186	172	165
	21.0	191	173	158	146	136	137	125	117	104
	28.0	152	135	127	115	106	107	95	87	86
	1.8	14.047	17.933	21.730	25.466	29.155	32.805	40.011	47.118	61.119
	3.5	7.117	9.086	11.010	12.903	14.772	16.621	20.272	23.873	30.967
L1	7.0	3.900	4.978	6.030	7.070	8.094	9.107	11.108	13.081	16.968
(μH)	14.0	1.984	2.533	3.069	3.597	4.118	4.633	5.651	6.655	8.632
	21.0	1.327	1.694	2.053	2.406	2.755	3.099	3.780	4.452	5.775
	28.0	0.959	1.224	1.483	1.738	1.989	2.238	2.730	3.215	4.171
	1.8	8.917	The value of L2 remains constant for all values of anode impedance.							
	3.5	4.518								
L2	7.0	2.476								
(μH)	14.0	1.259								
	21.0	0.843								
	28.0	0.609								

Tables 7.1 and 7.2 assume that source and load impedances are purely resistive; the values will have to be modified slightly to compensate for any reactance present in the circuits to be matched. Under certain circumstances matching may be compromised by high values of external capacitance, in which case a less-than-ideal value of *Q* may have to be accepted.

SINGLE SIDEBAND TRANSMISSION

Single sideband suppressed-carrier telephony transmission (commonly called *SSB*) is a specialised form of amplitude-modulated telephony, and a brief examination of basic AM theory is therefore an essential preliminary to any description of the more advanced system.

AM carrier and sideband relationships

The carrier of an AM transmission does not vary in amplitude: it is at all times of constant strength. The modulation introduced at the transmitter heterodynes the carrier and produces sum and difference frequencies. These are symmetrically disposed either side of the original carrier frequency and constitute two bands of side frequencies – those below the carrier form the lower sideband and those above the carrier form the upper sideband. The carrier by itself does not convey any intelligence; the intelligence is conveyed solely by the sideband frequencies.

Consider what happens when a carrier of 1000kHz is modulated 100% by a pure audio tone of 2kHz. The energy propagated from the antenna would be three entirely separate and individual outputs on 998, 1000 and 1002kHz (Fig 7.12). These three channels of RF energy would travel quite separately through the ionosphere and would eventually arrive at the receiving antenna. They would then be accepted by the receiver, heterodyned by the local

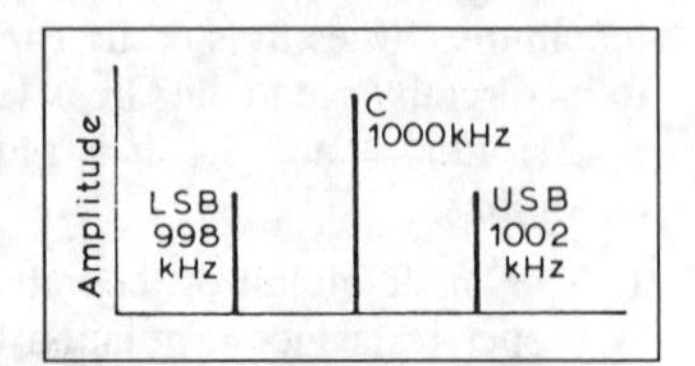

Fig 7.12. Amplitude frequency relationship of AM signal modulated with single tone of 2kHz

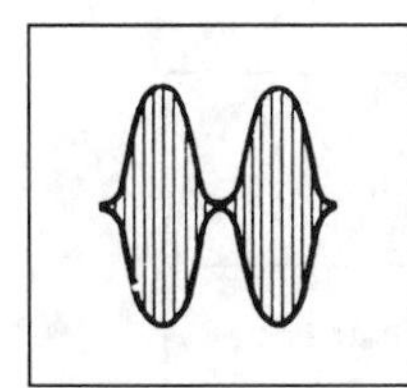

Fig 7.13. Modulation envelope at the detector

oscillator and converted to the final intermediate frequency of the receiver. If the IF passband was centred on 460kHz the IF amplifier would present three separate frequencies (462, 460 and 458kHz) to the detector. The combined effect of these three frequencies at the detector would in turn produce the modulation envelope of Fig 7.13.

Fig 7.14 illustrates in diagrammatic form the relationship of the carrier and its sidebands under speech modulation; it is this representation, rather than that of Fig 7.13, which forms the foundation stone on which to build a solid understanding of SSB. There is, however, one noteworthy point which is more immediately evident from Fig 7.13 than from the alternative presentation. At the modulation crest the carrier and sideband voltages are in phase and add together; the voltages at the anode and screen or collector of the modulated stage will also rise to twice the applied DC potentials, so the components in these circuits must be appropriately rated to avoid breakdown. As the transmitter peak envelope voltage (PEV) at the crest of modulation is double that of the unmodulated carrier, and as output power is proportional to voltage squared, the peak envelope power (PEP) will be increased four times. Assuming that the modulated stage is being operated at 150W DC input power (for many years the UK maximum amateur band power limit) with an overall efficiency of 66%, the unmodulated carrier output will be 100W and the peak envelope power output 400W (the current UK power limit).

Single sideband transmission

If the transmitter of Fig 7.12 were radiating a lower-sideband SSB signal, the carrier on 1000kHz and the upper sideband on 1002kHz would be suppressed at the transmitter, and the only output to the antenna would be continuous RF energy on 998kHz. This would be converted by the receiver to an IF of 462kHz and fed to the product detector. The carrier of 460kHz would be generated by a *carrier insertion oscillator* (CIO) and also fed to the product detector. A vector diagram would show the inserted carrier from the CIO as a vertical line with the sideband vector rotating around it at the modulating frequency; the modulation envelope would be recovered at the detector and the resultant audio output would be 2kHz. It will be seen that the original 2kHz tone input has been recovered, yet the SSB transmitter has radiated only one continuous RF output on a single frequency of 998kHz.

If the tone modulation at the transmitter had been changed from 2 to 3kHz the sideband radiated would change from 998 to 997kHz. This would be converted by the receiver to an IF of 463kHz, combined in the product detector with the local carrier on 460kHz and the resultant audio output would be 3kHz. Should the transmitter be modulated with both the 2 and 3kHz tones simultaneously, two RF outputs would be radiated from the antenna on 997 and 998kHz. These would be converted by the receiver to two IF outputs of 463 and 462kHz. The resultant output recovered by the detector would be 3 and 2kHz – the original modulating frequencies.

When the tone modulation is replaced by the output of a speech amplifier the transmitted sideband becomes essentially a band of frequencies undergoing continuous changes in amplitude and frequency as the voice changes in inflection and intensity. Provided that the receiver IF passband is sufficiently wide, all frequencies within the received sideband will be amplified equally and fed into the product detector. The locally inserted carrier

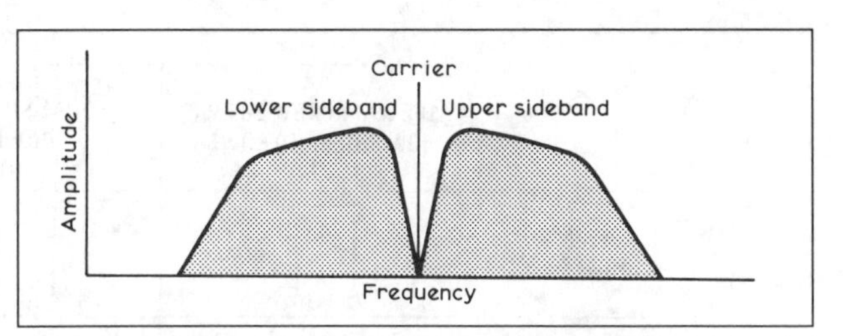

Fig 7.14. Amplitude/frequency relationships of carrier and sidebands with 100% speech modulation

merely serves as a datum line against which the sideband is demodulated – the resultant audio output is the original speech modulation.

Communications efficiency

In an AM transmitter, the amplitude of each sideband is limited to half that of the carrier, so its maximum power will be one-quarter of the unmodulated output power. In the example already quoted of the 150W (DC input power) amateur transmitter with an overall efficiency of 66%, the unmodulated carrier RF output would be 100W and the maximum power in one sideband would be 25W; both sidebands together would produce 50W. It is not, however, possible to specify the effectiveness of the transmission until something is known about the equipment on which it is being received. The sidebands necessarily occupy a band of frequencies 3kHz wide on either side of the carrier, so that if the receiver has a bandwidth of 6kHz they both add in phase and contribute to the total talk power. If a more selective receiver with a 3kHz bandwidth is used, only one sideband contributes to its output. Under the crowded conditions which obtain in the amateur bands, good selectivity is almost invariably a necessity so that only 25W of the signal actually conveys intelligence to the receiver. This is not a particularly profitable return for the generation of 100W of carrier power, not to mention the audio power required to effect high-level modulation.

As the carrier remains constant in frequency and amplitude it conveys no intelligence from transmitter to receiver, and serves merely as a datum line against which the sidebands are demodulated. All the intelligence is conveyed in the sidebands, so that the carrier may be omitted and the signal demodulated perfectly clearly if the reference function is transferred to a local oscillator at the receiving end. The attractiveness of substituting the signal from a low-power oscillator in the receiver for the 100W carrier is obvious.

The AM power amplifier which produces a 100W carrier must be capable of a peak output of 400W. With modifications, the full capacity of the amplifier may be utilised to radiate intelligence conveying only single sideband energy. This does not necessarily mean that the peak envelope power output of the hypothetical amplifier will be 400W because there are important differences between the operating conditions of high-level modulated Class C amplifiers and the Class AB or B linear amplifiers customarily employed in SSB output stages.

Double sideband transmission (DSB)

There is no difficulty at all in suppressing the carrier; a simple balanced modulator can be constructed from a pair of diodes or transistors. More commonly, purpose-made integrated circuits ensure precise balance and hence excellent carrier suppression. The problem is how best to remove one of the sidebands. While double sideband suppressed carrier (DSB) is a more efficient mode of transmission than AM, its reception is complex,

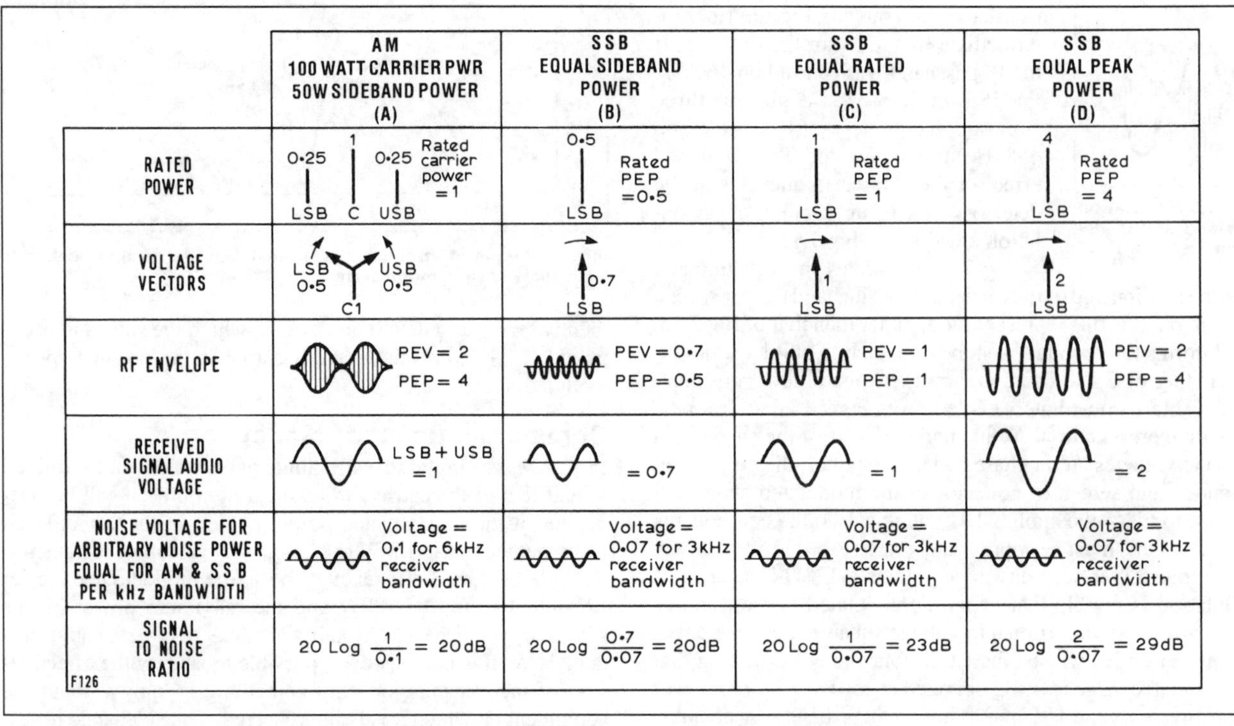

Fig 7.15. Power relationships for AM and SSB transmission. Single-tone sine-wave modulation

requiring a highly stable and phase-locked carrier insertion oscillator if the original modulating waveform is to be reproduced. Elimination of one sideband reduces the bandwidth by half to 3kHz or less and is much simpler to receive as SSB. The CIO may wander quite a few hertz away from the correct frequency before intelligibility suffers appreciably, and does not have to be locked in phase with the original carrier.

DSB transmissions can be converted to SSB to simplify reception by using a selective SSB receiver. The unwanted sideband is removed by the receiver filter, which has sufficient bandwidth to accommodate only one sideband. This is then demodulated as SSB, and either sideband may be selected.

Half of the power radiated by a DSB transmitter is not used for reception and merely serves to increase the general level of interference. In this respect it is no worse than conventional AM, but has the advantage of having no continuous carrier. The relative simplicity of construction of DSB transmitters compared to an AM transmitter of comparable power output accounts for their former popularity but they have now been superseded almost completely by SSB.

Effective power – AM versus SSB

Perhaps the clearest method (least open to misinterpretation) is to show the relative efficiency of AM and SSB systems in diagrammatical form where the powers and voltages concerned are to the same relative values. This method has been adopted in Fig 7.15. The basis of comparison given in column A is an AM transmission of 100W RF output power rating, 100% modulated with 50W of sideband power. (Assuming 66.6% efficiency this represents a DC input power of 150W, modulated 100% with 75W of audio.) At the crest of the modulation cycle the peak envelope power output (PEP) is four times the carrier power: 400W. The term *peak envelope power* (PEP) is defined as the average power during one RF energy cycle at the crest of the modulation envelope. (This should not be confused with the peak value of a sinusoidal waveform.)

The traditional method of power measurement used for AM transmitters was to measure the DC input power to the final amplifier. The measurement of the peak power by this method is impossible and initially caused considerable difficulty in accurately measuring the power of SSB transmissions. Peak RF output power can however be measured easily using an oscilloscope and a calibrated resistive load. This is now the accepted method of power measurement and the UK licence conditions have been modified accordingly.

The carrier power of one unit in value requires a half-unit of audio power for 100% modulation (this is the maximum power that can be used; any greater audio input would produce overmodulation and distortion) and this produces two sidebands each containing 0.25 units of power. As voltage is proportional to the square root of the power, the carrier voltage is 1 and the voltage of each sideband is 0.5. The RF envelope developed by the voltage vectors is shown, and for 100% modulation the peak envelope voltage (PEV) is the sum of the carrier and the two sideband voltages; this equals two units. This results in a PEP of four units of power.

The RF signal is demodulated in the receiver and the demodulator develops an audio output voltage that is equivalent to the sum of the upper and lower sideband voltages. The noise power per kilohertz is an arbitrary value equal for AM and SSB. For a 20dB signal-to-noise ratio, the noise voltage would be 0.1 units for the 6kHz receiver bandwidth and the signal-to-noise ratio is then

$$20 \log (\text{audio voltage/noise voltage})$$

Column B shows the power and voltage relationships for an SSB transmission of equal sideband power to the AM transmission. The audio output from the SSB transmission (recovered by heterodyning the received signal with a locally inserted carrier) is 0.707 units in value. This represents a loss in demodulator output voltage of 3dB due to the SSB power being in one sideband. However, the reduction in receiver bandwidth gives a

3dB advantage so consequently the signal-to-noise ratio is the same for the two modes of transmission.

Column C shows the relationship for an SSB transmission of equal rated power (100W carrier power AM transmitter and a 100W PEP SSB transmission). It is seen that the audio voltage developed at the output of the demodulator is equal to the audio voltage of the AM transmission. The reduction of receiver bandwidth gives a 3dB advantage and the signal-to-noise ratio is 23dB.

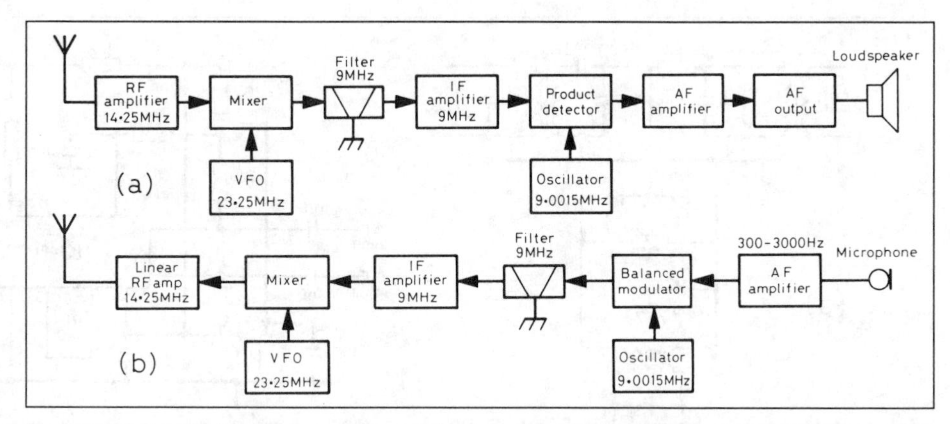

Fig 7.16. (a) A typical single-conversion superhet receiver. (b) A single-conversion SSB transmitter

Column D shows the relationship for an SSB transmission of equal PEP to the AM transmission. It is seen that the SSB transmission gives a gain in demodulator output voltage of 6dB. This, in addition to the 3dB improvement in receiver noise output, gives a total signal-to-noise ratio of 29dB, a system gain for SSB of 9dB.

As already mentioned, the carrier envelope voltage of the AM transmission serves no useful purpose other than to provide a reference frequency for demodulation. It should also be noted that for both modes of transmission the audio voltage recovered at the demodulator is directly proportional to the total sideband voltage, and that the two transmitters, AM and SSB of equal rated power, will produce an equal receiver audio voltage.

SSB advantages

There is no inherent difficulty in the construction of a SSB transmitter. In the filter type, the circuitry bears a remarkable resemblance to a single sideband receiver: Fig 7.16. The power supply is smaller than required for a comparable power AM transmitter due to the reduction in mean power and there is no requirement for a modulator and its associated power requirements. The SSB transmitter can be built and used initially to gain operating experience with a low power output – say 20W PEP. At a later date the power can be increased to a higher level by building a suitable linear amplifier and power supply and driving this with the existing transmitter.

Operating advantages over AM can be summarised as follows.

1. A power gain of 9dB with SSB operation (made up of 6dB gain at the transmitter and 3dB at the receiver) represents an equivalent power increase at the transmitter of eight times.
2. The bandwidth of a SSB signal is confined to the width of the modulating signal and is somewhat less than half the equivalent AM signal, thus utilising minimum spectrum space.
3. Only essential information is transmitted with no superfluous carrier, giving a considerable effective power gain.
4. SSB signals are effected far less adversely by the transmission disturbances inherent in ionospheric propagation.

The techniques used in SSB transmitters are closely allied to those found in SSB receivers: heterodyning and mixing have the same meaning and are both frequency translation processes. All superhet receivers translate the frequency from the required amateur band to the IF channel (or channels) and then to the final audio channel. In a SSB transmitter this process is reversed, converting an audio signal to a DSB signal at an intermediate frequency where the unwanted sideband is removed prior to translation to the desired amateur band.

BALANCED MODULATORS

Balanced modulators are essentially the same as balanced mixers, balanced demodulators and product detectors; they are tailored to suit different circuit applications especially with respect to the frequencies in use. The balanced modulator is a circuit which mixes or combines a low-frequency (audio) signal with a higher frequency (RF) signal in order to obtain the sum-and-difference frequencies (sidebands); the original RF frequency is considerably attenuated by the anti-phase or balancing action of the circuit. A *singly* balanced modulator is designed to balance out only one of the input frequencies, either f_1 or f_2, normally the higher frequency. In a *doubly* balanced modulator, both f_1 and f_2 are balanced out, leaving the sum and difference frequencies $f_1 + f_2$ and $f_1 - f_2$. In addition, intermodulation products (IMD) will appear in the form of spurious signals caused by the interaction and mixing of the various signals and their harmonics.

Balanced modulators come in many different forms, employing a wide variety of devices from a simple pair of diodes to complex ICs. In their simplest form they are an adaptation of the bridge circuit, but it should be noted that diodes connected in a modulator circuit are connected differently to those in a bridge rectifier. Simple diode balanced modulators can provide high performance at low cost. Early designs used point-contact germanium diodes while more recent designs use hot-carrier diodes (HCD). The HCD offers superior performance with lower noise, higher conversion efficiency, higher square law capability, higher breakdown voltage and lower reverse current combined with a lower capacitance. In practice, almost any diode can be used in a balanced modulator circuit, including the ubiquitous 1N914.

In the early days of SSB, simple diode balanced modulators were very popular, easy to adjust and capable of good results. Doubly balanced diode ring modulators have subsequently proved very popular because of their higher performance. However, they incur at least a 6dB signal loss while requiring a high level of oscillator drive. Doubly balanced modulators offer greater isolation between inputs as well as between input and output ports when compared to singly balanced types.

The introduction of integrated circuits has resulted in a multitude of ICs suitable for use in balanced modulator and mixer applications. Plessey have manufactured a number of devices dedicated to radio communication applications such as the SL640 and SL1640. These devices are expensive but have proved popular as they require few external components. The popular MC1496 doubly balanced mixer IC from Motorola is a considerably cheaper device, but requires numerous external components. The majority of IC mixers are based upon a doubly balanced *transistor tree* circuit, using six or more transistors on

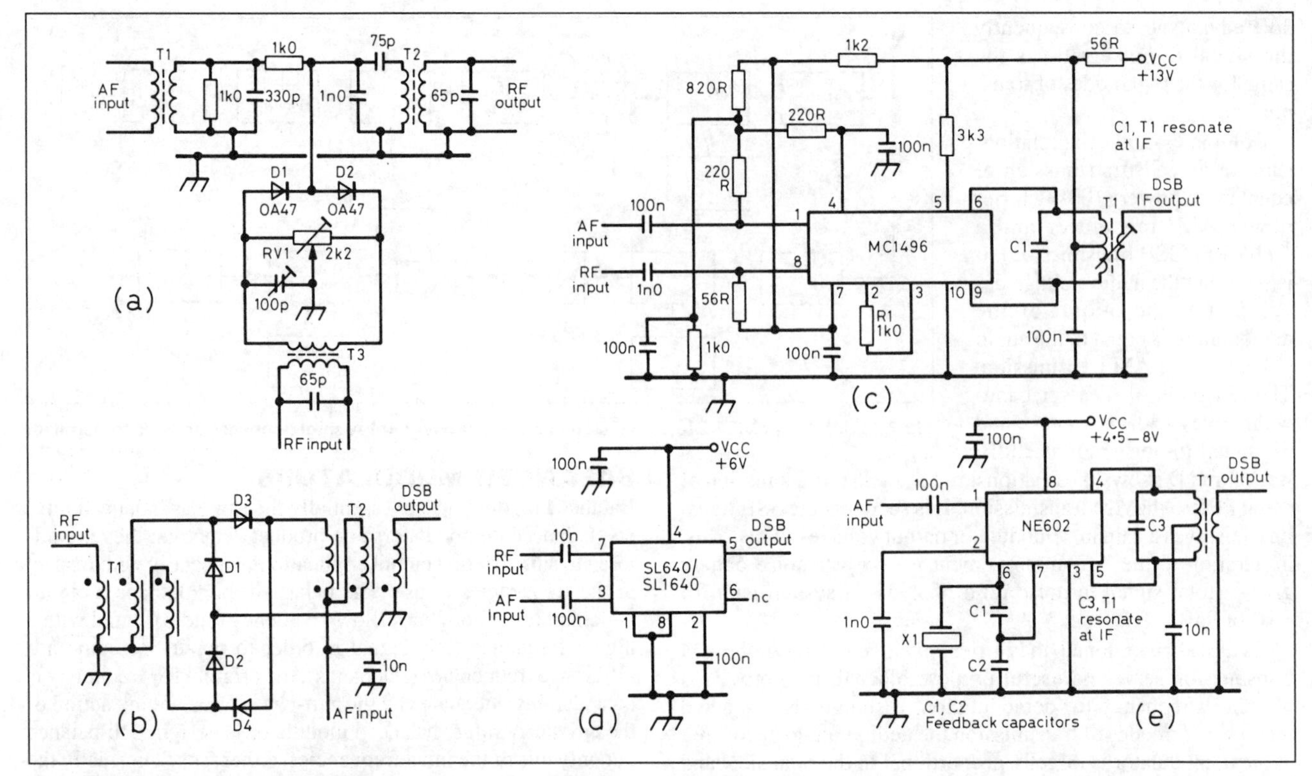

Fig 7.17. Balanced modulators. (a) Singly balanced diode modulator. (b) Doubly balanced diode ring modulator. (c) MC1496 doubly balanced modulator. (d) SL640 doubly balanced modulator. (e) NE602 doubly balanced modulator with internal oscillator

one IC. The major difference between different types of IC lies in the location of resistors which may be either internal or external to the IC.

IC mixers offer conversion gain, lower oscillator drive requirements and high levels of balance, but IMD performance can be inferior to that of diode ring modulators. Devices such as the Plessey SL6440 permit control of the current through part of the transistor tree, which in turn considerably improves the IMD characteristics.

Fig 7.17 illustrates a range of practical balanced modulator circuits. The shunt-type diode modulator, Fig 7.17(a), was common in early valve SSB transmitters, offering a superior balance to that achievable with conventional valve circuitry. Fig 7.17(b) shows a simple diode ring balanced modulator capable of very high performance – devices such as the MD108 and SBL1 are derivations of this design. The designs in Figs 7.17(c) and 7.17(d) illustrate the use of IC doubly balanced mixers; note the difference in external components between them! Fig 7.17(e) shows a very versatile IC, the NE602, primarily designed for very-low-power VHF receiver mixer applications. It offers simplicity of design, a low external component count and has the further advantage of an internal oscillator.

SIDEBAND ATTENUATION

The double sideband signal generated by the balanced modulator has to be turned into SSB by attenuating one of the sidebands. No matter what system may be used, the unwanted sideband is not eliminated completely – it is merely attenuated to the extent at which its nuisance effect becomes negligible. A filter attenuation of 30–35dB has come to be regarded as the minimum acceptable standard. With care, suppression of 50dB or more is attainable but it is debatable whether there is any practical advantage in striving after greater perfection.

The unwanted sideband may be attenuated either by phasing or filtering. The two methods are totally different in conception, and will be discussed in detail.

Although the filter method provides the classic way to generate an SSB signal and is now used almost exclusively, the phasing method will be described first because of its popularity in the early days of amateur SSB.

The phasing method

The phasing method of SSB generation can be simply explained with the aid of vector diagrams. Fig 7.18(a) shows two carriers A and B, of the same frequency and phase, one of which is modulated in a balanced modulator by an audio tone to produce

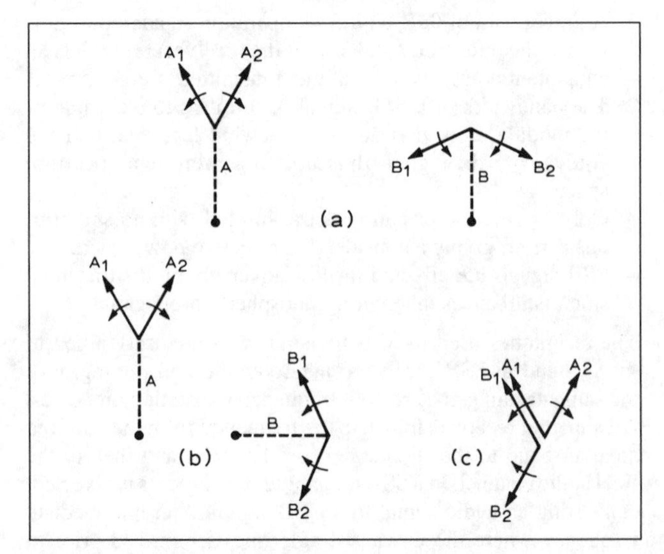

Fig 7.18. Phasing system vectors

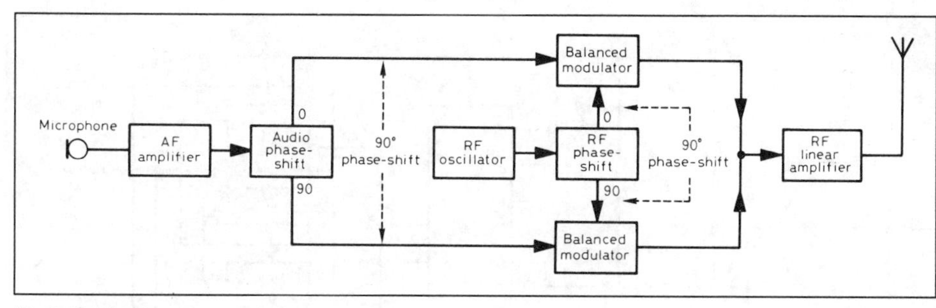

Fig 7.19. Phasing-type exciter

contra-rotating sidebands A1 and A2, and the other modulated by a 90° phase-shifted version of the same audio tone. This produces sidebands B1 and B2 which have a 90° phase relationship with their A counterparts. The vector presentation is similar to Fig 7.12 and the carrier vector is shown dotted since the carrier is absent from the output of the balanced modulators. Fig 7.18(b) shows the vector relationship if the carrier B is shifted in phase by 90° and Fig 7.18(c) shows the addition of these two signals. It is evident that sidebands A2 and B2 are in antiphase and therefore cancel whereas A1 and B1 are in phase and are additive. The result is that a single sideband is produced by this process.

A block diagram of a phasing-type transmitter is shown in Fig 7.19 from which will be seen that the output of an RF oscillator is fed into a network in which it is split into two separate components, equal in amplitude but differing in phase by 90°. Similarly, the output of an audio amplifier is split into two components of equal amplitude and 90° phase difference. One RF and one AF component are combined in each of two balanced modulators. The double-sideband, suppressed-carrier energy from the two balanced modulators is fed into a common tank circuit. The relative phases of the sidebands produced by the two balanced modulators are such that one sideband is balanced out while the other is reinforced. The resultant in the common tank circuit is an SSB signal. The main advantages of a phasing exciter are that sideband suppression may be accomplished at the operating frequency and that selection of the upper or lower sideband may be made by reversing the phase of the audio input to one of the balanced modulators. These facilities are denied to the user of the filter system.

If it were possible to arrange for absolute precision of phase shift in the RF and AF networks, and absolute equality in the amplitude of the outputs, the attenuation of the unwanted sideband would be infinite. In practice, perfection is impossible to achieve, and some degradation of performance is inevitable. Assuming that there is no error in the amplitude adjustment, a phase error of 1° in either the AF or the RF network will reduce the suppression to 40dB, while an error of 2° will produce 35dB, and 3.5° will result in 30dB suppression. If, on the other hand, phase adjustment is exact, a difference of amplitude between the two audio channels will similarly reduce the suppression. A difference between the two voltages of 1% would give 45dB, and 4% 35dB approximately. These figures are not given to discourage the intending constructor, but to stress the need for high precision workmanship and adjustment if a satisfactory phasing-type SSB transmitter is to be produced.

The early amateur phasing transmitters were designed for fundamental-frequency operation, driven directly from an existing VFO tuning the 80m band, and used a low-Q phase-shift network. This low-Q circuit has the ability to maintain the required 90° phase shift over a small frequency range, and this made the network suitable for use at the operating frequency in single-band exciters designed to cover only a portion of the chosen band. The RF phase-shift network is incapable of maintaining the required accuracy of phase shift for operation over ranges of 200kHz or more, and the available sideband suppression deteriorates to a point at which the exciter is virtually radiating a double sideband signal.

For amateur band operation a sideband suppression of 30–35dB and a carrier suppression of 50dB should be considered the minimum acceptable standard. Any operating method that is fundamentally incapable of maintaining this standard should not be used on the amateur bands. For this reason, the fundamental type of phasing unit is not recommended. For acceptable results, the RF phase shift must be operated at a fixed frequency outside the amateur bands. The SSB output from the balanced modulator is then heterodyned to the required bands by means of an external VFO.

Audio phase-shift network

Achieving the audio phase shift necessary for SSB generation in a phasing exciter traditionally required the use of high-tolerance components, often necessitating the use of a commercially made phase-shift network. Such devices can prove more costly than the crystal filter required for a filter-type exciter.

An alternative method of devising the required 90° phase shift using off the shelf values was devised by M J Gingell and is referred to as the *polyphase network* (Fig 7.20). Standard 10% tolerance resistors and capacitors are used in the construction of a six-pole network capable of providing four outputs of equal amplitude, all lagging one another by 90°. The network is designed to phase shift audio signals between 300Hz and 3000Hz, and it is therefore necessary to limit the bandwidth of the audio input using a filter or clipper circuit which can be either active or passive. Audio input to the polyphase network is derived using a simple phase splitter to provide the two phase inputs required by the network. Resistors used in the network are of one common value and Mylar audio-grade capacitors are suitable for the capacitive elements.

RF phase-shift network

Traditionally the most satisfactory way to produce a 90° RF phase shift was to employ a low-Q network comprising of two loosely coupled tuned circuits which exhibits a combination of inductance, resistance and capacitance. Fig 7.21(a) illustrates such a network in the anode circuit of a valve amplifier. The

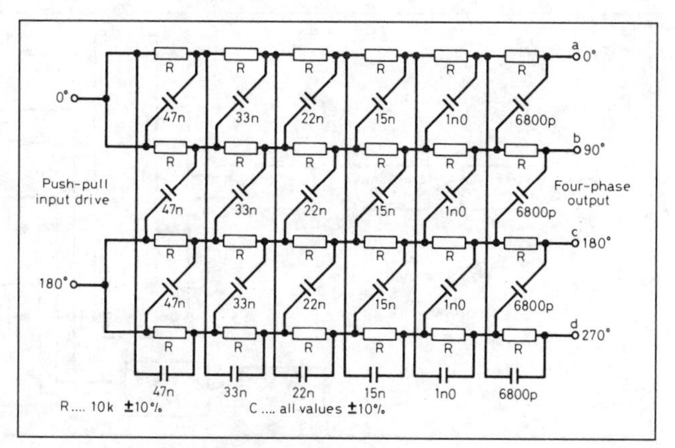

Fig 7.20. Gingell polyphase network

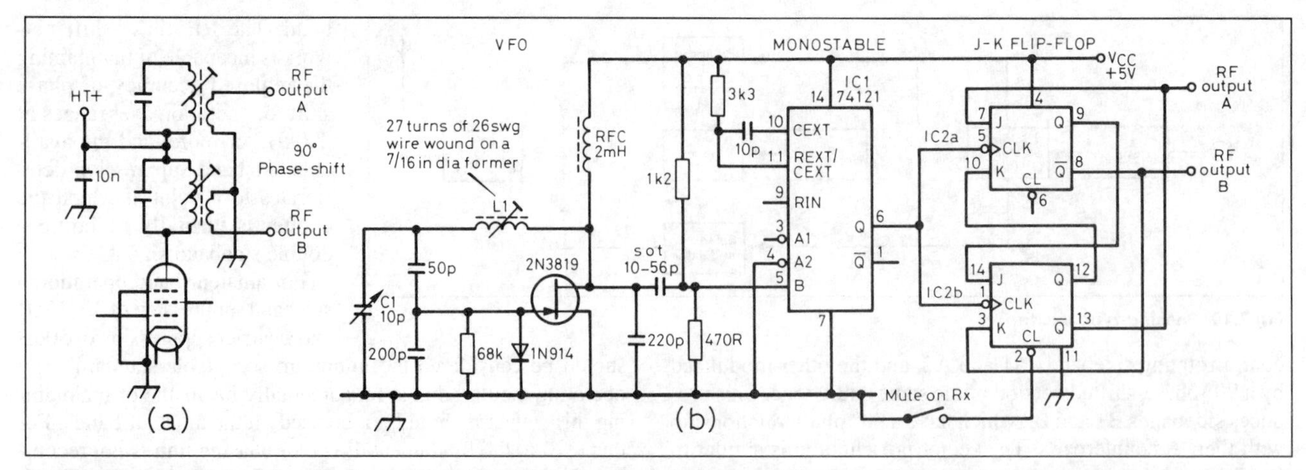

Fig 7.21. Methods of obtaining RF phase shift. (a) Traditional method of obtaining 90% phase shift using loosely coupled tuned circuits. (b) Active RF phase shifter 7.2MHz VFO providing 1.8MHz output with 90% shifts

primary coils are inductively coupled while the link couplings are connected in series. When both circuits are tuned to resonance there will be exactly 90° phase shift between them. Difficulties occur when the frequency is changed, and the network has to be retuned, restricting the bandwidth to no more than 200kHz.

With the advent of digital ICs, it has become relatively easy to obtain the required phase shift by dividing the output of an oscillator using a flip-flop IC. Fig 7.21(b) shows the RF circuitry for a 160m phasing exciter, in which the signal from a VFO tuning 7.2–8MHz is divided by 4 using a 7473 (J-K flip-flop), providing a square-wave output between 1.8 and 2MHz. A 74121 monostable IC is used to convert the sinusoidal waveform from the VFO to a square wave suitable for driving the flip-flop which provides both 0 and 90° outputs. For fixed frequency generation, it is possible to use a TTL IC as a square-wave oscillator directly driving the flip-flop. The fundamental square-wave signal will be phased out in the balanced modulator.

Four-way phasing method

The four-way phasing method is an adaptation of the conventional phasing method and can be simply described as a double two-way method. Fig 7.22 illustrates a four-way phasing exciter. The major requirement for acceptable carrier and sideband suppression is a good audio phase shifter. The polyphase network (Fig 7.20) is ideal and provides the required four output signals at 90° phase intervals. The RF output from the carrier generator must also provide four RF outputs phase-shifted by 90° from one another, and this is achieved by using a dual J-K flip-flop which also divides the input frequency by a factor of four. The phase-shifted AF and RF signals are fed to four modulators, the outputs of which are summed in a tuned adder, resulting in a SSB output signal.

A practical 9MHz four-way SSB generator is illustrated in Fig 7.23. The TTL oscillator is operated at 12MHz, providing a 3MHz signal at the output of the flip-flop. As this square wave signal is rich in odd-order harmonics, the tuned adder can be adjusted to tune to the third harmonic in preference to the fundamental signal, and the result is a 9MHz output SSB signal.

One major disadvantage of the digital phase shifter is the necessity to operate the oscillator on four times the output frequency. The technique used in Fig 7.23 represents one solution to the problem, but an alternative would be to heterodyne the output to the desired frequency using a VFO.

The filter method

Since the objective is to transmit only a single sideband, it is necessary to select the desired sideband and suppress the unwanted sideband. The relationship between the carrier and sidebands is shown in the diagram of Fig 7.12. Removing the unwanted sideband by the use of a selective filter has the advantage of simplicity and good stability. The unwanted sideband suppression is determined by the attenuation of the sideband filter, while the stability of this suppression is determined by the stability of the elements used in constructing the filter. High stability can be achieved by using materials that have a very low temperature coefficient of expansion. Commonly used materials are quartz, ceramic and metallic plates.

The filter method, because of its proven long-term stability, has become the most popular method used by amateurs. At present three types of selective sideband filters are in common use:

(a) the high-frequency crystal filter;
(b) the low-frequency mechanical filter;
(c) the low-frequency ceramic filter.

Crystal filters

The crystal filter is the most widely used type of filter found in SSB transmitters. In a transceiver, one common filter can be

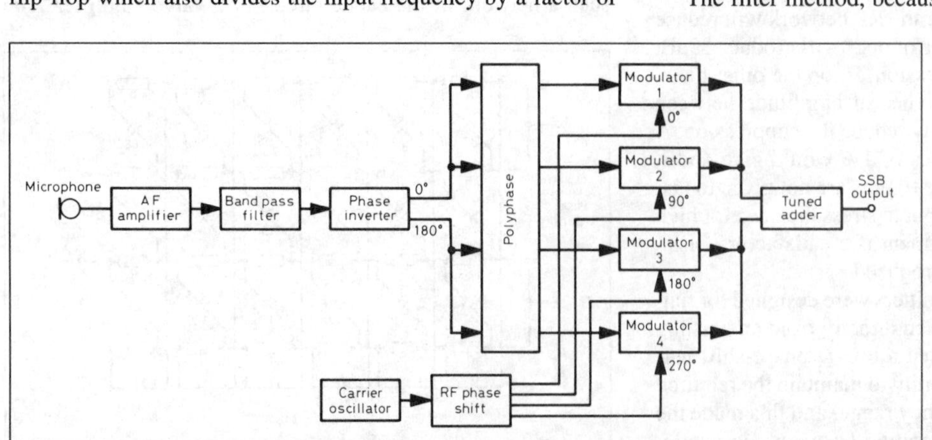

Fig 7.22. Four-phase SSB generator

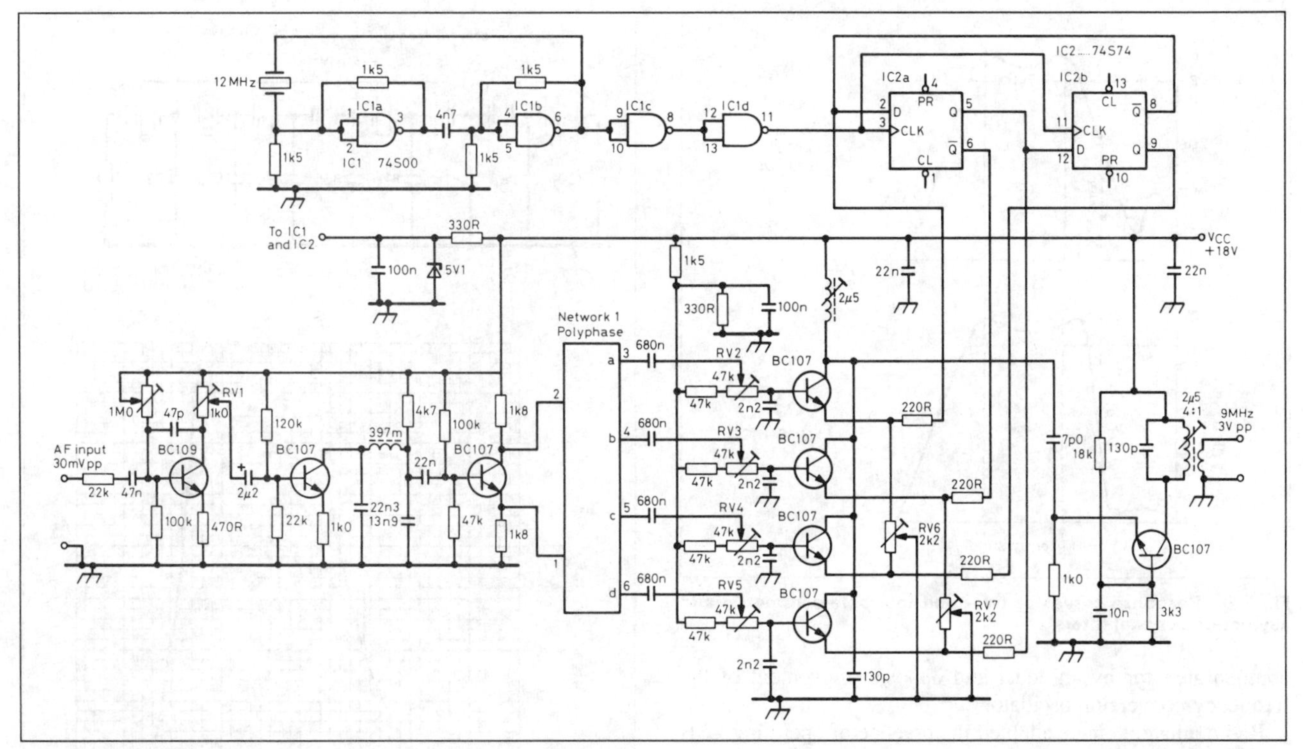

Fig 7.23. 9MHz phasing exciter

used for both transmit and receive functions. In the early days of SSB, filters were invariably home constructed using cheap and readily available surplus crystals in the 400–500kHz range. Commercially made low-frequency filters were available but were both bulky and relatively expensive. Low-frequency SSB generation requires two frequency conversions before operation on the higher-frequency bands is possible, if image responses are to be avoided. Generation of a SSB signal in the 9MHz range permits single-frequency conversion techniques to be employed to cover the entire HF spectrum, and for this reason 9MHz and 10.7MHz filters have virtually dominated the market. A number of other frequencies have been employed for filters, including 5.2MHz, 3.18MHz and 1.6MHz, the latter mainly for commercial applications.

The principle of operation of a crystal or quartz filter is based upon the piezo-electric effect. When the crystal is excited by an alternating electric current, it mechanically resonates at a frequency dependent upon its physical shape, size and thickness. A crystal will easily pass current at its natural resonant frequency but attenuates signals either side of this frequency. By cascading a number of crystals having the same, or very closely related, resonant frequencies it is possible to construct a filter having a high degree of attenuation either side of a band of wanted frequencies, typically 40–60dB with a six-pole filter, 60–80dB using a eight-pole filter, and 80–100dB with a 10-pole filter. The characteristic bandwidth of a SSB filter is selected to pass a communications-quality audio spectrum of typically 300–3000Hz. For SSB transmission the best-sounding results can be achieved using a 3kHz filter, but for receiver applications a slightly narrower filter is preferable and 2.4kHz has become the accepted compromise. SSB transmissions made using narrower filters have a very restricted audio sound when received. Filter bandwidths are normally quoted at the 6dB and 60dB attenuation levels, and the ratio of the two quoted bandwidths is referred to as the *shape factor*, 1:1 being the ideal but not realistically achievable. Anything better than 2.5:1 is regarded as acceptable.

The purpose of the crystal filter is to attenuate the unwanted sideband, but it has a secondary function: to attenuate further the already-suppressed carrier. A balanced modulator seldom attenuates the carrier by more than 40dB, which equates to 1mW of carrier from a 100W SSB transmitter. A further 20dB of carrier suppression is available from the sideband filter, making a carrier suppression of 60dB possible.

The passband of a crystal filter may be symmetrical in shape, Fig 7.24(a), or asymmetric as in Fig 7.24(b). Home-constructed filters are invariably asymmetric to some degree whereas commercially made filters will be designed to fall into one of the two categories. Assuming that we wish to generate a LSB signal with a carrier frequency of 9MHz, a SSB filter will be required to pass the frequency range 8.9975MHz to 9.0MHz, whereas if we wished to change to USB, the filter would be required to pass the frequency range 9.0MHz to 9.0025MHz. At first sight it would appear that two filters are required. In commercial equipment, the use of two filters is common practice, in which case the filters will most probably be asymmetric and annotated with the sideband that they are designed to generate and the intended carrier frequency. In amateur radio equipment a much cheaper technique is adopted: it is easier to use a symmetrical filter with a centre frequency of 9.000MHz and move the carrier frequency from one side of the filter to the other in order to change sidebands. Typically 8.9985MHz for USB and 9.0015MHz for LSB are used; note the filter frequency is above the carrier frequency to give USB and below the carrier frequency to give LSB. Asymmetric filters are invariably marked with the carrier frequency and have the advantage of higher attenuation of the unwanted carrier and sideband, due to the steeper characteristic of the filter on the carrier frequency side. This is the primary reason why they are used in commercial applications where they are required to meet a higher specification. One minor disadvantage of the symmetrical filter and switched carrier frequency method is that changing sideband causes a shift in frequency approximately equal to the bandwidth of the filter. This can be

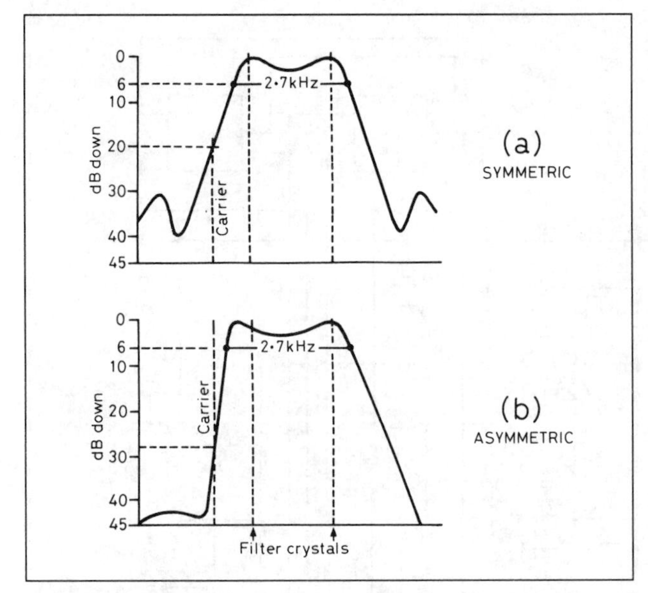

Fig 7.24. Response curves of two- and four-pole symmetric and asymmetric crystal filters

compensated for by an equal and opposite movement of the frequency-conversion oscillator.

Radio amateurs have adopted the practice of operating LSB on the low-frequency bands and USB on the high-frequency ones, and it is therefore necessary to be able to switch sidebands if operation on all bands is contemplated.

Home construction of crystal filters is only to be recommended if a supply of cheap crystals is available. Clock crystals for microprocessor applications are manufactured in enormous quantities and cost pence rather than pounds. Another source of crystals are those intended for TV colour-burst; the USA and continental frequency of 4.43MHz is the more suitable as the UK colour-burst frequency is in the 80m amateur band. Fig 7.25(a) shows an eight-pole ladder filter designed by G3UUR and constructed using colour-burst crystals . The frequency of individual crystals is found to vary by as much as 200Hz, but by careful selection of crystals it is possible to find those that are on frequency and those that are slightly above or below the nominal frequency. For optimum results it is recommended that the on-frequency crystals are located at either end of the filter while the centre four crystals should be slightly higher in frequency. Fig 7.25(b) shows the frequency response of the G3UUR filter constructed from TV colour-burst crystals.

Commercially available crystal filters are primarily confined to 9MHz and 10.7MHz types and, while the sources seem to come and go, the German KVG filters have remained available for several decades. Numerous other filters are available new, and surplus filters can often be found at bargain prices. Table 7.3 lists the popular KVG range of filters and alternative filters are listed in Table 7.4.

Mechanical filters

The mechanical filter was developed by the Collins Radio Company for low-frequency applications in the range 60–500kHz. The F455 FA-21 was designed specifically for the amateur radio market with a nominal centre frequency of 455kHz, a 6dB bandwidth of 2.1kHz and a 60dB bandwidth of 5.3kHz, providing a shape factor of just over 2:1.

The mechanical filter is made up from a number of metal discs joined by coupling rods. The discs are excited by magnetostrictive transducers employing polarised biasing magnets and must not

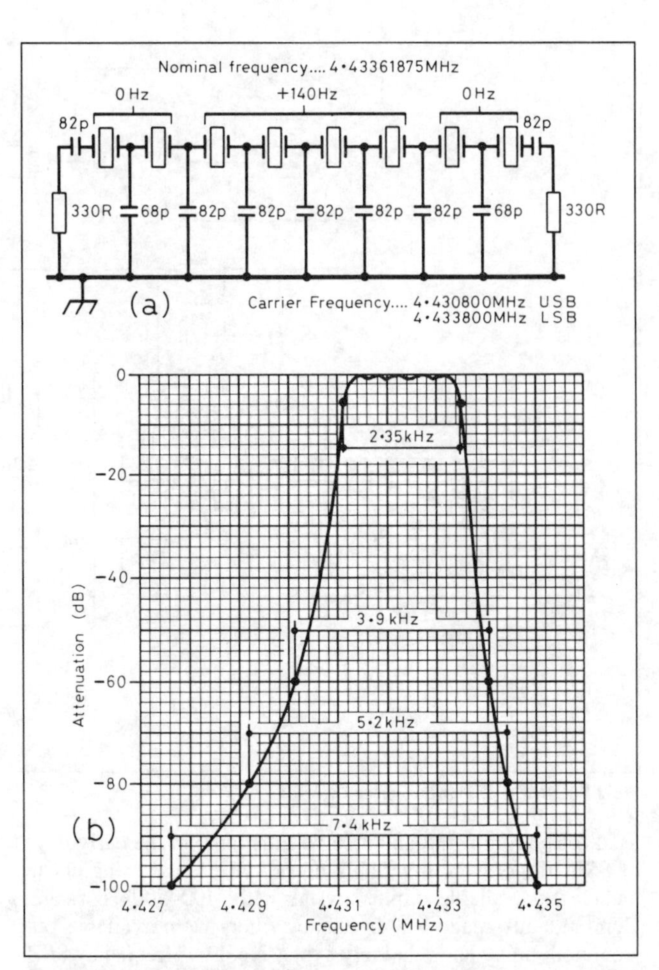

Fig 7.25. (a) Ladder crystal filter using 4.43MHz TV crystals. (b) Ladder crystal filter using P129 specification colour-TV 4.43MHz crystals. It provides a performance comparable with the ladder filter in the Atlas 180 and 215 transceivers. Insertion loss 4–5dB, shape factor (6/60dB) 1.66. Note that the rate of attenuation on the low-frequency side of the response is as good as an eight-crystal lattice design. On the HF side it is better

be used in circuits where DC is present. The input and output transducers are identical and are balanced to ground so the filter can be used in both directions. Mechanical filters have always been expensive but provide exceptional performance, and examples of the Collins filters can often be found in surplus equipment. Japanese Kokusai filters were available for a number of years and offered a more cost-effective alternative. Mechanical filters are now seldom used in new equipment.

Ceramic filters

Ceramic filters have been developed for broadcast radio applications; they are cheap, small in size and available in a wide range of frequencies and bandwidths. Narrow-bandwidth ceramic filters with a nominal centre frequency of 455kHz have been manufactured for use in SSB receiver IF applications and provide a level of performance making them ideally suited for use in SSB transmitters. Bandwidths of 2.4 and 3kHz are available with shape factors of 2.5:1 and having ultimate attenuation in excess of 90dB (Table 7.4).

SSB filter exciter

The SSB filter exciter is inherently simpler than the phasing-type exciter. Fig 7.26(a) illustrates a 455kHz SSB generator requiring just two ICs and using a ceramic filter. The NE602 doubly balanced modulator is provided with a 455kHz RF input signal

Table 7.3. KVG 9MHz crystal filters for SSB, AM, FM and CW applications

Filter type	XF-9A	XF-9B	XF-9C	XF-9D	XF-9E	XF-9M
Application	SSB TX	SSB TX/RX	AM	AM	FM	CW
Number of poles	5	8	8	8	8	4
6dB bandwidth (kHz)	2.5	2.4	3.75	5.0	12.0	0.5
Passband ripple (dB)	<1	<2	<2	<2	<2	<1
Insertion loss (dB)	<3	<3.5	<3.5	<3.5	<3	<5
Termination	500Ω/30pF	500Ω/30pF	500Ω/30pF	500Ω/30pF	1200Ω/30pF	500Ω/30pF
Shape factor	1.7 (6–50dB)	1.8 (6–60dB) 2.2 (6–80dB)	1.8 (6–60dB) 2.2 (6–80dB)	1.8 (6–60dB) 2.2 (6–80dB)	1.8 (6–60dB) 2.2 (6–80dB)	2.5 (6–40dB) 4.4 (6–60dB)
Ultimate attenuation (dB)	>45	>100	>100	>100	>90	>90

Table 7.4. Popular SSB filters

Filter type	90H2.4B	10M02DS	CFS455J	CFJ455K5	CFJ455K14	QC1246AX
Centre frequency (MHz)	9.0000	10.7000	0.455	0.455	0.455	9.0000
Number of poles	8	8	Ceramic	Ceramic	Ceramic	8
6dB bandwidth (kHz)	2.4	2.2	3	2.4	2.2	2.5
60dB bandwidth (kHz)	4.3	5	9 (80dB)	4.5	4.5	4.3
Insertion loss (dB)	<3.5	<4	8	6	6	<3
Termination	500Ω/30pF	600Ω/20pF	2kΩ	2kΩ	2kΩ	500Ω/30pF
Ultimate attenuation (dB)	>100	—	>60	>60	60	>90
Manufacturer/UK supplier	IQD	Cirkit	Cirkit	Bonex	Bonex	SEI

from its own internal oscillator. This can use a 455kHz crystal or the cheaper ceramic resonator, and the latter can be pulled in frequency to generate either a LSB or USB carrier. Microphone signals are amplified by a VOGAD (voice-operated gain-adjusting device) or any suitable AF amplifier, and fed to the input of the balanced modulator. A DSB signal at the output of the balanced modulator is fed directly to the ceramic filter for removal of the unwanted sideband, resulting in a 455kHz SSB output signal. This can be heterodyned directly to the lower-frequency amateur bands or via a second higher IF to the HF bands.

Fig 7.26(b) illustrates an alternative design for a 9MHz SSB generator where a separate JFET oscillator is used to provide the RF signal. Ceramic resonators are not readily available for 9MHz, so crystals are normally used in this circuit. Often two separate oscillators will be used to provide upper and lower sideband selection. The Plessey balanced modulator may be any of the SL640/SL1640 or SL641/SL1641 devices. A PNP transistor is DC coupled to the output of the balanced modulator to provide some gain to compensate for any losses in the filter, and it also aids impedance matching to the latter. The 560Ω collector load resistor provides the correct filter termination together with the 22pF shunt capacitor, which is in parallel with the transistor's internal capacitance. Output from the 9MHz SSB exciter can be heterodyned directly to any of the LF and HF amateur bands.

The only adjustment required in either circuit illustrated in Fig 7.26 is to adjust the oscillator to the correct frequency in relation to the filter; this frequency is normally located 20dB down the filter response curve. Quite often the oscillator can be adjusted for the best audio response at the receiver.

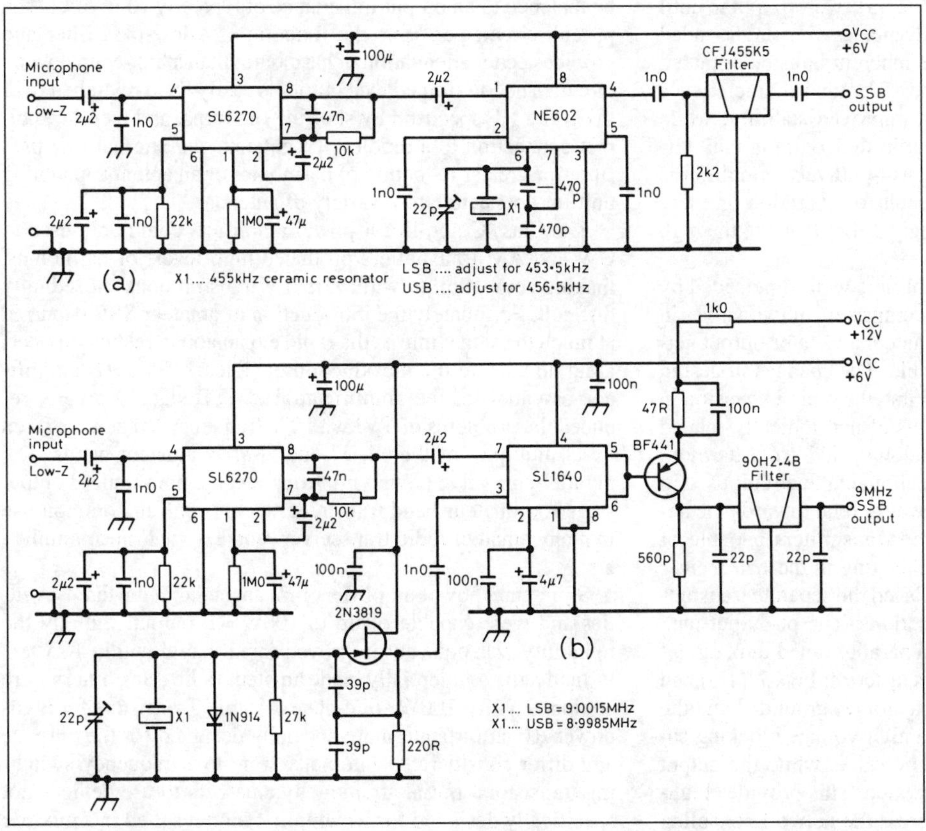

Fig 7.26. (a) 455kHz SSB generator; (b) 9MHz SSB exciter

POWER AMPLIFIERS

The RF power amplifier is normally considered to be that part of a transmitter which provides RF energy to the antenna. It may be a single valve or transistor, or a composite design embodying numerous devices to take low-level signals to the final output power level. RF amplifiers are also discussed in detail in Chapter 5.

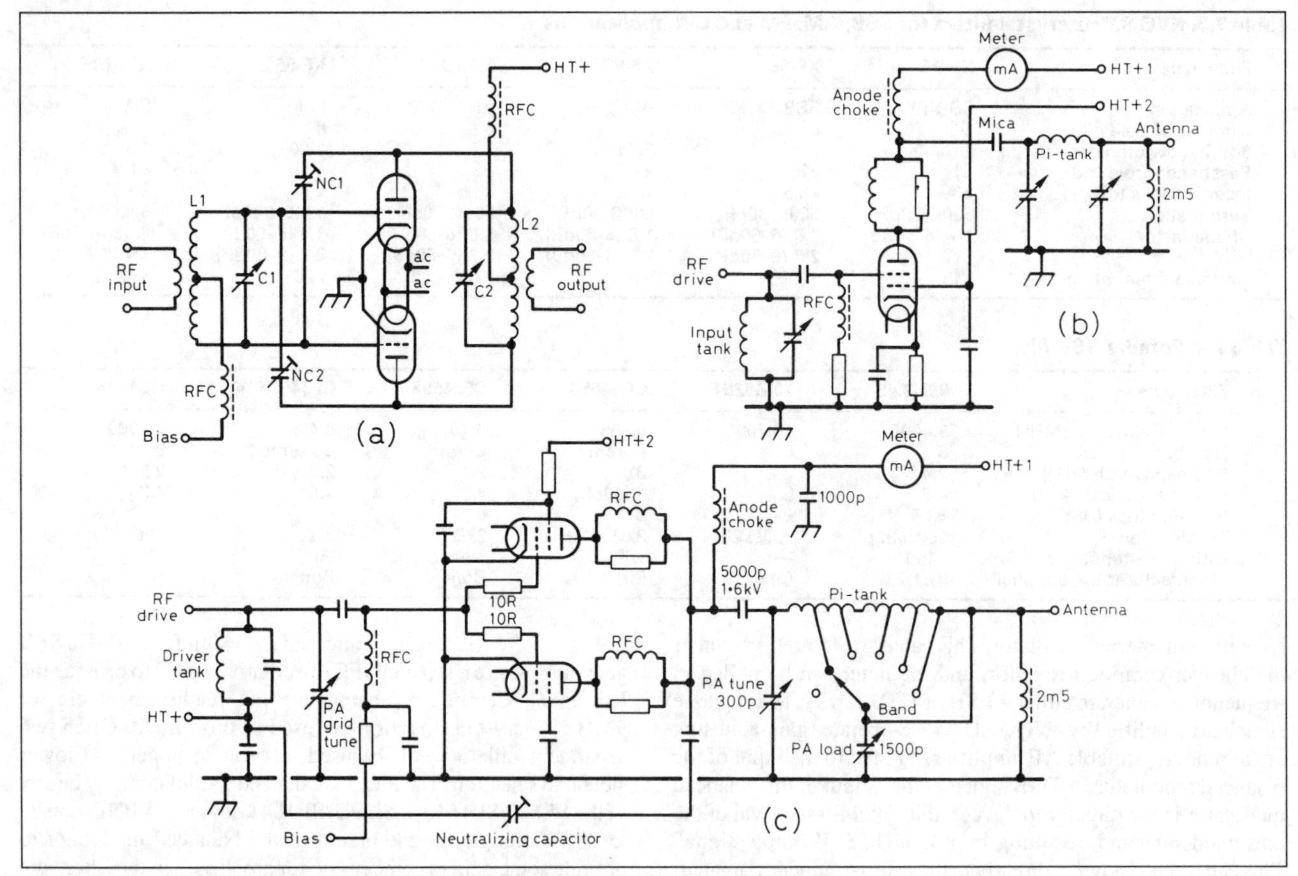

Fig 7.27. Valve power amplifiers. (a) Push-pull amplifier; (b) single-ended Class C amplifier; (c) parallel-pair RF amplifier

Push-pull valve amplifiers, Fig 7.27(a), were popular until the 'fifties and offered a number of advantages over single-ended output stages as in Fig 7.27(b). The inherent balance obtained when two similar valves having almost identical characteristics are operated in push-pull results in improved stability, while even-order harmonics are phased out in the common tank circuit. One major shortcoming of the push-pull valve amplifier is that switching of the output tank circuit for operation on more than one frequency band is exceedingly difficult due to the high RF voltages present.

During the 'sixties, push-pull amplifiers were superseded by the single-ended output stage, often comprising of two valves in parallel, Fig 7.27(c), coupled to the antenna via a pi-output network and low-impedance coaxial cable. The changes in design were brought about by two factors. First, the rapid expansion in TV broadcasting introduced problems of harmonically related TVI. This demanded greater attenuation of odd-order harmonics than was possible with the conventional tank circuit used in push-pull amplifiers. Second, there was a trend towards the development of smaller, self-contained transmitters capable of multiband operation, ultimately culminating in the *transceiver* concept which has almost totally replaced the separate transmitter and receiver in amateur radio stations. The pi-output tank circuit differs from the conventional parallel-tuned tank circuit in that it uses two variable capacitors in series: Figs 7.11(a) and 7.11(b). The junction of the two capacitors is grounded and the network is isolated from DC using a high-voltage blocking capacitor. The input capacitor has a low value, while the output capacitor has a considerably higher value. This provides high-to-low impedance transformation across the network, enabling the relatively high anode impedance of the power amplifier to be matched to an output impedance of typically 75 or 50Ω. The pi-tank circuit performs the function of a low-pass filter and provides better attenuation of harmonics than a link-coupled tank circuit. The low-impedance output is easily band-switched over the entire HF spectrum by shorting out turns, and facilitates direct connection to a dipole-type antenna, an external low-pass filter for greater reduction of harmonics or an antenna matching unit for connection to a variety of antennas.

The Class C non-linear power amplifiers commonly used in CW and AM transmitters produced high levels of harmonics, making compatibility with VHF TV transmissions exceedingly difficult. Fortunately the introduction of amateur SSB occurred at much the same time as the rapid expansion in television operating hours, and the introduction of 'linear' Class AB amplifiers, essential for the amplification of SSB signals, greatly reduced the problems of TVI with TV frequencies that were often exact multiples of the 14, 21 and 28MHz amateur bands. The parallel-pair valve power amplifier became the standard output stage for amateur band transmitters and is still in common use in many amateur radio transceivers and external linear amplifiers.

Solid-state power amplifiers first appeared in the late 'seventies and were capable of up to 100W RF output. Initially the reliability was poor but improved rapidly, and within 10 years virtually all commercially made amateur radio transmitters were equipped with a 100W solid-state PA unit. Transistors for high-power RF amplification are specially designed for the purpose and differ considerably internally from low-frequency switching transistors. Initial attempts by amateurs to use devices not specifically designed for RF amplification resulted in a mixture of success and failure, probably giving rise to initial claims that solid-state power amplifiers were less reliable than their valve counterparts.

Due to the very low output impedances of solid-state bipolar devices, it is almost impossible to match the RF output from a solid-state power amplifier to a resonant tank circuit with its characteristic high impedance. As a result, low-Q matching circuits (Fig 7.9 and 7.10) are used to transform the very low (typical 1 or 2Ω) output impedance encountered at the collector of an RF power amplifier to the now-standard output impedance of 50Ω. As the matching circuits are not resonant, they are broadband in nature, making it possible to operate the amplifier over a wide range of frequencies with no need for any form of tuning. Unfortunately, the broad-band amplifier also amplifies harmonics and other unwanted products, making it essential to use a low-pass filter immediately after the amplifier in order to achieve an adequate level of spectral purity. It is also essential that the amplifier itself should not contribute to the production of unwanted products. For this reason solid-state amplifier designs have reverted back to the balanced, push-pull mode of operation, with its inherent suppression of even-order harmonics. As band switching and tuning of the amplifier is neither possible or necessary, construction of the amplifier is relatively simple. Typically, HF broad-band amplifiers will provide an output over the entire HF spectrum from 1.5 to 30MHz.

For powers in excess of 100W, valves are still popular and are likely to remain in use for some time to come on the grounds of cost, simplicity and superior linearity.

Power amplifiers can be categorised into two basic types, valve and solid-state, and then further into sub-groups based upon the output power. Low-power amplifiers can be regarded as 10W or less, including QRP (usually regarded as less than 5W input power), medium power up to 100W, and high power in excess of 100W and up to the maximum permitted power.

The class of operation of a power amplifier is largely determined by its function. Class C amplifiers are commonly used for CW, AM and FM transmissions because of their high efficiency. It is also a pre-requisite for successful high-level modulation in an AM transmitter. Due to the non-linear operation of an amplifier operated in Class C, the harmonic content is high and must be adequately filtered to minimise interference.

Single sideband transmission demands linear amplification if distortion is to be avoided and amplifiers may be operated in either Class A or B. Class A operation is inefficient and normally confined to driver stages, the high standing current necessary to achieve this class of operation being usually unacceptable in amplifiers of any appreciable power rating. Most linear amplifiers for SSB amplification are operated in between Class A and B, in what are known as Classes AB1 and AB2, in order to achieve a compromise between efficiency and linearity.

Solid-state versus valve amplifiers

The standard RF power output for the majority of commercially produced amateur radio transmitters/transceivers is 100W, and at this power level solid-state amplifiers offer a number of advantages, as follows.

1. Compact design.
2. Simpler power supplies requiring only one voltage (normally 13.8V) for the entire equipment;
3. Broad-band, no tune-up operation, permitting ease of operation;
4. Long life, with no gradual deterioration due to loss of emission;
5. Ease of manufacture and reduced cost.

There are of course a number of disadvantages with solid-state amplifiers and it is for this reason that valves have not disappeared entirely.

1. Individual transistors are not capable of producing power outputs much in excess of 150W, so that a number of devices have to be connected together to raise the power level.
2. The relatively low voltages used require very high currents for the generation of any appreciable power, placing demands upon the devices and their associated power supplies. Heatsinking and voltage stabilisation become very important. Above 150W output, the use of 13.8V power supplies becomes impracticable.

While the construction of solid-state amplifiers up to 1kW is feasible using modern devices operated from a high-current 50V power supply, valve designs offer a simpler and more cost-effective alternative. However, the development of VMOS devices with much higher input and output impedances is beginning to bridge the gap. Quite possibly within a few years VMOS devices will offer a cheaper alternative to the valve power amplifier at the kilowatt level.

Impedance matching

All types of power amplifier have an internal impedance made up from a combination of the internal resistance, which dissipates power in the form of heat, and some complex reactances. For maximum transfer of power, the device resistance should equal the load resistance and consequently, when reactance is present, the load impedance must be equal to the source impedance. As both source and load impedances are fixed values and not liable to change, it is necessary to employ some form of impedance transformation or matching in order to obtain the maximum efficiency from an amplifier. The power may be expressed as:

$$P_{input} = P_{output} + P_{dissipated}$$

where P_{input} is the DC input power to the stage; P_{output} is the RF power delivered to the load and $P_{dissipated}$ is the power absorbed in the source resistance and dissipated as heat.

$$\text{Efficiency} = \frac{P_{output}}{P_{input}} \times 100\%$$

When the source impedance is equal to the load impedance, the current through either will be equal as they are in series, with the result that 50% of the power will be dissipated by the source and 50% will be supplied to the load. The object of a power amplifier is to provide maximum power to the load, and the highest possible efficiency is therefore essential. Design of a power amplifier must also take into account the maximum dissipation of the output device as specified by the manufacturer. An optimum load resistance is selected to ensure maximum output from a power amplifier while not exceeding the amplifying device's power dissipation. Efficiency increases as the load resistance to source resistance ratio increases and vice versa. The optimum load resistance is determined by the device's current transfer characteristics and for a solid-state device is given by:

$$R_L = \frac{V_{cc}^2}{2P_{out}}$$

Valves have more complex current transfer characteristics which differ for different classes of operation.

Essentially the optimum load resistance is proportional to the ratio of the DC anode voltage to the DC anode current divided by a constant which varies from 1.3 in Class A to approximately 2 in Class C.

$$R_L = \frac{V_a}{KI_a}$$

The output from a RF power amplifier is usually connected to an antenna system, neither of which are likely to be of the same impedance so a matching network must be employed. Two methods are commonly used: pi-tank circuit matching for valve circuits and transformer matching for solid-state amplifiers. The variable nature of a pi-tank circuit permits matching over a wide range of impedances, whereas the fixed nature of a matching transformer is dependent upon a nominal load impedance of typically 50Ω, and it is therefore almost essential to employ some form of antenna matching unit between the output and the final load impedance. Matching networks serve to equalise load and source resistances while providing inductance and capacitance to cancel any reactive elements.

Valve power amplifiers

Valve power amplifiers are commonly found in the output stages of older amateur radio transmitters. Usually two valves will be operated in parallel, providing twice the output power possible with a single valve. The output stage may be preceded by a valve driver amplifier or a solid-state driver stage. The valve amplifier is capable of high gain when operated in the tuned input, tuned output configuration often referred to as *TPTG* (tuned plate tuned grid). Two 6146 valves are capable of producing in excess of 100W RF output with as little as 500mV of RF drive signal at the input. Operation of two valves in parallel increases the inter-electrode capacitances by a factor of two and ultimately affects the upper frequency operating limit. Fig 7.27(c) illustrates a typical output stage found in amateur transceivers.

Neutralisation

The anode-to-grid capacitance of a valve provides a path for RF signals to feed back energy from the anode to the grid. If this is sufficiently high, oscillation will occur. This can be deliberately overcome by feeding back a similar level of signal, but of opposite phase, thus cancelling the internal feedback, and this process is referred to as *neutralisation*.

Once adjusted, neutralisation should not require adjusting unless the internal capacitance of the valve changes. However, if valves are replaced, the neutralisation will almost certainly require adjustment to compensate for the difference in internal capacitance.

While there are numerous ways of achieving neutralisation, the most common method still in use is series-capacitance neutralisation. A low-value ceramic variable capacitor is connected from the anode circuit to the earthy end of the grid input circuit as in Fig 7.27(c).

The simplest way to adjust the neutralising capacitor is to observe the anode current when the PA output tuning capacitor is adjusted through resonance. The current should reduce gradually to a minimum value and then rise smoothly again to its previous value. Any asymmetry of the current either side of the resonant point will indicate that the PA is not correctly neutralised. Neutralising capacitors should be adjusted with non-metallic trimming tools to avoid the introduction of additional capacitance.

Parasitic oscillation

It is not uncommon for a power amplifier to oscillate at some frequency other than one in the operating range of the amplifier. This can usually be detected by erratic tuning characteristics and a reduction in efficiency. The parasitics are often caused by the resonance of the connecting leads in the amplifier circuit being resonated with the circuit capacitance. To overcome problems at the design stage it is common to place low-value resistors in series with the grid, and low-value RF chokes, often wound on a resistor body, in series with the anode circuit. Ferrite beads may also be strategically placed in the circuit to damp out any tendency to oscillate at VHF.

HIGH-POWER AMPLIFICATION

Output powers in excess of 100W are invariably achieved using an add-on linear amplifier. In view of the relatively high level of drive power available, the amplifier can be operated at considerably lower gain, with the advantages of improved stability and no requirement for neutralisation. Input circuits are usually passive, with valves operated in either passive-grid or grounded-grid modes. The grounded-grid amplifier has a cathode impedance ideally suited to matching the pi-output circuit of a valve exciter. Pi-input networks are usually employed to provide the optimum 50Ω match for use with solid-state exciters. Passive grid amplifiers often employ a grid resistor of 200 to 300Ω, which is suitable for connecting to a valve exciter but will require a matching network such as a 4:1 auto-transformer for connection to a solid-state exciter.

Output matching

There are only two output circuits in common use in valve power amplifiers – the pi-output network is by far the most common and suitable values for a range of anode impedances is provided in (Table 7.1). More recently, and especially for applications in high-power linear amplifiers, an adaptation of the circuit has appeared called the *pi-L output network*, and here the conventional pi-network has been combined with an L-network to provide a matching network with a considerable improvement in attenuation of unwanted products. The simple addition of one extra inductor to the circuit provides a considerable improvement in performance. Suitable values for a pi-L network are given in Table 7.2 for a range of anode impedances.

Valve amplifiers employing pi-output networks require a suitable RF choke to isolate the RF at the anode of the power amplifier from the high voltage power supply. This choke must be capable of carrying the anode current as well as the high anode voltages likely to be encountered in such an amplifier. An essential requirement for this anode choke is that it must not have any resonances anywhere in the operating range of the amplifier or it will overheat with quite spectacular results. Chokes are often wound in sections to reduce the capacitance between turns and may employ sections in varying diameters. Ready-made chokes are available for high-power operation from Barker and Williamson (RF Engineering Ltd in the UK).

Solid-state power amplifiers

Solid-state power amplifiers are invariably designed for 50Ω input and output impedances and employ broad-band transformers to effect the correct matching to the devices. While single-ended amplifiers, Fig 7.28(a), are often shown in test circuits, their use is not recommended for the following reasons.

1. High levels of second harmonic are present and are difficult to attenuate using standard low-pass filters.
2. Multiband operation is more difficult to achieve due to the more complex filter requirements.

The broad-band nature of a solid-state amplifier with no requirement for bandswitching enables push-pull designs to be used,

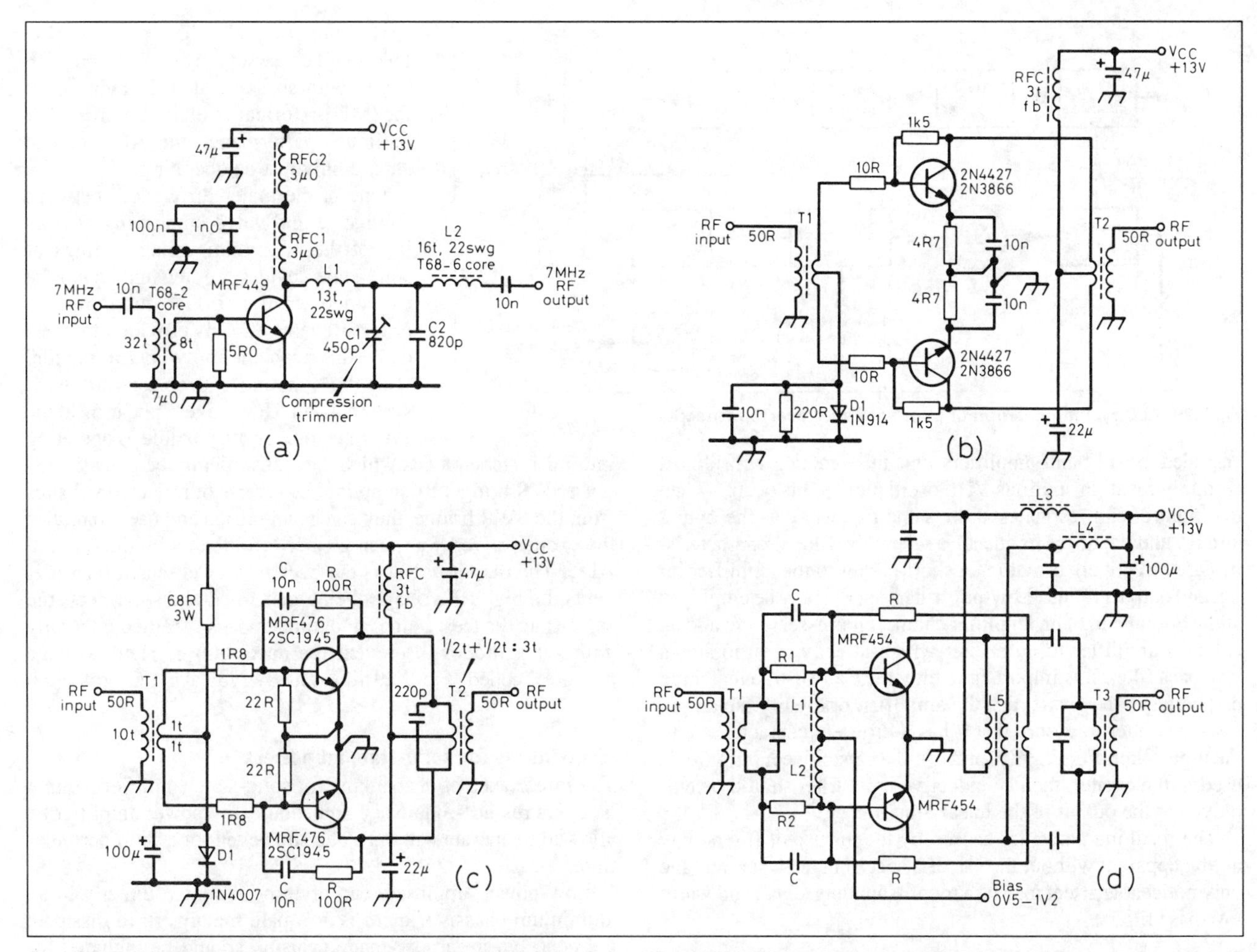

Fig 7.28. Solid-state power amplifiers. (a) Single-ended Class C amplifier (15W); (b) low-power amplifier driver (c) low-power push-pull amplifier (10–25W); (d) medium-power push-pull amplifier (100W)

taking advantage of their improved balance and natural suppression of even-order harmonics. The gain of a solid-state broadband amplifier is considerably lower than that of a valve power amplifier (typically 10dB) and may necessitate cascading a number of stages in order to achieve the desired power level. This is easily achieved using common input and output impedances. The reduced gain has the advantage of aiding stability, but the gain rises rapidly with a reduction in frequency and demands some form of frequency-compensated gain reduction. The latter is achieved using negative feedback with a series combination of R and C: Figs 7.28(b) and (c). The use of VHF power transistors is not recommended in the HF spectrum as instability can result even when high levels of negative feedback are employed.

Solid-state amplifiers can be operated in Class C for use in CW and FM transmitters, but their use for AM transmission should be treated very carefully for the following reasons.

1. It is almost impossible to achieve symmetrical amplitude modulation of a solid-state PA.
2. Device ratings must be capable of sustaining double the collector voltage and current on modulation peaks, ie four times the power of the carrier. Additional safety margins must be included to allow for high RF voltages generated by a mismatched load. Invariably suitable devices are not available for other than low-power operation.

For AM transmission it is recommended that the signal be generated at a low level and then amplified using a linear amplifier. Again, allowance should be made for the continuous carrier and the peak power on modulation peaks.

Output filters

Solid-state amplifiers must not be operated into an antenna without some form of harmonic filtering. The most common design is the *pi-section filter*, comprising typically of a double pi-section (five-element) and in some cases a triple pi-section (seven-element) filter. Common designs are based upon the Butterworth and Chebyshev filters and derivations of them. The purpose of the low-pass filter is to pass all frequencies below the cut off frequency (f_0), normally located just above the upper band edge, while providing a high level of attenuation to all frequencies above the cut-off frequency. Different filter designs provide differing attenuation characteristics versus frequency, and it is desirable to achieve a high level of attenuation by at least three times the cut-off frequency in order to attenuate the third harmonic. The second harmonic, which should be considerably lower in value due to the balancing action of the PA, will be further attenuated by the filter which should have achieved approximately 50% of its ultimate attenuation. Elliptic filters are designed to have tuned notches which can provide higher levels of attenuation at selected frequencies such as $2f$ and $3f$. Filters are discussed in detail in Chapter 5.

The desire to achieve high levels of signal purity may tempt constructors to place additional low-pass filters between

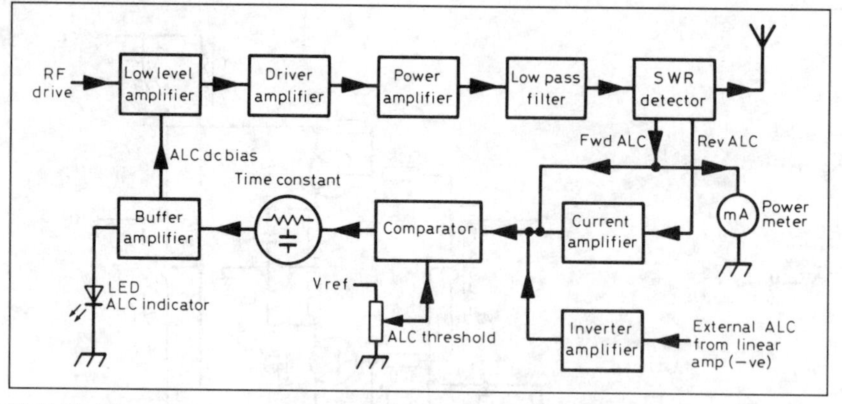

Fig 7.29. ALC system incorporating forward and reverse protection

cascaded broad-band amplifiers, but this practice will almost certainly result in spurious VHF oscillations. This occurs when the input circuit resonates at the same frequency as the output circuit, and occurs at frequencies where the filters' characteristics are effectively providing a short-circuit to the amplifier input and output circuits. Low-pass filters should only be employed at the output end of an amplifier chain. If it is essential to add an external amplifier to an exciter which already incorporates a low-pass filter, it is important to either use a resistive matching pad between the exciter and the amplifier, or modify the output low-pass filter to ensure that it has a different characteristic to the input filter: Figs 7.34(b) and (c). If a capacitive input filter is used in the exciter, then an inductive input filter should be employed at the output of the linear amplifier.

The resulting parasitics caused by the misuse of filters may not be apparent without the use of a spectrum analyser, and the only noticeable affect may be a rough-sounding signal and warm low-pass filters.

Amplifier matching

Broad-band transformers used in solid-state amplifiers consist of a small number of turns wound on a stacked high-permeability ferrite core. The secondary winding may be wound through the primary winding which may be constructed from either brass tube or copper braid. The grade of ferrite is very important and will normally have an initial permeability of at least 800 (Fairite 43 grade). Unfortunately, different manufacturers of ferrite materials use different numbering and grading systems. Too low a permeability will result in poor low-frequency performance and low efficiency. Some designs use conventional centre-tapped transformers for input and output functions, and carry the full DC bias and PA currents, while other designs, Fig 7.28(d), include phasing transformers to supply the collector current while the output transformer is blocked to DC by using series capacitors. The latter arrangement provides an improvement in IMD performance of several decibels, but is often omitted on the grounds of extra cost in commercial amateur radio equipment.

Amplifier protection

The unreliability of early solid-state PAs was largely due to a lack of suitable protection circuitry. ALC (automatic level control) has been used for controlling the output of valve amplifiers for many years – by sampling the grid current in a valve PA it is possible to provide a bias that can be fed back to the exciter to reduce the drive level. A similar system is used for solid-state amplifiers, but it is usually derived by sampling some of the RF output present in a SWR bridge circuit. ALC is very similar to the AGC system found in a receiver. One disadvantage of sampling the RF output signal is that excessive ALC levels will cause severe clipping of the RF signal with an associated degradation of the IMD performance of the amplifier. For optimum performance the ALC system should only just be operating.

One of the major differences between solid-state and valve amplifying devices is that the maximum voltage ratings of solid-state devices are low and cannot be exceeded without disastrous consequences. Reverse ALC is provided to overcome this problem and works in parallel with the conventional or forward ALC system. High RF voltages appearing at the PA collectors are attributable to operating into mismatched loads which can conveniently be detected using a SWR bridge. By sampling the reverse or reflected voltages from the SWR bridge, they can be amplified and used to reduce the exciter drive by a much greater level than with the forward ALC. The output power is cut back to a level which then prevents the high RF voltages being generated and so protects the solid-state devices. Some RF output devices are fitted internally with zener diodes to prevent the maximum collector voltage being exceeded. Fig 7.29 illustrates a typical ALC protection system.

Heatsinking for solid-state amplifiers

The importance of heatsinking for solid-state amplifiers cannot be overstressed – on no account should any power amplifier be allowed to operate without a heatsink, even for a short period of time.

Low-power amplifiers can often be mounted directly to an aluminium chassis if there is adequate metalwork to dissipate the heat, but a purpose-made heatsink should be included for power levels in excess of 10W. For SSB operation the heatsinking requirements are less stringent than for CW or data operation due to the much lower average power dissipation. Nevertheless, an adequate safety margin should be provided to allow for long periods of key-down operation.

Solid-state power amplifiers are in the order of 50% efficient, and therefore a 100W output amplifier will also have to dissipate 100W of heat. It can be seen that for very-high-power operation heatsinking becomes a major problem due to the compact nature of solid-state amplifiers. The use of copper spreaders is advisable for powers above 200W. The amplifier is bolted directly to a sheet of copper at least 0.25in thick, which in turn is bolted directly to the aluminium heatsink. The use of air blowers to circulate air across the heatsink should also be considered.

The actual mounting of power devices requires considerable care; the surface should ideally be milled to a flatness of plus or minus one half of one thousandth of an inch. For this reason, die-cast boxes must not be used for high-power amplifiers as the conductivity is poor and the flatness is nowhere near to being acceptable. Heatsinking compounds are also essential for aiding the conduction of heat rapidly away from the device. Motorola Application Note AN1041/D provides guidance in the mounting of power devices in the 200–600W range.

Power dividers and combiners

It is possible to increase the power output from a solid-state amplifier by combining the outputs of a number of smaller amplifiers so that the sum of the output powers is equal to the desired power level. It is practical to combine the outputs of four

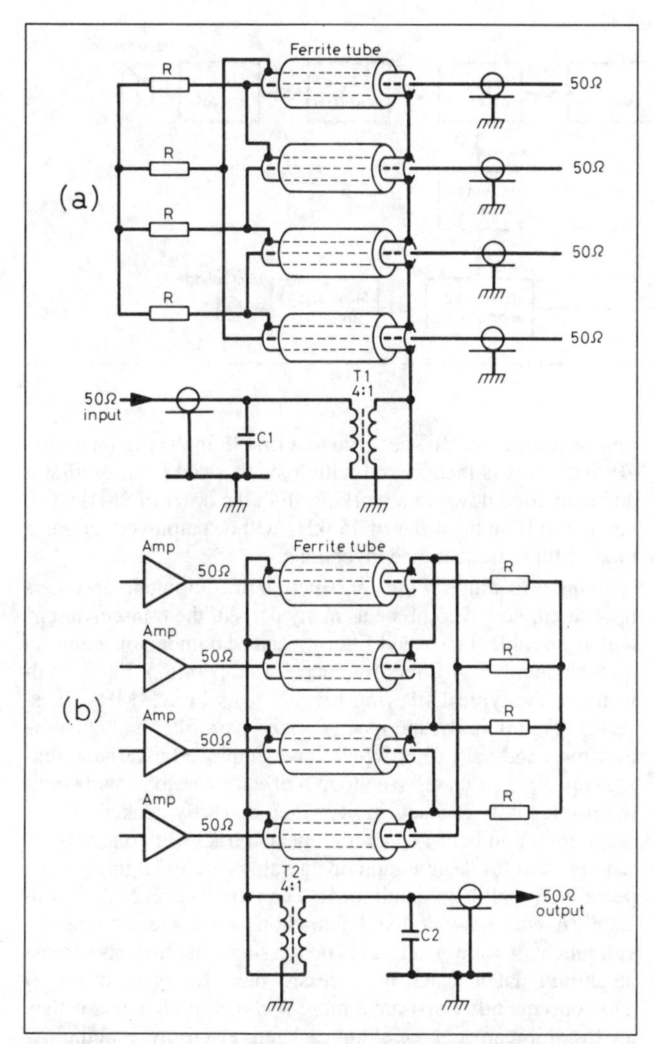

Fig 7.30. Combining multiple power amplifiers. (a) Four-port power divider. (b) Four-port output combiner

100W amplifiers to produce 400W output, or two 300W amplifiers to produce 600W output.

Initially the drive signal is split using a power divider and then fed to a number of amplifiers that are effectively operated in parallel. The outputs of the amplifiers are summed together in a power combiner that is virtually the reverse of the divider circuit. Fig 7.30 illustrates a four-way divider and combiner.

The purpose of the power divider is to divide the input power into four equal sources, providing an amount of isolation between each. The outputs are designed for 50Ω impedance, which sets the common input impedance to 12.5. A 4:1 step-down transformer provides a match to the 50Ω output of the driver amplifier. The phase shift between the input and output ports must be zero and is achieved by using 1:1 balun transformers. These are loaded with ferrite tubes to provide the desired low-frequency response without resort to increasing the physical length. In this type of transformer, the currents cancel, making it possible to employ high-permeability ferrite and a relatively short lengths of transmission line. In an ideally balanced situation, no power will be dissipated in the magnetic cores and the line loss will be low. The minimum inductance of the input transformer should be 16μH at 2MHz – a lower value will degrade the isolation characteristics between the output ports and this is important in the event of a change in input VSWR to one of the amplifiers. It is unlikely that the splitter will be subjected to an open- or short-circuit load at the amplifier input, due to the base-frequency compensation networks in the amplifier modules. The purpose of the balancing resistors R is to dissipate any excess power if the VSWR rises. The value of R is determined by the number of 50Ω sources assumed unbalanced at any one time. Except for a two-port divider, the resistor values can be calculated for an odd or even number of ports as:

$$R = \left(\frac{R_L - R_{IP}}{n + 1}\right) n$$

where R_L is the impedance of output ports (50Ω); R_{IP} is the impedance of input port (12.5Ω); n is the number of correctly terminated output ports.

Although the resistor values are not critical for the input divider, the same formula applies to the output combiner where mismatches have a larger effect on the total power output and linearity.

The power divider employs ferrite sleeves having a μ of 2500 and uses 1.2in lengths of RG-196 coaxial cable; the inductance is approximately 10μH. The input transformer is wound on a 63-grade ferrite toroid with RG-188 miniature coaxial cable. Seven turns are wound bifilar and the ends are connected inner to outer, outer to inner at both ends to form a 4:1 transformer.

The output combiner is a reverse of the input divider and performs in a similar manner. The ballast resistors R must be capable of dissipating large amounts of heat in the event that one of the sources becomes disabled, and for this reason they must be mounted to a heatsink and be non-inductive. For one source disabled in a four-port system, the heat dissipated will be approximately 15% of the total output power, and if phase differences occur between the sources, it may rise substantially. The resistors are not essential to the operation of the divider/combiner; their function is to provide a reduced output level in the event of an individual amplifier failure. If they are not included, failure of any one amplifier will result in zero output from the amplifier combination. The output transformer must be capable of carrying the combined output power and must have sufficient cross-sectional area. The output transformer is likely to run very warm during operation. High-frequency compensating capacitors C1 and C2 may be fitted to equalise the gain distribution of the amplifier, but they are not always necessary.

A two-port divider combiner is illustrated in Fig 7.31; operation is principally the same as in the four-port case but the input

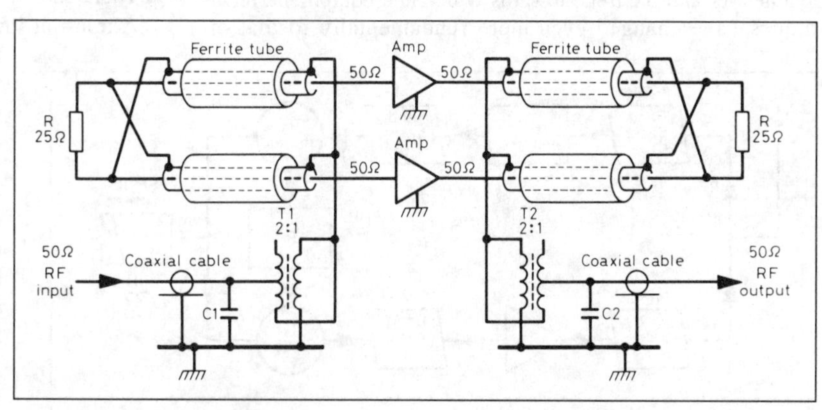

Fig 7.31. Two-port RF divider and combiner

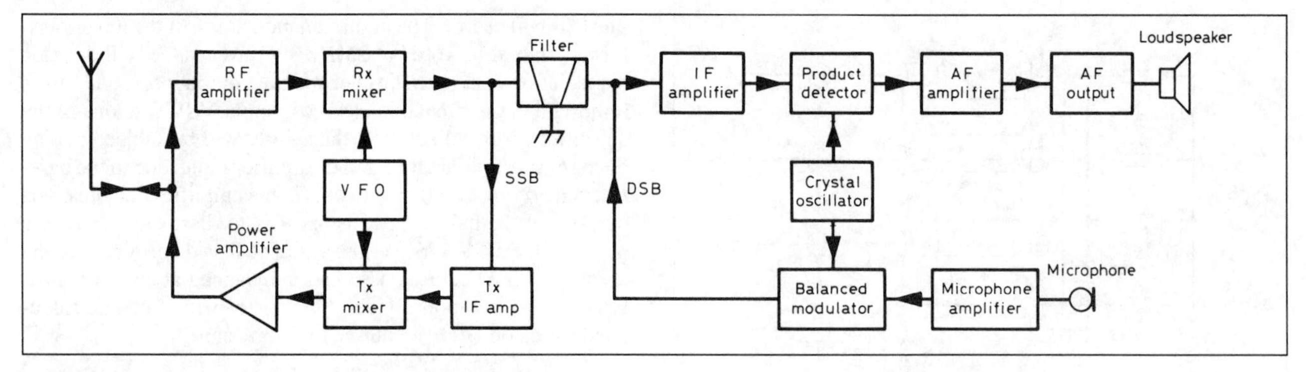

Fig 7.32. SSB transceiver block diagram

and output transformers are tapped to provide a 2:1 ratio. Detailed constructional notes of two- and four-port dividers and combiners are given in Motorola Application Notes AN749 and AN758 respectively.

TRANSCEIVERS

Separate transmitter and receiver combinations housed in one cabinet may be referred to as a *transceiver*. This is not strictly correct as a transceiver is a combined transmitter and receiver where specific parts of the circuit are common to both functions. Specifically, the oscillators and frequency-determining components are common and effectively synchronise both the transmitter and receiver to exactly the same frequency. This synchronisation is a pre-requisite for SSB operation and transceivers owe their existence to the development of SSB transmission. While many early attempts at SSB generation used the phasing method, the similarity of the filter-type SSB generator circuit to a superhet SSB receiver circuit (Fig 7.16) make interconnection of the two circuits an obvious development (Fig 7.32). True transceive operation is possible by simply using common oscillators, but it is also advantageous to use a common SSB filter in the IF amplifier stages, which provides similar audio characteristics on both transmit and receive as well as providing a considerable saving in cost.

Initially low-frequency SSB generation necessitated double-conversion designs, often employing a tunable second IF and a crystal-controlled oscillator for frequency conversion to the desired amateur bands. This technique was superseded by single-conversion designs using a high-frequency IF in the order of 9MHz, with a heterodyne-type local oscillator consisting of a medium-frequency VFO and a range of high-frequency crystal oscillators. With the advent of the phase-locked loop (PLL) synthesiser and the trend towards wide-band equipment, techniques have changed even more fundamentally to that of up-converting the HF spectrum to a first IF in the region of 40–70MHz. This is then mixed with a synthesised local oscillator and converted down to a working IF in the order of 9MHz. Often a third IF in the order of 455kHz will be employed, giving a total of three frequency conversions.

In order to simplify the construction of equipment, designers have attempted to combine as many parts of the transceiver circuit as possible. Front-end filtering can be cumbersome and requires a number of filters for successful operation. Fig 7.33 illustrates the typical filtering requirements in a 14MHz transceiver. Traditionally, the receiver band-pass filter and even the transmit band-pass filter would have employed a variable tuning capacitor, often referred to as a *preselector*, to provide optimum selectivity and sensitivity when correctly peaked. To arrange for a number of filters to tune and track with one another requires careful design and considerable care in alignment, especially in multiband equipment. The introduction of the solid-state PA with its wide-band characteristics has lead to the development of wide-band filters possessing a flat response across an entire amateur band. By necessity these filters are of low Q and consequently must have more sections or elements if they are to exhibit any degree of out-of-band selectivity. By employing low-Q, multi-section band-pass filters it is possible to eliminate one filter between the receiver RF amplifier and the receive mixer. The gain of the RF amplifier should be kept as low as possible and in most cases can be eliminated entirely for use below 21MHz. By providing the band-pass filters with low input and output impedances, typically 50Ω, switching of the filters can simplified to the extent that one filter can be used in both the transmit and receive paths, thus reducing the band-pass filter requirement to one per band.

Fig 7.34(a) illustrates a typical band-pass filter configuration for amateur band use, and suitable component values are listed in Table 7.5.

A transmit low-pass filter is essential for attenuation of all unwanted products and harmonics amplified by the broad-band amplifier chain. Suitable values for a typical Chebyshev filter, Fig 7.34(b), suitable for use at the output of a power amplifier chain providing up to 100W output, are given in Table 7.6. One disadvantage of the Chebyshev filter design is a slow rise time of the filter attenuation characteristic, with limited suppression of the second harmonic. The elliptic function filter, Fig 7.34(d), has a greatly improved characteristic. The addition of two capacitors placed in parallel with the two filter inductors results in the circuits L1, C4 and L2, C5 being resonant at approximately two and

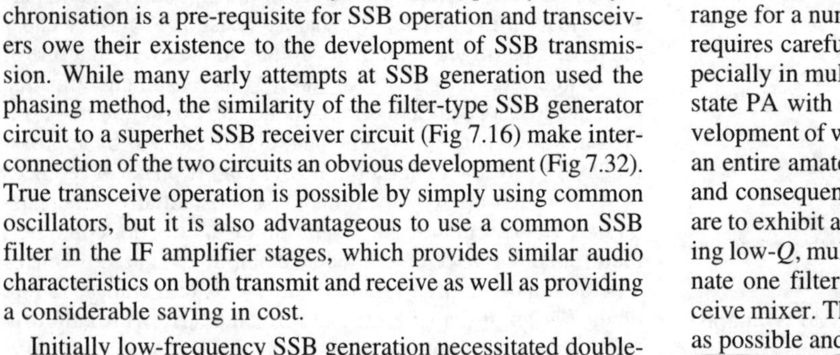

Fig 7.33. HF transceiver RF filtering

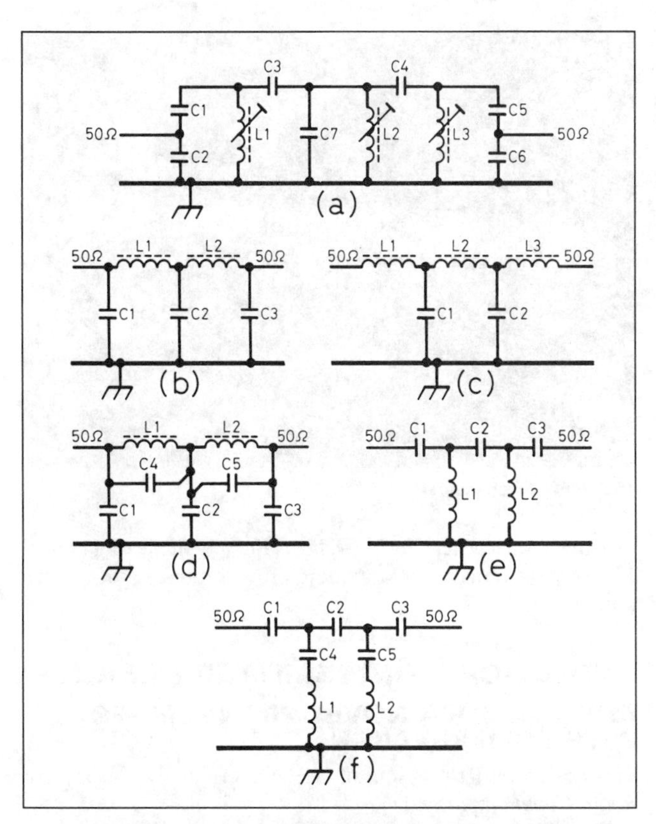

Fig 7.34. Band-pass, low-pass and high-pass filters. (a) Band-pass filter; (b) Chebyshev low-pass filter capacitive input/output; (c) Chebyshev low-pass filter inductive input/output; (d) elliptic function low-pass filter; (e) Chebyshev high-pass filter; (f) elliptic high-pass filter

three times the input frequency respectively to provide peaks of attenuation at the second and third harmonic frequencies. Where possible, the use of elliptic function filters is recommended – typical values for amateur band use are given in Table 7.7.

For an external solid-state amplifier being driven by an exciter that already contains a capacitive-input low-pass filter such as that in Fig 7.34(b), the inductive input design, Fig 7.34(c), may be necessary at the final amplifier output. The combination of the two different types of filter should eliminate parasitic oscillations which will almost certainly occur if two filters with similar characteristics are employed. Values for the inductive input filter are given in Table 7.8 for use at powers up to 300W.

The development of HF synthesisers has led to the development of HF transceivers providing general-coverage facilities and requiring yet further changes in the design of band-pass filters for transceiver front-ends. There is a finite limit to the bandwidth that can be achieved using conventional parallel-tuned circuit filters. For general-coverage operation from 1 to 30MHz, approximately 30–40 filters would be required, and this is obviously not a practical proposition. While returning to the mechanically tuned filter might reduce the total number of filters required, the complexity of electronic band-changing would be formidable. By combining the characteristics of both low-pass and high-pass filters, Figs 7.34(e) and (f), the simple action of cascading the two such filters will result in a *band-pass filter*, Fig 7.35(a), having a bandwidth equal to the difference between the two filter cut-off frequencies. The limiting bandwidth will be one octave, ie the highest frequency is double the lowest frequency. Practically, filter bandwidth is restricted to slightly less than one octave. Typical filters in a general-coverage HF transceiver might cover the following bands:

Table 7.5. Band-pass filter of Fig 7.34(a). 50Ω nominal input/output impedance

Band (m)	L1, L3 (μH)	L2 (μH)	L-type	C1, C5 (pF)	C2, C6 (pF)	C3, C4 (pF)	C7 (pF)
160	8.0	8.0	27t KANK3335R	1800	2700	180	750
80	5.8	5.8	KANK3334R	390	1800	47	270
40	2.8	2.8	KXNK4173AO	220	1000	10	150
30	1.3	1.3	KANK3335R	220	1000	10	180
20	1.2	1.2	KANK3335R	120	560	4.7	100
17	0.29	—	Toko S18 Blue	220	750	8.2	—
15	0.29	—	Toko S18 Blue	180	560	6.8	—
11	0.29	—	Toko S18 Blue	100	560	2.7	—
10	0.29	—	Toko S18 Blue	82	390	2.7	—

All capacitors are polystyrene except those less than 10pF which are ceramic. L2 and C7 are not used on the 10–17m bands.

Table 7.6. Chebyshev low-pass filter of Fig 7.34(b)

Band (m)	L1, L2	Core	C1, C3 (pF)	C2 (pF)
160	31t/24swg	T50-2	1200	2500
80	22t/20swg	T50-2	820	1500
40	18t/20swg	T50-6	360	680
30/20	12t/20swg	T50-6	220	360
21/17	10t/20swg	T50-6	100	220
11/10	9t/20swg	T50-6	75	160

All capacitors are silver mica or polystyrene – for 100W use 300VDC wkg; for <50W use 63VDC wkg. Cores are Amidon: T50-2 Red or T50-6 Yellow.

Table 7.7. Elliptic low-pass filter of Fig 7.34(d)

Band (m)	L1	L2	Core	C1 (pF)	C2 (pF)	C3 (pF)	C4 (pF)	C5 (pF)
160	28t/22swg	25t/22swg	T68-2	1200	2200	1000	180	470
80	22t/22swg	20t/22swg	T50-2	680	1200	560	90	250
40	18t/20swg	16t/20swg	T50-6	390	680	330	33	100
30/20	12t/20swg	11t/20swg	T50-6	180	330	150	27	75
17/15	10t/20swg	9t/20swg	T50-6	120	220	100	12	33
11/10	8t/20swg	7t/20swg	T50-6	82	150	68	12	39

Capacitors 300VDC wkg silver mica up to 200W. All cores Amidon.

Table 7.8. Chebyshev low-pass filter inductive input of Fig 7.34(c)

Band (m)	L1, L3	L2	Core	C1, C2 (pF)
160	8.1μH/24t	11.4μH	T106-2	1700
80	4.1μH/17t	5.8μH/21t	T106-2	860
40	2.3μH/13t	3.2μH/15t	T106-2	470
30/20	1.18μH/10t	1.65μH/12t	T106-6	240
17/15	0.79μH/8t	1.11μH/10t	T106-6	160
11/10	0.57μH/7t	0.8μH/8t	T106-6	120

This filter is for use with a high-power external amplifier, when capacitive input is fitted to exciter. Capacitors silver mica: 350VDC wkg up to 200W; 750VDC wkg above 300W. Use heaviest possible wire gauge for inductors. All cores Amidon.

(a) 1.5–2.5MHz
(b) 2.3–4.0MHz
(c) 3.9–7.5MHz
(d) 7.4–14.5MHz
(e) 14.0–26.0MHz
(f) 20.0–32.0MHz

It can be seen that six filters will permit operation on all the HF amateur bands as well as providing general coverage of all the in-between frequencies. Hybrid low/high-pass filters invariably use fixed-value components and require no alignment, thus simplifying construction. The transmit low-pass filter may also be left in circuit on receive in order to enhance the high-frequency rejection; it has no effect on low-frequency signals. The use of

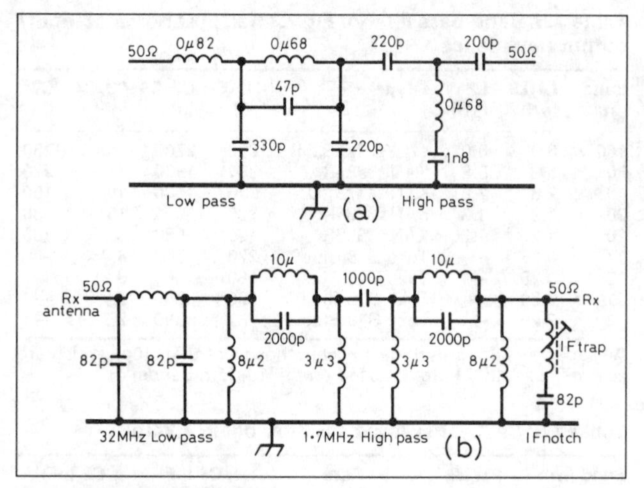

Fig 7.35. Multifunction filters. (a) Band-pass filter using high/low-pass filters; (b) composite receiver input filter

The G4ENA transceiver

separate high-pass filters in the receiver input circuit prior to the band-pass filter serves to eliminate low-frequency broadcast signals. Typically, a high-pass filter of the multi-pole elliptic type, Fig 7.35(b), having a cut-off of 1.7MHz, is fitted to most commercial amateur band equipment.

TRANSVERTERS

Transverters are transmit/receive converters that permit equipment to be operated on frequencies not covered by that equipment. Traditionally HF equipment was transverted to the VHF/UHF bands, however, with the increase in availability of 144MHz SSB equipment, down-conversion to the HF bands has become popular.

A transverter takes the output from a transmitter, attenuated to an acceptable level, heterodynes it with a crystal-controlled oscillator to the desired frequency and then amplifies it to the required level. The receive signal is converted by the same process in reverse to provide transceive capabilities on the new frequency band. The techniques employed in transverters are the same as those used in comparable frequency transmitting and receiving equipment. Where possible it is desirable to provide low-level RF output signals for transverting rather than having to attenuate the high-level output from a transmitter with its associated heatsinking requirements.

Occasionally transverters may be employed from HF to HF in order to include one of the new WARC bands on an older transceiver, or to provide 160m band facilities where they have been omitted. In some cases it may prove simpler to add an additional frequency band to existing equipment in preference to using a transverter. Fig 7.36 shows the schematic of a typical 144/14MHz transverter providing HF operation with a VHF transceiver.

PRACTICAL TRANSMITTER DESIGNS

QRP + QSK – A NOVEL TRANSCEIVER WITH FULL BREAK-IN

This design by Peter Asquith, G4ENA, originally appeared in *Radio Communication* [1].

Introduction

The past decade has seen significant advances in semiconductor development. One such area is that of digital devices. Their speed has steadily improved to the point which permits them to be used in LF transceiver designs. One attractive feature of these components is their relatively low cost.

The QSK QRP Transceiver (Fig 7.37) employs several digital components which, together with simple analogue circuits, provide a small, high-performance and low-cost rig. Many features have been incorporated in the design to make construction and operation simple.

One novel feature of this transceiver is the switching PA stage. The output transistor is a tiny IRFD110 power MOSFET.

This device has a very low 'ON' state resistance which means that very little power is dissipated in the package, hence no additional heatsinking is required. However, using this concept does mean that good harmonic filtering must be used.

The HC-type logic devices used in the rig are suitable for operation on the 160m and 80m bands. The transmitter efficiency on 40m is poor and could cause overheating problems. Future

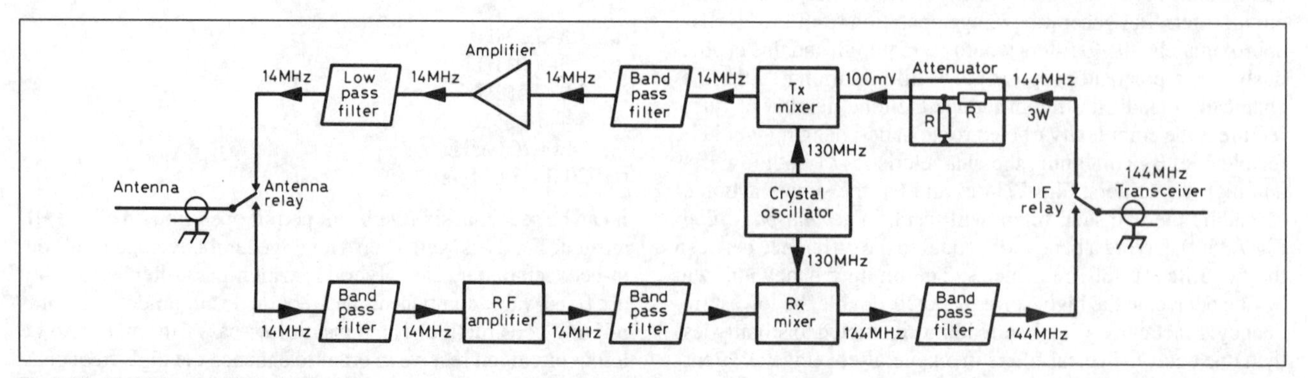

Fig 7.36. 144MHz to 14MHz transverter

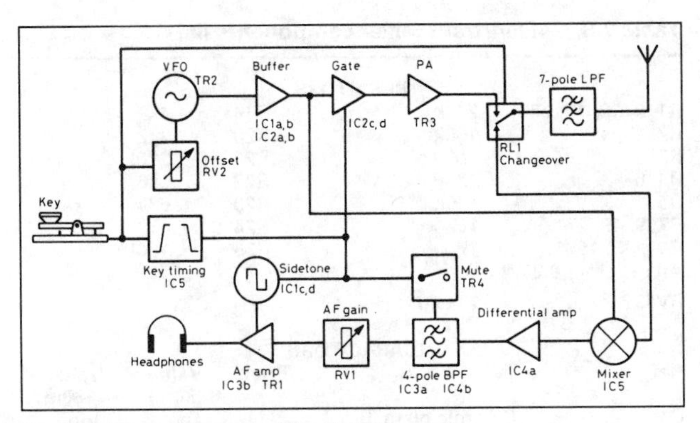

Fig 7.37. Block diagram of the QRP transceiver

advances in component design should raise the top operating frequency limit.

Circuit description

VFO

The circuit diagram of the transceiver is shown in Fig 7.38. TR2 is used in a Colpitts configuration to provide the oscillator for both receive and transmit.

The varicap diode D1 is switched via RV2/R1 by the key to offset the receive signal by up to 2kHz, such that when transmitting, the output will appear in the passband of modern transceivers operating in the USB mode. C1 controls the RIT range and C29/30 the band coverage. IC1a, IC1b and IC2a buffer the VFO. It is important that the mark/space ratio of the square wave at IC1b is about 50:50. Small variations will affect output power.

Transmitter

When the key is operated, RL1 will switch and, after a short delay provided by R19/C15, IC2c and IC2d will gate the buffered VFO to the output FET, TR3. TR3 operates in switch mode and is therefore very efficient. The seven-pole low-pass filter after the changeover relay removes unwanted harmonics, which are better than –40dB relative to the output.

Receiver

The VFO signal is taken from IC2a to control two changeover analogue switches in IC5, so forming a commutating mixer and providing direct conversion to audio of the incoming stations. IC4a is a low-noise, high-gain, differential amplifier whose output feeds the four-pole CW filter, IC4a and IC3a, before driving the volume attenuator, RV1. IC3b amplifies and TR1 buffers the audio to drive headphones or a small speaker. When the key is down TR4 mutes the receiver, and audio oscillator IC1a and b injects a sidetone into the audio output stage, IC3b. The value of R5 sets the sidetone level.

Constructional notes

The component layout is shown in Fig 7.39.

1. Check that all top-side solder connections are made.
2. Wind turns onto toroids tightly and fix to PCB with a spot of glue.
3. Do not use IC sockets. Observe anti-static handling precautions for all ICs and FETs.
4. All VFO components should be earthed close to VFO.
5. Fit a 1A fuse in the supply line.
6. Component suppliers: Bonex, Cirkit, Farnell Electronic Components etc.

Test and calibration

Before TR3 is fitted, the transceiver must be fully operational and calibrated. Prior to switch-on, undertake a full visual inspection for unsoldered joints and solder splashes. Proceed as follows.

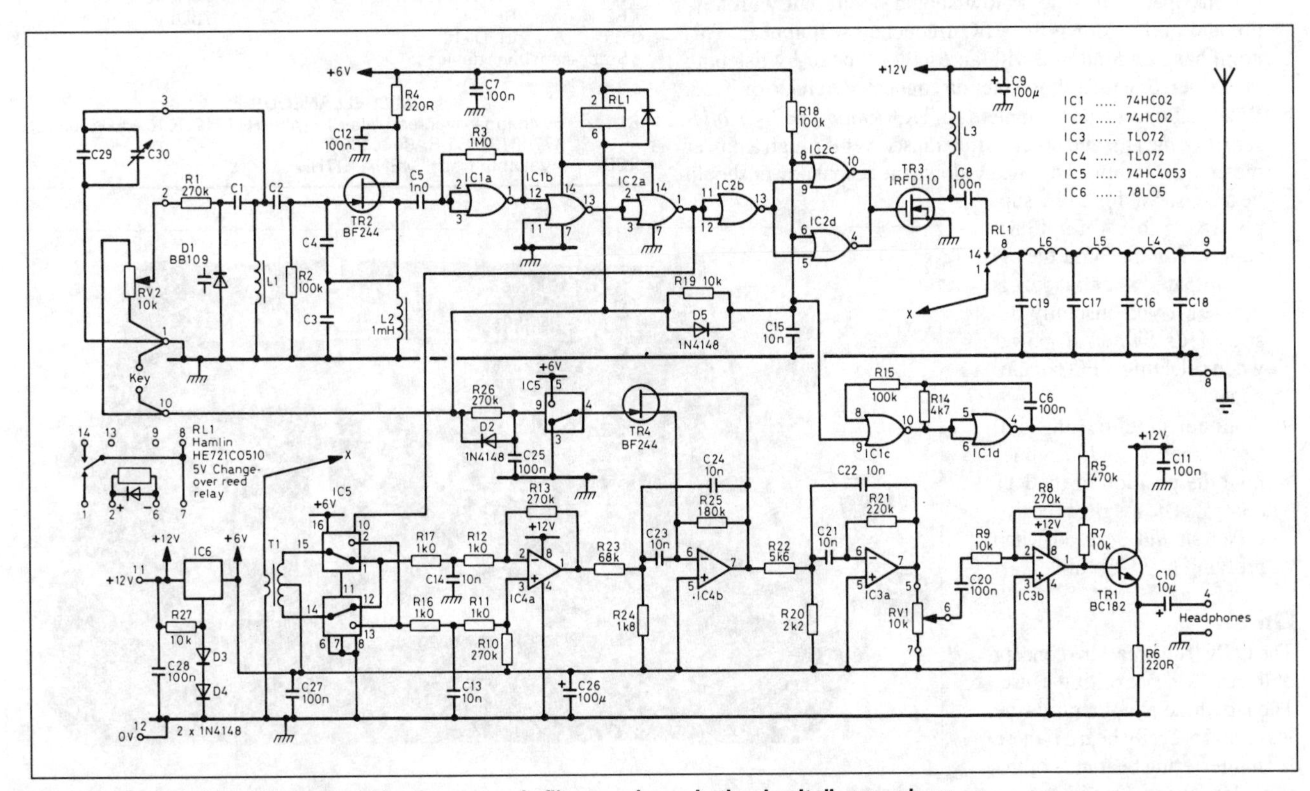

Fig 7.38. The switching PA stage requires a seven-pole filter, as shown in the circuit diagram above

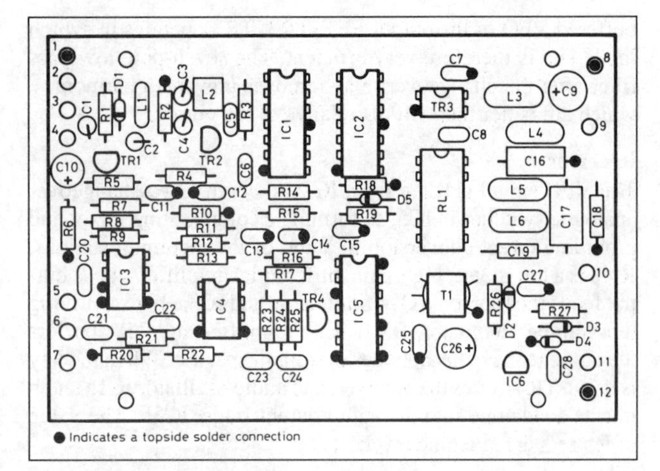

Fig 7.39. A neat component layout results in a compact unit suitable for portable operation. (The PCB supplied by G4ENA is slightly larger than this.)

1. Connect external components C29/30, RV1/2, headphones and power supply.
2. Switch on power supply. Current is approx 50mA.
3. Check +6V supply, terminal pin 7. Voltage is approx 6.3V.
4. Select values for C29 to bring oscillator frequency to CW portion of band (1.81–1.86MHz/3.50–3.58MHz). Coverage should be set to fall inside the band limits of 1.810/3.500MHz.
5. Connect an antenna or signal generator to terminal pin 9 and monitor the received signal on headphones. Tuning through the signal will test the response of the CW filter which will peak at about 500Hz.
6. Connect key and check operation of sidetone and antenna changeover relay. Sidetone level can be changed by selecting value of R5.
7. Monitor output of IC2c and IC2d (TR3 gate drive) and check correct operation. A logic low should be present with key-up, and on key-down the VFO frequency will appear. This point can be monitored with an oscilloscope or by listening on a receiver with a short antenna connected to IC2c or IC2d.
8. When all checks are complete fit TR3 *(important! – static-sensitive device)* and connect the transceiver through a power meter to a dummy load. On key-down the output power should be at least 5W for +12V supply, rising to 8W for 13.8V supply. *Note*: should the oscillator stop when the key is pressed it will instantly destroy TR3. Switch off power when selecting VFO components.
9. Connect antenna and call CQ. When a station replies note the position of the RIT control. The average receive offset should be used when replying to a CQ call.

On air

The QSK (full break-in) concept of the rig is very exciting in use. The side tone is not a pure sine wave and is easily heard if there is an interfering beat note of the same frequency. One important

Table 7.9. G4ENA transceiver components list

RESISTORS

Ref	Value	Ref	Value
R1, 8, 10, 13, 26	270k	R14	4k7
R2, 15, 18	100k	R20	2k2
R3	1M	R21	220k
R4, 6	220R	R22	5k6
R5	470k	R23	68k
R7, 9, 19, 27	10k	R24	1k8
R11, 12, 16, 17	1k	R25	180k

All resistors 0.25W 2%

RV1, 2	10k lin

CAPACITORS

Ref	Type	Pitch	Value (80m)	Value (160m)
C1	Ceramic plate 9	2.54	4p7	15p
C2	Polystyrene	—	47p	100p
C3, 4	Polystyrene	—	220p	470p
C5	Ceramic monolithic	2.54	1n	1n
C6, 7, 8, 11, 12, 20, 25, 27, 28	Ceramic monolithic	2.54	100n	100n
C9, 26	Aluminium radial 16V	2.5	100μ	100μ
C10	Aluminium radial 16V	2.0	10μ	10μ
C13, 14, 15, 21 22, 23, 24	Ceramic monolithic	2.54	10n 10%	10n 10%
C16, 17	Polystyrene	—	1n5	2n7
C18, 19	Polystyrene	—	470p	1n
C29*	Polystyrene	—	470p	820p
C30*	Air-spaced VFO	—	25p	75p

*Select on test component

INDUCTORS

Ref	Type	80m	160m
L1	T37-2 (Amidon)	31t 27swg (0.4mm)	41t 30swg (0.315mm)
L2	7BS (Toko)	1mH	1mH
L3	T37-2	2.2μH 23t 27swg	4.5μH 33t 30swg
L4, 6	T37-2	2.9μH 26t 27swg	5.45μH 36t 30swg
L5	T37-2	4.0μH 31t 27swg	6.9μH 41t 30swg
T1	Balun	2t primary, 5+5t secondary, 0.2mm 36swg (28-43002402)	

SEMICONDUCTORS

Ref	Type	Ref	Type
D1	BB109	IC1, 2	74HC02*
D2, 3, 4, 5	1N4148	IC3, 4	TL072*
TR1	BC182 (not 'L')	IC5	74HC4053*
TR2, 4	BF244	IC6	78L05
TR3	IRFD110*		

*Static-sensitive devices

MISCELLANEOUS

RL1	5V change-over reed relay, Hamlin, HE721CO510; PED/Electrol, 17708131551-RA30441051
PCB	Available from G4ENA, QTHR

Inside view of the G4ENA transceiver

note is to remember to tune the receiver into a station from the high-frequency side so that when replying your signal falls within his passband.

Both the 160 and 80m versions have proved very successful on-air. During the 1990 Low Power Contest the 80m model was operated into a half-wave dipole and powered from a small nicad battery pack. This simple arrangement produced the highest 80m single band score!

Its small size and high efficiency makes this rig ideal for portable operation. A 600mAh battery will give several hours of QRP pleasure – no problem hiding away a complete station in the holiday suitcase!

Front view of transceiver. On this prototype the TUNE switch is labelled TEST

A QRP TRANSCEIVER FOR 1.8MHz

This design by S E Hunt, G3TXQ, originally appeared in *Radio Communication* [2].

Rear view of transceiver

Introduction

This transceiver was developed as part of a 1.8MHz portable station, the other components being a QRP ATU, a battery-pack and a 200ft kite-supported antenna. It would be a good constructional project for the new Class A licensee or for anyone whose station lacks 1.8MHz coverage. The 2W output level may seem a little low, but it results in low battery drain and is adequate to give many 1.8MHz contacts.

The designer makes no claim for circuit originality. Much of the design was adapted from other published circuitry; however, he does claim that the design is repeatable – six transceivers have been built to this circuit and have worked first time. Repeatability is achieved by extensive use of negative feedback; this leads to lower gain-per-stage (and therefore the need for more stages) but makes performance largely independent of transistor parameter variations.

Circuit description

The transceiver (Fig 7.40) comprises a direct-conversion receiver together with a double-sideband (DSB) transmitter. This approach results in much simpler equipment than a superhet design, and is capable of surprisingly good performance, particularly if care is taken over the mixer circuitry.

During reception, signals are routed through the band-pass filter (L1, L2 and C25–C31) to a double-balanced mixer, M1, where they are translated down to baseband. It is vital for the mixer to be terminated properly over a wide range of frequencies, and this is achieved by a diplexer comprising R34, RFC2 and C32–R34. Unwanted RF products from the mixer, rejected by RFC2, pass through C32 to the 47Ω terminating resistor R34. The wanted audio products pass through RFC2 and C34 to a common-base amplifier stage which is biased such that it presents a 50Ω load impedance. The supply rail for this stage comes via an emitter-follower, TR5, which has a long time-constant (4s) RC circuit across its base. This helps to prevent any hum on the 12V rail reaching TR6 and being amplified by IC3.

The voltage gain of the common-base stage (about ×20) is controlled by R37 which also determines the source resistance for the following low-pass filter (L3, L4 and C39–C43). This filter is a Chebyshev design and it determines the overall selectivity of the receiver. The filter is followed by a single 741 op-amp stage which give adequate gain for headphone listening; however, an LM380 audio output stage can easily be added if you require loudspeaker operation.

On transmit, audio signals from the microphone are amplified in IC1 and IC2, and routed to the double balanced mixer

Top view of transceiver

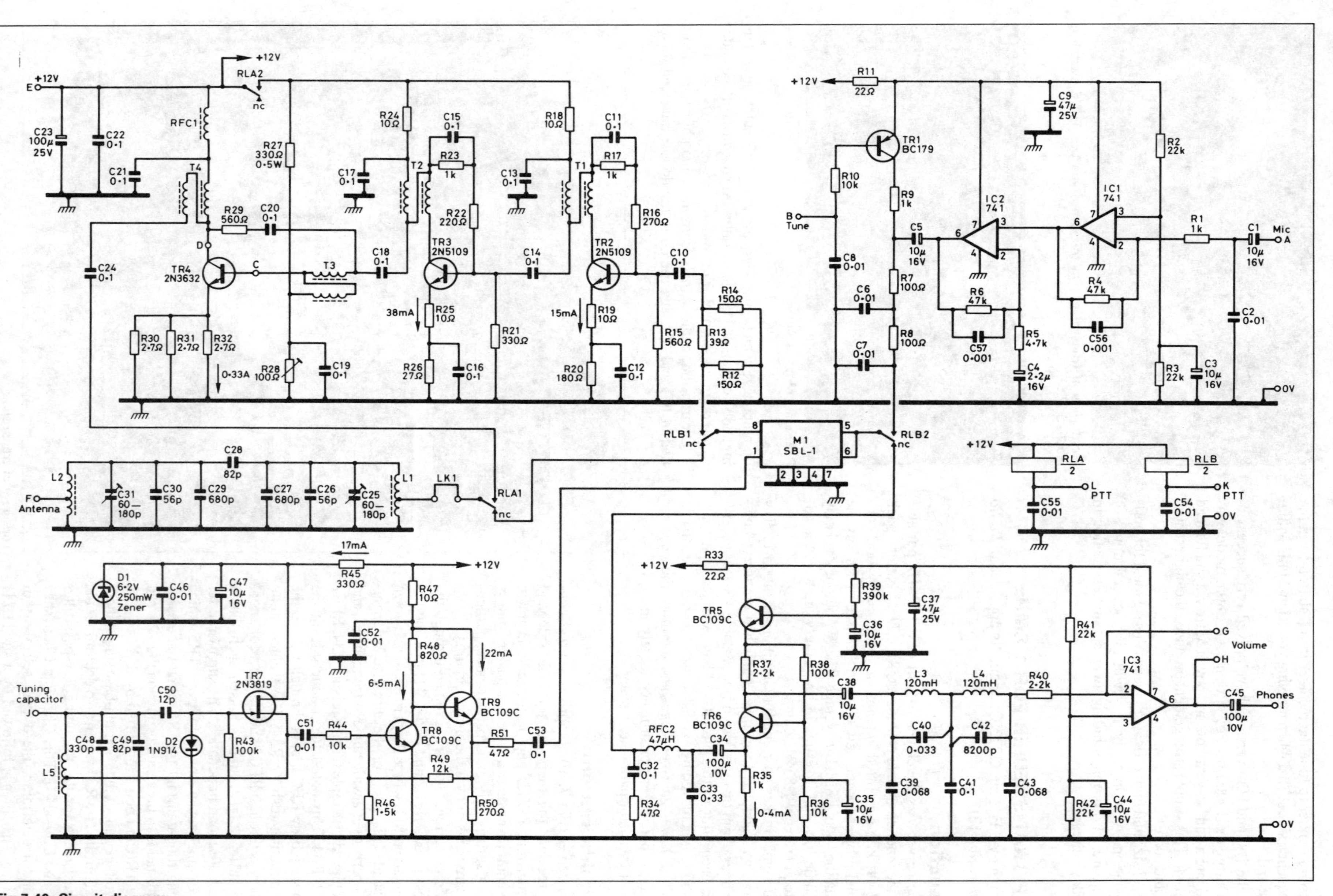

Fig 7.40. Circuit diagram

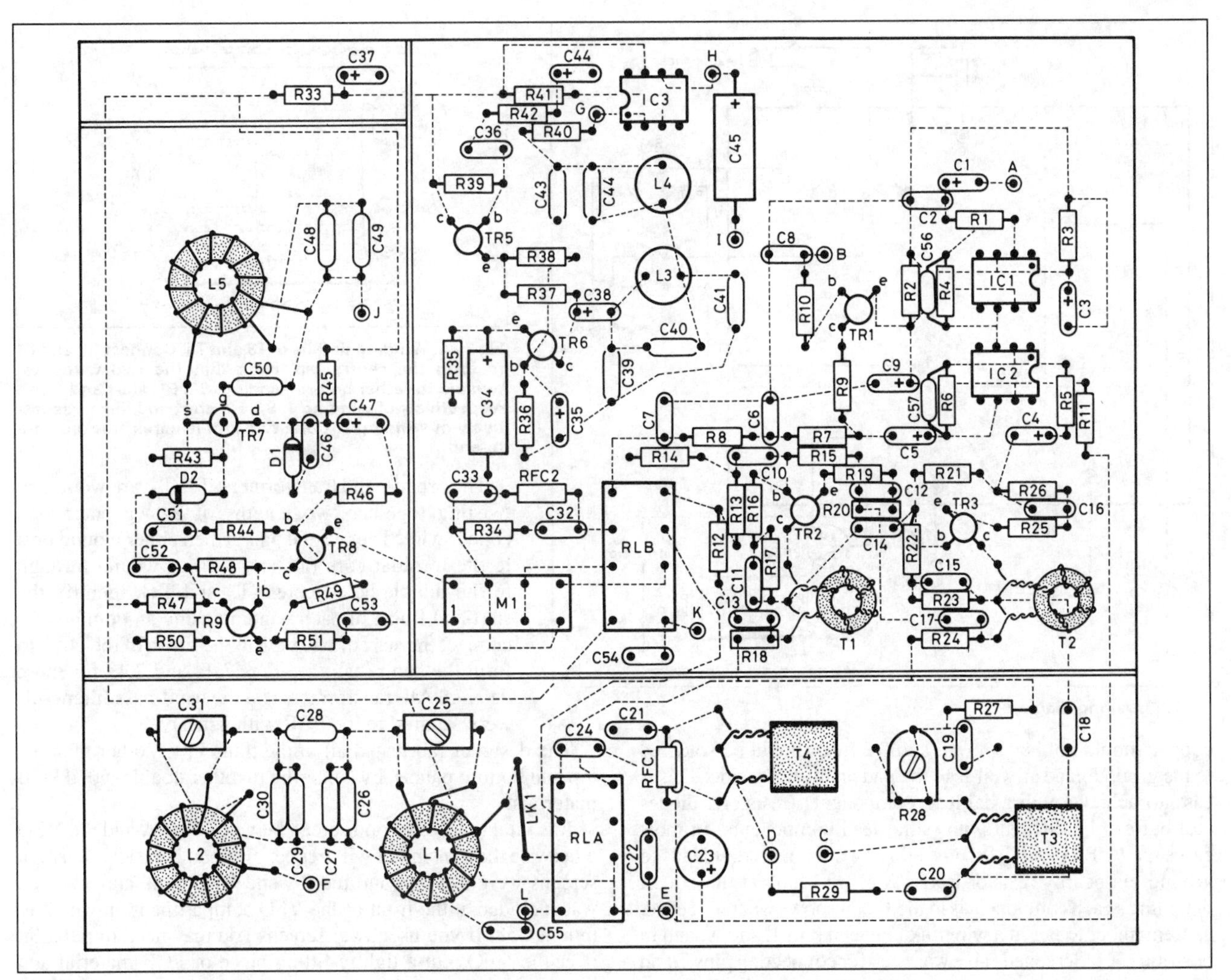

Fig 7.42. Component layout

where they are heterodyned up into the 1.8MHz band as a double-sideband, suppressed-carrier signal. Capacitors C56 and C57 cause some high-frequency roll-off of the audio signal and thereby restrict the transmitted bandwidth. A 6dB attenuator (R12–R14) provides a good 50Ω termination for the mixer. The DSB signal is amplified by two broad-band feedback amplifiers, TR2 and TR3, each having a gain of 15dB. TR3 is biased to a higher standing current to keep distortion products low.

The PA stage is a single-ended design by VE5FP [3]. The inclusion of unbypassed emitter resistors R30–R32 establishes the gain of the PA and also helps to prevent thermal runaway by stabilising the bias point. Additional RF negative feedback is provided by the shunt feedback resistor R29. The designer chose to run the PA at a moderately high standing current (330mA) in order to reduce distortion products, thinking that at some stage he might use the transceiver as a 'driver' for a 10–15W linear amplifier. The PA output (about 2W PEP) is routed through the band-pass filter to the antenna. The designer used a 2N3632 transistor in the PA because he happened to have one in the junk-box; the slightly less expensive 2N3375 would probably perform just all well. VE5FP used a 2N5590 transistor but this would need different mounting arrangements.

At the heart of the transceiver is a Hartley VFO comprising TR7 and associated components. The supply of this stage is stabilised at 6.2V by zener diode D1 and decoupled by C46 and C47. It is important for best stability that the Type 6 core material is used for L5 as this has the lowest temperature coefficient of permeability. Output from the VFO is taken from the low impedance tap on L5.

The VFO buffer is a feedback amplifier comprising TR8 and TR9. The input impedance of this buffer is well-defined by R44 and presents little loading of the VFO. Its gain is set by the ratio R49/R44 and R51 has been included to define the source resistance of this stage at approximately 50Ω.

Change-over between transmit and receive is accomplished by two DPDT relays which are energised when the PTT lines are grounded. A CW signal for tuning purposes can be generated by grounding the TUNE pin – this switches on TR1, which in turn unbalances the mixer, allowing carrier to leak through to the driver and PA stages.

Construction

The transceiver is constructed on a single 6 by 5in PCB. The artwork, component layout and wiring diagram are shown in Figs 7.41 (see Appendix 1), 7.42 and 7.43 respectively. The PCB is double-sided – the top (component) surface being a continuous groundplane of unetched copper.

Without the facility to plate-through holes, some care needs to be taken that components are grounded correctly. Where a component lead is not grounded, a small area of copper must be removed from the groundplane, using a spot-cutter or a small twist drill. Where a component lead needs to be grounded, the

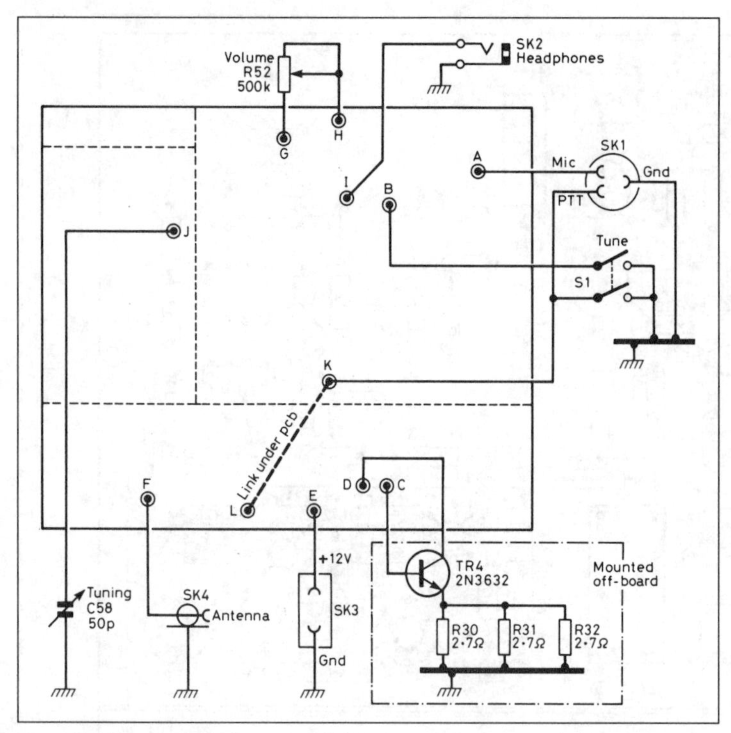

Fig 7.43. Wiring diagram

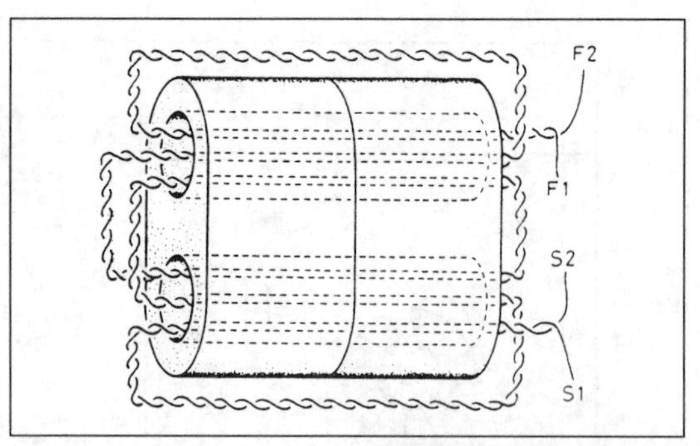

Fig 7.45. Winding details of T3 and T4. Connect S2 and F1 to form the centre tap. Note that the two wires are twisted together before winding. S1, F1: start and finish respectively of winding 1. S2, F2: start and finish respectively of winding 2. Core: two 2-hole cores stacked end-to-end

copper should not be removed and the lead should be soldered to the groundplane as well as to the pad on the underside. This is easy to achieve with axial-lead components (resistors, diodes etc) but can be difficult with radial-lead components. In most cases the PCB layout overcomes this by tracking radial leads to ground via nearby resistor leads. A careful look at the circuit diagram as each component is loaded soon shows what is needed.

Remember to put in a wire link between pins L and K, and in position LK1. Screened cable was used for connecting pins G and H to the volume control – connect the outer to pin H.

There are no PCB pads for C56 and C57, so these capacitors should be soldered directly across R4 and R6 respectively. TR4 must be adequately heatsinked as it dissipates almost 4W even under no-drive conditions. TR4 was boltedthrough the rear panel to a 1.5 by 2.5in finned heatsink. Resistors R30–R32 are soldered directly between the emitter of TR4 and the groundplane.

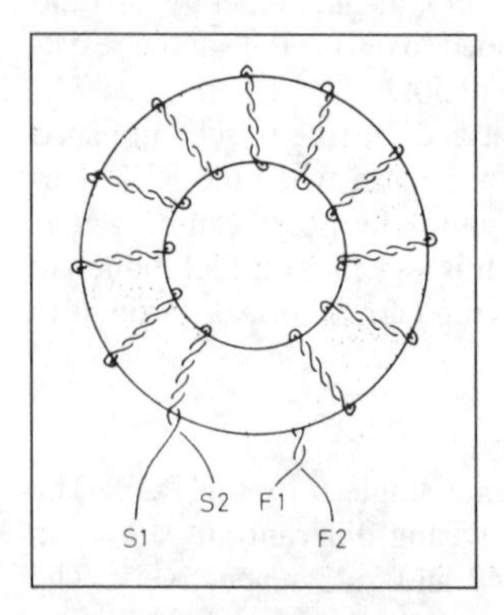

Fig 7.44. Winding details of T1 and T2. Connect S2 and F1 to form the centre tap. Note that the two wires are twisted together before winding. S1, F1: start and finish respectively of winding 1. S2, F2: start and finish respectively of winding 2. Core: 10mm OD ferrite toroid

It is important that the VFO coil L5 is mechanically stable. Ensure that it is wound tightly and fixed rigidly to the PCB; the coil was 'sandwiched' between two Perspex discs and bolted through the discs to the PCB. Also, be sure to use rigid heavy-gauge wire for connecting to C58.

The designer used a 6:1 vernier slow-motion drive which, with the limited tuning range of 100kHz, provides acceptable bandspread; the 0–100 vernier scale (0 = 1.900MHz, 100 = 2.000MHz) gives a surprisingly accurate read-out of frequency, the worst-case error being 1kHz across the tuning range.

The broad-band transformers, T1–T4, are wound by twisting together two lengths of 22swg enamelled copper wire. The twisted pair is then either wound on a ferrite toroidal core (T1 and T2), or wound through ferrite double-holed cores (T3 and T4). Identify the start and finish of each winding using an ohmmeter – connect the start of one wire to the finish of the other to form the centre tap (see Figs 7.44 and 7.45 for more details). All transformers and the band-pass filter coils were secured to the PCB with adhesive.

The designer fabricated all of the transceiver, other than the top and bottom panels, by soldering together double-sided PCB materials.

It is vital to have a good screen between the PA and the VFO otherwise the transmitter will frequency modulate badly. 2in high screens were used around the PA and VFO area, and a screen was included at the front of the VFO compartment on which to mount C58. If you use lower screens you may need to put a lid over the VFO; cut a tightly-fitting piece of PCB material and bolt it in position to four nuts soldered into the corners of the VFO compartment.

Alignment

Check the PCB thoroughly for correct placement of components and absence of solder bridges.

Turn the volume control fully counter-clockwise, the TUNE switch to the off position and R28 fully counter-clockwise. Connect the transceiver to a 12V supply and switch on. Check that the current drawn from the supply is about 50mA.

Check the frequency of the VFO either by using a frequency

Detail of top view with C58 removed to show mounting arrangement of L5

Table 7.10. 1.8MHz QRP transceiver components list

Resistor	Value
R1, 9, 17, 23, 35	1k
R2, 3, 41, 42	22k
R4, 6	47k
R5	4k7
R7, 8	100R
R10, 36, 44	10k
R11, 33	22R
R12, 14	150R
R13	39R
R15, 29	560R
R16, 50	270R
R18, 19, 24, 25, 47	10R
R20	180R
R21, 45	330R
R22	220R
R26	27R
R27	330R 0.5W
R28	100R preset
R30, 31, 32	2R7
R34, 51	47R
R37, 40	2k2
R38, 43	100k
R39	390k
R46	1k5
R48	820R
R49	12k
R52	500k log pot

Capacitor	Value
C1, 3, 5, 35, 36, 38, 44, 47	10µ 16V tant bead
C2, 6, 7, 8, 46, 51, 52, 54, 55	0.01µ ceramic
C4	2µ2 16V tant bead
C9, 37	47µ 25V tant bead
C10–22, 24, 32, 53	0.1µ ceramic
C23	100µ 25V elect
C25, 31	60–180p trimmer (Cirkit 06-18006)
C26, 30	56p silver mica
C27, 29	680p silver mica
C28	82p silver mica
C33	0.33µ
C34, 45	100µ 10V elect
C39, 43	0.068µ
C40	0.033µ
C41	0.1µ polystyrene
C42	8200p silver mica
C48	330p silver mica
C49	82p silver mica
C50	12p silver mica
C56, 57	0.001µ ceramic
C58	50p air-spaced variable, SLC law (Maplin FF45Y)

Component	Description
L1, 2	37t on T68-2 core tapped at 7t from ground
L3, 4	120mH (eg Cirkit 34-12402)
L5	57t on T68-6 core tapped at 14t from ground
RFC1	2t on small ferrite bead
RFC2	47µH choke
T1, 2	10t twisted wire on 10mm OD ferrite toroidal core Al = 1µH/t (eg SEI type MM622). See Fig 7.44.
T3, 4	4t twisted wire on two 2-hole ferrite cores. Al = 4µH/t (eg Mullard FX2754). See Fig 7.45.
TR1	BC179
TR2, 3	2N5109 or 2N3866
TR4	2N3632 (see text)
TR5, 6, 8, 9	BC109C
TR7	2N3819
D1	6.2V 250mW zener
D2	1N914
IC1, 2, 3	741 op-amp
M1	Mini-circuits SBL-1 double-balanced mixer
RLA, B	DPDT 12V relay (eg RS Electromail 346-845)
SK1	Microphone socket
SK2	Headphone socket
SK3	DC power socket (eg Maplin YX34M)
SK4	Antenna socket
S1	DPDT toggle switch

Miscellaneous

Slow-motion drive for C58 (eg Maplin RX40T)
Heatsink approx 1.5 by 2in
Knob for R52

Table 7.11. Bipolar transistor DC voltages (with 12.2V supply)

	Emitter	Base	Collector	Note
TR1	12.2	11.6	11.8	Tune switch operated
TR2	2.85	3.6	12	Transmit
TR3	1.4	2.15	11.6	Transmit
TR4	0.3	1	12.2	Transmit
TR5	11.2	11.8	12.2	
TR6	0.4	1	10.3	
TR8	0	0.65	6.75	
TR9	6	6.75	12	

Table 7.12. FET DC voltages (with 12.2V supply)

	Source	Gate	Drain
TR7	0	0	6.2

Table 7.13. AC voltages

Circuit node	AC voltage	Notes
TR7 source	2.6V p/p	1.8MHz RF
TR9 emitter	2.6V p/p	1.8MHz RF
Mic input	4mV p/p	Transmit audio
IC1 pin 6	200mV p/p	Transmit audio
IC2 pin 6	2.2V p/p	Transmit DSB RF
TR2 base	200mV p/p	Transmit DSB RF
TR4 collector	15V p/p	Transmit DSB RF
Ant (50Ω)	30V p/p	Transmit DSB RF

counter connected to the source of TR7, or by monitoring the VFO on another receiver. With C58 set to mid-position, the frequency should be about 1.95MHz; if it is very different, you can adjust L5 slightly by spreading or squeezing together the turns. Alternatively, major adjustments can be made by substituting alternative values for C49. Check that the range of the VFO is about 1.9 to 2.0MHz.

Plug in a pair of headphones and slowly advance the volume control; you should hear receiver noise (a hissing sound). If you have a signal generator, set it to 1.95MHz and connect it to the antenna socket; if not, you will have to connect the transceiver to an antenna and make the next adjustment using an off-air received signal. Tune to a signal at 1.95MHz and alternately adjust C25 and C31 for a peak in its level.

Connect the transceiver to a 50Ω power meter or through an SWR bridge to a 50Ω load. Plug in a low-impedance microphone and operate the PTT switch. Note the current drawn from the supply – it should be about 200mA. Slowly turn R28 clockwise and note that the supply current increases; adjust R28 until the supply current has increased by 330mA. Release the PTT switch and operate the TUNE switch; the power meter should indicate between 1 and 2W.

At this stage, final adjustments can be made to C25 and C31. Swing the VFO from end to end of its range and note the variation in output power. The desired response is a slight peak in power at either end of the VFO range with a slip dip at mid-range. It should be possible to achieve by successive adjustments to C25 and C31. For those of you lucky enough to have access to a spectrum analyser and tracking generator, LK1 was included to allow isolation of the band-pass filter.

If you have any problems, refer to Tables 7.11 to 7.13 which show typical AC and DC voltages around the circuit. If necessary, you can tailor the gain of IC2 to suit the sensitivity of your microphone by changing the value of R5.

Final thoughts

In retrospect it would have been useful to have included the low-pass filter (L3, L4, C39–C43) in the transmit audio path in order to further restrict the bandwidth. Normally the roll-off achieved by C57 and C56 combined with the low output power means that you are unlikely to cause problems for adjacent QSOs. However, when using a 200ft vertical antenna during portable operation, the transceiver puts out a potent signal and a reduction in bandwidth would then be more 'neighbourly'.

A CW facility could be added fairly easily using the TUNE pin as a keying point. You would need to add RIT (receive independent tune) facilities – probably by placing a varactor diode between TR7 source and ground. You might also consider changing to a band-pass audio filter rather than a low-pass audio filter in the receiver.

The transceiver can be adapted for other bands by changing the VFO components and the band-pass filter components – all other circuitry is broad-band. You will need to worry more about VFO stability as you increase frequency, and you may find the

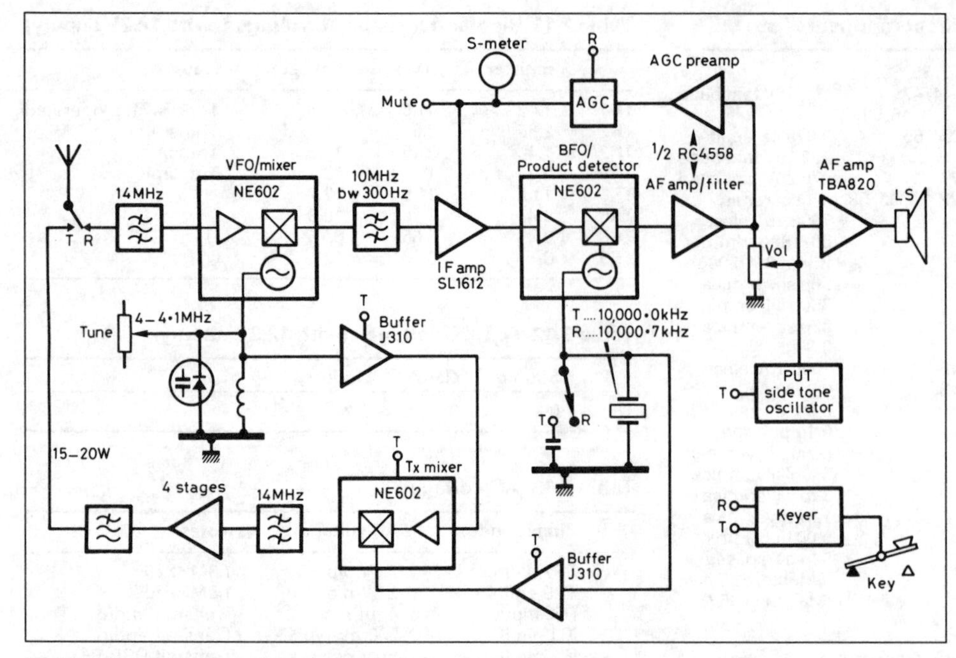

Fig 7.46. Block diagram of the complete transceiver

gain of the buffer falls – you can overcome this by decreasing the value of R44. The noise figure of the receiver is adequate for operation on the lower frequency bands but on 14MHz and above you will probably need a preamplifier. Those who enjoy experimentation might try changing the VFO to a VXO, adding a preamplifier to the receiver, and seeing if operation on 50MHz is possible!

Finally, it has been interesting to note that, despite theory, with careful tuning it is quite possible to resolve DSB signals on the direct-conversion receiver.

CW THE EASY WAY – A 14MHz CW TRANSCEIVER

This design by George Fare, G3OGQ, originally appeared in *Radio Communication* [4].

Introduction

There is no reason why it should be expensive to get on to the HF bands and work DX provided you are content to work CW. The transceiver presented here is capable of working worldwide DX under reasonable conditions and is quite small enough to be used portable.

It was conceived originally as a club project to enable new Class A licensees to get on the air easily and cheaply, giving them experience in construction at the same time.

Basic specification

As it is dedicated to CW, there are no compromises in the design as usually happens with multimode transceivers.

It includes a single-conversion superhet receiver with a home-made crystal filter and audio response tailored for CW. Sensitivity is 1μV for a S/N ratio of 12dB. The transmitter has an output power of 15 to 20W and incorporates semi-break-in.

Coverage is 14 to 14.1MHz, but this could quite easily be changed to most other HF bands by changing the VFO, the two band-pass filters and the output circuit.

Use of a superhet receiver does not necessarily introduce difficulties in construction compared with a direct conversion receiver. In fact it is much easier to produce good (ie single-signal) selectivity and there is a complete absence of microphonics which are sometimes difficult to suppress in a direct conversion design.

Incorporation of relatively cheap ICs and standard crystals make this design very cost effective and there is no need for expensive test equipment to align it.

Design in detail

A block diagram of the transceiver is given in Fig 7.46 and the circuit diagram in Fig 7.47.

It is based on the use of NE602 ICs which contain an oscillator, buffer and double balanced mixer in an eight-pin DIL package. They are readily obtainable and all three ICs cost about the same as a single SBL-1 mixer. The chip also contains an input amplifier which typically gives a gain of 20dB at 8V. The third order input intercept point is −12.5dBm (which is approximately +5dBm output intercept) and offers a conversion gain of 14dB minimum.

The IC is designed for optimum low-power performance and there is a limitation on its strong-signal performance owing to the presence of the RF amplifier. It was always intended during the design of the receiver to incorporate a switched attenuator in the front end but, in practice, this has not been found to be necessary and the idea was dropped.

On receive, the signal is passed through a band-pass filter to the combined mixer and VFO. The VFO is tuned by means of a varicap diode and has proved to be exceptionally stable. A variable capacitor could, of course, be substituted if desired. The VFO runs at 4MHz giving the IF of 10MHz – chosen because 10MHz crystals are usually accurate, readily available and cheap.

The filter following the mixer is a lower-sideband ladder filter consisting of two crystals with input and output matching and has a bandwidth of about 300Hz. Three crystals should be purchased at the same time, the third being used in the BFO.

After passing through the filter, the signal is amplified by a SL1612 which is AGC controlled. This amplifier is cut off on transmit by a voltage applied to the AGC pin.

After amplification, the signal is passed to the second NE602 which acts as a combined IF amplifier, product detector and BFO. The BFO operates at 10MHz on transmit and is shifted by about 700Hz on receive. In order to be compatible with commercial transceivers, it is necessary for the BFO to be above the transmit frequency on receive. As the filter is not symmetrical (the sharpest cut-off being in the high-frequency side) this is in any case preferable as we can arrange for the transmit frequency to be at the IF with the receiver BFO above the passband of the filter.

Audio from the NE602 is then amplified and filtered by one half of a dual op-amp (RC4558) which is configured to give a gain of 20, a Q of 10 and a centre frequency of 750Hz. The main purpose of this filter is to cut down the noise from the wide-band IF amplifier, as the main selectivity is governed by the crystal filter.

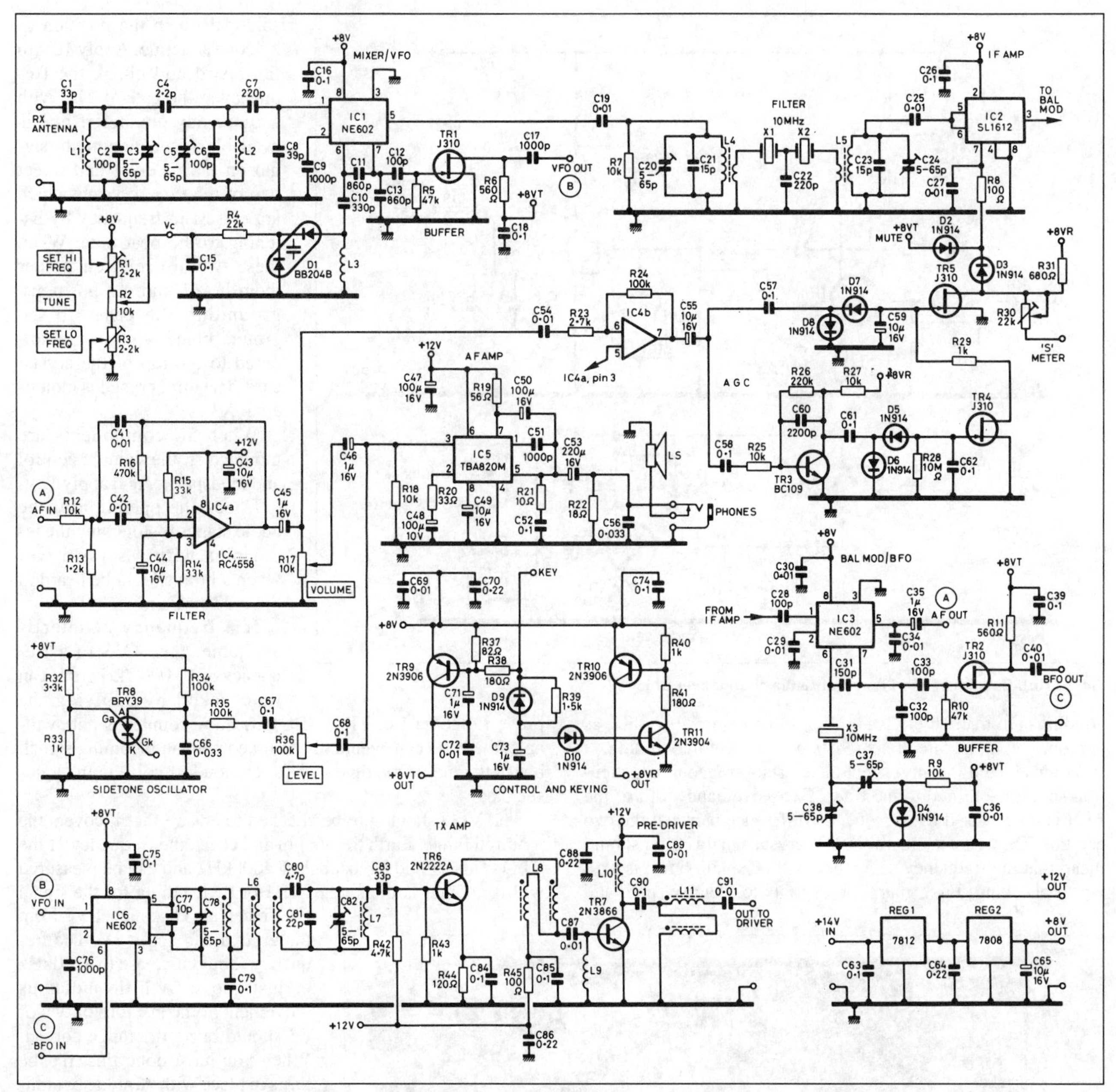

Fig 7.47(a). Circuit of the main board for the 14MHz CW transceiver

A further amplifier follows which increases the audio to loudspeaker level.

AGC is audio derived and is taken from the output of the first half of the dual op-amp and amplified by the second half. Time constants of the AGC system are tailored to CW and, although only one stage of amplification is controlled by the AGC, it is nevertheless quite effective.

On transmit, the VFO and BFO signals from the appropriate NE602 are buffered and fed to a third NE602 which acts only as a mixer to produce a 14MHz signal. A band-pass filter follows and then four stages of amplification boost the signal to the required output. The final amplifier has a T-section filter in the output and harmonic radiation is well within acceptable limits.

Keying is performed by switching on the VFO and BFO buffers and the transmit mixer, at the same time muting the receiver IF amplifier. A sidetone oscillator is also switched on by the transmit voltage and the output is fed to the audio amplifier. All other stages except the AGC remain on whether in transmit or receive, thus making break-in keying feasible. The only speed limitation is the use of a relay for antenna changeover which is separately keyed with a time constant which prevents the relay changing over between dots.

Construction

Except for the driver and PA stages, and antenna changeover, all the circuitry is contained on one main board measuring 5in × 4.5in. PCB layout is given in Fig 7.48 (see Appendix 1).

The board is double sided, the top surface being used as a ground plane with all grounded components soldered direct. Mount the components as shown in Fig 7.49 except for the three crystals. (Note also the three links between the tracks and ground plane.) When winding toroids, space the windings around the core to occupy about 270° and count the number of outside loops to ascertain how many turns have been applied. Bifilar and trifilar

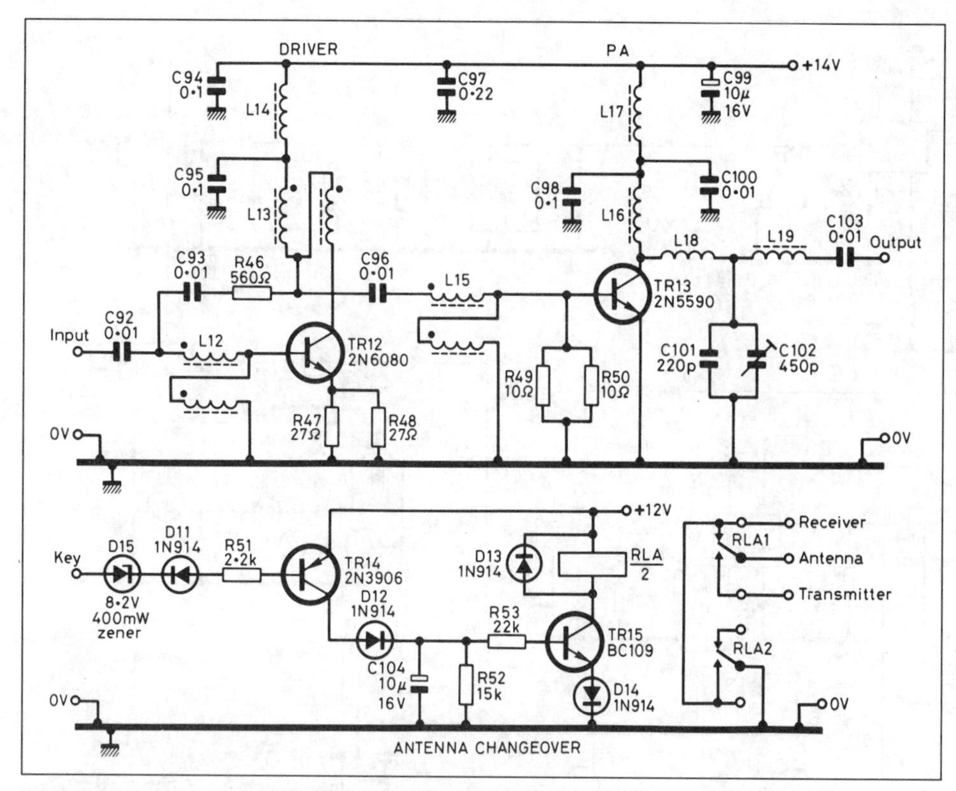

Fig 7.47(b). Circuits of the PA and antenna changeover unit

windings are made by twisting two or three wires together, as appropriate, at the rate of six twists per inch before winding.

When all components except the crystals are mounted, attention should be turned to the filter. Excessive bandwidth of the filter can be caused by a frequency difference between the two crystals. The use of standard 10MHz crystals in this filter should mean that the frequency difference will be small, but it is well worth checking. First, mark the crystals to identify them and solder them in the position of X3, one at a time. Apply 12V to the board and check the frequency at the gate of TR2 with a frequency counter if one is available. If not, monitor the signal on the receiver and select two of the three crystals which are closest in frequency by listening to the beat note. When these two are selected, solder them in X1 and X2 positions, grounding the cases to the ground plane with a wire soldered to the top of the crystal case. The third crystal is mounted at X3.

When all components are mounted fit the volume control and loudspeaker and apply 12V.

Check the operation of the audio stages by touching the input pins with a screwdriver when a hum should be heard in the loudspeaker.

If a frequency counter is available, tune C38 to a frequency of 10,000.7kHz (this can be checked by applying the probe to the gate of TR2). If a frequency counter is not available, the BFO can eventually be tuned with an incoming signal, but in the meantime tune C38 for the loudest noise in the loudspeaker.

The VFO should not be checked to ensure that it covers the required range. Earth the Vc pin and check the frequency of the VFO. This should be just below 4000kHz and can be measured with a counter at the gate of TR1 or by listening for the signal on a receiver. Disconnect Vc from the ground and apply 8V. The frequency should increase to 4100kHz or just above. A little judicious movement of the last turn or two of L3 should bring the range correct. When you have done this, fix the turns in place with Araldite or similar adhesive. The tuning potentiometer and its trimmers (R1 and R3) may now be wired in.

Attaching an antenna to the receiver input pin (a short length of wire will do) should produce audible signals. Peak C3 and C5 for the loudest signal and then tune the filter. In the absence of measuring equipment, the filter can be tuned by means of C20 and C24 for the loudest signal consistent with a low level of background noise. Another rather elegant method is to connect an antenna to pin 5 of IC1 and tune the filter for reception of 10MHz standard frequency signals. As already mentioned, one of the characteristics of a ladder crystal

Interior view of the transceiver. The PA board is mounted vertically on the right with its heatsink and the antenna changeover board is mounted vertically near the input band-pass filter. Note that the VFO is positioned at the opposite end of the cabinet from sources of heat such as the PA board and regulators. The photo is that of an early version. The AGC and tone oscillator stages have since been revised but the general layout is identical

filter is that the sharpest cut-off occurs on the high-frequency side, forming a lower sideband filter.

The shape of the passband is determined by the size of C22 and correct matching at the input and output, and the object is to make the passband above 10MHz as narrow as possible so as to produce only one signal. The trick therefore is to tune C20 and C24 until only one response is heard. This all sounds much more complicated than it really is and a couple of hours' patient work will be well rewarded. Final adjustment is probably best carried out when the whole receiver is complete, using 14MHz signals, to ensure that each signal is only heard once.

The AGC action should now be checked. With no signal, there should be about 0.3V on pin 7 of the SL1612. Injecting a strong signal or listening to a strong signal should increase the AGC voltage. You can then calibrate the S-meter by adjusting R30 either subjectively by listening to a signal which you judge to be S9, or by setting S9 for a 50µV input at the antenna terminal.

Attention should now be directed to the transmit side.

Fit a 47Ω resistor between the output pin and ground earth the KEY pin. The sidetone should be heard from the loudspeaker and the level can be adjusted by R36. Adjust C37 for a frequency of 10,000kHz, measured at the gate of TR2.

The RF output can be checked by means of an RF voltmeter or an oscilloscope; failing that, the signal can be monitored on a receiver.

Peak up trimmer C78 and C82 for maximum output which should be about 6V peak-to-peak or 2.1V RMS. The two trimmers interact slightly and the adjustment should be performed two or three times.

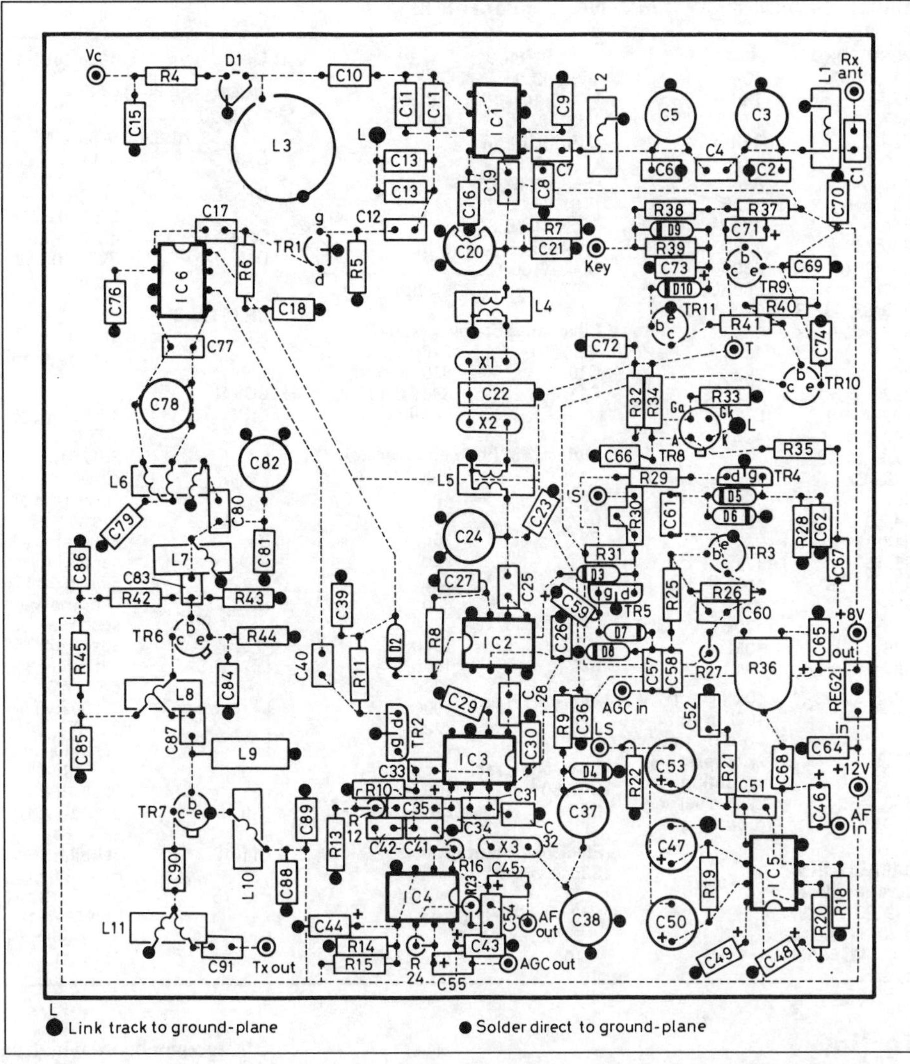

Fig 7.49. Main board component layout. L: track to ground plane; O: solder to ground plane

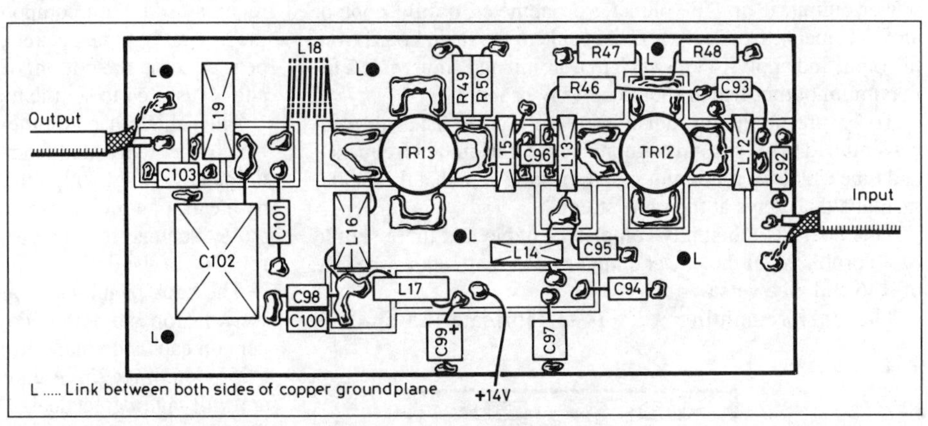

Fig 7.50. PA board component layout. L: link between both sides

The main board is now complete and the driver and PA board should be constructed next.

The components layout is given in Fig 7.50 and the PCB track layout in Fig 7.51 (see Appendix 1). Drill 10mm diameter holes in the transistor positions and link the two sides together by drilling six holes where shown and solder a wire link through.

After making the PCB, it should be mounted on its heatsink using the transistors to sandwich the board. The transistors should then be soldered in place making sure that there is no strain, particularly upwards, on the leads. Heatsink requirements are quite modest as the PA runs at 75% efficiency and a 6.2°C/W heatsink is sufficient.

Following that, all other components are mounted and soldered direct to the surface of the PCB. Keep all leads as short as possible.

Table 7.14. 14MHz CW transceiver components list

RESISTORS	
R21, 49, 50	10R
R22	18R
R47, 48	27R
R20	33R
R19	56R
R37	82R
R8, 45	100R
R44	120R
R38, 41	180R
R6, 11, 46	560R
R31	680R
R29, 33, 40, 43	1k
R13	1k2
R39	1k5
R51	2k2
R23	2k7
R32	3k3
R42	4k7
R7, 9, 12, 18, 25, 27	10k
R52	15k
R4, 53	22k
R14, 15	33k
R5, 10	47k
R24, 34, 35	100k
R26	220k
R16	470k
R28	10M
All fixed resistors ¼W.	
R2, 17	10k log pot
R36	100k preset, horiz mounting
R1, 3	2k2 preset, vert mounting
R30	22k preset vert mounting
CAPACITORS	
Ceramic plate	
C19, 25, 27, 29, 30, 34, 36, 40, 41, 42, 54, 69, 72, 87, 89, 91, 92, 93, 96, 100, 103	0.01µ
C56, 66	0.033µ
C15, 16, 18, 26, 39, 52, 57, 58, 61, 62, 63, 67, 68, 74, 75, 79, 84, 85, 90, 94, 95, 98	0.1µ
C64, 70, 86, 88, 97	0.22µ
Silver mica or polystyrene	
C22, 101	220p
C10	330p
C11, 13	390+470p in parallel
Mullard 682/630 series ceramic	
C4	2p2
C80	4p7
C77	10p
C21, 23	15p
C81	22p
C1, 83	33p
C8	39p
C2, 6, 12, 28, 32, 33	100p
C31	150p
C7	220p
C9, 17, 51, 76	1000p
C60	2200p
Electrolytic radial mount	
C47, 50	100µ 16V
C53	220µ 16V
Electrolytic tant bead	
C35, 45, 46, 71, 73	1µ 16V
C43, 44, 49, 55, 59, 65, 99, 104	10µ 16V
C48	100µ 10V
TRIMMERS	
C3, 5, 20, 24, 37, 38, 78, 82	5–65p
C102	450p
DIODES	
D1	BB204B
D2–D14	1N914 etc
D15	8V2 400mW zener
CRYSTALS	
X1, 2, 3	10MHz HC18/U 32p load cap
COILS	
L1, 2	15t 30swg on T37-6
L3	33t 26swg closewound 10mm dia on former
L4	27t primary 3t secondary 30swg on T50-6
L5	6t primary 27t secondary 30swg on T50-6
L6	12t trifilar 30swg on T37-6
L7	23t 30swg on T37-6
L8	10t bifilar 30swg on FT37-61
L9	22µH RFC
L10	20t 30swg on FT37-61
L11, 12	7t bifilar 30swg on FT37-61
L13, 15	6t bifilar 30swg on FT37-61
L14	9t 30swg on FT37-61
L16	7t 26swg on two FT37-61
L17	7t 26swg on FT37-61
L18	8t 24swg closewound 5/16in dia
L19	9t 24swg on T80-6
All coils except L3, L9 and L18 wound on Amidon toroidal cores.	
TRANSISTORS AND FETS	
TR1, 2, 4, 5	J310
TR3, 15	BC109
TR6	2N2222A
TR7	2N3866
TR8	BRY39
TR9, 10, 14	2N3906
TR11	2N3904
TR12	2N6080
TR13	2N5590
INTEGRATED CIRCUITS	
IC1, 3, 6	NE602
IC2	SL1612
IC4	RC4558
IC5	TBA820M
VOLTAGE REGULATORS	
REG1	7812
REG2	7808
MISCELLANEOUS	
2 jack sockets	
1 loudspeaker 8Ω	
1.6:1 slow-motion drive	
1 SO239 socket	
1 power plug and socket	
1 heatsink (Redpoint 2Y)	
1 meter, 100µA	
1 relay 12V (Radiospares 346-845) DPCO	

Testing

To test, fit a dummy load and some means of measuring the power output, ie an RF voltmeter or ammeter, oscilloscope or an SWR meter. Connect a coaxial cable to the main board from the input and apply 14V. Switch to transmit and adjust C102 for maximum output which should be 15W or more.

To ensure minimum radiation of harmonics, listen to a receiver tuned to the second harmonic of the transmitted frequency and tune C102 for minimum signal, making sure that the fundamental signal stays at the same strength.

This board is inherently completely stable and there should be no problems. If the power output is too great, reduce the size of R46 and vice versa.

The only remaining work is to fabricate the antenna changeover board which can be built on a piece of Veroboard or to the track layout in Fig 7.52 (see Appendix 1) if the specified relay is used, with component layout at Fig 7.53, and connect the key and phone sockets, tune control and trimmers. R3 can be used to set the bottom of the range to 14,000kHz on transmit and R1 is used to set the top of the range to 14,100kHz.

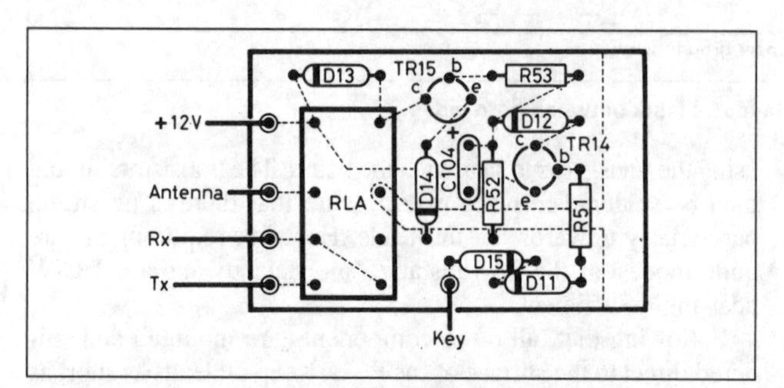

Fig 7.53. Component layout for the antenna changeover

The cabinet used on the prototype was home-made with the front panel fabricated from double-sided PCB. The tune potentiometer is fitted with a 6:1 reduction drive which is calibrated by means of a receiver or counter. The tuning is not linear but is quite adequate for this purpose as we are only tuning 100kHz for 270° of the dial.

The back panel was constructed from 16swg aluminium as were the top and bottom cover which are bolted to 1in high runners on each side made from PCB material. The overall size is 5.25in × 6.5in × 2in. A commercial case can, of course, be used and will save a great deal of time and effort, but it must be metal. Plastic cases are not suitable. For base station use, a larger case could be used incorporating a PSU.

Power requirements

Power requirements are quite modest, 110mA being needed on receive and 2.25A on transmit with key held down. 14V are needed to produce a well-regulated 12V from the 7812 regulator, although a 13.8V supply has been successfully used. The supply should be well regulated or the CW signal will be degraded.

Components

All components are readily available from a number of sources. The driver and PA transistors offer considerable latitude; almost any NPN transistor capable of 2W output can be used as a driver: BGQ34, 2N5589, 2N-5590 have all been successfully used instead of the 2N6080.

For the final amplifier MRF-449A, MRF450A, 2N6084, BLW60, PT9796A, ASO-12 devices have all been tried and work satisfactorily, the only problem being the method of mounting.

A BFY90 will perform as well as the 2N3866. Where a BC109 is specified, almost any NPN transistor will do and MPF102 can be used in place of J310. Any PNP transistor capable of passing 200mA may be used in place of the 2N3906.

Polystyrene capacitors may be used in place of silver mica.

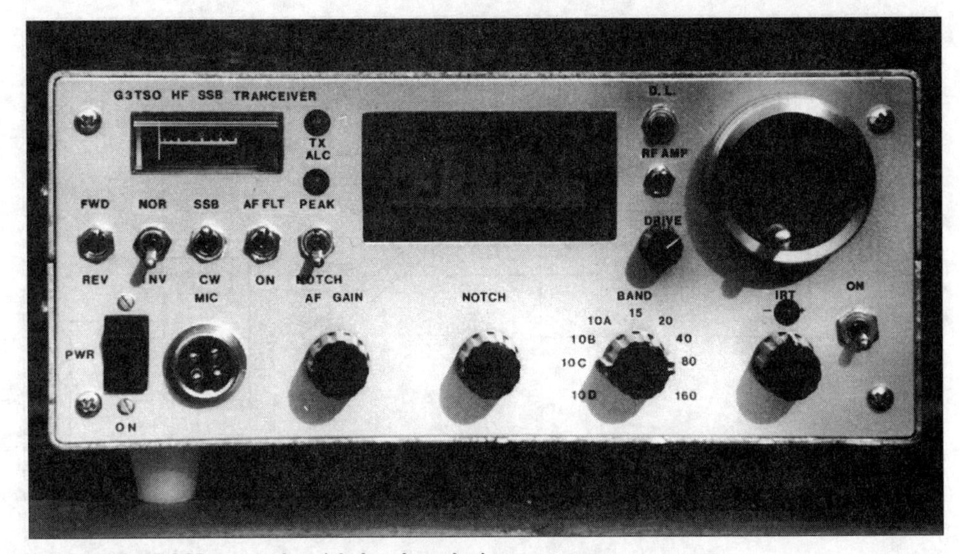

The G3TSO HF SSB transceiver (six-band version)

Performance

The CW waveform is a bit hard, but no clicks can be detected when monitored on a receiver nor have any been reported in QSO. This is due to exaggerated shaping of the rising waveform in the keying circuit which the Class C transmitter stages tend to straighten, the result being an acceptable waveform.

The presence of a 'built-in' RF amplifier as part of IC1, as already mentioned, can lead to some intermodulation distortion. If you have very strong local QRM, it might be necessary to introduce an attenuator (just like the TS940S) before the input band-pass filter.

If will, however, come as a very pleasant surprise to you to find how much easier it is to work DX with relatively low power using CW than it is with SSB. There is no point in listing countries worked but, as an example, WAC was made in one operating session. The sensitivity of the receiver is excellent – well up to the standard of much more expensive receivers – and operating convenience is very good. All that is required to go from receive to transmit is to press the key!

It should go without saying that an ATU or an antenna with an impedance of 50Ω should be used for best performance, or else the input band-pass filter will not perform properly. The transmitter won't be correctly matched either and so the harmonic output will rise. If it is impossible to correctly match the antenna, for example in a mobile or portable situation, then an additional low-pass filter is desirable between the output and the antenna.

A MODULAR MULTIBAND TRANSCEIVER

This section is an updated version of an article by Mike Grierson, G3TSO, in *Radio Communication* [5].

Introduction

Construction of a multiband SSB transceiver is a project likely to deter the most ardent constructor; however, by building a series of small modules the apparent complexity of such a project can be greatly reduced. With careful design, individual modules can be constructed, tested and aligned where necessary before being brought together as a project exhibiting a high degree of sophistication.

One of the major stumbling blocks encountered by anyone designing or building multiband equipment is the process of band changing. Traditionally, large rotary switches are used; these often reach from the front to the back of the equipment, with numerous wafers switching a multitude of different circuits. The switches are hard to find, difficult to wire up and impose serious limitations on the mechanical layout and construction of the equipment.

This switching problem can now largely be overcome by the use of diode and relay switching. Individual circuits can be switched remotely by a single DC voltage taken from a 12-way wafer switch. The layout of modules is thus independent of the switch location and so each module can be mechanically self-contained, making testing and alignment a simple process. Equipment can be constructed to suit the individual requirements of the constructor and may either be spread out into a large chassis or packed neatly into a high-density package in order to produce miniature equipment. Modules also permit considerable flexibility in future development or modification of equipment as well as making servicing somewhat easier.

With the increasing cost and complexity of commercial amateur radio equipment, home construction is once again becoming a viable proposition. Components are available from a number of mail order suppliers as well as at numerous rallies. Use of modern components, broad-band techniques and integrated circuits makes construction and alignment considerably easier than it was two decades ago when home construction was more commonplace.

The choice of a 9MHz IF enables a single VFO operating from 5.0 to 5.5MHz to provide coverage of both the 14MHz and 3.5MHz bands with a minimum of switching. Local oscillator injection for multiband operation is achieved by mixing the 5MHz VFO signal with the output from a switched crystal oscillator, and this is then filtered and amplified to provide the +7dBm required by the diode ring mixer on the exciter board.

The kit PA supplied by Cirkit Holdings is designed to give 10W continuous output over the frequency range 1.5 to 30MHz and employs a pair of 2SC1945 PA transistors. They are rated at 15W per device and are capable of providing up to 20W output for SSB or CW operation with less than 2mW of drive. This is suitable for low-power operation as a PA, or it can be used to

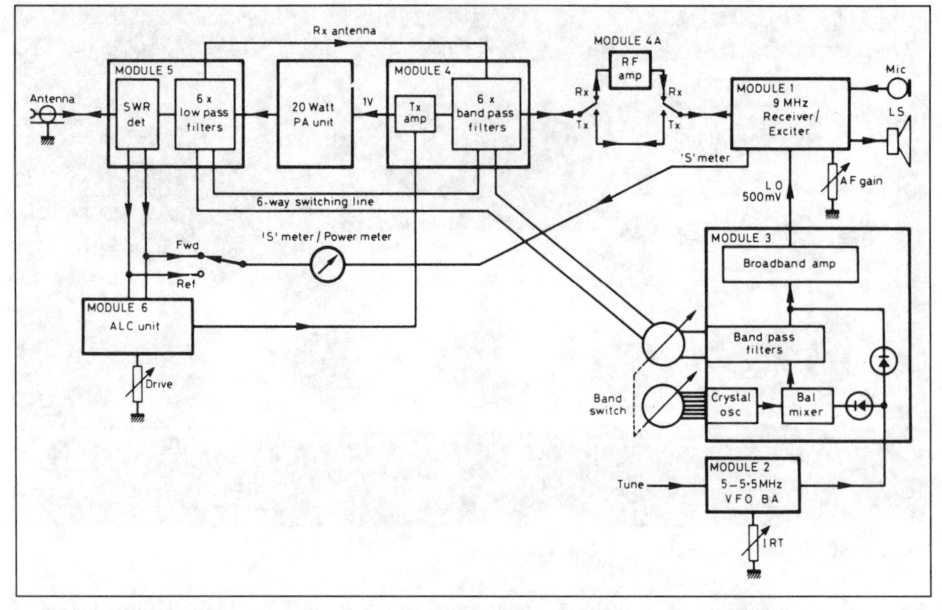

Fig 7.54. How the separate modules for the multiband transceiver link together. Each of the six basic units can be built and fully tested prior to final assembly.

drive a solid-state or valve linear amplifier to considerably higher output.

Bandswitching has been achieved by using low-capacitance switching diodes (any TV front-end switching diode will do); signals are able to pass through those diodes that are biased on from a 13V supply and are confronted by a high-impedance path by those diodes that are biased off. This is a technique used in almost all modern commercial equipment and enables band-switching to be effected with a single wafer bandswitch linked by ribbon cable to the various modules that require band-switching. In practice, a second wafer is used on the bandswitch to enable selection of the appropriate oscillator crystal. The low-pass filters used in the transmitter output path are switched using miniature relays as much larger currents are present in this part of the circuit.

A SWR detector is included in the design. This provides both forward and reverse ALC voltages used to control the output power on transmit and provide final protection under high SWR conditions. Meter indications of FWD and REF power are displayed on a combined S-meter.

The circuit

A block diagram of the basic multiband transceiver is illustrated in Fig 7.54.

For simplicity this will be broken down into a series of modules, each of which can be constructed and tested in its own right. All are constructed on relatively small PCBs designed to complement the dimensions of the exciter module.

Copyright notice

Please note that the annotated printed circuit layouts supplied for the G3TSO Modular Transceiver are copyright © 1992 by Mike Grierson, G3TSO. Individual radio amateurs may make copies of the boards for their own personal use, individually or for group projects subject to the following conditions:

The PCBs may not be made in commercial quantities, sold,

Table 7.15. G3TSO transceiver performance

GENERAL SPECIFICATION
Single-superhet transceiver.
Bandwidth: 2.4kHz @ –6dB, 4.3kHz @ –60dB.
Power output: variable from 100mW to 20W PEP (SSB), 20W (CW) 100% duty cycle.
Receiver sensitivity: 0.25µV for 10dB S+n/n (28MHz).
AGC: typically 3dB change of output for 80dB change of input.
RF amp gain: 15dB.
IF: 9MHz.
IF rejection: greater than 60dB.
AF output: 1W.

MEASURED PERFORMANCE
Local oscillator unit

Band (m)	LO freq (MHz)	2nd harmonic (dB)	Spurious responses
160	11	–45	Greater than –55dB
80	5	–30	Greater than –55dB
40	16	–32	1 @ –35dB (2 × xtal – VFO), others greater than –60dB
20	5	–30	Greater than –55dB
15	12	–35	Greater than –50dB
10A	19	–33	1 @ –42dB, others greater than –50dB
10B	19.5	–33	1 @ –47dB, others greater than –50dB
10C, 10D	20	–33	Greater than –50dB

Receiver performance
S-meter calibration using Cirkit 200µA S-meter adjusted for FSD at max AGC voltage:
FSD typically 100mV PD.
50µV PD gives S8 indication on 160, 80, 15 and 10m bands and S7½ indication on 40 and 20m.
Receiver sensitivity: 0.28µV for 10db S+n/n (14MHz); 0.25µV for 10dB S+n/n (28MHz).

Receiver spurious responses
The following internally generated spurious responses can be heard on the receiver:
80m: A weak response occurs at 3.6MHz when the VFO is operating at 5.4MHz. This also produces a similar out-of-band response on 20m at 14.4MHz.
15m: Band-edge birdie just out of band at 20.997MHz owing to 3rd harmonic of 7MHz xtal. 4th harmonic of VFO produces a response at 21.333MHz.
10m: Band-edge birdie just out of band at 27.997MHz owing to 2nd harmonic of 14MHz. Similar birdie from 14.460MHz xtal occurs at 28.92MHz. However, 10C range overlaps this frequency.

Second-channel interference
80m: Minimum detectable signal on 14MHz: 1mV PD.
20m: Minimum detectable signal on 3.5MHz: 300µV PD.

Each band was subjected to a 3mV signal across the HF spectrum and no spurious responses were found except for:
15m: 1 response at 23MHz which disappeared at 300µV PD.
10m: 3 responses at 26.5, 27.7 and 30MHz which disappeared at 300µV PD.

TRANSMITTER OUTPUT

Band (m)	2nd harmonic (dB)	3rd harmonic (dB)	Spurii (dB)
160	–33	–42	Greater than –50
80	–38	–42	Greater than –60
40	–42	–60	Greater than –50
20	–50	–32	Greater than –50
15	–42	–22	Greater than –50

Measured at 18W output into 50Ω.

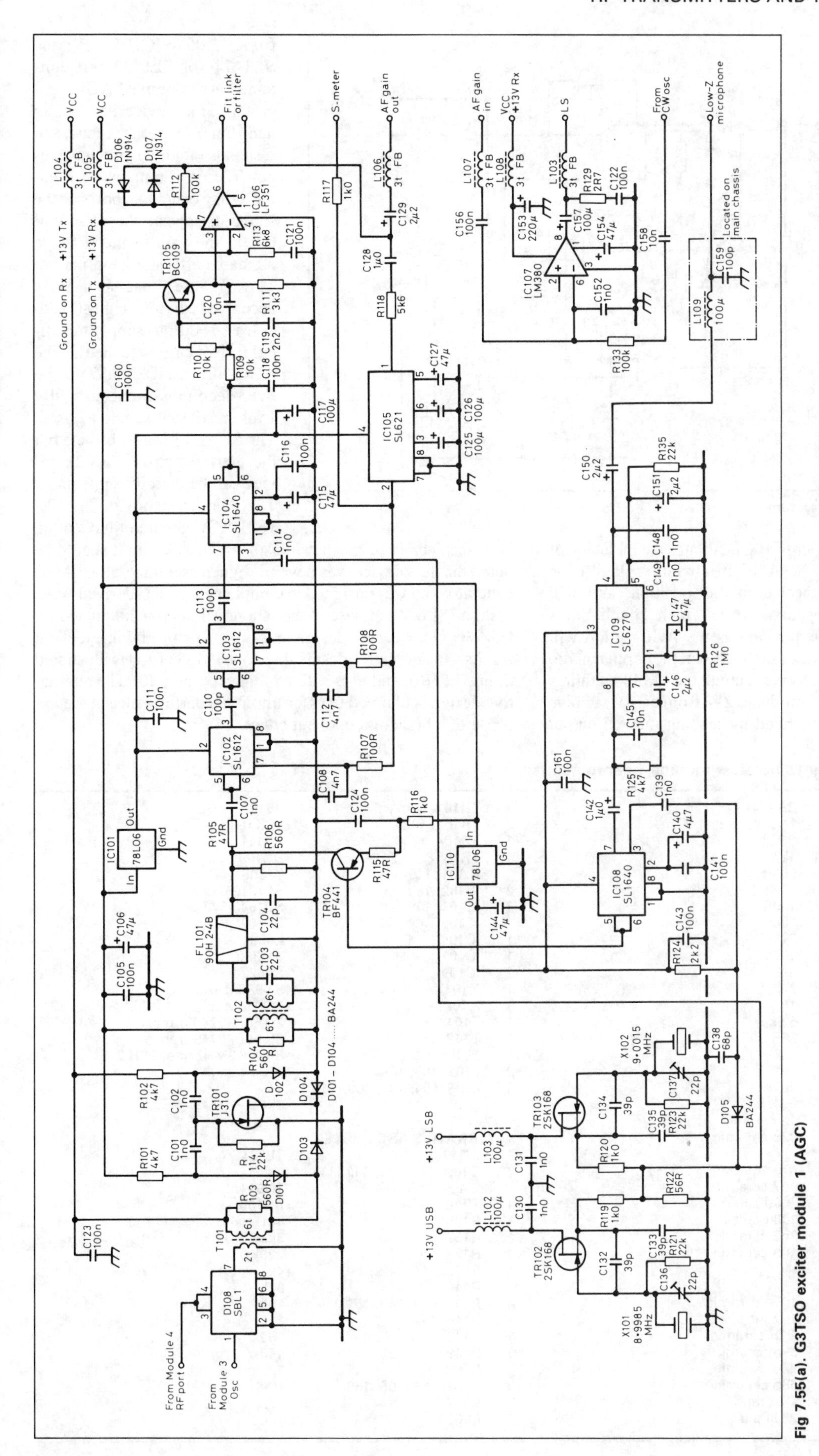

Fig 7.55(a). G3TSO exciter module 1 (AGC)

or advertised, either individually or as part of a kit of parts, without the written consent of the copyright owner.

Module 1: transceiver exciter board

The heart of the G3TSO Modular Transceiver is a self-contained 9MHz transceiver module based on a design by J Bryant, G4CLF, of Plessey Semiconductors. It utilises the famous Plessey SL1600 series communications ICs.

Fig 7.55(a) shows the circuit of the transceiver module. On receive, signals are fed to the input of an SBL1 ring mixer (MD-108) where they are mixed with a 500mV RMS local oscillator signal. Note: the RF and IF ports of the SBL1 are reversed in this design to permit operation below 1MHz. The resulting IF signal is amplified by TR101, a J310 bi-directional amplifier, before being filtered in an eight-pole 9MHz SSB filter. The J310 passes a heavy bias current in order to achieve a good intermodulation characteristic and may appear warm to the touch – some samples may suffer from thermal runaway. After filtering, the IF signal is amplified by a pair of cascaded SL1612 (or SL612) IC amplifiers having approximately 68dB gain.

IC104, an SL1640 or SL1641 product detector, heterodynes the IF with a local oscillator TR102 or TR103 depending upon which sideband is required. The resulting audio signal is fed to TR105, an active audio low-pass filter, and is then amplified further by IC106, an FET op-amp. Diodes D107 and 108 provide an element of noise or crash limiting. A small plug and socket inserted in

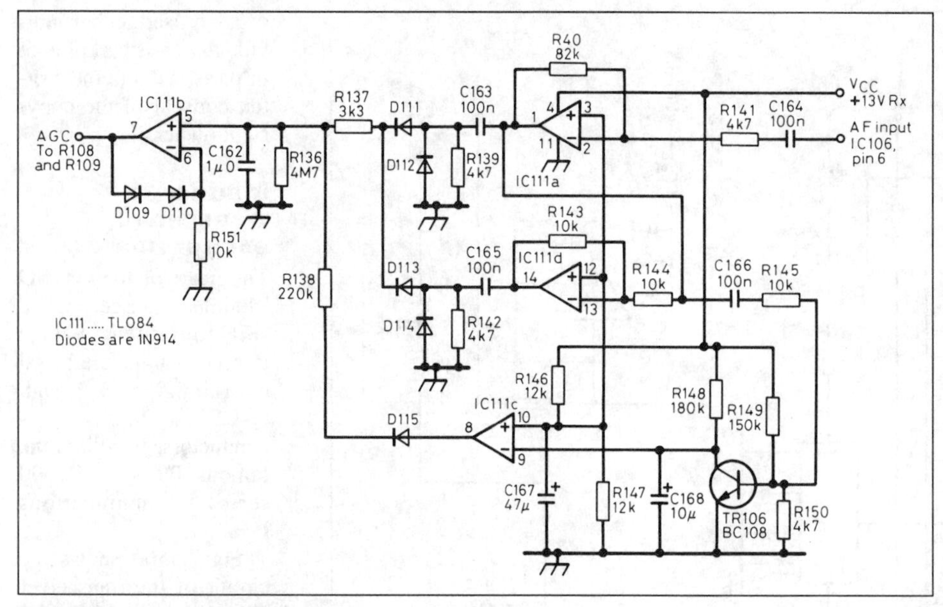

Fig 7.55(b). Alternative AGC system

the audio line immediately after IC106 facilitates the inclusion of an active audio CW filter; the AGC bandwidth will also be reduced with the filter connected in this position, and will characterise the sharp tuning of a crystal filter. Notch filters should not be located in this position as the AGC action will seriously degrade the depth of the notch. Most of the popular audio output ICs can be used in the receiver output; the LM380 requires few external components and produces 2W from a 13V supply.

Audio-derived AGC is generated by feeding the AF output from IC106 to IC105, a Plessey SL1621 (or SL621) self-contained audio-derived AGC generator. This device has a long hang time, about one second, and responds rapidly to an input signal. The AGC remains very stable during operation, but once the signal disappears the noise will rise as the gain of the IF stage increases after the hang period. The 6V line must be heavily decoupled adjacent to the IC with at least 100μF to supply the initial rush of current to the IC. The AGC line itself should not be over-decoupled, and normally 15nF maximum should be used. The SL1621 is now obsolete but the more expensive SL621 device can be directly substituted. The characteristics of the SL-1621 AGC generator have often been the subject of much criticism and discussion; it should be noted that this device was never designed as a stand-alone AGC generator and was designed to complement an RF-derived AGC system (SL624). Its use alone is a novel approach adopted by G3ZVC in his earlier design and is capable of giving excellent results. An alternative design developed by G3TXQ is illustrated in Fig 7.55(b), and uses a TL084 quad op-amp IC. The alternative design is claimed to offer improved performance at the expense of a higher component count.

Table 7.16. G3TSO modular transceiver module 1 components list

Qty	Components	Value
5	D101, 102, 103, 104, 105	BA244
2	D106, 107	1N914
1	D108	SBL1
3	R101, 102, 125	4k7
3	R103, 104, 106	560R
2	R105, 115	47R
2	R107, 108	100R
2	R109, 110	10k
1	R111	3k3
2	R112, 133	100k
1	R113	6k8
4	R114, 121, 123, 135	22k
4	R117, 116, 119, 120	1k
1	R118	5.6k
1	R122	56R
1	R124	2k2
1	R126	1M
1	R129	2R7
10	C101, 102, 107, 114, 130, 131, 139, 148, 149, 152	1n ceramic
3	C103, 104, 137	22p ceramic
6	C106, 115, 127, 144, 147, 154	47μ 16V tant
2	C108, 112	4n7 ceramic
3	C110, 113, 159	100p ceramic
4	C117, 125, 126, 157	100μ tant
1	C119	2n2 ceramic
3	C120, 145, 158	10n ceramic
13	C123, 105, 111, 116, 118, 121, 122, 141, 143, 156, 160, 161, 124	100n ceramic
2	C128, 142	1μ tant
4	C129, 146, 150, 151	2.2μ ceramic
4	C132, 133, 134, 135	39p ceramic
1	C136	22p ceramic
1	C138	68p ceramic
1	C140	4.7μ tant
1	C153	220μ alu
1	TR101	J310
2	TR102, 103	2SK168
1	TR104	BF441
1	TR105	BC109
2	IC101, 110	78L06
2	IC102, 103	SL1612/612
2	IC104, 108	SL1640/640
1	IC105	SL1621/621
1	IC106	LF351
1	IC107	LM380
1	IC109	SL6270
1	X101	8.9985MHz
1	X102	9.0015MHz
1	T101	2t+6t FX2249 transf
1	T102	6t+6t FX2249 transf
1	FL101	8-pole 9MHz filter 90H 2.4B
3	L102, 103, 104	100μH axial
6	L103, 104, 105, 106, 107, 108	3t FB FX1115
ALTERNATIVE AGC SYSTEM		
1	IC111	TL084
7	D109, 110, 111, 112, 113, 114, 115	1N914
1	TR106	BC108
1	R136	4M7
1	R137	3k3
1	R138	220k
4	R139, 141, 142, 150	4k7
1	R140	82k
3	R144, 143, 145, 151	10k
2	R146, 147	12k
1	R148	180k
1	R149	150k
1	C162	1μ
4	C163, 164, 165, 166	100n
1	C167	47μ
1	C168	10μ

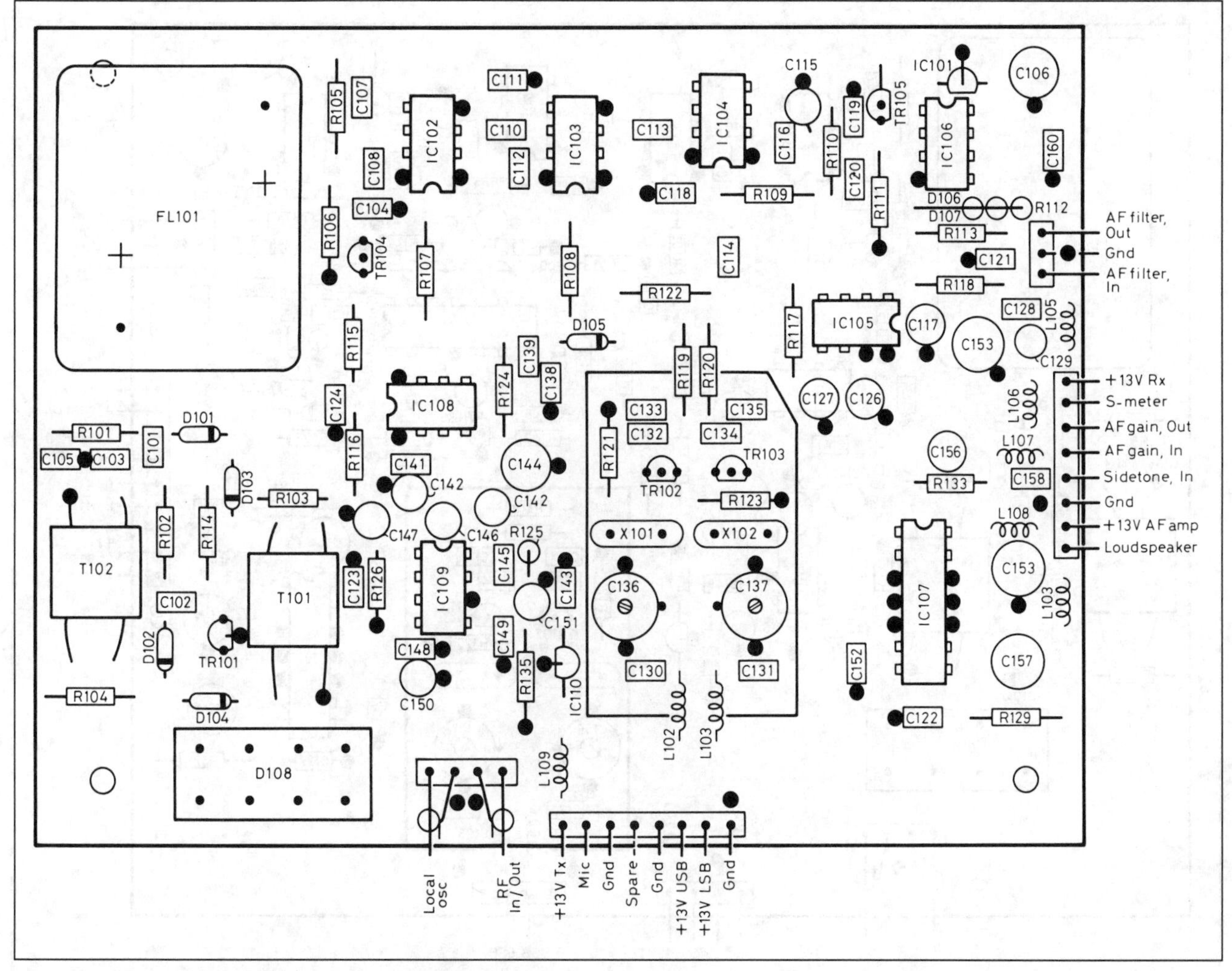

Fig 7.57(a). G3TSO exciter module 1 – SL1621 version component location

On transmit, microphone signals are amplified by IC109, an SL6270 VOGAD (voice-operated gain-adjusting device). This is designed for a balanced low-impedance input of 300Ω, or may be operated with an unbalanced input impedance of 150Ω by decoupling one input to ground; low-impedance microphones with impedances as high as 1kΩ can be fed directly to the SL6270. The SL6270 has 60dB gain; this is too much for most applications and should be reduced by placing a resistor between pins 7 and 8 of the device. 1kΩ reduces the gain to 40dB. The value of C145 should also be changed if R125 is added, typically C145 is 4.7nF without R125 and 47nF with 1kΩ.

Audio from the VOGAD is fed directly to the balanced modulator IC108, an SL1640 (or SL640), together with the local oscillator signal which is fed via a diode switch D105, this being used to limit the radiation of the oscillator signal while on receive. The local oscillator signal level can be adjusted by changing the value of R122 if required and should not be allowed to exceed 100mV RMS; 50mV is ideal. The resulting 9MHz double sideband signal is amplified in TR104, a PNP RF-type transistor, and then fed to the SSB filter for removal of the unwanted sideband. TR101, the J310 amplifier, has now been diode switched to pass the SSB signal to the SBL1 which acts as the transmit mixer. The resulting signal is heterodyned with the VFO or local oscillator to the desired output frequency, where it requires suitable filtering to remove the image response. The receive preselector can suitably perform this function. Output is in the order of 100mV RMS or slightly higher and can be adjusted if necessary by changing the value of R115 which sets the gain of TR104.

It is essential for the operation of the diode switching associated with TR101 to ground the +13V transmit line on receive and to ground the +13V receive line on transmit or the exciter module will not work.

Crystal filters for 9MHz are available from a number of sources and eight-pole types such as the XF-9B from KVG and the IQXF-90H2.4B filter, supplied by IQD Ltd of Crewkerne, are recommended and usually come complete with the two matching carrier crystals.

Transistors TR102 and 103 can be almost any high-gain JFET and the rare 78L06 regulator ICs can be replaced by the more common 78L05 type with a 390Ω series resistor in the earth return lead. If the SL1641 ICs are used, a 330Ω resistor must be connected between pin 5 and the 6V line as it is not fitted internally.

PCB track details and layouts for both versions of module 1 are shown in Fig 7.56 (see Appendix 1) and Fig 7.57.

Module 2: VFO

Fig 7.58 illustrates both modules 2 and 3: the VFO, the premix mixer, crystal oscillator, band-pass filters and broad-band amplifier. These are capable of producing the required local oscillator (LO) signal to the SBL1 mixer in module 1.

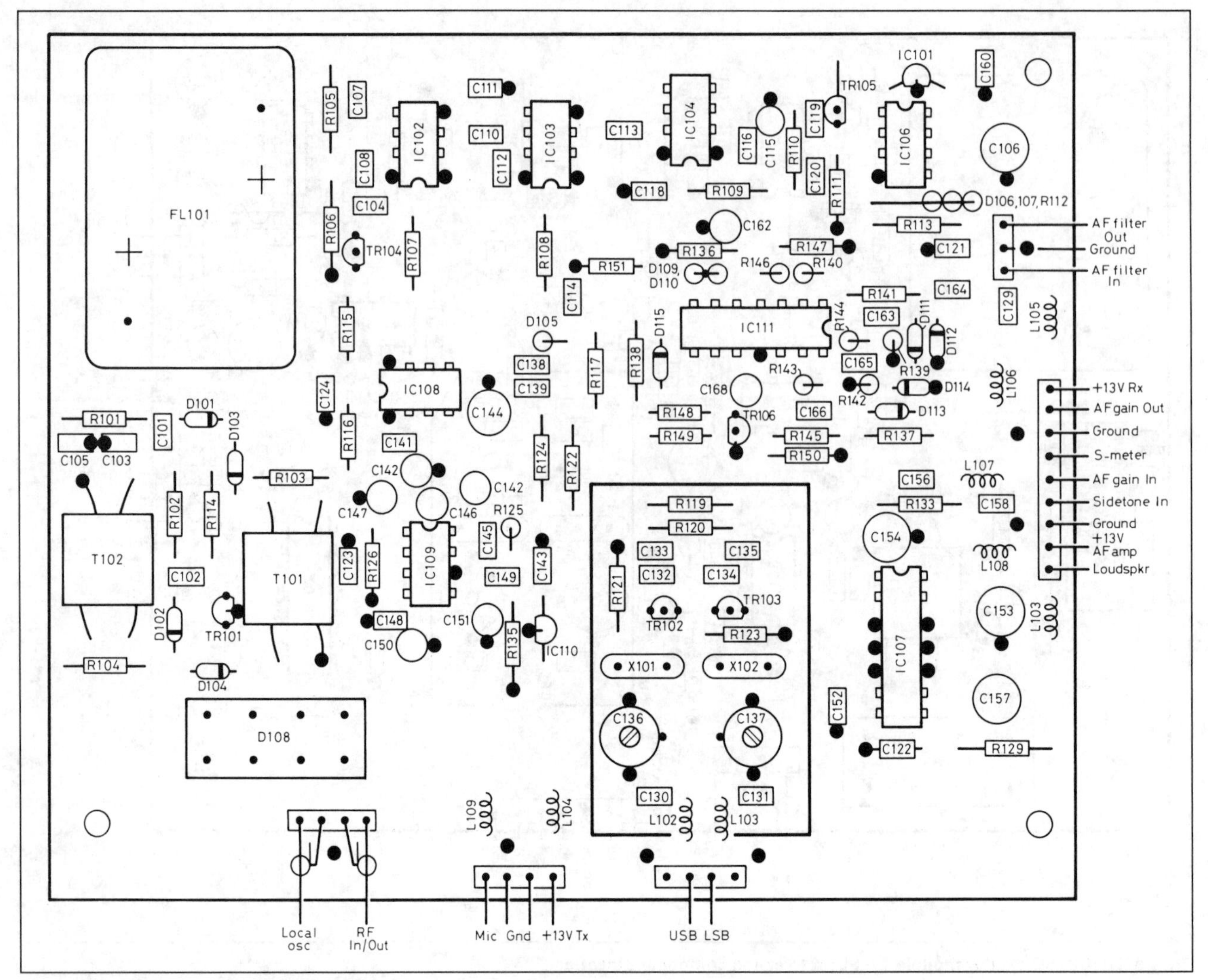

Fig 7.57(b). G3TSO exciter module 1 – TL084 version component location

The VFO is a conventional Clapp version of the Colpitts tuneable oscillator employing a 2N3819 FET. The type is not critical and almost any high-gain FET device will suffice.

The coil L201 is wound on a ceramic former provided with a powdered iron dust slug to vary the inductance. Special attention should be paid to the slug which must not be ferrite – and which should be securely mounted to prevent vibration. Most purpose-built ex-equipment coils have a lockable or tightly fitting slug and, if alternative coil formers such as the Neosid type are used, the cores must be tightened by taking a length of wool through the former to lock the core. It is better to use a coil with no slug in preference to a loose one, but alignment will be more difficult.

The coil is securely mounted to the side of the VFO enclosure with a clearance of at least 0.375in all round it. C205 is a plated brass capacitor with bearings at either end and with a capacitance of approximately 50pF.

Capacitors C206, C207 and C209 are polystyrene types while C203 is silver mica. This has the effect of providing an element of positive temperature coefficient to counter the negative coefficient of the polystyrene types. Any residual drift can usually be reduced by substituting C203 with a similar-value capacitor of different manufacture of design. If any form of frequency jumping occurs it can be caused by high RF currents present in C203. This can be eliminated by making C203 up from three lower-value capacitors wired in parallel.

D202, a silicon diode, serves to stabilise the gate voltage of the FET oscillator while D203 provides a regulated supply for the VFO and buffer amplifier. IRT or 'clarifier' operation is provided by varying the DC voltage on D201 and provides about 2.5kHz variation either side of the carrier frequency.

The VFO output is loosely coupled to the gate of TR202, another 2N3819 FET acting as a source follower to isolate the VFO from any marked changes in load. The output level from this stage is designed to feed IC301, the premix mixer, and is inadequate for direct injection to the SBL1 without further amplification.

The VFO is housed in an purpose-built aluminium box housing C205 and L201 (Fig 7.59) and a small PCB. Rigid construction techniques should be used throughout with heavy-gauge wire used for all off-board connections. All components should be firmly mounted to the PCB, which in turn should be secured to the enclosure with a minimum of four screws. C203 can be mounted above the PCB on pillars to enable component changes to be made without having to remove the PCB.

Mechanical stability of a VFO is vital if worthwhile electrical stability is to be achieved. Moving or vibrating components can give rise to FM, frequency jumping and instability.

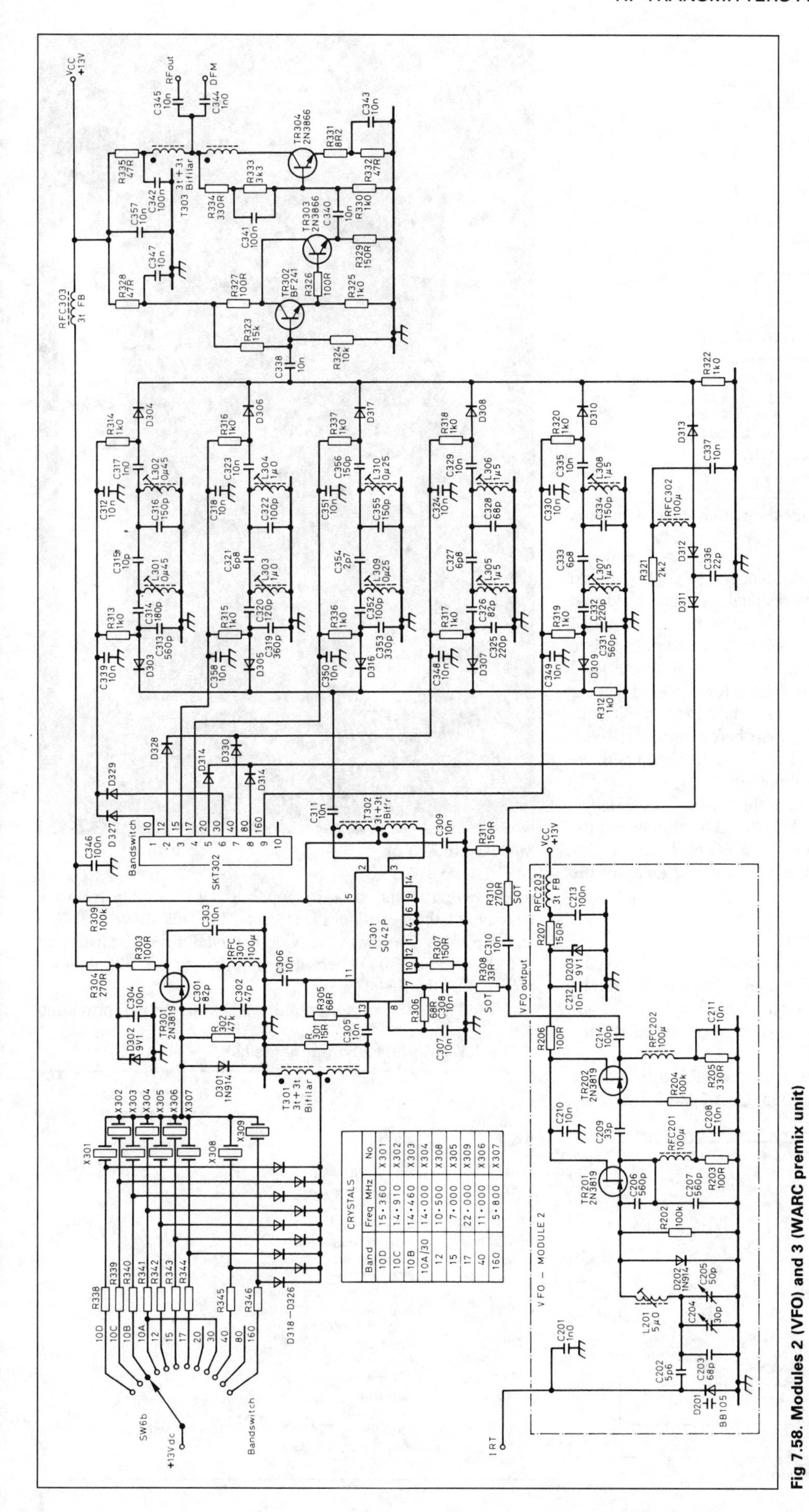

Fig 7.58. Modules 2 (VFO) and 3 (WARC premix unit)

The PCB layout (Fig 7.60, see Appendix 1) includes some unnecessary tracks which were originally intended to provide a higher level of output required to drive the SBL1. In this design it is left unused. The PCB component layout is illustrated in Fig 7.61.

VFO alignment

The VFO is aligned by adjusting L201, C204 and C205; the values indicated permit coverage of 5.0 to 5.5MHz ±20kHz at the band edges for the full range of C205.

To commence alignment, centre C205 and C204 and ensure that the DC feed point to the IRT circuit is grounded. With power applied to the VFO, adjust the slug in L201 until the VFO oscillates at 5.250MHz; this can be checked with a suitable receiver or a digital frequency meter. The tuning capacitor C205 should be swung through its entire range and coverage of the VFO checked. Typically the VFO may tune to one end of the desired range but not the other, in which case, if the tuning range is insufficient, reduce the value of C204, readjust L201 and check the coverage again. If the coverage is too great then increase the value of C204 and repeat the operation. After several adjustments of C204 and L201 it should be possible to tune the range 5.0 to 5.5MHz with a little over at each end, say, 10 to 20kHz. When the lid is placed onto the VFO unit the frequency coverage may change significantly owing to extra capacitance and some form of external adjustment of L201 and C204 is desirable.

The IRT may be checked by applying a DC bias to R201. The component values indicated on the chassis diagram (Fig 7.76) will allow a swing of the VFO frequency by about 2.5kHz either side of the nominal frequency. This is sufficient for most purposes but can be increased or decreased by changing the range of the bias voltage.

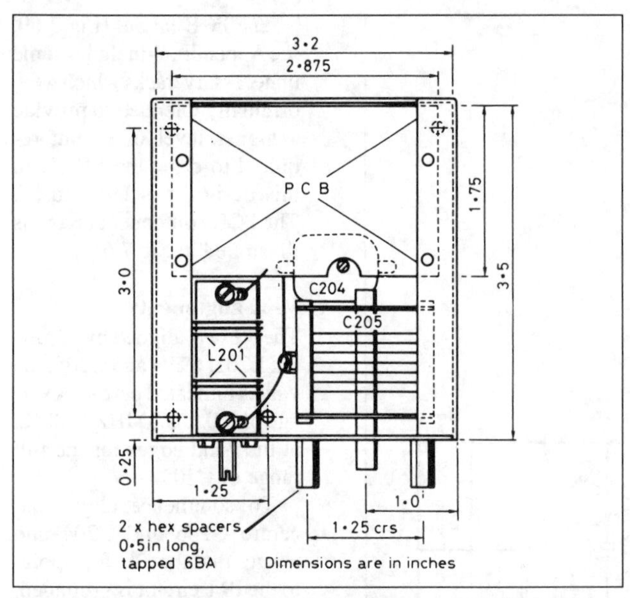

Fig 7.59. Layout of components inside the purpose-built aluminium box for Module 2 (the VFO)

When the VFO is complete and aligned, the stability should be checked for over one hour. Any residual drift can be reduced by replacing C203 with either another silver mica capacitor of the same value or a combination of silver mica and polystyrene types. This process is a little tiresome but will produce a VFO with very acceptable stability. It should be remembered that the application of a soldering iron introduces heat to the VFO components and must be allowed to dissipate before further stability checks are made. A drift of less than 100Hz in 30min is tolerable, but greater amounts can be annoying.

It is perhaps as well to mention at this stage that any slow-motion drive unit used with the VFO should be mounted in the same plane as the VFO or direct to the VFO enclosure to prevent any frequency movement being caused by flexing of the chassis and front panel.

Module 3: the premix unit

Output from the VFO unit is fed via miniature RG174 coaxial cable to IC301, a Siemens SO42P double-balanced mixer IC (Fig 7.58). R308 attenuates the signal slightly and may be adjusted to ensure that the VFO injection does not exceed 100mV RMS at the mixer port.

In addition, the VFO signal is routed via R310 to a diode switch comprising D311, 312 and 313. On 14 and 3.5MHz the diode switch is biased on by applying a 13V supply from the bandswitch through diodes D314 and 315. This allows the VFO signal to pass directly to the wide-band amplifier consisting of TR302, 303 and 304, where it is amplified to the 500mV level required by the SBL1 mixer.

TR302 and 303 are configured as a Darlington pair providing

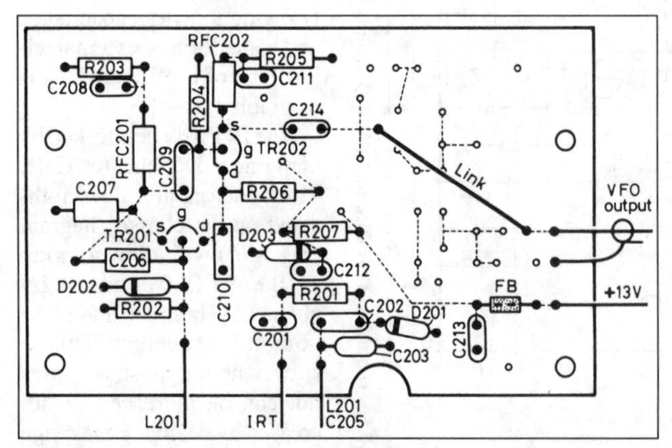

Fig 7.61. Module 2 component location

Module 2 (the VFO)

Table 7.17. Module 2 components list

R201	47k
R202, 204	100k
R203, 206	100R
R205	330R
R207	150R
D201	BB105Varicap
D202	1N914
D203	9V1 zener
TR201, 202	2N3819
RFC301, 302	1mH axial choke
C201	1n ceramic
C202	5p6 ceramic
C203	68p silver mica
C204	30p airspaced trimmer
C205	50p variable (see text)
C206, 207	560p poly
C208, 210, 211, 212, 213	10n ceramic
C209	33p poly
C214	100p poly
L201	29t 24swg 19mm ceramic former with slug adjustment. Approx 8µH
FB2	3t FX1115

Table 7.18. VFO mixing frequencies

Band (m)	Crystal (f_1) (MHz)	VFO (f_2) (MHz)	Output (f_1 + f_2) (MHz)
160	5.800	5.0–5.5	10.8–11.3
40	11.000	5.0–5.5	16.0–16.5*
30	14.000	5.0–5.5	19.0–19.5†
17	10.500	5.0–5.5	15.5–16.0
15	7.000	5.0–5.5	12.0–12.5
12	22.000	5.0–5.5	17.0–16.5*
10A	14.000	5.0–5.5	19.0–19.5†
10B	14.460	5.0–5.5	19.46–19.96
10C	14.910	5.0–5.5	19.91–20.41
10D	15.360	5.0–5.5	20.36–20.86

* and † indicate filters common to both bands

high gain with high input and low output impedance. They serve to drive TR304, a 2N3866 power amplifier operating in Class A and provided with negative feedback to improve linearity. A heatsink is required on this latter device. The output of TR304 is transformed to 50Ω by T303, a 4:1 transformer, and output should be at least 500mV RMS. Stage gain is set by R331, nominally 8.2Ω, and can be adjusted if necessary. The output is fed by RG174 coaxial cable to the SBL1 ring mixer.

TR301 is a JFET crystal oscillator employing a 2N3819; again most high-gain FETs will suffice. It is diode switched for multiband operation and the choice of crystal frequencies ensures that the oscillator is only required to operate over the range 5.8 to 22.0MHz. Crystals are HC18/U types specified for fundamental parallel resonance; they are soldered in and have their cans grounded. No provision is made for the adjustment of individual crystal frequencies. Any reluctance for crystals to oscillate in this circuit is usually attributable to insufficient feedback and may be cured by increasing the feedback capacitors C301, C302. In some cases a 30pF trimmer from gate to ground has effected a cure.

Output from the oscillator is taken by sampling the crystal current in T301, a 4:1 step-up transformer coupled directly to the second input port on IC301, the premix mixer. Output from the mixer is developed across T302, a balun transformer providing a 50Ω match to the following band-pass filters. On 3.5 and 14MHz no crystal is selected, the mixer ceases to function, and the VFO signal is routed directly to the broad-band amplifier. On the remaining bands the VFO is mixed with the crystal oscillator and the resulting signals are filtered in the band-pass filters before final amplification. The mixing process is shown in Table 7.18.

The VFO and crystal oscillator signals are suppressed to a low level in the double-balanced mixer; the sum and difference frequencies are present and must be filtered to provide the desired output. A series of diode-switched band-pass filters are used and are tuned to accept the $f_1 - f_2$ frequencies. Each of the five filters comprises of a pair of top-coupled, parallel-tuned circuits with a 50Ω input impedance and a higher output impedance. The LC ratio of the filters is chosen to provide a low Q in order to achieve adequate bandwidth over the entire band. One filter is used for all four segments of the 28 to 29.7MHz band.

Choice of crystal frequencies for the three HF bands results in the LO being on the low side of the signal; this has the effect of causing sideband inversion on the HF bands. This effect can be used advantageously to invert the LSB signal from the exciter to USB, required by convention on the HF bands. A complication occurs on 3.5MHz because the VFO frequency is subtracted from the IF; sideband inversion occurs and in addition the band tunes in reverse. The same phenomena also occurs on 12m. The sideband inversion can be overcome by reversing the sideband oscillators when 3.5MHz is selected and is achieved with the diode network comprising D8–D16 (Fig 7.76). The sideband selector switch is labelled NORMAL and INVERT and produces USB on the HF bands and LSB on the LF bands when in the NORMAL position. INVERT selects the opposite sideband on any band.

Table 7.19. Module 3 components list

Qty	Component	Value
1	R301	15R
2	R302	47k
6	R303, 309, 326, 327	100R
2	R304, 310	270R
2	R305, 306	68R
3	R307, 311, 329	150R
3	R308, 205, 334	330R
14	R313, 312, 314, 315, 316, 317, 318, 319, 320, 322, 325, 330, 336, 337	1k
1	R323	15k
1	R324	10k
1	R331	8R2
1	R333	3k3
3	R335, 328, 332	47R
9	R345, 321, 338, 339, 340, 341, 342, 343, 344, 346	2k2
1	SKT302	9-way header
1	SW6b	SW 12-way
1	X301	M451A IQD 15.360MHz
1	X302	M455A IQD 14.190MHz
1	X303	M459A IQD 14.460MHz
1	X304	A195A IQD 14.000MHz
1	X305	A136A IQD 7.000MHz
1	X306	A193A IQD 11.000MHz
1	X307	To order 5.800MHz
1	X308	(10.7MHz)? 10.500MHz
1	X309	Conv xtal 22.000MHz
8	D301, 314, 315, 327, 328, 329, 330	1N914
1	D302	9V1 zener
22	D318, 303, 304, 305, 306, 307, 308, 309, 310, 311, 312, 313, 316, 317, 319, 320, 321, 322, 323, 324, 325, 326	BA244 or similar
1	C301	82p ceramic plate
1	C302	47p ceramic plate
32	C303, 305, 306, 307, 308, 309, 310, 311, 312, 318, 323, 324, 329, 330, 335, 337, 338, 339, 340, 343, 345, 347, 348, 349, 350, 351, 357, 358	10n ceramic
5	C346, 213, 304, 341, 342	100n ceramic
4	C313, 331	560p poly
1	C314	180p poly
1	C315	10p ceramic
4	C316, 334, 355, 356	150p poly
3	C317, 344	1n ceramic
1	C319	360p poly
1	C320	120p poly
3	C321, 327, 333	6p8 ceramic
2	C322, 352	100p poly
1	C325, 332	220p poly
1	C326	82p poly
1	C328	68p poly
1	C336	22p ceramic
1	C353	330p poly
1	C354	2p7 ceramic
4	RFC301, 302	100μH axial
1	RFC303	3t FB FX1115
2	L301, 302	0.45μH Toko S18
2	L303, 304	1μH KANK3335R
4	L305, 306, 307, 308	1.5μH KXNSK4513B or KXNSK4172EK
2	L309, 310	0.25μH Toko S18
3	T301, 302, 303	3t+3t bifilar on Fairite 28-430002402 balun core
1	IC301	SO42P
3	TR301	2N3819
1	TR302	BF241
2	TR304, 303	2N3866

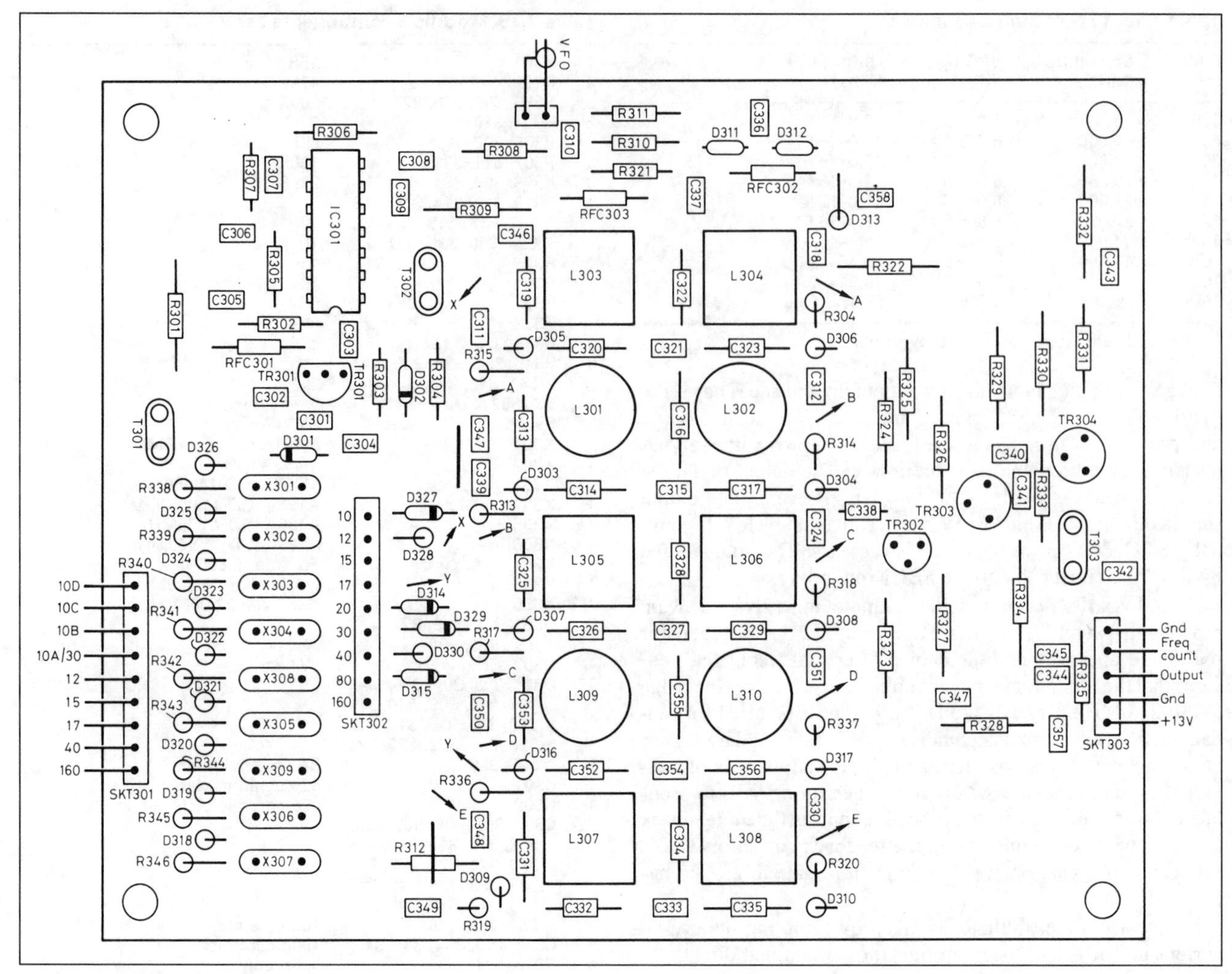

Fig 7.63. Module 3 component location

The choice of oscillator crystals is based upon a range of off-the-shelf crystals from IQD Ltd. They were cheap and provided the required band coverage, including complete but overlapping coverage of 28MHz. This may cause a few calibration problems with a mechanical dial but is of little consequence if a digital frequency display is used.

The choice of 7.00 and 14.0MHz crystals presents a slight problem in that the harmonics from both these crystals appear on the lower band edge of the 21MHz and 28MHz bands. They are quite weak, and in practice fall slightly below the lower band edge. A simple solution for the CW operator would be to use crystals for 6.990MHz and 13.990MHz, so ensuring that the unwanted signals are well out of band. The VFO tuning will be offset by 10kHz as a result of the frequency change, but can be compensated for by using an additional IRT circuit operated by the bandswitch if required. As the transceiver was primarily designed for SSB operation the use of the existing crystals has proved perfectly satisfactory.

Module 3 is constructed on a single-sided glassfibre PCB measuring 3.7 × 4.7in. The track layout is illustrated in Fig 7.62 (see Appendix 1) and the component layout in Fig 7.63. The board has a high component density and it is important to use miniature components. There are also a number of flying links located above the PCB. Toko coils are used and may be off-the-shelf types of appropriate inductance or rewound from surplus units.

Testing and alignment

No alignment of the crystal oscillators is required or even possible. In practice, crystals tend to oscillate slightly low in frequency. The mixer and broad-band amplifier has no adjustments other than the selection of R308 and R310 – the actual values used are indicated in Fig 7.58. R308 is chosen to set the VFO injection to IC301 at 100mV rms.

With 13V applied to either D314 or D315, the VFO signal should appear at the input to the broad-band amplifier. If power is applied to the amplifier then the VFO signal should be available at the output at around 500mV RMS. The output level is adjustable by selecting R310 which should be chosen to give a similar output level on 14MHz and 3.5MHz to that obtained when the mixer unit is operating.

The crystal oscillator may be tested by applying 13V and connecting a jumper lead across SKT301 to select the appropriate crystal. Each crystal should be checked in turn to ensure that the circuit oscillates and that the frequency is correct. The crystal oscillator injection to IC301 should not exceed 100mV RMS.

Band-pass filter alignment can be simple and does not require masses of expensive test equipment. As few amateurs will have access to a sweep generator an alternative method will be described. Each filter may be tested independently once it is wired up, but the low-impedance input should be terminated with a 50Ω resistor and a 1kΩ resistor connected across the output. Using a signal generator and a suitable measuring device such as an oscilloscope or VVM, inject a signal into the filter input while monitoring the level at the filter output. Peak the filter in the middle of the desired band so as to obtain maximum output. Each filter may then be adjusted to broaden the bandwidth by adjusting one coil at the LF end of the band and the other at the HF end of the band. Tuning of each coil in a filter will be interactive with the other inductors, and several adjustments will be necessary to achieve a flat response across the desired band. By sweeping the signal generator across and outside the band the output level should be observed to be as flat as possible up to the band edges and then rapid attenuation of the signal should occur as the generator moves further away from the band edge.

Further alignment of the filters is possible when both oscillators and the mixer are operating. With 13V connected to each filter switching in turn, and with the corresponding crystal selected, it should be possible to observe the local oscillator signal level at the output of module 3. As the VFO is tuned across its range the output level should be clean, on the correct frequency and fairly constant across the entire band. Minor adjustment of the band-pass filters can be made to achieve a level response. The output level from module 3 should be checked into a 50Ω load and should be approximately 500mV. The output level will vary slightly from band to band but nevertheless is fairly constant. Some adjustment of output level is possible by adjustment of R331, varying the gain of TR304.

Module 4: band-pass filter unit

This filter provides front-end selectivity on receive as well as filtering the transmit signal.

Module 4, Fig 7.64, comprises nine band-pass filters – one for each band. The filters are bi-directional and consist of a number of tightly coupled, parallel-tuned circuits, capacitively tapped to provide a 50Ω input and output impedance. The filters have a high *C* to *L* ratio and are designed to provide a compromise between bandwidth and adequate selectivity for receiver operation. On transmit only minimal attenuation of the low-power transmit signal occurs. No external preselector tuning is provided or necessary once the filters are correctly aligned.

Each filter is selected into circuit by forward biasing two BA244 low-capacitance switching diodes at either end of the filter. Input and output coupling is capacitive into a common line between the filters. Toko coils are used throughout and can be selected for their inductance value or rewound from old stock. No complicated taps or coupling windings are employed, each inductor being a straight solenoid. If ready-made coils are used be careful not to short out any secondary windings or taps as this will have adverse affects on filter operation.

D403 and D404 provide switching of the transmit and receive signal paths; D403 routes the receive signal from the antenna change-over relay to the filters.

On transmit, additional voltage amplification is required after the BPF and D404 routes the transmit signal to IC401, a Plessey SL610 RF amplifier IC whose gain is controlled both manually and by ALC action. Output from the SL610 is in the order of 1V RMS and is capable of providing adequate drive for the following PA unit. The SL610 operates from a 6V supply provided by D402 from the 13V transmit rail. ALC is applied as a DC bias to pin 7 of the IC which is shunted by D401, a 5.6V zener diode which prevents more than 6V appearing on the IC. If the SL610 is operated into a low-impedance load (50Ω) it is important to include a series resistor of approximately 100Ω in the output line to prevent parasitic oscillations occurring in the low VHF region. This resistor will limit the drive available and the use of a DC-coupled transistor (2N3866 or equivalent) is a preferable way of matching the SL610 to a low-impedance load. It is important not to overdrive the SL610 as it can limit severely, producing a flat-topped output which will introduce severe distortion when amplified by the following wide-band amplifier.

Table 7.20. Module 4 components list

1	R401	100R
1	R402	220R
1	R404	See note
23	R422, 405, 406, 407, 408, 409, 410, 411, 412, 413, 414, 415, 416, 417, 418, 419, 423, 424, 425, 426, 427, 428, 429	1k
1	R420	560R
1	R421	15k
28	C405, 402, 403, 406, 407, 408, 449, 450, 451, 452, 453, 454, 455, 456, 457, 458, 459, 460, 463, 464, 466, 468, 476, 477, 483, 484, 490, 491	10n ceramic
6	C409, 401, 404, 448, 465, 467	100n ceramic
2	C411, 415	1800p poly
2	C410, 416	2700p poly
5	C412, 413, 439, 441, 472	180p poly
3	C414, 478, 482	750p poly
4	C418, 422, 443, 447	390p poly
6	C417, 423, 424, 430, 469, 475	1000p poly
2	C419, 420	47p poly
1	C421	270p poly
6	C429, 425, 470, 474, 479, 481	220p poly
1	C498	150p poly
4	C427, 426, 471, 473	10p ceramic
2	C436, 432	120p poly
6	C437, 431, 438, 442, 485, 489	560p poly
2	C434, 433	5p6 ceramic
3	C435, 486, 488	100p poly
1	C480	8p2 ceramic
1	C440	6p8 ceramic
2	C487, 445	2p7 ceramic
2	C444, 446	82p poly
2	C461, 462	1n ceramic
2	D402, 401	5V6 zener diodes
17	D403, 404, 405, 406, 407, 408, 409, 410, 411, 412, 413, 414, 415, 416, 418, 419, 420, 421, 422, 423	BA244 or similar
1	D417	1N4001
2	RL401, 402	SPCO OUC Type
2	RFC401, 402	1mH
3	L403, 401, 402	27t 8μH. Use KANK3334R
1	T401	28-430002402 balun 3t+3t
3	L407, 408, 409	2.8μH KXNK4173AO
3	L406, 404, 405	5.8μH KANK3334R
8	L421, 413, 414, 415, 416, 420, 422, 423	0.25μH Toko S18
1	RFC403	3t FB FX1115
3	L419, 417, 418	1.3μH KANK3335R
3	L410, 411, 412	1.2μH KANK3335R
1	TR401	ZTX327
1	IC401	SL610
1	IC402	SL560

Polystyrene-type capacitors are used in the filter unit combining high stability with small size and low cost. Each filter is constructed independently on the single-sided glassfibre PCB (Fig 7.65, see Appendix 1), and all filters are connected to SKT401 by fly wires.

It is important to ensure that both ends of each filter are connected to the same pin on SKT401 if they are to switch correctly – these fly wires should be added under the PCB. A layout of module 4 is illustrated in Fig 7.66.

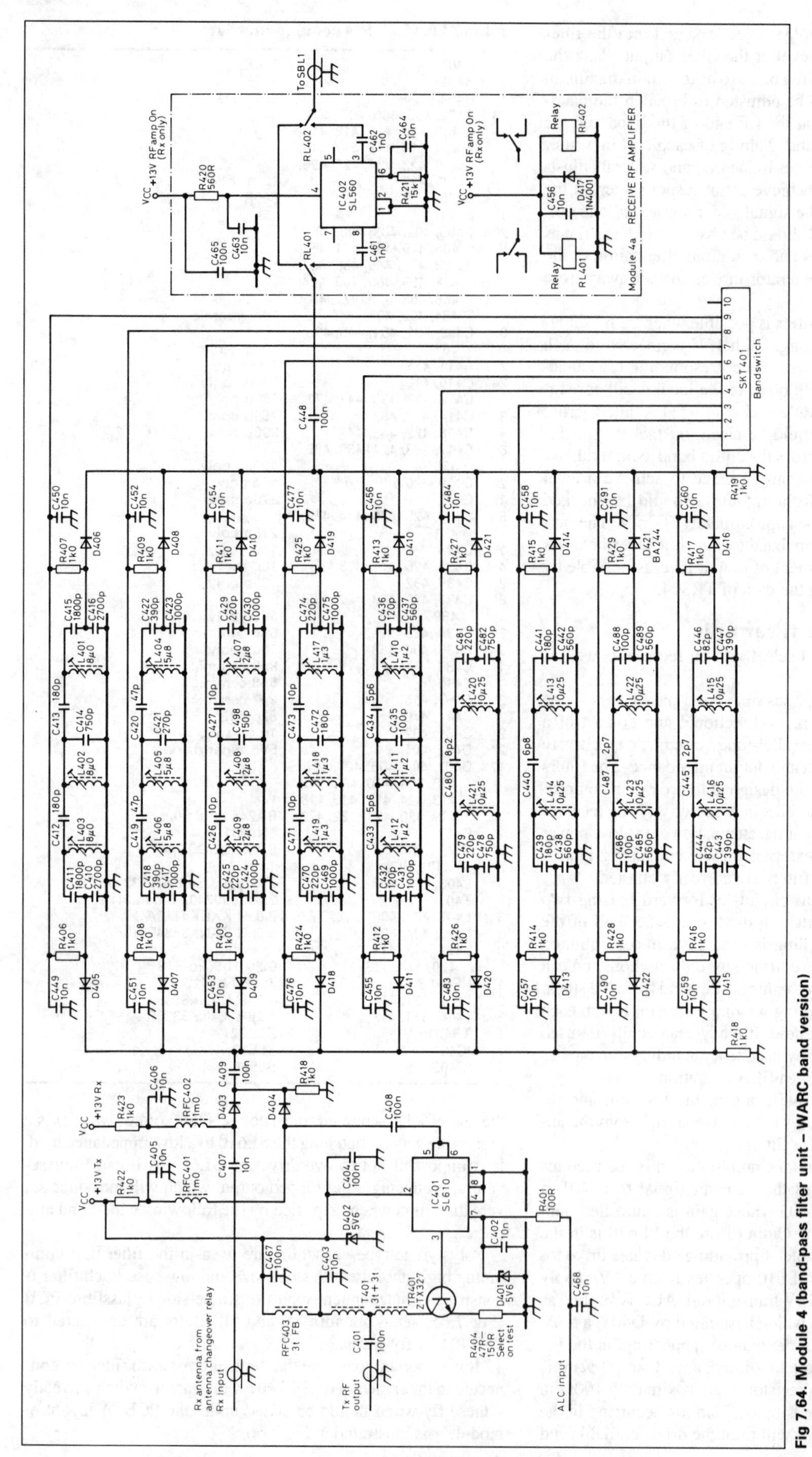

Fig 7.64. Module 4 (band-pass filter unit – WARC band version)

Module 4a: RF amplifier

While not actually part of module 4, the optional receiver RF amplifier is inserted between module 4 and the mixer to provide increased gain on 28 and 21MHz. It is not essential and certainly not required on 14MHz or below.

A Plessey SL560C is used as a 50Ω gain block with approximately 15dB gain and is switched into circuit on receive using subminiature relays. The RF amplifier and relays are switched from the 13V receive rail which ensures that the unit drops out on transmit. This is essential as the amplifier is located in a bi-directional signal path. It is also important to ensure the DC supply to the IC is removed when not in use as it is prone to oscillation if left unterminated.

Construction

Module 4 is constructed on a 2.9 × 7.2in single-sided PCB. The RF amplifier module 4A is housed on a separate double-sided PCB measuring 1.0 × 2.75in and is illustrated in Figs 7.67 (see Appendix 1) and 7.68.

Alignment of the band-pass filters can be achieved simply using a signal generator and a VVM or oscilloscope in the same manner as the filters in module 3, except that both ends of the filters should be terminated with a 50Ω resistor.

The 1.8 to 14MHz filters comprise three parallel-tuned circuits, while the 21 to 28MHz filters have only two tuned circuits. The triple-tuned filters have three distinct peaks during alignment and tuning of each circuit is interdependent. The simplest method of alignment is to peak each filter in the centre of the

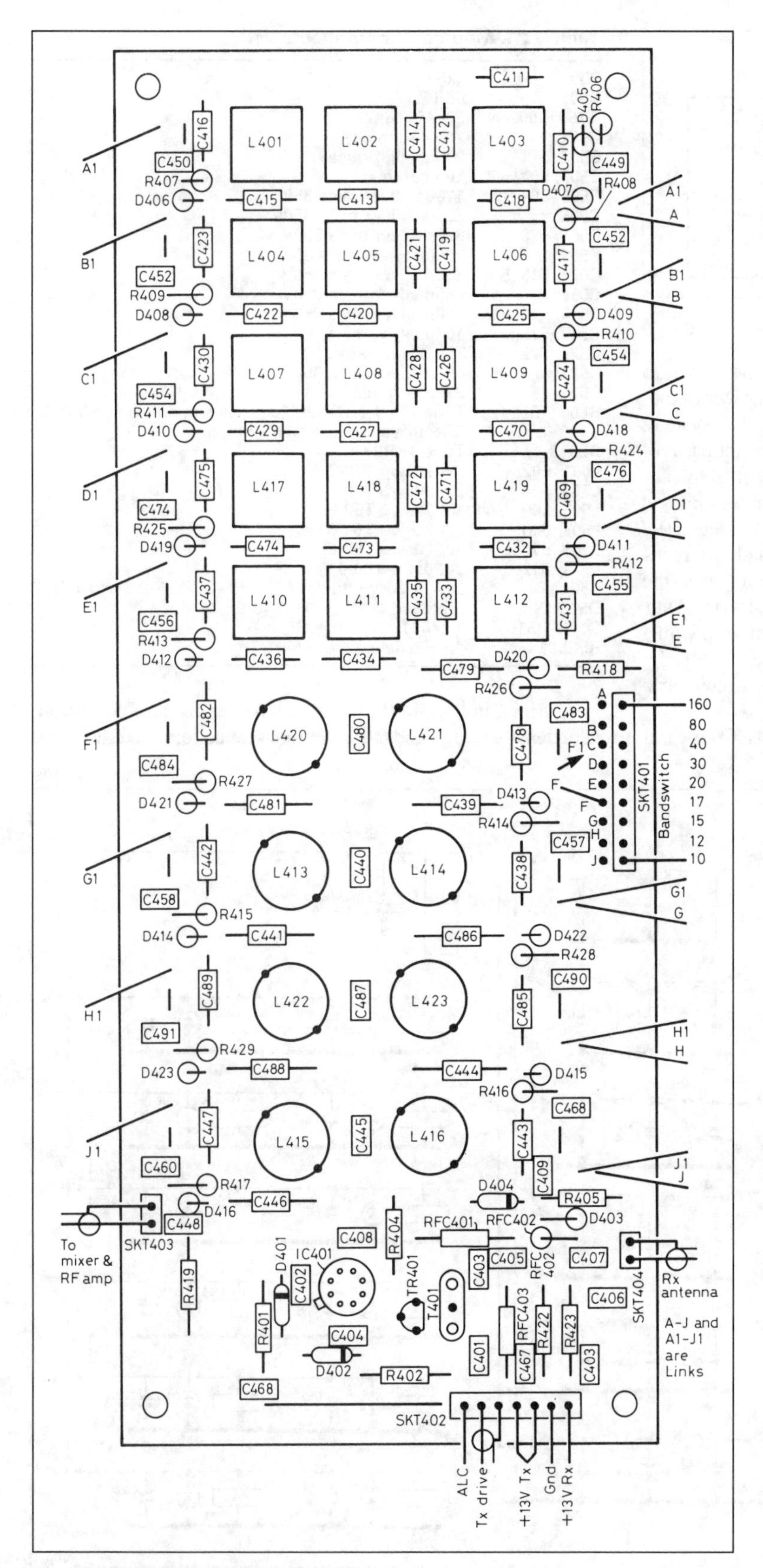

Fig 7.66. Module 4 component location

required passband and then increase the bandwidth by adjustment of the input and output tuned circuits towards the HF and LF ends of the band. The centre inductor can then be used to flatten the response across the band. Tuning is fairly critical and small adjustments should be made until a level response is achieved across the desired band. The signal generator should be swept slowly across the band during the alignment process and the output from the filter monitored on either the VVM or oscilloscope. When the generator is swept well away from the required band, the output signal should be rapidly attenuated. Check for the odd response that may appear some way out of band and which may be due to one of the tuned circuits being considerably off resonance. The 21 to 28MHz filters, with only two inductors, are considerably easier to align.

No adjustment of either the transmit or receive RF amplifiers is necessary or possible, but the gain of the transmit amplifier can be adjusted by means of the manual drive control.

During testing, if no DC bias is applied to pin 7 of the SL610, it will rise to the supply voltage and should be grounded for maximum gain.

Module 5: low-pass filter unit

The transmitter RF section employs a broad-band amplifier taking signals from the milliwatt region up to the final output of several watts. Before this signal can be fed to an antenna it is essential to remove any unwanted harmonics that may have been generated in the amplifier chain.

Module 5 (Fig 7.69) comprises of six Chebyshev low-pass filters having cut-off frequencies coincident with the top edge of each band segment on the transceiver. Even harmonics will be cancelled to a large extent by the push-pull PA and driver amplifiers, but any residual second harmonic will be further attenuated by the filter. The third harmonic and above will be attenuated by the filter by at least 50dB.

Filter switching is achieved using a series of miniature relays used to select the desired filter into circuit, and all filters not in use are grounded at both ends to prevent stray signal paths for higher-frequency products around the filter. All capacitors used in the filters should ideally be of the silver mica type and at least 350V DC working for use up to 200W. Polystyrene capacitors with a 125V rating can be used for powers up to about 50W. It should always be born in mind that the RF voltage at the antenna socket will rise with increase in power and SWR.

Capacitor values are critical to correct operation of the filters and must be adhered to – any values that cannot be obtained should be made up by paralleling two or more capacitors in order to achieve the desired value.

The low-pass filters are retained in circuit during receive to provide additional filtering with the result that the antenna changeover relay RL13 is located on the transmit side of the filter. This is a DPCO type relay and grounds the RX and TX ports when not in use. This is particularly important as the RX antenna line returns to the vicinity of the BPF which is passing low-power transmit signals. The use of a heavy-duty ferrite bead on this line is also advisable.

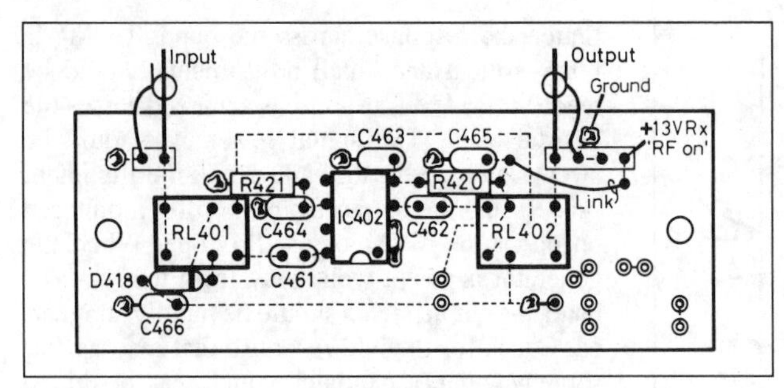

Fig 7.68. Module 4a component location

An SWR detector is included on the LPF unit and serves to provide meter indications of power and reflected power output as well as providing DC voltages for use by the ALC system. The bridge is a current-sampling type and therefore not particularly frequency conscious. T501 samples the current in the antenna line which develops a voltage across R502. This is summed with the RF voltage developed across the potential divider C501/502, producing forward and reflected voltages which are rectified by D501 and D502 before being fed as DC voltages to the ALC unit. Potentiometers are provided to scale the FWD and REF voltages for presentation to a panel meter. An IC op-amp with a gain of two is used as a buffer amplifier to prevent the panel meter loading the SWR bridge; it also serves as an S-meter amplifier on receive.

The LPF is constructed on a single-sided glassfibre PCB measuring 3in × 4.3in (Fig 7.70, see Appendix 1). The component density is high and the silver mica capacitors used should be

Table 7.21. Module 5 components list

R501	1k8
R502	68R ½W
RV501, 502	22–25k trim
C501	10p ceramic trim
C502	200p silver mica
C503, 504, 505	10n ceramic
C505, 507	1200p silver mica 350V
C506	2500p silver mica 350V
C508, 510	820p silver mica 350V
C509	1500p silver mica 350V
C511, 513, 515	360p silver mica 350V
C512	680p silver mica 350V
C514, 516, 518	220p silver mica 350V
C517, 519	100p silver mica
C520, 522	75p silver mica 350V
C521	160p silver mica 350V
C523–C534	332n ceramic
RL501–RL512	Type 211NA DOO9M20 surplus at rallies or SMR12 (Electrovalue) or YX94C (Maplin)
RL513	Type TKR22
L501, 502	31t 26swg T50-2
L503, 504	22t 20swg T50-2
L505, 506	18t 20swg T50-6
L507, 508	12t 20swg T50-6
L509, 510	10t 20swg T50-6
L511, 512	9t 20swg T50-6
T501	9t + 9t 26swg bifilar FT37-43 (GW3TMP Electronics)
D501, 502	Matched OA47
D503–D515	1N4001
RFC501	1mH axial choke

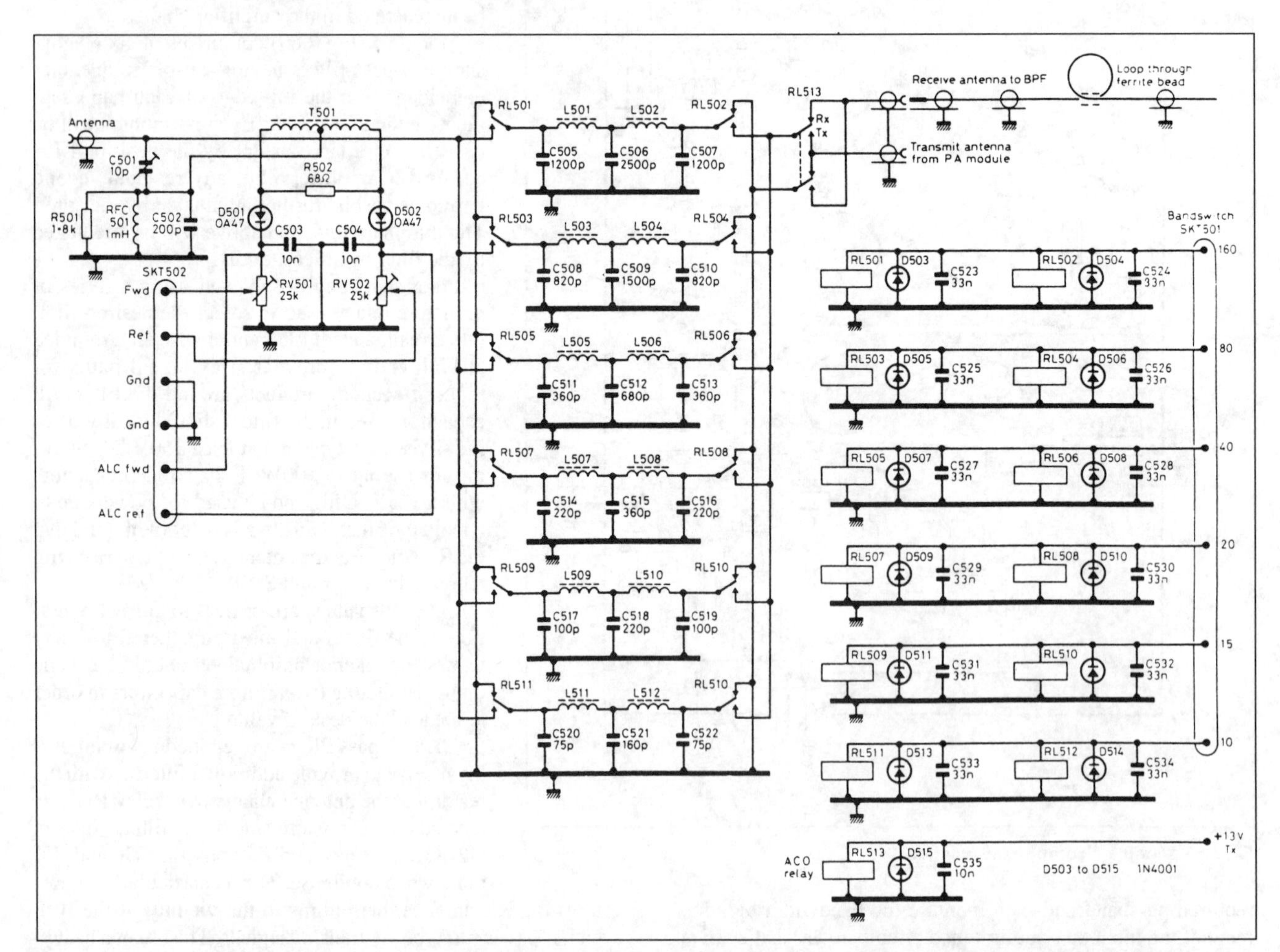

Fig 7.69. Circuit for the low-pass filter system (module 5) which buffers harmonics fo the transmitter output from the antenna

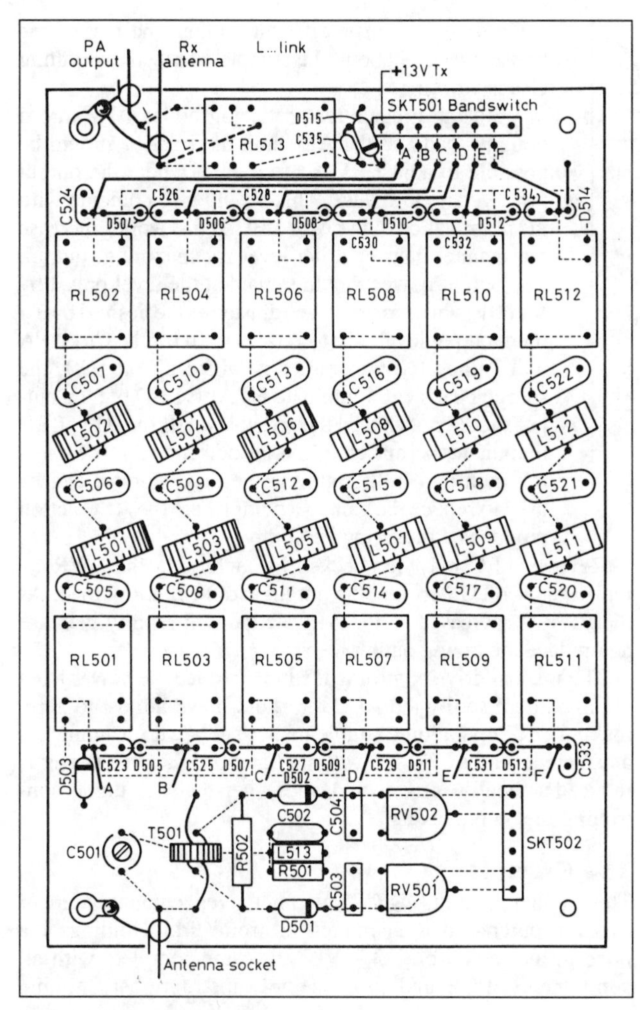

Fig 7.71. Module 5 component location

of the modern smaller design rather than the traditional large variety. There is no reason why the PCB layout should not be expanded a little to make more room if required. For operation on the WARC bands, the 14MHz filter is used on 10MHz, the 21MHz filter on 18MHz and the 28MHz filter on 24MHz; additional filters are not necessary. Component layout is given in Fig 7.71.

All relays on the LPF board should be capable of carrying the PA output current and a good rating to aim at is 2A. All relays must be shunted by diodes to suppress switching transients and all relay supplies and switching lines must be decoupled to prevent RF pick-up. The relays may be 6V types wired in series or 12V types in parallel.

Testing and alignment

No alignment of the filters is necessary if they are constructed correctly. A DC path should be checked through each filter when the appropriate relays are activated and the signal path should be checked through the antenna change-over relay. When connected to the bandswitch the LPF relays should operate on changing bands. If power is removed from the board all relays drop out and there is an open-circuit across the LPF unit.

The SWR unit must be balanced before it can be used – the antenna input should be terminated with a 50Ω dummy load and a transceiver capable of 10W output should be connected to the input of the bridge. C507 is adjusted to null out the voltage appearing at the REF ALC terminal.

In some cases it may be necessary to reverse the FWD and REF connections – this will become apparent if the voltage cannot be dipped or nulled. Calibration of the SWR meter is a matter of personal choice but if the meter used has a power scale below the S-meter scale it may be convenient to set the FWD scale for 25W FSD. The REV scale can be made more sensitive if required as the ALC system will not allow excessive reverse power to appear.

Module 6: ALC unit

ALC is essential on a multiband transceiver to ensure a constant output power on all bands and to prevent excessive overdrive on some bands. It can also provide SWR protection by reducing the drive level at a predetermined SWR.

The ALC unit is self-contained and uses a single IC, IC601, which can either be a Motorola MC3401 or the more common LM 3900 quad Norton current comparator. FWD current from the SWR unit is fed via R605 (Fig 7.72) to IC601b which it is compared with a reference current derived from RV601 through R607. The reference current will determine the ALC threshold or maximum gain of the transceiver. If the output level starts to exceed this level the comparator will produce an increasing voltage at the output of IC601b. This is fed via a time constant circuit which provides a rapid attack and slow decay to IC601c which operates as a buffer amp. The output from IC601c is a DC bias which can be used directly to control the gain of IC401 on module 4. An increasing voltage on the ALC line reduces the gain of the SL610 and holds the transceiver output power at the

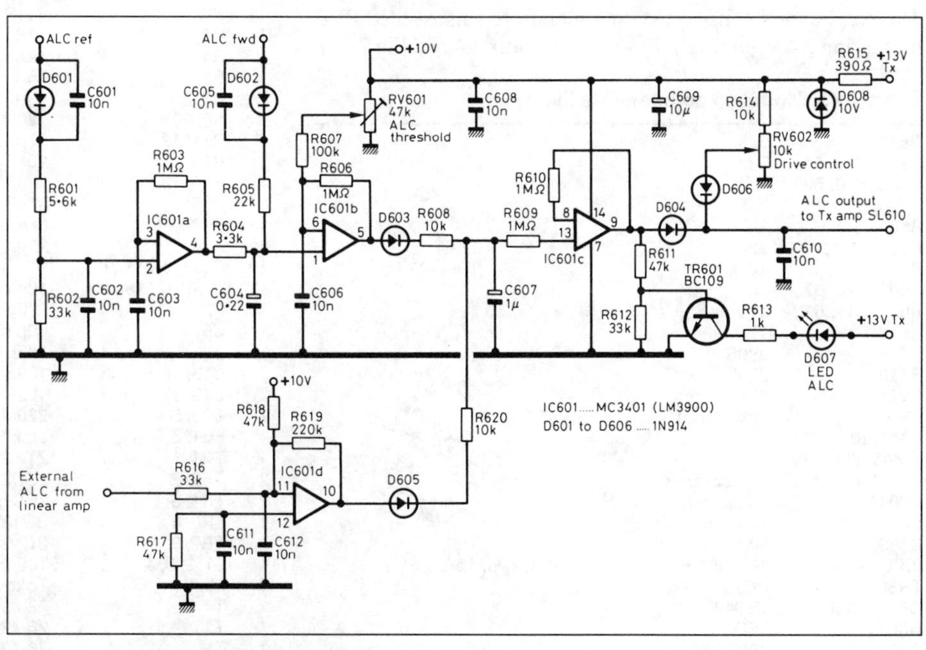

Fig 7.72. Module 6 (self-contained ALC unit)

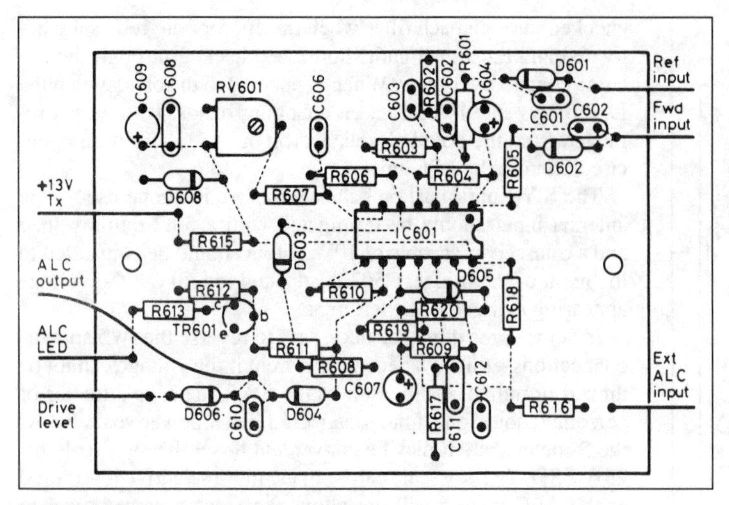

Fig 7.74. Module 6 component location

level set by RV601. Output from the ALC unit is summed with a DC bias obtained from the manual drive control by diodes D604, 606 and permits the drive control to override the ALC, in turn allowing power reduction from the maximum level set by the ALC threshold down to about 100mW. Reverse ALC is provided by IC601a which amplifies the REV current obtained from the SWR unit and sums it with the FWD current fed to IC601b. The result is that any reflected power will cause greater ALC action than that produced by forward power, and the output level will be reduced as long as the high SWR condition exists. The threshold level for reverse ALC action is controlled by the gain of IC601a which can be altered by changing R601.

The fourth stage of the Norton amp IC601d is used to feed ALC from an external linear amplifier into the ALC unit in order to control the total system gain. It is designed to take negative-going ALC voltages and is compatible with most valve linear amplifiers. If not required, this stage can be omitted. The ALC range required by the SL610 is from +2V to above +4.5V with maximum gain occurring at 2V or below.

Construction and testing

The ALC unit is constructed on a glassfibre single-sided PCB measuring 2.4 × 3in (Fig 7.73 – see Appendix 1 – and Fig 7.74).

Table 7.22. Module 6 components list

R601	5k6
R602, 616, 612	33k
R603, 606, 609, 610	1M
R604	3k3
R605	22k
R607	100k
R608, 614, 620	10k
R611, 617, 618	47k
R613	1k
R615	390R
R619	220k
RV601	47k preset
RV602	10k lin pot
C601, 602, 603, 605, 606, 608, 610, 611, 612	10n ceramic
C604	0.22μ tant
C607	1μ tant
C609	10μ tant
IC601	MC3401 (Watford Electronics) or LM3900
TR601	BC109
D601–606	1N914
D607	LED
D608	10V zener

Testing the unit is difficult without the rest of the transceiver and should be completed when everything else is working.

With the transmitter producing about 10W into a dummy load and the ALC unit in circuit it should be possible to adjust RV601 to a point where the output power starts to reduce. This should also coincide with the illumination of the ALC LED. If this works, increase the manual gain or drive level for maximum output. The output power should remain at the level preset by RV601, which can now be adjusted to the desired maximum output level – 18 to 20W is ideal for the Cirkit PA unit. When the transmitter is talked up on SSB, the speech peaks will illuminate the ALC LED but the output should not exceed the preset level. If the ALC LED illuminates permanently, it is an indication that the transmit gain is a little high and the drive control can be used to reduce the gain to a point where the ALC action produces a flickering LED.

Reverse ALC may be checked by increasing the SWR by inserting an ATU between the output and the dummy load. As the SWR is increased so the ALC action will rapidly increase and reduce the power output level.

The manual drive control will simply reduce the power from the maximum set by the ALC threshold down almost to zero, making QRP operation a simple matter and ideal for tuning up an antenna. The ALC LED will not function when the power is reduced manually as it only indicates when the ALC unit is controlling the output level.

The PA unit

The original PA unit, Fig 7.75, was not given a module number as it was purchased as a complete kit from Cirkit Holdings at a price in the region of £33 + VAT. It came complete with all components, PCB and full assembly instructions. Suitable

Table 7.23. 20W PA unit components list

2	RV1, 2	100R trimpot
2	R1, 5	100R
2	R13, 14	100R 1W
1	R2	680R
1	R3	12R
1	R4	4k7
1	R6	56R
2	R7, 8	33R
2	R9, 10	120R
2	R11, 12	22R
1	R15	10k
2	R16, 19	150R
1	R17	330R
1	R18	6k8
1	R20	220R
10	C1, 3, 4, 6, 7, 9, 10, 11, 12, 16	10n ceramic
1	C2	270p mica
1	C5	68p mica
1	C8	220p mica
2	C13, 15	100n ceramic
2	C16, 17	10μF tant
2	C20, 21	220μF alu
4	FB1, 2, 3, 4	Link 1, 2, 3, 4 ferrite beads 26-43000101 (8 off)
1	TR1	ZTX327 (2N3866)
2	TR2, 3	2SC2166
2	TR4, 5	2SC1945
2	TR6, 8	BD139
2	TR7, 9	BC108
4	L1, 2, 3, 4	100μH axial
2	L5, 6	Toko 283AS 1m8
1	T1	28-43002402 balun 6t:2t
1	T2	26-43006301 tube × 1, 1+1t:1t
1	T3	26-43006301 tube × 4, 1/2t+1/2t:2t

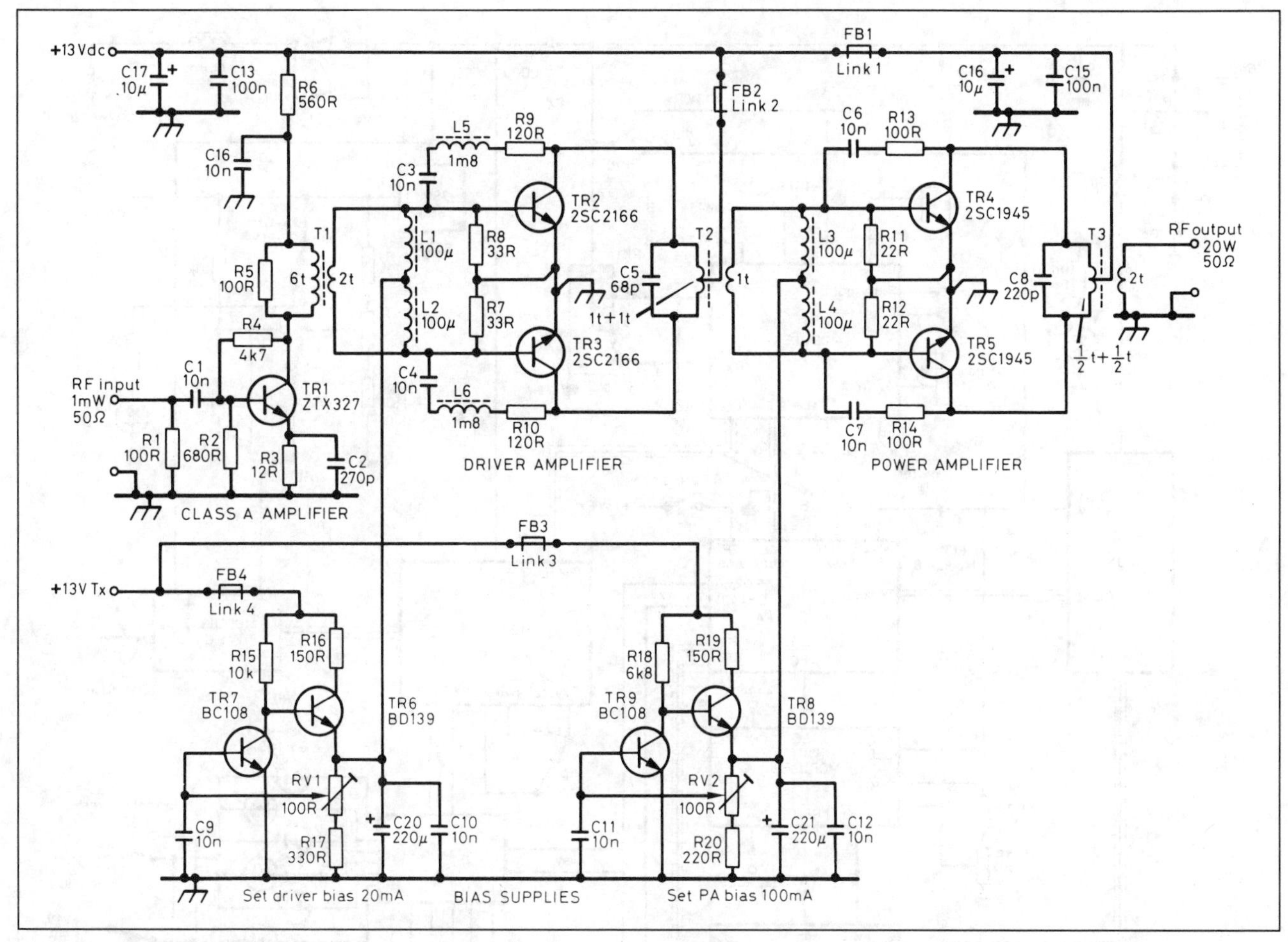

Fig 7.75. 20W power amplifier

alternative driver and PA circuits are illustrated in Figs 7.28(d) and 7.28(c).

Transmit control of the PA is achieved by switching the 13V bias rail to the two bias regulators while leaving the amplifier connected permanently to the +13V rail.

After some months of operation two minor modifications became necessary. The two 100Ω PA feedback resistors R13 and R14 were increased to 1W ratings as they were prone to overheating. It was also discovered that the PA bias was slow to stabilise when switching to transmit, with the result that the first few words were distorted. The PA was found to be almost in Class C until the bias supply stabilised. This was easily cured by changing R18 in the bias circuit from 10kΩ to 6.8kΩ, giving better regulation.

Main chassis

All modules have been housed in an aluminium chassis which serves as a convenient frame as well as providing an element of screening and heatsinking. Figs 7.76 and 7.77 illustrate the chassis wiring required to interconnect the modules, plus the associated switching functions.

Bandswitching is achieved with a two-wafer Yaxley-type switch employing two one-pole 12-way wafers with break-before-make contacts. The rear wafer switches the various crystals into circuit. Initially direct switching of crystals was employed, necessitating short lead lengths to minimise stray capacitance. By adopting diode switching, the layout becomes far less critical. The front wafer is used to switch the 13V rail to the various circuits that require bandswitching using multiway ribbon cable via a small junction PCB. This can be fabricated from a strip of Veroboard or be specially made.

All switching lines are decoupled to ground at the junction PCB and the various multiway cables are connected to the respective modules using a simple plug-and-socket system. Fig

Table 7.24. Main chassis components list

R1–4	10k histab
R5, 6	1k
R7, 8, 9	1M
R10	8k2
RVO1	10k log AF gain
RVO2	10k lin IRT
RVO3	25k preset
C1	1000µ 25V
C2–8, 11	10n ceramic
C9, 10	47n ceramic
C12	100p ceramic
C13	1n ceramic
C14	10µ tant
D1	1N4001
D2–13	1N914
LEDs × 3	
FBX	3t FX1115
FB	3t Fairite 26-43006301
S1	SPCO Fwd/Ref
S2	DPCO Nor/Inv
S3	SPCO RF amp on/off
S4	DPCO IRT on/off
S5	DPCO Power on/off (to carry 5A)
S6	Yaxley 2 wafers 1-pole 11-way
IC1	7808 reg
IC2	CA3130 meter amp
Meter	200µA S-meter (Cirkit)
Plugs and sockets as required.	

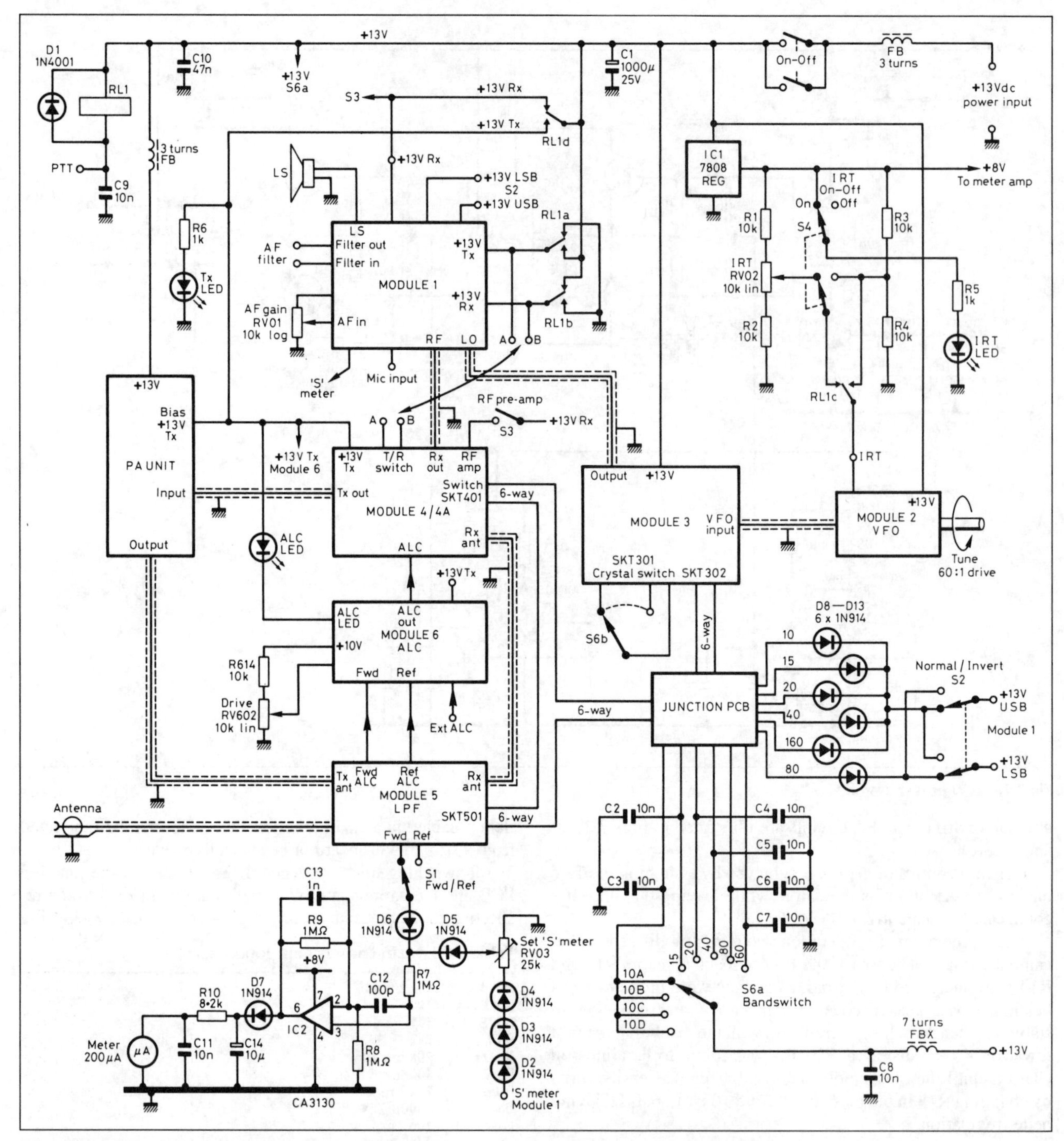

Fig 7.76. Complete chassis wiring system for the transceiver – all switching and the inter-module links

7.77 provides data for the interconnection of the various modules for nine-band operation.

The VFO is housed in a metal enclosure which must be securely fastened to the main chassis and tuned with a suitable slow-motion drive with a reduction ratio of at least 60:1 for smooth tuning. Either a mechanical dial or a digital frequency counter can be used to display the operating frequency.

An enormous range of possibilities exist for the layout of the transceiver. It can be large to accommodate additional modules and future modifications, or miniature if that takes your fancy. The original transceiver was made 9.5in wide, 3.5in high and 12.5in deep, making it ideal for portable working and still smaller than any multiband commercial transceiver.

Location of the individual modules is not likely to be critical provided a sensible approach is adopted. The PA unit ideally should be mounted on the rear panel where heatsinking is relatively easy to arrange. A 2.5 × 4 × 0.5in finned heatsink has proved more than adequate and barely runs warm in SSB operation. The LPF unit should also be located near the rear of the unit and should be screened from the remainder of the transceiver.

Conclusion

The modules described were originally built in an attempt to produce a home-constructed multiband amateur transceiver without undue complication. The results achieved have been far

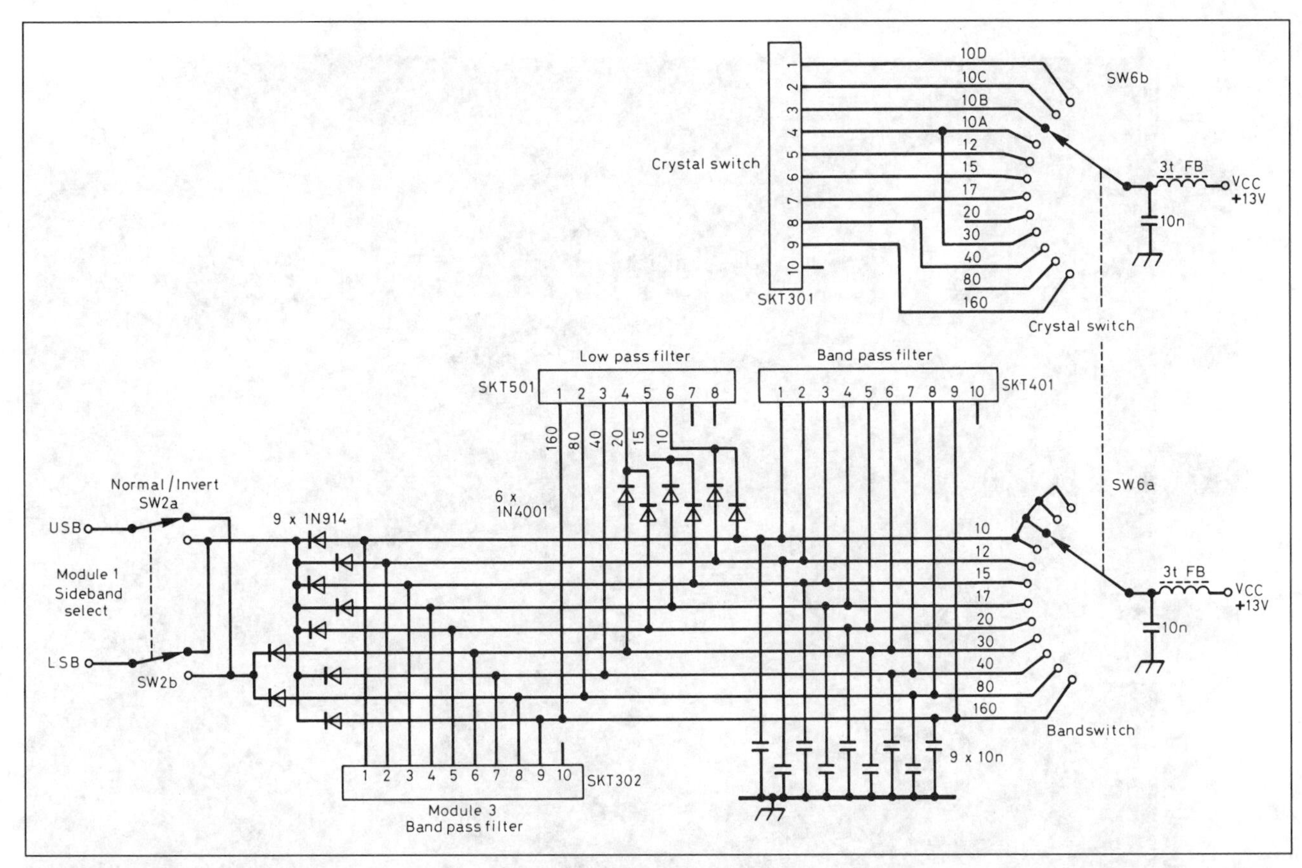

Fig 7.77. Band switching – WARC version

better than originally expected and the home-made transceiver is now used in preference to the author's commercial one. No sophisticated test equipment was used for testing or alignment, but access to a DFM, signal generator and a VVM or an oscilloscope is necessary.

Each module was constructed and tested independently before final assembly. This was initially done in breadboard form to simplify alignment and testing. The chassis was home-constructed to facilitate the mounting of the various modules.

Additional features may be added to the basic transceiver to suit personal preference and may include such items as a digital frequency counter, speech processor, audio filter, CW sidetone and keying unit and a variety of other add-ons. Since its original publication in *Radio Communication* in 1988, the WARC bands have been included in the design and have been successfully incorporated by many of the constructors of this project. The PCB layouts have been modified to include the additional components as well as some minor modifications, and consequently the boards have grown slightly in size from the originals. Two alternative versions of the exciter module are included in an attempt to thwart IC obsolescence.

Top view of the interior of the G3TSO HF SSB transceiver, showing VFO, exciter module, PA and LPF at rear, aux audio module and DFM

Bottom view of the interior of the G3TSO transceiver, showing band-pass filter, premix unit, PA and LPF units

MEDIUM AND HIGH-POWER SOLID-STATE HF LINEAR AMPLIFIERS

The construction of a solid-state linear amplifier now represents a realistic alternative to valve designs even up to the maximum UK power level of 400W. The construction of solid-state amplifiers has often been discouraged in the amateur press with claims that such projects can only be made with the use of expensive test equipment, and in particular, a spectrum analyser. The simple spectrum analyser described by Roger Blackwell, G4PMK, in *Radio Communication* November 1989 (see Chapter 15) is ideal for the purpose, and fully justifies the small amount of money necessary for its construction.

Motorola Applications Notes AN762, AN758 and EB104, available from Motorola in the UK, provide the designs for solid-state linear amplifiers with outputs in the range 140W to over 1kW. The applications notes (AN) and engineering bulletins (EB) describe the construction and operation of suitable amplifiers using Motorola devices and include printed circuit foil information, making construction relatively easy. It is interesting to note that the PA units fitted to virtually all of the currently available commercial amateur radio equipment are based upon these designs by Motorola, with only a few individual differences to the original design.

With the availability of suitable application notes it is perhaps surprising that few amateurs seem to have embarked upon such a project. High prices for solid-state power devices, combined with a lack of faith in their reliability, may well account for the reluctance to construct solid-state linear amplifiers. Components for the construction of solid-state linear amplifiers have been readily available in the USA for some time: Communications Concepts Inc of Xenia, Ohio have offered a complete range of kits and components for the Motorola designs, and many of these are now available in the UK from Mainline Electronics, PO Box 235, Leicester.

The kits include the PCB, solid-state devices, and all the components and the various ferrite transformers already wound, so that all you have to do is solder them together. A large heatsink is necessary and is not supplied.

Home construction of PCBs for an amplifier is possible from the foil patterns available in the application notes, but it should be borne in mind that the thickness of copper on much of the laminate available to amateurs is unknown and inadequate for the high current requirements of a linear amplifier. A ready-made PCB is therefore a very sensible purchase and the kit of parts as supplied by CCI represents a very cost effective way of building any of the Motorola designs.

Choosing an amplifier

Perhaps the most useful amplifier for the radio amateur is described in AN762. It operates from a 13V supply and is capable of providing up to 160W output with only 5W of drive. This design is the basis for the majority of commercial 100W PA units and lends itself to both mobile and fixed station operation using readily available power supplies. Table 7.27 illustrates a number of alternative designs.

AN758 describes the construction of a single 300W output amplifier operating from a 50V supply and further describes a method of using power combiners to sum the outputs of a number

Table 7.25. G3TSO transceiver voltages

MODULE 1

IC	1	2	3	4	5	6	7	8	Notes
102, 3	Gnd	6	2	Gnd	0.8	0.8	0	Gnd	Pin 7 rises to 5V with AGC
104	Gnd	2.7	0	6	5.2	5.2	2.8	Gnd	
105	1.2	0	1	0.4	1	nc	Gnd	Gnd	Pin 2 rises to 2V with ant noise, 2–5V with AGC
106	nc	4.8	4.8	Gnd	nc	4.8	13	nc	
107	1.4	Gnd	Gnd	6.6	13	Gnd	Gnd	0	
108	Gnd	2.7	0	6	6	6	2.6	Gnd	
109	0.1	3	6	1.6	0	Gnd	1.4	1.4	

RX measurements – no ant input. TX measurements drive max, no AF input.

TR	s	g	d
102, 3	7	0	13
101	0	0	13

TR	e	b	c
105	4.8	5	13
104	6	6	0

MODULES 2 and 3

TR	s	g	d
201	0.14	–2	9.1
202	1.2	0	9.1
301	0	0	8

TR	e	b	c
302	3.8	4.4*	9.6
303	3	3.8	9.6
304	1.5	2.2	12†

* RF input to TR302 200mV pp (80m).
† RF col TR304 3V pp.

IC301	1	2	3	4	5	6	7
	Gnd	13	13	Gnd	13	Gnd	3
	8	**9**	**10**	**11**	**12**	**13**	**14**
	3	Gnd	0.7	1.4	0.7	1.4	Gnd

MODULES 4 and 4A

	1	2	3	4	5	6	7	8
IC401	Gnd	5.6	1.8	Gnd	1	1	0*	Gnd
IC402	Gnd	Gnd	1.8	6.4	nc	1.8	nc	4.2

* Pin 7 varies with ALC and drive.

MODULE 6

IC601	1	2	3	4	5	6	7
	0.5	0.1	0.5	0.6	0.1	0.6	0
	8	**9**	**10**	**11**	**12**	**13**	**14**
	0	0	0.05	0.05	0.1	0	10

Measurements taken on TX were with no AF input and no ALC action.

of similar units to provide power outputs of 600 and 1200W respectively. Two such units would be ideal for a full-power linear amplifier for UK use, giving an output comparable to the FL2100 type of commercial valve amplifier. It is recommended that any prospective constructor reads the relevant applications note before embarking upon the purchase and construction of such an amplifier.

After constructing a number of low-power transceivers it was decided to commence the construction of the AN762 amplifier. The EB63 design is similar, but uses a slightly simpler bias circuit.

AN762 describes three amplifier variations: 100W, 140W and 180W using MRF453, MRF454 and MRF421 devices respectively, and the middle-of-the road MRF454 140W variant was chosen (Fig 7.78).

Table 7.26. G3TSO transceiver RF signal levels in prototype

Band:	160	80	40	20	15	10A	10B	10C	10D
Xtal osc o/p R301 (mV) 175		120	—	160	—	140	175	175	175
Base TR301 (mV)	250	250	350	250	300	200	200	250	400
LO o/p C344 (V)	2.5	2.0	3.75	2.0	3.0	2.0	2.5	2.75	3.75
VFO o/p pin 7 IC301	125mV. R308/C310 700mV								
Band-pass filt i/p C448	200mV approx								
Band-pass filt o/p C409 (mV)	150	200	150	120	175	175	175	175	175
SL610 o/p pin 3 (V)	1.5	1.1	1.0	0.8	0.6	0.6	0.6	0.6	0.6
RF o/p PA 50Ω load (W)	15	20	20	20	20	20	20	20	20

Note: all voltages are peak-to-peak measured on an oscilloscope. Convert to RMS by dividing by 2.8.

Other useful voltages:

IC109 AF o/p pin8	250mV
IC108 DSB output pins 5 and 6	400mV
Xtal filter input	400mV
TR101 gate	200mV
D108 output	400mV

All the voltages measured on the original transceiver should be fairly close to those built from the *Radio Communication* design. All measurements were made using a CDU150 with an ×10 probe. A VVM or multimeter with RF probe can be used and will give RMS readings.

Constructing the amplifier

Construction of the amplifier is very straightforward, especially to anyone who has already built up a solid-state amplifier. The PCB drawings (Fig 7.79, see Appendix 1) and layouts are very good and a copy of the relevant application note was included with the kit. Fixing the transformers to the PCB posed a slight problem: on previous amplifiers that the author has built they were soldered directly to the PCB, but not this one. Approximately 20 turret tags (not supplied) are required: they are riveted to the board and soldered on both sides. A modification is required if you wish to switch the PA bias supply to control the T/R switching: the bias supply must be brought out to a separate terminal rather than being connected to the main supply line. This is achieved by fitting a stand-off insulator to the PCB at a convenient point; the bias supply components are then soldered directly to this stand-off rather than directly to the PCB.

A number of small ceramic chip capacitors have to be soldered directly to the track on the underside of the PCB; it should be borne in mind when doing this that the clearance between the PCB and heatsink is slightly less than 1/8in and the capacitors must not touch the heatsink.

It is advisable to mount the PCB to the heatsink before attempting to make any solder connections to it, as it is necessary to mark the mounting holes accurately, ideally drilling and tapping them either 6BA or 3mm. The power transistor mounting is very critical in order to avoid stress on the ceramic casing of the devices. The devices mount directly onto the heatsink and should ideally be fitted by drilling and tapping it. The PCB is

Table 7.27. Motorola RF amplifiers

Number	Power out (W)	Supply voltage (V)
AN762	140	12–14
EB63	140	12–14
EB27A	300	28
AN758	300–1200	50
EB104	600	50

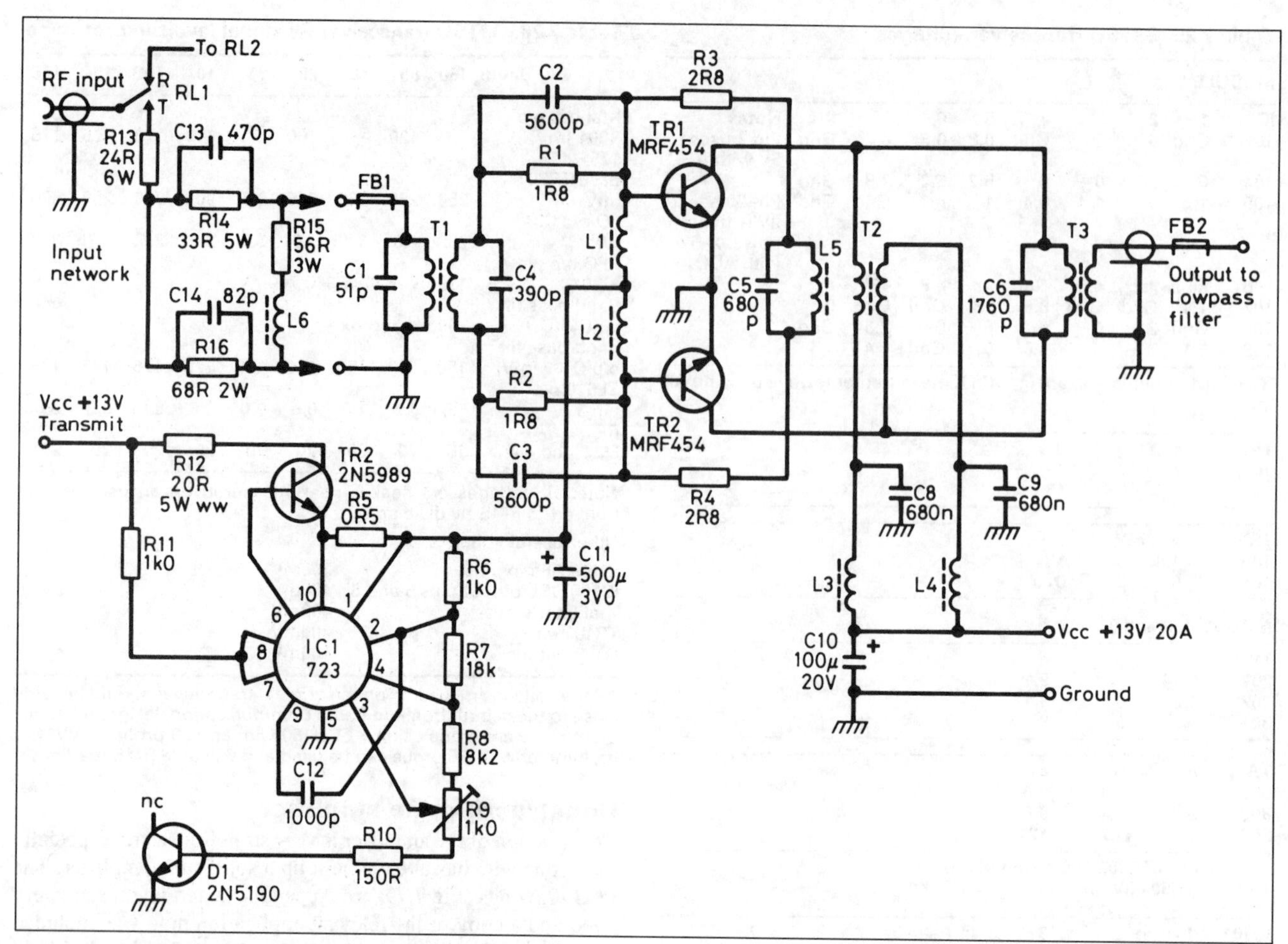

Fig 7.78. The HF SSB 140–300W linear amplifier circuit

raised above the heatsink on stand-offs made from either 6BA or 3mm nuts, so that the tabs on the transistors are flush with the

The HF linear amplifier

PCB – they must not be bent up or down. When the PCB and transistors have been mounted to the heatsink correctly, they may be removed for the board to be assembled. The transistors should *not* be soldered in at this stage. The nuts to be used as the stand-offs can be soldered to the PCB, if required, to make refitting the latter to the heatsink a little simpler, but ensure the alignment of the spacers is concentric with the holes.

Assembly should commence with the addition of the turret tags and stand-offs. Then the ceramic chip capacitors should be added under the PCB. D1, which is really a transistor, is also mounted under the PCB; only the emitter and base are connected – the collector lead is cut off and left floating. This transistor is mounted on a mica washer and forms a central stand-off when the PCB is finally screwed down to the heatsink. The mounting screw passes through the device which must be carefully aligned with the hole in the PCB. A number of holes on the PCB are plated through and connect the upper and lower ground planes together, and it is a good idea to solder through each of these holes. The upper-side components can be mounted starting with the resistors and capacitors, and finally the transformers can be soldered directly to the turret tags. Soldering should be to a high standard as some of the junctions will be carrying up to 10A or more.

When the board is complete it should be checked at least twice for errors and any long leads removed from the underside to ensure clearance from the heatsink. Mount the PCB to the heatsink and tighten it down. Now mount the power transistors which should fit flush with the upper surface of the PCB. Tighten them down, ensuring that there is no stress on the ceramic cases. If any of the connections need to be slightly trimmed to fit, cut

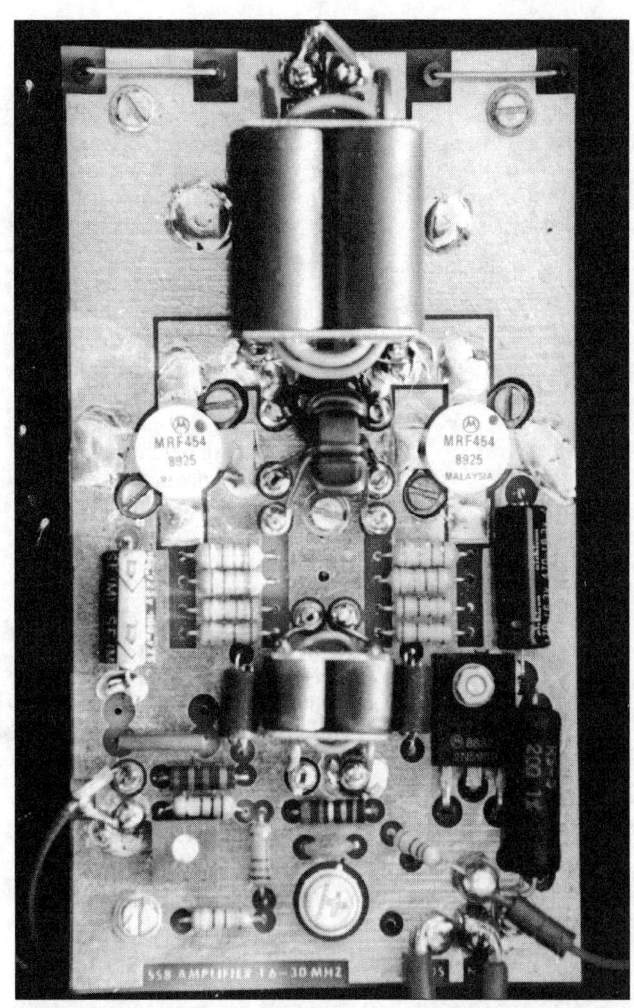

Circuit board of the linear amplifier

them with metal cutters. Ensure the collector tab is in the correct place. Once the transistors fit correctly, they can be removed again and very lightly tinned. The PCB should also be lightly tinned. The transistors can now be refitted and tightened down. *Now* they can be soldered in but, once in, they are very difficult to remove, so take great care at this stage. The amplifier board is now complete. It will need to be removed once more to allow the application of silicon heatsink compound to the devices in contact with the heatsink. Always unscrew the transistors first and then the PCB; refitting is a reverse of this process.

Construction of the amplifier takes very little time but requires considerable care to avoid damage to the output devices; the metal work may take a little longer. A large heatsink is required for 140W and an even larger one for 300W. It is recommended that the higher-power amplifiers are mounted onto a sheet of 0.25in copper which is in turn bolted to the main heatsink. (Note: on no account should the transistors be mounted onto a diecast box due to surface imperfections and poor thermal conductivity.) Blowing may be necessary at the higher powers or if a less-than-adequate heatsink is used.

Setting up and testing the amplifier

There is only one adjustment on the amplifier, making setting up relatively simple. Before connecting any power supplies, check and re-check the board for any possible errors. The first job is to test the bias supply. This must always be done before connecting the collector supply to the amplifier as a fault here

Solid-state amplifier – low-pass filter and ALC units

could destroy the devices instantly. With +13V connected to the bias supply only, it should be possible to vary the base bias from approximately 0.5V to 0.9V. Set it to the lowest setting, ie 0.5V. Disconnect the bias supply from the 13V line. When conducting any tests on the amplifier always ensure that it is correctly terminated in a 50Ω resistive dummy load. Apply +13V to the amplifier and observe the collector current on a suitable meter. It should not exceed a few milliamps – if it does something is wrong, so stop and check everything. Assuming that your amplifier only draws 3 to 4mA, connect the bias supply to the +13V supply and observe an increase in current, partly caused by the bias supply itself and also by the increased standing current in the output devices. The current can be checked individually in each of the output devices by unsoldering the wire links L3 and L4 on the PCB. Set the bias to 100mA per device by adjusting R9. The current should be approximately the same in each device: if it is not it could indicate a fault in either device or the bias circuitry to it. Increase the standing current to ensure that it rises smoothly before returning it to 100mA per device. Once the total standing current is set to 100 + 100 = 200mA, the amplifier is ready for operation.

With a power meter in series with the dummy load, apply a drive signal to the input, steadily increasing the level. The output should increase smoothly to a maximum of about 160W. It will go to 200W but will exceed the device specification.

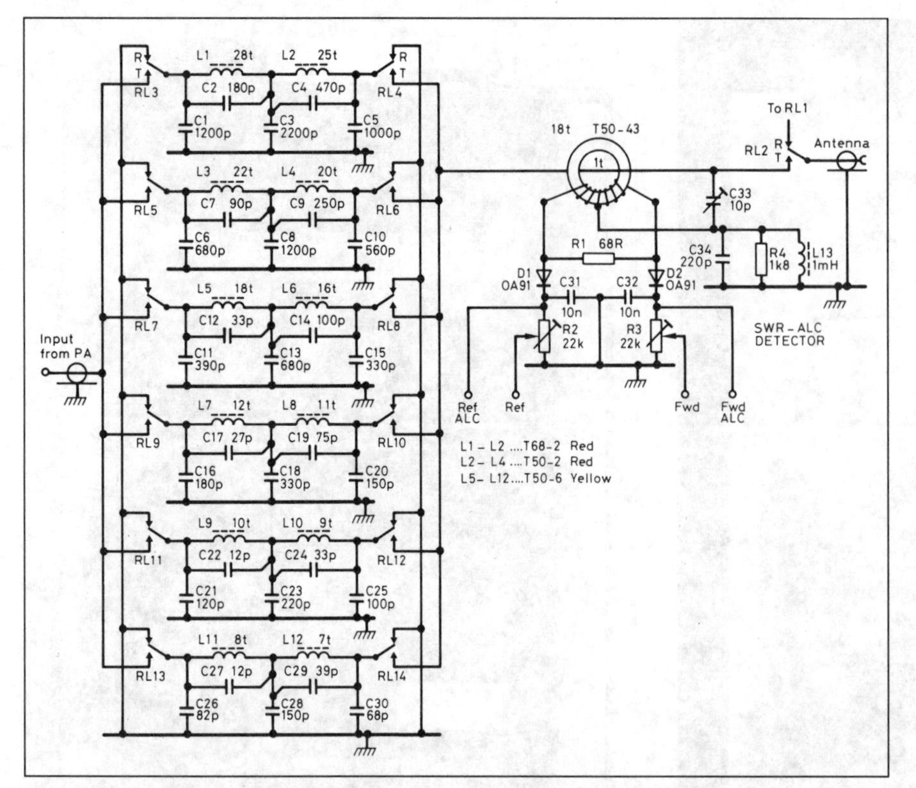

Fig 7.80. The amplifier is followed by six switched filters and a detector for SWR and ALC circuitry

Observing the output on a spectrum analyser should reveal the primary signal, together with its second, third and higher harmonics. Check that there are no other outputs. Removal of the input signal should cause the disappearance of the other signals displayed. It is helpful during initial setting up to monitor the current drawn by the amplifier. At full output, efficiency should be in the order of 50%, perhaps lowering slightly at the upper and lower frequency limits and increasing a little somewhere in the 20MHz range. The maximum current likely to be drawn by the 140W amplifier is in the order of 24A.

Putting the amplifier to use

Building and setting up the amplifier is undoubtedly a simple operation, and may lead one into a false sense of security. Before the amplifier can be used it must have a low-pass filter added to the output to remove the harmonics generated. For single-band operation only one filter would be required, but for operation on the HF amateur bands a range of filters is required with typically six switched filters covering the range 2 to 30MHz.

For most applications a five-pole Chebyshev filter will provide all the rejection required, but the majority of commercial designs now use the elliptic type of filter providing peaks of rejection centred around the second and third harmonics. Such filters can be tuned to maximise the rejection at specific frequencies. An elliptic function filter was decided upon as it only requires two extra components over and above the standard Chebyshev design and setting up is not critical.

The construction of a suitable low-pass filter (Fig 7.80) may take the form of the inductors and capacitors mounted around a suitable wafer switch, or they may be mounted on a PCB and switched in and out of circuit using small low-profile relays. This makes lead lengths shorter and minimises stray paths across the filter. Unused filters may be grounded easily using relays permitting only one filter path to be open at a time. The relays need only to be able to carry the output current; they are not required to switch it. 2A contacts are suitable in the 100–140W range. Amidon cores ensure the duplication of suitable inductors while silver mica capacitors should be used to tune the filters. The voltage working of the capacitors should be scaled

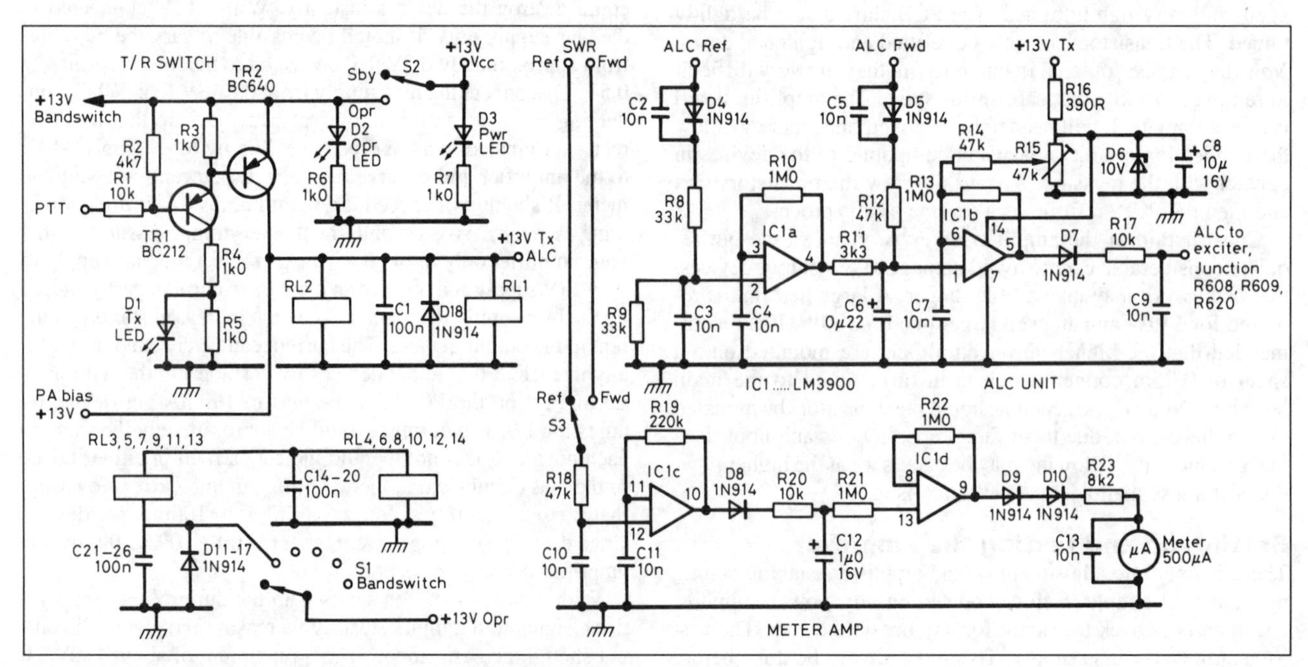

Fig 7.81. Transmit/receive switching, ALC and SWR circuits

Table 7.28. HF linear amplifier components list

HF LINEAR AMPLIFIER	
C1	51p chip
C2, 3	5600p chip
C4	390p chip
C5	680p chip
C6 (C7)	1760p (2 × 470p chips plus 820p silver mica in parallel)
C8, 9	0.68µ chip
C10	100µ 20V
C11	500µ 3V
C12	1000p disc
C13	470p silver mica
C14	82p silver mica
R1, 2	2 × 3.6R in parallel
R3, 4	2 × 5.6R in parallel
R5	0.5R
R6	1k
R7	18k
R8	8k2
R9	1k trimpot
R10	150R
R11	1k
R12	20R 5W WW
R13	24R carbon 6W*
R14	33R carbon 5W*
R15	56R carbon 3W*
R16	68R carbon 2W*
* Make up from several higher values in parallel.	
L1, 2	VK200 19/4B choke (6-hole ferrite beads)
L3, 4	Fairite beads × 2 (2673021801) on 16swg wire
L5	1t through T2
T1	2 × Fairite beads 0.375in × 0.2in × 0.4in, 3:1 turns
T2	6t 18swg ferrite 57-9322 toroid
T3	2 × 57-3238 ferrite cores (7d grade) 4:1 turns
FB1, 2	Fairite 26-43006301 cores
RL1	OUD type
L6	0.82mH (T50-6)
TR1, 2	MRF454
TR3	2N5989
D1	2N5190
IC1	723 regulator
LOW-PASS FILTER	
C1	1200p
C2, 16	180p
C3	2200p
C4	470p
C5	1000p
C6, 13	680p
C7	90p
C9	250p
C10	560p
C11	390p
C12, 24	33p
C14	100p
C15, 18	330p
C17	27p
C19	75p
C20, 28	150p
C21	120p
C22, 27	12p
C23	220p
C25	100p
C26	82p
C29	39p
C30	68p
C31, 32	10n ceramic
C33	10p trimmer
C34	220p silver mica
Note: C1–C30 silver mica 350VDC	
L1	28t 22swg T68-2
L2	25t 22swg T68-2
L3	22t 22swg T50-2
L4	20t 22swg T50-2
L5	18t 20swg T50-6
L6	16t 20swg T50-6
L7	12t 20swg T50-6
L8	11t 20swg T50-6
L9	10t 20swg T50-6
L10	9t 20swg T50-6
L11	8t 20swg T50-6
L12	7t 20swg T50-6
T1	18t bifilar T50-43 pri: 1t
R1	68R
R2, 3	22k trimpot
R4	1k8
L13	1mH RFC
RL2	OM1 type
RL3–14	2A SPCO PCB mtg, 6V coil
D1, 2	OA91 or OA47
SWITCHING AND ALC	
R1, 17, 20	10k
R2	4k7
R3–7	1k
R8, 9	33k
R10, 13, 21, 22	1M
R11	3k3
R12, 14, 18	47k
R15	47k trimpot
R16	390R
R19	220k
R23	8k2
TR1	BC212
TR2	BC640
IC1	LM3900
C1, 14–26	100n
C2–5, 7, 9–11, 13	10n
C6	0.22µ
C8	10µ 16V
C12	1µ 16V
D1–3	LEDs
D4, 5, 7–18	1N914
D6	10V zener
S1	1-pole 6-way
S2, 3	SPCO
Meter	500µA or similar

to suit the power level being used; ideally, 350V working should be used in the 100–150W range and, for powers in the region of 400–600W, 750V working capacitors should be used, the latter being available from CCI.

The antenna change-over relay may be situated at either end of the low-pass filter. If it is intended to use the filter on receive then the relay will be placed between the amplifier and the filter, but if the amplifier is an add-on unit then it may not be necessary to use the filter on receive and the relay may be located at the output end of the filter. The filter performance is enhanced if it is mounted in a screened box with all DC leads suitably decoupled.

SWR protection and ALC

One of the major shortcomings of early solid-state amplifiers was PA failure resulting from such abuse as overdriving, short-circuited output, open-circuited output and other situations causing a high SWR. A high SWR destroys transistors either by exceeding the collector-base breakdown voltage for the device or through overcurrent and dissipation.

ALC (automatic level control) serves two functions in a modern-day transmitter: it controls the output power to prevent overdriving and distortion and can be combined with a SWR detector to reduce the power if a high SWR is detected, which reduces the voltages that can appear across the output device and so protects it.

A conventional SWR detector provides indication of power (forward) and SWR (reflected power) which can be amplified and compared with a reference. If the forward power exceeds the preset reference an ALC voltage is fed back to the exciter to reduce the drive and hence hold the power at the preset level. A high SWR will produce a signal that is amplified more than the forward signal and will reach the reference level more quickly, again causing a reduction in the drive level.

The circuit shown in Fig 7.81 has been designed to work in conjunction with the ALC system installed in the G3TSO modular transceiver and produces a positive-going output voltage. The LM3900 IC used to generate the ALC voltage contains two unused current-sensing op-amps which have been used as a meter buffer with a sample-and-hold circuit providing a power meter with almost a peak reading capability. In practice it reads about 85% of the peak power compared to the 25% measured on a typical SWR meter.

T/R control

There are many ways of controlling the T/R function of an amplifier. It was decided to make this one operate from the PTT line but unfortunately direct connection resulted in a hang-up when relays in the main transceiver remained activated after the PTT line was released. The buffer circuit comprising TR4 and TR5 simply switch the input and output relays from receive to transmit and provide a PA bias supply on transmit.

Interfacing amplifier to exciter

The G3TSO modular transceiver was used as the drive source for the AN762 solid-state amplifier, a mere 5W of drive producing a solid 140W output from the linear. A little more drive and 200W came out. This was rapidly reduced by setting the ALC threshold. Initially the recently constructed spectrum analyser showed the primary signal, harmonics suitably reduced by the action of the elliptic low-pass filter, but alas a response at 26MHz and not many decibels down on the fundamental. A quick check with the general-coverage transceiver revealed that there really was something there, while a finger on the 80m low-pass filter showed quite a lot of heat being generated.

Investigations revealed quite clearly that this type of broadband amplifier cannot be operated with a capacitive-input, low-pass filter at either end without it going into oscillation at some frequency, usually well above the cut-off frequency of the filter. The filter input impedance decreases with frequency, and with

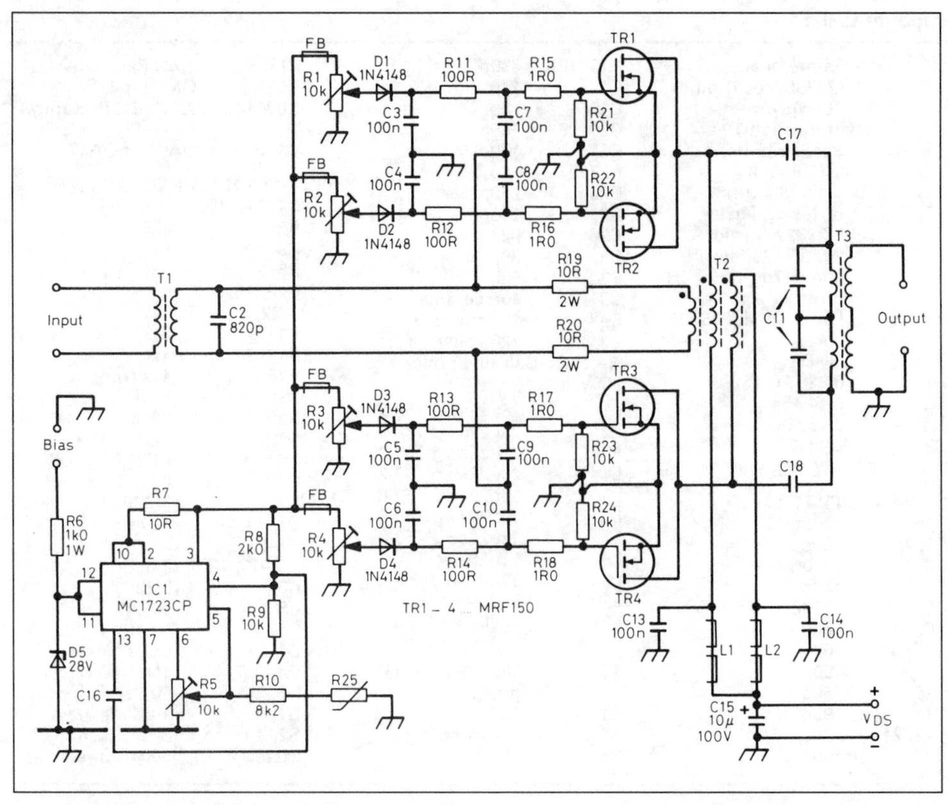

Fig 7.82. 600W output RF amplifier

two such filters located at either end of the amplifier there comes a point where the input circuit and output circuit resonate at the same frequency and a spurious oscillation occurs. Removal of either filter solves the problem.

A direct connection between the exciter and linear amplifier is the preferred solution, but is not always practicable in the case of add-on amplifiers where there is already a low-pass filter installed in the exciter. Another solution is to provide a resistive termination at the input of the amplifier; this is far simpler to effect and is used in a number of commercial designs. The network used comprises of a 50Ω carbon resistor placed across the input of the amplifier which effectively reduces the input impedance to 25Ω, so a 30Ω resistor is placed in series with the drive source to present a near-50Ω impedance to the exciter. Power from the exciter will be absorbed in these resistors which must be made up from a number of lower-wattage resistors in parallel, ie five 150Ω resistors make a 30Ω resistor with five times the power rating. In addition it was found necessary to add a ferrite bead to the input and output leads to the amplifier to effect a complete cure to the parasitic problem which was at its worst on 21MHz, the parasitics occurring above 40MHz. An alternative solution would be to use an inductive-input low-pass filter (Fig 7.34(c) and Table 7.8).

Table 7.29. Components list for 600W output RF amplifier

R1–R5	10k trimpot
R6	1k, 1W
R7	10R
R8	2k
R9, 21–24	10k
R10	8k2
R11–14	100R
R15–18	1R
R19, 20	10R, 2W carbon
R25	Thermistor, 10k (25°C), 2k5 (75°C)
All resistors ½W carbon or metal film unless otherwise noted.	
C1	Not used
C2	820p ceramic chip
C3–6, 13, 14	100n ceramic
C7–10	100n ceramic chip
C11	1200p each, 680p mica in parallel with an Arco 469 variable or three or more small mica capacitors in parallel
C12	Not used
C15	10µ, 100V elec
C16	1000p ceramic
C17, 18	Two 100n, 100V ceramic each (ATC 200/823 or equivalent)
D1–4	1N4148
D5	28V zener, 1N5362 or equivalent
L1, 2	Two Fair-Rite 2673021801 ferrite beads each or equivalent, 4µH
T1	9:1 ratio (3t:1t)
T2	2µH on balun core (1t line)
T3	See Fig 7.84
T1–T3 can be obtained from Mainline Electronics.	
TR1–4	MRF150
IC1	MC1723CP

Conclusion

The construction of a solid-state high-power amplifier is very simple, especially as the parts are obtainable in kit form in the UK and via international mail order from the USA. The use of a spectrum analyser, no matter how simple, greatly eases the

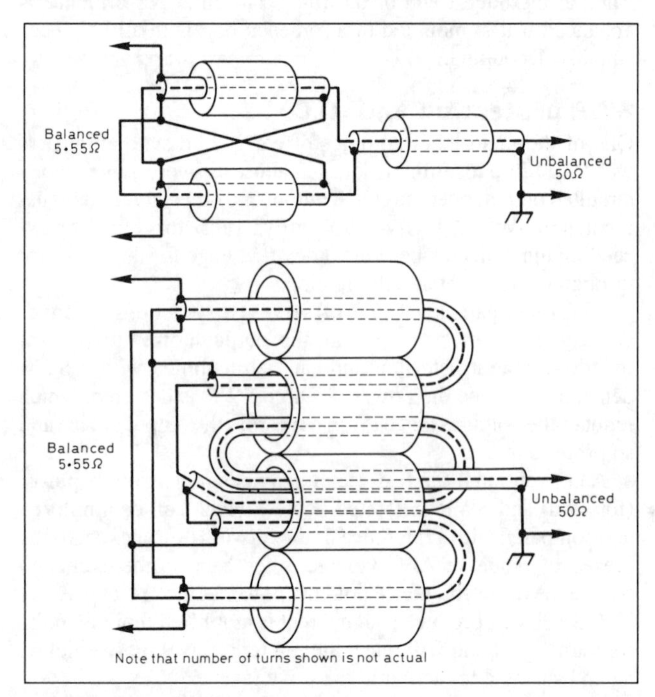

Fig 7.84. Winding the transformers

setting up of such an amplifier and ensures peace of mind when operating it.

Amplifiers are available for a number of different power levels, and can be combined to provide higher power levels. The use of 13V supplies practically limits powers to about 180W and below, while 28 or 50V facilitates higher-power operation without the need for stringent PSU regulation.

The full-power solid-state linear is now possible at a price showing a considerable saving on the cost of a commercially made unit.

There is a great similarity between all the Motorola designs: the 300W amplifier described in AN758 is virtually the same as the AN762 amplifier, the fundamental difference being the 28V DC supply voltage. The 600W MOSFET amplifier (Figs 7.82, 7.83 in Appendix 1, 7.84 and 7.85, plus Table 7.29) described in EB104 represents a real alternative to the valve linear amplifier and operates from a 50V supply, but considerably more attention must be paid to the dissipation of heat.

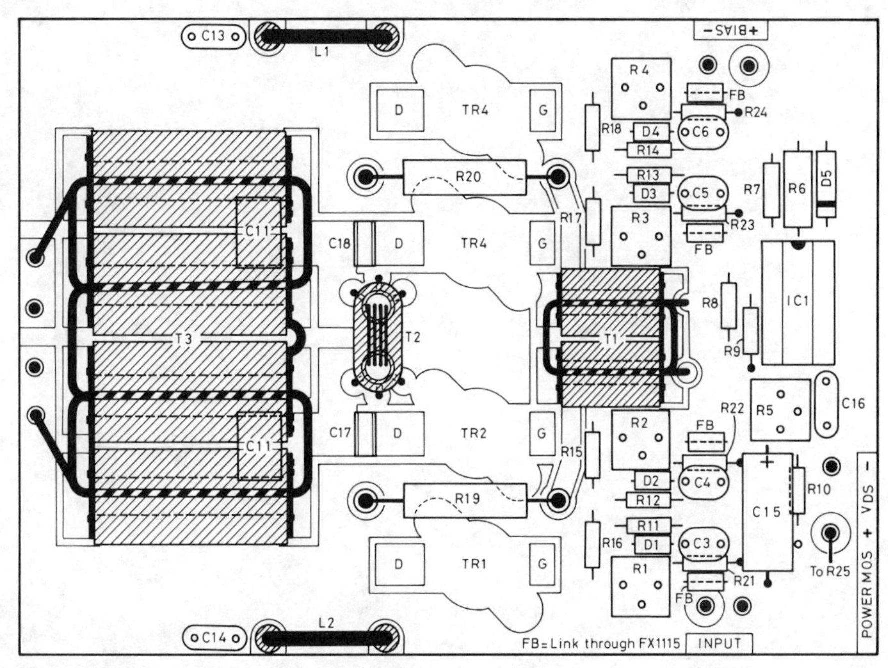

Fig 7.85. 600W amplifier component layout

Addresses

Motorola UK, 69 Fairfax House, Buckingham Road, Aylesbury, HP20 2NF.

Communications Concepts Inc, 508 Millstone Drive, Xenia, Ohio 45385, USA.
Tel (513) 426 8600.

REFERENCES

[1] 'QRP + QSK – a novel transceiver with full break-in', Peter Asquith, G4ENA, *Radio Communication* May 1992.

[2] 'A QRP transceiver for 1.8MHz', S E Hunt, G3TXQ, *Radio Communication* September 1987.

[3] 'Wideband linear amplifier', J A Koehler, VE5FP, *Ham Radio* January 1976.

[4] 'CW the easy way – a 14MHz CW transceiver', George Fare, G3OGQ, *Radio Communication* September 1988.

[5] 'A modular multiband transceiver', Mike Grierson, *Radio Communication* October/November 1988.

8 VHF/UHF receivers, transmitters and transceivers

THIS chapter will explore the theory and practice of designing receiving and transmitting equipment, including transceivers, for the 50, 70, 144 and 432MHz amateur bands. The 50MHz band will be new to many radio amateurs, but in terms of equipment design and construction, the principles are very similar to those for 70MHz equipment. 50, 70 and 144MHz (the 6m, 4m and 2m bands respectively) fall within the VHF spectrum, while 432MHz (the 70cm band) is within the UHF spectrum.

RECEIVERS

Standards for VHF/UHF receivers are strongly based on the performance expected from HF receivers, in particular the ability of the receiver to detect, without any deterioration in performance, a weak signal in the presence of one or more unwanted strong signals present at the same time.

Above 50MHz background noise is much lower, so with a good receiver it is possible to realise a performance superior in terms of sensitivity and signal-to-noise ratio. A HF signal of a few microvolts is often down in the noise but at VHF and UHF communication between stations can be achieved with signal levels as low as a few nanovolts ($1\text{nV} = 1\text{V} \times 10^{-9}$). On the HF bands a limit is imposed by both man-made and natural interference, beyond which any attempt to recover signals is fruitless. In VHF/UHF signal reception, there is no appreciable atmospheric noise with the exception of that caused by lightning discharges or from electrically charged rain drops. The limiting factor, when the receiver (and antenna) is in a good location, is extraterrestrial noise but the receiver can be designed to respond to signals only slightly above this level.

Definition of noise

Broadly, noise is unwanted signal of a more or less random nature within the pass-band of the receiver. It may be natural or man-made. Examples of natural noise are the radiation from the Sun or, as described earlier, that from electrical storms and charged rain drops. These can only be avoided by excluding the Sun or the electrical storms from the 'field of view' of the antenna. Also, there is the inescapable noise generated in a resistor at any temperature above absolute zero, and *shot noise* produced in semiconductors, caused by the random generation and recombination of electron-hole pairs in their operations.

Examples of man-made noise are the radiation from switches and thermostats when they break current, and the radiation from computers caused by their processing pulses with very fast rise and fall times.

In the design of VHF/UHF receivers only the inescapable natural noise needs to be considered. Resistors introduce thermal noise, due to the random motion of charge carriers which produce random voltages and currents in the resistive element. There is unfortunately no such resistor that will not produce these random products unless the receiver is operated at a temperature at absolute zero (0K). However, resistor noise generation can be minimised, particularly in the receiver front-end, by the correct choice of resistor. Metal film types are recommended. Thermal noise is also known as *Johnson* or *white noise*.

Shot noise in semiconductors is due to charge carriers of a particle-like nature having fluctuations at any one instance of time when direct current is flowing through the device. The random fluctuations cause random instantaneous current changes. Shot noise is also known as *Schottky noise*.

Noise factor and noise figure

The *noise factor* is the ratio of the input signal-to-noise ratio to the output signal-to-noise ratio. The *noise figure* is the noise factor expressed in decibels and is used as a figure of merit for VHF and UHF circuits.

$$f = \frac{\text{Input S/N}}{\text{Output S/N}}$$

$$\text{NF} = 10 \log_{10} f$$

It is measured as the noise power present at the receiver output assuming a conventional S/N ratio of 1 at the input. An ideal noiseless receiver does not produce any noise in any stage. Thus the equation becomes 1/1 or a noise factor of 1 or 0dB. The noise factor of a practical receiver which will generate noise in any stage, particularly the front-end, is the factor by which the receiver falls short of perfection.

Amateur communication receiver manufacturers usually rate the noise characteristics with respect to the signal input at the antenna socket. It is commonly expressed as (signal + noise)/noise or *signal-to-noise ratio*. The sensitivity is usually expressed as the voltage in microvolts at the antenna terminal required for a (signal + noise)/noise ratio of 10dB. Sensitivity can also be specified as the minimum discernible signal or *noise floor* of the receiver.

An important point to remember in VHF/UHF receivers is that the optimum noise figure of an RF amplifier does not necessarily coincide with the highest maximum usable gain from that stage.

Figs 8.1 and 8.2 illustrate this feature. The transistor is capable of operation up to 2.0GHz. As Fig 8.1 shows, however, the gain at noise figure falls but the actual noise figure increases with increasing frequency. This characteristic is also shown in Fig 8.2 but here the gain at noise figure and the actual noise figure are plotted for 500MHz and 1GHz against variations in collector current. Note that the maximum gain occurs with good input matching but minimum noise does not. For example, a GaAsFET preamplifier may have an input VSWR as high as 10:1 when tuned for the lowest noise figure.

The definition of noise figure and degradation of receiver performance due to noise implies that the front-end stages, namely the RF amplifier and mixer, must use active devices,

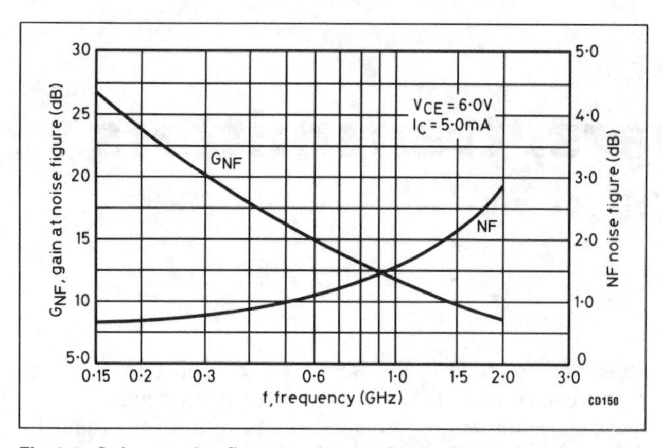

Fig 8.1. Gain at noise figure and noise figure versus frequency

either bipolar or field effect types, with a low inherent noise figure. The noise figure quoted for the transistor illustrated in Figs 8.1 and 8.2 applies only when the device is connected to a 50Ω source. As can be seen the noise figure increases by almost two times between 1 and 2GHz. This figure can be reduced by mismatching to the source at 1GHz and above.

Modern design theory and practice now employ *S-parameters* (scatter parameters) to obtain maximum performance from an RF amplifier and mixer while maintaining the noise figure at low levels. However, S-parameters require an advanced knowledge of design which includes the use of the Smith chart and the availability of some sophisticated equipment such as a network analyser. Manufacturers' data on RF devices includes tables of S-parameters.

This method of design of amplifiers and mixers is outside the scope of this handbook.

Intermodulation

Intermodulation occurs when two or more signals combine to produce addition (spurious) signals which were not originally present at the receiver input and possibly cause interference to a weak wanted signal. The receiver front-end is handling many incoming signals of different strengths but only these signals passing through the selective (IF) filters will eventually be detected. The RF circuits can have a bandwidth of several megahertz but the selective IF filters reduce the bandwidth to that required for adequate resolution of signals, dependent on the method of modulation of the wanted carrier. At low signal levels the front-end will have optimum linearity, ie there is no unwanted mixing between signals. However, as stated previously, very strong signals will cause the front-end to go into its non-linear region of operation, and then these signals will mix together and produce new signals which can appear in the IF pass-band. Second-order *intermodulation products* (IPs) are caused by two signals mixing, viz f_1 and f_2, and generating new frequencies which appear as $(f_1 + f_2)$ and $(f_1 - f_2)$ and the second harmonics of each signal ($2f_1$ and $2f_2$), generating the second-order IPs. However, if f_1 and f_2 are close spaced, their second-order IPs will be well spaced and can be easily filtered out by the selective IF filters.

However, if the f_1 and f_2 signals are increased in strength then another set of IPs is generated. These are third-order intermodulation products, due to the fact that mixing occurs between three signals. The three signals can be independent but the same products can be generated by f_1 and f_2 by themselves. These frequencies f_1 and f_2 can add or subtract to produce the following third-order IPs:

Third harmonics:	$(f_1 + f_1 + f_1)$ and $(f_2 + f_2 + f_2)$
Sum products:	$(f_1 + f_1 + f_2)$ and $(f_2 + f_2 + f_1)$
Difference products:	$(f_1 + f_1 - f_2)$ and $(f_2 + f_2 - f_1)$

It is clear that if f_1 and f_2 are equally spaced above and below the wanted frequency, interference will be severe. When $f_1 + f_2$ are close to the wanted frequency the third-order sum products will appear in the third harmonic area of this frequency and will be attenuated by the selective filters in the receiver. However, when the difference products containing a minus sign are close to f_1 and f_2 and are generated by the receiver, the filters, however selective, will not remove these spurious signals. These products could cause unwanted interference to a wanted weak signal. When third-order intermodulation products are generated in the receiver, they will increase in level by 3dB for every 1dB increase in the levels of f_1 and f_2. Thus the appearance of intermodulation products above the receiver noise floor is quite usual. When further levels of f_1 and f_2 occur, higher odd-order intermodulation products are generated, ie fifth, seventh etc, which can interfere with a weak wanted signal. These higher-order products will appear even quicker than third-order products but require stronger f_1 and f_2 signals, eg fifth-order products will be generated five times as fast as f_1 and f_2 as the level of f_1 and f_2 is increased.

Significant intermodulation products can only result from f_1 and f_2 when their strength is high. If either the f_1 or f_2 signal disappears, leaving only one signal, the intermodulation product will disappear.

Optimising linearity in the receiver front-end will minimise generations of these unwanted intermodulation products and hence interference to wanted signals.

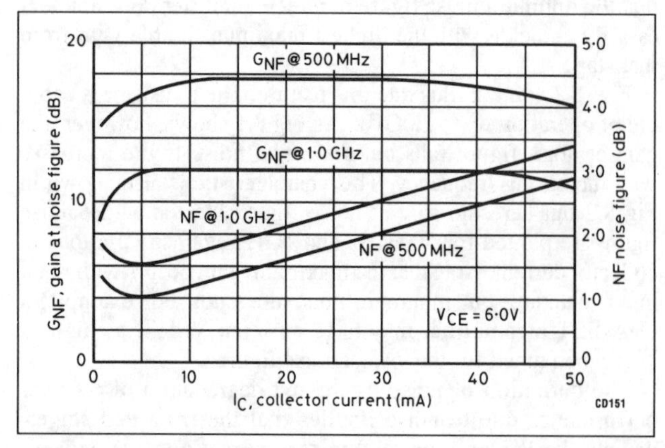

Fig 8.2. Gain at noise figure and noise figure versus collector current

Gain compression

This occurs when a strong incoming signal appearing at the antenna socket causes one (or more) stages in the front-end to drive into the non-linear region of its output characteristic. As an example, when an amplifier stage is operating in its linear region an increase by 3dB in signal level at its input will cause, by linear transfer, a 3dB increase in signal level at the output. However, a further increase in input signal level could cause non-linear transfer and limit the output level increase to 1dB. A very strong signal could drive the stage into extreme non-linearity, making the stage degenerative (gain less than 1) and desensitising the receiver. Background noise will decrease in level together with all other signals, including wanted weak signals.

As with intermodulation, optimising front-end linearity will minimise receiver desensitisation by strong signals.

Reciprocal mixing

This phenomenon occurs when the receiver local oscillator is allowed to produce excessive sideband noise on its carrier and a strong off-channel RF carrier mixes with this noise to produce the IF. Reciprocal mixing causes an increase in receiver noise level when a strong carrier appears, the opposite effect of gain compression. The receiver selectivity is not necessarily defined by front-end RF filters. The 'cleanest' local oscillator carriers are generated by LC (VFOs) and crystal-controlled oscillators. Some of the 'noisiest' oscillators are found in receivers employing a synthesised system, for example a phase-locked-loop/synthesised oscillator. Some earlier receivers suffered from reciprocal mixing effects, ie generation of an unwanted spurious signal in the IF pass-band, for example noise on the voltage-controlled oscillator control line, leading to FM noise sidebands. However, modern receivers now employ 'quiet' synthesised local oscillators which minimise reciprocal mixing.

Dynamic range

The main problems of front-end overload are gain compression, intermodulation and reciprocal mixing. Each phenomenon has its own characteristic and level at which strong unwanted signal(s) cause degradation in receiving wanted signals. Only one strong signal causes gain compression or reciprocal mixing, whereas two are required to cause intermodulation products. The gain of the front-end, ie the RF stage and the mixer, should be kept as low as possible consistent with good sensitivity and signal/noise performance. Gain compression and intermodulation are caused when either or both stages are driven beyond their linear transfer range. The front-end should be designed so it cannot be overloaded by even the strongest amateur-band signals. Intermodulation products will not be a problem if they are restricted to the level of background noise level, and gain compression and reciprocal mixing effects are not a problem if they do not significantly change the system noise level.

The lowest end of the dynamic range will be designed for the lowest-power audible signal, and conversely the highest end will be designed for the unwanted signal of the highest power level, ie signals without any overload effects degrading the front-end performance. This principle is called *spurious-free dynamic range* but the range will change according to the differing power levels of unwanted signals.

RECEIVER FRONT-END STAGES

The requirements then, for front-end stage design in a VHF/UHF receiver are (a) low noise figure, (b) large dynamic range and (c) power gain consistent with good sensitivity and signal-to-noise ratio.

The noise figure and dynamic range requirements have already been described in detail. However, 'power gain' must be brought into the equation to complete the design philosophy. 'Power gain' needs some explanation because sheer power gain is not sufficient in itself or even desirable. In a multistage receiver with, say, eight stages of gain, input noise originating in the first stage, normally an RF amplifier, will be amplified by the eight gain stages, that in the second by seven and so on. If the effective noise voltages are denoted by V_1, V_2, $V_3 \ldots V_8$ and the stage gains by G_1, G_2, $G_3 \ldots G_8$, the total noise present at the receiver detector will be:

$$V_1(G_1\, G_2\, G_3 \ldots G_8) + V_2(G_2\, G_3 \ldots G_8) + V_3(G_3 \ldots G_8)$$

and so on. If the voltage gain of the RF amplifier (G_1) is high, for example 20dB (10 times) or more, the important noise contribution is due to G_1. Provided the remaining gain stages are correctly designed and provide evenly distributed gain, the overall noise contribution from them will be very small.

Additional noise generated by any stage which is degenerative or is actually oscillating will degrade the overall receiver noise performance, and may actually cause receiver desensitisation and consequent poor weak-signal performance.

However, at the output of G_3 (eg the first IF stage) an amplified signal will be large compared to the noise contributed and $G_4 \ldots G_8$ should not degrade noise performance. The function of the RF stage is to provide just sufficient gain to overcome the noise contribution of the mixer stage. The mixer is by definition a non-linear device and normally contributes more noise than any other stage, no matter how well designed. The RF stage therefore considerably improves the receiver signal-to-noise ratio, improving weak-signal performance. If the RF stage gain is too high then the problems of strong signals, ie intermodulation, gain compression and reciprocal mixing products, will, as described, degrade receiver performance.

A system of gain control on the RF amplifier may appear to be the answer. However, the use of AGC must be carefully considered, otherwise the weak-signal performance will be degraded if the onset of AGC is not delayed. The RF stage linearity can itself be degraded by applying an AGC current (bipolar transistor) or voltage (field effect transistor) to reduce the amplifier gain. The noise factor of a transistor amplifier is also dependent on emitter (source) current. A large variation from the manufacturer's data given for noise factor $V\ I_E\ (I_s)$ will also degrade weak-signal performance. Circuit layout and correct shielding for VHF/UHF RF amplifiers is of paramount importance for stable operation. Instability and spurious oscillation can be produced via RF feedback through the amplifier transistor. Even the latest designs of transistor, both bipolar and field effect types, can give rise to these effects. Regeneration or actual oscillation can be prevented by neutralising the internal feedback by using an external circuit from output to input of the amplifier. This will feed back an equal amount of out-of-phase signal, neutralising the feedback, and, providing there is no other feedback path, the amplifier will be stable.

Examples of neutralising circuits are described in the sections on RF amplifiers. These circuits must be adjusted for optimum neutralising exactly at one frequency, but adjustment should be close enough for operation on any frequency within the VHF bands. Neutralising is not advised for UHF RF amplifiers.

Circuit noise

Noise due to devices other than transistors is produced solely by the resistive component; inductive or capacitive reactances do not produce noise. Inductors of any form have negligible resistance at VHF and UHF but the leakage resistance of capacitors and insulators is important. It is imperative to choose high-quality capacitors such as silver ceramic, polycarbonate, or polystyrene types with negligible leakage current and high-*Q* properties. Tantalum capacitors should be used where it is necessary to decouple at LF as well as VHF and UHF. Attention should be given to the use of low-noise resistors. Carbon or metal film types (of adequate power dissipation) must be used. The old-style carbon composition resistors are very good noise generators. 25dB noise difference between film and composition resistors has been observed when a direct current was passed through the two types, both being the same value.

Circuit noise due to regeneration has already been discussed. Common causes of regeneration are:

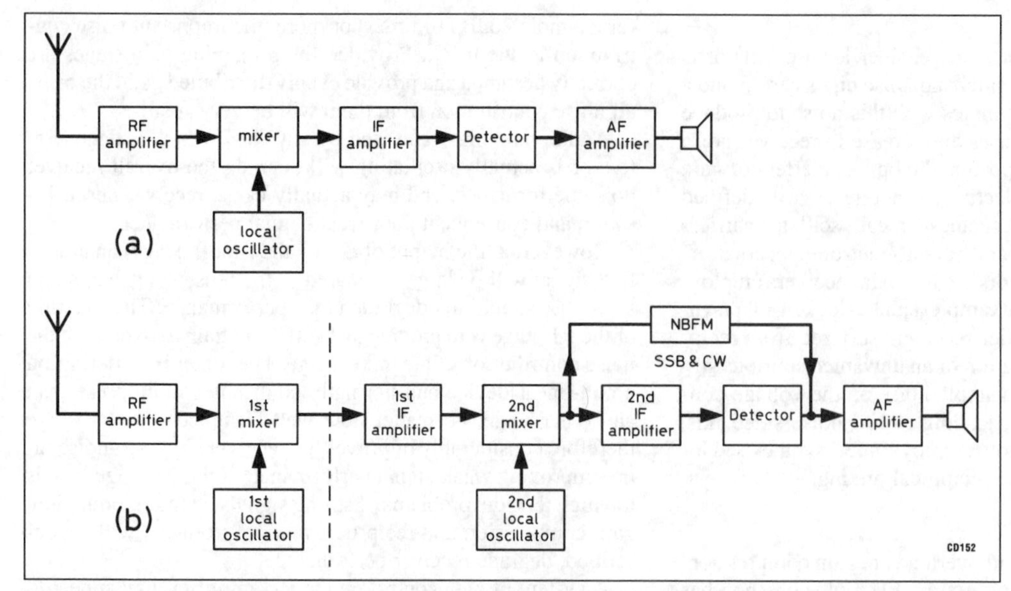

Fig 8.3. Two conventional superheterodyne configurations. (a) Self-contained single superhet with tuneable oscillator. (b) Self-contained double superhet or converter in front of a single superhet. Either or both oscillators may be tuneable

(a) *Insufficient decoupling of voltage supplies (at each stage) and particularly emitter (source) and collector (drain) circuits.* Suitable capacitor values and types are as follows:

10MHz	47nF (0.047μF) ceramic disc or plate
10–20MHz	22nF (0.22μF) ceramic disc or plate
20–30MHz	10nF (0.01μF) ceramic disc or plate
30–100MHz	4.7nF (0.0047μF) ceramic disc or plate
100–200MHz	2.2nF (0.0022μF) ceramic disc or plate
200–500MHz	1nF (0.001μF) ceramic disc or plate
500–1000MHz	220pF (0.00022μF) ceramic disc or plate

Leadless capacitors should be used above about 400MHz.

(b) *Closely sited input and output circuits.* Every attempt should be made to build the amplifier(s) in a straight line, and where possible to use shielded coils, particularly for 6m, 4m and 2m equipment.

(c) *Insufficient or wrongly placed screening between input and output circuits.* In transistor amplifiers where a screen is required, it should be mounted close fitting across the transistor with input and output circuits (normally base and collector) on opposite edges of the screen. The screen can conveniently be made of double-sided copper laminate and soldered to the main PCB.

(d) *Circulating IF currents in the PCB due to multipoint grounding.* Decoupling capacitors for each stage should be grounded at a single point very close to the emitter (source). It is preferable to use a double-sided PCB using one side as the ground plane.

Effect of bandwidth on noise

If the noise factor of a receiver is measured with a noise generator, it is independent of receiver bandwidth. Generator noise has the same characteristics as circuit noise so, for instance, if the bandwidth doubled, the overall noise is doubled. For the reception of a signal of finite bandwidth, however, the optimum signal-to-noise ratio is obtained when the bandwidth of the receiver is only just sufficient to accommodate the signal. Any further increase in bandwidth results in increased noise. The signal-to-noise ratio at the receiver detector therefore depends on the power per unit bandwidth of the transmitted signal.

As an example, a receiver may generate 0.25μV of noise for each 2.5kHz of bandwidth. Assume an SSB transmitter radiates a sideband signal of 2.5kHz bandwidth and produces 2.5μV of signal at the receiver detector. The signal-to-noise ratio is therefore 10 provided the receiver overall bandwidth is 2.5kHz. If the transmission bandwidth is reduced to 1.25kHz for a CW signal and the radiated power is unchanged, the receiver input will remain at 0.25μV but, if the bandwidth is also reduced to 1.25kHz, only 0.125μV of noise is detected by the receiver, and the signal-to-noise ratio will increase to 20.

Using receiver bandwidths which exceed transmission bandwidths is therefore undesirable when optimum signal-to-noise ratio is the prime factor. Transmitters with poor frequency stability will either require the receiver to be retuned or to use wider bandwidth, resulting in a degraded signal-to-noise ratio. Fortunately well-designed transmitters with PLL synthesised oscillators or crystal-controlled oscillators are now employed in the majority of the VHF/UHF bands.

CHOICE OF RECEIVER CONFIGURATION

Receivers using other than superheterodyne techniques are rare on VHF or UHF. Modern superheterodyne receivers may have one, two or three frequency changes before the final IF, each with its own oscillator which may be tuneable (by the receiver tuning control) or of fixed frequency. Receivers may have a variety of configurations; two are illustrated here in Fig 8.3. Fig 8.3(a) shows a conventional single superheterodyne for use on the HF bands. The local oscillator will be partially synthesised, ie use a pre-mixer driven by an HF crystal-controlled oscillator and a LF VFO to produce the local oscillator for the main mixer. Fig 8.3(b) shows a double superheterodyne. This can be a purpose-built receiver (or as illustrated in Fig 8.3(a)) to which is added (to the left of the dotted line) a VHF/UHF *converter*. The first local oscillator is crystal controlled and tuning is accomplished by the HF receiver (second) oscillator.

The main disadvantage of this method of using an HF receiver as a 'tuneable IF' amplifier preceded by a converter is again the problem of overloading the first amplifier and second mixer with strong signals. This will result in intermodulation and reciprocal mixing products, if not gain compression, particularly if the converter gain is high, say, 20 to 30dB (×10 to ×30).

A superior arrangement is to build a tuneable IF amplifier containing all the refinements of a normal HF receiver, including an NBFM IF amplifier and detector, and restrict the tuning range to a few megahertz to cover the VHF/UHF ranges of the converters. The HF receiver gain in front of the second mixer must be low. The first IF amplifier can be omitted but the premixer selectivity should be retained. The converter gain (from

VHF/UHF to first IF) should also be low: 10 to 14dB (×3 to ×4). This will result in a VHF/UHF receiver with a very good noise factor and dynamic range, and demodulation of NBFM in addition to CW and SSB signals.

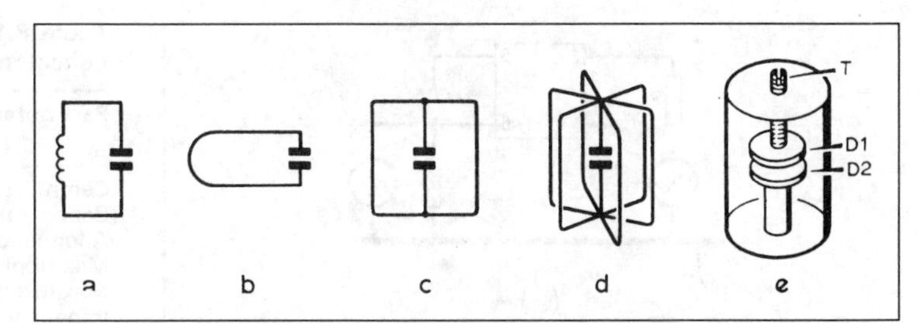

Fig 8.4. Progressive development of tuned circuits from a coil to a cavity as the frequency is increased

Choice of the first IF

In any superheterodyne receiver it is possible for two incoming frequencies to mix with the local oscillator to give the IF, these being the desired signal and the image frequency. A few figures should make the position clear. It will be assumed that the receiver is to cover the 144 to 146MHz band, and that the first IF is to be 4 to 6MHz. The crystal oscillator frequency must differ from that of the signal by this range of frequencies as the band is tuned and could therefore be 144 − 4MHz = 140MHz or, alternatively, 144 + 4MHz. Assuming that the lower of the two crystal frequencies is used, a signal on 136MHz would also produce a difference of 4MHz, and unless the RF and mixer stages are selective enough to discriminate against such a signal, it will be heard along with the desired signal on 144MHz. From the foregoing it will be appreciated that the image frequency is always removed from the signal frequency by twice the IF and is on the same side as the local oscillator.

It should be noted that even if no actual signal is present at the image frequency, there will be some contributed noise which will be added to that already present on the desired signal. It is usual to set the RF and mixer tuned circuits to the centre of the band in use so that on the 144MHz band they should be at least 2MHz wide in order to respond to signals anywhere in the band. This bandwidth only represents approximately 1.4% of the mid-band frequency and it is not surprising that appreciable response will be obtained over the image frequency range of 134 to 136MHz unless additional RF filtering is employed. Naturally, the higher the first IF, the greater the separation between desired and image frequencies. However, an IF as low as 4 to 6MHz is feasible, provided some attempt is made to restrict the bandwidth of the converter by, for example, employing two inductively-coupled tuned circuits between the RF and mixer stages, thus providing a band-pass effect.

The choice of the first IF is also conditioned by other factors. First, it is desirable that no harmonic of the oscillator in the main receiver should fall in the VHF band in use and, second, there should be no breakthrough from stations operating on the frequency or band of frequencies selected for the first IF.

Many HF receiver oscillators produce quite strong harmonics in the VHF bands and, although these are high-order harmonics and are therefore tuned through quickly, they can be distracting when searching for signals in the band in question. The problem only exists when the converter oscillator is crystal controlled, as freedom from harmonic interference is then required over a band equal in width to the VHF band to be covered. This also applies, of course, to IF breakthrough.

As it is practically impossible to find a band some hundreds of kilohertz wide which is unoccupied by at least some strong signals, it is necessary to take steps to ensure that the main receiver does not respond to them when an antenna is not connected. Frequencies in the range 20 to 30MHz are often chosen since fewer strong signals are normally found there than on the lower frequencies, but this state of affairs may well be reversed during periods of high sunspot activity.

With the greatly increased use of general-coverage receivers covering 100kHz to 30MHz, the best part of the spectrum for 6m, 4m and 2m is from 28 to 30MHz. Full coverage of the 70cm band will require the receiver to be tuned from 10 to 30MHz. IF breakthrough is minimised and frequency calibration is simple.

Tuned circuits

Tuning is readily achieved at HF by *lumped circuits*, ie those in which the inductor and capacitor are substantially discrete components. At VHF the two components are never wholly separate, the capacitance between the turns of the inductor being often a significant part of the total circuit capacitance. The self-inductance of the plates of the capacitor is similarly important. Often the capacitance required is equal to or less than the necessary minimum capacitance associated with the wiring and active devices, in which case no physical component identifiable as 'the capacitor' is present and the circuit is said to be tuned by the 'stray' circuit capacitance.

As the required frequency of a tuned circuit increases, obviously the physical sizes of the inductor and capacitor become smaller until they can no longer be manipulated with conventional tools. For amateur purposes the limits of physical coils and capacitors occur in the lower UHF bands: lumped circuits are often used in the 432MHz band but are rare in the 1.3GHz band.

Distributed circuits

Fig 8.4 illustrates how progressively lower inductances are used to tune a fixed capacitor to higher frequencies. In Fig 8.4(b) the 'coil' is reduced to a single *hairpin loop*, this configuration being commonly used at 432MHz. Two loops can be connected to the same capacitor as in Fig 8.4(c).

This halves the inductance and can be very convenient for filters. Fig 8.4(d) represents a multiplication of this structure and in Fig 8.4(e) there is in effect an infinite number of loops in parallel, ie a cylinder closed at both ends with a central rod in series with the capacitor. If the diameter of the structure is greater than its height it is termed a *rhumbatron*, otherwise it is a *coaxial cavity*.

The simple hairpin, shown at Fig 8.4(b), is a very convenient form of construction: it can be made of wide strip rather than wire and is especially suitable for push-pull circuits. It may be tuned by parallel capacitance at the open end, or by a series capacitance at the closed end.

In a modification of the hairpin loop, the loop can be produced from good-quality double-sided printed circuit board and such an arrangement is known as *microstripline*. The loop is formed on one side; the ground-plane side of the PCB, through the dielectric, makes the stripline. When the PCB is very thin, the result is called *microstrip* which is used in many commercial receivers.

Bandpass circuits

Tuning of antenna and RF circuits to maintain selectivity in the front-end of a VHF/UHF receiver cannot be undertaken with

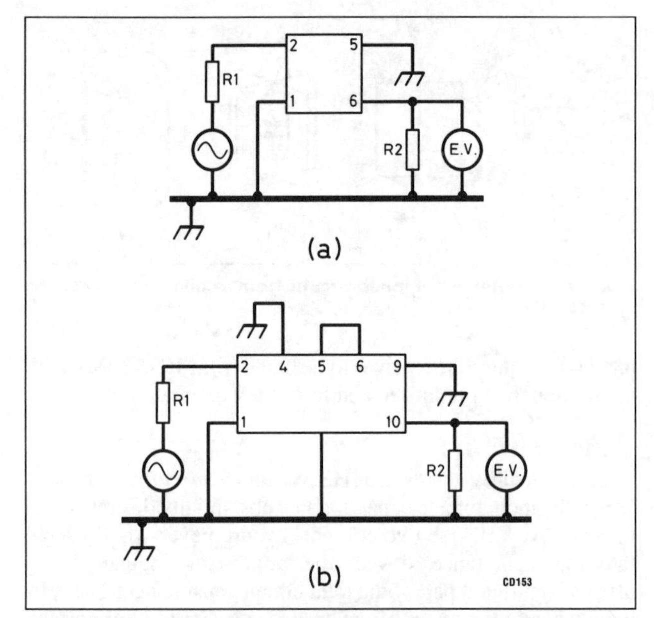

Fig 8.5. Test circuits for (a) Toko HRW and (b) Toko HRQ filters. The case lugs must be grounded. R1 = R2 = 50Ω

normal ganged tuning capacitors, not only due to the effects of strong capacitance coupling between circuits, which become prominent in these frequencies, but also due to the difficulty of procuring small-swing (say 20pF) multiganged capacitors. The varicap diode can be used to replace mechanical capacitors, but can degrade the receiver performance when strong signals are present by rectifying these signals and introducing intermodulation products into the mixer, thus seriously reducing the receiver's dynamic range.

Fortunately modern construction techniques have enabled coil manufacturers to introduce a band-pass circuit in a very small screened unit, namely the *helical filter*.

The helical resonator in simple terms is a coil within a shield. However, a more accurate description is a shielded, resonant section of helically wound transmission line having a relatively high characteristic impedance. The electrical length is approximately 94% of an axial quarter-wavelength. One lead is of the winding is connected to the shield, the other end is open-circuit. The Q_u of the resonator is dictated by the size of the shield, which can be round or square. Q_u is made higher by silver plating the shield. Resonance can be adjusted over a small range by opening or closing the turns of the helix. The adjustment is limited over the small frequency range to prevent degradation of Q_u. Modern miniature helical resonators can be obtained in a shield only 5mm square, but the minimum resonant frequency is normally 350MHz. Maximum F_r can be 1.5GHz. Large-size resonators (10mm square) will resonate down to 130MHz.

The band-pass filter is obtained by combining two to four resonators in one unit with slots cut in each resonator screen of defined shape to couple the resonators. This forms a high-selectivity tuned circuit with minimum in-band insertion loss and maximum out-of-band attenuation.

Helical filters can be cascaded to increase out-of-band attenuation. As an example, a quadruple filter with a centre frequency of 435MHz might have a 3dB bandwidth of 11MHz and 25dB attenuation at ±15MHz, with a ripple factor of 2dB and insertion loss of 4dB (see Table 8.1).

Tuning is accomplished by a brass screw in the top of the screen, one for each helix. The nominal input/output impedance is 50Ω, formed by placing a tap on the helix. This impedance is ideal for matching to antennas, and to RF amplifiers and mixers designed using S-parameters. Thus the helical filter replaces conventional tuned circuits in the receiver front-end, resulting in a considerable improvement in selectivity. One note of caution: the screening can lugs must be soldered perfectly to the PCB, otherwise the out-of-band attenuation characteristics of the filters will be degraded.

These filters can be used for 2m, 70cm and 23cm receiver front-ends.

Table 8.1. Electrical characteristics of Toko HRW and HRQ helical resonators

Parameter	HRW (231MT-10001A)	HRQ (232MT-1001A)
Centre frequency (MHz)	435	435
Bandwidth at 3dB (MHz)	12 min	11
Attenuation (dB)	20 min at ±30MHz	25 min at ±15MHz
Max ripple (dB)	1.5	2
Max insertion loss (dB)	2.5	4
Impedance (Ω)	50	50

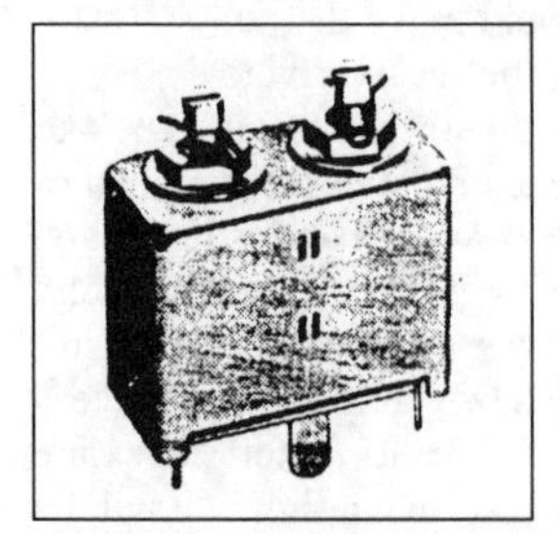

Fig 8.6. The Toko HRW resonator

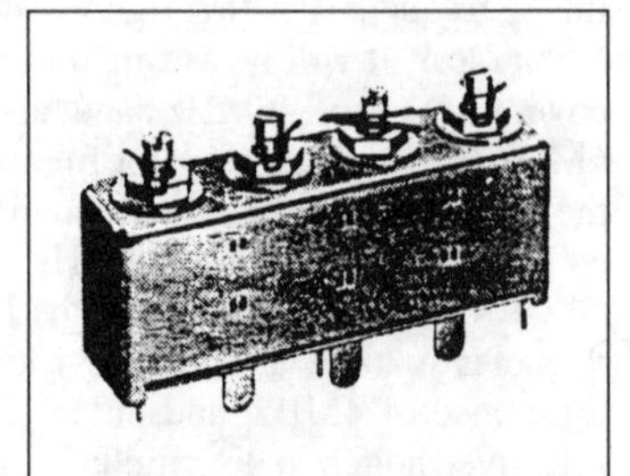

Fig 8.7. The Toko HRQ resonator

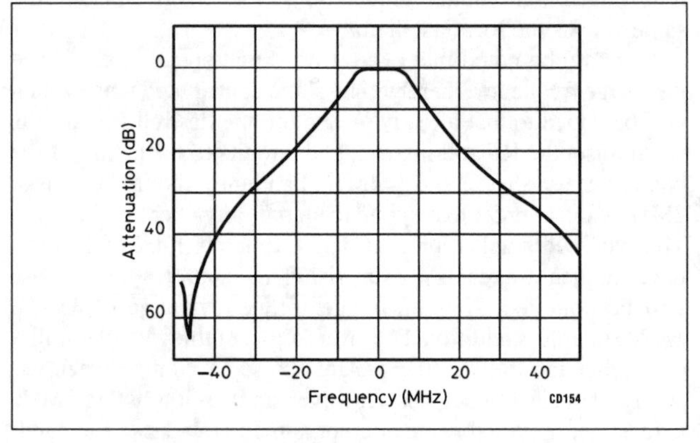

Fig 8.8. The Toko HRW frequency response

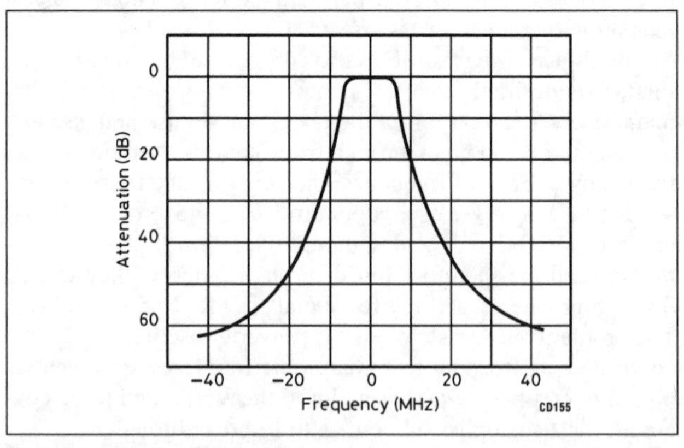

Fig 8.9. The Toko HRQ frequency response

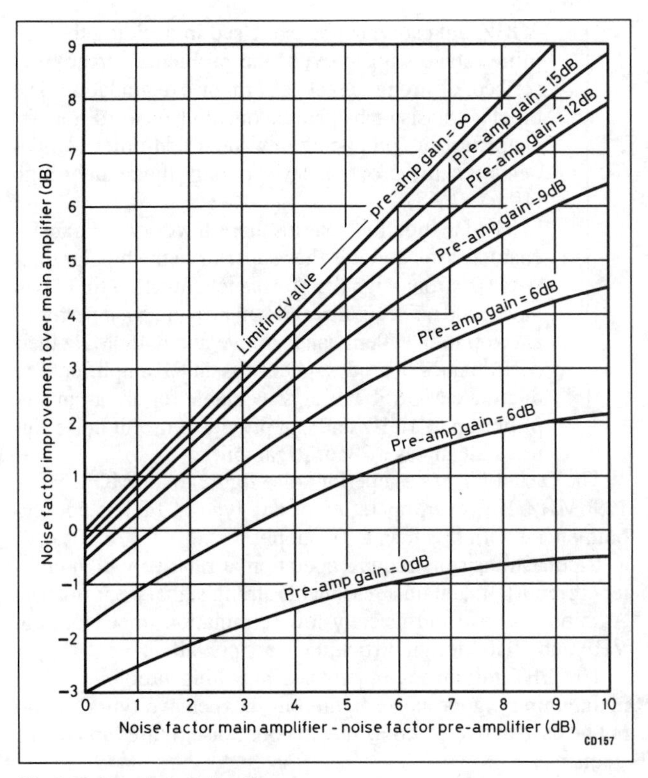

Fig 8.10. Receiver noise figures

VHF/UHF RF AMPLIFIERS

Preamplifiers

As available devices improve and new circuit designs are published, it will become apparent that a receiver which may have been considered a first-rate design when built is no longer as good as may be desired. Specifically, a receiver using early types of transistor may not be as sensitive as required, although the local oscillator may perform satisfactorily. The sensitivity of such a receiver can be improved without radical redesign by means of an additional separate RF amplifier, usually referred to as a *preamplifier*. Such an amplifier should have the lowest possible noise figure and just sufficient gain to ensure that the overall performance is satisfactory.

Fig 8.10 shows the improvement to be expected from a preamplifier given its gain and the noise factor of the preamplifier alone and the main receiver.

An example will suffice to show the application. An existing 145MHz receiver has a measured noise figure of 6dB and is connected to its antenna via a feeder with 3dB loss. It is desired to fit a preamplifier at the mast head; what is the performance required of the preamplifier?

Suppose a BF180 transistor is available; this has a specified maximum noise figure of 2.5dB at 200MHz and will be slightly better than this at 145MHz. The main receiver and feeder can be treated as having an overall noise figure of 3 + 6 = 9dB. From Fig 8.10, if the preamplifier has a gain of 10dB, the overall noise figure will be better than 4.1dB. Increasing the gain of the preamplifier to 15dB will only reduce the overall noise figure to 3.6dB, and may lead to difficulty due to the effect of varying temperatures on critical adjustments. The addition of so much gain in front of an existing receiver is also very likely to give rise to intermodulation products from strong local signals. If it is desired to operate under such conditions, it is essential that provision is made for disconnecting the preamplifier when a local station is transmitting.

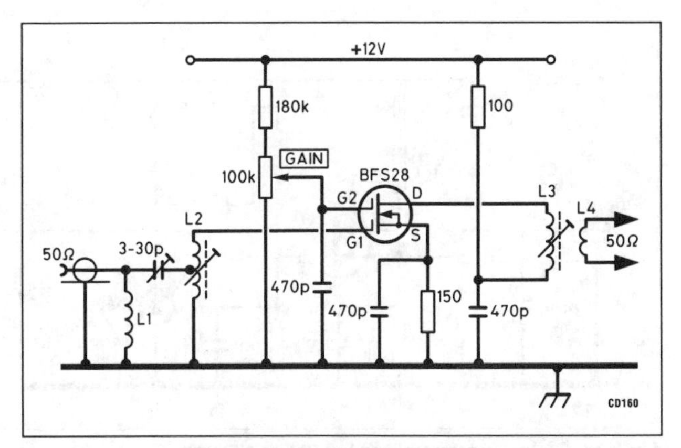

Fig 8.11. Dual-gate MOSFET RF amplifier with gain control. L1: 3 turns 1.0mm enamelled wire, 6.0mm ID, 8.0mm long. L2: 2 turns 1.0mm tinned copper wire on 8.0mm former 10mm long, tapped at 1½ turns and tuned with dust core. L3: 6 turns 1.0mm enamelled wire on 8mm former tuned with dust core and coupled to L4. L4: 2 turns 1.0mm enamelled wire on same former, close spaced to capacitor end of L3

RF amplifier design

The advent of the field effect transistor (FET) eased the design of VHF RF amplifiers in two ways. The relatively high input resistance at the gate permits reasonably high-Q tuned circuits, providing protection against strong out-of-band signals such as from broadcast or vehicle mobile stations. Also the drain current is quite exactly proportional to the square of the gate voltage; this form of non-linearity gives rise to harmonics (and the FET is a very efficient frequency doubler) but a very low level of intermodulation.

The use of a square-law RF stage is not, however, as straightforward as it first appears. Second-order products are still present, such as the sum of two strong signals from transmitters outside the band, typically Band II broadcasts, mixing to generate the unwanted signals on the 144MHz band. It follows, of course, that if two FET amplifiers are operated in cascade a band-pass filter is required between them to ensure that the distortion products generated in the first stage are not passed to the next stage where they will be re-mixed with the wanted signal.

A development of the FET was the metal oxide semiconductor FET (MOSFET), in which the gate is insulated by a very thin layer of silica. The gate therefore draws no current and a high input resistance is possible, limited only by the losses in the gate capacitance. These devices may be damaged by static charges, and must be protected against antenna pick-up during electrical storms, and also from RF from the transmitter feeding through an antenna changeover relay with an excessively high contact capacitance. MOSFETs now have protective diodes incorporated in the device which limit the input voltage to a safe level, and these devices are thus much more rugged.

In the dual-gate MOSFET the drain current is controlled by two gates and various useful circuit improvements result. If it is desired to apply gain control to a conventional transistor stage the control voltage is applied to the same electrode as the signal, but the result can be a reduction in power handling capability, showing up as a cross-modulation. The dual-gate FET avoids this problem, and automatic or manual gain control can be applied to gate 2 without reducing the signal-handling capability at gate 1. Fig 8.11 shows how such an RF stage for a converter is arranged. When a strong local station causes intermodulation at the mixer or an early stage of the main receiver, the RF gain can be reduced until interference-free reception is again possible.

The GaAsFET is similar to the MOSFET but is based on

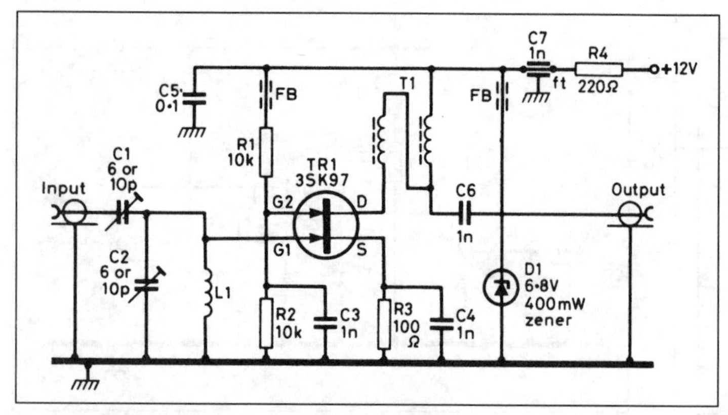

Fig 8.12. GaAsFET preamplifier for 144 or 432MHz

galliuum arsenide rather than silicon. Gallium arsenide has a larger electron mobility than silicon and therefore has a better performance at UHF (see Chapter 3). GaAsFETs, although primarily designed for use in television receiver tuners, are entirely suitable for 144MHz and 432MHz preamplifiers or as a replacement for an existing RF amplifier. These devices are now mass-produced by major semiconductor manufacturers and include the 3SK97, 3SK112, 3000 and the CF739 (Siemens).

In a correctly designed amplifier the GaAsFET will provide excellent performance at these frequencies. The quoted noise figure for the 3SK97 is 1dB at 900MHz. Silicon diodes are mounted in the GaAsFET chip to protect the gates against ESD (electrostatic discharge) breakdown.

The noise figure obtainable is related to the manufacturer's data sheet. Biasing is the same as for silicon MOSFETs, ensuring, however, that the gates are never positively biased with respect to the channel.

The circuit for a GaAsFET preamplifier [1] which incorporates a self-biasing arrangement is shown in Fig 8.12. Construction of L1 and T1 for 144MHz and 432MHz is given below.

L1 *144MHz*: 6 turns 2.0mm tinned copper wire 6mm ID 13mm long
432MHz: copper line 15mm wide 57mm log spaced 4mm above ground plane

T1 *144MHz*: 12 turns bifilar wound 0.5mm enamelled copper wire
432MHz: 2½ turns 0.5mm enamelled copper wire connected in series as 4:1 transformer on Amidon T-20-12 toroid core.

Double-sided copper laminate is used for mounting most of the components with a vertical screen of tinplate forming a screen between input and output circuits. Care must be taken against static when mounting the GaAsFET in the circuit. After careful checks for constructional errors the current should be checked before alignment. This should be between 25–30mA.

Power gain alignment will be in the order of 26dB at 144MHz and 23dB at 432MHz. Any tendency towards instability can be cured by fitting a ferrite head over the drain lead close to the FET.

An attenuator must be used between the preamplifier and input to an existing receiver or converter to prevent degradation of the strong-signal performance.

Designing and tuning RF amplifiers

The choice of the active device will have a significant effect on the weak and strong signal performance of the amplifier. Bipolar transistors are available with excellent noise figures up to 4GHz, but care must be exercised in deciding the amplifier stage gain to avoid the problems, already described, of strong-signal performance degradation. The bipolar transistor has an exponential forward transfer characteristic, producing unwanted odd-order components when the output level is only moderately high (10mV RMS).

The junction FET has a square-law forward transfer characteristic and it is thus superior to the bipolar transistor in strong-signal performance. Junction FETs will have good noise figures and will provide good performance up to the 70cm band. Above about 450MHz their performance degrades quickly as an RF amplifier. The dual-gate MOSFET is now available for use as an amplifier up to 1GHz and can provide a useful noise figure and gain as a 23cm RF amplifier.

The GaAsFET has a superior noise figure to the MOSFET at 1296MHz. However, operation of any type of FET is not recommended with less than a 9V supply.

Bipolar transistors require operation at relatively high emitter currents to maintain a reasonable strong signal performance. As bipolars have intrinsically low termination impedances at VHF and UHF, design difficulties may prevail.

The FET minimises impedance matching problems while maintaining a good noise figure but inexpensive types should not be used above 144MHz due to poor noise figures above this frequency.

Lumped circuits can be used up to 144MHz, but tuned lines are more convenient and ease design problems at 70cm and 23cm.

The input network to the RF amplifier should be as low in insertion loss as possible as any loss will be effectively added to the front-end noise figure. An L-network is preferred to a high-*Q* circuit as the latter can have significant loss at VHF. Additionally the L-network will provide optimum impedance matching. It is essential to employ high-quality components to minimise input circuit insertion losses.

An important point is that transistor terminating impedance matching for optimum noise figure is not necessarily the same as for optimum gain (power transfer). However, with careful design and consequent adjustment and measurement of the RF amplifier, an acceptable noise figure commensurate with required gain is quite possible.

The dual-gate MOSFET has the virtue of an almost identical noise and optimum gain match; therefore adjusting the RF amplifier for best noise figure will also result in maximum gain. It is not normally necessary to neutralise a correctly designed dual-gate MOSFET amplifier to achieve best noise figures and stability. However, a JFET amplifier operated in common-source mode usually requires neutralising to ensure it is stable under all conditions. A typical JFET amplifier is illustrated in Fig 8.13 with an inductance neutralising circuit. The cascode circuit of Fig 8.14 combines grounded-source and grounded-gate FETs to give high gain and low noise figures. Strong-signal levels will cause less susceptibility to overload and gain compression than a single-FET amplifier.

Although bipolar transistors are more easily overloaded by strong signals compared to FETs, both PNP and NPN types can successfully be used in VHF and UHF amplifiers. Choosing the correct transistor is of paramount importance. Parameters given in manufacturers' data must be studied before choosing a particular device. These include dynamic characteristics as well as DC operating conditions, eg gain-bandwidth product (F_T), collector base capacitance (C_{cb}), emitter base capacitance (C_{eb}), and

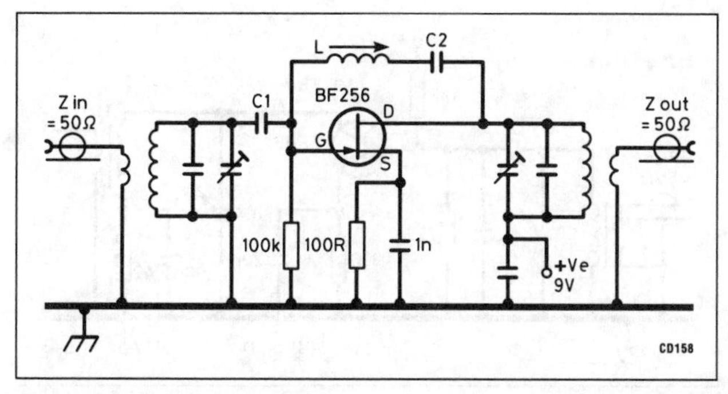

Fig 8.13. JFET VHF RF amplifier incorporating an inductive neutralisation circuit

noise figure (dB). F_T and NF are usually quoted against collector current, collector/emitter voltage and frequency, eg when I_c = 10mA, V_{ce} = 10V and f = 200MHz, the F_T is typically 1.5GHz and NF = 3.5dB. As a rule of thumb the F_T should be at least 3f. Thus the transistor could be used on all bands up to and including 70cm. However, a device with a quoted F_T of 4GHz would be required for an RF amplifier operating at 1296MHz. The noise figure for this UHF device would typically be 3dB. A typical bipolar RF amplifier circuit is shown in Fig 8.15.

Mixers

The common forms of mixers use diodes and most types of transistors.

Diode mixers

A diode operates non-linearly either around the bottom bend, where the current through the diode is proportional to the square of the applied voltage (see Fig 8.16(a)), or by the switching action between forward conduction and reverse cut-off, as shown in Fig 8.16(b). In the first case it is often necessary to forward-bias the diode with DC to obtain an optimum working point. A bias of 100 to 200mV is typical but will vary with the type of diode. The maker's data sheets should be consulted for the optimum working conditions.

The second type of mixer is used where a high overload level is required. Signals approaching one-tenth of the local oscillator power can be handled without distortion, and the local oscillator level is limited only by the power-handling capacity of the diodes. The noise generated in the mixer rises with increasing diode current, however, and this sets a limit to the usable overload level if maximum sensitivity is required, although it is

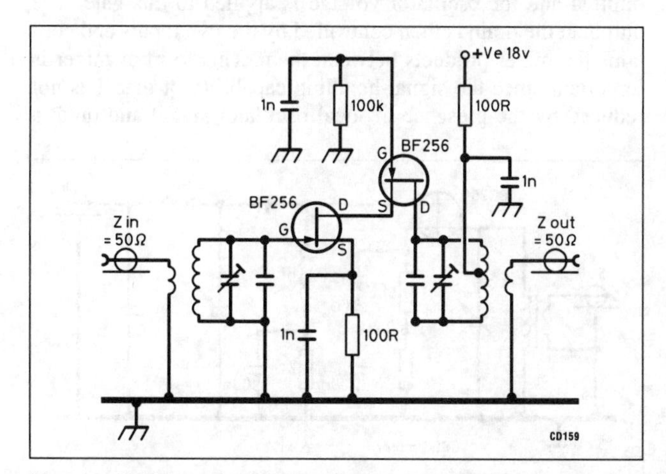

Fig 8.14. JFET cascode VHF RF amplifier suitable for use at 6m, 4m and 2m

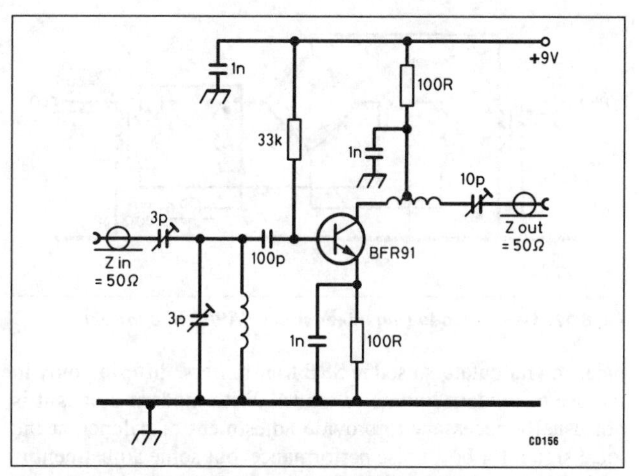

Fig 8.15. Typical bipolar VHF RF amplifier

possible to adjust the local oscillator power to select a compromise between sensitivity and overload capacity.

The diode mixer is necessarily 'lossy', and a loss of between 3 and 6dB may be obtained in practice. It is therefore essential that the stage following the mixer has the lowest possible noise figure.

Balanced diode mixers

Noise components from the local oscillator can be reduced to a low level by means of balanced mixers; these are similar to the

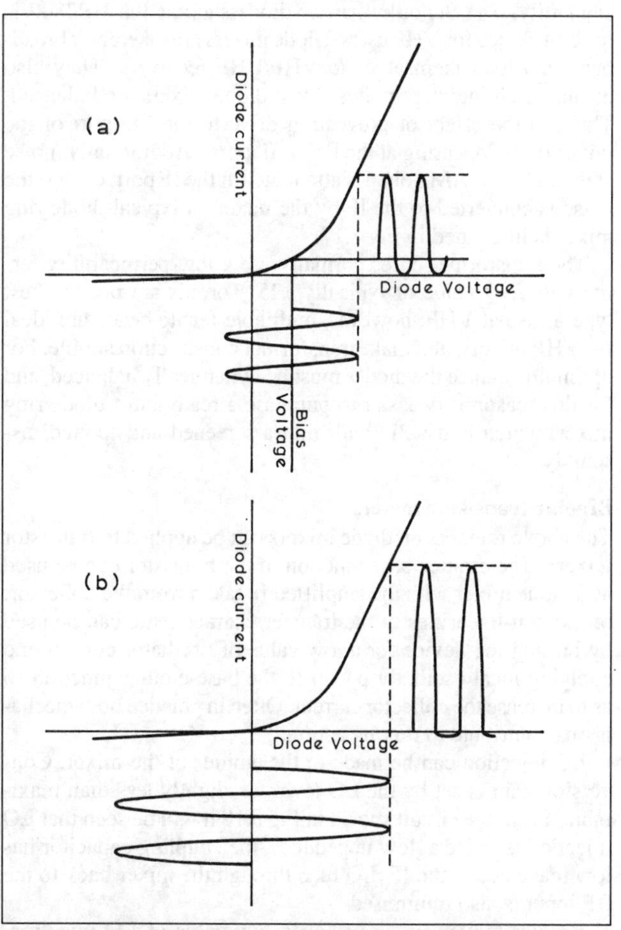

Fig 8.16. Working conditions of diode mixers. In (a) forward bias is required to provide an optimum working point. In (b) the local oscillator power is higher and no bias is required

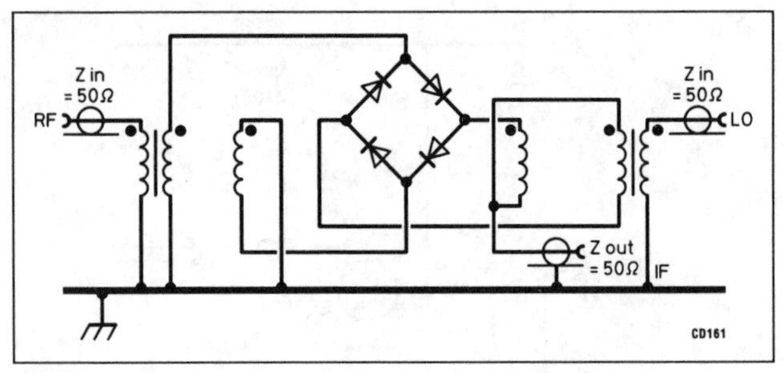

Fig 8.17. Typical diode ring mixer circuit *(ARRL Handbook)*

balanced modulators used in SSB transmitters, differing only in the use of modern low-noise diodes. With modern diodes it is not usually necessary to provide adjustment of balance at the local signal for best noise performance, but some adjustment is desirable to reduce local oscillator radiation from the antenna. In many cases it will be necessary to include a simple filter in the antenna lead to reduce this radiation; this can conveniently be the same filter as is used to reduce the second-channel response. Conversion loss is minimised with as little local oscillator (LO) injection as 1mW (0dBm), but IMD rejection is improved by increasing drive level to +7 to +10dBm, providing the diodes can handle the increase current.

Diodes suitable for receiver mixers are all semiconductor types, the most suitable being the hot-carrier, high-speed silicon switching or Schottky barrier diode. The latter are the only types used in mixers designed specifically to have a high overload capability. The Schottky barrier diode such as the HP2800 is the best choice for VHF users. Diode mixers are inherently broadband and lend themselves to VHF/UHF receivers. They also display high intercept points. Most diode mixers are balanced. This has the effect of preventing drive to the LO port of the mixer from appearing at the RF or IF ports. Additionally, noise at the LO port (AM noise) is attenuated at the IF port, even if the noise is converted to the IF by the mixer. A typical diode ring mixer is illustrated in Fig 8.17.

The transformer cores normally use a low-permeability ferrite with a Q value of typically 125. Toroids are not the best type to use at VHF; however, multihole ferrite beads are ideal for VHF mixers, and make transformer construction simple. For optimum balance the diodes must be dynamically balanced, and for this reason it is easier to purchase a ready-built diode ring mixer, which is usually built into a screened and 'potted' assembly.

Bipolar transistor mixers

The above remarks on diode mixers can be applied to transistor mixers. The emitter-base junction of the transistor can be used as a diode mixer with the amplified IF taken from the collector; or the non-linearities of the transfer characteristic can be used by biasing the device for a low value of oscillator current and applying local oscillator power to the base-emitter junction so as to increase the collector current. Often in practice both mechanisms contribute to mixing action.

LO injection can be made to the emitter of the mixer. Conversion gain is set by the LO level for slightly less than maximum. From the circuit shown in Fig 8.18 it will be seen that LO injection is made at low impedance, the coupling capacitor has low reactance at the IF. LO feed through the mixer back to the RF input is also minimised.

Bipolar mixers require a fairly low value of LO injection. However, the intrinsic exponential forward-transfer characteristic severely limits large-signal handling. Blocking and IM

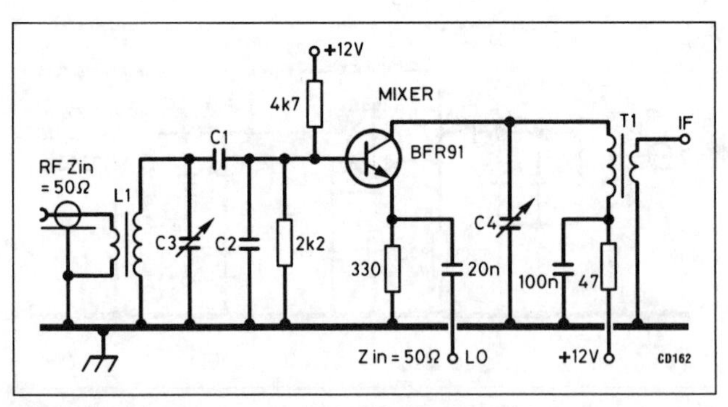

Fig 8.18. Bipolar mixer circuit *(ARRL Handbook)*

products are easily realised even with moderate RF gain being employed before the mixers.

Field effect transistor mixers

Junction FETs can provide exceptionally good performance as VHF mixers. The input impedance is high but the conversion gain is approximately 25% of the gain of the same device used as a VHF amplifier. Biasing is critical, and is normally chosen so the gate-source voltage is 50% of the pinch-off voltage of the device. The LO is applied to the source, the level being chosen to avoid the pinch-off region but not sufficient to cause the gate diode to be driven into conduction. Normally the peak-to-peak amplitude should be a little below the FET pinch-off voltage.

The JFET has a moderate output impedance typically around 10kΩ (resistive). This makes impedance matching for the IF filter easier normally via an inverting network (as shown in Fig 8.19) or a step-down transformer.

The JFET mixer can improve IM products reduction by up to 30dB compared to a bipolar mixer. However, it does not have an ideal square-law input characteristic due to the effect of bulk resistance associated with the source terminal.

Dual-gate MOSFET mixers

MOSFET mixers can provide considerable conversion gain, with a low noise figure and high intercept point. LO power requirement is also low, easing the design of oscillator/buffer amplifier circuits. Optimum conversion gain as obtained with about 5V peak-to-peak applied to gate 2. IM products at the IF are usually very low.

Fig 8.20 shows the use of a dual-gate MOSFET as a mixer. The signal input is applied to gate 1 as for an RF amplifier, but instead of a decoupled bias supply to gate 2, the decoupling is omitted and the oscillator voltage is applied to this gate. The output at the drain is then controlled by the two inputs and contains the mixer products between them. This form of mixer is important since the signal-handling capability of gate 1 is not reduced by the presence of local oscillator signal and quite a

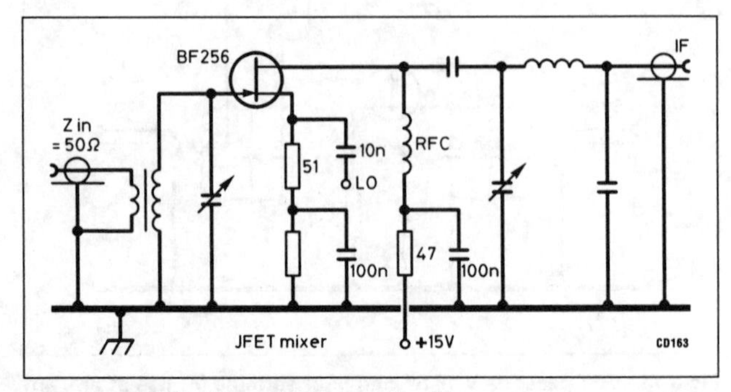

Fig 8.19. A JFET mixer *(ARRL Handbook)*

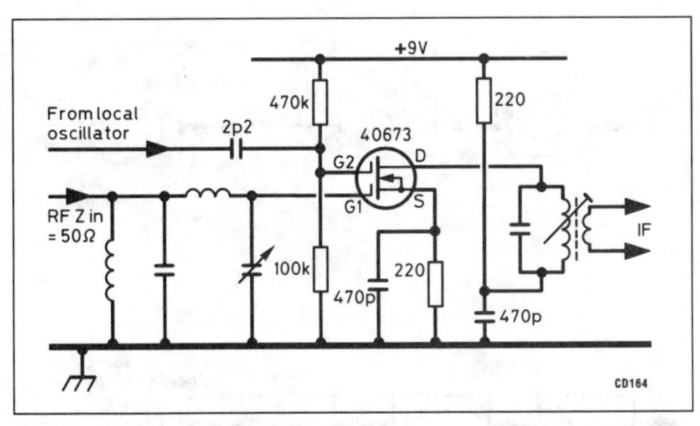

Fig 8.20. A dual-gate MOSFET mixer

high overload level results. Next to the diode mixer, the dual-gate MOSFET mixer has the highest overload level of any discrete semiconductor mixer.

IF filters

HF crystal filters

The performance of a purpose-built double-conversion VHF/UHF receiver with a 'tuneable' first local oscillator can be enhanced by using a crystal filter between the first mixer and IF amplifier. Commercial crystal filters are now readily available, not only for the well-known IFs of 10.7 and 21.4MHz, but also for 45 and 75MHz. Until recently, crystal filters above about 25MHz were only available with third-overtone-mode crystal elements. Now 45 and 75MHz filters are available with fundamental crystal elements. Recommended filters for VHF/UHF bands are:

6m and 4m	10.7MHz
2m	21.4MHz
70cm	45.0MHz
23cm	75.0MHz

These filters are available in two-, four- or six-pole versions; a six-pole filter will give the best attenuation to out-of-pass-band signals. Good selectivity must be achieved before the first IF amplifier, and the crystal filter will give superior selectivity compared to an LC filter. Crystal filters have low input and output impedances, the actual impedance being dependent on the number of poles. These impedances vary from 3kΩ/3pF at 10.7MHz to 500Ω/1pF at 75MHz and are given in the maker's literature.

The preferred matching mechanism to the first mixer is the pi-network as illustrated in the mixer circuits. This network has an impedance inversion property. The input impedance of the crystal filter appears as an open-circuit to out-of-pass-band signals while the load presented to the drain (of a MOSFET mixer) appears as a near-short-circuit. This will enhance the mixer IM product attenuation.

Bandwidth

Crystal filters must be correctly chosen for the type of modulation to be detected by the receiver. The –3dB bandwidth for SSB filters can be as little as 2.1kHz (500Hz for CW receivers). The –60dB bandwidth can be 4.2kHz, giving a shape factor of 2:1. The pass-band of the filter characteristic can possess ripple, and will not introduce pre-detector audio modulation distortion. However, filters for NBFM receivers must have a –3dB bandwidth of ±3.75kHz for 12.5kHz channel spacing and ±7.5kHz for 25kHz channel spacing, for linear 'detection' of NBFM transmitters using maximum frequency deviation of the modulator.

Pass-band ripple must be minimal (not more than 1dB) to avoid modulation distortion. The –60dB bandwidth for a typical six-pole NBFM crystal filter can be 30kHz for 12.5kHz channel spacing and 60kHz for 25kHz channel spacing, giving a shape factor of 4:1. All crystal filters have some insertion loss and it is usual to build a post-filter IF amplifier in the receiver to negate this loss.

RECEPTION OF FM SIGNALS

There are two principal features in receivers designed to receive FM signals, namely limiting rather than linear amplifiers precede the detector and the latter is designed to convert IF variations into AF signals of varying amplitude, dependent on the degree of frequency variation in the transmitter carrier. Frequency (and phase) modulation of transmitters is described later in this chapter.

The FM receiver

The block diagrams of an FM receiver and AM/SSB receiver are shown in Fig 8.21. The principal difference between the receivers are the IF filter bandwidths (see above) and the IF amplifier gains required before the detector.

It is necessary to provide sufficient gain between the antenna and detector of an FM receiver to ensure receiver quieting, ie optimum signal-to-noise ratio with the weakest signal. Usually this is less than 0.35μV PD or –116dBm (into 50Ω).

Thus it is necessary to use the double-superheterodyne principle to achieve the required gain, usually greater than 1 million or 120dB, whilst ensuring optimum stability independent of the input frequency.

Other receiver stages, particularly the RF amplifier, mixer, oscillator and audio stages, can be identical to those employed in AM/SSB/CW receivers.

In a multimode receiver designed for reception and detection of all principal methods of modulation, the difference in signal-to-noise ratio and effect of interference is very noticeable between FM and AM/SSB signals. The limiter and detector (discriminator) for FM signals reduce interference effects, usually impulse noise, to a very low level, thus achieving a high signal-to-noise ratio. However, it is necessary to align the detector correctly and in use tune the receiver accurately to achieve noise suppression.

An unusual effect peculiar only to FM receivers, and known as *capture effect*, occurs when a strong signal appears exactly on the frequency to which the receiver is tuned. If this strong signal has a carrier amplitude more than two to three times that of the wanted signal, the strong signal will be detected. This effect can be a problem in mobile operation, particularly in a geographical area between two repeater outputs on the same frequency.

Weak-signal reception in AM and FM receivers can be degraded by a much stronger carrier on or near the frequency of the weak carrier.

Bandwidth

As already stated in the previous section on filters, it is essential to choose the IF filters designed for NBFM reception. A narrow-bandwidth filter will introduce unwanted harmonic distortion. Too wide a bandwidth will degrade adjacent channel selectivity. Distortion effects are aggravated by transmitters exceeding the recommended limit on frequency deviation, though with modern NBFM transmitters this is less of a problem. Poor adjacent channel selectivity can cause receiver desensitisation,

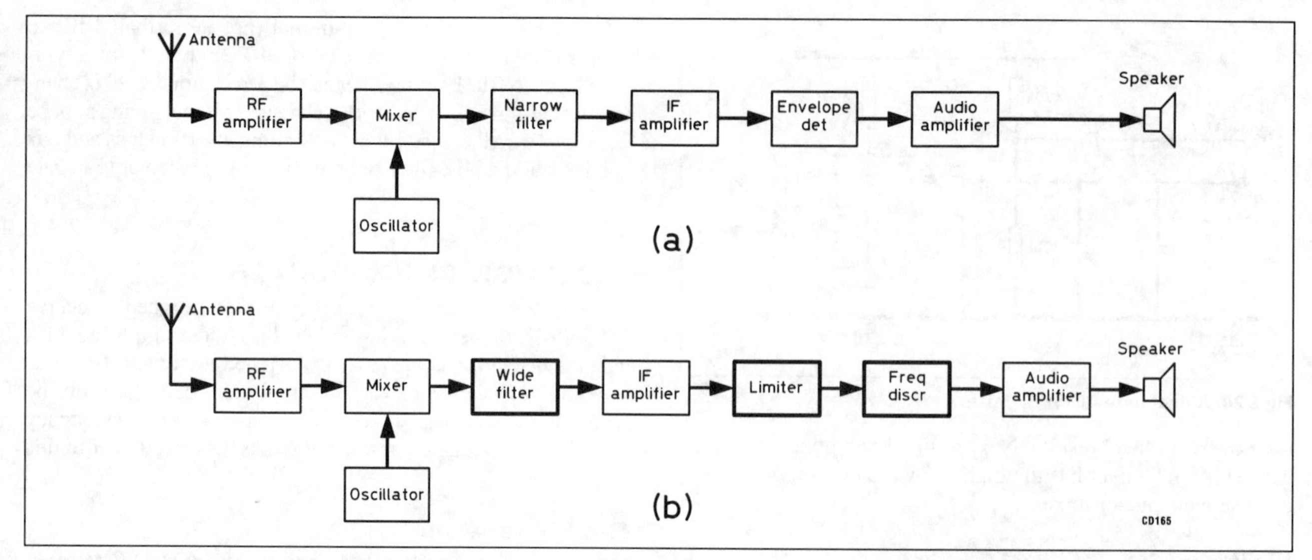

Fig 8.21. Block diagrams of (a) an AM and (b) an FM receiver. Dark borders outline the sections that are different from the FM set *(ARRL Handbook)*

particularly when strong signals are present on either or both adjacent channels. Weak-signal reception can also be degraded by a transmitter with poor adjacent channel power rejection. These effects again cause receiver desensitisation but the transmitter modulation will not appear on the wanted signal (cross-modulation cannot in theory occur in FM systems).

Limiters

Limiting amplifiers are specifically designed to introduce gain compression into their forward-transfer characteristics. Discrete HF transistors or multistage ICs can be used as limiting amplifiers. When an amplifier is driven into limiting, ie when its output signal level remains unchanged when the input signal level is varied, the amplifier effectively removes any sudden amplitude change, thus removing impulse noise and amplitude modulation prior to the detector.

The input signal to the limiter must be large to ensure full saturation, typically several hundred millivolts for both transistor and IC designs. With an RF input of 0.35μV or better, several stages of IF gain are required prior to the limiter. This can be obtained with discrete amplifiers but IC amplifiers now available which offer greater simplicity, require less peripheral components, and deliver consistently predictable results. When the RF signal input increases, limiting action starts and the receiver signal-to-ratio improves until, at a certain level and above, the noise disappears. This is known as the *receiver quieting* characteristic referred to earlier in this section, usually 20dB at 0.35μV or better for a well-designed NBFM receiver.

A good example of a typical IC is the Motorola MC3361, an IF subsystem which is described on p8.15.

Limiting amplifiers, both discrete and IC types, are illustrated in Fig 8.22.

It is convenient in discrete designs to provide transformer interstage coupling. In IC amplifiers, resistor collector loads for differential amplifier stages are employed. This limits the maximum IF to about 1MHz, hence the use of 455kHz as the second IF in modern receivers.

The high-gain Motorola MC1590 linear IC will saturate with 100mV of signal input and gives superior limiting action compared to a two-stage discrete design.

In the discrete design the base bias on either or both transistors can be varied to set the desired limiting input level. This will set the limiting knee characteristic of the transistors to a point at which for an increasing input level there will be no

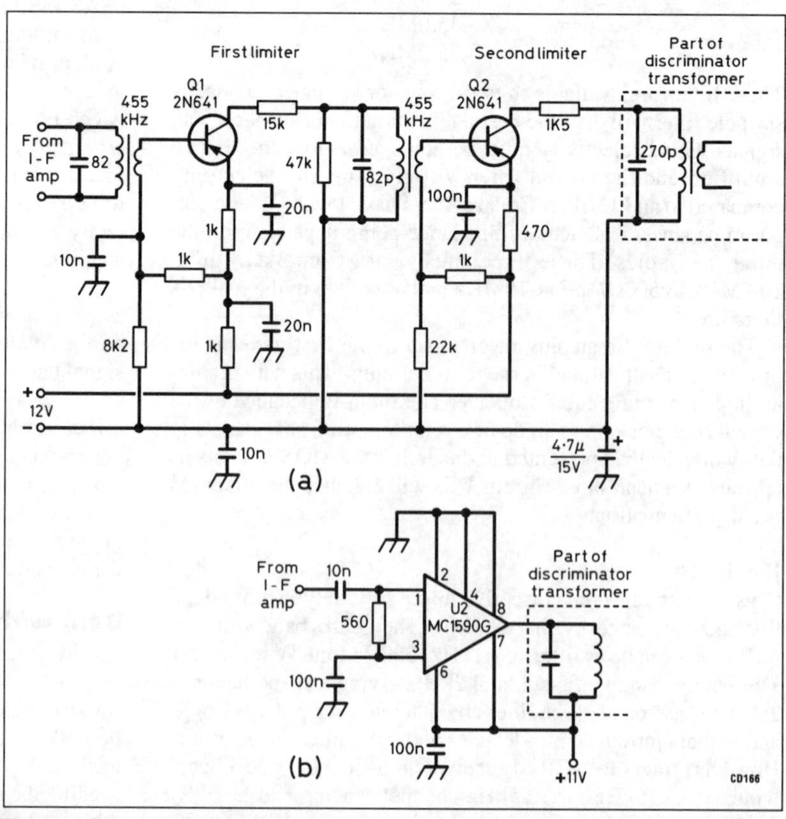

Fig 8.22. Typical limiter circuits using (a) transistors and (b) a high-gain linear IC *(ARRL Handbook)*

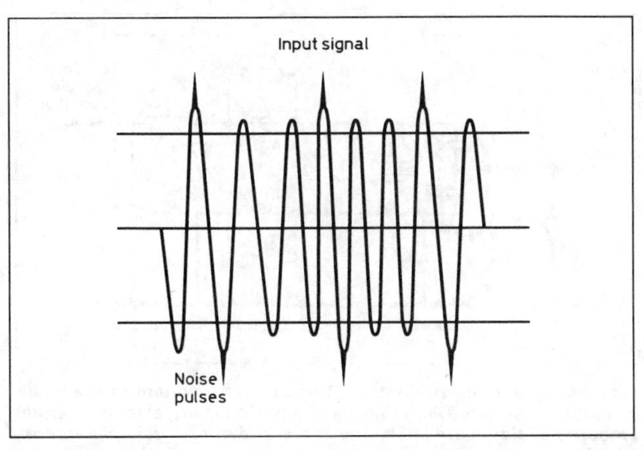

Fig 8.23. Input waveform to a limiter stage shows AM and noise *(ARRL Handbook)*

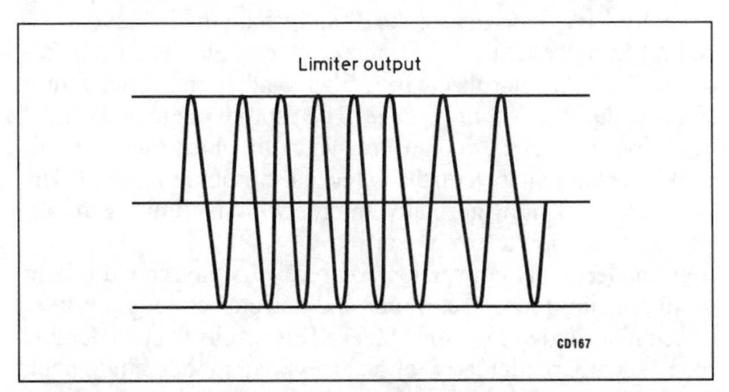

Fig 8.24. The same signal as in Fig 8.23, after passing through two limiter stages, is devoid of AM components *(ARRL Handbook)*

further increase in collector current. The amplifiers saturate, giving the required limiting action and consequent noise level reduction for good receiver quieting.

FM demodulators

The FM detector, or more correctly the FM *discriminator*, was evolved to respond only to frequency deviation from the centre carrier frequency. The degree of deviation corresponds to the amplitude of audio modulation applied to the transmitter modulator. For example, if the carrier is deviated by ±2.5kHz by the modulator, the receiver discriminator frequency will also be deviated by ±2.5kHz. This frequency 'shift' is detected linearly when the signal is both higher and lower than the centre frequency. This is shown clearly in Fig 8.25 which illustrates the 'S' curve characteristic of an FM discriminator. Provided the swing is in the linear (straight) portion of the line, the audio recovered from the carrier signal will be undistorted. Typical maximum frequency deviation in amateur FM equipment is ±5kHz compared to ±75kHz for broadcast FM transmitters and receivers. For this reason FM systems for voice and data communications are normally known as *narrow-band FM* (NBFM) systems. The discriminator curve is linear to about ±5kHz, and a relatively small percentage of distortion would be introduced into the recovered audio. Measures are usually taken in the transmitter to limit peak deviation to ±5kHz to avoid over-deviation.

A practical discriminator circuit, known as the *Foster-Seeley* circuit, is illustrated in Fig 8.26. T1 is the discriminator transformer. Voltage is developed across the primary from the final limiter stage. Primary to secondary coupling induces a current in the secondary 90° out of phase with the current in the primary. The carrier at the primary is also coupled to the centre tap on the secondary via a coupling capacitor. The voltages on the secondary are combined on either side of the secondary in such a way that these voltages lead and lag the primary voltage by equal amounts (degrees) when an unmodulated carrier is present. The resultant rectified voltages are of equal and opposite polarity, and hence no audio component is recovered. When the carrier is deviated the phase of the voltage between primary and secondary is changed, resulting in an increased output level on one side and decreased output level on the secondary. These level differences in the two voltages, after rectification by D1 and D2, represent the recovered audio signal. This discriminator, whilst giving good results, responds to variations in carrier level unless driven hard by the preceding limiter stages.

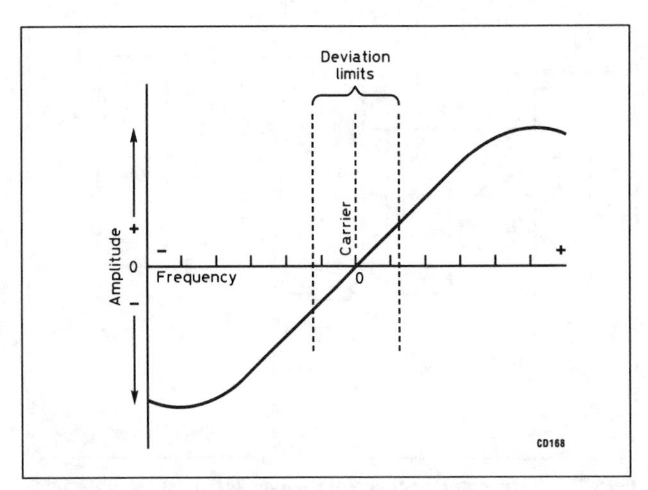

Fig 8.25. The characteristic of an FM discriminator

A modified form of FM detector, known as the *ratio detector*, became popular in broadcast FM receivers, and has been widely used in amateur receivers. The detector is far less susceptible to carrier amplitude variations and, although audio distortion products are greater, it is ideal for NBFM detection. The ratio detector is illustrated in Fig 8.27. T1 is a standard detector transformer. As its name implies, the ratio detector functions by dividing the rectified DC voltage from D1 and D2 into a ratio equal to the amplitude ratio present on each side of the transformer secondary. This detector will function satisfactorily with large changes in carrier amplitude, giving a constant level of audio recovery for a constant carrier deviation. As it only responds to ratios, only FM is detected. In Fig 8.27 the rectified DC voltage is developed across R5 and R6, and this voltage is maintained at a constant level by the shunt capacitor C6. Note

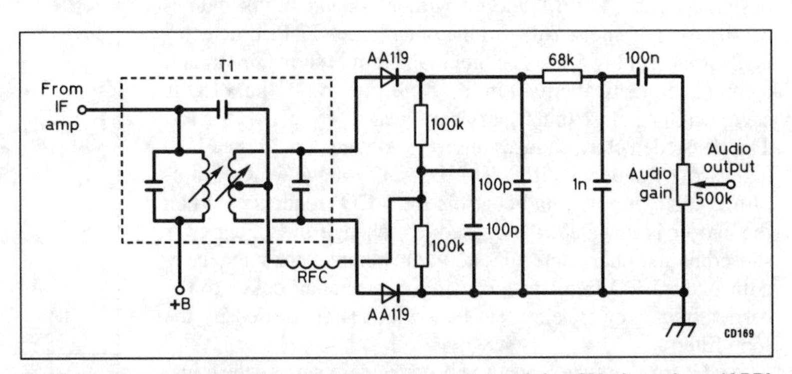

Fig 8.26. Typical frequency discriminator circuit used for FM detection *(ARRL Handbook)*

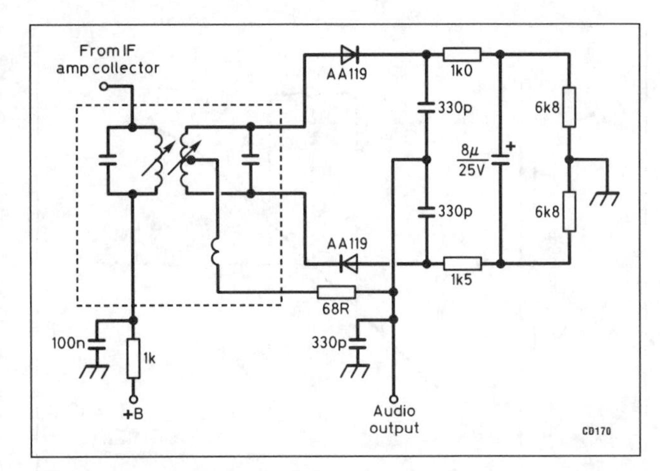

Fig 8.27. A ratio detector circuit of the type often used in entertainment radio and TV sets *(ARRL Handbook)*

that the diodes, usually a matched pair of germanium detectors, are connected in series rather than series-opposing as in the frequency discriminator circuit (Fig 8.26). The recovered audio is obtained from the tertiary winding of T1 (T1c). This winding is tightly coupled to the primary. The IF carrier is filtered out by R2 and C5. R5 and R6 are selected to present a relatively low impedance load to T1 secondary when the carrier is sufficient to drive D5 and D6 into full conduction. R4 can be adjusted for optimum AM rejection (at centre frequency) at a carrier level just sufficient to 'turn on' D5 and D6.

The ratio detector sensitivity is 6dB less than the frequency discriminator. The transformer primary to secondary coupling co-efficient (Q_k) should not exceed 0.5, otherwise the transformer may be difficult to align at centre frequency for maximum DC voltage across R5 and R6, and maximum audio recovery. The ratio detector can give excellent results, but does require at least one preceding limiter stage.

Due to the difficulties encountered in building and aligning frequency discriminators it is wise to consider an FM detector which requires minimal alignment. Fig 8.28 illustrates a discriminator where the transformer secondary is replaced by a crystal or ceramic resonator chosen to be resonant at the IF. The resonator is shunted by an inductor which is aligned for maximum audio recovery after aligning C1 and L1 to resonate at the IF. C2 and C3 are equal value and C4 is added (to either side of the resonator and L) to compensate for circuit imbalance and to ensure equal carrier amplitudes are applied to D1 and D2.

Phase-lock-loop detectors (PLL)

With the advent of the single-chip PLL it has been possible to design a reliable NBFM detector without tuned circuits and necessity for alignment. An example of a modern IC PLL detector is illustrated in Fig 8.29. The block diagram is shown in (a) and a complete practical circuit in (b). Referring to (a), the VCO is set to oscillate close to the carrier frequency, typically 455kHz. The phase detector produces an error voltage should the VCO and carrier frequency differ. This is a DC voltage which is amplified after filtering, and controls the VCO frequency. When the carrier is deviated, the consequent change in frequency is sensed by the phase detector and the resultant error voltage re-adjusts the VCO frequency, causing it to remain locked to the carrier frequency. The system bandwidth is controlled by the loop filter.

As the error voltage produced corresponds exactly to the transmitter frequency deviation during modulation, a PLL functions

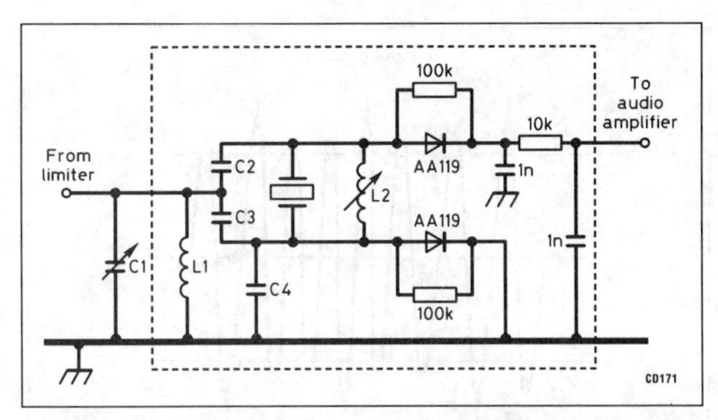

Fig 8.28. Crystal discriminator. C1 and L1 are resonant at the IF. C2 is equal in value to C3. C4 corrects any circuit imbalance so equal amounts of signal are fed to the detector diodes *(ARRL Handbook)*

as a precise FM detector. Detector sensitivity is high, requiring typically 1mV of carrier for the IC and circuit in Fig 8.29(b). R1 and C1 initially set the VCO frequency close to the carrier frequency and C2 controls the loop filter bandwidth, which in turn controls the PLL *capture range*. The capture range is the total deviation from carrier centre frequency to which the loop will gain and maintain lock on the unlocked carrier input signal. The PLL detector is now normally on-chip with the limiting amplifier, eg as in the Plessey SL6601.

In modern dual-conversion FM receivers the second mixer, oscillator, limiting IF amplifier and detector circuits are integrated in a single monolithic block. This results in considerable space saving, particularly for handheld and mobile equipment. Other advantages are a well-designed performance specification, low power consumption, and ease of setting up and alignment.

Additional circuitry normally integrated into the block includes an active filter, squelch, scan control and mute switch. The scan control is used in conjunction with a digital PLL tuning system

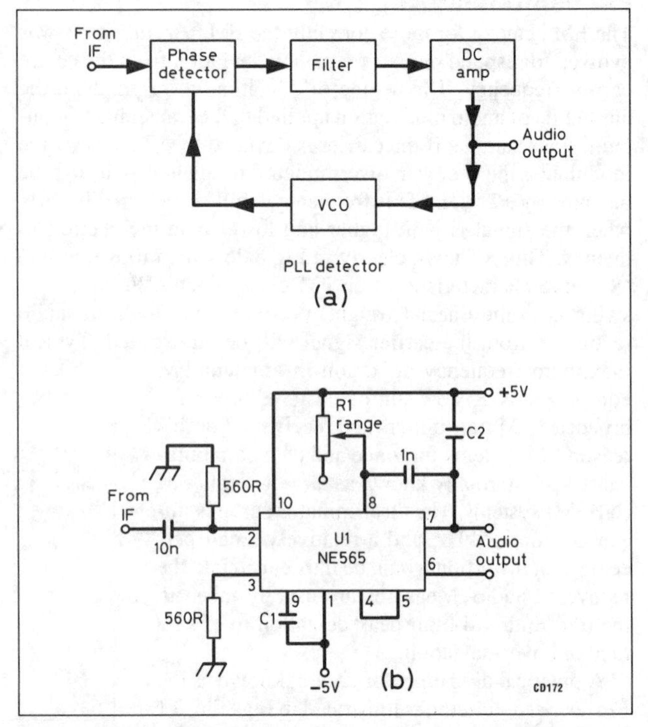

Fig 8.29. (a) Block diagram of a PLL demodulator. (b) Complete PLL circuit *(ARRL Handbook)*

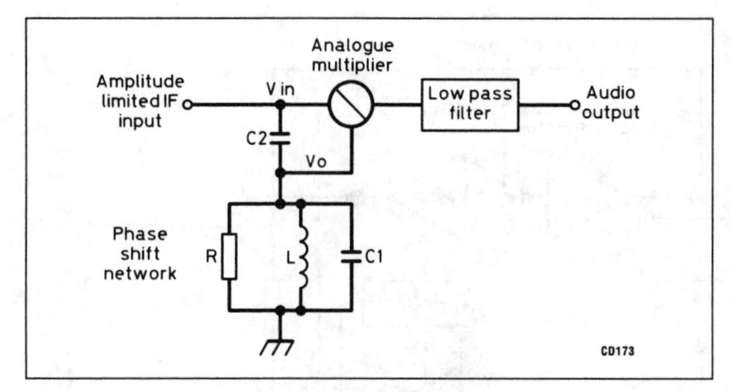

Fig 8.30. Symbolic circuit of a quadrature detector

for 'locking' onto channels where the receiver has scanning facilities.

The detector is known as a *quadrature discriminator* (sometimes known as a *coincidence detector*). The symbolic circuit of a quadrature detector is shown in Fig 8.30.

As its name implies this discriminator functions due to the IF signal being split into two parts. One part passes through a 90% phase-shift network, which is multiplied by the IF signal deviation from centre frequency. The other unshifted part is multiplied with the 90% shifted part to select the audio frequency portion of the multiplex output spectrum. For small deviations as in NBFM receivers the phase shift is sufficiently linear to give acceptable audio quality. The working *Q* of L in the detector can be controlled by shunting the inductor with a resistor to improve the linearity. This will increase the peak deviation capability of the detector at the expense of reduced audio recovery. C2 produces the 90% phase shift while C1 resonates with L at the IF centre frequency.

The functional block diagram of a typical low-power NBFM IC for use in dual-conversion receivers is shown in Fig 8.31.

In a typical application, the mixer-oscillator combination converts the first IF input down to 455kHz where, after external band-pass filtering, most of the IF amplification is done. The quadrature detector receives the audio from the IF signal (as described). In the absence of an input signal noise will be present above the normal voice audio frequencies. This noise 'band' is monitored by an active filter and a detector. The squelch trigger circuit identifies whether noise or an audio tone is present, its output is used to control scanning. At the same time the internal switch is operated which can be used to mute the audio signal in the absence of a signal carrier.

The 10pF on-chip capacitor produces the 90% phase shift between one output port of the limiting amplifier and the input port of the quadrature detector. Fig 8.32 illustrates the application of this IC and shows the external components required to complete a fully functional NBFM IF/detector system.

The oscillator is an internally biased Colpitts type. The application circuit shows a 10.245-MHz fundamental mode crystal oscillator calibrated for parallel resonance with a 32pF load capacitance. However, a 20.945MHz crystal can be used for a 21.4MHz IF input and a 44.545MHz third-overtone crystal for a 45MHz input (the IF input frequency can be as high as 60MHz). C1 and C2 form the load capacitance.

The mixer is doubly balanced to reduce spurious output

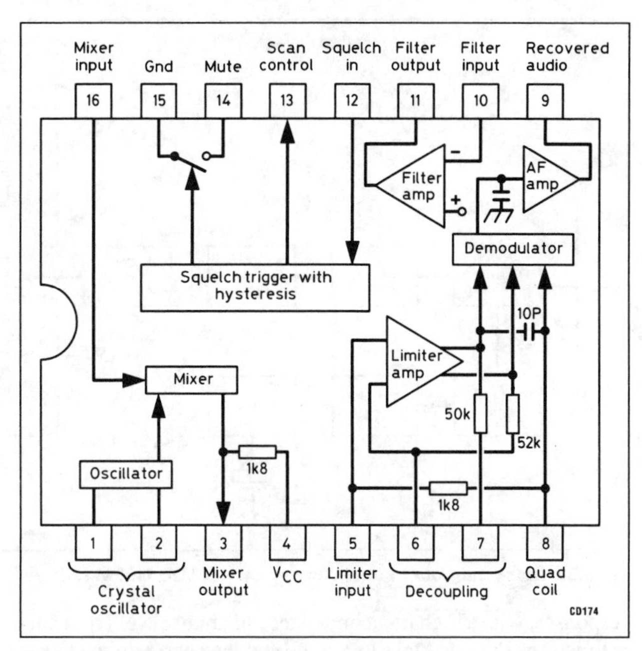

Fig 8.31. Functional block diagram of the Motorola MC3361 low-power NBFM receiver IC (Motorola)

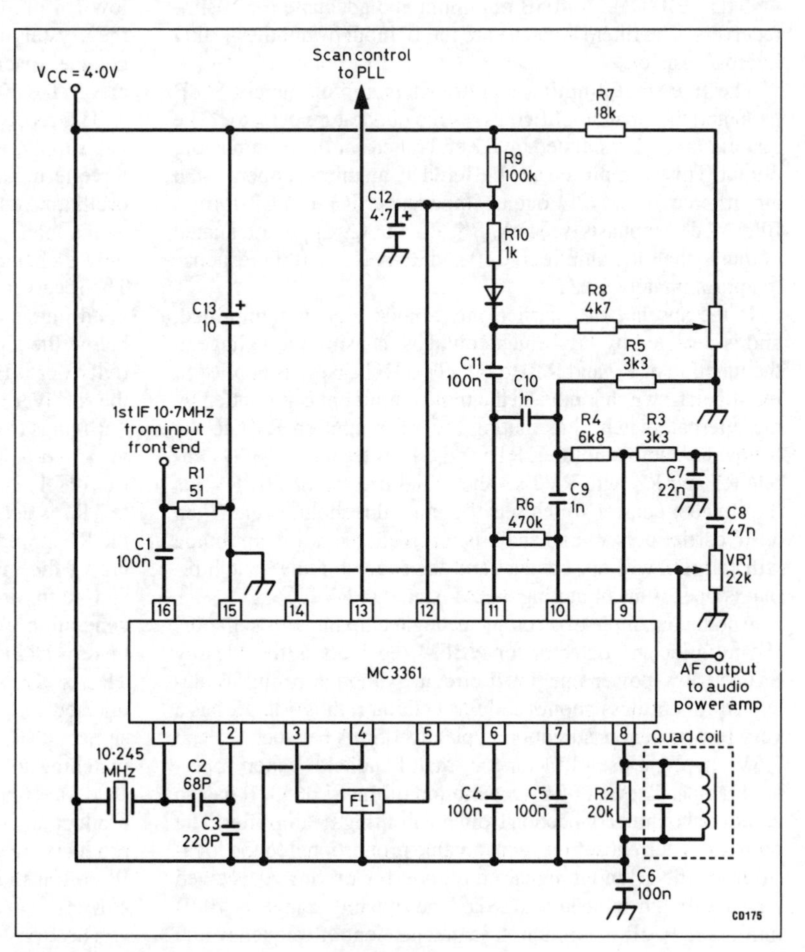

Fig 8.32. Practical circuit using the MC3361 (Motorola)

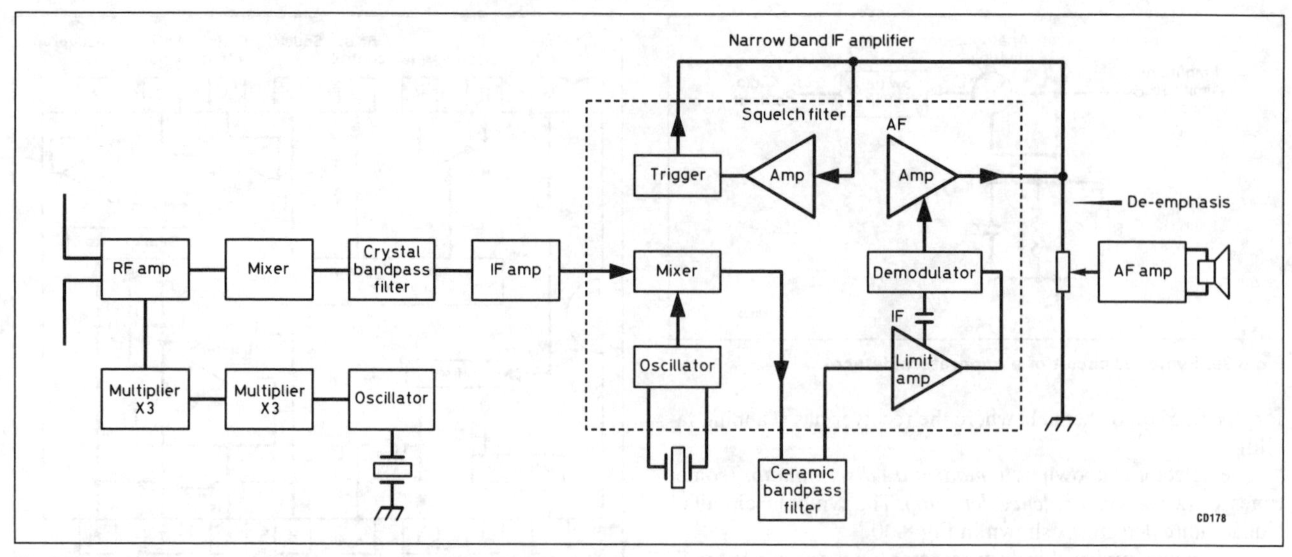

Fig 8.33. Block diagram of dual-conversion 432MHz NBFM receiver

responses, with a high input impedance of about 3kΩ. This characteristic enables crystal filters to be matched easily to the mixer input. Similarly the output impedance is fixed by the internal 1.8kΩ resistor, for good matching to the 455kHz ceramic band-pass filter (FL1).

The filter bandwidth is ±3.75kHz for 12.5kHz, and ±7.5kHz for 25kHz channel spacing respectively. Ultimate selectivity is a function of the filter stop-band attenuation. This measured at 4.55kHz ±100kHz is 40dB minimum and adequate for NBFM receivers. The filter is matched to the IF input by another 1.8kΩ internal resistor.

The five-stage limiting amplifier has approximately 80dB gain, and the final amplifier drives the quadrature detector. The quadrature coil is shunted by R2 and tuned for maximum audio output. This is amplified and buffered by an internal operational amplifier with a 470Ω output impedance. R3 and C7 form a filter or de-emphasis network. R4, R5 and C9 apply attenuated output to the filter amplifier. C10 and R6 peak the filter response to approximately 10kHz.

In the absence of a carrier only a noise signal is amplified, and is detected by D1, which conducts, causing the voltage at the junction of R9 and R10 to fall. This DC charge is applied to the squelch switch input and the mute output pin is grounded by the internal switch, thus muting the AF output on RV1 to the following audio amplifier. R10 and C12 filter the DC at D1 anode R7 and R8 with RV2 set the initial biasing of D1. RV2 is the squelch control which sets the mute threshold level. When an IF carrier present and audio is received, the noise amplitude will fall. D1 will not conduct and the squelch mute switch remains open, thus unmuting the AF signal at RV1.

Another example of a complete single chip mixer (oscillator) IF amplifier and detector for NBFM receivers is the Plessey SL6653 low-power integrated circuit. Although primarily designed for cordless phones and FM cellular radios, this IC has a very low power consumption, typically 1.5mA between 2.5 and 7.5V supply, a useful factor for small handheld, battery-operated radios. The SL6654 is a variation of the SL6653. It has an additional output connection from the limiting IF amplifier. The output is a current source, with a value proportional to the logarithm of the IF input signal amplitude for driving a received signal strength indicator (RSSI). The dynamic range is 70dB minimum, 92dB maximum. It is independent of frequency and has good temperature and supply voltage stability. The RSSI current change is typically 1.22μA/dB. Maximum output current ranges between 50 and 80μA.

432MHz NBFM receiver

This 70cm receiver embodies many of the design principles described in this chapter. It is based on the double-superheterodyne principle to provide adequate sensitivity and selectivity, combined with low front-end noise, a good dynamic range and low IM products. Both local oscillators are crystal controlled, the crystals in the first oscillator being chosen to give 25kHz channel spacing for reception of simplex and/or duplex frequencies in the 70cm band.

The receiver is designed to operate from a nominal +10.8V DC supply, for powering the front-end and audio amplifier. A three-terminal regulator supplies a stable 8V to the first local oscillator/multiplier chain and the limiting IF amplifier/FM detector integrated circuit. Receiver control is simple, requiring only a channel selector switch, AF gain and squelch controls. The receiver IFs are 45MHz and 455kHz. This feature ensures good image response rejection (the image response is 90MHz below the signal frequency) and good adjacent channel selectivity (±25kHz from the signal frequency). A block diagram of the receiver is shown in Fig 8.33.

Signals from the antenna appear at the RF amplifier input G1 via a two-pole helical band-pass filter (Fig 8.34). The RF amplifier (TR1) is a dual-gate MOSFET. Input and output matching to TR1 is designed for an RF gain of only 2 (6dB). This ensures the RF amplifier has just sufficient gain to negate the noise figure of the mixer and good large-signal handling of the front-end, ie minimum gain compression and minimum IM product generation. Amplified signals from TR1 drain arrive at the first mixer (TR2) input G1 via a three-pole helical band-pass filter. TR2 is also a dual-gate MOSFET. The first local oscillator is injected into the source of TR2. The resultant 45MHz first IF signal at TR2 drain is developed across IFT1. Input and output matching to TR2 is designed for a conversion gain of only 3 (10dB). This ensures optimum large-signal handling, low IM product and other spurious-signal generation by the mixer. IM products are also reduced by the low dynamic impedance of the IF coil at the drain. This impedance is approximately 3kΩ at 45MHz.

The first local oscillator final frequency ($F_s - F_{IF} = F_{osc}$ or 432 – 47 = 387MHz) is generated by the oscillator/multiplier

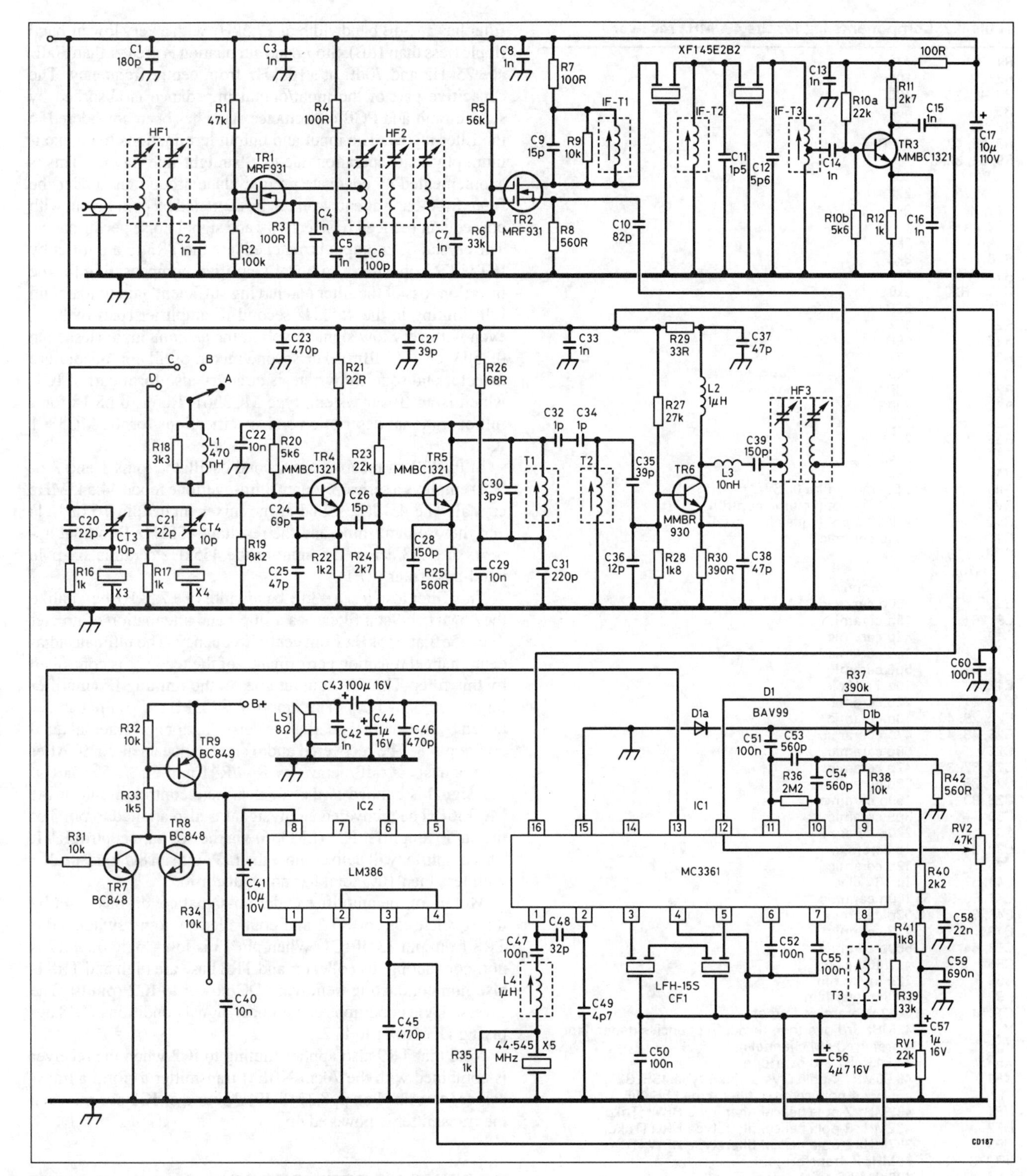

Fig 8.34. Circuit diagram of the 432MHz NBFM receiver

chain TR4, 5 and 6. TR4 is a modified Colpitts oscillator using a series-resonant third-overtone mode crystal as the frequency control element. TR5 and 6 are both triplers multiplying the oscillator frequency of 43MHz by nine for the final injection frequency of 387MHz. Band-pass filtering is used to attenuate unwanted oscillator harmonics (by T1 and T2) and multiplier harmonics (by helical filter HF3) to minimise spurious signal levels at the mixer source. This reduces production of spurious mixer outputs.

The circuit shows only a four-channel selector switch. Further channels may be added by increasing the number of 'ways' on the selector switch and the crystal complement. Fine tuning onto the specified channel frequencies is accomplished by adjustment of TC1–TC4.

The tap on IFT1 provides impedance matching to the 45MHz four-pole third-overtone crystal filter. The characteristic input and output impedances of the crystal filter is 560Ω/2.5pF and is supplied as a matched pair in UM1 style holders (HC49). The

Table 8.2. Components list for the 432MHz receiver

R1	47k
R2	100k
R3, 4, 13	100R
R5	56k
R6, 39	33k
R8, 25, 42	560R
R9, 31, 32, 34, 38	10k
R10a, 23	22k
R11, 24	2k7
R12, 14–17, 35	1k
R18	3k3
R19	8k2
R10b, R20	5k6
R21	22R
R22	1k2
R26	68R
R27	27k
R28, 41	1k8
R29	33R
R30	390R
R33	1k5
R36	2M2
R37	390k
R40	2k2
All resistors are 5% metal film 0.25W.	
RV1	22k 10% pot (volume control)
RV2	47k 10% pot (squelch control)
C1	180p ceramic
C2–5, 7, 8, 13–16, 33, 42	1n ceramic
C6	100p ceramic
C9, 26	15p ceramic
C10	82p ceramic
C11	1p5 ceramic
C12	5p6 ceramic
C17, 41	10μ 110V elec
C18–21	22p ceramic
C22, 29, 40	10n ceramic
C23, 45, 46	470p ceramic
C24	68p ceramic
C25, 37, 38	47p ceramic
C27, 35	39p ceramic
C28, 39	150p ceramic
C30	3p9 ceramic
C31	220p ceramic
C32, 34	1p ceramic
C36	12p ceramic
C44, 57	1μ 16V elec
C47, 50–52, 55	100n ceramic
C48, 60	33p ceramic
C49	4p7 ceramic
C53, 54	560p ceramic
C56	4μ7 16V elec
C58	22n mylar film
C59	680n mylar film
CT1–4	2–22p ceramic trimmer
X1–4	43MHz 3rd overtone (exact frequencies dependent on preferred FM channels)
X5	44.545MHz 3rd overtone
XF1	45.00MHz 4-pole crystal filter, type 4SE2B2
CF1	455kHz 4-pole ceramic filter, type LFH-155
HF1	432MHz 2-pole helical filter, type HRW (Toko)
HF2	432MHz 3-pole helical filter, type HRQ (Toko)
HF3	387MHz 2-pole helical filter, type HRW (Toko)
IFT1–3	45MHz IF transformer
L1	470nH RFC axial
L2	1μH RFC axial
L3, 4	10μH RFC axial
T1, 2	129MHz RF transformer
TR1, 2	MRF931 (Motorola)
TR3–5	MMBC1321 (Motorola)
TR6	MMBQ930 (Motorola)
TR7, 8	BC848 (Philips)
TR9	BC849 (Philips)
IC1	MC3361 (Motorola)
IC2	LM366 (National)
LS1	8R 500mW loudspeaker

All ceramic capacitors are 5%. All mylar film capacitors are 10%. All electrolytic capacitors are –20/+80% tolerance

filter has a –3dB bandwidth of ±7.5kHz with a very low in-band ripple (less than 1dB). Stop-band attenuation is greater than 30dB at ±25kHz and 70dB at ±100kHz from centre frequency. The capacitive part of the input/output impedance is absorbed by stray circuit and PCB capacitance. This has been considered in the filter layout. The input and output terminations have maximum physical separation due to a 'straight-line' layout. This is supplemented by adequate ground plane around and under the filter. Stop-band attenuation can be degraded by poor layout with consequent reduced IF selectivity and signal/noise performance. The crystal filter output is matched into the TR3 IF amplifier by IFT3. TR3 operates as a linear amplifier, compensating for the insertion loss of the filter and having sufficient gain to maintain full limiting in the 455kHz second IF amplifier (part of IC1), even with very low signal levels at the antenna input (less than 0.35μV or –116dBm). The second mixer, oscillator, quadrature detector and squelch/hysteresis circuitry also form part of IC1, which is an IF subsystem, type MC3361. Refer to p8.15 for a full operational description and specifications for the MC3361 IC.

In this 70cm receiver the second oscillator (pins 1 and 2 of IC2) employs a series-resonant third-overtone mode 44.545MHz crystal. The 45MHz signal at the mixer input (pin 16 of IC1) and the oscillator mix, and the resultant 455kHz IF output appears at pin 3 across the input of the 455kHz ceramic six-pole band-pass filter (CF1).

This filter also has a –3dB bandwidth of ±7.5kHz but, unlike the 45MHz crystal filter, has a stop-band attenuation of greater than 35dB at ±15kHz from centre frequency. The ultimate adjacent-channel rejection performance of the receiver is controlled by this filter. The filter output goes to the limiting IF amplifier input (pin 5 of IC1). The amplified 455kHz IF output is detected by the internal quadrature demodulator and external quad coil at pin 8. The recovered audio signal appears at pin 9. After de-emphasis at 6dB/octave by R40/R41 and C58/C59, part of this signal is applied to the squelch/mute control circuit of the MC3361. The recovered audio signal is also applied to pin 3 of the audio amplifier IC2 (LM386) via the AF gain control RV1. This amplifier will deliver more than 1W into an 8Ω loudspeaker with less than 10% total harmonic distortion.

Muting of the amplifier under no-signal conditions is set by the squelch control RV2 and controlled by logic switch TR7/TR8 from pin 13 of IC1. When pin 13 is low (logic 0), TR7 is non-conducting. Its collector and TR8 base are high and TR8 is also non-conducting, removing DC power to IC2 (pin 6). The reverse logic conditions occur under signal conditions, TR8 applying DC power to IC2.

Note that TR9 also applies muting to IC2 when the receiver is combined with the 70cm NBFM transmitter to form a transceiver (described on pp8.37–8.39). Muting of IC2 occurs when the transmitter is powered up.

RECEIVER OSCILLATORS

It has already been mentioned that the local oscillator may be either tuneable or fixed in frequency. Each system has its merits and these will now be considered (see also Chapter 5).

It is clear that if the final bandwidth of the complete receiver is to be the same as that used on the lower frequencies, the stability of the local oscillator must be as good as before, even though the frequency is much higher. For the 50 and 70MHz bands it is possible to build a local oscillator of reasonable stability to operate at frequencies either above or below the signal frequency (by the value of the first IF), but apart from some inevitable drift there is also a rapid small change in frequency known as *scintillation* which mitigates against a note better than about T8.

The same disadvantages apply, but with greater force, to the 144 and 432MHz bands, even if the circuit is designed to operate at one-half or one-third of the injection frequency and followed by frequency multipliers.

Crystal-controlled local oscillators can, however, be used in VHF and UHF converters or in a purpose-built receiver first oscillator.

The advantages and disadvantages of both types of oscillator are summarised below.

Tuneable oscillators

(a) Advantages

- (i) A directly calibrated dial is possible.
- (ii) The cost of a crystal is eliminated.
- (iii) The circuitry is simple.
- (iv) There is less likelihood of spurious-signal interference from the main receiver.
- (v) Only one clear channel for the first IF is required on the main receiver.

(b) Disadvantages

- (i) Long-term oscillator drift makes dial calibration unreliable.
- (ii) Warm-up drift can be troublesome.
- (iii) It is difficult to obtain a better than T8 note, but not impossible.
- (iv) Accurate tuning of SSB signals can prove difficult.
- (v) Oscillators cannot readily be used remotely with mechanical tuning.

Crystal-controlled oscillators

(a) Advantages

- (i) Accurate logging of stations on the main receiver dial is possible.
- (ii) A T9 note for CW is assured.
- (iii) Accurate SSB signal tuning is possible.
- (iv) Negligible short term or warm-up drift.
- (v) Absence of controls permits remote operation of a converter.

(b) Disadvantages

- (i) More expensive in transistors (for 432MHz and above) and crystal.
- (ii) Possibility of additional self-generated spuriae, particularly if frequency multiplication is necessary.
- (iii) More adjustment and therefore more difficult to align initially.

TRANSMITTER OSCILLATORS

For the VHF/UHF bands the variable frequency oscillator (VFO) and the crystal-controlled oscillator (CCO) may be used as the initial frequency generator. Their design and construction is identical to receiver tuneable and crystal-controlled oscillators, but optimised for a higher power output level to drive the following buffer, amplifier or frequency multiplier stages in the transmitter.

An additional design for transmitter use is the variable crystal oscillator (VXO). See Chapter 5 for further information.

The advantage and disadvantages of the VFO and CCO have been summarised in the receiver oscillator section, but additional comments are necessary on transmitter oscillator design.

VFOs

(a) Advantages

- (i) Choice of design to cover the whole of the communications of each VHF/UHF band.
- (ii) Easy to frequency or phase modulate for NBFM transmissions.

(b) Disadvantage

Great care must be exercised in the construction not only to achieve good frequency stability but also to generate a 'clean' waveform free from unwanted frequency or amplitude modulation.

CCOs

(a) Advantages

- (i) Reliable frequency location
- (ii) Ability to generate a 'clean' waveform for CW transmissions.

(b) Disadvantages

- (i) A crystal selection switch is required for frequency changes, eg channel changes in an FM transmitter.
- (ii) Care must be taken in the design to prevent unwanted frequency generation by the unused crystals connected to the selection switch.
- (iii) Not easy to frequency or phase modulate for NBFM transmissions.

VXOs

(a) Advantages

- (i) Small variable tuning range permits avoidance of co-channel interference, with reliable frequency setting, short term and warm-up drift.
- (ii) Moderately easy to frequency or phase modulate for NBFM transmissions.

(b) Disadvantage

In some designs of VXO it is not possible to deviate far from the nominal crystal frequency without degradation of the output waveform.

Each of these methods of frequency generation will now be described in more detail.

Tuneable or variable frequency oscillators

Three common types of tuneable or variable frequency oscillators normally found in LF/HF receivers and transmitters can be used, with buffer amplifiers and frequency multipliers in VHF equipment.

To maintain adequate frequency stability the multiplication factor should not exceed ×24. For the VHF bands this factor and the VFO frequencies are stated below.

6m 5.5555–5.7777MHz × 9
4m 5.833–5.875MHz × 12
2m 6.000–6.083MHz × 24

It is important to understand that the total change (drift) in frequency at the final multiplier output is the initial drift in the VFO multiplied by the frequency multiplication factor. As an example, if the frequency drift in the 2m VFO (with time and temperature) is 100Hz, the frequency change at the multiplier will be 2.4kHz. This change is not serious for NBFM transmission and reception but is not acceptable for SSB or CW equipment. The multiplication factor also limits the use of LF VFOs and multipliers for the three VHF bands.

These VFO types are illustrated in Fig 8.35.

The series-tuned Colpitts (a) and the Hartley (b) oscillators

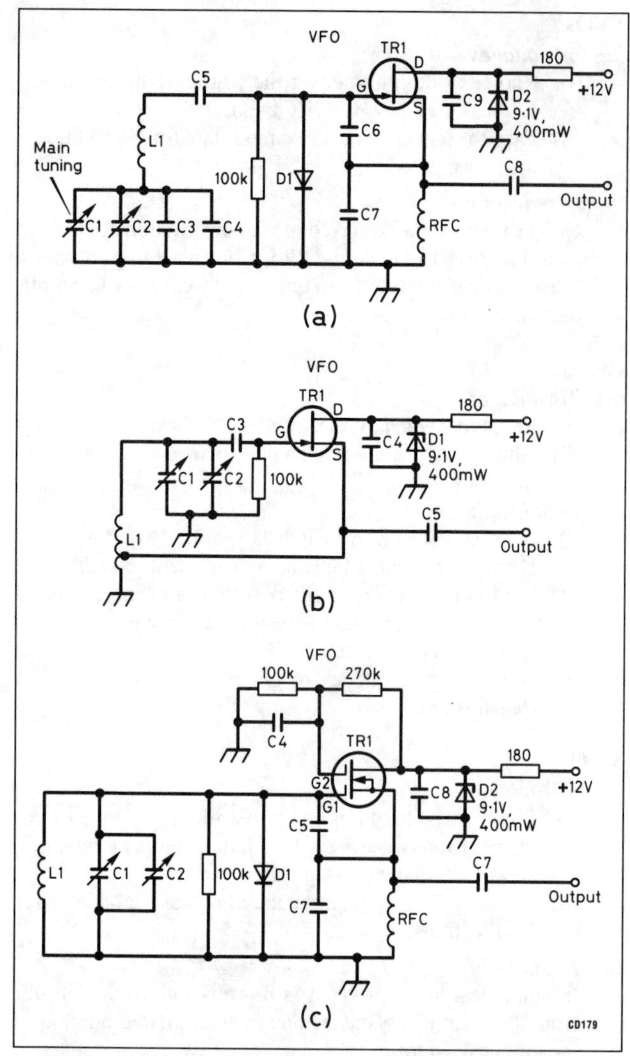

Fig 8.35. Three common types of VFOs for use in receivers and transmitters *(ARRL Handbook)*

offer good stability, both circuits using a JFET as the active component. The JFET G_m should be high. The dual-gate MOSFET in the parallel tuned Colpitts oscillator (c) also has a high G_m, eg the 3N211 or the 40673. This circuit also offers good stability. Gate 2 of TR1 should be bypassed by a high-quality, low-reactance capacitor. The shunt effect of C6 and C7 in Fig 8.35(a) and C5 and C6 in Fig 8.35(c) can be minimised by selecting a small value of coupling capacitor between the FET gate (G1 in Fig 8.35(c)) and C1. This will increase the VFO range but too small a value may prevent oscillation. The circuit in Fig 8.35(a) removes the shunt C problem as it requires a higher value L1 inductance than the circuit of Fig 8.35(c). Oscillator stability with temperature can initially be controlled by using polystyrene or polycarbonate capacitors. Drift compensation can then be added by employing disc ceramic capacitors with specific temperature characteristics.

Fig 8.36 illustrates a stable series-tuned Colpitts VFO with two buffers (one tuned) to increase the output amplitude and provide multiplier load isolation. This circuit can successfully be used for VFOs in VHF NBFM receivers and transmitters where $f_{osc(max)}$ is limited to 6MHz. The pi-network output circuit transforms the second buffer amplifier load impedance of 500Ω to the 50Ω output impedance. The 3300Ω shunt resistor across the pi-network inductor increases to the output. Long-term drift can be limited to 60Hz over a 3h period.

Crystal-controlled oscillators

The aim in a crystal-controlled oscillator/multiplier chain for a converter should be to produce the least number of frequencies in order to avoid unwanted heterodynes or 'birdies'. This means the use of overtone crystals and ideally no frequency-multiplier stages. Such an arrangement is quite feasible for the 50 and 70MHz bands where the crystal frequencies would be 22 and 42MHz respectively for a tuneable IF of 28–30MHz. However, for the 144MHz band, the local oscillator frequency is 116MHz for a tuneable IF at 28.30MHz. The cost of a fifth-overtone crystal may have to be considered.

In this case, and certainly for the 432MHz and 1296MHz allocations, some frequency multiplication is necessary; the aim should still be to start with as high a frequency as possible.

Many designs utilising crystals requiring high multiplication factors have been described in the literature and, provided certain precautions are taken to avoid frequencies which have harmonics falling inside the band concerned or frequencies which beat with harmonics of the oscillator in the main receiver, satisfactory results can be obtained.

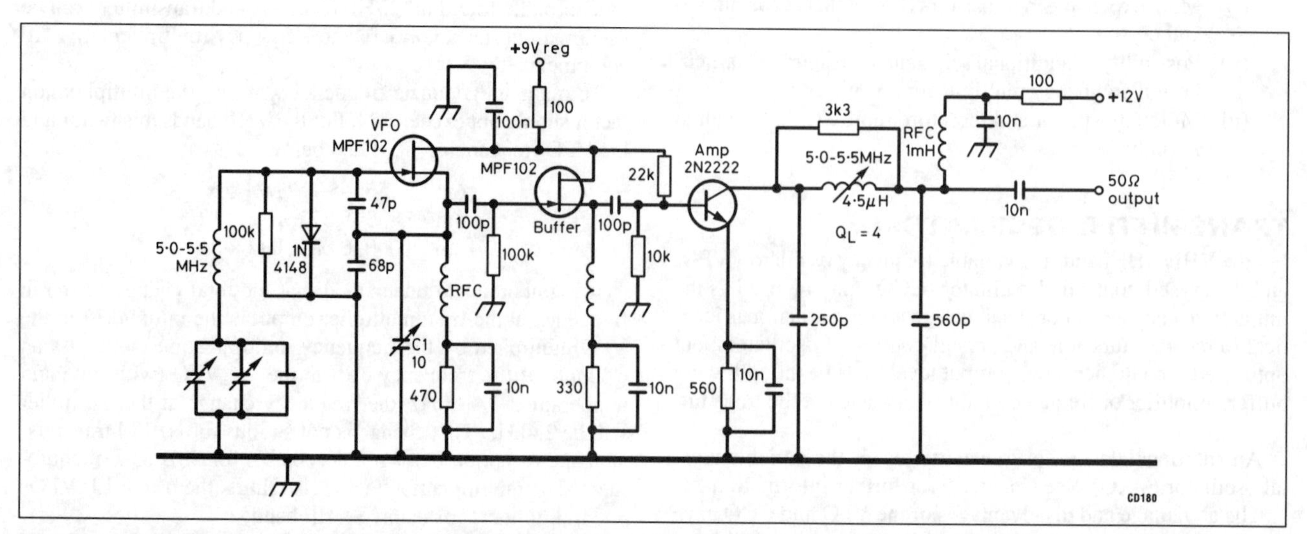

Fig 8.36. Suggested circuit for a stable series-tuned Colpitts VFO. Buffering follows the oscillator to increase the output level and provide load *(ARRL Handbook)*

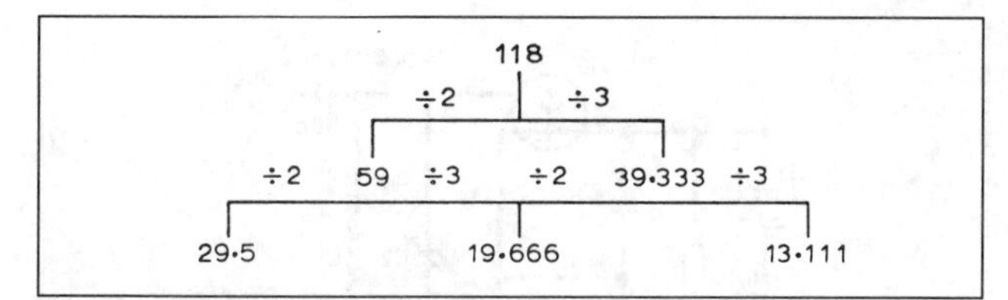

Fig 8.37. Type of chart for determining the fundamental crystal frequency required in an oscillator multiplier chain

Before deciding upon the frequency of a crystal for this service a chart such as that shown in Fig 8.37 should be prepared. As an example, assume that it is found that if the local oscillator in the main receiver covers the range 26.5–28.5MHz, no harmonics from it will fall into the 144MHz band. If the main receiver has an IF of 500kHz and its oscillator operates above the signal frequency, a first IF of 26–28MHz will be satisfactory. Since 26MHz will be the tuning position for a signal on 144MHz, the injection frequency for the converter will be 118MHz (ie 144 – 26) and reference to Fig 8.37 will show how this may be obtained.

Crystals for overtone oscillators are specifically cut to vibrate to approximately three or five times their fundamental frequency. The actual overtone frequency is stated on the crystal case. There is no output at the third or fifth sub-harmonic of the frequency. Third-overtone crystals should be used up to 60MHz operation. Above this frequency fifth-overtone crystals must be employed. Reliable operation and frequency stability is highly dependent on the oscillator circuit design, and specifically the choice of transistor. Typical circuits are shown in Figs 8.38 to 8.41. The cost of such crystals may seem high when compared with surplus fundamental-mode types but, unless the frequency of operation is very high or the specified tolerance very close (which is necessary for most amateur purposes), careful consideration should be given to the purchase of an overtone crystal. It should be borne in mind that such a crystal will give a greater certainty of trouble-free operation and, in the case of one of the high-frequency crystals oscillating at 60MHz or more, the greater cost may well be offset in the saving in transistors and other components, and the smaller space required for the converter.

Generally speaking it is preferable to employ crystals specifically intended for overtone operation: some essentially fundamental-mode crystals will be found perfectly satisfactory while others of similar type refuse to function or require an excessive amount of feedback, with the attendant possibility of oscillation uncontrolled by the crystal.

Oscillator circuits for converters, whether for fundamental-frequency or overtone crystals, are very similar to those employed in crystal-controlled transmitters, and reference should be made to the appropriate chapters for further information. For ease of reference, however, a selection of circuits is given here. Methods of optimising performance are described for each type of oscillator.

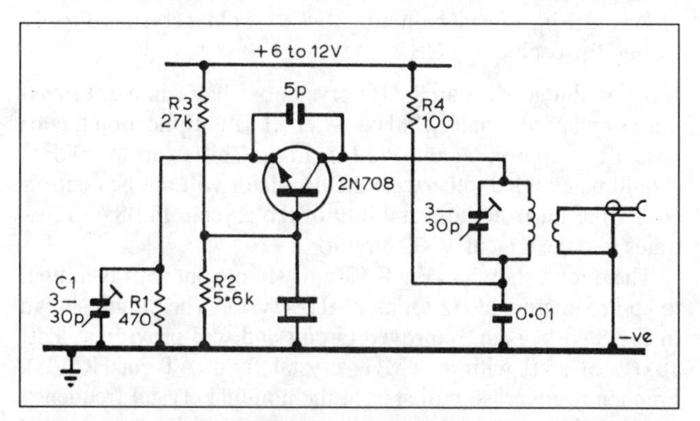

Fig 8.38. Transistor overtone oscillator circuit suitable for use with either third- or fifth-overtone crystals

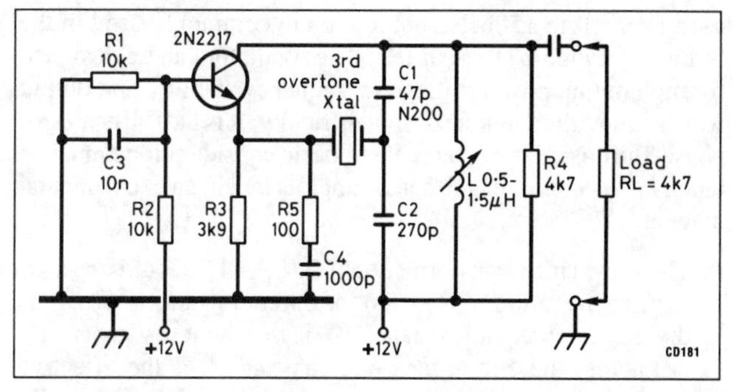

Fig 8.39. 22MHz grounded-base oscillator (decoupling not shown)

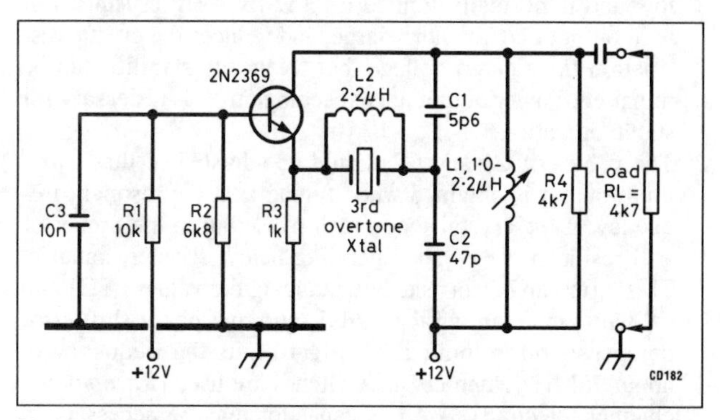

Fig 8.40. 42MHz grounded-base oscillator (decoupling not shown)

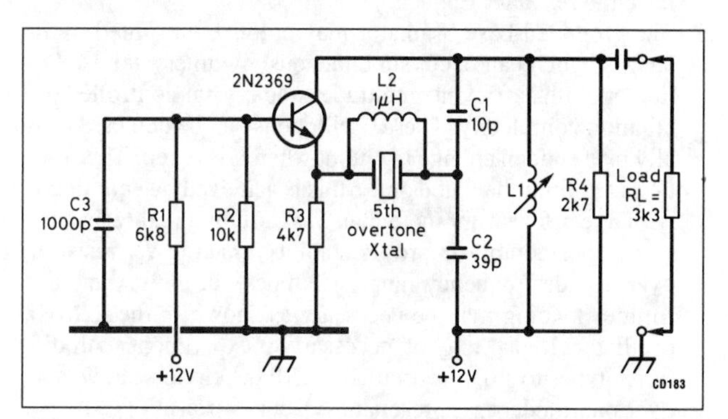

Fig 8.41. 70MHz grounded-base oscillator (decoupling not shown)

Crystal oscillator circuits may operate the crystal either in the parallel (high-impedance) mode or in a series-resonant circuit. In the latter the crystal forms part of the feedback loop so that, if it is replaced by a capacitor, oscillation will take place at a frequency determined by the inductance in the circuit and its associated capacitances. This can, in practice, happen even with the crystal in position as a result of the capacitance between the crystal electrodes, and spurious (uncontrolled) oscillations may be generated unless the circuit is properly adjusted.

A very satisfactory transistor overtone oscillator, suitable only for use with overtone crystals, is that shown in Fig 8.38. Almost any transistor capable of oscillating at a frequency of 100MHz or higher is suitable. The 3–30pF trimmer C1 is not essential but provides a ready means of controlling the amount of output. With other types or transistor it is worthwhile experimenting with other values of R1 and R3 for optimum results. There is little point in changing the value of R2.

The grounded-base oscillator (Figs 8.39–8.41) is an example of a design suitable for VHF equipment. This design can be used from 10 to 150MHz but it is more commonly used in the frequency range 20 to 200MHz. The oscillator can be designed for high output power with good frequency stability and output waveform without undue design difficulty. It is basically a zero-phase-shift oscillator, where three basic considerations must be taken into account, which can be approached in an experimental manner.

1. The impedance transformation $[(C_2/C_1) + 1]^2$ should approximate to the ratio $R_T/(R_{in} + R_E)$ for optimum gain, where R_{in} is the transistor input resistance, R_T is the shunt resistance (total) across the output tank circuitry and R_E is the effective series resistance (ESR) of the crystal. The ESR is usually low and is normally limited to 60Ω for VHF crystals. For high output C_2/C_1 is fairly large, and reduces the crystal dissipation for a given voltage, but frequency stability can be impaired. Power output not exceeding 5mW is necessary for stable operation.
2. The values of C1 and C2 should be adjusted so that maximum output is obtained when tuning L at the resonant frequency of the crystal. Too much phase lag in the transistor will result in the crystal operating below series resonance. This error can be corrected by reducing the values of C1 and C2 while maintaining the C2/C1 ratio, but phase-shift error compensation is limited. This fact limits the frequency to about 75MHz when crystal switches are used (for up/down channel selection). A series capacitor may be necessary to raise the resonant frequency.
3. The grounded-base oscillator may produce unwanted oscillations which can occur simultaneously with crystal oscillation or, with sufficient amplitude, stop crystal-controlled oscillation completely. These oscillations can be detected usually by a sudden change in output when L is tuned. The solution is to limit the value of R_T; this is achieved by shunting L with a resistor of not more than 5kΩ. Adding real resistance to R_T considerably improves stability, making R_T resistive over a wide frequency range. L2 tunes out the crystal C_O. Emitter loading may be necessary as shown in the 22MHz oscillator. It also may be necessary to experiment with different types to stop the oscillation. All the circuits shown use overtone-mode series-resonant AT cut crystals.

The general design of crystal-controlled oscillators has been discussed in Chapter 5. The Pierce, Colpitts and Clapp oscillators are actually the same circuit but with the ground point at one of three locations for each oscillator, ie at the emitter for the Pierce, at the collector for the Colpitts and at the base for the Clapp. In practical circuits stray capacitances and biasing resistors shunt different parts for each of the three configurations, resulting in performance differences for each circuit. At VHF it is necessary to use a different oscillator configuration, ie the zero-phase-shift oscillator, to ensure reliable starting-up, and frequency and output stability with supply and temperature variations.

Fig 8.42 shows two practical oscillator circuits of this type. C1 is included for adjusting the crystal to the frequency for which it has been ground. In circuits where considerable shunt capacitance is present (Fig 8.42(a)), the trimmer is usually connected in series with the crystal. When there is minimal parallel capacitance (approximately 6pF in the circuits at Fig 8.42(b)) the netting trimmer can be placed in parallel with the crystal. Whether a series or parallel trimmer is used will depend also on the type of crystal used (load capacitance and other factors).

A third-overtone crystal is illustrated in Fig 8.42(b). Satisfactory operation can be had by inserting the crystal as shown by the dashed lines. This method is especially useful when low-activity crystals are used in the overtone circuit. However, C1 will have little effect if the crystal is connected from gate to drain, as shown. C2-L1 is adjusted slightly above the desired overtone frequency to ensure fast starting of the oscillator. The circuits shown in Fig 8.42 can be used with dual-gate MOSFETs also, assuming that gate 2 is biased with a positive 3 to 4V.

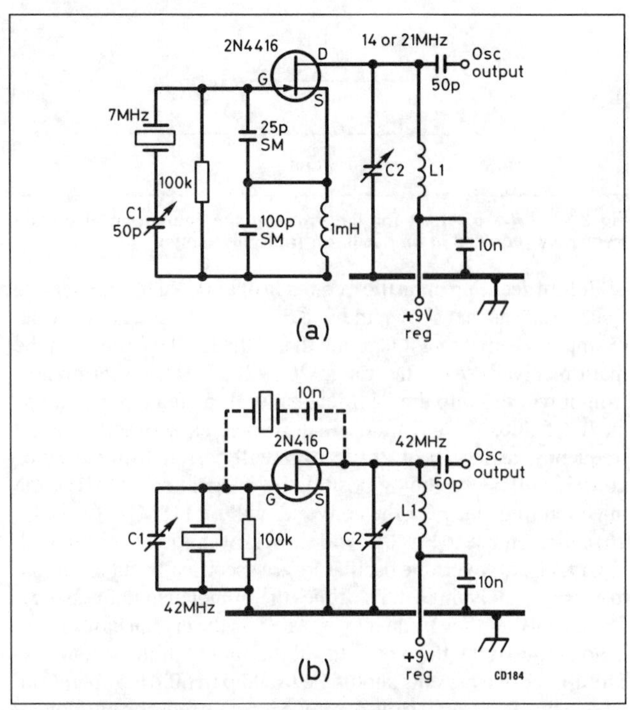

Fig 8.42. Practical examples of crystal-controlled oscillators that can be trimmed. (a) Harmonic oscillator. (b) Third-overtone oscillator *(ARRL Handbook)*

Variable-frequency crystal oscillators

Variable-frequency crystal oscillators (VXOs) provide a useful replacement of the conventional crystal oscillator when it is necessary to shift the oscillator by a few kilohertz. It is essential to ensure the oscillator stability is still controlled by the crystal when changing the frequency. Only a few 'cuts' of crystal can be 'pulled' from their nominal frequency. The best crystal is the fundamental mode AT-cut HC6/U type. A few rules must be observed when designing and building a VFO circuit as explained below.

1. Circuit stray capacitance must be reduced to a minimum.
2. Use only low-capacitance switches and low-minimum-capacitance variable capacitor.
3. Avoid using crystal mounting sockets. Solder crystals directly into the oscillator PCB.

It is possible to shift an 8MHz crystal oscillator in a 2m transmitter by approximately 8kHz. After ×18 multiplication the total shift is approximately 90kHz. Large shifts (up to 50kHz) should be avoided, otherwise the oscillator will not be controlled by the intrinsic high stability of the crystal. Fig 8.43 illustrates two designs of VXO circuit.

The circuit shown in Fig 8.43(a) is simple but shift is limited to approximately 5kHz for an 8MHz crystal. The circuit shown in Fig 8.43(b) is an improved circuit and will provide at least 10kHz of shift with an 8MHz crystal (both AT-cut HC6/U). Frequency coverage will start at the nominal crystal frequency and shift downwards. Bias stabilisation for the FET is controlled by D1 in both circuits. The diode reduces the FET junction

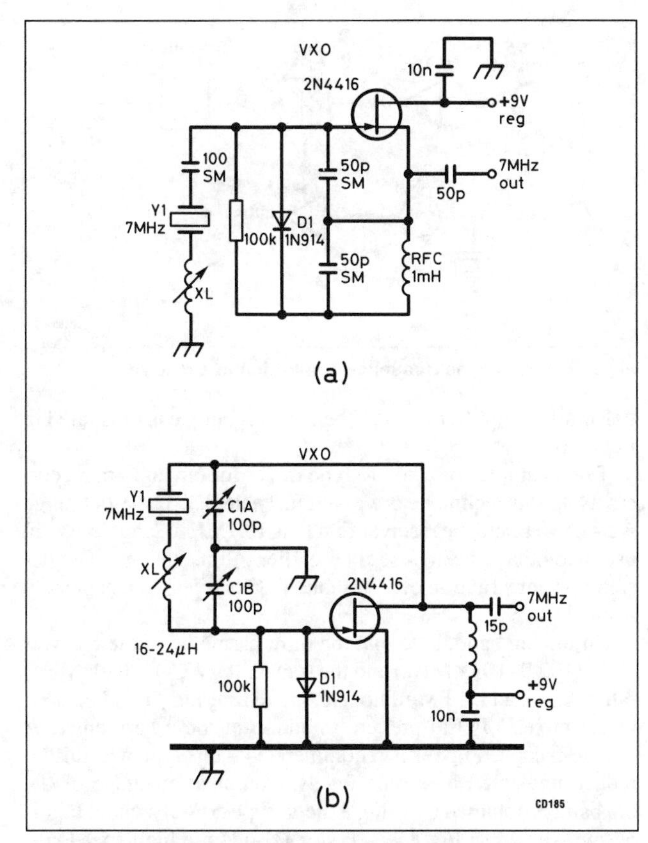

Fig 8.43. Two types of VXO circuit *(ARRL Handbook)*

capacitance at the peak of positive RF voltage swing by acting as a clamp, limiting the FET G_m during the time of these RF voltage swings. D1 also reduces the VXO harmonic amplitudes by controlling the non-linear change in the FET junction capacitance. In Fig 8.43(b), the reactance of XL is initially set to provide maximum possible frequency shift when C1 (a dual-gang capacitor) is tuned from minimum to maximum capacitance. Once set XL must not be readjusted. Frequency stability should be checked to ascertain the crystal is still controlling the VXO at maximum frequency shift. A buffer amplifier should be included between the VXO and first mixer in the receiver and multiplier/amplifier stages in the transmitter. VXOs are normally used in simple portable equipment where full band coverage is not required and frequency stability is not paramount, eg in NBFM equipment.

FREQUENCY MULTIPLIERS

It is usually necessary to use frequency multiplication after initial frequency generation, either from a VFO or crystal-controlled oscillator, particularly to reach the required LO frequency for the 70cm and 23cm bands.

It is always good practice to start with a reasonably high oscillator frequency requiring fewer stages of multiplication, rather than the reverse, because there is a distinct possibility that one of the unwanted multiples of the frequency oscillator might reach the mixer with sufficient power to convert (perhaps inefficiently) a strong local signal outside the amateur band and cause its appearance within the range of the tuneable IF.

Other spurious responses may appear at other harmonics of the crystal frequency. If, for example, a crystal frequency of 35MHz had been chosen, then the harmonics at 70 and 105MHz would give rise to a completely different set of possible spurious responses.

In addition to possible unwanted injection frequencies, all local oscillators generate noise. The magnitude of this noise depends on the operating conditions of the last frequency multiplier and the Q of the final multiplier tuned circuit. This final multiplier stage can be considered as an RF amplifier at the local oscillator frequency, and if its output tuned circuit has too low a Q, the stage will have appreciable gain at the signal frequency and so add to the noise input to the mixer. If, as is so often the case with UHF receivers, there is little or no RF gain in front of the mixer, oscillator noise can be quite serious.

Oscillator noise and unwanted injection frequencies can both be reduced to insignificance by interposing a filter between the local oscillator and the mixer. This filter may consist of a single high-Q tuned circuit called a *high-Q break* or two loosely coupled tuned circuits. Helical filters (described earlier in this chapter) provide an ideal solution to this problem. For example, a two-pole filter with a centre frequency of 1000MHz, while having a 1dB bandwidth of 18MHz, will attenuate at ±100MHz by at least 30dB. Two two-pole filters may be cascaded to give greater attenuation at ±100MHz.

The filter should be inserted between the final multiplier and the first mixer.

VHF transistors operated in Class C perform adequately as multipliers with efficiencies as follows: doubler 50%, tripler 33%, quadrupler 25% (Fig 8.44(a)). If a higher multiplication factor is required, eg ×9, it is desirable to use two or more multipliers. The transistor may be operated with forward bias but the stage must be driven hard enough to override the bias and operate in Class C. The VHF multiplier can be single-ended (Fig 8.44(a)) or with two transistors, configured as a push-push doubler (Fig 8.44(b)) or push-pull tripler (Fig 8.44(c)). The efficiency of the latter two multipliers is higher than for the single-ended types, and additionally the driving frequencies are well attenuated at the output. The push-pull tripler discriminates against even-order harmonics and thus reduces their level appearing at the LO drive to the mixer. R1 is used to establish electrical balance between TR1 and TR2.

Care must be taken in aligning a multiplier chain for the correct intermediate frequencies as illustrated in Fig 8.45. Here the output of a crystal oscillator at 46MHz is multiplied by three and three to generate 414MHz for a 432MHz converter. By error, the 138MHz tuned circuit is tuned to the fourth harmonic of the oscillator at 184MHz, but because of the low Q of this circuit, appreciable fundamental power from the oscillator appears at the input to the last stage. The tripler, designed as a non-linear stage, mixes harmonics of the drive frequencies and selects one of these, eg (184 × 2) + 46 = 414MHz which is the required output. The final multiplier is, however, very inefficient in this mode of operation, and requires an inordinate amount of RF drive. Much time can be wasted adjusting this apparently inefficient stage when the fault lies in an earlier one.

The cure for these and similar problems is to check all tuned circuits with an FET dip oscillator to verify that they are all operating at the required frequencies.

MIXER-TYPE VFOS OR PARTIAL FREQUENCY SYNTHESIS

The upper frequency limit for a VFO in HF equipment designed for good CW and SSB stability is about 10MHz. In VHF equipment, using multipliers after the VFO to reach the final frequency will degrade the oscillator stability, as described earlier. To overcome this problem a good solution is to use a heterodyne or mixer-type VFO. The low-frequency VFO output is mixed

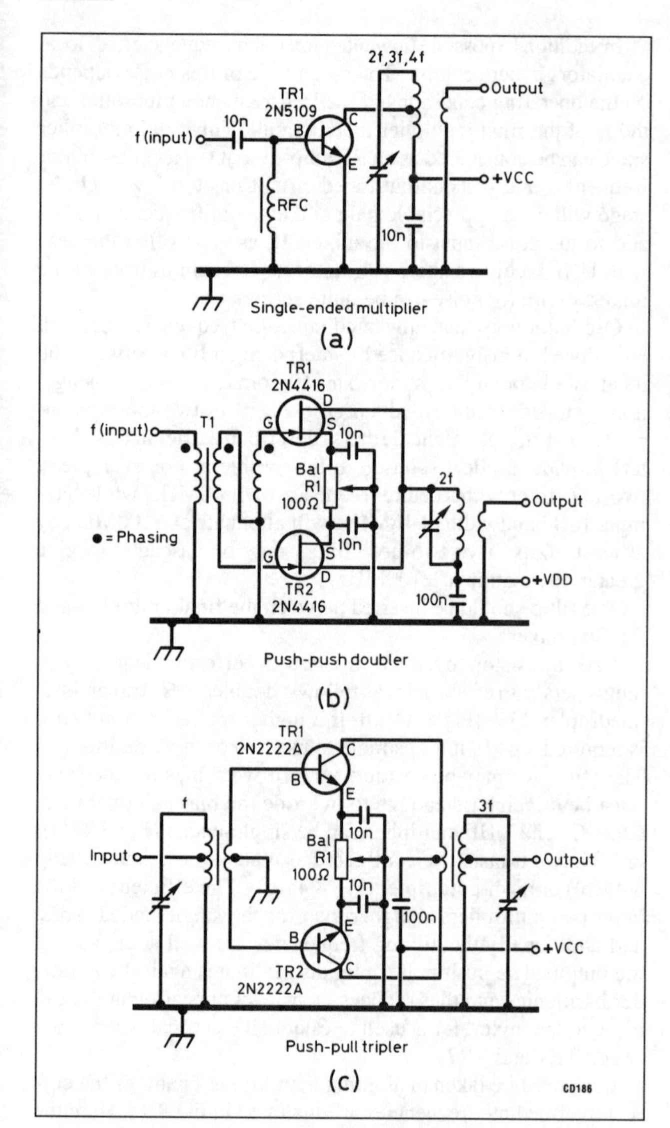

Fig 8.44. Frequency multipliers. (a) Single-ended. (b) Push-push doubler. (c) Push-pull tripler *(ARRL Handbook)*

with a high-frequency crystal-controlled oscillator output to obtain the LO for the receiver (plus or minus the IF) or the transmitter final output frequency. Both the oscillator outputs are buffered before injection into the heterodyne mixer, in practice through LC band-pass filters to reduce the amplitude of unwanted harmonics entering the mixer, which itself has a further band-pass filter at the output. This will reduce harmonic levels from the mixer output appearing at the receiver first mixer or the

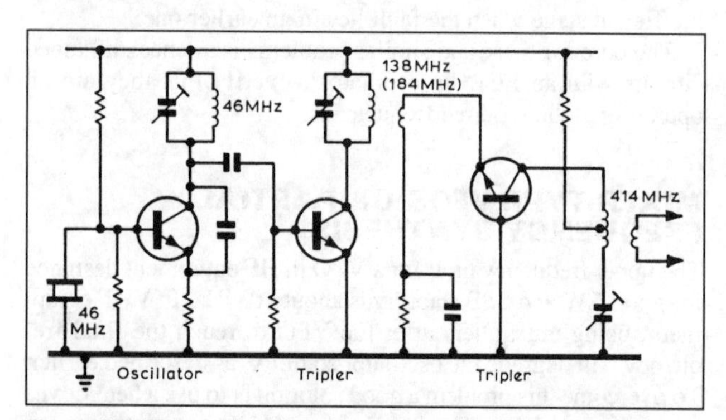

Fig 8.45. Typical local oscillator chain showing an error in alignment

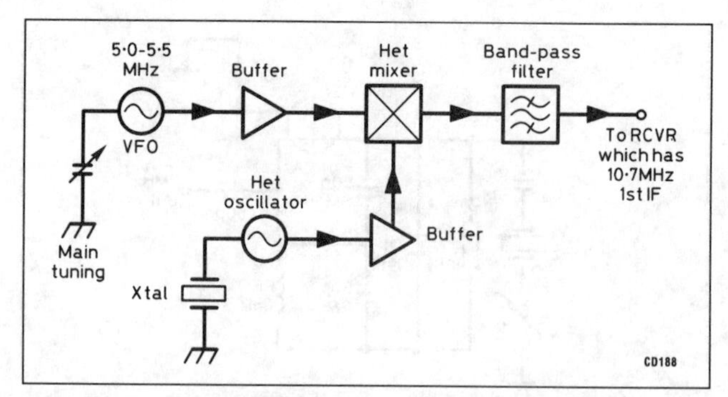

Fig 8.46. Heterodyne frequency generation in a receiver

transmitter amplifier inputs. The basic technique is illustrated in Fig 8.46.

Two examples of a heterodyne mixer for 6m and 4m receivers using this technique are given in Table 8.3. The VFO range is 5–5.5MHz and the receiver first IF is 10.7MHz. Suitable crystal oscillators have been described earlier in this chapter. The design and construction of a suitable VFO is given in Chapters 5, 6 and 7.

On the 6m band, note that the third harmonic of the receiver CCO (102.9–107.4MHz) and the transmitter CCO (90–93MHz) fall within Band II FM radio. Adequate filtering should be used in the mixer VFO to prevent radiation of these harmonics. A double-balanced mixer is recommended – this type will further reduce unwanted harmonic levels. Complete coverage of the 6m band is obtained by using a diode-type crystal switch. In this design, shown in Fig 8.47, diodes D1–D4 are high-speed silicon diodes. One of the four crystals are selected by applying via S1 +12V to one of the diodes, biasing it into full conduction and completing the circuit between the crystals and their respective trimmers. S1 can be located remotely.

FREQUENCY SYNTHESIS

This principle has already been described in Chapters 5 and 6. This section describes how the technique is applied to VHF equipment, where it is necessary to use a prescaler between the VCO and the phase comparator.

Digital phase-locked synthesisers

This is the most common type of frequency synthesiser found in modern amateur radio equipment. It uses a combination of high-speed digital dividers, a voltage-controlled oscillator and a technique analogous to a servo-controlled loop to produce a stable but programmable output frequency. This output frequency cannot be adjusted in a continuous manner as in a VFO but is tuned in a number of discrete steps. In FM equipment it is convenient to make these steps equal to the channel spacing, ie 25 or 12.5kHz. In the case of an SSB transceiver the step needs to be much smaller in order to ensure that the signal can be recovered correctly. 100Hz should be regarded as the maximum step size on SSB unless some other form of fine tune is available.

The actual programming of the synthesiser can be performed

Table 8.3. Examples of heterodyne mixers for 6m and 4m

Band (m)	Coverage (MHz)	RX CCO (MHz)	TX CCO (MHz)	RX LO (MHz)	TX carrier (MHz)
6	50.0–50.5	34.3	45.0	39.3–39.8	50.0–50.5
6	50.5–51.0	34.8	45.5	39.8–40.3	50.5–51.0
6	51.0–51.5	35.3	46.0	40.3–40.8	51.0–51.5
6	51.5–52.0	35.8	46.5	40.8–41.3	51.5–52.0
4	70.0–70.5	54.3	65.0	59.3–59.8	70.0–70.5

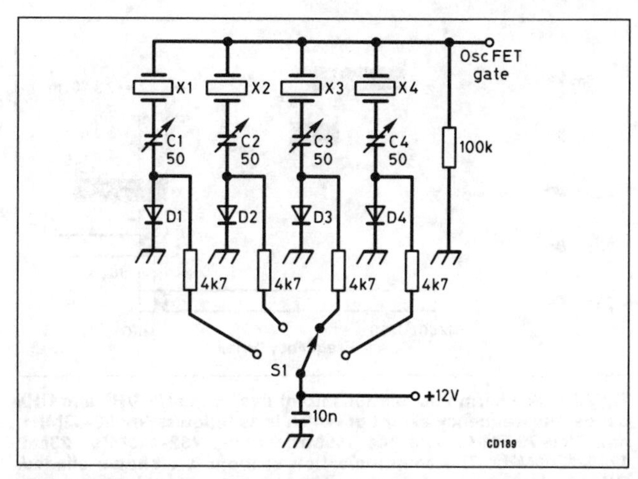

Fig 8.47. Methods for changing crystals by means of diode switching (ARRL Handbook)

in a number of ways, ranging from simple BCD-coded switches up to microcomputers. It is also possible to use tuning knobs coupled to shaft encoders or mechanical/optical switches to make it appear that the equipment is continuously tuned, the frequency being displayed on a digital display.

A block diagram of a basic PLL synthesiser is shown in Fig 8.48. The output from the synthesiser is provided by a voltage-controlled oscillator (VCO). This is normally a conventional LC oscillator which has all or part of its tuning capacitance replaced by a varicap diode. This means that its output frequency can be set by the voltage applied to the control input. Under normal circumstances this oscillator would be too unstable to use; the purpose of the synthesiser is to stabilise the frequency of the VCO. This done by dividing the VCO frequency down by means of a programmable divider.

The output from the programmable divider is applied to a phase-sensitive comparator which is also driven by a stable reference frequency. The output from the comparator is a voltage which is proportional to the phase and or frequency difference between its inputs. This error voltage is passed through a loop filter which cleans up the signal and also helps to determine the response time and stability of the synthesiser. It is then applied to the control input of the VCO. The phase and gain of the loop is arranged so that the two inputs to the comparator are brought into phase lock (ie on the same frequency, with zero or a static phase error) by adjusting the VCO frequency. Under these conditions the output from the programmable divider must be at frequency F_{ref}.

$$F_{out} = N \times F_{ref}$$

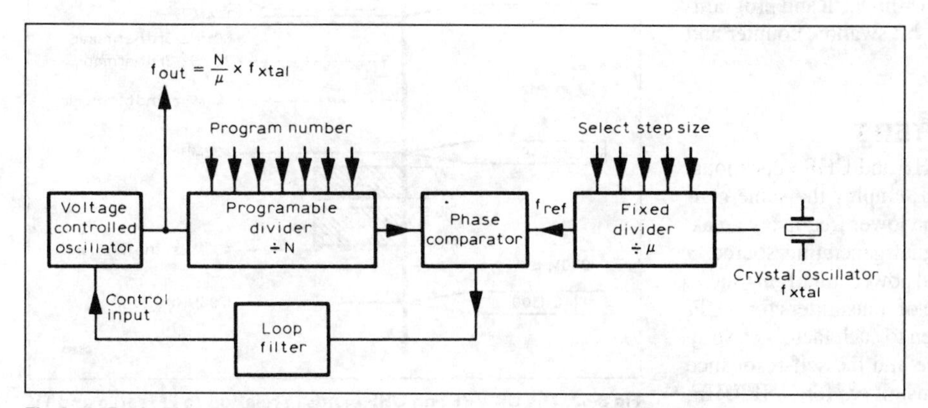

Fig 8.48. Basic PLL synthesiser

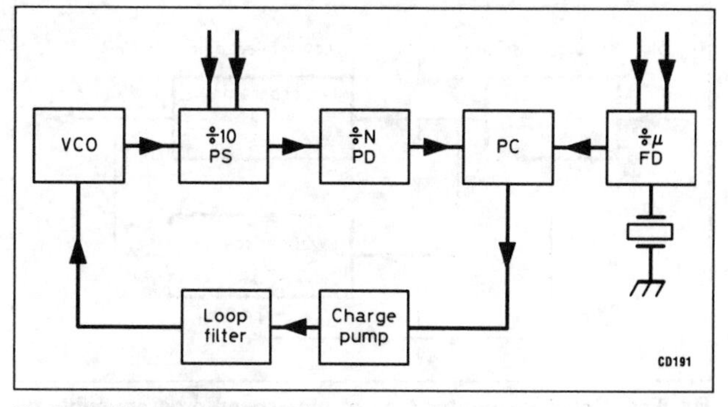

Fig 8.49. Use of prescaler

eg if F_{ref} = 1MHz and N = 145 then F_{out} = 145MHz. Now by selecting the division ratio N any output frequency can be achieved in multiples of F_{ref}. As F_{ref} is normally obtained from a crystal oscillator, as far as the long-term stability is concerned, the output is the same as a crystal-controlled source. If the reference divider ratio M is brought into the equation:

$$F_{out} = \frac{N}{M} F_{xtal}$$

By selecting N and M any output frequency can be obtained from a given reference crystal. For example, suppose a 145–146MHz FM synthesiser with 25kHz channels is required with a 5MHz reference crystal. M must be 200 and N will range from 5800 to 5839.

VHF/UHF prescalers

It is very difficult to make a fully programmable divider above about 30MHz, and it is therefore not practical to make a VHF synthesiser of the type described above. This problem is overcome by using a prescaler between the VCO output and the programmable divider input with the synthesiser loop as shown in Fig 8.49.

Swallow counting

A fixed ECL divider or prescaler could be used to reduce the VCO frequency to one that can be handled by conventional dividers. A fixed ÷10 prescaler before the programmable divider would mean that the reference frequency going to the phase comparator would also have to be reduced by a factor of 10. This would slow down the operation of the control loop and give longer locking times, together with less protection against VCO microphony.

An alternative to using a fixed prescaler is to use a dual-modulus device as typified by an ECL ÷10/11 device. It is possible to use this prescaler in conjunction with some low-frequency control logic to make a fully programmable UHF divider. The technique is called *swallow counting*.

The full programmable divider consists of three blocks as shown in Fig 8.50. Initially the swallow counter is loaded with value 5 and the 10/11 counter is in the ÷11 mode. When the swallow counter is empty the modulus of the prescaler is changed to ÷10. At this time there will have been 11 × S pulses at the input. The rest of the cycle, ie until the main divider is empty, is $(N - S)$ counts and

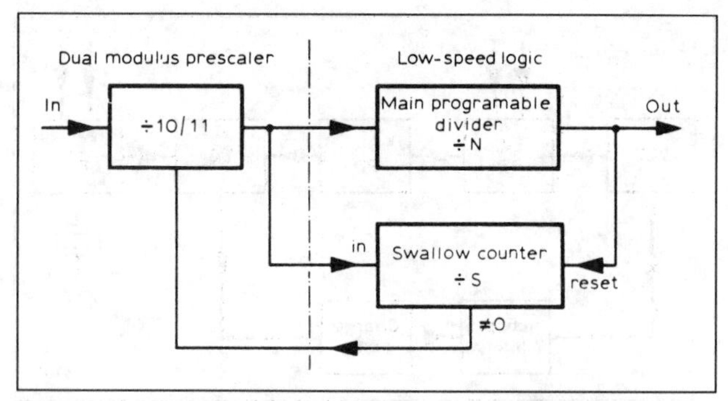

Fig 8.50. Swallow-counter type of high-speed programmable divider

as the prescaler is set to 10, corresponds to $10(N-S)$ pulses at the input. This gives a total division ratio N' where

$$N' = 11S + 10(N-S)$$
$$N' = 10N + S$$

The main divider N will have steps of 10 and single steps are provided by the swallow counter S. This type of divider can be implemented in SSI/MSI packages or in a number of LSI synthesiser systems that are on the market.

Design of frequency synthesisers

Synthesisers are notoriously difficult to design and to build with a low spurious output level. The design of the phase comparator and loop filter is very critical because any noise or AC components will modulate the VCO and produce unwanted sidebands on the output frequency. These can give rise to considerable interference on transmit. Also there is the possibility of the system going 'out of lock' – this will mean that the VCO could wander completely out of control up and down the band.

Many of these problems are a function of the phase comparator and fortunately there are LSI circuits available which use a very clean type of comparator called a *sample-and-hold phase comparator* which incorporates foolproof out-of-lock indication. The alignment of a synthesiser requires a spectrum analyser and great care is needed in the construction to prevent unwanted frequencies reaching the VCO. It is therefore recommended that synthesiser construction should only be undertaken when suitable experience and test equipment is available.

Modern PLL frequency synthesisers are now contained in single-package ICs. A typical package utilising bi-CMOS technology contains a 1.1GHz dual modulus prescaler, control signal generator, 16-bit shift register, 15-bit latch, programmable reference divider, 1-bit switch counter, phase comparator with phase conversion function, charge pump, reference oscillator (external crystal), 19-bit shift register, 18-bit latch and programmable divider consisting of a binary 7-bit swallow counter and binary 11-bit programmable counter.

VHF AND UHF TRANSMITTERS

Transmitters for use in the amateur VHF and UHF allocations, the frequency spectrum above 30MHz, employ the same general principles as those designed for the lower frequency areas, that is to say, they utilise a stable signal generating source, a chain of frequency multipliers and a final power amplifying stage. There the resemblance ends: the design of transmitters for VHF/UHF calls for great attention to be given to such factors as stray capacitance, excessive lead inductance and the self-resonance of certain components. Moreover, transmitters for VHF/UHF, unlike those intended for the HF spectrum, customarily cover

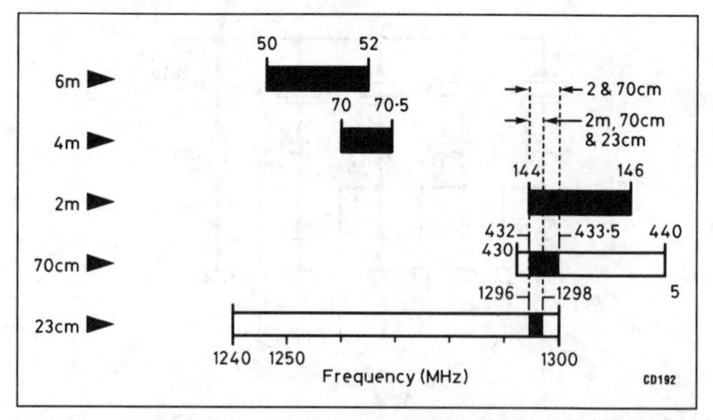

Fig 8.51. The harmonic relationship of five of the UK VHF and UHF bands, the frequency extent of which is as follows. 6m: 50–52MHz; 4m: 70.0–70.5MHz; 2m: 144–146MHz; 70cm: 432–440MHz; 23cm: 1240–1300MHz. The communication sections are shown shaded. Other amateur services occupy the remainder

one amateur allocation only in the interests of maximum efficiency.

In the UK the VHF/UHF bands are 6m, 4m, 2m, 70cm and 23cm, of which the last three are in partial harmonic relationship (Fig 8.51), a circumstance that permits the design of a three-band exciter capable of furnishing drive to multipliers for the next bands up. Fig 8.52 shows the proximity of these bands to broadcasting services.

In Fig 8.53 the block diagram illustrates the stages required to realise multiband output. The earlier stages of the exciter unit would need to be either tuneable or broad-banded enough to cover the whole range of output frequencies required.

Choice of frequency

The actual number of stages required in the exciter section of a VHF/UHF transmitter is governed by the starting frequency of the oscillator and the final operating frequency. In between a number of doubling or trebling stages are provided. At lower frequencies quadrupling may be used. Table 8.4 shows typical oscillator starting frequencies for four VHF/UHF bands, the multiplication factors required, and the final output frequencies. For convenience a 2MHz span is shown for the 70cm and 23cm

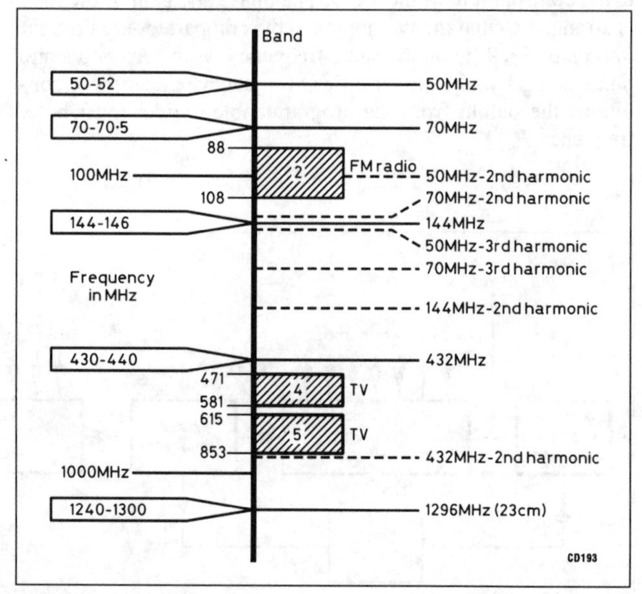

Fig 8.52. The UK VHF and UHF bands in relation to FM radio and TV broadcasting bands

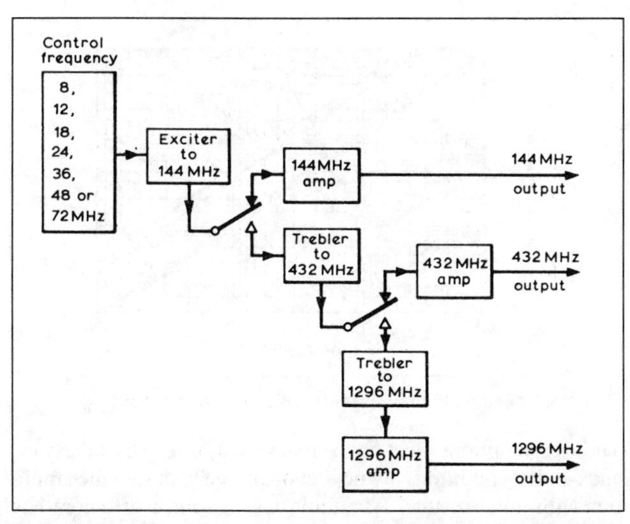

Fig 8.53. Block diagram to show how operation in the three harmonically related areas of the 2m, 70cm and 23cm bands may be achieved. Inter-unit switching may be by coaxial relays in the linking coaxial leads

bands, though in fact these bands extend considerably beyond these limits and in their upper reaches accommodate amateur television transmissions. Voice communication in the 70cm band in the UK is confined to the area 432.000 to 433.500MHz.

In addition to the multipliers in a VHF/UHF transmitter chain, additional stages are occasionally called for in such services as a buffer (isolation) amplifier between a variable frequency oscillator (VFO) and a subsequent stage, to prevent frequency pulling, or as an amplifier after the last multiplier to provide sufficient output to drive a large final amplifier.

The choice of frequencies to be used in the oscillator and following frequency multiplier stages will be governed by the crystals or VFO which the constructor has available. Each of these two sources of frequency control may require a different starting frequency; a high starting frequency may be used with crystals (provided they are operated within their ratings), but adequate stability in a VFO becomes more difficult to achieve as the starting frequency is raised.

Choice of multiplier frequency is governed by a further factor: the need to ensure that no undue radiation occurs on unwanted frequencies. For this reason it is desirable to operate exciter stages at low power levels even if this necessitates the addition of a driver-amplifier stage later in the design. Such an amplifier both raises the power level to the final stage and attenuates in its tuned circuits unwanted harmonics generated earlier.

Study of the figures given in Table 8.4 will assist in the selection of a starting frequency (crystal or VFO). For example, an oscillator at 72MHz will require no more than a doubler stage after it to reach the 144MHz (2m) band. 432MHz may be reached by adding a tripler stage. A further tripler will deliver output in the 1296MHz band. At such a high starting frequency separation of harmonics at the final frequencies is an easy matter.

Table 8.5. Comparison of various transistor parameters for small-signal and power operation

Parameter	Class A small-signal amplifier	Class C power amplifier P_{out} = 1W
V_{ce}	15V DC	13.6V DC
I_c	80mA	—
Input capacitance or inductance	12nH	21pF
Transistor output resistance	199Ω	92Ω
Output capacitance	4.6pF	5.0pF
G_{PE}	12.4dB	8.2dB

VHF/UHF POWER AMPLIFIER DESIGN

Early attempts to use small-signal RF design techniques in the design of power amplifiers led to poor results and amplifier performance, and difficulties in amplifier alignment. After these frustrating attempts, power amplifier designs were successfully pioneered using large-signal transistor input impedances. This principle has since been used by all of the major designers and manufacturers of communication equipment. The principle of large-signal impedances replaces the trial-and-error process, and is described here as it is applied to RF power amplifiers used in transmitters.

An example is shown here of the problem of trying to adopt small-signal parameters to power amplifier design (Table 8.5). The example shows the comparison of a VHF transistor used in a 300MHz Class A small-signal amplifier and the same transistor used in Class C power amplifier designed for an output power of 1W. In this table, resistances and reactances shown are parallel components, ie the large-signal input impedance is 38Ω in parallel with 21pF etc.

Table 8.4. Multiplication factors required for five VHF/UHF bands

Multiplication factor	6m band 50–52	4m band 70.0–70.5	2m band 144–146	70cm band 432–434*	23cm band 1296–1298†
×2	25–26	35.013–35.350	72–73		
×3	16.666–17.333	23.342–23.566	48–48.6	144–144.667	432–432.667
×4 (×2 ×2)	12.5–13.0	17.507–17.675	36–36.5	108–108.5	
×5	10.0–10.4	14.005–14.140	28.8–29.2	86.4–86.8	
×6 (×2 ×3)	8.333–8.666	11.671–11.783	24–24.3	72–72.3	
×8 (×2 ×2 ×2)	6.25–6.50	8.753–8.837	18–18.25	54–54.25	
×9 (×3 ×3)		7.781–7.855	14.4–14.6	48–48.2	144–144.2
×10 (×5 ×2)		7.025–7.070	16–16.2	43.2–43.4	129.6–129.8
×12 (×3 ×2 ×2)		5.836–5.891	12–12.16	36–36.165	108–108.16
×16 (×2 ×2 ×2 ×2)			9–9.125	27–27.125	81–81.125
×18 (×3 ×3 ×2)			8–8.11	24–24.1	72–72.1
×20 (×5 ×2 ×2)			7.2–7.3	21.6–21.7	64.8–64.9
×24 (×3 ×2 ×2 ×2)			6–6.083	18–18.0825	54–54.08
×32 (×2 ×2 ×2 ×2 ×2)				13.5–13.653	40.5–40.563
×36 (×3 ×3 ×2 ×2)				12–12.05	36–36.05
×40 (×5 ×2 ×2 ×2)				10.8–10.85	32.4–32.45
×48 (×3 ×4 ×4)				9–9.04125	27–27.04
×64 (×4 ×2 ×2 ×2 ×2)				6.75–6.7813	20.25–20.2813

* For convenience the band 432–434MHz is shown although in the UK there is little amateur voice communication above 433.5MHz.
† Communication section.

The largest differences in Table 8.5 relate to the transistor input impedance. As the operation of the device is changed from Class A to Class C, the complex input impedance changes by a large order of magnitude and actually changes from inductive to capacitive reactance. Additionally there is considerable difference in the transistor output resistance and power gains. This example shows clearly the errors introduced into a power amplifier design using the small-signal parameters of the transistor.

The network theory for power amplifier design has already been given in the HF transmitter chapter. Large-signal input and output impedance data for any given RF power transistor must be used in calculations of amplifier design. Large-signal impedance data, together with power output data, provide the essential information necessary to design the amplifier networks and a prediction of the expected performance to be achieved when the design is completed. It is also necessary to understand conditions of test and large-signal impedance data presentation. Large-signal input and output impedance refer to the impedance at the actual transmitter base and collector terminals when operating with the predicted output power for a given DC supply voltage in a matched amplifier. The term 'matched' refers to the condition where the input and output networks of the amplifier provide the best possible power transfer to and from the transistor. It is important to distinguish between large-signal impedances and small-signal two-port parameters normally measured at a Class A amplifier at low signal levels.

Consulting manufacturers' data will reveal that most RF transistors are measured in common-emitter Class C mode.

Much of the information in the transistor data sheets is presented in resistance and capacitance parallel-equivalent form. It may also appear in series-equivalent form. When designing a power amplifier the data should be studied to ascertain in which form it is presented. For convenience the series-parallel equivalent conversion equations are given at the end of this section in the section on complete designs (Fig 8.72).

An example of data presentation is given in Figs 8.63–8.65 later on. Studying the complex input impedance at 50MHz for the 2N5899 transistor with 20W power output indicates a 1.2Ω resistor in parallel with 1000pF capacitance. Similarly, at 70MHz for the same power output, the data indicate a 0.5Ω resistor in parallel with 0pF capacitance.

When a particular transistor has been selected for a required output power and input voltage, the data sheets for the device should be studied to determine the large-signal input and output impedances. This is simply a matter of reading off the complex impedances from the data sheets. It is possible that only the output capacitance may be quoted but the collector load resistance can be calculated as now described.

Large-signal impedance data for VHF and UHF transistors does not normally give the load resistance information, for the reason that this resistance is easily calculated. Conditions for load resistance derivation depend on certain assumptions, based on the theory that the collector voltage of the amplifier is a sine wave which swings from $2V_{CC}$ to zero in the tuned output network. V_{CC} is the DC supply voltage to the collector. The assumptions are given below.

1. $V_{CE(sat)}$ is equal to zero (oscillator emitter saturation voltage).
2. There is zero voltage drop in the DC collector supply.
3. The output network has the request value of loaded Q to reproduce the sine-wave voltage regardless of the transistor conduction angle, eg 180° for Class C.
4. The collector load presents zero impedance at all harmonics of the operating frequency.

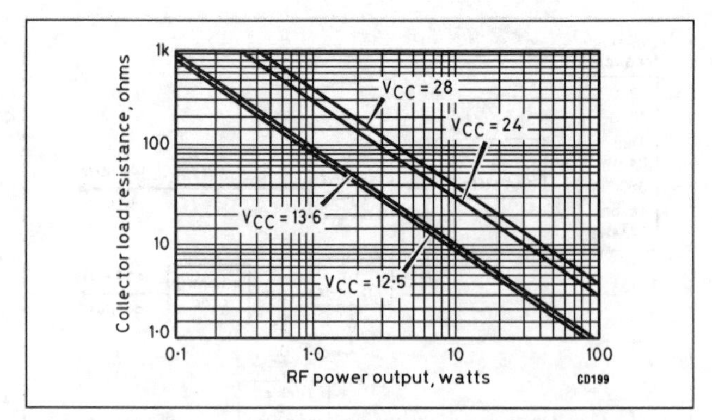

Fig 8.54. Collector load resistance versus power output

These assumptions must not be regarded as true, particularly (1) and (4), but conditions are near enough true to design the amplifier with good results. Assuming the existence of theoretical conditions, the parallel-equivalent collector resistance R_L is then a function of the required power output and V_{CC} only. The equation for solving R_L is readily derived thus:

$$R_L = \frac{V_{CC}^2}{2P}$$

where P is the RF power output. Therefore, in the power amplifier design, the complex collector load impedance is the conjugate of the equivalent and collector load resistance, calculated from the formula. Fig 8.54 illustrates a graphic solution for R_L at three familiar DC voltage levels of 13.6, 24 and 28V.

Power amplifiers using lumped components at VHF with direct voltage supplies varying from 7.5V to 28V, giving RF power levels from milliwatts to in excess of 100W, can be successfully built using this equation to calculate R_L. This indicates that lumped-component collector output networks are the best choice for proper impedance matching, ease of adjustment and tuning and low output loss. There is normally sufficient tuning and matching range to compensate for errors introduced by using this formula.

If $V_{CE(sat)}$ of the power transistor is known at the required frequency of operation, and also the collector current swings, the equation can be modified easily to:

$$R_L = \frac{(V_{CC} - V_{CE(sat)})^2}{2P}$$

In recent years many UHF power transistors have become available. Using these devices in power amplifiers requires a revised approach to design problems with the large-signal impedance matching technique. This is due to the use of microstrip matching networks and higher $V_{CE(sat)}$ values at UHF. Consulting the device data will reveal the output impedance presented in a full complex form instead of the parallel-equivalent output impedance to calculate the value of the collector load resistance. The data also shows how the UHF device should be evaluated in a microstrip circuit for its impedances, which is as close to the actual application as possible. The Smith chart is also used because of the convenience of designing the microstrip networks. The reader is referred to Chapter 12 for more information on the Smith chart.

Having designed a VHF or UHF RF power amplifier from the manufacturer's data, several factors must be taken into account before testing begins.

1. The source and load impedances must be an accurate 50 + j0Ω usually available on good RF generators and RF power load resistors.

Fig 8.55. Production test circuit for a 12V 70W VHF power transistor. (Courtesy of Semelab plc)

2. Deviation from a true 50Ω source will introduce gain measurement and input impedance errors.
3. The driver source should have minimum harmonic output level.
4. The amplifier should be 'tuned' for maximum output power in the load resistance.

At low and medium powers, simple tuning networks can match to a wide range of transistor impedances without many problems. As powers increase, impedances fall rapidly and matching becomes more difficult. Terminal impedances of 1–2Ω are not unusual for high-power transistors, and circulating currents of tens of amps can be encountered. Clearly, components carrying this level of current must be suitable for the purpose. As shown in Fig 8.55, layout and construction become critical: inductors typically become wide printed tracks or thick copper straps. Capacitors capable of carrying high current are normally either metal-clad mica or porcelain ceramic chips (ATC or equivalent), and very few other types are suitable in these situations. Using higher supply voltages brings several benefits: higher impedances (which makes matching much easier), higher gain and lower price per watt.

To some extent at 144MHz, and certainly at 432MHz, the parasitic inductance of capacitors must be taken into account. For both chip and metal mica types, this is about 0.5nH. At 432MHz this is a reactance of +1.3Ω. A final reactance of, say, –8Ω (46pF at 432MHz) actually requires a capacitor with a capacitive reactance of –9.3Ω. This is 40pF, 15% less than calculated. Fig 8.56 shows another example. A leaded capacitor, even with minimum lead length, will have an unpredictable parasitic inductance of several nanohenrys. This makes it impossible to swap capacitors in matching networks with any certainty, so

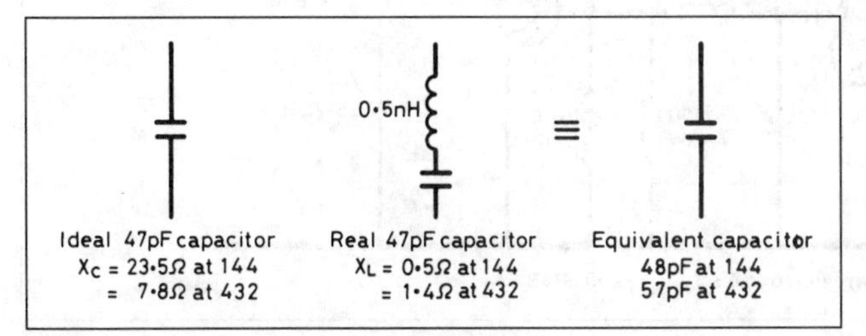

Fig 8.56. Parasitic inductance in capacitors

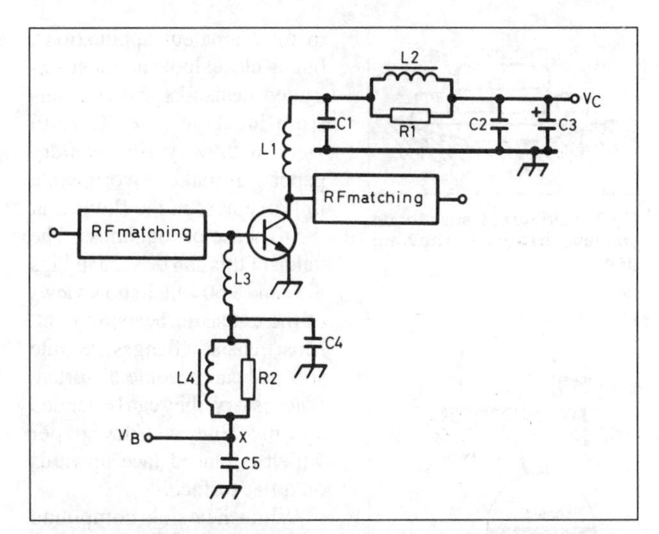

Fig 8.57. Transistor decoupling networks. L1, L3: RFC suitable for the operating frequency. L1 is usually an air-spaced inductor and L3 can be a moulded choke. L2, L4: ferrite-cored inductor, eg 2t on Siemens B62152 two-hole core, type A1X1 for L2, A4X1 for L4. C1, C4: decoupling at operating frequency; typically 1000pF. C2: 0.1μF ceramic. C3: 47μF. R1, R2: 10Ω 1W carbon or metal film.

designs need to be reproduced accurately to have a good chance of success. For example, Motorola application note EB46 [2] describes a VHF 80W amplifier which is fixed tuned in various bands by component selection. The difference between 143–170MHz and 155–175MHz is that in the first, a 500pF capacitor is made from 2 × 250pF and in the second it is made from 200 + 300pF.

Stability

It is important to remember that VHF and UHF power transistors have increasing gain at lower frequencies which can easily result in destructive oscillation. Many designs show no signs of instability, but a change to an 'equivalent' transistor or even a different batch of the original can change this dramatically. Some simple precautions can minimise the risk of such problems. The single most important one is to ensure that the transistor sees a resistive load at frequencies much below the operating frequency. Fig 8.57 shows networks which achieve this. At low frequencies, the RFC and RF decoupling are transparent and the transistor 'sees' the resistor and LF decoupling to ground. The LF choke provides the DC path for the supply current. In the base circuit of a Class C amplifier, point X is grounded and C4/5 are omitted. Another technique which improves LF stability is resistive feedback, shown in Fig 8.58. The RF choke isolates the resistor at the operating frequency and the capacitor provides DC blocking.

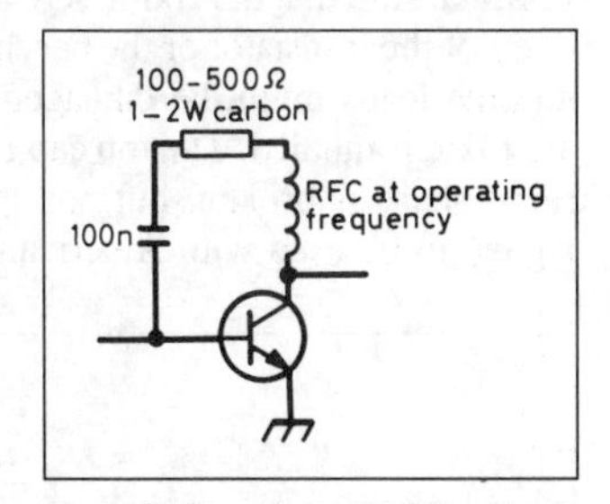

Fig 8.58. Transistor RLC feedback

Cooling

As with valves, provision of cooling is vital for high-power transistor amplifiers. A lot of heat has to be conducted through a small area with low temperature rise: around 20W/sq cm at the flange is typical. Philips' recommendation [3] for the heatsink is for a surface flatness of 0.02mm (0.001in) and a surface roughness of 0.5μm (almost mirror finish). These specifications are difficult to achieve without machining and probably unnecessary

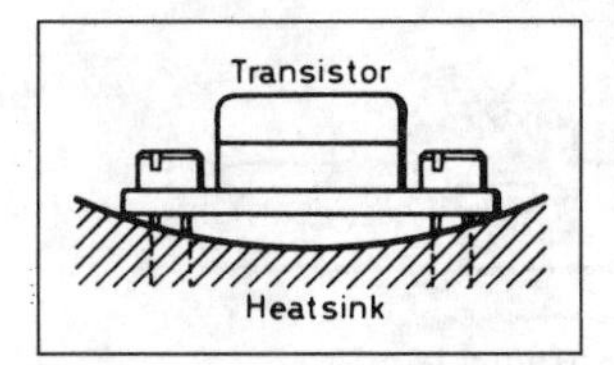

Fig 8.59. View of transistor on concave heatsink, showing gap

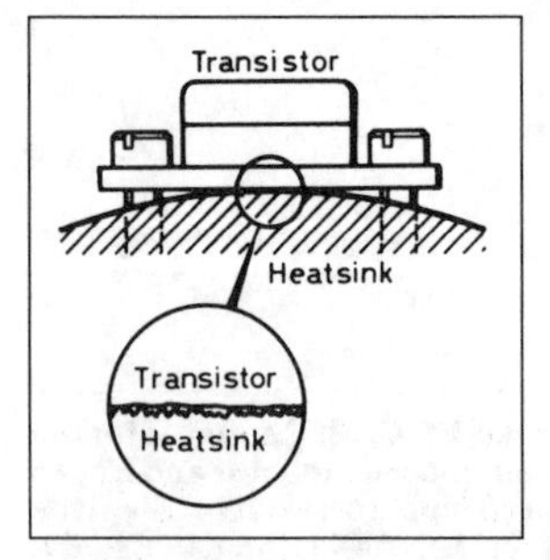

Fig 8.60. View of transistor on convex heatsink, showing gaps. Magnified view shows micro crevices in the transistor-heatsink interface

in most amateur applications, but a close look at most extruded heatsinks shows a far-from-ideal surface. Careful work with very fine wet/dry paper can make a worthwhile improvement in the flatness at the expense of roughness. The value of this can be seen in Figs 8.59 and 8.60 which show views of the transistor/heatsink joint. Most transistor flanges are quite soft and can become distorted. If necessary, they can be sanded against fine wet/dry paper which is placed face upwards on a flat surface.

White heatsink compound should be used in preference to clear silicone grease, and it is intended to fill the micro crevices in the enlarged view in Fig 8.60, rather than the gaps. The thermal resistance of heatsink compound is slightly better than air but a hundred times worse than metal, hence the emphasis on flatness and roughness to get maximum metal-to-metal contact at a microscopic level [4]. Before fitting the transistor, deburr the fixing holes well and clean all swarf, dust from sanding and solder/flux splashes from the heatsink and transistor using methylated spirits or similar; any debris under the transistor will undo all your work in making the heatsink flat. Using a thin layer of thermal compound, just enough to obscure the metal, press the transistor into place as hard as possible with a finger, then remove it – any voids in contact will be obvious. It should be possible to achieve contact over the whole flange with this amount of thermal compound; if this is not possible, then either the transistor or heatsink is not flat enough. *Warning*: thermal compound is waterproof and tenacious and will ruin your clothes; methylated spirits is a useful solvent to have on hand.

Make sure that the fixing screws do not foul against the main body of the transistor or the beryllium oxide between the flange and the leads might be damaged; particles of beryllium oxide are toxic if inhaled. The top cap is made from alumina which is razor sharp if broken, but non-toxic. Most transistors are designed to be used with American 4-40 UNC screws, but M2.5 or M3 can be used as an alternative. M2.5 is probably better as there is greater clearance for the head and less likelihood of overtightening, which reduces the thermal contact to the heatsink. Screws should be tightened to 0.7Nm, which is tight but well short of 'white knuckle' effort. Always use a plain washer between the screw and the flange.

Table 8.6. Performance data of the 40W amplifier

Power output at 50MHz	40W
Power input at 50MHz	20mW
Total current	5.4A
Overall efficiency	50%
2nd harmonic output	–25dBc
3rd harmonic output	–45dBc

TRANSMITTER DESIGN EXAMPLE

A typical example is illustrated for a 40W 50MHz transmitter designed by Motorola and utilising their 2N5847 and 2N5849 NPN silicon RF power transistors [5]. These transistors are constructed in the low-profile stripline package, providing a very-low-impedance emitter ground path which is vital for obtaining the highest possible power gain.

The transmitter is designed for CW or FM operation. All three stages require RF drive to forward-bias the base-emitter junctions during each half-cycle of the Class C mode. Therefore keying can be applied to the preceding exciter to remove drive during 'key-up'. Similarly NBFM can be achieved in the exciter as described earlier in this chapter. A schematic of the transmitter is shown in Fig 8.61 and performance details given in Table 8.6.

Design requirements

The first stage in designing this transmitter is to determine the necessary number of stages to increase the power input of 20mW to the final power level of 40W into 50Ω. The number of stages is determined by the power gain of each stage, and the block diagram in Fig 8.62 shows how this is determined by simple estimation. The 2N5849 will give a power output of 40W with a 12.5V supply from 6W input power. Allowing for 1dB circuit losses, the required drive level becomes 7W from the penultimate amplifier. The 2N5847 is rated at 7W power output for an input power of 0.5W. The 2N4073 will deliver 0.6W output from 0.02W input power and, again allowing for circuit losses, will adequately drive the penultimate amplifier.

Although these devices are old and may not be obtainable, the principles apply to all devices and the manufacturer's literature should be consulted to get the correct design parameters.

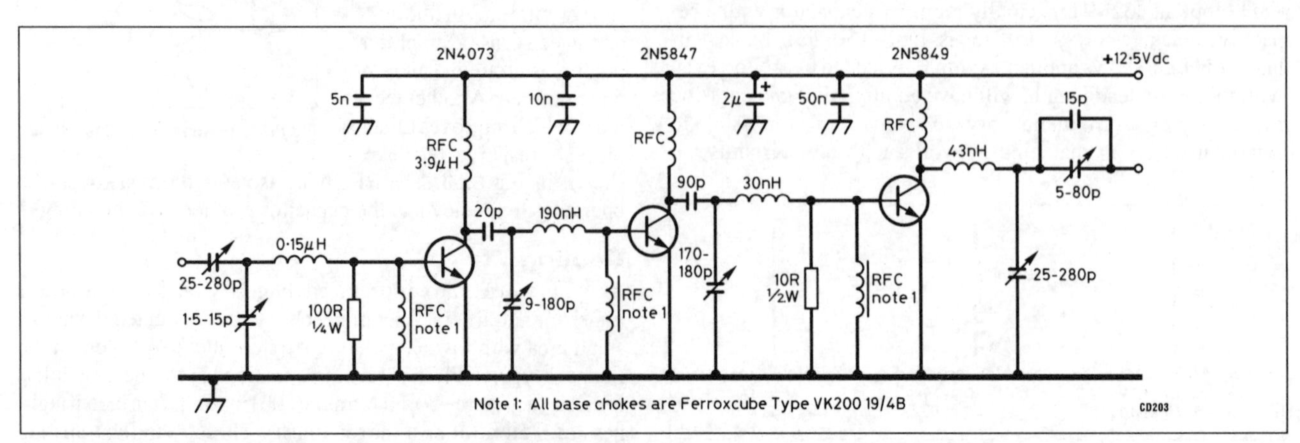

Fig 8.61. 50MHz 40W transmitter for 12.5V operation (Motorola)

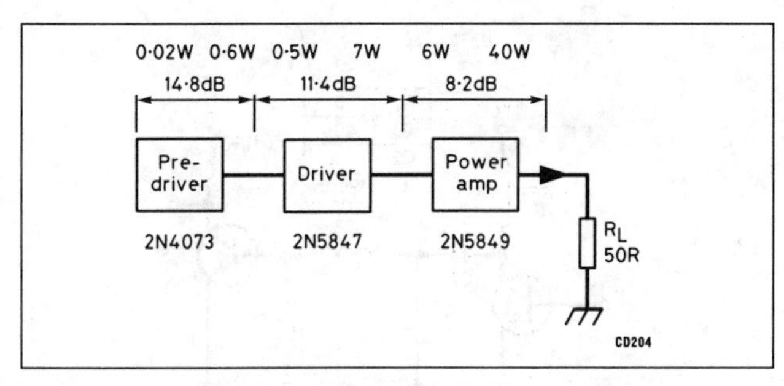

Fig 8.62. Transmitter block diagram

With regard to heatsinking, always use a larger heatsink than the calculations suggest.

High power output with good efficiency indicates the amplifier is operated in Class C, with each stage in common-emitter configuration.

As discussed earlier, in power amplifier design it is necessary to use the large-signal impedance data for each of the stages. These values are obtained from data sheet curves (Figs 8.63–8.65) and for the 2N5849 are as follows (for V_{cc} = 12.5V, f = 50MHz, P_{out} = 40W): R_{in} = 0.8Ω, C_{in} = 500pF, C_{out} = 625pF (all parallel equivalent).

The resistive portion of the collector load R_L is (assuming peak-to-peak collector voltage of $2V_{CC}$) is calculated with the formula:

$$R_L = \frac{(V_{CC} - V_{CE(sat)})^2}{2P_0}$$

where P_o is the RF output power and V_{CC} is the the collector DC voltage supply. For the 2N5849 transistor the collector load is:

$$R_L = \frac{(12.5)^2}{2 \times 40} = 1.95\Omega$$

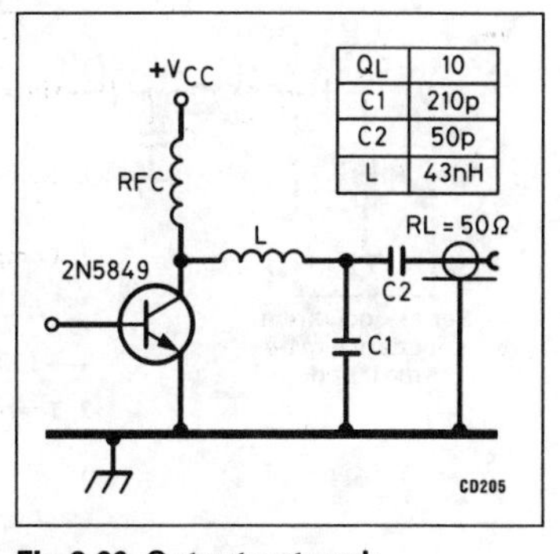

Fig 8.66. Output network

As the load impedance is small compared to the 50Ω termination impedance, the final stage network configuration is given in Fig 8.66 will realise optimum power transfer to the load (antenna) with good harmonic attenuation and ease of tuning. Comparative values based on a loaded Q of 10 are calculated from the formula presented in Fig 8.67.

The driver/final interstage network has to be matched to the input impedances of the 2N5849 and the 2N5847 calculated collector load impedance. With a drive power level of 7W this impedance is 11.2Ω in parallel with 160pF. The configuration is again the T-network as shown in Fig 8.68 and the values are calculated using the formula given in Fig 8.69. The 2N5849 is biased into Class C by the choke between base and emitter. The correct type of choke must be used, ie low Q at 50MHz, to avoid resonance with the base-emitter junction capacitance. The collector choke DC resistance must be very low as the 2N5849 collector current can peak to several amps. The resistance must not exceed 0.1Ω otherwise there will be significant output power loss.

As with the driver/final network, the pre-driver/driver interstage network must match – the pre-driver

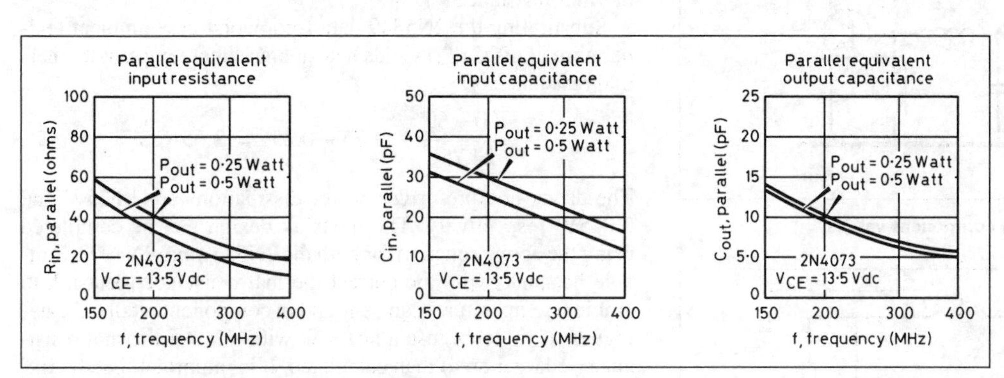

Fig 8.63. Large-signal impedances of 2N4073

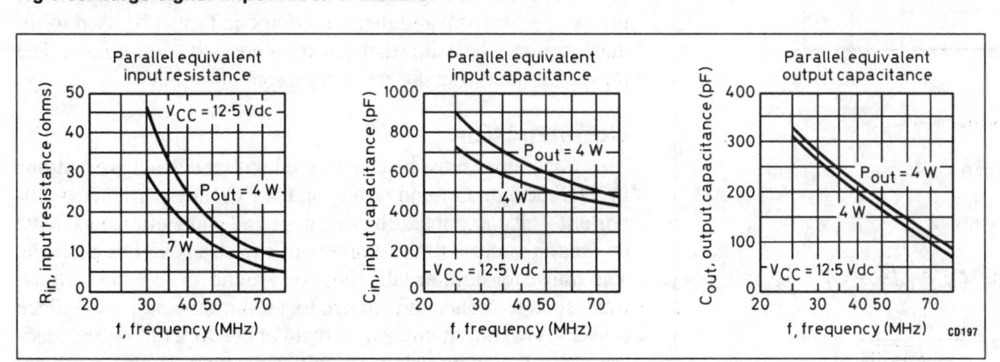

Fig 8.64. Large-signal impedances of 2N5847

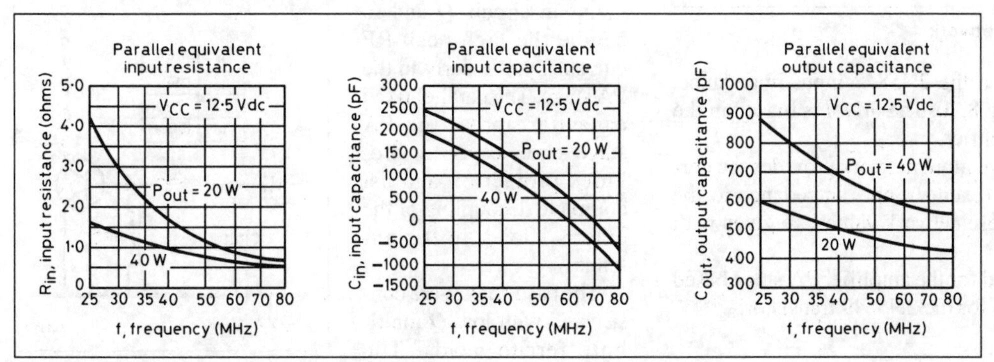

Fig 8.65. Large-signal impedances of 2N5849

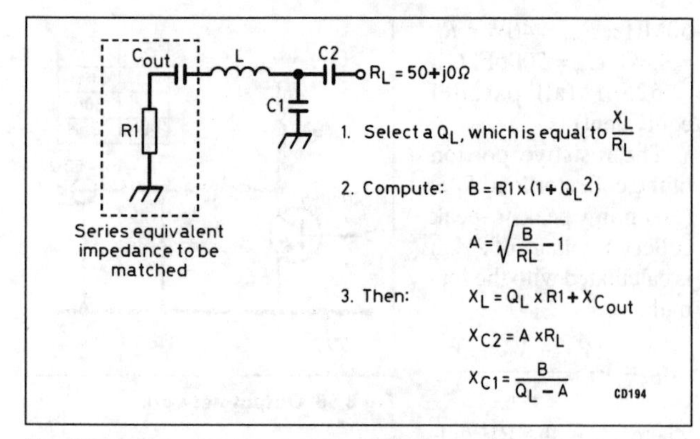

Fig 8.67. Input or output matching network

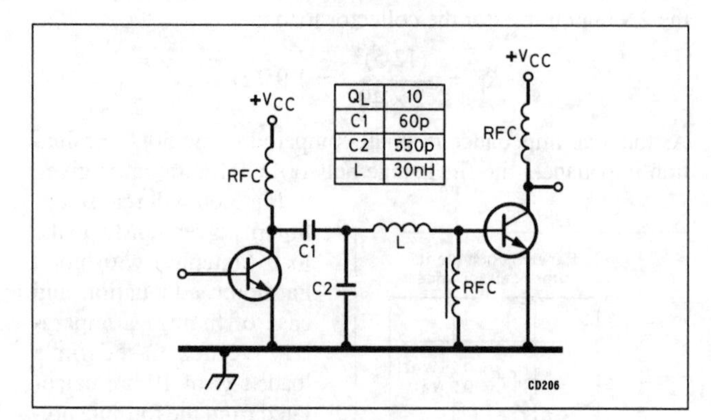

Fig 8.68. Driver/final interstage component values

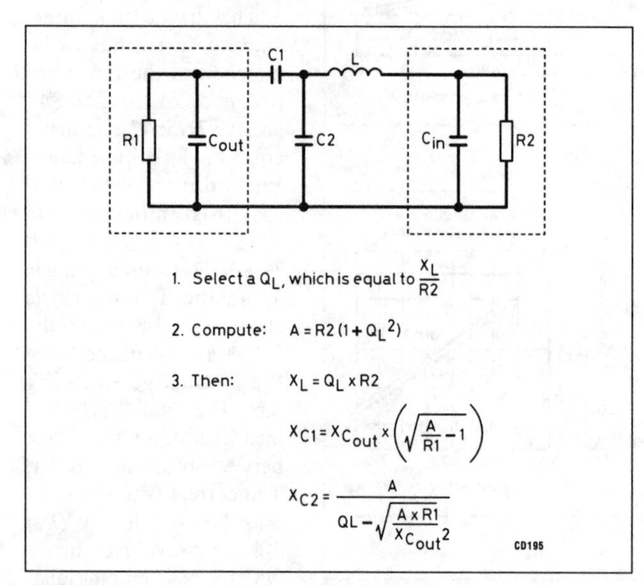

Fig 8.69. Interstage matching network

output impedance must match the 2N5847 input impedance. Circuit values are given in Fig 8.70. The driver is biased in the same manner as the final amplifier.

The 2N4073 pre-driver stage input network provides the correct matching from the 50Ω (exciter) input impedance to the 2N4073 input impedance. These network values are given with the circuit in Fig 8.71.

A heatsink must be provided for the amplifier PA stage based on the following thermal data for the 2N5849 transistor:

$T_{j(max)}$ = 200°C
P_D = 24.75W

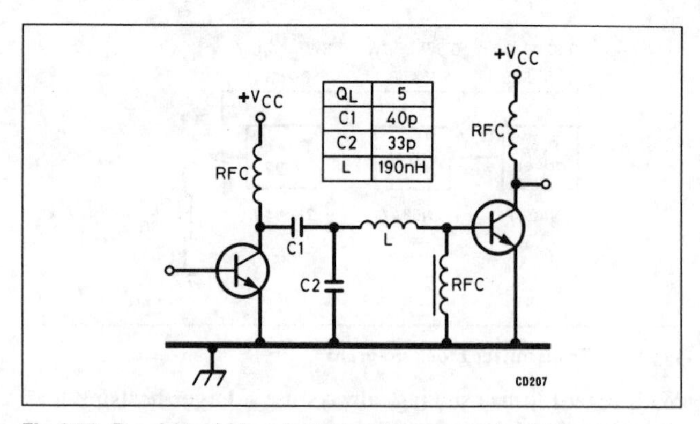

Fig 8.70. Pre-driver/driver interstage component values

θ_{JC} = 1.75°C/W
θ_{CS} = 0.25°C/W using thermal compound and mounting nut torque

where $T_{j(max)}$ is the maximum junction temperature, θ_{JC} is the junction/case thermal resistance, and θ_{CS} is the case/heatsink thermal resistance.

Substituting the 2N5849 data for a worst-case ambient temperature of 60°C (T_A) yields heatsink/ambient temperature thermal resistance of:

$$\frac{200 - 60}{24.75} - (1.75 + 0.25) = 3.65°\text{C/W}$$

The driver and pre-driver power dissipations are 1.625W and 0.045W respectively. The heatsink design can be completed using the above equation for both the 2N5847 and 2N4073. Suitable heatsinks with the correct thermal resistance (heatsink to ambient temperature) can be found in component retailers' catalogues. If in doubt, use a heatsink with a lower thermal resistance (ie larger area) than calculated. It is important that the fixing holes for the 2N5849 are free of burrs to ensure correct seating on the heatsink and thermal compound must be used to reduce the case/heatsink thermal resistance to a minimum. The torque on the fixing nut must not exceed 8in/lb.

Construction

This transmitter must be constructed with the usual precautions in VHF design. Ground points on the PCB are particularly important – all components having a ground connection must have this made as close to the emitters of each transistor as possible. The transistor emitter tabs must be grounded very close to the transistors to reduce inductance to a minimum, and also reduce power loss to minimum. Adjustment of network tuning and loading is facilitated by using variable capacitors. High-quality capacitors must be used to maintain circuit Q and to handle the high peak RF currents, particularly in the PA stage. The parallel fixed capacitor across the PA stage series capacitor provides protection against complete decoupling of the external load, ie the antenna.

The base chokes are constructed with low-Q multi-hole ferrite beads. The 100Ω parallel resistor

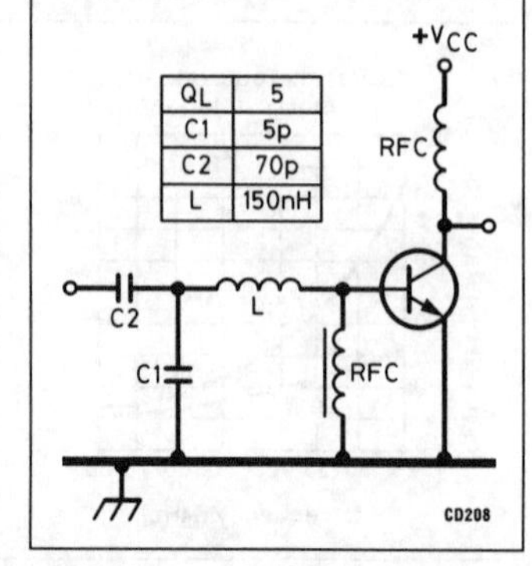

Fig 8.71. Input network

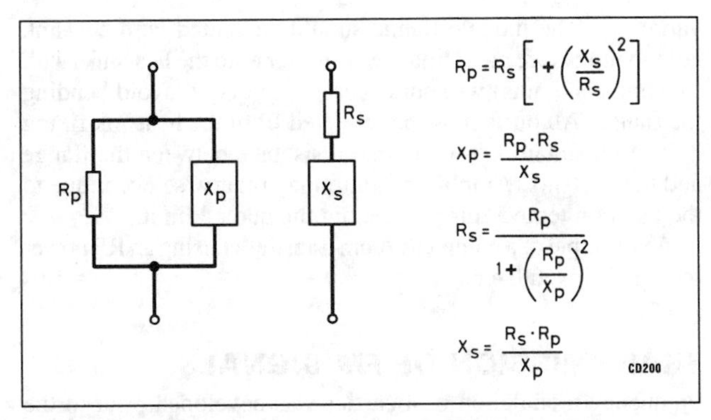

Fig 8.72. Parallel-to-series and series-to-parallel impedance conversion equations

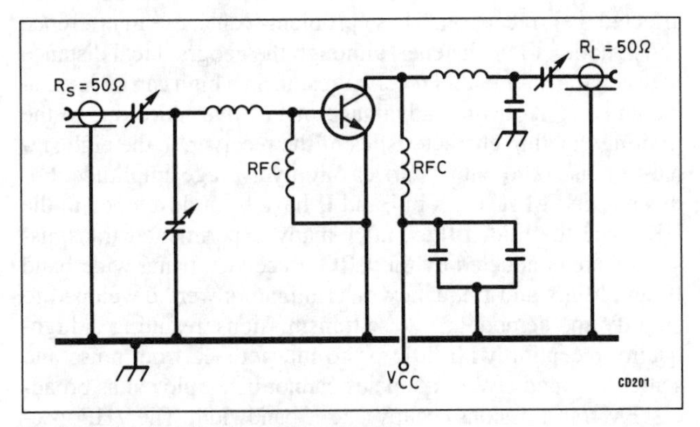

Fig 8.73. Typical test amplifier circuit

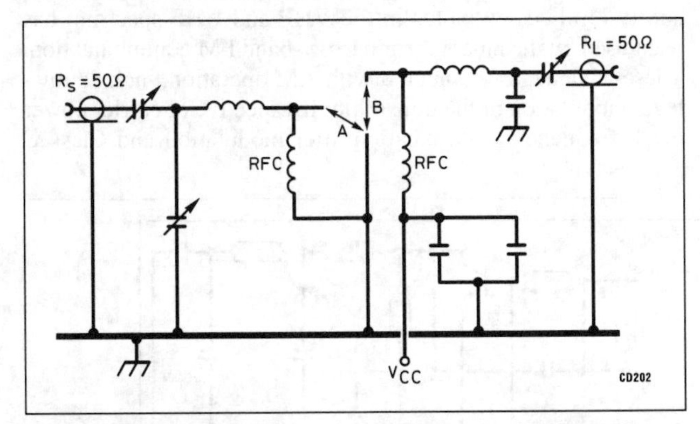

Fig 8.74. Test circuit with transistor removed. Complex impedances are measured at the base and collector circuit connections of the test transistor (points A and B respectively). Desired data will be the conjugates of these impedances

across the pre-driver choke and similarly the 10Ω parallel resistors across the PA stage choke ensure circuit stability at all drive levels.

As stated earlier, it is very important to align transmitters correctly. Suitable test equipment is a 50Ω VSWR/power meter, designed for the frequencies and power in use. The amplifier must be tuned for maximum power output with minimum reflected power. At VHF and UHF there is usually a small value of reflected power as it is not possible to null this out in practice unless the power load resistor is an accurate 50Ω. Since all output stages produce harmonics, they should always be fitted with a low-pass filter (see Chapter 22) before the antenna.

MODULAR VHF/UHF POWER AMPLIFIERS

With the advent of VHF/UHF power modules, many of the problems of designing and building medium-power PAs for transmitters have been largely overcome. Power outputs from 2W to more than 30W for CW and FM equipment are now common. The design of a three-stage amplifier for 50MHz has been described but it is not easy to test and align this amplifier without suitable calibrated test equipment, although the theoretical design has been achieved using transistor large-signal parameters, with the aid of the Smith chart for deducing complex input/output impedances. For example, power gain and power output for a given DC input into a 50Ω load can be predicted. Less easy to predict is power degradation over a wide temperature range, load mismatch with a high VSWR (to infinity), amplifier stability with varying power inputs and spurious outputs at frequencies below a particular VHF/UHF pass-band. These factors are taken into account by the designer and manufacturers of a power module. The electrical characteristics for a typical VHF modular 2W amplifier are given in Table 8.7.

Table 8.7. Electrical characteristics of the MHW401 series

Characteristic	Symbol	Min	Max	Unit
Frequency range MHW401-1 MHW401-2 MHW401-3	—	400 440 470	440 470 512	MHz
Output power* (50Ω load) (P_{in} = 50mW, V_S = 7.5V DC)	P_{out}	1.5	2.0	W
Power gain	G_p	15	—	dB
Efficiency* (P_{out} = 1.5W, V_S = 7.5V DC)	h	40	—	%
Harmonics* (P_{out} = 1.5W, Reference)	—	—	–45	dB
Input impedance*	Z_{in}	—	50	Ω
Power degradation* (P_{out} = 1.5W, T_C = 25°C) (T_C = 0°C to 60°C)	—	—	0.3	dB
Power degradation* (P_{out} = 1.5W, T_C = 25°C) (T_C = 0°C to 80°C)	—	—	0.7	dB
Load mismatch (VSWR = ∞, V_S = 11V DC, V_{SC} set for Pout = 2.0W)	—	No degradation in P_{out}		
Stability (P_{in} = 25 to 75mW, load mismatch 10:1, 50Ω reference, V_S = 4.0 to 11V DC, V_{SC} adjusted for P_{out} = 0.5 to 2.0W)	—	All spurious outputs more than 60dB below desired signal		

* P_{in} = 50mW, V_{SC} adjusted for 1.5W output

Power modules use NPN bipolar transistor 'dies' integrated with stripline tuned circuits techniques for input/interstage/output matching. The 'integrated circuit' is 'grown' onto a common substrate which itself is attached to a metal flange type base plate. This flange is the common ground for the amplifier and also acts as a low thermal resistance path for dissipating heat from the transistor junctions to a suitable heatsink. The heatsink should have sufficiently low thermal resistance to limit the junction temperature to safe limits, even with high ambient temperatures. The power module is usually encapsulated by a plastic cover, leaving the metal flange exposed for mounting to the heatsink. Construction pin-out data and dimensions are illustrated in Fig 8.75 for a typical device.

2W modules may also contain two transistors; three are normally used in modules with more than 10W output power. The internal circuit diagram of the 2W modular amplifier is illustrated in Fig 8.76.

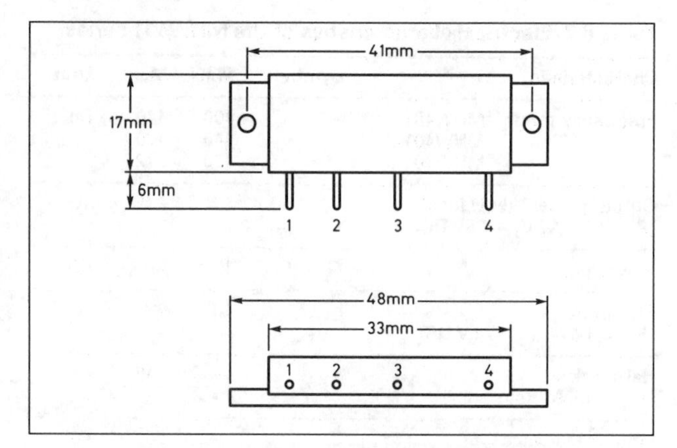

Fig 8.75. Pin-out details for the Motorola MHW401

Nominal operation is based on V_S and V_{SC} (controls the power input) given in the characteristics table. The quoted power output for a given DC input power into 50Ω will then be realised. This is the preferred method of operation – if necessary the RF power input should be adjusted to obtain the quoted power output. This will also result in the best performance over a wide ambient temperature range.

The input/interstage/output capacitor values have a low reactance at the pass-band of the module. This results in reduced power gain below the pass-band. In higher-power modules base return chokes are integrated into the amplifier as all the transistors are operated in Class C. While these chokes slightly degrade gain in the pass-band, they also have greater effect at lower frequencies. It is important to ensure the amplifier source and load impedances are true resistive 50Ω (a low-pass filter is integrated between the final amplifier and the output terminal), even though the amplifier is stable with a 2:1 VSWR source and load impedances, also with phase angle and input/output power variations. The module flange should be coated with heatsink compound before mounting the amplifier onto the heatsink (with two bolts and nuts) without excessive torque to avoid bending the flange. All burrs must be removed from the heatsink fixing holes to maintain a low thermal resistance between the flange and the heatsink. Amplifier failure may otherwise occur due to the junction temperature exceeding the quoted limit.

An example of a complete 70cm transmitter using an RF power module is given later.

TRANSMISSION OF FM SIGNALS

Frequency modulation of a carrier was developed prior to the second world war. AM broadcast transmissions suffered from the effects of man-made and natural noises (and static) which affected their reception. These problems can cause interference and irritation to the listener, although the geographical distance between transmitter and receiver results in a high carrier level at the antenna. A distinct advantage of FM over AM, due to the limiting/quieting characteristics of the receiver, is the high signal-to-noise ratio with a carrier of only average amplitude. For this reason FM services in Band II have been developed in the UK since the mid-'fifties, after many experimental transmissions were conducted by the BBC. Receivers using wide-band IF amplifiers and frequency discriminators were developed to amplify and demodulate these transmissions, resulting in high-quality reception with little or no interference from noise and static. MW and LW frequencies cannot be employed as broadcast FM transmissions occupy a large bandwidth. The VHF spectrum offers many advantages for FM operation and the availability of allocated bands in the VHF and UHF spectrum has been used by the amateur for narrow-band FM communication. These advantages, compared with AM operation, include low-level modulation of the transmitter irrespective of carrier power level, frequency multiplication after modulation and Class C

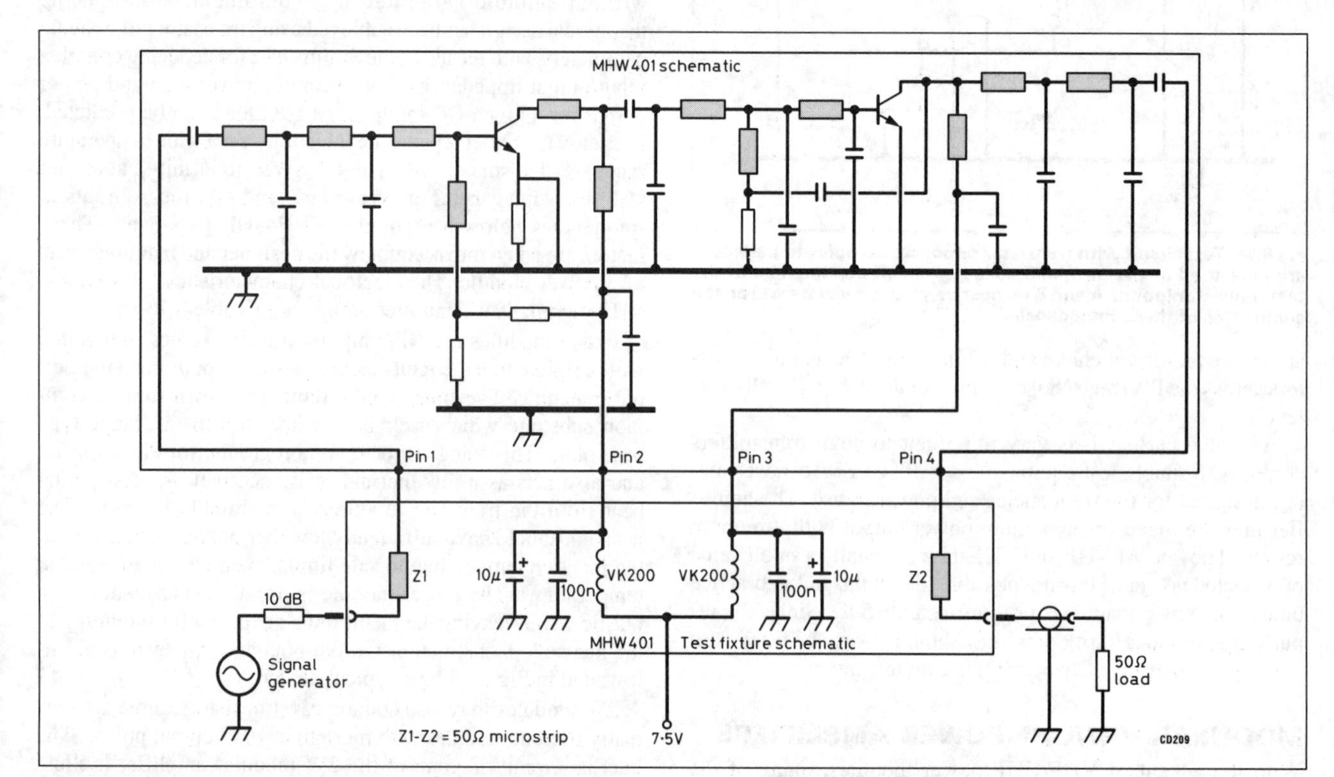

Fig 8.76. UHF power module MHW401 test set-up, showing its internal circuit (Motorola)

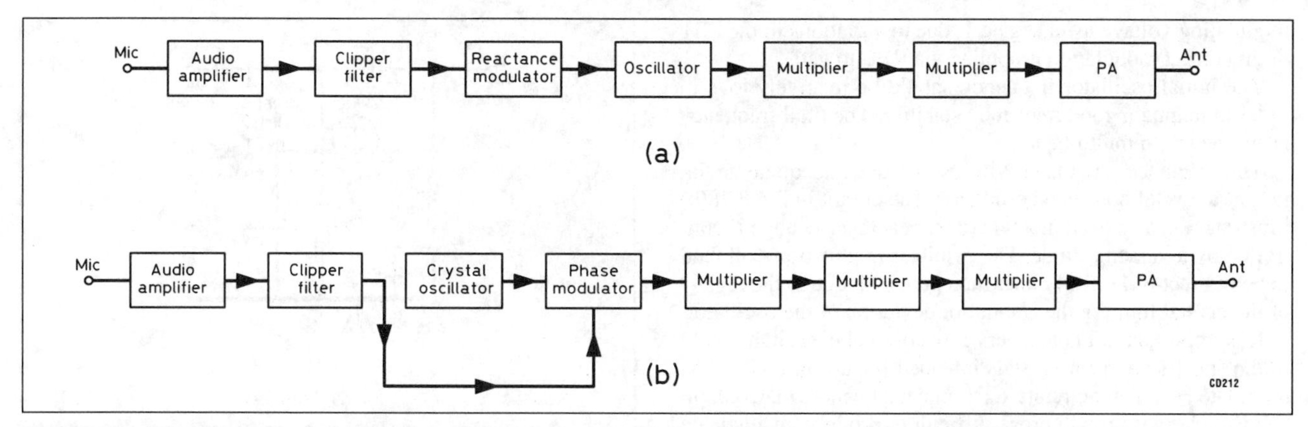

Fig 8.77. Block diagrams of typical FM exciters. (a) Frequency modulation. (b) Phase modulation *(ARRL Handbook)*

operation of the PA stage, ie non-linear, for highest efficiency. These advantages are particularly important in battery-operated portable and mobile equipment, reduced current consumption increasing the time intervals between battery charging.

Frequency and phase modulation

It is possible to modulate a carrier by introducing instantaneous frequency or phase changes to the carrier by a modulating signal. Frequency modulation results from direct variations of carrier frequency. Similarly phase modulation results from varying the phase of the carrier current. The two modulation processes are not independent as the carrier frequency and phase are varied at the same time by the modulating signal.

Frequency modulation

FM results from applying a modulation signal to the modulator, causing the carrier frequency to be increased on the half-cycle and decreased on the opposite half-cycle of modulating signal. The carrier frequency change, or deviation, from its nominal frequency is proportional to the amplitude of the modulating signal. This amplitude may be constant, eg modulation by a tone for repeater access, or instantaneous during voice transmissions. Deviation is thus small with low-level modulation and conversely large with high-level modulation. Peak deviation occurs at maximum instantaneous carrier frequency change, either plus or minus the carrier frequency.

Phase modulation

PM results from phase shift of the current flowing in a circuit, eg an oscillator, causing frequency change of the carrier. This process is known as *indirect FM* as shifting the phase gives the corresponding frequency shift. This shift or deviation is directly proportional to the degree of phase shift both in angle and velocity. The phase shift is thus directly proportional to the modulation signal frequency, and therefore the actual shift or deviation from nominal carrier frequency is directly proportional to the instantaneous modulation frequency *and* amplitude. In FM only the amplitude of the modulating frequency is proportional to frequency deviation.

Modulation circuits

For FM or direct modulation of the carrier, the best method is to use a reactance modulator. The RF tank circuit in a transmitter oscillator is made to act as a variable inductance or capacitance by a transistor connected to it.

Fig 8.78(a) illustrates a typical reactance modulator circuit employing a dual-gate MOSFET. The oscillator tank circuit is connected to gate 1 of the MOSFET by R1 and blocking capacitor C2. The transistor input capacitance is shown in dotted lines (C3). R1 resistance is large compared to the reactance of C3 (at the oscillator frequency). Therefore RF current flowing through R1 and C3 will virtually be in phase with the RF voltage present across L1 and C1. C3 will have a voltage-to-current lag of 90° across its terminals. The modulator drain RF current will be in phase with the gate 1 voltage but with a 90° phase lag to the current through C3 or 90° phase lag to the voltage across L1/C1. This current lag through L1/C1 produces the same effect as an inductance connected in parallel with the tank current.

The oscillator frequency is deviated from nominal in direct proportion to the amplitude of the drain current lag of the FET modulator. The transistor transconductance varies with the

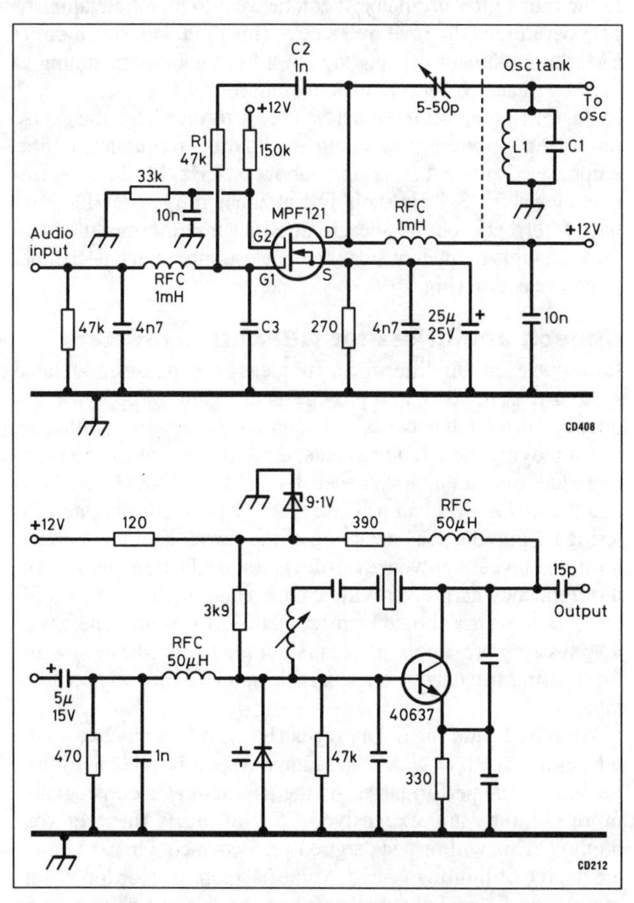

Fig 8.78. Reactance modulators using (a) a high-transconductance MOSFET and (b) a varactor *(ARRL Handbook)*

modulating voltage level at gate 1, due to variations in the FET drain current modulation is applied via RFC to gate 1.

The initial oscillator frequency should be relatively low, in order to maintain good frequency stability. The final frequency is attained with multipliers.

The reactance modulator will also produce adequate deviation of a crystal-controlled oscillator. The circuit in Fig 8.78(b) illustrates such an oscillator where the reactance change is controlled by a varactor diode. The resultant modulation will tend to be indirect FM or PM rather than pure FM, due to the high Q of the crystal limiting the amount of deviation of the oscillator.

It is important to quote a very low notional capacitance (C_0) in the specification for crystals intended for use in FM or PM modulated oscillator circuits. C_0 should not be more than 0.02pF (20fF), otherwise it will prove difficult to produce an adequate degree of positive and negative deviation of the oscillator irrespective of the modulation amplitude.

The degree of frequency deviation attained in the reactance modulator in Fig 8.78(a) depends directly on the FET transconductance, normally high for this application. Deviation is also dependent on the ratio of the resistance of R1 and the reactance of C3 at the oscillator frequency. $R_1 < X_{C3}$ for increased frequency shift. Additionally, increasing the L1/C1 ratio will give a higher degree of deviation, but oscillator frequency stability will be impaired by too high a ratio. Normally the capacitance value is a compromise between stability and adequate linear modulation. DC voltage change on the modulator and oscillator transistors must be eliminated by using a well-regulated and temperature-compensated power supply.

It is not strictly necessary to connect a reactance modulator to the transmitter oscillator; it can be used to modulate amplifiers operating at the final frequency. This modulation is indirect FM, the modulator causing the amplifier tank circuit tuning to vary by phase changes in the tank current.

An FM-compatible signal will be produced by the phase modulator providing the audio modulation frequency is pre-emphasised at 6dB/octave up to about 3.5kHz. Modulation frequencies about 3.5kHz should be attenuated by a suitable low-pass filter between the speech amplifier and the modulator to prevent high-frequency sidebands degrading adjacent channel power rejection (Fig 8.79).

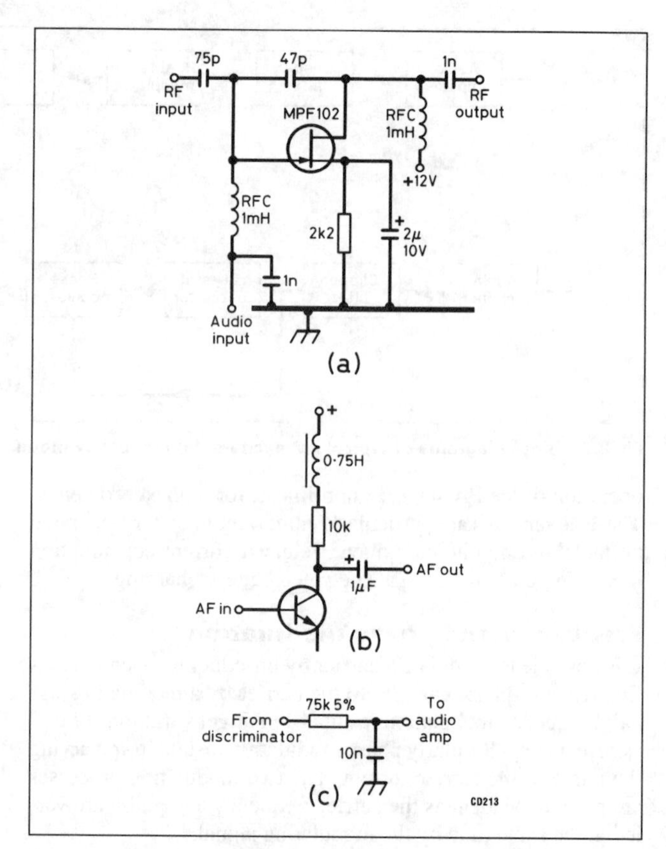

Fig 8.79. (a) The phase-shifter type of phase modulator. (b) Pre-emphasis and (c) de-emphasis circuits

Speech amplifiers for NBFM transmitters

In any speech amplifier for a frequency- or phase-modulated system it is necessary to provide audio gain, clipping, a frequency-correction network and band-pass or low-pass filtering.

Employing these requirements, the audio signal can be set to a predetermined maximum speech band limit (300Hz–3.5kHz) and the audio level can be controlled by peak clipping and filtering to limit the peak deviation of the carrier again to a predetermined level to prevent over-deviation of the transmitter. Audio frequency harmonics will be attenuated by the low-pass filter. Pre-emphasis should be used in an FM system, whereas in PM systems pre-emphasis occurs automatically, both systems introducing frequency compensation at approximately 6dB/octave.

Amplitude limiting (*clipping*) can be introduced by two methods, either at RF or at AF, and either system is capable of producing a good performance. As the RF method is considerably more complex and expensive, AF limiting is the preferred method. This will provide a good performance when a moderate degree of limiting is used. Some harmonic distortion is normally introduced, but with careful design this can be kept to an acceptable low level. The speech bandwidth should be limited, with a considerable attenuation of the response below 300Hz to limit the harmonics which will fall in the audio pass-band. Frequencies above 3.5kHz should be attenuated to prevent unwanted adjacent channel power radiation desensitising a receiver tuned to the next channel below or above the transmitter frequency. Ideally the frequencies above 3.5kHz should be attenuated at 12dB/octave and preferably 24dB/octave. Dynamic compression or automatic level control of the AF signal amplitude can be employed, and harmonic distortion, with careful design, is not introduced into the speech amplifier. However, rise-and-fall time-constants (fast attack, slow decay) are difficult to control on the normally complex audio waveform.

Fig 8.80 illustrates a comprehensive speech processor, which employs many of the principles just described. The processor includes a speech amplifier, an active filter, a balanced limiter and a further active filter. Low-noise transistors TR1 and TR2 function as the speech amplifier. TR3 attenuates frequencies below 300Hz in a high-pass filter configuration. TR4 and TR5 function as a differential amplifier operating as a balanced limiter. TR6 and TR7 in a low-pass filter configuration attenuate frequencies above 3kHz. The ultimate roll-off is 24dB per octave beyond the 3dB cut-off frequency. The filters combine to form an effective band-pass filter covering 300Hz to 3kHz. The overall frequency characteristic is shown in Fig 8.81. An output level of 3V peak-to-peak is available which will ensure adequate deviation in an FM exciter employing a varicap diode as the modulator. The maximum gain from the speech processor is 68dB (×2500), making it suitable for use with low-output dynamic microphones. The gain control RV1 should be adjusted for a 'just clipping' output when speaking normally into the microphone. The processor has been designed for a very good overall performance, with low harmonic distortion and controlled

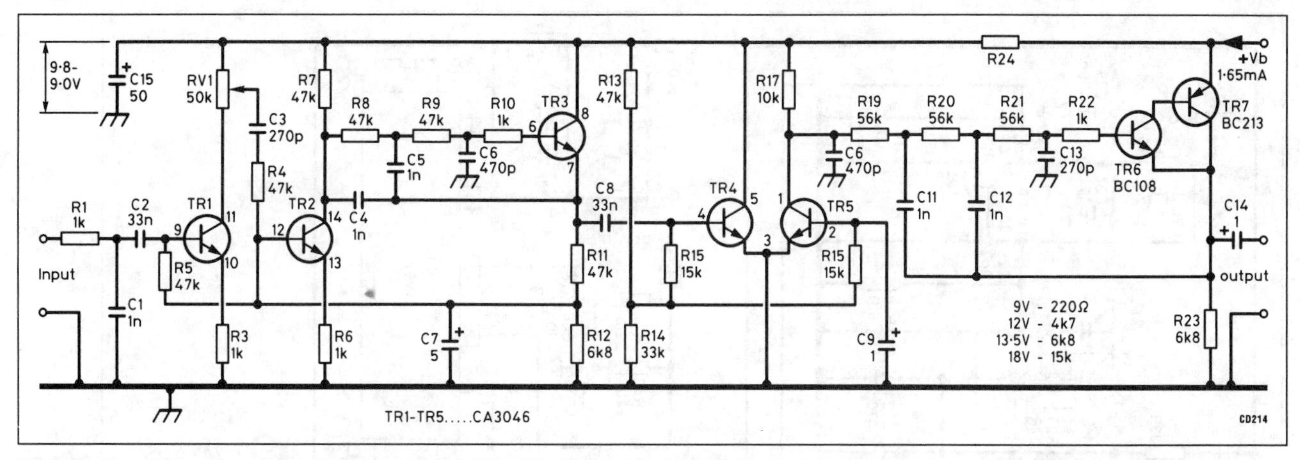

Fig 8.80. Circuit diagram of the speech processor. TR6: BC108, BC148, BC168, BC183, BC238, 2N3904 or similar NPN transistor. TR7: BC213, BC158, BC178, 2N3906 or similar PNP transistor

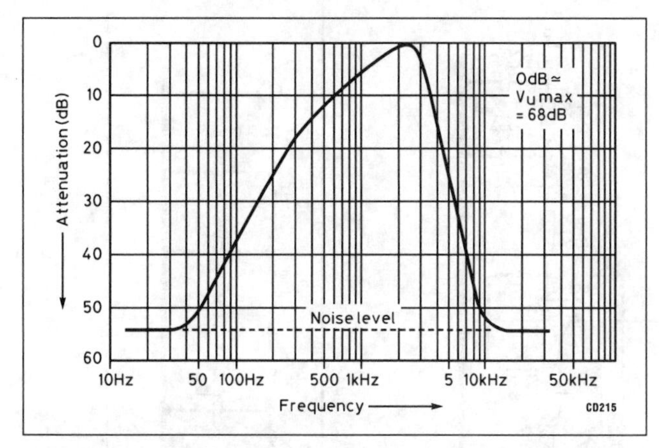

Fig 8.81. Overall frequency response of the circuit at low levels

deviation. These characteristics will ensure correct modulation of the exciter with no over-deviation and good adjacent-channel power attenuation. A speech amplifier employing op-amps and cascaded active low-pass filters is incorporated in the 432MHz NBFM transmitter described below.

432MHz NBFM transmitter

This complete 5W transmitter embodies all of the principles of transmitter design described in this chapter. The design includes a crystal-controlled oscillator, two multiplier stages, a driver amplifier and a power amplifier module. It also includes a modulator for NBFM transmissions, a microphone amplifier and active audio low-pass filters to drive the modulator. The circuit incorporates helical filters (at 432MHz), for attenuating in-band oscillator and unwanted multiplier harmonic levels, and a low-pass filter after the PA module for attenuating harmonics of 432MHz which fall in television Band V and in bands allocated to other services.

The transmitter block and circuit diagram are shown in Figs 8.82 and 8.83 respectively.

TR1 is configured as a modified Colpitts oscillator, using third-overtone crystals for frequency control and stability, at one-ninth of the final frequency. Fine frequency adjustment is controlled by trimmers TC1 to TC4. Only four channels are shown for simplicity. Further channels can be added by increasing the number of 'ways' on the selector switch. The crystals are chosen to oscillate at the initial frequency of 48MHz with 25kHz spacing between each channel (at 432MHz). The crystals are specified to operate with 32pF load capacitance, with a frequency tolerance of ±10ppm and a very low notional capacitance of less than 0.02pF. TR2 operates as a frequency multiplier tripling to 144MHz at its collector. T1 and T2 form a band-pass filter at 144MHz to attenuate unwanted oscillator harmonics (eg second and fourth). TR3 is a second frequency multiplier tripling to the final frequency of 432MHz at its collector. L3 is tuned by CT5 for maximum output at 432MHz (with C20 and C21). TR4 operates as a pre-driver, increasing the power level from the final tripler to approximately 1mW (0dBm) to drive fully amplifier TR5. Helical filter HF1 and HF2 provide a high degree of attenuation to unwanted multiplier harmonics, eg eighth and tenth. TR5 is a limiting amplifier/driver designed to provide a constant power output level to the PA irrespective of variation in drive level from the oscillator multiplier stages, compensating for crystals which may have differing activity levels. RV1 controls the limiting level, allowing TR4 collector to 'saturate', therefore no further power gain in TR4 is available once saturation level is reached. TR5 will give a constant output level of +10dBm (10mW) when its power input level is varying by up to 6dB (×4).

The PA module will provide a minimum of 5W output (+37dBm or 7dbW) into a 50Ω load at 432MHz with a 10.8V power supply from 10mW input. The internal circuit and

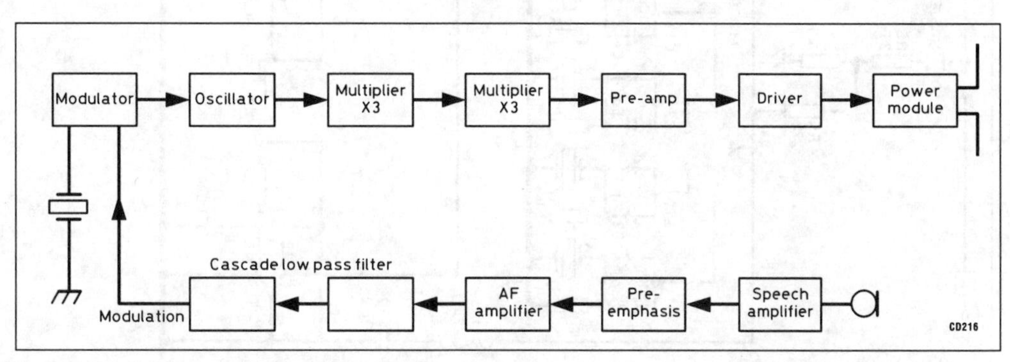

Fig 8.82. Block diagram of 5W 432MHz NBFM transmitter

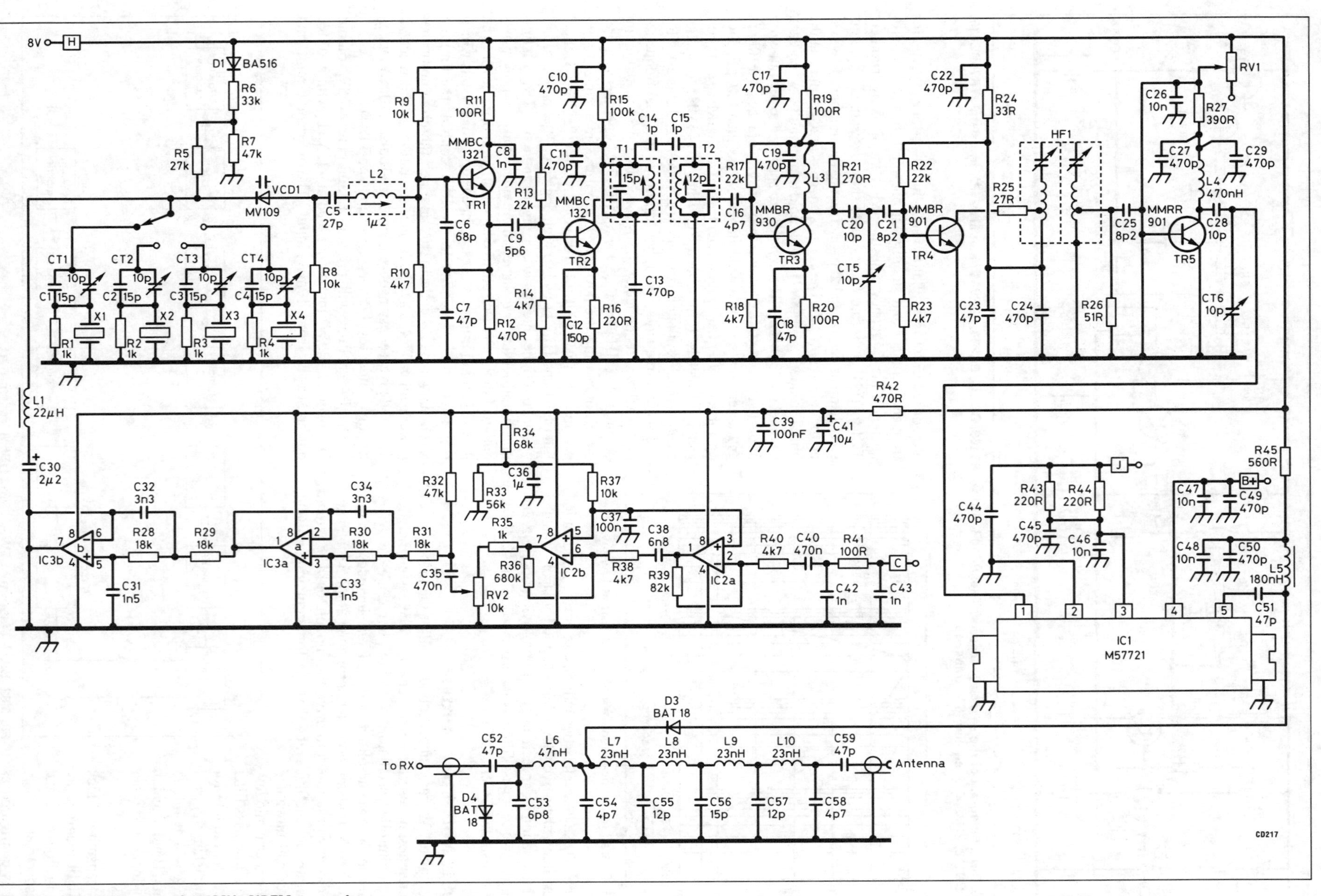

Fig 8.83. Circuit diagram of 432MHz NBFM transmitter

Table 8.8. Components list for the 432MHz NBFM transmitter

R1–4	1k
R5	27k
R6	33k
R7, 32	47k
R8, 9, 37	10k
R10, 14,18, 23, 38, 40	4k7
R11, 19, 20, 41	100R
R12, 42	470R
R13, 17, 22	22k
R15	100k
R16, 43, 44	220R
R21	270R
R24	33R
R25	27R
R26	51R
R27	390R
R28–31	18k
R33	56k
R36	680k
R39	82k
R45	560R
All resistors 5% metal film, 0.25W.	
RV1	1k 10% preset
C1–4, 56	15p ceramic
C5	27p ceramic
C6	68p ceramic
C7, 18, 23, 51, 52, 59	47p ceramic
C8, 42, 43	1n ceramic
C9	5p6 ceramic
C10, 11, 13, 17, 19, 22, 24, 27, 29, 44, 45, 49, 50	470p ceramic
C12	150p ceramic
C14, 15	1p ceramic
C16, 54, 58	4p7 ceramic
C20,28	10p ceramic
C21, 25	8p2 ceramic
C26, 46–48	10n ceramic
C30	2µ2 16V elec
C31, 33	1n5 mylar film
C32, 34	3n3 mylar film
C35, 40	470n mylar film
C36	1µ 16V elec
C37, 39	100n mylar film
C38	6n8 mylar film
C41	10µ 16V elec
C53	6p8 ceramic
C55, 57	12p ceramic
CT1–4	2–10p ceramic trimmer
X1–X4	48MHz 3rd overtone (exact frequencies dependent on preferred FM channels)
T1, 2	144MHz transformer
HF1	432MHz 2-pole helical filter, type HRW (Toko)
L1	22µH RFC axial
L2	1.2µH coil with iron-dust core
L3, 4	470nH RFC axial
L5	180nH RFC axial
L6	47nH air-wound, 4mm OD
L7–10	27nH air-wound coil, 4mm OD
TR1, 2	MMBC1321 (Motorola)
TR3	MMBR930 (Motorola)
TR4, 5	MMBR901 (Motorola)
IC1	M57721M (Mitsubishi)
IC2, 3	MC33172 (Motorola)
VCD1	MV109 (Motorola)
D1	BA516 (Motorola)
D3, 4	BAT18 (Philips)

specification of the module is given in the section on modular power amplifiers. DC power is permanently applied to pin 4 of the M57721M. Power is applied to pin 2 and bias to pin 3 when the transmitter is switched on. This technique reduces the loading on the DC power switch. Total transmitter current is 1.5A. The nine-element Chebyshev low-pass filter between the PA and the antenna socket provides high attenuation of 432MHz harmonics – better than 60dB (less than 10µW) at 864MHz and better than 70dB at the third and higher-order harmonics.

VCD1 is a varicap diode modulator incorporated in the oscillator circuit for direct frequency modulation of the transmitter. L2 is adjusted (with one of the crystals short-circuited across the 1kΩ resistor) for an approximate frequency of 432MHz. The short is then removed and each crystal can then be fine-tuned onto each channel. L1 prevents oscillator voltage being attenuated by the audio low-pass filter load resistor. The microphone amplifier (IC2a and IC2b) provides sufficient AF gain for an electret microphone to modulate the transmitter to a peak deviation of ±5kHz (F_{mod} = 1kHz). R40 and C40 introduce pre-emphasis of 6dB/octave to increase deviation at audio frequency from 2 to 3kHz for improved communication with a receiver with 6dB/octave de-emphasis. IC2b limits the AF output swing, preventing over-deviation of the transmitter. RV2 sets the nominal deviation to ±3kHz at 1kHz modulation frequency. IC1a and b form a cascaded, unity-gain, active, low-pass filter with an ultimate 24dB/octave cut-off above 3.4kHz. This filter minimises the possible desensitisation of a nearby 432MHz NBFM receiver tuned to one channel (±25kHz) away from the transmitter output frequency by high attenuation of the adjacent channel power (80dBc). IC1a output swing deviates the modulator to f_0 ± 555Hz. This deviation is multiplied nine times by the two triplers to ±5kHz at 432MHz.

The transmitter can be used with the 432MHz NBFM receiver to form a complete 70cm transceiver. The RX/TX antenna switch is shown in the transmitter circuit. VHF switching diodes D3 and D4 form this switch. When the transmitter is powered up DC is also applied to D3 and D4 via R45 and L5 which limits the diode current to a safe level. The diodes are therefore biased into full conduction. D4 completes the RF path from the PA output to the low-pass filter input. D3 effectively shorts the receiver input, giving additional protection against breakdown of the RF amplifier. D3 and D4 are non-conducting, effectively being open-circuit when the transmitter is powered down. This removes the receiver input short-circuit and open-circuits the PA from the low-pass filter. The filter then gives high attenuation of strong-signal TV transmissions which could cause receiver desensitisation (*blocking*).

TRANSCEIVERS AND TRANSVERTERS

VHF/UHF transceiver design and construction follows the same rules and concepts employed in their HF counterparts, described in Chapter 7. However, if a HF transceiver is already in use in the shack it is usually much more economical to use a VHF or UHF transverter in conjunction with the transceiver, rather than purchasing a dedicated VHF or UHF rig. The multimode types, particularly those with multimode facilities, are at least as expensive as an HF transceiver.

In most transverter designs, the receiver and transmitter conversions are achieved with the use of a common local oscillator, usually crystal controlled, for converting down the receiver frquencies and converting up the transmitter frequencies. The transmitter section is normally followed by a linear amplifier to raise the output power to the required level and to enable transmission of SSB as well as CW and FM signals.

The block diagram of Fig 8.84 illustrates the essential stages used in the design of a VHF/UHF transverter.

The preferred band on the HF transceiver when using a transverter is usually 10m as it will cover a 2MHz tuning range, this being ideal for the 6m, 2m and 70cm bands. The possibility

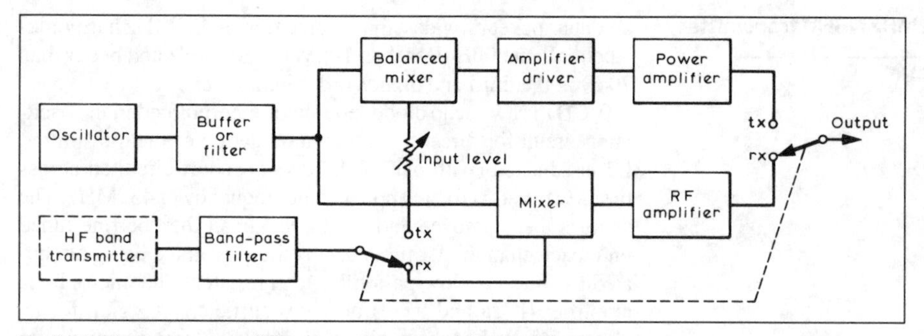

Fig 8.84. Block diagram of the essential stages of a complete transverter

of unwanted HF signals, including image signals, appearing at the transverter input in the 28 to 30MHz band is also much reduced.

The overall transversion gain of the receiver section must be minimised to usually no more than ×10 (20dB) to avoid strong-signal overload of the HF transceiver which would cause unwanted generation of spurious in-band signals.

For good performance with simplicity, a MOSFET or GaAsFET is usually the best choice for the RF amplifier, as described earlier in this chapter. The gain should not exceed ×2 to ×3 (6 to 10dB), thus mimising strong-signal overload of the mixer. The mixer should also employ a MOSFET, usually a dual-gate type for maximising strong-signal performance and ease of oscillator injection. Mixer conversion gain should be limited to ×3 (10dB), thus minimising odd-order IMD products and reciprocal mixing effects.

Another, more complex, solution is to use a higher-gain (say 20dB) amplifier followed by a hot-carrier diode ring mixer.

The choice of local oscillator frequency needs careful consideration to avoid both in-band and out-of-band signals, the former causing spurious responses to be detected in the receiver, the latter causing unwanted spurious signals to be transmitted which may cause interference with other services. Out-of-band signals are the most serious as far as the amateur is concerned.

The oscillator must be stabilised to mimimise frequency drift over short and long periods, particularly when using SSB reception and transmission. The frequency must be maintained within 100Hz from switching in the transverter and must not be affected by temperature changes, particularly in portable operation. Any of the oscillator circuits described in this chapter are suitable.

It is usual practice to include a buffer amplifier between the oscillator and mixer to increase the drive level to give adequate injection voltage. This technique enables the oscillator to be operated at a low level, reducing the possibility of crystal heating and consequent frequency drift to a minimum. The buffer amplifier output should be filtered by a high-*Q* break, in the form of a band-pass coupled pair (for VHF) or helical filter (for UHF) to attenuate oscillator harmonic levels at the mixer input.

The transmit mixer performance can be enhanced by using a balanced mixer as illustrated in Fig 8.85. The JFETs can be replaced by MOSFETs, giving a further reduction in the level of unwanted mixer spurious generation. Oscillator injection would normally be at G2 of the MOSFETs, similar to the receiver mixer configuration, with a ×3 (10dB) oscillator-to-signal input ratio.

Signal input from the HF transceiver would normally be at G1 to take advantage of the balanced symmetrical arrangement of the drain circuit being in push-pull compared with the push-push circuit at the gate inputs.

The mixer output level is raised to the required level by the following linear amplifiers; usually two or three stages are necessary for the correct overall power gain to produce required output power.

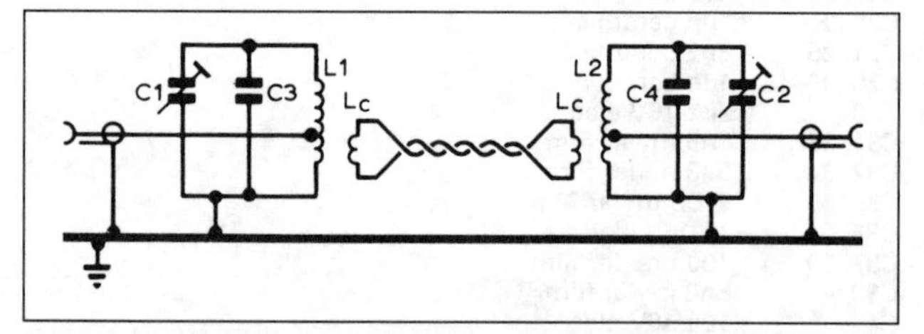

Fig 8.86. High-*Q* band-pass filter for 28MHz. L1, L2: 8t 1.0mm tinned copper wire 12mm OD, 12 to 14mmm long, tapped at 1t from ground end. L_c: 1t link coupling coil, 1.0mm enamelled wire, wound on ground of L1, L2. C1, C2: 50pF max air trimmer. C3, C4: 47pF silver mica or other suitable type

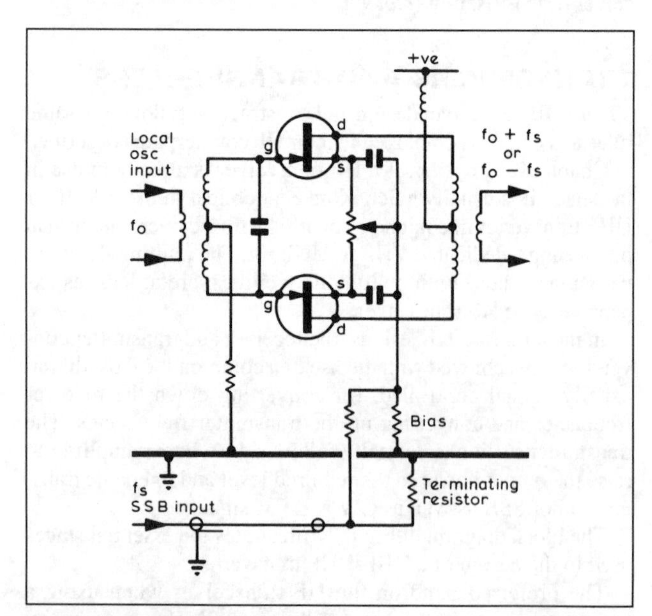

Fig 8.85. Typical balanced mixer using pair of JFETs. MOSFETs can be used as described in the text

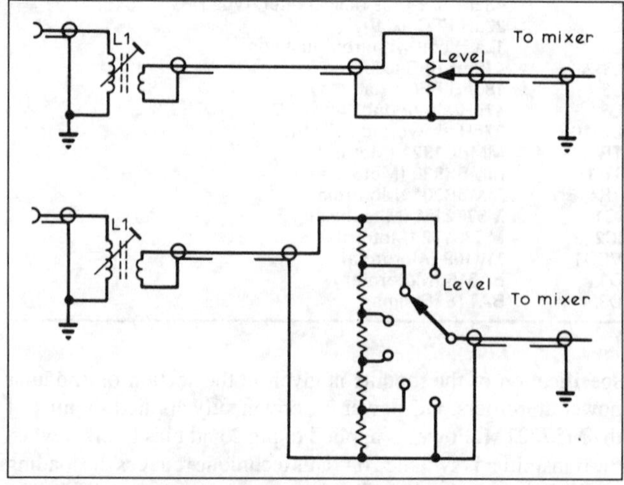

Fig 8.87. Methods of obtaining drive control between the SSB exciter and transverter mixer. L is part of the high-*Q* band-pass filter

Inductive coupling between the mixer output and each succeeding stage can be in the form of simple coupled circuits, with input and output tapping to match the low impedances of the power transistors in the linear amplifier.

A low-pass filter must be included after the final linear amplifier to attenuate out-of-band transmitter spurious outputs, both harmonic and non-harmonic, to minimise interference to VHF radio and UHF television as well as other services.

A correct interface between the HF transceiver and the transverter is necessary to attenuate spurious signals and control the input drive level to the transverter. A high-Q band-pass filter, as illustrated in Fig 8.86, will attenuate unwanted signals by up to 30dB depending on frequency. The filter is designed for 28MHz but component values can be scaled for other HF bands if required.

The transceiver output level has to be set for the correct low level for the transverter input. Modern HF transceivers have an output facility from the driver stage, but it is still necessary to have drive level control. Methods of obtaining this control are shown in Fig 8.87. The control can be conveniently fitted at the transverter input following the band-pass filter.

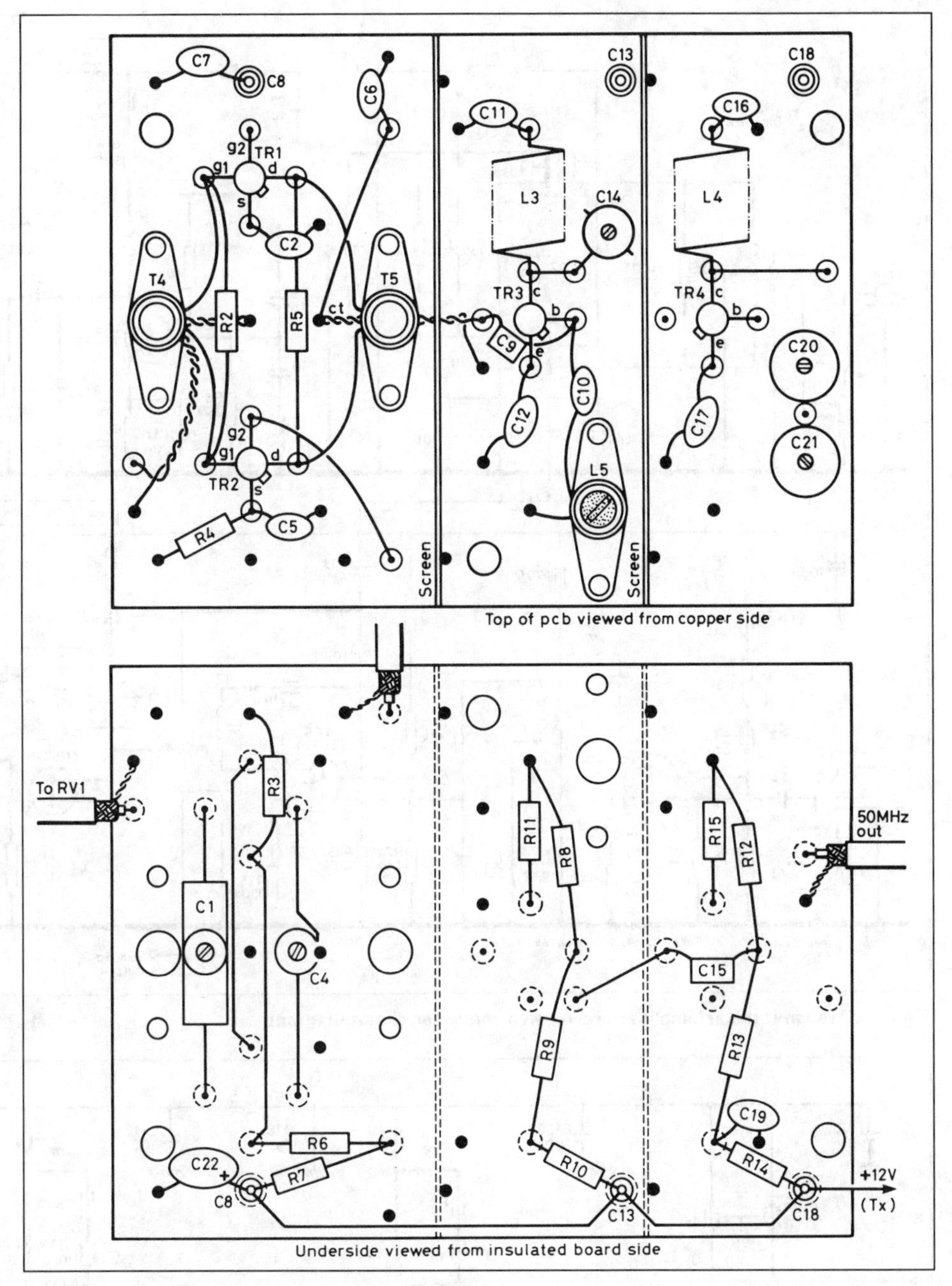

Fig 8.88. Transmit mixer/amplifier component layout

A 50MHz receive/transmit converter

This design was originally described by D S Jones, GW3XYW, in the April 1986 issue of *Radio Communication*. It utilises tried and tested circuits previously published [6, 7].

The receiver section in Fig 8.89 converts 50MHz signals down to 28MHz. It employs dual-gate MOSFETs in the RF amplifier and mixer stages TR5 and TR6, and a crystal-controlled oscillator TR8, followed by a JFET buffer amplifier TR7. It is constructed on its own PCB which allows the receive section to be constructed first, followed by the transmit section at a later date.

The transmit section in Fig 8.89 employs a balanced mixer TR1 and TR2, again using dual-gate MOSFETs. Input at 28MHz is converted to 50MHz in this stage; oscillator input to the mixer is taken from the receiver oscillator. A cascade buffer amplifier (TR3 and TR4) is tuned to 50MHz, providing adequate drive to the linear amplifier shown in Fig 8.90. The final stage TR103 incorporates a 2N6082 power amplifier.

The various units are constructed on a single-sided PCB which should not be etched. Make the anchor points by drilling 0.052in holes and inserting metal pins. Where the pins need to be insulated, remove the copper by lightly countersinking with a larger drill (0.125in). If the pins are earth points, then they should be soldered in place. Combinations of single-sided, earthed and insulated anchor points on which to mount the components are thus produced. Solder the feedthrough capacitors direct to the earthplane. This method of construction is clean, easy and suitable for RF up to VHF, and, if a mistake is made, the hole can be soldered over and a new one drilled. Wind the self-supporting coils on twist drills to obtain the correct internal diameter. Screens (of tinplate or PCB) should be mounted as shown in Figs 8.88 and 8.91. A substantial heatsink should be mounted on a spacer in good thermal contact with the 2N6082 output transistor.

Alignment is simple and follows normal practice. A GDO is very useful for setting up an approximate resonance although, if this is attempted with the FET stages, the drains must be

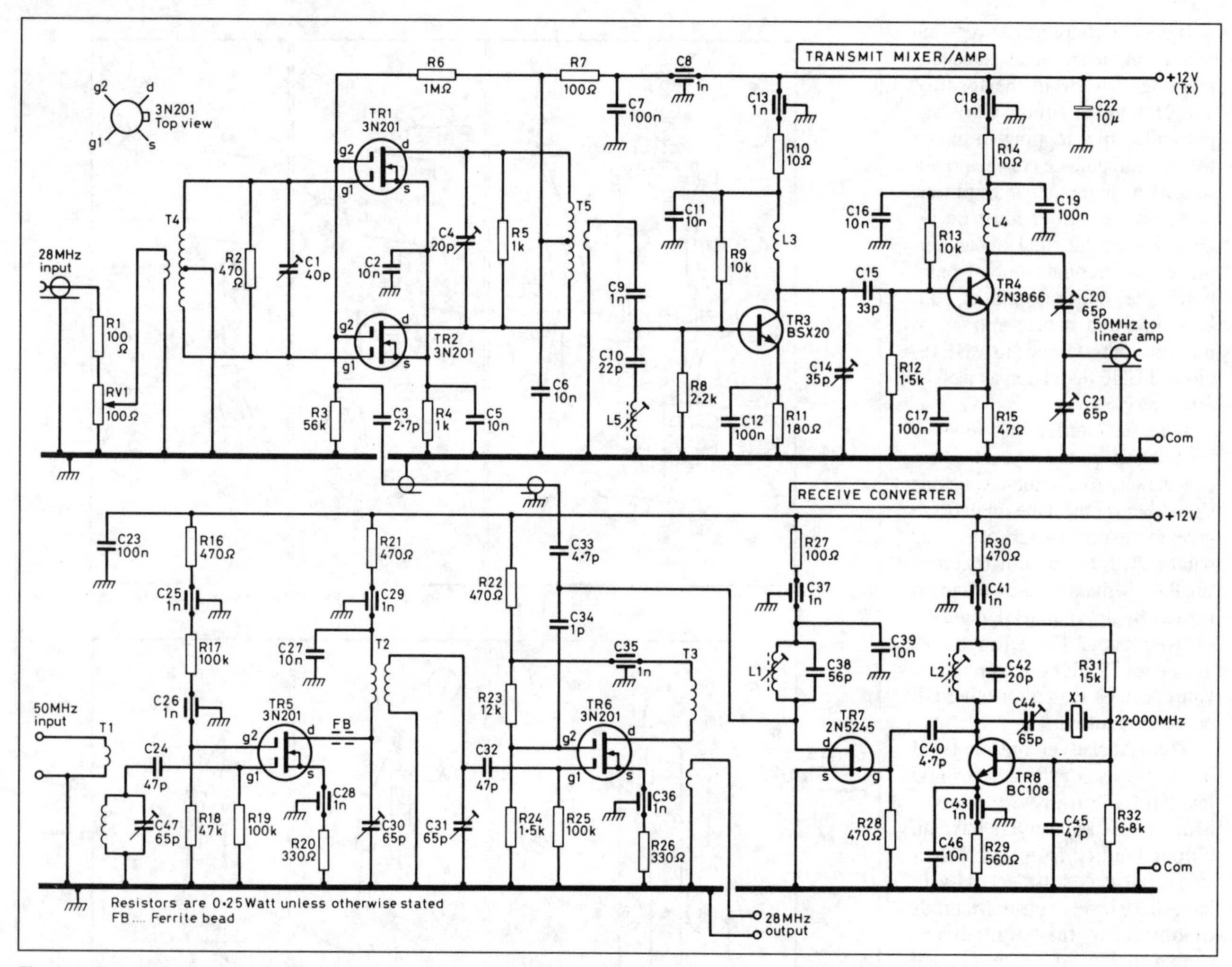

Fig 8.89. Transmit mixer/amplifier and receive converter circuit diagram

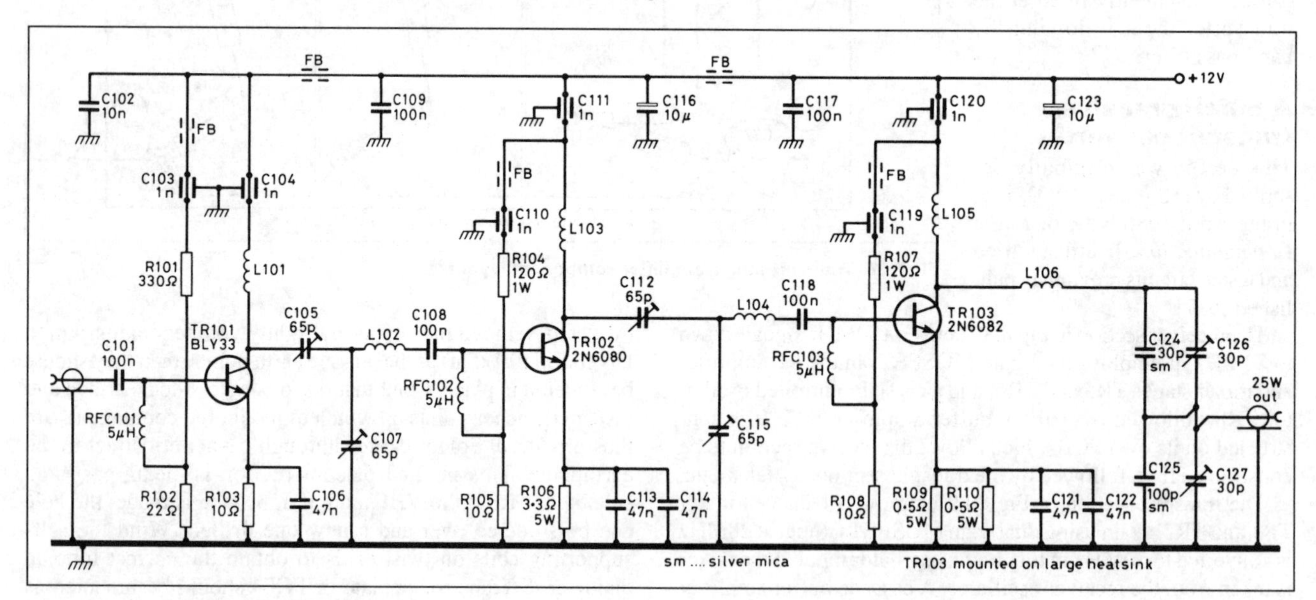

Fig 8.90. Linear amplifier circuit diagram

disconnected or the power applied, otherwise satisfactory 'dips' will not be observed.

The 22MHz crystal mixes with the 50MHz input to produce sum and difference frequencies at 72MHz and 28MHz. It is the latter frequency which is tuned and selected by LC circuits. The reverse applies on transmit mixing. Any instability encountered around the RF amplifier stage can be eliminated by connecting a 6.8kΩ 0.25W resistor across the T2 primary.

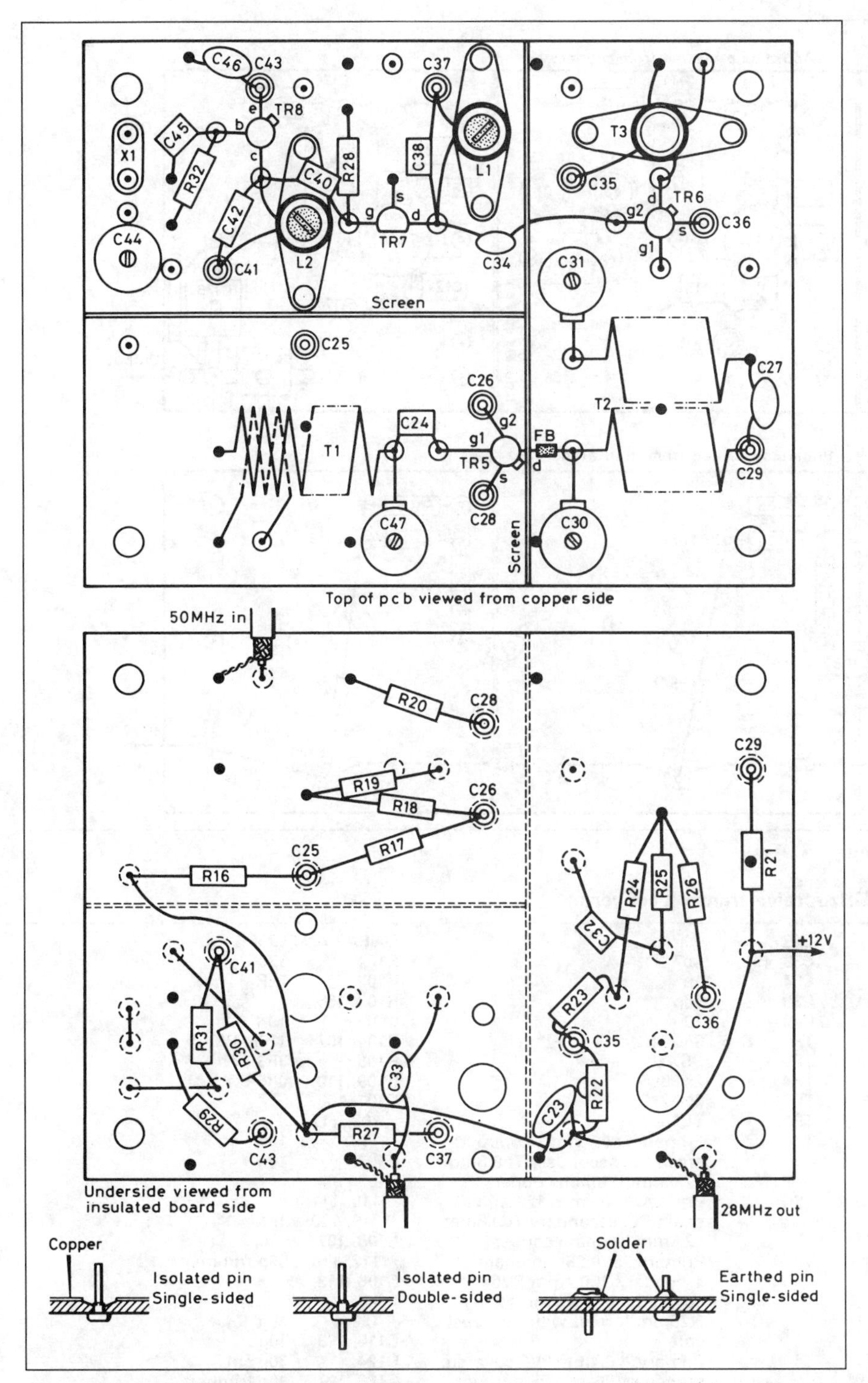

Fig 8.91. Receive board component layout

A series resonant trap (Ct and Lt) for 44MHz is included in the transmit amplifier chain – this helps to suppress any spurious output from the crystal oscillator second harmonic. It is recommended that the three finished panels are mounted either in separate boxes or in a single box with screens.

For the initial tune-up, a supply voltage of 12V is adequate, but in order to realise full power output the supply may be increased to 13.5V. An input of approximately 1mW is required for full output. Therefore, if the converter is to be used with a transceiver, a suitable attenuator should be included in the circuit.

HIGH-POWER VHF/UHF AMPLIFIERS

The key mechanism in high-power amplifiers of any type, especially at VHF and higher frequencies, is the low-loss transformation of impedance between the active device and the load, which is usually 50Ω.

The examples chosen not only represent working circuits, but also illustrate techniques and materials which result in reliable, high-performance amplifiers. This is not to imply that the use of alternative methods and components will result in any lack of success, but there situations in most high-power amplifiers where component ratings and characteristics are significant in achieving satisfactory performance. The random use of surplus or 'junkbox' components can lead to problems.

50MHz

A design by G3WZT [8] is shown in Figs 8.93 and 8.94. BLW96 is specified for HF SSB and Band II FM use, and so is well suited to 50 or 70MHz operation. Several other manufacturers offer transistors with similar specifications, although changes to the matching circuits might be necessary. The original article covers the design of matching networks in considerable detail and is recommended reading for anyone contemplating a high-power transistor amplifier. The amplifier produces a very linear 100W PEP from 6–7W PEP input.

Component details are given in Table 8.10.

The transistor input impedance is very low (0.4Ω in series with 0.15Ω inductive) which requires two stages of transformation to match to 50Ω with reasonable loss and bandwidth. The output impedance is much higher at 4.5 +j5Ω and a single matching section is needed. L105/6 and C118–20 are a low-pass filter to remove harmonics. Transistor matching circuits usually operate at low Q so harmonic output levels are often higher than with valves; a harmonic filter is should always be used between the amplifier and antenna.

Tuning uses mica compression trimmers. These are generally better than rotary film types as they avoid a moving contact in the high current path.

Bias is generated by TR103 and TR104. The base voltage of TR103 provides temperature compensation for the RF transistor and TR104 delivers the peak current demand. In this case, power to the whole amplifier is switched for transmit/receive. Switching the bias supply off is sufficient in most applications, but the RF transistor base must always have a DC path to ground.

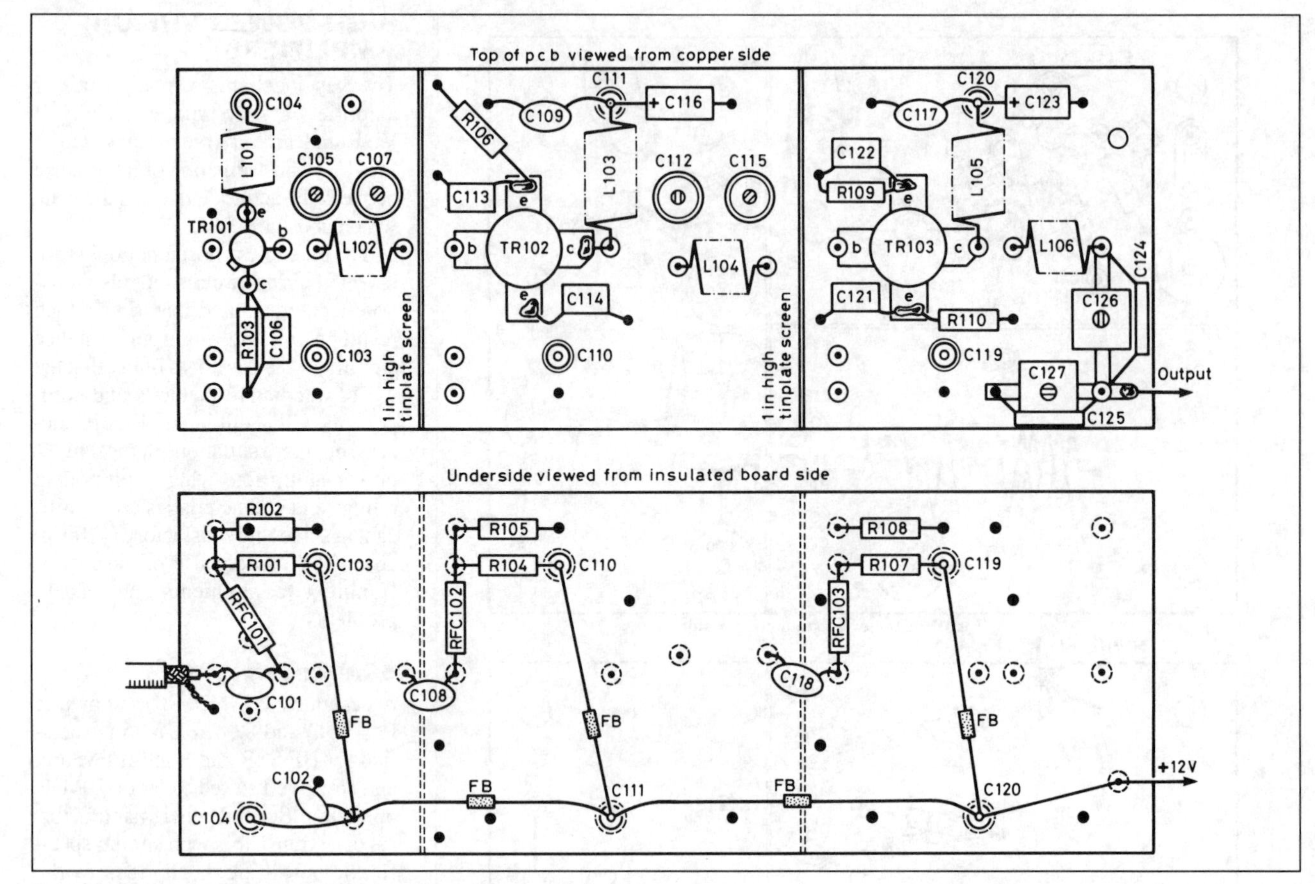

Fig 8.92. Linear amplifier component layout

Table 8.9. Components list for the 50MHz receive/transmit converter

TRANSMIT MIXER/AMP AND RECEIVE CONVERTER

R1, 7, 25	100R
R2, 16, 21, 22, 28, 30	470R
R3	56k
R4, 5	1k
R6	1M
R8	2k2
R9, 13	10k
R10, 14	10R
R11	180R
R12, 24	1k5
R15	47R
R17, 19, 25	100k
R18	47k
R20, 26	330R
R23	12k
R29	560R
R31	15k
R32	6k8
RV1	100R
C1	40p
C2, 5, 6, 11, 16, 18, 27, 39, 46	10n
C3	2p7
C4	10p
C7, 12, 17, 19, 23, 24	100n
C8, 9, 13, 25, 26, 28, 29, 35, 36, 37, 41, 43	1n
C10	22p
C14	35p trimmer
C15	33p
C20, 21, 30, 31, 44	65p trimmer
C22	10µ
C33, 40	4p7
C34	1p
C38	56p
C42	20p
TR1, 2, 5, 6	3N201
TR3	BSX20
TR4	2N3866
TR7	2N5245
TR8	BC108
T1	Primary 2t 0.8mm 12.5mm ID enam Cu, secondary 7t 0.8mm 12.5mm ID enam copper
T2	Primary 7t 0.8mm 12.5mm ID enam Cu, secondary 7t 0.8mm 12.5mm ID enam copper
T3	Primary 25t 0.25mm enam, secondary 2t 0.4mm PVC covered, both wound on 6.25mm former with iron-dust core
T4	Primary 2t 0.4mm PVC covered, secondary 8t + 8t 0.25mm enam, both wound on 6.25mm former – no core
T5	Primary 4t + 4t 0.4mm enam, secondary 2t 0.4mm PVC covered, both wound on 6.25mm former – no core
L1	12t 0.56mm enam wound on 6.25mm former with iron-dust core
L2	20t 0.4mm enam wound on 6.25mm former with iron-dust core
L3, 4	7t 0.8mm 9.5mm ID enam
L5	9t 0.56mm enam wound on 6.25mm former with iron-dust core

LINEAR AMPLIFIER

R101	330R
R102	22R
R103, 105, 108	10R
R104, 107	120R 1W
R106	3R3 5W
R109, 110	0R5 5W
C101, 108, 109, 117, 118	100n
C102	10n
C103, 104, 110, 111, 119, 120	1n
C105, 107, 112, 115	65p trimmer
C106, 113, 114, 121, 122	47n
C116, 123	10µ
C124	30p sm
C126, 127	30p trimmer
L101	7t 18swg 0.375in ID, tinned Cu
L102	5t 18swg 0.375in ID, tinned Cu
L103	7t 18swg 0.375in ID, tinned Cu
L104	3t 18swg 0.375in ID, tinned Cu
L105	8t 18swg 0.375in ID, tinned Cu
L106	7t 16swg 0.375in ID, tinned Cu
RFC101–103	5µ VHF choke, RS238-255
TR101	BLY33
TR102	2N6080
TR103	2N6082

Single-sided PCB pins, RS433-624
Double-sided PCB pins, RS433-630
Twist drill bits, 1.3mm (0.052in), RS549-195

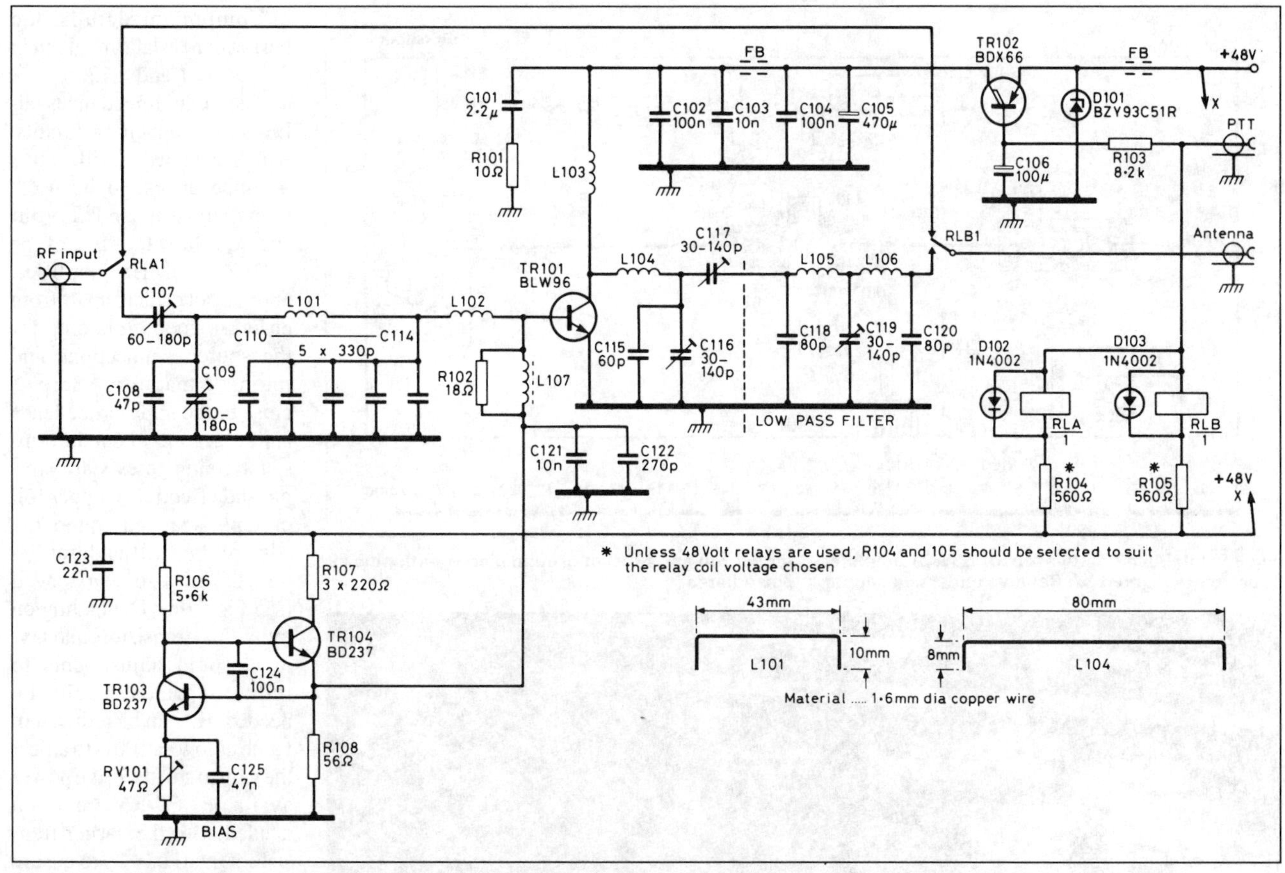

Fig 8.93(a). G3WZT 50MHz amplifier

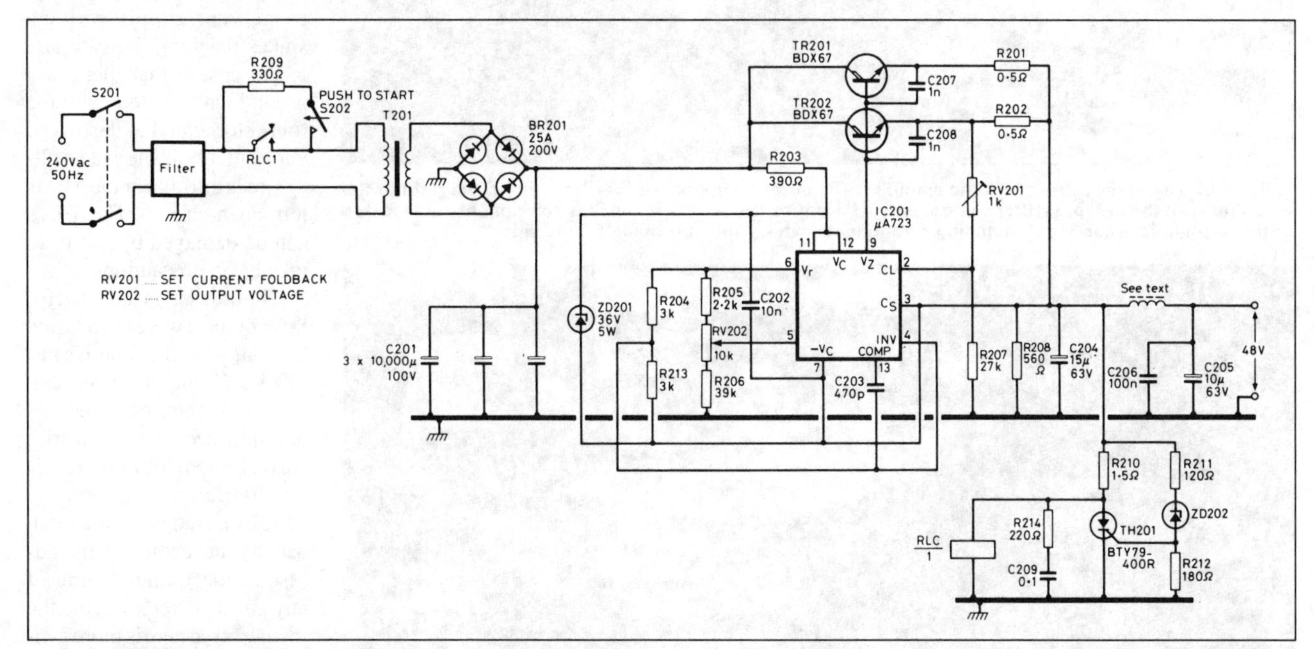

Fig 8.93(b). PSU for the G3WZT amplifier

C101/R101 ensure that the transistor has a resistive load at low frequencies and FB provides a lossy series component over a very wide frequency range.

144MHz

In recent years power MOSFETs have overtaken bipolar transistors in many applications up to 500MHz. For amateur service, MOSFETs offer the advantages of high gain, easy biasing, ruggedness and higher impedances. A simple, high-performance amplifier designed by G8GSQ is shown in Fig 8.95. Contrast the circuit for the 28V FET amplifier with Fig 8.55, a 12V amplifier for a similar output power. Also compare the gains; 10–20dB for the 28V FET amplifier, depending on the transistor, and 7dB for the 12V bipolar.

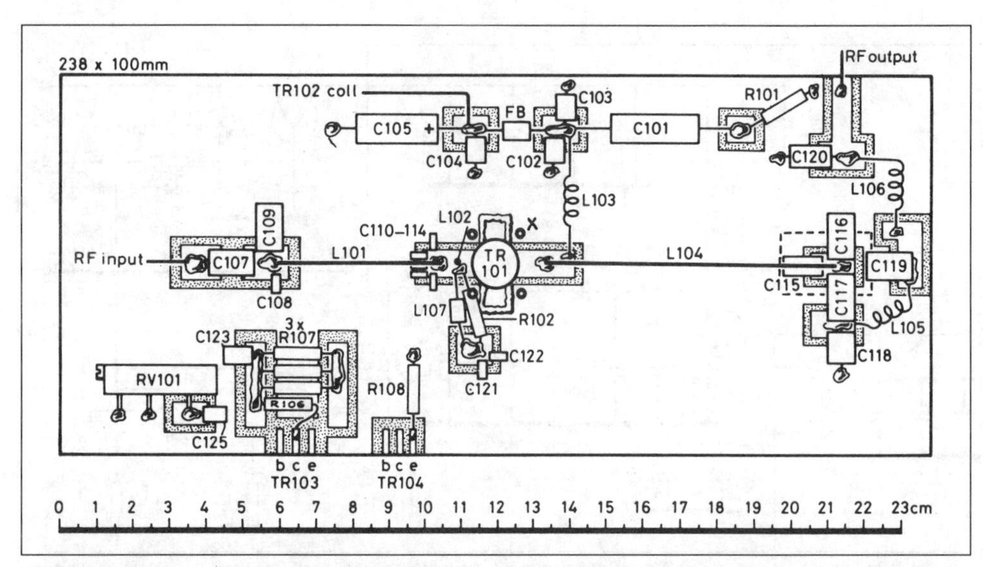

Fig 8.94(a). Main board layout of G3WZT amplifier. Connect top and bottom ground planes with pins at four points marked 'x'. Remove underside copper in dotted area.

Fig 8.94(b). The underside view of the amplifier. The output components can be seen on the right of the amplifier with the low-pass filter components adjacent to the coaxial relay. Bias components are located in the left-hand corner. Input matching components can be seen on the left-hand side

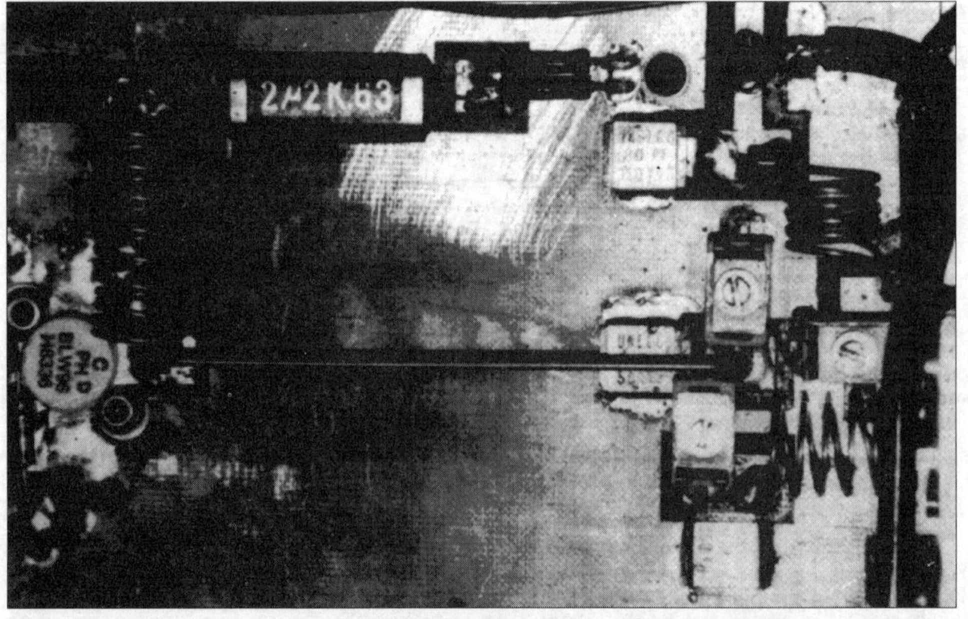

Fig 8.94(c). Close-up view of the output matching components and the low-pass filter

Component details and performance data are given in Tables 8.11 and 8.12.

The high impedances allow simple matching circuits, which can cover a wide range of impedances, to be used. Construction and layout should follow the style of the G3WZT 50MHz amplifier, and are not critical apart from ensuring good grounding for the source connections and tuning capacitors. Screws close to the appropriate leads and solder tags under the transistor fixing screws are simple and effective. Copper foil or tape wrapped round the edge of the PCB and transistor cut-out also works well (see Fig 8.96). Depending on individual transistors and layouts, slight adjustments to inductor sizes might be needed. R2 can be reduced in value as low as 100Ω to adjust the gain to suit the drive power available. C9/R6 are often needed with the earlier transistors (marked '*') to ensure stability. Always include D1 to prevent supply voltage spikes reaching the gate and R7 to ensure that there is a ground path for the gate; the transistor can be destroyed very quickly if the gate voltage goes too high or the gate is left open-circuit. The FETs can be damaged by static, so take basic precautions.

The transistors (all rated for 80W output) were each tested by tuning for maximum saturated CW output power, and then adjusting a two tone signal until acceptable linearity was achieved (fifth-order IMD < –40dBc, seventh-order < –55dBc). Better linearity can usually be achieved by adjusting the output tuning at any given power level, but the method used here is more generally applicable to amateur circumstances. Third-order IMD is often in the region of –25 to –30dBc in power FETs, even at low output power levels, while the high-order products drop very quickly. In practice, the high third-order products do not appear to affect

Table 8.10. Components for G3WZT 50MHz amplifier and PSU

AMPLIFIER	
R101	10R 0.5W carbon film
R102	18R 0.5W carbon film
R103	8k2 0.25W carbon film
R104, 105	Select to suit relays used
R106	5k6 0.5W carbon film
R107	3 × 220R 6W wire wound
R108	56R 0.5W carbon film
RV101	47R cermet trimpot
C101	2µ2 63V polycarbonate
C102, 104, 124	100n 100V monolithic ceramic
C103, 121	10n 100V monolithic ceramic
C105	470µ 63V tubular elec
C106	100µ 63V tubular elec
C107, 109	60–180p mica compression trimmer
C108	47p 50V ceramic chip
C110–114	330p 50V ceramic chip
C115	60p 250V Unelco mica or ATC
C116, 117, 119	30–140p mica compression trimmer
C118, 120	80p 250V Unelco mica or ATC
C122	270p 100V monolithic ceramic
C123	22n 100V monolithic ceramic
C125	47n 100V monolithic ceramic
D101	BZY93 C51R zener
D102, 103	1N4002
TR101	BLW96
TR102	BDX66 PNP Darlington
TR103, 104	BD237
FB	Suppression bead. Material, 3S2 (blue)
L101	See separate drawing
L102	15 × 7mm pad on PCB
L103	12t 1.2mm copper wire, 9mm ID, 28mm long
L104	See separate drawing
L105, 106	4½t 1.2mm copper wire, 10mm ID
L107	2½t 0.5mm enam copper wire wound through 6-hole ferrite bead
RLA, RLB	50R coaxial, type CX120P
POWER SUPPLY UNIT	
R201, 202	0R5 10W wire wound
R203	390R 2.5W wire wound
R204, 213	3k0 0.25W carbon film
R205	2k2 0.25W carbon film
R206	39k 0.25W carbon film
R207	27k 0.25W carbon film
R208	560R 5W wire wound
R209	330R 25W aluminium clad (see text)
R210	1R5 10W wire wound
R211	120R 0.25W carbon film
R212	180R 0.25W carbon film
R214	220R 0.5W carbon film
RV201	1k0 cermet trimpot
RV202	10k cermet trimpot
C201	10,000µ × 3, 100V
C202	10n 100V
C203	470p 100V
C204	15µ aluminium elec 63V
C205	10µ aluminium elec 63V
C206	100n 100V
C207, 208	1n ceramic 50V
C209	0µ1 polycarbonate 250V
TR201, 202	BDX67 NPN Darlington
IC201	µA723
ZD201	36V 5W zener
ZD202	See text
TH201	BTY79-400R
SW201	DPDT 250V AC at 5A
SW202	Momentary push 250V AC at 5A
RLC	48V DC coil, 5A contact rating
Mains transformer, 55V AC 300VA	Filter, 5A mains RFI
Rectifier, 25A 200V bridge module	PCB, glassfibre FR4

signal quality, and it is the high-order products which determine the apparent bandwidth of the signal.

AMPLIFIERS USING VALVES

The choice of valves for VHF and UHF power amplifiers is wide and varied, but supply and cost considerations at the time of writing and during the expected lifetime of this book focus attention on the 4CX250B and its many variants.

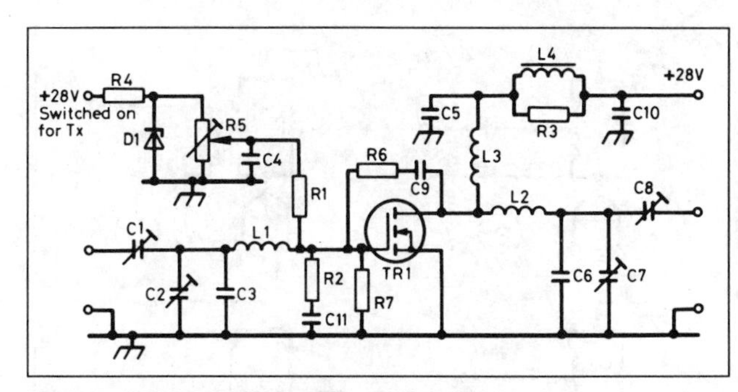

Fig 8.95. G8GSQ 144MHz FET amplifier circuit

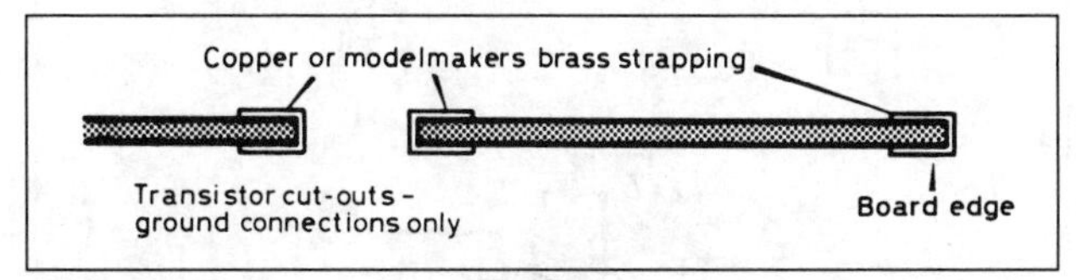

Fig 8.96. Copper foil ground connections

The step down in impedance from the anode (typically 1–4kΩ) to the output at 50Ω is usually carried out in a single tuned circuit where the output is capacitively or inductively coupled or tapped into the circuit. This large transformation in a single step demands a high Q in order to avoid excessive loss in the anode circuit. It is very important to use high-quality components and good construction techniques. At higher frequencies the inductive part can take the form of a transmission-line element, while at microwaves the whole tuned circuit can become a cavity. Most output circuits can be rearranged into one of the forms in Fig 8.97. Part (d) shows a rearranged pi network. The parallel-tuned circuit behaves similarly to a $\lambda/4$ transmission line. The tuned circuit can alternatively be series tuned, where it is configured to behave as a $\lambda/2$ transmission line.

Table 8.11. Components for G8GSQ 144MHz FET amplifier

R1	10k
R2	100R-5k range for gain adjust
R3	10R 1W metal film
R4	2k2
R5	10k preset potentiometer
R6	1k 1W carbon or metal film
R7	100k
C1, 2, 7	5–60p
C3	22p NPO ceramic
C4, 5, 11	1n ceramic
C6	30p metal clad mica or porcelain chip
C8	30p air-spaced trimmer
C9, 10	0.1µ 63V ceramic
L1, 2	1/2t 1.6mm wire, 20mm ID
L3	7t .7mm wire, 7mm ID
L4	1.5t 1mm wire on Siemens B62152A1X1 two-hole core
D1	12V 400mW zener diode
TR1	Siliconix DV2880* or DU2880* Motorola MRF172* or MRF173 Philips BLF246 Acrian VMIL80FT or AFT Polycore or Semelab D1004 or D1005

* See text.

Table 8.12. Performance of 144MHz FET amplifier

Transistor	Sat P_o at	P_i	Linear P_o at	P_i (PEP)
D1005	136	5	90	2.25
BLF246	107	8	80	4.1
VMIL80FT	106	5	80	2.5
MRF172	100	8	55	2.3

V_d = 28V, I_{dq} = 0.5A, R2 = 100kΩ in each case

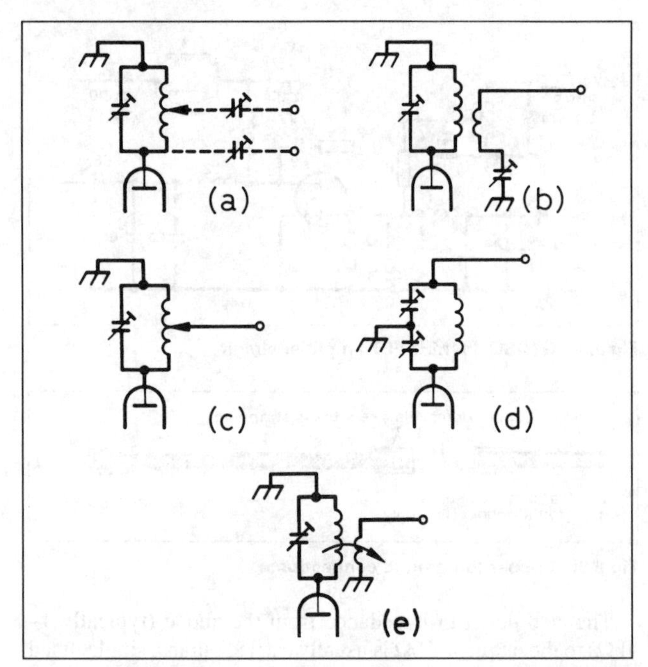

Fig 8.97. Anode circuit configurations excluding DC feeds and blocking

Cooling

Adequate cooling is essential and, within practical limits, it is impossible to overcool a valve. A number of designs have proposed the use of 120mm axial fans for single 4CX250 amplifiers. These fans are not designed to work against the back pressure of the anode cooler and, while they might prove adequate under individual circumstances, radial (squirrel cage) fans are much more suitable. Two methods of cooling are widely used; Fig 8.98 shows the original design principle where the grid compartment is pressurised and air is guided through the anode cooler by a chimney. The back pressure of the anode is increased by the socket and chimney, and the blower must overcome this.

Fig 8.99 shows a technique which has become popular since the publication of the K2RIW 70cm amplifier [9]. Here the anode compartment is pressurised and exhaust air is carried from the anodes to the outside by chimneys, which means that the full blower pressure is available at the anode. The chimney is made from thin PTFE sheet or Mylar film (plastic professional drawing film), rolled into a tube of a couple of turns, using the valve as a former, and held in shape with masking tape and DIY silicone sealing compound. It is very important that the chimney seals well onto the valve and into the lid as any gaps will allow the airflow to bypass the valve anode. With this method,

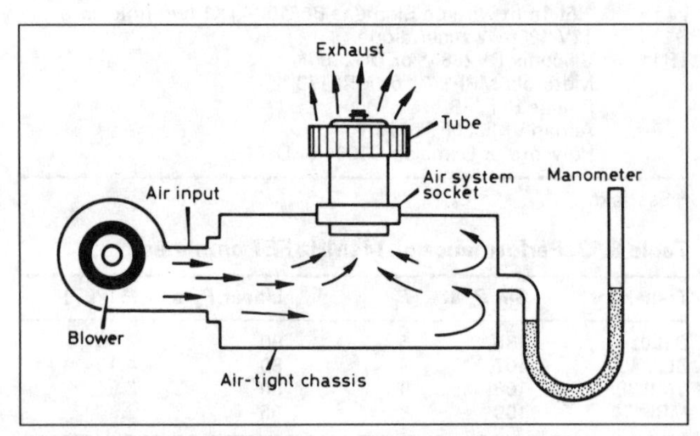

Fig 8.98. Grid blown cooling

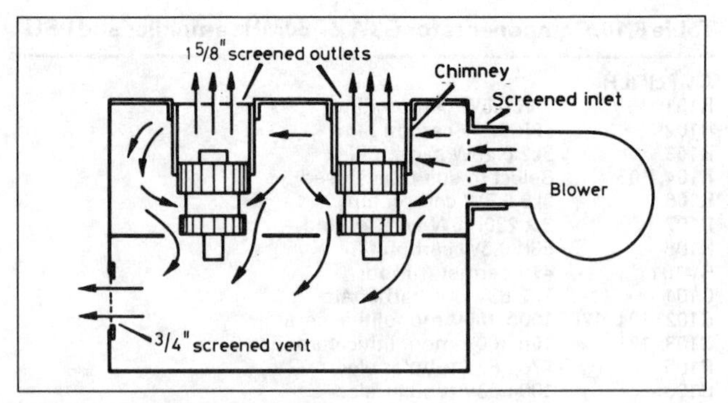

Fig 8.99. Anode blown cooling

it is also vital to provide a good flow of air through the socket to cool the heater and grid pins, otherwise the ceramic can crack. Around 40% of the anode airflow is a reasonable target [10]. As a simple check, use a jam thermometer and measure the anode exhaust air temperature: 90–95°C is an absolute maximum [11].

Very few blowers appearing on the surplus amateur market are truly capable of providing the air pressure needed at full power. The cost of a new blower designed to work at the correct pressure is not extreme (£50 in 1993), and represents a sound investment.

Power supplies

Never forget that the power supplies needed for valve amplifiers WILL KILL YOU if you let them.

An anode supply in the range 1500–2500V is optimum. Higher voltages give higher gain and better linearity, but also demand valves in good condition to avoid flashover. A current-limiting resistor of 10–50Ω and fuse will avoid the most serious damage in the event of a flashover, and interlocks to prevent trying to transmit without anode voltage are worthwhile, although a correctly designed screen supply should automatically avoid damage in this case.

Screen grid supplies have several demands placed on them. Good regulation should be maintained while sourcing and sinking current, and under dynamic conditions as the current varies with modulation. Current limiting is desirable to prevent excessive dissipation in the screen grid, and the supply needs to be protected against anode-to-screen flashover. A great many different ideas have been proposed over the years, but the most widely used solution uses power zener diodes. Electronic regulators must be treated with care, as they use feedback control loops which are potential oscillators. All circuit details, including load capacitance of the sockets, must be duplicated exactly. Gas-filled or metal oxide surge arrestors should be fitted at the socket screen contact to protect the screen supply and socket capacitors in the event of anode-to-screen flashover.

Screen grid voltage is typically in the range 325–375V. Higher values give higher gain, while lower values give better linearity.

Control grid supplies can be very simple, provided that the amplifier is not driven into grid current. This will always be the case where linear operation is needed for SSB, but maximum output and efficiency with FM or CW will be obtained (at the expense of gain) with Class C operation and grid current. In this case, the grid supply must maintain the bias voltage in the presence of grid current.

The bias voltage for any given valve can vary widely depending on the condition of the valve and the other supply voltages. For Class AB operation, –40 to –60V is typical. For Class C use, the valve is biased beyond cut-off: –90V is typical, depending on RF drive power.

Don't forget the heaters: longevity is enhanced by controlling the in-rush current and running at the correct voltage (6.0V for –B, 26.5V for –F, +0/–5%) measured at the valve base [12, 13]. Anode and screen supplies should not be applied until the heaters have warmed up for about a minute [14].

The circuits described here are suited to mains-powered operation. Many amplifiers perform poorly under portable conditions where the regulator circuits cannot cope with the variations in generator voltage. In this case, more complex regulators might be more appropriate and the reader is referred to other articles which detail such designs [15–17].

Bases

The choice of valve base is not generally critical except for use at 432MHz. Correct RF decoupling of the screen grid is, however, vital to correct operation. Common bases are Eimac SK600/610/612 for use up to 144MHz and SK620/630 for 432MHz. The latter two have modified low-inductance contacts and a lower value screen grid capacitor which ensure stable UHF operation. Numerous problems have been reported where VHF bases have been used in UHF amplifiers [18–20]. Surplus UHF bases are abundant and these are well suited to the VHF amplifiers where there is mechanical flexibility. 432MHz designs tend to rely on accurate mechanical reproduction around Eimac bases.

50/70MHz

Figs 8.100 and 8.101 show a 70MHz amplifier, designed by G8GSQ, which can readily be adapted to 50MHz by simple scaling of tuning component values. Mechanical construction, other than input to output screening, is not critical and the original was built in two die-cast boxes bolted back to back. The gain can be adjusted to suit individual circumstances by altering the value of R1; 500Ω suits about 2W drive for very clean legal limit output, while about 200Ω should suit 10W input. A maximum of about 2kΩ should be used, or the amplifier might need neutralising. Component details and performance are given in Tables 8.13–8.15.

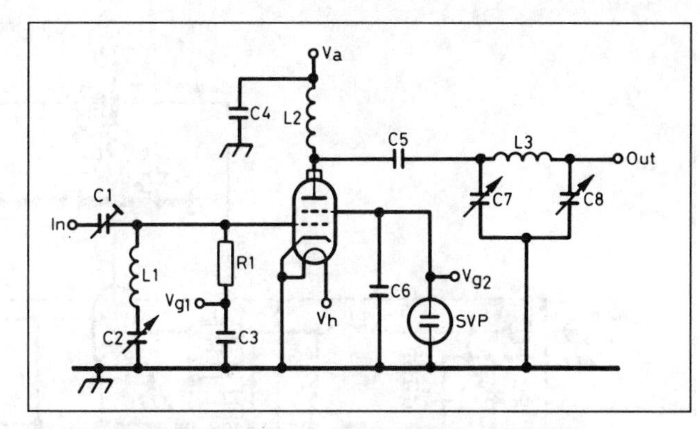

Fig 8.100. G8GSQ 4CX250 amplifier circuit for 50/70MHz

Table 8.15. Test results for 70MHz amplifier

P_i	P_o	I_a	V_a	IMD3	IMD5
0	0	75	2200		
2	150	140	2130	–30	–50
3.5	240	160	2100	–25	–40
5	290	180	2070	–25	–28

All powers are two-tone PEP. V_{g2} = 325V.

Table 8.13. Components for 70MHz amplifier

R1	470R 2W carbon or metal film. See text
C1	40p airspaced variable
C2	30p airspaced variable
C3	1n 500V ceramic disc
C4	300p 8kV ceramic
C5	500p 5kV 'transmitting' type
C6	1000p 1kV ceramic disc
C7	40p widespaced variable
C8	150p variable
L1	8t 1.6mm wire, 18mm ID
L2	20t 1mm wire, 25mm ID, 35mm long
L3	3t 1.6mm wire, 38mm ID, 12mm long
SVP	Siemens B2B470 surge arrestor or 275V(AC)/370V(DC) metal oxide varistor

Table 8.14. Components for 70MHz amplifier PSU

R1–12	330k 1W metal film
R13–19	220k 2W
R20	10k 5W
R21	220k 2W
R22	4k7 5W
R23	10k potentiometer
R24	4k7 5W
R25	Select on test for 6.0V at valve. Typ 0.1R
C1–6	200µ 500V
C7	100µ 500V
C8	68µ 250V
T1	1500V, 300mA
T2	(1) 300V/60mA, (2) 100V/50mA, (3) 6.3V/3A
D1–12	1N4007
D13	600V 1A bridge rectifier
D14–19	51V 5W zener diode
D20	68V 5W zener diode
D21	400V 1A bridge rectifier
D22, 23	1N4007
M1	250mA
M2	10/0/10mA centre zero
M3	10/0/10mA centre zero
RLA	DPCO, contact rated for 500V, coil to suit.

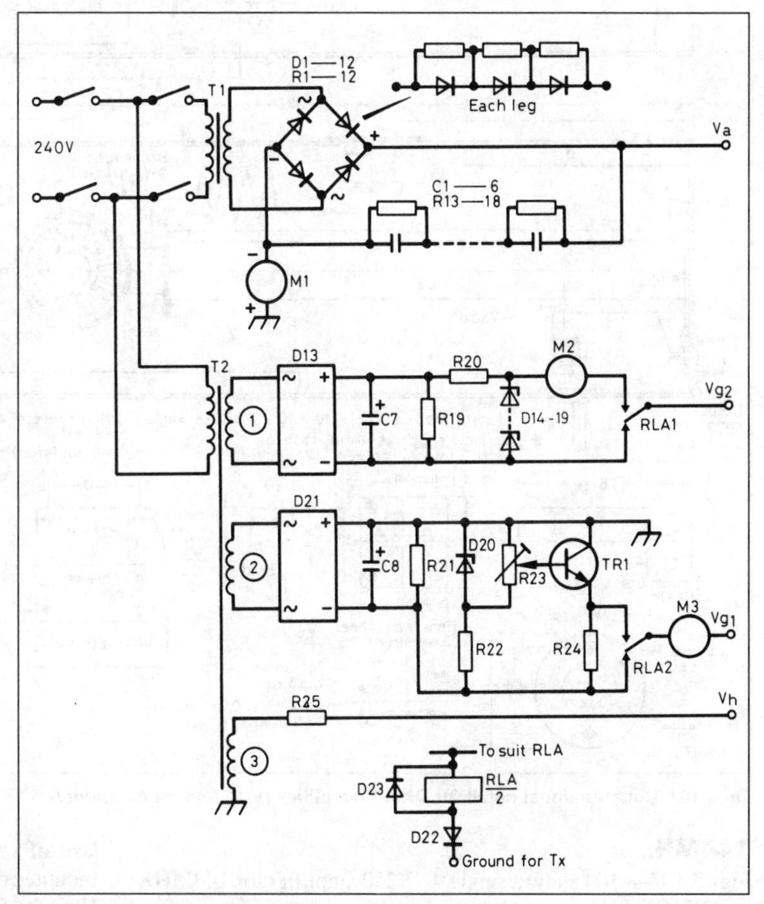

Fig 8.101. PSU circuit for 50/70MHz amplifier

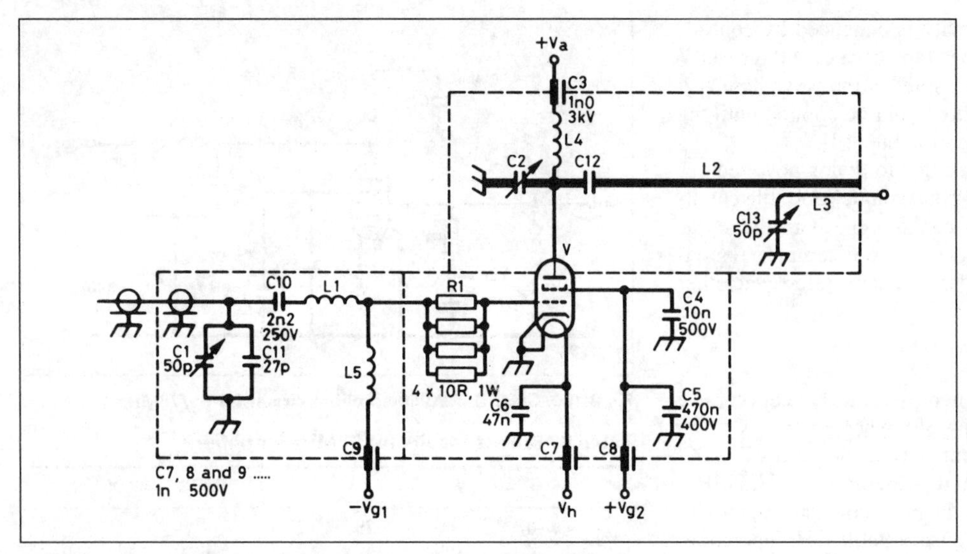

Fig 8.102. DK1OF 144MHz amplifier circuit *(VHF Communications)*

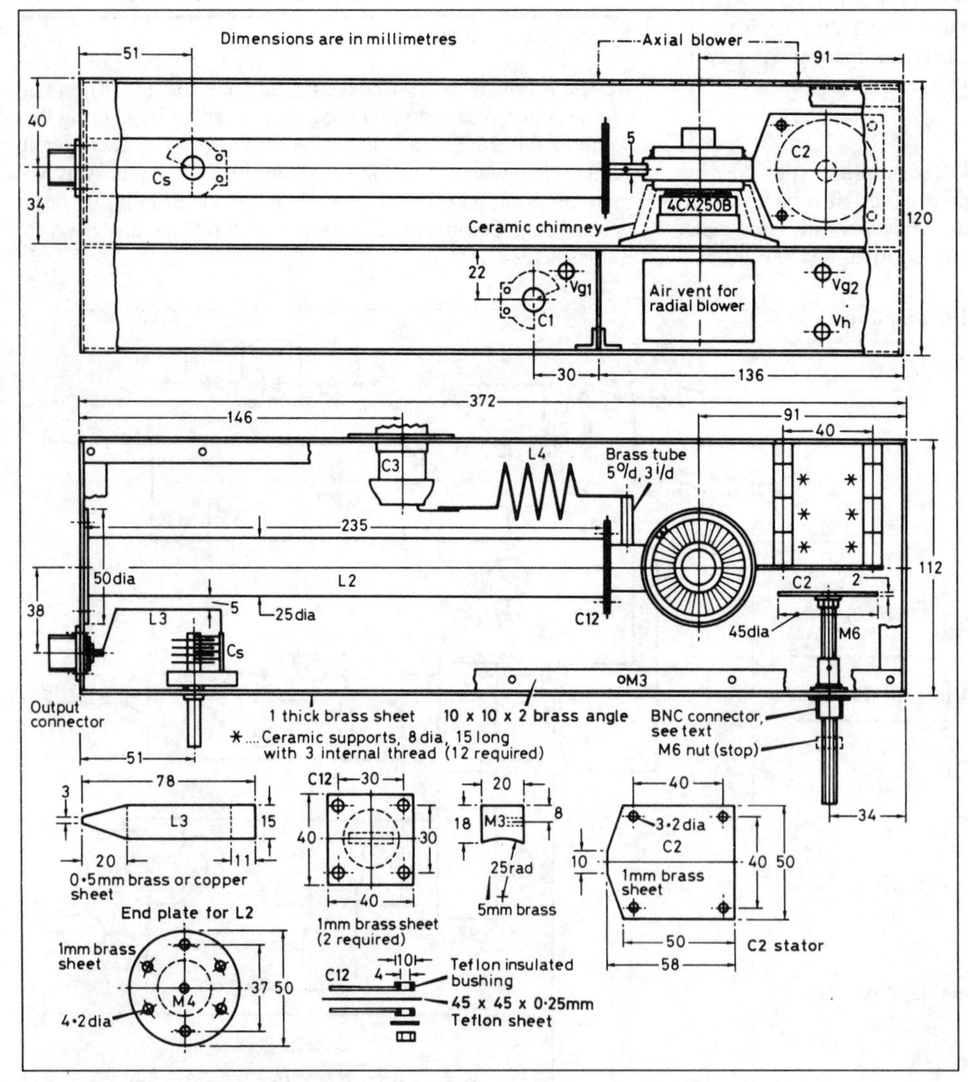

Fig 8.103. Constructional detail for DK1OF amplifier *(VHF Communications)*

144MHz

Figs 8.102–8.104 show a single 4CX250 amplifier for 144MHz by DK1OF [21] which demonstrates a number of interesting features. The grid circuit uses an L-match rather than the more usual tuned circuit arrangement. The inductance is provided by a wire length of about 15mm, and the leads of C10 can be used for this. The network Q is controlled by the 2.5Ω series resistance which ensures that the tuning is broad. The resistance also greatly reduces variations of input impedance with drive level. Finding the correct wire length for L1 can be fiddly but, once found, a good and stable input match is assured.

Component details and performance are given in Tables 8.16 and 8.17.

The anode circuit uses a transmission-line element as the inductive component, tuned by C2, and low loss is achieved by using a length of 25mm copper water pipe. Output coupling is through an inductive link made from brass strip.

C2 and C12 are noteworthy. C12, the DC blocking capacitor, is made from two 40mm square plates of copper with an oversize piece of 0.25mm PTFE sheet separating them. PTFE is soft, and extreme care should be taken to remove burrs and swarf from the metalwork to avoid the risk of puncturing the dielectric. Thin PTFE/glass PCB with the copper etched off both sides can be used as a stronger alternative but normal epoxy/glass PCB should not be used. The nuts and bolts which compress the sandwich are insulated using PTFE bushes. Nylon or hard plastic bushes usually supplied with power transistors should not be used. The finished component has excellent DC blocking and low loss RF characteristics.

A screw terminal capacitor of 100–200pF can be used as an alternative, in which case the increased inductance should be allowed for by shortening the copper pipe to maintain the overall distance from the centre of the valve to the end of the pipe.

The tuning capacitor C2 consists of a fixed plate connected to the anode and a moving plate mounted on a threaded spindle running in a panel-mounted bush. Here the RF current flows through the spindle and bush threads,

Table 8.16. Components for DK1OF 144MHz amplifier

R1–4	10R 1W carbon or metal film
C1	50p airspaced
C2	See text
C3	1000p 3kV feedthrough
C4	10n 500V ceramic disc
C5	0µ47 500V plastic film
C6	47n 100V ceramic disc
C7–9	1n feedthrough
C10	2n2 500V ceramic disc
C11	27p NPO ceramic or mica
L1	See text
L2	25mm copper pipe, 235mm long
L3	Brass strip, 15mm wide, 80mm long
L4	9t of 1.6mm wire, 25mm ID, 35mm long
L5	15t of 1.6mm wire, 15mm ID, 40mm long

Table 8.17. Performance of DK1OF 144MHz amplifier

P_i	P_o	I_a	V_a	I_{g2}	I_{g1}
0	0	100	2200	0	0
4	395	300	2000	5	0
8	520	420	1900	20	5

so a tight-fitting thread should be sought. In the original, the bush was made from a BNC connector with the centre and PTFE removed, and then tapped M6. Fig 8.104 shows an alternative arrangement where the moving plate is an earthed 'flapper', with the threaded spindle controlling its position. This is electrically superior to having a moving contact in the RF path, and so is preferred, especially at higher powers and frequencies. A split-stator ('butterfly') capacitor also achieves this, although the spindle is at half the peak RF voltage, so insulated mountings and spindle couplings are needed.

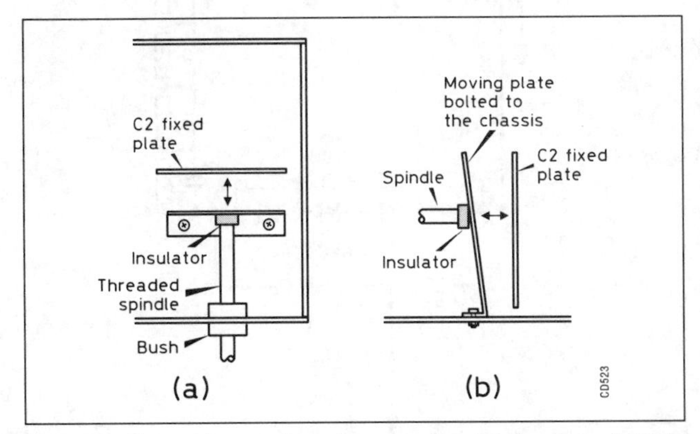

Fig 8.104. Alternative anode tuning capacitor. (a) Top view. (b) Side view

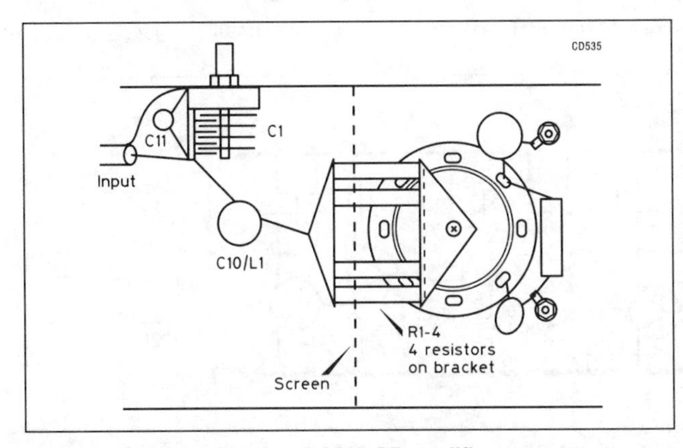

Fig 8.105. Grid circuit layout of DK1OF amplifier

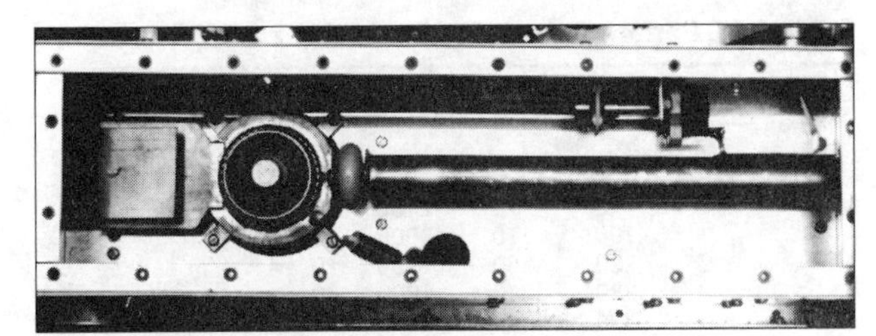

Fig 8.106. G4XZL 144MHz amplifier

Fig 8.107. Anode tuning detail

Fig 8.108. Output coupling detail

Fig 8.109. Grid circuit detail

Figs 8.106–8.108 show an amplifier based on this design built by G4XZL. A 'doorknob' transmitting capacitor is used instead

Table 8.18. Performance of G4XZL 144MHz amplifier

CW operation

P_i	P_o	I_a	V_a	I_{g1}
0	0	70	2100	0
1	106	170	2000	0
2	220	250	1950	0
3	330	310	1900	0.2
4	399	340	1850	0.8

Two-tone operation (powers are PEP, IMD is dBc)

P_i	P_o	IMD3	IMD5	IMD7	I_{g1}
1	110	–32	–60	–70	0
2	227	–30	–43	–55	0
3	349	–26	–38	–42	0
4	411	–22	–26	–35	1

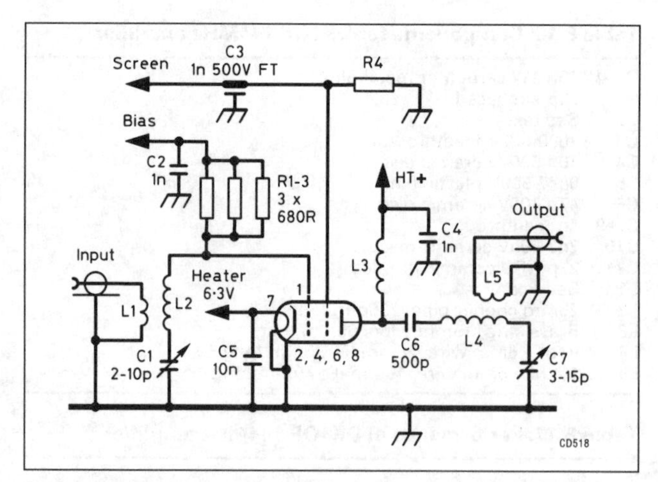

Fig 8.110. G4DHF 144MHz amplifier circuit

of the manufactured one, and the anode tuning capacitor shows another alternative mechanism whereby the flapper position is controlled by a string (actually the outer covering of some PTFE-covered wire), which passes through PTFE bushes in the fixed plate and chassis. It was found empirically for this valve and circuit that the output coupling needed to be tighter than in the original, and a piece of thin PTFE sheet is used to avoid any risk of shorting between the anode line and coupling loop.

Performance data are given in Table 8.18.

In the grid circuit, R1 is made from two 6Ω power terminations instead of the carbon resistors and C10 is a chip component. Both of these have less inherent inductance than the components used in the original, so L1 is made longer to compensate.

An alternative compact design by G4DHF, which has been proven in numerous contests and expeditions, is shown in Figs 8.110–8.114. Grid matching uses a series-tuned circuit with inductive coupling. Resistive damping helps to define the gain and also controls variations in input impedance. The anode circuit is a half-wave circuit, tuned at the end remote from the valve. The inductor is made from modelmaker's brass strip, wound as a coil. Output is taken from a two-turn link mounted on a rotating shaft which allows the coupling to be varied. Optimum output power was obtained after the coupling was increased by slightly compressing the anode coil turns around the coupling loop. The anode voltage is supplied through a RFC directly to the anode so that the tuned circuit can be isolated by C5. Linear output power is about 330W for 4W input.

Component details are given in Tables 8.19 and 8.20.

An axial fan is positioned to draw outside air over the anode tuning components. This improves reliability by reducing hot-spots and improves tuning stability during prolonged operating periods.

There are a couple of points to note in the power supply. Firstly, power is removed from the screen regulator diodes during receive periods in order to reduce heat and remove any slight chance of electrical noise. Secondly, anode and heater volts are applied together as a consequence of the EHT transformer also carrying the heater winding. This is contrary to the valve manufacturer's recommendations [14], but the simultaneous application of control grid cut-off bias and grounded screen grid have served to avoid any ill

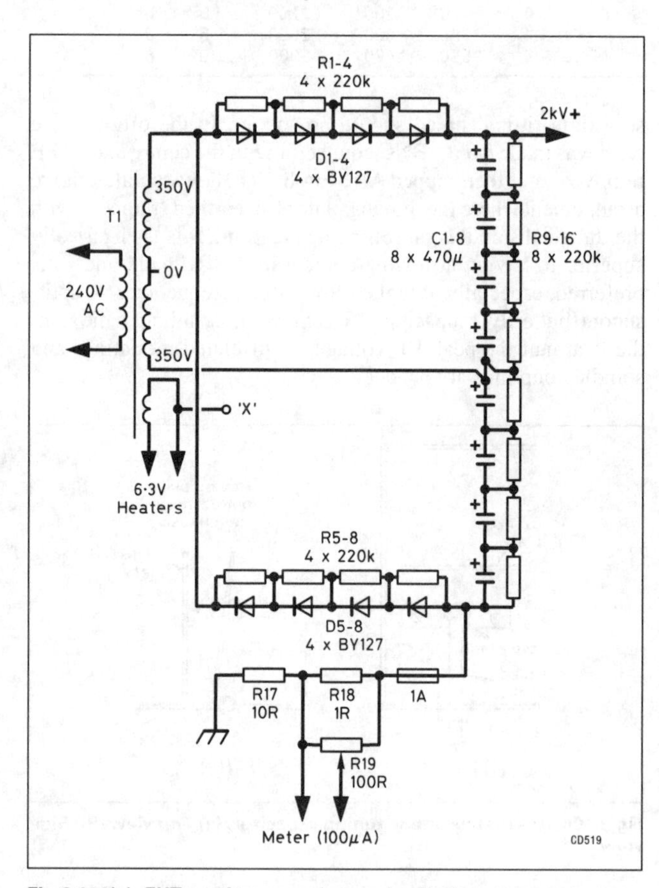

Fig 8.111(a). EHT section – anode supply circuit

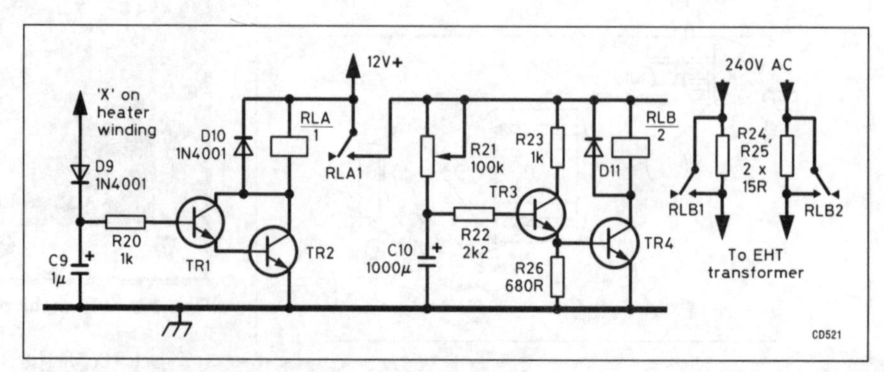

Fig 8.111(b). EHT section – sense and slow start circuits

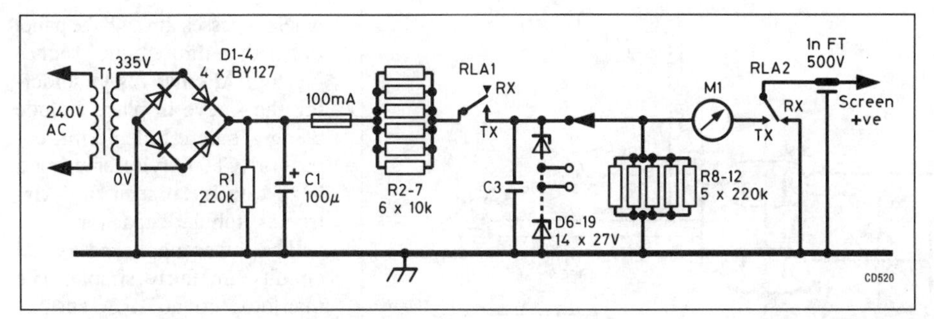

Fig 8.112(a). Grid section – g2 screen supply circuit

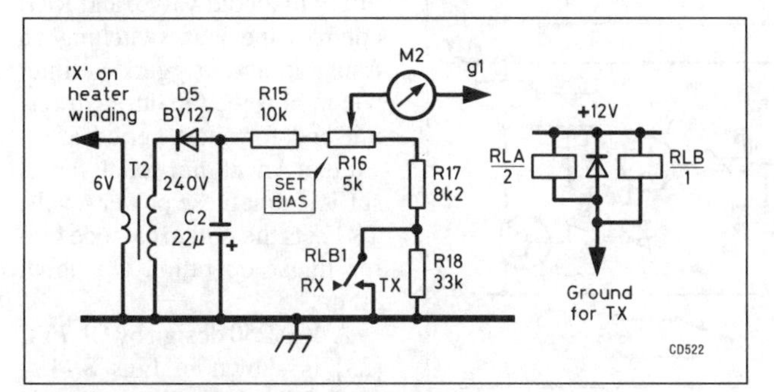

Fig 8.112(b). Grid section – g1 supply circuit

Fig 8.113. Overall view

Fig 8.114. Anode circuit layout

effects. TR1–4 provide a soft-start function, limiting the in-rush current into the EHT transformer.

There is no reason why anode and grid circuits should not be interchanged between designs.

432MHz

As an alternative to the complexity of tetrode circuits, a simple but effective 100W amplifier from G8PQG [22] is shown in Figs 8.115–8.118, using two 2C39 or similar valves in parallel, grounded grid. The input pi network and output half-wave line have enough tuning flexibility to cope with variations in valve types. Input power is typically 6W for 100W output.

Component details are given in Table 8.21.

Fingerstock can be difficult to obtain and an alternative is to make some from beryllium copper draught-excluder strip. A 6mm wide strip is pleated and then wrapped round the valve and the ends soldered together to form a ring. The same strip is

Table 8.19. Components for G4DHF 144MHz amplifier

R1–3	680R 1W, carbon or metal film
R4	275V(AC)/370V(DC) metal oxide varistor
C1	10p air-spaced variable
C2	1n 250V ceramic disc
C3	1n 500V feedthrough
C4	1n 3kV ceramic disc
C5	10n 50V ceramic disc
C6	500p 3kV 'transmitting' type
C7	15p wide airspaced variable
L1	1t 1.2mm wire, 12mm ID, coupled 1t from C1 end of L2
L2	5t 1.6mm wire, 10mm ID
L3	16t 1.2mm wire, 5mm ID
L4	3t 10mm × 0.5mm brass strip, 35mm ID
L5	2t 1.6mm wire, 15mm ID, coupled 1t from C7 end of L4

Table 8.20. Components for G4DHF 144MHz amplifier PSU

EHT SECTION	
R1–16	220k 2W
R17	10R 5W wirewound
R18	1R 3W wirewound
R19	100R pot
R20, 23	1k
R21	100k pot
R22	2k2
R24, 25	15R 5W wirewound
R26	680R
C1–8	470µ 300V
C9	1µ 16V
C10	1000µ 16V
T1	350–0–350V at 1A, 6.3V at 4A toroidal
D1–8	BY127
D9–11	1N4001
TR1–3	BC108
TR4	BFY51
RLA	Miniature 12V relay, SPCO contact
RLB	12V coil, 240V 6A contacts
M1	100µA
GRID SUPPLIES	
R1, R8–12	220k 2W
R2–7	10k 3W wirewound
R15	10k
R16	5k pot
R17	8k2
R18	33k
C1	100µ 450V
C2	22µ 450V
C3	10n 500V ceramic disc
T1	325V 100mA
T2	6.3V 100mA miniature, wired for step-up
D1–5	BY127
D6–19	27V 5W zener diode
RLA	12V coil, DPDT contacts rated for 350V
RLB	12V coil, SPDT contacts
M1	5mA, centre zero
M2	5mA, centre zero

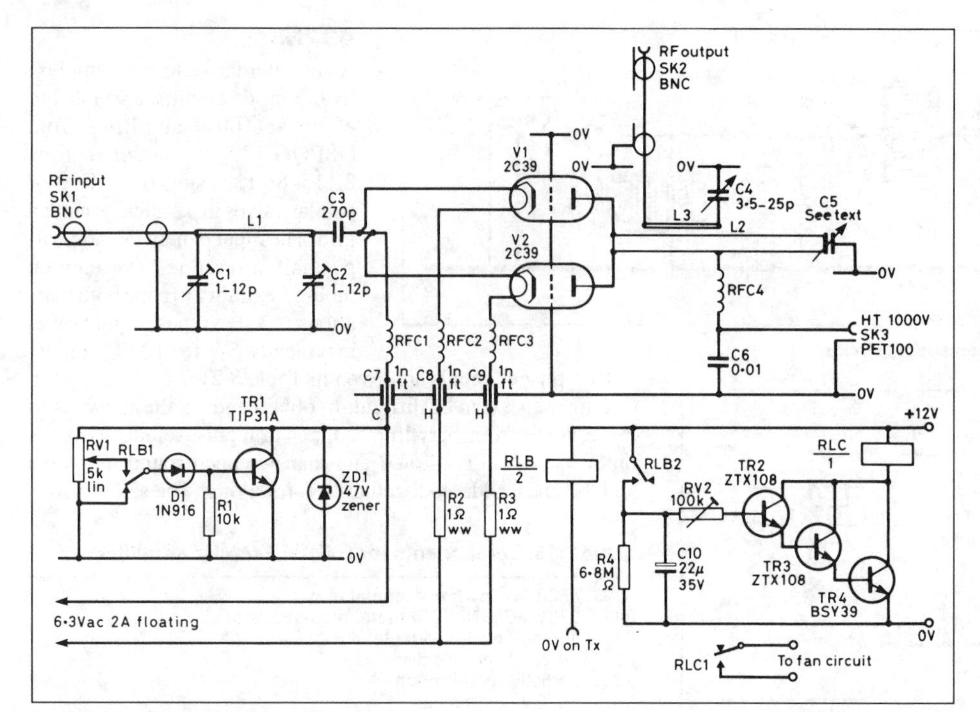

Fig 8.115(a). G8PQG 432MHz amplifier circuit

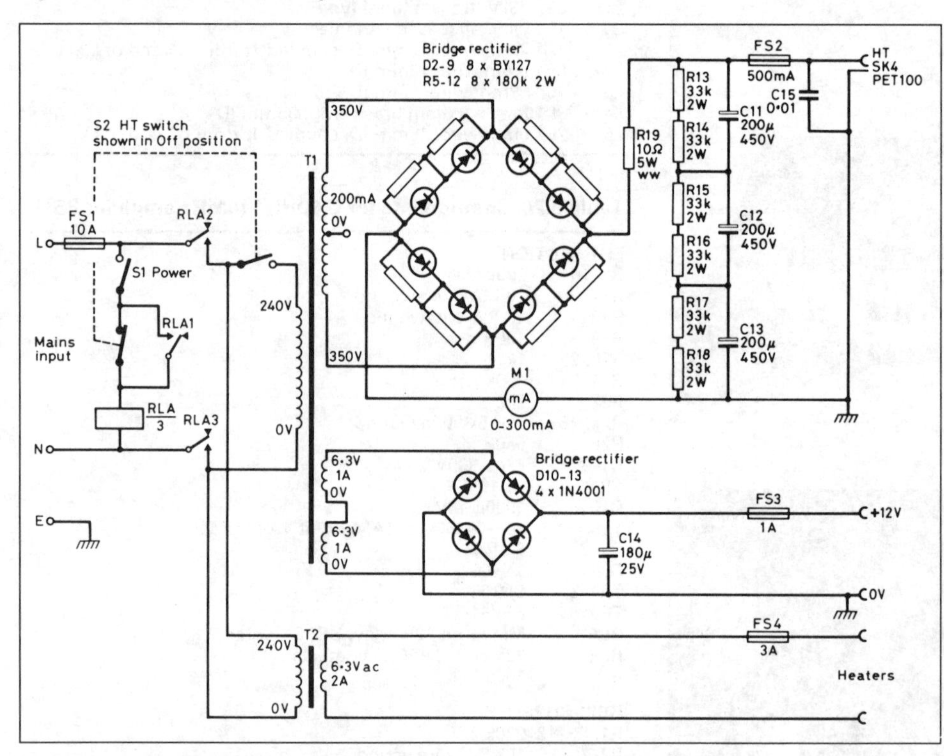

Fig 8.115(b). G8PQG 432MHz amplifier PSU circuit

used to fabricate the cathode contact and yet more is wound as a spiral spring for a tight fit in the heater contact.

The original chassis was constructed entirely from brass sheet with soldered joins, but alternatively, a chassis could be constructed from aluminium sheet and angle with small brass panels bolted in place where soldered connections were required (eg grid contacts). Those prepared to experiment might find that double-sided PCB could be used instead of brass throughout.

Minimum inductance in the ground connection of C4 is crucial and the spindle needs to be connected directly to the chassis where it passes through the panel rather than through the grounding lug. Alternatives to soldering the sleeve in place include using a single-hole fixing capacitor (C804 style) or using a high-quality piston or 'Airtronics' tubular capacitor.

The power supply and control circuits are fairly simple. The interlock around RLA ensures that HT cannot be applied unwittingly to a cold valve, and RLB controls the bias switching on transmit and triggers a timer which runs the fan during transmit and for short period afterwards. An 80mm axial fan is sufficient at these power levels. TR1 sets the grid to cathode bias for linear operation (I_{aq} about 30mA).

A 4CX250 design by DL7YC [23] is shown in Figs 8.119–8.121. This is similar to designs by G8AYN [24] and W2GN [25], and uses a single valve to produce up to 500W CW in Class C or 250W PEP linear.

Component details and performance data are given in Tables 8.22 and 8.23.

The overall dimensions of the amplifier are 307×192×152mm, with a dividing panel to give two sections, each 307×192×76mm. The upper one houses the anode circuitry and the grid circuits are contained in a smaller box of 180 × 98 × 76mm within the lower section. All dimensions should be followed closely. Even small variations could result in the amplifier not tuning correctly. It is also important that the lids should fit very tightly.

Both anode and grid circuits use half-wave transmission lines tuned at the end remote from the valve. Electrically, about half of the grid line is within the valve and socket and so the bias connection to the voltage node is at the grid terminal on the base. Both grid and anode circuits are tuned using flapper capacitors. In the anode, control is through PTFE ribbon (or PTFE-covered wire with the inner removed). Nylon and PVC are unsuitable in the high RF fields, although dial drive cord based on glass fibre will work well. The grid circuit has its own small enclosure to isolate the RF circuitry. Power supplies are connected through feedthrough capacitors and the RF input coaxial cable must be grounded as close as possible to the enclosure wall. In the grid circuit an eccentric cam moves the tuning flapper and RF is coupled in through a low-value airspaced trimmer.

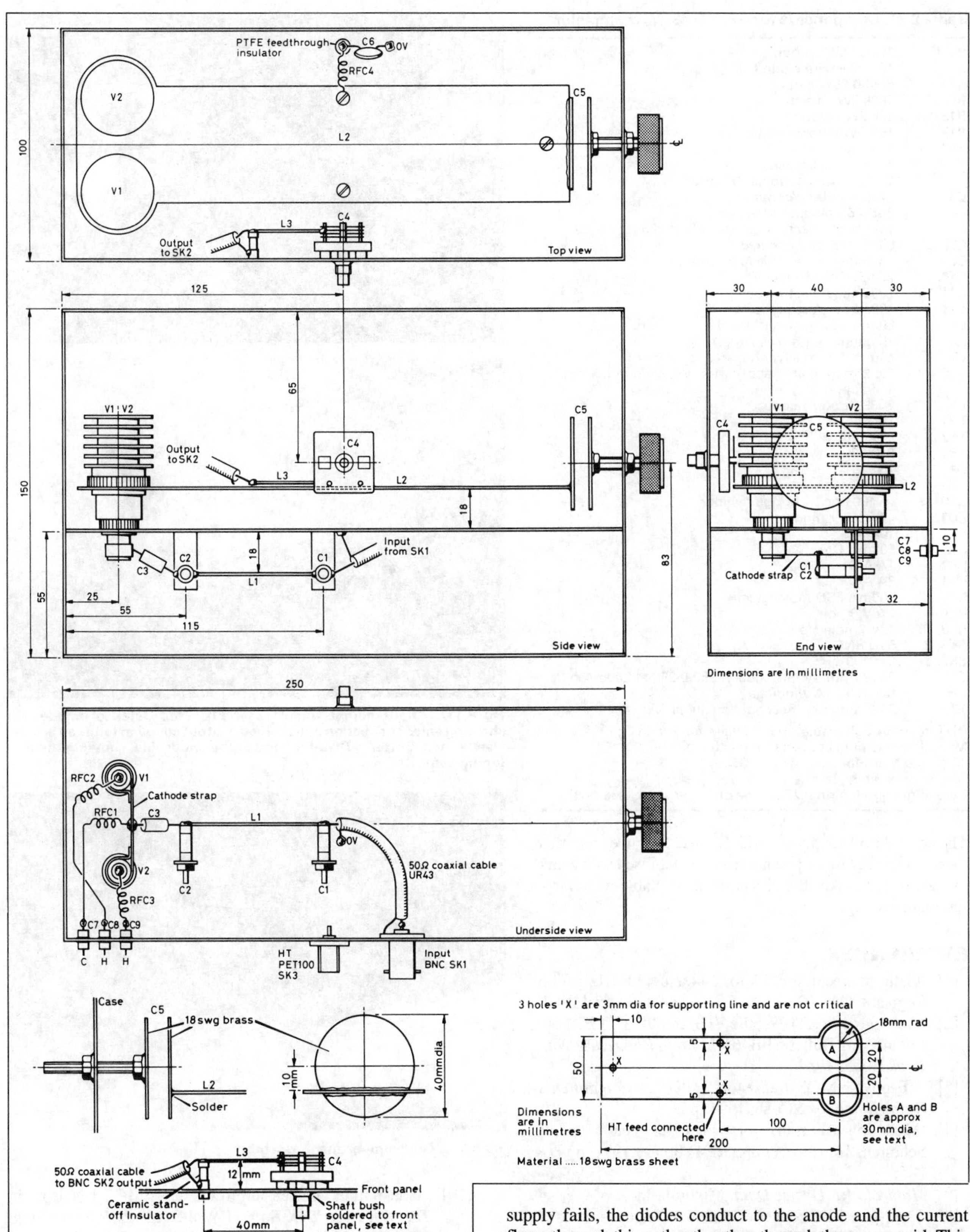

Fig 8.116. Constructional details for the G8PQG amplifier

The power supply circuits include an interesting screen grid protection scheme using D16–18. Normally, when anode voltage is present, the diodes are reverse biased but, if the anode supply fails, the diodes conduct to the anode and the current flows through this path rather than through the screen grid. This alternative to an anode voltage interlock is useful but not strictly essential, as R1 already limits the maximum screen current to keep the screen dissipation within the 12W limit. (*Author's note.* The screen grid circuitry has been changed slightly from the original in order to ensure correct operation under negative g2 current and provide protection against anode-screen flashover.)

Table 8.21. Components for G8PQG 432MHz amplifier

R1	10k 0.25W carbon
R2, 3	1R 2.5W wire wound
R4	6M8 0.25W carbon
R5–12	180k 2W carbon
R13–18	33k 2W carbon
R19	10R 5W wire wound
RV1	5k linear pot
RV2	100k carbon preset
C1, 2	1–12p ceramic tubular trimmer
C3	270p tubular ceramic
C4	3p5–25p air-spaced variable
C5	Two brass discs, 18 gauge, 40mm dia (see text)
C6, 15	0.01μ 2kV disc ceramic
C7–9	1n solder-in ceramic feedthrough
C10	22μ 35V electrolytic
C11–13	200μ 450V electrolytic
C14	180μ 25V electrolytic
L1	60mm 16swg silver-plated copper wire
L2	18-gauge brass (see Fig 8.116)
L3	40mm 16swg silver-plated copper wire
RFC1–4	10t 26swg enam copper wire, close wound, 5mm, self-supporting
V1, 2	2C39 or equivalent
TR1	TIP39A
TR2, 3	ZTX108
TR4	BSY39
D1	1N916
D2–9	BY127
D10–13	1N4001 or 1A bridge rectifier
ZD1	47V zener diode
FS1	10A A/S
FS2	500mA Q/A
FS3	1A A/S
FS4	3A A/S
M1	300mA FSD moving coil
RLA	240V 3-pole c/o
RLB, RLC	12V 2-pole c/o
SK1, 2	50Ω BNC
SK3, 4	Pet 100
T1	240V primary. Secondary: 350-0-350 at 200mA plus two 6.3V 1A windings
T2	240V primary. Secondary: 6.3V at 2A
PTFE feedthrough insulator and pillars for anode line	
Miscellaneous plugs/sockets for patch leads	
18-gauge brass for case, L2 and C5	
Fan, approx 3in square	
Beryllium copper strip 0.25in wide or finger stock (see text)	

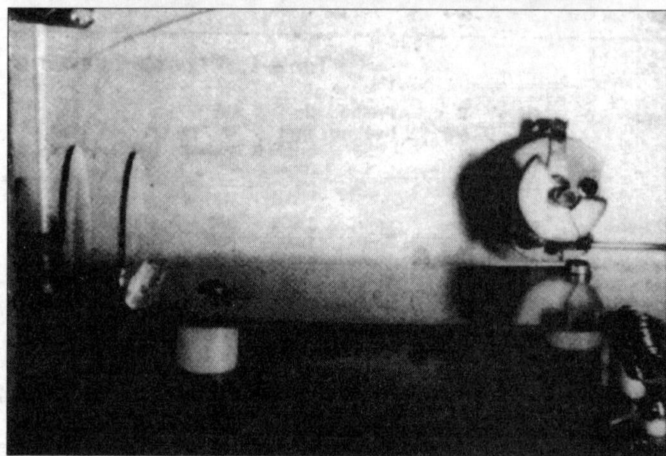

Fig 8.117. Anode compartment assembly. Top: Detail of anode line showing valve connection and HT feed. Output coupling loop can also be clearly seen. Bottom: The other end of the anode line, showing the disc tuning capacitor

Fig 8.118. Grid compartment assembly

The control grid supply is a widely used design, except that the input voltage to the potentiometer is stabilised with zener diodes, which makes the bias adjustment stable with varying mains voltage.

REFERENCES

[1] 'Gallium arsenide FETs for 144 and 432MHz', John Regnault, G4SWX, *Radio Communication* April 1984.

[2] 'A single device, 80W, 50Ω VHF amplifier', T Bishop, Engineering Bulletin EB46, *Motorola RF Device Data Book* B055, 1984.

[3] 'RF power MOS transistors', *Philips Components Data Handbook* SC08b, 1991, p29.

[4] 'Modeling RF transistors when the heat's on', J Scholten, *Microwaves and RF* February 1984, pp97–105.

[5] *Motorola RF Device Data*, Motorola Inc.

[6] 'Dual-gate FET converters for two and four metres', C W Westwood, G3VFD, *Radio Communication* October 1969.

[7] 'A 70MHz transistorised transmit/receive converter', D F Harvey, G3WOS, *Radio Communication* February 1977.

[8] 'A single stage linear amplifier for 50MHz', J Matthews, *Radio Communication*, June 1986, pp404–409.

[9] 'A stripline kilowatt amplifier for 432MHz', R Knadle, *QST* April 1972, pp49–55 and May 1972, pp59–62.

[10] 'Transmitters, power amplifiers and EMC', J Nelson, in *The VHF/UHF DX Book*, I White (ed), DIR Publishing, 1992, pp6-20 to 6-24.

[11] 'Stripline kilowatt for two meters', F Merry, *Ham Radio* October 1977, pp10–24.

[12] 'External anode tetrodes', W Orr, *Ham Radio* June 1969 pp23–27.

[13] 'Technical Topics', *Radio Communication* June 1982, p498.

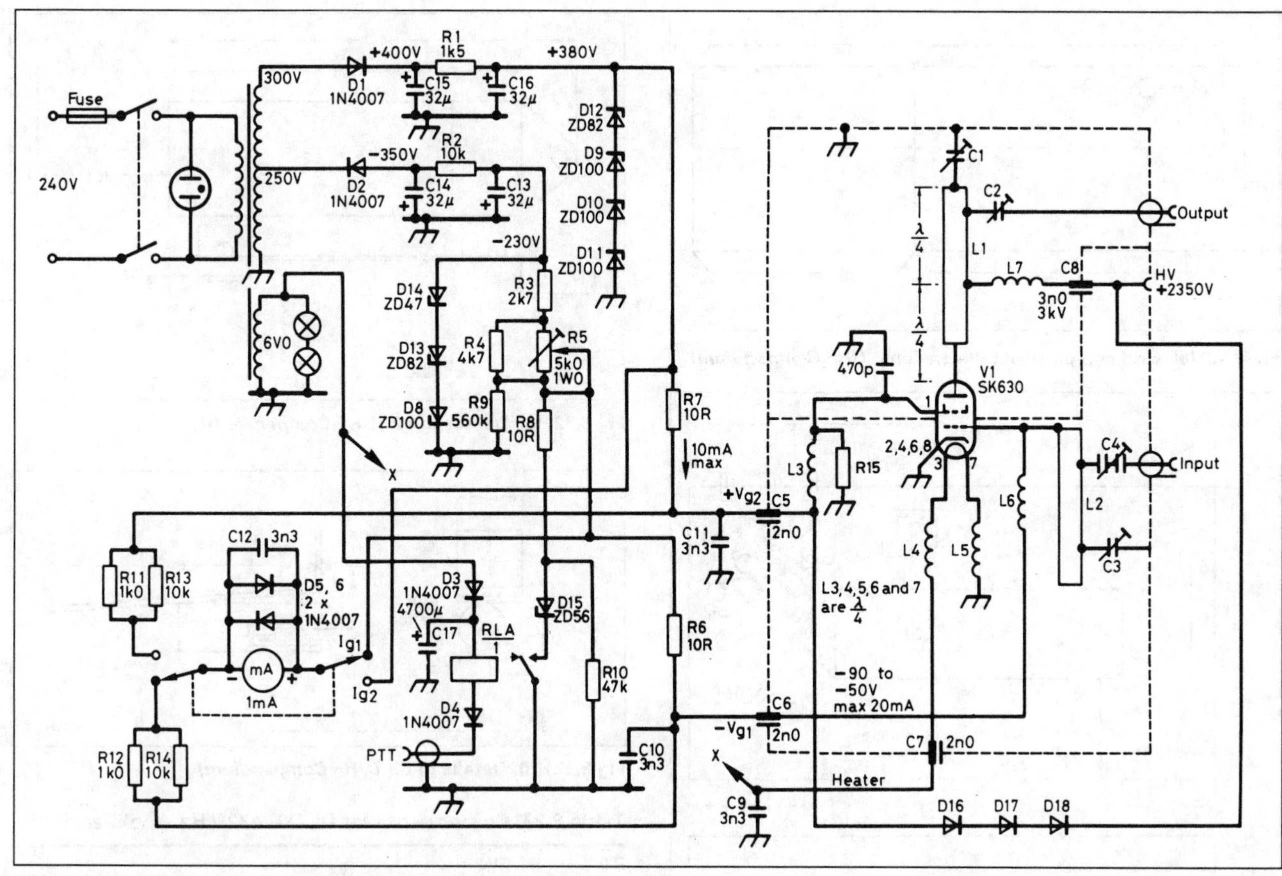

Fig 8.119. DL7YC 432MHz amplifier and PSU circuits *(UHF Compendium)*

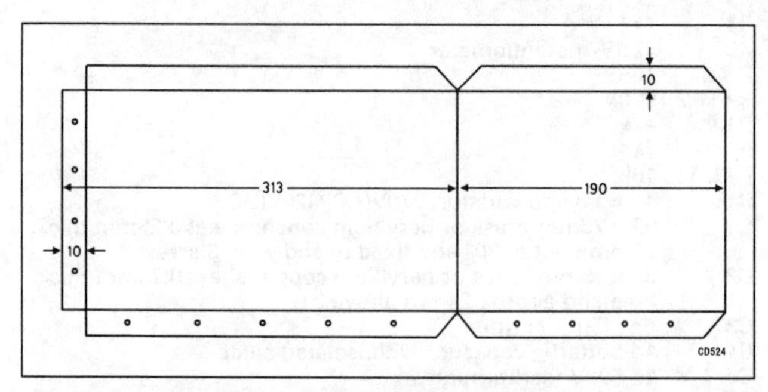

Fig 8.120(a). Anode compartment dimensions *(UHF Compendium)*

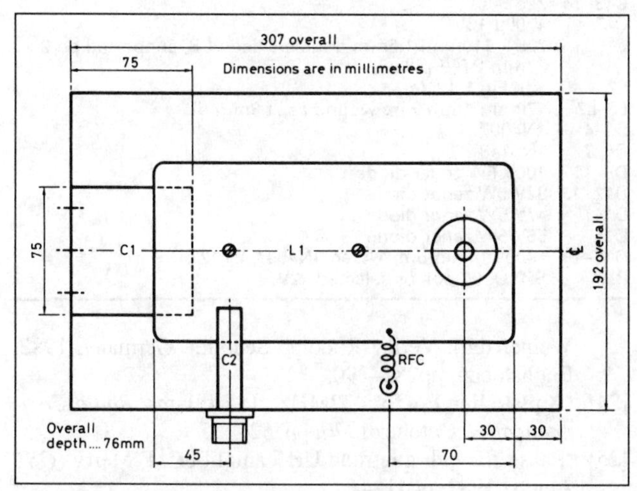

Fig 8.120(b). Anode compartment construction

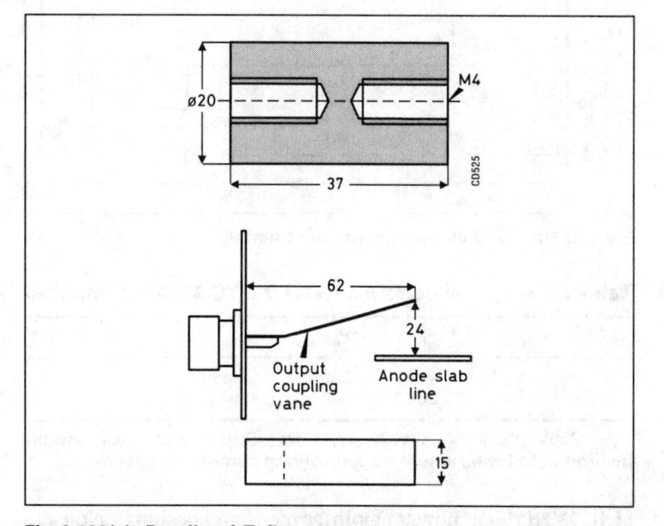

Fig 8.120(c). Details of Teflon spacers and output coupling vane. Two spacers are used to mount the anode line to the base plate *(UHF Compendium)*

[14] Technical data for 4CX250B, Eimac division of Varian.

[15] 'Power supply and control circuits for a 4CX250B amplifier', A Wade, *Radio Communication* October 1977, pp762–768.

[16] 'A power supply and control system for tetrode amplifiers', J Nelson and M Noakes, *Radio Communication* December 1987, pp902–907 and January 1988.

[17] 'Recipe for a longer life – keep the heaters under proper control', J Nelson and M Noakes, *Radio Communication* July 1988, pp529–532.

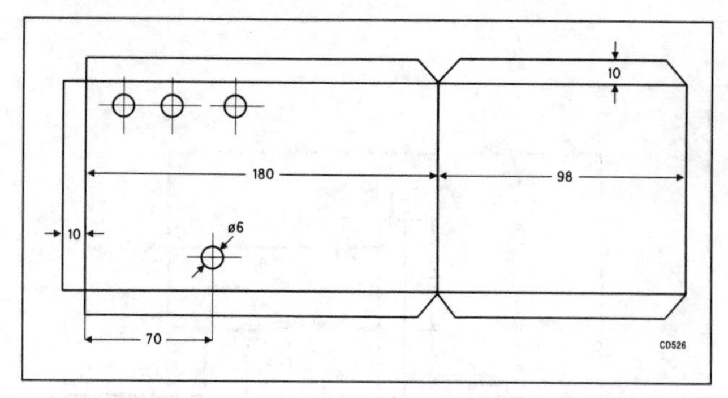

Fig 8.121(a). Grid compartment dimensions *(UHF Compendium)*

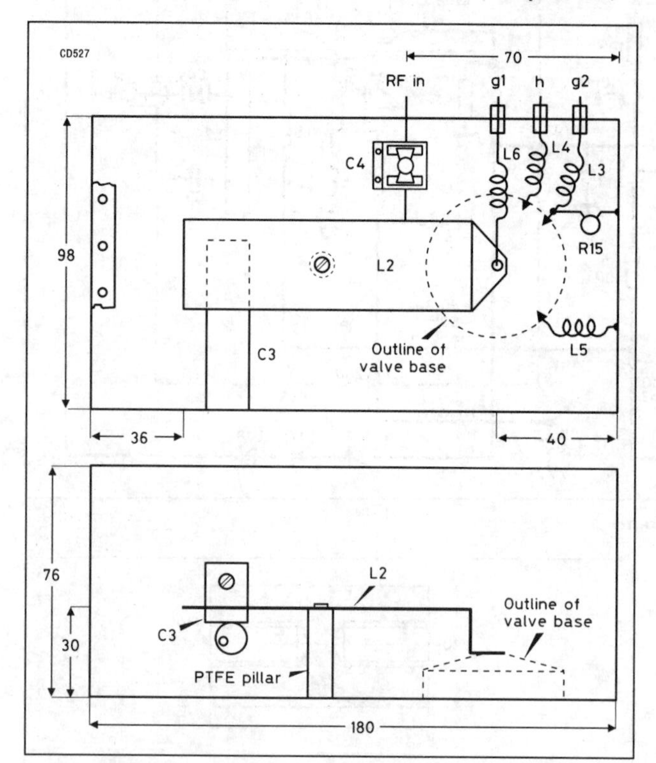

Fig 8.121(b). Grid compartment construction

Table 8.22. Typical performance of DL7YC 432MHz amplifier

P_i	P_o	I_{g1}
4	250	3 peak
20	500	20

V_a = 2350V and V_{g2} = 385V. Author's note: SSB output power should be limited to levels where no control-grid current is observed.

[18] 'VHF/UHF power amplifier mods and designs'. Notes to accompany high-power amplifier kits sold by G Brown, GJ4ICD.

[19] 'Technical Correspondence', *QST* July 1975, p47.

[20] 'Transmitters, power amplifiers and EMC', J Nelson, in *The VHF/UHF DX Book*, I White (ed), DIR Publishing, 1992, pp6-19 to 6-21.

[21] 'A 400W power amplifier equipped with the 4CX250B', J Kestler, *VHF Communications* 2/1978, pp100–113.

[22] 'The G8PQG 100W 432MHz linear amplifier', D Hewitt, *Radio Communication* November 1983, pp980–984.

[23] '"High power" 70cm power amplifier in half-wave technique', M Plotz, in *UHF Compendium Vols I and II*, K Weiner (ed), Verlag Rudolph Schmidt, Germany, 1982, English edn, pp282–290.

[24] 'A plate line PA for 432MHz', L Williams, *Radio Communication* October 1976, pp752–755.

[25] 'Phase III with a tetrode UHF amplifier', F Merry, *QST* August 1982, pp41–44.

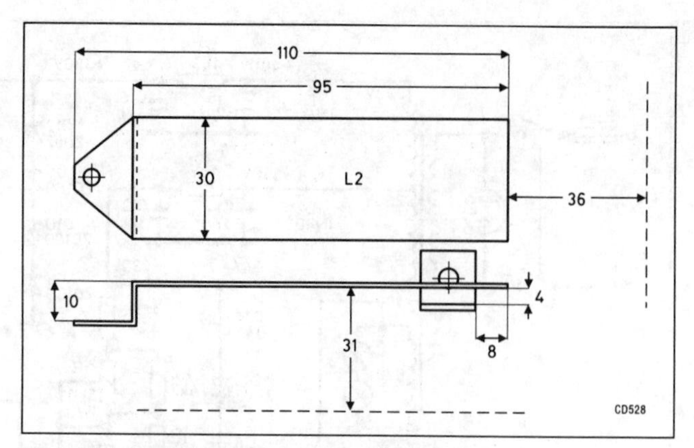

Fig 8.121(c). Details of L2 *(UHF Compendium)*

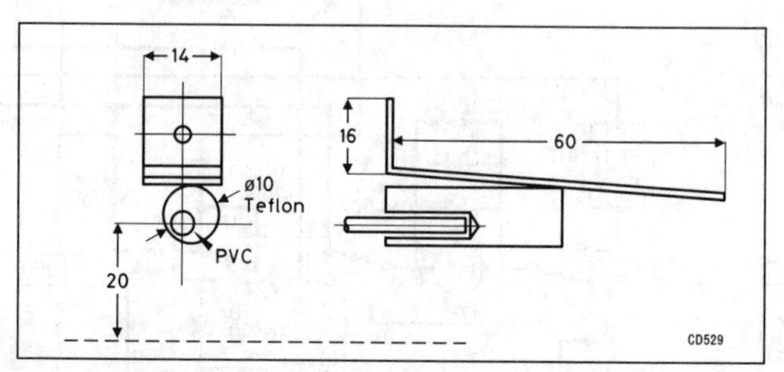

Fig 8.121(d). Details of C3 *(UHF Compendium)*

Table 8.23. Components for DL7YC 432MHz amplifier

R1	1k5 5W
R2	10k 2.5W
R3	2k7
R4	4k7
R5	5k 1W potentiometer
R6–8	10R
R9	560k
R10	47k
R11, 12	1k
R13, 14	10k
R15	Metal oxide varistor, 320V(AC)/420V(DC)
C1	95 × 75mm brass or beryllium copper sheet 0.25mm thick. 20mm bent at 90° and fixed to end wall (3 screws)
C2	62 × 15mm brass or beryllium copper sheet 0.25mm thick. Free end approx 24mm above L1
C3	See Fig 8.121(d)
C4	1p butterfly capacitor with isolated shaft.
C5–7	2n 500V feedthrough
C8	3n 3kV feedthrough
C9–12	3n3 500V ceramic
C13–16	32µ 500V
C17	4700µ 16V
L1	225 × 110mm, 1.6mm double-sided PCB, supported on 2 × 37mm PTFE pillars.
L2	See Fig 8.121(c)
L3–L7	170mm 1mm wire wound as 7t 8mm dia
D1–4	1N4007
D6, 7	1N4148
D8–11	100V 5W zener diode
D12, 13	82V 5W zener diode
D14	47V 5W zener diode
D15	56V 5W zener diode
D16–18	1000V minimum, 1A, eg 1N4007, BY127
RLA	SPNO. 9V coil or selected 12V.

9 Microwaves

ALTHOUGH a chapter devoted to microwave techniques is new to this handbook, such techniques have been in existence for a very long time. They have been largely neglected by the majority of amateurs, possibly because they were regarded as too difficult, or perhaps because the microwave bands were regarded as being suitable for use only over limited line-of-sight distances. Both these views are no longer tenable.

Many of the techniques for generating and receiving microwave frequencies were investigated and developed more than 60 years ago, in the 'thirties. Microwave usage was given added impetus by the development of radar and the advent of the second world war. Before 1940, the definition of the higher parts of the radio frequency spectrum [1, 2] read like this:

30 to 300Mc/s	Very high frequencies (VH/F)
300 to 3000Mc/s	Decimetre waves (dc/W)
3000 to 30,000Mc/s	Centimetre waves (cm/W)

Radio frequencies above 30,000Mc/s (now 30GHz) did not apparently exist! Various definitions have appeared in the intervening years. These have included terms such as *super-high frequencies* (SHF) and *extra-high frequencies* (EHF).

In the course of time, the unit of frequency *cycles per second* (c/s), its decimal multiples, *kilocycles per second* (kc/s) and *megacycles per second* (Mc/s), have been replaced by the unit *hertz* (Hz), its decimal multiples *kilohertz* (kHz), megahertz = 10^6Hz (MHz), gigahertz = 10^9Hz (GHz) and terahertz = 10^{12}Hz (THz).

Today, the term *microwave* means all radio frequencies above 1000MHz (1GHz). The division between radio frequencies and other electromagnetic frequencies, such as infra-red, visible (light) frequencies, ultra-violet and X-rays, is still not well defined since many of the techniques overlap, just as they do in the transition between HF and VHF or UHF and microwaves. To a large extent the divisions are artificial insofar as the electromagnetic spectrum is a frequency continuum, although there are, however, several good reasons for these divisions.

First, somewhere between about 400 and 1000MHz propagation via ionised media (E-layer or F-layer) ceases – there appear to be a few, largely unconfirmed, reports of auroral (E-layer) reflection in the amateur 432MHz band, but certainly no higher in frequency.

Second, and more importantly, around 1GHz (30cm wavelength) the *lumped-circuit* techniques used at lower frequencies are replaced by *distributed-circuit* techniques such as resonators and microstrip. Conventional components, such as resistors and capacitors, become a significant fraction of a wavelength in size and so they are nowadays replaced by very small, leadless, surface-mount devices (SMDs).

Conventional valves (vacuum tubes) and silicon bipolar solid-state devices are usable beyond 1GHz – perhaps to about 3.5GHz – and, as frequencies continue to increase, these devices are replaced by special valves such as klystrons, magnetrons, travelling-wave tubes, varactors and gallium arsenide field-effect transistors (GaAsFETs). Some of these devices, referred to elsewhere, are usable well into the millimetre-wave spectrum, ie above 30GHz.

However, many of these special microwave components were very expensive in amateur terms until they started to appear on the surplus market. Amateur equipment development, for this reason alone, has often lagged behind professional (particularly military) applications development.

The scene is changing! Recent consumer-market developments (for example, satellite TV broadcasting) have meant that many mass-produced, advanced devices are very competitively priced and many are becoming available on the surplus market at the present time. Amateur design, construction and operation are once again advancing very rapidly as a result of these developments, coupled with the mastery of these techniques by skilled and dedicated amateurs.

The range of the current amateur microwave allocations is such that all of these techniques are embraced: the lowest microwave frequency amateur allocation, the so-called *23cm band*, (1240MHz to 1325MHz in the UK), can thus be regarded as the transition point from 'conventional' radio techniques and components to the 'special' microwave techniques and components to be reviewed here.

In the space of a single chapter it will only be possible to summarise some of the practical techniques involved by outlining a few representative designs for most of the bands below 30GHz currently used by amateurs. Since most amateur microwave interest centres on the use of narrow-band modes to achieve long-distance, weak-signal communication, the majority of the designs outlined here will concentrate on such equipment.

In some instances construction and alignment procedures are described in some detail, again to illustrate the techniques used by amateurs in the absence of elaborate or costly test equipment, such as microwave noise sources, power meters, frequency counters or spectrum analysers. Most of the designs described are capable of being home-constructed without elaborate workshop facilities (most can be constructed using hand-tools, a generous helping of patience and some basic knowledge and skills!) and aligned with quite ordinary test equipment such as matched loads, directional couplers, attenuators, detectors, multimeters and calibrated absorption wavemeters. More details of components and techniques (including wide-band modes) than are possible to give here are available elsewhere [3–6].

AMATEUR MICROWAVE ALLOCATIONS

Most countries in the world have amateur microwave allocations extending far into the millimetre-wave region, ie above 30GHz. Many of these allocations are both *common* and *shared Secondary*, ie they are similar in frequency in many countries but are shared with professional (in this case *Primary*) users

Table 9.1. UK Amateur Service Allocations

Allocation (MHz)	Amateur status	Preferred (alternative) narrow-band segment (MHz)
1240–1325	Secondary	1296–1298
2310–2450	Secondary	2320–2322
3400–3475	Secondary	3456–3458 (3400–3402)
5650–5680	Secondary	5668–5670
5755–5765	Secondary	5760–5762
5820–5850	Secondary	
10,000–10,500	Secondary	10,368–10,370 (10,450–10,452)
24,000–24,050	Primary	24,048–24,050
24,150–24,250	Secondary	(24,192–24,194)
47.0–47.2GHz	Primary	Not defined
75.5–76.0GHz	Primary	Not defined
142–144GHz	Primary	Not defined
248–250GHz	Secondary	Not defined

who take preference. Amateur usage must, therefore, be such that interference to Primary users is avoided and amateurs must be prepared to accept interference from the Primary services, especially in those parts of the spectrum designated as *Industrial, Scientific and Medical* (ISM) bands. The current UK Amateur Service allocations are summarised in Table 9.1, and the UK Amateur Satellite Service allocations are shown in Table 9.2. All the familiar transmission modes are allowed under the terms of the amateur licence: in contrast to the lower frequency bands, most of the microwave bands are sufficiently wide to support such modes as full-definition fast-scan TV (FSTV) or very high-speed data transmissions as well as the more conventional amateur narrow-band modes, such as CW, NBFM and SSB.

Many of the bands are so wide (even though they may be Secondary allocations) that it may be impracticable for amateurs to produce equipment, particularly receivers, that cover a whole allocation without deterioration of performance over some part of the band. Most amateur operators do possess a high-performance multimode receiver (or transceiver) as part of their station equipment and this will frequently form the 'tuneable IF' for a microwave receiver or transverter: commonly used intermediate frequencies are 144–146MHz or 432–434MHz, either of which are spaced far enough away from the signal frequency to simplify the design of good image and local oscillator carrier sideband-noise rejection filters.

There are 'preferred' sub-bands in virtually all the amateur allocations where the majority of narrow-band (especially weak-signal DX) operation takes place. Typically 2MHz-wide sub-bands, often harmonically related to 144MHz as shown in Table 9.3, were formerly adopted for this purpose.

Some of these harmonic relationships are no longer universally available or usable because the lower microwave bands are rapidly filling up with Primary-user applications. Indeed,

Table 9.2. UK Amateur Satellite Service Allocations

Allocation	Amateur status	Comments
1260–1270MHz	Secondary	ETS only
2400–2450MHz	Secondary	ETS/STE. Users must accept interference from ISM users
5650–5670MHz	Secondary	ETS only
5830–5850MHz	Secondary	STE only
10,450–10,500MHz	Secondary	ETS/STE
24,000–24,050MHz	Primary	ETS/STE
47.0–47.2GHz	Primary	ETS/STE
75.5–76.0GHz	Primary	ETS/STE
142–144GHz	Primary	ETS/STE
248–250GHz	Primary	ETS/STE

ETS = Earth to space. STE = Space to earth. ISM = Industrial, Scientific and Medical applications.

Table 9.3. Some harmonic relationships for the microwave bands

Starting frequency (MHz)	Multiplication	Output frequency (MHz)
144	×3	432
	×9	1296
	×16	2304
	×24	3456
	×46	5760
	×72	10,368
	×108	24,192
432	×3	1296
	×8	3456
	×24	10,368
	×56	24,192
1152	+144*	1296
	×2	2304
	×3	3456
	×5	5760
	×9	10,368
	×21	24,192

* Note: additive mixing, not multiplication.

the position is changing particularly rapidly at present and the reader should refer to the latest ITU/IARU bandplans to get up-to-date information on current amateur usage, even though the current narrow-band segments are indicated in Table 9.1.

MICROWAVE LOCAL OSCILLATOR SOURCES AND LOW-POWER TRANSMITTERS

Microwave transmitters, like transmitters for any other frequencies, start with high-quality, stable local oscillator (LO) sources which are 'pure', ie free from unwanted harmonics, noise and spurious products. These attributes are, of course, also the main requirements for the LO source of a receiver or transverter, so that it seemed natural to start by considering microwave sources and low-power transmitter designs together before discussing microwave receivers, transceivers, and transverters.

For stability reasons, most LO sources intended for narrow-band operation in the microwave bands are crystal controlled – either multiplied from a lower frequency or phase/frequency locked to a lower-frequency, crystal-controlled reference oscillator. Generation directly at the final frequency is usually considered impractical due to the absence of suitable high-frequency crystals; therefore the crystal oscillator/multiplier approach is the accepted method of reaching the final frequency. *Self-excited* or *free-running* fundamental-frequency oscillators (whose frequency is determined by a high-Q tuned circuit), such as Gunn oscillators or dielectric-resonator stabilised oscillators, are used in the 10 and 24GHz bands where simple wide-band applications demand relatively uncritical short-term frequency stability.

In high-stability, narrow-band applications, the crystal oscillator/multiplier chain may be deliberately offset from the final operating frequency by some predetermined amount in order to allow generation of the final frequency by mixing with a second, intermediate frequency (IF) source. As mentioned earlier, it is common amateur practice to use a multimode 144MHz (or 432MHz) transceiver as the IF, enabling all the facilities of that transceiver to be translated into the microwave band of choice. The use of a 28 to 30MHz intermediate frequency, which is quite customary in the VHF and UHF bands, is not advisable in microwave equipment since it is almost impossible to adequately suppress the unwanted image product by means of filters. For higher microwave bands, the 1.3GHz amateur band (1240 to 1300MHz) may occasionally be used as an IF: there is no reason, apart from possible breakthrough, why any frequency above

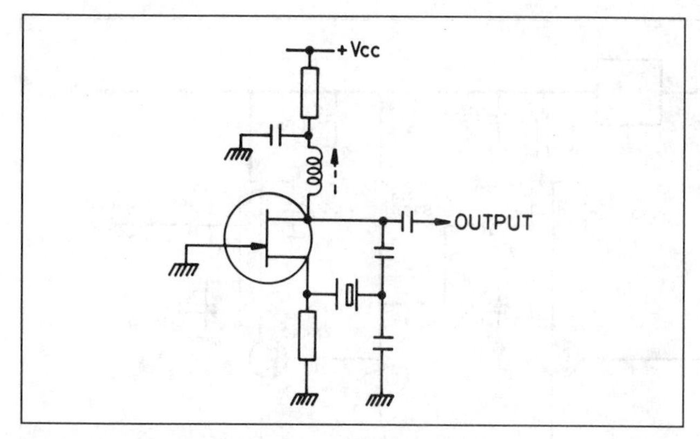

Fig 9.1. Circuit of JFET Colpitts oscillator circuit

about 144MHz should not be used as an IF, and the use of a high-performance multimode scanning receiver should not be discounted for receive IF signal processing.

The crystal frequency chosen to start the chain will obviously depend on the required final frequency. If too low a starting frequency is chosen, then a large order of multiplication will be needed which may cause problems with stability. It may also prove difficult to provide sufficient interstage filtering to reduce the level of unwanted products at the output. Too high a crystal frequency and the crystal becomes very expensive and fragile. The optimum starting frequency for most amateur microwave applications is therefore in the range 90 to 120MHz.

Crystal oscillators in the region of 100MHz use overtone crystals, usually operating on their fifth or seventh overtone. There are a number of suitable oscillator circuits which can be used but, for reasons of reproducibility, stability and noise performance, the choice is reduced to just a few. Perhaps the best known of these is the series-resonance crystal Colpitts oscillator using a power FET as the active element [7]. The circuit of Fig 9.1 has justifiably become the standard oscillator used in almost all current European amateur microwave designs. It does, however, suffer from one drawback. To achieve good phase-noise performance it is necessary to use a power FET such as the Texas Instruments P8000. This type of device is notoriously difficult to obtain. As a result there has been a tendency to use lower-power FETs such as the Siliconix J310. Although these work well in this circuit, the noise performance is not especially good and this may be of considerable concern in sources for 10GHz or higher.

Setting the crystal oscillator on exactly the required frequency can occasionally prove difficult due to excessive crystal-mount capacitance. This may be 'tuned out' by adding a small inductance across the crystal. The inductance needs to resonate with the crystal-mount capacitance at the crystal series resonant frequency.

An alternative oscillator circuit that is capable of phase-noise performance at least as good as the power FET Colpitts circuit is the Butler oscillator of Fig 9.2. In this circuit the function of amplitude limiting is performed by a second stage, leaving the function of the first stage as an amplifier. A further advantage of the Butler oscillator is the inherently high harmonic output. The limiter stage output circuit is usually tuned to the required harmonic of the overtone oscillator. For example, a 100MHz crystal oscillator stage may have its output tuned to 300MHz, at which frequency several milliwatts will be available, hence eliminating the need for one additional stage of multiplication after the oscillator stage.

The Butler oscillator is also particularly easy to modulate as

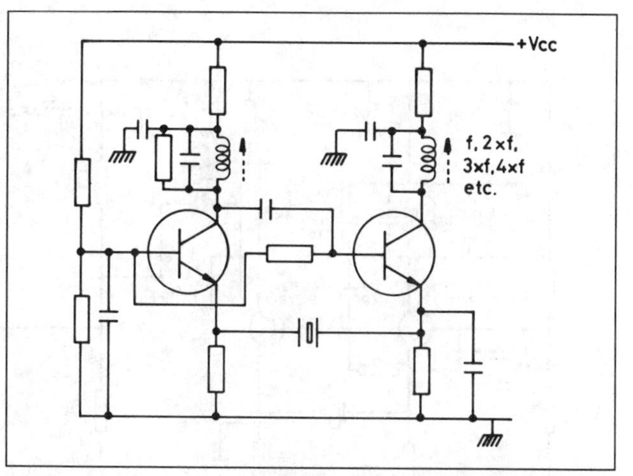

Fig 9.2. Circuit of Butler oscillator using bipolar junction transistors (BJT)

shown in a later section. However, the modulation produced by the simple circuit given there is a mixture of FM and phase modulation. For many applications this does not matter.

The bipolar transistor Colpitts overtone oscillator circuit, Fig 9.3, so often used in early amateur designs, has almost entirely fallen from favour as its relatively poor noise performance has now been widely recognised.

A high-quality UHF source for microwave use

A high-quality UHF source, based on these principles, was developed by members of the RSGB Microwave Committee from the Plessey AMETS transmitter [8]. The circuit uses the Butler oscillator, briefly described above, and was designed to have excellent oscillator noise sideband characteristics. Its output was intended to be multiplied up to the microwave region, and produced a good-quality note, even at 10GHz. The output frequency can be anywhere in the range 360 to 440MHz, depending on the intended application. Details of how this range can be extended are given later. Although alternative, more recent circuits which give output at much higher frequencies are described later, it is often more convenient or cost effective to provide amplification of the source at UHF, rather than in the microwave region. Thus, this UHF source, although designed some years ago, is still important to the design of many items of amateur equipment for the microwave bands.

The heart of the design is the Butler crystal oscillator which eliminates the need for frequency multiplier stages and, due to

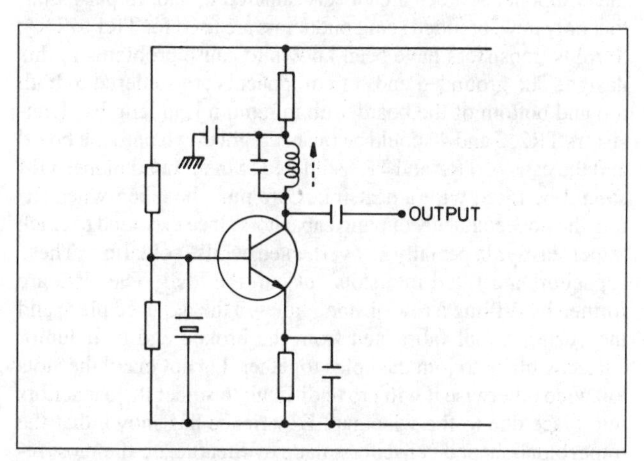

Fig 9.3. Circuit of a BJT Colpitts oscillator

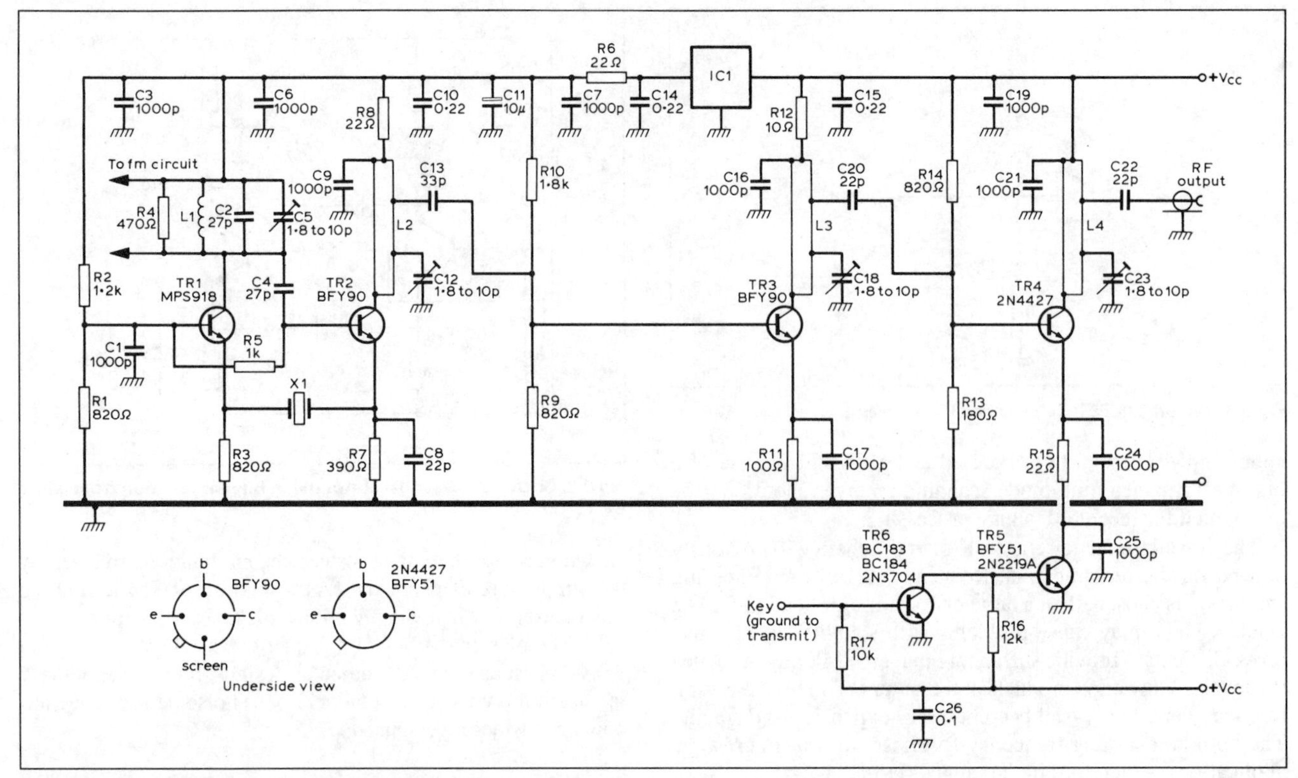

Fig 9.4. Schematic diagram of the high-quality UHF source

low crystal loading, gives a very low noise output compared with the commonly used single transistor circuits.

Two Class A amplifiers follow the oscillator stages to increase the output level to at least 100mW. The use of Class A amplifier stages contributes to the overall low-noise performance and makes the chances of instability less likely. The circuit of the source is shown in Fig 9.4 and the component values are given in Table 9.4.

The unit is constructed on 0.064in (1.6mm) thick double sided copper-clad glassfibre epoxy board (dielectric constant = 4.8) with an earth (ground) plane on the component side of the board. The use of materials with a different dielectric constant could result in incorrect resonant frequencies for the tuned circuits and should be avoided. The PCB artwork for the source is shown in Fig 9.5 (in Appendix 1).

Construction is generally straightforward and the layout of components on the PCB is shown in Fig 9.6. It is important that the component values are closely adhered to and, in particular, that only new, branded semiconductors are used for TR1 to TR4. Surplus transistors have been known to cause problems in this design. The grounded ends of components are soldered to both top and bottom of the board with minimum lead lengths. Transistors TR2, 3 and 4 should be pushed right down onto the board and the cans of TR2 and TR3 soldered to the ground plane. TR4 should be fitted with a heatsink. Care must be taken when fitting the trapezoidal decoupling capacitors since they tend to crack rather easily, especially if overheated whilst soldering. These capacitors are fitted into slots cut into the PCB. The slots are formed by drilling a row of small holes in the required place and then using a tool fabricated from the broken end of a Junior hacksaw blade to join the holes together. Do not make the slots too wide otherwise it will prove difficult to solder the capacitors into place due to the wide gap. Experience has shown that the trapezoidal-shaped capacitors made by Steatite are the most reliable type to use in this design.

The ground planes on both sides of the board are joined by wire 'worms' made from lengths of 20–24swg wire (eg scrap component leads) soldered in all places marked with an 'X' on the component layout.

Two wire links are required on the board to connect L1 and

Table 9.4. High-quality UHF source components list

R1, 3, 9, 14	820R
R2	1k2
R4	470R
R5	1k0
R6, 8	22R
R7	390R
R10	1k8
R11	100R
R12	10R
R13	180R
R15	22R
R16	12k0
R17	10k0
All resistors 0.25W TR4 2% (Electrosil, metal oxide)	
C1, 3, 6, 7, 9, 16, 17, 19, 21, 24, 25	1000p leadless disc (Steatite 1000 pF/80/20 TEFK7 400V)
C2, 4	27p subminiature ceramic disc
C5, 12, 18, 23	1.8–10p film trimmer (Mullard 809-05002)
C8, 20, 22	22p subminiature ceramic disc
C10, 14, 15	0.22μ polyester
C11	10μ tantalum
C13	33p subminiature ceramic disc
C26	0.1μ ceramic disc
TR1	MPS918
TR2, 3	BFY90
TR4	2N4427
TR5	BFY51 or 2N2219A
TR6	BC183 or BC184 or 2N3704
IC1	LM340T12 or 7812
X1	HC18/U
L1 to L4	Printed lines on board
Heatsink, TO5 (TR4)	
PCB mounting socket, SMA, SMB, or SMC	

L4 to the positive supply. Either two- or three-pin trimmer capacitors can be used to tune the inductors, but be careful to get the lead orientation correct. The rotor side of the trimmers should be connected to the earthy side of the tuned circuits to minimise stray capacitance effects when trimming tools are used.

A variety of methods can be used to extract the output signal from the board. A small-diameter coaxial lead such as RG174/U can be soldered direct to the output connection or alternatively a small socket such as SMA, SMB or SMC (CONHEX) can be fitted where indicated.

Provision has been made on the board for a TO220-style IC regulator for the oscillator supply to reduce chirp when the source is A1A keyed. Since these regulators generally require at least 2.5V 'head room', a 14.5V supply will be needed for 12V operation. Alternatively, an 8V regulator could be used with a 12V supply at the expense of power output. In many applications it is possible to eliminate the regulator altogether and wire a link across the two outermost regulator holes.

When the board has been wired correctly, it can be mounted in a die-cast box. The board may be fitted by means of bolts through its corners. A minimum spacing of 5mm should be maintained in order to avoid detuning the microstrip lines.

Alignment is easy, but needs a few items of simple test equipment. Connect a suitable power supply to the board (depends on the regulator used, see above) and check that the current is approximately 150mA. If the current is much in excess of this value it would be wise to check for the presence of short-circuits or incorrectly placed components. When all is well, the trimmers should be set initially to the following approximate positions:

C5 50% meshed
C12 40% meshed
C18 80% meshed
C23 50% meshed

These preset positions apply only if a 96MHz crystal is to be used. A low-power 50Ω detector connected to the output will probably indicate the presence of some output when the supply is connected. The trimmers can now be *slightly* readjusted for a peak in output. If no output is indicated, it will be necessary to use a wavemeter to investigate the tuning of each stage in order to find out what is wrong.

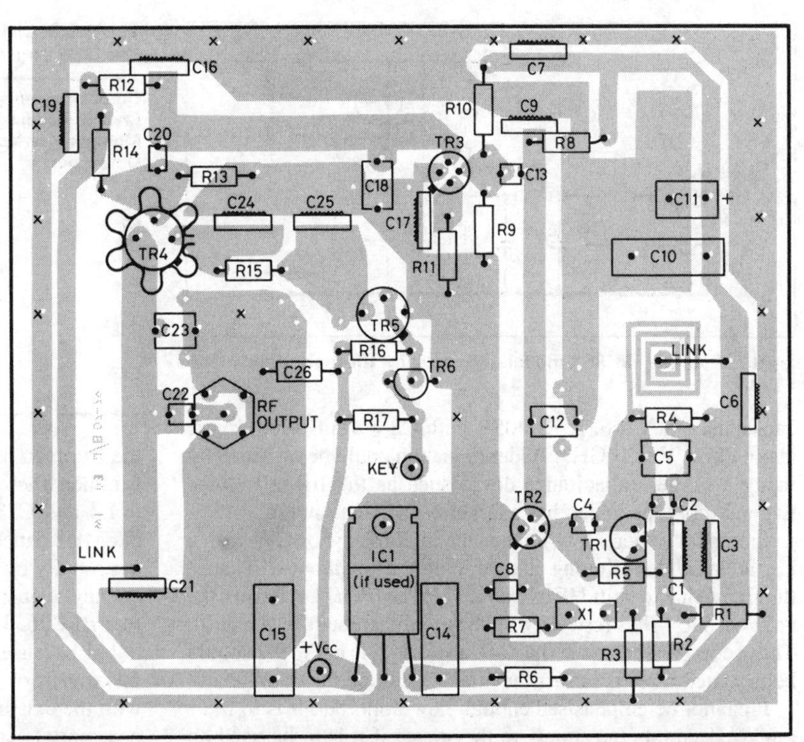

Fig 9.6. Top: layout of the components on the high-quality UHF source PCB. Bottom: photograph of the completed board

The position of C5 should not be too critical as regards output power and may be used to trim the frequency. If crystal frequencies outside the range 90 to 100MHz are used, it may be necessary to alter the value of C2 so that C5 is set to about mid-travel when the frequency is correct. If necessary the frequency can be raised slightly by placing a 5–20pF trimmer in series with the crystal. However, this may reduce power output or reduce stability. It may be preferable to put up with the small frequency offset and to calibrate it out at IF, rather than sacrifice stability.

The output stage can be on/off keyed by TR5 and TR6. However, this pulls the oscillator slightly and the keying chirp when multiplied to 10GHz is unacceptable unless the oscillator voltage regulator is used, and even then is still noticeable. Frequency shift keying (FSK) is the preferred method of keying. If the A1A keying facility is not required, TR5 and TR6 and their associated components should be omitted and the end of R15, previously connected to TR5, grounded directly.

FM or FSK can be produced using the circuit shown in Fig 9.7 in which a varicap diode varies the capacitance across L1. The value of C2 must then be reduced to maintain the total

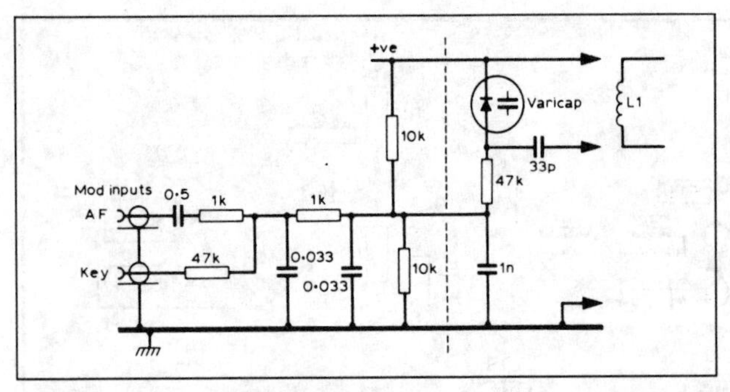

Fig 9.7. A simple FM/PSK modulator circuit for the high-quality UHF source

capacitance at 27pF. A BB405B with C2 = 18pF should give about 2kHz/V at 10GHz. Wider deviation could be obtained by using a higher-capacitance diode such as the BB110G or a hyperabrupt type which has a larger capacitance swing.

Inputs for either a low-impedance microphone or CW with a key are provided. Keying an earth onto the key input will cause the frequency to shift HF by about 1kHz at 10GHz. Altering the value of the 47kΩ resistor will change the amount of the shift. Those components to the right of the dotted line in Fig 9.7 should be mounted directly across the tuned circuit on the main board.

It cannot be emphasised enough how important it is to thoroughly decouple the supply to the source if it is to be used on the higher bands and particularly the modulation input if the source is to be modulated. Supply decoupling is best provided by the use of a Filtercon where the supply enters the box. In the event of a Filtercon not being available, an ordinary 1000pF feedthrough capacitor together with a small 10–100μH RFC (outside the box, in series with the supply lead) should be adequate. It is also a wise precaution to check with a wide-band oscilloscope that any voltage regulating circuits are not oscillating. The results shown in Table 9.5 were obtained with a typical board.

The major factor affecting frequency coverage is the maximum resonant frequency of the tuned circuits. As designed, 500MHz is about the highest frequency that can be obtained.

Referring to Fig 9.4, L1 and C2/C5 resonate at the crystal frequency. By removing C2 and replacing C5 with a trimmer that has lower minimum capacitance, the resonant frequency can be raised to over 200MHz. The effect of this change will also be seen as a reduction in the adjustment range of C5. Further increase in the maximum resonant frequency can be achieved by shorting out a portion of L1 at the centre of the coil. The three remaining tuned circuits can also be resonated at a higher frequency by replacing C12, C18 and C23, respectively, with variable capacitors of lower minimum value. Shortening of the tuned lines is not so easy to achieve because of the position of the coupling and decoupling components. When making adjustments to the tuned circuits there should be two positions where

Table 9.5. Performance of the high-quality UHF source

Supply voltage (V)	Output power (mW)
11	225
12	270
13.5	325
15 (without IC1)	350
15 (with IC1)	330

Frequency stability (multiplied to 10GHz):
without IC1 5kHz/V
with IC1 <100Hz/V
Temperature stability (at 10GHz): 2kHz/°C

Table 9.6. List of modifications for increased coverage from the UHF source

Output frequency (MHz)	384	404	464	524	594
Crystal frequency (MHz)	96	101	116	174.666	198
Multiplication factor	×4	×4	×4	×3	×3
Output power	+24dBm	+20	+22	+21	+17
L1	as is	as is	as is shorted	1/2t shorted	1t
C2	22p	22p	8p2	0	0
C5	1.8–10p	1.8–10	1.8–10p	1–3.5p	1–3.5p
	70%	30%	90%	60%	50%
C12	1.8–10p	1.8–10p	1.8–10p	1–3.5p	1–3.5p
	50%	50%	20%	60%	10%
C18	1.8–10p	1.8–10p	1.8–10p	1.8–10p	1–3.5p
	70%	70%	50%	10%	50%
C23	1.8–10p	1.8–10p	1.8–10p	1–3.5p	1–3.5p
	70%	40%	20%	40%	0%

the trimmers give a peak, since there are two positions of the trimmer where the capacitance is equal.

L4 and C23 have the lowest resonant frequency at about 580MHz. Since the performance of the 2N4427 transistor is falling rapidly by this frequency, any attempts to increase the frequency of operation should be accompanied by a change of TR4 to a BFR96. This transistor will require mounting beneath the board because of its package type. The lower capacitance of this transistor will allow a higher frequency of operation even with the existing components. Table 9.6 lists the modifications necessary to cover some higher frequencies.

The temperature coefficient of the UHF source has been measured at approximately ±0.2ppm/°C depending on the crystal used. This results in a frequency shift of about 40kHz at 10GHz when operated over the normal temperature range encountered in the shack of 10 to 30°C. Typical temperature coefficients for AT-cut crystals range from 0 to ±1ppm/°C. Thus, the main cause of drift when using standard-specification crystals seems to be the crystal itself, the rest of the circuit having much less effect. Because the crystal is very loosely coupled to the oscillator circuit, as it should be in a very-high-stability circuit, it is not possible to use conventional compensation employing capacitors with known temperature coefficients. A specially developed circuit is shown in Fig 9.8.

In this circuit, temperature compensation is achieved by sensing the crystal temperature with a thermistor and applying a correction to the crystal oscillator frequency by means of a varicap diode. This produces a capacitor with an effective temperature coefficient of up to ±20,000ppm. The thermistor is connected in a bridge circuit which produces an output voltage that varies linearly with temperature over the range 0–60°C. This is then connected to an operational amplifier with variable gain and offset; hence the centre frequency and rate of change of frequency with temperature can be independently adjusted.

As shown, the circuit compensates for oscillators with a negative temperature coefficient. To compensate for positive coefficients the thermistor and R3 should be swapped over.

The characteristic of the DKV6533A is fairly linear over the range 5–10V, so the output voltage should be set to give 7.5V across the varicap at room temperature. The gain should initially be set to unity.

An alternative to the DKV6533A is the BB110G. This diode, however, has a less linear characteristic over the same voltage range. This means that it will not be possible to obtain correction over the full range. The gain preset will need to be set up individually on each board if best results are to be obtained. This requires a means of accurately measuring small changes in frequency. As the frequency counter reference is not subjected to the same extremes of temperature as the board under test, and

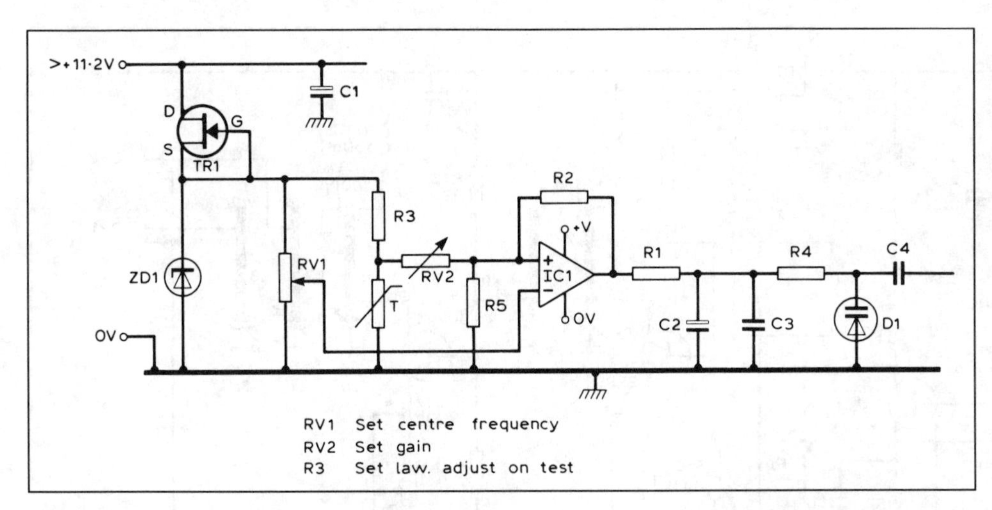

Fig 9.8. Temperature compensation circuit for the high-quality UHF source

Component values:

R1	4k7	C1	10µ
R2	100k	C2	1µ
R3	4k7	C3	1n
R4	47k	C4	100p
R5	100k	D1	DKV6533
RV1	100k	ZD1	7V5
RV2	220k	IC1	LF355
T	5k (25°C)	TR1	2N5457

the tests can be performed quickly, the demands on stability are not too great. With the gain set near unity, measure the frequency shift over a range of temperatures. These may conveniently be 5–10°C obtained at the bottom of a refrigerator, 20°C at room temperature and 40°C near to a room heater. Allow plenty of time (20–30min) for the temperature in the box to reach equilibrium. The change in gain can then be calculated. Once an adjustment has been made, the measuring process can be repeated. If it is found that the compensation required is non-linear, indicated by different error at high and low frequencies, the value of R3 can be changed. Table 9.7 gives various values for R3 together with the resulting value of voltage ratio at different temperatures.

Once the approximate value for RV2 is known, it is worth altering the value of R2 so that RV2 is near maximum resistance or replacing it with a fixed resistor and a smaller value preset in series to enable finer adjustments to be made. Any movement of the preset will affect the frequency of the oscillator, so it is worth using a better-quality ceramic trimming potentiometer.

Having gone to the trouble of temperature compensating the oscillator, it makes sense to ensure that the unit is rigidly housed in a draught-proof box. The oscillator must also be powered from a well-stabilised supply.

With temperature compensation and the other precautions, it should be possible to achieve a coefficient of 0.01ppm per degree or a drift of only 1kHz at 10GHz over the specified temperature range.

With the oscillator this stable, the only reliable way of accurately measuring the frequency is to use an off-air standard such as Droitwich, Rugby, or any other international frequency standard. Of course, the oscillator could also be locked to the same standard, to achieve really high stability!

Simpler, non-proportional, temperature control can be achieved by the use of a Murata Posistor crystal heater which simply clips onto the crystal case and is connected to the +12V supply rail. Type PTH507B01BM500N016 is suitable for the recommended HC18/U crystal. This device, after a short warm-up period, will maintain the crystal at a steady temperature regardless of ambient changes.

Table 9.7. Thermistor stabiliser for UHF source

Temperature (°C)	R3 = 2k	R3 = 5k	R3 = 10k	Thermistor resistance (kΩ)
0	0.13	0.23	0.43	13.0
10	0.18	0.31	0.53	9.0
20	0.26	0.41	0.64	5.7
30	0.34	0.51	0.72	3.9
40	0.42	0.60	0.79	2.7
50	0.54	0.70	0.85	1.7
60	0.67	0.80	0.91	1.0

A high-quality microwave source for 1.0 to 1.3GHz

An alternative source, with output in the range 1.0 to 1.3GHz, was originally designed [9] to provide two +10dBm (10mW) outputs at 1152MHz for use in a 1296MHz transverter with a 144MHz IF. Later work showed it was possible to use the same board, with a suitable crystal, anywhere in the range 1000 to 1400MHz with only minor changes in component values and output spectral purity. Versions of the board have since been produced to provide outputs between 700 and 1500MHz. At these two extremes it has been found necessary to change the length of some of the tuned microstrip lines in order to maintain resonance. PCBs are available from the Microwave Committee component service. There has been much recent interest in the board because of its use as a 1299MHz packet radio link transmitter when used with the companion amplifier design, described later, to form a low-cost 1W transmitter capable of being modulated at up to 9600 baud.

The circuit of the oscillator source is shown in Fig 9.9 and the component values in Table 9.8. The crystal oscillator section uses the Butler configuration used in the previous design. In the present design the oscillator operates with a crystal frequency between 90–110MHz, although by changing the value of C3 the circuit can be made to operate reliably between about 84–120MHz. Operation outside this range may require a change in the value of other components. A 9V integrated circuit regulator (78L09) stabilises the supply to the oscillator and limiter stages.

The third stage is a×2 multiplier. Input, in the frequency range 250–330MHz, is taken from the high-impedance end of the tuned circuit formed by L3 and C8 via C9. The output of TR3 is tuned by the coupled microstripline tuned circuits L4/C13 and L5/C14 to the range 500–660MHz.

A high-impedance tap on L5 couples the doubled signal to the base of the final multiplier stage, TR4. The output of this stage is tuned by three coupled tuned microstriplines; L6/C18, L7/C19 and L8/C20. The tuning range of this filter is highly dependent upon the type of trimmer capacitors chosen for C18, 19 and 20. Using the specified capacitors the filter will tune from 980 to 1400MHz, encompassing the range of second harmonics from the previous stage (1000 to 1320MHz).

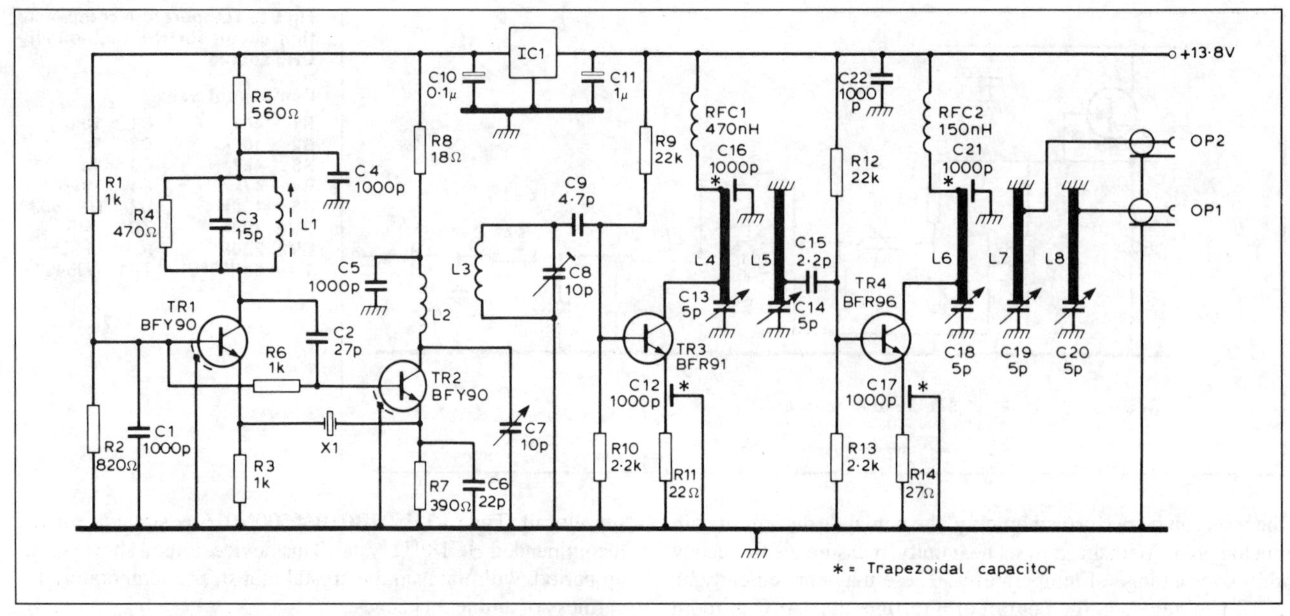

Fig 9.9. Schematic diagram of the microwave source G4DDK-001

Three stages of filtering are used to achieve a very clean spectrum at output 1. A slightly less clean output is available at output 2, since this output is taken from the second stage of the filter. Even so, the output here is more than adequate for use in a receive converter with output 1 being reserved for the transmit converter. Each output is at a level of +10dBm (10mW), but if only one output is required then this should be taken from output 1 and the track to output 2 cut where it leaves L7. A single output of +13dBm (20mW) should be available in this configuration. Fig 9.10(a) and (b) are analyser plots of the outputs of the board.

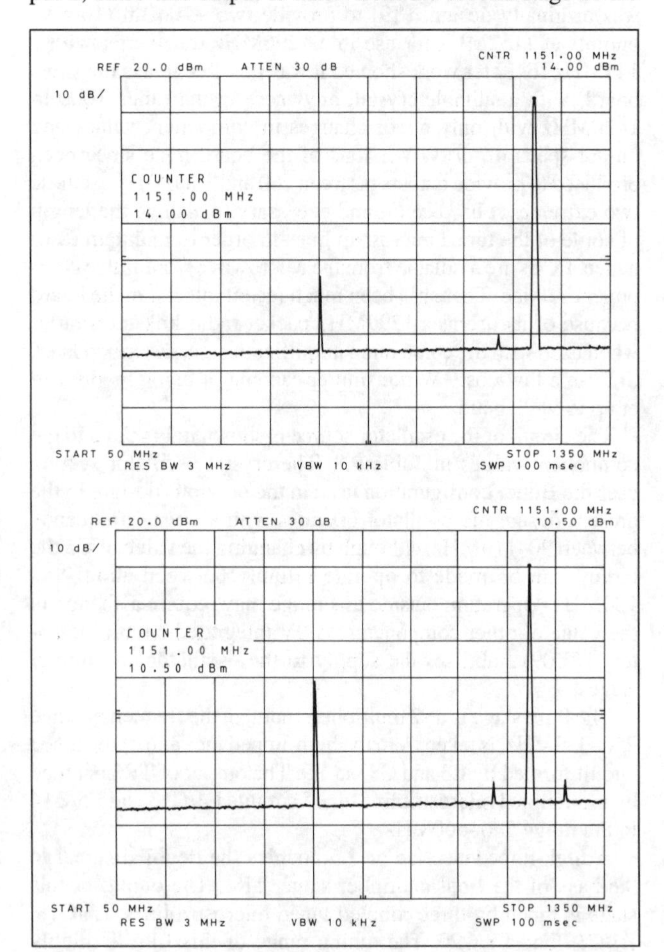

Fig 9.10. A spectrum analyser plot of the output from the microwave source G4DDK-001. (a) Output 1, output 2 terminated in 50Ω (b) Output 2, output 1 terminated in 50Ω. If output 2 is not required, cut the track near the line. Output 1 should then look like (a), but be at +13dBm

Table 9.8. Components list for the oscillator/multiplier unit G4DDK-001

RESISTORS	
R1, 3, 6	1k0
R2	820R
R4	470R
R5	560R
R7	390R
R8	18R
R9, 12	22k0
R10, 13	2k2
R11	22R
R14	27R
0.25W miniature carbon film or metal film	
CAPACITORS	
C1, 4, 5, 22	1000p high-*K* ceramic plate, eg Philips 629 series
C12, 16, 17, 21	1n trapezoidal capacitor from RSGB Component Service or Piper
C10	0.1μ tantalum bead, 16V working
C11	1μ tantalum bead, 16V working
C3	15p low-*K* ceramic plate, eg Philips 632 Series
C7, 8	5mm trimmer, 10p maximum
C13, 14	5mm trimmer, 5p maximum
C18, 19, 20	5mm trimmer, 5p maximum. Must be able to reach 0.9pF minimum, eg SKY (green) or Murata TZ03 (black)
INDUCTORS	
L1	TOKO S18 5.5t (green) with aluminium core
L2, 3	3t of 22swg tinned or enamelled copper wire, 3mm ID, turns spaced 1 wire diameter, height of coil 5mm above board.
L4–8	Printed on PCB
SEMICONDUCTORS	
TR1, 2	BFY90, available from Cirkit, Bonex, Piper etc
TR3	BFR91A, available from Cirkit, Bonex, Piper etc
TR4	BFR96, available from Cirkit, Bonex, Piper etc
IC1	78L08, Piper, STC Components etc
MISCELLANEOUS	
X1	5th overtone crystal in HC18/U case. Frequency of crystal = F_{out}/12

Fig 9.12. Photograph of the microwave source G4DDK-001 in the recommended-size die-cast box. The picture shows an early version of the source. L1 and L2 were laid out slightly differently in this version

PCB artwork for the UHF source is shown full size in Fig 9.11 (in Appendix 1). Fig 9.12 is a photograph of a finished unit. Construction is similar to the UHF source described earlier, with the trapezoidal capacitors being carefully soldered into the slots in the board as shown in the component overlay diagram Fig 9.13. Capacitors C18, 19 and 20 must be miniature 5mm diameter types such as SKY, Murata or Oxley. The use of larger 7mm trimmers such as the popular Dau or Philips types will inevitably lead to tuning problems. The circuit was designed to take the small types. This also applies to C7, 8, 13 and 14, where overcoupling and consequently tuning problems can also occur.

Again, alignment is straightforward. Connect DC power and check that the current drawn is no more than about 150mA. If significantly more, then check for short-circuits or wrongly placed components.

When all is well, proceed with alignment. Place an absorption wavemeter pick-up coil close to L1 and tune to the crystal frequency. A strong reading should be indicated on the meter. Peak the indication by turning the core of L1. Turn the oscillator on and off to check that it re-starts satisfactorily. If it doesn't, then turn the core of L1 about a quarter-turn and try again. Too close coupling of the wavemeter (ie too much power absorption) may also inhibit easy re-starting, so use the loosest coupling possible to give adequate indication.

Set an analogue (moving coil) multimeter to the 2.5V range (or nearest equivalent) and measure the voltage across R11. This should be no more than a few hundred millivolts. Peak the reading by tuning C7 and C8. Confirm that the frequency selected is three times the crystal frequency by placing the coil of the wavemeter close to L3. The reason for using an analogue meter is that digital meters do not measure in real time and therefore tend to show what you have just done, rather than what you are doing, whilst adjusting the trimmers: indeed, if a digital meter is used, it is possible to miss the increase in measured voltage which occurs as each circuit is brought to resonance – the tuning is sharp!

Transfer the meter leads across R14 and peak the reading by tuning C13 and C14. Again check the correct harmonic (twice the preceding stage) has been selected by using the wavemeter.

Finally connect a low power wattmeter (+10 to +20dBm full scale, ie 10 to 100mW) to the output and tune C18, 19 and 20 for a maximum reading. Confirm the correct harmonic (now the output frequency and twice the preceding stage) has been selected by using the wavemeter. It may now be necessary to go back and *slightly* re-peak the trimmers for an absolute maximum reading at the final output frequency.

Final setting of the frequency of the crystal oscillator can now be done by either using a known high-accuracy frequency counter or by connecting the source as the local oscillator of your 1296MHz converter and listening for a beacon whose frequency is known. L1 can then be adjusted to bring the signal onto the correct receiver dial calibration.

If difficulty is experienced in pulling the frequency to that marked on the crystal, then it is very likely you have a non-standard crystal. Pulling the frequency too far can result in the oscillator failing to restart after switching off and then on. The cure is to put a small-value ceramic plate capacitor, say 10–33pF, in series with the crystal by cutting the PCB track near the latter. If you have to do this modification, then keep the leads of the new capacitor short and use a zero temperature coefficient (NP0) capacitor, or frequency may drift unacceptably as the crystal oscillator warms up.

The source may be modulated in a similar manner to the previous design, although if the source is to be used as a transmitter in the 1.3GHz band then

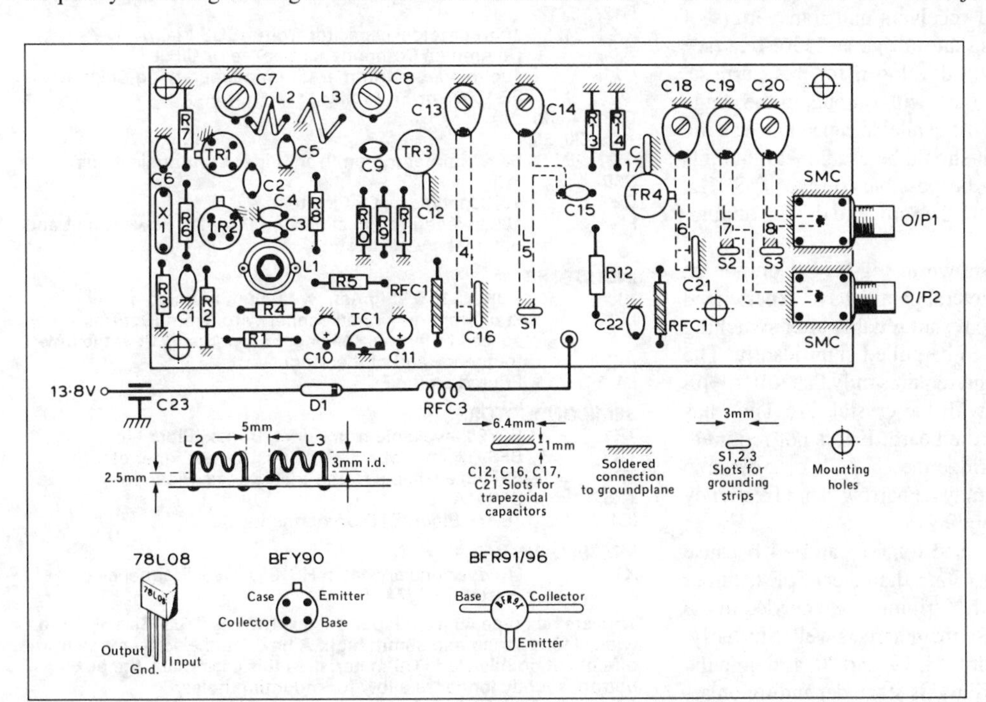

Fig 9.13. Layout diagram of the components for the microwave source G4DDK-001

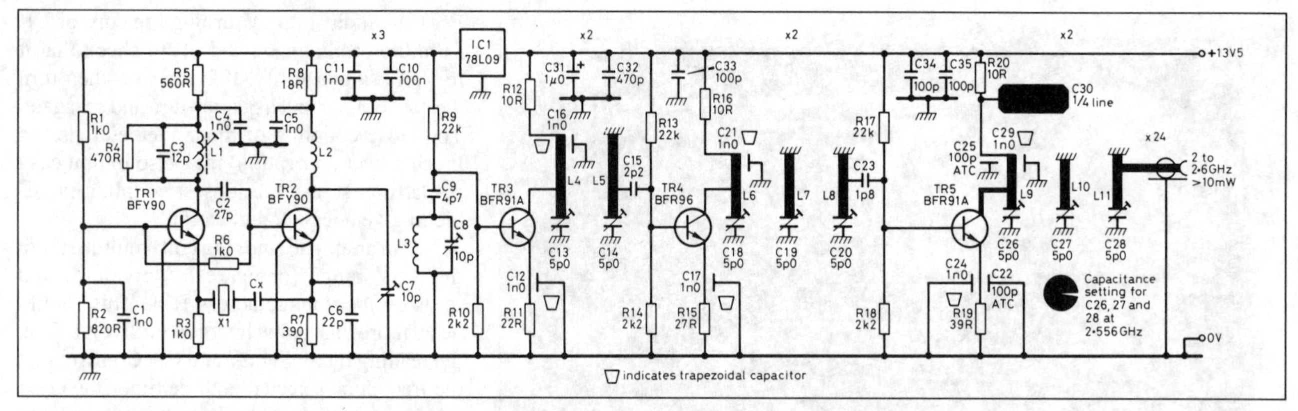

Fig 9.14. Schematic diagram of the 2.5GHz source G4DDK-004

greater deviation will be required than when multiplying to 10GHz. The modulator circuit shown in Fig 9.7 can be used.

If the source is to be used as a 1.3GHz beacon source, then the required 800Hz frequency shift can easily be obtained using this same circuit, although note that the 'sense' of the deviation should be such that the marker signal (carrier only) should be at the nominated beacon frequency and 'space' keys the beacon low in frequency by 800Hz, returning to 'mark' for each character element. With the circuit given, this means the keying voltage should be low for mark and high for space (conventional), ie mark corresponds to an earth condition on the keying lead.

A high-quality microwave source for 2.0 to 2.6GHz

This oscillator source [10] was a direct development of the previous design, but incorporates an additional multiplier stage to provide output in the 2–2.6GHz frequency range. This range encompasses such uses as the local oscillator in receive and transmit converters operating in the 2.3GHz amateur band. It could also be used as a low-power personal beacon or control transmitter operating in the same band under the terms of the UK licence.

Several 10GHz narrow-band receivers and transverters of modern design require local oscillator drive at 2556MHz (see later). This source was originally developed for this purpose. Versions of this unit have been built with outputs in the range 2176–2556MHz. On the basis of the available capacitance swing left in the tuned circuits, the unit should be usable with outputs from 2.0–2.6GHz. It may even be possible to reach 2.8GHz, opening up the possibility of using the source to drive a doubler to 5.6 or 5.7GHz.

The circuit of the source is shown in Fig 9.14. Component values are given in Table 9.9. Several people have experienced problems getting the oscillator to restart reliably after switching off if the crystal frequency has to be pulled significantly. The cure for this problem is simple and requires only that a 10–33pF capacitor be inserted in series with the crystal. Provision has been made for the capacitor on the board. If it is not required, then a *short* wire link should bridge the Cx pads or a ceramic 1nF capacitor used. The source may be narrow-band frequency modulated using the circuit in Fig 9.7.

SKY (green) trimmers were used for C13 and 14 because they were available. Any small (5mm diameter) 5pF trimmer could be used but, since green SKY trimmers are needed in the output filter, it makes sense to use them here as well. Similarly, SKY trimmers should be used for C18, 19, and 20, although the black Murata type could be used in this stage depending on required frequency.

Table 9.9. Components list for the oscillator/multiplier unit G4DDK-004

RESISTORS	
R1, 3, 6	1k0
R2	820R
R4	470R
R5	560R
R7	390R
R8	18R
R9, 13, 17	22k0
R10, 14, 18	2k2
R11	22R
R12, 16, 20	10R
R15	27R
R19	39R
0.25W miniature carbon film or metal film	
CAPACITORS	
C1, 4, 5, 11	1000p high-*K* ceramic plate, eg Philips 629 series
C2	27p low-*K* ceramic plate, eg Philips 632 Series
C3	12p low-*K* ceramic plate, eg Philips 632 Series
C6	22p low-*K* ceramic plate, eg Philips 632 Series
C9	4p7 low-*K* ceramic plate, eg Philips 632 Series
C15	2p2 low-*K* ceramic plate, eg Philips 632 Series
C23	1p8 low-*K* ceramic plate, eg Philips 632 Series
C31–33	470p medium-*K* ceramic plate, eg Philips 630 series
C22, 34, 35	100p low-*K* ceramic plate, eg Philips 632 series
C10	0.1µ tantalum bead 16V working
C12, 16, 17, 21, 24, 29	1n trapezoidal capacitor from RSGB Microwave Committee Components Service or Cirkit
C7, 8	10p miniature trimmer (5mm diameter), eg Cirkit 06-10008 or SKY (black)
C13, 14, 18, 19, 20, 26, 27, 28	SKY trimmer (green) from Piper Communications
C30	PCB track
Cx	10–33p, type as C3 (see text)
Ft	1000p feedthrough capacitor(s) for DC power input and optional crystal heater
INDUCTORS	
L1	TOKO S18 5.5t (green) with aluminium core
L2, 3	2t of 1mm dia tinned copper wire, 4mm ID, turns spaced to fit hole spacing. Exceptionally 3t at the low-frequency end of the range
L4–11	Printed on the PCB
SEMICONDUCTORS	
TR1, 2	BFY90, available from Cirkit, Bonex, Piper etc
TR3	BFR91A, available from Cirkit, Bonex, Piper etc
TR4	BFR96, available from Cirkit, Bonex, Piper etc
TR5	BFR91A
IC1	78L09, Piper, STC Components etc
MISCELLANEOUS	
X1	5th overtone crystal in HC18/U case. Frequency of crystal = $F_{out}/24$
Tinplate box type 45 from Piper. Also known as 7768. Size 55.5mm wide, 148mm long and 30mm high. A box could also be made from off-cuts of double-clad PCB material. In this case make the box bottom slightly longer to allow for mounting holes.	
Output socket, single-hole-mounting type SMA or SMB/C.	

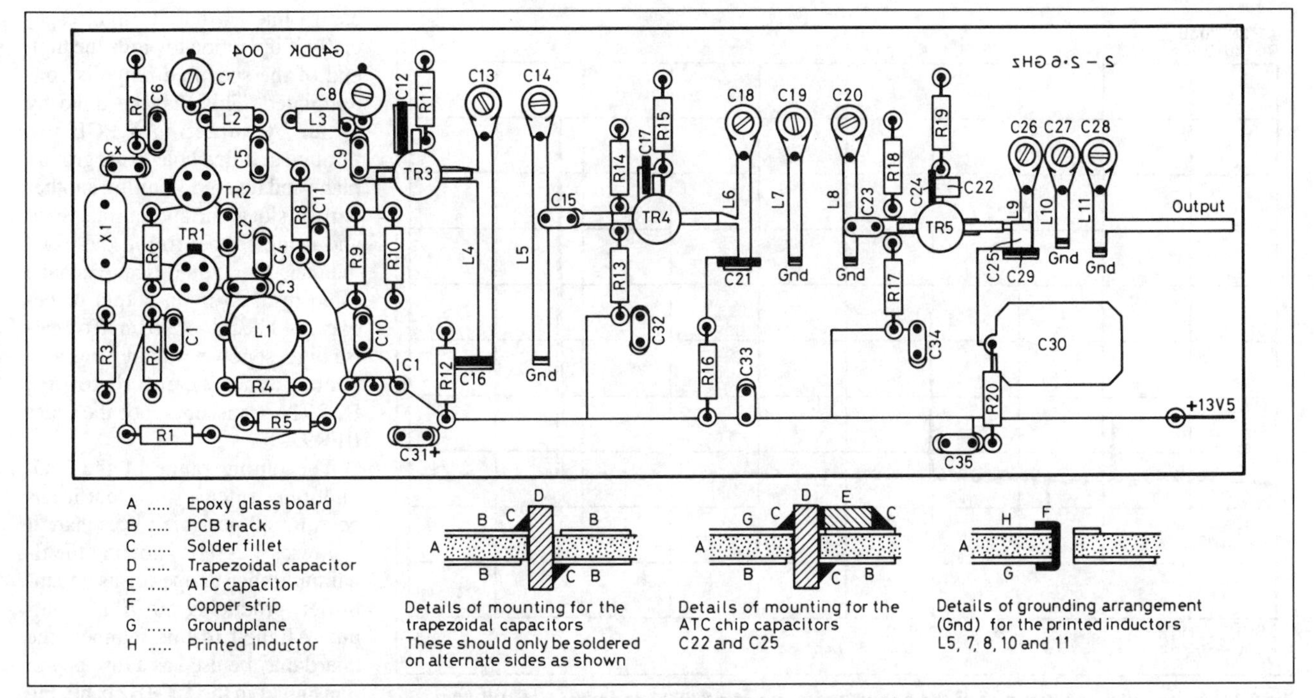

Fig 9.16. Layout diagram of the component for the 2.5GHz source G4DDK-004. Component side shown

The final multiplier uses a BFG91A and operates well at these frequencies, easily achieving the specified output. This type of transistor has two emitter connections and both must be thoroughly decoupled to ground to ensure stable operation; however, only one emitter has DC connection to ground through R19.

Additional decoupling has been provided in the collector supply circuit of this stage using a λ/4 open-circuit, low-impedance transmission line. The open-circuit at the end of the line is transformed to a short-circuit λ/4 away at the junction of R20 and the λ/4 choke line formed by the track from C29 to R20. Although this arrangement can only be optimum at one frequency, in practice the bandwidth of the line is such that it still remains effective over the entire range 2.0–2.6GHz.

The opportunity has been taken in this design to eliminate the collector supply chokes used in the 1152MHz unit. These chokes have caused problems in the past due to difficulty in obtaining the required axial-lead types specified. The 10Ω resistors are not as effective as chokes but the decoupling capacitor values have been carefully chosen to be as effective as possible in the frequency ranges encountered in their respective stages. The unit has been designed to fit into the type 45 tinplate box (also known as type 7768). This box is 55.5 × 148 × 30mm and retails at low cost compared with die-cast boxes of similar size. A full-size PCB layout is shown in Fig 9.15 (Appendix 1) and a component overlay in Fig 9.16.

The PCB material used for the oscillator is good-quality 1.6mm thick epoxy glass, double clad with 1oz copper. Slight differences in ε_r between board materials may affect performance slightly at the high-frequency end of the range. A high-quality PCB (G4DDK-004) is available, ready drilled, slotted and tinned from the Microwave Committee Components Service. Construction is similar to the designs described earlier, except that the board is soldered into a sheet metal (tinplate) box rather than being mounted in a die-cast box. It therefore does not require mounting holes.

The board is positioned so that the top of L1 will be 5mm below the rim of the box when soldered in place. Small areas of board need to filed away to clear the two overlapping corners of the tinplate box. The output socket mounting hole should be drilled to allow the spill of the connector to lie flush with the output track on the PCB. It is better to use a single-hole mounting socket and solder it flush to the outside wall of the box with its spill protruding into the box. The socket should be an SMA, SMB, or SMC (CONHEX) type. N types are too large and BNC connectors can prove unreliable at these frequencies. Also drill a hole in the same end of the box to take the feedthrough capacitor that will be used to bring DC power into the box. If a crystal heater is to be used then also drill holes for the power feed for this in the other end wall of the box.

If the unit is to be used as a transmitter then an additional feedthrough capacitor will be needed for the modulation input. The value of this capacitor will need to be carefully chosen not to unintentionally decouple the modulation.

Five short lengths of thin copper strip are used to ground the ends of L5, 7, 8, 10 and 11, as shown in the component overlay diagram Fig 9.16. Resistors and capacitors are fitted next, taking care to solder grounded leads both top and bottom of the board. These are followed by L2 and L3 and TR1 and 2 into place, making sure they are seated well down onto the board but leaving just enough room to solder the case lead of both transistors to the ground plane of the board. Trimmer capacitors are fitted next, using the shortest possible leads. The remaining semiconductors and IC1 are fitted and finally the crystal, seating it well down onto the board. It may be advisable to earth the case of the crystal, especially if a heater is to be used.

Alignment is quite straightforward provided it is done in a logical way, using similar methods to those outlined for the 1.0–1.3GHz source. It is possible to align the oscillator/multiplier with nothing more than a multimeter. However, this will give little information as to what frequency the unit is tuned to, or what output level has been achieved. As far as aligning the unit is concerned, the most essential items of test equipment are an analogue (moving-coil) multimeter and absorption wavemeter(s) to cover the range 106–2600MHz. A digital multimeter is not advised for tuning-up since it can give *very* misleading results,

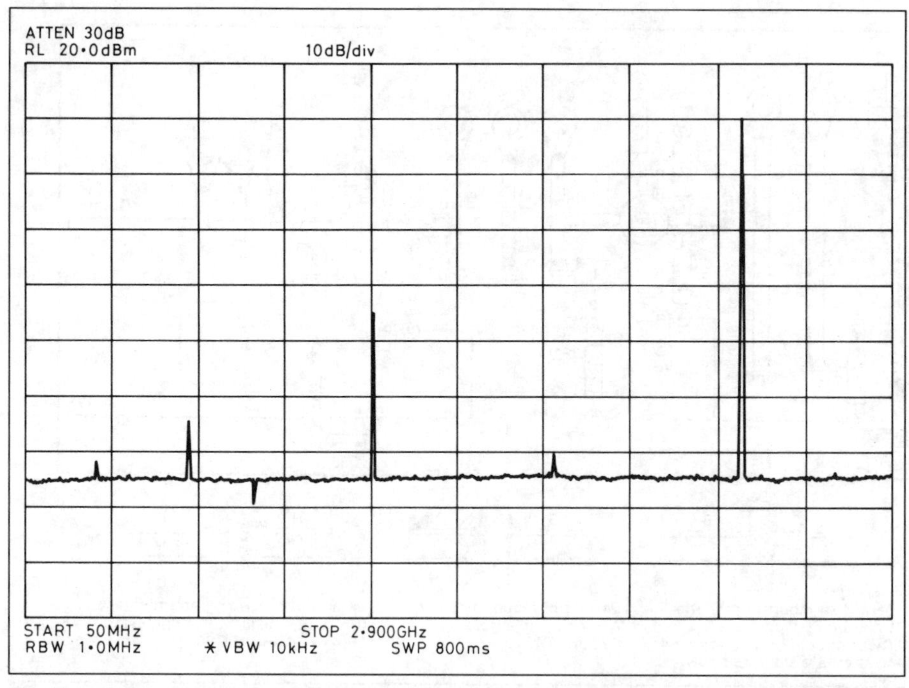

Fig 9.17. A spectrum analyser plot of the output spectrum of a 2.556GHz source, G4DDK-004

as discussed earlier. The wavemeter is also preferable to a frequency counter since it not only indicates the required output but also the presence of any unwanted signals which may not be 'seen' by a digital counter.

Due to component tolerances, there can sometimes be instability in the final multiplier stage, making the unit difficult to align: the remedy for this is to replace the BFG91A with a BFR91A and to add, on the upper surface of the board, additional decoupling in the form of two 100pF ATC (or similar) ceramic chip capacitors soldered across C17 and C24. It is possible with a little care to fit these extra capacitors between the ground plane and the silvering of the existing trapezoidal capacitor, making modification to the board unnecessary.

It has also been found that the centre trimming capacitor on the output filter (C27), which is shown 'staggered' in Fig 9.16 if turned round, so that all three capacitors are in line, will result in more output without significant degradation of spectral purity. It has sometimes been found advantageous to shorten the centre line by 1 to 1.5mm, especially if operation towards the high end of the specified range is contemplated. This may be done by either by soldering a PCB pin through a drilled hole in the ground plane and line or by cutting another narrow slot at the appropriate point and using copper foil in the usual manner. These modifications have often resulted in an output of between +11 and +13dBm. To realise highest power output it has also been found essential to use BFR91A transistors, not the older BFR91.

The tuning range of the final multiplier filter is such that it has been found possible to resonate it as low as 1.2GHz, allowing the final multiplier to operate as an amplifier, producing 50–70mW output. Aligned in this manner, the board may be used as a low-power transmitter in the 1.3MHz band. Fig 9.17 is a spectrum analyser trace of the output spectrum of this design when aligned for output at 2.56GHz.

A high-quality microwave source for 10.0 to 10.5GHz

This design by G3WDG [5, 11] requires +10dBm (10mW) or more drive from a 2.5–2.625GHz source [10] to provide a minimum of 50mW output (typically 80–100mW, dependent on the gain of individual GaAsFETs fitted) anywhere in the band 10.0–10.5GHz. The required level of drive can be supplied by the 2.0–2.6GHz source just described. The circuit is a good illustration of the use of modern surface-mount microwave components, MMIC and GaAsFET technology applied to microstrip PCB techniques. Employing the latest circuit concepts, it is nevertheless a design which is entirely practical for the home constructor. Although the components are small and a certain degree of care and attention to detail is required, none of the constructional methods are particularly difficult or out-of-the-ordinary. To make construction easier a special precision PCB, using PTFE-glass substrate, and other components are available as a 'short kit' from the Microwave Committee Components Service. The circuit, shown in Fig 9.18, comprises a MMIC amplifier at the drive frequency. This drives an active (GaAsFET) ×4 multiplier which is followed by a filter and two stages of amplification at the final frequency.

The output of the broad-band MMIC amplifier is matched into the impedance of the multiplier FET by a lumped-circuit network consisting of L2 and L3. L3 is also used to feed the negative gate bias to the FET

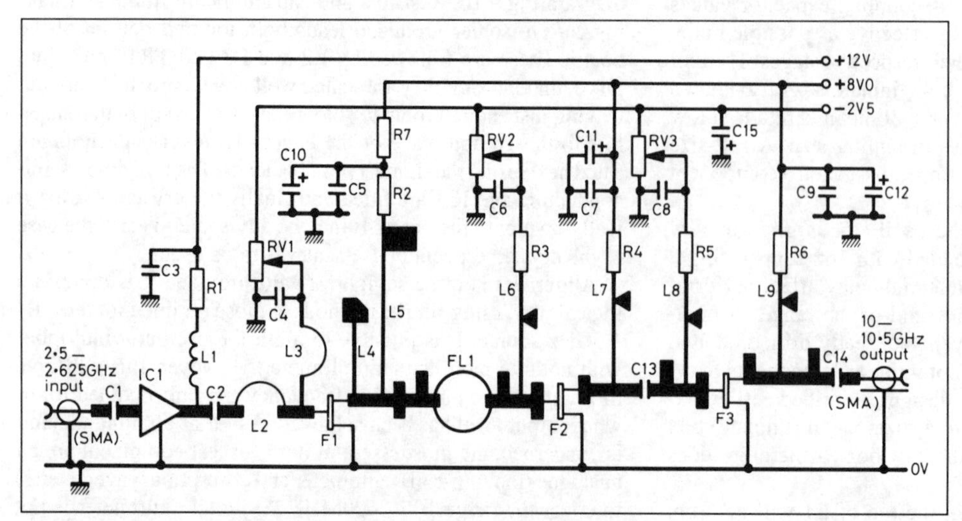

Fig 9.18. Circuit of the G3WDG-001 2.5 to 10GHz multiplier/amplifier

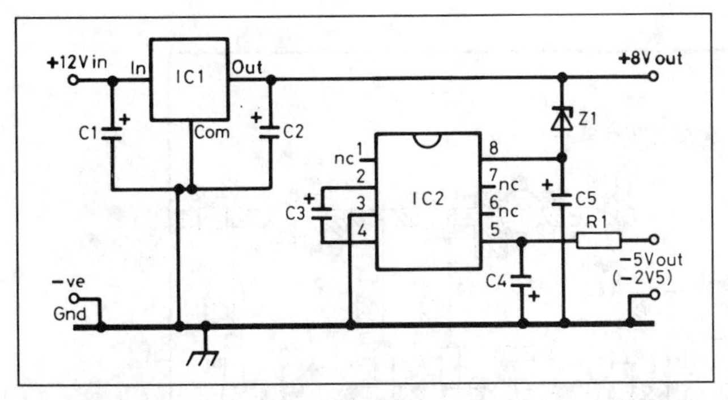

Fig 9.19. Positive and negative voltage regulator circuits for the G3WDG-001 module (G4FRE). Component values: IC1, μA7808; IC2, ICL7660PCA; Z1, 3V0 or 3V3, 400mW Zener diode; R1, 3k3 0.25W metal film; C1 1μF tantalum bead, 16V wkg; C2, 0.1μF tantalum bead, 10V wkg; C3, C4, 22μF tantalum bead, 10V wkg; C5 10μF tantalum bead, 10V wkg; PCB, G4FRE-023

from RV1, which sets the optimum operating bias for the multiplier. The output at the FET drain consists of a series resonant circuit at 2.5GHz, followed by some matching elements at 10GHz. The series resonant circuit is formed by L4 and the printed capacitor identified by the cut corner. The function of this circuit is to 'short-circuit' the input frequency (present at the drain) to ground. This improves multiplier efficiency considerably.

The drain bias circuit consists of a λ/4 choke, L5, connected to a λ/4 stub. The tip of the stub is at very low impedance at its resonant frequency (10GHz). This is transformed to a high impedance by L5. Configured this way, there is very little disturbance to signals on the microstrip line from the bias network. This type of bias network is very effective at its operating frequency but needs additional decoupling at lower frequencies. This is accomplished primarily by R2 and C5 which load the drain resistively at low frequencies, giving broad-band stability. C10 is used for further decoupling at very low frequencies. This configuration is used widely in this and other designs described later. In many locations the λ/4 stub is replaced by a small triangular element (known as a *radial stub*) which has the same properties as a conventional stub, ie very low impedance at the tip of the element. It is, however, very much smaller, resulting in a more compact layout. R7 is used to set the drain voltage to an optimum value for best multiplier performance.

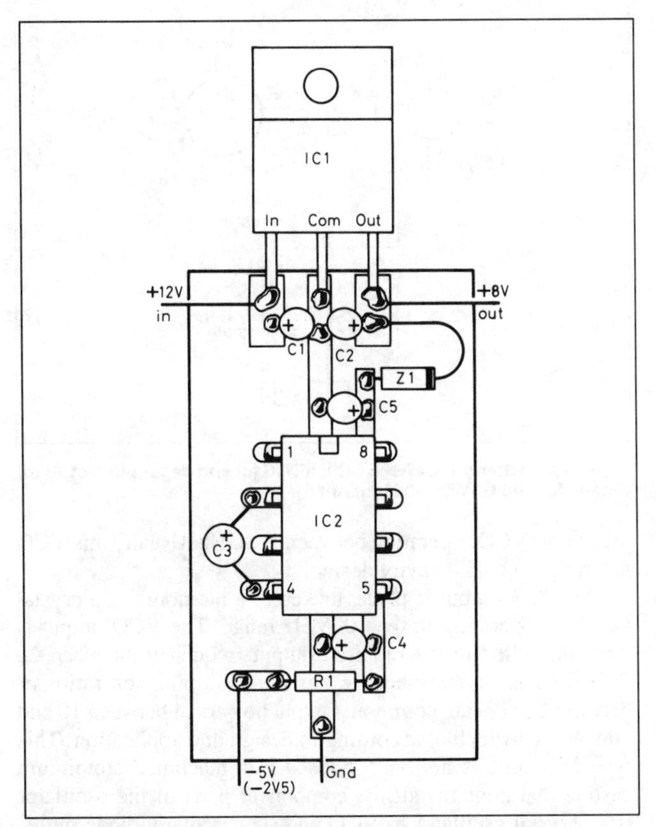

Fig 9.20. Layout for the G3WDG-001 regulator circuits (G4FRE). All components are mounted on the track side of the PCB. Board size approximately 32 × 13mm

The output from the multiplier consists of several harmonics, but principally the wanted fourth harmonic of the drive frequency. The output is fed into a resonator filter FL1 by means of a probe. The filter is tuned by means of an M4 screw, locked in place with a lock-nut. Filtered output is coupled into the gate of the first amplifier F2 by means of a second probe. The lengths and spacings of the probes have been adjusted for optimum coupling and *Q*-value.

GaAsFETs F2 and F3 form a cascaded two-stage amplifier yielding about 20dB gain. Each stage has gate and drain bias circuits similar to those of the multiplier stage. DC blocking and RF coupling are accomplished by means of special 2.2pF capacitors (ATC series 100 or 130): ordinary SMD capacitors used for coupling or decoupling at lower microwave frequencies are definitely unsuitable in these positions.

The drain supply to all three GaAsFETs should not exceed +8V and the operating point of each stage is set individually by means of a network of fixed and variable resistors across a –2.5V rail. Both the +8 and –2.5V supplies are derived from a nominal +12V supply, using a small auxiliary supply regulator PCB, the circuit of which is shown in Fig 9.19 and the layout in Fig 9.20.

The layout of the main PCB and the component positions are shown in Fig 9.21. Table 9.10 lists the components needed for the unit. The PCB was designed to fit into a standard tinplate box for immediate screening and it was intended that both the drive source and the multiplier/amplifier should be mounted inside a larger screening enclosure. The 'boxes-within-boxes' approach to screening is strongly recommended for all microwave equipment, regardless of its operating frequency. Not only does this practice provide additional screening to prevent noise pick-up and possibly instability, but it also provides an additional degree of thermal stability for the circuits concerned.

For more detailed methods of construction and alignment, see [11]. Fig 9.22(a) and (b) are spectrum analyser traces of the output of this design, showing the high spectral purity attained by the application of adequate filtering.

If the initial crystal frequency in the drive source is 106.5000MHz, then the final output frequency will be 10,224MHz which is suitable for either a 10,368MHz receive converter or transmit converter using a 144MHz IF. Alternatively, a crystal of 108.0000MHz will yield output at 10,368MHz which enables use as a transmitter or signal source (eg beacon) in the current narrow-band section of the 10GHz band. It should be stressed that the unit will, with a suitable drive frequency, cover the whole of the band, from 10.0GHz to 10.5GHz.

Voltage-controlled oscillators (VCOs)

Another type of oscillator source is starting to appear in amateur microwave designs. For many years professional microwave system engineers have used a versatile phase-locked loop technique for generating signals anywhere in the range from 1 to over 20GHz. The technique uses a voltage-controlled

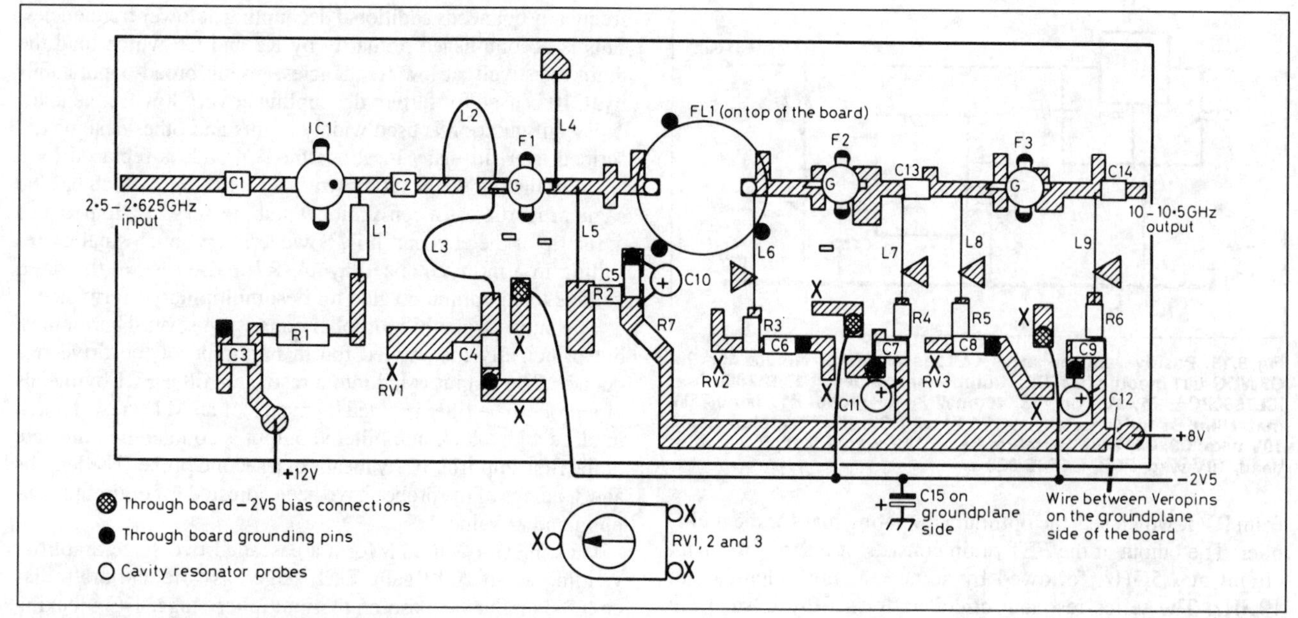

Fig 9.21(a). Layout of the G3WDG-001 circuit

Table 9.10. Components list for the G3WDG-001 2.5 to 10GHz multiplier/amplifier

RESISTORS

R1	39R ¼W carbon film
R2–6	47R SMD, 0805 size (*)
R7	220R (may need to be adjusted on test) SMD, size 0805 (*). Alternatively, use a wire-ended ¼W carbon or metal film resistor, stood on end
RV1–3	2k2 miniature horizontal preset with legs bent out at 90° and cut to fit board. These are fitted after all other parts except the GaAsFETs. This is necessary due to other parts being located below them. Connect to the points marked 'X' on the layout diagram. No tracking is provided between the bias connection points. This should be done with fine, insulated wire

CAPACITORS

C1–9	220p SMD, 0805 size (*)
C10–12	10µ tantalum bead or miniature electrolytic, 10V wkg
C13, 14	2.2p ATC chip capacitor, series 100 or 130 (*)

INDUCTORS

L1	8t 0.315mm diameter enamelled copper wire (ECW) (*), close wound and self-supporting
L2	16mm length of 0.315mm diameter ECW (*) formed into a hairpin shape and laid flat on the board. 1mm at each end used for soldered termination
L3	As L2, but 19.5mm long
L4, 5	Straight length of 0.315mm ECW (*) between stub and track
L6–9	Straight length of 0.2mm silver or tinned copper wire between radial stub and track as shown. A single strand of braid from a scrap length of RG214 cable is suitable. Mount flat to PCB

SEMICONDUCTORS

F1–3	P35-1108 GaAsFET (Birkett 'black spot') or similar
IC1	Avantek MSA0504

MISCELLANEOUS

FL1	Silver-plated brass cavity resonator (*). Tuning by means of an M4 screw with lock-nut. Two probes, each 4.7mm overall length as shown

PCB pins, RS Components 433–864, or Vero 1mm dia, 1.5mm head dia (*)
Two SMA sockets, flange fitting
Tinplate box type 7754 (37 × 111 × 30mm) from Piper Communications
Solder feedthrough capacitors, 1n to 10n or Filtercons
Positive/negative supplies: positive from 7808 IC, negative from ICL7660 voltage converter on G4FRE-023 PCB (*)

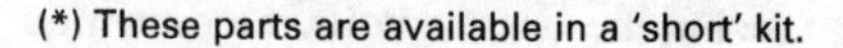

(*) These parts are available in a 'short' kit.

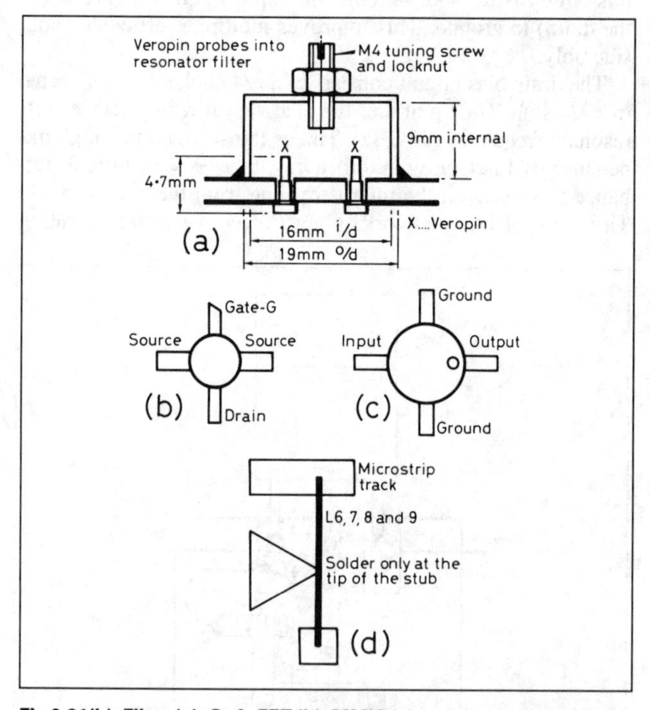

Fig 9.21(b). Filter (a), GaAsFET (b), MMIC (c) and decoupling stub (d) details for the G3WDG-001 circuit

oscillator (VCO) operating between 1–2GHz. Usually this VCO is a rugged coaxial cavity design.

The VCO output is phase locked to a harmonic of a crystal oscillator operating in the 100MHz range. The VCO output is then multiplied to the required output frequency, between 2–20GHz, using a step recovery diode. Multiplication ratios of between 2 and 9 are common. Output powers of between 10 and 50mW are available, according to design and application. This type of source is normally housed in a machined aluminium casting that contains all the component parts of the oscillator from crystal oscillator to VCO and step recovery diode multiplier. Several companies manufacture PLL sources, including GEC in England and Frequency West, CTI and California

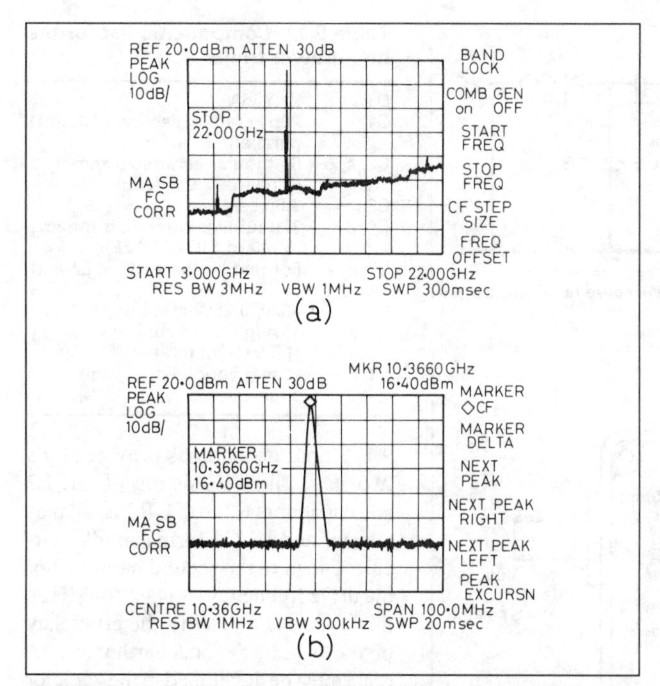

Fig 9.22. Spectrum analyser traces of the output from the G3WDG-001 multiplier/amplifier. (a) The output between 3GHz and 22GHz showing the remarkably low level of sub-harmonics and harmonics. The rising noise level is due to the spectrum analyser, not the multiplier! The drive source was the G4DDK-004 board at a level of +10dBm. (b) Close-in view of the output spectrum showing the clean output obtainable with this design

Microwave in the USA [12, 13]. Surplus sources of this type have occasionally appeared at amateur radio rallies in the last few years, especially in the USA. American radio amateurs, in particular, have been quick to use them in their microwave equipment designs. With some types, the phase noise on the output signal may be rather too high for low-noise applications. However, PLL sources are sure to become more popular as they become more easily available to radio amateurs. It seems unlikely that many amateurs would attempt to build their own source of this type due to the mechanical complexity of the design.

VARACTOR DIODE MULTIPLIERS

Three of the four microwave local oscillator sources described may be used directly in receivers or as low-power transmitters. In the case of the 1.3, 2.3GHz and 10GHz sources, considerable emphasis has been placed on the interstage band-pass tuning and the filter at the final frequency, so that the output from any of these sources is essentially free from sub-harmonics (frequencies below the final output frequency) and harmonics (multiples of the individual multiplier stage frequencies above the final output frequency), that is unwanted products. These have been reduced to at least –40dBc (40dB or 10,000 times down on the carrier level).

Whilst all the sources described give output of high spectral purity, the output may need to be multiplied further (or mixed) to be of use in either microwave transmitters or receivers. This can be carried out in one of two ways:

1. Low-level, active multiplication and/or mixing (transistor or FET) followed by amplification and filter(s). The use of active multipliers and filters has already been illustrated in the two microwave sources for 1.3 and 2.3GHz, whilst multiplication, amplification and filtering has also been amply illustrated by the 2.5–10GHz source, also described earlier.
2. Amplification at UHF, followed by a high-level, passive varactor multiplier (or mixer) and filter(s).

Notice that, in both cases, emphasis is placed on the most important feature of the chain: the filter. It must be stressed again that it is vitally important that adequate filtering is applied to all microwave sources to ensure removal of the unwanted products of multiplication or mixing.

The main disadvantage of the first method is that it may be difficult for amateurs to provide adequate, cost-effective power amplification at microwave frequencies, at or above 2.3GHz, and especially to power levels above a few tens of milliwatts. The first method, active multiplication followed by amplification and filtering, has only become feasible as far as amateurs are concerned by improved availability and falling prices of microwave bipolar transistors, microwave monolithic integrated circuits (MMICs) and GaAsFETs. This method is likely to gain popularity as suitable consumer devices become more commonplace.

The main advantage of the second method is that it is relatively easy and inexpensive to provide high-gain UHF amplification up to a level of tens of watts (if necessary), which can then be applied to a passive varactor multiplier to give a few watts output at the desired higher frequency. Note that the varactor *multiplier* does not require a power supply other than the drive power. It is essential that a good filter immediately follows the multiplier stage. A little additional suppression of unwanted products may be expected from the selectivity of an antenna and the input/output circuits of the multiplier itself, but most of the rejection of unwanted products *must* be provided by the filter(s).

An important disadvantage of the second method is that, unless carefully tuned and matched, varactors readily become unstable and their output breaks up into a 'comb' of broad-band spurious outputs, not necessarily related to the input frequency. Furthermore, varactors are two-port devices and their input and output tuning and matching are invariably interactive and power dependent. It is thus very difficult to cascade two or more varactor stages without the considerable risk of instability due to these interactions. Interaction may be reduced by using matched, resistive *pads* (attenuators) between the stages. This will of course reduce efficiency which, with high-order multiplication, is already low.

Power output is approximately proportional to $1/n$, where n is the multiplication factor. Thus, it can be seen that efficiencies of about 50% and 35% can be obtained from doublers and triplers respectively: at higher orders of multiplication, efficiency drops rapidly to low levels. As efficiency falls, so the varactor and its associated circuits dissipate more of the input energy as heat. Unless adequately heatsinked, the device is likely to fail early in its service life.

Lastly, it is becoming increasingly difficult for amateurs to obtain suitable varactor devices, many manufacturers having discontinued varactor manufacture (at least for the lower microwave bands) in favour of power GaAs devices which are much more efficient as active multipliers.

Despite all these disadvantages, passive varactor multipliers – even if only in the form of very-low-power Schottky diode multipliers in receiver local oscillator chains – still have a place in modern microwave technology. Varactors are still widely used professionally to generate frequencies high into the millimetre-wave range, certainly up to at least 100GHz. Briefly described next are some examples of practical amateur passive multipliers and mixers.

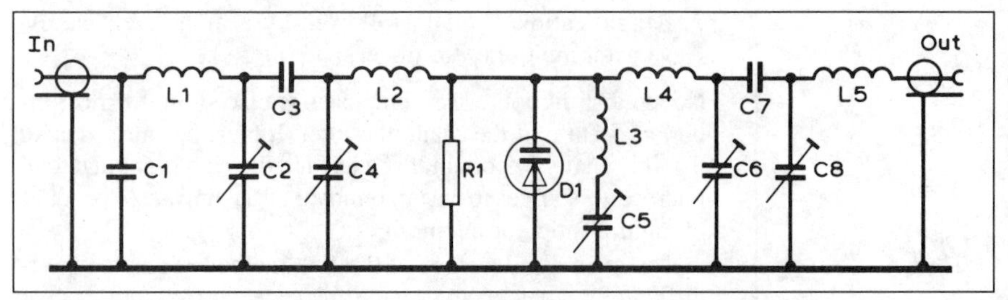

Fig 9.23. Schematic diagram of the 1152MHz/1296MHz tripler by Microwave Modules and G8AGN

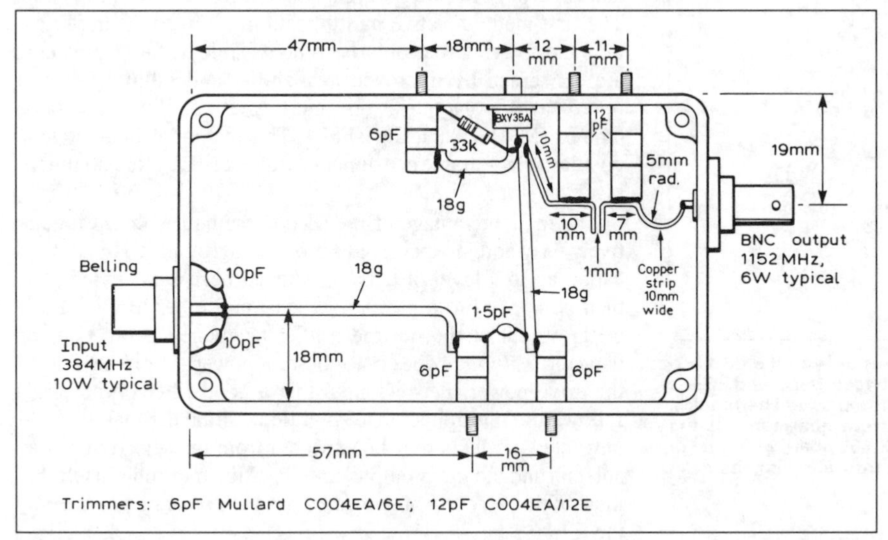

Fig 9.24. Layout of the 1152/1296MHz tripler using a simplified input circuit and Belling-Lee coaxial input socket

A high-power multiplier for 1296 or 1152MHz

The varactor tripler illustrated in Figs 9.23 and 9.24 was originally designed and manufactured in the UK by Microwave Modules for use as a multiplier from 432MHz to 1296MHz and is one of the most successful and reliable designs ever produced. It was also available, retuned, for 384MHz input and 1152MHz output. Commercial production has ceased but details were kindly provided by the former manufacturers.

The circuit is given in Fig 9.23 and the layout should be obvious from Figs 9.24 and 9.25. A list of parts is given in Table 9.11. Input matching is provided by a two-stage filter consisting of L1, L2 and capacitors C1 to C4. R1 is the bias resistor and L3/C5 forms an idler circuit. Tuned to the second harmonic of the drive frequency (864 or 768MHz), this circuit is crucial to the efficiency of the tripler. Second harmonic currents must be developed in the varactor so that mixing with the fundamental can take place to produce usable power at the third harmonic. L4, L5 and C6, 7 and 8 filter the output and provide matching to 50Ω.

Table 9.11. Components list for the high-power tripler

D1	BXY35A
C1	20p ceramic disc (two 10p in parallel)
C2, 4, 5	6p tubular ceramic trimmer (Mullard C004EA/6E)
C3	1.5p ceramic
C6, 8	12p tubular ceramic trimmer (Mullard C004EA/12E)
C7	Formed by the ends of L4 and L5
R1	33k0, 0.25W carbon
L1–3	18swg tinned copper wire (TCW) bent to fit
L4, 5	20swg copper strip 10mm wide

Ideally the tripler should be aligned using a swept frequency and swept power source. However, when the multiplier is to be used at a fixed drive level, as would be the case if used with a single frequency source, swept power can at least be dispensed with. With care, this tripler can be aligned without the swept frequency source as well.

Connect the tripler into a test set-up such as that shown in Fig 9.26. The coupler at the output ensures that the varactor only 'sees' the low VSWR of the 50Ω power meter. If the wavemeter were directly coupled into the varactor, there is a possibility that instability would be introduced when the wavemeter is tuned. First adjustments should be made with low power input to avoid excess dissipation in the varactor. Dissipation will be reduced as alignment proceeds. C2 and C4 are adjusted for minimum VSWR at the input. The wavemeter is tuned to the output frequency and C6 and C8 adjusted for peak reading on the wavemeter power indicator. The wavemeter is retuned to the second harmonic frequency and C5 is adjusted for minimum reading on the wavemeter power indicator. This should correspond to maximum output as indicated on the power meter.

Go back to the beginning and slightly readjust all variables for maximum output on the power meter. Increase the input power and go through the alignment again at the new power level. It should be possible to drive the multiplier with up to about 10W input, but only if at least 3W is measured at the output, otherwise the varactor may be damaged by excess dissipation.

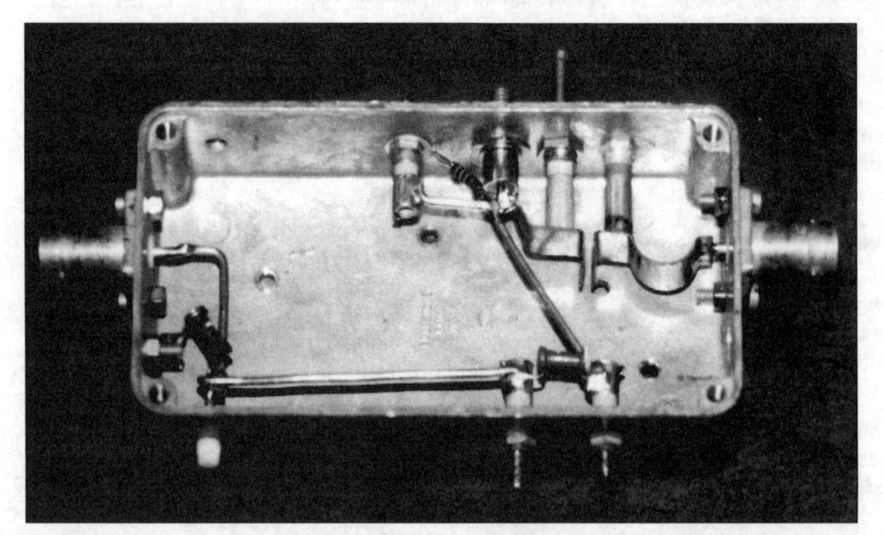

Fig 9.25. Photograph of the 1152/1296MHz tripler. Note that the input socket and the connection to C1/L1 are different to those shown in Fig 9.24; this is not significant

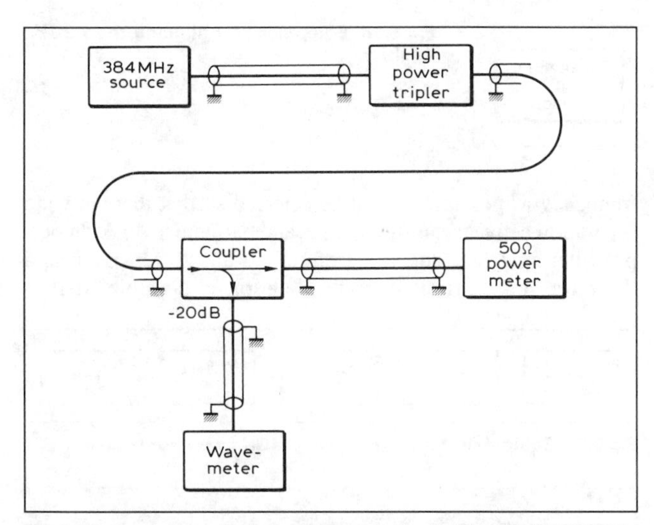

Fig 9.26. Recommended alignment set-up for the 1152/1296MHz tripler

A simple doubler for 2.3GHz

Fig 9.27 shows the layout and leading dimensions for a simple doubler from 1160 to 2320MHz (or 1152 to 2304MHz), designed by DF7QF [14]. It is capable of producing about 2.5W at 2320MHz for an input of 5W at 1160MHz. With better heatsinking, the varactor is capable of about 6W output for a maximum of 12W input.

The walls and bottom of the trough which house the doubler are made from 1.6mm double-sided glassfibre-epoxy PCB material and so can be easily made at home. The pieces are cut to size and the necessary holes drilled. Spacers C and D are made from the same material, but with the copper cladding removed. Part H is a 25mm-long piece of 7mm OD tube with two slits at each end, allowing it to be compressed slightly. Nuts E, F, G and tube H are soldered in place.

BNC1 and BNC2 are UG1094/U single-hole fixing sockets. BNC1 has the protruding PTFE insulation removed and kept for later use. The spill is then cut to leave only 3mm showing. A 6mm diameter disc is soldered centrally onto the cut-off spill. BNC2 has approximately 40mm of 2mm diameter silver-plated copper wire soldered to its spill. A 38mm-long piece of 6mm OD tube is soldered centrally to the body of the socket, the piece of PTFE salvaged from the other socket being inserted into the open end of the tube to maintain concentricity. The centre conductor is then cut to leave 4mm protruding from the tube. A 5mm diameter brass disc is soldered onto the end of the conductor to complete the input section, which is then inserted into part H.

The output connector mounting nut and the input and output tuning screw guide-nuts are soldered to the outer walls of the trough and the output socket and tuning screws, complete with their locknuts, are screwed into place.

The diode mounting screw is made from an M6 × 12 brass or copper screw with a 1.8mm hole drilled centrally in its end. Line L is made from a piece of 6mm OD brass or copper rod with a 1.75 to 1.8mm hole drilled through it in the position shown. Part K is a 10 × 10 × 1mm thick brass plate soldered inside the trough to provide a bearing support for the diode mounting screw. The diode mounting screw should be fitted into this plate and then passed through the 6mm hole in the trough wall. The plate is then soldered in place.

The spacers are fitted to the line L and the assembly placed in the trough. Using the shank of a 1.8mm drill as a jig, both diode mounting holes are aligned. Parts C, D and L are fixed in place using cyanoacrylate glue. The bias resistor is soldered into place, using the shortest possible lead length. The diode may now be mounted using the fingers to tighten the mount which is then locked into place with the locknut. The heatsink is fixed to the diode mount by means of another nut.

The alignment method requires a similar test set-up to that described for the previous multiplier, although alignment is simpler since there is no idler circuit to adjust. The suggested test and alignment set-up is shown in Fig 9.28. Note that this design does *not* have an integral filter in its construction, so that it is essential to connect a suitable band-pass filter close to, or directly at, the output socket.

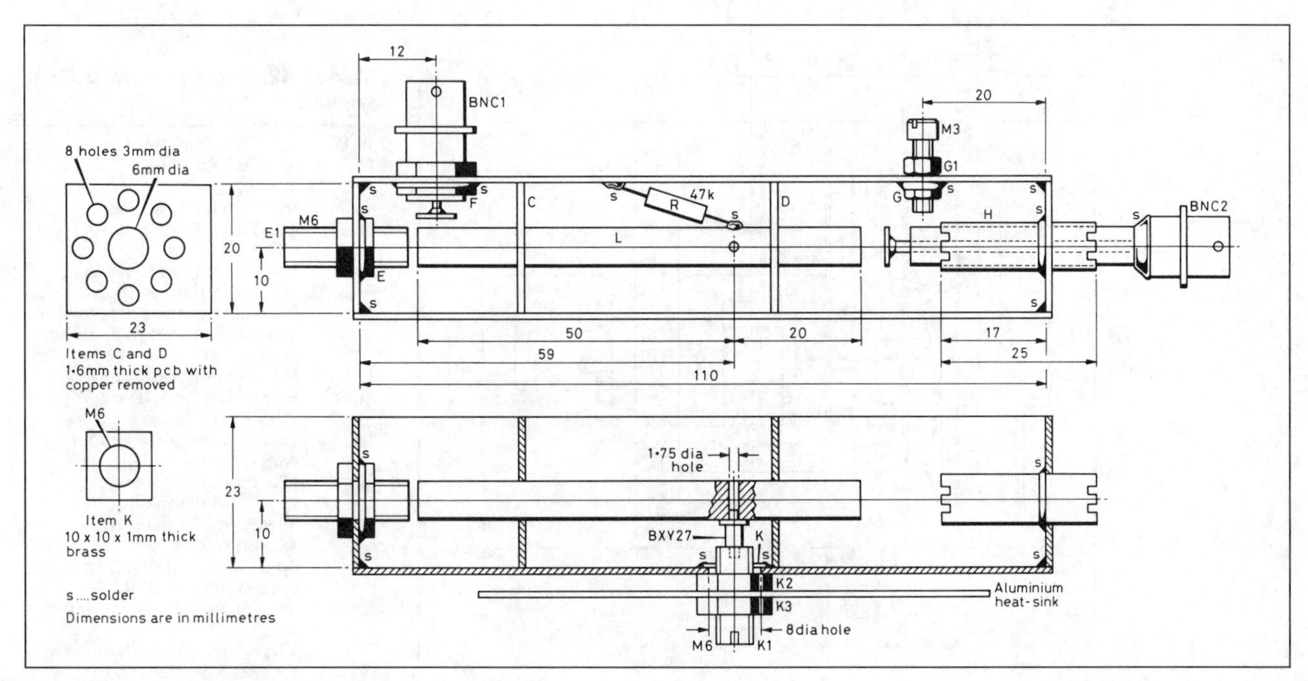

Fig 9.27. Construction of the 2.3GHz doubler

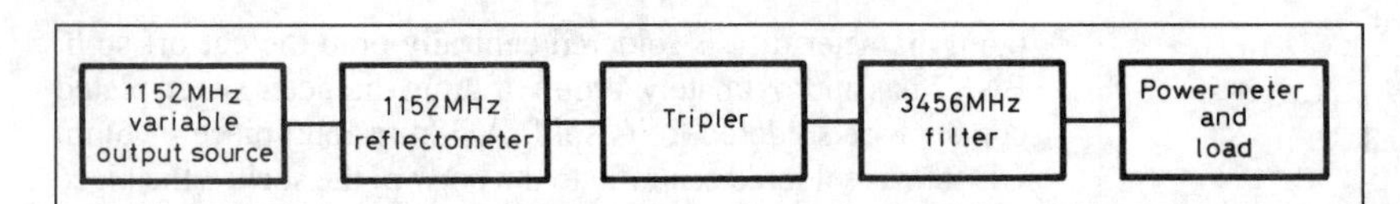

Fig 9.28. Equipment for aligning the 2.3GHz doubler

A varactor tripler for 3.4GHz

Similar principles have been used in this tripler design, developed by G0BPU from a design in reference [15]. The typical performance of a carefully constructed unit should be about 1.25W output at an efficiency of about 30% when using a BXY28e varactor diode. For full constructional details and alignment procedure, see reference [5]. An overall dimensioned diagram is shown in Fig 9.29, a suggested alignment set-up in Fig 9.30 and a photograph of the finished unit in Fig 9.31. The labelled parts in Fig 9.29 are listed in Table 9.12. Note that although a single-pole output filter is included in this design, the output should still be connected to a high-quality filter before being connected to an antenna.

Waveguide varactor multipliers

At higher frequencies, such as 5.7 or 10GHz, waveguide sizes are such that it becomes practical (and often convenient) to construct varactor multipliers and their associated filters actually within either a single section of waveguide or two short lengths bolted together, if it is more convenient to construct the multiplier and filter separately.

Waveguide has the added advantage that it will not support frequencies below its cut-off frequency: this can have the beneficial effect of reducing sub-harmonics which are below the cut-off frequency still further, leaving the output filter to clean up only those frequencies which lie above cut-off.

In designing and building such multipliers, as high a drive frequency as possible should be selected so that there is wide separation between sub-harmonics and harmonics (making output filtering easier and more effective) and lower-order, and therefore more efficient, multiplication is possible. If it is

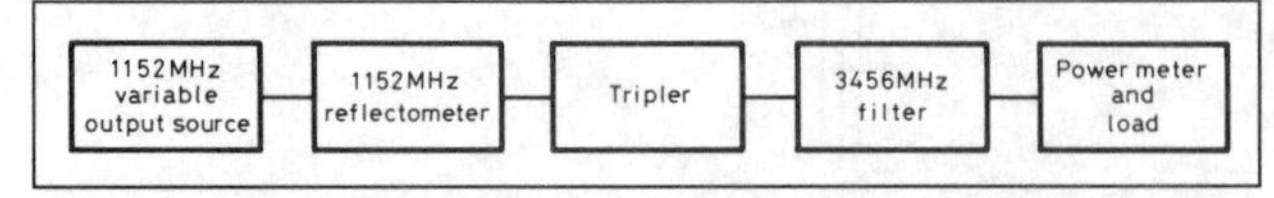

Fig 9.30. Equipment for aligning the 3.4GHz varactor tripler

Fig 9.31. Photograph of the completed 3.4GHz tripler

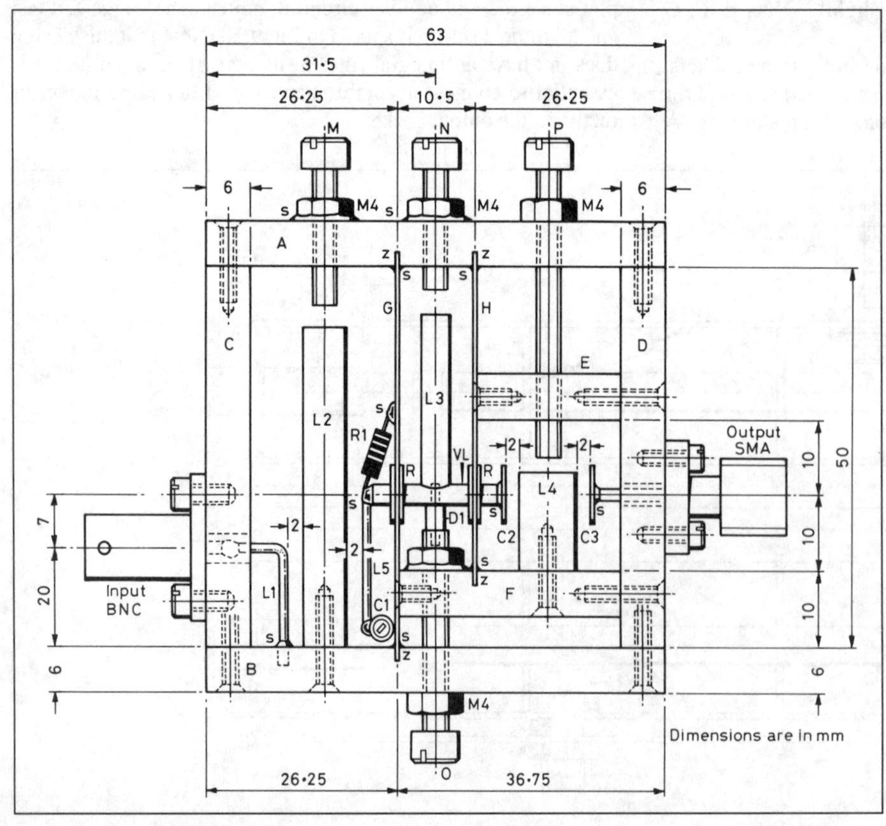

Fig 9.29. Overall dimensions of the 3.4GHz varactor tripler

Table 9.12. List of parts for 3.4GHz tripler

A, B	63 × 25 × 6mm bar
C, D	50 × 25 × 6mm bar
F	30.5 × 25 × 6mm bar
G	54 × 25 × 0.5mm sheet
H	44 × 25 × 0.5mm sheet
I, J	63 × 62 × 0.5mm sheet
M	M4 × 20mm brass screw
N	M4 × 20mm brass screw
O	M4 × 30mm brass screw with 1.6mm dia by 3mm deep hole in one end
P	M4 × 40mm brass screw
Q, R	8mm OD × 3mm ID × 2mm thick PTFE spacer
L1, 5	16swg silver plated copper wire
L2	42 × 6mm diameter rod
L3	23 × 4mm diameter rod
L4	13 × 8mm diameter rod
VL	18 × 3mm diameter rod
C1	6p ceramic trimmer 3mm maximum diameter
C2, 3	8mm diameter × 0.5mm thick disc
R1	47kΩ 0.5W carbon film resistor
D1	BXY28e varactor diode

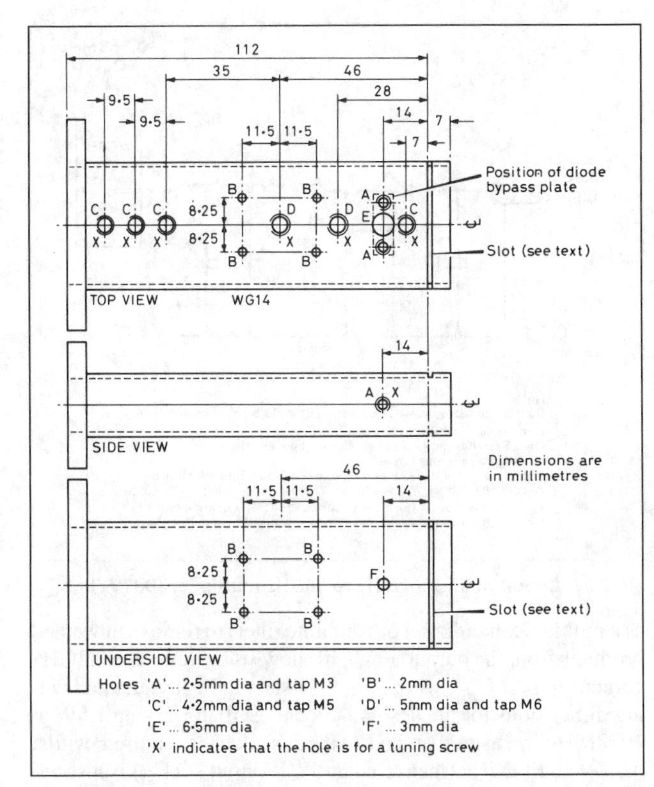

Fig 9.32. Mechanical details of the 5.7GHz multiplier waveguide assembly

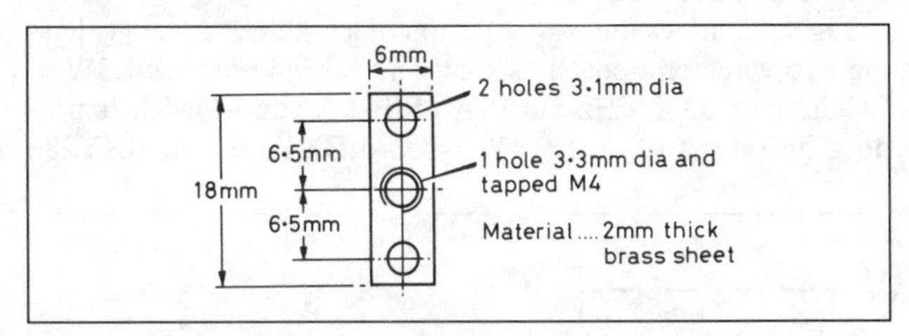

Fig 9.33. Diode bypass plate details, 5.7GHz multiplier

intended to use the harmonic relationships outlined earlier, then 1152MHz (nominal) is a good drive-frequency choice for transmitters for several amateur bands.

A waveguide multiplier for 5.7GHz

The dimensions for a simple ×5 multiplier capable of a power output of over 100mW at 5.7GHz are shown in Fig 9.32. It consists of a 120mm-long piece of copper or brass waveguide-14, terminated at one end by a flange for connection to the antenna feed (or other equipment) and terminated at the other by a short-circuit. A post-mounted varactor diode is spaced a critical distance from the short-circuit and the cavity is tuned by a screw between the varactor and the short-circuit. Output is matched into the waveguide filter, which consists of four posts and a tuning screw, by a matching screw immediately in front of the diode mount. Output matching is achieved by means of a three-screw tuner immediately adjacent to the output flange. Fig 9.33 gives details of the diode bypass plate, Fig 9.34 details the diode mounting post dimensions and Fig 9.35 gives detail of the diode mount assembly. The input matching circuit is shown in Fig 9.36.

Detailed construction and alignment were described in reference [5]. With an input of 1.5W at 1152MHz, an RF output at 5760MHz of at least 100mW should be obtained using a BXY28E diode. With 2.5W input, 150mW output should be

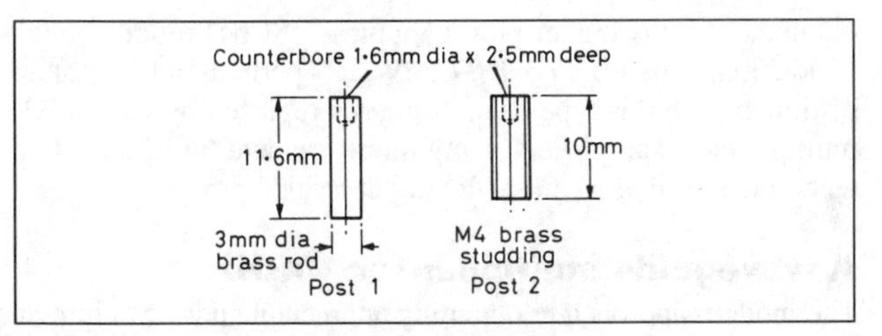

Fig 9.34. Diode post details, 5.7GHz multiplier

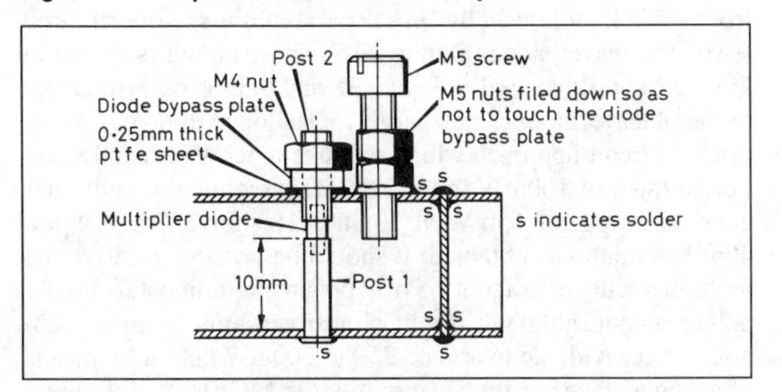

Fig 9.35. Diode mount details, 5.7GHz multiplier

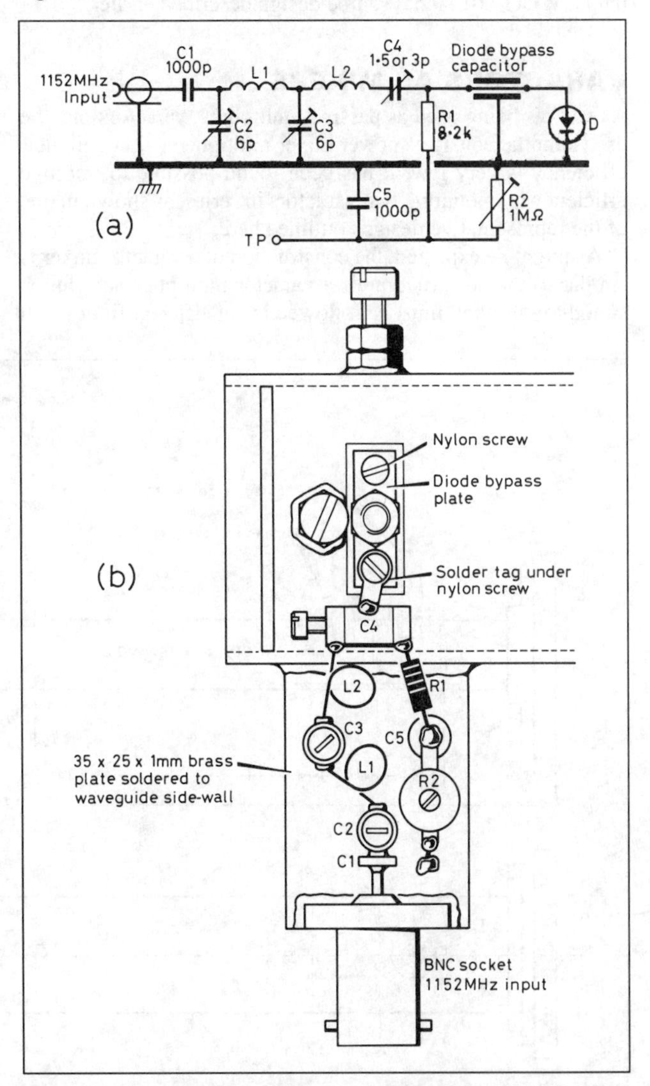

Fig 9.36. Input matching network, 5.7GHz multiplier. (a) Circuit diagram. (b) Layout. C1, C5, 1000pF leadless disc; C2, C3, 6pF tubular trimmer; C4, 1.5 or 3pF PTFE tubular trimmer; L1, 1t 1.6mm TCW, 6mm ID; L2, 1t 1.6mm TCW; R1, 8k2, 0.125W; R2, 1M0 preset; D, BXY28e, BXY39e or VSC64J varactor

obtained. At this higher power input, a VSC64J diode should yield 200mW or a BXY39E 275mW. It is perhaps only a matter of time before this type of multiplier is replaced by GaAsFET multipliers and amplifiers, using microstrip and 'pill-box' filter techniques similar to those already described.

A waveguide multiplier for 10GHz

The modern approach to obtaining appreciable power output at 10GHz is, as already described, to use an active ×4 multiplier from 2.5GHz, followed by GaAsFET amplifiers to the required level. The waveguide multiplier [16] whose circuit is shown in Fig 9.37 and illustrated in Figs 9.38 and 9.39 is representative of the older techniques now largely abandoned in favour of the printed circuit approach which requires much less metalwork! For an input of 100mW in the 1152MHz region, the multiplier gave an output of 15mW at 10,368MHz: for 250mW input, 40mW output was obtained. It should be possible to drive the multiplier with up to about 1.5W to obtain proportionately higher power output, although it will almost certainly be more economic to provide up to around 250 to 300mW output by means of a single-stage GaAsFET amplifier (eg MGF1801) following the G3WDG-001 signal source design described earlier.

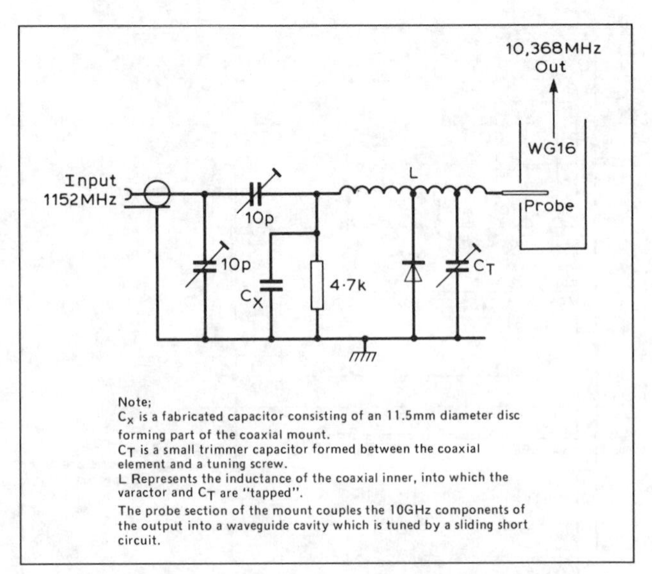

Fig 9.37. Circuit of an 1152MHz to 10GHz multiplier (DJ7VY)

VARACTORS AS MIXERS

As well as being used as passive multipliers, varactors may be used simultaneously as power mixer/multipliers, although their efficiency is very low. It has been found possible to improve efficiency by mounting two varactors in series, as shown in one of the representative designs outlined here.

As might be expected, the construction of a varactor mixer is similar to the construction of a varactor multiplier and, like a straight multiplier, must be followed by an efficient filter (such as a multipole interdigital or combline filter) to remove unwanted products from the output. Fig 9.40 shows the circuit of a 2.3GHz varactor mixer [17], whilst Fig 9.41 shows the layout and leading dimensions for the design. A local oscillator giving 1.5W at 1008MHz (measured at the LO input socket) is required, whilst 0.25W at 144MHz (measured at the IF input socket) is also required. This configuration provided about 0.35W linear (SSB) output at 2320MHz.

Fig 9.42 shows the configuration of a 3.4GHz mixer [18] using two varactor diodes in series. This design requires 1.5W of LO drive on 3312MHz and 10W of IF drive on 144MHz to produce an output of nearly 1W when a BXY39e and BXY28e

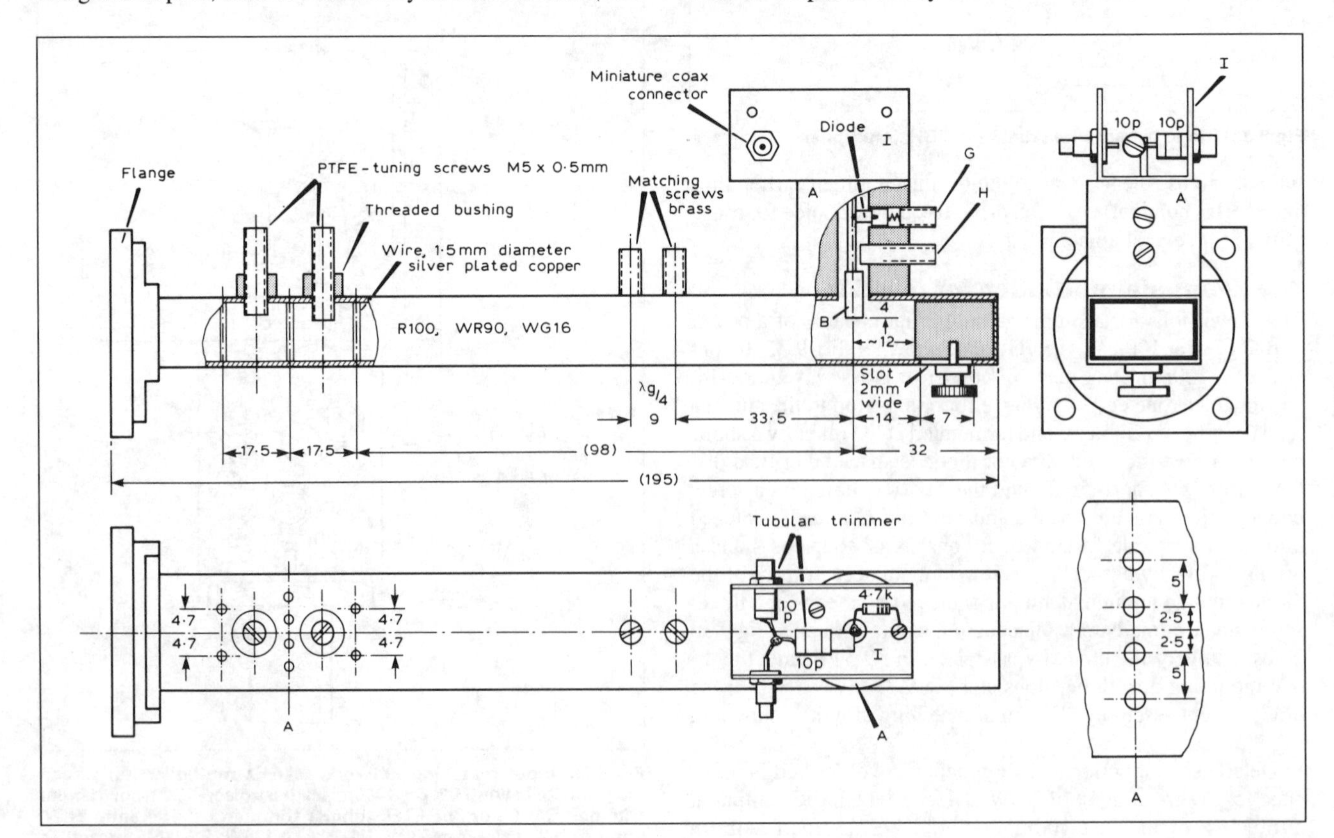

Fig 9.38. General layout of the DJ7VY 10GHz multiplier

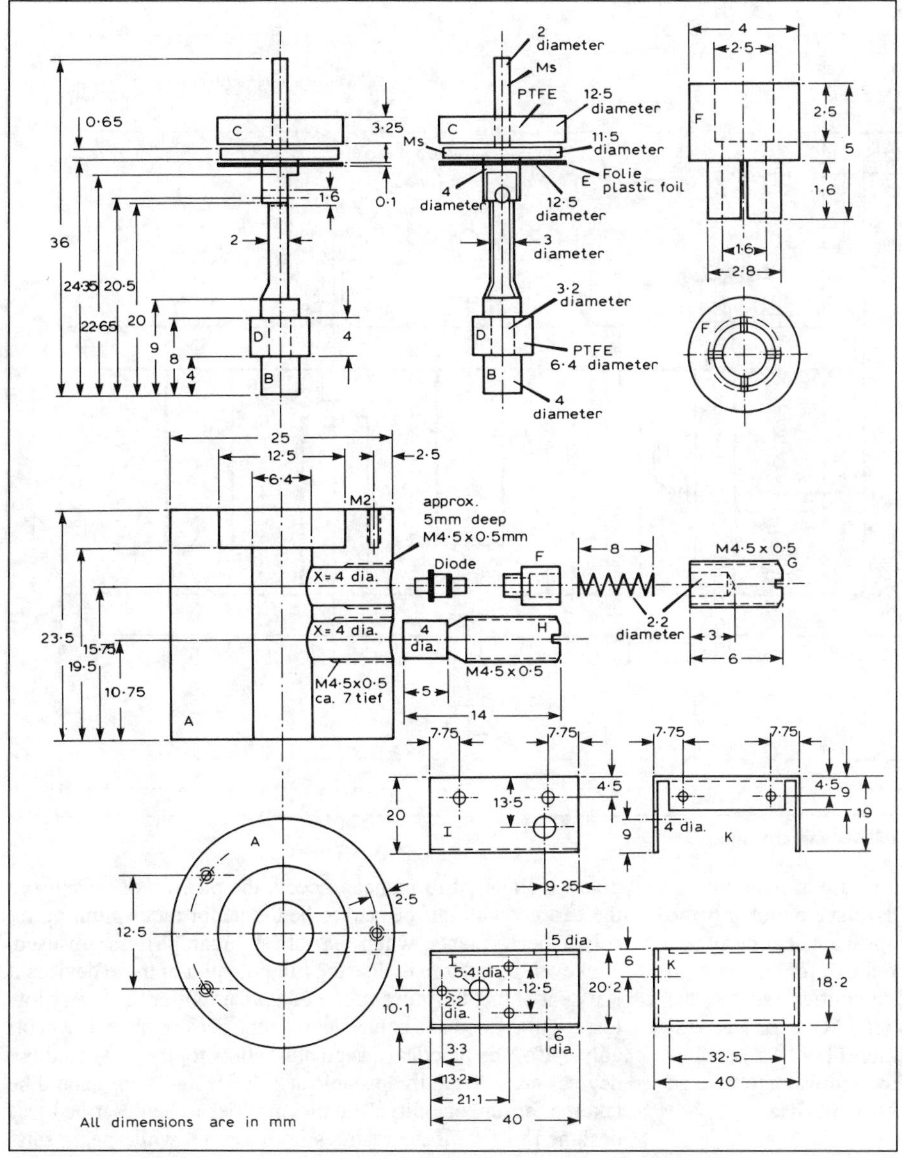

Fig 9.39. Parts detail for the DJ7VY 10GHz multiplier

combination was used. A pair of BXY27e (or BXY28e or BXY39e) diodes produced 0.5W whereas a single diode of the same type yielded only 0.1W.

MICROWAVE AMPLIFIERS

Microwave amplifiers can be divided roughly into two types: small-signal receive preamplifiers where a low noise figure is more important than gain, and large-signal transmit amplifiers where power gain is more important than low noise.

For ordinary receiver applications bipolar transistors (or some MMIC amplifiers) may be used, at least up to 3.4GHz, although the performance of GaAsFETs may be vastly superior. At 1.3GHz, commonly available GaAsFETs may have noise figures in the 0.1–0.3dB range but the amplifier circuit will need careful design and layout, since the devices exhibit high gain and are not unconditionally stable at these lower frequencies. GaAsFETs are used almost exclusively for low-noise applications above about 3.4GHz and, in a properly designed circuit, will perform with better stability than at lower frequencies in the microwave spectrum. At about 8GHz and above (up to at least 40GHz) GaAsFETs and their variants (HEMTs and PHEMTs) are used in low-noise receiver applications.

Low-power transmitter amplifiers may be based on either MMICs (again up to about 3–4GHz, with power output of tens of milliwatts), discrete component/bipolar transistor or GaAsFET amplifiers with outputs up to a few watts. Thick-film hybrid integrated amplifiers (at present up to about 2GHz) with power outputs of a few watts or tens of watts may also be used. Both types of integrated amplifier mentioned have the advantage of operating from a standard 12–14V power supply and being matched into 50Ω input and output impedances.

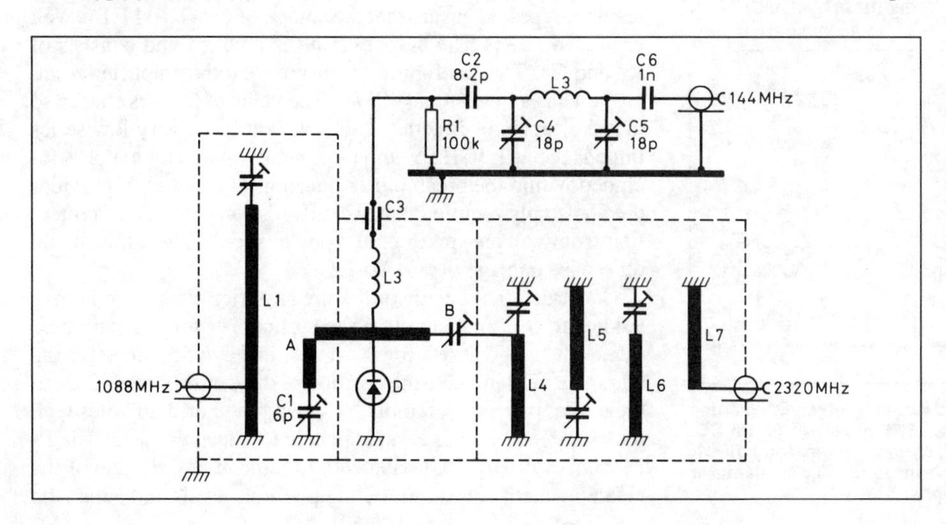

Fig 9.40. Circuit of a 2.3GHz varactor mixer. D1, BXY27; R1, 100k, carbon; L1, 54mm long, 10mm OD, 8mm ID copper tube; L2, 4t 0.4mm dia TCW, 3mm ID; L3, 6t 1mm silver-plated copper wire, 7mm ID; L4–7, 10mm OD, 8mm ID copper tube; C1, 6pF tubular ceramic trimmer, maximum 3mm diameter; piece A, 1.5mm long by 6mm wide, copper foil 0.2mm thick; piece B, 20mm × 8mm piece of copper foil; piece C, M4 studding with 1.6mm hole, 2mm deep in one end, to hold diode, fitted with locknut; piece D, 4mm diameter × 35mm long brass rod with 1.6mm hole drilled radially at its centre; also required two PTFE bushes 6mm OD, 4mm ID, used to support piece D

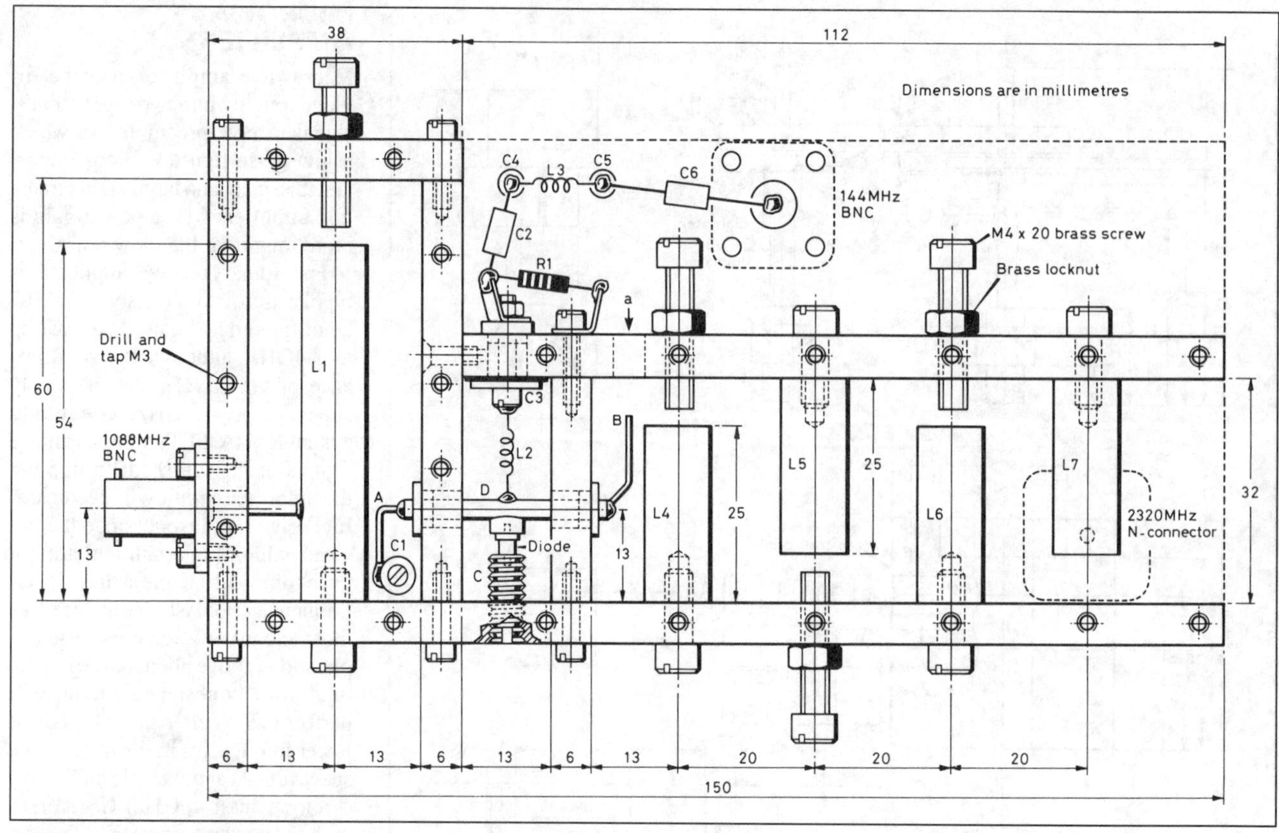

Fig 9.41. Mechanical construction of the 2.3GHz varactor mixer

GaAsFETs are available for transmitter use at most microwave frequencies up to about 18 or 20GHz but are probably too expensive, beyond about 1W output, for most amateur purposes. It is probably more feasible and cost effective for amateurs to use either grounded-grid disc-seal triode amplifiers (such as the 2C39A family) at frequencies up to about 3.5GHz or 'second-user' surplus travelling-wave tube amplifiers (TWTAs, see elsewhere) with a drive requirement of a few milliwatts to give an output of several watts at frequencies above 3.5GHz.

RECEIVE PREAMPLIFIERS

A high-performance GaAsFET preamplifier for 1.3GHz

Modern GaAsFETs are capable of producing very low noise figures at 1.3GHz, and preamplifiers using these transistors can give excellent performance. Indeed, the pick-up of noise from the earth can almost be said to be the major factor limiting receiver performance when GaAsFET preamplifiers are used. However, in order to realise the full potential of these devices it is essential that the input circuit of a preamplifier has a very low loss, or the added noise may degrade the performance unacceptably. Also, any practical design must allow for the fact that these devices are potentially unstable at 1.3GHz and steps should be taken to ensure stability. The preamplifier to be described was designed to take these features into account while being relatively easy to construct.

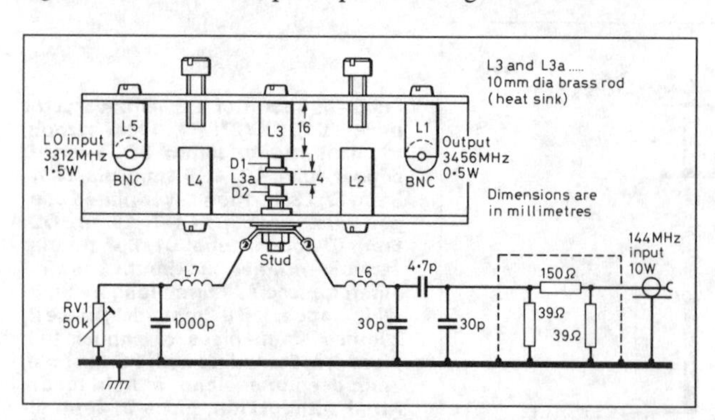

Fig 9.42. Schematic diagram of a 3.4GHz varactor mixer. D1, D2, see text; RV1, 50k carbon preset; L1, L2, L4, L5 16mm long, 10mm OD, 9mm ID copper tube; L6, 5t 18swg TCW, 8mm ID, 10mm long; RFC1, 3t 26swg ECW on FX1110 bead; L3A, 4mm long, 10mm diameter copper bar; L3B, 16mm long, 10mm copper bar

The circuit diagram of the preamplifier is shown in Fig 9.43. The low-loss input circuit consists of L1, C1 and C2. Source bias is used so that the preamplifier can be run off a single positive rail supply. The two source decoupling capacitors are leadless types, to ensure that the source of the GaAsFET is well grounded at RF. The output circuit is untuned and consists of R2 and C5. This configuration ensures that the amplifier is stable and has a low output VSWR. The value of C5 was chosen so that the capacitor is series resonant, ie it has a very low series impedance at 1.3GHz. A three-terminal voltage regulator is included within the preamplifier housing. This not only provides the 5V supply required, but also affords some degree of protection from voltage spikes on the power supply line which could otherwise damage the GaAsFET.

Constructional details of the preamplifier are shown in Figs 9.44(a) to (g). First cut out a sheet of copper for the main box, mark out the positions for the holes, make scribe lines for the corners and to indicate the position of the centre screen. Cut out the corner pieces (preferably by sawing) and drill all holes with the exception of holes E, and tap hole D. Clean the sheet using a PC rubber or Brasso, taking care to remove any residue of the cleaning agent at the end. Bend up all four walls and adjust the

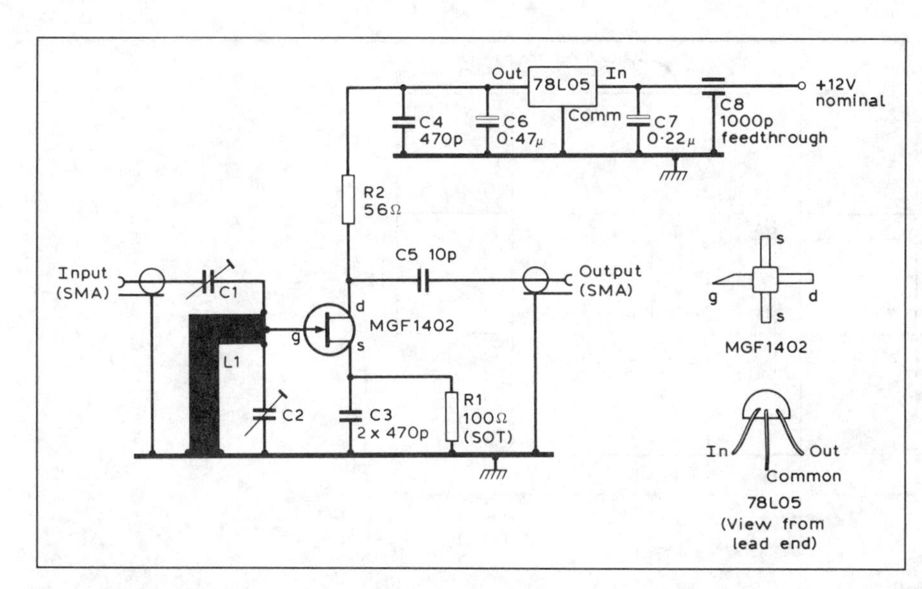

Fig 9.43. Circuit of high-performance 1.3GHz GaAsFET amplifier (G3WDG)

Component details:

- **L1** Stripline inductor, see Fig 9.44(e)
- **C1** Input capacitor, see Fig 9.44(f)
- **C2** Tuning capacitor (0BA or M6 screw)
- **C3** 2 × 470p trapezoidal leadless 'disc' capacitor
- **C4** 470p trapezoidal leadless 'disc' capacitor
- **C5** 10p miniature ceramic plate capacitor
- **C6** 0.47μ tantalum, 16V working
- **C7** 0.22μ tantalum, 16V working
- **C8** 1000p bolt-in feedthrough capacitor
- **R1** 100R (select on test)
- **R2** 56R

corners using pliers to ensure that the walls touch. The centre screen should be fabricated next. It is best to make this piece slightly oversize initially and then to file it to be a tight fit in the main box. The screen should be cleaned after drilling and deburring the hole. The leadless capacitors are then soldered to the screen, in the positions shown in Fig 9.44(c). The soldering is best done by clamping the sheet in a horizontal plane and applying a small flame from underneath. Tin each capacitor on one side using a soldering iron and small amount of solder, and tin the screen in three places, again using only a small amount of solder. Place the capacitors on the screen, tinned sides down, and reheat the screen under each capacitor in turn.

As soon as the solder melts, press down gently on the capacitor until it is flat on the screen. Using an ohmmeter, check that the capacitors are not shorted to the screen. Remove the excess solder by filing if shorts are found. Fabricate the retaining plate as shown in Fig 9.44(d) and place it on to the screen so that the bumps align with the capacitors. Fix it in position using an M2 screw and nut.

Locate the screen in the box using the scribe marks as a guide to correct alignment. Measure the distance between the screen and the inside of the end wall of the larger compartment, and check that this agrees with the dimension given in Fig 9.44(a). Move the screen if necessary. Clamp the screen in position using a toolmaker's clamp applied to the sidewalls of the main box around the screen. Jig a clean brass nut in position at hole D using a stainless steel or rusty screw to hold it in place. Tighten the nut slightly against the wall of the box. Mount the box in a vice so that the junctions of the sidewalls are horizontal, and solder along the junctions using a small flame from underneath each corner in turn, re-positioning the box each time. It is easiest to preheat the corner using the flame and to make the joints with a soldering iron. This reduces the chance of re-melting the joints already made. The screen should be soldered next, along all three sides using the same technique, with the box mounted so that the open side is uppermost. Finally, invert the box and solder the nut in position.

When the assembly is cool, remove the jigging screw and run a tap through the nut and sidewall. Remove the tap and check that the tuning screw runs freely in and out. If necessary, file away any excess solder around the nut so that the output connector fits into position correctly. Drill the fixing holes for the input and output connectors.

The lid is made next. Cut out the material for the lid and place the box symmetrically on it. Scribe around the outside of the box on the lid, cut out the corners and bend the lid over the box to form a tight fit. Using a 1.6mm bit, drill through the sidewalls of the lid and box (holes E). Tap the holes in the box and open out the holes in the lid to 2.1mm.

The next stage in the construction is to make the input line, details of which are given in Fig 9.44(e). The line should be cleaned before bending up the tabs. The overall length of the line is quite critical and since errors can occur during bending, it is best to make the line slightly longer initially and then file off excess material at the 'input' end of the line. Recheck all dimensions of the line using vernier callipers before proceeding. The line can then be fitted into the box, using M2 fixings.

Measure the distance between the top surface of the line and the bottom of the box, at the fixing end. This should be 4mm ±0.1mm. If necessary, remove the line and file the holes in the box so that this dimension can be achieved. Similarly, measure the height of the input end and bend the line up or down as necessary. The input tab is made next and should be bent so that when it is located on the input connector, the sides and top of the tab align with the tab on the input line. The tab can then be soldered to the input connector, leaving a 0.5–1mm gap between the tabs. The assembly is completed by fitting the components. The layout is shown in Fig 9.44(g). After mounting the grounding tag, solder R1 into position. When making the joint to the source decoupling capacitor, do not allow solder to run along the capacitor or it will be difficult to mount the GaAsFET correctly. Cut the leads of C5 to 2mm length and solder one end to the output connector. Bend the leads of C5 so that the free lead is in line with the hole in the screen.

Mount the feedthrough capacitor and connect up the voltage regulator circuitry. Apply power and check that +5V is available at the output of the regulator (connect a 1kΩ resistor between the output of the regulator and ground during this measurement). Cut the source leads of the GaAsFET to 2.5mm length. Using tweezers, twist the drain lead through 90° and, holding the end of the drain lead, mount the GaAsFET into position with its source leads touching the bypass capacitors. Unplug the soldering iron and solder the drain lead to the free end of C5. Check that the gate lead passes centrally through the hole in the screen and adjust the position of the device by careful bending if necessary. The source leads of the GaAsFET are soldered next: again, unplug the iron just before making the joints. Soldering the gate lead requires some care.

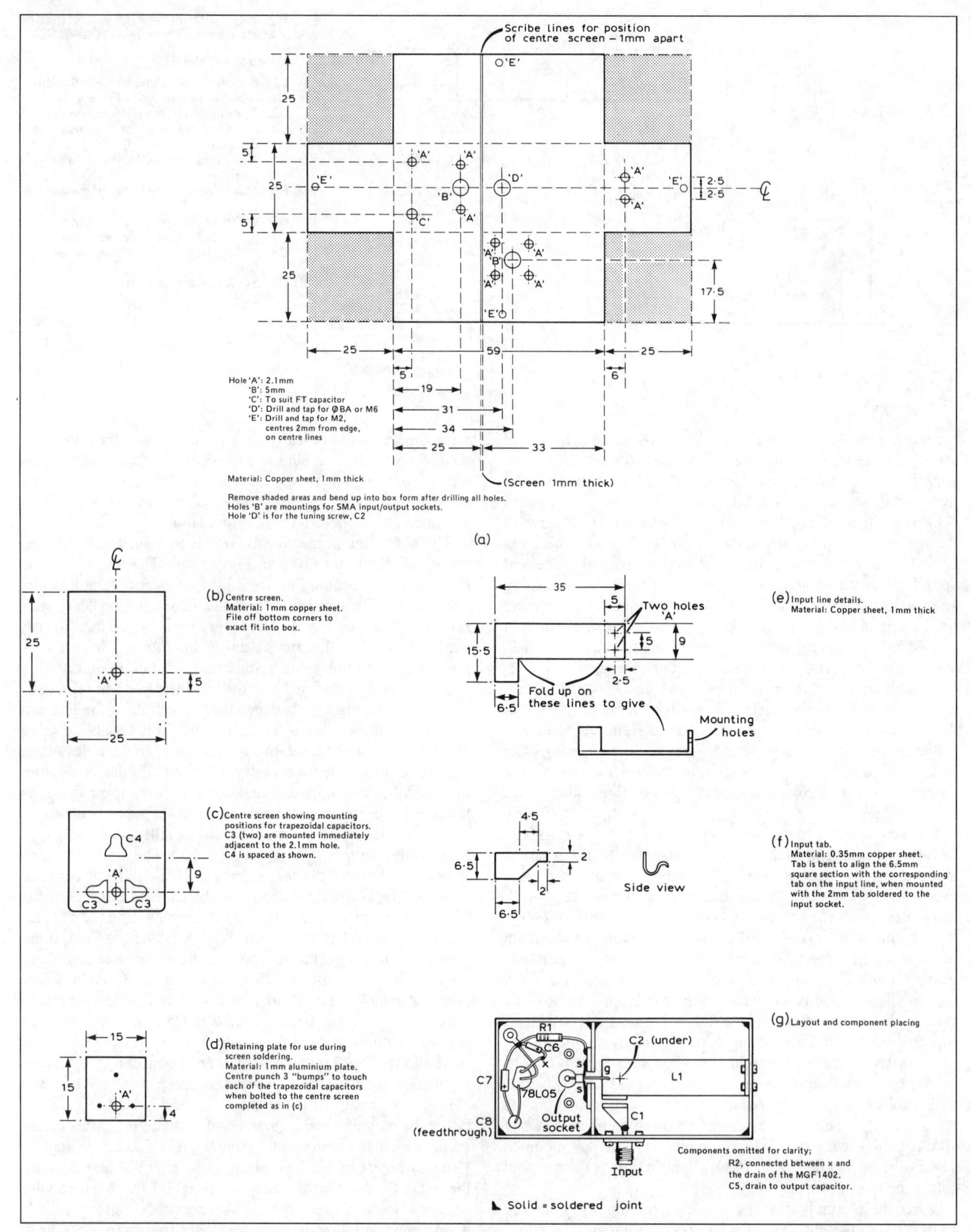

Fig 9.44. (a) Details of the main box for the GaAsFET amplifier. (b) Centre screen. (c) Centre screen showing mounting positions for trapezoidal capacitors. (d) Retaining plate for use during screen soldering. (e) Input line details. (f) Input tab, material 0.35mm copper sheet. (g) Layout and component placing

First, slightly loosen the fixings of the input line but not enough to allow the line to move. Using a minimum of a 45W iron, heat the end of the line at its edge until the solder flows. The body of the iron should be connected via a flying lead to the box

Table 9.13. GaAsFET preamplifier performance

Frequency (MHz)	Gain (dB)	Noise figure (dB)
1235	15.6	1.04
1245	16.4	0.90
1255	16.6	0.76
1265	16.4	0.65
1275	16.1	0.58
1285	15.6	0.56
1295	14.8	0.55
1305	13.9	0.56
1315	12.9	0.58
1325	12.2	0.62

during this operation. Move the iron to the centre of the line next to the gate lead. Then solder the gate lead to the line. Re-tighten the fixings of the input line. The final operation is to fit R2. Cut one lead to 2mm length and tin. With the iron unplugged, solder the other end of R2 to C4 such that the free end is touching the drain lead of the GaAsFET about 1mm from the package. Then solder R2 to the drain of the GaAsFET.

Adjustment is as follows: with the preamplifier connected in front of a suitable 1.3GHz receiver, apply power and check the current drawn. This should lie somewhere between 10 and 15mA. If not, the value of R1 will need to be altered. Next, fit the tuning screw and adjust for maximum noise. Tune in a weak signal and adjust the position of the input tab and the tuning screw for optimum signal-to-noise ratio. Alternatively, an automatic noise figure alignment aid can be used [19]. The performance of the preamplifier is shown in Table 9.13.

For applications where noise figure is at a premium, the MGF1402 device specified can be replaced by the MGF1412-09. This device will yield a lower noise figure, by approximately 0.2dB.

A GaAsFET preamplifier for 2.3GHz

This preamplifier was designed by OE9PMJ [20] for use on 2320MHz EME but the design is equally suitable for 2320MHz tropospheric working. The original design used two MGF1412-11-08 GaAsFETs and produced a noise figure of 0.6dB and a gain of 32dB. The version to be described here is much cheaper to build than the original design. It uses an MGF1412-11-09 and a MGF1402 which will produce a noise figure of 0.7dB and a gain of 29dB. The circuit diagram of the preamplifier is shown in Fig 9.45. The major mechanical details are shown in Figs 9.46 and 9.47. The case is folded from a single piece of 0.5mm brass sheet and soldered at the corners. The two holes at the ends of the box are drilled to clear the centre pins of the input and output connectors used, preferably N-type or SMA. The two partitions are made from 0.5mm brass sheet which are first cut to fit in the appropriate position, then drilled as shown in Fig 9.48. The dimensions of the inductors which are shown in Fig 9.49 are made of 0.5mm copper sheet. The 3mm by 5mm tabs on L1 and L3 are to enable the inductors to be bolted to the case using very short M2 screws. The appropriate-sized holes are drilled in the case to suit capacitors C1 and C7 and the 1nF feedthrough capacitors C5 and C10. In the authors' version, 0.3–3.5pF Johanson trimmers (type 5800) were used which required the holes to be 3mm. Alternatively, as shown in Fig 15.21, capacitors C1 and C7 may be replaced by M4 bolts running in M4 nuts soldered to the case. This method is especially useful if good-quality trimmers cannot be obtained.

The box, partitions and RF connectors are then soldered

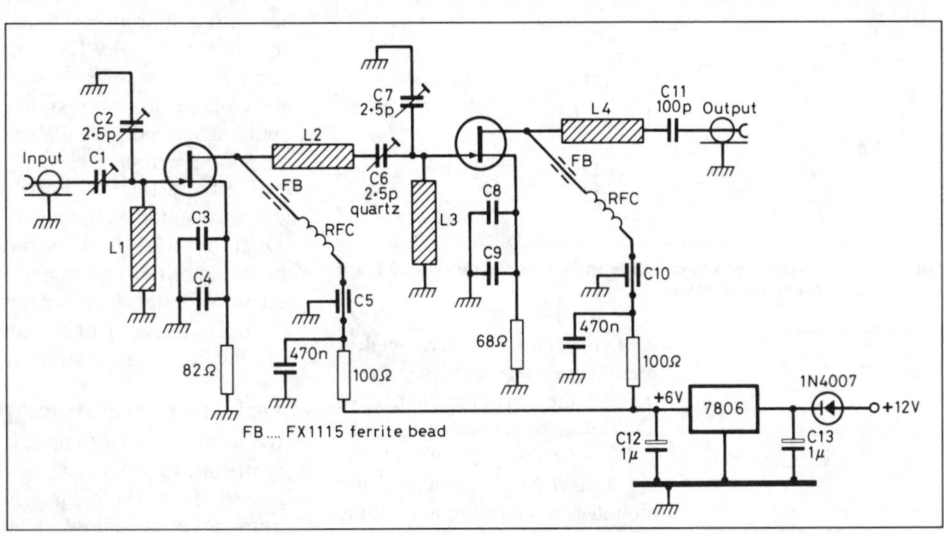

Fig 9.45. Circuit diagram of a two-stage 2.3GHz GaAsFET preamplifier

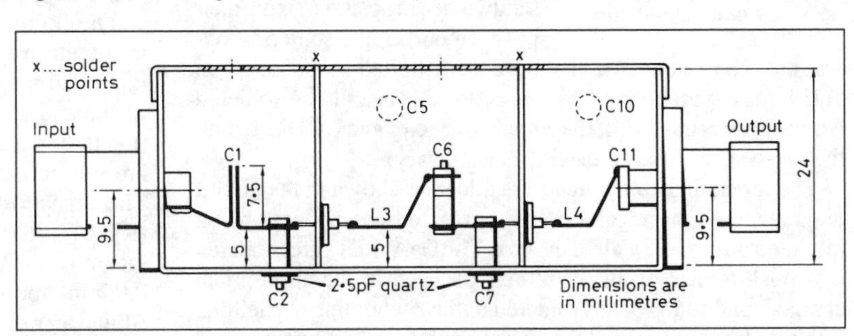

Fig 9.46. Side view of constructional layout of the 2.3GHz preamplifier

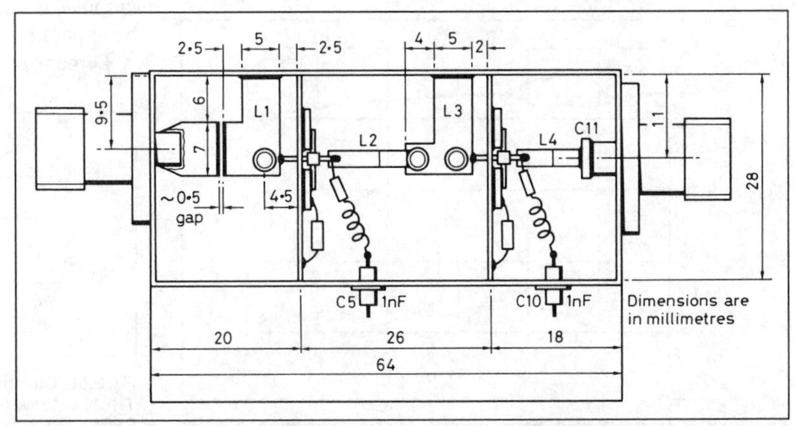

Fig 9.47. Top view of constructional layout of the 2.3GHz preamplifier

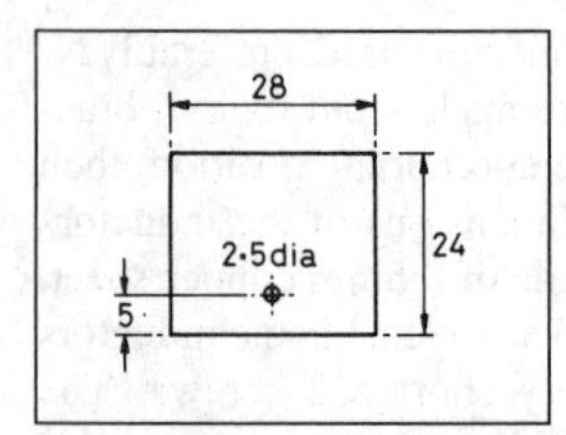

Fig 9.48. Preamplifier partitions, two required. Material is 0.5mm brass

together with a large soldering iron. Whilst the assembly is cooling, the 1nF feedthroughs (if solderable) and the leadless disc capacitors (C3, C4, C8 and C9) are carefully soldered into position using a small soldering iron.

The source decoupling capacitors are soldered as close as possible to the hole in the partition to ensure minimum source lead length on the FET. Direct heat should not be applied to the discs but the surface tinned whilst the reverse side of the copper is heated. The disc is then gently pressed in position and soldered.

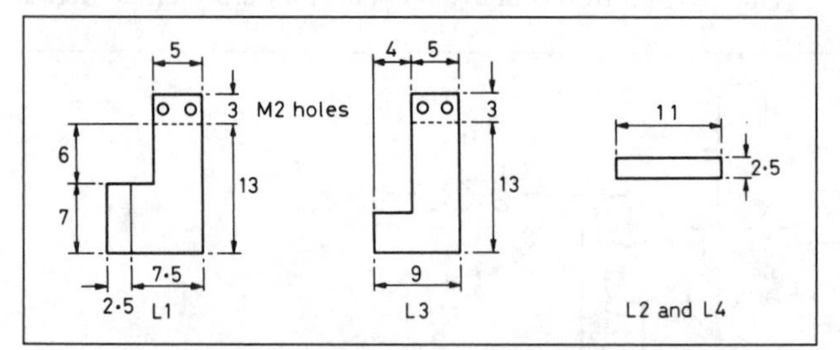

Fig 9.49. Dimensions of the stripline inductors. Material is 0.5mm copper, preferably silver plated

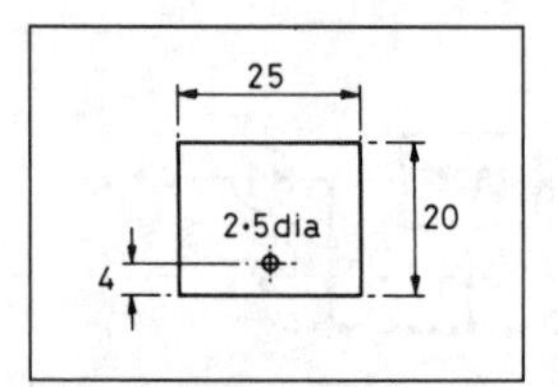

Fig 9.50. Leadless disc capacitor retaining plate. Material is 1.6mm aluminium

If difficulty is experienced in retaining the leadless disc capacitors on the partitions, a pair of aluminium plates made as shown in Fig 9.50 may prove useful. Using a steel M2.5 screw and nut mounted in the centre hole of the two plates, the leadless discs are sandwiched in position, fixing their position during the soldering operation. The tuning trimmers are next bolted or soldered in position as appropriate. Chip capacitor C11 should be soldered between the centre pin of the output connector and L4, shortening the centre pin of the connector as necessary.

The remaining components are added as shown in the circuit diagram, with the regulator and associated components being mounted on the side wall of the box. The GaAsFETs are mounted into position last of all, with their package in the appropriate chamber and source leads soldered to the decoupling capacitors. The gate leads are passed through to the hole in the partition and soldered to the gate inductor. Extreme care should be taken to prevent damaging the FETs with static charges.

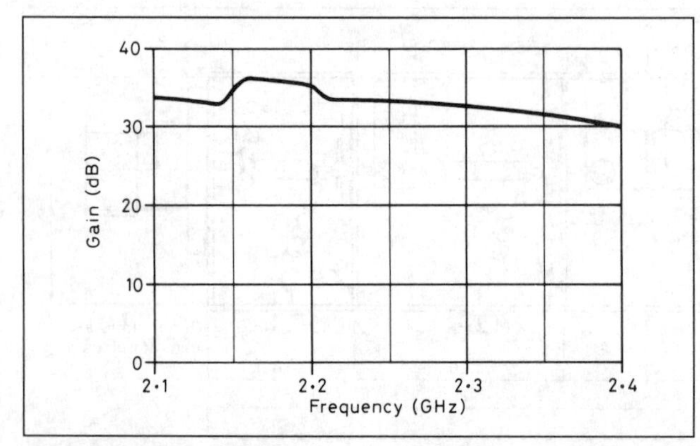

Fig 9.51. Measured gain performance of the prototype 2.3GHz preamplifier, using two MGF1412-11-08 devices

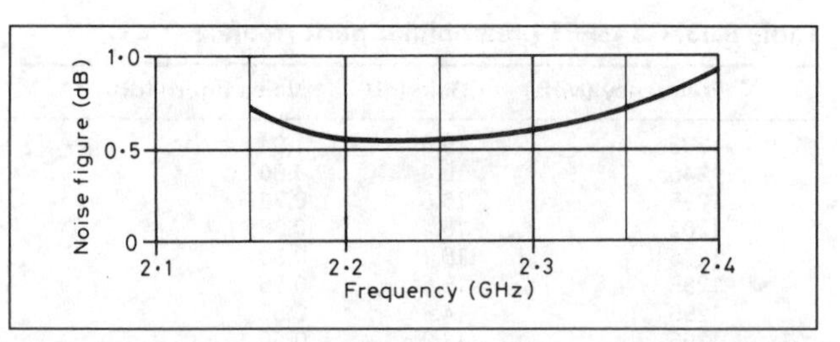

Fig 9.52. Measured noise performance of the prototype 2.3GHz preamplifier, using two MGF1412-11-08 devices

Alignment is quite straightforward: first, the DC bias conditions of each FET should be checked by connecting a voltmeter across the appropriate source resistor. For the MGF1402 the current is nominally 10mA and for the MGF1412 the current is 15mA, so the source resistor is adjusted to obtain the appropriate voltage drop across itself. If an automatic noise figure optimisation aid is available [19], this may be used to initially align the preamplifier.

Final adjustments (or the whole alignment process for those without a suitable alignment aid) should be made for maximum signal-to-noise ratio when monitoring a weak signal source with the preamplifier configured as it is will be used (ie the same length of feeder, same antenna etc). The gain against frequency and noise figure against frequency measurements made on the OE9PMJ original preamplifier are shown in Figs 9.51 and 9.52.

A bipolar transistor preamplifier for 3.4GHz

At 3.4GHz, bipolar transistors such as the NE64535 can give a significant improvement in receiver noise figure at relatively little expense. The preamplifier to be described here is a compromise between lowest noise figure and highest gain, as optimising one parameter usually degrades the other.

The circuit of a suitable preamplifier, developed by G3WDG, is shown in Fig 9.53. The function of the series capacitors on input and output is twofold. They act both as DC blocking elements and as part of the matching circuitry. In the latter function their values are fairly critical and, since suitable capacitors are not commercially available, they have to be home-made. The capacitors in the prototype were made from PCB material (RT/Duroid type 6010, 0.63mm (0.025in) thick). The material was cut, then filed to shape. The dimensions are critical to 0.05mm and should be checked with a micrometer during the filing operation. The required dimensions are 1.60 × 1.60mm for the input capacitor (0.36pF) and 1.84 × 1.84mm for the output capacitor (0.47pF). Other constructional details should be apparent from Fig 9.54.

Dimensions are given for the transmission lines on 0.5mm

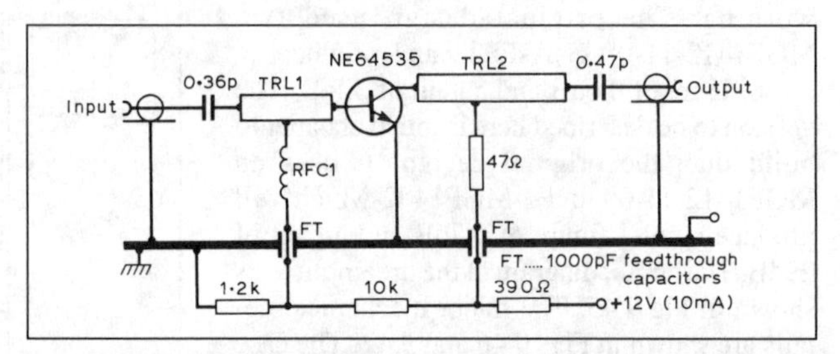

Fig 9.53. Circuit diagram of the G3WDG 3.4GHz bipolar preamplifier. TRL1, 0.128λ length of 50Ω transmission line, tapped midway for RFC1. TRL2, 0.336λ length of 50Ω transmission line, tapped 0.14λ from transistor end. RFC1, 5t of 0.4mm dia ECW, 1mm ID

(0.020in) thick RT/Duroid type D-5880. This material was chosen as it is a good medium for microstrip circuits at this frequency (it is usable to beyond 10GHz). Conventional 1.6mm Teflon board or 1.6mm epoxy glass fibre is *not* suitable for this application. The connector types shown (SMA) were chosen because of their excellent performance and size compatibility with the microstrip. The socket bodies are soldered to the reverse of the PCB and also to the 'chassis'.

It is possible to use N-type input and output connectors. The inner of the N-socket is filed down to form a small tab, as shown in Fig 9.55(a). This tab is soldered to the 50Ω microstrip lines on the PCB. In order to minimise the discontinuity at this junction with the microstrip, a cut-out is filed in the PCB to accommodate the bush on the connector. In order to allow for this, the length of the PCB must be increased by 6mm. The rear side metallisation on the PCB is soldered to the body of the N-socket as shown in Fig 9.55(b). BNC connectors are *not* suitable for this application. The cut-out for the transistor is made with a sharp blade such as a scalpel. After fitting the emitter grounding strips, made of 0.25mm copper foil, and soldering them to the reverse side of the board, they are flattened to lie flush to the board and filed slightly. This ensures that the transistor is located with a minimal gap between the base collector leads and the microstrip. When soldering in the transistor (which should be the final operation) cut all the leads to about 3mm in length. Ensure they are soldered to the microstrip directly at its ends, after pre-tinning the lines with a hot, clean soldering iron. Note the base connection has the chamfered end. Provided that the preamplifier has been constructed carefully, it should work immediately upon application of power. The only adjustment that may be necessary is to set the collector current to 10mA by making small changes to the 1.2kΩ or 10kΩ resistors, by adding a suitable resistor in parallel with one or the other. Do *not* alter the value of the 390Ω resistor. The performance of the prototype is shown in Fig 9.56. It can be seen that the preamplifier has low input and output VSWRs. The noise figure of the whole preamplifier is only 0.75dB higher than the minimum possible for the device when designed for lowest possible noise figure.

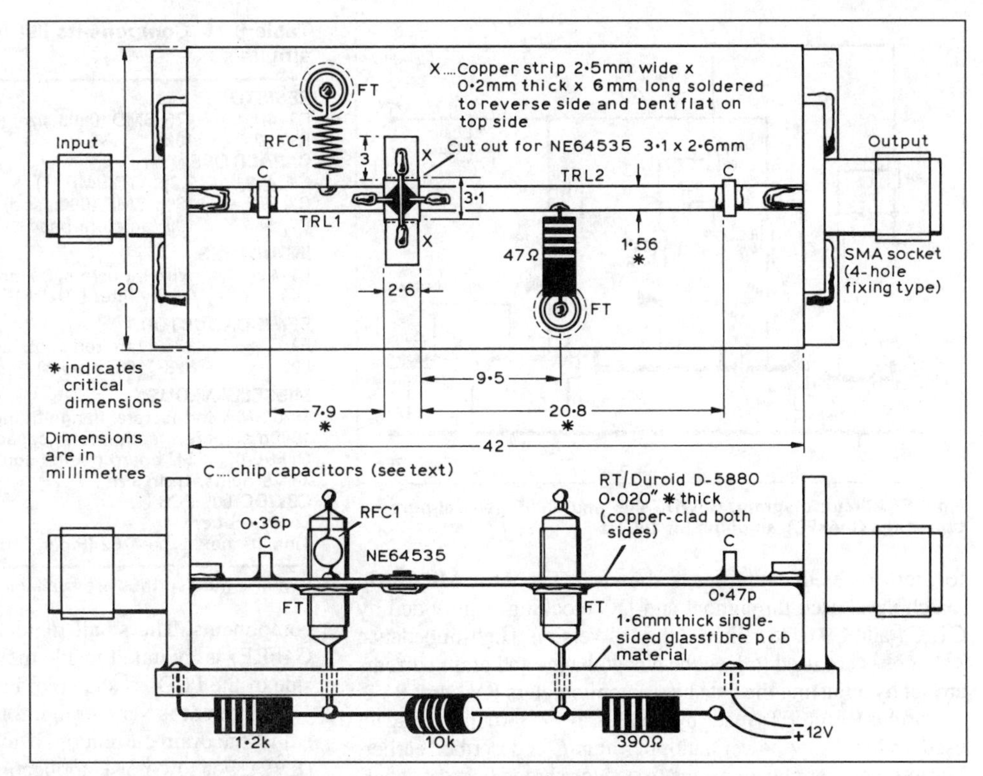

Fig 9.54. Constructional layout of the 3.4GHz bipolar preamplifier

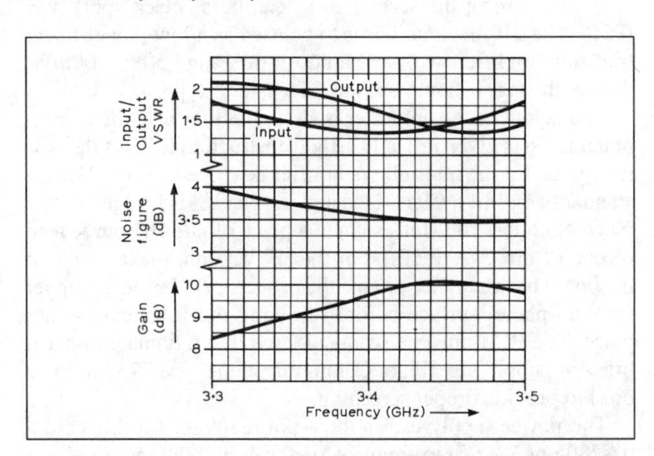

Fig 9.56. Measured performance of the prototype 3.4GHz preamplifier

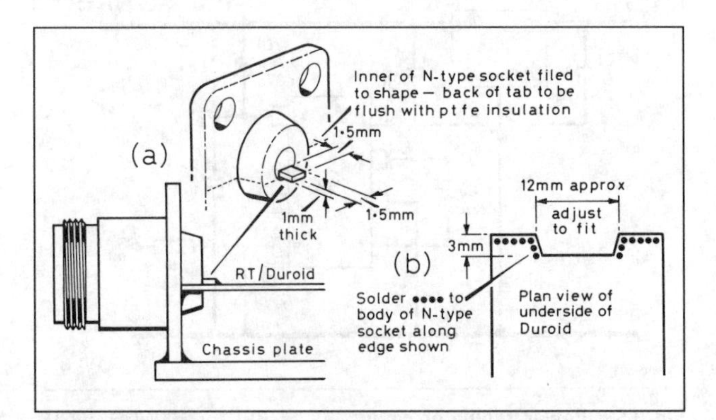

Fig 9.55. Details of the N-socket to RT-Duroid transition. (a) Modification of N-connector. (b) Modification of board. Note overall length of board must be increased by 6mm

When a single preamplifier stage is used with an interdigital converter using an MBD102 diode (see later), an overall noise figure of 6.2dB is obtained. When two of these preamplifiers are cascaded the overall noise figure is reduced to 4.2dB.

A general-purpose, two-stage GaAsFET amplifier for 10GHz

This general-purpose amplifier, designed as one of a series of modules by G3WDG [21] can be built in two forms, a standard version and an 'F' version which contains a narrow-band filter to provide rejection of image-channel noise when used ahead of a receiver with poor or no image rejection. The circuit diagram

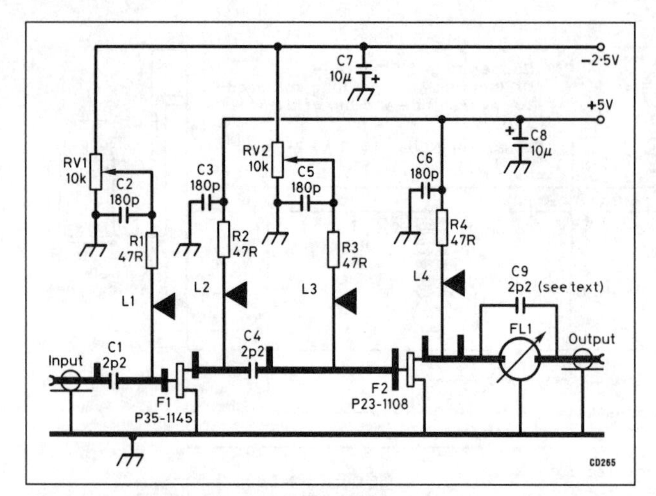

Fig 9.57. Circuit diagram, G3WDG-005 and -005F general-purpose two-stage GaAsFET amplifier for 10GHz

Table 9.14. Components list for the G3WDG-005 and -005F amplifiers

RESISTORS	
R1–4	47R SMD (0805 size) (*)
RV1, 2	10k
CAPACITORS	
C1, 4, 9	2.2p ATC SMD (*)
C2, 3, 5, 6	180p SMD (0805 size) (*)
C7, 8	10μ tantalum bead
INDUCTORS	
L1–4	Wire inductors, 0.2mm dia
FL1	Cavity filter ((*), WDG-005F only)
SEMICONDUCTORS	
F1	P35-1145 (red spot, surplus) or similar
F2	P23-1108 (black spot, surplus) or similar
MISCELLANEOUS	
Two SMA connectors, flange fitting	
1000p solder-in feedthrough capacitor	
G4FRE-023 PSU board (*) and components	
16 Veropins, 1mm size (*)	
G3WDG-005 PCB (*)	
Lossy rubber (*)	
Tinplate box, type 7752 (Piper Communications)	

(*) Parts marked thus are available in a 'short' kit.

for both types of amplifier is shown in Fig 9.57. Microstrip matching is used throughout and DC blocking is provided by C1, C4 and C9 or the filter in the 'F' version. High-impedance λ/4 chokes are used to feed the bias voltages and drain currents are set by adjusting the gate bias potentiometers RV1 and RV2.

Surplus GaAsFETs made by the Plessey 3-5 Group were again used, as in the low-power multiplier/amplifier described earlier. For low-noise applications, the first stage should be a 'red spot' type (P35-1145); the second stage can be a 'black spot' type (P35-1108). If the amplifier is to be used as a low-power linear transmit amplifier, with an output in the range 50 to 100mW, then both stages should use the 'black spot' type.

The layout of the amplifier is shown in Fig 9.58 and a components list is given in Table 9.14. Construction follows the outline given for the multiplier/amplifier described earlier. For the standard amplifier, C9 is fitted and the gap 'S' is bridged with a piece of copper foil (for example a piece of cut-off source lead of one of the GaAsFETs). For the 'F' version, make sure that the board has clearance around the probe-pin holes on the upper (ground-plane) surface before soldering the filter cavity into place. The PCB must be seam-soldered into a tinplate or sheet brass or copper box (Piper Communications type 7752 tinplate box) to provide proper screening.

The power supply uses a three-pin regulator (in this case a μA7805 or 78L05) to supply +5V for drain bias and a voltage inverter to supply −5V for gate bias, both supplies being derived from a nominal +12V input. The circuit is shown in Fig 9.59 and the layout in Fig 9.60. In this configuration, the zener diode is replaced by a link and the value of R1 is 4.7kΩ. Note that all the conventional components are used as surface-mount components. The small regulator/inverter PCB (designed by G4FRE) is mounted inside the enclosure on the ground-plane side of the PCB, as shown in Fig 9.61.

Alignment is very straightforward. For the standard version adjust the drain current of F1 to 15mA (RV1) and F2 to 20mA (RV2). For low-noise applications, these currents can be readjusted later for best gain/noise performance, assuming that noise measuring equipment is available. Where used as a transmit amplifier, the currents are individually adjusted for best gain/power output. The initial values given will be near enough optimal for most purposes.

The 'F' version must, of course, have its filter tuned to the required image rejection frequency: in a typical amateur receiver application, with the receiver LO at 10,224MHz and a 144MHz IF, the filter will be required to pass a signal frequency of 10,368MHz: simply tune the cavity for maximum receiver noise.

Typical performance of prototype amplifiers was 18–20dB gain, noise figure 1.9–2.0dB. If instability is encountered, it is recommended that a piece of absorptive rubber (RAM or *radar absorptive material*) is stuck to the lid of the box above the track side of the PCB, using contact adhesive.

Brief mention must also be made of the use of HEMT (high electron mobility transistor) devices as very-low-noise receiver preamplifiers. These devices, which are capable of noise figures

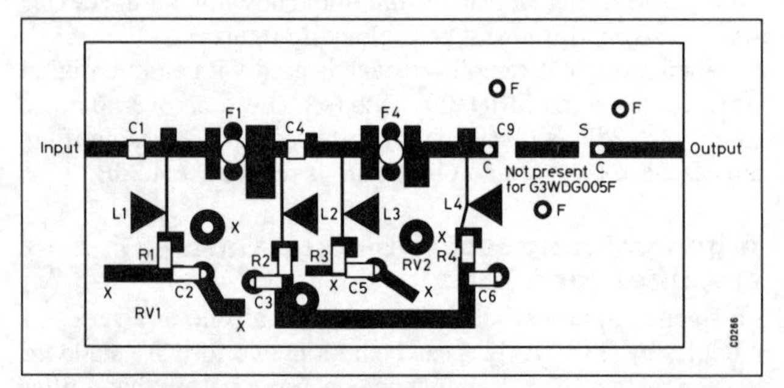

Fig 9.58. Layout of the G3WDG-005/(F) two-stage 10GHz amplifier (not to scale). C = filter probes (L = 3.4mm), F = filter locating pins, S = shorting link (see text), X = position of RV1, 2 connections

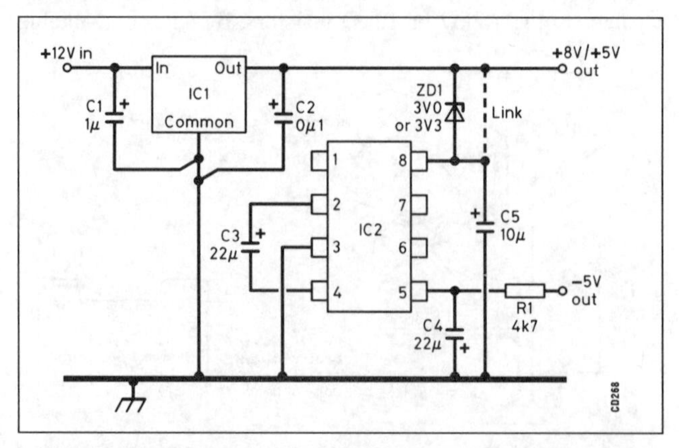

Fig 9.59. Power regulator circuit (5V or 8V) for G3WDG 10GHz amplifiers. IC1, μA7808 (8V) or μA7805/78L05 (5V). IC2, ICL7660PCA. ZD1, 3V0 or 3V3 400mW zener diode or shorting link (see text). C1–C5, tant bead. C1 is 16V wkg, rest are 10V wkg. PCB is G4FRE-023

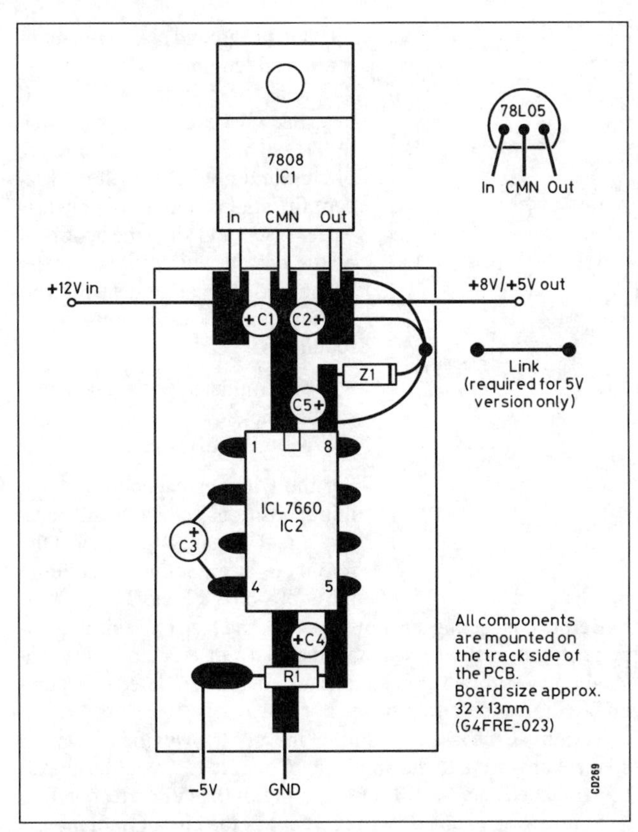

Fig 9.60. Layout of power regulator circuit for G3WDG 10GHz amplifiers (not to scale)

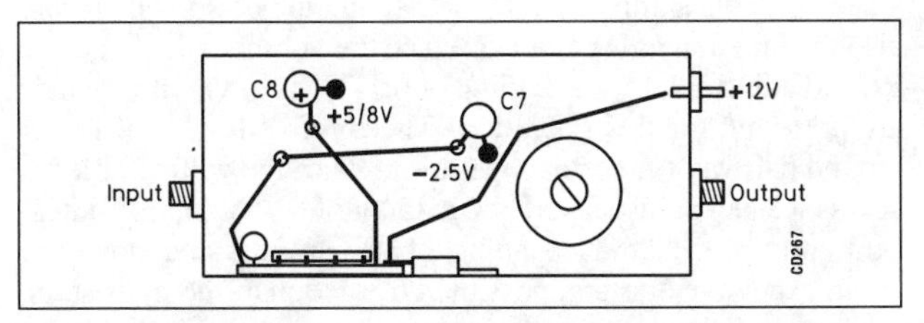

Fig 9.61. Positioning the power regulator circuit for the 10GHz amplifier (not to scale)

of 1dB or less at 10GHz, have only recently become available at prices which can be regarded as economic by amateurs. The circuit and layout for such an amplifier are not greatly different to the microstrip circuits, using 'ordinary' GaAsFETs, described above. A typical amateur-designed low-noise amplifier (LNA), using a Fujitsu FHX-06 HEMT, described in reference [59], is available in kit form from the Microwave Committee Components Service [40]. In use in many UK amateur stations, this amplifier is capable (with careful construction) of noise figures around 0.8 to 1dB, so low that the losses incurred in feeder and connectors must be taken into consideration when installing such an amplifier!

TRANSMIT (POWER) AMPLIFIERS

A 1W solid-state linear amplifier for 1.3GHz

The amplifier described in this section was designed to provide 1W output at 1152MHz when driven by 10mW from the G4DDK-001 [9] oscillator board, making it suitable for driving frequency multipliers to the higher amateur bands. The board may also be used at frequencies as high as 1300MHz, where it could be used in conjunction with a 1296MHz version of the oscillator board to make a 1W FM or CW transmitter.

Since the amplifier operates in the linear mode, it could also be used in conjunction with a low-level mixer/ModAmp (Avantek MMIC) amplifier to produce a cost-effective 1296 or 1269MHz (satellite) up-converter such as that described in reference [22] and later in this chapter. A Mitsubishi M57762 PA block requires just 1W drive for greater than 10W output across the whole of the 23cm band, so that the amplifier described is a suitable driver for this purpose. The 3dB bandwidth of the prototype 1W amplifier measured 58MHz, suggesting it could also be suitable for 24cm ATV operation. The circuit of the 1.3GHz amplifier is shown in Fig 9.62. It is possible to provide the required 20dB gain with just two stages, but previous experience has shown that keeping the gain in each stage to more manageable levels generally means greater stability and less dependence on selected devices. The first stage of the amplifier uses a BFR91A bipolar transistor, although slightly improved results were obtained in the prototype with an NEC NE021, since compression in the first stage tends to limit the final output to 1W. Operating the amplifier at greater than 1W output significantly increases dissipation in the output stage and makes the BFQ-34 more susceptible to mis-termination failure.

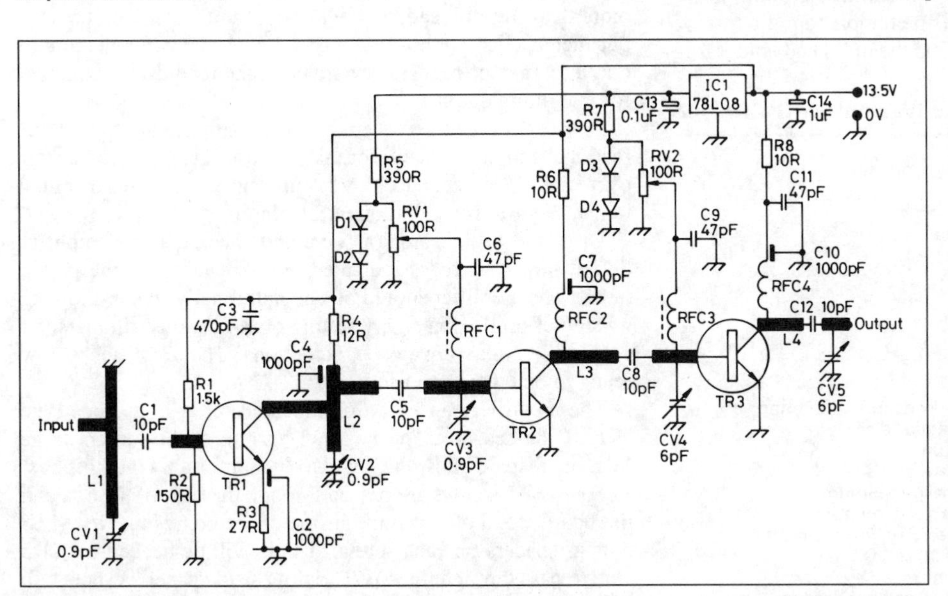

Fig 9.62. Schematic diagram of the G4DDK-002 1W 1.3GHz linear power amplifier. TR1, BFR91A; TR2, BFG34; TR3, BFQ34

The first stage operates in Class A at a collector current of 17mA. The second stage operates in Class AB, also at a quiescent bias current of 17mA. Substitution of an alternative device has proved difficult in this stage, the BFR96 being definitely a non-starter, although the BFR96S or BFG96 may be satisfactory. A BFQ34 is used in the output stage. Maximum output at 1152MHz may be as much as 1.5W with a selected device. Class AB operation at a quiescent bias of 35mA is used. Since the original article was published, G4LOJ has suggested the use of a Motorola MRF511 as a

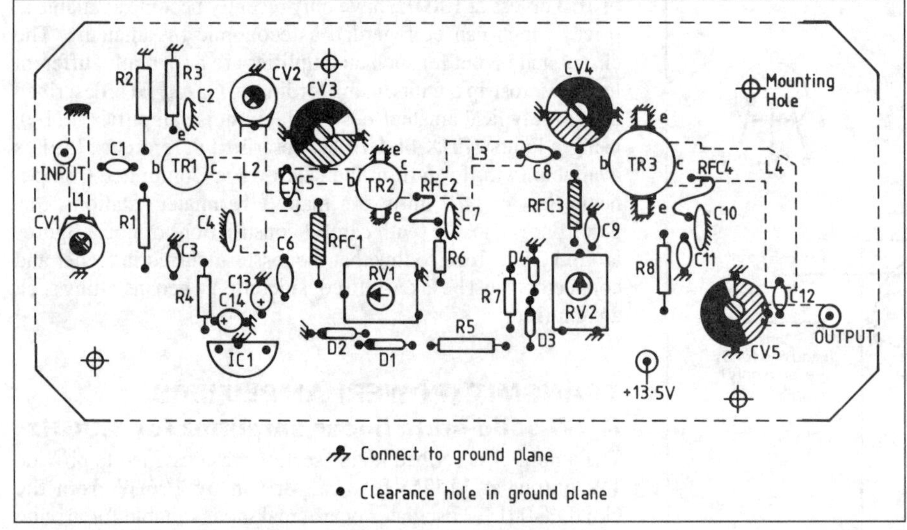

Fig 9.64. Component layout for the 1W amplifier G4DDK-002

replacement for the often-difficult-to-get BFQ34. However, it has been found that the addition of a 5–10pF trimmer between TR3 collector and ground is needed if the full gain of the transistor is to be realised.

If only 300 to 400mW are needed, it is possible to leave out the BFQ34 and associated bias components. In this case take the output from the base pad vacated by TR3. The match to 50Ω at this point is excellent.

The PCB artwork is shown in Fig 9.63 (Appendix 1), the layout in Fig 9.64 and component values in Table 9.15. As with the companion LO source, a PCB is available from the RSGB Microwave Components Service. Full constructional details are given in reference [5]. The order of construction should be to fit the board into its box, ensuring that it can be easily mounted and dismounted. Next fit any grounding foil or pins. Fit resistors and capacitors next, ensuring they are soldered to both sides of the board where appropriate. Semiconductors are fitted last and suitable short lengths of miniature coaxial cable are soldered to the input and output pads. The prototype used RG174 terminated with SMC plugs. Rather than form the braid of the cable into pigtails for soldering, it is preferable to put a turn of 24 to 26 swg tinned copper wire round the braid and solder this

Table 9.15. Components list for the 1W amplifier G4DDK-002

TR1	BFR91A, NE021
TR2	BFG34
TR3	BFQ34
D1–4	1N914,1N4148
IC1	78L08
R1	1.5k
R2	150R
R3	27R
R4	12R
R5, 7	390R
R6, 8	10R
All 0.25 or 0.3W miniature carbon film	
RV1, 2	100R 0.25W carbon track, horizontal mounting
C1, 5, 8, 12	10p miniature ceramic plate or disc
C2, 4, 7, 10	470 to 1n trapezoidal (coffin)
C6, 9	47p miniature ceramic plate or disc
C13	0.1μ tantalum bead, 10V wkg minimum
C14	1μ tantalum bead, 16V wkg minimum
CV1, 2	0.9p minimum foil trimmer, Sky, Oxley or Murata
CV3–5	6p foil trimmer, Mullard 808 series
RFC1, 3	1μH miniature moulded choke
RFC2, 4	1.5t 1mm dia TCW, 3mm ID, self supporting
L1–4	Printed lines on 1/16in double-sided glass epoxy board

direct to the ground plane with minimum lead length.

Even for alignment a small heatsink will be required on the stud of TR3. This can be a small block of aluminium suitably drilled to accept the stud of the transistor. It is easier to test and align the board out of the box in which it will finally be housed. For alignment the following items of test equipment are required:

- 50Ω terminated power meter
- Multimeter
- Wavemeter to cover 1.0–1.5GHz

Set the trimmer capacitors to the initial positions shown. Set the bias resistors RV1 and RV2 such that the rotors are at the ground end, ie no bias volts on TR2 and TR3.

Before connecting any power, check and then recheck all components are in the right place and that those components that should have ground connections are actually soldered to ground on both top and bottom of the board.

Terminate the amplifier output in a 50Ω power meter. Do not connect any drive to the amplifier. Connect 13.5V to the amplifier and check across R3 for a reading of 0.44V, corresponding to a current of 17mA. If wrong, recheck the circuit, looking especially for short-circuits or incorrectly placed components. When all is well, go on to check for 8V at the output of IC1. Check for approximately 1.4V at the junction of R5 and D1 and also the junction of R7 and D3. Place the voltmeter leads across R6 and adjust RV1 for a reading of 0.17V (17mA). Three hands are preferred for this operation! The rotor of RV1 will be set around halfway for normal operation. Place the voltmeter leads across R8 and adjust RV2 for a reading of 0.35V (35mA quiescent current). Make these adjustments carefully and slowly to avoid excessive bias and possible destruction of the transistors.

The oscillator input can now be connected. With the specified trimmers, set initially as shown, it should be unnecessary to more than slightly readjust CV1 and 2 for a reading on the output meter. Also slightly readjust CV3, 4 and 5 for maximum output. It may be necessary to go back and readjust all trimmers for maximum output.

Loosely couple the wavemeter to the amplifier output. A directional coupler is useful for this purpose. Tune the wavemeter over its entire range, checking for unwanted responses that may indicate spurious oscillation in the amplifier.

Disconnect the oscillator drive and check that the amplifier output falls to zero as indicated by no reading on the power meter. The amplifier should be unconditionally stable, showing no sign of oscillation at any setting of the various trimmers. Of course there may always be exceptions! The amplifier is now ready to be boxed and used.

The amplifier can be accommodated in an Eddystone type 27134P die-cast box. The diagonal cut-outs on the corners of the PCB are there to allow the board to fit comfortably into this box.

Four M2.5 screws are passed through the base of the box and the board stood off from the inside of the box using two M2.5 nuts as spacers on each screw. A hole will be needed for TR3 stud to pass through the box. Careful use of spacer washers will be needed to ensure the capstan of TR3 is just correctly spaced from the base of the box. The best approach here is to measure

Table 9.16. Thick-film hybrid power amplifiers: typical specifications

Type	Band (MHz)	Bandwidth (MHz)	P_{out} (min W)	P_{in} (mW)	Class
M67715	1240–1300	1220–1320	1.5	10	AB
M57787	1240–1300	1240–1300	3.0	10	C
M57762	1240–1300	1240–1300	20.0	1000	AB

the required spacing carefully and then machine a suitable-size aluminium block as the spacer. This, if large enough, will be a better heatsink than the base of the die-cast box.

The box will need drilling to accept two sockets for the amplifier input and output. A bolt-in feedthrough capacitor of 1–10nF should be used to carry the DC supply into the box.

Hybrid thick-film solid-state amplifiers

Several broad-band power-gain block amplifiers are available in the UK, suitable for either Class B (FM) or Class AB (linear) power amplification across the entire VHF and UHF spectrum. They have drive requirements ranging from a few milliwatts to several hundred milliwatts to give outputs in the range from 1W to about 50W. Higher output is available by driving two or more amplifiers via power splitters and combining their outputs using similar splitters as combiners. The modules are almost invariably designed for 12.5 to 14V DC input, 50Ω input and output matching, and all exhibit quite high power gain. Any module should, therefore, be mounted directly on a suitable heatsink and both DC and RF connections made via 50Ω microstripline circuits which incorporate suitable coupling and decoupling components. The manufacturer's recommendations should be followed closely to ensure stability and maximum reliability.

Three such units are presently commonly available for the amateur 1240 to 1300MHz band, all manufactured by Mitsubishi: their characteristics are summarised in Table 9.16. The principal precautions to be observed when using these modules are:

1. It is essential to use adequate heatsinking and, to ensure good thermal conductance, thermal compound must be used between the module's fin and the heatsink to which it is bolted.
2. Ensure that the heatsink has a flat, smooth surface and that there are no particles of foreign matter between the fin and the heatsink when mounting the module into place. If there are rough spots or if excessive stress is caused by overtightening the fixing screws, the substrate of the module may be cracked (an expensive error!).
3. Dropping the module onto a hard surface can cause similar expensive damage!
4. Avoid excessive soldering temperatures: 260°C for not longer than 10s or 350°C for not longer than 3s is quoted as acceptable.
5. Resin-cored solder should be used, ie use solder with a non-corrosive flux.
6. If solvents are used for flux removal, chlorinated solvents (eg trichlorethylene) must be avoided: ethanol, methylated spirit or surgical spirit are suitable.

A high-gain, single-valve 1.3GHz power amplifier

One of the problems confronting the 1.3GHz operator is how to raise the one to two watts available from a solid-state transverter *economically* to significant output levels of tens of watts. Solid-state devices for this task are available but are expensive and, due to low stage-gain available from most transistors, several stages may be needed to reach, say, 25W unless hybrid power-gain blocks such as those made by Mitsubishi are used. Even these dedicated units are quite expensive and susceptible to damage if not properly installed and heatsinked.

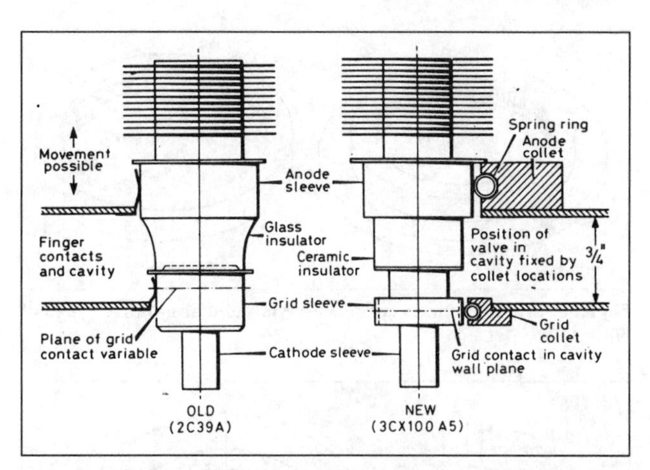

Fig 9.65. High-gain 1.3GHz amplifier: old and new methods of mounting valves in cavity (omitting anode supply details for clarity)

Most 1.3GHz power amplifiers are still designed around valves of the 2C39 family and one of the problems that beset early attempts at efficient amplifier design was achieving high gain. Most of the current amplifier designs have evolved from the box-cavity tripler first published by G2RD [23]. Although the use of power triplers is virtually obsolete in this day of transverters, the design found new life as the basis for amplifier design. Conversions of the G2RD tripler to both high-level mixers and power amplifiers were described by G3LTF and G3WDG [24]. These designs yielded a gain of no more than 8.5dB. References to gains of 10–12dB were reported [25, 26] from multi-valve amplifiers using the more recent ceramic-insulated versions of the 2C39. A brief reference [27] was made in *QST* magazine to a design by N6CA in which a single-valve amplifier, modified for water cooling, yielded a gain of 13dB at normal drive levels and up to 15dB at lower drive levels. The keys to success in seeking high stage gain lay in the use of the highest possible EHT (eg 1.3kV) and paying scrupulous attention to the RF grounding of the grid. Tests by N6CA showed that deliberately increasing the grid-to-ground inductance by raising only one in four of the contact fingers on the grid ring led to some 3dB loss in gain, due to the introduction of negative feedback. The design was more fully described in reference [28].

The present design by G4PMK and G3SEK [29], based on these principles, allows gains of 15dB or more to be realised. Note that all dimensions to be given in this section are in inches, since it is still easier for the UK constructor to obtain brass strip and tubing in Imperial sizes rather than metric. Examination of valve types and the professionally designed UPX4 amplifier (converted to amateur use by W2IMU), enabled the depth of the new version of the cavity to be fixed at 0.75in and a coarse tuning screw to be provided, as described by G3LTF and G3WDG. The differences in the valve mounting positions for the old and new designs are shown in Fig 9.65.

The greatest problem in amateur designs using the 2C39 series of valves has always been the contact rings for the grid and anode. To insist on an extremely low-inductance grid contact in the plane of the baseplate makes matters worse than ever!

Straight finger-stock (Fig 9.65) is ruled out because it projects either into or out of the anode cavity. Folded-over finger-stock is used in commercial preformed grid rings, but is not readily available; N6CA's experiments suggest that the inductance of the resulting contact is barely low enough.

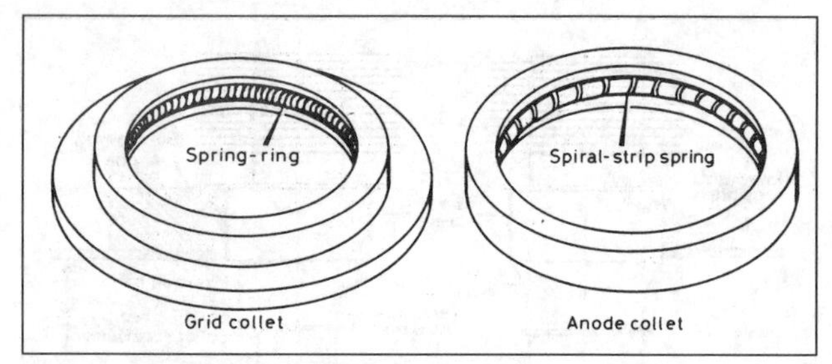

Fig 9.66. Grid and anode collets, showing two alternative types of spring material

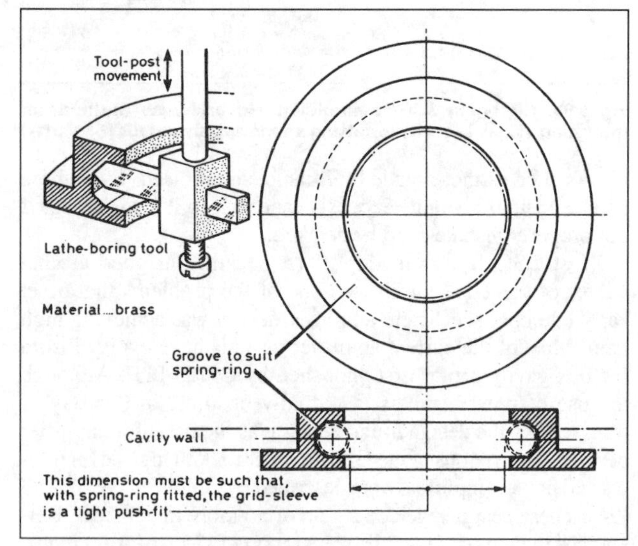

Fig 9.67. Details of grid collet

The solution is to use a ring of spiral spring to contact the valve, the spring-ring itself being held in a collet (Fig 9.66). In effect, the valve is contacted by several quarter-turns of the spring, all of which are electrically in parallel, combining to make a contact of extremely low inductance. The collet can be let into the base of the cavity so that the contact is made in the correct plane.

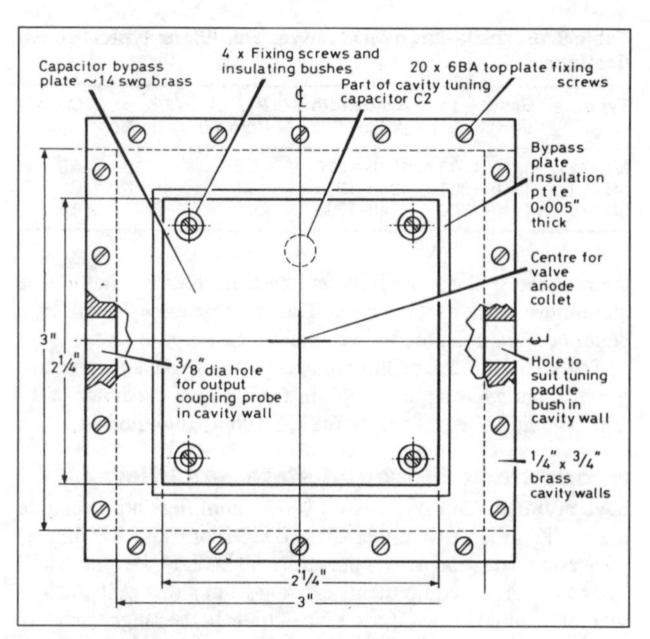

Fig 9.69. Top view of anode cavity assembly

The ideal spring-ring material is a loosely wound, silver-plated spring of about 0.25in diameter. A perfectly acceptable home-made substitute is a spiral wound from narrow (eg 0.1in wide) phosphor-bronze strip, such as draught-excluder; this gives fewer contacts to the valve but each turn of the strip has lower individual inductance. Fig 9.66 shows the two alternative types of spring-ring in their collets.

Precise dimensions of the collet depend to a large extent on the available spring-ring material, and the prototypes were turned by cut-and-try out of old brass vacuum fittings. The first step is to bore out the blank to just clear the grid sleeve of the valve. Then the internal groove is formed using a small boring tool (inset, Fig 9.67), repeatedly trying first the spring-ring alone for size and, in the later stages, both the spring-ring and valve. The fit of the valve can also be adjusted by pulling or squeezing the spring-ring. When all is well, the valve will be gripped gently but uniformly as it is twisted into place. Owing to the 'lay' of the turns of the spring-ring, the valve can only be twisted in one direction – the same for insertion and removal – so if spring-rings are used for both the grid and anode connectors they *must* be wound in the same sense.

Fig 9.68. General view of the G4PMK/G3SEK single-valve power amplifier

The anode is much more forgiving of stray inductance than the grid contact, so a ring of ordinary finger-stock would probably suffice. The prototypes used spring-ring anode connectors.

Many other features of the G2RD/G3LTF/G3WDG designs were retained. The cathode input circuit closely follows the original, the coarse tuning screw has already been mentioned and the fine tuning paddle and coupling loop are also

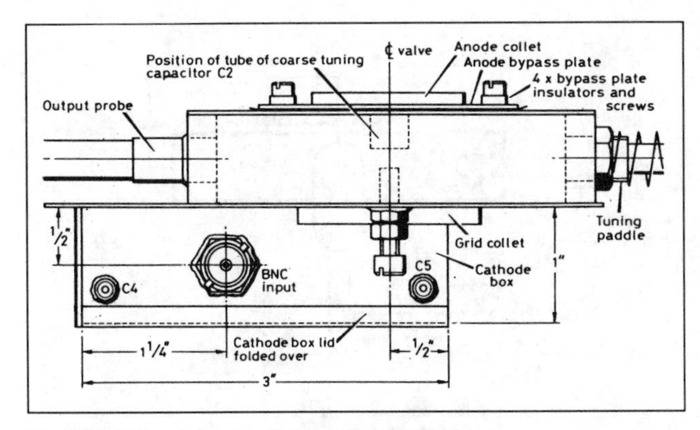

Fig 9.70. Side view of anode cavity assembly

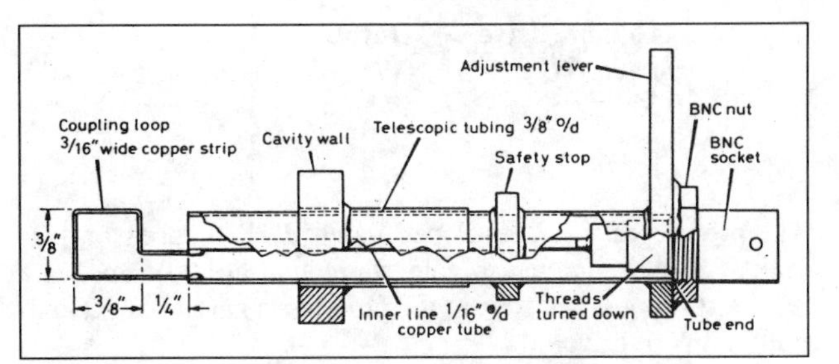

Fig 9.72. Output coupling probe

as before. Coupling with the magnetic field in the cavity is strongest when the loop is almost fully withdrawn to the cavity wall and at right-angles to the baseplate. Coarse loading adjustment is by sliding the loop in and out, and fine adjustment by rotating it.

A general view of the amplifier is shown in the photograph (Fig 9.68) and the leading dimensions are given in Figs 9.69 to 9.73. Non-critical dimensions are not given, being at the discretion of the constructor. As noted earlier, the dimensions of the anode and grid connectors are only critical in that they must be adjusted to provide a good fit to the valve. However, the two collets *must* be coaxial in order to avoid shear forces on the valve, and detailed assembly instructions are given at the end of this section.

The cathode circuitry below the baseplate (Figs 9.70 and 9.71) is assembled after the grid collet has been soldered into place. Rather than fabricating the RF bypass capacitor for the 'cold' end of the cathode stripline, the present design uses the entire end-wall to act as the capacitor by making it from double-sided glassfibre PCB (Fig 9.71), chamfering the copper from the inside edges to prevent a DC short-circuit.

Contrary to popular belief, grounded-grid amplifiers are not unconditionally stable and this high-gain design requires some attention to the possibility of stray feedback paths. Some stability problems were encountered when one of the prototypes was operated very close to the transverter driving it, the system gain at 1.3GHz being of the order of 40dB. The top of the cathode compartment was therefore covered with a close-fitting lid of perforated copper sheet. This, together with careful bypassing of the heater supplies, solved the problem completely.

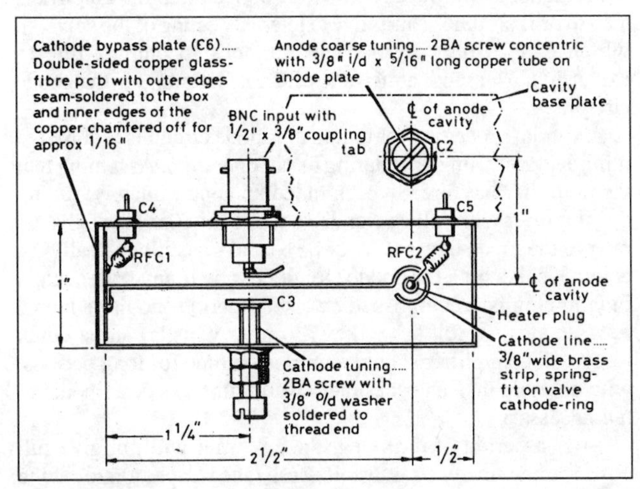

Fig 9.71. Details of cathode box and anode coarse tuning capacitor

The sliding loop coupling probe (Fig 9.72) is made using telescoping brass tubing available from good model shops. A safety stop *must* be provided to prevent the loop from touching the anode sleeve of the valve. The tuning paddle (Fig 9.73) needs to be well grounded to RF. This can be ensured by a strong compression spring over the shaft, which maintains a firm contact between the paddle and shaft bushing. It is helpful if the external controls indicate the true orientations of the loop and paddle within the cavity. As an optional refinement, all components can be silver-plated. The brass parts of the prototypes were given a thin but tenacious coating of silver by the method given in the constructional chapter.

The circuit diagram of the amplifier is very simple (Fig 9.74). For maximum gain, a fairly high standing current of the order of 50mA is required, ie DC efficiency has to be sacrificed. At low drive levels the amplifier will operate at virtually constant anode current, so simple cathode-resistor biasing will suffice. During development of the amplifier a 250Ω wire-wound potentiometer proved quite satisfactory, and a 22kΩ resistor connected from the cathode bypass to ground allows the valve to cut off safely during receive periods or if the bias resistor fails. At higher drive levels, constant voltage biasing must be used in order to maintain linearity on SSB, and an arrangement in which a single transistor acts as both bias regulator and T/R switch is shown in Fig 9.74. The zener diode sets the cut-off bias on receive and limits the transistors' collector voltage to below V_{ceo}.

The usual precautions regarding heater voltage should be observed when using the amplifier. At no time must the heater voltage exceed 6V. It is important that the cathode of the valve be allowed to reach full operating temperature before the anode voltage is applied. A delay of 60–90s is adequate.

The power gain achievable will depend on the type of valve and on its operating history, if it is second-hand. One of the prototype amplifiers, using a good but not remarkable 7289, gave the following measured performance with an EHT supply of 1kV, when the input and output matching were optimised to suit the available level of drive power.

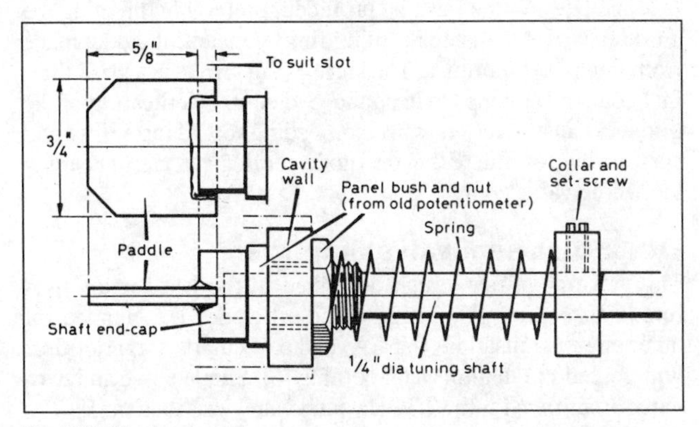

Fig 9.73. Tuning paddle

Available drive power (W)	Output (W)	Power gain (dB)
0.35	27	19
0.5	32	18
1.0	40	16

The prototypes were developed using EHT supplies of 1.0–1.1kV. Some reduction in gain was found at 800V and, in a brief test using 1.5kV, one of the prototypes gave 60W RF output for 1W of drive.

Assembly of the cavity is as follows:

1. Mark out the locations of the side walls and the centre of the cavity on the baseplate.
2. Solder the sidewalls in position. If the ends of the bars can be faced-off square (not impossible by hand or with a three-jaw lathe chuck) they may be pre-assembled into a square frame before soldering. After soldering, hone the top face of the sidewalls flat.
3. Mark out and pilot drill the 20 fixing holes in the cavity top-plate. Do *not* drill the centre hole yet. Tape the top plate accurately into position on the side walls and, on a drill press, drill two holes in diagonally opposite positions through the top plate and into the side walls. Tap these two holes and secure the top plate more firmly before drilling and tapping the rest of the fixing holes.
4. Again on a drill-press, drill square through the pilot hole in the base plate and through the top plate.
5. Use the pilot hole in the cavity top plate to locate the centre of the anode bypass plate when marking and drilling through the latter for the four retaining screws.
6. Open out all pilot holes to full size. Be careful to retain concentricity.
7. Drill the four holes in the anode bypass plate slightly oversize for the shoulders of the available insulating bushes. Leave the retaining screws slack until the valve has been fitted squarely into place for the first time, then tighten them.

Apply about 800V HT, and set the cathode current to about 50mA with no drive. Apply about 0.5W of drive and adjust the cathode tuning to maximise total cathode current. Then tune the anode cavity and adjust the output loop for maximum power output. The anode voltage can now be increased up to 1.1kV and the bias should be reset to give 50mA cathode current with no drive. All tuning adjustments should be re-optimised.

Drive power can be increased until the cathode current reaches 150mA. Higher drive levels than this will produce more power, but valve life will be shortened. At high drive levels, do not run the amplifier at full output continuously for more than a few seconds.

Adequate cooling must be provided, preferably directing the air-stream over the anode cooling fins by means of a duct made from Perspex or Formica. The spring-ring provides a good thermal contact and helps to keep the grid cool. Overheating of the grid can cause electron emission, leading to DC instability and shortened valve life. Efficient anode cooling can aid in keeping the whole valve cool.

Other disc-seal valve amplifiers

The valve amplifier design described in the last section uses forced-air cooling. Higher gain and efficiency are claimed for other circuit realisations (N6CA) and particularly for amplifiers with forced-circulation water cooling [6]. Single valve and two-valve amplifiers for the 2.3GHz band were described in [5]. The 3.4GHz band is more problematical, since the conventional

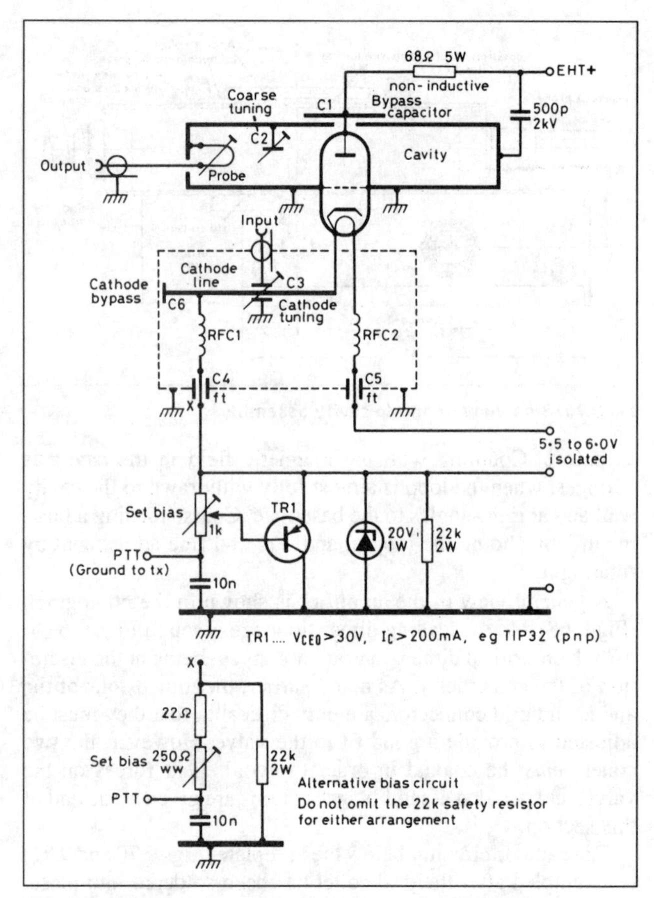

Fig 9.74. Circuit diagram of amplifier

2C39A type of disc-seal triode has lost gain and efficiency at this frequency. However, a few amateur designs have been published, such as that in reference [30].

Using 2C39 amplifiers

All amplifiers of this type can suffer from differential thermal effects during transmit and receive periods. This difficulty can largely be overcome by cooling the valve adequately on transmit and reducing or removing the airflow on receive, so that its temperature remains more nearly constant. A further improvement could be expected from the use of one of the modern temperature-compensated derivatives of the 2C39, eg the 7855.

At higher power levels, it will be noticed that the amplifiers tend to drift off tune. This is due to internal heating of the valve(s), which causes the internal capacitances to change due to expansion. The most noticeable effect is detuning of the anode circuit.

A special procedure for tuning the anode circuit is required to minimise the tuning drift during on-air operation. Assuming that the amplifier has already been tuned up approximately, run up the amplifier with full carrier drive for about 10s. Then, change over to the mode in use (CW or SSB) and transmit normally for about 30s, keeping the anode circuit on tune if any drift occurs. Final tuning is accomplished by a short period (no more than a few seconds) of full drive (key down, or whistle) and a quick retune. The amplifier should then stay on tune for long periods, with no more than an occasional 'warm tune', as described, being necessary.

After a period of receiving, the amplifier will not give full power immediately on transmit. Resist the temptation to retune but keep transmitting, possibly at full carrier to accelerate the

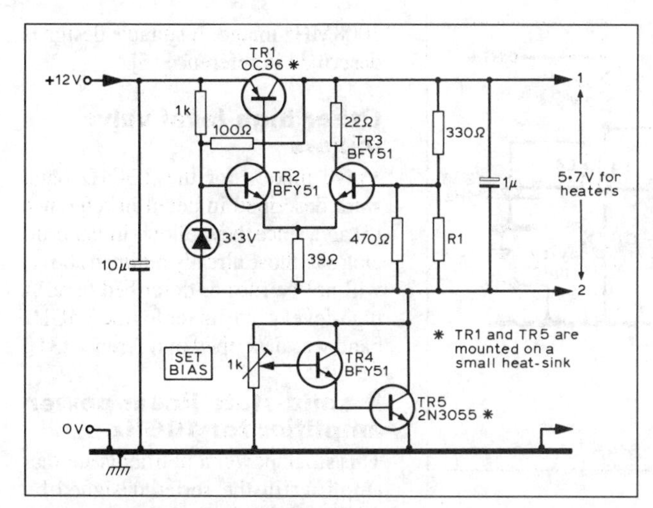

Fig 9.75. Circuit diagram of the G3ZUD heater/cathode bias supply. The zener diode is rated at 5W

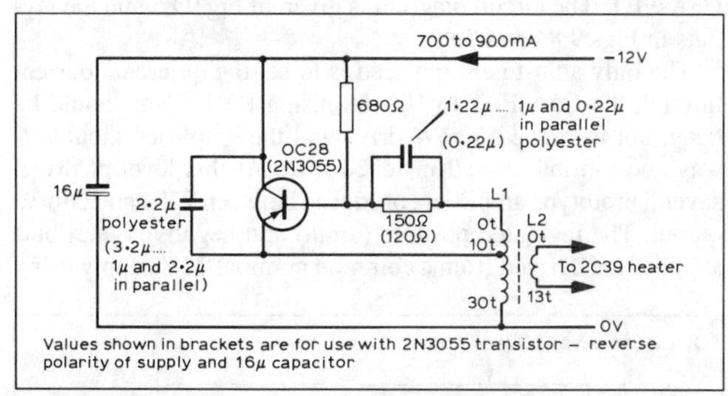

Fig 9.76. Circuit diagram of the GW8AAP 2C39 heater inverter power supply. L1 and L2 wound on LA1 or LA3 pot-core. L2 (1mm ECW) wound first, L1 (0.5mm ECW) wound second

process, until full power is reached. The time taken to reach full power can be reduced if blowers are switched off during receive periods.

Heater supply circuits for 2C39 amplifiers

Since the 2C39 family of valves is normally operated in grounded-grid and the cathode and heater are connected together, this means that an isolated heater supply is required or the valve cannot be biased correctly. When a mains supply is available, this is not a problem since a transformer can be used with a separate winding for each heater if more than one valve is used. However, when a single 12V DC supply is all that is available (as might be the case when operating portable), providing

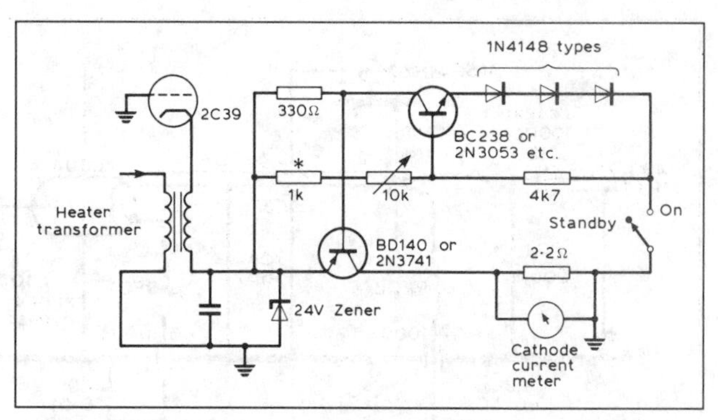

Fig 9.78. Bias circuit by LA8AK. Note: the value of the resistor marked '*' sets the lower bias voltage limit

isolated supplies is rather more difficult. Two circuits are given to overcome this problem.

The first, due to G3ZUD, is suitable for use when one 2C39 is to be powered from a 12V source and is shown in Fig 9.75. TR1, 2 and 3 form a stabilised supply for the heater; R1 should be selected on test to give the desired output voltage (5.7–6.0V). TR4 and 5 form a simple constant-voltage reference for the cathode bias. The voltage at output 2 can be varied from 1.5–6.0V and the corresponding output at output 1 varies between 7.5–12V. Thus, connecting the cathode to either output allows a wide range of cathode bias voltage to be used.

The second circuit was developed by GW8AAP with some assistance from G3AVJ and is shown in Fig 9.76. This uses a DC-AC inverter to produce the heater supply; the DC isolation between the windings of the transformer provides the necessary isolation of the heater/cathode from ground. One inverter circuit is required for each valve used. Efficiency is good and the inverter will start reliably from 'cold' under load. The values may need some adjustment, however, if different transistors or pot-cores are used. The transistors should be mounted on an adequate heatsink.

Bias circuits for 2C39 amplifiers

Many bias circuits have been published for this series of valves. Two selected circuits are illustrated in Figs 9.77 and 9.78. Neither is critical in construction and both are equally effective. The choice is left to the constructor.

Valve amplifiers as power mixers: a high-level linear transmit mixer for 1.3GHz

This high-level transmit mixer, capable of several watts output, uses a 2C39A valve in grounded-grid. The circuit diagram of the mixer is shown in Fig 9.79. The 144MHz drive and 1152MHz local oscillator signals are combined in the cathode circuitry and fed to the cathode of the valve. The circuit is designed to provide high isolation between the two input ports. Since the valve is operated in grounded-grid mode, an isolated heater supply must be used. The cathode circuitry, together with the component values, is shown in Fig 9.80. It is built in a trough-section box, mechanical details of which are shown in Figs 9.81 and 9.82. A tight-fitting lid is recommended to prevent radiation of the 144 and 1152MHz signals. A screen/cathode bypass capacitor fabricated from 1.6mm double copper-clad epoxy

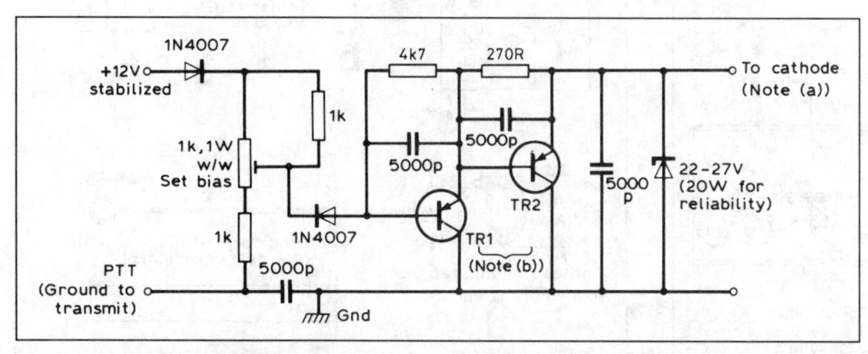

Fig 9.77. Bias circuit by G6CMS. Note: (a) maximum current dependent on H_{fe} product of TR1 and TR2 and maximum power dissipation limit of TR2. (b) TR1 V_{ceo} 50, I_c max 1A; TR2 V_{ceo} 50, I_c max 10A (use high-H_{fe} device for TR1)

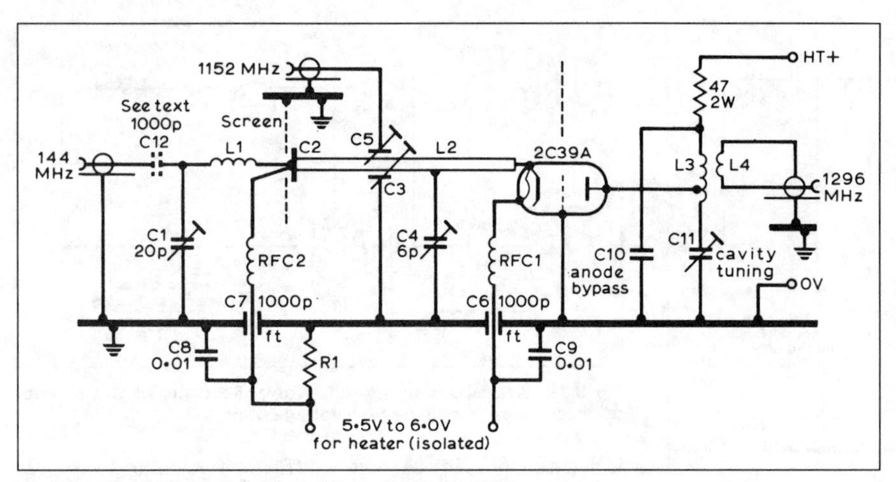

Fig 9.79. Circuit for a high-level valve mixer for 1.3GHz

PCB material is soldered into the trough. The cathode and heater connections are made from thin brass sheet. The anode cavity recommended is the same as that used in the single-valve power amplifier described before. The input circuit can be mounted on the anode cavity by a number of small L-shaped pieces of brass, soldered to the trough and screwed to the anode cavity (Fig 9.82).

Apply 5.5–6.0V to the heater and +400V to the anode, and ensure that there is adequate forced-air cooling applied to the latter. The cathode current should be set to approximately 20mA by varying R1. A value of 270W should be found suitable for initial tests.

Approximately 2–3W of 1152MHz drive should then be applied and the 6pF trimmer adjusted for maximum anode current. Fine tuning may be accomplished by adjusting the 2BA screw. The input matching can be optimised by bending the input tab to vary its spacing from the line and then retuning. When the anode current has been maximised it should be in the range 50–60mA. Leaving the 1152MHz drive connected, apply about 5W of 144MHz drive and adjust the 20pF capacitor for lowest input VSWR. The anode current should then be around 80–100mA. The anode tuning and loading should then be optimised for maximum power output.

When the unit is operating correctly, the anode voltage may be increased to 800–1000V. Since the mixer is only about 10% efficient, adequate forced-air cooling is essential to prolong the life of the valve. Power output at 1000V is about 10W and 3–4W if a 400V supply is used.

If it is intended to use the mixer alone as the transmitter, a filter should be used to reject the 1152MHz feedthrough and 1008MHz image. A suitable design is described in reference [5].

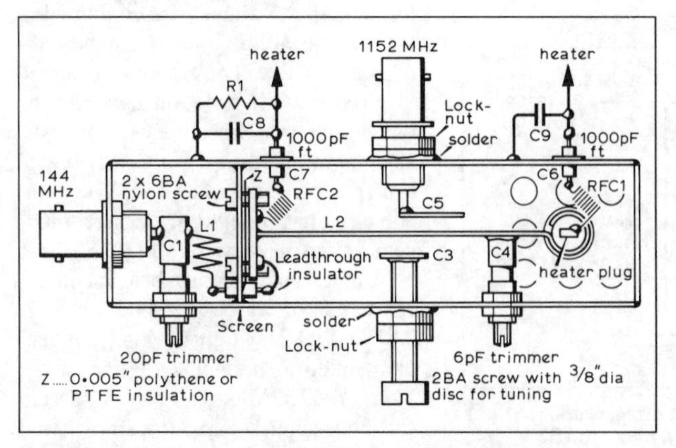

Fig 9.80. Layout of the cathode circuit, high-level mixer

Other high-level valve mixers

Valve mixers for the 2.3GHz band were described in detail in reference [5] and, since they follow similar principles to those already outlined above, will not be further described here. A high-level valve mixer for the 3.4GHz band was described in reference [31].

A solid-state linear power amplifier for 10GHz

This small power amplifier is another amplifier in the series designed by G3WDG [21] for home construction. It is a single-stage amplifier using a Mitsubishi MGF1801 GaAsFET. The circuit diagram is given in Fig 9.83 and the layouts in Figs 9.84 and 9.85.

The only adjustment required is to set the quiescent current through the GaAsFET to 100mA, using RV1. Care should be taken not to exceed 50mW drive and the amplifier should always be run into a well-matched load. At this level of drive, several prototype amplifiers produced between 275 and 325mW output. The regulated positive (drain) and negative (gate) bias supplies are derived from a common nominal 12V supply using

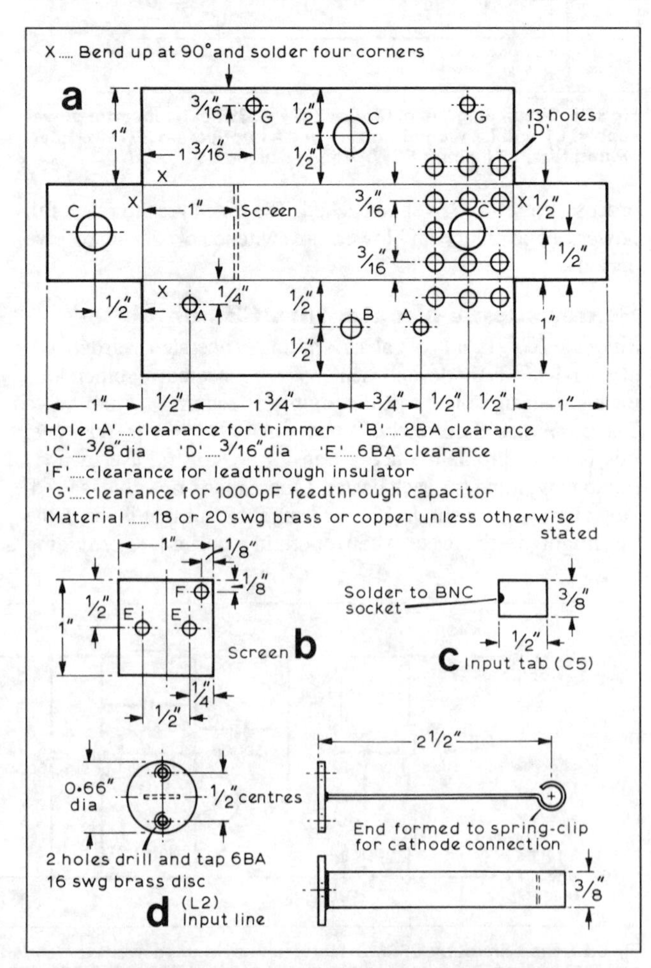

Fig 9.81. High-level mixer: details of construction of (a) the trough, (b) the screen, (c) the input tab, (d) the input line

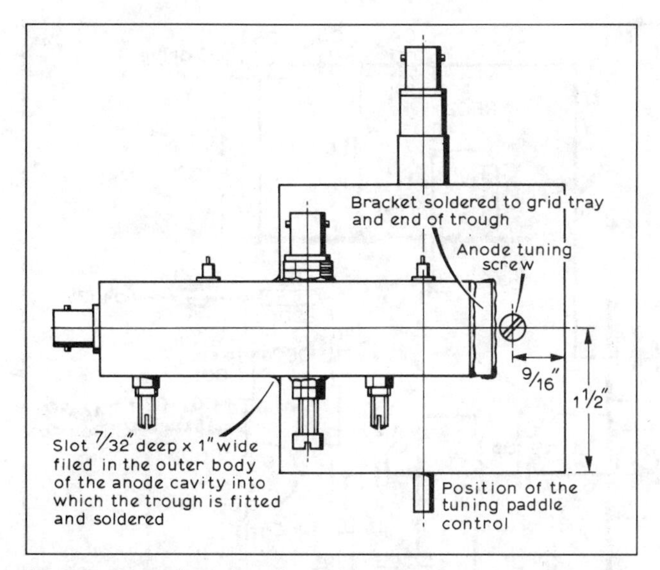

Fig 9.82. Approximate location of input circuitry on anode cavity

the circuit and layout of the regulator shown in Figs 9.19 and 9.20.

Using similar design principles, circuit and layout, a further linear power amplifier requiring a maximum of 230mW drive to produce 1W output (typically 1.1 to 1.2W, device dependent), designated the G3WDG-007, has recently been designed. This uses a Mitsubishi MGF2403A power GaAsFET and, like the other modules in this series, has proved reliable, reproducible and entirely suitable for home-construction. Full details are available in reference [60] and a kit of parts, including the essential heatsink block, is available from the Microwave Committee Components Service [40].

RECEIVERS

Nearly all amateur microwave receivers now employ crystal-controlled oscillator converters to mix incoming signals to a convenient intermediate frequency, usually 144MHz. The only exceptions are probably Gunn self-excited oscillator sources used

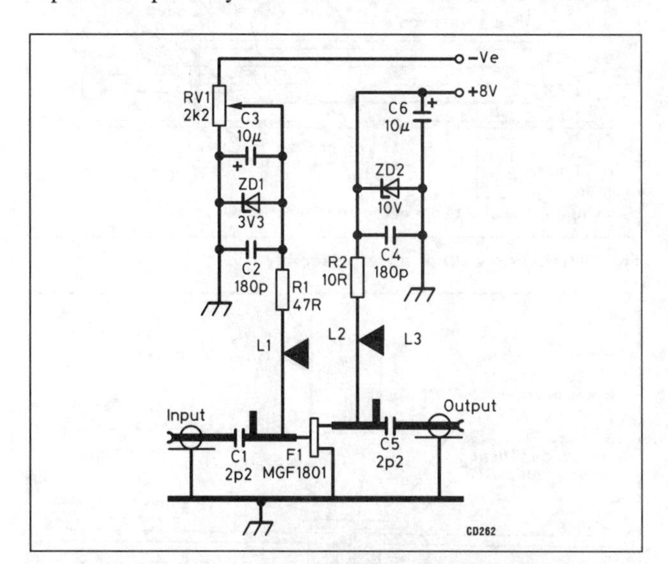

Fig 9.83. Circuit of the G3WDG-006 10GHz power amplifier. R1 and R2 are SMD (0805 size) chip type. C1 and C5 are ATC types. C2 and C4 are SMD (0805 size) chip type. C3 and C6 are 10V wkg tantalum bead type. L1 and L2 are 0.2mm wire inductors. ZD1 and ZD2 are 400mW types

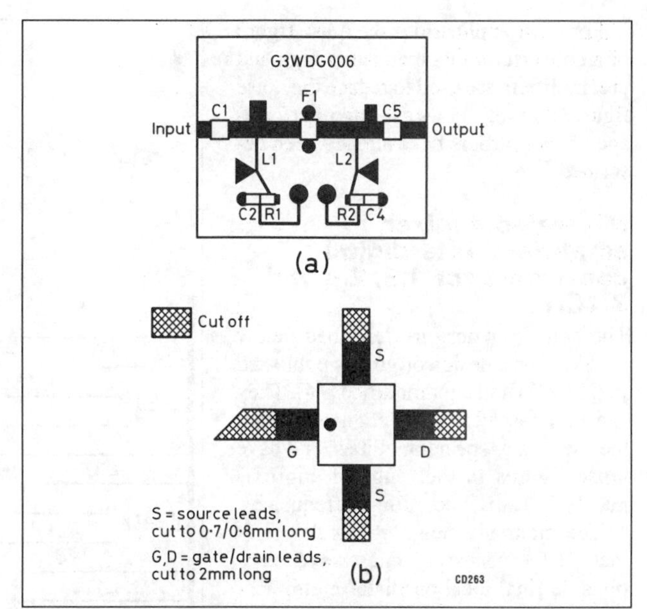

Fig 9.84. (a) Layout of RF components for the G3WDG-006 power amplifier. (b) MGF1801 connections and lead preparation. (Not to scale)

by amateurs for simple wide-band transmitters and receivers in the 10, 24 and 47GHz bands.

Converters may or may not employ integral pre-mixer low-noise amplifiers, but most employ an integral low-noise post-mixer amplifier to help present a good match and signal transfer to the following intermediate frequency receiver.

In this section converters are described for the lower microwave bands using mixers based on interdigital filters. They are recommended as simple, general-purpose converters. A second type of design uses broad-band amplifiers, microstrip, printed 'hairpin' filters and requires less mechanical engineering in its construction: these techniques are equally suitable for both transmit and receive purposes and so it is possible to combine a receive converter, a transmit converter and the final LO multiplier on a single PCB. This type of receive converter is described in the section on integrated equipment and transverters.

Both types of circuit, properly dimensioned, are suitable for the 1.3, 2.3 and 3.4GHz amateur bands. At frequencies above this (for instance 5.7, 10 and 24GHz), the smaller mechanical dimensions and closer tolerances required rules out the interdigital approach, and similarly the dimensional accuracy required of 'printed' filters is such that it is difficult to produce them with adequate Q, accurate coupling and low loss so that alternative arrangements must be sought. This may involve the use of coupled cavities constructed within waveguide [5, 6] or the use of small high-Q 'pill-box' cavities [32] probe-coupled into and out of impedance-matched microstrip transmission lines, such as those in the 10GHz source already described.

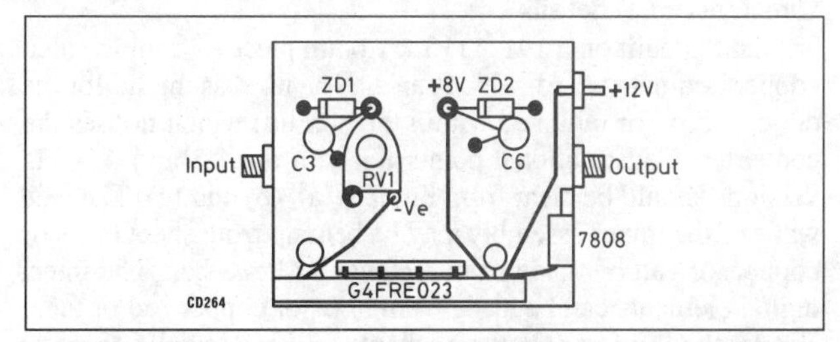

Fig 9.85. Layout of DC circuitry for the G3WDG-006 power amplifier (not to scale)

For most applications the noise figure of a converter on its own is too high and preamplifiers are used to reduce the noise figure. Suitable low-cost, high performance preamplifiers have already been described.

Mixers/post-mixer amplifiers: interdigital converters for 1.3, 2.3 and 3.4GHz

The converter designs described below are based on a design originally published in *QST* [33] and later modified [34]. They can be constructed and aligned without the use of any special facilities, and have noise figures in the range of approximately 8–12dB, according to frequency.

The main advantage of this design is that all the necessary microwave functions, ie final local oscillator multiplication, filtering, mixing and signal filtering, are all performed by one assembly which is based on an interdigital filter. This means that a separate final local oscillator multiplier does not have to be constructed, nor is any additional signal filtering necessary to suppress the image response. The interdigital design has been found very easy to duplicate and get working, and is in use by a large number of stations. Although a certain amount of metalwork is involved, it is especially recommended for beginners.

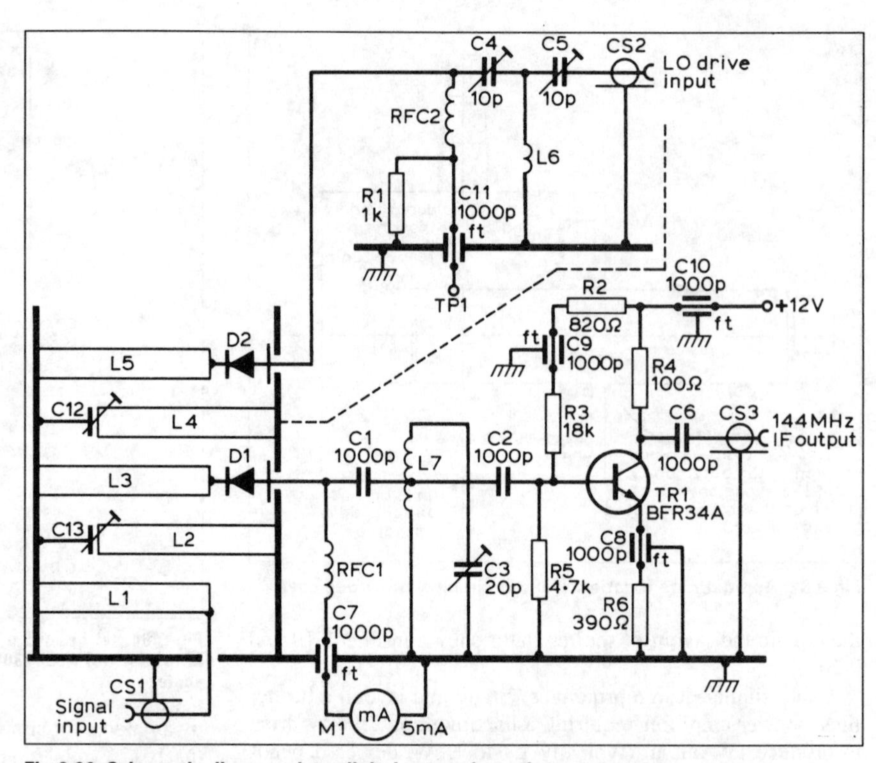

Fig 9.86. Schematic diagram, interdigital network receive converter

The heart of the unit is the interdigital network. Referring to Fig 9.86, this network consists of five rod elements L1–L5. L1, L3 and L5 are low-*Q* coupling elements, while L2 and L4 are resonant at the signal and local oscillator frequencies respectively, and perform the filtering functions. D2 acts as a multiplier producing a few milliwatts at the local oscillator frequency which varies, according to the application, as follows, assuming an IF of 144 to 146MHz:

Band (MHz)	Drive frequency (MHz)	LO required (MHz)	Multiplication factor
1296–1298	384	1152	×3
2320–2322	362.666	2176	×6
3456–3458	368	3312	×9

Drive at a level of 50 to 150mW at the LO input connection is sufficient for all these applications, producing a mixer current of several milliamps in all versions of the converter. A UHF LO design to provide this drive power was described early in this chapter. D1 is the mixer diode and the 144MHz IF output is fed to a low-noise amplifier stage using a BFR34A transistor.

Constructional details

The unit is built on a 191 × 115 × 1.6mm piece of double-sided copper laminate board, which can also be used as the lid for the die-cast box (or mounted within the box lid) which houses the converter. Constructional points for both the 1.3 and 2.3GHz versions should be clear from Fig 9.87(a), (b) and (c). The side walls of the mixer assembly may be bent up from sheet brass or copper, or can be made from rectangular brass bar. The interdigital elements can be made from brass or copper rod or tube. The method used to fix the elements to the sidewalls depends on whether rod or tube elements are used. Rod elements can be fixed by means of a tapped hole in one end or by soldering. Tube elements can be soldered, or held in place by means of a filed nut, soldered into one end of the element. A cover made from double-sided copper laminate is fitted to the mixer assembly as the final stage of construction. No screening is required at the ends of the interdigital unit. If the local oscillator source is built into the same box as the mixer assembly, the IF preamplifier will have to be screened by means of an enclosed box which can be fabricated using double-sided copper laminate. Fig 9.88 shows

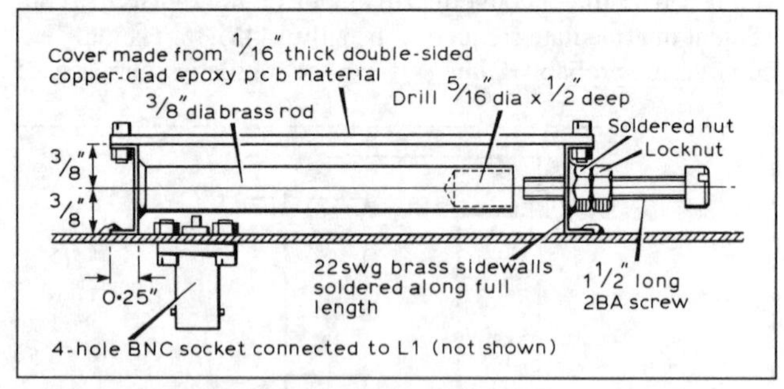

Fig 9.87(a). Side view of mixer assembly

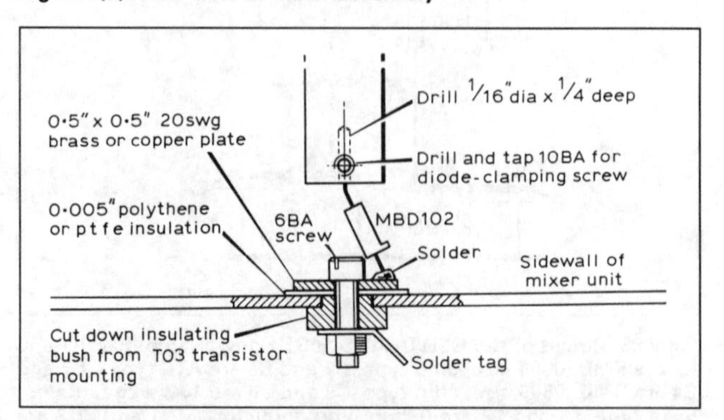

Fig 9.87(b). Method of fitting D1 and D2

the layout and dimensions for the 3.4GHz version of the converter.

Alignment

The 1.3GHz version of the converter can be aligned without any special equipment in the following way; alignment of the 2.3 or 3.4GHz versions is very similar although, obviously, the frequencies involved are different:

1. Apply 50 to 150mW drive at 384MHz (nominal) to the LO drive input socket, and connect a voltmeter between TP1 and earth. Carefully adjust the 10pF trimmers C4 and C5 using an insulated trimming tool until a maximum reading of at least 1.5V is obtained.
2. Monitor mixer current on the meter fitted for the purpose, insert the tuning screw C12 and screw in until maximum mixer current is obtained. This should be at least 1mA. With the specified dimensions the first peak observed will correspond to the correct tuning point.

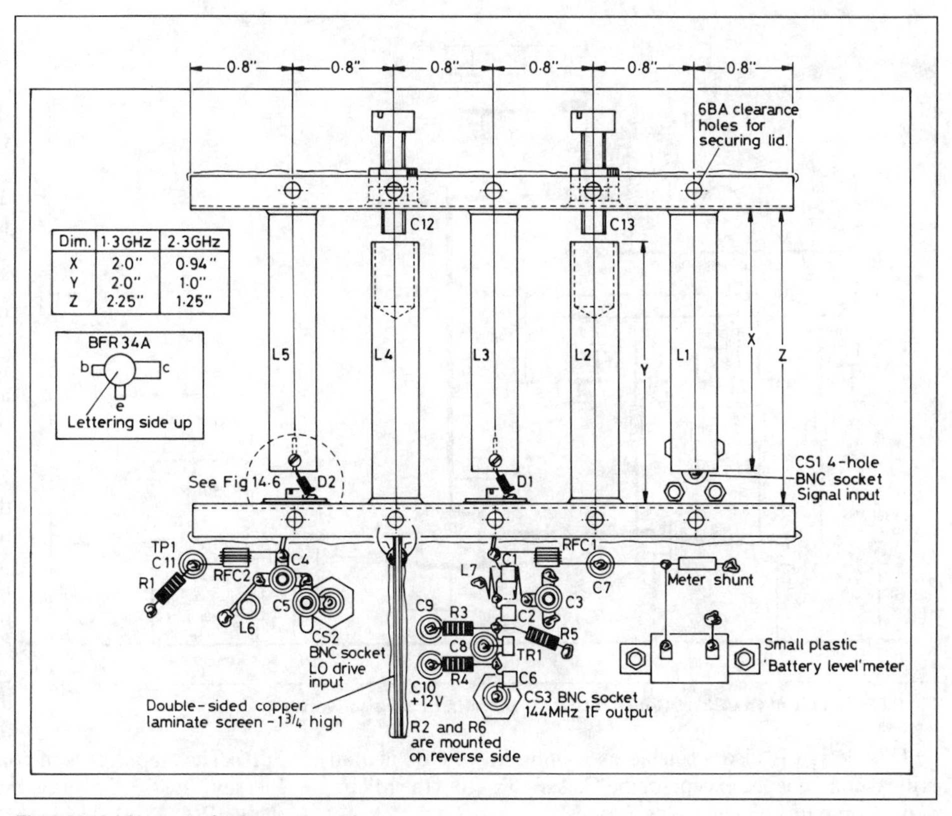

Fig 9.87. (c) Layout of mixer assembly

3. Connect a 144MHz receiver to the IF output socket and adjust the 20pF trimmer (C3) for maximum noise. If no peak is found, it will be necessary to experiment with the tapping points of the 1000pF capacitors on L7.
4. Connect a preamplifier or noise source to the antenna socket, insert tuning screw C13 and screw in until a noise peak is heard. Alternatively, an antenna can be connected, and C13 screwed in until maximum signal strength is obtained from a local station or beacon.
5. Carefully readjust C12, C13 and the 20pF trimmer C3 for best signal-to-noise ratio on a weak signal. An automatic noise figure optimisation aid may also be used. A description of suitable amateur equipment and its method of use was given in reference [19].

The post-mixer low-noise amplifier, using a BFR34A bipolar transistor tuned to the amateur 144–146MHz band, is typical of a 'universal' VHF amplifier of simple construction and good performance. The circuit was shown in Fig 9.86: layout is uncritical and may be that shown in Fig 9.87 or the slightly different version shown in Fig 9.88.

The G3WDG-002 10GHz to 144MHz receive converter

Like the 10GHz signal source and various 10GHz amplifiers already described, this converter is another module in the series designed by G3WDG [35]. This module requires a drive input of approximately 10mW in the 2.5–2.6GHz region. This can conveniently be provided by the G4DDK-004 module described earlier.

It incorporates a ×4 multiplier chain to generate the local oscillator signal, a dual-diode mixer and two stages of low-noise preamplification before the mixer. The design also incorporates the familiar BFR34A low-noise post-mixer amplifier at the intermediate frequency. The front-end noise figure of several prototypes has been measured at less than 3dB. It is possible to improve this figure by using an external preamplifier, for instance using a HEMT device which, in a properly configured circuit, can yield noise figures of 1dB or less.

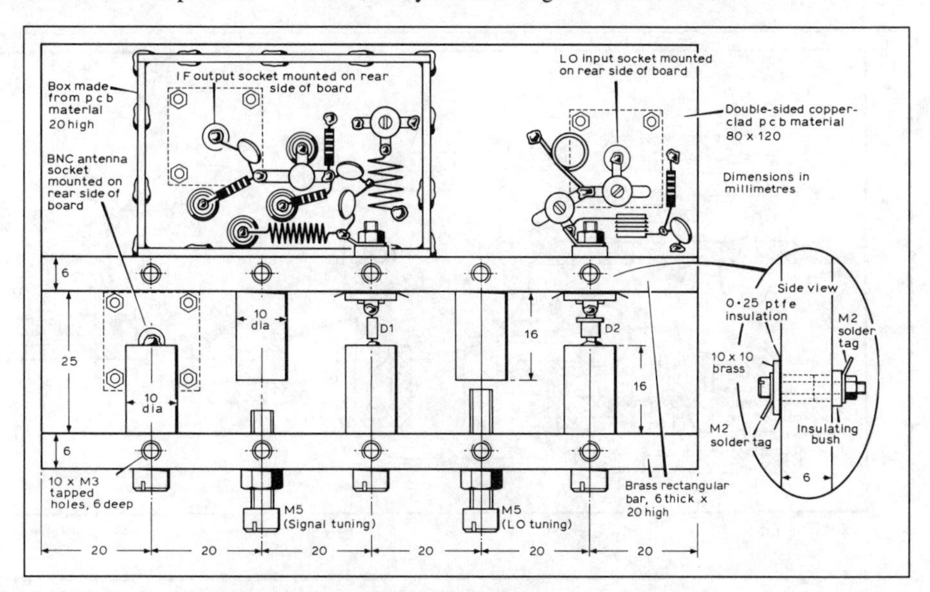

Fig 9.88. Layout of the interdigital converter for 3.4GHz

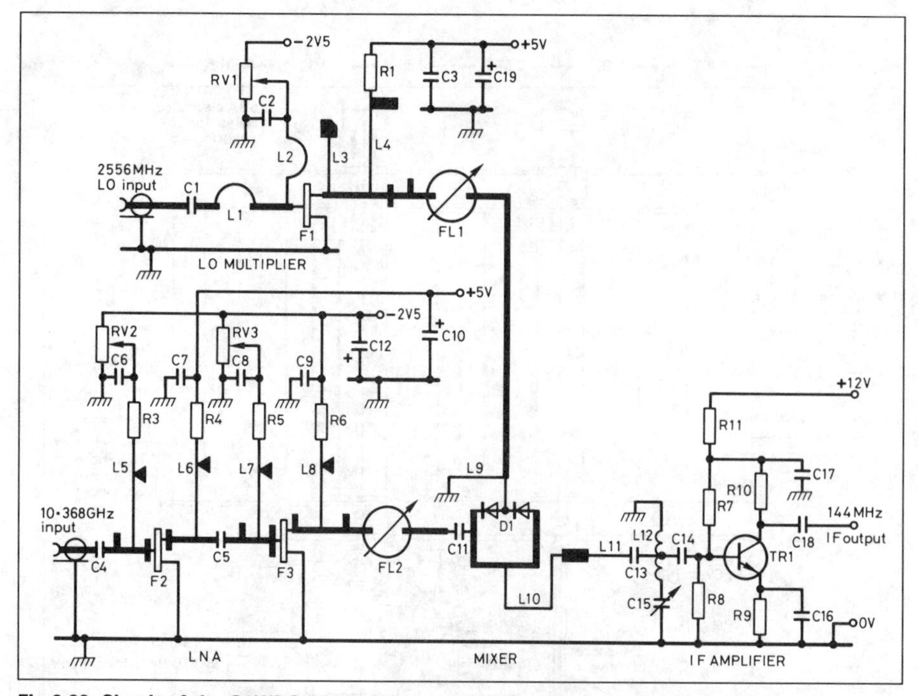

Fig 9.89. Circuit of the G3WDG-002 10GHz to 144MHz receive converter

This design is also available as a 'mini-kit', PCBs and all critical components, except for the GaAsFETs, from the Microwave Committee Components Service.

The circuit is shown in Fig 9.89, the layout of the board and components in Fig 9.90 and their values in Table 9.17. Referring to the circuit diagram, Fig 9.89, the 2556MHz LO input signal is fed to F1 which acts as a ×4 frequency multiplier producing a few milliwatts at 10,224MHz. The multiplier circuit is identical to that used in the G3WDG-001 module. The input signal is fed to the gate of the FET via a lumped-element matching network L1/L2. The cold end of L2 is decoupled via C2 and negative gate bias is applied via L2. The output of the FET is matched to 50Ω via microstrip elements, and drain bias is fed via a λ/4 line, L4, which is decoupled at 10GHz by a low-impedance λ/4 stub. L3 and the chamfered element (a microstrip shunt capacitor) form a series-resonant circuit at the input frequency (2556MHz) to improve the efficiency of the multiplier. Wide-band stability is provided by decoupling elements R1, C3 and C19. A number of harmonics are present in the output from the FET. The wanted fourth harmonic is selected by the cavity filter FL1 and passed to the LO port of the hybrid ring mixer via a microstrip matching network.

The mixer uses a series diode pair D1 connected between the ends of a folded 3λ/4 line. This configuration gives good rejection of the LO signal at the RF port and vice versa. L9 is a shorted λ/4 line to provide the required low-impedance IF return path, while having no effect at the LO frequency. The conversion loss of the mixer including the matching networks is about 6–7dB.

A two-stage low-noise RF preamplifier (LNA) is provided to reduce the noise figure of the unit to a more acceptable level (less than 3dB). The amplifier is of conventional design and uses surplus Plessey FETs. It is very similar to the power amplifier used in the G3WDG-001 design, except that a low-noise FET can be used in the input stage as an option. F2 can be either a 'red spot' or 'black spot' surplus GaAsFET according to the level of performance required. Overall noise figures of 2.6dB with a red-spot FET and 3.2dB with a black-spot FET are typical, but see the comments on stability later. The input circuit of F2 has an optional stub, which in some cases when connected can reduce the noise figure by a small amount. It is usually not required. The overall gain of the LNA is about 19–20dB and its own inherent noise figure in the region of 1.9–2.5dB.

The output from the LNA goes to the RF port of the mixer via a microstrip matching network and a high-pass filter (C11) after passing through filter FL2 which provides about 20dB of image rejection (with a 144MHz IF). Note that FL2 can easily be tuned

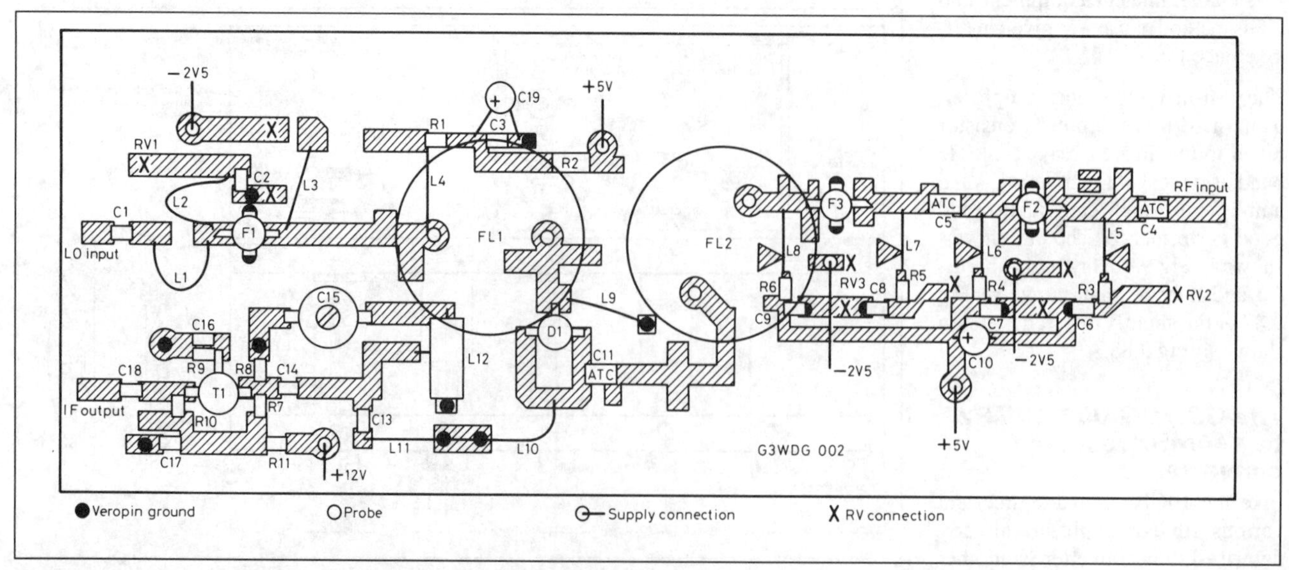

Fig 9.90. Layout of the RF components for the G3WDG-002 converter

Table 9.17. Components list for the G3WDG-002 10GHz to 144MHz receive converter

RESISTORS	
R1, 3–6	47R SMD, 0805 size (*)
R2	220R SMD, 1206 size, (or ¼W leaded)
R7	18k SMD, 0805 size (*)
R8	4k7 SMD, 0805 size (*)
R9	270R SMD, 0805 size (*)
R10	100R SMD, 0805 size (*)
R11	560R SMD, 0805 size (*)
RV1–3	2k2 horizontal preset, eg Allen-Bradley 90H, Bournes VA05H or Philips OCP10H etc
CAPACITORS	
C1–3, 6–9	220p SMD, 0805 size (*)
C13, 14, 16–18	1000p SMD, 0805 size (*)
C4, 5, 11	2p2 ATC chip capacitor, 100 or 130 series (*)
C12	22 to 47μ tantalum bead, 10V wkg
C10, 19	2μ2 to 10μ tantalum bead, 10V wkg
C15	30p trimmer, 5mm diameter, eg Murata TZ03Z300 (Green)
INDUCTORS	
L1	16mm length of 0.315mm dia ECW (*), formed into a hairpin. 1mm each end to be tinned and soldered to tracks as shown in layout
L2	As L1, but 19.5mm long
L3	Straight length of 0.315mm dia ECW (*), tinned 1mm each end, as above. The wire should be soldered to the edge of the stub and as close to the drain connection of F1 as possible
L4	Straight length of 0.315mm diameter ECW (*), tinned 1mm each end, as above. Solder between the stub edge and the probe connection to FL1
L5–8	Straight length of 0.2mm dia (not too critical) tinned or silver-plated copper wire. Solder between the track, stub point and terminal pad as shown in layout
L9	10mm length of 0.2mm diameter wire, as L5. Bend to fit between earth pin and mixer connection
L10 + L11	Single 20mm length of 0.2mm tinned or silver-plated copper wire, as L5. Solder between mixer centre, stub and C13, as shown in layout
L12	4t of 0.6mm diameter tinned or silver-plated copper wire. Wound to 5mm ID, turns spaced 1/2 wire diameter. Centre-tapped. Mount 1mm above the board.
FL1, 2	Cavity resonators as for G3WDG-001. See text and diagrams for details (*)
SEMICONDUCTORS	
F1, 3	P35-1108 GaAsFET (Birkett black spot)
F2	P35-1145 GaAsFET (Birkett red spot) or P35-1108 GaAsFET (Birkett black spot) – see text
D1	Alpha series dual diode (DMF 3909-99) (*)
TR1	BFR90/91
MISCELLANEOUS	
	Approx 33 PCB pins, RS Components 433-864 or Vero, 1mm dia, 1.5mm head dia (*)
	Three SMA sockets, flange fitting
	Tinplate box, type 7754 (37 × 111 × 30mm) from Piper Communications
	Feedthrough capacitors, solder-in, 1–10n, or Filtercons
	Regulated power supplies: positive from 7805 IC, negative from ICL7660 voltage converter, G4FRE-023 PCB (*) and components
	G3WDG-002 PCB (*)
	One piece of lossy rubber (*)

(*) Parts marked thus are available in a 'short' kit

to the wrong image (10,080MHz), so care should be taken when tuning up (see later). The IF output from the mixer is fed to a low-noise amplifier via a low-pass filter consisting of L10, a $\lambda/4$ line, and L11, to prevent 10GHz energy from reaching the IF amplifier. The latter is the well-proven BFR34A design from another application, constructed in surface-mount form to save space.

The negative bias generator used to supply the gate bias for the FETs uses the same PCB as that used in the G3WDG-001 module (G4FRE-023). Two modifications have been made for this application – the use of a 7805 regulator and the omission of the zener diode. The circuit is shown in Fig 9.59 and the board layout in Fig 9.60. Note that C12, the negative rail

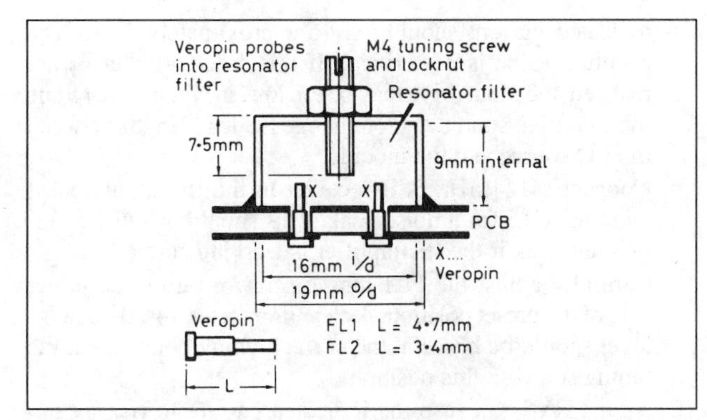

Fig 9.91. Tuning positions and probe lengths for filters for G3WDG-002 converter

decoupling capacitor, is not shown as it is fitted on the reverse side of the board.

Construction of the G3WDG-002 receive converter is quite straightforward. For more detailed methods of construction and alignment, see reference [35]. Suffice it to say that the order of construction is important and precautions need to be taken to avoid static damage during handling of the microwave diodes and FETs, or overvoltage/current damage when initially testing the converter. For this reason it is best to apply the supply voltage to the board *before* fitting the FETs, to check that both the +5V and –2.5V voltages are present and correct on the respective tracks/pins. On completion of this test, disconnect power and solder in the devices only if everything checks out correctly.

In this design there are two 'pill-box' filters to fit and note that the lengths of the filter probe-pins are significantly different for the LO and signal filters: make sure that the right pins are fitted in the right places!

Alignment with simple test-gear

Once completed, the PCB should be carefully examined for poor joints, accidental solder bridges and other forms of short-circuit. Once satisfied that all is well, the alignment procedure may begin. The constructor should already have checked before mounting the FETs and other semiconductors that the correct supply voltages will appear on the positive and negative supply rails when a 12V supply is connected to the input feedthrough capacitor.

1. Preset the FL1 and FL2 tuning screws as shown in Fig 9.91. Note that the 7.5mm dimension shown is the length of screw protruding from the locknut.
2. Turn the bias potentiometers RV1, RV2 and RV3 so that full negative bias will be applied to the gates of the three GaAsFETs when power is applied.
3. Insert a multimeter in series with the +5V supply between the regulator output and the +5V rail and set initially to, say, 500mA full-scale deflection.
4. Connect some form of matched load to the 10,368MHz input socket, such as a 10GHz-rated termination, attenuator or SMA-to-waveguide transition with a horn or similar well-matched (low VSWR) antenna connected.
5. With no oscillator drive applied, apply +12V to the power input feedthrough. The indicated current should be no more than a few microamps. Switch the range of the multimeter as necessary. If considerably more current is measured, look for short-circuits or misconnected components.
6. Adjust RV1 to give an indicated current of about 1mA on the meter.
7. Apply the LO signal (about 10mW at 2556MHz). The

indicated current should rise to approximately 7mA. The absolute value is not critical. If this order of value is not reached, the input drive may be too low, in which case retune the LO drive source and/or change the lead length between the LO source and the module.

8. Connect a 144MHz SSB receiver to the IF output socket and tune C15 for a noise peak. This should be quite a significant peak if the IF amplifier is working correctly.
9. Carefully adjust the FL1 tuning screw a turn or so either side of the preset position. A clear drop in the 144MHz noise level should be heard at the correct tuning point. Lock the tuning screw in this position.
10. Adjust RV2 to cause the indicated current to rise by approximately 12mA, and RV3 to cause a further increase of 15mA. The total current shown should now be in the order of 34mA. The LNA is now powered up at approximately the optimum bias currents for the two FETs.
11. Preliminary alignment is completed by tuning FL2 for a peak in the noise level. Careful tuning will show that two peaks can be heard. The correct one is with the tuning screw at the smaller penetration. Lock the screw in this position.
12. If a signal source is available (for example the G3WDG-001 source), check that this can be heard satisfactorily. It is worth reconfirming that FL2 is set to the correct image by tuning it for maximum signal. If no such signal source is available, try listening for other local signals on the band, perhaps harmonics from lower-frequency equipment.
13. Remove the test meter and make good the connection.
14. Final alignment can be done using a noise figure meter or a weak signal to optimise the performance. Adjust RV1, RV2, RV3, FL1, FL2 and C15 for best results. If a noise figure meter has been used, recheck with a signal source or generator that filter FL2 is on the correct frequency, as the noise figure meter will not reveal it is tuned to the wrong image!

After alignment, the converter should have a noise figure below 3.5dB. Prototypes have varied from 2.4 to 3.3dB. The overall gain should be in the region of 27–30dB.

During the initial tuning-up phase, the module should be operated with both top and bottom lids off. Under these conditions the module should be perfectly stable with either a red-spot or black-spot FET in the front end. However, problems with stability were encountered with some of the prototypes when the lid was put onto the component/microstrip side of the box. Note that the module works perfectly well without a lid and can be operated like this with no problems unless mounted close to another metal surface. In one of the prototypes, the impedance connected to the RF input socket had an effect too. Quality of construction, particularly how well the ground-plane side of the board is soldered to the box, may also have an effect. Prototypes using black-spot FETs seemed to be more stable then those using red-spot FETs.

A common cure for lid-induced oscillations (often used in commercial LNBs) is to mount a piece of lossy material in the lid above the LNA section of the unit. The choice of material is quite important and a lossy rubber, the same as that used professionally in many microwave applications, should be used. This is known as *radar absorptive material* (RAM). Black anti-static IC foam was tried initially, but was not nearly as effective as the proper material. The rubber sheet should be glued on the inside surface of the lid, as flat as possible, above the two-stage amplifier. Note that the lossy material on the lid does not degrade the performance: indeed the designers have often seen an improvement of up to 0.2dB with the lid in place!

As described above, the application is for narrow-band use at 10,368MHz. However, the design has been made sufficiently wide-band so that the unit can be used anywhere in the 10,000–10,500MHz band with virtually the same performance. All that has to be done is to choose the appropriate LO frequency and tune FL1 and FL2 to the desired LO and RF frequencies. Two particular applications might be in the 10,450–10,500MHz band for receiving future amateur satellites and (lower down in the band) for ATV.

A large amount of flexibility also exists with the choice of the IF, limited only by the tuning range of the '144MHz' tuned circuit. However, IFs below 144MHz are not recommended for high-performance applications as the image rejection will be insufficient and the noise performance will suffer. This will not be seen on a noise figure meter, though, so beware! Higher IFs should be possible by modifying the IF amplifier, although this has not been tried by the authors. However, the mixer on its own has good performance with IFs up to at least 1.3GHz. For ATV use, it should be possible to accommodate a standard amateur FM TV signal within the FL2 bandwidth, but the IF bandwidth might be too narrow. A damping resistor across L12 should increase the bandwidth but this has not been tried. A better solution would be to use a higher IF, eg 480, 612 or 1240MHz with a modified IF amplifier to suit. If a higher IF is used, the bandwidth of FL2 could be increased by using longer probes.

INTEGRATED EQUIPMENT: TRANSVERTERS

Amateurs will now seldom build a complete, self-contained microwave receiver or transmitter from scratch, preferring instead to design and build an integrated design that consists of a common LO source, linear transmit and receive converters and the appropriate DC and RF changeover/control system operating in conjunction with (and under the control of) a lower-frequency transceiver used as an IF. Such a combination of LO, transmit and receive converters and control circuits is commonly known as a *transverter*.

Most amateur operators have one or more multimode transceivers as part of their station equipment, often for one or more VHF or UHF bands such as 144MHz or 432MHz. As mentioned earlier, it is difficult to use an IF below 144MHz, since there can be image or oscillator noise problems caused by the inability to design filters to cope with the small (percentage) frequency separation when using an IF such as 28 to 30MHz. Such a low IF is suitable for many of the lower VHF/UHF bands, up to 144MHz or 432MHz, but is not suitable for bands above 1GHz for the reasons already stated.

Although 'integrated' in the sense that all the functions needed to provide complete microwave transmit and receive facilities are provided in a single transverter package, such equipment may, nevertheless, be made up from a number of discrete modules within the package, rather than the circuits being incorporated on a single board. This has the merit of allowing separate development of the circuits for each function and allows improvements to be made, either as the need arises or as improved devices and technologies become available, ie the modular approach is more flexible than would be a truly unitary, integrated approach using a single PCB. It can also avoid the possibility of feedback and instability caused by radiation from microstrip circuits. Even commercial multiband amateur equipment tends to adopt this general design philosophy, being built-up from converter or transverter modules working into a complex, multimode transceiver, itself being modular in construction.

The transmit converter *(up-converter)*, by mixing an IF

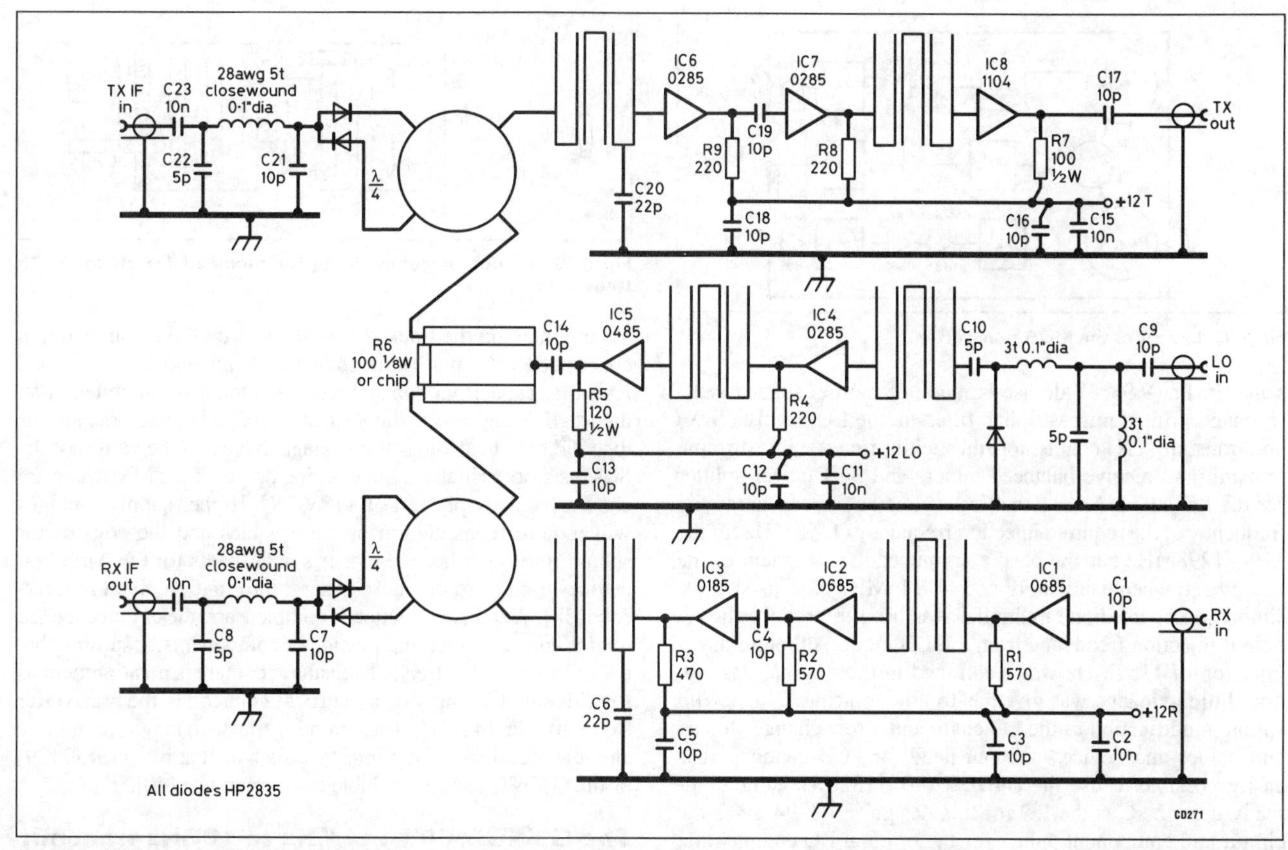

Fig 9.92. Circuit diagram, KK7B transverter. BA481 Schottky diodes can be used instead of the HP5082-2835 diodes. Other, more recent, MMICs can be substituted for the specified types, provided that the supply resistors are calculated (Ohm's Law) and adjusted in value to suit the devices chosen (see reference [31])

signal with the LO signal, followed by filters to remove unwanted products and amplifiers to the required power output level, will allow the lower frequency signal to be translated or *up-converted* to the required microwave band.

Similarly, the receive converter *(down-converter)* allows the microwave signal to be amplified and translated or *down-converted* to the intermediate frequency. In principle, the transmit and receive converters are not greatly different in concept, although they are of course designed to handle totally different signal levels. Both consist of a combination of linear mixers, linear amplifiers and suitable (usually band-pass) filters.

In theory, it is possible to use bilateral techniques, although this has been little practised by amateurs since there are the conflicting requirements for low-noise small-signal handling in a receive converter whereas good large-signal handling is required in the transmit converter. These conflicting requirements are difficult for amateurs to achieve in a bilateral circuit!

It should be obvious that both RF and DC input and output switching is necessary to accomplish the changeover from transmit to receive or vice versa, preferably controlled from the T/R switching line in the parent or 'host' transceiver used as the IF, so that some kind of interface is needed in most transverters. It should also be obvious that all the circuits in the transverter must be linear in order to fully utilise the multimode facilities of the parent transceiver. Some of the following designs illustrate these principles.

A single-board, no-tune 144MHz/1296MHz transverter

The comparatively recent development of economically priced and readily available microwave monolithic integrated circuits (MMICs) has allowed the development of a number of broadband (no-tune) low-power transverters from the 144MHz amateur band to the lower microwave bands, typically 1.3, 2.3 and 3.4GHz. Such designs use microstrip technology, including no-tune interstage band-pass filtering in the LO, receive and transmit chains.

A 144MHz to 1296MHz transverter circuit was described by KK7B in reference [22]. This circuit and layout, although it does not give 'ultimate' performance in terms of either receive noise figure or transmit output power (nor is it particularly compact in terms of board size), is probably one of the simplest and most cost-effective designs available at the time of writing. Its simplicity also makes it suitable for novice constructors. It is also flexible enough to allow the constructor to substitute new, improved MMICs as these become available, without major re-engineering.

The receive performance can be enhanced by means of an external (possibly mast-head) low-noise amplifier (LNA), such as the high-performance GaAsFET design already described. The transmit output level, at 13dBm (20mW), is ideal for driving the G4DDK-002 linear PA module, also already described. This, in turn, could drive either a solid-state power block amplifier or a valve linear amplifier.

Precision printed circuit boards for this design and a similar design for the 2.3GHz band [36] have been available for some time, produced and marketed by Down East Microwave in the USA. They are also available from a number of sources in the UK. A similar design concept was adopted for a transverter for the US 900MHz amateur band [37].

The original circuit diagram of the 1296MHz version is given in Fig 9.92 and the physical layout of the circuit is shown, not to

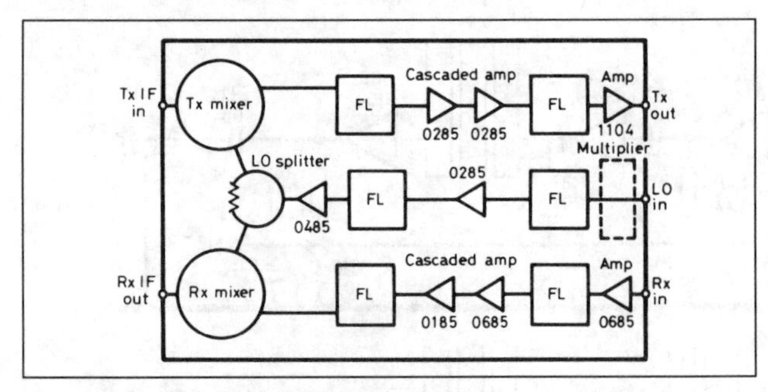

Fig 9.93. Layout of the KK7B transverter

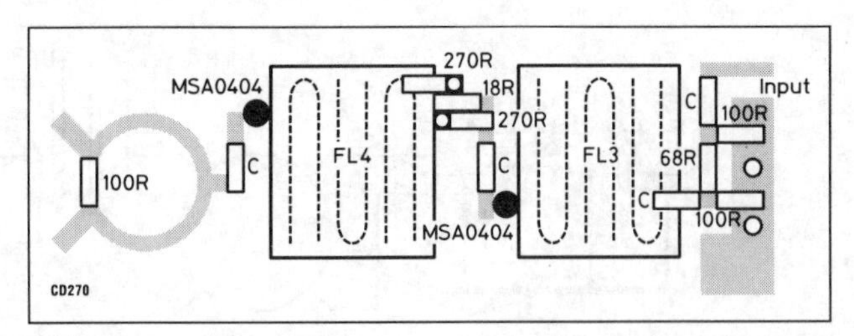

Fig 9.96. Layout of components for modified LO chain, KK7B transverter

scale, in Fig 9.93. Wide use is made of hairpin-shaped, self-resonant, printed microstripline filters in the LO, receive (RX) and transmit (TX) chains, together with printed microstripline transmit and receive balanced mixers and 3dB power splitter for the LO chain. An external LO source at any sub-harmonic frequency of the required injection frequency (1152MHz for the 1296–1298MHz narrow-band communications segment of the 23cm band when using an IF of 144–146MHz) was used and a simple, on-board diode multiplier was used to produce the required injection frequency from the LO input. Although direct injection of 1152MHz was mentioned in the original description, little guidance was given as to how to achieve this. With simple modification to the LO chain and a few changes to circuit values and devices, without need for PCB changes, it is easily possible to use the G4DDK-001 1152MHz source, already described, as the LO for this design. Fig 9.94 gives the circuit and component values for the original LO chain, while Fig 9.95 gives the modifications to allow the correct mixer injection levels to be attained when using the single +13dBm output option of the G4DDK-001 1152MHz source. Fig 9.96 shows the layout of the modified circuit using the existing PCB pads and tracks. Similar tactics could be adopted if it is desired to drive the 2.3GHz version of the transverter board from the G4DDK-004 2.0–2.6GHz LO source, also described earlier, although the attenuation values required and the gain required from each MMIC stage would have to be calculated for the new LO frequency. Construction is straightforward, using surface-mount techniques, ie all components, whether SMD or conventional, are mounted on the track side of the board unlike conventional construction. All non-semiconductor components – connectors, resistors, capacitors and inductors – should be mounted first, the MMICs and mixer diodes last, taking adequate precautions to avoid both heat and static damage. Note that the values of the bias resistors, which set the working points of the MMICs, were chosen for a supply rail of +12V DC. Higher supply voltages will require recalculation of these values and the constructor should refer to either the maker's data sheets for the particular devices used or to the more general information given in reference [38]. When construction is complete and the circuit checked out for correct values and placing of components, assuming that the LO source has already been aligned, there is no alignment as such! It should simply be a matter of connecting the transverter to a suitable 144MHz (multimode) transceiver via a suitable attenuator and switching interface such as that by G3SEK [39] or the G4JNT design available from reference [40].

The G3WDG-003 144MHz to 10GHz transmit converter

The G3WDG-003 transmit converter, together with the G3WDG-002 receive converter and the G4DDK-004 2.5GHz LO source, makes up a high-performance 10GHz amateur band transverter. When combined with a 144MHz transceiver, 10GHz antenna and suitable changeover relays, the system is capable of outstanding performance either from a fixed, home location or in portable use. Even better results may be obtained by adding a HEMT receiver preamplifier. Units of this prototype equipment, allied to a 30W travelling wave tube amplifier (TWTA) and 10ft solid dish antenna, allowed the designer, G3WDG, to conduct the first UK/USA EME contact with WA7CJO and the first UK/SM EME contact with SM4DHN early in 1993, just as this chapter was being prepared!

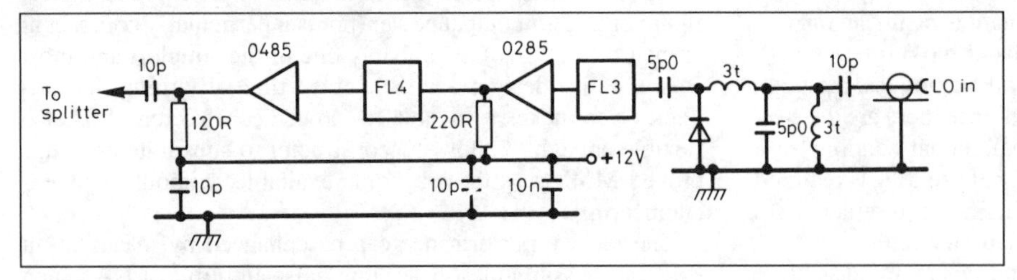

Fig 9.94. Original KK7B LO circuit

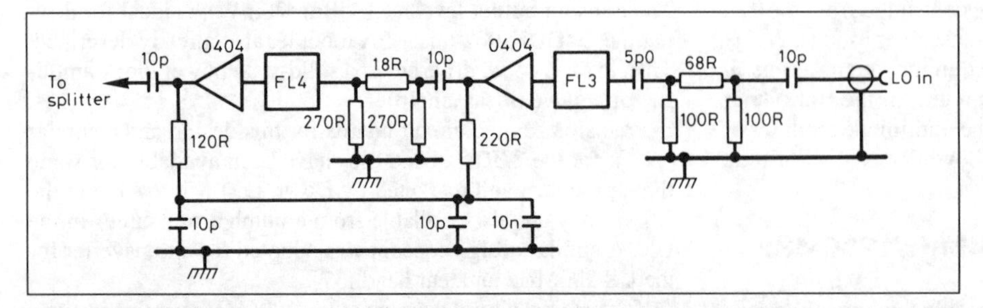

Fig 9.95. Modifications to KK7B LO circuit to allow use of G4DDK-001 1152MHz source and the higher output level available from the latter

The circuit is shown in Fig 9.97, the layout of the board and components in Fig 9.98 and their values in Table 9.18. Referring to the circuit diagram, Fig 9.97, the 2556MHz LO signal is fed to a MMIC amplifier IC1 that provides about 5.5dB gain. The output from this goes into a printed Wilkinson divider, which splits the signal into two equal, well isolated, outputs. One output is connected to J2, which is intended to provide the LO signal for the G3WDG-002 receive converter. The gain from J1 to J2 is about 2.5dB, allowing proper operation of the unit with LO powers in the range 5–15mW. The input circuit to the MMIC is modified from

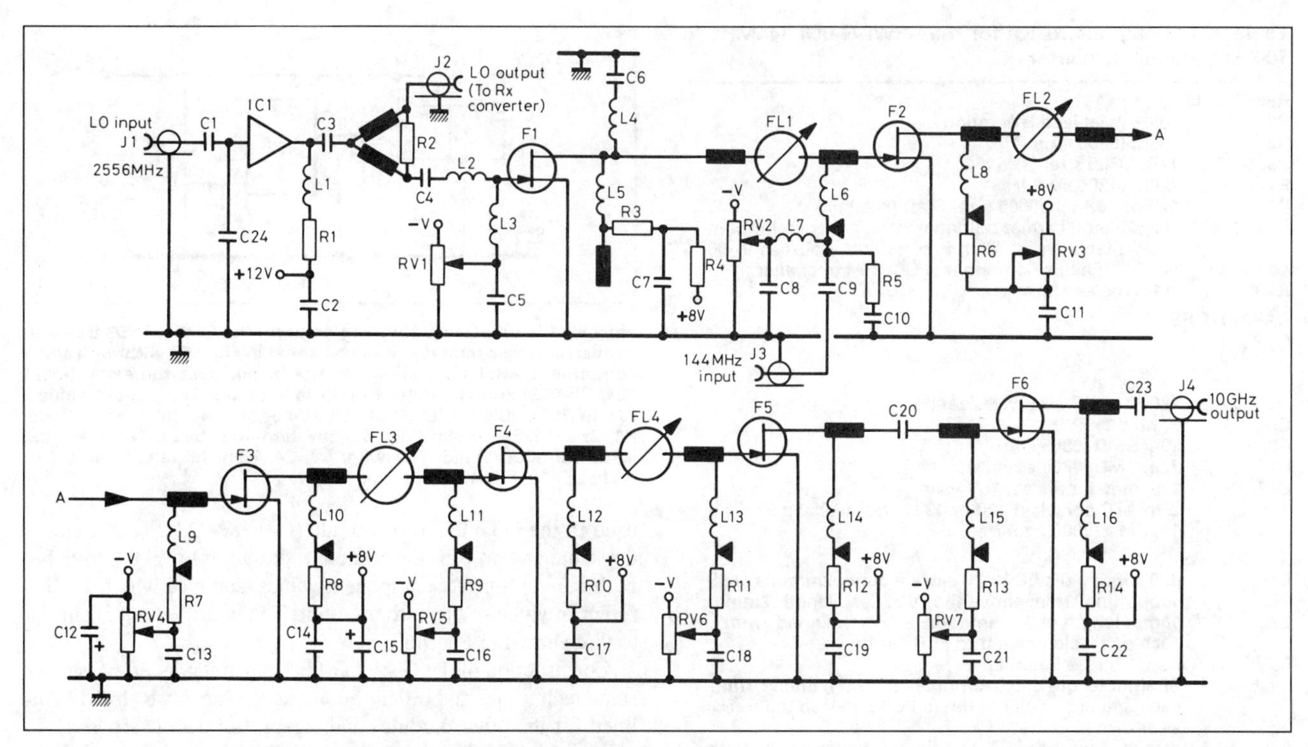

Fig 9.97. Circuit of the G3WDG-003 144MHz to 10GHz transmit converter

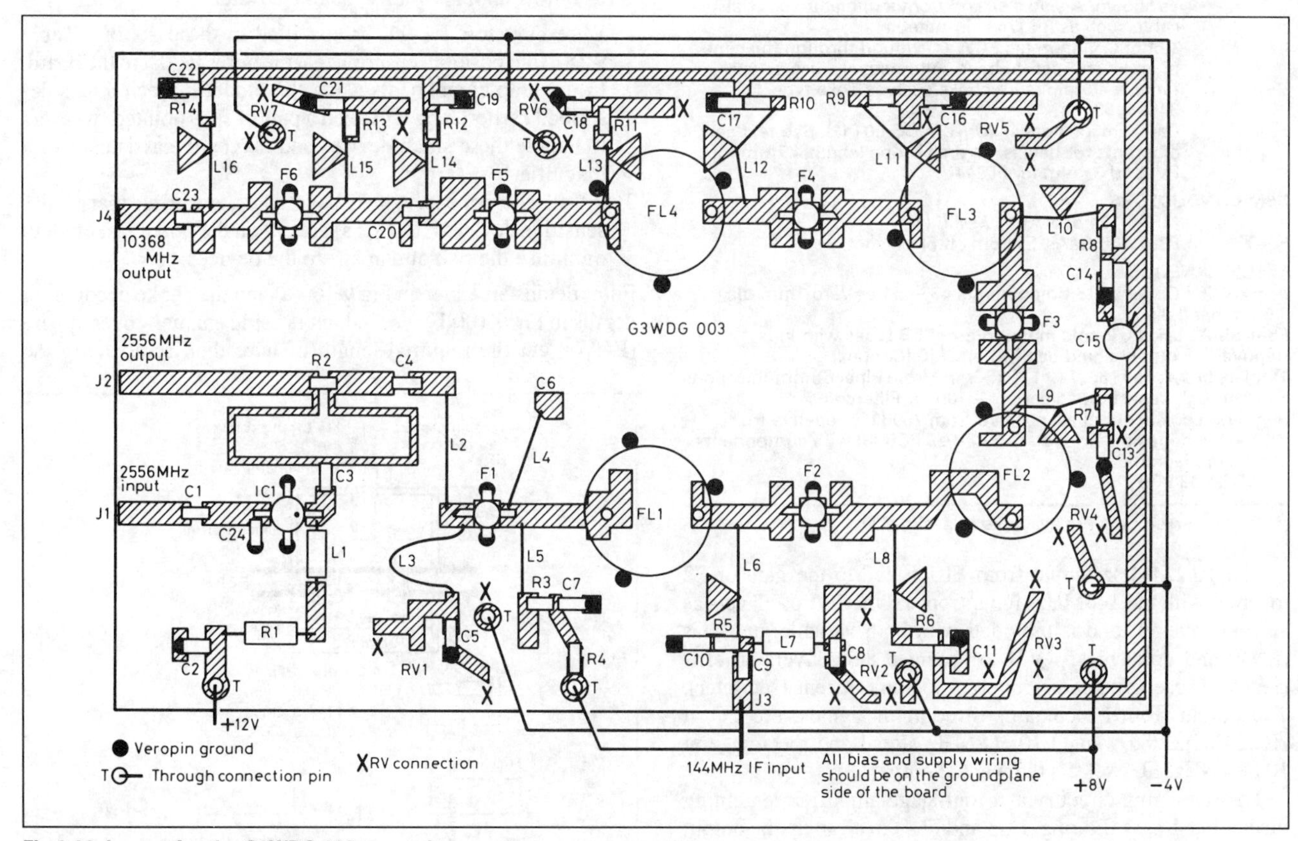

Fig 9.98. Layout for the G3WDG-003 transmit converter

that used in the G3WDG-001, in that a 1pF shunt capacitor has been added. This improves the input VSWR from about 5:1 to less than 1.5:1, and is a useful modification for the G3WDG-001 unit as described earlier. It decreases 2.5GHz drive requirement by about 2.5dB and stops resonant cable length effects from detuning the G4DDK-004 driver output circuit. The other output from the Wilkinson divider is fed to the ×4 multiplier stage F1 via a matching network L2/L3. The multiplier is identical to that used in the G3WDG-001 and -002 modules, and details may be found in earlier sections.

Table 9.18. Components list for the G3WDG-003 144MHz to 10GHz transmit converter

RESISTORS	
R1	100R, axial lead ½W rating
R2	100R, (0805 size) (*)
R3, 5–13	47R, (0805 size) (*)
R4	220R, (1206 size) (*)
R14	47R or 10R (*), (0805 size) SMD (see text)
RV1, 2, 4–7	10k skeleton cermet potentiometer, horizontal mounting. Suitable types: Allen Bradley 90H, Bournes VA05H or 3309, Philips Components OCP10H or similar.
RV3	2k2, type as above
CAPACITORS	
C1–5, 7, 11, 13, 14, 16–19, 21, 22	220p chip, SMD (0805 size) (*)
C6	Printed on PCB
C8, 9	10n, SMD (0805 size) (*)
C10	22p, SMD (0805 size) (*)
C12, 15	10μ, tantalum bead, 10V wkg
C20, 23	2.2p ATC porcelain 100 or 130 series (0.050in size) (*)
C24	1p, SMD (0805 size) (*)
INDUCTORS	
L1	6t 0.315mm dia ECW (*), close wound, 2mm dia, self-supporting, 1mm above the PCB. Lead length 2mm
L2	16mm length of 0.315mm dia ECW (*), tinned 1mm each end, soldered flat to PCB as shown.
L3	As L2, but 19.5mm long
L4, 5	Straight length of 0.315mm dia ECW (*), tinned 1mm each and and soldered flat to PCB between the tracks as shown
L6, 8–16	Straight lengths of 0.2mm dia (approx) silver plated or TCW, soldered between the tracks and the radial stubs as shown. A single strand from a miniature coaxial cable such as RG174/U is suitable.
L7	4.5t of 0.315mm dia ECW (*) wound through the centre of an FX1115 ferrite bead. Alternatively, use a 10μH radial lead miniature choke such as Toko type 348LS100
FL1–4	Cavity resonators as for G3WDG-001 (*). See text and diagrams for details. Overall probe length 4.7mm for FL1 and 3.4mm for FL2–4
SEMICONDUCTORS	
IC1	MSA 1104 or MAV11 MMIC (*)
F1–6	P35-1108 GaAsFET (Birkett black spot)

MISCELLANEOUS

Approx 58 PCB pins, RS Components 433-864 or Vero 1mm dia, 1.5mm head dia (*)
Four SMA sockets, 2-hole mounting, or PCB types with lugs removed. IF output could be SMB or SMC (Conhex)
Tinplate box, type 7760 (74×111×30mm) from Piper Communications
Feedthrough capacitor, solder-in, 1–10n, or Filtercon
Regulated power supplies: positive from 7808 IC, negative from ICL7660 voltage converter on G4FRE-023 PCB (8) with components shown in diagrams
G3WDG-003 PCB

(*) Parts marked thus are available in a 'short' kit

The 10,224MHz output from FL1 is fed to the gate of F2 together with the 144MHz signal from J3 via C9/L6. F2 acts as an up-converter, and is provided with both variable gate bias (RV2) and drain bias (RV3) to optimise the conversion efficiency. These adjustments interact to some extent (see later). The output from F2 contains three main signals, the LO at 10,224MHz, the wanted 10,368MHz signal and the image at 10,080MHz. FL2 selects the wanted output.

The remaining circuitry is a four-stage amplifier containing further band-pass filtering (FL3 and FL4) to clean up the output signal. The value of the drain resistor R14 is optional, 47 or 10Ω. The use of 10Ω may increase power output by a small amount in some cases.

The board layout is shown in Fig 9.97. Note that C12, the negative rail decoupling capacitor, is not shown as it is fitted on the reverse side of the board. The negative bias generator used to supply the gate bias for the FETs uses the same PCB as that used in the G3WDG-001 module (G4FRE-023). The circuit is the same as that of Fig 9.59 and is shown in Fig 9.99, together with the appropriate component values, some of which are different in value to the earlier circuit. The board layout is similar to that shown in Fig 9.60.

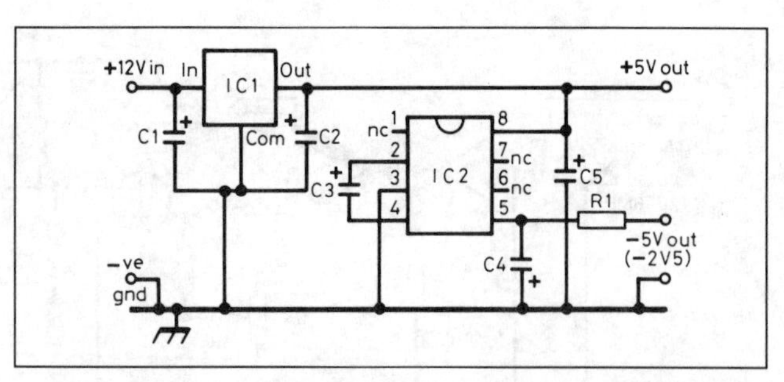

Fig 9.99. Circuit of the G4FRE regulator for the G3WDG-003 transmit converter. The circuit is the same as that in Fig 9.59, although some component values are different. The layout uses the same board (G4FRE-023) and is similar to that in Fig 9.60. Component values: IC1, μA7808; IC2, ICL7660PCA; Z1, 3V0 or 3V3, 400mW zener diode; R1, 1k5 0.125W metal film; C1, 1μF tantalum bead, 16V wkg; C2, 0.1μF tantalum bead, 10V wkg; C3, C4, 22μF tantalum bead, 10V wkg; C5, 10μF tantalum bead, 10V wkg

Construction of the G3WDG-003 transmit converter uses the same techniques and order of construction as those already outlined for the other modules and, again, full details are given in reference [41]. All parts are specified in the parts list of Table 9.18. Fitting should, briefly, be in the order:

1. Wire inductors L1–11, as specified in the parts list (Table 9.18), into position, ensuring that the wires lie flat to the board.
2. Fit all chip components using the mounting techniques described earlier. You will need a pair of fine-pointed tweezers to handle these small devices and maybe the assistance of a magnifier!
3. Fit all components which have leads, ensuring that static-sensitive devices (IC, FETs) are put on the board last of all to minimise the risk of damage to the devices.

Filter details are given in Fig 9.100(a) and the choke decoupling details in Fig 9.100(b). For inductors using enamel-covered wire (ECW), cut the required length of wire then scrape/chip the

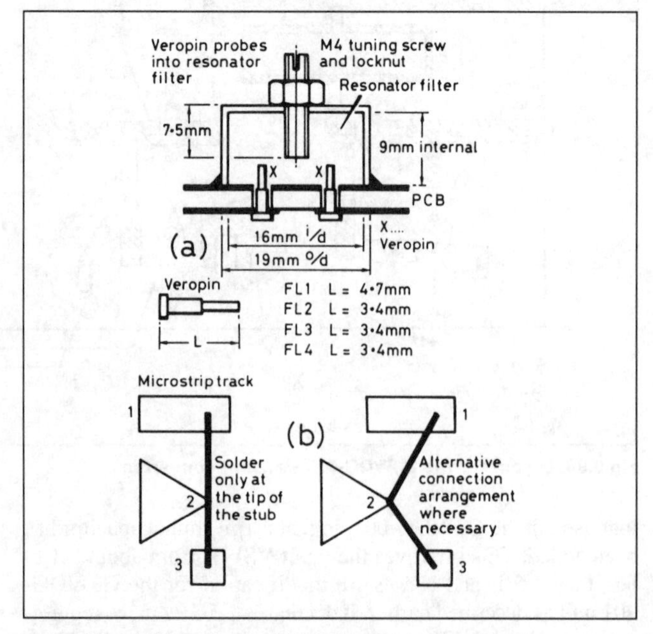

Fig 9.100. (a) Filter details, (b) choke decoupling details

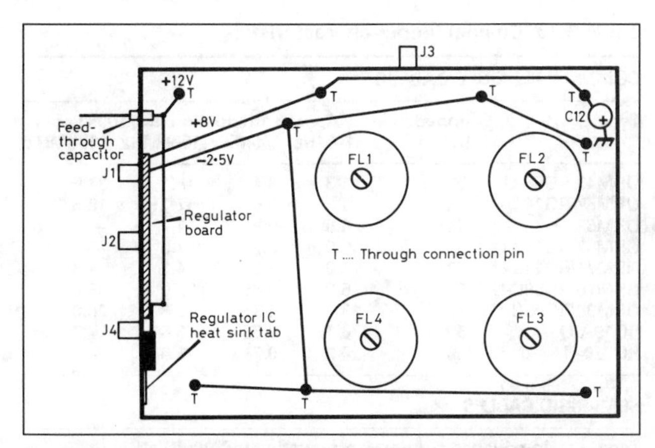

Fig 9.101. Positioning the regulator PCB and decoupling capacitor within the module enclosure, G3WDG-003 transmit converter

enamel from the last 1–2mm of each end using a scalpel blade. Tin each end. For inductors using one strand of a standard multistrand wire, tin one end and fit to the board as shown in Fig 9.100(b): solder first at position 1, then at position 2 as close as possible to the apex of the triangle, then at position 3. If any excess wire remains at 1 or 3, trim off carefully with a scalpel blade. Fig 9.101 shows the positioning of the regulator PCB and associated decoupling capacitor.

Alignment without laboratory equipment

The recommended way of tuning up the module without resorting to laboratory test equipment is to align a given stage while monitoring the drain current of the following stage, which acts as a power indicator. The gate bias of the monitor FET should be set to give a drain current in that FET of approximately 1mA. The procedure is to make temporary connections to the drain resistor of the following stage using thin flexible wires, which are then connected to an *analogue* multimeter. As noted earlier, digital multimeters are much more difficult to read and should be used *only* as a last resort, since they do not operate in real time! For example, when optimising the setting of RV1 and FL1 while aligning the multiplier stage, the multimeter is connected across R6 and RV2 is set to give a drain current of 1mA in F2. Adjustments are then made to R6 and FL1 for maximum current. In some cases the increase in current may be small and the adjustments need to be made carefully, especially the tuning screws. It helps if all tuning screws are preset to approximately the right penetration, ie with 7.5mm of thread below the bottom of the locknut. When adjusting the tuning screws, particularly FL2–4, the locknut should be kept reasonably tight with a spanner while turning the screw with a screwdriver (like setting the tappets in a car engine!), or the filter loss may be very high. The final tuning may be done using the locknut tightness as a fine adjustment, but take care that the correct tuning point is with the locknut quite tight, or again the loss may be higher than normal.

1. Connect a 50Ω load or >10dB attenuator to J2, and a 10GHz power indicator to J4 (eg coaxial to waveguide transition and waveguide diode detector).
2. Set the gate bias on all stages to cause all FETs to be pinched off (zero drain current).
3. Set gate bias of multiplier FET to 1mA using RV1.
4. Apply 2.5GHz drive to J1 (5mW minimum). The drain current of F1 should increase.
5. Using F2 as a power meter as described above, adjust FL1 and then optimise RV1.
6. Temporarily remove 2.5GHz drive and DC power, and transfer the test leads to R8. Set RV3 to its centre position and, after restoring DC power, set the current through F2 to 0.5mA with RV2. Reconnect 2.5GHz drive and apply 1mW of 144MHz to J3. The mixer current should increase incrementally on application of the drive signals. Next, using F3 as the power indicator, adjust FL2. This is the most difficult stage in the alignment, as there are three peaks since the filter resonates to 10,080, 10,224 and 10,368MHz. Finding the correct one may take a little time! The correct setting is with the tuning screw at the smallest penetration, and the power should disappear when 144MHz drive is removed. If the 10,080MHz peak is selected by mistake, it is possible to carry on with the alignment and all will appear normal except that the output will not be on the expected 10,368MHz!
7. Next optimise RV2 and RV3 for maximum power. The adjustments interact, so it is necessary to 'loop' through the adjustments until the best combination is found.
8. The same procedure should be followed to align the rest of the stages. Before starting optimisation, drain currents of the amplifier stages should be set to 15mA. After optimisation, the current can remain at whatever value resulted, with no risk of damage to the device. Finally, some power should be seen at the output, and a wavemeter should be used to confirm that the output is on 10,368MHz as desired.
9. The final stage in the alignment is to go around all the adjustments again for maximum output power at J4. In case the unit is operating at significant compression with 1mW 144MHz drive, adjustments will be easier if the 144MHz drive power is reduced. Prototype power outputs have been greater than 50mW, but it may be possible to achieve even more by tuning the microstrip lines with small pieces of copper or brass foil. If attempting this, start at the output and work backwards, as most will be gained (if any) around F6. Remove DC power before soldering the foil in position.
10. Check the level of output from J2. This should be 2–2.5dB higher than the drive power applied (below 10mW). At higher drive powers, IC1 will go into compression but this does not matter as plenty of drive will be available for the receive converter. Spectrum analyser traces of the output from the G3WDG-003 module are shown in Figs 9.102 and 9.103. Fig 9.104 plots power output from the module as a function of the level of 144MHz drive, clearly indicating the optimum drive level to be about 1mW.

ANTENNAS AND FEEDERS

Feeders

Feeders are used to transfer RF energy to and from an antenna. Ideal feeders are lossless: practical feeders have well-defined losses depending on their form, dimensions, construction and frequency!

A single-wire feeder with wave launchers (known as a *Goubier line* or *G-line* (or as a *G-string*!) has been little used because of the difficulty of installation and the fact that such a feeder must be kept several wavelengths away from metallic objects and as straight as possible to minimise radiation loss. Similarly, balanced twin feeder which may exhibit acceptably low theoretical losses is seldom used because exposure to the elements, particularly moisture, rapidly degrades the performance to unacceptable levels. This leaves two types of feeder in common use at microwave frequencies: coaxial feeders and waveguides.

As the frequency increases, so feeder losses increase, regardless of type. Typical losses at different frequencies are illustrated in Table 9.19 for *coaxial cable*. It can be seen that losses in

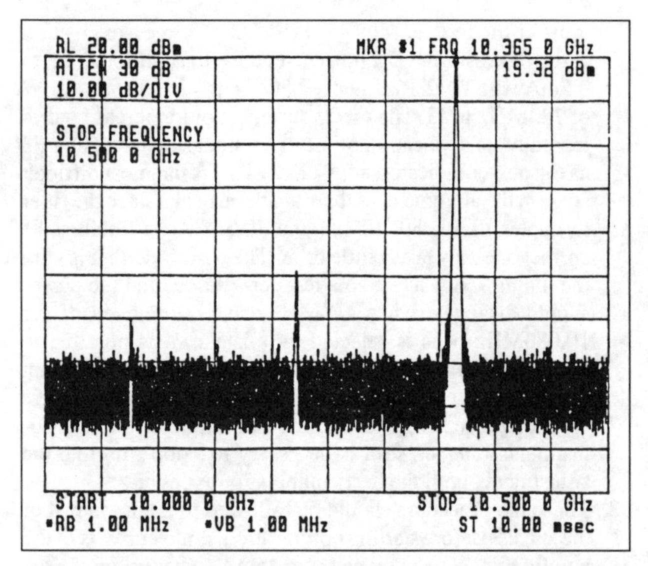

Fig 9.102. Output spectrum of the G3WDG-003 module from 10.0 to 10.5GHz, showing the wanted signal at 10,368MHz, the local oscillator signal at 10,224MHz and the image signal at 10,008MHz

Table 9.19. Coaxial feeder characteristics

COMMON FLEXIBLE CABLES

Type	Impedance (Ω)	Approx attenuation (dB/100ft) at 10MHz	100MHz	1000MHz	3000MHz
URM43/RG58-U	52	1.3	4.4	18.1	37.5
URM57/RG11A-U	75	1.9	3.5	7.1	16.5
URM63	75	0.5	0.9	1.7	—
URM67/RG213-U	50	2.0	3.7	7.5	—
URM74/RG218-U	51	1.0	1.9	4.2	—
URM76/RG58C-U	51	5.3	9.6	22.0	45.0
URM90/RG59B-U	75	3.5	6.3	12.3	26.0
RG19A-U	50	0.17	0.68	3.5	7.7
RG220-U	50	0.2	0.7	3.6	—

SEMI-RIGID CABLES

Type	Impedance (Ω)	Approx attenuation (dB/100ft) at 10MHz	100MHz	1000MHz	10GHz	18GHz
RG-405	50 ± 1.0	—	5.7	20.0	75.0	110.0
RG-402	50 ± 0.5	—	3.0	12.0	45.0	68.0
RG-401	50 ± 0.5	—	1.95	9.0	30.0	46.0
Heliax™						
LDF2-50	50	0.3	1.0	3.5	15.0	—
LDF4-50	50	0.2	0.7	2.7	13.0	—
LDF5-50	50	0.1	0.4	1.5	6.0 *	—

* Estimated, in absence of overmoding.

ordinary coaxial lines are too high for normal use, except in very short runs. The losses in miniature flexible or semi-rigid coaxial are even higher and their use is restricted to extremely short runs, for instance, in the interconnection of modules within equipment. The more expensive, larger-diameter, foam-filled or air-spaced helical membrane coaxial lines, of which the Andrew Corporation Heliax™ lines are representative, have sufficiently low losses to be used on moderate to long runs but require special connectors and detailed attention to waterproofing to prevent the ingress of water which causes rapid deterioration and greatly increased losses. It is possible to modify standard N-type coaxial connectors to allow proper fitting to some Heliax cables and, at the same time, to pressurise them with dry air to help keep out water [41].

Above about 5GHz, *waveguide* may be used, either conventional rigid, rectangular or circular waveguide or semi-flexible elliptical waveguide. All three types are hollow conductive tubes joined, where necessary, by means of metal-to-metal joints, typically flanges. The flanges usually have a groove to take a waterproofing and gas-tight gasket and, in professional use, the waveguide may be pressurised with dry nitrogen.

One characteristic of waveguide that is of considerable importance is the *cut-off frequency*. Waveguides act as high-pass filters and will only propagate frequencies above cut-off. The lowest frequency that will propagate is determined by the shape and dimensions of the waveguide and is roughly when the diameter (or the length of the larger dimension of a rectangular guide) is equal to $\lambda/2$. This means that the size of the waveguide determines the minimum frequency at which the guide can be used. The sizes and frequency ranges of waveguides usable in the amateur microwave bands are given in Table 9.20. For comparison with coaxial feeders, the loss (attenuation) in dB/100ft of waveguide-16 at 10GHz is approximately 5.5dB for brass guide or 2.2dB for copper guide. The main advantages of waveguide, therefore, are the relatively low loss per unit of length and the ability to construct many circuits (eg filters, couplers, detectors, multipliers) actually within short lengths of waveguide.

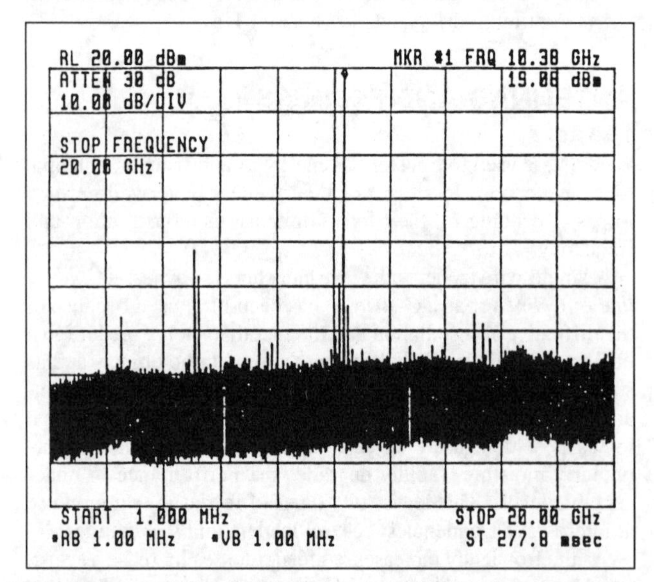

Fig 9.103. Wide-band spectrum analyser trace of the G3WDG-003 module from 1MHz to 20GHz

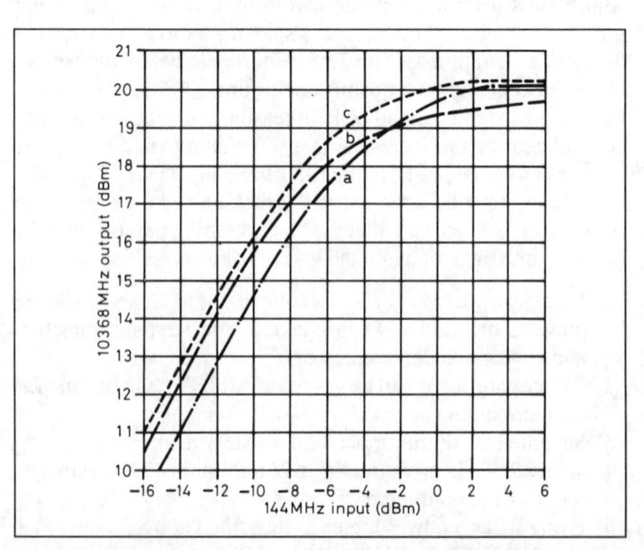

Fig 9.104. Power output of the G3WDG-003 at 10,368MHz vs drive power at 144MHz, for three prototypes

Table 9.20. Waveguide dimensions and characteristics

WG No	EIA desig	IEC desig	Recommended freq range (GHz)	Cut-off freq (GHz)	Internal dimensions (mm)		External dimensions (mm)		Wall (mm)	Aspect ratio
WG00	WR2300	R3	0.32–0.49	0.257	584.2	292.1	590.6	298.5	3.175	2.00
WG0	WR2100	R4	0.35–0.53	0.281	533.4	266.7	539.8	273.1	3.175	2.00
WG1	WR1800	R5	0.42–0.62	0.328	457.2	228.6	463.6	235.0	3.175	2.00
WG2	WR1500	R6	0.49–0.75	0.393	381.0	190.5	387.4	196.9	3.175	2.00
WG3	WR1150	R8	0.64–0.96	0.513	292.10	146.05	298.45	152.40	3.175	2.00
WG4	WR975	R9	0.75–1.12	0.605	247.65	123.82	254.00	130.18	3.175	2.00
WG5	WR770	R12	0.96–1.45	0.766	195.58	97.79	201.93	104.14	3.175	2.00
WG6	WR650	R14	1.12–1.70	0.908	165.10	82.55	169.16	86.61	2.032	2.00
WG7	WR510	R18	1.45–2.20	1.157	129.54	64.77	133.60	68.83	2.032	2.00
WG8	WR430	R22	1.70–2.60	1.372	109.22	54.61	113.28	58.67	2.032	2.00
WG9	—	—	—	1.686	88.90	44.45	92.96	48.45	2.032	2.00
WG9A	WR340	R26	2.20–3.30	1.736	86.36	43.18	90.42	47.24	2.032	2.00
WG10	WR284	R32	2.60–3.95	2.078	72.14	34.04	76.20	38.10	2.032	2.12
—	—	R35	—	2.258	66.37	29.50	70.37	33.50	2.000	2.25
WG11	—	—	—	2.448	60.25	28.50	63.50	31.75	1.626	2.11
WG11A	WR229	R40	3.30–4.90	2.577	58.17	29.08	61.42	32.33	1.626	2.00
—	—	R41	—	2.630	57.00	25.33	61.00	29.33	2.000	2.25
WG12	WR187	R48	3.95–5.85	3.152	47.55	22.15	50.80	25.50	1.626	2.15
WG12A	—	—	—	2.951	50.80	16.92	—	—	—	3.00
WG13	WR159	R58	4.90–7.05	3.711	40.39	20.19	43.64	23.44	1.626	2.00
WG14	WR137	R70	5.85–8.20	4.301	34.85	15.80	38.10	19.05	1.626	2.21
WG15	WR112	R84	7.05–10.0	5.260	28.50	12.62	31.75	15.88	1.626	2.26
Old English		—	—	5.902	25.40	12.70	—	—	—	2.00
WG16	WR90	R100	8.20–12.4	6.557	22.86	10.16	25.40	12.70	1.270	2.25
WG17	WR75	R120	10.0–15.0	7.869	19.05	9.525	21.59	12.07	1.270	2.00
WG18	WR62	R140	12.4–18.0	9.488	15.80	7.899	17.83	9.931	1.016	2.00
WG19	WR51	R180	15.0–22.0	11.57	12.95	6.477	14.99	8.509	1.016	2.00
WG20	WR42	R220	18.0–26.5	14.05	10.67	4.318	12.70	6.350	1.016	2.47
WG21	WR34	R260	22.0–33.0	17.36	8.636	4.318	10.67	6.350	1.016	2.00
WG22	WR28	R320	26.5–40.0	21.08	7.112	3.556	9.144	5.588	1.016	2.00
WG23	WR22	R400	33.0–50.0	26.34	5.690	2.845	7.722	4.877	1.016	2.00
WG24	WR19	R500	40.0–60.0	31.39	4.775	2.388	6.807	4.420	1.016	2.00
WG25	WR15	R620	50.0–75.0	39.88	3.759	1.880	5.791	3.912	1.016	2.00
WG26	WR12	R740	60.0–90.0	48.37	3.099	1.549	5.131	3.581	1.016	2.00
WG27	WR10	R900	75.0–110	59.01	2.540	1.270	4.572	3.302	1.016	2.00
WG28	WR8	R1200	90.0–140	73.77	2.032	1.016	4.064	3.048	1.016	2.00
WG29	WR7	R1400	110–170	90.79	1.651	0.826	3.175	2.350	0.762	2.00
WG30	WR5	R1800	140–220	115.75	1.295	0.648	2.819	2.172	0.762	2.00
WG31	WR4	R2200	170–260	137.27	1.092	0.546	2.616	2.070	0.762	2.00
WG32	WR3	R2600	220–325	173.49	0.864	0.432	2.388	1.956	0.762	2.00

Abbreviations: IEC – International Electrotechnical Commission, EIA – Electronic Industries Association (USA)

The principal disadvantages of waveguide are:

1. At low microwave frequencies the waveguide dimensions are too large to be practical.
2. Waveguide is a precision product and is, therefore, expensive compared with less precisely manufactured coaxial feeders.
3. Waveguide is rigid or, at best, semi-flexible. Connections between sections must be made with waterproof (and RF-proof) flanges which are also expensive.
4. Waveguide losses, due to mode changes, are minimised when the feeder run is as straight as possible. Bends are difficult to make, since the waveguide is rigid, and it is usual to use pre-formed bends. Rotary joints, necessary for antenna rotation, are available professionally but are difficult for amateurs to make!
5. The cut-off frequency is important: the choice of waveguide is dependent on the band to be used.

A typical amateur installation will invariably be a compromise: the main run of the feeder might be minimised by installation of the transmitter, receiver (or transverter) and the change-over arrangements close to the mast-head, possibly feeding both DC power up and IF up and down a common, ordinary coaxial feeder. For instance, the equipment might be housed in the loft space of a house, adjacent to the wall carrying the main antenna mast. The flexible section needed to rotate an antenna might then be a short length of suitable coaxial feeder. The use of a few feet or metres of a high-performance coaxial feeder will not be prohibitive, either in terms of cost or losses. Standard copper water pipe is sufficiently accurate to be used as a substitute for conventional circular waveguide, in some of the amateur microwave bands, when carefully chosen and installed [43]. Where mast-head or loft installation is impossible, such pipe might offer an inexpensive alternative to the use of rectangular waveguide.

ANTENNAS: GENERAL

Antennas for the amateur microwave bands are usually high gain, and therefore highly directional, in order to compensate to some degree for the weaker signals transmitted and received in these bands. For some applications, eg beacons or beacon/repeaters, lower-gain (wide beamwidth) or omnidirectional antennas may be used.

Where elevated, high-gain antennas are employed for general communications use, these are commonly either parasitic (Yagi or loop Yagi) type arrays or parabolic reflectors with suitable feeds. In the lower bands, up to and including 3.4GHz, arrays of stacked or bayed long Yagis or loop Yagis offer high gain, ease of feed, ease of mounting and rotation/tilting and good radiation patterns without the high *windage* (mechanical loading) and other mechanical complications of a dish of the size necessary to give the same gain. However, parasitic antennas are frequency-conscious (they have narrow bandwidth) and plane-polarised: only the feeds of parabolic dishes are frequency-conscious.

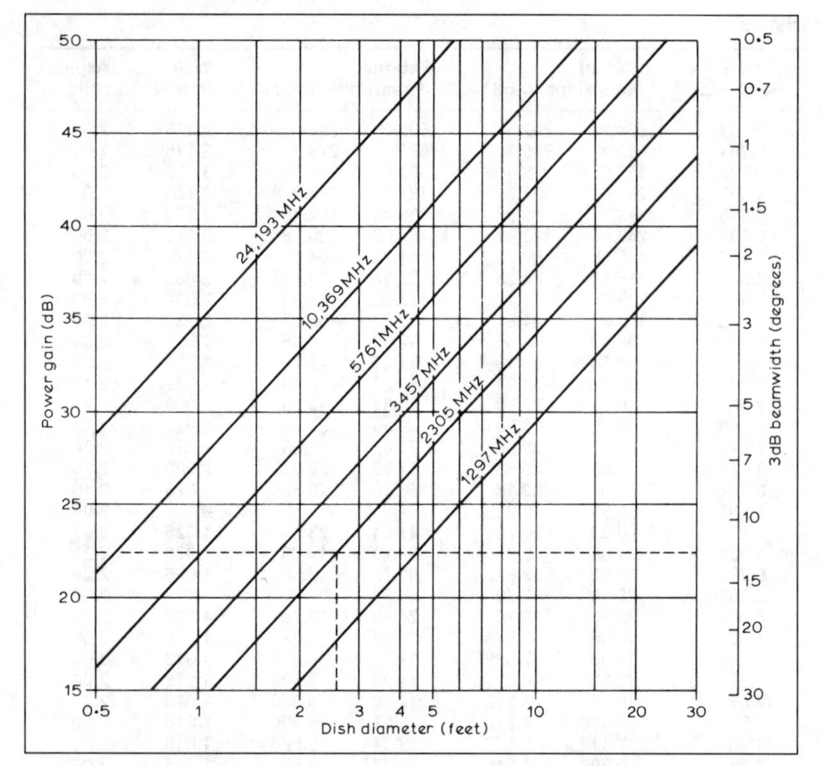

Fig 9.105. Relationship between the size of a dish, its gain and beamwidth as a function of frequency. An overall efficiency of 50% is assumed. As an example, a dish 2.5ft in diameter at 2305MHz will have a gain of about 22dB and a beamwidth of about 22°

To change the plane of polarisation of an array of Yagis, say from horizontal to vertical or circular, crossed-element arrays and potentially lossy phasing harnesses/switching or physical re-orientation of the antennas are required. The polarisation of a parabolic reflector, by contrast, can be easily changed (by altering only the feed) and circular polarisation, used for space communications, is easy to achieve. Furthermore, apart from the feed, parabolic reflectors are essentially aperiodic, ie can be used at any frequency, the gain being dependent mainly on the size (diameter) of the dish relative to the frequency at which it is to be used, assuming equal illumination efficiency at any particular frequency. See Fig 9.105.

For space communications use, possibly on more than one band, eg EME, a large skeleton dish [44], mounted at or near ground level and fully steerable in azimuth and elevation is a practical possibility in all but the smallest of gardens, although a solid dish may be preferable on the higher frequencies!

On the higher bands (5.7GHz and up), Yagi construction is difficult because of the closer tolerances necessary. Higher losses occur through the need for very accurate dimensioning of the antenna elements and their spacing. Smaller parabolic reflectors can be employed for the same (or more) gain, become more manageable and are an entirely practical proposition instead. On the 10GHz band, much use has recently been made of commercial offset-feed dishes designed for the satellite TV broadcasting bands around 11 to 12GHz.

Omnidirectional antennas commonly used include Alford slots for horizontally polarised signals or collinears for vertically polarised (mobile) signals on the lower frequency bands. Stacked slotted waveguide antennas are not uncommon in the higher frequency bands, even though the dimensions are quite critical.

For semidirectional applications (wide beamwidth) where maximum gain is not required, corner reflector, short Yagi or helical antennas have been used on the lower bands whilst optimal or sectoral horns can be effectively used on the higher bands. However, neither the corner reflector nor the helical antenna gives particularly good gain or radiation patterns in relation to the difficulty of their construction and their size and bulk, and so are now not much used by amateurs. Short Yagis or loop Yagis are easier to construct and give better gains and cleaner radiation patterns for a smaller size and bulk.

Conventional Yagis have been the subject of much experimental research by amateurs and probably the most notable amateur papers on long-Yagi design are those of DL6WU [45, 46]. The parallel development of loop-Yagi antennas by G3JVL shows both types of antenna to be very similar in performance: a 26-element DL6WU antenna, for instance, has a calculated gain of 18.5dBi and the comparable (but slightly longer) 27-element loop Yagi has a calculated gain of 20dBi. The choice of type may be largely determined by materials and fabrication facilities which the constructor may have available!

Loop Yagi antennas for 1.3, 2.3 and 3.4GHz

This type of antenna was first designed in 1974/5 by G3JVL and since then he has made a number of improvements, including longer versions with higher gain. The gain of the basic version is approximately 18dBi, longer versions correspondingly higher (see following tables).

The design of these antennas is shown in Fig 9.106. Construction is quite straightforward, but the dimensions given in Table 9.21 (1.3GHz), Table 9.22 (2.3GHz) and Table 9.23 (3.4GHz) must be closely adhered to. In drilling the boom, for example, measurements of the positions of the elements should be made from a single point by adding the appropriate lengths. If the individual gaps are marked out, then errors may accumulate to an excessive degree, resulting in poor performance. With the exception of the driven element, the elements are made from aluminium strip, the two holes in which are drilled before bending with a centre-to-centre spacing equal to the circumference specified in the appropriate table. The driven element should be made from copper strip. Commercial versions of these antennas may use all-stainless-steel construction, although this method of construction is not feasible for

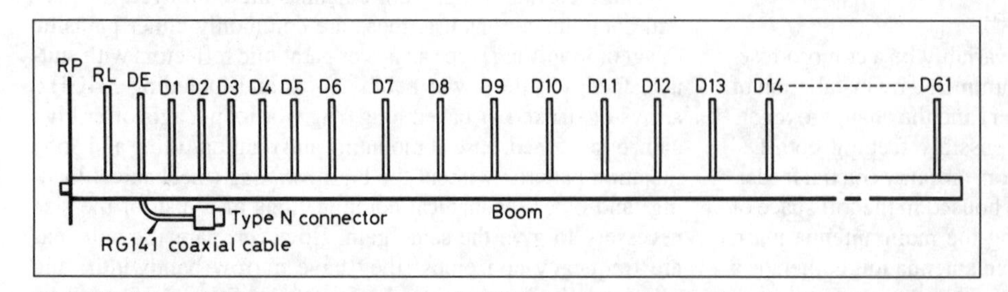

Fig 9.106. General diagram of construction of G3JVL loop-Yagi. Dimensions for 1.3, 2.3 and 3.4GHz versions are given in the tables

Table 9.21. Dimensions for 1.3GHz loop-Yagi antennas

Design frequency	1296MHz
Boom diameter	19mm OD
Boom length	5.5m
Boom material	Aluminium alloy
Driven element	2.5mm diameter welding or copper rod
All other elements	2.5mm diameter welding or brazing rod

Element sizes

Reflector plate	140 × 115mm
Element	**Length *L* (mm)**
Reflector loop	254
Driven element	232
Directors 1–12	215
Directors 13–20	208
Directors 21–30	200
Directors 31–40	195
Directors 41–55	192
Directors 56–63	191

Cumulative element spacings (mm)

RP	0	RL	79
DE	103	D1	131
D2	152	D3	198
D4	243	D5	275
D6	333	D7	424
D8	514	D9	605
D10	695	D11	785
D12	876	D13	966
D14	1057	D15	1147
D16	1238	D17	1328
D18	1418	D19	1509
D20	1599	D21	1690
D22	1780	D23	1871
D24	1961	D25	2051
D26	2142	D27	2232
D28	2323	D29	2413
D30	2503	D31	2594
D32	2684	D33	2775
D34	2865	D35	2959
D36	3046	D37	3136
D38	3227	D39	3317
D40	3408	D41	3498
D42	3589	D43	3679
D44	3769	D45	3860
D46	3950	D47	4041
D48	4131	D49	4222
D50	4312	D51	4402
D52	4493	D53	4583
D54	4674	D55	4764
D56	4855	D57	4945
D58	5035	D59	5126
D60	5216	D61	5307
D62	5397	D63	5487

Table 9.22. Dimensions for 2.3GHz loop-Yagi antennas

Design frequency	2320MHz
Boom diameter	12.7mm OD
Boom length	3.07m
Boom material	Aluminium alloy
Driven element	1.6mm diameter welding or copper rod
All other elements	1.6mm diameter welding or brazing rod

Element sizes

Reflector plate	80 × 65mm
Element	**Length *L* (mm)**
Reflector loop	143
Driven element	130
Directors 1–12	121
Directors 13–20	117
Directors 21–30	113
Directors 31–40	110
Directors 41–50	108
Directors 51–63	107

Cumulative element spacings (mm)

RP	0	RL	44
DE	58	D1	73
D2	85	D3	110
D4	136	D5	153
D6	186	D7	237
D8	287	D9	338
D10	388	D11	439
D12	489	D13	540
D14	590	D15	641
D16	691	D17	742
D18	792	D19	843
D20	893	D21	944
D22	994	D23	1045
D24	1095	D25	1146
D26	1196	D27	1247
D28	1297	D29	1348
D30	1399	D31	1449
D32	1500	D33	1550
D34	1601	D35	1651
D36	1702	D37	1752
D38	1803	D39	1853
D40	1904	D41	1954
D42	2005	D43	2055
D44	2106	D45	2156
D46	2207	D47	2257
D48	2308	D49	2358
D50	2409	D51	2459
D52	2510	D53	2560
D54	2611	D55	2661
D56	2712	D57	2762
D58	2813	D59	2863
D60	2914	D61	2964
D62	3015	D63	3064

the average home constructor. All elements, screws and soldered joints should be protected with polyurethane varnish after assembly, followed by a coat of paint on all surfaces. If inadequate attention is paid to this protection, then the performance of the antenna will deteriorate over a period of time as a result of corrosion.

If the specified materials are not available, or it is wished to use thicker or wider elements to increase the strength of the antenna, it is necessary to alter the lengths of all the elements to compensate.

The necessary data is given in Table 9.24. A general-purpose BASIC design program is given in reference [47].

Provided that the antenna is constructed carefully, its feed impedance will be close to 50Ω. If a VSWR or impedance bridge is available, the matching can be optimised by bending the reflector loop slightly toward or away from the driven element.

The antenna can be mounted using an element clamp from an old antenna. It is essential that the antenna be mounted on a vertical support, as horizontal metalwork in the vicinity of the antenna can cause severe degradation in its performance. It is usually best to mount the antenna with the loops pointing downwards, to reduce the likelihood of damage caused by perching birds.

An alternative, wide-band driven element (the G3JVL Looperiodic) was described in reference [48].

Stacking 1.3GHz Yagis

Yagi antennas are often combined together in an array to achieve higher gain. It is important that correct phasing, impedance matching etc is achieved. The antennas must be mounted a specific distance apart, known as the *stacking distance*. This is usually quoted by the manufacturer and may be different for the vertical and horizontal distances. For instance, the optimum stacking distance for the 1.3GHz G3JVL loop Yagi is 69cm in both planes. A slightly greater distance (about 75cm) is recommended for a DL6WU long Yagi.

At 1.3GHz it is common practice to use a matched power splitter to provide the feeds to the antennas in an array, instead of using specific lengths of mis-matched cables to feed the antennas from a common point. The power splitters shown in Fig 9.107 enable two or four antennas of a given impedance to be fed from a single coaxial cable of the same impedance. They

Table 9.23. Dimensions for 3.4GHz loop-Yagi antennas

Boom diameter	12.5mm OD
Boom length	2.0m
Boom material	aluminium alloy
Driven element	1.6mm diameter welding rod
All other elements	1.6mm diameter welding rod
Element sizes	
Reflector plate	52.4 × 42.9mm
Element	**Length *L* (mm)**
Reflector loop	99.2
Driven element	92.7
Directors 1–12	83.9
Directors 13–20	81.2
Directors 21–30	78.2
Directors 31–40	76.1
Directors 41–50	75.1
Directors 51–60	74.6

Cumulative element spacings (mm)

RP	0.0	RL	29.5
DE	38.6	D1	49.2
D2	57.2	D3	74.1
D4	91.1	D5	103.0
D6	125.0	D7	158.9
D8	192.8	D9	226.7
D10	260.6	D11	294.5
D12	328.4	D13	362.3
D14	396.2	D15	430.1
D16	464.1	D17	498.0
D18	531.9	D19	565.8
D20	599.7	D21	633.6
D22	667.5	D23	701.4
D24	735.3	D25	769.2
D26	803.1	D27	837.0
D28	871.0	D29	904.9
D30	938.8	D31	972.7
D32	1006.6	D33	1040.5
D34	1074.4	D35	1108.3
D36	1142.2	D37	1176.1
D38	1210.0	D39	1244.0
D40	1277.9	D41	1311.8
D42	1345.7	D43	1379.6
D44	1413.5	D45	1447.4
D46	1481.3	D47	1515.2
D48	1549.1	D49	1583.1
D50	1617.0	D51	1650.9
D52	1684.8	D53	1718.7
D54	1752.6	D55	1786.5
D56	1820.4	D57	1854.3
D58	1888.2	D59	1922.1
D60	1956.1	D61	1990.0

consist of a length of fabricated coaxial line that performs the appropriate impedance transformations. The inner is made exactly one half-wavelength long (λ/2) between the centres of the connectors, and the outer is made approximately 1.25in (37.15mm) longer. The outer can be made from square-section aluminium tubing, the ends of which and the access hole (for soldering the centre connector) being sealed with aluminium plates bonded with an epoxy adhesive.

Alternatively brass or copper tubing may be used, in which case the plates can be soldered. Other forms of materials can be configured to give the necessary impedance transformations and

Table 9.24. Correction factors for various loop-Yagi antenna materials

Thickness (mm)	(in)	CF × 100 (%)	Width (mm)	(in)	CF × 100 (%)	Boom dia (mm)	(in)	CF × 100 (%)
0.71	0.028	0.0	2.54	0.10	0.4	0.00	0.00	−0.7
1.63	0.0625	0.6	4.76	0.1875	0.0	12.7	0.50	0.0
*	*	*	6.35	0.25	−0.3	19.0	0.75	+0.9
*	*	*	9.53	0.375	−0.95	25.4	1.00	+2.1

* Other values are determined by plotting a straight line through the values given. Construct a graph to extend these values or use the BASIC design program given in reference [45].

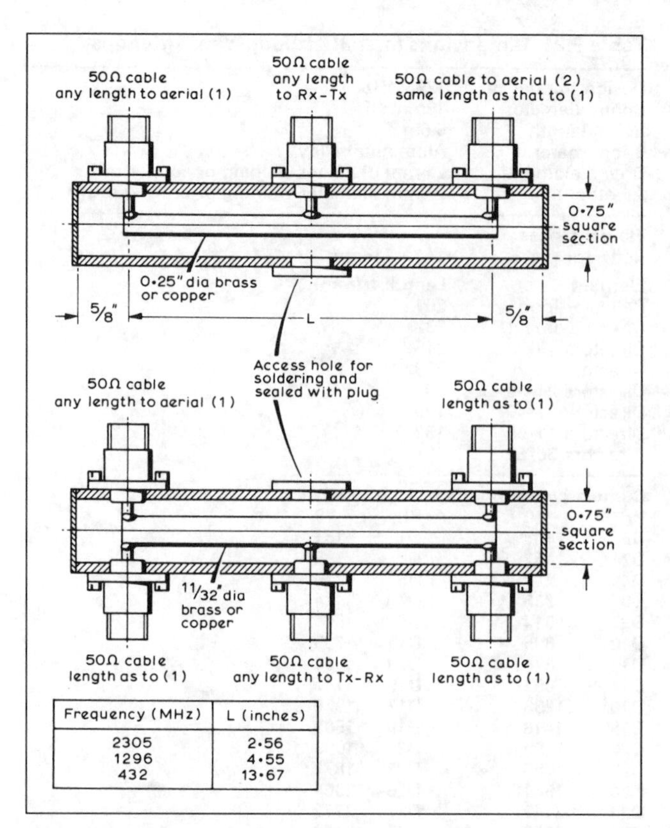

Fig 9.107. Power splitter/combiner for connecting two or four antennas to a single feeder

information to enable other designs to be realised is given in Fig 9.108.

For the two-way λ/2 splitter, using a 25.4mm square outer (23mm square internal dimension), the round centre conductor should be 7.59mm diameter: for the four-way λ/2 splitter, using the same outer, the inner diameter must be 10.81mm. It should be noted that these dimensions are quite critical and a 10% error will present a poor match. For this reason it is not satisfactory to use the nearest standard size of rod.

Alternative, less dimensionally critical, flat-strip inner conductor splitters were described in reference [49].

	50Ω system 2-way Zo = 72Ω	50Ω system 4-way Zo = 50Ω
d D	2·82	1·96
d D	3·32	2·31
d D	1·54	—
d D	1·66	—

Fig 9.108. Ratios of *d*/*D* for other useful coaxial configurations

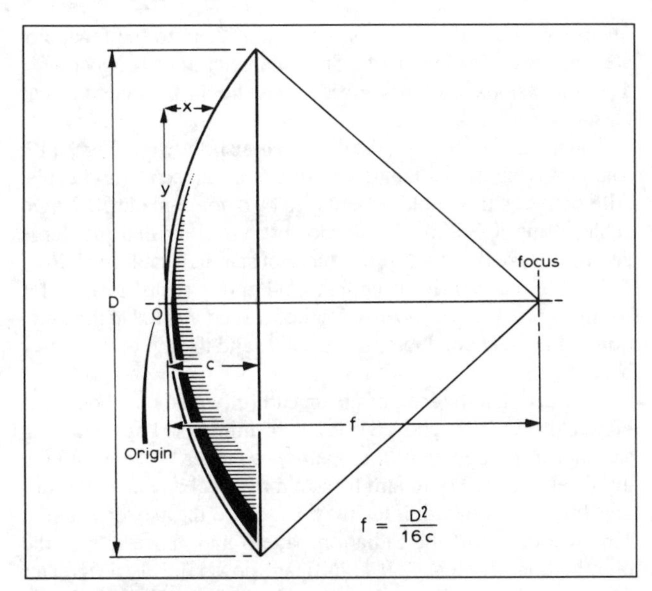

Fig 9.109. Basic parabolic dish geometry

Dish antennas

The basic geometry of a paraboloid is shown in Fig 9.109. The most important dimensions are the focal length *f*, the diameter *D* and the depth *c* at the centre. These are related by the formula:

$$f = \frac{D^2}{16c}$$

The feed is normally axial with the imaginary line joining the focus to the origin. The ratio *f/D*, using the notation of Fig 9.109, is the fundamental factor governing the design of the feed for a dish. The ratio is directly related to the angle subtended by the rim of the dish at its focus and also therefore to the beamwidth of the feed necessary to illuminate the dish effectively. Practical values of *f/D* range from about 0.2–1.0, with the value 0.25 corresponding to the common focal-plane dish in which the focus is in the same plane as the rim of the dish. Dishes deeper than this (*f/D* < 0.25) are used commercially where it may be important to minimise sidelobe response. However, such dishes are not easy to feed efficiently and gain begins to fall off as a consequence.

Amateurs may most frequently choose to use dishes with *f/D* ratios in the range 0.5–0.75 as these are comparatively easy to feed despite their disadvantages: these disadvantages include the need for a higher gain (and hence larger) feed, which can cause some aperture blocking, and the long focal length which requires the feed to be spaced further off the dish, leading to mechanical complications and an unbalanced structure. To avoid aperture blocking, the feed may be offset from the axis and the loss in gain, as a function of 3dB beamwidths for various *f/D* ratios, is shown in Fig 9.110. It can be clearly seen that the loss in gain for increasing feed offset is much less for a high *f/D* ratio: this principle has been exploited in the now-common offset-feed dishes used for satellite TV reception. Such dishes are eminently suitable for amateur use at 10 and 24GHz!

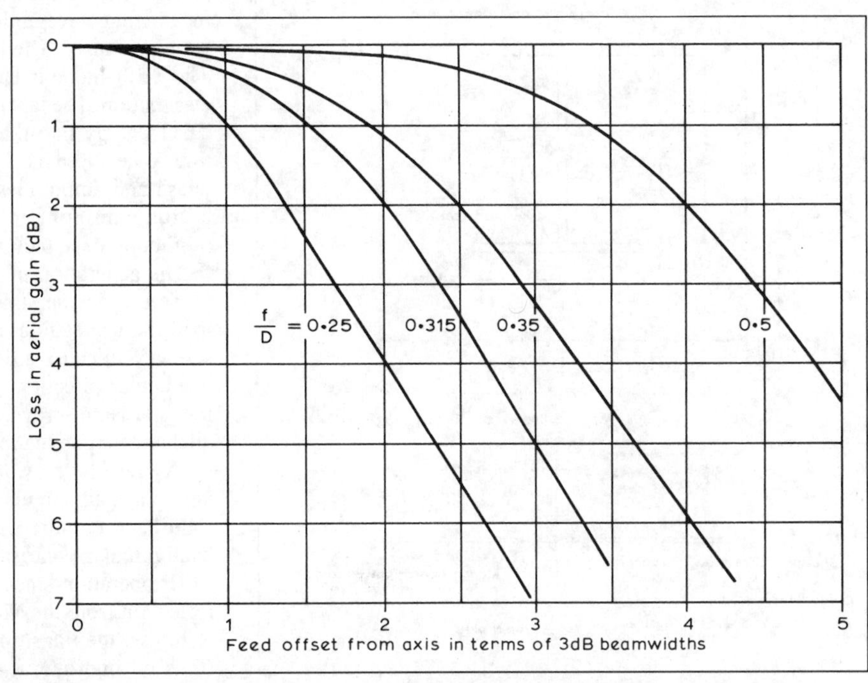

Fig 9.110. Loss in gain due to positioning feed off the axis of the dish

The gain of a dish antenna as a function of its diameter and frequency of use was shown in Fig 9.105, together with the 3° beamwidth. At 1.3GHz a dish of 5ft (1.5m) diameter has similar gain to an array of four Yagis or two long loop-Yagis (around 24dBi). Yagi arrays with eight or more antennas are very expensive if bought commercially and are also difficult to feed efficiently. A 2.1m diameter dish would have about the same gain. Dishes of this size are not too difficult to construct but may be difficult to mount and rotate.

A typical skeleton dish might have 10–20 ribs, with 13mm chicken wire as the reflecting surface. Constructional tolerances are not particularly severe at 1.3GHz, as deviations from a true parabola of up to 10mm would cause a maximum loss in gain of only 0.5dB. When designing a dish for 1.3GHz, an *f/D* ratio of 0.5 to 0.6 is recommended, as this gives a dish that is relatively easy to feed with high efficiency. Of course, a dish may already be to hand with a different *f/D* ratio and the feed should be chosen to match this. A number of different feed designs are described below.

Multiband feeds have been described in amateur literature, for instance [50–53]. Although such designs are invariably compromises, they may be useful where multiband operation is needed and the station is limited to using a single dish and feeder. Band-specific feeds will be outlined here.

Dipole/reflector feeds for 1.3 or 2.3GHz

The basic arrangement for this feed and dimensions for two bands are given in Fig 9.111. The radiation pattern makes it suitable for use with dishes with *f/D* ratios in the range 0.25–0.35. The feed is built around a length of fabricated rigid coaxial line, the inside dimensions of the outer and the diameter of the inner being chosen to produce a characteristic impedance of 50Ω.

The feed should be positioned in the dish so that the point halfway between the dipole and the reflector is located at the focus of the dish. The feed may then be moved inwards or

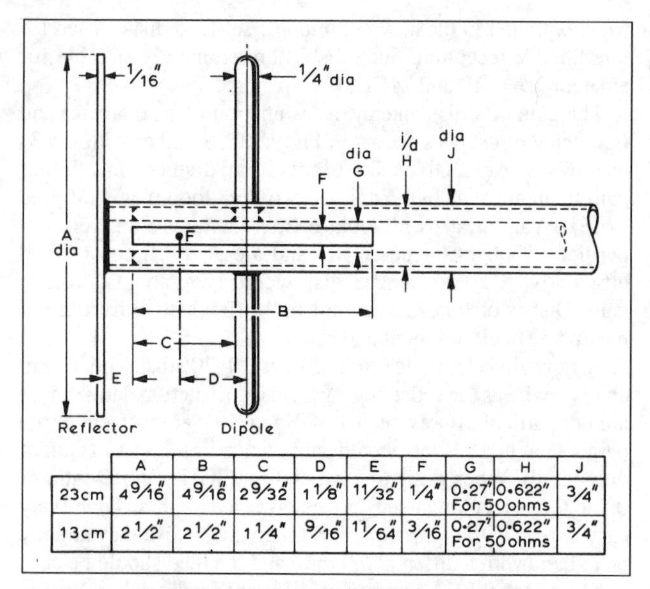

	A	B	C	D	E	F	G	H	J
23cm	4 9/16"	4 9/16"	2 9/32"	1 1/8"	11/32"	1/4"	0·27" For 50 ohms	0·622"	3/4"
13cm	2 1/2"	2 1/2"	1 1/4"	9/16"	11/64"	3/16"	0·27" For 50 ohms	0·622"	3/4"

Fig 9.111. Dipole and reflector feeds for 1.3 and 2.3GHz. The feed is positioned so that point F coincides with the focus of the dish

outwards slightly to optimise the gain. This design can be adapted to other frequencies by scaling the dimensions with frequency: the inside diameter of the outer and the outside diameter of the inner of the feed should remain the same, independent of frequency, since these dimensions determine the impedance of the coaxial section feeding the dipole.

A high-efficiency dish feed for 1.3GHz

The feed to be described is capable of feeding dish antennas with *f/D* ratios in the range 0.5–0.6 with high efficiency. The construction of the feed is shown in Fig 9.112. The design uses two folded dipoles $\lambda/2$ apart, driven in phase. The dipoles are positioned $\lambda/4$ above a square reflector. The dipoles and the $\lambda/4$ open-wire transmission lines which join them to the feedpoint are made from one length of 1.5mm diameter tinned copper wire. The dimensions are fairly critical and should be adhered to as closely as possible.

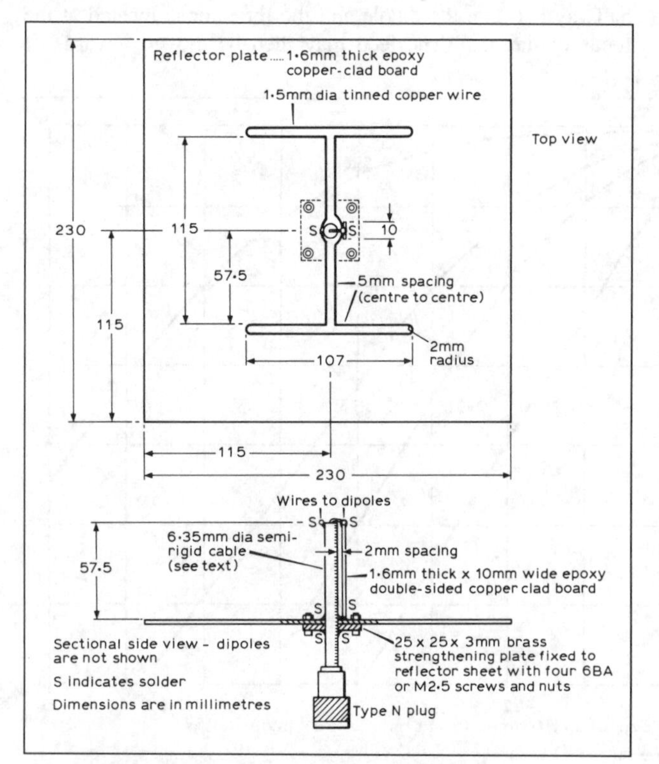

Fig 9.112. High-efficiency dish feed for 1.3GHz

The method of feeding the dipoles is also shown in Fig 9.112. The prototype used a length of 50Ω, 6.35mm semi-rigid cable. Alternatively, this could be replaced by a home-made 50Ω rigid cable, using a 6.35mm copper or brass rod, drilled just large enough to take the inner from a piece of standard cable, eg UR43. The cable and a strip of copper-clad epoxy board form a 1:1 balun. One side of the open-wire feeder is connected to the junction of the inner conductor of the cable and the top of the epoxy board.

Provided that the feed has been carefully made, it should exhibit a low VSWR. The VSWR can be minimised by squeezing together or pulling apart the open-wire feeder sections and the dipole elements. Try to aim for a symmetrical antenna after doing this, or the two dipoles may not share the power equally. This would distort the radiation pattern and hence reduce the illumination efficiency. If significant power levels, ie greater than about 1W, have to be used to measure the VSWR, do not make any adjustments with the power on!

In use, the feed should be located in the dish so that the focal point of the dish is midway between the plane of the dipoles and the reflector. The feed may then be moved inwards or outwards by a small amount to maximise the gain of the dish. The best way to mount the feed is to use a tripod or quadrupod support with the feed fixed between support members just in front of the apex of the support. Horizontal polarisation is obtained when the dipole elements are horizontal.

A high-efficiency circularly polarised feed horn for 1.3GHz EME

This dish feed was developed from an original design by W2IMU. It is primarily intended for optimum operation with dishes having *f/D* ratios in the range 0.5–0.6 but can be used, albeit with reduced efficiency, with deeper dishes, eg *f/D* ratio 0.4.

This feed horn differs from more conventional designs in that the circular waveguide operates in two modes, TE11 and TM11, simultaneously. The horn is termed *dual-mode* for this reason. The TE11 mode is launched by the probe in the smaller-diameter section. The tapered section converts some of the dominant TE11 energy into the TM11 mode in the larger-diameter section. When both modes reach the front of the horn, their relative phase and amplitudes give zero fields at the periphery of the aperture, resulting in very low rear and side radiation and thus minimum noise pick-up.

The generation of circular polarisation is performed by the 10 screws in the smaller-diameter section of the horn. These load the waveguide in the direction parallel to the axis of the screws and cause a delay of 90° but have no effect in the perpendicular direction. The feed probes are orientated at 45° to the polarising screws and thus generate equal components parallel and perpendicular to the screws. The parallel component is delayed in phase by 90° compared to the perpendicular component and thus circular polarisation results. One port produces right-hand circular polarisation while the other produces left-hand circular polarisation. This is exactly what is required for EME operation since the sense of polarisation is reversed on reflection from the Moon.

In use, the transmitter is permanently connected to one port (thus no high-power relay is required), while the receiver is connected to the other port via a small isolating relay (which is

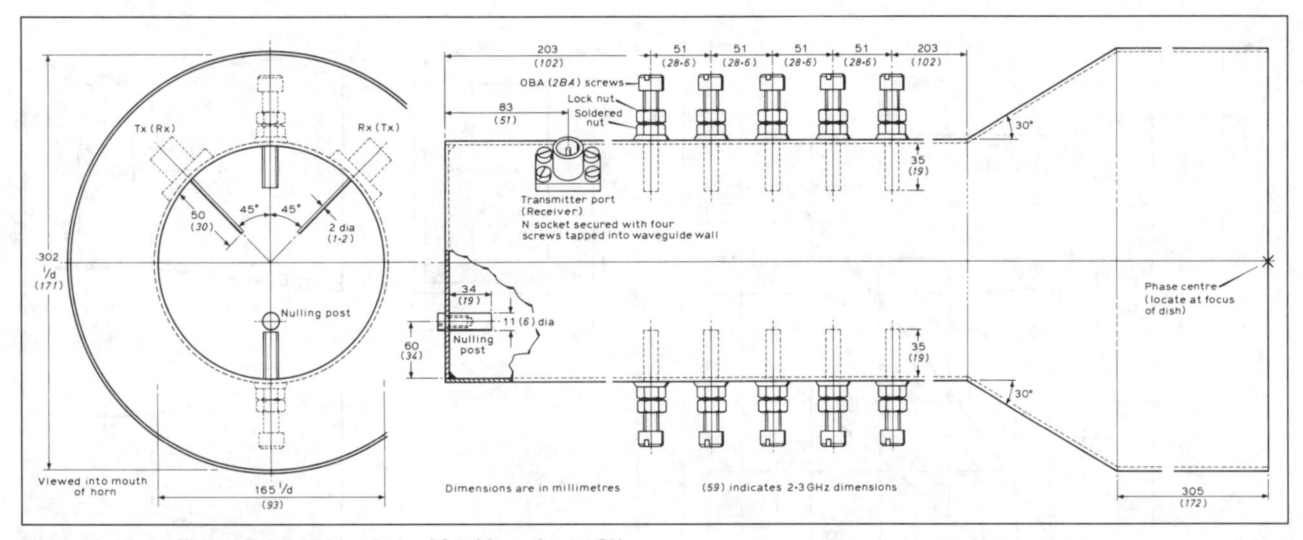

Fig 9.113. High-efficiency circularly polarised feed horn for 1.3GHz

required since the two probes are not totally isolated in practice). The degree of coupling between the probes depends on a number of factors, such as the exact orientation of them with respect to the polarising screws, the efficiency of the polariser and the focal length of the dish. A small post, mounted on the short-circuit back plate, is used to minimise the coupling between the probes.

Constructional details of the feed horn are given in Fig 9.113. The prototype version was constructed out of 0.8mm brass sheet, with the short-circuit plate cut from 3mm brass plate. The cone-to-cylinder joints were secured by means of many small tabs that overlapped the sections to form a butt joint. The rear plate was soldered in position.

Aluminium has also been used, with argon-arc welded joints. In this case, the polarising screws require a mounting bush. A suitable bush can be made from the screwed bushing and nut salvaged from an old panel-mounting variable resistor. The moulded body of the resistor is broken away from the bushing using a hammer and the shaft is replaced by a piece of brass rod, drilled and tapped to take the polarising screws, soldered into the bush. The polarising screw mounting holes are enlarged to take the bush which is held in place by the salvaged nut.

The adjustment of this feed is rather complicated and requires some care if optimum performance is to be attained. The recommended procedure is to optimise the feed probes individually for best VSWR, set up the polariser and then adjust the nulling post and polariser for best overall performance. The circularity of the radiation is sampled using two helix antennas of opposite sense. The full procedure is as follows:

1. Insert one feed probe, with the nulling post not fitted, and adjust the length of the probe for best VSWR (less than 1.2:1). The probe can also be bent slightly towards or away from the back of the horn to improve the match. When no further improvement can be obtained, remove the probe and repeat the procedure with the other probe.
2. Set up a test range as shown in Fig 9.114. Make sure that there are no obstructions nearby which could cause stray reflections. With one probe inserted and the nulling post not fitted, set the polariser screws to the nominal positions. Check that the probe is at 45° to the polariser screws. Apply a signal to the probe and, using one helix antenna at a time, measure the received signal level. A considerable difference in the signal level received by the two helix antennas should be noted. Select the helix with the lower signal and adjust the polariser screws (move each the same amount) for minimum signal, which corresponds to best circularity.
3. Insert the other probe (make sure it is exactly 90° to the first one) and fit the nulling post. Connect the detector to the probe just fitted. Then adjust the polariser screws, the position of the nulling post (and possibly the probe length/position) for minimum received signal, minimum reflected power and minimum coupling between the probes. It will be found that the adjustments interact to some extent, so the adjustments are a repetitive process! Also, it may happen that one of the

Fig 9.114. Test range set-up of OZ9CR for adjustment of the circularly polarised feed

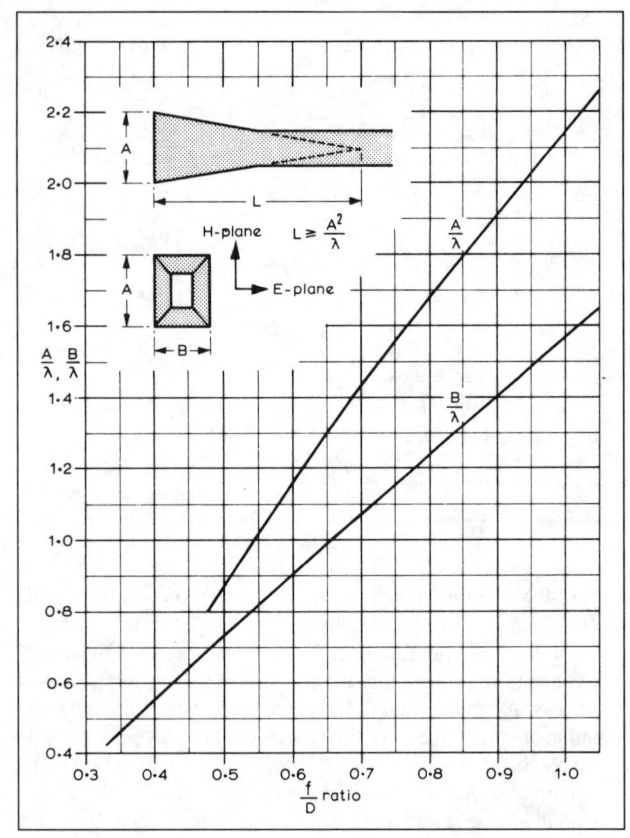

Fig 9.115. Form of horn antenna to feed dish and dimensions as a function of dish *f*/*D* ratio

polariser screws (eg the centre ones) can be used to optimise the VSWR without affecting the circularity too much.

4. Stop adjusting when the following specifications have been met: VSWR less than 1.2:1, difference in signal received by the two helix antennas greater than 20dB, coupling between probes less than 20dB.
5. Finally, check that the VSWR and polarisation discrimination are acceptable for the other feed probe. Continue adjusting if this is not the case.

In use, the feed horn should be positioned with its mouth located at the focus of the dish. It may then be moved inwards or outwards by a small amount to optimise the gain of the dish.

Pyramidal horn feeds for dishes

Pyramidal horns have significant advantages over most other types of feeds for dishes. This makes them especially suitable for use by amateurs. First, they offer a virtually perfect match over a wide range of frequencies and are therefore uncritical in their design and construction. Even quite large dimensional errors do not affect the quality of this match, but only the efficiency of illumination of the dish. A second advantage is that these horns are designed to produce optimum illumination of the dish in both planes. With other feeds, there may be little or no control of the ratio of beamwidths in each plane.

The form of the horn is shown in Fig 9.115. It consists of a length of waveguide which is flared in one or both planes to produce the beamwidth required. At higher frequencies, the horn will normally be fed by waveguide. The body of the feed will therefore usually consist of a length of waveguide that matches the rest of the system. On the lower microwave frequency bands (1.3, 2.3 and 3.4GHz) the horn is often fed by using coaxial

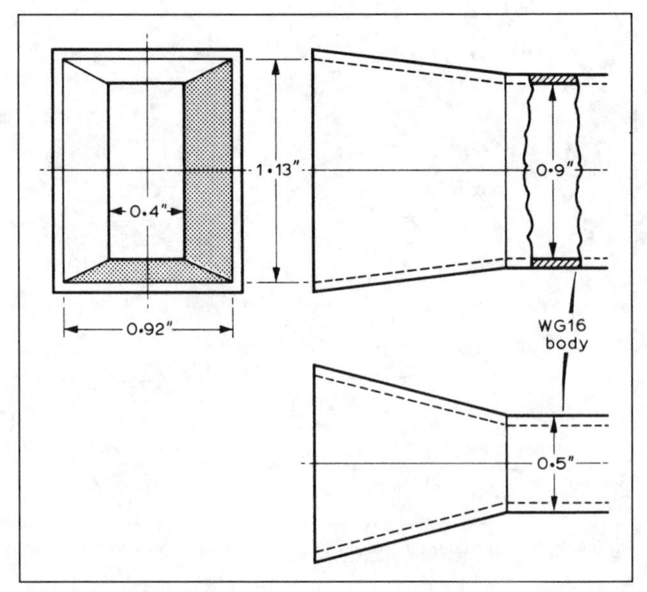

Fig 9.116. Typical pyramidal horn for a dish with *f*/*D* ratio of 0.53, for use at 10GHz

line. A waveguide to coaxial transition will be needed to serve as a method of launch. The horn design method is otherwise exactly the same and it is not necessary in this case to use standard waveguide dimensions.

The dimensions of the pyramidal horn as a function of the *f*/*D* ratio of the dish to be illuminated are given in Fig 9.115. They are based on an edge illumination of 10dB down on that at the centre of the dish. Due allowance has been made for space loss. The dimensions are given in terms of wavelength, so enabling a frequency of operation to be catered for. The actual dimensions are of course determined by multiplying the values of A/λ and B/λ by the wavelength in air at the design frequency.

It can be seen from Fig 9.115 that the aperture of the horn increases as the *f*/*D* ratio of the dish is reduced. The limit of the design is reached first with the 'A' dimension at a value of A/λ = 0.8 which corresponds to a minimum *f*/*D* ratio of 0.48. To feed dishes of a smaller *f*/*D* ratio, the Cassegrain sub-reflector system may be used [54].

Alternatively, the end of the waveguide may be suitably shaped to increase its beamwidth but, as this shaping has to be determined experimentally, much of the advantage of horn feeds is lost.

As an example of the design of a horn for a specific dish, consider a dish of diameter D = 0.914m (36in), which has a depth at its centre c = 108mm (4.26in). It is to be fed at 10,400MHz, for which λ = 28.846mm (1.136in). The focal length of the dish is given by $D^2/16c$ = 0.483m (19in) and the *f*/*D* ratio is therefore 0.53. From Fig 9.115, the values corresponding to this ratio are:

H-plane aperture	=	A/λ	=	0.96 (ratio)
A is broad-wall	=	28.846 × 0.96	=	27.69mm (1.09in)
E-plane aperture	=	B/λ	=	0.78 (ratio)
B is narrow wall	=	28.846 × 0.78	=	22.50mm (0.886in)
L = $A \times A/\lambda$	=	27.69 × 27.69/28.846	=	26.58mm (1.046in)

At this frequency a convenient waveguide is WG16, so a practical horn would have an aperture of 27.69 × 22.5mm (1.096 × 0.886in), tapering to 22.8 × 10.16mm (0.9 × 0.4in). The design is shown in Fig 9.116.

The same design procedure is applicable for any frequency to any dish with the same *f*/*D* ratio. For example, the feed horn

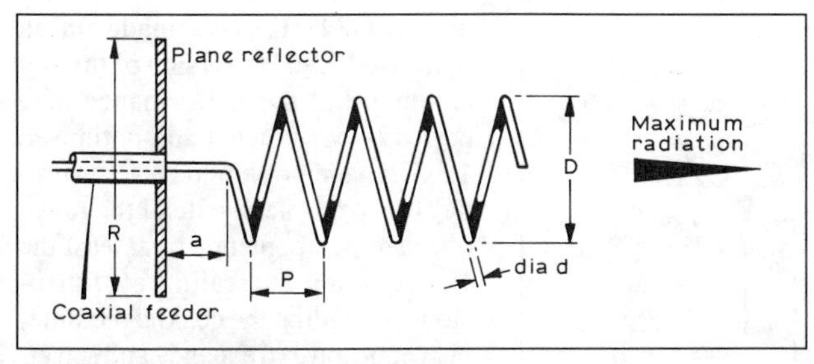

Fig 9.117. Main dimensions, axial helix antenna. Pitch angle 12°; R = at least $\lambda/2$; $D = \lambda/3$; $P = \lambda/4$; d = not critical; Impedance = 140Ω; use $\lambda/4$ transformer at feed point to match to 50Ω feeder

at 1296MHz would have an aperture of 222 × 180mm (8.74 × 7.10in), tapering to 165 × 82.5mm (6.5 × 3.25in) if WG6 were used. Similarly, one for use at 24GHz would have an aperture of 11.99 × 9.8mm (0.472 × 0.384in) tapering to 10.67 × 4.32mm (0.42 × 0.17in) for WG20.

Helix antennas for 1.3GHz

Helix antennas are useful both as test antennas for setting up circularly polarised antennas as described above, and also as general-purpose, medium-gain antennas of wide bandwidth. An eight-turn antenna will have about 10dB gain over an approximately 2:1 frequency range. Circular polarisation of either sense is obtained by winding the helix with a right-handed or left-handed thread. A simple $\lambda/4$ matching transformer is included and the VSWR can be optimised after construction by slightly shortening the final turn. No details of construction are given since there are now several inexpensive commercial sources of 1.3GHz helix antennas in the UK, although the general dimensions for an axial-mode helical antenna with a pitch angle of 12° are given in Fig 9.117.

Horizontally polarised, omnidirectional antennas for 1.3 and 2.3GHz

An omnidirectional, horizontally polarised, Alford slot antenna for 1.3GHz (and other bands) was developed by G3JVL to serve as a beacon or repeater antenna. Mechanical details of the antenna are shown diagrammatically in Figs 9.118 and 9.119. It consists of a length of slotted tubing which can be manufactured either by cutting a slot in a piece of suitable tubing, or by removing metal from a larger diameter piece of tubing and reforming the material left around a suitable former. The feed point can be made either in the centre of the slot or at the end by suitable design.

The width and length of the slot, the wall thickness and the diameter of the tubing are all related and the design data reproduced below was the result of much experimental work. Dimensions for end-fed and centre-fed antennas using different tubing sizes are given in Table 9.25.

The RF is fed to the antenna by a length of 3.6mm (0.141in) semi-rigid cable run inside the antenna to the centre of the slot or the end of the slot, as appropriate (Fig 9.118(a) and (b)), via a 4:1 balun, which is constructed in the end of the cable. Details of this are given in Fig 9.118(c) or Fig 9.119(b), with the balun slot dimensions for both bands listed in Table 9.25. It should be noted that the outer of the cable is slotted on both sides. The slots can be made, with care, using a broken blade from a small hacksaw. Connection is made between the balun and the slot by two tabs made from thin copper foil (Fig 9.119(b)) or by the use of two solder tags (Fig 9.119(a)). These may be fixed to the slot by soldering or using small screws. The base of the antenna can be terminated in an N-type bulkhead plug or socket as shown in

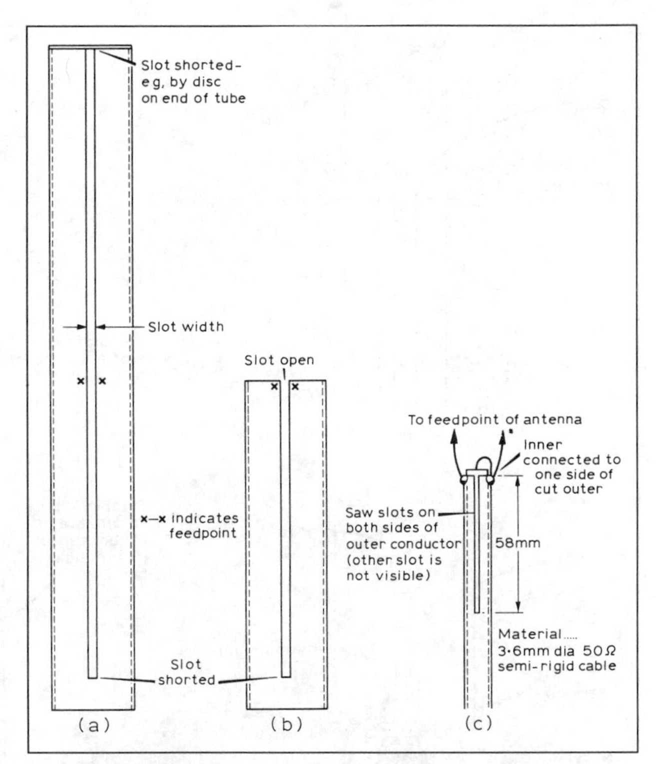

Fig 9.118. The G3JVL Alford slot antenna for 1.3 or 2.3GHz, showing mechanical details for (a) centre and (b) end-fed versions plus (c) construction of balun

Fig 9.119(c); the bulkhead fitting may need to be modified to allow correct fitting of the semi-rigid feeder. The length of tube above or below the slot is not critical and the lower end of the tube could be extended to provide a short mast section if this suits the constructor's needs. The feeder should be shaped so that it runs near the inside back face of the tubing as shown in Fig 9.119(d). Standard compression pipe-fittings and blanking plates can be used to close both ends of the centre-fed version of the antenna. A plastic cap can be used to close the open end of the end-fed version, if so desired. When built, the antenna should exhibit a low VSWR if the dimensions given have been followed closely.

If suitable test equipment is available, the match may be optimised by adjusting the width of the slot, either by squeezing the antenna in a vice, or by prising the slot apart. These operations should be done carefully! A small probe (Fig 9.119(e)) and sensitive diode detector/meter can be used to explore the voltage distribution along the slot. The probe should be held close to the tube, but not directly in front of the slot (hold it 20 or 30° round from the edge of the slot) and moved along its length. Patterns such as those in Fig 9.120 should be seen.

Table 9.25. Dimensions of Alford slot antennas for 1.3 and 2.3GHz

Antenna type	Tube dimensions OD (mm)	Wall (swg)	Slot width (mm)	Slot length (mm)	Balun slot Width (mm)	Length (mm)
1.3GHz						
End fed	38.1	16	11	254	1	58
Centre fed	38.1	16	11	509	1	58
End fed	31.75	20	4	254	1	58
Centre fed	31.75	20	4	509	1	58
2.3GHz						
End fed	19.0	18	3	140	1	26
Centre fed	19.0	18	3	280	1	26

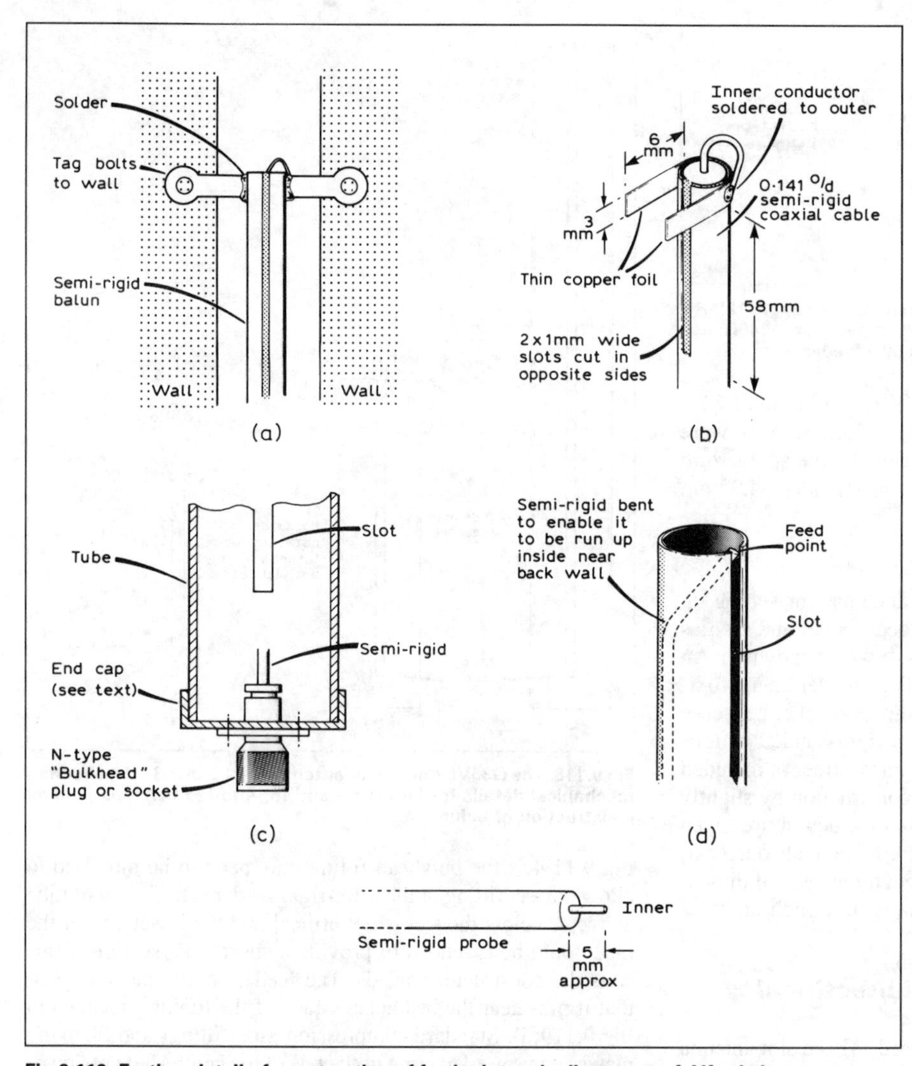

Fig 9.119. Further detail of construction of feedpoint and adjustment of Alford slot antenna

In use, the antenna should be mounted vertically as accurately as possible, since the vertical beamwidth is quite narrow. The length of the tubing below the slot is uncritical, so the same tube can be used for both mast and antenna. For weatherproofing, the antenna can be mounted inside a radome made from a length of 63mm diameter plastic drain pipe with 2mm wall thickness. If available, polypropylene tubing can be used with advantage, since it offers slightly lower loss than conventional PVC drain-pipe.

The performance of the slot antenna is illustrated in Figs 9.121, 9.122 and 9.123. It should be noted that these measurements,

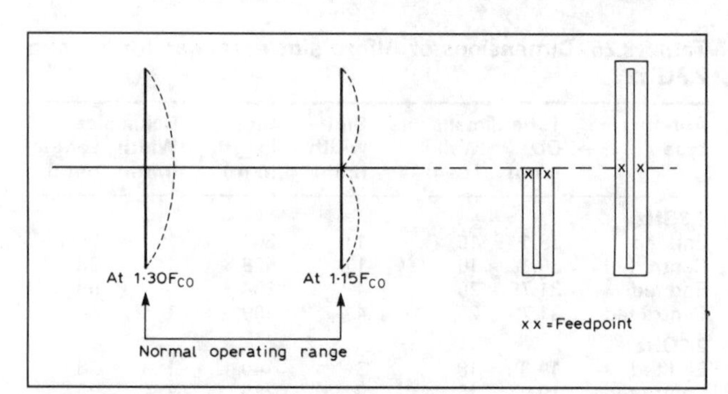

Fig 9.120. Typical field pattern using detector probe

due to G3TQF, were made on an optimised 2.3GHz version of the slot antenna, but the performance of a correctly constructed and optimised 1.3GHz version should be very similar. It is possible to extend the range of this type of antenna to at least the 10GHz band by scaling of dimensions (including the coaxial feedline) in proportion to frequency and selecting the optimum size of tube for the band concerned.

Horn antennas for the higher bands

Large pyramidal horns can be an attractive form of antenna for use at 10GHz and above. They are fundamentally broad-band devices which show virtually perfect match over a wide range of frequencies, certainly over an entire amateur band. They are simple to design, tolerant of dimensional inaccuracies during construction and need no adjustment. Horns are particularly suitable for use with transmitters and receivers employing free-running oscillators, the frequency of which can be very dependent on the match of their load (antenna).

Another advantage is that their gain can be predicted within a decibel or so (by simple measurement of the size of the aperture and length) which makes them useful for both the initial checking of the performance of systems and as references against which other antennas can be judged.

Their main disadvantage is that they are bulky compared with other antennas having the same gain. Large (long) horns, such as that illustrated in Fig 9.122, result in an emerging wave which is nearly planar and the gain of the horn is close to the theoretical value of $2\pi AB/\lambda^2$, where A and B are the dimensions of the aperture. For horns which are shorter than optimum for a given aperture, the field near the edge lags in relation to the field along the centre line of the horn and causes a loss in gain. For very short horns, this leads to the production of large minor lobes in the radiation pattern. Such short horns can, however, be used quite effectively as feeds for a dish.

The dimensions for an optimum horn for 10GHz can be calculated from the information given in Fig 9.124, and for a 20dB horn are typically:

$$A = 5.19\text{in (132mm)}$$
$$B = 4.25\text{in (108mm)}$$
$$L = 7.67\text{in (195mm)}$$

There is, inevitably, a trade-off between gain and physical size of the horn. At 10GHz this is in the region of 20dB or perhaps slightly higher. Beyond this point it is better to use a small dish. For instance, a 27dB horn at 10GHz would have an aperture of 11.8in (300mm) by 8.3in (210mm) and a length of 40.1in (1019mm) compared to a focal plane dish which would be 12in (305mm) in diameter and have a 'length' of 3in (76mm) for the same gain.

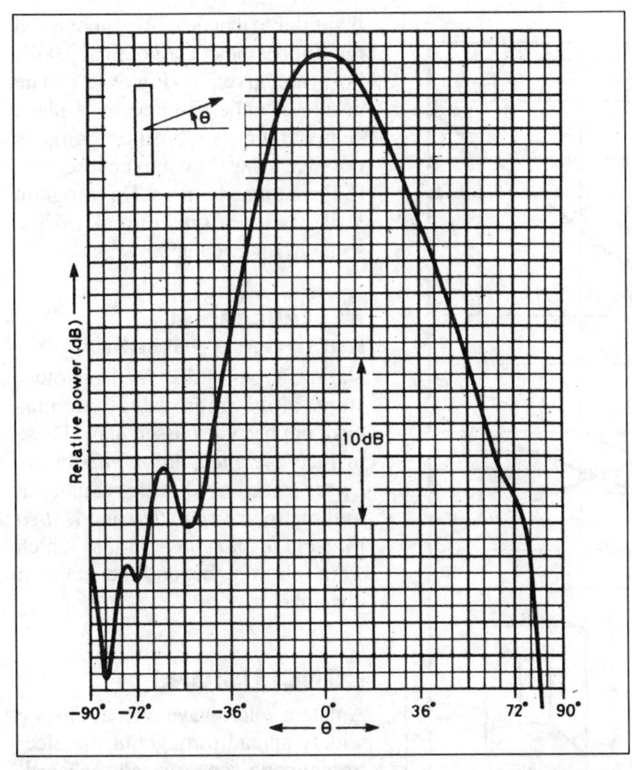

Fig 9.121. Performance of Alford slot antenna – vertical radiation pattern. (Note: measurements made on a 2.3GHz antenna; patterns given by a 1.3GHz version will be very similar)

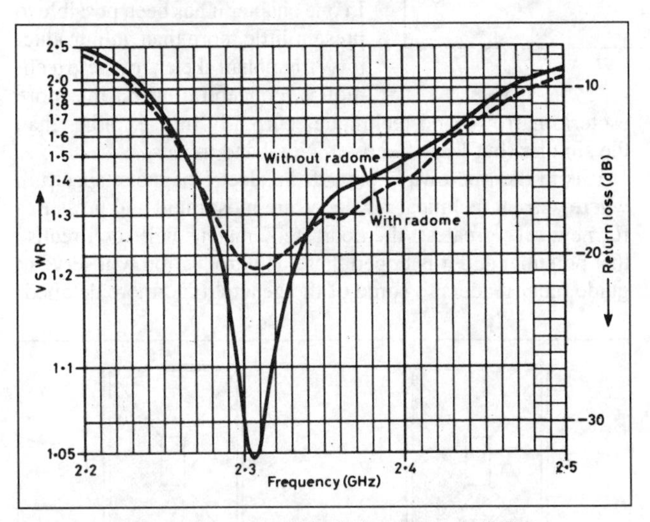

Fig 9.122. Performance of an Alford slot antenna: VSWR

A computer program for the design of both optimum and sectoral horns was been given in reference [55]. Either manual calculation or the program can be used.

Horns are usually fabricated from solid sheet metal such as brass, copper or tinplate. There is no reason why they should not be made from perforated or expanded metal mesh, provided that the size and spacing of the holes is kept below about λ/10. Construction is simplified if the thickness of the sheet metal is close to the wall thickness of WG16, ie 0.05in or approximately 1.3mm. This simplifies construction of the transition from the waveguide into the horn.

The geometry of the horn is not quite as simple as appears at first sight since it involves a taper from an aspect ratio of about 1:0.8 at the aperture to approximately 2:1 at the waveguide transition.

For a superficially rectangular object, a horn contains few right-angles, as shown in Fig 9.125 which is an approximately quarter-scale template for a nominal 20dB horn at 10.4GHz. If the constructor opts to use the 'one-piece' cut-and-fold method suggested by this figure, then it is strongly recommended that a full-sized template be drafted on stiff card which can be lightly scored to facilitate bending to final form. This will give the opportunity to correct errors in measurement before transfer onto sheet metal and to prove to the constructor that, on folding, a pyramidal horn *is* formed!

The sheet is best sawn (or guillotined) rather than cut with tin-snips, so that the metal remains flat and undistorted. If the constructor has difficulty in folding sheet metal, then the horn can be made in two or more pieces, although this will introduce more soldered seams which may need jigging during assembly and also strengthening by means of externally soldered angle pieces running along the length of each seam. Alternative methods of construction are suggested in Fig 9.126.

It is worth paying attention to the transition point which should present a smooth, stepless profile. The junction should also be mechanically strong, since this is the point where the mechanical stresses are greatest. For all but the smallest horns, some form of strengthening is necessary. One simple method of mounting is to take a short length of Old English (OE) waveguide, which has internal dimensions matching the external dimensions of WG16, and slit each corner for about half the length of the piece. The sides can then be bent (flared) out to suit the angles of the horn and soldered in place after carefully positioning the OE guide over the WG16 and inserting the horn in the flares. One single soldering operation will then fix both in place. After soldering, any excess of solder appearing inside the waveguide or 'throat' of the horn should be carefully removed by filing or scraping. The whole assembly can be given a protective coat of paint.

An alternative method would be to omit the WG16 section and to mount the horn directly into a modified WG16 flange. In this case the thickness of the horn material should be a close match with that of WG16 wall thickness, and the flange modified by filing a taper of suitable profile into the flange.

Whichever method of fabrication and assembly is used, good metallic contact at the corners is essential. Soldered joints are very satisfactory provided that the amount of solder in the horn is minimised. If sections of the horn are bolted or rivetted together, then it is essential that many, close-spaced bolts or rivets are used to ensure such contact. Spacing between adjacent fixing points should be less than a wavelength, ie less than 30mm.

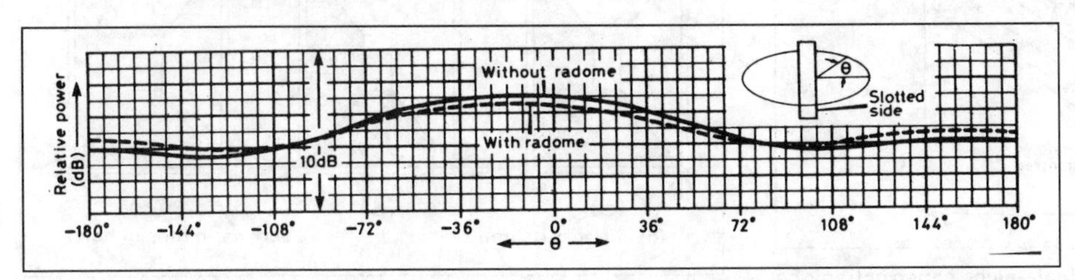

Fig 9.123. Performance of an Alford slot antenna: horizontal radiation pattern

Sectoral horns

The most useful type of sectoral horn is that with an H-plane flare.

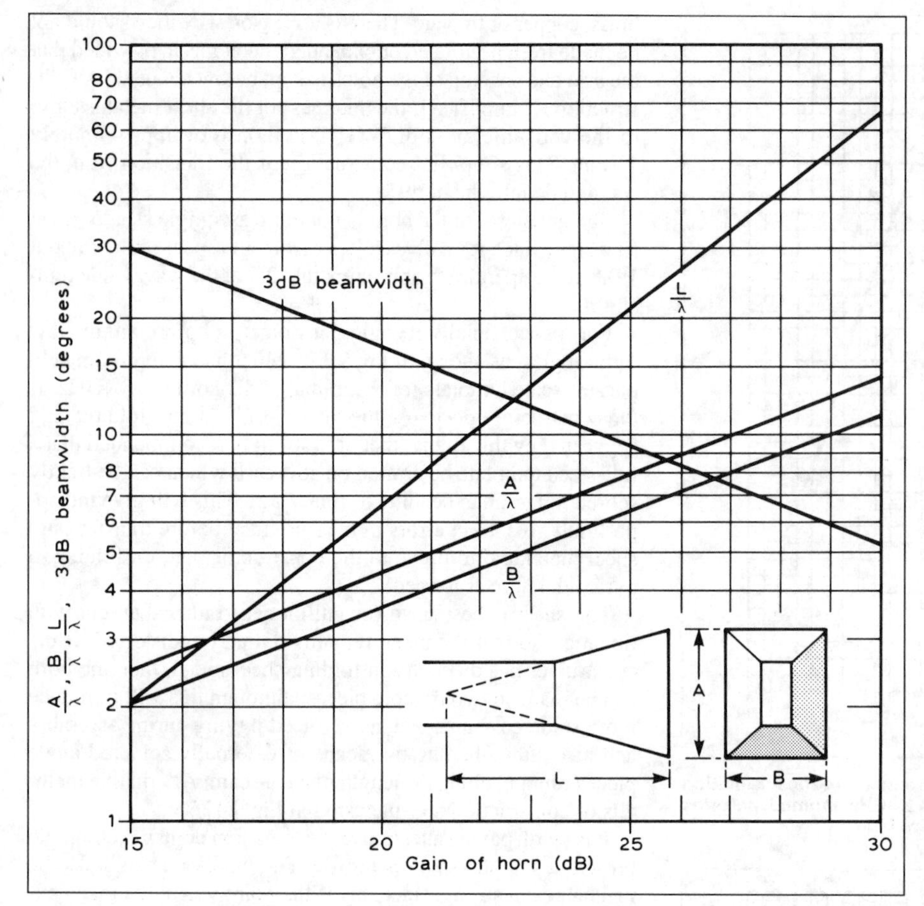

Fig 9.124. Horn design chart

With the broad faces of the guide vertical, a horizontally polarised field is produced which has an azimuth pattern of nearly 180° but a vertical pattern which is compressed into a few degrees (depending on the gain), thus making it useful for a beacon where semi-omnidirectional coverage is needed. Construction is similar to that described above and the dimensions of a nominal 10dBi horn are given in Fig 9.127. The design of either E-plane or H-plane sectoral horns of other gains is possible using the computer program [55] mentioned earlier. This program will also design conical horns, suitable for circular waveguide.

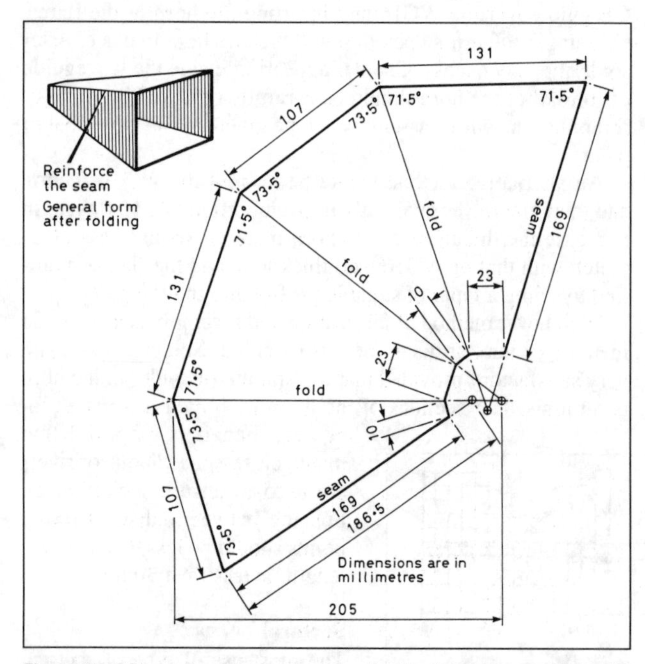

Fig 9.125. Dimensioned template for single-piece construction of a 20dB horn for 10GHz

Miscellaneous microwave antennas

Various types of dish feed and other stand-alone microwave antennas have not been discussed here. These include *biconical horns*, *dielectric lenses* (much used professionally at millimetre wavelengths) and *slotted waveguide antennas* [56–58] which were felt to be beyond the scope of this chapter.

CONCLUSIONS

Amateur microwave allocations are widely spread throughout the electromagnetic spectrum and, accordingly, require diverse techniques to exploit their unique characteristics. In one chapter it has been possible to present little more than an introductory 'thumbnail sketch' of these techniques, concentrating on the more esoteric narrow-band techniques currently in use, rather than the simpler (but less effective) wide-band modes.

It is in the microwave part of the electromagnetic spectrum where current technical advances are most rapid and pressures for new uses greatest: the dedicated amateur will soon realise that potential and it is hoped that this introduction will serve to guide other readers to some of the general (and more detailed)

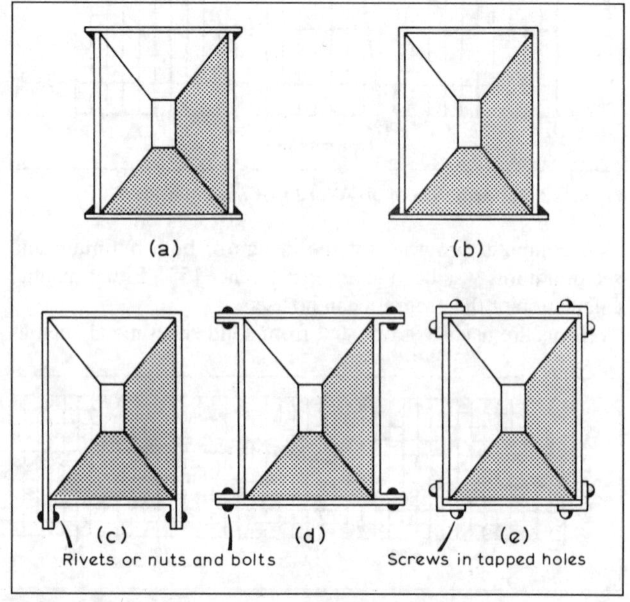

Fig 9.126. Alternative forms of horn construction

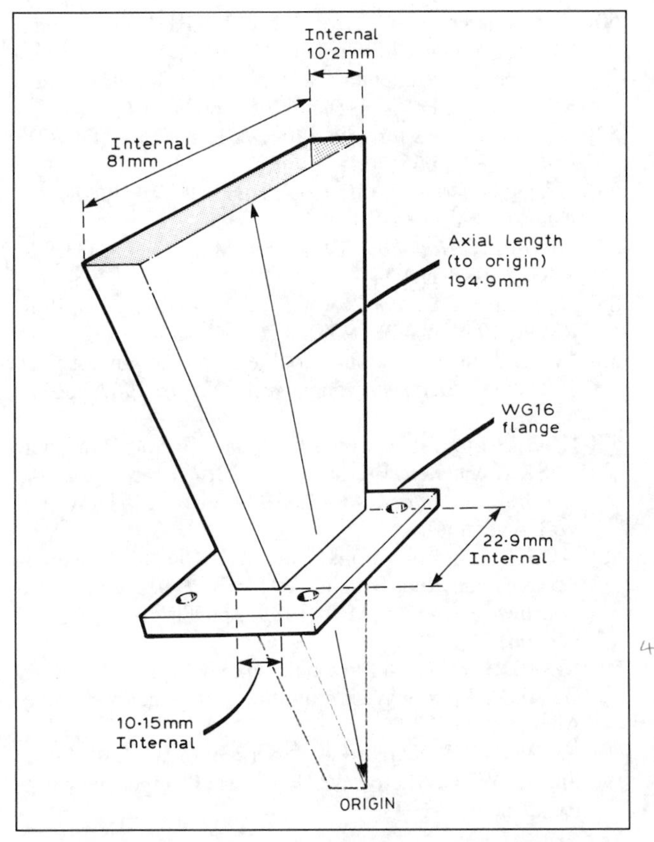

Fig 9.127. Typical sectoral horn dimensions. Both E-plane and H-plane sectoral horns can be computer designed (see text)

information available in many of the sources mentioned in the references below. In addition to the amateur press there are of course many professional source books, too numerous to list here. A library search will soon reveal those of most use to the experimenter!

REFERENCES

[1] CCIR, 1937.

[2] *Admiralty Handbook of Wireless Telegraphy*, HMSO, London, 1938.

[3] *Microwave Handbook*, Vol 1 – 'Components and operating techniques', ed M W Dixon, G3PFR, RSGB, 1989.

[4] *Microwave Handbook*, Vol 2 – 'Construction and testing', ed M W Dixon, G3PFR, RSGB, 1991.

[5] *Microwave Handbook*, Vol 3 – Bands and equipment', ed M W Dixon, G3PFR, RSGB, 1992.

[6] *UHF/Microwave Experimenter's Manual*, 'Antennas, components and design', ARRL.

[7] 'Overtone crystal oscillators in series and parallel resonance', H J Brandt, DJ1ZB, *VHF Communications* 1/77, pp38–43.

[8] 'A high-quality UHF source for microwave applications', RSGB Microwave Committee, *Radio Communication* October 1981, pp906–910.

[9] 'A local oscillator source for 1152MHz', S T Jewell, G4DDK, *Radio Communication* February 1987, p128 and March 1987, pp191–201.

[10] 'An oscillator/multiplier chain for the frequency range 2.0 to 2.6GHz', S T Jewell, G4DDK, *Radio Communication* August 1990, pp35–38.

[11] *Modern high-performance equipment for 10GHz*, 'Part 1, The G3WDG-001 multiplier/amplifier', C W Suckling, G3WDG, and G B Beech, RSGB Microwave Committee Components Service, August 1990.

[12] 'Phase-lock source update', K R Ericson, K0KE, *Proceedings of the Microwave Update*, ARRL, 1987.

[13] 'Phase-locked microwave sources', Greg McIntire, AA5C, *Proceedings of the Microwave Update*, ARRL, 1991.

[14] 'A frequency doubler for the 13cm band with 6W output power', O Frosinn, DK7QF, *VHF Communications* 3/79, pp141–143.

[15] '3456MHz tripler', C Neie, DL7QY, *Dubus Technik*, p215.

[16] R Griek, DJ7VF, and M Munich, DJ1CR, *VHF Communications* 2/79.

[17] 'Een Varactor mixer voor 13cm', H Schildt, PA0HEJ, *CQ-PA* 1982, pp423–429.

[18] D J Robinson, G4FRE, *RSGB Microwave Newsletter* 03/82, pp4–5.

[19] 'An alignment aid for VHF receivers', J R Compton, G4COM, *Radio Communication* January 1976.

[20] OE9PMJ correspondence, subsequently published in *Der SHF Amateur*, produced by J Dahms, DC0DA.

[21] *Modern high-performance equipment for 10GHz*, 'Part 4, Amplifiers', C W Suckling, G3WDG, RSGB Microwave Committee Components Service, 1992.

[22] 'A single-board, no-tuning 23cm transverter', Richard L Campbell, KK7B, *23rd Conference of the Central States VHF Society*, Rolling Meadows, Illinois, 1989, pp44–52 and *Engineering Notes*, *ibid*, pp53–55. Subsequently republished in the *ARRL Handbook for the Radio Amateur*, 69th edn, ARRL, 1992.

[23] R Dabbs, G2RD, *RSGB Bulletin* October 1965, p650. Also in the *VHF/UHF Manual*, 3rd and 4th edns, RSGB.

[24] C W Suckling, G3WDG, and P Blair, G3LTF, *Radio Communication* January 1976, p24.

[25] *Crawford Hills VHF Club (USA) Technical Report No 6*, July 1971.

[26] *Crawford Hills VHF Club (USA) Technical Report No 13*, December 1972.

[27] Chip Angle, N6CA, *QST* June 1981.

[28] 'A quarter-kilowatt 23cm amplifier', Chip Angle, N6CA. Part 1, *QST* March 1985; Part 2, *QST* April 1985.

[29] 'More gain from 1.3GHz amplifiers', Ian White, G3SEK, and Roger Blackwell, G4PMK, *Radio Communication* June 1983.

[30] '6W PA for 9cm', K D Broker, DK1UV, *Dubus Informationen* April 1981, pp257–261.

[31] 'A local oscillator, transmit mixer and linear amplifier for the 9cm band', H J Senckel, DF5QC, *VHF Communications* April 1980, pp236–245.

[32] '10GHz transverter in microstripline technique', P Vogl, DL1RQ, *Dubus Technical reports 2/86 and 4/86*.

[33] 'Interdigital converters for 1296 and 2304MHz', R E Fisher, W2CQH, *QST* January 1974.

[34] 'Interdigital converters for the GHz amateur bands', J Dahms, DC0DA, *VHF Communications* March 1978, pp154–168.

[35] *Modern high-performance narrow-band equipment for 10GHz*, 'Part 2, The G3WDG-002 receive converter', C W Suckling, G3WDG, Microwave Committee Components Service, 1991.

[36] Richard L Campbell, KK7B, *Proceedings of the Microwave Update*, ARRL, 1988.

[37] Richard L Campbell, KK7B, *Proceedings of the Microwave Update*, ARRL, 1989.

[38] 'VHF and microwave applications of monolithic microwave integrated circuits', Al Ward, WB5LUA, *ARRL UHF/Microwave Experimenter's Manual*, ARRL, 1990, pp7.32–7.47.

[39] *Microwave Handbook*, Vol 3, ed M W Dixon, G3PFR, RSGB, 1992, pp18.116–18.118.

[40] Microwave Committee Components Service, c/o Mrs P Suckling, G4KGC, 132A Newton Road, Rushden, Northants NN10 0SY, UK.

[41] *Modern high-performance narrow-band equipment for 10GHz*, 'Part 3, The G3WDG-003 transmit converter', C W Suckling, G3WDG, Microwave Committee Components Service, September 1991.

[42] *Microwave Handbook*, Vol 2, ed M W Dixon, G3PFR, RSGB, 1991, pp8.31–8.33.

[43] *Microwave Handbook*, Vol 3, ed M W Dixon, G3PFR, RSGB, 1992, pp18.5–18.6.

[44] *Microwave Handbook*, Vol 1, ed M W Dixon, G3PFR, RSGB, 1989, pp4.25–4.32.

[45] 'More gain from Yagi antennas', Gunter Hoch, DJ6WU, *VHF Communications*, Vol 9, March 1977, pp157–166.

[46] 'Extremely long Yagi antennas', Gunter Hoch, DJ6WU, *VHF Communications*, Vol 14, March 1982, pp131–138.

[47] *Microwave Handbook*, Vol 1, ed M W Dixon, G3PFR, RSGB, 1989, pp4.10–4.11, RSGB.

[48] *Microwave Handbook*, Vol 1, ed M W Dixon, G3PFR, RSGB, 1989, p4.10, RSGB.

[49] *Microwave Handbook*, Vol 1, ed M W Dixon, G3PFR, RSGB, 1989, pp4.16–4.17, RSGB.

[50] 'Wideband horn feed for 1.2 to 2.4GHz', Peter Riml, OE9PMJ, *Dubus* 2/86, pp110–111.

[51] 'Log periodic antenna feed for 1.0 to 3.5GHz', Hans Schinnerling, DC8DE, *Dubus* 2/83, pp99–101.

[52] 'Multiband feed for 1 to 12GHz', Klaus Neie, DL7QY, *Dubus* 2/80, pp66–76.

[53] 'Triband microwave dish feed', Tom Hill, WA3RMX, *QST* August 1990.

[54] *Microwave Handbook*, Vol 1, ed M W Dixon, G3PFR, RSGB, 1989, pp4.34–4.35.

[55] *Microwave Handbook*, Vol 1, ed M W Dixon, G3PFR, RSGB, 1989, pp4.38–4.39.

[56] 'X-band omnidirectional double-slot array antenna', T Takeshima, *Electronic Engineering* October 1967, pp617–621.

[57] '10GHz omni-directional antenna', Harold Reasoner, K5SXK, and Kent Britain, WA5VJB, *Proceedings of the Microwave Update*, ARRL, 1989, pp190–191 (WR90/WG16 version).

[58] '10GHz omnidirectional antenna', Harold Reasoner, K5SXK, and Kent Britain, WA5VJB, *Proceeding of the Microwave Update*, ARRL, 1991, pp308 (WR75/WG17 version).

[59] 'G3WDG-004 HEMT low-noise amplifier', C W Suckling, G3WDG, Microwave Committee Components Service, August 1993.

[60] 'G3WDG-007 10GHz 1W power amplifier', C W Suckling, G3WDG, Microwave Committee Components Service, December 1993.

10 Telegraphy and keying

CW TELEGRAPHY is the simplest form of amplitude modulation; intelligence is transmitted by switching on and off the carrier wave produced by the transmitter to form the dots and dashs of the morse code.

The bandwidth required by a properly keyed signal is small and should be directly related to the speed of sending involved. The optimum condition is not always easily attained, however, and the result of a maladjusted keying system is usually the radiation of spurious emissions over large parts of the frequency spectrum. It is also possible to obtain the opposite effect, however, through an attempt to avoid harmful keying transients; the result of this condition is to produce a very 'soft' characteristic which will prevent the reading of morse signals at high speeds.

Morse code characters are made up of dots and dashs, each character consisting of a unique combination of these elements.

The standard morse code (International Telegraph Code No 1) dictates that a space between the elements of a character shall be one dot length, and that the length of a dash and the interval between letters shall be three dot lengths. Fig 10.1, showing the length of a dot for a range of transmission speeds, should be used for reference when a keying waveform is being tailored, because it is important to provide the correct amount of softening of the key envelope in relation to the upper speed which is to be used: the rise or decay time should not exceed about one-third of a unit length at the highest speed, so that at 24 words per minute such times should not exceed about 15ms.

The only method of keying the transmitter to be discussed in this chapter consists of changing the amplitude of the carrier. An alternative method, known as *frequency shift keying*, is to vary the frequency, but this offers very little advantage for normal morse communication purposes, and is usually applied solely to operating teleprinters.

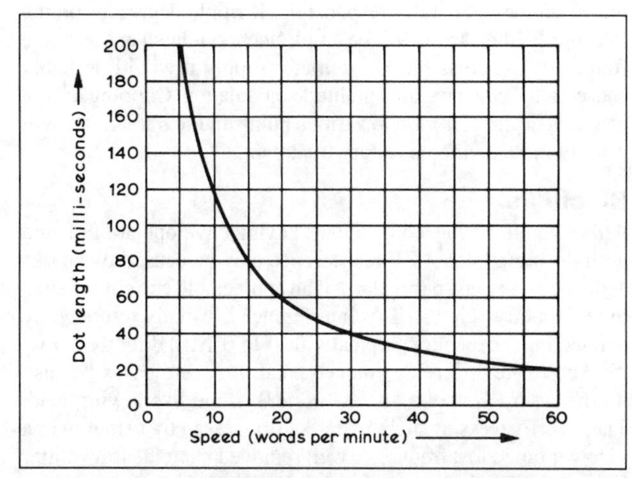

Fig 10.1. Graph showing dot lengths for a range of transmission speeds. Once the design figure of the maximum rate of sending for a transmitter has been decided, this graph can be used to select the appropriate time delays in keying filter circuits

KEYING CHARACTERISTICS

There are two main aspects which affect keying characteristics; envelope shape and frequency stability. Any keying trouble such as key clicks, ripple, chirp and spacer waves can be attributed to poor conditions in one of these areas.

Envelope shape

The envelope of a keyed signal is the outline of the pattern that the signal would display on an oscilloscope. It can be observed by feeding the keyed RF signal on to the Y-plates or vertical amplifier of a slow-scan oscilloscope, and setting the timebase in synchronism with the keying speed. The transmitter is best keyed by an automatic device producing a regular sequence of dots or dashs so as to obtain a steady display.

In general, if no precautions are taken, the pattern will be square and sharp, or 'hard' (Fig 10.2(a)) and will radiate interference over a wide range of frequencies. The rise and decay times of the carrier must therefore be lengthened until the interference is no longer objectionable, but without impairing the intelligibility at high speed. This requires rise and decay times of 5–20ms and the methods of achieving this will be described in the sections on different keying systems.

Key filter circuits rely for their operation on the time constant of a CR (or LR) circuit (see Chapter 1). Filter circuits are used to limit the rate of change of the bias voltage to the keyed amplifier stage, so that the operating conditions of the stage do not switch abruptly from conduction to cut-off.

The characteristics of the PA power supply may contribute to the envelope shape, as the voltage from a power unit with poor regulation will drop quickly each time the key is closed, and rise when it is released. This can lead to the shape in Fig 10.2(d). This effect is not necessarily undesirable but it can be lessened by use of a choke-input filter in the PA supply or by increasing the size of filter capacitors in the supply.

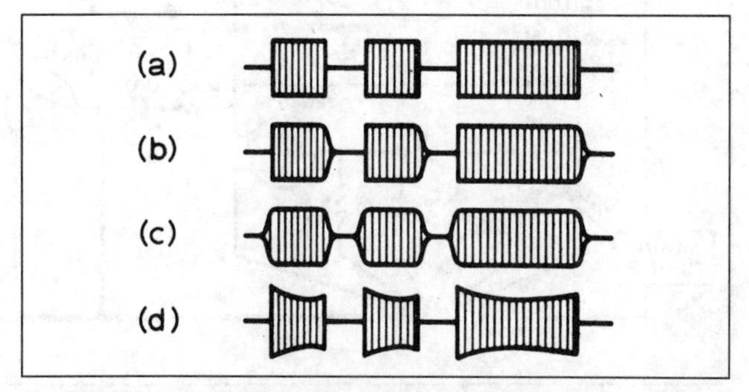

Fig 10.2. Keying envelope characteristics. (a) Click at make and break; (b) click at make with click at break suppressed; (c) ideal envelope with no key clicks; (d) effect on keying envelope of poor power supply regulation

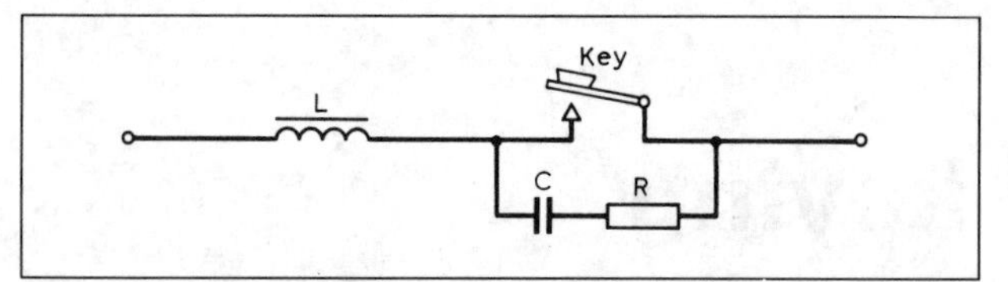

Fig 10.3. Typical key-click filter. L serves to prevent a rise of current. C, charging through R, serves to continue flow of current momentarily when key contacts open. Typical values: L = 0.01 to 0.1H, C = 0.01 to 0.1µF, R = 10 to 100Ω

Envelope ripple can be caused by RF feedback but poor power unit filters are generally responsible. If ripple is present on the PA supply, the carrier will be amplitude modulated at the ripple frequency, whereas on the oscillator supply it would probably cause both frequency and amplitude modulation. Clipping a large electrolytic capacitor on to various points in the transmitter will usually indicate the necessary treatment.

RF clicks

Although clicks caused by a hard keying envelope are radiated with the signal, local RF interference may be caused by sparks at the key contacts, particularly if an appreciable current is keyed in an inductive circuit. This interference is usually removed by connecting a capacitor (typically 0.01 to 0.1µF) directly across the key or keying relay contacts, and in severe cases by also inserting an RF choke (10 to 100mH) in the live keying lead. The effectiveness of such treatment is assessed by listening to a receiver tuned to a frequency well removed from the transmitter frequency. Clicks can also be caused by contact bounce in a keying relay or the contacts of a semi-automatic morse key. A typical key-click filter is shown in Fig 10.3.

Chirp

This is a form of frequency instability occurring each time the transmitter is keyed, and is recognised by a change in beat frequency at the beginning and end of each character when the signal is monitored on a receiver. A signal with chirp is undesirable, as it is less easy to copy, and is not suitable for reception by narrow-bandwidth receivers. It is usually more prevalent in transmitters controlled by a VFO, and there are three principal causes, which are:

1. *DC instability*. This occurs when a common power supply is used for the oscillator and the PA (or any circuit through which the current changes in sympathy with keying). No oscillator exhibits absolute stability under varying supply conditions, and therefore a voltage regulator should be incorporated in the oscillator supply. A separate oscillator supply might be needed to cure a difficult case, but improving the regulation of the common power unit and redesigning the oscillator to be less dependent on supply voltage variations should generally suffice.
2. *Pulling*. This refers to the effect on the oscillator frequency of one or more subsequent stages, the operating conditions of which change during the keying cycle. It can be expected if the stage following the oscillator draws input current (as would a Class C buffer or doubler), or if the early stages of the transmitter are tightly coupled. If the oscillator is on the same frequency as the PA, ie where no frequency multiplying or mixing is used, the likelihood of pulling is increased.

   Pulling can invariably be treated by improving the isolation of the oscillator, and an emitter follower is highly recommended for this. The oscillator should be loosely coupled to it, and by careful design it should be possible to produce a unit which can have its output short-circuited without shifting the frequency by more than a few hertz. Fig 10.4 shows a circuit combining a stable oscillator with an emitter follower.

   It may be simpler to replace the frequency-determining components in the oscillator with values to halve the original operating frequency and to use a subsequent stage as a doubler. This may, however, reduce the drive to the PA to such an extent that a further buffer stage would be necessary,
3. *RF feedback*. This is a stray signal leaking back from a high-level stage to a previous stage, particularly a variable frequency oscillator. It may have an appreciable effect on the frequency oscillation, depending on its strength and phase. The presence of feedback may be verified by noting the pulling which occurs on tuning the output circuits of the keyed stage through resonance. The feedback path may either be either internal, by virtue of valve or transistor capacitances, or external because of poor construction layout.

   Internal feedback may be treated by the methods outlined for pulling; isolation of the oscillator is of great importance. External feedback is only discovered after a transmitter has been built, and the commonest cause is the PA circuitry being close to the oscillator section. Here also pulling is most likely to occur when the PA runs on the same frequency as the oscillator, and the cure is either to resort to doubling, as before, or to screen both PA and oscillator. Sometimes it is sufficient to mount a metal plate between the two circuits.

   It is recommended that any long leads carrying RF power near the oscillator be screened. In addition, the HT line must be bypassed by means of series resistance and shunt capacitance.

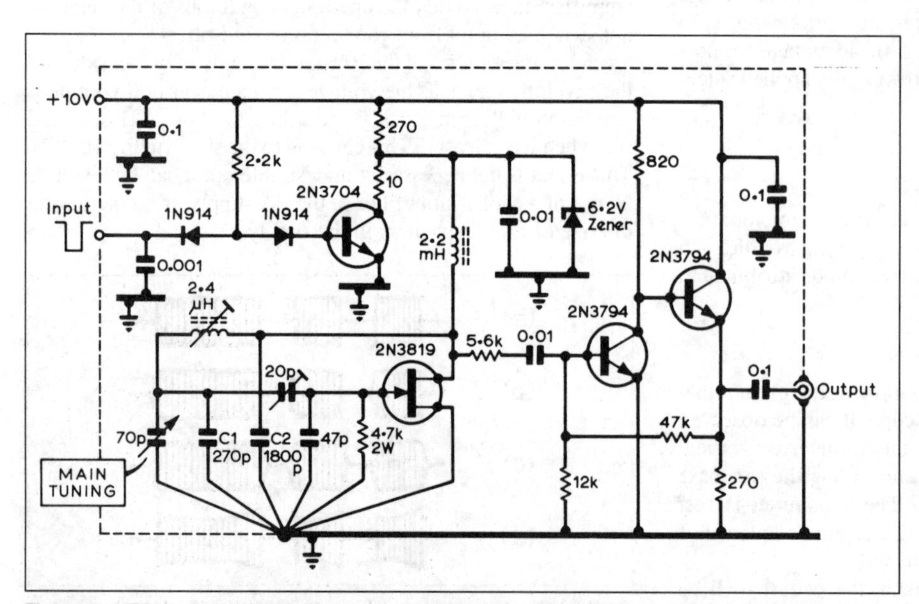

Fig 10.4. VFO and emitter follower designed to be stable against supply voltage and load fluctuations. Operating frequency with the component values shown is 5.88–6.38MHz, but this can be altered by changing L, C1 and C2 in inverse proportion to the desired frequency shift. The circuit around the oscillator permits it to be keyed for use in full break-in systems

Spacer waves or backwaves

The *spacer wave* is the small signal often radiated during key-up conditions. It is common for the spacer wave to be audible in a local receiver, but this does not mean that it is radiated far. If an appreciable spacer wave is radiated, it would make a signal difficult to copy. Causes are:

(a) the keyed transmitter stage not being completely cut off;

(b) leakage of RF through the PA valve capacitance.

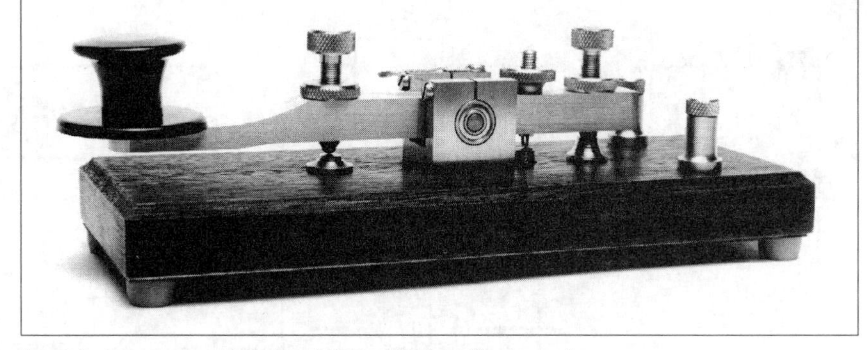

Fig 10.5. A modern straight key (photo: RA Kent (Engineers))

The first is often due to insufficient bias being applied to the keyed stage. If drive is applied to the PA continuously, the bias required for cut-off during key-up conditions will be the static value obtained from data sheets plus the peak value of the drive voltage, which is considerable. In a cathode-keying system, a leaky capacitor in the keying filter may cause a spacer wave.

In valve circuits, the second effect increases with the grid-to-anode capacitance of the PA: it is considerably worse with triodes than pentodes. It may be cured either by neutralising the stage in question or by keying more than one stage.

KEYING THE TRANSMITTER

Keying is the switching on and off of a transmitter to create the dots and dashs of the morse code. An ineffective system will most likely cause the deficiencies referred to earlier. Keying should take place at a point where the current is minimal to avoid sparking at the key contacts which may create interference. Likewise, the voltage should be low unless there is a relay between the key and the keying point.

It is generally considered to be bad practice to key a straight VFO, the collector/anode supply to which should be regulated. The VFO is normally followed by an isolating stage (emitter/cathode follower). The succeeding stage is a satisfactory keying point. When the key is up, there is no RF drive to the following stages, and the dependence of bias upon RF drive should be checked. Probably the only serious point here is the use of cathode bias on a tetrode valve output stage; this will not be cut off and may well radiate broad-band noise into a nearby receiver.

Transmit-receive switching

Associated with the keying process is the muting of the receiver when the key is down and the switching of the antenna from receive to transmit. This must be carried out in the correct sequence. This process can be done using relays to control relays, but it is not always satisfactory with a separate receiver and transmitter. In the latter case, a rotary switch to switch over the antenna (and to reduce the gain of the first stage of the receiver via a relay) is a reasonably satisfactory arrangement, unless one is interested in high-speed contest operation.

Modern amateur practice is the use of the commercial all-band transceiver for SSB and CW. With the transceiver, change-over from receive to transmit occurs automatically as the key is pressed and vice versa as the key is opened. The operator is therefore able to monitor the transmitting channel except when he or she is transmitting a morse character. The appearance of interference therefore becomes obvious, and the operator of the other station can indicate that interference is occurring by sending a series of dots. This is known as *break-in* (BK operation). Basically the same changeover switching system deals with telephony via the microphone press-to-talk (PTT) switch.

With the transceiver, keying deficiencies as defined earlier are indicative of a fault condition, in which case the equipment should be returned for investigation.

THE MORSE CODE AND MORSE KEYS

The morse code was published in America by Samuel Morse in 1834. By the end of the nineteenth century, telegraphic communication by land line using Morse's code covered most of America. With the invention of wireless telegraphy, the use of morse code soon became worldwide.

Literally thousands of different patterns of morse key have been produced. Fig 10.5 shows a modern straight key. The semi-automatic key, generally known as the *bug* after the trade mark of the original maker (Fig 10.6), has two pairs of contacts. Dashs are made singly by moving the knob to the left, thereby closing the front contacts. A train of dots is produced by similarly moving the paddle to the right against a stop. This causes the rear portion of the horizontal arm to vibrate and close the rear pair of contacts. A properly adjusted bug key will produce at least 25 dots.

Electronic keys are fully automatic and consist of two parts, the keyer containing the electronic circuits and a paddle key. The paddle key is moved from side to side, closing a separate pair of contacts on each side.

The simplest electronic key, electronically and to use, is the

Fig 10.6. A modern bug key (Vibroplex) (photo: J Hall, G3KVA)

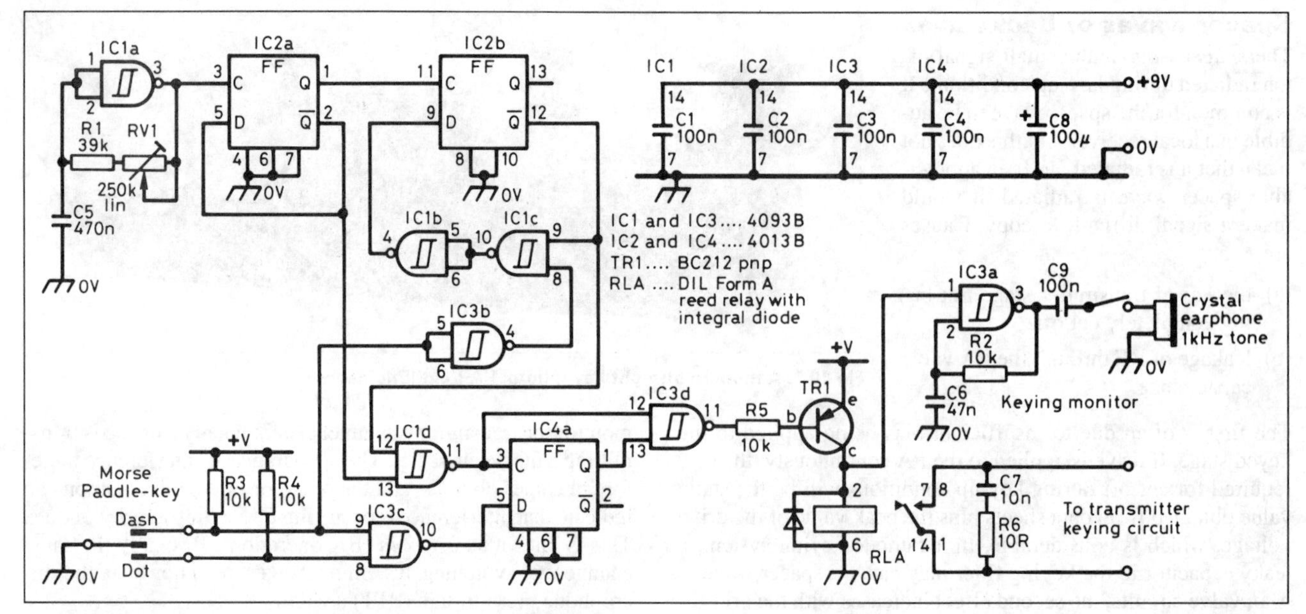

Fig 10.7. The G3BIK simple keyer uses four ICs for precision morse

el-bug in which movement of the paddle to the right produces a train of dots and to the left a train of dashs.

The *iambic keyer* is more complex electronically and requires a twin paddle key. There is a pair of contacts with each paddle, the fixed contacts being between the paddles. Thus in use the paddles are squeezed together (hence the other name *squeeze keyer*). According to which contact is closed, first a train of either 'dit-dah' or 'dah-dit' is produced.

These keys, particularly the iambic type, are complex and produce morse characters of the correct length and spacing. They contain a variable speed control and a speaker or similar for monitoring. Often the keyer contains the key, only the paddles protruding.

The weighting of a morse signal

The ratios of the lengths of the dot and dash and the spacing between them were established when the code was originated. It is now considered by some operators that in bad interference and at high speeds a morse signal is easier to read if the lengths of the dot and dash are increased slightly, ie if the *weighting* is increased. It is possible that these lengths are modified by the constants of the transmitter keying circuits, but this does not seem to have been a built-in adjustment.

The more complex electronic keyers are likely to include a weighting control.

A simple electronic keyer

This keyer [1], designed by E Chicken, G3BIK, is of basic design in that it uses only four integrated circuits and does not include the iambic facility. It has an built-in single paddle key which can be home-made from an open GPO type of relay.

According to which side a paddle is moved, a train of dots or dashs is produced. The speed range (adjustable) is approximately 5–35wpm. A small sounder is included as a side tone (keying monitor).

The circuit diagram, layout of the circuit board and details of the paddle key are given, together with a view of the complete unit (Figs 10.7–10.10). For full details, see reference [1].

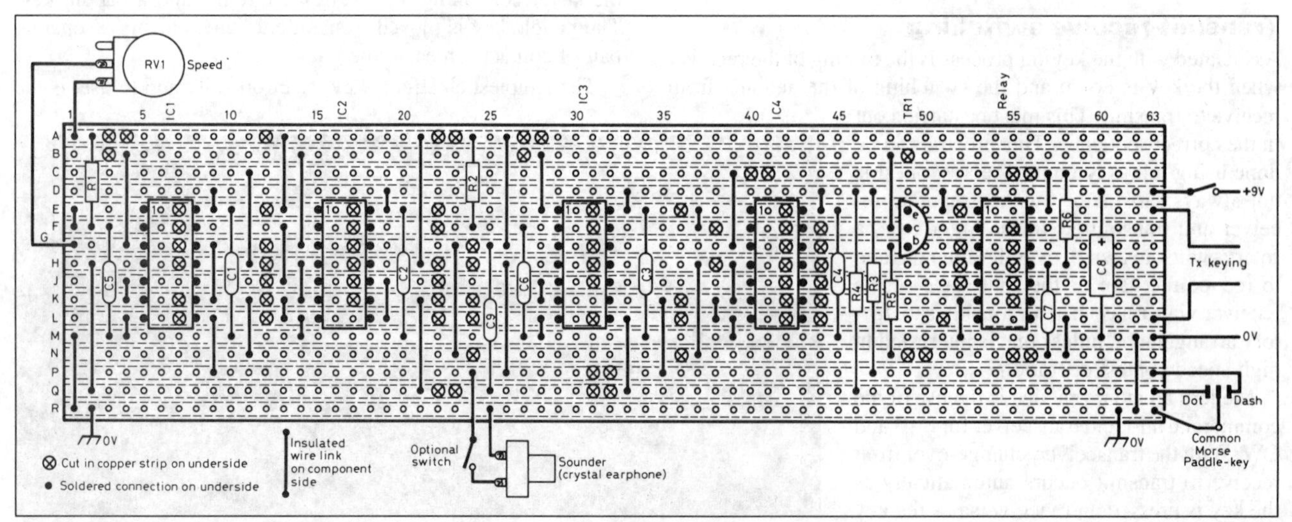

Fig 10.8. Construction is easy using strip board. The component side is shown

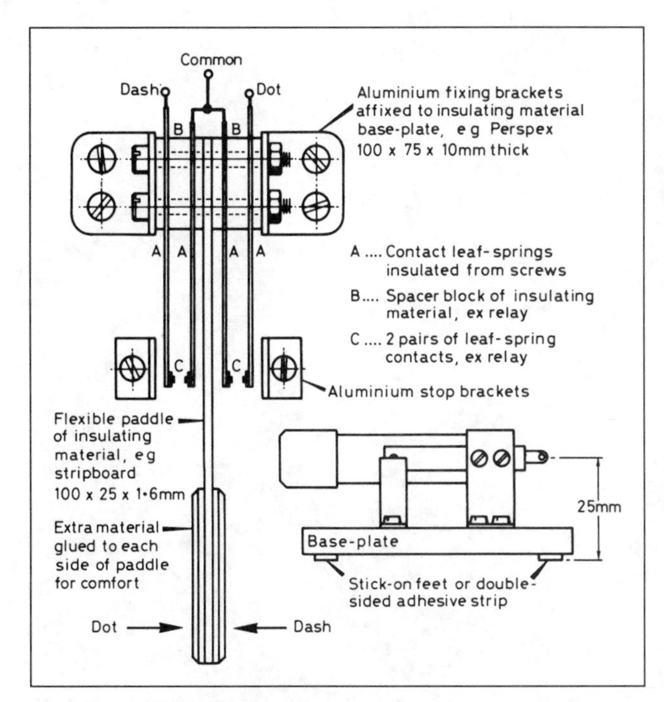

Fig 10.9. A simple paddle key can be made from an old relay

Fig 10.10. View of complete unit

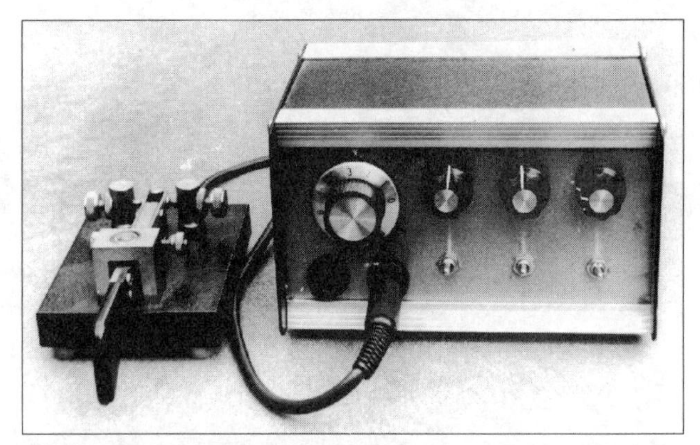

Fig 10.11. Front view of iambic keyer. Controls (left to right): speed, level, tone and volume

Fig 10.12. Internal layout of iambic keyer, showing the Curtis chip (centre)

A complex electronic keyer

This is a 'top-of-the-range' keyer which is based on a single CMOS integrated circuit developed specially for this application by Curtis Electro Devices Inc of California. The type number is 8044ABM.

The keyer is intended for home construction, but a reasonably high degree of experience is required. The facilities available are:

- Iambic modes (A and B) using a twin paddle key
- El-bug mode using single paddle key
- Speed range 8–50wpm
- Speed may be read on a low-range milliammeter
- Weight control (addition or subtraction)
- Internal single tone generator with miniature loudspeaker
- Dynamic dot and dash memories
- 'Tune' and straight key connections

Details of the 8044ABM (data sheet and application report) are available from:

Curtis Electro Devices Inc
4345 Pacific Street
Rocklin
CA 95677, USA
Tel (916) 632-0600

At the time of going to press (1994) the IC with the appropriate PCB was available from:

Mouser Electronics
2401 Highway 287 North
Mansfield
TX 76063, USA
Tel (817) 483-4422

together with full constructional details and the wiring necessary for the optional facilities which the constructor requires. Certain other components as indicated on the IC data sheet were also available.

Telephone orders with a credit card number are accepted by Mouser Electronics for immediate despatch by air mail.

The photographs and details of the keyer are by courtesy of T W Mansfield, G3ESH. This particular model was built into a Radio Spares box (Figs 10.11 and 10.12).

REFERENCE

[1] 'The BIK simple electronic key', E Chicken, G3BIK, *Radio Communication* August 1993.

11 Propagation

IT HAS been said that, without an ionosphere, wireless telegraphy might have remained an interesting but commercially unrewarding experiment in physics. Yet, would you believe, there *was* no ionosphere before 1932?

That has to be a trick question, of course. For the truth is that Balfour Stewart deduced from a study of geomagnetic storm data in 1878 (which was several years before Hertz begin his experiments with wireless transmission) that there had to be an electrically conducting layer high in the Earth's atmosphere.

In 1901, Marconi took a chance on it being so, and was rewarded for his faith by the achievement of the transatlantic 'first'. In the 'twenties, Appleton devised experiments to prove conclusively that such a layer existed; in fact, eventually he found more than one. And in 1932 Watson-Watt gave the region a general name – he called it the *ionosphere*.

Today, many people, including some who should know better, seem to think that we know all that we need to know about the ionosphere. But it preserves many a secret yet and there is still a place for careful experimentation by amateur operators.

This chapter can do no more than scratch the surface of a very wide-ranging topic. No attempt has been made in it to introduce divisions between HF, VHF, UHF etc because there are no such clear-cut divisions in the real world.

The story should really begin with a section about the Sun, because it is there that most of the direct influences on our signal paths have their origin. However, to appreciate that, we must first look at a few of the characteristics of our own atmosphere, and then run over some rather dull basic facts and figures which may nevertheless prove to be useful to you one day.

Don't be put off by the appearance of the mathematics. There is nothing very difficult here and a cheap scientific calculator can handle the tasks with ease. At a pinch, an even cheaper four-function calculator and a book of trig tables will see you through most of them.

As you read, compare the text with your own experiences on the bands. In all probability you know already much of the 'whats', 'wheres' and 'whens' of radio propagation. With any luck, this chapter will provide you with the associated 'whys' and 'hows'. Be warned, though. This subject is strongly addictive and there is room for more up there among the 'whos'.

THE EARTH'S ATMOSPHERE

There is a certain amount of confusion surrounding any attempt to identify various portions of the Earth's atmosphere, and this has come about as a result of there being no obvious natural boundaries as there are between land and sea. Workers in different disciplines have different ideas about which functions ought to be separated and it is particularly awkward for the radio engineer that he has to deal with the *troposphere* and *stratosphere*, which are terms from one set of divisions, and the *ionosphere*, which is from another.

Fig 11.1 shows the nomenclature favoured by meteorologists and now widely accepted among physicists generally. It is based on temperature variations, as might be expected from its origin. Note, incidentally, that the height scale of kilometres is a logarithmic one, and that the atmospheric pressure has been scaled in hectopascals (hPa), which are now replacing millibars in scientific work by international agreement. However, the two units are identical in magnitude; only the name has been changed, so the numbers remain the same. See *Weather* (Royal Meteorological Society) May 1986, p172, for an explanation.

In the *troposphere,* the part of the atmosphere nearest the ground, temperature tends to fall off with height. At the *tropopause,* around 10km (although all these heights vary from day to day and from place to place), it becomes fairly uniform at first and then begins to increase again in the region known as the *stratosphere.* This trend reverses at the *stratopause,* at an altitude of about 50km, to reach another minimum at the *mesopause*

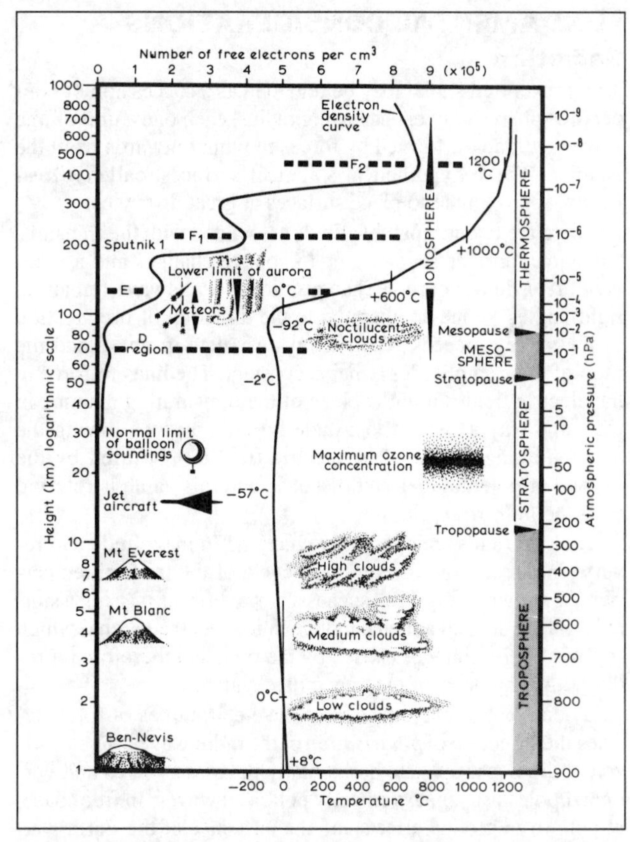

Fig 11.1. Some features of the Earth's atmosphere. The height scale is logarithmic, beginning at 1km above sea level. The equivalent pressure scale on the right is not regularly spaced because the relationship between pressure and height depends on temperature, which does not change uniformly with height

(ca 80km) after traversing the *mesosphere.* Above the mesopause temperatures begin to rise again in the *thermosphere,* soon surpassing anything encountered at lower altitudes, and levelling off at about +1200°C around 700km, where we must leave it in this survey.

The *ionosphere,* which has been defined as "the region above the Earth's surface in which ionisation takes place, with diurnal and annual variations which are regularly associated with ultraviolet radiation from the Sun, and sporadic variations arising from hydrogen bursts from sunspots" (*Chambers's Technical Dictionary*), overlaps the thermosphere, mesosphere and part of the stratosphere, but for practical purposes may be considered as lying between 60 and 700km. The name 'ionosphere' is perhaps misleading because it is the number of free electrons, rather than the ions they have left behind, which principally determines the electrical properties of the region. The electron density curve in the diagram shows a number of 'ledges', identified by the letters D, E, F1 and F2, which are the concentrations of free electrons described as 'layers' later on, when dealing with propagation in the ionosphere. They tend to act as mirrors to transmissions of certain wavelengths, while allowing others to pass through. The lowest ledge is generally referred to as the *D-region* rather than the *D-layer* because, as we shall see, its principal role is one of absorption rather than reflection, and its presence is usually easier to infer than it is to observe.

It may be found helpful to refer to this diagram again when features of the ionosphere and troposphere are dealt with in later sections of this chapter.

FUNDAMENTAL CONSIDERATIONS

Radiation

The transmitted signal may be regarded as a succession of concentric spheres of ever-increasing radius, each one a unit of one wavelength apart, formed by forces moving outwards from the antenna. These hypothetical spherical surfaces, called *wavefronts,* approximate to plane surfaces at great distances.

There are two inseparable fields associated with the transmitted signal, an *electric field* due to voltage changes and a *magnetic field* due to current changes, and these always remain at right-angles to one another and to the direction of propagation as the wave proceeds. They always oscillate in phase and the ratio of their amplitudes remains constant. The lines of force in the electric field run in the plane of the transmitting antenna in the same way as would longitude lines on a globe having the antenna along its axis. The electric field is measured by the change of potential per unit distance, and this value is referred to as the *field strength.*

The two fields are constantly changing in magnitude and reverse in direction with every half-cycle of the transmitted carrier. As shown in Fig 11.2, successive wavefronts passing a suitably placed second antenna induce in it a received signal which follows all the changes carried by the field and therefore reproduces the character of the transmitted signal.

By convention the direction of the electric lines of force defines the direction of *polarisation* of the radio waves. Thus horizontal dipoles propagate horizontally polarised waves and vertical dipoles propagate vertically polarised waves. In free space, remote from ground effects and the influence of the Earth's atmosphere, these senses remain constant and a suitably aligned receiving antenna would respond to the whole of the incident field.

When the advancing wavefront encounters the surface of the

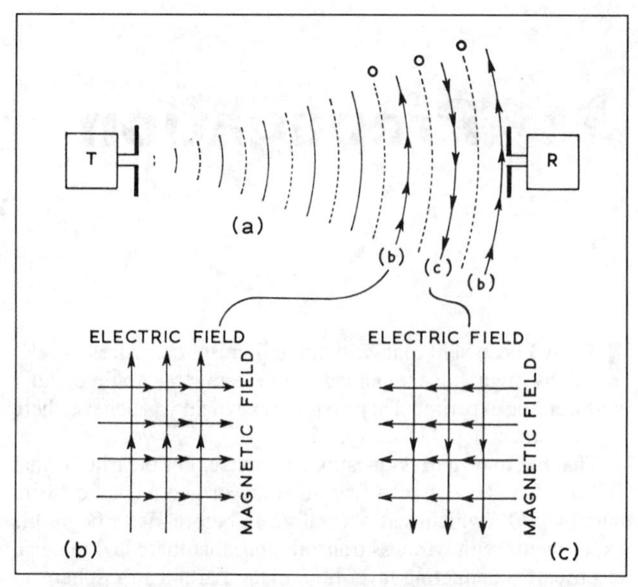

Fig 11.2. (a) A vertical section through the fields radiated from a vertically polarised transmitting antenna. The expanding spherical wavefront consists of alternate reversals of electric field, shown by the arrowed arcs; nulls are shown dotted. At rightangles to the plane of the paper, but not seen here, would be simultaneous alternate reversals of magnetic field (b) and (c). These two 'snapshots' must be rotated through 90° in imagination, so that the magnetic field lines run into and out of the plane of the paper. The view is from R towards T

Earth or becomes deflected by certain layers in the atmosphere, a degree of cross-polarisation may be introduced which results in signals arriving at the receiving antenna with both horizontal and vertical components present.

Circularly polarised signals, which contain equal components of both horizontal and vertical polarisation, are receivable on dipoles having any alignment in the plane of the wavefront, but the magnitude of the received signal will be only half that which would result from the use of an antenna correctly designed for such a form of polarisation (it must not be overlooked that there are two forms, differing only in their direction of rotation). Matters such as these are dealt with in detail in Chapter 13 – 'VHF/UHF antennas'.

Field strength

As the energy in the expanding wavefront has to cover an ever-increasing area the further it travels, the amplitude of the signal induced in the receiving antenna diminishes as a function of distance. Under *free-space* conditions an inverse square-law relationship applies, but in most cases the nature of the intervening medium has a profound, and often very variable, effect on the magnitude of the received signal.

The intensity of a radio wave at any point in space may be expressed in terms of the strength of its electric field at right-angles to the line of the transmission path. The units used indicate the difference of electric force between two points one metre apart, and Fig 11.3 (which should not be used for general calculations as it relates only to free-space conditions) has been included here to give an indication of the magnitudes of fields likely to be met with in the amateur service in cases where a more-or-less direct path exists between transmitting and receiving antennas and all other considerations may be ignored. In practice signal levels very much less than the free-space values may be expected because of various losses en route, so that field strengths down to about 1μV/m need to be considered.

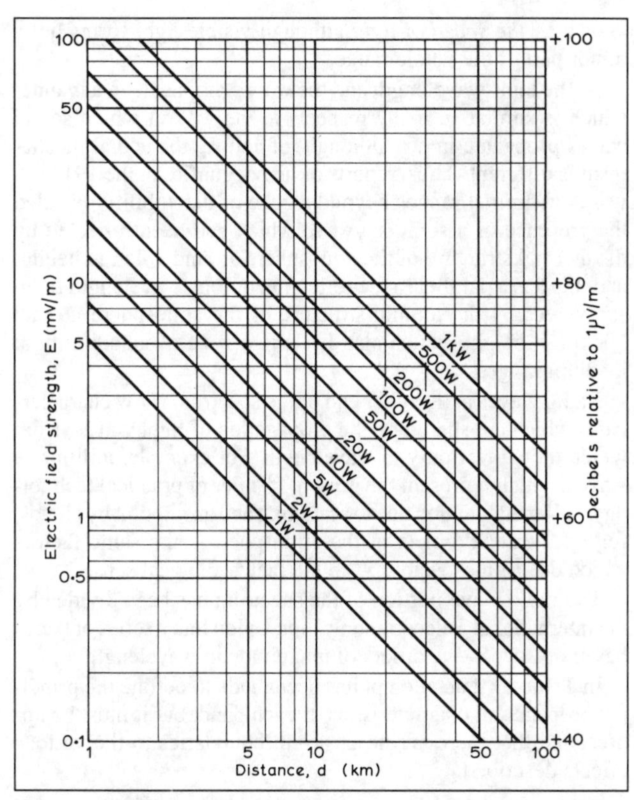

Fig 11.3. Field strength from an omnidirectional antenna radiating from 1kW to 1W into free space. This is an application of the *inverse distance law*

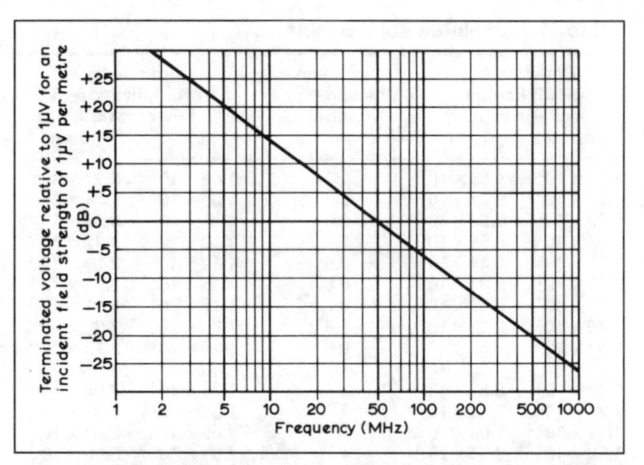

Fig 11.4. The relationship between the incident field strength on a half-wave dipole and the voltage at the receiver end of a correctly matched perfect feeder connected to it

Signal input

In many calculations dealing with propagation, the parameter of interest is not the field strength at a particular place but the voltage which is induced by it across the input of a receiver. If a half-wave dipole is introduced into a field and aligned for maximum signal pick-up, the open-circuit EMF induced at its centre is given by the expression:

$$e = \frac{E\lambda}{\pi} \qquad (1)$$

where e = EMF at the centre of the dipole, in volts
E = the incident field strength in volts/metre
λ = the wavelength of the transmitted signal in metres.

When connected to a matched feeder correctly terminated at the receiver the input voltage available will be half this, or $e/2$. By substituting frequency for wavelength the equation may be reduced to the more practical form:

$$V_r = \frac{47.8}{f} E \qquad (2)$$

where V_r = microvolts of signal across the receiver input
f = frequency of the transmitted signal in megahertz
E = the incident field strength in microvolts/metre.

It should be noted that for a given frequency the first term becomes a constant factor. By a further rearrangement:

$$E = \frac{f}{47.8} V_r \qquad (3)$$

which enables the field to be estimated from the magnitude of the received signal in cases where the receiver is fed from a perfectly matched half-wave dipole.

It is an advantage to work in decibels in calculations of this nature. If a standard level of 1µV/m is adopted for field strength and of 1µV for signal level it is a simple matter to take into account the various gains and losses in a practical receiving system. Fig 11.4 shows the terminated voltage at the receiver (V_0) in terms of decibels relative to 1µV for a normalised incident field strength of 1µV/m at the transmitter frequency, derived from expression (2). Then:

$$V_r(\text{dB}) = V_0(\text{dB}) + V_i(\text{dB}) + G_r(\text{dB}) - L_{fr}(\text{dB}) \qquad (4)$$

where V_r(dB) = input to receiver in decibels relative to 1µV
V_0(dB) = input in decibels relative to 1µV, for 1µV/m incident field strength (from graph)
V_i(dB) = the actual incident field strength in decibels relative to 1µV/m
G_r(dB) = the gain in decibels of the receiving antenna relative to a half-wave dipole
L_{fr}(dB) = loss in the feeder line between antenna and receiver

Example. The incident field strength of a 70MHz transmission is 100µV/m (or 40dB above 1µV/m). A three-element Yagi having a gain of 5dB over a half-wave dipole is connected to the receiver through 100ft of coaxial cable which introduces a loss of 2dB. From this, $V_0 = -3.5$, $V_i = +40$, $G_r = +5$ and $L_{fr} = +2$ (the fact that L_{fr} is a loss is allowed for in Eqn (4)), and $V_r = -3.5 + 40 + 5 - 2$dB = 39.5dB relative to 1µV, or a receiver input voltage of 94µV.

Table 11.1 may be found useful in converting the rather unfamiliar looking values of decibels met with in radio propagation work into their corresponding voltage or power ratios using only a cheap four-function calculator.

MODES OF PROPAGATION

Introduction

There are four principal modes by which radio waves are propagated. They are:

(a) *Free-space waves*, which are unaffected by any consideration other than distance;
(b) *Ionospheric waves*, which are influenced by the action of free electrons in the upper levels of the Earth's atmosphere;
(c) *Tropospheric waves*, which are subject to deflection in the lower levels by variations in the refractive index structure of the air through which they pass; and

Table 11.1. Skeleton decibel table

Combine by multiplication Voltage ratios up	down	Combine by addition (dB)	Combine by multiplication Power ratios up	down
1.01×10^0	9.89×10^{-1}	0.1	1.02×10^0	9.77×10^{-1}
1.02×10^0	9.77×10^{-1}	0.2	1.05×10^0	9.55×10^{-1}
1.03×10^0	9.66×10^{-1}	0.3	1.07×10^0	9.33×10^{-1}
1.05×10^0	9.55×10^{-1}	0.4	1.10×10^0	9.12×10^{-1}
1.06×10^0	9.44×10^{-1}	0.5	1.12×10^0	8.91×10^{-1}
1.07×10^0	9.33×10^{-1}	0.6	1.15×10^0	8.71×10^{-1}
1.08×10^0	9.23×10^{-1}	0.7	1.17×10^0	8.51×10^{-1}
1.10×10^0	9.12×10^{-1}	0.8	1.20×10^0	8.32×10^{-1}
1.11×10^0	9.02×10^{-1}	0.9	1.23×10^0	8.13×10^{-1}
1.12×10^0	8.91×10^{-1}	1	1.26×10^0	7.94×10^{-1}
1.26×10^0	7.94×10^{-1}	2	1.58×10^0	6.31×10^{-1}
1.41×10^0	7.08×10^{-1}	3	2.00×10^0	5.01×10^{-1}
1.58×10^0	6.31×10^{-1}	4	2.51×10^0	3.98×10^{-1}
1.78×10^0	5.62×10^{-1}	5	3.16×10^0	3.16×10^{-1}
2.00×10^0	5.01×10^{-1}	6	3.98×10^0	2.51×10^{-1}
2.24×10^0	4.47×10^{-1}	7	5.01×10^0	1.99×10^{-1}
2.51×10^0	3.98×10^{-1}	8	6.31×10^0	1.59×10^{-1}
2.82×10^0	3.55×10^{-1}	9	7.94×10^0	1.26×10^{-1}
3.16×10^0	3.16×10^{-1}	10	1.00×10^1	1.00×10^{-1}
1.00×10^1	1.00×10^{-1}	20	1.00×10^2	1.00×10^{-2}
3.16×10^1	3.16×10^{-2}	30	1.00×10^3	1.00×10^{-3}
1.00×10^2	1.00×10^{-2}	40	1.00×10^4	1.00×10^{-4}
3.16×10^2	3.16×10^{-3}	50	1.00×10^5	1.00×10^{-5}
1.00×10^3	1.00×10^{-3}	60	1.00×10^6	1.00×10^{-6}
3.16×10^3	3.16×10^{-4}	70	1.00×10^7	1.00×10^{-7}
1.00×10^4	1.00×10^{-4}	80	1.00×10^8	1.00×10^{-8}
3.16×10^4	3.16×10^{-5}	90	1.00×10^9	1.00×10^{-9}
1.00×10^5	1.00×10^{-5}	100	1.00×10^{10}	1.00×10^{-10}
1.00×10^{10}	1.00×10^{-10}	200	1.00×10^{20}	1.00×10^{-20}

Example: 39.5dB above 1μV (voltage ratio).
39.5dB = 30 + 9 + 0.5dB
Combining equivalents by multiplication
= $(3.16 \times 10^1) \times (2.82 \times 10^0) \times (1.06 \times 10^0)$ = 94 times 1μV, or 94μV.

(d) *Ground waves* which are modified by the nature of the terrain over which they travel.

Free-space waves propagate from point-to-point by the most direct path. Waves in the other three categories are influenced by factors which make them tend to overcome the curvature of the Earth, either by reflection as with ionospheric waves, refraction as with tropospheric waves, or diffraction at the surface of the Earth itself as with ground waves.

Wavelength is the chief consideration which determines the mode of propagation of Earth-based transmissions.

The spectrum of electromagnetic waves

The position of man-made radio waves in the electromagnetic wave spectrum is shown in Fig 11.5, where they can be seen to occupy an appreciable portion of a family of naturally-occurring radiations, all of which are characterised by inseparable oscillations of electric and magnetic fields and travel with the same velocity in free space. This velocity, 2.99790×10^8m/s (generally taken as 3×10^8m/s in calculations), is popularly known as *the speed of light* although visible light forms but a minor part of the whole range.

At the long-wavelength end the waves propagate in a manner which is similar in many respects to the way in which sound waves propagate in air, although, of course, the actual mechanism is different. Thus, reports of heavy gunfire in the 1914–18 war at abnormal ranges beyond a zone of inaudibility revealed the presence of a sonic skywave which had been reflected by the thermal structure of the atmosphere around 30km in height, and this has a parallel in the reflection of long wavelength radio sky-waves by the atomic structure of the atmosphere around 100km in height, which also leads to a zone of inaudibility at medium ranges.

Radio waves at the other end of the spectrum show characteristics which are shared by the propagation of light waves, from which they differ only in wavelength. For example, millimetre waves, which represent the present frontier of practical technology, suffer attenuation due to scattering and absorption by clouds, fog and water droplets in the atmosphere – the same factors which determine 'visibility' in the meteorological sense.

The radio wave portion of the spectrum has been divided by the International Telecommunication Union into a series of bands based on successive orders of magnitude in wavelength.

In Table 11.2 an attempt has been made to outline the principal propagation characteristics of each band, but it must be appreciated that there are no clear-cut boundaries to the various effects described.

Wave propagation in free space

The concept of *free-space* propagation, of a transmitter radiating without restraint into an infinite empty surrounding space, has been introduced briefly in the section on field strength where it was used to illustrate, in a general way, the relationship between the strength of the field due to a transmitter and the distance over which the waves have travelled.

It is only recently, with the advent of the Space Age, that we have acquired a practical opportunity to operate long-distance circuits under true free-space conditions as, for example, between spacecraft and orbiting satellites, and it is only recently that radio amateurs have had direct access to paths of that nature. Earth-moon-Earth contacts are becoming increasingly popular, however, and reception of satellite signals commonplace, and for these the free-space calculations apply with only relatively minor adjustments because such a large part of the transmission paths involved lies beyond the reach of terrestrial influences.

In many cases, and especially where wavelengths of less than about 10m are concerned, the free-space calculations are even applied to paths which are subject to relatively unpredictable perturbations in order to estimate a convenient (and often unobtainable) ideal – a bogey for the path – against which the other losses may be compared.

The basic transmission loss in free space is given by the expressions:

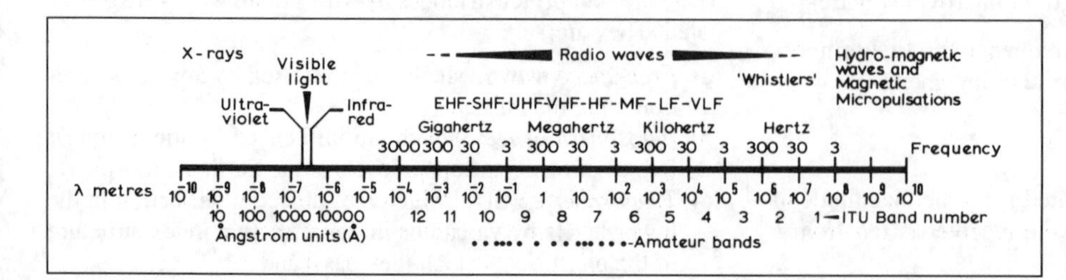

Fig 11.5. The spectrum of electromagnetic waves. This diagram shows on a logarithmic scale the relationship between X-rays, 'visible' and 'invisible' light, heat (infrared), radio waves and the very slow waves associated with geomagnetic pulsations, all of them similar in basic character

Table 11.2. A survey of the radio-frequency spectrum

ITU Band No	Metric name of band and limits by wavelength	Alternative name of band and limits by frequency	UK amateur bands by frequency (and usual description based on wavelength)	Principal propagation modes	Principal limitations
4	Myriametric 100,000–10,000m	Very low frequency (VLF) 3–30kHz	—	Extensive surface wave Ground to ionosphere space acts as a waveguide	Very high power and very large antennas required. Few channels.
5	Kilometric 10,000–1000m	Low frequency (LF) 30–300kHz	—	Surface wave and reflections from lower ionosphere	High power and large antennas required. Limited number of channels available. Subject to fading where surface wave and sky wave mix
6	Hectometric 1000–100m	Medium frequency (MF) 300–3000kHz	1810–2000kHz (160m band) (also known as *top band*)	Surface wave only during daylight. At night reflection from decaying E-layer	Strong D-region absorption during day. Long ranges possible at night but signals subject to fading and considerable co-channel interference
7	Decametric 100m–10m	High frequency (HF) 3–30MHz	3.50–3.80MHz (80m) 7.00–7.10MHz (40m) 10.10–10.15MHz (30m) 14.00–14.35MHz (20m) 18.068–18.168MHz (17m) 21.00–21.45MHz (14m) 24.89–24.99MHz (12m) 28.00–29.70MHz (10m)	Short-distance working via E-layer. Nearly all long-distance working via F2-layer	Daytime attenuation by D-region, E and F1-layer absorption. Signal strength subject to diurnal, seasonal, solar-cycle and irregular changes
8	Metric 10m–1m	Very high frequency (VHF) 30–300MHz	50.00–52.00MHz (6m) 70.00–70.50MHz (4m) 144.00–146.00MHz (2m)	F2 occasionally at LF end of band around sunspot maximum. Irregularly by sporadic-E and auroral-E. Otherwise maximum range determined by temperature and humidity structure of lower troposphere	Ranges generally only just beyond the horizon but enhancements due to anomalous propagation can exceed 2000km
9	Decimetric 1m–10cm	Ultra high frequency (UHF) 300–3000MHz	430–440MHz (70cm) 1240–1325MHz (23cm) 2310–2450MHz (13cm)	Line-of-sight modified by tropospheric effects	Atmospheric absorption effects noticeable at top of band
10	Centimetric 10cm–1cm	Super high frequency (SHF) 3–30GHz	3.400–3.475GHz (9cm) 5.650–5.850GHz (6cm) 10.00–10.50GHz (3cm) 24.00–24.25GHz (12mm)	Line-of-sight	Attenuation due to oxygen, water vapour and precipitation becomes increasingly important
11	Millimetric 1cm–1mm	Extra high frequency (EHF) 30–300GHz	47.00–47.20GHz (6mm) 75.50–76.00GHz (4mm) 142.00–144.00GHz (2mm) 248.00–250.00GHz (1.2mm)	Line-of-sight	Atmospheric propagation losses create pass and stop bands. Background noise sets a threshold
12	Decimillimetric (sub-millimetric)	— 300–3000GHz	—	Line-of-sight	Present limit of technology

$$L_{bf} = 32.45 + 20 \log f(\text{MHz}) + 20 \log r(\text{km}) \quad (5)$$

$$L_{br} = 36.6 + 20 \log f(\text{MHz}) + 20 \log r(\text{miles}) \quad (6)$$

where r is the straight line distance involved.

If transmitter and receiver levels are expressed in either dBW (relative to 1W) or dBm (relative to 1mW) – it does not matter which providing that the same units are used at both ends of the path – with other relevant parameters similarly given in terms of decibels, it is a relatively simple matter to determine the received power at any distance by adding to the transmitter level all the appropriate gains and subtracting all the losses. Thus:

$$P_r(\text{dBW}) = P_t(\text{dBW}) + G_t(\text{dB}) - L_{ft}(\text{dB}) - L_{bf}(\text{dB}) + G_r(\text{dB}) - L_{fr}(\text{dB}) \quad (7)$$

$$P_r(\text{dBm}) = P_t(\text{dBm}) + G_t(\text{dB}) - L_{ft}(\text{dB}) - L_{bf}(\text{dB}) + G_r(\text{dB}) - L_{fr}(\text{dB}) \quad (8)$$

where P_r = Received power level (dBm or dBW)
P_t = Transmitted power level (dBm or dBW)
G_t = Gain of the transmitting antenna in the direction of the path, relative to an isotropic radiator
L_{ft} = Transmitting feeder loss
L_{bf} = Free-space transmission loss
G_r = Gain of the receiving antenna in the direction of the path, relative to an isotropic radiator
L_{fr} = Receiving feeder loss

The free-space transmission loss may be estimated approximately from Fig 11.6 which perhaps conveys a better idea of its relationship to frequency (or wavelength) and distance than the nomogram generally provided. Unless a large number of calculations have to be made, it is no great hardship to use the formula for individual cases should greater accuracy be desired. It should be noted that antenna gains quoted with respect to a half-wave dipole need to be increased by 2dB to express them relative to an isotropic radiator.

Example. A 70MHz transmitter radiates 100W ERP in the direction of a receiver 550km away. The receiving antenna has a gain of 5dB over a half-wave dipole (2dB more over an isotropic radiator), and there is a 2dB loss in the feeder. In this case the effective radiated power is known, which takes the place of the terms P_t, G_t and L_{ft}.

P_{erp} = 100W = 10^5mW = 50dBm
L_{bf} = 124 from Fig 11.6 or by use of the expressions (5) or (6)
G_r = 7dB over an isotropic radiator
L_{fr} = 2dB

Then $P_t = P_{erp} - L_{bf} + G_r - L_{fr}$
= 50 – 124 + 7 – 2
= –69dBm, or 69dB below 1mW
= 12.6×10^{-6}mW

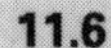

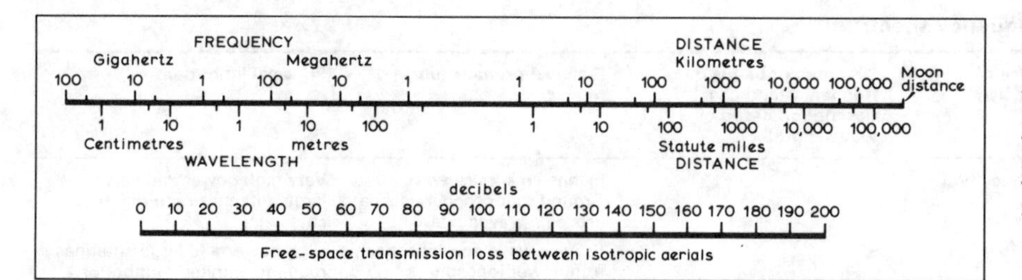

Fig 11.6. The effects of frequency (or wavelength) and distance on the free-space transmission loss between isotropic antennas. The length on the diagram between points corresponding to frequency and distance on the upper scale corresponds to the free-space transmission loss measured from the zero decibel mark on the lower scale

$= 12.6 \times 10^{-9}$W

If this power is dissipated in an input impedance of 70Ω the voltage appearing across the receiver V_r is $\sqrt{(P_r Z_{in})}$, in this case $\sqrt{(12.6 \times 10^{-9} \times 70)}$ which is 94μV. This example deals with the same situation which was considered earlier in connection with field strength and offers an alternative method of calculating signal input.

Wave propagation in the ionosphere

It has been shown that a transmitted signal may be considered as consisting of a succession of spherical wavefronts, each one a wavelength apart, and they approximate to plane surfaces at great distances. At certain heights in the upper atmosphere concentrations of negatively charged free electrons occur, and these are set into oscillatory motion by the oncoming waves, which causes them to emit secondary wavelets having a phase which is 90° in advance of the main wave. It is only in the forward direction that the original waves and their dependent wavelets combine coherently and their resultant consists of a wave in which the maxima and minima occur earlier than in the projection of the originating wave – to all intents and purposes the equivalent of the wave having travelled faster in order to arrive earlier. The amount of phase advancement is a function of the concentration of electrons and the change of speed is greatest at long wavelengths, decreasing therefore as the signal frequency increases.

The advancing wave-front, travelling, let us say, obliquely upwards from the ground, meets the layer containing the accumulation of free electrons in such a way that its upper portion passes through a greater concentration of charge than does a portion lower down. The top of the wavefront is therefore accelerated to a greater extent by the process just outlined than are the parts immediately below, which results in a gradual swing-round until the wavefront is being returned towards the ground as though it had experienced a reflection.

The nearer the wavefront is to being vertically above the transmitter the more quickly must the top accelerate relative to the bottom, and the more concentrated must be the charge of electrons. It may be that the density of electrons is sufficiently high to turn even wavefronts propagating vertically upwards (a condition known as *vertical incidence*), although it must be appreciated that deeper penetration into the layer will occur as the propagation angle becomes steeper.

Consider the circumstances outlined in Fig 11.7, where T indicates the site of a transmitting station and R1, R2 and R3 three receiving sites. For a given electron concentration there is a *critical frequency* f_0 which is the highest to return from radiation directly vertically upward. Frequencies higher than this will penetrate the layer completely and be lost in space, but their reflection may still be possible at *oblique incidence* where waves have to travel a greater distance within the electron concentration. This is not the case at point A in the diagram, so that reception by sky-wave is impossible at R1 under these circumstances, but at a certain angle of incidence to the layer (as at point B) the ray bending becomes just sufficient to return signals to the ground, making R2 the nearest location relative to the transmitter at which the sky-wave could be received. The range over which no signals are possible via the ionosphere is known as the *skip distance,* and the roughly circular area described by it is called the *skip zone*. Lower-angle radiation results in longer ranges, for example to point R3 from a reflection at C, and a second 'hop' may result from a further reflection from the ground. The longest ranges at HF are achieved this way – and it is possible for an HF signal to travel right round the world using a succession of hops.

If the angle φ is very large and conditions are favourable, the transmission path may lead from one point on the ionosphere to another without intermediate ground reflection. This is known as *chordal hop* propagation; signal strength is usually higher than normal because there is no ground reflection point to introduce losses.

From the point of view of the operator, it is θ, the angle of signal take-off relative to the horizon (Fig 11.8, inset diagram), that is mainly of interest in this application, not the angle of incidence at the ionosphere. The two curves in the main diagram show the distances covered for various values of θ at representative heights of 120km (E-layer) and 400km (F2-layer).

For oblique incidence on a particular path (eg the ray from T to R3 via point C in the ionosphere) there is a *maximum usable frequency* (MUF), generally much higher than the critical frequency was at vertical incidence, and this is approximately equal to $f_0/\cos\varphi$, where φ is the angle of incidence of the ray to the point of reflection, as is shown in the figure. The limiting angle which defines the point at which reflections first become possible is called the *critical wave angle,* and it is this function which determines the extent of the skip zone.

This mechanism is effective for signals in excess of about 100kHz, for which the concentration of electrons appears as a succession of layers of increasing electron density having the effect of progressively bending the rays as the region is penetrated.

Below 100kHz the change in concentration occurs within a distance which is small compared to the wavelength and which therefore appears as an almost perfect reflector. Waves are

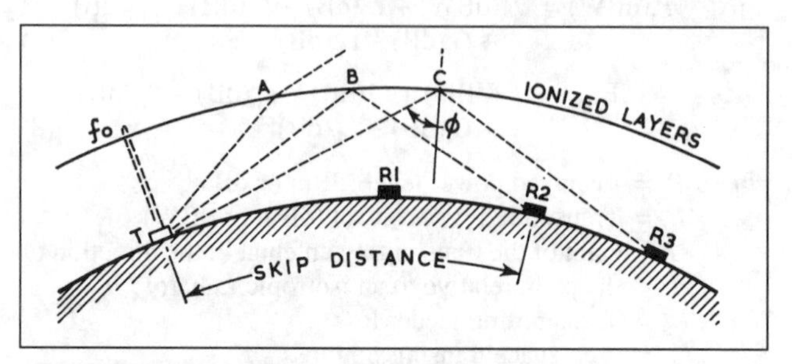

Fig 11.7. Wave propagation via the ionosphere. T is the site of a transmitter, R1, R2 and R3 are the sites of three receivers. The significance of the ray at vertical incidence (dotted) and the three oblique rays is explained in the text

propagated in that way over great distances by virtue of being confined between two concentric spheres, one being the lower edge of the layer and the other the surface of the Earth.

During the hours of daylight the quantity of free electrons in the lower ionosphere becomes so great that the oscillations set up by incident waves are heavily damped on account of energy lost by frequent collisions with the surrounding neutral air particles. Medium-wave broadcast band signals are so much affected by this as to have their sky-waves completely absorbed during the day, leading to the familiar rapid weakening of distant stations around dawn and their subsequent disappearance at the very time when the reflecting layers might otherwise be expected to be most effective. The shorter wavelengths are less severely affected, but suffer attenuation nevertheless.

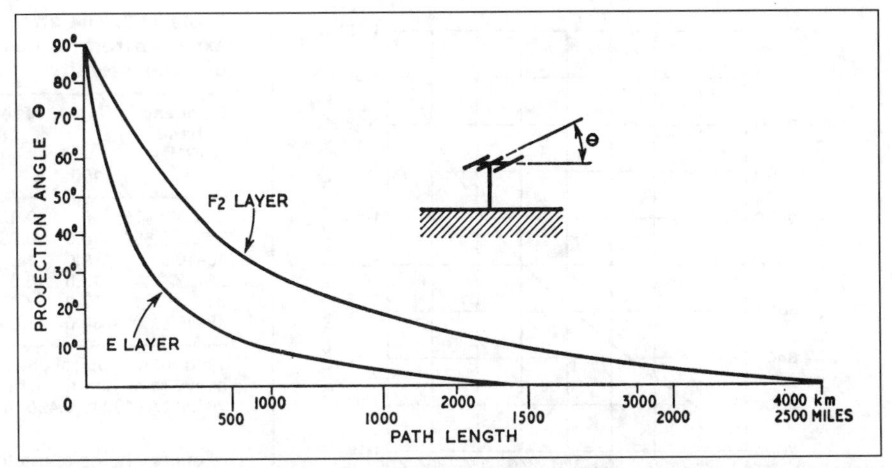

Fig 11.8. Relationship between angles of take-off and resulting path length. The two ionospheric layers are presumed to be at heights of 120km and 400km respectively

A certain amount of cross-polarisation occurs when ionospheric reflections take place so that the received signals generally contain a mixture of both horizontal and vertical components, irrespective of which predominated at the transmitting antenna.

Ionospheric scatter propagation does not make use of the regular layers of increased electron density. Instead forward scattering takes place from small irregularities in the ionosphere comparable in size with the wavelength in use (generally around 8m or about 35MHz). With high powers and very low angles of radiation, paths of some 2000km are possible and this mode has the advantage of being workable in auroral regions where conventional HF methods are often unreliable, but only a very small proportion of the transmitted power is able to find its way in the desired direction.

Wave propagation in the troposphere

The *troposphere* is that lower portion of the atmosphere in which the general tendency is for air temperature to decrease with height. It is separated from the *stratosphere,* the region immediately above, where the air temperature tends to remain invariant with height, by a boundary called the *tropopause* at around 10km. The troposphere contains all the well-known cloud forms and is responsible for nearly everything loosely grouped under the general heading of 'weather'.

Its effect upon radio waves is to bend them, generally in the same direction as that taken by the Earth's curvature, not by encounters with free electrons or layers of ionisation, but as a result of successive changes in the refractive index of the air through which the waves pass. In optical terms this is the mechanism responsible for the appearance of mirages, where objects beyond the horizon are brought into view by raybending, in that case resulting from temperature changes along the line-of-sight path. In the case of radio signals the distribution of water vapour also plays a part, often a major one where anomalous propagation events are concerned.

The refractive index, n, of a sample of air can be found from the expression $n = 1 + 10^{-6}N$, where N is the refractivity expressed by:

$$N = \frac{77.6P}{T} + \frac{4810e}{T^2}$$

where P = the atmospheric pressure in millibars or hectopascals
e = the water-vapour pressure, also in millibars or hectopascals
T = the temperature in kelvin or degrees absolute

Substituting successive values of P, e and T from a standard atmosphere table shows that there is a tendency for refractive index to decrease with height, from a value just above unity at the Earth's surface to unity itself in free space. This gradient is sufficient to create a condition in which rays are normally bent down towards the Earth, leading to a similar state of affairs as that which would result from the radio horizon being extended to beyond the optical line-of-sight limit by an average of about 15%.

The distance d to the horizon from an antenna of height h is approximately $\sqrt{(2ah)}$, where a is the radius of the Earth and h is small compared with it. The effect of refraction can be allowed for by increasing the true value of the Earth's radius until the ray paths, curved by the refractive index gradient, become straight again. This modified radius a' can be found from the expression:

$$\frac{1}{a'} = \frac{1}{a} + \frac{dn}{dh}$$

where dn/dh is the rate of decrease of refractive index with height. The ratio a'/a is known as the *effective Earth radius factor k*, so that the distance to the radio horizon becomes $d = \sqrt{(2kah)}$.

An average value for k, based on a standard atmosphere, is 1.33 or 4/3 (whence a common description of this convention, *the four-thirds Earth*). Notes relating to the construction of path profiles using this convention will be found in a later section of this chapter. When the four-thirds Earth concept is known to be inappropriate an estimate of a suitable value for k can be obtained from the surface refractivity N_s (obtained from Fig 11.9), using for N the value obtained from ground-level readings of pressure, temperature and vapour pressure but this method may be misleading if marked anomalies are present in the vertical refractive index structure.

Waves of widely separated length are liable to be disturbed by the troposphere in some way or other, but it is generally only those shorter than about 10m (over 30MHz) which need be considered. There are two reasons for this; one is that usually ionospheric effects are so pronounced at the longer wavelengths that attention is diverted from the comparatively minor enhancements due to the refractive index structure of the troposphere, and the other that anomalies, when they occur, are often of insufficient depth to accommodate waves as long as, or longer than, 10m.

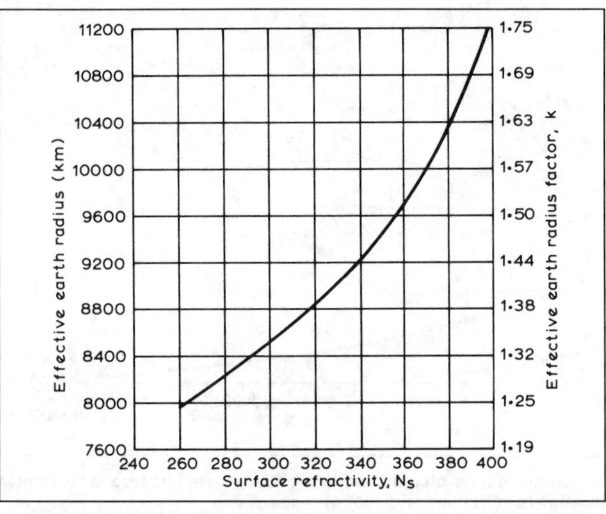

Fig 11.9. Effective Earth radius corresponding to various values of surface refractivity. A curved ray-path drawn over an Earth section between terminals may be rendered straight by exaggerating the Earth's curvature. An average value often used is equal to 1.33 times the actual radius; this is sometimes referred to as *the four-thirds Earth approximation.* (Based on *CCIR Report 244.*)

For example, it might be that the decrease of refractive index with height becomes so sharp in the lower 100m of the troposphere, that waves are trapped in an *atmospheric duct,* within which they remain confined for abnormally long distances. The maximum wavelength which can be trapped completely in a duct of 100m thickness is about 1m (corresponding to a frequency of 300MHz), for example, so that the most favourable conditions are generally found in the VHF and UHF bands or above. The relationship between maximum wavelength λ and duct thickness *t* is shown in the expression:

$$t = 500\ \lambda^{2/3}$$

where both *t* and λ are expressed in centimetres.

At centimetre wavelengths signals propagated through the troposphere suffer rapid fluctuations in amplitude and phase due to irregular small-scale variations in refractive index which give rise to continuous changes known as *scintillations,* and they are also attenuated by water in the form of precipitation (rain, snow, hail, etc) or as fog or cloud. As Table 11.3 shows, this effect increases both with radio frequency and with either the rate of rainfall or the concentration of water droplets. Precipitation causes losses by absorption and by random scattering from the liquid (or solid in the case of ice) surfaces and this scattering becomes so pronounced as to act as a 'target' for weather radars which use these *precipitation echoes* to detect rain areas.

At even shorter wavelengths resonances occur within the molecules of some of the gases which make up the atmosphere. Only one of these approximates to an amateur band, the attenuation at 22.23GHz due to water vapour. The principal oxygen resonances at around 60 and 120GHz are well beyond the reach of amateur activities, but are likely to prove to be limiting factors where professional work at millimetre wavelengths involves paths passing through the atmosphere, as opposed to outer space working.

Wave propagation near the ground

Diffraction is an alteration in direction of the propagation of a wave due to change in velocity over its wavefront. Radio waves meeting an obstacle tend to diffract around it, and the surface of the Earth is no exception to this. Bending comes about as a result of energy being extracted due to currents induced in the ground. These constitute an attenuation by absorption, having the effect of slowing down the lower parts of the wavefront, causing it to tilt forward in a way which follows the Earth's curvature. The amount of diffraction is dependent on the ratio of the wavelength to the radius of the Earth and so is greatest when the waves are longest. It also depends on the electrical characteristics of the surface, namely its relative permeability (generally regarded as unity for this purpose), its dielectric constant, ε (epsilon), and conductivity, σ (sigma). This diffracted wave is known as the *surface wave.*

Table 11.3. The attenuation in decibels per kilometre to be expected from various rates of rainfall and for various degrees of cloud intensity

Frequency band (GHz)	Precipitation (mm/h)					Fog or cloud water content (g/m³) (at 0°C)		
	100	50	25	10	1	2.35	0.42	0.043
3.400–3.475	0.1	0.02	0.01	—	—	—	—	—
5.650–5.850	0.6	0.25	0.1	0.02	—	0.09	—	—
10–10.5	3.0	1.5	0.6	0.2	0.01	0.23	0.04	—
21–22	13.0	6.0	2.5	1.0	0.1	0.94	0.17	0.02
24	17.0	8.0	3.8	1.5	0.1	1.41	0.25	0.03
48–49	30.0	17.0	9.0	4.0	0.6	4.70	0.84	0.09

100mm/h = tropical downpour; 50mm/h = very heavy rain; 25mm/h = heavy rain: 10mm/h = moderate rain, 1mm/h = light rain. 2.35g/m³ ≡ visibility of 30m; 0.42g/m³ ≡ 100m; 0.043g/m³ ≡ 500m.

Moisture content is probably the major factor in determining the electrical constants of the ground, which can vary considerably with the type of surface as can be seen from Table 11.4. The depth to which the wave penetrates is a function also of frequency, and the depth given by the δ (delta) value is that at which the wave has been attenuated to $1/e$ (or about 37%) of its surface magnitude.

At VHF and at higher frequencies the depth of penetration is relatively small and normal diffraction effects are slight. At all frequencies open to amateur use, however, the ground itself appears as a reflector, and the better the conductivity of the surface the more effective the reflection (Table 11.5). It is this effect which makes generalisations of wave propagation near the ground difficult to make for frequencies greater than about 10MHz where the received field is, more often than not, due to the resultant of waves which have travelled by different paths.

Further aspects of propagation near the ground will be dealt with in the section on multiple-path propagation where it will be necessary to consider the consequences of operating with antennas at heights of one to several wavelengths above the ground.

MULTIPLE-PATH EFFECTS

Introduction

The preceding descriptions of the various modes of propagation do not necessarily paint a very realistic picture of the way in which the signals received at a distant location depend on the radio frequency in use and the distance from the transmitter, excluding ionospheric components. The reason for this is that the wave incident on the receiving antenna is rarely only the one which has arrived by the most direct path but is more often the resultant of two or more waves which have travelled by different routes and have covered different distances in doing so. If these waves should eventually arrive in phase they would act to reinforce one another, but should they reach the receiving antenna in antiphase they would interfere with one another and, if they happened to be equal in amplitude, would cancel one another completely.

Table 11.4. Typical values of dielectric constant, conductivity and depth of wave penetration for various types of surface at various frequencies

Type of surface	Dielectric constant ε	Conductivity σ (mho/m)	Depth of penetration δ (m) 1MHz	10MHz	100MHz	1000MHz
Sea water (°C)	80	4–5	0.25	0.08	0.02	0.01
Fresh water (10°C)	84	1×10^{-3} to 1×10^{-2}	11	9	4	0.2
Very moist soil	30	5×10^{-3} to 2×10^{-2}	5.5	3	2	0.3
Average ground	15	5×10^{-4} to 5×10^{-3}	21	16	16	16
Very dry ground	3	5×10^{-5} to 1×10^{-4}	95	90	90	90

Table 11.5. The number of decibels to be subtracted from the calculated free-space field in order to take into account various combinations of ground conductivity and distance. The values are shown in each case for 1.8, 3.5 and 7.0MHz. Vertical polarisation is assumed

Distance (km)	Free-space field (dB) rel to 1μV/m for a 1kW transmitter	Sea, σ = 4			Land, σ = 3 × 10^{-2}			Land, σ = 3 × 10^{-3}			Land, σ = 10^{-3}		
		1.8	3.5	7.0	1.8	3.5	7.0	1.8	3.5	7.0	1.8	3.5	7.0
3	100	1	1	1	1	3	12	8	18	30	19	28	36
10	90	1	1	1	6	8	22	18	29	42	30	38	47
30	80	1	2	2	8	18	34	28	41	51	40	48	56
100	70	2	3	7	18	33	51	44	56	68	58	63	73
300	60	10	12	23	43	61	88	68	85	—	78	—	—
1000	50	48	55	—	—	—	—	—	—	—	—	—	—

These alternative paths may arise as a result of reflections in the horizontal plane (as in Fig 11.10(a), where a tall gasholder intercepts the oncoming waves and deflects them towards the receiving site) or in the vertical plane (as in Fig 11.10(b), where reflection occurs from a point on the ground in line-of-sight from both ends of the link). If the reflecting surface is stationary, as it ought to be in the two cases so far considered, the phase difference (whatever it may be) would be constant and a steady signal would result.

It may happen that the surface of reflection is in motion as it would be if it was part of an aeroplane flying along the transmission path. In that case the distance travelled by the reflected wave would be changing continually and the relative phases would progressively advance or retard through successive cycles (effectively an increase or decrease in frequency – the *Doppler effect*), leading to alternate enhancements and degradations as the two waves aid or oppose one another. This performance is one which is particularly noticeable on television receivers sited near an airport, and even non-technical viewers can instantly diagnose as *aircraft flutter* the fluctuations in picture brilliance which result. Any 'ghost' image on the television picture is evidence of a second transmission path, and the amount of its horizontal displacement from the main picture is a measure of the additional transmission distance involved. So, with the reflection from the moving aircraft, the displacement of the ghost picture will change as the path length changes, and its brilliance will reach a maximum every time the difference between the direct path and the reflected path is exactly a whole number of wavelengths. An analytical treatment of the appearance of aircraft reflections on pen recordings of distant signals has appeared in the pages of *Radio Communication*; see reference [1].

Because the waves along the reflected path repeat themselves after intervals of exactly one wavelength, it is only the portion of a wavelength 'left over' which determines the phase relationship in comparison with the direct-path wave. This suggests that relatively small changes in the position of a receiving antenna could have profound effects on the magnitude of the received signal when multiple paths are present, and this is indeed found to be the case, particularly where the point of reflection is near at hand.

Ground-wave propagation

It should now be evident why it was not possible to generalise on the relationship between distance and received signal strength when dealing with propagation near the ground. The surface wave, influenced by the diffraction effects considered earlier, is only one of the possible paths. If the spacing of the transmitting and receiving antennas is such that they are not hidden from one another by the curvature of the Earth there will also be a *space wave,* made up of a direct wave and a ground-reflected wave as suggested by Fig 11.11.

The combination of this space wave and the diffracted surface wave form what is called the *ground wave,* and it may sometimes be difficult (and often perhaps unnecessary) to try to separate this into its three components.

Beyond the radio horizon the direct and reflected rays are blocked by the bulge of the Earth, and the range attained is then determined by the surface wave alone. This diffracted wave is strong at low frequencies (including the amateur 160m band) but becomes less so as the carrier frequency increases and may be considered negligible at VHF and beyond. When occasional

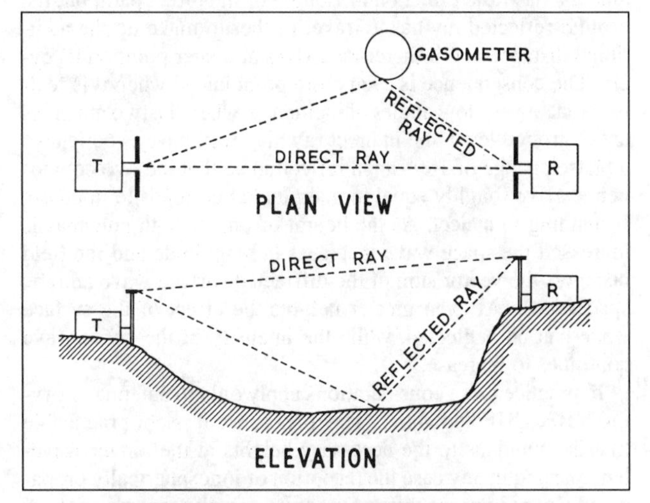

Fig 11.10. Multipath effects brought about by reflections in (a) the horizontal plane, and (b) the vertical plane

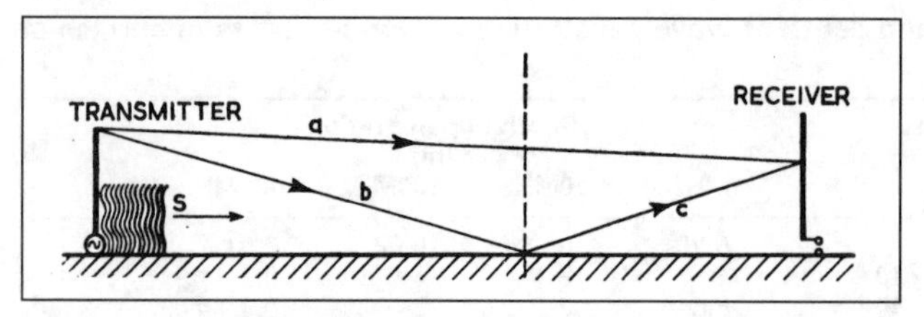

Fig 11.11. The diffracted surface wave S together with the space wave, composed in turn of a direct wave (a) and a reflected wave (b) and (c)

signals are received well beyond the horizon the dominant mechanism may be forward scatter.

The strength of the reflected component of the space wave depends largely on the conductivity and smoothness of the ground at the point of reflection, being greatest where oversea paths are involved, and least over dry ground and rock. An extensive treatment of the various factors concerned will be found in the Society's journal [2]. If perfect reflection is presumed, the received field strength due to the interaction of one reflected ray with the direct ray can be estimated from the expression:

$$E = \frac{2E_0}{d} \sin\left(2\pi \frac{h_t h_r}{\lambda d}\right)$$

where E is the resultant received signal strength, E_0 is the direct-ray field strength, h_t and h_r are the effective antenna heights above a plane tangential to the Earth at the point of reflection, d is the distance traversed by the direct ray, and λ the wavelength, all units being consistent (eg metres).

It can be seen that the magnitude of the received signal depends on the relative heights of the two antennas, the distance between them, and, of course, the frequency.

This relationship suggests that doubling the antenna height has the same effect on the received signal strength as halving the length of the path. In view of the respective distances involved it will be appreciated that an increase in the height of one of the antennas has a greater effect than a comparable horizontal movement towards the transmitter.

The effect of varying height

A few moments of experiment with two pieces of cotton representing the two alternative ray paths will provide a convincing demonstration of the effect of altering the height of one or both of the antennas. An increase in height of (say) the receiving antenna has little effect on the length of the direct path, but the ground-reflected ray has to travel further to make up the additional distance and it therefore arrives at a later point on its cycle. The consequence is even more pronounced when it is realised that at very low angles of incidence, when the two antennas are at ground level, the indirect ray may well have experienced a phase change of 180° upon reflection so that the two components arrive roughly equal in magnitude but opposite in phase, so tending to cancel. As the height of one or both antennas is increased the space wave increases in magnitude and the field becomes the vector sum of the diffracted surface wave and the space wave. At even greater heights the effect of the surface wave can be neglected, while the intensity of the space wave continues to increase.

In practice these considerations apply only to antennas carrying VHF, UHF and above. This is because it is not practicable to raise antennas to the necessary heights at the longer wavelengths, and in any case the reception of ionospherically propagated waves imposes different requirements as regards angle of arrival.

As with other functions which depend on Earth constants for their effectiveness there is a marked difference between overland and over-sea conditions. There is more to be gained by raising an antenna over land than over sea, for high frequencies than for low, and for horizontally polarised waves rather than vertically polarised ones.

The ratio of the received field at any given height above ground to the field at ground level due to the surface wave alone (presuming that the two components of the space wave have cancelled one another) is known as the *height-gain factor*. This can be expressed either as a multiplier, or as the corresponding equivalent in decibels, using the voltage scale of relationships.

Over flat ground there is little to be gained from raising antennas for frequencies below about 3MHz, unless it is to clear local obstacles, but it should not be overlooked that it may be desirable to raise antennas no matter what their frequency of operation for reasons unconnected with height-gain benefits – to increase the distance to the radio horizon, for example.

The result of changing the receiving antenna height is by no means as predictable as some authorities would have us believe, and the subject is still a matter thought worthy of further investigation at some research establishments. The following figures summarise the gains to be expected after raising a receiving antenna from a height of 3m to a height of 10m above the ground, according to a current CCIR report [3], primarily concerned with television broadcasting frequencies but relevant nevertheless:

50–100MHz	Median values of height-gain 9–10dB.
180–230MHz	Median values of height-gain 7dB in flat terrain and 4–6dB in urban or hilly areas
450–1000MHz	Median values of height-gain very dependent on terrain irregularity. In suburban areas the median is 6–7dB, and in areas with many tall buildings 4–5dB

A simple rule-of-thumb often adopted by radio amateurs is to reckon on a height-gain of 6dB for each time that the antenna height is doubled (eg if 12dB at 3m height, then expect 18dB at 6m, 24dB at 12m, etc), but the presence of more component waves than the two considered can lead to wide departures from this relationship, particularly in urban areas.

If the terrain is not flat the result of altering the antenna height depends largely on the position in the vertical plane of the reflection points relative to the two terminals. Thus in Fig 11.12 the situation shown in (a) corresponds to the one already considered. Should the two antennas be sited on hills, or separated by a valley, as at (b), there will be large differences in path length between the direct and ground-reflected rays which alterations in antenna height will do little to alter, so that elevating it is unlikely to have very much effect on the received signal strength. On the other hand, the presence of high ground between transmitter and receiver, as at (c), may make communication between them difficult at low antenna heights, and in that case there would be a great deal to be gained from raising them. In cases (b) and (c) the two antennas should be considered as having effective heights of h'_T and h'_R respectively when dealing with height-gain calculations.

If the intervening high ground has a relatively sharp and well-defined upper boundary, such as would be the case with a mountain ridge, the receiving antenna height at which signals cease might be much lower than would be expected from line-of-sight considerations, even when refractivity changes are taken into account. This is because of an effect known as *knife-edge diffraction,* which often enables 2m operators situated in the Scottish Highlands (to cite an instance) to receive signals from other

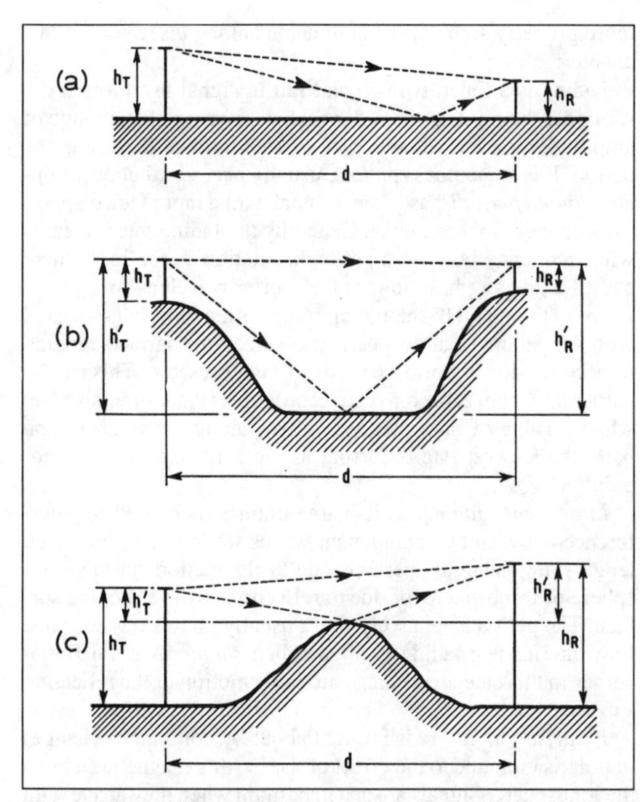

Fig 11.12. The effect of terrain on direct and indirect rays, showing the effective heights of the antennas when undertaking height-gain calculations.

stations which are apparently obscured from them by surrounding mountains.

The effect of varying distance

The effect on the field strength of varying the distance between transmitter and receiver is shown in Fig 11.13, where the result is again due to the interference between the direct wave and the ground-reflected wave passing through successive maxima and minima as the path difference becomes an exact odd or even number of halfwavelengths. (It must be remembered that a low-angle ground-reflection itself introduces a phase change of very nearly 180°). The spacing of the maxima (which are greater in magnitude than the free-space value) is closer the higher the frequency of operation and the shorter the path for a given frequency. The most distant maximum will occur when the path difference is down to one half wavelength; beyond that the difference tends towards zero and the two waves progressively oppose one another, the field rapidly falling below the free-space estimate. If the antenna heights are raised the patterns move outwards.

As with all these matters involving ground reflection there is a difference between the behaviour of horizontally and vertically polarised waves, and the foregoing description favours the former. The reflection coefficient and phase shift at the reflection point vary appreciably with the ground constants when vertical polarisation is employed. In practice, whichever is used, the measured field strength may vary considerably from the calculated value because of the presence of other components due to local reflections. A fairly reliable first estimate for VHF and UHF paths up to about 50km unobstructed length is just to allow for a possible increase or decrease of 10dB on the free-space figure.

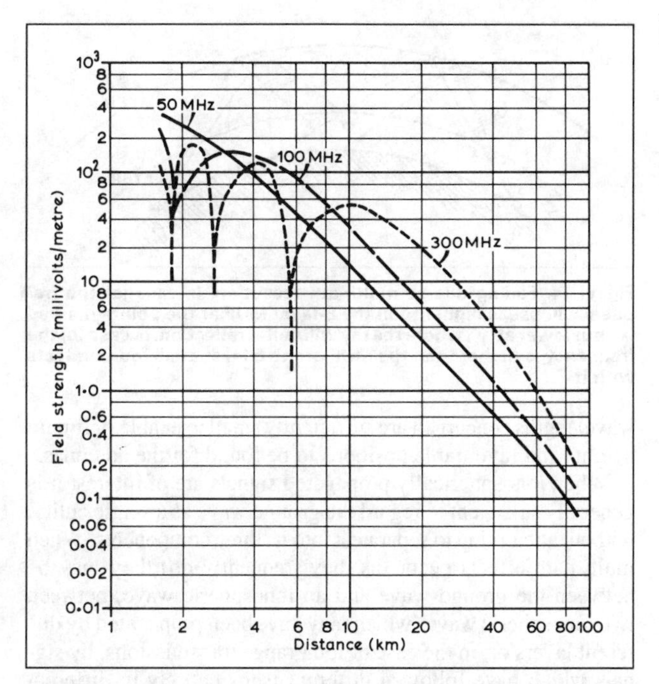

Fig 11.13. Relationship between field strength and distance at VHF and UHF

Fresnel zones

In all the explanations so far it has been presumed that reflection at a surface occurs at the point which enables the reflected ray to travel the shortest distance.

Because the surfaces considered in radio propagation work are neither plane nor perfect reflectors, the received waveform is the resultant of signals which have been reflected from an area, the size of which is determined by the frequency and by the separation of the terminal antennas, so that the individual reflected path lengths differ by no more than half a wavelength from one another. The locus of all the points surrounding the direct path which give exactly half a wavelength path difference is described by an ellipsoid of revolution having its foci at the transmitting and receiving antennas respectively. A cross-section of this volume on an intersecting plane of reflection encloses an area known as the *first Fresnel zone*.

The radius, R, of the first Fresnel zone at any point, P, is given by the expression:

$$R^2 = \frac{\lambda d_1 d_2}{d_1 + d_2}$$

where d_1 and d_2 are the two distances from P to the ends of the path, and λ is the wavelength, all quantities being in similar units. The maximum radius occurs when d_1 and d_2 are equal. Thus, on an 80km path at a wavelength of 2m, the radius of the first Fresnel zone at the midpoint is 200m, and this represents the clearance of the line-of-sight ray (corrected for refraction) necessary if the path is to be considered 'unobstructed'.

Higher orders of Fresnel zone surround the regions where similar relationships occur after separations of one wavelength, two wavelengths etc, but the conditions for reflection in the required direction rapidly become less favourable and Fresnel zones other than the first are rarely considered.

Ionospheric multi-path effects

The multiple-path effects so far considered have been mainly associated with VHF, UHF and SHF where very low angles of radiation and reception are generally involved, and where the

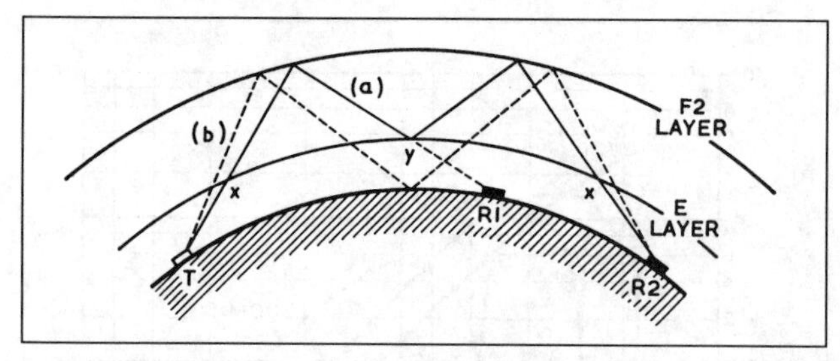

Fig 11.14. Fading due to multipath reception. In case (a) the frequency in use is higher than the E-layer MUF at the points marked 'x', but lower at 'y', where the ray will suffer reflection. In case (b) the frequency is higher than the MUF of the E-layer at all four contacts with it

wavelengths concerned are sufficiently small to enable optimum heights and favourable positions to be found for the antennas.

When ionospherically propagated signals are of interest it is generally sufficient to regard the ground wave as a single entity, without attempting to separate it into its three components. When multi-path effects occur (as they frequently do) they may be between the ground wave and an ionospheric wave, between two ionospheric waves which may have been propagated by different layers or, in the case of long range transmissions, by signals which have followed different paths entirely in different directions around the Earth's curvature, perhaps in several 'hops' between the ionosphere and the ground. Whatever the cause the result is inevitably a fading signal.

The ionosphere is not a perfect reflector; it has no definite boundaries and it is subject to frequent changes in form and intensity. These deficiencies appear on the received signal as continual small alterations in phase or frequency as the effective path length alters and, when only a single ionospheric wave is present, can pass almost unnoticed by the average HF listener, who is remarkably tolerant of imperfections on distant transmissions. However, when a second signal from the same source is present, which may be either the relatively steady ground wave or another ionospheric component, these phase or frequency changes become further emphasised by appearing as changes in amplitude as the waves alternately reinforce and interfere, and by distortion of modulated signals if the various sideband frequencies do not resolve back into their original form.

Pen recordings of signal strength generally show very clearly the period of fading which results when two modes of propagation begin to interact, continuing until the second predominates. This could occur as a result of circumstances similar to those shown in Fig 11.14, where the receiver at R2 may receive signals either by double-hop off the ground (as in path (b)) or double-hop off an intermediate layer, path (a), depending on the relationship between the signal frequency and the maximum usable frequency at point y. The transition between one propagation mode and another is generally accomplished within a relatively short time.

Occasionally very long-distance transmissions may be heard with a marked echo on their modulation. As the two signal components responsible have obviously travelled routes of markedly differing length they probably require very different azimuths at both ends of the path. This effect is most noticeable on omnidirectional broadcast transmissions and is minimised by the use of narrow-beam antennas for transmission and reception.

Fading

Fading is generally, though not exclusively, a consequence of the presence of multiple transmission paths. For that reason it is appropriate to include a summary of its causes here, although more properly some of the comments belong elsewhere in this chapter.

Fading is a repetitive rise and fall in signal level, often described as being *deep* or *shallow* when referring to the range of amplitudes concerned, and *slow* or *rapid* when discussing the period. It is sometimes *random,* usually *periodic,* but occasionally *double periodic,* as when a signal with a rapid fade displays slow changes in mean level. Generally the fading rate increases with frequency because a particular motion in the ionosphere causes a greater phase shift at the shorter wavelengths.

At VHF and UHF the fading rate is often closely associated with the pattern of atmospheric pressure at the surface, tending to become slower during periods of high pressure. This can be particularly noticeable on a pen recording of signal strength taken while a ridge of high pressure moves along the transmission path, the slowest rate occurring as the ridge crosses the mid-point.

Interference fading, as its name implies, is caused by interference between two component waves when one or both path lengths are changing, perhaps due to fluctuations in the ionosphere or troposphere, or due to reflections from a moving surface. The period is relatively short, usually up to a few seconds. Fast interference fading is often called *flutter.* Auroral flutter comes in this category, being caused by motion of the reflecting surfaces.

Polarisation fading is brought about by continuous changes in polarisation due to the effect of the Earth's magnetic field on the ionosphere. Signals are at a maximum when they arrive with the same polarisation as the receiving antenna. Period again up to a few seconds.

Absorption fading, generally of fairly long period, is caused by inhomogeneities in the troposphere or ionosphere. Period up to an hour, or longer.

Skip fading occurs when a receiver is on the edge of a skip zone and changes in MUF cause the skip distance to shorten and lengthen. Highly irregular as regards period of fade.

Selective fading is the name given to a form of fading characterised by severe modulation distortion in which the path length in the ionosphere varies with frequency to such an extent that the various sideband frequencies are differently affected. It is most severe when ground and sky waves are of comparable intensity.

Scintillations are rapid fluctuations in amplitude, phase and angle-of-arrival of tropospheric signals, produced by irregular small-scale variations in refractive index. The term is also used to describe irregular fluctuations on HF signals transmitted through the ionosphere from satellites and other sources outside the Earth.

Diversity reception

The effects of fading can be countered to a certain extent by the use of more than one receiving system coupled to a common network which selects at all times the strongest of the outputs available.

There are three principal versions in common use:

(a) *Space diversity,* obtained by using antennas which are so positioned as to receive different combinations of components in situations where multi-path conditions exist;

(b) *frequency diversity,* realised by combining signals which have been transmitted on different frequencies; and

(c) *polarisation diversity,* where two receivers are fed from antennas having different planes of polarisation.

It is unlikely that any of these systems have any application in normal amateur activities, but they may be of interest in connection with research projects relating to the amateur bands.

SOLAR AND MAGNETIC INFLUENCES

The Sun

Our Sun is at the centre of a complex system consisting of nine major planets (including ours), five of which have two or more attendant moons, of several thousand minor planets (or asteroids), and of an unknown number of lesser bodies variously classed as comets or meteoroid swarms.

It is a huge sphere of incandescence of a size which is equivalent to about double our moon's orbit around the Earth, but, despite appearances to the contrary, it has no true 'edge' because nearly the whole of the Sun is gaseous and the part we see with apparently sharp boundaries is merely a layer of the solar atmosphere called the *photosphere* which has the appearance of a bright surface, preventing us from seeing anything which lies beneath.

Beyond the photosphere is a relatively cooler, transparent layer called the *chromosphere,* so named because it has a bright rose tint when visible as a bright narrow ring during total eclipses of the Sun. From the chromosphere great fiery jets of gas, known as *prominences,* extend. Some are slow changing and remain suspended for weeks, while others, called *eruptive prominences,* are like narrow jets of fire moving at high speeds and for great distances.

Outside the chromosphere is the *corona,* extending a distance of several solar diameters before it becomes lost in the general near-vacuum of interplanetary space. At the moment of totality in a solar eclipse it has the appearance of a bright halo surrounding the Sun, and at certain times photographs of it clearly show it being influenced by the lines of force of the solar magnetic field.

The visible Sun is not entirely featureless – often relatively dark *sunspots* appear and are seen to move from east to west, changing in size, number and dimensions as they go. They are of interest for two reasons: one is that they provide reference marks by which the angular rotation of the Sun can be gauged, the other that by their variations in number they reveal that the solar activity waxes and wanes in fairly regular cycles. The Sun's rotation period has been found to vary with latitude, with its maximum angular speed at the equator. The mean rotation period with respect to the stars is 25.38 mean solar days, but a more useful figure is the *mean synodic rotation period* of 27.2753 days, which is the time required for the Sun to rotate until the same part faces the Earth taking into account the Earth's steady motion around it. We shall have occasion to note these periods again when dealing with the recurrence of abnormal magnetic activity on the Earth.

The Sun's rotations have been numbered since the year 1853 (rotation 1880 commenced in 1994 on 6 March), and observations of solar features are referred to an imaginary network of latitudes and longitudes which rotates from east to west as seen from the Earth. Remember, when looking at the mid-day Sun from the UK, north is at the top, the east limb is on the left side and the west limb on the right. A long-persistent feature which first appears on the east limb is visible for about 13½ days before it disappears from sight over the west limb.

For many purposes it is more convenient to refer the positions of noteworthy features to a related, but stationary, set of co-ordinates, the *heliocentric latitude and longitude,* in which locations are described with respect to the centre of the visible disc. An important statistic relating to sunspots is the time of their *central meridian passage* (CMP).

Radio telescopes detect features which are usually situated in the vicinity of the solar corona, and some of them reveal disturbances beyond the limbs of the visible disc. They often travel across the face of the Sun at a faster rate than any spots beneath.

The apparent diameter of the Sun varies with the choice of radio frequency. Because the lower frequencies come from the outer parts of the corona, the width of their Sun appears to be large. At high frequencies the sources are situated below the level which provides the visible disc so that the width of the Sun then appears to be small. The optical and radio frequency diameters are similar at a frequency of 2800MHz (10.7cm wavelength), so that was chosen to provide the daily solar flux measurements which provide an indication of activity which is independent of the amount of cloud cover in the vicinity of the observatory.

Sunspots

Sunspots are the visible manifestation of very powerful magnetic fields; adjacent spots often have opposing polarities. These intense magnetic fields also produce *solar flares,* which are emissions of hydrogen gas. They are responsible also for the ejection of streams of charged particles and X-rays.

Sunspot numbers have been recorded for over 200 years and it has been found that their totals vary over a fairly regular cycle occupying around 22 years where magnetic polarity is the criterion, or 11 years if only the magnitude of activity is considered. The 11-year peaks are known as *sunspot maxima*; the intervening troughs are *sunspot minima.*

The rise and fall times are not equal, though. Four years and seven years respectively are typical, although each cycle differs from the others in both timing and maximum value, as may be seen in Fig 11.15. At sunspot minimum the Sun may be completely spotless for weeks or even months (or once, in the 17th century, years).

Tables of daily *relative sunspot numbers* are prepared monthly at the Sunspot Index Data Centre (SIDC) in Brussels, from information supplied by a network of participating observatories. It is widely supposed by radio amateurs that those figures indicate the number of visible spots but that is not so. In accordance with a formula devised by Dr Wolf in Zurich (hence the description *Wolf number*, still used professionally) the relative sunspot number, R, is found from the expression:

$$R = k(10g + t)$$

where k is a regulating factor that keeps the series to a uniform standard, g is the number of spot groups, and t is the total number of spots.

Daily figures obtained from the Ursigram messages (see p11.20) use unity as the value of k, so, for example, Boulder figures obtained from that source record just 10 times the number of groups plus the total number of spots. Note also that those figures are provisional because they will have been prepared in haste to meet a deadline.

When the figures from the participating observatories have been combined in Brussels, a value of k which is less than unity will have been applied. That figure does not appear in the tables and it may well be subject to frequent variation, but it is currently around 0.7.

The Brussels figures are issued twice, first provisionally as soon as possible, then definitively after more careful scrutiny. From the monthly means of the definitive values a smoothed index R_{12} is obtained. This is the arithmetic mean of 12 successive monthly means, the result being ascribed to the period at the centre of the sample. In order to make that fall in the middle

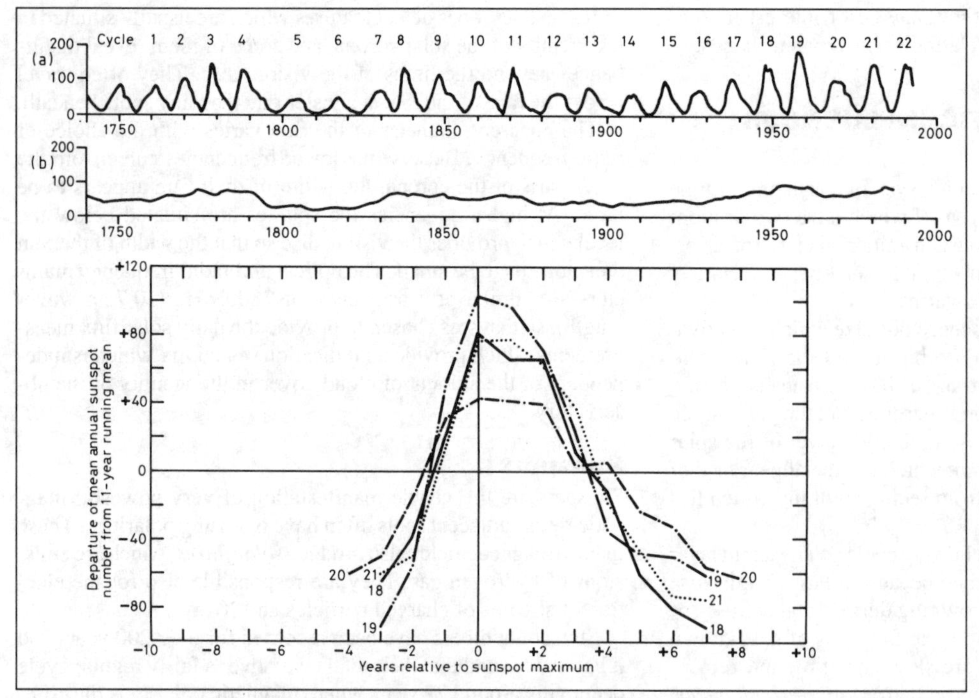

Fig 11.15. (a) Annual relative sunspot numbers, the means of monthly means of daily values. (b) The 11-year running means of the annual numbers. The succession of points reveals the underlying trend in solar activity. (c) A comparison of the four cycles most recently completed. The steep rise to maximum and the relatively slow decline thereafter are characteristic of all sunspot cycles

of the month, rather than between months, 13 months are taken but the first and the last are given only half weight in the calculations. From the nature of the 12-month running average – for that is what it is – it must be evident that R_{12} (which is the 'sunspot number' called for in ionospheric prediction programs) never reaches the peaks and troughs of the individual monthly means and it falls far short of the maxima of the daily values.

To put all these different versions of the 'sunspot number' into perspective we have only to look at a specific example, say the month which contained the peak of solar cycle 22, which was June 1989. The Boulder figures on the Ursigrams reached 401, and their monthly mean was 297. The SIDC Brussels definitive figure at maximum was 265, the monthly mean 196. However, the smoothed figure R_{12} for the month was only 158 and, remember, it is that one that you need for prediction programs, not the 401 from Boulder. You should be aware also that the latest smoothed figure available is always six months behind the current date, so a figure for this month has been a forecast six months ahead.

It is also of interest to observe that the peak Boulder figure was made up of 18 groups which, between them, contained a total of 221 spots. Put those figures in the formula and you come up with 401, the figure reported. The three largest groups accounted for 86, 53 and 26 spots respectively, and none of the others contained more than 7.

The solar wind

The solar corona was described in the last section as extending outwards until it becomes lost in interplanetary space. In fact it does not become lost at all, but turns into a tenuous flow of hydrogen which expands outwards through the solar system, taking with it gases evaporating from the planets, fine meteoric dust and cosmic rays. It becomes the *solar wind.*

Near the Sun the corona behaves as a static atmosphere, but once away it gradually accelerates with increasing distance to speeds of hundreds of kilometres per second. The gas particles take about nine days to travel the 150,000,000km to the Earth, carrying with them a magnetic field (because the gas is ionised) which assumes a spiral form due to the Sun's rotation. It is the solar wind, rather than light pressure alone, which is responsible for comets' tails flowing away from the Sun, causing them to take on the appearance of celestial wind-socks.

The existence of the solar wind was first detected and measured by space vehicles such as the Luniks, Mariner II and Explorer X, which showed that its speed and turbulence are related to solar activity. Regular measurements of solar wind velocity are now routine.

There is thus a direct connection between the atmosphere of the Sun and the atmosphere of the Earth. In the circumstances it is hardly surprising that solar events, remote though they may at first seem, soon make their effects felt here on Earth.

The Earth's magnetosphere

It is well known that the Earth possesses a magnetic field, for most of us have used a compass, at some time or another, to help us to get our 'bearings'. We know from such experiences that the field appears to be concentrated at a point somewhere near the north pole (and are prepared to believe that there is another point of opposing polarity somewhere near the south pole). Popular science articles have familiarised us with a picture of field lines surrounding the Earth like a section of a ring doughnut made up of onion-like layers.

Because the particles carried by the solar wind are charged their movement produces a magnetic field which interacts with the geomagnetic field. A blunt shock-wave is set up, called the *magnetosheath,* and the wind flows round it, rejoining behind where the field on the far side is stretched in the form of a long tail, the overall effect being reminiscent of the shape of a pear with its stalk pointing away from the Sun. The region within the magnetosheath, into which the wind does not pass, is called the *magnetosphere.*

On the Earthward side the magnetosphere merges into the ionosphere. Inside the magnetosphere there are regions where charged particles can become trapped by geomagnetic lines of force in a way which causes them to oscillate back and forth over great distances. Particles from the solar wind can enter these regions (often called *Van Allen belts* after their discoverer) in some way, as yet not perfectly understood.

The concentration of electrons in the magnetosphere can be gauged from the ground by observations on *whistlers,* naturally-occurring audio-frequency oscillations of descending pitch which are caused by waves radiated from the electric discharge in a lightning flash. These travel north and south through the ionosphere and magnetosphere from one hemisphere to another along the magnetic lines of force. The various component frequencies propagate at different speeds so that the original flash (which appears on an ordinary radio receiver as an *atmospheric*) arrives at the observer considerably spread out in time. The interval between the reception of the highest and lowest frequencies is a function of the concentration of electrons encountered along the way.

These trapped particles move backwards and forwards along the geomagnetic field lines within the Van Allen belts and some collide with atoms in the ionosphere near the poles where the belts approach the Earth most closely. Here they yield up energy either as ionisation or illumination and are said to have been *dumped.* These dumping regions surround the two poles, forming what are called the *auroral zones.* The radius of the circular motion in the spirals (they are of a similar form to that of a helical spring) is a function of the strength of the magnetic field, being small when the field is strong. Electrons and protons perform their circular motions in opposite senses, and the two kinds of spiralling columns drift sideways in opposite directions, the electrons eastward, the protons westward, around the world. Because of the different signs on the two charges these two drifts combine to give the equivalent of a current flowing in a ring around the Earth from east to west.

This ring current creates a magnetic field at the ground which combines with the more-or-less steady field produced from within the Earth. We shall see later the sort of effects that solar disturbances have on the magnetosphere, the ionosphere, and the total geomagnetic field.

The quiet ionosphere

With the stage set to follow the antics of the ions and electrons deposited by dumping we must pause again, this time to examine the normal day-to-day working of the ionosphere, which is dependent for its chemistry on another form of incoming solar radiation.

The gas molecules in the Earth's upper atmosphere are normally electrically neutral, that is to say the overall negative charges carried by their orbiting electrons exactly balance the overall positive charges of their nuclei. Under the influence of ultraviolet radiation from the Sun, however, some of the outer electrons can become detached from their parent atoms, leaving behind overall positive charges due to the resulting imbalance of the molecular structure. These ionised molecules are called *ions,* from which of course stems the word 'ionosphere'.

This process, called *disassociation,* tends to produce layers of free electrons brought about in the following manner. At the top of the atmosphere where the solar radiation is strong there are very few gas molecules and hence very few free electrons. At lower levels, as the numbers of molecules increase, more and more free electrons can be produced, but the action progressively weakens the strength of the radiation until it is unable to take full advantage of the increased availability of molecules and the electron density begins to decline. Because of this there is a tendency for a maximum (or peak) to occur in the production of electrons at the level where the increase in air density is matched by the decrease in the strength of radiation. A peak formed in this way is known as a *Chapman layer,* after the scientist who first outlined the process.

The height of the peak is determined not by the strength of the radiation but by the density/height distribution of the atmosphere and by its capability to absorb the solar radiation (which is a function of the UV wavelength), so that the layer is lower when the radiation is less readily absorbed. The strength of the radiation affects the rate of production of electrons at the peak, which is also dependent on the direction of arrival. The electron density is greatest when the radiation arrives vertically and it falls off as a function of zenith distance, being proportional to cos χ, where χ (the Greek letter chi) is the angle between the vertical and the direction of the incoming radiation. As cos χ decreases (ie when the Sun's altitude declines) a process of recombination sets in, whereby the free electrons attach themselves to nearby ions and the gas molecules revert to their normal neutral state.

Experimental results have led to the belief that the E-layer (at about 120km) and the F1-layer (at about 200km) are formed according to Chapman's theory as a result of two different kinds of radiation with perhaps two different atmospheric constituents involved.

The uppermost layer F2 (around 400km), which normally appears only during the day, does not follow the same pattern and is thought to be formed in a different way, perhaps by the diffusion of ions and electrons, but there are still a number of anomalies in its behaviour which are the subject of current investigation. These include the *diurnal anomaly,* when the peak occurs at an unexpected time during the day; the *night anomaly,* when the intensity of the layer increases during the hours of darkness when no radiation falls upon it; the *polar anomaly,* when peaks occur during the winter months at high latitudes, when no illumination reaches the layer at all; the *seasonal anomaly,* when magnetically quiet days in summer (with a high Sun) sometimes show lower penetration frequencies than quiet days in winter (with a low Sun); and a *geomagnetic anomaly* where, at the equinoxes, when the Sun is over the equator, the F2-layer is most intense at places to the north and south separated by a minimum along the magnetic dip equator. It is thought that topside sounding from satellites probing the ionosphere from above the active layers may help to resolve some of these anomalies in F2 behaviour.

The regular ionospheric layers

The various regular ionospheric layers were first defined by letters by Sir Edward Appleton who gave to the one previously known as the *Kennelly-Heaviside layer* the label 'E' because he had so marked it in an earlier paper denoting the electric field reflected from it, and to the one he had discovered himself the letter 'F', rather than call it the *Appleton layer,* as some had done. To the band of absorption below thus naturally fell the choice of the letter 'D', although this was generally referred to as a 'region' rather than a 'layer' because its limits are less easy to define.

From comments already made it will be appreciated that the regular ionospheric layers which these letters define exhibit changes which are basically a function of day and night, season and solar cycle.

Most of our knowledge of the ionosphere comes from regular soundings made at vertical incidence, using a specialised form of radar called an *ionosonde* which transmits short pulses upwards using a carrier which is continuously varied in frequency from the medium-wave broadcast band through the HF bands to an upper limit of about 20MHz, but beyond if conditions warrant. Reflections from the various layers are recorded photographically in the form of a graph called an *ionogram,* which displays *virtual height* (corresponding to radar range) as a function or signal frequency. *Critical frequencies,* where the signals pass straight through the layers, are read off directly. There are many such equipments in the world; the one serving the United Kingdom is located near Slough. It is under the control of the Rutherford Appleton Laboratory at Chilton, near Didcot, which houses one of a number of World Data Centres to which routine measurements of the ionosphere are sent from most parts of the world.

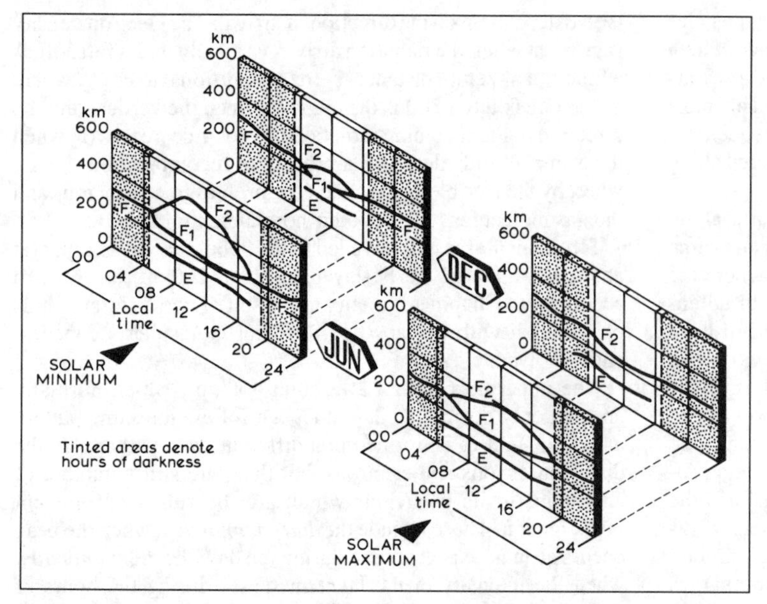

Fig 11.16. Typical diurnal variations of layer heights for summer and winter at minimum and maximum states of the solar cycle

The two sets of diagrams (Fig 11.16 and Fig 11.17) summarise the forms taken by the diurnal variations in height and critical frequency for two seasons of the year at both extremes of the sunspot cycle. The actual figures vary very considerably from one day to the next, but an estimate of the expected monthly median values of maximum usable frequency and optimum working frequency between two locations at any particular year, month and time of day can be obtained from predictions published several months in advance by a number of research establishments.

The critical frequencies of the E and F1-layers are a function of R_{12}, the smoothed SIDC Brussels (formerly Zurich) relative sunspot number (which is predicted six months in advance for this purpose), and the cosine of the zenith distance χ, the angle the Sun makes with the local vertical, and thus, to a first approximation:

$$f_0E = 0.9\,[(180 + 1.44R)\cos\chi]^{0.25}$$

(usually to within 0.2MHz of the observed values), and:

$$f_0F1 = (4.3 + 0.01R)\cos^{0.2}\chi$$

which is less accurate because of uncertainty in the value of the exponent which varies with location and season.

The F2-layer is the most important for HF communication at a distance, but is also the most variable. It is subject to geomagnetic control which impresses a marked longitudinal effect on the overall world pattern. causing it to lag behind the sub-solar point so as to give maximum values in critical frequency during the local afternoon.

The F2 critical frequency f_0F2 varies with the solar cycle, as shown in Table 11.6, which shows monthly median values for Slough, applicable to sunspot numbers of 0 and 150. In recent years it has been found possible to predict the behaviour of the F2-layer by extrapolation several months ahead, using an index known as *IF2*, which is based on observations made at about 10 observatories.

The *maximum usable frequency* (MUF) which can be used on a particular circuit may be calculated from the critical frequency of the appropriate layer by applying the relationship:

$$\text{MUF} = F/\sec\varphi$$

where φ is the angle that the incident ray makes with the vertical through the point of reflection at the layer. The factor ($\sec\varphi$) is called the *MUF factor*; it is a function of the path length if the height of the layer is known. Table 11.7 shows typical figures obtained by assuming representative heights.

To an operator, the *optimum working frequency* (OWF) is the highest (of those available) which does not exceed the MUF. As will be seen later, both MUF and OWF take on different meanings in the context of ionospheric predictions.

There is a lower limit to the band of frequencies which can be selected for a particular application. This is set by the *lowest useful frequency* (LUF), below which the circuit becomes either unworkable or uneconomical due to the effects of absorption and the level of radio noise. Its calculation is quite a complicated process beyond the scope of this survey.

It is often useful to be able to estimate the radiation angle involved in one- or two-hop paths via the E and F2-layers. Fig 11.18, also prepared for average heights, accomplishes this. It is a useful rule-of-thumb to remember that the maximum one-hop E range is 2000km and that the useful two-hop E range, twice that (4000km), is also the one-hop F2 range. Of course, all extreme ranges require very low angles of take-off.

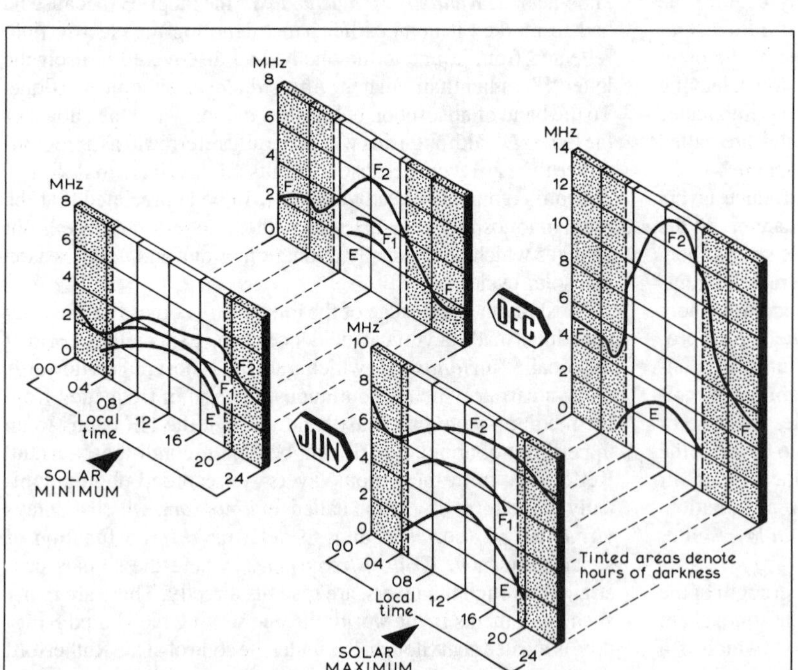

Fig 11.17. Typical diurnal variations of F-layer critical frequencies for summer and winter at the extremes of the solar cycle

Irregular ionisation

Besides the regular E, F1 and F2-layers there are often more localised occurrences of

Table 11.6. F2-layer critical frequencies at Slough. The two rows show mean median value of f_0F2 in megahertz for three month weighted mean sunspot numbers of 0 and 150

	Jan	Feb	Mar	Apr	May	Jun
$R_3 = 0$	5.3	5.1	4.75	4.80	5.10	5.03
$R_3 = 150$	12.17	12.42	11.67	9.88	8.23	7.70
	Jul	**Aug**	**Sep**	**Oct**	**Nov**	**Dec**
$R_3 = 0$	4.72	4.75	4.90	5.69	5.58	5.32
$R_3 = 150$	7.73	7.68	8.81	11.26	12.93	12.53

Table 11.7. MUF (maximum usable frequency) factors for various distances assuming representative heights for the principal layers

Layer	Distance 1000km	2000km	3000km	4000km
Sporadic E	4.0	5.2	—	—
E	3.2	4.8	—	—
F1	2.0	3.2	3.9	—
F2 winter	1.8	3.2	3.7	4.0
F2 summer	1.5	2.4	3.0	3.3

ionisation which make their contribution to radio propagation. They generally occur around the heights associated with the E-layer and often the effects extend well into VHF, although the regular E-layer can never be effective at frequencies of 30MHz or more.

Sporadic E (Es) has been observed at HF on ionospheric sounding apparatus since the early 'thirties. It has been shown to take the form of clouds of high density of ionisation, forming sheets perhaps a kilometre deep and some 100km across in a typical instance, and appearing at a height of 100–120km.

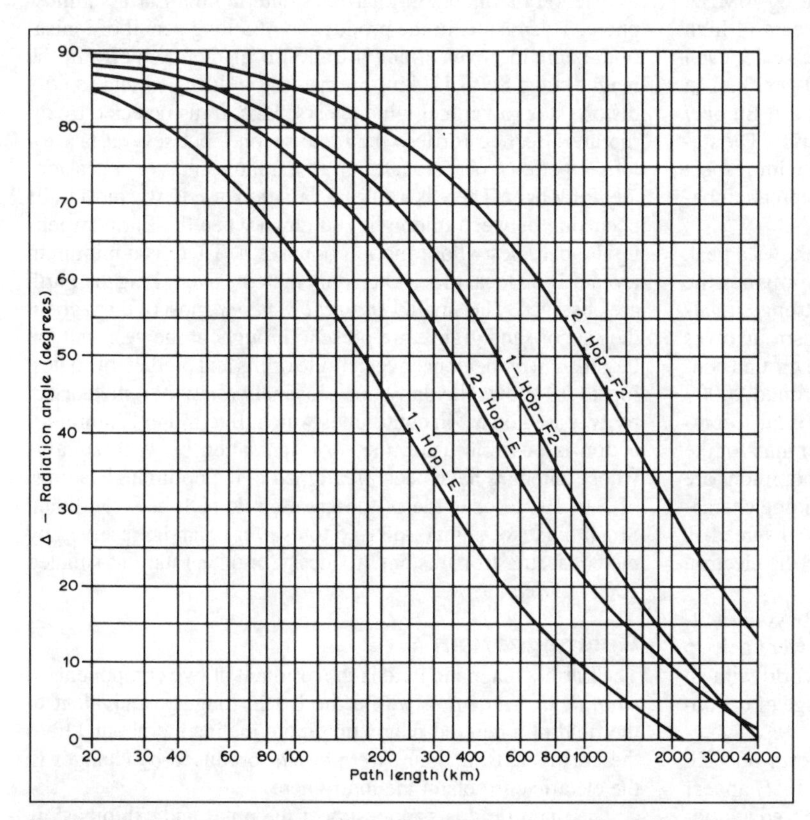

Fig 11.18. Radiation angle involved in one-hop and two-hop paths via E and F2-layers. (From NBS publication *Ionospheric Radio Propagation*, p191.)

Table 11.8. Comparative characteristics of tropospheric and VHF sporadic E propagation

Tropospheric sporadic-E	VHF sporadic-E
May occur at any season	Mainly May, June, July and August
Associated with high pressure, or with paths parallel to fronts	No obvious connection with weather patterns
Gradual improvement and decline of signals	Quite sudden appearance and disappearance
Onset and decay times similar over a wide range of frequencies	Begins later and ends earlier as radio frequency increases
Observed at VHF, UHF, SHF	Rarely above 200MHz
Area of enhancement relatively stable for several hours at a time	Area of enhancement moves appreciably in a few hours
May last a week or more	Duration minutes or hours, never days
Wide range of distances with enhanced signals at shorter ranges	Effects mainly at 1000–2000km. No associated enhancement at short ranges

However, strangely, over the years not one of the participating observatories has ever found evidence of a layer having sufficient electron density to support propagation at 144MHz – yet radio amateurs make use of something that behaves as though it was sporadic E many days of the year.

It is mainly a summertime phenomenon in May to August (the months without an 'r' in them) so far as the Northern Hemisphere is concerned, but there is also some activity during the latter part of the year. Distances worked are rarely less than 500km, mainly between 1000 and 2000km, and usually with good-quality, steady, strong signals. Two-hop Es has been observed on occasion, but that is more likely to involve two small clouds rather than one big one. Also on record are cases where 2000km has been exceeded by the aid of tropospheric enhancement at one end of the path. (Generally, VHF Es and tropospheric modes are quite separate and a list of their distinguishing features will be found in Table 11.8.)

The numbers and durations of sporadic-E events decrease with frequency. To date, the highest recorded frequency appears to be 218MHz but that may not necessarily be the absolute limit. Due to the mystery surrounding the mechanism of the mode at VHF, many observers make a long-term specific study of Es, and there has long been an International Amateur Radio Union co-ordinator whose task is to guide national societies into setting up useful co-operative research projects.

Auroral E (Ar) is closely connected with geomagnetic disturbances. At times of high activity (popularly known as *magnetic storms*), the regions around the north and south poles where visual aurora are commonplace (the *auroral ovals*) move towards the equator, taking with them the capability of returning VHF signals that have been directed towards them.

It used to be thought that antennas had to be turned towards the north (some even said the magnetic north) in order to take advantage of this mode but, thanks to careful observing by a group of dedicated amateurs over a period of many years, it is now known that optimum bearings can and do change considerably during an auroral event, and that from the UK a gradual swing towards the east may be expected as the activity develops. Radar measurements have shown that the reflecting regions are usually around 110km in height, but it should be noted that visual aurora extends very much beyond that. However, there is quite good general relationship between the radio and visual forms of aurora, although attempts to match details on pen recordings of received signals against observations of changes in the structure of auroral forms seen from the transmitting site have been disappointing.

The signal paths are of necessity angled, often with one leg much longer than the other, the two antennas being directed towards a common reflection point that is at a latitude higher than that of either station (forward scatter is unlikely). Stations in central Europe are able to work with beam headings considerably west of north; that sector is of little use to UK operators because there are no available contacts in the North Atlantic.

In Europe the signals have a raw, rasping tone that is readily recognised again, once identified. It is said to differ in character from the tone of auroral signals met with in North America.

A radio auroral event typically begins in the afternoon and may appear to be over by the traditional tea-time, but many events exhibit two distinct phases, and the evening phase is often the better in terms of DX worked, partly due perhaps to the greater number of stations likely to be active at that time. The event will frequently finish with dramatic suddenness, just as some of the longest paths are being achieved.

A book outlining the theory of auroral-E propagation, together with an analysis of observations made over much of the time since the second world war, will be found in the bibliography at the end of this chapter.

Trans-equatorial propagation (TEP) is one of the success stories of amateur radio research. Much of the pioneering work was carried out at 50MHz, which is a band not available to all radio amateurs even now. Higher-frequency working, eg 70MHz, is possible. Instances of TEP tend to favour the years of high solar activity, but it is present, on a reduced time scale, even near sunspot minimum. Paths are typically 3000–9000km in length, usually with a north/south bias: examples are Europe/Africa, Japan/Australia, North America/South America. The stations in contact are usually symmetrically located with respect to the magnetic dip equator, with the path in between them being perpendicular to it.

Two types of TEP have been recognised. One shows a peak in activity at around 1700–1900 local time. That provides the longest of the contacts (9000km or more) with strong signals and low fading rate. It is thought that the mechanism involves two reflections at the F region without intermediate ground contact. The other tends to peak in the evening at around 2000–2300 local time. Signal strengths are high but there is an accompaniment of deep and rapid flutter. Paths are shorter than for the afternoon type, perhaps no more than 6000km, and opinions are divided as to the mechanism involved. The rapid fading characteristic seems to connect in some way with equatorial spread-F, which is a diffuse effect caused by irregularities in the electron density of the F region.

Before television moved from VHF to UHF and beyond, suitably placed radio amateurs in places like Greece were able to entertain visitors from less-fortunate parts of the world with an impressive display of ultra-DX TV, taking advantage of the opportunities offered by TEP.

Operators in southern Europe make use of a form of VHF propagation in which *field-aligned irregularities* (FAI) appear to play an important part. The effect has been observed at mid-latitudes both on the Continent and in North America.

The FAI 'season' runs very closely parallel to that of VHF sporadic E, May to late August, and some instances have been known to follow conventional Es openings.

The mode is characterised by signals arriving away from the expected great-circle bearing. The scattering area responsible is apparently very small and some elevation of perhaps 10–15° has been required (in Italy) to find it. High-power transmitters and low-noise receivers are essential.

Italian amateurs recognise two distinct areas which appear to vary but little from one occasion to the next. One, located in the west of Switzerland, provides 2m contacts with Spain, southern France, Hungary and Yugoslavia. The other is located over Hungary itself and that provides Italian amateurs with openings to the Balkan peninsula.

Although FAI would appear to have very little direct application for UK amateurs at present, a nodding acquaintance with it might lead to fresh discoveries about its capabilities. One theory is that it is associated with anomalies in the Earth's magnetic field. Central Europe is not the only place with those.

The more-conventional *ionospheric scatter mode* is one that is developed commercially to provide communications over some 800–2000km paths, using ionisation irregularities at a height of about 85km, which is in the D-region. The frequencies used were between 30 and 60MHz but, at the top of the range especially, very high powers were necessary.

Received signals were weak, but they had the advantage of being there during periods of severe disruption on the HF bands. The systems that were set up have fallen from favour nowadays and satellite transponders are commonly used instead for much of the traffic. Any amateur involvement would have to be limited to the 50MHz band, but the high power involved suggests that this is now one for the history books.

Meteoric ionisation is caused by the heating to incandescence by friction of small solid particles entering the Earth's atmosphere. This results in the production of a long pencil of ionisation extending over a length of 15km or more, chiefly in the height range 80 to 120km. It expands by diffusion and rapidly distorts due to vertical wind shears. Most trails detected by radio are effective for less than one second, but several last for longer periods, occasionally up to a minute and very occasionally for longer. There is a diurnal variation in activity, most trails occurring between midnight and dawn when the Earth sweeps up the particles whose motion opposes it. There is a minimum around 1800 local time, when only meteors overtaking the Earth are observed. The smaller *sporadic meteors,* most of them about the size of sand grains, are present throughout the year, but the larger *shower* meteors have definite orbits and predictable dates. Fig 11.19 shows the daily and seasonal variation in meteor activity, based on a 24h continuous watch. Intermittent communication is possible using meteoric ionisation between stations whose antennas have been prealigned to the optimum headings of 5 to 10° to one side of the great-circle path between them. Small bursts of signal, referred to as *pings,* can be received by meteor scatter from distant broadcast (or other) stations situated 1000–1200km away.

Geomagnetism

The Earth's magnetic field is the resultant of two components, a *main field* originating within the Earth, roughly equivalent to the field of a centred magnetic dipole inclined at about 11° to the Earth's axis, and an *external field* produced by changes in the electric currents in the ionosphere.

The main field is strongest near the poles and exhibits slow secular changes of up to about 0.1% a year. It is believed to be

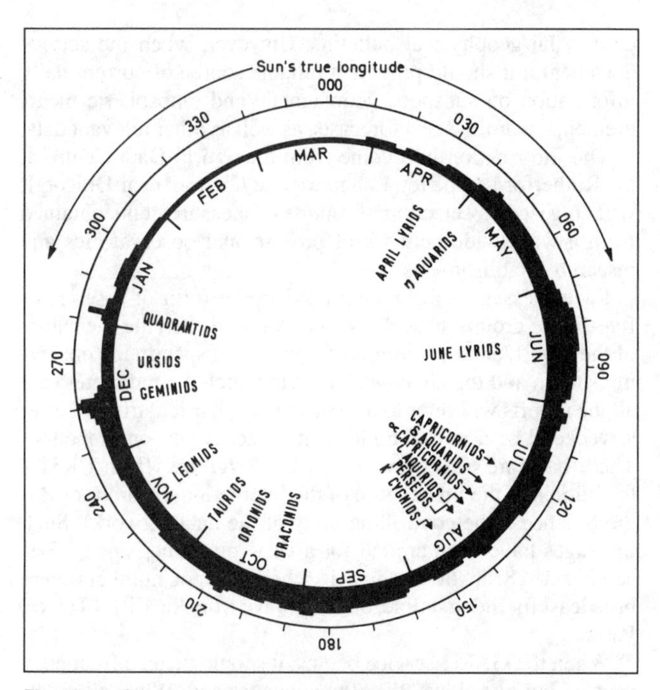

Fig 11.19. Seasonal variation of meteor activity, based on a daily relative index. Prepared from tables of 24h counts made by Dr Peter Millman, National Research Council, Ottawa. The maximum rate corresponds to an average of about 300 echoes per hour corresponding to an equivalent visual magnitude of 6 or greater

due to self-exciting dynamo action in the molten metallic core of the Earth.

The field originating outside the Earth is weaker and very variable, but it may amount to more than 5% of the main field in the auroral zones, where it is strongest. It fluctuates regularly in intensity according to annual, lunar and diurnal cycles, and irregularly with a complex pattern of components down to *micropulsations* of very short duration.

Certain observatories around the world are equipped with sensitive *magnetometers* which record changes in the field on at least three different axes, the total field being a vector quantity having both magnitude and direction. In the aspect of analysis which is of interest to us the daily records, called *magnetograms,* are read-off as eight *K-indices,* which are measures of the highest positive and negative departures from the 'normal' daily curve during successive threehourly periods, using a quasi-logarithmic scale ranging from 0 (quiet) to 9 (very disturbed). The various observatories do not all use the same scale factors in determining *K*-indices; the values are chosen so as to make the frequency distributions similar at all stations. Most large magnetic disturbances are global in nature and appear almost simultaneously all over the world. Another but somewhat similar indicator is the *planetary index* K_p which is graded on a finer scale in thirds thus: 0, 1–, 1o, 1+, 2–, 2o, 2+, . . . 9–, 9o, 9+. It is formed by a combination of *K*-figures from 12 selected observatories. Indices of 5 or more may be regarded as being indicative of magnetic storm conditions.

K- and K_p-indices are based on quasi-logarithmic scales which place more emphasis on small changes in low activity than high. For some purposes it is more convenient to work with a linear scale, particularly if the values are to be combined to derive averages, as of the day's activity, for example. This is often expressed as an *A-figure*, which records the *daily equivalent amplitude* on a linear scale which runs from 0 to 400.

There is a tendency for occasions of abnormally high geomagnetic activity (and in consequence auroral activity at VHF and UHF) to recur at intervals of approximately 27 days, linked to the *solar synodic rotation period.* The chart shown in Fig 11.20 clearly shows some long-persistent activity periods over the two-year interval 1974/75. A blank chart showing these co-ordinates to cover the current year with an overlap (known as a *solar rotation base map)* is published in the data section of the *RSGB Amateur Radio Call Book.*

The original diagram on which it is based was prepared by plotting the highest *K*-figure for each day on the spot determined by the longitude of the Sun's central meridian facing the Earth

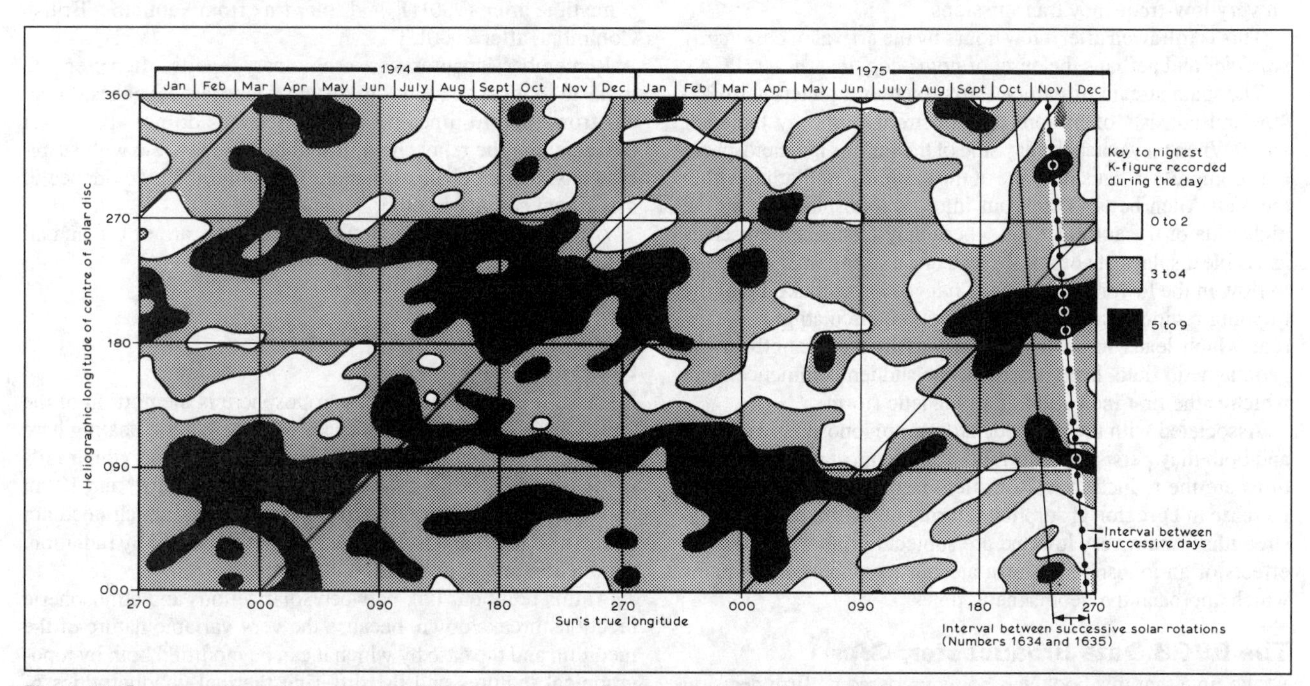

Fig 11.20. Geomagnetic activity diagram. The black areas indicate sequences where a *K*-figure of 5 or more was recorded at Lerwick Observatory. A horizontal line on this diagram denotes a recurrence period of 27 days, linked to the Sun's rotation period as seen from the Earth; the diagonal lines show the slope associated with a 25-day rotation period, such as the Sun has in relation to a fixed point in space

(thus a measure of the solar rotation), and a parameter called the *Sun's true longitude*, which indicates the position of the Earth around its orbit. Successive rotations build up a raster of daily figures in the way shown by the dots along the sloping right-hand edge, and the resulting chart should really be considered as being cylindrical, with the upper and lower edges brought together. The black areas surround the days when a *K*-figure of five or more was recorded – magnetic storm days – and the unshaded areas enclose relatively quiet days when the *K*-index was two or less. 27-day recurrences are clearly marked on this section of the record but occasionally there are periods when there is a marked tendency for storms to recur after an interval which appears to be linked to the Sun's rotation period relative to the stars – indicated by the slope of the diagonal across the diagram. A good example of this occurred in 1971/72.

Periods of high geomagnetic activity are usually centred around the time of the equinoxes, with relatively quiet periods near the solstices. However, a *sudden commencement* (SC) *is* likely to occur at any time, linked to an event on the Sun. There is an 11-year cycle in activity which tends to lag the solar cycle by a year or so.

Sudden ionospheric disturbances

We are now in a position to review the sequence of events which begins with the occurrence of a suitably positioned major flare on the Sun. When this takes place there is an emission of electromagnetic radiation covering a very wide range – X-rays, ultraviolet, visible light and radio waves between 3cm and 10m in length – all of which reaches the Earth in about eight minutes. The X-rays and UV light cause immediate increases in the D-layer ionisation, leading to *short-wave* (or *Dellinger*) *fade-outs* which may persist for anything up to two hours. They frequently affect the E-layer also and occasionally the F-layer. At this time some magnetometers in high latitudes may record a short fluctuation known as a *crotchet.* Other effects observed are a *sudden enhancement of atmospherics* (SEA), a *sudden absorption of cosmic noise* (SCNA), and *sudden phase anomalies* (SPA) on very low-frequency transmissions.

This is followed after a few hours by the arrival of *cosmic ray particles* and perhaps the onset of *polar-cap absorption* (PCA).

The main stream of particles arrives after an interval of 20–40h and consists of protons and electrons borne by the solar wind. When they reach the day side of the Earth's magnetosphere they compress it, causing some of the particles oscillating within the Van Allen belts to spill out into the ionosphere along the night side of the auroral zone, where they manifest themselves as visible displays of aurora, also causing a strong *polar electrojet* to flow in the lower ionosphere. Changes in the make-up of the trapping regions leads to variations in the circulating ring-current which leads to violent alterations in the strength of the geomagnetic field, bringing about the sudden commencement, which is the first indication of a magnetic storm.

Associated with the magnetic storms are ionospheric storms, and both may persist for several days. The most prominent features are the reduction in F2 critical frequencies (f_0F2) and an increase in D-region absorption. During the storm period signal strengths remain very low and are subject to flutter fading. The effects of an ionospheric storm are most pronounced on paths which approach the geomagnetic poles.

The RSGB Data Broadcaster, GAM1

At the time that this book was being prepared no firm decision had been made either as to the starting date for the RSGB Data Broadcaster transmissions on GAM1, or to the content of the daily solar/geophysical bulletins. However, when the service does begin, it should provide a reliable source of current daily information on sunspots, geomagnetic and ionospheric measurements, warnings and forecasts, as well as other relevant data.

The information will come from the World Data Centre at the Rutherford Appleton Laboratory, at Chilton, near Didcot. It will consist of a selection of reports of measurements obtained from a worldwide network of professional observatories and research establishments.

Each message will be circulated in the form of a series of five-figure groups, headed by a key word identifying the nature of the report, and two groups of figures that specify the observing station and the Greenwich date to which the data refer (not all the reports will refer to the current day, particularly just after a weekend or a public holiday, not necessarily one of ours.) These basic messages are known as *Ursigrams* (from URSI – the initials of the French title of the International Union for Radio Science – the controlling body of the data network). Such messages have been around for a very long time, since 1 December 1928, in fact, which is when sunspot numbers were broadcast for the first time, on long-wave from the Eiffel Tower, Paris.

When the GAM1 service begins, it should be on a frequency near to, but just above, the 80m amateur band. When all is settled, full details will be given in the Sunday morning RSGB news bulletins and the codes in use will be explained in a leaflet that will be available from the Society's HQ.

It will be important to recognise that the daily sunspot numbers from Boulder, Meudon and Catania will all be much higher than the final figures which will emerge, a month at a time, from the Sunspot Index Data Centre, Brussels, because no weighting factors will have been applied to them. As a very rough guide, however, taking two-thirds of the reported figures should provide something comparable.

Note also that sunspot numbers and solar flux measurements for a given day are always made at local noon. That means that the Boulder figure for today can only be available in the UK at some time after 1900UT, and solar flux from Penticton, British Columbia, after 2000UT.

Remember, Ursigrams are preliminary reports, often prepared in haste to meet deadlines in distant countries. Mistakes do occur from time to time and later corrections do not always get passed along the relay chain. It will be wise to view with suspicion any sudden changes that are not supported by evidence of associated activity in other parameters.

All times given in the Ursigram messages are in UT and all dates are Greenwich dates.

TROPOSPHERIC PROCESSES

Introduction

Because it is all around us the troposphere is the portion of the atmosphere which we ought to know best. We are dealing here with the Sun's output of electromagnetic radiation which falls in the infrared portion of the spectrum, between 10^{-6} and 10^{-5}m wavelength, is converted to heat (by processes which need not concern us here) and is distributed about the world by radiation, conduction and convection.

At this point our link between solar actions and atmospheric reactions breaks down, because the very variable nature of the medium, and the ease by which it can be modified both by topographical features and the differing thermal conductivities of land and sea, leads to the development of air masses having such widely contrasting properties that it becomes impossible to find a direct correlation between day-to-day climatic features and solar emissions. We must accept the fact that in meteorology 'chance' plays a powerful role and look to functions of the resulting weather pattern for any relationships with signal level, without enquiring too deeply into the way in which they may be connected with events on the Sun.

Pressure systems and fronts

The television weatherman provides such a regular insight into the appearance and progressions of surface pressure patterns that it would be wasteful of space to repeat it all here. Suffice it to record that there are two closed systems of isobars involved, known as *anticyclones* and *depressions* (or, less-commonly nowadays, *cyclones)* within or around which appear *ridges* of high pressure, *troughs* of low pressure (whose very names betray their kinship), and *cols,* which are slack regions of even pressure, bounded by two opposing anticyclones and two opposing depressions.

The most important consideration about these pressure systems, in so far as it affects radio propagation at VHF and above, is the direction of the vertical motion associated with them.

Depressions are closed systems with low pressure at the centre. They vary considerably in size, and so also in mobility, and frequently follow one another in quick succession across the North Atlantic. They are accompanied by circulating winds which tend to blow towards the centre of the system in an anticlockwise direction. The air so brought in has to find an outlet, so it rises, whereby its pressure falls, the air cools and in doing so causes the relative humidity to increase. When saturation is reached, cloud forms and further rising may cause water droplets to condense out and fall as rain. Point one: depressions are associated with rising air.

Anticyclones are generally large closed systems which have high pressure in the centre. Once established they tend to persist for a relatively long time, moving but slowly and effectively blocking the path of approaching depressions which are forced to go round them. Winds circulate clockwise, spreading outwards from the centre as they do, and to replace air lost from the system in this way there is a slow downflow called *subsidence* which brings air down from aloft over a very wide area. As the subsiding air descends its pressure increases, and this produces dynamical warming by the same process which makes a bicycle pump warm when the air inside it is compressed. The amount of water vapour which can be contained in a sample of air without saturating it is a function of temperature, and in this particular case if the air was originally near saturation to begin with, by the time the subsiding air has descended from, say, 5km to 2km, it arrives considerably warmer than its surroundings and by then contains much less than a saturating charge of moisture at the new, higher, temperature. In other words, it has become warm and dry compared to the air normally found at that level. Point two: anticyclones are associated with descending air.

In addition to pressure systems the weather map is complicated by the inclusion of *fronts,* which are the boundaries between two air masses having different characteristics. They generally arrive accompanied by some form of precipitation, and they come in three varieties: *warm, cold* and *occluded.*

Warm fronts (indicated on a chart by a line edged with rounded 'bumps' on the forward side) are regions where warm air is meeting cold air and being forced to rise above it, precipitating on the way.

Cold fronts (indicated by triangular 'spikes' on the forward side of a line) are regions where cold air is undercutting warm. The front itself is often accompanied by towering clouds and heavy rain (sometimes thundery), followed by the sort of weather described as 'showers and bright intervals'.

Occluded fronts (shown by alternate 'bumps' and 'spikes') are really the boundary between three air masses being, in effect, a cold front which has overtaken a warm front and one or the other has been lifted up above the ground.

It is perhaps unnecessary to add that there is rather more to meteorology than it has been possible to include in this brief survey.

Vertical motion

It is a simple matter of observation that there is some correlation between VHF signal levels and surface pressure readings, but it is generally found to be only a coarse indicator, sometimes showing little more than the fact that high signal levels accompany high pressure and low signal levels accompany low pressure. The reason that it correlates at all is due to the fact that high pressure generally indicates the presence of an anticyclone which, in turn, heralds the likelihood of descending air.

The reason that subsidence is so important stems from the fact that it causes dry air to be brought down to lower levels where it is likely to meet up with cool moist air which has been stirred up from the surface by turbulence. The result then is the appearance of a narrow boundary region in which refractive index falls off very rapidly with increasing height – the conditions needed to bring about the sharp bending of high-angle radiation which causes it to return to the ground many miles beyond the normal radio horizon. Whether you regard this in the light of being a benefit or a misfortune depends on whether you are more interested in long-range communication or in wanting to watch an interference-free television screen.

The essential part of the process is that the descending air must meet turbulent moist air before it can become effective as a boundary. If the degree of turbulence declines, the boundary descends along with the subsiding air above it, and when it reaches the ground all the abnormal conditions rapidly become subnormal – a sudden drop-out occurs. Occasionally this means that operators on a hill suffer the disappointment of hearing others below them still working DX they can no longer hear themselves. Note, however, that anticyclones are not uniformly distributed with descending air, nor is the necessary moist air always available lower down, but a situation such as a damp foggy night in the middle of an anticyclonic period is almost certain to be accompanied by a strong boundary layer. Ascending air on its own never leads to spectacular conditions. Depressions therefore result in situations in which the amount of ray bending is controlled by a fairly regular fall-off of refractive index. The passage of warm fronts is usually accompanied by declining signal strength, but occasionally cold fronts and some occlusions are preceded by a short period of enhancement.

To sum up, there is very little of value about propagation conditions which can be deduced from surface observations of atmospheric pressure. The only reliable indicator is a knowledge of the vertical refractive index structure in the neighbourhood of the transmission path.

Radio meteorological analysis

It remains now to consider how the vertical distribution of refractive index can be displayed in a way which gives emphasis to those features which are important in tropospheric propagation studies. Obviously the first choice would be the construction of atmospheric cross-sections along paths of interest, at times when anomalous conditions were present, using values calculated by the normal refractive index formula. The results are

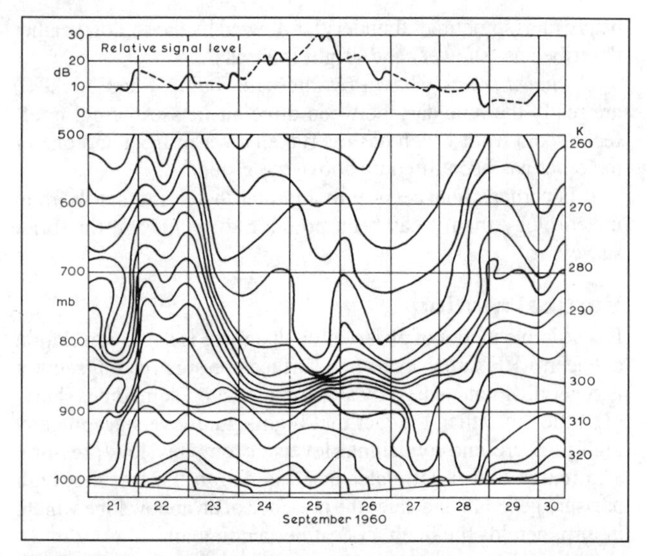

Fig 11.21. The relationship between variations in potential refractive index in the atmosphere and signal strengths over a long-distance VHF tropospheric path. (With acknowledgements to *J Atmos Terr Phys*, Pergamon Press.)

often disappointing, however, because the general decrease of refractive index with height is so great compared to the magnitude of the anomalies looked for that, although they are undoubtedly there, they do not strike the eye without a search.

A closely-related function of refractive index overcomes this difficulty, with the added attraction that it can be computed graphically and easily, directly from published data obtained from upper-air meteorological soundings. It is called *potential refractive index K* and may be defined as being the refractive index which a sample of air at any level would have if brought *adiabatically* (ie without gain or loss of heat or moisture) to a standard pressure of 1000mb; see references [4] and [5].

This adiabatic process is the one which governs (among other things) the increase of temperature in air which is descending in an anticyclone, so that, besides the benefits of the normalising process (which acts in a way similar to that whereby it is easy to compare different-sized samples of statistics when they have all been converted to percentages) there is the added attraction that the subsiding air tends to retain its original value of potential refractive index all the time it is progressing on its downward journey. This means that low values of *K* are carried down with the subsiding air, in sharp contrast to the values normally found there. A cross-section of the atmosphere during an anticyclonic period, drawn up using potential refractive index, gives an easily-recognisable impression of this.

Fig 11.21 is not a cross-section, but a time-section, showing the way in which the vertical potential refractive index distribution over Crawley, Sussex, varied during a 10-day period in September 1960. There is no mistaking the downcoming air from the anticyclone and the establishment of the boundary layer around 850mb (about 1.5km). Note how the signal strength of the Lille television transmission on 174MHz varied on a pen recording made near Reading, Berkshire, during the period, with peak amplitudes occurring around the time when the layering was low and well-defined, and observe also the marked decline which coincided with the end of the anticyclonic period. Time-sections such as these also show very clearly the ascending air in depressions (although the *K* value begins to alter when saturation is reached) and the passage of any fronts which happen to be in the vicinity of the radiosonde station at ascent time.

For anyone interested in carrying out radiometeorological analysis at home – and, be assured, it is a very rewarding exercise in understanding the processes involved – the propagation chapter in the fourth edition of the RSGB *VHF/UHF Manual* will provide full details (see the bibliography at the end of this chapter).

A visit to the National Meteorological Library, London Road, Bracknell, Berkshire, will provide you with any basic upper air data that you may need for your analysis, but you will of course have to do the conversions to refractive index yourself. If you need current data on a regular basis you can get it on your home computer, using a suitable interface between it and your HF receiver.

At the time of writing, meteorological data broadcasts in the UK are undergoing sweeping changes. In September 1993 the former GFA and GFE, Bracknell, facsimile services merged into a single broadcast on 3289.5, 9040 and 11,086.5kHz, with a split schedule on 4610 (1800–0600) and 14,582 (0600–1800). Other facsimile broadcasts originate from Northwood on 2374, 3652, 4307, 6446, 8331.5, 12,884.5 and 16,912kHz, and from Offenbach, Germany, on 134.2, 3855, 7880 and 13,882.5kHz. Some, or all, of the latter may be transferred to satellite as early as 1994.

A publication entitled *Meteorological Facsimile Broadcasts*, costing around £10 a copy and updated by a series of two-monthly supplements costing about £12 a year, may be obtained from the World Meterological Association, Case Postale 2300, CH-1211, Geneva, Switzerland, who will quote exact prices (in Swiss francs) on request.

The future of UK radio teleprinter broadcasts is uncertain after April 1994, when the Meteorological Office hands over responsibility for them to the Ministry of Defence. At present the transmissions are made from GFL on 4489, 10,551.3 and 14,356kHz, with a split service on 6835 (1800–0600) and 18,230kHz (0600–1800). Some cuts in the service are possible because the Meteorological Office no longer has an internal need for RTTY broadcasts.

If you are interested in meteorological broadcasts and have more than a passing interest in meteorology, you should consider becoming a member of the Royal Meteorological Society (104 Oxford Road, Reading, RG11 7LJ). Its monthly *Weather Log*, which comes as a regular supplement to the magazine *Weather*, will provide you with a miniature North Atlantic and European surface chart for every day of the year. For an additional nominal sum to cover copying and postal charges you can subscribe to a newsletter which provides details of the content of UK and other European meteorological broadcasts.

PRACTICAL CONSIDERATIONS

Map projections

Maps are very much a part of the life of a radio amateur, yet how few of us ever pause to wonder if we are using the right map for our particular purpose, or take the trouble to find out the reason why there are so many different forms of projection.

The cartographer is faced with a basic problem, namely that a piece of paper is flat and the Earth is not. For that reason his map, whatever the form it may take, can never succeed in being faithful in all respects – only a globe achieves that. The amount by which it departs from the truth depends not only on how big a portion of the globe has been displayed at one viewing, but on what quality the mapmaker has wanted to keep correct at the expense of all others.

Projections can be divided into three groups: those which show areas correctly, described, logically enough, as *equal-area projections*; those which show the shapes of small areas correctly, known as *orthomorphic* or *conformal projections*; and those which represent neither shape nor area correctly, but which have some other property which meets a particular need.

The conformal group, useful for atlas maps generally, weather charts, satellite tracks etc includes the following:

Stereographic, where latitudes and longitudes are all either straight lines or arcs of circles, formed by projection on to a plane surface tangent either to one of the poles *(polar),* the equator *(equatorial),* or somewhere intermediate *(oblique).* Small circles on the globe remain circles on the map but the scale increases with increasing radius from the centre of projection.

Lambert's conformal conic, where all meridians are straight lines and all parallels are circles. It is formed by projection on to a cone whose axis passes through the Earth's poles.

Mercator's, where meridians and parallels are straight lines intersecting at right-angles. The meridians are equidistant, but the parallels are spaced at intervals which rapidly increase with latitude. It is formed by projection on to a cylinder which touches the globe at the equator. Any straight line is a line of constant bearing (a *rhumb line,* not the same thing as a *great circle,* which is the path a radio wave takes between two given points on the Earth's surface). There is a scale distortion which gets progressively more severe away from the equator, to such an extent that it becomes impossible to show the poles, but most people accept these distortions as being normal, because this is the best-known of all the projections.

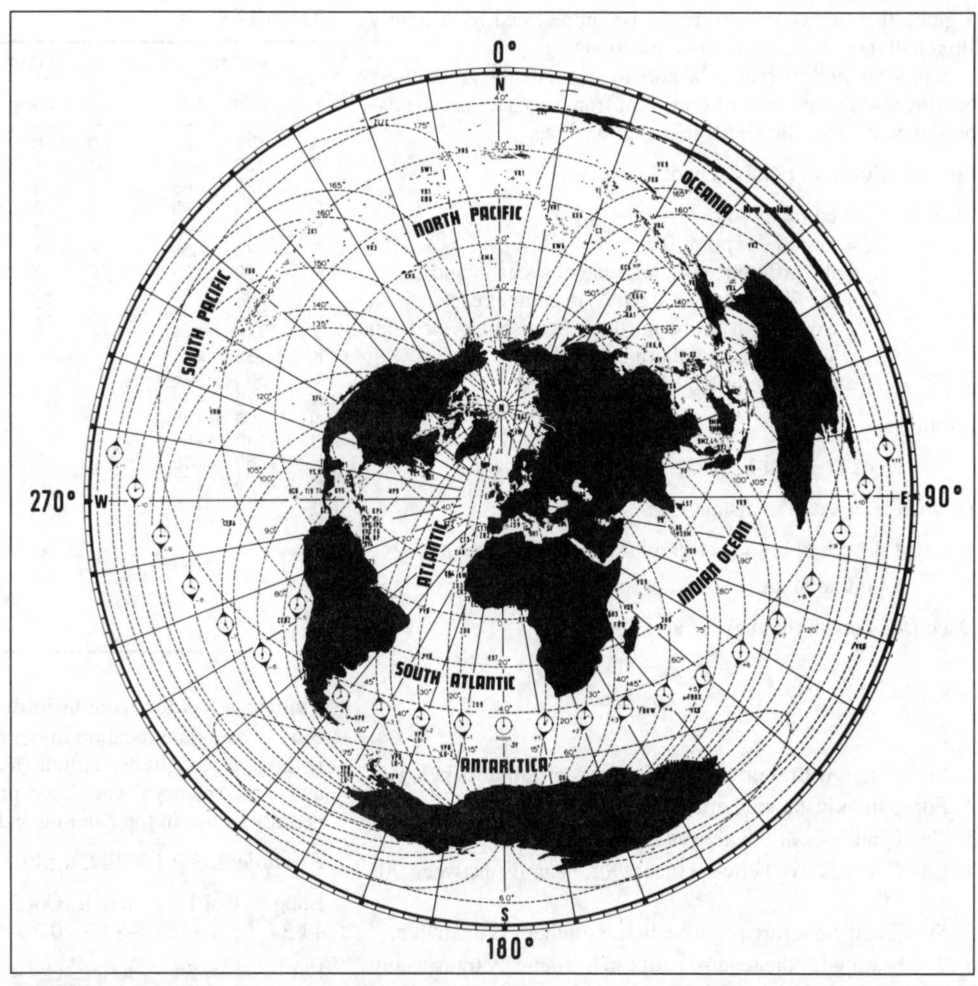

Fig 11.22. Example of an azimuthal equidistant (or *great-circle*) map. This map, available in a large size suitable for wall-mounting, shows the true bearing and distance from London of any place elsewhere in the world. (For magnetic bearings add 6° to the true bearing)

Transverse Mercator is a modification of the 'classical' system, and is formed by projection on to a cylinder which touches the globe along selected opposing meridians. It therefore corresponds to an ordinary Mercator turned through 90°, and is of value for displaying an area which is extensive in latitude but limited in longitude. A variant is the *universal transverse Mercator,* which forms the basis of a number of reference grids, including the one used on British Ordnance Survey maps.

The equal-area group is used when it is necessary to display the relative distribution of something, generally on a worldwide scale. It includes the following.

Azimuthal equal-area projection, having radial symmetry about the centre, which may be at either pole *(polar),* at the equator *(equatorial),* or intermediate *(oblique).* With this system the entire globe can be shown in a circular map, but there is severe distortion towards its periphery.

Mollweide's homolographic projection, where the central meridian is straight and the others elliptical.

Sinusoidal projection, where the central meridian is straight and the others parts of sine curves.

Homolosine projection, which is a combination of the previous two, with an irregular outline because of interruptions which are generally arranged to occur over ocean areas.

The final group includes the two following, which are of particular interest in propagation studies.

Azimuthal equidistant, centred on a particular place, from whence all straight lines are great circles at their true azimuths. The scale is constant and linear along any radius. Well known as a *great-circle map,* Fig 11.22.

Gnomonic, constructed by projection from a point at the Earth's centre on to a tangent plane touching the globe. Any straight line on the map is a great circle. Because the size of the map expands very rapidly with increasing distance from the centre they do not normally cover a large area. Often produced as a skeleton map on which a great circle can be drawn and used to provide a series of latitudes and longitudes by which the path can be replotted on a more detailed map based on a different projection.

Great-circle calculations

The shortest distance between two points on the surface of the Earth lies along the great circle which passes through them. On a globe this great-circle path can be represented by a tightly stretched thread joining the two locations.

It is sometimes useful to be able to calculate the great-circle bearing and the distance of one point from another, and the expressions which follow enable this to be done.

First label the two points A and B.

Then let L_a = latitude of point A
L_b = latitude of point B
L_0 = the difference in longitude between A and B
C = the direction of B from A, in degrees east or west from north in the northern hemisphere, or from south in the southern hemisphere.
D = the angle of arc between A and B.

It follows that:

$$\cos D = \sin L_a.\sin L_b + \cos L_a.\cos L_b.\cos L_0$$

D can be converted to distance, knowing that:

1 degree of arc = 111.2km or 69.06 miles

1 minute of arc = 1.853km or 1.151 miles

Once D is known (in angle of arc), then:

$$\cos C = \frac{\sin L_b - \sin L_a.\cos D}{\cos L_a.\sin D}$$

Note:

1. For stations in the northern hemisphere call latitudes positive.
2. For stations in the southern hemisphere call latitudes negative.
3. Cos L_a and cos L_b are always positive.
4. Cos L_0 is positive between 0 and 90°, negative between 90° and 180°.
5. Sin L_a and sin L_b are negative in the southern hemisphere.
6. The bearing for the reverse path can be found by transposing the letters on the two locations.

It is advisable to make estimates of the bearings on a globe, wherever possible, to ensure that they have been placed in the correct quadrant.

The IARU Locator

It is always nice to know where the other fellow lives, and radio amateurs have been asking the question ever since the earliest days of operating. But all too often the answer has been something like "16km south of (some place that your atlas does not regard as worthy of a mention)" and that is a *very* unsatisfactory state of affairs if points in a contest are riding on the result.

In the early 'sixties amateurs on the Continent came up with a suggestion for a locator code based on latitude and longitude; it found a great deal of favour, despite the fact that it had been put together unsatisfactorily.

Then, as time went on, the International Amateur Radio Union took charge of developing the idea for worldwide use and the result consists of a group of six characters that define position to 0.04° of latitude and 0.08° of longitude.

Maps showing the relationship between the code and its corresponding territory are available but Table 11.9 will do the job for you just as well. One thing, though – unless you are really happy about adding and subtracting a lot of numbers that may be either positive or negative, you might be well advised to let your calculator sort them out for you. The numbers themselves are no problem, of course, but a mistake somewhere with a sign can, quite literally, make a world of difference to the answer!

Table 11.9

	Letters 1 Long	2 Lat		Figures 3 Long	4 Lat		Letters 5 Long	6 Lat
A	−170	−85	0	−9	−4.5	A	−0.96	−0.48
B	−150	−75	1	−7	−3.5	B	−0.88	−0.44
C	−130	−65	2	−5	−2.5	C	−0.79	−0.40
D	−110	−55	3	−3	−1.5	D	−0.71	−0.35
E	−90	−45	4	−1	−0.5	E	−0.63	−0.31
F	−70	−35	5	+1	+0.5	F	−0.54	−0.27
G	−50	−25	6	+3	+1.5	G	−0.46	−0.23
H	−30	−15	7	+5	+2.5	H	−0.38	−0.19
I	−10	−5	8	+7	+3.5	I	−0.29	−0.15
J	+10	+5	9	+9	+4.5	J	−0.21	−0.10
K	+30	+15				K	−0.13	−0.06
L	+50	+25				L	−0.04	−0.02
M	+70	+35				M	+0.04	+0.02
N	+90	+45				N	+0.13	+0.06
O	+110	+55				O	+0.21	+0.10
P	+130	+65				P	+0.29	+0.15
Q	+150	+75				Q	+0.38	+0.19
R	+170	+85				R	+0.46	+0.23
						S	+0.54	+0.27
						T	+0.63	+0.31
						U	+0.71	+0.36
						V	+0.79	+0.40
						W	+0.88	+0.44
						X	+0.96	+0.48

Finding a locator from latitude and longitude

First express the location in degrees and decimals of a degree, in the order longitude, latitude (north and east are positive, south and west are negative). Then proceed as below, entering the nearest values in the columns indicated.

Example: Long 1.37E, Lat 51.03N

Long	Col 1	deficit	Col 3	deficit	Col 5	Check
+1.37	J = +10	−8.63	0 = −9	+0.37	Q = +0.38	+1.38

Lat	Col 2	deficit	Col 4	deficit	Col 6	Check
+51.03	O = +55	−3.97	1 = −3.5	−0.47	A = −0.48	+51.02

The checks are totals of the figures shown against the locator characters, following column numbers.

Taking the characters in column order, the required locator is JO01QA.

Finding latitude and longitude from a locator

Longitude is obtained from the first, third and fifth characters of the group, latitude from the second, fourth and sixth characters. Let us consider IO82XE.

Look up, in the table, the number corresponding to each of the characters in its appropriate column, listing them in pairs, then adding:

−10	+55
+7	−2.5
+0.96	−0.31
−2.04	+52.19

The first quantity is the required longitude, 2.04°W, the second is the latitude, 52.19°N.

The QTH Locator

The old European QTH Locator (originally known as the *QRA Locator*) still has its uses, despite the fact that it has been superseded by the IARU Locator which has a worldwide application. One reason for this is that its large squares (which are all that are used in most research projects) cover exactly the same ground as the similar-sized squares of the new locator. Another is that

Table 11.10

Longitude				Latitude			
U	11W	H	15E	A	40½N	N	53½N
V	09W	I	17E	B	41½N	O	54½N
W	07W	J	19E	C	42½N	P	55½N
X	05W	K	21E	D	43½N	Q	56½N
Y	03W	L	23E	E	44½N	R	57½N
Z	01W	M	25E	F	45½N	S	58½N
A	01E	N	27E	G	46½N	T	59½N
B	03E	O	29E	H	47½N	U	60½N
C	05E	P	31E	I	48½N	V	61½N
D	07E	Q	33E	J	49½N	W	62½N
E	09E	R	35E	K	50½N	X	63½N
F	11E	S	37E	L	51½N	Y	64½N
G	13E	T	39E	M	52½N	Z	65½N
First letter				Second letter			

those squares are identified by two characters instead of four, which leads to a great saving of space where extensive listings are concerned. But perhaps the most compelling reason is that the research people who use them most cannot be persuaded to give them up!

For those reasons the QTH Locator table has been preserved here, albeit now in a skeleton form (Table 11.10). The mid-square longitudes extend ±1° without change of letter, the mid-square latitudes ±0.5°. Some operators extend the latitude scale to cover the countries bordering on the Mediterranean coast of North Africa by repeating the alphabet from Z to Q, or even beyond; any resulting ambiguity is easily resolved by inspecting the callsigns, but this point needs to be watched if your computer is ever set the task of sorting locations in order of QTH locator squares.

Plotting path profiles for VHF/UHF working

For a tropospheric propagation study, it is standard practice to construct a path profile showing the curvature of the Earth as though its radius was four-thirds of its true value. This is so that, under standard conditions of refraction, the ray path may be shown as a straight line relative to the ups and downs of the intervening terrain.

It will be found convenient to construct the baseline of the chart using feet for height and miles for distance, for the reason that, by a happy coincidence, the various factors then cancel, leaving a very simple relationship that demands little more than mental arithmetic to handle.

Suppose that a profile is required for a given path 40 miles in length. First, decide on suitable scales for your type of graph paper (Fig 11.23 was drawn originally with 1in on paper representing 100ft, height, and 10 miles, distance). Then construct a sea-level datum curve, taking the centre of the path as zero and working downwards and outwards for half the overall distance in each direction, using the expression $h = D^2/2$ to calculate points on the curve (Table 11.11). Remember that this works only if h is in feet and D is in miles.

Then prepare a height scale. This will be the construction scale reversed if the contours of your map are in feet, or its metric equivalent if it is one of the more recent surveys. Usually it will be sufficient to limit this scale to just the contour values encountered along the path. Plot the heights, corresponding to the contours, vertically above the datum curve (a pair of dividers may be found helpful here) and add the antenna heights at the two terminal points. Draw a line through all the points and you are ready for business.

If you can draw a straight line from transmitting antenna to receiving antenna without meeting any obstacles along the way,

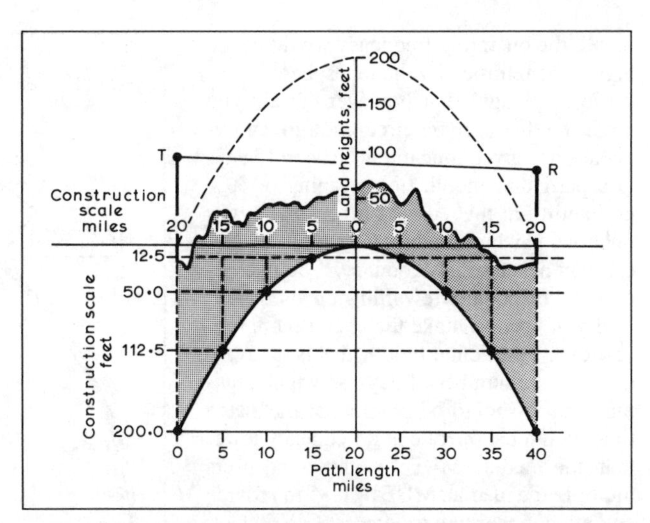

Fig 11.23. Construction of a 'four-thirds-earth' profile from data in Table 11.11 below. Land heights are measured upward from the lower curve. On this diagram, rays subjected to 'normal' variations of refractive index with height may be represented by straight lines

Table 11.11

D (miles)	Horizontal scale (inches)	D^2	$D^2/2 = h$	Vertical scale (inches)
5	0.5	25	12.5	0.125
10	1.0	100	50.0	0.500
15	1.5	225	112.5	1.125
20	2.0	400	200.0	2.000

then your path should be clear under normal conditions of refraction.

Ionospheric predictions

The quality of a radio circuit is highest when it is operated at a frequency just below the maximum usable frequency (MUF) for the path. Three regions, E, F1 and F2, are considered in the determination of MUF.

For the F2-layer, which is responsible for most HF long-distance contacts, the MUF for paths less than 4000km is taken as being the MUF that applies to the mid-point. For paths longer than 4000km, the MUF for the path is the lower of the MUFs at the two ends along the path direction. For this purpose the end point locations are not those of the terminals, but of their associated *control points* where low-angle radiation from a transmitter would reach the F2-layer. Those points are taken to be 2000km from each terminal, along the great circle joining them.

To put this into perspective, a control point for a station located in the midlands of England would lie somewhere above a circle passing close by Narvik, Leningrad, Minsk, Budapest, Sicily, Algiers, Tangier, mid-Atlantic and NW Iceland, the place depending on the direction of take-off. It means that (for example) the steep rise in MUF associated with UK-end sunrise will occur something like four hours earlier on a path to the east than on a path to the west.

Taken over a month, the day-by-day path MUF at a given time can vary over a considerable range. A ratio of 2:1, as between maximum and minimum, could be considered as typical. Monthly predictions are based on median values, that is to say, on those values which have as many cases above them as below. Therefore the median MUFs in predictions represent 50% probabilities because, by definition, for half of the days of the month the operating frequency would be too high to be returned by the ionosphere.

Note, though, that however certain you might be that a given circuit at a given frequency at a given time might be open 15 days in a particular month (the meaning of 50% probability in this context), there is no way of knowing which 15 days they might be. Nor can you be sure that 'good days' on one circuit will be equally rewarding on another.

If you want to make the circuit more reliable over a particular month, that is to say, to increase the number of days on which communication should be possible at the stated time, you must operate at a frequency lower than the median MUF . Commercial users multiply the median MUF by 0.85 to provide an *optimum working frequency* (OWF), having a theoretical reliability of 90%, representing 27 days in a month. At the other end of the scale, some amateur operators in search of rare DX would regard a 10% probability (three days a month) as worthy of interest. Their figure is the *highest path frequency* (HPF), obtained by dividing the median MUF by 0.85.

All this may seem to suggest that the lower the operating frequency, the greater would be the chance of success, but that is not true. The reason is that the further down from the median MUF that one operates, the greater become the losses due to absorption, and eventually these become the dominant factor. For any given path there is a *lowest useful frequency* (LUF) at which those losses become intolerable.

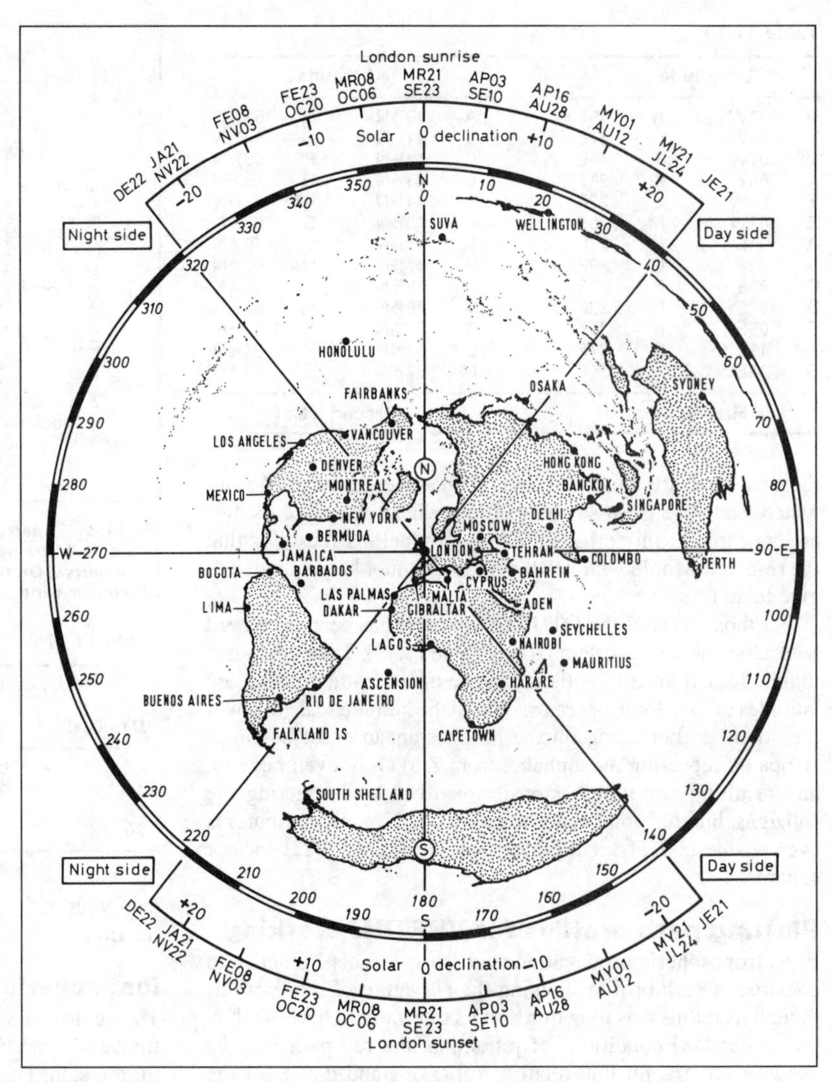

Fig 11.24. Grey-line propagation map

Unfortunately it is beyond the scope of this book to be able to provide detailed information about the preparation of monthly predictions and about how to relate the values to the amateur bands. However, each issue of *Radio Communication* contains a table giving current data in a form that will satisfy the needs of most operators based in the UK.

Those predictions are unique in that the information is given in the form of percentage probability for each of the HF amateur bands, taking both the HPF and the LUF into account. A figure of 1 represents 10% (or three days a month), a figure of 5 represents 50% (or 15 days a month), and so on. Multiply by three the figure given in the appropriate column and row and you will have the expected number of days in the month on which communication should be possible at a given time on a given band.

Many operators believe that if it is known that solar activity is higher (or lower) than was expected when the predictions were prepared, then they can raise (or lower) all the probabilities by a fixed amount to compensate for the changed circumstances. That is not so. The highest probability will always appear against the band closest to the OWF (as defined in the predictions sense). If the OWF is altered by fresh information, then the probabilities on one side of it will rise but on the other side will fall. The only sure way of establishing amended figures would be to enter the revised particulars into the computer program and to run off a complete set of new predictions.

Any operator who requires predictions tailored to particular path preferences is advised to obtain a reliable prediction program for his or her computer, preferably one based on CCIR Report 340. But follow the input instructions carefully. Monthly prediction programs usually require an index of solar activity comparable to the 12-month smoothed relative sunspot number R_{12}, or to IF2, which is derived from ionospheric data. There is nothing to be gained by substituting the latest unsmoothed monthly sunspot figure. Also, the use of raw daily figures, such as those from Boulder or Meudon, in the expectation that they will yield meaningful daily ionospheric predictions, is a misunderstanding of the highest degree. The ionosphere does not respond to fluctuations of daily sunspot numbers. On a daily time scale signal performance is much more responsive to changes in the level of geomagnetic activity, but that is something that is unlikely to be one of the input requirements of your computer program.

The locations of places for which predictions are given every month in *Radio Communication* may be seen against their relative beam headings in the great-circle outline map of Fig 11.24.

Grey-line propagation

The grey line is the ground-based boundary around the world that separates day from night, sunlight from shadow.

Many operators believe, and with some justification, that signals beamed along the grey line near sunrise or sunset will reach distant locations that are also experiencing sunrise or sunset for relatively short periods of time when conventional predictions may appear to be pessimistic.

Noon is a north-south phenomenon; all places having the same longitude encounter it at the same instant of time. But the grey line runs north and south only at the equinoxes, in March and September. At all other times it cuts across the entire range of meridians and time zones.

Sunrise and sunset are the periods when MUFs may be expected to rise or fall through their greatest range of the day. Couple that with the fact that the one grey line represents sunrise (rising MUF) on one side of the world and sunset (falling MUF) on the other, spring or summer on one side of the equator and autumn or winter elsewhere, and you have a very strong prospect that favourable circumstances for propagation will occur somewhere along the line.

This mode is for the lower frequency bands, 1.8, 3.5 and 7MHz, because the normal house rules apply: the operating frequency has to be below the MUF for the path or the signals will pass through the ionosphere somewhere instead of being reflected.

On the ground the grey line may be considered to be a great circle, but a rather badly defined one because the Sun is not a point source of light – hence twilight time, of course.

To work the grey line you need to know three things: your latitude (atlas), sunrise/sunset times for your area (daily newspaper), and the declination of the Sun for the day in question (*Whittaker's Almanack* or the *British Astronomical Association's Handbook*).

The ground azimuth at sunrise is given by the expression

$$\text{sin azimuth} = \frac{\text{sin declination:}}{\text{cos latitude}}$$

and the ground azimuth at sunset by:

$$\text{sin azimuth} = \frac{-\text{ sin declination}}{\text{cos latitude}}$$

and their reciprocals. Those equations are the basis for the sunrise and sunset scales above and below the skeleton great-circle map, Fig 11.24. Interpolate for the appropriate date and lay a straight-edge right across the map so that it passes through the centre (London, in this case). That is your grey line and it should work in either direction, if it is to work at all. Your line will show which areas ought to be accessible if fortune smiles upon you.

That is the good news. The bad news is that it is not the ground-based shadow that determines the state of the ionosphere because the Earth's shadow is shaped like a cone; the area of darkness at F2 heights is appreciably less than has been considered in the preceding paragraph. Sunrise comes earlier than on the ground, sunset later. In fact, in mid-summer the F2-layer is in sunlight for all 24 hours of the day over the whole of the UK. So, the ionospheric grey line (as opposed to the ground-based grey line) cannot be considered as a great circle and, therefore, cannot be represented by the straight line on the great-circle map.

That should not stop you from trying your luck, however. But there is no point in trying to calculate the true outline of the Earth's shadow, because your signals are going to take the great-circle route no matter what you come up with on your computer. You may as well make all your plans using your ground-based data, because it is easy to come by, and make up for its likely deficiencies by being generous with your timing. At the sort of frequency you will be using, the beamwidth of the antenna will be wide enough to take care of direction.

The beacon network

In various parts of this chapter you will have found references to amateur propagation research and perhaps one day you may feel sufficient interest to contribute some effort to it yourself.

You may feel the need then for a selection of reliable high-standard signal sources on which to base your studies. You will find them among the very extensive network of amateur beacon stations which operate on an international basis, most of them for 24 hours a day. However, apart from the research aspect, the beacon network provides every operator with reliable signals that may be used to assess band conditions at any time.

Unfortunately it is not practical to include a beacon list here. For one thing there are too many beacons – at the time of writing there are over 400 of them between 50MHz and 24.192GHz, without considering HF at all – but, more importantly, the list could not be kept up to date. However, the RSGB can supply at any time a current computer printout through their book sales department. Check a recent issue of *Radio Communication* for details.

For beacons from 50MHz and upwards an alternative source of information is the winter (Nr 4) issue each year of *DUBUS Magazine* (see the bibliography at the end of this chapter for the address of the publisher).

GB2RS News

Finally, do not forget that the weekly RSGB news bulletins, broadcast every Sunday, always contain items of interest concerning propagation, including a summary of solar/geophysical events that occurred during the previous week. Times and frequencies depend on your location, and your local radio club will have the details. Or you could write to RSGB Headquarters, enclosing a stamped envelope addressed to yourself. Ask for the current GB2RS schedule.

REFERENCES

[1] 'Flare spot', P W Sollom, *Radio Communication* December 1970, p820; January 1971, p20; February 1971, p92.
[2] 'The ground beneath us', R C Hills, *RSGB Bulletin* June 1966, p375.
[3] Report 239-6 in *Propagation in non-ionised media*, Study Group 5, CCIR, XVIth Plenary Assembly, Dubrovnik, 1986, Geneva, ITU, 1986.
[4] *VHF/UHF Manual*, 4th edn, G R Jessop (ed), RSGB, 1983, Chapter 2 (Propagation).
[5] 'Patterns in propagation', R G Flavell, *Journal of the IERE*, Vol 56 No 6 (Supplement), pp175–184.

BIBLIOGRAPHY

General works

Physics, Parts 1 and 2, 3rd edn, D Halliday and R Resnick, John Wiley and Sons, London and Toronto, 1978.

Sun, Earth and Radio, J A Ratcliffe, World University Library, London, 1970.

The Sun, Iain Nicholson, Mitchell Beazley Publishers, 1982.

Meteorology for Seafarers, R M Frampton and P A Uttridge, Brown Son & Ferguson, Glasgow, 1988.

Propagation topics

Ionospheric propagation, Study Group 6, CCIR, ITU, Geneva (updated every four years).

La propagation des ondes, S Canivenc, F8SH, Soracom editions, Paris, 1984 (in French).

'Recent work on mid-latitude and equatorial sporadic E', J D Whitehead, *Journal of Atmospheric and Terrestrial Physics*, Vol 51, No 5, pp401–424, 1989.

Radio Auroras, C Newton, G2FKZ, RSGB, 1991.

Propagation in Non-ionized Media, Study Group 5, CCIR, ITU, Geneva (updated every four years).

Records of propagation events

At HF – *DX News Sheet* (weekly, by subscription), RSGB.

At VHF, UHF, SHF – *DUBUS Magazine* (quarterly, by subscription), DUBUS Verlagsgesellschaft, Postfach 500368, W-2000, Hamburg 50, Germany.

12 HF antennas

IN SETTING UP a link for radio communication between two stations, certain specific items of equipment must be provided at each end of the circuit. At the sending end there must be a transmitter which imposes the signal intelligence upon a carrier wave at radio frequency and amplifies it to the required power level. At the other end a receiver is required which will again amplify the weak incoming signal, and then decode from it the original intelligence.

The signal passes from one station to the other as a wave propagating in the atmosphere, but in order to achieve this it is necessary to have at the sending end something which will take the power from the transmitter and launch it as a wave, and at the other end extract energy from the wave to feed the receiver. This is an *antenna* and, because the fundamental action of an antenna is reversible, similar antennas can be used at both ends. The antenna then is a means of converting power flowing in wires to energy flowing in a wave in space, or is simply considered as a coupling transformer between the wires and free space.

FUNDAMENTAL PROPERTIES

Many of the fundamental properties of antennas are common to their use in any part of the radio frequency spectrum and in free space, ie when the influence of the ground and surrounding objects can be neglected, a piece of wire which is one half-wavelength long will have the same directional radiation characteristics and appear as exactly the same load to a transmitter whether it be a half-wavelength at 1MHz or at 1000MHz. All that matters is that the wire should maintain a constant relationship between its physical length and the wavelength used. Strictly speaking, the wire diameter should also be scaled with the wavelength since the ratio of length to diameter determines the sharpness of resonances, but this does not affect the basic radiating or receiving properties. The classification into antennas for HF and antennas for VHF in this handbook is one based on the differing practical requirements for antennas for use in the two frequency ranges, taking account of physical size limitations and also the widely differing mechanisms of propagation which dictate different forms of antenna. Much of what is to be said in this chapter about the fundamental characteristics of antennas is equally applicable to both HF and VHF, and it is only later in the chapter that consideration is given to antennas specifically designed for the HF bands. At that point the reader with VHF interests should pick up the story of specialised VHF antennas in Chapter 13 – 'VHF/UHF antennas'.

Wave motion

An understanding of antenna behaviour is very closely linked with an understanding of basic propagation, and some aspects cannot be dissociated from it.

In Chapter 11 it is explained that radio waves are propagated as an expanding electromagnetic wavefront, the intensity of which decays as it moves further out from the source. The wavefront is formed of electric and magnetic fields which exist at right-angles to one another in the plane of the wavefront, and each field may be represented by parallel lines of force having lengths proportional to the strength of the field. The intensity of the wavefront is usually varying sinusoidally with time, as the waves expand like ripples on a pond, and the peak value of the wave decays as the wave moves further and further away from its source. The arrangement of one cycle of the wave along its direction of travel at a particular moment in time is represented in Fig 12.1, as it appears in general, and also as it appears to the observer whom it is approaching.

Some imagination is necessary to appreciate fully the structure of the wave as it propagates through space as a whole. The behaviour of the separate but related electric and magnetic fields is more easily visualised if one imagines them as oscillations travelling along a slack string which is being excited at one end. An observer at a point along the string will see approaching him a succession of crests and troughs which pass him along the string. A complete electromagnetic wave is then two such strings identically located but oscillating at right-angles and with the crests on each string passing the observer at the same instant.

The distance between successive peaks in the intensity of the electromagnetic wave as it moves along its direction of propagation is called the *wavelength* and is customarily measured in metres. The wave moves through space with the velocity of light, or more exactly light moves through space with the velocity of electromagnetic waves since light is really electromagnetic radiation in a narrow band of extremely high frequencies. This velocity is approximately 3×10^8 metres per second (186,000 miles per second), and is known as the *velocity of propagation.* As stated above, from the point of view of a fixed observer, the wave will pass by as a succession of peaks and troughs. The portion of a wave between successive peaks (or troughs) is called one *cycle,* and the number of such cycles which pass the observer in one second is known as the *frequency* of the wave, usually measured in hertz (Hz), where one hertz denotes one cycle per second.

Since the wave travels with a known velocity, it is easy to

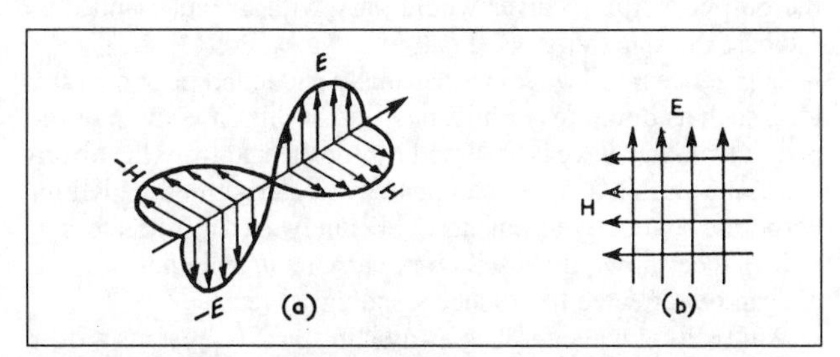

Fig 12.1. Instantaneous representation of a travelling radio wave: (a) along the path of travel and (b) for the wave approaching the observer

establish a relationship between the frequency and the wavelength, and if f is the frequency in hertz (Hz) and λ (lambda) the wavelength in metres, then

$$f \times \lambda = 3 \times 10^8$$

For example, a wavelength of 160 metres corresponds to a frequency of 1.87×10^6Hz or 1.87MHz.

Referring again to the wave illustrated in Fig 12.1, an observer facing the wave would 'see' the field as shown in Fig 12.1(b) which would be alternating at the frequency f. It is this alternating field that excites a receiving antenna into which it delivers some of its power. The arrows indicate the conventional directions of the electric and magnetic fields E and H, relative to the direction of motion. If the polarity of one component along either E or H in the diagram is reversed, then the wave is receding instead of approaching the observer. The components E and H cannot be separated; together they are the wave and together they represent indestructible energy. They wax and wane together and when the energy disappears momentarily from one point, it must reappear further along the track; in this way the wave is propagated.

Travelling and standing waves

The wave illustrated in Fig 12.1 is unrestricted in its motion and is called a *travelling wave*. As long as they are confined to the vicinity of the earth all waves must, however, eventually encounter obstacles. In order to understand what happens in these circumstances, it is convenient to imagine that the obstacle is a very large sheet of metal since metal is a good conductor of electricity and, as will be seen, an effective barrier to the wave (Fig 12.2). In this case, the electric field is 'short-circuited' by the metal and must therefore always be zero at the surface. It manifests itself in the form of a current in the sheet (shown dotted) like the current that is induced in a receiving antenna, and this current re-radiates the wave in the direction from whence it came.

In this way a wave is reflected from the metal but, due to the relative directions of field and propagation, the components of forward and reflected waves appear as in Fig 12.2(b). It will be seen that at the surface of the metal the two electric fields are of opposite polarity and thus cancel each other out, but the magnetic fields are additive.

If the two waves are combined the resultant wave of Fig 12.2(a) is obtained in which the electric and magnetic maxima no longer coincide, but are separated by quarter-wave intervals. Such a combined wave does not move, because the energy can alternate in form between electric field in one sector and magnetic field in an adjacent one. This type of wave is called a *standing wave* and exists near reflecting objects comparable in size with the length of the wave. If a small receiving antenna and detector were moved outwards from the front of the reflector the output of the receiver would vary with distance and thus indicate the stationary wave pattern.

In practice radio waves more usually encounter poor conductors, such as buildings or hillsides, and in this case some of the power from the wave is absorbed by or reflected from the obstacle. Later in this chapter, antennas will be described which incorporate conductive elements deliberately used as reflectors. It will also be shown that oscillatory currents in antennas or feeders can be reflected to produce standing waves.

A travelling wave moving away into space represents a flow of energy from the source. The transmitter can be regarded as being at the centre of a large sphere, with the radiated power passing out through its surface. The larger the sphere, the more

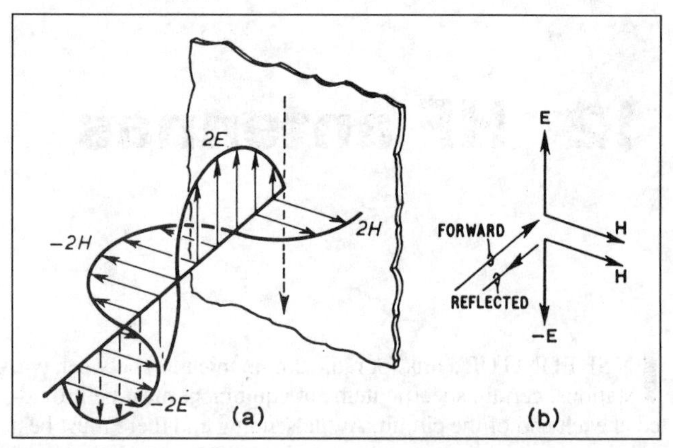

Fig 12.2. Standing waves due to short-circuit reflection at a metal surface, with analysis into forward and reflected wave components

thinly will it be spread, so that the further the receiver recedes from the transmitter, the smaller will be the signal extracted by means of a receiving antenna of given size from a given area of the wave front.

Resonant antennas

If an oscillatory current is passed along a wire, the electric and magnetic fields associated with it can be considered as a wave attached to the wire and travelling along it as far as it continues, and if the wire finally terminates in, say, an insulator, the wave cannot proceed but is reflected. This reflection is an open-circuit reflection and the wire produces standing-wave fields complementary to those generated by a short-circuit reflection as in Fig 12.2, with corresponding voltages and currents. Fig 12.3 shows two typical cases where the wire is of such a length that a number of complete cycles of the standing wave can exist along it. Since the end of the wire is an open-circuit, the current at that point must be zero and the voltage a maximum. Therefore at a point one quarter-wavelength from the end, the current must be a maximum and the voltage will be zero. The positions of maxima are usually known as *current* (or *voltage*) *antinodes* or *loops* and the intermediate positions as *nodes* or *zeros*. At positions of current loops, the current-to-voltage ratio is high and the wire will behave as a low-impedance circuit. At voltage loops the condition is reversed and the wire will behave as a high-impedance circuit. A wire carrying a standing wave as illustrated in Fig 12.3 exhibits similar properties to a resonant circuit and is also an efficient radiator of energy, most of the damping of the circuit being attributable to this radiation. This is a *resonant* or *standing wave antenna* and the majority of the antennas met with in practice are of this general type. The wire need not be of resonant length but it is then equivalent to a detuned circuit and, unless brought into resonance by additional capacitance or inductance, it cannot be energised efficiently.

The length for true resonance is not quite an exact multiple of the half-wavelength because the effect of radiation causes a slight retardation of the wave on the wire and also because the

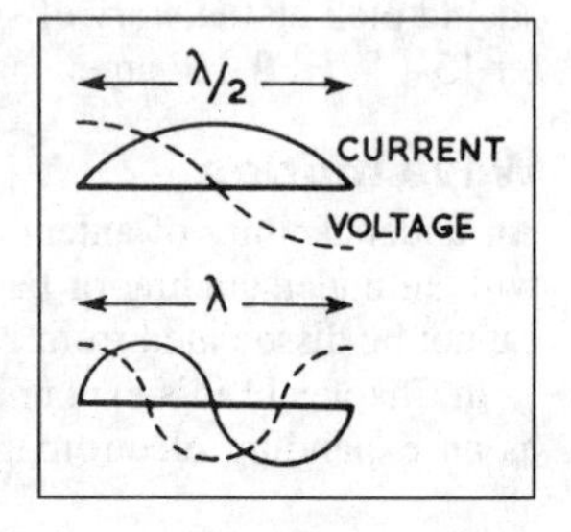

Fig 12.3. Standing waves on resonant antennas, showing voltage and current variations along the wire. The upper antenna is $\lambda/2$ long and is working at the fundamental frequency; the lower is a 1λ or second harmonic antenna

supporting insulators may introduce a little extra capacitance at the ends. An approximate formula suitable for wire antennas is:

$$\text{Length (feet)} = \frac{492(n - 0.05)}{f}$$

where n is the number of complete half-waves in the antenna and f is the frequency in megahertz.

Radiation

The actual physics of the radiation of energy from the wire is an involved matter, which is best left to theoretical textbooks. It suffices to say here that the current flowing up and down the wire gives rise to a magnetic field around the wire, while the charges in motion (which constitute the current) carry with them an electric field. Due to the reversing nature of the current, the two fields are mutually supporting and expand outwards from the wire, carrying with them energy from the exciting current. There exists in the immediate vicinity of the wire an oscillating field known as the *induction* field (similar to that surrounding an induction coil or a magnet), but this decays in strength rapidly as the distance from the wire increases. At a distance of $\lambda/2\pi$, or approximately one-sixth of a wavelength, it is equal in strength to the *radiation* field, but beyond one or two wavelengths has fallen to a negligible level. Radiation takes place from any elevated wire carrying a radio frequency current unless prevented by screening or cancelled by an opposing field of equal magnitude. At any given frequency and for a given direction relative to the wire the field strength produced at a distant point is proportional to the current multiplied by the length of wire, as it appears to the observer, through which it flows.

Directivity

The radiation field which surrounds the wire is not uniformly strong in all directions. It is strongest in directions at right-angles to the current flow in the wire and falls in intensity to zero along the axis of the wire; in other words the wire exhibits *directivity* in its radiation pattern, the energy being concentrated in some directions at the expense of others. Later in this chapter it will be explained how directivity may be increased by using numbers of wires in arrays. These antenna arrays are called *beams* because they concentrate radiation in the desired direction like a beam of light from a torch. Because a number of wires or elements are needed to create the additional directivity, beam antennas usually require more space than simple ones, and this limits the extent to which they can usefully be employed at the longer wavelengths. When space is limited to an average garden, useful beam action tends to be obtainable only with antennas operating on 14MHz or higher bands, those for 1.8MHz, 3.5MHz and 7MHz being generally limited to simple arrangements involving single wires. In the VHF bands, the wavelength is sufficiently short that the various elements of the antenna may be made rigid and self-supporting, and the antenna system may therefore be quite elaborate.

Dipoles

One of the most commonly used words in antenna work is *dipole.* Basically a dipole is simply some device (in the present context an antenna) which has two 'poles' or terminals into which radiation-producing currents flow. The two elements may be of any length, and a certain amount of confusion sometimes arises from the failure to state the length involved. In practice it is usually safe to assume that when the word 'dipole' is used by itself, it is intended to describe a half-wavelength antenna, ie a radiator of electrical length $\lambda/2$ fed by a balanced connection at the centre. Any reference to gain over a dipole is assumed to refer to this $\lambda/2$ dipole. When reference to another form of dipole is intended, it is usual to state the overall length, eg a *full-wave dipole, short dipole* etc. A short dipole is less than $\lambda/2$ in length but needs to be tuned to resonance by the addition of inductance, usually in the centre, or some form of capacitive end-loading as discussed later. Shortening has little effect on radiation pattern but, if carried too far, leads to poor efficiency and excessively narrow bandwidth.

Loops containing between two-thirds and one-and-a-half-wavelengths of wire have radiation patterns very similar to those of half-wave dipoles.

A further reference sometimes encountered is to the *monopole* or *unipole.* This is an unbalanced radiator, fed against an earth plane, and a common example is the ground-plane vertical described later in this chapter.

Gain

If one antenna system can be made to concentrate more radiation in a certain direction than another antenna, for the same total power supplied, then it is said to exhibit *gain* over the second antenna in that direction. In other words, more power would have to be supplied to the reference antenna to give the same radiated signal in the direction under consideration, and hence the better antenna has effectively gained in power over the other. Gain can be expressed either as a ratio of the powers required to be supplied to each antenna to give equal signals at a distant point, or as the ratio of the signals received at that point from the two antennas when they are driven with the same power input. Gain is usually expressed in decibels: a table of conversion from voltage and power ratios will be found in the preceding chapter (Table 11.1). Gain is of course closely related to directivity, but an antenna can be highly directive and yet have a power loss as a result of energy dissipated in antenna wires and surrounding objects. As we shall see later, this is the main reason why it is impossible to obtain high gains from very small antennas.

It is important to note that in specifying gain for an antenna, some reference to direction must be included, for no antenna can exhibit gain simultaneously in all directions relative to another antenna. The distribution of radiated energy from an antenna may be likened to the shape of a balloon filled with incompressible gas, with the antenna at the centre. The amount of gas represents the power fed to the antenna, and the volume of the balloon can only be increased by putting in more gas. The balloon may be distorted to many shapes, and elongated greatly in some directions so that the amount of gas squeezed in those directions is increased, but this can only be achieved by reducing the amount of gas in some other part of the balloon: the total volume must remain unaltered. Likewise the antenna can only direct extra energy in some required direction by radiating less in others.

The gain of an antenna is expressed in terms of its performance relative to some agreed standard. This enables any two antennas to be directly compared, since if two antennas have for example a gain relative to the standard of 6dB and 4dB respectively, the first has a gain of 6 – 4 = 2dB relative to the second. It is unfortunate that two standards exist side by side and will be encountered in other references to antennas. One standard often used is the theoretical *isotropic* radiator, which radiates equal power in all directions, ie its solid polar diagram is a sphere. This is a strictly non-practical device which cannot be constructed or used, but has the advantage that the comparison is not complicated by the directional properties of the reference antenna.

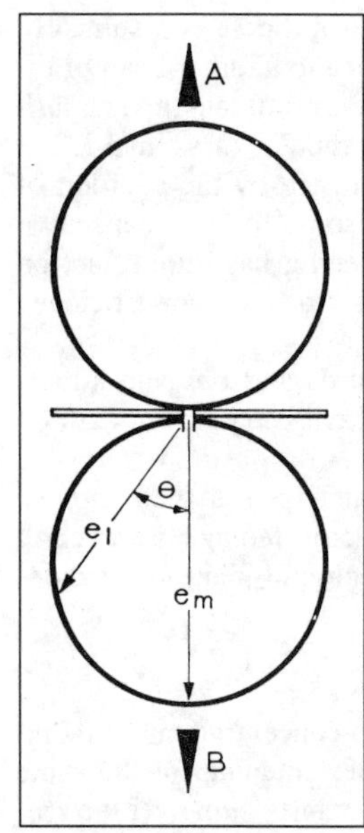

Fig 12.4. Directive polar diagram of a λ/2 (or shorter) dipole shown in section: e_l and e_m represent the relative field strength at any given distance in the directions indicated. The circles and the dipole all lie in the same plane and an observer moving round either circle experiences a constant field strength which, in free space, is inversely proportional to distance

The other standard is the *half-wave dipole* which has in its own right a directional pattern as illustrated in Fig 12.4. This is a practical antenna which can be built and is therefore a more realistic basis for comparison, but it should be noted that gain (dBd) expressed *relative to a λ/2 dipole* means by inference relative to the maximum radiation from the dipole, ie in directions A or B of Fig 12.4. The polar diagram consists of two circles and the relative gain in any direction relative to the wire, such as that of e_l, may be found by describing a circle, filling in e_l and e_m as shown, and using a rule to measure their relative lengths. This is known mathematically as a *cos θ pattern* because the ratio of e_l and e_m is given by the cosine of the angle θ between e_l and e_m. The radiated power moves outwards in straight lines so that in free space a given package of energy occupies four times the volume at twice the distance and the field strength (being measured in volts) is halved. This is one aspect of a more general rule according to which field strength is directly proportional to the apparent length of a given dipole whatever the distance or direction to the observer, so that by imagining what would happen to it visually as one moves through the surrounding space one can get a good idea of what happens in the case of the radio signal, though due allowance should be made for polarisation.

Gain relative to an isotropic radiator should be designated as 'dBi' but the reader is advised to beware of a common tendency to express it in decibels without reference to the standard employed. This can lead to a disparity of 2.15dB in claimed results, being the relative gain of the two standards employed (the gain of a λ/2 dipole relative to the isotropic source). In such cases it is safer to assume the more conservative figure when comparing different antenna performance unless one is sure that the same standard antenna has been assumed in each case.

Because direction is inevitably associated with a statement of gain, it is usually assumed in the absence of any qualifying statement that the gain quoted for any antenna is its gain in the direction of its own maximum radiation. Where the antenna system can be rotated, as is often the case on 14MHz and higher frequencies, this is not so important, but when the antenna is fixed in position the superiority it exhibits in one direction over another antenna will not hold in other directions, because of the different shapes of the two directivity patterns. Antenna A may have a quoted gain of 6dB over antenna B but only in the directions which favour the shape of its radiation pattern relative to that of antenna B. There is an important distinction between transmitting gain and effective receiving gain. In the first case it is required to maximise the power transmitted and in the second case we have to maximise the signal-to-noise ratio, and the two gains will be the same only if there are no power losses and noise is isotropic, ie arriving equally from all directions. In the

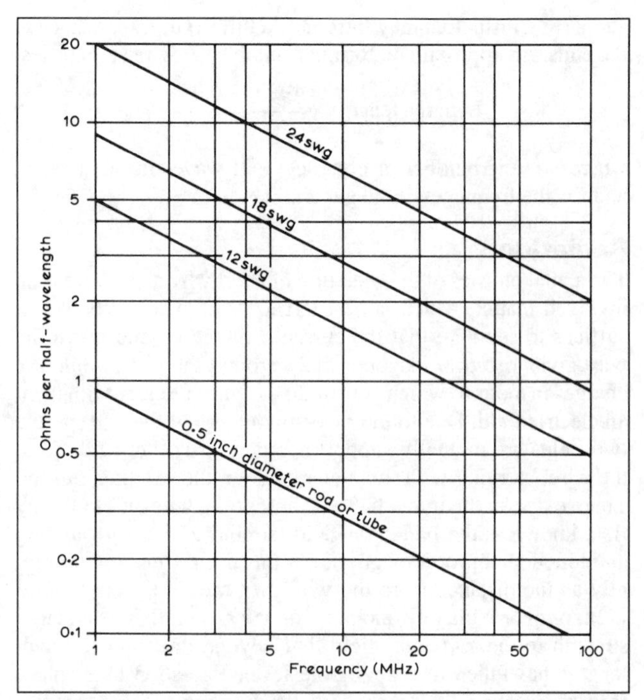

Fig 12.5. RF resistance of copper conductors. Resistance per unit length is proportional to the circumference of the conductor and the square root of the frequency

HF bands, the useful receiver sensitivity is limited by external noise which is usually well above the receiver noise level, and, so long as this remains true, signal-to-noise ratio is unaffected by losses in the antenna system. Typically, with a low-noise receiver, antenna losses could reduce the power transmitted by up to 10dB or more before starting to affect the performance adversely when the same antenna is used for reception.

Radiation resistance and antenna impedance

When power is delivered from the transmitter into the antenna, some small part will be lost as heat, since the material of which the antenna is made will have a finite resistance, albeit small, and a current flowing in it will dissipate some power. The bulk of the power will usually be radiated and, since power can only be consumed by a resistance, it is convenient to consider the radiated power as dissipated in a fictitious resistance which is called the *radiation resistance* of the antenna. Using ordinary circuit relations, if a current I is flowing into the radiation resistance R, then a power of I^2R watts is being radiated. As depicted in Fig 12.3 the RMS current distribution along a resonant antenna or indeed any standing wave antenna is not uniform but is approximately sinusoidal. It is therefore necessary to specify the point of reference for the current when formulating the value of the radiation resistance, and it is usual to assume the value of current at the anti-node or maximum point. This is known as the *current loop*, and hence the value of R given by this current is known as the *loop radiation resistance*: in practice the word 'loop' is omitted but inferred.

A λ/2 dipole has a radiation resistance of about 73Ω. If it is made of highly conductive material such as copper or aluminium, the loss resistance may be less than 1Ω. The conductor loss is thus relatively small and the antenna provides an efficient coupling between the transmitter and free space. The effective loss resistance per half-wavelength of copper conductor varies with frequency and with size of conductor as shown in Fig 12.5; for comparison with the radiation resistance of a dipole these figures must be divided by two, since the current distribution along the dipole is more or less sinusoidal and the mean-square current is therefore only half the current in the centre. Approximately, for non-magnetic materials, the resistance varies as the square root of the resistivity and inversely as the surface area. Values obtained with the aid of Fig 12.5 should be increased by about 25% for aluminium and 65% for aluminium alloy. For magnetic materials the losses are much higher.

When the antenna is not a resonant length, it behaves like a resistance in series with a positive (inductive) or negative (capacitive) reactance and requires the addition of an equal but opposing reactance to bring it to resonance, so that it may be effectively supplied with power by the transmitter. The combination of resistance and reactance, which would be measured at the antenna terminals with an impedance meter, is referred to in general terms as the *antenna input impedance.* This impedance is only a pure resistance when the antenna is at one of its resonant lengths.

Fig 12.6 shows, by means of equivalent circuits, how the impedance of a dipole varies according to the length in wavelengths. It will be seen that the components of impedance vary over a wide range.

The input impedance of the antenna is related specifically to the input terminals, whereas the radiation resistance is usually related to the current at its loop position. It is possible to feed power into an antenna at any point along its length so that the input impedance and the loop radiation resistance even of a resonant antenna may be very different in value, although in this case both are pure resistances. Only when the feed point of the antenna coincides with the position of the current loop on a single wire will the two be approximately equal (Fig 12.7(a)). If the feed point occurs at a position of current minimum and voltage maximum, the input impedance will be very high, but the loop radiation resistance remains unaltered (Fig 12.7(b)). For a given power fed into the antenna, the actual feed-point current measured on an RF ammeter will be very low, but because the input impedance is high, the power delivered to the antenna is the same. Such an antenna is described as *voltage fed,* because the feed point coincides with a point of maximum voltage in the distribution along the antenna. Conversely an antenna fed at a low-impedance point, usually a current maximum, is described as *current fed.*

The input impedance of a current-fed λ/2 dipole consisting of a single straight conductor is approximately equal to 73Ω, though somewhat dependent on height (Fig 12.86), and will be much the same value irrespective of the size of wire or rod used to fabricate the dipole. The input impedance of a voltage-fed λ/2 dipole is very high, and its precise value depends not only upon the loop radiation resistance of the dipole, which is independent of the method of feed, but also the physical size of the wire

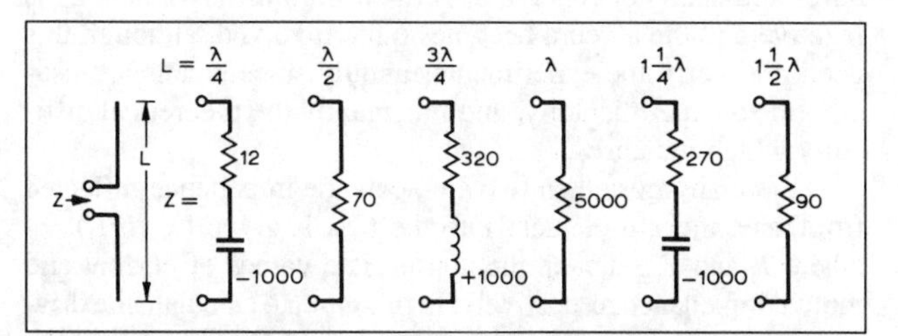

Fig 12.6. Typical input impedance (Z_i) value for dipoles of various lengths. The values for L = λ/2 and 3λ/2 are always approximately as shown but values for other lengths vary considerably according to the length/diameter ratio of the antenna. The values given are typical for wire HF antennas. Note how the reactance changes sign for each multiple of λ/4

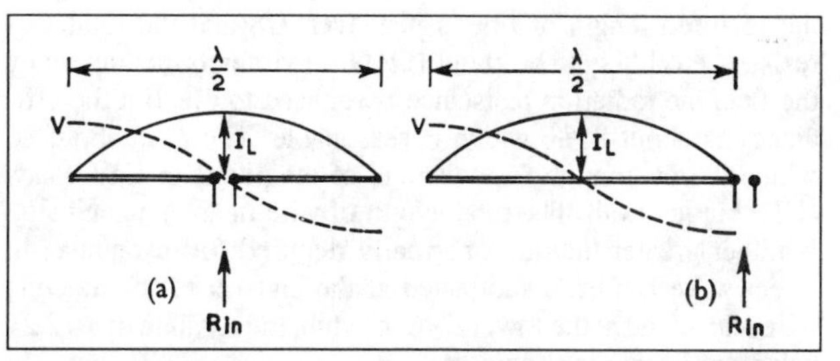

Fig 12.7. The input resistance of a λ/2 dipole is low at the centre and high at the end, although the loop radiation resistance is the same in each case. In (b) R_{in} is twice the R_e of Table 12.5

used. The dipole wire between the current loop and the current zero at the ends may be considered to act as a λ/4 transformer between the loop radiation resistance at the centre and the input impedance at the end. As a transformer, the wire must exhibit a certain characteristic impedance Z_0, and the transformer ratio will depend upon the value of Z_0, which in turn depends upon the ratio of the conductor diameter to its length. If the end impedance is R_e, and the radiation resistance R_r, then $R_e = Z_0^2/R_r$ approximately: the value of R_r is fixed by the current distribution, and hence the value of R_e will change as Z_0 or the wire length/diameter ratio changes. Typical values of R_e for different ratios are given in Table 12.5; the behaviour of transformers is dealt with in more detail in the section on transmission lines.

In certain cases, even though a λ/2 dipole is centre-fed at the current loop position, the input impedance may differ considerably from the value of 73Ω. If the λ/2 dipole is folded (see p12.33), a transformation again occurs to produce a different input impedance. If the dipole is the driven element of a multi-element array such as a Yagi, then the input impedance will also be modified (usually reduced) because of the presence of the other elements. If a wire is made up to be 33ft long and centre-fed, it would be resonant (λ/2) at 14MHz, three-quarter wavelength (3λ/4) at 21MHz and full wave (λ) at 28MHz. At 14MHz it is easy to supply power into the 70Ω input resistance, but at full-wave resonance, ie 28MHz, the high value of 5000Ω requires the use of some form of transformer. Again at 21MHz where the impedance is complex and evidently high, special arrangements are necessary. These problems are discussed later in connection with tuned feeders and antenna couplers and in relation to specific antennas.

This same antenna on 7MHz would only be λ/4 long, with a low resistance of 12Ω and a high capacitive reactance. It would require a *loading coil* of about 23μH inductance (1000Ω reactance) for resonance at 7MHz. This loading coil will have appreciable loss resistance, typically about 6Ω, so that only two-thirds of the power from the transmitter reaches the antenna. The remainder is expended as heat in the coil so that the efficiency is 67% and signal reports will be degraded by about one-third of an S-unit. With 18Ω total input resistance and a reactance of 1000Ω, the Q of the circuit (ratio of reactance to resistance) would be approximately 50, which is just sufficient to permit reasonable coverage of the band without retuning. On 3.5MHz the same antenna would have a resistance of only 3Ω and a reactance of 2000Ω. The loss resistance of the loading coil would be 12Ω, the efficiency only 20%, and it would be impossible to shift frequency over more than about 25kHz without having to re-tune. The losses could be at least halved by using big coils, probably needed in any case for dissipating the heat, but the bandwidth would then be reduced to a mere 15kHz.

Greater efficiency may be achieved by using a length of open-wire line (or tuning stub) in place of the loading coil. For 7MHz the required length of line is just over λ/8 and the total loss resistance is likely to be about 1Ω. Due to transformer action by the line, the radiation resistance is reduced to 6Ω, but the efficiency is about 85%, which is reasonable. The Q is about 86 which is low enough for coverage of the phone or CW bands (UK) but not both. The total length of wire in antenna and stub is rather greater than that normally required for resonance; in effect, wire has been subtracted at the high Z_0 of the antenna wire and added at the lower Z_0 of the stub, the formula on p12.27 being applicable in each case.

Starting from the known value for a λ/2 dipole, the radiation resistance of other antennas may be deduced from knowledge of the gain and current distribution. For example, knowing that the directivity of a λ/4 dipole is very similar to that of a λ/2 dipole, it follows that the fields produced must be similar for the same power radiated but, since the length is halved, the current must be doubled and one might therefore expect the radiation resistance to be reduced to 17.5Ω or rather more than the 12Ω indicated in Fig 12.6. We are, however, concerned with the average current, which for a half-sine-wave as depicted in Fig 12.7 is 2/π times the current in the centre, whereas for a short dipole the distribution represents the two ends of a sine wave which are more or less triangular so that the average current is half the centre current. The radiation resistance is thus reduced by $(\pi/4)^2$ to 10.8Ω.

The remaining slight error lies in the assumption that the two antennas have identical gain, which is nearly true because in both cases the antennas are a close approximation to point sources of radiation. This means that, to reach any given point in space, the contributions from all parts of the radiator have to travel nearly the same distance, arrive in almost the same phase, and are directly added. Nevertheless, if we look at a λ/2 dipole from a nearly end-on position, we find contributions from near its two ends arriving almost half a wavelength apart and therefore cancelling each other. These components of the field are small and can for most purposes be ignored, but they account for a reduction of about 4% in the relative field strength from the shorter antenna. This is equivalent to a 4% reduction of current and allowance for this brings the estimated radiation resistance up to 12Ω as illustrated. The variation of radiation resistance with dipole length, for different types of loading, is plotted in Fig 12.8.

The possibility of approximating each element by a point source is particularly helpful towards understanding the operation of beam antennas as described later in this chapter.

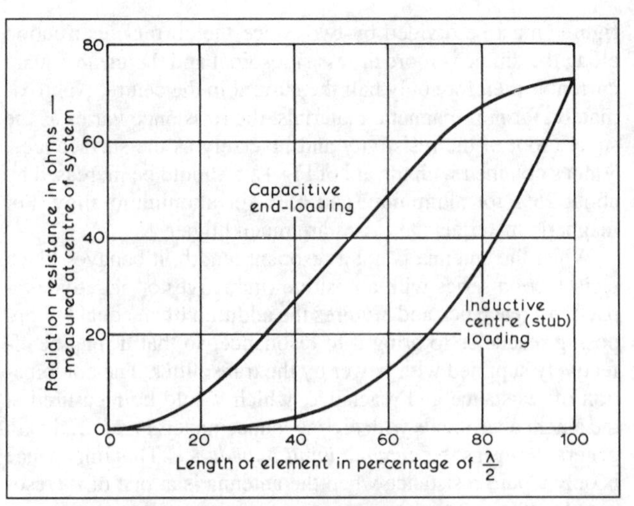

Fig 12.8. Radiation resistance of short dipoles with alternative methods of loading. For beams designed on the guidelines of Fig 12.15, the radiation resistance varies with length of element in the same ratio as that of single dipoles

Mutual impedance

If two antennas are in proximity there is mutual coupling between them just as there would be between a pair of tuned circuits, except that instead of mutual reactance there is a *mutual impedance* which usually includes both resistance and reactance. It plays an important role in the operation of small beam antennas such as those normally used for the amateur HF bands, causing a big reduction of input impedance so that relatively large currents flow in the elements. The following example illustrates the calculation of input impedance for closely-spaced two-element arrays and provide some insight into a problem which becomes extremely complex for practical systems involving additional elements.

Mutual impedance Z_m is defined as the ratio $-E_2/I_1$ where E_2 is the voltage induced in one element by a current I_1 flowing in another. It should be noted that E_2 is the voltage which would be required to produce the current I_2 (which is the current actually flowing) if the driven element was removed, and the minus sign arises from the nature of magnetic induction. If Z_2 is the impedance of the second element we have $I_2 = E_2/Z_2$, but since $E_2 = -I_1Z_m$ the ratio of the currents, $(-I_2/I_1)$ is given by Z_m/Z_2. The first element has, in addition to the driving voltage E_1, a voltage $-I_2Z_m$ induced from the second element, and putting these facts together we find that the driving point impedance (E_1/I_1) is given by $Z_1 - Z_m^2/Z_2$, where Z_1 is the self-impedance of the first element and usually equal to the radiation resistance R, any reactive component being tuned out.

In the case of Z_2, the situation is slightly more complicated because the element is usually detuned slightly, which (as explained later) increases the gain and favours one direction of propagation rather than the other so that the antenna system is now said to possess a *front-to-back ratio.* For best performance a phase shift of about 30° from the antiphase condition is required and for the simple case where Z_m is a pure resistance R_m (which happens for a convenient spacing of about λ/8 between dipoles) the phase-shift in radians is roughly equal to X/R where X is the reactance inserted in series with the radiation resistance R. (Instead of using a coil, the detuning effect can be achieved by altering the length of the element slightly, which is usually more convenient). In this situation, which (though relatively simple) is of considerable practical importance, the driving point impedance is given by:

$$\frac{E_1}{I_1} = R\left(1 - \frac{R_m^2}{R^2 + X^2}\right)$$

This is the resistive part of what is really a complex impedance, but the reactive portion is neutralised by normal tuning procedures and need not concern us further. For other spacings Z_m is reactive and the algebra becomes quite involved, although this need not worry the experimenter unduly since beams are usually tuned experimentally, and it is mainly the theoretical difficulty which increases.

If there are more than two elements the impedance reflected from, say, the nth element into the first is given by $(I_n/I_1).Z_{mn}$ where I_n and Z_{mn} denote the appropriate values of current and mutual impedance respectively. In this case, to calculate the driving point impedance the above expression would have to be elaborated to take into account the relative phases of the currents, and the mutual impedance of all the elements and their images in the ground, Fig 12.24 on p12.15. This would have to be repeated for each element, leading to a complex set of

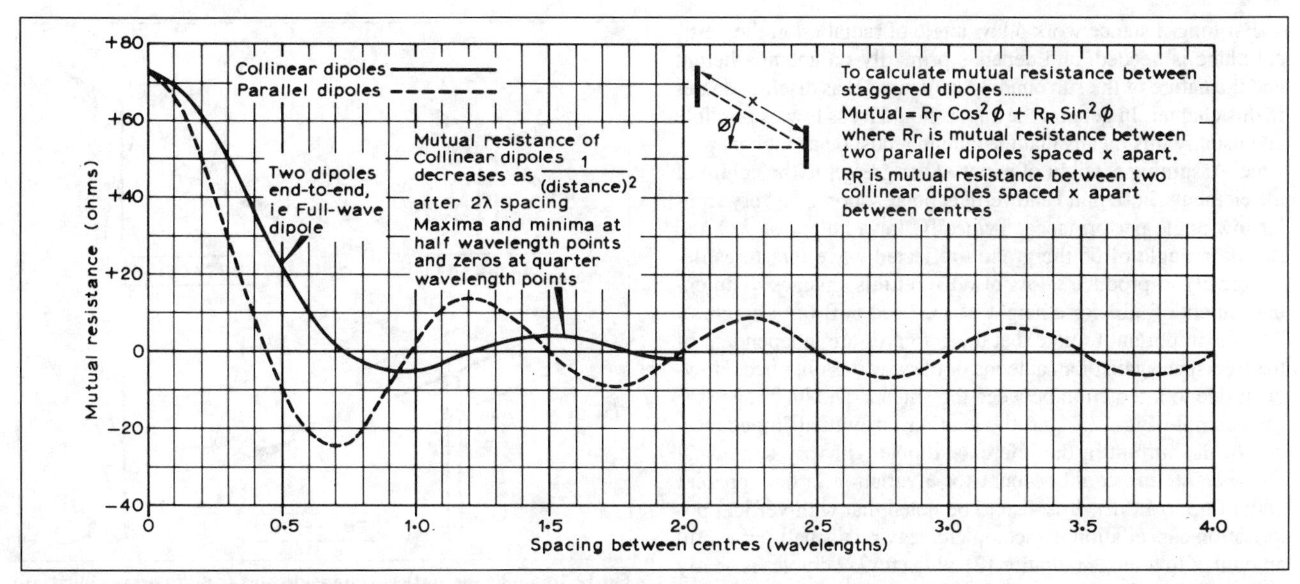

Fig 12.9. Variation of mutual resistance between parallel and collinear λ/2 dipoles as the spacing is altered

simultaneous equations soluble only by computer. We can, however, effect a remarkable simplification by assuming that all the currents are of equal amplitude and either equal or opposite in phase, in which case all the coefficients I_n/I_1 are equal to 1 or −1 and we have:

$$\frac{E_1}{I_1} = R \pm R_{m2} \pm R_{m3} \pm R_{m4} \pm \text{etc}$$

the reactance terms (which add in the same way) being eliminated by the normal tuning process. The mutual resistances of parallel or collinear λ/2 dipoles may be obtained from Fig 12.9 and added or subtracted as required; thus for two collinear dipoles with their ends almost touching (centres λ/2 apart) the input resistance of each is 73 + 22 = 95Ω which reduces the current in each, and therefore the field at a distant point, by 1.2dB compared with what it would be with no interaction. On the other hand, for similarly phased parallel dipoles spaced by 0.7λ we have 73 − 24 = 49Ω and the field is increased by 1.8dB. This procedure is particularly useful in dealing with arrays of widely spaced elements; thus in the absence of mutual resistance the gain obtainable from stacking any number n of identical radiators (regardless of whether these are dipoles or elaborate beam antennas) is precisely equal to n, as explained on p12.16. For two elements, the basic gain of 3dB is modified in the case of the above examples to 3 − 1.2 = 1.8dB and 3 + 1.8 = 4.8dB respectively. A more dramatic illustration of the relation between gain and mutual impedance is provided by the 8JK beam, Fig 12.14 on p12.10, in which close-spaced parallel elements are fed in antiphase. For a spacing of 0.15λ we have R_m = 61Ω so that, allowing for the phase reversal, the radiation resistance of each becomes 73 − 61 = 12Ω and the total power radiated for a current of I in each element is $24I^2$ compared with $73I_0^2$ for a single dipole. Equating these powers, we find that $I = 1.71I_0$. Along a line at right-angles to the elements the fields radiated do not quite cancel out because one has a start of 0.15λ over the other so that there is a 54° phase difference. The resultant field is therefore 2 × 1.71 sin (54/2)° times that of a dipole, ie 1.55 times, which implies a gain of 3.84dB.

In practice, simple addition and subtraction of mutual resistances can usefully be applied only to symmetrical arrangements in which each element 'sees' an identical pattern of other elements and their ground images. Otherwise different elements will experience different reflected impedances, and currents will be unequal unless the phases and amplitudes are independently adjusted in every case. In the case of close-spaced beams at reasonable heights, ground images, being relatively remote, have negligible effect on impedances compared with those due to interactions between the elements.

PRINCIPLES OF DIRECTIVE ARRAYS

All practical antennas are directive to some extent since, as we have seen, there is no radiation from a wire directly along its axis. When there is more than one half-wavelength in an antenna or if an antenna system is built up from a number of separate radiators, it can be represented as a number of *point sources,* each representing one standing-wave current loop. The total radiation can then be regarded as the resultant of a number of components, one from each point source. In any given direction these components may have to travel different distances, so that they do not arrive at a distant point in the same relative phases as they had in the wire. They can therefore augment or oppose each other. It is, however, possible to combine a number of elementary antennas so that their radiation accumulates in some favoured direction, at the expense of radiation in other directions. This gives a much stronger signal than a single antenna would give in the required direction for the same power input. For these reasons antennas are said to be *directive* and to have *gain.*

The concept of gain was discussed on p12.3 and, as we have seen, directivity and gain are closely interrelated. There are many ways of constructing directive arrays and the choice between them depends largely on the space and height available as well as basic requirements such as whether it is desired to achieve equally good performance in all directions or concentrate on one or more particular directions. Rotary beams with two or three elements meet the requirements for coverage of all directions with a gain of about 5–6dB [1] and comparable results may be obtained by switching between two or three fixed reversible beams. However, higher gains require larger beams, resulting in a rapid escalation of practical difficulties, and the azimuthal patterns of fixed arrays become narrow so that a large number of separate arrays are necessary if it is desired to achieve uniform coverage of all compass directions.

For long-distance work a low angle of radiation in the vertical plane is needed, and depends primarily on antenna height and the nature of the surrounding contours [2] as discussed later in this chapter. In general the angle should be as low as possible and usually this means that the antenna must be as high as possible. Assuming a height of about λ/2 or greater to the centre of the elements, horizontal and vertical polarisation give very similar low-angle performance. Typically, for a height of λ/2 and radiation angle of 5° the ground reflected wave interferes destructively to produce a loss of 6dB but this changes to an enhancement of 3dB for a height of 3λ/2 and 6dB for a height of 3λ. It is important to note that these figures are independent of the free-space gain of an antenna or the way it is obtained. However, due to interaction between the antenna and its image (see for example Fig 12.9 and the section on mutual impedance, p12.6) the current in the reference dipole will vary somewhat with height, and could account for a variation in the apparent gain of up to 1.2dB. It is also to be noted that with vertical polarisation cancellation is incomplete, leaving a small but useful amount of low-angle radiation [2] which for λ/2 dipoles is nearly independent of height. For short radiators close to ground there is an additional basic loss of 3dB (p12.61).

Long-wire antennas have very distinctive radiation patterns, as will be seen later, but these patterns have several main beams or lobes and hence the gain is not so great as when most of the transmitter power is concentrated into one main beam. In order to achieve the latter condition, two or more elements, such as λ/2 dipoles, are combined in special ways and the combination is called an *array*.

There are three general classes of array – the *broadside array* in which the main beam is at right-angles to a plane containing the elements, the *end-fire array* in which the main beam is in the same direction as the row of elements and the *collinear array* which employs co-phased elements arranged end-to-end, radiation being a maximum at right-angles to the line of the elements; the three classes are illustrated in Fig 12.10. It should be noted that the term 'end-fire' refers to the layout of the array and not to the ends of the wires, although the term is sometimes applied to a long wire because its direction of maximum radiation tends towards the direction of the wire.

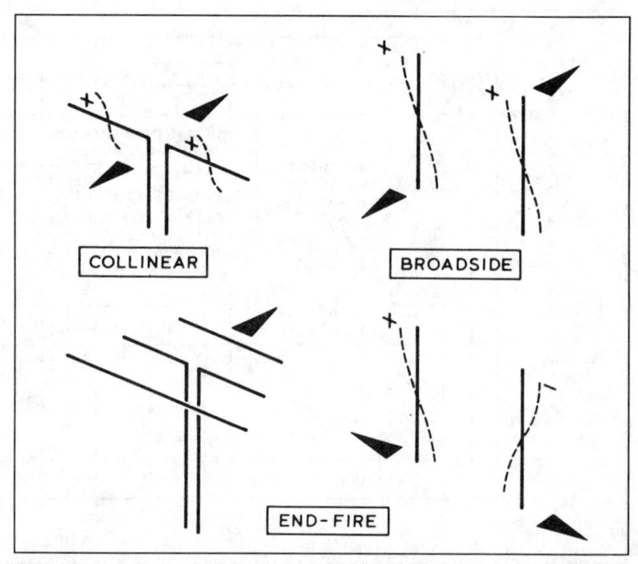

Fig 12.10. Collinear vertical broadside and end-fire arrays illustrating the terms. The arrows indicate directions of maximum radiation and the voltage standing waves on three of the antennas show their relative polarity or phase. The lower left antenna is a three-element parasitic array and is unidirectional

Radiation patterns and polar diagrams

From the point of view of effective gain it is immaterial whether this comes from horizontal directivity, vertical directivity or both, but the practical usefulness of a fixed array using horizontal collinear elements is restricted by the relatively narrow beamwidth in the horizontal plane. To illustrate the radiation pattern of an antenna, *polar diagrams* are used in the form of curves, the radius of which in any direction represents the relative strength of signals in that direction.

The radiation from an antenna occurs in three dimensions and therefore the radiation pattern is best represented by the surface of a solid object. A polar diagram is any section of the solid shape, and a large number of sections may be necessary to reduce the antenna radiation pattern to two dimensions. In practice it is necessary to be content with two polar diagrams taken in the principal planes, usually the horizontal and vertical, and giving the two cross-sections of the main beam.

Where the polar diagram has a definite directional form, the angle between the directions where the power radiated is half the value at the point of maximum gain (–3dB) is called the *beamwidth*. These points are marked on Fig 12.124 on p12.75.

To avoid confusion when discussing radiation, directions in the horizontal plane are referred to as *azimuth;* angles above the horizontal, in the vertical plane, are called *wave angles* or directions in *elevation*. Confusion often arises additionally when the expression horizontal (or vertical) polar diagram is used, unless it is made clear by a statement of the polarisation of the antenna with respect to the Earth's surface. When reference is made to the polar diagram of an antenna in free space, the terms 'horizontal' and 'vertical' have no meaning, and the more precise descriptions of *E-plane* and *H-plane* polar diagrams are to be preferred. These are unambiguous, since the direction of the electric and magnetic fields around the antenna is a function only of the direction of current flow. The electric field (or E-plane) is parallel to the direction of current and therefore usually parallel to the radiating wire. The magnetic field (or H-plane) is at right-angles to the current and therefore normal to the radiating wire. The polar diagram of the λ/2 dipole illustrated in Fig 12.5 is then an E-plane diagram: the H-plane diagram of the dipole will be a circle. Such an antenna is then said to possess *E-plane directivity*, and is omnidirectional in the H-plane.

No matter what name is used to describe the polar diagram, it should be remembered that these radiation patterns are for long distances and cannot be measured accurately at distances less than several wavelengths from the antenna. The greater the gain, the greater the distance required.

Construction of polar diagrams

Fig 12.11 is a plan of two vertical antennas A and B, spaced by λ/2, and carrying equal in-phase currents. A receiver at a large distance in the direction a at right-angles to the line of the antennas (ie *broadside*) receives equally from both A and B. Since the two paths are equal the received components are in phase and directly additive, giving a relative signal strength of two units. In the direction b the signal from A travels λ/2 further than that from B. The two received components are therefore one half-cycle different, or in antiphase, and so cancel to give zero signal. In an intermediate direction c the two paths are effectively parallel for the very distant receiver, but one component travels farther by the distance x. In this case the components have a relative phase $\psi = 360x/\lambda$ degrees. The addition for this direction must be made *vectorially* using the 'triangle of forces' also illustrated in Fig 12.11, in which the two equal lines

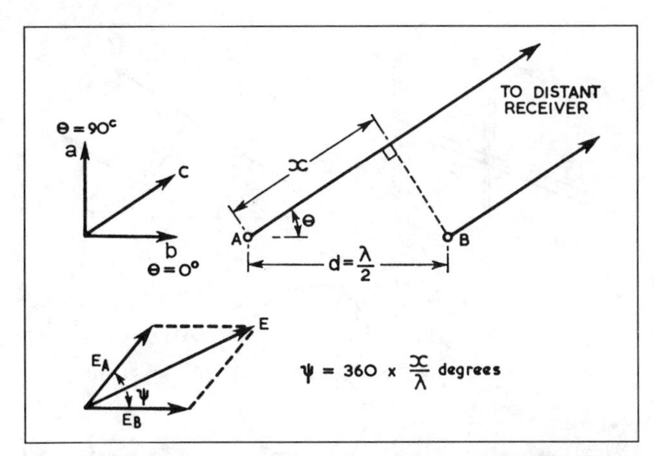

Fig 12.11. The radiation pattern of two sources can be calculated by vector addition of the respective field contributions at various angles of azimuth. For simple arrays this can be converted into a trigonometrical formula

representing signal strength from A and B are set at the phase angle ψ which corresponds to the path difference x; the total or resultant signal is equal to the relative length of the line E which completes the triangle. This process could be carried out for all directions, but is tedious and it is usual therefore to convert the operations to a trigonometrical formula from which the result can be calculated more quickly.

In the general case of two antennas carrying equal but dephased currents and spaced a distance d apart, the appropriate formula for the H-plane polar diagram is:

$$E = 2E_0 \cos\left(\frac{\pi d}{\lambda} \cos\theta + \frac{\phi}{2}\right) \qquad (1)$$

where θ is the angle measured as shown in Fig 12.11, ϕ is the electrical phase difference between the currents in the two antennas, and E_0 is the value of the field from one antenna alone in the direction θ. In the particular case illustrated in Fig 12.12, θ = 0, E_0 is constant for all directions of θ, and $d = \lambda/2$, so that the shape of the polar diagram can be plotted from the expression:

$$\cos\left(\frac{\pi}{2} \cos\theta\right)$$

The absolute magnitude of the field E will depend upon the actual value of the current flowing in each antenna and is of no direct importance when calculating the shape of the pattern. It comes into account only when the relative patterns of two antenna systems are being considered and the question of gain arises.

The polar diagram found in this way is a figure-of-eight, the curve marked λ/2 in Fig 12.12(a). If, on the other hand, the two antennas had been in antiphase, the resultant signal would have been zero in the direction a and maximum in direction b, and the polar diagram as in Fig 12.12(b).

In that case, ϕ = 180° (or π) in equation (1) above, and the polar diagram is given by:

$$\cos\left(\frac{\pi}{2} \cos\theta + \frac{\pi}{2}\right)$$

$$= \sin\left(\frac{\pi}{2} \cos\theta\right)$$

This expression for the antiphase antennas is very similar in form to that for the in-phase antennas, and the change from cosine to sine reveals that the general pattern is turned through 90° of azimuth. This is confirmed by the dotted curves of Fig 12.12.

A similar procedure is used if the antennas do not carry equal currents, or if they have an arbitrary phase relation. In this case the sides of the triangle would be drawn in lengths proportional to the two antenna currents and the relative phase of these currents would be added to the angle ψ which arises from the path differences.

In Fig 12.11 the path difference x and corresponding phase shift ψ are of course proportional to the spacing between the antennas, which in the example above was λ/2. If the spacing is one wavelength then the phase difference changes twice as rapidly with change of direction, and so there are twice as many maxima and minima to the pattern, giving the curves marked 'λ' in Fig 12.12. These patterns are said to have four *lobes*.

With increased spacing between the two antennas, more lobes appear, two for each half-wave of spacing, and the pattern becomes like a flower with many petals. This type of pattern is not very useful, but if the intervening space is filled with antennas spaced λ/2, one pair of lobes grows at the expense of all the others, giving a sharp main beam with a number of relatively small *minor lobes.* This is the basis of stacked beam arrays. It should be noted, however, that the beam is only developed in the plane in which the array is extended; a broadside array of vertical antennas or collinear array of horizontal antennas all in phase produces a beam which is narrow in the horizontal plane, but its vertical pattern is the same as that of a single antenna. It is therefore necessary to extend a stacked array in the vertical direction if a sharp vertical pattern is required. In an end-fire array where the antennas are spaced and phased to give the main lobe along the line of antennas, the vertical and horizontal patterns are developed simultaneously by the single row of radiators because the array is extended simultaneously in both of these

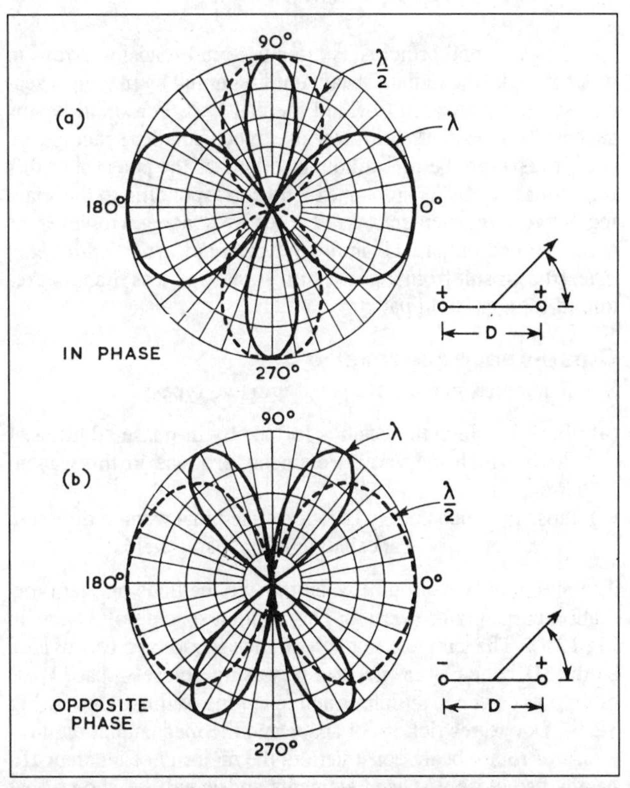

Fig 12.12. Horizontal polar diagrams of two vertical antennas spaced λ/2 and λ. The upper diagram is for antennas in phase, the lower diagram for antiphase connection. If separate feeders are used the pattern can be changed by reversing polarity of one feeder. Directions 90°, 270° are broadside: 0°, 180° end-fire. Note that the end-fire diagrams are broader than the broadside ones

planes. The azimuth pattern of an end-fire array is always broader than that of a broadside array of the same length; this effect can be noted in the patterns of Fig 12.12. Note that for the purpose of this discussion the interaction of horizontal and vertical planes is assumed to be along the direction of propagation.

Although the trigonometrical formula for the polar diagram of a large array may be very complex, and therefore laborious to plot in full, it is always readily possible to find the directions of minimum radiation or *nulls* by solving the equation for θ when *E* is made equal to zero.

Unidirectional patterns

The patterns so far considered are symmetrical: a somewhat different combination of antennas gives a unidirectional pattern. The two dipoles in Fig 12.13 are connected a quarter-wave apart along a common feeder, with a wave entering from the left. In the direction away to the right it does not matter whether radiation leaves via the first antenna or the second; it takes the same time to reach its destination. Thus to the right the components of radiation are additive because they are in phase. This is shown by the upper set of broken lines. To the left, however, the wave from the end dipole has to travel further than that from the first dipole by a distance equal to twice the spacing between them. This extra journey is λ/2 and hence the two components differ in phase by 180° and cancel. Because of the λ/4 spacing the currents in the two antennas have a phase difference of 90° and are said to be in *quadrature*.

The actual shape of the pattern can again be determined by substituting the appropriate values for *d* and ϕ in equation (1) given earlier, to give:

$$\cos\left(\frac{\pi}{4}\cos\theta + \frac{\pi}{4}\right)$$

This fundamental principle is often used in broadside arrays in order to make the radiated beam unidirectional by placing a second set of antennas λ/4 behind the main set. For small beams such as those suitable for rotary use, much smaller spacings are used, the pattern being cardioidal so long as the phase shift differs from 180° by an amount exactly corresponding to the spacing between the elements, eg 135° for λ/8 spacing. However, as demonstrated on p12.11, improved gain and *directional characteristics* result from a phase shift somewhat less than that required for a cardioid pattern.

Close-spaced arrays

Beam antennas may be classified into two types:

(a) those in which the beam is formed by in-phase addition of fields which the various elements produce in the wanted direction; and

(b) those in which energy is concentrated in a wanted direction by less-effective cancellation of opposing fields.

The region of transition between the two methods of beam formation corresponds (very roughly) with a spacing of λ/4, as in Fig 12.13. The simplest example of the second type is provided by the 8JK which is an end-fire array using a closely spaced pair of dipoles driven in antiphase and operating as illustrated by Fig 12.14. Due to restrictions of space and the mechanical requirements of rotary beam construction, the majority of amateur HF beams use closely spaced elements and it will be shown that those with two elements can be evolved from the 8JK by the introduction of a small phase shift.

The nearly-antiphase excitation of elements is an inherent feature of all close-spaced beams. If such elements are fed in

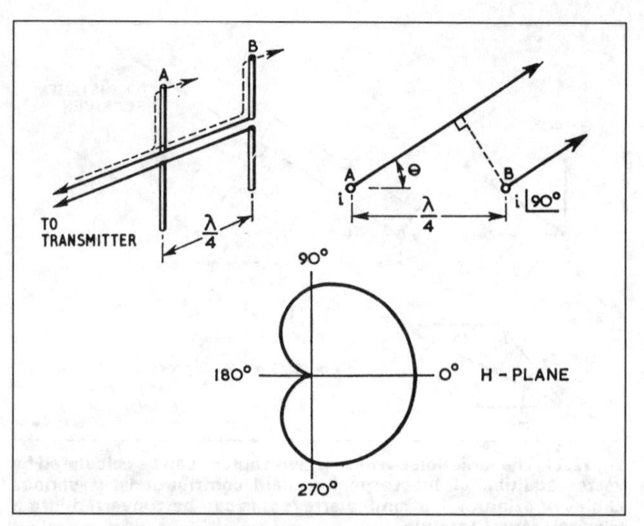

Fig 12.13. Illustrating the fundamental principle of the driven reflector. The two wave components have equal path lengths in one direction, and therefore add, but in the other direction the paths differ by λ/2 and the components cancel. The polar diagram is the geometrical figure called a cardioid (heart shape)

the same phase the radiation resistances are additive so that the current in each is reduced, the total current tends to stay constant, and the radiation pattern and gain is the same as that of a single dipole. This follows also from the fact that for any given point in space, signals from each element arrive in nearly the same phase, a difference of, say, 45° being insignificant when the fields are adding. On the other hand, if the elements are fed in antiphase a narrow beam can be formed because (as shown in Fig 12.14) the extent to which cancellation fails to take place in any given direction is proportional to the apparent spacing of the elements as viewed from that direction. This varies as cos θ, so that a second cos θ pattern is superimposed on the basic dipole pattern, giving $\cos^2\theta$ in the end-fire direction and cos θ in the broadside direction, a concentration of energy which produces a gain of 4dB and is independent of the dimensions of the beam.

Arrays characterised by this property are known as *supergain arrays* since for a given gain they are much smaller than 'normal' arrays using the additive principle. Unfortunately, since the increased field strength is attributable to the difference between opposing fields, each of these must be relatively large. This implies large currents in the element so that for a given

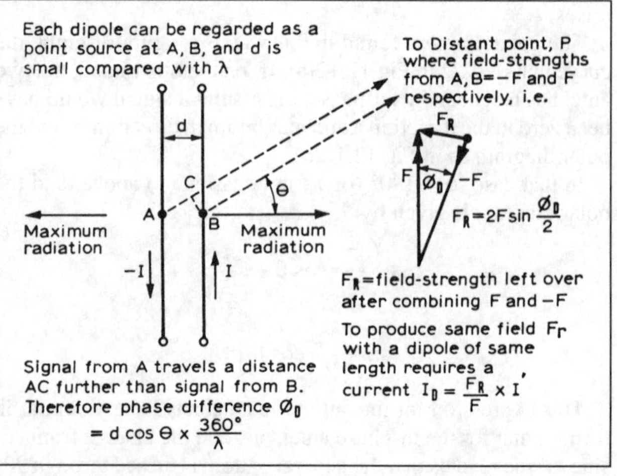

Fig 12.14. Mechanism of 8JK antenna

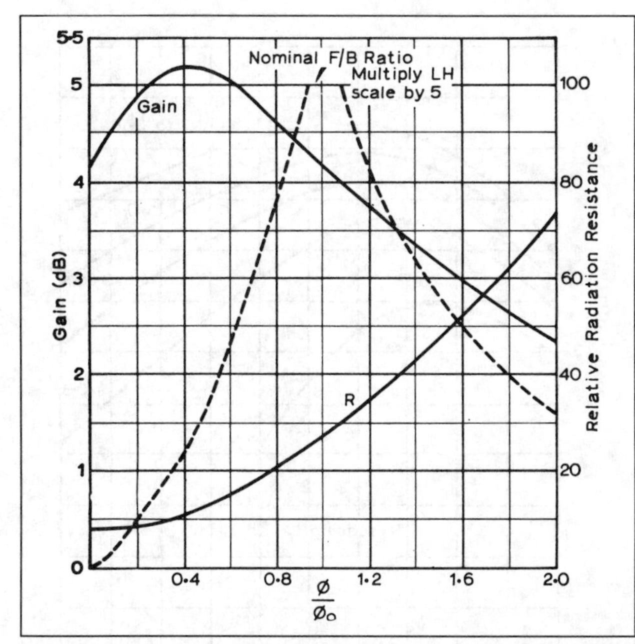

Fig 12.15. Variation of gain, radiation resistance and front-to-back ratio with ϕ/ϕ_0 where ϕ_0 is the phase shift corresponding to the distances between the elements and ϕ is the electrical phase shift relative to the antiphase condition. Resistance scale is correct in ohms for each of a pair of dipoles spaced λ/8 (equal currents are assumed)

radiated power the radiation resistance is relatively low; in effect resistance has been subtracted from each element by the presence of the other, the difference being defined as the mutual resistance R_m. It follows that without R_m there could be no gain and, since R_m is closely linked to gain and hence to the radiation pattern (p12.7), it is outside the control of the designer. The low value of resistance means that bandwidth is therefore decreased and larger-diameter conductors are needed to prevent energy from being dissipated as heat instead of radiation.

This subtractive method of beam forming can only be applied with close spacing, since otherwise there must be some direction in which the field is additive and thus relatively large.

If the phase of the current in one element is advanced by an amount equal to the spacing, ie 45° for a spacing of λ/8, the total phase shift is doubled for the direction of advancement and zero for the opposite direction so that a cardioid pattern results exactly as described in the previous section for λ/4 separation and 90° phase shift, Fig 12.13, though better ways of achieving equal currents together with the desired phasing will be found later (Figs 12.130 and 12.131). One highly desirable consequence of this is an increase of four times in radiation resistance, with the gain remaining the same as for an 8JK system.

Better performance is obtained with somewhat smaller values of phase shift (ϕ) and Fig 12.15 is a set of universal curves showing how gain, front-to-back ratio (nominal) and radiation resistance vary with phasing. It should be noted that as phase shift is reduced the single null of Fig 12.13 splits into two in different directions as shown by Fig 12.16 so that, despite the reduction of nominal front-to-back ratio, the average discrimination against interference tends to be improved. In fact, if interference were equally likely to arrive from all directions in space, adjustment for maximum gain would by definition ensure minimum interference. The variation of backward field strength with bearing and phase shift is plotted in Fig 12.17.

In the discussion so far both elements have been treated as if driven with equal currents, though more often only one element

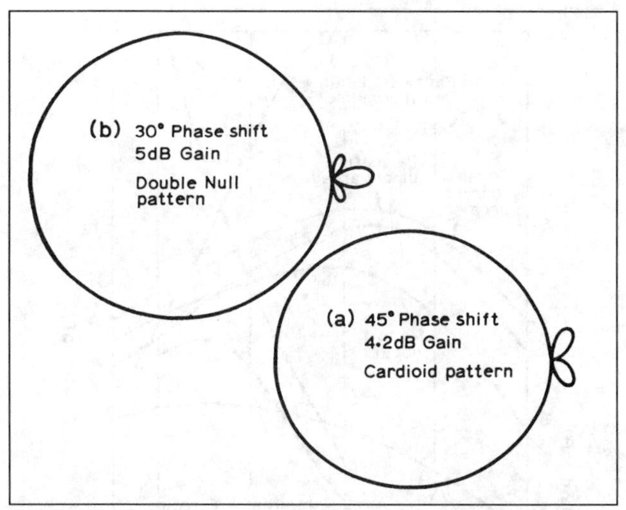

Fig 12.16. Variation of E-plane polar diagram with phase angle

is driven, the other being excited *parasitically* from the driven element via the mutual impedance, phase being adjusted by tuning the parasitic element which may be operated either as a *director* or *reflector.* In this case equalisation of currents is impossible without mutual reactance which is readily available with loops and bent elements, whereas standard practice favours straight elements; this is a special case in so far as, at normal spacings, a fairly accurate balance exists between the centre to centre inductive (+ve) and end-to-end capacitive (–ve) couplings. In consequence deep nulls are impossible, the effect on directivity being catastrophic.

The obvious remedy is to add some reactive coupling by bending the elements. Directors require an increase of inductive coupling [1, 20, 28–30] but this is usually not recommended because bringing the centres together decreases the effective spacing and so lowers the radiation resistance, whereas folding the ends inwards maintains the effective spacing besides favouring simple and compact forms of construction such as the VK2ABQ antenna. In this case, as also with most loop arrays, the parasitic element must be operated as a reflector. In addition to establishing current equality the adjustment of mutual reactance X_m also determines the phasing, resulting in a pattern similar to Fig 12.16(b) with nulls at 128° relative to the beam heading

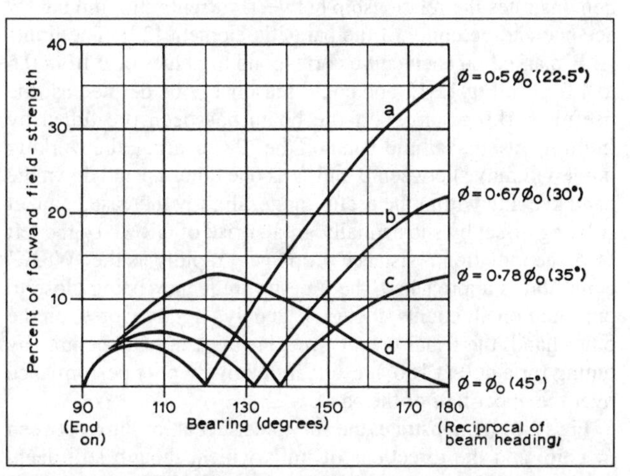

Fig 12.17. Variation of backward field strength with bearing and phase shift. Angles in brackets apply for $S = \lambda/8$. Curve (a) corresponds to maximum gain and curve (d) to maximum nominal back-to-front ratio. Curve (b) provides near-maximum gain and better rejection of interference than (d) for all angles less than 147°

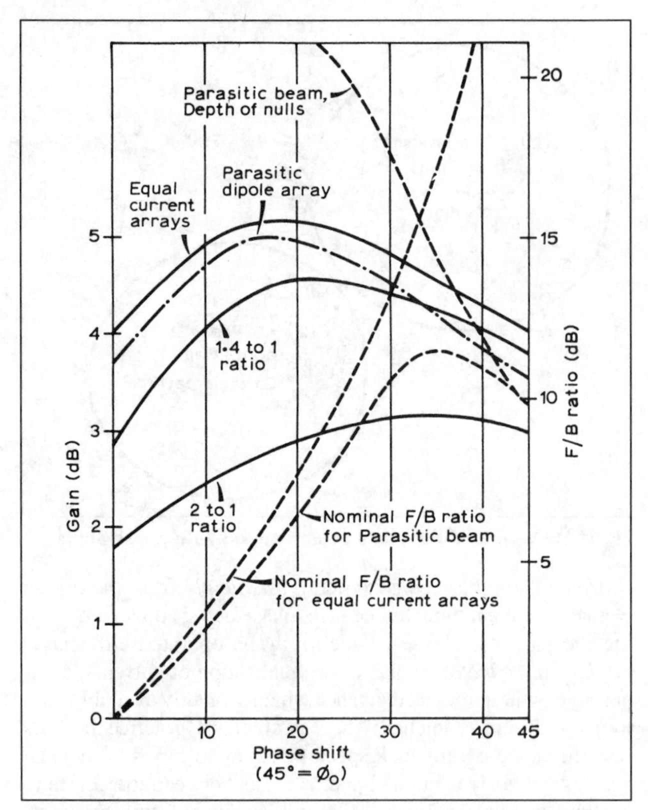

Fig 12.18. Gain and front-to-back ratio for two-element arrays with unequal currents, compared with equal current arrays. Spacing is λ/8

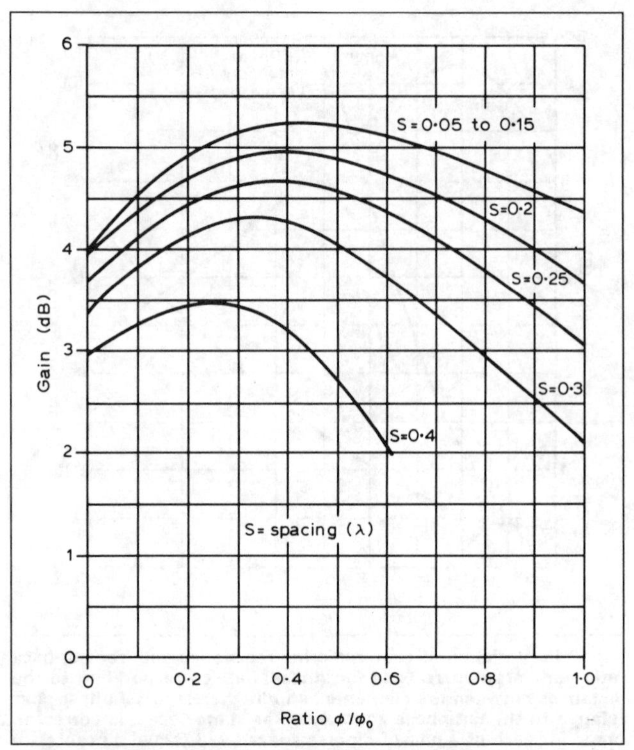

Fig 12.19. Variation of gain with phase shift at large spacings. (Top curve is reproduced from Fig 12.15 for comparison). For 'straight' arrays without current equalisation, maximum gains are only slightly reduced (eg 4.0dB for $S = \lambda/4$) but front-to-back ratios are much worse

in the event of the reflector being self-resonant. In marked contrast the most usual method of phasing, ie detuning the reflector, further degrades the current ratio.

The phase angle ϕ is given by ($\tan^{-1} X_m/R_m$) in the case of a self-resonant reflector and ($\tan^{-1} X/R$) if $X_m = 0$ and the reflector detuned by a reactance X.

Practical aspects of current equalisation

Fig 12.18 includes the effect of current ratio on performance in some typical cases, Fig 12.19 shows the extent to which the available gain decreases with increase of spacing, and Fig 12.20 demonstrates the relationship between current ratio, mutual reactance and detuning of the parasitic element [29]. The limits A, B marked on each curve correspond to values of ϕ from 0.6 to 1.0 ϕ_0 in Fig 12.15 and could reasonably be defined as 'the useful working range'. If the beam has been pre-tuned by 'nulling' a signal at mid-band on the 128° bearing, the working point will move between B and A as one tunes up and down the band keeping within the useful range, slightly increased gain at A being offset by substantially greater risk of QRM. To the left of A the radiation resistance drops very rapidly as the 'W8JK' condition is approached, the tune-up frequency being close to optimum on all counts, though B is only slightly worse; on the other hand, the usual practice of placing it at band-centre (by tuning for a null at 180°) results in relatively poor performance over the upper half of the band.

Fig 12.21 demonstrates the very precise relationships between R, gain, and the directions of nulls which, though still deep, move round as one tunes through the band. If full advantage is taken of constructional opportunities arising from small size and light weight, such antennas can be competitive with much larger arrays. Their main virtue, however, lies in the possibility,

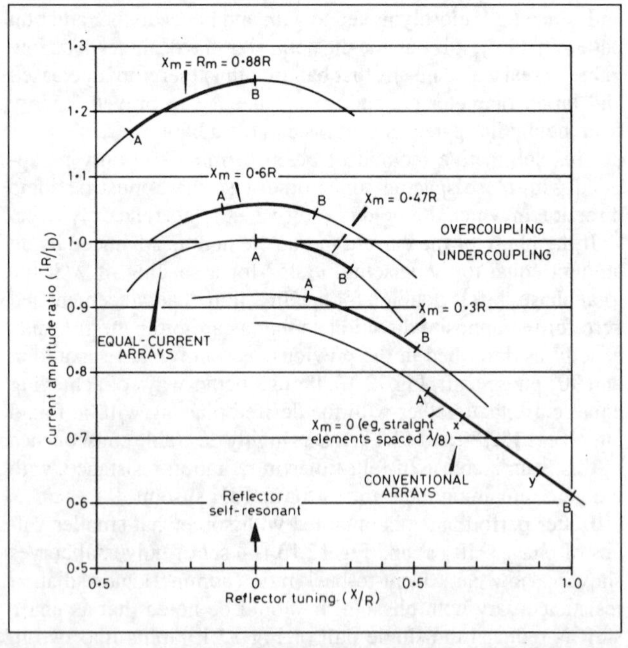

Fig 12.20. Current ratio for pairs of elements spaced λ/8 in terms of reflector tuning as X_m is varied. AB is an advisory working range in conformity with Figs 12.15 and 12.17. At A, $\phi/\phi_0 = 0.6$, nulls at ±128°. At B, $\phi/\phi_0 = 1.0$, null at 180° relative to beam heading. R is the radiation resistance of a single element, and reflectors require negative values of X_m or positive values of X; the reverse applies in the case of directors. Coupling depends on element shape and spacing, but lengths are not critical. To assist perspective, points x, y on the lower curve correspond roughly to 5% detuning of tubular and wire dipoles respectively. Variation of reflector tuning allows nulls to be moved in either direction from the mean position which is determined by the ratio of X_m to R_m

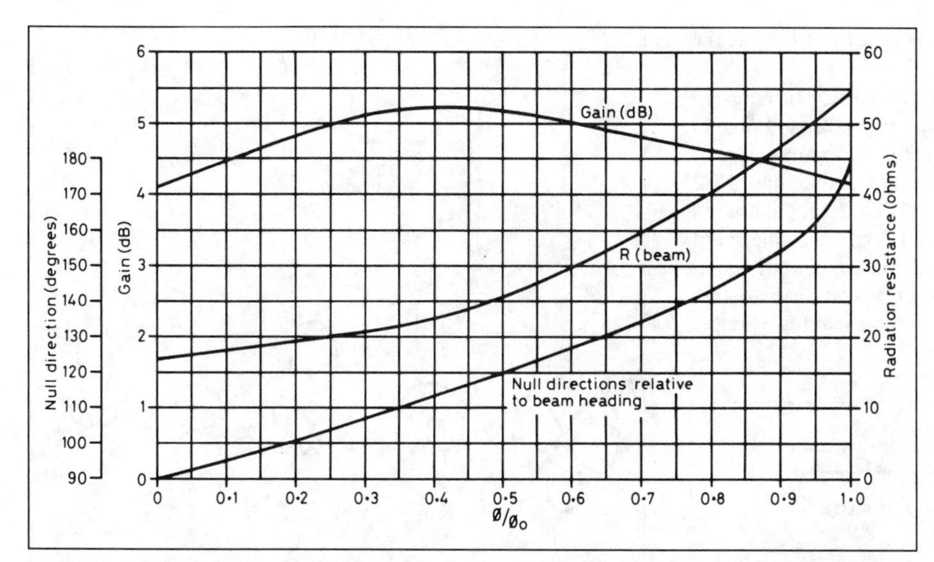

Fig 12.21. Relationship between phase angle, gain, null directions, and radiation resistance for all close-spaced, two-element beams with equal-current drive. Note: losses not included; gain figures valid within 0.3dB for null depths less than 10dB (approx). Radiation resistance (in ohms) for λ/2 elements spaced λ/8 (see text for other beams) (*Ham Radio*)

subject to the provision of separate feeders to each element, of remotely fine-tuning the reflector. It should then be a simple matter to optimise performance at any point in the band, nulls being moved around as required, and the ability to reverse beam direction instantaneously is a further important bonus; subject to correct coupling (ie if currents are equal), deep nulls are guaranteed apart from such fault conditions as feeder radiation or undue proximity to another antenna.

With two feeders fine adjustment of coupling can if necessary be carried out in the shack, otherwise strict adherence to adjustment procedures outlined on pp12.80–12.84 is essential; in particular, reliance on tune-up at ground level is unsafe, though aiming for about 25% undercoupling should reduce the margin of error and hence the risk of having to lower the antenna down again for further adjustments.

The importance of current equalisation in the case of two-element arrays has been widely recognised but methods of achieving it have for the most part not been entirely successful, tending to work well enough in some cases but not in others. The main reason for failure in the case of driven arrays lies in the fact that (as can be inferred from a study of Fig 12.37) phasing lines require to be suitably matched at both ends which is difficult or impossible with close-spaced elements, one of which may have a zero or negative value of radiation resistance, $X_m \sin \phi$ ohms being added to one element and subtracted from the other. Even if $X_m = 0$, as happens for straight dipoles spaced λ/8, problems remain since any adjustments to either element affect the matching of both.

The method explained on p12.79 based on establishing resonance at the feed point and then detuning the elements by equal amounts in opposite directions overcomes the basic problem but is essentially a 'two knob' system, requiring independent adjustment of phasing and current ratio if its potential is to be fully exploited. After long years of experience with systems of this sort the change to parasitic operation in line with Fig 12.20 effected an improvement sufficiently dramatic to ensure that in most cases driven operation is no longer a recommended option.

The choice of director or reflector operation is immaterial if the mutual impedance is purely resistive, which occurs in the case of dipoles for a spacing of about λ/8; at larger spacings, or in the case of loop elements, the mutual impedance includes capacitive reactance which increases the current in a reflector and reduces it in a director so that directors are then relatively inefficient. This is particularly marked with loop antennas (eg those using quad or delta loops, Figs 12.118 and 141), with reflector operation providing almost equal currents and deep nulls, whereas directors provide much reduced gain and back-to-front ratio.

Parasitic beams frequently employ a reflector and one or more directors but, as shown on p12.85, the increase of gain obtainable in practice with close spacing is limited to about 1dB, although the theory of supergain antennas predicts a possible gain proportional to the square of the number of elements. This is unfortunately subject to very precise adjustment of current amplitude and phases, and, as the size is reduced or the number of elements increased, bandwidth decreases and losses increase at an astronomical rate.

Point sources

We have already made use of the fact that antenna elements can be represented by point sources and, although it is always admissible to use more than one point per element, the need only arises if this radiates in more than one plane or if it contains separate concentrations of current with a distance of more than about λ/4 *between* them. In the latter case radiation from different parts of the element could arrive in phase for one direction in space and yet be appreciably different in phase for another direction, in which case the radiation pattern will differ from that of a point source, and if the reduction of radiation in one direction accounts for a significant amount of energy there may be enough gain in another to invalidate the point-source assumption. An important borderline case is provided by the quad loop (Fig 12.141), the top and bottom of which could be regarded as individual point sources separated by λ/4; with a single loop this slightly reduces the radiation in the vertical plane and provides a small stacking gain, about 1dB. With loops assembled into a beam, the stack approximates to a pair of Yagis which, as discussed later, require a separation of the order of λ/2 instead of λ/4 before they can be regarded as stacked effectively, so that even in this case no appreciable error is to be expected from the point source approximation.

Array factor

The radiation pattern of a large antenna array can often be calculated by breaking down the array into units of which the individual pattern is known, and then combining the patterns of the units together. For the purposes of combining the unit patterns, each is assumed to be a point source of radiation, located at its physical centre, such as the antennas A and B in Fig 12.22, and an expression F_1 derived for the pattern of the individual sources. The final pattern F of the whole antenna is then obtained by multiplying F_1 by the pattern F_2 of two antiphase point sources spaced λ/2 so that:

$$F = F_1 \times F_2$$

The expression F_1 is known as the *unit pattern* and F_2 as the *array factor.*

For very large arrays of dipoles this technique can be repeated by successive breaking down of the array into smaller and smaller units. The application of the array factor to the determination of the E-plane polar diagram for a pair of collinear antiphased λ/2 dipoles is shown in Fig 12.22, where F is the final pattern, F_1 that of the dipole and F_2 the array factor or pattern of the set of point sources. F_1 is unity for the azimuth pattern of vertical antennas. If each antenna were fitted with a reflector, then another multiplying factor would be included to represent for example the cardioid (heart shape) of Fig 12.13.

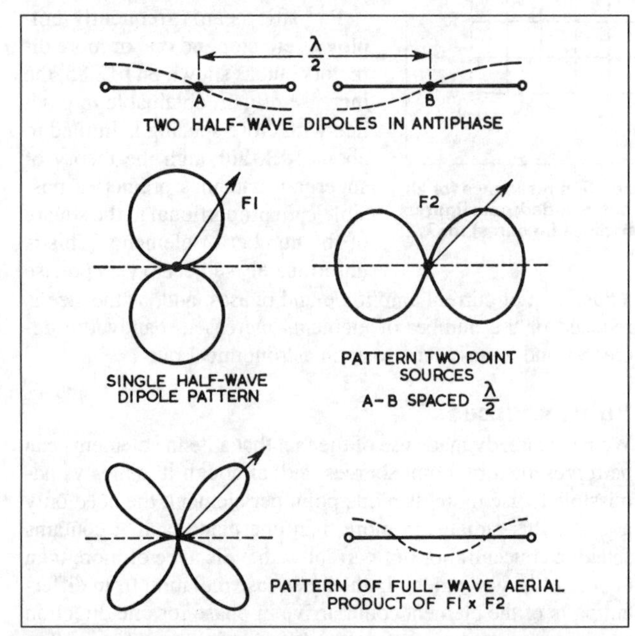

Fig 12.22. The array factor. Each dipole has a pattern like F1. Two sources in opposite phase at A and B would have a pattern like F1. For any direction (eg the arrows) the chord F1 is multiplied by that of F2 for the same angle. When this is done for all angles the pattern shown at the bottom is obtained for the complete antenna

A particular application of array factor is the conversion of H-plane to E-plane polar diagrams for arrays made up of parallel λ/2 dipoles. Once the H-plane pattern has been established by measurement or calculation, the E-plane pattern is obtained merely by multiplying the H-plane pattern by the dipole factor F_d (the E-plane pattern of a single λ/2 dipole) so that:

$$\text{E-plane} = \text{H-plane} \times F_d$$

Since the λ/2 dipole possesses E-plane directivity normal to its axis, the effect of converting the H-plane diagram is to sharpen it in the directions at right-angles to the axis of the dipoles at the expense of radiation in the directions along the axis of the dipoles. This is illustrated in Fig 12.23 which shows the E- and H-plane diagrams for an array of two parallel dipoles with reflectors.

Vertical plane diagrams are found in the same way, by treating the antenna and its image in the ground as an array of two sources spaced by twice the height of the real antenna.

It is important to realise that unless F2 is narrower than F1 in at least one plane its effect on the overall pattern will be negligible so there will be no stacking gain; this means that in general the higher the unit gain, the greater must be the stacking distance, though there are exceptions such as a broadside stack of collinear elements in which case the two patterns are orthogonal

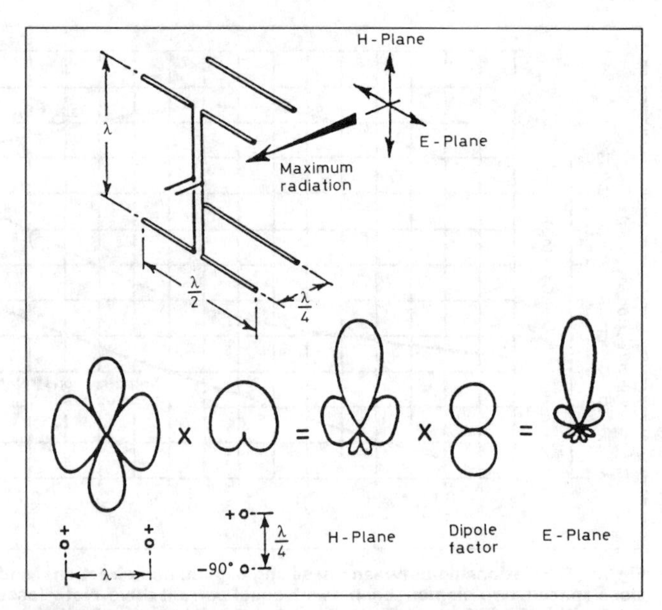

Fig 12.23. Conversion of H-plane to E-plane polar diagrams using the principle of array factor

and therefore independent; in this case gains can be directly added. An interesting example of the same principle arises in the case of arrays of short elements which have gains relative to single elements of the same type up to 0.4dB greater than their gain in dBd; this is because the λ/2 reference dipole itself has a gain of 0.4dB due to radiating less energy in nearly-endwise directions which are outside any array pattern. It follows that in the case of an array the overall pattern (and therefore gain) is to a close approximation independent of element length; typical examples include quad loops (which closely resemble shortened folded dipoles), the inverted-V, the VK2ABQ array, and the small delta-loop arrays featured later in this chapter, all of which can be expected to provide virtually the same gain in dBd as the same number of λ/2 elements.

Vertical radiation patterns over earth

The relationship between the sign of the antenna and its image for the separate cases of horizontal and vertical polarisation is shown in Fig 12.24. In the case of the horizontally polarised antenna, the image is of opposite sign or phase, and the resultant field along the surface of the reflecting plane will always be zero. In the case of vertically polarised antennas over perfect ground the image would be in phase and the field a maximum at low angles.

For horizontal polarisation, there are maxima in the vertical plane whenever $2h \sin \theta$ is equal to an odd number of half-wavelengths, thus producing a phase reversal, and at these angles the direct wave and ground-reflected waves reinforce each other, giving 6dB gain compared with the free-space pattern. On the other hand, there is almost-zero radiation when $2h \sin \theta$ is equal to zero or an even number of half-wavelengths. Typical vertical plane polar diagrams are given in Fig 12.25(c) and (d), and it will be evident from the above discussion that these represent the array factor for any two sources fed in antiphase and spaced by $2h$.

Vertical polar diagrams for any horizontal array may therefore be obtained by multiplying its free-space polar diagram by the appropriate pattern from Fig 12.25, with the nose of each lobe corresponding to a voltage multiplication of two, and it is important to note that this multiplying factor is entirely independent of the type of antenna. If the array has a narrow

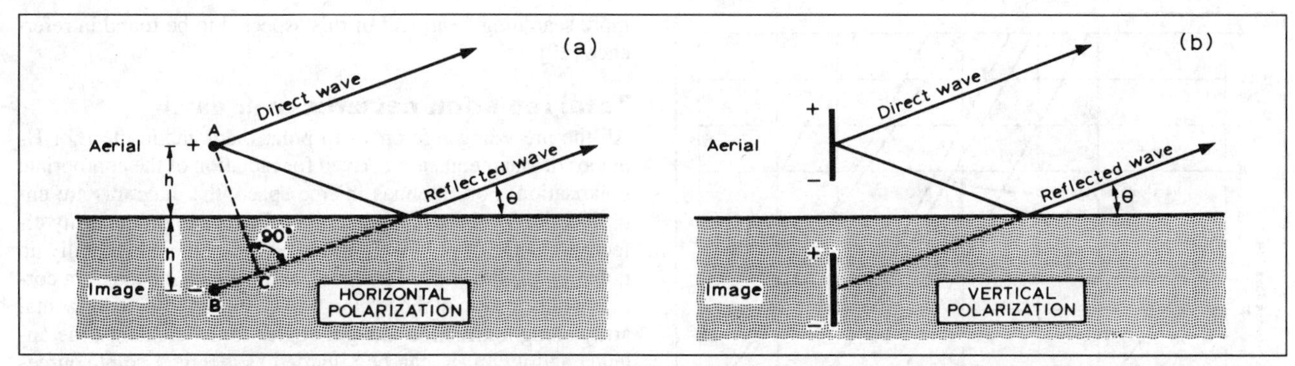

Fig 12.24. Images of antennas above a perfectly conducting ground plane. The vertically polarised antenna produces an image which is in phase and supports radiation along the surface. The horizontally polarised image is in antiphase and cancels the radiation along the earth's surface. The path length from the image is greater by the distance $2h \sin \theta$, resulting in a phase difference of $4\pi h \sin \theta/\lambda$ radians

free-space radiation pattern in the direction corresponding to the vertical plane, this would suggest that it derives its free-space gain from restriction of radiation in this plane rather than by a narrow azimuthal beamwidth, and the narrow vertical pattern does not imply any additional gain due to a lowering of the angle of radiation in the presence of the ground. The effect of the ground is almost fully accounted for by the appropriate array factor which is identical for all horizontal antennas at the same height, but there is another small factor resulting from the variation of radiation resistance with height, Fig 12.101. This can increase or decrease the current, causing a loss of 1.2dB for a dipole at a height of 0.4λ or a gain of 0.9dB if the height is 0.55λ etc. It is to be expected that this effect will be neutralised in the case of close-spaced beams with antiphase elements such as the 8JK and its derivatives since each element 'sees' two images of opposite phase and almost equidistant.

Effect of earth conductivity

So far the ground has been considered as a perfect conductor which is a useful approximation in the case of horizontal polarisation but far from true for vertical polarisation except for the special case of sea water when it is valid for elevation angles greater than about 3°.

Due to a phenomenon which is known as the *Brewster effect* there is an almost complete sign reversal of the image at the angles of elevation most important for DX, ie a few degrees, so that vertical and horizontal antennas centred at the same height tend to give very similar performance.

At the *Brewster angle*, which is of the order of 10–20° depending largely on ground conductivity and dielectric constant, the reflection coefficient falls to a low value and at higher angles the performance starts to decrease on account of the dipole factor. Even this is, however, an oversimplification since the

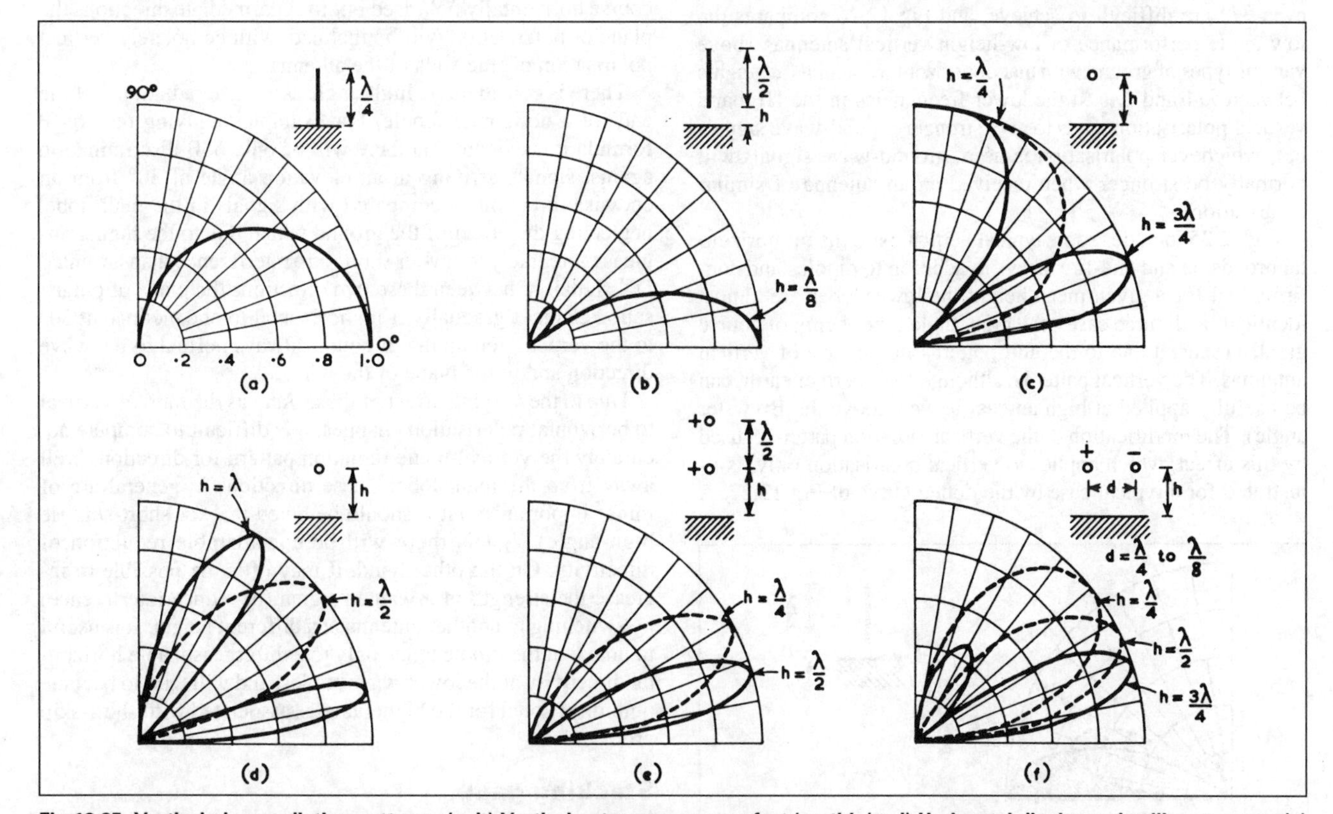

Fig 12.25. Vertical plane radiation patterns. (a, b) Vertical antenna over a perfect 'earth'. (c, d) Horizontal dipoles and collinear arrays. (e) Broadside horizontal arrays. (f) End-fire horizontal systems, eg W8JK. Only half of each pattern is shown: they are symmetrical about the vertical axis unless reflectors are also used. Diagrams (a) and (b) hold for all azimuth directions: the remainder are for broadside direction only. The beamwidth in azimuth depends on the length of the arrays as shown in Fig 12.124

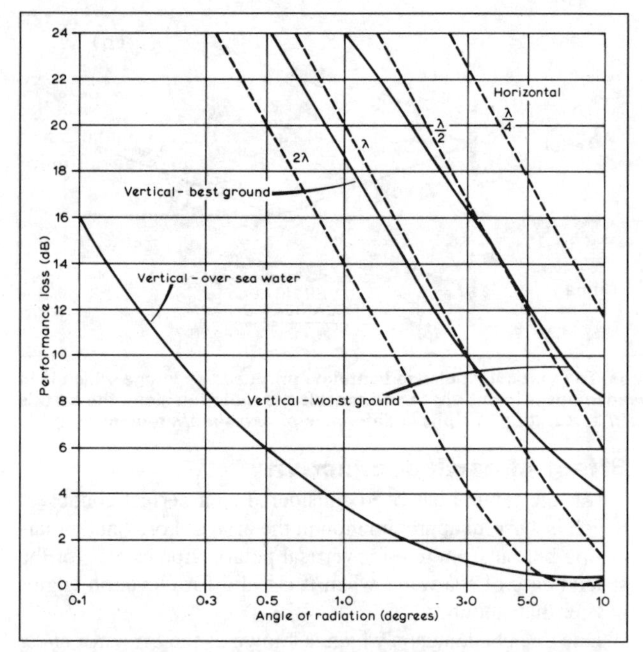

Fig 12.26. Comparison of short horizontal and vertical radiators at HF, assuming flat open country. 'Zero loss' occurs with in-phase addition of the direct wave and a reflected wave of equal amplitude. Antenna heights are indicated in wavelengths for horizontal polarisation (dotted curves). Vertical polarisation curves are calculated for low height and a frequency of 7MHz; performance deteriorates slightly as frequency increases

reflection coefficient at low angles is appreciably less than unity, and leaves some useful radiation however low the antenna. This can be extremely useful at low frequencies where heights of even λ/2 are difficult to achieve, and Fig 12.26 compares the low-angle performance of low-height vertical antennas above various types of ground with that of horizontal antennas at heights between λ/4 and 2λ. At the lower frequencies in the HF band vertical polarisation tends to give stronger ground-wave signals but, whichever polarisation is used, ground-wave signals will normally be stronger when received on an antenna of similar polarisation.

Fig 12.25 includes some typical vertical patterns for horizontal broadside and end-fire arrays in addition to dipoles and confirms that for a given mean height the lower lobes are almost identical in all three cases, the high-angle lobes being of course greatly reduced due to the unit patterns in the case of vertical antennas. The vertical patterns, although for a perfect earth, can be usefully applied at high angles (ie well above the Brewster angle). The modification of the vertical radiation pattern caused by this effect, which applies to vertical polarisation only, is illustrated for a typical case by the dotted curve of Fig 12.27. A more searching treatment of this aspect is to be found in reference [20].

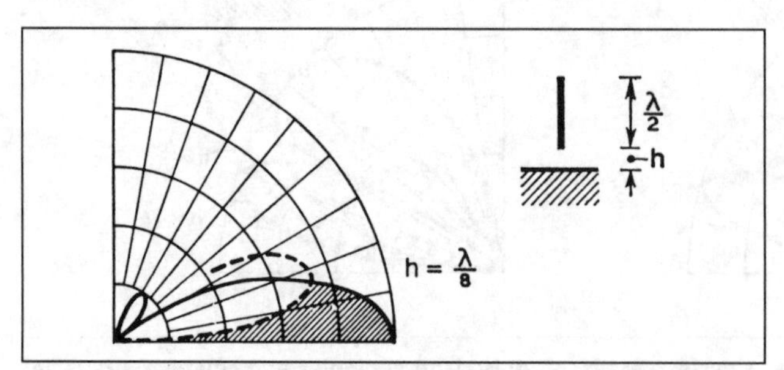

Fig 12.27. Effect of ground conductivity and Brewster reflection on vertical radiation pattern of a vertically polarised antenna over typical ground

Total radiation patterns over earth

All the preceding references to polar diagrams in the E or H-planes of an antenna are derived for radiation of the appropriate polarisation. For antennas in free space, the diagrams are entirely valid for there is no external influence to produce any effective change of polarisation. In free space there is literally no radiation off the ends of a dipole, and the E-plane diagram correctly falls to zero in those directions. Similarly, for a horizontal array along a direction at right-angles to the elements, the antenna and its image can be regarded as identical point sources and the polar diagrams computed from the appropriate array factor as described. However, it should be appreciated that the diagram so obtained is relative to the centre of the system, ie a point midway between the antenna and its image, zero elevation being relative to this point and not to the actual antenna, ie zero field coincides with ground level.

If a dipole is disposed horizontally above an earth plane acting as a reflector, there will then be some radiation off the end even at zero elevation, as may be appreciated by holding a pencil parallel to the surface of a mirror and observing the image from the end-on direction at various heights above the mirror. Closing one eye, the pencil shrinks to a dot but the image, though perhaps greatly reduced in length, remains visible. Neglecting ground losses, the amount of radiation is determined by the usual cos θ formula and is therefore proportional to the apparent length of the image as viewed in this way. Since this image appears to be in the vertical plane if the mirror is horizontal, the radiation is vertically polarised.

Ground-wave propagation normal to the antenna wire is of course horizontally polarised but for intermediate directions the plane of polarisation will be inclined, with completely vertical polarisation off the ends of the antenna.

There is considerable high-angle sky-wave radiation 'off the end' of a horizontal dipole. For example, applying the cos θ formula it is evident that there will be only 6dB discrimination against signals arriving at an elevation angle of 30° from an endwise direction as compared with signals in the main lobe, neglecting the effect of the ground which due to the higher angle is more likely to favour short-skip interference than a wanted DX signal. In between these two directions the plane of polarisation changes gradually from horizontal to one inclined at 30° to the vertical, remaining of course always normal to the wave direction and in the plane of the wire.

Due to the varying effect of the ground as the ratio of vertical to horizontal polarisation changes, it is difficult to compute accurately the vertical-plane radiation pattern for directions well away from the main lobe. These directions in general are of minor importance but it should be noted that for short-skip (ie high-angle) signals there will be considerable reduction of directivity. On the other hand, it may often be possible to increase the strength of a wanted signal (or reduce interference) by switching to another antenna at a different height. It is useful to note that the dipole tends only to exhibit reasonable horizontal directivity at the lower wave angles, and will tend to become omnidirectional for the higher angles associated with short-skip propagation.

Stacking gain

The general relationship between gain and directivity was explained on p12.3, the effect of mutual coupling on antenna currents on p12.6, and the procedure for building up the directional

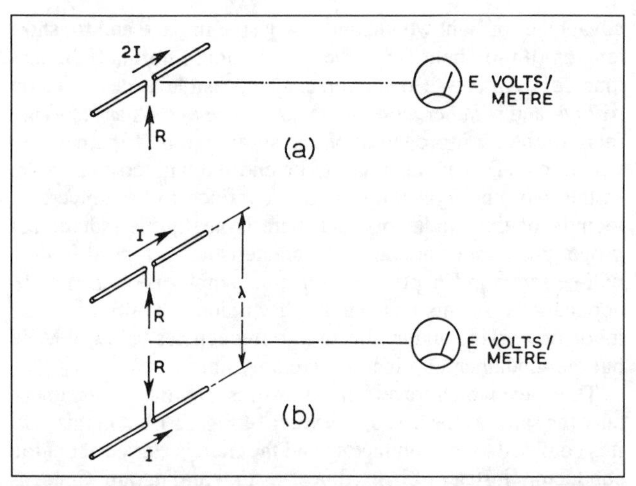

Fig 12.28. Stacking two dipoles gives an effective gain of 3dB (ignoring mutual effects) because the input power is proportional to the square of the current and the distant field to the total current

pattern from individual elements has now also been covered. It remains to assemble these factors together to establish the gain of arrays of various types. The achievement of gain with collinear, broadside and widely spaced end-fire arrays depends upon the fact that the distant field, which excites the receiving antenna, is directly related to the radiation current, whereas the power fed to the antenna is proportional to the square of the radiation current. Fig 12.28 shows two λ/2 dipoles stacked one wavelength apart, each with a radiation resistance *R.* They are centre fed so that the radiation resistance is nominally the same as the input impedance, and each carries a loop radiation current *I.* For one dipole, the distant field is *E* due to a radiation current 2*I,* and the power required to be fed to the dipole to achieve this is $(2I)^2 \times R = 4I^2R$. If the second dipole is now introduced and the current is split between them so that each has a current *I* and gives a distant field *E*/2, the fields will be in phase broadside to the dipoles and will add to give the same field strength *E* as before, but the power delivered to each dipole is now I^2R, so the total power is $2I^2R$. For the same distant field, the power required is only half that of a single dipole, so a power gain of 3dB has resulted from stacking the two dipoles. What has happened is that the fields radiated by the individual dipoles have added in phase in the desired directions but have either cancelled or added less effectively in others, which results in a sharpening of the radiation pattern in the desired direction, and hence gain.

By simple extension this argument leads to the result that *n* elements provide a gain of *n* times, but this is slightly in error since it fails to take account of mutual impedance which, as demonstrated earlier, may increase or decrease the element currents somewhat even with wide spacing. Typically a reduction of 1.2dB is found for a collinear pair and an increase of 1.8dB for an optimally spaced broadside pair so that the effective gains in these two cases will be 1.8dB and 4.8dB respectively. The latter figure is the maximum gain obtainable from two elements fed in phase, and the required spacing (0.7λ) is known as the *optimum stacking distance.* As shown on p12.11, however, considerably greater gain is obtained from close-spaced elements with nearly-antiphase excitation. In the example of the two dipoles spaced one wavelength apart, the mutual resistance is very nearly zero, and the simple figure of 3dB is correct.

The principle of stacking gain may be extended to more than two radiators, and to radiators each of which is highly directive in its own right, but wherever many elements are involved, consideration must be given to the mutual coupling between each element and all other elements in turn. The derivation of the optimum stacking distance for such systems becomes very complicated and varies with the type of array, but in general the higher the directivity and intrinsic gain of each unit in the array, or the larger the number of elements, the greater is the optimum stacking distance between them. Two broadside dipoles require a spacing of 0.7λ, four are better spaced at 0.8λ intervals, eight at 0.9λ intervals and so on. On the other hand, to get maximum gain from a fixed length of array, instead of a fixed number of elements, the optimum stacking distance tends to remain constant regardless of the length. Typically, a broadside or end-fire array 4λ in length made up of λ/2 dipoles would provide about 12.5dB gain in each case, the number of elements required being 7 and 13 respectively [4]. It will be noted from the broadside examples that there is little or no change in optimum stacking distance as the array is lengthened.

When a fixed number of beam arrays are stacked, the optimum distance increases roughly as the square root of the power gain and for small Yagi arrays with 6dB gain (p12.75) it is equal to about 1.2λ, though not unduly critical.

This aspect is covered more fully in Chapter 13 – 'VHF/UHF antennas'.

TRANSMISSION LINES

Three separate parts are involved in an antenna system: the radiator, the feed line between transmitter and radiator, and the coupling arrangements to the transmitter. Wherever possible, the antenna itself should be placed in the most advantageous position, and a feed line used to connect it to the transmitter or receiver with a minimum of loss due to resistance or radiation. In some circumstances (eg in temporary installations or to simplify multiband operation) the feed line is omitted and the end of the antenna is brought into the station and connected directly to the apparatus.

By the use of transmission lines or *feeders,* the power of the transmitter can be carried appreciable distances without much loss due to conductor resistance, insulator losses or radiation. It is thus possible to place the antenna in an advantageous position without having to suffer the effects of radiation from the connecting wires. For example, a 14MHz dipole 33ft in length can be raised 60ft high and fed with power without incurring appreciable loss. If, on the other hand, the antenna wire itself were brought down from this height to a transmitter at ground level, most of the radiation would be propagated from the down-lead in a high angular direction. An arrangement of this nature would be relatively poor for long-distance communication.

Types of line

There are three main types of transmission line:

(a) The single wire feed arranged so that there is a true travelling wave on it, ie the line is terminated in its characteristic impedance (Z_0);

(b) The concentric line in which the outer conductor (or sheath) encloses the wave (*coaxial feeder*);

(c) The parallel wire line with two conductors carrying equal but oppositely directed currents and voltages, ie balanced with respect to earth (*twin line*).

Single-wire feeders are usually connected to a point on a resonant antenna where the impedance formed by the left- and right-hand portions in parallel matches the Z_0 of the wire (see Fig 12.103(a) on p12.64). This is now rarely used, being basically

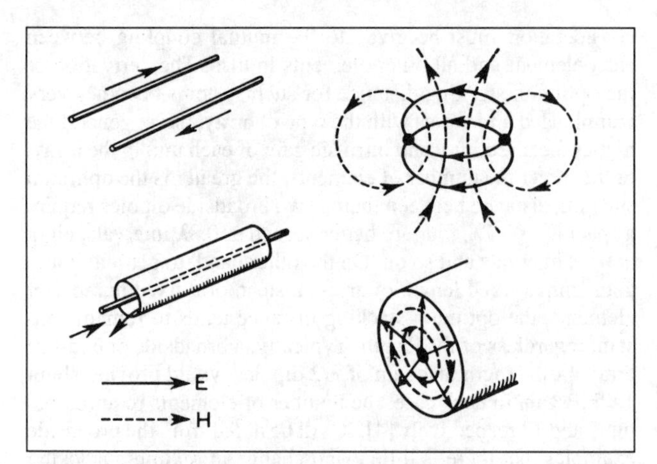

Fig 12.29. Coaxial and two-wire transmission lines, with cross-sections of their wave fields. The field directions correspond to waves entering the page

inefficient due to losses in the return path, which is via the ground, and to radiation from the feeder which acts also to some extent as a terminated long-wire antenna; against this it offers the advantages of light weight and low visual impact, and for short lengths (up to about 0.5λ) the losses should normally be less than 1dB. An end-feed version is also possible as described on p12.64, and is rather more interesting since as well as resolving more problems it proved to be the starting point for a new development allowing antennas to be end-fed with coaxial cable. In the two other types, the field is confined to the immediate vicinity of the conductors and there is negligible radiation if proper precautions are taken. Characteristics of typical feeders will be found in Chapter 13 with some emphasis on their VHF applications. In this chapter the discussion is confined to the mode of working and application to frequencies below 30MHz but the fundamental principles are the same.

The wave which travels on a transmission line is fundamentally the same as the free-space wave of Fig 12.1 but in this case it is confined to the conductors and the field is curved about the conductors instead of being linear as in that diagram. Concentric and two-wire lines and their associated fields are illustrated in Fig 12.29. In the concentric line the current passes along the centre conductor and returns along the inside of the sheath. Due to *skin effect* at high frequencies the currents do not penetrate more than a few thousandths of an inch into the metal; hence with any practical thickness of the sheath there is no current on the outside. The fields are thus held inside the cable and there is no radiation, provided that current is not allowed to flow on the outside of the cable.

In the twin line the two wires carry 'forward and return' currents producing equal and opposite fields which effectively neutralise each other away from the immediate vicinity of the wires. When the spacing between wires is a very small fraction of the wavelength, the radiation is negligible provided the line is accurately balanced. In the HF range, spacings of several inches may be employed, but in the VHF range small spacing is important.

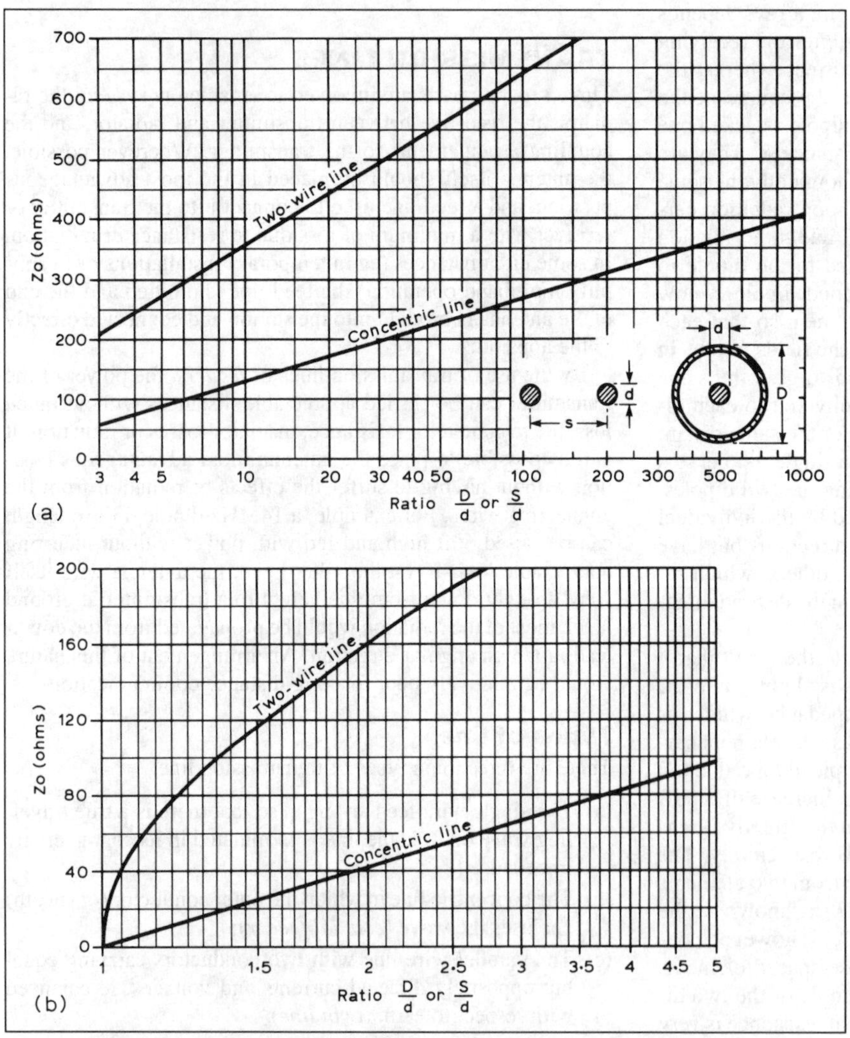

Fig 12.30. Chart giving characteristic impedances of concentric (coaxial) and two-wire lines in terms of their dimensional ratios, assuming air insulation. When the space around the wires is filled with insulation, the impedance given by the chart must be divided by the square root of its dielectric constant (permittivity). This ratio is called the *velocity factor* because the wave velocity is reduced in the same proportion

Characteristic impedance

In the earlier section on the behaviour of currents in radiating wires, it was stated that a travelling wave is one which moves along a wire in a certain direction without suffering any reflection at a discontinuity. The same applies to transmission lines, although in this case the presence of reflections and therefore standing waves does not cause radiation if the line remains balanced or shielded. If the line were infinitely long and free from losses a signal applied to the input end would travel on for ever, energy being drawn away from the source of signal just as if a resistance had been connected instead of the infinite line. In both cases there is no storage of energy such as there would be if the load included inductance or capacitance and the line, so far as concerns the generator of the signal, is strictly equivalent to a pure resistance. This resistance is known as the

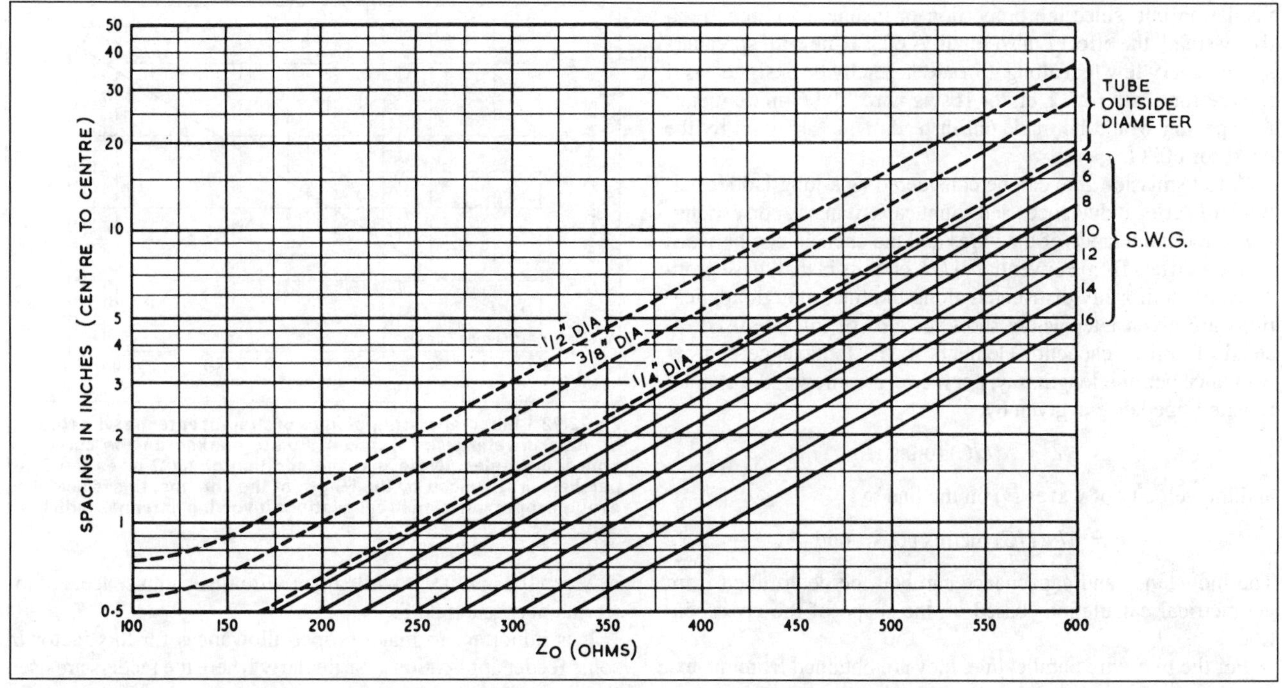

Fig 12.31. Characteristic impedance of balanced lines for different wire and tube sizes and spacings, for the range 200–600Ω. The curves for tubes are extended down to 100Ω to cover the design of Q-bar transformers. Air spacing is assumed

characteristic impedance of the line and usually denoted by the symbol Z_0. Suppose now that at some distance from the source we cut the line; what has been removed is still an infinitely long line and equivalent to a resistance Z_0 so if we replace it by an actual resistance of this value the generator will not be aware of any change. There is still no reflection, all the power applied to the input end of the line is absorbed in the terminating resistance, and the line is said to be *matched*.

Again because no reflections occur at the end of a correctly matched line, the ratio of the travelling waves of voltage and current, V/I, is Z_0. This enables the load presented to the feeder by the antenna to be in turn presented to the transmitter, without any change in the process. This is irrespective of the length of line employed, since the value of the characteristic impedance Z_0 is independent of the length of the line. In order to achieve maximum efficiency from a transmission line, it should be operated as close to a matched condition as possible, ie the load presented by the antenna should be arranged, either directly or by means of some impedance transformer, to present a good match to the line. However, the degree to which the load impedance can be permitted to depart from the characteristic impedance without introducing appreciable extra losses is quite large, as discussed in the section on attenuation. On the other hand, bandwidth considerations or the need to avoid load variations which could damage transmitters may impose stringent matching requirements.

The characteristic impedance is determined by the dimensional ratios of the cross-section of the line, and not by its absolute size. A concentric line or cable with diameter ratio 2.3:1 and air dielectric is always a 50Ω line, whatever its actual diameter may be. If it is connected to an antenna of 50Ω radiation resistance, then all the power available at the far end of the line will pass into the antenna and the impedance at the sending end of the line will also appear to be 50Ω. (Coaxial cables can conveniently be constructed with characteristic impedance values between about 50 and 120Ω. Twin lines have higher impedances: in practice, between 80 and 600Ω. Figs 12.30 and 12.31 show the characteristic impedance of coaxial and two-wire lines in terms of the dimensional ratios, assuming air between the conductors. The formula for concentric lines of inner and outer diameters d and D respectively is:

$$Z_0 \text{ (ohms)} = 138 \log (D/d)$$

Thus if the diameter ratio D/d is 2.3:1 the logarithm of 2.3 is 0.362 and this multiplied by the constant 138 gives $Z_0 = 50\Omega$. If the space between conductors is filled with insulating material with a *dielectric constant* ε (permittivity) greater than unity, the above value of Z_0 must be divided by the square root of the dielectric constant.

The usual material for insulation is polythene, which has a permittivity of 2.25. The square root of 2.25 is 1.5 so that a 'solid' polythene cable has a characteristic impedance two-thirds of the value given by the formula. Many cables have a mixed air/polythene dielectric, and for these it is necessary to estimate the effect of the dielectric; this can be carried out by measuring the *velocity factor* as described later.

The characteristic impedance of two-wire lines of wire diameter d and centre spacing S is given by the approximate formula:

$$Z_0 \text{ (ohms)} = 276 \log_{10} (2S/d)$$

For greater accuracy at small spacings use should be made of Fig 12.31. Appreciable errors may be introduced by bending and, during use, such lines should not be coiled with the turns touching.

As with concentric lines, an allowance must be made for the effect of the insulation. A thin coat of enamel or even a PVC covering will not produce any material change in lines with an impedance of 300Ω or less, but in the moulded feeder commonly known as *80Ω flat twin* the electric field between the wires is substantially enclosed in the polythene insulator, which is thus effective in reducing the impedance to two-thirds of the value given by the formula.

When an open-wire line is constructed with wooden or polythene spacers it is again difficult to estimate the effect of the insulation but, since the proportion of insulator to air is relatively small, the effect is also small. A 600Ω line with spreaders spaced every few feet along it would normally be designed as if it were for about 625Ω, eg for 16swg wires (0.064in diameter) the spacing would be made 6in instead of 5in as given by the chart for 600Ω.

A transmission line can be considered as a long ladder network of series inductances and shunt capacitances, corresponding to the inductance of the wires and the capacitance between them. It differs from conventional *L/C* circuits in that these properties are uniformly distributed along the line, though applications are given later where short sections of line are used instead of coil or capacitor elements. If the inductance and capacitance per unit length, say, per metre, are known, the characteristic impedance is given by:

$$Z = \sqrt{(L/C)} \text{ (ohms)}$$

and the velocity of waves (v) on the line by:

$$v = 1/\sqrt{(L/C)} \text{ metres per second}$$

The inductance and capacitance can both be determined from geometrical calculations based on the shape of the cross-section.

For the two-wire parallel line, they are obtained from the expressions:

$$L = 0.921 \log_{10} (2S/d) \text{ microhenrys/metre}$$

$$C = \frac{12.05\varepsilon}{\log_{10} (2S/d)} \text{ picofarads/metre}$$

where ε is the dielectric constant of the material in the space between the conductors (= 1.0 for air). Similarly for concentric lines:

$$L = 0.46 \log_{10} (D/d) \text{ microhenrys/metre}$$

$$C = \frac{24.1\varepsilon}{\log_{10} (D/d)} \text{ picofarads/metre}$$

Characteristic impedance is an important property also of antennas: a λ/2 dipole can be treated for this purpose as a λ/4 line which has been opened out, thereby reducing capacitance and increasing inductance so that it acquires a higher value of Z_0 as plotted in Fig 12.32, based on reference [31]. For a single wire with ground return Z_0 is given by:

$$138 \log (2h/r)$$

where h is the height and r is the radius of the conductor.

Velocity factor

When the medium between the conductors of a transmission line is air, the travelling waves will propagate along it at the same speed as waves in free space. If a dielectric material is introduced between the conductors, for insulation or support purposes, the waves will be slowed down and will no longer travel at the free-space velocity. The velocity of the waves along any line is equal to $1/\sqrt{(LC)}$; from the expressions for L and C given above, the value of C depends upon the dielectric constant of the insulating material. The introduction of such material increases the capacitance without increasing the inductance, and consequently the characteristic impedance and the velocity are both reduced by the same factor $\sqrt{\varepsilon}$. The ratio of the velocity of waves on the line to the velocity in free space is known as the *velocity factor.* It is as low as 0.5 for mineral or PVC insulated lines and is roughly 0.66 for solid polythene cables (ε = 2.25). Semi-airspaced lines have a factor which varies between 0.8 and 0.95, while open-wire lines with spacers at intervals may reach 0.98.

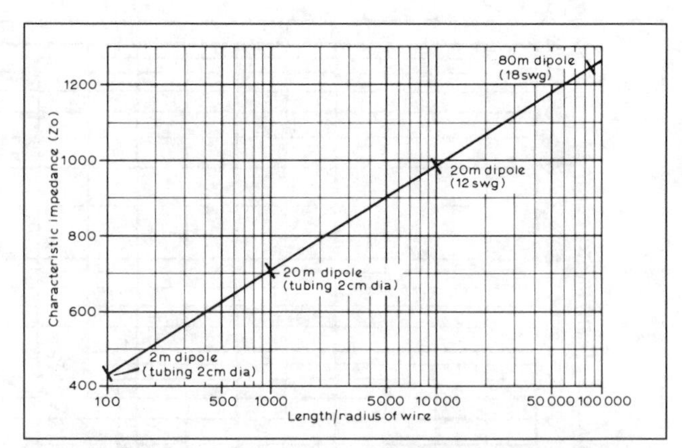

Fig 12.32. Characteristic impedance of straight antenna wires (based on data in reference [4], p864). Points marked on the curve are typical examples. (Note that the addition of 120Ω gives the Z_0 of parallel wires spaced by the length of the antenna; this is useful as a rough approximation in calculations involving antennas with bent ends)

It is important to make proper allowances for this factor in some feeder applications, particularly where the feeders are used as tuning elements or interconnecting lines in antenna arrays, or as chokes for achieving EMC. For example, if $v = 2/3$, then a λ/4 line would be physically λ/6 long (2/3 × 1/4 = 1/6).

In practice, the velocity factor v can be found by short-circuiting a length of cable with about 1in of wire formed into a loop and then coupling the loop to a grid dip oscillator. The *lowest* frequency at which the cable shows resonance corresponds to an *electrical* length of λ/4; then

$$v = \frac{f\text{(MHz)} \times \text{Length (feet)}}{246}$$

and should have a value between 0.5 and unity.

Standing waves

It has already been stated that when a transmission line is terminated by a resistance equal in value to its characteristic impedance, there is no reflection and the line carries a pure travelling wave. When the line is not correctly terminated, the voltage-to-current ratio is not the same for the load as for the line and the power fed along the line cannot all be absorbed – some of it is reflected in the form of a second travelling wave, which must return along the line. These two waves, forward and reflected, interact all along the line to set up a *standing wave.*

The flow of power along the line can be interpreted as the progress along the line of a voltage wave and a current wave which are in phase, the product of which is the value of the power flowing.

If the voltage V at a point on the line is given by the expression:

$$V = V_0 \sin \omega t$$

and the current at the same point by:

$$I = I_0 \sin \omega t$$

then the amount of power flowing is the product of the RMS voltage and current:

$$P = 0.707\, V_0 \times 0.707\, I_0 = 0.5\, V_0 I_0$$

and is independent of the time or position on the line, varying only as the peak amplitude of the voltage or current wave is altered. Since the voltage and current waves can be expressed in identical terms, it is convenient to consider the current wave, bearing in mind that the discussion is equally applicable to the voltage wave.

The I and V waves are both of sinusoidal form, varying in amplitude at an angular rate $\omega = 2\pi f$ where f is the frequency at which the transmitter is generating the RF power. Such a wave may be represented graphically by a vector of constant magnitude (or length) rotating at an angular speed ω. The actual instantaneous value of the wave at any moment is obtained by projecting the length of the vector on to a line passing through the origin (Fig 12.33). The nature and use of vectors is fully covered in any textbook dealing with AC waves but for the present purpose it is sufficient to regard a vector as a line with its length representing amplitude and its angle indicating relative phase.

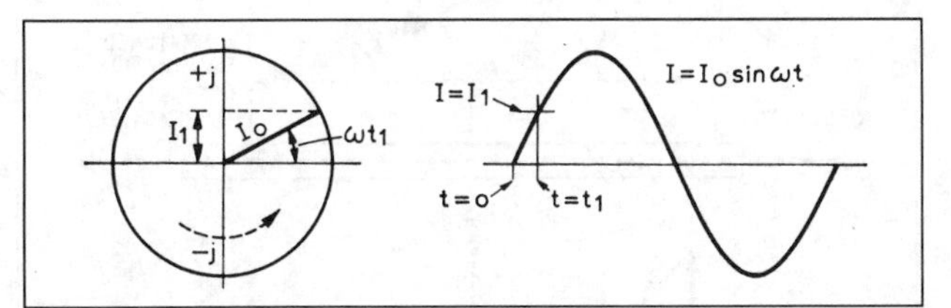

Fig 12.33. Graphical interpretation of a sinusoidal wave represented by a vector. A cosine wave is similar but displaced by quarter of a cycle

Thus the distribution of current along the line *at a particular moment of time* due to the passage of the current wave is also sinusoidal, with the phase of the current lagging more and more with respect to the generator as the distance from it increases. It can therefore be represented by a whole series of vectors, each appropriate to a particular physical point on the line. This illustrates the physical meaning of the statement that a piece of line is 'one wavelength' long, since this is the distance between adjacent points along the line at which the current is equal in amplitude and phase, ie the vectors are identical (Fig 12.34).

The vectors are drawn for a loss-less line, in which the current wave suffers no attenuation during its passage along the line. It is convenient to explain the principles of generation of standing waves by reference to the behaviour of such a line and subsequently to examine what happens when the line exhibits some amount of attenuation, as is the case in practice to a greater or lesser degree.

The current wave moving along the line will eventually reach the far end, and will then be influenced by the termination it meets. In order to establish the magnitude of this effect we must now look at the various loads that can exist. These may be divided broadly into three groups depending upon whether all, some, or none of the incident power is reflected:

1. A resistive termination equal in value to the characteristic impedance of the transmission line. By definition there will be no reflection from such a termination and all the current will flow into this load. Hence all the power delivered by the generator will be dissipated in this load (Fig 12.35(a)).
2. A resistive termination of value other than in (a) above. This may or may not have a reactance associated with it. At such a termination reflection will occur, the actual amount of the reflected current being dependent upon the relative values of the resistive part of the load and the characteristic impedance of the line. In general, the greater the difference between these two, the larger the proportion of current reflected. Also, if the load resistance is greater than the characteristic impedance, a phase reversal occurs, ie the reflected current wave is

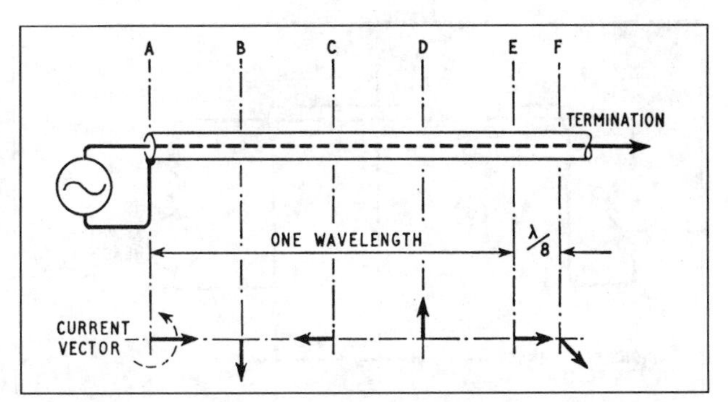

Fig 12.34. Initial variation of incident current vector, according to position along the line at any instant of time

180° out of phase with the incident wave. In the most general case where a reactance is also involved, a further phase shift will occur, the amount depending upon the ratio of load resistance to reactance. Thus some of the generator power is dissipated in the resistive part of the load and some is reflected back along the line (Fig 12.35(b)).
3. An entirely reactive termination. This includes the two extreme cases of a short-circuit and an open-circuit. Since there is no resistive component in the load, no power can be dissipated and therefore all must be reflected. This means that the whole of the incident current wave is reflected back down the line. There will be a phase change relative to the incident wave, the actual value of which will depend upon the reactance of the load.

In the limiting case of a short-circuit, the current in the short-circuit will have a maximum value and because there is no discontinuity the phase change is zero; in the case of an open-circuit, no current will flow in the load, and this implies cancellation of the forward by the reflected current wave so that there is a phase change of 180° and the reflection coefficient has a negative sign. These are both special cases of the more general condition covered in (b).

In every case, apart from that of a perfect termination, some proportion of the incident current is reflected at the far end of the transmission line, and commences to flow back along the line at the same rate as the incident current flowing towards the end. This reflected current wave will, depending upon the circumstances of the termination, commence with an amplitude and phase both differing from the incident current wave. However, since the reflected wave is travelling back along the line, its value at any one moment in time may also be expressed in terms of a current vector, which is rotating in a clockwise direction, opposite to the incident current vector (since the waves are travelling in opposite directions) but nevertheless rotating at the same angular rate, since the frequency of the wave remains unaltered. This is illustrated in Fig 12.35(b)–(d) where the incident and reflected current wave vectors are shown at a precise moment of time for both the end of the line and a point $\lambda/4$ away from the end.

The reflected current wave is thus travelling back along the line towards the generator supplying the power. At any point on the line the net current is represented in amplitude and phase relative to other points by the vector sum of the two waves as illustrated in Fig 12.36. At any point on the line the RF current is of course varying sinusoidally with time at the signal frequency and it is the maximum amplitude of this which is represented by the solid vectors in the lower line of the figure. The instants of time at which these amplitudes are reached for different points on the line are separated by fractions of a cycle which are

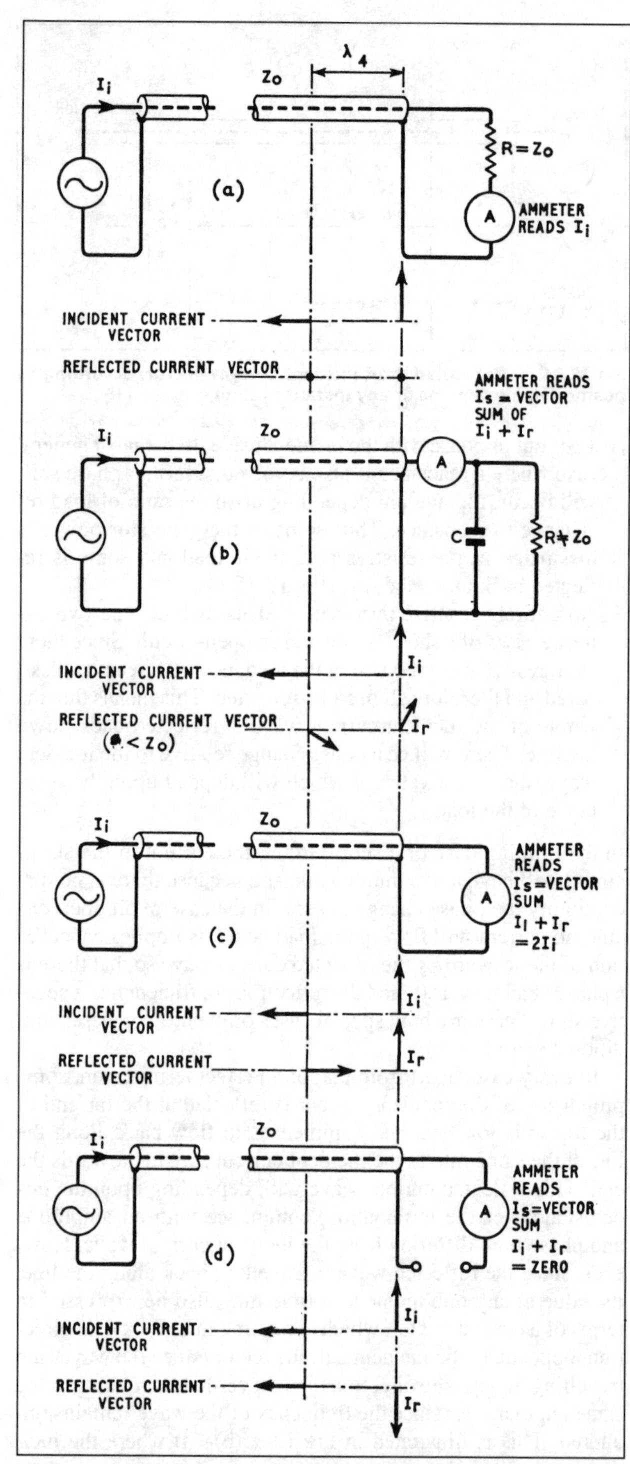

Fig 12.35. (a) Arrangement of incident and reflected current vectors in the vicinity of a correctly terminated load. (b) Arrangement of incident and reflected current vectors in the vicinity of a mismatch load of complex impedance. (c) Arrangement of incident and reflected current vectors in the vicinity of a short-circuit. (d) Arrangement of incident and reflected current vectors in the vicinity of an open-circuit

represented by the relative angles of the vectors, and the variation of phase along the line is non-linear unless the reflected wave is zero.

The standing wave pattern may be observed by voltage or current meters attached to the line, or moved along the line, since the indications correspond to averages over many cycles.

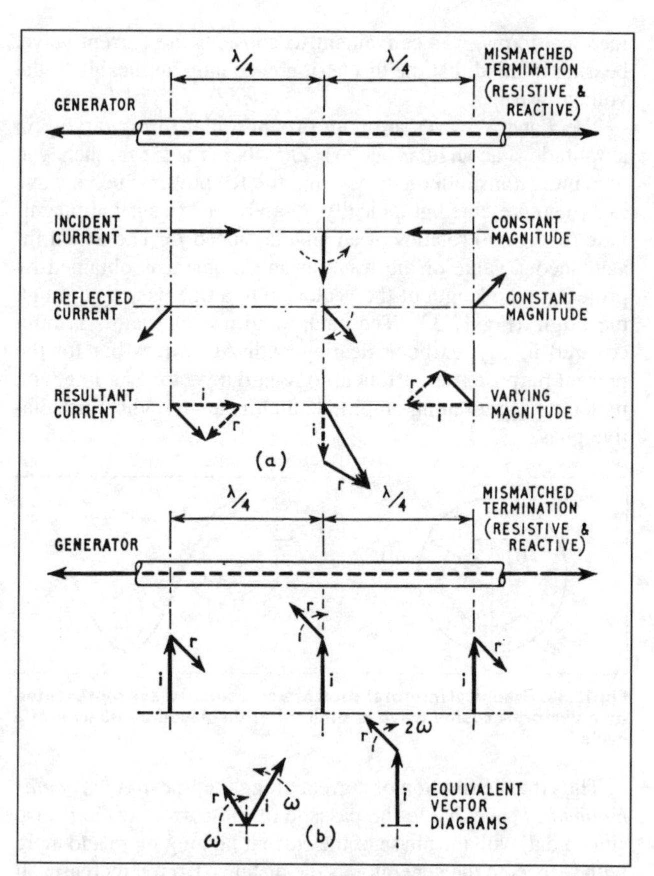

Fig 12.36. (a) Combination of incident and reflected currents along a typical section of the line. (b) Effect on (a) of maintaining the incident current vector stationary. The reflected current vector effectively rotates at twice the speed to produce maxima and minima of current which have a crest-to-crest distance of λ/2

If on the other hand we were to measure the instantaneous values at one particular moment of time, the readings would be influenced as much by the phase variations as by the amplitude variations and would tell us little.

In Fig 12.36 the *i* and *r* vectors from the previous figure have been redrawn to show the phase variation of the reflected wave relative to the incident wave along the line, from which it will be seen that the same relative phases exist at any two points along the line separated by half a cycle, which means that maxima and minima occur twice per wavelength along the line. It follows that, as the incident wave moves along the line at the velocity of propagation, the reflected wave vector rotates relative to it at twice the signal frequency. Fortunately this harmonic frequency does not exist as such, the situation being analogous to that of two trains approaching each other at 50mph and passing therefore at a *relative* velocity of 100mph; in the same way the incident wave travelling outwards along the line 'sees' the reflected wave rushing back past it at what appears to be twice its own velocity. The interaction of forward and reflected waves produces the familiar standing-wave pattern as typified by Fig 12.37.

The standing wave pattern is not a sine wave, although it is approximately sinusoidal for low values of SWR; it departs increasingly from this form as the SWR increases, to an extent where in the limit of an open- or short-circuit termination it becomes a plot of the *amplitude* of a sine wave, ie it resembles a succession of half-sine waves (Fig 12.38(b)). Because of the fact that in space the reflected current wave vector is effectively rotating at twice the generator frequency, the positions of successive maxima and minima occur at λ/4 intervals along the

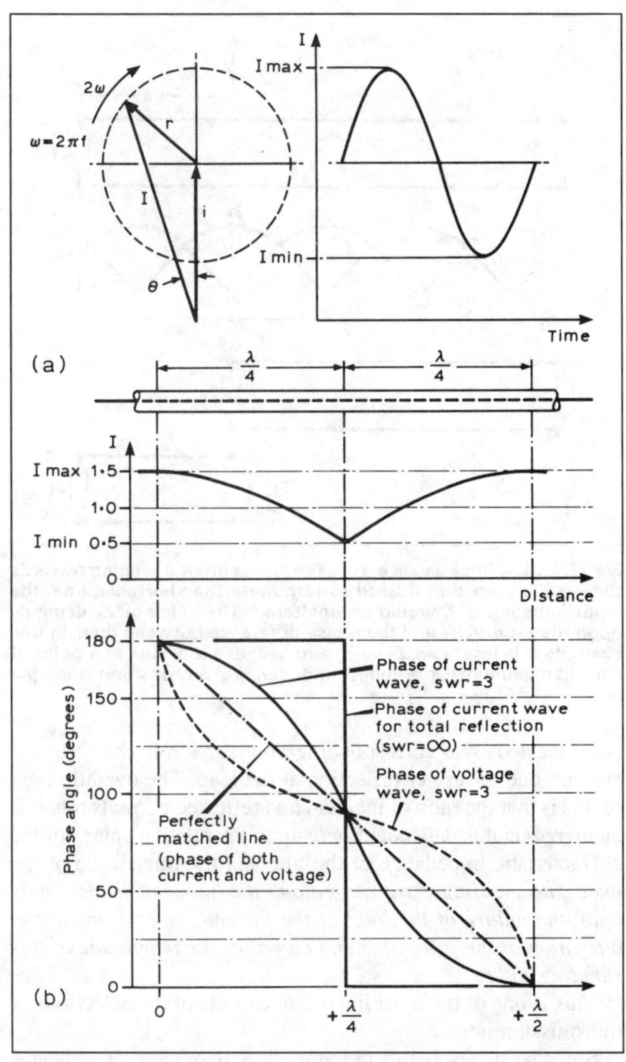

Fig 12.37. (a) Development of the standing wave pattern due to interaction of the incident and reflected current vectors. Variation of line current with distance is derived from the variation with time of the two vectors. At any given point on the line the relative phase of the vectors is fixed but, as the incident wave travels along the line, the relative phase changes with time at the rate of 2ωt. (b) Variation of phase along the line corresponding to the current wave in (a). The dotted line shows the phase variation for the voltage wave. Phase variation is also shown for a perfectly matched line and a completely mismatched line

line. It is important to note that the voltage pattern, though similar in other respects to the current distribution, is displaced from it by λ/4 so that voltage maxima coincide with current minima and vice versa. This follows from the facts that voltage must be zero at a short-circuit, where current is a maximum, and current must be zero at an open-circuit. The voltage standing waves are shown in Fig 12.38 by dotted lines.

The distance from crest to crest of the standing wave is λ/2 at carrier frequency, this being also the distance between points of maximum amplitude of a travelling wave, and in both cases the voltages or currents at successive maxima are opposite in phase. In the case of a travelling wave, however, a meter inserted in the line 'sees' a succession of peaks and troughs, and records an average over many cycles, which is the same for any point in the line if losses are neglected. In general, for a standing wave, the conditions of voltage, current and therefore impedance repeat themselves every half-wavelength along the line, and use is made of this property for impedance matching.

Table 12.1. VSWR in terms of reflection coefficient

Z/Z_0 or Z_0/Z = VSWR	Reflection coefficient percentage
1.0	0
1.5	20
2	33
3	50
5	67
9	80
∞	100

The *standing wave ratio (SWR) k* is the ratio of the maximum and minimum values of the standing wave (*I* or *V*) existing along the line. Also, by definition, the *reflection coefficient r* is the ratio of the reflected current vector to the incident current vector. Thus the maximum value of the standing wave will be $(1 + r)$ and the minimum value of the standing wave will be $(1 - r)$, the SWR and reflection coefficient being related by the expression:

$$k = \frac{1+r}{1-r}$$

The value of *r* used in the above expression lies between zero (matched line) and unity (complete mistermination) but it is sometimes expressed as a percentage. Typical values are given in Table 12.1.

As defined above, the standing wave ratio will have a range of values from unity (matched line) to infinity (complete mistermination). Occasional references will be found to values of SWR in the range zero to unity. These refer to a scale of values obtained by inverting the expression given, and are exactly reciprocal to the more generally used figures, ie an SWR of 0.5:1 is exactly the same as an SWR of 2.0:1. This system is more frequently encountered in microwave work. Sometimes the designation 'VSWR' is used in place of 'SWR' to emphasise the normal convention of expressing it in terms of voltage or current, not power ratio.

Fig 12.38(a) shows the current standing wave pattern on a line which is terminated in a purely resistive load smaller than Z_0. The standing wave is a maximum at the load and therefore a minimum at a point λ/4 back from the load.

If a line is terminated in a mismatch R_L which is entirely resistive, then the value of the SWR is conventionally given by the simple relationship:

$$\text{SWR} = R_L/Z_0 \text{ or } Z_0/R_L\text{, whichever is larger}$$

This follows from a consideration of the power distribution at the resistive load. Considering the balance of power flow at the point of connection of the load to the line, the incident power, ie that associated with the forward travelling wave, is given by:

$$P_{in} = I_i^2 \times Z_0$$

and the reflected power, ie that associated with the backward travelling wave, is given by:

$$P_{ref} = I_r^2 \times Z_0$$

both from the earlier definition of $Z_0 = V/I$ along the line.

Assuming a positive value of reflection coefficient, ie R_L less than Z_0 so that I_i and I_r are in phase, the power in the load resistance R_L will be:

$$P_L = (I_i + I_r)^2 \times R_L$$

and must equal the difference between the incident and reflected power, so that:

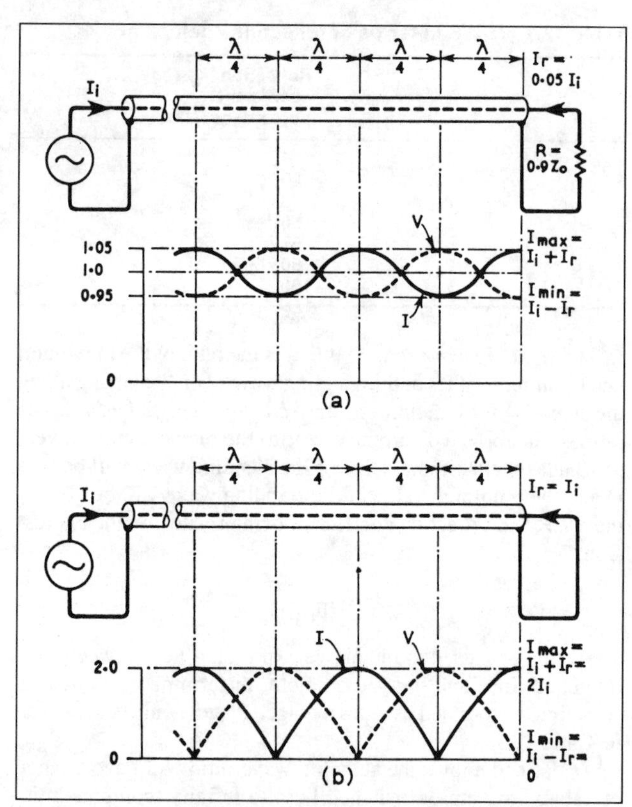

Fig 12.38. (a) Standing wave pattern for a near-matched line ($R = 0.9Z_0$). (b) Standing wave pattern for a completely mismatched line (R = short-circuit). The curve of Fig 12.37 is typical of the transition at medium SWR from the near-sinusoid of Fig 12.38(a) to the rectified sinusoid of Fig 12.38(b)

$$Z_0(I_i^2 - I_r^2) = R_L(I_i + I_r)^2$$

and

$$\frac{R_L}{Z_0} = \frac{I_i + I_r}{I_i - I_r} = \frac{1 + r}{1 - r} = k\,(\text{SWR})$$

If r is negative (ie R_L greater than Z_0) the expression inverts so that $k = Z_0/R_L$.

This relationship between R_L, Z_0 and the SWR is a basic one which is exploited to good effect when using transmission lines as impedance transformers as discussed later.

The reflected current wave travelling back along the line from the mistermination at the load end will ultimately return to the generator which supplies the power. This generator will itself possess an internal impedance which may or may not match the Z_0 of the line. Subject to a suitable mismatch, the reflected wave will be re-reflected at the generator and will travel forward again towards the load as a new incident wave which will in turn be reflected at the load end to exactly the same extent as was the original incident wave. This process of reflection at the load and re-reflection at the generator will continue until all the power is absorbed in the load, provided there are no line losses and the mismatch at the load is countered by the appropriate 'mismatch' between line and generator. Part of the power in the load will have been delayed by several passages along the transmission line but the time interval, though sometimes long enough to cause ghost images in television reception, is of no significance for the narrow-band modulation systems used in HF communication.

The net effect of the passage up and down the line of travelling waves of current is to modify in turn the value of the incident and reflected currents. However, each contribution to the incident current due to re-reflection at the generator is accompanied by a corresponding contribution to the reflected current, due to further reflection at the load. The overall effect of this is that the ratio of incident and reflected currents remains unaltered, and is dependent *only* upon the relative values of the characteristic impedance of the line and the impedance of the load. *The standing wave ratio along the line is dependent only upon the nature of the load at the far end, and no amount of alteration at the generator end can alter the magnitude of this standing wave.*

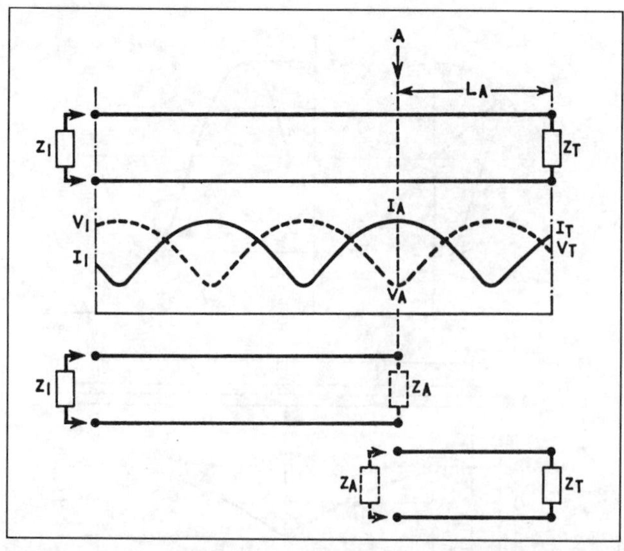

Fig 12.39. The impedance across the line at point A looking towards the load is such that if used to terminate the shortened line, the input impedance Z_I would be unaltered. The value of Z_A depends upon the ratio V_A/I_A and the phase difference between then: in this example it is less than Z_T or Z_I, and because it occurs at a point of current maximum is the lowest impedance at any position along the line

This is one of the most important aspects of the behaviour of transmission lines.

Since no power is lost in the lossless line, the current waves are not attenuated during their passage along the line, and hence the SWR remains at a constant value for the whole length of the line.

Input impedance

When a transmission line is operated in a mismatched condition, it may be necessary to know what value of impedance is seen across the input to the line. This is the load which is presented to the transmitter, and is termed the *input impedance* of the line.

The voltage and current along a misterminated line vary in the manner indicated by Fig 12.38 and depend as already described on the nature of the load impedance. At the far end of the line the ratio of voltage to current is of course equal to the load impedance and similarly the ratio at any other point along the line describes an impedance which can be said to exist across the line at that point. If the line were cut at a point A and this value of impedance connected, the standing wave pattern between this point and the transmitter would be unaffected, and the input impedance would remain the same as before. Equally, the impedance across the line at the point of cutting is the input impedance seen looking back into the section of line which has been cut off. This is illustrated by Fig 12.39. The actual value of the impedance at any point along the line is determined by the ratio of the actual line voltage and current at this point, taking due account of phase differences. Fig 12.38 shows that the

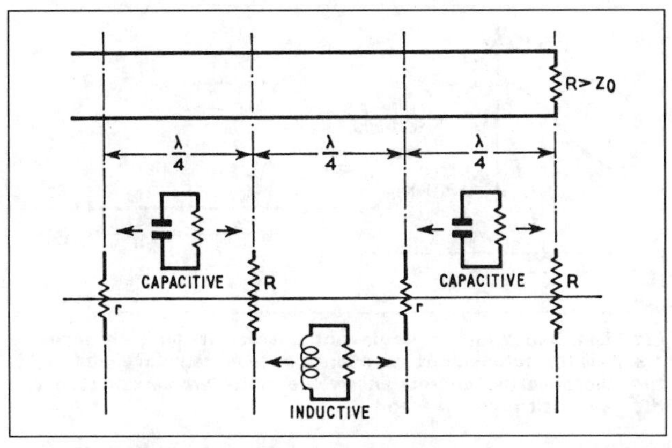

Fig 12.40. Variation of input impedance along a misterminated line with $R > Z_0$. The impedance is alternately inductive and capacitive for succeeding λ/4 sections; at the precise λ/4 points it is purely resistive

apparent load impedance as observed at a voltage maximum is purely resistive and greater than Z_0. Equally, at current maxima the impedance will also be resistive, although of a lower value. At these points current and voltage are in phase, and in between them the current leads and lags alternately. This causes the impedance to become in turn capacitive and inductive along successive quarter-wavelengths of line, changing sign at the positions where the reactive term becomes zero in Fig 12.40. Note that if R is less than Z_0, the capacitive and inductive sections are interchanged.

The input impedance is dependent not only upon the load impedance, but also the length of line involved. It is a common fallacy to believe that by altering the length of a line, the match can be improved. The adjustment of line length may result in the presentation to the transmitter of a more acceptable load impedance so that it delivers more power but this indicates merely the provision of an insufficient range of adjustment at the transmitter. The standing wave pattern remains completely unaltered.

The case of a purely resistive termination is illustrated in Fig 12.38. In practice the far-end termination may be inductive or capacitive. The effect of this is to cause the positions of voltage maxima to move along the line towards the generator by an amount determined by the nature of the reactive termination. If this is capacitive, as in Fig 12.41, there will then exist along the line a short region AB of capacitive impedance before the first voltage minimum. The reactive termination could be replaced by the input impedance of a further short length of line BC terminated in a suitable pure resistance: the length AB is then such as to make up the quarter-wavelength AC. The lower the capacitive reactance of the termination, the closer is the point of voltage minimum.

An important aspect of the input impedance of a mismatched line is its repetitive nature along the line. The conditions of voltage and current are repeated at intervals of λ/2 along the line, and the impedance across the line repeats similarly. *The input impedance is always equal to the load impedance for a length of line any exact number of half-wavelengths long (neglecting line losses), irrespective of its characteristic impedance.* Such a length of line can be used to connect any two impedances together without introducing unwanted variations, and without regard to the Z_0 of the line being used. This property is the basis of the operation of 'tuned' lines as antenna feeders, and its application is discussed later. Also, any existing feeder arrangement can be extended in length by any number of half-wavelengths without altering the SWR or the load as seen at the input

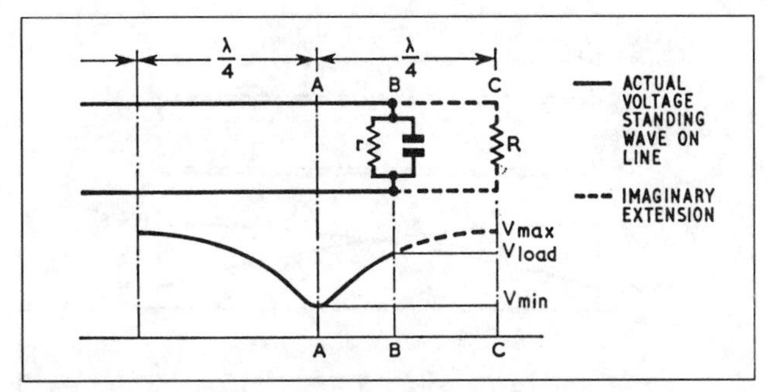

Fig 12.41. Derivation of the standing wave pattern on a line terminated in a capacitive load impedance. The identical pattern is provided by a hypothetical resistive load at a particular distance beyond the actual line load

end. Because the repetitive nature of the line impedance is a function of frequency, being tied to intervals of λ/4, the precise duplication of the load impedance at the input can only occur at a number of discrete frequencies for any specific length of line, for which the line is 1, 2, 3 . . . n half-wavelengths long.

The extent to which the input impedance varies when the frequency is varied slightly determines the bandwidth over which the antenna system may be used without readjustment of the transmitter. For a given percentage change in frequency, there will be a greater variation of input impedance the longer the physical length of line, the change in electrical length (expressed in fractions of a wavelength) being greater, leading to a larger change in the voltage and current at the input to the line. This of course involves properties not only of the line itself but also of the load connected to it. Particular attention should be paid to the bandwidth aspects of lines when they are operating with high SWR such as matching or EMC stubs, or as 'tuned' feeders. In the case of EMC stubs, which operate with a nearly infinite SWR, the effective Q is quite high and their useful bandwidths as filters is strictly limited, often to the extent of ±2% of nominal design frequency.

Line transformers

It is possible to use short lengths of line with values of Z_0 different from that of the main feeder as impedance transformers for matching a wide range of load impedances to different values of line impedance. It has already been shown that the standing wave ratio:

$$k = Z_0/R_L \text{ or } R_L/Z_0$$

whichever is greater, so that translating this into the symbols of Fig 12.40 we have:

$$k = R/Z_0$$

so that

$$r = Z_0/R \text{ and } Z_0 = \sqrt{R_r}$$

If the quarter-wavelength of line adjacent to the load in Fig 12.40 is replaced by a different line having a characteristic impedance Z_T the main transmission line will 'see' a load impedance Z_T^2/R and by suitable choice of Z_T this can be made equal to Z_0 so that it looks like a matched load. The impedance transformation ratio Z_0/R *is* therefore given by Z_T^2/R^2 and the required value of Z_T by $\sqrt{Z_0R}$. The matching line is known in this case as a *quarter-wave transformer* and its operation is shown pictorially by Fig 12.42. It should be noted that R1 and R2 are interchangeable so that a λ/4 section of 150Ω line will match a resistance of 450Ω to a 50Ω line, or alternatively for example a 50Ω impedance to

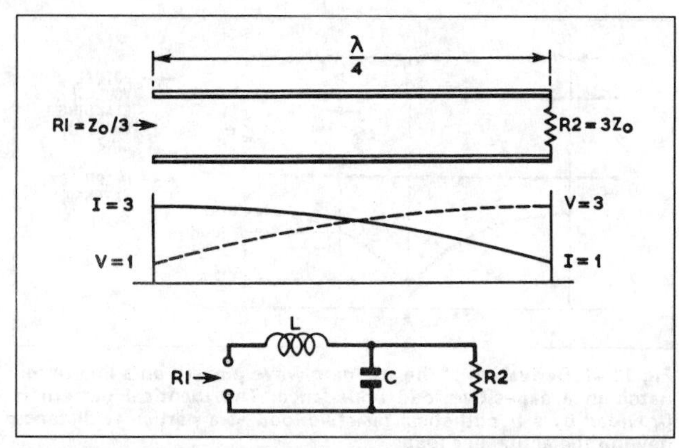

Fig 12.42. Illustrating the principle of the λ/4 transformer in the case of a 9:1 impedance ratio. The *V*/*I* ratio is inverted by the standing wave on the line with VSWR = 1, giving $R_1 R_2 = Z_0^2$. This corresponds to the series-parallel action of a resonant circuit in which $L/C = Z_0^2$. Note that R_2 is the parallel resistance and R_1 the equivalent series resistance

a 450Ω open-wire line. This property is known as *impedance inversion*, an impedance of, say, one tenth of the matching line impedance Z_T being transformed to 10 times Z_T and vice versa. This relationship holds good for any ratio of impedances.

A transmission line can always be approximated by one or more inductances and capacitances, and the simple *LC* circuit shown in Fig 12.42 provides a close equivalent to the λ/4 transformer; *L* and *C* are in resonance and ω*L* replaces the Z_T of the transmission line formula.

Apart from questions of possible convenience it is not always possible to find or construct matching lines with suitable characteristics, and lumped constants provide an important alternative.

As an example of the use of line transformers, Fig 12.43 illustrates the matching of a broadside array of two dipoles into a 70Ω line. The ratio of the higher impedance to the Z_T of the transformer is closely analogous to the *Q* of a tuned circuit and the ratio in this case, 1.4, implies a large bandwidth. For large transformation ratios the bandwidth may be improved by the use of two λ/4 sections in series, transforming to an intermediate value at their junction, which should ideally be the geometric mean of the end impedances. In such an arrangement the effective ratio of each transformer is reduced, which itself improves their bandwidth, and in addition the two tend to be mutually compensating with change of frequency to produce a further improvement, Fig 12.44. Use of this technique is somewhat restricted by the limited choice of cable impedances available.

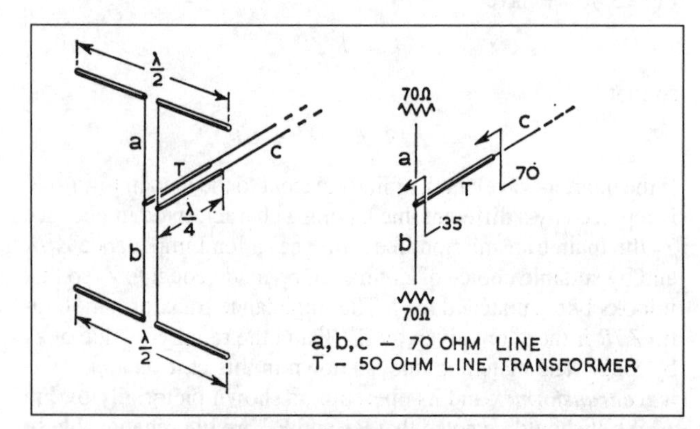

Fig 12.43. Use of a λ/4 50Ω line transformer to match two co-phased dipoles into a single 70Ω main feeder. The impedance at the junction of a and b is 70/2 = 35Ω, which is transformed up to 70Ω again

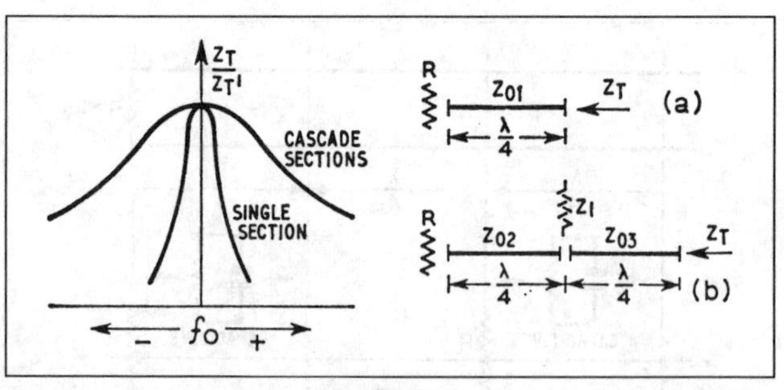

Fig 12.44. Bandwidth of single- and double-section λ/4 transformers. Z_T is the actual input impedance at each frequency while Z_T' is the theoretical design frequency value. In the two-section case, Z_i is the geometric mean of *R* and Z_T

The inverting properties of λ/4 of line can also be used to relate the voltage at the input to the current in the load (and vice versa). In Fig 12.42, the SWR on the line can be defined in a number of ways:

$$\text{SWR} = \frac{V_2}{V_1} = \frac{Z_0}{R_1} = \frac{Z_0 I_1}{V_1}$$

so that $V_2 = I_1 \times Z_0$ and similarly $V_1 = I_2 \times Z_0$.

The voltage at the input end is determined *only* by the characteristic line impedance and the current at the other end, and does not depend upon the value of the load in which the current flows. Any change of load impedance alters the value of impedance across which this constant voltage is developed, and hence the power which must be supplied by the transmitter in order to maintain the current.

Stubs

A line is completely mismatched when the far end is terminated by either a short- or open-circuit. There is then no load resistance to dissipate any power, and a 100% reflection of voltage and current takes place. The standing wave pattern is that of Fig 12.38(b). Except at resonance, sections of line with open- or short-circuit terminations approximate to pure capacitance or inductance, although there is usually some resistance arising from losses in the conductors or dielectric material and these can be considerable (as explained later) unless the lines are air-spaced or very short. Such pieces of line are called *stubs* and can be used to compensate for, or match out, unwanted reactive impedance terms at or close to a mismatched termination, and allow the major part of the feed line to operate in a matched condition. To minimise losses, and also to preserve the bandwidth of the antenna system, the lengths are usually restricted to λ/4 or less.

The input reactance of a loss-free short-circuited line is given by the expression:

$$X_{in} = Z_0 \tan\theta \text{ (inductive)}$$

($\approx Z_0 \theta$ when θ is small) and for an open-circuited line by:

$$X_{in} = -Z_0 \cot\theta \text{ (capacitive)}$$

where θ is the electrical line length and Z_0 its characteristic impedance as given for example by Fig 12.31.

Figs 12.45 and 12.46 illustrate the manner in which the input reactance varies with line length. For sections shorter than λ/4, the short-circuited line is inductive, and the open-circuited line capacitive, as may be appreciated from visualising very short lengths of line which bear a close physical resemblance to inductors or capacitors.

A short-circuited line can also be made to behave like an open-circuited line of the same electrical length (and vice versa) by

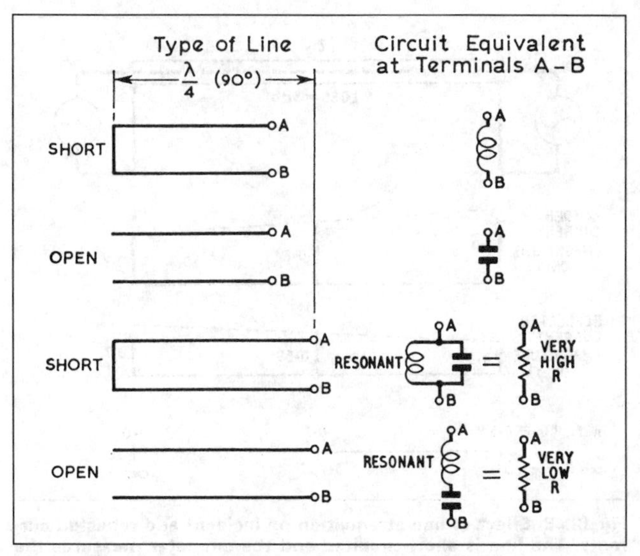

Fig 12.45. Circuit equivalents of open-circuited and short-circuited lines

adding a further λ/4 of line: this is another application of the inverting properties of a λ/4 section of line described previously. The behaviour of the two types of stub is completely complementary, and if two are joined together at their input ends they will make up between them precisely one quarter-wavelength when their respective reactances are equal and opposite in sign. This is closely analogous to a parallel-tuned circuit, and a short-circuited λ/4 line is said to be a *resonant section*, possessing all the properties of the parallel tuned circuit. Near resonance the reactance/frequency slope of the resonant line is similar in shape to that of the lumped-constant circuit, and a change of Z_0 is equivalent to a change in the *L*/*C* ratio. Such sections of line may be used as shunt chokes across other lines without affecting the latter, and may also at VHF be used as tuned circuits for

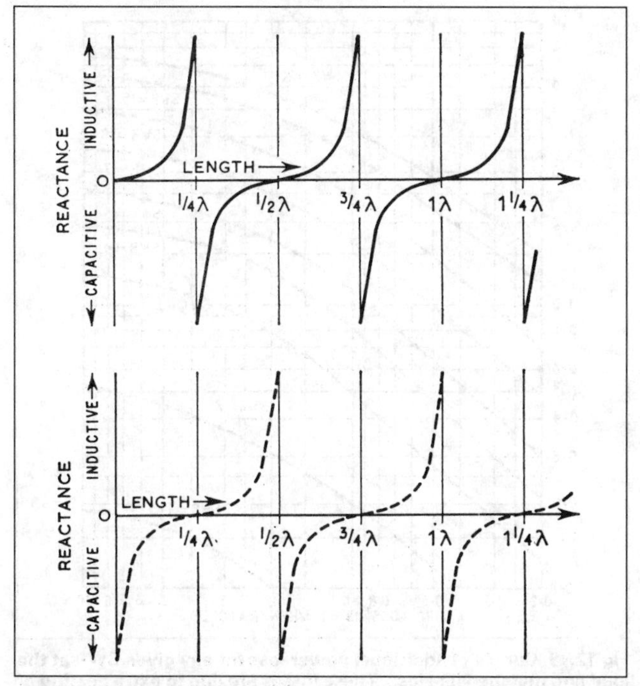

Fig 12.46. Variation of input reactance with length for short-circuited (solid) and open-circuited (broken) lines. The two sets of curves are identical and displaced by λ/4

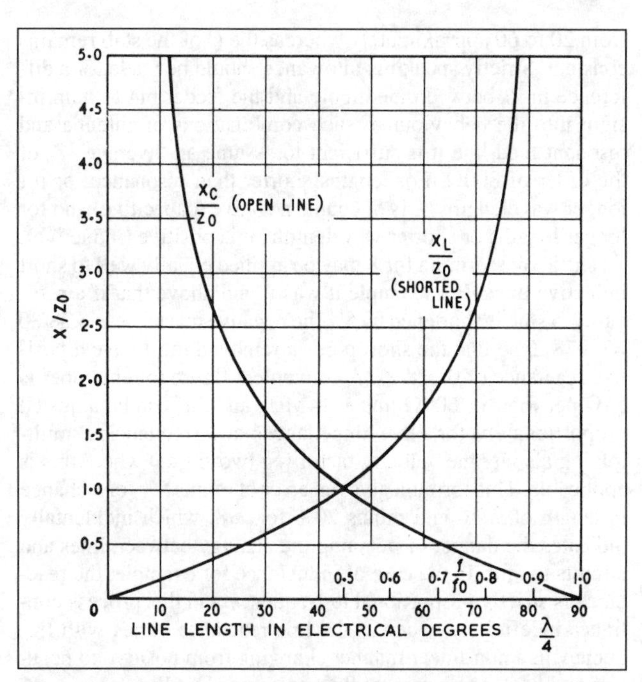

Fig 12.47. Reactance curves for open-circuited and short-circuited lines. The quantity X/Z_0 multiplied by the characteristic impedance of the line equals the reactance at the input terminals. Longer lines alternate as capacitance or inductance according to Fig 12.46. The curves also indicate the manner in which input reactance varies with frequency for nominal λ/4 sections

receivers and transmitters, the physical lengths being then conveniently short.

In complementary fashion, a λ/4 open-circuit line section behaves as a series resonant circuit with a very low input resistance at the centre frequency. In the ideal case this is a perfect short-circuit: in practice the line losses degrade this to a finite but low value of resistance, which for all practical purposes can be considered a short-circuit. The reactance/frequency slope of both open- and short-circuit resonant sections is shown in Fig 12.47 which illustrates the effect of variations in the Z_0. For the open-circuited lines of Fig 12.45 operating near resonance but slightly detuned the reactance is given by the simplified formula:

$$X = Z_0\,\theta = 6.28\,Z_0 \cdot \frac{\Delta l}{\lambda} \text{ ohms}$$

where $\Delta l/\lambda$ is the fraction of a wavelength by which the line falls short of or exceeds the resonant length. This formula is valid for any length of line *or antenna system* observed at or near a current loop and, given the radiation resistance (if any) and loss resistance, it enables the bandwidth of the system to be roughly estimated. Thus for a λ/4 line (or λ/2 dipole which can be thought of as the λ/4 line 'opened out'), 5% change of frequency produces a length error of λ/80. Z_0 for the λ/2 dipole may be as much as 1000Ω so that X = 78Ω which is comparable with the radiation resistance and therefore indicates a half-bandwidth of the order of 5%. For an open-wire line Z_0 is typically 600Ω and from Fig 12.5 the loss resistance at 14MHz using 18swg is 2.4/2 = 1.2Ω. The reactance equals this resistance for an error in line length or a change in frequency of 5 × 1.2/68%, ie the total bandwidth is 12/62.8% and the Q is therefore 62.8/12 × 100 = 525.

If an additional λ/2 is added to the stub or a λ/2 dipole is fed with λ/2 of open wire line, the reactance changes three times as fast, the loss resistance is multiplied by three, but radiation resistance stays the same so that for the dipole case the Q increases from 20 to 60 approximately whereas the Q of the stub remains constant. Strictly speaking, allowance should be made for a difference in Z_0 between the dipole and the feeder but to gain insight into the behaviour of such combinations of antenna and resonant feed line it is sufficient to assume an 'average' Z_0 of the order of 800Ω. For lengths shorter than resonance, or for longer wavelengths, X is of course negative (capacitive) and for longer lengths or shorter wavelengths it is positive (inductive).

The above formula for X may be applied equally well to short inductive lines. For example it was found above that if a resonant λ/4 stub is shortened by 5% the negative reactance is 6.28/80 = 0.078 Z_0 so that the short portion removed must have a positive reactance of 0.078 Z_0. A convenient figure to remember is 55Ω per foot for 600Ω line at 14MHz and this can be adjusted proportionately for other impedances and frequencies, multiplying also by the velocity factor (eg two-thirds) where this is applicable. This formula gives an error of minus 5% for a change in length of λ/16 and minus 20% for λ/8, which incidentally indicates the danger of pursuing the analogy between lines and circuits too far. In the case of inductance for example, the reactance is strictly proportional to frequency and this process continues indefinitely, whereas the line reactance varies with frequency in a non-linear manner changing from positive to negative and back again repeatedly as discussed earlier.

Some practical applications of matching stubs are considered on p12.32 in the section dealing with impedance matching.

Attenuation and loss

In practice all transmission lines have some loss associated with them. This loss may be due to radiation, resistive losses in the conductors, and leakage losses in the insulators; but however it arises, it is a function of the actual construction of the line and the materials employed. With the best of the moulded twin or coaxial lines supplied for amateur radio and television use this attenuation is less than 1dB per 100ft (when correctly matched) at frequencies below 30MHz. 1dB loss in 100ft means that at the far end of the feeder the amplitude of voltage or current is just under 90% of that at the input end, and about 80% of the power is delivered to the load. A second 100ft extension would deliver 80% of the power left at the end of the first 100ft or 64% of the original output. For each extra 100ft the factor is multiplied again by 0.8 but in decibel ratios 1.0dB loss is added. Open-wire lines can be remarkably efficient: a 600Ω two-wire line made from 16swg copper wire spaced about 6in, carefully insulated and supported, has a loss of only a few decibels *per mile* in the HF range.

The *matched loss* of the line is quoted in the manufacturers' published information usually as *n*dB per 100ft. It increases more or less in proportion to the square root of the frequency at which it is being used, ie a line having a quoted matched loss of 0.5dB/100ft at 10MHz will have a matched loss of approximately 1.5dB/100ft at 90MHz. One effect of this loss is to improve the SWR as measured at the input end of the line since at this point the reflected wave has suffered attenuation due to the outward and return journeys along the line, whereas the outgoing wave is unattenuated. This can on occasion conceal the existence of a considerable mismatch at the load end of the line. This effect is illustrated in Fig 12.48 for the extreme case of a short-circuited line. Losses increase with increasing SWR since they are proportional to the square of current or voltage. This means for example that the increased loss due to heating of the conductors at a current maximum is less than counterbalanced by reduced heating at the adjacent minimum, and similar considerations apply to the dielectric losses. Extra losses arising in this way depend only on the matched line loss and the SWR at the load, not on the length of line, and may be obtained from Fig 12.49.

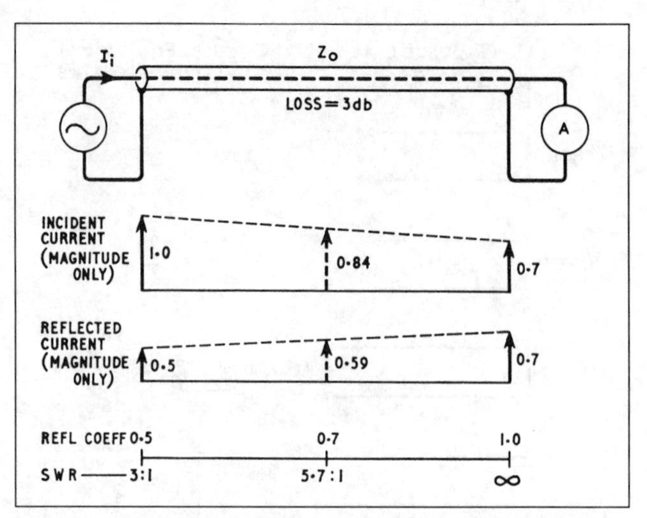

Fig 12.48. Effect of line attenuation on incident and reflected currents. The line is short-circuited and the ammeter measures the current in the short-circuit

When a signal is fed into a lossy line terminated by a short-circuit, it arrives at the far end with reduced amplitude but in almost the same phase as if there were no losses. The reflected wave is initially of the same amplitude as the incident wave but in turn experiences attenuation on its way back to the generator as illustrated in Fig 12.48. In this case an infinite SWR at the termination appears as a ratio of only 3:1 at the transmitter. If sufficient cable is available it is possible to use it in this way as a dummy load of reasonable accuracy and relatively low cost for use in the adjustment of transmitters; thus, referring to Fig 12.49, a line with a matched loss of 3dB will have a total loss of at least 10dB when completely mismatched as in this example,

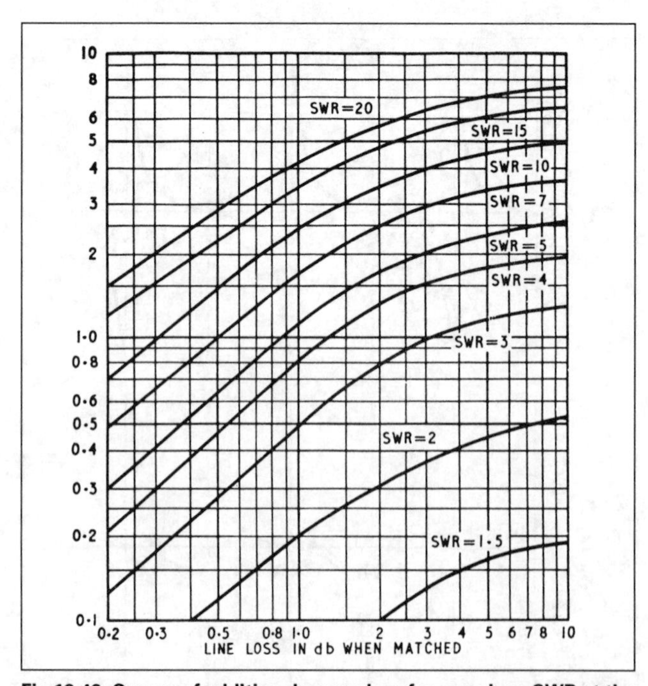

Fig 12.49. Curves of additional power loss for any given SWR at the load and matched line loss. These losses are due to extra heating of the line with high values of SWR and will be incurred in addition to the matched line loss regardless of whether or not the generator matches the line

the input SWR will be only 1.22:1 and the load impedance presented to the equipment will be equal to $Z_0 \pm 20\%$. The cable, because of its considerable bulk, is able to dissipate a considerable amount of heat. This technique is particularly useful at VHF and UHF where the smaller diameter cables tend to be quite lossy. It should be noted that Fig 12.49 has in one respect been over-simplified, since the losses are not distributed uniformly along the line but are about three times as great per unit of length near the termination, where the SWR is large.

In typical practical installations the improvement in SWR at the transmitter due to line losses can be very large, leading to an over-optimistic assessment of performance as illustrated by the following example. In this case the matched-line loss (theoretical and measured) was 1.5dB and the SWR at the transmitter was 2. The extra line loss due to this was estimated from Fig 12.49 to be only 0.25dB and from DX signal reports the antenna appeared to be working well until another antenna became available for comparison, when it became obvious that something was amiss. Measurements at the load end of the feeder revealed an SWR of 5, thus increasing the total loss to an estimated 3.2dB and accounting, along the lines of Fig 12.48, for most of the 'improvement' in SWR at the transmitting end. (Further losses were attributable to feeder radiation as discussed later.)

As already indicated, any losses in a transmission line are increased by the presence of standing waves, whereas in the case of the loss-less line all the power delivered to the line is eventually dissipated in the load, although, as we have seen, a certain proportion is delayed in time due to multiple reflections. When the mistermination of the line is not gross, the amount of power returned from the first reflection at the load is not large.

Assuming that the generator completely misterminates the line at the sending end then the power from this first load reflection will reappear at the load after a journey back down the line and up again during which time it is attenuated by twice the matched line loss (go and return). At this stage the major proportion of the attenuated reflected power is delivered to the load, and a small proportion is re-reflected; thus if 20% is reflected on the first occasion only 4% (less some attenuation) is reflected a second time and the net power delivered to the load after only one reflection represents most of that delivered by the generator.

In practice, any reactive term present in the load offered to the transmitter by the line is removed by retuning of the output stage, and any variation in the resistive term is compensated by readjustment of whatever coupling device is used between the transmitter and the line. These adjustments ensure that the effective loading of the transmitter and therefore the power delivered to the antenna system remains constant despite variations of load impedance.

In other words, the transmitter presents virtually a complete mistermination to the reflected power in the line, and the additional power lost due to the presence of reflections from the load is relatively small, arising only from the attenuation of the reflected wave. This point is frequently misunderstood and it is a common belief that once power has been reflected from a misterminating load this reflected power is lost. This case *only* occurs when the variation of line input impedance caused by reflections at the load is not compensated by tuning and matching adjustments at the transmitter. This is a situation of limited practical importance but can arise for example due to an insufficient range of adjustment or if attempts are made to obtain instantaneous comparisons (ie without retuning) between antennas which are not equally well matched. If the transmitter has been adjusted with a matched load and its output is then switched

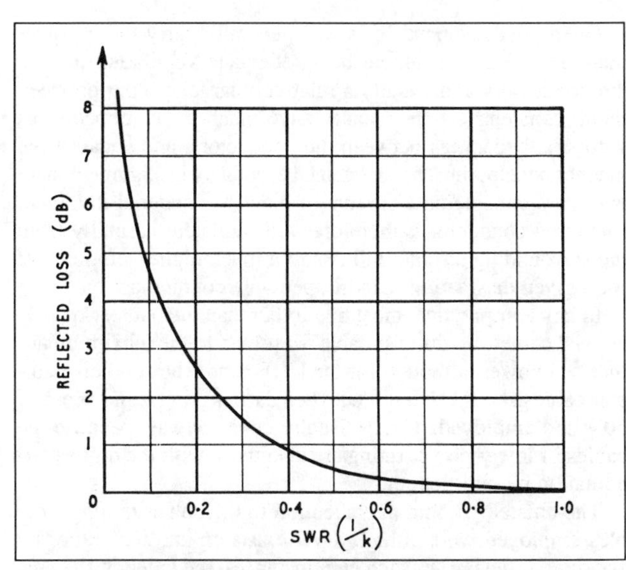

Fig 12.50. Curve of mismatch loss, ie the amount of power reflected from a mismatch for a given SWR. This is also the curve of absolute power loss (when the generator matches the line, and therefore absorbs the reflected power) for lines of zero matched loss

to another load with a known SWR, the loss may be obtained from Fig 12.50; for example, with an SWR of 3 there will be a loss of only 1.5dB or less than half an S-unit, though in the case of a linear amplifier the incorrect loading could lead to distortion and 'splatter'.

Matching adjustments are not usually provided in the case of receivers and it is sometimes erroneously thought that reception can be degraded by antenna mismatch. Maximum signal-to-noise ratio does not in general require a good match but a certain optimum degree of mismatch, though this consideration is usually of no importance for HF reception in view of the high external noise levels from galactic and other sources. Because these external noise levels tend to be a long way above the internal receiver noise it is usually permissible to insert considerable attenuation in front of the receiver and this may indeed be beneficial in view of the prevalence of intermodulation and similar effects which contribute to interference and effective noise levels.

As already noted, the additional line loss at high values of SWR for the normal transmitting situation is dependent only on the total matched line loss and the SWR, and not influenced by the length of the line. From Fig 12.49 it will be noticed also that when the matched line loss is very high, the additional loss due to the presence of a standing wave tends to become independent of the matched line loss. Cable loss may be determined very simply by measurement of SWR at the input end with the far end short-circuited. If for example the loss is 3dB, the reflected wave arrives back at the input with an attenuation of 6dB so that the measured SWR is $(1 + 0.5)/(1 - 0.5)$, ie 3.0. For monitoring the condition of a feeder cable it should be possible to obtain the necessary mismatch by operating at a frequency for which the antenna is not designed, taking care of course not to overload the cable.

Power handling

There is always a definite limit to the amount of RF power which can safely be transmitted along any form of transmission line. The limitation is set either by excessive heating of the conductors, which can lead to deformation or destruction of the insulation between them, or by voltage breakdown between the conductors, either in air or in the insulating material.

Open-wire balanced lines will generally carry more power than any other form, there being adequate ventilation around the conductors and usually a relatively large spacing between them. Concentric lines usually have much lower impedances with small spacings between the conductors, and because the inner conductor must be maintained accurately in alignment there is often a considerable amount of dielectric material involved. The inner conductor is therefore well insulated thermally from the surrounding air and will rise to a much higher temperature for a given dissipation than its open wire counterpart.

In amateur practice, most lines other than the poorest-quality coaxial cables are quite suitable for use up to the maximum authorised power, although for the UHF bands the smaller-diameter cables should be avoided when carrier powers in excess of 50W are employed. This is usually done anyway, because the cables of lower power rating are also those with the highest attenuation per unit length.

The amateur should pay attention to the power rating of cables employed when a high SWR exists on the line, either by accident or design. In such circumstances, the value of the current at maxima of the standing wave may rise to such an extent that local overheating will occur, with consequent damage. The voltage maxima may also approach the limiting value for flashover. These conditions are unlikely to arise in a well-designed installation except perhaps under fault conditions, a possibility which should not be overlooked.

Practical feeders

The majority of commercially available coaxial cables have impedances of the order of 50Ω or 70–75Ω and twin feeder is available for impedances of 75, 150 and 300Ω. Two- or four-wire open line having impedances in the range 250–750Ω, though not as a rule commercially available, is readily constructed, providing the advantages of much lower cost and greatly reduced losses, typically 0.4dB per 100yd at 14MHz compared with 1.8 to 4dB for coaxial cable. Against this, coaxial cable and low-impedance twin-lead is often more convenient to use, with the latter providing a good match to the centre of a $\lambda/2$ dipole without the need for transformers or baluns. In general, however, practical beam antennas have relatively low impedances so that some form of transformer is needed between the antenna and line. In addition, if unbalanced feeders are used to feed balanced antennas or vice versa, a suitable converter (balun) should be used as discussed later, and these devices can also be used as impedance transformers so that the choice of feeder is mainly dictated by personal preferences, practical convenience in individual cases or, in the case of commercial antennas, the instructions of the manufacturer. If very long feeders are required, the use of open-wire line in conjunction with some form of impedance transformation at the antenna is normally essential.

Coaxial cables are usually made with a solid or stranded inner conductor and a braided wire sheath. Two-wire lines of 75–300Ω impedance are generally made with moulded polythene insulation and are reasonably flexible. The properties of typical cables are listed in Chapter 22 – 'General data'.

Almost any copper wire may be used for the construction of open-wire lines and although 12–18swg is advisable for very long lines it can be inferred from Fig 12.5 on p12.5 that even with 24swg the losses only rise to 0.9dB per 100yd at 14MHz. These figures do not take account of loss by radiation, commonly thought to be large in the case of open-wire lines but typically (as discussed later) less than 0.1% for lines with 6in spacing at 14MHz. Spacers are needed at intervals and subject to using the minimum number (depending on wire gauge,

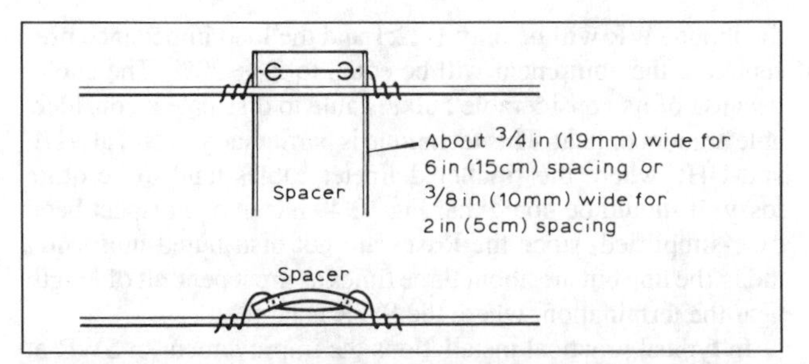

Fig 12.51. Simple and inexpensive 600Ω line construction. Plastic piping of 2–3in (5–8cm) diameter is cut into short lengths and then sawn lengthwise into strips, a pair of small holes being drilled near each end of each strip. Spacing may be 2–6in (5–15cm), with insulation at intervals of some 12–15 times the line spacing. With wide spacing, losses are reduced and construction time is much quicker, but symmetry is more easily upset. Note that the spacers are slightly curved and this helps to prevent slippage

spacing, length of span, exposure to wind etc) almost any insulating material can be used provided it does not absorb moisture. Polythene rod is particularly recommended whereas it is inadvisable to use lossy materials such as bakelite for long lengths of line. Narrow spacings are inconvenient in so far as they require more spacers, and there is more risk of trouble from snow and ice. Rod spacers may be fixed as shown in Fig 12.51. Nylon fishing line has excellent insulating and mechanical properties and may be used for supporting open-wire line, guiding it round obstacles etc, and also for supporting antennas without additional insulators. Polythene line, which exhibits much less stretch, is also suitable and inexpensive lines, suitable for lengths of 100ft or more, may be constructed by using tensioned polythene cord as mechanical support for 22swg or finer wire gauges.

Although the above procedures are recommended, considerable liberties may be taken in the design of open-wire lines without appreciably closing the performance gap, which is strongly in their favour relative to low-impedance lines. The lead-in for example may use plastic insulated wire taken through holes drilled in a window frame without incurring measurable losses, but it is possible to go too far in this direction and in one instance a lead-in using twin plastic lighting flex pushed through a single hole in a wooden frame gave rise to a spectacular fireworks display; the situation was somewhat extreme as there was an accidental feeder short and 400W PEP applied to the line, but this incident suggests a serious fire risk which could easily be overlooked. It is therefore recommended that good insulation, preferably some sort of ceramic bushing, be used at this point, though short lengths of coaxial cable having 4 or 5kV voltage ratings provide a simpler alternative and, despite much use by the author (including some extreme fault conditions), have given no trouble at UK power levels. Soft-drawn enamelled instrument wire can be used, although when the line is to be strained between insulators, it is better to use hard-drawn or cadmium-copper wire. Prestretching to about 10% extra length by means of a steady pull will, however, appreciably harden soft-drawn copper wire and render it less liable to stretching in service.

Both 75 and 300Ω twin lines suffer severely from the effect of moisture in damp weather. This is most noticeable in the case of tuned lines, because water has a very high dielectric constant and affects the velocity factor. The 300Ω feeder in which the central web is removed in large slots is less affected by rain than the type with a continuous web, typical increases being 5% and 20% respectively, though in most cases losses remain negligible. In general there are few problems in the case of single driven elements operating at low SWR, and in other cases possibilities include the use of 300Ω open-wire line which is not difficult to construct if wires can be tightly stretched between two points, 16swg spaced 1cm being suitable. The slotted plastic line can then be used for any portions of the feeder run which need to be flexible, up to a total length which decreases with SWR but is typically at least λ/6 for an SWR of 6.

Soot deposited on the surface will tend to retain moisture and accentuate these troubles. The remedy is to clean the line periodically and give it a dressing of silicone wax polish to repel any moisture. Tubular 300Ω line should be sealed so that water cannot enter the interior. Although the PVC outer jacket of a concentric cable is good enough to allow the cable to be buried, the open end of the cable is very vulnerable for it can 'breathe in' moisture which does irreparable damage to the line. There is a variety of satisfactory sealing materials available, such as Bostik cement, Bostik sealing strip (putty) and Sylglas tape which is loaded with a silicone putty.

Particular attention should be paid to the adequate sealing of plugs and sockets used out of doors on coaxial cables. Even those intended for professional use (eg BICC Teleconnectors, US type C and BNC) are not entirely waterproof, and the usual Belling-Lee coaxial plug is virtually unsealed in its own right. Such fittings should be securely taped for at least 1in along the cable with a water-repellent tape.

Balanced lines should be arranged so that as far as possible they are kept clear of masts, buildings and other obstructions. Failure to do this may sometimes result in unbalanced currents in the two conductors, which then become a source of undesirable radiation and adversely affect antenna performance. Unbalance can also be caused by abrupt changes of direction of the feeder, and to a lesser degree by slackness in the line which can allow it to swing about in the wind and to twist.

Coaxial cables lend themselves to less obtrusive installation. If they are correctly terminated in an unbalanced load, no currents will flow on the outside of the outer conductor and they may therefore be laid close to any other surface or object (brickwork, tubular masts, etc) without affecting their behaviour in any way. It is only necessary to ensure that any bend is made on a sufficiently large radius to avoid distortion of the cable itself. Typically, bends in ¼in diameter TV-type cables may have as little as ½in radius, whereas those in larger cables such as UR67 should be of not less than 2in radius. The recommended minimum bending radius for cables can often be found in manufacturers' literature, but a useful rule of thumb is to use a radius equal to at least twice and preferably four times the cable diameter. It is easy to damage coaxial cables by excessively tight fastenings, such as staples driven hard into the cable support, and care should be exercised when making such fixings to avoid bruising the cable.

Radiation from feeders

Radiation from feeders can arise in several different ways. Fig 12.52 illustrates a typical antenna fed at the centre with a feeder which may be of any type, and it is assumed for the moment that the system is energised by inductive coupling into an antenna tuning unit with no earthing at the transmitter. If the feeder has an effective electrical length which (allowing for top-loading by the dipole) is an even number of quarter-wavelengths, it will resonate as an antenna and can be used as such, in fact a 14MHz dipole with about 44ft of vertical feeder, voltage fed at the lower end, performs well as a low-angle radiator for 7MHz DX. On the other hand, a length of 17 or 51ft would enable it to radiate efficiently on 14MHz, though probably in an unwanted direction or angle of elevation. With the arrangement shown the unwanted mode could be excited by capacitive coupling at the

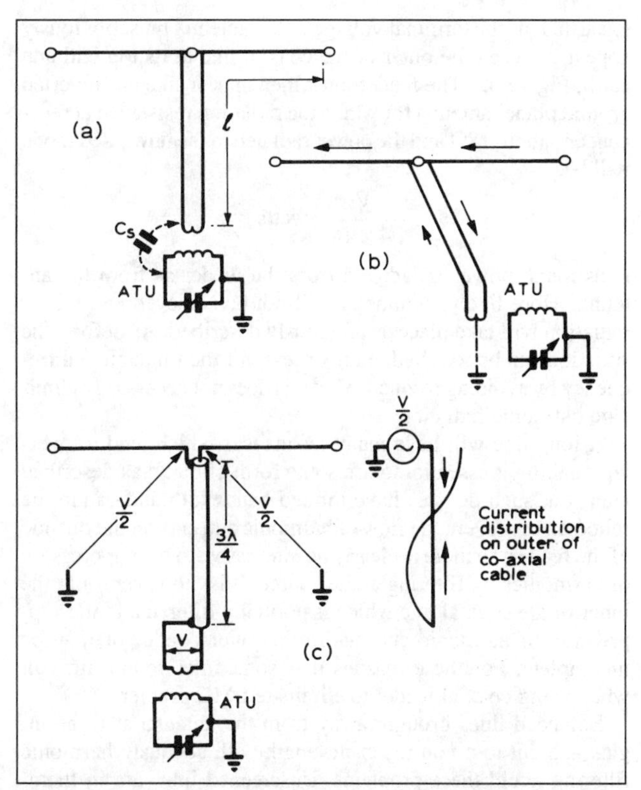

Fig 12.52. (a) If *l* is an odd number of wavelengths the feeder is energised as a vertical radiator via the stray capacitance C_s. (b) If not brought away at a right-angle the feeder is energised by unequal inductive or capacitive couplings as indicated by arrows. (c) Excitation of standing waves in coaxial feeder if current distribution is symmetrical. If feeder is at earth potential excitation can take place as at (a) or (b)

transmitter unless the coupling coil is balanced to earth and shielded. The two halves of the dipole can also induce currents in the vertical feeder (the outer in the case of a coaxial line) but, given perfect symmetry, these are equal and opposite and therefore cancel. On the other hand, any asymmetry due to failure to bring the feeder away from the dipole at right-angles or not connecting it to the exact centre of the dipole, or proximity of one half of the dipole to a tree or building, will lead to a net current in the down lead and hence to radiation in an unwanted and probably lossy mode with considerable potential for EMC problems.

Earthing the feeder at the transmitter alters the length required for resonance and leads to a rather less predictable situation as the earth resistance and effective length of the earth connection are introduced into the problem. The safest course is probably to avoid earthing and keep the electrical length of the system as far from resonance as possible. If necessary, the resonance can be shifted by a short counterpoise or earth connected to the centre point of the coupling coil as shown dotted or a short-circuited λ/4 stub. Avoiding resonance minimises but does not necessarily eliminate the unwanted radiation at the fundamental frequency, and the problem, though basically similar, may prove much more intractable at the harmonic frequencies responsible for EMC troubles.

In the case of balanced lines, the unwanted current is induced in the same direction in both conductors, subtracting from the transmission line current in one conductor, adding in the other, and displacing the SWR pattern in one relative to the other; this makes it impossible to obtain reliable measurements of SWR.

If a dipole fed with coaxial line is symmetrical with respect to earth, half the terminal voltage of the antenna must obviously appear between the outer of the coaxial line at its top end and earth, Fig 12.52. The feeder must then appear like an 'inverted ground plane' antenna for which the radiation resistance at resonance is about 18Ω and the power radiated in the unwanted mode will be:

$$\frac{V^2}{4 \times 18} \text{ watts}$$

ie as much power is radiated from the feeder as from the antenna. More likely, symmetry will not be achieved and feeder radiation will take place as previously described. As before, the situation can be rectified, more or less, for the fundamental frequency by avoiding resonance but this does not necessarily eliminate harmonic radiation.

A long line will have many resonances which tend to overlap, making it essential to use some form of balun as described later, but such devices have limited bandwidth and cannot be relied on to prevent the flow of harmonic currents on the outside of the feeder, so there is clearly no alternative to the suppression of harmonics by filtering at the source. It is, however, only the inner of the coaxial line which is normally filtered and the suppression of harmonic currents in the outer conductor may be incomplete. For these reasons it is sometimes found difficult when using coaxial feeder to eliminate EMC problems.

Balanced lines brought away from the antenna at right-angles and cut to a non-resonant length with adequate harmonic filtering avoid these problems. Unscreened lines are in themselves potential radiators but, subject to accurate balancing, this should be negligible for frequencies in the HF band; thus the power radiated for an RMS line current of I amps and a spacing D is given by:

$$P_r = 160 \left(\frac{\pi D}{\lambda}\right)^2 I^2$$

D and λ being measured in the same units so that D/λ is the line spacing measured in wavelengths. This is the same power as would be radiated from the same current in a short dipole of length equal to the line spacing. For 600Ω line with 6in spacing and a current of 1A, the line radiation at 14MHz is therefore 160 $(\pi/34)^2$ = 0.09W for 600W supplied to the antenna. The proportion of any harmonic power radiated by the line increases of course as the square of the harmonic number, but seems likely to remain insignificant.

The figures given by the above formula must be multiplied by four to account for radiation from the terminal connectors, but even so, for a 75Ω line with 0.08in spacing only about one millionth of the incident power is radiated as compared with the possibility of up to one-half as previously discussed for an unbalanced line.

This is still not the full story. A major advantage of coaxial line is that it can be laid anywhere, even buried underground; in contrast to this, low-impedance twin line generates a field which can be detected externally and, though it spreads only a short distance, makes it advisable to keep the line clear of trees, buildings etc, at least to the extent of mounting it on stand-off insulators. It is a disadvantage of open-wire lines that the balance is very easily disturbed and the presence of objects near the line can cause asymmetry and hence radiation; the use of open-circuited matching stubs is particularly dangerous and if their use is unavoidable they should be brought away from the line at right-angles with the separate wires carefully equalised in length, and balance is particularly sensitive to proximity of the open ends of such stubs to surrounding objects. Even if the layout

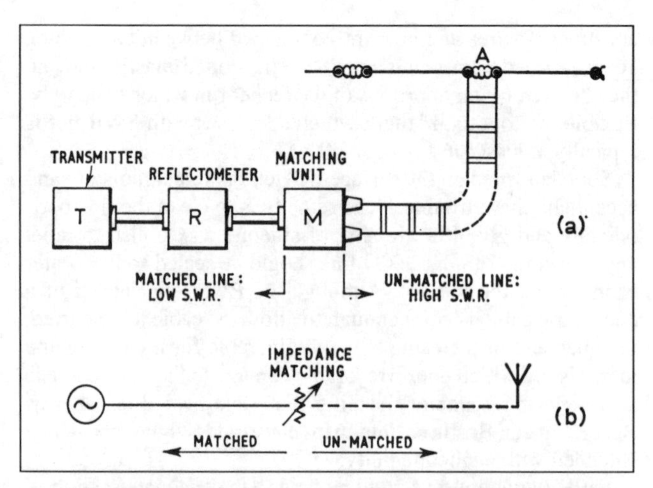

Fig 12.53. Application of impedance matching. Introduction of the matching unit can improve matters only on the section M-T. The feeder on the antenna side of the matching unit M-A, remains 'unmatched' with a high SWR. It should therefore be as short as possible and of low attenuation (see text). The matching unit, often called an *antenna coupler* in this application, can be adjusted using a reflectometer to indicate minimum reflected power in the 'matched' section of the line to the transmitter

appears to be perfect, balance should be carefully checked, eg with simple indicators such as that illustrated in Fig 12.86.

IMPEDANCE MATCHING

In nearly every arrangement of antenna and feeder system there is a requirement to provide some means of impedance transformation, or *matching,* at a particular point between the antenna and the transmitter. It may take the form of the antenna coupler mentioned above, which transforms the varying input impedance for a tuned line system down to a suitable load value for the transmitter and helps to provide discrimination against harmonics or other spurious frequencies. It may be a network at or near the antenna terminals which transforms the antenna input impedance to the correct value to match a length of flat line back to the transmitter. In each case the function of the transformer is to alter the impedance on the antenna side of it to a suitable value for presentation to the transmitter or the line connecting it to the transmitter. The matching network, whatever its construction, has no effect on the impedance of the antenna and therefore cannot alter the SWR on the line which connects it back to the antenna terminals. It is important to understand that the process of matching results in the elimination of standing waves only on the line between the point of application of the matching and the transmitter delivering the power (Fig 12.53).

It will be clear from the discussion of tuned-line feeder systems that they are acceptable, carrying as they do a high standing wave, only when, as in the case of open-wire line, the inherent line loss and hence the additional loss due to the standing wave is very low. It is also evident (as explained on p12.25) that the bandwidth of the antenna system reduces rapidly when the length of the tuned line is increased, the rate of change of reactance with change of frequency being proportional to the length of line whereas the radiation resistance is fixed, so that even in the narrow span of an amateur band retuning of the coupling circuits may be necessary as the frequency is changed. Any necessary matching should therefore be carried out as near as possible to the antenna. In a similar way, a reduction of bandwidth occurs if a large ratio of impedance is to be transformed, because high-ratio transformers have smaller bandwidths. It is therefore best to avoid any large ratios between the impedances of antennas and feeders, and this also helps to reduce losses. This cannot always be done, especially if a long line is necessary, because it would then be preferable to employ a high-impedance line in order to minimise losses. The importance of avoiding feeder radiation (sometimes referred to as the *Marconi effect*) has been stressed in the previous section and in many cases the balun required to prevent this in the case of coaxial feeders can also act as a matching unit.

The technique employed for matching will depend upon the nature of the unmatched impedance and the physical details of the antenna. In some cases it is possible to modify the antenna itself to achieve the required match to the feed line, without significantly altering its characteristics as a radiating element. Examples of this are the use of folded wires, and the adjustment of reflector spacing and length in Yagi-type beam antennas. Resonant antennas fed with high-impedance balanced lines may usually be matched by tapping the line across a suitable portion of the antenna wire, eg by means of the T or delta matches described later, but in some cases it is necessary, or easier, to use a network consisting of lumped-constant components or transmission-line stubs interposed between the antenna terminals and the main feeder.

Folded dipoles

The λ/2 dipole is a balanced antenna and requires a balanced feeder. Normally this is a 70Ω line, but it is sometimes necessary to step up the antenna input impedance to a higher value; for example, where the feeder is very long and 300 or 600Ω line is employed to reduce feeder loss. Another case is that of a beam antenna in which the dipole input impedance is too low to match 70Ω line. A separate transformer to step the antenna impedance up to these values would reduce the bandwidth, but it has been found that by folding the antenna the step-up can often be accomplished without reduction of bandwidth. Fig 12.54(a) shows a 1λ wire folded into a λ/2 dipole; this bears some resemblance to a loop antenna and has some useful properties in common with the quad and delta loops into which it can be formed by pulling the wire out to form respectively a square or triangular shape.

The relative direction of currents, at current maximum, is shown by arrows in the diagram. In a straight 1λ wire they would be of opposite polarity, but folding the full-wave has 'turned over' one current direction; thus the arrows in the diagram all point the same way. The folded dipole is therefore equivalent to two single-wire dipoles in parallel.

Since half the current flows in each wire, the radiation resistance referred to the centre of either is four times that of a simple dipole, ie 300Ω approximately, which is particularly convenient for matching to standard ribbon feeder, though the higher impedance and wider bandwidth of an open-wire line greatly eases the problems of resonant feed lines, and this can be exploited to good effect for the development of multiband beams as discussed later. Wider bandwidth is obtained with folded dipoles because the reactance change with varying frequency is approximately that of two λ/4 stubs connected in series and is therefore rather less than twice that of a simple dipole, whereas the radiation resistance, as explained above, is increased by four times. Typically, a simple dipole which is well matched to low-impedance line at the centre of the 14MHz band could show an SWR as high as 14 at the band edges, whereas the corresponding figure for a folded dipole fed with 300Ω line would be better than 1.2.

If the antenna is made three-fold (Fig 12.54(b)), it is equivalent to three radiators in parallel, the feed-point resistance is multiplied nine times and it would be used with a 600Ω line.

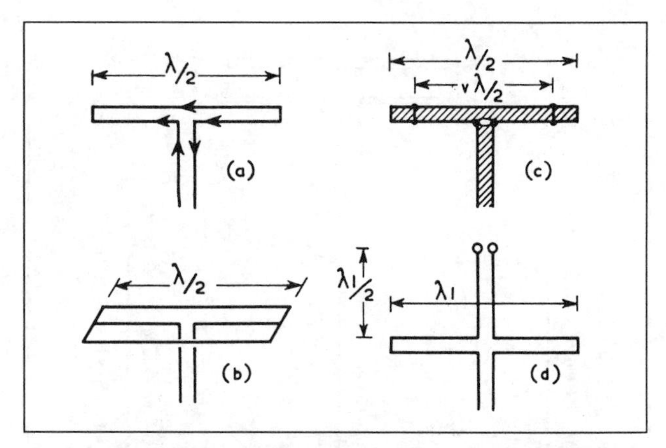

Fig 12.54. Folded dipoles including (a) simple fold; (b) three-fold; (c) folded dipole made from 300Ω ribbon feeder; (d) arrangement for working on half- and full-wave resonance

In some multi-element arrays the radiation resistance is very low, and folding the driven element is a useful aid to correct matching since with N wires carrying equal currents the resistance is raised N^2 times.

The folded dipole is usually regarded as a one-band antenna, and will not normally operate on even harmonics of the frequency for which it is cut because the 'folded' currents cancel each other. However, the alignment of currents is correct for the odd harmonics, so that a folded dipole for 7MHz can also be used on 21MHz, although the radiation pattern will have multiple lobes. On the other hand, a 14MHz folded dipole with a λ/2 resonant feeder will be nearly resonant on 21MHz and will provide some gain at the higher frequency, the radiation pattern being intermediate between those of λ/2 and 1λ dipoles. With suitably tuned feeders, a 21MHz folded dipole may be used on 28MHz and 14MHz.

A folded dipole one wavelength long can be made to radiate but only if the centre of the second conductor is broken; in such a case the input impedance is *divided* by four. The centre impedance of a 1λ dipole is 4000 to 6000Ω; that of the folded version 1000 to 1500Ω. Fig 12.54(d) shows how such an antenna can be made to work at both fundamental and second harmonic frequencies by means of a length of open-circuit twin line (called a *stub*) used as a 'frequency switch'. At the frequency at which the antenna is λ/2 long, the stub is one electrical quarter-wave long and, since its other end is open, the input end behaves as a short-circuit and effectively closes the gap in the antenna. At the full-wave frequency the stub is one half-wave long and behaves as an open-circuit.

The folded dipole can be conveniently made from 300Ω flat twin feeder as in Fig 12.54(c) because the spacing is not critical and the electrical length of the system, as a radiator, is not affected by the insulating material since the voltages have the same magnitude and phase in both wires and no current flows between them. The two halves of the dipole can, however, also be regarded as a pair of short-circuited stubs connected in series, and in this role the normal velocity factor applies so that each stub is about 0.05λ longer than resonance. The capacitance being roughly 6pF per foot, this is equivalent at 14MHz to a reactance of about 1200Ω across the feeder which increases the SWR to 1.3. The velocity factor of ribbon feeder being about 0.8, the SWR may be improved by placing the short-circuits between the wires at this fraction of a quarter-wavelength from the centre rather than at the end, eg 27ft apart in a 33ft long 14MHz antenna.

Other ratios of transformation than four or nine can be

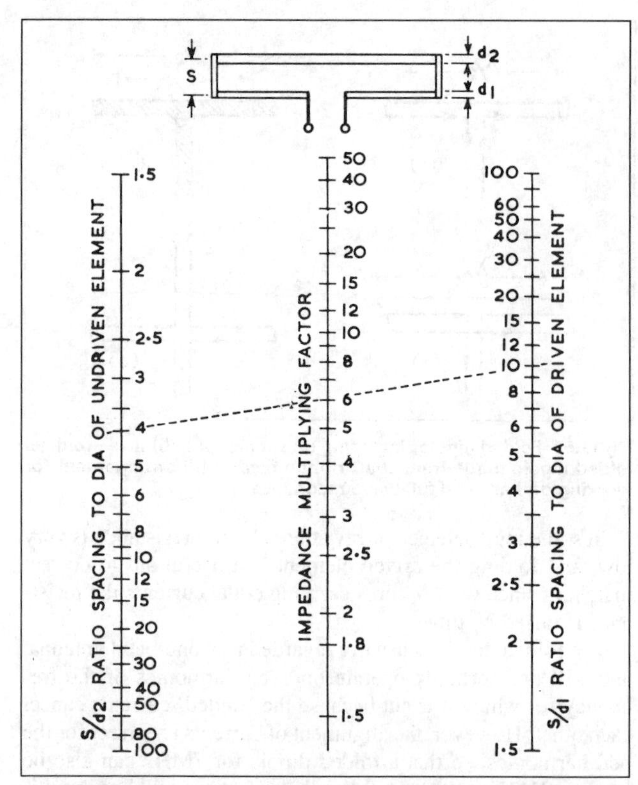

Fig 12.55. Nomogram for folded dipole calculations. The impedance multiplying factor depends on the two ratios of conductor diameter to spacing between centres, and is always 4:1 when the diameters are equal. A ruler laid across the scales will give pairs of spacing/diameter ratio for any required multiplier. In the example shown the driven element diameter is one-tenth of the spacing and the other element diameter one quarter of the spacing, resulting in a step-up of 6:1. There is an unlimited number of solutions for any given ratio. The chart may also be used to find the step-up ratio of an antenna of given dimensions

obtained by using different conductor diameters for the elements of the radiator. When this is done, the spacing between the conductors is important and can be varied to alter the transformation ratio. The relative size and spacings can be determined with the aid of the nomogram in Fig 12.55.

These variations of the basic folded dipole do not lend themselves readily to multiband operation.

Antenna transformers

In antenna systems comprising 1λ (or end-fed λ/2) radiators, it is possible to obtain some control over the high input impedance by utilising the radiator itself as a λ/4 transformer, making use of its similarity to a λ/4 open-ended stub which has been unfolded. The principle was explained on p12.26. The *loop* radiation resistance of the radiator is principally a function of its length, and is independent of the conductor size. By varying the latter to alter the effective *L/d* ratio of the radiator, the Z_0 of the 'transformer' can be altered, and hence the input impedance at the voltage feed point. This is not practicable as an empirical adjustment after erection, but does permit some elementary matching to be built-in to the antenna at the design stage to restrict the extent of the initial mismatch to the open-wire main feeder.

The effective Z_0 of the radiator as a transmission line being a function of the length/diameter ratio, and the length determined by the operating frequency, the diameter must be adjusted to control Z_0. This may be achieved by using a thicker wire, or more effectively by using several wires connected in parallel

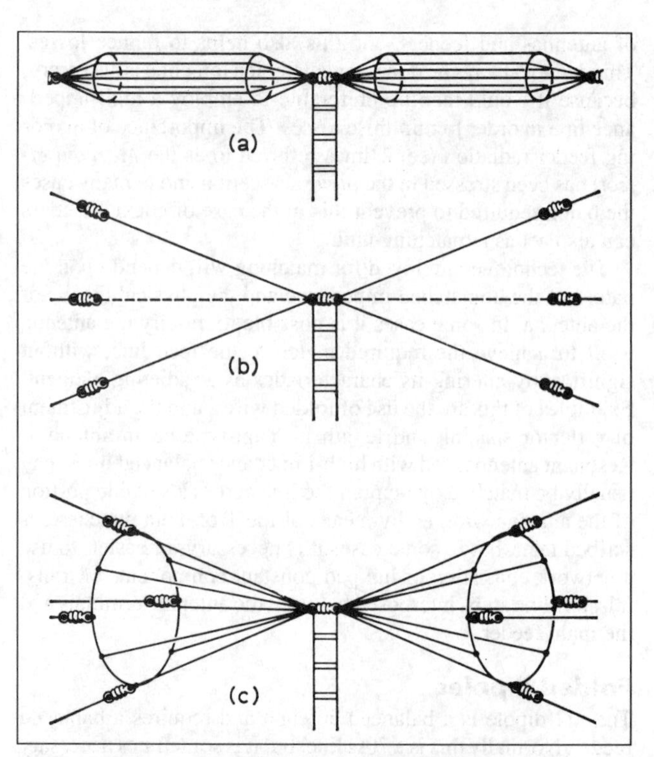

Fig 12.56. Broad-band dipoles. (a) Cage. (b) Fan (two or three wire). (c) Conical. The semi-conical form is often used for broad-band vertical monopoles over an earth plane

and spaced out to form a cage. Another alternative is to use several wires which are spread out from their common end at the feed point, to form a fan or semicone. These methods are illustrated in Fig 12.56. The curve in Fig 12.57 shows the variation of input impedance of a 1λ biconical dipole in terms of the cone angle: it is correct for solid cones and indicative of the general tendency for elemental cones comprising several wires. The values of input impedance for a fan dipole will be 50–100% higher for the same angle.

The use of large-diameter conductors or fans of wires improves the bandwidth of λ/2 and λ dipoles equally, since in accordance with line transformer theory, the *Q* is approximately equal to R/Z_0 or Z_0/R, whichever is the larger. The need for a reduction of Z_0 arises with λ/2 dipoles if it is required to combine large bandwidth with matching to a low-impedance line (ie one matching the radiation resistance), otherwise the folded dipole makes better use of the extra conductors up to a maximum of three or four, beyond which further matching problems may arise. With full-wave dipoles, however, there are two separate

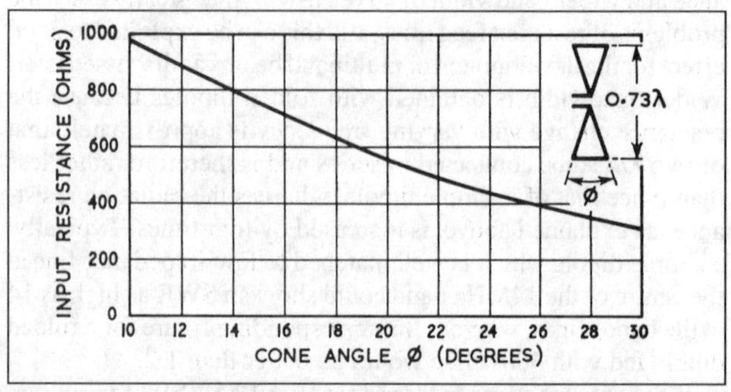

Fig 12.57. Input resistance of a 1λ bi-conical dipole as a function of the cone angle. The overall length should be 0.73λ at the mid-frequency, independent of the cone angle

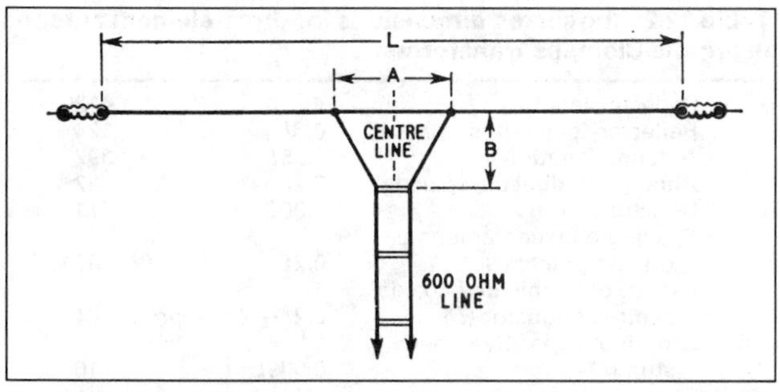

Fig 12.58. Use of the delta match with a $\lambda/2$ antenna fed with 600Ω open line

problems, the bandwidth of the dipole itself and the constraints imposed by the high feed-point impedance relative to the feeder Z_0 which may further restrict the bandwidth unless some form of cage or fan is adopted; for the amateur HF bands, however, this problem is unlikely to be acute unless the dipole forms part of a close-spaced beam system.

When fans of wires are used, the length required for resonance is reduced, typically by a factor of about 0.8. A simple explanation of this is implied by the $6.28\, Z_0\, \Delta l/\lambda$ formula for the reactance, p12.27. Thus, if a low impedance (Z_{0L}) resonant line is shortened by an amount Δl, this subtracts a reactance proportional to $Z_{0L}.\Delta l$. If this length is now restored, using a line of higher impedance Z_{0H}, the added reactance is proportional to $Z_{0H}.\Delta l$ which is more than the amount removed, so that some shortening must be applied to one or other sections of the line, and the overall length for resonance is less than it was originally. In the case, for example, of a $\lambda/2$ fan dipole, Z_0 is higher where the wires are closer together, ie in the centre, which is the reference point for application of the formula. This demonstrates an important property of all antenna or antenna-plus-feeder systems involving a change of Z_0, enabling us to predict, for example, the well-known fact that for resonance a quad loop requires slightly more wire than a folded dipole, and other facts which appear to be less well known, eg when a non-resonant dipole is loaded by means of a stub, the total length of wire required will be longer for a short dipole but shorter for lengths between $\lambda/2$ and λ, and harmonic resonances will not be quite coincident. Similarly a 14MHz quad or delta loop with a $\lambda/2$ resonant stub *nearly* resonates at 21MHz, the length of the whole system then being $3\lambda/2$, but to obtain exact resonance the stub must be shortened by about 2ft.

Tapped antenna matching: the delta match

Pursuing the analogy between a resonant antenna and a tuned circuit, any point on it will have an impedance *relative to ground* varying from a low value (one-quarter of the radiation resistance) at points of maximum current, up to a very high value at points of maximum voltage. A single-wire feed line (p12.17), if connected to the centre, would cause a current to flow outwards in both directions which would of course be quite wrong, but when displaced to one side the line 'sees' an inductive reactance one way and a capacitive reactance the other so that equal but opposite currents flow in the two directions and the required current distribution is set up. This is exactly analogous to connecting a wire to a tap on a tuned circuit and, by choosing the right tapping point, the impedance of the line can be matched. This is the principle of the single-wire feed, which (as discussed on p12.17) is now largely obsolescent due to wastage of power by radiation from the feeder and losses in the ground return path. These disadvantages are completely overcome by using *two*

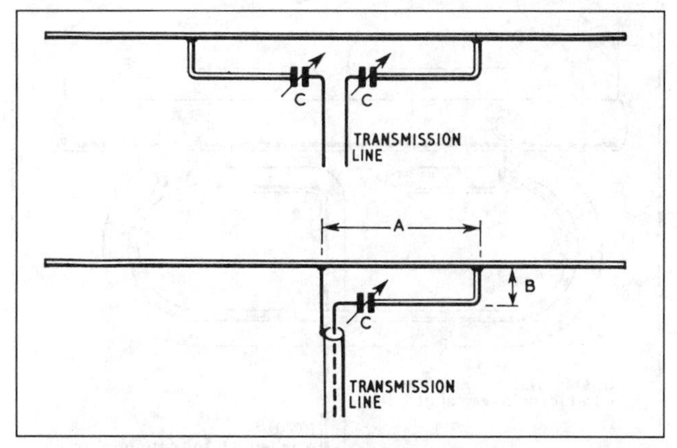

Fig 12.59. Tuning out residual reactance in the tee- and gamma match systems with series capacitors. Capacitors with a maximum capacitance of 150pF are suitable at 14MHz. For higher frequencies proportionately smaller capacitances may be used

wires, tapped each side of centre as shown in Fig 12.58. This is called the *delta match*, and is a convenient way of connecting an open-wire balanced line to a $\lambda/2$ dipole or beam element. The precise dimensions *A* and *B* of the delta section may be obtained experimentally, but for the case of a 600Ω feed line are given approximately by:

$$A = \frac{118}{f\,(\text{MHz})} \qquad B = \frac{148}{f\,(\text{MHz})}$$

where *A*, *B* are measured in feet.

It might be thought that the delta section would radiate but, using its resemblance to an 8JK antenna, it is possible to arrive at an approximate figure of 0.002dB for the power loss by radiation. The previous formula is not applicable as the ratio of feeder length to spacing is less than specified and in any case most of the radiation from the delta is in the wanted mode. There is no need to make direct connection to the dipole, and the end of the feeder may be fanned out into a delta loop which provides inductive coupling to the antenna. Ordinary PVC insulated mains cable secured against the dipole with plastic tape has been found satisfactory, the dimension A being increased by about 10% as compared with direct connection. It is believed, though not fully established, that inductive coupling may reduce the risk of line imbalance, but either method provides a simple and satisfactory method of matching between a dipole (or beam element) and a 600Ω line. These methods cannot be used in their simple form with low-impedance lines, as the reactance of the loop is then excessive relative to Z_0 and must be removed by tuning. In the case of a close-spaced two- or three-element beam the dimensions *A* and *B* will be reduced by some 30 to 50% compared with those for a dipole.

The tee and gamma match

The dimensions *A* and *B* for the delta match are slightly interdependent but *B* is not critical and the whole of the delta may be folded upwards to form a *tee-match* as in Fig 12.59.

The tee-match is particularly suitable for feeding close-spaced beam antennas using low-impedance twin-lead and the *gamma match*, also illustrated, is evolved from it for use with unbalanced lines. In each case there is some mismatch caused by the inductance of the coupling loop unless this is tuned out, eg as shown in Fig 12.59. In the case of a delta match, as in Fig 12.58, the loop can be regarded primarily as an extension of the transmission line and the inductive effect is small, but tuning is usually essential in the case of tee and gamma matches, particularly

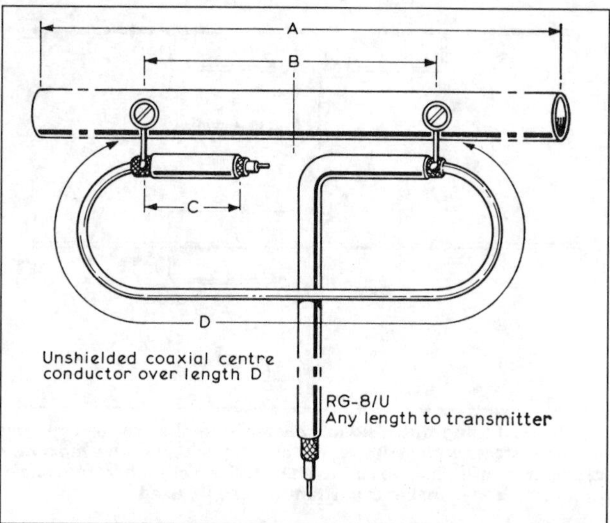

Fig 12.60. The Clemens transformer, a flexible arrangement which will match a wide range of dipole impedances into a concentric 50 or 70Ω line, and at the same time convert from balanced to unbalanced circuits. The suggested dimensions for a three-element antenna using this transformer are given in Table 12.2

with low-impedance lines and low antenna impedances. The capacitors will require to be adjusted empirically for minimum SWR on the main feed line: they will also need to be sealed adequately against weather and dirt accumulation, which would otherwise make them liable to voltage breakdown. The physical size and rating should be adequate to withstand the full line voltage with an adequate factor of safety.

As in the delta match, the exact dimensions of the matching section are a matter of experimental adjustment for minimum SWR on the main feed line. For a λ/2 dipole fed with 600Ω line, the approximate dimensions are given by:

$$A = \frac{180.5}{f\text{(MHz)}} \qquad B = \frac{114}{f\text{(MHz)}}$$

where A is in feet and B in inches.

These formulae apply only when the extra conductor is of the same diameter as the antenna conductor. If a different size of conductor is used for the matching section, an impedance transformation is obtained analogous to that achieved by varying the diameter ratio in a folded dipole.

The Clemens match

With the gamma match, power is fed into one half only of the radiator, the other being excited by induction. This is inherently an unbalanced arrangement and does not fully meet the requirements for elimination of feeder radiation. A further disadvantage is the need to adjust the length of the matching section. An improved version has been developed which enables the tapping points to be fixed and also incorporates balance-to-unbalance transformation. It is known as the *Clemens match* and, as shown in Fig 12.60, can be constructed entirely of coaxial cable, the cable shield being attached to the metal antenna element using self-tapping screws.

This arrangement will cater for a very wide range of impedance transformations. The coaxial feeder cable is taken to the centre 'neutral' point of the antenna, carried along one side to a distance of 0.05λ where the outer conductor is bonded to the antenna, the inner conductor being looped back to an equal distance the other side of centre and there joined to the driven element via a capacitor which is formed from the cable. The spacing between D and the driven element is variable from 0.01 to 0.02λ which adjusts the ratio, while the capacitor tunes out the transformer reactance and also helps with the tuning of the radiator.

Table 12.2. Suggested dimensions for three-element antenna using the Clemens transformer

Reflector length	L	412
Reflector-to-antenna spacing	$0.3L$	124
Antenna length (A)	$0.95L$	392
Antenna-to-director spacing	$0.2L$	82
Director length	$0.90L$	371
Spacing between antenna driving points (B)	$0.2L$	82
Length of unshielded coaxial centre conductor (D)	$0.2L$ + 2in approx	84
Length of capacitive open stub (C)	$0.039L$	16

L is the length of a λ/2 in free space (5900/f inches at frequency f MHz). The third column gives the actual dimensions in inches for a 14.2MHz antenna.

The dimensions given are those recommended for matching to a particular design of three-element Yagi antenna and may need modifying for other types of antenna by trial and error along the lines indicated above. The cable should be sealed against damp, Bostik sealing strip or Sylglas tape both being suitable for the purpose.

Line transformers

Often the most convenient way of performing impedance matching between the antenna and the main feeder is by employing *line transformers.* These are made up as described on p12.27 from short sections of transmission line and are hence physically convenient to assemble and connect at the antenna terminals. They are also lighter than the equivalent coil and capacitor networks and more dependable when exposed to the weather.

The behaviour of short sections of transmission line with various terminations was explained on p12.26 where it was shown that λ/4 sections can be considered equivalent to parallel and series tuned circuits. Shorter line sections behave like inductance or capacitance depending upon their length and whether they are open- or short-circuit at the far end.

Stub matching

If the main feeder is connected directly to the antenna terminals the antenna will usually constitute an unmatched load, and a system of standing waves will be set up along the feed line as explained on p12.22. Moving away from the antenna, the impedance along the line will alter in sympathy with the pattern of voltage and current waves, as explained and illustrated in Fig 12.39. There will occur a position along the line *not more than λ/2 from the antenna* at which the resistive component of the impedance will equal the line characteristic impedance, but will have associated with it some degree of reactance. By connecting in parallel with the main line at that point a transmission-line stub, which is either open- or short-circuited at the far end to give an equal but opposite reactance at its input end, the unwanted reactance on the main line can be cancelled or 'tuned out', and the only term left is the resistive one which then presents a correct match to the remainder of the line. This is the principle of *stub matching.*

It is possible to calculate the length and position of the stub from a knowledge of the SWR on the unmatched line and the position of a voltage maximum or minimum. The charts shown in Figs 12.61 and 12.62 give the length of the stub and its distance from the point of maximum or minimum voltage *in the direction of the transmitter.* They are drawn specifically in terms

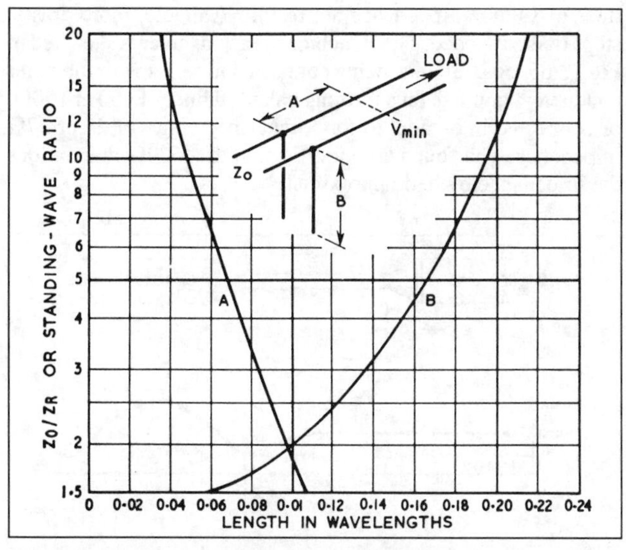

Fig 12.61. Position and length of an open stub as a function of Z_0/Z_R or SWR when reference is at a voltage minimum

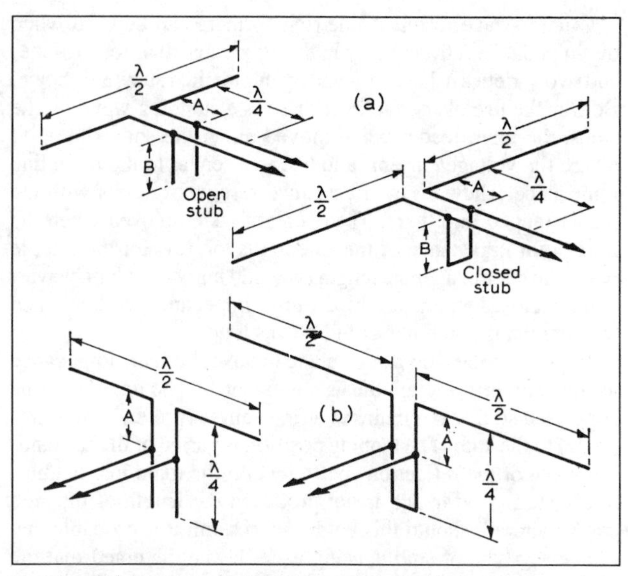

Fig 12.63. Use of matching stubs with various antennas

of a stub of the same impedance Z_{01} as the main feed line. However, a line of different impedance Z_{02} may be used to provide the stub by determining the ratio X/Z_{01} from Fig 12.47, and thence X for the stub length given by the stub-matching charts. The new ratio X/Z_{02} may then be used to find a new stub length again from Fig 12.47. It may sometimes be advantageous to use a stub of lower Z_0 than that of a main feed line when the latter is relatively high, since the bandwidth of the match will then be improved. However, particular care must be taken in using low-impedance lines as stubs in view of the possibility of high losses; for example a 75Ω line with a loss rating of 1dB per 100ft will have a 2.5% loss per 10ft which is equivalent to a resistance of 2Ω in series with a λ/4 stub, or a Q of 40. If used as a line transformer with a voltage step-up ratio greater than two, there will be a power loss exceeding 5%. Attention is also drawn to the warning on p12.32 regarding the added danger of line imbalance when using open stubs.

The curves of Fig 12.61 apply when the point of reference is a voltage minimum, and those of Fig 12.62 when the reference is a voltage maximum. Since these voltage conditions along the line alternate at intervals of λ/4, either short-circuit or open-circuit stubs may be employed at the appropriate points on the line to achieve exactly the same match on the remainder of the line. Also, because of the inverting properties of a λ/4 section of line, the open-circuit stub may be replaced by a short-circuit stub which is λ/4 longer, and vice versa. This system of stub matching is extremely flexible, and within certain limits permits the matching to be carried out at a point which is physically convenient, and with a stub which can be either open- or short-circuit to suit the particular mechanical layout. Generally speaking, the use of short-circuit stubs is preferable; apart from the problem of balance mentioned above, the electric field at the remote end of an open-circuit stub is very high, and the 'end effects' of the supporting insulator are difficult to assess accurately. Their action is to modify the electrical length of the stub and hence to 'de-tune' it. In the case of short-circuit stubs, the remote end is accurately defined by the shortening bar and the length is not subject to the same end effects. The voltage at the end is practically zero, and they can be tensioned or otherwise supported without the need for a strain insulator. There is also no objection to a DC connection from the centre of the shorting bar to earth to provide lightning protection for the antenna and feeder system.

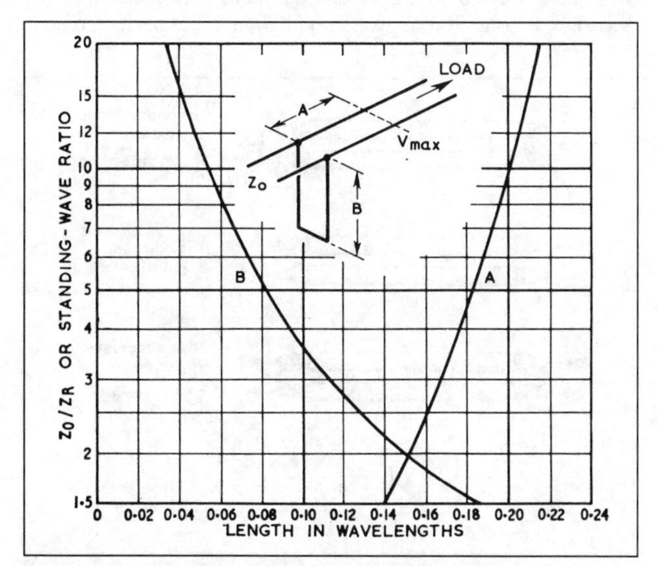

Fig 12.62. Position and length of a closed stub as a function of Z_R/Z_0 or SWR when reference is at a voltage maximum

Another useful feature of stub matching is that it is not necessary for the antenna to be resonant, ie cut to an exact number of quarter-wavelengths at the operating frequency. If the antenna is of an arbitrary length, the input impedance will be complex, and the series reactance necessary to tune the antenna is effectively provided by the short piece of line between the antenna and the first voltage maximum or minimum.

Quarter-wave resonant sections

When the standing wave ratio is high, the length of the stub together with its distance from the reference point add up to λ/4, and because of this, such stubs are sometimes referred to as *quarter-wave resonant line transformers*. Typical stub matching arrangements are shown in Fig 12.63(a) and rearranged in Fig 12.63(b) to illustrate the quarter-wave feature of this system of matching. The action is electrically identical, but it may sometimes be physically more convenient to mount a fixed λ/4 stub across the antenna input, and to adjust the tapping point of the main feeder along this stub for minimum SWR on the main line.

Quarter-wave resonant line transformers can be used when the impedance ratio is fairly high (eg greater than four-to-one) and two varieties using short- and open-circuited stubs are shown. Both make use of the fact that there is a standing wave on the stubs, the main feeder being moved along the stub to a point where the voltage/current ratio becomes equal to the main line impedance. They are in effect tuned resonant circuits with the feeder tapped into them. The open stub is employed when the feed-point impedance of the antenna is too low for the line, ie when the overall antenna length is an odd number of half-waves, and the closed stub is used for a high-impedance feed, ie when the antenna is a number of full waves long.

In practice the stub can be made to have the same impedance as the main line, though this is not essential. The line should be supported so that it remains at a fixed angle (preferably a right-angle) to the stub. The tapping position is varied until the standing wave or the reflection coefficient on the main line is minimised. If a good match is not produced the length of the stub can be altered; should this not be successful, it is probable that the wrong type of stub is being used. It may be noted that the antenna itself need not be a resonant length, though the antenna-plus-stub must be; also, the stubs may be increased in length by one or more quarter-waves in order to bring the tap position nearer to ground level, changing from 'open' to 'short-circuit' or vice versa with each quarter-wavelength addition. One method of adjustment requires a simple current indicator such as the one illustrated in Fig 12.86. If the current in a closed stub is less than the current on the antenna side of the junction, the length of antenna plus stub needs to be shortened, and vice versa, the SWR in the main feeder being unimportant for this part of the procedure. When the length of antenna plus stub is correct, the tapping point for the main feeder can be moved up and down the stub to obtain the minimum SWR.

The technique of stub matching is applicable both to balanced and unbalanced lines, and may be used with advantage on coaxial lines at VHF where the physical length of the stub is not great. Again because of the 'end effect' of the open-circuit stub, a short-circuit version is to be preferred, which then completes the outer screen of the coaxial system and eliminates the possibility of unwanted signals being picked up on the inner conductor of an open-circuit stub.

Series quarter-wave transformers

The impedance inverting properties of a $\lambda/4$ section of line were explained earlier in the section dealing with transmission lines and it was shown that a resistive load may be transformed to provide a match to some other impedance by one or more $\lambda/4$ sections of line having appropriate values of Z_0. If the load is partly reactive, it may be supplied with power from any convenient feed line and, moving back along this line from the antenna, a position of voltage minimum may be located. Viewed from this point the load will appear to be a low value of resistance with no reactive component, and the next $\lambda/4$ section of line (moving towards the transmitter) may be designed to provide the required transformation. This may be merely a matter of 'pinching in' the conductors of an open wire to reduce the Z_0, but it may also require a larger diameter of conductor. Such matching sections are often referred to as *Q-bar transformers,* and their action is illustrated by Fig 12.64. By measuring the SWR in a line of known impedance it should be possible to deduce the value of matching line Z_0 required and design it according to the data given on pp12.18–12.19. As will be apparent from Fig 12.31, it is difficult to effect much change of Z_0 merely by reducing the spacing of a 600Ω line, so that if a large ratio is required and no lines of suitable impedance are readily available, more drastic steps will be needed. These include the arrangements sketched in Fig 12.65, coaxial lines being connected in series with the outer conductors bonded, thus forming balanced lines of 100 or 150Ω, or connected in parallel to form unbalanced lines of 25 or 37Ω impedance. The four-wire modification of a 600Ω line divides the impedance by half approximately.

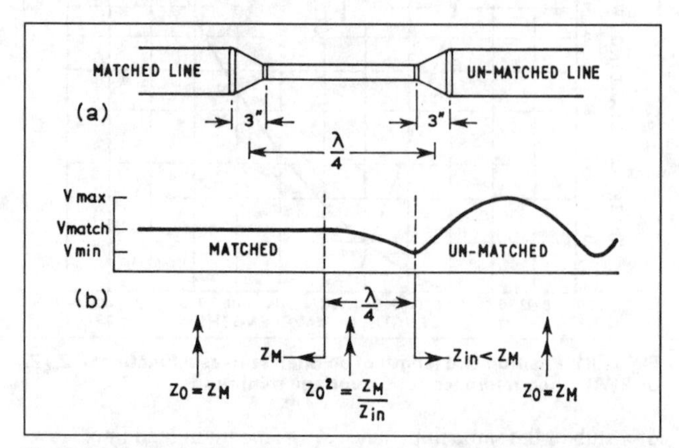

Fig 12.64. Series λ/4 transformer (Q-bar) using 'pinch' matching. The voltage standing wave is a minimum at the point of commencing the λ/4 matching section, which is pinched together until the standing wave disappears on the following length of line. The 'pinch' can be determined experimentally using string to hold the wires together. The string can then be replaced by permanent insulators

The bandwidth of an amateur-band HF antenna is usually determined by the antenna itself in connection with any resonant matching stubs, the ratio and therefore the effective loaded Q of $\lambda/4$ matching transformers being relatively low so that they do not introduce any further appreciable restriction of bandwidth.

Tapered lines

When the required transformer ratio is not high, for example when an 800Ω rhombic antenna is to be matched to a 600 or 300Ω line, it is practicable to use a simple tapered line. At the antenna end, the spacing is correct for 800Ω and gradually decreases to that for the lower impedance. This arrangement gradually converts the impedance from one value to the other. It is a wide-band device and, provided the taper is not less than one wavelength long, it will give a good, though not perfect, match.

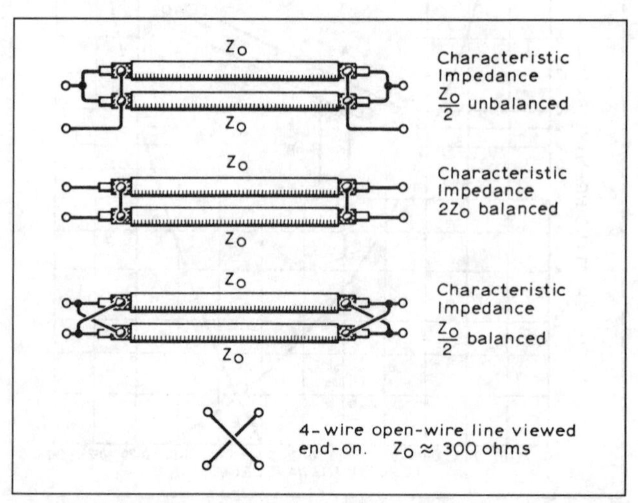

Fig 12.65. Methods of obtaining non-standard line impedances, eg for matching sections

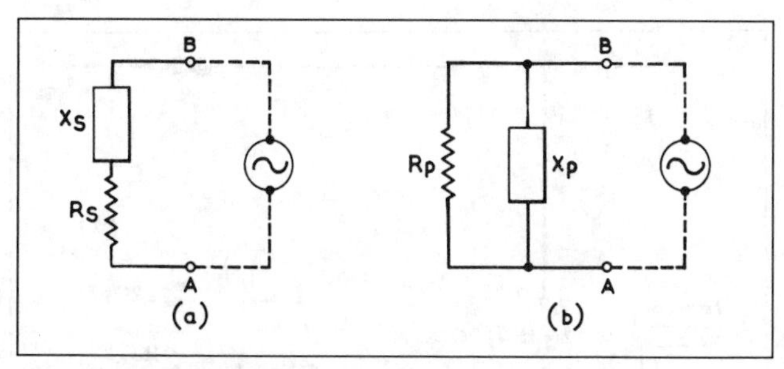

Fig 12.66. Equivalent impedance circuits

Network transformers

At ground level, particularly with coaxial feed lines, it is often more convenient to employ a coil and capacitor matching network. This could be similar to any of the couplers used near the transmitter (though probably simpler, eg as shown in Fig 12.42).

The ability to transform one value of resistance to another without loss is of course a common requirement in radio frequency circuits generally, not only in antenna systems. This general problem may be studied in terms of a 'generator' supplying power to a load consisting of resistance and reactance, as in Fig 12.66. The generator may be a valve supplying power to an anode load, or it can be *any device from which power comes,* regardless of what may intervene in the form of tuned circuits or lines with or without losses. The important properties of generators are the available power and the source impedance. The available power, ie the maximum power obtainable from a generator, is expressed in terms of the open-circuit voltage E_0 and the effective internal series resistance R_G, and is equal to $E_0^2/4R_G$, any internal reactance being ignored for this purpose. The requirements for maximum power transfer are that the generator should see a 'conjugate match', or in other words if the load consists of a resistance R_S in series with a reactance X_S and the generator has a series reactance X_G, we require $R_G = R$ and $X_G = -X_S$.

Suppose now that the transmitter is intended to work into a 75Ω load subject to proper adjustment of tuning and matching controls. When the adjustment is completed successfully with a 'dummy load', this will automatically ensure a conjugate match and the reader who is interested only in getting as much power as possible into the antenna need not be concerned further with what is happening inside the transmitter, provided he can make the antenna look like 75Ω and connect it in place of the dummy load. The problem next to be considered is how to make *any* impedance look like 75Ω using convenient 'lumped' constants, ie capacitors and inductors. There is one further constraint: this must be achieved without losses, and if the components are loss-free (no internal resistance) all the available power must then go into the load for there is nothing else to absorb it. Usually there is a choice of many alternative ways of effecting the required transformations, and because in general the required components are reasonably efficient devices, low-loss transformation can usually be achieved, though it is advisable to avoid large currents in large inductors, or excessive voltages which produce leakage losses in capacitors. Losses are proportional to the square of these currents and voltages, and the loss resistance of a coil tends to be proportional to the inductance, ie Q values do not vary much assuming reasonable design practice.

Consider now two elementary networks involving series and shunt elements respectively as shown in Fig 12.64; for these two networks to be identical, that is present the same load to a generator connected across the terminals AB, the admittance seen at these terminals must be the same in each case, ie:

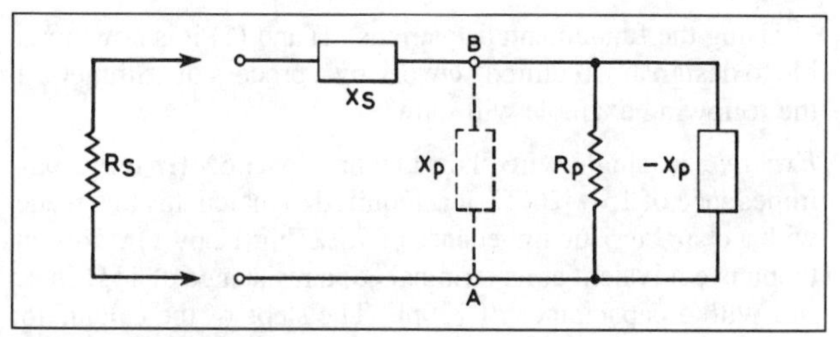

Fig 12.67. Transformation: stage 1

$$\frac{1}{R_p} + \frac{1}{jX_p} = \frac{1}{R_s + jX_s} \quad (1)$$

From this it is simple to derive the following two identities:

$$R_p = R_s\left(1 + \frac{X_s^2}{R_s^2}\right) \quad (2)$$

$$X_p = X_s\left(1 + \frac{R_s^2}{X_s^2}\right) \quad (3)$$

The 'operator' j, (which is mathematically the square root of minus one) merely denotes that the quantity after it is a reactance and that, if added to a resistance, the addition must be done vectorially. This can be done by drawing a right-angled triangle with the two shorter sides proportional to the two quantities, the magnitude of the sum then being represented by the length of the longer side.

Using these basic formulae it is possible to convert a network of series elements to its equivalent network of parallel elements (and conversely). Consider next the circuit shown in Fig 12.67. From an inspection of Fig 12.66 and equation (2) above, it can be seen that it is possible to select a value of X_s to put in series with the resistance R_s, such that the shunt equivalent resistance is equal to R_p, but has in parallel with it a residual reactance X_p. If now a reactance of value $-X_p$ is placed across the terminals AB, the net effect will be to produce at those terminals a pure resistance equal to R_p, which is the requirement of the transformer. Then the basic form of the transformer is an L-network comprising only reactive elements as shown in Fig 12.68. This also meets the terms of the original specification. If the ratio $R_2/R_1 = p$, it can be shown that:

$$X_1 = \pm R_1\sqrt{(p-1)} \quad (4)$$

$$X_2 = \frac{\mp pR_1}{\sqrt{(p-1)}} \quad (5)$$

X_1 and X_2 must have opposite signs in all cases or, in more general terms, the network must consist of one inductive and one capacitive element. It is more usual in practice to make the series element inductive and the shunt element capacitive for reasons which vary with individual circumstances (eg a series DC path to the anode of a valve), but there is no basic objection to the reversal of this practice in cases where it may be more suitable.

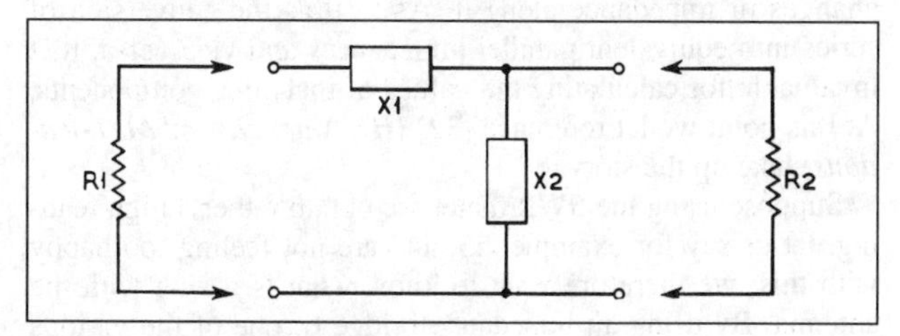

Fig 12.68. Transformation: stage 2

Using the fundamental equations (4) and (5) it is now possible to design any required network by a process of arithmetic as the following example will show:

Example. A single wire short antenna for 3.6MHz has a base impedance of 15 – j200Ω. It is required to match this to a feeder with a characteristic impedance of 75Ω (Fig 12.69(a)). This antenna is equivalent at its terminals to a resistance of 15Ω in series with a capacitance of 220pF. The steps of the calculation are as follows:

(a) Connect an equal and opposite reactance +j200 in series to tune out the antenna capacitance and leave only the resistive term of 15Ω: the transformation required is then from 15Ω resistive to 75Ω resistive (Fig 12.69(b))
(b) From equation (4), X_1 (series term) = ± 15 √5 – 1 = ± 30Ω
(c) From equation (4), X_2 (shunt term) = ∓ (5 × 15)/√(5 – 1) = ∓ 37.5Ω

Thus the transformer section required is as shown in Fig 12.69(c) and the complete arrangement in Fig 12.69(d). To convert the reactance into physical components, since $X_L = 2\pi fL$ and $X_C = 1/2\pi fC$:

+j230 is the reactance of 10µH at 3.6MHz.
–j37.5 is the reactance of 1250pF at 3.6MHz.

If the coil has a *Q* of 200, ie a series loss resistance of 1Ω, this will be equivalent to an antenna loss resistance of 1Ω. Strictly speaking, the matching network should be modified slightly as if the antenna had in fact a resistance of 16Ω, and one-sixteenth of the transmitter power will be unavoidably lost as heat in the coil. This, however, is a loss in signal of 0.3dB only, and unlikely to have much impact on reports of signal strength.

The complete network is shown in Fig 12.69(e)). By making each element variable over a convenient range, it is possible to correct for the coil resistance and any inaccuracies in the original figure taken for the antenna base impedance, and also for variations in the working frequency.

A suitable protective cover for the network is a large airtight can, used upside down, with components mounted on the lid. A metallised lead-through seal from an old capacitor or transformer can be soldered into the lid for the antenna lead. A sealed RF connector is preferable for the cable connection. The can should have a coat of paint, and Bostik putty can be used to seal all the joints; a small bag of silica-gel may be placed in the can to guard against condensation. The components of the network should be capable of handling the transmitter power; if fixed capacitors are used they should be of the foil-and-mica type.

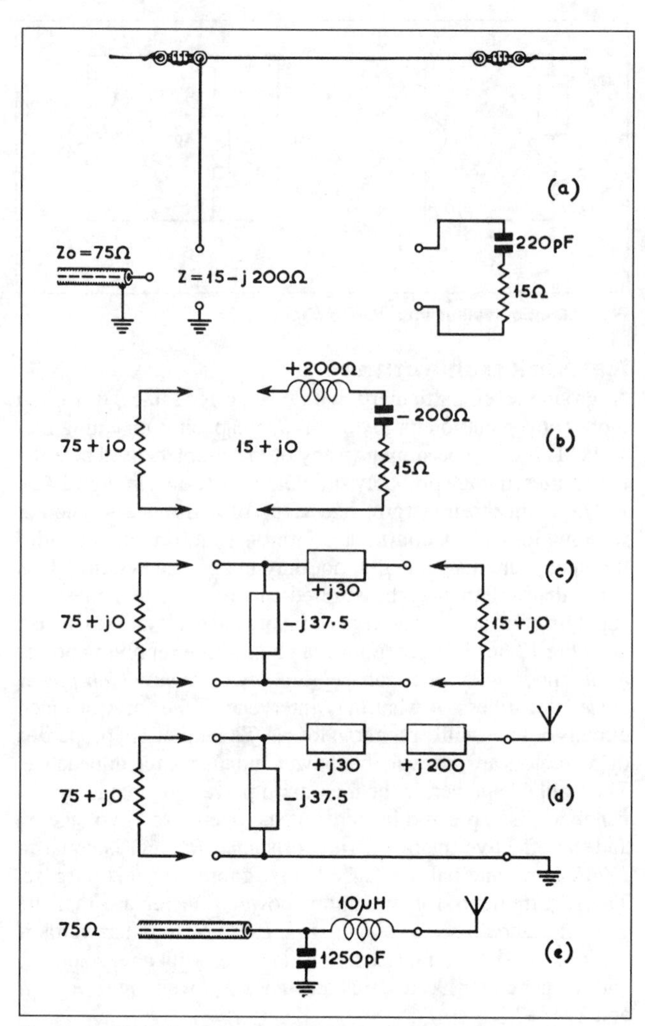

Fig 12.69. Matching of two arbitrary impedances

The Smith chart and why it is useful

Calculations such as the above can be much simplified by using the Smith chart; this is a highly versatile instrument which, despite the handicap of a somewhat fearsome appearance, provides a simple means for carrying out basic transmission line (ie antenna and feeder) calculations such as establishing the impedance and SWR at any point on a feeder even if there are several changes of impedance along it. As well as the conversion of series into equivalent parallel impedances and vice versa, it is invaluable for calculating the values of matching components. At this point we let reference [32] (*HF Antennas for All Locations*) take up the story.

Suppose using the SWR meter we obtain rather a high reading, let us say for example 2.5, and are not feeling too happy with this; we therefore want to know what is wrong with the antenna. By using an impedance bridge or one of the various tricks described in this chapter it should be possible to discover the impedance 'seen' by the transmitter, and before erecting the antenna the reader will have remembered of course to measure the electrical length of the feeder by coupling it into a GDO. You didn't? Sorry, but it is 'back to square one'; given this information a mere glance at the Smith chart would have revealed exactly what was wrong and what to do about it!

What it is

As explained in reference [32] (which should be consulted if the reader is looking for insight rather than an instruction book) the chart is simply a sheet of graph paper specially designed for the plotting of reactance against resistance. As one moves along a mismatched transmission line both of these quantities vary, and the unique feature of the Smith chart is the 'cooking' of the scales in such a way that, provided there is no attenuation in the line, *all the points lie on the same circle.*

The reader should now look at Fig 12.70 which shows the chart, a point (marked X) with its circle, and two dotted lines which will be discussed later. The point where the circle crosses the SWR scale indicates the SWR in the line and a set of concentric circles can be drawn representing various values of SWR. The chart owes its somewhat fearsome appearance mainly to professional users who like having a lot of lines drawn in; this means that answers can be read off to better than one place of decimals which does not usually help much but looks more impressive. Before being too critical of this version, however, the reader should count his blessings – most of them look a lot worse.

Copies of the chart may be obtained from various suppliers, and will be found to differ slightly from Fig 12.70. The author has deleted scales which the average amateur does not need and may find slightly confusing, and the resistance scale has also been used as the SWR scale which is what it is anyway. References to susceptance and conductance have also been deleted since these may confuse some readers and are merely reactance and resistance turned upside down, a useful trick which the chart is rather good at. This does not mean they have to be given new names but a lot of people think it helps and you may agree with them.

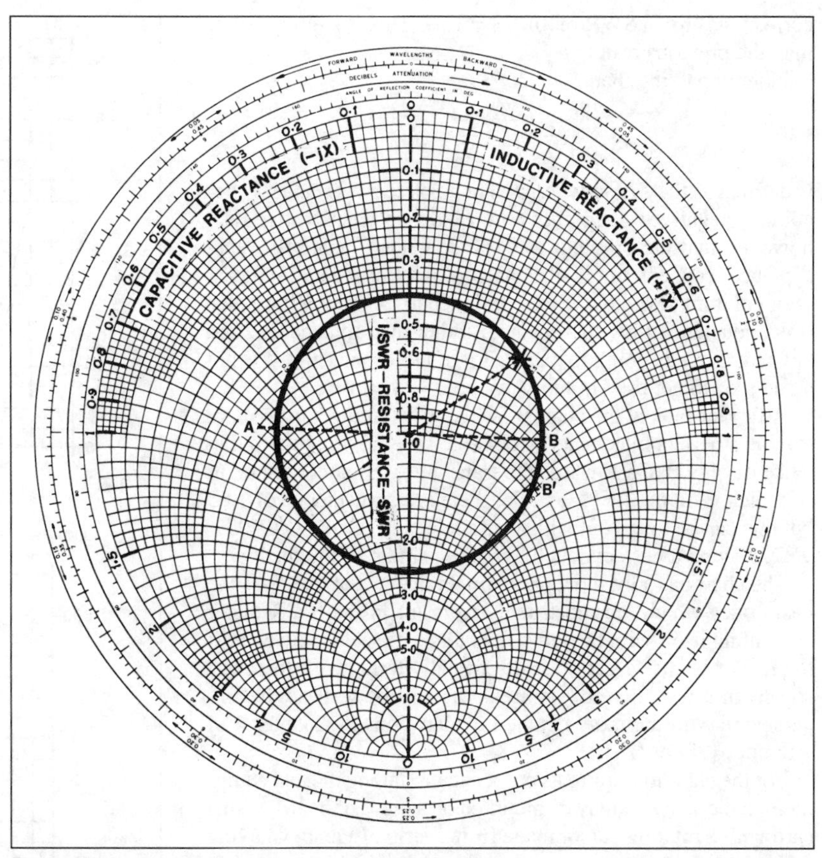

Fig 12.70. Smith chart, showing its use for finding the cause and cure for a high value of SWR. A measured impedance (point X) is plotted on the chart and its circle drawn in. By moving along the transmission line for the appropriate distance the impedance at the antenna end is given by a point such as A or its reciprocal B. By moving back slightly B moves round to the point B′ on the $R = 1.0$ circle and a perfect match is obtainable, subject to tuning out the reactance $0.95Z_0$ with the appropriate variable capacitor

Using the chart

We must now return to the plight of the operator whose measurement of a SWR of 2.5 has caused him to appeal to the chart for help and who may by now be getting impatient. To simplify the example it will be assumed that the line has a characteristic impedance (Z_0) of 100Ω, in which case the scale markings can be read as 'hundreds of ohms' and the measured impedance ought to be a point such as 50 + 43j lying on the 2.5 SWR circle. If after checking the measurements it still fails to do so, one may need a new SWR bridge, a new impedance bridge or both but agreement is unlikely to be exact – it will probably be necessary to cheat slightly by coaxing the impedance measurement onto the nearest point on the circle.

The next step, and this is 'what the chart is all about', is to discover the impedance at the antenna end of the line by moving along the line the correct number of wavelengths as indicated by the scale round the outside of the chart. Each time round the circle is a distance of λ/2 and if, for example, the line length is 1.2λ we have to make 2.4 circuits of the chart 'towards the load'. This brings us to the point A which is an impedance of (67 – 70j)Ω, and means that by connecting an inductance of +j70Ω in series we would be left with a resistive impedance of two-thirds the line impedance, ie an SWR of 1.5. This might be acceptable but the example also tells us that a much better match can be obtained by changing to a feeder impedance of 75Ω, which has the further merit of being nearly a standard value.

There is still the possibility that we can do even better by using a shunt inductance as explained below, but first the reader must be introduced to another important virtue of the chart, namely the fact that with the scale markings arranged as they are so that '1.0' comes in the centre of the chart, it can be used for any value of line impedance Z_0. For this purpose an impedance must be divided by Z_0 before entering it on the chart, and the final answer must be multiplied by Z_0 to get it 'back into ohms' (see also p12.99). This is known as *normalisation* and, following from the fact that the reciprocal of one is also one, it has the further virtue of allowing the same scales to be used for conductance/susceptance as for resistance/reactance which is its reciprocal.

Words like 'normalisation' have a deterrent effect but the reader has just been taken through this process quite painlessly (it is hoped) by leaving him free to think that the omission of a couple of noughts from the scale figures was due to laziness on the part of the chart designer. Reverting to more popular values of line impedance, one must of course be prepared to face up to such added difficulties as dividing by 50 instead of 100.

Series-to-parallel conversion

Thus far, the chart enables us to find the impedance at *any* point along a line of any impedance if we know what is happening at any other given point in the line. It will also be recalled that impedance inversion takes place along a λ/4 line so that a normalised impedance Z at one end appears like $1/Z$ at the other. So far we have in effect been putting $Z = R_s + jX_s$ but we know also that for the equivalent parallel circuit:

$$\frac{1}{Z} = \frac{1}{R_p} + \frac{1}{jX_p}$$

because this is the way one 'adds' impedances in parallel. $1/R_p$ and $1/X_p$ are by definition the conductance and susceptance.

Let us therefore refer back to the impedance $R_s + jX_s = (67 - 70j)\Omega$ which has just been obtained from the chart. In its normalised form this was 0.67 – 0.7j and to invert it all that is necessary is to go half-way round the circle. This pushes point A across to B at the opposite side of the circle where it may be read off as 0.73 + 0.76j, the first of these terms being $1/R_p$ and the second one $1/jX_p$ which means that $X_p = -1/0.76$. This, after de-normalising, turns into a capacitive reactance of 132Ω and can be removed by connecting 132Ω of inductive reactance across the line, leaving only the conductance 0.73 which corresponds to an SWR of 1/0.73 = 1.37. This is slightly better than the previous result.

Note in passing that:

$$\frac{1}{jX_p} = -j\frac{1}{X_p}$$

is defined as the *susceptance* so that a positive reactance X_p becomes a negative susceptance ($1/X_p$) and vice versa. Notice also, however, that the problem was solved without using the 'susceptance' concept, but if a bridge was used for the initial measurement it may well have been designed to measure admittance (ie conductance plus susceptance) rather than impedance, and anyway if one is going to make a lot of use of the chart it is not a bad idea to get used to thinking 'both ways up'.

A further look at the chart shows how by taking a little more trouble a perfect match can be obtained. By moving back a distance of only 0.03λ towards the transmitter, B moves round to B′ which is on the R = 1.0 circle; there is a susceptance of 1.05 but this can be tuned out with a parallel reactance of 100/1.05 = 95Ω, leaving a perfect match in the remainder of the feeder.

The chart has a lot more tricks up its sleeve, but these are the basic ones which are most likely to result in the SWR meter contenting itself with the left-hand half of its scale. However, if the reader wants to get on really friendly terms with the chart and learn a few more of its tricks, he will be well advised to persevere with the more detailed explanations in Chapter 4 of reference [28] or [29].

For the chart to acquire its proper status among the test equipment in the shack, it may be mounted on a piece of board or stiff card with a rotating cursor made from a strip of Perspex having scale markings corresponding to values of SWR from, say, 1 to 10. An alternative is to inscribe a set of SWR circles directly on the chart using coloured ink. The chart may be modified to line up with Fig 12.70 with the aid of typing correction paint and a pair of scissors, but there is no need for this once the user is master of the situation and no longer concerned by appearances.

Values for the feed-point impedances of dipoles, suitable for entry into the chart, are provided by Fig 12.71.

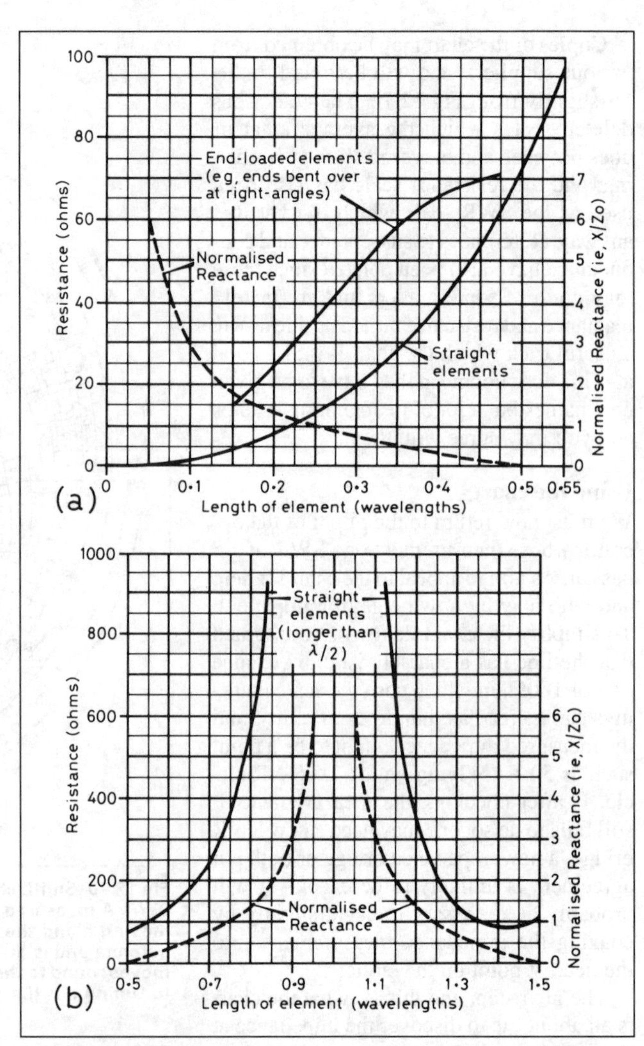

Fig 12.71. Variation of impedance with length at the centre of dipole elements. These are the figures required when working with the chart in its impedance form as described here. Radiation resistance is expressed as an equivalent series resistance at the centre of the element. Unlike the equivalent parallel resistances they are independent of wire diameter. The reactances are in normalised form suitable for direct entry into the chart, but can be converted into ohms by using them as multipliers for the appropriate Z_0 values, obtainable from Fig 12.32. The resistance figures for straight elements, based on a computer program, were supplied by W0JF

Baluns

The majority of HF antennas are inherently balanced devices and equal voltages exist or should exist to earth from each input terminal. Exceptions include long-wire end-fed antennas often used on the lower frequency bands and verticals which are driven against either a ground plane or true earth.

It is not possible to connect an unbalanced feeder such as coaxial line to a balanced antenna and maintain zero potential on the outside of the line. In such cases currents are forced, as explained on p12.32, to flow down the *outside* of the outer conductor by the voltages appearing at the antenna terminal to which it is connected. These currents give rise to unwanted radiation from the line itself since the field due to them cannot be cancelled out by the field arising from the current flowing in the inner conductor, which is contained entirely within the outer conductor of the line and cannot penetrate beyond it.

The existence of the unwanted radiation field around the feeder will modify the pattern of the antenna, and possibly also its input impedance due to the coupling now existing between feeder and antenna, and will represent a loss of energy by radiation in undesired directions. As with an inadequately balanced twin wire line, it is also a potential source of EMC and RF feedback problems within the station itself.

If it is required to drive a balanced antenna directly from an unbalanced line, then a choke or transformer can be employed to prevent the unwanted currents flowing back down the outside of the line. Such a device is known as a *balance-to-unbalance transformer* (or *balun*) and often replaces the gamma match described earlier. The action of transmission line baluns depends principally upon the use of a λ/4 section of line as a parallel resonant circuit to present a high-impedance path to the unwanted currents. A number of balun designs based on this principle are described in Chapter 13 and, although they will all work in theory at HF as well, their physical length precludes their convenient use below the VHF bands. A suitable balun for the longer wavelengths consists of a pair of bifilar-wound choke coils connected as shown in Fig 12.72. These coils may be sealed up in a suitable waterproof housing mounted at the antenna feed point and taking the place of the usual centre insulator. The coaxial feeder may be soldered in place or taken into the housing through a weatherproof plug and socket, with a clamp or other device for taking the mechanical strain of the feeder.

The coils act as a choke for asymmetrical currents while having negligible effect on symmetrical currents, and there is a 4:1

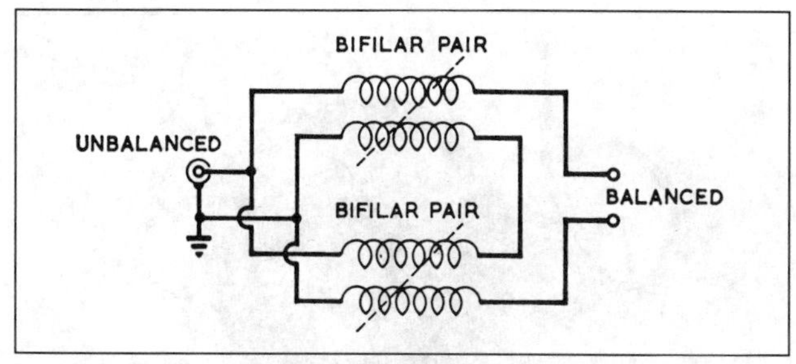

Fig 12.72. Pairing of bifilar choke coils

impedance transformation, with the higher value on the balanced side.

Transformer-type baluns using ferrite or iron powder cores can be designed to operated over a wide frequency range such as 3 to 30MHz. The principle is the same as that of low-frequency transformers except that, due to the relatively low permeability of cores having sufficiently small RF losses, special precautions are needed to maintain a high coefficient of coupling. It is not possible to rely on the core alone to provide the necessary coupling, and use must be made of the fact that if two wires of equal length are laid side by side in close contact, and if the insulation between them has negligible thickness, then all the lines of magnetic force surrounding one wire must also go round the other. This by definition is unity coupling and a voltage applied across the ends of one wire will produce an identical voltage across the other. For efficient transformer action the inductance must be large enough to ensure that connecting the transformer has no effect on the input other than that due to whatever load may be connected to the secondary. In practice there will always be some leakage inductance which can be regarded as in series with the load, and for a given coupling coefficient the leakage inductance increases in proportion to the self-inductance so that, if this is adequate for 3.5MHz operation, it may sometimes be difficult to get a low SWR on 28MHz. As with mains transformers there are losses in the windings and core, and reducing the inductance to improve SWR at 28MHz can lead to unacceptable core losses at 3.5MHz. In addition self-capacitance may have unpredictable effects on performance at the higher frequencies.

Despite these complexities it is not difficult to design small broad-band baluns with losses of only a small fraction of a decibel, and acceptable SWR. With efficient design, baluns wound on toroid cores of only 2in diameter are capable of handling powers up to 1kW PEP, and rod-type baluns can be equally compact and efficient. The impedance ratios obtainable are somewhat restricted by the necessity for complete overlap between the windings but Fig 12.73 shows how a trifilar winding may be used to obtain balun action with a 1:1 impedance ratio. Comparing this with a double-wound transformer (bifilar) it is found that the use of three wires instead of two tends to increase the leakage, but this is more than compensated as a result of the auto-transformer connection whereby half the turns of each winding are shared with the other.

For winding onto a toroid core the three wires may be laid side by side and bound tightly together with plastic tape so that they form a flat strip, or they may be bunched together and lightly twisted to keep them in position without air gaps. In the case of ferrite rods, the three wires may be wound side by side as a single-layer winding without air gaps. Typically five to eight turns have been found suitable for ferrite cores, including 1in diameter 'ferrite antenna' rod and also 2in diameter toroid rings. Although the wires must be wound as a tight bunch without

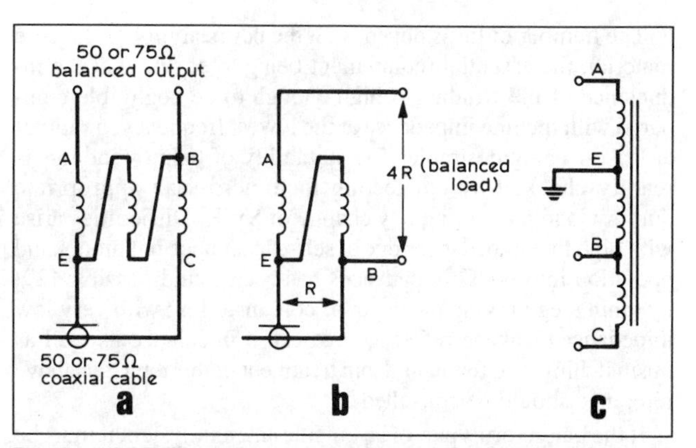

Fig 12.73. (a) Trifilar balun, 1:1 impedance ratio. (b) Bifilar balun, 4:1 ratio. (c) Trifilar balun redrawn as an auto-transformer to illustrate principle of operation. In bifilar case BC is omitted. The three (or two) windings are wound as one with the least possible spacing between wires but individual turns may be spaced out on the core

spacing there is no objection to spaces between turns. Powder cores require rather more turns, a minimum of 14 being recommended in the case of the 1kW kits obtainable from Ferromagnetics, PO Box 577, Mold, Clywd, CH7 1AH.

In the case of 1:1 trifilar baluns there is a tendency for SWR to increase appreciably with frequency, due to the leakage inductance, but considerable improvement can usually be obtained by use of a suitable parallel capacitance, typically about 50pF, though the actual value must be found by trial and error. For powers up to the legal limit a voltage rating of 500 should be adequate for this capacitor except under extreme fault conditions which might in any case result in damage to the core. Alternatively, if the balun is used in, for example, the centre of a dipole, leakage inductance together with any lead inductances may be compensated by adjustment of dipole length for minimum SWR.

Much smaller cores may be used for low-power and receiver applications, and a 1:1 balun using 17 turns, on a 0.69in core type T-68-6 obtainable from the above source, had measured losses averaging less than 0.1dB, an SWR better than 1.35 without capacitive compensation, and handled a power of 20W CW with very little temperature rise. Ferrite cores suitable for powers up to the legal limit include a pair of Ferroxcube FX1588 toroid rings which should be stacked and bound with acetate, polyester or silk tape before winding with six trifilar turns. Ferrite antenna rod of 5/8in to 1in diameter is usually suitable, and rod-type baluns may also be purchased ready made from various sources. In the majority of cases the measured losses have been found to vary with frequency, from less than 0.1dB to a maximum of 0.2 to 0.4dB.

Bifilar winding auto-transformers connected as in Fig 12.73(b) provide 4:1 impedance transformation, eg from 75Ω coaxial feeder to 300Ω twin lead. Because of lower leakage inductance it is much easier in this case to maintain low values of SWR over the entire frequency range; conversely if multifilar windings are used to obtain impedance ratios other than 1:1 or 4:1 it is rather difficult to achieve good performance over a large range of frequencies. Despite recommendations in some quarters, the author has had no success with tapped windings and this result is to be expected since the use of taps conflicts with the overlap requirement for minimising leakage inductance, although in some cases this may be removed by tuning.

The transformer-type baluns described above are all suitable for use with 50 or 75Ω coaxial feeder over at least the frequency range 3.5 to 30MHz. Subject to reasonable care in winding to ensure symmetry, balance should be within a few per cent.

The number of turns depends on the permeability of the core material, the essential requirement being to ensure that the inductance of the winding is high enough to be negligible compared with the line impedance at the lowest frequency, a ratio of at least 4 being desirable. The suitability of balun windings is readily checked by connecting them across an appropriate dummy load and noting any change of SWR. Difficulties arise with high line impedances due to self-capacitance but monoband operation into 600Ω impedances has been found possible [29] by tuning, eg varying the depth of core insertion; with very low impedances leakage reactance can cause imbalance as well as mismatching, and for an in-depth treatment of this topic the above reference should be consulted.

If the balun forms part of an outside antenna system it must be protected from the weather, and two methods are illustrated. Perhaps the simplest method is to attach the toroid to a triangular sheet of Perspex, and completely envelop the assembly in Araldite. The coaxial line can be readily secured with a sheet metal clamp bolted to the Perspex; the toroid is held in position by the combined action of the mounting wires and Araldite. The dimensions and drilling points are shown in Fig 12.74; the inner holes X retain the balun leadout wires, and the holes Y secure the antenna wires. The jumper between holes X and Y should preferably be a length of copper braid to prevent damage caused by flexing of the assembly in high winds. With auto-transformer windings all connections to the balun will usually be close together and holes marked X may need repositioning.

Another simple method of mounting is to enclose the balun in a short length of Marley plastic drainpipe, the ends being covered with discs of paxolin. The discs can be cut with a flycutter, the pilot holes being used to allow a brass screw to clamp the ends together, or alternatively used as fixing points to tie the unit to a mast. The coaxial cable is held tight by a rubber grommet in the centre of the tube wall, while the balanced winding can either be connected to the dipole by braid fly leads, or alternatively a small strain insulator cut from Perspex sheet can be mounted within the unit; this possesses the added advantage that all joints are totally enclosed. The unit should be sealed with Evo-stik or Twinpack Araldite. This method is readily adaptable for use with rod-type baluns.

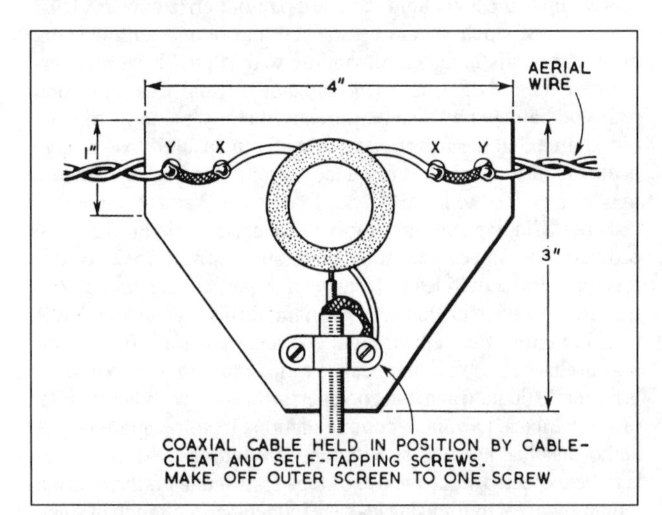

Fig 12.74. A triangular block of Perspex makes a strong mounting for a balun positioned at the centre of a λ/2 dipole. The balun is anchored with its connecting wires and a coating of Araldite, the coaxial cable is held with a cable cleat, and the antenna wires are passed through holes Y. Braid is used to connect the output of the balun to the antenna wires to avoid strain on the antenna connections caused by flexing in the wind

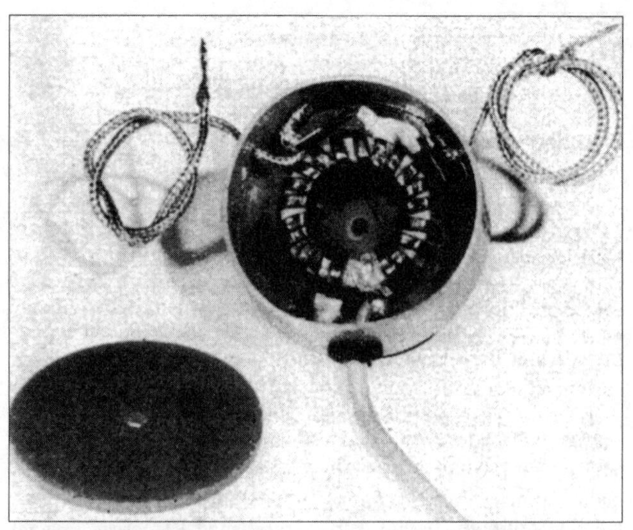

An inexpensive mount using a short length of plastic drainpipe

The fixed impedance ratios provided by baluns designed according to the principles set out above are unlikely, without further assistance, to provide the exact match demanded by many transistor output stages; this can usually be achieved for mismatches typical of coaxial feeder systems by means of a simple pi-network (eg Fig 12.78).

METHODS OF FEEDING AN ANTENNA

End-feeding

Any random length of antenna wire, whether resonant or not, may be energised by bringing one end of it into the station and connecting it to an earth or counterpoise through a tuned circuit, link coupled to the transmitter.

This method has the advantage of simplicity, and has often been used with great success, particularly in the early days of amateur radio, but cannot be generally recommended as it involves bringing the antenna right into the station. This means that radiation takes place in the vicinity of house wiring, brickwork etc, leading to unpredictable losses and added risk of EMC problems. The local radiated field in and around the shack may also give rise to feedback problems in audio and RF circuits, and will contribute little or nothing to the useful radiation, being generally absorbed in the fabric of the building in which the station is housed. If it is necessary to use this method, it is advisable to use a counterpoise rather than rely on an earth connection; earth wires, though in this case just as much part of the radiating system as the antenna wire, are usually attached to the building, thereby introducing losses additional to those of the earth itself. Typically, a counterpoise consists of λ/4 of wire at any convenient height, though preferably out of doors, but if the antenna is fed at or near a voltage point almost any random short length of wire is adequate. On no account should RF be allowed to find its own way to earth via housing wiring or other random paths.

If the antenna is being fed near a low-impedance point a quarter-wavelength is not in fact recommended in view of the power lost by radiation from it in an unwanted mode. For monoband operation much shorter lengths, inductively loaded, are generally suitable [33] subject to ensuring that losses in the loading coil are small compared with the feed-point resistance; in some cases there may be a reduction in bandwidth, and it is advisable if possible to use more than one wire. A suggested prescription is to use a λ/4 wire but take the ground connection

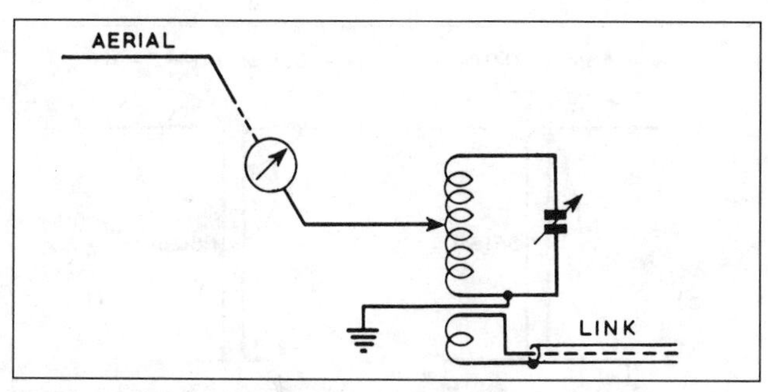

Fig 12.75. Antenna coupler for end-fed antennas. For some lengths of wire a series tuning capacitor and perhaps a separate coupling loop instead of a tap may be required

to its centre through a loading coil, the required reactance being about 250Ω. In the case of multiband antennas either retuning or the use of separate counterpoises is recommended except that in some cases, due to much larger values of feed-point resistance at the higher frequencies, this may not be essential. A further useful trick to remember, if the antenna is fed near a voltage point, is the use of a short length of matched single wire line to 'push the high voltage outside the window' as described on p12.64.

Valve transmitters usually employ a pi-circuit for the PA tank, designed to work into an output impedance within a limited range of low impedances, typically 40–150Ω. Since the end impedance of a current-fed antenna, ie one which is an odd number of quarter-wavelengths long, is also low, and of the same order as the figures quoted, it would be quite possible to couple the antenna straight into the output socket of the transmitter and achieve a reasonable match by means of adjustments to the variables in the PA tank circuit. For the reasons which have been outlined above the use of this technique is deprecated, tempting as it may be to the amateur beginning on 1.8MHz with a simple transmitter and a short wire antenna. It is perhaps permissible in field-day conditions, when the transmitter is in the open and can be connected in the direct path between the antenna and earth. A separate tuned circuit should be employed as an antenna coupling network in *all* other circumstances involving use of a random length of wire as an antenna. This network may be located in any suitable position and link coupled to the transmitter tank circuit by a low-impedance coaxial line to achieve adequate separation between the radiating part of the antenna and the building housing the station.

If the total antenna length is such as to cause the antenna section nearest to the transmitter to carry a high RF voltage, there may be considerable loss of power should it have to pass through or near to surrounding structures. On the other hand, an antenna tuned to an odd number of quarter-wavelengths and consequently of low input impedance can also be inefficient as a result of the comparatively high resistance of the earth connection. Unless a counterpoise is employed, it is generally better therefore to use an intermediate length, for example an odd number of eighth-wavelengths, so that the impedance will have some intermediate value, thereby reducing the current at the earth connection.

A suitable coupling circuit is illustrated in Fig 12.75. The inductor and capacitor may be similar to those in the transmitter output tank circuit. The link coil may vary from about four turns at 3.5MHz to one turn at 28MHz and may be tightly coupled into the earthed end of the transmitter tank coil, the antenna tap then being adjusted for correct loading of the transmitter. For good efficiency the resonance of the antenna coupler should be fairly broad, correct loading combined with high Q being an indication of insufficiently tight coupling and excessive dissipation of power in the tuned circuit. Alternatively a similar link coil may be used at the transmitter or, better still, the link may be fed from the output of a pi-network. With a non-resonant length of antenna it will be found that, as the tap position is moved up the coil (from the 'earthy' end) in order to draw more power, the coupler circuit will need retuning.

Most low-power transmitters (up to 100W or so) now employ transistor output stages which require a very accurate impedance match but a tuner such as Fig 12.78 (see later) (or, subject possibly to some adjustment of feeder length, based on Fig 12.80) should be able to take care of this situation.

Earth connections

When an antenna system requires an earth connection the use of counterpoises as described above, though otherwise recommended, do not always improve the appearance of a garden and may be neither possible nor convenient. There is no direct relation between the DC and RF properties of an earth connection, though in both cases it may be acceptable to take it to a copper spike or tube several feet long, or to the nearest large earthed conductor such as a main water pipe. The joint should be heavily coated with bituminous paint to avoid corrosion. In this case *the earth connection of the electricity supply is dangerous* since RF voltages could lead to an insulation breakdown; also it could introduce noise into the receiver or be responsible for spreading interference to nearby television receivers. If the station is at the top of a building, the antenna coupler may fail to function properly on some frequencies because the earth lead is long enough to exhibit resonance. In such cases a 'shortening' capacitor of 50–100pF may be inserted into the RF earth lead to de-tune it. This capacitor should be a variable air-spaced type. On the other hand, if the shack is at ground level a direct earth connection may be preferred on the score that, unlike a counterpoise, it does not need retuning for multiband operation. For best RF performance a number of earth wires laid in shallow trenches can be generally recommended.

Further information on earth connections will be found later in the section dealing with low frequency antennas.

Important. Special rules about earthing apply in the case of domestic installations fitted with PME, as set out in Chapter 21.

Matched versus resonant feeders

Some feeders are described as matched lines and others as resonant lines. It is sometimes thought that these are two different kinds of line, resonant lines being usable with any load impedance whereas matched lines have to be operated at an SWR better than some low value such as 1.4 to 1. A resonant line is in fact merely a badly mismatched line, a point being reached with increasing SWR at which it becomes convenient to regard the antenna plus feeder as a resonant system rather than as a matched (or mismatched) resistive load.

If an antenna is to be used on one frequency only, there is rarely a good reason for not attempting to match it, eg by using a $\lambda/4$ resonant line or equivalent lumped-constant network to transform a high impedance down to a lower value. Referring to Fig 12.49, matching is not critical from the point of view of power transfer and even an SWR of 3.0 only increases the feeder loss by a fraction of one decibel unless the feeder is extremely lossy, but this degree of mismatch may have considerable nuisance value with a long feeder; for example a 2% frequency change with a feeder three wavelengths long is equivalent to a change in line length of $\lambda/16$ and this would convert an impedance of $3Z_0$ into one of $(1.4 + 1.3j)Z_0$. Adjustments to obtain a conjugate match at the transmitter in the second case would have little effect on SWR in the first case so that Fig 12.50 would be applicable and there would be a reflected power loss of 1.4dB, probably combined with flat-topping of the transmitter unless the tuning and matching are both readjusted.

When an antenna is required to work on several wavebands the matching problem usually requires a different solution for each band and the difficulties tend to escalate rapidly with the number of wavebands to be covered, except that an antenna fed with matched line will work on odd harmonics, eg a 7MHz antenna will work also on 21MHz. Since the feed point is rarely accessible, it is often the lesser evil to forego any attempt to match the line at the antenna and use resonant lines, accepting a large SWR on all bands and relying on adjustment of the ATU to provide a match at the transmitter. If, for example, a 14MHz dipole is directly fed at the centre with 600Ω line, the SWR will be about 8.0. Because the line impedance is roughly the geometric mean of the current loop and voltage loop impedances, the same antenna will work as a 1λ dipole on 28MHz with a comparable SWR, and will perform rather similarly on 21MHz also. It will even work on 7MHz, though in this case the radiation resistance is very low (8Ω) and the SWR will be very high, about 80 to 1.

The loss in 100ft of matched 600Ω line is typically 0.07dB at 14MHz and, extrapolating from Fig 12.45, a resonant feeder loss of the order of only about 0.3dB would be expected at the higher frequencies. The loss on 7MHz can be worked out with the aid of Fig 12.5 and for the same long lines comes to about 1.3dB assuming use of 12swg copper wire, the bandwidth being about 35kHz. On the other hand, with low-impedance feeder the SWR would be about 67 at 28MHz and, extrapolating from Fig 12.49, the actual line loss will be at least 6dB, assuming a matched line loss of 1dB; moreover, unless the power is very low, breakdown of the cable due to high voltages and overheating is to be expected. These examples underline the necessity of using open-wire line for resonant feeders if low-loss operation is required on even harmonics of the fundamental frequency or in any other circumstances involving large SWRs or high powers.

Tuned lines

The input impedance at the lower end of a resonant line will depend upon the actual load impedance presented at the far end, and upon the length of feeder involved. In general, it will be quite arbitrary and the feeder is employed solely as a means of connecting the antenna to the transmitter without attempting to transform it to any particular value. It is therefore necessary to terminate the tuned line in a matching network which will transform the impedance at the lower end of the line into the optimum load resistance for the transmitter.

One type of antenna coupler for use with tuned open-wire lines is basically similar to the end-feed coupler shown in Fig 12.75 but differs in detail, and the tuned circuit will require to be of a series or parallel form, depending on whether the impedance at the lower end of the line is lower or higher than the Z_0 of the open-wire line, as illustrated in Fig 12.76.

Without a knowledge of the exact antenna impedance, it is not possible to predict the precise value of the input impedance to the feeder or to determine the values of capacitance and inductance in the coupler. It is, however, possible to determine whether a series or parallel coupler is required with the aid of Fig 12.77 which is based on the total length of the system from the far end of the antenna to the near end of the feeder.

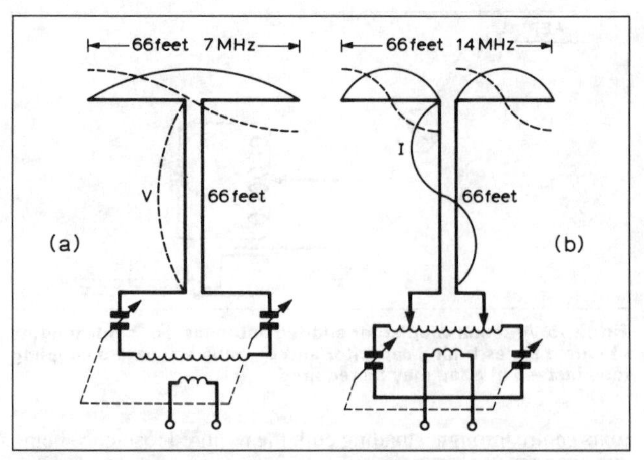

Fig 12.76. Centre-fed antennas using tuned feeders. The antenna to the left is λ/2 long and has a low input impedance; the feeder is also λ/2 long and thus repeats the low impedance so that series tuning could be used. On the right is a similar antenna operating at twice the frequency of the other (full wave). The input impedance is high, and the feeder one wavelength long; parallel tuning would be best. The effect of feeder length is discussed in the text and illustrated in Fig 12.77. The tuning capacitor sections should each be 50pF maximum for 14MHz and higher frequencies; proportionately larger values are needed for lower frequencies. The voltage rating of the capacitors should be the same as for those in the transmitter output tank circuit

The impedance presented to the coupler will contain both resistive and reactive terms. The reactive term is neutralised by adjustment of the tuning control on the coupler (usually a variable capacitor) but provision must also be made for obtaining the correct transformation of the resistive term; this may be done by providing a range of tappings on the coil of the coupler unit, in order that the feeder may be tapped in at a position along the coil best suited to the value of the input resistance. A wide range of adjustment is desirable to accommodate very low impedances on taps near the coil centre, and very high impedances towards the outer ends of the coil. When the impedance is very low, corresponding to a current maximum on the line, it is possible to connect the feeder to a separate coupling coil of one or two turns and retain the parallel circuit, the coupling being adjusted to obtain the equivalent of adjustable taps, and many other combinations of coupling and tuning methods may be devised to suit particular circumstances, some examples being found later in this chapter. In general, if the feeder input is of low impedance but largely reactive, series tuning is advisable, though it may be combined with parallel tuning which can be retained as the main variable.

When the feeder input resistance is of the same order as the feeder Z_0, a parallel coupler is necessary and the feeder should be tapped down the coil to obtain the best match to the transmitter. This situation implies that the feeder is more or less matched to the antenna and alternative couplers for this case are discussed in the next section.

In general, very high impedances at the bottom of the feeder should be avoided by judicious selection of the length of feeder employed, using the chart of Fig 12.77. High impedances are often difficult to match into the parallel tuned circuit because of the tendency to 'run out of coil' when selecting the correct tappings. The high impedance also has associated with it a high voltage which will appear across the components of the tuned circuit and requires disproportionately large components to avoid the danger of flash-over. This is particularly likely to occur between the vanes of the air-spaced tuning capacitor during peaks of modulation if the vane spacing is inadequate for the high

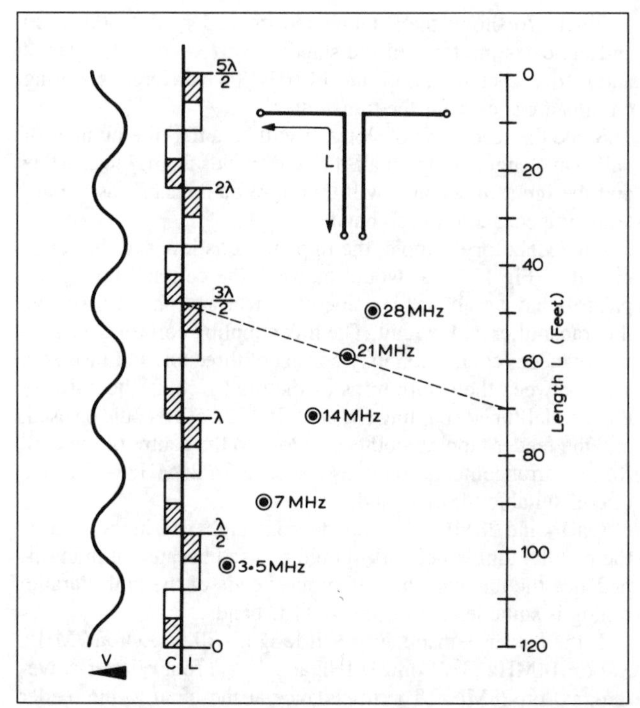

Fig 12.77. Standing wave chart for tuned feeders. A line through the length *L* of feeder plus half-antenna, and the frequency point, will show on the wavelength scale the nature of the input impedance. Rectangles to the left of the line are regions of capacitive impedance, those to the right inductive. The shaded areas correspond to high-impedance input (high voltage) and call for parallel-tuned couplers (Fig 12.76(b)) and the blank areas to low impedances (high current) which require close taps on a parallel-tuned coupler, or a series circuit (Fig 12.76(a)). The chart may also be used for feeders alone in which case, as coded, it applies for the use of high-resistance loads. When the load is lower than Z_0 the code must be reversed; the shaded areas indicate low input impedance, inductive to the left. The velocity factor of the feeder is not allowed for in the chart, and the physical length of the feeder must be divided by this factor when computing length *L*

voltages developed. If the tuned line has to be brought into the shack, regions of current maxima are preferred because RF voltages 'getting loose' can cause feedback troubles, as well as losses in walls and window frames etc. Unfortunately, achievement of this condition at the lowest frequency ensures high voltages at all the even harmonics, and the best compromise will probably be found by aiming for a current maximum at the highest frequency, where the above troubles, if allowed to occur, are likely to be more serious.

It is desirable to reduce the length of resonant line to a minimum because of the bandwidth problem discussed earlier, and it is good practice if the tuner can be located, say, at the base of the mast or in a nearby garden shed, continuing from there to the shack with matched line. If this avoids the need for re-matching when altering frequency within a band, and allows reasonable access for band changing, it should be the method adopted. Otherwise the tuner should if possible be located at the point of entry into the building. It is inadvisable to extend widely spaced open-wire lines carrying high RF voltages for any appreciable distance inside a building, and particularly into the station itself, because of their tendency to aggravate the troubles mentioned above if any current imbalance exists; if such a course proves unavoidable, particular care must be taken to maintain balance and this will be assisted by reducing the spacing of the line.

The change of Z_0 will usually be unimportant with a resonant open-wire line but it can be avoided by reducing the wire

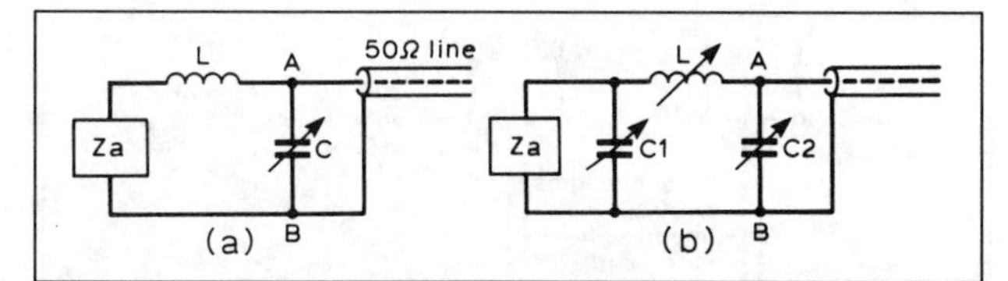

Fig 12.78. L and pi-matching networks for 50Ω lines are shown at (a) and (b) respectively. The antenna impedance Z_a may be expressed in the form $R_a + jX_a$, and X_a, if capacitive, tuned out by part of the reactance *X*. If R_a is less than 50Ω it can be made to appear as 50Ω across AB by increasing the value of *X*, the inductive reactance resulting from this being tuned out by C. If R_a exceeds 50Ω the input and output terminals may be reversed. If X_a is inductive it can be tuned to resonance by the left-hand capacitor of the pi-circuit, (b), and over a limited range the additional control in the case of the pi-circuit can obviate the need for a variable inductance

diameter proportionately, a course strongly recommended in the case of a matched line which also needs to be carefully balanced, though the situation is much worse with a large SWR; this is mainly on account of the regions of high voltage, the field from which may extend unevenly into adjacent brickwork etc which has poor insulating properties. If, however, a high-current section of line passes near another conductor and the layout is asymmetrical, this also may cause unbalance and additional losses, and both of these effects may aggravate EMC problems.

The stress laid on the importance of balanced lines being truly balanced must not be taken as implying that coaxial line is any better in this respect since it is inherently unbalanced and baluns, even when fitted, may not help in the event of unbalanced coupling between the antenna and the outer of the feeder cable. Moreover the author has not been alone in finding balanced lines less troublesome [6]. One arrangement which has proved very satisfactory is to resonate open-wire lines by means of 300Ω extensions from their point of entry into the shack as illustrated by Figs 12.79 and 12.80. This allows fine-tuning and matching to be carried out at a point of low impedance adjacent to the transmitter using low-rated components, aided in some cases by judicious use of series or parallel reactances at the point of entry [29, 38]. In the case of a pair of beam elements, fine adjustment of coupling is usually possible by overlapping the lower ends of the extension lines.

For multiband operation, which is envisaged as the most

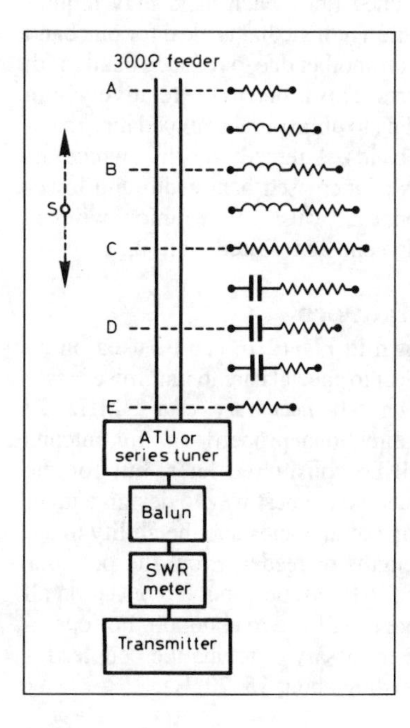

Fig 12.79. Determination of line lengths. AE represents λ/2 of 300Ω line adjacent to the transmitter, though we are concerned only with the existence of SE, of which the line length must be determined separately for each band to provide a resistive (low) impedance at E. The line is divided into eight sections for which the nature and magnitudes of impedances seen looking outwards are denoted by symbols. S, the point of entry of the feeder into the shack, can in principle be anywhere on AE but the need for excessive lengthening of SE can be avoided by parallel tuning with capacitance at C

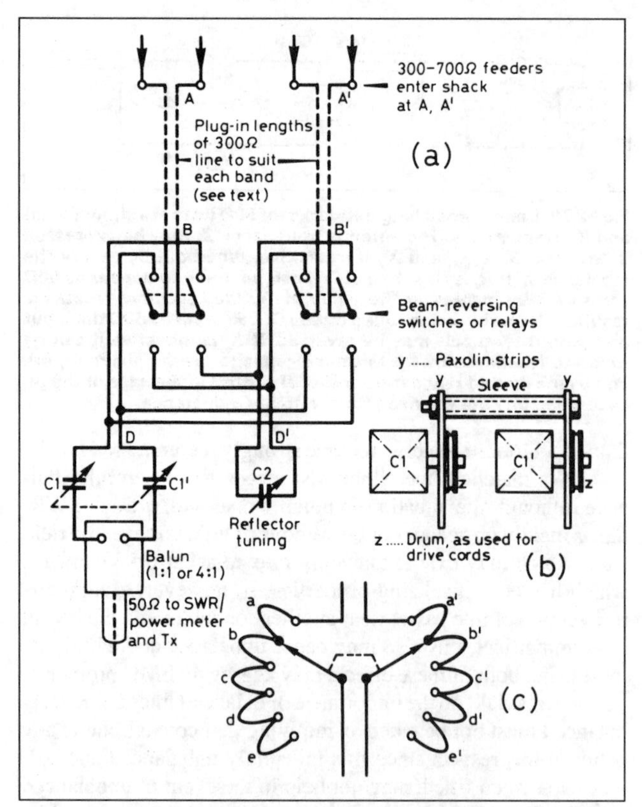

Fig 12.80. (a) Shows the tuning and matching system used with a multiband beam based on a pair of small close-spaced delta loops (the 'Claw', p12.81). With 600Ω feeder its length should if possible be arranged to avoid high RF voltage at A, A' on the lowest frequency band. Extension line lengths in excess of about 0.3λ can also be avoided by series capacitance at A, A'. Plug-in baluns were used but could be switched. An ATU can be substituted for C1. For components below the B-B' line, 250V ratings (500V for C2) should be adequate. This applies also to any switches used as a replacement for plugs at B, B'. (b) Method of ganging C1, C1' if only standard single-ended capacitors are available. (c) Line stretchers: aa', bb' etc are 4 to 6in lengths of 300Ω line. They may be inserted at B, B' or, with some modification of AB, A'B', used instead of C1, C2, though the tuning range is more restricted. Note: full power must not be applied with the wrong extension leads in place

important application for tuned lines, each band may require different treatment and an arrangement that is ideal for one band may be less 'user-friendly' on another due to reduced bandwidth and more-critical adjustments. This is due to the relatively high antenna Z_0 (Fig 12.32), leading to unfavourable impedance transformation on one or more bands. A remedy for this, which offers the further advantages of increased bandwidth and lower losses, is to reduce the antenna Z_0 by using two or more wires in parallel for the radiating element, as featured on p12.72.

Centre-fed horizontal wire

The 7MHz λ/2 dipole shown in Fig 12.76 can be used on all bands from 10 to 80m, subject to special care to minimise losses on 80m as was earlier shown to be necessary for a 14MHz λ/2 dipole used on 7MHz. The directional properties of this antenna on the various bands will be considered later, but for the moment the problem is to decide the best way to design a tuner unit which provides coverage of all bands and the ability to accommodate any desired lengths of feeder. From the previous example it follows that the SWR will be about 8.0 or 9.0 on all bands except 3.5MHz where it will rise to about 60. For operation on this band it will be necessary to retune the coupler for any frequency change exceeding about 15–20kHz.

Fig 12.76 shows the antenna and its feeder (also 66ft long) and indicates the form of the standing waves on both antenna and feeder when used on 7 and 14MHz. Two different coupling circuits are shown for the transmitter.

Since the feeder is 66ft long, it will be a definite number of half-wavelengths long on all the 'old' bands from 7 to 28MHz and the input impedance will therefore be the same as the antenna impedance for each band.

On 7MHz for example, the impedance is low and the series circuit of Fig 12.76(a) would be used, the coil and tuning capacitors having about the same electrical and physical size as the transmitter tank circuit. The link coupling coil for the transmitter feeder connection may consist of three turns of insulated wire between the centre turns of the tuning coil. Alternatively the parallel-tuned coupling circuit of Fig 12.76(b)) could be used, and the feeder tapping points set close to the centre of the coil. Similar arrangements could also be used for 21MHz with a size of coil suitable for that band.

On 14 and 28MHz the impedance is high, and in these cases the parallel tuning network should be used because at high impedance the taps will be at the outer ends of the coil. Parallel tuning is suitable also for the WARC bands.

If the feeder is made only 33ft long it will be λ/4 on 7MHz, λ/2 on 14MHz, 3λ/4 on 21MHz and 1λ (ie four quarter-wavelengths) on 28MHz. The impedance at the input to the feeder will be the same as with the 66ft feeder on 14 and 28MHz but on 7 and 21MHz, where the antenna impedance is low, the odd λ/4 feeder will transform it to a high value on these bands also. The parallel-tuned coupler would therefore be used suitable in each of these cases, also for 18MHz, though 10 and 25MHz require series tuning.

With the antennas described the SWR is high and consequently the coupling circuit will carry large voltages and currents. It is essential therefore to use good-quality components. A suitable arrangement for multiband working consists of a plug-in coil base with six contacts, two for the coil, two for the feeder and two for the transmitter link coil with a separate coil unit for each band. The tuning capacitor must have two sections in order to preserve feeder balance, as discussed later under 'antenna couplers'. For 100W of RF power the feeder current may not exceed about 0.25A in each line when the impedance is high, but it may well exceed 1A for the low-impedance cases, eg 66ft feeder on 7 and 21MHz. This is a disadvantage in that the same RF ammeters cannot be used in all cases but, when the current is low and the voltage high, a small neon lamp may be used instead as a voltage indicator.

Judicious choice of feeder lengths will avoid a high-voltage feed point on most bands. A suitable length may be found with the aid of Fig 12.77. On this standing wave chart the total lengths may (if desired) include one half of the antenna so that the voltage wave illustrated starts from one antenna insulator. It will be seen that total lengths of 45 or 90ft bring the feeder input into the high-current sections of the chart in nearly every case when using tuned feeders. It is not essential to make the antenna itself a resonant length; with the top increased to 84ft length, this arrangement is known as the *extended double Zepp* and provides a gain of 3dB on 14MHz as compared with 2dB for the length shown.

The increased length has the advantage of doubling the radiation resistance on 3.5MHz, making it much more attractive for this band, but in both cases it may often be found preferable to connect the feeders in parallel and use the system as a top-loaded vertical operating against a counterpoise or another antenna, as discussed later.

Flat lines

An important feature of flat lines lies in the fact that the input impedance of the line is equal to the load impedance and by definition to the characteristic impedance Z_0, *irrespective of the line length.* It follows from this that the route and length of feeder employed may be decided primarily by the physical layout of the antenna system and its position relative to the transmitter, though losses increase of course in proportion to the length. Because the matched loss of coaxial cables is considerably greater than for open-wire lines, such cables are usually employed as nominally flat lines, with a minimum standing wave and hence minimum additional loss. It is also of course advantageous wherever possible to operate open-wire lines in a matched condition as already discussed. The advantages of this are wider bandwidth and the associated convenience of operation, better balance and therefore less risk of feeder radiation or other possible complications, and the ability if required to use very long lines with negligible loss so that the antenna can be placed in the best location available. Though not common practice, presumably due to other difficulties, there is no technical reason to prevent someone living in a valley from installing an antenna on an adjacent hilltop, anything up to a mile or so distant. The loss could be held down to 3dB at 14MHz and the advantage could be many S-units.

It was shown in the section on transmission lines that any given line can be matched only by adjustment of the load impedance so that it equals the Z_0 of the line. Thus, unlike the tuned line method of antenna feeding, the matching arrangements must be incorporated *at the antenna end* of the flat line to obtain correct operation.

If the line is to be operated in a matched condition for its entire length, some means of matching must be incorporated into the actual antenna design and, if the line employed is of the coaxial or unbalanced type, usually some form of unbalance-to-balance transformer (balun) will also be required at the antenna terminals. Techniques of matching at the antenna are discussed in the earlier section on impedance matching, which also considers the design of baluns.

Because the use of flat lines dictates the need for impedance-matching facilities at the antenna end of the line, it is possible to standardise on a characteristic impedance for flat lines for all applications, and arrange to match any particular antenna impedance which is encountered to that standard line impedance. The commonest coaxial line impedances are 75Ω and 50Ω. The former is used in Europe for domestic TV purposes, but the majority of amateur commercial equipment is designed for the lower impedance in conformity with American and Japanese domestic practice.

It is advisable to develop a station where possible around one or other of these standard values, rather than to attempt the simultaneous use of both. Transmitters and all auxiliary equipment, such as harmonic filters, SWR indicators etc, can then be constructed to the selected standard impedance, thus avoiding the need for duplication of components and simplifying the operation of the station.

Fig 12.81 illustrates some of the basic differences between tuned and flat-line operation, the matching network in the ease of a flat line being located at the antenna. In the ease of the tuned line, the matching network is normally accessible so that it can incorporate tuning and matching adjustments, and it is use of a tuned line which usually makes this both possible and necessary. With a flat line the matching network, probably a balun with suitable impedance transformation, is not accessible and must therefore be a broad-band device if more than one

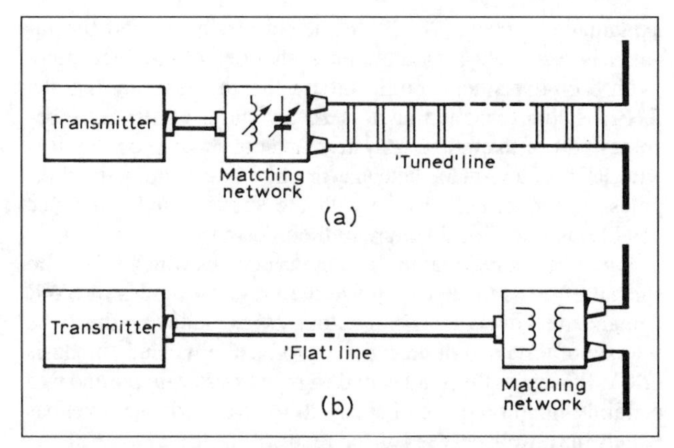

Fig 12.81. Transmission lines. (a) Tuned line. (b) 'Flat' line

band has to be covered. Modern practice tends rightly or wrongly towards the use of matched unbalanced coaxial line wherever possible, though the relative merits of balance and unbalanced line are subject to some controversy and for further discussion the reader is referred to p12.32 and reference [6]. The use of tuned lines tends to be restricted, as previously discussed, to multiband antenna designs, a popular example being the G5RV multiband dipole described on p12.70. Methods of achieving multiband operation with matched-line systems include remote switching or tuning based on relays or servo-controlled motors in the antenna (pp12.69 and 12.83), log-periodic antennas (not an easy option for the home constructor or those unable to erect a suitable tower), and the use of traps or stacking which in the case of beams restrict operation to two or three bands only.

It has been found [29, 43] that with suitable transformers an antenna can often be end-fed with coaxial cable against an artificial ground such as a short inductively loaded counterpoise; for example, moving outwards a short distance from one end of a λ/2 dipole one finds an impedance of 450Ω in series with a large inductive reactance which can be tuned out with a series capacitor. The transformer of Fig 12.73 can be rearranged to give the required 9:1 impedance ratio though it has been found necessary to tune out its self-capacitance, eg by varying the amount of core-insertion. The shortening of the antenna has little effect on radiation resistance or pattern provided it does not exceed about λ/8. Examples of antennas featuring this arrangement will be found on pp12.59 and 12.64. As in all cases of asymmetrical feeding, strict attention must be paid to the rules on p12.31 for preventing the flow of current on the outer of the coaxial feeder, the linear trap (Fig 12.82) tending to be effective in all but the worst cases.

A recent proliferation of ideas according to which dipoles are fed off-centre with coaxial or twin line is a cause for concern; as

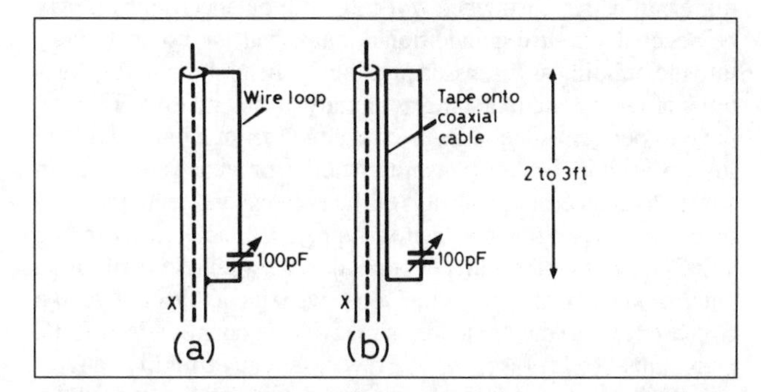

Fig 12.82. Traps for eliminating antenna current on outers of coaxial lines. Values and dimensions are for 14MHz but are not critical

explained elsewhere [28, 29] this is permissible only if the antenna is resonated separately on both sides of the feed point (with L on one side, C on the other), thereby ensuring that the feeder is connected at a point of zero voltage. Otherwise it becomes part of the radiating system and the shack probably fills with RF. Ground-plane antennas are likewise asymmetrical dipoles, and strict compliance with the above-stated resonance conditions is a critical feature of their design.

It will be noticed that in the antenna designs which follow no special effort is made to achieve the very low values of SWR demanded by many transistor output stages, believing this to be a waste of time and often counter-productive. Undue emphasis on SWR restricts the number of design options available and will certainly inhibit the use of small, thin-wire, invisible antennas which may well be needed for reconciling the desires of the operator with good neighbourly relations. Feeders can tolerate an SWR of at least 2:1 and to meet the needs of output transistors a simple π network ATU (or, for those who can afford one, an automatic ATU) is more than adequate. Balun transformers such as Fig 12.73 can also be connected as matching transformers with various ratios, ie 9:1, 9:4, 4:9, 1:9 and, subject to the tuning out of reactances, this allows most antennas to be centre-fed with coaxial cable without exceeding an SWR limit of 2:1.

ANTENNA COUPLERS

In Fig 12.81 the flat line has been shown going directly to the transmitter but coupling such a line directly to the output circuit provides insufficient discrimination against harmonics or other possible spurious signals, and is an obsolete practice unsuited to modern conditions. At the very least a harmonic filter is required, and this may either be built into the transmitter or fitted externally but in close proximity. The normal output from a transmitter is through a coupling network and harmonic filter into a coaxial RF connector. In some cases, where the antenna is already fed by coaxial lines, it can be connected directly to the output of the filter, but more generally it is necessary to use an intermediate coupler or transformer. Even if the line is of the same impedance as the transmitter output (say 70Ω), it may be a balanced twin feeder and cannot therefore be connected to the unbalanced concentric transmitter output. It is therefore common practice to use an *antenna coupler* or *antenna tuning unit* (often abbreviated to 'ATU'), such as those illustrated in Figs 12.76 and 12.83, even when the transmitter is working into a flat line as in Fig 12.81(b). An antenna coupler not only provides additional rejection of harmonics but also some protection against radiation of local oscillator, intermediate frequency and other spurious signals which because of their relatively low frequency, are not suppressed by the harmonic filter.

The antenna coupler is sometimes built as part of the transmitter, in which case when working with balanced lines it may be necessary to use an additional coupler at the point of entry into the building as discussed previously. In addition to the functions of tuning and matching as already discussed the ATU may have to perform such tasks as conversion from a balanced feed line to an unbalanced transmitter outlet or vice versa, and in some cases electrical isolation of the antenna and feeder system from the transmitter and mains supply. Connection as in Fig 12.83(a) or any other direct connection between the earth of an antenna system and that of the house wiring (in this case via the rig), though common practice, is inadvisable on grounds of EMC and mains-borne noise, and in a few cases can be highly dangerous. This situation arises when the house wiring is protected by one of the older (voltage-operated) varieties of earth-leakage trip, in which case any ground connection to the antenna or feeder, or even leakage to ground, will short-circuit the trip and leave the entire mains installation unprotected. A similar situation arises in the case of PME (Chapter 21), otherwise modern installations rely on an RCD (residual current device) which is exempt from this problem.

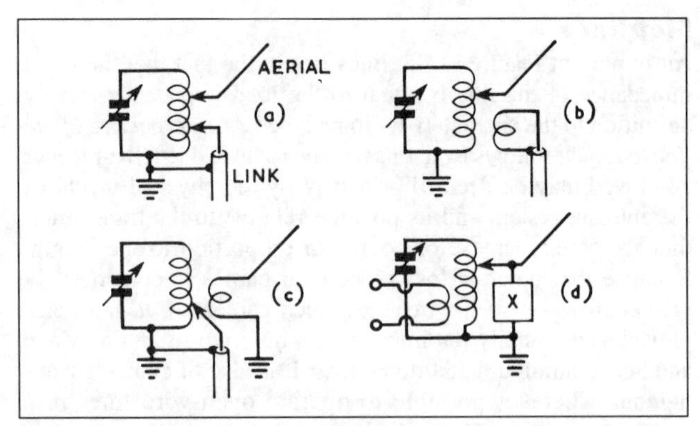

Fig 12.83. Various forms of antenna coupler for end-fed antennas. The use of reactance *X* may be needed in some cases for tuning out extreme values of antenna reactance

The general design principles of couplers have already been discussed but as stated earlier, many practical variations are possible, and the following notes are intended to assist in choosing the optimum configuration for any particular situation, and suitable design parameters.

Unbalanced antennas

Fig 12.83 shows variations of an antenna coupler intended for unbalanced antennas or feeders. In (b) to (d) the inductance acts as a two-winding transformer, the capacitor serving to resonate the coil to ensure a high degree of coupling, to help the network to discriminate against unwanted frequencies and to assist in harmonic suppression. The auto-transformer arrangement of Fig 12.83(a) is deprecated on various grounds as discussed above, although with a high-impedance antenna any safety requirement can be met by isolating it for DC with a small series capacitor not exceeding a few hundred picofarads. The link may have either a direct tap connection or a separate coil of a few turns interwound with the main inductance at the earthed end.

The tuning components may conveniently have a reactance of about 500Ω at the operating frequency. This may most conveniently be translated into practical figures by reckoning the capacitance as 1pF per metre of wavelength and the inductance as 0.25μH per metre. For example, at 7MHz, approximately 40pF and 12μH could be used. There is nothing critical about these values, though for a very-high-impedance antenna load (as may occur when using tuned feeders) it is advisable to increase the inductance and use less capacitance. Coils of about 3in diameter wound with 14–26swg wire on a ceramic former will handle powers up to the UK legal limit without overheating. The tuning capacitor should be of good quality, have a plate spacing of 1/16in or more and have adequate capacitance to tune to resonance at about half its maximum value. The link coil may be wound outside the tuning coil with about ¼in clearance and should be either self-supporting or rely on spacers of good-quality insulating material such as ceramic or polystyrene rod. Polystyrene rod may be notched to fit over the turns of the tuning coil, its ends being secured with tape or Araldite to the ends of the ceramic former, and the coupling winding may be embedded in the polystyrene by an application of heat from a soldering iron at the points of contact.

Couplers for 14–28MHz may use a self-supporting tuning coil wound with 1/8in to 3/16in diameter copper tubing. If a variable coupling is required it is necessary to use either a self-supporting tuning coil and a well-insulated coupling winding which can be slid in between the turns of the tuner, or alternatively split the tuning coil into two closely coupled halves wound on separate formers. The link may vary from four or five turns at 3.5MHz to one turn at 28MHz.

In adjusting the coupler the aim should be to achieve the sharpest possible tuning consistent with negligible loss of efficiency. This not only minimises the radiation of harmonics or other spurious signals but gives valuable additional protection to the receiver against various forms of spurious signal and particularly intermodulation interference, which is usually caused by powerful signals in adjacent broadcast bands. With good construction the unloaded coil Q should be at least 300 which means that a loaded Q of 30 would result in a power loss of 0.4dB and an improvement in efficiency of only 0.2dB would result from loading twice as heavily, ie reducing the Q to 15. A loaded Q of 30 allows coverage of, say, all or most of the phone portion of the 14MHz band without retuning, whereas to achieve equal convenience of operation on 3.5MHz the Q must be reduced to 7 or 8 with correspondingly less rejection of harmonics and intermodulation interference. It is suggested that adjustment may be made by starting with maximum coupling and if necessary reducing the antenna tap until it is just possible to obtain adequate loading. The tap is then reduced, readjusting the ATU and PA as necessary to keep the loading constant until the current or power going into the antenna just starts to decrease; if, however, there is no EMC problem and a larger bandwidth is required (as is likely on 3.5MHz), the tap position may be increased.

Balanced antennas

When the antenna or the feeder is balanced it is necessary to revert to a balanced circuit such as Fig 12.76(b) which has already been discussed in some detail. Similar components and methods of construction are suitable, the main differences being the provision of symmetrically arranged pairs of taps and the advisability of a two-gang capacitor with the two portions in series to ensure that the stray capacitances to ground are symmetrical.

Marconi effect

To reduce the risk of feeder radiation due to the feeder wires being excited in parallel as a vertical radiator, as discussed on p12.32 and often known as *Marconi effect,* the capacitance between the windings must be kept as small as possible and it is advisable to screen the coupling loop. This may be done by forming it from coaxial cable as illustrated in Fig 12.84. Trouble may still be experienced if the length of feeder plus antenna is near resonance, and the easiest way to rectify this situation is likely to be the connection of an earth or counterpoise to the frame (ie centre) of the split tuning capacitor. It is sometimes helpful to 'damp' the unwanted resonance, eg by a 200Ω 2W resistor in series with the earth connection.

Matched lines

When supplying a matched line the design requirements for the coupler are modified to the extent that no need arises for feeding into a high value of impedance, and much smaller values of inductance may be used.

Advantage should be taken of this, as it reduces the magnitude of RF voltages and it may often be possible to use a two-gang receiving-type capacitor even at the UK legal power limit.

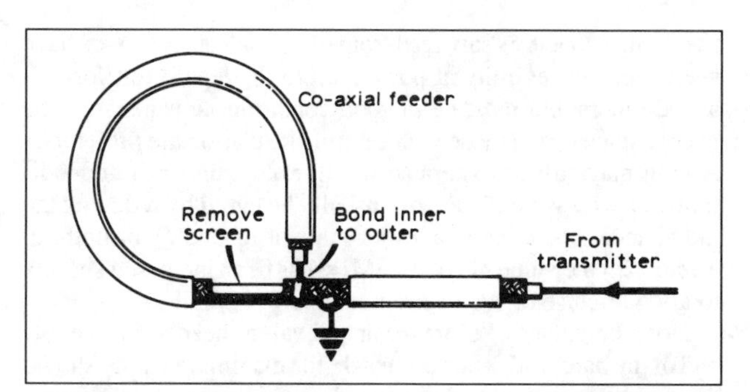

Fig 12.84. Screened coupling loop. The outer is sometimes partly removed as shown, to ensure that the current return path is inside rather than outside the sheath

Apart from this the risks of (a) feedback troubles and (b) excitation of the Marconi effect are reduced.

By way of example, a balanced resonant feeder connected to a dipole can present an impedance as high as 6000Ω; if this is connected across a reactance of 500Ω having a Q of 300 there will be a 4% wastage of power. In contrast, for a 600Ω line with an SWR of 2:1 the reactance can be reduced in the ratio 6000/1200, ie to 100Ω for the same loss in the worst case, and the voltages generated will be less than half. In practice, when working with experimental antennas, a value of 150Ω has been found a good compromise, providing somewhat greater flexibility, but this is not critical and a reactance of 150Ω at 21MHz provides a coil that can be tuned over the range 14–28MHz, a second coil being required to cover 3.5–7MHz.

For feeding low-impedance balanced line a second coupling coil (same number of turns) may be used; this can consist of two coils in parallel to assist in achieving symmetry. The arrangement of Fig 12.83(c) may also be used by omitting the earth connection from the coupling coil, which should be located close to the earthy end of the tuning coil. If Marconi effect occurs it may be reduced by connecting an earth, not that of the transmitter casing, to the centre point of the coupling coil.

Fig 12.83(c) is also suitable for coaxial feeders, the coupling coil being connected to the line and the earth if required being connected to the outer of the line.

Table 12.3 gives suggested L and C values for feeding a balanced matched open-wire line. These should be found suitable for an SWR up to at least 3. The turns are based on the assumption of 3in diameter and a length/diameter ratio of unity. For other diameters the turns must be adjusted inversely as the square root of the ratio of diameters, smaller diameters giving slightly larger losses which for a given coil are roughly proportional to the *loaded Q.* With so many variables depending on individual circumstances it is difficult to give any fixed rules, but couplers handling the full legal power limit should preferably not be less than 2in diameter and need not be more than 3in. For the purpose of the table the reactance has been allowed to vary slightly around 150Ω in order to arrive at a whole number of turns, but considerable variation in dimensions is admissible to suit available components and formers etc. The more robust varieties of

Table 12.3. Suggested L and C values for feeding a balanced open-wire line

Band (MHz)	L (μH)	C (pF)	Number of turns
3.5	10.20	200	14
7.0	4.05	128	9
14.0	1.80	70	6
21.0	1.25	45	5
28.0	0.80	40	4

two-gang capacitors salvaged from old broadcast receivers have been used successfully at powers up to the legal limit for balanced tuners, but must be in good condition, ie vanes must be evenly spaced, and those with ceramic insulation are preferable. Arcing may still be experienced but, except under fault conditions (eg excessive SWR), can usually be cured by reducing the inductance, operating at a lower value of loaded Q, or both. In Table 12.3 the value of L for 3.5MHz has been increased slightly to allow the use of such capacitors.

Note the values of C are the actual values needed for the low end of the band with a perfect match; the maximum values should preferably be about double to provide a wide range of adjustment.

These values of capacitance are of course the effective totals, ie with a balanced circuit 200pF is obtained by connecting two 400pF capacitors in series. With four-gang capacitors a total of about 500pF can be obtained. For tuners feeding resonant lines at voltage maxima the number of turns should be increased by a factor of 1.7 to 2.0 and the capacitances reduced to a third or quarter respectively with provision for a much larger variation, say up to twice the design values, and receiver capacitors are not suitable.

If the need arises the coupling may be increased somewhat in link circuits by tuning out the inductance of the coupling coil, reckoning this as about 0.25µH for a one-turn coil of 4in diameter, and directly proportional to diameter. This additional resonance is extremely flat and no adjustment is needed except from one band to another. When more turns are needed (up to three or four at 3.5MHz or two at 7MHz), they may be wound tightly together, the inductance increasing almost as the square of the turns. The ideal value of inductance for the coupling coil depends on the loaded Q of the ATU as well as on the line impedance and, for values of 10 or more, single-turn coupling coils have been found satisfactory for low-impedance lines at 14MHz and higher frequencies.

The coils are concentric and single layer with a spacing of ¼in or less, and may be constructed as previously described with the inner coil on a ceramic former and the outer coil wound over it using polystyrene rods located in end-plates of any convenient insulating material, with a clearance of at least ¼in between the end turns and the end plates if these consist of a lossy material such as Perspex or bakelite.

Tuning compensation

When the antenna or feeder is coupled to the antenna tuning unit it is sometimes impossible to adjust the circuit for reasonably high loading. This occurs with tuned feeders having a high standing wave ratio, in which case a low-reactance load may be presented by the line. In such cases it is better to compensate the feeder reactance separately before transforming the resistance component, using a coil or capacitor connected in parallel with the feeder (see Fig 12.83(d)). If more capacitance than is available appears to be required for resonance, then capacitance loading is necessary; if the converse, inductive loading is the solution. A loading component of the same value as the coil or capacitor of the tuning circuit may be tried first.

Adjusting matching units and couplers

The recommended way of adjusting the units just described requires an SWR bridge or reflectometer. The layout of a typical set-up is shown in Fig 12.53. The matching unit may be that of Fig 12.76(b), and a suitable reflectometer is described in Chapter 15.

The procedure is quite simple and should present no difficulties if followed carefully and in the correct sequence. The reflectometer should be capable of handling the power output of the transmitter and should be of the type which reads both forward and reflected power. For maximum convenience the instrument should employ two meters, enabling both powers to be observed simultaneously.

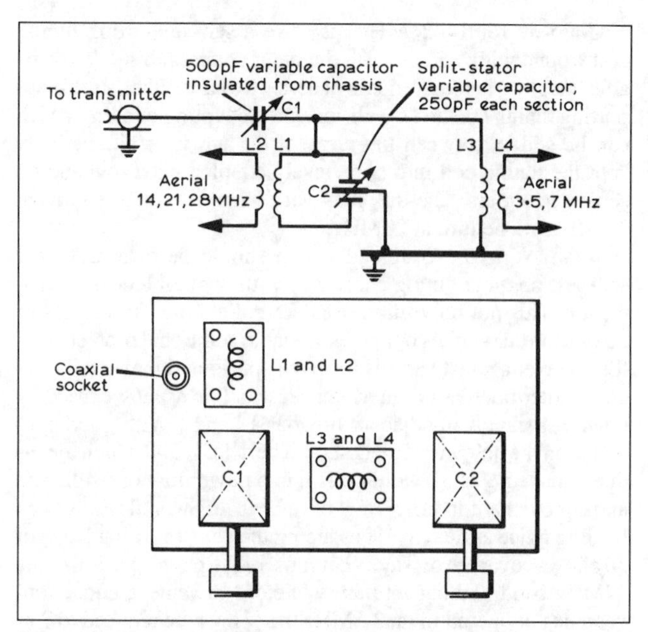

Fig 12.85. Circuit diagram of the Z-match coupler and suggested layout. C1 is the series capacitor and C2 the split-stator capacitor in the multiband tuning circuit

Connect up as shown in Fig 12.53 and adjust the transmitter power or the reflectometer sensitivity so that in the forward reading position the meter indicates well up the scale with the taps set at a trial position equidistant from the centre of the coil.

Now read the reflected power. Vary the tuning of the matching unit for the minimum possible reading on the meter.

If a low reading is not possible it will be necessary to adjust the tap positions, making certain that they are symmetrical about the centre of the coil.

If the coupling between the coils is variable then after tuning for lowest reflected power the coupling should be varied to obtain an even lower reading, repeating the tuning adjustment. Once again, if a low reading is not possible the position of the taps should be changed and the whole procedure repeated.

Multiband couplers

There are many variations of the antenna couplers described. For example, a pi-network similar to those described in Chapter 7 ('HF transmitters') for transmitter output circuits is often attractive because it can be arranged to cover as many as four amateur bands with the same set of components.

The couplers of Figs 12.76 and 12.83 can usually be arranged to cover two adjacent amateur bands but where complete coverage is needed it may be necessary to make up a coil together with its link and taps as a plug-in unit for each band, though there may still be some difficulty if the capacitor has too large a minimum capacitance for 28MHz or insufficient maximum capacitance for 3.5MHz.

A popular multiband antenna tuning unit for 3.5–28MHz is shown in Fig 12.85 and is known as a *Z-match coupler.* It is a compound network using two pairs of windings and is capable of matching the wide range of impedances which may be presented by a tuned antenna line. Coils L1 and L2 may each be 5 turns tightly coupled, while L3 and L4 are 8 and 6 turns respectively. L1 and L3 may be about 2.5in diameter and L2 and L4 about 3in diameter, with 0.25in spacing between turns.

The series capacitor C1 is 500pF maximum and it should be noted that it is 'live' on both sides; the frame should be connected to the transmitter link cable and an insulated extension shaft provided. The other capacitor, C2, is the split-stator type, 250pF per section. C2 tunes the coupler and C1 adjusts the load to the transmitter. A standing wave indicator (reflectometer) is again an aid to tuning. C1 and C2 are adjusted for minimum reflection and the transmitter can then, if necessary, be trimmed for maximum output.

Power indicators

It is necessary to be cautious in interpreting the power radiated in terms of meter readings, because the current and voltage vary from point to point in an antenna system. Some suggestions as to what to expect may therefore be helpful. A 100W DC input transmitter will deliver up to 1A into an 80Ω matched line, but only about 0.35A into a matched 600Ω feeder. In both cases this can be taken to represent true power. In a tuned line, however, the current may vary over a wide range and does not reveal the real radiated power, though it still serves as a guide for tuning and matching adjustments since the current at any point beyond that at which adjustments are made will be proportional to the square root of the true power, provided the adjustment is not of a nature to upset the balance of the system. RF thermo-ammeters are available with maximum readings of 0.5 to 5A, but usually have rather cramped and limited scale at low readings so that they cannot be used for widely differing currents. Care is necessary in cases where the current is greatly different in the same antenna for different bands, since these meters are easily damaged by overload. Such overloads are particularly prone to occur in the event of a fault developing in the antenna or feeder system.

Low-wattage lamps provide slightly more sensitive current indication and are cheaper to replace when destroyed by overload, but both devices are inconvenient to use as they usually require the line to be cut for their insertion, though if the current is high enough (eg at the centre of a dipole) they may be tapped across a few inches of conductor which then acts as an inductive shunt. Even this may be inconvenient and, if the voltage is high enough, a neon lamp is a more handy though less precise indicator.

A sensitive current meter which is portable and requires no direct connection to the line is illustrated in Fig 12.86; the line current induces a voltage in the loop which is detected by a silicon or germanium diode, and with a 4in square loop the sensitive microammeter provides a useful indication from a power level of only a few milliwatts in a flat open-wire line, enabling measurements to be made with very little radiation from the antenna. It is, however, advisable to use a little more power and space the loop at least 0.25in away from the feeder in order to ensure a true current reading since the meter, acting as an antenna, can respond slightly to line voltage and there is also a risk of unbalancing the line appreciably by its use. The loop can be used to indicate the current in each line separately, or it can be held so that it couples to both lines, in which condition full-scale deflection is obtainable even on an insensitive meter (3mA FSD) for about 500mW radiated power, ie a line current of only 30mA. The loop dimensions are not critical but should be comparable with the line spacing. These figures are of course only a rough guide and the instrument is not suitable for use as an accurate power meter. It can, however, be calibrated accurately in terms of *relative* power or voltage by coupling it loosely to the PA and plotting meter reading versus RF oscilloscope deflection, or against antenna current measured with an accurate meter. At very low currents readings are square-law but for induced voltages of about 0–5V upwards a reasonably linear law is obtainable. The device should be mounted at the end of an insulated handle, the longer the better to minimise body capacitance effects.

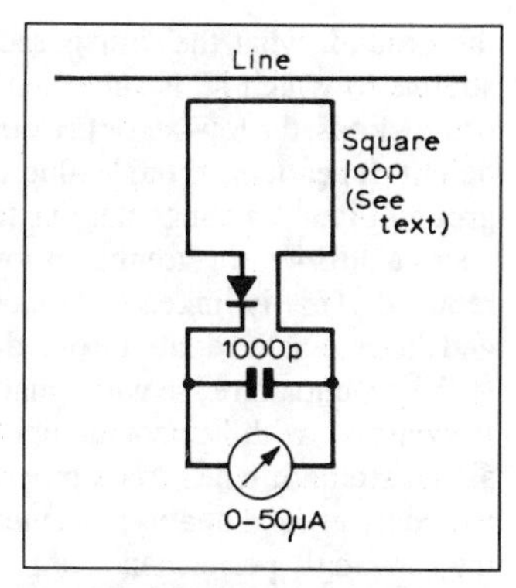

Fig 12.86. Sensitive current meter using germanium or silicon diode

The current in a low-impedance twin feeder may also be detected by an external pick-up device, eg a few inches of feeder taped to the main feeder, with one end shorted and the meter and rectifier of Fig 12.86 applied to the other end. As it stands the instrument can also be used as a sensitive detector of unwanted current flowing on the outside of a coaxial line, but measurements of currents in the line are not possible without damage to it. For measurements on coaxial lines it is however almost as simple to construct a reflectometer on the lines of Fig 15.77. The reflectometer will be found much more useful as it provides an actual SWR measurement for comparison with the SWR on the transmitter side of the ATU. It is an easy and inexpensive matter to construct two similar instruments, one for use on each side of the ATU, using a single meter with a two-pole, four-position switch, but the meter should be as sensitive as possible to allow adequate RF filtering of the leads by the use of series resistances. at least 10kΩ in each lead being recommended.

DESIGNS FOR PRACTICAL ANTENNAS

The first sections of this chapter have been devoted to the general aspects of antennas and transmission lines without any specific attempt to relate these to the particular requirements of any one amateur frequency band. The rest of the chapter deals mainly with practical antenna designs and has been divided into those primarily suitable for the low-frequency bands (1.8–7MHz), those which are more likely to have practical application for the higher-frequency bands (7–28MHz), and various types of beam antenna suitable mainly for the higher frequencies, though considerable overlap exists between these categories, with vertical antennas tending to find their most important role at the lower frequencies where their relatively modest requirements for horizontal area is a major advantage. The distinction, though convenient for the grouping of antennas, is often difficult to maintain since most LF antennas have their HF counterparts and vice versa. The so-called 'ground-plane' featured in the LF section is equally important at HF and enjoys a popularity which seems remarkable in view of the misunderstandings and controversy by which it has been surrounded; much of this can now be dispelled [29, 33] by recognising it as a dipole, one pole of which is non-radiating so that the radials can be shortened by loading with little or no reduction in efficiency, and a better understanding of ground losses which follows from a study of the behaviour of horizontal wires close to ground as featured in the HF section, pp12.58 and 12.62. The use of horizontal wires at the lower frequencies is largely for top-loading, ground-plane radials, and multiband systems such as the G5RV.

On the higher bands the choice between horizontal and vertical polarisation depends in a complicated way on properties of the ground, what the user is seeking to achieve and the constraints to which he is subjected. In general, in the absence of obstructions, the low-angle performance of verticals is not greatly height-dependent, though (due to its image in the ground) a ground-plane antenna, starting from a height of 0.2λ, experiences a 3dB loss in strength of low-angle signals as its height is reduced. This is linked to broadening of the radiation pattern and the associated doubling of radiation resistance, but very close to the ground there are additional losses due to ground currents. In contrast, with horizontal antennas over flat ground the DX signal strength tends to be proportional to height and little affected by ground quality; this means that with increasing height the low-angle performance of the horizontal antenna overtakes that of the vertical at heights which can be as low as λ/4 but may be much more depending on the ground quality. If it is not possible to put up a good horizontal antenna the vertical may well do better, and before making a choice it is advisable if possible to make comparative tests with temporary antennas.

LOW-FREQUENCY ANTENNAS

Choice of polarisation

One feature of operation on the lower frequency bands is the widespread use of vertical antennas. Reasons include ease of erection and also the important role of ground-wave propagation at these frequencies, particularly on 1.8–2.0MHz. Due to the low height in relation to the wavelength, ground-wave propagation is very inefficient with horizontal polarisation, although better ground-wave signals will usually be obtained when both antennas have the same polarisation, regardless of whether this is horizontal or vertical. Vertical polarisation is advantageous also for DX, since even at 3.5MHz it is usually not possible to erect a horizontal antenna at sufficient height to achieve good low-angle radiation.

Against this, the 3.5MHz band is particularly suitable for short- and medium-range communication via the E-layer in the daytime and the F-layer at night, including distances just beyond the ground-wave range which can usually be reached in daylight hours by nearly vertical incidence reflection from the E-layer, whereas for much of the time long-distance low-angle paths are useless due to excessive absorption in the ionosphere. Vertical antennas are relatively poor as high-angle radiators so that horizontal antennas, even at low heights, are usually better for short-range sky-wave working and reasonably satisfactory over medium distances. The directive properties of the horizontal antenna have comparatively little influence at the high angles appropriate to these distances.

Marconi antennas

Most of the discussion so far of simple types of antenna has assumed horizontal polarisation and, although in principle these antennas can be erected vertically, this is rather difficult in the case of, say, a λ/2 dipole at 3.5MHz which would require a mast about 140ft high! The usual procedure if vertical polarisation is required is to use as long a vertical wire as possible and tune it as a monopole resonator working against some form of earth connection. The antenna, in conjunction with its image in the ground, then bears some resemblance to a shortened dipole. If the wire is λ/4 long with its base at ground level it can be thought of as a bisected λ/2 antenna and its radiation resistance is about 35Ω, ie half that of a λ/2 dipole. Such a height is not usually possible, even for 3.5MHz where it would be 66ft, and hence it is necessary to use an electrically short antenna and to load it in such a way as to bring it to resonance, accepting the inevitably low value of radiation resistance. For example an antenna 0.1λ high (50ft for 1.8MHz) has a radiation resistance of 4Ω, which would not matter if the earth were a perfect conductor, but this is far from being the case, and the resistance due to even a very elaborate earth system, together with the loss due to the current returning through the surface, may contribute 20Ω or more. The efficiency is thus low. The relative efficiency of various combinations of earth resistance and antenna height are considered in more detail in the following section on earth systems.

Above-ground systems of short loaded radials can often be used to reduce or eliminate earth losses, particularly at higher frequencies; at the lower frequencies, to be equally effective, radials must be longer, higher, and in general more difficult to dispose of so that if, as frequently happens, both height and space are at a premium, additional methods of improving efficiency may need to be sought. Let us consider first a short vertical wire and assume access to the base only; this wire can only be brought to resonance by means of a series inductance which has some loss resistance, this being added to the other system losses. Apart from this, the current distribution in the wire, being the tip of a sine wave, is triangular and the mean current is only half the base current. If, on the other hand, the antenna can be tuned by capacitance between the top end and ground, and if this provides the main current path, the current distribution on the vertical wire will be nearly uniform; the mean current will then be almost equal to the base current, the required value of this being therefore halved for a given field strength. This means that the radiation resistance is multiplied by four, and in addition the loading coil is eliminated. Capacitive end-loading may be achieved in practice by various combinations of horizontal wires, and even when it is not possible to tune the vertical radiator fully by this means the amount of inductive base loading required may be greatly reduced. On the other hand, it is sometimes possible in this way to increase the total effective length of the antenna to more than λ/4, thereby raising the impedance at the feed point, and hence even further reducing the current and the power loss in the earth connection.

The simplest form of top end-loading is to add a horizontal top to make a T or inverted-L antenna (Fig 12.87(a)). This loading should ideally be arranged to bring the current maximum into about the centre of the vertical portion – though it may sometimes be advantageous (eg if there are local obstructions) to bring it as high as possible when the loading is of T-form.

With L-type construction it is undesirable to have too much current at the top where it is liable to result in wastage of power by useless radiation from the horizontal section, whereas with the T-configuration the radiation from the two halves of the top is in antiphase and tends to cancel. For an L-antenna the electrical length is given approximately by the usual formula, p12.3, the distance from the end of the wire to the centre of the current loop being about 130ft at 1.8MHz. For a T-antenna the equivalent electrical length of the top portion can be reckoned as about two-thirds of the actual length for lengths up to λ/4, decreasing to half for a length of λ/2 when the centre of the T becomes the centre of a λ/2 dipole.

If the length obtainable by these means is insufficient, the next alternative is to increase the loading effect of the top by making it into a cage or 'flat top' of two or three wires joined in parallel; this increases the capacitance and effectively lengthens the antenna. If this procedure is not practicable, the antenna may be loaded by including a coil (see Fig 12.87(b)) near the free end of the antenna. In this case the principle of operation is as follows: if the length of wire beyond the coil has a capacitive reactance to ground of $-X$, an inductive reactance of $X/2$ gives a

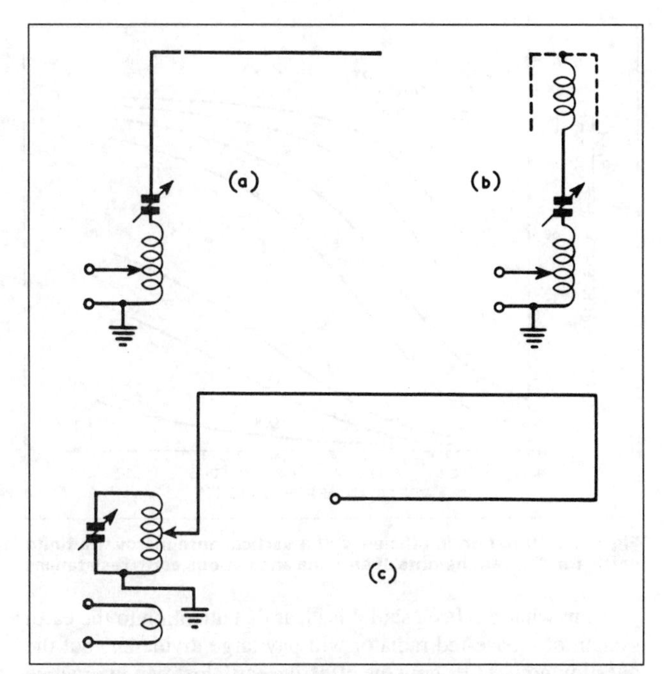

Fig 12.87. Antennas for low frequency bands. (a) λ/4 (inverted-L) antenna with series tuning. (b) End-loading to raise the efficiency of a short antenna. (c) Special extended antenna with high efficiency, detailed in text. The input current in antenna (a) and (b) may be 0.5A for 10W input; that in (c) is low, say 0.1A for 10W

net reactance of $-X/2$ which is equivalent to doubling the capacitance or, approximately, doubling the length of wire if this is shorter than about λ/16. The coil also has a resistance $X/2Q$, part of which appears as a loss resistance in series with the radiation resistance and thus limits the amount of this type of loading which can be usefully employed.

As an example, a 34ft length of vertical wire top-loaded to resonate on 3.6MHz will have a radiation resistance of about 18Ω. The vertical wire has an inductive reactance of about 300Ω which can be tuned to resonance by 1500Ω of capacitance in series with an additional 1200Ω of inductance which will have about 4Ω loss resistance. This would require a coil of 53μH in conjunction with a capacitance of 32pF, which could be provided by 16ft of horizontal wire in place of the 70ft or more required without the coil. Being in series with the radiation resistance plus the earth resistance, the loss in the coil has relatively little effect on performance, though the process cannot be carried very much further without loss of efficiency since halving the capacitance and doubling the inductance will roughly double the coil loss. On the other hand, if the capacitive top is long compared with the vertical wire, it is immaterial electrically whether the coil is placed at the top or, more conveniently, at the base.

In the above instance, whichever method of capacitive loading is used improves the efficiency by three to four times if the earth resistance is large. This example has been slightly oversimplified by ignoring some aspects of the current distribution but serves to illustrate the basic principle as well as the main practical aspects, and it will be obvious to the reader that there are many possible variations of this technique which find particular application in the field of mobile and indoor antennas.

Other methods of end-loading may be devised; for example, folding the antenna back on itself with a spacing of, say, 6in.

Antennas for 1.8MHz with relatively high input currents are preferably series-tuned at the base, using about 250pF capacitance with say 30μH inductance. A suitable coil for 1.8MHz could be made with 20 turns of 16swg wire on a 3in diameter former, spaced to occupy a length of 3–4in. The transmitter link could be tapped across a few turns at the earthed end of the coil or the wire itself coupled to the transmitter output tank coil. About half the above values should be used for 3.5MHz.

To reduce the deleterious effect of an imperfect earth connection the antenna may have a total length of about 3λ/8 or 5λ/8, the current in the vertical part being high in both cases, though the earth current is considerably reduced. A very effective antenna is shown in Fig 12.87(c) where the main half-wave is folded into a U with one leg near the ground and adjusted in length until maximum current occurs at a point half-way up the right-hand vertical section. The down-lead makes the total length up to about 5λ/8, so that the impedance at the feed point is a few hundred ohms and capacitive, while the earth current is relatively low. The coupling circuit for this antenna should be parallel tuned but can use the same components as the previous examples.

The 1.8MHz antennas of Figs 12.87(a) and 12.87(c) can also be used effectively on 3.5MHz and higher frequencies, though their performance will not be accurately predictable.

Earth systems

The above examples have demonstrated the importance of a low earth resistance when the ground acts as the return path for the flow of RF currents. As will be evident from this discussion, earth resistance tends to present a major problem on the low frequencies, particularly on the 1.8MHz band, as it is often difficult to dispose of a sufficient length of wire to achieve a radiation resistance larger than the earth resistance which is in series with it and represents wasted power. There are two lines of attack on this problem; increasing radiation resistance where possible by one or more of the methods outlined above, and decreasing the resistance of the earth connection. To illustrate the problems of earth resistance, Fig 12.88 shows the current paths when a short vertical radiator is driven by a 'generator' (ie transmitter) connected between the bottom of the radiator and ground.

The current distribution along the radiator possesses a maximum value at or near the ground, decreasing approximately sinusoidally (or in the case of a very short radiator, linearly) to the end of the radiator at which point it must be zero. Because of the capacitance of the radiator to ground, charges are induced in the earth surrounding it, giving rise to a circulating current which flows back to the generator. This current flows in the ground at a penetration which varies with frequency and near-surface conductivity, decreasing as both increase. The net effect is to include in the series circuit representing the antenna system a resistive loss term through which the antenna current flows, and

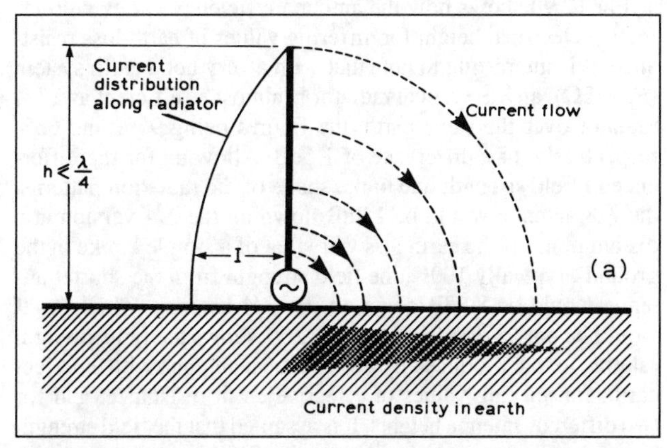

Fig 12.88. Current distribution on a short vertical radiator over a plane earth

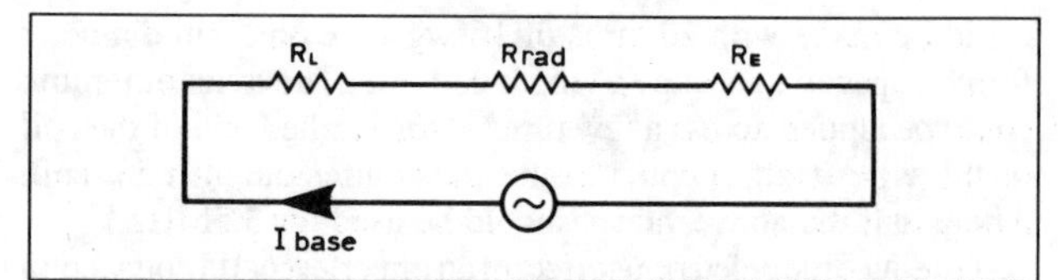

Fig 12.89. Equivalent circuit for a short vertical antenna over a plane earth, R_L, ohmic losses; R_{rad}, radiation resistance; R_E, effective earth resistance

which is dependent upon the near-surface conductivity in the neighbourhood of the antenna.

In order to determine the antenna efficiency we need to compare the power dissipated in the various resistive elements of the circuit; as illustrated by Fig 12.89 these comprise the ohmic loss R_L of the antenna tuning components and the antenna conductor itself, the radiation resistance R_{rad}, and the earth loss resistance R_E.

Radiation resistance is a fictitious resistance which is included to account for the power dissipated by radiation, ie its value is determined by the useful radiated power and the current flowing in the antenna at the feed point. It does not exist as a physical resistance in the same way that R_L and R_E do. Of the three resistive elements dissipating power, only that in R_{rad} can be considered useful; that in R_L and R_E is wasted as heat, although it still has to be supplied by the transmitter. Since the same current flows through all resistances, the efficiency of the antenna system is given by:

$$\text{Efficiency } \eta = \frac{\text{Power radiated}}{\text{Power supplied}} = \frac{R_{rad}}{R_L + R_E + R_{rad}}$$

From Fig 12.5 the antenna wire resistance will typically contribute about 1Ω per quarter-wavelength and a tuning coil of 30μH with a Q of 200 will add a further 2Ω, making a total of 3Ω. The form of loading illustrated in Fig 12.87(b) if used will increase this figure, the extent to which it can be exploited being determined as already discussed by the point at which R_L becomes comparable with $R_E + R_{rad}$. The value of R_{rad} is fixed for a particular antenna by its physical size and shape, being typically 35Ω for a vertical λ/4 wire, and as low as 9Ω for a vertical λ/6 wire. Hence, the shorter the antenna (electrically), the better is the earth system required to maintain a given efficiency. This can also be expressed by saying that for a given input power, the shorter antenna with lower radiation resistance gives rise to a larger feed current, and hence the earth loss must be reduced in proportion. The value of R_E varies widely, from as much as 300Ω for a simple earth in sandy soil, to as low as 2–3Ω for a comprehensive earth system in good soil.

Fig 12.90 shows how the antenna efficiency varies with effective electrical height for differing values of earth loss resistance. It is interesting to note that with a very good earth system ($R_E < 2\Omega$), an λ/8 vertical radiator is almost as efficient as a λ/4 radiator over the same earth, the figures being 95% and 66% respectively, or a difference of 1.6dB. Allowing for the difference in field strength due to the shape of the radiation patterns, the λ/8 antenna would be 1.7dB down on the λ/4 version at a distant point. If the earth loss were that of a simple 'spike in the ground', typically 100Ω, the field strength from the shorter antenna would be 8.5dB down on the λ/4 antenna, which itself would be only 27% efficient. Fig 12.91 shows the field strength relative to that produced by a λ/4 vertical antenna with a perfect earth, for varying values of effective earth resistance, and for two different antenna heights. It is assumed that the field strength is measured at ground level, and the rate of attenuation with distance the same in both cases.

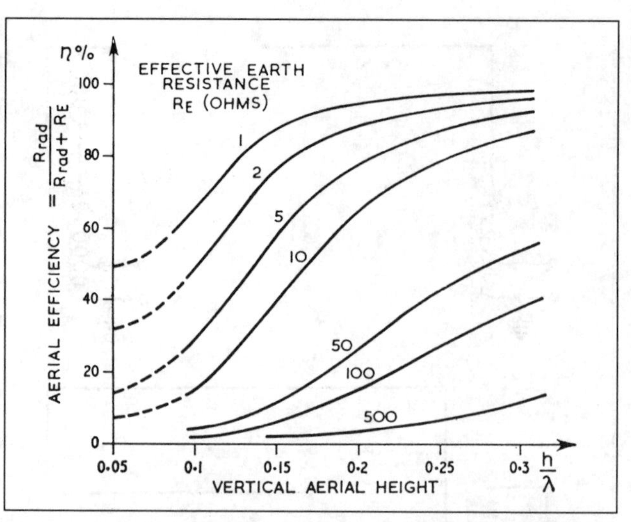

Fig 12.90. Variation in efficiency of a vertical antenna over a finite earth, for different heights of antenna and various earth resistances

From what has been said it is clear that attention to the earth system of a base-fed radiator will pay large dividends, but the question now arises how the effective earth loss can in practice be reduced. The figure of 100Ω for a large spike in sandy soil is not unrealistic, and to reduce this figure to the 2–3Ω of a near-perfect earth is beyond the scope of most amateurs, requiring as it would a massive system of radial earth wires, in excess of 120 in number and extending far out up to a wavelength from the base of the antenna – and all this in best-quality agricultural ground as well! In practice, the best rule of thumb is to get as much copper wire into the ground as possible in the immediate vicinity of the antenna base, concentrating it near the antenna at the expense of the edges of the garden. The radial wires should be at least 16swg, and buried as near the surface as possible consistent with their remaining undisturbed by gardening activities since the depth of current penetration is a function of the near-surface resistivity. The inner ends of the wires should be joined to a heavy copper circular bus-bar (say 6swg wire in a 12in diameter loop). Brazing is to be preferred for resistance to corrosion, although soft soldering is adequate provided the joints are painted with a bitumen paint to seal out moisture. It is

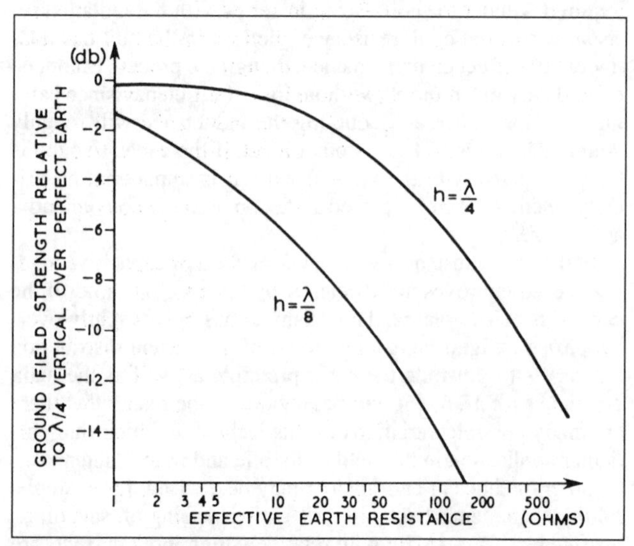

Fig 12.91. Comparative ground-wave field strength of λ/8 and λ/4 vertical antennas over an imperfect ground (neglecting ground-wave attenuation)

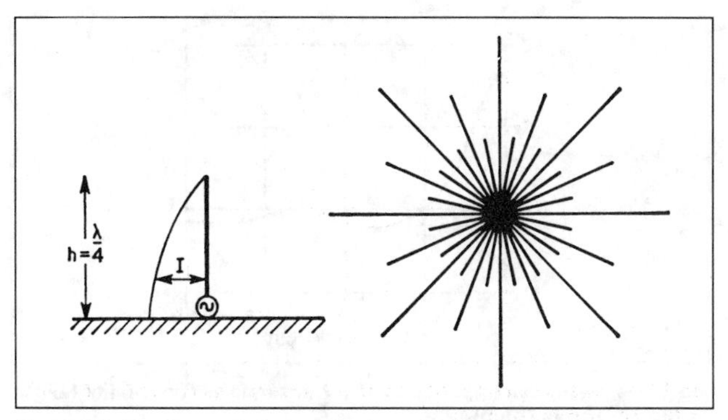

Fig 12.92. Typical radial earth system for a short vertical antenna. The earth currents are concentrated near the base

permissible to join the outer ends of the wires, but too much interconnection will result in large circulating currents being induced in the buried loops with a consequent increase in losses – the opposite in fact of the desired result (Fig 12.92).

The remarks in the previous paragraph apply mainly to amateur antennas for the LF bands having lengths of a quarter-wave or less and the current maximum at the base of the antenna. For those in the position to erect verticals (natural or loaded) with an effective length approaching a half-wave there is no need for an earth connection, and Fig 12.149 on p12.91 illustrates a 7MHz beam antenna based on the use of short end-loaded vertical dipoles. Used singly at the lowest possible height, these elements have a radiation resistance of about 24Ω but this figure can be expected to vary somewhat with height and ground conductivity. With direct scaling for 3.5MHz a height of over 50ft would be required but if a relatively longer span is accepted this could be considerably reduced for single elements, particularly if broadband operation is not essential.

It will be obvious from the above examples that every possible effort should be made to avoid reliance on earth return, and this is the main advantage of the various forms of ground-plane antenna. Even if the radials are reduced to two (a reasonable compromise), a run of 133ft is required to accommodate a full-size ground plane on 3.5MHz, but the vertical can be end-loaded by the use of a T-configuration, the radiation resistance being halved when the height is reduced from λ/4 to λ/8, and (as explained below) shorter inductively loaded radials are in any case to be preferred to full-sized ones [33]. In general, however, these systems, like the previous example, tend to have too narrow a bandwidth for matched coverage of the whole band, so that access for tuning may be essential.

It has been pointed out that a particular virtue of the 3.5MHz band lies in its suitability for short- and medium-distance (high-angle) sky-wave communication for which horizontal polarisation tends to be more suitable, but in both cases there is some value in being able to select either polarisation at will, and it is worth noting that the horizontal dipoles described below can frequently also be used by connecting the feeder wires in parallel to act as a vertical radiator, with the dipole providing top-loading or acting as an 'inverted ground plane'. This mode will usually be found better for DX except in the case of hilltop or hillside sites. Though not generally ideal for ground-wave propagation at the lower frequencies, horizontal polarisation may give better ground-wave signals in those cases where the antenna at the other end is also horizontal; also, being a balanced load such antennas do not require an earth return, and even for ground-wave communication they offer a better compromise in those cases where the earth system is very poor.

Finally, a word of warning about ground connections. It is sometimes suggested both in print and over the air that a rising water pipe is a good earth. This is not always true in practice since the amount of pipe in contact with the earth *in the vicinity of the antenna* may not be large, and the contact resistance will be indefinite because of corrosion deposits on the outer of the pipe. There will probably be indifferent contact between lengths of pipe at the screwed unions because of the sealing compounds used, and in any case modern practice is often to use plastic tubing which will negate the whole exercise. There is an obvious temptation in some cases to use the earth connection of the electricity supply for LF antenna systems but, as previously noted, *this may inject RF voltages into the supply and create a dangerous situation.* It may also introduce noise into the receiver or be responsible for spreading interference to nearby TV receivers.

It is a mistake to bring a long 'earth' lead into the radio room, which is often some distance from the point of connection to true earth, being in the limit on the upper floors of a building. The result of this is that the long earth lead will necessarily radiate since it carries the antenna feed current, and in consequence of the lead impedances the equipment in the shack will be up-in-the-air to RF with many consequent problems of filtering and feedback. A better arrangement is to install the ATU in a box at ground level immediately adjacent to the earth mat, and connect it back to the shack with a low-impedance coaxial line matched into the ATU. This will not only isolate the shack from the radiating part of the antenna system, but will permit the vital vertical section of the antenna to be installed clear of obstructions which would otherwise degrade its performance.

Ground planes and radials

The *ground-plane antenna* (Fig 12.93) is essentially a dipole, one pole of which has been prevented from radiating by arranging it in the form of two or more radial wires carrying currents which oppose each other, its radiation pattern being exactly the same as that of any other short dipole. This means that since the length is halved, the current is doubled and the radiation resistance *R* divided by 4, which brings it down to 72/4 = 18Ω plus the small gain correction explained on p12.6 which brings it up again slightly to 20Ω. Advantages of arranging one pole in this way include the possibility of bringing it down close to ground level where it takes over the role of the relatively lossy ground connections discussed above, with the radiation resistance rising to the usual 'grounded λ/4 vertical' value of 36Ω.

As well as providing a return path for the antenna current, the radials exercise another important function of the ground connection, that of tying the outer conductor of the coaxial feeder firmly down to ground potential, thereby preventing the flow of current on the 'outer' of the feeder which could result in undesirable forms of radiation or additional ground losses. A major virtue of the ground-plane antenna is that it can be erected anywhere instead of being tied to ground. Its portability and the ability to erect it in a very small apace are further important assets, a view not encouraged by insistence in most of the literature on λ/4 radials. In fact, as explained in references [29] and [33], recognition of the non-radiating status of the radials means that they can be shortened without affecting radiation efficiency so long as the losses in the essential loading coil remain small compared with the radiation resistance. With very short radials, it becomes more difficult to prevent the feeder becoming an undesirable addition to the radial system, but with care lengths down to λ/20 and even less have been used successfully [29, 33]. It will, moreover, be obvious by inference from Fig 12.20 that, far from being mandatory, λ/4 is the one length to be

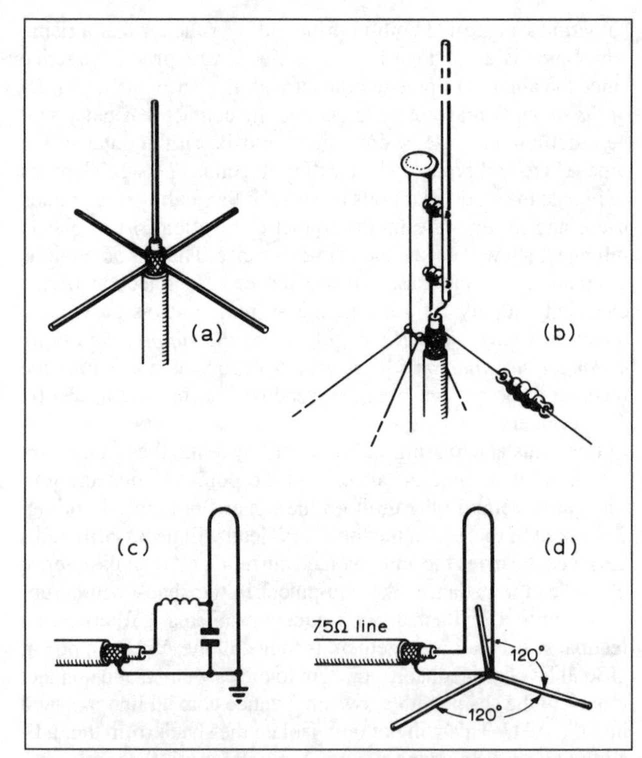

Fig 12.93. λ/4 vertical antennas. (a) Ground plane. (b) Ground plane with sloping wires which may be used as guy wires. (c) Folded wire with transformer. (d) Folded ground plane. The cable size is exaggerated for clarity. See text for recommendations regarding radials

avoided since close to resonance phase is changing very rapidly and, allowing for finite length tolerances, the equalisation of radial currents becomes highly problematic. In view of this it is essential for short radials to share a common loading coil; the inductance of this is essentially a replacement for that of the inner portions of the radials, though distributed capacitance has to be taken increasingly into account as radial length is reduced. With lengths below about λ/8 or λ/12, bandwidth decreases significantly and, due to high RF voltages, it becomes even more important to protect the ends of radials from accidental contact.

Fig 12.94 shows how R varies with the height of the radials, this being about 0.1λ less than the effective mean height which corresponds to the 'centre of gravity' of the current distribution and is the appropriate figure for comparison with other vertical systems; these figures are not quite the full story since at low heights part of the antenna current returns via the ground as illustrated by Fig 12.95 which explains the occurrence of ground losses in the case of horizontal wires (including radials) at low height. The resulting losses decrease very rapidly with height, typically becoming negligible in the case of radials at heights above about 0.05λ. They can be rendered negligible by means of a wire mesh screen placed on the ground underneath the radials [35] or by using very large numbers of radials [33], but it is usually easier to increase the height slightly. Better still, if suitable supports are available, is replacement by end-loaded dipoles as used in Fig 12.126 or inverted ground planes such as the original G3VA version [45], Fig 12.96(a), or Fig 12.96(b) where the separation of wires at the top overcomes the problem of unequal current division described earlier with reference to Fig 12.20.

Incidentally, Fig 12.96 provided a convenient test-bed for checking-out the new method of end-feeding described on p12.64, results being almost identical with Zepp feed based on the G6CJ balancing stub (p12.67) or, in the absence of the stub,

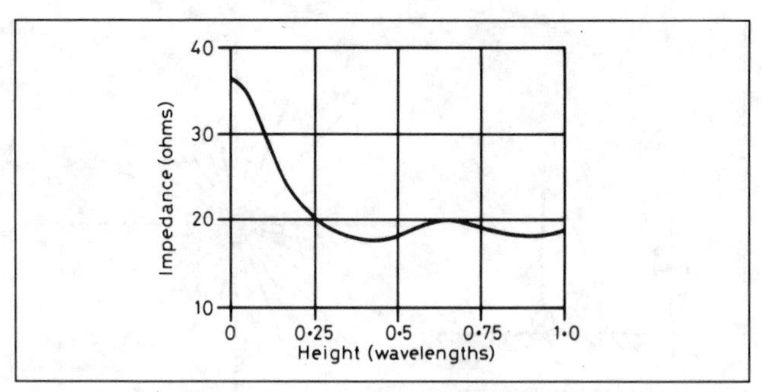

Fig 12.94. Base feed impedance of λ/4 antenna as function of height of radials above the ground

subject to applying a linear trap to the coaxial cable outer as shown in the figure.

An improved ground-plane radiator which is particularly suitable for use as a beam element is shown in Fig 12.97. As the radials are short, the imbalance is not important and at 14MHz a pair mounted on fence posts has proved to be a very effective beam antenna. Being at ground level they are easy to tune, coupling being adjustable by swivelling them round so that ends of the radials approach each other. Observed features of this arrangement have included the absence of ground losses in the case of parasitic elements, even at base heights of only a few inches; this is attributed to the distributed nature of the signal-injection which permits current flow without ground involvement.

It must be emphasised that difficulties in preventing feeder radiation escalate with increasing height and could perhaps explain much conflicting information to be found in the literature; increasing the number of radials should be helpful in this respect, and for more detailed discussions covering aspects which may still be open to controversy the references [29, 33] should be consulted. Losses in the ground adjacent to the antenna are further discussed in a later section (p12.61).

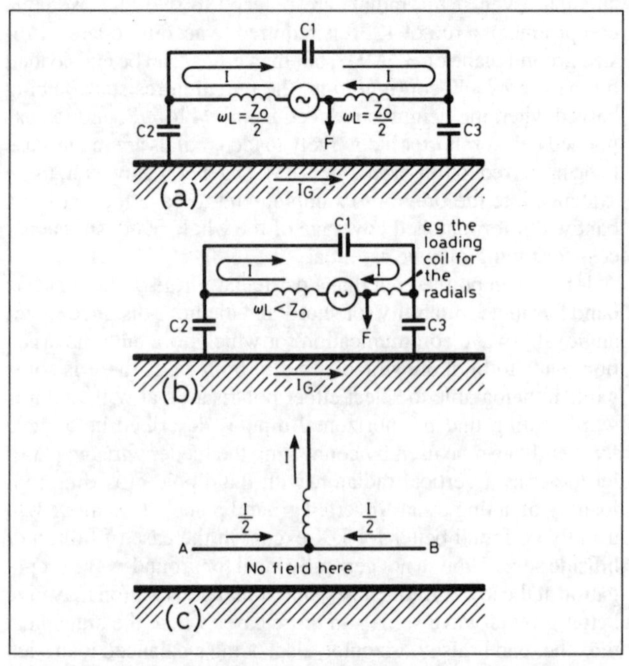

Fig 12.95. Equivalent circuits for dipoles close to ground: (a) symmetrical and (b) asymmetrical. The feeder (at F) will radiate unless maintained close to ground potential. (c) shows why the ground current I_G might be reasonably expected to disappear in the case of a ground-plane antenna (ARRL)

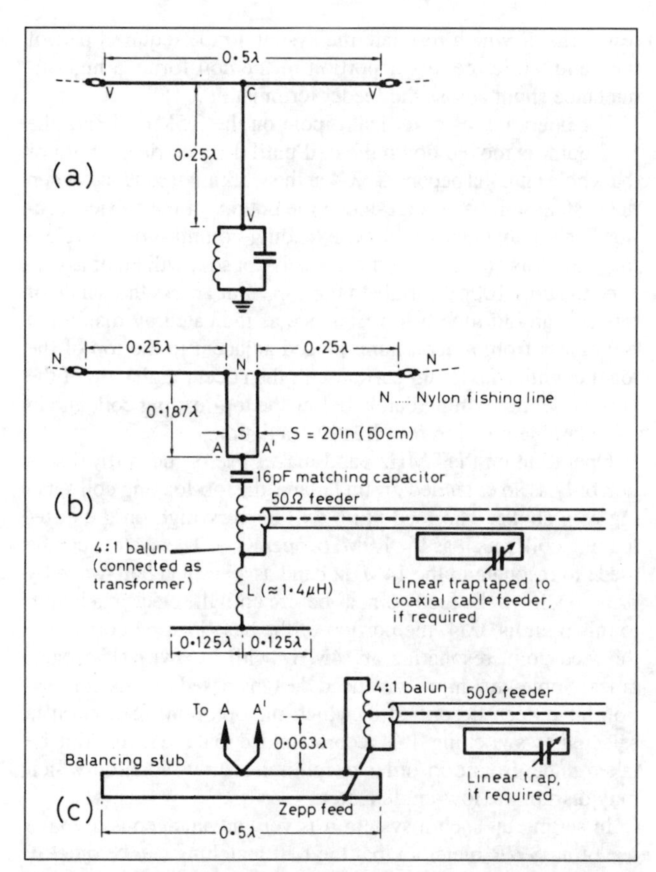

Fig 12.96. The G3VA inverted ground-plane antenna; inversion increases effective height and eliminates ground losses but makes it even more important for the radials to be non-resonant, at least 5% increase or decrease in top length being suggested. (b) Improved version based on Fig 12.103 ensures equal currents, improves bandwidth and avoids the limitations of the Zepp feed. Values of L and the matching capacitor are for 21MHz. The counterpoise length can be varied between wide limits subject to alteration of L. (c) The Zepp alternative. Note restoration of radiator length to the usual λ/4

Other ground effects are more properly considered under the heading of propagation, but it needs to be strongly emphasised that the 'ground plane' is in no sense a reflecting plane, has no effect on angles of propagation, and no functions other than those described above. An entirely different situation would arise in the case of a 'true' ground plane, ie a plane of high conductivity which (to comply with the Fresnel zone requirements specified in reference [44]) would need to extend for several wavelengths along the required line of propagation and have a width of the order of six times the antenna height, in which case considerable enhancement of the signal could reasonably be expected.

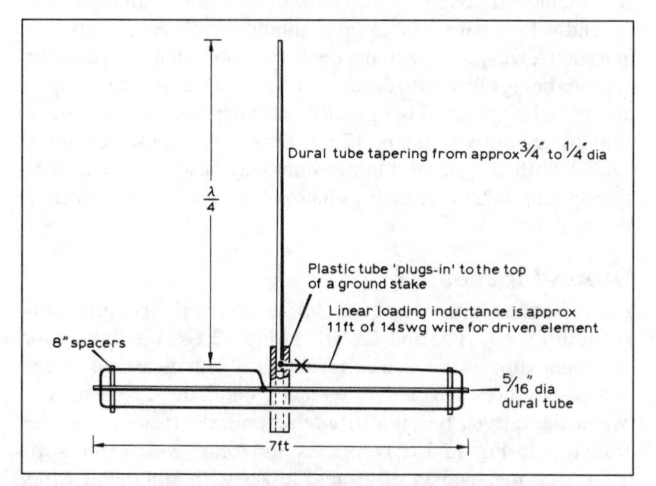

Fig 12.97. Constructional details of a 14MHz monopole with short, inductively loaded counterpoise. The antenna may be fed with 50Ω coaxial cable at point X (outer to counterpoise) but do not expect an SWR of less than about two unless additional impedance transformation is used. Subject to maintaining resonance the counterpoise dimensions are not critical. This design can be readily adapted for LF use

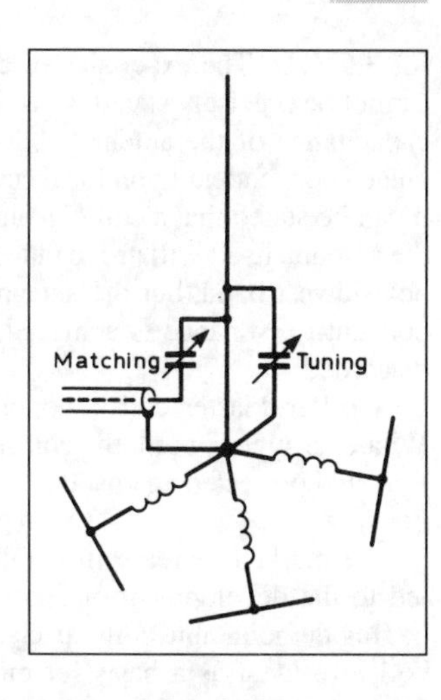

Fig 12.98. Multiband vertical – length λ/4 approx at lowest frequency. Capacitors must be accessible or remotely controlled for adjustment to suit the band in use. Separate counterpoises are shown for each band but alternatively series tuning of the lowest-frequency counterpoise should be acceptable. Coverage of at least one octave can be expected

Multiband ground plane

The principle of the all-band dipole, Fig 12.111, may be extended to the ground plane by using a set of verticals, one for each band joined in parallel at the base, and a separate set of radials for each band. Alternatively it may be found possible to use traps (p12.68) in the vertical portion, in conjunction either with separate sets of radials or trapped radials.

A more efficient method of achieving multiband operation is shown in Fig 12.98. If the components are accessible five-band (14–28MHz) operation should be feasible, otherwise replacement of capacitors by fixed values with bandswitching may be worth considering.

The loaded vertical

Vertical or ground-plane antennas present little problem at the higher frequency bands where a self-supporting radiator up to 0.375λ long or a wire supported by a mast can be erected. Not infrequently a metal mast or lattice tower insulated from earth at its base can be used effectively, as is common practice at broadcast frequencies. The loading of vertical antennas to permit the use of relatively short radiators has been discussed on p12.54 and antennas designed for the higher frequencies can often be pressed into service on the lower frequencies by the use of one or other of the loading devices described. Loading is common practice in mobile installations where the length of whip antenna is strictly limited by practical considerations.

Base-loading is clearly the most convenient since the loading coils are accessible for changing bands, and can also be used as a coupling transformer on the lines shown in Fig 12.87. Because of relatively large coil losses, base-loading is generally confined to adapting an antenna for use on the next lowest frequency band, because the inductance will not then be large enough to upset the current distribution seriously or to introduce excessive losses. Thus a vertical antenna designed for λ/4 or 3λ/8 on 28 or 21MHz can be simply adapted for use on 14MHz; and more commonly a 14MHz vertical may be loaded for use on the 7MHz band. A 32ft vertical resonant at 7MHz is not unduly difficult to erect, and may be base-loaded effectively for 3.5MHz. The exact size of a base-loading inductance coil cannot be precisely stated, since in effect it is tuned by the self-capacitance of the antenna above it to earth, and this capacitance is dependent upon local conditions. As a guide, however, it can be stated that a coil containing an equal length of wire to the antenna itself will prove rather too large for resonance in the next lowest band, but the antenna can be tapped down such a coil until resonance is achieved, and the unused portion then discarded.

A roller-coaster coil is popular for base-loading purposes. Whatever type is used, the coil should be of low-loss construction, and protected against the weather by a suitable waterproof housing.

The marked increase in mobile activity in recent years has led to the development of a number of loaded whip antennas having demountable coils specially suitable for this work. Such coils are ideal as a basis for more permanent loaded vertical antennas for the HF bands. Mobile antennas are in many instances centre-loaded, which implies that the coil will be located part-way up the mobile system, and may be designed for mounting upon a section of 1in diameter aluminium tubing. The section of whip above this coil determines its resonance, and often takes the form of a 4ft to 5ft section of 5/8in diameter tubing. Coils are available in interchangeable form for the various amateur bands.

Since in a mobile system of this kind maximum current is carried in the section below the loading coil, there will be a very large reduction in loss resistance if this section is greatly lengthened. Making this change, however, has only a slight effect upon the resonant frequency, which tends to rise as the capacitance of the top section to earth is reduced, but is simultaneously lowered by the added inductance of the lengthened lower section. Excellent loaded verticals can therefore be constructed by mounting a standard coil and top whip section for the band required upon a mast of 1in diameter tubing, which may itself be of resonant length for a higher amateur band. The mast must be insulated at its base, either by the aid of a mounting insulator or a short section of bakelite rod or tube, and will be fed as already described. There is no objection to a small trimming-inductance at the base, since the frequency in the loaded-mode is likely to be a little high; should it be otherwise, the top section must be shortened with caution as this length is very critical. An antenna of this kind will require to be guyed at a point just below the loading coil, and at other points according to the actual construction and overall height chosen.

The antenna described is effectively top-loaded for the lowest frequency band, and will give useful low-angle radiation. Bearing in mind that a vertical radiator need not necessarily be quarter-wave, it is possible to excite such a system on several bands by altering the feeder termination. Where space is very limited, the feeder line reasonably short, and the base of the antenna accessible for adjustment, a vertical system on these lines can be a useful compromise, providing a DX capability on several bands.

As a practical example, the vertical antenna may consist of from 32ft to 40ft of 1in diameter tubing, terminating at the base in a coil similar to that described in the context of Fig 12.87(b) with provision for tapping the feeder up or down the inductance, but without the series capacitor shown in that illustration. At the top this 1in tubing carries a loading coil of the type used in 1.9MHz mobile installations, and carrying above it the usual whip section which will resonate towards the HF end of the 1.8 to 2.0MHz band. To operate on this band, the feeder is taken very nearly to the top of the base-loading-coil, the remaining few turns of which resonate the system to the required part of the band while the larger portion of the coil forms a high-inductance shunt across the feeder termination.

For operation as a vertical dipole on the 3.5MHz band, the feed point is moved down the coil until the electrical length of the whole antenna becomes 3λ/4 at the working frequency, when the system will be voltage-fed at the bottom of the vertical section. The inductance of the base loading-coil must be sufficient to ensure this condition. Where this is not so it will be of assistance to add a 100pF variable tank capacitor across the whole of this coil and to tune it to resonance as indicated by maximum brightness from a neon lamp placed adjacent to the top of the loading coil. Maximum current will then occur at the top of the vertical section, immediately below the top-loading coil, and in a favourable position for effective radiation.

Operation on the 7MHz band makes use of the vertical section only as an extended ground plane, the top-loading coil serving as a choke. The feed point will be very high on the base-loading coil, much as for 1.9MHz operation. The system can be made to resonate in the 14MHz band as a vertical half-wave by bringing down the feed point as before until the electrical length to this point is 3λ/4, the portion of the base loading coil above the feed point resonating at 14MHz with its own self-capacitance. Somewhat more advanced designs based on this conception have been described in which the optimum feed point is selected by switching. It is recommended that the feeder line be taken underground in order to minimise stray radiation which may disrupt the low-angle pattern.

In setting up such a system it is very advantageous to make use of an SWR meter so that the best matching can be quickly found but, providing the feeder length is not excessive and that the transmitter loading is satisfactory, results are likely to prove reasonably acceptable.

Horizontal dipoles

In an earlier section it was explained that the frequent need for high-angle propagation experienced on the LF bands is best met by the use of a horizontally polarised antenna. The appropriate lengths for λ/2 dipoles for the 1.8MHz and 3.5MHz bands are given in Fig 12.99.

The current distribution along the dipole is roughly sinusoidal and is concentrated in the middle with little or no radiation from the ends of the wire. The antenna should therefore be supported in a such a way as to keep the centre region as high as possible, the ends being allowed to droop or to hang down, depending upon the available space. Two possible arrangements for restricted gardens are shown in Fig 12.99. Shortened dipoles centrally loaded with a coil or tuning stub may also be used, with appropriate matching, their radiation resistance being given in Fig 12.8.

Ground losses

The radiation resistance of λ/2 dipoles above perfect ground is given in Figs 12.100 and 12.101. In Fig 12.100 the figures for horizontal dipoles are also compared [35] with measured values of feed-point resistance over typical ground, the difference between the curves being attributable entirely to losses in the ground, and Fig 12.101 compares horizontal with vertical dipoles. The mechanism of ground losses with horizontal wires close to ground was illustrated in Fig 12.95(a) and (b) and Fig 12.102 shows the derivation of a similar circuit for ground-plane antennas.

Ground losses are mainly associated with horizontal wires including radials and, as demonstrated by Fig 12.100, they are

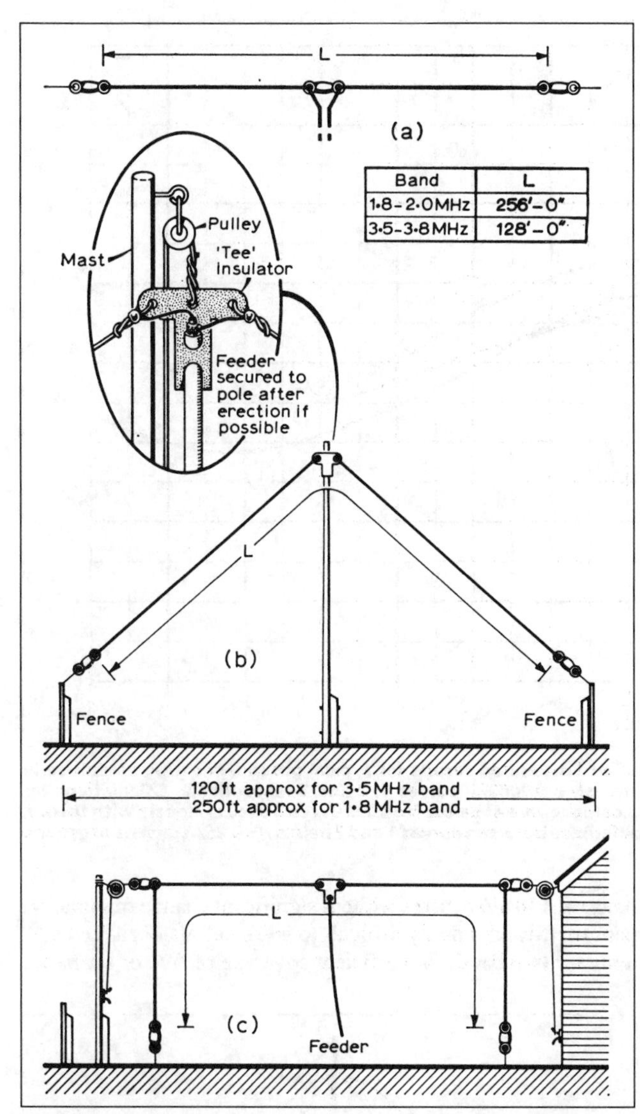

Fig 12.99. Dipole horizontal antennas for 1.8 and 3.5MHz. The current is low at the ends and contributes little to the radiated signal. If alternative use as a vertical radiator is required, the feeder must be brought down well away from the pole which should be non-metallic

the major factor in restricting the performance of horizontal antennas at low heights. It will be seen that as the height of a horizontal $\lambda/2$ dipole is reduced, the feed-point resistance departs significantly from the 'perfect ground' curve below about 0.17λ, levelling out at a typical value of 50Ω, and then starts to rise again. It will be seen that the rise is quite small down to a height of about 0.05λ but becomes increasingly steep with the approach to ground in line with theoretical predictions [35], reaching in practice a typical maximum value of 130Ω at heights of a few inches, whereas the radiation resistance (which in this case is equal to the feed-point resistance if there are no losses) tends to zero at zero height.

For matching one is concerned only with the feed-point impedance and the distinction has no relevance, but the effect of the ground losses on radiated power at antenna heights such as a few feet is catastrophic. The loss in terms of power effectively radiated can be estimated by comparing the upper and lower solid-line curves in Fig 12.100.

It is of particular interest to note that removal of losses should render the low-angle DX performance almost independent of height below about 0.2λ, since halving the height divides the radiation resistance by four so that the current is doubled and the low-angle field remains constant. This could be within the bounds of possibility, given that Fig 12.100 also includes a measured curve showing the effect of placing a ground screen under the antenna, the result being almost indistinguishable from the 'perfect earth' curve. For an antenna height of $\lambda/8$ the expected gain would be in the region of 6dB, a mesh size of 0.01λ being more than adequate [35] and 0.025λ only about 1dB worse. The size of screen is not stated but reference [36], also claiming that the rise in resistance can be prevented by a wire-mesh ground-screen, advises a distance of $\lambda/2$ in all directions. In the HF band the provision of such large screens would be no mean undertaking and the cost-effectiveness would be in some doubt. Nevertheless, in the absence of factual data relating size to efficacy, given also that such screens rely for their effect solely on the reduction of near-field losses in the ground directly under the antenna, these fields reducing rapidly in extent as the height tends to zero, much smaller screens might be worth trying. Evidence for this exists in the case of small loop antennas and radial systems so that it could be a fruitful field for the experimenter.

In the absence of such a ground screen, as most amateurs are all too well aware, horizontal antennas, even if they are at a height of only, say, 0.3λ, are much more likely to bring in the DX than when they are lying on the ground! This is consistent with the fact that although the ground reflection coefficient for horizontal polarisation approaches unity for *low-angle* propagation, it drops to 0.6–0.8 for vertical incidence, suggesting appreciable absorption.

Ground losses with vertical antennas

Ground losses are much less in the case of vertical antennas since the radiator is at right-angles to the ground surface and, comparing Fig 12.102 with Fig 12.95(a), it can be seen that C2 in the vertical case is likely to be smaller and much less height-dependent. Even in the case of a grounded monopole, the loss estimated from typical ground characteristics came to a mere 1.6dB; the starting point was a measured value of 130Ω (as in Fig 12.100) for the feed-point resistance of a $\lambda/2$ dipole at a height of an inch or so. Since as the height approaches zero there ceases to be any coupling between the two arms, the RF resistance to ground from its centre should be in the region of $130/4 = 32.5\Omega$, resulting in a loss of 2.7dB which can be roughly halved by using two dipoles at right-angles. The 32.5Ω figure was checked, and appeared to be little changed by burial of the 'dipole' although the resonance ceased to be measurable. The ground was light, sandy soil of high resistivity, and the experiment provided a good example of the absence of correlation between RF and DC resistance. In the case of above-ground radials, losses decrease very rapidly with height and the main height-determining factor is likely to be head clearance! These results were found to be independent of the length or number of radials which must be well-insulated and clear of any surrounding obstructions.

HIGH-FREQUENCY ANTENNAS

There is a wide choice of antennas for the higher frequency bands, ranging from simple horizontal or vertical wire types to various kinds of small fixed or rotary beams and, for the fortunate few with plenty of space, large arrays which enable high gains to be obtained in one or more fixed directions. The choice depends on individual circumstances and preferences including the desire for single or multiband operation, and the space available.

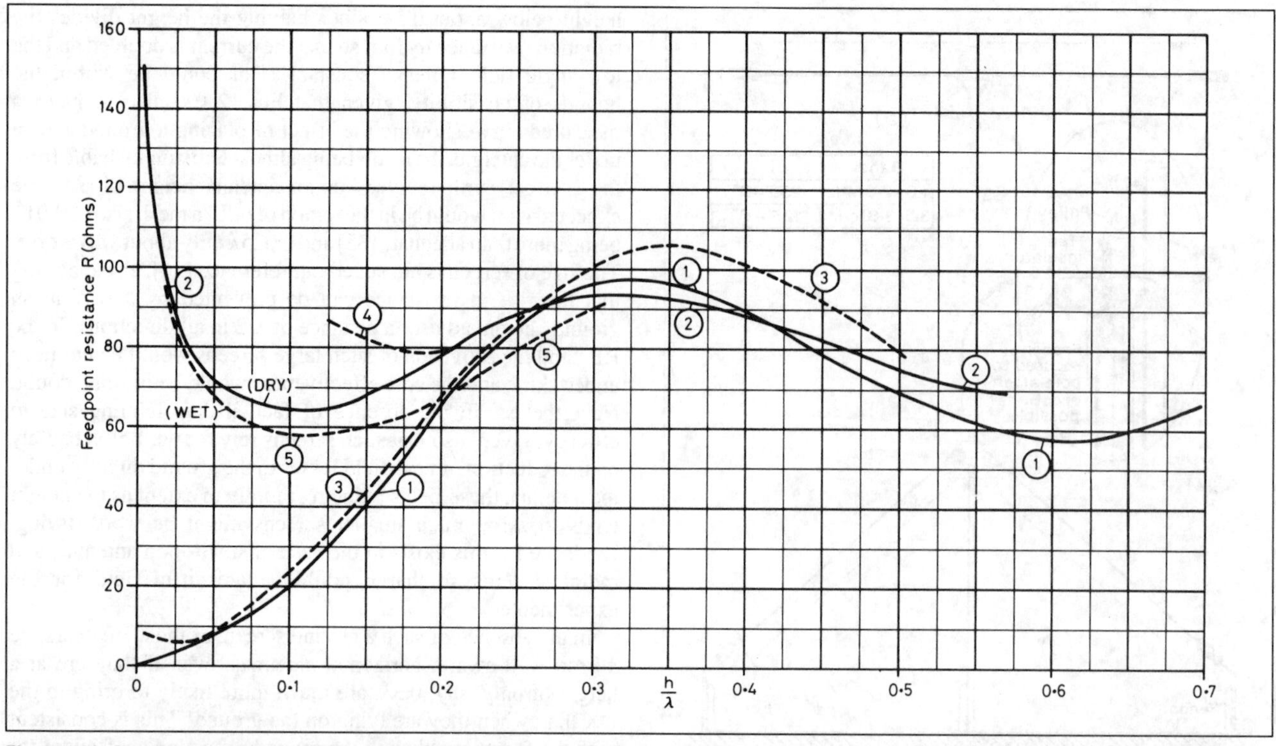

Fig 12.100. Comparison based on reference [35] between perfect ground and measured results for typical ground, showing steep rise in resistance at very low heights and the effect of a ground screen mesh (mesh = 0.0003λ): curves 1, 2 and 3 respectively. Calculations for dielectric ground (K = 5) show good agreement with curve 2 except as indicated (curve 4) below h = 0.2λ. Curve 3 checks closely with theory, putting K = ∞. Curve 5 illustrates the effect of wet weather. Note that the difference between curves 1 and 2 below h = 0.25λ consists of ground losses *(ARRL Antenna Compendium Vol 3)*

Simple wire antennas

Generally speaking, almost any length of wire can be tuned to work as an antenna on any band although, as we have seen, it is difficult to make short lengths radiate efficiently on the lower frequencies. Longer wires, on the other hand, tend to have rather complex radiation patterns on the higher frequencies but even in the best directions their gain performance for lengths up to 2λ rarely exceeds that of a dipole at the same mean height. The simplest forms of antenna include horizontal dipoles and loops, and these are capable of efficient multiband operation over a range of 1 to 1½ octaves without significant change of radiation pattern. It is not unduly difficult to work out matched feed systems for two bands, but efficient coverage of five or six bands

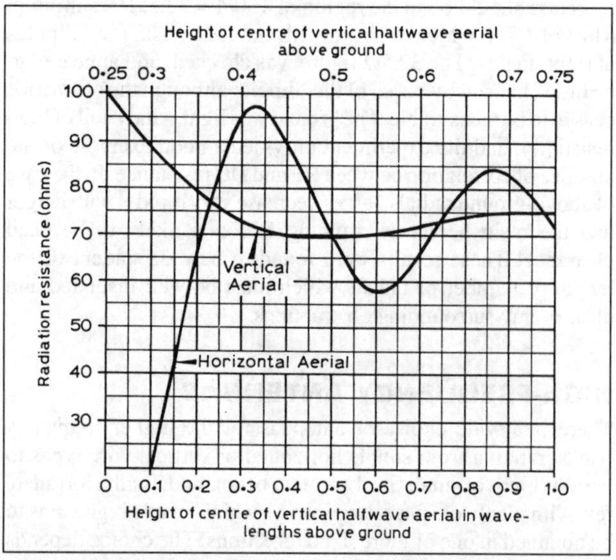

Height of centre of vertical halfwave aerial above ground

Vertical Aerial

Horizontal Aerial

Radiation resistance (ohms)

Height of centre of vertical halfwave aerial in wavelengths above ground

Fig 12.101. Radiation resistance of λ/2 horizontal and vertical dipoles as a function of height above a perfect earth

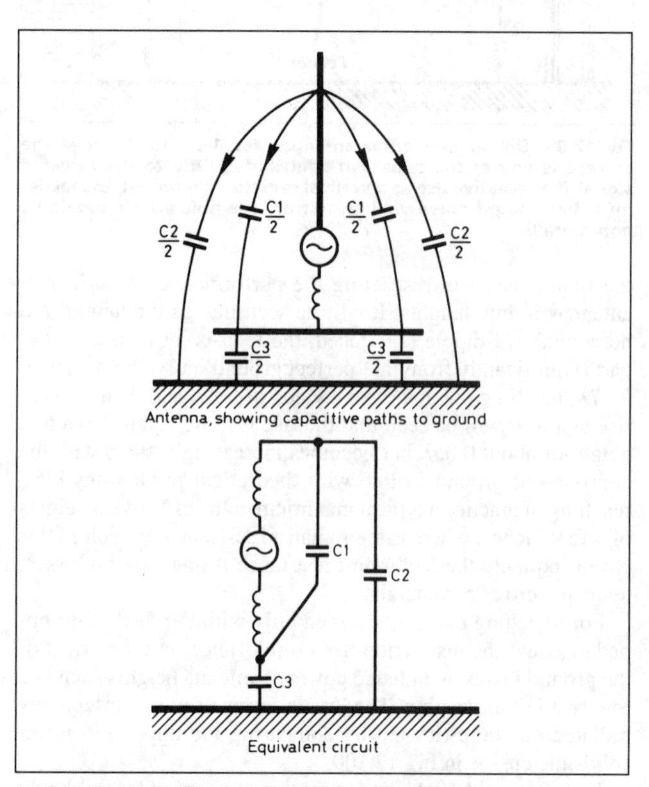

Fig 12.102. Equivalent circuit for ground-plane antenna with inductively loaded radials close to ground level

without retuning, rematching or some form of compromise is hardly conceivable. The great virtue, as we shall see, of resonant feeder systems is that all necessary operations can be carried out by the operator from the comfort of the shack; even in this case there is often a price to be paid in terms of bandwidth, leading to the need for more frequent retuning, though efficiency is much less likely to be affected.

When a simple wire antenna is long enough to cover lower frequency bands as well there is then a break-up of the radiation pattern at higher frequencies (p12.65) and the direction of maximum radiation on one band coincides with a null or poor efficiency on another, and if it is not desired to use resonant open-wire feeders with the possible disadvantages of high RF voltages around the shack, some more-or-less lossy compromise may have to be accepted.

Ideally all radiators, however simple, should be suspended as high as possible and in the clear, using heavy-gauge wire and good insulation, but usually there are practical constraints such as the need to use existing supports, eg trees or chimneys. The antenna may have to be more or less invisible, or the span available may be much less than $\lambda/2$ at the frequency in use. Fortunately many 'liberties' can be taken without compromising performance but a lot of conflicting ideas are in circulation as to what is or is not permissible. The following guidelines are aimed at providing optimum DX performance and, embody a lot of practical experience inspired or supported by theory, but use has also been made of experimental results which do not lend themselves readily to theoretical treatment.

Height should be as great as possible unless the ground is sloping and the antenna must obviously be orientated for the wanted direction. For efficient operation the length of the radiator must not be less than about two-thirds the normal resonant dimensions if loading is by inductance at a current loop, or one-third if end-loading by capacitance is used. When electrically short radiators are used, greatly increased care must be taken to keep voltage maxima (eg the ends of radiators) away from lossy insulating material such as bricks and trees etc, thicker conductor must be used and extreme precautions to avoid high-resistance joints at or near a current loop are advisable. As a rough guide it can be taken that the above figures for permissible shortening are valid for wire sizes of 12swg or thicker and spacings of not less than 3–4ft between the ends of the radiator and objects such as tree branches. There must be no other conductor within several feet of, and parallel to, the radiator. Regardless of antenna dimensions, no other resonant conductor should be within a distance of about 0.7λ or if it is in front, similarly oriented and in the direct line of propagation, at least 2λ and possibly 3λ; there appears, however, to be no available data on the effect of metal supporting poles, which should be regarded with obvious suspicion unless the antenna is symmetrical in the sense that fields due to any one part of it in the vicinity of the mast are cancelled out by an equal and opposite field.

The higher the radiation resistance, the smaller the permissible wire gauge and the less important the proximity of trees etc. A quad loop (or even beam) can be 'buried' in tree branches, and it has been found that twigs, even if wet, can make contact with a dipole at least half-way out from the centre without producing noticeable deterioration as judged by signal reports involving instantaneous comparisons between two antennas. Vertical radiators are, however, liable to be adversely affected by the close proximity of tree trunks and supports etc. Dipoles and full-size quad or delta loops may be constructed from very thin wire (as evidenced by Fig 12.5), provided the mechanical requirements are met. Insulation between points of high voltage must be good. Polythene cord or polystyrene rod are good insulators and adequate for most situations, but bakelite or Perspex are not.

Considerable departure from 'straight line' shape is usually admissible *provided* feeder balance is maintained and the *inverted-V dipole* (Fig 12.99(b)) is a particularly important example, being probably the easiest of all antennas to support in an average situation, since it needs only one main support (though if only a short span is available additional poles of lower height will be required) and the two ends can be allowed to 'droop' so that very little tension need be applied to the wire. It provides an excellent antenna for portable work since the centre can be supported from a single tree branch or by a 'clothes prop' constructed from bamboo garden canes, the ends being strung out to points as far away as possible using light polythene cord. 24swg or thicker wire is satisfactory for this application.

It is commonly thought that the inverted-V dipole is a compromise since it must produce vertically polarised radiation in an endwise direction, but although quite a strong ground-wave field can be measured in this direction it represents a negligible amount of wasted energy and the radiation pattern is hardly affected for apex angles down to 90°. This may be demonstrated by resolving each arm of the V into a pair of point sources of radiation as explained on p12.9, one horizontal, the other vertical, noting the resemblance of the two vertical sources to an '8JK' system, Fig 12.14. Intuitively one might expect the 'centres of gravity' of the current distributions to be about one-third of the way out from centre, but for greater precision it is not difficult to do a rough graphical integration to take account of the phase differences between the endwise fields produced by different portions of the current distribution. This gives the result that for currents I in the horizontal sources the 8JK system comprises currents $0.97I$ with a physical separation of 48°, and the field strength produced is 0.35 times that of the normal mode. Allowing 4dB for the gain in the 8JK mode, the power wasted by it is 0.2dB only. Except in this inefficient endwise direction, the polar diagram is almost identical with that of a horizontal dipole, and the assembly of two such elements into a beam will almost completely cancel the '8JK' radiation. At an apex angle of 90° the radiation resistance is halved so that the minimum satisfactory wire diameter is doubled. The bandwidth, however, is almost halved, whereas for 120° apex angle the reduction is only about 25%, so it is useful to keep the angle as large as possible.

If a horizontal dipole is suspended between two supports, the required wire gauge is determined by mechanical requirements and therefore dependent on the total span and on the weight of the feeder so that no general rules can be given, though 16swg is usually satisfactory assuming open-wire or other lightweight feeder. It should be noted that this method of construction requires two supports, the ends of the wire carry high RF voltages and should not be too close to these points, and there will be some sag in the centre where height is most needed. For comparison the inverted-V dipole loses effective height to the extent of about one third of the total 'droop', ie 4ft and 2ft 9in respectively for apex angles of 90° and 120° at 14MHz. Approximate resonant lengths for straight horizontal dipoles centred on the higher frequency bands are as follows:

Band (MHz)	Length
7	66ft 0in
14	33ft 0in
21	22ft 0in
28	16ft 0in

Resonant lengths are influenced by various proximity effects and also by bends in the wire. Fig 12.99 indicates considerable shortening in the case of an inverted-V; this is attributable to a change in Z_0 which, at low heights, occurs towards the ends of the wires due to the fact that they are much closer to ground than the wire centre. This effect is expected to be negligible for end-heights in excess of λ/4. It is probably best for the length to be determined experimentally, eg starting with an extra 5% the ends of the dipole may be folded back on themselves symmetrically until minimum SWR is obtained in the feeder.

Alternative methods of feeding are described on p12.36. To avoid Marconi effect (p12.49) it is usually reckoned that the feeder should be allowed to hang vertically from the feed point connection for at least a length equal to a quarter-wave in free space but, as will be appreciated from earlier discussion, p12.32, the incidence of this effect depends not only on the disposition of the feeder but also on its length and the connections at the transmitting end.

The Windom antenna

The principles of single-wire-fed (*Windom*) antennas, Fig 12.103, and the reasons for their obsolescence have been explained on p12.17. These objections can be overcome by balancing one Windom against another, but rearrangement of the wires as a collinear pair or extended double Zepp (p12.67) is thought to offer more advantages, being in general equally convenient and an ideal multiband radiator, whereas the multiband properties of the Windom are limited to even harmonics, and the double Windom requires a span of not less than λ at the lowest frequency.

Attention should however be drawn to the 'end-feed' version of the Windom based on Fig 12.103(b) as an alternative option; at a distance *l* from one end of a dipole one finds a resistance equal to the Z_0 of the feeder, the inductive reactance in series with it being tuned out by a small capacitor. Though suffering from the same basic defects (losses due to feeder radiation and ground currents, and possible EMC problems) it has important mechanical and visual advantages compared with centre-feed or its chief competitor, the Zepp feed (p12.67). Note also that the feeder can be any length including zero, in which case the antenna becomes end-fed with coaxial cable (Fig 12.103(c)) and at reasonable heights antenna current returns directly to the counterpoise, thus eliminating ground losses. An identical result can, alternatively, be arrived at by taking, say, a ground-plane antenna, lengthening the radiator by the appropriate amount, and tuning out the reactance.

It is interesting to note that short lengths of single-wire feeder can often be used to push relatively high RF voltages away from insulators, lead-ins, and switches as in the case of Fig 12.152. Fig 12.104 provides design data covering all of these applications.

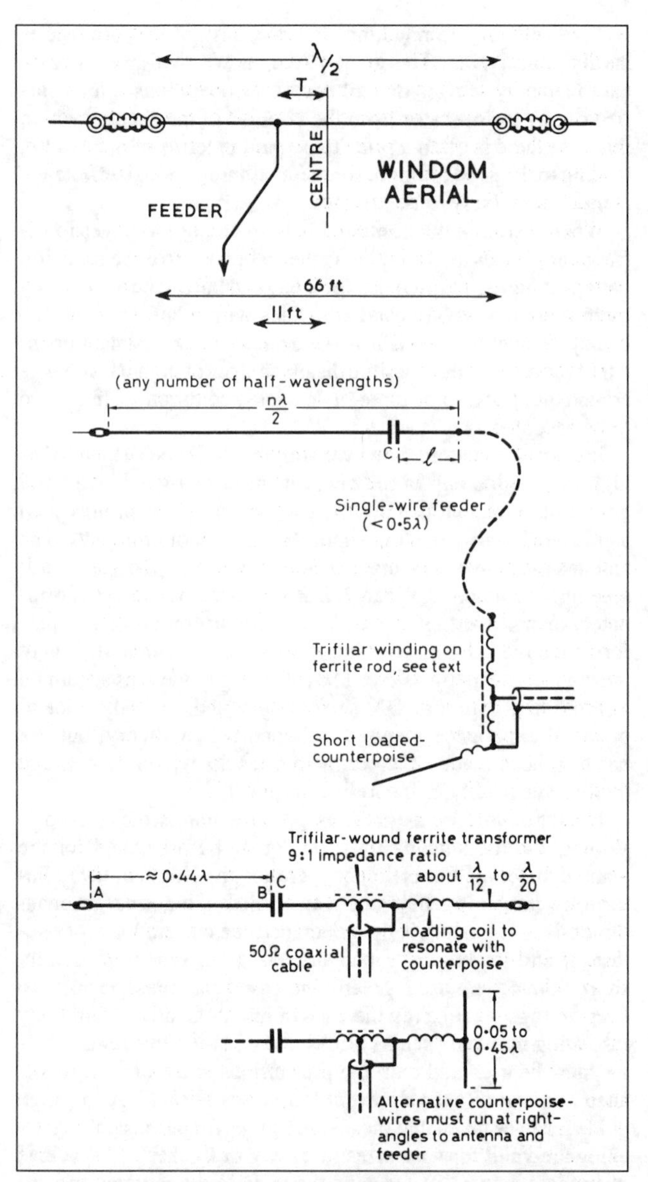

Fig 12.103. Single-wire feeders. The original version of the Windom antenna is shown at (a) with typical dimensions. (b) shows the end-feed equivalent, values of *l* and *C* being obtainable from Fig 12.104, and (c) shows two slightly different versions of the same antenna with the feeder length greatly shortened or reduced to zero, resulting in a 'dipole end-fed with coaxial cable'. For other applications of the same principle see Figs 12.96 and 12.155. Note that Fig 12.104 provides rough guidelines only, being based on a typical 'average' value of 500Ω for the Z_0 of single wires

Asymmetrical twin feed

It will be realised that a long-wire antenna, ie one which is two or more half-wavelengths long, can be fed with low-impedance twin line at any current-maximum position, such as the point a quarter-wave from one end (Fig 12.105). In this case 70–80Ω twin line can be used and matches the antenna well enough, though the impedance at such a point in the antenna is somewhat greater than that of a single half-wave. This can be done with a 7MHz antenna, and at 21MHz the feeder is again at a current maximum position. Other bands have an even harmonic relationship, so that the feeder is badly matched.

Long-wire horizontal antennas

The basic patterns of antennas of various harmonic lengths are shown in Fig 12.106 and further information is given in Table 12.4. It will be seen that there are two lobes in the diagram for each half-wavelength of wire, and that those nearest the end-on direction are always strongest. It should be remembered that these diagrams are sections, and the reader should try to visualise the lobes as sections of cones about the antenna. The reason for this concentration of energy in the nearly end-on direction is immediately obvious if the antenna is visualised as a string of point sources alternating in sign. In the direction of the wire there is of course no radiation, whereas at a small angle to the wire the radiation from all the sources adds up in phase. Viewed from a larger angle to the wire, for every current loop of one sign there is another of opposite sign and almost at the same distance, so that radiation tends to be cancelled. At a small angle to the wire each source is of course a very weak radiator and, although this is compensated by the large number of sources,

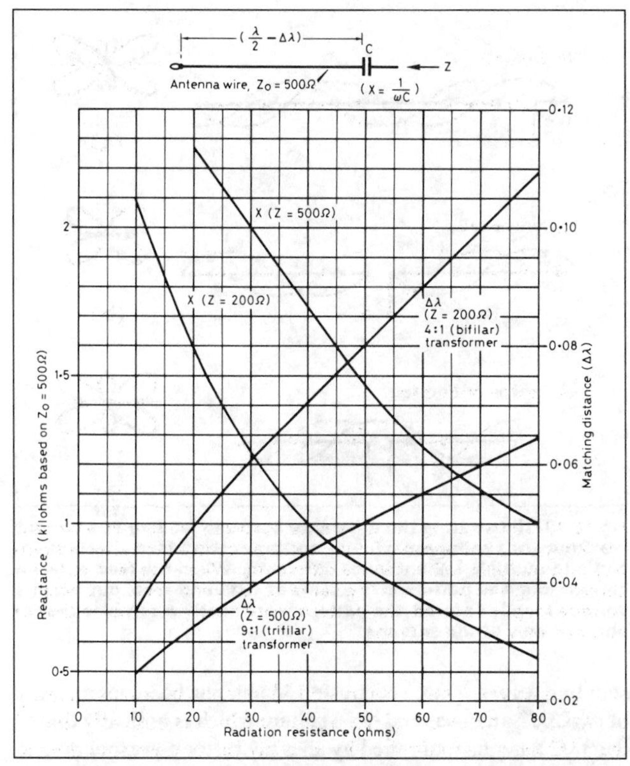

Fig 12.104. Matching data for single wires end-fed with coaxial cable. Radiation resistance refers to a current maximum. *l* is given in wavelengths and the reactance of C in ohms for two values of impedance, providing matches to 50Ω through 4:1 or 9:1 balun-type impedance transformers. Tuning is required in the case of the higher ratio, eg by varying the amount of core insertion (ARRL)

it follows that for a given gain the length of wire required is very much greater than if all the sources are assembled in a fully co-operative manner as they are with other types of array. The same applies to a wire terminated in a suitable resistance at the far end so that there are no reflections and no standing-wave pattern, except that in this case there is a travelling-wave and no radiation takes place in the backwards direction because radiation from any point is cancelled by radiation of opposite sign which started out half a cycle earlier but has half a wavelength further to travel.

Polar diagrams represent effectively the azimuth directivity of these antennas, but their general assessment must include the effect of height, and this can be done by considering them in relation to the vertical plane diagrams of Fig 12.25. There is, however, one very important feature: namely the existence of a major lobe of radiation at moderate angles of elevation in the endwise direction; thus the lobe at 36° to the wire in the 2λ pattern applies equally to the horizontal and vertical patterns and at this angle, depending on height, the ground reflection

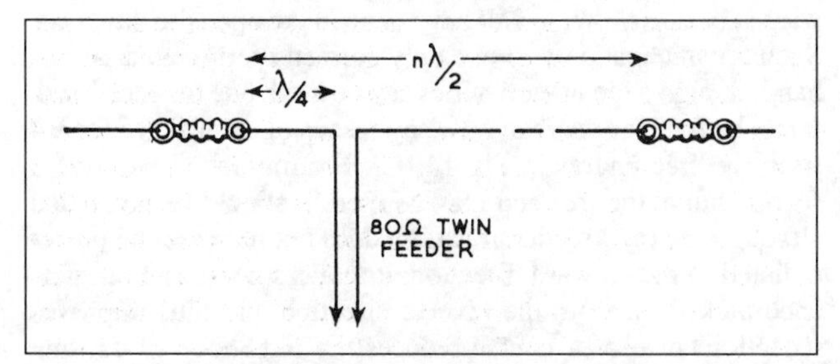

Fig 12.105. Asymmetrical twin line feed for a harmonic antenna

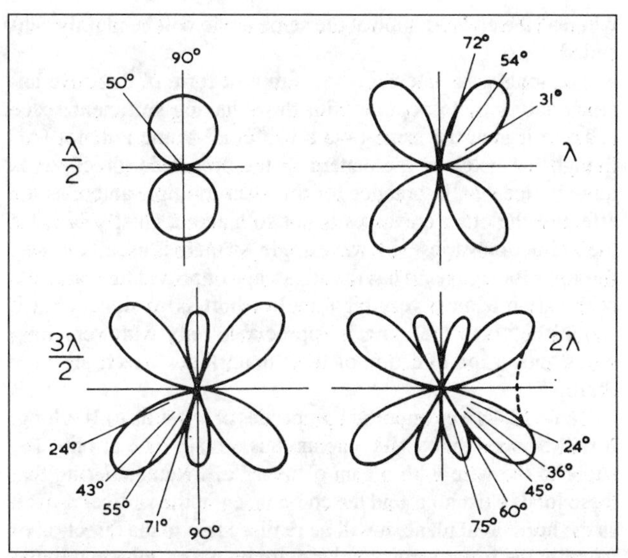

Fig 12.106. Theoretical polar diagrams for wire antennas up to 2λ in length. The angles of the main lobes and crevasses are shown, also the angles at which the loss is 3dB in the main lobe. The lobes should be visualised as cones about the wire. When the antennas are horizontal, radiation off the end can take place at useful wave angles, especially from very long wires, eg the broken line on the antenna. Details of long-wire antennas are given in Table 12.4

may reinforce the sky wave or possibly produce cancellation, causing the lobe to split into two.

For a dipole, assuming comparable ground reflection in both cases, the end signal will be 6dB (one S-point) less than the broadside at 30° elevation, but this effect is more marked with longer radiators, where the end pattern almost fills in for elevations of 15–30°, as shown for the 2λ case in Fig 12.106. In the case of a 3λ antenna (which would be about 100ft for 28MHz) the end radiation at a 15° wave angle may be much greater than the broadside signal from a dipole at the same height. It should be noted, however, that in going from broadside to high-angle end radiation the sign of the ground reflection coefficient is reversed, the latter radiation being mainly of vertical polarisation and the angle of maximum radiation being well above the Brewster angle except for very long wires. This means, for example, that for a height of λ/2 the broadside radiation at 30° angle of elevation will gain 6dB from the ground reflection

Table 12.4. Properties of long-wire radiators

Length (λ)	Angle of main lobe to wire	Gain of main lobe over half-wave dipole (dB)	Radiation resistance (Ω)
1	54° (90°)	0.4	90
1.5	42°	1.0	100
2	36° (58°)	1.5	110
2.5	33°	1.8	115
3	30° (46°)	2.3	120
4	26° (39°)	3.3	130
5	22° (35°)	4.2	140
6	20° (31°)	5.0	147
8	18° (26°)	6.4	153
10	16° (23°)	7.4	160

The number of complete conical lobes (see Fig 12.107) is equal to the number of half-waves in the antenna. The main lobe is the one nearest to the direction of the wire, and the figures in this table give its direction and gain. When a multiple full-wave antenna is centre-fed the pattern is like that of one half, but with more gain in the main lobe. The angles in brackets correspond to this case. When the antenna is terminated, or self-terminating, the radiation resistance is 30 to 50% greater, and the main lobe slightly nearer to the wire.

whereas the end radiation at the same angle will be mainly cancelled.

The long-wire antenna is the simplest form of directive antenna and is quite popular with those having sufficient space because it is at the same time a useful all-round radiator. Although the notch in the pattern in the broadside directions is quite noticeable, in practice for the even-multiple antennas the effect of the other crevasses is not so marked, chiefly because their direction varies with wave angle, so there is usually a 'way through' for signals. It has the advantage of providing some discrimination against very-high-angle (short-skip) signals but it should be noted that gain is appreciable only with very long wires and in the direction of their main lobes which are very sharp.

Table 12.4 gives important properties of wires up to 10λ long. It will be seen that a 10λ antenna has a main lobe at only 16° angle to the wire with a gain of over 7dB. Remembering that these lobes exist all round the antenna, eg in the vertical as well as the horizontal plane, it will be realised that in the direction of the wire there is a vertical polar diagram with a lobe maximum at 16° elevation. This angle of radiation may be quite close to the Brewster angle, in which case the reflection coefficient is small and, taking into account the gain and the vertical polarisation, a long-wire antenna is likely to give more radiation 'off its end' than a dipole at the same height. (It is this effect which sometimes causes confusion over the use of the term *end-fire* which refers to a particular type of array and not to the direction of a main lobe with respect to a wire.) It should, however, be noted that unless a wire is long enough to have appreciable gain it cannot differ much in main-lobe performance from a dipole having the same height and polarisation; this is because (as we have seen on p12.15) the effect of the ground is accounted for by an array factor which, though dependent on ground constants, polarisation, antenna height and angle of elevation, is independent of the number and disposition of the point sources of radiation comprising the antenna.

The radiation resistance (R_D) figures in Table 12.4 are the resistance at any one current maximum, say λ/4 from the end, and are representative free-space values, fluctuating somewhat with height. However, it is highly improbable that a long wire will behave in the same way as a dipole (Fig 12.101) since each current loop 'sees' many image loops, some of one polarity and some of opposite polarity, instead of seeing only one loop of opposite sign. From Table 12.4 it will be seen that the resistance increases steadily as the size of the radiating system increases; this is a general rule for all large directive antenna systems.

Effect of feed position

The general subject of how to feed energy to a long wire has already been discussed and it was indicated that it could be fed with a single wire near one end or preferably with balanced feeders in the centre. In each case the system is capable of multiband operation. A balanced line can be connected at any voltage or current maximum, for example λ/4 from one end, but in this arrangement the antenna works only on its fundamental and odd harmonics.

It should be noted that shifting the feed point from a current to a voltage maximum may produce an entirely different radiation pattern. The long wires described above were continuous with alternate positive and negative current loops along them. This is the situation with end-feed, single-wire feed, and feeding at any current loop, but if the antenna is fed with balanced line (Fig 12.107) at a voltage maximum an extra phase reversal is introduced at this point, as can be seen by sketching the

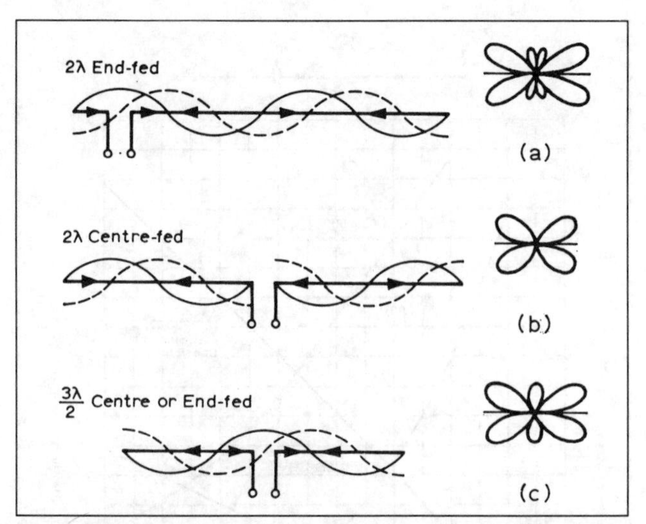

Fig 12.107. Illustrating the difference between feeding at a current maximum or a voltage maximum, and how centre-feed affects even- and odd-multiple λ/2 antennas differently. When the feed enters a current loop the pattern is the same as with end-feed, but when a voltage loop is entered, the pattern more nearly resembles that of one half only of the antenna

standing waves. Thus a centre-fed 3λ antenna becomes an array of two 3λ/2 antennas and has a pattern which is basically that of the 3λ/2 antenna multiplied by an array factor corresponding to the 3λ/2 spacing between the centre of the two halves. Since this particular array factor is very nearly the same as the pattern of each half, the final diagram is very like that of a single 3λ/2 antenna, but with emphasis on the broadside lobe. By comparison the 3λ end-fed antenna has a null in the broadside direction.

It should be noted that any asymmetrical feed arrangement will tend to be imperfectly balanced. The presence of Marconi effect should therefore be suspected and precautions taken as discussed on p12.31. Feeding elsewhere along the radiator is admissible only if it is separately resonated in each direction from the feed point and in addition carefully checked for balance [28, 29].

In practice a long wire fed at one end tends to radiate best from the opposite end, the pattern of a 2λ antenna tending to become like that of Fig 12.108 (these patterns are ideal patterns). This tendency is greater the longer the wire, and occurs because the lobes to the right (in Fig 12.108) are due to the radiation from the forward wave along the wire, while those toward the left can only be due to the wave reflected from the far end. Because of loss by radiation, the reflected wave is weaker than the forward wave; thus a long wire tends to behave like a lossy transmission line. The one-way effect can be enhanced by joining the far end of the antenna to a λ/4 'artificial earth' wire through a 500Ω (matching) resistor. This resistor must be able to absorb about 25% of the transmitter power if the antenna is only 2λ long, but only about 10% if it is very much longer. The power lost in this resistor is not wasted because it would all have gone in the opposite direction. Such an antenna is of course only correctly terminated on one band, though a fan of earth wires can be used, one for each band. It can be fed (for one-band use) by means of an 80Ω feeder λ/4 from the free end as in Fig 12.108. For multiband working a second fan at the free end may be used. It should be noted that although the unidirectional pattern does not increase the power radiated in the forward direction, it reduces noise and interference picked up from the reverse direction and thus improves reception. It also of course reduces the likelihood of causing interference to other stations during transmission.

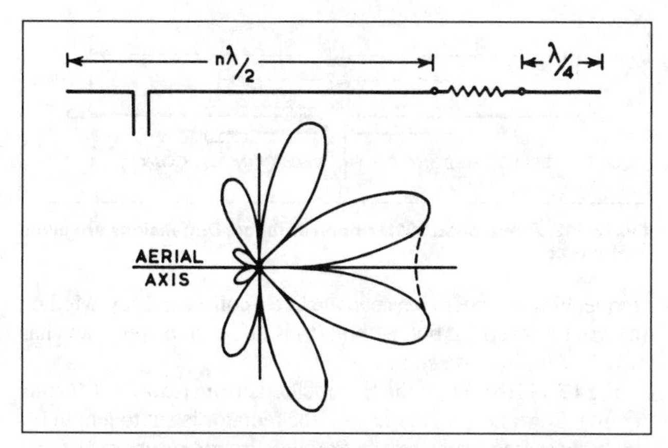

Fig 12.108. A 2λ terminated antenna and its radiation pattern, showing unidirectional effect. The end-to-end ratio depends on several factors but may be 3 to 5dB for a 1λ unterminated antenna and more than 10dB for a terminated antenna. The 500Ω terminating resistor is earthed to a λ/4 artificial earth. A multiple 'earth' fan may be used to cover two or three frequency bands. These antennas do not need critical length adjustment

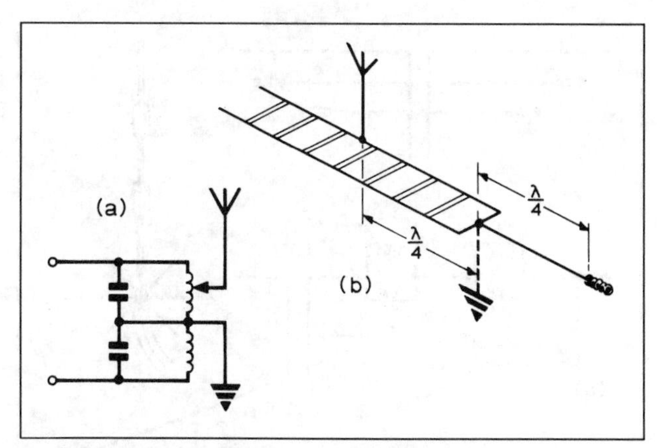

Fig 12.109. Unbalanced antenna – balanced line

The Zepp antenna

A simple wire antenna can be made to work on a number of harmonically related bands by feeding it at one end. This will always present a high impedance and can be connected to open-wire tuned lines. The most usual arrangement is similar to Fig 12.76 but with one half of the antenna disconnected at the top, creating imbalance. The feeder may be any length with series or parallel tuning in accordance with Fig 12.77 and various other matching options are available. This is known as *Zepp feeding* and featured in most of the textbooks but, despite its time honoured status, this arrangement as illustrated is very uncertain in its behaviour [8]. Put bluntly it usually *does not work,* and the reason is the same one that we have met before in a different guise. In going from a balanced to an unbalanced system *or vice versa* a balun is essential. Without one, there is no guarantee that the antenna will work better than a random length of wire nor, to be fair, any certainty that it will not work, but the Zepp feed has been found particularly uncooperative in this regard. A suitable balun consists of a λ/4 short-circuited stub, Fig 12.109. The end-fed antenna may now be seen clearly for what it is, namely a high impedance to ground connected across one half only of a resonant system which experiences relatively little damping so that the SWR is extremely high. The fact that it does not appear so with the conventional Zepp feed is a measure of the inefficiency of the system and, as reference [8] states, simple connection of the antenna to one side of the line will not work – it is necessary to add a 'transformer winding' in the form of a stub to 'tell' the line it is balanced.

An example of the balanced Zepp in use was provided by Fig 12.96(c): in this case there was no need for the stub as well as the trap and the Zepp feed appeared to be satisfactory, but apart from its adverse effects on bandwidth the balanced Zepp is somewhat cumbersome and the method of end-feeding with coaxial cable featured in Figs 12.96 and 12.103 is usually more attractive; the bit of radiator 'lost' in the matching process is sufficiently short, and the current in it low enough, to ensure negligible effect on the radiation.

The extended double Zepp

This antenna system is simple to erect and adjust and gives a gain of approximately 3dB over a λ/2 dipole, at the price of a rather narrow azimuthal beamwidth. The horizontal polar diagram, like that of a 1λ dipole, has main lobes which are at 90° to the run of the wire. In addition there are four minor lobes at 30° to the wire. At twice the frequency or more the pattern becomes similar to that of 1λ and 2λ antennas. The system is a type of collinear antenna (see p12.8).

The layout is shown diagrammatically in Fig 12.110 and comprises the two lengths of wire each 0.64λ long, fed in the centre with an open-wire line. Extension improves the SWR which is reduced to about 6. The table of Fig 12.110 gives the design length for the centre of different bands. A feeder length of 45ft is again a good compromise and the coupling arrangements described for the ordinary Zepp are equally suitable, but as this is a balanced system there is no need for the additional stub. For other lengths the form of coupling (series or parallel) can be found from Fig 12.77. In this case the length *L* should include the half-top of the antenna.

MULTIBAND ANTENNAS

In the average location, an amateur will not have space for many antennas, and thus the problem of making one array work usefully on more than one amateur band is an important one. Throughout this chapter, care has been taken to indicate the multiband possibilities and limitations of the various types of antenna, but compromises of one sort or another are inevitable unless the operator is content with monoband operation, and in planning an installation it is useful to have a wide range of options available; the antennas to be described in this section all include multiband operation as a main objective, and many of them are essential building blocks for multiband arrays to be described later.

Multiband dipoles and ground planes

One method of constructing a multiband dipole is shown in Fig 12.111(a). Dipoles are cut for the various bands and supported about one foot apart in the same plane, and then joined to a single

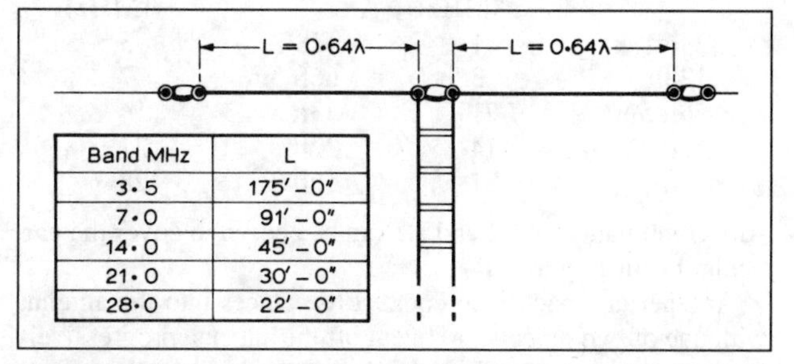

Band MHz	L
3·5	175′ – 0″
7·0	91′ – 0″
14·0	45′ – 0″
21·0	30′ – 0″
28·0	22′ – 0″

Fig 12.110. The extended double Zepp antenna employs centre-feed with tuned lines

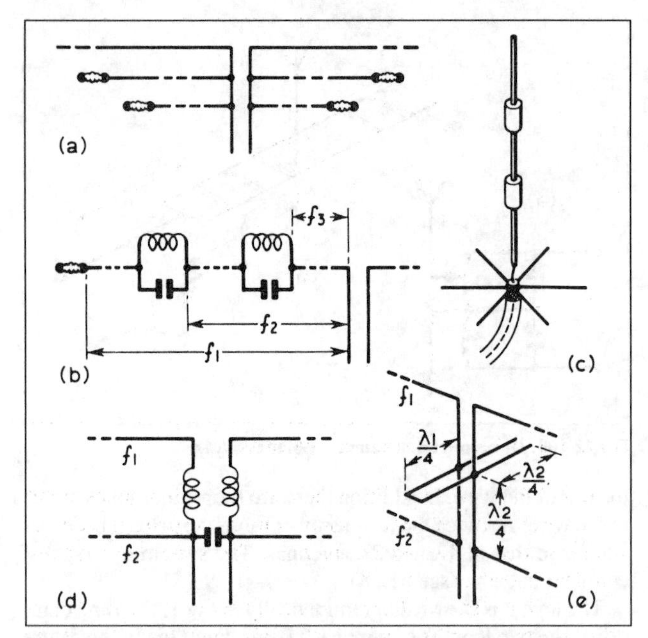

Fig 12.111. Multiband arrangements. (a) Parallel dipoles. (b) Multiband dipole. (c) Ground plane equivalent of (b). (d, e) Two 'frequency switch' filters for feeding two antennas over one transmission line

80Ω twin line. The theory is that the antenna which is in tune at λ/2 resonance takes all the power; just as the coupling between resonant circuits is relatively small when they are tuned to widely different frequencies so the coupling between the dipoles can be ignored as a first approximation and the power radiated from the 'wrong' dipoles will be a small fraction of the total, except in the case of a 7MHz dipole which will tend to produce a long-wire type of radiation pattern when the system is driven on 21MHz. Applications of this method are subject to the limitation that interactions between closely spaced wires, even when resonant frequencies are separated by half an octave or more, can be severe and unpredictable so that the use of a single wire tuned by a 'resonant' open-wire feeder is much to be preferred; this is the bases of the very popular G5RV antenna described below, Arrangements based on Fig 12.111(a), though not normally encountered as beam elements, have nevertheless achieved some popularity among those looking for a 'simple' antenna.

The 'ribbon antenna' shown in Fig 12.112 is a good example, capable of being fed with 75Ω flat twin or 75Ω coaxial cable. L1 acts as a λ/2 dipole on 7MHz and as three half-waves on 21MHz; L2 functions as a λ/2 dipole on 14MHz. The lengths of L1 and L2 can be calculated from the formula 468/*f*(MHz), the answer being in feet. Dimensions for L1 and L2 are as follows when cut for mid-band operation:

L1 length	Band (MHz)	L2 length	Band (MHz)
246ft	1.8	128ft	3.5
128ft	3.5	66ft 5in	7
66ft 5in	7/21	33ft	14
33ft	14	22ft	21
22ft	21	16ft 3in	28

Any combination of L1 and L2 can be chosen to cover the particular bands required.

Another approach is to connect reactances into the antenna with one of two objects: (a) to cut off the antenna progressively for each frequency band, so that it is a dipole for each band, or (b) to use the reactance as a phase changer, so that at the higher

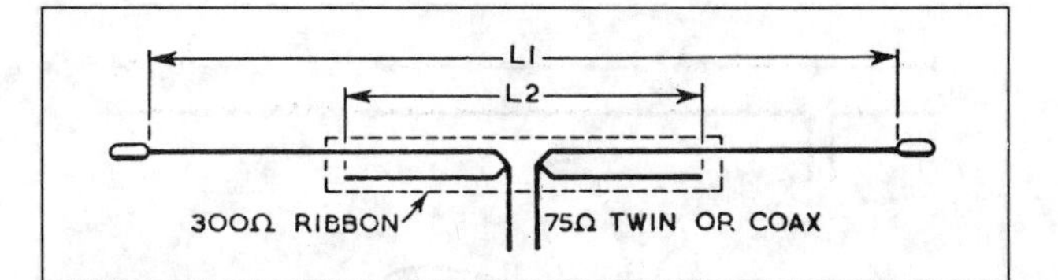

Fig 12.112. A two-band 300Ω ribbon antenna. Dimensions are given in the text

frequencies the extensions behave like a collinear array. Method (a) will be described but method (b) is difficult to apply, and has only had limited use so far.

Fig 12.111(b) shows the first method, using resonant *LC* units (*traps*). Starting from the feeder, the radiator is cut to length for the highest frequency, say f_3. Parallel circuits resonant at f_3 are then inserted, one for a ground-plane antenna, two for a dipole, and the antenna is then extended until it resonates at f_2, the next lower frequency. The procedure is repeated for a third frequency if required. Finally the lengths of the sections are re-trimmed for each band in the same order. For a dipole working on 7 and 14MHz, *L* and *C* values of 2μH and 50pF respectively are claimed to be suitable for the 14MHz traps. Taking just-acceptable losses as the criterion, these values should be more or less proportional to wavelength and to the characteristic impedance Z_a of the antenna, assuming the basic element to be a λ/2 dipole.

Table 12.5 gives examples of Z_a and the effective impedance R_e viewed between the *ends* of the dipole, this being equal of course to Z_a^2/R_r. Assuming a coil *Q* of 200 the two traps in series will have an impedance at 14MHz of 2 × 200 × 177 = 70,800Ω and $2R_e$ for 1¼in tubing is 5850Ω so that only 7.6% of power (ie 0.35dB) is lost in the traps. At band edges the traps present a reactive shunt of 14,000Ω and the SWR can be expected to rise to about 1.5 which is quite acceptable. In the case of a typical beam antenna, however, $2R_e$ is increased to about 24,000Ω and with the same traps the SWR would rise to about 6, causing considerable inconvenience even if the feeder cable and ATU are able to cope with the situation.

Sometimes damping is added to the traps to improve the SWR but, for the above example, this would cause a loss of about 4dB for an SWR of 2 at band edges. It is obvious therefore that coil *Q* must be kept high and the *L/C* ratio increased until the SWR is acceptable. At lower frequencies than their resonance, the traps provide an inductive reactance which operates like the coil in Fig 12.87(b) to reduce the length of conductor required for resonance and, for the above example, the overall length required to obtain resonance on 7MHz will be about 52ft.

This example may be adapted for other frequencies along the lines indicated, bearing in mind that the increased *L/C* ratios needed for beams will require additional shortening of the elements. In view of the complex electrical and mechanical requirements, the design of trapped beam antennas is not a task to be undertaken lightly and, if traps are purchased, it is important

Table 12.5. Characteristic and anti-resonant dipole impedances

Ratio *L/D*	Char imp Z_a	End R_e (λ/2)	Centre R_C (λ)	Typical antenna
16,600	1130	8750	13,500	66ft of 18swg
2500	920	5750	8900	16ft of 14swg
320	655	2925	4530	33ft of 1¼in dia tubing

This table gives the λ/4 characteristic impedance, Z_a, the end-impedance R_e of λ/2 dipoles and the centre impedance R_C of λ antennas in terms of conductor length/diameter (*L/D*) ratio. R_e is based on a radiation resistance of 73Ω, and the full-wave R_C on 95Ω per half-wave. Values may vary by 20% in practice due to environment.

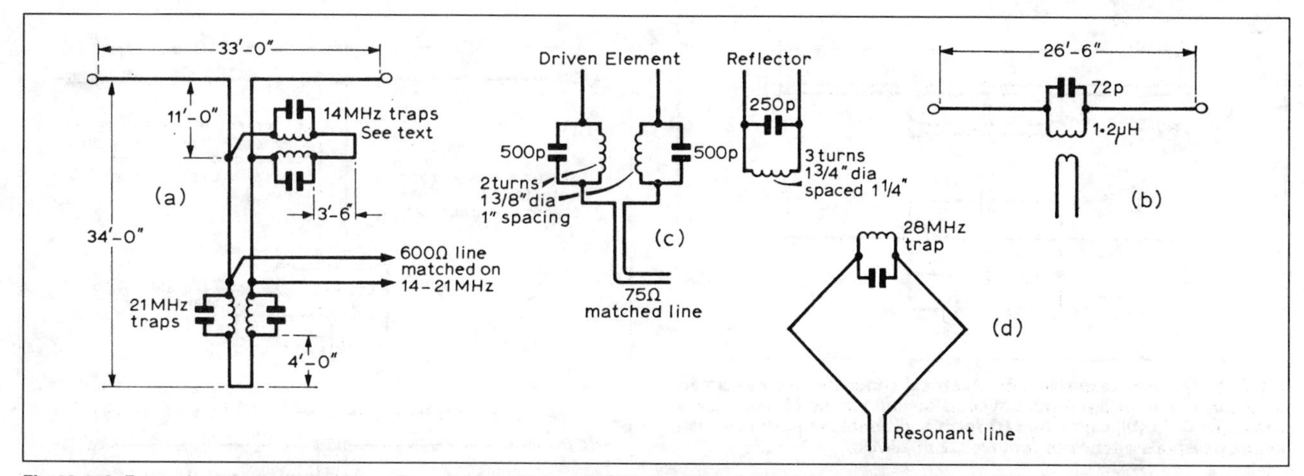

Fig 12.113. Examples of tuned circuits as frequency switches. (a) Illustration of principle enabling a dipole to be matched into an open-wire line on several wavelengths. (b) Shortened dipole which is resonant and matched on 14 and 21MHz. Component values depend on Z_a, being roughly correct for tubing of 0.7in diameter. (c) Driven element and reflector terminations for loops as Fig 12.146(a), 16ft 8in square, L = 70ft. This avoids the need for adjustment of stubs when changing bands. (d) Use of 28MHz trap in top corners allows a 14MHz quad to operate as a bi-square at 28MHz

that their design should be related to the particular application and their suitability verified along the lines indicated above.

Replacing the tubing assumed for the above dipole example by thick wire will roughly double the losses, but these still remain quite small. A suitable coil would comprise eight turns of 16swg wire 2in diameter and 2in long supported by polystyrene strips cemented in position. The coil should be mounted over a long insulator and the capacitor may be mounted along the insulator. The whole assembly should be tuned to resonance by adjusting the coil with a grid dip oscillator, and then sealed into a polythene bag. The antenna described above will also work quite well on 21MHz. The required voltage rating of the capacitor can be obtained from the anticipated power level and the antiresonant impedance which in turn depends of course on the antenna wire gauge. For this example a rating of 1000V should cover all eventualities up to the 'legal limit' of power.

The theory of operation described is approximate, and does not allow for the coupling which exists between antenna sections. It is therefore advisable to adjust the antenna during construction.

When the process is applied to a ground-plane antenna, Fig 12.111(c), the physical construction is simpler, because the antenna can be made of self-supporting tube. The tube can be broken and supported by rod insulators over which the coil is wound; the trap circuits can then be well protected by tape, or even encased in glassfibre. A 'whip' can be used for the top section. Separate ground-plane radials may be provided for each band.

When it is required to use two antennas on different bands with a common feeder, this can be done by inserting a low-pass filter immediately after the high-frequency antenna (Fig 12.111(d)). The filter has a cut-off midway between the two frequencies: at this cut-off frequency the reactance of the capacitor is equal to the impedance (Z_0) of the line, while each inductor has half this reactance. An alternative form of filter (Fig 12.111(e)) uses stubs connected a distance $\lambda/4$ from the first (f_2) antenna. One stub is $\lambda_2/2$ (ie $2 \times \lambda_2/4$) and is open-circuit; this therefore presents a short-circuit on to the line at f_2 and thus the second stub has no effect on the first. The short-circuit becomes a high impedance at the second antenna (f_1) by transformation. The second (shorted) stub tunes out the first at f_1, when the total length of the two stubs is $\lambda/4$. The length of line from the second stub to the first antenna is not important.

Fig 12.113 illustrates the use of tuned circuits as frequency switches. In Fig 12.113(a) a $\lambda/2$ 14MHz dipole is fed with open-wire line, and a matching stub for this band is adjusted in the usual way, after which 21MHz traps are inserted as shown. The stub will then require to be shortened slightly. The process is repeated for 21MHz, the stub for this band requiring to be lengthened slightly when the 14MHz traps are in place. The process may be extended to achieve a matched line on any three bands but thereafter the method gets too cumbersome to be recommended. Because the traps are bridged across 600Ω line instead of the high impedance of the previous example, losses in them tend to be negligible and their design is greatly simplified. At 21MHz the length of the dipole is of course $3\lambda/4$ and it has a slight gain, about 1dB.

Fig 12.113(b) shows a simple way of obtaining a compact two-band dipole; circulating currents are quite large, leading to increased losses and reduced bandwidth, but efficiency tends not to be seriously degraded except in the case of beam elements. At (c) resonators tuned to a geometric-mean frequency are used to bring resonances into coincidence. (This has been found particularly useful in multibanding by means of harmonic resonances which tend to be slightly separated as in the case of Fig 12.146). At (d) the tuned circuit is used to open-circuit the top of the loop at 28MHz.

The linear resonator

The tuned circuits in Fig 12.113(b) to (d) can be replaced by linear resonators [28, 29, 46], Fig 12.114. Efficient operation is possible on three bands because the values of capacitance required for 21 and 28MHz are too small to have any appreciable effect at 14MHz, the effective value of C at 21MHz being enhanced by the inductance of its leads. A smaller value is required for 28MHz so that for triband operation this needs to be switched.

Passive switching as shown at (b) has been used successfully but the linear resonator is no exception to the 'two bands easy, three bands difficult' rule which applies to most multiband systems other than those based on mechanical switching or resonant lines. The spacing between the parallel conductors is critical since it should not exceed about 10in for correct operation as described above and, if less than 5in, losses due to circulating currents may start to become appreciable and bandwidth is reduced. For the same reasons only 'full-size' elements are recommended, and spacing within the resonators must be accurately maintained by rigid construction or plenty of spacers.

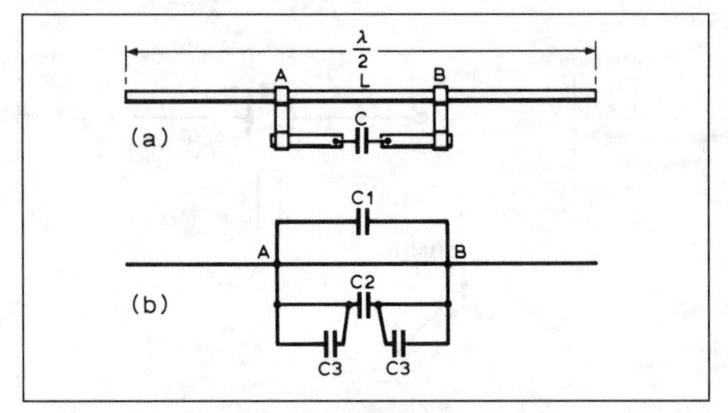

Fig 12.114. Linear resonator. An electrical conductor AB having an inductance *L* is bridged by a second conductor which includes a capacitor C. At (b), capacitors C3 form additional resonators which act as passive switches to remove C2 at 28MHz

For feeding driven elements, the methods shown in Fig 12.59 have been used successfully, with SWR values better than 2 in most cases, and better compromises may be possible. Based on Fig 12.113(d) a 14MHz quad was successfully used as a bi-square beam on 28MHz, giving additional gain though the bandwidth was rather narrow: about 300kHz. Advantages of linear resonators include the fact that they can be applied to any conductor without having to cut it, making them particularly suitable for use as traps to keep RF current away from places where it is not wanted, eg as in Fig 12.96.

The G5RV 102ft dipole

The 102ft dipole has been found an excellent compromise suitable for all HF bands and can be fed in either of the ways illustrated in Fig 12.115. In the upper diagram, tuned feeders (300 to 600Ω) are used all the way: in the lower version the high-impedance feeder is 34ft long and is connected into a 72Ω twin or coaxial cable. At this junction, the antenna impedance is low on most bands, as can be checked with the aid of Fig 12.77 for a length of 34 + 51 = 85ft.

It has previously been suggested that there is an optimum height for this antenna (ie λ/2 or 1λ above ground) but, although in some cases these heights may provide better matching, the G5RV is no exception to the general rule that antennas should normally be erected as high as possible.

On 1.8MHz the two feeder wires at the transmitter end are connected together or the inner and outer of the coaxial cable joined and the top plus 'feeder' used as a Marconi antenna with a series-tuned coupling circuit and a good earth connection.

On the 3.5MHz band, the electrical centre of the antenna commences about 15ft down the open line (in other words, the middle 30ft of the dipole is folded up). The antenna functions as two half-waves in phase on 7MHz with a portion 'folded' at the centre. On these bands the termination is highly reactive and the ATU must of course be able to take care of this if the antenna is to load satisfactorily and radiate effectively.

At 14MHz the antenna functions as a 3λ/2 antenna. Since the impedance at the centre is about 100Ω, a satisfactory match to the 72Ω feeder is obtained via the 34ft of λ/2 stub. By making the height a half-wave or a full-wave above ground at 14MHz and then raising and lowering the antenna slightly while observing the standing wave ratio on the 72Ω twin-lead or coaxial feeder by means of an SWR bridge, an excellent impedance match may be obtained on this band. If, however, low-angle radiation is required, height is all important and, as most cables will withstand an SWR of 2 or greater, any temptation to improve the SWR by lowering the antenna should be resisted.

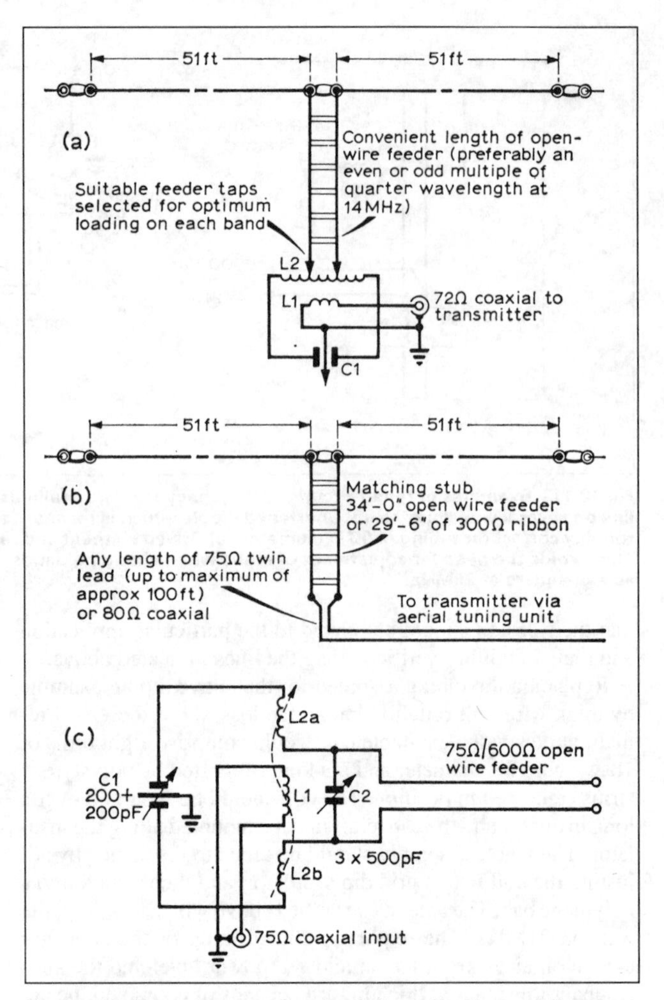

Fig 12.115. Two versions of a simple but effective multiband antenna for 1.8–30MHz. L1 is the coupling coil and C1, L2 form a resonant circuit at the operating frequency

On 21MHz, the antenna works as a slightly extended 2λ system or two full-waves in phase, and is capable of very good results, especially if open-wire feeders are used to reduce loss. On 28MHz it consists of two 3λ/2 in-line antennas fed in phase. Here again, results are better with a tuned feeder to minimise losses although satisfactory results have been claimed for the 34ft stub and 72Ω feeder.

When using tuned feeders, it is recommended that the feeder taps should be adjusted experimentally to obtain optimum loading on each band using separate plug-in or switched coils. Connection from the ATU to the transmitter should be made with 72Ω coaxial cable in which a harmonic suppression (low-pass) filter may be inserted.

With tuned feeders operation on the WARC bands and use of lengths other than 102ft is greatly facilitated, though some relatively high values of SWR can be expected unless multiple wires are used for the radiator as advocated on p12.72. It should also be noted that the radiation pattern is of the general long-wire type and the position of lobes and nulls will vary with length and frequency.

Loops as antennas and beam elements

Loops have a number of advantages over dipoles but also some disadvantages so that in choosing between them each case has to be considered on its merits. A number of different loops are featured in Fig 12.116 of which the quad will be the most familiar.

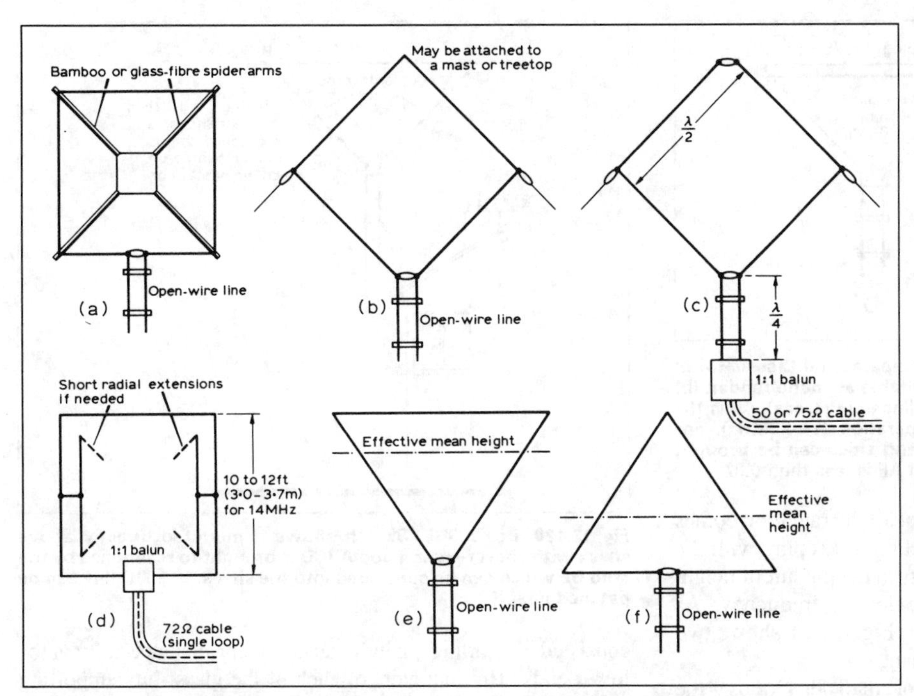

Fig 12.116. Loop antennas. (a), (b) are quad loops, (a) being most suitable for rotary beams and (b) more convenient for fixed arrays. At (c) the top of the loop is open-circuited to provide operation at twice the frequency with 3dB gain. At (d) loading by means of capacitive 'hats' allows considerable reduction in size. Delta loops are shown at (e) and (f) with dotted lines to illustrate the effective mean height; this is much lower in case (f)

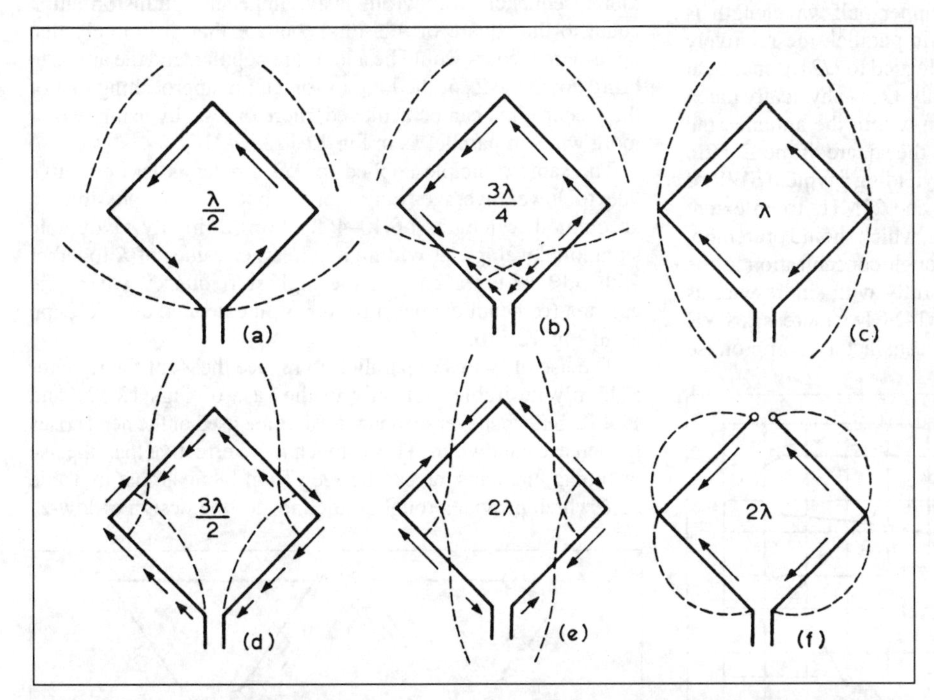

Fig 12.117. Current distribution in loops of various sizes at different frequencies. The arrow lengths indicate very roughly the difference in the relative contribution to the total field strength. Illustrations (a), (c), (d) and (e) represent a 14MHz quad loop excited at 7, 14, 21 and 28MHz respectively, while (b) and (d) correspond to slightly oversized 21MHz loop used at 14 and 28MHz. Illustration (f) shows the effect at 28MHz of open-circuiting the top corner of a 14MHz loop, thereby turning it into a bi-square element as in Fig 12.113(d); by using a linear resonator based on Fig 12.114 this effect can be achieved without disturbance of operation at 14 or 21MHz

The tendency to regard it as a stacked pair of dipoles is misleading, though a single loop does have a gain of about 0.7dB compared with a λ/2 dipole – see reference [29], p112. The possibility of fitting loops into much narrower spaces than dipoles is perhaps their biggest advantage, the width of the quad at its fundamental resonance being only λ/4. This is an important feature, though it is worth noting that only about 0.25dB of gain is lost by bending over the ends of a dipole to give it the same width. The quad has a larger value of *R*, which makes it more suitable for use with an open-wire feeder, but in taking advantage of this for multiband operation the useful frequency range is found to be more limited owing to pattern break-up at the higher frequencies, though this may not be important unless it is required for use as a beam element. At a circumference of 2λ there is no radiation normal to a quad loop, though in practice this has been found a highly effective radiator in the 45° directions, and even at 3λ/2 most of the radiation is still in the normal direction despite substantial sidelobes which indicate this as a desirable upper size limit for operation as a beam element.

The radiation pattern of quad loops can be visualised from the current distributions shown in Fig 12.117, and Fig 12.119 shows the variation of *R* with loop size and therefore frequency for the quad and several other shapes of loop. Fig 12.118 contrasts the classic form of delta loop antenna with the right-angled inverted delta, Fig 12.118(b), which is particularly interesting as the horizontal mode of radiation prevails over a relatively wide frequency range (well in excess of one octave) and for a given circumference it provides the largest values of *R*. The rapid decrease in *R* below a circumference of about 3λ/4 sets an approximate lower frequency limit for full efficiency as a beam element and for those of us unable to erect a 'big beam' small delta loops with circumferences of about 0.75λ at 14MHz such as Figs 12.120-12.122 have emerged as recommended options; in particular, used as beam elements they can provide coverage of all five bands from 14 to 28MHz with competitive performance and a minimum of environmental problems. With the same loop sizes efficiencies of over 50% were obtained at 10MHz by using heavier conductors and short feeders or special matching arrangements; likewise, scaling down the loop size by a factor of two-thirds resulted in a loss of only half an S-unit at 14MHz [28, 29, 38].

The small size of these loops offers a number of further advantages since it places current nulls in the middle of the sides

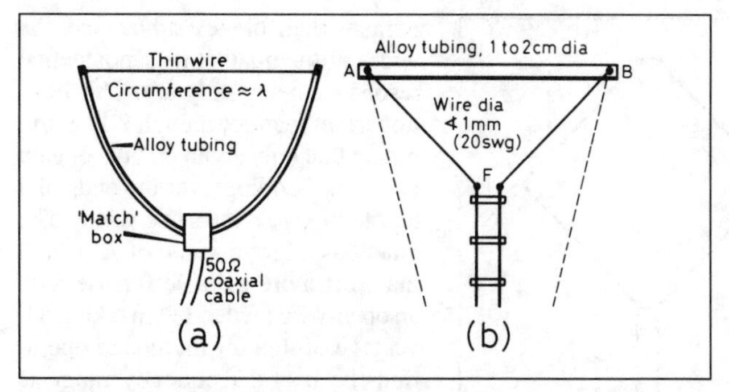

Fig 12.118. Delta loop – two versions compared. (a) Classic form: features include extra height but only suitable as monobander. (b) Small version for 'pushing up anywhere' (including trees). AB is 18–25ft, ABFA is 40–60ft for multiband operation (14–28MHz). For monoband operation, AB is 0.25–0.35λ and sides can be brought down steeply as shown dotted, provided AF is less than 0.3λ

at 14MHz (see Fig 12.120) ensuring that all the radiation comes from the top, and lightweight construction in keeping with the small size of the elements further assists in the pursuit of height which is particularly valuable at the lowest frequency. As alternatives to the 'stub substitute' DE, Fig 12.121 shows two ways of making small loops self-resonant.

As noted above, an important aspect of small delta loops is the relatively large value of R on the higher frequency bands which means that acceptable values of SWR should be obtainable on all bands subject to special provisions for the lowest band only. This can take two forms; in the case of the impedance-transforming loop (Fig 12.122(a)) the Z_0 of the upper half-wavelength is reduced by using two or more wires in parallel, the relatively short lower portion being inductively loaded to λ/2 resonance at 14MHz. Such a loop, though electrically 1λ, is physically much smaller but, due to transformer action within the antenna, the feed-point resistance is stepped up by the square of the Z_0 ratio for presentation to the open-wire line, yielding a typical SWR of only 3 or 4. SWR is degraded on 18 and 21MHz to an extent depending on the manner of the loading, which should preferably be distributed as in Fig 12.121(a), though concentration at the bottom corner has been used successfully with single-wire as well as ITL elements. Though ideal for 14MHz it increases SWR on 18 and 21MHz, and in one case a satisfactory compromise

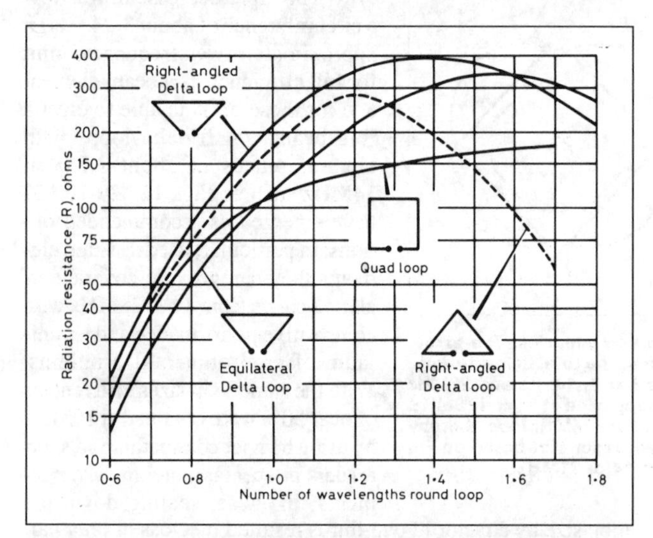

Fig 12.119. Variation of R with loop size. R is radiation resistance as perceived at top centre, and size is measured in wavelengths of circumference. Values are approximate, based on graphical integration as described in Chapter 2 of references [28] or [29]

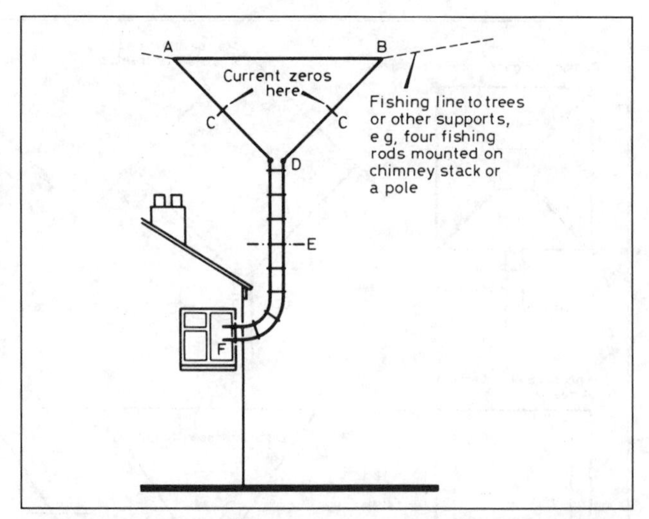

Fig 12.120. Basic SDL. Sketch shows it mounted directly above shack with short feeder. Loop ABCD is brought to resonance by the stub DE which can be continued into the shack, or 300Ω line can be patched in at E

consisted of winding the inductances over a distance of 2ft at the lower ends (1in diameter) of each of the glassfibre supporting arms.

Fig 12.122(b) represents a more basic approach to this problem. There is usually an impedance discontinuity at the feed point in the region of 2:1 and this can be relied on to produce, at one or more frequencies, a 'wrong way' impedance transformation equal to the square of this ratio; but for this, it is likely that open-wire feeders would be a lot more popular, and the aim is to bring down the Z_0 of the loop to something approaching that of the feeder. This can be achieved, more or less, by using two or more wires in parallel as in Fig 12.122(b).

The same principle applied to λ/2 dipoles is less effective due to lower average values of R but, if it is possible to accommodate lengths of 40–45ft (13m), a highly favourable situation develops, providing much larger values of R together with 3dB of extra gain at the highest frequency where the antenna (or beam element) becomes an extended double Zepp as in Fig 12.110.

The use of 'wires in parallel' to reduce the Z_0 of the radiator is highly desirable, not only in the case of Figs 12.121 and 12.122, but whenever resonant feeders are used or the need arises to increase bandwidth. This is much more effective than the use of tubing elements, as can be seen from the example in Table 12.6 which provides rough guidelines for the design of low-Z_0

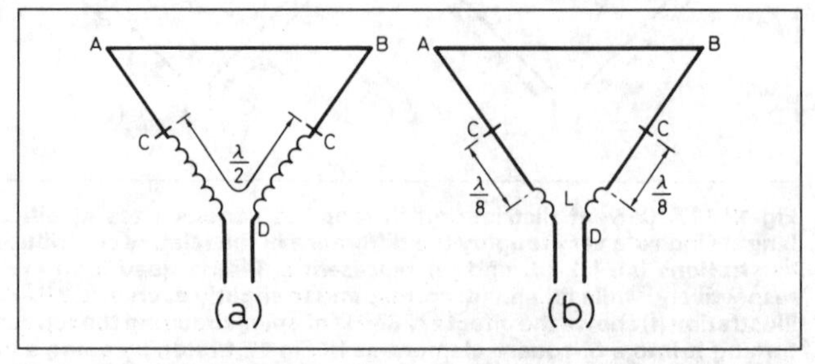

Fig 12.121. Two methods of making small loops self-resonant. (a) Helically loaded λ/2 dipole CDC has roughly the same Z_0 as CABC so that the impedance seen by the feeder is roughly equal to radiation resistance R. (b) λ/8 line CD results in an impedance $2R\,jZ_0$ at lower corner so that, putting $\omega L = Z_0$, the feeder sees a resistance $2R$. This method provides a useful impedance step-up of 2, but L can cause problems at 18 or 21MHz where the loop is nearly self-resonant. It helps to wind it in the form of short helices forming part of CD

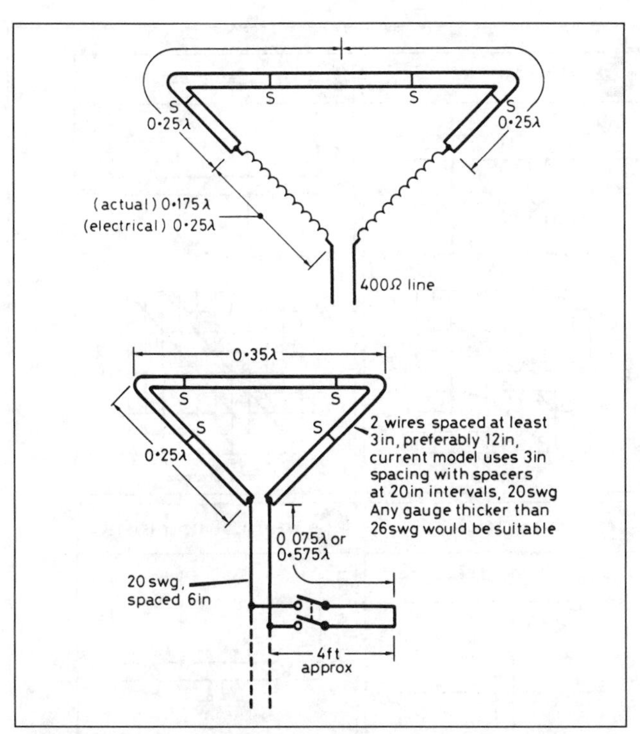

Fig 12.122. (a) Impedance-transforming loop beam element. Z_0 data for wires in parallel can be obtained from Table 12.6. For 14MHz about 60ft of 20swg copper will be required for helices. For the 400Ω line, 16swg spaced 1in is suitable. Shorts at S to be spaced no more than 13ft. Wire spacing can be reduced to 3in and any feeder impedance from 300–600Ω used with little adverse effect. (b) Constant-Z_0 version of (a). Switched stub improves bandwidth on 14MHz (the MDF), eg if system length above the stub equals the feeder length below it, bandwidth is doubled. (Not needed with short feeders.)

elements. However, no great accuracy can be expected, especially when wires are bent as in the case of loops; even so, designs based on this data have worked well in practice.

BEAM ANTENNAS

Beam antennas may be divided into two main classes: physically small beams using close-spaced elements and wide-spaced beams or long-wire systems. The small beams can usually, if desired, be built as rotary systems and are particularly suitable for general-purpose amateur use, whereas the larger beams, by virtue of narrow beamwidths and requirements for large areas of land, find their main application in commercial point-to-point circuits. The smaller types of rhombic, Vee, and multiple-dipole arrays nevertheless play a substantial role in amateur communication and provide outstanding performance when there is sufficient space available combined with a primary interest in a small number of fixed directions.

There are other ways of classifying beams, eg as collinear, broadside, end-fire (Fig 12.10) and long-wire. Small rotary beams are a special case of the end-fire category although, as explained on p12.10, there is a radical change in the mechanism of beam formation from additive to subtractive as the spacing of an end-fire array is reduced. Each of these types can further be subdivided into *parasitic* and *driven* arrays.

In broadside and collinear arrays the elements are connected together by phasing lines, so that they are all in one phase, but in the end-fire antennas the elements may be all connected to give a progressive phase change along the array (driven arrays) or there may be one driven dipole together with a number of

Table 12.6. Characteristic impedance of multiwire and helically loaded dipole elements

Number of conductors or, for helices, the approx no of turns per inch	Diameter (in)	(cm)	Spacing (in/cm) or, for helices, the resonant length (ft/m)		Z_0 (Ω)
1	0.04	0.1			1000
1	0.08	0.2			900
1	0.8	2			650
2	0.04	0.1	3	7.5	715
2	0.04	0.1	12	30	630
2	0.04	0.1	24	60	590
2	0.08	0.2	12	30	550
3	0.04	0.1	3	7.5	630
3	0.04	0.1	12	30	515
4	0.04	0.1	4	10	490
Helix, 4.8	1	2.5	7	2.1	2350
Helix, 3.5	1	2.5	7	2.1	1650
Helix, 2.5	1	2.5	15	4.5	1100

nearly resonant free parasitic elements which modify the local field of the radiator so that a unidirectional end-fire pattern is produced (parasitic arrays).

Parasitic reflectors can also be added to broadside and collinear arrays to produce a unidirectional beam instead of the fore-and-aft pattern of a single row of elements.

Driven arrays in general use resonant elements interconnected by tuned lines and are usually single-band, though by using traps or tuned feeders multiband arrangements are possible. Arrangement of elements (broadside or collinear) at the same height and along a line at right-angles to the direction of propagation produces a narrow azimuthal pattern, whereas if sufficient height is available vertical stacking can be used to reduce high-angle radiation without affecting directivity in the horizontal plane, though due to the reduction of mean height for a given mast height it may also reduce very low-angle radiation. On the other hand, end-fire systems provide gain by virtue of reduced beamwidth in both horizontal and vertical planes.

Simple arrays suitable for fixed-direction amateur use include the lazy-H described below and the two-element collinear array (which is in fact the same thing as a 1λ dipole), preferably with the addition of reflectors, directors or both. Two or more close-spaced beams, if widely spaced from each other, may also be connected together to form high-gain arrays.

Long-wire antennas as described on p12.65 can also be assembled into arrays such as rhombics and V-beams, which, despite the disadvantage of requiring a relatively large area of land for a given gain, have many advantages, being easier to erect and adjust besides operating into a matched line over a wide frequency range without readjustment. Coverage can include as many as four amateur bands, though the optimum performance is limited to a 2:1 frequency range. Their patterns are not so well defined as those of dipole arrays, a difference very noticeable in reception, and, like the end-fire arrays, their vertical and azimuthal patterns are determined simultaneously by the height and length. Long-wire arrays do not normally use reflectors but can be made unidirectional by terminating them with matching resistances.

Collinear and long-wire arrays have the disadvantage of producing a rather narrow azimuthal pattern which restricts the geographical coverage, a disadvantage in amateur work where even. a small country like New Zealand may cover nearly 90° of bearing, and over near-antipodal distances there is no guarantee that signals will follow a great-circle path. This disadvantage is, however, somewhat mitigated in the case of rhombic antennas by

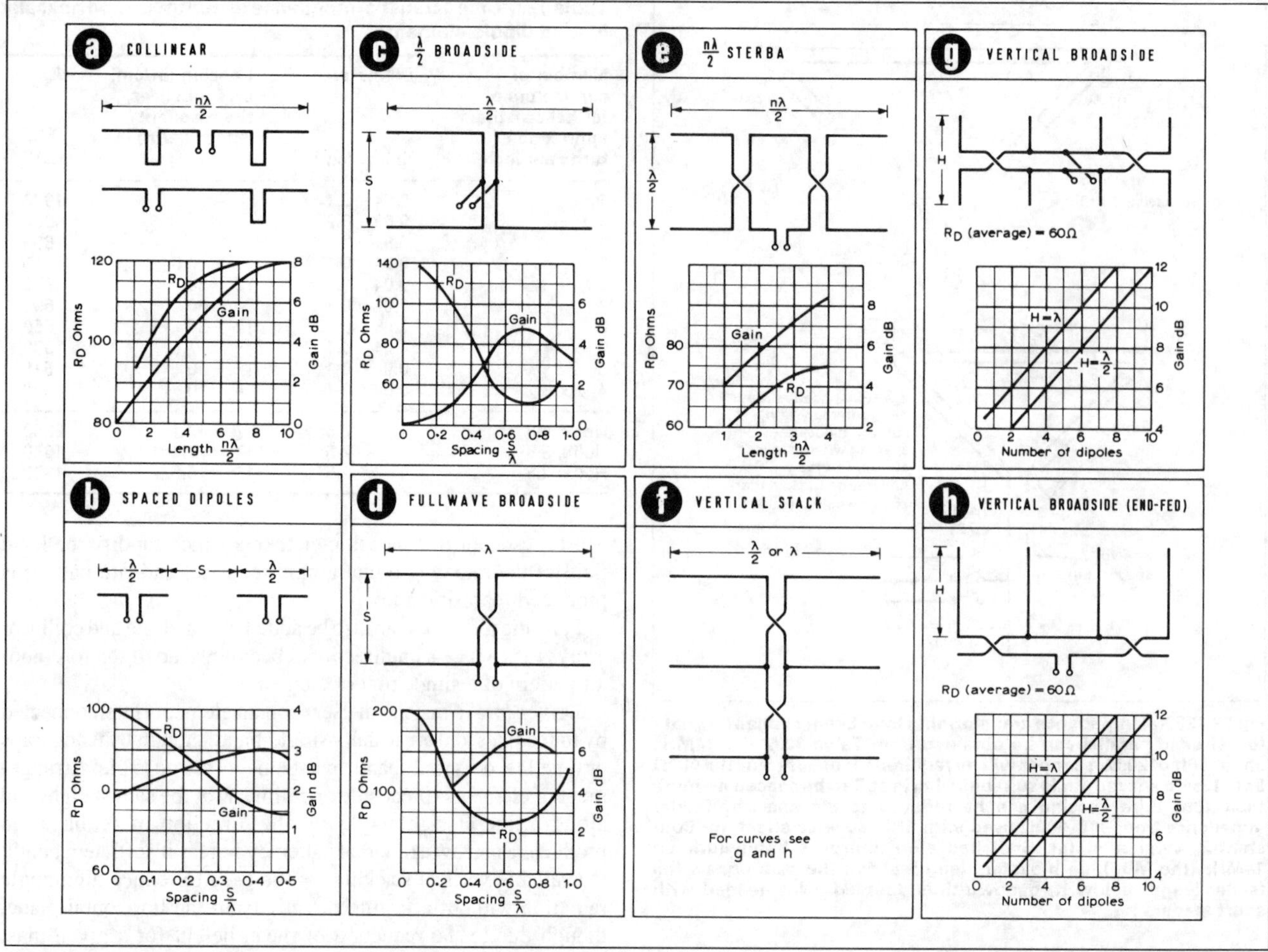

Fig 12.123. Four general types of broadside array. (a) Collinear arrays. (b) End-spaced dipoles. (c, d, e) Two-tier Sterba or Barrage arrays. (f) Pine tree or Koomans stacked horizontal λ/2 or λ dipoles. (g, h) Vertically polarised broadside arrays. Gain figures are with reference to a free-space dipole, in terms of spacing or total length in half-wavelengths. Resistance figures are average over the array, and are added in series or parallel according to the feed arrangements as described in the text. Various feed positions are shown, and details are also given in the text. The antenna in (c) can be arranged to give a broadside beam over a 2:1 frequency range, eg all bands from 14 to 28MHz. In (g, h) two half-waves in phase (*H* = λ) are reckoned as 'one dipole'

rather high sidelobe levels which often produce good signals in directions other than those of the main beam. This in turn is offset by the disadvantages that interference may be received from or caused in the directions of the sidelobes.

ARRAYS WITH WIDELY SPACED ELEMENTS

Broadside arrays – general

Fig 12.123 gives a selection of simple broadside arrays together with gain and average radiation resistance figures. A variety of feed connections are shown, which are interchangeable as discussed below, and it is apparent how to extend the arrays beyond the number of elements illustrated.

For all these arrays the azimuth patterns can be estimated as a function of the length of the array from Fig 12.124 and, in most cases, the broadside vertical patterns as a function of height from Fig 12.25.

In broadside arrays, the elements are all in phase and the interconnecting *phasing lines* or stubs must be adjusted to secure this condition. The spacing between elements and between the centre of the elements need not be one half-wave but can be varied up to about 3λ/4, beyond which minor lobes in the pattern become too large to ignore. The choice of spacing is, however, in practice dependent on the type of phasing line used. The position of the feed point depends on the input impedance required, or on requirements for multiband operation. A centre-feed position should be used if possible, especially in long arrays, because power is being radiated as the currents travel along the array, and the more distant elements may not receive their proper share. Uneven power distribution can cause the beam to broaden and 'squint'.

Collinear arrays

The collinear array is the simplest method of obtaining a sharp azimuthal pattern and is simply a row of λ/2 radiators strung end-to-end. To bring all elements into phase it is necessary to provide a phase reversing stub between the high-voltage ends of each pair, except where this position is occupied by the feeder.

The simplest form of collinear array is the centre-fed full-wave or *double Zepp* with a high-impedance tuned line feed at the centre and, as we have seen (p12.66), this can be used on other bands; when the total length is only one half-wave the impedance is between 60 and 100Ω but in the full-wave condition it is 5000Ω or more. When the frequency is raised to the value giving three half-waves the impedance is about 100Ω, while at two full-waves it is about 3000Ω. The high impedances can be lowered to between 1000 and 2000Ω by using a flat top of twin wires 2–3ft apart, joined in parallel, in order to improve the SWR or permit the use of a 300Ω line; this does not alter the low-impedance value.

Radiation is broadside for lengths up to about 1.3λ. At higher frequencies it becomes multi-lobed as shown in Fig 12.107(b).

Tuning and matching collinear arrays

In order to match a 1λ dipole, a λ/4 stub may be added at the centre and low-impedance line connected into the end of the stub, or 300 or 600Ω line tapped on to the stub (Fig 12.63(b)). The array needs tuning when a stub is used, and for this purpose the stub is made a little too long and a moving short-circuit provided. It may be possible to couple a grid dip oscillator into the bottom of the stub to find resonance. Approximate dimensions (in feet) are given by 470/*f*(MHz) for the radiators and 240/*f* for the stub.

If a low-impedance line is used the stub is left open, the line then being connected in place of the short-circuit and moved along to find the position for minimum SWR. Fig 12.123(b) is a variation in which the two halves are separated and is useful because the two equal-length, low-impedance feed lines can be connected in parallel, in or out of phase by means of a plug connector in the station, giving broadside or full-wave patterns at will. When the gap is λ/2 the pattern may be found by multiplying the λ-patterns of Fig 12.12 by sin ϕ where ϕ is the angle to the wire (see the section dealing with array factors).

Longer arrays may use λ/2 elements with tuning stubs and a feed point either at the centre of one element (current maximum) or at a phase-reversal point (voltage maximum). The impedance at the centre of any element rises rapidly with the number of elements but falls rapidly at the phase-reversal position, the two meeting at about 1200Ω which is the characteristic impedance of a typical antenna. These longer antennas should be tuned up section by section as described above, the total length of wire in feet between centres of shorting bars being 950/*f*(MHz).

Horizontal broadside arrays

These antennas, consisting of two horizontal arrays one above the other, give more gain than single arrays for the same width provided the height of the lowest dipole is not less than about half the height of the upper one. At low wave angles their azimuthal patterns are the same as for collinear arrays, but with λ/2 spacing the vertical pattern has only one main lobe whatever the height. In planning broadside arrays it is important to realise that at angles well below the main lobe of the upper dipole the lower element produces less field strength than the upper one by an amount which, assuming flat ground, is *proportional to the ratio of the heights*. If therefore it receives an equal share of the power the field strength will be less than if the upper unit is used alone unless the above 'half-height' rule is enforced. The benefit from the lower element is therefore unlikely to be appreciable, except at fairly high angles, unless the height of the lower element is at least λ/2, and decreases very rapidly with spacing between the elements, being only 1dB for λ/4 spacing even at the most favourable height.

Wide-band array

The simple array using two λ/2 dipoles cut for the lowest frequency (Fig 12.123(c)) can be used as a broadside array over a 2:1 frequency band, eg 14 to 28MHz. For this purpose the vertical spacing at the lowest frequency should be at least 3λ/8 wavelength. The phasing line should be 600Ω (not crossed) with the feed point at its centre. The SWR on 600Ω main feeder will be between 6 and 10 but can be improved by using two or more well-spaced wires in parallel for the radiating elements as

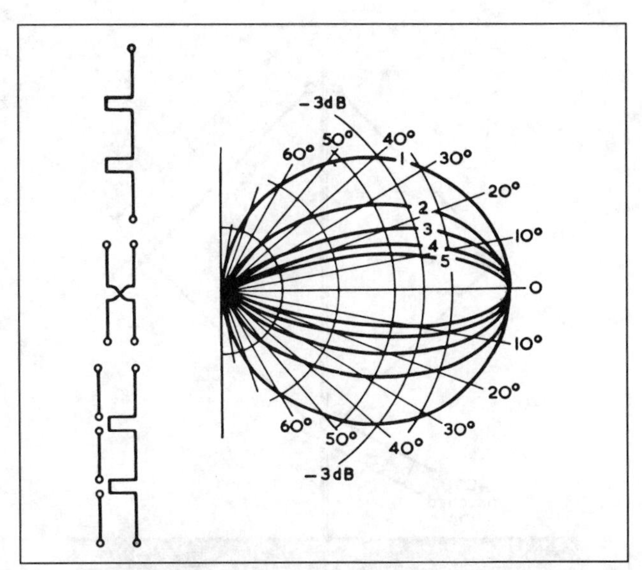

Fig 12.124. Horizontal polar diagrams of collinear and similar arrays for one to five λ/2 overall length, showing half-power points (–3dB). Without a reflector the patterns are bidirectional. With a reflector the patterns are only very slightly sharper, because the forward pattern of a dipole and reflector in this plane is very little different from that of a dipole. Minor lobes are not shown but should be –14dB or lower in a well-adjusted array. The pattern of a 1.25λ dipole is slightly sharper than that of a 1λ dipole but has minor lobes at –10dB level at about 60°. Vertical patterns of vertically stacked dipoles are also represented by the upper half of the above curves

explained above, There is nothing critical about any dimensions on this antenna. Methods of matching it to a line on any one frequency are considered later in this section.

Lazy-H

When the above array is one wavelength long it is called a *lazy-H*, the effective gain then being about 2dB higher. With the phasing line connected as in Fig 12.123(d) it must be electrically λ/2 long and crossed to restore phase. The impedance across the bottom end of the phasing line is then of the order of 3000Ω and a λ/4 stub transformer using 500Ω line provides a good match to a low-impedance twin feeder for single-band operation. On the other hand, if it is connected as in Fig 12.125 with a centre tap feed, there is no need to cross the line as the two branches are always in phase. In this case, if $S = \lambda/2$, the 600Ω line connecting the dipoles becomes a pair of λ/4 transformers which reduce the feed-point resistance to around 70Ω so that for single-band operation a low-impedance feeder can be used without additional transformation.

When a λ/2 broadside array for 14MHz, Fig 12.123(c), is used on 28MHz it becomes a lazy-H on the higher frequency

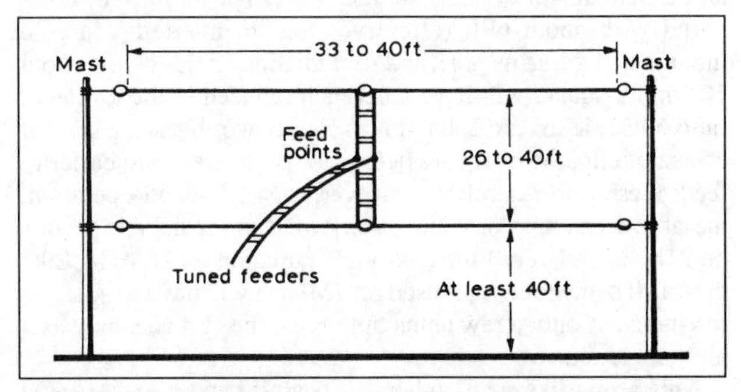

Fig 12.125. A lazy-H array for 14 to 28MHz. The dimensions given are a minimum for useful low-angle gain at the lowest frequency over flat ground

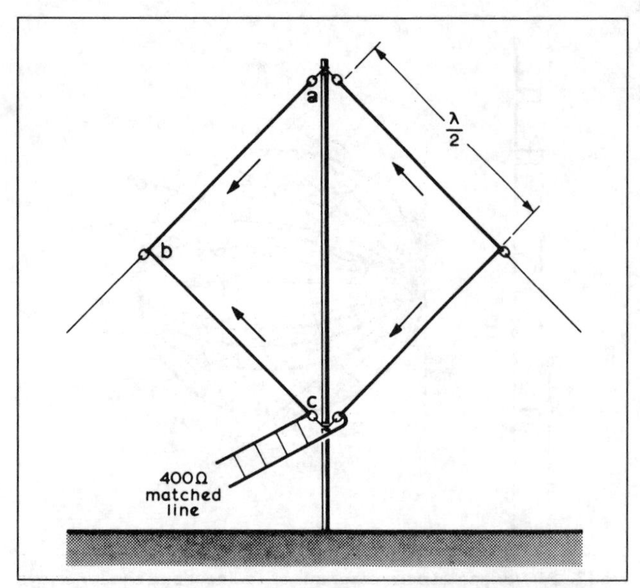

Fig 12.126. Bi-square antenna. Length of wire each side of the feed point is given by 960/*f* feet where *f* is the frequency in megahertz

and the gain is appropriately increased, ie from a matter of 2–4dB at 14MHz for the spacings suggested in Fig 12.100 to 6dB at 28MHz with an intermediate value at 21MHz. If the recommended spacings are not realisable, eg if the spacing is reduced to 15ft and the height of the lower element to 30ft, the array approximates to a single dipole at 37ft on 14MHz but the arrangement may still be considered worthwhile since most of the gain on 28MHz is retained.

The bi-square antenna

A development of the lazy-H known as the *bi-square,* Fig 12.126, is particularly attractive as it can be supported from a single pole, although the gain is somewhat less than that of the original. Two can be mounted at right-angles and switched to provide omnidirectional coverage, the antenna wires in this case acting also in part as guy wires. The radiation resistance is 300Ω so that it can be fed with either 300 or 600Ω line. The gain usually claimed is 4dB, but the radiation resistance in fact implies a gain of 3dB which is also the figure to be expected by extrapolation from the gain curves of Fig 12.123, taking into account the rather close spacing of the elements. The gain may be increased by a reflector or director, though problems then arise if it is desired to suspend two beams from the same pole. The gain, like the figures quoted for the lazy-H and other broadside arrays, is relative to a dipole at the same *mean* height, not the height of the top element; thus for a frequency of 14MHz and a pole height of 75ft, the mean height will be 52ft, compared with about 69ft (effective) for an inverted-V dipole mounted on the same pole, and for radiation angles below about 10° the bi-square will have a net gain reduced in the (voltage) ratio 69/52, ie to 0.5dB, but this is clearly worth having since it costs nothing, the bi-square being as easy to erect and easier to feed; these remarks relate, however, to single-band operation, the above example providing somewhat unpredictable results on 21MHz and very little low-angle radiation on 28MHz. Like the 14MHz inverted-V, if used on 7MHz it will have a radiation resistance of only a few ohms but unlike the V it acts only as a high-angle radiator.

This array, like many other HF beams, can be excited as a very efficient vertical radiator on the LF bands if the feed point is accessible or the feeder is brought down to, say, an accessible point near the base of the mast, where it can be short-circuited and energised via an appropriate ATU with suitable provision for bandswitching.

Sterba curtain

Longer broadside arrays have, of course, sharper patterns and greater gain, generally up to 4dB more than a collinear array of the same length. A six-element array with series feed (Fig 12.123(e)) has effectively the input impedance of six dipoles in series, say 500Ω, while an eight-element array fed at the base of the centre phasing line would resemble four full-wave centre-fed antennas in parallel, about 800Ω. In either case, the VSWR on a 600Ω line would be low enough without extra matching. The element length (feet) should be about 470/*f*(MHz) and the phasing line on these larger antennas may be 600Ω open wire or 300Ω ribbon of resonant length.

Phasing lines

When the phasing lines are part of a series connection (eg Figs 12.123(d) and (e)) they must be electrically one or more complete half-wavelengths long. When the length is λ/2 they must be transposed to offset the phase reversal due to the wave travelling along the length of line, but if they are one wavelength long the phase is restored and the lines are not crossed. However, if the feeder is tapped into the middle of the phasing line as in Fig 12.123(c) the current divides in phase regardless of its length and no cross-over is needed in that phasing line.

The velocity factor of open-wire lines is nearly unity and the length factor for a half-wave is 0.48–0.49λ; thus it is usually necessary to space the upper and lower rows by λ/2 in order to achieve a practical construction. When 300Ω ribbon is used advantage can be taken of its velocity factor of 0.8 and the antennas spaced vertically by a 3λ/8 wave with an electrical λ/2 of ribbon (365/*f*) or 3λ/4 with a full-wave of ribbon. On the other hand, 300Ω ribbon is not so good for small arrays because in such applications it carries a high SWR. The greater spacing gives about twice as much effective stacking gain for *the same mean height.*

Broadside verticals and stacks

Vertical patterns of these arrays are given in Fig 12.25 and horizontal patterns for two elements in Fig 12.12. The patterns are broader than those of the equivalent collinear arrays. The impedance of the individual dipoles has an average value of about 60Ω.

The stack of horizontal dipoles is really the same antenna rotated. Its azimuth pattern is that of a single horizontal element; the vertical pattern is not illustrated, but improves relative to that given in Fig 12.25(e) as the number of elements increases. In these arrays advantage can be taken of the velocity factor of 80Ω line (0.67) or 300Ω lines (0.8) to make the phasing lines one wavelength long (uncrossed) and increase the spacing of the elements to 2λ/3 or 3λ/4, but the 80Ω line should not be used with 1λ dipoles because of the high VSWR which would result.

End-fire arrays

The vertical broadside array, Fig 12.123(g), may be converted into an end-fire array by reversal of alternate feeder connections so that antennas λ/2 apart are excited in opposite phases. Due to the space separation between the elements their fields then add up in phase along the line of the array instead of broadside to it. This has the valuable feature of allowing the beam to be rotated 90° by a simple switching operation although, if there are many elements, the beams will be too sharp to provide all-round coverage without elaborate phase-shift networks. The end-fire connection offers the important advantage of broader azimuthal coverage, enabling the beam to be used for communication with much larger geographical areas, and it also allows the beam to be mounted horizontally though it cannot then be switched back to operate in the broadside mode. The gain of a *horizontal* end-fire array is virtually identical with that of a *vertical* broadside array but the vertical end-fire array with λ/2 spacing has considerably less gain, by about one-third for up to four elements. The same gain as before may be obtained from a given length, but only by reducing the spacing to 3λ/8 or less, ie increasing the number of elements.

When the spacing is less than λ/2 the phase shift between elements must also be reduced and this has the advantage of producing a unidirectional pattern but, being no longer a simple matter of phase reversal, gives rise to practical difficulties. This problem has been analysed in detail (pp12.10–12.13) for the relatively simple case of two close-spaced elements, leading to the conclusion that the best and easiest solution is to use 'parasitic' excitation of one element from the other. The simplest way out of this problem is to use parasitic excitation of the radiators, this arrangement being known as a *Yagi* after one of its inventors. Any number of parasitic elements can be used, though one or two are more usual; these can be supported from a single mast and rotated to produce some 5–6dB of gain in any desired direction.

LONG-WIRE BEAM ANTENNAS

The V-beam antenna

A long-wire antenna two wavelengths long has a lobe of maximum radiation at an angle of 36° to the wire. If two such antennas are erected horizontally in the form of a V with an included angle of 72°, and if the phasing between them is correct, the two pairs of lobes will add fore and aft along a line in the plane of the antenna and bisecting the V. Remaining lobes do not act in this way and so this provides what is essentially a bidirectional beam, although minor lobes will occur away from the main beam.

Fig 12.127 illustrates the principle. If the waves on the wires were visible, they would be seen to flow in the directions of the arrows, to appear in phase from the front of the array. Hence an anti-phase excitation is necessary, eg a balanced feed line at the apex.

The directivity and gain of V-beams depend on the length of the legs and the angle at the apex of the V. The correct choice of this angle also depends on the length of the legs which are likely to be the limiting factor in most amateur installations, and this is the first point to be considered in designing a V-beam.

The correct angle and the gain to be expected in the most favourable direction is given in Table 12.7.

The horizontal polar diagram varies with the leg length and is virtually bidirectional along a line bisecting the apex. The shorter lengths show quite a broad lobe which becomes rapidly narrower as the number of wavelengths in each leg is increased.

The layout of a typical V-beam is shown in Fig 12.127(b).

Probably the best way of feeding a V-beam is by the use of open-wire line as the impedance is high. If a V-beam is designed for a particular frequency it can then be used successfully at higher frequencies.

This antenna is too large for use by the majority of amateurs but where space is available it is attractive in view of its simplicity. Its other main limitation is the narrowness of the main lobes which restricts all-round coverage unless a sufficient number can be erected to cover all the main land masses.

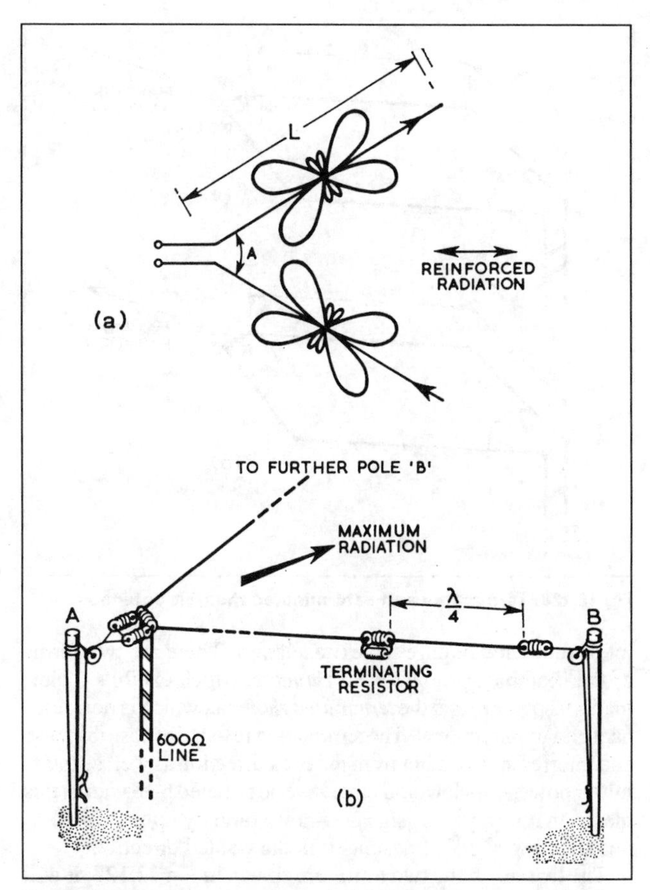

Fig 12.127. The V-beam antenna derived from two long wires at an acute angle *A*. Addition of resistors as shown in (b) results in a unidirectional pattern

The V can be made unidirectional if it is terminated, for example, with the artificial earth described in connection with Fig 12.108. A suitable value of resistor would be 500Ω for each leg. The input impedance, in the resonant condition, may rise to 2000Ω in a short V but will be between 800 and 1000Ω in a longer or terminated antenna and thus 600Ω feed lines can be used.

The rhombic

Early difficulties of terminating a wire high in the air led to the development of the *rhombic* antenna in which a second V is added, so that the ends can be brought together. The same lobe addition principle is used but there is an additional complication, because the lobes from the front and rear halves must also add in phase at the required elevation angle. This introduces an extra degree of control in the design so that considerable variation of pattern can be obtained by choosing various apex angles and heights above ground.

The rhombic antenna gives an increased gain but takes up a

Table 12.7. V-beam antennas

Leg length (λ)	Gain (dB)	Apex angle
1	3	108°
2	4.5	70°
3	5.5	57°
4	6.5	47°
5	7.5	43°
6	8.5	37°
7	9.3	34°
8	10.0	32°

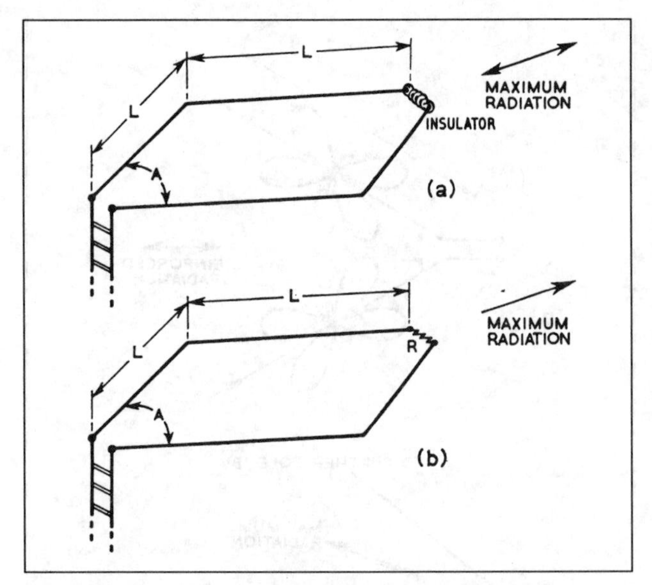

Fig 12.128. Terminated and unterminated rhombic antennas

lot of room and requires an extra support. There are two forms of the rhombic – the *resonant rhombic*, which exhibits a bidirectional pattern, and the *terminated rhombic*, which is non-resonant and unidirectional. The terminating resistance absorbs noise and interference coming from the back direction as well as transmitter power which would otherwise be radiated backwards; this means that it improves signal-to-noise ratio by up to 3dB without affecting signals transmitted in the wanted direction.

The layouts of the two forms are shown in Fig 12.128. It will be evident that the resonant rhombic can be considered as two acute-angle V-beams placed end-on to each other. Advantages of the rhombic over a V-beam are that it gives about 1–2dB greater gain for the same total wire length and its directional pattern is less dependent on frequency. It also requires less space and is easier to terminate.

The use of tuned feeders enables the rhombic, like the V-beam, to be used on several amateur bands.

The non-resonant rhombic differs from the resonant type in being terminated at the far end by a non-inductive resistor comparable in value with the characteristic impedance, the optimum value being influenced by energy loss through radiation as the wave travels outwards. An average termination will have a value of approximately 800Ω. It is essential that the terminating resistor be as near a pure resistance as possible, ie without inductance or capacitance – this rules out the use of wire-wound resistors. The power rating of the terminating resistor should not be less than one-third of the mean power input to the antenna. For medium powers suitable loads can be assembled from series or parallel combinations of say, 5W carbon resistors. Higher-power loads may be constructed from a number of Morganite Type 702 resistors.

The terminating resistor may be mounted at the extreme ends of the rhombic at the top of the supporting mast or an open line of 800Ω can be brought down from the top and the resistor connected across this at near ground level. The impedance at the feed point of a terminated rhombic is 700–800Ω and a suitable feeder to match this can be made up of 16swg wire spaced 12in apart. Heavier gauge wire will need somewhat wider spacing, which can be determined by extrapolation from the curves of Fig 12.31.

The design of the rhombic antennas can be based on Table 12.7, considering them to be two V-beams joined at the free ends. These figures will give a wave angle of the main lobe of approximately 15° in all cases, when the antenna height is one wavelength.

The design of V and rhombic antennas is quite flexible and both types will work over a 2:1 frequency range or even more, provided the legs are at least 2λ at the lowest frequency. For such wide-band use the angle is chosen to suit the length *L* at the mid-range frequency. Generally the beamwidth and wave angle increase at the lower frequency and decrease at the upper frequency, even though the apex angle is not quite optimum over the whole range. In general, leg lengths exceeding 6λ should not be used because the beam is then too narrow to cater for random variations in the direction of propagation which does not always stick precisely to the great circle or a particular angle of elevation. The vertical radiation pattern of rhombic antennas is modified by the presence of the ground in exactly the same way as that of other horizontal arrays, and for maximum very-low-angle radiation the height should be as great as possible.

A comparison with the figures given for tuned arrays in Fig 12.123 will show that, for the same total length of radiator, V and rhombic antennas give somewhat lower gain, but against this must be set the simplicity of construction, wide-band properties and ease of feeding. They are applicable to amateur communication primarily in those situations where there is special interest in a particular geographical area extending over not more than 10–20° on a great circle map, and where there is room to accommodate not less than 2λ per leg at the lowest frequency which provides a gain of 8–10dB on three bands, compared with 5–6dB for, say, a triband rotary beam of considerably greater complexity. The gain increases in proportion to leg length and the coverage angle decreases roughly as the inverse square root of the leg length.

CLOSE-SPACED TWO-ELEMENT ARRAYS

'Flat top' or W8JK end-fire array

These antennas use two anti-phase radiators in end-fire formation (Fig 12.129) and the theory has been explained on p12.11. Briefly, in any given direction θ the fields produced do not quite cancel because the distance from the two elements is slightly different by an amount proportional to d/λ and also to cos θ. The radiation pattern is therefore similar to that of a dipole multiplied by cos θ in both planes. From this there arises a theoretical gain of 4dB but unfortunately the extent to which cancellation takes place is quite impressive and a gain of 4dB, representing as it does the difference between two large quantities, implies very large circulating currents. This translates into a radiation resistance of only 8.5Ω per element or just over 4Ω for the two in parallel, assuming λ/8 spacing. Values for other spacings and for 1λ elements will be found in Table 12.8.

The high currents in turn imply high voltages and consequent losses in insulators and in the surroundings generally so that the theoretical gain is never reached. It has been claimed that gains up to 3.5dB can be obtained with suitable precautions which should include a somewhat larger spacing (say 0.15–0.20λ), use of heavy-gauge conductors, avoiding insulators as far as possible except near voltage nodes (though use of four end-insulators is unavoidable unless self-supporting elements are used), mounting in the open away from houses and trees, and using an open-wire matching stub as close as possible to the antenna so that high currents are confined to the elements themselves and the minimum amount of additional conductor necessary for matching.

The use of resonant feeders for multiband operation, though attractive in principle, is therefore not possible without large

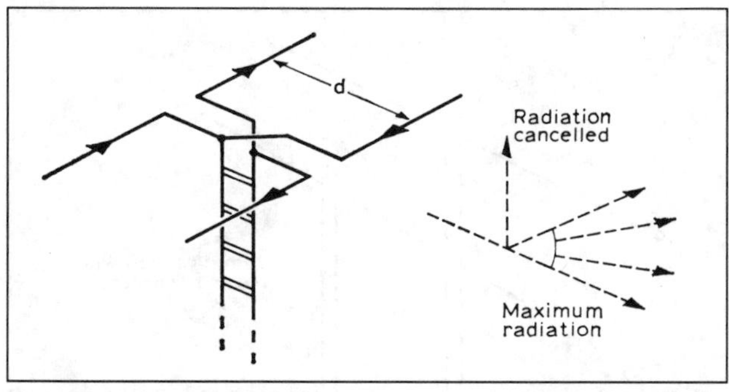

Fig 12.129. The basic W8JK flat-top array. The antiphase currents in the elements cancel any vertical radiation and reinforce in the plane of the array

losses unless the feeder length can be kept very short or the beam dimensions increased. The position may be eased considerably by use of folded dipole elements, accepting some restriction of multiband capabilities, though two-band 14/21 or 21/28MHz operation is feasible. For single-band operation the element lengths may be anywhere between about 0.5λ and 1.25λ, and are therefore not critical; shorter lengths involve even lower radiation resistances, whereas the longer length corresponds to the extended double Zepp, giving higher gain and increased radiation resistance. Even longer lengths, or harmonic operation of the longer length, are feasible subject to the acceptability of long-wire type radiation patterns. Radiation resistance increases nearly as the square of the spacing which may be increased to 0.25λ, accepting about 0.5dB drop in theoretical gain. At 0.5λ spacing the 8JK becomes a wide-spaced end-fire array with 2.5dB gain and a feed-point impedance in the region of 2400Ω for 1λ elements.

Phased arrays

As explained on p12.11 the performance of any close-spaced two-element beam may in general be derived by taking the 8JK as a starting point and assuming a phase shift between the elements comparable with that corresponding to their spatial separation. Such phased arrays may be realised in practice by driving both elements, or one element only with parasitic excitation of the other. First the driven versions will be considered, and Fig 12.130 shows how a generator may be connected to two equal resistances so that a phase shift is introduced without disturbing the equality of the currents. The impedance of the two arms is of equal magnitude, therefore the currents are equal, but the phases are shifted in opposite directions, each by an angle given by $\tan \phi = X/R$. For small phase shifts ϕ (in radians) is equal to X/R. In the case of a close-spaced beam the situation is in general greatly complicated by the mutual coupling but for

Table 12.8. Impedance values for W8JK antennas

Spacing S (λ)	$L = \lambda/2$ R_d	R_e	$L = \lambda$ R_d	R_e
0.1	6	40,000	10	50,000
0.15	12	20,000	20	25,000
0.2	20	12,000	30	16,000
0.25	33	7500	50	10,000
0.3	46	5500	65	8000
0.4	64	4000	100	5000
0.5	85	3000	125	4000

Approximate theoretical impedances in ohms at two points of a W8JK array. R_d is the impedance at the centre of any λ/2 element and R_e the impedance to earth of any free end. Figures are based on an antenna characteristic impedance of 700Ω. For typical wire elements R_e will be roughly doubled.

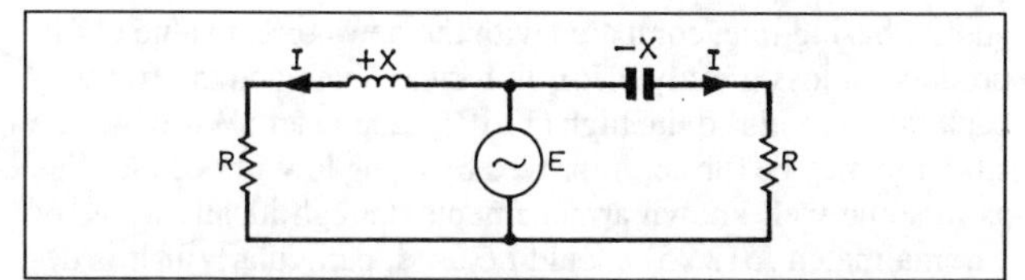

Fig 12.130. The ±X method of phasing. Total phase shift ϕ is given by $\tan(\phi/2) = X/R$ or approximately by $\phi = 2X/R$ radians if this is less than unity. For parallel λ/2 dipoles spaced λ/8 and fed out of phase, R = 137.5Ω, the mutual resistance being added to the radiation resistance. (In-phase connection is unusable because the resistances are then subtracted and X becomes too critical.) Note that the currents are equal only if the reactances are equal

the relatively simple case of λ/8 spacing the mutual coupling is a pure resistance R_m and the phase shift for each element is given by $X/(R + R_m)$ if the dipoles are connected in opposition, ie '8JK fashion', and $X/(R - R_m)$ if they are connected in parallel without phase reversal, ie if to get the required nearly antiphase condition we start from an in-phase connection. R is the intrinsic radiation resistance of each dipole considered in isolation.

These formulae demonstrate the need to use the antiphase condition, adjustments being more critical in the ratio $(R + R_m)/(R - R_m)$ or about 15 times if the in-phase connection is used. Fig 12.15 on p12.11 shows the relation between phase shift, gain, radiation resistance and nominal back-to-front ratio and we may decide from this to aim for example at a phase angle of 151°, ie starting from the 8JK, 14.5° or 0.2 radian each side. Therefore, $X - 0.2/(73 + 64) = 27.4\Omega$ which from the formula on p12.27 is equal to the reactance of 6in of open-wire line at 14MHz. In other words, to obtain the required phase shift we must first make sure the system consisting of the two resonators and the feeder between them is exactly resonant, and then shift the feed point *6in only* off-centre which subtracts the required reactance from one side and adds it to the other.

There are many types of driven array which use a phasing line, or unequal feeder lengths, to obtain the required phase shift, and are based on the assumption that phase shift is equal to the length of line expressed in electrical degrees. This is true only if each feeder is accurately matched at the feed point which is extremely difficult to achieve, particularly as the mutual coupling includes reactance (X_m) except in the case of straight elements spaced by λ/8; this reactance causes the radiation resistances to be unequal, a resistance $X_m(\cos \phi)$ being subtracted from one and added to the other. (It will be recalled from p12.11 that ϕ is the spacing expressed in electrical degrees, ie ϕ (in degrees) is given by $360\,(s/\lambda)$.)

The method described above is in fact identical with the phasing-line method with the addition of a rigorous procedure for determining the correct length *and position* of phasing line, the length for this particular example being only 1ft. For 45° phase shift this becomes 18in instead of the usual 8ft, and must also be disposed symmetrically with respect to resonance. Phase shift corresponds to line length *only* in the special case of perfect matching *at the antenna*, and matching of course depends in turn on the phase shift, so that any adjustment of one requires readjustment of the other.

To account for the occasional success of phasing-line methods, Fig 12.15 shows the critical dependence of radiation resistance on phasing which could therefore be used primarily as a matching adjustment; element currents will then be equal, there will be a phase shift, and therefore some gain and front-to-back ratio.

By using folded-dipole elements the situation is greatly improved and with the aid of Fig 12.15 it can be shown that use of an 8ft phasing line, compared with the new correct value of 6ft, produces a loss of only 0.7dB in forward gain and the front-to-back ratio remains quite high (17dB), despite an SWR of 4. The situation may be further improved by using low-impedance line as in some well-known arrangements, though ideally a Tee or gamma match (p12.35) should be used, particularly if it is desired to work at the point of maximum gain. The $\pm X$ method in conjunction with resonant feed lines permits simple and precise adjustment at any one frequency from ground level, or even in the shack, and also allows multiband operation but, like other resonant multiband feeder systems, has the disadvantage of being frequency-sensitive. However, acceptable band coverage can be obtained on 14/21MHz or 21/28MHz using folded dipole elements designed for 14 and 21MHz respectively, provided the feeder length does not greatly exceed $\lambda/2$ at the lowest frequency. Three-band coverage is feasible using single-wire 14MHz dipoles or 21MHz dipoles (single-wire or folded) but these designs are compromises involving bandwidth restriction or excessive spacing with some reduction in gain, or both.

There appears to be no reason to maximise the 'nominal' back-to-front ratio, since this does not give the best average discrimination against unwanted signals [28, 29], taking into account all directions, and it is better to operate closer to the condition for maximum gain; on the other hand, larger phase shifts produce greater bandwidth, sometimes assist matching, and are helpful in the design of miniature beams.

The $\pm X$ method of phasing has been applied in a number of different ways to quad and dipole elements, including the Swiss quad (see below) which uses loops of slightly different sizes. The simplest arrangement is almost identical in appearance with the 8JK, Fig 12.129, the resonance requirement being met by shortening the elements to 27ft 6in with 8ft 4in spacing. The shortening of the elements unfortunately almost halves the radiation resistance, making it advisable to operate with phase shifts well in excess of those required for maximum gain, optimum results being obtained with the main feeder displaced from the centre of the system by about 4½in in the required direction of fire, corresponding to about 135° phasing. A 168pF capacitor bridged across the 600Ω line at 2ft from the feed point acts as a suitable 'matching stub' but an SWR of at least three must be expected at band edges, narrow bandwidth being of course a disadvantage of all centre-loaded short-dipole systems. If the antenna is not correctly resonated, displacing the feed point causes more current to flow to the shortened (capacitive) side if the overall length of the elements and connecting feeder is too great, and the longer (inductive) side if the length is too short. Tuning can therefore be checked with any meter capable of indicating relative current, eg Fig 12.86).

An improved version results in lower losses, wider bandwidth, and operation on any pair of frequencies separated by about half an octave, such as 14 and 21MHz. It uses folded dipole elements of standard length for the lowest frequency, each fed with $\lambda/2$ of resonant feeder, the feeders being connected in parallel antiphase. For 14MHz the elements may be spaced about 9ft and fed by connecting a 50Ω line (balun essential if using coaxial cable) to points about 2ft either side of centre, permitting beam reversal by means of a switch or relay as in Fig 12.131. For operation on 21MHz the feeders require shortening by 4ft in the case of 6in spacing between the wires of the dipoles, or 3ft 6in with the dipoles 'opened out' to an average spacing of 3ft. The same feeder can be used, connected 14in off-centre. Shortening may be carried out electrically by series tuning with two centrally located capacitors of 20–25pF, and in this case it has been found possible to obtain points for feeder connection

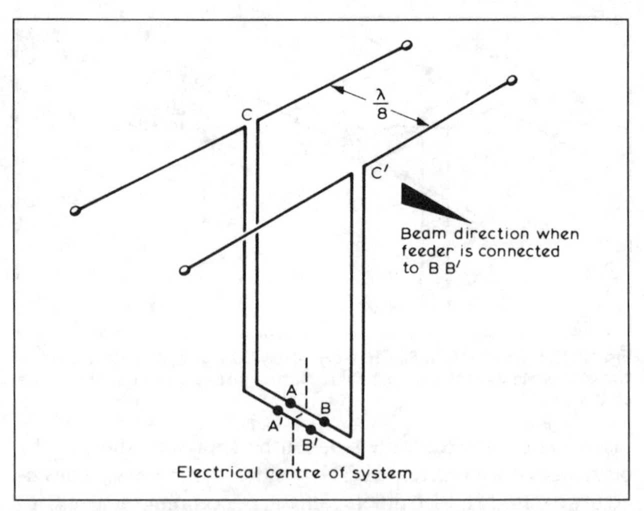

Fig 12.131. Two-element array with resonant (or mismatched) feeders. If the elements are resonant the length CABC' must be an even number of half-wavelengths. AB can be regarded as 'the phasing line' but is much less than $\lambda/8$ if open-wire lines are used. Note that the feeders are crossed over as for the W8JK antenna

which are a reasonable compromise for both bands, although the beam now fires in the opposite direction on 21MHz because the physically longer side has become electrically shorter. Many variations on this theme are possible.

Parasitic arrays

The broadside arrays of an earlier section (pp12.8–12.9) radiate equally fore and aft. In order to make them unidirectional, reflectors are often added at a distance of $\lambda/4$ behind each dipole of the array, and this has the further advantage of increasing the gain by 3dB. The reflector may be driven but, as discussed above in the context of phased arrays, there are problems in adjusting arrays with driven reflectors so that the currents are equal in amplitude but different in phase, because one effect of the reactive part of the mutual impedance (p12.13) is to produce unequal reactances and radiation resistances in antenna and reflector. As we have seen, this difficulty can be overcome by using the $\pm X$ method, illustrated by Fig 12.130, and as a practical example by Fig 12.131 which gave the author excellent service over many years prior to discovering as already related (Fig 12.20) the superior virtues of parasitic operation in which the element currents are equalised by adjustment of mutual reactance between nearly self-resonant elements.

For this purpose some enhancement of coupling relative to that provided by straight elements is essential and Fig 12.132 shows a number of ways by which this may be achieved. At (a) and (d) the ends of dipoles are brought closer together to increase capacitive coupling and at (b) the end-loads also provide increased coupling. Loops (c) couple more tightly than dipoles. Fig 12.132(a) may be recognised as a rectangular version, as previously recommended by the author [28, 29], of the VK2ABQ antenna; this can be adjusted for deep nulls in any back direction and, in line with Fig 12.19, up to 5dB gain can be expected as compared with only 3dB for the original square shape [39] adjusted for a single null at 180°. With suitable tuning the dimensions can be varied within wide limits, 20ft by 10ft being a recommended size for 14MHz. To accommodate the required wire length (about 35ft but check with a dip oscillator) the ends may be folded back as shown or the corners tucked in (see photo). Centre-feed with 50Ω coaxial cable and 1:1 balun is recommended, each element having its own feeder to permit beam reversal and some degree of remote fine-tuning.

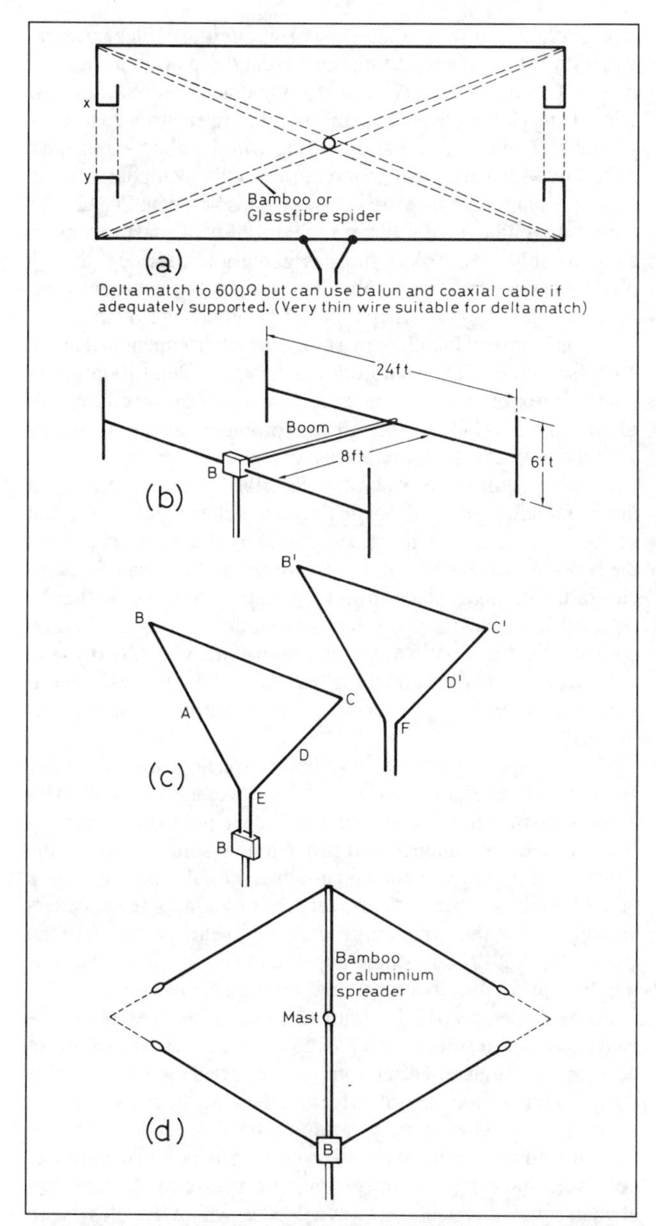

Fig 12.132. Two-element horizontal beams with reduced length and enhanced coupling. Reflector should preferably be a duplicate of the driven element. Otherwise, if currents equal (ie if deep nulls are obtainable), tune reflector to the low edge of desired band. (a) Bent dipole elements (20 × 10ft suggested, though dimensions are not critical), xy (nylon fishing line) approx 30in. (b) End-loading by vertical rods. Single elements have been used successfully with the dimensions shown. Coupling may need augmenting. (c) Small delta loops (ABCD) should be just over λ/2. Spacing EF may be reduced to increase coupling. BB' may be 0.12–0.2λ; EF < BB'/2. (d) Erect between posts or as inverted V. Spreader (or boom) may be 9–12ft for 14MHz *(Ham Radio)*

For triband operation Fig 12.133 shows two alternative configurations and Fig 12.134 gives further details of one successful design but the need, as noted, for 28MHz traps in the 21MHz elements should serve as a warning against the hidden perils of the close-stacking approach to multibanding. No problems have been encountered with 14/21MHz operation but arrays based on small loops with resonant feed-lines such as Figs 12.120–12.122 are much more useful since the operator can adjust them from the shack to obtain 'best possible' performance on five bands subject only to the need for a certain amount of patience in coping with QSY problems (helped by Figs 12.79 and 12.80)

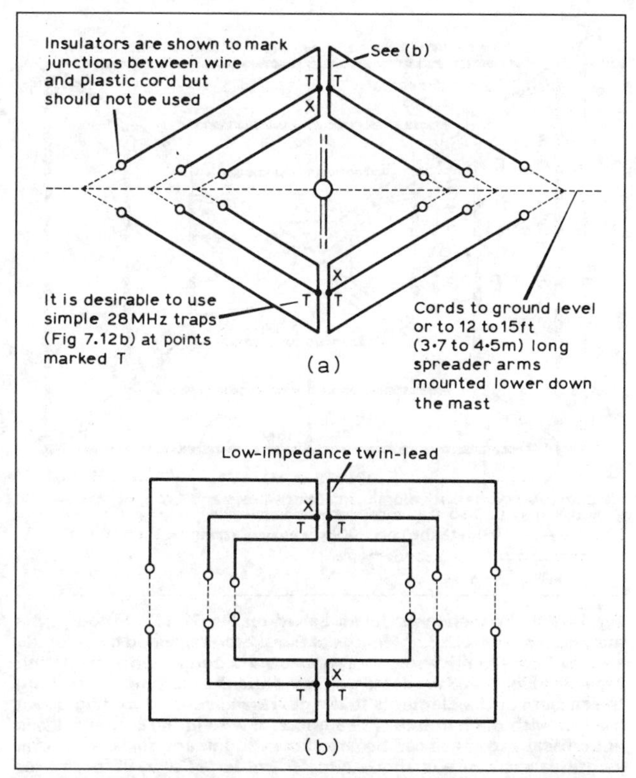

Fig 12.133. Multiband dipole arrays. Elements are constructed in parallel using low-impedance twin-lead, and spacing between adjacent ends is adjusted to ensure equal currents in conjunction with correct phasing; (a) can be erected as an inverted-V and (b) is basically similar to the VK2ABQ array [39]. Total wire lengths for 14MHz are approximately 35ft 6in (10.8m) with a spacing between ends of 12–15in (30–38cm), these dimensions being scaled down for higher frequencies in the appropriate ratio. Both may be fed at X with 50Ω cable and 1:1 balun. Use similar feeders for each set of elements. Preferred system uses 1:4 bal-bal transformer into 300 or 600Ω line (not plastic); open-wire line is used in multiples of 20m, terminating in a 4:1 balun plus coaxial line for the remainder of the feeder run

if he 'wants the lot'! It should perhaps be added that conflicting claims which sometimes encourage the use of stacked arrays are in line with the author's own experiences, but this course is not advised as it relies on the delicate balance between positive and negative mutual reactances when elements tuned to different frequencies are very closely spaced [29, 30].

Arrays of small delta loops such as Figs 12.120–12.122 may be constructed in many different ways to suit varying circumstances; one recommended method is to support the loops from glassfibre fishing-rod blanks angled upwards and outwards from the masthead, taping the sides of the loops onto the supporting arms, an arrangement known from its appearance as the *Claw* (see photo). To avoid feeder radiation and achieve maximum operating convenience (eg beam-reversal without retuning) symmetry is important, and as far as possible beam elements should be identical, but dimensions are not otherwise critical since with the recommended use of two feeders all tuning can usually be carried out in the shack. In some cases, however, especially with long feeders, it may be helpful to connect matching stubs closer to the antenna for one or more bands. The Claw construction has been used successfully with several other types of loop and further details will be found in references [29] and [30]. An important aim in the case of multiple-wire loops is to equalise tension so that no wire is allowed to go slack, in which case it is liable to flexing and eventual breakage; a method of equalising tension which has proved to be successful is shown in Fig 12.135.

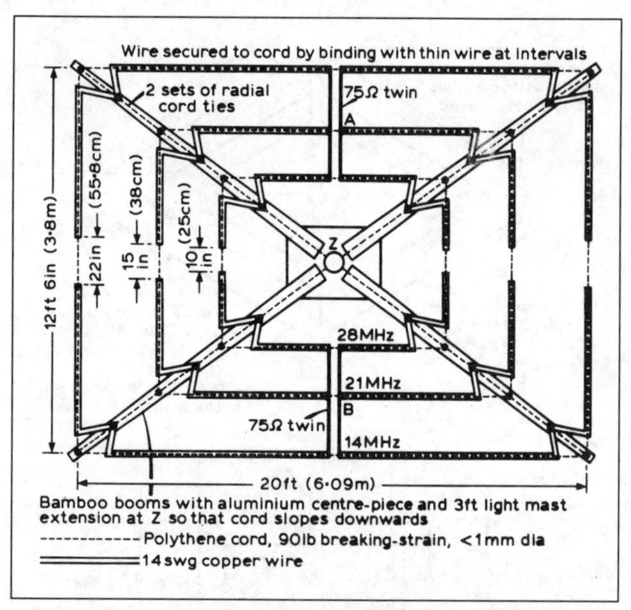

Fig 12.134. Two-element beam based on Fig 12.133. Spider arms may consist of 4ft (1.2m) lengths of Dural tube extended by carefully selected 8ft (2.4m) bamboo garden canes bound with insulating tape. See Fig 12.133 for details of feed. Note the method of attaching the corners of the elements to the polythene cords, avoiding direct contact with the bamboo. The amount of wire in the corner folds is not critical and these can be used for taking in any slack. The total length of each wire is about 35ft (10.7m) for 14MHz, 23ft (7m) for 21MHz and 17ft (5.2m) for 28MHz but can vary with the size and shape of the corner folds. To optimise performance on 28MHz, linear traps [28, 29] may be needed at A, B, tuned to 28MHz, for suppression of current in 21MHz elements when operating on 28MHz. See text for alternative options

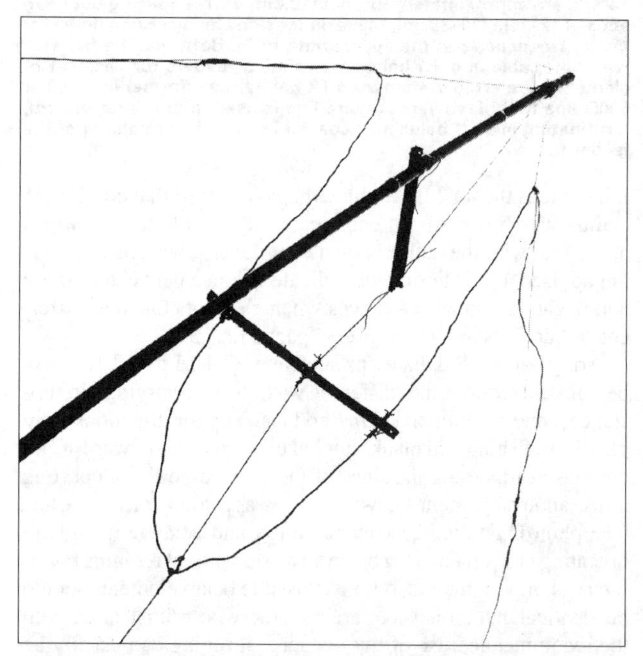

The tucking in of the corners of the 14MHz elements of the antenna shown in Fig 12.134. The folds provide loading and are readily adjustable, thus providing considerable freedom as regards element sizes, shapes and spacings. Spacers obtained by lengthwise cutting of short lengths of plastic pipe are used to maintain the shape of the folds which (as may be inferred from the picture) is not critical! This method allows the wire to be kept well away from wet bamboo but it remains nevertheless advisable to give the bamboo several coats of polyurethane varnish. The same method has many other applications; in particular it allows the tubing elements of typical HF beams to support folded-wire elements for lower-frequency bands

Fig 12.136 shows an alternative, though less elegant, construction based on supporting the top centres as well as the corners of the loops. This is easier to implement and, because the wires hang down clear of them, the glassfibre arms can be replaced by bamboo or by alloy tubing with bamboo extensions, materials which are cheaper and more readily available. Used to support a pair of single-wire (20swg) loops based on Fig 12.120, the total weight for the beam including 8ft of mast extension came to only 8½lb (4kg) and has given good service, though plenty of scope remains for the experimenter, particularly in respect of multiple-wire loops. A major advantage of this construction is ease of handling in a cluttered environment in marked contrast to the Claw or traditional forms of delta loop; these tend to be extremely top-heavy, making them dangerous to handle on top of a ladder, though this problem can be overcome with the help of a temporary mast extension.

In the case of the SDL arrays described above, operation at the fundamental or *main design frequency* closely resembles that of the VK2ABQ with the sides providing the same loading as the bent-over ends of the latter. However, at the higher frequencies radiation takes place from the whole of the loop so that by tapering the spacing of the loops inwards towards the bottom corners as in Fig 12.136 the effective spacing, typically 0.15λ at 14MHz, rises only to something less than 0.25λ at 28MHz, consistent with the goal of near-optimum performance throughout the range.

Tapered spacing with bottom-corner spacings of 3 to 6ft has been found generally satisfactory for coverage of 14–28MHz more often than not, though achieving deep nulls on all bands is likely to require patience and provision of some means in the shack for fine adjustment of coupling. Possibilities include a variable link between separate tuners or overlapping of feeders as suggested in the context of Fig 12.80. In early versions of the Claw [29, 30, 40] adjustable capacitors (in or out of phase as required) were used between feed lines inside the shack.

As explained on p12.11, critical coupling between self-resonant elements results in very deep nulls at 128° to the beam heading, although null directions can be varied by fine-tuning of the reflector without greatly affecting their depth, and this is also what happens when tuning through a band in the event of the reflector tuning being fixed. Fixed tuning is not a first choice, however, since it is usually a simple matter with only two elements to allow each its own feeder which enables beam reversal and remote fine-tuning of the reflector to be carried out in the shack. Advantages of this include the ability to vary null directions from the operating position, avoid or reduce delays associated with beam rotation, in-shack compensation of small errors in tuning or coupling (or even large ones if resonant feeders are used) by in-shack adjustments and, by use of the beam-reversing switch which interchanges the feeders, nearly real-time monitoring of beam effectiveness. Further to this, by taking advantage of the relatively lightweight construction to increase height, arrays such as those just described can be expected in many cases to outperform much larger beams.

Recommended elements such as Figs 12.120–12.122 may be assembled in many other ways including almost-invisible fixed beams [29] suspended spiders-web fashion from trees or other supports by means of nylon cord which also provides the insulation. As an alternative to the SDL, if a sufficient span is available, reference has already been made to the extended double Zepp, Fig 12.95; designed for 28MHz this becomes a slightly oversized λ/2 dipole at 14MHz and can likewise be used with resonant feeders for two-element five- or six-band beams, the ends being bent inwards VK2ABQ-fashion to increase coupling.

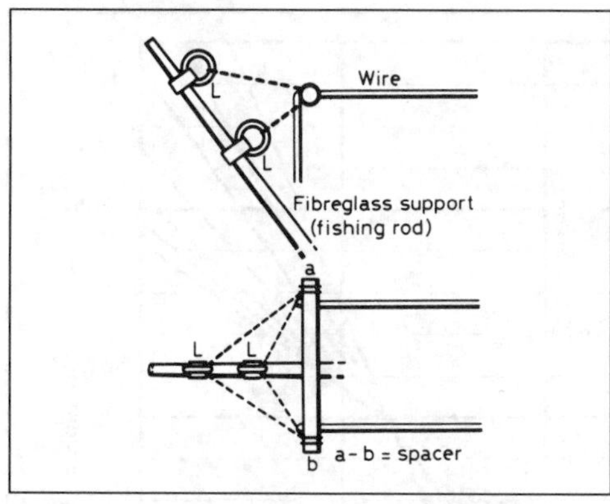

Fig 12.135. Equalising tension in paired wires of ITL or CIL. Nylon line, two looped lengths as shown dotted, is free to slide through loops, rings or insulators L

Apart from the extra gain at the highest frequency this can be particularly recommended [29] for three-element arrays, straight elements being preferred in this case as equal currents are not needed; moreover, by using only the outer pair on the lowest bands, near-optimum spacings can be achieved throughout a range of two octaves.

Those with more modest objectives or a dislike of open-wire lines should note that most types of element other than very small ones can be adapted for two-band operation with little or no adverse effect on gain, and efficient triband operation may be feasible [28, 29] though opportunities are more restricted.

As a further practical point, long lengths of plastic-insulated twin feeder, regardless of impedance or quality, should not be used in the open for remote tuning of reflectors or for driven arrays other than the W8JK; this is because of phase changes resulting from the change in velocity factor (5% to 20%) which occurs when wet, though short lengths may be used, eg for the bypassing of a beam rotator [28, 29].

It is clear from experience with the above arrays and theoretical considerations presented earlier that resonant feeders or remote tuning (including band-switching) devices located in the antenna are capable of providing gain and null depths on five or six bands comparable with those of monoband beams, whereas the erection of five or six monoband beams all enjoying equal pride of place would be an extremely rare achievement.

Adjustments are made sometimes for best nominal front-to-back ratio and sometimes for maximum forward gain. Referring again to Fig 12.15, the first condition tends to give slightly less gain but higher radiation resistance which may be desirable to preserve bandwidth. However, it is important to realise that maximum gain *by definition* implies minimum interference if, as is true for the average case, interference is no more likely to be coming from one direction than another. This follows from the fact that maximum gain is obtainable only by achieving the maximum concentration in one direction at the expense of the average of all other directions. To select one particular direction which happens to be at −180° and adjust for minimum interference from that one direction, neglecting the possibility that this

The Claw antenna. Small delta-loop elements are held up by glassfibre arms angled upwards and outwards from the masthead. Bottom corners are brought down to a short boom where they are attached to open-wire feeders. Spacing of the top wires is set to 10ft by cord ties (not visible). The span is 20ft, and with all the radiation coming from the top the height gain (14MHz) is 13ft. Loop types used have included single wire (Fig 12.120), ITL (Fig 12.121(b)) and constant-Z_0 (Fig 12.122(b)) as in the photo

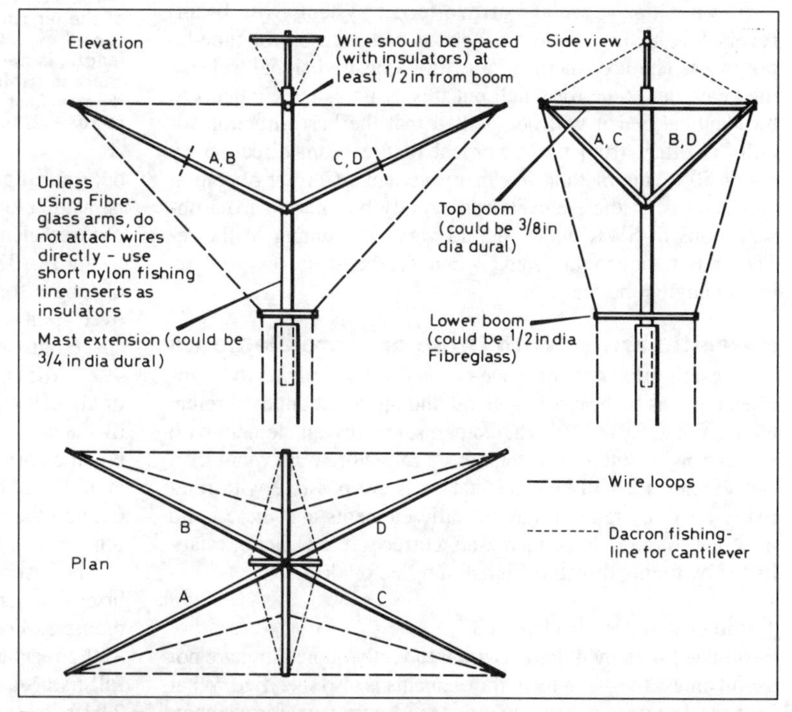

Fig 12.136. Constructional details of SDL array. A, B, C, D are spider arms, eg glassfibre fishing-rod blanks or suitably weatherproofed bamboo (these can be partly metal but only inwards from the dotted lines marked on elevation). Details of attachment of spider arms to mast not shown (could be secured to short lengths of angle held to mast by U-clamps)

could increase the signals picked up from or radiated in other unwanted directions is, to say the least, devoid of logic; on the other hand it is easier to adjust for a minimum than a maximum and it is evident from Figs 12.17 and 12.18 that if this course is adopted, the signal chosen should be on a bearing in the region of –120° to –150°.

A field strength indicator may help in the adjustment of these arrays, but this does not always give the best result, because the local field pattern of any antenna is different from the distant field, especially near the ground. If a field strength meter is used it should be connected to a short dipole, set up parallel to the array, as far away as possible and preferably at the same height. The greater the gain of the array, the greater the distance at which the field strength meter must be placed. With small beams (5–6dB gain) field-strength indicators at a distance of λ and the same height as the beam, or at more or less any height if the distance exceeds about 2λ, have been found to give an indication of optimum performance in fair agreement with reports from distant stations. Particular sources of error observed in such measurements have included the use of too short a dipole for the field strength meter, so that the harmonic sensitivity is many times greater than the fundamental sensitivity, and pick-up on insufficiently-filtered leads used for wiring the meter back to the point where the adjustments are being made. Polarisation of the field strength measuring antenna must be correct, meters with short vertical rod antennas being useless for measurements on horizontal antennas. An important point is that the impedance of the driven antenna changes when the parasitic element is tuned, so that the power drawn from the transmitter also varies with this tuning. Thus, unless a power-flow monitor, not an ammeter but a reflectometer, is used the optimum antenna gain adjustment may not be obtained.

It is a major advantage of two-feeder systems with beam-reversal switching that other stations can be used for tune-up purposes. The lazy method is to find a desired signal, reverse the beam, and tune for a null but this assumes the elements to be identical and it will be recalled that the best direction for nulls is offset from the reciprocal of the beam direction by about 50°. This method has many variants. Correct operation and identity of the elements can usually be inferred from observations of SWR which should vary with tuning of the reflector but remain the same when the beam is reversed by interchanging the feeders.

Parasitic arrays with three or more elements

The description *Yagi*, after one of the inventors, tends to be applied to parasitic beams in general, though traditionally it refers to end-fire arrays of straight elements, ie a driven element with usually one or more directors and a reflector. With more than two elements it will be clear that nulls are possible with relatively small currents in the parasitic elements and there is no need for coupling to be increased, correct phasing being established by tuning directors high and reflectors low.

Tuning the parasitic elements

Formulae for element lengths are frequently quoted, but are not useful unless the diameter of the elements is also specified. What controls the performance is the reactance of the parasitic element; that is, the amount by which the element is detuned from resonance, and this, in terms of length, depends on the length/diameter ratio. The reactance chart, Fig 12.137, can be used to determine for a given conductor diameter the amount of detuning required to produce any desired value of X in accordance with Fig 12.20, for calculation of antenna bandwidth, and for

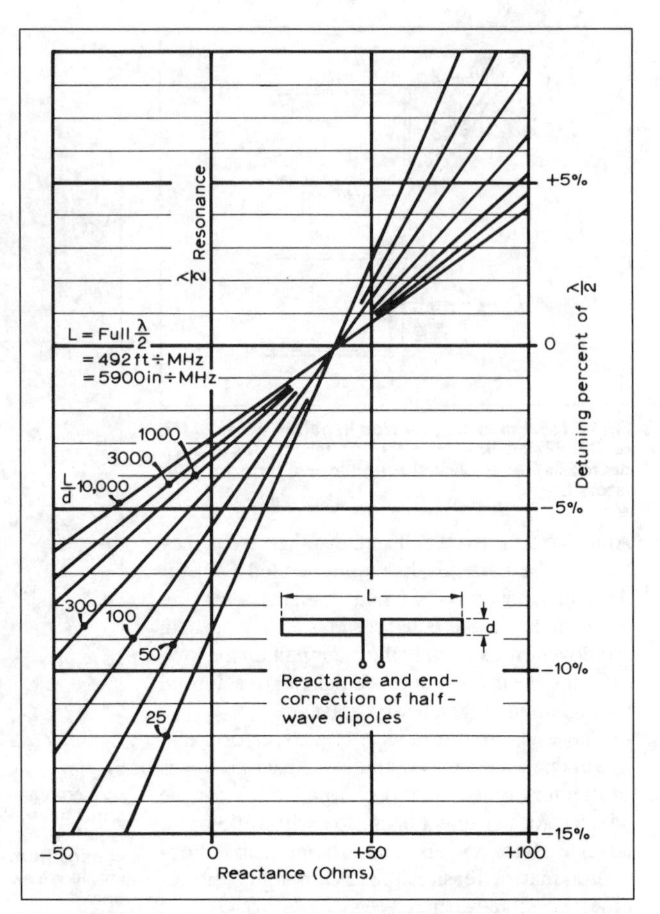

Fig 12.137. Tuning and reactance chart for λ/2 dipoles, as a function of the length/diameter ratio. A radiator exactly λ/2 long is 'over-tuned' by 42Ω and 'end-correction', given as a percentage of the length, is necessary to bring it to resonance (zero reactance). The chart is useful for the construction of parasitic arrays and VHF dipoles. Each one per cent of length corresponds to five units in the factor 492/*f* or 60 in 5900/*f* (*f* is in megahertz)

determining the length changes needed for generating the parasitic element reactances required for optimising the design of three-element beams in line with Figs 12.138–12.140.

Fig 12.137 shows reactance as a function of percentage change of length for various relative conductor diameters. The cross-over point of the curves occurs when the radiator is a true free-space half-wavelength long, the length in feet being given by 492/*f*(MHz). At this length the conductors all have a reactance of about +40Ω, and to obtain resonance must be *end corrected* by the percentage which brings the reactance to zero. For example, the correction for a relatively 'fat' antenna, with $L/d = 100$, is 5.5% so that for 28.0MHz the full half-wave would be 17ft 6in and the end correction 1ft, making the resonant length 16ft 6in.

The best reflector has a positive reactance of about 40Ω, which brings its length to almost a true half-wave, independently of diameter. For 0.1λ spacing, best director action occurs at –10 to –40Ω reactance with a length some 5 to 10% shorter than the full half-wavelength. For example at 21.0MHz, $\lambda/2 = 492/f = 23.4$ft. A 1in diameter λ/2 radiator would have a L/d ratio of about 300, and an end correction of –44%, giving a resonant length of 22ft 4in, while a director of –30Ω reactance would be 7.4% short of true half-wave, or 21ft 7in.

In practice it is always advisable to find the best adjustment experimentally, using the above theory as a guide, because the environment of the antenna has some effect. The parasitic elements can be tuned first, independently of the driven element, which may then be adjusted for optimum tuning and impedance match, as described later.

Three-element arrays

Theoretically, the gain is proportional to the square of the number of elements, but this requires very precise adjustment of all currents and phases together with acceptance of very narrow bandwidths and low radiation resistances which lead to severe losses. A four-element driven array on these lines which gave a measured gain of 8.7dB out of a theoretical 10.1dB has been described [3, 28, 29] but the interdependence of amplitudes and phases would probably rule out achieving comparable results with Yagi arrays. There is no advantage in using more than one reflector in an in-line array, though many directors may be used with advantage if they are widely spaced. For HF work one or, at the most, two directors are the limit because of the size of the array, but in the VHF bands longer arrays with more than three elements may be built, and are described in Chapter 13.

The characteristics of two-element arrays can easily be calculated (p12.10) but the analysis of arrays with three or more elements is difficult because of the much larger number of variables. A vast amount of experimentation, as well as computer studies, by amateurs and others has resulted in an infinite variety of prescriptions for element lengths and spacings without any clearly recognisable 'best' design emerging.

The vertical and horizontal radiation patterns of a typical three-element Yagi array are given in Fig 12.138.

The maximum gain theoretically obtainable for parasitic arrays with equally spaced elements is 7.5dB for three elements. These patterns have their half-power points at a beamwidth of about 55° in the E-plane (the plane of the elements) and 65–75° in the H-plane (at right-angles to the elements). These high gains are, unfortunately, associated with a feed-point resistance of only a few ohms, so that in practice the loss due to conductor resistance and environment will be relatively large. In addition, the tuning is very sharp, so the array will only work usefully over a small part of, say, the 14MHz band, while the line matching for a low VSWR is very difficult.

The power loss due to conductor resistance, or high feeder VSWR, is such that these high gains are never realised in practice. Better general performance is obtainable by designing the antenna for less gain, as may be appreciated with the aid of Fig 12.139. This is a 'contour' map of gain and impedance as a function of parasitic tuning, for a Yagi with element spacings of 0.15λ. The lower right-hand corner of the chart is clearly the region to aim for, where the gain is still over 6dB while the feed resistance approaches 20Ω. This region of the chart corresponds to rather long reflectors (+40 to +50Ω reactance) and rather shorter than optimum directors (–30 to –40Ω), and also is found to be the region of best front-to-back ratio.

Although the chart is for antennas with both spacings equal to 0.15λ, the resistance and bandwidth are both somewhat improved, with the gain remaining over 6dB, if the reflector spacing is increased to 0.2λ and the director spacing reduced to 0.1λ. Such an array would operate satisfactorily over at least half the 28–29MHz band or the whole of any other HF band for which it could be constructed.

The overall array length of 0.3λ is practicable on all bands from 14MHz upwards, though the 20ft boom required for a 14MHz array is rather heavy for a rotary array, and there is a temptation to shorten it. Spacings should not be reduced below 0.1λ for both reflector and director as the tuning will again become too critical and the transmitter load unstable as the

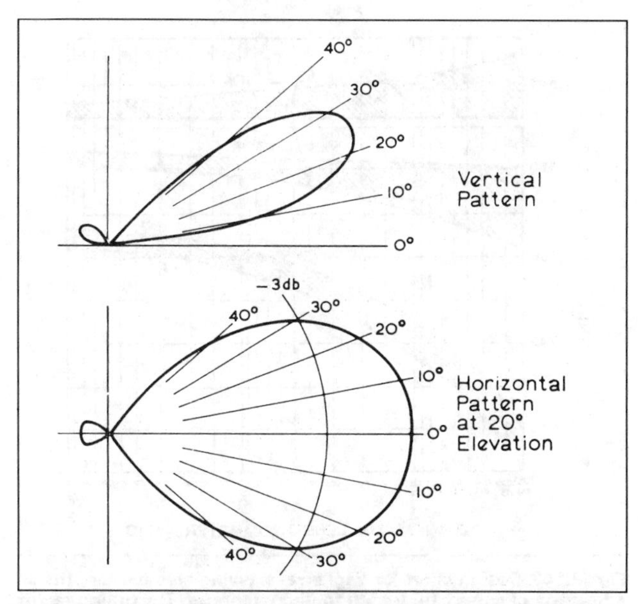

Fig 12.138. Vertical and horizontal radiation patterns of a three-element Yagi array with 0.15λ spacings, and erected at a height of λ/2. These patterns, which were obtained from a scale model, do not vary much with other spacings of the elements. Note that the front-to-back ratio along the ground may be very different from that at 20° elevation. The vertical pattern of which only the lower lobe is shown in the figure, depends on height, which should be as great as possible

elements or the feeder move in the wind. It should not be overlooked that in general for a given cost and degree of practical difficulty, the lighter the beam, the higher it can be erected; in terms of practical results this may be worth a lot more than additional elements.

The optimum tuning does not vary appreciably for the different spacings, and it is therefore possible to construct a practical design chart (Fig 12.140). An array made with its element lengths falling in the shaded regions of the diagram will give good

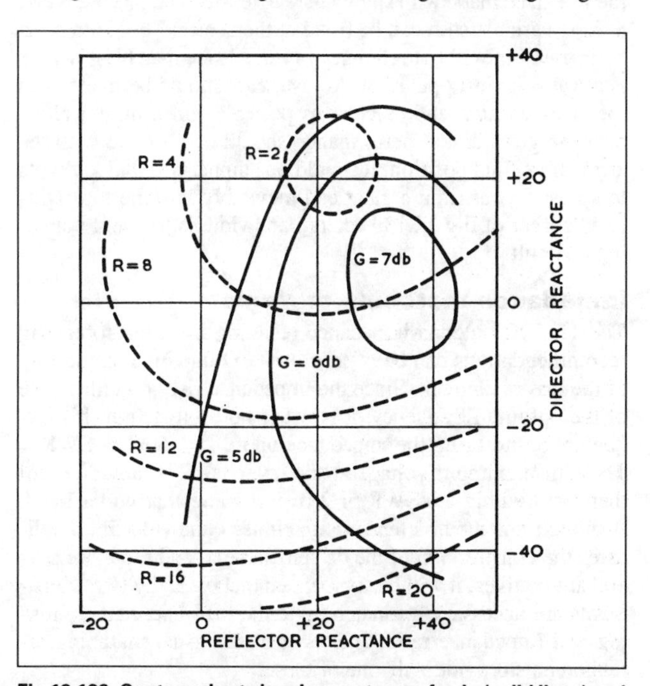

Fig 12.139. Contour chart showing contours of gain (solid lines) and input resistance (broken lines) as a function of the tuning of the parasitic elements. This chart is for a spacing of 0.15λ between elements, but is also typical of arrays using 0.2 + 0.1λ spacing

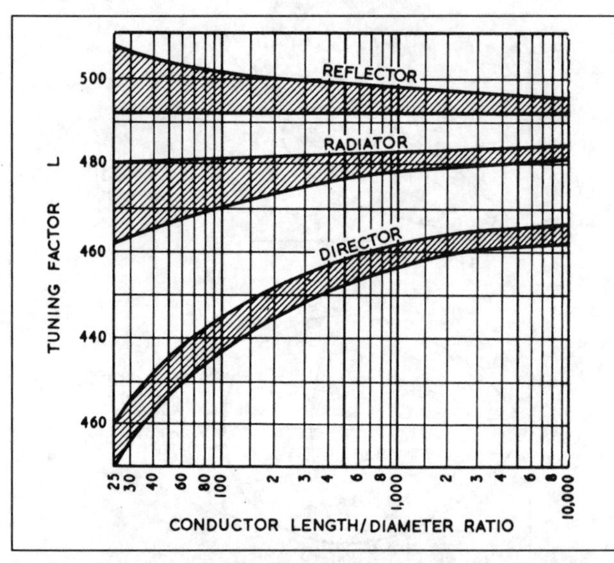

Fig 12.140. Design chart for Yagi arrays, giving element lengths as a function of conductor length-to-diameter ratio. The tuning factor *L* is divided by the frequency in megahertz to give the lengths in feet. These curves are for arrays of overall length 0.3λ with reflector reactance +40 to +60Ω and director –30 to –40Ω, and give arrays of input impedance between 15 and 20Ω. Element lengths which fall within the shaded areas will give an array which can be used without further adjustment, though the front-to-back ratio may be improved by adjusting the reflector

performance without further adjustment. The length of the director may be decided in advance. The reflector may then be adjusted experimentally to improve the front-to-back ratio. It will be seen that the radiator is somewhat longer than a normal dipole; this is because the parasitic elements have a detuning effect on it. The addition of a second director also spaced 0.1 to 0.15λ will not materially affect the above recommendations.

The adjustment of the array can be carried out using the nominal front-to-back ratio as the criterion. For this purpose the elements can be made with sliding tube extensions or, alternatively, a short variable stub can be fitted at the centre. The tuning can be carried out without reference to impedance matching, which is always the last operation. Adjustments should be made with the array as high and as clear as possible, since tuning alters near the ground, and performance should be checked with the array in its final position. It should be emphasised that attempts to adjust for maximum gain lead inevitably into the top right-hand corner of Fig 12.139 where bandwidth is low and matching difficult.

Impedance matching of Yagis

The 15–20Ω antenna impedance resulting from the foregoing recommendations can be matched to 80Ω line by a single fold of the driven element. Since the impedance rises on either side of the optimum frequency, it is better not to match at this frequency but to bring the impedance up to, say, 60Ω (VSWR = 1.3 with maximum voltage at the antenna). The antenna will then hold within a VSWR of 2 over a somewhat wider band. Although this method tends to maximise bandwidth it complicates the construction of the driven element, and there are several alternatives; it will be appreciated that basically two adjustments are necessary for matching to the feed line, namely tuning and impedance ratio, and most of the usual matching arrangements provide only one of these.

The operation of adjusting the length of the radiator at the same time as the impedance ratio can be tedious and difficult. If, however, the dimensions are carefully established in accordance with Fig 12.140 it should only be necessary to carry out the matching adjustment to obtain reasonable performance, the T and gamma transformers of Fig 12.59 being suitable for this purpose. Alternatively high-impedance feeder may be used with a delta match, but this can cause difficulty with disposal of the feeder which must be kept well clear of any metal mast. Provided insulated wire is used and the main run is well clear of the mast, accidental contact with a wooden mast at one or two points is not serious with a well-matched open-wire line. Details of an alternative design for a three-element beam centred around the Clemens match, which also provides balance-unbalance conversion, will be found on p12.36.

The cubical quad

This antenna, though largely bypassed in the earlier discussion of loop antennas, remains very popular with DX operators on the 14, 21 and 28MHz bands and calls for further discussion.

In its basic monoband form it usually consists of a driven element in the shape of a square, each side of which is λ/4 long. Behind this driven element is placed a closed square of wire to act as a parasitic reflector. The layout is shown in Fig 12.141 and in this configuration it is horizontally polarised. Feeding in the middle of one of the vertical sides would change the polarisation to vertical. The loops can alternatively be mounted with their diagonals vertical, the feed point being at a current loop in both cases and the performance identical.

The currents in the top and bottom wires of any one loop are in phase and add to give broadside horizontally polarised radiation. Those in the vertical sides are in antiphase and cancel in the broadside direction, giving rise only to a small amount of vertically polarised radiation off the side.

The radiation resistance and other properties of the cubical quad vary with the spacing between the driven element and the parasitic reflector in the same way as other small beams, and Fig 12.15 is applicable subject to use of the appropriate scale factors, ie 1.7 for the radiation resistance of each element if both are driven or 2.5 for the more usual arrangements using a driven element and reflector. These figures are derived from those for folded dipoles, dividing by two because the end portions of each half-wavelength of wire (which account for 30% of the total field strength from a dipole) have been prevented from radiating, and reduced by a further 10% to account for the fact that with only two elements some residual stacking gain can be expected, though this will vary with phasing. The usual scale factor S^2 for the radiation resistance, where S is the spacing in eighths of a wavelength, also applies.

The cubical quad presents a good match to 75Ω cable with an element spacing of 0.15λ and the reflector tuning adjusted for

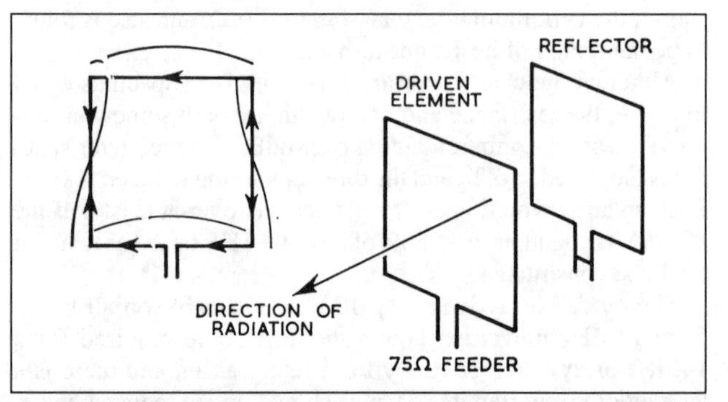

Fig 12.141. The basic cubical quad showing the current distribution and direction around the driven element. Radiation from the vertical sides is mainly cancelled as the currents are in opposition

$\phi/\phi_0 = 0.8$ (p12.11) but by shortening the reflector it should be possible to obtain a match to 50Ω cable together with an increase of 0.6dB in gain. Up to 30% of overcoupling with consequent loss of directivity has been observed with a spacing of λ/8 but this is critically dependent on phasing and the presence or otherwise of any losses. The reader will appreciate however that many variations are possible.

There is *no* justification for claims that the quad provides gain figures significantly greater than those of dipole arrays, or that it achieves enhanced DX signal gain by concentration of energy at lower angles of elevation. The latter belief apparently arises from the fact that a quad element radiates less energy vertically upwards than a dipole, but this is in fact the origin of the very slight superiority of quads over dipole arrays and claimed gains in excess of 6dB are at variance with the laws of nature; practical antennas consist of point sources located above a reflecting plane, the distribution of energy in space and therefore the gain being determined entirely by the basic geometry of the situation. The relative phases and amplitudes of currents are part of this geometry and need to be 'right', also the sources themselves have to comply with a few obvious rules in the interests of efficiency, but for a given geometry, polarisation, and number of point sources no type of antenna can be electrically superior to another. In some cases, however, the quad may provide a further advantage in the form of a 'periscope effect', due to the ability of radiation from its top half to clear local obstructions. In such cases the various small-loop arrays (pp12.81–12.84) should fare even better as their top half does all the radiating so that no energy is wasted.

To appreciate why a quad loop approximates to a single point source it is convenient to imagine the top and bottom horizontal wires replaced by separate λ/2 dipoles, ie two sources spaced by λ/4. From Fig 12.9 the mutual resistance between them is 40Ω so that each dipole has a radiation resistance of 73 + 40 = 113Ω. For equal power radiated therefore the current in each is 0.565 times that in a single dipole, the total field is increased by 13% or 1.1dB only, and the loop is much closer in performance to one than two separate sources of radiation. Even this slight advantage is eroded when allowance is made for the 'short dipole' effect or two elements are arranged as a beam, since an increase of gain increases the optimum stacking distance (p12.17).

In its favour the quad is particularly easy to adjust for good results, the small span of the elements simplifies construction of rotary beams, and the near-equality of currents in the driven element and parasitic reflector produces sharp nulls in the radiation pattern which can often be used by rotation of the beam to suppress interference. Though this feature is shared with other equal-current arrays, it has the advantages in this case of occurring naturally at convenient values of spacing without the need for any other adjustment.

Another property of quad loops is the possibility of using them with resonant feeders as multiband radiators in various ways as discussed below.

There are certain points of importance to note in the electrical construction of a cubical quad. Because of the variation of Z_0 round the loop, the wire-length formulae used for dipoles are not applicable and the overall length is slightly longer than it would be for an antenna in free space. For the driven element, each side of the square may be 248/*f*(MHz) and fed with 75Ω flat twin, or coaxial with a balun. If an unbalanced coaxial cable feed is used there will, as with other antennas, be a distortion of the lobes caused by the imbalance. However, if suitable tuning stubs are used neither the dimensions nor the shape of the loop are in any way critical.

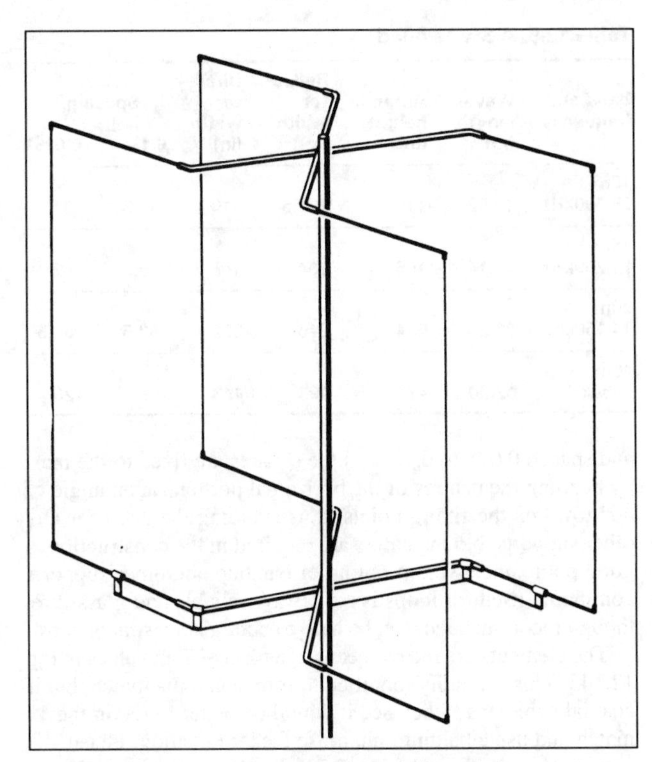

Fig 12.142. General arrangement of the Swiss quad

The size of a parasitic reflector element is usually made physically the same as that of the driven element but electrically longer by means of a short-circuited stub which can be adjusted for maximum gain or front-to-back ratio. It will usually be found quite easy to adjust the stub for a nominal front-to-back ratio better than 20dB, though the idea that reflectors need to be tuned lower than the driven element is quite wrong as was demonstrated by Fig 12.20. Unfortunately this has been overlooked in published descriptions of quads which in place of an adjustable stub often specify a reflector size some 3% larger than the driven element and in one case of poor performance investigated by the author the situation was rectified by removing exactly this amount of wire from the reflector. Referring back to p12.11, the reasons why the 3% figure is wrong can be simply stated as follows. If deep nulls are obtainable this is proof that element currents are nearly equal, which is impossible without reactive coupling; if the reflector is self-resonant this automatically provides near-optimum phasing with nulls occurring at 128° relative to the beam heading. Tuning the reflector lower moves the nulls round towards the usual 180° position (ie points B in Fig 12.20) where they coincide, a highly undesirable procedure when carried out as usual at band centre because tuning higher in the band then carries one over into the right-hand half of Fig 12.15 where gain and front-to-back ratio are rapidly decreasing. It is, however, even more important to ensure that one does not go much beyond points A to the left of Fig 12.20 which results at first in a rapid rise of SWR after which the reflector changes into a director, albeit a rather inefficient one due to the sign of the mutual reactance being negative with the usual forms of construction.

The Swiss quad

Quads may of course also use pairs of driven elements and the Swiss quad (Fig 12.142) described below provides improved mechanical construction without sacrificing performance [12]. The elements consist of two parallel squares having λ/4 sides

Table 12.9. A Swiss quad

Band and frequency	Wave-length (m)	Antenna height (in)	Reflec-tor width (in)	Direc-tor width (in)	Spacing (in) 0.1λ	0.075λ
10m 28,500kHz	10.52	116	121.5	110	41.3	31
15m 21,200kHz	14.14	156	164	148	55.5	42
20m 14,150kHz	21.20	234	246	222	83.5	62.5
40m 7050kHz	42.60	470	493	443	168	126

and spaced 0.075 to 0.1λ, and the squares are fixed to the mast by bending the centres of the horizontal portions at an angle of 45° towards the fixing points, thus avoiding the need for any other supports. No insulators are required in the construction, a good point since waterproofing of bamboo and other supports commonly used for loops is not always reliable and glassfibre, though recommended, can be hard to obtain at reasonable cost.

The elements are fed by means of a form of T-match as in Fig 12.143. This is readily modified to form a gamma match, but if coaxial cable has to be used it would be better to retain the T-match and use a balun to minimise feeder radiation. Phasing is obtained by the $\pm X$ method (p12.79), making one element shorter and the other longer than the resonant length. Despite the pinching-in, it has been possible to make the elements more or less full size overall; the pinch merely fulfils the general requirement for the 'lengthening' of loop elements to establish resonance which, as explained on p12.80, is an essential feature of the method of phasing.

Fig 12.15 can be used as a rough design guide also for the Swiss quad but the narrow spacing reduces the radiation resistance so that a phase shift of 30° between the elements is likely to be a reasonable compromise, giving near-maximum gain. Adjusting the figures for a spacing of 0.1λ and the above scale factor of 1.6 gives a required phase shift of 24° and a radiation resistance of 21Ω. From the explanation of the 8JK antenna (p12.69) it follows that $(R - R_m)$ must be proportional to R and to S^2, from which R_m = 104Ω and the required reactance given by $(R + R_m)\phi/2$ is equal to 46Ω. Assuming Z_0 = 800Ω for the loop, this requires ±2% detuning.

The original article specifies ±2.5% and quotes figures upwards of 30Ω for the radiation resistance, both figures being consistent with working slightly nearer to the condition of maximum nominal back-to-front ratio which could be helpful in view of the rather low radiation resistance.

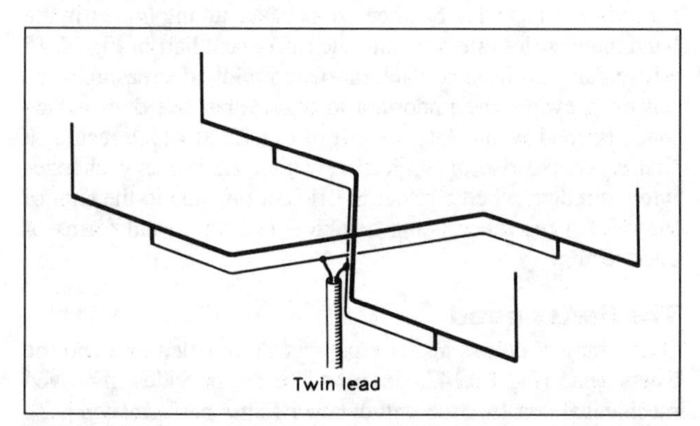

Fig 12.143. Balanced feed and matching system

There is clearly considerable latitude in the design of this type of antenna but Table 12.9 represents a proven design and is the obvious starting point for further experiments, being taken from reference [12] which may be consulted for general guidance on constructional methods. It should be noted that the 'X' section in the middle contributes little to the wanted radiation because the wires are inclined at 45° to the direction of propagation and their average spacing is only half that of the main wires. The radiation from them is a clover-leaf pattern with some radiation at right-angles, and indeed all four petals would become equal with '8JK' phasing. At 0.1λ spacing the 'X' section will have negligible effect on the performance or the calculations, but if the spacing is increased the importance of the 'X' increases rapidly, not only because of its own greater dimensions, but also because the width of the loops will shrink. This tends to reduce the radiation resistance as the inverse square of the width, thus offsetting and ultimately reversing the usual trend whereby an increase of spacing increases the radiation resistance.

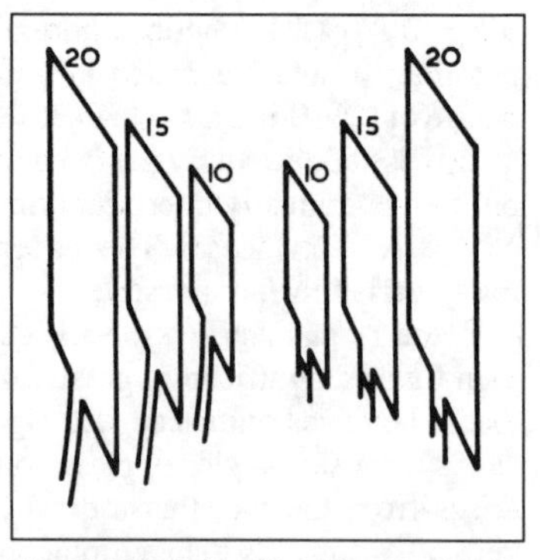

Fig 12.144. A three-band nest of cubical quads maintaining typical spacing for each band

In the case of driven arrays using the reactance method of phase shifting it is important to distinguish between the radiation resistances and the feed-point impedance, since the reactances form a tuned circuit which acts as a transformer. But for this and the T-match, 'looking' into the antenna from the feeder one would 'see' the two radiation resistances in parallel, ie 10Ω, but this will be transformed by the reactances to 40 or 50Ω and the T-match provides further adjustment.

Multiband quads

Fig 12.144 illustrates the most usual form of multiband quad, the driven elements being connected in parallel to a single feeder. Considerable interaction can be expected and on these grounds alone the DJ4VM quad [47], Fig 12.145, is much to be preferred; in addition it can provide efficient operation on at least five bands with 3dB of additional gain at 28MHz where it functions as a 'bi-square'. Compared with the small loop arrays featured earlier, which also use resonant feeders, it has a similar gain advantage, broader bandwidth and a wider operating frequency range helped by freedom from the vertically polarised endwise mode of radiation which determines the upper frequency limit in the case of conventional quad or SDL arrays. Disadvantages, shared with other quads, include high windage and visual impact. In contrast the SDL arrays are much easier to support and handle, for a given mast height they provide more effective height, and their light construction entails fewer hazards.

Fig 12.146 shows two simple ways of obtaining a multiband resonance. If the length is equal to two half-wavelengths on 14MHz it will be three on 21MHz, four on 28MHz and one on 7MHz, ie nominally resonant in every case, though slight adjustments are needed for the reasons given on p12.35. Of the two alternatives, the closed loop (a) is more generally useful, because the lower end is a point of current maximum on each band, and therefore in many cases accessible for tuning and matching into any non-resonant line. This is true in principle of the open loop; the end of the resonant feeder can be coupled

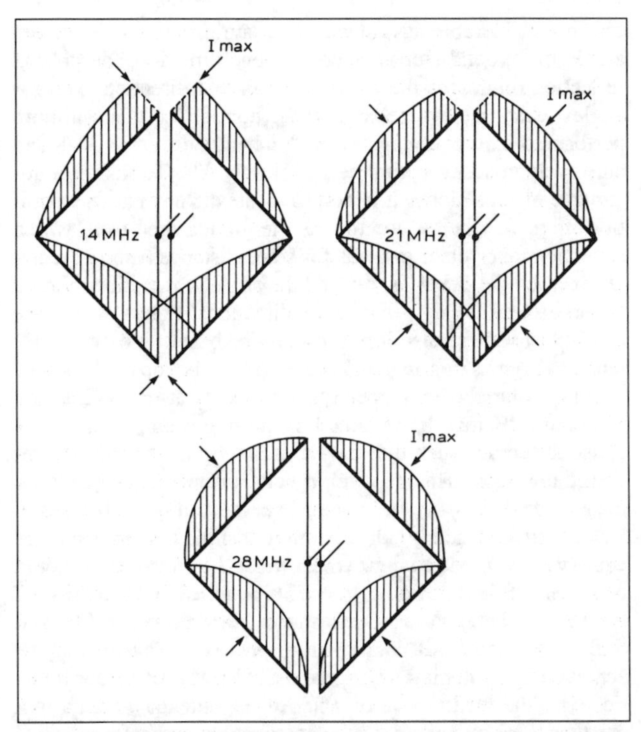

Fig 12.145. Current distribution in centre-driven quad elements [47]. The current loops are always symmetrical. The feed-point impedance is variable and tends to be highly reactive so that resonant feeders are needed for multiband operation *(Ham Radio)*

loosely through a very small capacitance (eg by proximity) so that it presents a match into open-wire line, but practical difficulties of adjustment are considerable. The open loop can be replaced by a full-wave dipole.

In the case of a 14MHz quad loop used on 21MHz or a 21MHz loop used on 28MHz, the current node slides up the side to a point about 11ft from the top centre, and vertically polarised radiation off the ends is less fully cancelled, but it has been estimated that loss from this cause is more than offset by higher intrinsic gain arising from the 'oversize' elements, and as observed on sky-wave signals the radiation patterns obtained from a driven-element plus reflector are roughly comparable on the two frequencies. For two-band operation the loops should preferably be mounted with the diagonals vertical [11]. The loop length needs reducing slightly on 21MHz and this may be done

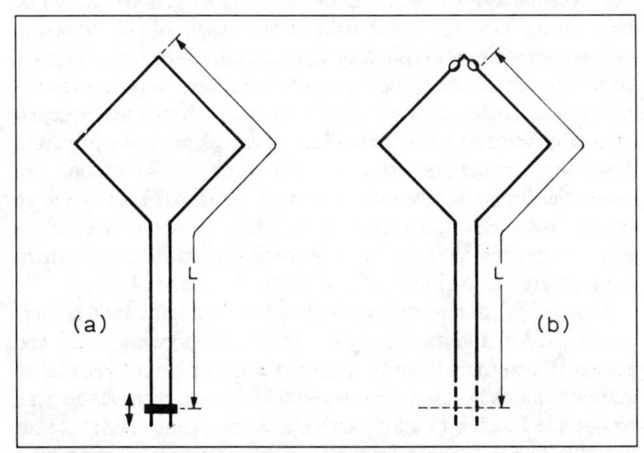

Fig 12.146. Examples of multiband resonators. If $L = \lambda/2$ at 7MHz, resonance occurs also at 14, 21 and 28MHz approximately, but small adjustments of *L* are needed. This applies to any size of loop

for a reflector by connecting a $\lambda/4$ open-ended stub about 2ft from the end; this has negligible effect on 14MHz. Alternatively, frequency switches may be used on the lines of Fig 12.113(c). In this case the traps have to effect only a small adjustment of line length and the values are not critical. The 14MHz quad also works, but not as a beam, on 7MHz and 28MHz but for lower frequencies it is probably better to excite one or more of the loops as a vertical, eg Zepp-fed (Fig 12.109) at the lower end as a $\lambda/2$ vertical on 7MHz or fed with coaxial cable against a ground-plane as a $\lambda/4$ vertical antenna on 3.5MHz.

An example of a 14MHz quad used very successfully on 3.5 and 7MHz for long-haul DX has been described by G8PO [17] and, in general, any vertical bunch of wires provided it is not 'nailed to the mast' is likely to be a useful candidate for consideration as a low-frequency antenna.

On its second harmonic frequency a conventional quad or delta loop can be resolved into two end-fire two-element arrays, one of which fires vertically upwards so that half the energy is wasted, thereby reducing the gain approximately to that of a dipole. A single 14MHz quad loop can therefore be expected to work more or less equally well on 28MHz except that the direction of propagation and the polarisation are both switched through 90° so that radiation is vertically polarised and 'off-the-end'. Very little gain is obtainable on 28MHz by exciting both elements of a close-spaced 14MHz quad, though a two- or four-lobed pattern with useful gain should be obtainable by feeding wide-spaced ($\lambda/4$ on 14MHz) elements with suitable phasing. A more efficient triband beam can be achieved by open-circuiting the top of the loop so that it turns into a bi-square (Fig 12.126) on the higher frequency, and this may be done by means of a frequency switch (Fig 12.113(d)) if narrow band operation is acceptable. If the loop is required to work on 14, 21 and 28MHz the *L/C* ratio must be kept low in order not to upset the 21MHz radiation pattern, values of 0.5µH and 62pF being suitable if the loop size is reduced to 16ft square. Identical treatment can be applied to a 14MHz folded dipole to obtain triband operation. Yet another alternative arises from the fact that a 21MHz loop retains a reasonable amount of radiation resistance, about 36Ω for a 12ft 6in square loop when driven on 14MHz; this is coupled with the advantage of increased effective height since most of the radiation comes from the top half of the loop.

With all these arrangements the simplest method of feeding is to use resonant feeders or stubs as in Fig 12.146(a) which can be matched at their lower ends into low-impedance feeder, this being much easier with loops in view of the higher radiation resistances and the near-coincidence of resonances if the length is suitably chosen. It is possible even without the use of tuned feeders to devise elements with simultaneous resonances on two or three bands by using a variety of stubs and other loading devices [11, 18]. Triple resonance, however, combined with efficient matching to an untuned line on all bands, is difficult to achieve without compromises of one sort or another, and the only systems currently enjoying much popularity employ trapped elements.

Arrays with directional switching

As an alternative to rotary beams, two or preferably three fixed reversible beams may be used to provide all-round coverage, with instant switching of beam direction being a major advantage. If only two horizontal beams are used, mutually at right-angles, performance will be within about 1dB of maximum gain over four 60° segments but in between these it will fall off roughly to equality with a dipole. Two beams may be mounted on the same pole with a small vertical separation for mechanical clearance,

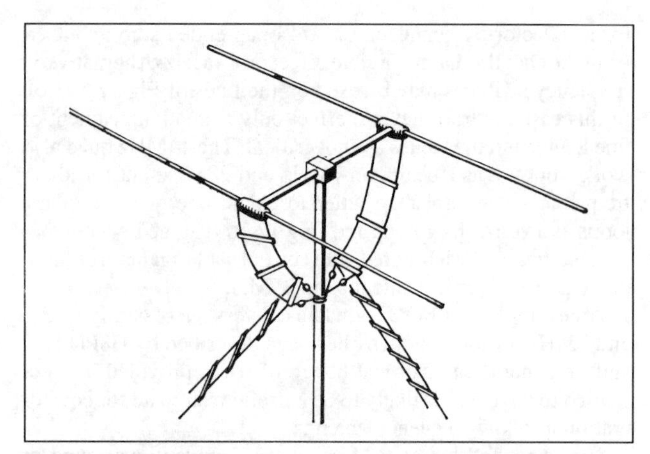

Fig 12.147. Arrangement to permit rotation of reversible beam with open-wire feeders. The top guy wires should be attached at or below the anchorage point of feeders. About 120° to 140° of rotation is sufficient in most cases

provided the one not in use is detuned. This is achieved most easily with the aid of resonant feeders, the ends of which are accessible from ground level, and such an arrangement also facilitates multiband operation.

Beam reversal may be achieved in various ways. Switching a director to act as a reflector is one possibility but may have considerable effect on the SWR. This may be overcome in the case of a three-element beam by using a switch or relay to shorten one or the other of the parasitic elements, an arrangement particularly recommended with loop or folded dipole elements having $\lambda/2$ resonant feeders which may enable both tuning and switching to be carried out at or near ground level. Any type of two-element beam can be made reversible by attaching feeders to both elements so that either can be used as a driven element, the other being tuned via the feeder as a parasitic element.

From this and earlier discussion it will be clear that the feeding of beams with pairs of open-wire feeders brings many advantages including multiband operation, instantaneous reversal of beam direction, and nulling of interference or optimisation of gain on any frequency from the operating position; against this it might be thought to create problems with beam rotation, or feeder radiation resulting in EMC problems, but if anything the reverse is true [6] in respect of EMC, and beam reversal means that only about 140° of rotation is needed. A suitable feeder layout is shown in Fig 12.147 and in some cases this enables a pair of cords to be substituted for the beam rotator.

Multiple arrays

It has been shown that unless considerable space or height is available, the effective gain of high-frequency beams cannot exceed (though it should equal) a total of 5 or 6dB. On the other hand, if enough height is available it may pay to stack two beams vertically, though it needs to be pointed out that the results of vertical stacking will be disappointing in terms of DX signals unless large heights are available; this is because the lower antenna drags down the mean height of the system relative to that of the top antenna alone, thus tending to wipe out any extra gain. As an example the low-angle gain for a beam at a height of λ and another one at half the height stacked below it is only 0.6dB greater than that of the top one alone. On the other hand, assuming a large garden, farm or access to neighbouring fields a number of other possibilities arise; it is quite likely that due for example to uneven ground contours or local obstructions different antenna positions will favour different directions, and by using a number of switched beams it may be possible to obtain considerable advantage in certain directions. If for example the ground slopes steeply in one direction, the ground reflection reinforces the direct wave in that direction even for angles which are low relative to the horizon, so that optimum performance may be achieved with a height as low as $3\lambda/8$ and closely approached with a height of only $\lambda/4$. To take full advantage of such slopes it is best to locate the antenna in a position some way below the top, as the ground area from which reflection takes place extends for some distance behind the antenna when the height is low and the angle of radiation relative to the ground is large. For other directions, unless slopes are available for these also it will usually be best to mount the antenna as high as possible and preferably at the top of the slope.

If two rotary beams can be spaced apart by at least $\lambda/2$, a gain of about 3dB may be obtained by driving them both, with a phase difference such that the fields add up in phase in the required direction. This does not involve critical adjustments, as the gain drops by only 1dB for a power inequality of 10 times or a phase error of nearly 60°, assuming that the two antennas are equally good. If one antenna is xdB worse than the other it should be given xdB less power, the benefits obtained from addition of the inferior antenna, for performance differences of 3dB and 6dB, then being 1.8dB and 1.1dB respectively. Phasing may be achieved by switching additional short lengths of feeder in series with the feed to one or other of the antennas, which will produce a 2:1 mismatch if both antennas are perfectly matched initially; this can be corrected, eg by a $\lambda/4$ transformer (p12.25), if it is outside the range of adjustment provided on the ATU. In principle the $\pm$X method of phasing can also be used but, unless corrected, the matching will then vary with the amount of phase shift. With balanced feeder a single $\lambda/6$ section of phasing line can be switched in as required to give either antenna a 60° lag compared with the other, and this with reversals gives six phases spaced 60°. The maximum phasing error is therefore 30° or, translating this into performance, 0.3dB loss. A six-pole six-way switch is required. With coaxial cable the arrangement shown in Fig 12.148 on p12.84 is suitable and can be used with any wide-spaced pair of antennas.

VERTICAL BEAMS

It is difficult to give any definite rules governing the choice of horizontal or vertical polarisation, because of wide variations in ground constants as well as individual circumstances, but some guidance [2] is provided by Fig 12.23 from which it may be inferred that a low vertical antenna over poor ground should be about equal to a horizontal antenna at a height of $\lambda/2$ for radiation angles of 2–8°, typical of those required for long-distance communication. Over good ground the vertical may be appreciably better. It should be noted that whereas a horizontal antenna is greatly helped by foreground reflections from a steep ground slope the vertical does so only to a much smaller extent, because the image is tilted back into the ground. Fresh water or marshy ground is likely to be 'good', but it is not an alternative to sea-water which owes its excellence to a conductivity which is much greater than that of any likely alternative.

Most of the horizontal beams described earlier can in principle be used in a vertical position. Other things being equal, the pattern of a vertical beam is narrower in elevation but broader in azimuth than a horizontal beam so that all-round coverage may be obtained with a smaller number of fixed beams, and it is also possible to use a single driven element surrounded by parasitic elements suitably switched to give omnidirectional coverage.

Despite these attractive features, vertical beams have not

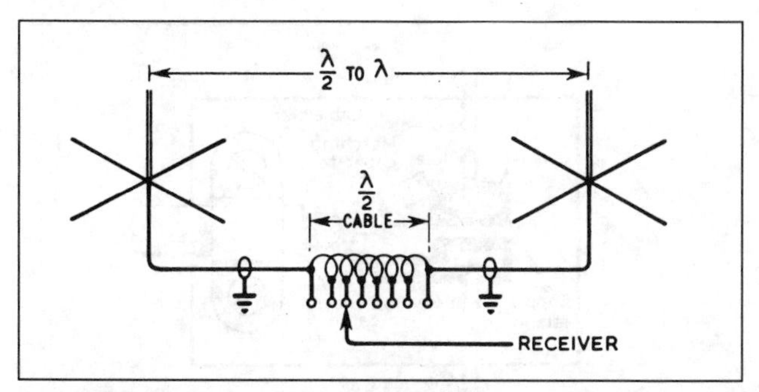

Fig 12.148. Arrangement of two ground-plane antennas providing electrical steering of the patterns given in Fig 12.12. This gives a gain of 3dB and is often a great help for interference reduction in a receiver as the nulls in the radiation pattern are steerable and the vertical polarisation tends to reject high-angle noise and interference. Elements based on Fig 12.97 are suitable and can be used also to form close-spaced arrays as described on p12.58

achieved much popularity. It is usually easier mechanically to erect a horizontal beam than a vertical beam at the same mean height which would of course be a greater maximum height, serious inefficiency may result from the proximity of masts, trees or other 'lossy' vertical objects, and radiation from feeders is more difficult to prevent due to the inherent asymmetry of vertical antennas. The use of lower-end Zepp feed as in Fig 12.109 is usually successful but cumbersome and restricts bandwidth. Figs 12.96(b) and 12.103(c) illustrate a preferred alternative.

From Fig 12.6 the impedance at the centre of a 1λ dipole is typically 5000Ω and, but for the mutual impedance between the two halves, Fig 12.9, this would rise to 8500Ω. The Zepp feed 'sees' in effect one half of this across one half of the feeder, equivalent to 17,000Ω across the whole feeder, and this is multiplied by about 2–4 times in the case of a close-spaced beam antenna with likely values of spacing and phasing, so that the effective Q will be typically of the order of 80 and the bandwidth much less than the width of, say, the 14MHz band.

If the maximum height is severely restricted, centre-feed is less of a problem, the height of the feed point being probably comparable with the height of an upstairs window, allowing the feeder to be brought away at right-angles all the way to the shack or at least for a considerable distance. Assuming sufficient ground area, vertical beams can in these circumstances be very attractive, even for the lower bands, and a simple array using two ground-plane radiators and providing all-round coverage with a gain of 3dB is illustrated in Fig 12.148. With λ/2 spacing and in-phase feed, the beam is bidirectional at right-angles to the line joining the antennas whereas with antiphase feed the pattern is rotated through 90° and the array becomes end-fire with a broader radiation pattern. This arrangement is suitable for use also on the low-frequency bands. There are of course many possible ways of combining the radiation from as many vertical radiators as space permits and in general a gain of n times can be realised from n radiators spaced upwards of half a wavelength. A particularly interesting arrangement suggested by GW3NJY [16] uses two pairs of grounded λ/4 verticals located around a common centre, ie at the corners of a half-wavelength square. By feeding one diagonally opposite pair in-phase and the other out-of-phase so that the two patterns coincide, then combining the two pairs with a phase shift of 90°, a unidirectional pattern is obtained which may be switched to provide all-round coverage. This idea should be equally applicable not only to ground-plane elements but also to dipoles, thereby avoiding earth losses.

Short end-loaded dipole elements can be assembled to form long Yagi arrays, as in Fig 12.149, or added to other arrays to provide increased gain in a particular direction (or more than one direction provided elements not in use are thrown out of resonance). Consistently good results have been obtained on 7MHz over the long path from the UK to Australia using up to six elements suspended from one or two catenaries, the lower ends being at times only a foot or so and never more than 6ft (1.8m) from the ground. The basic element size was 25ft (7.6m) for the vertical portion and 30ft (9.1m) for the horizontal loading sections, one or both of the lower arms being bent back or the vertical portion reefed in to obtain director action. Best results were obtained with the director elements shortened so that the currents decreased by about 30% from each element to the one in front, the spacing being λ/4.

It was the role of this antenna to be used exclusively for maintaining good long-path communication with VK during three sunspot minimum periods, being rolled up and stored away as soon as 14MHz conditions returned to normal. Each time the configuration was different, though favouring the use of directors, and with two or more of these the reflector if used was almost 'dead' so that it made little or no contribution to the forward field. The design was empirical, bearing little resemblance to one produced by G0GSF who used a computer to take the guesswork out of it [37], ending up with the two-element design illustrated in Fig 12.150 and the computed relationship between element dimensions shown in Fig 12.151. This indicates considerable latitude in respect of element shape, inclusive of the dimensions shown in Fig 12.149. For Fig 12.150, the computed gain and front-to-back ratio is given by Fig 12.152, and Fig 12.153 shows his recommended matching circuit, though this might have to be modified somewhat to suit a different set of dimensions.

The antenna is envisaged as a "simple wire-element Yagi with shortened elements hanging vertically from light nylon line", the elements being "light, almost invisible, and their wind

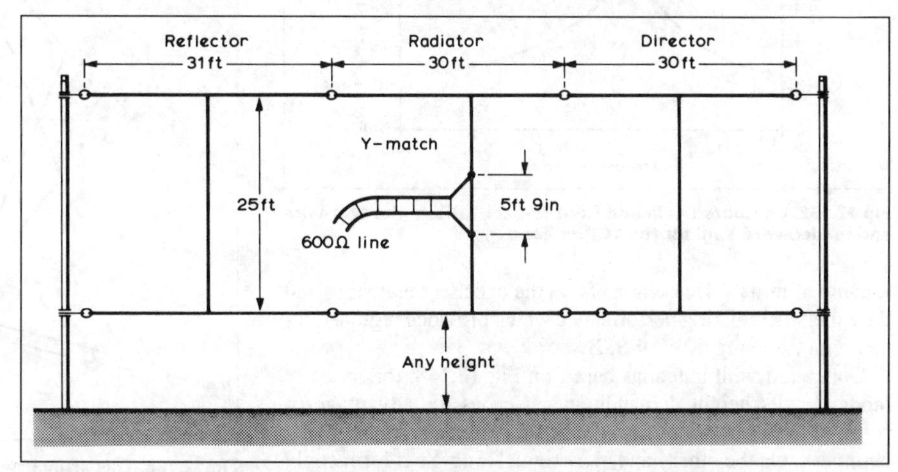

Fig 12.149. Vertical Yagi array using short end-loaded elements. Height is not critical, no drop-off in performance being noticed with 'sag' to within 1ft (30cm) of the ground. The reflector and director are tuneable by adjusting the length of the verticals or the span of the lower horizontal wires. Additional directors may be used to fill whatever space is available; the optimum spacing for such additions is probably about 0.35λ

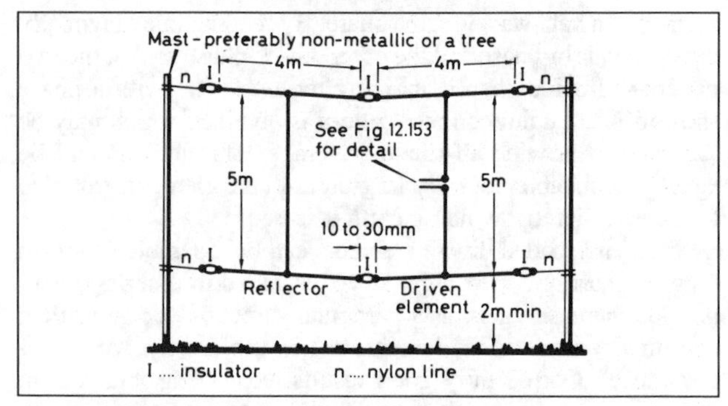

Fig 12.150. Two-element vertical end-loaded wire Yagi for the 14MHz band

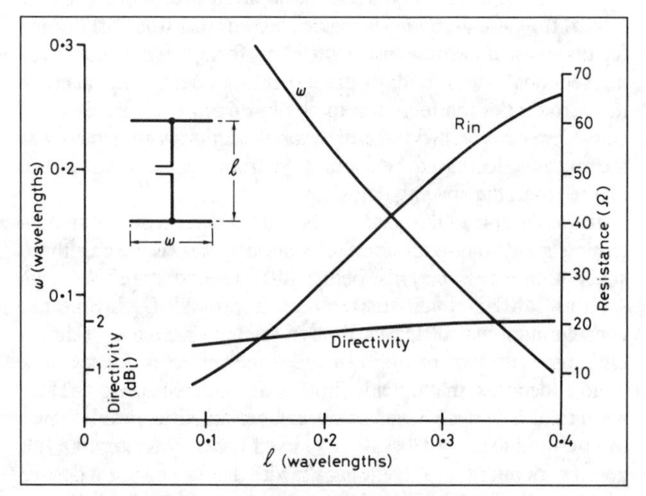

Fig 12.151. Computed relationship between *l* and *w* at resonance as well as the variation in the input resistance and directivity of the end-loaded dipole

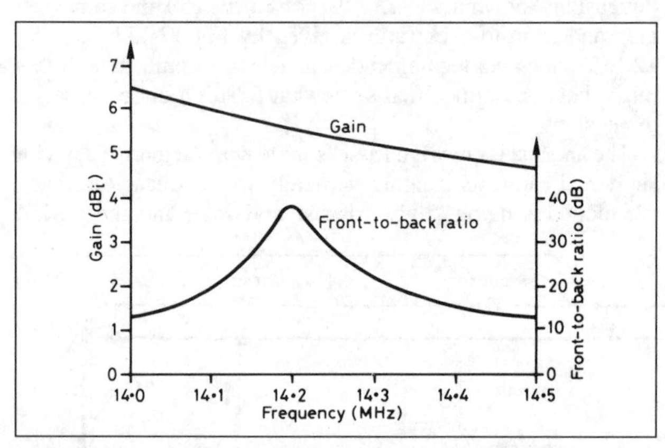

Fig 12.152. Computed gain and front-to-back ratio of the two-wire end-loaded wire Yagi for the 14MHz band

loading minimal'. He comments on the excellent matching and the excellent radiating performance which provided regular contact with ZS using 50W of SSB.

Compared with antennas based on Fig 12.148, this is more modest in its height demands and also has the advantage of being able to use widely available supports such as trees and chimneys. On the other hand, the original long-Yagi form could be particularly useful to someone with a long narrow strip of ground and severe height restrictions, though a seven-element 14MHz array scaled from the 7MHz design was disappointing to the extent of being 2–3 S-points down (at the good location) into VK compared with a typical horizontal array. Into South America the drop was only one S-point, however, and a computer study would no doubt improve matters.

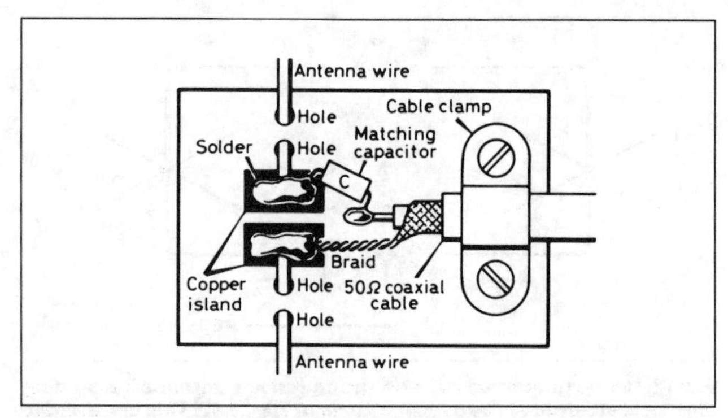

Fig 12.153. Single-sided glassfibre PCB at the feed point, potted in silicone rubber. C is 220pF for the 14MHz band

The maypole array

Derived from the quad, the *maypole array* [28–30, 33] provides similar gain using four λ/4 wires hanging down from a single pole as shown in Fig 12.154 with their top halves acting also as part of the guy-wire system. The wires are bent back at right-angles towards the base of the mast which in the case of a 21MHz array consisted of a fishing rod sitting on a fence post. Two or (as shown) three-element options are available, four-quadrant directional switching being readily achieved by means of plugs and sockets, but with the original version, as shown, remote switching is a matter of some difficulty due to high RF voltages at the feed point. An alternative version in which the elements are centre-fed is feasible [28] but rather cumbersome and the preferred solution is based on the new method of end-feeding shown in Fig 12.100. Fig 12.155 shows how this can be implemented in the case of the maypole array and Fig 12.156 illustrates one of several switching options; this was successfully implemented using a pair of 'ordinary' sealed relays out of the

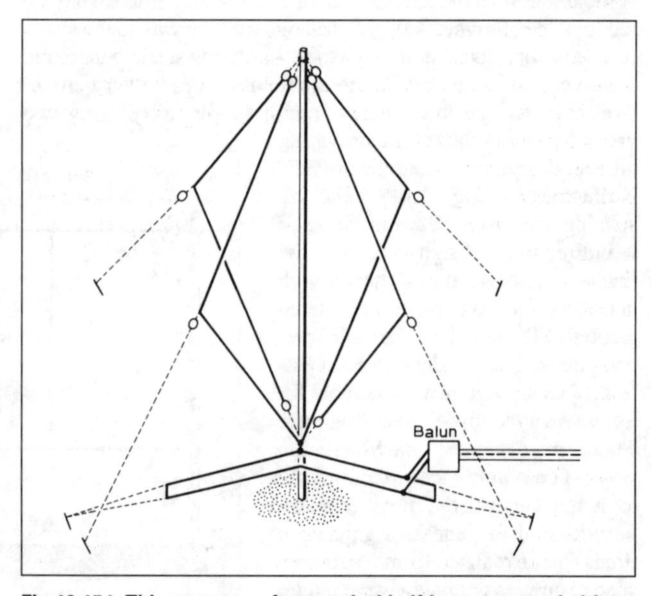

Fig 12.154. This array uses four vertical half-loops suspended from a single pole, with the top portions of the loops forming part of the guy-wire system. As shown, diagonally opposite pairs are fed in phase as the driven element, the other two elements being tuned as director and director. The equivalent quad configuration has some important advantages; see text

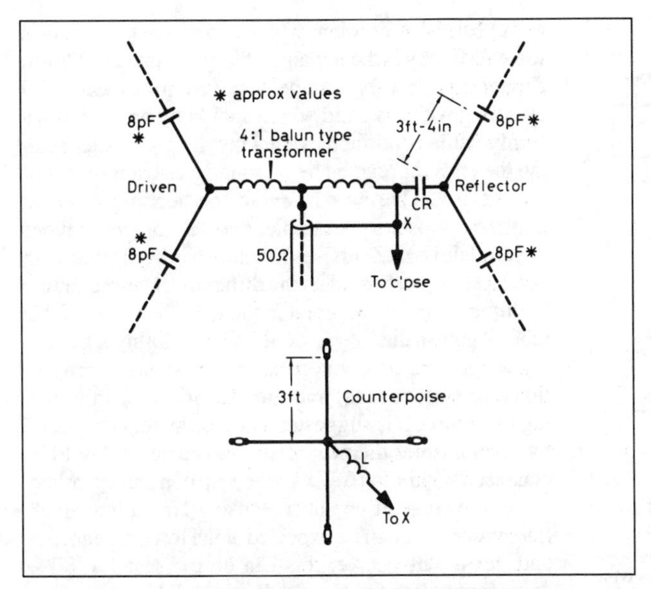

Fig 12.155. (a) Application of coaxial end-feed, Fig 12.103(c), to a four-quadrant switchable 21MHz array based on Fig 12.154 and supported by a single fishing rod. It uses short, matched single-wire lines to keep high RF voltages away from the switching. CR tunes reflector by resonating with the single-wire feeders (which in the reflector case are 'inductive to ground' so the inductance has to be removed). (b) Counterpoise details. Four radials (each 3ft long) are used for reasons of symmetry. L is eight turns, 2.75in diameter, 6in long, wound over fishing rod extension. This arrangement can be greatly simplified by substituting plugs and sockets or clips for the switching, making it ideal for portable operation

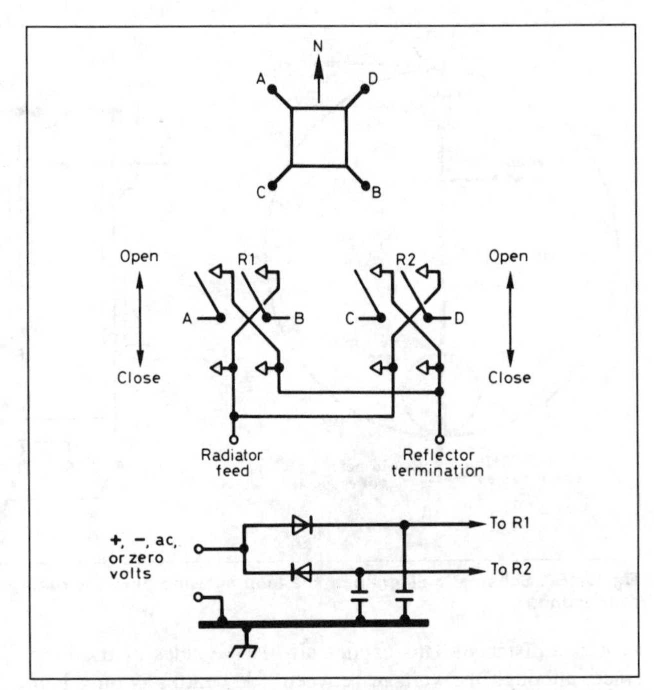

Fig 12.156. The switching sequence is as follows:

R1, R2 both open, BD driven, so beam heading EAST

R1 closed, R2 open, AD driven, so beam heading NORTH

R1 open, R2 closed, BC driven, so beam heading SOUTH

R1, R2 both closed, AC driven, so beam heading WEST

Note that the relay wires need to be well bypassed and filtered; also that there is the possibility of two-wire control for four directions as shown in the lower diagram

junk box but in the absence of components designed for the job it would be safer to use open relays or motor-driven switches mounted in plastic food-boxes for easy access. With these changes it has been found much better to use adjacent pairs of wires connected in parallel and arranged as a two-element beam, thus taking advantage of the wide bandwidth which is an inherent feature of this type of element; previously this had been masked by the narrow bandwidth of the Zepp feed. Adjustments are easier with two elements, and gain roughly the same. Multiple 'maypoles', phased along the lines of Fig 12.148, are a further interesting possibility.

SMALL LOOP ANTENNAS

The need for smaller transmitting antennas has led to an upsurge of interest in small loops; though not a new idea, these have only recently emerged as a best choice for this purpose, aided by the development of new components [29, 42, 43].

In seeking to reduce the size of antennas the essential aims are to keep the radiation resistance R as high and the losses R_l as low as possible. R is strictly governed by the dimensions, varying as the square of length in the case of short dipoles and, for small loops, the fourth power of loop diameter, ie inversely as the second and fourth powers of λ respectively. In the case of dipoles (or monopoles), size reduction, when pressed to the limit, demands a large amount of inductive loading, resulting in an adverse factor roughly proportional to λ into the comparison with loops. On the other hand, loops are able to overcome their initial handicap by using conductors of large diameter, resulting in very low values of R_l. As the inductance is also fairly low this implies large values of Q and correspondingly narrow bandwidths which in extreme cases may be no more than sufficient for coverage of a single SSB channel; this is a far cry from the full-band coverage commonly demanded, a typical case of mistaking the means for the end which can equally well be achieved by making the antenna remotely tuneable from the operating position. To achieve this result there are a number of problems to be resolved; firstly the tuning mechanism must conform to high standards of mechanical precision which tend to be expensive, and adding further to the design problems it is important to ensure that any added losses are negligible compared with the R_l of the loop.

Loops are commonly considered as responding to magnetic fields and dipoles to electric fields, but this can be highly misleading since all antennas respond to both fields, and it is interesting to note that the radiation resistance of a square loop can be worked out correctly either by the usual loop formula:

$$R = 31{,}200\,(A/\lambda^2)^2$$

or [29] by treating the loop as two pairs of antiphase-connected dipoles at right-angles (in effect, as a pair of very small W8JK beams – nothing could be more 'electric' than these), adding the radiation resistances. In the loop formula R is in ohms, A and λ^2 in square metres. After obtaining R from the above formula, the next thing we need to know is the loss resistance from Fig 12.5; this is proportional to $\sqrt{f}$ so that with $R \propto f^4$ we find on nearing the useful limit that efficiency is decreasing as $f^{3.5}$.

The need to minimise RF resistance makes it necessary to use copper tube of the largest possible diameter, in practice about 1in because of difficulty in bending larger sizes added to the law of diminishing returns. There should be no joins except (unavoidably) at the ends since solder, especially when corroded, is a likely source of RF resistance, and further to reduce losses the capacitor must be well constructed with no moving contacts. Also, because of very high voltages, 'lossy objects' of all kinds (including the ground and, not least, the operator) must be

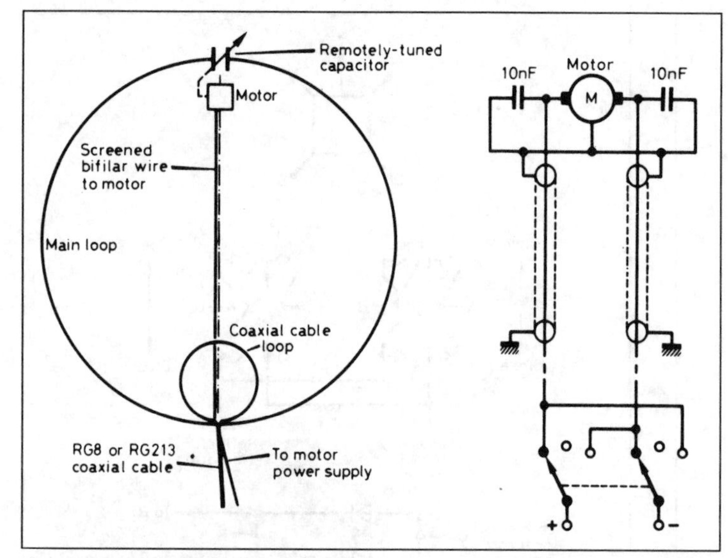

Fig 12.157. Schematic diagram of the loop antenna plus the tuning motor connections

kept at a distance! This applies also to the sides of the loop since, although the voltage between side-centres is only half that across the capacitor, it will still be large.

The feed normally uses a small coupling loop at the bottom of the main loop, and the tuning motor cables can be taken up the centre of the supporting mast which can be plastic or preferably glassfibre.

A loop circumference of 0.25–0.3λ at the highest frequency is recommended, except that at 29MHz loops larger than 0.25λ have been found difficult to tune. A circumference of 0.25λ at 28MHz translates into a loss of only 3dB at 14MHz and about 7dB at 10MHz, but in view of the rapid decrease of bandwidth and efficiency, added to escalation of voltage across an expensive capacitor, restriction of frequency range to one octave [42] would appear to be good advice.

To this must be added the importance of ensuring symmetry and checking for possible RF currents on the feeder or motor control wires (which must be screened as well as effectively bypassed at the motor). If helpful the braids could be grounded by means of short loaded counterpoises (p12.58) at right-angles to the plane of the loop. They should also be bonded to any metal supporting mast.

Figs 12.157–12.160 give the main constructional details [42] of a loop for 14 to 29MHz, apart from the tuning capacitor and motor which will depend on the contents of the reader's junk box, his success in tracking down 'surplus' components, the size of his bank balance, or whether he really needs to cover five (or more) bands! A problem which appears to have caused some difficulty is the bending of large-diameter (22mm) copper pipe, but this can be achieved quite easily [42] by filling with dry sand which should be pressed down firmly while continuously shaking the pipe. After sealing the ends this can be bent around a circular object of correct size. One way to avoid the bending problem might be to use, for example, two ½in diameter tubes in parallel; I1ARZ advises that this has been tried with good results, but considers the difficulty of co-ordinating multiple loops to be greater than the bending of 1in tube. Against this, there could be possibilities of saving weight and cost (given that RF resistance is proportional to surface area) and the idea seems worth placing on record; it is suggested that conductors be spaced by several times their diameter, and currents should be equalised within 5–10% with the help of a current probe.

For coverage of one octave (using 1in tubing) an efficiency of 50% can be expected at the lowest frequency and, for a half-octave, this can be increased to 80%; these figures are for circular loops and, because of the reduced area, decrease to 34% and 70% respectively for the same length of tubing formed into a square loop. For a monoband loop 5ft in diameter an efficiency in excess of 90% at 14MHz can be expected, and Fig 12.161 shows how this varies with frequency. Fig 12.162 shows how efficiencies corresponding to those in Fig 12.161 vary with the conductor diameter, and in line with this I1ARZ reports favourably on loops formed from RG-213 coaxial cable using the outer braid only. The author is also indebted to I1ARZ for information that in the monoband case fine tuning is possible by means of a coil wound on a toroid core which is threaded onto the bottom of the loop; this coil is connected by coaxial cable to a similar coil in the shack, series-tuned by a large variable capacitor. In this way it is possible for narrow-band operation to avoid the problems of remote tuning with very-high-voltage variable capacitors.

For further advice and suggestions the references should be consulted. It is encouraging to note that the small loop seems now to be firmly established as a 'no compromise' transmitting antenna. Further to this, reference [41] points out that it performs well indoors, but also draws attention to the potential hazard of prolonged exposure to strong magnetic RF fields, though the degree of risk is still an unresolved question.

Small loops seem to be acquiring an enviable reputation for working well close to ground and the author's experience of zero losses in the case of parasitic ground-plane elements [29, 33] suggests a possible explanation since in both cases current is able to flow between the terminals of the 'equivalent generator' without necessarily flowing through the ground. In the case of the loop the path is a series-resonant circuit of extremely high Q and therefore low impedance. On the other hand, if the top and bottom of the loop are visualised as very short horizontal dipoles, some ground losses are to be expected (p12.61) in which case they should be reducible by means of a ground screen under the loop.

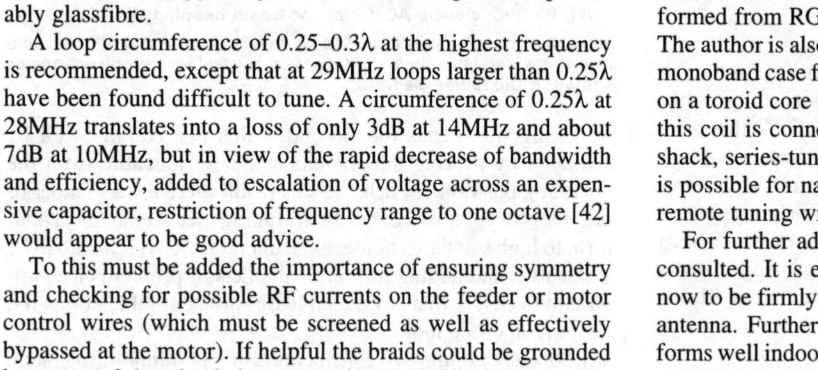

Indirect confirmation of this can be inferred from claims that such screens act as 'reflectors', though in real life reflections occur elsewhere and involve much larger areas. A more likely explanation (assuming low height) would be the reduction of ground losses.

As an example, a loss figure for a λ/16

Copper saddle bracket soldered to loop with torch-flame
'U-bolts' for fixing loop
Supporting plastic mast
Loop
Loop
Metal mast sleeved over plastic mast
Coaxial socket
Motor feed connector

Fig 12.158. Details of the bottom of the loop – from the front

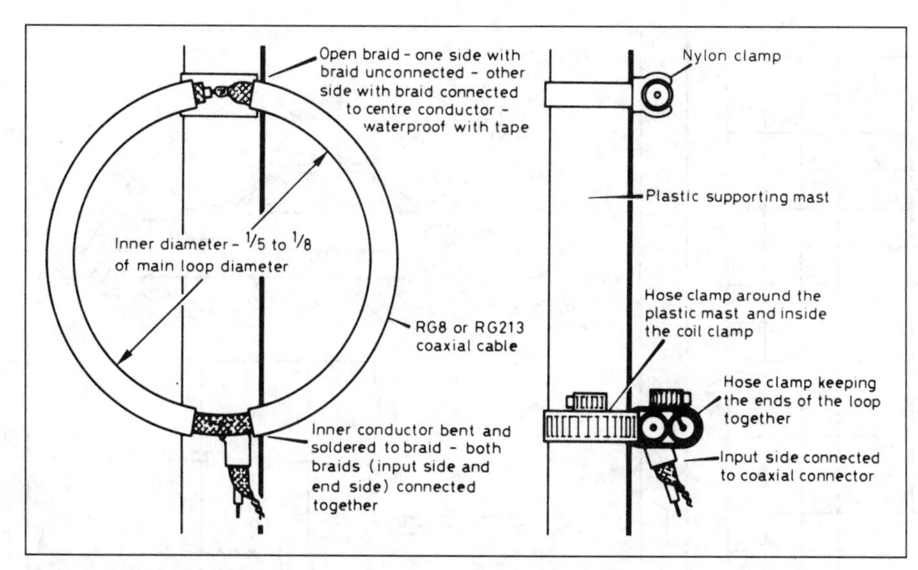

Fig 12.159. Construction of the coaxial feed loop

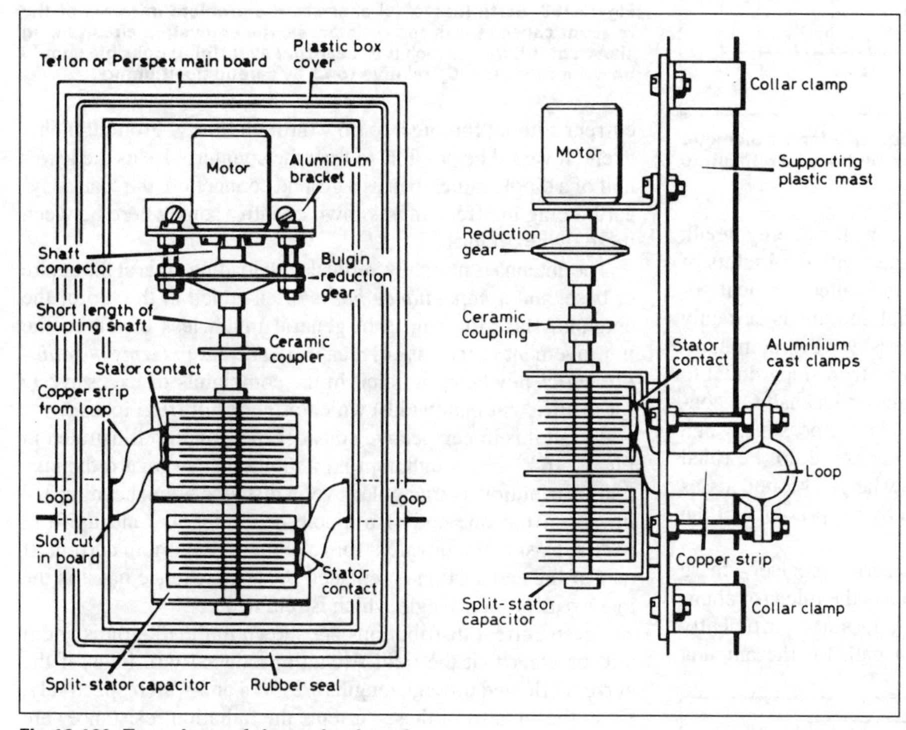

Fig 12.160. Two views of the tuning board

square loop at ground level can be estimated from the relative lengths of conductor at zero height, recalling that a $\lambda/2$ dipole is equivalent to a λ/π dipole carrying a uniform current. This yields a resistance of $(\pi/16)^2 \times 80 = 3\Omega$. For a loop of this area $R = 0.48\Omega$ and for the loop away from ground at 14MHz we have $R_l = 0.13\Omega$. It follows that if all the loss can be removed by using a conductive mat to provide a low-resistance path for currents induced in the adjacent ground, efficiency can be increased from 13% to 79%.

Figures for different heights and circular loops are difficult to calculate, but the loss will decrease very rapidly with height, and division by a factor somewhere between 10 and 100 can be expected for a circular loop at a height equal to its diameter.

There appears to be some confusion also about directivity and gain, in which respects it will by now come as no surprise to readers that 'small loops' (at the upper end of the recommended frequency range) differ little in respect of gain from other small radiators of sizes up to and including that of a quad element, though they are *not* usable as close-spaced beam elements; it will also be noticed that the radiation pattern is omnidirectional in the plane of a loop but 'cos θ' at right-angles, whereas the opposite applies in the case of dipoles and ground-plane antennas. In free space it would be impossible to determine from the radiation pattern which type of antenna is being used to produce it, and at a height of 0.3λ the antenna with its image constitute a collinear pair providing 3dB of gain in both cases. At ground level the loop, especially if rotatable, should be more effective for discriminating against local noise sources, while the lack of vertical directivity makes it equally suitable for DX and short-skip communication; on the debit side, short-skip QRM from the wanted direction or its reciprocal will be worse.

If despite all the obvious precautions the expected performance is not achieved, radiation from the feeder (at the wrong angle, or absorbed by lossy surroundings) is the most likely culprit; the risk of this is reduced by feeding the loop at the bottom, placing the capacitor at the top in line with the above references, and contrary advice (currently prevalent) should be treated with caution. Loops can give a good account of themselves at low heights and even in some cluttered environments, but claims that they do so underground must be rejected, being probably accounted for by feeder radiation picked up on mains leads and thence from a very large antenna! Though mostly used in a vertical position, the possibilities of horizontal mounting should not be overlooked as the resulting cancellation of fields in the vertical plane should tend to eliminate ground losses at low heights; losses in adjacent walls should also be reduced and in theory the DX performance of a small horizontal loop near the top of a high-rise block should almost equal that of a three- or four-element beam at half the height, given the linear relation [28, 29] between height and low-angle field strength. Attenuation through the building should tend to result in good front-to-back ratios. At the least, this would appear to be an interesting field for experimenters.

MOBILE ANTENNAS

Earlier in this chapter the reader was introduced to the basic theory and practical aspects of fixed antenna installations. Much of this applies to antennas of all types but there are a number of

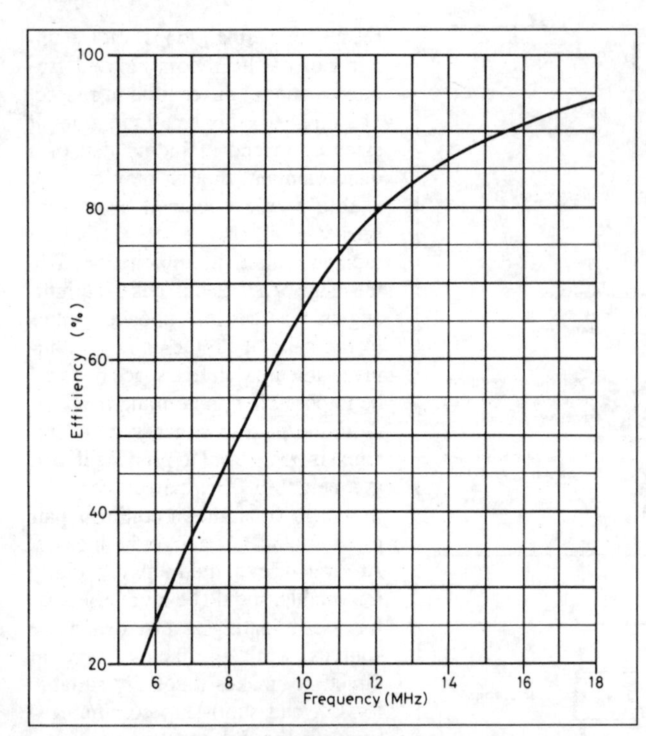

Fig 12.161. Variation in efficiency with frequency for 5ft diameter circular copper loop. Conductor diameter is 1in (may be difficult to tune above 14.5MHz)

features peculiar to mobile operation. As well as being small, mobile antennas must be rigidly constructed with road safety of primary importance, and height is strictly limited; vertical polarisation is normally considered essential and this is certainly true for ground-wave and medium-range sky-wave communication but, parked by the roadside with a steep slope down to the sea and an inverted-V antenna, the author has had Q5 contacts with VK using only 0.25W of SSB, and the possibility of a nearly-as-good antenna attached to the car could not be ruled out solely on technical grounds. Due to low height, ground losses play a more important role, complicated by the presence of the car body as illustrated in Fig 12.163.

Though difficult to analyse or even describe precisely, there must be some loss since the car body breaks the rules for counterpoise earths, ie that they must be low loss and sufficiently clear of the ground to provide a return path for the antenna

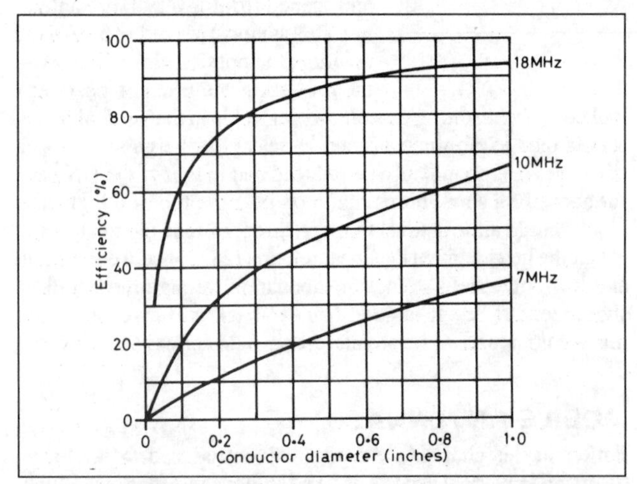

Fig 12.162. Variation of efficiency with conductor diameter for 5ft diameter circular copper loop

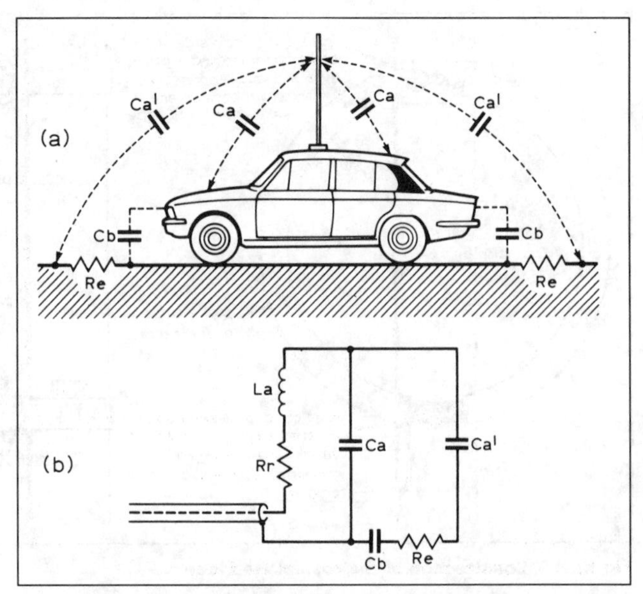

Fig 12.163. Earth losses. (a) analyses the problem in terms of the relevant capacitances and resistances, the equivalent circuit being shown at (b), from which it is clear that everything possible should be done to reduce C_a' relative to C_a by careful positioning

current without forcing it to flow through 'lossy' ground. In this event, it would be possible to treat the counterpoise as the lower half of a dipole rather than as a ground connection, the 'car body' earth being in effect an unknown quantity somewhere between these two extremes.

The antenna is normally vertical with loading coils at the centre or base, and a *capacitance hat* is often added at the top of the antenna. Base-loading is in general much less efficient, the improvement in the case of a capacitance hat *in terms of radiation efficiency* being implicit in the complaints of excessive Q (or insufficient bandwidth) which often result from its use!

To get this in perspective consider the situation illustrated in Fig 12.164. As a rough approximation one can regard the current distribution as triangular over a distance not exceeding $\lambda/8$ inwards from one end of a dipole or monopole, and uniform over at least $\lambda/16$ outwards from a point of maximum current. If part of the end section is replaced by a capacitance hat it is the inner part of the triangle which is retained.

These current distributions are shown in the diagrams and it will be clear that the field strength produced will be as if the current I flowed through lengths $h/2$, $3h/4$ and $7h/8$ respectively. From the squares of these currents the radiation resistances are in the ratio 1, 2.25 and 3.063. It will be found that at the lower frequencies the inductance of the antenna below the coil is negligible compared with the required loading inductance. If this equals L in case (a), we require $2L$ in case (b) and in case (c), since eb = h, it is again equal to L. Since the value of L has little effect on the best obtainable coil Q, which is likely to be of the order of 300 in each case, the coil loss resistances R_c are given respectively by $\omega L/300$, $\omega L/150$, and $\omega L/300$. It is necessary also to assume an earth-loss resistance (unknown) of R_e which is the same in all three cases.

From inspection of Fig 12.165 it is found that the proportion of power radiated is given by $R/(R + R_c + R_e)$ but at low frequencies R is only a fraction of an ohm and very much less than R_c or R_e so that the efficiency becomes $R/(R_c + R_e)$ approximately. The capacitance of an 8ft (2.4m) whip is in the region of 30pF so at 3.5MHz we have for case (a) $\omega L = 1/\omega C = 1600\Omega$, $R_c = 5.3\Omega$ and a value of 10Ω is often assumed for R_e. For case (b), R

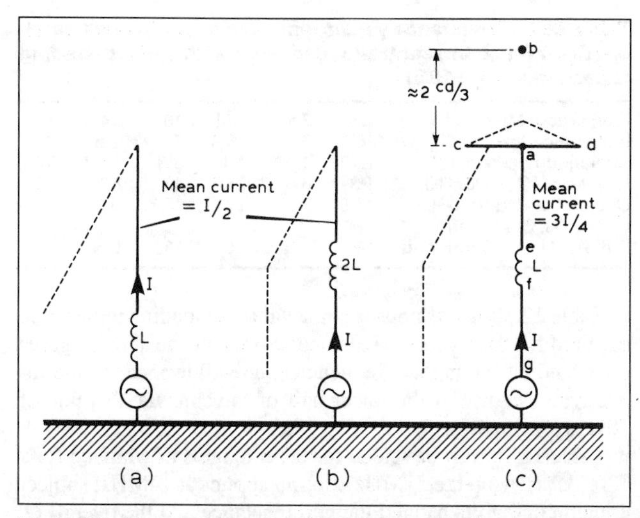

Fig 12.164. Alternative methods of loading a mobile antenna. The height *h* of the radiator is assumed to be the same in each case, and the shape of the current distribution is shown by dotted lines. At (c) a hypothetical increase in height ab = ae has been replaced by horizontal loading cd which leaves the current distribution below a unaffected. The length of cd is roughly equal to 1.5 ab. The value of *L* is doubled in case (b) since the length of wire above it, and hence the capacitance, is halved

is multiplied by 2.25 and R_c by 2, and for case (c) R is multiplied by 3.06. The efficiencies (relative) are 0.06, 0.11 and 0.2 so that centre-loading is 2.6dB better than base-loading, and the top hat gives a further improvement of 2.6dB.

It may be of interest to note that with $R_e = 0$ these ratios become 0.5dB and 4.35dB respectively so that the improvement due to centre-loading comes about mainly because of the earth resistance. Without this, the rise in R_c largely offsets the improvement in R. The overall Q is given by $\omega L/(R_c + R_e)$, ie 105, 78 and 105 respectively, the 2.6dB improvement in case (c) compared with (b) being obtained at the price of a 26% reduction in bandwidth. However, it is important to note that this denotes an improvement in efficiency due to use of the hat and (with so little in hand) very much a feature to be encouraged.

The situation is slightly different at 14MHz because the inductance of the lower portion of the antenna is no longer negligible compared with L and the current in the centre of the antenna is some 13% less than at the base. The reactance above the coil is 400, 800 and 400Ω respectively but the lower half of the antenna contributes about 200Ω of inductive reactance in cases (b) and (c), so that ωL = 400, 600 and 200Ω respectively. Corresponding values of R_c (ohms) are 1.33, $2.0 \times (0.87)^2$ and $0.67 \times (0.87)^2$, these being the values referred to the feed point for comparison with R, the factors in brackets serving to take account of the sinusoidal current distribution. The radiation resistance in case (a), from Fig 12.166, is 5.5Ω and the factors for cases (b) and (c) are approximately as before, so that (again taking R_e as 10Ω) the efficiency is 33, 52 and 62% respectively. The improvement from centre-loading is 2dB and the further advantage from the capacitance hat is only 0.8dB. With R_e removed, the top hat would reduce the coil losses by four but these are small enough for the improvement in overall performance to be negligible.

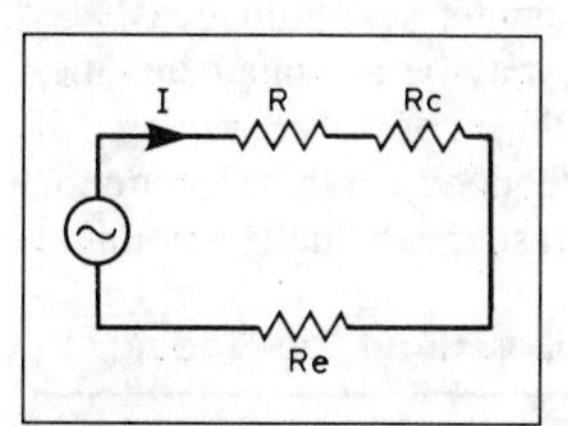

Fig 12.165. Estimation of efficiency of mobile antenna. *R* = radiation resistance, R_c = loading coil resistance, R_e = earth losses. The antenna is assumed to be resonant

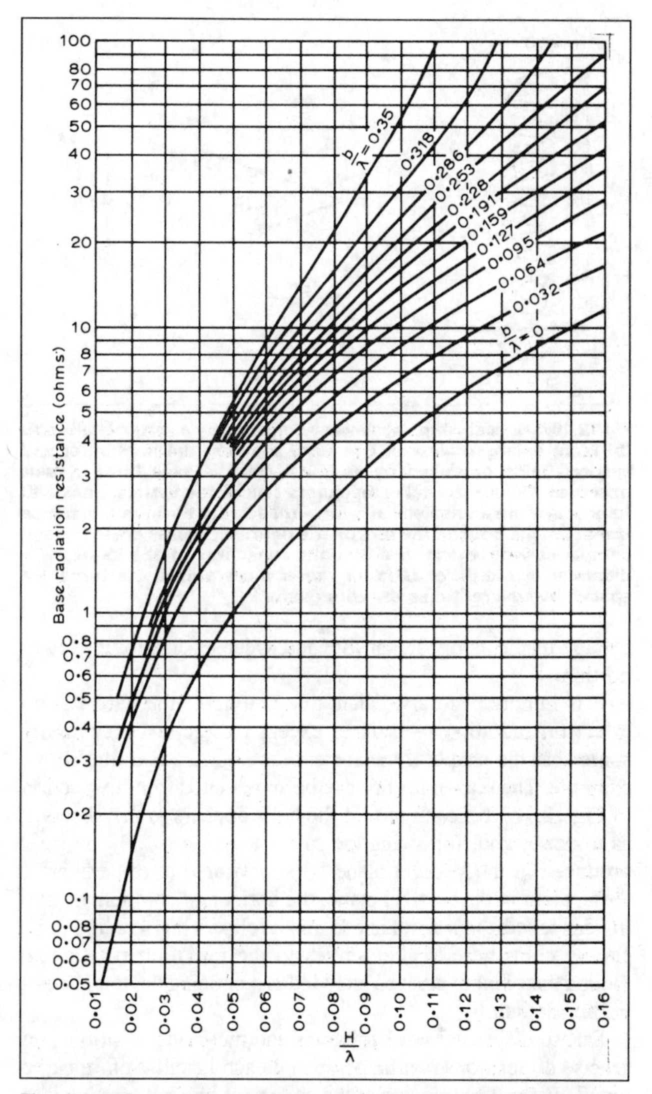

Fig 12.166. Base radiation resistance (Ω) of vertical antennas up to 0.16λ long with non-radiating top termination. Height of antenna = *H*, length of top = *b*. The special case of *b* = 0 represents an unloaded short vertical antenna

At 28MHz the antenna is self-resonant and losses are even less. One might expect this to be reflected in impressive performance but, for DX communication as compared with 14MHz, an average fixed station will be obtaining an extra 6dB or so of height gain from its horizontal antenna and the mobile station can expect to be relatively down by some 9–10dB.

Radiation resistance is proportional to f^2 and, for heights of λ/8 or less, also to h^2. The 8ft (2.4m) base-loaded whip has a radiation resistance at 3.5MHz of only 5.5/16 = 0.34Ω and the efficiency is therefore 0.34/15.64 or 2.2% only, rising to 6.6% for case (c).

Fig 12.167 shows another common form of mobile antenna, the *helical whip*. A 3.5MHz helix might consist of about λ/4 of 14swg (2mm) wire wound on an 8ft (2.4m) rod of 1in (25mm) diameter. From Fig 12.5 the HF resistance is 0.8Ω and, assuming a sinusoidal current distribution, the radiation resistance may be obtained directly from the height of the whip relative to that of a λ/4 monopole, ie it is equal to $(4h/\lambda)^2 \times 36\Omega$, giving an efficiency of 3.6% compared with 4% for the centre-loaded whip without a capacitance hat.

Both systems can be improved by a hat, and there seems little to choose between them. The helix can be somewhat improved

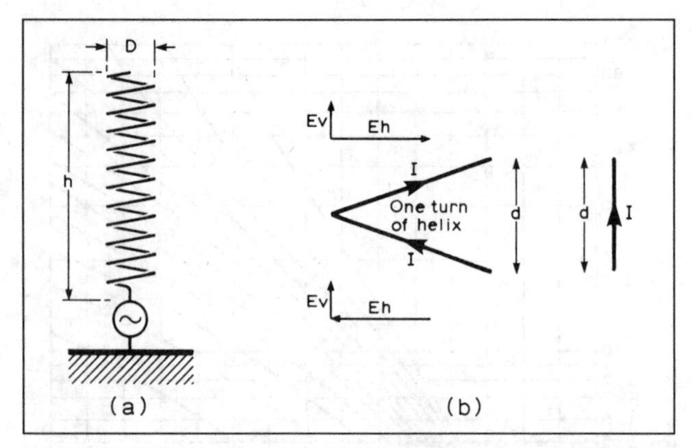

Fig 12.167. Helical whip. (a) shows schematically a vertical helix and (b) is an enlarged view of one turn, showing the horizontal and vertical fields produced by each half-turn as seen by a distant observer. The horizontal components cancel, the vertical ones add to produce the equivalent of a current *I* flowing through a distance *d* equal to the pitch of the turn on the right. The losses are increased compared with those in a straight conductor of the same wire diameter in the ratio $\pi D/d$, or rather more unless the turns are spaced by several times the wire diameter

by tapering, ie using thicker wire and wider spacing for the lower portion.

It is important to appreciate that coiling of the wire has no effect on radiation resistance, except indirectly by changing somewhat the shape of the current distribution as in the above example. The reason for this can be appreciated from inspection of Fig 12.167(b); each turn of the helix appears to the observer as a zigzag and, following the rule set out on p12.3, the field produced in the vertical plane is proportional to the extent of these wires in the vertical plane, the horizontal components being cancelled. As a corollary to this it follows that loading coils cannot contribute in themselves to the radiation resistance, though their length must be included in reckoning the total height of the antenna.

Due to differences in car bodies and mounting positions, no precise dimensions can be given and each installation must be tuned *in situ*. The position of the coil is not critical; moving it to the middle of the top half at 3.5MHz requires L and therefore R_c to be doubled, but increases R in the ratio $(7/6)^2$ and reduces the efficiency by 10%. There is therefore a loss in signal strength of 0.4dB, though there could be a slight additional loss due to the self-capacitance of the coil which could be approaching self-resonance. Assuming a stray capacitance of 1.5pF, which is probably equal to about 20% of the total tuning capacitance, this increases the current in L by 20% but reduces the required value of L by the same amount, so that there is a net increase of 20% in the effective value of R_c. This results in a signal loss of 0.3dB.

The capacitance of a short whip antenna ($<\lambda/8$) can be reckoned as roughly 3.4pF/ft (11.2pF/m). This assumes an average diameter of 0.5in (13mm) and is a bit on the low side for longer lengths but, subject to normal practical constraints, differences of more than 15% are unlikely. Such figures can in any case be used only as rough guidelines in view of the many other variables as already indicated.

Table 12.11. Inductance values, efficiencies and overall bandwidth for 8ft (2.4m) centre-loaded whip with 7pF top-loading capacitance (R_e = 10Ω)

Frequency (MHz)	1.8	3.6	7.05	10	14.2	18
Inductance (μH)	287	67	16.8	7	2.6	0
Radiation resistance (Ω)	0.27	1.09	4.1	8.3	16.3	27
Efficiency (Q = 200) (%)	1.3	6.8	24.7	43	60	72
Overall bandwidth (kHz)	11.5	37	141	450	934	
Coil turns, 3in (7.6cm) diam, 6in (15.2cm) long	98	47	24	15	9	

Table 12.10 shows approximate values of loading inductance required for the various bands on the basis of the above figures for C and, at the higher frequencies, due allowance for the inductive reactance of the lower half of the antenna. A value of 10Ω is assumed for R_c.

Table 12.10 may be scaled for other lengths of whip, eg for a 12ft (3.7m) whip the 21MHz column applies at 14MHz, subject to maintaining unchanged the coil reactance and the overall Q. In other words, the inductance for 14MHz becomes 2.1×1.5 = 3.15μH and the bandwidth (Q = 100) becomes 1420/1.5 = 947kHz. For maximum Q the coil size should be as large as possible but it is not critical. Weather protection tends to degrade the Q but is necessary if there is any chance of moisture being trapped between turns. The use of a long coil with self-supporting turns should avoid this but it is possible only at the high frequencies; too long a winding degrades the Q and a diameter of 3in (7.5cm) with a length of 6in (15cm) is a reasonable compromise. Coil turns may be obtained from Table 12.11, using the fact that inductance is proportional to the square of the number of turns.

For multiband operation the best method is to change the coil since it would appear that any attempt at variable tuning must result in some loss, particularly if a capacitor is used as this 'works' by increasing the current in the coil; typically, if I is doubled then L is halved and losses doubled.

To obtain maximum benefit from use of a hat it should be as large as possible, so that its removal or reduction as a means of band changing is not acceptable, except perhaps at the HF end of the range where efficiency is still moderately high even without the hat.

Top-loading

A horizontal bar as in Fig 12.164(c) is not the most convenient method of top-loading, a 'skeleton' disc, Fig 12.168, being generally favoured [1]. For a solid disc the capacitance in picofarads is roughly equal to 0.9 times the diameter in inches (0.36 times the diameter in centimetres) and the skeleton disc is only slightly less. To obtain the required capacitance of 13.5pF a disc diameter of 15in (38cm) with at least four spokes should be satisfactory. Alternatively, for a cylinder having a length equal to the diameter [2] the capacitance in picofarads is equal to twice the diameter in inches (0.8 times the diameter in centimetres).

One effect of top-loading which may be advantageous mechanically is that it allows the loading coil to be mounted if necessary in a lower position. The tendency in this direction is evident if one thinks of an extreme case in which the amount of

Table 12.10. Inductance values, efficiencies, and overall bandwidth for 8ft (2.4m) centre-loaded whip without top-loading

Frequency (MHz)	1.8	3.6	7.05	10	14.2	18	21.3	25	28.5
Inductance (μH)	582	142	35.5	17	7.24	4	2.1	0	
Radiation resistance (Ω)	0.2	0.8	3	6	12	17	21	28	36
Efficiency (Q = 100) (%)	0.26	1.9	10	22	42	62			
Efficiency (Q = 300) (%)	0.62	3.7	16	30	50	60	66	73	78
Overall bandwidth (Q = 100) (kHz)	21	48	124	490	1420				
Overall bandwidth (Q = 300) (kHz)	9	24.5	78	417	1290				

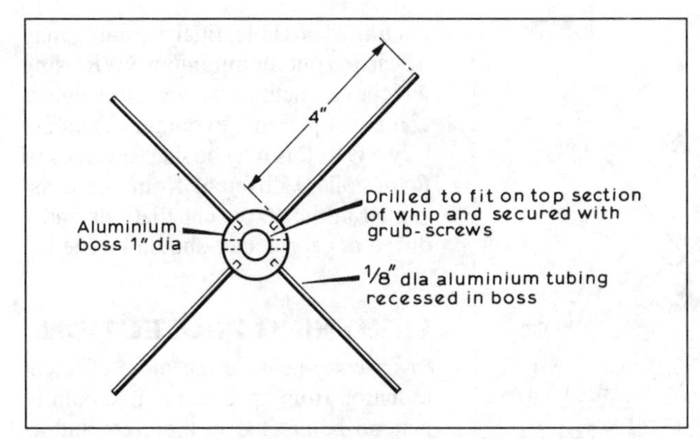

Fig 12.168. Capacitance hat suitable for vertical whip antenna

top-loading is so large that it results in uniform current in the vertical portion of the antenna. Further loading, even if placed at the base, then has no effect on the radiation resistance.

To take a more practical example, assume an 8ft (2.4m) whip with the loading coil at the top and surrounded by a cylinder of 8in (20cm) length and diameter; the current distribution is virtually uniform, thus increasing the radiation resistance by four compared with Fig 12.164(a), ie to 1.36Ω at 3.5MHz. The capacitance is 16pF so that L = 129μH (total) or 125μH for the coil. The coil resistance (Q = 300) comes to 9.16Ω and the radiation efficiency is 1.36/20.52 or 6.6%, ie very slightly worse than the previous example in which the coil was placed halfway down. Next consider base-loading, but since the coil no longer has to be supported, let it be supposed that the cylinder can be increased to 14in (35.6cm) length and diameter; this doubles C, thereby halving L and R, but the current is now nonuniform to the extent that the radiation resistance factor is 2.25 (as for Fig 12.164(b)) instead of 4. The efficiency is therefore 0.8/15.4 or 5.2%, which is slightly worse, being 1.5dB down in terms of signal level compared with the original figure of 6.8% efficiency for centre- plus top-loading, but still 3.74dB up compared with simple base-loading.

There is of course an exact optimum position for the loading coil, but the optimum is extremely flat, especially (as we have just seen) if top-loading is also used, and there is unlikely to be much advantage in departing from the centre position for the loading coil. Table 12.11 provides design data for top-loading in accordance with Fig 12.164(c). The whip will be roughly self-resonant at 21MHz and for higher frequencies it will be necessary to reduce the amount of top-loading.

Coil construction

Nylon or PTFE is recommended for coil formers and the wire gauge should be chosen so that the spacing between turns is not less than 0.6 times the wire diameter. Self-supporting air-spaced coils are preferable at the higher frequencies and can be wound with 3/16in (5mm) or ¼in (6mm) diameter copper tubing. Mechanical fixings can be devised using, for example, wing nuts to allow interchange of coils plus the top sections; however, tapped coils may be used for greater convenience if some slight drop in efficiency is accepted at the higher frequencies. One method of coil assembly [1] is shown in Fig 12.169. Loading coils should be protected from moisture and this can be done by means of shields cut from old polythene bottles.

Earth resistance

Despite the crucial role of earth resistance and the capacitance to ground of the car body, there are few data available for these

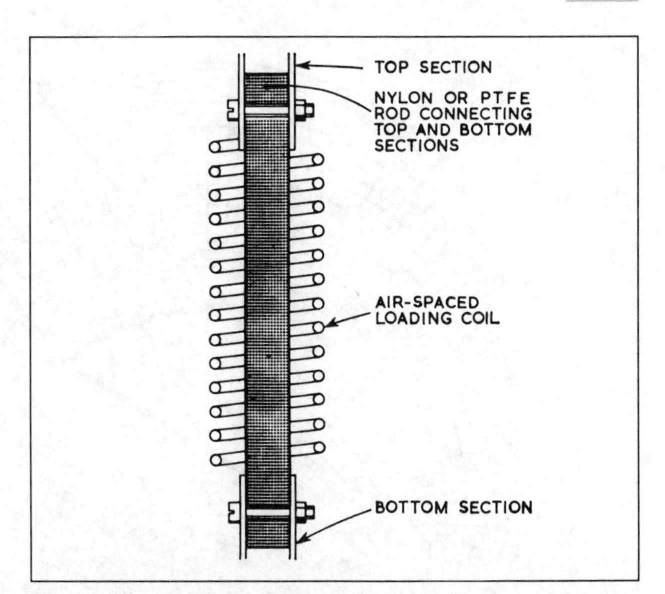

Fig 12.169. Loading coil assembly. An insulated rod is inserted into the bore of the whip above and below the coil, and runs through the centre of the latter. The rod must have good insulating properties and mechanical strength, PTFE or nylon being the most suitable materials

quantities, the 10Ω resistance assumed in the above calculations being a value which happens to be consistent with some measurements at 4MHz [48].

A possible source of earth resistance is poor electrical contact between different portions of the car body and, if for example the antenna is mounted on a wing or bumper, it is advisable to check the DC resistance between this, the chassis and other large areas of metal such as the roof.

Matching

Matching between the mobile antenna and the transmitter is probably best achieved by means of an L-network, Fig 12.170. For the examples given earlier the total resistance R of the antenna is very roughly 20Ω in all cases, since reduction in radiation resistance is accompanied by increases in coil resistance and a constant figure of 10Ω has been assumed for R_e.

By tuning the antenna 'slightly long' a reactance $+X_s$ is caused to appear in series with R_a and this is equivalent to a parallel combination of resistance R_a' and reactance X_p. For matching, X_s is chosen so that $R_a' = Z_0$ and the resulting value of X_p is then tuned out by a parallel reactance $-X_p$. This can be done very easily with the Smith chart; referring to Fig 12.171, moving halfway round the circle from any point P corresponding to (R_a + jX_s) brings us to a point P′, the co-ordinates of which are equal to $1/R_a'$ and $1/X_p$. For this exercise one uses the 'normalised' values of R_a and X_s obtained by dividing Z_0 into the actual quantities, and the answer is interpreted via the reverse process of multiplying by Z_0 (p12.40).

(p12.40)

For matching to 50Ω the normalised R_a is 0.4, and going from P to P′ must be done so that, starting from a point on the R = 0.4 circle and proceeding on a straight line via O, we arrive on the R = 1.0 circle in such a way that OP is equal to OP′. Laying a transparent ruler across the chart, it is easy

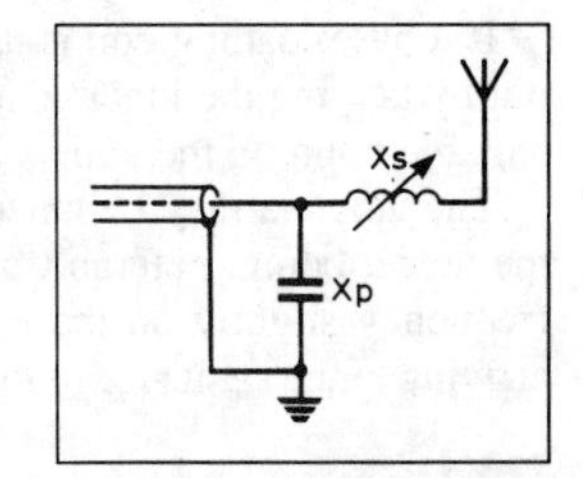

Fig 12.170. L-match. X_p is determined as shown in Fig 12.171; use of this value ensures that a match can be achieved by tuning the antenna slightly low

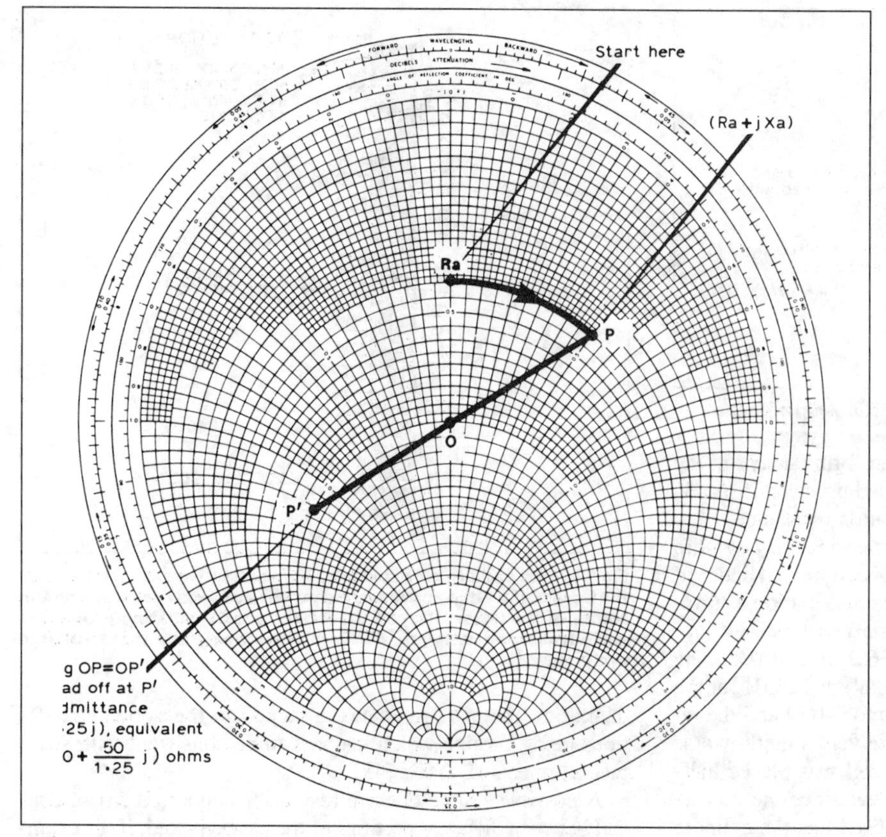

Fig 12.171. Use of Smith chart to determine X_p. Given that $R = 0.4Z_0$, points are found on the 0.4 and 1.0 resistance circles such that OP = OP′, all three points being in a straight line. This gives the value of X_p and also of X_s, though this latter value is found in practice by tuning for minimum SWR after connecting the right value of X_p

to find a position such that intercepts with the $R = 0.4$ and $R = 1.0$ circles are equidistant from the centre as shown. Corresponding reactance values read from scales round the edge are $X_s = 0.5$ and $1/X_p = -1.25$. The last figure is the most interesting one since it tells us that if $R_a = 20\Omega$ a capacitive reactance of 50/125 (ie 40Ω) will provide a perfect match subject to adjustment of antenna length, size of capacitance hat or loading inductance, whichever is most convenient. Since X_s is slightly greater than R_a, the amount of detuning required to effect the matching will be just over half the bandwidth figures given in the tables.

The corresponding capacitance values are given in Table 12.12. It will usually be sufficient to select the nearest convenient values. In cases where the antenna is nearly self-resonant, eg for 24 or 28MHz, the radiation resistance plus the earth resistance should provide a reasonable match to 50Ω feeder without the use of a capacitor, though the antenna length or loading must be adjusted for resonance.

If a base-loading coil is used, matching may also be carried out by tapping the inner conductor of the coaxial feeder onto a suitable point on the coil.

The antenna may be tuned approximately by using a GDO, the feeder being preferably short-circuited which will place the frequency slightly on the low side as indicated earlier. After arriving at the right size of loading coil with as much capacitive loading as possible, final trimming may be carried out for minimum SWR using whichever method is most convenient. If it is not possible to obtain a satisfactory SWR, this may be due to values of R_e or coil Q different from those assumed in the above calculations and a different capacitance should be tried.

Table 12.12. Values of matching capacitance assuming R_a = 20Ω

Frequencies (MHz)	1.8	3.6	7.05	10	14.2	21.3
Capacitance (pF)	2212	1106	564	400	280	186

LIGHTNING PROTECTION

In order to achieve the most efficient radiation from an antenna, it should in general be erected in the clear and as high as possible. Its potential as a lightning hazard is thereby increased and serious consideration should be given to this problem which unfortunately is one of considerable complexity, so that no one set of simple rules can be devised to suit all situations. Some understanding of the general principles involved and, incidentally, the weakness of many common practices, is therefore desirable.

The object of lightning protection systems is achieved by providing a *safe* conducting path between ground and the atmosphere above the structure to be protected [24]. Radio antenna installations usually embody the main ingredients of a protective system but may tend to increase the hazard instead of reducing it, because of failure to meet certain minimum specifications for such systems. The majority of lightning strikes involve currents in the region of 2000A to 100,000A, with an absolute maximum in the region of 220,000A. These are of short duration, eg a rise time of a few microseconds and decay time of a millisecond or less, though a complete lightning discharge may comprise a sequence of such strokes following the same path and lasting up to one second or more. Despite the short duration these currents cause intense heating if they pass through a bad joint in a metal conductor or poor insulators such as trees or brickwork where they may cause sudden generation of steam; in each of these cases the effect may be explosive, fires may be started, and there is danger to any individual near the path of the discharge. The current can pass safely to ground only if the following conditions are satisfied:

1. The current path is of adequate conductivity and cross-section.
2. The earth resistance is low enough.
3. Other conductors in the vicinity of the lightning conductor are adequately isolated or bonded to it.
4. There are no people or animals in the vicinity of the earth termination, where large potential gradients exist at the ground surface during a lightning strike.

The most frequent cause of damage to equipment is not a direct strike but voltage or current induced in antennas and mains wiring by lightning strikes in the vicinity, at distances up to several hundred yards. Static charges due to an accentuation of normal atmospheric stresses also come into this category. Protection against these effects may be obtained even with a high-resistance earth-connection but the presence of an HF antenna, despite adequate protection against static electricity, could in some cases increase the likelihood of a direct strike without providing the means for dissipating it.

Lightning danger varies enormously between different areas of the world and this must obviously enter into the assessment of what precautions should be taken. The BSI Code of Practice [25] puts forward a points system for deciding whether a building needs protection and the majority of private dwelling houses in the British Isles would appear to be exempt, though many of them might be put at risk by erection of an HF beam or ground-plane antenna on the roof. Many antennas mounted on towers or tall masts would also appear to be in need of protection; on the other hand, some antennas can be designed so that they themselves act as efficient protective systems. Short of full lightning protection it is recommended [26] that TV antennas etc should be protected against atmospheric electricity by earthing with a conductor of not less than 1.5mm^2 cross-section, the outer conductor of a coaxial cable being regarded as suitable. An HF transmitting antenna should obviously receive at least this degree of protection. Antennas and masts may be earthed for this purpose by connection to an existing system of earthed metal work, eg suitable water pipes, the point of connection being as high as possible in the system. In the case of VHF or UHF antennas parasitic elements need not be earthed, but this may be advisable with the much larger elements used for HF beams.

Whether or not full protection is provided, the earth resistance should be as low as possible. For full protection the BSI Code of Practice [25] recommends a value not exceeding 10Ω and conductor cross-sections of the order of 60mm^2, though opinions differ on the latter figure; the official standards of some countries run as low as 28mm^2, and values as low as 5mm^2 have been described as adequate despite a small number of instances of damage to 60mm^2 conductors [24]. An earth resistance as low as 10Ω is often difficult to achieve but in such cases an annual dose of a solution of rock salt can be very effective. Low earth resistances are sometimes obtained by laying conductors in trenches near the surface, but it seems reasonable to suppose that this might tend, for a given resistance, to create dangerous potential gradients over a much larger area in the event of a strike.

The bonding of other conductors to the lightning protective system can raise complex problems and, if it appears necessary, the references should be consulted. Isolation is the alternative, and for full protection requires separations of 1ft per ohm of earth resistance, plus 1ft for every 15ft of structure height to allow for the inductive voltage drop in the down conductor; however, for a slight increase in risk these distances may be halved. A rough idea of the zone of protection of a lightning conductor system is obtained by imagining a cone with a 90° apex angle extending down from the tip of the conductor – this may be extended by the efficient bonding to the conductor of, say, a 'plumbers delight' HF beam erected over the roof. However, most HF transmitting antennas lie wholly or partly outside the protected zone and, because of inadequate conductor sizes, are unsuitable for its extension. In such cases the feeder (eg outer of coaxial cable), metal masts and parasitic elements should be connected to the highest convenient point on the lightning conductor. If there is no lightning protection system they should be *earthed outside the building.*

In the case of an antenna mounted on a tower or tall metal mast, all components including parasitic elements should be adequately earthed and unless a separate lightning conductor is fitted the tower must be free from high-resistance joints (eg fully welded). Even if resistance is initially low, possibilities of corrosion following exposure to the weather must not be overlooked.

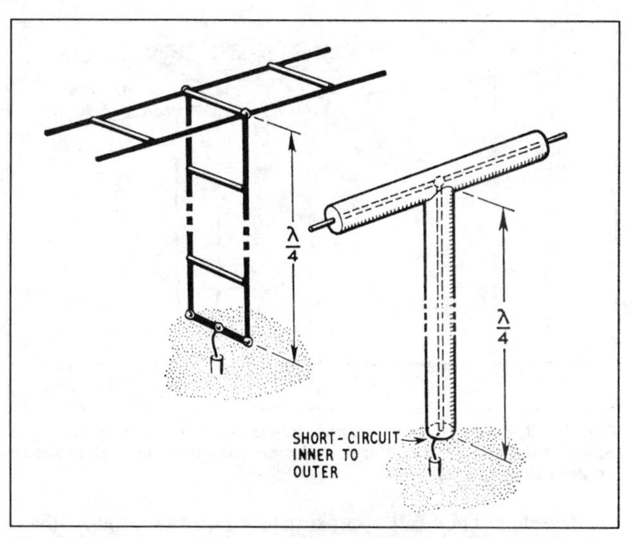

Fig 12.172. λ/4 lightning protection. The balanced line is approximately λ/4 in free space: the coaxial line is λ/4*V* where *V* is the velocity factor of the cable employed to make the stub

Otherwise, if the lightning current is allowed to pass to earth via the tower, explosive disintegration and collapse could occur. The earth resistance must be as low as possible and a single 3ft earth spike, as illustrated in earlier editions of this handbook, though adequate for the discharge of static electricity, is insufficient for lightning protection. Balanced feeders cannot be directly earthed except via resonant stubs as illustrated in Fig 12.172. Various spark-gap type lightning-arresters are suitable as an alternative [27].

The usual change-over switch for earthing an antenna and, worse still, other 'executive' arrangements likely to require the handling of a feeder during a thunderstorm, such as removing it from the rig and plugging it into an earth connection, are not merely useless for lightning protection but could prove highly dangerous. Even in the event that an operator remembers to 'switch to earth' before the outset of a storm, "any idea that they provide safety is illusory" [24]. This is because the switch is not capable of carrying lightning currents, nor is the gap sufficient for providing the required isolation which needs to be several feet as already indicated. It should be remembered that in the event of a lightning strike, even if the earth resistance is as low as 10Ω, voltages as high as two million could be lying around in the shack!

The BSI Code of Practice recommends where possible a common earth electrode for lightning protection and *all other services,* the lightning conductor being of course routed outside the building, and the installation must comply with regulations for the other services. It should be noted that an earth resistance of 10Ω is not low enough to provide in itself the fault protection required by wiring regulations and, if this is the best figure available, an earth-leakage trip will be fitted. Some older types are voltage-operated with a coil connected between true earth and the 'earth connection' of the wiring system. If an earthed coaxial feeder is plugged into a transmitter or receiver to which the usual three-pin plug is fitted, the trip coil will be wholly or partly short-circuited and a dangerous situation involving the whole of the house wiring could be created.

CONSTRUCTING ANTENNAS

All the antennas described in this chapter can be constructed at home without much difficulty provided that suitable materials

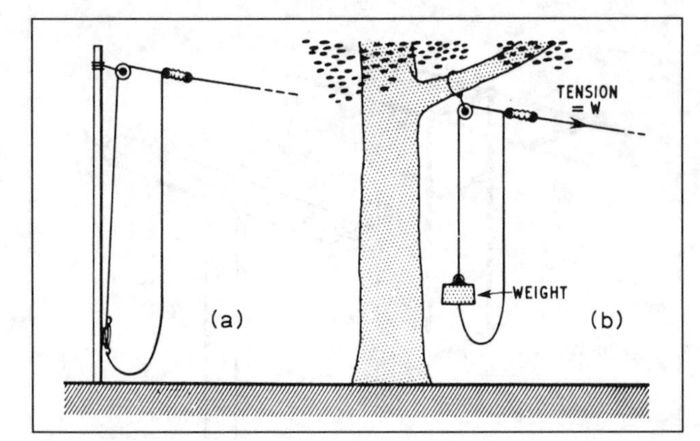

Fig 12.173. Halyard connections to poles and trees. The weight *W* is equal to the tension *T* required in the antenna wire. In both cases an endless loop is employed

are selected. They fall approximately into two groups, those which are formed from wires under tension between fixed anchoring points and those which employ rigid or semi-rigid conductors or light frames which act as supports for flexible radiators. All rotatable beam antennas come in the second group.

Wire antennas

The cardinal points of design of a fixed wire-type antenna are:

(a) choice of conductor type and gauge;
(b) insulators;
(c) end anchorages for tensioning; and
(d) method of support.

The size of conductor to be employed depends upon the working tensions involved. The longer the span and the heavier the insulators, the greater will be the tension for a given amount of sag. Single copper conductors are cheapest, the choice of wire gauge being determined by the need to keep the loss resistance small compared with the radiation resistance and also by mechanical considerations. Ideally, for 'short dipoles' or dipole beam elements, 14swg or thicker should be used in order to reduce losses and maintain bandwidth. However, stranded wire, eg 7/029, is slightly easier to handle. Thinner wires may be used for loops and folded dipoles even when these are part of a beam and, given a rough idea of the likely radiation resistance, Fig 12.5 provides the necessary guidance for meeting the electrical requirements. Wire should always be tensioned to avoid breakage due to flexing, but if anything more than light tension is required it is essential to employ hard-drawn copper to prevent stretch. For a 14MHz dipole the wire size can be as low as 30swg for a loss of only 0.5dB and, though liable to mechanical breakage, such an antenna has the advantage of being much less conspicuous in areas where bye-laws or aesthetic requirements call for an invisible antenna. Wire antennas may be suspended between any available supports, using polythene line without additional insulation, lightly tensioned as in Fig 12.173. If only one support is available, or if a few feet of extra height can be obtained in this way, the dipole may be erected as an inverted-V.

A solid insulator is likely to be required in the centre of the dipole to provide mechanical anchorage for the feed line, and in the case of $\lambda/2$ dipoles almost any insulating material is suitable. Losses at this point can, however, have a serious effect on the performance of 1λ dipoles or 'short dipoles'. Insulators can be obtained as proprietary items in glazed china or glass, or can be manufactured at home from polythene rod. It is possible to cast individual tension and centre feed-point insulators in Araldite

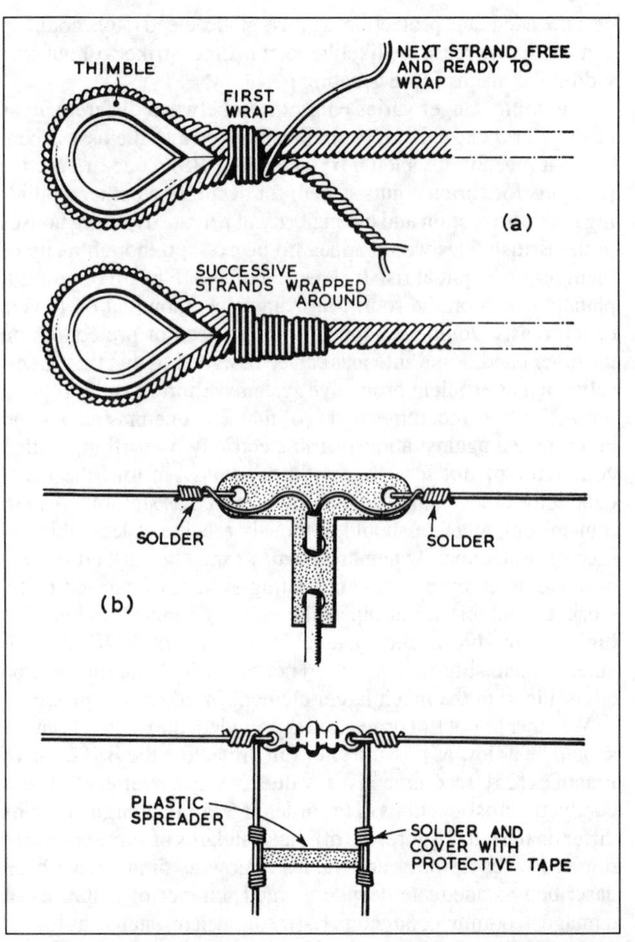

Fig 12.174. (a) The PO splice for securing stranded wire around a thimble or insulator. (b, c) Methods of supporting feeders from dipole centre insulators. The exposed end of the coaxial cable in (b) must be taped to exclude moisture. The 'tee' insulator is described in the text

resin, which has an acceptable power factor and low dielectric loss up to 30MHz. In all cases where a rigid insulator is not required for mechanical reasons, polythene cord will be found satisfactory.

The correct method of attachment of halyards and conductor wires to a thimble or insulator is shown in Fig 12.174(a). To form the PO splice in the stranded conductor, one strand of the free end of wire is untwisted back to the insulator, and then wrapped tightly around both wires to bind them together. A second strand is then untwisted back to the end of the first wrapping, and wrapped to continue the binding action. The process is repeated until all the strands of the free end have been used. This neat splice may also be used on galvanised stay wires for masts, and is very strong. It is aesthetically better and inherently safer as a permanent splice than the familiar Bulldog wire rope grips sometimes employed.

One method of securing the feed-point insulator for coaxial and for open-wire line is shown in Fig 12.174(b) and (c). It is important to ensure that the weight of low-impedance line is supported firmly by the insulator and not by the connections made to the antenna conductor. The latter should be wrapped to form a sound mechanical joint before soldering.

Antenna halyards should be of polythene, nylon, or other plastic material since conventional rope or cord tends to rot and in the meantime is much affected by weather conditions. They should always be reeved as an endless loop to avoid a climb to the top of the pole if an antenna or halyard connection fails: Fig 12.173(a). Some plastic cords are reinforced by a fine strand of wire and these must be avoided in view of possible effects on the antenna radiation pattern, apart from the fatal consequences likely to ensue if they come into accidental contact with overhead electric cables.

Wooden or metal poles, towers, or the walls or eaves of buildings are all suitable tension anchorages. In the latter case a long free end to the halyard is preferable to keep the antenna away from lossy brickwork and metal guttering. The branches or trunks of trees have disadvantages as halyard anchors as they move excessively in the wind and cause large variations of sag in antenna wires, and possibly failure under extreme tension. If they must be used, then some form of counterbalance on the lower end of the halyard is essential: Fig 12.173(b). The weight should be free to move up and down, and is equal to the tension in the antenna wire.

Beam antennas

Various forms of rotary beam antennas can be constructed either from wires tensioned on a framework or from self-supporting tubes, remembering that wires should be of the heaviest convenient gauge, though 18 or 20swg is electrically satisfactory for quad or delta loops of normal size and at least $\lambda/8$ spacing. The framework can be made up from suitable timber which can be screwed together, or from bamboo poles clamped to end fixing plates. When timber is used, particularly for rotary Yagi-type antennas, a hardwood is to be preferred. Softwoods should always be primed and painted for normal outdoor protection, or given a thorough creosote or polyurethane treatment. It is advisable, particularly if short elements are used, to keep any wood or bamboo supports well away from wires to ensure satisfactory performance in wet weather.

The most elegant form of Yagi antenna construction is the *'plumber's delight'*, so called because all the elements and the supporting boom are made from tube sections. Since all elements in a Yagi are at zero RF potential at their centres, they may be joined to and supported by a metallic boom. This form of construction is usual in commercially available antennas, but often presents problems for the home constructor who may like to consider the alternative of using two or more fixed but reversible wire beams to provide all-round coverage. Apart from the important operational advantage of instantaneous switching between directions, such arrays can often use lighter supports and are relatively cheap. Sometimes they can even be concealed among tree branches without incurring appreciable losses. Wire beams using inverted-V dipole elements may be supported from a single mast and spreader.

Many ideas for the construction of quad antennas will be found in reference [21].

Vertical antennas

Vertical antennas should as far as possible be kept well clear of supporting structures. They may be suspended from a catenary which can also provide end-loading, eg as in Fig 12.149, or may be self-supporting. If stay wires are required polythene cord is suitable in view of its excellent insulating properties. Sometimes a metal mast is used in conjunction with a base insulator as a single-band radiator. The base insulator can be made from a soft drink or wine bottle, clamped around the lower part of the outside, and with the bottom end of the vertical tubing either inside or over the neck of the bottle. A hock bottle is probably the most suitable shape for this purpose. In all such cases it is essential to prevent the bottle filling up with rainwater.

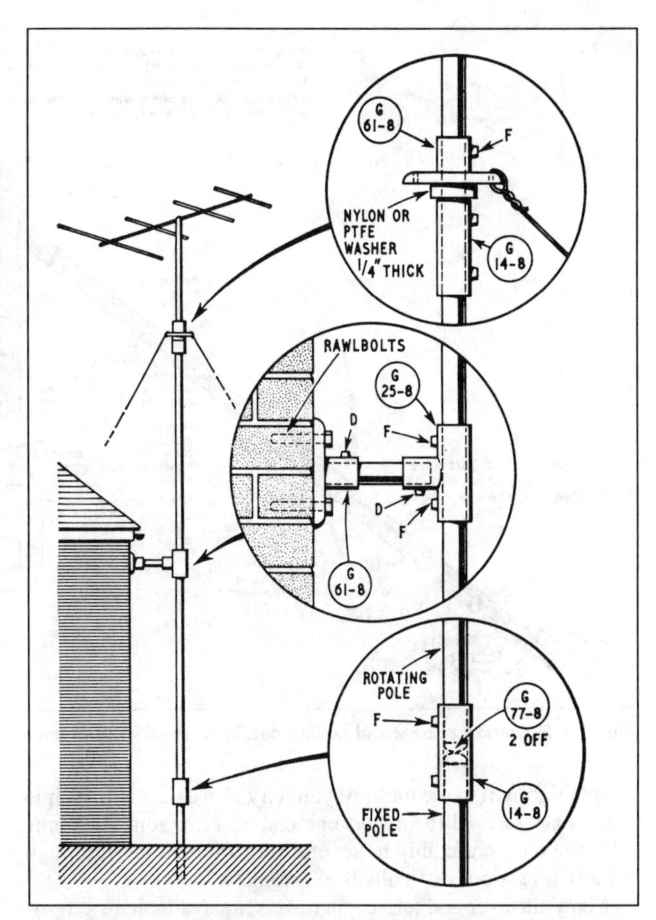

Fig 12.175. A light rotating mast using Kee Klamp scaffold fittings. The clamping screws at F should be left slack and grease forced into the fitting at those positions. For extra security the stub pole can be drilled ½in clearance at points D and ½in BSW bolt substituted for the Allen screw, in order to penetrate the wall of the tube and form a lock against possible pull-out. A simpler form of construction can be used for fixed poles

MASTS AND RIGGING

Traditionally wire antennas have been supported by stayed or self-supporting wooden poles, usually painted or creosoted for weather protection, but the use of steel or aluminium tubing with suitable coupling and stay attachment fittings is now more usual and many kits are available. With 2in diameter aluminium scaffold poles readily available and suitable for supporting the smaller types of HF beam, arrangements such as Fig 12.175 can be recommended. Guy rings and couplers for joining mast sections are available from various suppliers, polypropylene rope suitable for guys is available from most hardware and DIY stores, and problems with beam rotation are greatly simplified by the reversible beams described earlier in this chapter which, as explained, have other important advantages. Crank-up and tiltover towers are available from several suppliers and, though relatively expensive, can be used for supporting much larger beams.

Metal guy wires if used should be broken up by insulators into non-resonant sections. It should be noted that metal masts, unlike guy wires, cannot be split up in this way and may have unpredictable resonances which can be excited by any asymmetry in the antenna system, so that the measures recommended (p12.32) for the prevention of feeder radiation may assume added importance; they should not be used to support vertical antennas with the possible exception of ground planes mounted at the top, although in this case the radials, if allowed to droop, could

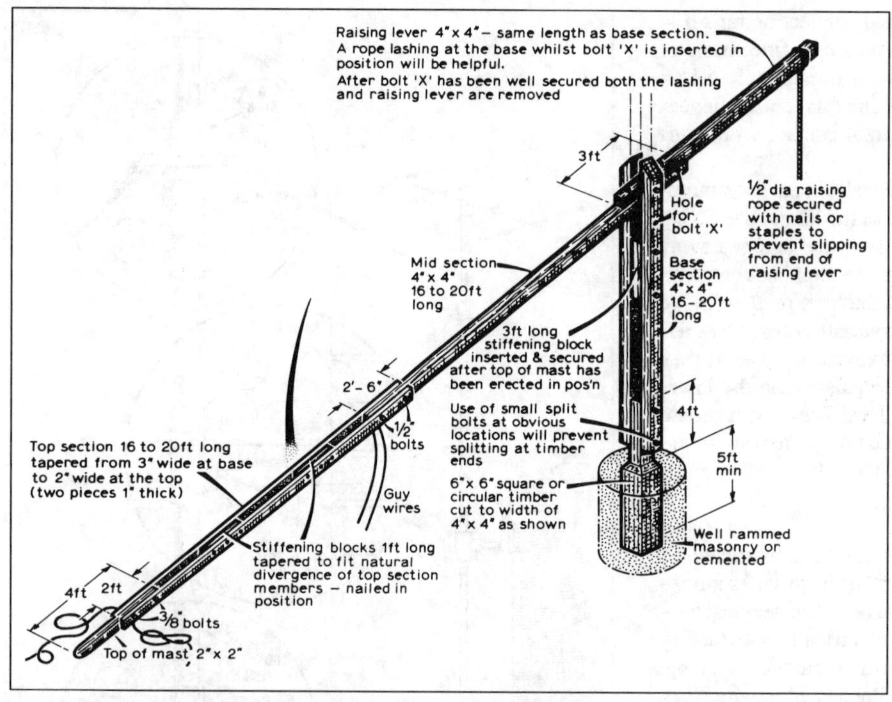

Fig 12.176. Constructional and raising details of the 55ft wooden mast *(ARRL Antenna Book)*

couple tightly into the mast. Asymmetry also exists in principle if the mast is used to support one end of a horizontal antenna which should preferably be several feet from the mast, though usually no serious difficulty is experienced.

Many tilt-over and telescopic towers are available as proprietary articles and the telescopic variety can be motor operated. On the other hand, wooden masts can be constructed from timber at low cost as in Fig 12.176 and lend themselves readily to tilt-over methods of erection for easy replacement or maintenance of antennas.

Excellent articles on antenna masts and rigging have appeared in *Radio Communication* [22, 23] and will help the reader wishing to adopt a 'professional' approach to antenna erection, or to benefit from the experience of others in evolving cheap and simple designs of antenna support while avoiding some of the more serious pitfalls. Every installation presents a different set of problems, and solutions have few features in common. Careful consideration must be given to any dangers to persons or property which could result from the collapse of a mast and beam during severe weather conditions and it is essential to err on the side of safety, even though this usually leads towards rather massive construction. However, there are cases where collapse would be of minor consequence and repairs easily effected, for example where a light support is used for an inverted-V dipole or ground plane, and materials such as the PVC tubing available from builders' merchants have been used successfully in these cases.

Guys are usually required, but the stranded galvanised clothes line often used has only limited life, particularly when the galvanising has been damaged with pliers. Polypropylene rope (monofilament) is an excellent general-purpose material for guys and halyards, Terylene being slightly better but considerably more expensive. Nylon and polythene have been used successfully but stretch under tension and this can be disastrous when they are used as guys. Guy anchorages for light masts may consist of 2ft lengths of 2in angle-iron driven into the ground approximately at right-angles to the line of pull of the guys. Particular attention should be paid to safety during the erection of masts, avoiding any danger of contact with power wires. Pairs of guy wires, carefully measured and attached to points at right-angles to the line of erection will prevent the mast falling sideways. The 'gin-pole' method permits the single-handed erection of masts up to 40ft or more in length. The gin-pole, which may conveniently be a ladder, is held vertically at the foot of the mast which is on the ground, and the free guy or guys from the top of the mast are passed over the top of the ladder and extended to a lower position where they can be easily secured. For single-handed erection a tackle may be rigged between the top of the mast and a suitable anchorage. A temporary crossbar should be clamped to the base of the ladder so that the mast sits on it during erection, being afterwards lifted off into a suitable socket [22]. In working out the required strength of guy-wires, it is convenient to estimate the effective surface area presented to the wind by the antenna, and add on the area of the top one-third of the mast. The wind drag is then given by $v^2/400$ where v is the wind velocity in mph. Depending on the location and other factors, v is usually taken as 80–100mph, and a factor of safety of six should be allowed for the breaking strain of the guys. Typical wind drag (100mph) is 5.81b per foot length for 2in tubing, 0.17lb per foot for 14swg wire, 135lb for a three-element 14MHz Yagi and 260lb for a cubical quad [23].

General considerations of safety are dealt with in Chapter 21.

ANTENNA MEASUREMENTS

The most important property of HF antennas is usually the effective DX performance which is difficult to measure direct]y as it involves not only the antenna but its entire surroundings and particularly the ground contours which may have markedly different effects even for antennas at say, opposite ends of a typical suburban garden. Even if the area is flat, power lines and trees can have marked effects in the case of horizontally and vertically polarised antennas respectively. Data on these effects are few, but in one case vertical antennas suspended from, on either side of and barely clear of, a row of trees were found to be ineffective when 'looking through' the trees. Local measurements of impedances and polar diagrams can be used to establish that an antenna is working approximately in the correct manner but when evaluating a new antenna it is advisable to obtain a considerable number of comparative signal reports before and after the change.

The yardstick for such comparisons may be provided by another amateur whose operating conditions are maintained constant, or another antenna (eg a dipole) at least $3\lambda/4$ away from the antenna under test and end-on to it, although these requirements may be relaxed if the antenna not in use *including any parasitic elements* can be completely detuned, eg by the open-circuiting of resonant feeders. It will be difficult to get consistent results unless both antennas have similar vertical-plane radiation patterns, which for horizontal antennas and low angles are a function only of height and ground contours. More consistent results have been obtained over long transequatorial paths such as G/VK or G/ZS, probably due to the prevalence of single modes of propagation (eg chordal hop) whereas transatlantic paths can sustain many different multihop modes, some of which may involve quite high angles of radiation. The 'yardstick' station need not be particularly close, geographical separation up to 100 miles or more being usually less significant than a difference of, say, 2:1 in antenna height, though correlation may break down completely at times, particularly if band conditions are changing. This 'rule' also breaks down for near-antipodal paths since places which are close together geographically may be on completely different great circle bearings, as will be obvious from looking at New Zealand on a great circle map centred on the UK.

Local measurement of antenna performance usually involves three particular properties of the antenna:

(a) impedance;
(b) polar diagram (in both planes); and
(c) gain relative to a reference standard.

These will now be discussed in turn; see Chapter 15 for details of suitable test equipment.

Impedance

It is desirable to know the approximate value of input impedance of an antenna for two reasons:

(a) to determine a suitable form of matching transformer at the antenna;
(b) to verify as discussed below that the radiation resistance is of the right order, since if too low this indicates faulty design or tuning, and if too high suggests the presence of excessive losses.

In the case of an antenna fed with matched open-wire line the position and length of matching stubs after following normal adjustment procedures should give the information required (Figs 12.61 and 12.62). With low-impedance feeder and correct tuning of the antenna, SWR should be a minimum around the middle of the band and, if the antenna is fed without a transformer, or with a transformer of known ratio, radiation resistance may be deduced from the minimum SWR. An SWR of 2.0 in 50Ω cable could result from an antenna impedance of 25 or 100Ω, but such ambiguities may be resolved by connecting, say, a 100Ω resistance across the antenna terminals. This would have little effect in the first case, but would improve the SWR to unity in the second case.

Matching may be checked with an impedance bridge, noise bridge or reflectometer which should if possible in the first instance be connected close to the antenna, otherwise what is interpreted as a good match may in fact represent only a high value of feeder loss. Even if the feeder loss is low, as measured into a resistive load, additional loss and measurement errors may be caused by line imbalance or currents flowing on the outer surface of a coaxial feeder. To illustrate this by a practical example, a two-element beam fed with 120ft of coaxial feeder, without a balun, had an SWR of 1.4 measured at the transmitter, and the feeder loss into a matched load was 1.5dB. Despite this the performance was very badly down and an improvement of about 6dB, as estimated from DX signal reports, was obtained by changing over to open-wire feeder with a matching section. Investigation revealed an SWR in the first case of 4.0 *at the antenna.* This gave [19] a predicted SWR at the transmitter, allowing for line losses, of 2.0. Further reduction of SWR and at least

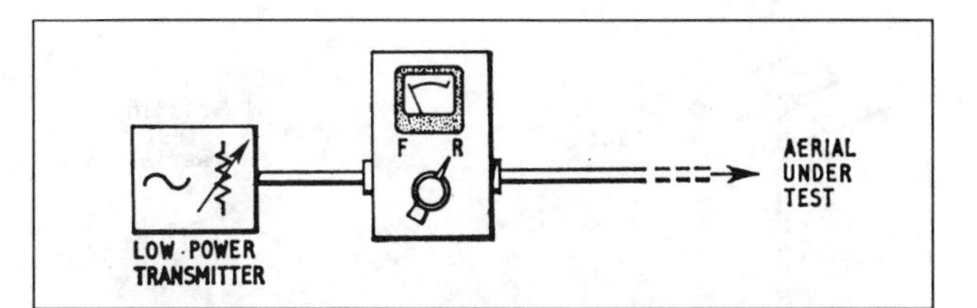

Fig 12.177. Adjustment of antenna impedance using a reflectometer. The output of the transmitter is set to maintain a constant indicated forward power. The antenna is then adjusted for minimum reflected power to match its impedance to the cable employed

half the total loss was attributed to loss by radiation from the feeder.

After the performance of an antenna has been established, the reflectometer may be installed in the shack for permanent monitoring, but it should not be regarded as giving an infallible indication of correct operation and losses; for example, moisture getting into the feeder will improve rather than degrade the SWR. Fig 12.177 illustrates a typical test set-up using a reflectometer to indicate correct adjustment of the antenna for minimum reflected power in the line. When carrying out such adjustments it is important to maintain a constant forward power from the test oscillator and this too can be verified by means of the reflectometer.

Adjustment of antenna impedance should ideally be carried out with the antenna in its operating position, since it is influenced by height above ground. Often this is not feasible and it may be necessary for reasons of accessibility to make adjustments at a height of 10ft or so with a final performance check at full height.

It is permissible, and usual, to leave the reflectometer connected in the transmission line during normal operation (with a reduction of sensitivity appropriate to the transmitter power employed). It is then possible to maintain a continuous check on the SWR and detect antenna faults as they arise.

When antennas using tuned feeders are employed, the adjustment of impedance is transferred to the antenna coupler (see p12.48). Again a reflectometer may be used in a short piece of line of correct impedance for the transmitter output load, and the antenna coupler adjusted to give minimum SWR on this piece of line (see Fig 12.81).

Polar diagram

The horizontal polar diagram of an antenna can be measured to a reasonable degree of accuracy without having to raise the antenna to an excessive height above ground. A figure of 0.2–0.3λ may be considered adequate, with a minimum of 8ft, to ensure adequate general clearance.

The horizontal polar diagram of small beam antennas depends on the number of elements and the relative phase and amplitudes of currents. In the case of a pair of loop or dipole elements a good idea what to expect can be gleaned from Figs 12.15, 12.17 and 12.18. With more elements the pattern is less predictable but a slightly narrower main lobe and reduction of other lobes can be expected. In both cases measurement of polar diagrams is useful for providing a rough check that the beam is operating correctly but it is important to realise that these diagrams are only rather loosely related to sky-wave performance. In particular, discrimination against short- and medium-range sky-wave signals 'off the end' is quite poor because at an angle of incidence of 30°, for example, the vertical polar diagram is only 6dB down. Also, for directions 'off the back' the effective element spacing for high-angle signals is different from what it

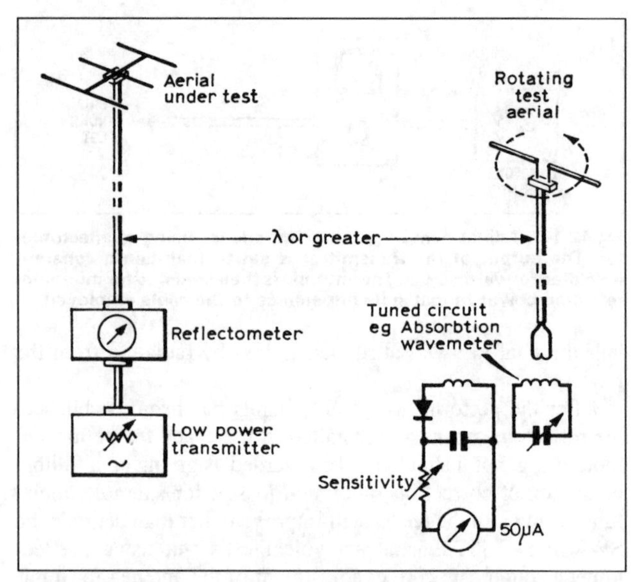

Fig 12.178. Test site for measurement of antenna radiation patterns. The site must be as free as possible from obstructions. The reflectometer is used in the forward position to monitor the transmitter output which must remain constant. The S-meter should first be calibrated using a signal generator into the receiver input

is in the horizontal plane so that nulls will occur, if at all, in directions slightly different from those observed in the horizontal plane.

The test antenna must have the same polarisation as the one being tested or, in some cases, it may be useful to make a measurement of both vertical and horizontal fields. This may reveal, for example, the presence of feeder radiation, but great care is needed in interpreting results if both polarisations are present. For example, an inverted-V dipole radiates a very small amount of power in a vertically-polarised '8JK' mode, enough to produce a relatively strong ground signal 'off the end', although this has little effect on sky-wave performance.

Fig 12.178 illustrates a suitable test set-up for measuring horizontal polar diagrams. The distance between antennas should be as great as possible, otherwise the signal from one element travels significantly further than that from another and is thus subject to greater attenuation. A null which would be observable at greater distances is therefore partly filled in, and at the low heights usually favoured for such measurements the inverse distance-squared law tends to apply, so that if one element produced a field of 1V at a distance of λ an identical element spaced λ/8 behind it produces a field of 64/81V, and a null will be filled in to the extent of being only 14dB down. Observations under these conditions, though not useless, are not accurate either, and need careful interpretation. The greater the gain of the antenna being tested, the greater is the separation needed. At short distances accuracy is also affected by the presence of capacitive and inductive fields in addition to the radiation field. In the context of Fig 12.178 height is not critical but should be considerably greater than the spacing between the elements; even so, there could be a significant change of tuning and radiation pattern when the antenna is raised to its full height.

Depending on sensitivity the field-strength meter may be linear or square-law, and this can be checked against transmitter output as observed on an oscilloscope or other linear voltage indicator.

The possibility of harmonic resonances which greatly increase the sensitivity of the test meter to harmonics should not be overlooked, and harmonic radiation from the transmitter must be prevented. This is particularly important if a short antenna is used for the test set.

Gain

Due to the complex role played by ground reflections it is not possible using ground-wave signals to make accurate measurements of effective DX gain of HF antennas. Although for many purposes flat ground can be regarded as a nearly perfect mirror with reflection taking place at a point as in Fig 12.24, the 'point' is in fact a Fresnel zone of considerable extent depending on antenna height and the angle of reflection. For the reflection to be fully effective both antennas must 'see' the whole of the relevant zone, which extends in the case of short ranges and low angles (as for ground-wave propagation) from the vicinity of the transmitting antenna to a long way beyond the measuring antenna. The allowance to be made for ground reflection must therefore depend on the polar diagram of the antenna under test and will be different for a beam and for the reference antenna. Errors will also occur because the mutual impedance between the antenna and its image will exert a different influence in the two cases and there will probably also be absorption or field distortion by nearby objects which will also be different for different antennas. These problems are all much less acute in the case of measurements on VHF and UHF antennas.

Even if ground-reflection can be regarded as taking place at a point there is *no limit to the possible error* of measurement, as the following example illustrates. A lazy-H with a vertical separation between elements of 0.7λ is mounted at a fairly large mean height $n\lambda$ and a test antenna is placed at a distance $2n\lambda$ so that ground reflection takes place at an angle of 45°. Radiation from the two elements cancels at the point of reflection so that the test antenna experiences the true free-space gain of the lazy-H. If the lazy-H is replaced by a reference dipole, the ground reflection may (depending on the height) add or subtract from the direct signal, giving a resultant field anywhere between zero and 6dB greater than the free space value, so that the lazy-H with its true free-space or DX signal gain of 4dB may have in this case an *apparent* gain anywhere between 'plus infinity' and –2dB with respect to a dipole! Effects of this nature seem to be the most likely explanation of the large but impossible gain figures so often claimed and conversely perhaps the premature demise of some promising new ideas! At the very least this example emphasises the need for extreme caution when interpreting gain figures, both measured and claimed. With this proviso, gain measurements may be attempted with the arrangement of Fig 12.178 by replacing the antenna under test with a reference dipole and comparing the signal levels for the same value of radiated power. In the case of beams with tuned feeders to each element it is a simple matter to open-circuit all but one element which can then be used as the reference dipole.

Tested in this way, close-spaced beams can be expected to show a gain of at least 3dB and sometimes as much as 6dB, but usually rather less than the theoretical value; if, however, the radiation resistance and polar diagram of a two-element beam are correct and loss resistances from Fig 12.5 relatively small, the gain must also be correct and any measurements which indicate otherwise should be ignored. In particular, it should be noted from Fig 12.21 that with two elements (in the absence of losses) there is an *exact* relationship between gain and the direction of nulls.

Comparisons between antennas based on the average of a sufficient number of signal reports must also be interpreted carefully, remembering that the comparison is not merely between a quad and a Yagi (or whatever the antennas may be) but between two different combinations of *antenna plus environment*. If, as usual, these 'environments' include the other antenna, it becomes important to ensure that one antenna is not in the 'line of fire' of the other, as the following example illustrates. A 14MHz fixed reversible beam for the north-south directions, which appeared to be working well, was swung round through 90° for comparison with another beam fixed on the long path to VK and found to be down by two S-units, the second beam being 80ft in front and having a good front-to-back ratio so that interaction was at first considered unlikely; it was suspected initially as a result of gain measurement on the lines indicated above, then confirmed by signal reports with the second antenna lowered to the ground, and finally substantiated on theoretical grounds. Any nearly resonant element in the line of fire tends to generate an antiphase field, and the relative magnitude of this as a function of separation distance can be inferred from the mutual impedance or from the maxima of the mutual resistance curves, Fig 12.9, with due allowance for ground images. In this 'parasitic' role a beam does not operate as such and has no front-to-back ratio, which exists only in respect of signals applied to (or taken from) its proper terminals. For an accuracy of ±1dB in comparisons based on average DX signal reports a separation of about 2.0 or 3.0λ appears to be required, but for antennas end-on to each other useful (though not highly accurate) comparisons can usually be made with quite small separations, eg 0.6λ between centres. However, the difficulties may be aggravated or eased by environmental differences.

CHOOSING ANTENNAS

This chapter has attempted to explain antennas and the principles underlying their design: a large number of antennas have been illustrated together with many accessory devices. The beginner may well be confused by the range of information, and therefore need advice on the best way to start.

In particular he may well be discouraged by the complex and expensive installations often considered to be necessary for good results, and if he attempts without previous experience to build, say, one of the more exotic varieties of multiband rotary beam this is likely to breed further discouragement. It is likely that most pleasure will be obtained and more will be learnt by starting with a simple antenna which is bound to work, such as a one-band inverted-V dipole which can be 'hitched up' in the centre to the highest point available and the ends allowed to droop (with an apex angle which should not be less than 90°) towards lower anchoring points. Alternatively, if two high points are available the dipole may be suspended between them, or two inverted-V dipoles may be mounted at right-angles with separate feed lines to provide switched coverage in all directions. If sloping ground is available this should be exploited, good results being obtainable 'down the slope' with antennas at comparatively low heights [2]. If space is too limited for a dipole, a ground plane may be tried but results can be very disappointing on the higher frequencies and it is important that the base should be as high as possible, eg mounted on a chimney or rooftop.

Simple antennas such as these are often capable of providing daily SSB contacts with all continents and, in favourable cases, performance may be within one or two S-units of the best obtainable. For multiband operation the use of dipoles with resonant feeders should present no serious problems, the G5RV antenna being recommended, space permitting, for coverage of all bands from 3.5 to 28MHz; otherwise note should be taken of the particular advantages of quad or delta loops as described on pp12.71 and 12.72, and many beginners might find these to be an ideal starting point.

The next step could be the erection of a fixed reversible monoband beam using a pair of small loop or dipole (including inverted-V) elements, each with its own 50Ω feeder.

Alternatively for rotary operation the modified VK2ABQ array (Fig 12.134) can be recommended. It should, however, be noted that multiband operation over an octave or more is very easily achievable by the use of resonant feeders, delta loops having circumferences of about 0.8λ or dipoles with a length of 5/8λ on the lowest band being particularly suitable, as described on pp12.67, 12.72 and 12.83. Correct coupling between elements is very important and should be adequately realisable by following the guidelines provided. Such beams provide deep steerable nulls and are fully tuneable without leaving the shack. Being instanteously reversible, only about 120° is needed (readily achievable with two cords), and their small size and weight often enables them to be erected at a greater height than conventional Yagis such as Fig 12.140 which may be a preferred choice for those with better-than-average facilities at their disposal, such as the ability to erect one or more 50ft towers with separate arrays for the 'old' and WARC bands..

There are of course excellent multiband beams available from commercial sources, as well as some that are less good, and the beginner who thinks he may eventually want to purchase one of these would be well advised to take note of the ones which seem most consistently to produce the biggest signals from long distances. However, of the options available, the log-periodic [28, 29, 49] seems to be the only rotatable array capable of covering all bands from 10 or 14 up to 28MHz.

Low-frequency antennas have not been separately considered in the above discussion owing to the complex variety of individual problems and solutions but the reader will have appreciated that many of the HF antenna systems can also be energised in various ways as LF radiators, as discussed on pp12.82 and 12.87.

Those fortunate enough to be able to choose a site should pick one with plenty of clear space, preferably on a hilltop. With such a site, success comes more easily. The ability to achieve low-angle propagation will increase the hours per day during which stations in, say, USA or Australia can be worked. It should be borne in mind that if the ground is sloping, particularly if it is a steep slope, very good results will be obtained very easily down the slope but it may be difficult or impossible to work against the slope. With more than one antenna, or even for multiband work, preset antenna couplers will be found a great advantage.

RECEIVING ANTENNAS

At the beginning of this chapter it was stated that the receiving problem is not quite the same as the transmitting one. The reciprocity theorem as applied to antennas is true in a limited sense but assumes that interference is isotropic, ie equally likely from all directions, whereas atmospheric noise arrives mainly at low angles, galactic noise from high angles, and man-made noise occurs in local concentrations. Prevailing galactic noise levels ensure that useful sensitivity in the HF band is normally limited by external noise so that losses in antennas or feeder systems, up to a point, attenuate noise and signal equally and are likely to be acceptable for reception, though certainly not for transmission. The rule concerning the use of the transmitting antenna for reception is usually a good one; beam arrays, especially the unidirectional ones, can be a great help for reducing interference from directions not in the beam of the antenna. Long-wire types are not so good in this respect, and often bring in more cosmic or atmospheric noise than other types; vertical antennas normally reduce the galactic background noise level, and help to discriminate against short-range interference – say from 500 to 1000 miles, because of their selective vertical patterns, though they may well be inferior for transmission. A properly constructed antenna with a good transmission line always helps to minimise man-made interference radiated by house wiring etc.

REFERENCES AND BIBLIOGRAPHY

[1] 'Supergain aerials', L A Moxon, G6XN, *Radio Communication,* September 1972.

[2] 'Low-angle radiation', L A Moxon, *Wireless World,* April 1970.

[3] 'A new approach to the design of super-directive aerial arrays', A Bloch *et al, Proc IEE,* Part III, September 1953.

[4] *Radio Engineers Handbook,* F E Terman, McGraw-Hill, 1943, pp799–802.

[5] *Radio Engineers Handbook,* F E Terman, McGraw-Hill, 1943, p193.

[6] 'A 14 Mc co-ax fed dipole and TVI', F G Rayer, G3OGR, *RSGB Bulletin* March 1965. See also 'Why coax?', E M Wagner, G3BID, *73* November 1971.

[7] 'The determination of the direction of arrival of short radio waves', H T Friis, C B Feldman, W M Sharpless, *Proc IRE,* Vol 22, 1934, pp47–78.

[8] 'Aerial reflections', F Charman, BEM, G6CJ, *RSGB Bulletin* December 1955.

[9] 'Two-element driven arrays', L A Moxon, G6XN, *QST* July 1952.

[10] 'Theoretical treatment of short Yagi aerials', W Walkinshaw, *JIEE,* Part IIIA, Volume 93, 1946, p564.

[11] 'Multiband quads', L A Moxon, G6XN, *CQ* November 1962.

[12] 'The Swiss Quad Beam Aerial', R A Baumgartner, HB9CV, *RSGB Bulletin* June 1964.

[13] 'The HRH Delta-loop Beam', H R Habig, K8ANV and Lew McCoy, WIICP, *QST* January 1969. See also 'Technical Topics', *Radio Communication* May 1969 and May 1973.

[14] 'Evaluating aerial performance', L A Moxon, *Wireless World* February and March 1959.

[15] 'The 9M2CP Z-beam', Technical Topics, *Radio Communication* August 1971.

[16] 'Unidirectional antenna for the LF bands', Malcolm M Bibby, GW3NJY, *Ham Radio* January 1970.

[17] 'The G8PO 'Guy Wire' array', J E Ironmonger, G8PO, *RSGB Bulletin* July 1962.

[18] 'More about the Minibeam', G A Bird, G4ZU, *RSGB Bulletin* October 1957.

[19] 'Measuring cable loss by SWR', O J Russell, *RSGB Bulletin* November 1961.

[20] 'Gains and losses in aerials', L A Moxon, G6XN, *Radio Communication,* December 1973 and January 1974.

[21] *All about Cubical Quad Antennas,* William I Orr, W6SAI, Radio Publications Inc, Wilton, Conn, USA.

[22] 'Ropes and rigging for amateurs', J Michael Gale, G3JMG, *Radio Communication* March 1970.

[23] 'Aerial masts and rotation systems', R Thornton, GM3PKV, and W H Allen, G2UJ, *Radio Communication* August and September 1972.

[24] 'The protection of structures against lightning', J F Shipley, *JIEE,* Part I, December 1943.

[25] *The Protection of Structures against Lightning*, British Standard Code of Practice, BSI, CP326, 1965.

[26] *The Reception of Sound and Television Broadcasting*, British Standard Code of Practice, BSI, CP327.201, 1960.

[27] 'Lightning and your aerial', G R Jessop, G6JP, *Radio Communication* January 1972.

[28] L A Moxon, *HF Antennas for All Locations*, RSGB, 1982.

[29] Les Moxon, *HF Antennas for All Locations*, 2nd edn, RSGB, 1993.

[30] 'Two-element HF beams', Les Moxon, *Ham Radio* May 1987, p8.

[31] F E Terman, *Radio Engineers Handbook*, McGraw-Hill, 1943, p864.

[32] See [28] pp56, 242, or [29], pp65, 284.

[33] 'Ground-planes, radial systems, and asymmetric dipoles', L A Moxon, G6XN, *ARRL Antenna Compendium* Vol 3, 1993.

[34] 'The feed impedance of an elevated vertical antenna', Guy Fletcher, VK2BBF, *Amateur Radio (Australia)* August 1984.

[35] 'Input impedance of horizontal antennas at low heights above ground', R F Proctor, *Proc IEE*, Part 3, Vol 07, May 1950.

[36] *The ARRL Antenna Book*, 1988 edn, p3-11.

[37] 'Designing end-loaded HF wire Yagis', Brian Austin, G0GSF/ZS6BKW, *Radio Communication* September 1989, p44.

[38] 'All-band beam antennas', Les Moxon, *Radio Communication* August 1991.

[39] Fred Caton, VK2ABQ/G3ONC, *Electronics Australia* October 1973.

[40] 'The Claw Mark 4 Antenna', Pat Hawker, G3VA, *Radio Communication* January 1987, p28. 'Two-element HF beams', Les Moxon, *Ham Radio* May 1987.

[41] 'Waterproofing', 'In Practice', *Radio Communication* January 1989.

[42] 'Electrically tuneable HF loop', Roberto Craighero I1ARZ, *Radio Communication* February 1989.

[43] The *ARRL Antenna Book*, 15th edn, ARRL, 1988, p3-11.

[44] 'Siting criteria for HF communication centres', W F Utlaut, *NBS Technical Note 139*, US Dept of Commerce, April 1962.

[45] *Amateur Radio Techniques*, 7th edn, Pat Hawker, G3VA, RSGB, 1980, p270.

[46] 'The 'disappearing inductance' – a new trick and some better beams', L Moxon, *Radio Communication* April/May 1977.

[47] 'A new multiband quad antenna', W Boldt, DJ4VM, *Ham Radio* August 1969.

[48] *The ARRL Antenna Book*, 12th edn, ARRL, 1970, p293.

[49] 'The log-periodic dipole array', Peter D Rhodes, K4EWG, *QST* November 1973.

13 VHF/UHF antennas

THE antenna is the essential link between free space and the transmitter or receiver. As such, it plays an essential part in determining the characteristics of the complete system. The design of the antenna and its working environment will decide its effectiveness in any particular system.

An antenna is defined as: "A structure that transforms electromagnetic energy contained in a guided wave to that of free space propagation or vice versa."

The parameters that describe the properties of an antenna in accomplishing this transformation are:

1. Input impedance
2. Directivity
3. Gain
4. Radiation efficiency
5. Radiation pattern
6. Polarisation
7. Bandwidth

Regardless how one uses a system, all antennas have these basic interactive properties. Characteristics that are of most interest to the amateur and professional designers are the input impedance (usually called the *match*) or SWR, the antenna gain, the antenna radiation pattern and the polarisation.

RADIATION PATTERN

The radiation pattern of an antenna is generally its most basic requirement. It decides the spatial distribution of the radiated energy, which is the direction in which the signal is transmitted or from where it is received.

It is measured and recorded as a graph representing the distribution of radiation about an antenna, usually in spherical co-ordinates.

Radiation patterns are in terms of voltage field strength or power density using a decibel scale. The patterns are referred to an absolute value or compared with some reference level. They can be displayed in rectangular or polar format using spherical co-ordinates θ (theta) and φ (phi). Spherical co-ordinates are angles in two principal planes that position any point on a sphere.

The simplest presentation, most common to radio amateurs, is a polar plot in one or preferably two principal planes with a decibel amplitude scale. Amplitude is relative to a dipole (dBD) or an 'isotropic radiator' (dBi).

Fig 13.1(a) is an example of a rectangular pattern plot. Figs 13.1(b) and 13.1(c) are examples of the common forms of polar pattern plots.

These polar diagrams are plotted on a logarithmic (decibel, dB) scale and represent reality. Occasionally they are plotted on a linear scale which makes the main lobe look sharper and makes the side lobes look much smaller. The type of plot should always be made clear.

In many amateur systems the requirement is for a highly directive pattern to achieve increased gain and interference

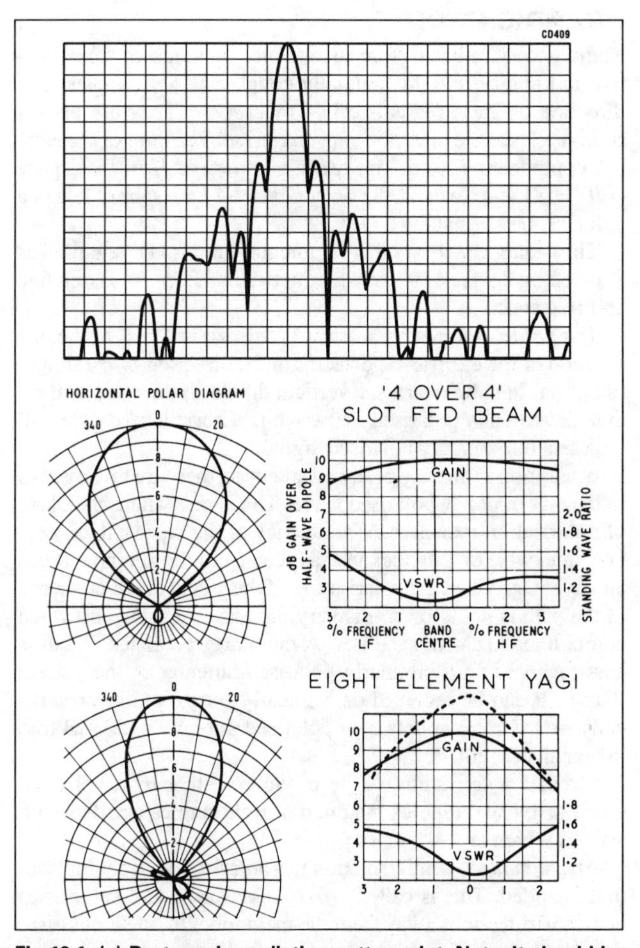

Fig 13.1. (a) Rectangular radiation-pattern plot. Note: It should be symmetrical about the main lobe but symmetry is affected by the presence of conductors near the antenna and by asymmetry in the feeder. (b) Polar diagram of a 4-over-4 slot-fed beam. (c) Polar diagram of an 8-element Yagi. Note: These have the same forward gain but the 4-over-4 has a wider beamwidth (about 40° versus about 25° at 3dB down.) This may be an advantage (aiming has to be less accurate) or a disadvantage (less discrimination against strong signals on the same frequency but from a different direction). Note also that the bandwidth of the 4-over-4 is greater than that of the Yagi

reduction. However, for such systems as repeaters and mobile communication it is often necessary to radiate energy in an omnidirectional pattern to provide a broadcast type of coverage.

Amateur antennas normally have radiation patterns with a simple shape. The important characteristics of the patterns can be specified by the *beamwidth* and *sidelobe level* in the two principal planes. The usual reference planes taken are the E (or vertical) plane and the H (or horizontal) plane.

The 'beamwidth' in a principal plane of the radiation pattern is defined by the pattern's angular width at a level that is 3dB down from the beam maximum. This is also known as the *half-power beamwidth*.

The term 'sidelobe' usually refers to radiation in any direction other than that required. The front-to-back ratio with directional beam antennas is specifically the ratio of the energy radiated in the required direction to that radiated in the diametrically opposite direction.

Both are expressed as the number of decibels that the unwanted signal is (hopefully) below the maximum radiation of the main beam.

POLARISATION

Radio waves, radiated from an antenna, are made up from electric and magnetic fields mutually coupled at right-angles. The direction of these fields is called the *vector*. They are also at right-angles to the direction of propagation. The ratio of the electric component E to the magnetic component H in free space ($E/H = Z$) is known as the *impedance of free space*. It has a value of approximately 377Ω.

This statement shows that an antenna can also be regarded as a matching device or transformer between the transmission line and free space.

The *polarisation* of an antenna is defined in terms of the orientation of the electric-field vector in the direction of maximum radiation. In simple terms, a vertical dipole in the clear will radiate a vertically polarised signal while a horizontal dipole will radiate a horizontally polarised signal.

In addition to linearly polarised antennas, there are those which radiate a *circularly* polarised wave. The is one where the plane of polarisation rotates at the frequency of the signal. It may rotate clockwise or anticlockwise as seen from the antenna. It has an advantage when the polarisation of the signal at the other end of the path is not known or is varying, and is most useful when communicating with satellites. A circularly polarised signal is best received on a circularly polarised antenna of the correct 'hand'. It can be received on a linearly (horizontally or vertically or anything in between) polarised antenna with 3dB loss in signal strength.

Circular polarisation can be produced either by a helix antenna or by two dipoles mounted at right-angles and fed with signals which are 90° out of phase.

An antenna may emit radiation in a polarisation different from that intended. This is called *cross polarisation* and sometimes occurs in directions away from the main lobe. In linear polarisation, the cross polarisation is at right-angles to the intended polarisation. In circular polarisation it is the component that has the opposite hand to that intended. If it is attempted to receive a circularly polarised signal with an antenna of the wrong hand, the signal will be many decibels down and may not even be heard at all.

GAIN

The gain of an antenna is a basic property that is frequently used as a figure of merit. Gain is closely associated with directivity, which in turn is dependent upon the radiation patterns of an antenna.

The gain (power gain) is defined as 4π (pi) times the ratio of the radiation intensity in a given direction to the net power accepted from the accompanying transmitter.

The gain can also be defined in practical terms as the ratio of the maximum radiation signal in a given direction to the maximum radiation signal produced in the same direction from a reference antenna with the same power input.

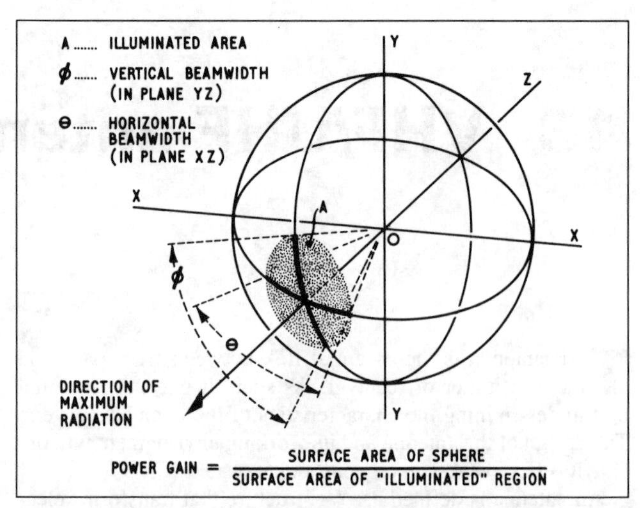

Fig 13.2. Radiation from an antenna. An isotropic radiator at point O will give uniform illumination over the inner surface of the sphere. A directional radiator will concentrate the energy into a beam which will illuminate oniy a portion of the sphere, as shown shaded

Due to the law of reciprocity, the same principle is true for a receiving system.

The reference antenna is normally an *isotropic* radiator. This is a hypothetical, loss-less, antenna that radiates uniformly in all directions. One way to appreciate the meaning of antenna gain is to imagine the radiator to be totally enclosed in a hollow sphere, as indicated in Fig 13.2. If the radiation is distributed uniformly over the interior surface of the sphere the radiator is said to be *isotropic*.

An antenna that causes the radiation to be concentrated into any particular area of the inside surface of the sphere, and produces a greater intensity than that produced by an isotropic radiator fed with equal power, is said to have *gain*. This gain is inversely proportional to the fraction of the total, interior, surface area that receives the concentrated radiation.

The gain of an antenna is usually expressed as a power ratio, either as a multiple of so many 'times', or in decibel units. For example, a power gain of 20 times could be represented as 13dB (ie $10 \log_{10}20$).

The truly isotropic radiator is a purely theoretical concept, and in practice the gain of beam antennas is usually compared with the radiation from a single $\lambda/2$ dipole fed with an equal amount of power. The radiation pattern of even a single $\lambda/2$ dipole is markedly non-uniform, having the form of a torus (like a ring doughnut where the centre hole has been reduced to nothing). In consequence the power gain of such an antenna compared with the hypothetical isotropic radiator is 1.64 times or 2.15dB.

The area of 'illumination' is not sharply defined as shown by the shaded region in Fig 13.2 but falls away gradually from the centre of the area. The boundaries of the illuminated area are determined by joining together all points where the radiation intensity has fallen by half (ie 3dB, the half-power points). The gain of the antenna can then be determined by dividing the total surface area of the sphere by the illuminated area, eg if the total surface area were $100cm^2$ and the illuminated area bounded by the half-power points was $20cm^2$, the gain of the antenna would be five times or 7dB.

This concept of gain measurement can be rationalised to a simple formula and compared to either the isotropic radiator or the more practical $\lambda/2$ dipole to give a close estimate of antenna gain (normally within 1dB if the lobe structure is simple and of a low level). If the 3dB (half-power) points are measured in both of the principal planes the following relationships can be used.

$$\text{Gain relative to } \lambda/2 \text{ dipole} = \frac{27{,}000}{\theta_E \theta_H}$$

where θ_E is the angular width in degrees at the half-power point in the E or vertical plane and θ_H is the angular width in degrees, at the half-power points in the H or horizontal plane. So:

$$\text{Gain in decibels relative to dipole (dBD)} = 10 \log_{10} \frac{27{,}000}{\theta_E \theta_H}$$

A more accurate calculation, with gain relative to the isotropic level is:

$$\text{Gain isotropic} = \frac{41{,}252}{\theta_E \theta_H}$$

The calculated figure for antennas with gain greater than 10dBi and sidelobes below –10dB relative to the peak gain can be within 0.4dB of the true value. With all sidelobes below –17dB the error is reduced to 0.1dB. The level for the sidelobe figure assumes a 'solid sphere' without nulls for simplicity of calculation. Taking the sidelobe level at the peak value or even a mean usually produces results better than that calculated. To obtain good results the beamwidth –3dB points need to be carefully measured. However, 1 or 2° error in beamwidth can produce less than 1dB error. This may be acceptable for most purposes.

Besides examining the action of a transmitting antenna in concentrating the radiated power into a beam, it is also useful to examine the way in which the antenna structure changes the reception of an incoming signal. In this study it is convenient to introduce the concept of *capture area* or *aperture* of the antenna. This idea is frequently misunderstood, probably because it may appear to relate to the cross-sectional area of the beam (as represented by 'A' in Fig 13.2). It is related to the inverse of the cross-sectional area of the beam. An antenna that has a high gain usually has sharply focused 'beam' with a small physical cross-sectional area. At the same time the capture area of the antenna is large. The larger the capture area, the more effective is the antenna compared with a simple dipole.

The actual size of the antenna system does not always give a reliable indication of the capture area. A high-gain array may have a capture area considerably greater than is determined by its physical dimensions. The fundamental relationship between the capture area and the power gain of antenna system is:

$$A = \frac{G_i \lambda^2}{4\pi}$$

where A is the capture area and λ is the wavelength (measured in the same units as A) and G_i is the power gain relative to an isotropic radiator. A $\lambda/2$ dipole has a gain of 1.64 relative to the isotropic radiator.

The formula can be modified to give the capture area in terms of the gain of a $\lambda/2$ dipole, instead of G_i, simply by introducing the factor 1.64, thus:

$$A = \frac{1.64 \times G_D \lambda^2}{4\pi} = 0.13\, G_D \lambda^2$$

Note that for a dipole, since $G_D = 1$, the capture area is approximately $\lambda^2/8$.

This formula shows that if the wavelength is kept constant, the capture area of an antenna is proportional to its gain. Therefore, if an increase in gain results in a narrower beamwidth there is a corresponding greater capture area. 'Beamwidth' is used

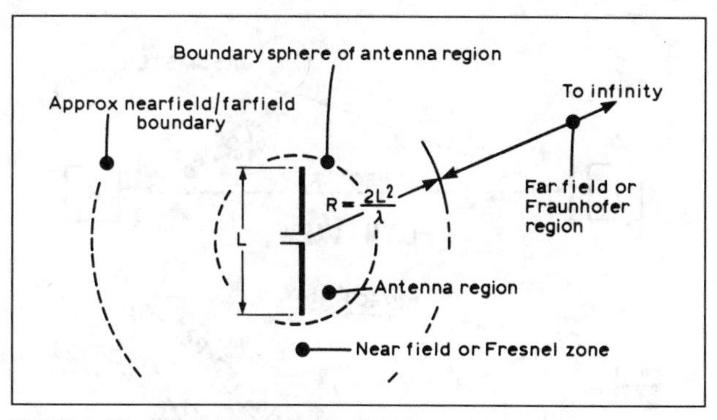

Fig 13.3. The near and far fields of an antenna

here to signify both horizontal and vertical dimensions, ie in effect the cross-sectional area.

The formula also shows that for any given power gain the capture area is proportional to the square of the wavelength. An antenna with a power gain of 10 times, relative to a dipole at 600MHz (0.5m), would have a capture area one-sixteenth of that of an antenna having a similar power gain at 150MHz (2m). To achieve an equal capture area for the 600MHz antenna it would need a 16 times greater gain than the 150MHz antenna, ie 160 times more than a dipole.

This is unfortunate because it is the capture area of an antenna that determines its effectiveness in absorbing the incoming radiation. It means that as the wavelength is reduced it becomes increasingly important to design an antenna to have greater gain.

ANTENNA HEIGHT GAIN

A very important factor, which is often a 'grey area' for many amateurs and professionals alike, is the effect of an antenna's height on its performance above ground. Similarly of interest are the effects that may be apparent due to surrounding terrain or nearby structures.

Apart from the changes of pattern due to ground reflections, adding or cancelling signals at various wavelength distances from the antenna, there is one other important mechanism to consider. It was previously mentioned that an antenna has an electric and magnetic field radiating from it. The electric field is the one normally considered the link between antennas as the magnetic field strength decreases with distance more rapidly than the electric field. Any object in the radiated fields will disrupt them, even the distant receiving antennas. However, there is a region of maximum electromagnetic intensity called the *near field* where an object has the most effect. Nearby structures, including the antenna mounting itself, will have an effect on the radiation pattern and often the impedance of the antenna.

A simple formula that gives an approximate distance outside which measurements of the radiation patterns will have some sense is:

$$R = \frac{2L^2}{\lambda}$$

where R is the distance, L is the largest linear dimension of the receiving or transmitting antenna and λ is the wavelength. All dimensions should be in the same units of measurements.

Under ideal conditions in the Fraunhofer region measured field components are transverse. The shape of the field pattern is independent of the distance at which it is measured (refer to Fig 13.3). However, in the Fresnel region the radial field may be appreciable. The near field or Fresnel zone is bounded by an

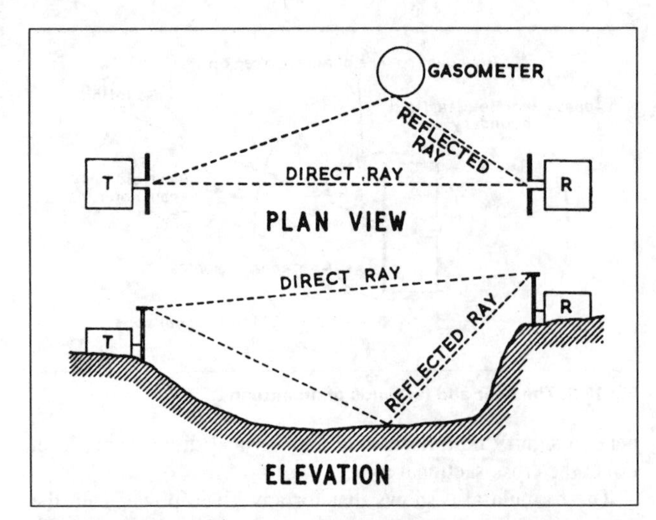

Fig 13.4. Interference between direct and indirect (reflected) rays. The strength of the received signal depends on the phase difference between the two rays when they arrive at the receiver, and on their respective intensities

imaginary sphere of radius $R = 2L^2/\lambda$, where L is the overall length of the antenna, and the far field or Fraunhofer zone is outside that. These are arbitrarily chosen but work. If the antenna is a half-wave dipole, the sphere has a radius of $\lambda/2$. The shape of the field pattern changes with the distance from the antenna and is modified by the presence of the measuring device and other objects.

Antenna height above ground and disruption of the near field pattern will give an apparent or actual change of the received or transmitted signal. Assuming the antenna is mounted with minimal change to the near field, the main consideration will be addition or cancellation of the field pattern due to ground reflections.

Several points should be considered for optimum performance, when deciding the height to mount the antenna above ground.

1. The antenna should be above local screening from buildings and other obstacles.
2. The rule-of-thumb figure of approximately 12m (40ft) is worth considering. At this level the antenna can be above the layer of electrical interference and signal variations caused by the heat layer above buildings.
3. If there is no screening, the antenna mast is on a ground area that is flat for several wavelengths and with low mast heights, the radiation from the antenna will be raised in the vertical plane.
4. As the height increases above ground the pattern will level off, giving the main radiation in the required horizontal plane. However, secondary lobe structures will still be present and vary due to several factors. It is rare that the antenna manufacturer's quoted pattern will be exactly obtained unless it is mounted under the same conditions as those used by the manufacturer.
5. In general, as height increases, the pattern directivity improves and an additional gain of 6dB is obtained each time the mast height is doubled.

Fig 13.5 gives approximate height gains obtained at various frequencies for various heights above ground. Over 12m (40ft) above ground, assuming all obstacles have been cleared, a 24m (80ft) mast will be required to increase the gain a further 6dB. The additional expense for the mast is rarely justified by the 6dB gain improvement.

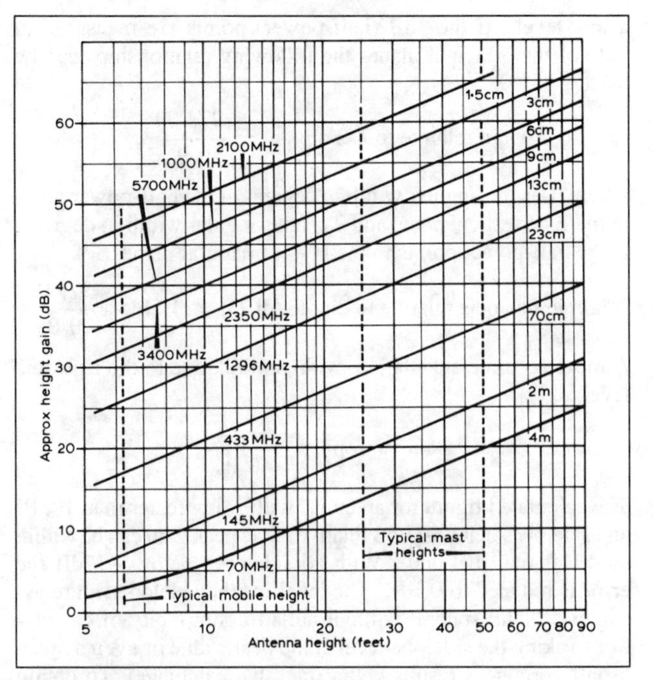

Fig 13.5. Antenna height gain correction factor

Should the station be well sited, on a hill for instance, doubling the mast height may give little or no improvement. The effective height above ground can relate to a point at the bottom of the hill, not the base of the mast.

Conversely, a station in a valley or behind a hill may find that over a certain mast height there is a considerable increase in gain. It can be in excess of the gain expected for the height increase. This is often due to a more favourable propagation angle to the hilltop (*crest factor*) or by obtaining a line-of-sight over the hill. Often a change of the antenna's vertical mounting angle above the horizontal plane or polarisation in this situation can give a gain improvement.

The three basic arrangements for receiving and transmitting antennas, and the intervening ground, are illustrated in Fig 13.6.

A classic plane-earth case is shown in Fig 13.6(a), and under such conditions the signal received at the distant antenna is given by the relationship:

$$e = \text{constant} \times \frac{h_T\, h_R}{\lambda d^2}$$

where: h_T = height of transmitting antenna
h_R = height of receiving antenna
λ = wavelength
d = distance between antenna

In this expression h_T, h_R and d must all be in the same units. d must be much larger than either h_T or h_R (by a factor of at least 10). This is usually the case in practice.

From this expression it is clear that an increase in either h_T or h_R will result in a corresponding increase in e. Doubling the height will give an increase of 6dB. This is the *6dB height gain rule*. Although each doubling of the mast gives 6dB improvement, there soon arises a practical limit beyond which the added complexity of raising the antenna does not pay sufficient dividends.

The 'classic' case of Fig 13.6(a) may be considered that of the average VHF station. In the same terms the case of Fig 13.6(b) may be considered that of the 'portable' VHF operator who has selected a good site. Here, the antenna height above immediate

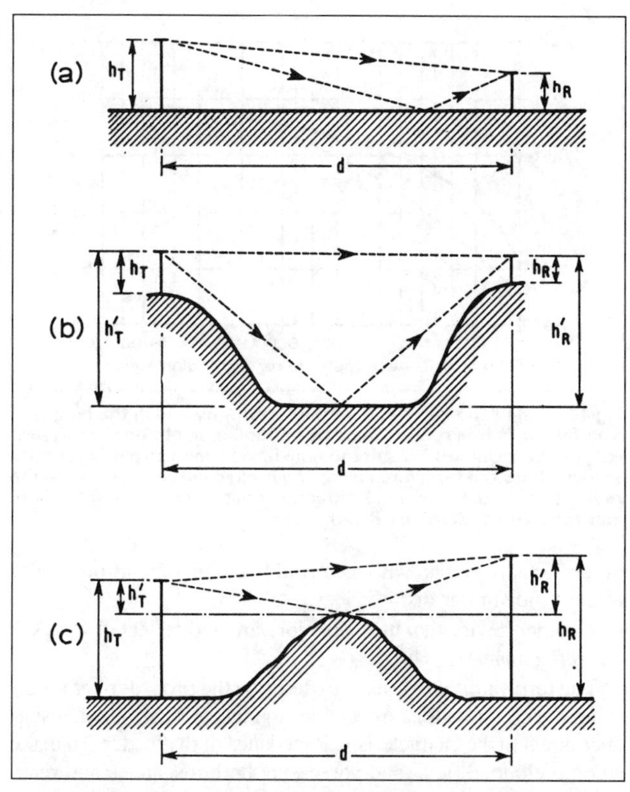

Fig 13.6. The effect of ground profile on direct and indirect rays. (a) Linear height gain. (b) Marginal height gain. (c) Rapid height gain. (The vertical scale has been exaggerated in each case.)

ground is small compared with the effective height above the ground level at the point where the indirect ray is reflected from the intervening ground.

$$e = \text{constant} \times \frac{h'_T\, h'_R}{d^2}$$

where h'_T and h'_R are the *effective* heights of the two antennas.

There is still height gain achievable by increasing antenna height locally, but not at the same rate as in the first example. To obtain a gain of 6dB it is necessary to double h'_T and this will require a many-fold increase in h_T. In the limit it clearly becomes not worthwhile seeking any great antenna height. This is often the case for portable stations on hilltops. Additional signal strength obtained by raising the antenna can be cancelled by the increased loss in extended feeder cables.

In the third case, the poorly sited station whose antenna is just able to see over the surrounding higher ground, the reverse of case (b) applies. The effective height h'_R is much less than h_R and a small increase in the height of the antenna is required to bring massive improvements in signal level.

BANDWIDTH

Unlike the properties previously discussed, the bandwidth of an antenna or antenna system does not have a unique definition. Depending upon the operational requirement of the system in which the antenna is to be used, its functional bandwidth may be limited by any one of several parameters. These can be:

1. Change of pattern shape or direction.
2. Increase in sidelobe level.
3. Loss of gain.
4. Change of polarisation.
5. Deterioration of impedance.

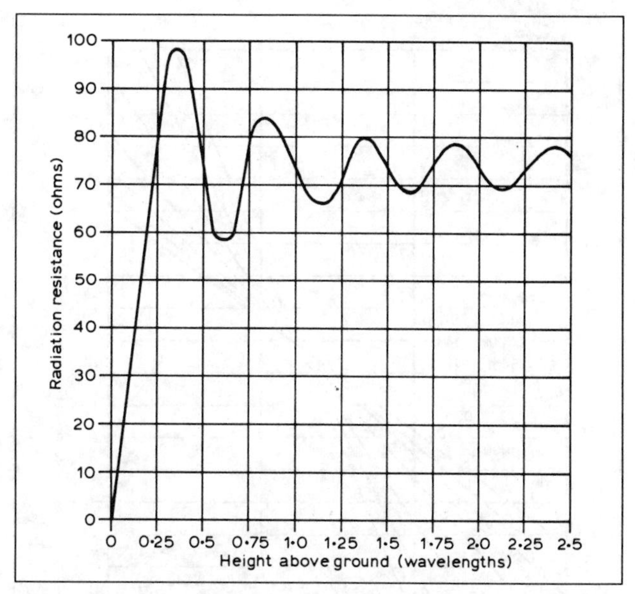

Fig 13.7. Variation in radiation resistance of a horizontal λ/2 antenna with height above a perfectly conducting ground (*ARRL Antenna Book*)

Amateurs usually consider the impedance match and, as a secondary consideration, gain, as the main criteria when quoting bandwidth. With the relatively limited frequency range of the amateur VHF and UHF bands the gain normally does not change too radically, so the impedance bandwidth is normally quoted.

However, this is not always the case with high-gain antenna systems where the gain and the pattern shape or direction of radiation may change quite considerably with bandwidth.

For antennas of small dimensions (ie when the linear dimensions are the order of λ/2 or less) the limiting factor is normally the impedance performance. With circular polarisation antennas the change of the polarisation characteristics is the limiting factor on bandwidth. For end-fire linear arrays, collinears and the like, the pattern direction followed by its shape can deviate considerably before significant deterioration of gain and more so, impedance characteristic.

The bandwidth of the antenna is very dependent on its value of Q. The higher the Q, the smaller the bandwidth.

IMPEDANCE

The input impedance of an antenna system is important since it directly affects the efficiency of energy transfer to or from the antenna.

Radiation from transmission lines and cables can also modify the antenna's pattern and in particular can also lead to interference between co-located systems.

The input impedance of an antenna system depends not only on the individual antenna elements but on the mutual impedance between elements. In addition, transmission lines and transmission-line components that are connected between antennas and from the antenna to the transmitter or receiver can modify the *effective* impedance.

Design of a complex antenna system will be governed as much by the interconnections of the transmission lines as by the characteristics of the individual antenna elements.

Impedance matching

For a feeder to deliver power to the antenna with minimum loss, it is necessary for the antenna to behave as a pure resistance. It

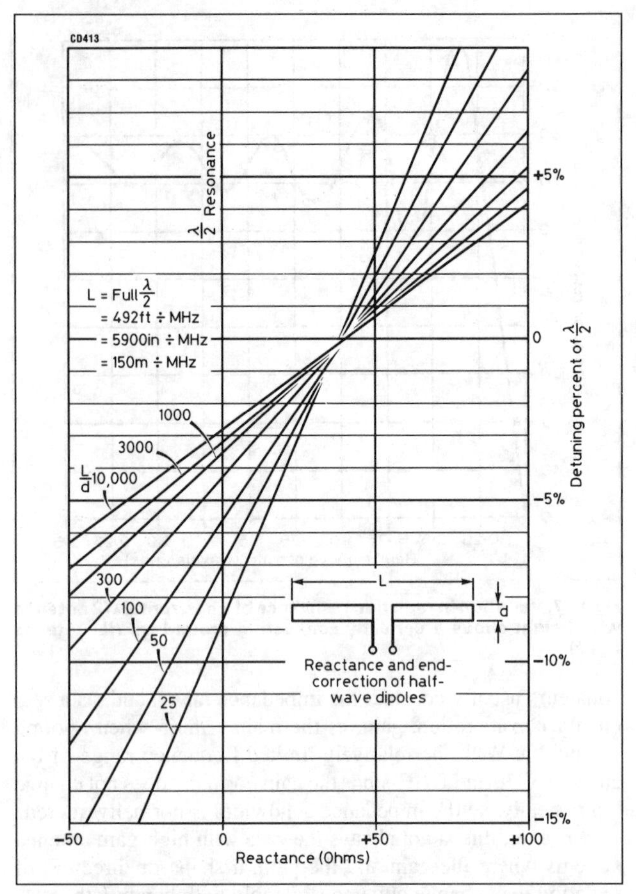

Fig 13.8. Tuning and reactance chart for λ/2 dipoles as a function of the length/diameter ratio. A radiator exactly λ/2 long is 'overtuned' by 42Ω, and 'end correction', given as a percentage of the length, is necessary to bring it to resonance (zero reactance) The chart is useful for the construction of parasitic arrays and VHF dipoles. Each 1% of length corresponds to 1.52 units in the factor 150/*f* (metres), 5 units in 492/*f* (feet) or 60 in 5900/*f* (inches)

also needs to be equal in value to the characteristic impedance of the feeder. Under these conditions no energy is reflected from the point where the feeder is joined to the antenna, and in consequence no adverse standing waves appear on the line.

When the correct terminating resistance is connected to any feeder, the voltage and current distribution along the line will be uniform. This may be checked by using a device to explore either the magnetic field (H) or the electric field (E) along the line.

One such device, suitable for use with a coaxial feeder, is a section of coaxial line with a longitudinal slot cut in the wall, ie parallel to the line. A moveable probe connected to an RF voltmeter is inserted through the slot. This samples the electric field at any point, and the standing wave ratio may be determined by moving the probe along the line and noting the maximum and minimum readings. The distance between adjacent maxima or between adjacent minima is λ/2.

A radiator exactly λ/2 long is 'overtuned' by 42Ω, and 'end correction', given as a percentage of the length, is necessary to bring it to resonance (zero reactance). Fig 13.8 is useful for the construction of parasitic arrays and VHF dipoles. Each 1% of length corresponds to 1.52 units in the factor 150/*f* (metres), 5 units in the factor 492/*f* (feet) or 60 in 5900/*f* (inches). *f* is in megahertz (MHz).

The fields surrounding an open line may be explored by means of an RF voltmeter. However, it is much more difficult to obtain

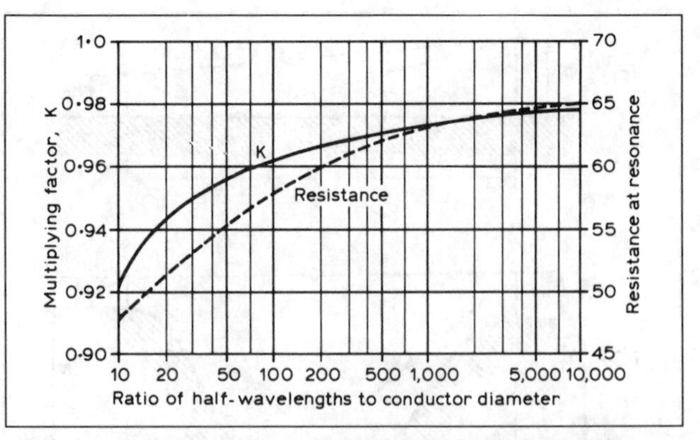

Fig 13.9. The solid curve shows the factor *K* by which the length of a half-wave in free space should be multiplied to obtain the physical length ot a resonant λ/2 antenna having the length/diameter ratio shown along the horizontal axis. The broken curve shows how the radlation resistance of a λ/2 antenna varies with the length/diameter ratio (*ARRL Antenna Book*)

precise readings than with a coaxial line due to hand-proximity effects and similar disturbances.

Another device that measures forward and reflected waves is the *reflectometer*.

The term *matching* is used to describe the procedure of modifying the effective load impedance to make it behave as a resistance equal to the characteristic impedance of the feeder. To make a complete load (ie a load possessing both resistance and reactance) behave as a resistance, it is necessary to introduce across the load a reactance of equal value and opposite 'sign' to that of the load, so that the reactance is effectively 'tuned out'.

A very convenient device that can theoretically give reactance values from minus infinity to plus infinity (ie pure capacitance to pure inductance) is a section of transmission line. This can be a length variable between zero and λ/2 with an open-circuited end. Alternatively a length a little greater than λ/2 can be used with a 'short-circuit' capable of being adjusted over 1λ. The short-circuited stub is to be preferred since it is easier to construct.

Although there is no need to make the characteristic impedance of a stub equal to that of the transmission line, it may be desirable to do so for practical reasons.

In addition to tuning out the reactance, a match still has to be made to the transmission line.

The impedance at any point along the length of a λ/4 resonant stub varies from zero at the short-circuit to a very high impedance at the open end. If a load is connected to the open end and the power is fed into the stub at some point along its length, the stub may be used as an auto-transformer to give various values of impedance according to the position of the feed point.

This is shown in Fig 13.10(a). The distance *L* is adjusted to tune the antenna to resonance and will be λ/4 long if the antenna is already resonant. The distance *L* is adjusted to obtain a match to the line. It is usually more convenient to have a stub with an adjustable short-circuit which can slide along the transmission line (see Fig 13.10(b)).

In practice, matching can be achieved entirely by the cut-and-try method of adjusting the stub length and position until no standing waves can be detected. The feeder line is then said to be *flat*.

The frequency range over which any single stub matching device is effective is quite small and, where wide-band matching is required, another matching system must be used.

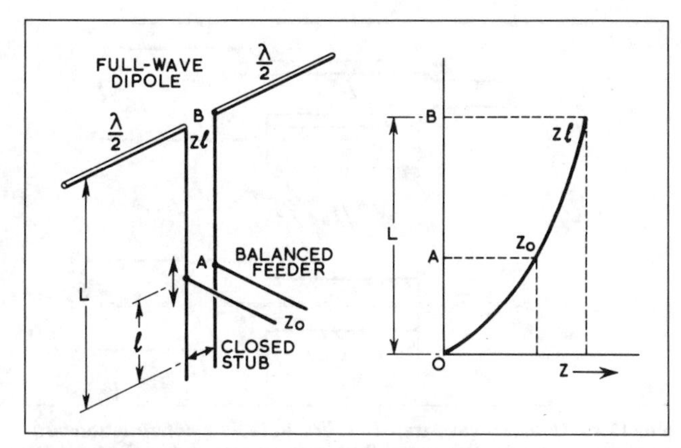

Fig 13.10(a). Stub matching applied to a λ/2 dipole

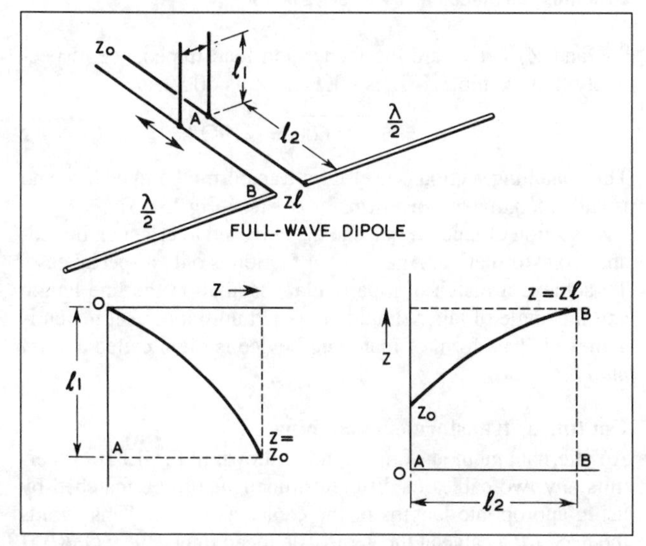

Fig 13.10(b). Stub matching with a movable short-circuited stub

Stub tuners

On a coaxial line it is impracticable to construct a stub with an adjustable position. However, two fixed stubs spaced by a certain fraction of a wavelength can be used for matching purposes (see Fig 13.12). Spacing usually employed is λ/8 or odd multiples thereof. With this spacing, independent adjustment of the short-circuiting plungers gives a matching range from 0.5 times the characteristic impedance (Z_0) of the transmission line upwards. As the spacing is increased towards λ/2 or decreased towards zero, the matching range increases, but the adjustments then become extremely critical and the bandwidth very narrow.

The theoretical limit of matching range cannot be achieved owing to the resistance of the conductors and the dielectric loss, ie the Q is limited. To obtain the highest Q the ratio of outer-to-inner conductor diameters should be in the range 2:1 to 4:1 (as for coaxial baluns). An important mechanical detail is the provision of reliable short-circuiting plungers that will have negligible inductance and ensure low-resistance contact. These can be constructed of short lengths of thin-walled brass tubing. Diameters should be chosen so when they are slotted and sprung they make a smooth sliding contact between both inner and outer conductors.

The two-stub tuner may be applied to open transmission lines if it is inconvenient to have a movable stub. Stubs must be mounted laterally opposite to each other to prevent mutual coupling (see Fig 13.12).

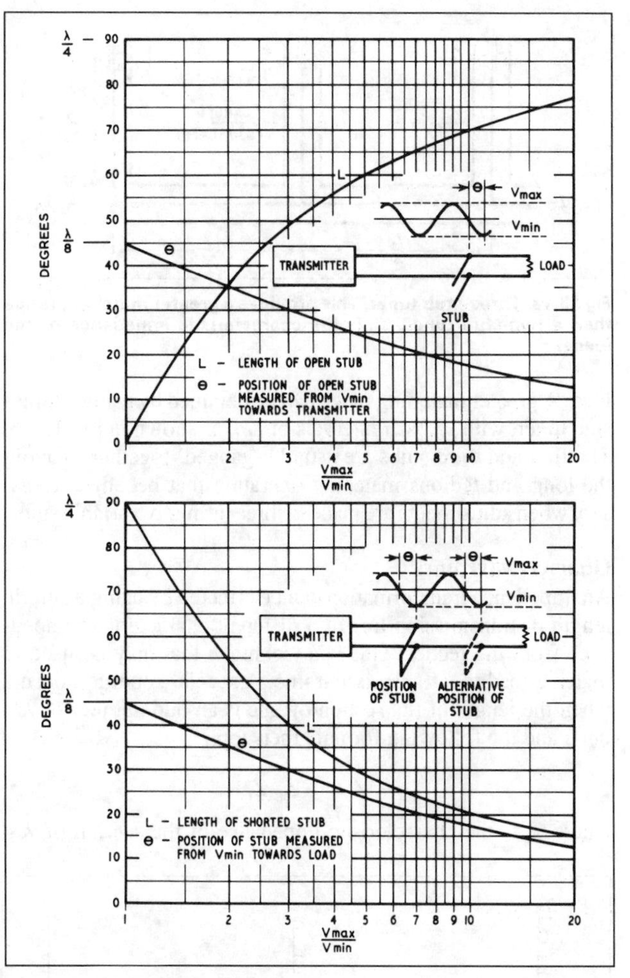

Fig 13.11. Impedance matching charts

This type of tuner may, of course, be used for other purposes than to feed an antenna. For example, it will serve to match an antenna feeder into a receiver, or a transmitter into a dummy

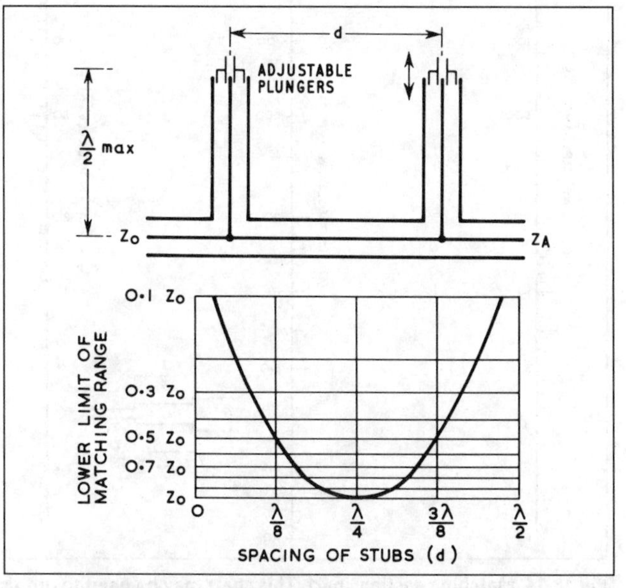

Fig 13.12. Two-stub coaxial tuner. The graph shows the lower limit of the matching range: the upper limit is determined by the Q of the stubs (ie it is dependent on the losses in the stubs). Z_0 is the characteristic impedance of the feeder

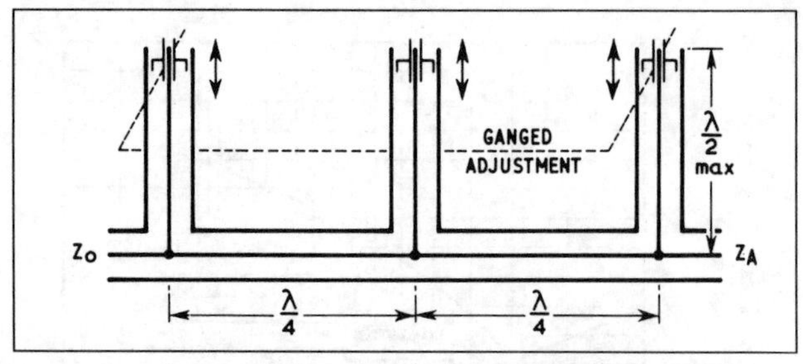

Fig 13.13. Three-stub tuner. This provides a greater matching range than a two-stub tuner. Z_0 is the characteristic impedance of the feeder

load. A greater matching range can be obtained by using a three-stub tuner, with stubs at intervals of λ/4, as shown in Fig 13.13. The first and third stubs are usually 'ganged' together to avoid the long and tedious matching operation that becomes necessary when adjustments are made to three infinitely variable stubs.

Quarter-wave lines

An impedance transformation can be effected by using a certain length of transmission line of a different characteristic impedance from the feeder. This is a technique that may be used to match a load to a transmission line. A special condition occurs when the length of the section of line is an odd number of λ/4 units and the following formula then applies:

$$Z_t = \sqrt{Z_0 . Z_l}$$

where Z_t is the characteristic impedance of the section of λ/4

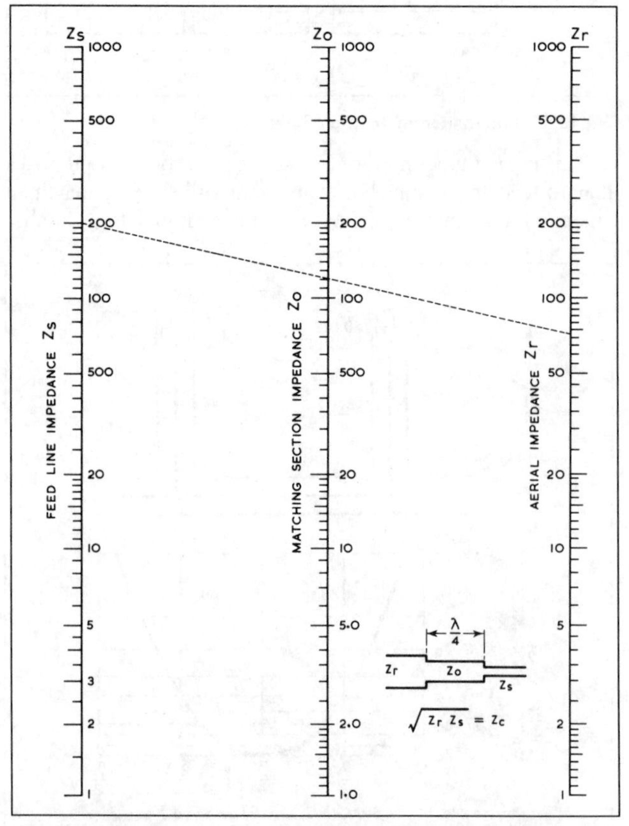

Fig 13.14. Matching section chart. This chart may be used to obtain the surge impedance of a λ/4 matching section used as an impedance transformer from one real impedance to another. In the example shown, Z_r is 72Ω and Z_s is 200Ω, indicating that a λ/4 matching section of 120Ω is needed

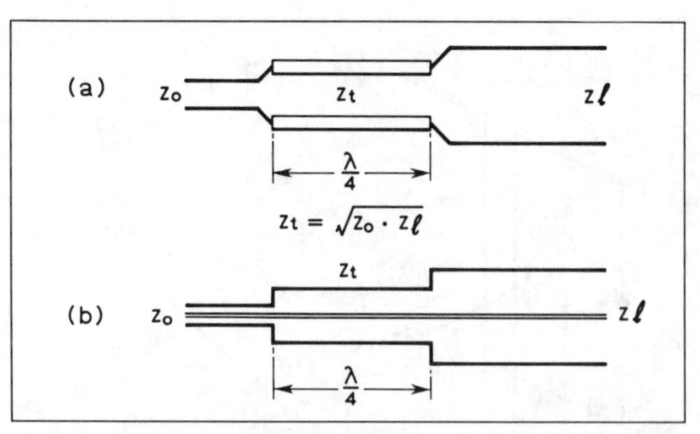

Fig 13.15. Quarter-wave transformers. In (a) is a construction suitable for open-wire lines, and (b) is the corresponding method for coaxial cables. Where a solid dielectric section is used, due allowance must be made for the velocity factor

line and Z_0 and Z_l are the feeder and load impedance respectively. For example, if Z_0 is 80Ω and Z_l is 600Ω:

$$Z_t = \sqrt{80 \times 600} = 219\Omega$$

This matching section is useful for transforming impedance and is called a *quarter-wave transformer* (see Fig 13.15).

A section of tapered line can also be used to effect an impedance transformation. Again, a λ/4 section is only a special case. To achieve a match in a particular installation the line length and the angle of taper should be varied until a perfect match is achieved. This form of matching device is often called a *delta match*.

Cot (linear transformer) matching

An alternate method of matching is shown in Fig 13.16. It permits any two cables of different impedance to be matched by using appropriate lengths of the cables as shown. This avoids the need for a cable at the geometric mean impedance. G3KYH simplified the original formula to that shown. He stated that "for a 50/75Ω transformer this works out to an electrical length of 29.3° for each section of cable. The physical length must of course take into account the velocity factor of the cables (typically 0.66 to 0.86)."

Balance-to-unbalance transformers

In most cases an antenna requires a balanced feed with respect to ground. Therefore, it is necessary to use a device that converts the unbalanced output of a coaxial cable to the balanced output required by the antenna. This device also prevents the radio wave contained within the cable from tending to 'spill over'

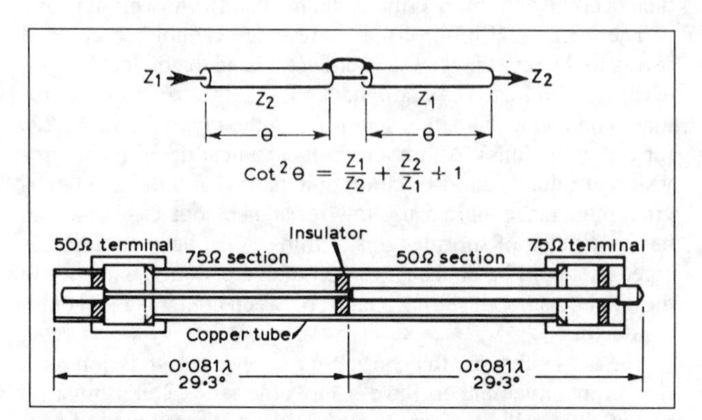

Fig 13.16. Transmission-line transformers which provide a simple way of matching 50 and 70Ω coaxial cables

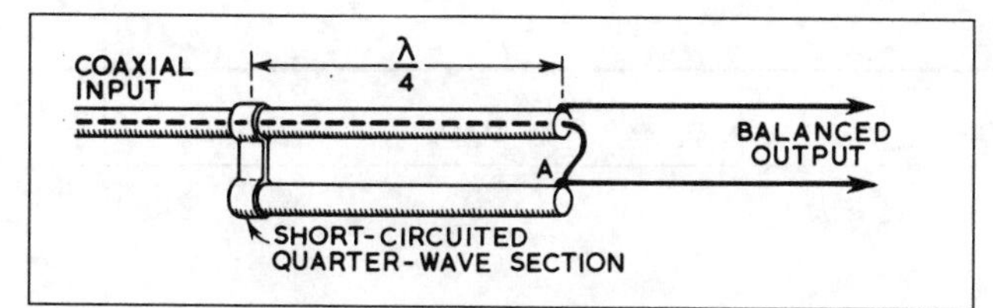

Fig 13.17. Quarter-wave open balun or Pawsey stub

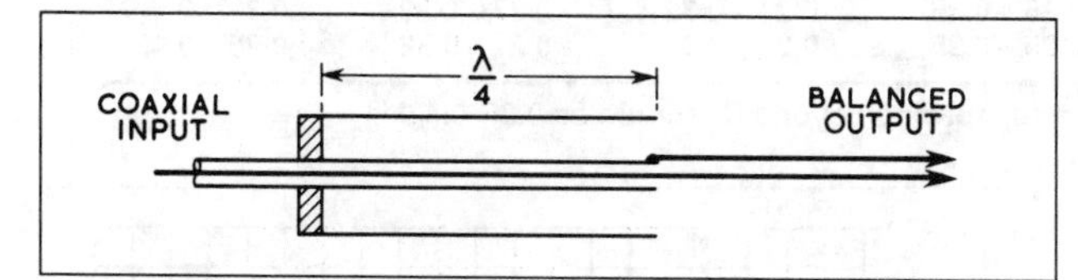

Fig 13.18. Coaxial sleeve balun

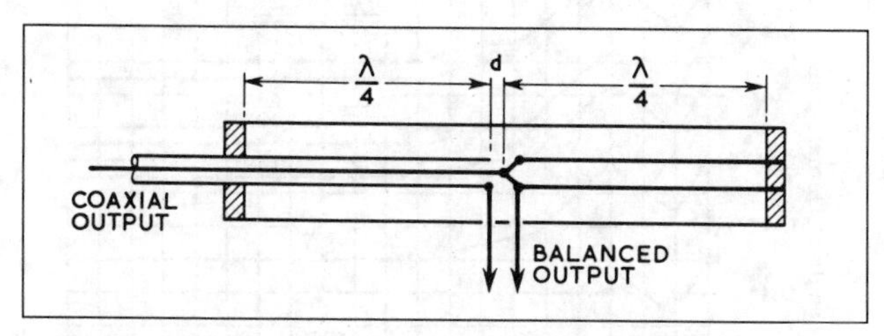

Fig 13.19. Totally enclosed coaxial balun. The right-hand section acts as a metal insulator

the extreme end and travel back over the surface of the cable. Whenever this occurs there are two important undesirable effects; first, the re-radiated wave modifies the polar diagram of the attached antenna, and secondly the outer surface of the cable is found to have an RF voltage on it.

To prevent this, a *balance-to-unbalance transformer* (abbreviated to *balun*) is connected between the feeder cable and the antenna. The simplest balun consists of a short-circuited λ/4 section of transmission line attached to the outer braiding of the cable as shown in Fig 13.17. This is often known as a *Pawsey stub*. At the point A the λ/4 section presents a very high impedance that prevents the wave from travelling over the surface. The performance of this device is, of course, dependent upon frequency, and its bandwidth may have to be considered in the design.

Several modifications of the simple balun are possible. For example, the single λ/4 element may be replaced by a λ/4

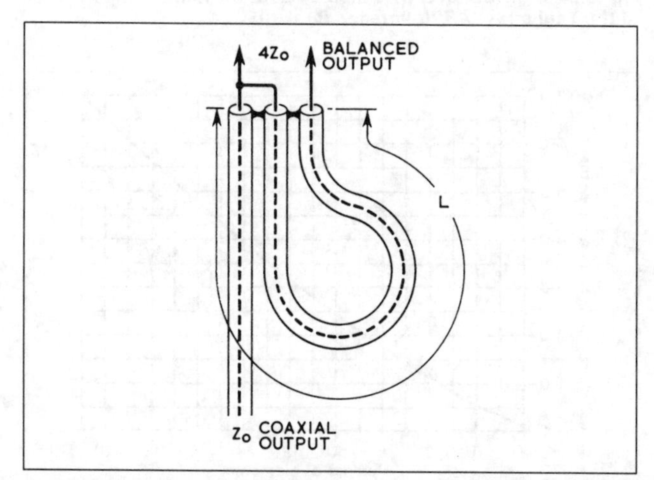

Fig 13.20. A coaxial balun giving a 4:1 impedance step-up. The length *L* should be λ/2, allowing for the velocity factor ot the cable. The outer braiding may be joined at the points indicated

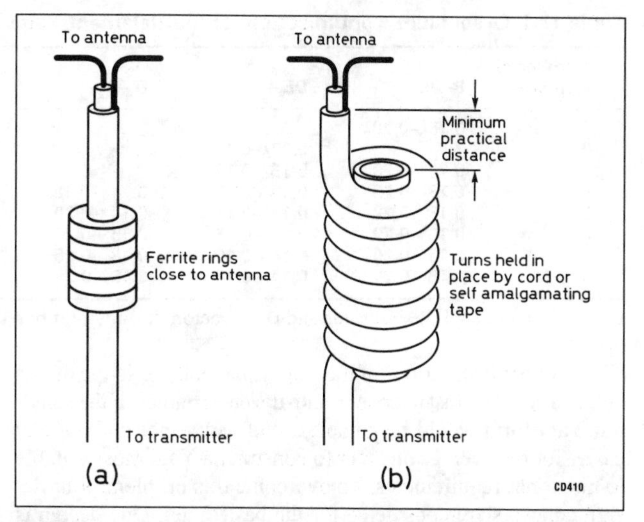

Fig 13.21. Two alternative baluns which are inherently wide band. In each case, the balun should be as close to the antenna as possible. The number of ferrite rings or the number of turns depends on the operating frequency. At 144MHz, 5–8 rings or 8–10 turns on 25mm dia should be satisfactory

coaxial sleeve. This can reduce radiation: see Fig 13.18. To prevent the ingress of water and to improve the mechanical arrangement, the centre conductor may be connected to a short-circuited λ/4 line. The line acts as a 'metallic insulator' as shown in Fig 13.19. The distance *d* should be kept small. Capacitance between the sections should be kept small otherwise the λ/4 section will not be resonant at the desired frequency. A satisfactory compromise is to taper the end of the λ/4 line, although this is by no means essential. In practice, at a frequency of 435MHz about 1/8in (3mm) is a suitable spacing. The whole balun is totally enclosed with the output taken through two insulators mounted in the wall.

A useful variation is that shown in Fig 13.20 that gives a 4:1 step-up of impedance. The λ/2 loop is usually made from flexible coaxial cable, and allowance must be made for the velocity factor of the cable when calculating a half-wavelength.

Coaxial-sleeve baluns should have an outer-to-inner diameter ratio of between 2:1 and 4:1.

The type of balun illustrated in Fig 13.19 has a larger effective bandwidth than the other types described here. However, they are all suitable for the VHF amateur bands.

A balun with a wide frequency range can be constructed by passing the antenna end of the coaxial cable through a series of ferrite rings (Fig 13.21). These raise the impedance to energy trying to flow down the outer of the coaxial cable. Suitable ferrites are type 43. An alternative is to form the coaxial cable into a 'choke' by winding it on to a former of about 25mm diameter to have 10 to 20 turns. Obviously this can only be done with small-diameter coaxial cable because there is a minimum radius through which any coaxial cable may be bent. It is specified by the makers.

THE YAGI AND ITS DERIVATIVES

It has been said that if you take a dozen variables and fit them in line along a supporting boom you will produce a Yagi antenna. This system was first conceived by Uda and constructed by Yagi in a form similar to that shown in Fig 13.22. It consists of a driven element combined with an in-line parasitic array. There have since been many variations of the basic concept, including its combination with log-periodic and backward-wave techniques.

Table 13.1. Greenblum's optimisation for multielement Yagis [1]

Number of elements	R–DE	DE–D_1	D_1–D_2	D_2–D_3	D_3–D_4	D_4–D_5	D_5–D_6
2	0.15λ–0.20λ						
2		0.07λ–0.11λ					
3	0.16 –0.23	0.16 –0.19					
4	0.18 –0.22	0.13 –0.17	0.14λ–0.18λ				
5	0.18 –0.22	0.14 –0.17	0.14 –0.20	0.17λ–0.23λ			
6	0.16 –0.20	0.14 –0.17	0.16 –0.25	0.22 –0.30	0.25λ–0.32λ		
8	0.16 –0.20	0.14 –0.16	0.18 –0.25	0.25 –0.35	0.27 –0.32	0.27λ–0.33λ	0.30λ–0.40λ
8 to *N*	0.16 –0.20	0.14 –0.16	0.18 –0.25	0.25 –0.35	0.27 –0.32	0.27 –0.32	0.35 –0.42

DE = driven element, R = reflector and D = director. *N* = any number. Director spacing beyond D_6 should be 0.35–0.42λ

To cover all variations of the Yagi is beyond the scope of this antenna chapter. Just to cover a half-dozen variables of the standard Yagi form would be complex, and would not make it any easier for the average amateur to construct a Yagi most suitable to his or her requirements. To overcome this problem, four design concepts are considered for the basic Yagi. One design is from a study by Greenblum in the USA and re-confirmed by G8CKN in the UK in 1978. The second study is from Chen and Cheng who produced an analytical study to maximise the directivity of the Yagi-Uda array by adjustment of the dipole lengths and spacing. The third is a measurement study by Viezbicke for the US Department of Commerce and National Bureau of Standards. Many independent investigations of multi-element Yagi antennas have shown that the gain of a Yagi is directly proportional to the array length. The fourth is the double-optimisation method for long Yagis due to Günter Hoch, DL6WU.

There is a certain amount of latitude in the position of the elements along the array. However, the optimum resonance of each element will vary with the spacing chosen. With Greenblum's dimensions [1], in Table 13.1, the gain will not vary more than 1dB. The most critical are the reflector and first director as they decide the spacing for all other directors and most noticeably affect the matching.

The optimum director lengths are normally greater the closer the particular director is to the driven element. (The increase of capacitance between elements is balanced by an increase of inductance, ie length.) However, the length does not decrease uniformly with increasing distance from the driven element.

Fig 13.23 shows experimentally derived element lengths for various material diameters. Elements are mounted through a cylindrical metal boom that is two or three diameters larger than the elements. Some variation in element lengths will occur using different materials or sizes for the support booms. This will be increasingly critical as frequency increases. The water absorbency of insulating materials will also affect the element lengths, particularly when in use.

Fig 13.24 shows the expected gain for various numbers of elements if the array length complies with Fig 13.25.

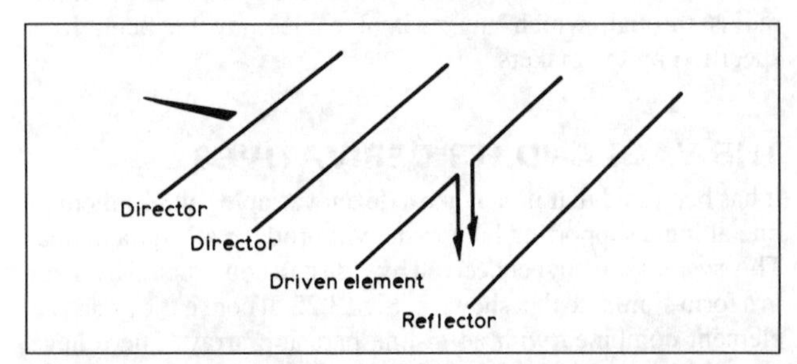

Fig 13.22. 'Four-element' antenna system, using two directors and one reflector in conjunction with a driven element

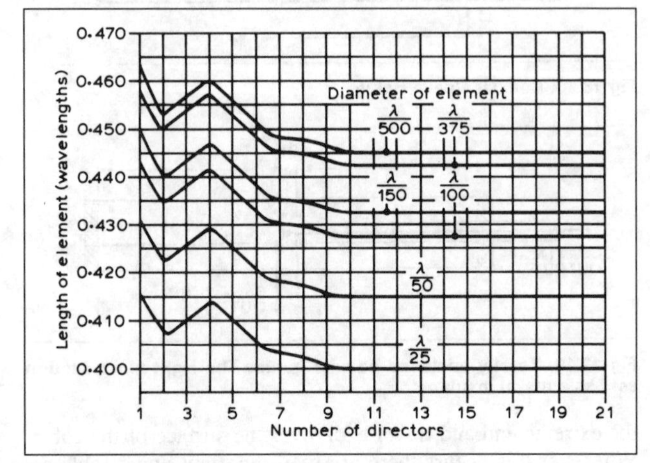

Fig 13.23. Length of director versus its position in the array for various element thicknesses (*ARRL Antenna Book*)

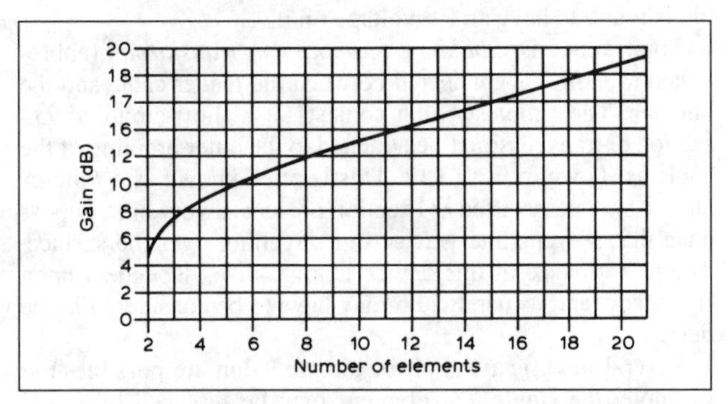

Fig 13.24. Gain (dB) over a λ/2 dipole versus the number of elements of the Yagi array (*ARRL Antenna Book*)

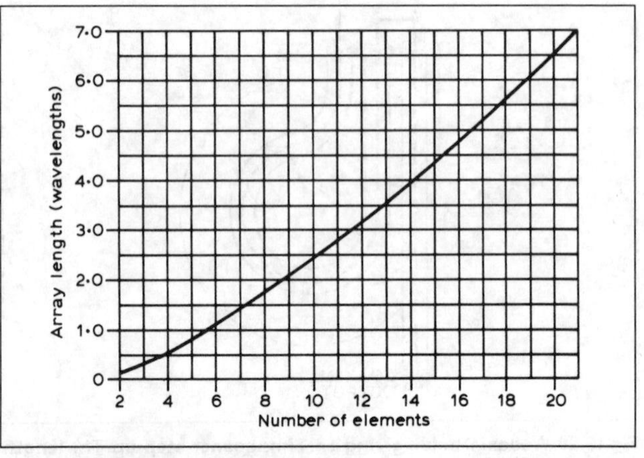

Fig 13.25. Optimum length of Yagi antenna as a function of number of elements (*ARRL Antenna Book*)

Table 13.2. Directivity optimisation for six-element Yagi-Uda array (perturbation of element lengths)

	h_1/λ	h_2/λ	h_3/λ	h_4/λ	h_5/λ	h_6/λ	Directivity (referring to $\lambda/2$ dipole)	Gain (dBD)
Initial array	0.255	0.245	0.215	0.215	0.215	0.215	7.544	8.78
Length-perturbed array	0.236	0.228	0.219	0.222	0.216	0.202	10.012	10.00

$b_{i1} = 0.250\lambda$, $b_{i2} = 0.310\lambda$ (i = 3,4,5,6), $a = 0.003369\lambda$.

Table 13.3. Directivity optimisation for six-element Yagi-Uda array (perturbation of element spacings and element lengths)

	h_1/λ	h_2/λ	h_3/λ	h_4/λ	h_5/λ	h_6/λ	b_{21}/λ	b_{22}/λ	b_{43}/λ	b_{34}/λ	b_{35}/λ	Directivity (referring to $\lambda/2$ dipole)	Gain (dBD)
Initial array	0.255	0.245	0.215	0.215	0.215	0.215	0.250	0.310	0.310	0.310	0.310	7.544	8.78
Array after spacing perturbation	0.255	0.245	0.215	0.215	0.215	0.215	0.250	0.289	0.406	0.323	0.422	11.687	10.68
Optimum array after spacing and length perturbations	0.238	0.226	0.218	0.215	0.217	0.215	0.250	0.289	0.406	0.323	0.422	13.356	11.26

The results obtained by G8CKN using the 'centre spacing' of Greenblum's optimum dimensions shown in Table 13.1 produced identical gains to those shown in Fig 13.26. Almost identical radiation patterns were obtained for both the E and H planes (V or H polarisation). Sidelobes were at a minimum and a fair front-to-back ratio was obtained.

At frequencies below 1000MHz ground reflections, and in particular differences between H and V polarisation, give a false indication of antenna performance. Most professional antenna measurement ranges require special techniques to remove these errors. However, with an optimum pattern shape for the measurement receiving antenna, results at 100MHz are sufficiently close to the expected theoretical figures as to be acceptable for most pattern measurement. Fig 13.26 shows radiation patterns for such an optimised vertically polarised antenna using Greenblum's dimensions.

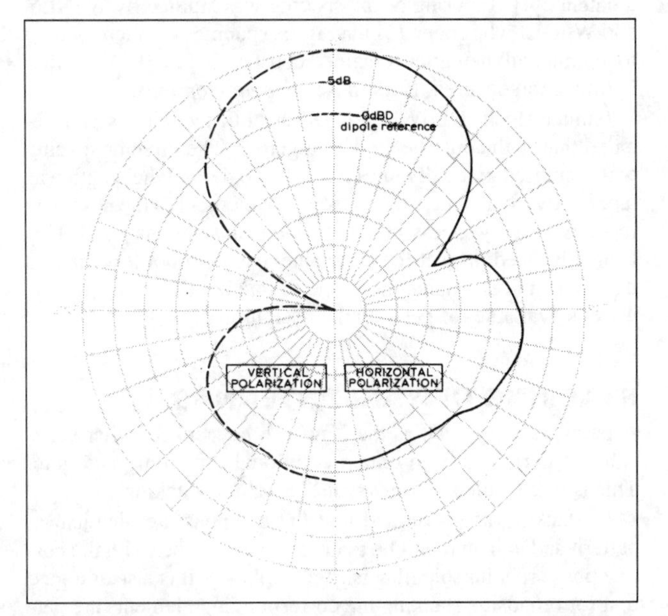

Fig 13.26. Decibel radiation patterns for a four-element Yagi optimised for minimum ground and mast distortion and minimum sidelobes configured for vertical polarisation using Greenblum's dimensions. The use of a decibel scale emphasises the lobe structure, giving more definition at the lower radiation levels

The antenna is mounted such that the null is in line with the supporting mast and the ground. The antenna supporting stub on the top of the mast was made of an insulating material such as glassfibre or resin-impregnated wood and was approximately λ/2 in size. A technique where measurements are conducted in the vertical plane, instead of horizontally, has produced good results as low as 30MHz.

Considerable work has been carried out by Chen and Cheng on the optimising of Yagis [2]. Tables 13.2 and 13.3 show their results obtained in 1974, by optimising both spacing and resonant lengths of the elements. Figs 13.27 and 13.28 show comparative radiation patterns for the various stages of optimisation.

The example is for a six-element Yagi with conventional shortening of the elements. The gain figure produced was 8.77dB relative to a λ/2 dipole. Optimising the element lengths produced a forward gain of 10dBD. Returning to the original element lengths and optimising the element spacing produced a forward gain of 10.68dBD. This is identical to the gain shown for a six-element Yagi in Fig 13.25. Using a combination of spacing and element length adjustment obtained a further 0.57dBD gain, giving 11.25dBD as the final forward gain.

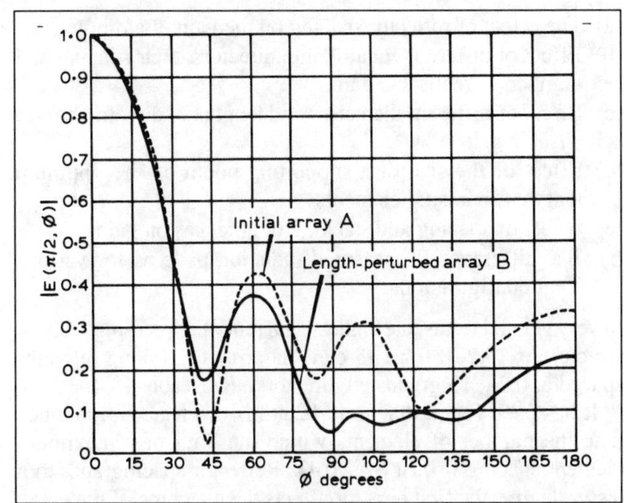

Fig 13.27. Normalised patterns ot six-element Yagi-Uda arrays from Table 13.2. The gain improvement of B over A is +1.2dB (*Proc IEEE*)

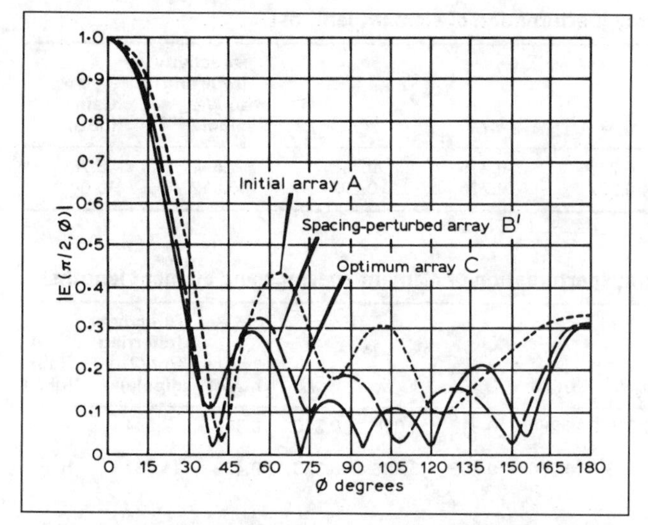

Fig 13.28. Normalised patterns of six-element Yagi-Uda arrays from Table 13.3. The gain improvement of B over A is +1.9dB, while that of C over A is +2.5dB (*Proc IEEE*)

Comparing the Greenblum results with those of Chen and Cheng shows, from the comparison of number of elements, that optimum element spacing has been achieved with different dimensions. A different reflector or first director spacing has been used to give good matching with a folded active element. An investigation of Greenblum's expected gain figures for optimum, total array length is shown in Table 13.4. The gain figures are in close agreement with Chen and Cheng's results. Typical values are 11.9dBD for a six-element Yagi compared with 11.25dBD.

A publication of the US Department of Commerce and National Bureau of Standards [3] provides very detailed information on Yagi dimensions. Results have been obtained from experimental measurements to optimise designs at 400MHz using a model antenna range.

The information, presented largely in graphical form, shows very vividly the effect of different antenna parameters on realisable gain. For example, it shows the extra gain that can be achieved by optimising the lengths of the different directors, rather than making them all of uniform length. It also shows just what extra gain can be achieved by stacking two elements, or from a 'two-over-two' array.

In fact it shows:

(a) The effect of reflector spacing on the gain of a dipole.
(b) Effect of different equal-length directors, their spacing and number on realisable gain.
(c) Effect of different diameters and lengths of directors on realisable gain.
(d) Effect of the size of a supporting boom on the optimum length of parasitic elements.
(e) Effect of spacing and stacking of antennas on gain.
(f) The difference in measured radiation patterns for various Yagi configurations.

In very general terms, the highest gain reported for a single boom structure is 14.2dB for a 15-element array (4.2λ long reflector spaced at 0.2λ, 13 graduated directors). See Table 13.4.

It has been found that array length is of greater importance than the number of elements within the limit of a maximum element spacing of just over 0.4λ. Reflector spacing and, to a lesser degree, the first director affects the matching of the Yagi. Optimum tuning of the elements, and therefore gain and pattern shape, varies with different element spacing.

Table 13.4. Optimised lengths of parasitic elements for Yagi antennas of six different boom lengths

Length of Yagi (λ)	0.4	0.8	1.20	2.2	3.2	4.2
Length of reflector (λ)	0.482	0.482	0.482	0.482	0.482	0.475
Length of directors (λ)						
1st	0.424	0.428	0.428	0.432	0.428	0.424
2nd	—	0.424	0.420	0.415	0.420	0.424
3rd	—	0.428	0.420	0.407	0.407	0.420
4th	—	—	0.428	0.398	0.398	0.407
5th	—	—	—	0.390	0.394	0.403
6th	—	—	—	0.390	0.390	0.398
7th	—	—	—	0.390	0.386	0.394
8th	—	—	—	0.390	0.386	0.390
9th	—	—	—	0.398	0.386	0.390
10th	—	—	—	0.407	0.386	0.390
11th	—	—	—	—	0.386	0.390
12th	—	—	—	—	0.386	0.390
13th	—	—	—	—	0.386	0.390
14th	—	—	—	—	0.386	—
15th	—	—	—	—	0.386	—
Director spacing (λ)	0.20	0.20	0.25	0.20	0.20	0.308
Gain (dBD)	7.1	9.2	10.2	12.25	13.4	14.2

Element diameter 0.0085λ. Reflector spaced 0.2λ behind driven element. Measurements are for 400MHz by P P Viezbicke.

Rationalising, near-optimum patterns and gain can be obtained using Greenblum's dimensions for up to six elements. Good results for a Yagi in excess of six elements can still be obtained where ground reflections need to be minimised.

Chen and Cheng have employed what is commonly called the *long Yagi technique*. Yagis with more than six elements start to show an improvement in gain with fewer elements when this technique is employed.

Typical dimensions of Yagi antennas for three bands are given in Table 13.5.

Long Yagi antennas

The NBS optimisation above has been extended by American amateurs [4]. Tapering of the spacing was studied by W2NLY and W6QKI who found [5] that, if the spacing was increased up to a point and thereafter remained constant at 0.3–0.4λ, another optimisation occured. Both these are *single optimisation.*

Günter Hoch, DL6WU, looked at both techniques and decided that both could be applied together. The director spacing was increased gradually until it reached 0.4λ and the length was tapered by a constant *fraction* from one element to the next. The result is a highly sucessful *doubly optimised* antenna [6, 7]. This is an advanced project for large antennas with not less than 10 directors, and is outside the scope of this book. However, software is available to speed up the design process [8].

STACKING OF YAGI ANTENNAS

A parasitic array such as the Yagi can be stacked either vertically or horizontally to obtain additional directivity and gain. This is often called *collinear and broadside* stacking.

In stacking it is assumed that the antennas are identical in pattern and gain and will be matched to each other with the correct phase relationship, that is, 'fed in phase'. It is also assumed that for broadside stacking the corresponding elements are parallel and in planes perpendicular to the axis of the individual arrays. With vertical stacking it is assumed the corresponding elements are collinear and all elements of the individual arrays are in the same plane.

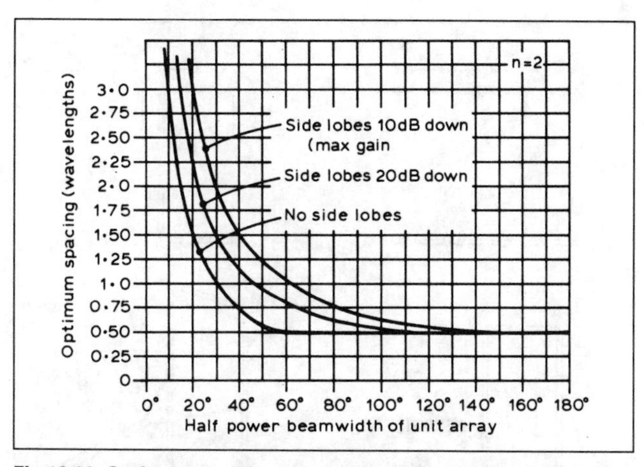

Fig 13.29. Optimum stacking spacing for two-unit arrays. The spacing for no sidelobes, especially for small beamwidths, may result in no gain improvement with stacking (*ARRL Antenna Book*)

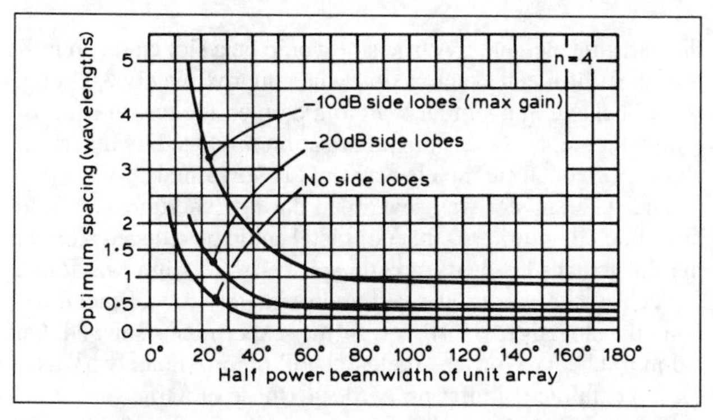

Fig 13.30. Optimum stacking spacing for four-unit arrays (*ARRL Antenna Book*)

The combination of the radiation patterns can add but can also cancel. The phase relationships, particularly from the side of the Yagi, are very complex. Because of this complexity the spacing to obtain maximum forward gain does not coincide with the best sidelobe structure. Usually maximum gain is less important than reducing signals to the sides or behind the array.

This being the case, optimum spacing is one that gives as much forward gain as possible as long as the sidelobe structure does not exceed a specific amplitude compared with the main lobe.

Fig 13.29 gives typical optimum spacing for two arrays under three conditions: (a) optimum forward gain with sidelobe down 10dB, (b) sidelobe 20dB down and (c) virtually no sidelobe.

The no-sidelobe case can correspond to no additional forward gain over a single antenna.

The maximum forward gain of two stacked arrays is theoretically +3dB with +6dB for four stacked arrays. More complex arrays could produce higher gain but losses in the matching and phasing links between the individual arrays can outweigh this improvement. When stacking two arrays, the extra achievable gain is reduced at close spacing due to high mutual impedance effects. With two seven-element arrays a maximum gain of about 2.5dB can be achieved with 1.6λ spacing; with two 15-element arrays it was also possible to achieve the extra 2.5dB but the spacing needed to be 2λ.

The use of four arrays, in correctly phased two-over-two systems, can increase the realisable gain by about 5.2dB. Using seven-element Yagis this produced a total gain of 14.2dB. With 15-element optimised Yagis a total gain of 19.6dB was obtained.

Table 13.5. Typical dimensions of Yagi array components

	Length		
	70.3MHz	**145MHz**	**433MHz**
Driven elements			
Dipole (for use with gamma match)	79 (2000)	38 (960)	12¾ (320)
Diameter range for length given	½–¾ (12.7–19.0)	¼–3/8 (6.35–9.5)	1/8–¼ (3.17–6.35)
Folded dipole 70Ω feed			
l length centre/centre	77½ (1970)	38½ (980)	12½ (318)
d spacing centre/centre	2½ (64)	7/8 (22)	½ (13)
Diameter of element	½ (12.7)	¼ (6.35)	1/8 (3.17)
a centre/centre	32 (810)	15 (390)	5 1/8 (132)
b centre/centre	96 (2440)	46 (1180)	152 (395)
Delta feed sections (length for 70Ω feed)	22½ (570)	12 (300)	42 (110)
Diameter of slot and delta feed material	¼ (6.35)	3/8 (9.5)	3/8 (9.5)
Parasitic elements			
Element			
Reflector	85½ (2170)	40 (1010)	13¼ (337)
Director D1	74 (1880)	35½ (902)	11¼ (286)
Director D2	73 (1854)	35¼ (895)	11 1/8 (282)
Director D3	72 (1830)	35 (890)	11 (279)
Succeeding directors	1in less (25)	½in less (13)	1/8in less
Final director	2in less (50)	1in less (25)	¾in less
One wavelength (for reference)	168¾ (4286)	81½ (2069)	27¼ (693)
Diameter range for length given	½–¾ (12.7–19.0)	¼–3/8 (6.35–9.5)	1/8–¾ (3.17–6.35)
Spacing between elements			
Reflector to radiator	22½ (572)	17½ (445)	5½ (140)
Radiator to director 1	29 (737)	17½ (445)	5½ (140)
Director 1 to director 2	29 (737)	17½ (445)	7 (178)
Director 2 to director 3, etc	29 (737)	17½ (445)	7 (178)

Dimensions are in inches with millimetre equivalents in brackets.

(This was the highest gain measured during the experiments by Viezbicke.) The effects of stacking in combination with the physical and electrical phase relationship can be used in two ways to reduce directional interference.

An improvement in front-to-back ratio can be accomplished in vertical stacking by placing the top Yagi λ/4 in front of the lower Yagi as shown in Fig 13.31. The top antenna is fed 90° later than the bottom antenna.

If the interference is largely from the forward direction, feeding the antennas in phase (as shown in Fig 13.32) and varying the spacing between the arrays will give a primary null in the combined patterns at specific angles. Results for five-element Yagis are shown in Table 13.6. This can be useful where beams are set in a fixed direction.

Table 13.6. Spacing of Yagi antennas to obtain desired null angles as described in the text

Null angle (°)	10	15	20	25	30	35	40	45°	50	55	60
Spacing (λ)	2.5	1.75	1.5	1.25	1.0	0.85	0.75	2.25	1.9	1.7	1.65

An improvement in front-to-back ratio can be accomplished in vertical stacking by placing the top Yagi λ/4 in front of the lower Yagi as shown in Fig 13.31. The top antenna is fed 90° 'later' than the bottom antenna.

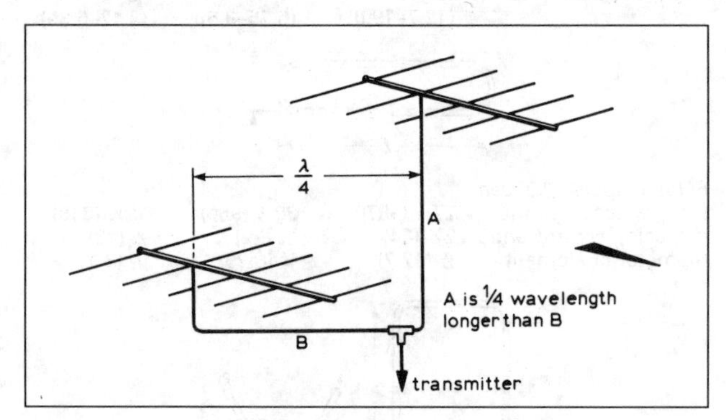

Fig 13.31. Improving the front-to-back ratio of stacked Yagi antennas with offset vertical mounting

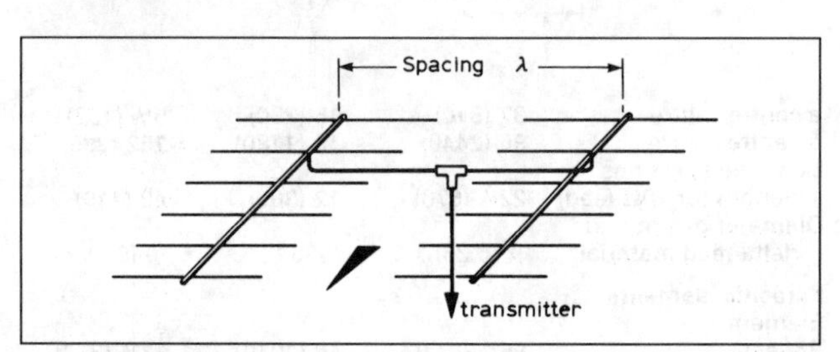

Fig 13.32. Desired null angle as shown in Table 13.6 is determined by the spacing between the two Yagi antennas

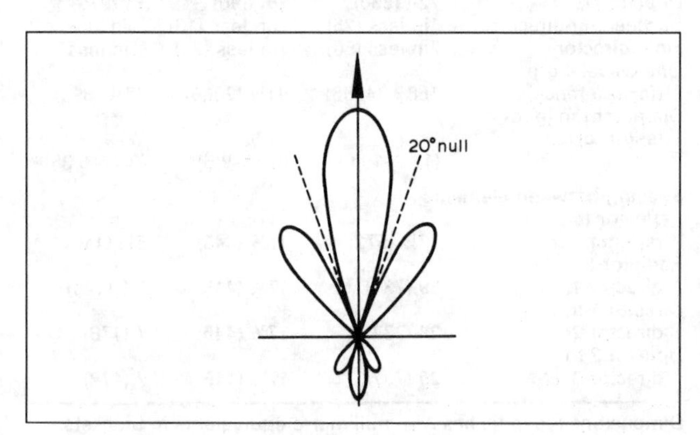

Fig 13.33. Response of two five-element Yagi antennas positioned for a 20° pattern null

Disadvantage of conventional Yagi arrays

Perhaps the most important disadvantage of the Yagi array is that variation of the element lengths and spacing causes interrelated changes in the feed impedance. To obtain the maximum possible forward gain experimentally is extremely difficult. For each change of element length it is necessary to readjust the matching either by moving the reflector or by resetting a matching device.

However, one method has been devised for overcoming these practical disadvantages. It involves the use of a radiating element in the form of a *skeleton slot*. This is far less susceptible to the changes in impedance caused by changes in the length of

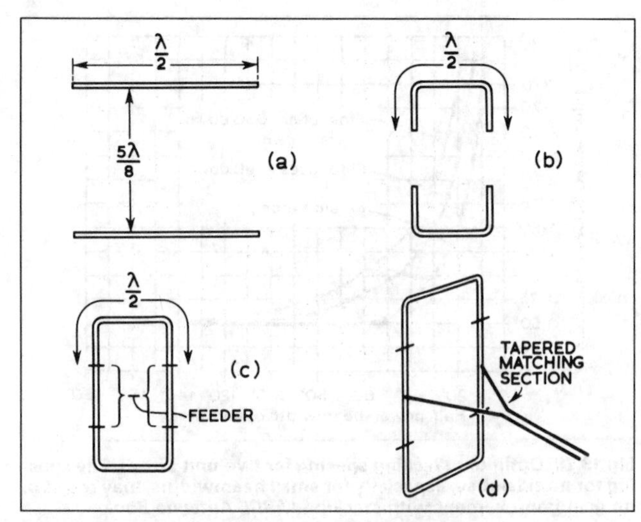

Fig 13.34. Development of a skeleton-slot radiator

the parasitic elements. A true slot would be a slot cut in an infinite sheet of metal. Such a slot, when approximately λ/2 long, would behave in a similar way to a dipole radiator. In contrast with a dipole, however, the polarisation produced by a vertical slot is horizontal (ie the electric field is horizontal).

The skeleton slot was developed during experiments to find how much the 'infinite' sheet of metal could be reduced without the slot antenna losing its radiating property. The limit was found to occur when there remained approximately λ/2 of metal beyond the slot edges. However, further experiments showed that a thin rod bent to form a 'skeleton slot' (approximately 5λ/8 by 5λ/24) exhibited similar properties to those of a true slot.

The way a skeleton slot works is shown in Fig 13.34. Consider two λ/2 dipoles spaced vertically by 5λ/8. Since the greater part of the radiation from each dipole takes place at the current antinode (ie the centre) the ends of the dipoles may be bent without serious effect. These 'ends' are joined together with a high-impedance feeder, so that 'end feeding' can be applied to the bent dipoles. To radiate in phase, the power should be fed midway between the two dipoles. The high impedance at this point may be transformed down to that of the feeder cable with a tapered matching section/transmission line (ie a delta match). Practical dimensions of a skeleton-slot radiator are given in Fig 13.35.

It is important to note that two sets of parasitic elements are required with a skeleton-slot radiator and not one set as with a true slot. One further property of the skeleton slot is that its bandwidth is greater than a pair of stacked dipoles.

Skeleton-slot Yagi arrays in a stack

Skeleton-slot Yagi arrays may be stacked to increase the gain but the same considerations of optimum stacking distance as previously discussed apply. The centre-to-centre spacing of a pair of skeleton-slot Yagi arrays should typically vary between 1λ and 3λ depending on the number of elements in each Yagi array.

Each skeleton-slot Yagi may be fed by 72Ω coaxial cable with equal lengths of feeder to a convenient common feed point. It is desirable to use a balun at the point where the cable is attached to each array. A coaxial λ/4 transformer can be used to transform the impedance to that of the main feeder.

As an example, if a pair of skeleton-slot Yagi arrays, each of 72Ω feed impedance, is stacked, the combined impedance will be half 72Ω, ie 36Ω. This may be transformed to 72Ω with a λ/4 section of 52Ω coaxial cable, allowance being made for its

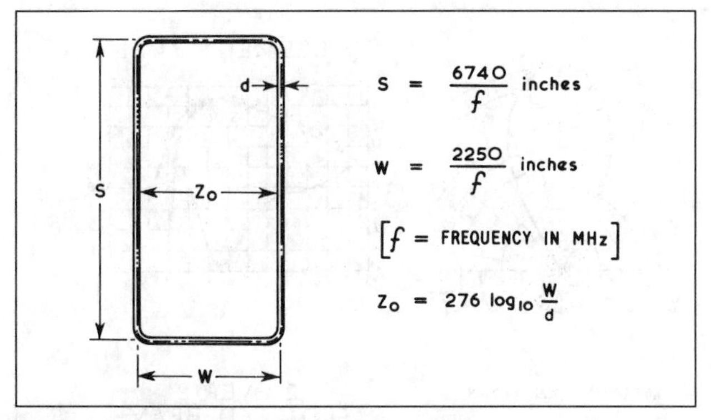

Fig 13.35. Dimensional relationships of a skeleton-slot radiator. Both *S* and *W* may be varied experimentally from the values indicated by these formulae. For small variations the radiation characteristics of the slot will not change greatly, but the feed impedance will undergo appreciable change and therefore the length of the delta matching section should always be adjusted to give a perfect match to the transmission line

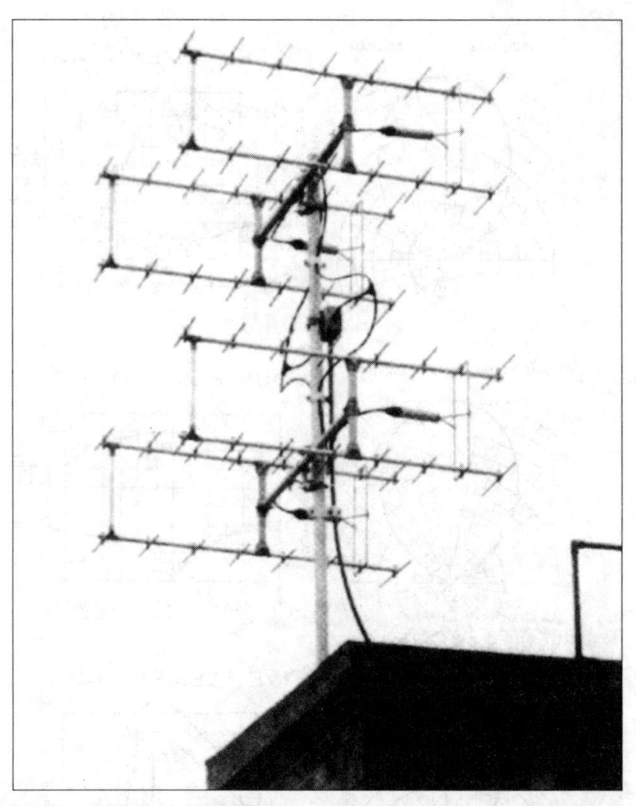

Fig 13.37. A high-gain 432MHz antenna consisting of four 8-over-8 fed Yagi units arranged in a square formation

velocity factor. Larger assemblies of Yagi arrays can be fed in a similar manner by joining pairs and introducing λ/4 transformers until a single feed remains for the whole system.

STACKED DIPOLE ARRAYS

Both horizontal and vertical beamwidth can be reduced and gain increased by building up arrays of driven dipoles. This arrangement is usually referred to simply as a *stack*, sometimes as a *broadside* array. Since this type of array is made up from a number of radiating dipoles, the feed impedance would be extremely low if the dipoles were centre-fed.

However, the impedance to earth of a dipole at its end is high, the precise value depending upon the length-to-diameter ratio of the dipole. It is more convenient to use a balanced high-impedance feeder to 'end-feed' a pair of collinear half-wave dipoles (a 'full-wave' dipole)

The element length for resonance and the feed impedance in terms of wavelength/diameter ratio is shown in Table 13.7.

The full-wave (1λ) dipoles are usually mounted with a centre-to-centre spacing, horizontally and vertically, of λ/2, and are fed in phase.

Typical arrangements for stacks of 1λ dipoles are shown in Fig 13.39. Note that the feed wires between dipoles are λ/2 long and are crossed so that all the dipoles in each bay are fed in phase.

The impedance of these phasing sections is unimportant provided any separators used are made of low-loss dielectric material. Also, there must be sufficiently wide separation at the cross-over points to prevent contact.

To obtain the radiation pattern expected, all dipoles should be fed with equal amounts of power. This cannot be achieved in practice because the dipoles that are farthest from the feeder have a greater feeder loss than the nearest. However, by locating the main feed point as symmetrically as possible these effects are reduced. It would be preferable for the antenna shown in Fig 13.39(a) to be fed in the centre of each bay of dipoles.

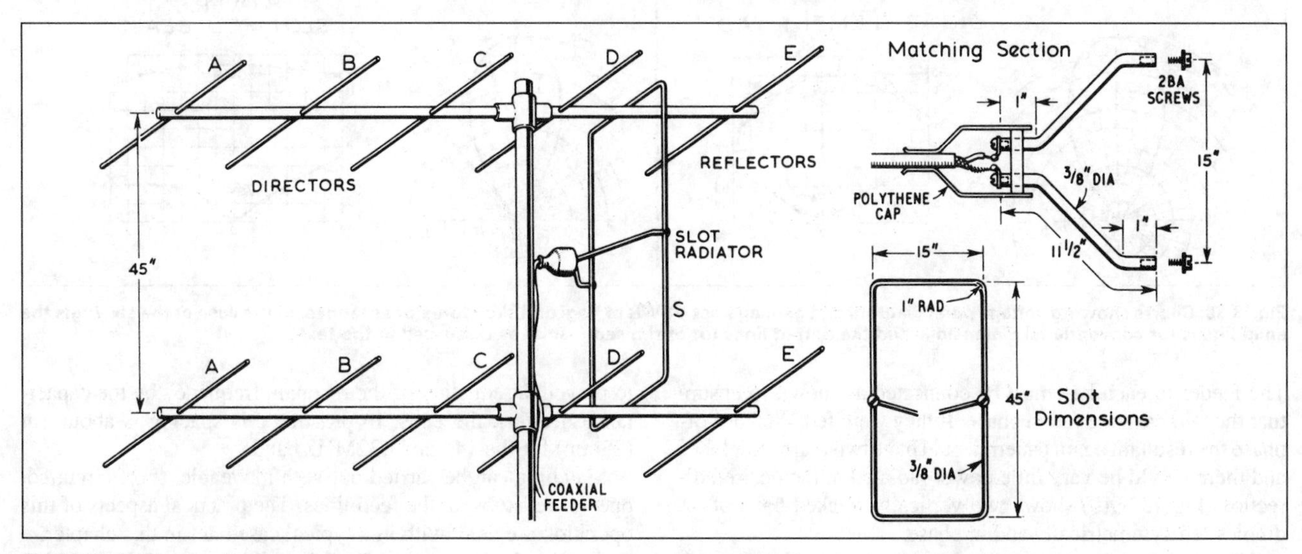

Fig 13.36. A six-over-six slot antenna

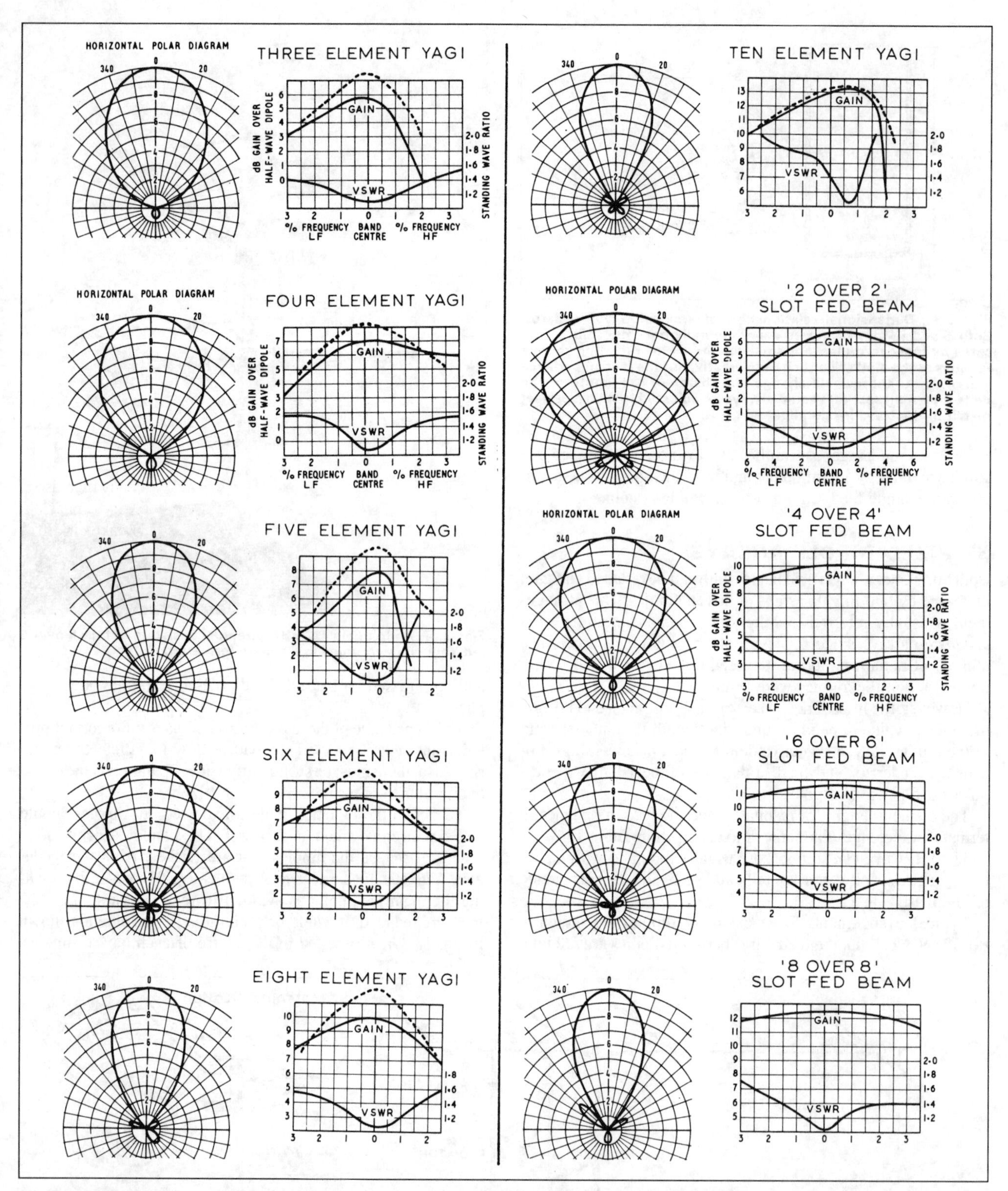

Fig 13.38. Charts showing voltage polar diagram and gain against VSWR of Yagi and skeleton-slot antennas. In the case of the six Yagis the solid line is for conventional dimensions and the dotted lines for optimised results as discussed in the text

The feeder to each bay must be connected as shown to ensure that the two bays are fed in phase. If they were fed 180° out-of-phase the resultant beam pattern would have two major sidelobes and there would be very little power radiated in the desired direction. Fig 13.39(b) shows two vertically stacked bays of 1λ dipoles fed symmetrically and in phase.

The spacing at the centre of each 1λ dipole should be enough to prevent a reduction of the resonant frequency by the capacitance between the ends. In practice this spacing is about 1in (25mm) for the 144 and 432MHz bands.

Matching can be carried out with moveable, short-circuited, open-wire, stubs on the feed lines. The practical aspects of this operation are dealt with in a typical case later in this chapter.

As with the Yagi array, the gain will be increased by placing

Table 13.7. Resonant lengths of full-wave dipoles

Value of wavelength / Diameter	Dipole length / Wavelength for resonance	Feed impedance (Ω)
50	0.85	500
100	0.87	900
150	0.88	1100
200	0.896	1300
300	0.906	1500
400	0.916	1700
700	0.926	2000
1000	0.937	2400
2000	0.945	3000
4000	0.951	3600
10,000	0.958	4600

The dimensions used in calculating the ratios must be in the same units (eg both in metres or both in centimetres). From *Aerials for Metre and Decimetre Wavelengths* by R A Smith.

a reflector behind the radiating elements at a spacing 0.1 to 0.25λ. A figure of 0.25λ is frequently chosen. For the 432MHz band and for higher frequencies, a plane reflector made up of mesh wire netting stretched on a frame can replace the resonant reflector. The hole size for the mesh must be less than 0.05λ, ie 1in (25mm) for 432MHz. The mesh of the wire should be so orientated that the interlocking twists are parallel to the dipole. The wire netting should extend at least λ/2 beyond the extremities of the dipoles in order to ensure a high front-to-back ratio.

The λ/2 sections of the 1λ dipole should be supported at the current antinodes, ie at their centres, either on small insulators or in suitably drilled wooden vertical members. Supports should not be mounted parallel to the elements because of possible influence on the properties of the antenna.

The bandwidth of this type of antenna is exceptionally large and its adjustments generally are far less critical than those of Yagi arrays.

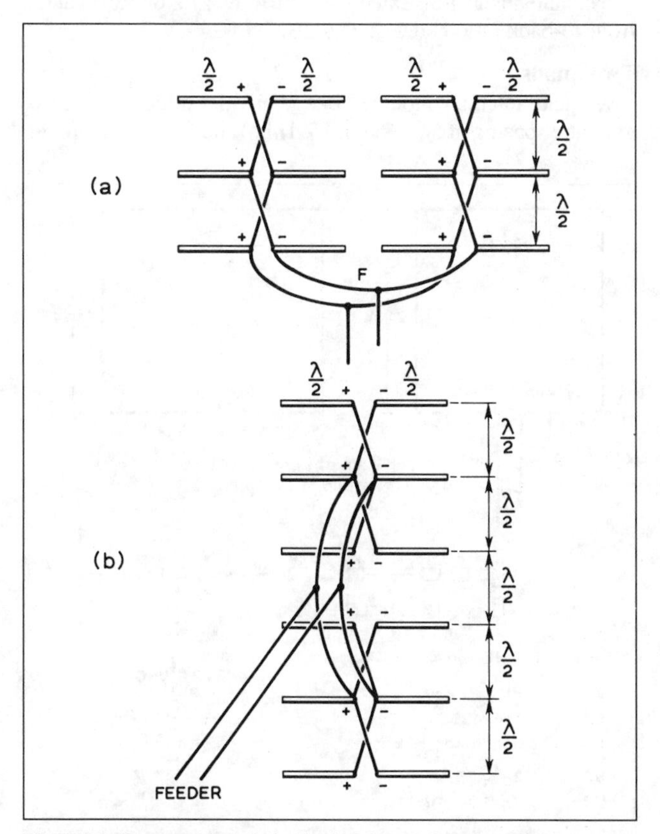

Fig 13.39. Two stacks of 1λ dipoles. Note that the feedpoint F is equidistant from each bay of dipoles. For examples of element lengths, see Table 13.7

For a stack having an adequate wire-net reflector, the horizontal beamwidth θ, vertical beamwidth φ, and power gain *G* (compared with an isotropic radiator) can be calculated approximately from the following formulae:

$$\theta = \frac{51\lambda}{a} \qquad \varphi = \frac{51\lambda}{b} \qquad G = \frac{4\pi ab}{\lambda^2}$$

where *a* and *b* are the horizontal and vertical dimensions of the reflector respectively. Both *a* and *b* are in the same units as the wavelength. These formulae are true only for an array that is large compared with the wavelength. They are suitable as a criterion for judging antennas of any type, provided the equivalent aperture or capture area is known.

Disadvantage of multi-element arrays

As the frequency becomes higher and the wavelength becomes shorter, it is possible to construct high-gain arrays, although, as already described, much of the gain advantage is offset by the reduction in capture area. If many driven or parasitic elements, either in line or in stack, are adopted, the complication of numerous feeder cables rapidly increases. As frequency increases, the radiation and resistive loss from open-wire lines, phasing lines and matching sections also increase. This makes it difficult to ensure an equal power feed to all radiators.

Preferably the antenna or antenna array should have a few separately 'fed' elements and other critical parts, ie phasing and matching components. Feed cable resistive losses and impedance mismatch should be kept as small as practical. Damaged cables must be avoided.

12-element stack for the 432MHz band

Fig 13.40 illustrates the assembly of a 12-element stack arranged in two horizontal bays of three 1λ dipoles with a gain of 13dBD. With the dimensions shown the centre frequency is 432MHz. Radiating elements are constructed from thin-walled 3/8in (9.5mm) diameter brass tubing to which the feeder and phasing wires are soldered. The centres are supported on ½in (13mm) insulators mounted on vertical 1in by 1in (25mm) members. The reflecting screen is made from thin-mesh galvanised-wire netting mounted on a framework of aluminium-alloy angle. Each bay of three 1λ dipoles is fed at the centre. Feed lines are taken straight through the reflector. Each of the feed lines is matched to the 300Ω open-wire feed line by a moveable, short-circuited, stub. The two 300Ω feeders, one from each bay, are joined together to the output of a totally enclosed coaxial balun and a two-stub tuner. The antenna is fed with 72Ω coaxial cable.

To match the antenna to the feeder, first one bay of the beam is disconnected and a 300Ω 1W carbon resistor is substituted. The stub on this side should be set to exactly λ/4 long.

The remaining bay is then matched approximately to the open-wire feed line. This may be determined by using a very-low-wattage bulb (eg 6V, 0.06A), the screwed body of which is held in the hand. The presence of standing waves is seen from variation in the glow in the bulb as it is slid along the feed line with its centre connection in contact with one of the feed-line conductors. (*WARNING – possible RF hazard.*) To continue, the procedure is reversed with the resistor being placed to represent the bay previously matched. Both bays are then reconnected and the final matching carried out by means of the two-stub tuner using a slotted line or other matching device. (Subject to it having the appropriate impedance, a VSWR bridge can be used to set up this antenna system.) If no special apparatus is available the array should be tuned for maximum gain as already described.

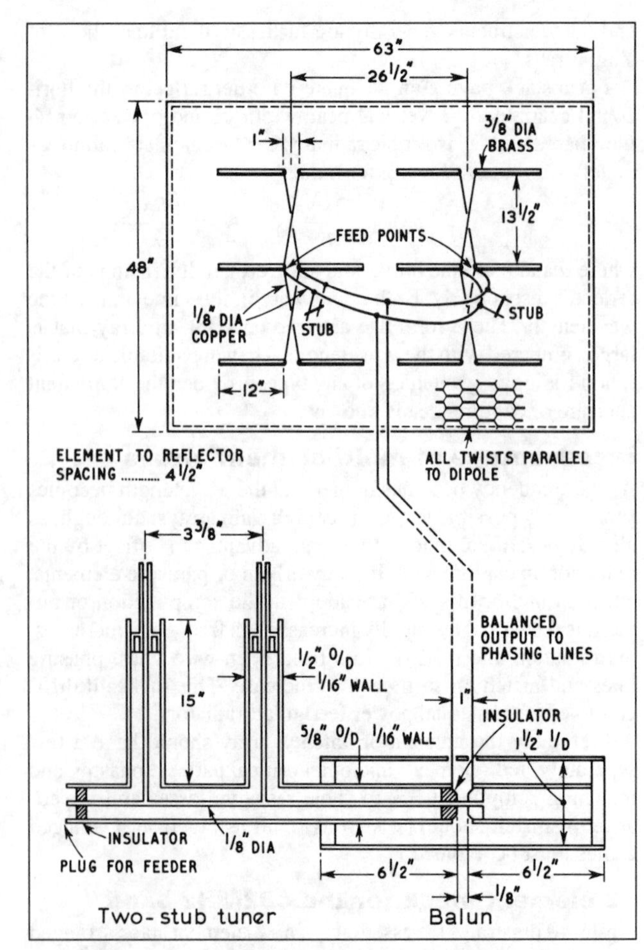

Fig 13.40. A 12-element stack for 432MHz, complete with two-stub tuner and enclosed coaxial balun

THE QUAD

In spite of its relatively small physical size a forward gain of 5.5 to 6dB can be obtained with good front-to-back ratio. Additional quad or single element directors can be added to the basic two-element array in the same manner as the Yagi.

Typical dimensions for both 70MHz and 144MHz are given in Table 13.8. The actual spacing should initially be made adjustable. For a 144MHz antenna the spacing between the radiator and reflector will be between 7in (178mm) and 9in (229mm) for a 72Ω input impedance. A balun should be used to connect the cable to the radiator. This has been dispensed with by some users but is advisable even when "the feeder is short and low loss."

An alternative to a balanced input would be the use of a gamma

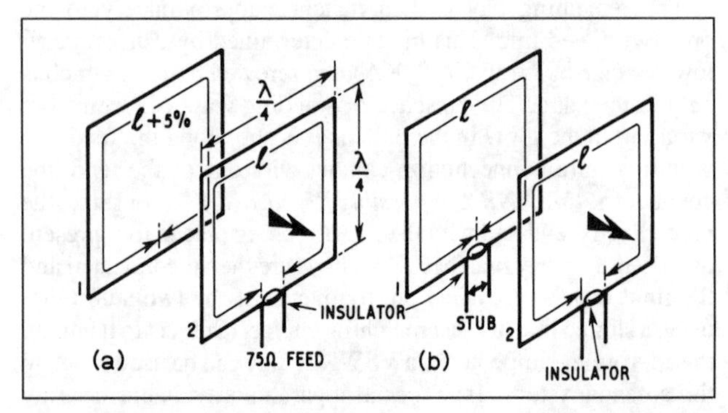

Fig 13.41. Quad antenna dimensions

Table 13.8. Design dimensions for 70 and 144MHz quad antenna

Band (MHz)	Reflector 1 total length	Radiator 2 total length	Director (if used)	Approx length of stubs if used Reflector s/c	Director s/c
70 (a)	173 (4390)	165 (4190)	157 (3990)	—	—
70 (b)	165 (4190)	165 (4190)	165 (4190)	8 (203)	8 (203)
144 (a)	84 (2130)	80 (2030)	76 (1930)	—	—
144 (b)	80 (2030)	80 (2030)	80 (2030)	4 (101)	4 (101)

Dimensions are in inches with millimetre equivalents in brackets.

match with the driven quad element 'joined' at the original feed point.

The elements may be made of 1/8in to ¼in diameter aluminium, preferably solid so that the corners can be easily bent. If the vertical dimensions of both elements are made the same, then two short cross-pieces can be used to mount the antenna to the mast. These cross-pieces may be made of metal. This will produce a very solidly built antenna that can withstand high winds. Additional protection may be required for the feed point and stub (if used). Quads may be stacked or built into a four-square assembly in the same way as the basic Yagi.

A four-square quad antenna for 144MHz

The dimensions of the basic quad are shown in Fig 13.42. Taking the 'squared up' figures of 201in (5105mm) for each side of the driven element is approximately 0.255λ in 'free space' at 144MHz. The spacing between the two elements to match a 72Ω feeder cable was found to be 7in, approximately 0.08λ. The spacing between the elements has a markedly critical effect on the VSWR. Forward gain of the antenna is not changed in so drastic a manner.

The antenna as illustrated has a forward gain of 5dBD and a front-to-back ratio better than 20dB.

Two quads

Two quad antennas mounted one above the other at a centre-to-centre spacing of 5λ/8 (65in, 1651mm) and combined through

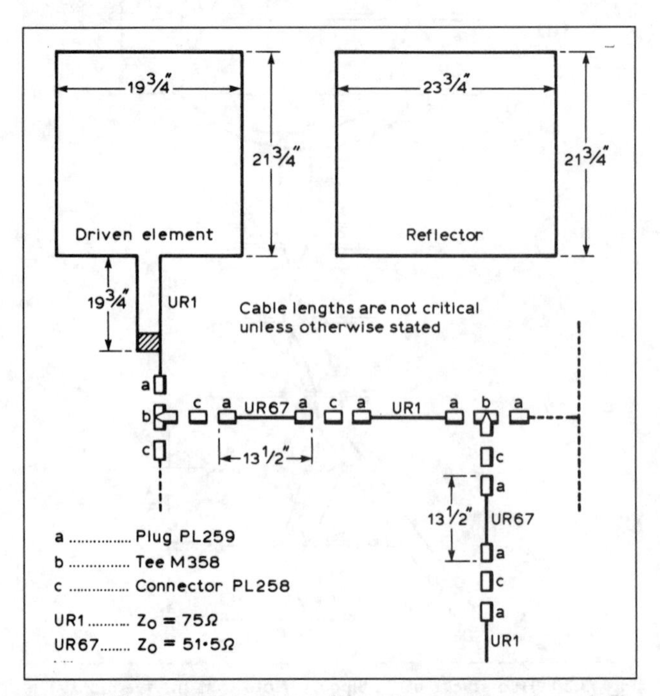

Fig 13.42. Physical arrangement of the antenna system shown in Fig 13.43 with details of matching sections of coaxial cable

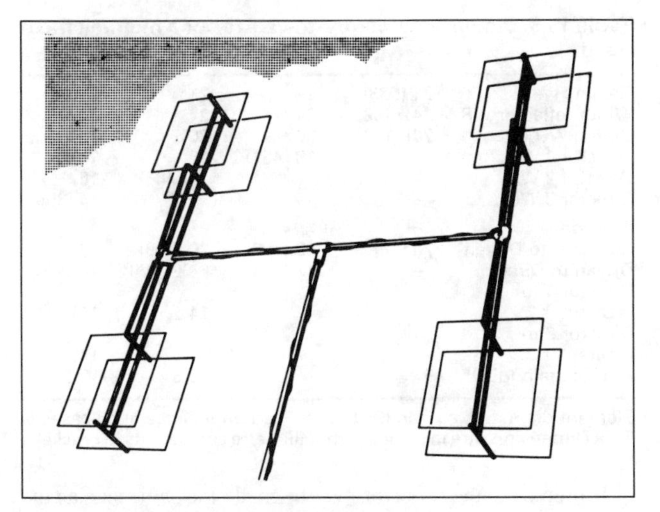

Fig 13.43. A 144MHz cubical quad array

a single λ/4, 51.5Ω transformer. Performance figures for this type of antenna are typically a forward gain of 8.2dB over a λ/2 dipole with a front-to-back ratio better than 10dB at 144MHz. To obtain the 72Ω input impedance for the double system it may be necessary to increase the spacing of the reflectors to driven element on both quads to 9in (229mm). This need to change the spacing is due to a reduction in input impedance of the individual quad caused by mutual coupling of the pair.

Four quads

The final form of the antenna is shown in Figs 13.43 and 13.44. The layout is determined by the best 'run' for the feeder cable and to avoid a long length of unguyed mast. The spacing between the elements is 9in (229mm), as for the double quad. Vertical spacing between quads is 65in (1651mm). The horizontal spacing is set at 78in (2m) centre-to-centre, ie one wavelength in 'free space'.

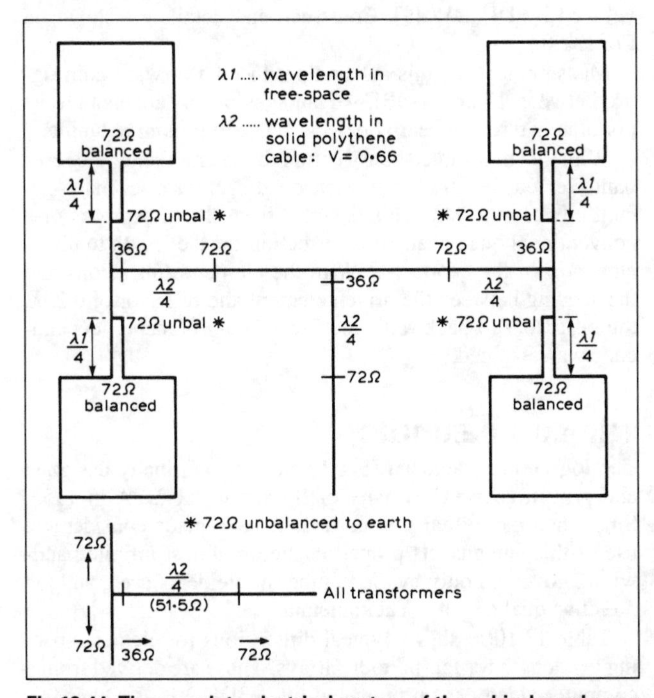

Fig 13.44. The complete electrical system of the cubical quad array described

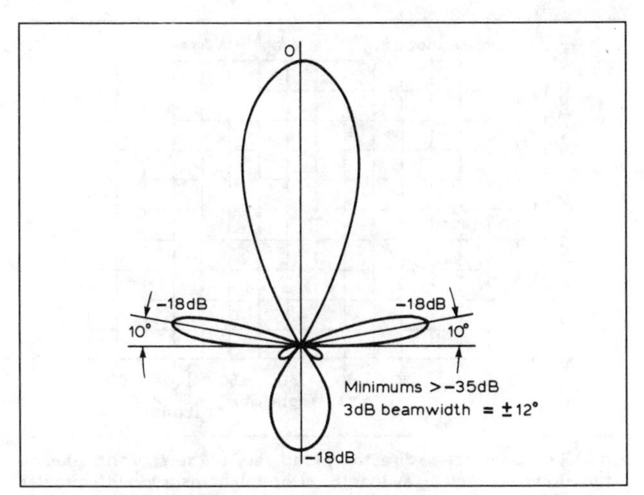

Fig 13.45. Horizontal polar diagram of the complete array

The gain of this arrangement is 13.5dBD with a front-to-back ratio of 18dB. The polar diagram is shown in Fig 13.45.

Multi-element quad antennas

Interest in multi-element quad antennas has increased, mainly because they can offer a better performance with reduced sidelobes compared with the average simple Yagi.

Generally the maximum number of elements used is five. Where more gain is needed, a pair may be stacked vertically or horizontally. (For maximum mechanical strength the vertical arrangement is to be preferred.)

The whole structure may be made up of aluminium tube (or solid rod for the elements). The only insulator necessary is the feed point of the driven element.

In construction, it is an advantage to make each element from one piece of material. A 3/8in aluminium rod will bend to form corners much more readily than tube that would also need a 'filler'. The corner radius should be kept small. Large or small, allowance must be made for the resultant 'shortening' of element length, ie side of the quad element.

For mechanical simplicity (and appearance) it is an advantage to arrange for all the element heights to be the same, and vary the width.

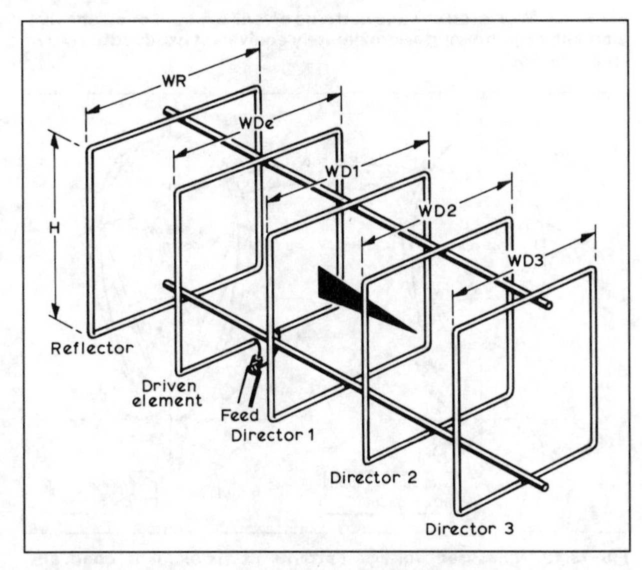

Fig 13.46. General arrangement of a multi-element quad

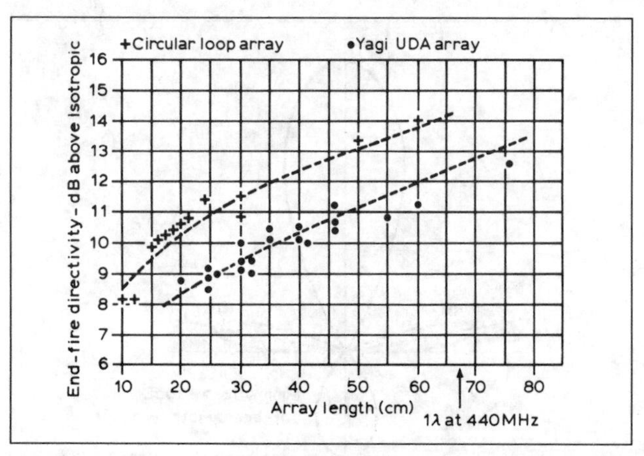

Fig 13.47. Comparative directivity (and gain) of the Yagi and quad as a function of overall array length. Although measured with circular loops, the performance with the square loops used in the quad is comparable. The measurements were made at 440MHz by W0HTH but also apply at lower frequencies (*ARRL Antenna Book*)

Fixing the elements to the boom and the boom to the mast is conveniently done with standard TV antenna fittings. Although suitable blocks or clamping arrangements can be made by the constructor, they often tend to be unnecessarily heavy. Purchased TV fittings can be more cost-effective than obtaining raw materials and there is also much less effort involved in construction.

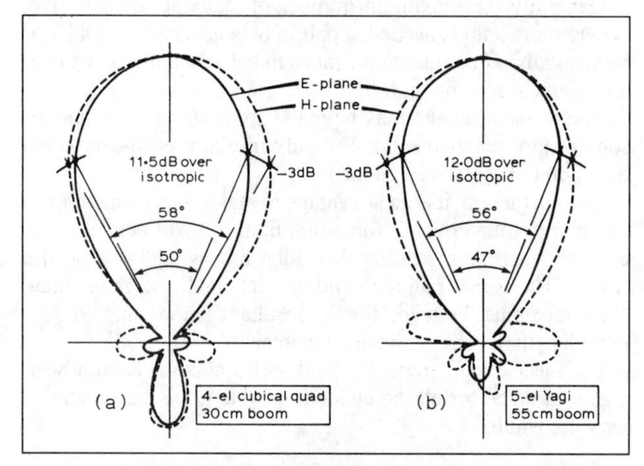

Fig 13.48. Measured voltage patterns of four-element quad and five-element Yagi showing approximately equivalent bandwidths (*ARRL Antenna Book*)

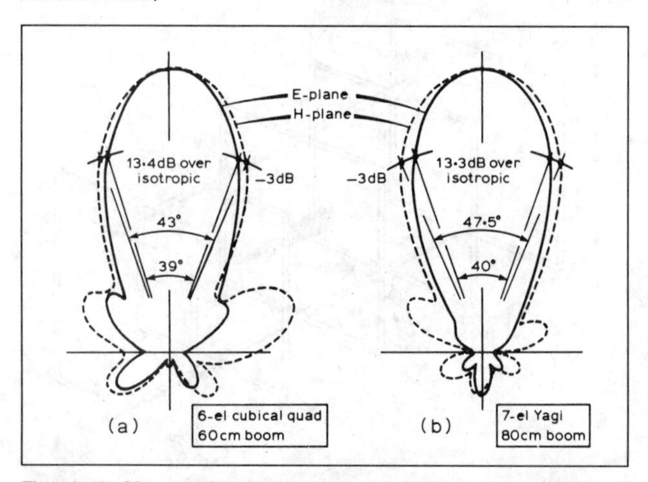

Fig 13.49. Measured voltage patterns of six-element quad and seven-element Yagi (*ARRL Antenna Book*)

Table 13.9. Dimensions, centre-to-centre, for a multi-element quad

Height H	21 (533)	21	21	21
Width reflector WR	24½ (622)	24½	24½	24½
Driven WD_e	20½ (520)	20½	20½	20½
Director 1 WD_1	—	18 (457)	18	18
Director 2 WD_2	—	—	16 (406)	16
Director 3 WD_3	—	—	—	14 (356)
Spacing				
Reflector to Driven	7 (178)	19 (483)	20 (508)	20
Driven to Director 1	—	12 (305)	14½ (368)	14½
Director 1 to Director 2	—	—	14½	14½
Director 2 to Director 3	—	—	—	14½
Approx gain (dBD)	5	7	10.5	12.5

Element diameters all ³/₈in (9.35mm). Feed impedance in all cases is 75Ω. Dimensions are in inches with millimetre equivalents in brackets.

If preferred, the reflector may be made the same size as the driven element, and tuned with a suitable stub. If vertical polarisation is required instead of horizontal then the feeder can be attached to the centre of one of the vertical sides of the driven element. (The same 'side' must be used for correct phase relationship with stacked arrays.)

All quad loops are the same height and the antenna looks similar to a pair of stacked Yagis. The centre bottom section of the driven element is left open for the feed line.

All joints between coaxial cables should be as short and of as low resistance as possible. Coaxial 'T' pieces and connectors should be used if available. Waterproofing is essential at all joints or connectors. The coaxial harness must be secured to the mast by cleats or tape to prevent damage to joints in high winds.

The velocity factor of coaxial cable is 0.66 for solid polythene or 0.86 for semi-air-spaced cable.

A quadruple quad antenna

An alternative to the multi-element quad is the stacked multi-element quad using parasitic coupling instead of a phasing harness. A four-stacked, two-element or 'quadruple quad' was developed by DL6DW [9]. Constructional details are shown in Fig 13.50.

Measurements of this antenna produced a forward gain figure between 10 and 11dBi. Advantages of this antenna are its portable nature and nearby objects cause minimum detuning.

With this multi-quad construction the interaction of the resonant element lengths gives each quad section a resonant circumference of 1.04λ. This deviates from the dimensions of a conventional quad even after correction for the length to diameter ratio of the conductor. With the element dimensions and the spacing between the driven element and reflector of 0.25λ, the best front-to-back ratio and lowest environmental dependence was obtained.

THE LOG PERIODIC

The log-periodic antenna (Fig 13.51) was originally designed and proven at the University of Illinois in the USA in 1955. Since then the military, in particular, have made considerable use of this antenna. Its properties are an almost infinite bandwidth, governed only by the number of elements used, and the directive qualities of a Yagi antenna.

Table 13.10(a) shows typical dimensions for element spacing and length for log-periodic arrays. These are derived from a computer-aided design produced by W3DUQ [10]. Other frequency bands can be produced by scaling all dimensions.

The tabulated parameters have a 5% overshoot of the working frequency range at the low end and a 45% overshoot at the high-frequency end. This is done to maintain logarithmic response over the complete frequency range specified as the log-periodic active over approximately four elements at any one specific frequency. The alpha or logarithmic element taper is 28° for all three antennas. They have a forward gain of 6.55dBD, with a front-to-back ratio of typically 15dB and a VSWR better than 1.8:1 over the specified frequency range.

The construction can be straightforward. It should be noted that the element lengths for the highest-frequency antenna were calculated for the elements to be inserted completely through the boom, ie flush with the far wall. The two lower-frequency antennas have element lengths calculated to butt flush against the element side of the boom. If the elements are inserted through the boom on the 21–55MHz and 50–150MHz antennas, a length correction factor must be added to each element.

The supporting booms are also the transmission line between the elements for a log-periodic antenna. They must be supported with a dielectric spacing from the mast of at least twice the boom-to-boom spacing. Feed-line connection and the arrangement to produce an 'infinite balun' is shown in Fig 13.52. Any change in the boom diameters will require a change in the boom-to-boom spacing to maintain the feed impedance. The formula to achieve this is:

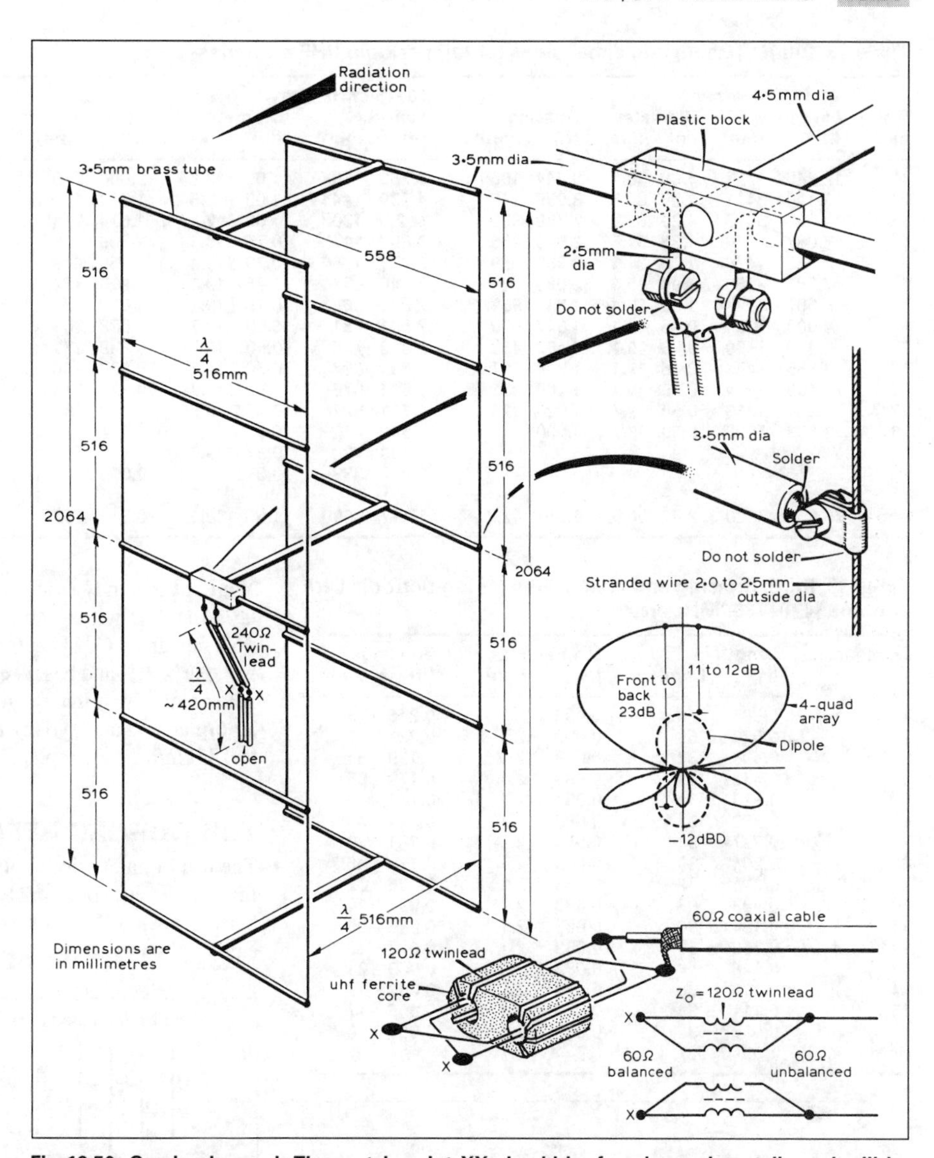

Fig 13.50. Quadruple quad. The match point XX should be found experimentally and will be approximately 200mm from the open end (*VHF Communications*)

$$Z_0 = 273 \log_{10} D/d$$

where D is the distance between boom centres and d the diameter of the booms. The antenna can be oriented for either horizontal or vertical polarisation if a non-metal mast section is used. The horizontal half-power beamwidth will be typically 60° with a vertical half-power beamwidth of typically 100°.

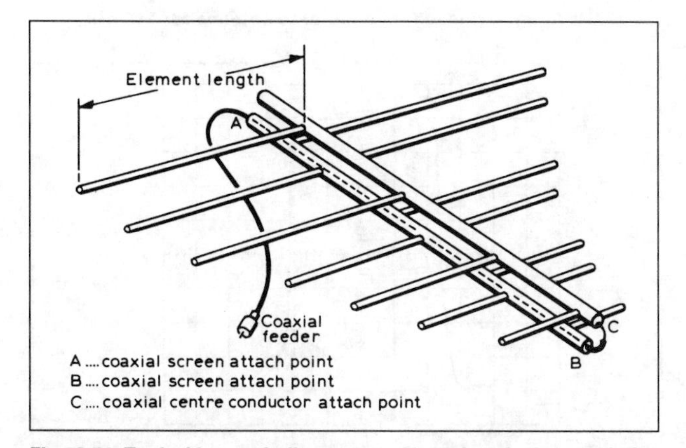

Fig 13.51. Typical log-periodic antenna. Note that the bottom is fed from the coaxial outer while the top boom is fed from the centre conductor (*Ham Radio*)

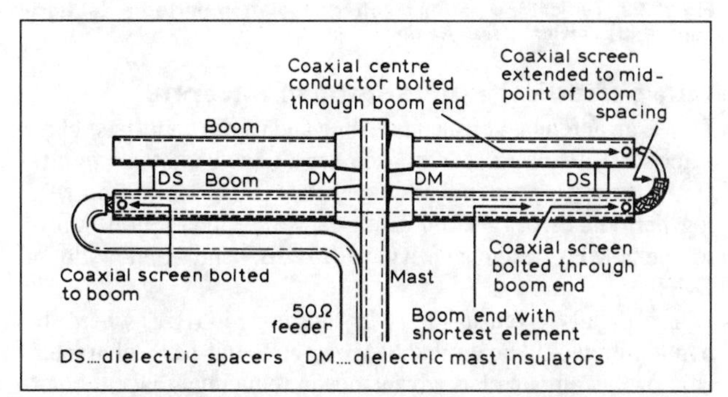

Fig 13.52. Feeding the log periodic is relatively simple. Remove the outer plastic jacket from the feedline for the entire length of the boom, so that the coaxial outer is permitted to short itself inside the boom as well as the solid electrical connections at each end of the boom (*Ham Radio*)

Table 13.10(a). Spacing and dimensions for log-periodic VHF antennas

	21–55MHz array						50–150MHz array						140–450MHz array					
Ele-	Length		Diameter		Spacing		Length		Diameter		Spacing		Length		Diameter		Spacing	
ment	(ft)	(mm)	(in)	(mm)	(ft)	(mm)	(ft)	(mm)	(in)	(mm)	(ft)	(mm)	(ft)	(mm)	(in)	(mm)	(ft)	(mm)
1	12.240	3731	1.50	38.1	3.444	1050	5.256	1602	1.00	2.54	2.066	630	1.755	535	0.25	6.7	0.738	225
2	11.190	3411	1.25	31.8	3.099	945	4.739	1444	1.00	2.54	1.860	567	1.570	479	0.25	6.7	0.664	202
3	10.083	3073	1.25	31.8	2.789	850	4. 274	1303	1.00	2.54	1.674	510	1.304	397	0.25	6.7	0.598	182
4	9.087	2770	1.25	31.8	2.510	765	3.856	1175	0.75	19.1	1.506	459	1.255	383	0.25	6.7	0.538	164
5	8.190	2496	1.25	31.8	2.259	689	3.479	1060	0.75	19.1	1.356	413	1.120	341	0.25	6.7	0.484	148
6	7.383	2250	1.00	25.4	2.033	620	3.140	957	0.75	19.1	1.220	372	0.999	304	0.25	6.7	0.436	133
7	6.657	2029	1.00	25.4	1.830	558	2.835	864	0.75	19.1	1.098	335	0.890	271	0.25	6.7	0.392	119
8	6.003	1830	0.75	19.1	1.647	500	2.561	781	0.50	12.7	0.988	301	0.792	241	0.25	6.7	0.353	108
9	5.414	1650	0.75	19.1	1.482	452	2.313	705	0.50	12.7	0.889	271	0.704	215	0.25	6.7	0.318	97
10	4.885	1489	0.75	19.1	1.334	407	2.091	637	0.50	12.7	0.800	244	0.624	190	0.25	6.7	0.286	87
11	4.409	1344	0.75	19.1	1.200	366	1.891	576	0.50	12.7	0.720	219	0.553	169	0.25	6.7	0.257	78
12	3.980	1213	0.50	12.7	1.080	329	1.711	522	0.375	9.5	0.648	198	0.489	149	0.25	6.7	0.231	70
13	3.593	1095	0.50	12.7	0.000		1.549	472	0.375	9.5	0.584	178	0.431	131	0.25	6.7	0.208	63
14							1.403	428	0.375	9.5	0.525		0.378	115	0.25	6.7	0.187	57
15							1.272	388	0.375	9.5	0.000		0.332	101	0.25	6.7	0.169	52
16													0.290	88	0.25	6.7	0.000	
Boom	25.0	7620	2.0	50.8	0.5	12.7	16.17	5090	1.5	38.1	0.5	152	5.98	1823	1.5	38.1	0.5	152

Table 13.10(b). Spacing and dimensions for log-periodic UHF antenna (420–1350MHz array)

Element	Length		Diameter		Spacing	
	(ft)	(mm)	(ft)	(mm)	(ft)	(mm)
1	0.585	178	0.083	2.1	0.246	75
2	0.523	159	0.083	2.1	0.221	67
3	0.435	133	0.083	2.1	0.199	61
4	0.418	127	0.083	2.1	0.179	55
5	0.373	114	0.083	2.1	0.161	49
6	0.333	101	0.083	2.1	0.145	44
7	0.297	91	0.083	2.1	0.131	40
8	0.264	80	0.083	2.1	0.118	36
9	0.235	72	0.083	2.1	0.106	32
10	0.208	63	0.083	2.1	0.095	29
11	0.184	56	0.083	2.1	0.086	26
12	0.163	50	0.083	2.1	0.077	23
13	0.144	44	0.083	2.1	0.069	21
14	0.126	38	0.083	2.1	0.062	19
15	0.111	34	0.083	2.1	0.056	17
16	0.097	30	0.083	2.1	0.000	0
Boom	1.99	607	0.5	12.7		

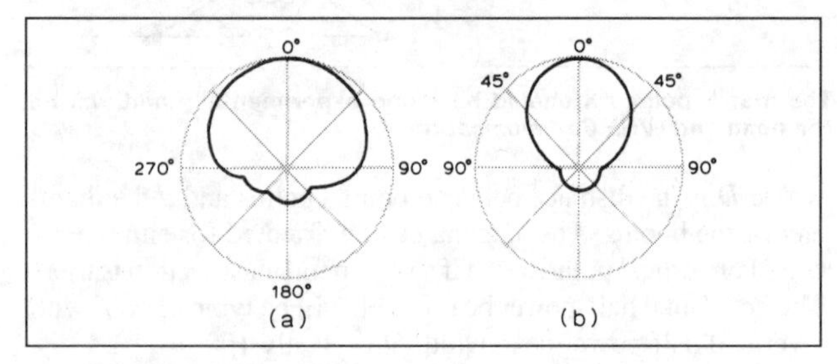

Fig 13.53. Typical log-periodic voltage radiation patterns: (a) horizontal, (b) vertical (*Ham Radio*)

Log-periodic Yagi band-pass antenna

This is an antenna with an interesting and useful band-pass characteristic, giving a flat response over a wide band, and significant attenuation outside. It consists of a combination of a log-periodic driven section with a parasitic Yagi section. A prototype 50MHz design gave a gain of 12dBD and a bandwidth of 2MHz.

The details given in Fig 13.54 are for 144MHz, where the bandwidth would be around 6MHz. The 'band-pass' characteristic of this antenna has advantages in reducing adjacent channel interference. It also gives a constant performance over the whole 144 to 148MHz band.

The simple Yagi, by comparison, is essentially a narrow-band antenna. (Wide-band operation can be obtained with the simple Yagi but this may result in lower forward gain. This is particularly the case with a multi-element, high-gain, Yagi).

A design by G3FDW for a log-periodic Yagi antenna to cover just one VHF band has recently been published [11]. The claimed advantage is that it has a good gain (11dbD for a boom length of 3.66m or 12ft at 70MHz) and is robust to withstand winds of up to 185km/h or 115mph.

THE CORNER REFLECTOR

The use of an aperiodic plane reflector, spaced behind a radiating dipole, has previously been discussed. If this reflector is bent to form a 'V', as shown in Fig 13.56, a much higher gain can be obtained.

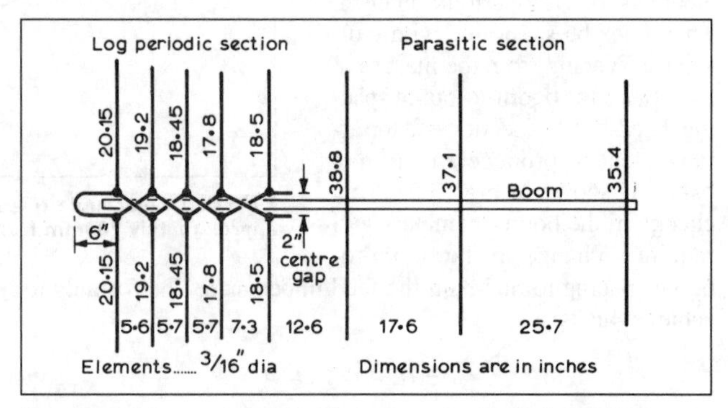

Fig 13.54. A log-periodic Yagi band-pass antenna for 145MHz

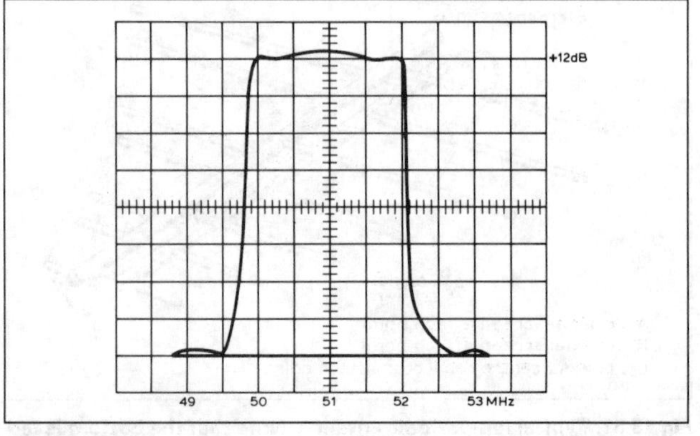

Fig 13.55. Gain versus frequency characteristic of the 50MHz log-periodic Yagi

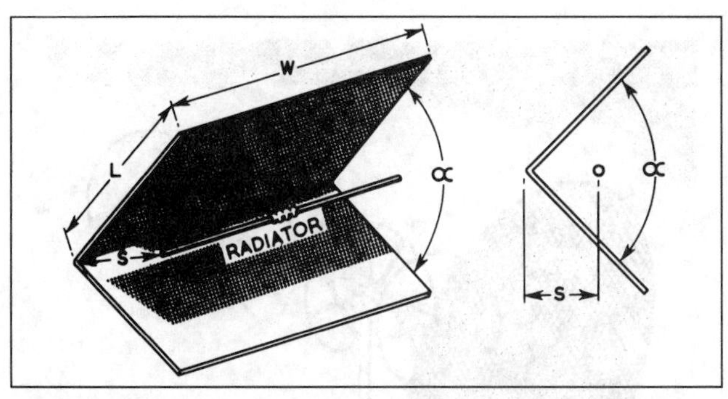

Fig 13.56. Corner reflector. The λ/2 dipole radiator is spaced parallel with the vertex of the reflector at distance S; its characteristics are shown in Figs 13.57 and 13.58

The critical factors in the design of such an antenna array are the corner angle α and the dipole/vertex spacing *S*. The curves in Fig 13.57 show that as α is reduced, the gain theoretically obtainable becomes progressively greater. However, simultaneously the feed impedance of the dipole radiator falls to a very low value, as seen from Fig 13.58. Matching becomes difficult and a compromise has to be reached. In practice the angle is usually made 90° or 60°. Match adjustment in a 60° corner reflector is more critical but maximum obtainable gain is higher. The final matching of the radiator to the line may be carried out by adjusting the distance *S*. It does not greatly affect the gain over a useful range of variation but causes a considerable change in radiation resistance. A two-stub tuner can be used for final adjustments.

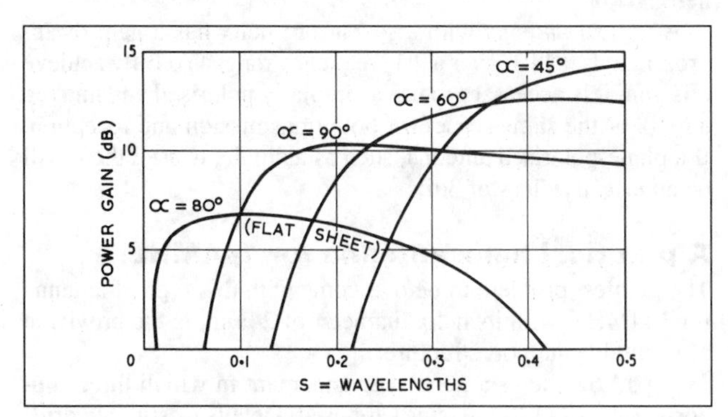

Fig 13.57. Theoretical power gain obtained by using a corner reflector with a λ/2 dipole radiator

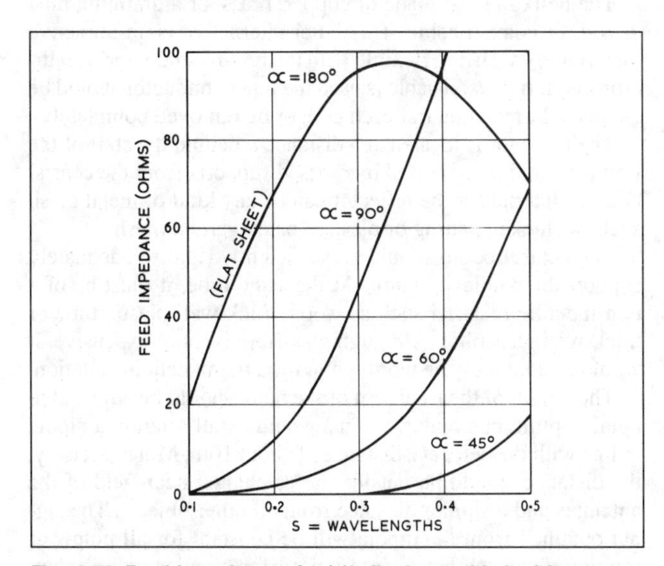

Fig 13.58. Feed impedance of a λ/2 dipole provided with a corner reflector: see Fig 13.56

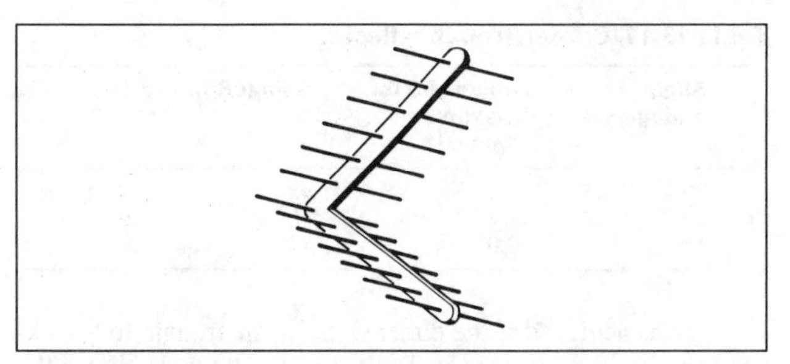

Fig 13.59. The corner reflector can be modified by using a set of metal spines arranged in V-formation to replace the sheet metal or wire-netting reflector

The length *L* of the sides of the reflector should exceed 2λ to secure the characteristics indicated by Fig 13.57 and 57. Reflector width *W* should be greater than 1λ for a λ/2 dipole radiator.

The reflecting sheet may be constructed of wire netting as described previously or alternatively may be fabricated from metal spines arranged in a V-formation, all of them being parallel to the radiator. See Fig 13.59. The spacing between adjacent rods should not exceed 0.1λ.

A useful approximation for the power gain *G* referred to a λ/2 dipole is

$$G = 300/\alpha$$

where α is the angle between the sides measured in degrees.

Maximum dipole/vertex spacing *S*, included in the curves shown, is λ/2. Spacing greater than this would require cumbersome constructions at lower frequencies. At higher frequencies larger spacing becomes practicable with higher gain than suggested by Fig 13.57. See Table 13.11. This suggests that the corner reflector can be attractive for the 1.3GHz band. The distance across the 'opening' should be more than 4λ to give the results shown.

THE TROUGH REFLECTOR

To reduce the length of a large corner reflector the vertex can be cut off and replaced with a plane reflector. This arrangement is called a *trough reflector*. Similar performance to a large corner

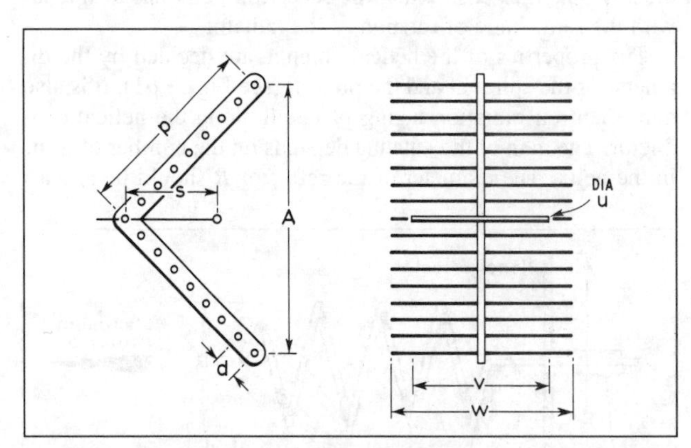

Fig 13.60. Dimensions for a 60° corner reflector antenna system giving a gain of about 13dBD. The feed impedance of the dipole radiator is 75Ω. The apex may be hinged for portable work

	Dimensions in millimetres							
Band	*p*	*s*	*d*	*v*	*w*	*A*	*u*	λ
144	2540	1016	152	965	1270	2540	9.5	2083
433	889	337	38	324	508	889	6.4	692
1296	305	114	12.7	102	203	305	3.2	232

Table 13.11. Corner/trough reflector

Angle α (degrees)	Value of *S* for maximum gain (λ)	Gain (dBi)	*T* (λ)
90	1.5	13	1–1.25
60	1.25	15	1.0
45	2.0	17	1.9

reflector is achieved if the dimensions of the trough do not exceed the limits suggested in Table 13.11. The table shows the gain obtainable for greater values of *S* than those covered by Fig 13.57 when the reflector is of adequate size. *T* is the distance between the plane reflector and the theoretical apex of the vertex.

The resulting antenna has a performance very little different from a full-size corner reflector. It presents fewer mechanical problems: sides are significantly shorter and the plane centre portion is easier to mount to a mast.

The gain of both corner reflectors and trough reflectors may be increased still further by stacking two or more in phase. Alternatively, collinear dipoles can be combined in a wider reflector and fed in phase. Not more than two or three radiating units should be used if the virtue of the simple feed arrangement is to be retained.

AXIAL-MODE HELIX

The helix antenna is a simple means of obtaining high gain and a wide-band frequency characteristic. When the circumference of the helix is of the order of 1λ axial radiation occurs, ie the maximum field strength is found to lie along the axis of the helix. This radiation is circularly polarised, the sense of the polarisation depending on whether the helix has a right or left-hand thread.

A pick-up dipole used to explore the field, in the direction of maximum radiation, will show no change of amplitude as it is rotated through 360°. This shows true circular polarisation. At any point to the side of the helix the wave will be elliptically polarised, ie the horizontal and vertical components will be of unequal strength.

A helix may be used to receive plane or circularly polarised waves. When signals are received from a transmitting helix care must be taken to ensure that the receiving helix has a 'thread' with the same hand of rotation as the radiator.

The properties of the helical antenna are decided by the diameter of the spiral *D* and the pitch *P* (see Fig 13.61). It is also dependent on radiation taking place all along the helical conductor. The gain of the antenna depends on the number of turns in the helix. The diameter of the reflector *R* should be at least

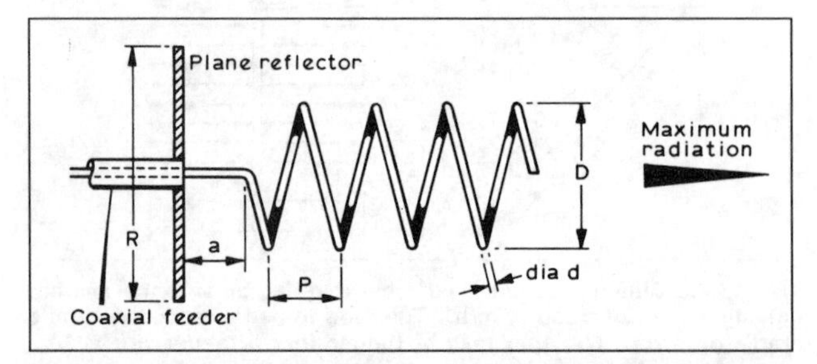

Fig 13.61. The helix antenna. The plane reflector may take the form of a dartboard type of wire grid. The dimensions given in Table 13.12 are based on a pitch angle of 12°. The helix, which may be wound of copper tube or wire, the actual diameter of which is not critical, must be supported by low-loss insulators

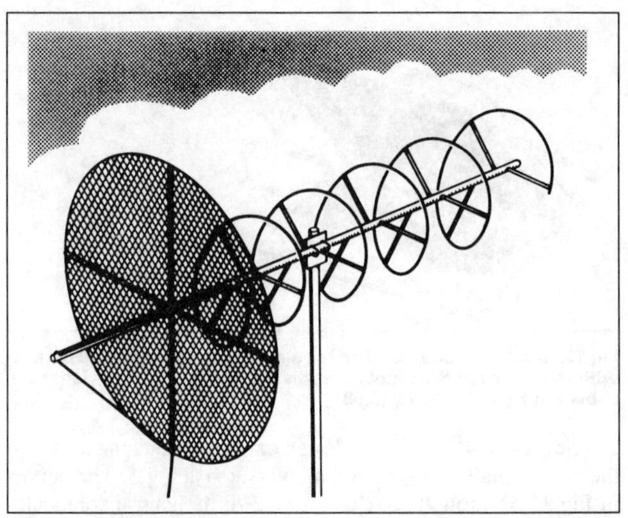

Fig 13.62. General arrangement ot support structure for a five-turn helical antenna for 144MHz

λ/2, with the diameter of the helix *D* about λ/3 and the pitch *P* about λ/4.

A helix of this design will have a termination/feed impedance of about 140Ω. This may be transformed to the feeder impedance by means of a suitable λ/4 transformer. Gain of the antenna is directly proportional to the number of turns in the helix.

A typical antenna with a seven-turn helix has a gain of approximately 12dBi over a 2:1 frequency range. To fully achieve this gain it is necessary to use a circularly polarised antenna (eg a helix of the same sense) for both transmission and reception. If a plane-polarised antenna, such as a dipole, is used there will be an effective loss of 3dB.

A practical helix antenna for 144MHz

The greatest problem to be overcome with this type of antenna for 144MHz, with its helix diameter of 24½in, is the provision of a suitable support structure.

Fig 13.62 shows a general arrangement in which three supports per turn (120° spacing) are used. Details of suitable drilling of the centre boom are given in Fig 13.63.

The helix may be made of copper, brass, or aluminium tube or rod, or coaxial cable. This latter alternative is an attractive material to use, being flexible with the braid 'conductor' weatherproofed. If coaxial cable is used the inner conductor should be connected to the outer at each end, or be removed completely.

The reflector is located at a distance *a* behind the start of the first turn, and is supported by crossed supports from the central boom. Material for the reflector can be any kind of metal mesh such as chicken netting or plastic-coated garden mesh.

The central boom should be sufficiently rigid to adequately support the whole structure. At the same time, it must be of a non-metallic material such as wood, thick-wall plastic tube or thick-wall glassfibre. Although glassfibre is more expensive, it would undoubtedly be worthwhile for a permanent installation.

The length of the final turn of the helix should be adjusted to obtain optimum circularity. This would entail rotating a dipole in line with the helix at a distance of about 10m. More precisely, the distance should be outside the calculated near-field of the antennas and a similar distance from all other objects. The signal obtained from the dipole will be constant for all points of rotation when the helix is optimised for circular polarisation. Any variation of the signal is known as the *polarisation axial*

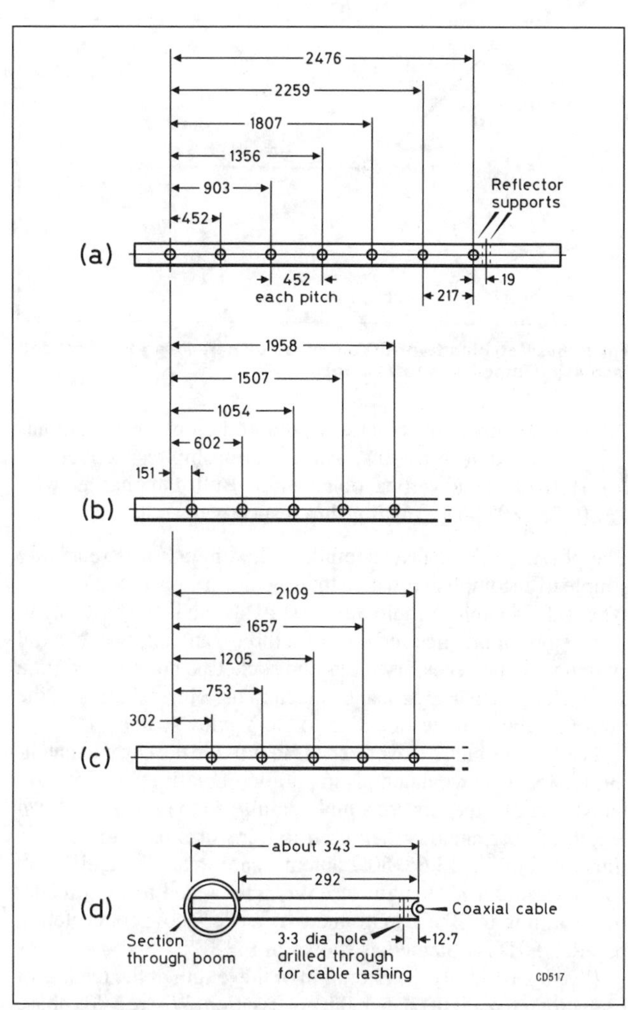

Fig 13.63. (a) First side drilling dimensions, reflector support holes are drilled at right-angles; (b) and (c) are drilled at intervals of 120° and 240° respectively from (a). (d) Cutting and filing dimensions for the element stand-offs

ratio or *boresight ellipticity*, and is expressed as a ratio or decibel figure.

Control of antenna polarisation

Vertical polarisation is the most popular for mobile operation in the UK. This is because it is easier to obtain omnidirectional radiation and moderate gain with a vertical polarised antenna compared with a horizontally polarised antenna. Vertical gain antennas are far less obtrusive than their horizontally polarised counterparts. This is important on a vehicle, where the mechanical simplicity of a short vertical rod considerably outweighs the complexity of a halo or crossed dipole. In addition, a horizontal polarised antenna must be at least $\lambda/2$ above the vehicle surface to ensure low-angle radiation.

The establishment of repeaters with mobile communicators using vertical polarisation means that a fixed station can only operate satisfactorily if a means of changing polarisation is available. It is possible to use two antennas of different polarisations. Ideally antennas capable of receiving all variation of polarisation are required. This becomes complex and expensive.

Space communication, where control of polarisation is difficult or impossible, has forced the use of circular polarisation. It is surprising that it is not used more between fixed stations for long-distance terrestrial work. The fundamental advantage of this type of polarisation is that, since reflections change the direction of polarisation, there can be far less fading and aircraft 'flutter'.

Table 13.12. General dimensions for 144, 433 and 1296MHz helix antennas

Band	Dimensions D	R	P	a	d
General	0.32λ	0.8λ	0.22λ	0.12λ	
144MHz	25½ (648)	64 (1626)	17¾ (450)	8¾ (222)	3/16–½ (4.8–12.7)
433MHz	8¾ (222)	22 (559)	6 (152)	3 (76)	½ (12.7)
1296MHz	3 (76)	7 (178)	2 (50)	1 1/8 (28)	¼–1/8 (6.4–3.2)
Turns	6	8	10	12	20
Gain	12dB	14dB	15dB	16dB	17dB
Beamwidth	47°	41°	36°	31°	24°

Dimensions in inches, millimetres are given in brackets. The gain and beamwidth of the helical antenna are dependent upon the total number of turns as shown above.

Bandwidth = 0.75 to 1.3λ

$$\text{Feed impedance} = 140 \times \frac{\text{circumference}}{\lambda} \text{ ohms}$$

(Note: λ and circumference must be in the same units.)

$$\text{Beamwidth (degrees)} = \sqrt{\frac{12{,}300}{\text{No of turns}}}$$

The use of circular polarisation at one end only, with normal horizontal or vertical at the other end of the link, will result in a 3dB loss. To achieve the full advantage of circular polarisation it is necessary for all stations to use it. However, changes of polarisation caused by propagation from plane-polarised stations are often received stronger with a circular-polarised antenna.

The usual practice when using circular polarisation is to standardise on clockwise rotation in the northern hemisphere. The direction of polarisation/rotation is as seen looking forward from the rear of the antenna.

Changing all VHF operations to circular polarisation is obviously not practical. If a system of switching polarisations is used it would soon become evident that circular polarisation is the best compromise. If you use a system of polarisation switching, big variations are found in polarisation from stations, in particular mobiles. Quite often a mobile using a vertical antenna will be equal strength on all polarisations or the opposite to that expected.

To obtain circular polarisation would indicate the use of the helix antenna. This normally produces 'modes' of circularity dependent upon whether the thread of the antenna element is wound clockwise or anti-clockwise. Horizontal or vertical polarisation is possible from helix antennas by using two helices and suitable phasing, for example.

One compromise for receiving circular-polarisation signals is to use *slant polarisation*. To obtain this a single Yagi is set at an angle of 45°. This enables horizontal and vertical signals to be received almost equally. At first sight one would expect a loss of 3dB for H and V polarised signals compared with an appropriately aligned Yagi. However, long-term practical measurements have shown when averaged that this arrangement gives a 6dB improvement with typical mixed polarised signals. In addition this arrangement it is little affected by the mounting mast, unlike a vertical polarised Yagi.

The simplest way of being able to select polarisation is to mount a horizontal Yagi and a vertical Yagi on the same boom, giving the well-known *crossed Yagi* (Fig 13.64). Separate feed to each section of the Yagi brought down to the operating position will enable the user to switch to either horizontal or vertical. It is perhaps not generally realised that it is quite simple to alter the phasing of the two Yagis in the shack and

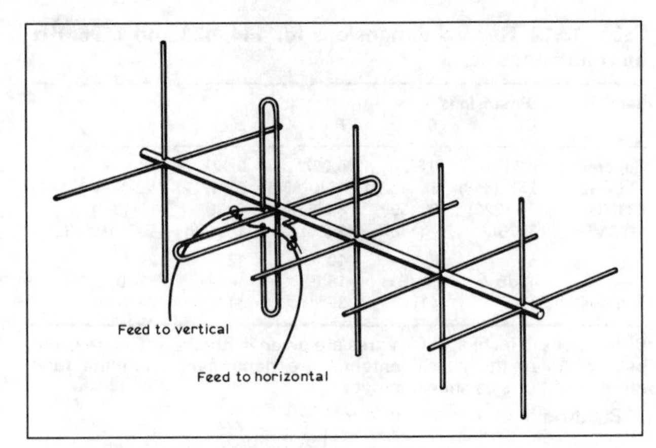

Fig 13.64. General arrangement of a crossed Yagi antenna

obtain six polarisation options. These are two slant positions (45° and 135°), two circular positions (clockwise and anti-clockwise) and the original horizontal and vertical polarisation.

The presence of the mast in the same plane as the vertical elements on a Yagi considerably detracts from performance as already stated, so a simple solution is to use a crossed Yagi configuration, mounting the antennas at 45° relative to the vertical mast. The mast then has little or a balanced effect on the antennas. With appropriate phasing, vertical and horizontal polarised radiation patterns can be obtained that are unchanged by the presence of the mast.

If a crossed Yagi is mounted at 45°, with individual feeders to the operating position, the polarisation available and the phasing required is as follows:

(a) Slant position 45° and 135°. Antennas fed individually.

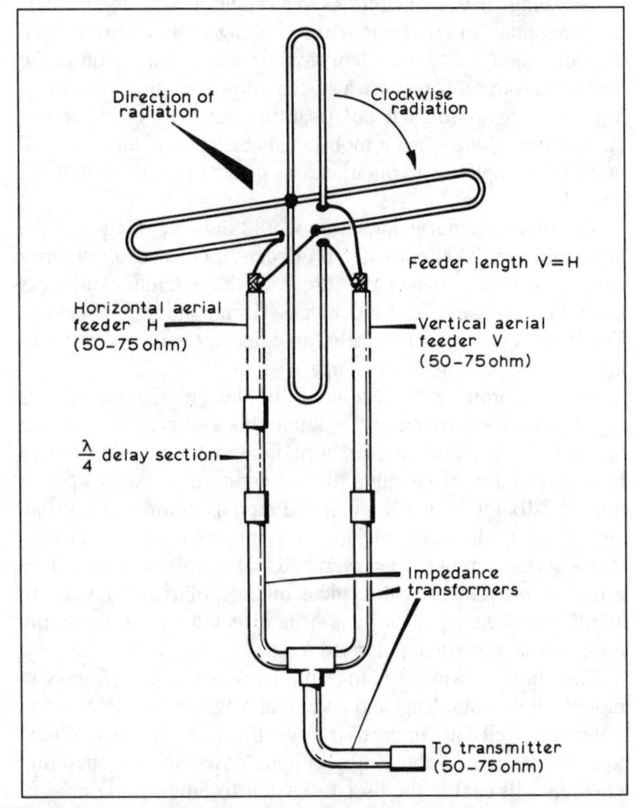

Fig 13.65. General arrangement of feeders with delay line (phasing) for clockwise radiation

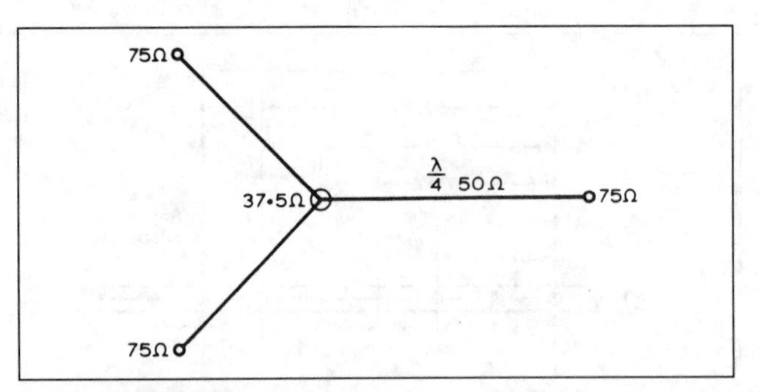

Fig 13.66. Matching two 75Ω antennas by paralleling to 37.5Ω and increasing impedance to 75Ω again

(b) Circular positions clockwise and anti-clockwise. Both antennas fed 90°+ or 90°– phase relationship respectively.

(c) Horizontal and vertical polarisation. Both antennas fed with 0° or 180° phase relationship respectively.

The phasing may appear complicated, although it is relatively simple to accomplish it with a three-gang, six-position, Yaxley-type wafer switch. A multiway coaxial switch is the best way to do the job. In practice the cost of a three-gang six-way coaxial switch with the necessary plugs and sockets is not justified for a 144MHz system. Care must be taken in the dimensioning of the phasing cables and termination to the Yaxley switch.

The first problem to overcome is to provide the correct matching for feeding two antennas in parallel. Briefly, with 75Ω antennas the two feeders are simply paralleled to give 37Ω. A λ/4 length of 50Ω cable is then used to transform back to 75Ω, as illustrated in Fig 13.66. 50Ω antennas are treated in a different way by using a λ/4 length of 75Ω cable added to each feeder transforming to 100Ω. With the two feeders now connected in parallel, 50Ω is obtained as shown in Fig 13.67.

Phasing is simply the alteration of the length of the feeders of each crossed Yagi to change the polarisation. Where a 90° phase shift is required, λ/4 of cable is inserted and, where a 180° phase shift is required, λ/2 cable is inserted. The polarisation switch must switch in the appropriate λ/4 'impedance transformer' and correct phasing by connecting the appropriate length(s) of cable.

There is an added complication. Not all antenna systems are 50Ω. However, antenna termination of 50Ω is the international standard for VHF and above. As some amateurs still use 75Ω coaxial cables and low-loss TV cables with this impedance are readily available, consideration has been given to systems with a characteristic impedance of 75Ω.

Figs 13.68 and 13.69 show the necessary switching arrangements for 75Ω and 50Ω antennas respectively. The normal drawing of a switch makes the illustration of the 75Ω system complicated. The alternative, Fig 13.69, is drawn as a side view of the

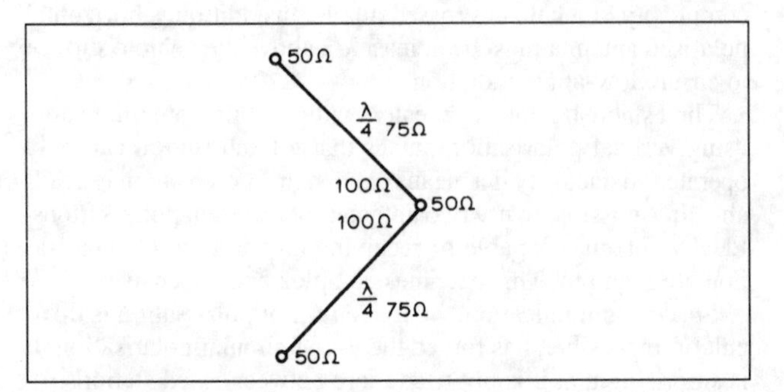

Fig 13.67. Matching two 50Ω antennas by increasing impedance to 100Ω and paralleling to 50Ω again

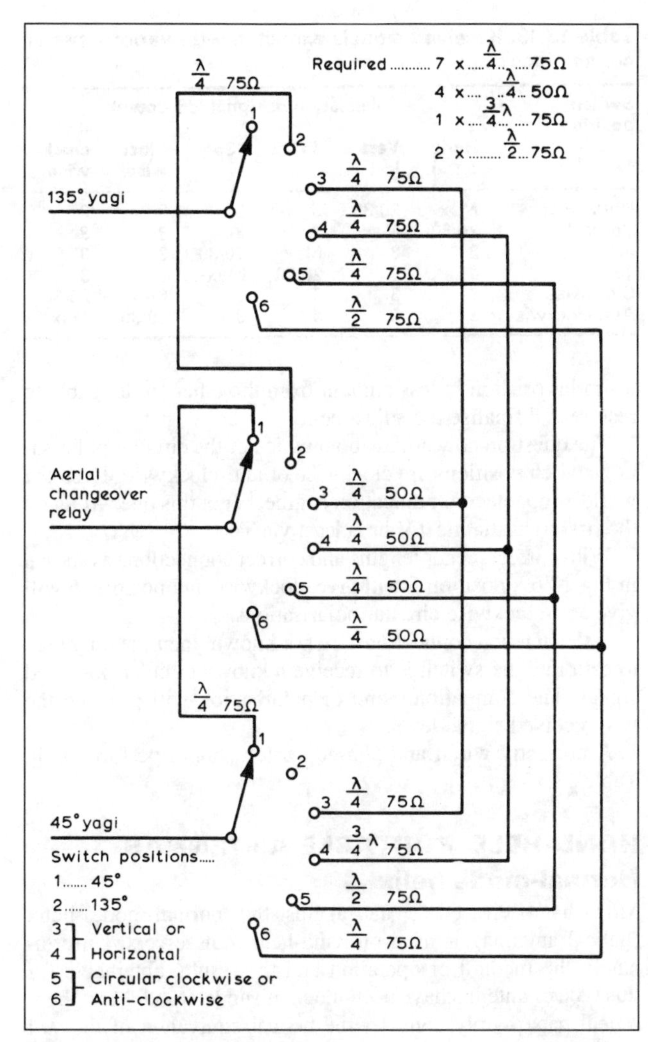

Fig 13.68. 75Ω phasing and matching switch

Yaxley switch with the six contacts visible in a vertical line. The moving contact is not shown. Note that the 50Ω version is much simpler as there is no need to manufacture 'T' junctions in the cables.

An alternative to accurate cutting of the phasing cables is to

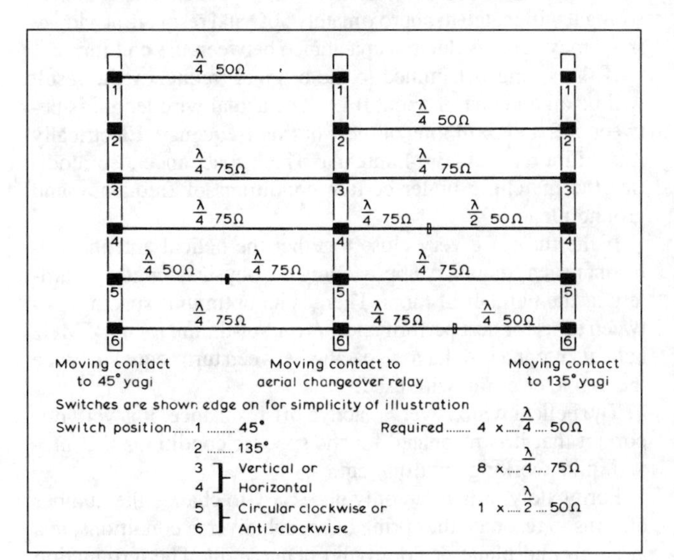

Fig 13.69. 50Ω phasing and matching switch

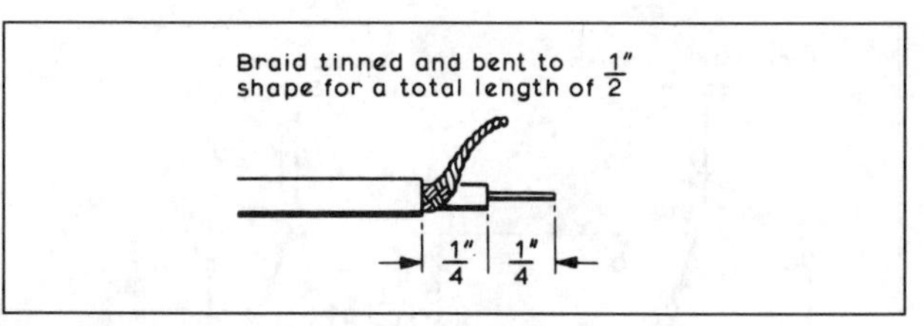

Fig 13.70. Method of 'tailing' coaxial cable

finally set lengths with a grid dip oscillator (GDO). First, use the smallest possible diameter cable to reduce the mechanical problems of connection to the contacts of the switch. Type UR43 for 50Ω and UR70 for 75Ω with a solid dielectric is preferred.

To obtain λ/4 of cable, cut off more than the calculated length, typically 13–15in (33–38cm) of solid dielectric cable at 144MHz. Leave one end open-circuit and join the other end with the shortest possible loop that will produce a dip on the GDO. It is surprising just how small that loop can be and still give a reading with a sensitive GDO. A virtual short-circuit may still 'couple'.

Check the dip frequency, which will probably be around 120MHz, then carefully clip pieces off the open end of the cable until the dip occurs at 145MHz. If cable of the same type has been used throughout the switch, there is no need to 'dip' each length. Uniformity of new cable is sufficient for the dimensions of the 'dipped' λ/4 section to be doubled or trebled where a λ/2 or 3λ/2 section is required.

The slight shortening of the cables when they are prepared for connection is compensated by the length in the switch contacts.

Remember when wiring the switch that all cable ends should be made as short as possible to maintain impedance. The configuration shown in Fig 13.70 must be followed. All outer braids on each wafer of the switch must be joined by the shortest possible route and not connected to the frame of the switch.

Miniature rotary switches with small-diameter cable makes for a neat assembly. However, great care is needed to deal with the many coaxial connections to a switch of this small size.

The joining of a length of 50Ω and 75Ω is important. Every effort should be made to maintain the coaxial form of the cable by pushing the braid back away from the inner and carefully making the inner connections. Cover the inner connection with polythene tape, to avoid any possible short-circuit. Bring the braids back over the tape and bind securely with fine wire. Any attempt to solder the braid is likely to be disastrous as the polythene easily melts.

Further protection may be given by a layer of tape over the entire joint. The 'T' junctions required on the '75Ω' switch may be made by cutting small triangular sections of tinplate and carefully soldering the outer of each cable to the triangle. (Small pieces of single- or double-sided PCB is an alternative that requires less heating time). Fig 13.71 illustrates the 'triangle' method.

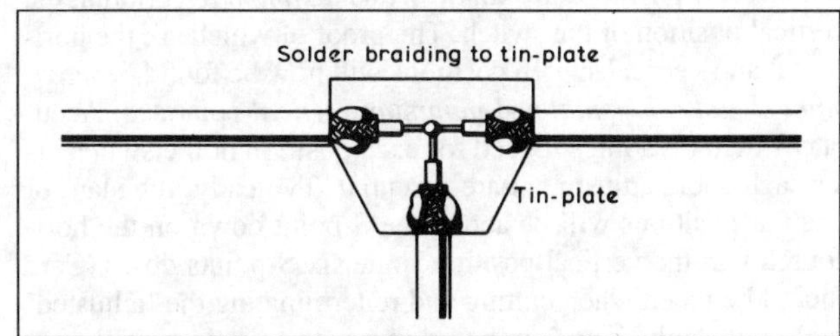

Fig 13.71. Method of joining three cables

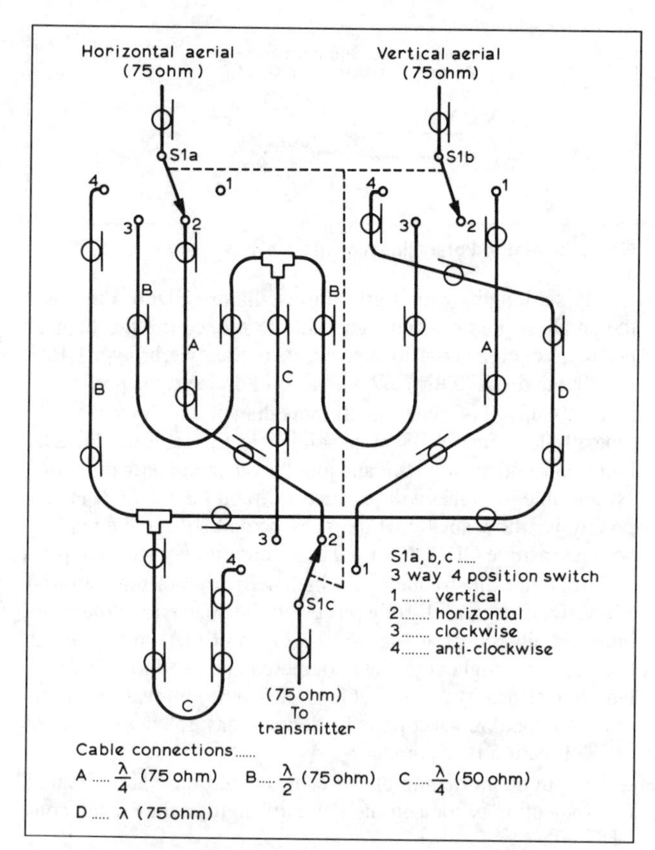

Fig 13.72. An alternative arrangement for feeding crossed Yagi antennas which provides various polarisations at the click of a switch

After ensuring the switch unit is correctly built the feeder cables from each 'half' of the crossed Yagi must be made the same length. Ideally, these feeders should be cut mechanically and electrically to equal length before installation. The two Yagi sections should be in the same place on the boom. It is mechanically difficult to have the two sets of elements in the same position along the mounting boom, and it is therefore necessary to correct this mechanical displacement of phase by an 'equal' change of one of the feeder cables. (The cable to the 'rear-most' fed Yagi will need to be *shortened* to compensate for the 'positional phase delay'. The length compensation *must take into consideration the velocity factor of the cable in use.*

A method of adjustment, and a means of checking if a compensated cable is correct, is to receive a horizontal-polarised signal (of constant amplitude) from a local station. It is important to ensure the transmitting and receiving antennas are 'beamed' directly at each other as radiation from other than the main lobe will not necessarily be of the correct polarisation. This fact will become very evident in subsequent use of the switch. The feeder lengths should be adjusted so that all slant and circular positions are *equal*, with *maximum* rejection in the vertical position of the switch. The proof of which are the horizontal and vertical switch positions will now be found, ie *maximum signal horizontal* and *minimum vertical* polarised. Accurate S-meter readings logged for each position of the switch after each cable adjustment are essential. Typically, the slant or circular positions will be about one S-point down on the horizontal with the vertical position some six S-points down. Care should be taken when cutting and re terminating the 'adjusted' cable with only 1 or 2cm being removed at a time. If the recorded readings indicate that the last cut was one too many, cut

Table 13.13. Received signals expected with various switch connections

Switch position	Polarisation of signal (dB down)					
	Horizontal	Vertical	45°	135°	Clockwise	Anti-clockwise
Horizontal	Max	20/30	3	3	3	3
Vertical	20/30	Max	3	3	3	3
45°	3	3	Max	20/30	3	3
135°	3	3	20/30	Max	3	3
Clockwise	3	3	3	3	Max	20/30
Anti-clockwise	3	3	3	3	20/30	Max

a similar or slightly less amount from the other feeder cable to restore and finalise the adjustment.

The question now arises about which of the circular polarisation switch positions are clockwise or anti-clockwise. Even the world's top telecommunication engineers got this one wrong on the first transatlantic TV broadcast via Telstar.

With correct feeder lengths and correct connections as shown in Fig 13.69, position 5 will give clockwise and position 6 will give anti-clockwise circular polarisations.

If the antenna connections are not known, then the only way to calibrate the switch is to receive a known circular polarised signal. The compatible hand of polarisation will produce the best received signal level.

A correctly wired and phased switch should perform as in Table 13.13.

HAND-HELD PORTABLE ANTENNAS

Normal-mode helix

Much has been said for and against the 'normal-mode' helix (helical antenna) as used on hand-held transceivers. Unfortunately the method of operation and the results obtainable for this type of antenna have been much misunderstood. Most theoretical papers only consider the helical equivalent of the $\lambda/4$ whip while most of the antennas in use are the equivalent of a physically reduced $3\lambda/4$ whip.

A helix will work in the normal mode when the diameter and pitch of the helix is less than 0.1λ. When working in this mode, the radiation is from the side of the helix.

A $3\lambda/4$ whip over a moderate ground plane has a resistive match very close to 50Ω. If this whip is coiled into a helical spring it will match to approximately 50Ω and resonate at a lower frequency. This is due to capacitance between the coil turns.

If the spring is trimmed to the original frequency the result will be an antenna of about 0.1λ. The actual wire length is between $\lambda/2$ and $5\lambda/8$ long at the working frequency. Electrically it is still a resonant $3\lambda/4$ antenna. This capacitance also modifies the matching under certain conditions of frequency and ground plane.

If the turns are very close together the helical antenna will resonate at a frequency approaching the physical length, regardless of the number of turns. There is an optimum 'spacing' between turns for best performance. A rule-of-thumb for a 145MHz helical antenna is to have a spacing between turns equal to twice the diameter of the wire used.

The helical whip is very reactive off-resonance. It is very important that it is resonated for the specific conditions that prevail in its working environment.

Fortunately, it is often only necessary to change the number of turns to resonate the spring over such diverse conditions, ie a large ground plane or no ground plane at all. The termination impedance can vary between 30 and 150Ω at the extremities.

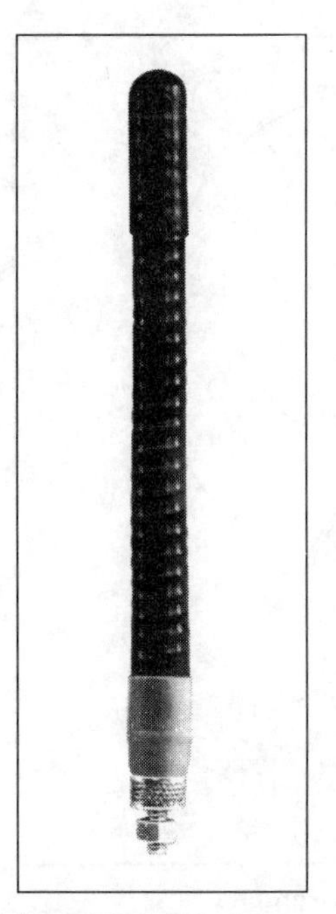

Fig 13.73. A typical commercial helical antenna with screw mounting facility

Under typical 'hand-held' conditions, although to a small extent depending on the frequency of operation, the spring will be fairly close to a 50Ω impedance match. Fig 13.74 shows the number of turns required for a typical 9mm diameter helix for 3λ/4 resonance.

As the helical is a reduced size and aperture antenna two factors arise. First, the radiation resistance is lower than the equivalent linear whip so the choice of a good conducting material is important to remove resistive losses. A plain steel spring compared with a brass or copper-plated helical can waste 3dB of power as heat. The physical aperture of the helical whip is a third that of a λ/4 whip. This would imply a loss of 4.77dB.

Results obtained from copper-plated, Neoprene-sheathed helical antennas, correctly matched to a hand-held transmitter at 145MHz, gave at worst –3dB and at best +1dB compared with a λ/4 whip. (A λ/4 whip with minimal ground plane would be about –6dB compared to a λ/2 dipole.) One thing observed was that the top of the helical often needed to be raised to the same position as the top of the equivalent λ/4 whip to receive maximum signal strength.

A helical, resonant and matched, on a λ/2 square ground plane can give results 2–3dB below a λ/2 dipole. An alternative arrangement using a bifilar-wound helix gives identical results (within 0.2dB) to a λ/2 dipole.

The helical antenna has an interesting operating characteristic when supported close to the body, particularly at the higher frequencies. Fig 13.75 shows typical results of a 145MHz and a 70MHz helical as it is brought closer to the body. The interesting effect occurs at several centimetres from the body, where the resonant frequency of the helical surprisingly increases.

Closer than 2cm the operating frequency suddenly decreased, due to body capacitance as one would expect. Unfortunately this very changeable area occurs at the typical mounting distance of a body-worn transceiver. However, many transceivers are raised to the mouth when transmitting and this puts the antenna back to its best operating position.

HB9CV mini-beam

An antenna that falls into the categories horizontal- or vertical-polarised; portable rather than mobile; and for base station use; is the HB9CV mini-beam. Similar antennas are the lazy-H and ZL Special often used on the HF bands. The HB9CV version has one or two mechanical advantages that make it particularly suitable for VHF portable use.

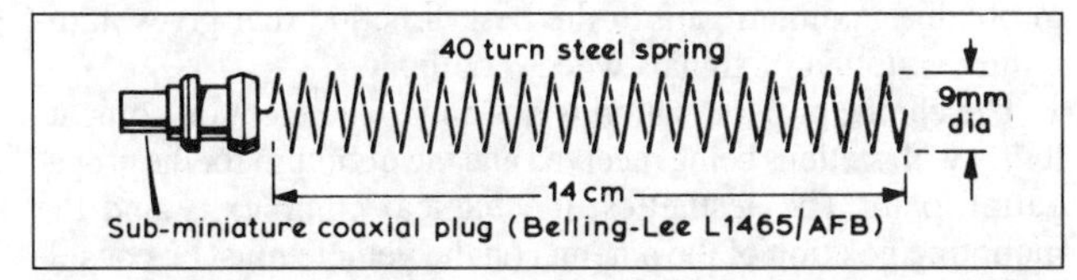

Fig 13.74. Details of a home-made helical whip for 145MHz. A BNC plug could also be used

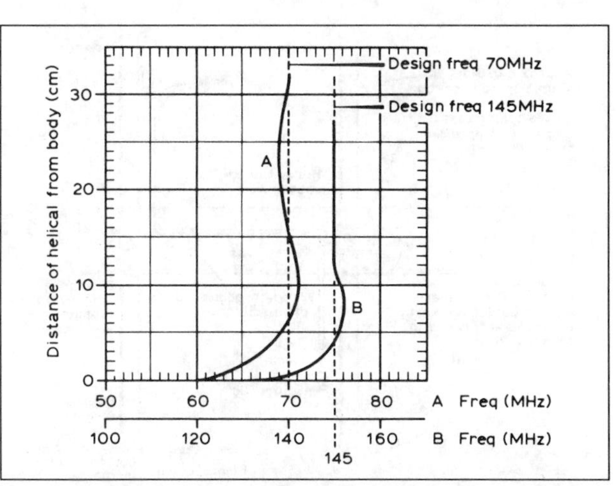

Fig 13.75. Frequency shift of a helical antenna on a typical hand-held transceiver for various distances from the body

Figs 13.76(a) (taken from reference [12]) and Fig 13.76(b) show two methods of construction for the HB9CV antenna. A point that should be stressed is that a series capacitor of 3–15pF is required to adjust the gamma match/phasing combination to a VSWR of about 1.3:1 referred to 50Ω. The element spacing, and in particularly the transmission line spacing (5mm), is critical

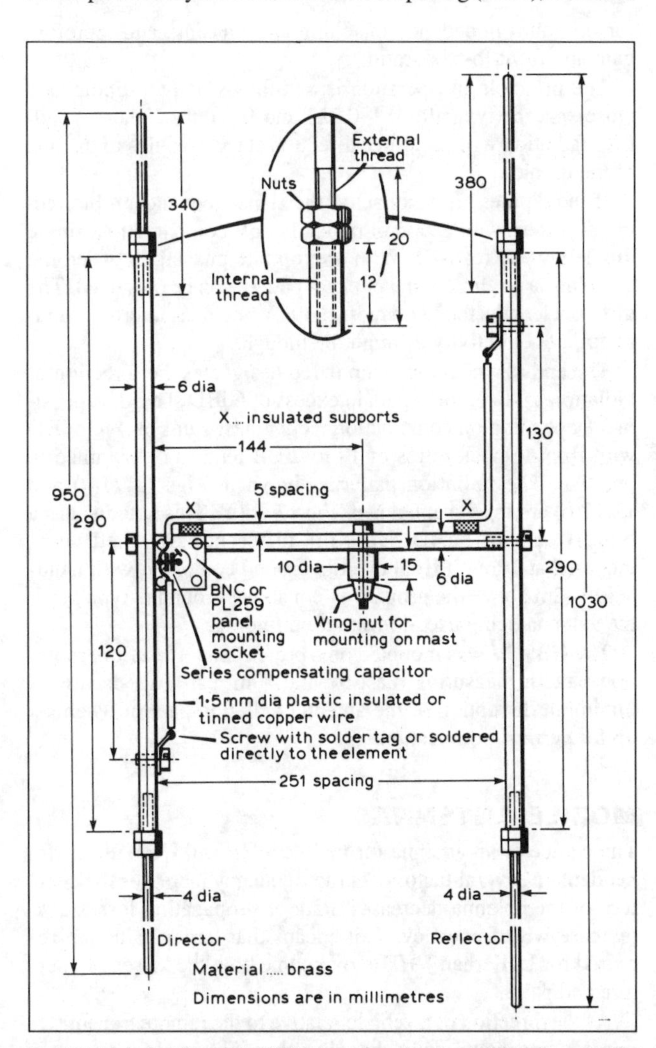

Fig 13.76(a). A collapsible HB9CV antenna for the 144MHz band (*VHF Communications*)

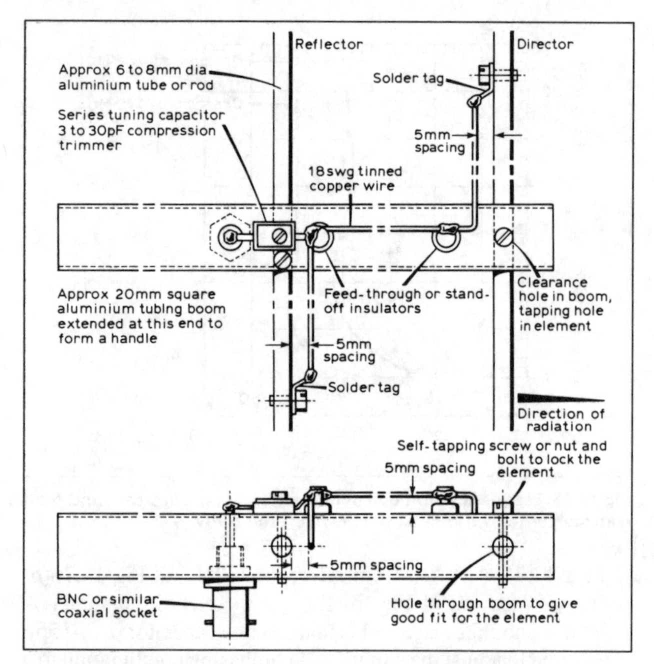

Fig 13.76(b). Alternative construction of the HB9CV. Dimensions as per Fig 13.76(a)

for optimum impedance matching and phasing, and therefore gain and front-to-back ratio.

The principle of operation is as follows. If two dipoles are close spaced (typically 0.1–0.2λ) and fed out of phase, 'end-fire' radiation will occur in a direction at right-angles to the line of the dipoles.

If the dipoles are resonant at the same frequency, a bidirectional pattern with a gain of typically 3dB compared to a single dipole will be realised. With appropriate phasing between the elements, a unidirectional or beam pattern can be produced. The different lengths found on most HB9CV antennas improve bandwidth, not directivity as might be thought.

The end at which the beam is fed designates the direction of radiation. A theoretical gain in excess of 6dBD should be possible. Depending on construction techniques, gains of 4 to 5dBD with front-to-back ratios of 10 to 20dB tend to be obtained in practice. The radiation patterns shown in Figs 13.77(a) and 13.77(b) are for the antenna of Fig 13.76(a). This antenna has a typical gain of 5dBD. Note the difference obtained when mounted at 10m (30ft) above the ground compared with hand-held 1–2m above the ground. 1–2m above ground is typical for the antenna being used for direction finding.

The HB9CV was mounted on a professional glassfibre radiation pattern measuring mast for the 10m test. This ensured a minimum disruption of the antenna radiation pattern when set up for vertical polarisation.

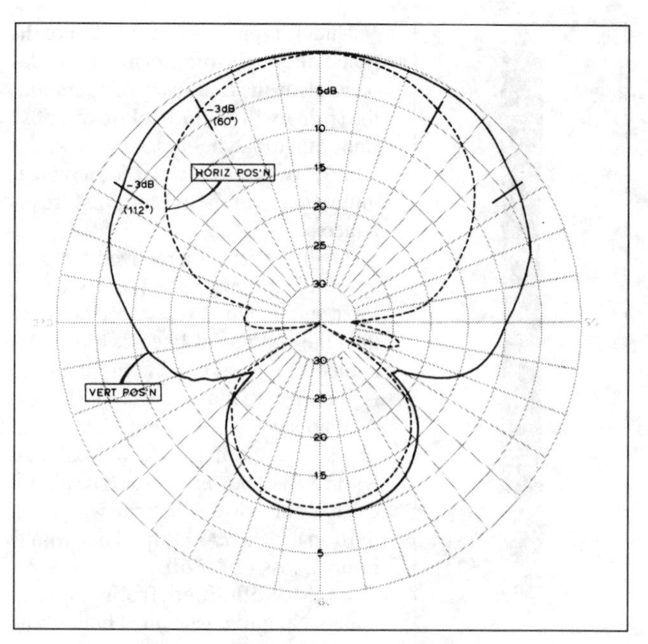

Fig 13.77(a). HB9CV antenna at 10m above ground

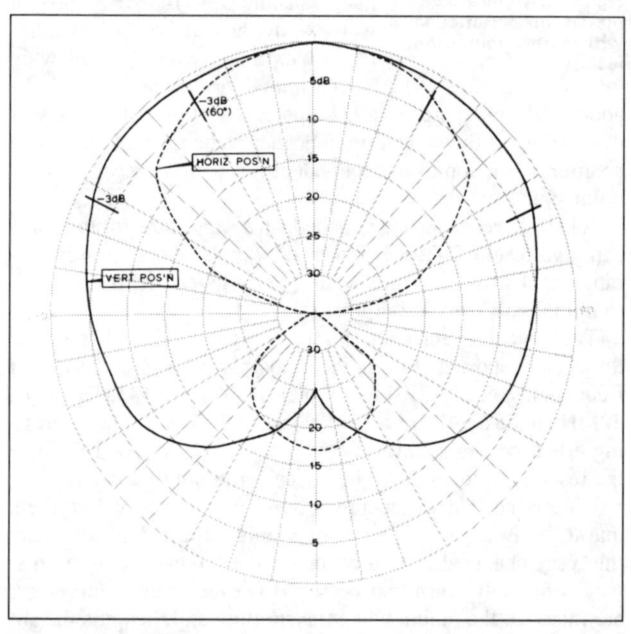

Fig 13.77(b). HB9CV antenna at 1–2m above ground, hand held

MOBILE ANTENNAS

The choice of an antenna for mobile VHF and UHF use is dependent on several factors. As the frequency increases the aperture of the antenna decreases. Path or propagation losses also increase with frequency. This means that larger gains are required for UHF than VHF to overcome both the losses of aperture and path.

As the direction of a vehicle relative to the remote transmitter or receiver is continually changing, there is a need for an omnidirectional antenna system. To achieve gain in the horizontal plane, while retaining an omnidirectional pattern, will require considerable reduction of the pattern in the vertical plane.

For example an omnidirectional antenna of 6dBD gain will have a typical half-power point (–3dB) of under 30°. The narrow beam or disc that is produced can result in considerable variation in transmitted and received signal strength as the vehicle or antenna tilts. This is particularly the case where signals are reflected from nearby objects. A compromise has to be made to obtain maximum gain in the best direction that gives minimum disruption of signals when mobile.

The choice of polarisation is not only dependent on compatibility with stations being received and the optimum for the propagation path. The aesthetics, mechanical complexity and the mounting position of the antenna on the vehicle must be considered.

Antennas, particularly when vehicle mounted, must always

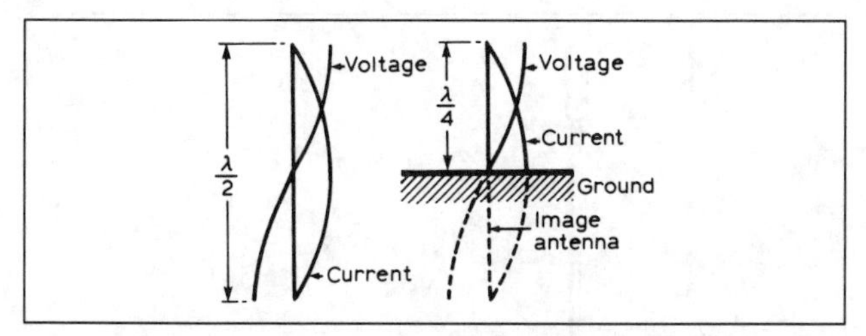

Fig 13.78. The λ/2 antenna and its grounded λ/4 counterpart. The missing λ/4 can be considered to be supplied by the image in ground of good conductivity

be considered as an integral part of the environment in which they are to be used. Radiation patterns quoted by manufacturers can be completely different when the antenna is installed. Antenna measurements are rarely made with the conditions that prevail in typical usage.

High-gain, relatively large, antennas can lose gain with probable loss of omnidirectivity if the antenna is mounted at a lower point to enable access to a garage, for example. The difference in mounting an antenna on the wing or boot of a car compared with mounting it on the top dead centre of the car roof can be at least 3dB. Variation of the radiation pattern can occur due to close-in reflections and surface-wave effect across the vehicle, as well as restriction of the 'line of sight'.

There are several antennas in current use that are worth considering. One or two specialised antennas are available, or can be readily fabricated by the radio amateur, that also merit consideration. Mobile antennas can be considered in three basic groups:

1. Vertically polarised antennas more often used for FM and repeaters.
2. Horizontally polarised antennas normally used for SSB and long distance communication.
3. Circularly polarised antennas.

There is also a sub-group of low-profile antennas for vertical and horizontal polarisation with physical heights below 0.1λ.

Vertical antennas

λ/4 whip

This is the simplest and most basic mobile antenna. It is derived from the doublet or λ/2 dipole. Marconi replaced half of the doublet with a ground plane as shown in Fig 13.78. He found that the image of the vertical λ/4 section was 'reflected' in the ground plane. This produced an antenna that was substantially the same as the original dipole.

The theory of operation showed that if the ground plane was infinitely large and made of a perfectly conducting material, all of the radiation associated with the lower half of the dipole was radiated by the top half. In fact, a 3dB improvement over the dipole would be achieved.

In practice the size of the ground plane and its resistive losses modify the pattern. The full 3dB is never realised. (Measurement by the author of a 5GHz monopole on an aluminium ground plane of 40λ diameter showed a gain of 2.63dBD). Figs 13.79 and 13.80 show optimum patterns of a λ/4 whip measured on a ground plane of λ/2 sides and 1λ sides. Although the pattern is raised from the horizontal, on a medium ground plane the loss of horizontal gain is relatively small (20° and 1dB at 0° in Fig 13.79, but 40° and 6dB at 0° in Fig 13.80).

However, as the ground-plane size increases the main lobe continues to rise until the situation of Fig 13.81 occurs. When a vertical radiator is mounted over a ground plane as described, the input impedance is typically halved. For the λ/4 whip or monopole, the input impedance is typically 36Ω + j, that is to say approximately half the resistance of the dipole but with an additional reactive component.

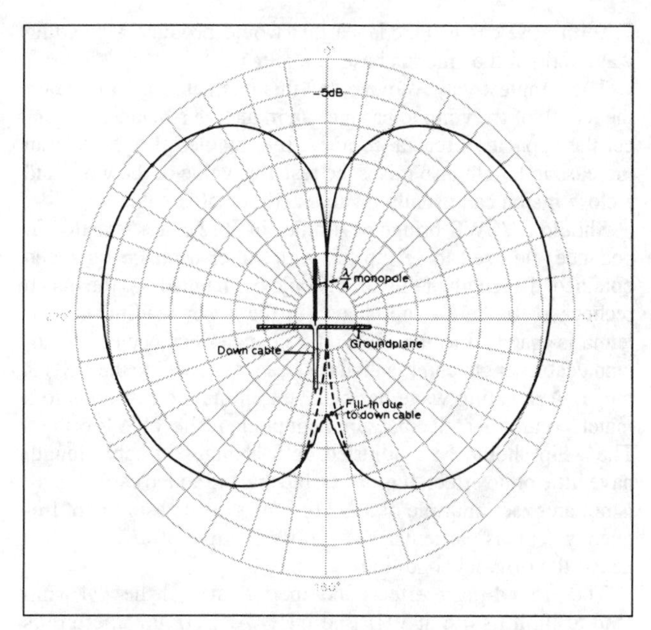

Fig 13.79. Decibel radiation pattern of a λ/4 monopole over a λ/2 square ground plane at 145MHz

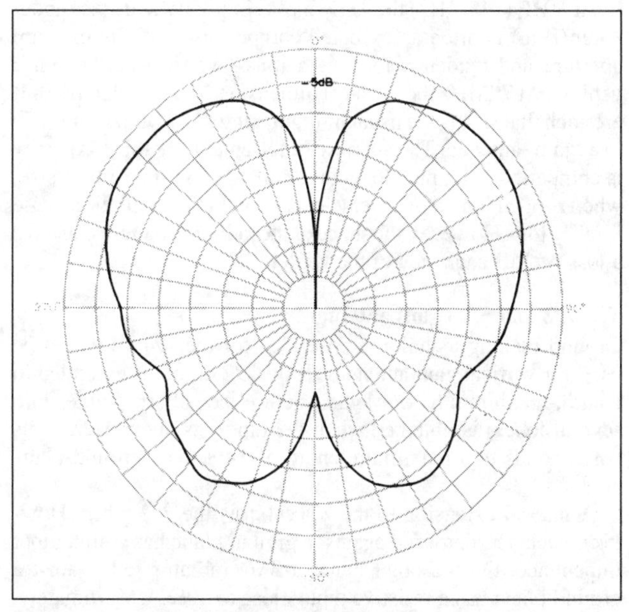

Fig 13.80. Decibel radiation pattern of a λ/4 monopole over a 1λ square ground plane at 145MHz

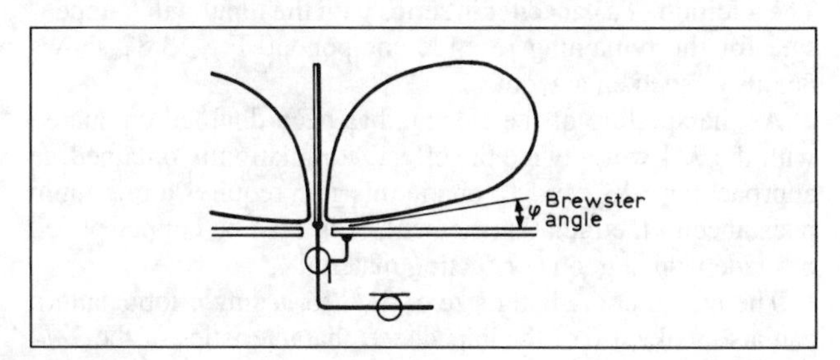

Fig 13.81

With 50Ω cable impedance this would produce a standing wave ratio at the antenna base of about 1.5:1.

The simplest way to overcome this mismatch is to increase the length of the whip to produce an inductive reactance to cancel the capacitive reactance normally obtained. In practice an increase in length also raises the resistive value of the whip and a close match can usually be obtained to a 50Ω cable.

Should a VSWR bridge or similar (of 50Ω characteristic impedance) be used to set up the whip, after a match has been obtained, the length of the cable should be changed and the match rechecked. If there is no change in the meter reading the antenna is matched to the cable. If a change does occur, the antenna/cable length combination has been 'matched' to the VSWR meter. A standing wave has been set up that corresponds to a match condition at the measurement point, ie the VSWR bridge. The whip should be readjusted until changes in cable length have little or no effect. It is preferable that the added cable length is not an exact multiple of a λ/2 or λ/4 at the measurement frequency. This is particularly so with a multiple of λ/2, as it 're-peats' the original condition.

The ground-plane effects and aperture size of the λ/4 whip tend to limit its use at VHF and UHF. At UHF the aperture is small and the pattern is raised in the vertical plane by a large ground-plane area. It has therefore limited use at UHF.

At VHF (145MHz) the λ/4 whip's simplicity and size (about 49cm/19in) is often an accepted compromise with its medium aperture and tendency to have a raised vertical pattern on a vehicle. At 70MHz the physical dimensions (about 102cm/40in) are such that size is the usual limit, making a λ/4 whip preferable to a 'gain' antenna. The aperture of the antenna at this frequency is compatible with path loss conditions. The ground-plane size, when roof-mounted on a vehicle, is such that the radiation angle is fairly low. However, the shape of the radiation pattern can have a loss of 3dB each side of the vehicle.

The λ/2 and 5λ/8 gain antenna

Ground-planing techniques described for the λ/4 whip can be used for vertical gain antennas. If the λ/2 dipole is extended in length, maximum forward gain (before the pattern divides into several lobes) is obtained when the dipole is about 1.2λ. This corresponds to a maximum length of 5λ/8 for a ground-plane antenna.

A natural extension to the λ/4 whip is the λ/2 whip. However, such a radiator fed against a ground plane has a high input impedance. On the other hand, a 3λ/4 radiator fed against a ground plane has a resistive input close to 50Ω. Unfortunately, it is above the optimum length for a reasonable pattern shape.

If the λ/2 whip could be made to look like a 3λ/4 radiator then it would be possible to obtain a 50λ resistive input. A series coil at the ground-plane end of a λ/2 radiator can be used to resonate at 3λ/4, but the input is still of fairly high impedance and reactive. If, however, the coil is shorted to the ground plane, tapping up the coil will provide the required match/input point. The addition of a capacitor in series with the input will compensate for the remaining reactive component. Fig 13.82 shows details of such an antenna.

As the aperture of the antenna has been doubled compared with the λ/4 whip, twice the effective radiation is obtained, ie approaching 3dB gain. To obtain this gain requires a minimum resistance in the radiating element, ie it must be copper-plated or made from a good conducting material.

The maximum radiator size of 5λ/8 for a single-lobe pattern can also make use of the impedance characteristics of the 3λ/4 radiator.

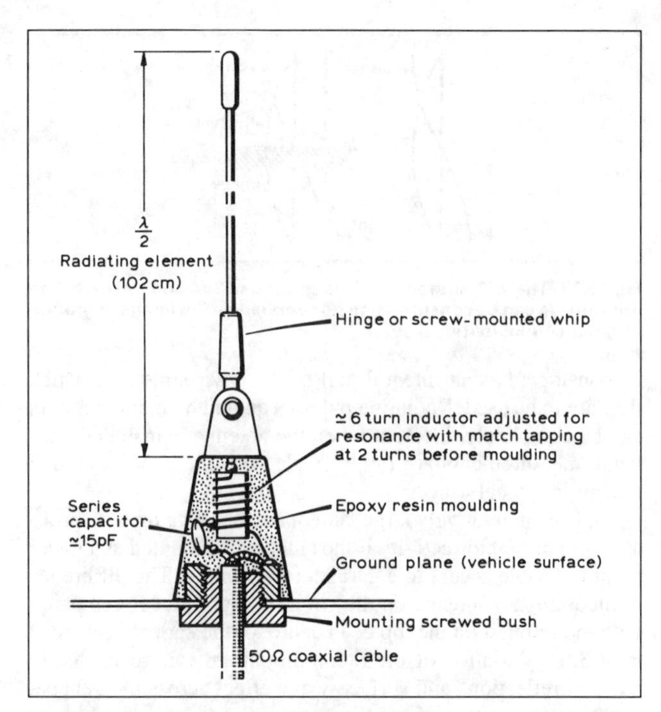

Fig 13.82. A typical home-built λ/2 mobile antenna and mount

Construction is simpler than for a λ/2 antenna. If the radiating element is made 5λ/8 with a series coil, equivalent to λ/8, at the ground plane end, an input impedance very close to 50Ω is obtained. With correct materials a gain close to 4dBD can be achieved from the further increase in aperture. The radiation pattern is raised more than that of a λ/2 antenna, so the improved gain of the 5λ/8 may not always be realised. However, the simplicity of construction is an advantage.

Fig 13.83 gives details of the series 5λ/8 whip. One other advantage of this antenna is that over a wide range of mounting and ground-plane conditions it will self-compensate for impedance and resonance changes. It is preferable for both the λ/2 and 5λ/8 antennas to be 'hinged', particularly if roof-mounted, to enable folding or 'knock' down by obstructions like trees and garages.

Various gain figures have been reported for the 'five-eighth wave' whip antenna. Unfortunately not all antennas use optimum materials. Resistive steel wires or rods produce heating loss. The use of a glassfibre-covered wire changes the resonant length by as much as 20%. The radiator therefore has to be cut shorter than 5λ/8, with an accompanying loss of aperture.

The construction of the series coil is important. Movement of the coil turns will change the antenna's resonance, giving apparent flutter. Some transceivers with VSWR-activated transmitter close-down will be affected by change of resonance of the antenna. This can make the power output of the transmitter continually turn down or be switched off, producing what appears as extremely severe 'flutter' on the transmission.

Several of the '5λ/8 ground-plane antennas' discussed in various articles are in fact antennas not truly of this nature.

One of these devices worth considering for its own merits is that shown in Fig 13.84. It consists of a 5λ/8 vertical element with a reactive sleeve of 0.2λ at the ground-plane end. The gain obtained from this antenna is typically 1.8dBD. As can be seen, the actual radiating element A–A is shorter than a λ/2 antenna.

Another antenna family, with similar properties but different in construction, includes the 'J' and Slim Jim. These are described later in this chapter.

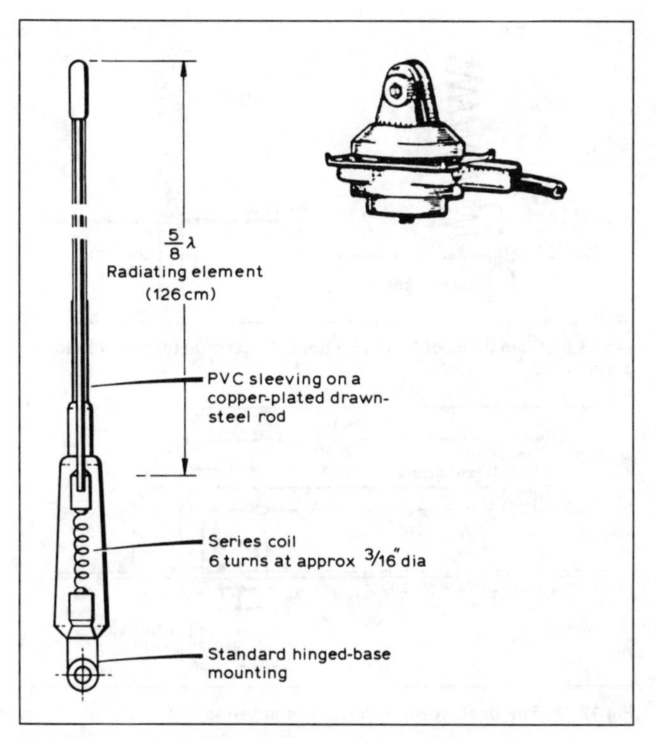

Fig 13.83. A typical commercial 5λ/8 mobile antenna and mount (top right)

7λ/8 whip

This mobile antenna is derived from the Franklin collinear of Fig 13.104(c). It consists of two λ/2 elements coupled by a series 'phasing' capacitor. One effect of the capacitor is to resonate the combined elements at a lower frequency than that of a single λ/2 element. However, reducing the length of the top element resonates the arrangement back to the original frequency.

The base impedance above a perfect ground plane is 300–400Ω with some capacitive reactance. A series loading coil in combination with an L-matching section gives a good match to 50Ω coaxial feeder. The match is maintained with quite modest ground plane size (λ/4 radials or λ/2 diameter metal surface). This makes the 7λ/8 whip suitable for vehicle mounting or for use as a base-station antenna.

The final length of the two radiator elements is somewhat dependent on their diameters and the design of the series capacitor and matching unit. Fig 13.85 shows the general appearance and dimensions of a commercial version of the 7λ/8 whip, together with typical circuit components and current distribution in each element.

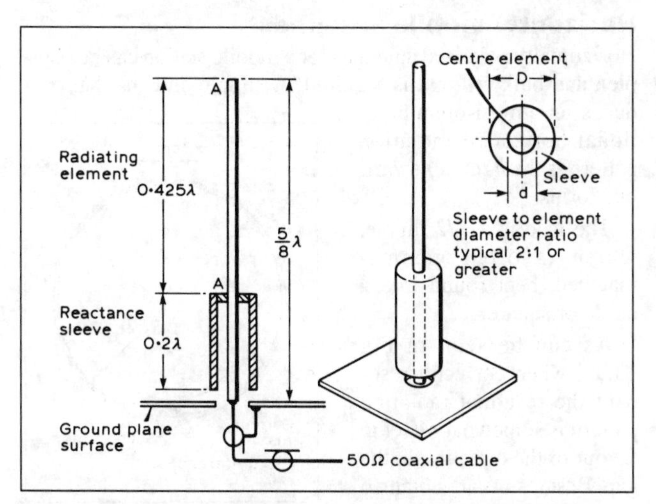

Fig 13.84. The reactance-fed 5λ/8 monopole. Typical gain is +1.8dBD (*Ham Radio*)

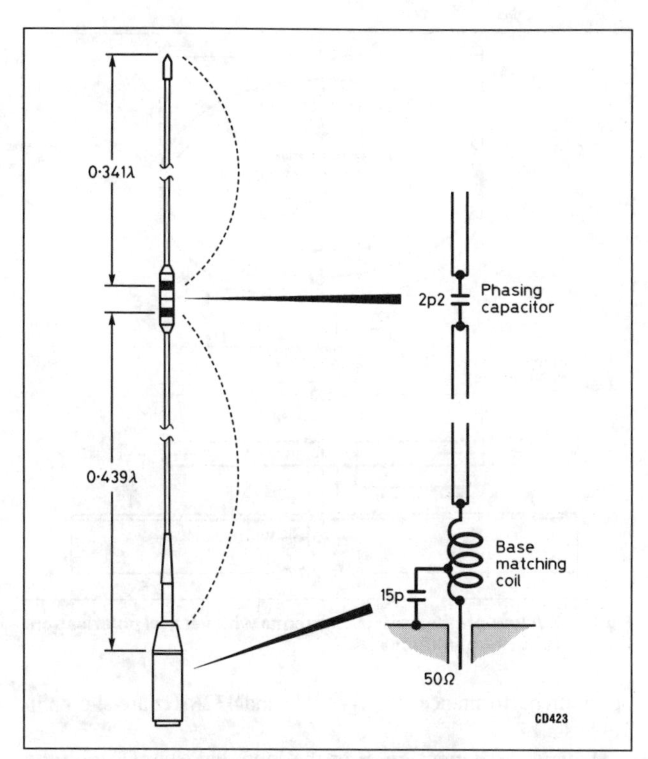

Fig 13.85. The 7λ/8 whip antenna. This is effectively two half-waves in series with a series coupling/phasing capacitor. The overall assembly is that of a commercial form of the antenna. Also shown are typical electrical dimensions and components. The dotted lines show the approximate current distribution

The theoretical gain of this antenna with the dimensions shown in Fig 13.85 is 4.95dBi (2.8dBD) over a perfect ground plane. The professionally measured gain, with the whip on a 1m ground plane, was slightly over 4.7dBi for the full 144MHz band. The radiation pattern in the E (vertical) plane was predominantly a single lobe (torus/doughnut) peaking at 4° above the horizon and with a 3dB beamwidth of 38.5°.

Low-profile antennas

An alternative to vertical ground-plane antennas is a device to reduce the size of the system. The reduction of size normally implies loss of aperture and therefore gain. However, of the three antennas discussed in this section, one can produce a gain of +1dBD.

The λ/2 ring radiator

Although called a 'ring' radiator, radiation is produced by the 'slot' formed between the horizontal λ/2 ring and the ground plane. Consider a λ/2 slot in a metal sheet. If the sheet is rolled into a cylinder such that the two ends of the slot come together, an omnidirectional vertically polarised radiator is produced. As with the conventional λ/2 slot, an impedance match can be obtained by tapping along from one end. If the slot is just under λ/2, a capacitor across the centre of the slot will resonate it to λ/2.

As with the skeleton slot developed by G2HCG, the ground-plane sheet at the top of the slot is reduced to produce a ring. The lower ground-plane section is 'bent' into the horizontal plane. The outcome is the low-profile antenna of Fig 13.86.

The dimensions in 'electrical degrees' and specific sizes for

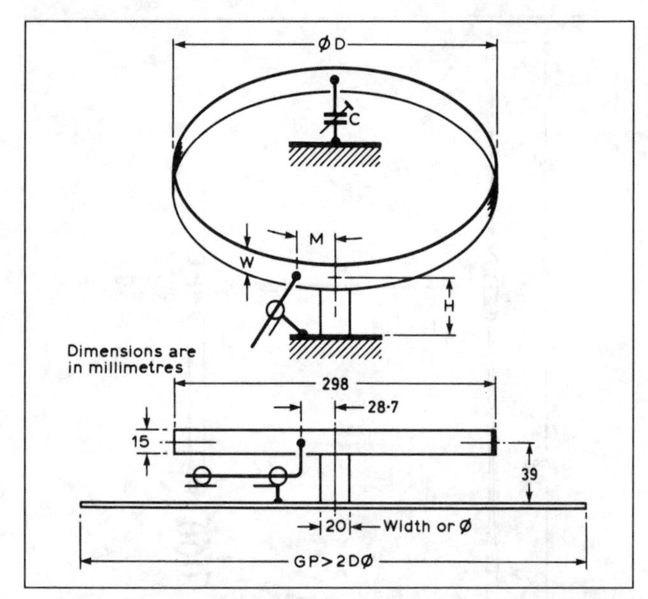

Fig 13.86. A low-profile vehicular antenna with vertical polarisation. Gain is 1dBD, termination 50Ω

optimum performances for 144MHz and 432MHz are shown in Table 13.14.

Halving the dimension *H* or the loop diameter *D* (with the necessary increase of capacitance to re-tune and change of match point) will halve the radiation capability.

The λ/2 ring radiator is a high-*Q* antenna and has therefore a reduced bandwidth compared to a dipole (typically 1–2% compared to 10% for a monopole).

If the ground plane is completely reduced, as was the top section as previously described, a 'double ring' radiator is produced. Both ring radiators lend themselves to discreet fixed antennas.

The ground-planed normal helical

The normal-mode helical (spring) antenna, when vehicle mounted, is approximately 2 to 3dB down on a dipole and the order of 0.1λ in height. Often an acceptable match to 50Ω can be obtained by simply adjusting its resonant length.

Alternatively, a small inductance or capacitor across the base or an input tapping on an 'earthed' helical, as shown in Fig 13.87, will provide the required matching.

The blade antenna

If an earthed λ/4 vertical is 'bent over' and an 'input' is tapped at a point from the earthed end, as shown in Fig 13.88, a predominantly vertical, low-profile, antenna is produced. Being so

Table 13.14. λ/2 ring radiator dimensions

	Theoretical	VHF measurement antenna	UHF measurement antenna
Frequency	*f* MHz	145MHz*	433MHz
Diameter *D*	52°	298mm	100mm
Height *H*	8°	39mm	15.5mm
Diameter *d*	nom. 1–2°	15mm 20swg strip	10mm 20swg strip
Match *M*	5° for 50Ω	28.7mm	9.7mm
Tuning capacitor C	To give capacitive reactance, nominally 250–500Ω	2–5pF	0.5–2pF

* Tuneable 137–148MHz.

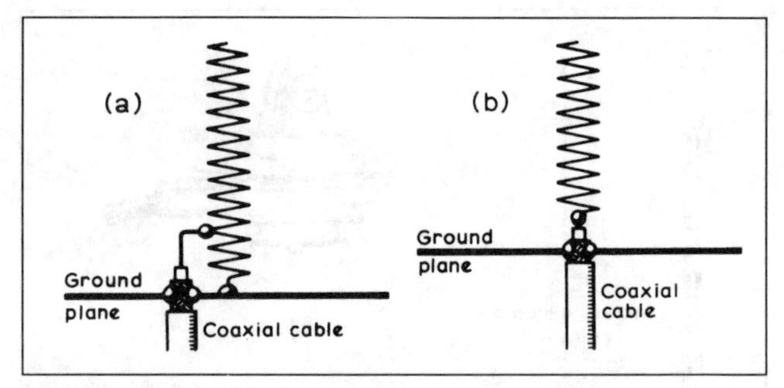

Fig 13.87. Two ways of feeding a helical antenna: (a) shunt feed, (b) series feed

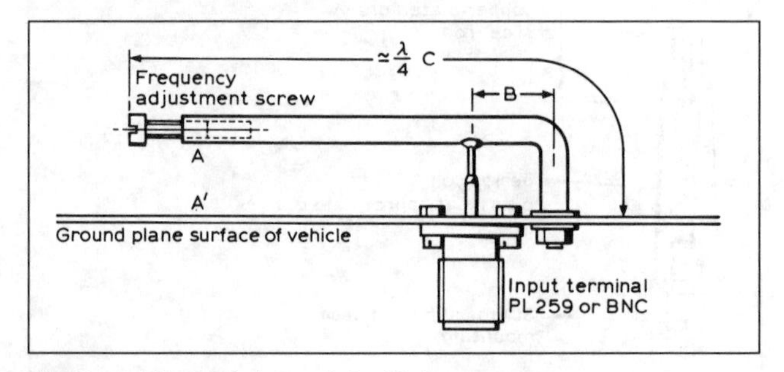

Fig 13.88. The basic form of the blade antenna

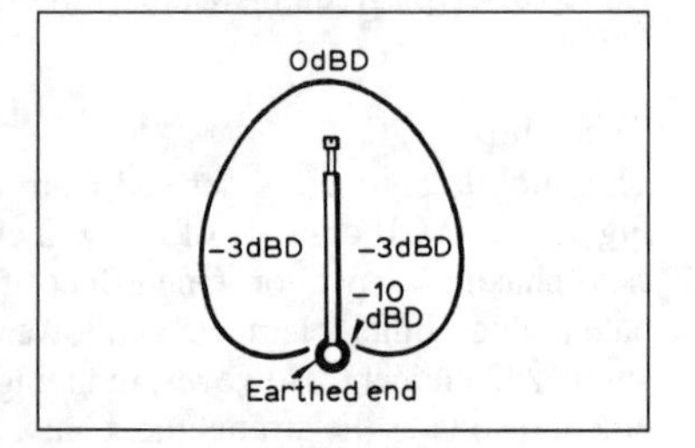

Fig 13.89. Typical radiation pattern of the blade antenna. Dimension C should be less than λ/4 such that an inductor formed by an extended screw at A will resonate the antenna. Dimension B is about λ/20 of a wavelength at the operating frequency

close to the ground plane it produces vertical polarisation with a pattern typically that of Fig 13.89.

This antenna, often with a small capacitor across A–A' to give a range of tuning, is called a *blade antenna*. It is sometimes known as a *dog leg* antenna when used without a cover. This name comes from its 'cocked-over leg shape'. The 'blade antenna' name is due to its original use on aircraft where it was moulded into a blade shape to produce streamlining.

Horizontal mobile antennas

Horizontally polarised antennas for a mobile station can be complex and bulky if gain is required. A simple antenna that produces an almost-omnidirectional horizontal radiation pattern is the *halo* in its various forms.

The halo is a λ/2 dipole, which is often gamma-matched, bent round into a circle or square.

As can be seen in Fig 13.90, when correctly resonant the resultant radiation pattern is somewhat offset in favour of the direction of the gap. Best results are obtained when it is mounted at a minimum height of 70cm (0.34λ

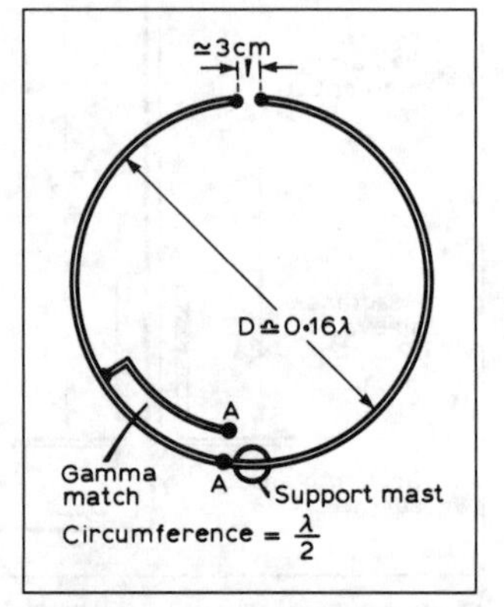

Fig 13.90. Dimensions of the λ/2 halo

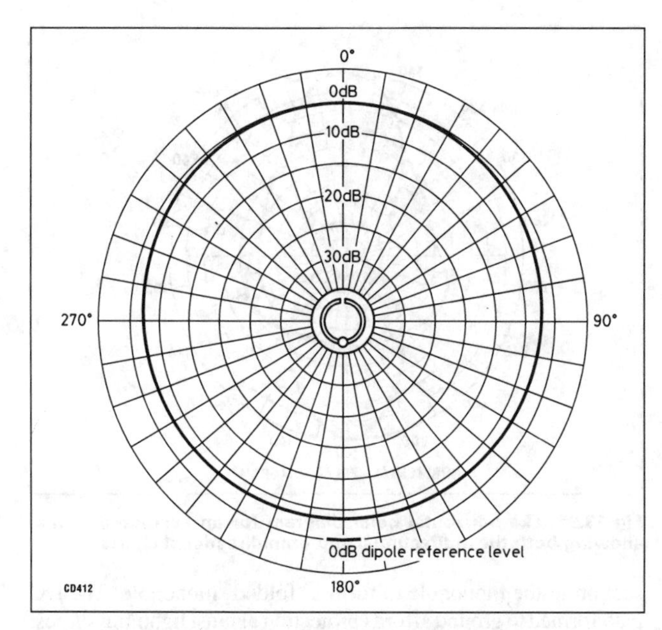

Fig 13.91. Decibel radiation pattern of halo antenna

at 144MHz) above the ground plane produced by the vehicle roof.

Simple crossed dipoles

The *turnstile*, produced from crossed dipoles, provides a simple yet very effective horizontally polarised, omnidirectional antenna. It consists of two horizontal dipoles mounted at right-angles and fed with equal power, but with a 90° phase difference (Fig 13.92). Matching to 75Ω is quite simple, and with a little adjustment a very low VSWR can be obtained. The radiation pattern produced tends to be square. The 'corners', at element ends, are about 1dB up on a dipole's maximum radiation.

The 90° phase difference is readily obtained with a resonant λ/4 cable between the dipoles. A further resonant, series, λ/4 is then added as a matching transformer to the characteristic line impedance. Because the vehicle ground plane becomes a reflector there is a predominance of radiation upwards. This upward radiation is reduced if the crossed dipoles are about 0.5λ above the surface of the vehicle.

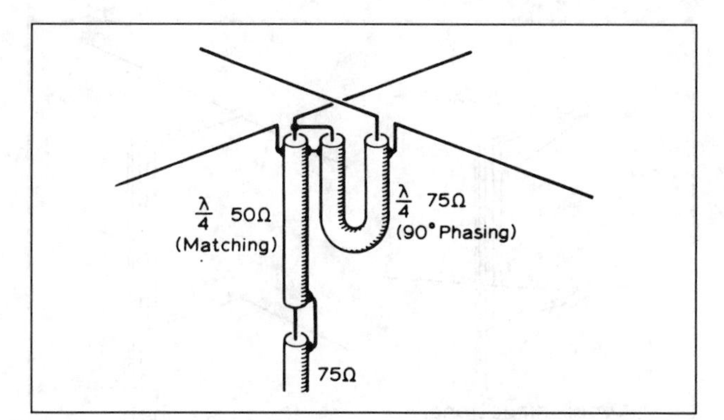

Fig 13.92(a). Phasing and matching arrangement of crossed dipoles

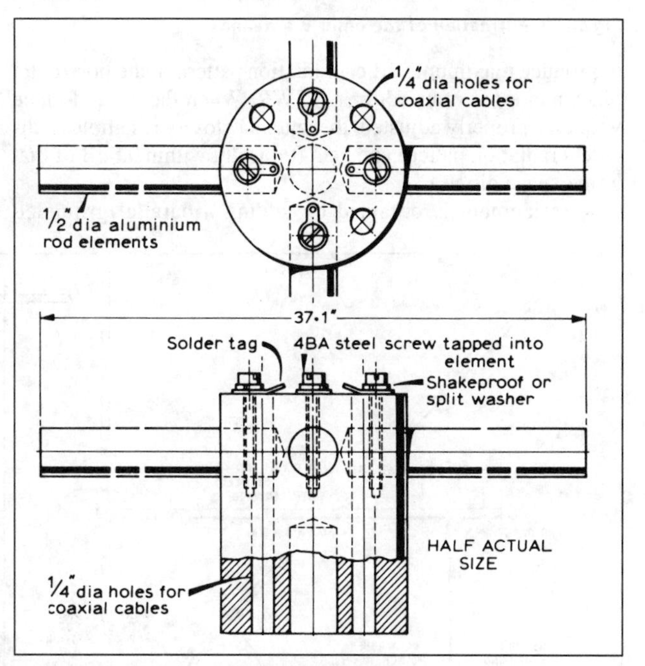

Fig 13.92(b). Details of central insulator

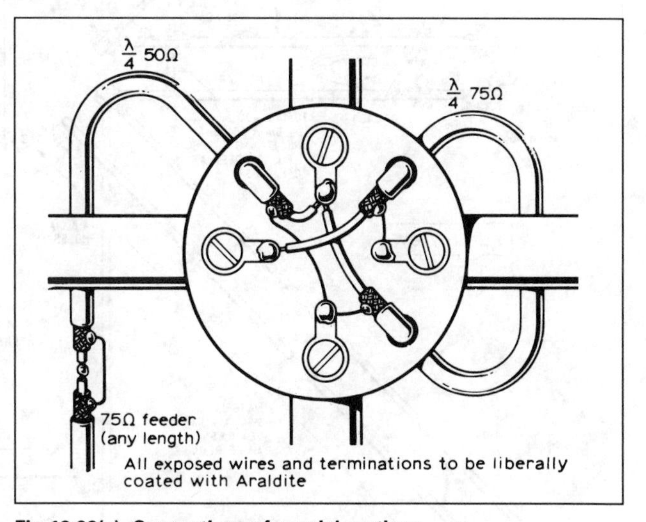

Fig 13.92(c). Connections of coaxial sections

OMNIDIRECTIONAL BASE STATION ANTENNAS

The horizontally polarised omni-V for 144MHz

This antenna consists of a pair of λ/2 dipoles. The centres of the dipoles are physically displaced to produce 'quadrature' radiation with the ends of each dipole 'supported' on a λ/4 shorted stub. A pair of Q-bars are tapped down the stubs to a point where the impedance is 600Ω. When the two units are fed in parallel, they produce an impedance of 300Ω at the centre. A 4:1 balance-to-balance coaxial transformer is fitted to the centre point of the Q-bars to enable a 75Ω coaxial feeder cable to be used. (*Note*: adjustment of the Q-bars to give 400Ω would set the final impedance for a 50Ω feeder cable.) The general arrangement is shown in Fig 13.93(a). Fig 13.93(b) shows how the antenna may be arranged to give a bidirectional radiation pattern.

The λ/4 ground plane

Since base-station monopole antennas are usually mounted on masts atop buildings, a perfect ground plane is not present. Often the ground plane is simulated by wires extending horizontally from the base of the monopole. Two popular forms of such antennas are shown in Figs 13.97(a) and (b). The ground-plane wires are usually four in number and from 0.28λ to 0.30λ in length. The optimum length is often selected experimentally

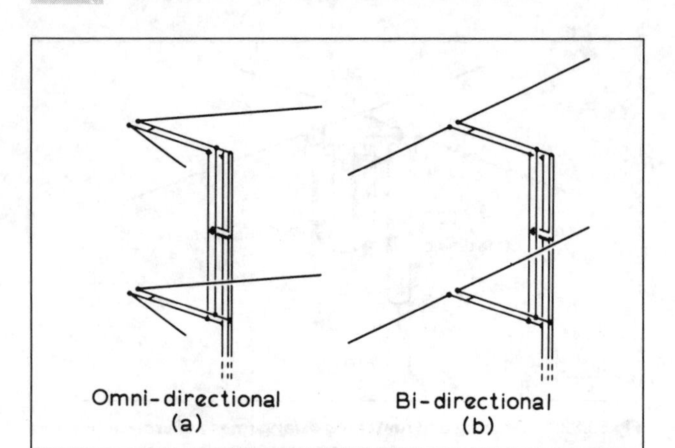

Fig 13.93. Formation of the omni-V antenna

to produce maximum vertical radiation pattern in the horizontal plane and more often for best VSWR. When the ground-plane wires are properly adjusted, the antenna closely resembles a dipole's radiation pattern and gain (typically within 0.5dB of that a free-space dipole).

A refinement is obtained by adding a parallel grounded

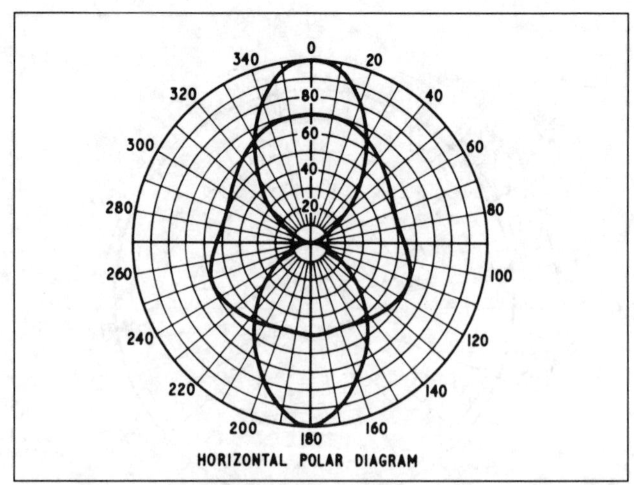

Fig 13.95. The horizontal polar diagram for an average antenna, showing both the bidirectional and omnidirectional charts

section to the monopole to form a 'folded' monopole. The DC path formed to ground affords protection against lightning surges. A 'folded' configuration is inherently more broad-band than the

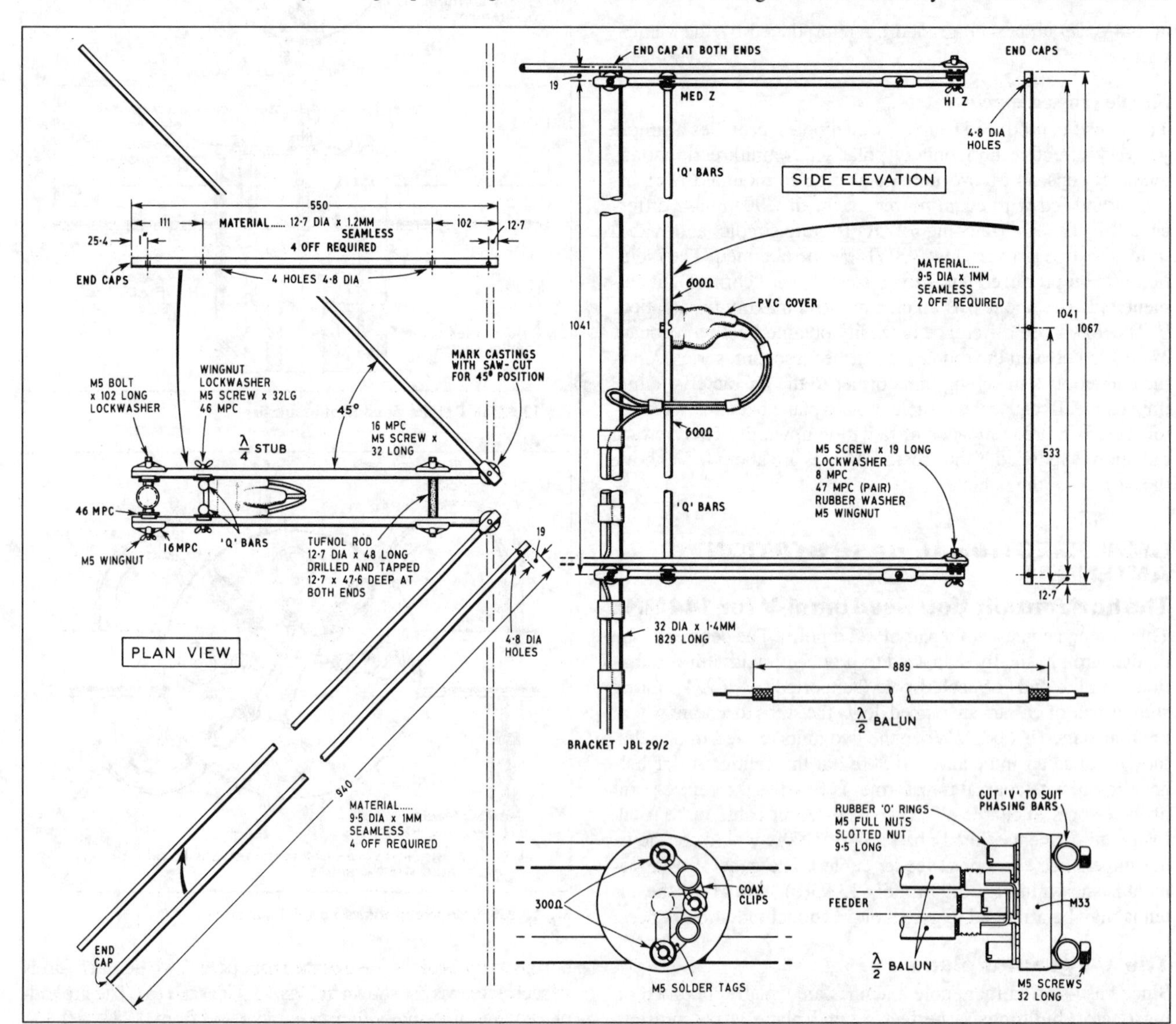

Fig 13.94. The mechanical details of the construction of the omni-V (dimensions are in millimetres)

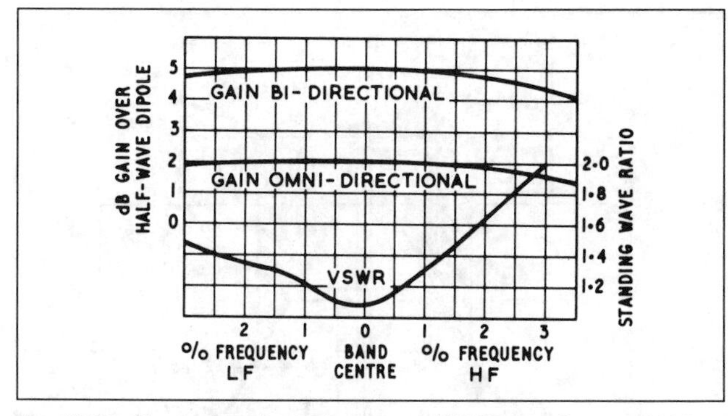

Fig 13.96. Chart showing gain versus VSWR on the omni-V just described

single element. For equal diameters in the two parts of the 'fold', the radiation resistance would be quadrupled (nominally 4 × 37.5 = 150λ). The grounded side of the folded element is made smaller in diameter to lower the resistance to 50Ω. This provides a much better impedance match to the feeder cable than the simple monopole.

An alternative feed is shown in Fig 13.97(c). The short-circuit λ/4 line formed in the upper end of the mast supports the monopole rigidly and, due to its high impedance, does not load the feed point. A λ/4 matching section is sometimes placed in the transmission line at the feed point to improve the impedance match between the antenna and the line.

The skirted antenna

Although the λ/4 monopole over a ground plane behaves electrically as a λ/2 dipole in free space, the radial projections of the ground rods are sometimes physically inconvenient. Furthermore, the ground rods 'shield' the mast and prevent the excitation of parasitic elements to form high-gain arrays. The coaxial *skirted antenna* (Fig 13.98) behaves as a λ/2 dipole in free space by using a 'skirt' to form the bottom half of the dipole and a 'choke' to isolate the antenna from the mast.

The gain sleeve dipole (vertically polarised)

The *gain sleeve dipole* is derived from the 1.8dBD shunt-feed 5λ/8 previously mentioned in the mobile antenna section.

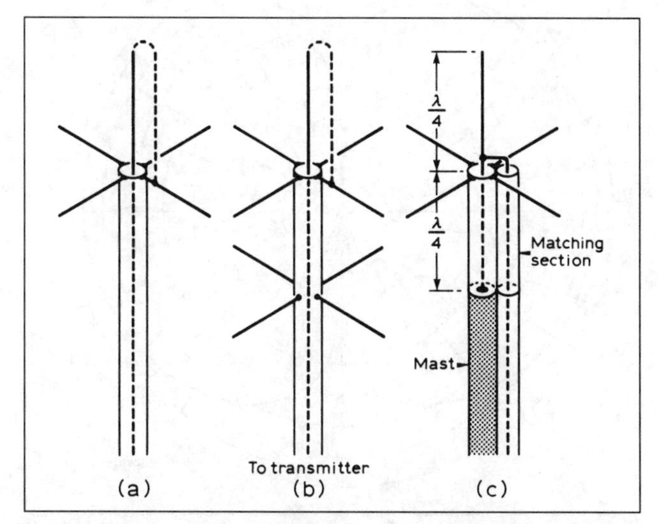

Fig 13.97. Ground-plane simulation for the mast-mounted vertical antenna. The folded section is shown in dotted lines, and a modified feed is shown in (c). (From *Antenna Engineering Handbook* by Jasic. © 1961 McGraw-Hill. Used with the permission of McGraw-Hill Book Co)

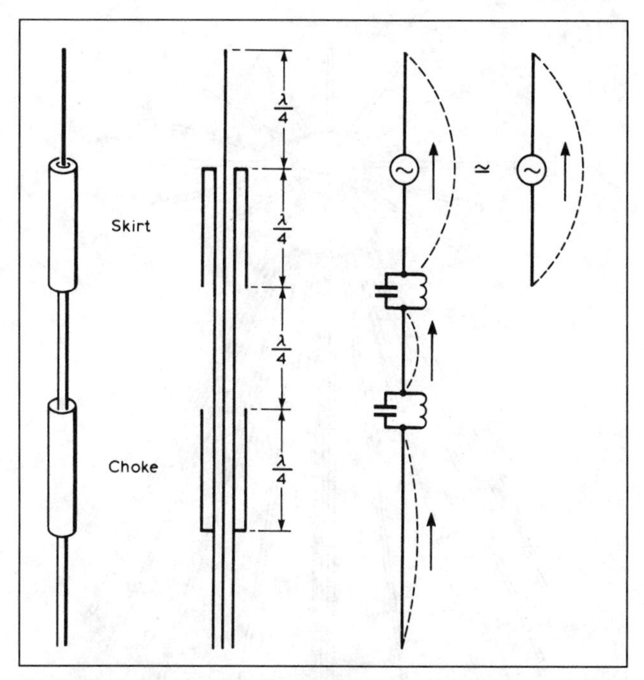

Fig 13.98. Coaxial-fed λ/2 dipole with choke. (From *Antenna Engineering Handbook* by Jasic. © 1961 McGraw-Hill. Used with the permission of McGraw-Hill Book Co)

The radiating element B–B is a centre-fed 1λ element but it is fed coaxially to make it an end-fed one. Having approximately twice the aperture of the λ/2 dipole, a gain of typically 2.5–3dB is achieved.

Mechanical construction is open to interpretation but a 'beer can' or plastic water-pipe formats are two solutions. Note that the mounting point should be at A–A and not on the 0.25λ sleeve.

The discone

This is increasingly being used by amateurs to provide a single antenna covering several VHF/UHF bands. It has been used for many years for commercial and military purposes. It is not only omnidirectional and vertical polarised but has wide-band characteristics. A single antenna is capable of covering the 70, 144 and 432MHz bands or 144, 432 and 1296MHz. However, as the antenna can operate over roughly a 10:1 frequency range, it will more readily radiate harmonics present in the transmitter output, so it is therefore important to use a suitable filter to adequately attenuate them. The radiation angle tends to rise after the first octave of frequency and this is the normal acceptable working range. If correctly constructed, a VSWR of less than 2:1 can be obtained over the octave range. One characteristic of the discone is a very sharp deterioration of the VSWR at the lowest frequency of operation.

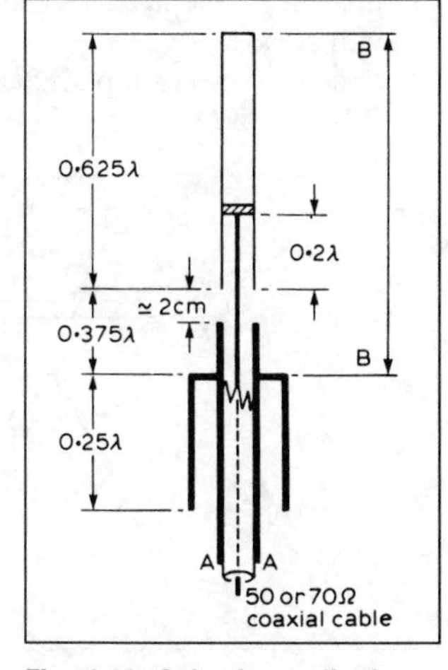

Fig 13.99. Gain sleeve dipole

The discone consists of a disc mounted above a cone, and ideally should be constructed from sheet material. This can be difficult to realise without sheet metal facilities, and appearance or wind

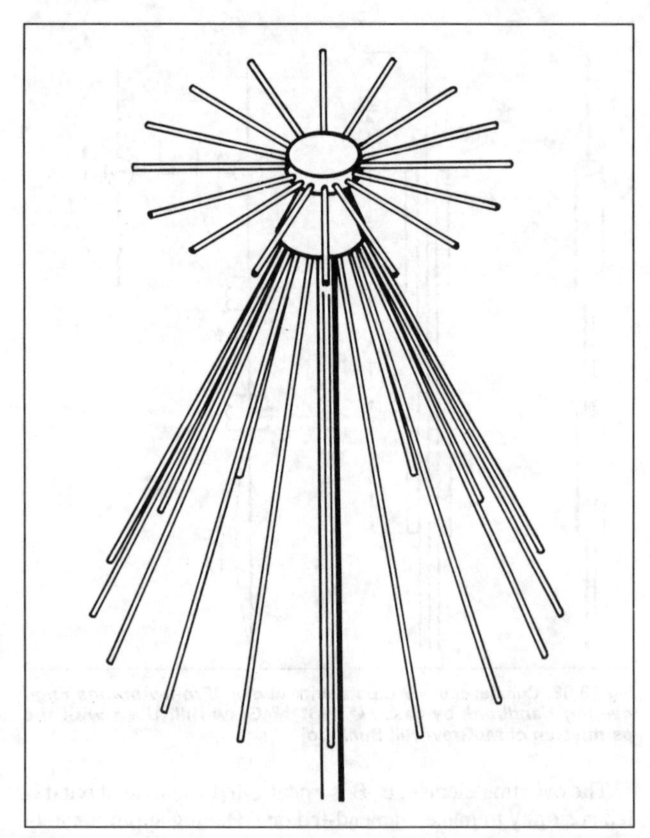

Fig 13.100. General arrangement of skeleton form of discone antenna

effects add to the problem. However, with only a little loss of performance the components may be made of rods or tubes as illustrated in Fig 13.100. At least eight or preferably 16 rods are required for the 'disk' and 'cone' for reasonable results. Open mesh may be used as an alternative to sheet metal or rods.

The important dimensions are the end diameter of the cone and the spacing of this from the centre of the disc. These are instrumental in obtaining the best termination impedance, ie 50Ω.

The primary parameters are shown in Fig 13.101 with the following dimensions:

A The length of the cone elements are λ/4 at the *lowest* operating frequency (2952/*f* MHz inches).

B The overall disc diameter should be 70% of λ/4.

C The diameter of the top of the cone is mainly decided by the diameter of the coaxial feeder cable. For most purposes 0.5in will be suitable.

D Spacing between top disc and the cone top is 20% of *C* or 0.1in for 50Ω.

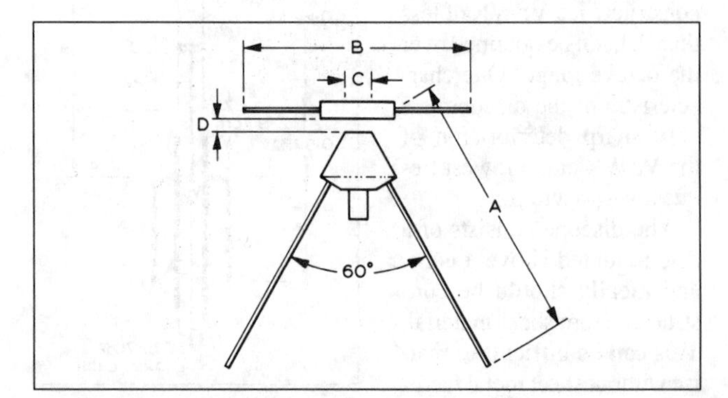

Fig 13.101. Primary dimensions of discone antenna

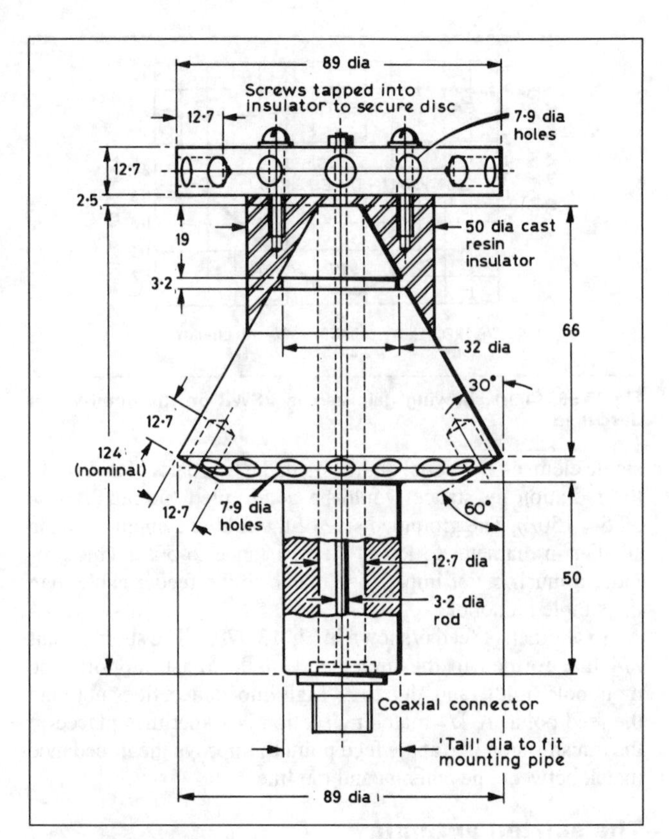

Fig 13.102. Details of a hub assembly

The detail given in Fig 13.102 of the 'hub' construction will be suitable for any design using a 50Ω feeder cable and may be taken as an example. A suitable insulator can be made with a potting resin or turned from nylon, PTFE or other stable low-loss material.

An extension of the discone is the *helicone*. The elements of

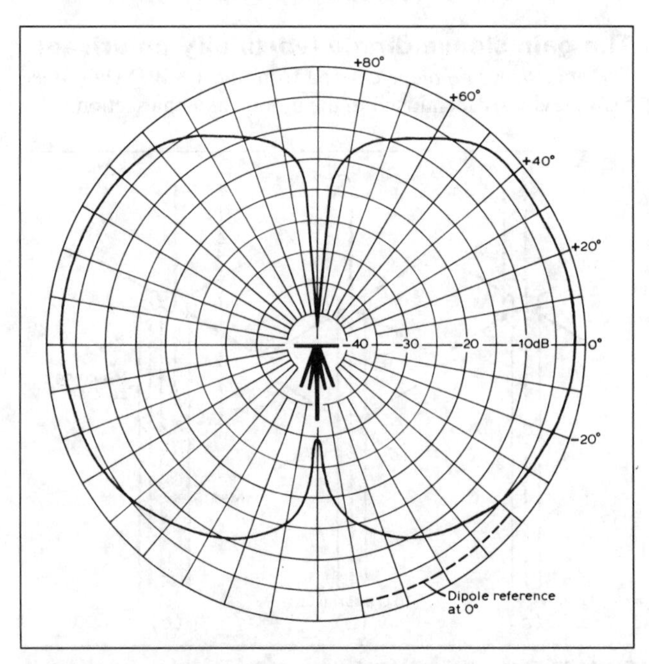

Fig 13.103. Typical discone and helicone decibel radiation pattern over the first 2:1 frequency range. As the frequency increases above 2:1 the pattern tends to rise above the horizontal level until at about 5:1 in frequency the main direction of radiation is above 45° from the horizontal

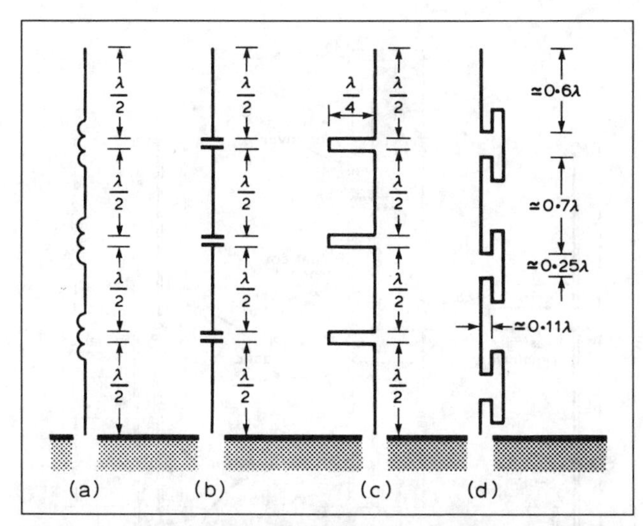

Fig 13.104. The collinear as developed by Marconi and Franklin. (From *Antenna Engineering Handbook* by Jasic. © 1961 McGraw-Hill. Used with the permission of McGraw-Hill Book Co)

the conventional discone can be replaced with helical elements working in the normal mode as previously discussed.

In its simplest form only eight elements are required for the disc and for the cone. Gain and the radiation pattern are essentially the same for both the discone and helicone, but for the helicone the usable bandwidth is reduced to approximately one-third.

Collinears (vertical polarisation)

As previously mentioned with vertical mobile antennas, increasing the radiator length produces an increase in gain. This is true until the centre-fed dipole is over 1.2λ long or the ground-planed form is over 5λ/8 long. Any further addition in length leads to the radiation pattern breaking into several lobes. An investigation into why this occurs shows that the phase relationship along the radiator is such that radiation from various parts are in opposition, thus cancelling radiation in certain directions. If some mechanism could be employed to change this phase relationship or radiating elements up to 5λ/8 could be added together, then the increase in gain could be obtained.

Use of the collinear antenna is a solution to this problem. One arrangement is an end-to-end series of co-phased, λ/2 dipoles over a ground plane. The centre-fed form, ie extending each arm of the dipole with co-phased elements, is twice the length of the ground-planed form. It also has the problem, when used vertically, of the feed cable modifying the pattern of the lower half. Correct phase relationship can be obtained by reversing the phase of alternate λ/2 sections. This is done by inserting anti-resonant networks every λ/2. One form is a λ/4 non-radiating transmission line (shorted stub). This can be wound into a coil to reduce its profile.

If the stub is an open transmission line, as opposed to coaxial cable, the stub will need to be shortened to compensate for the increased reactance produced by the coil turns. Marconi used series coils as shown in Fig 13.104(a). This was extended to the use of series capacitors, λ/4 lines and a fold-back or zig-zag technique by Franklin.

Practical collinears in radio amateur use tend to be variations of Fig 13.104(a) and (c). The radiating elements are not limited to λ/2 sections; λ/4, λ/2 and up to 5λ/8 radiators can be combined with various degrees of success. Several collinears omit the ground-plane and replaced it with variations of the λ/4 matching transformer. This leads to a reduction in gain as the antenna

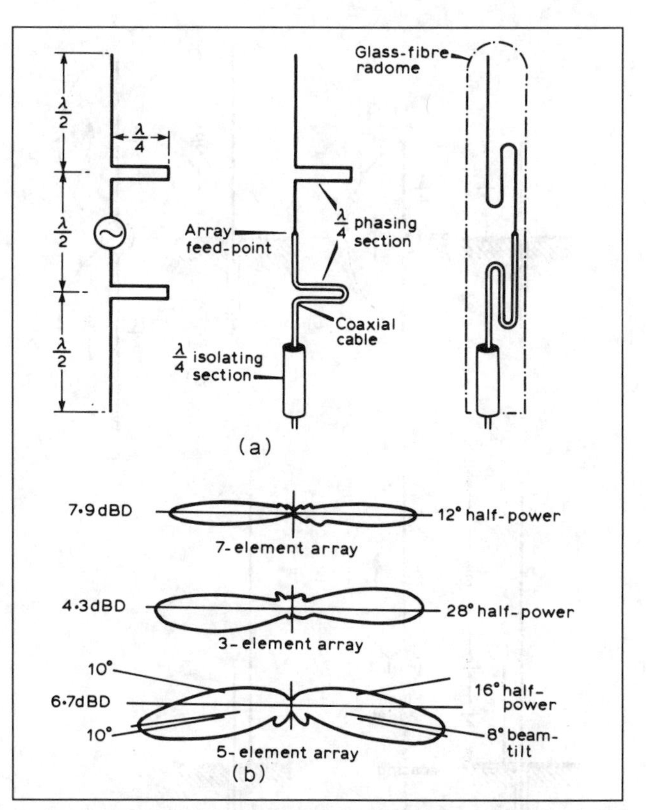

Fig 13.105. (a) Development of modified Franklin array. (b) Vertical radiation patterns. (From *Antenna Engineering Handbook* by Jasic. © 1961 McGraw-Hill. Used with the permission of McGraw-Hill Book Co)

no longer has an image reflected in the ground plane. However, when mounted under its normal working conditions an 'image' is reflected in the mast supporting the antenna and almost the maximum theoretical gain is obtained. Variation of the ground-plane structure will also modify the input impedance and radiation angle of an antenna.

Referring an earlier section on stub matching, a centre-fed 1λ or a λ/2 end-fed dipole can be matched by connecting to the 'open' end of a short-circuited λ/4 line. Similarly, a λ/2 dipole can be end fed with an open-circuit λ/4 line, where it is used as a series impedance transformer. This arrangement is shown in Fig 13.106(c) and (d). It is called the *J antenna* due to the shape, particularly when used without a ground plane.

Fig 13.106(c) and (d) has an unbalanced feed and can be connected directly to coaxial cable without a balun. The radiating λ/2 section should be set to resonance. This will require compensation for the length-to-diameter ratio of the radiating element. Fig 13.106(c) and (d) then require the λ/4 transformer adjusted for correct match, with any length of feeder cable.

Unfortunately, in both cases there is a reactive component which can upset matching. A very useful addition, developed by G2BCX for the *Slim Jim*, overcomes this problem [13]. Fig 13.107 shows the addition that consists of a folded open-circuit, λ/2 dipole, matched with a λ/4 transformer. When correctly set up, the Slim Jim produces low-angle radiation (about 10° above the horizon). The Slim Jim requires a 1:1 balun for correct operation but reasonable results have been obtained without one. An alternative is to use the feed arrangement of Fig 13.106(c) and (d) without the ground plane. A further extension to the J antenna is to use λ/4 sections, as suggested by Franklin, to produce a collinear. Fig 13.108(b) is one variation using a coaxial short-circuited, λ/4 transformer to give an unbalanced input.

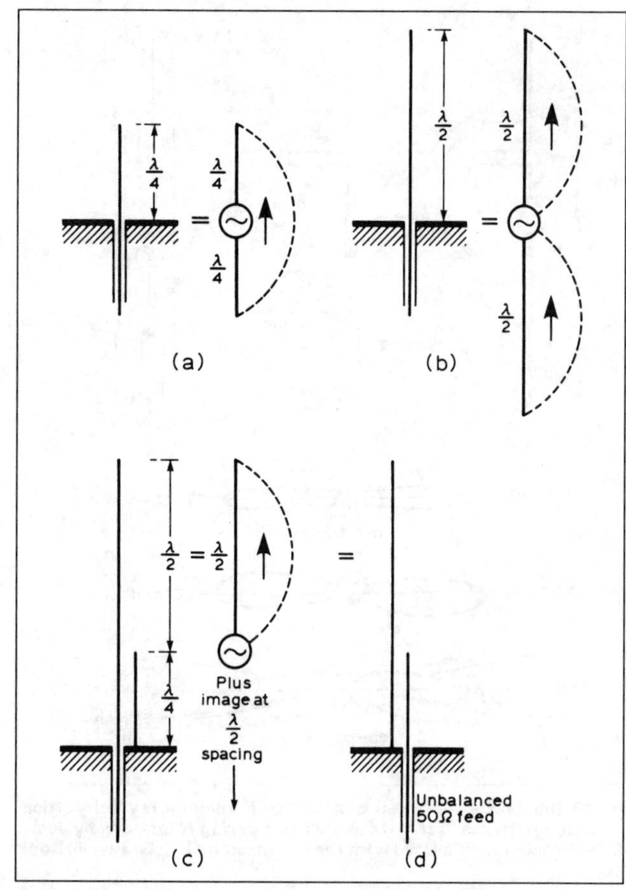

Fig 13.106. The derivation of the grounded J antenna. (From *Antenna Engineering Handbook* by Jasic. © 1961 McGraw-Hill. Used with the permission of McGraw-Hill Book Co)

This antenna and that of Fig 13.108(c) have both been used successfully to produce low-angle radiation for the GB3SN 144MHz repeater.

With both (b) and (c) the λ/4 phasing transformer has been 'wound up' as a bifilar coil (each coil being wound in the opposite hand). While the inductive component is cancelled, the mutual capacitance on the windings makes them physically shorter than λ/4.

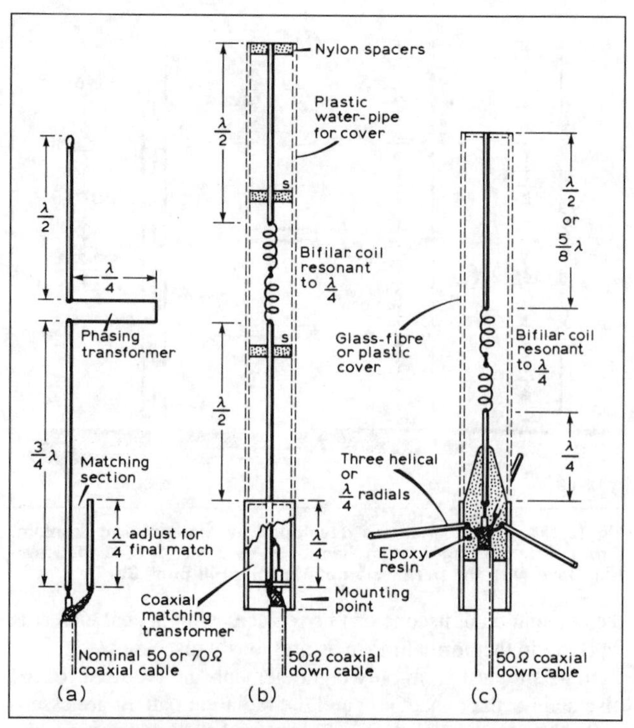

Fig 13.108. A collinear form of the J antenna. (a) The addition of λ/4 sections as suggested by Franklin. (b) Use of a coaxial short-circuit λ/4 transformer to give an unbalanced input. The tapping point in the matching transformer is approximately 0.15λ from the 'earthy' end. (c) A variant of (b) with radials. With both (b) and (c) the λ/4 phasing transformer has been 'wound up' as a bifilar coil (each coil being wound in the opposite hand). While the inductive component is cancelled, the mutual capacitance on the windings makes them physically shorter than λ/4

The series array in Fig 13.109 employs periodic annular slots in the outer conductor of a coaxial line to excite λ/2 dipoles formed by adding symmetrical 'skirts' about the slots. The dipole centres are spaced at 0.7λ. This keeps the sidelobes of the

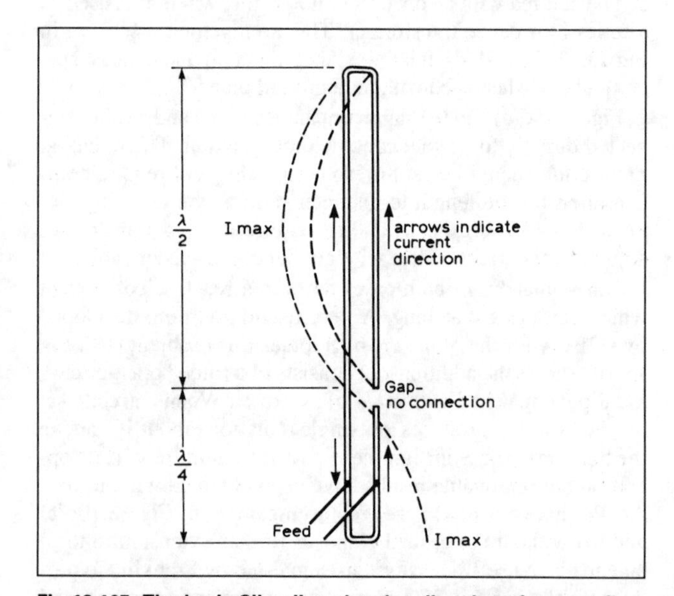

Fig 13.107. The basic Slim Jim, showing direction of current flow and phase reversal in matching stub (*Practical Wireless*)

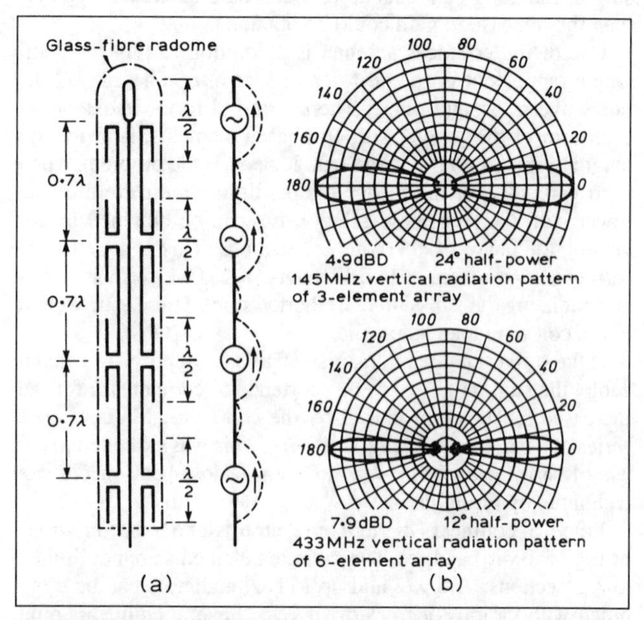

Fig 13.109. A series-fed collinear array of λ/2 dipoles. (From *Antenna Engineering Handbook* by Jasic. © 1961 McGraw-Hill. Used with the permission of McGraw-Hill Book Co)

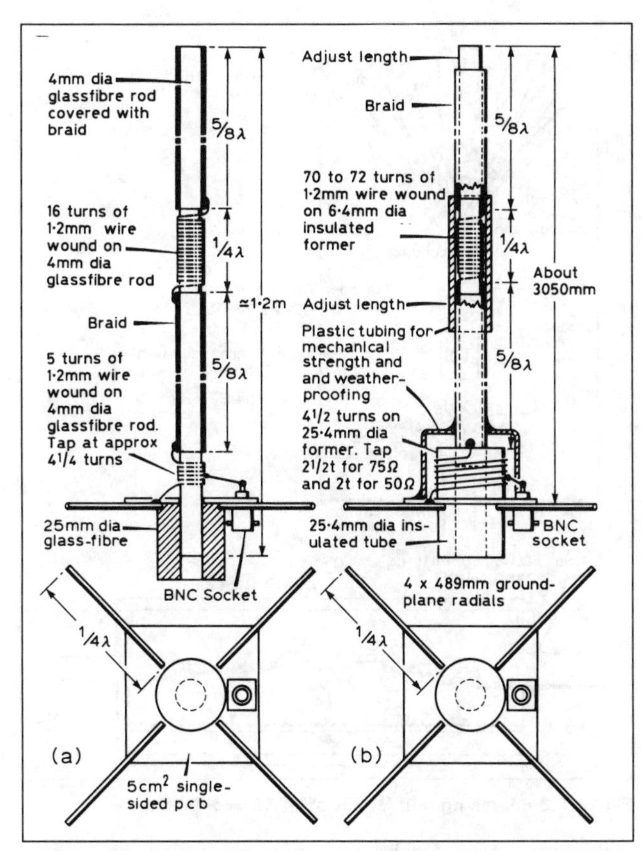

Fig 13.110. (a) A 432MHz collinear. (b) A 144MHz collinear (*UK Southern FM Journal*)

individual sections 15 to 20dB down on the main radiation. Dielectric loading of the feed line produces the full wavelength required for correct phasing between elements. Typical voltage radiation patterns are shown in Fig 13.109(b).

A variation of the techniques described, using coils as with the original Marconi concept, is shown in Fig 13.110(a) for 432MHz and Fig 13.110(b) for 144MHz. The expected gain is between 6 and 7dBD.

Materials required for Fig 13.110(a) are as follows:

- One 2.5cm diameter 10cm long glassfibre tube.
- One 4.0mm diameter 1.2m long glassfibre rod.
- Four 2.0mm diameter 20cm long glassfibre rods.
- Length of braiding from 'junk' large coaxial or multicore cable.
- Length of 1.2mm wire for matching coils.
- Approximately 5cm square of singled-sided PCB.

First, adjust the bottom 5λ/8 element to give minimum VSWR by moving the tapping point on the bottom coil (approx four turns). A fine adjustment can be made by altering the length of the first 5λ/8 element.

Next fit the centre matching coil and the top element. To obtain the best results, both elements should be the same length and approximately 5λ/8. Further improvement in VSWR is obtained by adjusting the centre matching coil (the coil is spread over λ/4).

The matching coil provides the phase change necessary to feed the top element and so adjustment is quite critical. If the matching coil has to be 'squeezed up' to obtain a good VSWR, then the coil has too many turns. The opposite is true if the coil has to be greater than λ/2 for a good VSWR.

To prevent the collinear going 'off tune' once set up, the elements are secured to the centre glassfibre rod and the matching coil taped with self-amalgamating tape. Provided care is taken in setting up, a VSWR close to 1.1:1 can be obtained.

Materials required for Fig 13.110(b) are as follows:

- Two 12.7mm diameter by 1206 ±12mm, 5λ/8 elements (adjustable).
- Four 495mm rods for the ground plane.
- One 6.4mm diameter by 762mm insulated rod.
- One 25mm diameter insulated tube (a cotton reel can be used).
- 1.6mm wire for matching and phasing coils.

The diagram shows extra insulated tubing over the matching and phasing coils to give more mechanical strength and weatherproofing.

Setting up is carried out as follows. First, adjust the length of the bottom 5λ/8 element to give minimum VSWR.

Secondly, fit the phasing-coil and the top element. The top element must be the same length as the set-up, bottom element. Next obtain the best VSWR by 'adjusting' the turns of the phasing-coil.

The coil provides the phase change necessary to 'feed' the top element. It consists of a length of 1.6mm wire, (about 1λ), coiled up to give 70–72 turns on a 6.4mm diameter former. The λ/4 spacing between the two elements is more critical than the number of turns. 68 turns gave a satisfactory VSWR with the prototype.

Some difficulty may occur in setting up the phasing coil. If more than seven turns have to be removed, go back to the first adjustment stage to ensure the bottom 5λ/8 element is correctly matched. If the bottom element is not correctly set up the collinear will not tune up. Careful adjustment should produce a VSWR of 1.1:1 at the chosen operating frequency.

A technique, widely used for commercial systems, combines λ/2 dipoles in phase from a single source. Alternatively, combination with an appropriate 'phase length' for the feeder cables between the dipoles. There is a disadvantage with this form of antenna array that some interaction occurs between cables and radiating elements. However, the disadvantage is balanced with the ability to modify the radiation pattern shape by adjustment of dipole spacing or cable phase length. The example given in Figs 13.111, 13.112 and 13.113 is probably the simplest to set up and was devised for the GB3ER 432MHz band repeater.

SATELLITE ANTENNAS

For the radio amateur, satellite antennas fall into two groups. Both groups consist of *ground-station antennas*, that is, those on the ground rather than on the satellite itself. The first group comprises *steerable antennas*, which enable the passage of the satellite to be tracked across the sky. The second group consists of *fixed antennas*, which have a hemispherical radiation pattern to receive the satellite signals equally in any direction. These antennas do not therefore require to track the satellite's passage. The tracking antennas are usually of high gain, while the fixed antennas are usually of low gain, due to their hemispherical coverage. Fortunately, signal losses between ground and the line-of-sight satellite are low. With no obstructions, low-gain antennas of the fixed variety are often acceptable for reception of amateur or weather satellites.

As many satellites rotate or change their relative position with respect to the ground station, both groups of antennas are normally circularly polarised. Right-hand circular polarisation is the convention for earth/space communication.

Of the higher-gain tracking antennas, the crossed Yagi and the helix antenna are the main ones used. The crossed Yagi is

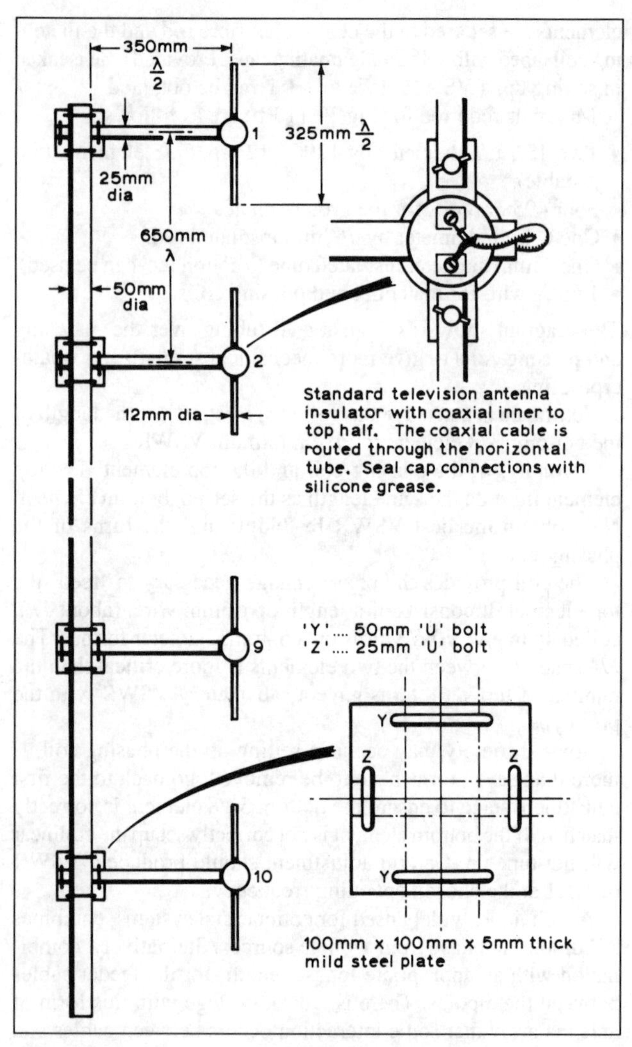

Fig 13.111. Mechanical details of GB3ER collinear

probably the easiest to construct and most readily available commercially. The sections on pp13.26 and 13.24 give details of the crossed Yagi and the helix.

For fixed or low-gain steerable antennas, several variations of crossed dipoles can be used. The *volute*, which is a fractional turn, four-element helix, can be used to give either directional

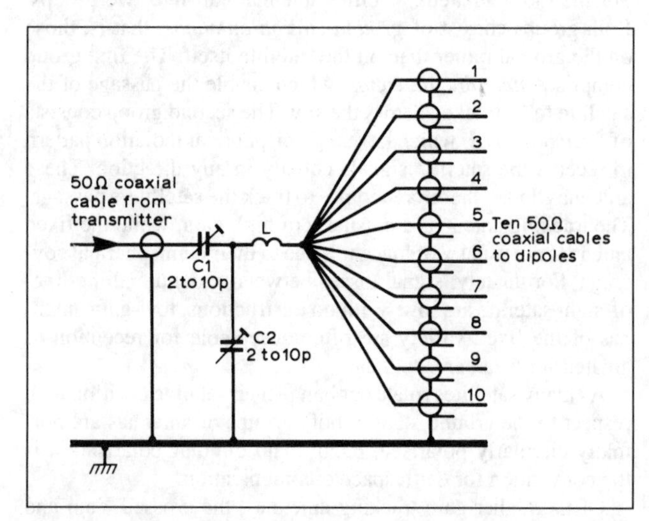

Fig 13.112. Matching unit of GB3ER collinear

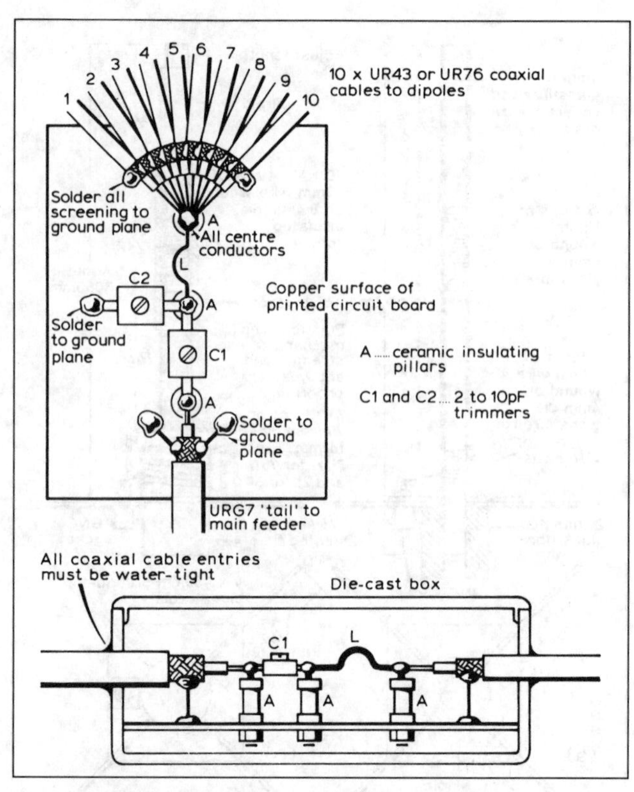

Fig 13.113. Matching unit layout of GB3ER collinear

gain or hemispherical, circular polarised, coverage. (A conventional single element helix requires two or more 'turns' to obtain circular polarisation.)

Fig 13.114 shows a simple arrangement of *crossed dipoles* above a ground plane for 145MHz. This type of antenna can be scaled for use at 29, 145 or 432MHz. Mechanical problems may make the reflectors inadvisable in a 29MHz version. The height above ground can be about 2m for 145MHz and 3m for 29MHz.

Typical dimensions are:

29MHz	driven elements (λ/2)	188in	4775mm
145MHz	driven elements (λ/2)	38in	965mm
	reflectors	40.5in	1030mm
	spacing (0.3λ)	24.5in	622mm

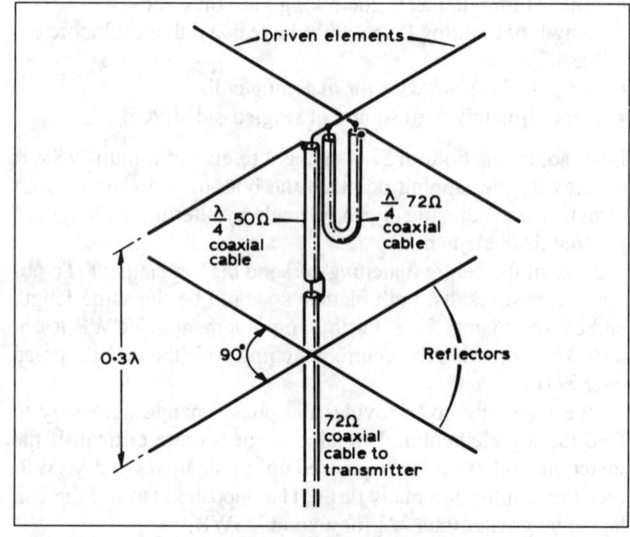

Fig 13.114. A crossed-dipole antenna for 145MHz

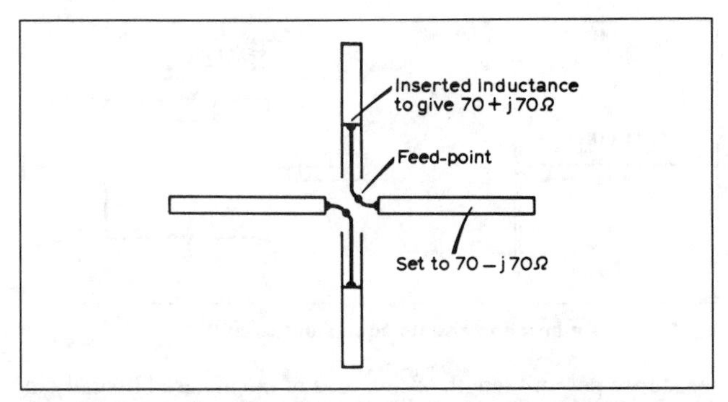

Fig 13.115. Achieving phase quadrature by introducing a reactance in one arm

The phasing line comprises λ/4 of 72Ω coaxial cable. The matching section is λ/4 of 50Ω cable. When calculating the length of the λ/4 sections, the velocity factor of the cable must be taken into account. Typically this is 0.8 for cellular and semi-airspaced cables, and 0.66 for solid dielectric cables, but verification of the figure for the particular cable used should be obtained. As an example, a matching section of RG59/U would be 13in (330mm) in length.

For a centre-fed crossed dipole, it is advisable to have a 1:1 balun to ensure a consistent pattern through 360° of azimuth. Dependent on the spacing between the dipoles and ground plane, the radiation pattern can be made to be predominantly to the side, for satellites low on the horizon, or up for overhead passes.

By drooping the dipole elements at 45°, with a spacing of approximately 0.4λ between the dipole mounting boss and the ground plane/reflectors, a hemispherical radiation pattern can be obtained. As ground reflections affect horizontal and vertical polarisation differently, low-to-horizon flight paths will not produce circular polarisation. This is due to 'ground scatter' of the satellite signal when it is low on the horizon, and ground reflections locally at the ground-based antenna.

Circular polarisation is normally produced by feeding one dipole 90° out of phase to the second dipole. This is achieved by having an extra λ/4 on one side of a combining 'harness'.

An alternate approach to this method of phasing is to use the phase properties of a capacitive or inductive reactance.

Suppose, for example, that the length and diameter of the dipoles are set to give a terminal impedance of 70 – j70Ω (capacitive). If a second, crossed dipole is set to be 70 + j70Ω (inductive) when combined (Fig 13.115) the terminal impedance of the arrangement becomes 70Ω ± j0, ie 70Ω resistive. As the two dipoles are connected in 'parallel', the current in each dipole is equal in magnitude. However, due to the opposite phase differences of 45° produced by the capacitive and inductive reactance, phase quadrature (a 90° phase difference) is produced which gives circular polarisation.

The addition of a 1:1 balun provides a direct match to a 70Ω coaxial line. If the impedance of the balun is correctly proportioned a match can be made with a 50Ω coaxial cable. The dipole elements can be 'drooped', as previously described, to improve the hemispherical coverage. One way of providing the opposing reactances is simply to make one dipole long (inductive at the working frequency), and one dipole short (capacitive at the working frequency).

Fig 13.116 shows a working example of the 'starpole' turnstile arrangement. The value of the reactive component was chosen as ±25Ω. Dimensions were based on the reactive of a dipole shown earlier in Fig 13.8.

The volute can also make use of a phasing line or the

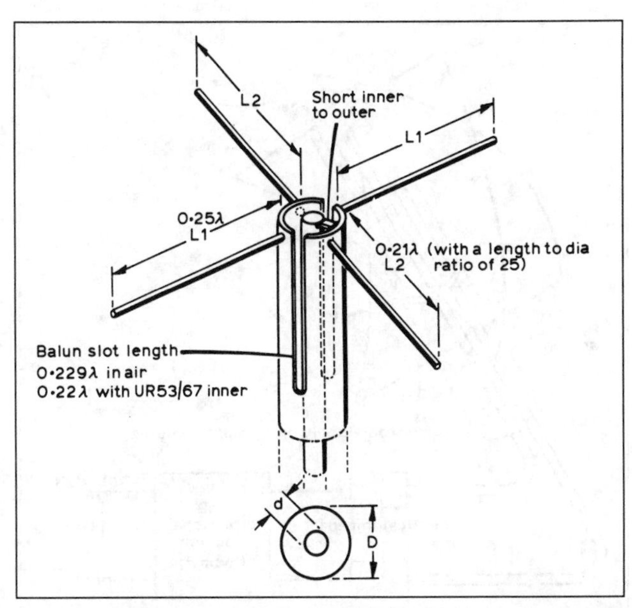

Fig 13.116. A starpole turnstile. D/d = 1.86 for 75Ω and 1.5 for 50Ω

reactance method to produce circular polarisation. The number of 'turns' or part turns of the radiating elements combined with their length can be used to produce various radiation patterns. Radiation patterns produced for several combinations of turns and resonant lengths are shown in Figs 13.119(a) to (d). General details of the volute are shown in Figs 13.117 and 13.118 taken from reference [14]. Elements that are multiples of λ/4 have open-circuit ends, while the elements that are multiples of λ/2 can be short-circuited to the mounting structure.

MATERIALS FOR ANTENNA CONSTRUCTION

The radiating efficiency of an antenna may be defined as the ratio of the power radiated to the power input to the antenna. The difference between the radiated power P_r and the input power P_{in} is the power lost in dissipation by the antenna itself, P_d.

An antenna can be represented by the equivalent circuit shown in Fig 13.120. Power in such a circuit can only be 'lost' in a resistive element. Both the power dissipating elements are therefore shown as R_r and R_d. R_r is proportional to the power radiated

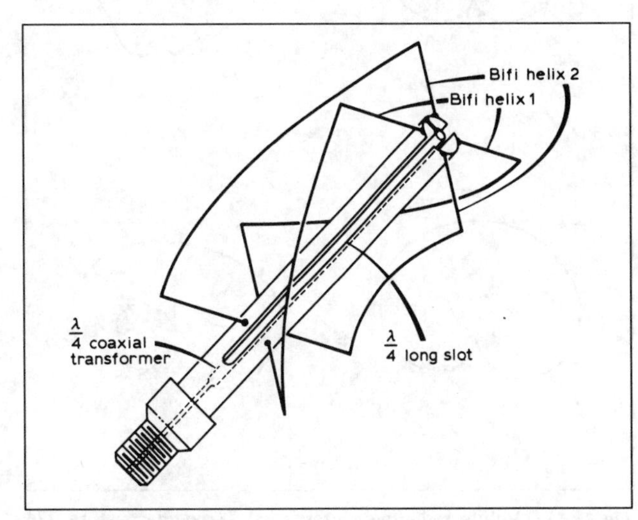

Fig 13.117. A quarter-turn volute with split sheath or slot balun (*Microwave Journal*)

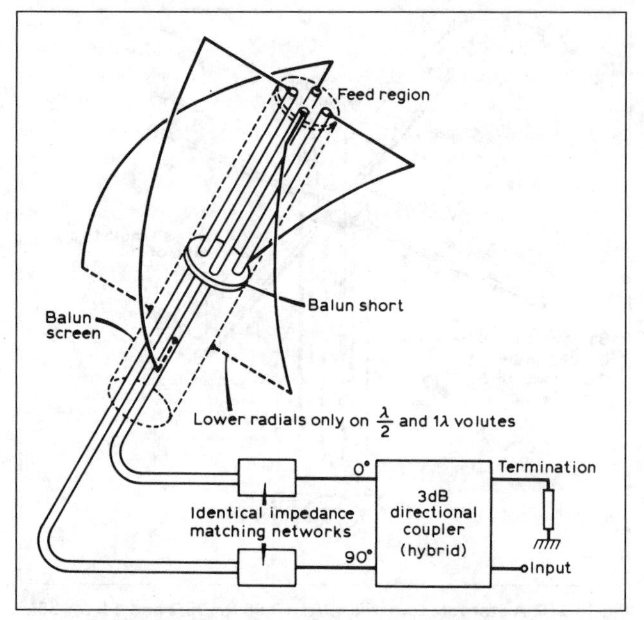

Fig 13.118. The general arrangement using Pawsey stub baluns. A half-hybrid or λ/4 phasing harness as used for the crossed dipoles can be used in place of the directional coupler (*Microwave Journal*)

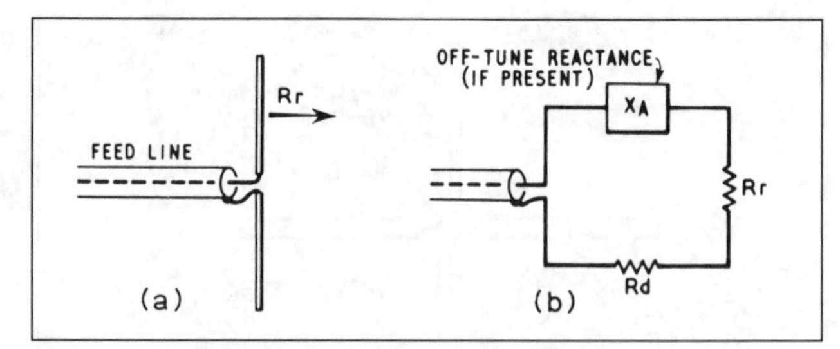

Fig 13.120. An antenna and its equivalent circuit

(effective radiation resistance). R_d is proportional to the power dissipated in the antenna itself (effective loss resistance). When considering the relative efficiencies of various materials for antenna elements, it is the value of the effective loss resistance R_d that is important. Since most VHF and UHF antennas are made up elements of uniform cross-sectional area, the DC resistance may be calculated from:

$$R_{DC} = \frac{L}{A} \times \text{resistivity}$$

where L is the length, A is the area, and 'resistivity' is the resistance per unit length or unit area of the material (usually in ohm-metres). For example, taking L as λ/4 at 144MHz, and A as the area of a 1/8in (3mm) diameter rod, then typical values of R_{DC} at 20°C are:

Metal	Resistivity (Ω-m × 10^{-8})	Resistance (R_{DC}) (Ω)
Copper	1.72	0.0023
Aluminium	2.83	0.0037
Zinc	5.90	0.0077
Brass	7.50	0.0098

All these are negligible with respect to the radiation resistance R_r of a λ/2 dipole, which for the chosen element diameter is about 65Ω. This, however, is only part of the effective loss resistance. Note that ferrous metals such as steel have much greater losses unless heavily copper plated.

As the frequency of the current flowing in the material is increased from zero frequency (or DC) another factor, *skin effect*, modifies the current distribution in the cross-section of the conductor. Current is concentrated more in the outer 'skin' as frequency increases.

This reduces the 'working area' of the conductor and increases its effective resistance. At VHF the skin carrying most of the current becomes quite thin, about 0.0005in, and is proportional to the square root of the conductor resistivity. This shows that the skin will be thicker for metals with higher resistivity and RF resistance will be less (compared with the DC resistance), than for good conductors. RF resistance, R'_{RF}, for a current constant along the length of a λ/2 element is approximately as follows at 145MHz:

Metal	Skin depth (in)	Skin depth (mm)	RF resistance (R'_{RF}) (Ω)	Ratio R'_{RF} to R_{DC}
Copper	0.00022	0.0056	0.34	148
Aluminium	0.00028	0.0071	0.41	110
Zinc	0.00040	0.0102	0.60	78
Brass	0.00045	0.0114	0.68	69

The radiation efficiency and antenna loss in decibels for the metals considered are as shown below:

Metal	R_L (Ω)	Radiation efficiency (%)	Loss due to antenna elements (dB)
Copper	0.17	99.75	0.011
Aluminium	0.20	99.70	0.013
Zinc	0.30	99.50	0.022
Brass	0.34	99.40	0.026

These figures are for elements of 1/8in diameter; for elements of the more popular diameters ¼in to ½in the RF resistance will be lower, although for very thin-walled tubes the DC resistance may be greater.

The following simple formulae can be used to calculate the skin depth of R'_{RF} resistance.

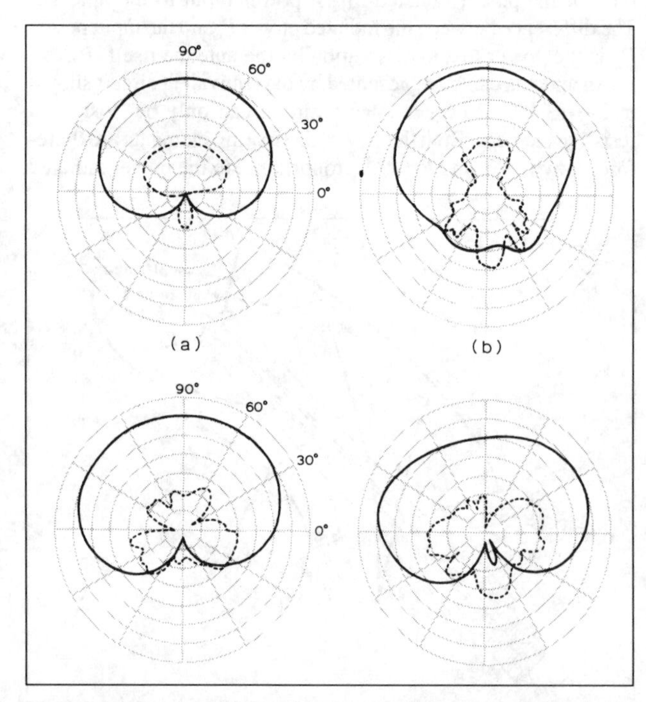

Fig 13.119. Volute radiation patterns. (a) Three-quarter-turn λ/4 volute. (b) Three-quarter-turn λ/2 volute. (c) Three-quarter-turn 3λ/4 volute. (d) Three-quarter-turn 1λ volute (*Microwave Journal*)

Table 13.15. Electrochemical potentials of metals

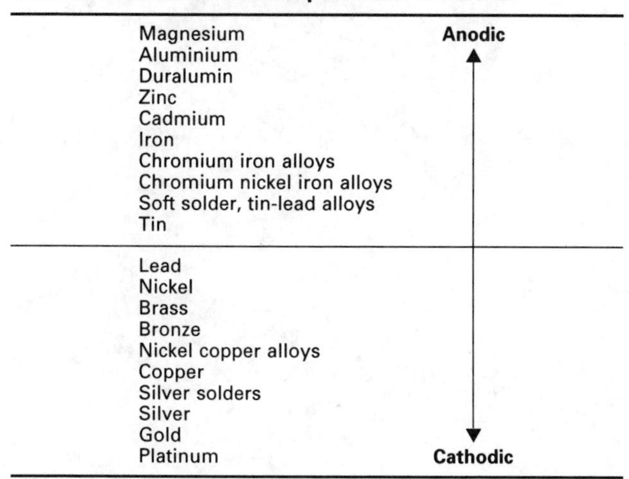

Metal	
Magnesium	Anodic ↑
Aluminium	
Duralumin	
Zinc	
Cadmium	
Iron	
Chromium iron alloys	
Chromium nickel iron alloys	
Soft solder, tin-lead alloys	
Tin	
Lead	
Nickel	
Brass	
Bronze	
Nickel copper alloys	
Copper	
Silver solders	
Silver	
Gold	
Platinum	↓ Cathodic

For solid rod elements:

$$R'_{RF} = R_{DC} \times \frac{\text{Diameter}}{4 \times \text{skin depth}}$$

Common dimensions must be used, ie millimetres or inches, but not mixed. The *ratio* is the important factor. Resistance is in ohms (Ω).

For thin-walled tubes:

$$R'_{RF} = R_{DC} \times \frac{\text{Wall thickness}}{\text{skin depth}}$$

$$\text{Skin depth at VHF (m)} = 2\sqrt{\frac{\sigma}{\mu f}}$$

where σ is resistivity in ohm-metres, μ is permeability (taken as 1 for non-ferrous metals) and f is frequency in hertz.

Electrolytic corrosion

The use of dissimilar metals in contact is likely to cause corrosion in one of them. Table 13.15 lists metals and alloys in order of their electrochemical potential. The higher up the list the metal is, the more likely it is to corrode, except that some, such as aluminium, form an adherent film of oxide which protects the metal from further corrosion. This will be disrupted by the presence of, for example, chorides from pollution or sea spray, and corrosion will continue. If a metal high up the list is in contact with one lower down, the former will be corroded and will protect the latter. This is the basis of *galvanising* steel (coating it with zinc) to protect it.

As mentioned, aluminium alloys are particularly susceptible to corrosion, especially if in contact with copper. Stainless steel in contact with aluminium does not cause much corrosion and should be used (as a washer) between aluminium-alloy elements and copper-wire feeders. Joints should also be protected from moisture and air by means of a coating of non-acidic grease or silicone rubber.

REFERENCES

[1] 'Notes on the development of Yagi arrays', C Greenblum, QST and other ARRL publications, edited by Ed Tilton, W1HDQ.
[2] 'Yagi-Uda arrays', Chen and Cheng, Proc IEEE, 1975.
[3] 'Yagi antenna design', P P Viezbecke, *NBS Technical Note 688*, December 1976.
[4] 'How to design Yagis', W1JR, *Ham Radio* August 1977, pp22–30.
[5] 'Long Yagis', W2NLY and W6KQI, *QST* January 1956.
[6] 'Extremely long Yagi antennas', Günter Hoch, DL6WU, *VHF Communications* March 1982, pp131–138; *VHF Communications* April 1977, pp204–211.
[7] 'DL6WU Yagi Designs' in *The VHF/UHF DX Book*, ed Ian White, G3SEK, Dir Publishing, 1993.
[8] Software is available from the RSGB Microwave Committee on receipt of a disc plus return postage and packing.
[9] DL6DW, *VHF Communications* May 1971.
[10] W3DUQ, *Ham Radio* August 1970.
[11] 'The log-periodic Yagi', G3FDW, *Radio Communication* July 1994, pp13–17.
[12] 'The HB9CV antenna for VHF and UHF', H J Franke, DK1PN, *VHF Communications* February 1969.
[13] F Judd, G2BCX, *Practical Wireless* April 1978.
[14] 'Resonant quadrifilar helix design', CC Kilgus, *Microwave Journal* December 1970.

14 Power supplies

AMATEUR radio equipment normally derives its power from one of three sources:

(a) The public AC mains, which are 240V at 50Hz in the UK.
(b) Batteries, which are either primary or secondary.
(c) Petrol-driven or diesel-driven generators or alternators.

Occasionally wind generators or solar panels are used but these are rare and are always used in conjunction with a secondary battery. They will not be considered further.

For fixed stations the AC mains is readily available, is cheap and is almost always used. It can be converted by transformers, rectifiers and smoothing circuits to the wide range of direct voltages and currents necessary for amateur equipment use.

Batteries, primary and secondary, have always provided a convenient (although expensive, with primary batteries) source of power, especially for low-powered transistorised equipment.

Engine-driven generators can give an output of DC or AC. However, the most popular give an output of 240V at 50Hz, ie an exact match for the domestic mains.

Wind generators and solar panels are rarer and usually give a DC output suitable for charging a 12 or 24V secondary battery.

Safety

The operation of all power supplies (except perhaps low-voltage primary batteries) can be dangerous if proper precautions are not taken. The domestic mains can be *lethal*. High-voltage power supplies for valve equipment are also *lethal* and great care must be taken with them. There is more on safety in Chapter 21.

SUPPLIES FROM THE DOMESTIC MAINS

Note that in all the following, it is assumed that the mains is 240V at 50Hz.

A *rectifier* (or rectifiers) is needed to convert AC to DC. Rectifiers are nearly always silicon diodes (see Chapter 3) which are highly efficient in that they have a very low forward resistance and a very high reverse resistance. Germanium diodes have been used in the past for low-voltage applications but are now obsolete. The diode has a conventional symbol (Fig 14.1) in which the arrow points in the direction of the *conventional* current flow (*not* the electron flow!) and thus the arrow is the anode and the plate is the cathode.

Fig 14.2 shows three types of rectifier circuits which cover nearly all the applications in amateur radio, together with diagrams indicating the shape of the current wave supplied by the rectifier. In all cases this can be analysed into a DC component and an AC component. The latter is called the *ripple*. It is evident that the half-wave circuit (Fig 14.2(a)) has a worse ripple than the full-wave circuits and it has its lowest frequency component at 50Hz. Also, it will be seen that DC flows through the transformer which may cause magnetic saturation and overheating if it is not designed for this service.

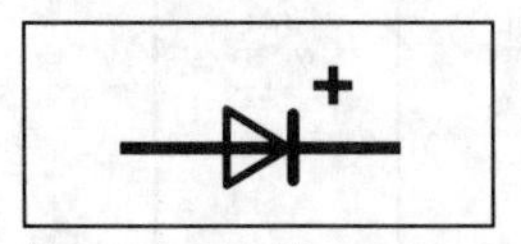

Fig 14.1. Rectifier symbol

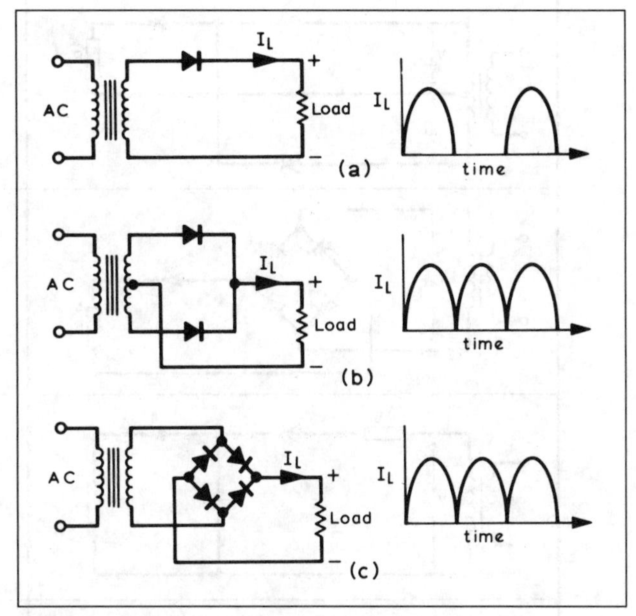

Fig 14.2. Rectifier circuits showing the output current waveforms with resistive loads: (a) half-wave, (b) full-wave or bi-phase half-wave, (c) bridge

The full-wave circuit needs a centre-tapped transformer while the bridge circuit does not. In the latter case, there are always two diodes in series.

In most cases the diodes (rectifiers) feed a large capacitor (the *smoothing capacitor*) which stores energy during the part of the cycle when the instantaneous voltage is low. In a few cases, the diodes feed an inductor (*choke*) followed by a capacitor.

A rectifier diode has three important parameters:

(a) Maximum mean forward current.
(b) Maximum peak forward current.
(c) The *peak inverse voltage* which is the voltage across the diode when it is *not* conducting. This is made up of the instantaneous voltage of the transformer, ie when at its negative peak, added to the voltage of the smoothing capacitor. The voltages and currents associated with these configurations are shown in Table 14.1 and the parameters for some common diodes are shown in Table 14.2.

There are many hundreds of types of rectifier diodes. Those shown in Table 14.2 are only a small selection of 'normal' types. Other categories are *fast recovery diodes* which are used in switch-mode power supplies (see later) and *soft recovery*

Table 14.1. Operating conditions of single-phase rectifier circuits

Circuit	DC output voltage	PIV across diode	Diode DC current	Diode peak current	Secondary RMS current
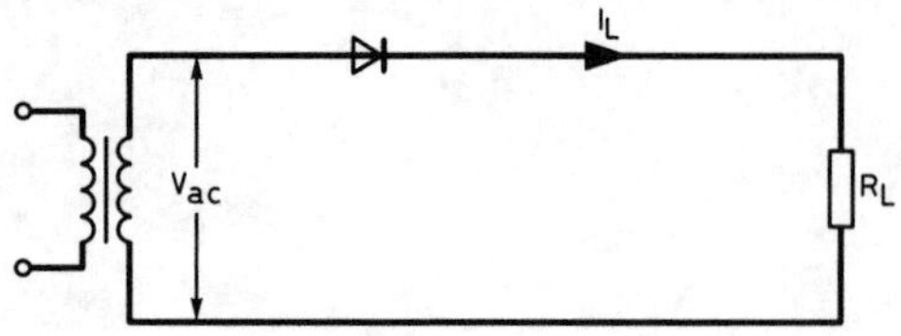	$0{\cdot}45V_{ac}$	$1{\cdot}4V_{ac}$	I_L	$3{\cdot}14I_L$	$1{\cdot}57I_L$
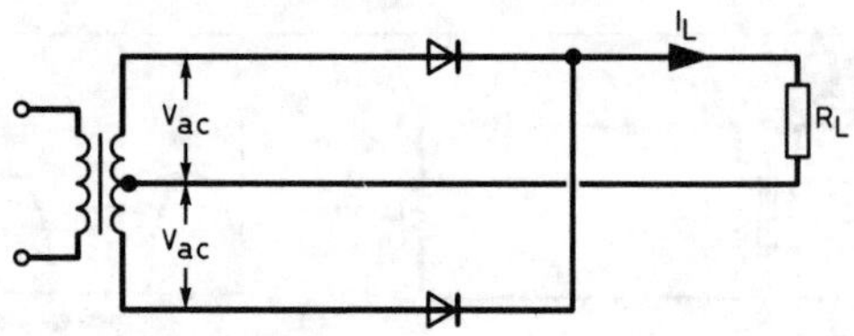	$0{\cdot}9V_{ac}$	$2{\cdot}8V_{ac}$	$0{\cdot}5I_L$	$1{\cdot}57I_L$	$0{\cdot}785I_L$
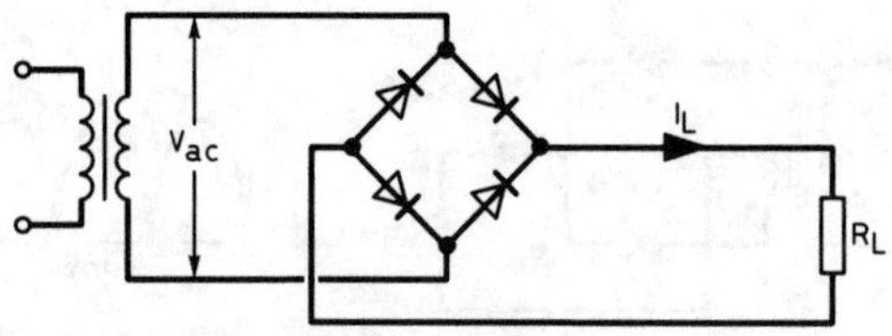	$0{\cdot}9V_{ac}$	$1{\cdot}4V_{ac}$	$0{\cdot}5I_L$	$1{\cdot}57I_L$	$1{\cdot}11I_L$
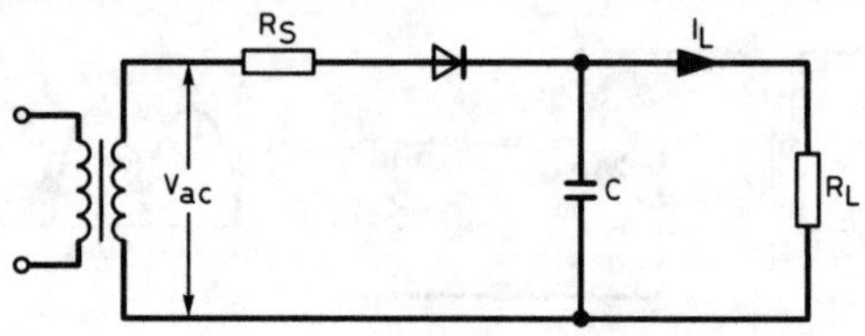	$1{\cdot}4V_{ac}$ (no load)	$2{\cdot}8\ V_{ac}$ maximum	I_L	See Fig 14.8	= Diode RMS current See Fig 14.7
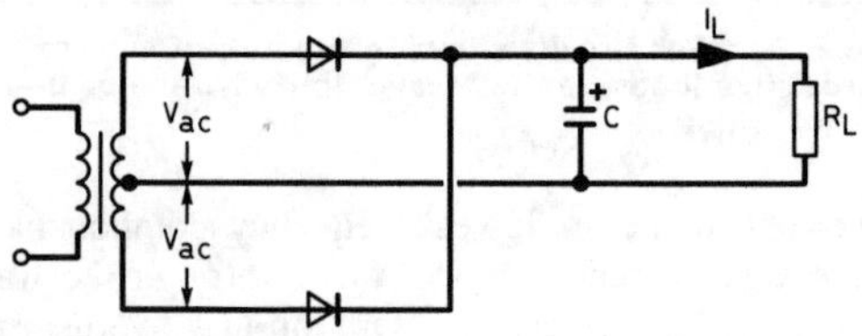	$1{\cdot}41V_{ac}$	$2{\cdot}82V_{ac}$	$0{\cdot}5I_L$	See Fig 14.6	I_L
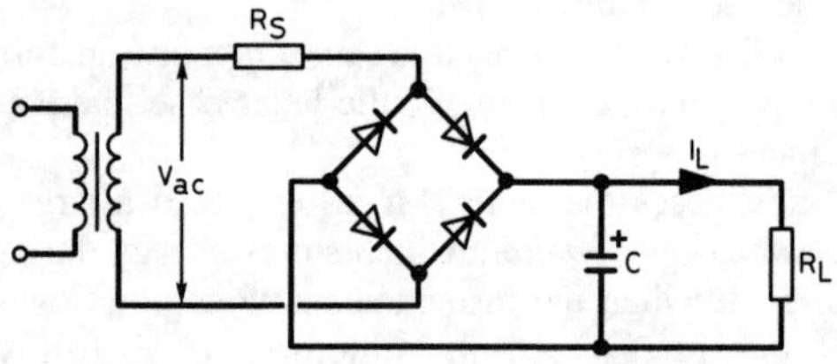	$1{\cdot}4V_{ac}$ (no load) See Fig 14.6	$1{\cdot}4V_{ac}$ maximum	$0{\cdot}5I_L$	See Fig 14.8	= Diode RMS current x 1·4 See Fig 14·7
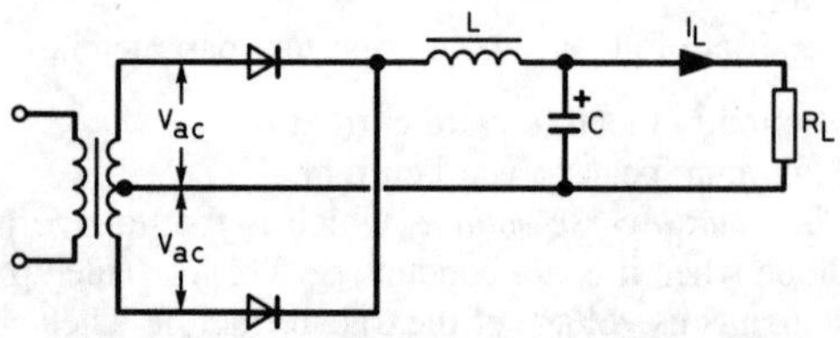	$0{\cdot}9V_{ac}$	$1{\cdot}41V_{ac}$	$0{\cdot}5I_L$	$2I_L$ when $L = L_C$	$0{\cdot}65I_L$
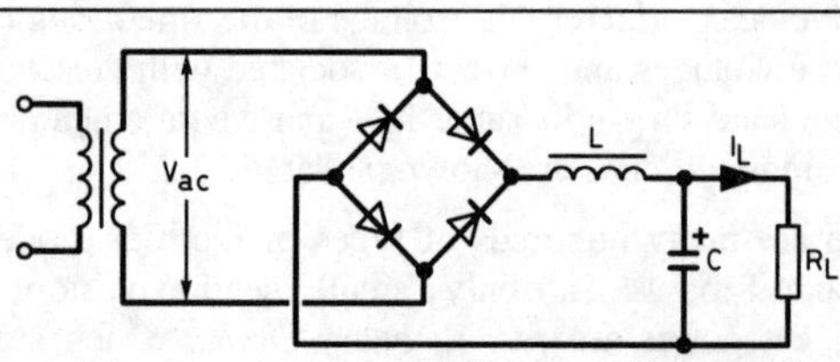	$0{\cdot}9V_{ac}$	$1{\cdot}4V_{ac}$	$0{\cdot}5I_L$	$2I_L$ when $L = L_C$	$1{\cdot}22I_L$ when $L = L_C$

CD219

Table 14.2. Electrical characteristics of some common diodes and bridge rectifiers

Type	V_{RRM}	I_{av}	I_{FRM}	I_{FSM}
Diodes				
1N4001*	50	1.0	10	20
1N4007*	1000	1.0	10	20
1N5401†	100	3.0	20	60
1N5408†	1000	3.0	20	60
BY98-300	300	10	50	100
BY98-1200	1200	10	50	100
BY96-300	300	30	100	200
BY96-1200	1200	30	100	200
Bridge rectifiers				
1KAB10E	100	1.2	25	50
1KAB100E	1000	1.2	25	50
MB151	100	15	150	300
MB156	600	15	150	300
GBPC3502	200	35	200	400
GBPC3506	600	35	200	400

Note: The diodes marked * and † are wire ended – the rest are mounted on screwed studs. V_{RRM} is the maximum reverse voltage or peak inverse voltage, I_{av} is the average output current in amps, I_{FRM} is the maximum repetitive peak current in amps, I_{FSM} is the maximum non-repetitive peak forward current with a maximum duration of 5ms.

diodes which, because they switch relatively slowly, cause less RF interference than ordinary types.

There are higher current rectifier diodes available with ratings of over 1000A but these are not really of use to the amateur! Furthermore, there are many packages of four diodes in bridge formation ready to mount on a heatsink. These are available at relatively low cost with ratings up to 35A and 500V. A selection of these is listed in Table 14.2.

Voltage multipliers

When a DC voltage greater than the peak of the alternating voltage is needed, a *voltage multiplier* can be used. These can give large multiplications but generally have poor *regulation*, ie the output voltage droops significantly when current is drawn. Their operation can be visualised roughly by thinking of the diodes as 'ratchets'. The mechanical ratchet passes motion in one direction only and in the multiplier circuit each stage 'jacks up' the following stages. Representative circuits are shown in Fig 14.3.

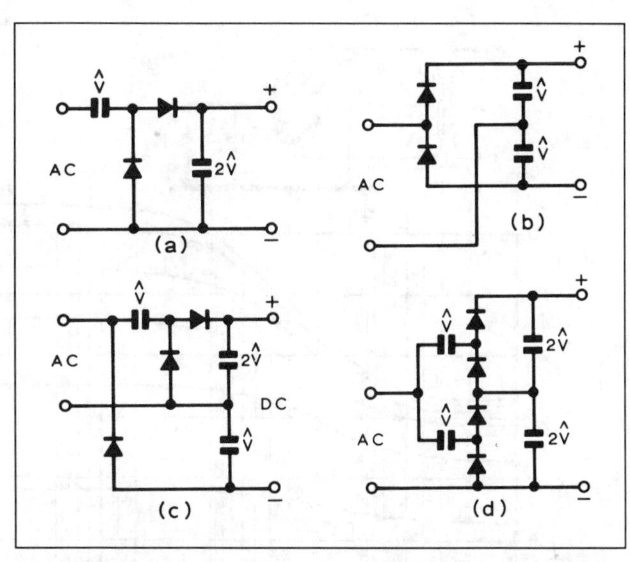

Fig 14.3. Voltage multiplier circuits. (a) Half-wave voltage doubler; (b) full-wave voltage doubler; (c) voltage tripler; (d) voltage quadrupler. V̂ = peak value of the AC input voltage. The working voltages of the capacitors should not be less than the values shown

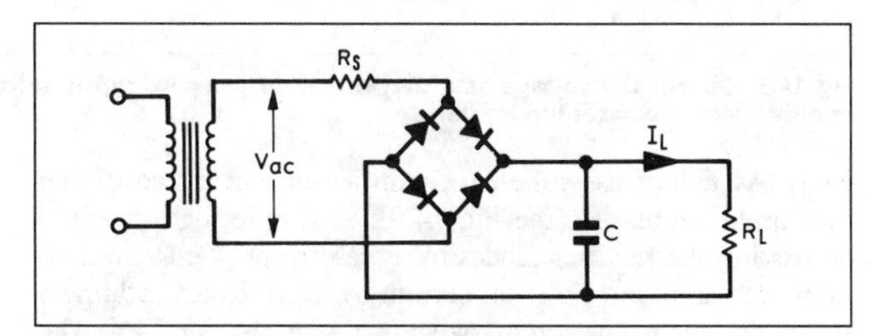

Fig 14.4. Bridge rectifier with capacitor input filter

Fig 14.6 shows this relationship graphically. Note that ω is equal to $2\pi f$ where f is the frequency.

The rectifier peak current is considerably higher than the average current. As the effective resistance of the transformer becomes a smaller and smaller fraction of the load resistance, the peak current of the rectifiers also increases (see Figs 14.7 and

Smoothing circuits

These are just low-pass filters which follow the rectifier diodes and the behaviour of the rectifier circuit depends on the input element of the filter. This can be a capacitor or an inductor and either will generally be followed by further components to give even greater reduction in the ripple.

Capacitor input

An example is the bridge rectifier (Fig 14.4) in which R_s is the effective resistance of the transformer (the resistance of the secondary plus N^2 times the resistance of the primary, where N is the turns ratio).

The capacitor (the smoothing capacitor) charges up during part of the cycle and discharges when the input voltage falls during the rest of the cycle (see Fig 14.5). This means that the diodes conduct during only part of the cycle. The ratio of output voltage to peak input voltage depends on the size of the smoothing capacitor, the load and the effective resistance in series with the rectifier.

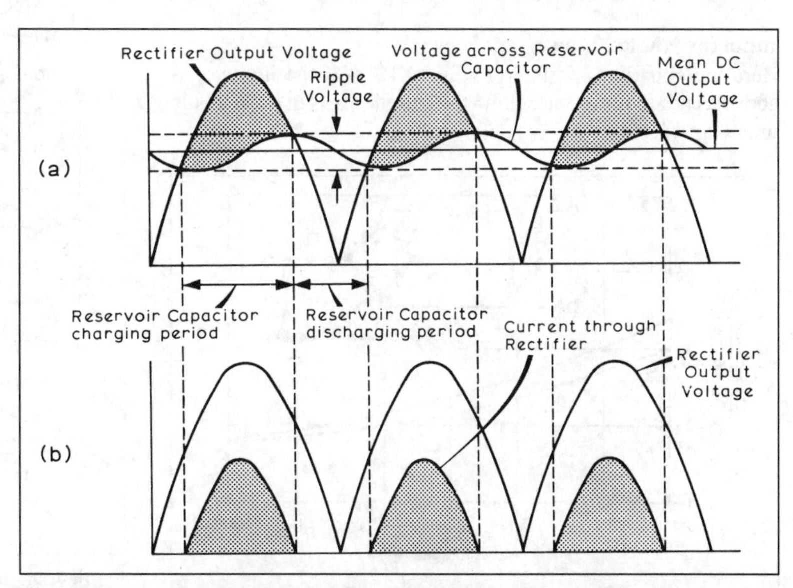

Fig 14.5. Curves illustrating the output voltage and current waveforms from a full-wave rectifier with capacitor input filter. The shaded portions in (a) represent periods during which the rectifier input voltage exceeds the voltage across the reservoir capacitor, causing charging current to flow into it from the rectifier

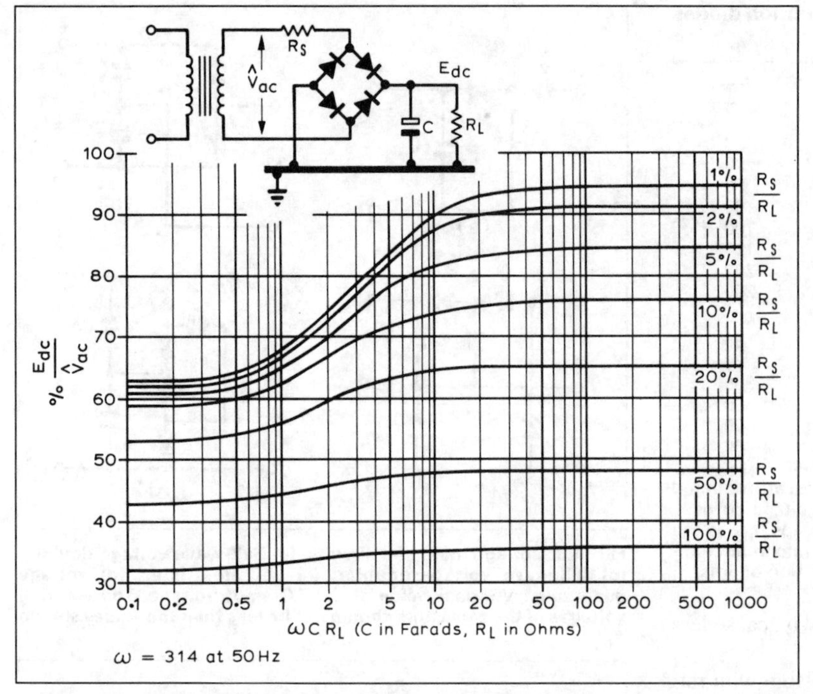

Fig 14.6. Output DC voltage as a percentage of peak AC input voltage for a bridge rectifier with capacitor input filter

14.8). As C increases, the peak rectifier current increases and the ripple decreases (see Fig 14.9). A simple idea to avoid destroying the rectifier diodes by excessive peak current is to limit the maximum 'in-rush' current by, if necessary, adding a series resistor to the circuit as shown as R_S in Fig 14.4. The minimum value of this is given by:

$$R_S = V/I_{FRM} \quad (1)$$

where V is the RMS voltage output of the transformer.

Then calculate the equivalent resistance of the transformer. If it is more than the value calculated above, no further added resistance is necessary. If it is less, add a series resistor to make up the difference, bearing in mind the power rating necessary.

Inductor (choke) input

Here the situation is different (Fig 14.10) – if the inductor is above a certain limit (see below) the current flows in it the whole time (Fig 14.11).

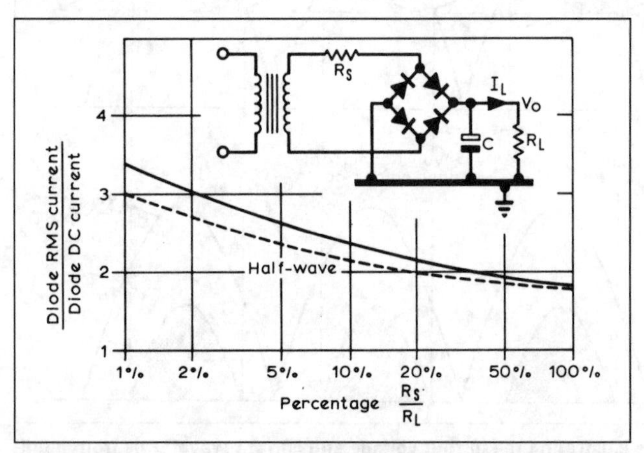

Fig 14.7. Relationship between diode RMS current and percentage R_S/R_L for values of ωCR_L greater than 10. (ω = 314 for 50Hz mains). The dotted line applies to half-wave rectifiers

The critical value for the inductor is given by:

$$L_c = \frac{R_s + R_L}{3\omega} \quad (2)$$

where L_c is in henrys and R_s and R_L are in ohms. If R_s is much less than R_L this reduces to:

$$L_c = \frac{R_L}{940}, \text{ assuming 50Hz supply} \quad (3)$$

It will be seen that the inductance required increases as the load current decreases (the load resistance increases) so it is necessary to provide a minimum current by means of a *bleeder* if the inductor input is to remain effective. The value of this bleeder is 940 times the maximum inductance value (in henrys).

Choice of components

The ideas given here are for power supplies which will always work but may not be the most economical in components. The reason for this is that, in general, components are cheap but time is important and trouble-shooting can be difficult.

Components should always be chosen on a 'worst-case' basis. That is, assume that:

1. The mains voltage can fluctuate by ±10% from its nominal value. (This doesn't allow for drastic 'load shedding' during a very hard winter!)
2. Electrolytic smoothing capacitors can have a tolerance of +50% to −20%, ie one marked '100µF' can have any value between 150µF and 80µF.
3. Rectifier diodes should have a peak reverse voltage rating of at least 2.83 times the output voltage of the transformer (if one is used) *when the mains is 10% higher than normal.*
4. Choose a diode or diodes with an I_{av} at least twice the required value.

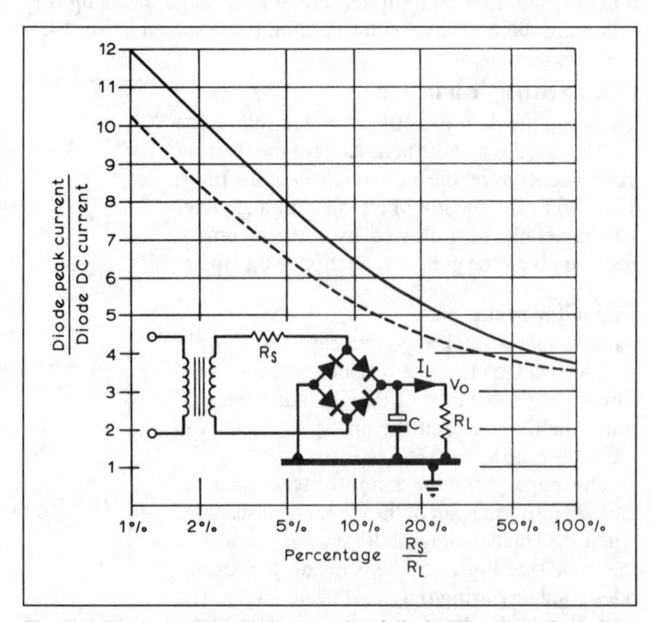

Fig 14.8. Diode peak current as a ratio of diode DC current for values of ωCR_L greater than 10. Note: in a bridge rectifier circuit, diode DC current is half the load current. The dotted line applies to half-wave rectifiers; in this case the diode DC current is equal to the load current

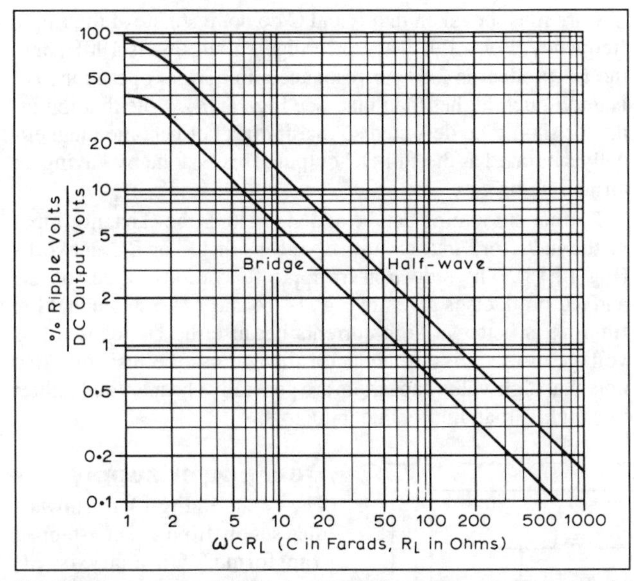

Fig 14.9. Percentage ripple voltage (RMS) against values of ωCR_L. ($\omega = 2\pi f$ where *f* is the mains supply frequency)

Capacitor input

Consider the circuit in Fig 14.12 where the output is 13.8V at 5A. (This peculiar voltage is that of a 12V lead acid battery on charge and is often specified for mobile equipment.) This corresponds to a load resistance of 13.8/5 = 2.76Ω.

Consider the following:

(a) The secondary resistance of the transformer is 0.1Ω and the primary resistance is 5Ω. The turns ratio (see later) is 0.0729:1 so the effective resistance of the primary at the secondary is 0.027Ω and the total effective resistance of the transformer is 0.127Ω.

(b) The bridge rectifier is a type SKB25/02 which has an I_O of 10A and an I_{FRM} of 350A. The RMS transformer output voltage is 17.5; see below. The minimum value of R_S is 17.5/350

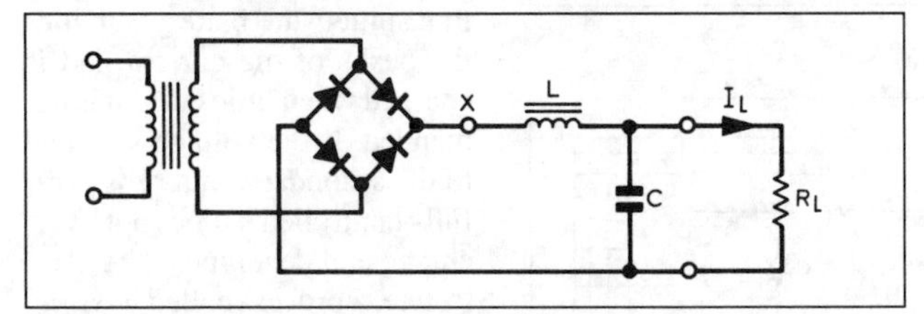

Fig 14.10. Bridge rectifier with choke input filter

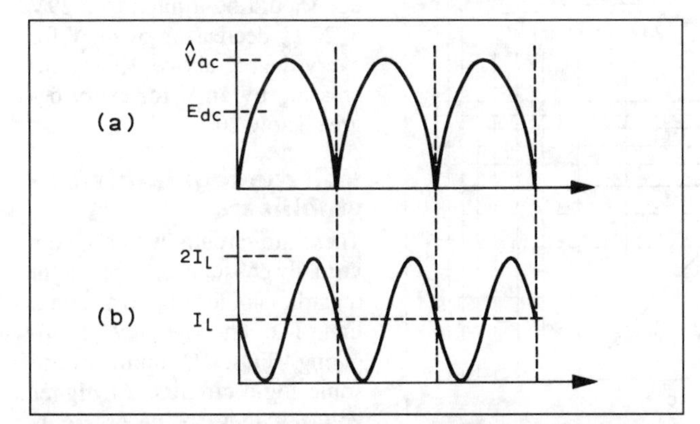

Fig 14.11. Waveforms at rectifier output (point X in Fig 14.10) in a choke input circuit. (a) Voltage waveform. (b) Current waveform ($L = L_C$)

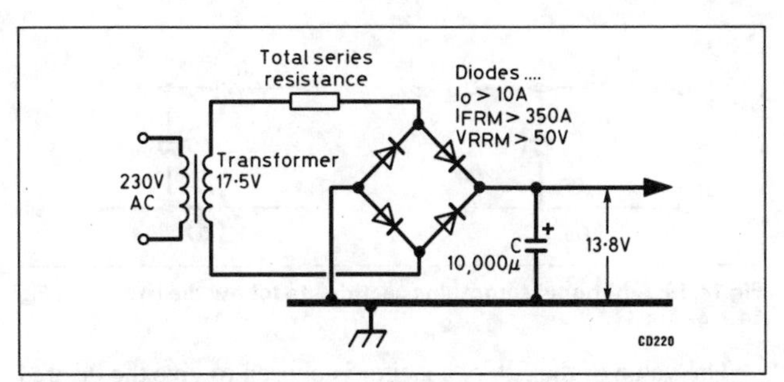

Fig 14.12. Circuit diagram for a power supply for 300V at 100mA

= 0.047Ω which is well below the effective transformer resistance so no added series resistor is necessary.

(c) $R_S/R_L = 0.047/2.76 = 0.017$ or 1.7%. The average diode current is 2.5A (two diodes for the 5A).

(d) Referring to Fig 14.8, the peak diode current is 12 times this, ie 30A, and the RMS current from Fig 14.7 is 3.2 times the mean, ie 8.0A.

(e) Assume $\omega CR_L = 100$ (this is an arbitrary choice based on the need for a low ripple voltage – see later) so $E_{DC}/E_{peak} = 0.95$ (Fig 14.6). Therefore secondary volts = 13.8/0.95 = 14.5V. However, this does not include the voltage drop of about 1.5V per diode, ie a total of 3V, so the secondary voltage required is 17.5.

(f) R_L is 2.76Ω so C is $2.76/2\pi f$ which equals 8800μF where *f*, the mains frequency, is 50Hz. However, electrolytics have a tolerance of –20 to +50% so this should be scaled up to 11,000μF. It is probable that a 10,000μF capacitor would be suitable.

(g) Referring to Fig 14.9, the ripple would be about 0.6%.

The design of all other power supplies, no matter what voltage or current, follows the same rules with two exceptions:

1. Where an inductor (choke) input filter is used (see below).
2. Where a 12 or 24V secondary battery (usually a car starter battery) is 'floated' across a DC supply and takes the place of the smoothing capacitor.

Inductor input

Here the inductor is connected directly to the rectifier diodes (Fig 14.10) and is followed by a smoothing capacitor. Note that it is equally valid to use a centre-tapped transformer and a full-wave rectifier with just two diodes.

As mentioned above in equations (2) and (3), the inductor must have a certain minimum value for a given load. This must be calculated for the smallest current expected, ie that of the bleeder if fitted.

The voltage and current values and waveforms for a circuit using at least this value of inductor are shown in Fig 14.11. It is clear that the peak rectifier current is only double the mean current. Under these conditions, the transformer RMS current is 1.22 times the load current and the average current per diode is half the load current.

The output voltage is given by:

$$E_0 = 0.9\,V_{AC} - I_L(R_C + R_T) - E_D \qquad (4)$$

where E_0 is the output voltage, I_L is the output (load) current, R_C is the resistance of the inductor, R_T is the effective resistance of the transformer, and E_D is the forward voltage drop of the rectifier diode(s) (one for the full-wave circuit, two for the bridge).

E_D can be neglected in high-voltage (say, greater than 100V) power supplies and can be estimated from makers' figures for the others.

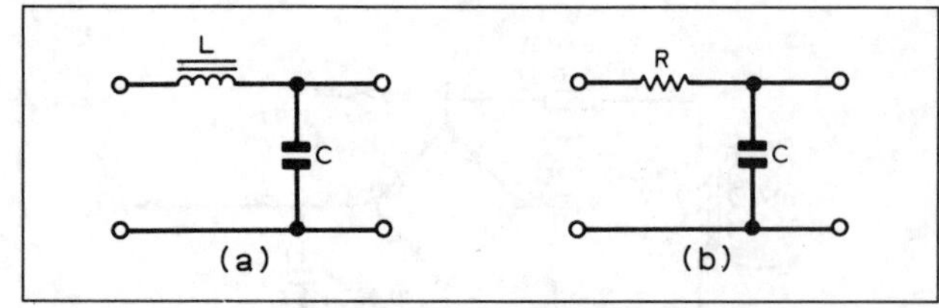

Fig 14.13. Additional smoothing sections to follow the circuits of Fig 14.4 or Fig 14.10

The value of the filter capacitor is chosen to give the desired ripple voltage E_R according to:

$$E_R = \frac{E_0}{0.8LC} \qquad (5)$$

where L is in henrys and C is in microfarads.

Care must be taken that L and C do not resonate at the ripple frequency. For a full-wave or bridge circuit this is 100Hz and the *LC* product is 2.53 for resonance. In normal operations *LC* is very much higher than that (see Fig 14.14). Note that the inductor should be designed so that it does not become magnetically saturated at the full DC output. This is done by having an air gap in the core.

Further smoothing can be added to either the capacitor input or the inductor input circuits by means of LC or RC networks (Fig 14.13). The reduction in ripple for various values of *LC* and *RC* products is given in Figs 14.14 and 14.15. Note that LC filtering is better at high currents because the DC causes less voltage drop across the inductor than across the resistor. Also note that high-value capacitors are generally cheaper (and lighter in weight!) than high-value inductors.

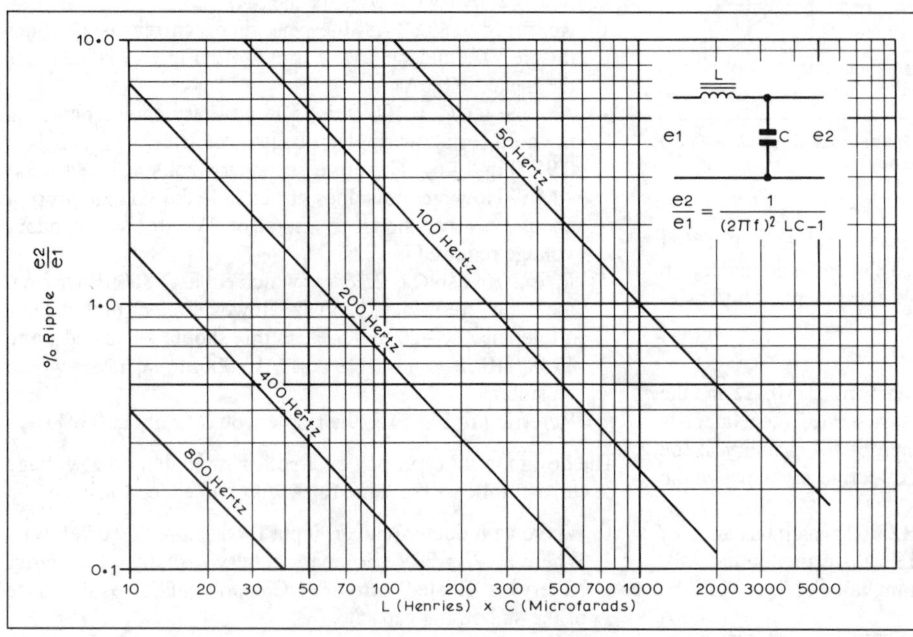

Fig 14.14. Relationship between percentage ripple and product of LC

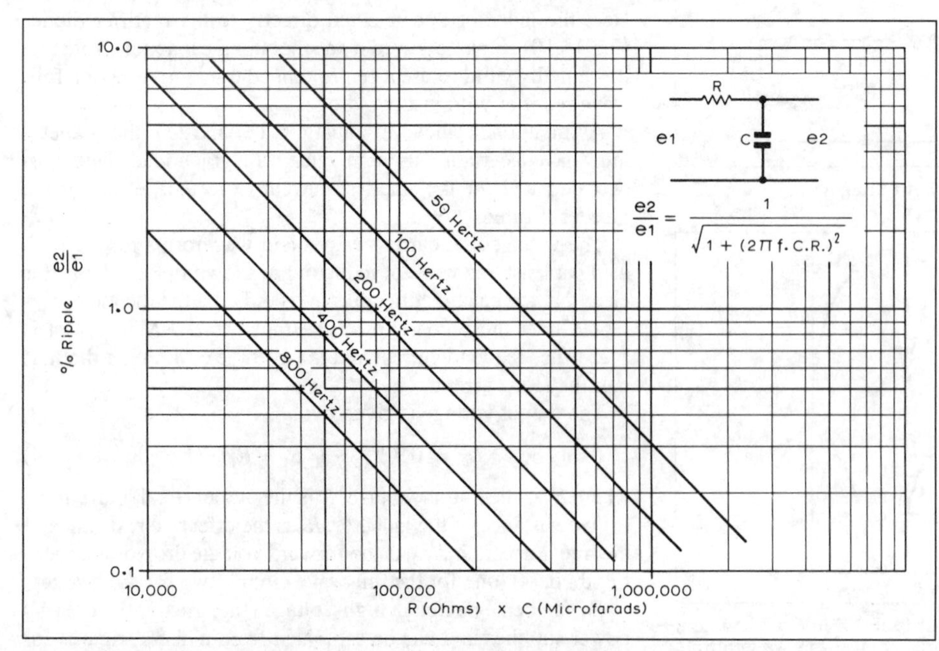

Fig 14.15. Relationship between percentage ripple and product of RC

Dual power supply

The circuit in Fig 14.16 shows a dual supply from a centre-tapped transformer. On analysis, although it appears to use a bridge rectifier, the two halves of the bridge feed the two supplies separately. This emphasises the advantage of the silicon diode, which does not need a heater supply, over a thermionic valve which does. The latter would need *three* separate heater supplies.

Float charging of secondary cells

In this the secondary battery (usually a lead-acid car starter battery) is connected permanently to the mains-driven DC power supply. In use, say for a transmitter, the battery supplies the peaks of the current and is charged when little current is demanded. If the voltage supplied to the secondary battery is carefully controlled, it does not overcharge and decompose the electrolyte, a process called *gassing*. To achieve this, the supply voltage should be limited to 2.29V at 20°C, decreasing by 4mV for every degree above 20 and increasing by 4mV for every degree below 20.

Voltage regulators or stabilisers

These are circuits which give a virtually constant voltage output regardless of load up to a certain limit. They are necessary for supplying VFOs, DC amplifiers and some logic circuits. At higher voltages, they are necessary as screen and/or grid bias supplies for valve power amplifiers.

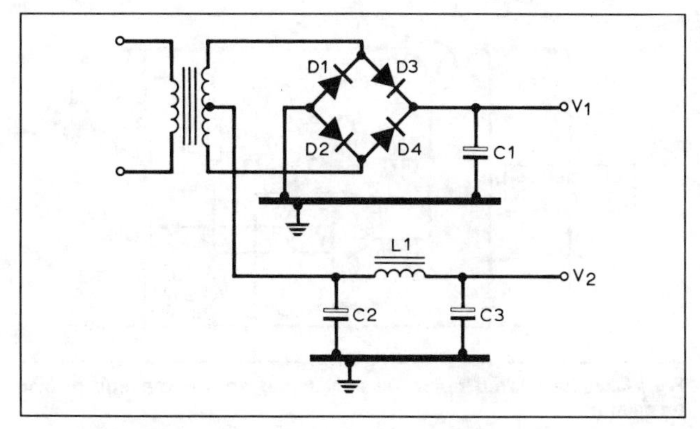

Fig 14.16. Circuit of a dual-voltage power supply

Zener diodes

These have a sharp breakdown voltage (see Chapter 3) and, if supplied from an unregulated source by a resistor (Fig 14.17), can form a simple voltage regulator which is adequate for small, more-or-less constant, loads.

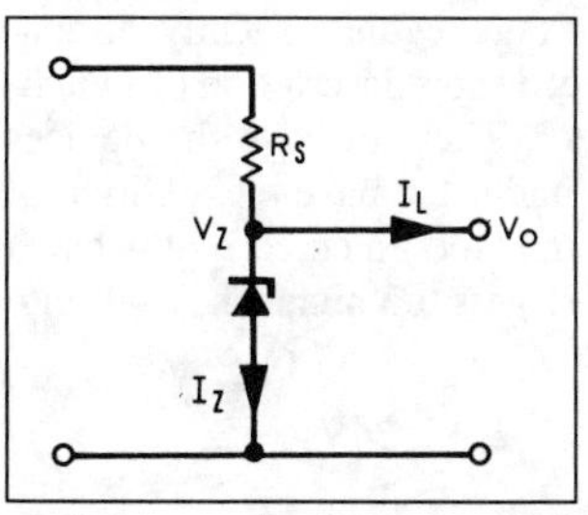

Fig 14.17. A simple zener diode regulator

When conducting, the zener diode exhibits a very small resistance so the voltage will rise slightly as the current through the diode rises; hence the need for a 'more-or-less constant' load. This resistance is called the *zener slope resistance*. Zener diodes also dissipate power – this is the zener voltage multiplied by the diode current. The maximum dissipation depends on the size of the active element, the packaging and the capacity of the heatsink, if any, to which it is attached. Finally, the zener voltage varies somewhat with temperature. In general, low-voltage zener diodes have a negative temperature coefficient (TC) and the higher voltage ones a positive one. At around 5–6V, the TC is near zero.

Table 14.3 gives some figures for typical zener diodes manufactured by Philips.

The series resistor (R_S) is calculated in such a way that the zener diode provides regulation when its input voltage (V_S) is at its minimum and the load current (I_L) is maximum. It is important to check that the zener diode's maximum dissipation is not exceeded when these are reversed, ie when V_S is at maximum and I_L is at its minimum. If there is a chance that the load can be disconnected with the power on, this must also be considered. The expression for the series resistor is:

$$R_S = \frac{V_{S(min)} - V_Z}{I_{L(max)} + I_{Z(min)}}$$

The dissipation of the resistor is given by :

$$W_R = \frac{(V_{S(max)} - V_Z)^2}{R_S}$$

Table 14.3. Electrical and thermal characteristics of some zener diodes

Type	Zener voltage (V)	Normal operating current (A)	Zener slope resistance (Ω)	Maximum dissipation (W)	Temp coeff (mV/°C)
BZX79C2V4	2.4	0.005	100	0.4	−3.5
BZX79C6V2	6.2	0.005	10	0.4	+2.0
BZX79C75	75	0.002	255	0.4	+80
BZT03C7V5	7.5	0.1	2	3.25	+2.2
BZT03C270	270	0.002	1000	3.25	+300
BZY91C10	10	5.0	NA	75*	NA
BZY91C75	75	0.5	NA	75*	NA

* On a heatsink.
'NA' means 'not available'.
While these are all Philips devices, all semiconductor manufacturers make zener diodes in one or more sizes. The data on Philips devices were used for this table because they were to hand and not because they are recommended above other makes. This table represents the extremes of size, power dissipation and voltage.

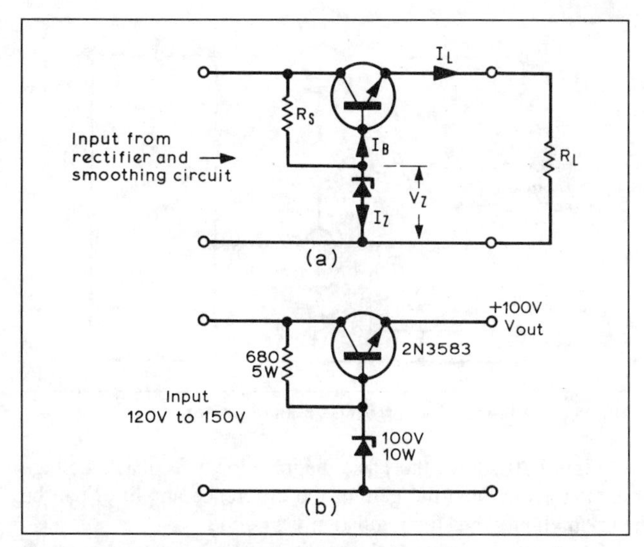

Fig 14.18. A series transistor regulator

Series regulators

In these, an active device is placed in series with the supply and negative feedback applied in such a way that the output voltage remains constant in spite of a varying load. The output voltage is always compared with a reference, and the difference amplified and used to control the series element in such a way as to minimise that difference. The greater the gain of the amplifier, the better is the final result, although a single transistor can be good enough for loads which only vary a little.

The series element can be a power transistor, a power FET or a valve. The last is only possibly used for the higher voltages (say, greater than 100V).

Figs 14.18 and 14.19 show the simplest type of regulator which has only two active devices, a series transistor (sometimes called the *pass transistor*) which may be simple (Fig 14.18) or a *Darlington pair* (Fig 14.19) and a zener diode. Since this has only a small gain, it is only suitable for a constant load.

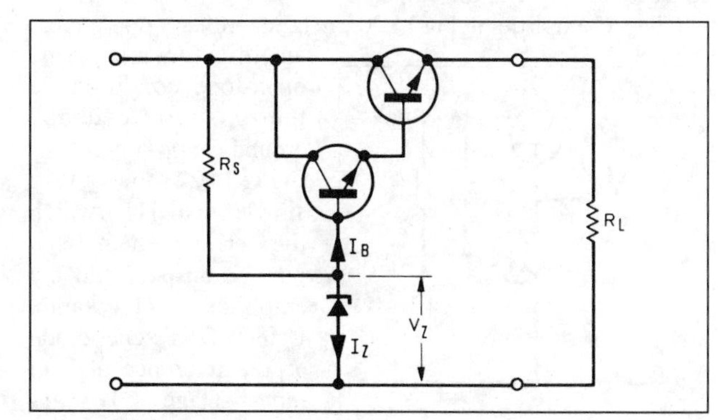

Fig 14.19. A series regulator using two compounded transistors as the series element

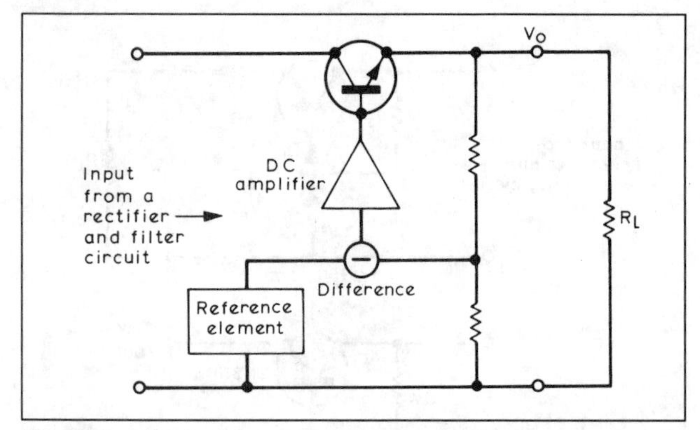

Fig 14.20. A feedback-controlled series regulator

Fig 14.20 shows the basic diagram for a feedback voltage regulator in which the gain of the difference amplifier can be very high and thus the regulation very good.

There are two parts to the quality of regulation of the output voltage. These are:

$$\text{Stabilisation factor} = \frac{\text{Change in input voltage}}{\text{Change in output voltage}}$$

and

$$\text{Output resistance} = \frac{\text{Change in output voltage}}{\text{Change in output current}}$$

With a good stabilised supply, the first should be 5000–50,000, and the second, 0.01–0.001Ω, ie the stabilisation factor should be as high and the output resistance as low as possible.

Basic circuits

Fig 14.21 shows a circuit with three active devices where a proportion of the output voltage determined by the ratio of R_2 to R_3 is compared with the reference voltage, that of the zener diode. If the voltage is low, the feedback makes the series transistor, TR1, conduct harder and vice-versa.

The values of the resistors are calculated as follows:

$$\frac{V_{S(min)} - V_O}{R_1} = \frac{I_{L(max)}}{h_{FE}}$$

so

$$R_1 = \frac{(V_{S(min)} - V_O)\, h_{FE}}{I_{L(max)}}$$

where h_{FE} is the 'worst case', ie minimum value of the gain of the transistor.

It is normal to connect a capacitor (C) between the emitter and base of TR2. This has the effect of bypassing R2 for ripple components and increasing the gain at ripple frequency. This reduces the ripple in the output.

The DC amplifier in Fig 14.20 can be an ordinary operational amplifier (op-amp) in the open-loop condition, ie there is no feedback around the op-amp.

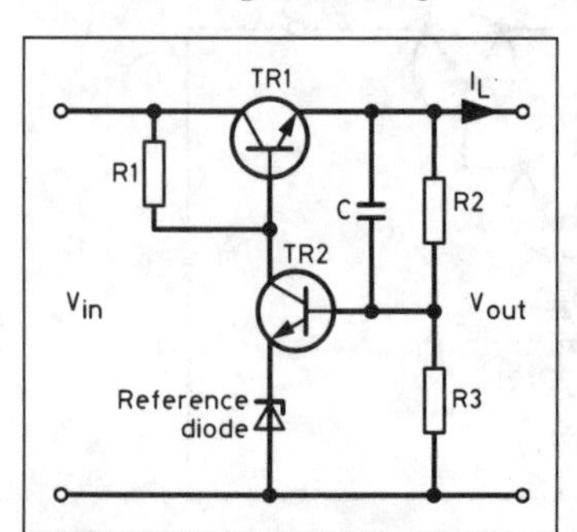

Fig 14.21. Basic regulator circuit with single transistor amplifier

Fig 14.22 shows a very simple circuit [1] in which the series element is a power transistor and the amplifier a 741 op-amp. Using a 5.6V zener diode as the reference and an input voltage of 18V approximately, the output can be varied from 6–12V.

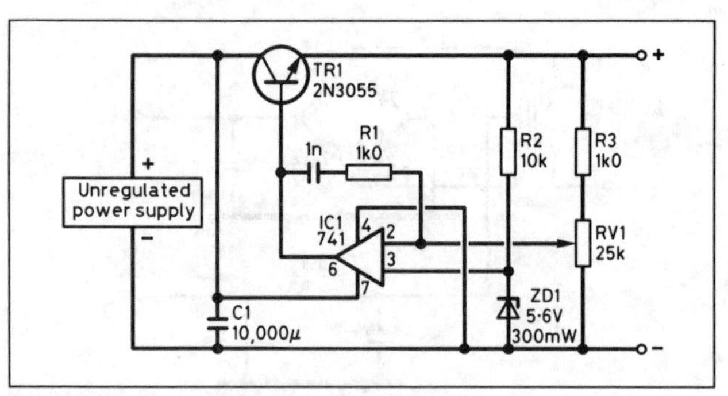

Fig 14.22. Regulated power supply using an op-amp and power transistor

The output resistance is about 0.0001Ω but this is only realised at the terminals since connecting wires will have a greater resistance than this.

IC voltage regulators or stabilisers

There are a large number of IC voltage regulators. Only the analogue types will be described here; a short description of switch-mode systems will be given later. They all have the devices above built onto the single chip and often have the addition of circuits that shut down the system if too much current is taken or if the temperature of the device gets too high. They fall into four main classes:

1. Positive fixed voltage (eg 5, 9, 12, 15, 24V)
2. Positive adjustable voltage.
3. Negative fixed voltage (as for positive types).
4. Negative adjustable voltage.

All need capacitors connected (see Fig 14.23) close to the input and output terminals to prevent spurious oscillation and the adjustable devices need one or two resistors to set up the output voltage.

Table 14.4 gives some typical examples. Circuits employed for adjustable types vary from manufacturer to manufacturer. Fig 14.24 shows one variant. Most of these have internal circuitry for short-circuit protection. This takes the form of a *fold-back* in which the output voltage falls sharply as the current is increased above the rated value (Fig 14.25).

This current limitation can be added to the discrete component types by adding a current-sensing resistor in the output circuit and using the voltage drop across this, after it reaches the rated value, to cut down the output voltage.

Fig 14.26 shows a typical circuit due to G3PCJ [2] which has current limitation using the above principles but with a FET (see below) control element. The power FET, TR1, is controlled by a common-base NPN error amplifier, TR2, the emitter of which is connected to a portion of the output voltage. Its base is held at

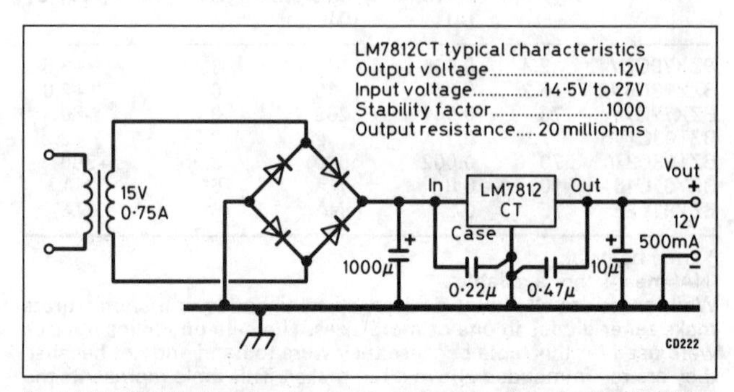

Fig 14.23. A 12V 500mA power unit using a regulator type LM7812CT

Table 14.4. IC voltage regulators

Type	Voltage	Current	Polarity	$V_{in(min)}$	$V_{in(max)}$
Fixed voltage					
MC78L05APC	5.0	0.1	+	6.9	30
MC79L05APC	−5.0	0.1	−	−6.9	−30
LM78 12CT	12.0	1.0	+	14.6	35
LM79 12CT	−12.0	1.0	−	−14.5	−35
LT1086CT12	12.0	1.5	+	13.5	25
MC78T 15CT	15.0	3.0	+	17.8	40
78P 05SC	5.0	10.0	+	8.25	40
LM2931A	5.0	0.4	+	5.65	25*
Adjustable voltage					
LM317LZ	1.2–37	0.1	+	NA	40
TL783C	1.25–120	0.7	+	NA	125
LM317T	1.2–37	1.5	+	$V_o + 3$	$V_o + 40$
LT1086CT	1.2–29	1.5	+	$V_o + 1.5$	$V_o + 30$
79HGSC	−2.1–24	5.0	−	NA	−35

* Low 'drop-out' type, ie it has a low voltage drop across the series transistor.

Notes. There are many other different voltage and current rated stabilisers and they are made by many different manufacturers. All the high-powered devices must be fixed to a suitable heatsink. All 'fixed voltage' devices can have their voltage adjusted upwards by adding a resistor, a diode or a zener diode in series with their 'common' lead. The value of resistor varies with the device and the manufacturer, and the latter's literature should be consulted.

+5V by the low-power three-terminal regulator IC1. The output voltage is given by:

$$V_{out} = \frac{(V_{ref} - 0.6) \times (R_1 + R_2)}{R_2}$$

The output voltage may be varied by varying the ratio of R_1 to R_2 but their sum should not be more than 10kΩ.

Current limiting is provided by R3 and TR3 which removes the drive from TR1 when the voltage across R3 exceeds about 0.6V. The use of remote sensing via the leads marked '*' and connected directly to the load improves the stability since it cancels the effect of the power lead resistance. The two 10Ω resistors are needed to prevent the system going 'haywire' if the sensing leads are disconnected from the output leads.

The pass transistor, TR1, can be a 2SJ174 which is capable of 20A output on a good heatsink.

Hybrid circuits are possible where an IC voltage regulator controls a high-current series transistor. One such was designed by G3UPA [3] (see Fig 14.27) for an output of 13.8V at 10A. The circuit was slightly modified by G1OSC [4] to include a resistor between the base and the source of TR2. It should be noted that the IC is supplying all the base current for TR2 and that, if TR2 has a low h_{FE}, the IC will be over-run. The use of a Darlington pair transistor (either a monolithic pair or two separate transistors) will remove this problem. Setting up is simple:

(a) R3 is set to the negative end of its travel.
(b) R1 is adjusted to give an output of 14.5V at the IC1 output.
(c) R3 is adjusted to give an output of 13.8V.

FETs

Instead of bipolar transistors, insulated-gate field effect transistors (IGFETs or MOSFETs) may be used as the controlling elements. They have only recently become cheap enough for this purpose and offer some advantages over bipolars. Two of these are that they are not so subject to 'thermal run-away' and they have a constant gain irrespective of drain current. (The h_{FE} of a power bipolar drops sharply as the collector current rises.) The only disadvantage is that they need a low-power auxiliary power supply since the gate needs to be 6–8V above the source for conduction to occur.

Fig 14.28 shows a modern circuit for such a stabiliser, due to G3LBA [5], which also has over-voltage protection (see later). The zener diode ZD2 provides protection for TR1 from too high a drain-to-source voltage at switch-on when all the capacitors are discharged. With an 'adequate' heatsink, it is capable of a continuous output of 9A, and of 20A for short periods such as for transmission with a transceiver. In this context, an 'adequate' heatsink would probably be one with a dissipation of about 1°C/W.

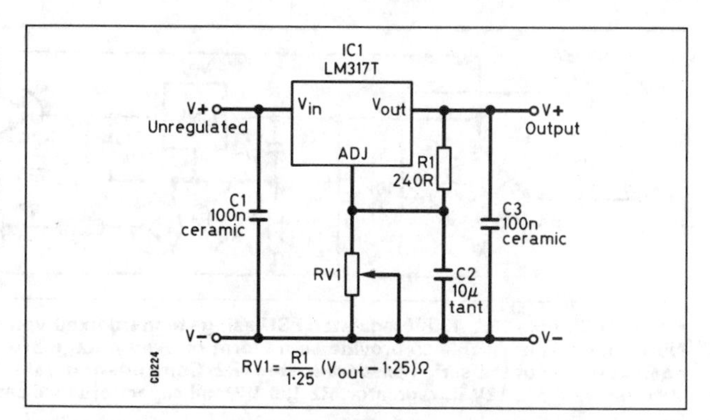

Fig 14.24. Adjustable stabiliser using an IC. The value of RV1 is $R_1(V_{out} - 1.25)/1.25\Omega$

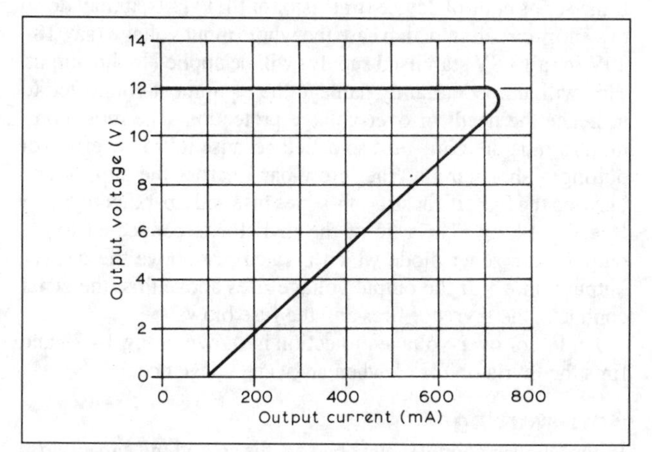

Fig 14.25. Voltage current characteristic showing 'fold-back'

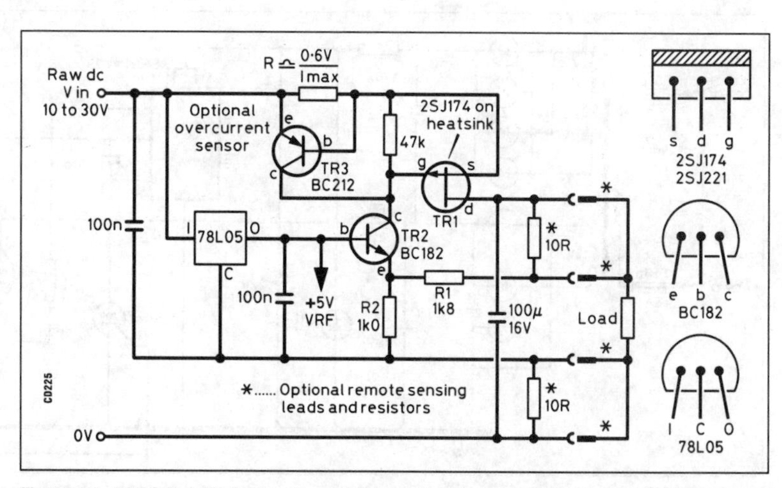

Fig 14.26. G3PCJ's efficient voltage limiting regulator drops only 0.4V for low input voltages

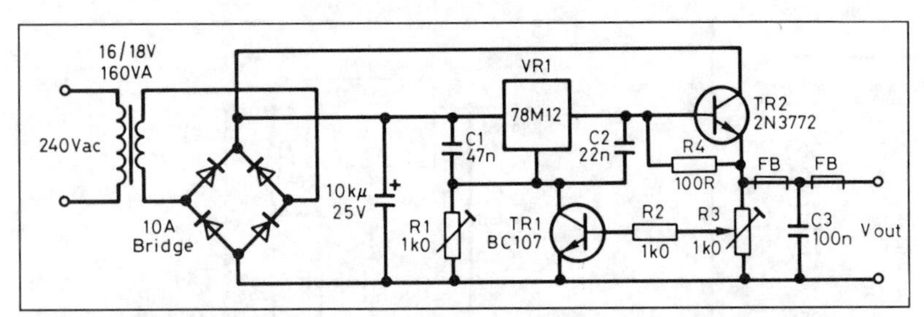

Fig 14.27. Simple 10A, 13.8V regulated PSU using standard fixed-voltage IC regulator. Note that it is advisable to provide some form of over-voltage protection to guard against failure of the series-pass transistor TR2. Component details: TR1 BC107/8/9, VR1 78M12 0.5A, 12V IC regulator, R2 1k0 ¼W, all capacitors disc ceramic

Fig 14.29. Regulated PSU with crow-bar protection. C1, smoothing capacitor; F1, suitable fuse – fusing current 10% above normal rated output; TR1 BTY79-400R for up to 10A supply; ZD, voltage 10% above rated output voltage

Over-voltage protection

If the series control device (transistor or FET) fails, it may do so to a short-circuit in which case the whole input voltage (say 18–20V for a 13.8V stabilised supply) will be applied to the output. This will almost certainly damage the equipment connected to it, hence the need for over-voltage protection. The most common form is the *crow-bar*, so called because it has the effect of putting a short-circuit (the 'crow-bar') across the supply and blowing the fuse. It consists of a series fuse and a parallel thyristor (see Fig 14.29). The gate of the thyristor is connected to the output via a zener diode which is rated just above the desired output voltage. If the output voltage rises above this, the zener conducts, the thyristor fires and the fuse blows.

An IC for over-voltage protection is shown in Fig 14.28 and fires the thyristor THY1 when an over-voltage occurs.

Soft-starting

When a power supply is switched on, the smoothing capacitor(s) is (are) uncharged and there is a large surge in current to charge it (them). With high-current supplies which must have large capacitors, this can be enough to damage the rectifier or blow the input fuse. *Soft starting* is a method of reducing this surge.

One method is to put a thermistor in series with the supply to the smoothing capacitor (Fig 14.30) This has a cold resistance of, say, 5Ω, and a hot one of less than 0.1Ω. One such device is the Siemens S2344R7.

High-voltage supplies using valves

At much above 100V, valves are probably better for stabilised power supplies than semiconductors, although high-voltage FETs and bipolar transistors are now available. The same principles apply to valve circuits as to semiconductor ones. A series (or pass) device, usually a triode or a triode-connected pentode or tetrode, controls the output. A portion of the output voltage is compared with a reference and the difference amplified and used to control the series valve so that the difference is minimised. The reference is usually a neon-gas-filled reference diode, although high-voltage zener diodes can be used.

It is not proposed to go into detail since the demand for such circuits is limited to their use as the screen supply in high-power linear amplifiers.

Fig 14.31 shows a typical circuit for a stabiliser with an output of 450V at 50mA. A higher current could be obtained by either using a valve with a greater capability or by putting a second valve in parallel with the first. V1 is the pass device and V2 the amplifier which is a cascode and has a high gain. The cathode of V2a is held at 85V positive to the 0V line by V3, and a part of the output is tapped from the potentiometer comprising R3, R4, RV1 and R5. The difference is amplified by V2a and V2b and the result applied directly to the grid of V1. C3 applies any ripple directly to the grid of V2a and so reduces this substantially. R6 and R7 are grid 'stoppers' which, together with the input capacitance of the valves, form a low-pass filter and prevent HF oscillation. RV1 is used to adjust the output voltage to exactly the desired value.

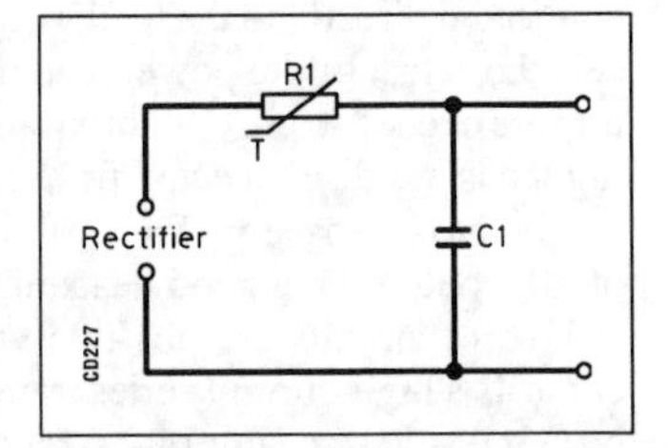

Fig 14.30. Soft-starting circuit. C1, smoothing capacitor; R1, Siemens S2344R7 thermistor

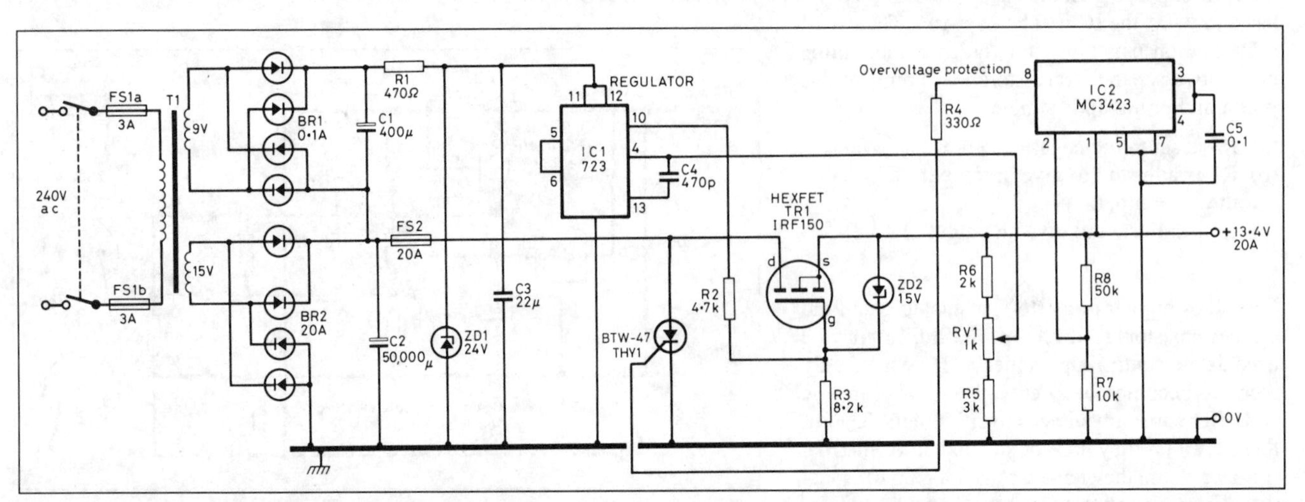

Fig 14.28. High-current PSU providing 13.4V at up to 20A (peak) used by G3LBA/PA3, showing the simplicity made possible by the use of a HEXFET (six power FETs in a single package) pass device

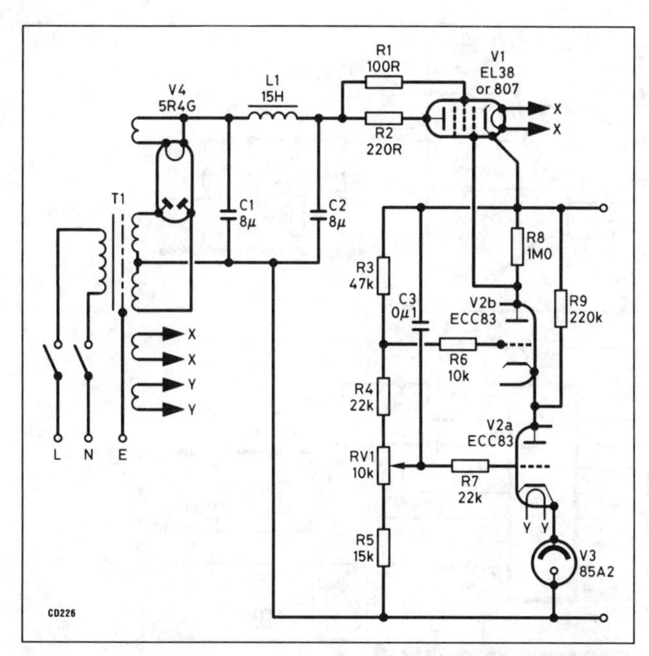

Fig 14.31. Regulated PSU using valves. T1, 650-0-650V 60mA, 5V 2A, 6.3V 3A, 6.3V 2A

Constant-current circuits

From time to time constant-current circuits are needed. For example, if a constant *voltage* is required from a zener diode, it is best achieved by feeding it with a constant current. This can be done by using a FET with a bias resistor in its source circuit; see Fig 14.32. A similar circuit with a bipolar transistor can be used and, by using high-powered devices, it can work at up to a few amps.

Switched-mode power supplies

In all the linear power supplies mentioned above, conversion to direct voltage occurs after a transformer has changed the alternating voltage of the mains to that which is required – see Fig 14.2. Thus the mains transformer has to carry the full load at mains frequency, 50Hz. The laws of physics dictate that the transformer must be large and heavy. Table 14.5 gives the weight of several of the most efficient transformers, ie those with grain-oriented toroidal cores.

If the transformation could be done at a much higher frequency, the size and weight of the transformer would be greatly reduced since the core could be made of soft ferrite which has a low density, and the windings would need many fewer turns than at 50Hz.

This is the principle of switched-mode power supplies. First, the mains is rectified directly and fed to a smoothing capacitor. This unregulated DC supply is fed to a power MOSFET or bipolar transistor switch which is operated at 50–100kHz or even higher. Bipolar transistors are rarely used since power FETs are now as cheap as bipolars and much easier to use. The output of this switch is connected to a ferrite-cored transformer whose secondary (or secondaries) (Fig 14.33) is (are) rectified to provide the required outputs. The rectifiers must be high-speed or 'fast recovery' types.

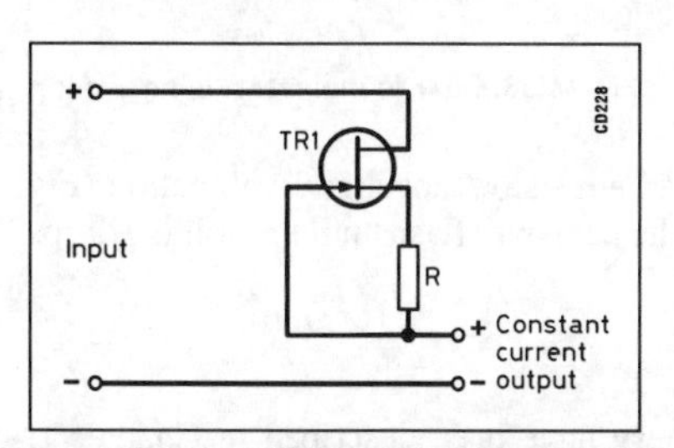

Fig 14.32. Principle of the constant-current circuit. TR1, small JFET; R, to set current – depends on G_m of FET

Table 14.5. Transformer ratings

Rating (VA)	Weight (kg)	Weight (lb)
30	0.42	0.9
60	0.85	1.9
100	1.2	2.6
160	1.9	4.2
330	3.3	7.3
530	5.0	11.0

The high-frequency drive to the FET switches is generated by a dedicated IC and is in the form of a rectangular wave with 'dead time' of a few microseconds (Fig 14.34). This allows control of the output by varying the 'on' time of each drive and it can be done within the IC. The dead time ensures that both switches cannot be on at the same time.

Fig 14.35 shows a circuit of a switched-mode power supply [6] where the mains power is supplied through a fuse, rectified by BR1 and fed via a resistor to limit in-rush current to the smoothing capacitor made up from two units in series with equalising resistors. The only reason why there are two capacitors in series is so that the circuit can easily be modified to work on 110V. The resistors R2 and R3 make sure that the voltage is equally divided between the two capacitors. They also act as bleeders to discharge the capacitors when the unit is switched off. Unregulated DC appears across the combination of C1 and C2.

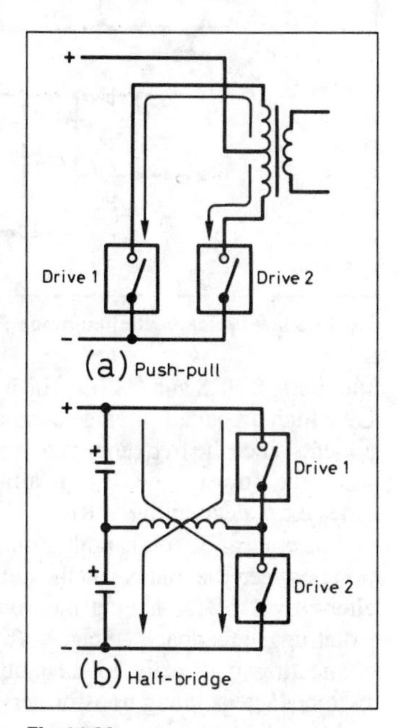

Fig 14.33

The basic power switch is of the half-bridge type and is essentially TR1, TR2, C3, C4 and T3. Other components in this area function as follows. R4 and R5 do the same job as R2 and R3, and the combinations C5/R6 and C6/R7 are *snubbers*. With fast switching waveforms in inductive circuits, there is always the risk of high-voltage spikes occurring at the switching edges, especially if the magnetic circuit is not perfect. The snubbers, together with the inherent drain-to-source diodes (ie ones present as part of the manufacturing process) of TR1 and TR2 keep these spikes within acceptable limits. T5 is a current transformer and forms part of a dynamic current limiting circuit feeding R16 and the full-wave rectifier D3 and D4.

The power transformer T3 uses a centre-tapped secondary and D1a and D1b are the fast rectifier diodes. They have

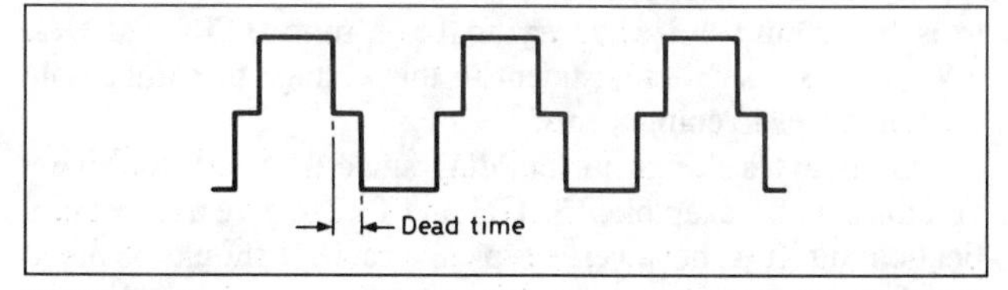

Fig 14.34

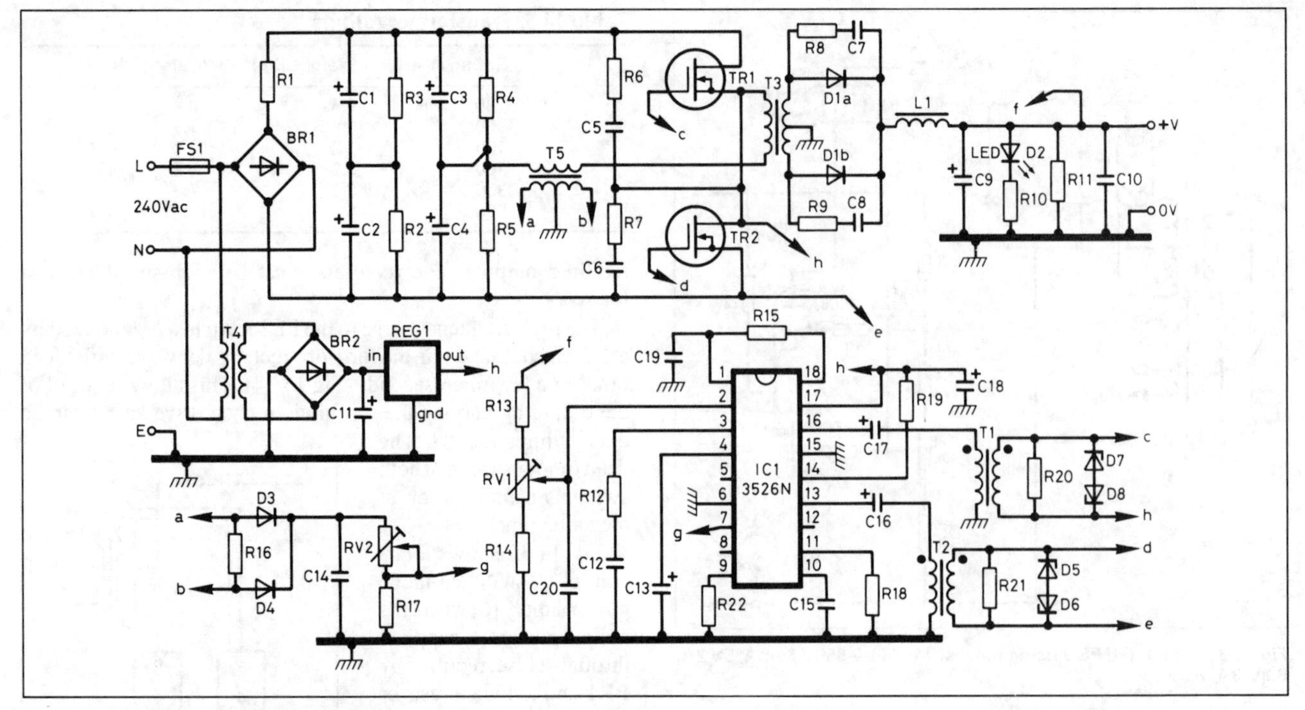

Fig 14.35. A typical switched-mode PSU

snubbers, C7/R8 and C8/R9. The output filter comprises L1 and C9 which are small in value compared with 50Hz supply components since the frequency is so much higher. D2 and R10 show when the power is on, R11 is a bleeder resistor and C10 provides extra decoupling at RF.

The control circuit is built around a SG3526 IC. Since isolation between the mains and the output is essential, there are the alternatives of (a) having the control on the mains side and isolating the feedback voltage or (b) isolating the control circuit, connecting it directly to the output, and driving the power switches via isolating transformers. The second course is better and used here because it enables testing with an oscilloscope to be carried out on the control circuit with safety. It does mean that a small auxiliary power supply is needed and provided by T4, BR2, REG1 etc. These give an output of 12V at a low current.

IC1 is the most important component. It produces antiphase outputs to drive the power switches via T1 and T2, which are small ferrite-cored devices with bifilar windings, giving a 1:1 ratio. R20 and R21 ensure that the gates of TR1 and TR2 are always biased to earth to prevent them from accidentally switching on at the wrong time and D5–D8 clip any dangerous spikes to limit the voltage applied to the gates.

R22 and C15 are the oscillator timing components, with R18 defining the dead time. C13 provides a 'soft start'. Dynamic current limiting comes from the current transformer, T5, via a full-wave rectifier and smoothing capacitor which gives an output voltage proportional to the AC flowing in T5. A proportion of this voltage is fed to IC1 and reduces the main output if the current rises too high. Feedback from the main output goes to the potential divider comprising R13, RV1 and R14, and controls the output voltage by varying the on-time to TR1 and TR2. RV1 allows a small adjustment to this voltage to compensate for tolerances in components.

This is not a design for building since there are too many variables in, for example, T3, TR1 and TR2 to give a fully practical circuit. It is, however, a typical circuit. It should be noted that there are many other manufacturers of control ICs and, while they all do the same thing, they have different details in their circuitry. Makers' data sheets will often give basic practical circuits.

EMC aspects

As might be expected, switched-mode power supplies can generate radio frequency interference (RFI) since they contain devices which are switched sharply at high frequencies. When building one, it should be contained in an earthed metal case using proper RFI filters on the mains input and suitable filters on all the outputs. If ventilation is necessary, the vents should be covered with copper gauze, either soldered in place or bolted with several bolts. An alternative is to make part of the case from perforated sheet metal.

Safety

These power supplies are hazardous in that there is no isolation of the input part from the supply mains such as is achieved with the transformer in the linear type. The direct voltage to the power switch is 1.414 times the mains voltage, ie 340V. The smoothing capacitors must be provided with a good bleeder resistor *and* with something to indicate when they are still charged. A neon and a resistor/capacitor combination (Fig 14.36) is suitable since the neon will flash until the voltage drops to about 80V.

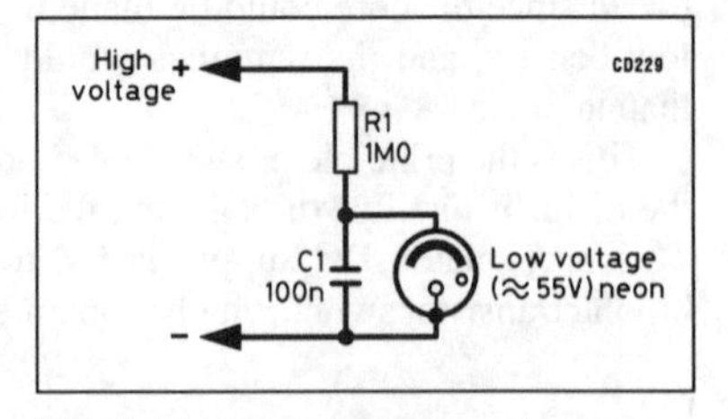

Fig 14.36. Charge indicator using a neon

BATTERIES

The basic types of battery have been described in Chapter 1, 'Principles'. They are of two types: primary or *one shot* and secondary or *rechargeable*. Strictly speaking, a 'battery' is an

Table 14.6. Characteristics of typical zinc-carbon cells and batteries

Type	Voltage (V)	Weight (g)	Maximum size (mm)	Current range (mA)
AAA	1.5	9	45 × 10.5	0–25
AA	1.5	18	50.5 × 14.5	0–40
C	1.5	48	50 × 26.2	20–60
D	1.5	110	61.5 × 34.2	25–100
PP3	9	38	48.5 × 17.5 × 26.5	1–10
PP9	9	410	81 × 52 × 66	5 × 50
C (HP)	1.5	as 'C' above	0–1000	
D (HP)	1.5	as 'D' above	0–2000	

Note. Where there are two dimensions, the first is the length and the second the diameter. The current range is that which gives a reasonable life. Manufacturers do not often give capacities in ampere-hours. The shelf life of either type of cell is about a year although it can be improved by keeping it cold. 'HP' is the high-powered type.

assembly of two or more cells but a single cell has also come to be known as a battery.

Primary batteries and cells

These fall into two main groups, depending on the metals used as their electrodes. The cell derives its energy from the metal of the negative electrode. This can be zinc or lithium. The positive electrode has less effect on the cell voltage but it is surrounded by a *depolariser* which reacts with the hydrogen gas which would otherwise coat and insulate that electrode. The nature of this depolariser has a large effect on the cell's ability to produce current over a period. The electrolyte also has some effect on this.

Zinc-carbon

These form the oldest and cheapest cells and are commonly called *dry batteries* because the electrolyte, although *not* dry, is immobilised so that it cannot spill. Three different electrolytes are used: ammonium chloride in ordinary cells, zinc chloride in 'high-power' cells and sodium hydroxide in 'manganese-alkaline' cells. Table 14.6 gives the parameters of some typical cells and batteries.

Manganese-alkaline cells and batteries are made in the same sizes as the above and here the manufacturers do give capacities. Table 14.7 gives some typical values.

Zinc-air cells

These are similar to the above but use the oxygen of the air as the depolariser. They are bought sealed and only 'come to life' when the seal is broken and air given access to the electrode. They must be used in a well-ventilated housing. They have a higher *energy density* than the above cells, ie they pack more power into a given weight or volume.

Lithium cells

Here the negative electrode is metallic lithium, which is a very reactive metal, and the electrolyte contains no water. They have a higher voltage than the above cells, and have a very high energy density and a low weight. Depending on the electrolyte and the positive electrode, they have cell voltages from 1.5 to nearly 4V. They have a shelf life of at least 10 years. These are relatively new and expensive. A typical cell is an AA size which has a capacity of 1.9Ah and a voltage of 3.7. It is used to back up the memory in a computer.

Table 14.7. Characteristics of typical manganese-alkaline batteries

Type	Voltage	Weight (g)	Capacity (Ah)
AAA	1.5	11	1.2
AA	1.5	22	2.7
C	1.5	67	7.8
D	1.5	141	18.4
PP3	9	45	0.55

Note. The dimensions are as above and the capacity is in ampere-hours. They have a shelf life of several years.

Table 14.8. Characteristics of typical nicad cells and batteries

Size	Voltage (V)	Capacity (Ah)	Weight (g)
AAA	1.2	0.18	10
AA	1.2	0.5	22
C	1.2	2.2	70
D	1.2	4.0	135
PP3	8.4	0.11	46
PP9	8.4	1.2	377

Note. The dimensions are the same as those for the zinc-carbon cells/batteries above.

Other systems

There are several other systems which are used only for specialised purposes. Very small *button cells* use either mercury oxide or silver oxide as depolariser. They have a very flat discharge curve, ie the voltage stays almost constant during discharge. They are expensive and used for in cameras, hearing aids and pocket calculators. There are so many types that there is no point in listing them here.

Safety note

All spent cells or batteries should be disposed of with care. *They must not be incinerated* as this may cause an explosion. This is especially true of the lithium cells since some of these operate at an internal pressure of several bar and will explode with violence if incinerated.

Secondary cells and batteries

There are three main types: nickel-cadmium (often called *nicad*), nickel-metal hydride and lead-acid.

Nicads

These are the most common in amateur equipment and come in many sizes. At 1.2V, they have a lower voltage than primary cells but can be recharged many times. Table 14.8 gives a selection.

The disadvantage is that they contain a highly toxic element, cadmium, and should be disposed of with care. If necessary, consult the local environmental health authority for guidance. Some authorities have containers specially for them at their disposal points.

Nickel-metal hydride cells

These are relatively new and there is currently little commercial information available about them. They have advantages over nicads in that they have a higher energy density and they do not contain such a toxic material as cadmium. A simple comparison between one of these and the same size in nicad is given in Table 14.9.

Lead-acid cells and batteries

These are typified by the motor car starter battery although smaller and larger versions are made. Most of the ones used by radio amateurs will be batteries of six cells connected in series to make a nominal 12V battery. Commonly used capacities are from 1 to 100Ah. Many modern types are 'maintenance-free' in that, if slightly over-charged, the gases (hydrogen and oxygen) liberated recombine within each cell to form water. Table 14.10 gives a brief selection of 12V batteries. It should be noted that there are many intermediate sizes and that different manufacturers may make them in different shapes.

Table 14.9. Comparison of metal hydride and nicad 'C' size cells [7]

	Metal-hydride	Nicad
Capacity (Ah)	3.5	2.0
Voltage (V)	1.2	1.2
'Memory'*	None	Severe
Toxics	None	Cadmium
Discharge rate (A)	<12–15	<15
Overcharge capability	Cont. at C/5	Cont. at C/5

* 'Memory' is an alleged effect [8] which shows up if a cell is only partially discharged before recharging. It is said to reduce the capacity of the cell. This has been disputed [9].

Charging secondary batteries

Nicads

Normally these are charged using a *constant-current* charger. The charge rate is determined by dividing the capacity in ampere hours by 10 and charging at this rate for 14 hours. This is the so-called *C/10 rate* and results in an overcharge of 40% which is safe for all types of cell. Some are specially designed to be charged at much higher rates but this must be much more carefully controlled.

Lead-acid batteries

These are normally charged with a *constant-voltage* charger with current limiting to avoid a large current flowing if the battery is heavily discharged. If the voltage is limited to approximately 2.4V per cell, the current virtually ceases when the cell is fully charged. The state of charge is indicated by the specific gravity of the electrolyte which is dilute (about 30%) sulphuric acid if this is accessible. When fully charged this is about 1.27 and when discharged it will be about 1.10.

Discharging

Secondary cells have a very low internal resistance and can produce a very high current if short-circuited, so caution is needed to avoid this.

The rated capacity in ampere hours refers to the one-hour rate, ie a 5Ah cell should last for an hour at a discharge rate of 5A. The voltage will fall, in the case of nicads, from 1.2 to 1.0V per cell. At lower rates, the capacity is somewhat higher and at higher rates, it is lower.

Lead-acid batteries should not be discharged below 1.8V per cell.

Safety

As mentioned above, the electrolyte of the lead-acid battery is dilute sulphuric acid. It should be treated with caution and not allowed to touch the skin. If it does, it should be washed off with running water as soon as possible. In particular, the eyes *must* be protected from it.

The electrolyte of both the nicad and the metal-hydride cell is potassium hydroxide (also known as *caustic potash*) which is a powerful alkali. It, too, is corrosive to skin and should be washed off as soon as possible if spilt on it. Again, the eyes are very vulnerable to damage from this.

Table 14.10. 12V lead-acid batteries

Capacity (Ah)	Height (mm)	Width (mm)	Depth (mm)	Weight (kg)
1.2	57.5	97	48	0.6
6.0	103	151	65	2.4
24	125	166	175	8.5

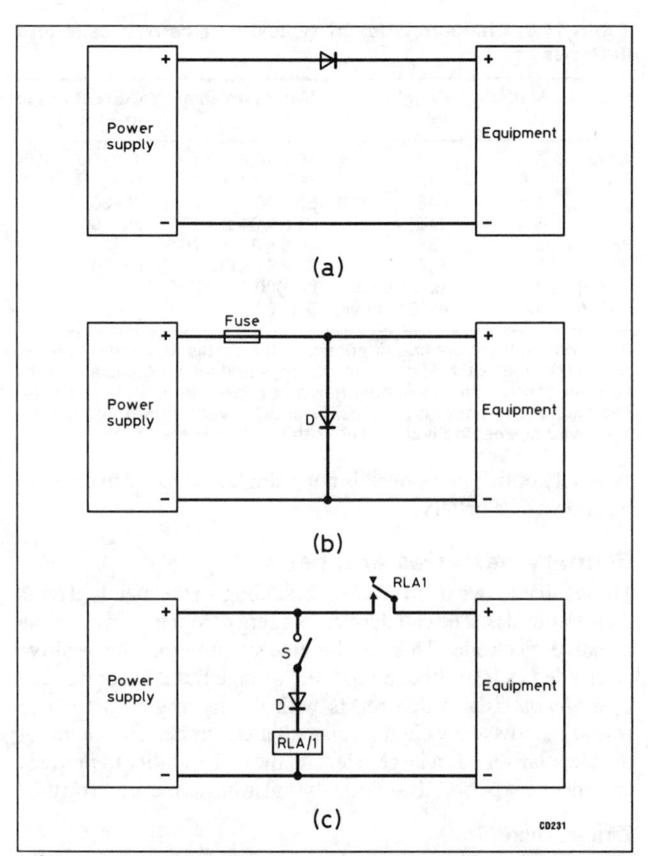

Fig 14.37. Three methods of reverse-polarity protection. D1 is a silicon diode; in (a) it must carry the whole current. In (c), S is the on-off switch, A/1 is the relay operating coil and A1 is the normally open relay contact

REVERSE POLARITY PROTECTION

Applying power to a piece of equipment with the wrong polarity can cause serious damage. There are three simple methods for preventing this:

(a) Put a power diode in series with the supply. This has the disadvantage of wasting 0.7V across the diode (see Fig 14.37(a)).

(b) Put a power diode in parallel with the device and a fuse in series. If the power is connected incorrectly, the diode conducts and the fuse blows (see Fig 14.37(b)).

(c) Use relay switching with a diode in series with the relay coil so that it will only operate if the power is connected correctly (see Fig 14.37(c)).

REFERENCES

[1] 'Technical Topics', *Radio Communication* 1973, p402.
[2] 'Technical Topics', *Radio Communication* April 1993, p52.
[3] 'Technical Topics', *Radio Communication* October 1991, p33.
[4] 'Technical Topics', *Radio Communication* March 1992, p39.
[5] 'Technical Topics', *Radio Communication* October 1975, p785.
[6] *Radio Communication* October 1989, p40.
[7] *Encyclopedia of Chemical Technology,* Vol 3, Kirk-Othmer, 4th edn, John Wiley, New York, 1992, p1070.
[8] *Art of Electronics*, Horowitz and Hall, 2nd edn, Cambridge University Press, 1989, p927.
[9] 'Technical Topics', *Radio Communication* June 1989, p34.

15 Measurements and test gear

CORRECT operation of amateur radio equipment involves measurements to ensure optimum performance in order to comply with the licence conditions and to avoid interference to other users. This will involve the use of test equipment as will the repair and maintenance of equipment. Some professional test gear is very expensive but it is the intent of this chapter to show how some of the cheaper (and maybe home-built equipment) can be used to good effect.

For further information, equipment and a more detailed discussion of some of the topics the reader should consult reference [1].

CURRENT AND VOLTAGE MEASUREMENTS

Most electrical measurements rely on the measurement of voltage and current. To this end many types of instrument have been developed such as meters, oscilloscopes, spectrum analysers etc. These are all examined in this chapter.

The ubiquitous multimeter (Fig 15.1) tends to be used for many voltage and current measurements nowadays. The units are either analogue or digital, they are relatively cheap and usually provide resistance measurement as well. Because these units are so cheap it is usually not worth making one except for the experience. However, in making power supplies, amplifiers etc one needs to have meters which are dedicated to a single function or a group of functions. This is the aspect that is considered in the following sections and how the meters can be adapted to the ranges that need to be used.

One problem with all measuring instruments is how they affect the circuit they are measuring, due to the power they require to provide the input signal.

Fig 15.1. Photo of typical analogue and digital multimeters

Fig 15.2. Photo of various moving-coil and moving-iron meters

ANALOGUE METERS

These are of electromechanical design and consist (amongst others) of the moving-coil and moving-iron type meters (Fig 15.2). The moving coil meter has a linear scale while the moving iron meter is non-linear, the scale being very cramped at the lower end. The moving-coil meter is the most sensitive and the most accurate. The moving-coil meter will respond to DC only, while the moving-iron instrument is AC/DC with a response up to about 60Hz. The modern analogue meter tends to be rectangular in appearance, older types usually being round.

The sensitivity of analogue meters is defined by the current that must flow through them in order to provide full-scale deflection (FSD). The moving-coil range starts at about 50µA FSD while the moving iron range is about 100mA FSD. These analogue meters do draw current from the circuit under test to operate and, because the coil is made from wire, the coil has resistance (R_m). When used for measuring current (Fig 15.3(a)) the meter is placed in series with the circuit and so there is a voltage drop – typically 100 to 200mV. When used as a voltmeter (Fig

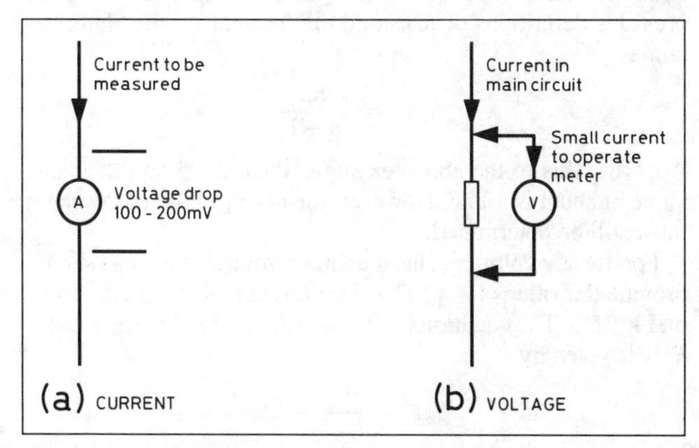

Fig 15.3. Use of meters for measurement

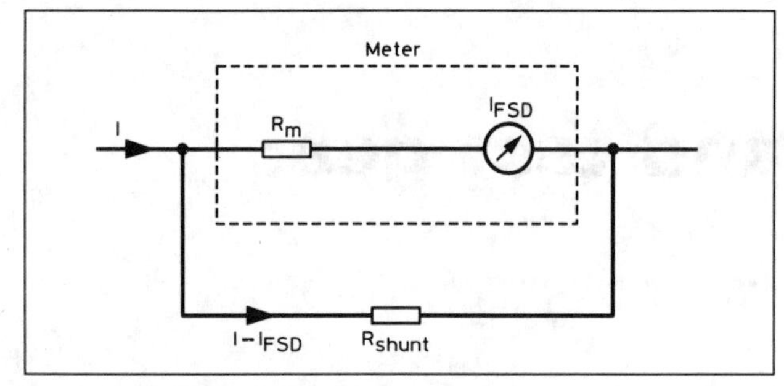

Fig 15.4. Arrangement for current shunts

15.3(b)) the current drawn depends on the basic meter movement and, if this takes more than 10% of what is flowing in the circuit, then the circuit conditions are being progressively affected.

Extending the range of analogue meters

The meters referred to above come in various fixed arrangements and may not suit the ranges it is desired to measure. It is possible, by the addition of resistors, to extend the range of meters, possibly still using the original scaling. There is no reason why the scale should not be redrawn – it is possible to use transfers or, if one is skilful, to redraw it. The scale plate can often be removed.

An analogue meter basically requires current to operate; consider the measurement of current initially. The FSD of the meter cannot be changed so it is necessary to shunt some of the current to be measured around the meter, the typical circuit being shown on Fig 15.4.

Here, assuming the maximum current to be measured is I, the shunt resistance is given by:

$$R_{shunt} = \frac{R_m I_{FSD}}{I - I_{FSD}}$$

where I_{FSD} is the current for full-scale deflection of the meter and R_m the resistance of the latter. It is normal to choose I so that only a multiplying factor is required of the scale reading. The power rating of the shunt can be calculated and is $(I - I_{FSD})^2 R_{shunt}$.

Example: It is desired to use a 100µA FSD meter to measure a maximum current of 500µA. The resistance of the basic movement is 2000Ω. Substituting these values in the above formula gives:

$$R_{shunt} = 500\Omega \text{ with a power rating of } 80\mu\text{W}$$

An alternative way of considering this problem is to consider what the multiplying factor (n) of the scale must be. Using the previous definitions of resistors, the formula for the shunt becomes

$$R_{shunt} = \frac{R_m}{n - 1}$$

Applying this to the above example, then $n = 5$ and the same value of shunt is found. However, the power rating of the shunt must still be determined.

For use as a voltmeter, the maximum voltage to be read should provide the value of I_{FSD}. The circuit used in this case is shown on Fig 15.5. The equation for the resistance of the series resistor R_{mult} is given by

$$R_{mult} = \frac{V}{I_{FSD}} - R_m$$

The power rating for the resistor is given by $I_{FSD}^2 R_{mult}$.

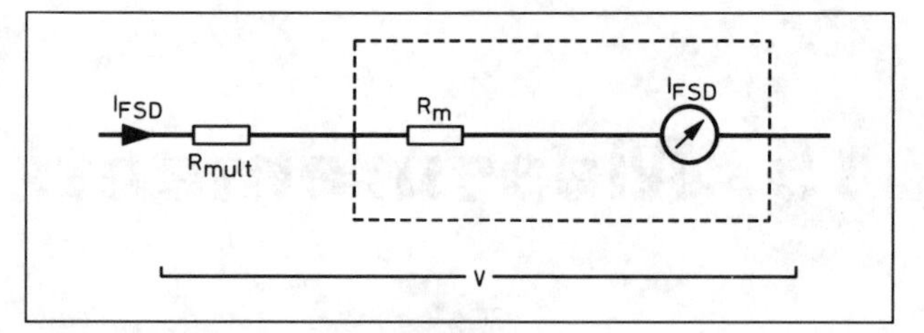

Fig 15.5. Arrangement for voltage multipliers

Example: A 50µA movement meter with a coil resistance of 3000Ω is required to measure voltages up to 30V. Calculate the multiplier resistor.

$$R_{mult} = 597\text{k}\Omega \text{ with a power rating of } 1.5\text{mW}$$

These simple calculations show the basis on which the familiar multimeter is based and how they are designed. The switch on the multimeter merely switches in different shunt and multiplier resistors. Remember, these calculations only apply to DC for the moving-coil meter.

Meter sensitivity

The sensitivity of a voltmeter is usually expressed in ohms/volt. This is merely the reciprocal of the full-scale current sensitivity I_{FSD} of the basic meter. Hence, a 1mA meter used as a voltmeter would be described as 1000Ω/V and a 50µA meter as 20,000Ω/V.

Effect on circuit readings

Putting a voltmeter across a resistor may upset the circuit conditions, and the loading effect of a meter has to be considered. For example, putting a meter which requires 50µA across a resistor through which only 100µA flows will disturb the circuit significantly. Putting the same meter across a resistor through which 10mA flows will have little effect. How can one gauge this or guard against it?

Consider a 20,000Ω/V meter. Set on the 10V range this will have a resistance of 10 × 20,000 = 200kΩ. It is suggested that any resistance across which this voltmeter is placed should have a maximum value of one-tenth of this, eg 20kΩ. Hence, for any range one can use this rule-of-thumb method. The smaller the percentage, the more accurate will be the reading.

For ammeters the point that must be considered is the voltage drop across the ammeter in relatively low voltage circuits (ie $I_{FSD} \times R_m$). For example, a 0.5V drop across an ammeter is unacceptable in a 12V circuit but it is immaterial in a 100V circuit. One must therefore choose a meter that has as low a coil resistance as possible. This reduces the in-circuit voltage drop and keeps any shunt resistance value as high as possible. If possible, use an ammeter of I_{FSD} equal or just greater than the range required.

For mains circuits of 100V or above the moving-iron meter represents a more viable alternative and tends to be cheaper.

Meter switching

In order to save cost (and sometimes panel space), it may be worthwhile for a meter to serve several functions. This is more likely to be used in valve circuits for measuring grid and anode voltages and currents. These normally require different ranges for the various parameters being measured. For convenience, two meters would be used – a voltmeter and an ammeter. *In all instances a break-before-make switch should be used.* Care should also be exercised in selecting the switch when used in high-voltage circuits.

When measuring current, the resistance of wire and switch

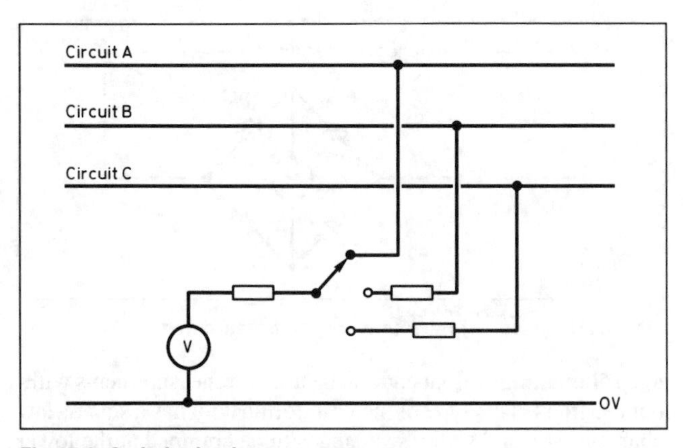

Fig 15.6. Switched voltage measurements

contacts may affect the value of low-value shunts. The switching in of different multiplier resistors is of little consequence as these tend to have high-value resistances.

Voltage measurements are normally measured with respect to 0V or earth. This means that one end of the voltmeter is fixed – see Fig 15.6. Knowing the characteristics of the meter, the various values of series resistance can be calculated. It is suggested that the lowest value is usually wired directly in series with the meter and then the other values chosen such that this value plus the additional one equals the value calculated. Assuming that circuit A has the lowest voltage to be measured (Fig 15.6), then some current limiting always exists in series with the meter.

For current measurements the problem is overcoming contact and wire resistance when low-value shunts are used. For the purposes of this discussion a meter is assumed to have 1mA FSD and coil resistance of 100Ω. Fig 15.7 shows how switching could be arranged for the measurement of current on three ranges. Switching/conductor resistance is unlikely to be a problem with circuit A, but it may be a problem on circuit B and certainly will be on circuit C.

One solution is to use a non-switchable meter for any current range which requires a low shunt value, typically less than about 0.5Ω. A different approach is to consider the meter as measuring volts across a resistor. The problems of measuring a voltage and the current taken must then be considered as previously discussed. If a 50μA meter was used then it must be possible to develop a minimum voltage drop of about 150mV; for a 1mA

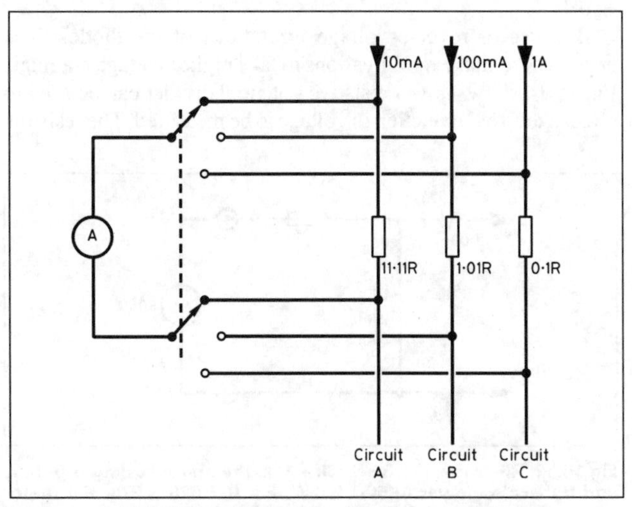

Fig 15.7. Switched current measurements

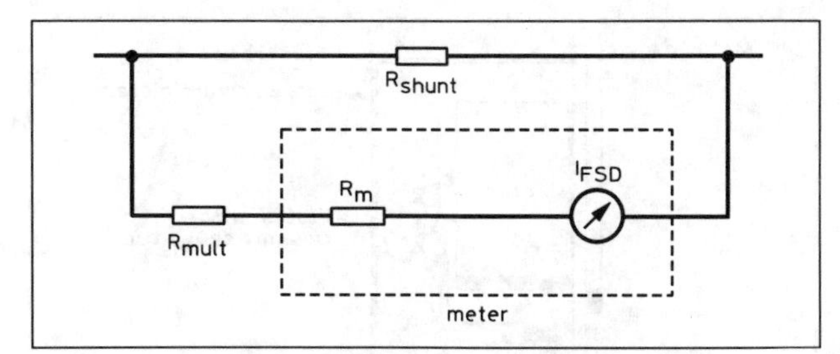

Fig 15.8. Current measurement by volt-drop method

movement it should be about 100mV. The voltage drop should be equal to or greater than $I_{FSD} \times R_m$. The typical circuit used for this arrangement is given on Fig 15.8.

Meter protection

Meters are relatively expensive and easily damaged if subjected to excessive current. Damage can be prevented simply and cheaply by connecting two silicon diodes in parallel (anode to cathode) across the meter terminals as in Fig 15.9, and this should be regarded as standard practice. No perceptible change of sensitivity or scale shape need occur.

A characteristic of silicon diodes is that they remain of very high resistance until the anode is some 400mV above the cathode, at which point they start to conduct and the resistance falls to a low value. Since the voltage drop across the average meter is around 200mV, it follows that a silicon diode connected across the meter will have no effect even when the meter shows full-scale deflection. If, however, the meter is overloaded to twice the FSD and the voltage across the meter rises to 400mV, the diodes will begin to conduct and shunt the meter against further increase of fault current.

Most meters will stand an overload of at least twice the FSD without damage but it is wise to include a series resistor as shown on Fig 15.9 to ensure the protection afforded by parallel diodes without affecting the meter. The series resistance ensures that the voltage drop across the meter/resistor combination is 200mV minimum. Parallel diodes are used because excessive current in either direction can damage the meter.

Example: What series resistor should be added to a 1mA FSD meter with a coil resistance of 100Ω?

At 1mA FSD the voltage drop across the meter is 1mA × 100Ω = 100mV. Thus the drop across the series resistor should also be 100mV, and this requires a resistance of 100Ω. This then means that the meter is protected for currents in excess of 400mV/200Ω = 2mA.

If an additional series resistor is to be included then any shunts to be included to increase the current range should be placed across this combination and the series resistance taken into account when making the calculations.

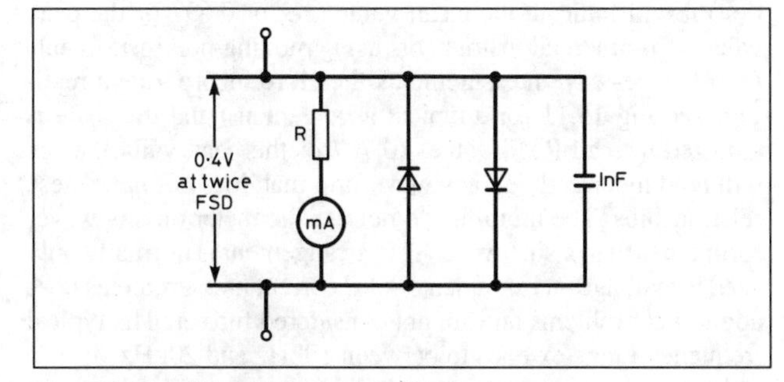

Fig 15.9. Meter protection using diodes

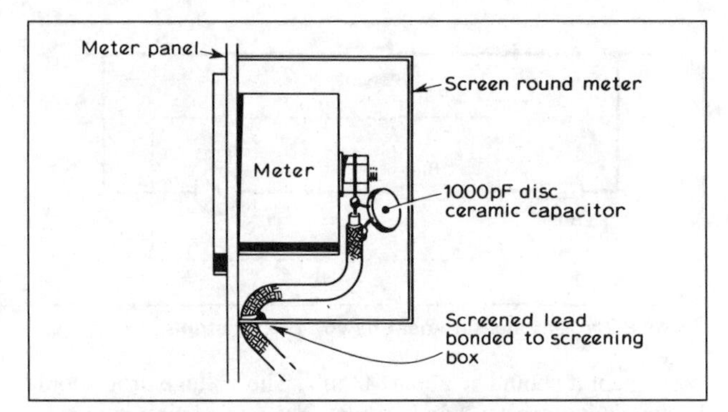

Fig 15.10. Screening and bypassing a meter in a transmitter

For most cases, small-signal silicon diodes such as the OA202, 1N914 or 1N4148 are satisfactory – they have the advantage of having an inherently high reverse resistance – ie a low reverse leakage current is required as this shunts the meter circuit. However, it is important that under the worst fault conditions the diode will not fail and go open-circuit, thus affording no protection. An example of this with small-signal diodes would be in a high-voltage supply where a large current could flow in the event of a short-circuit of the power supply. In these cases a rectifier diode should be used, such as the 1N400X or 1N540X series. The reverse current of these diodes may be a few microamps and, depending on the current to be measured, may have a slight effect on the sensitivity of the meter circuit.

Although diode protection should be applied as routine in order to safeguard instruments, it can cause some unusual effects if measurements are made with an AC signal imposed on a DC signal. This AC component, providing it is symmetrical, should not normally introduce any error but, if the AC is large enough to bring the diodes into conduction at the peak of the cycle, it introduces a dynamic shunt on the meter. This can be partly confusing when back-to-back diodes are used as the meter sensitivity will drop without any offset reading to warn what is happening. These effects are most likely to occur when measuring rectified mains or when RF is present.

Whenever a meter is to be used when RF may be present (this includes even a power supply output voltmeter) it is wise to shunt the meter with a capacitor, typically a 1000pF ceramic type – see Fig 15.10. In addition, if strong RF fields are likely to be present, eg in a transmitter, it would also be wise to shield the meter and possibly feed it via screened cable.

AC measurements

If an alternating current is passed through a moving-coil meter there will normally be no deflection since the meter will indicate the mean value and, in the case of a waveform symmetrical about zero, this is zero. If, however, the AC is rectified so that the meter sees a series of half-sine pulses (full-wave rectification) it will indicate the mean value ($2/\pi$ or 0.637 of the peak value). Commercial instruments using moving-coil instruments for AC sine-wave measurements therefore incorporate a rectifier (see Fig 15.11 for a typical arrangement) and the scale is adjusted to read RMS values (0.707 of the peak value). They will read incorrectly on any waveform that does not have these relationships. The moral is: do not use the meter on any waveform other than a sine wave. This arrangement is normally only used for voltage measurements – AC current measurements pose additional problems and are not considered further. The typical frequency range extends to between 10kHz and 20kHz.

Moving-iron instruments, as previously mentioned, do respond

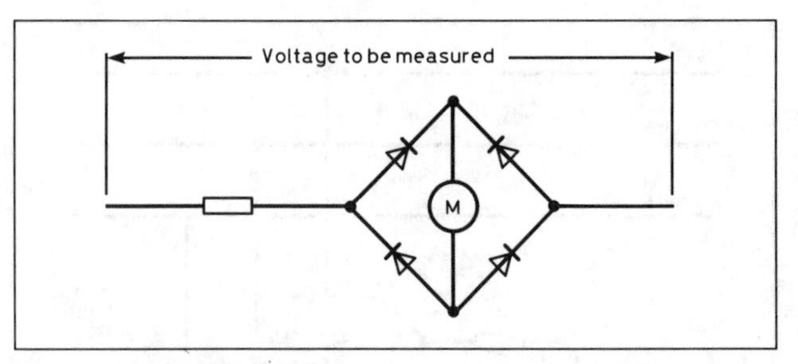

Fig 15.11. Typical arrangement for AC measurements

to an alternating current and can be used for measurements without rectifiers. This type of meter unfortunately has a square-law characteristic and so the scale tends to be cramped at the lower end. Moving-iron meters normally have a full-scale reading of about 20% more than the normal value to be displayed. They are not used for multimeters.

Other AC measurements can be accomplished by means of electronic voltmeters or oscilloscopes.

ICs do exist (eg AD536, 636, 736, 737, SSM2110) which will provide the RMS of any waveform but their frequency range is limited.

RF measurements

These probably pose the biggest problem: the circuit under test should not be loaded, capacitance has an increasing effect as frequency rises and the diodes used for rectification must handle the frequencies concerned. The diode characteristics required mean they have a relatively low reverse-voltage rating (1N914 is 100V, OA202 is 150V with slightly poorer RF capabilities) and the forward diode voltage drop. The approach in measuring RF voltages is to rectify as soon as possible and then use DC measuring circuits.

Fig 15.12 shows a typical probe for measuring RF voltages. Capacitor C1 provides DC isolation, D1 rectifies the signal and the resistor is used to convert what is essentially a peak reading to an RMS reading on the meter. For the 50µA meter it is possible to use an individual meter or the most sensitive range on many multimeters. If possible use the precautions for the meter as depicted on Fig 15.10. Fig 15.13 shows the typical construction of a probe, the exact method being left to the ingenuity of the constructor. A scrap length of 15mm central heating piping may make a good tube. The probe should be useful for frequencies from 50kHz to about 150MHz with an accuracy of about ±10%.

Because of reverse-voltage limitations of the diodes, it is necessary to make modifications to take higher voltage readings. Fig 15.14 shows how a resistive potential divider can be used to effect a ten-fold reduction in voltage to be measured. The resistors

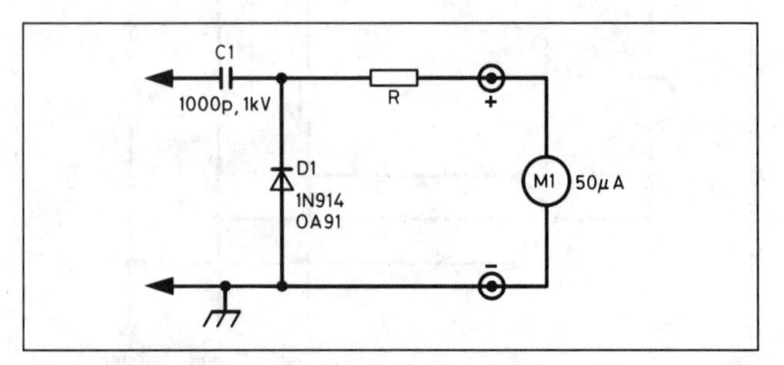

Fig 15.12. RF probe. For R = 270k + 12k, the meter scaling is 0–10V, and full-scale, power in 50Ω is 2W. For R = 820 + 27k, the meter scaling is 0–30V, and full-scale, power in 50Ω is 18W

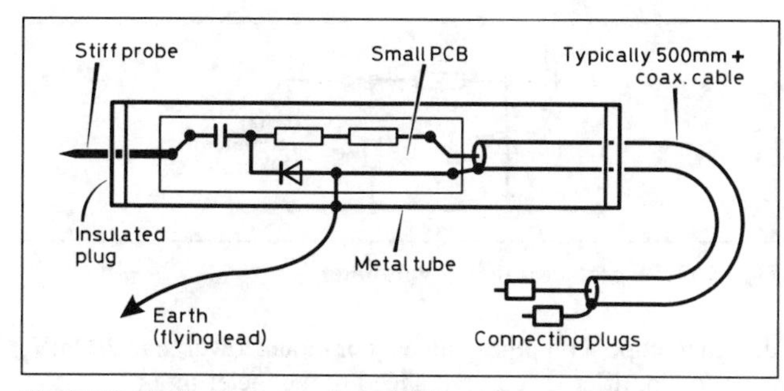

Fig 15.13. Typical construction of RF probe

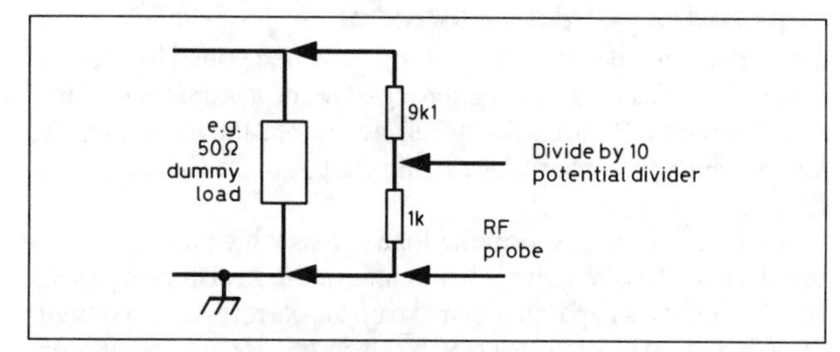

Fig 15.14. Suggested method for higher voltages

should of course be suitable for RF and of adequate power rating. An alternative approach is to use several diodes in series but they will need equalising resistors across them.

DIGITAL METERS

The digital meter is fast becoming more common than analogue types and its price is now comparable in most instances. It provides a very accurate meter at reasonable price. Its disadvantages are that the smallest digit can only jump in discrete steps (hence digital) and that it requires a battery.

The digital meter (Fig 15.15) works by converting an input analogue voltage to a digital signal that can be used to drive either an LED (light-emitting diode) or LCD (liquid crystal display). The conversion technique used is either an analogue-to-digital (A-D) converter or the dual ramp technique. A digital meter is often quoted as having, for example, a 3½ digit display. This means that it will display three digits 0–9, with the most significant being only a 0 (normally suppressed) or a 1, ie a maximum display showing 1999 as well as + or – signs.

There are quite a few ICs made by various manufacturers that provide a basic digital voltmeter, external components being required for extending the range, over-voltage protection and displays. These ICs provide outputs suitable for driving LEDs, LCDs or providing BCD outputs for further processing.

The digital meter is basically a DC voltage measuring device (as opposed to the moving-coil meter which is current controlled).

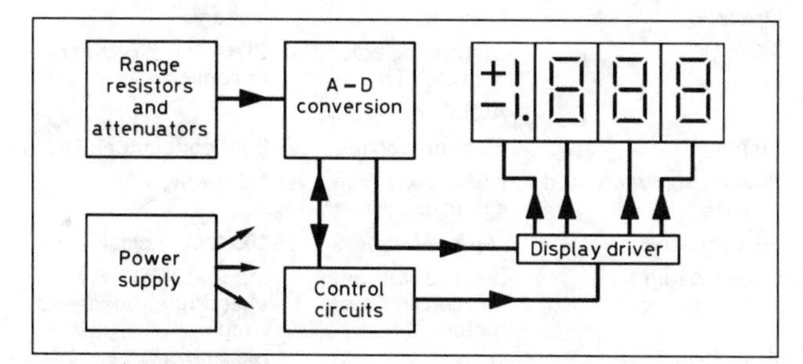

Fig 15.15. Block diagram of digital meter (3½-digit display)

Fig 15.16. Photo of typical digital panel meter

Hence all measurements to be made must be converted to a voltage.

The digital form of the multimeter is readily available at reasonable cost and it is not worth the exercise of making one of these meters. The approach here, as for the analogue meter, is to understand the basic principles and how to apply them to specific situations.

The digital panel meter

The best approach for a digital display is to use a panel meter module (which includes the above ICs) and comes with a 3½ or 4½ digit display. These are relatively cheap and provide a good basis for making various types of metering system.

It is normally a module based on an LCD display and either plugs into a DIL socket or is on a small PCB. They have (typically) a full scale reading for a 199.9mV DC input, work from 5V or 9V DC supplies (depends on model), consume very low current (eg 150–300μA on a 9V supply) and have an input resistance of at least 100MΩ. Because of this high input resistance they present virtually no loading on the circuit under test.

The panel meter itself will provide an accuracy of 0.1% or better but this does not take into account any external signal conditioning circuits such as amplifiers or attenuators. In addition to these parameters, some of the displays will also show units or prefixes such as μ, m, V, A, Ω, Hz etc (referred to as *annunciators*). They can be purchased with and without backlighting; a typical unit is shown on Fig 15.16. The main design consideration in using these units is to get the parameter to be measured to a DC voltage in the range 0–199.9mV. This can include amplifiers, attenuators and rectifiers.

The following designs are based on the Anders OEM22 module which is readily available. The panel meter consists of a liquid crystal display driven by a 7136 IC which contains an A-D converter and LCD drivers. The unit can be driven from 5V (5mA) providing two links are made on the board or direct from 9V (300μA). It comes with a leaflet containing technical details [2]. The principles explained can, however, be applied to modules available from other manufacturers.

A practical digital voltmeter

Fig 15.17 shows the arrangement for a digital voltmeter for DC voltage ranges of 200mV, 2V, 20V and 200V. The unit requires a 9V DC supply. Components are listed in Table 15.1.

Resistors R1 to R5 form a potential divider network with switch S1a selecting the correct input, ie the maximum voltage to the panel meter is to be 199.9mV. Resistor R6 and diodes D1/D2 provide protection for the panel meter should the wrong range

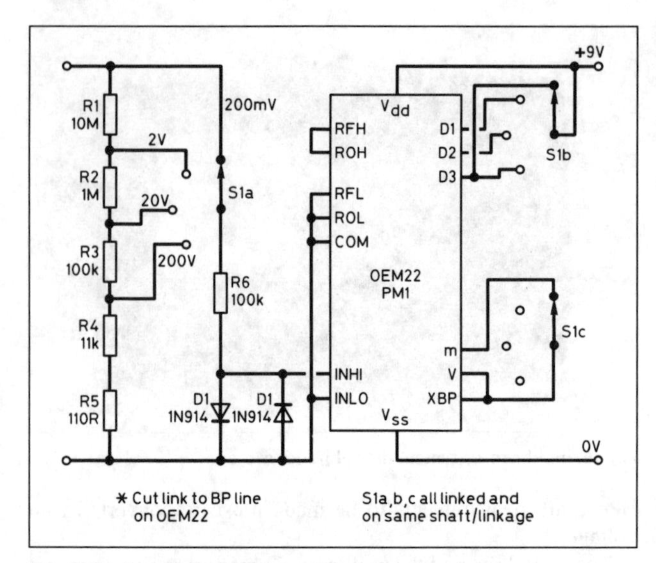

Fig 15.17. A practical digital voltmeter

Table 15.1. Practical digital voltmeter component list

PM1	Anders panel meter type OEM22	R3	100k, 0.5W, 1%
S1	Rotary switch 3-pole, 4-way	R4	11k, 0.5W, 1%
D1, 2	1N914 or similar	R5	110R, 0.5W, 1%
R1	10M, 0.6W, 1%	R6	100k, 0.5W, 5%
R2	1M, 0.5W, 1%		

(S1a) be inadvertently selected and introduce an error of less than 0.1%. S1b selects the position of the decimal point while S1c selects the annotation to be shown. The link MUST be cut to the BP line on the panel meter. Because the input resistance of the meter module is of the order of 100MΩ it represents negligible loading on the potential divider chain. The overall input resistance of the meter is about 10MΩ.

Sufficient information is provided for the reader to adapt this

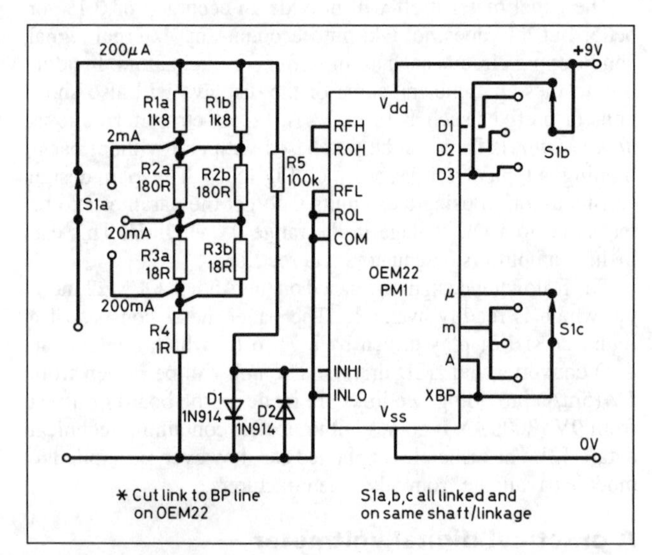

Fig 15.18. A practical digital ammeter

Table 15.2. Practical digital ammeter components list

R1a, 1b	1k8, 0.5W, 1%	R5	100k, 0.5W, 5%
R2a, 2b	180R, 0.5W, 1%	PM1	Anders panel meter type OEM22
R3a, 3b	18R, 0.5W, 1%	S1	Rotary switch, 3-pole, 4-way
R4	1R, 0.5W, 1%	D1, 2	1N914 or similar

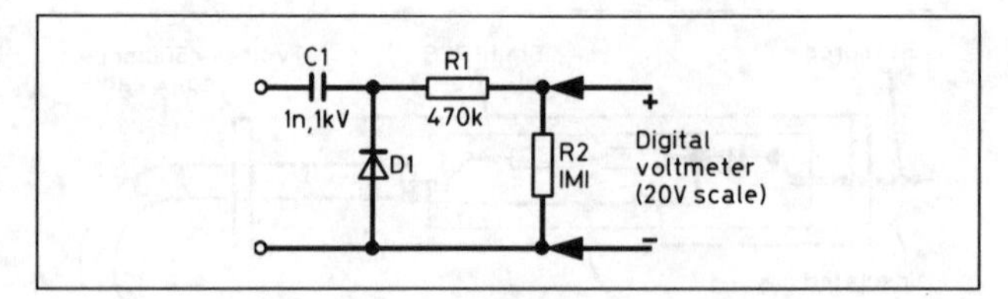

Fig 15.19. RF probe for digital voltmeter

design to cope with other ranges. For inputs lower than 200mV, then an amplifier is required ahead of the meter input.

A practical digital ammeter

This relies on measuring the voltage drop developed by the current to be measured passing through the measurement resistor, and it must be 200mV for full scale. Hence the circuit of Fig 15.18 results for a meter measuring 200μA to 200mA in decade ranges.

Resistors R1 to R4 form the load across which the voltage is developed from the current being measured. Resistors R1 to R3 involve resistors in parallel to make up the correct value required. Switch S1a selects the input, S1b selects the decimal point positions and S1c selects the correct annotation for the range being used. The combination R5/D1/D2 provides protection for the panel meter input. The unit can be powered from a PP3 battery or equivalent. Table 15.2 lists the components used.

RF measurements

Similar problems arise for the digital meter as explained earlier for the analogue meter. A slight modification is made to the RF probe circuit and this is shown on Fig 15.19. This assumes the meter has a scale with 20V full scale and an input impedance in excess of 10MΩ. The resistors provide scaling from peak to RMS for a sine-wave input. The construction should be similar to that shown on Fig 15.13.

Comparison of analogue and digital meters

Table 15.3 assumes that the analogue meter has no electronic circuit associated with it as this may alter its characteristics. It should also be borne in mind that the input to a digital meter may be affected by input amplifiers and attenuators.

THE OSCILLOSCOPE

The oscilloscope (Fig 15.20) is a general-purpose instrument for examining electrical waveforms. It can be used for various

Table 15.3. Comparison of analogue and digital meters

Feature	Analogue meter	Digital meter
Operation	Current	Voltage
AC or DC	DC moving-coil (AC with rectifiers) AC/DC moving-iron	DC (AC with rectifiers or converters)
Display	Electromechanical	Semiconductor
Power supply required	None (taken from circuit under test)	DC supply
Best sensitivity	50μA FSD typical	199.9mV typical
Circuit loading	Depends on circuit and sensitivity of meter	Input >10MΩ, may affect high-impedance circuit
RF interference	None	Possible due to internal oscillator

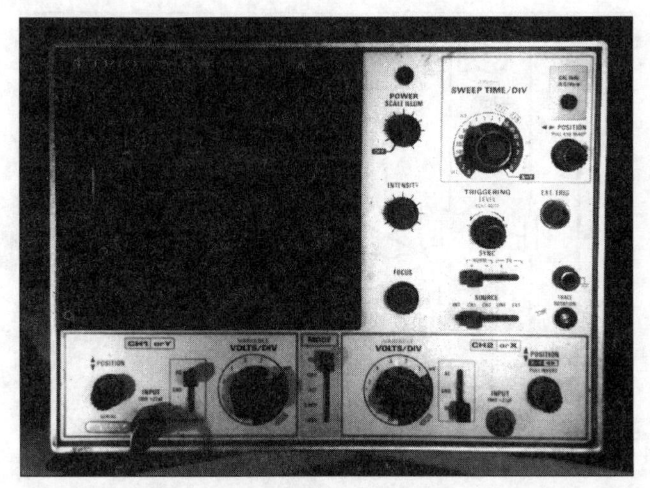

Fig 15.20. Photo of a dual-trace oscilloscope

sets of measurements depending on how it is has been set up. It is the intention of this section to explain briefly how an oscilloscope works and how it can be used for taking various measurements.

The basic oscilloscope

Some oscilloscopes can display a single trace while others can display two traces or even more with adapters. The single-trace oscilloscope has a cathode-ray tube with a single electron gun firing at the phosphor. The two-trace oscilloscopes fall into two categories – the *dual-trace* and the *dual-beam*. In the dual-trace oscilloscope there is a single electron gun but the control of it is split between the two traces to be shown – first one and then the other etc but using the same timebase. In the dual-beam type there are two electron guns in the same cathode-ray tube which are independent of each other, hence the two beams can use different timebase settings.

Fig 15.21 shows the absolute basics for an oscilloscope. An oscilloscope consists of a display (usually a cathode-ray tube) which displays an electrical waveform. These signals have been processed in some way (eg amplified) for them to be suitable for display. The oscilloscope also contains an oscillator (or timebase) which causes the display beam to traverse the display face in the horizontal plane.

In addition a power supply is required to supply the amplifier and timebase with low-voltage supplies and the tube with a high voltage (usually in the kilovolt region). The screen is split into two directions, the X (or horizontal) and Y (vertical direction).

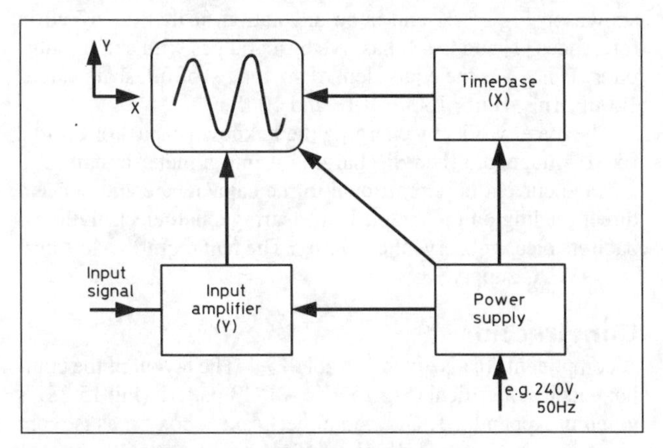

Fig 15.21. Basic block diagram of an oscilloscope

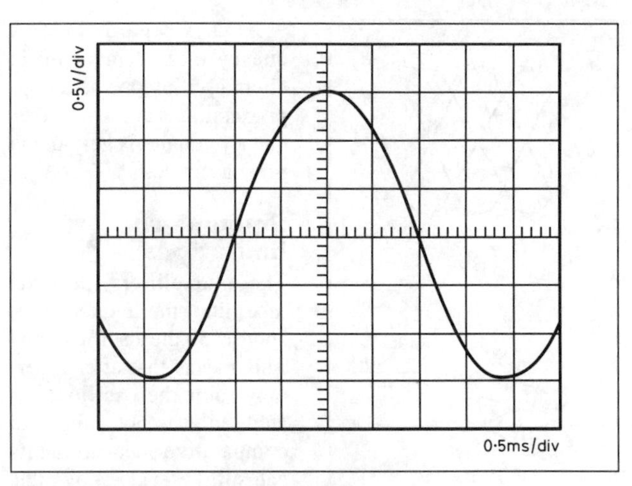

Fig 15.22. Typical displayed waveform

Voltage measurements

Consider the oscilloscope screen display as depicted on Fig 15.22. This is obviously a sine wave but what is its voltage? What voltage is one talking about? The easiest voltage measurement to take is the *peak-to-peak* value. The vertical displacement (Y) is 6 divisions. One must also take into account the setting of the Y-controls, say these are at 0.5V/div. The peak-to-peak voltage is therefore $6 \times 0.5 = 3$V. The peak value is half of this, ie 1.5V; the RMS value is 0.7071 times this value, ie 1.06V.

Frequency measurements

There are two possible methods that can be used here. The first method for making a frequency measurement is similar to the above, except that now the horizontal (X) axis is used with its setting. One problem with something like a sine wave is estimating a point on a curve and so it is an estimate only. Measure from like point to like point, eg the two negative peaks. The distance between the two negative peaks is 8 units. If the X or timebase setting is 0.5ms/div, then this represents a period of $8 \times 0.5 = 4$ms. The frequency is the reciprocal of this, ie 250Hz.

It should be noted from this exercise that the period of a rectangular waveform is easier to estimate than that of a sine wave.

The second method requires the oscilloscope/equipment to be set up as shown in Fig 15.23. The known source should be adjusted until a stationary pattern is obtained on the oscilloscope screen: Fig 15.24(a). The number of loops in both horizontal (N_H) and vertical (N_V) planes should be counted. The unknown frequency can then be calculated by:

$$f_{\text{unknown}} = \frac{N_V}{N_H} \times f_{\text{known}}$$

Fig 15.24(b) shows some typical Lissajou figures for some common ratios.

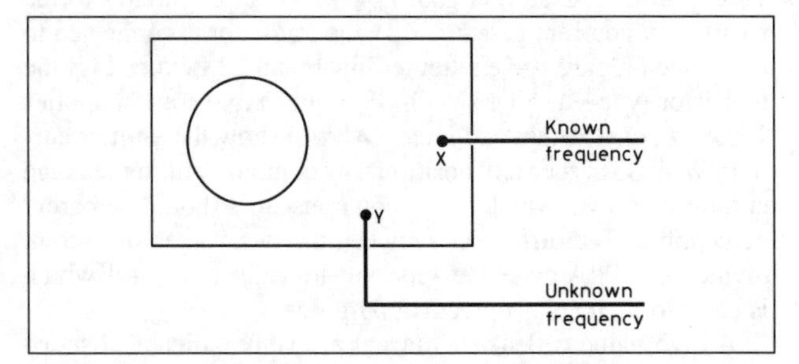

Fig 15.23. Second method of frequency determination

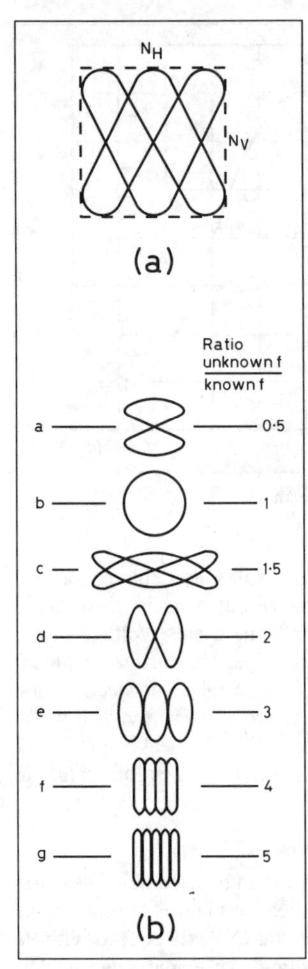

Fig 15.24. Typical Lissajou figures

Always regard the frequency measurement made on an oscilloscope as an approximate value – a frequency counter is very much more accurate.

Equipment limitations

The Y-amplifiers (plus the tube) limit the frequency response of the oscilloscope. This means that after a certain point the oscilloscope calibration is not valid, but comparative measurements can still be made above this point providing the frequency is not changed.

The capacitance of the oscilloscope and/or its probe may affect the circuit under test if the capacitance in the latter is of the same order as the oscilloscope input (20–40pF), eg in a tuned circuit.

The input voltage on an oscilloscope is normally quoted as x volts DC plus peak AC. Typical of these figures are 400V DC plus the peak AC signal that can be displayed. Exceeding this will damage internal components of the oscilloscope. Although a divide-by-10 probe can be used to extend the voltage range, these have a voltage limit (typically 600V DC) but may have to be derated as frequency rises – check the specification. The divide-by-10 probe will also reduce any loading effects on the circuit.

For most semiconductor applications the voltage limit never causes a problem, but with high-voltage valve circuits one must pay due regard to the limitations.

A DIODE AND TRANSISTOR TESTER

The circuit of Fig 15.25 shows a simple tester which will identify the polarity and measure the leakage and small-signal gain of transistors plus the forward resistance of diodes.

Testing transistors

To check the DC current gain h_{FE} (which approximates to the small signal current gain h_{fe} or β), the transistor is connected to the collector, base and emitter terminals and S2 switched for the transistor type. Moving switch S3 to the GAIN position applies 10μA of base current and meter M1 will show the emitter current. With S3 at the LEAK position, any common-emitter leakage current is shown, which for silicon transistors should be barely perceptible. The difference between the two values of current divided by 10μA gives the approximate value of h_{FE} + 1 which is close to h_{fe} for most practical purposes.

A high value of leakage current probably indicates a short-circuited transistor, while absence of current in the GAIN position

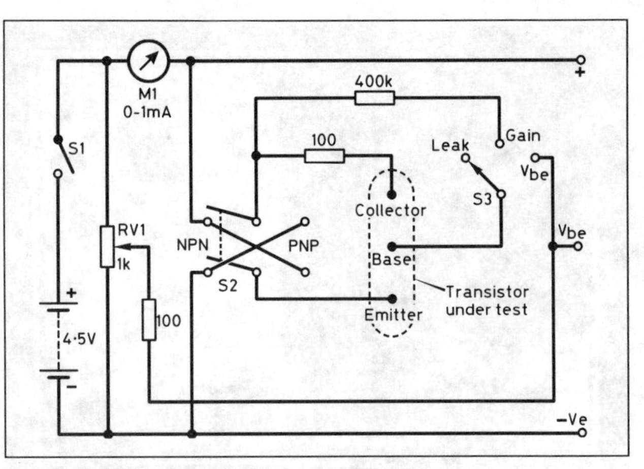

Fig 15.25. Simple transistor and diode tester

indicates either an open-circuited transistor or one of reversed polarity. No damage is done by reverse connection, and PNP and NPN transistors may be identified by finding the polarity which gives normal gain.

With S3 in the V_{be} position, the base-emitter voltage is controlled by RV1 which should be near the negative end for NPN and near the positive end for PNP. V_{be} may be measured by a voltmeter connected between the terminal marked 'V_{be}' and either the positive or negative rail depending on the polarity of the device. This test position may be used for FETs but only positive or zero bias is possible.

Testing diodes

The forward voltage drop across a diode may be measured by connecting it across the terminals marked '+' and 'V_{be}' with a voltmeter in parallel. The forward current is set by RV1.

Diodes may be matched for forward resistance and, by reversing the diode, the reverse leakage can be seen (which for silicon diodes should be barely perceptible). The value of forward voltage drop can be used to differentiate between germanium and silicon diodes.

The unit can also be used to check the polarity of LEDs as the maximum reverse voltage of 4.5V is hardly likely to damage the device (note: the reverse voltage applied to an LED should not exceed 5V). For this test RV1 should be set to about mid-position.

A LINEAR-SCALE CAPACITANCE METER

This instrument is based on the familiar 555 timer. The circuit is shown on Fig 15.26 and is an adaptation of that described in references [1] and [3]. It has five basic ranges with a ×10 multiplier. This gives the equivalent of six ranges of full scale values 100pF, 1nF, 10nF, 100nF, 1μF and 10μF.

The meter works by charging the unknown capacitor Cx to a fixed voltage and then discharging it into a meter circuit. The average current is proportional to the capacitance and hence a direct reading on the meter. If measuring small electrolytic capacitors please observe the polarity. The unit requires a low current 9V DC supply.

Construction

A components list is given in Table 15.4. The layout of the components is not critical (Fig 15.27). A PCB pattern (Fig 15.28) is given in Appendix 1. One can either make a box or, as is more usual, purchase one of the cheaper plastic types.

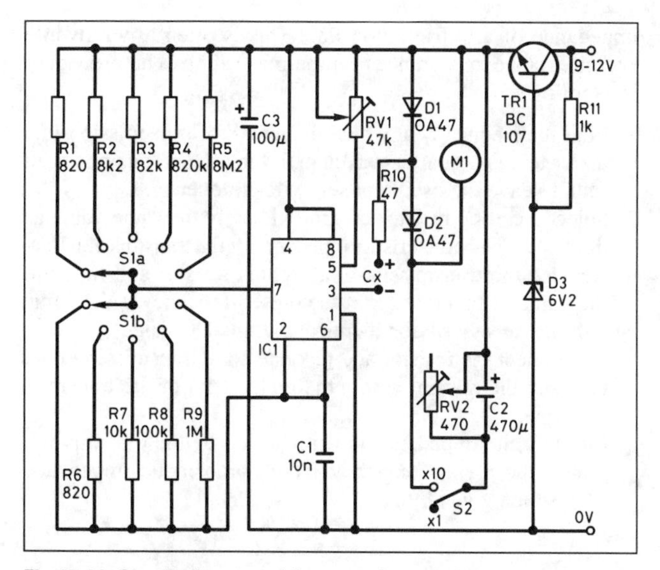

Fig 15.26. Circuit diagram of linear-scale capacitance meter

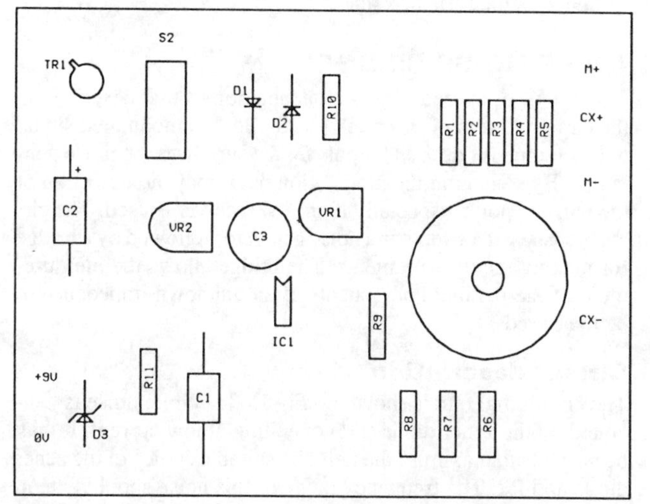

Fig 15.27. Layout of components in the linear-scale capacitance meter

Calibration

Calibration may be carried out on any range; if possible obtain 100pF, 1nF and 10nF capacitors with ±1% tolerance. With the range switch set to position 2 and multiplier switch S2 in the ×1 position, connect the 1nF capacitor. Adjust RV1 for full-scale deflection. Switch to the ×10 position of S2 and adjust RV2 for a meter reading of 0.1. Use the other capacitors to just check the other ranges. Calibration is now complete.

Table 15.4. Linear-scale capacitance meter components list

R1, 6	820	C1	10n polystyrene or silver mica
R2	8k2	C2	470μ, 16V
R3	82k	C3	100μ, 16V
R4	820k	D1, 2	OA47
R5	8M2	D3	6V2, 400mW zener
R7	10k	TR1	BC107 or similar NPN
R8	100k	IC1	555 timer
R9	1M	M1	50μA FSD meter
R10	47	S1	2-pole, 6-way rotary switch
R11	1k	S2	SPCO switch
RV1	47k preset		
RV2	470 preset		

All resistors are 0.25W, 5% unless specified otherwise.

Warning: If a large-value capacitor is to be measured, the meter will be overloaded.

AN RF IMPEDANCE BRIDGE

The need for an instrument which will measure impedance is felt at some time or other by every experimenting amateur. The instrument normally used is the full RF bridge, but commercial RF bridges are elaborate and expensive. On the other hand it is possible to build a simple RF bridge which, provided the limitations are appreciated, can be inexpensive and a most useful adjunct in the amateur workshop. In fact it is essential if experiments with antennas are undertaken.

The instrument described here will measure impedances from 0 to 400Ω at frequencies up to 30MHz. It does not measure reactance or indicate if the impedance is capacitive or inductive. A good indication of the reactance present can be obtained from the fact that any reactance will mean a higher minimum meter reading.

Circuit description

There are many possible circuits, some using potentiometers as the variable arm and others variable capacitors, but a typical circuit is shown on Fig 15.29. The capacitors have to be differential in action, mounted in such a way that as the capacitance of one decreases the capacitance of the other increases. The capacitors should be the type which have a spindle protruding at either end so that they can be connected together by a shaft coupler. To avoid hand-capacitance effects, the control knob on the outside of the instrument should be connected to the nearest capacitor by a short length of plastic coupling rod. These capacitors form two arms of the bridge, the third arm being the 100Ω non-inductive resistor and the fourth arm the impedance to be measured. Balance of the bridge is indicated by a minimum reading on the meter M1.

Construction

Construction is straightforward, but keep all leads as short as possible. The unit should be built into a metal box and screening provided as shown on Fig 15.29.

Signal source and calibration

The instrument can be calibrated by placing across the load terminals various non-reactive resistors (ie not wire-wound) of

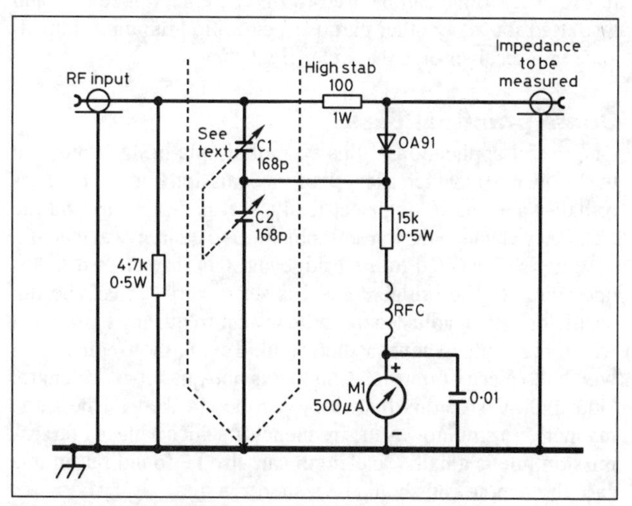

Fig 15.29. Simple RF bridge

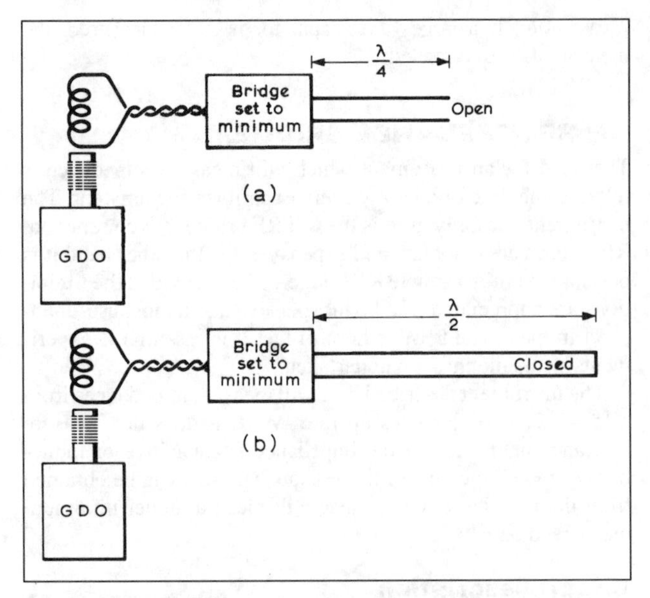

Fig 15.30. Use of the RF bridge with a dip oscillator

known value. The calibration should preferably be made at a low frequency where stray capacitance effects are at a minimum but calibration holds good throughout the frequency range. In using the instrument, it should be remembered that an exact null will only be obtained on the meter when the instrument has a purely resistive load. When reactance is present, however, it becomes obvious from the behaviour of the meter; adjusting the control knob will give a minimum reading but a complete null cannot be obtained.

The RF input to drive the bridge can be obtained from a dip oscillator, signal generator or low-power transmitter capable of giving up to about 1W of signal power. The signal source can be coupled to the bridge by a short length of coaxial cable directly or via a link coil of about four turns as shown on Fig 15.30.

If using the dip oscillator, care should be exercised in order to not overcouple with it as it may pull the frequency or, in the worst circumstances, stop oscillating. As the coupling is increased it will be seen that the meter reading of the bridge increases up to a certain point, after which further increase in coupling causes the meter reading to fall. A little less coupling than that which gives the maximum bridge meter reading is the best to use. The bridge can be used to find antenna impedance and also used for many other purposes, eg to find the input impedance of a receiver on a particular frequency.

Some practical uses

One useful application of this type of simple bridge is to find the frequency at which a length of transmission line is a quarter- or half-wavelength long electrically. If it is desired to find the frequency at which the transmission line is a quarter-wavelength, the line is connected to the bridge and the far end of it is left open-circuit. The bridge control is set to zero ohms. The dip oscillator is then adjusted until the lowest frequency is found at which the bridge shows a sharp null. This is the frequency at which the piece of transmission line is one quarter-wavelength. Odd multiples of this frequency can be checked in the same manner. In a similar way the frequency at which a piece of transmission line is a half-wavelength can also be found but in this case the remote end should be a short-circuit.

The bridge can also be used to check the characteristic impedance of a transmission line. This is often a worthwhile exercise, since appearances can be misleading. The procedure is as follows.

1. Find the frequency at which the length of transmission line under test is a quarter-wavelength long. Once this has been found, leave the oscillator set to this frequency.
2. Select a carbon resistor of approximately the same value as the probable characteristic impedance of the transmission line. Replace the transmission line by this resistor and measure the value of this resistor at the preset frequency. (*Note*: this will not necessarily be identical with its DC value).
3. Disconnect the resistor and reconnect the transmission line. Connect the resistor across the remote end of the transmission line.
4. Measure the impedance now presented by the transmission line at the preset frequency. The characteristic impedance (Z_0) is then given by:

$$Z_0 = \sqrt{Z_s \times Z_r}$$

where Z_s is the impedance presented by the line plus load and Z_r is the resistor value.

AN RF NOISE BRIDGE

The circuit described here is an adaptation of that described in the earlier edition of this book [3]. It allows a modulated signal to be obtained if desired by pulsing the supply to the noise generator. By switching the modulation on, it may aid detection of the balance point, especially if an AM receiver is used. The circuit consists of a wide-band noise generator followed by a bridge for making the measurements. The bridge allows the measurement of the parallel components of an unknown impedance to be measured.

Circuit description

The circuit diagram is shown on Fig 15.31. White noise is generated by the zener diode (D1) operating at low current. It may be possible to maximise the noise by suitable choice of the zener diode and R2. The frequency range of this noise should extend up to at least 200MHz. The noise source is followed by a three-stage wide-band amplifier (TR1, TR2 and TR3) to raise the noise level to the order of 100µV. This enables a receiver to be used as a null indicator.

The noise output from the amplifier is applied to a quadrifilar-wound toroid which forms a transformer T1. This provides two arms of a bridge circuit which has a variable resistor and capacitor in the third arm to obtain a balance against the antenna in the fourth arm. The bridge circuit is shown diagrammatically on Fig 15.32.

When the noise across the potentiometer/capacitor arm equals the noise across the antenna/capacitor combination the bridge is said to be balanced, and this occurs when the received noise signal is at a minimum. The impedance values can be obtained from the settings of RV1 and VC1. The inclusion of C7 allows an offset to be used so that inductive reactance can be measured. The mid-point setting of VC1 is equal to zero reactance. If a noise bridge is only required to measure the resistive part of the antenna impedance, omit C7 and VC1.

Timer IC1 is in astable mode and runs at about 850Hz with 50% duty cycle; this can be used to provide current supply for the zener circuit via S1 and thus modulate the noise source. The zener diode can be alternatively fed from the constant-voltage power supply line.

The unit requires about 25mA at 9V.

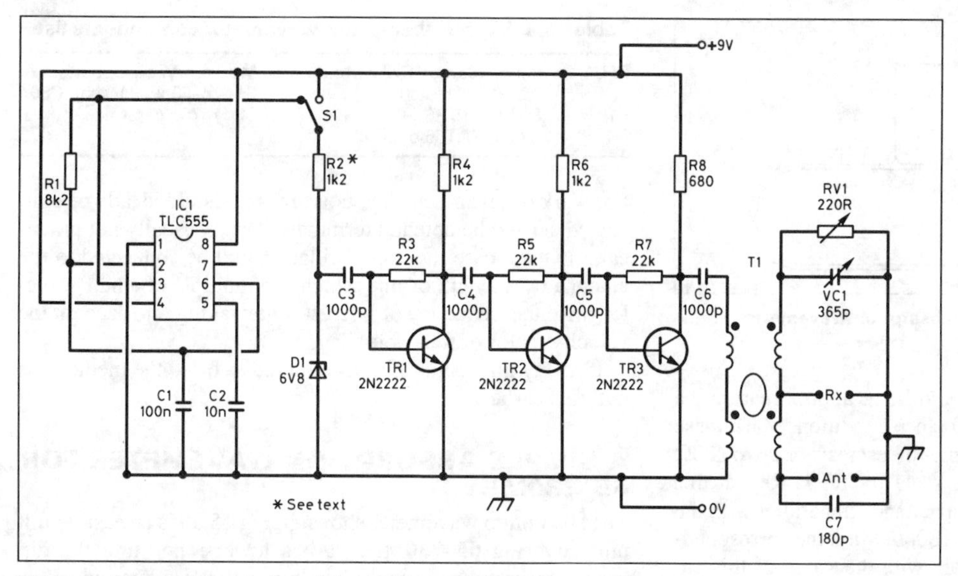

Fig 15.31. Circuit diagram of modulated RF noise bridge

Construction

A components list is given in Table 15.5. A PCB track layout is given in Fig 15.33 in Appendix 1 and corresponding component layout in Fig 15.34. Two types of core can be used for the toroid transformer (T1). If a dust-iron core is used (T50-6), then it should be wound with four windings of 14t each. If a ferrite-type core is used then it requires four windings of 6t each. The dust-iron core will provide a slightly lower frequency range. The ferrite core will have a slightly higher range, especially as the reduced number of windings has lower inter-winding capacitance.

Cut four lengths of 26swg (or similar) enamelled copper wire, twist them together and then thread them through the toroid to give the requisite number of turns and evenly spaced to cover the circumference. Divide the turns into two pairs, each pair consisting of two windings connected in series, the end of one winding connecting to the start of the other – be careful. Check that the two pairs are insulated from each other and endeavour to keep the lead lengths in the bridge as short as possible and symmetrical.

The variable resistor RV1 should be of high quality and with a carbon track – *not wire-wound!*

When constructing the circuits ensure that the noise generator and amplifiers are well screened from the bridge transformer and measuring circuit. The potentiometer case should not be earthed, and if it has a metal spindle this should not contact ground and be isolated from the user – it is easiest to use one with a plastic spindle.

The complete circuit should be mounted in a screened box such as a die-cast type with appropriate connection – eg UHF type or BNC. Keep any leads to the connectors to an absolute minimum with minimum capacitance. In order to avoid coupling into the measuring circuit of noise by way of current in earth loops, the earthed side of the noise source should not be joined to the general chassis earth of the bridge but should be taken by an insulated lead to the frame of the variable capacitor.

As in all high-frequency measuring circuits, lead inductances should be kept to an absolute minimum and where any lead length more than a fraction of an inch is unavoidable, copper foil at least 6mm wide should be used. All earth returns should be taken to the capacitor frame. Capacitor C7, which should be silver mica, can be soldered directly across the 'unknown' socket.

Calibration

Connect a test resistor (of a carbon type) across the unknown socket with the receiver tuned to 3.5MHz. Adjust RV1 and VC1 to give a null. The value of RV1 is at the position equal to the

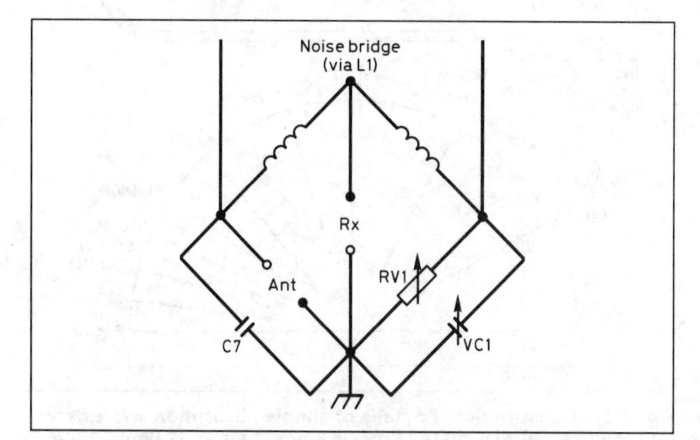

Fig 15.32. Diagrammatic representation of noise bridge

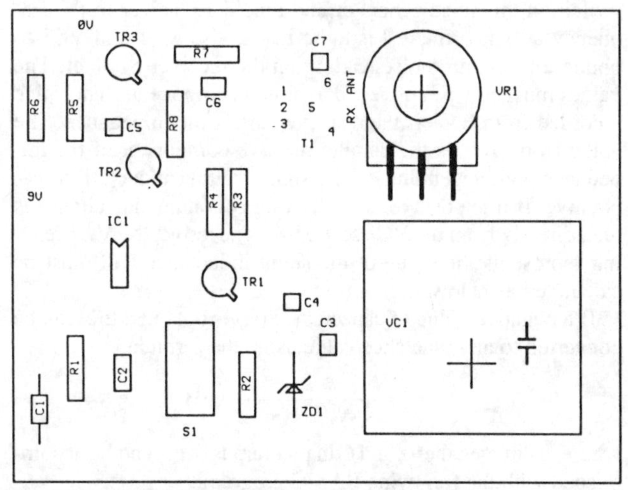

Fig 15.34. Component layout for the RF noise bridge

Table 15.5. RF noise bridge components list

R1	8k2	C1	100n, 50V, ceramic
R2*	1k2	C2	10n, 50V, ceramic
R3, 5, 7	22k	C3–6	1000p, 50V, ceramic
R4, 6	1k2	C7	180p silver mica
R8	680	VC1	365p, Jackson, type 01 Gang
RV1	220R pot, carbon	S1	SPCO switch
TR1, 2, 3	2N2222 or similar	T1	FT50-6 dust iron core
D1	6V8 zener, 400mW		4 windings, each 14t, or
IC1	TLC555		596100001 ferrite core
			4 windings, 6t

Resistors are 0.25W, 5% unless specified otherwise.

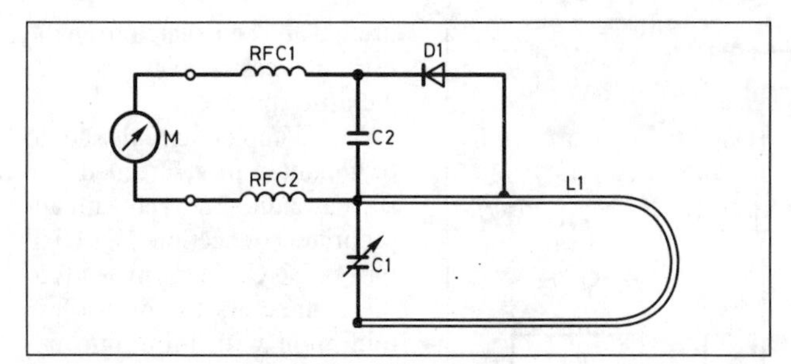

Fig 15.35. Circuit diagram of simple absorption wavemeter for 65–230MHz

test resistor and the capacitor should be at approximately the mid-mesh position or the zero reactance condition. Mark these positions. Repeat with different values of test resistor up to 220Ω in order to provide a calibrations scale for RV1. Repeat this operation with known values of capacitance in parallel with the resistor, up to a maximum value of 180pF. Mark the corresponding null positions on the VC1 scale with the value of this capacitance. Repeat this procedure at 28MHz to check the accuracy of the bridge. If the layout has been carefully attended to there should be little difference in the null positions.

To calibrate VC1 for negative capacitance values (ie inductance) it is necessary to temporarily place given values of capacitance in parallel with VC1. Gradually decrease the value of these capacitors (CT) from 150pF towards zero, obtaining null positions and marking the VC1 scale with the value of – (180 – CT) pF, ie if 100pF is substituted then the negative C value is 80pF.

Using the noise bridge

Connect the impedance to be measured to the 'unknown' socket, switch on the noise generator and tune the receiver to the frequency at which the test is to be made. Use RV1 and VC1 to obtain a minimum noise reading on the receiver S-meter. The values must now be converted to circuit components. The value recorded from RV1 is the resistive part of the impedance. The value from VC1 is the parallel reactive component of the impedance and, depending on the sign, is either inductive or capacitive. If it is positive, then the value of shunt capacitance is read directly from the VC1 scale. If it is negative, the VC1 reading represents the value of the shunt inductance and must be calculated as below.

If a negative value of capacitance (C) is obtained this can be converted to an inductance value using the formula:

$$L = \frac{1}{4\pi^2 f^2 (180 - C)} \ \mu\text{H}$$

where f is in megahertz and C in picofarads. This can be accomplished with the following BASIC program:

```
10 REM Noise bridge inductance calculation
20 CLS
30 C=180
40 INPUT "Negative C value in picofarads ",CV:
   CV=ABS(CV)
50 INPUT "Working frequency in megahertz ",F:
   F=F*1000000
60 K=1/(4*3.14159^2*F^2)
70 L=K*1/((C-CV)*1E-12)
80 L=L*1000000
90 PRINT "Inductance in microhenrys is ",L
100 END
```

Table 15.6. Simple absorption wavemeter components list

RFC1, 2	80t of 40swg ECW wound on 10k, 0.5W resistor	M1	1mA FSD or better
L1	See Fig 15.36(a)	C1	4–50pF, Jackson C804
D1	OA91, BAT85 or similar	C2	470p ceramic

For work on an antenna, the noise bridge should ideally be connected across the antenna terminals. This is usually not practical, in which case the noise bridge should be connected to the antenna by a length of line which is a multiple of a half-wavelength at the frequency of interest (*note:* taking into account the velocity factor of the cable).

If at any time calibration is required, refer to the methods as outlined earlier.

A SIMPLE ABSORPTION WAVEMETER FOR 65–230MHz

The absorption wavemeter shown in Fig 15.35 is an easily built unit covering 65–230MHz. For a lower-range unit the dip oscillator described in the following section can be used.

Construction is straightforward and all the components, apart from the meter, are mounted on a Perspex plate of thickness 3 or 4mm and measuring 190 × 75mm. Details of the tuned circuit are shown on Fig 15.36(a) and should be closely followed. The layout of the other components is not critical provided they are kept away from the inductor. A components list is given in Table 15.6.

For accurate calibration a signal generator should be required but, provided the inductance loop is carefully constructed and the knob and scale are non-metallic, the dial markings can be determined from Fig 15.36(b).

In operation the unit should be loosely coupled to the circuit under test and the capacitor tuned until the meter indicates resonance (a maximum). For low-power oscillators etc a more sensitive meter should be used (eg 50μA or 100μA).

The wavemeter can also be used as a field strength indicator when making adjustments to VHF antennas. A single-turn coil should be loosely coupled to the wavemeter loop and connected via a low-impedance feeder to a dipole directed towards the antenna under test.

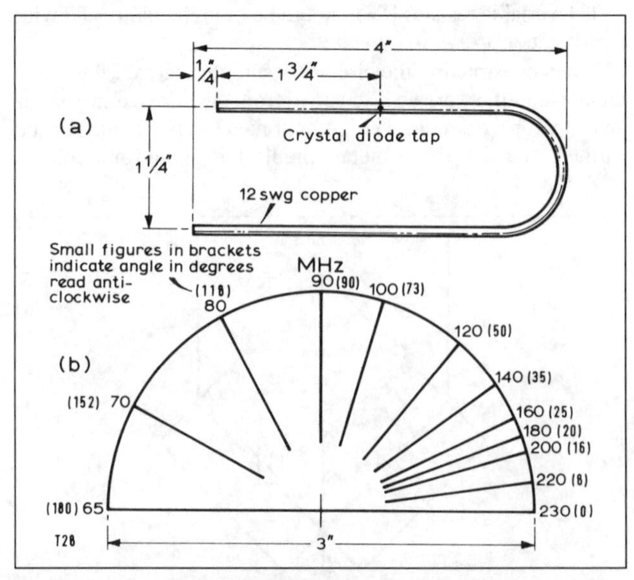

Fig 15.36. Constructional details of simple absorption wavemeter: (a) inductor; (b) dial plate. ¼in = 6.3mm, 1¼in = 31.8mm, 1¾in = 44.5mm, 3in = 75mm, 4in = 101.6mm

A FET DIP OSCILLATOR FOR 0.8–170MHz

The original version of this design by Tony Bailey, G3WPO, appeared in *Radio Communication* in November 1981 but the revised Mk2 version which is described here (Fig 15.37) is from 1987 [4]. The specification is:

- Dip and wavemeter function
- Coverage of 0.8–170MHz
- Audio and meter indications
- Battery operation

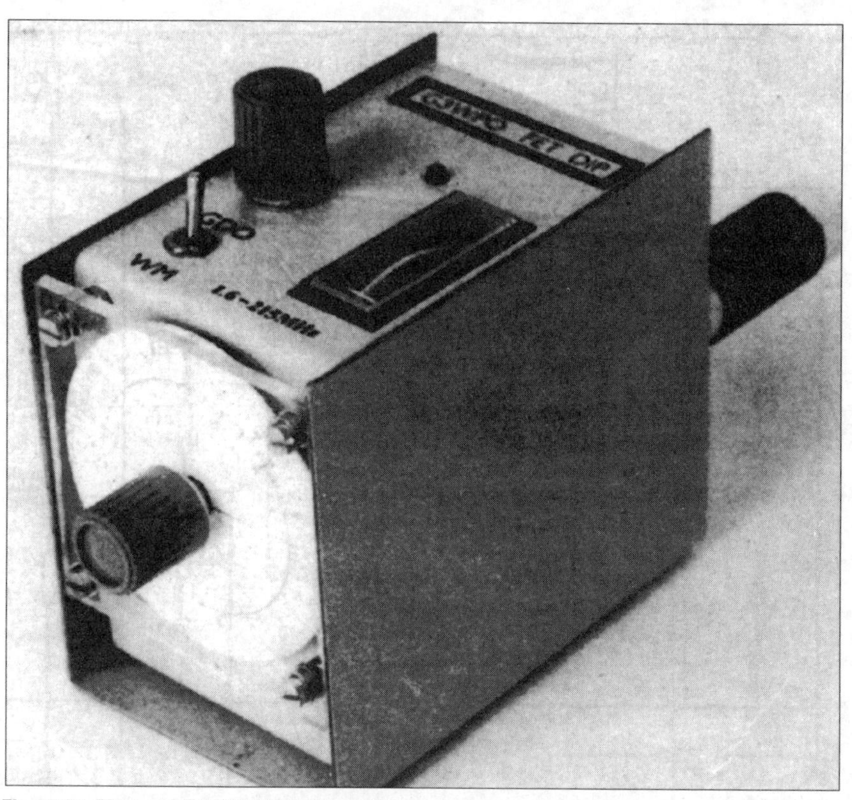

Fig 15.37. Photo of G3WPO Mk2 dip oscillator

Circuit description

The circuit diagram is given on Fig 15.38. The instrument is based on a Kalitron oscillator which is formed by the two MOSFETs TR1 and TR2.

The frequency-determining components are C1 and L1 (the plug-in coil); resistor Rs is included as part of the plug-in coil and is a gain-setting resistor.

The RF from the oscillator is detected by D2 and D3 which are Schottky barrier-type diodes with a good frequency response and lower forward voltage drop compared with devices such as the ubiquitous 1N914.

The detected DC is applied to the base of amplifier TR5 which controls the current flowing through the meter M and the audio multivibrator formed by TR3 and TR4. The frequency of the multivibrator is determined by C11/R19 and C10/R18. As the current through TR5 increases, the note

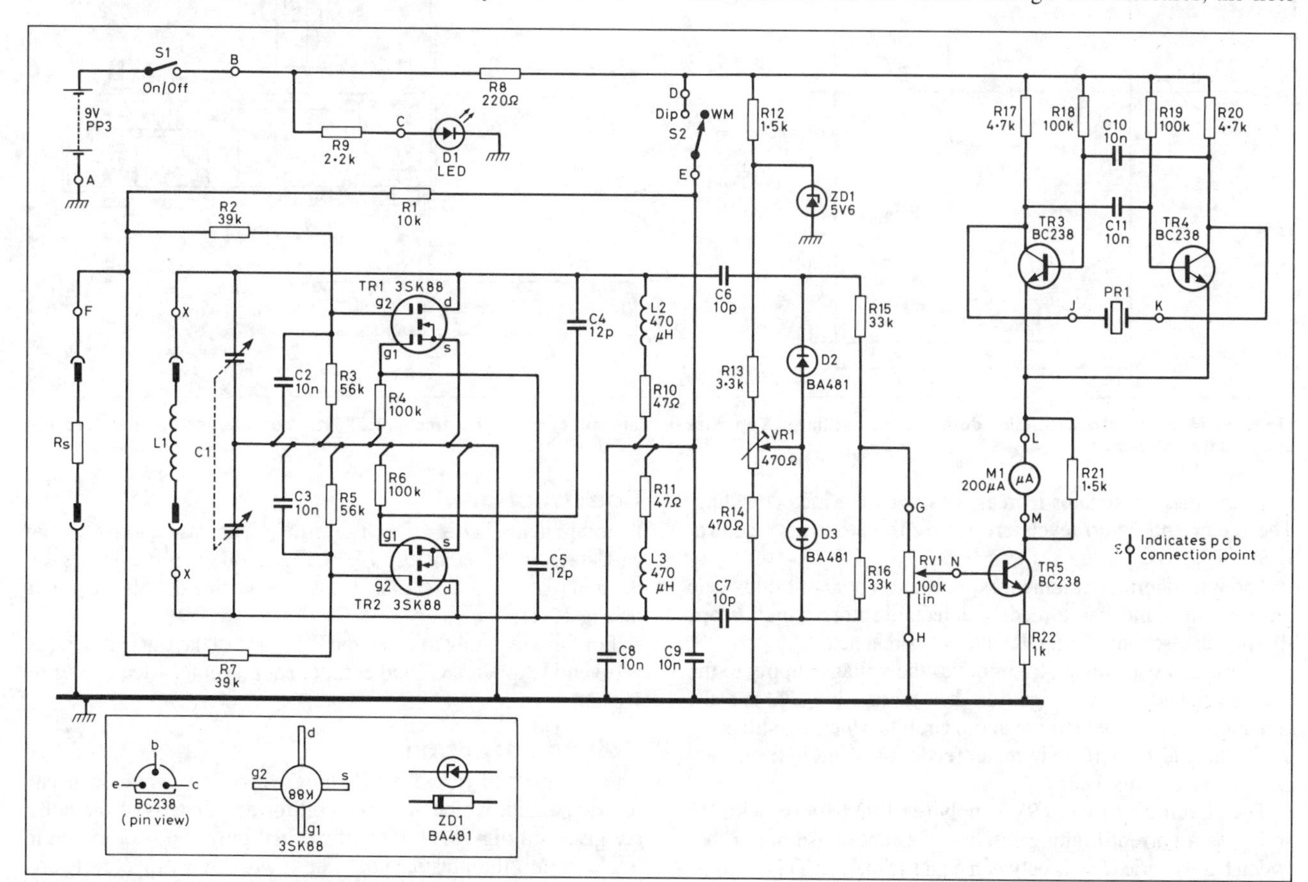

Fig 15.38. Circuit diagram of G3WPO Mk2 dip oscillator

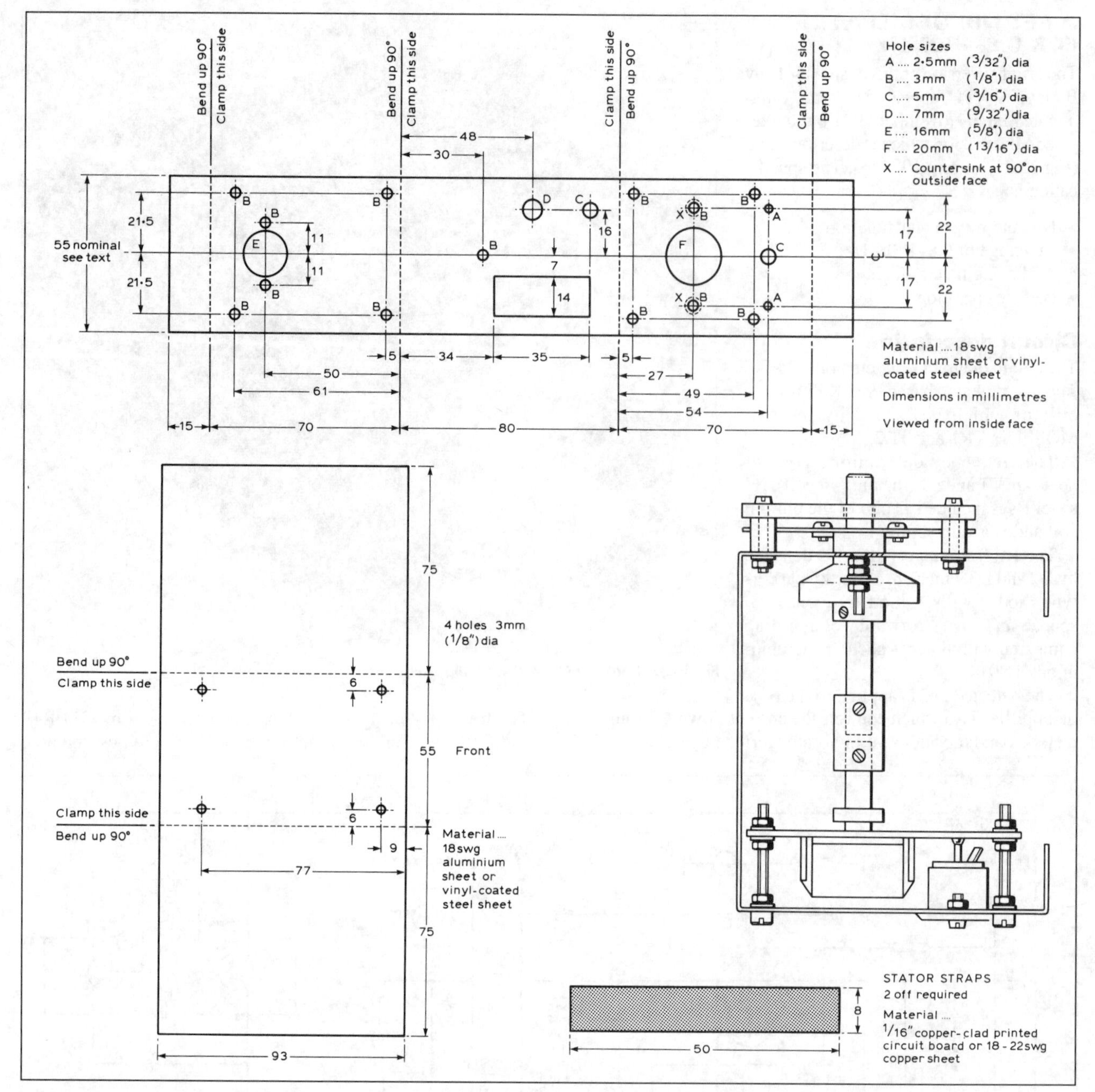

Fig 15.39. Mechanical construction details of dip oscillator. Top: chassis plan, left: cover, right: mechanical details of case and drive with stator straps detailed below

from the piezo-resonator increases as does the meter reading. The meter and audio levels are set by the sensitivity control RV1.

The multivibrator commences oscillation at about mid-scale on the meter and has a readily detectable note which drops sharply as resonance of the RF circuit is reached.

In use as a wavemeter, S2 removes the voltage supply to the oscillator. The received signal is then rectified by D2 and D3 and applied to the meter drive/audio circuits which are still powered. There is an increase in meter reading and audio frequency as resonance is reached.

The circuit runs from a 9V supply (eg PP3 battery) with R8 acting as a current-limiting resistor. Current consumption between bands varies and is between 5 and 15mA. LED D1 acts as a reminder of the equipment being ON.

Construction

A components list is given Table 15.7. The case and mechanical details are given on Fig 15.39. Fig 15.40 gives the details for the dial plate, dial cursor and scale. The wiring details are given on Fig 15.41.

For those wishing to make the PCB, a track layout (Fig 15.42) is given in Appendix 1 and component placement details in Fig 15.43).

Coil construction

These consist of five-pin DIN plugs fitted into rigid electrical plastic conduit which acts as a coil former. Details of the coils are given on Fig 15.44. Only the actual plug end is used and it is glued into the plastic tube. The lowest four range coils are wound directly onto the formers, and the two highest ranges are

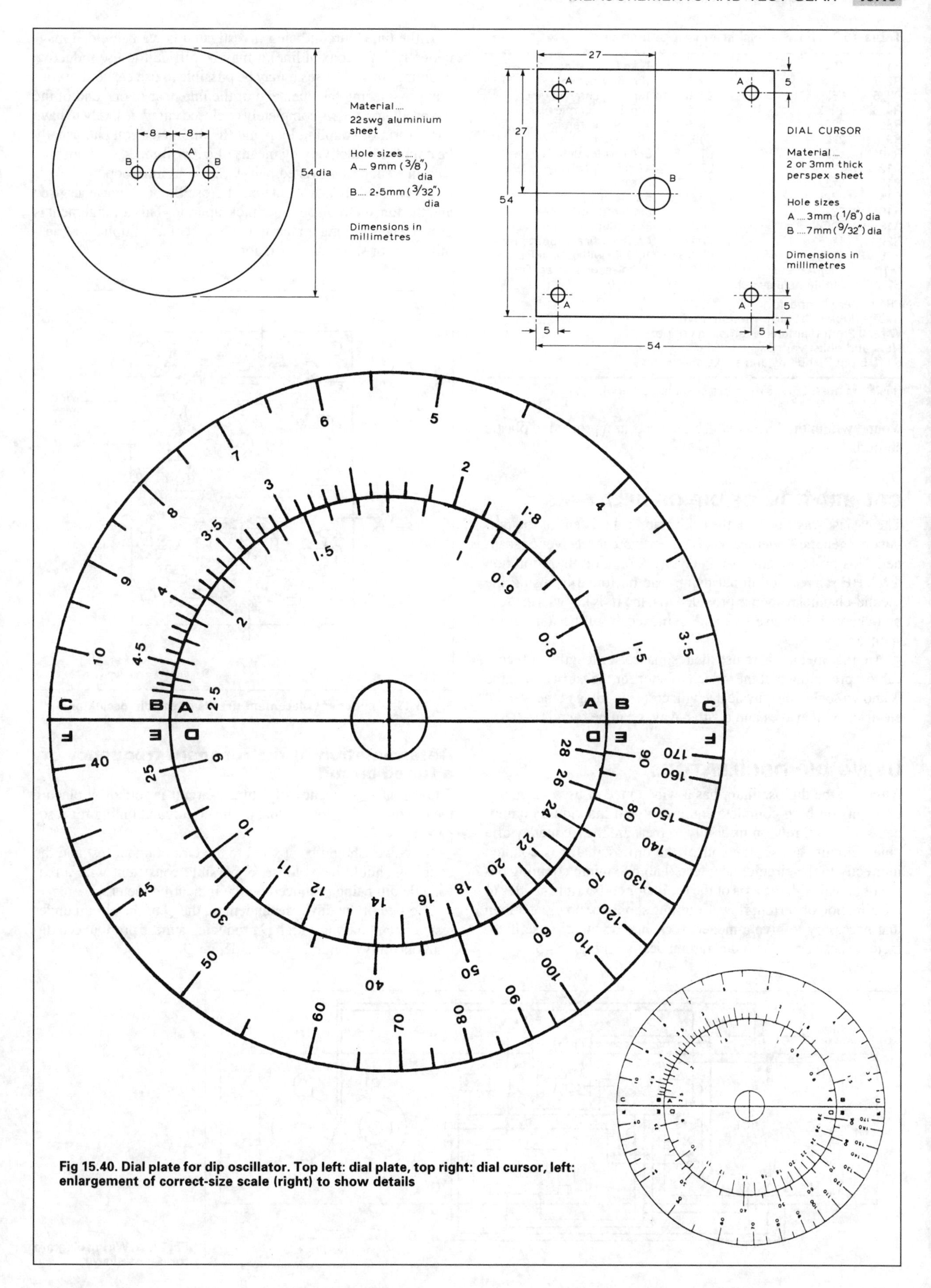

Fig 15.40. Dial plate for dip oscillator. Top left: dial plate, top right: dial cursor, left: enlargement of correct-size scale (right) to show details

Table 15.7. FET dip oscillator components list

R1	10k	C1	Toko Polyvaricon, 2 × 266p
R2, 7	39k	C2, 3	10n, 50V ceramic
R3, 5	56k	C4, 5	12p, 5%, ceramic disc
R4, 6	100k	C6, 7	10p, 5% ceramic disc
R8	220R	C8, 9	10n, 50V ceramic
R9	2k2	C10, 11	10n mylar
R10, 11	47R	L1	See text for details
R12, 21	1k5	L2, 3	470μH, eg Toko type 7BS
R13	3k3	TR1, 2	3SK88
R14	470R	TR3, 4, 5	BC238 or similar npn
R15, 16	33k	D1	3mm red LED
R18, 19	100k	ZD1	5V6, 400mW zener
R22	1k	VR1	470R, vertical mount preset
R18, 19	100k	RV1/S1	100k lin with ON/OFF switch
M1	200μA meter	PR1	Piezo-resonator, eg Toko PB2720
S2	Single-pole switch		

PP3 battery connector
Shaft coupler: 6:1 slow motion drive
Wire: 0.2mm diameter enamelled copper
Nuts and bolts, various
6 off, 5-pin DIN plugs and sockets – see text

Resistors are 0.25W, 5% unless specified otherwise

wound within the former which then acts as a protective plastic shroud.

CALIBRATION OF DIP OSCILLATORS

The easiest way to check the calibration is to listen for the output on a general-coverage receiver, an amateur receiver or scanner. This probably allows a good check on the calibration into the VHF range. Additional points can be found by using the second-channel response provided that the IF is known (the second-channel response is 2 × IF removed from the normal response).

Another method is to use the resonances of lengths of feeder cables, providing that the velocity factor for the particular cable is known so that the physical length corresponding to the wanted electrical half-waves and quarter-waves can be found.

USING DIP OSCILLATORS

Although the dip oscillator has a wide range of uses for measurements on both complete equipment and individual components, these all rely on its ability to measure the frequency of a tuned circuit. In use, the coil of the dip oscillator is coupled indirectly to the circuit under test, with maximum coupling being obtained with the axis of the oscillator coil at right-angles to the direction of current flow. Coupling should be no greater than that necessary to give a moderate change on the dip oscillator meter. These are shown diagrammatically on Fig 15.45.

If the tuned circuit being investigated is well shielded magnetically (eg a coaxial line) it may be difficult to use inductive coupling. In such cases it may be possible to use capacitive coupling by placing the open end of the line near to one end of the dip oscillator coil. A completely enclosed cavity is likely to have some form of coupling loop and the dip meter coil can usually be coupled inductively by means of a low-impedance transmission line such as a twisted pair with a coupling loop.

When used as a wavemeter, the oscillator is not energised and the tuned circuit acts as a pick-up loop. This arrangement is useful when looking for harmonic output of a multiplier or transmitter or for spurious oscillations.

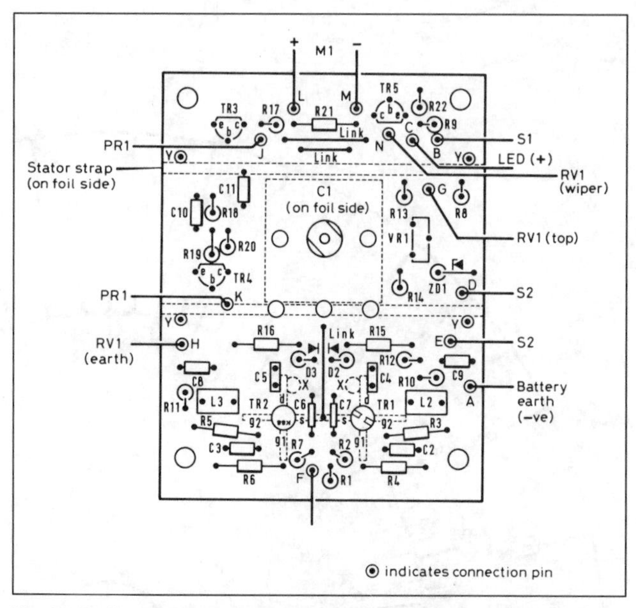

Fig 15.43. Component placement details of the dip oscillator

Determination of the resonant frequency of a tuned circuit

The resonant frequency of a tuned circuit is found by placing the dip oscillator close to that of the circuit and tuning for resonance.

No power should be applied to the circuit under test and the coupling should be as loose as possible consistent with a reasonable dip being produced on the indicating meter.

The size of the dip is dependent on the Q of the circuit under test, a circuit having a high Q producing a more pronounced dip than one only having low or moderate Q.

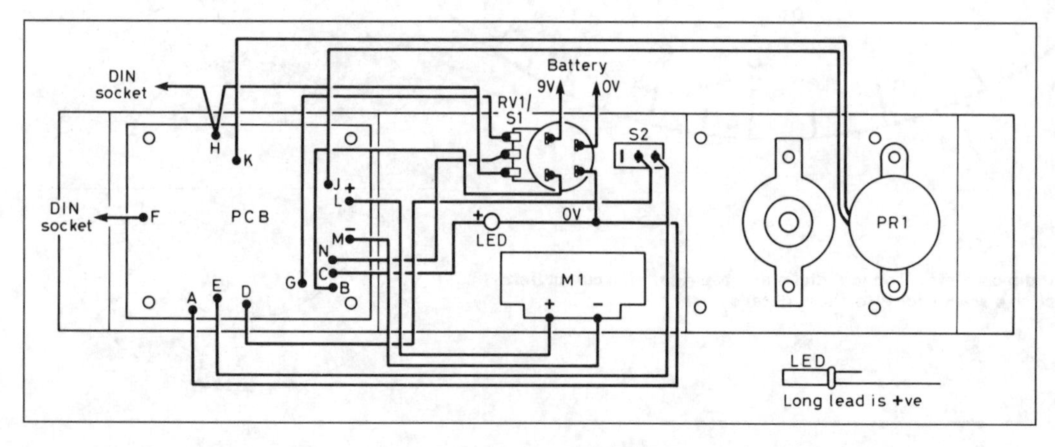

Fig 15.41. Wiring diagram for dip oscillator

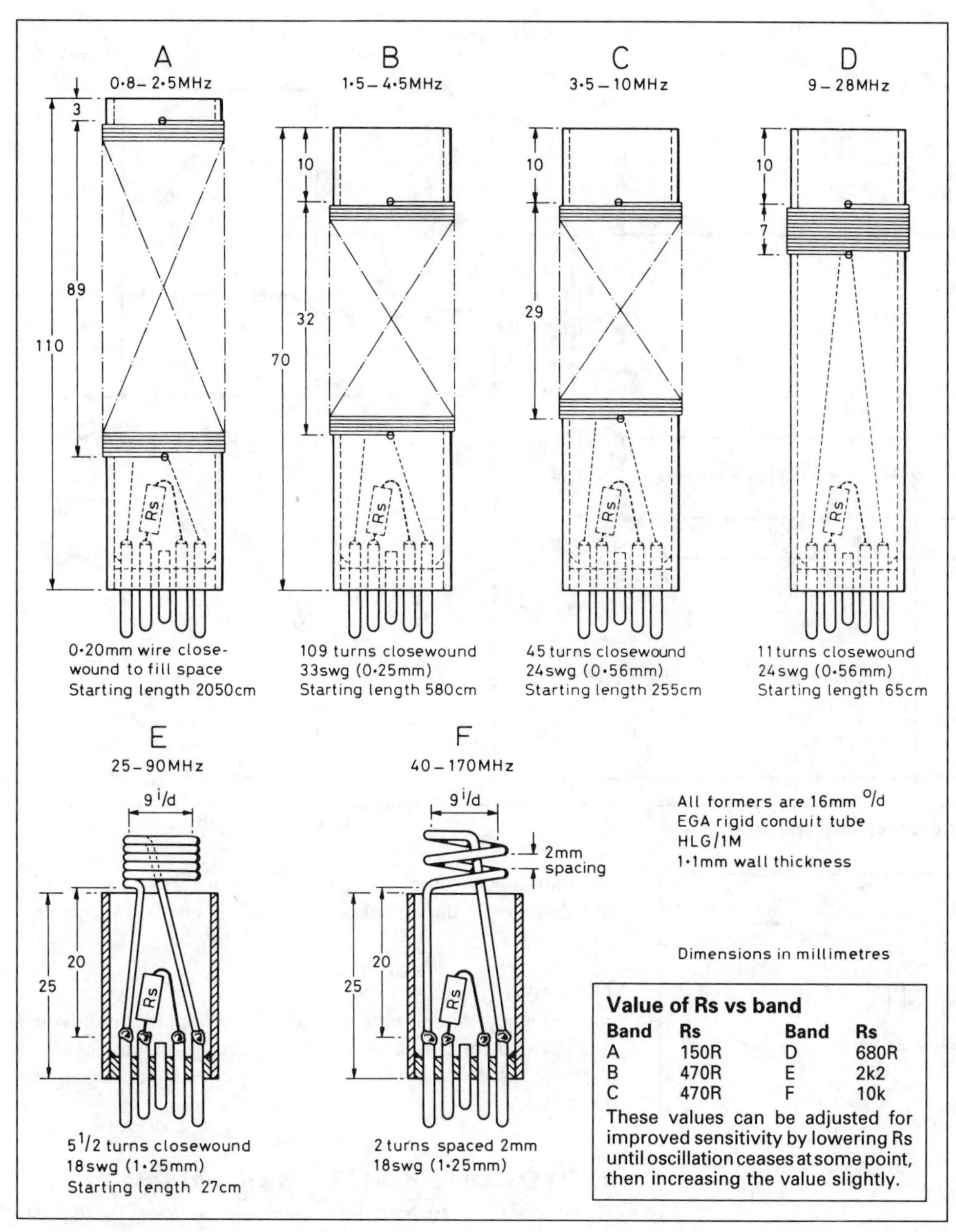

Value of Rs vs band

Band	Rs	Band	Rs
A	150R	D	680R
B	470R	E	2k2
C	470R	F	10k

These values can be adjusted for improved sensitivity by lowering Rs until oscillation ceases at some point, then increasing the value slightly.

Fig 15.44. Coil details for dip oscillator

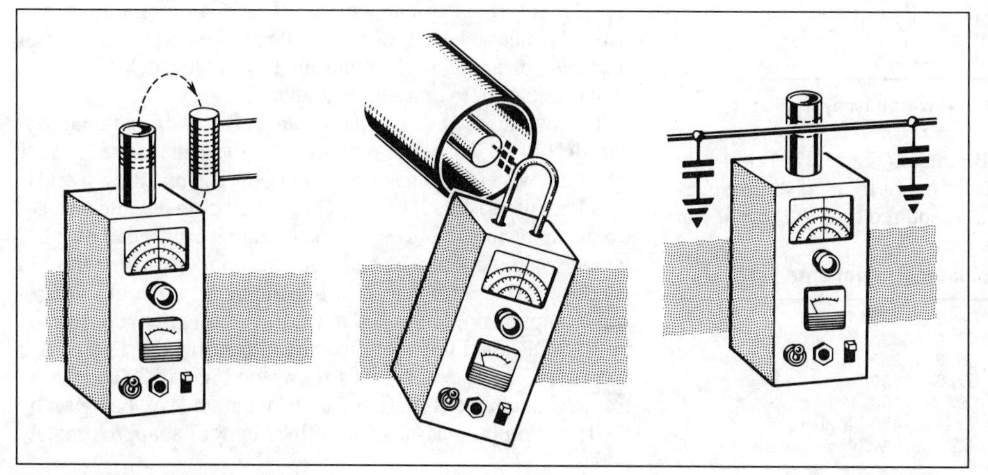

Fig 15.45. Using a dip oscillator

Use as absorption wavemeter

The instrument may be used as an absorption wavemeter by switching off the power supply in the dip oscillator and then using in the normal way. In this case power has to be applied to the circuit under test. Resonance is detected by a deflection on the meter due to rectified RF received. It should be noted that an absorption wavemeter will respond to a harmonic if the wavemeter is tuned to its frequency.

A LOW-FREQUENCY OSCILLATOR FOR 10Hz–100kHz

The circuit diagram for this oscillator is shown on Fig 15.46. It is based on a Wien bridge oscillator formed around IC1a and buffered by IC1b. The main frequency-determining components are R1/R2 and RV2 with capacitors C1 to C8. In the configuration shown, stable oscillation can occur only if the loop gain remains at unity at the oscillation frequency. The circuit achieves this control by using the positive temperature coefficient of a small lamp to regulate the gain as the oscillator varies its output. Potentiometer RV3 forms the output level control, with R4 giving a defined output resistance of approximately 600Ω and C11 providing DC isolation. Capacitors C9 and C10 provide power supply line decoupling.

The approximate ranges provided are:

Range 1	10Hz–100Hz
Range 2	100Hz–1kHz
Range 3	1kHz–10kHz
Range 4	10kHz–100kHz

The exact range is dependent on the tolerance of the components used and ambient temperature variations.

The circuit requires a symmetrical plus and minus supply between 9 and 15V.

Construction

A components list is given in Table 15.8. The layout of the

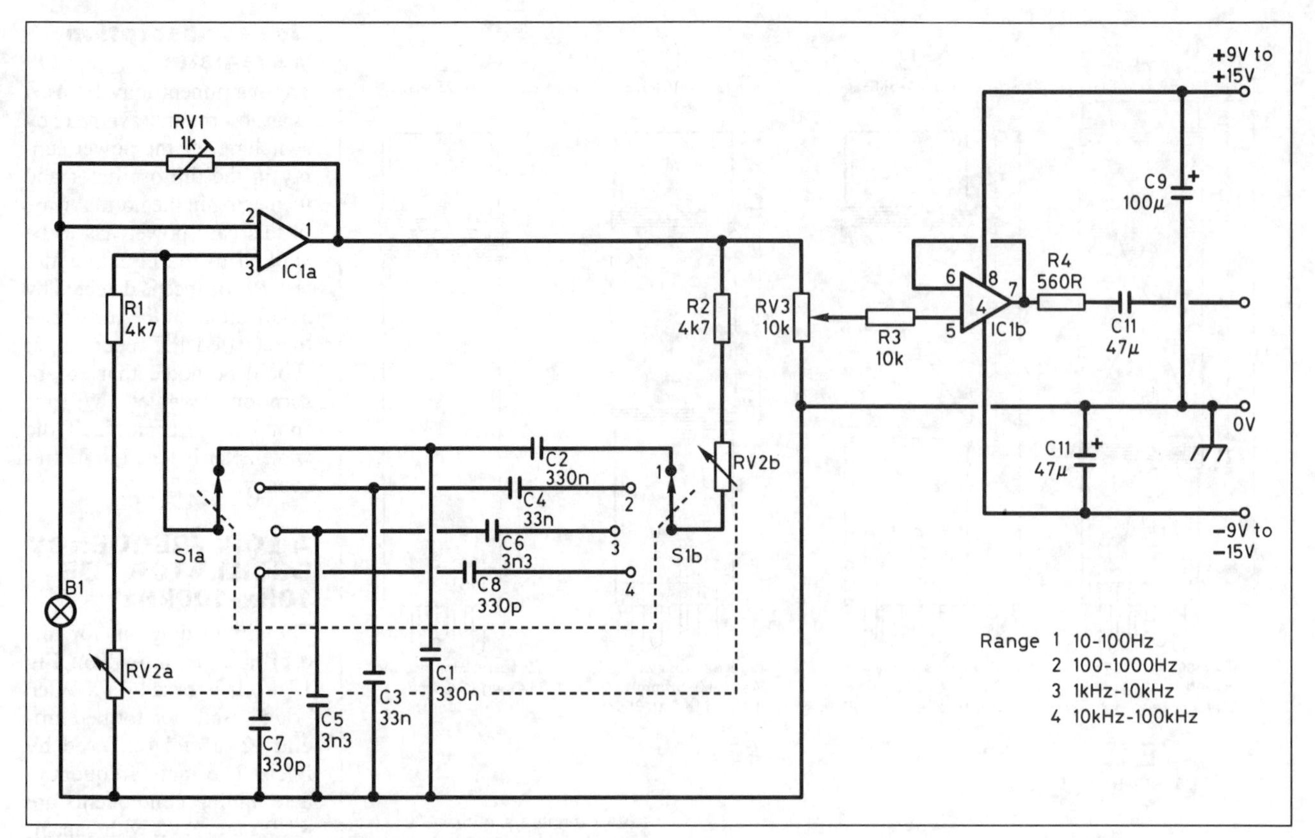

Fig 15.46. Circuit diagram of low-frequency oscillator

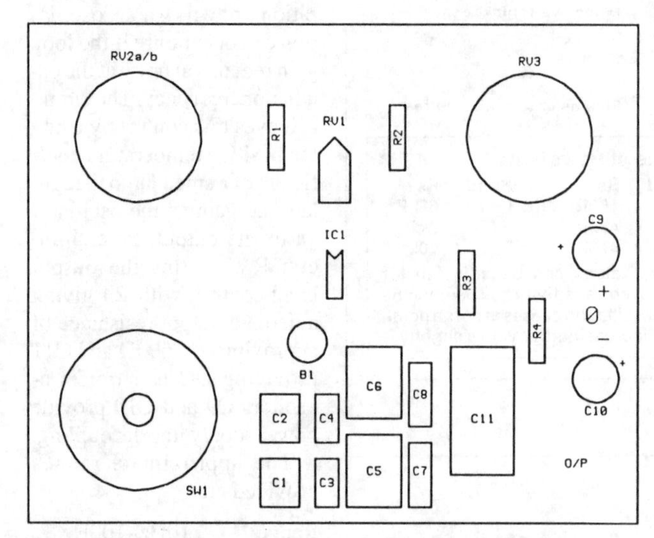

Fig 15.48. Component layout of the low-frequency oscillator

circuit is not critical, and a PCB pattern (Fig 15.47) is given in Appendix 1 and a component layout in Fig 15.48. If some ranges or the output level control are not required then the layout can be tailored accordingly. The feedback resistor RV1 should be adjusted so that the output on all ranges is just below the clipping level.

Table 15.8. Low-frequency oscillator components list

R1, 2	4k7	C1, 2	330n
R3	10k	C3, 4	33n
R4	560R	C5, 6	3n3
RV1	1k trimmer	C7, 8	330p
RV2	47k dual gang pot	C9, 10	100μ, 25V
RV3	10k lin pot	C11	47μ bipolar
B1	28V, 40mA bulb	IC1	LM358

All resistors are 0.25W, 5% unless specified otherwise.

Testing

No frequency calibration is required of this circuit but it would be wise to check with a frequency counter that the ranges are as suggested. An oscilloscope is required for setting up the adjustment of RV1.

A TWO-TONE BURST OSCILLATOR

This is a self-contained unit for producing two tones for the testing of sideband transmitters. The frequencies produced are approximately 1.4kHz and 2kHz, ie they are not harmonically related. It is formed around a circuit (Fig 15.49) previously published [1] but which has been modified to give a burst output (if required) of about 10Hz, controlled by S2. Switch S1 allows either tone or both tones to be output.

The two sine-wave oscillators are formed around quad op-amp IC1. The sine-wave outputs from these are routed via S1 which allows either tone through or both. The resulting signal is summed by resistors R13 and R14 before being fed into an electronic attenuator IC2. The attenuator gain is controlled by a DC voltage on pin 2. For the condition of continuous output this pin is connected to –6V via R15. For the burst mode the square-wave output from IC3 is fed to the integrating network formed by R15/C13, and this modified signal controls the gain of IC2 and provides a burst output as shown on Fig 15.52. The output from the attenuator is buffered by IC4, output level being set by RV1. The output resistance is defined by R17 at approximately 600Ω.

A +6, 0, –6V DC supply is required.

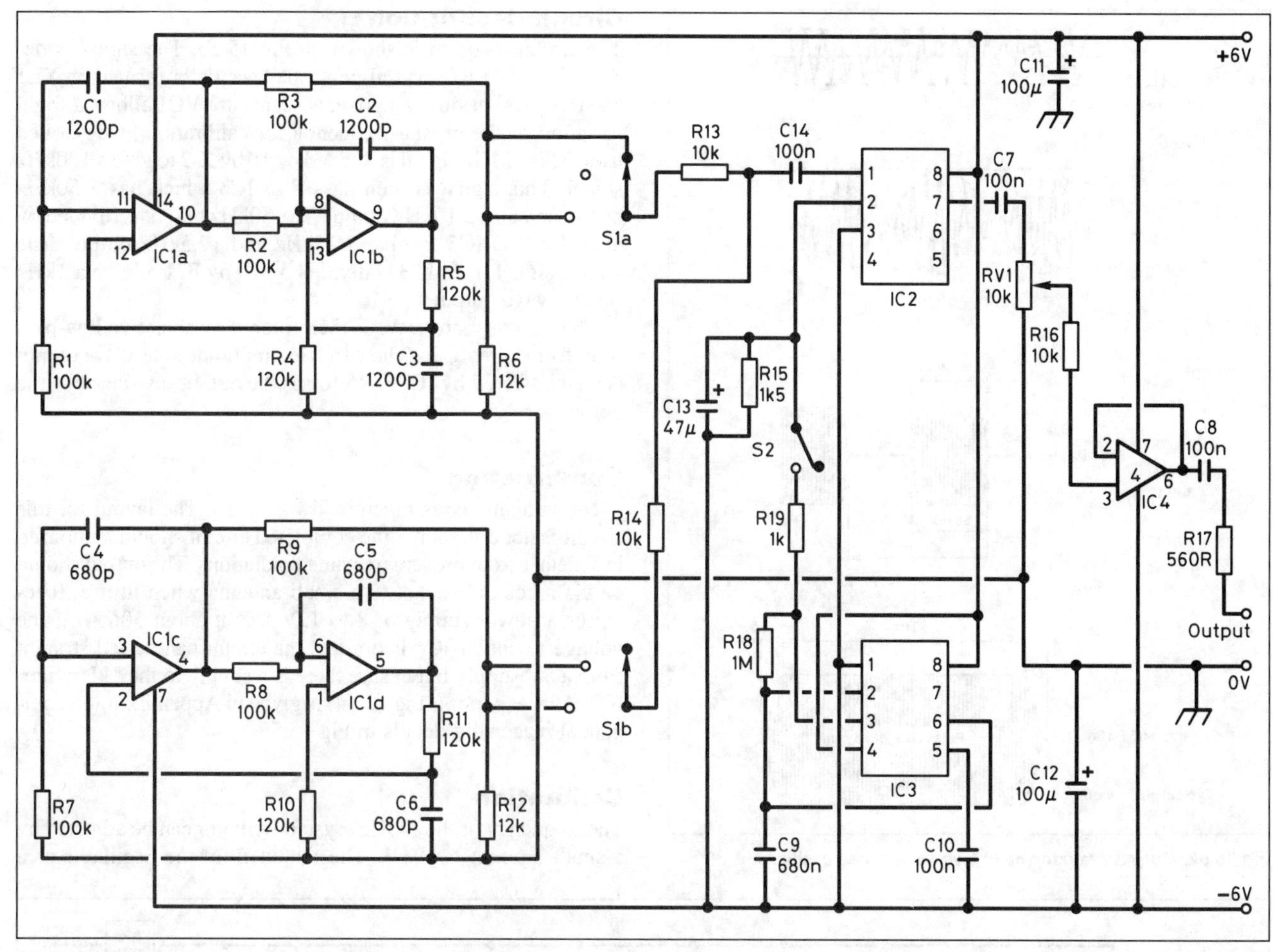

Fig 15.49. Circuit diagram of two-tone burst oscillator

Construction

A components list is given in Table 15.9. Construction is straightforward and requires no special techniques. The burst option can be omitted if required but ensure that pin 2 of IC2 is connected to –6V via R15; C13 can be omitted in this case as it only helps shape the envelope of the pulsed output. A PCB pattern (Fig 15.50) is given in Appendix 1 and corresponding component placement details in Fig 15.51.

Testing

No calibration of the circuit is necessary but the operation should be checked. Fig 15.52 gives typical outputs for the tone output, continuous or pulsed.

Table 15.9. Two-tone burst oscillator components list

R1, 2	100k	C1, 2, 3	1200p polystyrene
R3	100k	C4, 5, 6	680p polystyrene
R4, 5	120k	C7, 8	100n
R6	12k	C9	680n
R7, 8	100k	C10	100n ceramic
R9	100k	C11, 12	100μ
R10, 11	120k	C13	47μ, 16V
R12	12k	C14	100n ceramic
R13, 14, 16	10k	IC1	LM3900
R15	1k5	IC2	MC3340P
R17	560R	IC3	TLC555M
R18	1M	IC4	741
R19	1k	S1	4-pole, 3-way rotary switch
RV1	10k lin pot	S2	SPST PCB mounting

All resistors are 0.25W, 5% unless specified otherwise.

A CRYSTAL-BASED FREQUENCY MARKER

The purpose of this unit is to produce a 'comb' of output frequencies which are all based on a crystal. The unit described here gives outputs at harmonics of 1MHz, 100kHz, 25kHz, 12.5kHz and 10kHz with an additional output of a sine wave at 1kHz which may be useful as an accurate modulation signal. The sine wave output has an output resistance of approximately 600Ω and maximum amplitude of approximately 2.5V peak to peak.

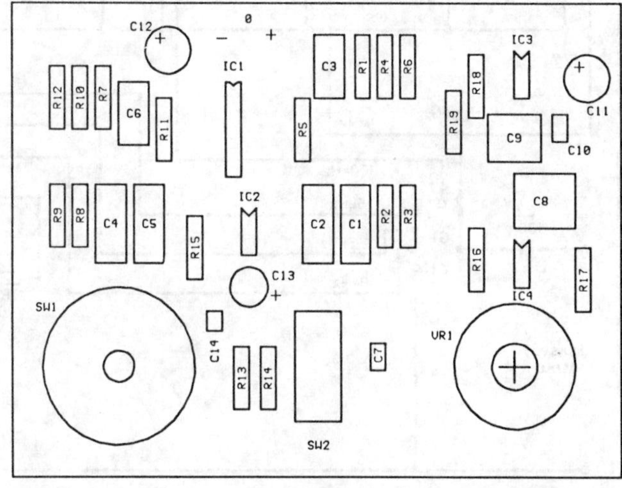

Fig 15.51. Component placement details for the two-tone burst oscillator

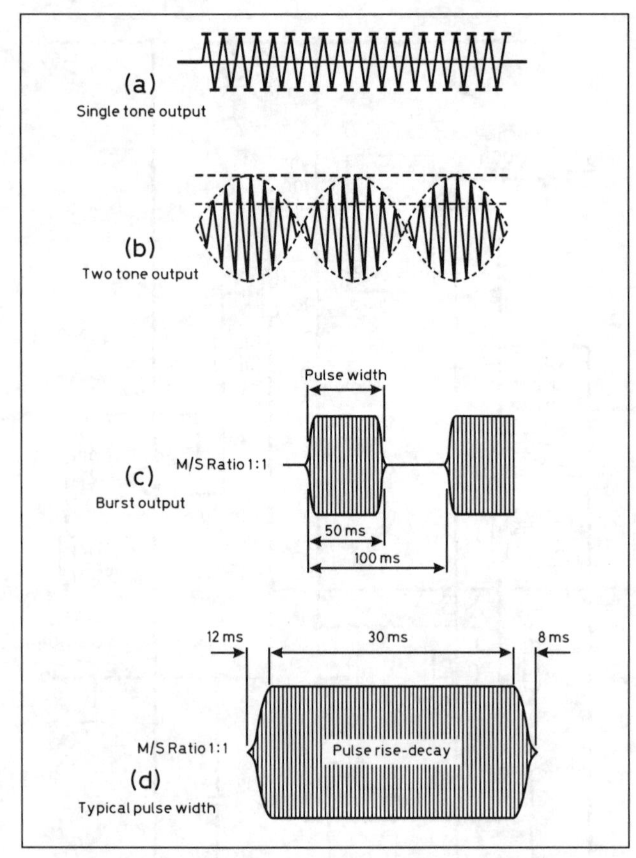

Fig 15.52. Output waveforms of two-tone burst oscillator

Circuit description

The circuit diagram is shown on Fig 15.53. The signal is derived from a 1MHz crystal-controlled oscillator formed by XL1 and IC1 plus various components. Capacitor VC1 allows a slight variation of the crystal frequency for calibration as described later. This 1MHz signal is divided by 10 by IC2 to give a 100kHz signal. This signal is then passed to IC3 which has a 50kHz output and also a 10kHz output. The 50kHz output is divided by dual flip-flop IC5 to give a 25kHz and 12.5kHz output. The 10kHz signal from IC3 is divided by 10 by IC4 to give a 1kHz square-wave output.

The 1kHz square wave is then filtered by an active low-pass filter formed by IC7a. The variable-amplitude sine-wave output is then buffered by IC7b. R5 forms the output resistance of the buffer.

Construction

A components list is given in Table 15.10. The layout for this circuit is not critical but the completed circuit should be housed in a metal box to prevent unwanted radiations. The output should be via a coaxial socket to a small antenna when in use. It requires a power supply of 8 to 12V DC at about 50mA. If the voltage regulator IC6 is omitted the circuit can be fed straight from a 5V supply but ensure there is a supply to the 1kHz filter IC7. A PCB pattern (Fig 15.54) is given in Appendix 1 and component placement details in Fig 15.55.

Calibration

The frequency of the 1MHz crystal oscillator can be adjusted by a small amount by VC1. The output from the oscillator or a

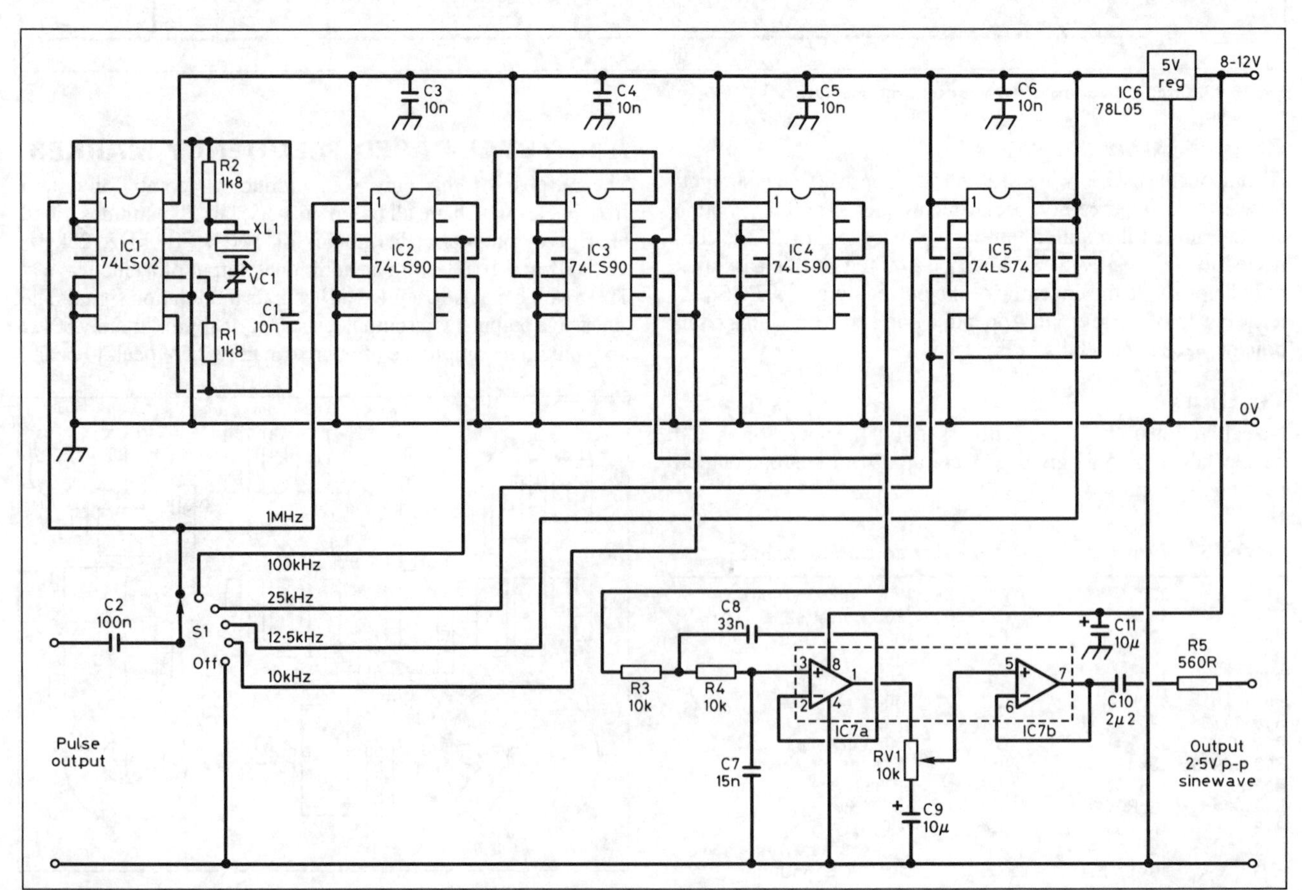

Fig 15.53. Circuit diagram for crystal-based frequency marker

Table 15.10. Crystal-based frequency marker components list

R1, 2	1k8	IC1	74LS02
R3, 4	10k	IC2, 3, 4	74LS90
R5	560R	IC5	74LS74
RV1	10k lin pot	IC6	78L05
C1	10n ceramic	IC7	LM358
C2	100n ceramic	S1	2-pole, 6-way rotary switch
C3–6	10n ceramic	XL1	1MHz crystal, HC6U
C7	15n	VC1	30p trimmer
C8	33n	IC sockets	14-pin 5 off
C9, 11	10μ, 25V tant bead		8-pin 1 off
C10	2μ2		

Resistors are 0.25W, 5% unless specified otherwise.

harmonic should be checked against an accurate frequency source.

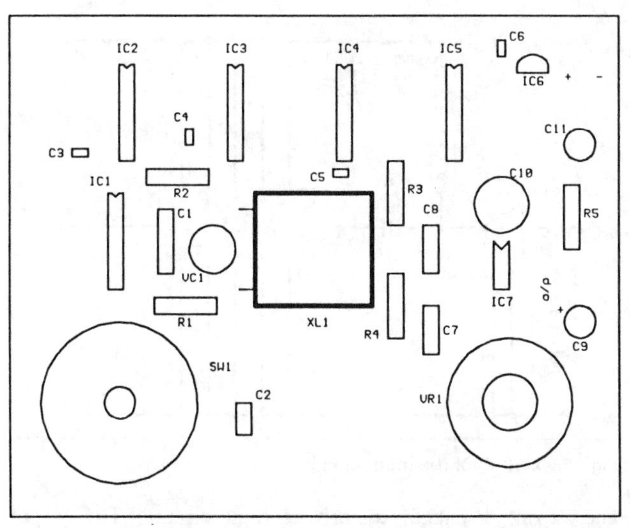

Fig 15.55. Component placement details for the crystal-based frequency marker

A 600MHz DIGITAL FREQUENCY COUNTER

This frequency counter is capable of operating up to about 600MHz with a sensitivity of 50mV or better. It is based on an eight-digit frequency counter IC which will count to 10MHz. The range is increased by the use of prescalers which divide the incoming frequency down to the 10MHz range. Fig 15.56 shows a block diagram of the system. It offers three input ranges, 0–6MHz (typically 10MHz), 0–60MHz and 0–600MHz. It is possible to build a 0–60MHz counter and then add the additional components later to increase the capability to about 600MHz.

The *gate time*, ie the *count period*, can be switched between 1s and 10s. At the 1s position the counter offers a seven-digit display, while at the 10s position the counter provides an eight-digit display. The decimal point is switched to denote the megahertz position, all digits to the left of it representing megahertz. The supply requirement is 5V at approximately 300mA.

References [5], [6] and [7] provided the basic information for the design; for extension in frequency ranges and sensitivity the reader is referred to [8].

Input circuit up to 60MHz

The circuit diagram is shown on Fig 15.57. The input impedance is approximately 1MΩ, with junction FET TR1 acting as a buffer.

The signal is then fed to IC3 which contains a group of transistors capable of operation up to 1GHz. This provides amplification

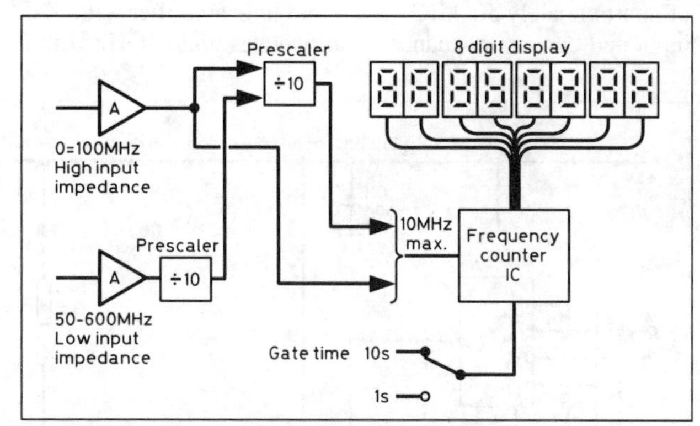

Fig 15.56. Block diagram of frequency counter

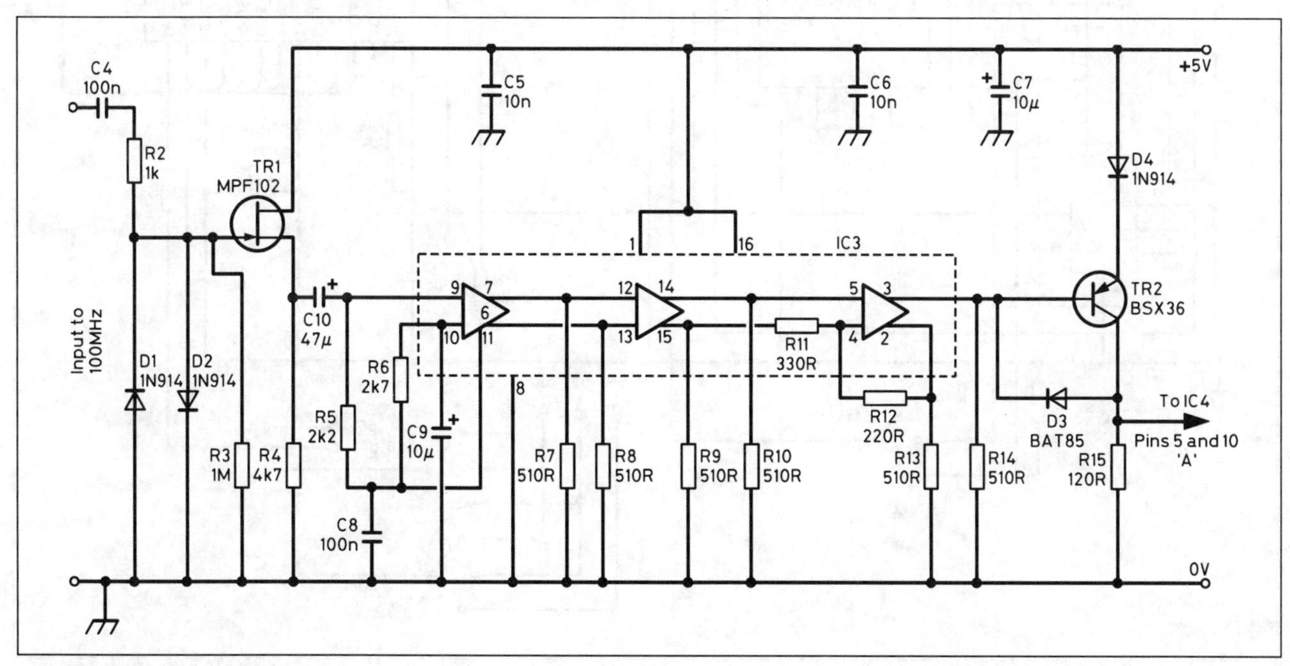

Fig 15.57. Up to 60MHz input circuit

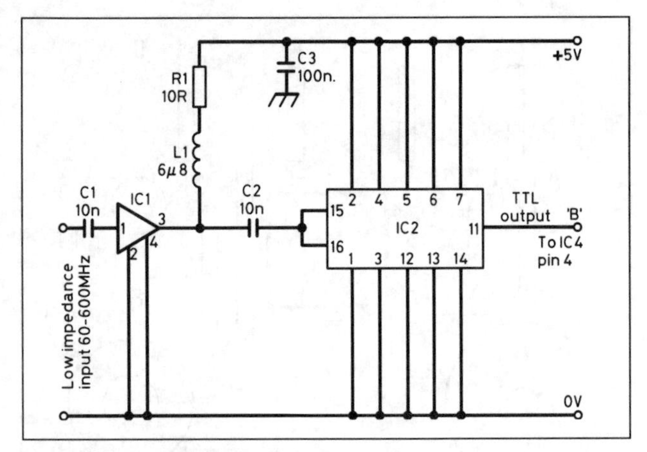

Fig 15.58. 10–600MHz input circuit

and an output capable of driving logic circuits. The output transistor TR2 is a high-speed PNP switch which translates the ECL output logic level of IC3 to a TTL-compatible level.

Input circuit 10–600MHz

This circuit is shown on Fig 15.58 and has an input impedance of approximately 50Ω. IC1 is a monolithic amplifier with 50Ω input and output impedances, and operates up to 1GHz with a

Table 15.11. Digital frequency counter components list

R1	10R	C1, 2	10n ceramic disc
R2	1k	C3, 4	100n ceramic disc
R3	1M	C5, 6	10n ceramic disc
R4	4k7	C7, 9	10μ 16V tant
R5	2k2	C8	100n ceramic disc
R6	2k7	C10	47μ 6V tant
R7, 8, 9	510R	C11	100n ceramic disc
R10, 13, 14	510R	C12	47p polystyrene/ceramic
R11	330R	C13	100p polystyrene/ceramic
R12	220R	C14	100n ceramic disc
R15	120R	C15	100n ceramic disc
R16, 17	1k	VC1	65p trimmer
R18, 20	10k	L1	6μ8 RF inductor
R19	22M	XL1	10MHz crystal
R21	100k	PB1	Pushbutton switch
R22	3k3	D1, 2, 4	1N914, 1N4148
IC1	MAR-1	D3	BAT85
IC2	SP8680B	TR1	MPF102
IC3	MC10116	TR2	BSX36, 2N5771, fast PNP switch
IC4	74F153	DP0–7	Double 7-segment LED display (4 off), common cathode, 0.5in
IC5	74F160		
IC6	ICM7216D	S1	4-pole, 3-way rotary switch
		S2	DPDT toggle, PCB mounting

Resistors are 0.25W, 5%

gain of some 16dB. The resulting output is fed to IC2 which is a high-speed divide-by-10 counter capable of operating up to at least 575MHz with a sine-wave input. The output selected is TTL compatible. The circuit will operate without IC1 but is not

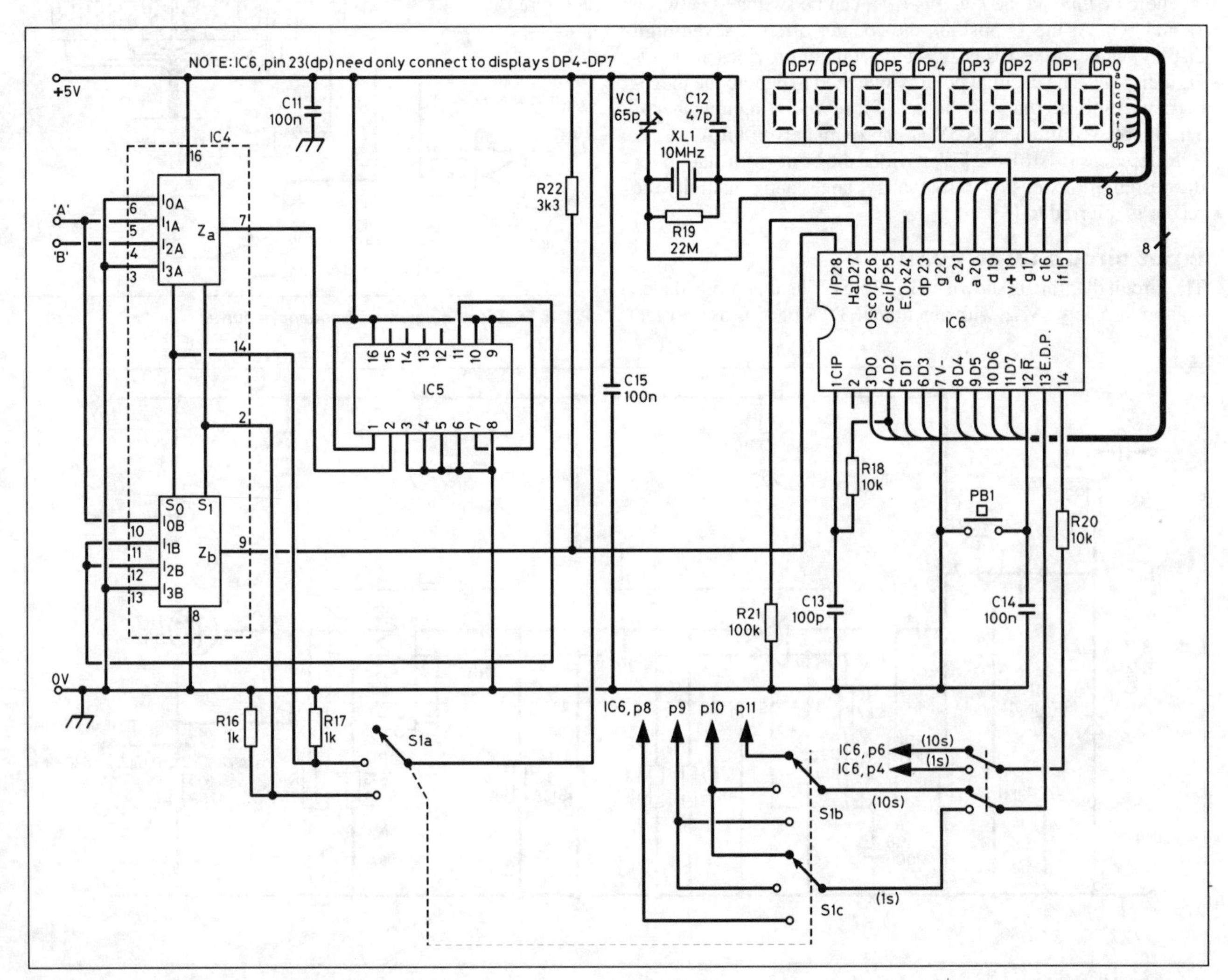

Fig 15.59. Control, counter and display circuit

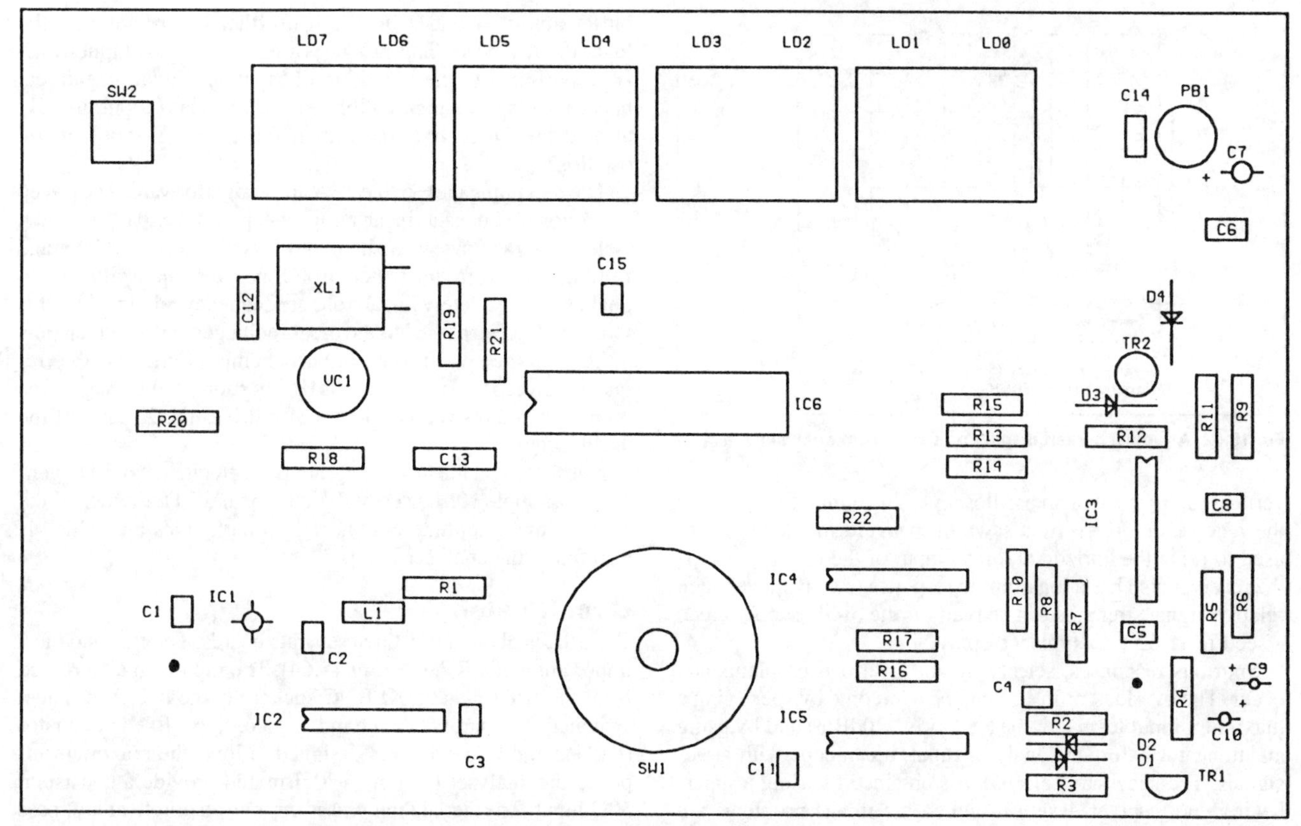

Fig 15.61. Component placement details for the digital frequency counter

as sensitive. Although a Mini-circuits monolithic broad-band amplifier is used, it is possible to use those from other manufacturers such as Avantek.

Control, counter and display circuit

Fig 15.59 shows all of the remaining digital side of the frequency counter and displays. The initial stages of the frequency counter consist of the input selection and prescaling circuits.

IC4 and IC5 are both 74F 'fast' TTL in order to cope with the speed. IC4 is a multiplexer which is controlled by DC signals supplied via S1a and determines the route of signals from the input amplifiers to the counter IC. IC5 is a divide-by-10 circuit which provides prescaling for all signals above 10MHz. IC6 is the heart of the digital frequency meter and provides all the count circuits, decoders and drivers for the eight-digit LED display. Timing is derived from a 10MHz crystal oscillator.

Switches S1b/c determine the position of the decimal point in conjunction with the gate time selection switch (S2) for the various inputs. PB1 is a reset button for the counter.

Construction

A double-sided PCB layout (Fig 15.60) is provided in Appendix 1, and component placement details in Fig 15.61. If one wishes to make one's own, it is essential that an earth plane is provided on the copper side for the two input circuits on Figs 15.57 and 15.58. It is wise to use good layout practice for the digital circuits, especially where they can run up to 100MHz. Adequate decoupling is also essential.

Do not use IC sockets for IC1, IC2, IC3, IC4 and IC5. The dual LED displays should be mounted on single-row sockets. The PCB is double-sided, and through connections are made mainly via component leads, including IC sockets. The sockets may therefore have to stand off the board slightly in order that they can be soldered on the top side. It is suggested that turned pin sockets are used.

There are three through connections to be made with wire links. Where necessary, clearance must be made around holes on the earth plane where connection is not made to the earth plane – use a sharp drill or cutter for this.

If a counter is only required up to 10MHz, then the output of the circuit of Fig 15.57 can be fed straight to pin 28 of IC6, omitting IC4 and IC5 and the range switch S2. If an input is required only up to 60MHz, then omit the components associated with the 600MHz input amplifier and alter the stop on S1 for only two positions.

Calibration

Using a highly accurate frequency counter, which has been calibrated against a standard, feed the same input signal into both counters and adjust VC1 for the same reading.

A SPECTRUM ANALYSER

This Simple Spectrum Analyser (SSA), designed by Roger Blackwell, G4PMK, offers reasonable performance over the approximate range 1–90MHz. It is fairly cheap to build and utilises almost any oscilloscope for its display. It has selectable, calibrated frequency sweep-width ranges, logarithmic signal-strength calibration, a dynamic range of over 50dB and a built-in frequency marker generator. A suitable oscilloscope should have a DC-coupled Y-amplifier offering a 100mV/div sensitivity and an external input to the X-amplifier. See reference [9] for the original article.

What is the purpose of a spectrum analyser and how does it operate? It is essentially no more than an electronically tuneable receiver, the S-meter output of which is connected to the

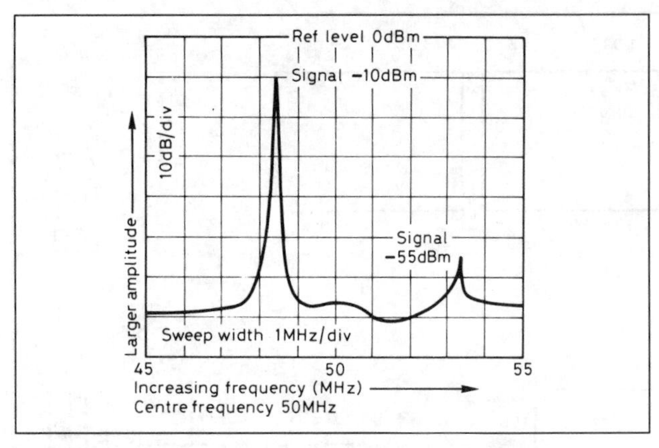

Fig 15.62. A typical screen display of a spectrum analyser

vertical (or Y) input of an oscilloscope. The tuning frequency of the receiver is driven by a sawtooth waveform which is also used to drive the horizontal (or X) input of the oscilloscope. If the receiver also has a logarithmic response to signal level, then relative signal strengths can be read off the oscilloscope screen – see Fig 15.62 for a typical display.

Spurious responses, selectivity and overload problems can occur. The overload problems can be overcome by specifying a maximum input level (for the SSA it is –20dBm) and by using an attenuator before the analyser input, it can cope with larger signals. The necessary selectivity is obtained by using a superheterodyne receiver design in which the image problems are minimised by using an intermediate frequency (IF) which is higher than the maximum frequency of the analyser – in this case 145MHz – which allows a readily available helical filter to be used.

The constructor can use his own methods of construction but there are PCB layouts in Appendix 1 and corresponding component layouts in this section.

Overview

Fig 15.63 shows a block diagram of the SSA. After attenuation (if required), the input signal is fed via a low-pass filter to the first (up-conversion) mixer, where the input frequency range of 0–90MHz is mixed with the varicap-tuned local oscillator which operates over the range 145–235MHz, giving a first IF of 145MHz. The signal is then passed through the helical filter to a second mixer and local oscillator, where it is down-converted to the second IF of 10.7MHz.

The signal next passes through wide or narrow IF filters, a buffer amplifier and a further wide filter before entering the logarithmic IF strip. This produces a signal strength output which is proportional to the logarithm of the input signal magnitude, hence the display can be calibrated in dBm. This output (usually termed the *video output*) is then fed to the Y-input of the oscilloscope.

The remaining parts of the SSA are straightforward. The sweep generator produces a linear ramp sweep voltage, part of which (selected by the sweep width control) is added to a DC voltage from the centre frequency control. Since varicap oscillators do not have a completely linear voltage/frequency relationship, this sweep voltage is passed to the break-point generator, which puts a 'kink' in the sweep where it will attempt to linearise the frequency sweep over the 70–90MHz portion of the range. The output from the sweep oscillator also drives the X-input of the oscilloscope.

Not shown on the block diagram is the frequency marker generator, a simple 10MHz crystal oscillator and TTL divider which gives a low-amplitude output, rich in harmonics and which is also fed to the analyser input.

Circuit detail

The SSA is divided into three separate boards. The first and most important is the RF unit (Fig 15.64). The input signal is routed from the front panel 50Ω BNC socket via the two front-panel switched attenuators (shown in Fig 15.65), to a fixed attenuator (R1, R2 and R3), which is designed to limit the maximum input to the analyser to about –20dBm and provide a consistent 50Ω input. The signal then passes through an elliptical low-pass filter (C3, C4, C5 and L1) to the first mixer which is part of IC1 (an MC3356). The MC3356 is intended to be used as a single-chip FSK receiver and has some special features which are exploited in the SSA. First, the IF amplifier has a signal strength output which is proportional to the logarithm of the input voltage; second, the local oscillator and mixer will work up to at least 250MHz. The local oscillator is tuned by varicap diode D1 using the sweep voltage from the sweep/video board. Note that two 1nF capacitors (C8 and C30) are fitted at the anode end of D1 because a low-impedance path is vital here to enable the highest frequency to be reached. Adding C30 to one of the prototypes increased the upper frequency limit by 5MHz!

The 145MHz output from IC1a goes to the first IF filter FL1 (a three-chamber helical type). The IF output from the filter is then down converted to the second IF of 10.7MHz in IC2 (an NE602), the local oscillator frequency of about 134.3MHz being set by L4, C2 and C13. Setting the second LO below the first IF removes the 21.4MHz (2 × 2nd IF) spurious response.

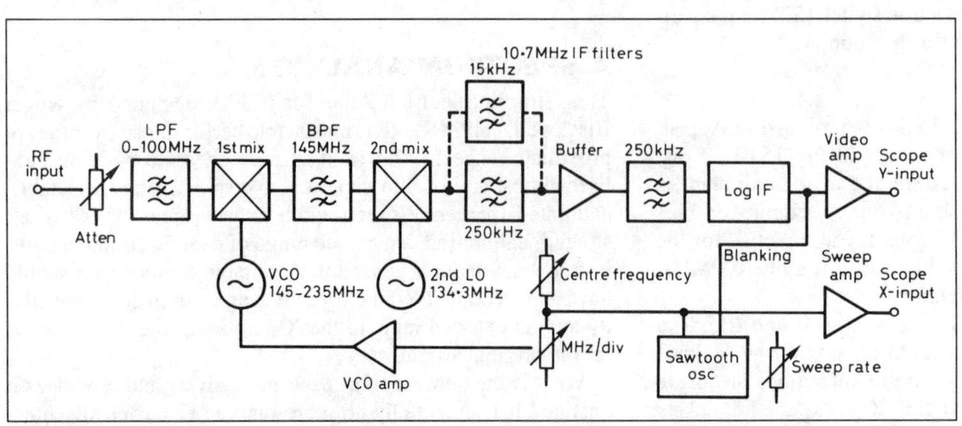

Fig 15.63. Block diagram of the Simple Spectrum Analyser

The NE602 requires a lower supply voltage than the MC3356 and this is obtained from a 5V regulator IC3.

Narrow (15kHz) or wide (250kHz) first IF filters (FL2 and FL3) are selected by means of miniature relays RL1 and RL2 and the front-panel switch S1 (IF bandwidth). After filtering, the signal is amplified by TR1 which is run at a relatively high standing current so as to provide a good dynamic range. Although the stage does not provide the

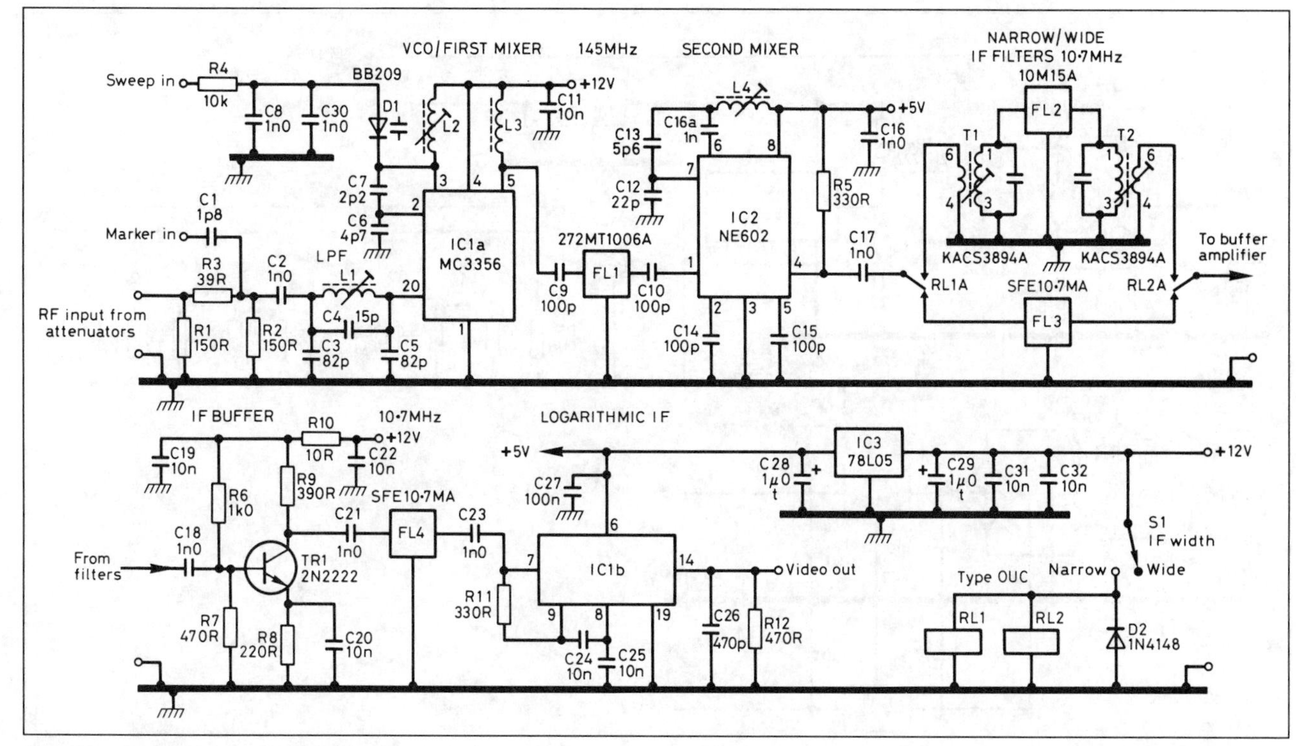

Fig 15.64. Circuit diagram of RF board

correct impedance terminations for the filters, in practice this is of little consequence. Removing C20 would improve the matching but with the consequential loss of over 20dB of sensitivity. The signal is then passed via a second filter (FL4) to the main IF signal processing circuit IC1b. This does one of the most difficult jobs in the analyser – it provides a DC output which is proportional to the logarithm of the IF input voltage. Here, in one circuit, the 10dB/division Y-axis calibration is achieved, with the (video) output being taken via a screened lead to the sweep/video board. The output from the FM discriminator is not used in this application.

Sweep generator

The sweep generator is shown on Fig 15.65. The ubiquitous 741 op-amp has been used throughout as they are cheap enough so that one can be liberal with them. The sweep ramp is generated by IC4 (a 555 timer), IC5 and current source TR2, with the sweep rate being controlled by a front-panel potentiometer RV1. The

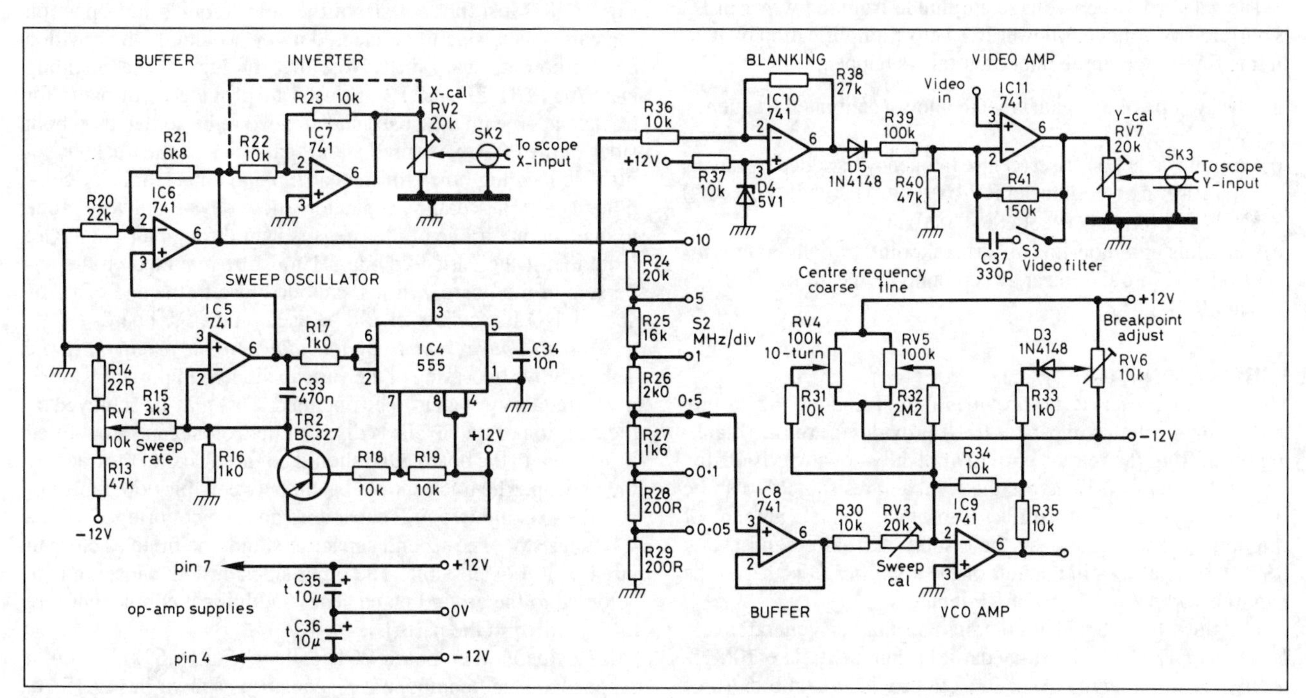

Fig 15.65. Circuit diagram of video/sweep board

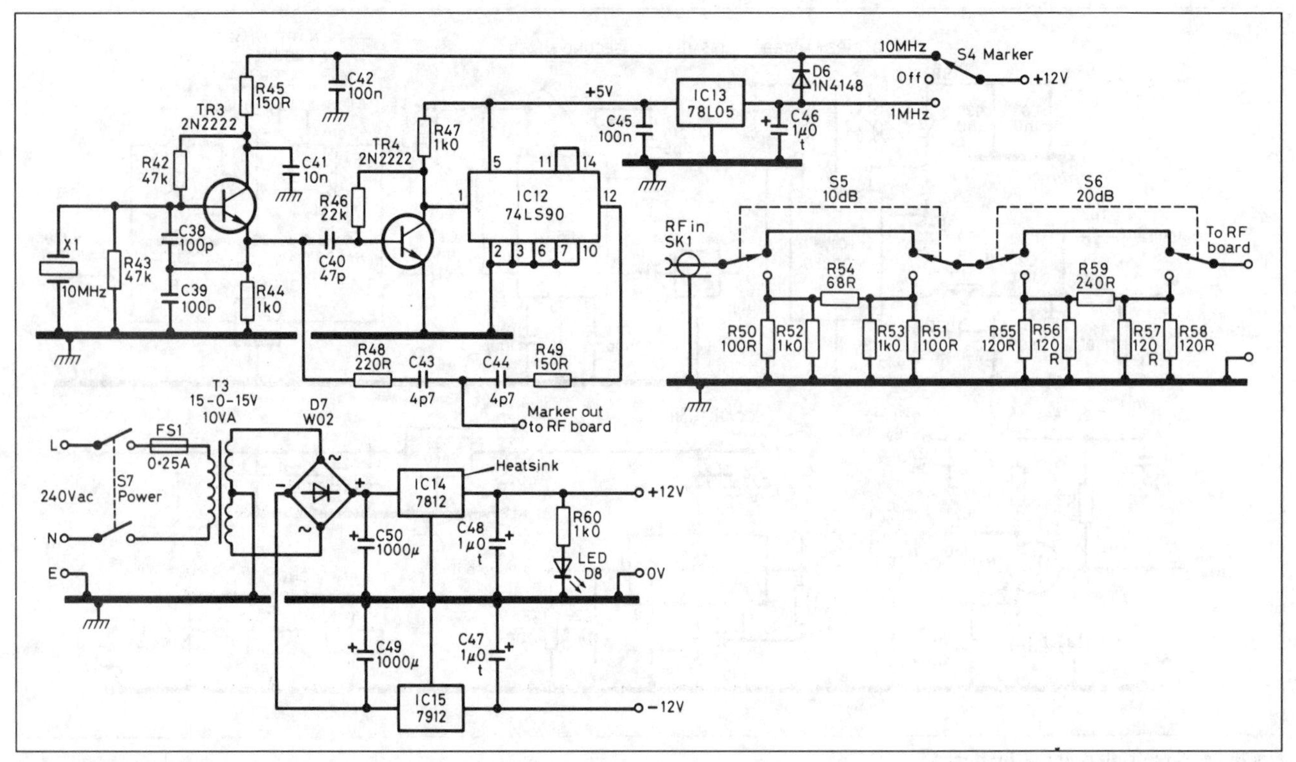

Fig 15.66. Circuit diagram of marker generator, PSU and front-panel attenuators

555 timer also provides a fast blanking pulse output for the video amplifier. The sweep output is buffered by IC6 before being fed to the sweep-width front panel control switch S2 (MHz/div) and the X-output at SK2. Depending on the oscilloscope used, the inverting unity gain buffer IC7 may not be needed. *Note*: if a positive voltage applied to the oscilloscope X-input deflects the spot to the right, then IC7 can be omitted; in this case omit R22 and R23 and connect IC6 (pin 6) to the top of X-CAL preset RV2 via link LK1 shown dotted.

The selected sweep voltage amplitude from the wiper of S2 is buffered by voltage follower IC8 before amplification by IC9. In this final stage, three important things happen:

(a) the sweep voltage gain is set to allow a calibrated frequency sweep;
(b) an adjustable DC offset (centre frequency) is added by means of a 10-turn potentiometer RV4 (CENTRE FREQUENCY COARSE) and RV5 (CENTRE FREQUENCY FINE); and
(c) an adjustable non-linearity (breakpoint) is deliberately introduced into the linear sweep ramp by means of RV6, D3 and R33.

Video amplifier

The video amplifier (IC11) provides a small amount of gain and, in conjunction with comparator IC10, provides the retrace blanking by shifting the retrace portion vertically downwards off the screen. Capacitor C37 across the feedback resistor R41 can be switched by S3 (VIDEO FILTER) to provide a small amount of smoothing of the 'grass' on the display if required and preset RV7 (Y-CAL) allows the output of the amplifier to be set to the required 100mV per 10dB of RF input.

The third board, which contains the marker generator and power supply (Fig 15.66) needs little comment. The 10MHz oscillator is divided by 10 in IC12 to provide a comb of 10 or 1MHz markers which can be added to the input signal to allow an easy method of frequency calibration. The power supply uses standard components and is conventional.

Construction

Components lists are given in Table 15.12. The RF board is constructed on a double-sided glass epoxy PCB, one side of which is not etched and is used as an earth plane. The PCB pattern (Fig 15.67) is given in Appendix 1 and a component placement diagram (together with drilling details) is shown in Fig 15.68. Most (but not all) of the holes require the copper on the earth-plane side to be cleared away around the hole with a counterbore or sharp drill. Note that the lugs of the shielding cans for FL1, T1 and T2 are used to provide earth paths for tracks underneath the board and so need to be soldered on both sides of the PCB. The small additional holes on the track layout provide locating holes for the earth plane connections of components – small ceramic capacitors these days seem very prone to disintegrate if one of their legs is bent through a right angle! Note that if the varicap diode D1 must be mounted on the underside of the board with its cathode close to the end of L2 as shown. NO IC sockets should be used on the RF board.

Note: this paragraph is an amendment to the original article. The effective bypassing of the varicap diode is important if maximum frequency span is to be obtained. This may be achieved by the addition of two leadless disc or trapezoidal capacitors fitted in slots cut in the board. One should be fitted next to the track by the varicap. The track should be cut between this new capacitor and the existing two 1nF capacitors and the cut bridged with a 47kΩ resistor. The second capacitor should be fitted as close to pin 4 of IC1 as possible. The other sides of both capacitors are soldered to the ground plane. These additional components are *not* specified in the parts list.

The video/sweep board PCB pattern (Fig 15.69) is given in Appendix 1 and component placement is shown in Fig 15.70. This is a single-sided PCB where the optional link LK1 (shown

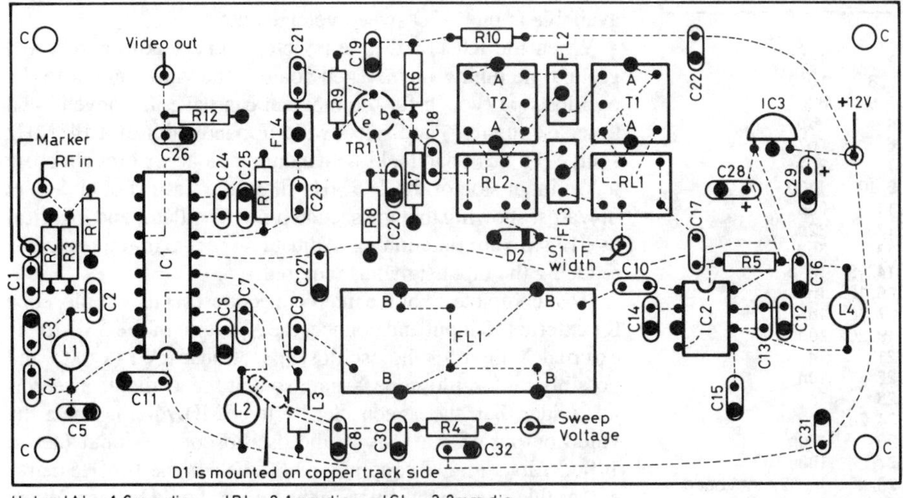

Fig 15.68. RF board layout. Note that the BB209 D1 is mounted on the underside of the board

dotted) should be fitted as explained earlier if the X-signal does not need to be inverted. Note that resistors R24–R29 are mounted on the rotary switch S2.

The third board, containing the marker generator and power supply, is also a single-sided PCB. The PCB pattern (Fig 15.71) is given in Appendix 1 and the component overlay is shown on Fig 15.72. Sufficient space for a small heatsink for IC14 has been allowed for.

The analyser needs to be housed in a metal case in order to provide the necessary shielding from stray signals but, when choosing or making a case, remember that access to the presets on both the RF and sweep boards is required. This means that the boards should be mounted so that there is easy access to both boards. The input attenuators are constructed on slide switches S5 and S6 using short leads and could well be fitted with a grounded screening box made of double-sided PCB.

Choose a good 10-turn potentiometer for the CENTRE FREQUENCY COARSE control and fit it with a large knob which has a cranked handle; this will save a lot of wear and tear on your fingers! The CENTRE FREQUENCY FINE control needs a good-quality, single-turn, carbon-track potentiometer. The connections to the RF board from the front-panel attenuators and the marker generator should be with miniature coaxial cable such as RG174. The video and sweep connections, as well as those to S2 and the X- and Y-output sockets, should be made with a small-diameter screened (audio) cable.

Test the boards on the bench before finally assembling them in the case. Final setting up of the RF board should be carried out when it is fixed in the case.

Alignment

Start by testing the power supply and marker generator. The latter can be easily checked by listening to its harmonics on an HF receiver, or use an oscilloscope on the input and output of the decade divider IC12. Next test the sweep and video board. It

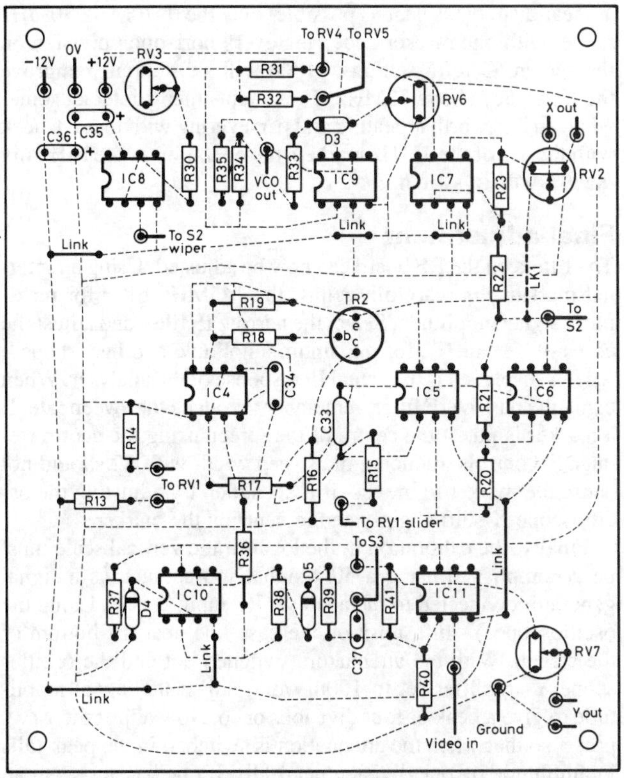

Fig 15.70. Component placement video/sweep board

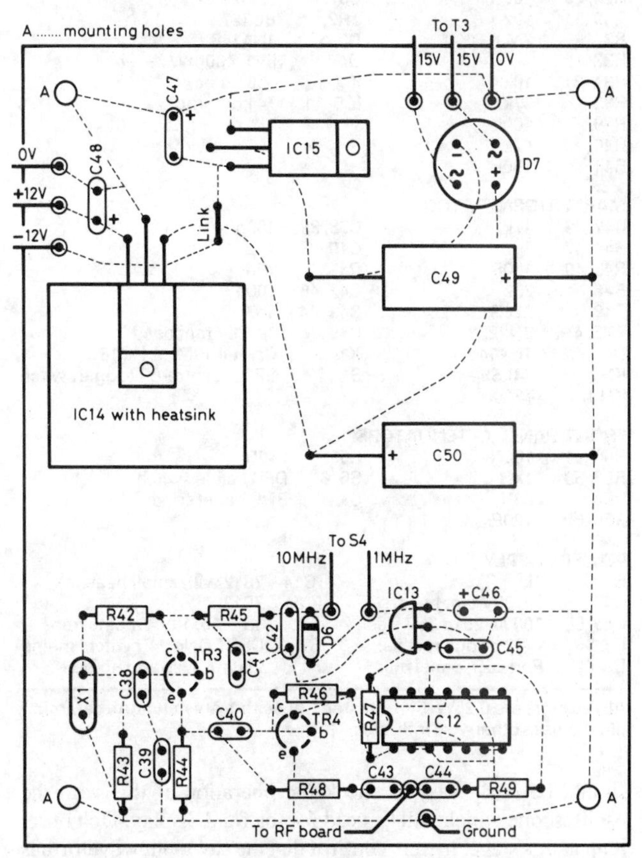

Fig 15.72. Component overlay marker generator/power supply

Table 15.12. Spectrum analyser components list

RF BOARD			
R1, 2	150R	C1	1p8
R3	39R	C2, 8	1n
R4	10k	C3, 5	82p
R5, 11	330R	C4	15p
R6	1k	C6	4p7
R7, 12	470R	C7	2p2
R8	220R	C9, 10	100p
R9	330R	C11	10n
R10	10R	C12	22p
D1	BB209 (or MV209, BB809)	C13	5p6
D2	1N4148	C14, 15	100p
TR1	2N2222	C16, 16a	1n
IC1	MC3356	C17, 18	1n
IC2	NE602	C19, 20	10n
IC3	78L05	C21	1n
FL1	272MT1006A CBT 145MHz helical filter	C22	10n
		C23	1n
FL2	10M15A 2-pole 10.7MHz crystal	C24, 25	10n
FL3, 4	SFE10.7MA 10.7MHz ceramic filter	C26	470p
		C27	100n
L1, 2	Toko S18 Coil 301SS0100	C28, 29	1μ, 35V tant bead
L3	Toko FL4 RF choke 348LS4R7	C30	1n
		C31, 32	10n
L4	Toko S18 coil 301SS0400	RL1, 2	Min relay type OUC
T1, 2	Toko transformer KACS3894A	S1	SPDT switch
SWEEP AND VIDEO BOARD			
R13	47k	RV1	10k lin pot
R14	22R	RV2, 3	20k cermet preset
R15	3k3	RV4	100k 10-turn
R16, 17	1k	RV5	100k lin carbon
R18, 19	10k	RV6	10k cermet preset
R20	22k	RV7	20k cermet preset
R21	6k8	S2	1-pole, 6-way rotary switch
R22, 23	10k	S3	SPST or SPDT toggle switch
R24	20k	SK2, 3	BNC panel sockets
R25	16k	C33	470n polyester layer
R26	2k	C34	10n
R27	1k6	C35, 36	10μ, 25V tant bead
R28, 29	200R	C37	330p
R30, 31	10k	TR2	BC327
R32	2M2	D3, 5	1N4148
R33	1k	D4	5V1, 400mW zener
R34–37	10k	IC4	555 timer
R38	27k	IC5–11	741 op-amps
R39	100k		
R40	47k		
R41	150k		
MARKER/GENERATOR			
R42, 43	47k	C38, 39	100p
R44, 47	1k	C40	47p
R45, 49	150R	C41	10n
R46	22k	C42, 45	100n
R48	220R	C43, 44	4p7
TR3, 4	2N2222	C46	1μ, 35V tant bead
D6	1N4148	X1	Crystal 10MHz, HC18U
IC12	74LS90	S4	SPDT, centre-off toggle switch
IC13	78L05		
FRONT-PANEL ATTENUATORS			
R50, 51	100R	R59	240R
R52, 53	1k	S5, 6	DPDT slide switch
R54	68R	SK1	BNC panel socket
R55–58	120R		
POWER SUPPLY			
R60	1k	IC14	7812 with small heatsink
C47, 48	1μ, 35V tant bead	IC5	7912
C49, 50	1000μ, 25 or 35V elec	T3	15-0-15, 10VA mains trans
D7	W02 bridge rectifier	S7	DPST ON/OFF switch, mains
D8	Red LED, panel mtg	F1	250mA fuse and holder

All resistors are 0.25W, 5% or better. Capacitors are miniature ceramic plate unless otherwise indicated.

should be fairly easy to check the operation of this with the oscilloscope which will be used for the final display. Do not set it up at this stage; merely confirm that the sawtooth waveform is available at the analyser X-output and that an attenuated version (with DC offset dependent on the centre frequency controls) is available at the VCO sweep voltage output.

When the RF board is complete, connect power to it and ground the tuning input. Then connect the video output to the oscilloscope (which for the moment can have its conventional timebase running) and select wide IF bandwidth. If a 145MHz source (eg a 2m handheld with a dummy load) is brought close to the input side of the 145MHz filter, the trace should deflect upwards, showing that the second mixer/oscillator and logarithmic IF strip are functioning. Adjust L4 for maximum response, reducing the input signal as required.

Now complete all of the interconnections, set the oscilloscope for external X-input and connect the X-output of the SSA to the external X-input of the oscilloscope. Adjust RV2 (X-CAL) and possibly the oscilloscope X-gain so that the available sweep is just wider than the screen. Set S2 to 10MHz/div, connect the video output of the SSA to the oscilloscope Y-input (set to 100mV/div with DC coupling) and switch on the 10MHz markers. At this stage, a few blips on the screen should be seen. When the VCO is correctly aligned, one of the blips will not disappear when the markers are switched off. This is the lower limit of the coverage – in other words 0MHz.

The next stage requires patience! Set RV4 (CENTRE FREQUENCY COARSE) to about mid-travel and unscrew the core of L2 so that it is about half way out of the coil – by now a few marker blips should be seen as well. Adjust L4 for maximum amplitude of the blips, noting that there will be two positions where this occurs – choose the position where the core is further inside the coils, as the other corresponds to the local oscillator being on the high side of the first IF. By careful adjustment of the VCO coil L2 it should be possible to see marker blips every 10MHz up to 90MHz, while still keeping the 0MHz blip. If necessary adjust L4 slightly. Coil L1 does not need adjustment – leave the core as supplied.

Adjust RV3 (SWEEP CAL) and RV6 (SET BREAKPOINT) to give a linear display (as near as possible) over the 0–70MHz/80MHz range, with one marker appearing every horizontal division on the screen. Careful setting of RV6 will substantially improve the frequency above 70MHz. These adjustments interact somewhat, so it is worth repeating and persevering with them. Check with the aid of the 1MHz markers the operation of the MHz/div (SWEEP WIDTH) switch.

Final adjustment

The filters on the RF board can now be adjusted. Using an internal marker blip, carefully adjust the 145MHz filter for maximum signal amplitude. Select the narrow IF filter and adjust the cores of T1 and T2 for maximum amplitude and best shape – what is displayed is the actual IF response of the analyser. When using the narrow IF filter, remember to reduce the sweep rate. If a marker is put at the centre of the screen using the centre frequency controls, reducing the sweep width with S2 should not cause the marker to move – if it does then try adjusting the oscilloscope X-shift slightly and re-centring the marker.

Finally, the calibration of the logarithmic vertical scale must be accomplished using a 50Ω signal source such as a signal generator connected to the analyser RF input socket. Using the oscilloscope Y-shift, position the base line near the bottom of the screen. With the attenuators switched out and the oscilloscope Y-amplifier set to 100mV/div, adjust the signal amplitude to give a peak of four divisions or so. Now adjust the Y-CAL preset so that when the attenuation is switched in the peak falls in amplitude by one division per 10dB. If one has access to an accurate signal source, set the oscilloscope Y-shift so that the top of the screen corresponds to –20dBm (in a commercial instrument this is termed the *reference level*). The noise floor of the analyser is about –85dBm but note that the lowest vertical division does not quite correspond to the 10dB/div calibration of the rest of the screen.

Practical hints

Bearing in mind the practical limitations of the analyser, ie the maximum input of –20dBm (+10dBm with both attenuators in), it will show its own shortcomings if it is overloaded. The dynamic range of the analyser is over 50dB. While the absolute sensitivity will vary across its range by 6dB or so, the relative calibration of 10dB/div remains unchanged for any given frequency. When using the narrow IF setting, slow the sweep down – watching the display as this is carried out will reveal why this is necessary. While not shown on any circuit diagram, a good RF filter is recommended on the mains input in order to keep the entry of RF purely to the front-panel socket.

While the SSA is not really suitable for making intermodulation distortion measurements or looking at oscillator noise, many useful and interesting tasks await it. By connecting a few feet of wire to the input, a fascinating picture of the HF spectrum emerges – try it during the day time and then at night when the 7MHz broadcast stations are in full swing. Use the 10MHz and 1MHz markers to find the way about the spectrum.

The upper limit of the analyser should be greater than 95MHz and, if living in a good signal-strength area, Band II VHF radio signals are visible. Connecting a good antenna should enable a good watch to be kept for 28MHz and 50MHz openings. If your rig has a mixer-type VFO, try looking at the output and be prepared for a shock.

While this unit does not have quite the performance of a commercial unit (or the price tag!), when used with a modicum of care it is a very useful tool.

POWER OUTPUT MEASUREMENTS

The UK Amateur Licence requires that one should be able to measure output power in order to comply with the licence conditions. The following information is taken from the *Amateur Radio Licence Terms and Limitations Booklet* BR68. Only the relevant paragraphs have been included.

Notes to the Schedule

(a) *Maximum power* refers to the RF power supplied to the antenna. Maximum power levels will usually be specified by carrier power. For emissions having a suppressed, variable or reduced carrier, the power will be specified by the peak envelope power (PEP) under linear conditions.

(e) Interpretation

(i) *Carrier power:* the average power supplied to the antenna by a transmitter during one radio frequency cycle taken under the condition of no modulation.

(iv) *Mean power:* the average power supplied to the antenna by a transmitter during an interval of time which is sufficiently long relative to the lowest frequency encountered in the modulation taken under normal operating conditions.

(v) *Peak envelope power (PEP):* the average power supplied to the antenna by a transmitter during one radio frequency cycle at the crest of the modulation envelope taken under normal operating conditions.

The oscilloscope can be used up to about 30MHz to monitor modulated waveforms and measure output power but above this it becomes an expensive item and may provide unwanted loading effects on the equipment being monitored. The familiar VSWR meter monitors forward and reflected signals and the scale can be made to represent power in a 50Ω line. It is possible to use an RF voltmeter across a given load to measure power. The higher one goes in frequency, the more difficult it becomes to measure the modulation and power with relatively cheap equipment. Yet it is a condition of the licence that these parameters can be monitored. It is more difficult to measure PEP than average carrier power.

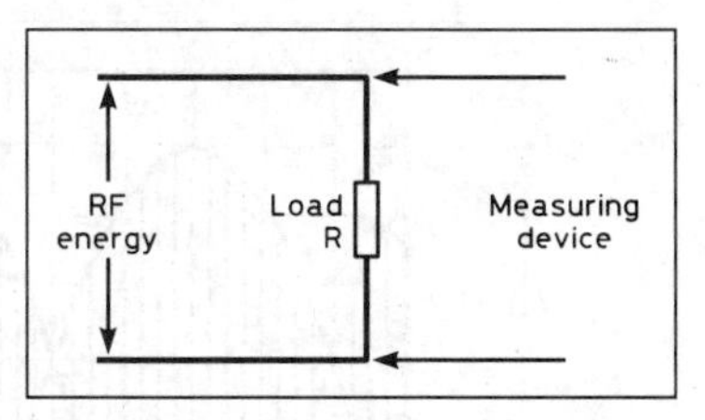

Fig 15.73. Power measurement arrangement

Constant-amplitude signals

In a carrier-wave situation (CW, FM or unmodulated AM), the output is of constant amplitude and so it is relatively easy to measure the output power. To measure these signals using the circuit as shown on Fig 15.73, the power output is given by:

$$P_{out} = V^2/R$$

where *V must be the RMS value of the voltage.*

This voltage measurement can be carried out using an oscilloscope or RF voltage probe. The SWR meter described later can also provide this value.

Amplitude-modulated signals

These pose more of a problem and two cases are dealt with below.

Amplitude modulation (A3E)

With no modulation the problem reverts to the measurement of power of a constant carrier as described above. If the carrier is amplitude modulated (A3E) then the overall output power increases. The power is divided between the sidebands and the carrier component. With 100% modulation the output power increases to 1.5 times the unmodulated condition – the power contained in each of the two sidebands being one-quarter that in the carrier. It is suggested that for this form of modulation the carrier power is measured (ie no modulation) and multiplied by 1.5 to give the maximum output power.

If an exact value for the output power is required it is necessary to determine the *modulation index*. This can be carried out using an oscilloscope of adequate frequency response. Set the oscilloscope as shown on Fig 15.74 and calculate the modulation depth *m*. The output power is then given by:

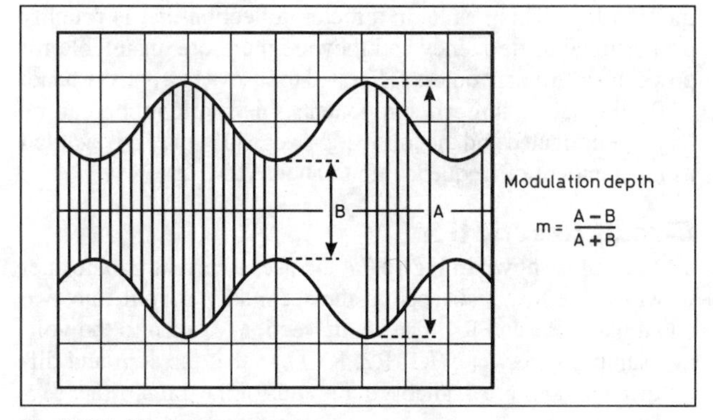

Fig 15.74. Modulation depth measurement

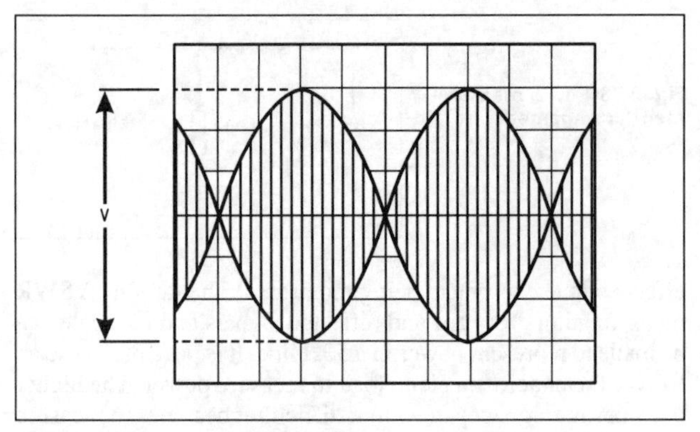

Fig 15.75. Two-tone test display

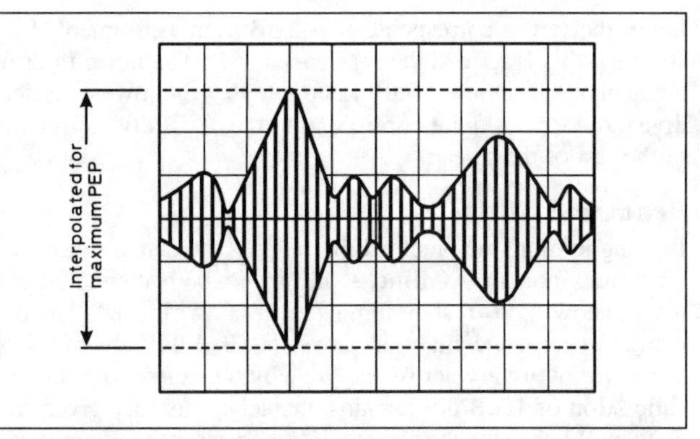

Fig 15.76. Speech waveform and interpolated maximum PEP level

$$P_{out} = \frac{V^2}{R}(1 + \tfrac{1}{2}m^2)$$

where V is the RMS value of the unmodulated carrier and R the load.

For test purposes an audio signal can be fed in at the microphone socket of the transmitter using the low-frequency oscillator and two-tone burst oscillator described earlier.

Single sideband (J3E)

With single sideband, no power is output until modulation is applied. The output envelope is non-sinusoidal in appearance. The normal method for measuring output power is by observation of the modulation envelope and determination of the peak envelope power – this is the parameter defined by the UK licensing authority. This can be accomplished using an oscilloscope of suitable frequency response as described below or using a SWR meter that will respond to peak envelope power – see reference [1].

Fig 15.75 shows the display when a two-tone test signal generated by the oscillator described earlier is fed in via the microphone socket. If the peak-to-peak voltage V at the crest of the envelope is measured across a load of value R, then the PEP is given by:

$$P_{out} = V^2/8R$$

The equivalent peak-to-peak voltage reading can then be interpolated for the maximum allowable PEP and the position noted on the display. Fig 15.76 shows a typical display for a speech waveform and the interpolated maximum PEP level.

AN SWR METER

Reflectometers designed as VSWR indicators have normally used sampling loops capacitively coupled to a length of transmission line. This results in a meter deflection that is roughly proportional to frequency and they are therefore unsuitable for power measurement unless calibrated for use over a narrow band.

By the use of lumped components this shortcoming can be largely eliminated and the following design [10] may be regarded as independent of frequency up to about 70MHz.

Circuit description

The circuit is shown in Fig 15.77 and uses a current transformer in which the low resistance at the secondary is split into two equal parts, R3 and R4. The centre section is taken to the voltage-sampling network (R1, R2, RV1) so that the sum and difference voltages are available at the ends of the transformer secondary winding.

Layout of the sampling circuit is fairly critical. The input and output sockets should be a few inches apart and connected together with a short length of coaxial cable. The coaxial cable outer must be earthed at one end only so that it acts as an electrostatic screen between the primary and secondary of the toroidal transformer. The layout of the sensing circuits in a similar instrument is shown on Fig 15.78.

The primary of the toroidal transformer is formed by threading a ferrite ring on to the coaxial cable. Twelve turns of 24swg (0.56mm) enamelled copper wire are equally spaced around the entire circumference of the ring to form the secondary winding. The ferrite material should maintain a high permeability over the frequency range to be used: the original used a Mullard FX1596 which is no longer available but suggested alternatives are Philips FX3852 or 432202097180 and Fair-rite 5961000301.

Other components in the sampling circuits should have the shortest possible leads. R1 and R2 should be non-inductive carbon types. For powers above about 100W, R1 can consist of several 2W carbon resistors in parallel. RV1 should be a miniature skeleton potentiometer in order to keep stray reactance to a minimum. The detector diodes D1 and D2 should be matched point-contact germanium types with a PIV rating of about 50V; OA91 diodes are suitable (with OA79 an alternative if available). The resistors R3 and R4 should be matched to 5% or better.

The ratio of the sampling resistors R1 and R2 is determined by the sensitivity of the current sensing circuit. As the two sampling voltages must be equal in magnitude under matched conditions, RV1 provides a fine adjustment of the ratio.

Germanium diodes as specified are essential if an instrument is to be used at low power levels, otherwise silicon diodes such as 1N914 may be substituted, a more recent alternative being Schottky diodes such as the BAT85. To increase the sensitivity at low power levels, eg 1W, then the feed line could be looped through the toroid. It may then be necessary to use a large toroid or smaller coaxial cable (but this will not cope with high powers!).

A components list for this project is given in Table 15.13 below.

Table 15.13. SWR meter components list

R1	5k carbon (see text)	C1, 2	10n ceramic
R2	390R carbon	T1	Philips FX3852, 4332202097180
R3, 4	27R, 2W carbon		Fair-rite 5961000301
R5, 9	4k7		with 12 turns 24swg ECW
R6, 10	33k	D1, 2	OA91 (matched)
R7, 11	100k	M1, 2	50μA FSD meters
R8, 12	330k	RV1	1k, skeleton pot, 0.5W

Resistors are 0.25W, 5% unless specified otherwise.

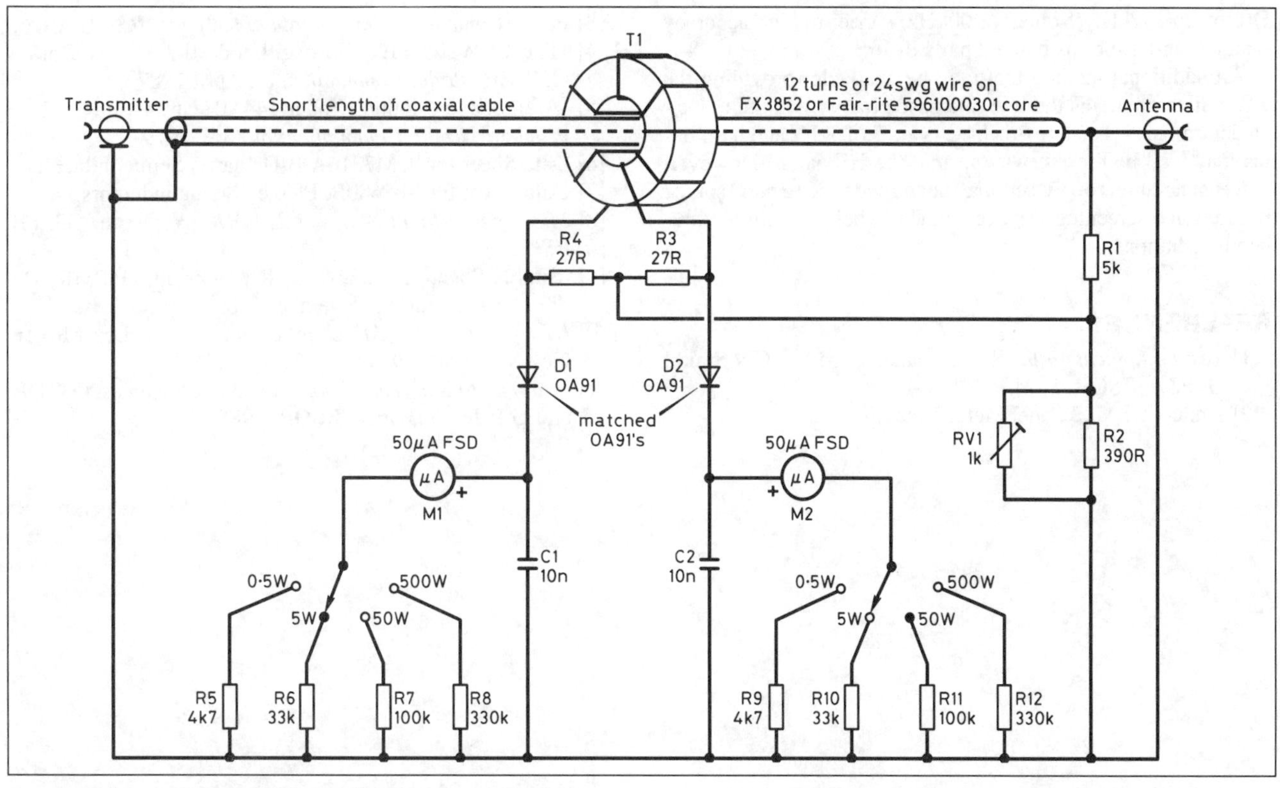

Fig 15.77. Circuit diagram of frequency-independent SWR meter

Calibration

Accurate calibration requires a transmitter and an RF voltmeter or possibly an oscilloscope. The wattmeter is calibrated by feeding power through the meter into a dummy load of 50Ω. RV1 is adjusted for minimum reflected power indication and the power scale calibrated according to the RF voltage appearing across the load. The reflected power meter is calibrated by reversing the connections to the coaxial line.

The instrument has full-scale deflections of 0.5, 5, 50 and 500W which is selected by the range switch. These should not normally be ganged since the reverse power will normally be much less than the forward power.

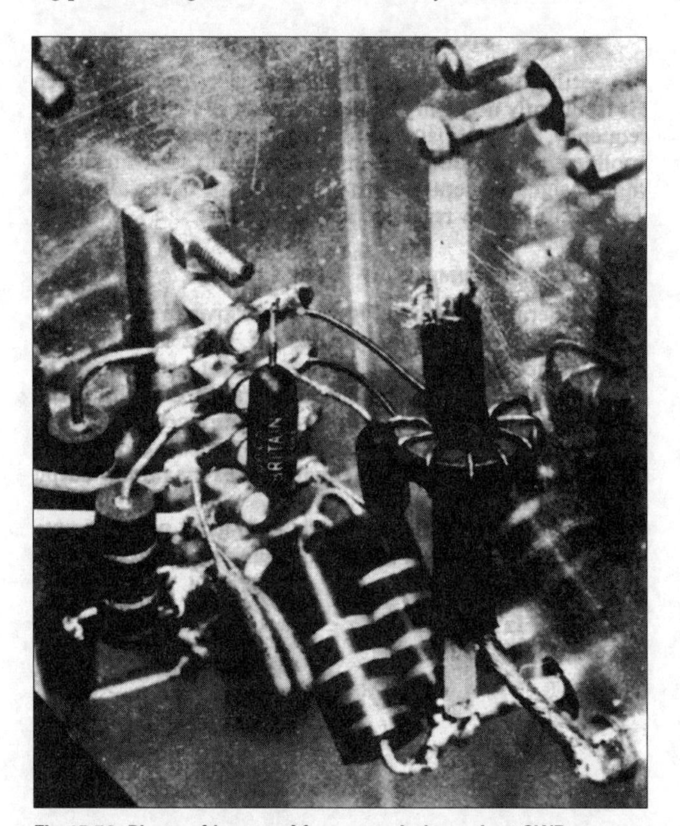

Fig 15.78. Photo of layout of frequency-independent SWR meter

STANDARD FREQUENCY SERVICES

There are various standard frequencies transmitted throughout the world and these can be harnessed in order to check other equipment against them. Typical of these transmissions [11] are those shown in Table 15.14.

These standard frequencies are maintained to an accuracy of typically one part in 10^{11}. However, if the sky-wave is used there could be a large error in reception due to Doppler shift and there will be fading of the signal. These problems can be avoided by using a low-frequency transmission such as those from MSF or WWVB. Timing information is also impressed on the signals in either GMT or UTC.

In addition in the UK the BBC maintains the accuracy of the

Table 15.14. Standard frequency transmissions

CHU	Canada	3330, 7335, 14,670kHz
FFH	France	2500kHz
IBF	Italy	5000kHz
JJY	Japan	2500, 5000, 10,000, 15,000kHz
MSF	UK	60kHz
LOL2	Argentina	5000, 10,000, 15,000kHz
RWM	Russia	4996, 9996, 14,996kHz
VNG	Australia	4500, 7500, 12,000kHz
WWV	USA	2500, 5000, 10,000, 15,000, 20,000, 25,000kHz
WWVB	USA	60kHz
WWVH	USA	2500, 5000, 10,000, 15,000, 20,000, 25,000kHz

Droitwich 198kHz (formerly 200kHz) carrier to high accuracy – on a long-term basis being 2 parts in 10^{11}.

An additional method that may be available (providing the TV radiates well!) is the line timebase of a TV and the associated harmonics. The line timebase is at 15.625kHz and harmonics may well be present way up into the HF bands. However, newer generations of TV sets may not provide such a good source if additional screening has been added to help minimise extraneous radiations.

REFERENCES

[1] *Test Equipment for the Radio Amateur*, 3rd edn, C V Smith, G4FZH, RSGB, 1994.

[2] Anders OEM22 Panel Meter Data Sheet.

[3] *Radio Communication Handbook*, 5th edn, RSGB, 1976.

[4] 'The G3WHO FET dip oscillator Mk2', Tony Bailey, G3WHO, *Radio Communication* April 1987.

[5] 'A frequency counter for a 144MHz transmitter', N B Pritchard, *Radio Communication* May 1979.

[6] Data Sheet for ICM7216A/B/D, Harris Semiconductors.

[7] Data Sheet for SP8680B, Plessey Semiconductors.

[8] *Microwave Handbook*, Vol 2, Ed M W Dixon, G3PFR, RSGB, 1991.

[9] 'Simple Spectrum Analyser', R Blackwell, G4PMK, *Radio Communication* November 1989.

[10] *HF Antennas for All Locations*, 2nd edn, Les Moxon, G6XN, RSGB, 1993.

[11] *Radio Data Reference Book*, 5th edn, R S Hewes, G3TDR, and G R Jessop, G6JP, RSGB, 1985.

16 Construction and workshop practice

IN SPITE of the now worldwide commercial market of amateur radio there are still many items of equipment that are worth making for yourself. Cost is an important factor and often things may be made very cheaply by utilising ex-equipment components. For the newcomer it is worthwhile starting the 'junk box' (usually many boxes) of a variety of useful bits and pieces collected from junk sales, rallies and the like. If you belong to a radio club, many of its members may have items to exchange or give away which will just 'do nicely' for your project. The creatively satisfying work of home construction requires many skills, all of which make this an interesting and rewarding facet of the amateur radio hobby. The skills learnt and the pleasures gained are well worth the effort. It all takes time, and for the newcomer it is better to start on some simple project, just to get the 'feel' of things, rather than to dive in on some marathon project which usually finishes up as a complete deterrent to construction or as a collection of bits in someone else's junk box!

Most amateurs deciding to build some piece of equipment start collecting components, information on circuits, source of components and other technical data, but often forget to enquire about the manufacturing techniques required. Materials, tools and how to use them; methods of component assembly; and making the finished job look good are just some of these techniques.

The first requirement is an elementary knowledge of the materials useful and normally available to constructors.

METALS

Aluminium and aluminium alloys

These are good electrical conductors and are the lightest in weight of the normal metals available. They are non-magnetic, of medium to high cost, and available in sheet, rod, tube and other forms. The annealed quality usually bends and machines easily but soldering requires special fluxes and solders, and often the soldered joint is not good electrically. With the unknown quality of aluminium that most amateurs encounter, soldering is best left alone or experimented with before committing the project to this method of fastening. Long-term corrosion is also a problem with soldered joints in aluminium unless specialist methods are applied to prevent this. Adhesive bonding is very good but such joints are not usually reliable electrically. Aluminium is non-corroding in normal use, but direct contact with brass or copper should be avoided as these react and encourage corrosion with the resultant troubles of poor or non-existent electrical contact. (See the section on corrosion.) Nickel-plated or stainless-steel screws, nuts and washers, rather than brass or steel, should be used to reduce the chances of corrosion around this type of fastening.

Typical uses: cabinets, boxes, panels, masts, beam antennas, heatsinks.

A metal that looks similar to aluminium, and which may come into the hands of the unwitting, is *magnesium alloy*. Filings and chippings from this metal are highly inflammable. They can burst into a glaring flame with the heat generated by filing and drilling, and trying to put out such a fire with water only makes matters worse. (Factories where this alloy is machined use a chemical fire extinguisher, one type of which goes under the title of *DX powder*!) It is difficult to tell the difference by just looking, but size-for-size magnesium alloy sheet weighs less than aluminium and usually on bending there is a 'granulating-cum-scrunching' feel.

Brass

This is an expensive but good non-magnetic electrical conductor, and is available in sheet, rod, tube and other forms. It can be soldered easily but adhesive bonding can prove difficult. For work involving bending or forming, the most suitable grade is *ductile brass*. For panels and non-formed parts, the *half-hard* and *engraving brasses* are adequate.

It can be annealed like copper, though care is necessary as brass is nearing its melting point when heated to bright red. Non-corroding in normal use but reacts with aluminium and zinc.

Typical uses: morse keys, terminal posts, weatherproof boxes, extension spindles, waveguides and other microwave components.

Copper

This is a very good though expensive electrical conductor. It is non-magnetic and is available in sheet, rod, tube and other forms. Before work of a forming or bending nature is attempted, this metal should be annealed by heating as uniformly as possible to a bright red heat and air or water cooled. If considerable bending is required, this annealing should be repeated as soon as the metal begins to resist the bending action. In the annealed state, copper bends very easily. Soft or hard soldering present no problems but adhesive bonding can be troublesome. It is non-corroding in normal use but does react with aluminium and zinc, especially in an out-of-doors environment. Cuts on the hands etc caused by copper should be cleansed and treated immediately they occur, for such cuts can turn septic very quickly.

Typical uses: heatsinks, coils, antenna wires, tuned lines, earthing stakes and straps.

Steels

These are electrical conductors, which are magnetic except for the expensive stainless types. They are cheap and available in numerous forms and qualities. The common grades are called *mild steel* or *GCQ (good commercial quality) steel*. The *black quality steels* are usually cheaper and used for such things as the stakes for mast guys, or similar 'rough' work. *Silver steel* is a special grade suitable for making tools, pivot pins and other items which require the parts to be tough or hardened and tempered. Most steel sheet forms commonly available will bend, solder and machine easily, and can be annealed by heating to bright red and allowing to cool slowly in air. Do not quench in water or in any way cool rapidly, for this may cause some steels to harden. Corrosion is a problem unless plated or well painted. For outdoor use the commercial process of galvanising is perhaps the best form of protection. The next best thing is a few coats of paint.

Typical uses: Masts and tabernacles, screw fixings, guy wire stakes and parts subject to high wear or heavy loads.

Tin plate

This is a good electrical conductor. It is magnetic, cheap and available in sheet form up to 0.5mm thick. It can be easily soldered, bonded, bent and machined, and is non-corroding in normal use. Cut edges should be re-tinned with solder if the full benefits of the non-corroding properties are required. It can be annealed but this will destroy the tin coating. This is a 'friendly' metal to use and is normally readily available in the form of biscuit tins and similar containers, hence the old timers' expression 'an OXO tin special'.

Typical uses: Boxes, screening cans and plates, light-duty brackets, retaining clips, spacers.

General comments on metals

All of the above metals work-harden and will break if repeatedly flexed at the same point. Annealing removes the effects of work-hardening, providing it is carried out before the part is over-stressed. There are professional standards which classify the above metals and each is given a specific identifier code. Fortunately, amateur constructors do not normally need to enter this maze of professional standards, and metal suppliers usually understand that to most of us steel is steel and brass is brass!

PLASTICS

Plastics are electrical and thermal insulators, and are not suitable as RF screens unless the plastic surface is specially metal coated. The insulation can be considered as excellent for most amateur purposes. The following is a brief description of the more commonly available and useful plastics, and is a very small selection of the many plastics in use today.

Laminates

Various base fabrics such as paper, cotton, glass, asbestos etc, are bonded together by selected resins, and usually compacted and cured under pressure. The combination of the resin and the base fabrics produce laminates which may be used for many applications. Most are available in sheet, rod and tube form. The cured laminate cannot be easily formed. Normal machining is possible, particularly if attention is paid to the lay of the base material. Drilled or tapped holes should be arranged so that they go through at right-angles to, and not in the same plane as, the laminations.

Where components made from this material are exposed to the elements and expected to insulate, the glass-based laminates should be preferred. This also applies where the dielectric properties are important (VHF converters, RF amplifiers etc). The normal heat generated by valves and similar electronic components will not harm these laminates but the glass or asbestos fabric-based laminates should be used for higher temperatures (100–140°C). Costs range from expensive for the paper and cotton bases to very expensive for the nylon and glass bases. Glassfibre repair kits for cars are a useful laminate for weatherproofing antenna loading coils and making special covers or insulators. The filler putty supplied with these kits may contain metal.

Typical uses: PCBs, matrix boards, coil formers, insulating spacers.

ABS (acrylonitrile butadiene styrene)

This is expensive and is available in natural white coloured sheet. It machines easily and can be formed by heating, similar to Perspex. Bonding requires proprietary adhesives. A tough plastic and a good insulator.

Typical uses: antenna insulators, coil formers, handles.

Acetal copolymer

This is a medium- to low-cost plastic and is available in white or black rod and sheet. It machines very easily without specialist tools and is a useful plastic to have available in the workshop. It cannot be formed easily. Bonding requires proprietary adhesives and such joints are usually the weakest part of any assembly.

Typical uses: bearing bushes for rotating beams, morse key paddles or electrical parts such as insulators or feeder spacers.

Acrylics (Perspex)

A medium-cost plastic which is available in clear or coloured sheet, rod or tube. It is non-flexible and can shatter or crack under shock or excessive loads, although it is often used for see-through machine guards. The clear sheet is ideal for covering and protecting the front panels of equipment.

Perspex may be formed by heating but not with a flame as this plastic is combustible and gives off toxic fumes if burnt. If placed in a pan of water and simmered (or in an oven) at around 95°C, the plastic softens and can be formed or bent very easily. Forming should be stopped and the work re-heated the moment hardening or resistance to bending is felt, otherwise breakage will occur.

Bonding requires proprietary adhesives. A properly made bond is structurally sound and can be transparent. Normal drilling, sawing and filing are straightforward, providing the work is adequately supported. Most sheet forms are supplied covered with protective paper and it should be worked with this left in place. Not very heat resistant and should not be placed in direct contact with any heat source such as lamps, heatsinks and valves. Where transparency is not required a better plastic to use would be acetal or, for particularly tough applications, ABS.

Typical uses: decorative and protective panels, dials, morse key paddles, insulated fabricated boxes and covers.

Nylon 66

This is a cheap- to medium-cost plastic, and is available in sheet, rod or tube. It is usually supplied in its natural creamy white colour. It can be machined but it does tend to spring away from any cutting edge, making tapping or threading difficult. It is not easily formed. Proprietary adhesives are available which claim to bond nylon successfully. There are other types of nylon but most of these are expensive and intended for special applications, such as bearings, gears etc. For most amateur purposes the acetal copolymer, mentioned previously, is generally an easier material to use.

Polyethylene (ultra high molecular weight)

A medium- to high-cost plastic, normally available in sheet and rod. It is usually supplied in its natural white colour and it can be machined easily but not formed. Proprietary adhesives are available for bonding. This is an ideal plastic for outdoor components such as insulators, feeder separators etc as it is virtually rot-proof.

Polycarbonate

This is an expensive material, normally available in transparent sheet form. It can usually be bought at builders' merchants where it is sold as vandal-proof glazing. A very tough plastic, virtually unbreakable (bullet proof!) and, though it can be machined, it will wear out normal tools very quickly. It is ideal for making an insulated base for a vertical mast/antenna and in other areas where impact, high loads and temperature changes would rule out other less-durable plastics.

Polypropylene

This is an expensive plastic, normally available in opaque-coloured rod form. It can be machined and formed but not bonded. It can be sensitive to prolonged frictional contact with metals, particularly copper, and disintegration can occur in these circumstances. Because of its strength and resistance to atmospheric attack, it is usually used by amateurs in its rope form for halyards, mast guys etc.

The twisted strands are normally melt-welded together to prevent fraying, using a soldering iron. A naked flame should not be used as a substitute for the soldering iron as this plastic burns and melts, and burning droplets can go anywhere, even on the hands!

Polystyrene

This is a relative cheap plastic and is available in a variety of types, shapes and colours from black to transparent. It can be formed, machined, painted and bonded very easily – a model maker's delight! Usually used by radio amateurs for coil formers, insulated extender spindles and in other areas requiring insulation. It is a particularly tough flexible material, although some transparent types can be brittle. Some forms are also heat sensitive. A paper/card laminate of this plastic is available and this is very useful for making mock-ups of cabinets, boxes etc. Model-making suppliers usually stock extruded polystyrene sections, some shapes of which can be utilised in making bezels and other cabinet embellishments.

PTFE

This is an expensive material, noted for its excellent dielectric performance and low frictional properties. It is available in sheet, tube and rod, and is normally supplied in its natural off-white colour. Not easy to machine and cannot normally be formed or bonded. Usually used for low-friction bearings, insulators (up to UHF), capacitor dielectric and the nozzles of de-soldering guns. The fumes from overheated PTFE are very toxic.

PVC

This cheap material is available in many forms including rod, tube and sheet. It is usually grey or black in colour, and can be easily machined, formed and bonded. Proprietary adhesives should be used (although hot-air welding with a filler rod is also possible) but skill is required to produce structurally sound joints. Certain of the building types of PVC encountered seem to have some conducting capabilities which can lead to problems if used in electrical or RF applications. A suggested test for this is to place a small sample in a microwave oven, and if no metal is present it should stay cool to just warm.

PVC insulating tapes are strong, cheap and normally self-adhesive, and are supplied in a variety of colours and widths intended for wrapped insulation. Some of the poorer-quality tapes do not weather very well and suffer adhesion failure with the first frost.

ADHESIVES

Many modern adhesives are hazardous and it is essential to follow meticulously the manufacturer's instructions when mixing, using and curing them. Most are insulators and unsuitable for electrical joints.

Three general rules should be applied for bonding:

1. Degrease the parts thoroughly; even finger marks impair results.
2. Roughen the joint faces unless a transparent joint is required.
3. Do not place bonded joints under a peeling type of load.

Epoxides

A group of medium to expensively priced heat-setting resins (usually self-generated heat), that can be used for bonding, surface coating, laminating or encapsulation. Air and gas bubbles are the biggest problem with encapsulation (this work is carried out professionally under vacuum). The problem can be minimised by warming the work and the resin to around 40°C and providing a generous shrinkage allowance with a large pouring area which can be cut off from the cured encapsulation. Careful thought should be given to the necessity of encapsulation, for once completed, the encapsulated module cannot be altered or repaired. Encapsulation is usually used when circuits are subjected to harsh vibrational and environmental conditions. (See also 'silicone sealants').

These are usually two-part adhesives and require careful mixing just prior to use. A structural joint should not be over-clamped during bonding, and a bonding gap of typically around 0.05mm is required for the joint to be made properly. In other words, don't squeeze out all of the glue!

Surface coatings can be applied by dipping, spraying or brushing. Flexible resins are usually used for this type of work and are ideal for protecting beams, traps etc.

Cyanoacrylates ('superglue')

These expensive adhesives are available in various grades, each intended for bonding a particular set of materials. The low- to medium-viscosity grades are suitable for most amateur work. They are scientific marvels of bonding and as such require correct and proper application to ensure success. Releasing or debonding agents are available and it is a wise precaution to keep some of this handy in case of accidental bonding of fingers etc. These adhesives should be used and cured in well-ventilated conditions. It is advisable to wear protective glasses when using this adhesive.

Toughened acrylics

These are expensive, fast-curing adhesives intended for structural joints. Various types are available and are usually supplied in two parts – the 'glue' and the primer/activator. The glue is applied to one side and the primer to the other. They are suitable for use on most of the materials already mentioned, but some may not be used with certain plastics as they dissolve the material and eventually the joint fails. They are usually easier to use than the cyanoacrylates.

OTHER MATERIALS

Silicone rubber compounds

These are medium-priced materials, available as paste in squeeze tubes and as a liquid in tins. When cured, they normally set to give a white or translucent silicone rubber finish. They are ideal for encapsulation and the sealing or weatherproofing of antenna connecting boxes and similar out-of-door items. The electrical insulating characteristics are excellent and can be used to prevent parts from vibrating in equipment used for mobile or portable work. One type of this compound emits acetic acid during curing and this may damage some insulators.

Though not normally sold as such, one type of this compound has been used successfully as a resilient adhesive for structural and pressure-sealing joints on metal, plastic and glass.

Self-amalgamating tapes

These are a form of insulating tape which, when stretched and overlap-wrapped around cables, coaxial plugs etc, will amalgamate or flow together as one. They are reasonably priced and available in widths up to 50mm either with or without a self-adhesive face. Excellent weather-resistant properties. The self-adhesive form of this tape is ideal for waterproofing antenna traps, joints and connectors.

CORROSION

There are two main processes of metal corrosion. The first relies on environmental conditions such as rain or condensation which results in an acidic electrolytic liquid being formed on the surface of the corroding metal. As the metal corrodes, the acidity of the liquid increases until the electrolytic process of corrosion becomes almost self-sustaining. The second occurs due to the electrolytic action occurring between dissimilar metals in contact, and is referred to as *galvanic action*. Both processes change the metal into a different form, which in the case of steel or iron we know as 'rust', and often refer to the process as *oxidisation*.

Table 16.1 shows the galvanic relationship between metals. The numbers are item numbers only to show the position of each metal in the galvanic series. The actual values depend on several factors such as temperature, radiation and acidity etc. This list enables metals to be selected which will have the minimum galvanic corrosion effect on one another. The greater the list separation, the greater will be the possibility of corrosion.

Table 16.1. Galvanic series

1. Magnesium alloys	8. Lead
2. Zinc	9. Tin
3. Aluminium alloys	10. Nickel
4. Cadmium	11. Brass
5. Mild steel	12. Copper
6. Stainless steel 18/8	13. Silver
7. Lead-tin solders	14. Gold

Note: the numbers are item numbers only and are not values!

For example, brass and copper are adjacent and would therefore not cause problems if in direct contact with each other. Brass and aluminium alloy are widely separated and corrosion occurs if these metals are in direct contact with each other. This state can be reduced by tinning the brass with lead-tin solder, which falls about halfway between the other two metals (*galvanic interleaving*). The higher item-numbered metal normally corrodes the lower item-numbered one.

Dissimilar-metal galvanic action corrosion is avoided by ensuring that the galvanic series separation is minimal. If widely separated metals do need to contact each other, then a suitable interleaving material should be used to reduce the galvanic separation level.

Corrosion by moisture or rain is more difficult to combat effectively. The commonly accepted anti-corrosion treatment consists of protection by paint and, providing the paint coat remains intact or is renewed regularly, this is a very effective treatment. Steel and iron can also be coated by protective metals such as zinc (*galvanising*) or nickel (*plating*), both methods being normally outside the range of the home workshop. Aluminium is slightly different, for this metal forms its own protective oxide barrier which, providing it is not disturbed, will prevent further corrosion. It is this oxide barrier which makes soldering aluminium difficult. Unfortunately, this natural protective oxide layer can be disturbed by stress or galvanic action, and the corrosion process bites deeper into the metal. Anodising is a process on aluminium which forms a controlled layer of oxide on the surface and presents a toughened surface finish which can be coloured by dyeing. Brass and copper can be considered as corrosion resistant for most amateur purposes. However, their surfaces do oxidise and this can impair good electrical contact. Certain platings, such as silver, nickel or chromium, can reduce this. The platings themselves can also corrode but usually at a much slower rate than the parent metal.

If not adequately protected, corroding metal will gradually lose strength and the device from which it is made will fall apart. This is usually seen as collapsed masts, broken antennas and similar expensive disasters.

Corrosion also affects RF and electrical connectors, particularly feeder-to-antenna connections, and causes a gradual decline in the overall performance of the system. Signals become weaker and calls to DX stations which used to be answered are ignored. Most observed lowering of performance in this area is usually (or eventually), discovered to be due to corrosion. This can be practically eliminated by first ensuring that no dissimilar metals are in contact to cause galvanic corrosion, and second that water is excluded from all connections. Copper coaxial-cable inners should never be connected directly to the aluminium elements of a beam but should be tinned first as previously stated. Coaxial plugs and sockets should be fitted with heat-shrink sleeving and wrapped with amalgamating tape to prevent the ingress of water. Connector boxes can be filled with silicone-rubber compound for a like purpose. Stainless steel antenna fittings are the least affected by corrosion but even these would benefit from a coating of protective lacquer, particularly on screwed fittings. The position of stainless steel in the above list is interesting, for sometimes the metal in contact with the stainless steel will corrode due to galvanic action, particularly in wet, smoggy or salt-laden conditions, and some stainless steel fittings are supplied with plastic interleaving spacers and washers to prevent this.

Condensation is also a problem with outdoor enclosed or boxed-in items such as rotators, remote tuners, antenna traps, coaxial cable and the like – where there is condensation, there will be corrosion. The simplest, but not necessarily the easiest, solution is to allow the enclosed area to 'breathe' by introducing suitable weather-resistant holes as drainage vents, while ensuring that these are at the lowest point of the enclosed area and cannot be inlets for rain or the run-off water from the outside of the enclosure. Complete sealing usually makes matters worse, for a fully water and air-tight enclosure still produces internal moisture by condensation. Full hermetic sealing is difficult to apply for it normally requires the ability to pressurise the enclosure with an inert gas, as with some military or maritime equipment. If the items within the enclosure can be protected as if they were exposed to the elements, much of the corrosive effect of condensation is reduced.

In the case of rotators, attempts at filling the voids with grease does not help, for the grease forms small pockets which can hold water. It is better to lacquer or wax oil the moving parts and the inside of the housing, and to use grease for its intended

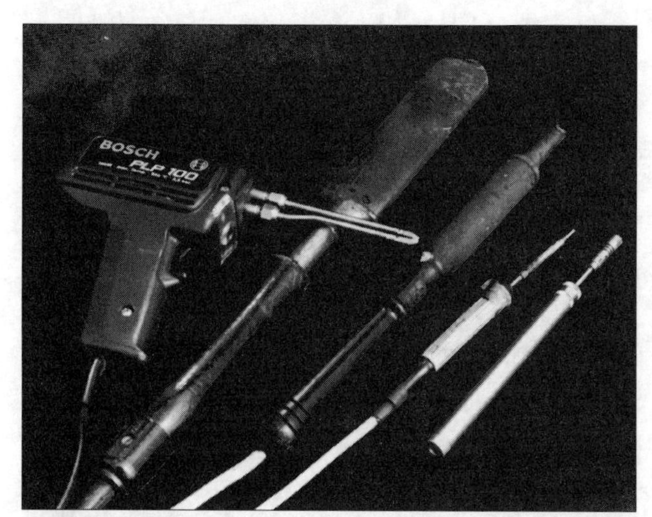

Fig 16.1. Soldering tools

purpose of lubrication. Any electrical items within the rotator should be sprayed with a commercial, non-insulating, waterproofing liquid.

The threads, screws, nuts, bolts etc should be given a light coating of anti-seize compound on assembly and, after assembly, sprayed over with lacquer or wax oil, or coated with a water-repellent grease such as lanolin. This makes for easier maintenance and reduces the possibilities of corrosion.

Corrosion is not limited to outdoor items. Corrosion of connector pins of microphones, plug-in PCBs, computers etc is not uncommon. Careful selection of mating materials to avoid galvanic action, combined with appropriate painting, plating etc and regular maintenance, will reduce the effects of corrosion.

TOOLS

Any tool bought with reliable use in mind should be the best you can afford, for cheap tools usually lead to frustration, like a cheap pair of wire-cutters that has its cutting edges notched by copper wire the first time used, and from then on will not cut but only fold the wire! Retiring toolmakers and fitters often sell their 'kit', and some very good but used tools can be obtained this way at reasonable prices. Tools also make very good presents!

Most amateurs have a shack, room or some place with a bench, hacksaw and a vice of one type or another. Accepting these, the range and type of useful tools available is virtually limitless, unlike most pockets. Basic tools and a few extras – some home-made – are listed here.

Soldering irons (Fig 16.1): 15W instrument: 50W electrician's (and a useful extra, a 200W heavy-duty iron) all mains powered.

The choice of type of irons in the above selection is a personal one, bearing in mind the cost. Temperature-controlled irons are excellent but expensive. The main factors to consider are the availability of the replacement parts (bits, heating elements, handles) and the ease with which these may be fitted. The anti-static properties of an iron should also be considered.

Battery and gas-powered irons are worth considering if on-site or outdoor work is likely. Similarly, the rapid heat soldering gun is a worthwhile addition.

De-soldering 'gun'. The miniature anti-static, hand-operated type is an effective and value-for-money device (Fig 16.2).

Electrician's pliers (also known as *combination pliers*) (180mm) (Fig 16.2).

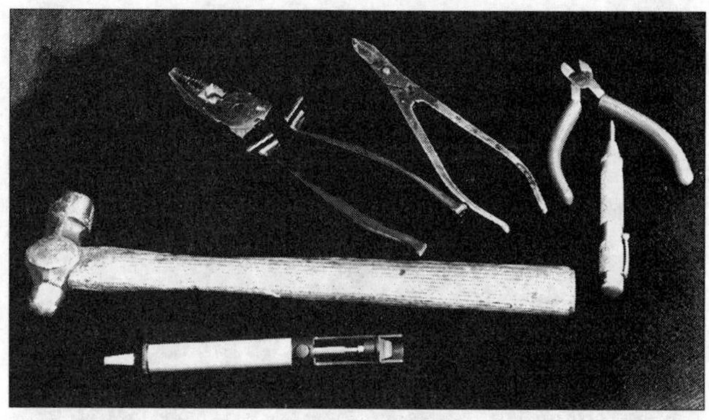
Fig 16.2. Selection of hand tools

Side cutters (also referred to as *diagonal cutters*) (120–180mm) (Fig 16.2). There are a variety of wire-cutters and the type specified here is suggested as the most suitable for general use. They will cut most wires up to 2mm diameter (barbed or fencing wire excepted!)

Watch-maker's shears (Fig 16.2).

Long-nosed pliers (also known as *Snipe nose*) (120–180mm).

8oz ball pein hammer (Fig 16.2).

8oz soft-faced hammer with replaceable heads.

Twist drills. These should be of high-speed steel in at least 1.0–12mm diameter (fractional sizes from 1/16–1/2in diameter). Carbon-steel drills are cheaper but require sharpening much more often.

A *centre drill* (BS1 size) is a worthwhile addition (Fig 16.3). Unfortunately, it is very easy to break the tips of these centre drills and it is worth considering keeping a few 'in stock'!

Drills need regular sharpening and some drill-sharpening device should be considered. There are many types available and the main points to consider when buying are that:

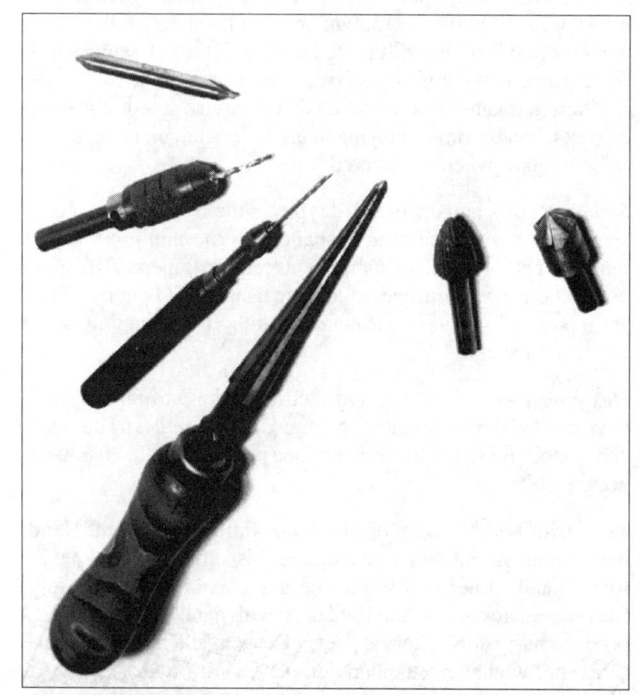
Fig 16.3. Anti-clockwise from the top left: centre drill, pin chuck, pin vice, taper reamer, file burr, countersink

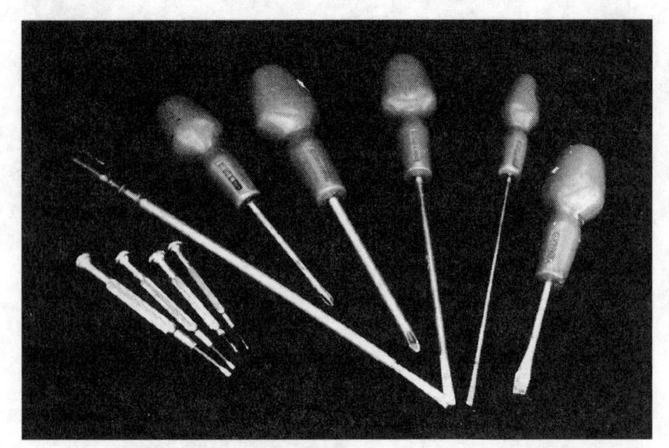

Fig 16.4. Selection of screwdrivers, including screw-holding type

(a) it will sharpen the range of drills in use;

(b) spare grinding stones can be purchased;

(c) little or no skill is required to obtain the correct drill point; and

(d) the speed is not so high that it will soften instead of grind the drills.

Electric drill. An electric drill should preferably be of the continuously variable speed type and should have a drill chuck capacity to at least match the range of drills on-hand. A drill stand is a valuable addition. The model-makers' variable speed, hand-held or stand-mounted drilling and grinding unit is a very useful extra, especially for the finer work of PCB making.

Pin chuck (Fig 16.3). This device enables small-diameter drills (1mm downwards) to be held in the normal drill chuck. Most pin chucks come with interchangeable collets to cover a range of small-diameter drills. This pin chuck should not be confused with the pin vice which is intended to hold small components during filing or fitting, and not drills.

Screwdrivers (Fig 16.4). Minimum requirement: parallel flat blade 3.2 × 100mm, 5 × 150mm and similar sizes for the cross-head types of screw (Phillips™, Pozidriv™ etc). A set of watch or instrument makers' screwdrivers are very useful.

The interchangeable-bit screwdriver is also worth considering but it can become annoying to use – the right bit never seems to be in place when it's wanted!

Spanners. Box and open-ended types in the BA and ISO metric sizes, plus a small adjustable spanner as the minimum starter requirement. Many older components require imperial/BA spanners but the newer European parts are usually ISO metric. Pliers are a poor substitute for spanners, leading to mangled nuts and scratched panels!

Hexagon socket keys. BA and metric sizes up to 8mm. These keys are available mounted in screwdriver handles. The interchangeable-bit screwdriver mentioned previously often has hexagon key bits.

Files with handles (Fig 16.5). 150–200mm second cut. Hand, half-round, round and three-square, also the same shapes in round-handled needle files. In the interests of personal safety, files should always be fitted and used with handles, for file tangs produce nasty cuts. There are sets of files available in the above selection, which are supplied complete with fitted plastic handles. The woodworker's shaper plane/file (Surform™) is very useful for the rapid trimming of aluminium sheet.

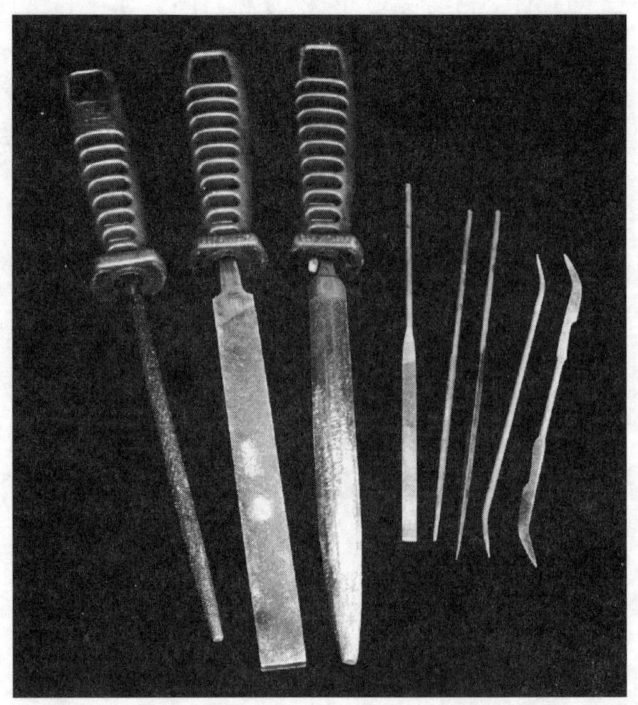

Fig 16.5. Selection of files. The two on the right are Reifler files, and are useful for filing awkward shapes

Tapered reamer (see Fig 16.3). Sometimes known as *repairman's reamer*. The 3–12mm size with handle will suit most requirements.

Hand countersink (Fig 16.3). The 12mm size with five or more cutting edges will cover most work and, mounted in a plastic handle, it is very useful for deburring holes.

The above list of tools can be considered as a starter kit for the newcomer to construction. With these few tools it is possible to attempt many of the jobs encountered in making or repairing something but, as construction skills grow, so will the tool chest! A tap and die set is extremely useful and should be placed high on the 'tools required next' list.

Useful extras

One of the problems most of us have in metalwork and the like is cutting or filing things square and straight. Also, marking out to ensure that parts will fit together correctly is sometimes not as easy as it appears. The few extras described here should simplify things.

Bending bars

See Fig 16.6. These are normally a home-made item, and the things to check when obtaining the steel angle are its straightness and squareness. If these are not 'true' the corrective actions needed will require skills which may be beyond those so far acquired. An old bed-frame angle, provided it has not rusted too badly, usually makes very good bending bars. One bed frame can be used to make several bending bars of different lengths. The choice of length and distance between the clamping bolts of the bars is related to the maximum width of metal or other sheet purchased and the maximum size of panel worked. Aluminium sheets, for example, usually come in 1.8 × 1.2m size and trying to cut this is rather difficult. It would be better, if buying in these sort of quantities, to have the sheets guillotined by the supplier to widths which will fit between the bending bar clamp bolts. (Suggest 482mm.) The length of the sheet will not matter.

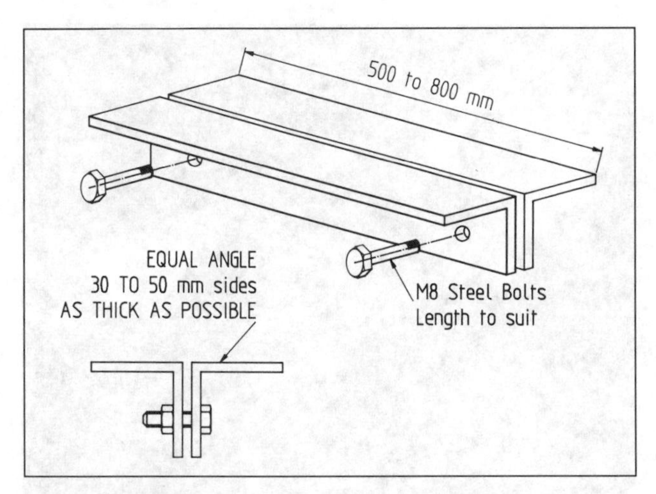

Fig 16.6. Bending bars

Hole embossing tool

See Fig 16.7. This very simple home-made tool forms holes along the edges of metal panels to accept self-tapping screws of about 2.5mm diameter (depending on metal thickness), and with an engagement length of about 12mm. It facilitates the rapid manufacture of cabinets, boxes and screens, eliminating much of the accurate metal bending normally associated with this type of work.

The working parts and the two side plates of the tool are made from 1in wide by 0.125in thick-ground gauge plate, which is a tough but workable steel capable of being hardened and tempered. It is not necessary to harden the finished tool, unless it is

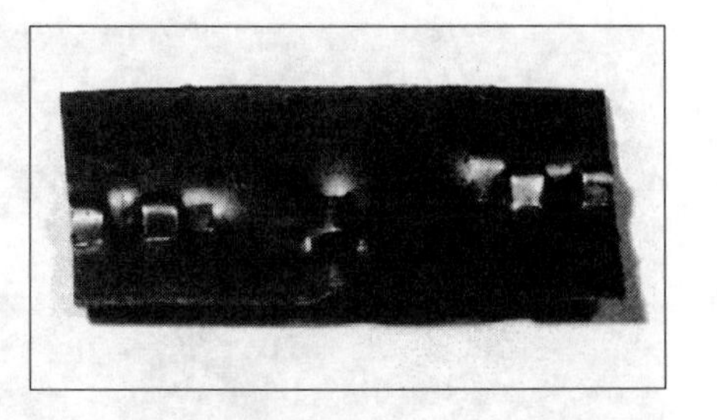

Fig 16.7(b). Holes made by embossing tool

intended to form mild steel but, if facilities are available to do this, it will improve the durability. The base and top plates are made from 1in wide by 0.375in thick mild steel.

The vee points must be on the centre-line of the tool. The vee notches are not so critical for they play no part in the shape of the hole produced, but they should allow sufficient clearance for the formed metal to flow into. The point and notch sizes shown are suitable for most metal thicknesses up to 1.6mm (16swg). The vee points can be made as a set, with the four plates clamped together. The fully shaped points should be polished with an oilstone to obtain square, burr-free, sharp edges along each face of the point. The edges of each point are the working parts of the tool and should be made with care. It is essential that the flat faces on both sides of the notch and point ends of each set of plates are square, parallel and level with each other.

Tapping-size holes should be marked out and drilled on each top plate only, and these used as a template to drill the remaining holes on each plate. The holes in each bottom plate should be tapped 2BA and the holes in the remaining plates opened up to 0.1875in diameter – a good fit for the *high tensile steel* 2BA screws. ISO M5 may be used in place of the 2BA screws and tapped holes. The outside faces of the holes in the top plates should be countersunk deep enough to ensure that the screw heads are slightly under-flush. It is good practice to lightly countersink both sides of every hole to remove burrs and facilitate assembly.

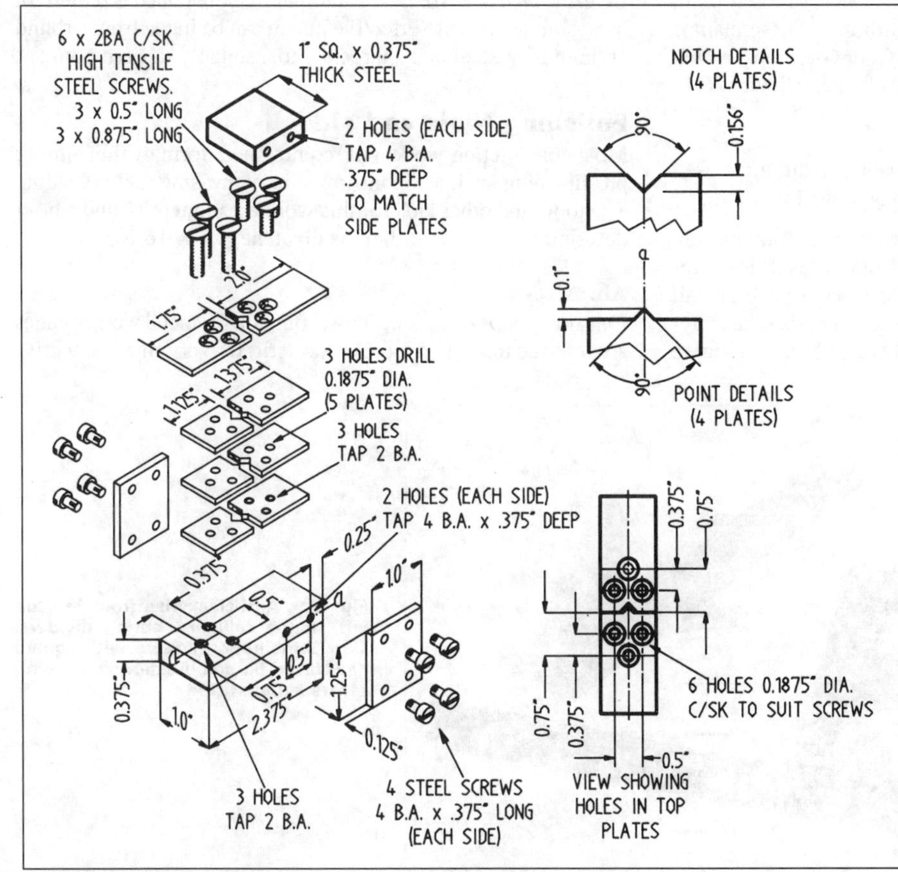

Fig 16.7(a). Hole embossing tool

The back edges should also be square and level to ensure that the vice pressure is applied evenly during forming. The 0.125in thick side plates and the 0.375in thick top and bottom plates form the bearing for the sliding part of the tool, which must slide easily with the minimum of play in any direction. Failure to get this right will ruin the action of the tool, for each point must just slide over the opposing one or two points with the minimum of clearance.

To operate, the tool is placed between the jaws of a 4in vice and supported by the extended top plates. The metal to be holed is rested on the base of the tool and the marked out hole position is aligned with the tip of the visible point. The vice is then closed with the minimum of force until each face of the working ends of the tool just contacts the metal sheet. *It is pointless trying to go beyond this, for squeezing the tool and the*

Fig 16.8. Toolmaker's clamp and joiner's G-cramp

work by excessive pressure from the vice will only ruin the tool and the work. The vice is then opened and the metal sheet gently prised away from the points. This method of releasing the formed sheet could be improved upon, but the tool would become more complex to make and, as a spot of oil on the working faces eases the problem, this extra complexity is not worthwhile.

The tool performs very well on aluminium, tin plate and annealed brass or copper in thicknesses up to 0.0625in (16swg). It is not recommended for use on sheet steel unless the points have been hardened and tempered.

Tool-maker's clamps

See Fig 16.8. These are available in several sizes. Two 100mm clamps are suitable for most amateur purposes. These clamps should not be confused with the joiner's C- or G-cramps, which are also useful and certainly better than no clamping device at all.

Measuring and marking-out tools

After the soldering iron, these will probably be the most used tools and buying the best possible will pay in the long term.

Engineers' combination square (Fig 16.9). The 300mm/12in ruler with square head only is preferred from the usefulness-to-cost viewpoint. Most of these combination squares have a small scriber, housed in a hole on the square head, and this saves buying a scriber separately. A complete quality combination square is expensive and, as well as the square head, it has a centre square and a clinometer head, neither of which are essential for normal amateur work. Beware of the cheap bazaar-type combination squares, for they are a waste of money and lead to corners which are not square and inaccurate measurements.

Fig 16.10. Finishing and polishing tools

Spring dividers (100mm size) (Fig 16.9).

Jenny callipers (100–200mm size) (Fig 16.9). These are also known as *odd-legs*.

Centre punch (Fig 16.2). An automatic centre punch is preferred, but a simple and cost-effective punch can be made from a round 100mm long steel nail sharpened to a suitably tapered point.

Finishing tools and aids

Most construction work will require some form of finishing or pre-finishing such as deburring, emery dressing and polishing. The tools and other aids for this work are numerous, and a brief description of some of them is given here (Fig 16.10).

Abrasives

Emery paper or cloth is the most common for metal work. Grades are referred to as 'very coarse' (80 grit) to 'very fine' (800 grit).

Fig 16.9. Anti-clockwise from the top left: Jenny callipers, spring dividers, 12in combination square with square head fitted, 6in combination square with centre square fitted

The 240 (medium) to 400 (fine) grades are suitable for most amateur uses.

Flap wheels for mounting in the electric drill come in various grit sizes and can save a great deal of work. The suggested grades are 60 grit to 120 grit, or finer.

Wet and dry paper is useful for the final smoothing before painting, or between coats of paint. Fine wire wool or plastic pan scouring pads are good substitutes for wet and dry paper.

Blocks of abrasive mounted in a rubber-like material (rather like erasers for paper) are also extremely useful and are available in fine to coarse grades. One of each grade is suggested. The super-fine grade block containing non-metallic abrasive is ideal for the cleaning and polishing of PCBs. (It cleans without removing too much copper.)

A *glassfibre brush pencil* is useful for cleaning the smaller areas prior to soldering. A brass wire brush insert is also available for this pencil.

An electric drill fitted with a *hard felt disc* or a *calico polishing wheel* used with polishing compound will save considerable effort when trying to polish a component to a bright and scratch-free finish. Polishing aluminium by this method is not recommended, for aluminium tends to 'pick-up' on the wheel and produce deep score marks on the panel being polished. The fine-grit flap wheel or the abrasive eraser are more suitable for aluminium, and the final polish can be made using a rotary bristle brush and thinned-down polishing compound.

Steel or brass wire brushes, either hand or machine types of various grades (coarse to fine), are very useful where paint, rust or other corrosion needs to be removed. These brushes can also be used (in skilled hands) to produce decorative effects on the surface of aluminium, brass, or stainless steel.

Warning

The dust and fumes created whilst polishing or grinding can be harmful and a face mask and protective glasses should be worn. Always ensure that the work place is well ventilated. Some polishing compounds can cause skin irritation and the use of barrier cream on the hands is advised. The wearing of protective gloves can save the hands getting filthy, but take care when working with powered tools to ensure that the gloves do not become entangled with the rotating parts. Even a DIY electric drill can break a finger!

USING THE TOOLS

Marking out and measuring

This is the important bit of construction. If not done carefully and accurately, failure is certain, hence the saying "cut once, measure twice". All measurements and squareness checks should be made from one or two datum edges.

Consider marking out the box used for the dip oscillator shown in Fig 16.11. The first operation is to obtain two edges straight and at right-angles to each other. The method of achieving this should be clear from the illustrations (Fig 16.12). These two edges are now the datum edges for vertical and horizontal measurements, and should be clearly marked. The overall size of sheet is marked out and cut to size. This results in a sheet with all corners square and all edges straight. With the bending bars as a guide it is really very simple! In this case the sides of the box are the same depth. Adjust the Jenny callipers to the required measurement and, locating from each edge in turn, scribe round

Fig 16.11. Dip oscillator using home-made box described in the text

the sheet. Check with the combination square from the datum edges that these lines are parallel and square with their respective edges. If Jenny callipers are not available, the combination square may be used by setting the ruler to the required measurement and scribing round the sheet with the scriber held against the end of the ruler. These lines show where the bending is to occur and are used to align the sheet in the bending bars (Fig 16.13). It is good practice to lightly centre-pop along the lines at

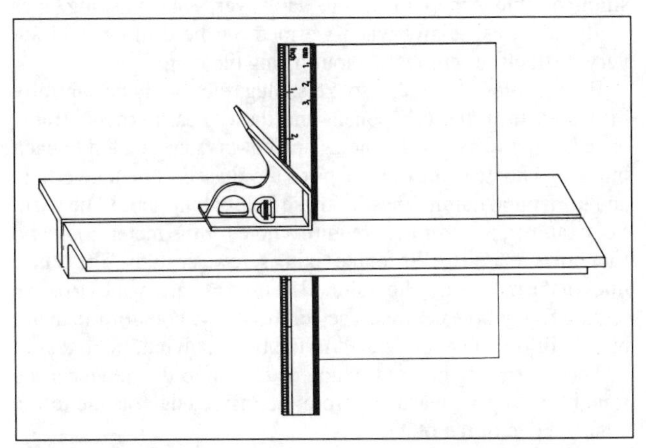

Fig 16.12. Squaring up the work prior to bending or cutting

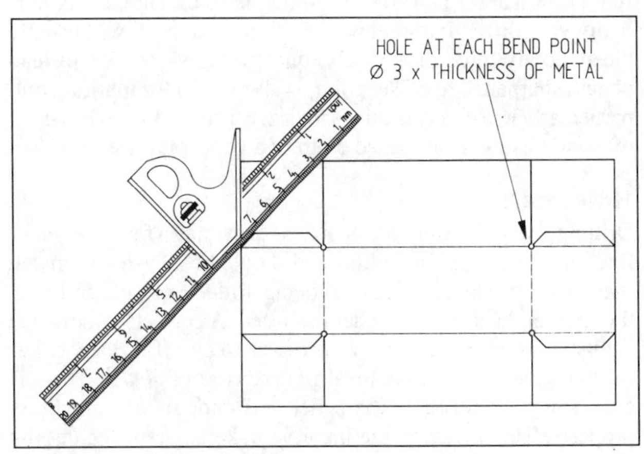

Fig 16.13. Marking out the box

about 20mm intervals to make 'sighting-up' in the bending bars easier – half the centre pop should be visible and the other half invisible inside the bending bars.

Existing boxes, panels etc, can be similarly marked out using two datum edges that are straight and at right-angles. If a truly square corner is not available, then the longest and straightest side should be used as a datum and the end of the combination ruler used to mark out the lines at right-angles. A suitably placed line at right-angles to this single datum can be used to mark off the measurements in this plane.

Lines marking the centres of holes and cut-outs should be scribed in, for the centre punch can then be located by feel into the notch formed by their respective intersections. It is worthwhile using the dividers to scribe-in each hole diameter around its centre-popped centre. This helps to ensure that the correct size hole is drilled and in checking that holes do not foul. If screw-up chassis punches are used, these scribed outlines can be used to locate the punch accurately.

When a gleaming or ready-painted cabinet, box or panel is purchased, marking out directly on to it would ruin the finish. Instead mark out in pencil on to a piece of draughting film cut to suit and then secure this to the required face with draughting tape. Check that everything is where it should be and, when satisfied, centre-pop through the hole positions and the outline of any cut-outs. Remove the film, scribe in the hole diameters and join up any cut-out outline dots. When scribing holes on to a ready painted surface, it is worthwhile gently scoring through the paint to reveal the metal, for this helps prevent the paint chipping off around the holes during drilling and reaming. Some constructors drill through masking tape stuck over the hole position to achieve the same results. However, some masking tapes really stick, especially when warmed by the drilling, and are very difficult to remove without lifting the paint.

Fixing holes for meters, sockets, plugs etc should be carefully measured from the component and these measurements transferred to the work. Where holes and cut-outs are related to each other, and not to some other edge of the chassis or panel, vertical and horizontal datum lines for these holes alone should be used. For example, a meter may require a hole for the meter body and four holes related to this for the fixing screws or studs. The centre lines of the meter body hole should be used as datum lines to mark out the fixing hole centres. The centres of the hole for the meter body will of course be related to the two main datum edges.

The centres of holes or studs of the same diameter can be found by simply measuring from the inside edge of one to the outside edge of the other.

It is tempting to use, and easier to see, markings made with a fine-point marker pen. Beware – the marking ink used is normally very difficult to remove, and if not removed, will usually bleed slowly through any subsequent paint work. A soft lead pencil, sharpened to chisel point, is also useful for marking out, particularly when 'trying things for size'. The marks can be wiped off easily using a dampened cloth or a very soft eraser.

Hole making

Drilling is a straightforward operation providing a few key guidelines are observed. The drill speed should be adjusted to match the size of drill and the material being drilled. As a general rule, the smaller the drill, the higher the speed. A correctly sharpened drill should not require heavy pressure to cut. If it does, either the material is too hard or the drill needs to run slower. There is a tendency for normally sharpened drills not to cut some brass properly. This is due to the incorrect rake angle of the cutting edges of the drill. The problem can be solved by sharpening the helix edge on the face of the flute to give a rake angle of about 15°, ie leaning backwards away from the normal cutting edge (Fig 16.14). Drilling some plastics can also present problems, such as chipping of the edges and breakaway of the material as the drill breaks through. Some improvement is possible by sharpening the drill to produce an included point angle of around 80°. The standard included point angle is 118°.

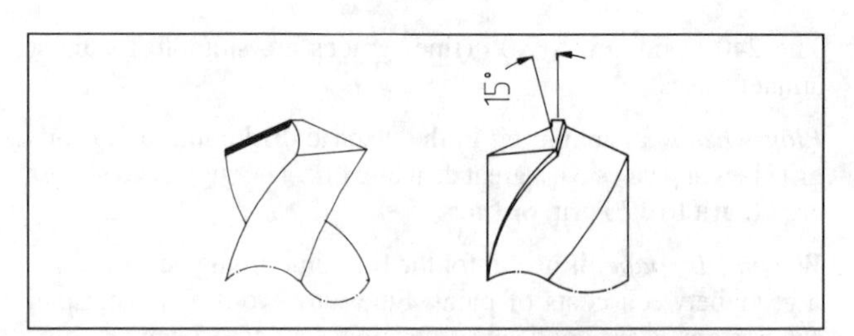

Fig 16.14. Modified drill rake angle to improve the drilling of brass

When drilling steel, it is advisable to use a coolant such as soluble cutting oil to keep the drill and the work cool. It also saves having to re-sharpen the drill so often. Paraffin is a good coolant for aluminium and copper. Brass and most plastics do not normally need a coolant, providing the drilling speed is correct, but treat as steel if necessary. These coolants may be applied either by an oilcan, or a brush. (Old liquid soap squeeze bottles make good coolant dispensers and any left-in detergent is not detrimental.) Soapy water is also a good coolant and certainly better than no coolant at all!

Holes should always be centre-popped before drilling and, if the BS1 centre drill referred to previously is available, this should be used next to provide an accurate location for the drill.

The holes drilled in thin sheet are often anything but round because the drill does not have enough depth of metal to round the hole properly before it breaks through. There are at least two ways round this. The easiest is to drill the hole undersize and bring to size using the taper reamer, which must be allowed to cut without forcing, otherwise another fancy-shaped hole will be produced. (Another reason for scribing-in the holes beforehand.) Another way is to alter the cutting angle of the drill to an included angle of about 140° and thin the chisel edge of the drill to a point. It is worthwhile keeping a set of drills sharpened in this manner especially for thin sheet drilling. These sheet metal drills normally require a much slower speed and the holes should also be pre-drilled using the centre drill.

Whatever is being drilled should be well supported and clamped to prevent rotation and lifting. Failure in this direction can lead at best to a broken drill and at worst to serious personal injury, for drills have a habit of picking-up just as they are about to break through the hole. (Plastic and copper are particularly susceptible to this.) A panel whirling round on the end of a 10mm diameter drill is a frightening sight!

All drilling will produce burrs around the hole edges and it is good engineering practice to remove these using the hand-held countersink bit referred to previously. A file will only scratch the rest of the surrounding surface and bend the burrs into the hole. The use of a large drill for hole de-burring is not recommended, particularly on the softer metals, unless the 'touch' for this method has been acquired.

It is usual to step drill holes larger than 10mm diameter, that is a hole of smaller diameter (3mm) is drilled first, then another slightly larger (+1mm), and so on until the finished size is reached. Step drilling is unsuitable for sheet material. There are special, though expensive, stepped drills available for drilling holes up to 40mm diameter in sheet material.

Fig 16.15. Large round hole making tools. Clockwise: deluxe home-made tank cutter, commercial tank cutter, hole cutter, screw-up punch, hole saw

Making large round holes can be tackled in at least two ways. The first requires a washer or tank cutter (Fig 16.15) and an extra slow speed drill (a joiners' brace is effective). The biggest snag with this method is trying to obtain an even cut around the full circle. By clamping the work to a block of wood and drilling through into the wood for the centre pilot of the tank cutter, a guide is provided which improves things a little. The main thing with this method is not to be in a hurry. The second method can be applied equally well to non-round holes. Contiguous holes of about 5mm diameter are drilled on the waste side of the hole or cut-out and 1 to 2mm away from the finished size markings. The waste is removed, using tin snips. The hole is then carefully filed to size, using the bending bars for support, and as a guide for any straight portions.

To de-burr large holes, a small half-round needle file can be used in the 'draw' fashion. The file is held at both ends and drawn round the edges to be de-burred in a manner similar to using a spoke shave. The file should be held at an angle to produce a small 45° chamfer around each edge of the hole.

Machine countersinking of holes requires a very slow speed drill (60rpm). They can be produced using the handle-mounted countersink bit referred to previously. Even the multi-toothed rose-countersink bit will chatter and leave a very unsightly surface if too high a speed is used. Countersunk screws will not sit properly on such holes. There are countersink file burrs available which do a similar job and can be used in the DIY drill, providing the work is held securely with the burr square to the work. They do have a tendency to skid over the surface, leaving a trail of deep scratches, if not located properly.

Drilling accurately positioned holes can be a problem and templates can be used with success. Keeping the small holes for multi-pin DIL ICs in-line and at the correct spacing is difficult, but the job is made easy by using a piece of the correct pitch matrix board as a drilling template. One of the holes (pin 1 is suggested) is drilled first in the required position. The template is located from this by passing a drill through the hole and the matrix board. The matrix board is then aligned with the rest of the marked-out hole positions and clamped using the toolmakers' clamps. The remaining holes are drilled through using the matrix board as a template. It helps to mark each row of holes on the matrix board template. If a considerable amount of such accurate drilling is to be done, it is worth making a set of metal templates for each size of IC using this method.

Sometimes the component itself can be used as a template to reduce the risk of error, eg slide switches, coaxial sockets, dial fixing holes etc.

Filing

Accurate filing is a skill which can only be acquired by practice. General rules are:

1. Always use a file handle. This eliminates the risk of running a file tang into the wrist and enables the file to be guided properly.
2. Use a sharp file. It is normal practice to use new files for brass and, as their sharpness wears off, use them for filing steel or aluminium. New files should be kept separately or otherwise identified.
3. Do not force the file to cut. Only a light relaxed pressure is required, which also aids the accuracy of filing.
4. Keep the file clean by brushing with a file card or by rubbing a piece of soft brass or copper along the teeth grooves.
5. Support the work properly. Trying to file with the work held in one hand and the file in the other is a guarantee of failure.

The height of the vice should be such that when the filer's arm is bent to place the fisted hand under the chin, the point of the elbow should just rest on the top of the vice jaws. An old but effective form of ergonomics!

A strip of emery cloth wrapped along the file can be used to obtain a better surface finish. This method was frowned upon in apprentice schools but it works and saves time. The draw filing technique referred to previously imparts the final touches to the filed edges and removes the filing burrs. The bending bars can be used as a filing guide (Fig 16.16) and a set of bars can be made especially for this purpose, checking them regularly to see that the guiding surfaces are not bowed by too much one-spot filing.

Bending and forming

The first essential is to ensure that the material can be bent. This sounds obvious but it is better to check first than to find out after all but the bending work has been done. Annealing can be applied as mentioned previously under the section on materials.

A metal hammer should never be used directly on the metal when bending or forming. Either use a soft-faced hammer or a block of wood or plastic as a buffer for the hammer blows (Fig 16.17). This prevents all the humps and hollows which would otherwise occur. (Do not use any metal as a buffer block as there is a danger of the metal chipping and flying into the face or eyes.) The block should be kept into and near the point of

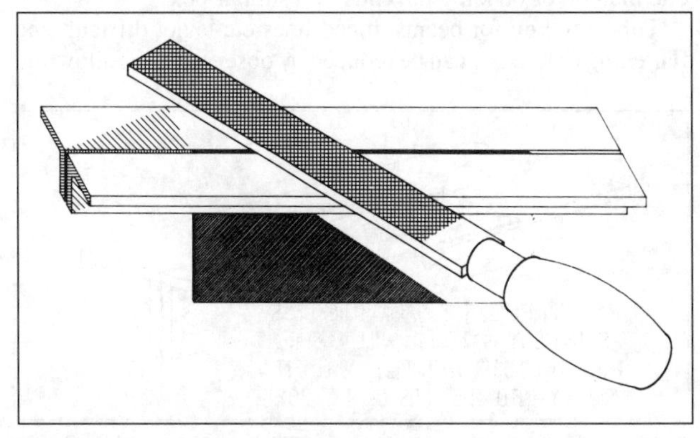

Fig 16.16. Bending bars as filing guide

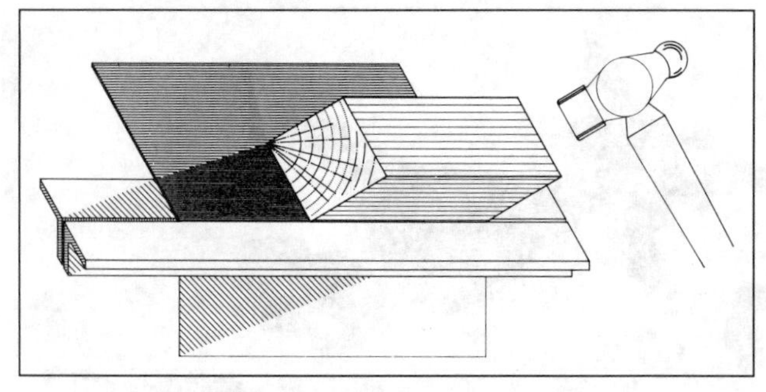

Fig 16.17. Method of bending in the bending bars. Note that the buffer block is held into the point of bend

bending. It seems easier and quicker to try to use the sheet as a lever and hammer as far away from the point of bending as possible. This will only produce a bend which curves up and back again. Where three sides are to be formed by bending, the point of intersection of the marked-out bend lines should be drilled before bending, with a hole diameter three times the thickness of the metal. This prevents corner bulge (Fig 16.13).

For reliable bending of sheet aluminium up to 1.5mm thick, it should be possible to bend a sample strip back on itself and hammer the fold flat without breakage. Annealing aluminium is possible, but under specialised conditions. A workshop 'dodge' which sometimes works, and is certainly worth a try if bending or forming proves difficult, is to gently heat the aluminium to such a temperature that a red-headed match (Swan Vesta), when placed on the heated surface, just softens but does not ignite. Care needs to be taken to avoid overheating, for this will destroy the aluminium's properties. This method of annealing has proved effective when bending tube or rod for the folded dipole elements of beams. In this case the area to be annealed is small and the 'dodge' can be applied easily. Any bending or forming should be carried out as soon as the annealed piece has cooled, for most grades of aluminium alloys start to re-harden, and may actually improve in strength after such heating. Heat treatment of aluminium and its alloys is a specialised process used to improve a particular characteristic of the material, and is carried out in controlled conditions not usually available in the home workshop. The above 'dodge' is about as specialised as possible and is more reliable than the older 'spit and sizzle' method, or 'is the clothes iron hot enough?' technique.

Another method is to rub soap on the surface and warm the aluminium until the soap turns brown. However, it is essential to ensure that it is the aluminium which is warmed and not the soap. This is ensured by applying the heat to the non-soaped face.

Fig 16.18 shows how to ensure that the end lugs of the box shown in Fig 16.11 fit snugly inside the box. Fig 16.19 shows one method of bending the ends of a similar box.

Tube bending for beams, tuned lines etc is not difficult and flattening or kinking can be reduced by observing the following:

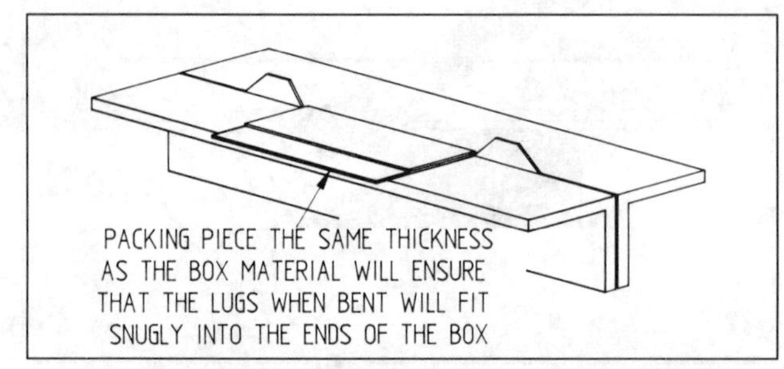

Fig 16.18. Making allowance for bending the end lugs of the box

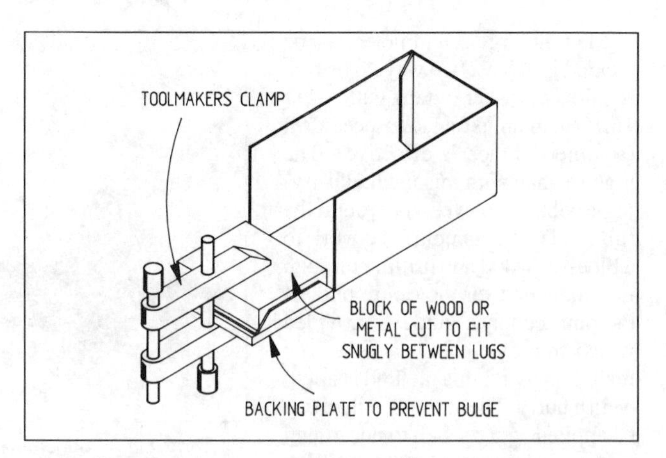

Fig 16.19. Bending the ends of the box using a toolmaker's clamp to secure the forming block

1. Ensure that the tube is suitable for bending. Anneal as necessary and re-anneal as soon as resistance to bending is felt.
2. Unless skilled or equipped with specialised tooling, do not attempt to bend to a radius of less than three times the outside diameter of the tube, eg 12mm OD tube should have a 36mm minimum bend *radius*.
3. Always bend round a former shaped to the required radius (Fig 16.20).
4. Pack the tube tightly with fine sand (birdcage sand is ideal, preferably unused!). Wet the sand and cork both ends. This will minimise the risk of kinking during bending, and the sand can be washed out afterwards. The tube can be re-annealed if necessary with the sand left in place but the corks should be removed to let the hot gases escape.

A low-melting-point, lead-like material is available which can be used in place of sand, and is known as *bending metal*. It is poured molten into the blocked, ended tube, allowed to set and, when the bending has been completed, is melted out to be used again. This process should be carried out with the tube at an angle to allow the hot air to escape. The molten bending metal should be poured very slowly for it has a tendency to 'blow-back'. Normal lead is not a substitute for bending metal and should not be used, particularly with aluminium tubes, for the melting temperature required for lead may destroy the aluminium's properties.

Modelling wax (also known as *American wax*) can be used in place of sand or the bending metal. The wax is heated, not boiled, until molten and then poured into the tube in the same manner as the bending metal. The wax should be allowed to set hard before bending. Unfortunately it is not possible to re-anneal with this wax left in place.

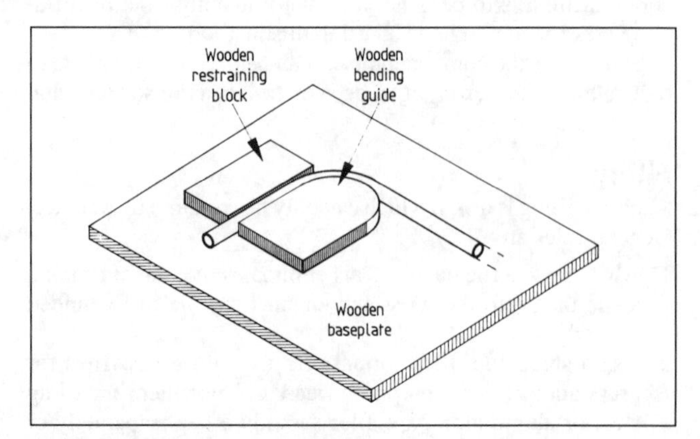

Fig 16.20. Tube or rod bending set-up

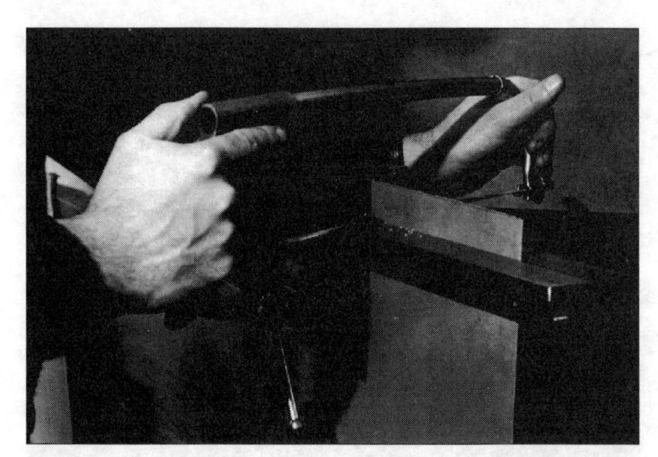

Fig 16.21. Bending bars as a sawing guide

Another tube-bending tool is the *bending spring*. This is often used by plumbers and pipe fitters. Each tube size should have a matching bending spring, for it is all too easy to use the wrongly matched spring and have it permanently trapped inside the bent tube!

Cutting

Cutting long strips of metal with tin snips or shears is an expert's job. The cut edges usually produced by non-experts are anything but straight and they require flattening to remove the cutting curl. Tin snips are best used where a one-snip cut will remove the required amount of metal, such as 45° corners, or the trimming to length of narrow strips. Snips should not be used on plastics, laminates or copper-clad board (PCBs), for the cutting action can cause de-lamination or shattering of the edges.

A guillotine is by far the quickest way of cutting most sheet material but it is not a normal home-workshop tool. For the home workshop the cutting of nearly all materials is best done by sawing, using a hacksaw and the bending bars as a guide and support (Fig 16.21). The hacksaw should be fitted with a 24 to 32 teeth per inch blade for most sheet work, and coarser blades for cutting blocks of material. The blades known as *ding-dong style*, ie with their cutting teeth arranged along a wavy edge, are very good for most work. When sawing copper-clad laminates such as PCBs, it is advisable to score through the copper beforehand using a sharp chisel. This will bevel the edge of the copper and prevent delamination due to sawing. The coolants recommended for drilling can be used if required. A fine-toothed *roofers' saw* is very useful to cut sheets which are too big to pass inside the hacksaw frame. This is similar to a joiners' saw but shorter with a stiffer fine-toothed blade. A *pad saw* or fine-toothed machine hacksaw blade which is handle mounted is equally effective in these circumstances. Some constructors use a fretsaw with either a metal-cutting blade or an Abrafile to overcome the problem, particularly when making large holes.

Hand-operated *nibblers* are available for cutting most thin sheet materials, including laminates, although they are sometimes difficult to guide properly. The power-operated nibblers are not normally a home-workshop tool. The DIY jig-saw fitted with a fine-toothed, metal-cutting blade is very good but it is essential to support the work very well and even then it is a noisy process. Similarly the DIY router-cutter is extremely useful, especially for large holes, but it can be a dangerous tool to use.

Cutting and sawing is one of those areas where gadgets are forever appearing, each claiming to save time and produce a better job! The hacksaw has not yet been replaced as a good general-purpose cutting tool.

Soldering

Learning to solder properly is not difficult and it is an essential skill for the home-constructor.

There are two main types of soldering: soft and hard. For most amateur purposes *soft soldering* is the norm. The type of soft solder most commonly used in electronic work is that known as *60/40*, ie 60% tin and 40% lead. A flux is required to enable the solder to 'wet' the surfaces to be joined, and in the case of electronic work it must be non-corrosive. To make things easier this type of solder is usually supplied with a built-in flux, in the form of core or cores surrounded by the solder alloy, and is known as *60/40 multi-cored solder*. This multi-cored solder comes in various diameters and the 22swg size is suitable for most electronic work. The melting point of this solder is around 185°C, which is low enough to reduce the risk of heat damage to most of the items being soldered. There are numerous other soft-solder alloys with differing characteristics, such as *low-melting-point*, *low-residue*, *high-melting-point* and others. The low-melting-point and low-residue solders are normally used with surface-mounted components. The high-melting-point solders are used for joints requiring high mechanical strength. Another type of solder is known as *universal* or *all-purpose*, and this can be used for soldering aluminium. Its main claim to fame is that it will solder most metals, but it requires too much heat for normal electronic work and is therefore best used for structural applications or work which is less sensitive to heat.

Structural parts may be joined by *Tinman's solder*. This has a high melting point, around 200°C, and will therefore require a hotter iron or even a gas torch. It is normally supplied in 10mm square bars, 300mm long, and not usually in multi-cored form. A suitable flux for this type of solder is a liquid known as *Baker's fluid*, which comes in various grades; the number 3 grade is suitable for most jobs. (It is sometimes referred to as *killed spirits*, from the method used to make it – a hazardous process.) *This flux is corrosive and should not be used for electronic or PCB work.* Soldered joints made using this flux may be neutralised by a good washing and brushing with hot soapy water. Commercially this neutralisation is carried out chemically but these chemicals are not too friendly to have around the home workshop. Paste fluxes are available which can be wiped along the faces before joining. Most paste fluxes are susceptible to overheating, which can nullify their fluxing action during structural soldering.

Hard or *silver soldering* produces very strong structural joints, which are also sound electrically, on brass, copper and most steels. The solder is available in various grades, shapes and sizes to suit the type of work to be joined. Typical applications are self-supporting waveguide assemblies and in areas where lower-temperature soft-soldering is required at a later stage of assembly. A silver-soldered component can be heated to accept soft solder without falling apart. The parts to be silver soldered need to be heated to around 600 to 700°C, and a gas torch is essential. The process is thus unsuitable for PCBs. The fluxes used are related to the melting temperature of the silver solder and the type of work – brass, copper, stainless steel etc. The standard Easy-flo™ flux powder and No 2 silver solder in 22swg wire form should cover most requirements. It is difficult to silver solder joints which were previously made with soft solder. No amount of cleaning of the previously soft soldered faces will alter this because the lead/tin in the former soft-soldered joint will not allow the silver solder to wet the surface properly. Similarly some metals with a high lead content will not silver solder.

Another process known as *brazing* is about the nearest most home workshops can come to welding. A workshop equipped for brazing is more for model making than amateur radio construction, although it is a useful process to have available.

Rules for soldering

1. All surfaces must be *scrupulously* clean. This applies equally well to the bit of the soldering iron. These bits are often iron plated to prevent rotting, and this plating should not be removed but wiped clean using a damp cloth or sponge. The bit should be kept 'tinned' with the solder and flux in use.

   In the case of PCBs, the copper cladding should be cleaned using the fine abrasive eraser block referred to previously. Component wires and pins etc should be similarly cleaned for, although many brand-new parts come with the wires dipped in a solder-through or flux coating, it is still worthwhile just making certain. There must be no dirty or greasy patches on any of the surfaces to be soldered and corrosion spots must be removed. Even finger marks can spoil the soldering, so "clean it and when you think it's clean, clean it again!"
2. Ensure that there is just sufficient heat to make the joint. Too much heat is harmful to electronic work and, in the case of structural work, it causes distortion. Not enough and the solder will not bond correctly.

Soldering technique

After cleaning the parts, the following technique should be used to make a sound mechanical and electrical joint on PCBs.

Apply the clean, tinned hot iron to the pad adjacent to the component wire to be soldered, and tin the copper surface by applying just sufficient *multi-cored* solder to the point where the tip of the iron is in contact with the copper. When the solder and flux starts to wet the copper surface, slide the iron into contact with the wire and apply a little more solder. As soon as both the wire and the copper surface are blended together in a *small* pool of molten solder, draw the iron up the wire away from the work. This whole process should take seconds, not minutes. The less time any component is under this form of heat stress the better. In the case of particularly heat-sensitive components, it is advisable to heatsink the component's leads during soldering. A good soldering heatsink can be made from a crocodile clip, on to each jaw of which is soldered a small strip of copper. A small set of metal forceps will also serve as a clamp-on heatsink. The clamp-on action is required so that both hands are left free to do the soldering. Commercial soldering heatsink clips are available.

This method of soldering applies equally well to the joining of wires to components such as switches, potentiometers etc. In this instance, good practice demands that the wires should be wrapped and secured mechanically before soldering, but bear in mind that the joint may need to be unsoldered later.

Good soft-soldered joints should look neat, with no draw-off points, and be continuous around or along the joint. Large blobs of solder do not ensure that a joint is well-made. The cored solder should always be applied to the work and the iron at the same time, not carried on the iron to the work.

To join structural parts together by soft soldering, each mating face should be *pre-tinned*, ie given a thin coat of solder using the appropriate flux to ensure proper 'wetting'. The tinned faces must be clean and shiny with no burnt or blackened flux spots. The two tinned faces are then painted with flux, brought together, heated to just melt the solder, and clamped together until the solder has cooled. There is no need to apply any more solder. If the job has been done properly, a neat, clean solder fillet will be seen along the mating edges. The flux residue should be washed away immediately after soldering to prevent long-term corrosion.

It is not good practice to have long soldered joints on PCB laminates. The expansion caused by the necessary heat will delaminate the copper. In these circumstances it is better to tack solder along the required structural joint, placing 2mm long solder-tacks at about 10mm intervals. Any such joints on copper laminates will be as mechanically strong as the laminate-to-copper bond. It is better practice to secure mechanical parts such as angle brackets, shaft housings etc by screws or clips which pass right through the laminate, and to use solder only to make a sound electrical connection.

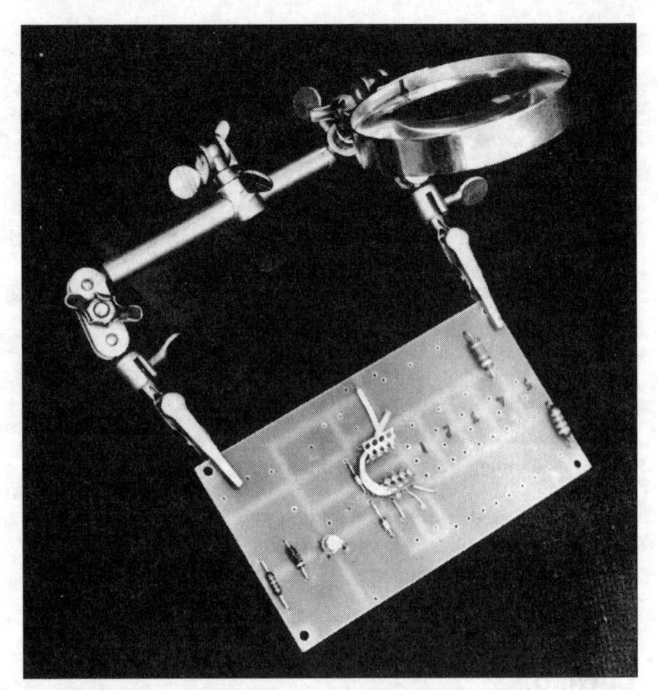

Fig 16.22. Work-holding fixture (the 'third hand')

General comments on soldering

All soldering should be carried out in a well-ventilated room. The fumes from flux and molten solder are irritants, especially to asthmatics. In the interests of safety, it is advisable to wear safety glasses when soldering. Molten solder and flux have a nasty habit of spitting and minute spots can fly anywhere, even into the eyes!

All work being soldered should be supported to leave the hands free to hold the soldering iron and apply the solder (Fig 16.22).

The greater the area to be soldered, the larger the soldering iron bit needs to be to transfer the heat to the work. A hotter iron is not a substitute for the correct bit size. The quantity of heat is related to the wattage of an electric soldering iron, and the larger the area to be soldered, the greater the wattage required. The soldering irons of various wattage specified previously should enable most soft-soldering jobs to be carried out.

Figs 16.23 and 16.24 show one type of soldering iron bit-saver. When the iron is placed on the holder arm, the diode is switched in series with the heater of the iron, approximately halving the voltage. This considerably reduces the 'bit rot' referred to previously, but at the same time allows the iron to remain hot – only a small delay occurs on picking it up again before soldering temperature is regained.

HEATSINKS

Heat levels for semiconductors are usually defined as 'not to be exceeded' junction temperatures. The power dissipation at these junction temperatures is specified in watts, and can range from

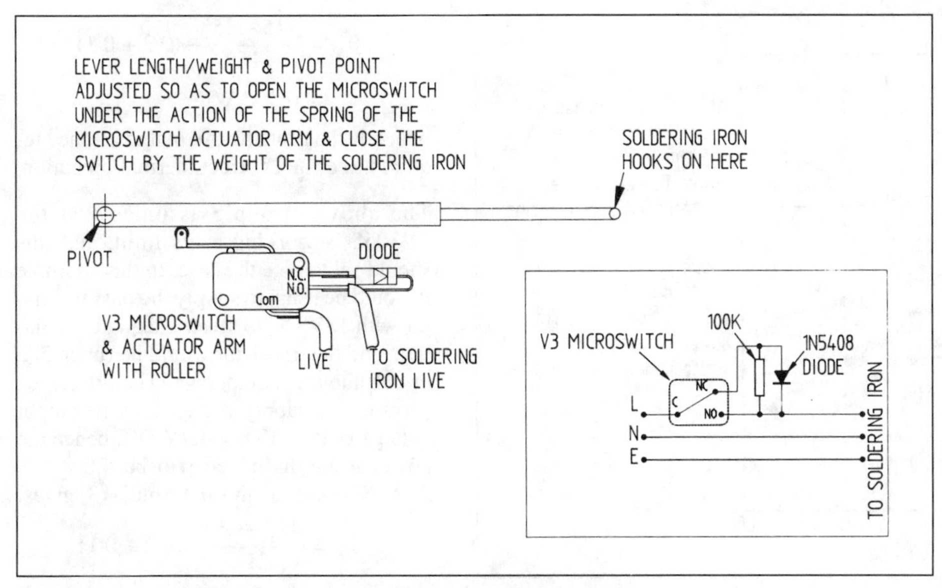

Fig 16.23. Circuit and mechanical details of a soldering iron saver

milliwatts to several hundred watts. Whenever the device is operating, heat is being generated. Unless some means of heat dissipation is used, the recommended operational limits will be exceeded very rapidly and the device will fail. A semiconductor junction takes microseconds to reach its operating temperature and only a few more microseconds to destroy itself if heat is not dissipated. Electronic protection circuits can help safeguard the device but will not eliminate the requirement for heat dissipation. Semiconductor devices generate heat very rapidly and it is essential to ensure proper cooling.

The heat must be dissipated at the same rate as it is being generated in order to ensure that the device remains within its specified temperature operating limits. The usual system consists of a heat-conducting material which takes the heat from the device and transfers it to the atmosphere. The materials used for heatsinks should have high thermal conductivity to conduct the heat away from the heat source, and high emissivity to radiate this heat to the atmosphere. Emissivity depends on the surface finish and texture of the radiating elements, rather than the type of material, but thermal conductivity depends on the type of material. Most commercial heatsinks are made from aluminium,

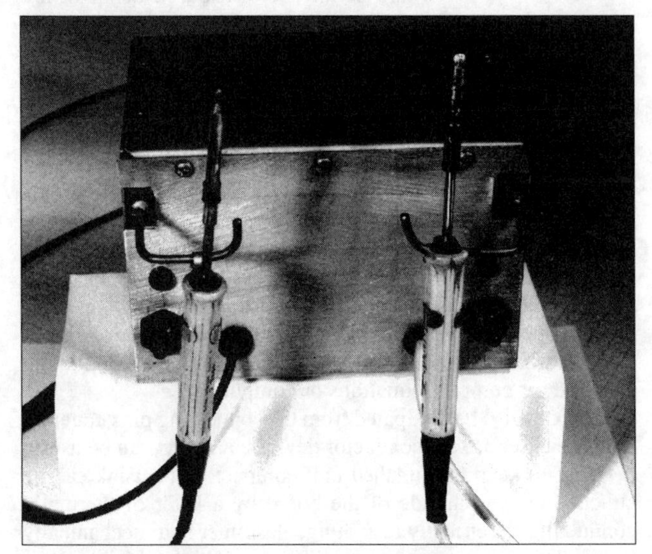

Fig 16.24. Dual soldering iron saver based on Fig 16.23

for it is lighter in weight, cheaper and can be extruded or cast into the complex shapes required for compact, efficient heatsinks. Aluminium can also be anodised and dyed black to increase its emissivity.

Heatsinks should be made and positioned such that the fins radiate to atmosphere. Fan cooling enables heatsinks to be smaller for the same thermal resistance, and as a guide, a 5 to 10 litres per second airflow over the fins can reduce the existing thermal resistance by up to 40%, depending on the number of fins receiving air. Painting a natural metallic finished heatsink with a very thin coating of matt black paint will also produce a reduction in overall thermal resistance by increasing the emissivity. The actual amount of reduction depends on the mattness of the black paint and the thickness of the coating but a 10 to 25% improvement could be expected.

Fig 16.25 shows the thermal resistance possible at various volumes for different metals, finishes and fin positioning. The bands indicate the effects of a matt black finish and the vertical or horizontal positioning of the fins. The top edge of each band represents the thermal resistance of natural-finished horizontal fins. The bottom edge represents matt-black-finished vertical fins. The band for aluminium or brass is an approximation, for the thermal conductivity of either metal varies widely depending on the alloying metals, but it can be considered reasonably accurate for most amateur purposes.

The heat transfer path is considered as flowing from the semiconductor (heat source) to ambient air, with each junction or transitional point treated as a thermal resistance (Fig 16.26). The thermal resistances of fins and heatsink to ambient are considered in parallel to give the total heatsink value. Similarly the thermal resistance of the device junction to air is considered to be in parallel with the total thermal resistance.

Heatsink formulae:

$$P_D = \frac{T_j - T_a}{\theta_{jc} + \theta_{cs} + \theta_{sa}} \qquad (1)$$

and

Total thermal resistance × Power dissipation
= Temperature rise in °C above ambient of the transistor junction (2)

from which:

$$\theta_{sa} = \frac{T_j - T_a}{P_D} - (\theta_{jc} + \theta_{cs}) \qquad (3)$$

also:

$$T_j - T_a = P_D \times \theta_{ja} \qquad (4)$$

where:

T_j = Maximum allowable junction temperature (°C)
T_a = Ambient temperature (°C)
θ_{ja} = Thermal resistance, junction to air (°C/W)
θ_{jc} = Thermal resistance, junction to case (°C/W)

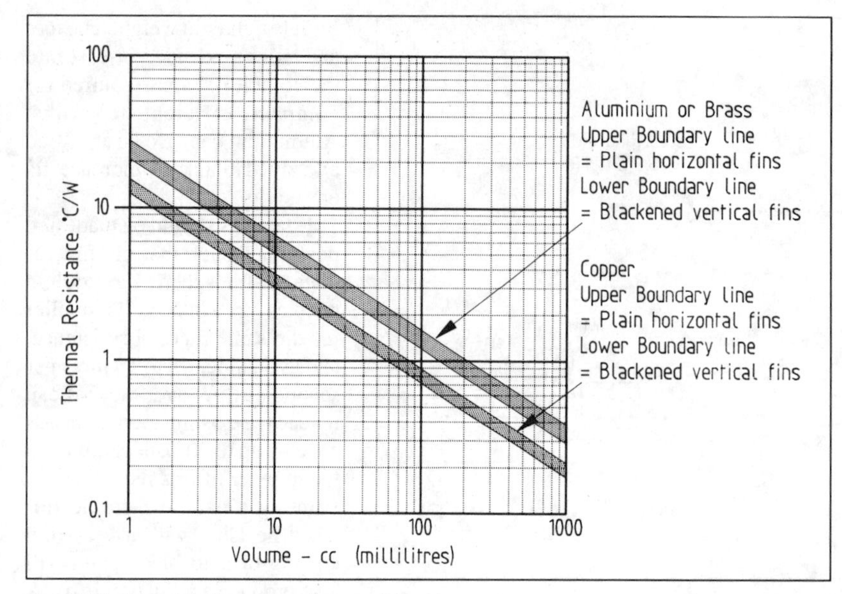

Fig 16.25. Heatsink thermal resistance/size chart

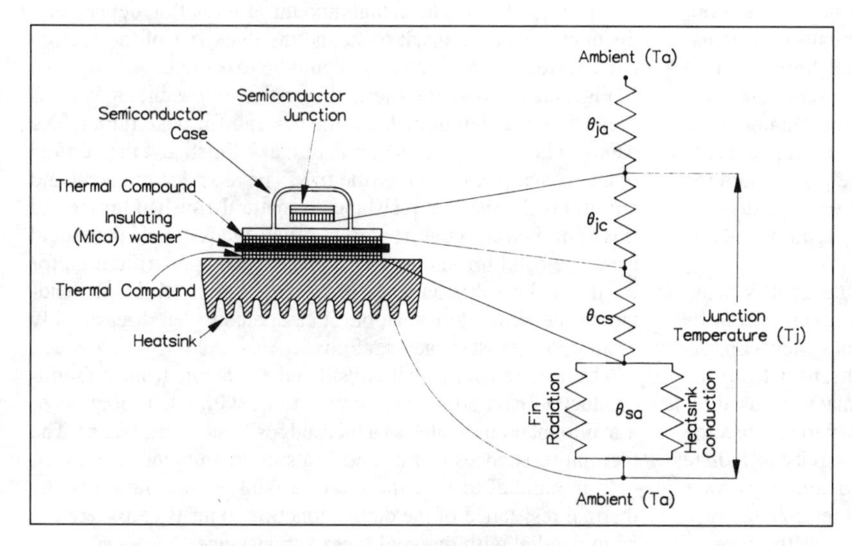

Fig 16.26. Heat transfer path and thermal resistance zones

θ_{cs} = Thermal resistance, case to heatsink, plus any insulating washer and heat-conducting compound (°C/W). (Can be assumed to be between 0.05 and 0.2°C/W).

θ_{sa} = Thermal resistance, heatsink to ambient air (°C/W).

P_D = Power dissipation (W)

Readily available semiconductor data do not always specify all of the above details but approximations may be used to design a heatsink suitable for most amateur applications. The usual data includes P_{tot} in watts (P_D) at a case temperature of usually 25°C. This can be used to derive the junction temperature (T_j) for substitution in formula (1). Because $\theta_{ja} = 25°C/P_{tot}$ and $T_j = 25°C/\theta_{ja}$ then $T_j = P_{tot}$°C.

Example: A 2N3055 is to be used as the pass transistor in a 13.8V DC regulated PSU. The available data shows P_{tot} as 115W at 25°C, thus T_j = 115°C and θ_{jc} = 0.2°C/W (25°C/P_{tot}). It is estimated that, for most applications, $\theta_{cs} \approx$ 0.1°C/W, assuming thermally conducting compound and a mica insulating washer are used. (If no insulating washer is used, $\theta_{cs} \approx$ 0.05°C/W but the heatsink may be 'live'.) Substituting in formula (3):

$$\theta_{sa} = \frac{115 - 25}{115} - (0.2 + 0.1)$$

$$= 0.3°C/W$$

which is the heatsink thermal resistance required to ensure correct thermal operation.

The above example assumes that the 2N3055 is working at its limit, and this should seldom be the case. In this example it would be usual to supply the pass transistor with 18V DC to allow adequate regulation, and the transistor would be run at 5A, or half the rated amperage. The difference between the supply voltage and the output voltage is 18 – 13.8 = 4.2V DC, hence the power to be dissipated will be 4.2 × 5 = 21W. Substituting this in formula (3) gives:

$$\theta_{sa} = \frac{115 - 25}{21} - (0.2 + 0.1)$$

$$= 4.0°C/W$$

This is the thermal resistance required for this power supply. If necessary a further allowance could be made for the inefficiency of the semiconductor and the calculated thermal resistance decreased to allow for this. In this example this factor has been ignored.

From Fig 16.25 it can be seen that about 6cm^3 of blackened or 9cm^3 of natural finished copper (vertical fins) will be a suitable minimum size. Fig 16.27 shows a typical home-made heatsink based on the above calculations. Ideally the two bent plates should be soldered together to minimise any thermal resistance at this junction. Thermal conducting compound should be used if this joint is not soldered. The copper should be annealed before bending.

Using a heatsink with a lower-than-calculated thermal resistance will not affect the operation of the cooled semiconductor. It would be unwise to use a heatsink with a higher-than-calculated thermal resistance. In use the heatsink will become warm to the touch but it should not become untouchable. If it does, then the semiconductor providing the heat is probably being overworked. It would be worthwhile checking the calculations and ensuring that each mating joint is made properly. Some semiconductor devices are designed to run at high temperatures and a hot heatsink would be expected. This heat represents waste energy, and the necessity of large cooling systems, though seeming to be the norm, indicate inefficiency.

Professional designers use very sophisticated computer programs to design their heatsinks. Even then they rely on trials and measurements to tune-up their results! The above formulae and chart are accurate for most amateur purposes, eliminating the need for complex equations or computers.

Heat can also be dissipated from the connecting pins, together with the case, of semiconductor devices, and this can be useful if trying to keep the finished unit compact. A heatsink can be attached to the pin side of the board by a layer of thermally conducting, electrically insulating, elastomer and mechanically fastened to the board with screws. This method can provide about

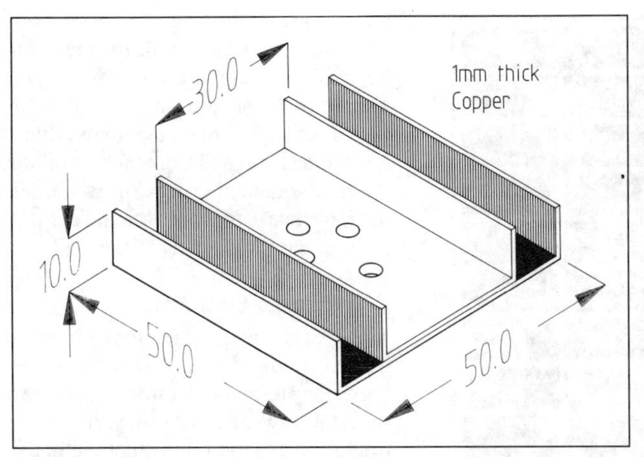

Fig 16.27. Heatsink based on the calculations of the example in the text

a 5 to 10% reduction in the required overall size of the main heatsink. Large soldering pads for the pin connections also act as heatsinks, and these can be used in conjunction with a suitable heatsink for such devices as audio amplifiers, power regulators, rectifiers etc.

The duty cycle of the device also affects the size of heatsink required. A 50% duty cycle can allow about a 20% reduction in the size of heatsink, depending on the heatsink design. During the OFF cycle, the heatsink must be able to dissipate all of the heat generated during the ON cycle, and it is usual in these circumstance to provide forced-air (or water) cooling to ensure that this is achieved. Many solid-state transceivers specify ON and OFF times for continuous full-power carrier (FSK, AM etc) operation. The small handheld transceivers start to get very warm if they are transmitting for longer than they are receiving. In the case of handheld transceivers, the batteries usually limit the transmit time but, if such transceivers were operated from a mains PSU, it would be essential to ensure that the heatsink arrangements were adequate.

MAKING CIRCUITS

Translation from diagram to working circuit is another accomplishment. The systems available are numerous and varied. Before making any circuit, experimental or permanent, it is worth considering what is expected of the finished project and relate this to the type of circuit construction available. All of the components should be collected together, tested and identified (see Chapter 2). A well-organised storage system, good test equipment (see Chapter 15) and a magnifying glass are invaluable. It is worth trying to understand how the proposed circuit is supposed to work and what function each component performs.

If it is intended for personal use only, then the circuit board needs to be functional rather than reproducible. If it is to be portable in use, as distinct from transportable, then weight, power consumption and compactness are important. If more than a few of the same circuits are needed, then it will be worth considering setting up to design and make printed circuit boards.

The following list suggests some of the factors worth considering before starting on any circuit design and making:

1. Permanent or experimental?
2. How will the circuit/s be housed and mounted?
3. What power supplies are required (battery, mains, internal, external etc)?
4. What types of inputs/outputs are required?

Fig 16.28. Example of a plug-in prototype board

5. What controls are required?
6. What safeguards are required (eg accidental switch-on, wrong polarity, over-voltage etc)?
7. Methods of construction. (Can you make it?)

In the rush to make the circuit, it is all too easy to discover afterwards that some of the above factors were important!

Many solid-state devices are susceptible to damage by static discharge. Handling and soldering such components requires care. It is safer to assume that all semiconductor devices are prone to static discharge damage and treat accordingly. Assembly and repair should be carried out using the normal anti-static precaution of connecting the soldering iron, PCB and operator to a common point. This equalises the static level and is further improved if the common point is properly grounded. Commercial equipment is available, consisting of an anti-static mat, wrist strap and connecting leads for the other tools. Semiconductors which are very sensitive to static are usually supplied mounted in a metal strip or wire clip which shorts together every pin of the device. This anti-static protection should be left in place until the device is plugged or soldered in position, and then carefully removed. A suitable size 'Bulldog' clip connected to the ground mat by a length of flexible wire makes an excellent anti-static connector for ICs.

Methods for experimental or temporary circuits

The main factor with this sort of work is adaptability. It should be possible to change components, and even the whole circuit, with the minimum of effort. The variety of methods devised over the years by fellow amateurs are ingenious and effective.

An old-time experimental circuit construction method was to use pins nailed into a wooden board and wrap or solder the wires and components to these. This was further improved upon by using drawing pins in place of pins and trapping the wires under the heads. If brass-headed drawing pins were used the wires could be soldered in place. Several of the subsequent methods of experimental circuit construction can trace their origins to these early techniques.

Many experimental circuits may be made using plug-in matrix boards of the type shown in Fig 16.28 which are also known as *prototyping boards*.) The RF characteristics of these boards are not of the best, but for digital, AF and other work they are ideal. This board can also be used to do trial component layouts during the designing of a printed circuit or a strip board layout.

Copper-clad laminates have become the accepted base for most circuit assembly. The types of laminates available have been mentioned previously in the section on materials. The

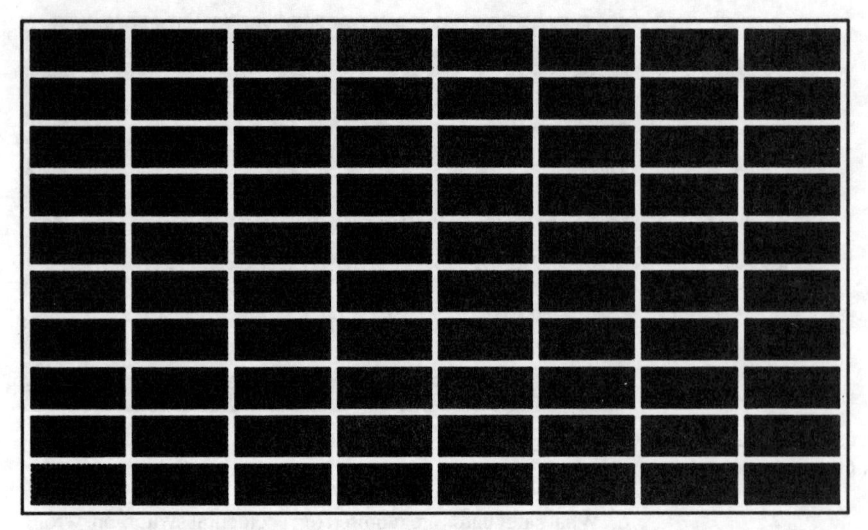

Fig 16.29. Grid pattern PCB for experimental circuits

glassfibre boards are preferred, whether single or double-sided, for they are kinder to repeated soldering and de-soldering than other laminates, reducing the chances of copper de-lamination.

For some experimental work, PCBs are made consisting of small copper pads in a grid pattern on the board (Fig 16.29). Components are soldered directly on to these pads without drilling holes (Fig 16.30). Commercial boards of this type are available for the experimenter who does not wish to go to the trouble and hazards of etching.

An allied technique is that of sticking the main components such as transistors and ICs etc on to the board with their wires uppermost and interconnecting as required with wire, resistors or capacitors, using the copper or one of the supply rails as ground. Tag strips or insulated connection pillars can be fastened to the board to provide more support for components and connections. This is a very quick way of making a circuit and has the advantage that changes may be made with the minimum of trouble. However, it's not a 'pretty sight', and it has been referred to as *the ugly system*, although it works very well, particularly for RF circuits.

Some of the simpler circuits can be made using only tag strips, eliminating the need for any copper laminate board.

Fig 16.31 shows a commercial system which provides a PCB with the same conductor layout as its matching plug-in prototype board.

Drilled matrix board without copper cladding, using special pins which are press-fitted into the appropriate holes for component mounting, is yet another method.

A combination of the above methods may be used to achieve a working circuit. A set of counter-boring drills are very useful, for they can be used to remove circles of copper around holes or to form small annular islands of copper to secure items which need to be insulated from the rest of the board. A counterboring drill is similar to an ordinary drill but with the cutting faces square to the axis of the drill.

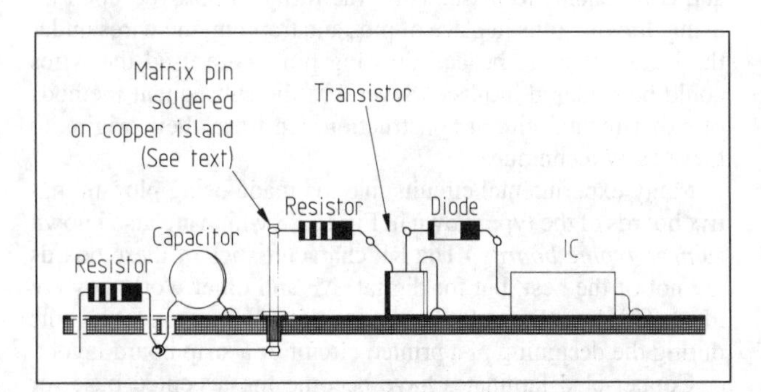

Fig 16.30. Various component mounting and connecting methods for experimental circuit construction

Permanent circuits

Many of the above-mentioned experimental circuit assembly systems can be used for permanent circuits, but for repeatability and more robust construction it is usual to turn to other methods.

One form of permanent circuit assembly which requires no home etching of boards uses the 0.1in pitch matrix of holes and copper strips, commonly known as *strip board* (Fig 16.32). The techniques of using this sort of board are relatively straightforward and it is very easy to translate from the experimental plug-in board assembly mentioned earlier. The main consideration is the number of holes required per connection and this is derived by careful scrutiny of the circuit diagram. For the more complicated circuits, it is possible to use a grid reference system; the copper strips lettered 'A' to 'Z' and the holes numbered '1' to '*n*'. The connections per component may then be identified on the circuit diagram by its appropriate grid reference. A drawing of the layout beforehand can often show up better ways of placement.

The unwanted strips of copper may be cut away using either a sharp knife or a special strip board cutter. The finished assembly can be made to look neat and tidy, particularly if any track

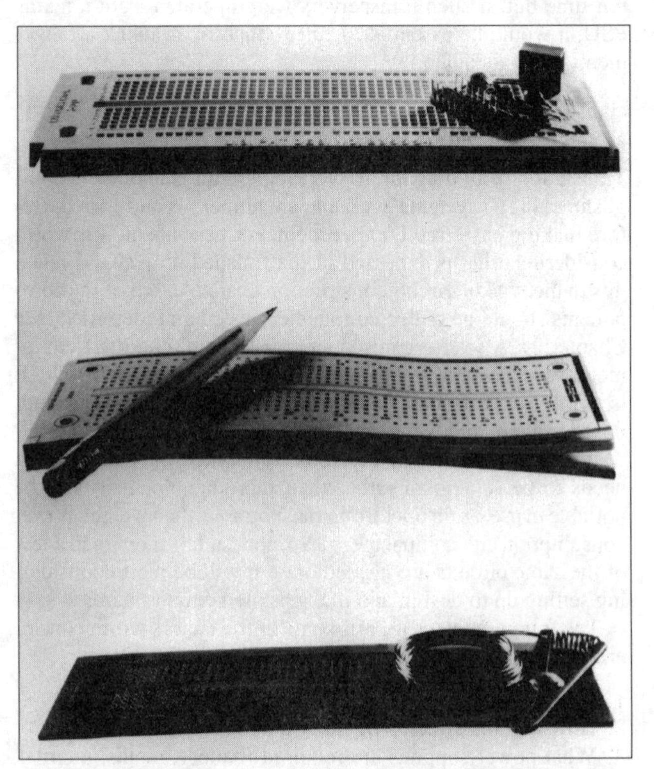

Fig 16.31. Another prototype-to-PCB system – plug-in board, sketch pad, PCB

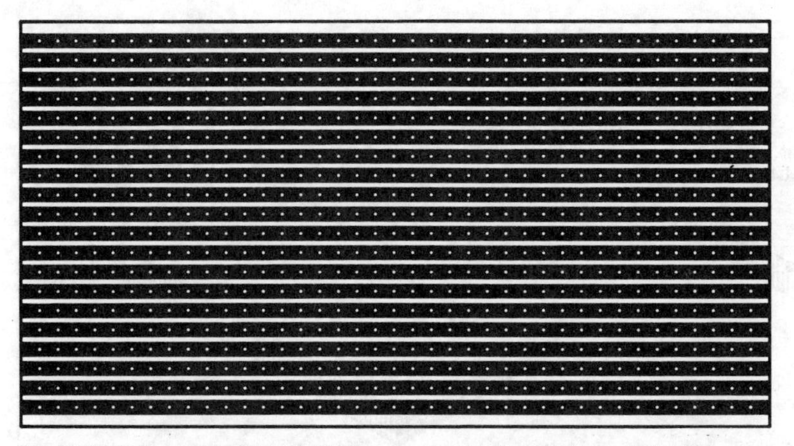

Fig 16.32. Strip board PCB suitable for experimental or permanent circuits

interconnecting wires are dressed against the board. These track interconnecting wires do sometimes present the problem of the insulation melting during soldering. To avoid this it is possible to use either bare wire or sleeve the wire with heat-resisting sleeving. Connecting wires for power supplies or controls should be wrapped to pins soldered to the board. This eliminates the problem of wire fracture that often occurs when these wires are passed through the holes and soldered to the tracks.

Similar boards are available for digital circuits, with holes and copper pads suitable for dual-in-line ICs. Many of these digital boards are available for wire wrap, or point-to-point wire and solder methods. These normally require expensive tools, components and wires but are favoured by digital equipment experimenters. They can save considerable time for things like memory boards, microprocessors etc and save the effort of designing and making special one-off PCBs for such circuits.

Etching

The next set of techniques involves etching. A brief outline of the equipment, materials and methods required for this are given as a guide.

The usual etchant is ferric chloride which is corrosive. It should not be allowed to contact any part of the body, particularly the mouth and eyes. Work should be carried out in a well-ventilated room and protective glasses, gloves and overalls should be worn. No child or pet should be allowed near when using or preparing etchants. The initial action in the event of accidental skin contact is to wash immediately with running water. Splashes to the eyes should be first-aided using a proprietary clinical eye wash and medical advice sought immediately.

This etchant is made by dissolving ferric chloride (hexahydrate) crystals in water. The proportions are around 1kg of crystals to 2 litres of boiled water. It is difficult to mix ferric chloride crystals and warm, *not boiling*, water helps. The solution should be mixed in a non-metallic vessel using a plastic or glass stirring rod. Nothing metallic should be used in the preparation and storage of this substance; plastic containers with plastic screw-top caps are ideal. Child-proof caps are recommended where there is the slightest chance of exploring little fingers. Most chemists will be pleased to advise and possibly supply suitably safe containers.

The container should be clearly labelled 'POISON – FERRIC CHLORIDE'. A permanent marker pen can be used to write on the plastic container, eliminating the risk of a sticky label dropping-off or becoming obliterated.

A small quantity (10–20ml for the above) of hydrochloric acid may be added to improve the solution.

Fig 16.33 shows a suggested arrangement for an etching bath. The essential requirements are that the etchant can be warmed to around 50°C and agitated continuously during etching. The etching bath should be large enough for the work to be fully immersed. The aquarium aerator block is glued to the bottom of the bath and air from an aquarium electric aerator pump is fed through a small bore plastic pipe passed over the lip of the container. The pump should be placed clear of the bath to prevent splashes of the etchant ruining the pump.

The PCB to be etched should be clamped in a plastic carrier and placed in the warm (60–80°C) foaming etchant. It usually takes under a minute with this set-up to completely etch a PCB. The shorter the etch time, the less the risk of under-cut and of the etch-resistant materials being washed or dissolved away. The board should be washed and scrubbed in water immediately after etching to remove all traces of the etchant. Failure to do this results in continued erosion even after the board has been assembled. Some of the faults of early commercial PCBs were found to be caused by inadequate cleaning after etching.

The cooled etchant can be poured back into its container and saved for future use. The partially used etchant can be kept for many months, particularly if hydrochloric acid has been added. It does not deteriorate but becomes saturated with copper and it is this which prevents further etching. It is interesting to place a small iron bar or plate into the seemingly spent etchant. The copper attaches itself to the bar and forms artistic surface patterns depending on the cleanliness of the bar – it might even be a method of reclamation!

Check before disposing of ferric chloride for it may not be acceptable to just pour it down the local drain.

Designing printed circuit boards

Designing a PCB requires skill, time and a great deal of patience. The process is intended for quantity reproduction of circuits and its value to the amateur constructor is worth careful thought. For the individual constructor, as distinct from a club project or a published design, it is better to consider 'modularisation' rather than an 'all-on-one-board' approach to printed circuits. This is the exact opposite of some commercial philosophy but it gives the home constructor much more flexibility, as well as being

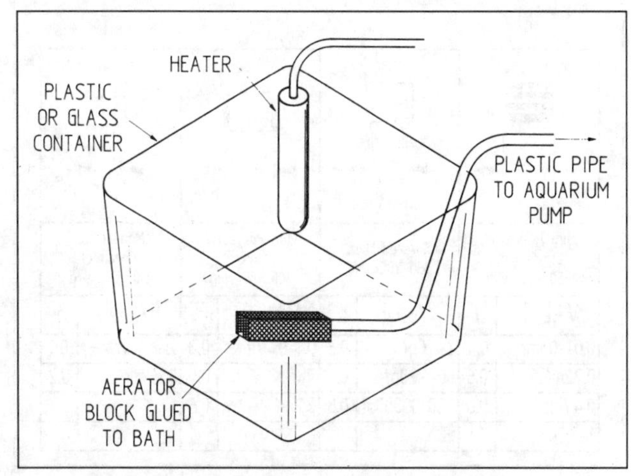

Fig 16.33. Etching bath using aquarium devices

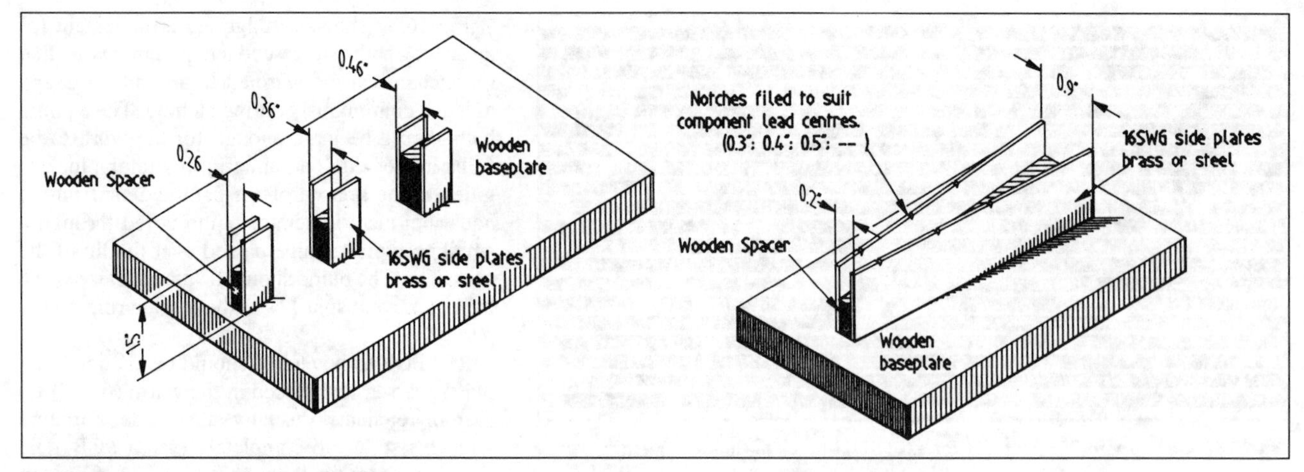

Fig 16.34. Component wire-bending jigs

less prone to error. Standardisation of board size helps during the design and making stages, and later when fitting the assemblies into a cabinet.

The standardising of the fixing centres of resistors is also helpful. Fig 16.34 shows two types of component wire-bending jigs which can be used to bend the leads to accurate centres for resistors of various wattage. It is particularly necessary to consider all of the previously mentioned design criteria, for PCBs are very restricting, especially in the size of components used and in the ability to make changes. For example, it is difficult to use, say, 0.5W resistors in place of the specified 0.125W ones or to change from a discrete transistor to an integrated circuit. With this in mind, the first essential before designing a PCB is to have either the components on-hand or to have accurate details of their sizes and fixings: it is pointless guessing (see Fig 16.35). It is possible to design the circuit such that more than one size of component may be used (Fig 16.36), but this may detract from the intended compactness and repeatability of the PCB.

Commercial PCB designers use computers to speed up the very tedious task of track routing and component layout. The computer even produces data for controlling such things as hole drilling, board shaping and the mechanised placing of components on the board. There are some very good programs available for the home computer which facilitate the designing of printed and schematic circuits, but it is usually a problem to output the results because most home computer printers will not produce accurate or good enough quality drawings. The drawings really require plotting but plotters are expensive. However, it is usually possible to produce plot files to disc and have these plotted out commercially.

Most home constructors use less expensive methods, the most popular being a sheet of 0.1in pitch graph paper as a guide, placed under tracing paper.

The approach to designing the printed circuit, by any method, consists of four main areas, as follows.

1. Positioning the active devices (semiconductors, ICs etc).

Wattage \ Size	A mm	B inches	TYPE
0.125	4.1	0.5	CARBON
0.25	8	0.6	HIGH STABILITY
0.5	11	0.7	HIGH STABILITY
1.0	16	1.0	HIGH STABILITY
2.0	24	1.3	HIGH STABILITY

Mini Polyester Dip Coated 250 VDC Wkg		Silvered Mica 350 VDC Wkg		Mini Polyester Layer 100 VDC Wkg		Monolithic Ceramic 100 VDC Wkg	
VALUE	D"	VALUE	E"	VALUE	F"	VALUE	G"
0.01-0.1mf	0.4	2-70pf	0.3	0.1-0.47mf	0.3	10pf - 0.047mf	0.1
0.22mf	0.6	80-220pf	0.5	1mf	0.4	0.1-1mf	0.2
0.47mf	0.8	220-10000pf	0.8	2.2mf	0.6		
1-2.2mF	1.1						

Fig 16.35. Some component fixing centres (note: 'mf' = 'µF')

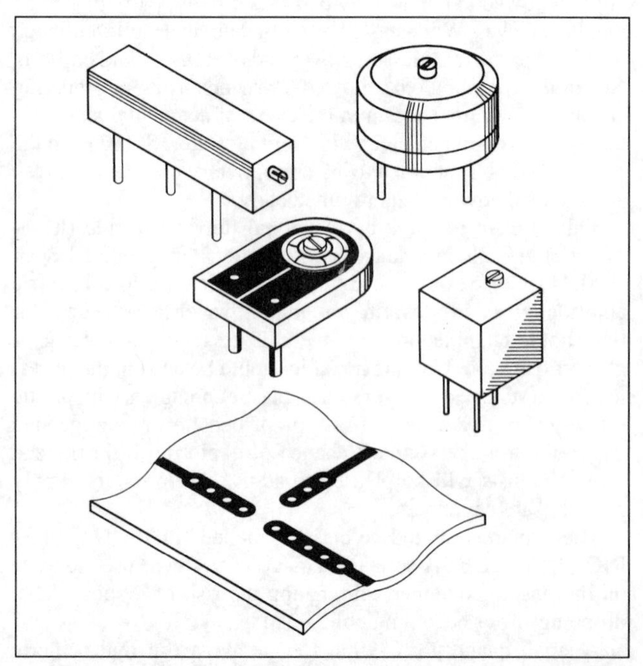

Fig 16.36. Accommodating different sizes of components on a PCB

2. Positioning the passive components such as resistors, capacitors etc.
3. Positioning inputs and outputs (connections).
4. Linking together the tracks.

A different coloured pen or pencil and a separate sheet of tracing paper may be used for each of the above areas. Draw the active devices, say, in red on sheet one and resistors etc in blue on sheet two and so on. This way it is possible to try various combinations without having to redraw everything. The more complicated the circuit, the greater the likelihood of many, many trials. A multiplicity of circuit jumper links usually indicates either a poor layout design or that the designer has had enough!

It is also essential to adopt some standard form of viewing, to avoid the all-too-easy pitfall of producing a beautiful but mirror-imaged board. Most designers seem to adopt the 'as viewed from the component side' standard even for double-sided boards.

Eventually, an acceptable circuit layout will be produced. At this stage a thorough check should be made to ensure that there are no crossed tracks, mirror-imaged components or other mistakes which would be annoying to discover when the board has been made. If you don't find any errors, look again! Some designers tick off each item on a copy of the schematic diagram as they verify that it is on the PCB layout. As with the previously mentioned strip board construction, it is worthwhile marking the number of components connected to each interconnecting track. It is all too easy to either miss one or to find that connections are made to the wrong track.

Guidelines for RF circuits, including high-speed digital

1. Use double-sided board with one face, usually the component side, as a ground plane.
2. Keep the input and output of each stage as separate as possible consistent with (3).
3. Keep tracks short.
4. Decouple all supply leads at the point of entry to the board as well as at the points specified by the circuit. Small ferrite beads placed over each supply lead usually eliminate any possibility of circuit interaction occurring through these leads. These ferrite beads can sometimes be used on signal leads in digital circuits to reduce the possibility of interaction or external RFI, and if these sort of problems occur it is worth trying.

General PCB guidelines

1. Keep mains and high-voltage circuits well away from other circuits, preferably by using separate boards. In the interests of safety, any tracks etc carrying high voltages should be covered by an insulating material to prevent accidental contact.
2. Avoid high current density points. Rapid changes of line width should be avoided, particularly in circuits carrying several amperes. The recommended current density I in amps is:

   $$I = 3(\sqrt[3]{w^2})$$

   where w is the width of the copper track in millimetres. Circuit board tracks make excellent fuses if incorrectly designed!
3. The capacitance C in picofarads between each side of a double-sided board is:

   $$C = 0.0885 \times K \times (A/h).$$

   where A is the area (sq cm) of the smallest side; h is the thickness (cm) of the laminate excluding copper, and K is the dielectric constant of the laminate material, which for glassfibre is about 4.5.
4. The recommended line or track width varies with wattage, but for most home production a 1mm width is about the smallest possible, with a minimum gap between tracks of not less than 1mm unless you are good at drawing. Commercial artwork for PCBs is often drawn four times full size and photographically reduced.
5. An 8mm minimum margin should be allowed around the board for handling during working and etching. This margin can be removed later.

Making printed circuits

There are several methods of transferring a home-designed or a published layout to the copper laminate. Which one to use depends on the number of boards required and the facilities available.

Suggested methods for a one-off PCB

The traced PCB layout is attached to the appropriate side of the laminate; which side depends as previously mentioned on the drawing system used. All the hole positions are carefully marked through using either an automatic centre punch set at its lowest level, or an ordinary centre punch which should be very lightly pressed to mark the surface. Heavy thumping centre popping will cause de-lamination and should be avoided. The tracing paper is removed and all the holes drilled using a 0.8–1.0mm diameter drill. In the case of dual-in-line IC holes, the matrix template mentioned previously can be used to assist accuracy. The holes are deburred and the board cleaned. Three different styles are shown in Fig 16.37.

Fig 16.37(a) is the usual type of line track and pads. This can be hand drawn using a special fibre-tipped pen or a tubular draughting pen. Fibre pens with a special etch-resistant ink are available especially for drawing directly onto copper. The fibre pens used for drawing overhead transparencies may also be used, providing they are of the etching or permanent ink types. Special ink is required for use in tubular drawing pens and is known as 'K' or 'P' ink. *These inks etch into plastic and must be used in special tubular pens, which should be cleaned out thoroughly immediately after use with the matching solvent for the ink.* The etching ink dissolves the plastic of ordinary tubular drawing pens.

An alternative to drawing with ink is to use special etch-resistant tapes and rub-on transfers. These must be very well burnished on to the clean and grease-free copper to ensure adhesion during the subsequent etching process. Most normal rub-on lettering will not withstand the rigours of etching and is easily washed away.

Fig 16.37(b) shows the simple lines and pads system, which can be drawn with either the pens mentioned previously or which can be painted using an artist's fine brush and cellulose paint or the special 'P' ink.

Fig 16.37(c) shows the scratching system, which requires no draughting skills. The copper side of the drilled board is painted all over with either cellulose paint or the 'P' ink; allowed to dry thoroughly, and the gaps between the tracks scratched away using a 0.5mm wide chisel-pointed piece of metal. It may be helpful to first mark out the scratching lines using a soft lead or wax pencil. The circuits produced by this method perform very well at all frequencies from AF up to UHF, and are very useful in providing extra heatsinking when heat-producing components are used. As a result, soldering requires more heat due to the larger areas of copper.

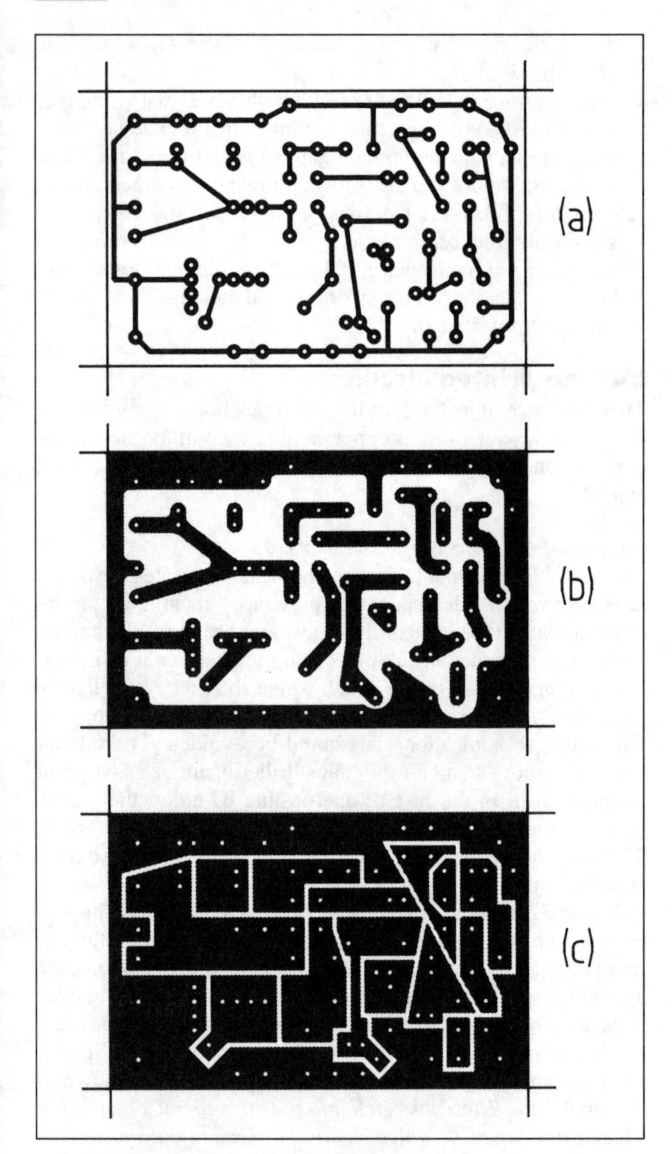

Fig 16.37. Three layout modes for PCBs

The circuit may also be transferred to the copper by using carbon paper and going over the tracing paper circuit with the carbon paper in contact with the copper. However, the transferred image needs to be inked-in and the transferred carbon often prevents this being done successfully.

Suggested method for multiple board production

This method uses photographic techniques to copy the circuit layout on to the copper. It is an expensive process and, as well as the etching bath, a further similar bath and agitator is required for developing the exposed board.

A source of ultra-violet light is also necessary, preferably with some form of accurate automatic timer. As UV light is harmful, this light source needs to be totally enclosed and incapable of being switched on until fully covered. Commercial high-intensity UV light exposure boxes are available which meet all of these requirements and, unless fully conversant with the hazards of UV light, it is safer to use one of these commercial units rather than attempt to make your own. High-intensity UV light produces serious permanent eye damage with effects similar to that experienced by the accidental viewing of an electric welding arc. Lower-intensity UV light boxes are available but exposure times usually become very long and the safety factor is not much improved.

Full darkroom facilities are not required, only protection from direct and reflected sunlight. A normally darkened room is usually adequate with subdued tungsten or non-UV lighting.

The copper-clad board is coated with a photosensitive etch-resistant coating, the thickness and smoothness of which controls the quality of the finished PCB. The positive type of photo-resist is the easiest to use for it does not require the making of a negative 'master'. With positive photo-resists, the unexposed parts remain in place during developing and the exposed parts are washed away. All photo-resists are expensive and usually have a limited shelf life.

Ready-photosensitised copper laminates are available and are usually supplied with an opaque plastic covering which is peeled away immediately prior to exposure. The board is usually cut with this plastic left in place to protect the sensitised coating and to ensure that no fogging occurs. The margins already referred to should be used, for sometimes light does fog the edges of these sensitised boards. However, if required, it is possible to 'roll your own' using either a spray or a paint-on photo-resist. Full instructions are usually supplied with either and these should be followed. It is essential that during and after coating no UV light strikes the coated surface, for this will cause 'fogging' and attendant poor quality of the finished circuit.

The coating must be of even thickness and free from dust. It is this which makes DIY coating very difficult, especially if repeatable results are required. Spin coating is about the best and an old variable-speed gramophone turntable set at about 100rpm makes an ideal spinner. The immaculately clean, grease-free and smooth board to be coated is secured as centrally as possible, copper side up, to the turntable. Allow the turntable to spin for a few minutes just to ensure that the board is securely fastened and, using a glass squeeze-dropper, gently apply drops of the photo-resist to the centre of the spinning board until it is seen that the whole board is covered. Photo-resist flies everywhere and it is preferable to place the turntable in a high-sided box. Usually by the time the turntable has come to rest the coating is dry enough to touch but only by the margins.

Normally the board is then dried by heat at the temperature specified by the manufacturer or else it is just air dried. Baking the coating *at the correct temperature* hardens it and improves the potential quality of the finished board. A clean and light-proof metal box (biscuit tin) can be used to house the freshly coated board, reducing the risk of dust settling on the surface and preventing UV light from pre-exposing the board. The same spin method can be used with the spray-on photo-resists for it does ensure fairly repeatable even coatings. If the coatings vary in thickness, the exposure times will vary, which makes repeat production difficult. It is similar to having a camera loaded with film with each frame having a different and unknown exposure rating.

Double-sided boards may be coated in a similar manner, coating and drying each side in turn. The double-sided board should be attached to the spinner using the 'margins' to ensure that the already-coated side is not scratched or marred by the fastenings. Obviously, UV light must not be allowed to fog the existing coating.

Using the previously generated circuit board design, a further drawing is required which must be on fully UV-transparent material (usually clear plastic film), and the tracks and pads etc must be completely opaque to UV light or failure is certain. The previously mentioned pens, inks, tapes and rub-on transfers can be used to make this 'master' transparency. Ordinary rub-on transfers can be used, for there is no etchant to resist. If the 'as viewed from the components side' system of drawing was used, this should be continued for single-sided boards when applying the ink or the transfers etc. These ought to come into direct contact with the photosensitised copper surface of the PCB during the exposure stage to ensure the best possible clarity of lines etc. The master for the component side of a double-sided board should also be produced to ensure this.

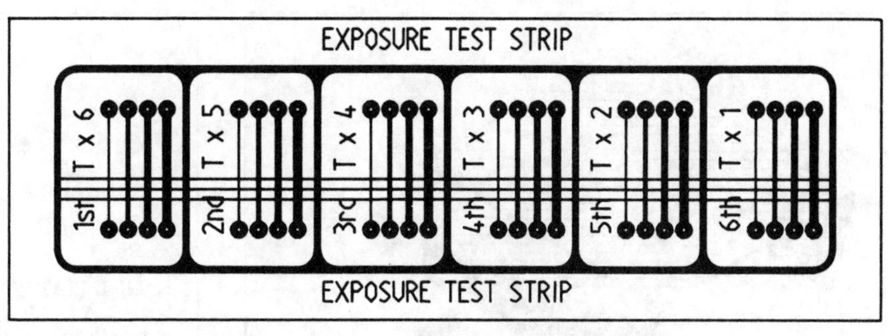

Fig 16.38. Exposure test strip for photosensitised PCB

A strip of the photosensitised board should be used to assess the exposure time required. Fig 16.38 shows a suggested exposure testing strip, which should be made of the same materials as those used for the master transparency. Each section of the test strip is progressively uncovered until the complete strip has been exposed. For example, if the timer is set to, say, 20s then each step would be 20s greater than the previous one. This is normal photographic practice. The fully exposed strip is now developed. The image of the test strip should be visible, at least in sections, and at this stage the board is washed thoroughly in running water, dried and then etched. It is possible to assess the correct exposure without etching but etching is more reliable and much easier to see. The developing is usually carried out by using sodium hydroxide dissolved in water in the ratio of 500g of sodium hydroxide to 5 litres of water. Cooled boiled water is preferred. During mixing, the sodium hydroxide tends to produce a boiling effect and great care should be taken to prevent contact with the eyes etc. *The usual safety precautions should be observed: glasses, gloves, no children or pets etc.* The mixed solution has a limited life and a fresh solution should be made for each batch of developing.

Some types of photosensitive resists use developers other than sodium hydroxide and most need careful handling, particularly in the home workshop.

Having established an exposure time, the set of PCBs can now be made. The master transparency is attached to the photosensitised board by one margin edge only, using either masking or double-sided tape. The assembly is placed face down towards the light source in the UV light box and the lid closed, ensuring that the master is in full contact with the photosensitised face of the board. Some exposure devices use a vacuum hold-down system to ensure this but, providing the board is not too big and the master transparency is allowed to roll into contact with the board, all should be well, with no air bubbles trapped to distort the image. Expose for the time determined by the test strip, develop, wash thoroughly and then allow the board to dry thoroughly before etching. If the coating has been hardened by baking, it is possible to go straight into the etching process, for the coating should not have softened too much during development.

To give some idea of the time involved, providing the boards are correctly coated, it is possible to produce a board ready for etching about every five minutes using the set-up and methods described. In the case of double-sided boards, each side is exposed before developing, and great care is needed to ensure that the previously exposed side is not fogged during the second exposure.

Registration is a problem with double-sided boards. The simplest method is to drill a few of the holes beforehand in a manner similar to that described in the one-off method and locate each 'master' to these holes. The holes are best drilled after the board is coated, otherwise the holes will cause variations in coating thickness. In the case of the ready-coated boards, the plastic protection is left in place during drilling. This is a technique commonly used in commercial prototype work and is referred to as *spotting through*. In this case all of the holes are drilled using computer control. Spotting through is made simple by holding the work against a light source.

Some constructors make a master transparency from either a published circuit or their own layout, using a xerographic copying machine. Unfortunately, some of these copiers do not produce opaque enough images for use with UV light. Also, a slight, non-uniform, scale change usually occurs, which makes hole centres inaccurate. This is particularly noticeable on the multi-pin dual-in-line ICs. A xerographic transparent copy of the exposure test strip shown in Fig 16.38 may be used to find out the quality of transparency produced, and this could save a considerable amount of wasted effort.

The completed boards made by either of the above methods should be washed thoroughly to remove all traces of etchant, and the resist, ink, transfers or paint should be removed using the appropriate solvent. Proprietary brush-cleaning solvent usually works. Some photosensitive resists need not be removed for they can be soldered through. Final drilling should be carried out and the last stage should be the removal of the handling margins. Commercial boards are usually tinned using either chemical plating or hot dip tinning. Tinning by running the soldering iron around the tracks is unnecessary and can cause delamination.

CABINETS

Housing the finished project is another important area. The cabinet is always on view and therefore should reflect the care and time devoted to the project. Commercially made cabinets appear to be the easy way out but often these cabinets do not meet the requirements exactly. They are either too big or too small. Also, they do not always allow for easy changes in layout. These restrictions, combined with the expense of commercial cabinets, are a great encouragement to make your own. This is a relatively easy thing to do, especially if the techniques and tools previously mentioned are acquired. It is also possible to re-use old cabinets from obsolete equipment, modifying these to suit your requirements.

Ideally any box or cabinet should be able to be altered with the minimum of effort and waste. To this end, the base and the front and rear panels should be separate items. The two cabinets shown in Figs 16.39 and 16.40 are very easy to make and meet this requirement. That shown in Fig 16.39 is intended for the smaller projects, such as keyers, QRP ATUs and active filters etc. Typical sizes are 130mm wide, 80mm high and 150mm deep. It can be sized for larger work but the cover becomes difficult to make. The cover of the larger version really needs

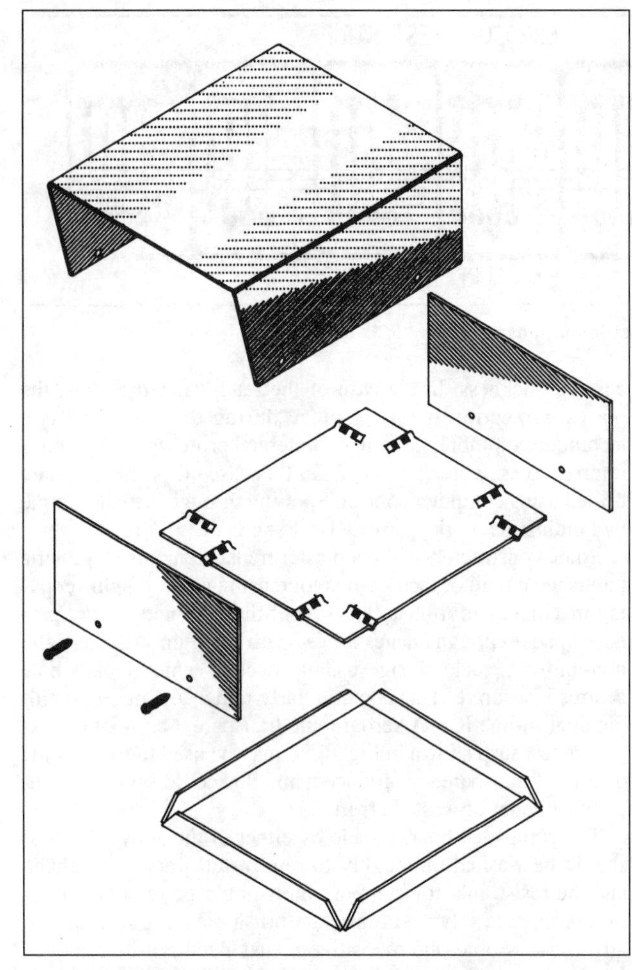

Fig 16.39. Small cabinet design with alternative base plates

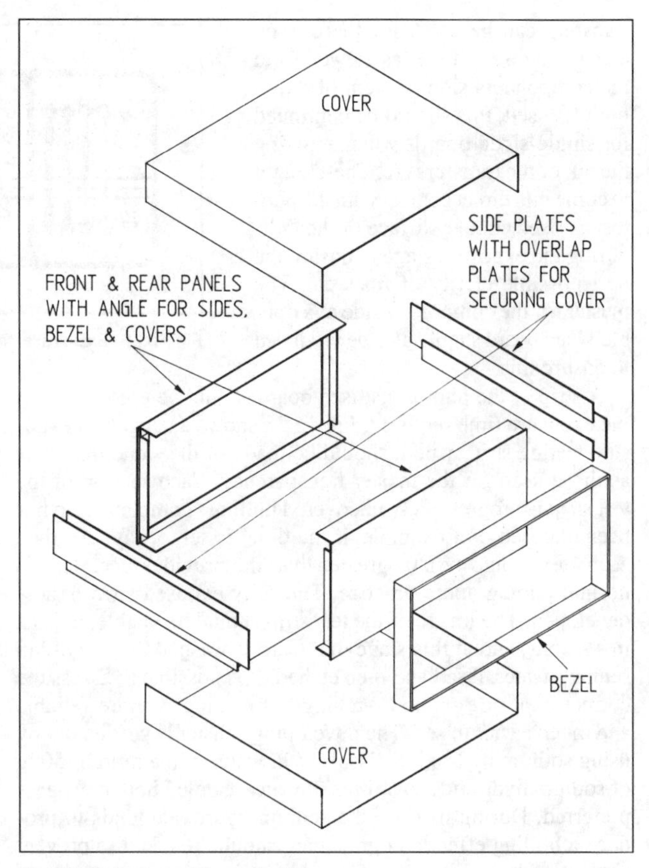

Fig 16.40. Large cabinet design

making from heavy-gauge material to reduce flexing, and bending becomes difficult. The cabinet shown in Fig 16.40 overcomes this problem and is suitable for such things as high-wattage ATUs and linears etc. The cabinets can be made from steel or aluminium, and a small version of Fig 16.39 can be made from tin plate.

Fig 16.39 shows two possible base plates. The flat base plate utilises the hole-embossing tool, and the edges of the cover should extend beyond the base plate to form feet. These feet can be edged with plastic strip of the type that is used to bind papers together. This plastic binding is available in various colours and, if used correctly, the finished job is very professional looking. The alternative bent base plate can be made using a former. The finished base plate and the former can then be used as a former to bend the cover, thus ensuring a good fit. The sides of the cover do not need extending if using the bent base plate.

In the transmitting radio shack, it is advisable to use metal boxes or cabinets to house electronic devices. This gives some measure of screening, which helps to reduce mutual RF interference. Plastic, wood and other non-conducting materials, though often making things easier to construct, do not provide such protection unless lined with metal foil which is grounded. These materials are best used to provide cabinet embellishments, such as bezels, feet, handles etc.

Fig 16.41 shows home-constructed equipment using the small cabinet or box. Fig 16.42 shows home-constructed equipment fitted into commercially made boxes. Also, the front and right-hand side pieces of equipment shown in this diagram utilise the paper panel methods referred to later.

FINISHING

Having built the equipment, the urge to use it may be so great that there is no time to finish it off. This is a shame, because the

Fig 16.41. Home-made equipment housed in the small cabinet design

Fig 16.42. Home-made equipment housed in commercial boxes. The items in the foreground and on the right use the paper/Perspex panel system described in the text

finish affects the appearance, which in turn adds to the pleasure of using home-made equipment. The time spent on construction may be considerable but that spent on finishing is undoubtedly the most rewarding.

Painting

This is probably the most common form of finish used. The combination of two or more colours can enhance the equipment. The choice is limitless and it is usual to try to adopt some standard colour scheme. If commercial equipment is already in the shack, it is possible to try to match or blend with its colour scheme. Brush painting very rarely gives a good surface over large areas and a considerable amount of work is required to burnish this form of painting into anything like a smooth finish. Brush-applied hammer or crinkle finish paints are useful for covering the outsides of cabinets but front panels look far superior and workmanlike with a monotone finish.

Spray painting has been available for sometime in the form of aerosol sprays and these are a reasonably cost-effective form of painting. A durable and pleasing appearance can be obtained, once the simple technique of using them has been learnt. The directions, usually given with product, for the method of application should be followed. The main area of difficulty usually arises when spraying the three-sided corners of a box. It often happens that too much paint goes on to one surface and runs; the other surfaces are left unpainted or just lightly coated. The trick is to spray each corner face in turn, masking-off the other faces, which may or may not be freshly painted, by holding a piece of cardboard in the path of the unwanted spray. A few practice runs on an unwanted piece of metal enables the technique of aerosol spray painting to be acquired quickly.

Before any painting is done, the surfaces should be suitably prepared, for the resultant ease of obtaining a smooth finish depends on the quality of the unpainted surface. The surface must be smooth and burr-free. This can be achieved quickly by using the flapwheel and drill mentioned previously or by rubbing down with emery cloth. In the case of aluminium, ensure that the emery cloth does not pick up the aluminium dust and form an abrasive lump which will score the surface. Any scratches or score marks should be smoothed out or, after priming, filled with one of the proprietary fillers which should be blended-in completely to leave a smooth flat surface. The abrasive block should be used for the final rub-down. The surfaces should then be washed and degreased thoroughly with scouring powder and the scouring pad. The now clean and grease-free surface should not be handled, for any finger marks will affect the finish.

Surfaces can also be prepared and sometimes finished by etching but, as most of the substances used for this technique are hazardous, this method is not recommended for the home workshop. A solution of ferric chloride will etch to produce a reasonable ready-to-paint surface on copper, brass, mild steel and aluminium. The surfaces must be clean and grease-free otherwise uneven etching will occur.

The next stage in painting is priming. Most aerosol paints have details of the recommended primer to use printed on the can, and these instructions should be followed because the manufacturer is as anxious as you that the painting works properly. If no instructions are given then the following types of primer should be used:

- For *steels* use the oxide primers to suit the type of paint that will be used for the final finish.
- For *aluminium* use the etch-type primers.
- If it really is necessary to paint *brass* or *copper* then the oil-bound undercoat paints should be used. Zinc or aluminium-based primers do not always work well on brass or copper and tend to lift after a short period of time.

After priming, and when the work has dried thoroughly, carry out any filler work necessary. When this has dried a light rub-down with very fine wet and dry paper will smooth off ready for the finishing coat or coats in the colour of your choice.

Many of the modern enamel spray paints require neither primers nor undercoats and still produce a durable pleasing finish, providing the surface of application has been well prepared. There are so many types of paint to choose from that a little personal experimenting is advisable. Most paint manufacturers will supply literature on their products, giving full details of the methods of application, durability etc. A few types are listed here.

Acrylic paints

A good gloss finish is possible with these reasonably priced paints but they are susceptible to some cleaning solvents. They are suitable for use on most metals or plastics and are available in aerosol cans. These paints are probably the best all-round paint for the finishing of home-constructed projects. Some of the hammer finish paints are of this type.

Bitumen paint

A useful and cheap paint. It is protective rather than decorative and is suitable for the protection of masts and other outdoor metal work. Usually it gives a black semi-gloss finish and is normally brush applied. White spirit can be used to clean brushes etc.

Cellulose dope

This is a fast-drying liquid which, like shellac, can be used as an adhesive or a coating to secure and waterproof. Unlike shellac, the dope may dissolve some plastics. It is usually available from model aircraft shops in clear or coloured form and is reasonably priced. The coloured dopes contain pigments and in some cases metal powder. These may provide an electrical conductive coating. The non-shrinking clear dope weathers very well. It is a good protective coating for beam elements and does not appear to upset the beam's performance. Normally it is brush applied, although it can be thinned with cellulose thinners for use in DIY-type paint sprayers.

Cellulose lacquer

A reasonably priced, quick-drying lacquer, giving a high and reasonably durable gloss which can be improved by polishing. It can be affected by some cleaning solvents. The rapid drying properties considerably reduce the risk of dust-marred work. It is available in aerosol cans and as normal paint. Special thinners are required for cleaning brushes etc.

Stoving paints

These are usually of the acrylic or alkyd types, which rely on stoving to complete the chemical changes of drying. They are usually applied by spray and the resultant finishes are hard, durable and can be very glossy. This is about the best paint finish possible but the stoving and spraying requirements usually present a problem in the home workshop. However, some commercial concerns do offer a small-quantity painting and stoving service, which is worth considering for that extra-special job. Some car body repair shops have these facilities also.

The number of coats of paint and the method of application is entirely a matter of choice. All painting should be carried out in an atmosphere as warm, dry and dust-free as possible, and in very well-ventilated conditions. If the painted work can be placed in an electrically heated oven at around 70°C, drying will be greatly assisted. Putting the work in a tin with a lid and placing this on top of a stove or central heating boiler is the next best thing, for it will keep dust from marring the work during drying. Rubbing down and cutting-in compounds supplied for car paint retouching are also very useful in obtaining that final smooth finish.

Other finishing materials

Shellac lacquer

A lacquer which is made by dissolving shellac flakes in methylated spirits or similar solvents. This is a well-tried, inexpensive waterproofing, insulating and light-duty adhesive-cum-lacquer. It can be used to protect antenna elements and connectors, and to secure wires, coils and components. The lacquer dries to a hard brown translucent finish. The drying process can be speeded up by the application of heat. (An old dodge, on coils, was to set fire to the applied lacquer, to 'flash-off' the methylated spirit solvent. The technique required careful timing to avoid burning the coated item but, when successful, the result was a hard and durable coating.)

The shellac flakes and methylated spirits are normally available from dispensing chemists, or it can be bought ready made as *French polish* or knotting compound. The ready-made lacquers are usually very watery and may contain solvents other than methylated spirits.

Self-adhesive plastic sheets

These are an alternative finish that can be used to good effect on the outsides of cabinets and boxes, saving a great deal of painting and rubbing down. Leather-effect cloth provides a very professional touch, particularly on the larger surfaces such as covers of ATUs, PSUs etc. This cloth, if not self-adhesive, may be glued down using contact adhesive very evenly applied to both surfaces. The previously mentioned panel cleaning should be carried out before applying either of these types of sheet finishings. The main problem with using these sheets is that of shrinkage and the edges curling with age. This can be eliminated or reduced by heating the panel to around 60°C (just too hot to hold) immediately before applying the sheet. Also, wrapping the sheet around the edges can sometimes improve the problem. With wrapping, edge bulge can become obtrusive. Trapped air is the bane of this process (a bit like wall-papering), and the knack is to peel-on the sheet using a straight edge or a rod to peel against from the non-adhesive side.

Self-adhesive coloured or metallic finish tapes can be used to line-in features of the panel or cabinet. There is a strong tendency, due to the ease of application, to overdo this and finish up with a cross between a juke box and a 'sixties car radiator grill.

Lettering panels

This is now very easy and, though machine engraving is still considered the best, it is little used because of cost. Most modern commercial equipment has the lettering screen printed on to panels, knobs and push-buttons etc. Screen printing can be another home workshop process. It is fascinating and worth looking at if you ever get fed up with amateur radio!

The wet slide-on transfers are effective and can be complementary to the more popular rub-on lettering such as Letraset™, Chartpak™ and others. These are expensive and it is worth considering carefully what style and sizes suit your tastes. The Helvetica™ Medium style seems to match the lettering on most commercial equipment, and is available in black or white in various point sizes. The range 10, 12 and 14 point should cover most requirements. A sheet of numbers only in the same range is also worthwhile. Variations of importance can be made using a combination of upper and lower case, with different point sizes of the same style. Mixing styles of lettering is not usually aesthetically effective.

Sometimes these rub-on letters can be difficult to apply, usually because the surface is uneven, cold, damp, greasy or too glossy. A very light rub-down with wet and dry paper should be adequate preparation for the surface. The paint will have a matt finish but this is taken care of at a later stage after lettering. The pre-release method will usually ensure success, particularly on the smaller characters. To pre-release the required character, the lettering sheet should be placed on to its non-stick backing paper and, with the two sheets resting on a hard surface, the area of the character should be rubbed over as normal. A change of texture can be seen through the carrier sheet when the character has released. In this state, the character, still lightly attached to its carrier sheet, may be applied where required, and usually a very light pressure will allow the carrier sheet to be withdrawn, leaving the character in place. The newly placed character must be burnished immediately into place through the non-stick paper, to ensure that it will not be lifted off during the next character application.

Alignment of the lettering during application is facilitated by using the guide markings on the lettering sheet. Some have the guides under each letter and some at the ends of each row of lettering. A strip of card can be arranged as a ruler guide and its alignment checked using the combination square. Lining-up the bottom of each letter as the lettering progresses is difficult and will not produce correctly aligned work. Also, for some reason, there is a tendency to gradually curve downwards, particularly if referencing from the immediately previous character. Spacing the characters equally, say around the centre line of a switch or volume control, is best carried out by working outwards from the centre letter or space. Typesetting is an art and a few experiments with the chosen style can help in getting things right on the finished work. The guide-bar under each letter on some lettering sheets is also a guide to spacing. Each bar should be positioned such that a continuous straight line is produced, and then the lettering will be at the correct spacing for that style. Rubbing on characters is a slow process, but this serves to ensure that only essentials are given, which in turn produces a good workmanlike labelled panel, not an instruction book! For interest, using the smallest letters available, it is worthwhile to date the work, say, in the bottom right-hand corner – this way you can see how your workmanship has improved over the years!

Unfortunately, transfers are easily rubbed off and require some form of protective covering. The quickest way is to spray the completed panel with clear lacquer but some lacquers react with the surface of application and this should be checked beforehand. Another effective way is to face-up the work using a thin sheet of transparent Perspex, which can be secured to the panel using screws and/or the existing component fixings. Self-adhesive clear sheets can also be used but this has the problem that they must be applied right first time, for any attempt to peel off and start again will remove all of the lettering. This technique can be used to remove unwanted or misplaced letters at the time of lettering and is superior to scraping. A quick dab with the sticky side of masking tape on the unwanted letter usually does the trick.

If clear plastic sheet is used to face the panel, other possibilities occur, which can eliminate painting the latter. Paper of the coloured art sort (eg Letraset) can be cut to size with all holes cut to match the panel, and the lettering etc applied to this. The lettered paper or card can be secured to the panel by the Perspex sheet. It should not be glued to the panel, for this will cause the paper or card to buckle and not sit flat. If a computer with a desktop publishing (DTP) package and laser printer is available, this can be used to produce paper/card panels very easily. Most laser printers produce only black printing which does restrict the possible colour schemes. Coloured card (which will pass through the printer) has been found very effective with this system. When properly applied, this paper or card system of panel marking and covering is very effective. A very pleasing effect can be produced by placing the coloured or metal finish narrow, self-adhesive tape, as previously mentioned, along the edges of the card/paper. Do not try to stick the card to the panel in this manner, as the differential expansion and contraction will cause the card to buckle even under the Perspex.

Etch engraving is also very effective and the same techniques used in the production of printed circuits are applied. It is a system worthwhile experimenting with in those lulls between projects.

Self-adhesive strip labels made on hand-operated lettering machines are a useful way of marking controls but do not add much to the appearance of the equipment. However, this type of strip is very useful in making Braille labels for equipment used by a blind operator. Matrix board (0.1in pitch) can be used as a guide for the Braille characters, for it is about the correct spacing for the Braille dots. These are formed through the guide and from the peel-off backing side of the tape, using a suitably thinned centre-punch, with the tape resting on a strip of rubber or soft plastic.

Home-made dials for dip oscillators and field strength meters etc can be made by attaching card on to a metal or plastic supporting disc and then calibrating and lettering as required. A final spray with clear lacquer and a lasting job is produced. Using these lettering techniques, it is also possible to recalibrate meter dials and so utilise functional meters which others would have thrown away.

Other finishes

Plating, anodising and colouring etc which have to be carried out by an appropriate firm are well worthwhile but a few words of advice may be helpful. Most finishing firms do not pre-polish or dress the work prior to finishing. With plating, the article is usually plated 'as received', scratch marks and all. The techniques described in the sections on materials and tools should be applied therefore as required to obtain a smooth and scratch-free surface before any item is sent for plating. Aluminium is usually anodised and coloured and does not require a highly polished surface, only a scratch-free one. The contractor may, on request, carry out some form of etch treatment prior to anodising. It is possible for lettering and other characters to be colour dyed on to aluminium during anodising and a master 'see-through' layout of this, to size, should be supplied to the finisher if this is required. These finishes are expensive but do provide corrosion protection as well as enhancing appearance.

If none of the aforementioned methods of finishing appeal, an old-fashioned method can be used to produce whirl patterns on the surface of the work. This is done be sticking a disc of emery cloth on the end of a piece of 12mm diameter dowel rod mounted in the electric drill. The patterns can be generated, either randomly or sequentially to suit, by bringing the emery/dowel straight down on to the work, holding it there briefly, and repeating as required. The same problem of metal pick-up with aluminium can occur. This process, derived from the days of hand scraping, is a quick way of visually achieving what was then considered a quality finish. Nowadays, it is considered wasteful and uneconomical! Yet another alternative is to 'grain' the work, using either fine emery cloth or an abrasive eraser, retaining and enlivening this surface by a spray of clear lacquer.

The polishing of cut edges or the removal of scratch marks on Perspex and most plastics can be done by sequentially filing, emerying (using first the medium grade and working down to the Crocus paper grade) and finishing off with a mixture of metal polish and powder cleaner, the proportions of which are reduced until the final polish is achieved with metal polish alone. Commercial compounds are available to do the same job. A hard felt pad mounted on the end of a drill can also be used with the normal buffing compounds to polish and remove the sharp edges. However, care should be taken with this method of polishing to avoid overheating which produces an 'ocean wave' effect on the surface being polished!

DESIGN THOUGHTS

Home construction allows individual ideas to be designed into the project being made. With commercial equipment 'you pays your money' but you don't always get exactly what you would like. Some controls you find unnecessary and others which you want are not there or are hidden inside the case. These factors are often overlooked until you make something for yourself and then aspects of commercial designs take on a different meaning. The quality or lack of it becomes apparent and it is very difficult

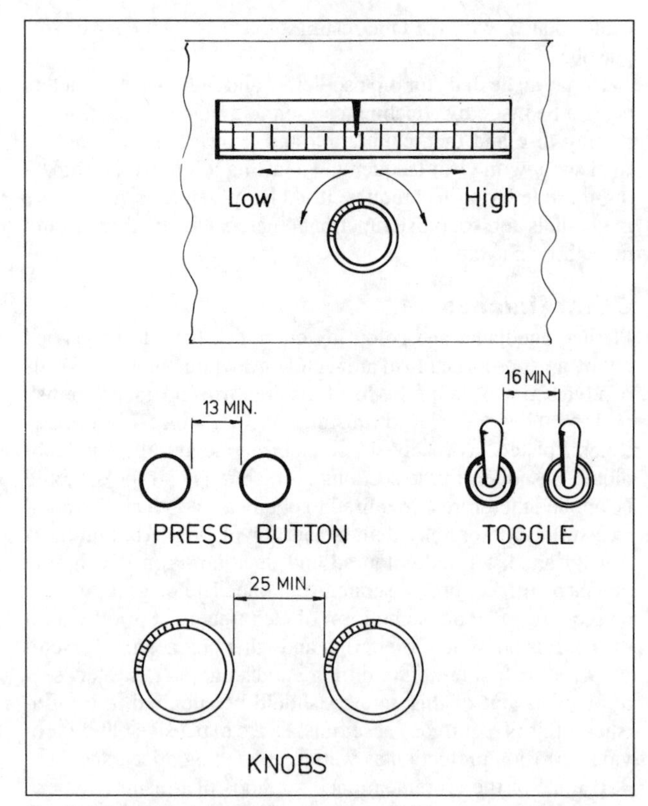

Fig 16.43. Suggested ergonomic strategy for some controls

to acquire this appraisal ability without having made, or having tried to make, something for yourself.

Over the years there has become an awareness of the value of ergonomics (see Fig 16.43). Most of us now know that controls should be positioned according to their purpose and that this purpose should be self-evident from either the type of control or the area in which it is placed. For example, most transceiver tuning knobs appear adjacent to the tuning display and, when rotated clockwise, the frequency increases. Usually, it is evident by looking at most transceivers which is the tuning knob.

Unfortunately the same cannot be said about press buttons and other controls. The use of press buttons needs great care, especially where accidental operation could cause damage. If correctly designed, it should be impossible to accidentally press two or more buttons at the same time. Also, if such an action is done accidentally or deliberately, it should not destroy the device, or send it into an unusual operating or locked-up state. Ideally, any button-type control should be accompanied by an indicator which shows what mode the button is in. Power supply switches are particularly important in this respect, especially with portable equipment, where it is all too easy to switch it on and not be aware of this fact.

Rotary switches are less susceptible to accidental switching but they must be positioned and have suitable style knobs to facilitate switching. Most rotary switches occupy more space than push buttons and it is this which usually restricts their use on the compact rigs of today. Rotary switches are ideal for such controls as mode and band selection. One look at the switch position shows immediately what mode or band is selected. It is impossible to attempt to select two modes or two bands and there is no need for any other form of indicator: a power-saving factor also. Toggle and slide switches are an equally effective form of self-indicating switches but toggle switches are prone to accidental operation.

Potentiometric rotary controls, such as AF volume, morse speed and RF gain etc should all be self-indicating. Also, clockwise rotation should increase the function. Rotary concentric controls should give each control a related and easily identified function.

Knob styles can also affect the quality of presentation of the finished work. Unfortunately, knobs are expensive but ex-equipment knobs can often be used as a cheaper alternative. The size of knob should relate to the accuracy required from the function it controls. For example, a tuning knob would not feel right if it was less than say 40mm diameter, but the same diameter for a volume control would be unsuitable and usually unnecessary. Slider controls are very good, especially when simultaneous operation is required on such as faders and mixers, but this facility is not usually required on radio equipment. These controls also take up space if they are to be accessed easily. Some commercial amateur radio equipment use the smaller type of this control for the less-utilised or preset controls.

Many more factors can be found, and it is discovering these which adds to the pleasure of home construction.

17 Electromagnetic compatibility

ELECTROMAGNETIC compatibility (usually abbreviated to 'EMC') is the ability of various pieces of electrical and electronic equipment to operate without mutual interference. So far as amateur radio is concerned, the object is to achieve good EMC performance: that is, not to suffer from received interference or to cause interference to others. This is, of course, not a simple matter since it involves not only the station but also any radio and electronic equipment operating in the vicinity. It is said that it takes two to make peace, but only one to make war, and a similar relationship exists between EMC and interference.

In practice, the amateur must endeavour to achieve peaceful co-existence by minimising interference caused by his (or her) station, and, where appropriate, increasing the immunity of susceptible local domestic radio and electronic equipment. Complementary to this is interference to amateur reception. This has always been a problem, but the proliferation of computers and control electronics in modern homes has brought it into greater prominence in recent years.

Most national administrations have enacted legislation defining minimum EMC standards which equipment on sale to the public must meet. These standards lay down maximum permitted emitted interference, and also the minimum immunity which equipment must have to unwanted signals. In the UK the standards are issued by the British Standards Institute, and are harmonised to the common standards of the European Community (EC). In general, EMC standards are framed round a normal domestic or industrial radio environment, and fall short of what would be ideal from the radio amateur's point of view. Amateurs tend to generate high field strengths and (as discussed later in the chapter) attempt to receive smaller signals than other radio users in a typical residential area.

Interference between stations jockeying for space in crowded amateur bands is part and parcel of amateur radio operation. Avoidance of this type of interference is mainly a matter of good operating practice, coupled with good receiver design, and is not specifically dealt with in this chapter. When the word 'interference' is used here, it is intended to mean incidental interference caused by transmitters to nearby radio and electronic equipment, or interference to reception from poorly suppressed equipment operating in the vicinity.

Interference, then, falls into two major categories:

1. Interference caused by operation of the transmitter.
2. Interference to amateur reception.

INTERFERENCE CAUSED BY THE TRANSMITTER

This is such a common occurrence that most amateurs have suffered from it at some time or other, and in all too many cases it has been the cause of bad feeling with neighbours. It also breaks down into two major categories:

1. Interference due to unwanted emissions from the amateur station. The general name for unwanted emissions is *spurious emissions* – often shortened to *spurious* or *spurii.*
2. Interference caused by the legitimate amateur signal on the normal operating frequency breaking through into some piece of susceptible equipment. This is usually called *breakthrough* to emphasise that it is not really a transmitter fault, but rather a defect in the equipment which is being interfered with.

When investigating interference it is essential to determine which of these is causing the problem – attempting to solve the wrong problem rarely results in success! Find out what sort of equipment is affected. If it is non-radio equipment such as a burglar alarm, then the cause must be breakthrough of some sort. If the susceptible equipment makes use of radio in some way (usually broadcast radio or TV but it could be some form of communication or control device) then the cause could be either breakthrough or a spurious emission from the transmitter.

The next step is to find out if the interference only occurs when the equipment being interfered with is tuned to specific frequencies. If so, check whether similar equipment in your own house (or in neighbours' houses) is affected when tuned to the same frequencies. In effect, if your transmitter is radiating a spurious signal, then anything tuned to that frequency is likely to be affected. Nowadays breakthrough is a much more likely cause of interference than spurious emissions, and this will be dealt with first.

BREAKTHROUGH

Breakthrough is by far the most important problem in amateur radio today. Over the years the number of electronic devices in the typical home have increased and nowadays almost every appliance has some form of electronic control. All electronic equipment is to some extent vulnerable to strong radio frequency fields and is a potential EMC threat. Any amateur who lives in close proximity to his neighbours will have to give some thought to the avoidance of breakthrough. This can be tackled in two ways:

1. By installing and operating the station so as to reduce the amount of radio frequency energy reaching neighbouring domestic equipment. The term *good radio housekeeping* has been coined to cover all aspects of this activity.
2. By increasing the immunity of the affected installation to the amateur signal.

Good radio housekeeping

The essence of good radio housekeeping is keeping your RF under reasonable control, putting as much as possible where it is wanted (in the direction of the distant station) and as little as possible into the local environment.

It is fortunate that installations designed to achieve this are

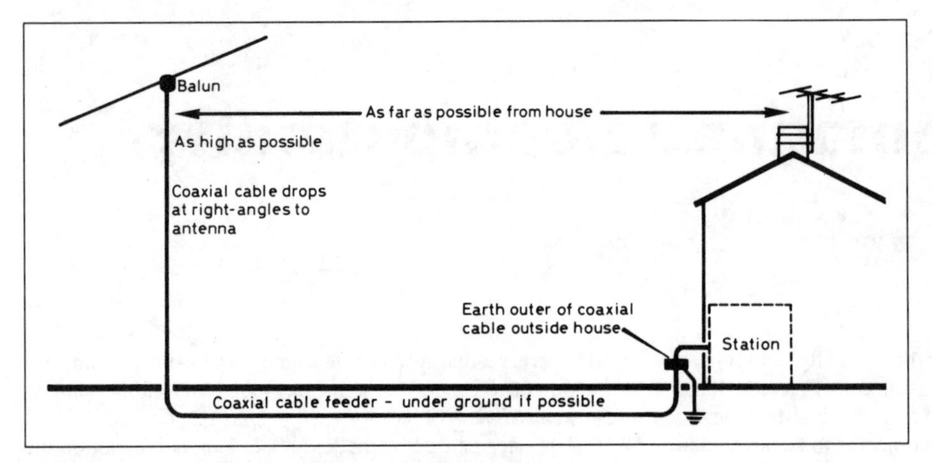

Fig 17.1. Antenna and feeder system with EMC in mind

also likely to minimise the pick-up of locally generated interference – adding a bonus to good neighbourliness.

It is essential to plan with EMC in mind where a station is to be installed in a typical urban environment in close proximity to neighbours.

Antennas

It is always good practice to erect any antenna as far from houses as possible, and as high as practical, but for HF operation the relatively long wavelengths in use give rise to special problems. In locations where breakthrough is likely to be a problem, HF antennas should be:

1. *Horizontally polarised.* House wiring and other leads tend to look like antennas working against ground, and hence are more susceptible to vertically polarised signals.
2. *Balanced,* to minimise out-of-balance currents, which can be injected into the house wiring, particularly in situations where a good earth is not practical. These currents will also give rise to unwanted radiation.
3. *Compact,* so that the whole of the radiating part of the antenna can be kept as far from house wiring as possible. Try to avoid antennas where one end is close to the house while the other end is relatively far away. This encourages direct coupling between the near end and the house wiring, inducing RF currents which will be greater than would otherwise be the case.

The sort of thing to aim for is a dipole, or small beam, located at least 15m from the house (and neighbouring houses), and fed with coaxial cable via a balun (Fig 17.1). This is not too difficult to achieve above 10MHz, but at lower frequencies a compromise is usually necessary. A balanced feeder can be used, but coaxial feeder is convenient and allows the use of ferrite chokes to reduce any unwanted currents which may find their way on to the braid.

Earths

From the EMC point of view, the purpose of an earth is to provide a low-impedance path for RF currents which would otherwise find their way into household wiring, and hence into susceptible electronic equipment in the vicinity. In effect, the RF earth is in parallel with the mains earth path as in Fig 17.2. Good EMC practice dictates that any earth currents should be reduced to a minimum by making sure that antennas are balanced as well as possible. An inductively coupled ATU can be used to improve the isolation between the antenna/RF earth system and the mains earth [1]. The impedance of the mains earth path can be increased by winding the mains lead supplying the transceiver and its ancillaries onto a stack of ferrite cores as described below (p17.4) for breakthrough reduction.

Antennas which use the earth as part of the radiating system – ie antennas tuned against earth – should be avoided since these inevitably involve large RF currents flowing in the earth system. If this type of antenna must be used, arrange for it to be fed through coaxial cable so that the earth, or better some form of counterpoise, can be arranged at some distance from the house.

The minimum requirement for an RF earth is several copper pipes 1.5m long or more, driven into the ground at least 1m apart and connected together by thick cable. The connection to the station should be as short as possible using thick cable or alternatively flat copper strip or braid.

Where the shack is installed in an upstairs room, the provision of a satisfactory RF earth is a difficult problem, and sometimes it may found that connecting an RF earth makes interference problems worse. In such cases it is probably best to avoid the need for an RF earth by using a well-balanced antenna system – but don't forget to provide lightning protection – see Chapter 21).

Operating in difficult conditions

What can be done if your garden is small or non-existent, or other domestic conditions make the antenna arrangements such as shown in Fig 17.1 impossible? The most important thing is not to get discouraged; many amateurs operate very well from amazingly unpromising locations. If there is no choice but to have antennas very close to the house or even in the loft, then it will almost certainly be necessary to restrict the transmitted

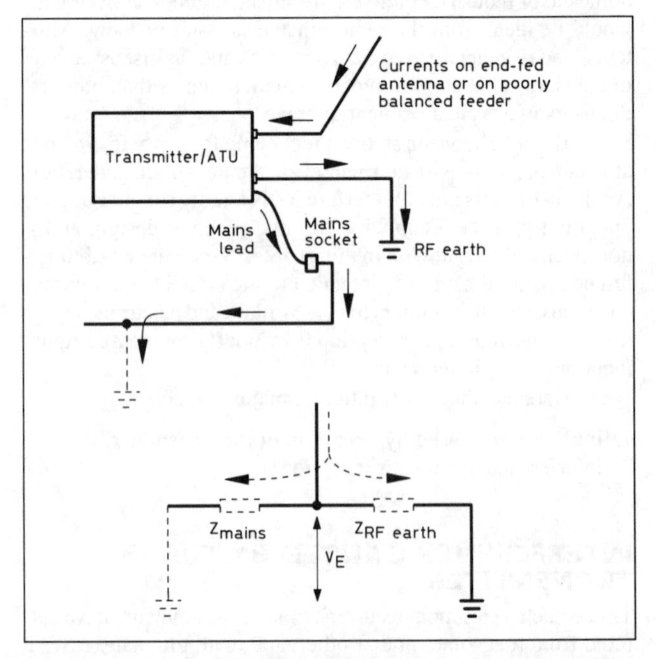

Fig 17.2. Earth current divides between RF earth and mains. The current down each path will depend on the impedances. The transmitter earth terminal will be at V_E relative to 'true' earth potential

power. Not all modes are equally 'EMC friendly', and it is worth looking at some of the more frequently used modes from this point of view.

SSB is a popular mode, but also the least EMC friendly, particularly where audio breakthrough is concerned.

FM is a very EMC friendly mode, mainly because in most cases the susceptible equipment sees only a constant carrier turned on and off every so often.

CW has two big advantages. First, providing the keying waveform is well shaped, the rectified carrier is not such a problem to audio equipment as SSB. The slow rise and fall gives relatively soft clicks which cause less annoyance than SSB. The second advantage is that it is possible to use lower power for a given contact than with SSB or FM.

Data modes used by amateurs are generally based on frequency-shift keying (FSK), and should be EMC friendly. All data systems involve the carrier being keyed on and off – when going from receive to transmit, and vice versa – and consideration should be given the carrier rise and fall times just as in CW.

Passive intermodulation products (PIPs)

This phenomenon is a fairly unusual aspect of good radio housekeeping but one which has turned up from time to time since the early days of radio. Traditionally it is known as the *rusty bolt effect*. It occasionally causes harmonic interference to be generated by amateur transmissions, but far more often it simply degrades receiver performance without being identified as a problem.

All mixing and harmonic generating circuits use non-linear elements such as diodes to distort the current waveform and hence to generate the required frequency components. A similar effect will be produced whenever the naturally produced semiconductor layer in a corroded metal joint forms an unwanted diode. These unwanted diodes are usually most troublesome in the antenna system itself, particularly in corroded connectors. In the case of a single transmitter the effect simply causes excessive harmonic radiation, but where two or more transmitters are operating in close proximity the result can be spectacular intermodulation product generation. On receive, the result is much-reduced receiver intermodulation performance, and in severe cases there will be a noticeably high noise floor which consists of a mishmash of unwanted signals. The best way to avoid troubles of this sort is to keep the antenna system in good repair and to examine all connections every few months. PIPs can be generated in corroded metal gutters and similar structures not directly associated with the antenna system. The solution is to clean up or remove the corroded metalwork or, where this is not possible, to short-circuit the corroded junctions with a conductive path. It is obviously good practice to keep antennas away from doubtful metalwork.

Dealing with breakthrough at the receiving end

Breakthrough is simply unwanted reception and the basic mechanism by which signals are picked up is the same as for any other reception. Breakthrough can occur to such a wide range of equipment that for the sake of general discussion it is simpler to assume some non-specific device, in other words a *black box*.

For signals to get into our black box, they must be picked up on a wire which is a significant fraction of a wavelength long, which means that on HF the external leads are the most common mode of entry and direct pick-up by the equipment itself is

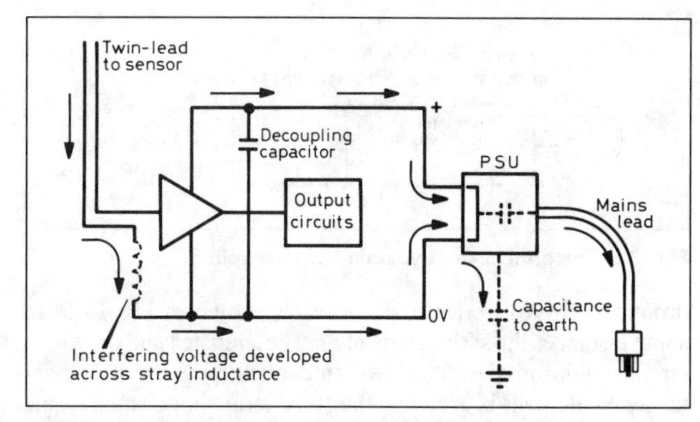

Fig 17.3. Path of RF signal in a typical sensor/alarm device

unlikely unless the transmitter field strength is very high. At frequencies above about 50MHz pick-up by wiring inside the black box becomes more likely but only if the box is made of non-conducting material such as wood or plastic.

Fig 17.3 shows how unwanted RF signals might get into the black box – in this case, some sort of alarm circuit is assumed with a sensor connected by twin cable, and an amplifier and power supply inside the box. The sensor lead acts as a crude earthed antenna so that electromagnetic energy from the transmitter causes currents to flow in the antenna formed by the sensor lead, through any stray impedance between the input connection and the 0V rail, and through the power supply to earth. These currents are called *common-mode currents* because they flow on both conductors in the cable in phase – in effect they act as if the pair of wires were one conductor. The wanted signals are *differential*, flowing in one direction on one wire, and the reverse direction in the other. Common-mode and differential currents are illustrated in Fig 17.4.

The key to avoiding breakthrough is to prevent the unwanted signals picked up on the external lead from getting into the circuits inside the box. There are two basic ways of doing this:

1. Bypass the unwanted RF by providing a low-impedance path across the vulnerable input circuit.
2. Use a ferrite choke to increase the impedance of the unwanted antenna where it enters the black box, effectively reducing the currents getting into the sensitive internal circuitry.

Bypassing the unwanted RF

The principle of bypassing is to arrange for a potential divider to be formed in which the majority of the unwanted signal is

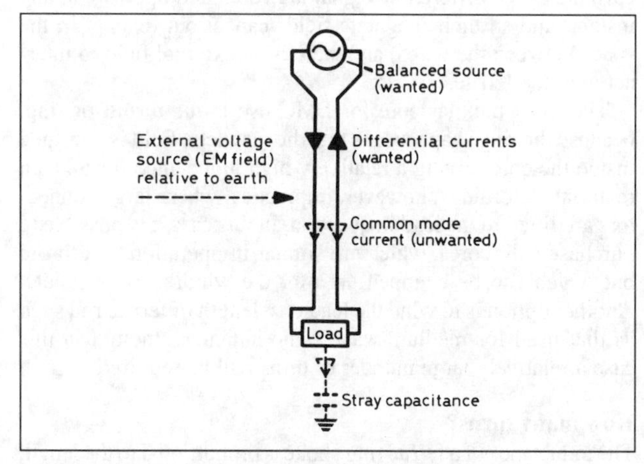

Fig 17.4. Differential-mode and common-mode currents

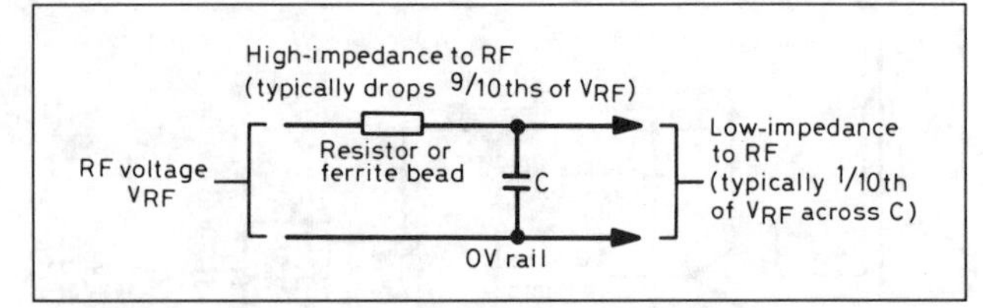

Fig 17.5. The principle of bypassing or decoupling

dropped across a series impedance, as shown in Fig 17.5. In some instances the series element may be a ferrite bead or, where circuit conditions permit, a low-value resistor, but in many cases no series element is used and the stray series impedance of the lead provides the series element. A ceramic capacitor with a value in the region of 1 to 10nF would be typical. It is important to keep the leads as short as possible, and to connect the 'earthy' end of the capacitor to the correct 0V point – usually the point to which the amplifier 0V is connected, and to which the inputs are referred. ('0V' in this context has much the same meaning as 'ground', but avoids any confusion with other meanings of 'ground' used in radio discussions such as in 'capacitance to ground' etc.)

Generally, bypassing is not very practical in domestic situations and is included as food for thought for anyone building their own equipment. It is inadvisable for the amateur to attempt to modify commercial equipment unless he (or she) has expert knowledge – in particular, *never attempt to modify equipment belonging to someone else.*

Ferrite chokes

Ferrite chokes are also used to form a potential divider to the unwanted RF currents, though in this case the series impedance is increased outside the black box, and the stray capacitance or resistance inside the box forms the shunt element (Fig 17.6). The great benefit of this technique is that it does not involve any modification inside the box. All that is required is to wind the susceptible lead on to a suitable ferrite core. Ferrite cores which are designed for EMC purposes have a reasonably high permeability, which is combined with a high loss resistance at the frequencies of interest. This enables a high impedance to be achieved without resonance effects becoming dominant. Further information on core types will be found in Appendix 3 of reference [1] and in reference [2].

A very important feature of this type of ferrite choke is that it acts only on the common-mode interfering currents – differential-mode wanted signals will not be affected. The go and return currents of the differential signal are equal and opposite at any instant, and so their magnetic fields cancel out (except in the space between the wires) and there is no external field to interact with the ferrite.

The most popular core for EMC use is the toroid or ring, because the ring shape means that the magnetic field is confined inside the core, giving a relatively high inductance for a given material. A toroid is, however, impractical where large connectors are fitted to the lead in question. In this case it is possible to purchase split cores, which are similar in operation to a toroid but which can be clamped up after the winding is complete. Another option is to wind the lead on a length of ferrite rod such as that used for medium-wave radio antennas, though in this case a relatively large number of turns will be required.

How many turns?

The inductance of a ferrite ring choke is proportional to the length of ferrite through which each turn passes (ie the thickness or

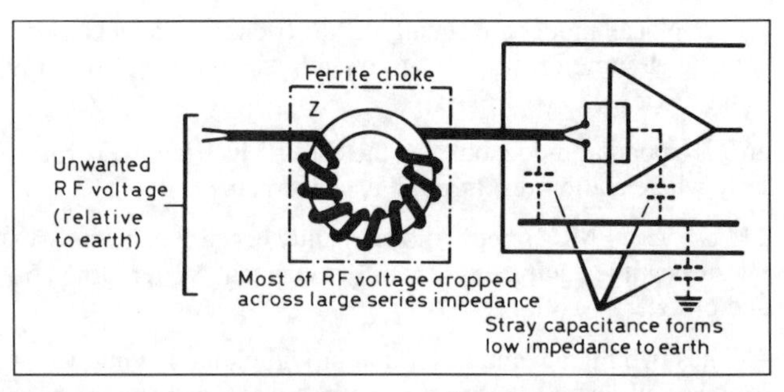

Fig 17.6. The series choke. The impedance of Z forms a potential divider with stray capacitance

'depth' of the ring) and also to the square of the number of turns. Traditionally cores about 6.5mm thick have been used, two of these being stacked to give an effective thickness of 13mm or so. This gave rise to the rule of thumb that, for choking HF signals, the number of rings in the stack multiplied by the square of the number of turns should be about 400. Typically we might have 14 turns on a two-ring stack. For VHF fewer turns should be used – about seven turns on a single ring will be effective at 144MHz. Where no other information is available the same rules should be used for split cores. Ferrite rods require more turns to make an effective choke: 20 or more at lower HF and somewhat fewer above 10MHz – see Appendix 3 of reference [1].

Recent experimental work [2] has shown that a very effective choke can be made by winding about 12 turns on a single (but relatively thick) ring of suitable ferrite material.

Breakthrough to TV and video equipment

In a TV installation, the most vulnerable point is the antenna input, and in the majority of cases eliminating unwanted signals at this point will clear up the problem. The most common route for the unwanted signals to get in to the set is via the braid of the coaxial feeder. The small size of UHF TV antennas makes it unlikely that large interfering signals will be picked up on the antenna itself, unless the frequency is relatively high – above about 100MHz or so.

First have a look at the general installation. Are the coaxial connectors correctly fitted with the braid firmly gripped and making good electrical contact with the outer connection of the plug? Make sure that the centre pin is properly soldered. Corrosion of the connector at the end of the cable coming down from the antenna may indicate that water has got into the cable; the commonest cause is chaffing or splitting of the sheath, though it could be a failure of the seal at the antenna itself.

TV antenna amplifiers

A well-designed masthead amplifier which is correctly installed, close to the antenna, is unlikely to give problems at HF because the length of unwanted antenna will be limited, but at VHF the conditions could be very different. Problems arise if the unwanted signal reaching the input to the amplifier is large enough to drive it into non-linearity and cause cross-modulation. Masthead amplifiers get their DC power supplies through the coaxial feeder, so that any filters or braid-breakers must pass (and not short-circuit) the required DC supply. Unless specific information is available, it is best to use only ferrite-ring chokes in the down leads from such amplifiers.

Indoor amplifiers are much more likely to cause problems, especially where HF is concerned, but fortunately they can be easily removed from the circuit for test purposes. Some amplifiers have several outputs, allowing more than one set to be operated from the same TV antenna. If the amplifier has to be retained in the installation it should be considered as part of the TV set (or sets), and the standard techniques, using chokes and filters, on the antenna side of the amplifier will still apply.

Chokes and filters for TV installations

The simplest way of choking-off signals travelling down the braid is to use a ferrite ring choke; used in this way it is often called a ferrite-ring *braid-breaker*. The low-loss cable commonly used for UHF TV is not suitable for winding on to the core. It is best to make the choke separately, using a length of smaller-diameter coaxial cable, with suitable TV connectors at each end. The photograph shows a choke for HF breakthrough. The coaxial cable used for the choke should, of course, be 75Ω, but there is no need to use low-loss cable since the length involved is quite short.

Braid-breakers and filters are available from a number of commercial manufacturers, and these are usually very effective provided that the correct type is used. Some devices are simply braid-breakers, with no filtering of signals travelling down the inside of the coaxial cable, while others include high-pass or band-stop filters optimised for specific bands, so it is important to study the specification before making a decision. In some instances it may be necessary to use two devices (for instance, a braid-breaker in series with a high-pass filter) to achieve the desired rejection. Further information on braid-breakers and filters can be found in Appendix 3 of reference [1] and in reference [2].

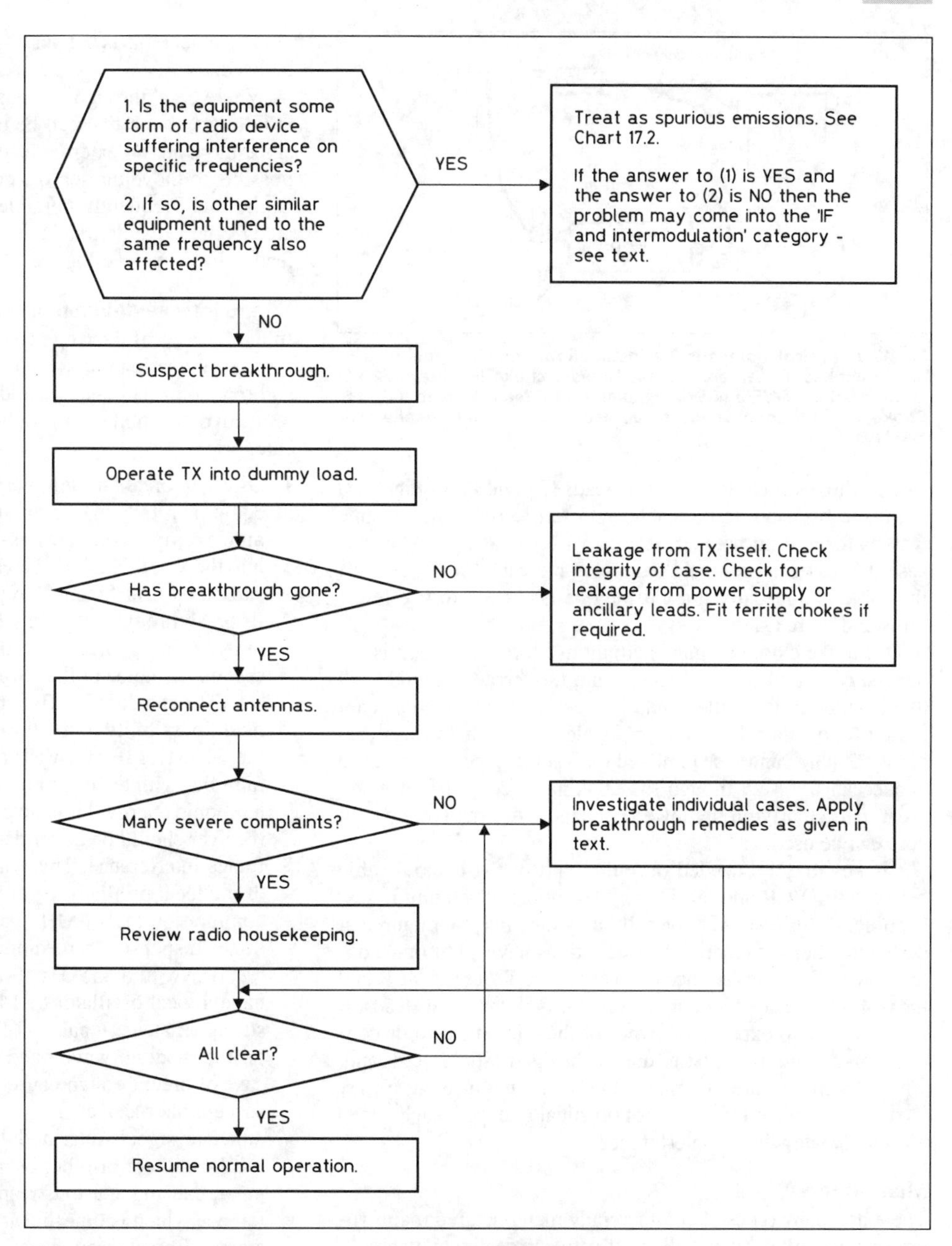

Chart 17.1. Breakthrough

At the time of going to press, the RSGB holds a range of suitable braid-breakers and filters. Further information and prices can be found in the society's journal *Radio Communication*.

Fitting the chokes and filters

Try the chokes and filters in order, starting with the antenna choke (F1 in Fig 17.7). Do not remove a device if it does not seem to work – in many cases of breakthrough, the unwanted signals enter the installation by more than one route, and the observed interference is the result of the signals adding or subtracting. If a ferrite choke is only partially successful, another should be tried in series to increase the rejection, but it is usually not worthwhile to go beyond two. When the sources have been identified, chokes which are not needed can be removed.

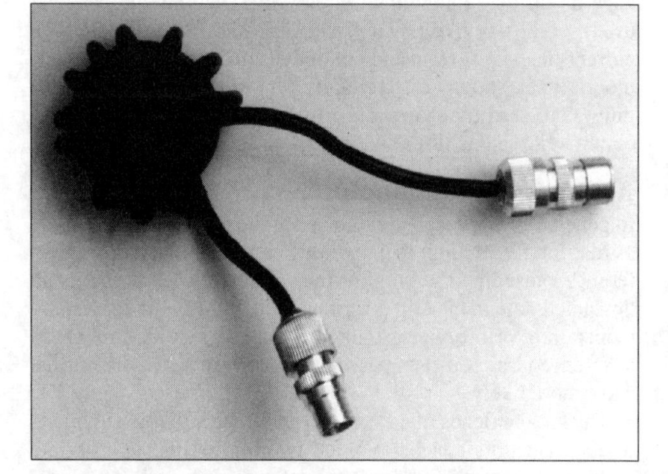

A ferrite ring choke suitable for TV coaxial down lead

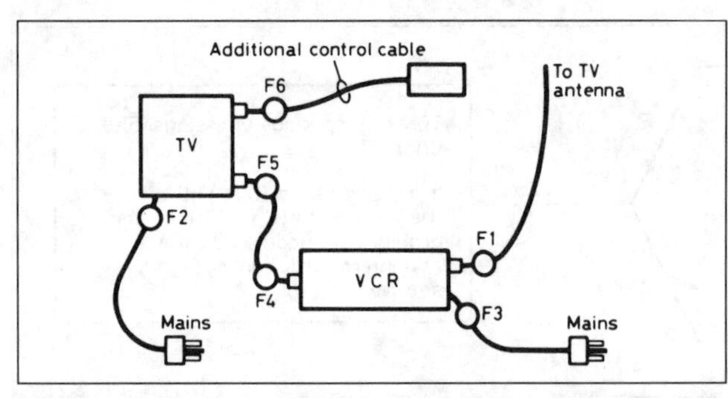

Fig 17.7. Typical domestic TV installation, showing positions of ferrite chokes. F1: antenna coaxial feeder choke (or filter). F2/F3: mains chokes. F4/F5: coaxial chokes on RF lead (VCR to TV). F6: Choke on additional leads to controls or external speakers or displays

The chokes fitted to the mains leads (F2 and F3 in Fig 17.7) are made by winding the cable onto ferrite rings in the same way as for a braid-breaker. It goes without saying that in this case it is essential to avoid damaging the cable in any way. If the cable is unusually thick it may be necessary to use fewer turns and more rings.

Due to the closed magnetic circuit in a toroidal core, it is not necessary to pull the cable tight onto the ferrite material – all that is required is for the winding to be secure and the ends not too close together. If it is not possible to use ferrite rings (because the plug cannot be removed for instance) then a ferrite rod choke can be used, though the choking effect will not be so great, particularly on the lower HF bands. Alternatively a split core can be used.

Chokes may be needed on either end of the coaxial cable between the VCR and the TV set, depending which unit is susceptible. If this lead is abnormally long they may be required at both ends. Some modern TV installations have additional control and audio cables connected to either the TV or the VCR and these should be dealt with in the same way as the mains leads. It is reasonable to expect the owner of the equipment to disconnect non-essential ancillaries during the diagnostic process and, where ancillaries are gimmicky and poorly engineered, it may have to be accepted that it is not practical to expect such a system to operate without interference.

Mains filters

There are many types of commercially manufactured mains filters on the market, but not all are effective in dealing with breakthrough. Generally filters which do not provide a high impedance on all the mains input wires – live, neutral and earth (if used) – will not give much protection, since the unwanted RF currents usually flow in and out of the equipment as a common-mode signal, and not differentially between the wires. The best thing is to start by trying a ferrite-ring choke and not to use a commercial mains filter unless a special need arises.

Radio and audio equipment

In general the approach to radio and audio problems is the same as that for TV, though there are differences of emphasis. Interference caused by the braid of the antenna feeder acting as an unwanted antenna is still a serious problem, but the larger elements on a VHF broadcast antenna make direct pick-up of interference (which is then passed down the feeder in the normal way) more likely.

The speaker leads of a stereo system frequently pick up interference and this can be avoided by winding the appropriate number of turns on to a ferrite ring (or pair of rings) to form a choke. Where this is not possible, a split core or a ferrite rod can be used.

Where breakthrough to a portable radio is experienced, there is not very much that can be done from the outside of the set since there are no external leads to choke. However, it is often possible to move the set to a position where the interference is negligible. Frequently portables only give trouble when operated from the mains, and in this case a ferrite choke in the mains lead is likely to be effective.

IF and intermodulation breakthrough

In many cases of interference to radio and TV, breakthrough is caused by overloading RF circuits or by rectified RF affecting audio or control circuits. In addition, radio and TV sets can suffer from more subtle forms of breakthrough. These fall into four categories:

1. Direct IF breakthrough, where the IF of some piece of radio equipment falls in or near an amateur band. The most common example is a transmission on the 10MHz band getting into the 10.7MHz IF of a VHF receiver. This is breakthrough caused by insufficient IF rejection in the receiver.
2. IF breakthrough caused by harmonics of the amateur transmitter being picked up in the receiver IF, for example, the second harmonic of the 18MHz band entering the IF of a TV set. (The standard TV IF in the UK includes 36MHz.) A less likely possibility is the third harmonic of 3.5MHz entering the 10.7MHz IF of a VHF receiver. In cases of this sort the fault lies with the transmitter which is be radiating too much harmonic energy – it is no excuse to say that the susceptible receiver should be better designed.
3. Image interference. This where a signal on the 'wrong side' of the local oscillator beats with it to give the IF. It is fairly common on the 1.8MHz band, where amateur signals give image responses on medium-wave receivers. For instance, a receiver with a 455kHz IF tuned to 990kHz (303m) would have a local oscillator of 1445kHz. This would beat with a strong amateur signal on 1.9MHz, which would be tuned in on the medium-wave band like any other signal. This is a case of breakthrough caused by poor image rejection of the susceptible receiver.
4. Amateur signals intermodulating with the harmonics of the local oscillator or other oscillators in the susceptible equipment, causing spurious responses. These give rise to interference which is tuneable at the receiver, but is nevertheless a case of breakthrough.

Interference to alarms and control electronics

There are so many possible types of equipment which might be affected, with new designs appearing almost every day, that it is only possible to deal with general principles. (The best source of specific up-to-date information is the EMC column in *Radio Communication*.) The usual procedure is to try to find out where the RF is getting into the system and to fit ferrite chokes to the offending leads. Bypassing techniques can be used, as discussed above, but are not practical for general amateur use since they involve internal modifications to the equipment.

If you are installing an alarm system yourself, make sure you buy the equipment from a reputable supplier (and that it has the appropriate BS number or the CE logo), and make sure that good installation practice is used.

Breakthrough to telephones

The main problem with modern telephones is breakthrough, caused by rectification and amplification of RF currents, and the solution is to avoid the RF getting into the equipment in the first place. There are two courses of action:

1. First check your radio housekeeping to see that you are not inviting trouble by poor antenna siting.
2. Prevent common-mode interfering currents getting into vulnerable circuits by fitting ferrite chokes as close to the susceptible units as possible. Winding the lead on to a ferrite ring is the most popular method of making a choke, but split-core or rod chokes can also be used.

Many households have extension telephones connected by plug and socket to the 'line jack' provided by the telephone company, and these may involve quite long lengths of interconnecting cable. At the start of an investigation, unplug any extension units and their cables, leaving only the instrument near the line jack connected. Clear any breakthrough on this, and then reconnect the extension leads and telephones one by one, dealing with breakthrough problems as they arise. In some cases it may be necessary to re-route vulnerable extension leads, but in most cases liberal use of ferrite chokes will prove effective.

In general it is unwise to attempt to carry out any internal modifications to telephone equipment, even if it is your own property; there are rules about equipment being approved for connection to telephone lines, and most telephone authorities (including those in the UK) do not permit unofficial modifications to anything connected to their lines. If privately purchased telephone equipment is abnormally susceptible to interference, then the supplier should be contacted. Where equipment is rented from the telephone company or administration, the problem should reported to the appropriate engineering department. Many companies undertake to correct breakthrough to rented telephones free of charge or alternatively they may exchange the telephone for a type which has better immunity. Breakthrough is not a very common problem (from the telephone company's point of view) so make sure that the person you report it to understands the problem and is aware of the company's policy.

Occasionally breakthrough may be caused by abnormal conditions on the lines before they come into the house. This should be reported to the authority responsible for the lines – but check your radio housekeeping first.

Intractable cases

Where a case of breakthrough fails to yield to the procedures that have been discussed in this section, it is time to call for some help, and in most cases this means contacting your national radio society. In the UK, the RSGB's EMC Committee has set up a countrywide network of EMC co-ordinators; their addresses and telephone numbers are published at frequent intervals in *Radio Communication* and also in reference [3].

INTERFERENCE CAUSED BY UNWANTED EMISSIONS FROM THE TRANSMITTER

As mentioned above, where interference is caused by spurious emissions from a transmitter, the interference can only be to equipment which uses radio in some way, and will be evident on similar equipment in the vicinity which is tuned to the same frequency. This is in contrast to breakthrough which is dependent on the susceptibility of the 'victim' equipment rather than any defect in the transmitter.

Spurious emissions fall into two major categories:

1. Predictable spurious emissions which are generated as part of the process of carrier generation, and hence are (at least in principle) predictable.

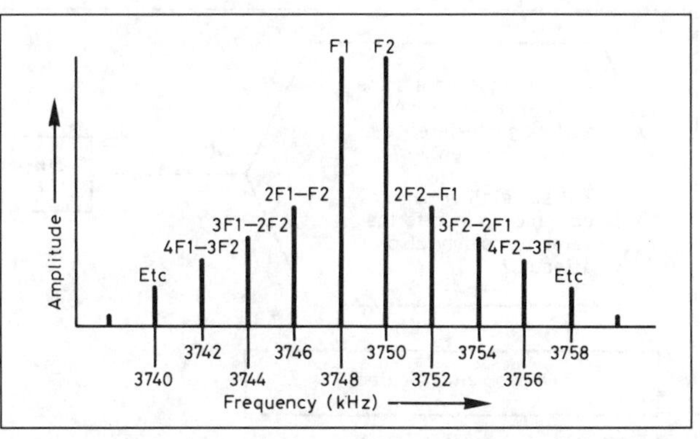

Fig 17.8. Odd-order 'intermods' from a two-tone SSB signal on the 3.5MHz band

2. Oscillations caused by faults in the design or construction of a transmitter. These give rise to unexpected emissions which can occur on almost any frequency. This is known as *instability*.

Predictable spurious emissions

The commonest predictable spurious emissions are harmonics of the carrier. These harmonics are simply a measure of the distortion of the sine wave which constitutes the carrier; an absolutely pure sine wave would have no harmonic content. In practice the object is to generate a carrier which is as undistorted as possible, and where necessary to filter it so that harmonics are kept to a negligible level.

The other major class of spurious emissions is the products caused by multiplication and mixing processes in the transmitter. In almost all modern transmitters the carrier is generated by mixing the outputs of oscillators (fixed and variable) to produce the final frequency. In most cases there are several mixer stages and at each stage unwanted products will be generated as well as the wanted signals.

The secret of good design is to make sure that unwanted products are at frequencies widely separated from the wanted signals so that they can be readily filtered out. As with harmonics, the object is to ensure that everything except the wanted output is attenuated to a negligible level.

How small 'negligible' is depends on the frequency bands in use and the power of the transmitter. As a general rule on HF, harmonics should be at least 50dB, and other spurii at least 60dB, below the carrier. On VHF the figures should be 60dB and 80dB respectively.

There is another class of predictable spurious emissions, and this is *splatter* – the generation of unwanted intermodulation products by the modulation process in an SSB transmitter. The important intermodulation products are the odd-order products because they appear close to the carrier and hence cannot be filtered out. (See also Chapter 1.) There will always be some intermodulation products generated in any SSB transmitter, and the object is to design and operate the transmitter in such a way that they are kept within reasonable bounds. About 35 or 40dB down on the carrier is a reasonable target to aim for (see Fig 17.8).

Identifying and rectifying predictable spurious emissions

Interference caused by harmonics of a transmitter can usually be identified by consideration of the frequencies on which interference is evident and the frequency on which the transmitter is operating. For instance, if the problem is interference to a VHF broadcast signal on 100.5MHz when the transmitter is operating on 50.25MHz, then it is very likely that second-harmonic radiation

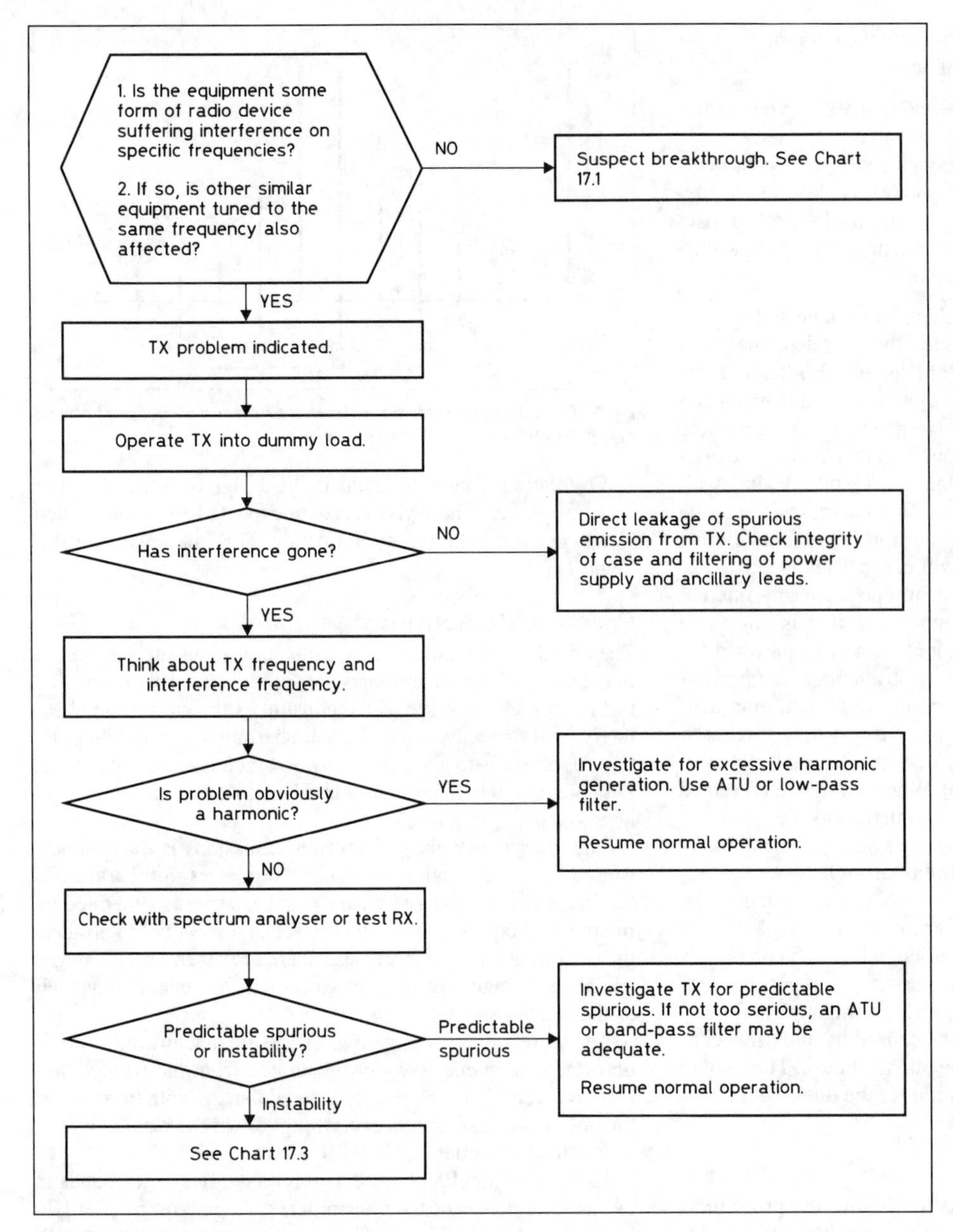

Chart 17.2. 'Predictable' spurious emissions

is the culprit. Where excessive harmonic radiation is suspected, first check the transmitter and its adjustment and, if the problem persists, fit a low-pass filter as in Fig 17.9. Note that the filter comes after the VSWR meter and, where an ATU is used, it should come after the filter. In mild cases an ATU alone may be sufficient.

With non-harmonic spurious emissions (which are often just called *spurious* or *spurii*) the situation is more difficult. The best procedure is to set up some way of monitoring the output from the transmitter to see if any excessive spurious signals are present. If a spectrum analyser or measuring receiver is available the task will be simple, but in most cases resort will have to be made to a general-coverage communications receiver. This can be used to tune round the frequencies on which interference is experienced while the transmitter is operated.

All receivers have spurious responses which can be difficult to differentiate from transmitted spurious emissions, and these will become much worse if the receiver is overloaded. Make sure that the signal from the transmitter as received on the test receiver is not greater than about S9 plus 20dB; this will avoid overloading and at the same time enable fairly weak spurious signals to be detected. Once the spurious signal has been identified, the transmitter should be investigated; the frequency of the spurious signal will often indicate the stage which is likely to be at fault. Where the problem is not too severe, using an ATU on HF or a suitable band-pass filter on VHF may be all that is required.

If a spectrum analyser or a measuring receiver is available it should be connected up as in Fig 17.10. It is good practice to start with more attenuation than is expected to be required and to remove it to establish the desired level. Remember that the attenuator connected to the transmitter must be capable of handling the full transmitter power. More information on suitable high-power attenuators will be found in Chapter 6 of reference [1].

Where a communication receiver has to be pressed into service, it will probably be simpler to set up the receiver with a short antenna some distance from the transmitter. In such an arrangement the magnitude of any spurious signals detected will be dictated by the antenna characteristics at the receiver, but the technique will none the less be adequate to show up serious spurious emissions.

Spurious emissions due to instability

This is caused by unintentional feedback in some stage of a transmitter causing oscillation. Instability also falls into two categories:

Fig 17.9. Position of the filter. The ATU ensures that the filter sees a 50Ω load

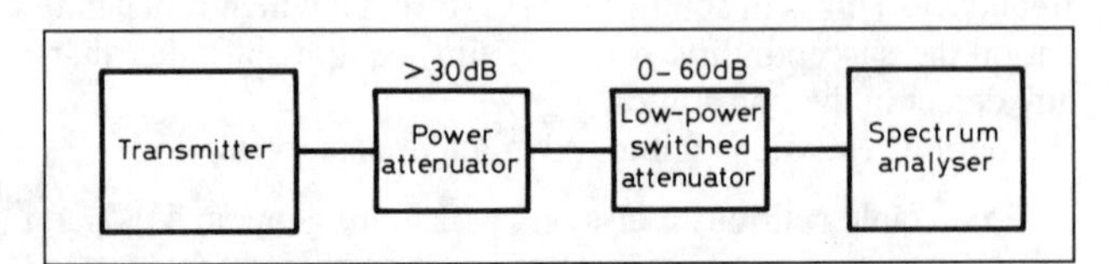

Fig 17.10. Connecting a spectrum analyser to a transmitter

1. Direct instability where feedback in one of the amplifying stages of the transmitter causes oscillation on or near the normal operating frequency.
2. Parasitic oscillation, where incidental inductance and capacitance cause oscillation at a frequency far removed from that at which the circuit would normally operate.

Both these effects are serious but direct instability is perhaps the most dangerous because it can result in very large spurious signals being generated at a frequency where the antenna can be expected to be a good radiator. In very bad cases the spurious signal can be comparable with the normal output power, causing widespread interference. Fortunately both types of instability are rare in modern, well-designed, transceivers but care should be exercised when repairs have been carried out. This is particularly true where power transistors have been replaced with 'near-equivalents'. The way to avoid instability is by employing good design and construction, with particular reference to layout, screening and decoupling. Where appropriate, *parasitic stoppers* should be included. See Chapter 5.

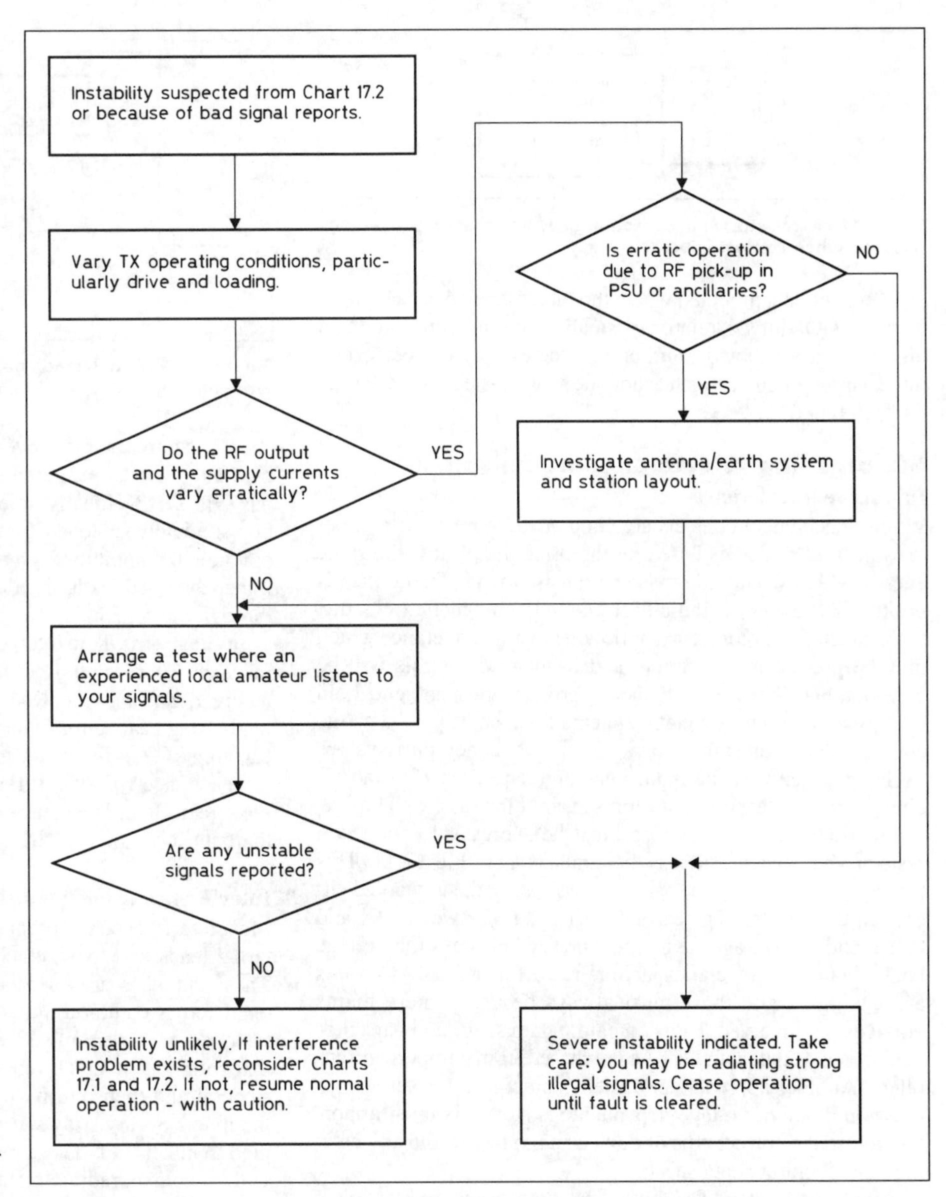

Chart 17.3. Instability

Evidence of instability

The two main indicators of instability are bad signal reports and erratic operation of the transmitter. If instability is suspected and sophisticated test gear, such as a spectrum analyser, is not available, the best procedure is to arrange for a report from someone who can receive you at good strength, and who knows what they are looking for. Ask them to look as far either side of your signal as possible, looking for unstable signals coming and going in sympathy with your main signal. Vary the drive and loading conditions of the transmitter – though in some modern transceiver the ability to this may be limited. It is important to remember that a transmitter may be stable when operating into a good 50Ω load but unstable when connected to an antenna.

Erratic operation usually means that the RF output and power supply current varies unpredictably as the drive is varied or as the ATU is tuned but, before jumping to the conclusion that the transmitter is faulty, make sure that the problem is not caused by poor station lay-out. It is not uncommon for the RF output of a transmitter to find its way into the power supply, or into ancillaries. This can cause very erratic operation. In a modern commercial transceiver, which is normally well behaved, this is more likely to be the cause of erratic operation than instability in the transceiver itself. Check the antenna/earth arrangements, and review your radio housekeeping – see Chapter 3 of reference [1].

RECEIVED INTERFERENCE

Background interference level

On HF there will always be noise picked up by the antenna, however well it is sited – this is usually called the *ambient noise*. This noise may be either man made or natural, and in most cases is much greater than the thermal noise generated in the front-end of the receiver. The 'bottom line' of this noise is natural static from thunderstorms and other atmospheric discharges which are always taking place somewhere in the world and are propagated over very long distances by the ionosphere. Added to this is the noise from electrical and electronic machinery of various sorts. As frequencies rise into the VHF region, the external noise reduces and becomes 'whiter', until above about 100MHz the thermal noise in the receiver front-end predominates, even when low-noise design techniques are used.

It is a general rule of radio that good transmitting antennas are also good receiving antennas but, so far as HF is concerned, this is modified by the fact that the ambient noise sets a limit to small-signal reception. For HF reception it may be better to mount a relatively poor antenna in a good location than to have an

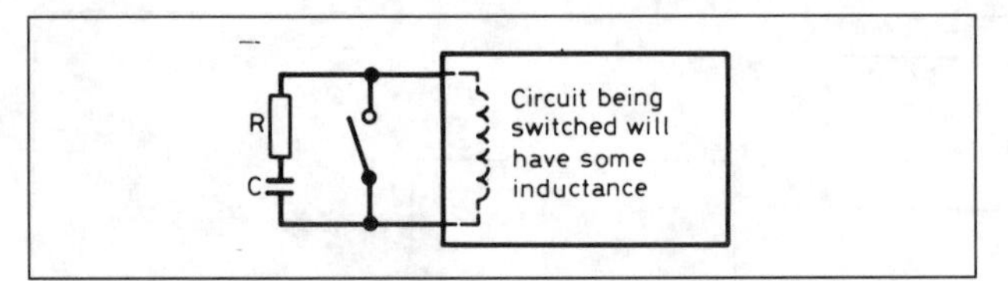

Fig 17.11. A resistor and capacitor used to absorb the energy released when contact is broken

efficient antenna mounted where the man-made noise is high. In difficult situations a relatively small active antenna mounted high up and well away from noise sources may out-perform a much larger antenna which cannot be so well sited – see Chapter 7 of reference [1].

Suppressing interference at source

Impulsive interference

Where mechanical contacts are concerned, there are a number of well-tried remedies based on the principle of absorbing the energy which would otherwise be released when the contact is broken. The energy is initially stored in the magnetic field, due to the normal operating current flowing in any inductance which may be present in the circuit, and in many cases this will be considerable. When the contact is broken, the magnetic field collapses and a large voltage appears for a short period as the contacts open, causing a spark. Radio frequency currents are exchanged between the inductance and capacitance in the vicinity of the contact, using the ionised air of the spark as a bridge.

The traditional way of absorbing the energy is to connect a resistor and capacitor across the contacts as in Fig 17.11. This effectively *quenches* the spark, by dissipating the unwanted energy in the resistor. The capacitor should be between 0.01 and 0.1μF and the voltage rating must be several times the voltage being switched. Special capacitors rated for use on AC mains are available, and these must always be used where mains voltages are involved. Units containing a resistor and capacitor in one encapsulated unit can be purchased from component suppliers. Another approach is to use a semiconductor surge suppression diode or voltage-dependent resistor. It is most important to use the correct type of device and to follow the advice in the manufacturers data sheet.

Small, low-voltage DC motors can be suppressed by using a shunt capacitor of between 0.05 and 1μF, and series ferrite cored chokes of 10 to 30μH, as in Fig 17.12. The chokes are more effective at higher frequencies, and may not be required if only low frequencies are involved. Mains motors are best dealt with by using one of the many mains filters which are available in the suppliers' catalogues. This should be installed as close to the machine as practical.

Interference from digital equipment

It is relatively easy to reduce the leakage of interference from digital equipment at the design stage – it is really a matter of good engineering practice. Good decoupling and the provision of a substantial ground plane for the common 0V rail is a good

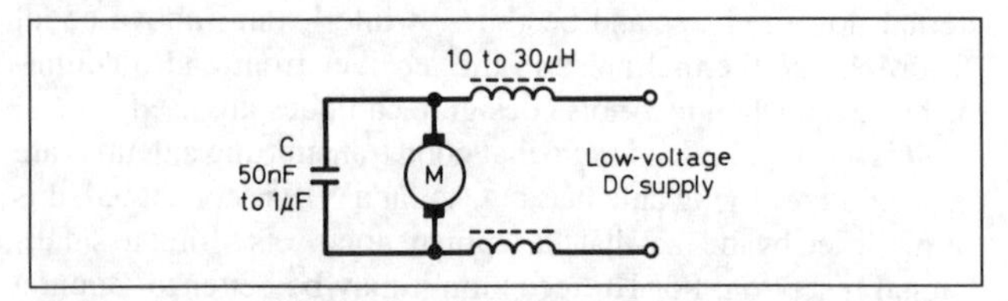

Fig 17.12. Suppressing a small low-voltage DC motor

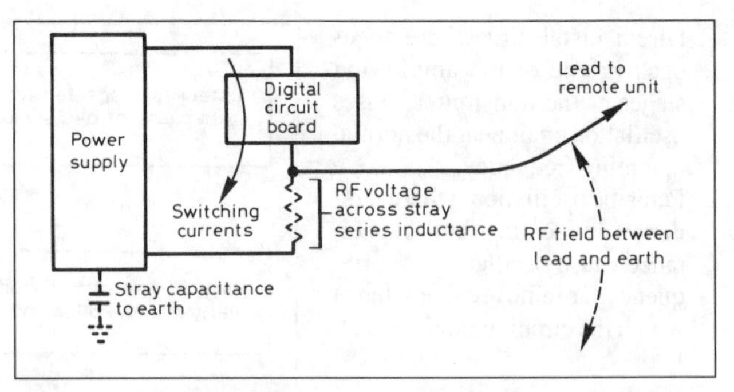

Fig 17.13. External lead, nominally at 0V, acting as unwanted antenna

start. It is important to prevent external leads having energy coupled into them from shared return paths, and so acting as antennas (Fig 17.13). Ideally, interference-generating circuits should be completely screened. The screen should be connected to the common 0V point through a path which has the lowest possible impedance. All leads should be decoupled where they pass out through the screen.

In situations where interference reduction is a major factor, for instance where digital control circuits are actually part of the receiver, special attention should be paid to screening and feedthrough capacitors should be used on all leads except those carrying RF. It is important to choose the correct value – too large a capacitance will distort fast digital signals. So far as possible, the screen should be continuous and, where there are any joints, there must be good electrical contact along the mating surfaces.

Interference from computer installations

Most interference in computer installations is caused by the external leads and their attached peripherals acting as antennas. These antennas are energised by interference voltages developed across common impedances, generally along the lines illustrated, rather simplisticly, in Fig 17.13. The overall effect is that common-mode currents flow from the computer to the peripheral and then return via any available path. This could be the mains or any stray wiring which may be capacitively coupled to either unit. These currents wandering all over the place will, of course, radiate interference.

Common-mode chokes

The simplest approach is to increase the impedance to common-mode currents by using ferrite chokes in exactly the same way as described above for dealing with breakthrough. Fig 17.14 illustrates the how the chokes reduce the flow of common-mode currents in a typical computer installation.

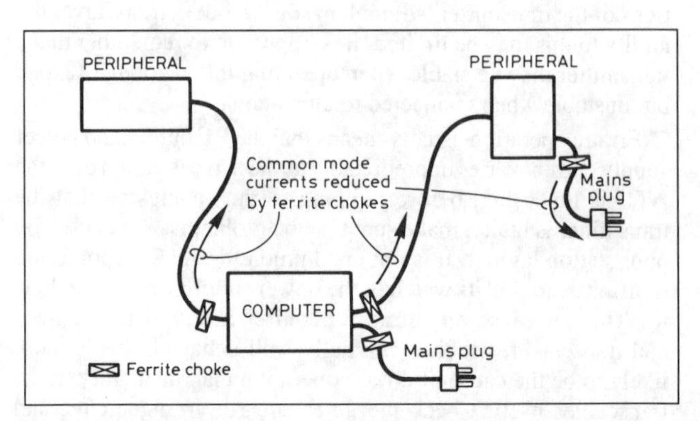

Fig 17.14. Reducing common-mode currents using ferrite chokes

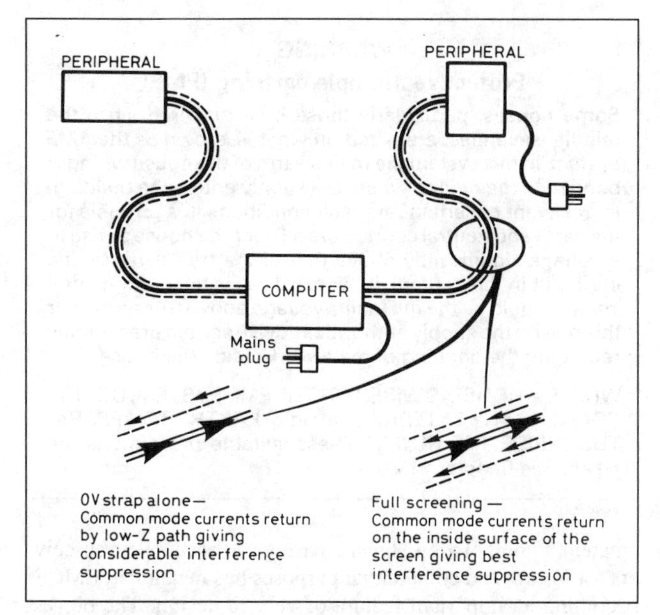

Fig 17.15. Confining common-mode current by 0V rail strapping and screening

The problem with computer installations is that the cables are usually unsuitable for winding on to ferrite rings. Clamp-up ferrite cores or ferrite rods are a possibility, but neither of these is very practical for use on ribbon cable. It is possible to purchase special clamp-on ferrite devices for ribbon cable, but at most amateur frequencies their effectiveness is limited. To achieve a reasonable degree of attenuation, the cable must form several turns on the core. Since inductance is directly proportional to the length of ferrite through which the wire passes, and proportional to the square of the number of turns, you get more attenuation by winding several turns on a relatively narrow ring than by using one turn on a thicker ring. (The rule is that every time the wire passes through the ring counts as one turn, so a ferrite ring simply slipped on to a wire counts as one turn.)

Screening leads

Where a screened lead is used to connect a peripheral to a computer, the screen has two functions. As might be expected, it does reduce radiation by forming an electrostatic screen around the conductors, but more importantly it provides a low-impedance path back to the computer for the common-mode currents which would otherwise leak back by devious routes, introducing interference into other units (Fig 17.15).

Ideally the screen should be good-quality braid, properly terminated in the correct connectors, but fortunately, provided the underlying principles are kept in mind, it is possible to achieve good results using low cost materials and a little ingenuity. The important thing about screening is not so much the nature of the screen, but where it is connected. It must be taken to a point which is as near as possible at true 0V potential at each end. Many modern computers and peripherals have external metalwork at earth potential or an external 'ground' terminal to which screens can be connected. In such cases there will be no problem finding somewhere to connect the screen. In simpler equipment which has no external metalwork at 0V/ground potential and a two-core mains lead, it must be appreciated that the manufacturer did not anticipate 0V connections being brought out from the case and this will be reflected in the design of the power supply. If in doubt, seek expert advice.

Under no circumstances connect anything to the chassis/0V rail of equipment which is not isolated from the mains by a suitable double-wound transformer. In particular remember that domestic TV sets, sometimes used as monitors, often have the chassis live to earth.

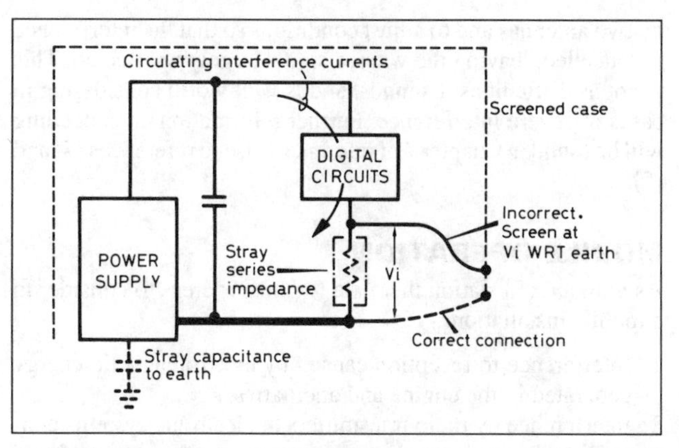

Fig 17.16. Incorrect connection of screened case. Screen is at interference voltage V_i with respect to earth

Before fitting screening to the leads, try strapping the 0V rails on the computer and the peripheral together with a short length of braid, about 5mm wide. If this reduces the interference, then you are on the right track. Improvements may not be noticeable if other peripherals are still connected. They should be disconnected or similarly strapped to the computer.

In some cases strapping may cure the problem but, if screening is required, there is a choice of using either special screening materials, available from component suppliers, or household aluminium cooking foil. The easiest way to make a contact with an aluminium screen is to include an uninsulated 'drain wire' inside the screen. In effect, wrap the screen over the cable leaving the strap, which we have just been discussing, inside.

Screening the case

In the early days of home computers, spraying the inside of the case with conductive paint to reduce interference was a popular measure. Nowadays, most computers have metal cases, or plastic cases with some built-in screening, and this practice has become less common. Before attempting any action of this sort, make sure that the problem is really radiation from the computer itself, and not the leads and peripherals. It goes without saying that spraying with conductive paint should be carried with great care, following the manufacturers instructions, and making sure the paint only goes where you want it to. It is essential to connect the screen to the correct 0V point, otherwise the screen could act as an unwanted antenna and make the interference worse (Fig 17.16).

Reducing the effects of interference at the receiver

Where it is not possible to suppress the interference at source, the only solution is to attempt to mitigate its effects at the receiver. There are two basic techniques available: the first is some form of *noise limiter* or *blanker* and the second is by cancellation.

The noise limiter has been around for a long time, but in recent years the it has been displaced by the blanker, which does a similar job in a rather more complex way, and generally much more effectively. More information on this topic will be found in Chapter 6.

Cancellation is a very powerful tool for dealing with received interference and deserves to be much more widely known. The principle is to receive both the wanted signal and the interference on two antennas and to adjust conditions so that the interference is cancelled, leaving the wanted signal largely unaffected. This is not as difficult as it sounds, and is well worth considering in cases of severe interference. Further information on cancelling will be found in Chapter 7 of reference [1], and references [4] and [5].

MOBILE OPERATION

As with a fixed station there are two major areas to consider in a mobile installation:

1. Interference to reception caused by electromagnetic energy generated in the engine and ancillaries.
2. Interference by radio transmitters to electronic systems controlling the engine and ancillaries.

At one time the most severe difficulty in mobile operation was interference generated by the ignition and by various motors and switches. In recent years this has become less of a problem for a number of reasons. One is that many modern cars have a broadcast radio fitted as standard, so that at least basic interference suppression is included. Another factor is that the majority of amateur mobile communication now takes place on VHF FM, which is inherently less susceptible to impulsive interference than SSB or AM. The immunity of an FM receiver to impulses (or any other AM signal) is due to the action of the limiter. Transceivers vary in the degree of limiting designed into the system. If a particular radio seems to suffer unduly from interference, suspect a lack of signal at the limiter, caused by a fault or a design weakness.

On SSB and AM the effect of impulsive interference can be greatly reduced by the use of a good noise blanker. A blanker operates by using the very short impulses of interference to operate a gate in the receiver IF, and its performance is dependent on a number of factors including the ability of the receiver front-end to handle large impulses. In practice the effectiveness of blankers varies considerably, so it is worthwhile trying the noise blanker in a practical test before deciding on a transceiver for mobile use.

RF interference to electronic control systems has become a major factor in mobile radio; not least because the possible safety aspects attract publicity. Basically the problem is caused by RF energy being picked up in the wiring of the car and entering the logic of the control circuits. The effects can vary from specific faults, such as failure of door locks or erratic flashers, to complete failure of the engine control system. Fortunately (from the amateur's point of view) the increasing use of mobile transceivers for leisure and commercial purposes has meant that EMC is becoming an important feature of vehicle design. The biggest problem is HF operation, where the cable harnesses can form quite efficient antennas.

The various metal assemblies in a car may not be in good RF contact, and bonding can result in a reduction of received noise and in improved immunity. Be careful to avoid accidentally short-circuiting the 12V supply; large currents can cause nasty burns, apart from being a fire risk.

High-power operation (particularly at HF) may not be possible in every make of vehicle and, bearing in mind all the relevant factors (including safety), this is probably a situation where moderation will pay dividends. Keeping the power down to reasonable levels (perhaps 25W mean power) on both HF and VHF would seem a sensible compromise.

WARNING

Protective multiple earthing (PME)

Some houses, particularly those built or wired since the middle 'seventies, are wired on what is known as the *PME system.* In this system the mains earth of the house wiring is bonded to the neutral where the supply enters the building. In the event of certain rare fault conditions it is possible for the earth and neutral conductors all over the house to rise to a voltage significantly above that of the true earth (ie the earth out in the garden). In extreme cases the earth neutral voltage could be the full mains voltage above true earth. For this reason the supply authorities advise certain precautions regarding the bonding of metalwork inside the house.

WHERE A HOUSE IS WIRED ON THE PME SYSTEM, DO NOT CONNECT ANY EXTERNAL (ie radio) EARTHS TO APPARATUS INSIDE THE HOUSE unless suitable precautions are taken. See Chapter 21.

REFERENCES

[1] *The Radio Amateur's Guide to EMC*, Robin Page-Jones, G3JWI, RSGB, 1992.

[2] 'Filters and ferrites in EMC', D Lauder, G1OSC, *Radio Communication* December 1993/January 1994.

[3] *RSGB Amateur Radio Call Book*, RSGB, published annually.

[4] 'Technical Topics', *Radio Communication* March 1993.

[5] '2m Interference Reduction System', T S Day, *Radio Communication* April 1992.

18 Amateur satellites and space communications

SPUTNIK 1, the first Earth satellite, was launched from the former USSR in October 1957. It carried a simple radio beacon which was heard all around the world but, though it reached a height of 939km at its farthest point from the Earth, its orbital low point was at 215km. The resulting atmospheric drag brought it down in altitude quite quickly and 92 days after launch it entered the thicker regions of the atmosphere around 100km where it burnt up.

On 12 April 1961, the Soviet Air Force pilot Yuri Gagarin was successfully launched into orbit and recovered after one orbit. Gagarin was a radio amateur, callsign UA1LO, but he was far too busy to use amateur radio during his classic flight. Meanwhile the pace of unmanned launches was rising rapidly, up towards 100 or more satellite launches a year, as various military groups took advantage of the excellent viewpoint offered by an Earth satellite for communications, navigation, weather observation, and photo-reconnaissance.

In December 1961, a few months after the Gagarin flight, the first amateur radio satellite (Oscar 1) went into orbit ('Oscar' stands for 'Orbiting Satellite Carrying Amateur Radio'). Oscar 1 was built by American radio amateurs and it was launched as ballast aboard a military rocket. As it orbited, its onboard beacon bleeped out the message that amateur radio had arrived in space. Since then amateur satellites have gone aloft at an average rate of slightly greater than one per year. As January 1994 ended, 41 launches had been made, with 18 of these launches still operational in various roles. Fig 18.1 shows the orbits of the Oscars operational in 1994, and the orbit of the Russian Mir space station which has recreational amateur radio equipment in regular use on-board.

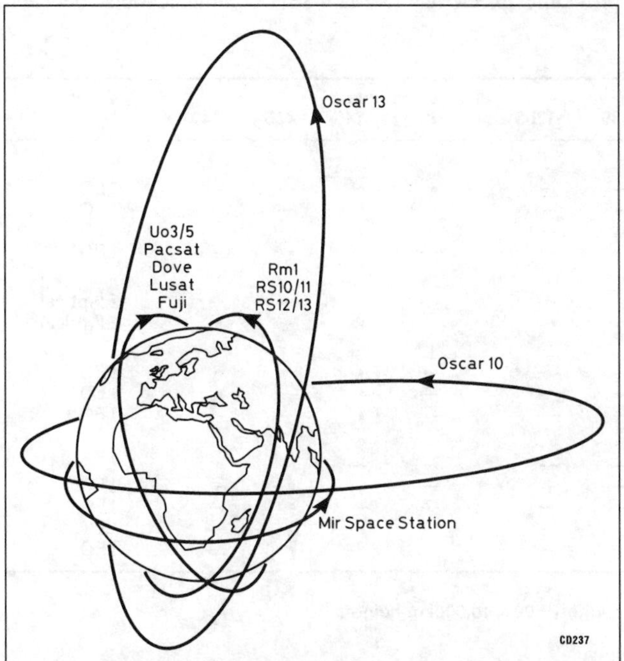

Fig 18.1. Typical amateur radio satellite orbits. Note that Mir, Uosats, Pacsats, RM1 etc are in LEOs (low Earth orbits). Oscars 10 and 13 have high elliptical orbits. Scales are approximate only

AMATEUR RADIO SATELLITES: 1961–1994

Oscars 1 and 2 launched in December 1961 and June 1962. These first radio amateur satellites were short-lived, primary-battery powered beacons. They worked well until their batteries discharged.

Oscar 3, launched in March 1965, was designed as a communications relay. You talked up to it on 146MHz and it repeated your signal and any adjacent signals on a 144MHz downlink. It allowed stations in Europe to talk briefly to stations in the USA while the satellite was mutually visible from both continents. Oscar 3 only lasted for 18 days but it clearly proved the concept of the satellite as an amateur radio international relay.

Oscar 4 (December 1965) was also a communications test. Unfortunately the rocket failed, leaving it in a very precarious short-duration orbit. While it lived it worked well.

Oscar 5 (January 1970) which followed was a much more complex design built in Australia. It tested magnetic stabilisation and other features which could be controlled from the ground.

Oscar 6 (October 1972), incorporated all the features of Oscar 5 and used them to provide extremely successful international communications links, eg UK to USA, every day for four years before it succumbed to battery collapse.

Long before Oscar 6 expired it had been joined in space in November 1974 by Oscar 7, a complex VHF/HF or UHF/VHF communications relay. Oscar 7 could duplicate Oscar 6's 145MHz up/29MHz down (Mode A) system or it could use the much clearer, less propagation-susceptible 432MHz up, 145MHz down (Mode B) system. Fig 18.2 shows how a wide-band transponder of the Oscar 7 type works. Oscar 7 provided many years of excellent service.

Oscar 8, intended as a stop-gap replacement for Oscar 7, was launched in March 1978. It used Mode J (145MHz up, 435MHz down) for its communications link, thereby providing a much stiffer test for would-be users, considering that at that time no commercial 435MHz receivers were available. Initially very unpopular because of reception difficulties, Mode J is now the favourite mode for many amateur satellite communicators, being removed from the very heavy interference which has gradually appeared on the Mode B systems 145MHz downlink frequencies.

In October 1978, the world awoke to find the first Russian amateur radio communications satellites were in orbit. 'RS1' and 'RS2', as they were called, suffered like the earlier Oscar satellites from limited battery performance, but in December 1981 the Russians orbited a group of six amateur radio satellites.

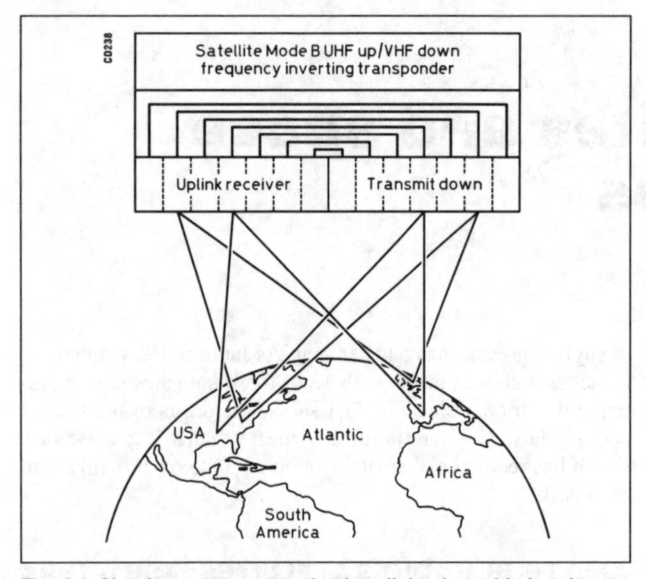

Fig 18.2. Simultaneous communications links via a wide-band transponder – Spain to/from Texas, UK to/from Kentucky

RS3 and RS4 were simple 29MHz beacons, but RS5 to RS8 inclusive were powerful Mode A communications relays which were to provide many years of excellent service. Simple to access and operate, these RS satellites and their modern replacements RS10/11 and RS 12/13 are superb vehicles for radio amateur satellite newcomers to start on.

This good news from Russia compensated slightly for the major disappointment which occurred in May 1980, when the highly ambitious Amsat Phase 3A high-altitude elliptical orbiter suffered a launch failure and went to a watery grave in the Atlantic Ocean.

Meanwhile a team at the University of Surrey in England was building a special, downlink-only, experimental satellite, Uosat 1 (Oscar 9), which was to be the forerunner of a steady stream of useful experimental and educational satellites coming from this stable. Equally useful were Russian satellites Iskra 2 and Iskra 3, small suitcase-sized satellites released into the ionosphere at 300km altitude via the airlock of the Salyut 7 space station. The Iskras provided amateur radio's first glimpse of the fascinating experiments which could be done with a satellite transmitting on 29MHz from inside the Earth's ionosphere.

Table 18.1. Frequency modes used by Amsat satellites

Mode	A	B	J	K	L	S	T
Uplink (MHz)	145	435	145	21	1269	435	21
Downlink (MHz)	29	145	435	29	435	2401	145

Note: Some RS satellites can operate simultaneously on two modes, eg KT, featuring modes K and T together, or KA, featuring K and A together.

During the period 1992–1994 there have been 26 further launches of amateur radio satellites, of which 18 still survive and are available for operations. They will be described in the next paragraphs. The two orbiters which have not survived were both special packet radio test satellites. Fuji 1 (Oscar 12) was taken out of service when its successor Fuji 2 (Oscar 20) was launched, and Uosat 3 (Oscar 14) suffered a similar fate when its duties as a packet store-and-forward international relay were taken over in 1992 by the more powerful Uosat 5 (Oscar 22).

OPERATIONAL AMATEUR SATELLITES

Table 18.2 lists radio amateur satellites operating in 1994. The satellites in the table can be separated into four distinct groups. Groups 1 and 2 comprise the communications satellites and Groups 3 and 4 the digital data satellites.

Group 1 low-Earth-orbit (LEO) communications satellites include RS10/11 and RS12/13, which provide very simple shortwave (29MHz) communications links via CW or SSB between stations up to 5000km apart, ie UK to Eastern North America. Also included in this group are RM1 (Oscar 21) and Fuji 2 (Oscar 20), each of which has a sophisticated VHF/UHF link facility providing excellent CW and SSB communications between

Table 18.2. Amateur radio satellites operational in 1994

Name	Oscar	Description	Uplinks 21	29	145	435	1269	Downlinks 29	145	435	2401	Orbit
Group 1												
RS10/11	—	Communications	•	—	•	—	—	•	•	—	—	LEO
RS12/13	—	Communications	•	—	•	—	—	•	•	—	—	LEO
RM1	21	Comms + digi	—	—	—	•	—	—	•	—	—	LEO
Fuji 2	20	Comms S+F	—	—	•	—	—	—	—	•	—	LEO
Group 2												
Oscar	10	Communications	—	—	—	•	—	—	•	—	—	Elliptical
Oscar	13	Communications	—	—	•	•	•	—	•	—	•	Elliptical
Group 3												
Uosat 2	11	Experimental	—	—	—	—	—	—	•	•	•	LEO
Dove	17	Educational	—	—	—	—	—	—	•	—	•	LEO
Sara	—	Experimental	—	—	—	—	—	—	•	—	—	LEO
Group 4												
Pacsat	16	Store-and-forward	—	—	•	—	—	—	—	•	•	LEO
Weber	18	Earth image	—	—	•	—	—	—	—	•	—	LEO
Lusat	19	Store-and-forward	—	—	•	—	—	—	—	•	—	LEO
Uosat 5	22	Image S+F	—	—	•	—	—	—	—	•	—	LEO
Kitsat	23	Store-and-forward	—	—	•	—	—	—	—	•	—	LEO

Notes:
'LEO' – low Earth orbit, below 1600km height; 'Elliptical' – elliptical orbit roughly 500 × 40,000km height.
Communications use CW or SSB.
Store-and-forward (S+F) is 1200 FSK up/1200 PSK down or 9600 FSK up/down.
Sara monitors natural HF from the planet Jupiter.
Fuji 2 (Oscar 20) is in either communications or store-and-forward mode, not both together.
Uosat 5 (Oscar 22) operates both image and store-and-forward modes together.
Four S+F and Earth imaging satellites (Eyesat, Itamsat, Posat and Kitsat 25) are in space being tested in January 1994.

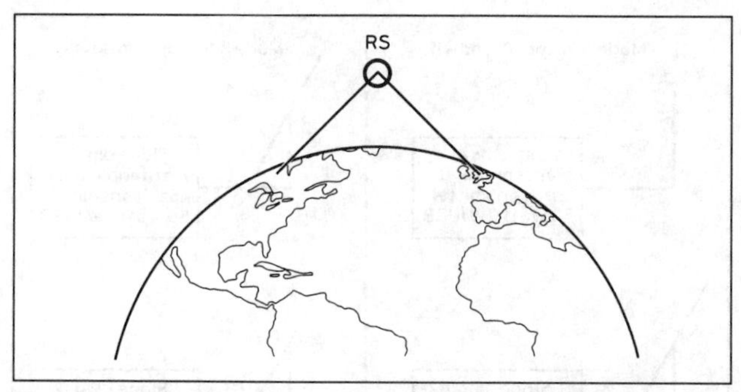

Fig 18.3. A RadioSport communications satellite connects Canada to the UK

continents. Both RM1 and Fuji have two separate transponders. They can operate in either the analogue voice/morse mode described above or they can provide digital radio links. Only one mode is in operation at any one time. Fig 18.3 shows how RS10/11 can be used as a simple communications link between Canada and the UK.

Group 2 high-altitude elliptical orbiters comprise Oscar 10 and 13. Both execute roughly two orbits each day, appearing almost stationary for up to six hours while at their *apogee* (highest altitude) as they provide multiband contacts across almost a complete hemisphere of the Earth. Oscar 10 is getting old and is now restricted to a single 435/145MHz link which can handle up to 40 or more conversations at a time. Oscar 13 provides two choices of operating frequencies, permitting a broad range of communications experiments. Fig 18.4 shows the Oscar 13 orbit, with the rough times taken to go around the different parts of the orbit. Note how the satellite appears to travel much more slowly when it is farthest from earth, where it gives greatest communications range.

Group 3 includes simple LEO experimental data satellites such as Uosat 2 (Oscar 11), Brasilian Dove (Oscar 17) and Amsat Belgium's Sara Jupiter monitor.

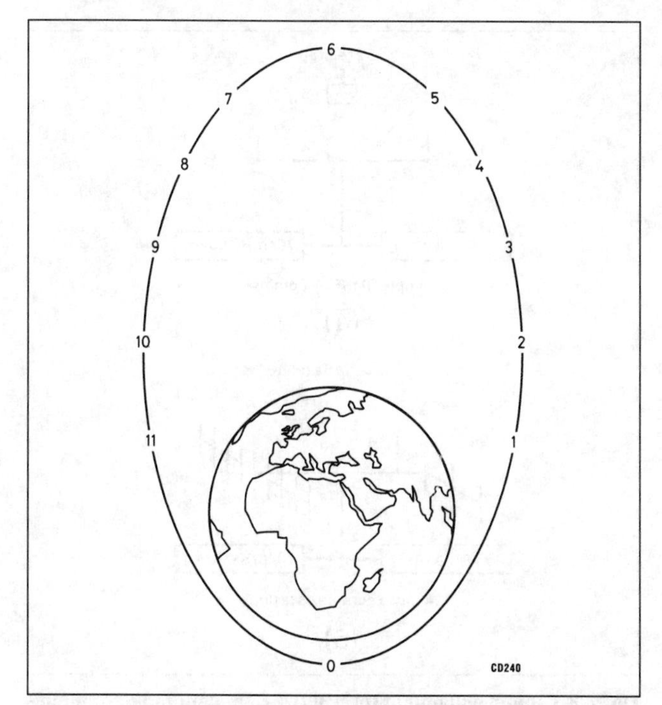

Fig 18.4. The Oscar 13 high-altitude elliptical orbit as seen from above its orbit plane. The numbers on the orbit track show where the satellite is each hour after perigee passage

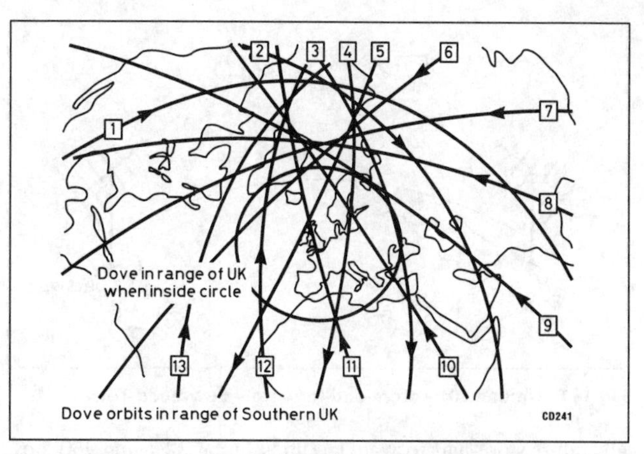

Fig 18.5. Dove educational satellite: orbits near the UK

Fig 18.5 shows the Dove daily orbit pattern near the UK. Uosat 2 has a similar pattern. Whenever Dove or Uosat 2 come within roughly 2800km of UK they are visible above UK station horizons inside the footprint circle shown on Fig 18.5. They can be received on a VHF FM receiver as they send *telemetry* – automatic digital telegraphy which reports the satellite's operational state and the conditions in its local space environment. Table 18.3 shows a frame of typical Amsat telemetry in the raw form, and reports a decoded example of one of its boxes. Decoded telemetry figures provide the material for hundreds of different experiments in schools and universities. Most Amsat satellites broadcast a roughly similar set of telemetry, excepting Sara – its telemetry reports levels of natural radio signals generated in the region between the planet Jupiter and its volcanic satellite Io.

The Dove telemetry is designed for direct on-line decoding by your computer. Software to do this is available from Amsat. The most interesting parameters are blocks 26 to 2B which change very rapidly as the satellite spins in the sunlight. Other Amsats use very similar telemetry.

Group 4 store-and-forward packet radio satellites comprise a recent addition to the Amsat series. Amsat pioneered the method shown in Fig 18.6 whereby a digital station in the UK can uplink a message to the satellite as it passes over the UK. The satellite stores this message in its large memory until it gets over the message addressee's station, when the latter can ask the satellite to download the message to him or her.

This system provides a very rapid way of distributing international messages, bulletins, and general news. Pacsat (Oscar 16), Lusat (Oscar 19), Fuji 2 (Oscar 20) and Uosat 5 (Oscar 22)

Table 18.3. Dove telemetry – raw and decoded

DOVE-1>TLM
00:59 01:5A 02:8E 03:35 04:5C 05:59 06:6E 07:4A 08:70 09:6A 0A:45 0B:EF 0C:E8 0D:DC 0E:10 0F:23 10:DC 11:A0 12:02 13:EE 14:BF 15:B5 16:76 17:76 18:78 19:77 1A:76 1B:07 1C:7D 1D:79 1E:DE 1F:68 20:D1

DOVE-1>TLM
21:C8 22:6E 23:30 24:2A 25:24 26:70 27:01 28:4D 29:01 2A:03 2B:02 2C:01 2D:7A 2E:5F 2F:AA 30:D8 31:AB 32:10 33:B8 34:AB 35:A7 36:AC 37:AF 38:C8

Decoding an individual box on the above table is performed as follows: Each box contains two hexadecimal numbers, eg box 0 has 00 and 59. The first number, hex 00, signifies 'Receiver E/F Audio' reported here. The second number, 59, is translated to decimal = $5 \times 16 + 9 = 89$. This decimal number is then applied to an equation of the form $A \times N^2 + B \times N + C$ where A, B and C are variables listed in the decode for this box (in this case A = 0, B = 0.0246 and C = 0). So Receiver E/F Audio = 89 × 0.0246 = 2.1894 volts.

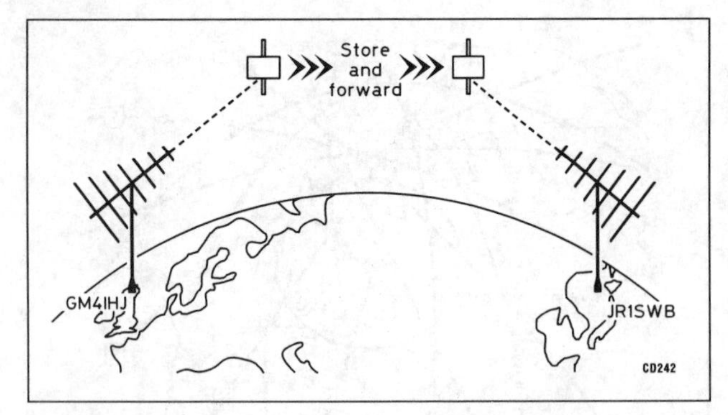

Fig 18.6. Upload UK – store-and-forward – download Tokyo

all feature store-and-forward facilities. Oscar 22, using 9600bps, is eight times faster than the others so it is presently heavily used for international traffic.

Up in space in January 1994 are four new satellites: Eyesat, Itamsat, Posat and Kitsat KO25. They feature a wide range of digital store-and-forward modes, and in two cases carry Earth imaging cameras.

Future satellites

Scheduled to launch in 1994/95 is a Russian Mode A (145MHz up, 29MHz down) communications satellite, which is presently awaiting launch as RadioSport 15 (RS15). This may well be the last Mode A satellite launched by Russian radio amateurs.

Beyond 1994 several satellites are being prepared for launch. One is by Israel – a store-and-forward satellite of the Fuji/Pacsat type, and another of similar type is coming from Mexico, possibly carrying a meteor radar. Further in the future is Sedsat, a satellite assembly provisionally planned to be launched with a 'tethered package' into a roughly circular high-latitude orbit at an altitude of about 1600km. Then a year or two later will come Phase 3D, a high-altitude elliptical orbiter development of the Oscar 13 type. The 3D project is being led by Amsat Germany.

Still on the drawing board, or at the discussion stages, are further amateur radio space projects involving a probe to the planet Mars, and an amateur beacon or communications package aboard a lunar or interplanetary solar sail experiment.

GROUND STATION EQUIPMENT

Group 1: LEO communications satellites

RS10/11 and RS12/13 provide the best point for newcomers to learn satellite operating techniques. Most of the equipment required is already present in an amateur radio station with ordinary HF communications facilities, and the antennas used can be simple HF types or home constructed from wire. In fact any HF beam, vertical or dipole will give good results on HF and a standard 2m omnidirectional antenna will suffice for the RS Mode A uplink. Fig 18.7 shows this Mode A (2m up, 10m down) configuration normally used on RS10. Using 25 or 50W to the 2m antenna and receiving on a simple 10m sloped dipole can produce good results with CW or USB communications into Asia or USA.

For Mode K operating, the usual mode for RS12, both uplink and downlink are at HF. So simple antennas and a 50W 15m transmitter can produce a good uplink. With a dipole and a standard HF communications receiver tuning 29.4MHz, CW or USB gives good results on the downlink.

For VHF/UHF Mode B (70cm up, 2m down) (satellite RM1) or 2m up and 70cm down (satellite Fuji/Oscar 20 in its analogue

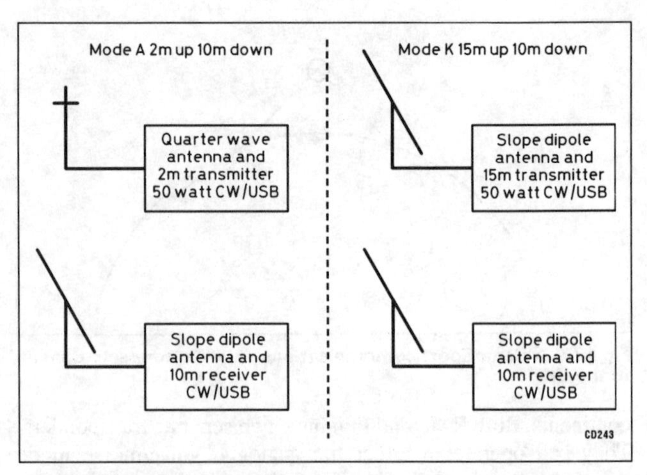

Fig 18.7. RadioSport 10/11/12/13 ground station equipment

JA mode), a more complex station may be required. You can use simple omnidirectional antennas as shown in Fig 18.8(a) but if you require long-range communication into America or Asia you will need better antennas. These need not be elevatable. All good long-distance communication occurs when the satellite is on your horizon. So simple, hand-trained, relatively wide-beam, five-element 2m Yagi and, say, 10-element 70cm Yagi antennas are more than adequate. Fig 18.8(b) shows this configuration.

Group 2: high-altitude elliptical-orbit communications satellites

Oscars 10 and 13 are 35,000km from the Earth (see Figs 18.1 and 18.2) for a large part of each orbit (as opposed to a maximum of 2500km for the LEO satellites), so much better antennas are needed. As Fig 18.9 shows, these should be long, multi-element, narrow-beam Yagis, and because they have a narrow beam they need to be accurately trained on the satellite by

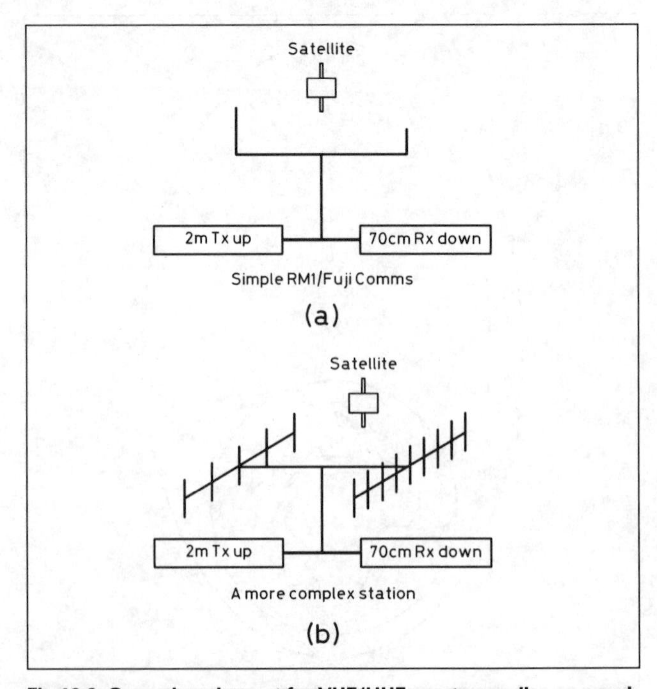

Fig 18.8. Ground equipment for VHF/UHF amateur radio communications satellites. (a) Simple RM1/Fuji communications. As shown, correct for Fuji; RM1 is 70cm up, 2m down. (b) A more complex station

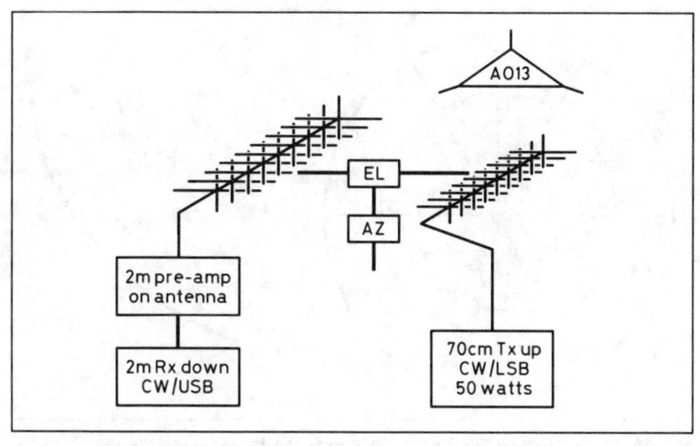

Fig 18.9. Oscar 10/13 Mode B ground equipment

azimuth and elevation rotators. It also helps to mount a low-noise preamplifier directly onto the output from the downlink Yagis. Some operators recommend circularly polarised helical antennas but these are awkward to build and the satellite signal polarisation can vary greatly from the designed nominal. To get more consistency in conditions where polarisation is changing, many operators use *XY Yagis* (antennas with elements in both the vertical and horizontal planes) and they switch from one set of elements to the other as polarisation changes require. Fig 18.9 shows the equipment required for Oscar 10/13 mode B (70cm up, 2m down) and a transmitter on 70cm will be suitable for the Oscar 13 Mode S uplink. Please note that a failure of Oscar 13's 435MHz downlink has resulted in loss of Mode J and L facilities. For Oscar 13 Mode S downlink reception, this author uses a 2401/144MHz converter fed by a home-made 20-turn helical antenna, and feeding a 2m receiver. See Table 18.1 for a list of modes and frequencies.

Group 3: telemetry/educational digital satellites

Dove (Oscar 17) is the newcomer's entry to reception of digital packet signals from space. Fig 18.10 shows a simple Dove reception station with an omnidirectional antenna feeding any 2m FM receiver or scanner tuned to 145.825MHz. Any terrestrial packet radio station can receive Dove simply by tuning to 145.825 when Dove is near the station (mid-morning or late evening). Uosat 2 (Oscar 11) telemetry can be received on this same frequency by a station using simple FM reception, but the downlink telemetry requires a special low-speed demodulator.

Group 4: packet store-and-forward station equipment

Fuji Oscar 20 is the best target for starting on two-way packet satellite work. The station shown in Fig 18.11, using ordinary simple protocol terrestrial packet software, is quite adequate for Fuji orbit passes coming over the UK. However, a more sophisticated station is desirable for more dedicated work using orbits at longer range or using Pacsat (Oscar 16) or Lusat (Oscar 18). This is shown in Fig 18.12, but please note that for Pacsat/Lusat you need to use a Fuji 1200bps phase-shift keying modem in place of the 9600bps modem shown in this diagram. Equally important, while connections direct to the transmitter varactor and from the receiver discriminator are not necessary for Pacsat/Lusat, an automatic frequency control feed from the Fuji modem to the 70cm receiver is a 'must'.

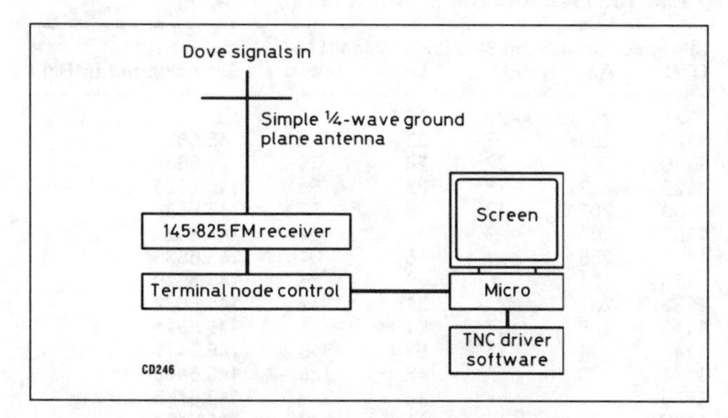

Fig 18.10. A Dove (Oscar 17) packet telemetry station

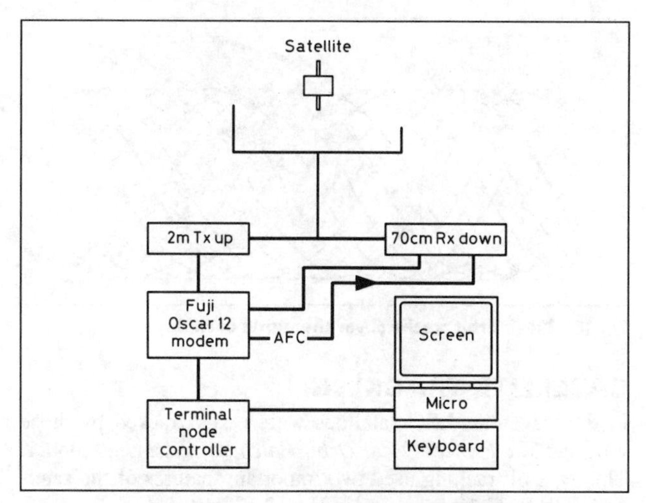

Fig 18.11. A simple station equipped for Fuji 2 (Oscar 20)

The station shown in Fig 18.12 is very complex. Indeed this sort of specialised satellite operation should not be attempted unless the station builder/operator has a year or two's experience of ordinary RS or Fuji communications operating. As shown in Fig 18.12, the Uosat 5 (Oscar 22) packet station requires special feeds to the transmitter varactor and from the receiver discriminator in order to avoid distortion of the wide-band 9600bps signal. Equally important, the operating protocol for accessing Oscar 22 (and Pacsat/Lusat) is very complex, requiring special software. So heavy are the software requirements of full-time Oscar 22/Pacsat/Lusat operation that many operators use special terminate-and-stay-resident software or, better still, multi-tasking software running under Desqview in order to concurrently access antenna control software, doppler tuning correction software, and last but not least the actual operating protocol software.

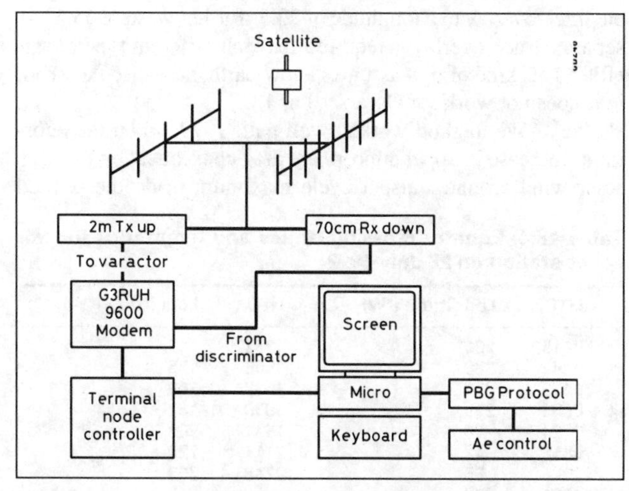

Fig 18.12. Complex 9600bps Uosat 5 (Oscar 22) packet station

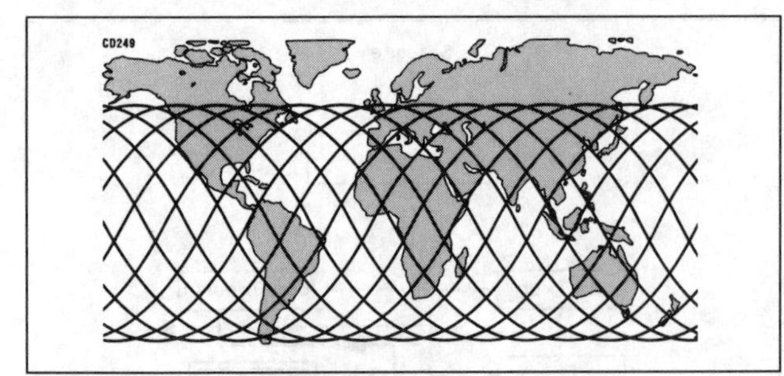

Fig 18.13. Mir orbit tracks cover the world each day

SATELLITE TRACKING

In the 'seventies LEO satellites were easily tracked by those who had *The Project Oscar Orbit Calendar* and a hand plotter. This type of tracking used two important features of the satellite's orbit which are shown in Fig 18.13.

As Fig 18.13 shows, one day's orbits form a distinct pattern of sine waves, each wave showing the ground track of a single orbit separated in time by the orbit period. In addition each sine-wave track is separated from the next by the approximate amount the Earth has turned while the satellite was doing one orbit. It is therefore possible to make a calendar based on the time and the longitude when the south-to-north-going orbit crosses the Equator. Table 18.4 lists Mir's Equator crossing times (EQXs) on a typical day.

Note the spacing pattern – roughly 92min and 23.5°. This can change slightly as Mir's height varies between 410 and 350km, as the thin atmosphere exerts a drag on the space station and brings it gradually lower. When Mir gets down around 350km the cosmonauts fire the engine and push it back to about 410km. The longer orbit time period and the greater longitude change per orbit occurs when Mir has the highest altitude.

Equipped with this information and a hand plotter as shown in Fig 18.14, the satellite operator aligns the transparency carrying the satellite track over the centre of the polar great circle map on the plotter to, say, an Equator crossing of longitude 42°W as shown in Table 18.4 for 0856 UTC, and reads off the 2min steps after the Equator crossing and the rough azimuth bearing as the satellite emerges above the ground station horizon. For example, as shown in Fig 18.14 the satellite is acquired about 12min after EQX, just after 0908 UTC at azimuth 194°, tracked through azimuth 180° at just after 0910 UTC, and lost at the horizon at azimuth 90° just after 0916 UTC. To follow the next orbit, allow for an EQX about 92min later and turn the transparent track overlay to a longitude of 23.5° further west, ie 65.5°. A separate track overlay is required for each different type of satellite. This kind of plot assumes a low-earth, near-circular orbit, ie it does not work for Oscars 10 or 13.

The above method worked well until 1979, when the enormous increase in upper atmosphere drag consequent on the high solar wind around sunspot cycle maximum made the printed

Table 18.4. Equator crossing, times and longitudes for Mir space station on 28 July 1992

UTC	Longitude (°W)	UTC	Longitude (°W)
0115	284	1332	112
0247	308	1505	135
0419	331	1637	159
0552	355	1810	182
0724	18	1941	206
0856	42	2114	229
1028	65	2246	253
1201	88		

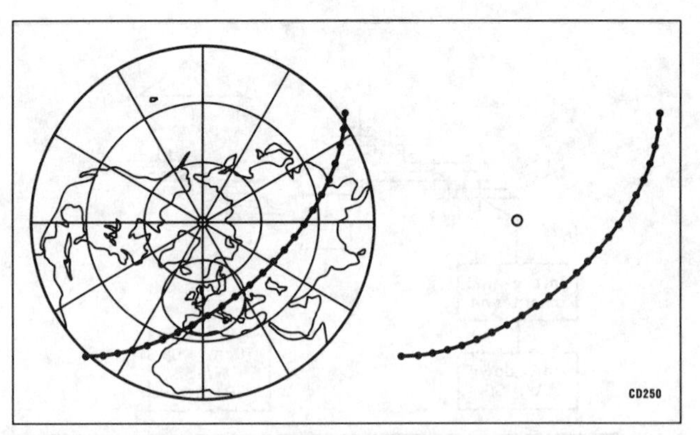

Fig 18.14. A plotter for Mir with separate track overlay

calendar long-term prediction hopelessly inaccurate. Help was at hand, however, in the form of the home computer which began to arrive in some radio amateur stations at this time. Table 18.5 shows a computer printout of Mir tracking data.

In this example Mir comes above the UK horizon a little after 1134 UTC. By 1136 it is at azimuth 251°, ie 251° clockwise measured from north in the horizontal plane, 19° south of west and at an elevation of 6° above the station horizon (very few stations have clear zero-elevation horizons, so satellite arrival time must be gauged with respect to actual station horizon elevation as seen from the antenna).

At 1140 Mir has already passed due south of the station. Then, just after 1144, Mir drops below the station horizon near azimuth 103°.

Latitude and longitude columns show the satellite earth sub-point location (Mir is actually about 400km above this point). The doppler column gives a coarse readout (due to single-precision calculation) showing the anticipated receiver tuning point for the Mir signal on the GM4IHJ station receiver.

Keplerian orbital elements

This data is derived from what are known as *Keplerian elements*. Roughly every two weeks Amsat packet bulletins list the elements which originate from the North American radar which tracks all satellites. NORAD passes on the data to the American space organisation NASA and it publishes the elements, which computer users then load into the tracking software in their computers. Table 18.6 shows a set of Keplerian elements for the Mir space station.

Elements are normally issued by NASA in this form, usually described as *two-line elements*. The sequences of numbers in this example have the following meanings:

Table 18.5. A computer printout of Mir tracking

Mir space station on 30/7/92 at GM4IHJ

UTC	Az	El	Lat	Long	Sig + doppler (MHz)
1122	263	–30	16	74	na
1124	263	–27	22	70	145.5527
1126	263	–22	28	64	145.5534
1128	262	–18	33	58	145.5528
1130	261	–13	38	51	145.5535
1132	260	–8	42	44	145.553
1134	256	–1	46	34	145.5536
1136	251	6	49	25	145.5529
1138	232	23	51	12	145.5532
1140	166	37	51	1	145.5511
1142	114	14	50	348	145.5477
1144	103	2	48	336	145.5466
1146	100	–4	45	328	145.5465
1148	97	–10	41	318	145.5464

Table 18.6. An example of Keplerian orbital elements

```
1 16609U 86 17 A 92203.46053722 .00018304 00000-0 29540-3 0 4465
2 16609 51.6235 075.3135 0002936 331.3180 028.7615 15.52944027367610
```

(a) The first number '1' is the line number.
(b) The sequence '16609U 86 17 A' relates to official designation number of Mir, object in space 16609, launched in 1986 as the 17th international launch of that year and 'A' which signifies it was the first item of that particular launch.
(c) The sequence '92203.46053722' reports that all this data applies to Mir at epoch year 92, day of the year 203 and time UTC (decimal day) 0.46053722.
(d) The next term '.00018304' reports orbit decay rate in revolutions per day squared.
(e) The sequence of zeroes indicates that a second-order decay term is not reported for Mir.
(f) The sequence '29540-3' reports either Bstar drag or radiation pressure term.
(g) The last four-number sequence '4465' reports element set number 446, checksum 5.
(h) The number at the beginning of the second line, ie '2', is the line number.
(i) Next we have a repeat of the satellite object in space number 16609.
(j) Then comes 51.6235 orbit *inclination* degrees with respect to the Equator, followed by 75.3135 degrees, the *right ascension* of the orbit ascending node.
(k) '0002936' reports the *orbital eccentricity* of 0.0002936 (circular = 0.000)
(l) Next comes *argument of perigee* of 331.318°.
(m) Position of Mir around its orbit at this epoch time is reported as *mean anomaly* of 28.7615°.
(n) Number of orbits/day, *mean motion*, is 15.52944027 (always eight decimal places).
(o) '36761' is the number of orbits Mir has made since its launch.
(p) The last number in the second line is the checksum.

Most Amsat software requires:

1. *Epoch* – year and day of year.decimal day
2. *Inclination* – degrees anticlockwise between equatorial and orbit planes
3. *Right ascension* (RAAN) – degrees between ascending node and First Point of Aries
4. *Eccentricity* – a ratio. Circle = 0.0000 but true circular orbits are very rare
5. *Argument of perigee* – degrees round orbit from ascending node to perigee
6. *Mean anomaly* – degrees round orbit from perigee to satellite at epoch time
7. *Mean motion* – number of orbits satellite makes in 24h day

The other elements are only used in very complex programs. Fig 18.15 shows the orbit features the elements refer to.

All this computer hardware and publication of Keplerian elements had been available for some time when Amsat introduced the Phase 3 series of high elliptical orbiters. This was fortunate because it is nearly physically impossible to track the complex long-duration orbits of Oscar 10 and Oscar 13, the first Phase 3 satellites, by any sort of manual method. Today almost all radio amateurs use computer prediction of satellite orbits, and many amateurs use sophisticated automatic computer-driven antenna systems for 'hands off' tracking – something which is particularly important if the operator is simultaneously trying to tune

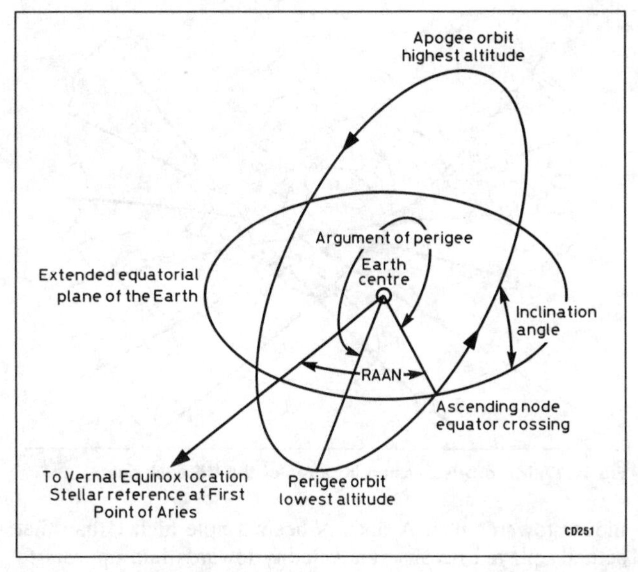

Fig 18.15. Keplerian elements

satellite signal doppler shift and use keyboard inputs to a complex digital data protocol.

Software for satellite tracking is available for some but not all types of computer. Do make sure that software is available before you buy a particular type of computer. The specialised amateur radio software market is supplied almost exclusively by radio amateurs selling their work through and in support of Amsat societies. Software can be obtained from AMSAT UK and AMSAT NA (see later for addresses).

DOPPLER SHIFT

Consider man A standing by the railway in Fig 18.16. Some distance from him is a bugler on a stationary wagon. Man A hears the exact note that the bugler plays. However, if the wagon

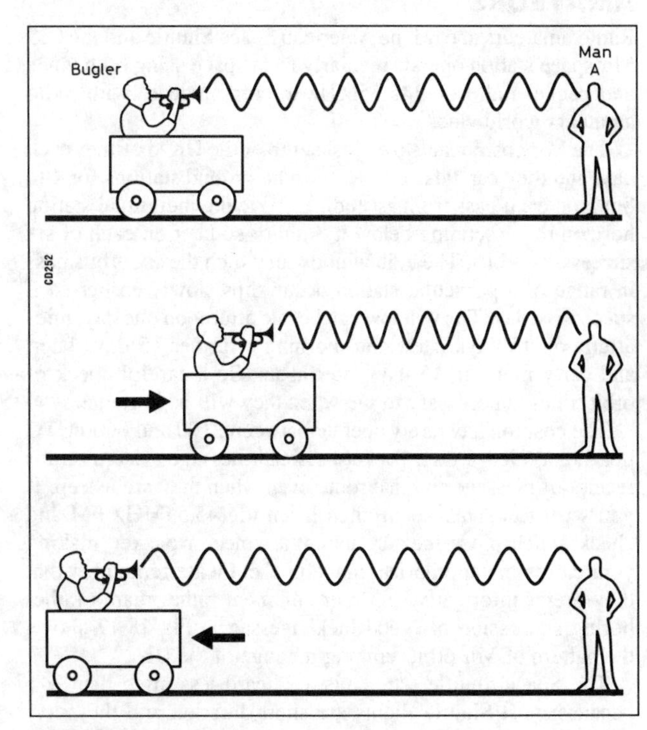

Fig 18.16. Doppler shift

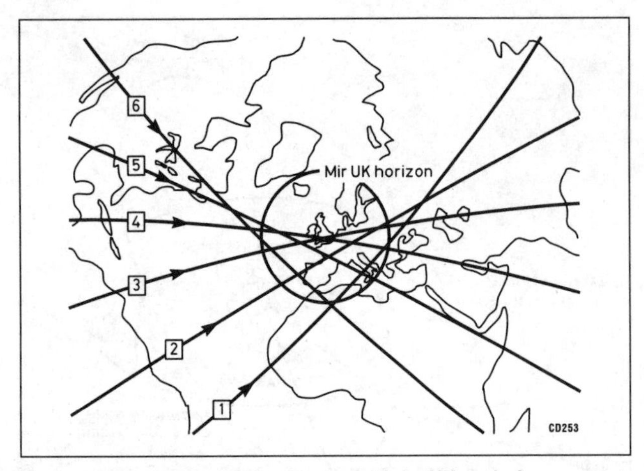

Fig 18.17. Mir orbits passing in range of the UK each day

moves towards man A, then A hears a note higher than that actually played because the velocity towards him appears to shorten the bugle note wavelength and to increase the bugle note frequency. Conversely, if the bugler moves away from man A, the latter hears a lower note than the one actually played. This effect is called the *doppler shift*, and the frequency shift of any sound or radio signal coming from an object moving relative to the observer is proportional to the velocity difference between the signal transmitter/bugle and observer and the normal speed of transmission in the medium (which in the case of radio waves is assumed to be the speed of light).

A satellite coming towards you is heard at a higher frequency (its transmission plus the doppler). A satellite going away from you is heard at a lower frequency (its transmission minus the doppler). So signals from high-speed satellites require careful tuning to match the received dopplered signal.

COSMONAUT/ASTRONAUT RADIO AMATEURS

Radio amateurs aboard the American Space Shuttle and the CIS Mir space station operate regularly from space using both voice and packet radio on 145.55MHz to communicate with radio amateurs worldwide.

The Mir cosmonauts pass just south of the UK six times each day, and they can talk to amateur radio ground stations for 4 to 8min on each pass from ascending above the individual station horizon to descending below it 8min or so later on each of six successive orbits. The 7.5h window in which the six orbits pass in range of a particular station occur slips slowly earlier each successive day. The window can be late afternoon one day, mid-afternoon 15 days later, mid-morning a further 15 days later, and early morning 15 days later again. So a careful check of orbit times is necessary to see when they will be in range.

The cosmonauts rarely operate between 2100 and 0600 UTC but they do leave their packet personal message system on for receipt of messages by that route even when they are asleep. If you wish to contact them, then listen to 145.55MHz FM and check which mode they are using, ie voice or packet, making your call in the appropriate mode in use. Please remember that they prefer informative news or comment rather than a rather boring succession of 'good luck' messages. Fig 18.17 shows the pattern of Mir orbits coming in range of the UK.

The Space Shuttle astronauts are heard less often than Mir because most Shuttle flights are short duration and they normally take the 'fuel economy near-equatorial' launch track, inclined at 28° to the Equator, which never brings them into range of the UK. Only about one flight in four takes a higher orbit inclination – which, depending on height, must be above 37° inclination if it is to bring Shuttle signals to UK radio amateurs. Shuttle flights over the UK, though rare, do provide an excellent mix of voice communications, packet radio, and slow-scan TV pictures. Shuttle flights carrying radio amateurs are usually described as *SAREX flights* (Shuttle amateur radio experiments).

Fig 18.18. Helen Sharman

The FM capture problem

Both Mir and Shuttle communications suffer from two distinct problems. First, the spacecraft operator can see over, and hear from, almost half a continent, whereas the ground station only hears the spacecraft and perhaps a few local stations inside his or her VHF horizon. So your signal may be one of a hundred all trying to talk to the space station at any one moment. This results in a phenomenon called *FM capture* whereby only the very strongest of signals gets through the interference and is heard in space. Secondly 145.55MHz FM is normally a ground station 'chat frequency'. This can produce unintentional interference. Most operators are prepared to wait for 8min to give Mir/Shuttle a clear frequency *if they know what is going on* and are asked politely to stop transmitting briefly.

Helen Sharman, GB1MIR

Helen Sharman, GB1MIR, gave schools across the UK excellent voice contacts from Mir in May 1991. Other contacts with Mir occur almost daily, while pre-arranged schools contacts with the Space Shuttle when it was near the UK have been a great success.

Satellite telemetry

Since the launch of the very first amateur radio satellites, it has been common practice to have the satellite transmit details of its internal 'health' back to Earth. These telemetered reports included details of battery charge, solar cell currents, system operating temperatures and local space environment conditions, thereby allowing ground station controllers to keep track of the satellite's health and welfare, and be appraised when it became necessary to send up commands to the satellite to alter one or more of its operational parameters. A typical example of this is *eclipse safety*. Many LEO satellites such as RS12 annually experience a month or so of orbits in which the satellite spends up to half the orbit in Earth shadow. To allow transponder operation to continue at these times when its solar panels are not seeing any sunlight would rapidly exhaust the satellite's power back-up batteries. Telemetry readings of battery data regularly inform controllers of the development of situations of this kind, allowing action to switch off the transponder before the batteries are totally exhausted. During the 'eighties, as controllers and designers got more experience, satellites were fitted with passive magnetic torqueing. When they noticed that solar cells were not seeing much sunlight, controllers could order current flow in coils inside the satellite such that the interaction between the magnetic fields thus generated and the magnetic field of the Earth produced a turning moment at the satellite, reorienting the solar panels to a better Sun angle.

Automatic telegraphy

Early telemetry signals used morse code. Automatic telegraphy was introduced for telemetry downlinks in the 'seventies and by the mid-'eighties a variety of automatic formats were being used. This development has continued in the 'nineties with complex auto-telegraphy and digital data transmissions being used, not only to report satellite state but also to report details of the satellites external environment and, in the case of the Sara satellite, long-range detection of radio signals from as far away as the planet Jupiter. Some satellites such as Dove (Oscar 17) and Uosat 2 (Oscar 11) are in everyday use as educational experiments, providing environmental readings from space which can be used in schools and universities for a wide range of quite complex experiments. Table 18.3 shows a sample block of raw data plus an example of the decoding of one of the 57 boxes reported by Dove using 1200bps AX25 packet protocol. This type of telemetry is used on most modern Amsat Oscars.

Software for decoding/displaying telemetry

Some excellent software for IBM PC compatibles is available from AMSAT UK and AMSAT NA (see below for addresses).

GETTING STARTED ON SATELLITES

The first requirement is an HF receiver and simple antenna. Tune around 29.357 or 29.408MHz until you hear a morse code beacon signing 'HI HI RS10 ' or ' HI HI RS12'. Then tune through the 40kHz above the beacon listening for CW or USB from stations communicating through the satellite.

If you have a scanner or FM receiver tuning 144/146MHz, try listening on 145.825MHz FM to see if you hear the telemetry from Dove or the digital voice and telemetry from Uosat 2. You only require a very simple 2m antenna for their signals. Equally strong is Mir on 145.55 using packet radio or voice, but you will need a good narrow-beam antenna in order to hear the more-distant Oscar 10 and Oscar 13 – look first for their beacons around 145.81MHz, then tune up the band slowly from the beacons, looking for CW up to 145.9MHz and voice from 145.9MHz to 145.96MHz approximately

To go beyond this very simple approach, you will need information on satellite orbits and times of passes in range of your station. Your local Amsat branch is the best place to get this.

AMSAT

'Amsat' is the name used by the group of volunteer national bodies who co-ordinate the activities of amateur radio satellite builders, controllers and users around the world. These national agencies are the logical point of contact for all would be satellite users:

Amsat UK, 94 Herongate Rd, Wanstead Park, London, E12 5EQ

Amsat NA, 850 Sligo Ave, Silver Spring, Maryland 20910, USA

Amsat DL, Holderstrauch 10, 3550 Marburg 1, Germany

Amsat Belgium, Thier des Critchions 2, B-4600 Chenee, Belgium

Amsat SA, PO Box 13273, Northmead, 1551, South Africa

Send them a large SASE or several International Reply Coupons for details of membership and available services.

Fig 18.19. Photo of GM4JJJ EME antenna

Equally useful for Amsat information is packet radio. UK packet bulletin boards carry regular reports of all operational amateur satellites. Beginners who are puzzled by the strange language and terms used in satellite work are welcome to address their packet radio queries to GM4IHJ @ GB7SAN (ie through the Glasgow bulletin board).

EME: BOUNCING SIGNALS OFF THE MOON

In 1946 a US Army radar successfully received echoes of its signals from the Moon. By 1960 radio amateurs were using the Moon to reflect messages from one continent to another and, by the 'nineties, *EME* (*Earth-Moon-Earth*, also known as *moonbounce*) facilities were in daily use at amateur radio stations using modest transmitter powers (below 1kW) and single long Yagi antennas.

Any time the Moon is above your horizon, you can receive lunar reflected signals by tuning the bottom-end frequencies of most VHF, UHF and microwave amateur radio bands. A typical station at GM4JJJ near Dunfermline, Scotland, shown in Fig 18.19, uses 450W of CW to the 4 by 12 long Yagis to communicate with stations in Japan, America, Africa and Europe.

Seen from the Earth, the Moon is only half a degree wide. Given a typical EME antenna comprising four long Yagis at 144MHz, the half-power points of the antenna beam will produce a cone of about 10° angle at its apex. Hence less than 1/400th of the radiated signal will hit the ½° circle of the Moon. In addition, the Moon reflects only 6% of the signal which hits it and the path from earth to Moon is a very long one, varying between 350,000 and 420,000km one way. Path losses average 252dB at 144MHz to 277dB at 2304MHz. So high-power transmitters and big antennas are necessary for even narrow-band CW signal echoes. Worse still, the signal's transit through the Earth's ionosphere results in unpredictable Faraday rotation of its polarisation. Vertically polarised transmissions rarely come back as vertically polarised echoes. So hearing your own echoes can be a matter of luck. Equally important, the uplink signal can be degraded by scintillation, then further degraded by irregular reflections from the Moon's surface. This irregular reflection is further complicated because it continuously changes as the aspect of the Moon from your station changes by a phenomenon known as *libration*. Libration can be both a help and a hindrance, producing deep fades when you hear nothing, followed by periods of marked signal enhancement.

Despite the difficulties mentioned above, EME is becoming increasingly popular. You can try EME for yourself even if you have only a single long Yagi with no elevation. Pick one of the EME Weekends which are scheduled every year (see the ARRL magazine *QST* for early warning of EME Weekends). Catch the Moon as it comes above your horizon so that the received signal may be enhanced by ground reflection and tune the lower 30kHz of the band just above 144.000MHz. You should hear EME from one of 'the big guns'. These are stations using very high power and massive arrays of 40 or more long Yagis.

Useful EME software is available from Amsat sources. Fig 18.20 shows a typical map produced by one of GM4IHJ's programs. The time is 0600 UTC on 22 July 1992. The Moon is over West Africa, just north of the Equator, and its footprint edge goes through Siberia, India, down to Antarctica, then west of South America, and ends by just missing California, British Columbia and Alaska. All Europe and Africa, all western Asia, South America and North America east of the Rocky Mountains are potential targets for the UK moonbouncer who aims his or her antenna at azimuth 191° and elevation 46°. The line at the bottom showing local time indicates that stations east of the UK are awake but west of the UK most of America is still asleep, although in places like Arizona the Moon will be on the eastern horizon at about 2200 local time, so good signals may be obtained from there.

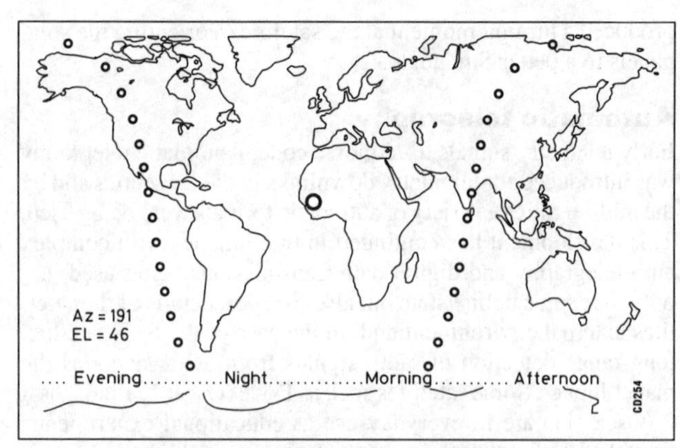

Fig 18.20. Lunar footprint at 0600 UTC on 22 July 1992

As the Moon moves across the UK sky, an alert EME operator may catch New Zealand as the Moon comes above the UK eastern horizon, and Hawaii as the Moon sets on the UK western horizon, with almost all stations in between at the appropriate times at which they enter the *lunar footprint* – that part of the Earth's surface seen from the Moon.

FURTHER READING

Space Radio Handbook, John Branegan, GM4IHJ, published by the RSGB.

Satellite Experimenters' Handbook, Martin Davidoff, K2UBC, published by the ARRL.

These books are available from RSGB Books, Lambda House, Cranborne Rd, Potters Bar, Herts, EN6 3JE.

19 Image techniques

THE TERM *image techniques* as applied to amateur radio encompasses quite a wide selection of differing modes, which include in the main *fast-scan television* as in commercial TV (FSTV), *slow-scan television* (SSTV) and *facsimile* (fax). Another mode which is fast becoming popular is *remote imaging*, which is essentially the reception of weather satellite images. However, this mode is more to do with the reception of extremely low-level signals, rather than with the technicalities of television or image techniques themselves.

SSTV and fax are narrow-band modes, which means that signals in these modes can be transmitted on normal voice channels. You can thus reach with SSTV and fax signals wherever you can reach by voice mode; in other words, on appropriate bands they are worldwide communication modes.

FSTV, on the other hand, is a wide-band mode, and as such is essentially a 'local' communication system although, when propagation conditions permit, relatively long-range contacts can be established over several hundreds of kilometres. However, it must be said that the quality of received pictures may not always be up to broadcast TV standards.

SLOW-SCAN TELEVISION

Slow-scan television has changed greatly since it was invented by Copthorne Macdonald and first used by radio amateurs many years ago. Nowadays, we have a plethora of systems within the mode and a variety of equipment that can be used to transmit and receive SSTV pictures. The original idea behind SSTV was to find a method by which a normal wide-band television picture could have its bandwidth reduced, so as to allow its transmission over a single-channel voice communication system. This meant that a typical (at that time) 3MHz wide television signal had to be reduced to around 3kHz – around a 1000 to 1 reduction in bandwidth! Nowadays, this reduction is even greater if a colour picture source is used, as this generally has a bandwidth around 5.5MHz. Because of this severe narrowing of the bandwidth, the system is only suitable for the transmission of still pictures – moving images are the preserve of FSTV, to be described later in this chapter.

To reduce the bandwidth of a television signal, both the horizontal (line) and vertical (frame or field) scanning rates must be reduced to as low a frequency as possible. At the outset, it was decided that both the line and frame frequencies could be conveniently derived from the domestic AC mains supply (50Hz in the UK). The basic frequencies and parameters of the system are shown in Table 19.1. It can be seen that the line frequency of 16.6Hz is obtained by dividing the mains frequency by three (in countries with 50Hz), and the frame frequency of 1/7.2Hz by dividing the mains frequency 50Hz by 360. In countries using a 60Hz mains supply different division ratios are used to arrive at the same standards.

The above refers to the 'original' SSTV concept but, in view of the fact that SSTV is increasingly using digital techniques and computer systems, there is a tendency to increase the number of lines to 128, which is a convenient binary number (10000000). Most SSTV monitors and receivers of the 'old' standard will decode pictures using 128 lines with no noticeable effect other than the picture being slightly larger on the screen. A full review of all the current SSTV modes can be found in reference [4].

Table 19.1. SSTV standards

Parameter	50Hz mains	60Hz mains
Line speed	16.6Hz (60ms)	15Hz (66ms)
Lines per frame	120 or 128	120 or 128
Frame speed	7.2s or 7.68s	8s or 8.53s
Aspect ratio	1 to 1	1 to 1
Scanning direction:		
Horizontal	left to right	left to right
Vertical	top to bottom	top to bottom
Sync pulse duration:		
Horizontal	5ms	5ms
Vertical	30ms	30ms
Subcarrier frequency:		
Sync	1200Hz	1200Hz
Black	1500Hz	1500Hz
White	2300Hz	2300Hz
Required transmission bandwidth	1.0 to 2.5kHz	1.0 to 2.5kHz

The basics of a slow-scan picture

The composition of a single SSTV picture line of the above original standard is shown in Fig 19.1. In order to separate the spectrum of the synchronisation (sync) pulse as much as possible, the line sync pulse length is made 5ms. (The sync pulses tell the receiver when to start the picture; the line pulses start each line and the frame pulses start each complete 'frame' of the picture.) Analysis shows that such a pulse width has a base video bandwidth of 200Hz. The frame sync pulse is made much wider than the line to make it easier to separate the two pulses in an integrating circuit. The frame pulse is thus 30ms long, which is approximately the length of one horizontal line.

The aspect ratio of the picture of 1:1 is rather an inherited standard, which was originally chosen to suit the ex-radar, long-persistence, cathode-ray tubes first utilised in SSTV display

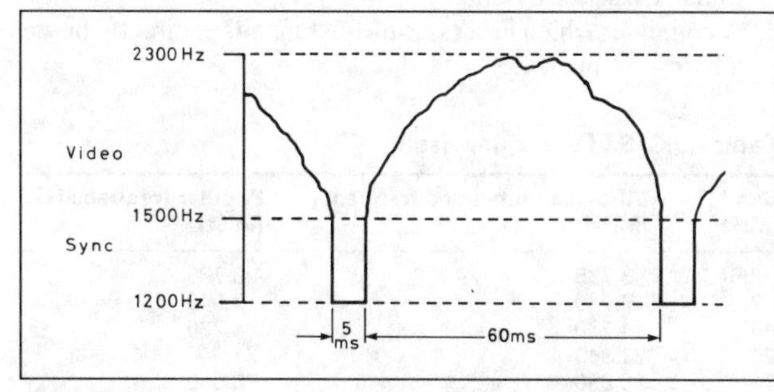

Fig 19.1. Frequency composition of a single slow-scan line

monitors. These tubes had round faces rather than the 4:3 rectangular faces of modern tubes, and thus a square picture rather than a rectangular one was more appropriate.

In order to avoid phase shift and drift problems within SSTV demodulators, the video information is modulated onto a subcarrier placed within the 3kHz SSTV spectrum. The subcarrier is frequency modulated by both video and synchronising signals. The basic video frequency at black level is 1500Hz, which rises to 2300Hz at peak white. The sync frequency of 1200Hz represents 'blacker-than-black', so that the visible raster is blanked out during *retrace*. This is when the spot returns from the end of the line on the right-hand side of the screen back to the beginning of the next picture line on the left-hand side. At the beginning of each frame the first 5ms line-sync pulse is replaced by the 30ms frame-sync pulse, during which the scanning spot resets from the bottom right of the displayed picture to the top left. Once again this flyback is visually suppressed. The placement of the three key modulating signals is shown in Fig 19.1.

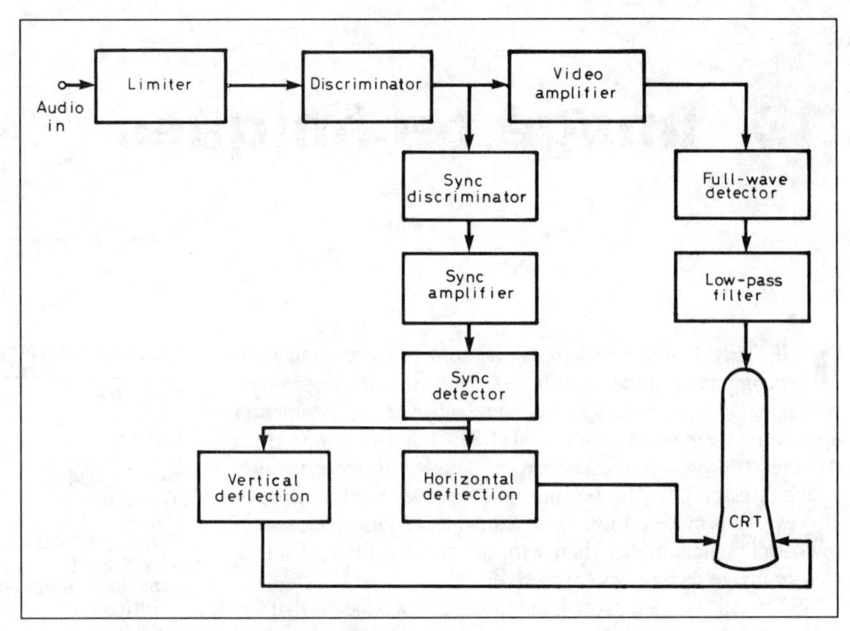

Fig 19.2. Block diagram of a slow-scan TV monitor

Receiving slow-scan television

SSTV pictures can be received using an ordinary communications receiver or transceiver covering the popular amateur bands. No modifications are required to the receiver, although the internal IF filter should be not less than 2.5kHz wide and 3kHz for preference. The SSTV signal is extracted either from an audio line output or from a headphone jack. These signals can then be either fed to the SSTV decoder or saved onto an ordinary domestic cassette or tape recorder for decoding at a later date.

Table 19.2 shows the amateur bands used for SSTV as well as the IARU recommended working frequencies and those frequencies which are perhaps most often used. The mode of transmission may be either SSB or FM.

Equipment

Broadly speaking, there are three methods by which a slow-scan picture can be displayed:

1. A conventional monitor containing an integral long-persistence cathode-ray tube (CRT), together with signal processing and deflection circuits.
2. A digital scan converter in which the received signal is digitised and stored in a memory. The memory is then scanned at fast-scan (625 or 525-lines) rate for displaying on a conventional TV set or monitor.
3. A computer which processes the picture either directly or via a hardware interface.

Table 19.2. SSTV frequencies

Band (MHz)	IARU recommended frequency (MHz)	Popular frequency (MHz)
3.5	3.735	3.730
7	7.040	7.040
14	14.230	14.230
21	21.340	21.340
28	28.680	28.680
144	144.500	144.500

The SSTV monitor

The basic principle of all SSTV monitors is the same. The audio signal from a receiver is a frequency-modulated subcarrier, and therefore FM detection is required. As with conventional FM receivers, a good limiter is required ahead of the discriminator to help eliminate AM noise etc and to present a constant amplitude to the demodulator.

The discriminator changes the FM signal to an AM one, but it is important to realise that the waveform still consists of a subcarrier centred around 1500Hz. The sync signal is recovered by using a tuned-circuit sync discriminator, which can be adjusted to accentuate its amplitude in order that both vertical and horizontal pulses can be recovered by threshold detectors.

The AM subcarrier signals will need to be detected to recover the original baseband video and sync signals. Full-wave rectification is most often used because the design of the necessary post-detection filter is eased.

The deflection circuits are controlled by the sync pulses to provide a raster on the screen in the same way as a conventional TV set. However, in simpler designs no internal generator is provided to scan out a raster in the absence of received syncs. When using this type of monitor one must take care that a stationary scanning spot does not burn the CRT phosphor, and ideally some form of spot suppression should be employed. The baseband video signal itself is filtered and used to bright up the CRT in the normal manner.

A block diagram of a slow-scan monitor is shown in Fig 19.2 and a design may be found in reference [1].

A digital scan converter

It is only in recent years that digital techniques and devices, especially memories, have come within the reach of amateurs. The main reason for adopting this technique in a receiver is to enable a domestic TV set (or monitor) to display the slow-scan pictures. The benefits will be realised by those who have experienced conventional slow-scan monitors as described in the previous section. With these the picture starts to fade at the top of the frame as it is being traced out, and because it is not possible to easily increase the overall brightness and persistence of the image, it is necessary to view pictures in subdued room lighting.

Also, the reception of colour transmissions is very difficult on such a monitor and necessitates several frames of each of the three primary colours, each viewed through an appropriate colour filter and recorded on film to photographically build up the full colour picture – this is hardly practical.

A block diagram of a simplified SSTV scan converter is shown in Fig 19.3. As with a conventional monitor, the incoming signal is limited to provide a constant amplitude signal. This is then passed to an analogue-to-digital (A-D) converter, the purpose of which is to digitise each line of information into 128 four-bit binary words. Two 512-bit shift registers act as buffer stores, the digital information for the first slow-scan line being stored in the odd-line buffer, which is controlled by a slow-speed clock. The storage operation takes 60ms, the length of a line. The second line is stored in the even-line buffer but, while this is going on, the first line is being loaded into the main memory bank.

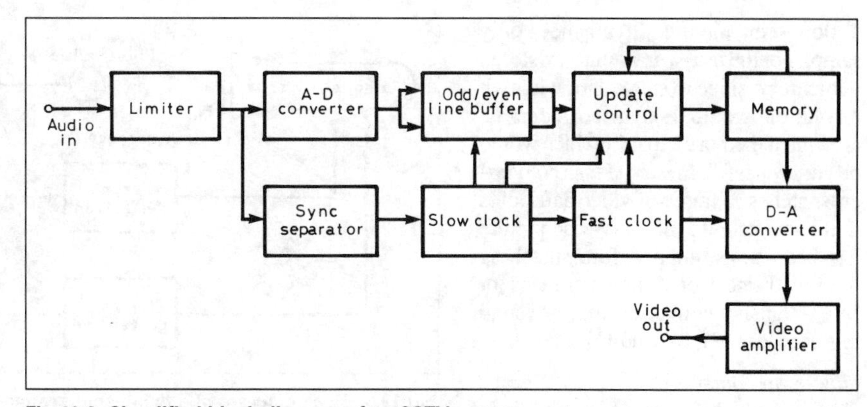

Fig 19.3. Simplified block diagram of an SSTV scan converter

The main memory consists of a set of four shift registers, each of which handles one bit of the four-bit word. Because the whole TV frame is made up of 128 lines, each containing 128 pixels, there will be a total of 16,384 four-bit words, and thus each shift register must have a capacity of 16k bits. The memory is continually scanned at fast-scan rate and the information passed to a digital-to-analogue (D-A) converter to bring it to a baseband video signal. The video is mixed with fast-scan syncs and either fed out to a monitor or used to drive a UHF modulator so that it may be viewed on a domestic TV set. A continuous picture is displayed on the TV screen and it stays there until a new SSTV frame is received, which slowly replaces the first picture as it progresses down the screen. Of course, any picture may also be stored on the screen for as long as you wish by disabling the input to the scan converter.

Naturally there is a difference in aspect ratio between the two TV systems: SSTV has a 1:1 ratio while FSTV has a 4:3 ratio. If this were left uncorrected it would result in distortion of the displayed picture, so arrangements are made within the scan converter to blank out the first and last eighths of each fast-scan line, resulting in a square SSTV picture framed by a completely black border on either side. This does not detract from the presentation of the picture on the screen.

Designs for SSTV receive converters appear in references [2], [3] and [4].

Computerised SSTV

Most personal computers, including of course the ubiquitous PC, are capable of being employed as SSTV receivers and picture generators. Most systems presently in use require some form of hardware interface between the computer and the radio receiver, apart from systems running on the Sinclair Spectrum, which are generally software-only SSTV receive systems. One of the main reasons for using a computer is to make use of the relatively large amount of memory available, as well as taking advantage of the excellent display facilities. Of course, once the information is in the computer's memory it is relatively simple to manipulate it in various ways to provide extra facilities and effects.

Transmitting slow-scan television

Sending SSTV pictures over the air is quite straightforward, and an ordinary HF or VHF transmitter or transceiver can be used – the combined SSTV signal is simply fed into the microphone socket. The band in which you are operating largely dictates the mode of emission and sideband convention, although both SSB and FM are used on VHF.

No modifications to the transmitter are necessary, but when using SSB there is one very important factor to be borne in mind. Transmitting ordinary speech using SSB means that the maximum power output is only reached at speech peaks, and therefore the duty cycle for the power amplifier is fairly low, enabling it to be run harder while not overheating. However, an SSTV signal transmitted in the same way produces a 100% duty cycle, due to the presence of the subcarrier. When transmitting slow-scan via SSB you must therefore turn down the audio gain so that the transmitter output stages are operating within their recommended limits.

Picture sources

The most popular slow-scan picture sources are computers, electronic pattern generators, keyboards, digital scan converters with fast-scan cameras, sampling cameras and flying spot scanners.

Computers

Computers can be used to generate both graphic pictures and text, which can of course be saved to disc for later use. As will be discussed later, computers can also form the basis of the SSTV system, rather than just being the picture generator.

Pattern generators

It is quite easy to make small logic circuits to produce such patterns as grey scale, chequer board, horizontal and vertical bars, grille etc but in their simple form these patterns are of only limited use because personalisation cannot easily be added to them. Such patterns nevertheless do have considerable value in providing test signals to help in aligning equipment.

Keyboards

The keyboard is an electronic typewriter, on which you can type a message which will be output as a combined slow-scan TV picture ready for transmission. These units were very popular some years ago and are still quite widely used. However, SSTV is a visual medium and it is generally considered bad practice to conduct an entire QSO using keyboards. They are, though, an excellent way of titling and captioning within a QSO.

Digital scan converters

The digital scan converter for transmission is similar in principle to the one described in the earlier section dealing with receiving SSTV. The transmitting system has to accept a fast-scan picture, store it and then scan the memory at SSTV rate in order to provide a slow-scan picture. In practice, only samples of the fast-scan picture are stored in memory, since to store all of it in high resolution would take a considerable amount of memory capacity which would be very wasteful. In use the scan converter 'snatches' a frame of video and stores it, then waits until the slow-scan picture has been transmitted before snatching the next frame. Construction projects for SSTV transmit converters may be found in references [3], [4] and [5].

Slow-scan cameras

Cameras which actually scan a vidicon tube at slow-scan rates are not often used these days. The tube essentially stores a latent image, so that a mechanical shutter may be used to 'freeze' the action at the beginning of a scanning period, with the image being read off during the remainder of the scan.

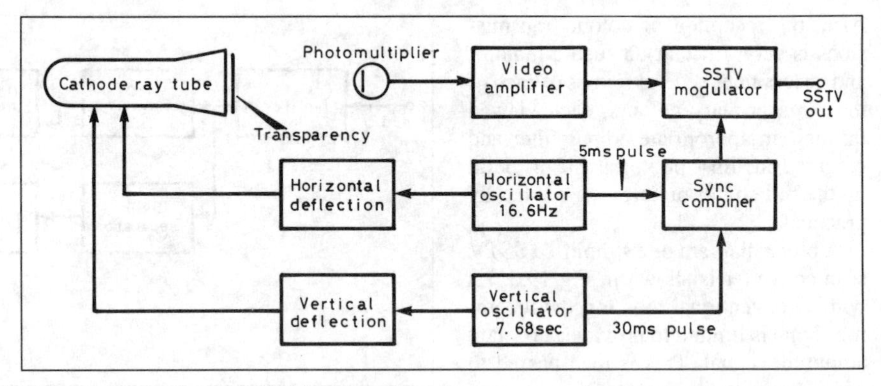

Fig 19.5. Block diagram of a flying-spot scanner

Sampling cameras

A sampling camera is basically a conventional vidicon camera, operating in a near-conventional manner, but having the actual video information sampled in order to produce a slow-scan picture. The normal (UK) fast-scan standards call for a line frequency of 15.625kHz and a frame speed of 50Hz. If one turns the camera on to its right side, then the 50Hz scan now becomes horizontal, and dividing this rate by three results in the correct slow-scan line speed. The modification required is normally to either drive the camera's frame sync circuit with external slow-scan line syncs or, in the case of a camera with internal synchronisation, alter the value of the timing circuit of the oscillator (usually by raising the capacitor value) to obtain the correct speed. A block diagram of a sampling camera is shown in Fig 19.4.

The fast-scan line (now running along the vertical axis) is sampled many times during the slow-scan frame period, and the resulting video is used to produce the slow-scan image. Of course, since a part of every fast-scan frame is sampled, it follows that in order to produce a complete slow-scan picture the subject must remain stationary for the duration of the slow-scan frame. The output from such a camera is fed direct to the transmitter microphone socket.

Flying-spot scanners

It must be said that equipping oneself with a live slow-scan camera is neither straightforward nor cheap. There is, however, one method of generating slow-scan pictures from photographs, slides or pictures which may appeal to the home constructor. This is by using a flying-spot scanner, and again the output of this unit is fed direct to the transmitter microphone socket.

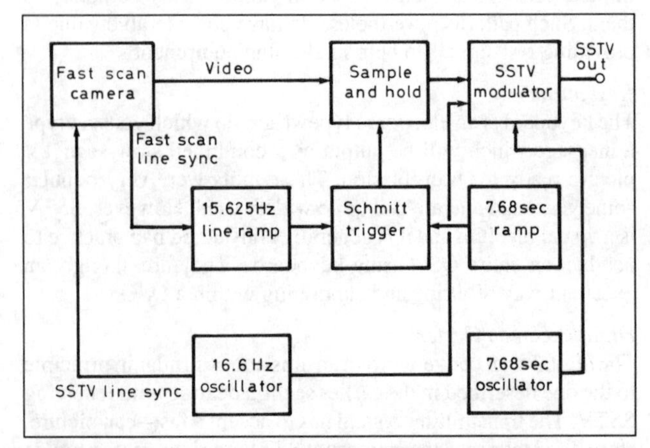

Fig 19.4. Block diagram of a sampling camera system

The principle of the flying-spot scanner is shown in the block diagram in Fig 19.5. With the aid of slow-scan deflection circuitry a raster is produced on the face of a small magnetic or electrostatic CRT. The raster is actually a fast-moving spot of light, which is used to scan the picture to be transmitted. This can be accomplished by placing a photographic transparency onto the face of the tube, allowing the light to shine through it and be picked up by a photo-multiplier. Another method is to use a lens system to focus the raster spot onto a photograph or drawing and pick up the reflected light with a photo-multiplier. In both methods the brightness of the light produced fluctuates depending upon the part of the picture being scanned.

A sensitive photo-multiplier tube, often a 931A, is used to pick up the light and convert it to a voltage which is proportional to the amount of light falling on its sensor. A photo-multiplier is generally used because it is so constructed that it provides a considerable amount of internal amplification, and thus the following amplification stages are kept to a minimum. It does, however, need a rather high voltage to operate (up to 1000V), and therefore many constructors prefer to use modern solid-state image sensors. A design for a flying-spot scanner may be found in reference [7].

Commercial systems

So far we have been discussing mainly home-brew equipment, but there are several commercially built items of SSTV equipment that are available both new and second hand. Manufacturers include Robot Research Inc, Davtrend Ltd and Wraase Elektronik, although other manufacturers continue to appear. Further information on commercial systems appears in references [3] and [4].

Software systems

Finally, we must mention the numerous software and computer-based SSTV systems available, especially as this method of receiving and transmitting SSTV is perhaps becoming the most widely used as time goes by.

Just about every home computer ever made has had a software or software/hardware SSTV receive or transceive system written and designed for it. However, during the past 10 years computer manufacturers have come and gone, leaving the stalwart few models whose pedigree has stood the test of time. The most notable computers for SSTV use in the UK are the Spectrum, BBC, Atari, Amiga and, rapidly coming up from behind, the PC and all its derivatives.

The Sinclair Spectrum, although rather limited in this application, can provide perhaps the simplest (and cheapest) SSTV station of all, especially since it may not need any hardware interfacing between the computer and the radio equipment. Commercial software packages available for this computer take advantage of this.

The Atari and Amiga computers are of course in a totally different league to the Spectrum, BBC etc. For a start, they are 16-bit machines, with greatly enhanced operating systems and vast (by comparison) amounts of memory available. As a consequence there are quite a few SSTV systems available for each. The Amiga computer in particular has superb graphics capabilities and excellent results should be obtained.

Last, but by no means least, the PC and its clones are now being supported with SSTV hardware and software, and this may yet become the most important computer for SSTV.

Further information on software systems is given in reference [4].

FACSIMILE

Facsimile (fax) is a process by which graphic information is converted into electrical signals which are transmitted by cable or radio and then reproduced exactly as the original. Until a few years ago the only method of operating fax available to the radio amateur was by converting ex-commercial equipment, which was essentially electromechanical in nature. While it is still possible to obtain, modify and use this equipment today, most amateur fax operation is conducted using computer-based systems. Software packages are available for most of the popular makes of personal computer, some of which have been covered above.

Much of the amateur activity with fax is actually in the reception of the various weather and news fax services around the world, rather than by reception and transmission of this mode between amateur stations.

FAST-SCAN TELEVISION

When TV amateurs started exhibiting at local and national shows many years ago, most of the general public had never seen a television camera, and the experience of actually viewing themselves and their friends on a TV screen caught the imagination. Nowadays, television has become part of our lives and is generally accepted without a second thought.

One of the main reasons why today's TV is taken for granted is that the technical side seems near to perfection. We have superb colours, hi-fi sound, world satellite links, slick programming and spectacular special effects. The techniques of programming and presentation are so polished that they are accepted without thought, which is how it should be.

Amateur fast-scan television (ATV) is perhaps one of the more fascinating and self-construction oriented modes in amateur radio today. While there is an amount of commercially made equipment for ATV this is mainly in the form of kits or units for the microwave bands. The starting place for all ATVers is 70cm, where relatively simple equipment can be built and contacts can be easily made. It must be mentioned here that it is not possible to use existing radio equipment for fast-scan television. Transmitters and receivers must be specially commissioned for this mode, whether it be for AM television on 70cm or FM television on 24cm or the microwave bands.

Fast-scan television is essentially a short DX mode, in that the average contact distance at 70cm is probably in the order of 30 to 50 miles or so for good-quality received pictures. It must be noted here that for a received picture to be of good, relatively

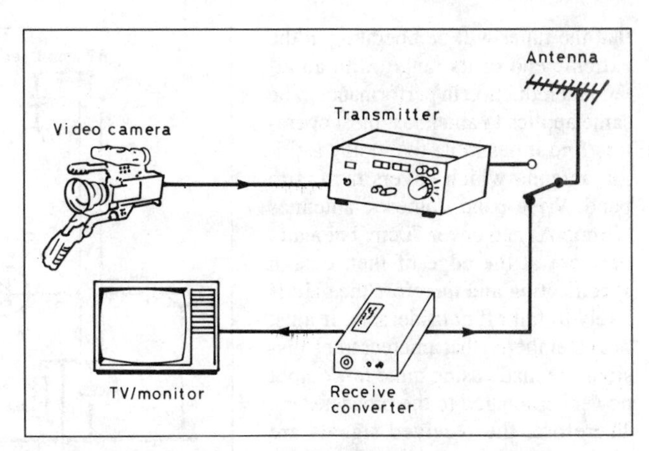

Fig 19.6. Block diagram of a basic FSTV station

noise-free quality, the received signal strength must be in excess of 20dB over S9 – much higher signal strength than for a noise-free phone contact!

In the UK, amateur television transmissions may be found on several of the UHF and microwave bands, the principal one traditionally being the amateur 70cm allocation (amplitude modulation around 436MHz – channel 17). The relatively recent introduction of ATV repeaters operating in the 1.3GHz band using FM has, however, prompted many stations to use that band as well for both repeater and simplex operation.

Intercarrier sound may be used as part of the vision transmission but, since the available bandwidth in the UK 70cm allocation is so restricted, it should never used on that band. (Similarly, transmissions should be restricted to monochrome only on the 70cm band). Instead, when operating 70cm ATV, 'talkback' is usually found in the 2m amateur band. There is an internationally agreed ATV calling channel on 144.750MHz but, owing to its popularity, stations usually move in multiples of 25kHz either side of the calling channel once a contact has been established. It is well worth monitoring this frequency in order to assess TV activity in your area. Another recognised 'talkback' and calling frequency, especially in mainland Europe, is 144.270MHz SSB. This method of talkback is also used with 1.3GHz transmissions and above, but there is sufficient space available on these bands to allow the use of full intercarrier sound, and this is becoming more widely used these days.

The station

An amateur TV station need be only as complex as the individual desires. Many stations employ no more equipment than a camera, a transmitter and a receiver. This basic system is adequate for normal communication and has the advantage of being easy to use and maintain.

The block diagram of such a basic amateur TV station is shown in Fig 19.6.

Amateurs who don't wish to put together a complete TV station often set up a simple receiving system in order to 'eavesdrop' on ATV pictures in their area. This can be very rewarding, especially if they live in a high activity area, and particularly when lifts in propagation conditions occur. Looking-in on a local ATV repeater is becoming a regular activity as the repeater network expands.

Receiving fast-scan ATV

There are several ways of receiving amateur TV on 70cm. Some domestic sets (particularly imported portable ones) will tune directly to 436MHz without modification. The only problem is that the tuner will be operating at the extreme end of its range, with an attendant reduction in performance. The same applies to antennas: most operators find it desirable to employ a single antenna which covers the entire band. Wide-band domestic antennas (Group A) can cover 70cm, but again they are at the edge of their design specification and therefore the gain is likely to fall off considerably. It must be remembered that amateur transmissions are made using quite low output powers compared to the broadcasters. Therefore, the received signals are likely to be weaker than broadcast ones and it makes sense to install the most sensitive equipment available.

If a domestic TV or VCR is to be used as a TV tuner, then purchase (or build) a good low-noise preamplifier to put in the antenna feeder. This will raise the level of the received signal and overcome the low gain in the domestic tuner.

The most effective way of receiving ATV pictures is to use a custom-built frequency converter. This changes the incoming signal to a frequency within the broadcast UHF TV spectrum, so that a normal, unmodified, domestic TV set can be used. ATV converters usually plug into the TV's antenna socket and a spare tuning button is used to tune in the amateur signals. There are a number of converters available on the amateur market, or alternatively the one described below can be built at home.

There are several commercially built 70cm antennas suitable for TV. The Jaybeam range (typically the 8/8 slot, the 18-element Parabeam or the range of Multibeams) are particularly suitable. Don't forget also to use good-quality coaxial cable. It is very easy to lose many decibels of signal in the cable, especially when some of the so-called 'low-loss' domestic TV cable is used (although there are some which are acceptable – especially if they are in very short lengths). Westflex 103 or Pope H100 cables are particularly suitable for more serious work, although these necessitate the use of large-style connectors.

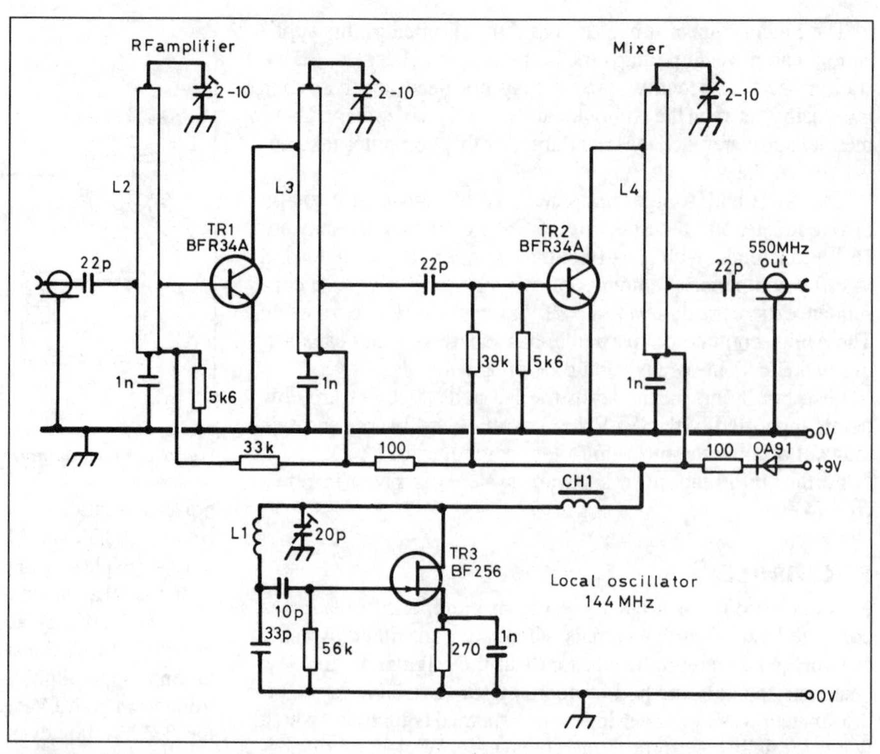

Fig 19.7. FSTV up-converter circuit

A 70cm ATV receive converter

The receive converter, as its name implies, converts the signals being received at 70cm to a frequency suitable for input to a standard 625-line UHF television set at approximately channel 36. Thus in this case we need an up-converter, converting our frequency around 435MHz to one at around 600MHz.

The converter described here was originally designed by G4DYP and has proved very popular over the years. A printed circuit board and crystal are available from the British Amateur Television Club, although the unit can be adequately built onto unetched copper laminate board, using small pieces of PCB material as mounting pads and connection points for the components. The design is simplicity itself, utilising no special techniques and not requiring any special test equipment to align.

The circuit diagram of the converter is shown in Fig 19.7 and the component layout in Fig 19.8. The 70cm antenna is coupled to the input tuned line L2 via a 22pF capacitor. The line is tuned to 70cm by the 2–10pF trimmer capacitor at one end. A second tuned circuit is used at the amplifier (TR1) output to further improve the selectivity of the converter and to provide a low-impedance feed to the mixer.

The mixer (TR2) collector circuit is tuned to the chosen IF and, although the tuned line L4 is the same size as L2 and L3, there is sufficient range on the 2–10pF trimmer capacitor to accommodate the higher frequency.

The local oscillator (TR3) employs a standard L/C circuit tuneable over an approximate range of 100 to 150MHz. Other frequencies could be used by varying the coil L1 and/or the 2–20pF trimmer capacitor. Local oscillator injection is somewhat unconventional, in that it relies on stray coupling into the mixer. This is achieved by the proximity of the components, especially L1, and for this reason the layout shown in Fig 19.8 should be closely followed if that method of construction is chosen.

All capacitors should be good-quality small disc or tubular ceramic types and resistors should be low-noise, quarter watt. Trimmer capacitors may be good-quality film dielectric or ceramic types. Choke CH1 is made by close winding as many turns of 34swg enamelled copper wire as will fit in a single layer wound onto a 1MΩ ¼W resistor. L1 is 6 turns of 20swg enamelled copper wire, 3/16in inside diameter and 3/8in long, and self-supporting.

If you are not using the printed circuit board available from the BATC, then take a piece of single-sided copper laminate board 2 3/16 × 1 11/16in and place it copper side uppermost. Now cut the pieces indicated by the shaded areas in Fig 19.8 from another piece of similar copper-clad board using a small saw. Glue these pieces copper side up onto the main board in the positions indicated, so that the copper is insulated from the earth plane. The small pads are ¼in square. Their positions on the board are not too critical, but should be close to those shown in Fig 19.8.

Once the circuit has been built and checked for short circuits etc, apply 12V DC and check that the oscillator stage is working correctly and will tune to the frequency required. This can be

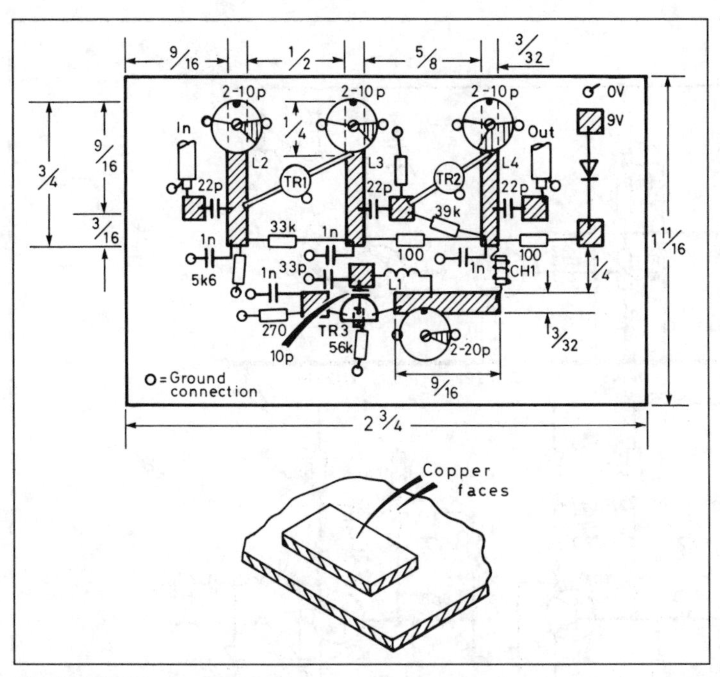

Fig 19.8. FSTV up-converter component layout

accomplished by closely coupling a wire loop to L1 and connecting it to a frequency counter. Alternatively, if a frequency counter is not available, closely couple L1 to the antenna socket of a 2m radio tuned to 145.000MHz and slowly adjust the oscillator trimmer until a signal is received. The oscillator is then set to around 145MHz. Connect the output of the converter to the antenna socket of a domestic TV set adjusted to a convenient channel around number 34 to 36. Switch on the converter and gently readjust the oscillator and mixer trimmer capacitors for maximum noise on the TV screen. The TV set may need to be slightly retuned to locate the signal from the converter.

Connect a signal generator set to 435MHz to the 70cm input of the converter and tune the input trimmer capacitors for maximum signal on the TV set, loosely coupling the converter to the TV set to reduce the signal input to the TV as necessary. Alternatively, if a signal generator is not available, a local amateur 70cm TV signal can be used. If a signal is not available, tune a 2m radio to 145.000MHz, key the radio (preferably into a dummy load!) and adjust the converter input trimmer capacitors for the third harmonic from the radio set, which will appear as a lessening of the noise on the TV set, eventually becoming a noise-free white screen as the signal is maximised.

Final tuning should be accomplished using a received amateur TV signal, reducing the received signal strength as necessary (by rotating the antenna off the transmitting station perhaps) until the converter is giving maximum conversion gain.

The finished unit should be housed in a screened metal enclosure fitted with good-quality (BNC or N-type) connectors.

The 1.3GHz band and above

The 1.3GHz amateur allocation (known as '24cm' by TVers) is being increasingly used in the UK, largely as a result of amateur television repeaters licensed and regulated by the Society. Although it is permitted to use normal AM transmissions on 24cm, for technical reasons it is more usual, and indeed advantageous, to use FM. There are a few commercially available receivers and transmitters available for 24cm and one or two in kit form for home building. Also, there are some very good designs and printed circuit boards available for receivers and transmitters from the British Amateur Television Club, full details of which can be found in references [7] and [8]. The main ATV activity in the UK above 24cm is at present on 10GHz (3cm), although there is interest growing in the little-used band at 2.4GHz (13cm).

Building ATV gear for the 3cm band is more akin to plumbing in many aspects, as we tend to use waveguide a great deal. A basic 15mW transmitter (it does not sound much, but the results that can be achieved are amazing – the world ATV 3cm DX record at present is 142km with just 15mW!) can be built from an ex-surveillance microwave head of the Solfan or Plessey variety, seen at most rallies for only a few pounds. The modifications are extremely easy and a suitable video and audio modulator is extremely simple to build. A receiver can also be fashioned using similar methods, but a specialised IF and FM demodulator must also be built.

Suitable designs and instructions for modifying a microwave head can be found in reference [7].

Transmitting fast-scan ATV

A special transmitter is required in order to be able to transmit fast-scan television. The reasoning for this is that television is a very-wide-band, spectrum-hungry mode. The average, well-filtered and controlled, 70cm ATV transmission will occupy around 4–5MHz of bandwidth, and a 24cm FM ATV transmission upwards of 16MHz, both being rather more than the less than 10kHz of conventional phone transmissions! Therefore, transmitters have to be designed which have very wide-band IF and PA stages. This is not particularly difficult and many such circuits abound. There are also commercially built ATV transmitters and kits available, as are projects and printed circuit boards from the British Amateur Television Club. Designs for a 70cm station can be found in references [7] and [8] along with the 24cm and 3cm projects as mentioned above.

When transmitting ATV on any band care must be taken so as not to cause any problems to other users due to the wide-band nature of the transmissions. This is especially significant on 70cm, where the UK allocation of 434–440MHz is just above the top of the repeater inputs, and any extra spread of a badly controlled ATV signal downwards could cause severe problems to local 70cm repeaters. Similarly, the amateur satellite communication area is to be found at the upper end of the 70cm allocation.

Picture sources for amateur television can be any form of fast-scan camera (such as the any one of the many ex-surveillance types seen at rallies), camcorder, VCR, computer, caption generator etc.

A 70cm FSTV transmitter

This transmitter design, although it has been around for some years, has been, and still is, very simple yet very effective. It is both easy to construct and align, and is capable of delivering up to 150mW of RF output in the 70cm band. It is intended as a low-power, self-contained transmitter or as a driver for subsequent linear amplifiers. A video modulator is included on the printed circuit board and requires only the addition of a 100Ω potentiometer to provide adjustment of the video input level.

A printed circuit board and crystal for this unit is available from the BATC. The circuit diagram and inductor details are shown in Fig 19.9 and Table 19.3. TR1 forms a crystal-controlled oscillator which operates at 108.875MHz. In order to ensure

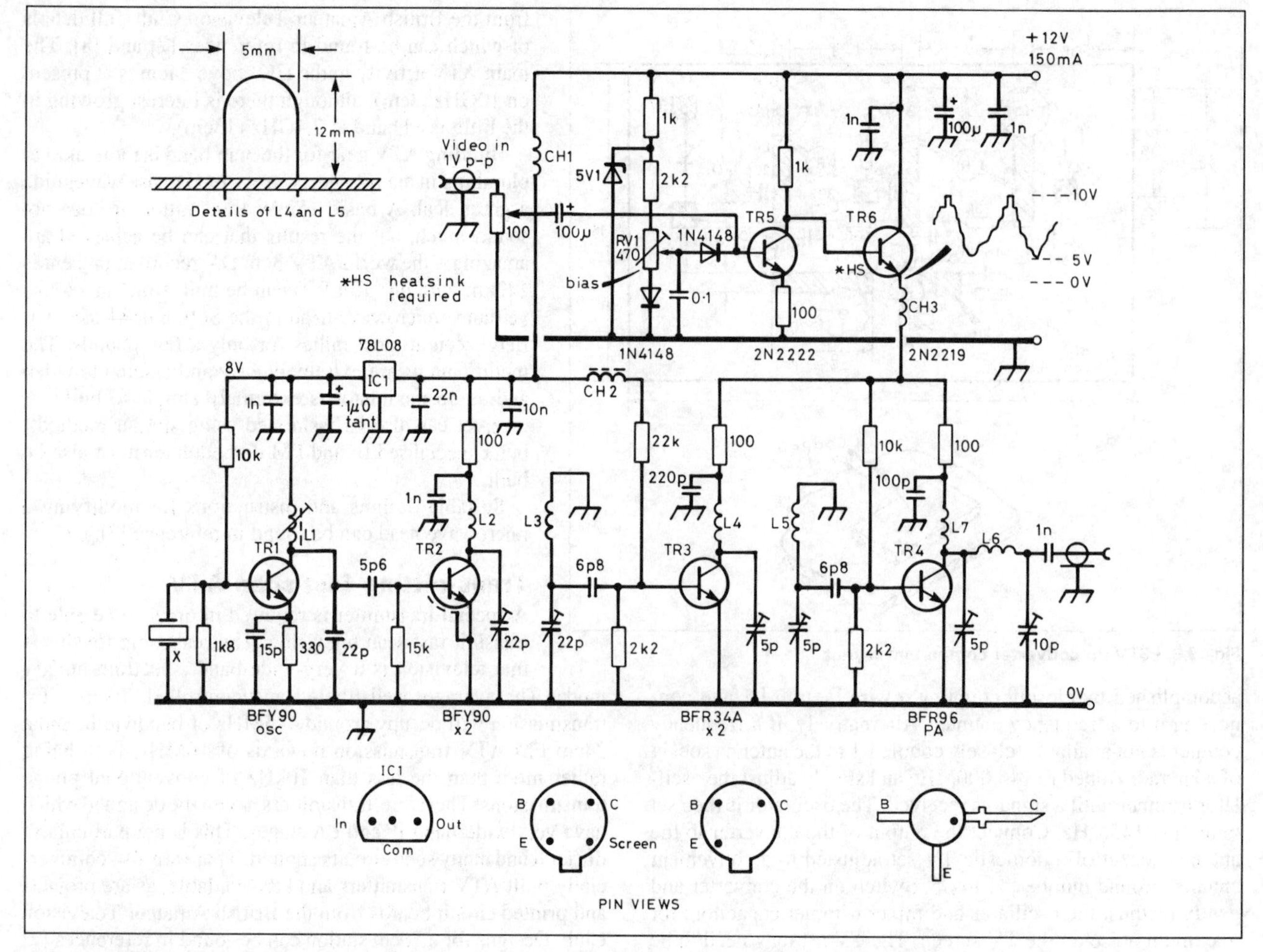

Fig 19.9. FSTV 70cm transmitter circuit, showing details of L5 and L6 plus transistor pin views

maximum stability and spectral purity the oscillator is powered from a three-terminal voltage regulator (IC1). The output is coupled directly to the base of TR2 which operates as a frequency doubler. The collector tuned circuit (L2) resonates at a frequency of 217.750MHz and, together with L3, forms a simple band-pass filter. TR3 is another doubler stage and brings the signal to its final frequency of 435.500MHz.

The collector of TR3 also connects to one half of a band-pass filter (L4) but derives its supply from the video modulator.

TR4 is the output amplifier and is also powered by the modulated rail. The collector connects to a simple pi-output stage which provides a low-impedance output suitable for matching into 50Ω.

Video modulation is applied to the base of amplifier TR5 via a panel-mounted 100Ω variable potentiometer which provides adjustment of the actual video level. TR5 base is biased from a potentiometer circuit fed from a zener diode stabilised voltage source. DC restoration is provided by a 1N4148 diode. TR6 acts as an emitter follower and delivers up to approximately 10V (modulated) to TR3 and TR4.

Table 19.3. Inductor details for the FSTV 70cm transmitter

L1	5½t 26swg enam Cu on 4.5mm former with core
L2, 3	1¾t 20swg enam Cu, 5mm ID, spaced 2mm from board
L4, 5	¾t 20swg enam Cu bent to shape as illustrated in Fig 19.9
L6	1¾t 20swg enam Cu or silver plated, spaced 2mm from board
L7	3t 26swg enam Cu, 3mm ID, close wound
CH1	10t 26swg enam Cu, wound on 3mm drill, self-supporting
CH2	3t thin wire on ferrite bead
CH3	8t 26swg enam Cu wound on 3mm drill, self-supporting

The component layout is shown in Fig 19.10. The double-sided PCB features a ground plane on the component side and, where possible, component leads which are connected to ground should be soldered to both sides of the board.

Although HC18 or HC25-U crystal packages are most often used for frequencies over 100MHz, provision has been made on the BATC circuit board for the use of the larger HC6-U crystal packages as well.

Trimmer capacitors should be good-quality PTFE film types. Try to use Mullard or DUA makes, as these are among the high-quality ones available. However, please note that this type of trimmer capacitor is not intended for lots of 'twiddling', and may become unserviceable if subjected to too many adjustments.

All lower-value capacitors are miniature plate ceramic types. One 100μF electrolytic is axial mounted, while the other is a vertical radial type. A small heatsink should be fitted to TR6.

L2 and L3 are wound in opposing directions and should be wound so as to fit the holes provided. The screen lead of TR1 may be either clipped off or soldered to ground on top of the board.

A suitably sized hole should be drilled to accept the former used for L1, which should be glued into position. Mounting holes ³/₁₆in in diameter should be drilled in the board at the locations for TR3 and TR4. These transistors are mounted on the print side of the board and carefully soldered to the tracks provided. Care should be taken to ensure that the devices are installed the

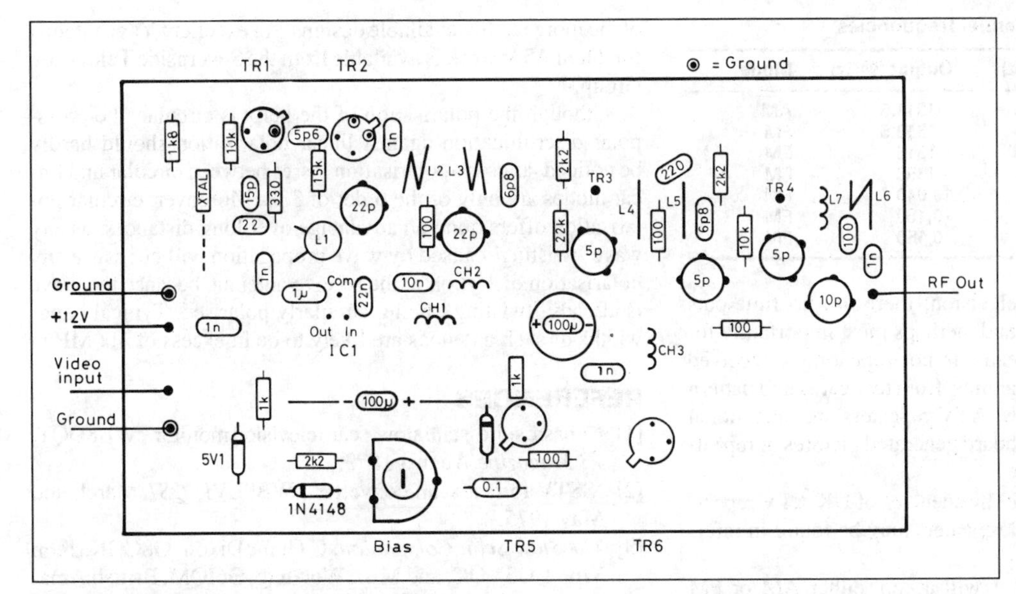

Fig 19.10. FSTV 70cm transmitter component layout

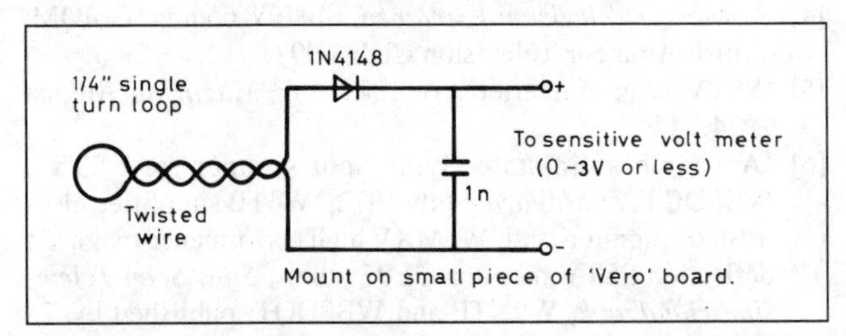

Fig 19.11. An RF 'sniffer' probe

correct way round. The printing on the transistor package should face downwards.

The completed unit should be housed in a screened metal box fitted with a BNC or N-type socket for the RF output. Power is fed in through a 1000pF feedthrough capacitor.

Alignment of the unit is straightforward and may be carried out using the minimum of equipment. Temporarily up-end CH3 and apply power to the unit. Check that there is +8V at the collector of TR1. Using the RF 'sniffer' probe shown in Fig 19.11, place the loop over the oscillator coil former and adjust the slug until the oscillator starts, indicated by a reading on the test meter.

Now place the probe near to L2 and adjust its trimmer capacitor for maximum indication on the test meter. While in this position re-peak the oscillator coil for maximum output, then withdraw the slug about a quarter of a turn. This should ensure that the oscillator starts readily. Switch the unit on and off several times to check that it does.

Remove the sniffer probe, reconnect CH3 and turn the BIAS control (RV1) fully clockwise. Set the VIDEO GAIN control to minimum. Make a test load/detector circuit as shown in Fig 19.12

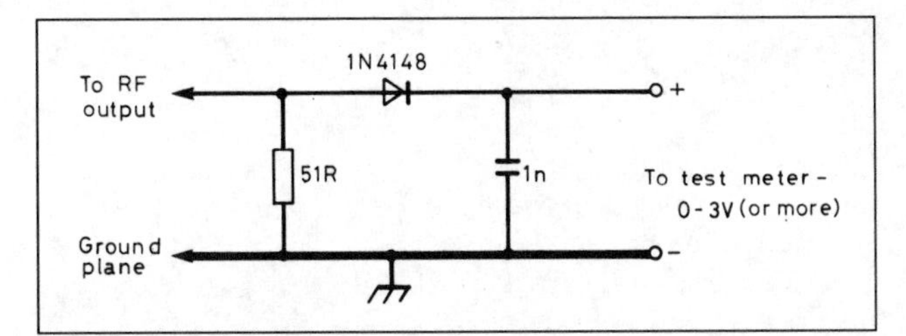

Fig 19.12. A test load and detector circuit

and connect it to the RF output socket.

Apply power and adjust L2, L3, L4 and L5 tuned circuits for maximum indication on the test meter. Adjust the pi-output tuned circuit for maximum output, by playing one capacitor off against the other until the final peak is reached.

At this stage variation in the output power should be noticed if RV1 is adjusted. If all is well, apply a video signal and turn up the VIDEO GAIN control. Turn RV1 slowly anti-clockwise, noting as you do that the output power falls. This indicates that video modulation is present. Do not be troubled if, when modulation is applied, the indicated power falls considerably. The power output indicated is the average video power and not the actual output power of the transmitter (or any AM transmitter). The true output power in the case of a video-modulated AM transmitter occurs when the sync tips occur, and can be measured by removing the video source. (Actually the true output power is approximately 95% of the power indicated with no video input and is referred to as the *peak sync power* or PSP).

The unit should now be producing around 150mW of PSP with a well-modulated video signal. The final setting of the VIDEO GAIN control is best set on-air, by increasing the video gain until the receiving station indicates that the white areas of the picture are being 'crushed', which indicates that the amplifier is being over-driven. Similarly, the BIAS control should be adjusted for best sync/video ratio, which is usually found to be with the control almost at the fully anti-clockwise position.

The transmitter has been carefully designed so that, even when adjusted as above without the aid of a spectrum analyser, all harmonics are better than 30dB below the carrier. However, if equipment is available the transmitter should be aligned for minimum harmonic content. Table 19.4 shows the typical harmonic output levels from this unit after correct alignment, where *f* is the carrier frequency.

UK ATV repeaters

As mentioned previously, there is a growing network of 24cm FM ATV repeaters in the UK. These repeaters allow much lower powered stations to operate without the necessity for complex antenna systems, large transmitter PAs or linear amplifiers. Essentially, ATV repeaters are much the same as their phone counterparts. However, there are three very important differences: to access an ATV repeater all you need is a fast-scan FM TV signal with the video conforming to the standard CCIR

Table 19.4. FSTV transmitter harmonic content

Harmonic (× fundamental *f*)	Level below carrier (dB)
0.5	–42
2	–44
3	–38
4	–42
5	–38
6	–41

Table 19.5. ATV repeater channel frequencies

Channel	Input (MHz)	Output (MHz)	Mode
RT1	1276	1311.5	AM
RT2	1249	1318.5	FM
RT2R	1249	1316	FM
RT3	1248	1308	FM
RT101	10,200	10,040	FM
RT102	10,255	10,150	FM
RT103	10,250	10,150	FM

specifications for 625-line television; there are no time-outs operating on ATV repeaters; and, perhaps most importantly for the constructor, when a repeater is not repeating a received picture it broadcasts its own pictures, from test-cards and pattern generators etc. In other words, ATV repeaters are operational 24h a day, either radiating on-board generated pictures or repeating incoming received pictures.

Table 19.5 gives the channel frequencies of UK ATV repeaters, and details of operational repeaters may be found in reference [9].

Repeater stations on RMT-1 will accept either AM or FM signals, while inputs on RMT-2 and RMT-3 are in FM only. Repeater output on RMT-1 is AM only, and the outputs on RMT-2 and RMT-3 are FM only.

The repeater video standard is 625-line CCIR fast-scan TV. A morse code identification giving callsign and location is present on the 6MHz audio carrier. FM transmissions can employ CCIR pre-emphasis as an option. Repeater access is by the presence of a standard 625-line video signal at the input.

Although the actual coverage of each repeater will depend on local geography, and in some cases antenna directivity, the average area covered is in the order of a 30km radius from the repeater. Apart from one or two exceptions, antennas are omnidirectional. *All* ATV antennas operate horizontally, regardless of band.

The widest frequency difference between an input and an output signal is 69.4MHz. A check on the specification of commercial antennas will show that most of them lack the bandwidth necessary for single-antenna ATV operation. Fortunately, there are one or two which are suitable for TV applications: the helix or helical beam and the quad-loop Yagi are two of the more readily available designs. An excellent Yagi antenna for 24cm ATV work is available from the Severnside Television Group.

Although the polarisation of the helix is circular, the cross-polar discrimination against linear polarisation should hardly be noticed, as cross-polarisation losses between circular and linear modes are only of the order of 3dB. Moreover, circular polarisation offers quite an advantage over long distances, as any wave 'twisting' caused by wave propagation will not cause any polarisation distortion at the receive antenna, because the wave is already 'twisting', being circularly polarised. Typical bandwidths of such antennas are likely to be in excess of 100MHz.

REFERENCES

[1] 'A basic solid-state slow-scan television monitor', WB8DQT, *73 Magazine* August 1973.

[2] 'SSTV to fast-scan converter', WB9LVI, *QST* March and May 1975.

[3] *The Slow Scan Companion*, C Grant Dixon, G8CGK, John Wood, G3YQC, and Mike Wooding, G6IQM, British Amateur Television Club, 1987.

[4] *Slow-Scan Television Explained*, Mike Wooding, G6IQM, British Amateur Television Club, 1992.

[5] 'SSTV scan converter', W0LMD, *73 Magazine* August 1974.

[6] 'A simple solid-state flying spot scanner for SSTV', WB8DQT, *73 Magazine* July 1972; 'W4TB simplified electrostatic monitor' and 'W6MXV high performance magnetic deflection SSTV monitor (5FP7 tube)', *Slow Scan Television Handbook*, W9NTP and WB8DQT, published by *73 Magazine*, 1972.

[7] *The ATV Compendium*, Mike Wooding, G6IQM, British Amateur Television Club, 1989.

[8] *An Introduction to Amateur Television*, Mike Wooding, G6IQM, and Trevor Brown, G8CJS, British Amateur Television Club, 1992.

[9] *RSGB Amateur Radio Call Book* published annually by the RSGB.

Note: the current address of the British Amateur Television Club can be found in reference [9].

20 Data communications

IT IS GENERALLY ASSUMED that data communication is a very modern aspect of amateur radio but the truth is that CW can be regarded as the very first amateur use of the data mode. However, the mode of operation first associated with data on the amateur bands is radio teletype (RTTY) which began to be used just after the second world war, using commercial or military surplus teleprinters. Very little then changed until the mid-'seventies when computers for personal use first became available and radio amateurs began to replace their noisy teleprinters with keyboards and TV screens. The computer revolution had begun.

At about this time a British amateur, Peter Martinez, G3PLX, worked on a new system for transmitting data over radio, and AMTOR (from 'amateur teletype over radio') came into being. The system is based on the commercial SITOR system and was the first type of amateur data communications to make use of error detection techniques, albeit in a very simple form.

Towards the end of the 'seventies a number of groups of amateurs in North America began experimenting with microprocessor and digital techniques to generate and process the data information. Up to the middle of 1982 no real co-ordination took place and a wide variety of protocols existed. (Some amateurs in the UK were also experimenting with what was to become packet radio, devising modes such as 'Cambridge' packet.) A standard was finally thrashed out in the USA based on the CCITT X.25 public packet switched-network protocol. Although not universally accepted immediately, Amateur X.25 (AX.25) became the standard for amateur packet radio. This new protocol and mode was destined to become the growth area in amateur radio in the 'eighties and 'nineties. It is easy to understand the attraction when a global amateur integrated network is in prospect to provide:

- Two-way terminal-to-terminal connections between your station and any other on the network.
- Computer file transfers between stations.
- The ability to leave messages for other stations anywhere in the world to be collected at their convenience.
- A variety of services such as databases available on host computers anywhere throughout the network.
- Automatic handling of third-party traffic through the network.
- Bulletins of matters of amateur radio interest available at all times.

Developments on data modes did not stop with packet radio. New protocols such as PACTOR and CLOVER, along with enhancements to existing protocols such as AMTOR, have continued. This, coupled with the advent of the DxCluster DX-spotting systems, has meant that data communications will continue to be one of the most exciting and developing modes in amateur radio.

Table 20.1. ITA2 and CCIR 476 codes, with AMTOR service information signals

Letters case	Figures case	ITA2 code Bit no 43210	CCIR 476 code Bit no 6543210
A	-	00011	1000111
B	?	11001	1110010
C	:	01110	0011101
D	WRU	01001	1010010
E	3	00001	1010110
F	ua	01101	0011011
G	ua	11010	0110101
H	ua	10100	1101001
I	8	00110	1001101
J	BELL	01011	0010111
K	(	01111	0011110
L	)	10010	1100101
M	.	11100	0111001
N	,	01100	1011001
O	9	11000	1110001
P	0	10110	0101101
Q	1	10111	0101110
R	4	01010	1010101
S	'	00101	1001011
T	5	10000	1110100
U	7	00111	1001110
V	=	11110	0111100
W	2	10011	0100111
X	/	11101	0111010
Y	6	10101	0101011
Z	+	10001	1100011
CR (carriage return)		01000	1111000
LF (line feed)		00010	1101000
LTRS		11111	1011010
FIGS		11011	0110110
SP (space)		00100	1011100
BLK (blank)		00000	1101010
AMTOR service information signals			
Control Signal 1 CS1			1100101
Control Signal 2 CS2			1101010
Control Signal 3 CS3			1001101
Idle Signal Alpha			0001111
Idle Signal Beta			0110011
Signal Repetition			1100110

RTTY

RTTY activity has developed from the surplus teletype or teleprinter type of system with all output on paper to systems using computers and VDUs, but the actual signalling system has changed very little.

Codes and signalling speeds

RTTY uses five pulses to present the letters, figures, symbols and machine functions. This five pulse code is commonly known as the *Murray code* in the UK and *Baudot code* in North America. The International Telegraph and Consultative Committee (CCITT) recognises these codes as the International Telegraph Alphabet No 2 (ITA2) which allows for national variations. The ITA coded character set is shown in Table 20.1.

Table 20.2. Characteristics of various common telegraphy systems

Speed	Baud rate	Units per second	Speed (wpm) (approx)	Start pulse length (ms)	Signal pulse length (ms)	Stop pulse length (ms)
60 speed (USA)	45.45	7.42	61.33	22	22	31
European/ CCIT standard	50.00	7.50	66.67	20	20	20
75 speed (USA)	56.88	7.42	76.67	17.57	17.57	25
100 speed (USA)	74.20	7.42	100.00	13.47	13.47	19.18
75 baud (European/ CCIT)	75.00	7.50	100.00	13.33	13.33	20

As the ITA2 is a five-pulse code with only two conditions for each pulse (binary 0 or 1, space or mark) this allows for 32 different combinations. Because it is necessary to provide 26 letters, 10 figures and punctuation marks, the 32 combinations are not enough. This problem is resolved by using the codes twice; once in letters (LTRS) case and again in figures (FIGS) case. Two special characters are used, LTRS and FIGS, to indicate which case is in use. The receiving station continues to use the last received case command character until it receives a different one. Control functions such as LTRS, FIGS, CR, LF, space and blank are available in either case. The remaining 26 have different meanings depending on whether the LTRS or FIGS case is selected.

As mentioned earlier, most teleprinter equipment came from either the American or British surplus markets and this highlighted a major problem associated with signalling speeds. Table 20.2 shows a comparison of the differing speeds. As is often the case, the actual speed used depended on the types of equipment available and this resulted in 45.45 bauds becoming the default standard because of the wider availability of surplus US equipment. In 1975 the IARU Region 1 conference in Warsaw voted in favour of using 45.45 bauds for all HF, VHF and UHF working.

Terminal units or modems

In order to transmit by radio the output of the teleprinter or computer and then to display this transmitted signal on a computer or teleprinter, it is necessary to use a piece of equipment to convert the appropriate signals. This equipment is usually referred to as a *terminal unit* or *modem* (modulator/demodulator). The normal methods of transmitting RTTY is *audio frequency shift keying* (AFSK) on VHF and above or *frequency shift keying* (FSK) on the HF bands. FSK is also occasionally used at VHF for long-distance working.

The tone pairs used were adopted at the IARU Region 1 Conference in Warsaw in 1975 and are 1445Hz as the mark and 1275Hz as the space, giving a shift of 170Hz.

A simple modem uses two audio oscillators set at 1445 and 1275Hz for the transmit side and two phase-locked loops (PLLs) set to operate at 1445 and 1275Hz on the receive side. The oscillators are connected to the microphone input of either an FM or SSB transmitter, giving AFSK or FSK respectively. The phase-locked loops are connected to the audio output of the FM or SSB receiver. On SSB tuning becomes critical and so some form of tuning indicator is built in. The modem described above usually accepts two different DC levels on transmit and gives out two different DC levels on receive. The display device (teleprinter or computer) converts the DC levels to a display on the paper or VDU. Fig 20.1 shows a block diagram of typical RTTY stations using either a teleprinter or a computer.

Experiments have been conducted using ASCII rather than Murray or Baudot code but this never became very widely used. The ASCII code is the basis for all types of digital communications and is shown in Table 20.3.

AMTOR

RTTY suffers from problems such as fading and noise which makes successful decoding difficult or impossible on occasions. In order to try and overcome these problems it is useful to try simultaneously to receive two or more versions of the transmission to compare. This can be achieved by using frequency, polarisation or space diversity. Another method of getting more than one version of the same transmission is to use *time diversity*. This is the basis of AMTOR.

AMTOR uses two forms of time diversity in either Mode A (ARQ – automatic repeat request) or Mode B (FEC – forward error correction). In Mode A, a repeat is only sent when requested by the receiving station but in Mode B each character is sent twice. In both cases the second type of time diversity is supplied by the redundancy of the AMTOR code.

As mentioned earlier, AMTOR was developed by Peter Martinez, G3PLX, from the commercial SITOR system. SITOR was devised to improve the communication between teleprinters using the ITA2 code. The system converts the five-pulse code to a seven-pulse code for transmission. There is an exact correspondence between the ITA2 five-pulse code and the AMTOR seven-pulse code. Under normal circumstances the seven-pulse code can have 128 possible combinations but the AMTOR code chooses to have a constant ratio of four marks to three spaces. This provides the simple form of error detection since any character which does not have this 4:3 ratio is known to be in error and can be rejected. There are 35 combinations which allow this constant ratio, and this allows compatibility with the 32 codes of ITA2 and three service information signals (Idle Signal Alpha, Idle Signal Beta and Repeat request (RQ)). The initial AMTOR system used only upper-case characters but system enhancements have recently been implemented by G3PLX to allow the transmission of lower-case characters as well. The original full AMTOR code is shown in Table 20.1.

Mode A operation

This is a synchronous system which transmits blocks of three characters from the transmitting or *information sending station* (ISS) to the receiving or *information receiving station* (IRS).

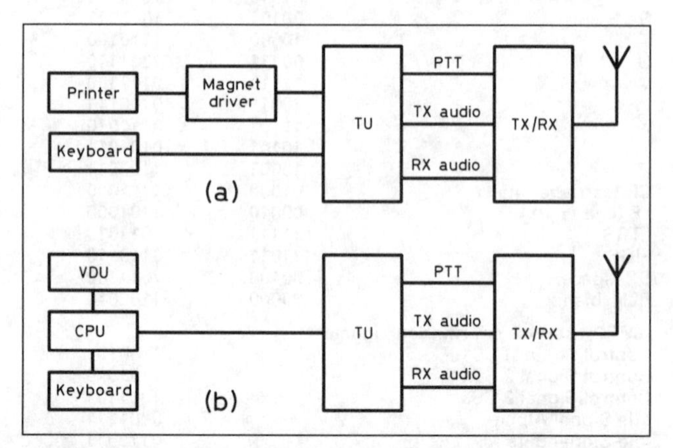

Fig 20.1. Block diagrams of RTTY stations using (a) a teleprinter, (b) a computer

Table 20.3. ASCII codes

ASCII code 6543210	Character	ASCII code 6543210	Character
0000000	NUL	1000000	@
0000001	SOH	1000001	A
0000010	STX	1000010	B
0000011	ETX	1000011	C
0000100	EOT	1000100	D
0000101	ENQ	1000101	E
0000110	ACK	1000110	F
0000111	BEL	1000111	G
0001000	BS	1001000	H
0001001	HT	1001001	I
0001010	LF	1001010	J
0001011	VT	1001011	K
0001100	FF	1001100	L
0001101	CR	1001101	M
0001110	SO	1001110	N
0001111	SI	1001111	O
0010000	DLE	1010000	P
0010001	DC1	1010001	Q
0010010	DC2	1010010	R
0010011	DC3	1010011	S
0010100	DC4	1010100	T
0010101	NAK	1010101	U
0010110	SYN	1010110	V
0010111	ETB	1010111	W
0011000	CAN	1011000	X
0011001	EM	1011001	Y
0011010	SUB	1011010	Z
0011011	ESC	1011011	[
0011100	FS	1011100	\
0011101	GS	1011101	]
0011110	RS	1011110	^
0011111	US	1011111	_
0100000	SP	1100000	£
0100001	!	1100001	a
0100010	’	1100010	b
0100011	#	1100011	c
0100100	$	1100100	d
0100101	%	1100101	e
0100110	&	1100110	f
0100111	’	1100111	g
0101000	(	1101000	h
0101001	)	1101001	i
0101010	*	1101010	j
0101011	+	1101011	k
0101100	,	1101100	l
0101101	-	1101101	m
0101110	.	1101110	n
0101111	/	1101111	o
0110000	0	1110000	p
0110001	1	1110001	q
0110010	2	1110010	r
0110011	3	1110011	s
0110100	4	1110100	t
0110101	5	1110101	u
0110110	6	1110110	v
0110111	7	1110111	w
0111000	8	1111000	x
0111001	9	1111001	y
0111010	:	1111010	z
0111011	;	1111011	{
0111100	<	1111100	|
0111101	=	1111101	}
0111110	>	1111110	~
0111111	?	1111111	DEL

Characters with two or three letters such as NUL or BS are control codes.

The station which initiates a QSO is referred to as the *master station* (MS). The MS first sends the selective call (*selcal*) of the called station in blocks of three characters, listening between blocks. The *slave station* recognises its selcal and answers that it is ready. The selcal is a four-letter group and is normally derived from the first character and last three letters of the station callsign. As an example, the selcal of GM4AUP would be GAUP.

At this point in the contact the MS has become the ISS and will send traffic as soon as the IRS says it is ready. As soon as contact is established the ISS sends its message in groups of three characters, pausing between groups for a reply from the IRS. The signalling rate is 100 bauds with each character accounting for 70ms and a three-character block occupying 210ms. The block repetition time is 450ms, so there is 240ms in each cycle when the ISS is not transmitting. This period is taken up by the propagation time from the ISS to the IRS, 70ms for the IRS to send its service information and the return journey back to the ISS. Also included in this 240ms period is the switching time from transmission to reception. This time should be less than 20ms. It is the 450ms block repetition cycle which limits the distance over which a Mode A QSO can take place.

Transfer from ISS to IRS is initiated by sending the +? sequence. A station may end the contact by sending an 'end of communication signal' consisting of three Idle Signal Alphas.

On the air Mode A signals have a characteristic chirp-chirp sound to them. The 210/240ms duty cycle allows some transmitters to operate at 100% power levels.

Mode B operation

When it is required to transmit to no particular station (for example when calling CQ or transmitting a bulletin) there is no one station to act as IRS. Mode B therefore uses a simple *forward error correction* (FEC) technique by sending each character twice. In order to allow time diversity each character is repeated after four other characters have been transmitted, thus eliminating a error associated with bursts of noise. At 70ms per character, this leaves 280ms between the end of the first transmission and the start of the second.

The receiving station tests for the constant four-mark to three-space ratio and prints only correct first or second transmission characters. If neither is correct then an error symbol is displayed.

Since there is no 50% duty cycle it may be necessary to reduce the transmitter power level to 25 to 50% of full rating.

PACKET RADIO

Packet radio is the first true amateur digital, as opposed to analogue, transmission system. This makes the relaying of signals much more efficient since the data is reconstituted at each stage of the link and any end-to-end noise and distortion is simply that of the digitising process and not the transmission of the digital information. One of the other main benefits of this mode of operation was always assumed to be that the channel could be shared by many users. Unfortunately the radio-based systems are different to computer networks in that not all stations can receive each other, thus making it more difficult for channel sharing.

As with all the above methods of data communications, packet radio makes use of a terminal unit. This terminal unit is rather more sophisticated that an ordinary RTTY terminal unit and usually contains a microprocessor and other associated circuitry. It is normally referred to as a *terminal node controller* (TNC) although the other name of *packet assembler disassembler* (PAD) gives a better clue as to the task the unit performs. Very simply, the function of the TNC is to take the data arriving from the computer, usually in ASCII form, and assemble it into packets which are then passed to the modem for conversion into audio tones. The receive side of the TNC performs the reverse of the tasks outlined. On VHF frequencies the transmission speed for most end-user access is 1200 baud with tone frequencies of 1200Hz (mark) and 2200Hz (space), with 300 baud and 200Hz shift being employed for HF applications. These standards coincide with Bell 202 and 103 modems for VHF and HF respectively. It is not necessary to use a TNC, as packet systems have

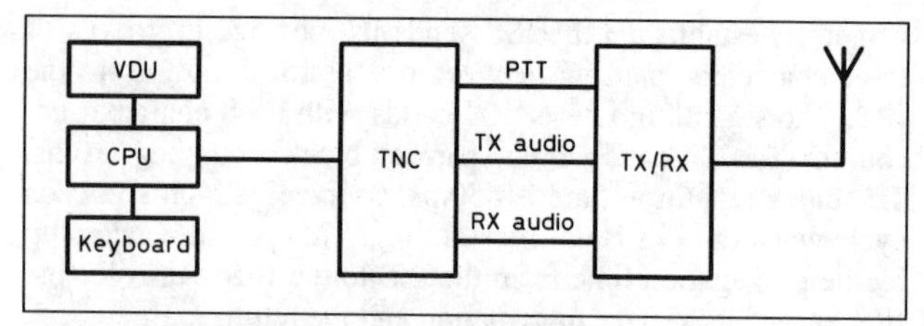

Fig 20.2. Block diagram of typical packet radio station

been designed using a simple modem with all the packet assembling/disassembling being done within the host computer.

A block diagram of a typical packet station is shown in Fig 20.2. Although the drawing shows a computer, a simple dumb terminal can be used; however, to make use of the full facilities for file transfer etc a computer is essential.

Channel access

The basis of a packet radio contact is that each station transmits some information and receives an acknowledgement. If no acknowledgement is received then the information is retransmitted. One of the main causes of non-receipt of acknowledgement is collision with another transmission of either the main transmission or the acknowledgement.

Early packet radio experiments made use of a channel access system in which a station transmitted without checking if the channel was free. If the transmission was not acknowledged within the correct time slot, the TNC waited a random length of time before retrying. Current packet systems make use of *data carrier detect* (DCD) – they listen for an empty channel before transmitting. This is not a guarantee against collisions, because two stations may decide to transmit at the same time, but it is an improvement.

AX.25 Level 2 link layer protocol

Version 2 of the AX.25 Level 2 protocol was adopted by the ARRL in October 1984. This protocol follows that of CCITT Recommendation X.25 except that the address field has been extended to accommodate amateur callsigns and an Unnumbered Information (UI) frame has been added. This protocol formally specifies the format of a packet radio frame and the action a station must take when it transmits or receives such a frame.

At this link layer, data is sent in blocks called *frames*. As well as carrying data, each frame carries addressing, error checking and control information. The addressing information carries details of the who sent the frame, who it is intended for and which station should relay it. This forms the basis of many stations sharing the channel since any station can be set up to monitor all frames on the channel, through various stages to monitor only those intended for it and ignore any others. The error-checking information allows the intended recipient to determine if the frame has been received free of errors. If this is the case and the two stations have previously established a connection, an acknowledgement is generated by the receiving station. If errors are detected the frame is ignored and some time later the sending station resends the frame.

AX.25 format

Packet radio transmissions are sent in frames with each frame divided into fields. Each frame consists of a start flag, address field, control field, network protocol identifier, information field, frame check sum (FCS), and an end flag. Fig 20.3 shows the format of a frame and Fig 20.4 shows a typical address field.

Flag field

Each frame starts and ends with a flag which has a particular bit pattern: 01111110. This pattern appears only at the beginning and end of frames. If five 1 bits show up elsewhere in the frame, a procedure called *zero insertion* (more commonly called *bit stuffing*) takes place and a 0 is inserted by the sending station and deleted by the receiving station. The receiver will therefore delete any 0 bit which follows five consecutive 1 bits that occur between the flag fields.

Address field

The address field consists of the destination field, source field and up to eight optional relay or *digipeat* stations. These fields usually contain callsigns and space is available for up to six characters per callsign with a seventh available as a *secondary station identifier* (SSID). This allows up to 16 different packet radio stations to operate with one callsign. The default is an SSID of 0. For example, GM4AUP-0 could be the real-time station, GM4AUP-2 could be a personal message system (PMS) and GM4AUP-4 could be a node station. The SSID byte in the digipeater address also contains information as to whether it is repeating a frame or not.

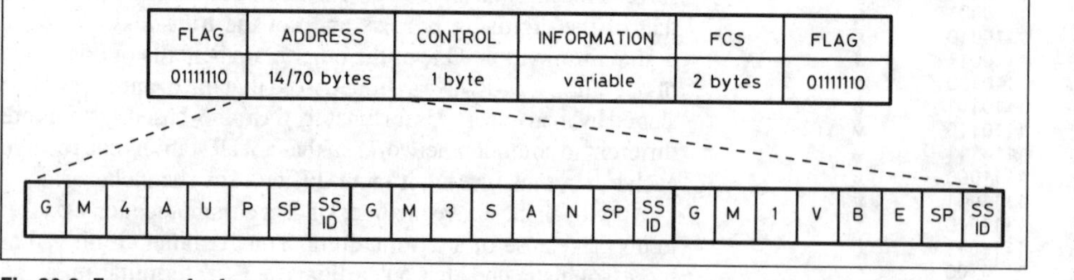

Fig 20.3. Format of a frame

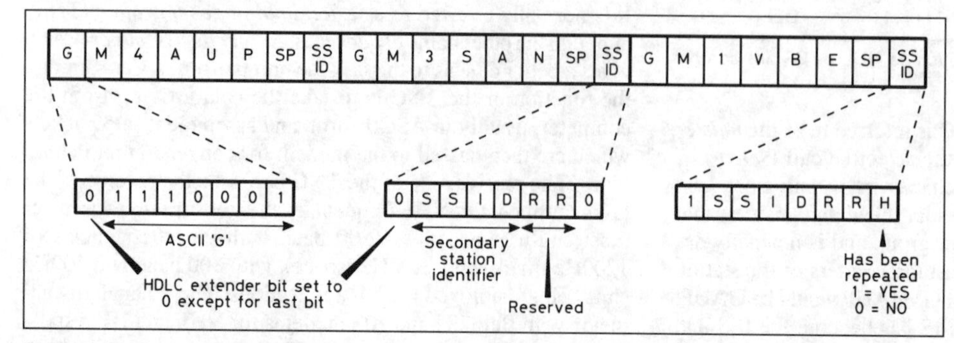

Fig 20.4. Typical address field

Control field

The control field is used to identify the type of frame being transmitted and the frame number.

Protocol identifier field

This field is contained within the information field and identifies what, if any, network layer protocol is being used.

Information field

The information field contains the data to be transmitted and can contain any number of bytes, up to a maximum of 256, of information.

Frame checksum field

The FCS is a 16-bit number calculated by the sender. On receipt of a frame the receiving station calculates a FCS and compares it with that received in the FCS field. If the two match then the receiving station acknowledges the frame.

AX.25 operation

As previously described, the TNC is the device which assembles the data into frames as above. When first powered up, the TNC is in a disconnected state and is monitoring traffic on the appropriate radio channel.

In order to communicate with another station it is necessary to enter the *connected state*. This is done by issuing a *connect frame* which contains the callsign it is requesting connect status with as the addressee. If the other station is on the air it responds with an *acknowledgement frame* and the stations become connected. If no acknowledge frame is received the requesting station re-issues the command a pre-determined time later and continues to do so until a preset number of tries has taken place. If no connection is established the requesting TNC issues a failure notification.

Once a link is established the TNCs enter the connected or information transfer state and exchange information and supervisory frames. The control field contains information about the number of the frame being sent and the number of the last one received (0 to 7). This allows both TNCs to know the current link status and which to repeat if necessary.

When in the connected state either station may request a disconnection which occurs after an acknowledgement is received or if no response is received after several attempts.

Packet operation

Packet operation currently makes use of the HF, VHF, UHF and SHF parts of the spectrum with both terrestrial and satellite links being utilised. In the early days, much packet operation was real-time person-to-person operation, either direct or through a *digipeater*. Most TNCs were capable of digipeat operation and this enabled stations who could not contact each other direct to do so by the on-frequency retransmission of the digipeater. As packet became more popular the real-time operation tended to be replaced by store-and-forward mail handling via special stations operating as mailboxes or *bulletin board systems* (BBS). The network also developed and the digipeater was replaced by the *multi-port node*.

Digipeaters

A digipeater was the first method of retransmitting information from one packet station to another. Most TNCs can be used as a digipeater as this function is usually contained within the AX.25 Level 2 firmware.

Fig 20.5 shows how two stations A and D can connect to each other using digipeaters B and C. In order for information to be passed from station A to station D via the digipeaters B and C, the information frame must be received by station D and the acknowledgement frame received by station A before a frame can be said to be successfully sent. Digipeaters B and C play no part in the process; they merely retransmit any frames that contain their callsigns in the digipeat portion of the address field. If the acknowledgement is not received by station A then the frame

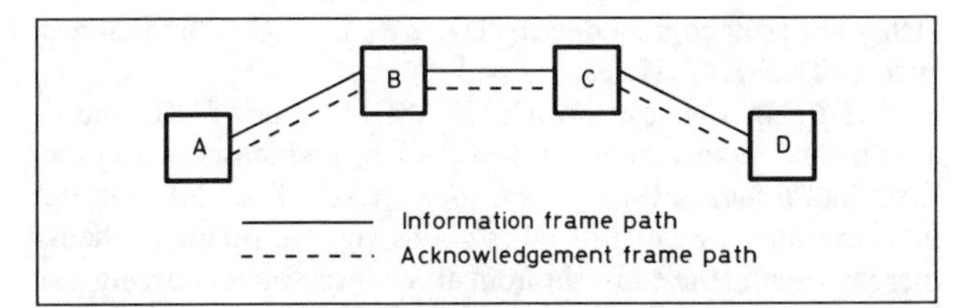

Fig 20.5. How two stations can connect to each other using digipeaters

is retried over the whole path. The use of digipeaters has reduced dramatically in recent years with the advent of the network nodes.

Network nodes

The advent of the network node significantly improved the packet radio system as a means of communicating with between packet-equipped stations in both real time and by the use of mailboxes. The major advantage of a network node over a digipeater is that any frame which is being transmitted is separately acknowledged between each individual element rather along the whole chain.

Fig 20.6 shows a system with station A trying to communicate with station D via the nodes B and C. In trying to communicate with each other the information is sent from station A to node B and acknowledged back to station A. Node B then passes the frame on to node C and receives an acknowledgement back. Node C then passes the frame to station D who acknowledges it back to Node C. If anywhere in the path no acknowledgement is received then the frame is retried only over the part of the path for which no acknowledgement has been received.

There are two types of network protocol in use – *virtual-circuit* and *datagram*. In the virtual-circuit protocol the appearance of a direct connection between the two stations is provided. In order to establish communications a 'call set-up' packet is sent through the network to make a path to the other station. Once this path is established information is sent through the circuit. Any packets sent do not have the full address of the required path because the network attempts to maintain this path for the duration of the contact. After the contact is completed the virtual circuit is cleared by removing the information on the path along the network. An example of a virtual-circuit protocol is the RATS Open System Environment (ROSE) developed by the Radio Amateur Telecommunications Society (RATS) of New Jersey. ROSE is a firmware replacement for TNC2 clones. The virtual-circuit protocol is not very common in the UK and most networking is done using the datagram protocol.

In the datagram protocol each packet contains full network addressing and routing information. This enables a packet to reach its destination via any route still open, regardless of how reliable the network may be. The network overhead is greater in this protocol but it has much greater flexibility and the end user does not need to know the route, only the node nearest him and the node nearest the station with which he desires to connect.

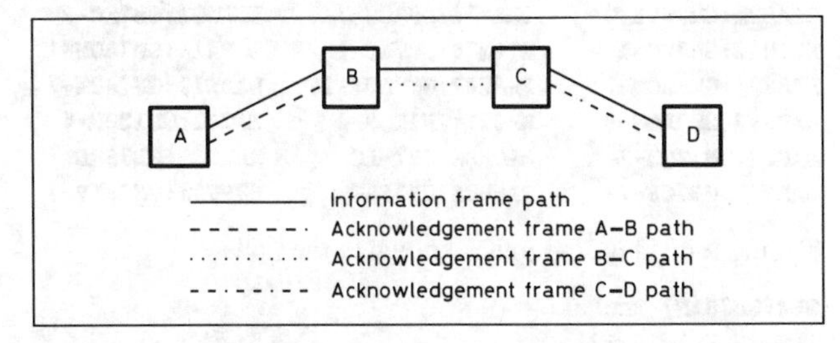

Fig 20.6. How two stations can connect to each other using network nodes

Datagram protocols used in the UK are NET/ROM (and clones such as TheNET), TheNODE and Internet.

NET/ROM was developed by WA8DED and W6IXU and is a firmware replacement for TNC2 clones which converts the TNC into a *network node controller* (NNC). Each NET/ROM node maintains a table of other nodes and the routes to them, thereby freeing the end-user from all worries about knowing the best route from one node to another. NET/ROM nodes can be connected together via their serial ports, allowing multi-port nodes to be built on the same site with cross-band or cross-frequency capability.

TheNODE, developed by G8BPQ, is a software implementation of a datagram protocol and is fully compatible with NET/ROM protocols, enabling their respective nodes to appear in each others' nodes table. TheNODE is commonly installed in PC clones and may be used as front-end software to enhance the capability of a mailbox.

The *Internet* protocol software was written by Phil Karn, KA9Q, and is more commonly known as *TCP/IP* which is an acronym for two protocols, the Internet Protocol (IP) and the Transmission Control Protocol (TCP). In reality KA9Q's TCP/IP consists of a suite of individual protocols, Address Resolution Protocol (ARP), File Transfer Protocol (FTP), Serial Line Transfer Protocol (SLIP), Simple Mail Transfer Protocol (SMTP), Telnet Protocol, User Datagram Protocol (UDP) as well as TCP and IP. Each station using TCP/IP is a network node with a unique IP address that has been assigned by the local IP address co-ordinator. The amateur TCP/IP network has been assigned the network name *AMPRNET* and all amateur addresses commence with the two digits 44, followed by three digits indicating the country code (as an example of a full address '44.131.5.2' is assigned to G3NRW). TCP/IP is becoming very popular in the UK and is said to offer many advantages over 'ordinary' AX.25.

Using network nodes

Network nodes are connected to in exactly the same way as a normal station by issuing the following command:

```
C GLW
*** CONNECTED to GLW
Welcome to GM3SAN's Packet Switch in Glasgow
Type ? for list of available commands.
```

By typing **?** the commands are displayed:

```
GLW:GM3SAN} BBS CONNECT BYE INFO NODES PORTS ROUTES
USERS
```

Typing **N** gives a list of the known nodes in the table:

```
GLW:GM3SAN} Nodes:
GB7AY-2              GB7AY-4              AUP:GM4AUP
BBSJSC:GB7JSC        BOTH16:GM1VBE-1      BOTH21:GM1VBE-2
BOTH42:GM1VBE-4      BOTH74:GM1VBE-7      BOTH99:GM1VBE-9
CLYDE:GM1VBE-10      EDIN21:GB7PH-2       EDIN70:GB7PH-7
GLAS21:GM0MVZ-2      GLAS74:GB7GN-7       KILN21:GB7KC-2
KILN42:GB7KC-4       KILN70:GB7KC-7       LARK16:GM7AOM-1
LARK21:GM7AOM-2      LARK42:GM7AOM-4      LARK72:GM7AOM-7
LARK99:GM7AOM-9      MAC21:GM1ZQM-2       MAC42:GM1ZQM-4
MAC70:GM1ZQM-7       MACBBS:GB7MAC        RAYNET:GB7SRL
SALS70:GB7CS-7       SANBBS:GB7SAN        SEPU21:GM0SEP-2
```

Typing **R** displays the routes known to the node:

```
GLW:GM3SAN} Routes:
4 GM1VBE-1 20 12!
3 GM1VBE-4 15 1!
> 3 GM4AUP 20 2!
4 GM7AOM-1 20 12!
3 GM7AOM-4 15 5!
1 GM1ZQM-2 10 1
3 GM1ZQM-4 10 1
1 GM1VBE-2 10 4
> 2 GB7KC-7 10 3
2 GB7CS-7 10 1
1 GM0MVZ-2 10 2
2 GM1ZQM-7 10 1
> 1 GM0SEP-2 10 1
2 GB7PH-7 10 1
> 3 GB7KC-4 10 1
1 GB7SRL 10 1
1 GB7KC-2 10 1
1 GB7PH-2 10 1
```

In the above list for GB7KC-4 the '>' indicates the route is in use, is available through port 3, callsign is GB7KC-4, with a route quality of 10 (default), with the number of direct routes of 1 and if the line is ended with a '!' then it indicates the route is locked and will not disappear from the table.

Typing **P** lists the available ports:

```
GLW:GM3SAN} Ports:
1  144.650 MHz   1200 Baud (DRSI)
2  432.625 MHz   1200 Baud (DRSI)
3  70.4875 MHz   1200 Baud (DRSI)
4  1299.00 MHz   1200 Baud (DRSI)
```

To connect to another node or station it is necessary to issue a further connect command:

```
c sals70
```

On successful connection the following typical message appears:

```
GLW:GM3SAN} Connected to SALS70:GB7CS-7
```

To disconnect the circuit the command is **B** and the following response is given:

```
*** DISCONNECTED: GLW
```

The above commands are typical of a TheNODE system but NET/ROM commands are very similar.

Mailboxes

A mailbox or *bulletin board system* (BBS) is one of the most important parts of the packet radio network. Messages can be sent from one amateur to another and from one side of the world to another with the knowledge that the recipient can read it in his or her own good time rather than when both parties are active at the same time. Each message is passed from mailbox to mailbox automatically until the destination mailbox is reached. The sending party only needs to know the mailbox which the receiving party uses for the message to be successfully delivered. All mailboxes in the UK are easily recognisable using callsigns in the GB7xxx series and there are approximately 250 such callsigns currently issued.

The requirements of a mailbox are a *system operator* (SysOp) with a significant amount of patience and time, a PC clone with large amounts of hard storage and memory and multiple TNCs, transceivers and antennas.

Using a mailbox

Mailbox operation is simple, although it is well worth downloading the appropriate help file soon after logging on for the first time. On first logging on to a mailbox the following is a typical response:

```
*** CONNECTED to GB7SAN
[FBB-5.14-ABFHM$]
GB7SAN BBS Glasgow. IO75WU.
Channel 3 - 4 active channel(s).
Ports available on 144.650 and 432.625 MHz
Hello IAN, welcome to the Glasgow MacPAC Mailbox.
You last logged in at 14:50GMT on 30-Mar-93.
1033 Active messages - The last message is # 92220.
The L command will list new messages from # 92192.
*** For MacPac Information Type 'D INFO' ***

3:GB7SAN BBS (A,B,C,D,F,G,H,I,J,K,L,N,O,R,S,T,U,V,W,X,Y,?) >
```

The help file can be received by typing **?H** and the response will be similar to this:

```
List of available commands :
A : Abort         - Abort listing.
B : Bye           - Log off the BBS.
C : Conference    - Access to conference.
D : DOS           - Access to FBBDOS, or to download a
                    file.
F : FBB           - Access to server mode.
G : Gateway       - Access to other frequencies by
                    'gateway'.
? : Help          - Help.
I : Info          - Information about the system.
J : Jheard        - List of the last few connected
                    stations.
K : Kill          - Kill messages.
L : List          - List messages.
M : Make          - Copy a message to a file.
N : Name          - Change your name.
NZ: Zip           - State your zip-code.
NH: homeBBS       - Type your home-BBS.
O : Option        - Select options (paging and
                    language).
PS: Servers       - Show which servers are available in
                    this BBS.
PG: Program       - Run (show) certain DOS-programs.
R : Read          - Read messages.
S : Send          - Send messages.
T : Talk          - Talk to SysOp.
U : Upload        - Upload a file to the BBS.
V : Verbose       - Verbose read of messages (like R,
                    but with forwarding headers).
W : What          - Which files are available.
X : Expert        - Change between Normal and Expert.
Y : Yapp          - Transfer binary files with the Yapp
                    transfer protocol.
Z : Delete        - Delete a file.
+ : Append        - Only for sysops: Append a file after
                    a message.
> : Send text     - To send text to another station
                    connected to the BBS.
= : Connect       - To connect to another station
                    connected to the BBS.
! : Info          - Short version of the I-command.
$ : Forwarded     - Only for SysOp: Show what BBSs that
                    will receive, or already have
                    received a message.
* : Wildcard      - Many possibilities, like @,?,#,=,*
For detailed help on each command, type ? [command].

3:GB7SAN BBS (A,B,C,D,F,G,H,I,J,K,L,N,O,R,S,T,U,V,W,X,Y,?) >
```

Three typical further help messages are shown below:

```
?S

S-command lets you send messages.
- Type S[type][space][callsign] to send a message.
  [type] can be P for Personal or B for Bulletin.
- Type S[type][space][callsign] @ [callsign] to send a
  message to a station at another BBS.
- Type SB ALL to send a bulletin to ALL.
- Type SR[space][messages#] to reply to a message.
  (Send Reply).
- Type SC[space][messages#][space][callsign] to send a
  copy of a message to another callsign. (Send Copy).
  You can write a message before sending the CTRL-Z,
  then the copy will be appended to your text. If you
  don't want to do this, just send CTRL-Z instead of
  any text.

3:GB7SAN BBS (A,B,C,D,F,G,H,I,J,K,L,N,O,R,S,T,U,V,W,X,Y,?) >

?R

R-command is for reading messages (not files).
Type R[space][message#] to read one message.
You can chain several message#, separated by a space.
Type RM to read ALL messages addressed to you [Read
Mine].
Type RN to read all NEW messages to you [Read New].
Type RU to read all UNREAD messages (status N) to or
from you.
Type R< to read messages from a certain callsign.
Type R> to read messages to a certain callsign.

3:GB7SAN BBS (A,B,C,D,F,G,H,I,J,K,L,N,O,R,S,T,U,V,W,X,Y,?) >

?D

D-command has 2 different functions.
1) If you send D [filename], the mailbox will send you
   the file with that name. This file must be in the
   root-directory for users. If the file is in a
   subdirectory, you must use full path, like this: D
   VHF/VHF.DX
   See also help with the W-command (? W).
2) If you send the letter D alone, you will enter
   FBBDOS. The commands in FBBDOS are very similar to
   those of MS-DOS. Users have access to a part of the
   BBS's hard-disk in FBBDOS.
   Following commands are allowed here:
   ? HELP            >= Use one of these to get help.
   DIR               >= List files in this directory.
   MD MKDIR [name]   >= Use one of these to make a
                        new directory.
   RD RMDIR [name]   >= Use one of these to remove a
                        directory. You cannot be in
                        the directory you want to
                        remove, and the directory
                        must be empty.
   CD [name]         >= Use CD to change directory.
   COPY [from] [to]  >= Copy a file named [from] to a
                        file named [to]. Can also be
                        used to copy between
                        directories.
   DEL [filename]    >= Deletes the file [filename].
   TYPE [filename]   >= Download the ASCII-file
                        [filename] from the BBS. The
```

```
                        file is divided into pages,
                        if you use paging when in
                        BBS-mode. The file must be in
                        your current directory.
GET [filename]      >=  Identical to TYPE, but always
                        without paging, and the file
                        is always ended with a
                        CTRL-Z.
PUT [filename]      >=  Use this command to upload an
                        ASCII-file to this directory.
YPUT and YGET       >=  These are identical to PUT
                        and GET, but uses the
                        YAPP-protocol for binary
                        transfer.
XPUT and XGET       >=  These are identical to YPUT
                        and YGET, but uses the
                        XMODEM-protocol for binary
                        transfer. Only for
                        telephone-modem.
EDIT [filename]     >=  This is a small editor for
                        editing texts and files in
                        the BBS. You can use this
                        editor ONLY on files that YOU
                        have uploaded to the BBS. Get
                        more info with H EDIT
F B EXIT QUIT       >=  Use one of these to go back
                        from FBBDOS to BBS-mode.
O                   >=  Use this to set 'options'
                        like in BBS-mode.
LIST                >=  Same as DIR, but shows also
                        descriptions of files, if
                        users have provided such a
                        description.
PRIV                >=  Gives special users access to
                        a special directory.
VIEW                >=  See the contents of an
                        archived file like *.ZIP,
                        *.LZH etc. VIEW filename.ext
NEW                 >=  List all new files since you
                        last sent NEW

3:GB7SAN BBS (A,B,C,D,F,G,H,I,J,K,L,N,O,R,S,T,U,V,W,X,Y,?) >
```

At all times the response:

```
3:GB7SAN BBS (A,B,C,D,F,G,H,I,J,K,L,N,O,R,S,T,U,V,W,X,Y,?) >
```

indicates that the BBS is waiting for an input.

To disconnect from a mailbox the command **B** is typed and the following response is typical:

```
Connect time : 2mn 21s - Compute time : 20s
Goodbye IAN, thanks for using the MacPAC mailbox.
*** DISCONNECTED: GB7SAN
```

The above commands are from a BBS using F6FBB software but most systems are similar in format. BBS software commonly in use in the UK has been written by W0RLI, WA7MBL, AA4RE, G1NNA, GW3TMH, G4YFB and F6FBB, with G1NNA and F6FBB being the most common in 1993.

DxClusters

A DxCluster is one of the more recent developments within packet radio which provides information on DX stations being worked/heard along with information on QSL managers, WWV propagation and prefixes for example. The operation is not dissimilar to mailbox operation but users stay connected to their local DxCluster for as long as they wish to receive announcements. The type of announcement the user receives is customised to suit his own needs and can be used to select prefix information, band information, mode information or a combination of all those.

Each cluster is generally referred to as a *cluster node* and these nodes can be connected together to each other via the network. This enables an item of DX information (commonly referred to as a *spot*) to propagated to all other cluster nodes in the network, thereby in theory enabling all connected users to see this spot in a matter of minutes.

Operating guidelines

Packet radio has become so successful as a message-passing service that it was necessary to draw up guidelines for the use of the network. These guidelines which were produced in conjunction with the Radiocommunications Agency of the DTI and are reproduced on the following pages.

CURRENT DEVELOPMENTS IN DATA COMMUNICATIONS

Two new modes have recently appeared on the scene which have attempted to enhance the ability of data communications to operate in noisy and varying radio channels.

PACTOR

PACTOR was developed in Germany by DF4KV and DL6MAA and they describe it thus: “PACTOR is an improved half-duplex synchronous ARQ system combining the reliability of packet radio with the fixed AMTOR time frame, and is specially designed for operation in noisy and fluctuating channels”.

The system is, as stated above, synchronous and the transceiver is switched between transmit and receive at fixed intervals. The total cycle time is 1.25s which is two and a half times the length of the AMTOR cycle time. There is also a ‘long path’ mode which overcomes one of the limitations of AMTOR. As with AMTOR the receiving station acknowledges the transmitted block with a short Control Signal block.

The tone shift is 200Hz, slightly wider than that used for RTTY and AMTOR, but existing modems or terminal units can handle it without any problems. If you listen to PACTOR it sounds very much like AMTOR except for the longer cycle time.

The major developments of PACTOR over AMTOR are the error checking and the baud rates. PACTOR uses cyclic redundancy checks (CRCs) and the tone polarity is reversed with every cycle, enabling the system to attempt to reconstruct one good packet from a corrupted one plus a corrupted repeat. The baud rate is toggled between 100 baud and 200 baud, depending on the link quality and conditions. This enables the speed to reduce if conditions change or interference increases.

The system supports the full ASCII character set as shown in Table 20.3 but throughput is increased by using a Huffman compression coding technique.

CLOVER

CLOVER™ was conceived by Ray Petit, W7GHM, and development started in 1987. However, the real basis of CLOVER began in 1975 when experiments with *coherent CW* (CCW) took place. This is a method of transmitting CW in a narrow bandwidth: 12 to 15Hz at 12wpm. This involves generating code with extremely accurate timings and transmitting on a very precise and stable frequency. The mode never proved popular although some amateurs are still working with it.

GUIDELINES FOR SYSTEM OPERATORS (SYSOPS) OF PACKET RADIO MAILBOXES

1. In addition to these guidelines please read carefully, and comply with, the 'Guidelines For the Use of the Packet Radio Network'.
2. Always ensure that the TNC parameters are adjusted correctly to suit the conditions on your Local Area Network (LAN), especially PACLEN, MAXFRAME, and CSMA parameters.
3. Ensure that you have a second SysOp (or more than one) who is familiar with the mailbox so that the service can be continued whilst you are away on holiday etc. The second SysOp (who must be a licensed amateur) need not have regular access to the equipment, but must know how to close down your station in an emergency, and should be able to do basic remote 'housekeeping' on the mailbox on a daily basis. Ensure that the RSGB Mailbox Co-ordinator knows who your additional SysOp is in case of an emergency.
4. Review all messages at least once per day to ensure that nothing contravenes the conditions of your licence. If in doubt about a message consult 'The Guidelines For the Use of the Packet Radio Network' and act appropriately. SysOps are advised to err on the side of caution; it is also your licence which could be at risk. During the daily review, check why any messages have remained unforwarded and look for any duplicate messages.
5. Ensure that the mailbox will auto-start after any power failure.
6. Where possible limit the downloading of large files to off-peak times or less-congested frequencies. SysOps should advise their users accordingly.
7. IARU Region 1 has advised that there should be no formal network links on the 144MHz band. You are strongly recommended, therefore, to link on other bands covered by your Notice of Variation.
8. Forwarding should only take place on fast efficient routes, preferably on a dedicated forwarding frequency. If it is essential to use the 144MHz band, or frequencies which are shared with users, then forwarding should on take place during the less-busy off-peak periods in your area.
9. Do not forward to unlicensed mailboxes or use unlicensed stations for forwarding. It is, however, permissible to forward personal mail automatically to the licensed amateur who is the intended recipient of that mail.
10. If you have any problems ask the Data Communications Committee's Mailbox Co-ordinator to help. He will probably have encountered your problem before, or will be able to put you in touch with someone who has.
11. Remember that at all times you are providing a service which should be as reliable and efficient as possible for your users while at the same time causing as little channel congestion as possible for other packet radio stations.
12. The RA have advised that the use of frequencies in the band 1298 to 1300MHz for mailbox linking is likely to be subject to occasional severe interference from the primary users of the band. It is likely that different parts of the country will experience problems on different frequencies and at different times of the day. The radio frequency environment is likely to be less hostile outside normal working hours.
13. For those providing an end-user service it is permissible to restrict the message categories available to the users, provided this is pre-advertised and even-handed. ALL trunk traffic must be carried until forwarded unless it contravenes the 'Guidelines For the Use of the Packet Radio Network'.
14. If a user persists in sending messages that contravene the 'Guidelines For the Use of the Packet Radio Network' after having been warned, he may be excluded. Users should not be excluded for any other reason.

The basis for developing CLOVER was a feeling that HF packet was very inefficient but AMTOR maintained a link in very poor conditions. The name 'CLOVER' was derived from one of the displays used by W7GHM for signal analysis which produced a clover leaf pattern. W7GHM continued his development of CLOVER through CLOVER-I which sent data at modest rates through very narrow bandwidths to CLOVER-II™ which is capable of much higher data rates and is compatible with most modern transceivers. In 1990 W7GHM joined forces with HAL Communications to bring out early versions of the software in late 1992.

The mode makes use of an completely new method of modulation which utilises a series of pulses rather than frequency shift keying. The pulses are 8ms long and transmitted at intervals of 32ms, and are shaped in amplitude to produce a narrow spectrum. This spectrum is shown in Fig 20.7 and the comparison with other HF data modes is shown in Fig 20.8.

The modulation technique is highly complex and makes use of phase and amplitude modulation of the pulses. The system can vary the amount of phase and/or amplitude modulation depending on band conditions. The data blocks in this modulation technique can also be varied and sent in blocks of 17, 51, 85 or 255 bytes length depending on conditions.

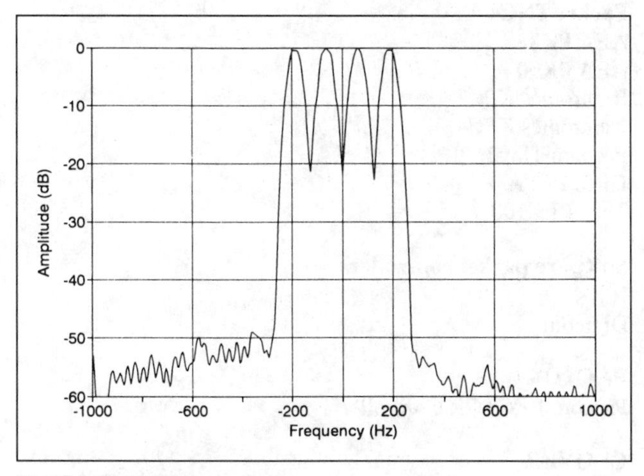

Fig 20.7. CLOVER-II spectrum

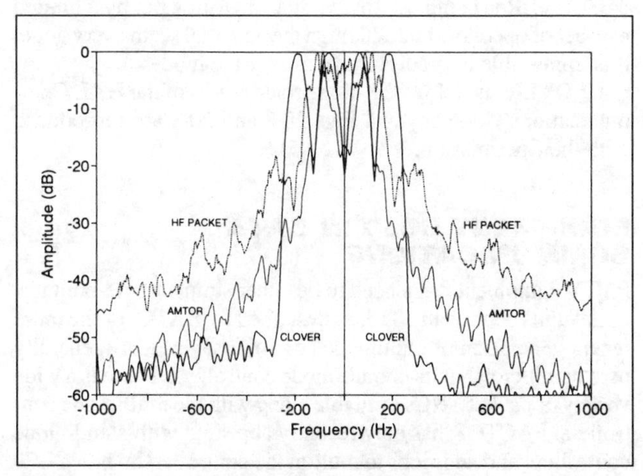

Fig 20.8. Comparison between CLOVER-II, AMTOR and HF packet spectra

GUIDELINES FOR THE USE OF THE PACKET RADIO NETWORK

The packet radio network in the UK and throughout the world is an immensely useful tool for the dissemination of information, the seeking of help and advice and the publication of amateur radio related news. It is not uncommon to find messages giving information on Amsat, RAYNET or other similar activities. The GB2RS news is also available on the network, as is local club news in the area of a particular mailbox. This use of the network is what was in many operator's minds when they spent large amounts of time and money in developing it.

Unfortunately the very success of the network has resulted in messages appearing which are of doubtful legality under the terms of the UK licence.

The RSGB Data Communications Committee, in consultation with the Radiocommunications Agency, has devised the following guidelines with which all operators are urged to comply.

These guidelines have been split into four sections in order to reflect:

A. The need for messages to be within the terms of the licence conditions and the implications if they are not.
B. Messages which could result in legal action being taken by other amateurs or outside bodies.
C. Actions to be taken when amateurs identify cases of abuse.
D. Other appropriate items.

Section A

1. All messages should reflect the purposes of the amateur licence, in particular "self training in the use of communications by wireless telegraphy".
2. Any messages which clearly infringe licence conditions could result in prosecution, or revocation, or variation of a licence. The Secretary of State has the power to vary or revoke licences if an amateur's actions call into question whether he is a fit and proper person to hold an amateur licence. An example of this could be unreasonable behaviour by using the packet network to carry on a dispute or to deliberately antagonise other amateurs.
3. The Radiocommunications Agency has advised that the Amateur Radio Licence prohibits any form of advertising, whether money is involved or not.
4. Messages broadcast to ALL are considered acceptable but should only be used when of real value in order to avoid overloading the network.
5. Do not send anything which could be interpreted as being for the purpose of business or propaganda. This includes messages of, or on behalf of, any social, political, religious or commercial organisation. However, our licence specifically allows news of activities of non-profit-making organisations formed for the furtherance of amateur radio.

Section B

1. Do not send any message which is libellous, defamatory, racist or abusive.
2. Do not infringe any copyright or contravene the Data Protection Act.
3. Do not publish any information which infringes personal or corporate privacy, eg ex-directory telephone numbers or addresses withheld from the *Call Book*.

Section C

1. Any cases of abuse noted should be referred in the first instance to AROS (the RSGB Amateur Radio Observation Service), which will take the appropriate action.
2. It is worth noting that any transmissions which are considered grossly offensive, indecent, obscene or menacing should be dealt with by the police. This action should also be co-ordinated by AROS initially.
3. Mailbox SysOps have been reminded by the Radiocommunications Agency that they have an obligation to review messages daily and that they should not hesitate to delete those that they deem unacceptable. It is worth remembering that their licence is also at risk as well as your own.

Section D

1. Do not send 'open letters' to individuals.
2. Do not write in the heat of the moment. Word-process your bulletin first, then reread it. You may feel differently after a few minutes.
3. Obey the Golden Rule – If you would not say it on voice, do not send it on packet.

Error correction is also present using Reed-Solomon coding which contains redundant information which can be used to detect and correct errors.

As can be seen, the modulation method and variable block length, coupled with the error-correction techniques, make CLOVER a potentially very robust data communication method.

CLOVER is being used at the time of writing but by a limited number of operators and, although the potential seems very good, it is impossible to predict the future for the mode.

CLOVER and CLOVER-II are trade marks of the HAL Communications Company and Figs 20.7 and 20.8 are reproduced with their permission.

EQUIPMENT USED IN DATA COMMUNICATIONS

RTTY equipment has tended to be either simple home-built terminal units such as the ST5, as described by BATC, or the more recent generation of multimode controllers. AMTOR generally uses commercial units or multimode controllers. There is a wide variety of packet TNCs available along with the multimode controllers. PACTOR has only recently appeared with stand-alone controllers and additions to multimode controllers, while CLOVER only currently has the PCI-4000 controller from HAL Communications.

Multimode controllers

AEA PK232
AEA PK900
Kantronics KAM
MFJ 1278
DSP1232/2232

Packet TNCs

AEA PK88
AEA PK90
Kantronics KPC3
Kantronics KPC4
Paccom Tiny2
DRSI PCPA
Thor RLC100

Software packet controllers

BMK
Digicom

PACTOR

Paccom PACTOR Controller

CLOVER

HAL PCI-4000

THE FUTURE

What the future holds is, as always, very difficult to predict. It depends on a number of factors such as licensing conditions as allowed by the Radiocommunications Agency and the other worldwide PTTs, the development of much faster networks, an integrated network which carries all traffic regardless of the type of use, and many other reasons. It is likely that satellites will be used much more as support to traditional international forwarding on HF.

One thing is certain – to anyone remotely interested in data communications it will be exciting.

BIBILOGRAPHY

Further information can found from the following books and periodicals:

Packet Radio Primer, Dave Coomber, G8UYZ, and Martyn Croft, G8NZU, RSGB, 1991.
NOSIntro, Ian Wade, G3NRW.
AX.25 Link Layer Protocol, ARRL.
ARRL Handbook 1993, ARRL.
Radio Communication, RSGB.

Information on licensing and policy matters with regard to data communications is available from the chairman of the Data Communications Committee c/o RSGB, Lambda House, Potters Bar, Hertfordshire EN6 3JE.

GLOSSARY OF TERMS

AFSK	Audio frequency shift keying
AMTOR	Amateur teleprinter over radio
ANSI	American National Standards Institute
ASCII	American Standard Code for Information Interchange
BBS	Bulletin board system or mailbox
CCITT	International Telegraph and Consultative Committee
CR	Carriage return
Digipeater	Digital repeater with single-frequency retransmission capability
FCS	Frame checksum
FSK	Frequency shift keying
LF	Line feed
MBX	See 'BBS'
Modem	Modulator/demodulator
Node	Digital repeater with multi-frequency retransmission capability
PAD	Packet assembler disassembler
TU	Terminal unit
TNC	Terminal node controller
RTTY	Radio teletype

21 Operating technique and station layout

HOWEVER much time and money have been spent on the construction or purchase of equipment, results are certain to depend to a large degree on the way in which the station is operated. A listening session on any of the amateur bands at almost any time will reveal many examples of bad behaviour and faulty operating techniques which are more usually the result of ignorance than a deliberate attempt to be a nuisance to other band users.

The two basic essentials of successful operating on any mode are judgement and courtesy. A third and most important factor is adequate knowledge of the various bands and their propagation habits. This can only come as a result of observation and practice, and is the one great advantage which the former keen short-wave listener has over the pure technician when first coming on the air. The newcomer may wonder exactly how he should begin his first contact: the wisest course for him to take is to first of all listen around carefully and establish exactly what signals are to be heard and the general level of activity. He can then decide whether to put out a 'CQ' call or call a specific station at the appropriate time – this is when that station has finished his previous contact or has himself put out a 'CQ' call. It is not good manners to try to 'break in' on his frequency while he is still in contact. Having established contact it is usual to exchange signal strength reports, names, locations, and details of equipment being used. Further general discussion should concern "matters of a personal nature in which the licensee, or the person with whom he is in communication, has been directly concerned . . ." (Wireless Telegraphy Act). It should be made a golden rule never to discuss politics, religion, or any matter which may offend the person to whom one is talking or anyone who may be listening.

Telephony operation

Always speak clearly and not too quickly, especially when talking to someone whose native tongue is not one's own. Remember that many overseas amateurs deserve great credit for the trouble that they have taken to learn a foreign language such as English in order to be able to talk to those who do not speak their own language. Their ability to converse may well be limited to the more standard subjects involved in the usual simple contact and the use of basic words or even of some of the Q-code (see Table 21.2) is helpful under such circumstances. The latter should never be used on phone under any other circumstances as nothing sounds more stupid to the casual listener than two local amateurs talking to each other by spelling out morse code abbreviations.

Morse operation

In general, the good morse code operator is the one whose copy is easy to read and who does not try to impress by sending faster than he is really capable of doing properly. The speed of sending should partly depend on circumstances – when conditions are poor it is sensible to send more slowly than when signals are loud and clear. *A good principle to follow is to send at the same speed as the operator at the other end of the contact* – and to slow down if he indicates that he is having difficulty. Morse from an electronic keyer can be a joy to the ear if the device is being operated correctly, and it should be the aim of the semi-automatic 'bug' or straight key owner to send out signals just as readable. It is highly desirable to be able to listen to one's own transmission to know how one's style is developing. There are five different ways of ending a morse code transmission and each indicates to a listener exactly what the intentions of the sender are. These are as follows:

1. '. . . $\overline{\text{AR}}$.' These letters are sent at the end of a call to a specific station before contact has been established, eg 'G3AAA de G3ZZZ $\overline{\text{AR}}$'.
2. '. . . K.' The letter 'K' is sent at the end of a 'CQ' call, or at the end of a transmission during an established contact where there is no objection to others joining in.
3. '. . . $\overline{\text{KN}}$.' This is used when a call is made to a specific station and replies are not desired for any other station, or at the end of a transmission when already in contact where other calls are not wanted.
4. '. . . $\overline{\text{SK}}$.' This combination is used at the end of a final transmission in a contact, immediately before the callsigns are given. It means that the sender is ready to receive other calls.
5. '. . . CL.' Denotes that the sender is closing down and will not answer any more calls.

In the above, a bar over two letters indicates they are sent as one character.

Note that it is not correct to call 'QRZ?' after finishing a transmission unless someone has already been heard calling – the right thing to do is to call 'CQ'. It should be borne in mind, however, that it is courteous to move off a frequency at the end of a contact if the station contacted was originally operating there. It is polite to move away from a frequency on which one has been fortunate enough to have been called by a rare DX station and leave it for others to use – there may be many others who would like a contact with the latter and it is not difficult to find a fresh frequency. Remember that no one has an absolute right to any frequency under any circumstances whatsoever and that the policy of 'live and let live' is by far the best to adopt.

The 'CQ' call

There are many occasions when it is a good idea to put out a 'CQ' call. One of these is when a band appears to be 'dead'. This can happen particularly on 21 and 28MHz when there can be suitable long-distance propagation but the level of activity is low, most having left in the belief (after a quick listen around) that it is a waste of time to transmit. There are important rules to be observed when putting out a 'CQ' call. Some of the more vital ones are as follows:

1. Listen on the chosen frequency to make sure that it is quite clear of other signals before starting to transmit.
2. When using telephony confirm step 1 by asking "Is anyone using this frequency please?" several times, and then listen very carefully for replies before commencing the call.
3. Keep calls short and listen frequently for replies. Many contacts have never taken place because potential callers have become bored while waiting for an opportunity to identify themselves and have moved away. The ideal arrangement is to use full break-in on morse and VOX when on SSB so that it is possible to 'listen through' one's calls. Not more than three 'CQs' followed by one's own callsign given twice – this being repeated until a contact is established – is perhaps the best procedure to be followed. If no answers are obtained it may be because the frequency chosen is subject to interference elsewhere in the world and a change to another one is advisable.

If contact with a specific area is desired it is a good idea to put out a directional call. For example, 'CQ VK, CQ VK, CQ VK de G' etc (in morse code) or "CQ Australia" on phone should produce replies from that part of the world only if conditions are suitable, and all other callers should be ignored. Likewise, if another station's directional CQ is heard and this does not apply to one's own area it should not be answered. When answering a 'CQ' call it is good practice to make transmissions short and to emphasise one's own call. The other man knows his own callsign well but not the caller's! Short calls mean minimum interference; if the contacts being made by the station being called are of the 'rubber stamp' variety (RST, name, location, promise to QSL) it is quite possible that a persistent caller may find that a contact has been established and is half over by the time he finishes his transmission. It is a very good idea to listen to any station for a short while before calling; it will then be possible to estimate the speed at which contacts are being made and also *where* callers should transmit.

DX working

This is a branch of amateur radio which holds a fascination of its own and commands a considerable following. The interpretation of the expression 'DX' varies according to the frequency being used, the power and equipment of the stations involved, and even the scarcity of licensed amateurs in the country in which the DX station is located. Without doubt the most important attribute which those looking for DX contacts need is patience. The more elusive stations are often running low power, may have simple equipment, and may be situated a very long way away. They may also not be particularly anxious to talk to yet another station in the British Isles! It is therefore all the more important to obey any calling instructions which they may give, and not to cause any interference problems.

On the whole it is more rewarding to seek out a station and call it at the correct time than to call 'CQ' and expect a reply from a rare station. Many of the operators in DX locations have set operating habits and tend to appear at regular times and on regular frequencies, and the good listener-amateur will try to find out about these and then lie in wait for his 'quarry'. This author considers that DX working is very similar to fishing and requires similar study.

There are a number of news sheets published throughout the world at regular intervals which contain a great deal of valuable information on the latest expeditions and also on frequencies and times of operation of more unusual stations.

Many expedition stations make a practice of never listening on their own frequencies but indicate where they will seek replies, and the more practised mean exactly what they say. On telephony, for example, they may say "Tuning 14,190 to 14,200kHz" and under these circumstances it is often better not to choose a calling frequency at the limits of the range indicated but to find a clear spot in between and call there. Some of the less-experienced operators keep tuning to stations using the same frequency as that of the previous station contacted and if this is the case it is advisable to zero beat with the latter. A call just off the frequency of the last station contacted has often produced results for the writer. It is common practice when using morse code to indicate tuning frequencies by sending 'U 5' or 'L 5' (for example) to indicate that the listening area is 5kHz higher or lower than the station's own transmitting frequency.

Under expedition conditions the DX station is usually trying to make as many contacts as possible in limited time and therefore tends to restrict exchanges to reports and callsigns only. Once again, adequate preparatory listening should alert the caller to what is required in order for him to avoid offending others who may be waiting for their turn.

Table 21.1. The phonetic alphabet

A	Alfa	J	Juliett	S	Sierra
B	Bravo	K	Kilo	T	Tango
C	Charlie	L	Lima	U	Uniform
D	Delta	M	Mike	V	Victor
E	Echo	N	November	W	Whiskey
F	Foxtrot	O	Oscar	X	X-ray
G	Golf	P	Papa	Y	Yankee
H	Hotel	Q	Quebec	Z	Zulu
I	India	R	Romeo		

Tail-ending

The use of this procedure, which consists of dropping one's own callsign in almost zero beat with a station just completing his last transmission, is accepted by some good operators but is open to grave abuse if used unskilfully and is usually best avoided. Likewise the temptation to break into a contact already established on SSB should be resisted unless one of the participants is well known to the caller.

It should be remembered that joining into a conversation between two complete strangers is similar to joining a conversation anywhere else and that the rules of common courtesy should be exercised.

Phonetics

The use of internationally recognised words as phonetics is most useful under certain circumstances and the alphabet suggested in Appendix 16 of the Radio Regulations, Geneva, 1959 (as described on amateur licences in the UK) is shown in Table 21.1. Remember that "words used in this manner shall not be of a facetious or objectionable character" (British Amateur Licence).

The Q-code

The most common abbreviations employed by amateurs using morse code are Q-codes (Table 21.2). Where a Q-code is followed by a question mark an answer is required and where appropriate this should be qualified by the addition of a number according to the following classification: 1 – very slight; 2 – slight; 3 – moderate; 4 – severe; 5 – extreme. For example 'QRM?' means 'is there any interference' and 'QRM 5' means 'there is extremely severe interference'. A great deal of information can be transmitted in a short time using morse code if full use is made of these standard abbreviations and codes. Their use may also be of great assistance at times during contacts with

Table 21.2. The Q-code

QAV	I am calling . . .
QCM	There seems to be a defect in your transmission.
QIF	. . . (callsign of station) is using . . . (frequency).
QRA	The name of my station is . . .
QRB	The distance between our stations is . . .
QRG	Your exact frequency in kilohertz is . . .
QRH	Your frequency varies.
QRI	Your note varies.
QRJ	Your signals are very weak.
QRK	The intelligibility of your signals is (1 to 5)
QRM	There is interference.
QRN	I am being troubled by atmospheric noise.
QRO	Increase power.
QRP	Reduce power.
QRQ	Send faster (. . . words per minute).
QRS	Send more slowly (. . . words per minute).
QRT	Stop sending.
QRU	I have nothing for you.
QRV	I am ready.
QRW	Please tell . . . that I am calling him.
QRX	I will call you again.
QRZ	You are being called by . . .
QSK	I can hear between my signals (ie I am using break-in).
QSL	I give you acknowledgement of receipt.
QSM	Repeat the last message.
QSP	I will relay to . . .
QSV	Please send a series of Vs.
QSW	I will transmit on . . . kHz.
QSY	Move to . . . kHz.
QSZ	Send each word or group twice.
QTH	My location is . . .
QTR	The exact time is . . .

amateurs who are in difficulty on telephony because of language differences (but not otherwise).

Other morse code abbreviations

Table 21.3 shows other abbreviations commonly used by amateurs.

The RST code

Originally the only code available for reporting signals was the one used to denote loudness and that was graded in the arbitrary scale R1 to R9. Later, when interference became a problem, it was necessary to indicate the readability of the signal and the scale QSA1 to QSA5 was used to denote this; at the same time the signal strength was given as QRK to QRK9. Later, when attention began to be paid to the quality of morse code signals another code was evolved – this is the now universally used RST code (Table 21.4). This shows Readability (R1 to R5), Signal strength (S1 to S9), and Tone (T1 to T9). On telephony it is usual to give the report as Readability (1 to 5) and Strength (1 to 9). A telegraphy note which is chirpy is indicated by the addition of a 'c' after the tone number.

Learning morse

It will be noticed that in Table 21.5 the various morse code characters are described in dits and dahs. It is most important that anyone wishing to learn the code should not attempt to do so by memorising first of all an image of dots and dashes which then have to be translated into sounds. It is far better to learn each character as a sound entity from the very beginning.

Many amateurs look upon the task of preparing themselves for a morse test as a great ordeal and never attempt it, thereby denying themselves the pleasure of owning a higher class licence with all its associated privileges. With determination and assistance from the various morse instruction tapes which are now on the market (and available from RSGB) a very useful degree of skill can be acquired. Morse signals can often be copied over long distances even if very low power is all that is available, and under

Table 21.3. Amateur morse code abbreviations

ABT	about
AGN	again
ANT	antenna
BK	signal used to interrupt a transmission in progress
CPY	copy
CQ	general call to any station
CUL	see you later
CW	morse code
DE	from (callsign)
DR	dear
EL	element
ES	and
FB	fine business – good
FER	for
FM	from
GA	good afternoon
GD	good day
GE	good evening
GM	good morning
HPE	hope
HR	here
HVE	have
HW	how
MNI	many
MSG	message
NW	now
OM	old man
OP	operator
PSE	please
PWR	power
R	receive(d)
RPRT	report
RST	readability, signal strength and tone report
RX	receiver
SIG	signal
SRI	sorry
TEMP	temperature
TNX	thanks
TU	thank you
TX	transmitter
TXR	transceiver
UR	your
VERT	vertical
VY	very
WID	with
WX	weather
XYL	wife
YL	young lady
73	best wishes
88	love and kisses

Table 21.4. The RST system

Readability	
1	Unreadable.
2	Barely readable, some words distinguishable.
3	Readable with considerable difficulty.
4	Readable with practically no difficulty.
5	Fully readable.
Signal strength	
1	Faint signals, barely perceptible.
2	Very weak signals.
3	Weak signals.
4	Fair signals.
5	Fairly good signals.
6	Good signals.
7	Moderately strong signals.
8	Strong signals.
9	Extremely strong signals.
Tone	
1	Extremely rough hissing note.
2	Very rough AC note.
3	Rough low pitched AC note, slightly musical.
4	Rough AC note moderately musical.
5	Musically modulated note.
6	Modulated note trace of whistle.
7	Near DC note, smooth ripple.
8	Good DC note, trace of ripple.
9	Pure DC note.

Table 21.5. Morse code and sound equivalents

A	di-dah	S	di-di-dit
B	dah-di-di-dit	T	dah
C	dah-di-dah-dit	U	di-di-dah
D	dah-di-dit	V	di-di-di-dah
E	dit	W	di-dah-dah
F	di-di-dah-dit	X	dah-di-di-dah
G	dah-dah-dit	Y	dah-di-dah-dah
H	di-di-di-dit	Z	dah-dah-di-dit
I	di-dit	1	di-dah-dah-dah-dah
J	di-dah-dah-dah	2	di-di-dah-dah-dah
K	dah-di-dah	3	di-di-di-dah-dah
L	di-dah-di-dit	4	di-di-di-di-dah
M	dah-dah	5	di-di-di-di-dit
N	dah-dit	6	dah-di-di-di-dit
O	dah-dah-dah	7	dah-dah-di-di-dit
P	di-dah-dah-dit	8	dah-dah-dah-di-dit
Q	dah-dah-di-dah	9	dah-dah-dah-dah-dit
R	di-dah-dit	0	dah-dah-dah-dah-dah

An abbreviated form of '0' (zero) is sometimes used and consists of one long dash. Similarly an abbreviated '9' in the form of an 'N' (dah-dit) is in quite frequent use and can save considerable time during contest work where many numbers are being transmitted.

Punctuation	
Question mark	di-di-dah-dah-di-dit
Full stop	di-dah-di-dah-di-dah
Comma	dah-dah-di-di-dah-dah

The comma is frequently used to indicate an exclamation mark.

Procedure signals	
Stroke (/)	dah-di-di-dah-dit
Break sign (=)	dah-di-di-di-dah
End of message (+ or $\overline{AR}$)	di-dah-di-dah-dit
End of work ($\overline{SK}$)	di-di-di-dah-di-dah
Wait ($\overline{AS}$)	di-dah-di-di-dit
Preliminary call ($\overline{CT}$)	dah-di-dah-di-dah
Error	di-di-di-di-di-di-di-dit
Invitation to transmit (K)	dah-di-dah.

One dah is equal to three di's (dits). The space between part of the same letter should equal one dit. The space between two letters should equal three dits. The space between two words should equal between five and seven dits.

today's crowded band conditions they take up far less space and are less subject to being rendered unreadable by interference.

QSL cards

Most amateurs and keen listeners experience pleasure when they receive written confirmation of enjoyable contacts or verification of reception of distant stations, and in former times the QSL card was considered to be the final courtesy of the contact. Present-day costs have meant that many can no longer follow a policy of confirming every contact but it should be the aim of all those who make promises to send out QSL cards to do so, or to state the fact that they do not quite clearly at the time. In many countries the national radio society runs a central bureau where members may send their cards for worldwide distribution in bulk at reduced cost. They are also able to deposit envelopes for collection of their incoming cards from other QSL bureaux.

One of the uses of QSL cards is proof of contact for the purpose of claiming the various certificates of operating proficiency which are offered by many of the world's national societies. It is therefore important to realise that the information given on cards should be accurate, and should contain the following data as an essential minimum:

1. The fact that a contact took place must be mentioned.
2. The date and time of the contact.
3. The frequency band and mode used.
4. The signal strength of the station contacted.

It is absolutely vital that no alterations whatever are made on cards which are likely to be used for the purposes of certain award applications (eg ARRL's DX Century Club). This applies whether such alterations are made by either the sender or the recipient.

It is not necessary to use elaborate cards, and oversized ones or those weighing more than four grams should not be sent via bureaux. A confirmation written in the form of a note or letter is equally acceptable and some enthusiasts have had blank cards printed with a suitable layout for the distant station to complete and return to the sender. A rubber stamping (suitably worded and applied to the applicant's own card) is also quite satisfactory.

Amateur bands

The up-to-date frequencies on which operation is permitted and band plans are listed in the *RSGB Amateur Radio Call Book* (published by the Society every year) and at the beginning of the year in *Radio Communication*, the monthly journal of the Society.

1.8MHz (160m) operation

Special care must be exercised when using 1.8MHz (often known as 'top band') as the amateur service is only permitted to use it as a secondary user. This means that amateur privileges depend entirely on good behaviour and the utmost care being used to avoid interference to the primary users of the band. It is absolutely vital that such interference should not occur and any request from, for example, a coastal station, to close down or move frequency must be instantly obeyed. It is more important than ever to listen carefully before transmitting. Coastal stations on SSB use upper sideband; amateurs use lower sideband.

The frequencies available to amateurs in various countries on 1.8MHz vary considerably and in some are very restricted. Different parts of the United States of America are allowed to use different parts of the band, eg stations in the Hawaiian Islands are allowed the 1900–2000kHz section whereas the rest of the USA uses the lower half of the band. Users of 1.8MHz who are looking for long-distance contacts follow a special transmitting and listening pattern. For example, it is the custom for European stations wishing to make transatlantic contacts to transmit in the 1825–1830kHz region and to listen for replies at the very low end of the band – this area is relatively free of other signals in Europe and the 1825–1830kHz band (known as the 'DX window') is relatively clear in America. During the relatively few hours when propagation permits long-distance contacts on 1.8MHz, other stations are considered to be courteous if they avoid this part of the band.

3.5MHz (80m)

The Region 1 allocation on the band covers 3500–3800kHz, but in Region 2 (which includes the American continent) the upper limit is 4000kHz. Australian amateurs are restricted to the portion below 3700kHz, and the portions which may be used by some other countries (eg India and Japan) are very small indeed. The whole band is available on a shared basis with other services. In order to facilitate intercontinental working it has been recommended by Region 1 IARU that 3500–3510kHz and 3790–3800kHz should be reserved for such purposes when propagation is favourable and that the segment 3635–3650kHz be reserved for use by stations in the CIS for intercontinental contacts.

7MHz (40m)

Amateurs in Region 1 (which includes Britain) have the poorest slice of this band and are limited to the portion between 7000

Table 21.6. Region 1 HF Band Plan

Band (MHz)	Type of emission
1.810–1.838	CW only
1.838–1.840	Digimode (RTTY), CW
1.840–1.842	Digimode (RTTY), phone, CW
1.842–2.000	Phone, CW
3.500–3.510	Intercontinental DX CW
3.500–3.560	CW only, contests CW preferred segment
3.560–3.580	CW only
3.580–3.590	Digimode, CW
3.590–3.600	Digimode (packet), CW
3.600–3.620	Phone, digimode, CW
3.600–3.650	Phone, contests preferred phone segment, CW
3.650–3.775	Phone, CW
3.700–3.800	Phone, contests preferred phone segment, CW
3.730–3.740	SSTV, fax, phone, CW
3.775–3.800	Intercontinental DX phone, CW
7.000–7.035	CW only
7.035–7.040	Digimode, SSTV, fax, CW
7.040–7.045	Digimode, SSTV, fax, phone, CW
7.045–7.100	Phone, CW
10.100–10.140	CW only
10.140–10.150	Digimode, CW
14.000–14.070	CW only
14.000–14.060	CW only, contests CW preferred segment
14.070–14.089	Digimode, CW
14.089–14.099	Digimode (packet), CW
14.099–14.101	International Beacon Project (IBP)
14.101–14.112	Digimode (packet), phone, CW
14.112–14.125	Phone, CW
14.125–14.300	Phone, contests phone preferred segment, CW
14.225–14.235	SSTV, fax, phone, CW
14.300–14.350	Phone, CW
18.068–18.100	CW only
18.101–18.109	Digimode, CW
18.109–18.111	IBP
18.111–18.168	Phone, CW
21.000–21.080	CW only
21.080–21.100	Digimode, CW
21.100–21.120	Digimode (packet), CW
21.120–21.149	CW only
21.149–21.151	IBP
21.151–21.335	Phone, CW
21.335–21.345	SSTV, fax, phone, CW
21.345–21.450	Phone, CW
24.890–24.920	CW only
24.920–24.929	Digimode, CW
24.929–24.931	IBP
24.931–24.990	Phone, CW
28.000–28.050	CW only
28.050–28.120	Digimode, CW
28.120–28.150	Digimode (packet), CW
28.150–28.190	CW only
28.190–28.199	IBP time shared, CW
28.199–28.201	IBP time shared
28.201–28.225	IBP continous duty, phone, CW
28.225–28.675	Phone, CW
28.675–28.685	SSTV, fax, phone, CW
28.685–29.200	Phone, CW
29.200–29.300	Digimode (packet NBFM), phone, CW
29.300–29.550	Downlink satellites
29.550–29.700	Phone, CW

Modes. Where several modes are shown in the sub-bands the first has priority, but this has to be understood to be on a NIB (non-interference basis, according to the Radio Regulations). A mode appearing in brackets means 'preferred area of activity'. The term 'phone' means all modes of this form of transmission. Up to 10MHz LSB should be used, and above that USB on HF bands. The term 'digimode' means all forms of this mode of transmission, eg RTTY, AMTOR, PACTOR, CLOVER, ASCII and packet radio.

1.8MHz band. Those Societies which have an existing SSB allocation below 1.840MHz may continue to use it. However, they are requested to take all steps with their licensing administration to adjust the phone allocation in accordance with the Region 1 Band Plan.

3.5MHz band. 3.500–3.510 and 3.775–3.800MHz – intercontinental operation should be given priority in these segments.

Contest preferred segments. Where no DX traffic is involved, the contest preferred segments should not include 3.500–3.510MHz or 3.775–3.800MHz. Member Societies will be permitted to set other (lower) limits for national contests (within these limits). This recommendation does not apply to digimode stations.

Satellite operation frequencies. FM (and other) operators should not transmit on frequencies between 29.3 and 29.55MHz in order to avoid interference to amateur satellite downlinks.

10MHz band. SSB may be used during emergencies involving the immediate safety of life and property, and only by stations actually involved in the handling of emergency traffic. It is recommended that unmanned stations using digital modes shall avoid the use of the 10MHz band. (The 10MHz band may be used for automatic packet forwarding during local daylight hours in the areas Africa and Middle East.) News bulletins on any mode shall not be transmitted in the 10MHz band.

Unmanned transmitting stations. It is recommended that any unmanned transmitting station on HF shall only be activated under operator control except for IARU approved beacons or specially licensed transmitting stations.

Transmitting frequencies. The frequencies given in the Band Plan are understood to be 'transmitting frequencies (not those of the suppressed carrier)'.

Experimentation with NBFM packet radio on the 29MHz band. Preferred operating frequencies on each 10kHz from 29.210 to 29.290MHz inclusive should be used. A deviation of ±2.5kHz should be used with 2.5kHz as maximum modulating frequency.

and 7100kHz. Elsewhere the upper limit is 7300kHz. It should be noted that contacts with stations in other Regions (eg those in the USA) who are transmitting above 7100kHz should not be made as they come within the definition of 'cross-band' contacts according to ITU regulations.

Band plans

On all bands there are recommended sections set aside for use by each mode. In some parts of the world (eg the USA) observance of these band sub-divisions is mandatory and their use also depends on the class of licence held by the operator. An outline of the Region 1 HF Band Plan (which is supported by all IARU member societies in Europe and Africa) is set out in Table 21.6 and should be observed at all times even though its recommendations are only advisory as far as British amateurs are concerned. Band plan details are subject to change and a more detailed version of Table 21.6, as well as plans for other bands, is given in the *RSGB Amateur Radio Call Book*.

Single sideband transmissions

This mode of operation enables its users to carry on telephone-type conversations if both stations are making full use of voice control (VOX). This is most convenient as questions can be answered as they arise, interference detected as soon as it occurs, and long monologues avoided. Regulations in the UK provide that stations operating in this way should identify by giving their callsigns at intervals of not more than 15min. Stations in the USA must identify every 10min.

When transmitting on SSB (J3E) it is very important to avoid driving the final amplifier too hard and causing serious deterioration of the signal quality and severe interference to other band users. A good general rule to follow is to see that on speech

Table 21.7. Allocation of international callsign series

Series	Allocated to
A2A-A2Z	Botswana
A3A-A3Z	Tonga
A4A-A4Z	Oman
A5A-A5Z	Bhutan
A6A-A6Z	United Arab Emirates
A7A-A7Z	Qatar
A8A-A8Z	Liberia
A9A-A9Z	Bahrain
AAA-ALZ	USA
AMA-AOZ	Spain
APA-ASZ	Pakistan
ATA-AWZ	India
AXA-AXZ	Australia
AYA-AZZ	Argentina
BAA-BZZ	China
C2A-C2Z	Nauru
C3A-C3Z	Andorra
C4A-C4Z	Cyprus
C5A-C5Z	Gambia
C6A-C6Z	Bahamas
C7A-C7Z	World Meteorological Service
C8A-C9Z	Mozambique
CAA-CEZ	Chile
CFA-CKZ	Canada
CLA-CMZ	Cuba
CPA-CPZ	Bolivia
CQA-CUZ	Portugal
CVA-CXZ	Uruguay
CYA-CZZ	Canada
D2A-D3Z	Angola
D4A-D4Z	Cape Verde Republic
D5A-D5Z	Liberia
D6A-D6Z	State of the Comoros
D7A-D7Z	Korea (RK)
DAA-DRZ	Germany
DSA-DTZ	Korea (RK)
DUA-DZZ	Philippines
E2A-E2Z	Thailand
E3A-E3Z	Eritrea
EAA-EHZ	Spain
EIA-EJZ	Eire
EKA-EKZ	Armenia
ELA-ELZ	Liberia
EMA-EMZ	Ukraine
EPA-EQZ	Iran
ERA-ERZ	Moldova
ESA-ESZ	Estonia
ETA-ETZ	Ethiopia
EUA-EWZ	Bielorussia
EXA-EXZ	Russian Federation
EYA-EYZ	Tadjikstan
EZA-EZZ	Turkmenistan
FAA-FZZ	France
GAA-GZZ	United Kingdom
H2A-H2Z	Cyprus
H3A-H3Z	Panama
H4A-H4Z	Solomon Is
H6A-H7Z	Nicaragua
H8A-H9Z	Panama
HAA-HAZ	Hungary
HBA-HBZ	Switzerland
HCA-HDZ	Ecuador
HEA-HEZ	Switzerland
HGA-HGZ	Hungary
HHA-HHZ	Haiti
HIA-HIZ	Dominican Republic
HJA-HKZ	Colombia
HLA-HLZ	Korea (RK)
HMA-HMX	Korea (DPRK)
HNA-HNZ	Iraq
HOA-HPZ	Panama
HQA-HRZ	Honduras
HSA-HSZ	Thailand
HTA-HTZ	Nicaragua
HUA-HUZ	El Salvador
HVA-HVZ	Vatican City
HWA-HYZ	France
HZA-HZZ	Saudi Arabia
IAA-IZZ	Italy
J2A-J2Z	Djibouti
J3A-J3Z	Grenada and Dependencies
J4A-J4Z	Greece
J5A-J5Z	Guinea-Bissau
J6A-J6Z	St Lucia
J7A-J7Z	Dominica
J8A-J8Z	St Vincent and Dependencies
JAA-JSZ	Japan
JTA-JVZ	Mongolia
JWA-JXZ	Norway
JYA-JYZ	Jordan
JZA-JZZ	Indonesia
KAA-KZZ	USA
L2A-L9Z	Argentina
LAA-LNZ	Norway
LOA-LWZ	Argentina
LXA-LXZ	Luxembourg
LYA-LYZ	Lithuania
LZA-LZZ	Bulgaria
MAA-MZZ	United Kingdom
NAA-NZZ	USA
OAA-OCZ	Peru
ODA-ODZ	Lebanon
OEA-OEZ	Austria
OFA-OJZ	Finland
OKA-OLZ	Czech Republic
OMA-OMZ	Slovak Republic
ONA-OTZ	Belgium
OUA-OZZ	Denmark
P2A-P2Z	Papua New Guinea
P3A-P3Z	Cyprus
P4A-P4Z	Aruba
P5A-P9Z	Korea (DPRK)
PAA-PIZ	Netherlands
PJA-PJZ	Netherlands Antilles
PKA-POZ	Indonesia
PPA-PYZ	Brazil
PZA-PZZ	Surinam
RAA-RZZ	CIS
S2A-S3Z	Bangladesh
S5A-S5Z	Slovenia
S6A-S6Z	Singapore
S7A-S7Z	Republic of the Seychelles
S9A-S9Z	Sao Tome, Principe
SAA-SMZ	Sweden
SNA-SRZ	Poland
SSA-SSM	Egypt
SSN-STZ	Sudan
SUA-SUZ	Egypt
SVA-SZZ	Greece
T2A-T2Z	Tuvalu
T3A-T3Z	Kiribati
T4A-T4Z	Cuba
T5A-T5Z	Somali Republic
T6A-T6Z	Afghanistan
T7A-T7Z	San Marino
T9A-T9Z	Bosnia Herzegovina
TAA-TCZ	Turkey
TDA-TDZ	Guatemala
TEA-TEZ	Costa Rica
TFA-TFZ	Iceland
TGA-TGZ	Guatemala
THA-THZ	France
TIA-TIZ	Costa Rica
TJA-TJZ	Cameroon
TKA-TKZ	France
TLA-TLZ	Central African Republic
TMA-TMZ	France
TNA-TNZ	Congo
TOA-TQZ	France
TRA-TRZ	Gabon
TSA-TSZ	Tunisia
TTA-TTZ	Chad
TUA-TUZ	Ivory Coast
TVA-TXZ	France
TYA-TYZ	Benin
TZA-TZZ	Mali
UAA-UIZ	Russian Federation
UJA-UMZ	Uzbekistan
UNA-UQZ	Russian Federation
URA-UZZ	Ukraine
V2A-V2Z	Antigua
V3A-V3Z	Belize
V4A-V4Z	St Kitts Nevis
V5A-V5Z	Namibia
V6A-V6Z	Micronesia
V7A-V7Z	Marshall Is
V8A-V8Z	Brunei
VAA-VGZ	Canada
VHA-VNZ	Australia
VOA-VOZ	Canada
VPA-VSZ	United Kingdom (Overseas)
VTA-VWZ	India
VXA-VYZ	Canada
VZA-VZZ	Australia
WAA-WZZ	USA
XAA-XIZ	Mexico
XJA-XOZ	Canada
XPA-XPZ	Denmark
XQA-XRZ	Chile
XSA-XSZ	China
XTA-XTZ	Burkina-Faso
XUA-XUZ	Cambodia
XVA-XVZ	Vietnam
XWA-XWZ	Laos
XXA-XXZ	Portugal
XYA-XZZ	Union of Myammar
YAA-YAZ	Afghanistan
YBA-YHZ	Indonesia
YIA-YIZ	Iraq
YJA-YJZ	Vanuatu
YKA-YKZ	Syria
YLA-YLZ	Latvia
YMA-YMZ	Turkey
YNA-YNZ	Nicaragua
YOA-YRZ	Romania
YSA-YSZ	El Salvador
YTA-YUZ	Yugoslavia
YVA-YVZ	Venezuela
YZA-YZZ	Yugoslavia
Z2A-Z2Z	Zimbabwe
Z3A-Z3Z	Macedonia
ZAA-ZAZ	Albania
ZBA-ZJZ	United Kingdom (Overseas)
ZKA-ZMZ	New Zealand
ZNA-ZOZ	United Kingdom (Overseas)
ZPA-ZPZ	Paraguay
ZQA-ZQZ	United Kingdom (Overseas)
ZRA-ZUZ	Republic of South Africa
ZVA-ZZZ	Brazil
2AA-2ZZ	United Kingdom
3AA-3AZ	Monaco
3BA-3BZ	Mauritius & Dependencies
3CA-3CZ	Equatorial Guinea
3DA-3DM	Swaziland
3DN-3DZ	Fiji
3EA-3FZ	Panama
3GA-3GZ	Chile
3HA-3UZ	China
3VA-3VZ	Tunisia
3WA-3WZ	Vietnam
3XA-3XZ	Guinea
3YA-3YZ	Norway
3ZA-3ZZ	Poland
4AA-4CZ	Mexico
4DA-4IZ	Philippines
4JA-4KZ	Azerbaijan
4MA-4MZ	Venezuela
4NA-4OZ	Yugoslavia
4PA-4SZ	Sri Lanka Republic
4TA-4TZ	Peru
4UA-4UZ	United Nations
4VA-4VZ	Haiti
4XA-4XZ	Israel
4YA-4YZ	ICAO
4ZA-4ZZ	Israel
5AA-5AZ	Libya
5BA-5BZ	Cyprus
5CA-5GZ	Morocco
5HA-5IZ	Tanzania
5JA-5KZ	Colombia
5LA-5MZ	Liberia
5NA-5OZ	Nigeria
5PA-5QZ	Denmark
5RA-5SZ	Madagascar
5TA-5TZ	Mauritania
5UA-5UZ	Niger
5VA-5VZ	Togo
5WA-5WZ	Western Samoa
5XA-5XZ	Uganda
5YA-5ZZ	Kenya
6AA-6BZ	Egypt
6CA-6CZ	Syria
6DA-6JZ	Mexico
6KA-6NZ	Korea
6OA-6OZ	Somalia
6PA-6SZ	Pakistan
6TA-6UZ	Sudan
6VA-6WZ	Senegal
6XA-6XZ	Malagasy Republic
6YA-6YZ	Jamaica
6ZA-6ZZ	Liberia
7AA-7IZ	Indonesia
7JA-7NZ	Japan
7OA-7OZ	Yemen
7PA-7PZ	Lesotho
7QA-7QZ	Malawi
7RA-7RZ	Algeria
7SA-7SZ	Sweden
7TA-7YZ	Algeria
7ZA-7ZZ	Saudi Arabia
8AA-8IZ	Indonesia
8JA-8NZ	Japan
8OA-8OZ	Botswana
8PA-8PZ	Barbados
8QA-8QZ	Maldives
8RA-8RZ	Guyana
8SA-8SZ	Sweden
8TA-8YZ	India
8ZA-8ZZ	Saudi Arabia
9AA-9AZ	Croatia
9BA-9DZ	Iran
9EA-9FZ	Ethiopia
9GA-9GZ	Ghana
9HA-9HZ	Malta
9IA-9JZ	Zambia
9LA-9LZ	Sierra Leone
9MA-9MZ	Malaysia
9NA-9NZ	Nepal
9OA-9TZ	Zaire
9UA-9UZ	Burundi
9VA-9VZ	Singapore
9WA-9WZ	Malaysia
9XA-9XZ	Rwanda
9YA-9ZZ	Trinidad & Tobago

peaks the automatic level control (ALC) only just operates. By far the best plan of all is to arrange for a sample of the transmitted signal to be viewed on a monitor oscilloscope at all times – in this instance 'flat-topping' becomes immediately obvious.

Logkeeping

The UK Radiocommunications Agency requires all amateurs to keep a log book recording full details of all transmissions. A well-kept log is essential, not only because it may be required for inspection by the authorities, but also because it provides a permanent record of events, results and observations which may be interesting or useful later on. Logs may now be kept on computer using a suitable data base, and such software can be obtained from several sources.

Suitable log books which comply with regulations are readily available and may be purchased from RSGB Headquarters.

It should be made a practice to carry out regular checks for interference with other users, eg television viewers, and such episodes should be duly entered into the log book. Few amateurs seem to realise that such tests are in fact stipulated in the Amateur Licence (section entitled 'Limitations on Use', section 4.).

Power

The transmitting licence details the various power inputs permitted on the various bands, and the maximum power output permissible in the case of SSB transmitters. The correct power to use for any contact is the lowest that will ensure satisfactory reception by the distant station and, except where the transmitter is designed only for low power, it is desirable to have some easy means of reducing power input whenever conditions permit. Cross-town contacts may be carried on with very low power indeed and a great deal of interference to other stations avoided. Local contacts should be arranged to take place on frequencies which are not open and in use for DX working, and 28MHz is an ideal band to use for the purpose during sunspot minima.

STATION LAYOUT

Before starting to construct equipment it is desirable to have some idea of the ultimate layout of the completed station, for the physical size of the various units will, to some extent, depend on whether a table-top, bureau-bookcase, cupboard, or console assembly will be used. This in turn will depend on the amount of space that can be devoted to the station. Nowadays it is relatively easy to buy or construct gear, capable of being operated at the full permissible power, to occupy very little space. This is the result of the availability of semiconductor devices and miniaturised components, and the need for effective screening of the modern transmitter as a precaution against EMC problems.

Choosing a site

Amateur stations have been set up in many different places in and around the home, the location obviously depending on domestic circumstances, the available accommodation, and the operator's ambitions. Ideally the best arrangement is for a small room to be set aside for the station. Not only does this provide the maximum comfort and quietness for operating, but also affords complete safety from danger for other members of the family. However, there are other quite suitable places in most homes.

Very efficient installations have been set up in cupboards under staircases, built into bureaux or cupboards in downstairs rooms or in sheds in gardens. All these places suffer from some drawback that does not exist for the lucky person who can devote a whole room to his station. The site under the staircase will inevitably be small, dark, and difficult to ventilate (a most important factor when extended operating periods are anticipated), and the station in the downstairs room shared with other members of the family will often be noisy, while the garden shed may be too cold and damp in the winter and too hot in the summer, besides being less accessible. The choice must be made by each individual, preferably in consultation with the rest of the family, after weighing the pros and cons of each alternative. The decision should take into account the siting of antennas – which should be as near to the transmitter as possible and conversely as far from television antennas as can be arranged. The accessibility of power points should also be borne in mind, and also the availability of suitable earthing for the equipment; this is *essential*, see below. The aim should be to arrange that operating, even over long periods, is a pleasure so that maximum efficiency is obtained.

Station wiring

The care that has gone into the construction of equipment should be continued when linking it together. Connecting cables should be short and concealed wherever possible. It is advisable to use connecting leads of differing colours since not only does this facilitate rapid servicing but it is also a safety precaution. A most important point to be watched where several plugs and sockets are to be used to carry different voltages is never to use identical components because it is only too easy to make an expensive mistake and put a wrong plug in a socket. In these days of the existence of some imported equipment, wired for 110V AC input, it is a good plan to feed the mains from a Variac or auto-transformer to a distribution board with three-pin sockets of a type different from all the 230V power sources. The 110V gear is then fitted with plugs which will only enter the 110V sockets and what would be an expensive disaster cannot occur!

Circuit diagrams of all equipment should be kept readily available, as this is an obvious aid to rapid and efficient servicing. Diagrams of all the interconnecting wiring will also be found to be invaluable.

Switching

The ideal number of switches to be used to change the equipment from the receive to the send position is none! The most convenient arrangement is to have VOX operation when operating on SSB, and full break-in when using morse. If the portion of the circuit where the transmitter is keyed carries a high voltage, it is vital to key it through a relay. In any case a keying relay should be used whenever the leads from the keyed circuit to the key are of any appreciable length (see Chapter 10 for further details).

Arranging the equipment

A typical arrangement of the equipment that goes to make up the average station is shown in Fig 21.1, but there are, of course, many other ways of assembling the various items according to personal preference. One of the most important features of the station is the operator's chair, which should be as comfortable as possible. From the operating position, all the main controls should be within easy reach, and all meters clearly in view. The transmitters should be sited so that antenna feeders can be connected with the minimum of bends and of length within the room. If spaced feeder is being used this should be kept as far as possible from other objects. Ideally feeders should pass through a small window. Replacement of the glass with Perspex for easier drilling is not recommended (see Chapter 12).

For the right-handed, the morse key is normally placed on the

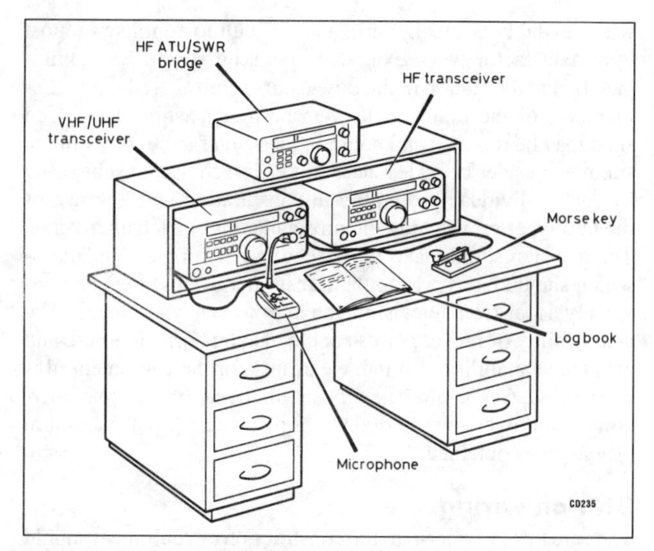

Fig 21.1. Typical table-top arrangement of an amateur station. A suitable operating table may be made from a standard flush door, polished and supported on two cupboards or two-drawer filing cabinets

right side of the operating surface but should be located for maximum comfort during long periods of use. Small pieces of insulating tape folded beneath the feet of the key will help to stop any tendency for it to wander around on the bench. The microphone can be located in any convenient position on the desk. A good flat area, with sufficient space to accommodate the key, microphone, log, and a scribbling pad is essential for comfortable operating.

Safety precautions

Safety is of paramount importance and every precaution should be taken to ensure that the equipment is perfectly safe, not only for the operator himself but also for the other members of the household, or visitors. Double-pole switches should be used for all AC supply circuits, and interconnected switches should be fitted to home-constructed equipment so that no part of it can have high voltage applied to it until the valve heaters and low-power stages have been switched on. This precaution may not only save the life of the operator, but also protects the transmitter against damage.

It should be possible to turn off power to the entire station by operating one master switch, located in a very prominent position, and all members of the household should know that in the event of an emergency this must be switched off before anything is touched. All antennas should be protected against lightning, either by the provision of a lightning conductor in the immediate vicinity, or by the use of lightning arresters. The construction of a suitable spark-gap arrester for use with open-wire feeders is shown in Fig 21.2. Arresters for coaxial cable are also available. Great care should be exercised before touching feeders which have been disconnected during a thunderstorm as they may hold a dangerous charge for a considerable time afterwards. Further information on lightning protection is given in Chapter 12. An often-overlooked precaution concerns the current-carrying capability of the mains supply to the station. An amateur station rated for the full legal power and fully equipped with ancillary apparatus can draw quite a heavy current from the supply, and when assembling the equipment it is important to calculate the current that will be drawn when everything is in use and to check that the house wiring will carry this amount. If there is any doubt new wiring should be installed. It is most

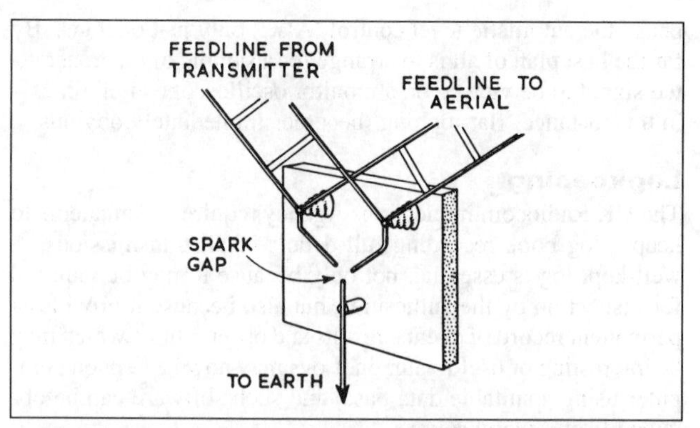

Fig 21.2. Lightning arrester. This arrangement is suitable for all types of twin-feeder systems. The mounting board should be of fireproof insulating material. Metal can be used provided that the spacing between the feeder and earth gaps is considerably smaller than the distance between the feeder gaps and the metal plate. Similar arrangements must be made for coaxial feeders

important that every amateur should develop a strict code of safety discipline for use when handling his equipment. It should be the rule never to work on equipment which is plugged into the AC supply if this can possibly be avoided. However, there are occasions when this is unavoidable and under these circumstances the following precautions should be followed:

1. Keep one hand in a pocket.
2. Never wear headphones.
3. Be certain that no part of the body is touching an object which is earthed and use a rubber or similar non-conductive covering over concrete floors.
4. Use insulated tools.
5. On transistorised equipment which has *low-voltage* power supplies which are capable of *high current*, don't wear a wrist watch with a metal strap because it can become red hot if it short-circuits the supply.

Before working on equipment of any kind, plugged into the mains or not, it is vital to make sure that all filter capacitors are fully discharged – these are capable of retaining what could be a lethal charge for a considerable time.

The vast majority of shocks sustained from electrical apparatus are derived from the 240V mains line lead. Every year there are 100 or more deaths in the UK due to electrocution, mostly as a result of accidental contact with mains voltage. Bear in mind, too, that voltages as low as 50V can cause a nasty shock in the right circumstances, eg if the hands are hot and sweaty. There is evidence to suggest that because of the different physiological effects, those who receive shocks from voltages of more than 1000V have a better chance of survival than those subjected to severe medium-voltage shocks. Voltages as low as 32V have been known to cause death – as the jingle says: "It's volts that jolts, but mils that kills."

The danger of electrocution is increased where the victim's skin resistance is lowered by dampness or perspiration, or where he grips an extensive area of 'live' metal while in good contact with earth. It is against this second possibility that particular care is needed in amateur stations. A particular hazard is equipment which has a mains-connected chassis which is being used under conditions for which it was not intended. All modern British television sets and a fair proportion of domestic broadcast receivers fall into this category; not only the 'AC/DC' sets but also – and this is not always appreciated – a large proportion of 'AC only' models. These sets are built to comply with the British Standards safety specification (BS 415) which, among other

SAFETY RECOMMENDATIONS FOR THE AMATEUR RADIO STATION

1. All equipment should be controlled by one master switch, the position of which should be well known to others in the house or club.
2. All equipment should be properly connected to a good and permanent earth. *Note*: Protective multiple earthing (PME) is now used in the majority of dwellings in Great Britain. With this the neutral is bonded to the earth at the electricity sub-station and all the metalwork (such as gas and water pipes, the latter including central heating pipes and radiators) in the house is also bonded to earth and therefore to neutral. Although it is unlikely, there is a small chance that a fault occurs in the neutral at the sub-station. If this happens, all the metalwork becomes LIVE. If an outside (RF) earth is in place, the full mains voltage will be present between the metalwork and the outside earth – DANGER! One solution is to disconnect the mains earth from all amateur equipment and rely on the outside earth. Alternatively, all the amateur equipment should be placed far enough from any metal connected to the mains earth so that the two cannot be touched at the same time. Another idea is to fit a heavy cable to the outside (RF) earth, feed it through some ferrite rings (toroids) and bond it to the mains earth. The equipment is connected to the earth rod side of the toroids. The purpose of the toroids is to prevent RF interference from the mains earth from getting to the equipment. Four turns on five toroids using 10mm^2 insulated wire is recommended. (Note A and reference [1])
3. Wiring should be adequately insulated, especially where voltages greater than 500V are used. Terminals should be suitably protected.
4. Transformers operating at more than 100V RMS should be fitted with an earthed screen between the primary and secondary windings or have them in separate slots in the bobbin.
5. Capacitors of more than 0.01μF capacitance operating in power packs, modulators, etc (other than for RF bypass or coupling) should have a bleeder resistor connected directly across their terminals. The value of the bleeder resistor should be low enough to ensure rapid discharge. A value of $1/C$ megohms (where C is in microfarads) is recommended. The use of earthed probe leads for discharging capacitors in case the bleeder resistor is defective is also recommended. (Note B). Low-leakage capacitors, such as paper and oil-filled types, should be stored with their terminals short-circuited to prevent static charging.
6. Indicator lamps should be installed showing that the equipment is live. These should be clearly visible at the operating and test position. Faulty indicator lamps should be replaced immediately. Gas-filled (neon) lamps and LEDs are more reliable than filament types.
7. Double-pole switches should be used for breaking mains circuits on equipment. Fuses of correct rating should be connected to the equipment side of each switch in the live lead only. (Note C.) Always switch off before changing a fuse. The use of AC/DC equipment should be avoided.
8. In metal-enclosed equipment install primary circuit breakers, such as micro-switches, which operate when the door or lid is opened. Check their operation frequently.
9. Test prods and test lamps should be of the insulated pattern.
10. A rubber mat should be used when the equipment is installed on a floor that is likely to become damp.
11. Switch off before making any adjustments. If adjustments must be made while the equipment is live, use one hand only and keep the other in your pocket. Never attempt two-handed work without switching off first. Use good-quality insulated tools for adjustments.
12. Do not wear headphones while making internal adjustments on live equipment.
13. Ensure that the metal cases of microphones, morse keys etc are properly connected to the chassis.
14. Do not use meters with metal zero-adjusting screws in high-voltage circuits. Beware of live shafts projecting through panels, particularly when metal grub screws are used in control knobs.
15. Antennas should not, under any circumstances, be connected to the mains or other HT source. Where feeders are connected through a capacitor, which may have HT on the other side, a low resistance DC path to earth should be provided (RF choke).
16. Antennas must be designed with due allowance for wind loading. For this, guidance from the antenna manufacturer is necessary and British Standard (BS) CP3 Chapter 5 for guyed masts and BS 8100 for self-supporting masts should be consulted [2].
17. Certain chemicals occur in electronic devices which are harmful. Notable amongst these are the *polychlorinated biphenyls* (PCBs) which have been used in the past to fill transformers and high-voltage capacitors and *beryllium oxide* (BeO) which is used as an insulator *inside* the case of some high-power semiconductors. In the case of PCBs, the names to look out for on capacitors are: ARACLOR, PYROCHLOR, PYRANOL, ASBESTOL, NO-FLAMOL, SAF-T-KUL and others [3]. If one of these is present in a device, it must be disposed of carefully. The local Health and Safety Authority will advise. In the case of beryllium oxide, the simple rule is DON'T OPEN ANY DEVICE THAT *MAY* CONTAIN IT.

Note A. – Owing to the common use of plastic water main and sections of plastic pipe in effecting repairs, it is no longer safe to assume that a mains water pipe is effectively connected to earth. Steps must be taken, therefore, to ensure that the earth connection is of sufficiently low resistance to provide safety in the event of a fault. Checks should be made whenever repairs are made to the mains water system in the building.

Note B. – A 'wandering earth lead' or an 'insulated earthed probe lead' is an insulated lead permanently connected via a high-power 1kΩ resistor or a 15W 250V lamp at one end to the chassis of the equipment; at the other end a suitable length of bare wire with an insulated handle is provided for touch contacting the high-potential terminals to be discharged.

Note C. – Where necessary, surge-proof fuses can be used.

precautions, lays down that it should be impossible for the little finger to touch any part of the chassis, and also specifies that double pole on-off switches are used. Old radio sets often do not comply with these standards, or may have damaged backs, or be used with exposed control spindles or grub screws. If there is any such equipment in your station, the exposed parts should be checked with a neon to make certain that they are not 'live' with the on/off switch in either position. Remember that a single-pole switch in the neutral lead will leave the chassis 'live' even with the set apparently turned off. After such checks have been carried out non-reversible plugs should be fitted to AC supply leads.

It is wise to check all three-pin supply sockets in the house to see whether they have been correctly wired; all too often this is not the case. A three-pin socket with the 'earth' socket at the top should have the 'neutral' socket on the bottom left, and the live 'line' socket on the bottom right – these directions apply when looking into the socket and must of course be reversed when

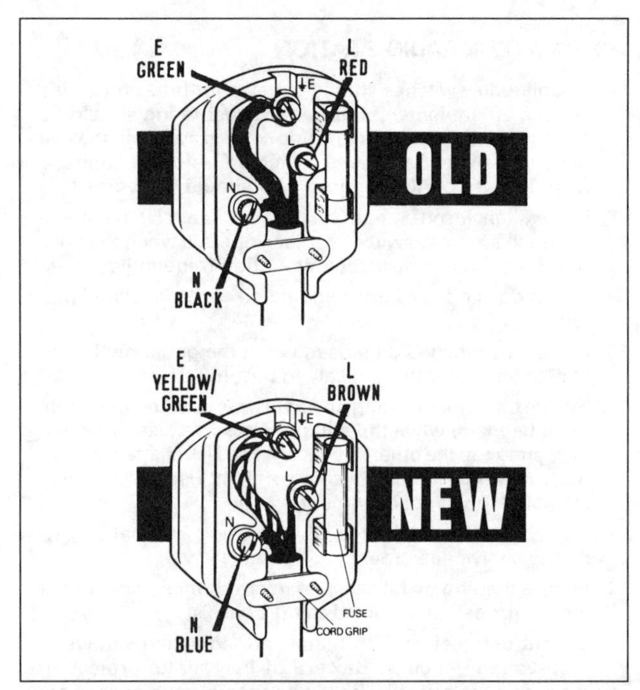

Fig 21.3. The correct wiring for three-pin plugs and sockets. To test that a socket is correctly wired, a lamp should light if connected between 'L' and 'N' or 'L' and 'E' but not when connected between 'N' and 'E'. A neon bulb will glow when touched against 'L'

looking at the back of the socket for wiring purposes (see Fig 21.3). Correct colour coding of leads is: 'Live', brown; 'Neutral', blue; 'Earth', yellow and green. It is very important to note that this coding may not apply to the wiring on some older or imported equipment, and the manufacturer's instructions should be very carefully studied before plugging into the supply. The use of modern fused plugs is recommended with fuses selected to just carry the necessary current.

An even greater hazard, because it is seldom anticipated, can arise under fault conditions on equipment fitted with a double wound (ie 'isolating') transformer of the type so often used in amateur equipment. It is by no means unusual or unknown for the primary winding to short-circuit to the screening plate between the primary and secondary, the core, or to one of the secondary windings, so that the chassis of the equipment becomes 'live'. Such equipment will often continue to operate quite normally and can thus represent a very real danger over a considerable period. The best safeguard against this danger is to ensure that the screening plate between the primary and the other windings, the core and the chassis, are all effectively earthed. The earth connection must be of very low resistance otherwise the supply fuses may not blow. These fuses should be of the minimum practicable rating – it is no use having a 50Ω resistance to earth and a 10A fuse – if this should be the case, the size of the electricity bills may be surprising but the hazard is likely to remain undetected!

Another source of danger is the electric tool which has developed a fault and which has a 'live' casing. This can happen, for example, with soldering irons and electric drills. The modern electric drill is designed with this danger in mind, but even so it should be remembered that in industry it is recommended that such tools should only be used in 'earth free' areas – which is far from the state of affairs in the average amateur installation. A very careful check should be kept on the leads to all such tools, and any 'tingles' felt when they are in use must be investigated immediately.

Many amateurs fit extra power sockets in their stations and the control arrangements may call for quite a lot of semi-permanent AC wiring and switching. In the UK these should always conform with the high standards laid down in the IEE Wiring Regulations. These are rather formidable reading for the non-professional but a number of books giving sound advice on modern wiring practice, based on the IEE recommendations, have been published and can often be obtained from local libraries. Advice can also be obtained from the offices of local electricity boards. In most countries overseas, similar regulations exist and operators in these areas are recommended to obtain copies or seek the advice of the supply authorities. Finally, taking the worst possible event into consideration, the operator and members of his household are advised to familiarise themselves with the procedures for the treatment of electric shock. While every amateur will try to construct and maintain his station so that it is completely safe for himself and any others who may visit it, there is always the possibility that an accident may occur. Owing to a component failure a visitor may receive an electric shock, or an antenna or mast may fall and injure someone or damage property. Such an occurrence can result in a legal action, and in these days can result in the award of very substantial damages against the person held to be responsible for the accident. This risk can, and should, be insured against, either by an extension to the existing Householder's Comprehensive policy or by taking out a separate public liability policy if only fire insurance is held. The annual premium for this will only be quite a small amount and readers cannot be too strongly urged to consult their insurance advisers over this matter. It is a wise precaution to have a fire extinguisher of the type suitable for use on electrical equipment in the radio room. The best type is that which directs a stream of carbon dioxide gas on to the burning area; the powder and Halon types may be used but the output from the former is liable to cause further damage to such electrical equipment with which it comes into contact.

Planning permission for amateur radio masts and antennas

Town and Country Planning legislation is aimed at securing 'the proper use of land' and, as part of this, to safeguard amenities, natural and otherwise, enjoyed by local inhabitants. In the UK, permission is been required before undertaking 'development' which is defined as 'carrying out of building, engineering, mining or other operations in, on, over or under land, or the making of any material change in the use of any buildings or other land'.

Such applications are considered by Local Planning Authorities (usually the District Council in the case of amateur radio masts or antennas) and will entail an assessment of the pros and cons, balancing the issues involved and making a final judgement upon the merits of the proposal. The Secretary of State for the Environment and his opposite numbers for Wales, Scotland and Northern Ireland have appellate roles over these decisions and can also issue guidance on procedures. Councils also may publish guidance policies, either in the form of 'Local Plans' relating to all or part of their areas or on specific topics. Look out for them.

The overall definition of 'development' set out in the first paragraph is subject to some exclusions set out in the legislation (currently in Section 55 of the Town and Country Planning Act 1990 as amended) and *two* are of relevance to radio amateurs. The first is that the carrying out for the maintenance, improvement or other alteration of any building of works which affect only the interior of the building or do not materially affect the external appearance of the building (the word 'materially' being very relevant here). The second is the use of any buildings or other land within the curtilage (the area attached to the dwelling) of a dwelling house for any purpose incidental to the enjoyment of the dwelling house as such. Amateur radio, as a hobby, falls within this and thus the 'transmitting' component of the hobby cannot be challenged in the way that, say, a PMR base station might be.

In addition to the exclusions set out above, the Act provides for General Development Orders (GDOs), to be prepared which in effect represent automatic permission for things which may not warrant developers or local authorities having to go to the trouble and cost of preparing and processing applications of little moment to the community or where the 'benefits', in a national context, will override any possible costs. Such GDO permissions can be qualified in a variety of ways not practicable in general legislation. This Secondary Statutory Regulation system has a more simple Parliamentary procedure and thus can be revised more easily. The provisions which allow the extensions to a dwelling by 10 or 15% of its cubic capacity, subject to not coming in front of the forward-most part, nor exceeding the maximum height of the roof of the original dwelling, for many years allowed substantial dormer windows to be built on the front and/or back faces of dwelling roofs and so providing a very cheap way of enlarging dwellings. However, in 1988 this provision was changed to preclude such dormers on roofs facing a public highway and thus reducing some unsightly alterations to street scenes.

Satellite antennas on houses have also caused much criticism, more so on blocks of flats (where in fact the GDO permissions do not apply), but also because the 90cm size constraint produced difficulties in those northern parts of the British Isles where the satellite footprints were less strong. Adverse amenity impacts from the numerous dishes in street schemes and the adverse picture effects were dealt with in July 1991 by changes which varied the size according to the location in the country, allowing smaller (45cm) dishes on the chimney and 70cm elsewhere on the building but retained the 90cm below the roof top in the South West and the area North of the Humber.

There are also two other parts of the GDO which relate specifically to telecommunications; one is the Telecom Code Systems (BT and Mercury) and the other to other commercial microwave systems. Neither is relevant to radio amateurs as, apart from these limitations to commercial systems, they specifically exclude siting on dwelling houses.

The 'exclusions', set out in the third paragraph above, together with the automatic permissions referred to in the fourth paragraph may well meet the needs of many amateurs. Remember that neighbours and others may not regard a mast and antenna as a thing of beauty!

In other cases it will be necessary to make an application for permission. *A study of the brochure 'Planning Permission – Advice to Members', which is available free to members from RSGB Headquarters, is strongly recommended.* Think carefully what you really *need* – not what you would like because a local amateur has received permission for a pair of 17m telescopic masts with a variety of HF, VHF and UHF antennas. Don't forget that his house is set in a pocket of woodland and that your corner bungalow on an estate will pose a very different set of 'costs' when your application is being considered. Do make sure that your application sets out fully why you *need* whatever it is that you show on the application forms and plans. Think about the siting and design of your proposal. Slimline (box-section or tubular) masts are more acceptable than square or triangular lattice masts although they won't carry so high wind loads at such heights. Some antennas are of better *visual* design than others – they don't have so many excrescences or breaks in their general lines which draw attention to them in an irritating way. Look at your proposal as if through a neighbour's eyes and show them where and why you want it. If you want a high mast to get antennas up to reduce EMC problems already existing, can you get help to have a report prepared to illustrate how the new set-up will reduce close-in field strengths? Don't try to mislead with antenna details. The Planning Authority is fully entitled to require details and a failure to supply them may result in permission subject to a condition which requires approval of detail drawings of the antenna before the mast is erected! At least one inspector endorsed this course of action; others have dismissed appeals because of the absence of clear information about antennas.

Make sure the mast, any guys and the antenna elements are *entirely* within your own land and do not trespass onto next door's airspace; if you do encroach, you must get his permission in writing and refer to this in supporting material, saying why the space is needed. Remember that an antenna which cannot be accommodated on your own land invites the comment "It's too big for this site".

Make good plans and fair copies of any explanatory material and show them to your neighbours; the Council has to consult your neighbours and it is better if you get to them first and explain things. Ask the Case Officer in the Planning Department if he or his chief has any query. If so, offer to come into the office and explain things.

Some amateurs believe in erecting things first and hoping that nobody will notice them! If you are lucky, you may get away with this for four years after which the Council cannot serve an enforcement notice. It often happens, however, that a new neighbour arrives after three years and can cause problems. In any event, if you have altered the antenna at all within the four years, you may well have put the mast and all at risk. This is currently a grey area and the enforcement law is being steadily tightened. The four-year rule now only applies to building or engineering works and changes of use can now be challenged for longer periods from the change taking place.

One final point. You will know, if you have got the planning permission brochure mentioned above, that the RSGB has a panel of advisers who may be able to help you, perhaps to avoid problems or to overcome them by negotiation with the authority or to appeal to the Secretary of State if you have been refused permission.

REFERENCES

[1] 'The killing ground', Peter Chadwick, G3RZP, *Radio Communication* 1987, p404.

[2] 'Wind loading', D J Reynolds, G3ZPF, *Radio Communication* 1988, pp252-255, 340-344, 629.

[3] 'Technical Topics', Pat Hawker, G3VA, *Radio Communication* 1984, p295

22 General data

Capacitance

The capacitance of a parallel-plate capacitor is:

$$C = \frac{0.224\,KA}{d} \text{ picofarads}$$

where K is the dielectric constant (air = 1.0), A is the area of dielectric (sq in), and d is the thickness of dielectric (in).

If A is expressed in centimetres squared and d in centimetres, then:

$$C = \frac{0.0885\,KA}{d} \text{ picofarads}$$

For multi-plate capacitors, multiply by the number of dielectric thicknesses.

The capacitance of a coaxial cylinder is:

$$C = \frac{0.242}{\log_{10}(D/d)} \text{ picofarads per centimetre length}$$

where D is the inside diameter of the outer and d is the outside diameter of the inner.

Capacitors in series or parallel

The effective capacitance of a number of capacitors in *series* is:

$$C = \cfrac{1}{\cfrac{1}{C_1} + \cfrac{1}{C_2} + \cfrac{1}{C_3} + \text{etc}}$$

The effective capacitance of a number of capacitors in *parallel* is:

$$C = C_1 + C_2 + C_3 + \text{etc}$$

Characteristic impedance

The characteristic impedance Z_0 of a feeder or transmission line depends on its cross-sectional dimensions.

(i) *Open-wire line*:

$$Z_0 = 276 \log_{10} \frac{2D}{d} \text{ ohms}$$

where D is the centre-to-centre spacing of wires (in) and d is the wire diameter (in).

(ii) *Coaxial line*:

$$Z_0 = \frac{138}{\sqrt{K}} \log_{10} \frac{d_o}{d_i} \text{ ohms}$$

where K is the dielectric constant of insulation between the conductors (eg 2.3 for polythene, 1.0 for air), d_o is the inside diameter of the outer conductor and d_i is the dameter of the inner conductor.

Decibel

The *decibel* is the unit commonly used for expressing the relationship between two power levels (or between two voltages or two currents). A decibel (dB) is one-tenth of a *bel* (B). The number of decibels N representing the ratio of two power levels P_1 and P_2 is 10 times the common logarithm of the power ratio, thus:

$$\text{The } ratio\ N = 10 \log_{10} \frac{P_2}{P_1} \text{ decibels}$$

If it is required to express *voltage* (or *current*) ratios in this way, they must relate to identical impedance values, ie the two different voltages must appear across equal impedances (or the two different currents must flow through equal impedances). Under such conditions the *power* ratio is proportional to the square of the *voltage* (or the *current*) ratio, and hence:

$$N = 20 \log_{10} \frac{V_2}{V_1} \text{ decibels}$$

$$N = 20 \log_{10} \frac{I_2}{I_1} \text{ decibels}$$

Dynamic resistance

In a parallel-tuned circuit at resonance the dynamic resistance is:

$$R_D = \frac{L}{Cr} = Q\omega L = \frac{Q}{\omega C} \text{ ohms}$$

where L is the inductance (henrys), C is the capacitance (farads), r is the effective series resistance (ohms), Q is the Q-value of the coil and $\omega = 2\pi \times$ frequency (hertz).

Frequency – wavelength – velocity

The velocity of propagation of a wave is:

$$v = f\lambda \text{ centimetres per second}$$

where f is the frequency (hertz) and λ is the wavelength (centimetres).

For electromagnetic waves in free space the velocity of propagation v is approximately 3×10^8m/s and, if f is expressed in kilohertz and λ in metres:

$$f = \frac{300{,}000}{\lambda} \text{ kilohertz}$$

$$\lambda = \frac{300{,}000}{f} \text{ metres}$$

$$\text{Free space } \frac{\lambda}{2} = \frac{492}{\text{MHz}} \text{ feet}$$

$$\text{Free space } \frac{\lambda}{4} = \frac{246}{\text{MHz}} \text{ feet}$$

Note that the true value of v is 2.99776×10^8m/s.

Impedance

The impedance of a circuit comprising inductance, capacitance and resistance in series is:

$$Z = \sqrt{R^2 + \left(\omega L - \frac{1}{\omega C}\right)^2}$$

where R is the resistance (ohms), L is the inductance (henrys), C is the capacitance (farads) and $\omega = 2\pi \times$ frequency (farads).

Inductors in series or parallel

The total effective value of a number of inductors connected in *series* (assuming that there is no mutual coupling) is given by:

$$L = L_1 + L_2 + L_3 + \text{etc}$$

If they are connected in *parallel*, the total effective value is:

$$L = \frac{1}{\frac{1}{L_1} + \frac{1}{L_2} + \frac{1}{L_3} + \text{etc}}$$

When there is mutual coupling M, the total effective value of two inductors connected in series is:

$$L = L_1 + L_2 + 2M \text{ (windings aiding)}$$

$$\text{or } L = L_1 + L_2 - 2M \text{ (windings opposing)}$$

Ohm's Law

For a unidirectional current of constant magnitude flowing in a metallic conductor:

$$I = \frac{E}{R} \qquad E = IR \qquad R = \frac{E}{I}$$

where I is the current (amperes), E is the voltage (volts) and R is the resistance (ohms).

Power

In a DC circuit, the power developed is given by:

$$W = EI = \frac{E^2}{R} = I^2R \text{ watts}$$

where E is the voltage (volts), I is the current (amperes) and R is the resistance (ohms).

Q

The Q-value of an inductance is given by:

$$Q = \frac{\omega L}{R}$$

where L is the inductance (henrys), R is the effective resistance (ohms) and $\omega = 2\pi \times$ frequency (hertz).

Reactance

The reactance of an inductance and a capacitance respectively is given by:

$$X_L = \omega L \text{ ohms}$$

$$X_C = \frac{1}{\omega C} \text{ ohms}$$

where L is the inductance in henrys, C is the capacitance in farads and $\omega = 2\pi \times$ frequency (hertz).

The total reactance of an inductance and a capacitance in series is $X_L - X_C$.

Resistors in series or parallel

The effective value of several resistors connected in *series* is:

$$R = R_1 + R_2 + R_3 + \text{etc}$$

When several resistors are connected in *parallel* the effective total resistance is:

$$R = \frac{1}{\frac{1}{R_1} + \frac{1}{R_2} + \frac{1}{R_3} + \text{etc}}$$

Resonance

The resonant frequency of a tuned circuit is given by:

$$f = \frac{1}{2\pi\sqrt{LC}} \text{ hertz}$$

where L is the inductance (henrys) and C is the capacitance (farads).

If L is in microhenrys (μH) and C is picofarads (pF), this formula becomes:

$$f = \frac{10^3}{2\pi\sqrt{LC}} \text{ megahertz}$$

The basic formula can be rearranged thus:

$$L = \frac{1}{4\pi^2 f^2 C} \text{ henrys}$$

$$C = \frac{1}{4\pi^2 f^2 L} \text{ farads}$$

Since $2\pi f$ is commonly represented by ω, these expressions can be written as:

$$L = \frac{1}{\omega^2 C} \text{ henrys}$$

$$C = \frac{1}{\omega^2 L} \text{ farads}$$

See Figs 22.1 and 22.2.

Time constant

For a combination of inductance and resistance in series the time constant (ie the time required for the current to reach $1/\varepsilon$ or 63% of its final value) is given by:

$$t = \frac{L}{R} \text{ seconds}$$

where L is the inductance (henrys) and R is the resistance (ohms).

For a combination of capacitance and resistance in series, the time constant (ie the time required for the voltage across the capacitance to reach $1/\varepsilon$ or 63% of its final value) is given by:

$$t = CR \text{ seconds}$$

where C is the capacitance (farads) and R is the resistance (ohms).

Transformer ratios

The ratio of a transformer refers to the ratio of the number of turns in one winding to the number of turns in the other winding. To avoid confusion it is always desirable to state in which sense the ratio is being expressed, eg the 'primary-to-secondary' ratio n_p/n_s. The turns ratio is related to the impedance ratio thus:

$$\frac{n_p}{n_s} = \sqrt{\frac{Z_p}{Z_s}}$$

where n_p is the number of primary turns, n_s is the number of secondary turns, Z_p is the impedance of the primary circuit (ohms) and Z_s is the impedance of the secondary circuit (ohms).

COIL WINDING

Most inductors for tuning in the HF bands are single-layer coils and they are designed as follows. Multilayer coils will not be dealt with here.

The inductance of a single-layer coil is given by:

$$L\,(\mu H) = \frac{D^2 \times T^2}{457.2 \times D + 1016 \times L}$$

where D is the diameter of the coil (millimetres), T is the number of turns and L is the length (millimetres). Alternatively:

$$L\,(\mu H) = \frac{R^2 \times T^2}{9 \times R + 10 \times L}$$

where R is the radius of the coil (inches), T is the number of turns and L is the length (inches).

Note that when a ferrite or iron dust core is used, the inductance will be increased by up to twice the value without the core. The choice of which to use depends on frequency. Generally, ferrite cores are used at the lower HF bands and iron dust cores at the higher. At VHF, the iron dust cores are usually coloured purple. Cores need to be moveable for tuning but fixed thereafter and this can be done with a variety of fixatives. A strip of flexible polyurethane foam will do.

Designing inductors with ferrite pot cores

This is a simple matter of taking the *factor* given by the makers and multiplying it by the square of the number of turns.

Example

A RM6-S pot core in 3H1 grade ferrite has a 'factor' of 1900 nanohenrys for one turn. Therefore 100 turns will give an inductance of:

$$100^2 \times 1900\text{nH} = 10000 \times 1900\text{nH} = 19\text{mH}$$

There are a large number of different grades of ferrite; for example, the same pot as above is also available in grade 3E4 with a 'factor' of 3300. Manufacturers' literature should be consulted to find these 'factors'.

Table 22.1

Diameter	Approx SWG	Turns/cm	Turns/in
1.5	16–17	6.6	16.8
1.25	18	7.9	20.7
1.0	19	9.9	25
0.8	21	12.3	31
0.71	22	13.9	35
0.56	24	17.5	45
0.50	25	19.6	50
0.40	27	24.4	62
0.315	30	30.8	78
0.25	33	38.5	97
0.224	34–35	42.7	108
0.20	35–36	47.6	121

Note: SWG is Imperial standard wire gauge. The diameters listed are those which appear to be most popular; ie they are listed in distributor's catalogues. The 'turns/cm' and 'turns/in' are for enamelled wire.

Table 22.3. Wire table

Diameter (mm)	Approx SWG	Max current (A)	Fusing current (A)	Resistance at 20°C (Ω/km)
2.5	12	7.6	325	3.5
2.0	14	4.9	225	5.4
1.5	16–17	2.7	147	9.7
1.0	19	1.2	81	22
0.71	22	0.61	46	43
0.5	26	0.30	28	87
0.25	32	0.076	10	351
0.20	36	0.049	7.1	541

'Max current' is the carrying capacity at 1.55A/mm^2. This is a very conservative figure and can usually be doubled. The 'fusing current' is approximate since it depends also on thermal conditions, ie if the wire is thermally insulated, it will fuse at a lower current.

Table 22.2. Coaxial cables

Type	Nominal impedance (Ω)	Outside diameter (mm)	Velocity factor	Capacitance (pF/m)	Maximum RF voltage (kV)	Attenuation per 10m of cable 10MHz (dB)	100MHz (dB)	1000MHz (dB)
UR43	50	5.0	0.66	100	2.6	0.4	1.3	4.5
UR67	50*	10.3	0.66	100	6.5	0.2	0.68	2.5
UR70	75*	5.8	0.66	67	1.8	0.5	1.5	5.2
UR76	50*	5.0	0.66	100	2.6	0.5	1.6	5.3
UR95	50	2.3	0.66	100	1.3	0.9	2.7	6.9
UR202	75*	5.1	0.84	56	—	0.4	1.1	4.2
UR203	75	7.25	0.84	56	—	0.2	0.8	2.7
CT100	75	6.65	0.84	56	—	0.2	0.6	2.1
RG58BU	50	4.95	0.66	100	3.5	0.5	1.7	5.6
RG58CU	50*	4.95	0.66	100	2.5	0.5	1.7	5.6
RG59BU	75	6.15	0.66	68	3.5	0.5	1.5	4.6
Min RG59	75	3.7	0.84	51	—	0.4	1.2	3.9
RG62AU	95	6.15	0.84	44	—	0.3	0.9	2.9
Min RG62	95	3.8	0.84	41	—	0.4	1.4	4.5
RG174AU	50*	2.8	0.66	101	2.1	0.3	0.9	2.9
RG178PE	50*	1.83	0.85	99	—	1.5	4.8	16
RG179PE	75*	2.54	0.85	69	—	1.2	4.0	13
RG402U	50	3.58†	—	—	—	—	—	—
RG405U	50	2.20†	—	—	—	—	—	—

* Indicates cable with flexible core.
† Indicates cable with solid drawn outer, ie rigid.
There are many further types of coaxial cable but these are the most popular ones; ie those listed in distributors' catalogues. The characteristics of others are listed in earlier editions of this handbook.

Table 22.4. Basic SI units

Quantity	Name of unit	Unit symbol
Electric current	ampere	A
Length	metre	m
Luminous intensity	candela	cd
Mass	kilogramme	kg
Thermodynamic temperature	kelvin	K
Time	second	s

Table 22.5. Derived SI units

Physical quantity	SI unit	Unit symbol
Electric capacitance	farad	F = A s/V
Electric charge	coulomb	C = A s
Electric potential	volt	V = W/A
Electric resistance	ohm	Ω = V/A
Force	newton	N = kg m/s^2
Frequency	hertz*	Hz = s^{-1}
Illumination	lux	lx = lm/m^2
Inductance	henry	H = V s/A
Luminous flux	lumen	lm = cd sr
Magnetic flux	weber	Wb = V s
Magnetic flux density	tesla†	T = Wb/m^2
Power	watt	W = J/s
Work, energy, quantity of heat	joule	J = N m

* Hertz is equivalent to cycle per second.
† Tesla is equivalent to weber per square centimetre.

Table 22.6. Comparison of Centigrade and Fahrenheit thermometer scales

Centigrade	Fahrenheit	Centigrade	Fahrenheit
−50	−58	+80	+176
−45	−49	+85	+185
−40	−40	+90	+194
−35	−31	+95	+203
−30	−22	+100	+212
−25	−13	+105	+221
−20	−4	+110	+230
−15	+5	+115	+239
−10	+14	+120	+248
−5	+23	+125	+257
0	+32	+130	+266
+5	+41	+135	+275
+10	+50	+140	+284
+15	+59	+145	+293
+20	+68	+150	+302
+25	+77	+155	+311
+30	+86	+160	+320
+35	+95	+165	+329
+40	+104	+170	+338
+45	+113	+175	+347
+50	+122	+180	+356
+55	+131	+185	+365
+60	+140	+190	+374
+65	+149	+195	+383
+70	+158	+200	+392
+75	+167		

Table 22.7. Colour coding for glass fuses

Colour	Rating (mA)	Colour	Rating (A)
Green/yellow	10	Green	0.75
Red/turquoise	15	Blue	1.0
Eau-de-Nil	25	Light blue	1.5
Salmon pink	50	Purple	2.0
Black	60	Yellow and purple	2.5
Grey	100	White	3.0
Red	150	Black and white	5.0
Brown	250	Orange	10.0
Yellow	500		

Note that this coding does not apply to the ceramic-bodied fuse commonly found in 13A plugs etc.

Table 22.8. Conversion factors

To convert	into	multiply by	Conversely
Amp hours	Coulombs	3600	2.778×10^{-4}
Atmospheres	Lb/sq in	14.70	0.068
Centigrade	Kelvin	°C + 273 = K	K − 273 = °C
Cubic inches	Cubic feet	5.787×10^{-4}	1728
Cubic inches	Cubic metres	1.639×10^{-5}	6.102×10^{4}
Degrees (angular)	Radians	1.745×10^{-2}	57.3
Dynes	Pounds	2.248×10^{-6}	4.448×10^{5}
Ergs	Foot pounds	7.376×10^{-8}	1.356×10^{7}
Feet	Centimetres	30.48	3.281×10^{-2}
Foot pounds	Kilowatt hours	3.766×10^{-7}	2.655×10^{6}
Gausses	Lines per sq in	6.452	0.155
Grams	Dynes	980.7	1.02×10^{-3}
Grams per cm	Pounds per in	5.6×10^{-3}	178.6
Horse power	Kilowatts	0.746	1.341
Inches	Centimentres	2.54	0.3937
Kilograms	Pounds (lb)	2.205	0.454
Kilometres	Feet	3281	3.048×10^{-4}
Kilometres	Nautical miles	0.540	1.853
Kilometres	Statute miles	0.621	1.609
Kilowatt hours	Joules	3.6×10^{6}	2.778×10^{-7}
Kilowatt hours	HP hours	1.341	0.7457
Knots	Miles per hour	1.1508	0.869
Lamberts	Candles per sq cm	0.3183	3.142
Lamberts	Candles per sq in	2.054	0.4869
Lumens per sq ft	Foot candles	1	1
Lux	Foot candles	0.0929	10.764
Metres	Feet	3.28	0.3048
Metres	Yards	1.094	0.9144
Miles per hour	Feet per second	1.467	0.68182
Nepers	Decibels	8.686	0.1151
Tons	Pounds	2240	4.464×10^{-4}
Watts	Ergs per second	10^{7}	10^{-7}

Table 22.9. Greek alphabet

Capital letters	Small letters	Greek name	English equivalent
Α	α	Alpha	a
Β	β	Beta	b
Γ	γ	Gamma	g
Δ	δ	Delta	d
Ε	ε	Epsilon	e
Ζ	ζ	Zeta	z
Η	η	Eta	é
Θ	θ	Theta	th
Ι	ι	Iota	i
Κ	κ	Kappa	k
Λ	λ	Lambda	l
Μ	μ	Mu	m
Ν	ν	Nu	n
Ξ	ξ	Xi	x
Ο	ο	Omicron	o
Π	π	Pi	p
Ρ	ρ	Rho	r
Σ	σ	Sigma	s
Τ	τ	Tau	t
Υ	υ	Upsilon	u
Φ	φ	Phi	ph
Χ	χ	Chi	ch
Ψ	ψ	Psi	ps
Ω	ω	Omega	ō

Table 22.10. Useful twist drill sizes

Screw size	2	3	4	5	6
Clearance drill	2.10	3.10	4.10	5.10	6.10
Tapping drill	1.55	2.65	3.50	4.50	5.20

Where all sizes are in millimetres.

Twist drill no	¼in 0.250 in	1 0.228 in	9 0.196 in	17 0.173 in	24 0.152 in	32 0.116 in	43 0.089 in	50 0.070 in
Clearance for woodscrew no	14	12	10	8	6	4	2	0
Clearance for BA	0	1	2	3	4	6	8	10
Tapping size for BA	—	—	0	1	2	4	6	8

Table 22.11. Component colour codes

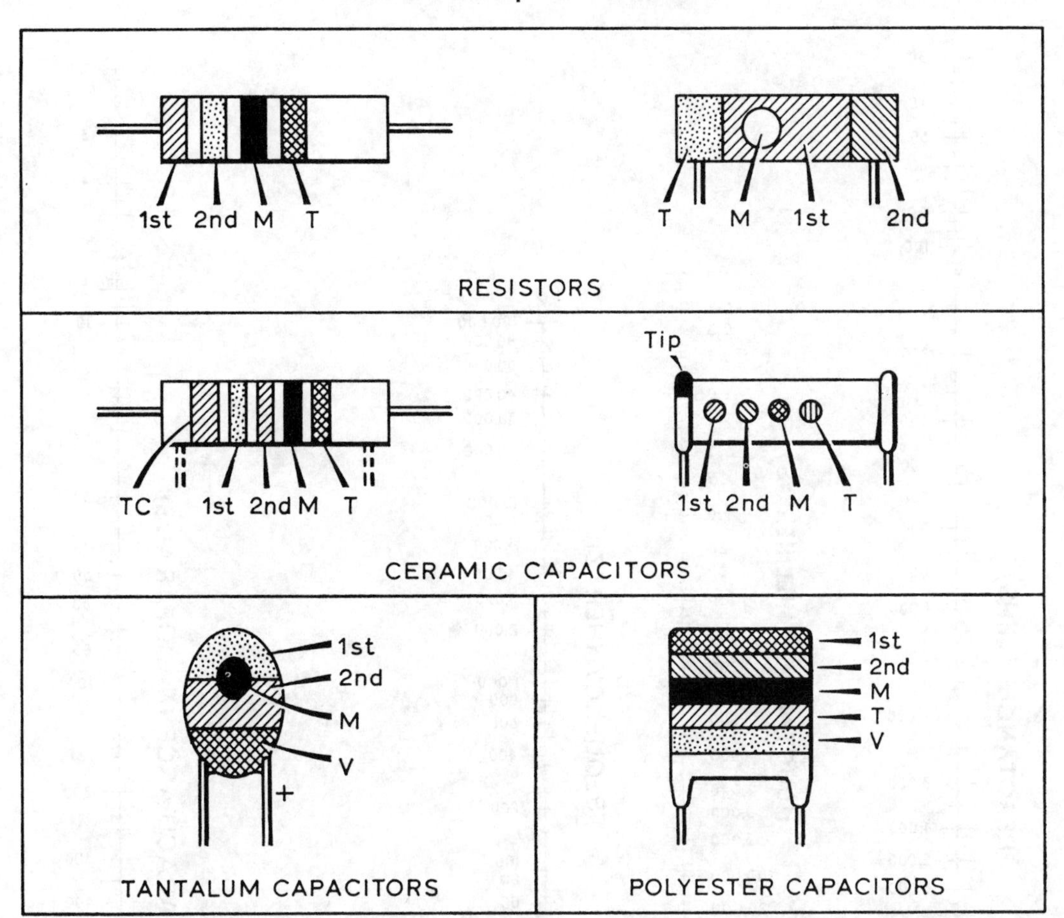

Colour	Significant figure (1st, 2nd)	Decimal multiplier (M)	Tolerance (T) (per cent)	Temp coeff (TC) (parts/10^6/°C)	Voltage (V) (tantalum cap)	Voltage (V) (polyester cap)
Black	0	1	±20	0	10	—
Brown	1	10	±1	−30	—	100
Red	2	100	±2	−80	—	250
Orange	3	1000	±3	−150	—	—
Yellow	4	10,000	+100, −0	−220	6.3	400
Green	5	100,000	±5	−330	16	—
Blue	6	1,000,000	±6	−470	20	—
Violet	7	10,000,000	—	−750	—	—
Grey	8	100,000,000	—	+30	25	—
White	9	1,000,000,000	±10	+100 to −750	3	—
Gold	—	—	±5	—	—	—
Silver	—	—	±10	—	—	—
Pink	—	—	—	—	35	—
No colour	—	—	±20	—	—	—

Units used are ohms for resistors, picofarads for ceramic and polyester capacitors, and microfarads for tantalum capacitors.

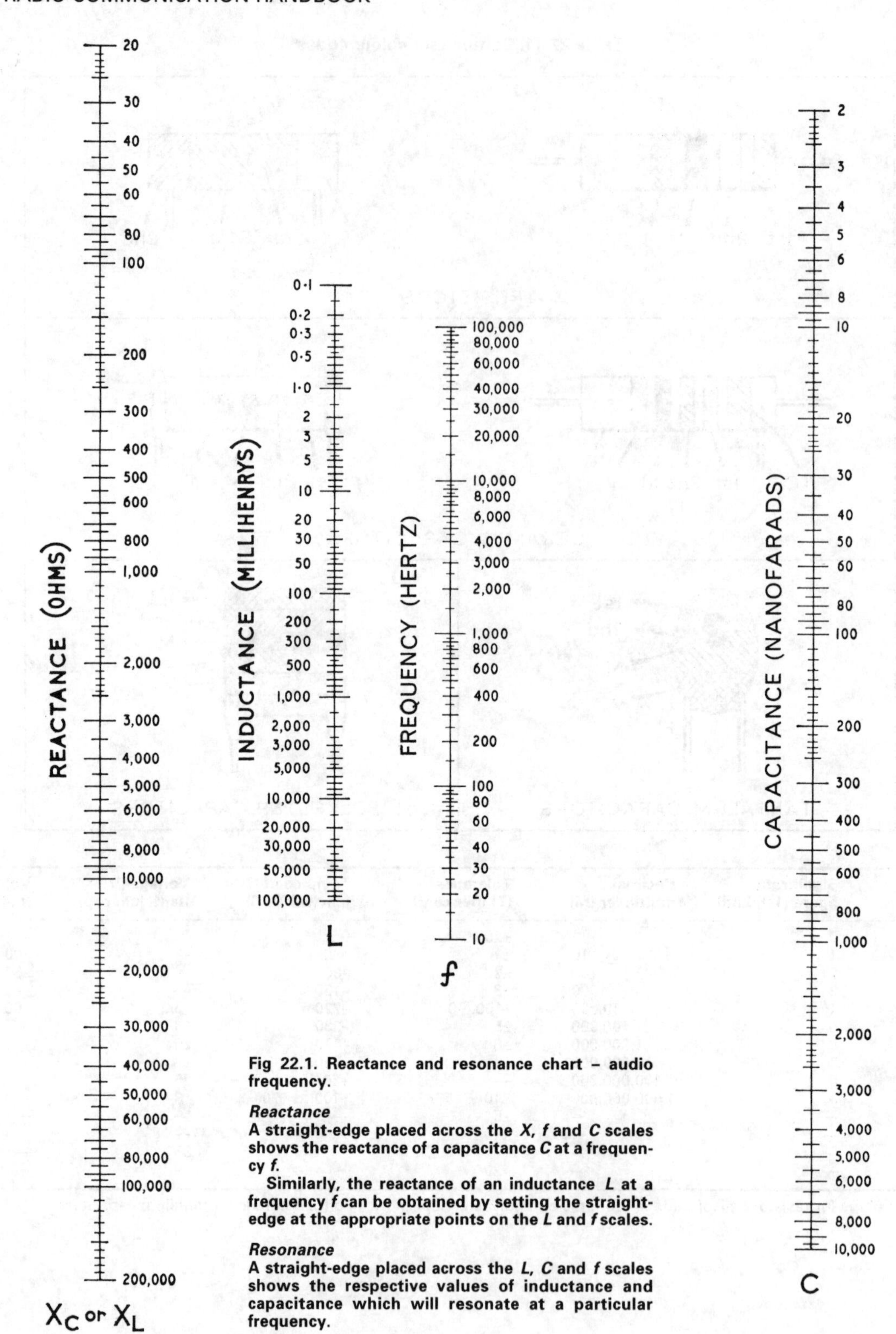

Fig 22.1. Reactance and resonance chart – audio frequency.

Reactance

A straight-edge placed across the *X*, *f* and *C* scales shows the reactance of a capacitance *C* at a frequency *f*.

Similarly, the reactance of an inductance *L* at a frequency *f* can be obtained by setting the straight-edge at the appropriate points on the *L* and *f* scales.

Resonance

A straight-edge placed across the *L*, *C* and *f* scales shows the respective values of inductance and capacitance which will resonate at a particular frequency.

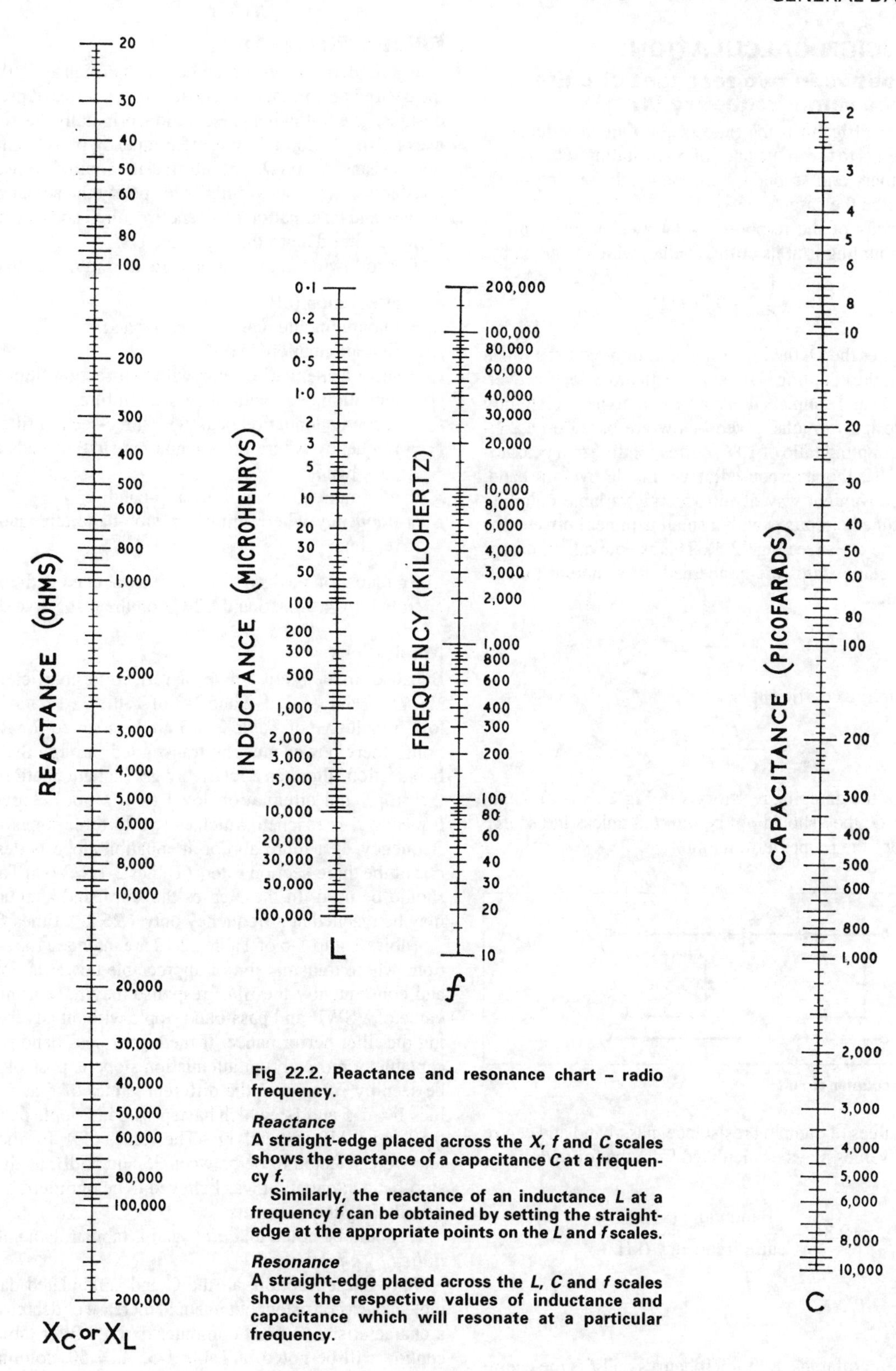

Fig 22.2. Reactance and resonance chart – radio frequency.

Reactance
A straight-edge placed across the *X*, *f* and *C* scales shows the reactance of a capacitance *C* at a frequency *f*.

Similarly, the reactance of an inductance *L* at a frequency *f* can be obtained by setting the straight-edge at the appropriate points on the *L* and *f* scales.

Resonance
A straight-edge placed across the *L*, *C* and *f* scales shows the respective values of inductance and capacitance which will resonate at a particular frequency.

FILTER DESIGN CALCULATIONS

Coupling between two resonant circuits tuned to the same frequency [1]

The coupling coefficient is the ratio of the mutual inductance between windings to the inductance of one winding. This is true where the primary and secondary are identical; for simplicity, this is taken to be the case.

When the peak of the response is flat and on the point of splitting, the coupling is at its critical value, which is given by:

$$k_c = \frac{1}{Q} \quad (Q_p = Q_s)$$

Hence, the higher the Q, the lower the coupling required. In an IF transformer, the coupling is set at the critical value; however, for use in wide-band couplers it is convenient to have it slightly higher. The design formulae given below are based on a coupling/critical coupling ratio of 1.86, corresponding to a peak-to-trough ratio of 1.2:1, or a response flat within 2dB over the band.

The most convenient way of introducing variable coupling between two tuned circuits is with a small trimmer between the 'hot' ends of the coils (see Fig 22.3). This is equivalent, except where phase relationships are concerned, to a mutual inductance of the value:

$$M = \frac{C_1}{C_1 + C} L$$

Hence the coupling coefficient is:

$$k = \frac{C_1}{C_1 + C}$$

The purpose of the damping resistors R in Fig 22.3 is to obtain correct circuit Q; they should not be omitted unless the source or load provides the proper termination.

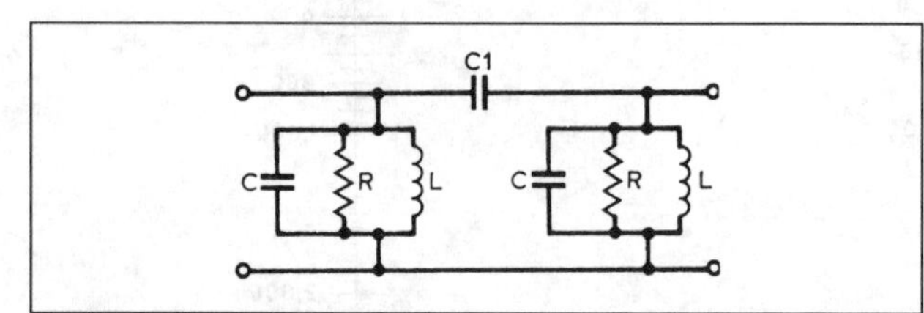

Fig 22.3. Basic coupler circuit

Given set values of damping resistance, pass-band and centre frequency, all values may be calculated from the following formulae:

$$k = 0.84 \frac{\text{Bandwidth (kHz)}}{\text{Centre frequency (kHz)}}$$

$$Q = \frac{1.86}{k} \qquad L = \frac{R}{2\pi f Q} \qquad C = \frac{1}{L}\left(\frac{1}{2\pi f}\right)^2$$

where C is in microfarads, L in microhenrys and f is the centre frequency in megahertz. R is in ohms.

Note that C includes all strays: if the calculated value of C is less than the estmated strays on any band, a lower value of R should be used. The bandswitch can increase the strays to 20pF or more.

Coupling capacitance C_1 is given by:

$$C_1 = \frac{k}{1-k} C$$

Elliptic filters [1]

Using modern design procedure, a 'normalised' filter having the desired performance is chosen from a series of precalculated designs. The following presentation, originally due to W3NQN, uses normalisation to a cut-off frequency of 1Hz and termination resistance of 1Ω, and all that is required to ascertain the constants of a practical filter is to specify the actual cut-off frequency and termination resistance required and to scale the normalised filter data to those parameters.

The following abbreviations are used in these curves:

A = attenuation (dB),
A_p = maximum attenuation in pass-band,
f_4 = first attenuation peak,
f_2 = second attenuation peak with two-section filter or third attenuation peak with three-section filter,
f_6 = second attenuation peak with three-section filter,
f_{co} = frequency where the attenuation first exceeds that in the pass-band,
A_s = minimum attenuation in stop-band,
f_s = frequency where minimum stop-band attenuation is first reached.

The attenuation peaks f_4, f_6 or f_2 are associated with the resonant circuits L4/C4, L6,C6 and L2/C2 on the respective diagrams.

Applications

Because of their low value of reflection coefficient (P) and VSWR, Tables 1-1, 1-2 and 1-3 of Table 22.12 and Tables 2-1 to 2-6 inclusive of Table 22.13 are best suited for RF applications where power must be transmitted through the filter. The two-section filter has a relatively gradual attenuation slope and the stop-band attenuation level (A_s) is not achieved until a frquency f_s is reached which is two to three times the cut-off frequency. If a more abrupt attenuation slope is desired, then one of the three-section filters (Tables 2-1 to 2-6 in Table 22.13) should be used. In these cases the stop-band attenuation level may be reached at a frequency only 1.25 to 2 times f_{co}.

Tables 1-4 to 1-6 of Table 22.12 are intended for AF applications where transmission of appreciable power is not required, and consequently the filter response may have a much higher value of VSWR and pass-band ripple without adversely affecting the filter performance. If the higher pass-band ripple is acceptable, a more abrupt attenuation slope is possible. This can be seen by comparing the different values of f_s at 50dB in Tables 1-4, 1-5 and 1-6 which have pass-band ripple peaks of 0.28, 0.50 and 1.0dB respectively. The values of A_s for the audio filters were selected to be between 35 and 55dB, as this range of stop-band attenuation was believed to be optimum for most audio filtering requirements.

It should be noted that *all* C and L tabular data *must be multiplied by a factor of* 10^{-3}.

With one exception, all the C and L tabulated data of each table have a consistent but unequal increase or decrease in value, a characteristic of most computer-derived filter tables. An exception will be noted in Table 1-5, A_s = 50, column C1. The original author points out that this is not an error but arose from a minor change necessitated in the original computer program to eliminate unrealisable component values.

How to use the filter tables

After the desired cut-off frequency has been chosen, the frequencies of f_s and the attenuation peaks may be calculated by multiplying their corresponding tabular values by the required cut-off frequency (f_{co}). The component values of the desired

Table 22.12. Two-section elliptic-function filters normalised for a cut-off frequency of 1Hz and terminations of 1Ω

Reflection coefficient, VSWR & Ap	As dB	fs Hz	f4 Hz	f2 Hz	C1 Farad	C3 Farad	C5 Farad	C2 Farad	L2 Henry	C4 Farad	L4 Henry
Table 1–1	70	3·24	3·39	5·42	110·4	235	103·5	4·34	199·0	11·72	187·5
p = 4%	65	2·92	3·07	4·88	109·6	233	101·0	5·39	197·9	14·67	183·7
VSWR = 1·08	60	2·56	2·68	4·24	108·2	229	96·9	7·20	195·8	19·88	177·3
Ap = 0·0069 dB	55	2·37	2·48	3·90	107·2	227	93·8	8·57	194·3	23·9	172·7
	50	2·13	2·23	3·48	105·5	223	88·6	10·88	192·0	31·0	164·7
Table 1–2	70	3·07	3·22	5·13	118·3	243	110·8	4·73	203	12·78	191·0
p = 5%	65	2·79	2·92	4·64	117·4	241	108·3	5·82	202	15·82	187·2
VSWR = 1·11	60	2·46	2·57	4·06	116·0	237	104·0	7·67	200	21·2	180·7
Ap = 0·011 dB	55	2·28	2·39	3·75	115·0	234	100·8	9·07	198·5	25·3	175·9
	50	2·06	2·16	3·36	113·2	230	95·6	11·43	196·0	32·4	168·1
Table 1–3	70	2·79	2·92	4·64	138·4	262	129·6	5·59	210	15·09	196·4
p = 8%	65	2·56	2·68	4·24	137·4	259	126·9	6·75	208	18·32	192·4
VSWR = 1·17	60	2·28	2·39	3·75	135·9	255	122·4	8·72	206	23·9	185·7
Ap = 0·028 dB	55	2·06	2·16	3·36	134·2	251	117·4	10·98	204	30·6	178·4
	50	1·887	1·970	3·05	132·2	245	111·8	13·55	201	38·4	170·3
Table 1–4	55	1·701	1·773	2·71	217	317	190·8	18·03	191·5	49·7	162·3
p = 25%	50	1·556	1·617	2·44	213	306	181·3	22·8	187·3	63·8	151·9
VSWR = 1·67	45	1·440	1·493	2·22	209	295	170·6	28·3	182·7	80·9	140·5
Ap = 0·28 dB	40	1·325	1·369	1·988	203	279	155·8	36·4	176·0	108·0	125·1
	35	1·236	1·273	1·802	195·9	262	139·2	46·4	168·2	144·3	108·3
Table 1–5	55	1·618	1·690	2·56	248	348	214	21·3	181·4	58·7	151·0
p = 33%	50	1·481	1·540	2·30	249	336	210	27·4	174·9	76·7	139·3
VSWR = 2·00	45	1·369	1·416	2·08	244	318	197·5	34·7	169·2	99·8	126·5
Ap = 0·50 dB	40	1·270	1·308	1·878	238	299	177·3	44·4	161·7	133·7	110·8
	35	1·186	1·222	1·700	229	280	163·3	57·0	153·9	177·6	95·5
Table 1–6	55	1·528	1·591	2·39	314	401	276	28·3	156·9	77·5	129·1
p = 45%	50	1·407	1·459	2·16	308	381	260	35·5	153·3	99·6	119·4
VSWR = 2·67	45	1·245	1·313	1·898	306	365	247	46·6	150·7	135·0	108·9
Ap = 1·00 dB	40	1·217	1·250	1·755	296	341	227	59·2	138·9	176·2	92·0
	35	1·145	1·174	1·597	284	315	203	75·4	131·6	237	77·7
	As dB	fs Hz	f4 Hz	f2 Hz	L1 Henry	L3 Henry	L5 Henry	L2 Henry	C2 Farad	L4 Henry	C4 Farad

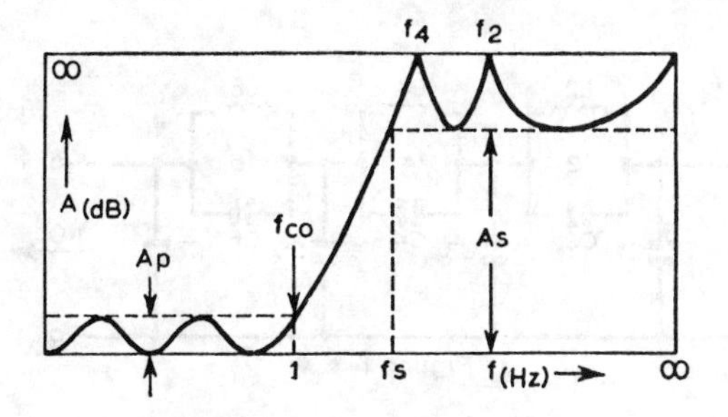

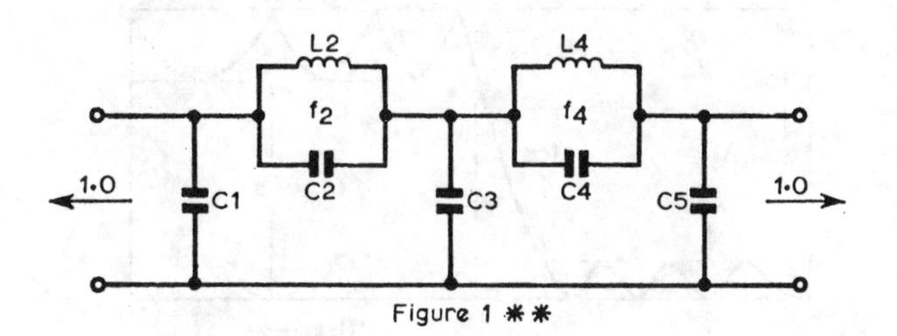

Figure 1 **

* All tabulated data of C and L must be multiplied by 10^{-3}; for example, in Table 1–1, the normalized value of C1 is 110·4 x 10^{-3}, for As = 70dB

** In the above tabulation, the top column headings pertain to Figure 1 while the bottom column headings pertain to Figure 2

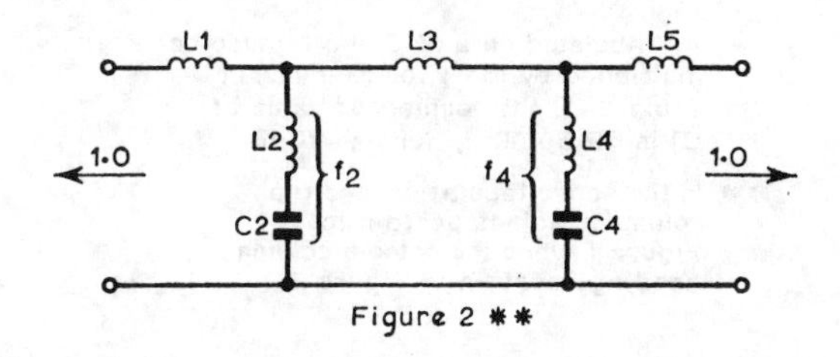

Figure 2 **

filter are then found by multiplying C and L values in the tables by $1/Rf_{co}$ and R/f_{co} respectively.

Example 1

A low-pass audio filter to attenuate speech frequencies above 3kHz with a minimum attenuation of 40dB for all frequencies above 3.8kHz, and to be terminated in resistive loads of 1.63kΩ. (This odd value has been chosen merely for convenience in demonstrating the design procedure.)

The circuit of Fig 1 in the Tables is chosen because this has the minimum number of inductors, which are both more expensive and have higher losses than do capacitors. The parameters are:

$$A_s = 40\text{dB} \qquad f_{co} = 3\text{kHz} \qquad R = 1.63\text{k}\Omega$$

From Table 1-5 of Table 22.12, A_s = 40dB, calculate f'_s, f'_4 and f'_2. (Numbers with the prime (′) are the frequency and component values of the final design: numbers without the prime are from the filter catalogue.)

(1) $f'_s = f_s(f_{co}) = 1.270 \times 3 = 3.81\text{kHz}$.

$f'_4 = f_4(f_{co}) = 1.308 \times 3 = 3.92\text{kHz}$

$f'_2 = f_2(f_{co}) = 1.878 \times 3 = 5.63\text{kHz}$

Table 22.13. Three-section elliptic-function filters normalised for a cut-off frequency of 1Hz and terminations of 1Ω

REFLECTION COEFFICIENT, VSWR & Ap	As dB	fs Hz	f4 Hz	f6 Hz	f2 Hz	C1 Farad	C3 Farad	C5 Farad	C7 Farad	C2 Farad	L2 Henry	C4 Farad	L4 Henry	C6 Farad	L6 Henry
Table 2–1	70	2·00	2·04	2·49	4·35	79·6	209	201	63·1	7·42	180·2	30·9	196·4	26·3	155·2
p = 1%	64	1·836	1·876	2·27	3·95	78·3	204	194·8	58·2	9·10	178·4	38·4	187·6	33·0	148·3
VSWR = 1·02	60	1·743	1·780	2·15	3·72	77·3	200	190·3	54·5	10·35	177·1	44·1	181·4	38·2	143·5
	55	1·624	1·657	1·990	3·41	75·8	194·2	183·5	48·5	12·42	175·2	53·8	171·4	47·2	135·6
Ap = 0·43 x 10⁻³dB	50	1·524	1·554	1·854	3·15	74·1	187·8	176·3	41·8	14·75	172·8	65·3	160·7	58·0	127·1
Table 2–2	70	1·836	1·876	2·27	3·95	93·8	222	212	75·7	8·34	194·8	35·8	201	29·4	167·0
p = 2%	64	1·701	1·737	2·09	3·61	92·5	216	205	70·7	10·08	193·1	43·8	191·6	36·2	160·0
	60	1·624	1·657	1·990	3·41	91·5	212	200	67·1	11·35	191·6	49·8	185·1	41·3	154·8
VSWR = 1·04	55	1·524	1·554	1·854	3·15	89·9	206	192·7	61·1	13·47	189·4	60·0	174·8	50·2	146·7
Ap = 1·7 x 10⁻³dB	50	1·414	1·440	1·702	2·86	87·5	196·9	182·1	52·2	16·70	186·1	76·4	160·0	64·8	135·0
Table 2–3	70	1·743	1·780	2·15	3·72	104·2	230	219	84·7	9·06	203	39·7	201	31·8	172·5
p = 3%	65	1·624	1·657	1·990	3·41	102·8	224	211	79·7	10·84	201	48·1	191·8	38·7	165·4
	60	1·524	1·554	1·854	3·15	101·2	217	203	74·1	12·86	198·3	57·8	181·6	46·8	157·5
VSWR = 1·06	55	1·440	1·466	1·737	2·92	99·5	211	194·8	67·9	15·12	195·9	69·0	170·8	56·3	149·1
Ap = 3·9 x 10⁻³dB	50	1·367	1·391	1·636	2·73	97·6	203	186·2	61·2	17·65	193·1	82·2	159·2	67·5	140·1
Table 2–4	70	1·701	1·737	2·09	3·61	113·0	236	224	93·0	9·37	208	41·6	202	32·7	177·0
p = 4%	65	1·589	1·621	1·942	3·32	111·6	230	217	88·0	11·18	205	50·2	192·3	39·6	170·0
	60	1·494	1·523	1·813	3·07	110·0	224	208	82·4	13·20	203	60·0	181·9	47·6	161·9
VSWR = 1·08	55	1·414	1·440	1·702	2·86	108·3	217	199·6	76·3	15·47	201	71·4	171·1	57·0	153·4
Ap = 6·9 x 10⁻³dB	50	1·325	1·347	1·576	2·61	105·6	206	187·5	67·3	18·94	196·9	89·7	155·6	72·2	141·3
Table 2–5	70	1·662	1·696	2·04	3·51	120·6	242	229	99·9	9·77	211	43·9	201	33·9	179·4
p = 5%	65	1·556	1·586	1·897	3·23	119·2	235	221	94·9	11·61	209	52·7	191·1	40·9	172·0
	60	1·466	1·494	1·774	3·00	117·6	228	212	89·3	13·67	206	62·8	180·8	49·0	164·1
VSWR = 1·11	55	1·367	1·391	1·636	2·73	115·2	219	199·7	81·0	16·81	203	78·8	166·2	61·9	152·7
Ap = 11 x 10⁻³dB	51·5	1·325	1·347	1·576	2·61	113·8	213	193·4	76·5	18·57	201	88·2	158·3	69·5	146·6
	50	1·305	1·327	1·548	2·55	113·1	211	190·2	74·1	19·51	199·7	93·2	154·4	73·7	143·5
Table 2–6	70	1·556	1·586	1·897	3·23	139·7	252	237	116·2	11·30	214	52·0	193·4	39·1	180·0
p = 8%	65	1·466	1·494	1·774	3·00	138·1	245	228	110·9	13·30	212	61·9	183·5	46·6	172·5
	60	1·390	1·415	1·668	2·79	136·3	238	218	105·0	15·54	210	73·2	173·0	55·3	164·4
VSWR = 1·17	55	1·325	1·347	1·576	2·61	134·4	230	208	98·6	18·05	207	86·3	161·9	65·4	155·8
Ap = 28 x 10⁻³dB	50	1·252	1·271	1·471	2·39	131·4	218	193·9	89·2	21·9	202	107·3	146·1	81·6	143·4
	As dB	fs Hz	f4 Hz	f6 Hz	f2 Hz	L1 Henry	L3 Henry	L5 Henry	L7 Henry	L2 Henry	C2 Farad	L4 Henry	C4 Farad	L6 Henry	C6 Farad

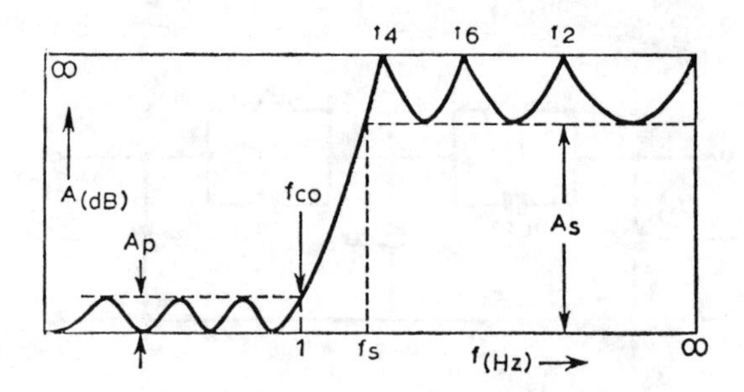

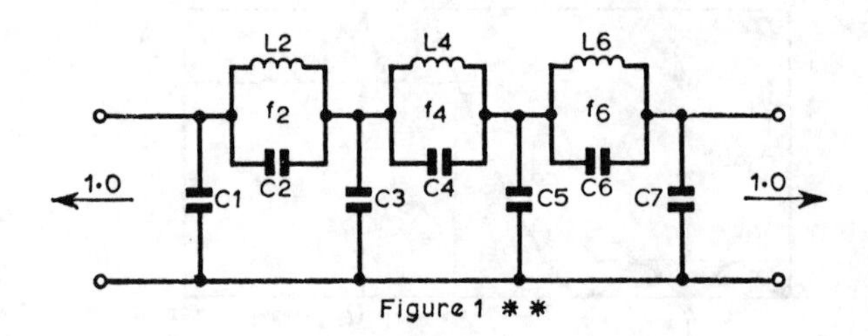

Figure 1 * *

* All tabulated data of C and L must be multiplied by 10^{-3}; for example, in Table 2-1, the normalized value of C1 is 79·6 x 10^{-3}, for As = 70dB

** In the above tabulation, the top column headings pertain to Figure 1 while the bottom column headings pertain to Figure 2

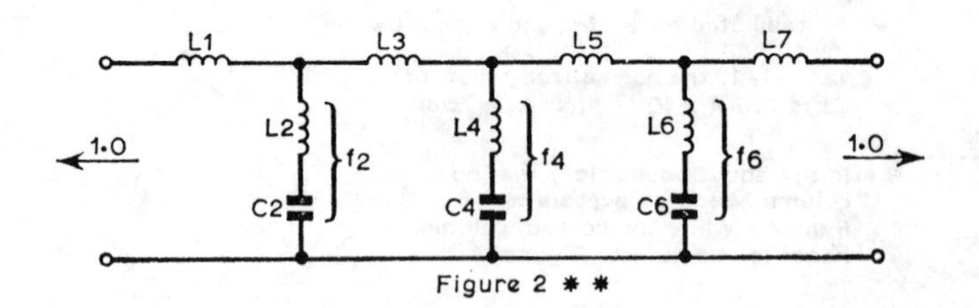

Figure 2 * *

(2) Calculate factors $1/Rf_{co}$ and R/f_{co} to determine the capacitor and inductor values.

$$1/Rf_{co} = 1/(1.63 \times 10^3)(3 \times 10^3)$$
$$= 1/(4.89 \times 10^6)$$
$$= 0.2045 \times 10^{-6}$$
$$R/f_{co} = (1.63 \times 10^3)/(3 \times 10^3) = 0.543$$

(3) Calculate the component values of the desired filter by multiplying all the catalogue tabular values of C by $1/Rf_{co}$ and L by R/f_{co} as shown below:

$$C'1 = C1(1/Rf_{co}) = (238 \times 10^{-3})(0.2045)10^{-6} = 0.0487\mu F$$
$$C'3 = C3(1/Rf_{co}) = (299 \times 10^{-3})(0.2045)10^{-6} = 0.0612\mu F$$
$$C'5 = C5(1/Rf_{co}) = (177.3 \times 10^{-3})(0.2045)10^{-6} = 0.0363\mu F$$
$$C'2 = C2(1/Rf_{co}) = (44.4 \times 10^{-3})(0.2045)10^{-6} = 0.00908\mu F$$
$$C'4 = C4(1/Rf_{co}) = (133.7 \times 10^{-3})(0.2045)10^{-6} = 0.00273\mu F$$
$$L'2 = L2(R/f_{co}) = (161.7 \times 10^{-3})(0.543) = 87.8mH$$
$$L'4 = L4(R/f_{co}) = (110.8 \times 10^{-3})(0.543) = 60.1mH$$

These calculations, which may conveniently be performed with a pocket calculator, complete the design of the filter.

It should be noted that all the elliptic-function data is based

Table 22.14. Butterworth filters

K	C_1 L_1	C_2 L_2	C_3 L_3	C_4 L_4	C_5 L_5	C_6 L_6	C_7 L_7	C_8 L_8	C_9 L_9	C_{10} L_{10}
1	2.000	—	—	—	—	—	—	—	—	—
2	1.4142	1.4142	—	—	—	—	—	—	—	—
3	1.000	2.000	1.000	—	—	—	—	—	—	—
4	0.7654	1.8478	1.8478	0.7654	—	—	—	—	—	—
5	0.6180	1.6180	2.000	1.6180	0.6180	—	—	—	—	—
6	0.5176	1.4142	1.9319	1.9319	1.4142	0.5176	—	—	—	—
7	0.4450	1.2470	1.8019	2.000	1.8019	1.2470	0.4450	—	—	—
8	0.3902	1.1111	1.6629	1.9616	1.9616	1.6629	1.1111	0.3902	—	—
9	0.3473	1.000	1.5321	1.8794	2.000	1.8794	1.5321	1.000	0.3473	—
10	0.3129	0.9080	1.4142	1.7820	1.9754	1.9754	1.7820	1.4142	0.9080	0.3129

on the use of lossless components and purely resistive terminations. Therefore components of the highest possible Q should be used and precautions taken to ensure that the filter is properly terminated.

It will be noticed that some rather curious values of both capacitance and inductance may emerge from the calculations but these may be rationalised to the extent that the tolerance on the values of components need not be closer than some ±3%.

Example 2

A three-section low-pass filter to suppress harmonics at the output of a transmitter covering the HF bands up to a frequency of 30MHz with a matching impedance of 50Ω and a minimum attenuation in the stop-band of 50dB.

The parameters are, from Table 2-2 (circuit Fig 2) of Table 12.13:

$$A_s = 50\text{dB} \qquad f_{co} = 30\text{MHz} \qquad R = 50\Omega$$

From Table 2-2 (bottom line) of Table 22.13, calculate f'_s, f'_4, f'_6 and f'_2.

(1) $f'_s = f_s(f_{co}) = 1.414 \times 30 = 42.4\text{MHz}$.
$f'_4 = f_4(f_{co}) = 1.440 \times 30 = 43.2\text{MHz}$
$f'_6 = f_6(f_{co}) = 1.702 \times 30 = 51\text{MHz}$
$f'_2 = f_2(f_{co}) = 2.860 \times 30 = 85.8\text{MHz}$

(2) Calculate factors $1/Rf_{co}$ and R/f_{co} to determine the capacitor and inductor values respectively.

$1/Rf_{co} = 1/50(30 \times 10^{-6}) = 66 \times 10^{-11}$
$R/f_{co} = 50/(30 \times 10^6) = 1.67 \times 10^{-6}$

(3) Calculate component values of the desired filter by multiplying all tabular values of C by $1/Rf_{co}$ and L by R/f_{co}, remembering to multiply *all* values in the tables by 10^{-3}.

$C'2 = C2(66 \times 10^{-11})$
$= (186.1 \times 10^{-3})(66 \times 10^{-11})$
$= 12{,}286.6 \times 10^{-14}\text{F}$
$= 12{,}282.6 \times 10^{-2}\text{pF}$
$= 122.8\text{pF}$

$L'1 = L1(1.67 \times 10^{-6})$
$= (87.5 \times 10^{-3})(1.67 \times 10^{-6})$
$= 146.1 \times 10^{-9}\text{H}$
$= 0.15\mu\text{H}$

$C'4 = (160 \times 10^{-3})(66 \times 10^{-11})$
$= 105.6\text{pF}$
$C'6 = 89.1\text{pF}$

$L'2 = 0.03\mu\text{H}$ $\quad L'3 = 0.33\mu\text{H}$
$L'4 = 0.13\mu\text{H}$ $\quad L'5 = 0.30\mu\text{H}$
$L'6 = 0.11\mu\text{H}$ $\quad L'7 = 0.09\mu\text{H}$

As a check, it will be found that the combination C4, L4 tunes to 43.2MHz and that the other two series-tuned circuits tune to the other two points of maximum attenuation previously specified.

In order to convert the values in the filter just designed to match an impedance of 75Ω it is only necessary to multiply all values of capacitance by 2/3 and all values of inductance by 3/2. Thus C6 and L6 in a 75Ω filter become approximately 59.4pF and 0.17μH respectively.

Butterworth filters [1]

Frequency response curve:

$$A = 10\log_{10}\left[1 + \left(\frac{f}{f_c}\right)^{2K}\right]$$

where A is the attenuation, f is the frequency for an insertion loss of 3.01dB, and K is the number of circuit elements.

Low- and high-pass filters

Table 22.14 is for normalised element values of K from 1 to 10 (number of sections) reduced to 1Ω source and load resistance (zero reactance) and a 3.01dB cut-off frequency of 1 radian/s (0.1592Hz). In both low-pass and high-pass filters:

$$L = \frac{R}{2\pi f_c} = L\,(1\Omega/\text{radian}) \qquad C = \frac{1}{2\pi f_c R} = C\,(1\Omega/\text{radian})$$

where R is the load resistance in ohms and f_c is the desired 3.01dB frequency (Hz).

An example of a Butterworth low-pass filter is given in Fig 22.4 (see Table 22.14 for element values). In these examples of five-element filters (a) has a shunt element next to the load and (b) has a series element next to the load. Either filter will have the same response. In the examples of five-element filters given in Fig 22.5, (a) has a series element next to the load and (b) has a shunt element next to the load. Either filter will have the same response.

Butterworth band-pass filters

Centre frequency $\qquad f_0 = \sqrt{f_1 f_2}$

Bandwidth $\qquad BW = f_2 - f_1$

If the bandwidth specified is not the 3.01dB bandwidth (BW_c), the latter can be determined from:

$$BW_c = \frac{BW}{(10^{0.1A} - 1)/2K}$$

where A is the required attenuation at cut-off frequencies.

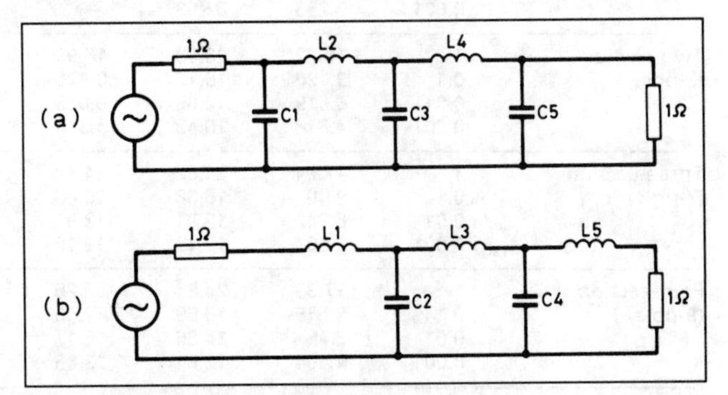

Fig 22.4. Butterworth low-pass filter

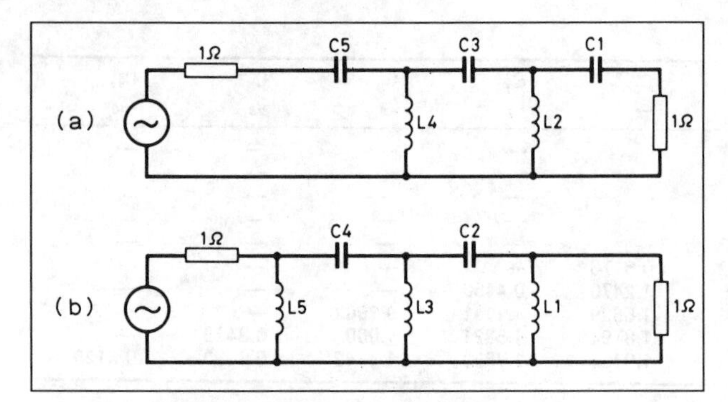

Fig 22.5. Butterworth high-pass filter

Lower cut-off frequency:

$$f_{cl} = \frac{-BW_c + \sqrt{(BW_c)^2 + 4f_0^2}}{2}$$

Upper cut-off frequency: $f_{cu} = f_{cl} + BW_c$

An alternative, more-convenient method, is to choose a 3.01dB bandwidth (as wide as possible) around the desired centre frequency and compute the attenuation at other frequencies of interest by using the transformation:

$$\frac{f}{f_c} = \left[\left(\frac{f}{f_0} - \frac{f_0}{f}\right)\frac{f_0}{BW_c}\right]$$

Chebyshev filters [1]

Tables 22.15 to 22.18 provide the essential information for both high-pass and low-pass filters of T and π form. Figures are given for pass-band ripples of 1, 0.1, 0.01, and 0.001dB which respectively correspond to VSWR of 2.66, 1.36, 1.10 and 1.03.

The filters in this case are normalised to a frequency of 1MHz and an input and output impedance of 50Ω. This means that for any particular desired frequency the component values simply have to be divided by the required frequency in megahertz.

The 1MHz is the cut-off frequency; attenuation increases rapidly above the frequency for a low-pass filter and correspondly below for a high-pass type.

The filter data is also dependent on the impedance which as given is for 50Ω. For other impedances the component values need to be modified by the following:

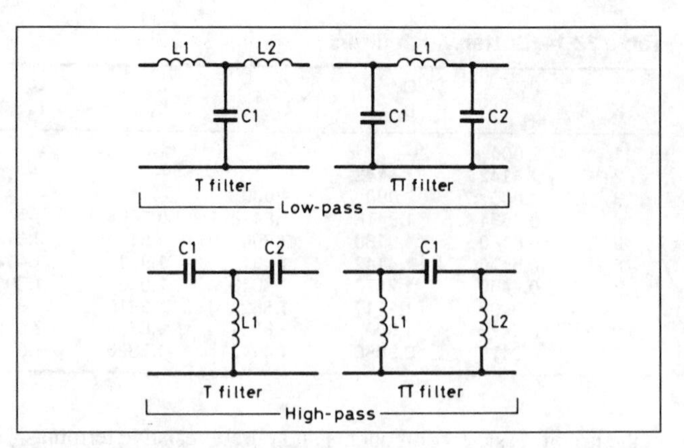

Fig 22.6. Single-section three-pole filter elements

$$\frac{Z_n}{50} \text{ for inductors} \qquad \frac{50}{Z_n} \text{ for capacitors}$$

where Z_n is the required impedance.

There is an advantage in using toroidal-form inductors due to their self-screening (confined-field) properties. Mica or silver mica capacitors are superior to other types for filter applications.

Practical filters for the amateur HF bands are given in Table 22.19.

Constant-*k* and *m*-derived filters [2]

The filter sections shown in Fig 22.7 can be used alone or, if greater attenuation and sharper cut-off is required, several sections can be connected in series. In the low-pass and high-pass filters, f_c represents the cut-off frequency, the highest (for the low-pass) or the lowest (for the high-pass) frequency transmitted without attenuation. In the band-pass filter designs, f_1 is the low-frequency cut-off and f_2 the high-frequency cut-off. The units for L, C, R and f are henrys, farads, ohms and hertz respectively.

All the types shown are for use in an unbalanced line (one side grounded), and thus they are suitable for use in coaxial line or any other unbalanced circuit. To transform them for balanced lines (eg 300Ω transmission line or push-pull audio circuits), the series reactances should be equally divided between the two legs. Thus the balanced constant-*k* π-section low-pass filter

Table 22.15. Chebyshev low-pass filter ('T' configuration)

	Ripple (dB)	L_1	L_2	L_3	L_4	L_5	C_1	C_2	C_3	C_4
Single section (3-pole)	1	16.10	16.10	—	—	—	3164.3	—	—	—
	0.1	8.209	8.209	—	—	—	3652.3	—	—	—
	0.01	5.007	5.007	—	—	—	3088.5	—	—	—
	0.001	3.253	3.253	—	—	—	2312.6	—	—	—
Two-section (5-pole)	1	16.99	23.88	16.99	—	—	3473.1	3473.1	—	—
	0.1	9.126	15.72	9.126	—	—	4364.7	4364.7	—	—
	0.01	6.019	12.55	6.019	—	—	4153.7	4153.7	—	—
	0.001	4.318	10.43	4.318	—	—	3571.1	3571.1	—	—
Three-section (7-pole)	1	17.24	24.62	24.62	17.24	—	3538.0	3735.4	3538.0	—
	0.1	9.40	16.68	16.68	9.40	—	4528.9	5008.3	4528.9	—
	0.01	6.342	13.91	13.91	6.342	—	4432.2	5198.4	4432.2	—
	0.001	4.69	12.19	12.19	4.69	—	3951.5	4924.1	3981.5	—
Four-section (9-pole)	1	17.35	24.84	25.26	24.84	17.35	3562.5	3786.9	3786.9	3562.5
	0.1	9.515	16.99	17.55	16.99	9.515	4591.9	5146.2	5146.2	4591.9
	0.01	6.481	14.36	15.17	14.36	6.481	4542.5	5451.2	5451.2	4542.5
	0.001	4.854	12.81	13.88	12.81	4.854	4108.2	5299.0	5299.0	4108.2

Inductance in microhenrys, capacitance in picofarads. Component values normalised to 1MHz and 50Ω.

Table 22.16. Chebyshev low-pass filter (Pi configuration)

	Ripple (dB)	C_1	C_2	C_3	C_4	C_5	L_1	L_2	L_3	L_4
Single section	1	6441.3	6441.3	—	—	—	7.911	—	—	—
(3-pole)	0.1	3283.6	3283.6	—	—	—	9.131	—	—	—
	0.01	2007.7	2007.7	—	—	—	7.721	—	—	—
	0.001	1301.2	1301.2	—	—	—	5.781	—	—	—
Two-section	1	6795.5	9552.2	6795.5	—	—	8.683	8.683	—	—
(5-pole)	0.1	3650.4	6286.6	3650.4	—	—	10.91	10.91	—	—
	0.01	2407.5	5020.7	2407.5	—	—	10.38	10.38	—	—
	0.001	1727.3	4170.5	1727.3	—	—	8.928	8.928	—	—
Three-section	1	3538	5052	5052	3538	—	17.24	18.20	17.24	—
(7-pole)	0.1	3759.8	6673.9	6673.9	3759.8	—	11.32	12.52	11.32	—
	0.01	2536.8	5564.5	5564.5	2536.8	—	11.08	13.00	11.08	—
	0.001	1875.7	4875.9	4875.9	1875.7	—	9.879	12.31	9.879	—
Four-section	1	6938.3	9935.8	10,105	9935.8	6938.3	8.906	9.467	9.467	8.906
(9-pole)	0.1	3805.9	6794.5	7019.9	6794.5	3805.9	11.48	12.87	12.87	11.48
	0.01	2592.5	5743.5	6066.3	5743.5	2592.5	11.36	13.63	13.63	11.36
	0.001	1941.7	5124.6	5553.2	5124.6	1941.7	10.27	13.25	13.25	10.27

Inductance in microhenrys, capacitance in picofarads. Component values normalised to 1MHz and 50Ω.

Table 22.17. Chebyshev high-pass filter ('T' configuration)

	Ripple (dB)	C_1	C_2	C_3	C_4	C_5	L_1	L_2	L_3	L_4
Single section	1	1573	1573	—	—	—	8.005	—	—	—
(3-pole)	0.1	3085.7	3085.7	—	—	—	6.935	—	—	—
	0.01	5059.1	5059.1	—	—	—	8.201	—	—	—
	0.001	7786.9	7786.9	—	—	—	10.95	—	—	—
Two-section	1	1491	1060.7	1491	—	—	7.293	7.293	—	—
(5-pole)	0.1	2775.6	1611.7	2775.6	—	—	5.803	5.803	—	—
	0.01	4208.6	2018.6	4208.6	—	—	6.098	6.098	—	—
	0.001	5865.7	2429.5	5865.7	—	—	7.093	7.093	—	—
Three-section	1	1469.2	1028.9	1028.9	1469.2	—	7.160	6.781	7.160	—
(7-pole)	0.1	2694.9	1518.2	1518.2	2694.9	—	5.593	5.058	5.593	—
	0.01	3994.1	1820.9	1820.9	3994.1	—	5.715	4.873	5.715	—
	0.001	5401.7	2078	2078	5401.7	—	6.410	5.144	6.410	—
Four-section	1	1460.3	1019.8	1002.7	1019.8	1460.3	7.110	6.689	6.689	7.110
(9-pole)	0.1	2662.2	1491.2	1443.3	1491.2	2662.2	5.516	4.922	4.922	5.516
	0.01	3908.2	1764.1	1670.2	1764.1	3908.2	5.578	4.647	4.647	5.578
	0.001	5216.3	1977.1	1824.6	1977.1	5216.3	6.657	4.780	4.780	6.657

Inductance in microhenrys, capacitance in picofarads. Component values normalised to 1MHz and 50Ω.

Table 22.18. Chebyshev high-pass filter (Pi configuration)

	Ripple (dB)	L_1	L_2	L_3	L_4	L_5	C_1	C_2	C_3	C_4
Single section	1	3.932	3.932	—	—	—	3201.7	—	—	—
(3-pole)	0.1	7.714	7.714	—	—	—	2774.2	—	—	—
	0.01	12.65	12.65	—	—	—	3280.5	—	—	—
	0.001	19.47	19.47	—	—	—	4381.4	—	—	—
Two-section	1	3.727	2.652	3.727	—	—	2917.3	2917.3	—	—
(5-pole)	0.1	6.939	4.029	6.939	—	—	2321.4	2321.4	—	—
	0.01	10.52	5.045	10.52	—	—	2439.3	2439.3	—	—
	0.001	14.66	6.074	14.66	—	—	2837.3	2837.3	—	—
Three-section	1	7.159	5.014	5.014	7.159	—	1469.2	1391.6	1469.2	—
(7-pole)	0.1	8.737	3.795	3.795	8.737	—	2237.2	2023.1	2237.2	—
	0.01	9.985	4.552	4.552	9.985	—	2286.0	1949.1	2286.0	—
	0.001	13.50	5.195	5.195	13.50	—	2584.1	2057.7	2584.1	—
Four-section	1	3.651	2.549	2.507	2.549	3.651	2844.1	2675.6	2675.6	2844.1
(9-pole)	0.1	6.656	3.728	3.608	3.728	6.656	2206.5	1968.9	1968.9	2206.5
	0.01	9.772	4.410	4.176	4.410	9.772	2230.5	1858.7	1858.7	2230.5
	0.001	13.05	4.943	4.561	4.943	13.05	2466.3	1911.8	1911.8	2466.3

Inductance in microhenrys, capacitance in picofarads. Component values normalised to 1MHz and 50Ω.

would use two inductances of a value equal to $L_k/2$, while the balanced constant-k π-section high-pass filter would use two capacitors of a value equal to $2C_k$.

If several low- (or high-) pass sections are to be used, it is advisable to use m-derived end sections on either side of a constant-k section, although an m-derived centre section can be used.

Table 22.19. Practical Chebyshev low-pass filters (3-section, 7-pole)

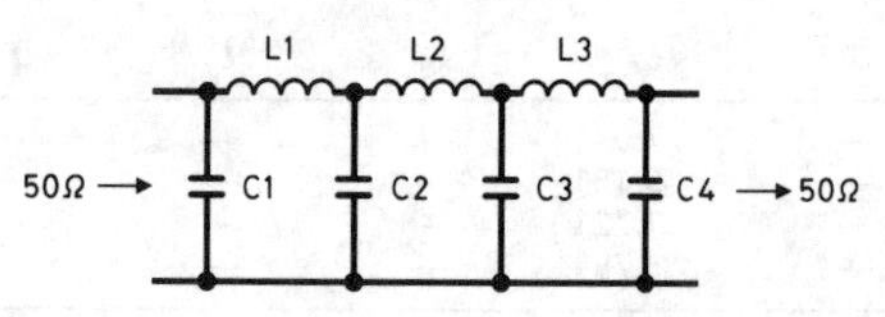

Amateur band	28	21	14	7	3.5	1.8	MHz
F_c	30.9	21.69	15.16	7.98	4.11	2.05	MHz
VSWR	1.10	1.06	1.09	1.08	1.07	1.09	
C1, C4	82	100	160	300	560	1200	pF
C2, C3	180	240	360	680	1300	2700	pF
L1, L3	0.36	0.49	0.72	1.37	2.62	5.42	µH
L2	0.42	0.59	0.85	1.62	3.13	6.41	µH

Ripple is 0.01dB.

The factor m relates the ratio of the cut-off frequency and f_∞, a frequency of high attenuation. Where only one m-derived section is used, a value of 0.6 is generally used for m, although a deviation of 10 or 15% from this value is not too serious in amateur work. For a value of $m = 0.6$, f will be $1.25f_c$ for the low-pass filter and $0.8f_c$ for the high-pass filter. Other values can be found from:

$$m = \sqrt{1 - \left(\frac{f_c}{f_\infty}\right)^2}$$

for the low-pass filter and:

$$m = \sqrt{1 - \left(\frac{f_\infty}{f_c}\right)^2}$$

for the high-pass filter.

The filters shown should be terminated in a resistance R, and there should be little or no reactive component in the termination.

Microstrip circuit elements [3, 4]

In the calculation of microstrip circuit elements it is necessary to establish the dielectric constant for the material. This can be done by measuring the capacitance of a typical sample.

$$\text{Dielectric constant } e = \frac{113 \times C \times h}{a}$$

where C is in picofarads, h is the thickness in millimetres and a is the area in square millimetres. This should be done with a sample about 25mm square to minimise the effects of the edges. Having found the dielectric constant, it is now necessary to calculate the characteristic impedance (Z_0) of the microstrip. There are many approximations for this but the following is simple and accurate enough (±5%) for amateur use since microstrip is fairly forgiving of small errors:

$$Z_0 = \frac{131}{\sqrt{(e + 0.47)}} \times \log_{10}\left(\frac{13.5h}{w}\right)$$

where e is the dielectric constant, h is the dielectric thickness (see Fig 22.8), and w is the conductor width. Note that h and w *must* be in the same units, eg both in centimetres. The formula assumes that the conductor is thin relative to the dielectric.

An accurate plot of Z_0 against w/h for dielectric constants between 2 (approximately that of PTFE) through 4 (approx that of epoxy-glassfibre) to 6 is given in Fig 22.9. The next operation is to determine the *velocity factor*, the ratio of the velocity of electromagnetic waves in the dielectric to that in free space. Here, too, the equations are complex but the factor only changes slowly as w/h changes. Fig 22.10 gives figures for the above range of dielectric constants.

A microstrip resonator has the form of a strip on one side of a double-sided PCB with the other side as a ground plane. It is usually a quarter-wavelength long. The length is calculated from the free space length multiplied by the velocity factor as estimated above. The line width depends on the required Z_0 and a starting point for experiment would be 50–100Ω. If it is necessary to tune the resonator accurately, it should be made shorter

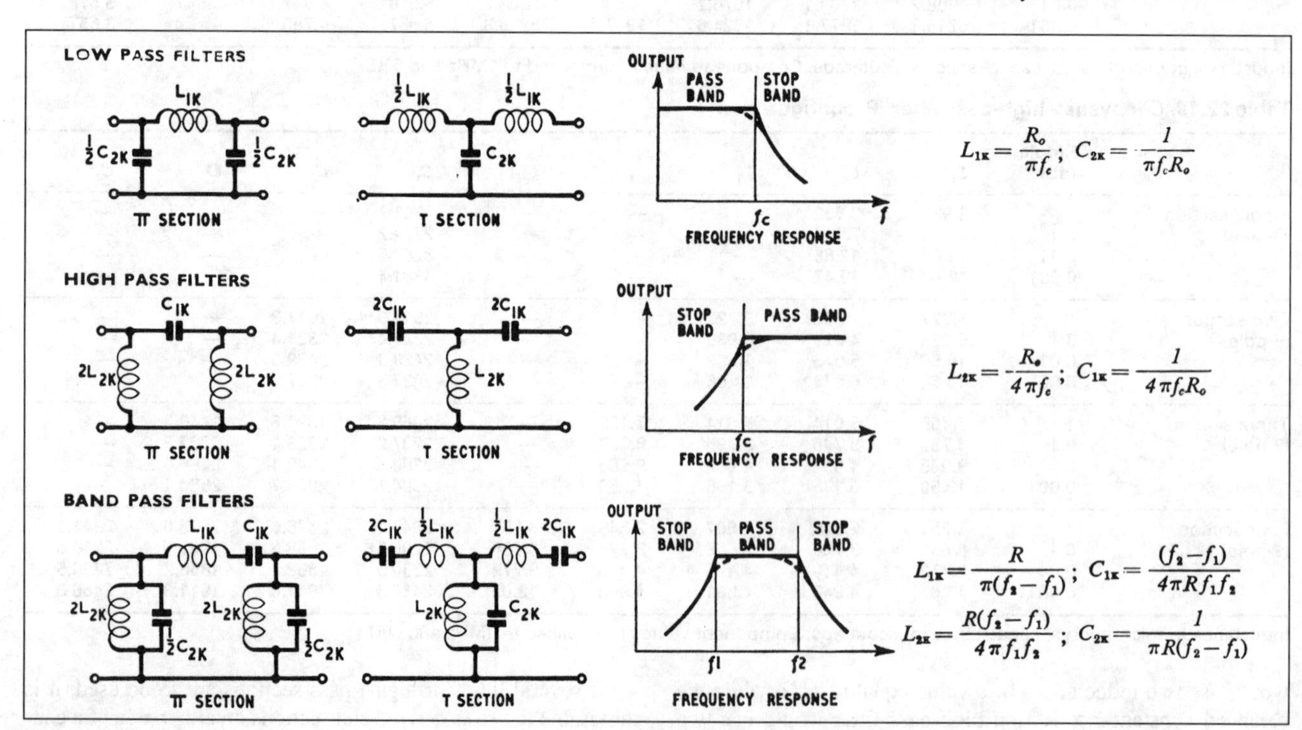

Fig 22.7(a). Constant-*k* filters

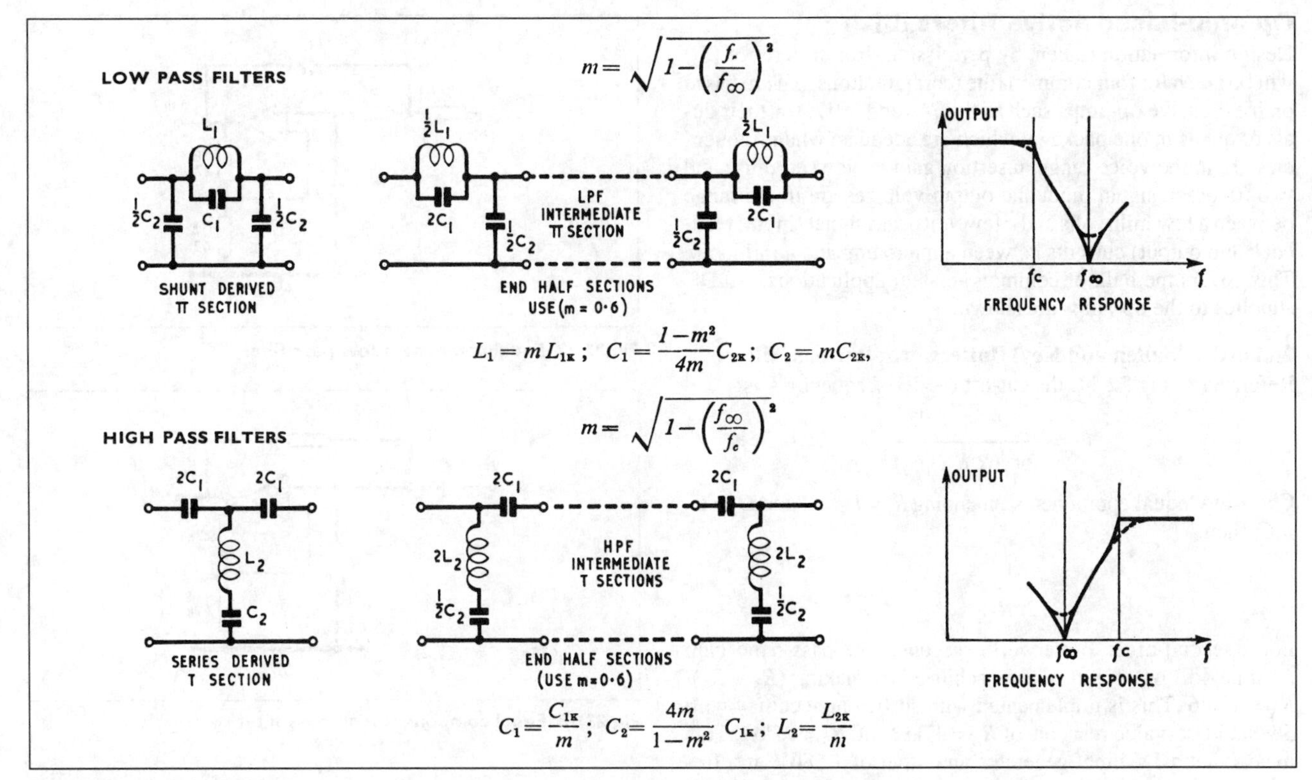

Fig 22.7(b). ***m*-derived filters**

than calculated above and a trimmer capacitor connected between the 'hot' end and the ground plane. The new length can be calculated from:

$$l = 0.0028\lambda \times \tan^{-1}\left(\frac{\lambda}{0.188CZ_0}\right)$$

where l is the length in centimetres, λ is the wavelength in centimetres, C is the capacitance (say at half maximum) in picofarads, Z_0 is the characteristic impedance in ohms and $\tan^{-1}(*)$ is the angle in degrees of which * is the tangent. * represents the figures in the bracket.

Coupling into and out of the line may be directly via tapping(s) or by additional line(s) placed close to the tuned line. A spacing of one line width and a length of 10–20% of the tuned line would be a starting point for experiment.

Materials

The most-used material for amateur purposes is glassfibre reinforced epoxy double-sided PCB. It has a dielectric constant of 4.0–4.5, depending on the resin used. Accurate lines may be made by scoring through the copper carefully with a scalpel or modelling knife and lifting the unwanted copper foil after heating it with a soldering iron to weaken the bond to the plastic.

For microwave use, glassfibre-reinforced PTFE with a dielectric constant close to 2.5 is the preferred material. Further information is given in the *Microwave Handbook* [4].

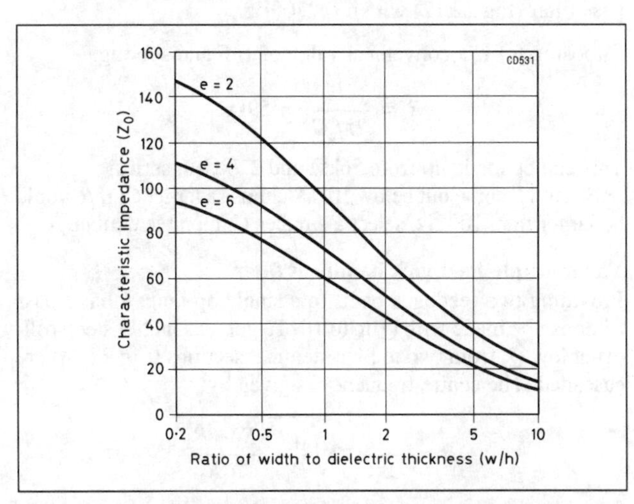

Fig 22.9. Characteristic impedance versus *w/h* for dielectric constants between 2 and 6

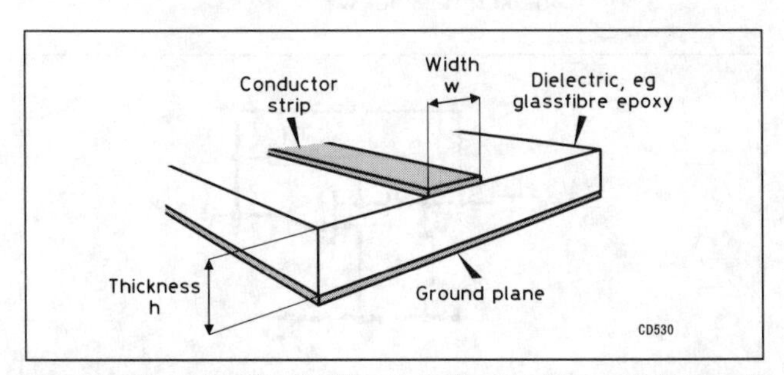

Fig 22.8. Dimensions involed in calculating the characteristic impedance of microstrip

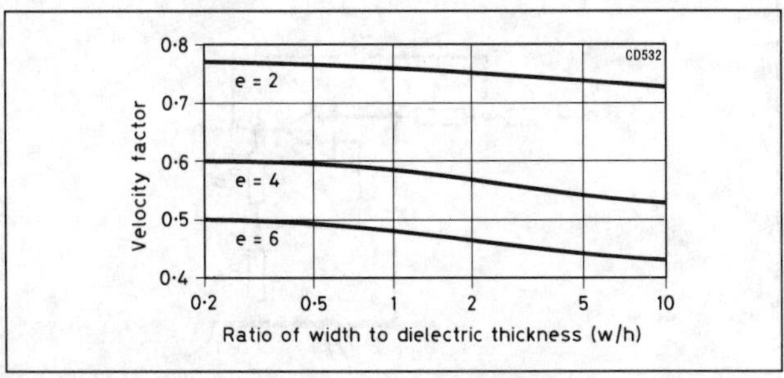

Fig 22.10. Velocity factor versus *w/h* for dielectric constants between 2 and 6

Op-amp-based active filters [5]

Design information (taken, by permission, from reference [5]) will be given for four common filter configurations. All are based on inexpensive op-amps such as the 741 and 301A (or their duals or quads in one package) which are adequate when frequencies are in the voice range, insertion gain is between unity and two (0–6dB), signal input and output voltages are in the range between a few millivolts and a few volts, and signal (input, feedback and output) currents between a microamp and a milliamp. This covers the bulk of common amateur applications. No DC supplies to the op-amps are shown.

2nd order 'Sallen and Key' Butterworth low-pass filter

Referring to Fig 22.11, the cut-off (–3dB) frequency f_c is:

$$f_c = \frac{1}{2\pi\sqrt{(R_1 R_2 C_1 C_2)}}$$

Choosing 'equal components', meaning $R_1 = R_2 = R$ and $C_1 = C_2 = C$, then:

$$f_c = \frac{1}{2\pi RC}$$

For a second-order Butterworth response, the pass-band gain *must* be 4dB or ×1.586. This is achieved by making $(R_A + R_B)/R_A = 1.586$. This is implemented with sufficient accuracy with 5% standard-value resistors of $R_A = 47\text{k}\Omega$ and $R_B = 27\text{k}\Omega$. This means that a 1V input generates an output of 1.586V at a frequency in the pass-band and 0.707 × 1.586 = 1.12V at the –3dB cut-off frequency; the roll-off above f_c is 12dB/octave or 20dB/decade.

Example. Design a two-pole 'equal component' Butterworth low-pass filter (Fig 22.12) with f_c = 2700Hz.

Choosing for C a convenient value of 1nF and solving:

$$R = \frac{1}{2\pi f_c C} = 59\text{k}\Omega$$

This can be made up from 56kΩ and 2.7kΩ in series.

Should R come out below 10kΩ, choose a larger C; if R would be larger than 100kΩ, select a smaller C; then recalculate R.

The multiple-feedback bandpass filter

Providing two feedback paths to a single op-amp, a band-pass filter can be made with Q up to 10. To get reasonably steep roll-off at low Q, from two to four identical sections (Fig 22.14) are cascaded. The centre frequency is given by:

$$f_0 = \frac{1}{2\pi C}\sqrt{\frac{1}{R_3} \cdot \frac{R_1 + R_2}{R_1 R_2}}$$

for which the three resistors can be calculated from:

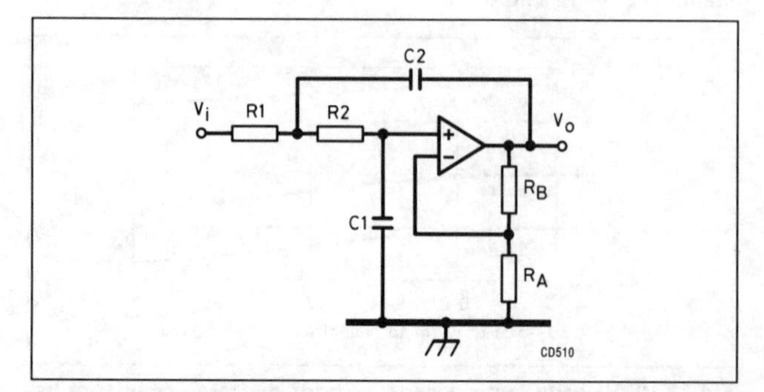

Fig 22.11. 'Sallen & Key' Butterworth low-pass filter

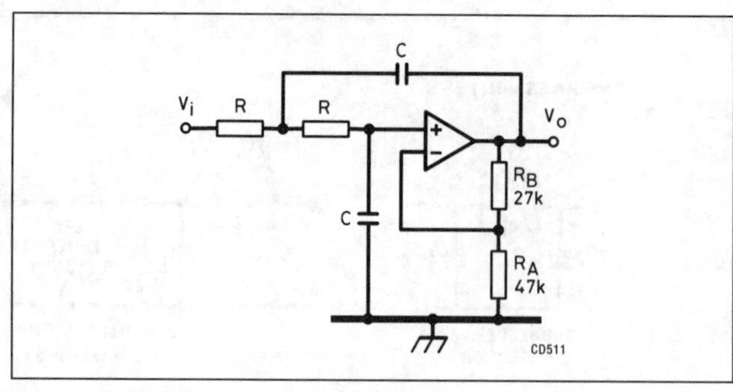

Fig 22.12. Equal-component low-pass filter

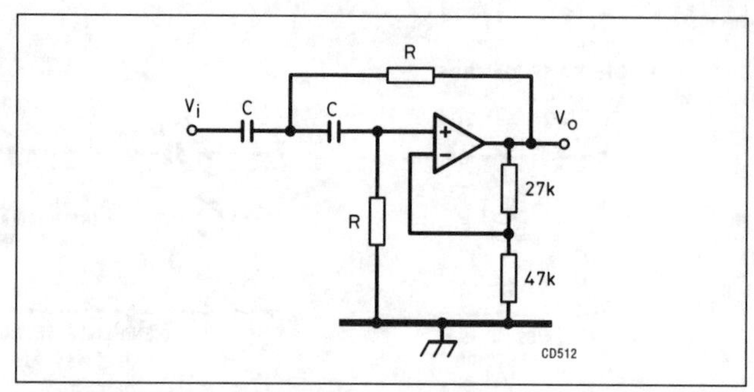

Fig 22.13. Equal component high-pass filter

$$R_1 = \frac{Q}{2\pi f_0 G_0 C}$$

$$R_2 = \frac{Q}{2\pi f_0 C(2Q^2 - G_0)}$$

$$R_3 = \frac{Q}{\pi f_0 C}$$

The equations for R_1 and R_3 combine into:

$$G_0 = R_3/2R_1$$

Also, the denominator in the formula for R_2 yields:

$$Q > \sqrt{(G_0/2)}$$

Example. Design a band-pass filter with centre frequency 800Hz, –6dB bandwidth of 200Hz and centre-frequency gain of 2.

A two-section filter, with each section having a 200Hz –3dB bandwidth, is indicated.

$$Q = 800/200 = 4$$

$$G_0 = \sqrt{2} = 1.4$$

Select a convenient C, say 10nF.

$$R_1 = \frac{4}{6.28 \times 800 \times 1.4 \times 10^{-8}}$$
$$= 56.9\text{k}\Omega, \text{ (use 56k}\Omega)$$

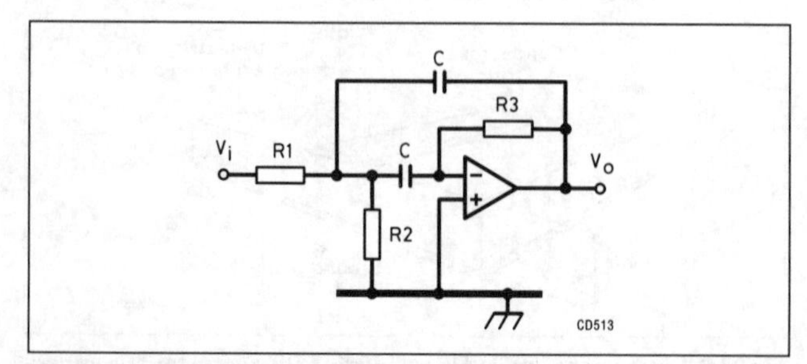

Fig 22.14. Multiple-feedback band-pass filter

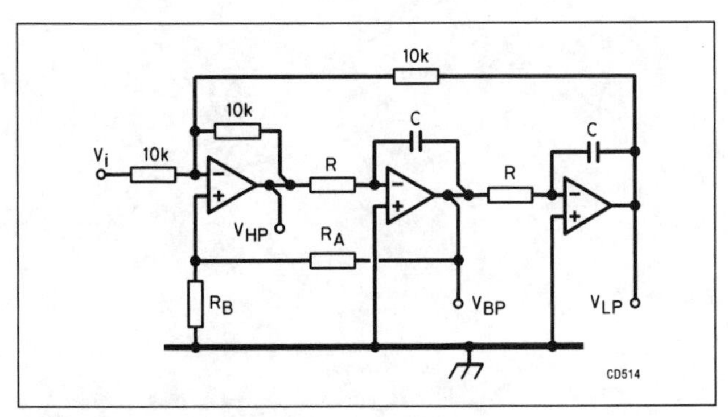

Fig 22.15. State-variable filter

$$R_3 = 2 \times 56.9 \times 1.4$$
$$= 159\text{k}\Omega \text{ (use } 100\text{k}\Omega + 56\text{k}\Omega)$$
$$R_2 = \frac{4}{6.28 \times 800 \times 10^{-8} \times (2 \times 4^2 - 1.4)}$$
$$= 2.60\text{k}\Omega \text{ (use } 5.1\text{k}\Omega \text{ in parallel with } 5.1\text{k}\Omega)$$

Note that the centre frequency can be shifted up or down at constant bandwidth and centre frequency gain by changing R_2 only, using ganged variable resistors for cascaded sections:

$$R_2' = R_2 \left(\frac{f_0}{f_0'}\right)^2$$

The state-variable or 'universal' filter

Three op-amps, connected as shown in Fig 22.15, can simultaneously provide second-order high-pass, low-pass and band-pass responses. The filter is composed of a difference amplifier and two integrators. The common cut-off/centre frequency is given by:

$$f_{cL} = f_{cH} = f_0 = \frac{1}{2\pi RC}$$

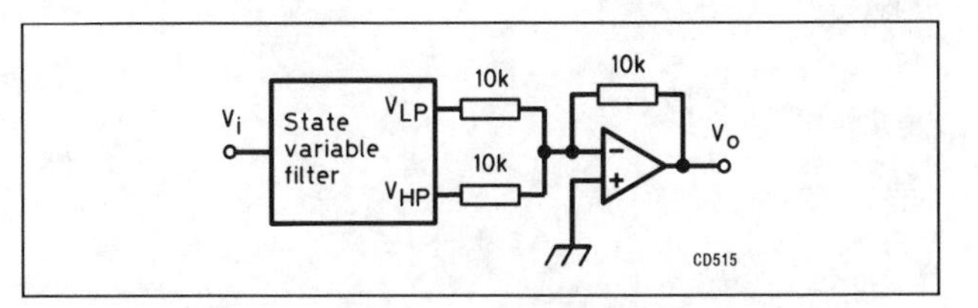

Fig 22.16. Using the state-variable filter to obtain a notch response

The filter's Q depends only on R_A and R_B:

$$R_A = (3Q - 1)R_B$$

There is no way to simultaneously optimise the performance of high-/low-pass and band-pass performance. For a Butterworth response, Q must be 0.7 and even for a second-order 3dB-ripple Chebyshev response the Q is no more than 1.3, obviously too low for good band-pass response. No DC voltage should be applied to the input of this filter and there should be no significant DC load on its outputs.

By adding the low-pass and high-pass outputs from a variable-state filter in a summing amplifier, a notch response is obtained. See Fig 22.16.

For an application of the variable-state filter refer to the active filters section in Chapter 5 and Fig 5.108 in particular.

REFERENCES

[1] *Radio Data Reference Book*, 5th edn, G R Jessop, G6JP, RSGB, 1985.

[2] *ARRL Radio Amateur's Handbook*, 1953 edn, ARRL, p542.

[3] *VHF/UHF Manual*, 4th edn, ed G R Jessop, G6JP, RSGB, 1983, p3.10.

[4] *Microwave Handbook*, Vol 1, ed M W Dixon, G3PFR, RSGB, 1989, p5.14*ff*.

[5] *The Design of Operational Amplifier Circuits, with Experiments*, Howard M Berlin, W3HB, E & L Instruments Inc, Derby, Conn, USA, 1977.

APPENDIX

1 Printed circuit board artwork

Fig 5.59(b). The SL560 50Ω line driver and test circuit PCB *(GEC-Plessey Professional Products Handbook 1991)*

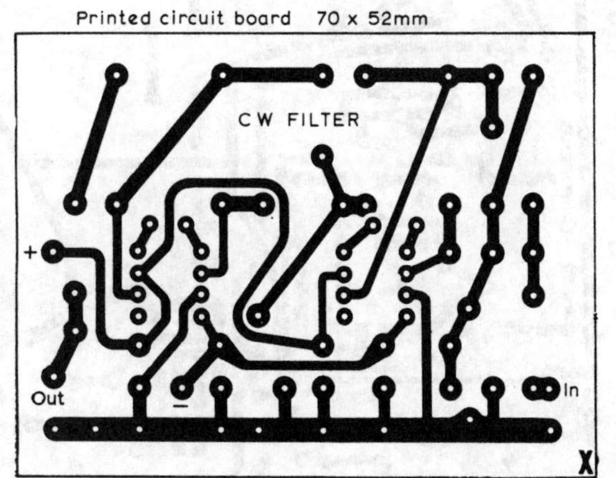

Fig 5.107(c). LA2IJ/LA4HK CW filter PCB layout

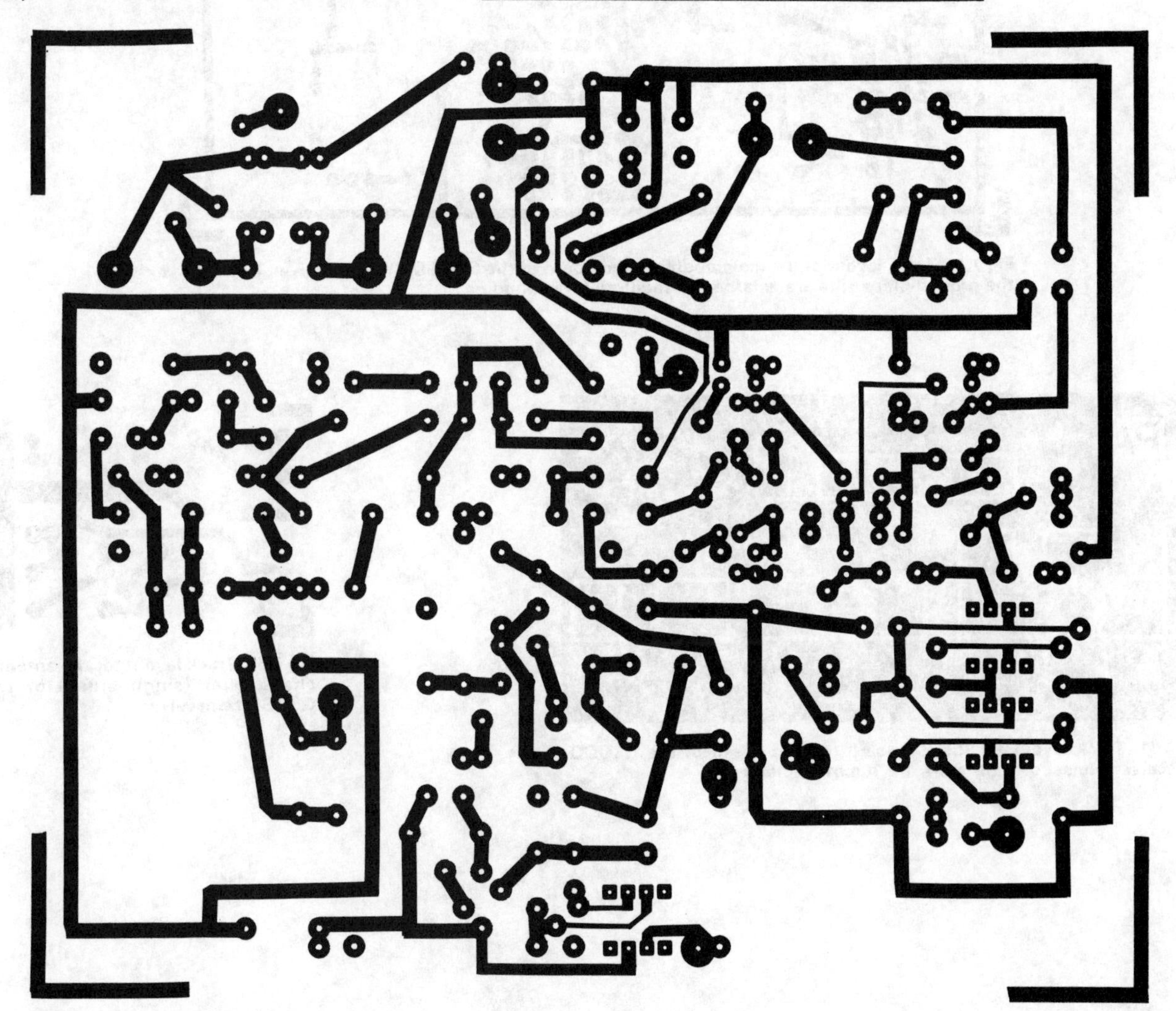

Fig 7.41. 1.8MHz QRP transceiver PCB layout

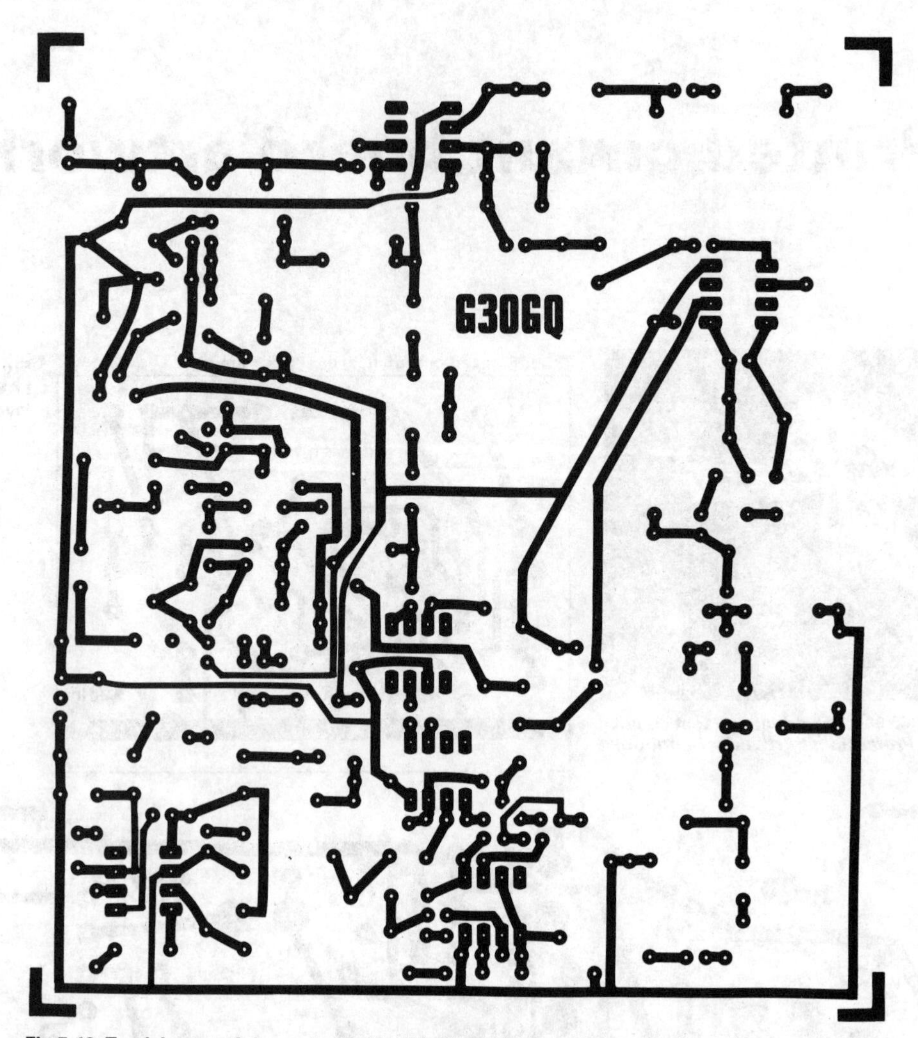

Fig 7.48. Track layout of the main double-sided board of the G3OGQ CW transceiver. Holes on the ground-plane side are enlarged as required for grounding

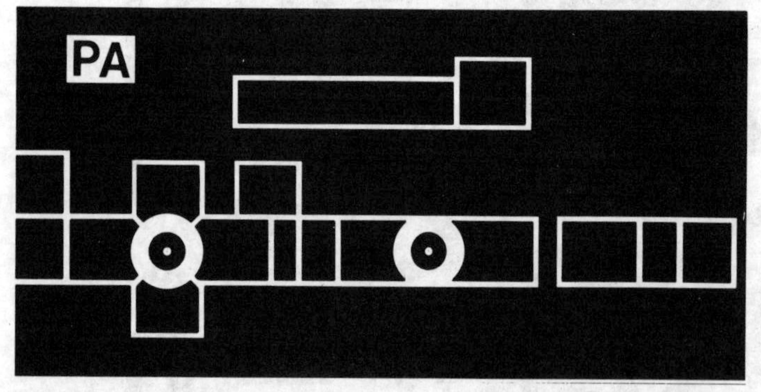

Fig 7.51. Track layout for the PA board (double-sided) of the G3OGQ transceiver. Holes for transistors are 10mm diameter

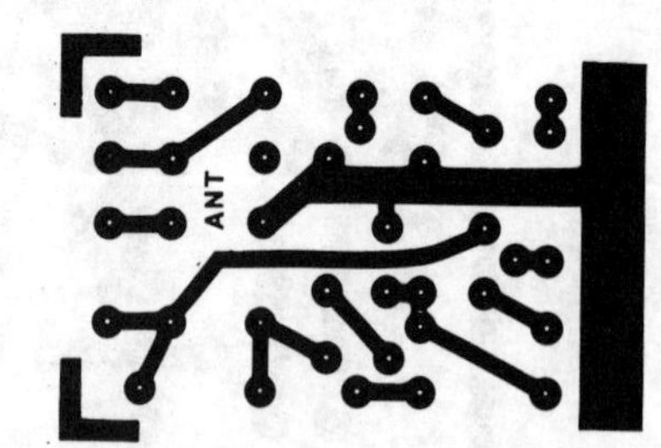

Fig 7.52. Track layout for the antenna changeover (single-sided) of the G3OGQ transceiver

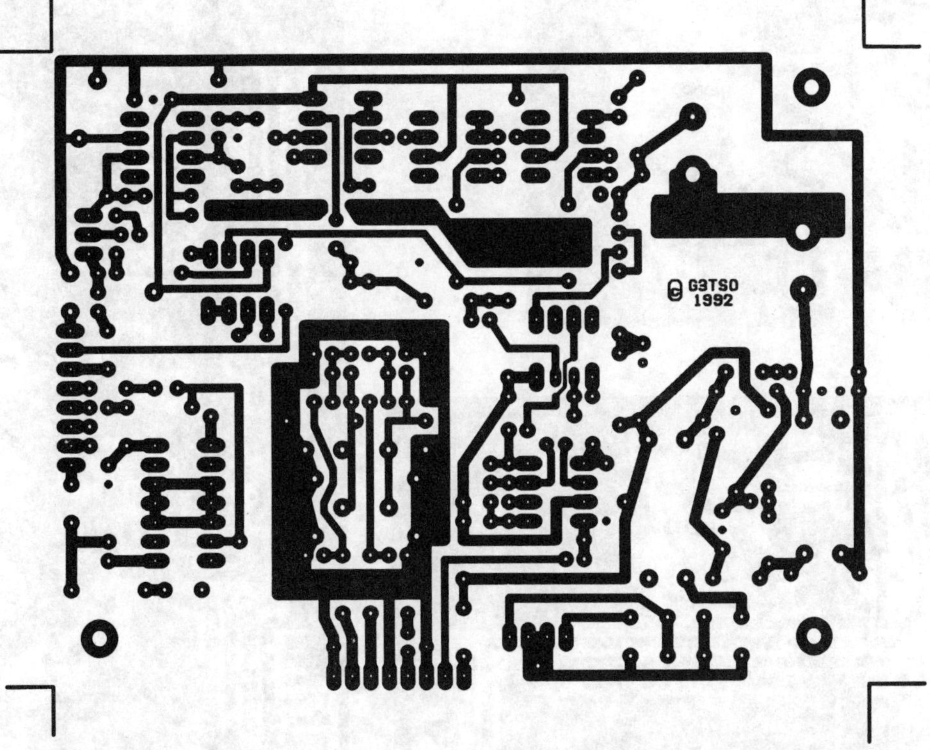

Fig 7.56(a). G3TSO exciter module 1 – SL1621 version PCB. Material: double-sided glassfibre. Upper surface is ground plane

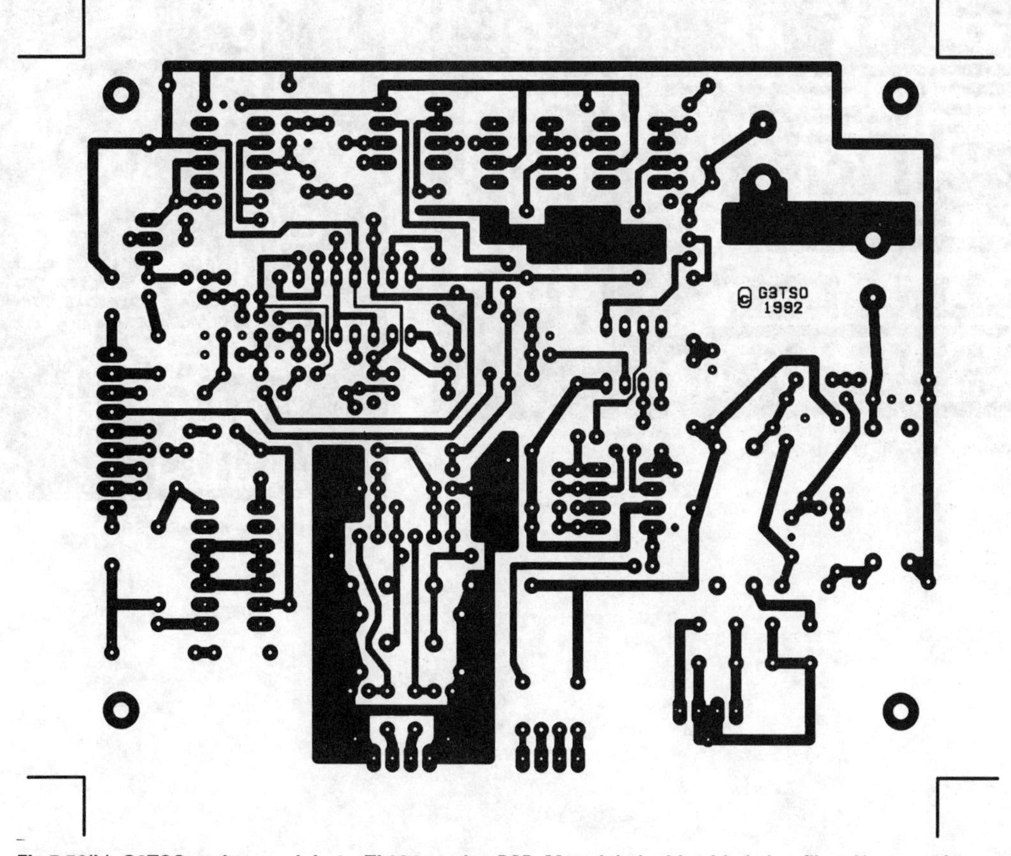

Fig 7.56(b). G3TSO exciter module 1 – TL084 version PCB. Material: double-sided glassfibre. Upper surface is ground plane

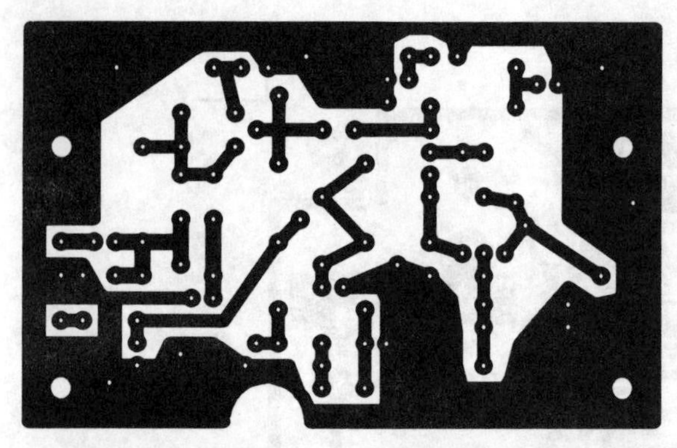

Fig 7.60. Module 2 PCB

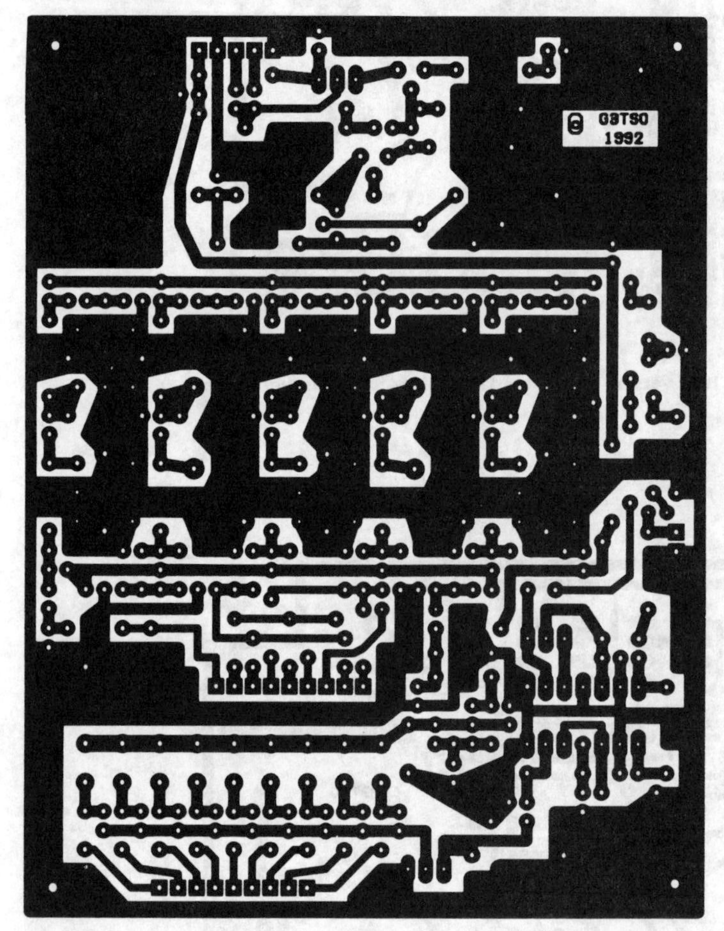

Fig 7.62. Module 3 (WARC premix) PCB

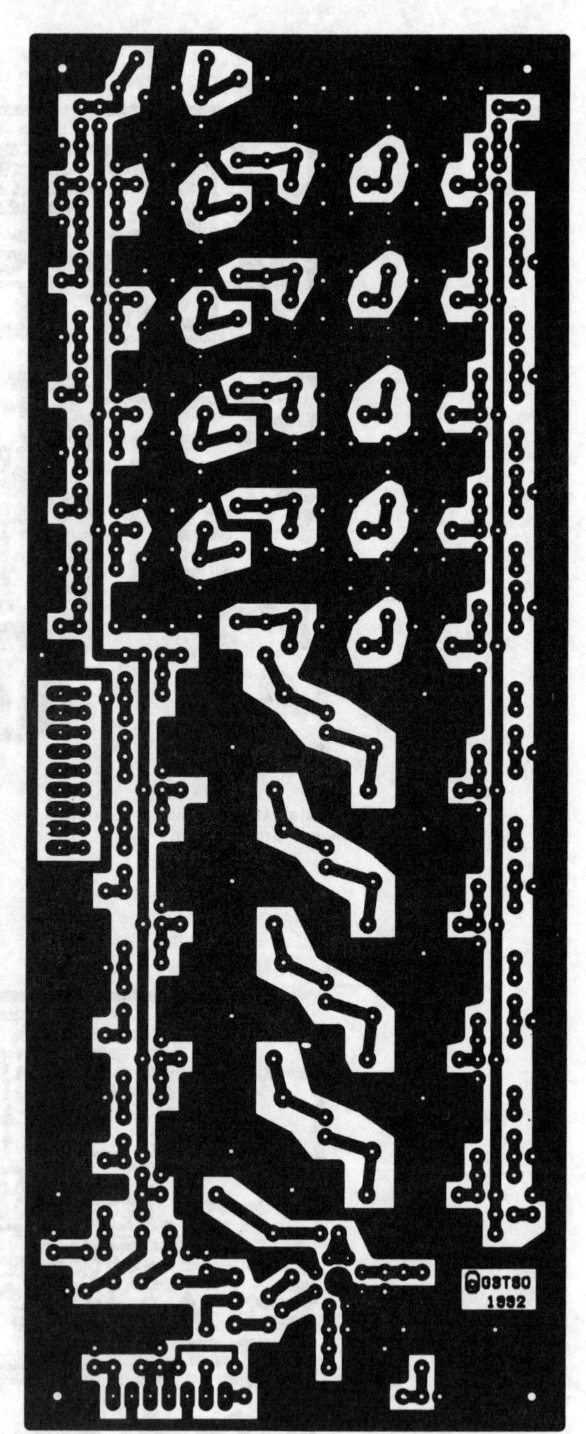

Fig 7.65. Module 4 (BPF) PCB

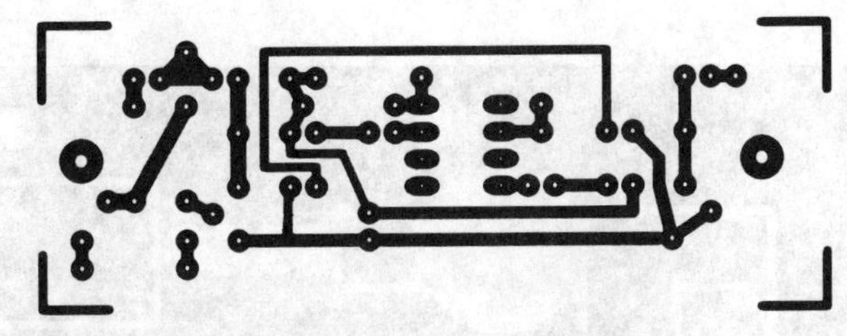

Fig 7.67. Module 4a (RF amp) double-sided PCB

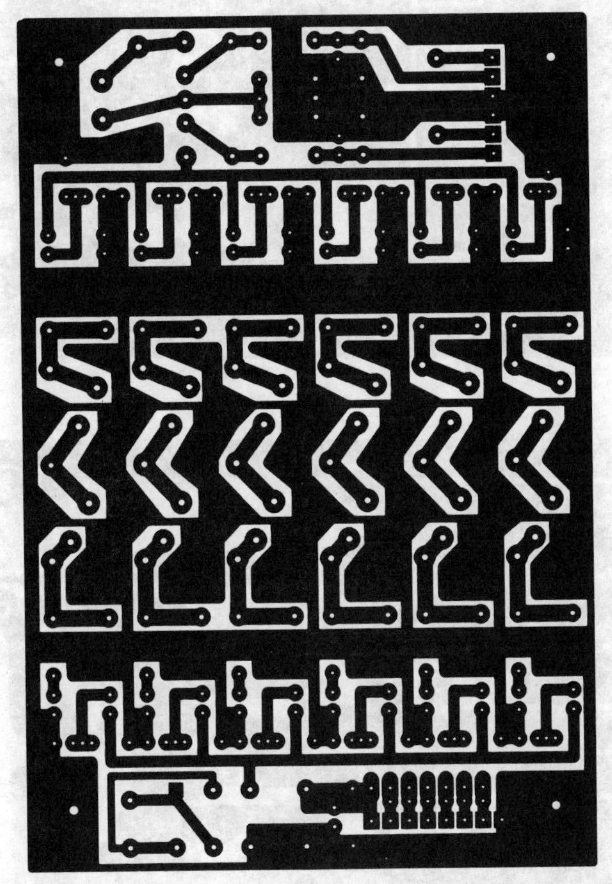

Fig 7.70. Module 5 (low-pass filter) PCB

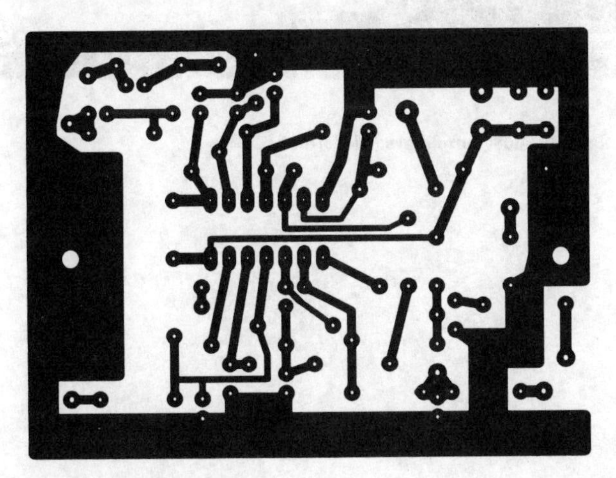

Fig 7.73. Module 6 PCB

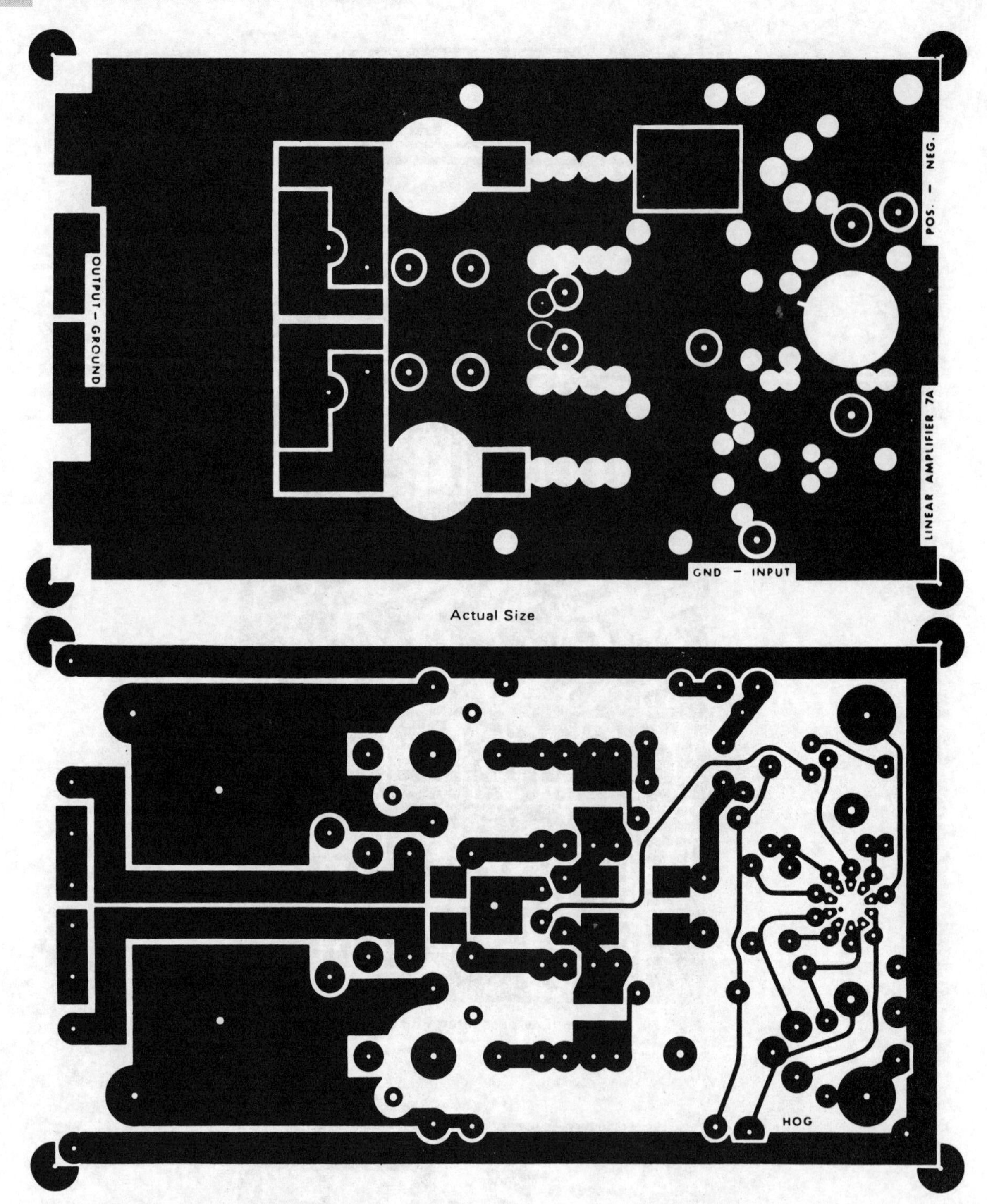

Fig 7.79. PCB layouts for the 140–300W amplifiers (Motorola)

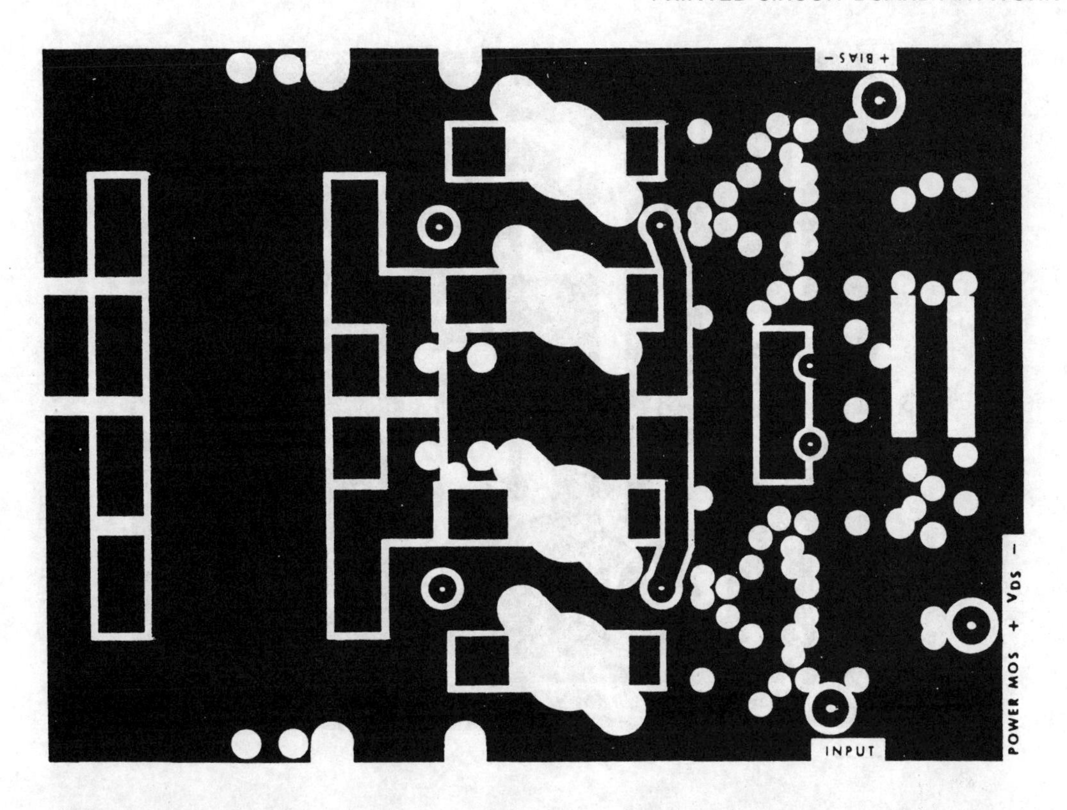

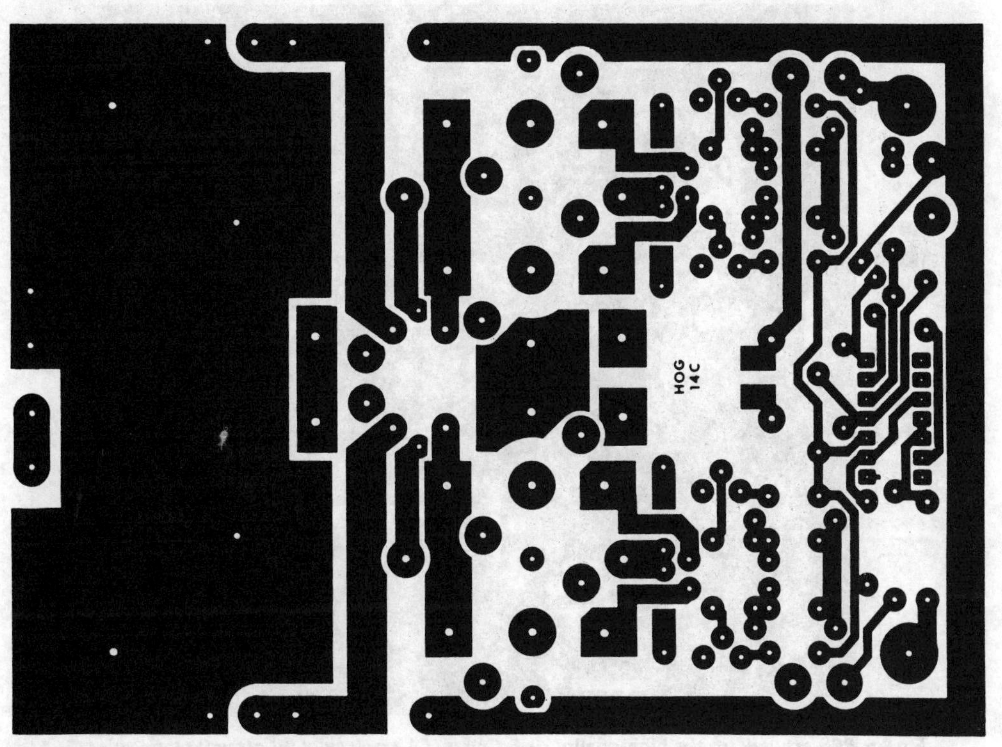

Fig 7.83. PCB layouts for the 600W amplifier (Motorola)

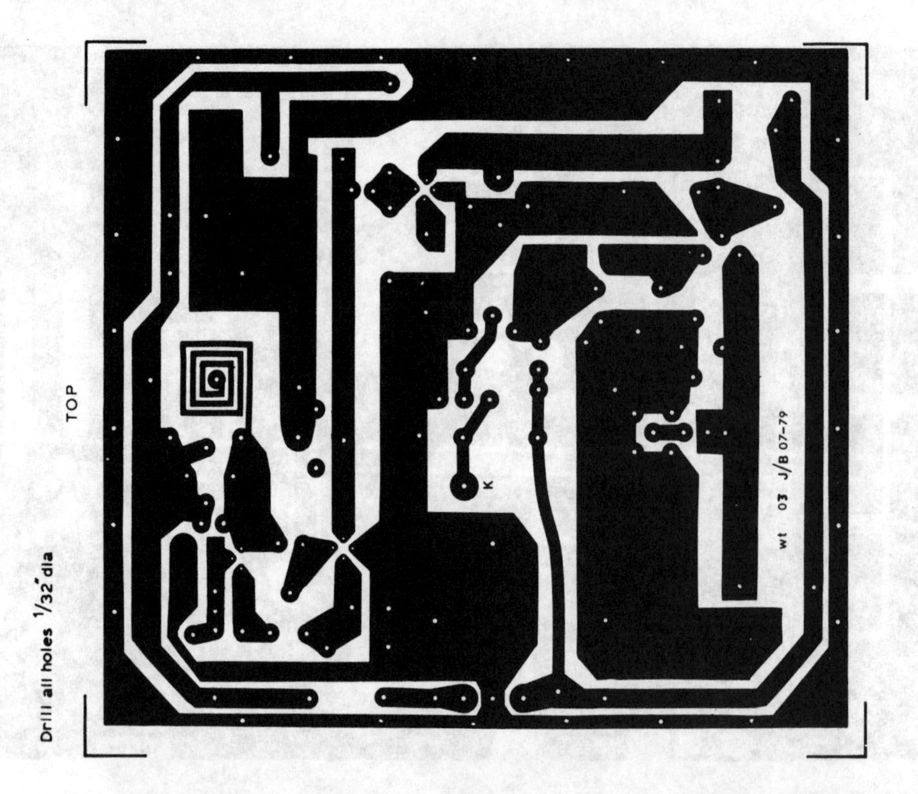

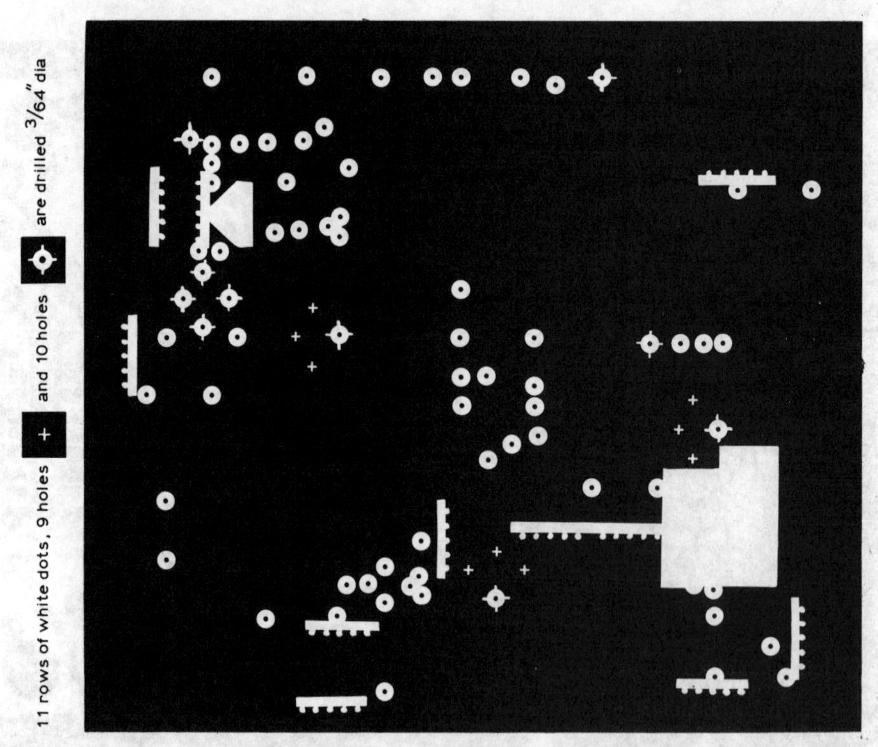

Fig 9.5. PCB artwork for the high-quality UHF source. (a) Track side; (b) ground-plane side

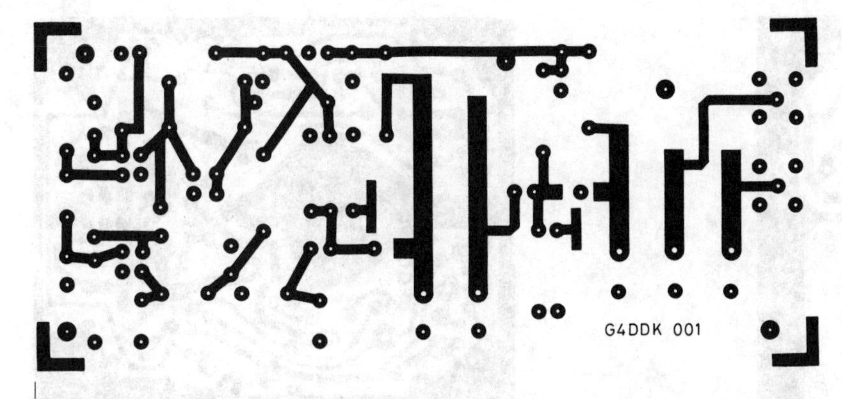

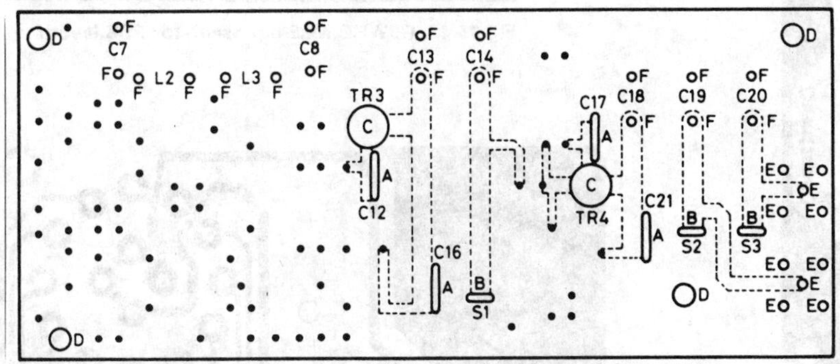

Slots 'A'.... 6·4mm long x 1·2mm wide   Slots 'B'.... 3mm long x 0·8mm wide
Holes 'C'.... 5mm dia   Holes 'D'.... 2·5mm dia   Holes 'E'.... 1·2mm dia   Holes 'F'.... 1mm dia
Holes marked • are 0·8mm dia although 1mm dia is permissible if more convenient

Fig 9.11. Printed circuit board artwork and drilling pattern for the microwave source G4DDK-001

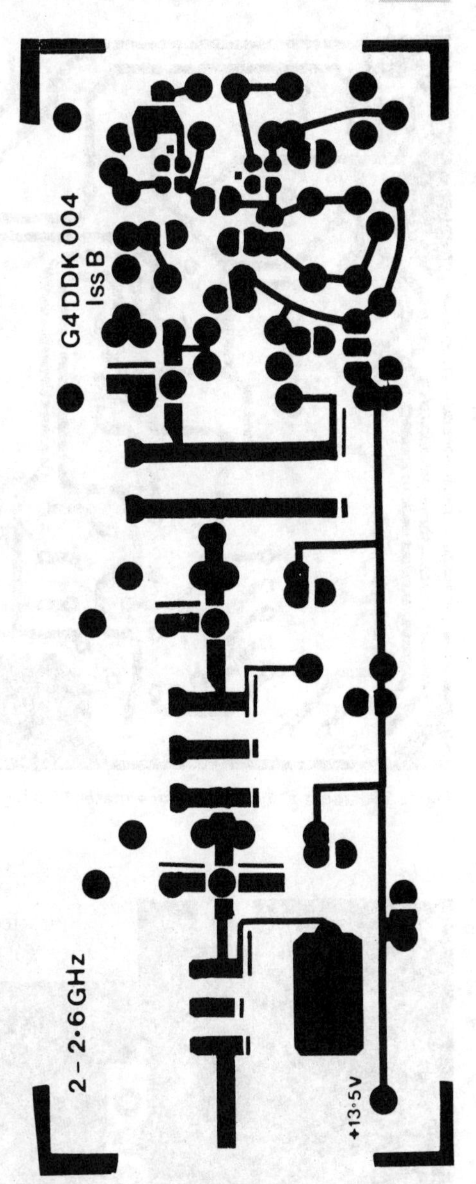

Fig 9.15. Printed circuit board artwork for the 2.5GHz source G4DDK-004

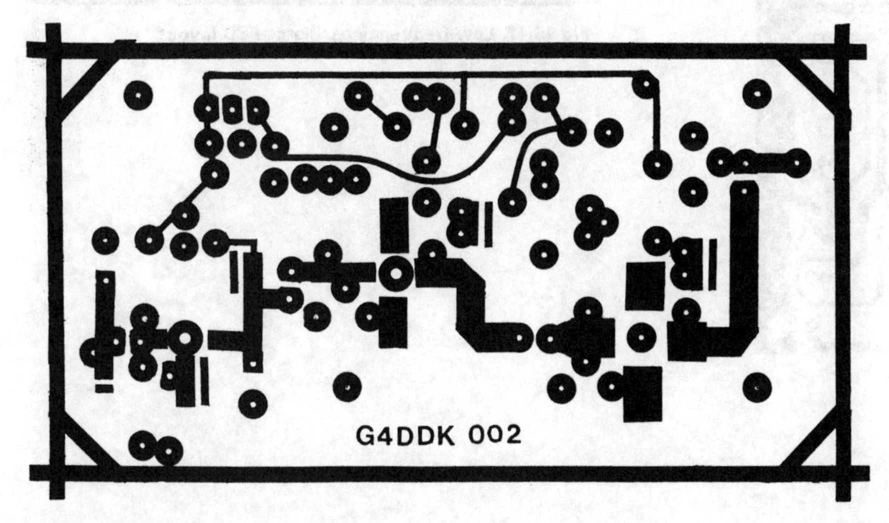

Fig 9.63. PCB artwork for the 1W G4DDK-002

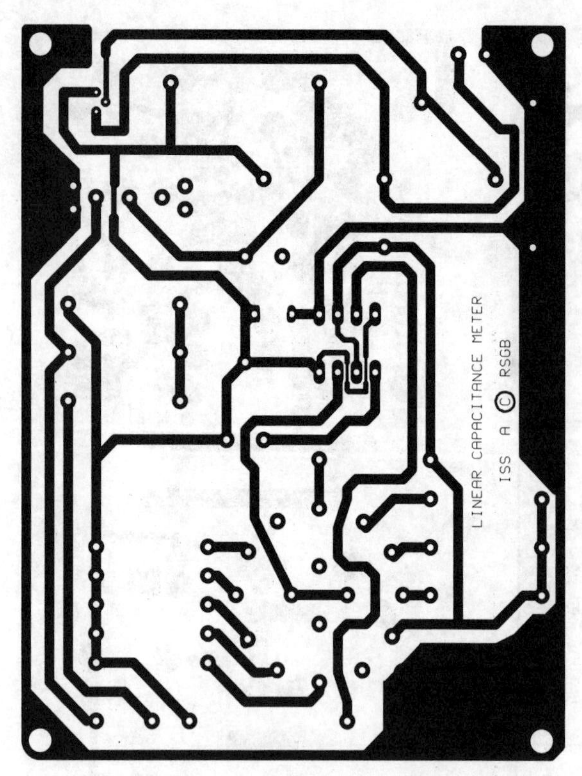

Fig 15.28. Linear-scale capacitance meter PCB layout

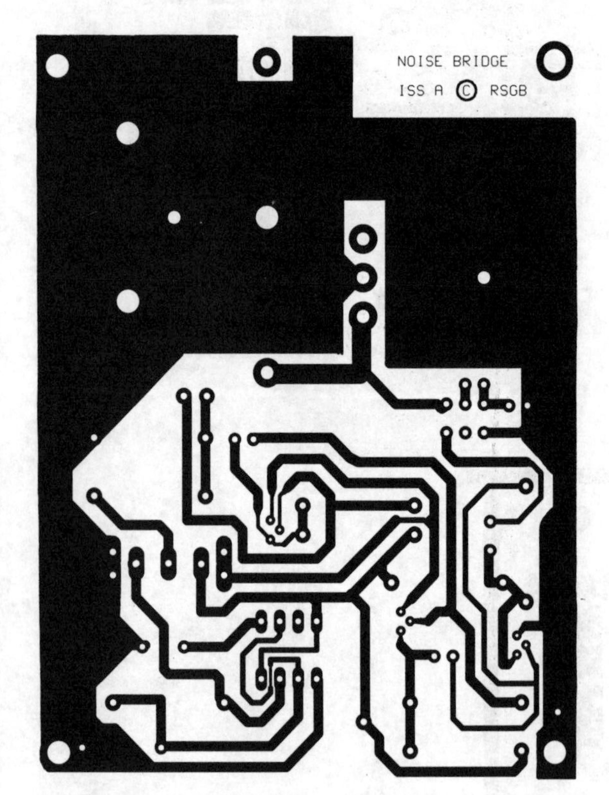

Fig 15.33. RF noise bridge PCB layout

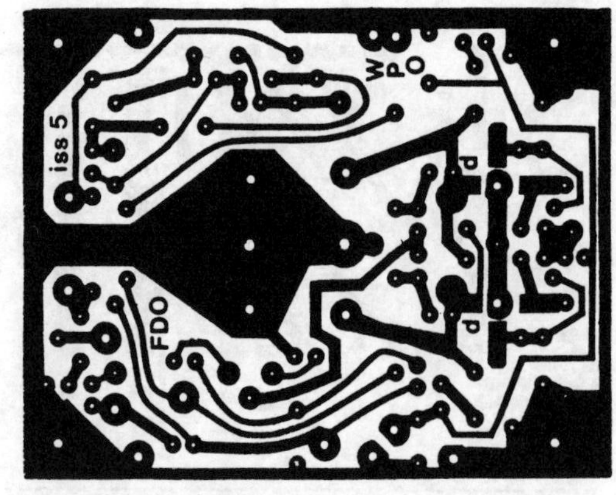

Fig 15.42. G3WHO Mk2 dip oscillator PCB layout

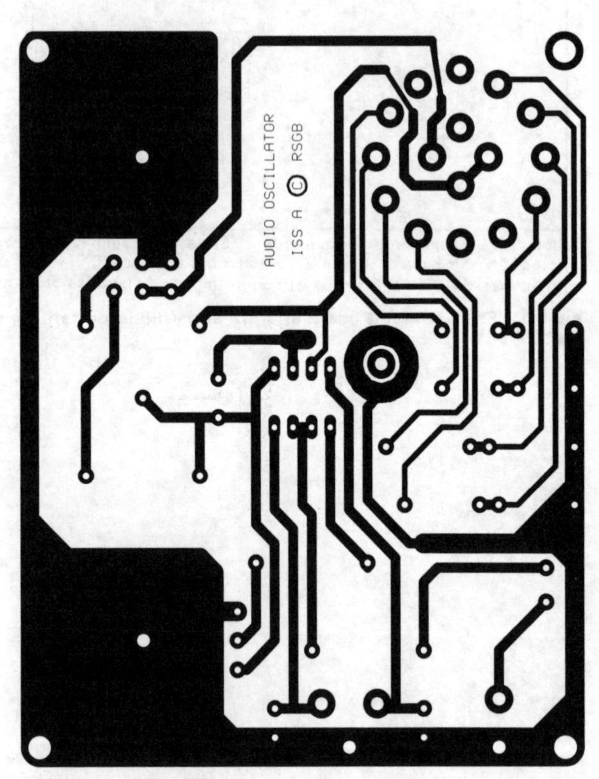

Fig 15.47. Low-frequency oscillator PCB layout

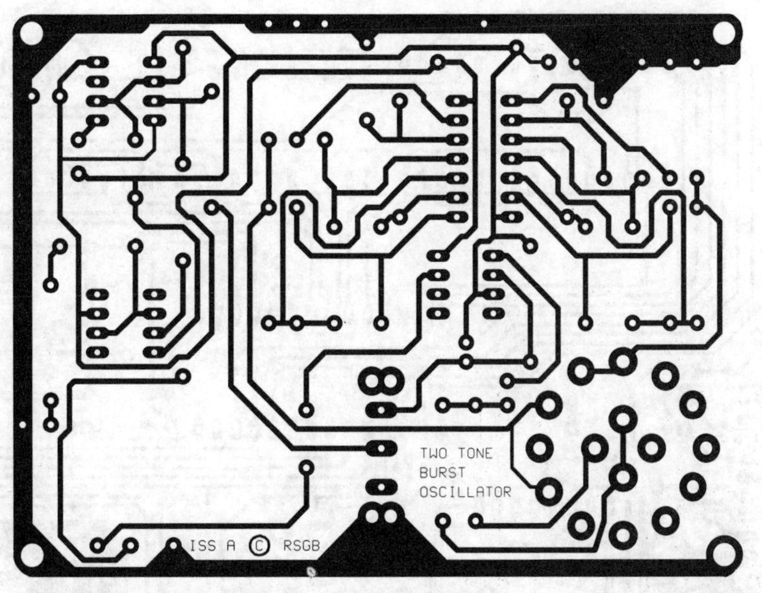

Fig 15.50. Two-tone burst oscillator PCB layout

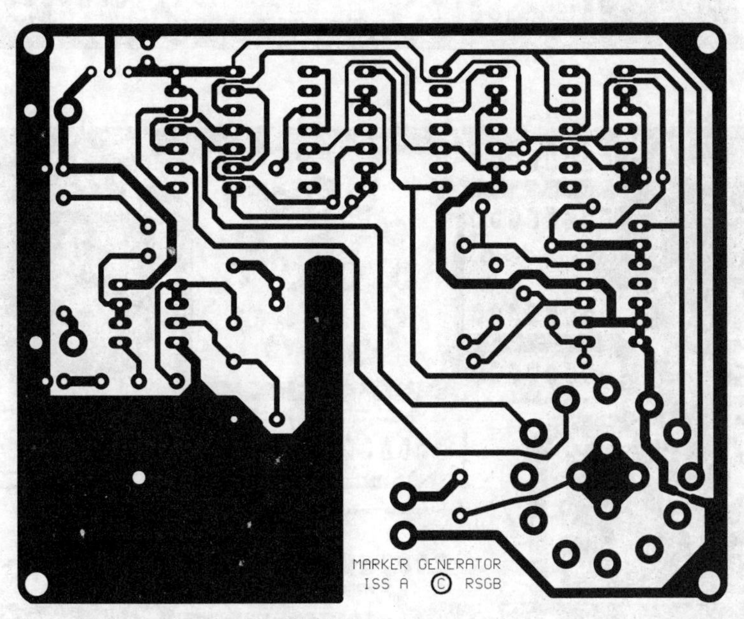

Fig 15.54. Frequency marker PCB layout

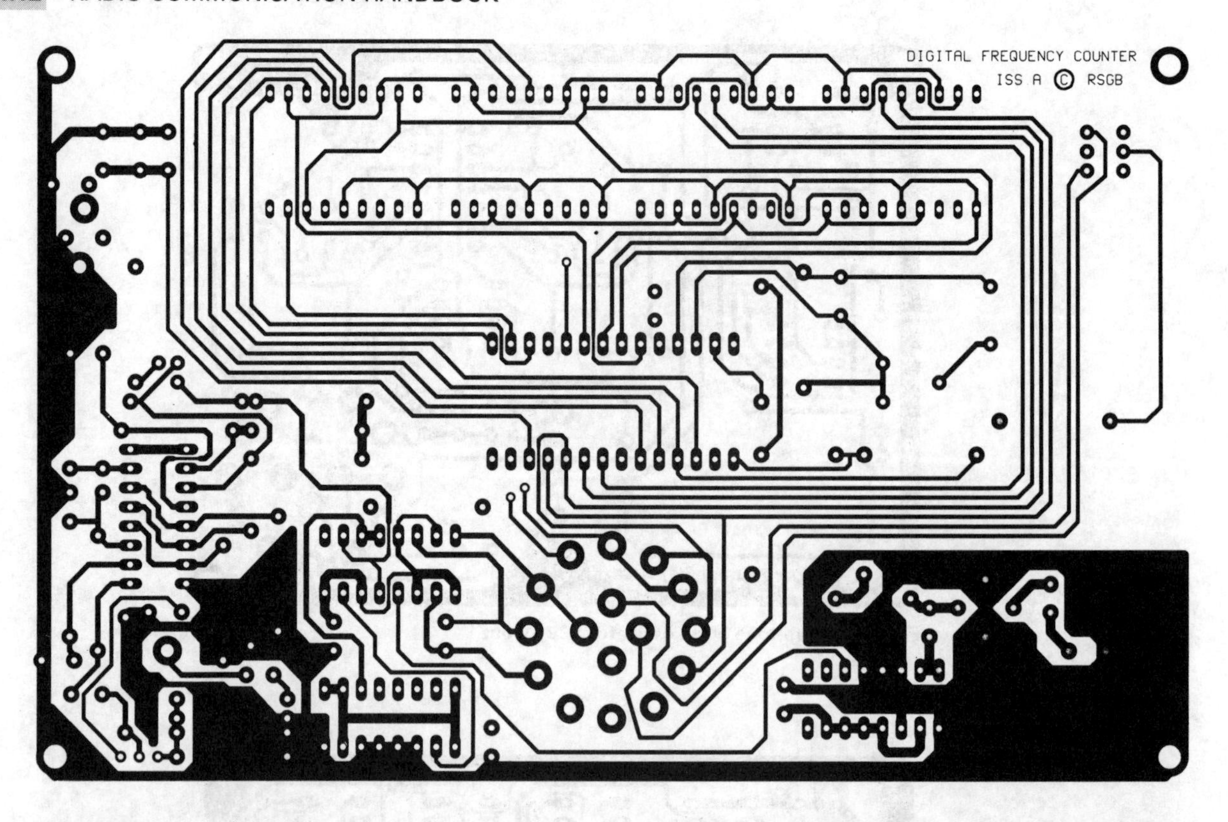

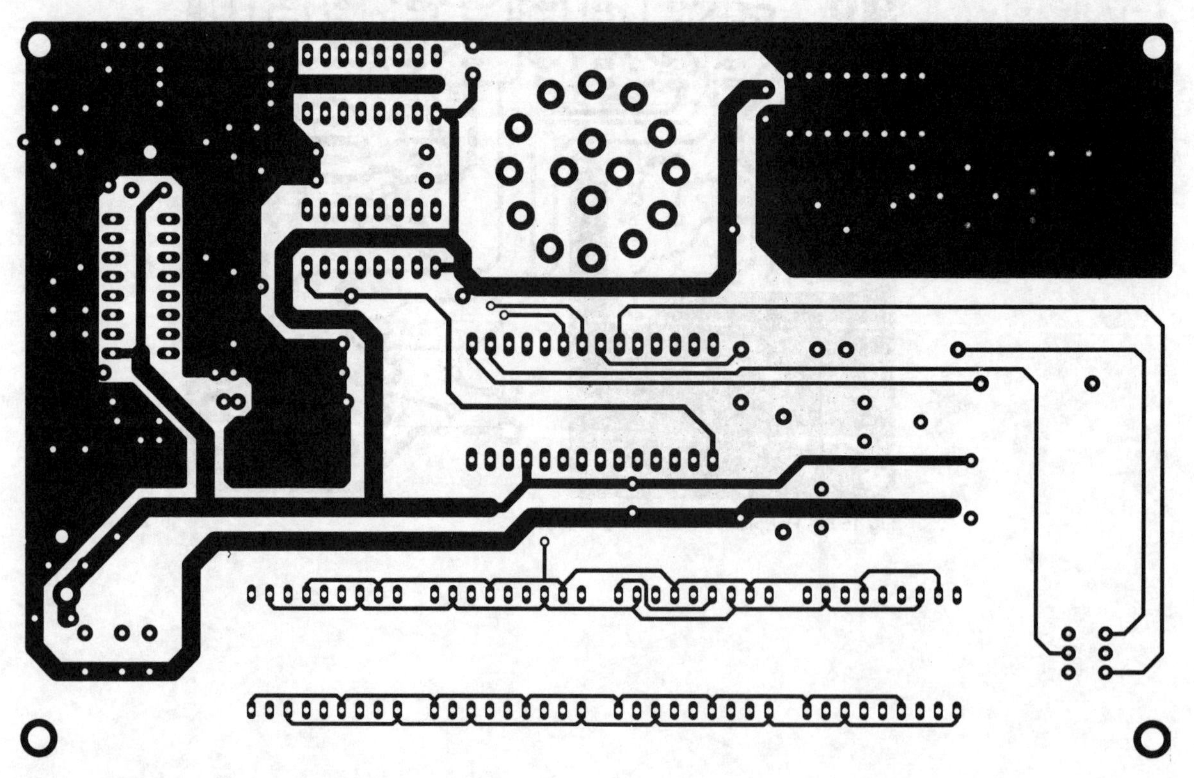

Fig 15.60. Digital frequency counter double-sided PCB layout

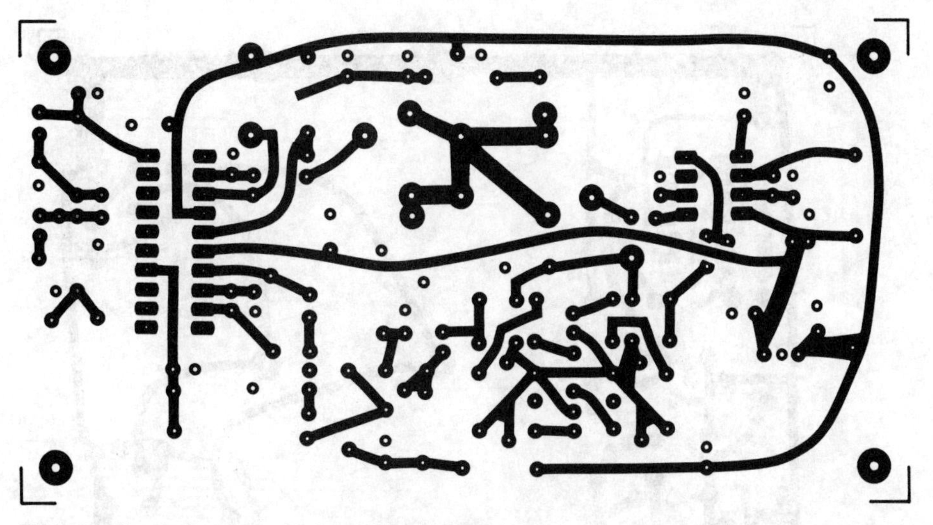

Fig 15.67. Simple Spectrum Analyser RF board PCB layout

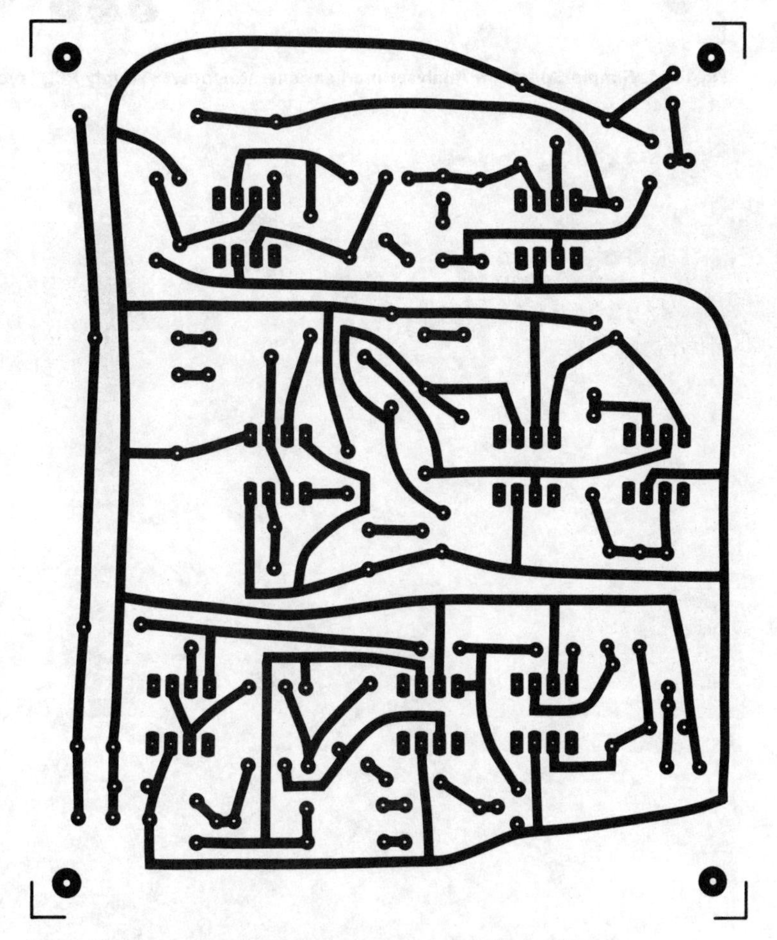

Fig 15.69. Simple Spectrum Analyser video/sweep board PCB layout

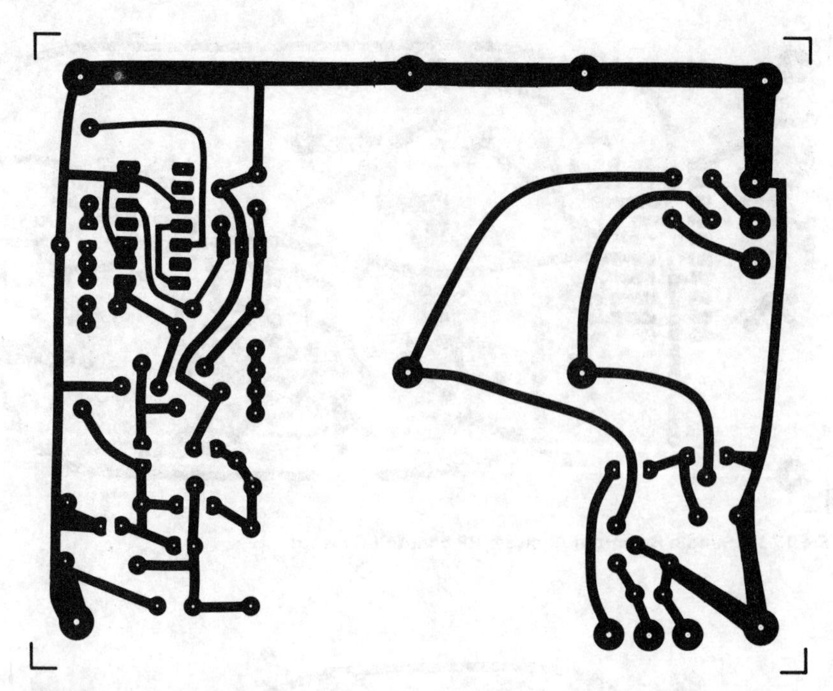

Fig 15.71. Simple Spectrum Analyser marker generator/power supply PCB layout

Index

D

E

F

W

Y

Z

Becoming Aware

TENTH EDITION

A TEXT/WORKBOOK FOR HUMAN RELATIONS AND PERSONAL ADJUSTMENT

Velma Walker

Tarrant County College

KENDALL/HUNT PUBLISHING COMPANY
4050 Westmark Drive Dubuque, Iowa 52002

You Can Be Whatever You Want to Be

There is inside you all of the potential to be whatever you want to be—all of the energy to do whatever you want to do.

Imagine yourself as you would like to be, doing what you want to do, and each day, take one step . . . toward your dream.

And though at times it may seem too difficult to continue, hold on to your dream.

One morning you will awake to find that you are the person you dreamed of—doing what you wanted to do—simply because you had the courage to believe in your potential and to hold on to your dream.

Book Team

Chairman and Chief Executive Officer Mark C. Falb
Senior Vice President, College Division Thomas W. Gantz
Director of National Book Program Paul B. Carty
Editorial Development Manager Georgia Botsford
Developmental Editor Angela Willenbring
Vice President, Production and Manufacturing Alfred C. Grisanti
Assistant Vice President, Production Services Christine E. O'Brien
Prepress Editor Angela Puls
Permissions Editor Colleen Zelinsky
Designer Suzanne Millius

Cover images courtesy of Jupiter Images and Photodisc

ISBN 13: 978-0-7575-2882-8
ISBN 10: 0-7575-2882-1

Printed in the United States of America.
10 9 8 7 6 5 4 3 2 1